1st International Conference on Sustainable Energy and Environment Development (SEED 2016)

E3S Web of Conferences Volume 10 (2016)

Krakow, Poland
17 - 19 May 2016

Editors:

M. Filipowicz
M. Dudek
T. Olkuski
K. Styszko

ISBN: 978-1-5108-3113-1

Printed from e-media with permission by:

Curran Associates, Inc.
57 Morehouse Lane
Red Hook, NY 12571

Some format issues inherent in the e-media version may also appear in this print version.

Printed by Curran Associates, Inc. (2016)

For additional information, please contact EDP Sciences – Web of Conferences at the address below.

EDP Sciences – Web of Conferences
17, Avenue du Hoggar
Parc d'Activité de Courtabœuf
BP 112
F-91944 Les Ulis Cedex A
France

Phone: +33 (0) 1 69 18 75 75
Fax: +33 (0) 1 69 28 84 91

contact-edps@webofconferences.org

Additional copies of this publication are available from:

Curran Associates, Inc.
57 Morehouse Lane
Red Hook, NY 12571 USA
Phone: 845-758-0400
Fax: 845-758-2633
Email: curran@proceedings.com
Web: www.proceedings.com

TABLE OF CONTENTS

ISSN (electronic edition): 2267-1242

Scientific Committee of the International Conference on the Sustainable Energy and Environment Development (SEED'16)

Chair of the Scientific Committee:
Assist. Prof. Mariusz FILIPOWICZ (AGH University of Science and Technology, Poland)

Scientific Committee members:

- Prof. Vincenzo BIANCO (DIME/TEC – Università degli Studi di Genova, Italy)
- Prof. Vyacheslav M. BYSTRITSKY (Joint Institute for Nuclear Research, Russia)
- Prof. Tadeusz CHMIELNIAK (Silesian University of Technology, Poland)
- Prof. Dorota CHWIEDUK (Warsaw University of Technology, Poland)
- Prof. Ewa CUKROWSKA (Witwatersrand University Johannesburg, Republic of South Africa)
- Prof. Bartłomiej GŁOWACKI (MA University of Cambridge, United Kingdom)
- Prof. Agata GODULA – JOPEK (Airbus Group, Germany)
- Prof. Janusz GOŁAŚ (AGH University of Science and Technology, Poland)
- Prof. Wacław GUDOWSKI (KTH Royal Institute of Technology, Sweden)
- Prof. Waldemar KAMRAT (Gdańsk University of Technology, Poland)
- Prof. Henryk KAPROŃ (Warsaw University of Technology, Poland)
- Prof. Anneliese KASPER-GIEBL (Technische Universitat Wien, Austria)
- Prof. Janusz KOTOWICZ (Silesian University of Technology, Poland)
- Prof. Wojciech NOWAK (AGH University of Science and Technology, Centre of Energy, Poland)
- Prof. Jozef M. PACYNA (Norwegian Institute for Air Research, Norway)
- Prof. Janusz SOWIŃSKI (Czestochowa University of Technology, Poland)
- Prof. Wojciech SUWAŁA (AGH University of Science and Technology, Poland)
- Prof. Marek ŚCIĄŻKO (Institute for Chemical Processing of Coal, Poland)
- Prof. Gareth TAYLOR (Brunel University London, United Kingdom)
- Prof. Janusz TRAWCZYŃSKI (Wroclaw University of Technology, Poland)
- Assist. Prof. Magdalena DUDEK (AGH University of Science and Technology, Poland)
- Assist. Prof. Anna GORCZYCA (Univeristy of Aglicullture in Krakow, Poland)
- Assist. Prof. Anita LEWANDOWSKA (University of Gdansk, Poland)
- Assist. Prof. Wojciech LIPINSKI (The Australian National University, Australia)
- Assist. Prof. Jarosław MARKOWSKI (Poznań University of Technology, Poland)
- Assist. Prof. Tadeusz OLKUSKI (AGH University of Science and Technology, Poland)
- Assist. Prof. Paweł PURGAŁ (Kielce University of Technology, Poland)
- Assist. Prof. Barbara TORA (AGH University of Science and Technology, Poland)
- Ph.D. Elizabeth CASTILLO-MARTINEZ (University of Cambridge, United Kingdom)
- Ph.D. Andrius GARBARAS (Center for Psychical Sciences and Technology Lithuania)
- Ph.D. Magdalena KISTLER (Technische Universitat Wien, Austria)
- Ph.D. Volker LENZ (Das Deutsche Biomasseforschungszentrum, Germany)
- Ph.D. Andreas ORTWEIN (Das Deutsche Biomasseforschungszentrum, Germany)
- Ph.D. Mariusz RUSZEL (Rzeszow University of Technology, Poland)
- Ph.D. Katarzyna STYSZKO (AGH University of Science and Technology, Poland)

Partners of the International Conference on the Sustainable Energy and Environment Development (SEED'16)

Conference was organized under the patronage of:

The Head of the Ministry of Science and Higher Education
Jarosław Gowin

Ministry of Energy
Krzysztof Tchórzewski

Ministry of the Environment
prof. Jan Szyszko

National Fund for Environmental Protection and Water Management

Energy Regulatory Office

The National Centre for Research and Development

The Head of Małopolska Provincial Office in Krakow
Józef Pilch

Marshal of Małopolska Region
Jacek Krupa

Municipality of Krakow – partnership

National Contact Point in Poland

AGH UST Rector
Prof. Tadeusz Słomka

Conference was sponsored by:

EDF Polska
Diamond Sponsor

Ciepło dla Krakowa
Diamond Sponsor

CC Poland Plus Platinum Sponsor	MetalERG. Platinum Sponsor	RWE Polska Platinum Sponsor

Media Partners:

Radio Kraków	Dziennik Polski	Ciepłownictwo, Ogrzewnictwo, Wentylacja	Rynek Instalacyjny
Instal	Energetyka Cieplna i Zawodowa	Energia+	Energetyka Wodna
Radio17	CIRE.PL	kierunekenergetyka.pl	Wrota Małopolski
Magiczny Kraków	odnawialnezrodlaenergii.pl	OZE.PL	GRAMwZIELONE.pl
WysokieNapiecie.pl	Studentnews.pl	Świat Kominków	Biuletyn AGH

Organizers of the International Conference on the Sustainable Energy and Environment Development (SEED'16)

Conference was organized by:

IZE

Foundation Institute
for Sustainable Energy

Faculty of Energy and Fuels,
AGH UST in Krakow

Statement of Peer review

In submitting conference proceedings to E3S Web of Conferences, I certify to the Publisher that I adhere to the **Policy on Publishing Integrity** of the journal in order to safeguard good scientific practice in publishing.

1. All articles have been subjected to peer review administered by the proceedings editors.
2. Reviews have been conducted by expert referees, who have been requested to provide unbiased and constructive comments aimed, whenever possible, at improving the work.
3. Proceedings editors have taken all reasonable steps to ensure the quality of the materials they publish and their decision to accept or reject a paper for publication has been based only on the merits of the work and the relevance to the journal.

Title, date and place of the conference

International Conference on the Sustainable Energy and Environment Development
May 17th - 19th, 2016
Kraków (Poland)

Proceedings editor(s):

Prof. Mariusz Filipowicz
Prof. Magdalena Dudek
Prof. Tadeusz Olkuski
Dr. Katarzyna Styszko

Date and editor's signature

Preface

The 1st International Conference on the Sustainable Energy and Environment Development (SEED) was held in the Center of Energy at the AGH University of Science and Technology in Krakow (Poland) on May 17th – 19th, 2016.

The organizers hosted over 150 participants from 13 countries (England, Austria, Czech Republic, Iraq, Ireland, Lithuania, Germany, Norway, Poland, Republic of South Africa, Turkey and Italy) representing 23 scientific and research institutions. There were 118 excellent talks and 84 excellent posters during the 14 plenary sessions and 2 poster sessions in the areas of: energy, sustainable development, cogeneration, fuel cells, environment protection and wastes management, sustainable urban architecture, levels of pollution from transportation, possibilities to reduce the emission of exhausts from transportation, energy production with an application renewable and non-renewable energy sources and nuclear energy.

The conference was under the patronage of Ministry of Science and Higher Education of Poland, Ministry of Energy of Poland, Ministry of the Environment of Poland, National Fund for Environmental Protection and Water Management, Energy Regulatory Office, The National Centre for Research and Development, Małopolska Provincial Office in Krakow, Marshal of Małopolska Region, Municipality of Krakow (partnership), National Contact Point in Poland and AGH-UST Rector. The conference was financially supported by EDF Poland, CC Poland Plus Sp. z o. o., MetalERG, RWE Polska and Małopolska Fund for Environmental Protection and Water Management (WFOŚiGW in Krakow). The event was also medially supported by 18 media institutions.

A subset of the presenters opted to write full conference papers. These peer-reviewed proceedings are the collection of those full papers. The papers well represent the range of topics presented in the conferences and help to describe better the term of sustainability in varied areas of research. The papers were peer-reviewed by 186 reviewers representing 54 research institutions from Poland and abroad.

The Conference was organized by Institute for Sustainable Energy (ISE) in cooperation with AGH University of Science and Technology in Krakow. The members of Organizing Committee were: Tadeusz Olkuski, Mariusz Filipowicz, Magdalena Dudek, Katarzyna Szramowiat, Krzysztof Sornek, Kamila Rzepka, Andrzej Raźniak and Marta Marczak. The Scientific Committee composed of 36 specialists from Poland and abroad was driven by Prof. Mariusz Filipowicz.

On behalf of the Institute of Sustainable Energy, we have a great pleasure to invite you for the participation in the second edition of the International Conference on the Sustainable Energy and Environment Development in 2017 (follow the updates in the website www.seedconference.pl and www.ize.org.pl).

Kraków, Poland
September 2016

On behalf of the Organizing Committee:

Katarzyna Szramowiat
SEED Conference Secretary

The conference was financially supported by Małopolska Fund for Environmental Protection and Water Management (WFOŚiGW in Krakow)

Qualitative and chemical characteristics of mineral matter in the selected lignite deposits in light of their suitability for clean coal technologies

Barbara Bielowicz[1,a]

[1] *AGH University of Science and Technology, Poland, Faculty of Geology, Geophysics and Environmental Protection, Al. Mickiewicza 30, 30-059 Kraków*

Abstract. An analysis of the variability of ash content and chemical composition of coal ash is crucial when assessing the possibility of using this raw material in clean coal technologies (CCS). The paper focuses on chemical, technological and quality characteristics of ash from the combustion of lignite from Polish deposits. Ash content on a dry ash-free basis for the samples from 9 deposits varies from 1 to 50%. The most commonly observed ash content in the studied samples of lignite is in the range between 10-20%. The measured ash melting temperatures from the collected lignite samples have shown that the sintering, softening, melting and flow temperatures, determined in an oxidizing atmosphere (air), are higher than those determined in a reducing atmosphere. In the majority of Polish lignite deposits, the dominant component of ash was SiO_2, followed by CaO. When evaluating the suitability of coal in energy production, with particular emphasis on the combustion and gasification processes, it is also necessary to study the behavior of ash at high temperatures and the impact of ash on the gasification and combustion facilities. Ash fouling and slagging processes, which are heavily dependent on the ash content, composition, melting point and the content of chlorine, phosphorus and sulfur in coal, have a negative effect on the combustion process.

1 Introduction

Lignite is one of the most important energy sources in the world. However, the introduction of the new EU energy and environmental policy has resulted in advances in innovative and more environmentally friendly technologies using coal, known collectively as clean coal technologies (CCT). Full implementation of the above mentioned technologies will open new applications for this energy raw material, which could strengthen its position in the market.

The direct combustion of lignite is the most common form of producing energy from this raw material (municipal and industrial heat and electricity production).

Currently, the combustion of coal is carried out mainly in fluidized bed boilers, pulverized coal-fired boilers and, to a lesser extent, in grate boilers. However, in order to meet the expected coal quality standards, related mainly to the environmental protection, particularly emissions limits for sulfur and nitrogen oxides, particulate matter and mercury [1], it is necessary to introduce increasingly better and cleaner technologies. In recent years, major advances in new coal technologies have been done. New coal technologies are being introduced while those already known are being improved. One of them is the gasification of coal.

The gasification of coal can be carried out via surface and underground gasification technologies. Coal gasification is a key technology for the production of energy from coal while taking care of the environment. The versatility and flexibility of gasification technologies enables the production of a wide range of products, including mainly liquid fuels, chemical fuels, gas fuels and many others. As opposed to conventional combustion, this technology has a positive impact on the environment because of its ability to capture CO_2 before the gasification process. The capture of carbon dioxide allows obtaining a gas producing much less harmful substances during the combustion process when compared to the conventional combustion of coal. In this process, 99% of sulfur and other harmful substances can be removed and processed for industrial use. Currently, there are about 200 coal gasification plants in the world, while 17 are under construction and another 50 are being planned [2]. Lignite is the main raw material used in the gasification process. According to the forecasts of the US Department of Energy, it will maintain its dominant position in the coming years [2].

In Poland, there are 90 documented lignite deposits, mainly in the Paleogene and Neogene sediments in the Polish Lowlands. The total proved recoverable reserves of lignite amount to 23 510.59 million tonnes.

The proved recoverable reserves in currently exploited deposits amount to 1 482.69 million tonnes, representing 6.3% of the total proved recoverable reserves. Lignite from these deposits is extracted in 5 mines, namely: Bełchatów, Turów, Adamów, Konin

[a] Corresponding author: bbiel@agh.edu.pl

and Sieniawa lignite mines. Given the current level of consumption, these reserves should last for another several decades. As of today, lignite is mainly used in large plants such as power stations.

2 The impact of the mineral matter on the suitability of coal for CCT

However, it should be underlined that coal is a mixture of carbonized organic matter represented by macerals and mineral matter. Therefore, the ash content (content of non-combustible mineral matter) is an extremely important parameter when assessing the possibility of using coal for combustion purposes. The term ash content applies to the residues (unburned mineral matter) produced after the complete combustion of coal under certain conditions. The content and composition of the non-combustible mineral matter in coal affect the quality of coal, mainly by reducing the gross calorific value [3], thus reducing its usefulness for energy purposes. The ash content of coal determines the possibility of its use in gasification and liquefaction processes.

The content and the type of mineral matter contained in the combusted coal also have an impact on this process. According to the current guidelines, the ash content (dry basis) of coal subjected to gasification process should be less than 40%. According to Shirazi et al. [3], the amount of the mineral and chemical matter and their composition can significantly reduce the amount of energy obtained due to the need to use part of it in order to heat the non-combustible materials. It has been observed that the energy loss was more significant in pulverized coal-fired boilers. Ash components react with each other at high temperatures and - forming new mineral phases and volatile compounds - affect the combustion process [4, 5, 6]. The most troublesome is the formation of fly ash, slag and chemical compounds inside the furnace, resulting in corrosion of boilers. The knowledge of the ash melting point is crucial when estimating the impact of mineral matter on a given combustion technology. The individual minerals have different melting points. The low ash melting point is the reason for many negative phenomena during combustion processes. The vitrified ash damages grate of the boiler and its housing material hinders the circulation of air and sticks to the unburned coal, which contributes to a decrease in thermal efficiency of the furnace. It can also - reacting with the material of the furnace walls - form low melting slag, which can lead to serious accidents. The ash melting point depends on its chemical composition. Generally, the more silicates and fewer oxide connections of iron and alkali, the higher the ash melting temperature. Both softening and melting of ash are very complex process. These reactions run parallel to the other chemical reactions, which are highly dependent on process time, fragmentation and arrangement of the ash components. Ash components react with each other and with their decomposition products both before melting and during the reaction [7]. The least troublesome in this respect is quartz, which has higher melting point (about 1800 °C) than the temperatures occurring in furnaces [6] and therefore does it does not pass into the liquid state.

Ash fouling and slagging processes, which are heavily dependent on the ash content, composition, melting point and the content of chlorine, phosphorus and sulfur in coal, have a negative effect on the combustion process [8]. In order to determine the tendency to form unfavorable residues in the furnace, the alkalinity/acidity, slagging and SiO_2 ratios, as well as the tendency to contamination of heating surfaces (fouling), are determined [9]. According to Crossley [10], coals with chlorine content above 0.3%, phosphorus content greater than 0.03% and sulfur content over - 1.8% have the greatest tendency to slagging and fouling. According to Ataköl et al. [11] this tendency is directly proportional to the Na_2O content and inversely proportional to the Al_2O_3 content. On the other hand, Róg [9] suggests that the higher the aluminum silicate content and the lower the iron and alkali content, the higher the melting point of ash and lower tendency to fouling.

Kaolinite decomposes already at 450 °C, losing OH groups in the crystal structure, while clay minerals (mainly illite and smectite) begin to decompose at about 500 °C to form new mineral phases, which can also contribute to the plastic deformation of ash and the formation of slag, which is difficult to remove.

Calcite and dolomite decompose at about 900°C, resulting in the formation of calcium oxide (CaO), which - upon reaction with water vapor - forms artificial minerals like portlandite ($Ca(OH)_2$) and interacts with aluminosilicates to form gehlenite ($Ca_2Al_2SiO_7$) and anorthite ($CaAl_2Si_2O_8$). Carbonates, especially calcite, can react with sulfur contained in coal and form the artificial mineral phase chemical composition typical for anhydrite, thereby permanently binding the sulphur [12]. This process has been used in one of the methods of dry flue gas desulfurization (flue gas resulting from the combustion).

Despite the fact that combustion technologies are constantly being improved, the gasification of coal is considered as one of the cleanest coal technologies. Gasification can be carried out in several types of gasifiers. The choice of the technology used depends largely on the quality of coal, defined by the following parameters: the maximum ash content, maximum moisture of fuel, ash melting temperature, reactivity, and the rank of coal. The rank of coal and reactivity of the fuel allow to determine whether the coal can be used in all types of gasifiers. The maximum moisture content below 35% or 5% depending on the type of gasifier can prevent the use of a given lignite type. In the case of fixed bed gasifiers, the maximum moisture content below 35% is good enough. Fluidized bed and entrained flow gasifiers require the complete drying of coal. The ash content and composition have significant impact on the gasification process because some minerals can act as catalysts and others as inhibitors, while the high content of mineral matter causes a decrease in system efficiency [13, 14] Therefore, the maximum permissible content of mineral matter in coal depends on the gasification technology used. Different ash melting points determine the choice of gasification technology and the removal of residues resulting from the process. Therefore, the knowledge of this parameter for a given fuel is necessary before designing the gasifier. In this context, the ash content, oxide composition and the ash melting point are of crucial importance when it comes to determining the suitability of minerals for gasification process.

According to Sekine et al. [15] the Ca, K and Na content improves the reaction rate. On the other hand, Johnson [16] suggests that dispersed alkaline substances can act as a catalyst and the reactivity is independent from the coal rank of the gasified coal, solely depending on the type of mineral matter. On the contrary, silicon (Si) and aluminum (Al) slow down the gasification of coal and tend to suppress the aforementioned process [15].

Skodras and Sakalloropoulos [17] have correlated the alkalinity/acidity index - hereafter referred to as alkali index, characterized by Sakawa et al. [18] – to the reactivity of coal during the gasification process. They have found that there is a clear relationship between the Ca content and the reactivity of coal, unlike in the case of Mg, K, Na, Fe and the reactivity. In addition, it has been found that oxides, mainly SiO_2, CaO and iron oxide can adversely affect refractory materials of the reactor, contributing to their corrosion [19].

The ash melting point, which determines the technology used (Table 1) is another parameter crucial for the gasification process [13, 20]. This parameter can be correlated as a function of the SiO_2, $Al2O_3$, TiO_2, Fe_2O_3, CaO, MgO, Na_2O and K_2O content in ash [21, 22]. The alkalinity/acidity index of ash can be used to correlate the ash melting point and its chemical composition. For example, the high alkali content in Ca, Fe, K_2O and Na_2O oxides in coal suggest low ash melting point. In the case of entrained flow gasifiers, the melting point of coal ash should be lower than their operating temperature in the range 1400-1600 °C. In the case of fluidized bed gasifiers it is necessary to use coal with ash melting point higher than the operational temperature of the reactor (1000 °C) in order to prevent ash slagging processes, favored by high pyrite and sodium silicate content in the gasified coal.

3 Research methodology

The study included field work and laboratory work. Field work was aimed at the collection of lignite samples, which were subjected to a series of laboratory tests. In total, about 150 lignite samples have been collected and used for proximate analysis. In addition, a large database consisting of archival results from mines and scientific literature has been collected. All the results obtained were used in the statistical analysis. The studies have used the lignite samples collected from the currently exploited and perspective deposits. The distribution of these deposits is shown in Figure1.

Research on the properties and usability of different coal types involves technical, physico-chemical and petrographic analysis of coal. Laboratory tests of the collected lignite samples included:

a-Technical and physical analysis of coal,

b-Chemical analysis of coal and coal ash. After the determination of free moisture content, the samples collected from lignite deposits were dried and grinded. The chemical and technical analysis used coal with a grain size of 0.02 mm. Technical analysis of coal has been carried out in accordance with the applicable standards for determining:

- Total and analytical moisture
- The ash content
- The ash melting point
- The sand content
- The oxide composition of ash.

Table 1. The characteristics of gasification reactors [20, 23; supplemented].

Reactor type	**Fixed bed gasification**		**Fluidized bed gasification**		**Entrained flow gasification**
Ash collection conditions	Dry at the bottom	Slag	Dry ash	Agglomerated ash	Slag
Typical designs	Lurgi	BGL	Winkler, HTW, KBR, CFB, HRL	KRW, U-Gas	KT, Shell, GEE, E-Gas, Siemens, MHI, PWR
Requirements for fuel					
Preferred fuel	lignite, hard coal, anthracite, waste	hard coal, anthracite, petroleum coke, waste	lignite, hard coal, anthracite, waste	lignite, hard coal, anthracite, waste biomass, petroleum coke	lignite, hard coal, anthracite, petroleum coke
Particle size	6-50 mm	6-50 mm	6-10 mm	6-10 mm	<100um
The maximum ash content%	unlimited	<25 (preferred)	unlimited	unlimited	<25 (preferred)
The maximum moisture%	<35	<35	<5	<5	<5
Ash melting temperature °C	>1200	<1300	>1100	>1100	<1300
The preferred rank of coal	All	High	Low	All	All
Reactivity	Low	Low	High	High	High

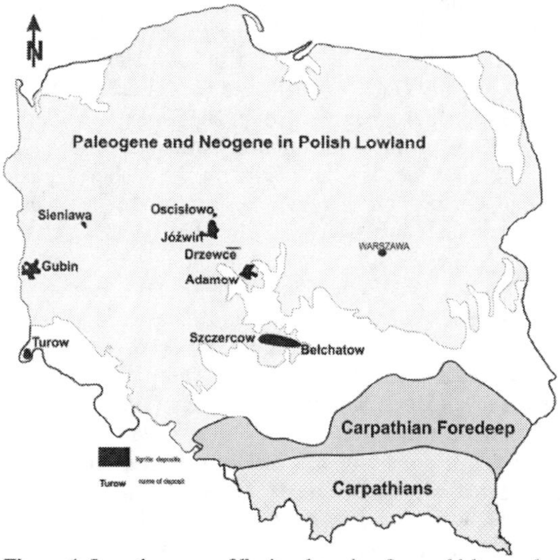

Figure 1. Location map of lignite deposits, from which samples for testing were collected [24; completed].

In order to determine the tendency for the formation of unfavorable by-products (residues) in the furnace, the following ratios were determined:

1) The alkalinity/acidity index (*base/acid (B/A) ratio*)

B/A = %(Fe_2O_3 + CaO + MgO + Na_2O + K_2O) / [%(SiO_2 + Al_2O_3 +TiO_3)] (1)

SiO_2, Al_2O_3,TiO_3, Fe_2O_3, CaO, MgO, Na_2O, K_2O, – the percentage of individual oxides in the ash.

2) The slagging index *(Rs)*

Rs = B/A ·S_t^d or Rs = B/A·Fe_2O_3 (2)

S_t^d – total sulfur content in coal (dry basis) Depending on the Rs value, the studied coal is characterized by:

— Low tendency to slagging – Rs < 0.6,
— Medium tendency to slagging – 0.6 < Rs < 2.0,
— High tendency to slagging – 2.0 < Rs < 2.6,
— Very high tendency to slagging – Rs > 2.6

3) The silica value, *(SV)*

SV= (SiO_2·100) / (SiO_2+Fe_2O_3+CaO+MgO) (3)

Depending on the SV value, the studied coal is characterized by:

— Low tendency to slagging – SV > 72,
— Medium tendency to slagging – 65 < SV < 72,
— High tendency to slagging – SV < 65.

4) The tendency to fouling of the heating surfaces (*Fouling Index, Rf)*

Rf=B/A · (Na_2O +K_2O) (4)

Depending on the Rf value, the studied coal is characterized by:

— No tendency to fouling – Rf < 0.6,
— High tendency to fouling – 0.6 < Rf < 40.0,
— Very high tendency to fouling – Rf > 40.0.

5) The alkali index AK

AK=Na_2O + 0.96559K_2O (A^d /100) (5)

A^d – the percentage of ash content in coal (dry basis) Depending on the AK value, the studied coal is characterized by:

— Low tendency to fouling with alkali components – *AK* < 0.3,
— Medium tendency to fouling with alkali components – 0.3 < *AK* < 0.45,
— High tendency to fouling with alkali components – 0.45 < *AK* < 0.6,
— Very high tendency to fouling with alkali components – *AK* > 0.6 [25].

4 Research results

The average results of the analysis for the individual deposits are shown in Table 2.

In the case of the examined lignite deposits, the ash content (A^d) on a dry basis is very variable parameter. For the aforementioned deposits, the average ash content on a dry basis ranges from 8.0% in the Turów deposit to 25.9% in the Szczerców deposit (Table 2). In the individual samples, this parameter is in the range from 4.9% in the Turów deposit to 68.7% in the Gubin deposit, with an average value of 20.3%. The asymmetrical distribution with a positive skewness can suggest a tendency for higher ash content in the lignite, which is clearly visible. The coefficient of variation for the ash content on a dry basis is 66.4%, which is fairly strong. The most frequently observed ash content in the studied lignite samples is in the range between 10-20%.

Sand, a granular fraction consisting of hard mineral grains, is a harmful element in the mineral matter of coal in ash resulting from the combustion of lignite. The main component of sand is quartz, followed by feldspar and mica. In the studied deposits, the sand content on a dry basis (p^d) is very variable – the highest in the Ościsłowo deposit, reaching on average to 8.3%, while the lowest in the Adamów deposit – 1.4% (Table 2). The above mentioned variability is very strong also

Table 2 The average chemical and technological parameters of domestic lignite deposits.

Deposit	Ash content A^d [%]	Sand content p^d [%]	Ash melting point t_B [°C]	CaO content [%]	Alkalinity/ acidity B/A [%]	Slagging index Rs	Silica value SV	Fouling index Rf	Alkali index AK
Adamów	8.6	1.4	1375	2.91	1.69	0.43	34.53	0.07	0.03
Bełchatów	18.0	4.4	1285	3.82	0.37	0.43	67.78	0.11	0.22
Drzewce	12.7	3.2	1310	3.73	0.49	0.75	66.34	0.06	0.05
Gubin	19.2	2.7	1300	2.52	0.38	0.79	73.28	0.31	0.65
Józwin	16.8	5.0	1365	3.80	0.55	1.04	64.34	0.07	0.10
Ościsłowo	25.4	8.3	>1500	2.93	0.25	0.44	79.73	0.01	0.03
Sieniawa	7.5	2.7	1390	2.03	-	-	-	-	-
Szczerców	25.9	4.4	1356	3.90	0.76	1.48	53.29	0.13	0.12
Turów	8.0	5.7	1345	0.65	0.34	0.15	76.70	2.48	6.52

in the case of individual samples of lignite, with a coefficient of variation amounting to 158% and an average sand content of 4.3%. While there is practically no sand in some of lignite samples from the Szczerców deposit, the highest sand content, reaching up to 47.7%, has been shown in the Gubin deposit. Generally, the examined samples contained up to 5.0% sand. In the case of Polish lignite deposits, a very strong variability of this parameter, with the average sand content of about 7.4%, can be observed.

The calcium content of coal, expressed by the CaO content, is associated with the mineral matter present in coal, especially - calcite ($CaCO_3$), partially aragonite ($CaCO_3$) and - occasionally occurring in coal deposits - calcium aluminum hydroxycarbonate e.g. para-alumohydrocalcite ($CaAl_2(CO_3)_2(OH)_4{\cdot}6H_2O$) [26]. However, it should be noted that part of the calcium, along with Fe, Mg, K and Na, reacts with the mineral matter (coal) to form forms dopplerite, a calcium salt [27], The calcium oxide content in the tested lignite ranges from 0.65% in the Turów deposit to 3.9% in the Szczerców deposit, while the average calcium content in the lignite is determined to be 2.85%. The coefficient of variation for this parameter is 82.39%, which indicates a strong variation of this parameter and at the same case suggests that it is difficult to estimate using statistical methods. The highest CaO content in domestic lignite deposits in Poland is observed in the Szczerców deposit [28]. In other deposits, this value usually does not exceed 5.00% [29]. This finding has also been confirmed in the presented study, as the CaO content in the majority of the examined lignite is in the range of 2-4%.

The ash melting temperatures of the collected lignite samples have shown that the sintering, softening, melting and flow temperatures, determined in an oxidizing atmosphere (air) are higher than those determined in a reducing atmosphere (a mixture of CO and CO_2 in the ratio 3:2). The melting point (t_B) in an oxidizing atmosphere ranges from 1285 °C (the Bełchatów deposit) to over 1500 °C (the Ościsłowo deposit) while a reducing atmosphere can range from 1170 °C (the Szczerców deposit) to > 1500 °C. This parameter is also highly variable for individual deposits.

The average chemical composition of ash, resulting from slow incineration of lignite, in the examined deposits indicates that in all of them, with the exception of the Adamów deposit, the dominant component of ash is SiO_2, ranging from 30.23% in the Adamów deposit to 67.7% in the Ościsłowo deposit. In the former, the dominant component of ash is CaO (40.39%), while the average content of this component in other deposits ranges from 8.9% in the Turów deposit to 25.5% in the Szczerców deposit. The share of Na_2O is usually low, in the range between 0.03% in the Adamów deposit and 6.5% in the Turów deposit. The content of magnesium oxide (MgO) is also limited, ranging from 0.20% to 5.02%. The aluminum oxide content (Al_2O_3) is highly variable and ranges from 1.01% in the Ościsłowo deposit to 17.23% in the Turów deposit. The average P_2O_5 content in ash of the analyzed raw material is below 0.31%, with the exception of one sample from the Jóźwin deposit, where it has reached 1.09%. Similar low levels have been determined for potassium oxide (K_2O), with the highest value, 2.77%, in the Gubin deposit, and the lowest - 0.01% - in deposits located in the Konin and Turek area. Sulfur oxide (SO_3) in the tested lignite ash ranges from 5.7% in the Turów deposit to 14.8% in the Gubin deposit. A high content of this component can be associated with the fact that sulphates in the lignite are not decomposing in the combustion process, as opposed to sulphides and organic sulfur [30]. The iron oxide (Fe_2O_3) has a significant share in the tested lignite ash, ranging from 2% in the Bełchatów deposit to 14.1% in the Adamów deposit. The ashes also contain manganese (MnO) and titanium oxides (TiO_3), but their share in the tested samples does not exceed 1%.

When it comes to the analyzed oxides, the content of Na_2O, P_2O_5, K_2O and MnO in the individual lignite ash samples is highly variable, which could also suggest that the estimation of their share in the ash using the ashes statistical methods will be difficult. The low variability can be observed only for the SiO_2 content in ash.

Among the tested lignite samples, the least favorable parameters regarding the tendency to slagging have been shown for the Adamów and Szczerców deposits, where a very high tendency to slagging has been observed.

The high and very high tendency to fouling of the heating surfaces with alkali components has been shown for ashes from the Gubin and Turów deposits.

5 Conclusions

The analysis has shown that the quantity and quality of mineral matter in lignite has a large impact on its suitability for clean coal technologies. According to a preliminary analysis of the selected parameters from several deposits in Poland, it can be stated that the ash content is a very variable parameter for both individual deposits and all of the tested lignite. The ash content is a parameter limiting the possibility of use of coal for the gasification processes. It is assumed that the ash content for the most promising technologies should not exceed 25%. Therefore, a significant portion of the resources has been classified as unsuitable for gasification purposes.

The analysis of ash melting temperatures allows concluding that the tested lignite can be successfully used for fluidized bed gasification. The ash melting point in all of the tested deposits is higher than 1100 °C, required for the fluidized bed gasification. The $CaCO_3$ content in some of the tested deposits is unfavorable for the gasification process. The aforementioned deposits occur mainly in the Bełchatów area, where large amounts of chalk lake - accompanying the lignite seams - can be observed. However, the majority of the tested lignite ash shows no tendency to fouling and slagging.

On the basis of the presented analysis it can be concluded that the Polish lignite is an excellent energy raw material for advanced coal technologies. However, due to the high variability of the key technological and chemical parameters, its quality must be constantly monitored.

Acknowledgements

This project was supported by the Polish National Science Centre under research project awarded by decision No. DEC-2013/09/D/ST10/04045 and Statutory Research No 11.11.140.320.

References

1. U.S. Environmental Protection Agency, *Clean Air Interstate Rule* www.epa.gov (2006)
2. U.S. Department of Energy, http://fossil.energy.gov/programs/powersystems/gasification/ (2010)
3. A. R. Shirazi, O. Börtina, L. Eklund, O. Lindqvist, Fuel **74** (1995)
4. J. C. Nankervis, R. B. Furlong, Fuel **55**, 4–8 (1980)
5. S. V. Vassilev, K. Kitano, S. Takeda, T. Tsurure, Fuel Processing Technology **45**, (1995)
6. A. P. Reifenstein, Kahraman H., C. D. A. Coin, N. J. Calos, G. Miller, P. Uwins, Fuel **78**, (1999)
7. B. Roga, L. Wnękowska, *Analiza węgla i koksu*. Wydawnictwo Naukowo-Techniczne, Warszawa (1966)
8. S. Su, J. H. Pohl, D. Holcombe, Fuel **82**, (2003)
9. L. Róg, *Wpływ budowy petrograficznej i chemicznej węgla kamiennego na temperaturę topliwości popiołu*. Prace Naukowe GIG, Górnictwo i środowisko, Kwartalnik nr **1** (2003)
10. H. E. Crossley, Journal of the Institute of Fuel **40**, (1967)
11. H. Ataköl, B. Hilmioğlu, E. Ekinci, Fuel Processing Technology **86**, (2005)
12. A. Filippidis, A. Georgakopoulos, A. Kassoli-Fournaraki, International Journal of Coal Geology **30**, (1996)
13. M. J. Keyser, M. Conradie, M. Coertzen, J. C. van Dyk, Fuel **85**, (2006)
14. T. F. Wall, G. S. Liu, H. W. Wu, D. G. Roberts, K. E. Benfell, Gupta S., J. A. Lucas, D. J. Harris, Progress in Energy and Combustion Sciences **28**, (2002)
15. Y. Sekine, K. Ishikawa, E. Kikuchi, M. Matsukata, A. Akimoto, Fuel **85**, (2006)
16. L. Johnson, *Relationship Between the Gasification Reactivities of Coal Char and the Physical and Chemical Properties of Coal and Coal Char.* http://www.anl.gov/PCS/acsfuel/preprint%20archive/Files/20_4_CHICAGO_08-75_0085.pdf. Institute, Palo Alto, California (1975)
17. G. Skodras, G. P. Sakellaropoulos, Fuel Processing Technology, **77-78**, (2002)
18. M. Sakawa, Y. Sakurai, Y. Hara, Fuel **61**, (1982)
19. A. G. Collot, International Journal of Coal Geology, **65**, (2006)
20. C. Higman, M. van der Burgt, *Gasification*. Second edition. Elsevier (2008)
21. M. Seggiani, Fuel **78**, (1999)
22. B. Alpern, J. Nahuys, L. Martinez, Communicacoes dos Servicos Geologicos de Portugal, Vol. **70**, (1984)
23. D. Simbeck, N. Korens, F. E. Biasca, S. Vejtasa, R. L. Dickenson, *Coal Gasification Guidebook Status, Applications, and Technologies.- Final report TR-102034.* Electric Power Research Institute, Palo Alto, California (1993)
24. J. R. Kasiński; M. Piwocki, Special Papers - Polish Geological Institute **7**, (2002)
25. B. Bielowicz, Fuel **96**, (2012)
26. M. Wagner, Węgiel Brunatny **4/57** (2006)
27. M. Wagner, International Journal of Coal Geology **2**, (1982)
28. M. Wagner, *Zmienność petrologiczno-sedymentologiczna i własności technologiczne kredy jeziornej w osadach neogenu typu wapiennego zapadliska tektonicznego na przykładzie złoża węgla brunatnego „Szczerców”,* AGH Uczelniane Wydawnictwa Naukowo-Dydaktyczne, Kraków (2007)
29. Z. Fąfara, K. Twardowski, J. Traple, K. Krochmal, *Statystyczne modele prognostyczne oparte na danych geofizyki wiertniczej służące do oceny wybranych własności węgla.* In: K. Matl, K. Twardowski (ed.)) Rozkład własności fizykochemicznych i technologicznych węgla brunatnego z uwzględnieniem domieszek szkodliwych dla środowiska w południowo-zachodniej części Niżu Polskiego (pokład łużycki II). Wyd. Centrum PPGSMiE PAN, Kraków (1996)
30. E. Stachura, T. Ratajczak, *Substancja mineralna w węglu brunatnym ze złoża Bełchatów (pole Bełchatów)*, Prace Geologiczne nr **153**, Wydawnictwo Instytutu Gospodarki Surowcami Mineralnymi i Energią PAN, Kraków (2005)

The Analysis of Heat Exchangers Geometry in Thermoelectric Generators for Waste Heat Utilization

Marcin Borcuch[1,a], Stanisław Gumuła[1], Michał Musiał[1] and Krzysztof Wojciechowski[2]

[1] *Department of Thermal Engineering and Fluid Flow, Faculty of Energy and Fuels*
[2] *Thermoelectric Technology Laboratory, Faculty of Materials Science and Ceramics, AGH University of Science and Technology, Al. A. Mickiewicza 30, 30-059 Cracow, Poland*

Abstract. The paper presents results of the analysis and comparison of the hot-side heat exchangers (HHXs) dedicated for the thermoelectric generators (TEGs). Efficient operation of TEG depends on, i.a. proper design of the HHX. Six geometries of the heat exchangers' cross-section have been investigated and analysed in view of heat transfer effectiveness (η_{TH}) and pressure drop (ΔP). As an assumption, useful heat exchange surface has been set up as 2400 cm^2, maintaining heat exchanger (HX) length as 30 cm, which is enough for the placement of the 32 thermoelectric modules able to generate at least 160 W of the electrical power. The source of waste heat are flue gases, in the analysis approximate as an air. Cold-side heat exchanger (CHX) has been simplified and calculated as a water flow around the casing of the HHX to achieve comparable results. As a base, circular profile has been presented. Numerical calculations provide results suggesting which shape is most suitable for specified application. Results could be the first guidelines for selecting and designing the HX for the TEG. Further investigation will focus on optimization of the chosen HX in view of increasing η_{TH} and minimizing ΔP.

1 Introduction

Increasing awareness of waste energy losses, especially waste heat, causes constant development in the recovery solutions. Among a wide range of waste heat harvesting solutions, thermoelectric generation seems to be competitive and with current progress in thermoelectric materials, TEG in close future could be a significant rival not only in waste heat recovery area but also in Renewables, between photovoltaic and wind energy. TEG is a device used for heat conversion into electrical energy. As a result of Seebeck phenomenon, due to temperature difference between hot and cold sides of thermoelectric module, heat flux is converted into electrical energy with the defined efficiency. The efficiency of a TEG as a whole device depends on both, the efficiency of the thermoelectric modules and the effectiveness of the HXs on the cold and on the hot sides. Big effort is put in development of TEGs to achieve more efficient constructions for different applications [1-6]. Location of the TEG varies from its usage for supplying residential boilers in electrical energy [1], projects of wood stoves able to generate heat, electricity and hot water [2], through cement rotary kilns with energy recovery unit [3], to automotive applications [4-6]. Especially the possibility of replacing the alternator in the cars seems to be one of the best economically justified locations of the TEG.

Despite big benefits from clean heat-into-electricity conversion in TEGs, there still is a big need to optimize the constructions of HXs and thermoelectric modules. HX in TEG can be treated as two separate elements, each for the hot and the cold side of thermoelectric module (Figure 1).

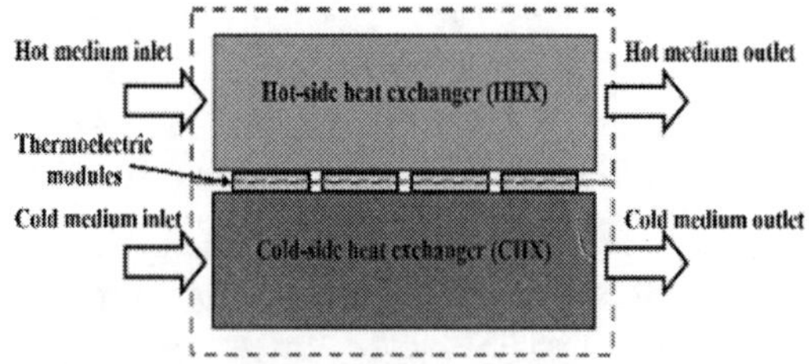

Figure 1. Scheme of the TEG.

In that approach, HHX is a crucial part, which is responsible for effective heat removal from the source. Most common applications of the TEG focus on waste heat harvesting from exhaust and flue gases, so the HHX should effectively remove heat from the gas and transport it to the module connected with the HX surface. High η_{TH} and minimal ΔP are the two most important parameters in choosing and designing of the HHX. Rectangular [7], square [8], hexagonal [9] as well as octagonal [10] shapes of HHX have been analysed so far by many research teams. Moreover, cost of manufacturing such HXs has to be taken into account in the designing process. HXs are often equipped with elements for increasing the heat transfer, like fins and insertions [11,12], however

[a] Corresponding author: borcuch@agh.edu.pl

comparison of pure geometries is necessary to relate exchangers to each other.

The paper presents the comparison and analysis of six HHX' profiles in view of heat utilization and pressure losses. Numerical analysis has been carried out with defined assumptions and obtained results have been interpreted. Square, rectangular, pentagonal, hexagonal, octagonal and as a reference circular shapes of HXs' cross sections were analysed and compared in view of η_{TH} and ΔP for varied mass flow rates.

2 Methodology

Commonly considered TEG's application is generation of electrical energy from waste heat contained in flue gases from domestic gas boiler [8]. Creation of self-powered heating unit, independent of the grid is a promising opportunity, especially for regions with unstable grid and frequently occurring blackouts. Decreasing energy demand of pumps and other boiler's equipment leads to lower of electrical energy consumption. For typical purposes, demanded electrical power of a gas boiler is in a range from 80 W to 160 W. Thus, TEG dedicated for this application should generate such power. Furthermore, usage of waste heat from flue gases will improve the overall efficiency of the system. Proper choose and design of the HHX is crucial for the effective heat utilization from flue gases. The numerical analysis of selected geometries of HHXs have been performed in ANSYS CFX. To obtain comparable results, pure aluminium profiles of 2 mm thickness were set as HXs. Flue gases were simulated by hot air, whole lateral surface has been cooled by water flow around the casing to avoid dissimilarities caused by the design and mismatching of the CHX.

2.1 Assumptions

TEG consists of the HHX, the CHX and the thermoelectric modules. To generate 160 W, 32 modules are required in the TEG. Modules are placed between HHX and CHX. Assuming dimensions of the module as a 5 cm x 5 cm square, 800 cm² of the exchanger will be covered by the modules. Including spacing and insulations, the overall heat exchange area, which is equal to the lateral surface of the HHX is set as 2400 cm² (Table 1). Taking into account the length of the device, which should not be too high because of the possibility to mount it (30 cm), the cross-section surface areas can be calculated for all shapes.

Table 1. Analysed cross-sections of heat exchangers.

Profile	Cross-section shape and the lengths of the sides/diameter	Inlet or outlet surface area[1]	Heat transfer surface area[2]
Square	20 (cm)	400 cm²	2400 cm²
Rectangle	5, 40	200 cm²	2400 cm²
Pentagon	16	440.4 cm²	2400 cm²
Hexagon	13.35	463 cm²	2403 cm²
Octagon	10	482.8 cm²	2400 cm²
Circle	25.5	510.5 cm²	2402.1 cm²

[1] Cross-section surface area, [2] Lateral surface of the exchanger.

2.2 Mesh generation

All meshes were created in the same way. Water surrounding the casing and air filled it were inflated near the boundary to obtain better solution. Average orthogonal quality of the mesh was in the range from 0.84 to 0.89 (Table 2), which is a very good result. Generated mesh for each case consists of three main parts (Figure 2):

- aluminium HX,
- air inside the HX,
- cooling water around the HX.

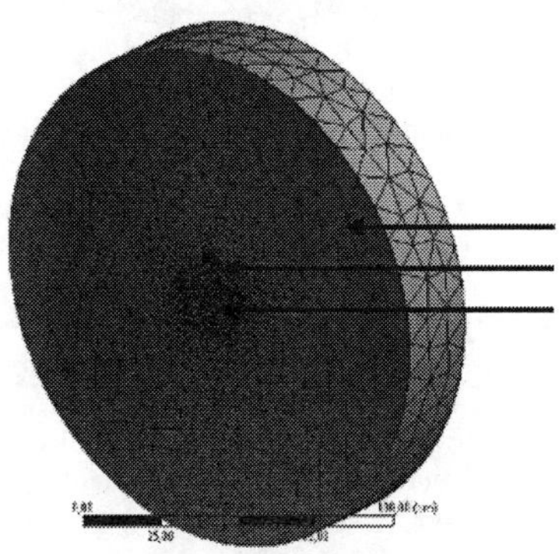

Figure 2. Main parts of generated mesh.

Table 2. Mesh quality.

Profile	Nodes	Elements	Average orthogonal quality
Square	50536	151943	0.85
Pentagon	54138	162428	0.87
Hexagon	66545	241658	0.89
Rectangle	44932	124483	0.84
Circle	62664	211330	0.89
Octagon	56081	190280	0.88

 DOI: 10.1051/e3sconf/20161000003

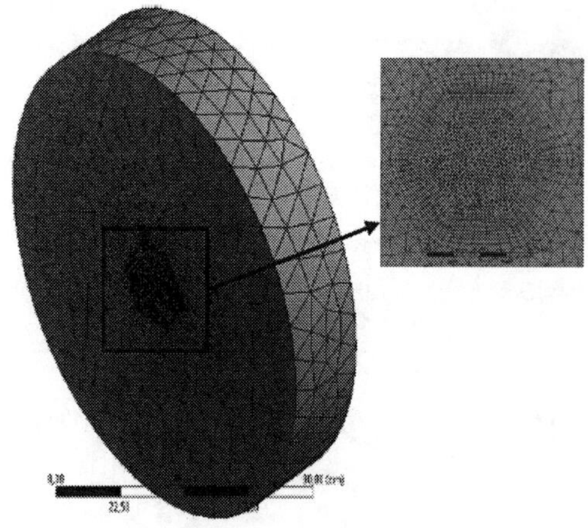

Figure 3. Generated mesh for analysis of hexagonal shape HX.

Mesh close to HX, both, on the air and the water sides has been refined to obtain more accurate results in this volume (Figure 3). Departing from the aluminum profile, temperature gradient is lower, so the density of the mesh also decreases.

2.3 Solver settings and calculated parameters

All analysed cases have been set with the same settings and boundary conditions:

- air ideal gas,
- inlet air temperature = 200°C,
- inlet water speed = 1 m/s,
- inlet water temperature = 25°C.

Spatially averaged governing equations for mass, momentum and energy were solved on control volumes to obtain temperatures and ΔP for fluid domain. Moreover, energy equation was solved for casing due to heat conduction in solid domain. Turbulence modelling was performed with Shear Stress Transport RANS model. The residual target was set as 10^{-5} to obtain accurate results. The same simulation was performed for each profile for varied mass flow rate, due to investigate their performance within typical and high flue gases mass flow rate values. It is related to domestic gas boilers and industrial devices to expand analysed scope. In the range from 0.01 kg/s to 0.1 kg/s steady-state simulation has been calculated with the interval of 0.01 kg/s, for the range from 0.1 kg/s to 1 kg/s, interval was equal to 0.1 kg/s.

Calculated parameters in all simulations were as following:

- area average ΔP between the inlet and the outlet of air flow,
- air outlet temperature (decrease due to water cooling),
- heat flux through lateral surface of the HX.

3 Results

3.1. Heat transfer effectiveness

The η_{TH} is estimated by two approaches:

- by defining the hot gas outlet temperature (Figure 4), where the difference between inlet and outlet temperatures is caused by the energy transfer from the air to the cooling water.
- By calculating the average heat flux through the heat exchange surface (Figure 5, Figure 6). Obtained values are area averaged.

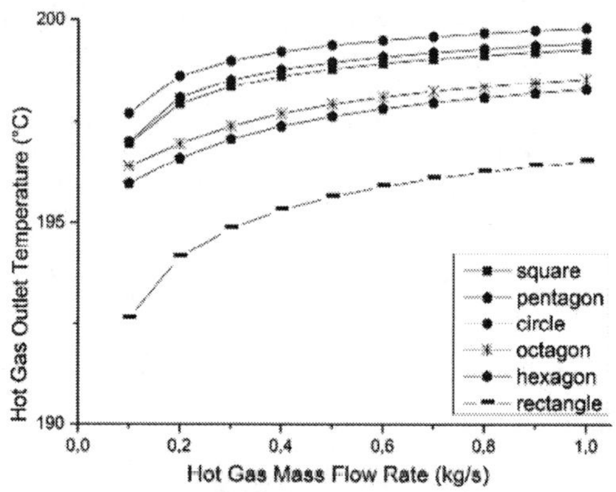

Figure 4. Hot gas outlet temperatures.

The dependency between mass flow rate of hot gas and average temperature at the outlet surface is similar for all profiles. The most effective heat transfer occurs in the rectangle HX. With increasing mass flow rate, the temperature difference between the inlet and the outlet of HX becomes smaller, the hot fluid is not able to cool off on 30 cm length of the HX. Although, the average heat flux increases with the increment of mass flow rate (Figure 5, Figure 6).

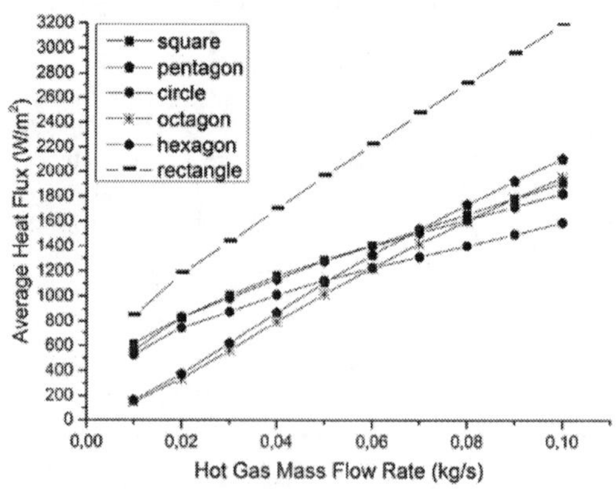

Figure 5. Area averaged heat fluxes.

The dependency between average heat flux and mass flow rate is similar for flows in all ranges of mass flow rates. Water temperature is stable, so the temperature difference is increasing. The best η_{TH} is obtained

 DOI: 10.1051/e3sconf/20161000003

for rectangular HX. Its domination is clearly seen is all range of flows. Omitting rectangular profile, for small flow rates, between 0.01 to 0.05 kg/s, square and hexagonal HX present the highest values of energy transferred from the flue gases to the coolant (Figure 5). For higher flow rates, pentagonal and octagonal HX have better heat transfer characteristics. Circular HX, calculated as a reference case, presents the worst results in view of heat transfer possibilities.

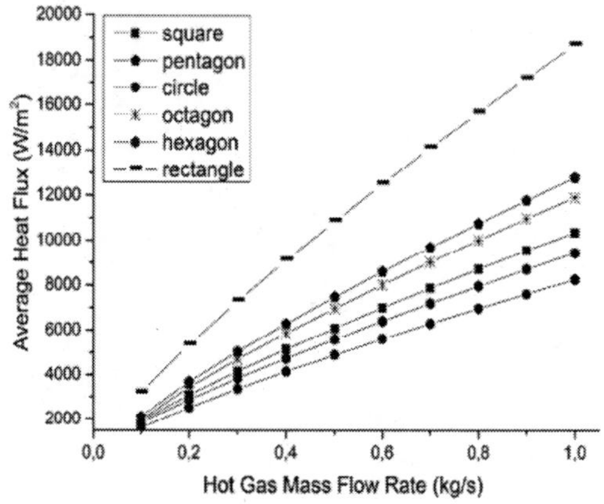

Figure 6. Area averaged heat fluxes.

Comparing outlet temperatures and average heat fluxes, the order of profiles is the same, starting from the most efficient HX: rectangular, pentagonal, octagonal, square, hexagonal and circular.

3.2 Pressure drop

Application of the TEG in the installation with the gas boiler will cause the occurence of the additional pressure losses. HX with the lenght equal to 30 cm can be treated as a local resistance with specfied ΔP, dependent on gas flow rate. Calculated charasteristics (Figure 7, Figure 8) presents the connection between ΔP and hot gas mass flow rate for all profiles.

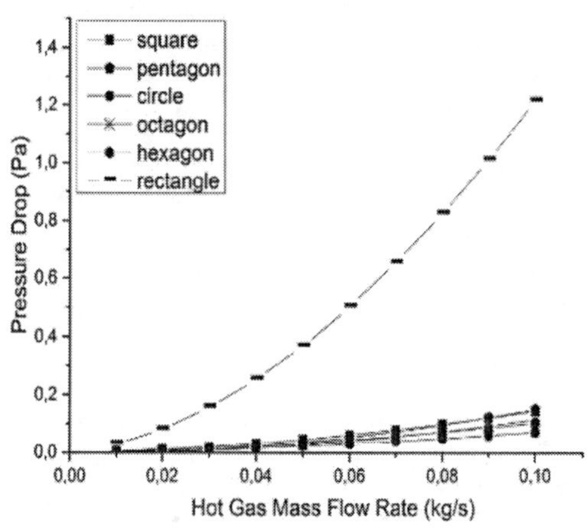

Figure 7. Pressure drop in the HX.

Pressure losses increase with the increment of the mass of the gas flowing through the HX. Big difference is visible between ΔP in rectangular HX and the others. Much larger losses are caused in the installation with the application of such element. This dependency is the same for all flow rates. The smallest ΔP is caused by the circular HX, which simulaniously validates presented data, because circular pipes have the best flow characteristics. For square, pentagonal, hexagonal and octagonal HX, pressure losses are relatively small.

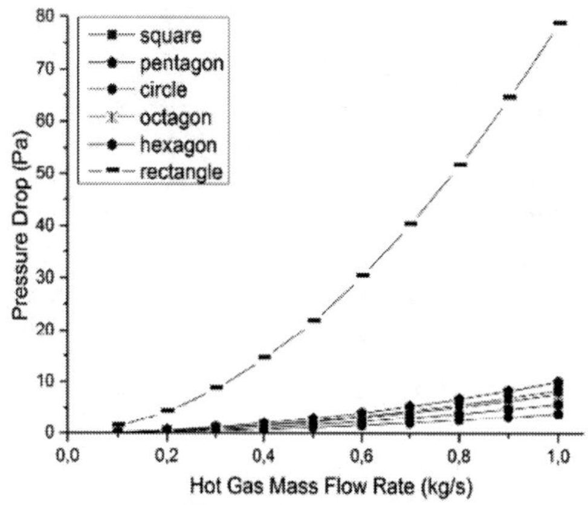

Figure 8. Pressure drop in the HX.

The highest heat transfer effectiveness but also the highest pressure losses characterise the rectangular HX. To compare all shapes, taking into account both factors, an additional analysis is necessary.

The heat transfer surface area is estimated as 2400 cm^2. The area, covered by 32 thermoelectric modules is equal to 800 cm^2. The heat flux, flowing through the thermoelectric modules will be converted into electrical energy with specified efficiency, according to the type and model of the module. The only input is the additional power, needed to overcome the power losses caused by ΔP. The net power, generated in each HX can be calculated as:

$$P_{TEG} = P_{HF} \cdot \eta_{TE} - P_{\Delta P} \quad (1)$$

$$P_{\Delta P} = \Delta P \cdot \dot{V} \quad (2)$$

P_{TEG} net power generated in the TEG [W]
P_{HF} thermal power transferred from the hot gas to the coolant through the thermoelectric modules [W]
η_{TE} efficiency of thermoelectric modules [-]
$P_{\Delta P}$ power losses due to pressure drop [W]
$\dot{V}$ volumetric flow rate [m^3/s]

Usage of commercially available modules with the efficiency about 5% for analysed HXs leads to obtaining not satisfactory results in the net power (Figure 9). High flow rates are necessary to generate more than 30 W of electrical power. In the case with rectangular HX, in some point, losses become greater than generated energy and application of that kind of TEG results only in net losses.

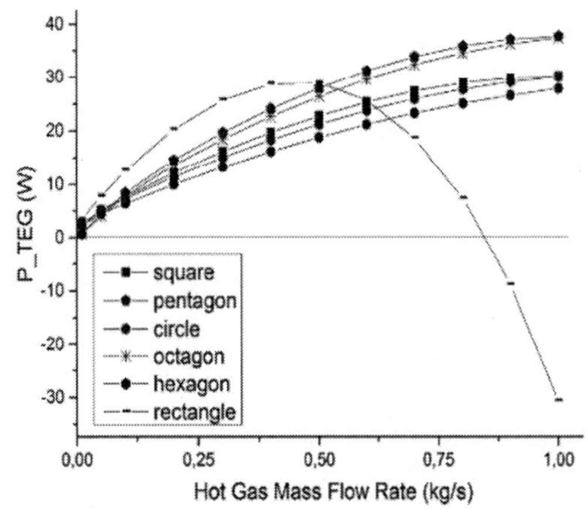

Figure 9. Net power generated in the TEG, η_{TE}=5%, surface covered by thermoelectric modules in 33%.

Usage of the modules with newly developed materials, with the average efficiency about 10% benefits in the increase of net generated power (Figure 10). Surface equal to 800 cm^2, covered by the thermoelectric modules allows to supply equipment with small energy demand.

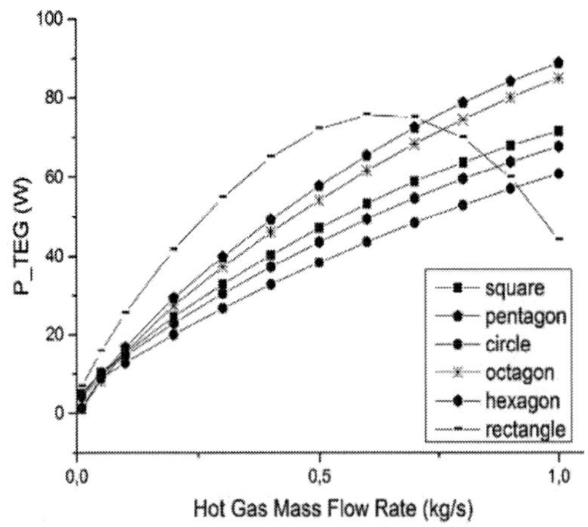

Figure 10. Net power generated in the TEG, η_{TE}=10%, surface covered by thermoelectric modules in 33%.

Covering the entire surface with thermoelectric modules significantly increases net generated power (Figure 11, Figure 12).

Usage of the modules with the efficiency about 10% allows to generate more than 300 W of electrical energy, which is sufficient to supply gas boiler as well as another electrical devices.

Besides the rectangular HX, in which despite the highest heat transfer effectiveness, the power losses have to be taken into account, the pentagonal and octagonal profiles present the best net generated power characteristics.

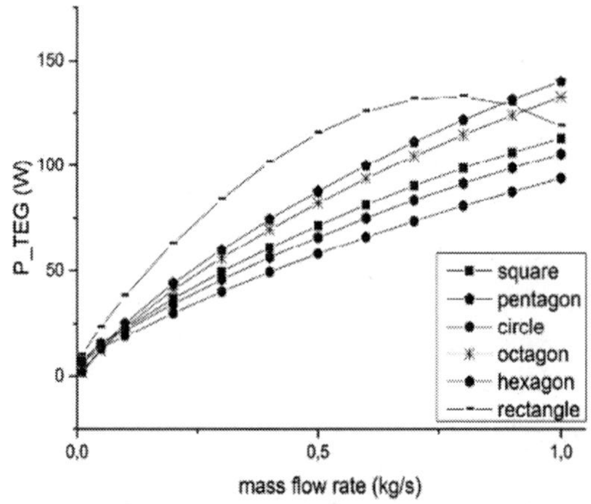

Figure 11. Net power generated in the TEG, η_{TE}=5%,surface covered by thermoelectric modules in 100%.

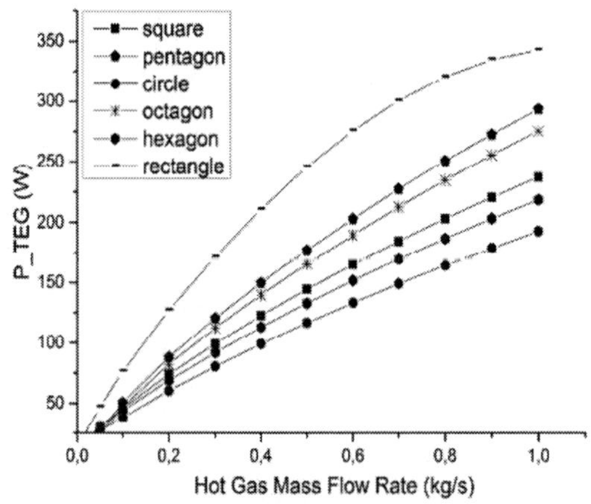

Figure 12. Net power generated in the TEG, η_{TE}=10%, surface covered by thermoelectric modules in 100%.

Starting of the outlet temperature, through the pressure drop, to area averaged heat transfer, the order of profiles remains the same. Rectangular HX presents the highest heat transfer effectiveness and the biggest pressure losses. It is followed by pentagonal, octagonal, square, hexagonal and circular profiles. The order of profiles obtained in the analysis is not consistent with the order of profiles determined by increasing or decreasing the number of sides. Presented data in next step will be verified experimentally. Moreover, the CHX will be included into analysis.

4 Conclusions

Numerical analysis of six HXs profiles has been carried out. The net generated power, taking into account losses due to pressure drop is presented for varied mass flow rates. The highest η_{TH} is obtained for the rectangular HX. Although, taking into account pressure losses, this profile is applicable only for high mass flow rates of hot medium. The circular HX, which has been calculated as a reference, presents the worst characteristics. Besides the others profiles, pentagonal HX seems to be optimal in view of η_{TH} and ΔP.

Analysed profiles were clear, without any insertions and fins. For increasing the η_{TH}, some modifications are required. Usage of fins will increase ΔP and simultaneously improve η_{TH}, so the optimal configuration should be specified and verified. Also other modifications will be implemented, maintaining ΔP as low as possible.

Acknowledgments

This study has been supported by the statutory funds of AGH UST, No. 11.11.160.438 and grant No. 15.11.210.325.

References

1. K.Qiu, A.C.S Hayden, J Electron Mater, **40**, 5, 606-610 (2011)
2. A. M. Goudarzi, P. Mazandarani, R. Panahi, H. Behsaz, A. Rezania, L. A. Rosendahl, J Electron Mater, **42**, 7, 2127-2133 (2013)
3. Q. Luo, P. Li, L. Cai, P. Zhou, D. Tang, P. Zhai, Q. Zhang, J Electron Mater, **44**, 6, 1750-1762 (2015)
4. Y. Y. Hsiao, W. C. Chang, S. L. Chen, Energy, **35**, 3, 1447–1454 (2010)
5. Y. Wang, C. Dai, S. Wang, Appl Energ, **112**, 1171–1180 (2013)
6. J. LaGrandeur, *Vehicle fuel economy improvement through thermoelectric waste heat recovery*. DEER conference, Chicago, 2005
7. X. Liu, Y. D. Deng, Z. Li, C. Q. Su, Energ Convers Manage, **90**, 121–127 (2015)
8. A. Martinez, J. G. Vian, D. Astrain, A. R. Guez, I. Berrio, J Electron Mater, **39**, 9, 1463-1468 (2010)
9. K. T. Wojciechowski, J. Merkisz, P. Fuć, J. Tomankiewicz, R. Zybała, J. Leszczyński, P. Lijewski, P. Nieroda, Combustion Engines, **154** (3), 60-71 (2013)
10. X. Wang, S. Wang, L. Zhou, X. Li, Adv Mat Res, V. **986-987**, p.848 (2014)
11. M. Musiał, M. Borcuch, K. Wojciechowski, J Electron Mater, **45**, 3, 1517-1522 (2016)
12. X. Liu, Y.D. Deng, K. Zhang, M. Xu, Y. Xu, C.Q. Su, Appl Therm Eng, **71**, 364-370 (2014)
13. J. Yu, H. Zhao, J Power Sources, **172**, 428–434 (2007)
14. C. Lu, S. Wang, C.Chen, Y. Li, Appl Therm Eng, **89**, 270-279 (2015)
15. L, Chen, J. Gong, F. Sun, C. Wu, Int J Therm Sci, **41**, 95–99 (2002)
16. H. Lu, T. Wu, S. Bai, K. Xu, Y. Huang, W. Gao, X. Yin, L. Chen, Energy, **54**, 372-377 (2013)
17. J. Esartea, G. Minb, D. M. Rowe, J Power Sources, **93**, 72-76 (2001)
18. M. Hatami, D. D. Ganji, M. Gorji-Bandpy, Renew Sust Energ Rev, **37**, 168–181 (2014)
19. W. He, G. Zhang, X. Zhang, J. Ji, G. Li, X. Zhao, Appl Energ, **143**, 1–25 (2015)

Performance analysis of the Single-Phase Grid-Connected Inverter of a photovoltaic system in water and wind applications

Dariusz Borkowski[1,a]

[1]*Institute of Electromechanical Energy Conversion, Cracow University of Technology, Krakow 31-155, Poland*

Abstract. Single-phase grid connected inverters are nowadays broadly developed and tested in various types of applications especially in photovoltaic systems. The main aim of the inverter control strategy is to extract the maximum energy from the PV system which corresponds to the maximum power at certain conditions. However, the MPPT methods are also important in other renewable energy conversion systems. This paper analyses the performance of a commercially available photovoltaic inverter in water and wind systems. Presented models are implemented in a laboratory test bench in the form of torque characteristics realised by an induction motor fed by the inverter with vector control. The parameters are scaled into relative variables to provide a proper performance comparison. Presented tests include step response to assess the performance of a system dynamic. The dynamic tests showed a fast response of the investigated systems. The MPPT tracking accuracy tested under realistic profiles is similar for both cases: 98% and 96% respectively for the wind and water systems. These results prove the satisfactory performance of the MPPT of the PV microinverter in these applications.

1 Introduction

Nowadays photovoltaic (PV) system technologies become decentralized where each PV module has its own inverter. This concept provides the optimal operation of an individual module and supports the easy change of the system topology [1]. The low power single-phase inverter which is connected to the grid is called the ac-module microinverter [2].

The topology of the microinverter is categorized firstly depending on the power processing stages into single- and multiple stages what is defined by the place of the decoupling capacitor. Next grouping is distinguished by the isolation transformer which can work at line frequency (LFT) or high frequency (HFT). In practice the HFT is used due to the low size and costs. From the grid side, the inverter target is to inject the sinusoidal current thus it operates as a current source [3].

The main aim of the inverter control strategy is to extract the maximum energy from the PV system which correspond to the maximum power at certain conditions. Due to the environmental variations the control techniques have to track the maximum power point (MPP) with a high efficiency [4]. In general two types of the MPPT strategies may be defined: indirect and direct control. The indirect strategies use the database with characteristic curves as a function of irradiances and temperatures for the specific PV type. The direct methods base on the various types of seeking methods without any prior knowledge of the PV features [5]. Usually PV microinverters need the simple and cheap MPPT algorithms therefore the most popular are direct techniques ex.: the hill climbing method or the perturb and observe (P&O) techniques.

However, the MPPT methods are important also in other renewable energy conversion systems. Wind turbines use these techniques to optimize the turbine speed under the various wind velocities. The indirect methods called the lookup table use the database or pre-known relationship to track the maximum power or maximum torque. Some variants ex. the tip speed ratio method, requires a wind speed sensor to provide the fast control action. The direct methods are similar to the MPPT of PV systems. The most popular is the hill climbing search (HCS) control which may be implemented by perturbing the control parameters and observe the output power (P&O) [6]. Extended versions engage the artificial neural network (ANN) effective after comprehensive training and/or the fuzzy logic controller which should be designed to the specific system. The wind speed fluctuations may result in inefficiency caused by slow response of the P&O methods. This problem may be solved by using the various control parameters for perturbing [7].

Another renewable energy source which may gain advantages from MPPT techniques is a water turbine. Variable speed techniques which become popular in hydropower solutions [8-10] create opportunity to implement the adaptive methods [11]. The scientific literature is poor and limited mainly to the "run of the river" small hydropower plants (SHP). Here the controller has two main functions: provide the possible highest conversion efficiency and maintain upper water on the constant level, what results in a maximal energy

[a] Corresponding author: dborkowski@pk.edu.pl

 DOI: 10.1051/e3sconf/20161000004

production. These tasks may be integrated in the one controller [12] or realized by the two separate governors [13]. However, especially attractive types of small hydropower are the reservoir plants working with the variable speed where the MPPT technique may be easily implemented.

The presented three types of renewable energy sources differ significantly due to the control variables and time constants. In spite of this the similar control techniques may be used. The MPPT direct methods seem to be the most appropriate.

This paper presents the possibility and analysis of the energy conversion effectiveness of the low power wind turbine and the reservoir SHP that use the single-phase grid-connected inverters designed for photovoltaic systems. The wind and water turbines are modelled by the torque characteristics and are implemented in the laboratory test bench of 1 kW nominal power. The commercially available PV inverter is used.

2 Wind turbine model

The mechanical torque produced by a wind turbine may be defined as follows [7].

$$T_w = \frac{1}{2\omega_w} \cdot \rho_w \cdot C_p \cdot A \cdot v_w^3 \tag{1}$$

where:

ω_w - turbine angular velocity (rad/s),

ρ_w - air density (kg/m³),

C_p - power coefficient (-),

A - wind turbine swept area (m²),

v_w - wind speed (m/s).

The power coefficient is a nonlinear function of tip speed ratio λ as well as the turbine pitch angle β and can be approximated by expression (2) [14].

$$C_p = 0.5176 \cdot \left(\frac{116}{\lambda_i} - 0.4 \cdot \beta - 5 \right) \cdot e^{\frac{21}{\lambda_i}} + 0.0068 \cdot \lambda \tag{2}$$

$$\lambda_i = \left(\frac{1}{\lambda + 0.08 \cdot \beta} - \frac{0.035}{\beta^3 + 1} \right), \quad \lambda = \frac{R \cdot \omega_w}{v_w}$$

where: R - wind turbine radius (m).

To simplify the analysis let assume that the pitch angle is constant and equals 0. The power coefficient curve is presented in Fig. 1a. Taking the reference values: $C_{p\max}$ - maximal C_p coefficient, λ_{opt} - optimal tip speed ratio as relative variables, the power coefficient curve is scaled as presented in Fig. 1b

The relative torque of the wind turbine can be calculated from formula (3) and it is shown in Fig. 2a at different wind speeds.

$$T_w^r = \frac{1}{\omega_w^r} \cdot C_p^r \cdot v_w^{r3}, \tag{3}$$

where: $C_p^r = \frac{C_p}{C_{pN}}$, $\omega_w^r = \frac{\omega_w}{\omega_{wN}}$, $v_w^r = \frac{v_w}{v_{wN}}$.

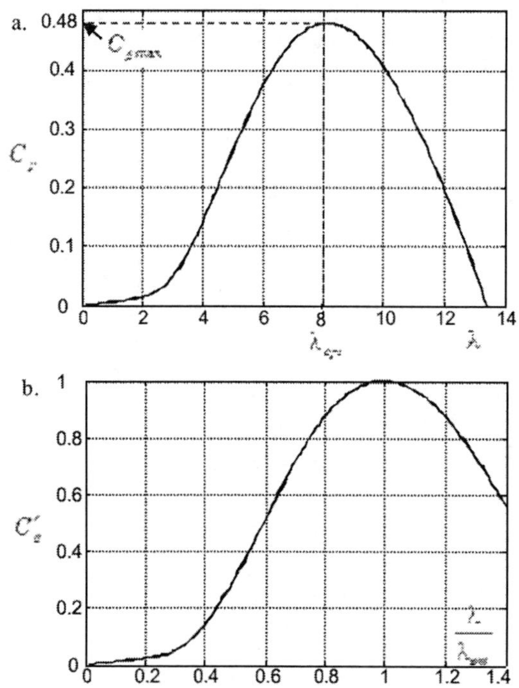

Figure 1. Power coefficient curve: a.) natural units b.) relative variables.

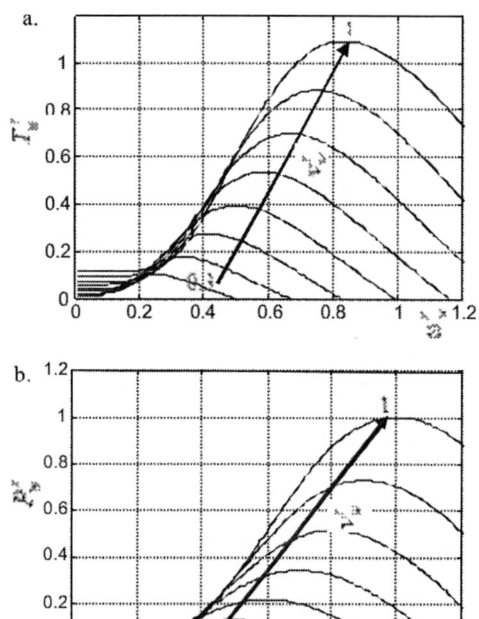

Figure 2. Wind turbine characteristics: a.) torque b.) power, versus relative angular velocity curves at different wind speeds.

3 Water turbine model

The water turbine features depend significantly on its construction. It performance is visualised usually on the water flow – turbine velocity plane (known as a hill chart) which presents isolines connecting the points that have the same efficiency. The example hill chart of the propeller turbine which is often used in SHPs is presented in Fig. 3a. Basing on this relation the torque function may be determined by (4) and approximated by the function (5) (Fig. 3b) [13].

$$T_h = \frac{9.81 \cdot \rho_h \cdot Q \cdot H \cdot \eta}{\omega_h} \tag{4}$$

$$T_h = C_H \cdot \left(a(\alpha) \cdot \omega_h^2 + b(\alpha) \cdot \omega_h + c(\alpha) \right) \tag{5}$$

where: ρ_h - water density (kg/m^3),
Q - water flow (turbine discharge) (m^3/s),
H - difference of water level between the upstream level and the downstream level (water head) (m),
η - turbine efficiency (-),
ω_h - water turbine angular velocity (rad/s),
C_H - water level coefficient (-),
$a(\alpha), b(\alpha), c(\alpha)$ - function coefficients (-),
α - guide vane angle (-).

Taking the nominal parameters: Q_N, H_N, η_N, α_N, ω_{hN} as reference values the function coefficients may be approximated by a polynomial function of the guide vane angle (6).

$$\begin{aligned} a(\alpha) &= -0.44423\alpha^3 + 2.32343\alpha^2 - 3.12\alpha \\ b(\alpha) &= 3.6548\alpha^5 - 13.58\alpha^4 + 15.256\alpha^3 - 5.552\alpha^2 + 2.094\alpha \\ c(\alpha) &= -2.4484\alpha^5 + 10.09\alpha^4 - 13.493\alpha^3 + 5.872\alpha^2 + 0.347\alpha \end{aligned} \tag{6}$$

The analysis presented in this paper concerns only the reservoir SHP under the changeable water head. To simplify the analysis the guide vane angle will be set to 100%. The water head influences the turbine torque through the water level coefficient and the turbine velocity which relative formula is given in (7).

$$T_h^r = C_H^r \cdot \left(a(\alpha) \cdot \omega_h^{r2} + b(\alpha) \cdot \omega_h^r + c(\alpha) \right) \tag{7}$$

$$C_H^r = \left(\frac{H}{H_N} \right)^{\frac{3}{2}}, \quad \omega_h^r = \frac{\omega_h}{\omega_{hN}} \sqrt{\frac{H}{H_N}}$$

The curves of turbine torque in the relative values are shown in Fig. 4a

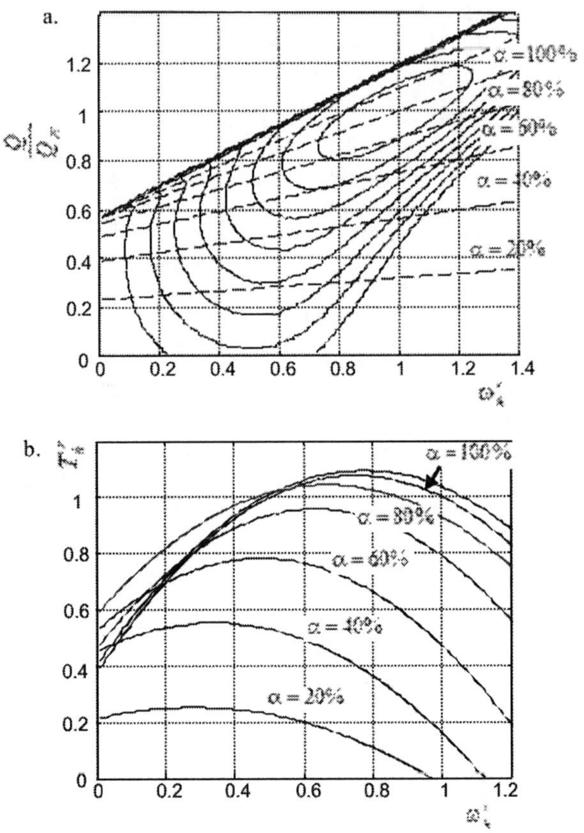

Figure 3. Propeller turbine characteristics: a.) hill chart (solid – efficiency isolines, dashed – guide vane angles), b.) turbine torque curves at different guide vane angles and nominal head.

4 Tests

4.1 Inverter description and MPPT dynamic test

The investigated single-phase microinverter is the commercially available Soladin 600 with parameters listed in table 1. The analysis of the topology showed that it is a dual-stage topology without inherent

Table 1. Chosen inverter parameters given by the Mastervolt

Parameter of input (solar)	Value	Parameter of output (grid)	Value
Nominal power at 25°C	535 W	Net voltage	230 V
MPP voltage	40-125 VDC	Maximum current	2.25 A
MPP tracker	> 99% MPP	Power factor	0.99
Nom. rated current	8 A	European efficiency	91%

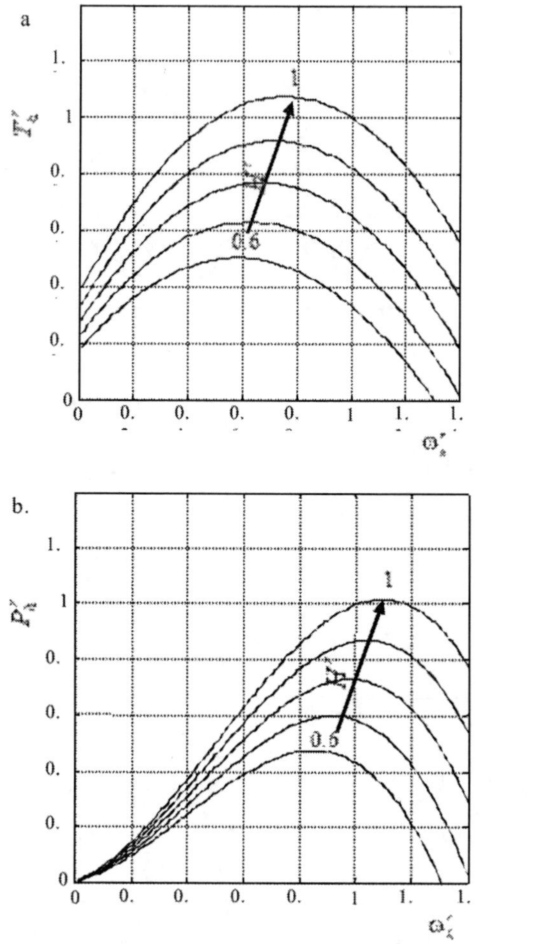

Figure 4. Water turbine characteristics: a.) relative torque b.) relative power, versus relative turbine angular velocity at different water heads.

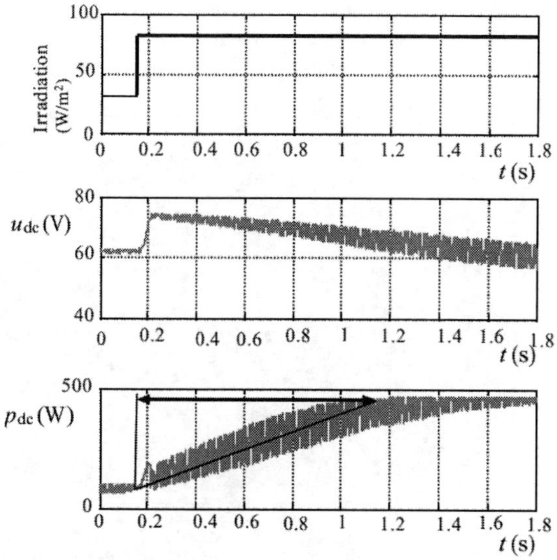

Figure 6. Dynamic test of inverter MPPT: a.) solar irradiation, b.) voltage across the PV modules, c.) generated DC power, for step-change of the irradiation.

power decoupling, which is similar to the Soladin 120 described already in the literature [3,15] (Fig. 5). The capacitor connected in parallel to the PV system work as an energy buffer.

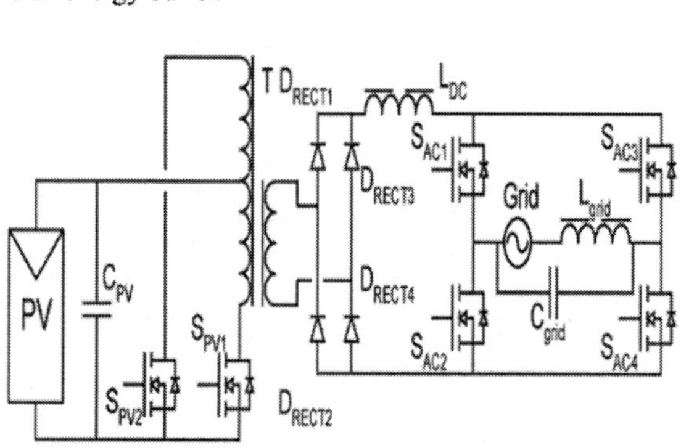

Figure 5. Topology of the microinverter Soladin 120 [3,14].

The dynamic tests of the inverter coupled with the PV panels (two Q-cells 250W connected in series) showed its fast MPPT response to the step change of the irradiance (Fig. 6). The time constant may be estimated at about 1second.

4.2 Test bench description

The modelled torque of the wind turbine and the water turbine are realised by an induction motor (IM) fed by the inverter with the vector control. The PLC controller calculates the actual torque from relation (3) or (7) as a function of the angular generator velocity using reference values listed in Table 2. The control signals (velocity and torque) are transmitted using analogue voltage signals what increase the system dynamic. The longest time constant (about 1s) is introduced by the inverter controller and the induction motor inertia. The motor is coupled by the elastic clutch with the permanent magnet synchronous generator (PMSG). The nominal parameters of the machines are shown in Table 2. In order to combine the PMSG with the microinverter the diode rectifier bridge is used (Fig. 7).

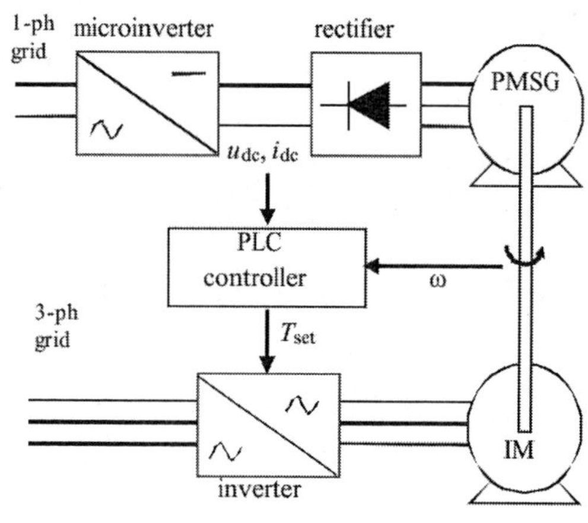

Figure 7. Energy conversion system structure of the laboratory test bench.

In this topology the DC current is proportional to the PMSG torque while the DC voltage is a linear relation of the generator velocity in a steady state operation [16].

Figure 8. Laboratory machines – permanent magnet synchronous generator coupled with induction motor of 0.8 kW.

The diode bridge introduces additional frequencies into the dc current (Fig. 9). Besides the 100Hz component generated by the microinverter MPPT control [17] which is visible in Fig. 9a, the current spectrum contains components at sixth multiples of the motor voltage frequency (Fig. 9b). This may affect the MPP tracking accuracy The current disturbances caused by the diode rectifier do not affect the key parameters of the microinverter. The conversion efficiency is above the 91% in the whole range of measured power (Fig. 10). power conditions become significant thus the power factor exceeds value 0.95 above the 40% of the nominal power.

Table 2. Nominal machine parameters and reference values

PMSG parameters		IM parameters		Reference values	
P_N	3.5k W	P_N	0.8 kW	ω	104,7 rpm
I_N	11 A	I_N	2.5 A	P	500 W
eff	111V/1000rpm	n_N	1420 rpm	I_{dc}	6 A
T_N	12 Nm	U_N	400 V		

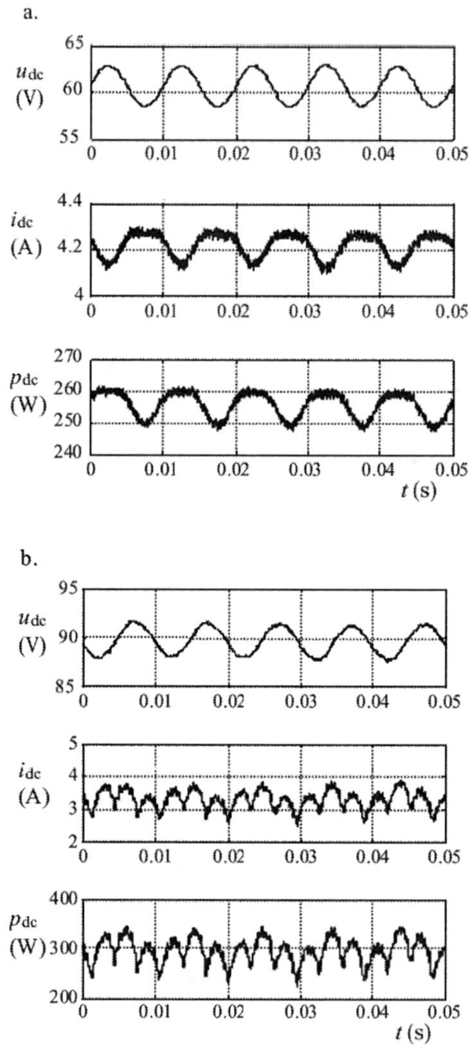

Figure 9. Time domain waveforms of DC signals of microinverter supplied by: a.) PV system b.) PMSG with diode rectifier.

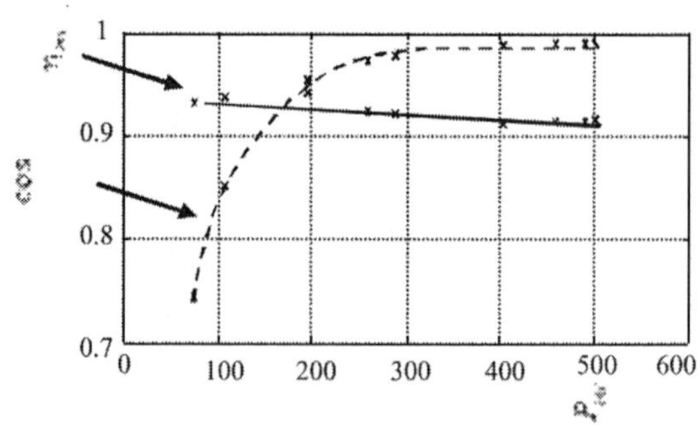

Figure 10. Performace of microinverter supplied by PMSG through diode bridge rectifier: a.) efficiency (solid line), b.) power factor (dashed line)

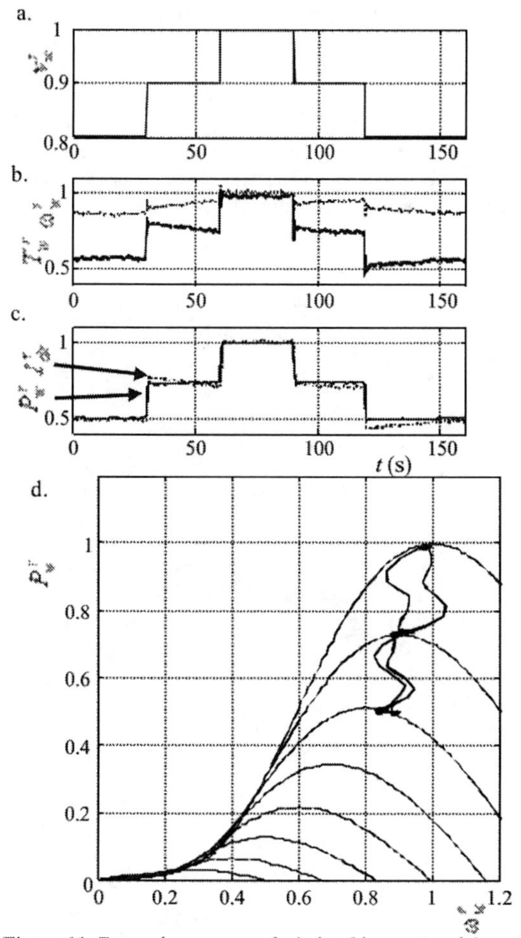

Figure 11. Dynamic response of wind turbine system: b.) angular velocity and turbine torque, c.) dc current and output power, d.) output power – angular velocity plane, to the step changes of wind speed a.).

4.3 Tests of wind turbine system

The operation effectiveness of the PV microinverter in the wind turbine system is assessed basing on the two tests. The dynamic performance is checked by stepping up and down the wind speed (Fig. 11).

The two stages of the MPPT algorithm are visible in every step change. Firstly, the step change of the wind v_w^r causes the fast turbine torque T_w^r modification what results in the speed ω_w^r variation. The tracking algorithm reacts immediately by the dc current I_{dc}^r modification to the voltage gradient of the synchronous generator which is proportional to the turbine speed. The tracking speed is satisfactory under the wind speed fast changes. The second stage is the seeking process of the MPP basing on the P&O method. The power curves near the MPP are plane for the constant wind speed thus the control process results in the small power changes in these areas (Fig. 11c).

The system efficiency is validated under a realistic wind profile (Fig. 12a). The power coefficient C_p^r (Fig. 12b) is affected by the wind speed changes especially at low wind speed values where the power curves are flatten.

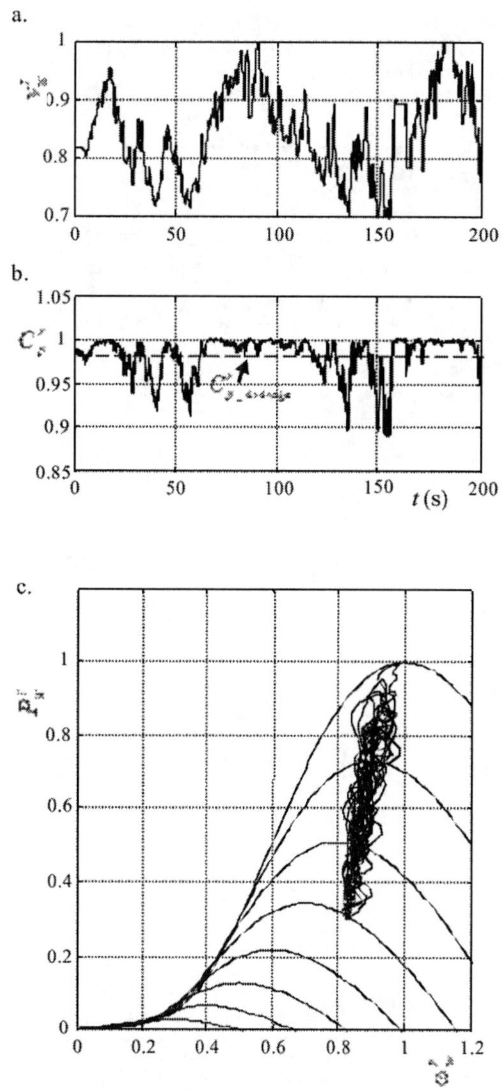

Figure 12. System performance under realistic wind profile: a.) wind speed, b.) power coefficient, c.) power versus angular velocity, in relative values.

The average power coefficient in a relative scale corresponding to the MPP tracking accuracy (8) [18] equals 0.983. The value of the average power coefficient equals 0.472 in a natural unit.

$$\eta_{MPPT} = \frac{P^r}{P_{MPP}^r} = C_p^r \tag{8}$$

where:

P^r - relative output power

P_{MPP}^r - relative power for maximum power point

4.4 Tests of water turbine system

The performance analysis of the PV microinverter in the water turbine system is evaluated similarly to the previous case. Here the water head H' is the variable parameter. Its step changes cause the two stages reaction of the MPPT algorithm. The dynamic response to the stepping up and down of the input variable is presented in Fig. 13.

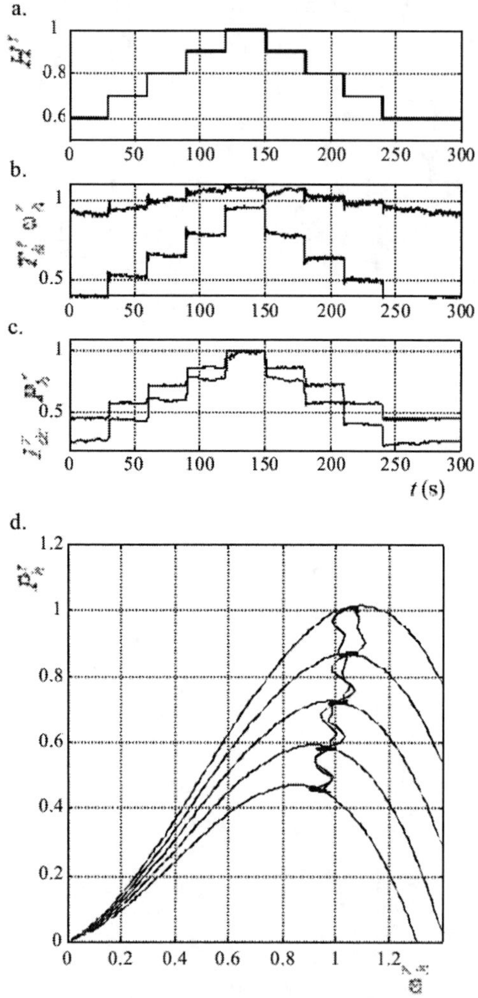

Figure 13. Dynamic response of water turbine system: b.) angular velocity and turbine torque, c.) dc current and output power, d.) output power – angular velocity plane, to the step changes of water head a.).

The propeller water turbine is characterized by the relatively small influence of the water head variation into the output power defined by (7). That is way the same step changes of the variable parameter (each by 10% of the relative value) results in the smaller power modifications than in the wind system. This decreases the MPPT seeking time.

The realistic water profile used for the system test (Fig. 14a) presents the water head changes in the reservoir SHP. The gradient of the water head depends on the tank capacity, turbine discharge and tank supply volume. Some fluctuations result from the water surface waving and the sensor accuracy. Slow changes and low amplitude of the variations improves the operation curve on the output power – angular velocity plane (Fig. 14c). The average MPP tracking accuracy defined by (8) equals 0.957 (Fig. 14b). This result is worse than for the wind system about 2.5%. The difference of the MPPT effectiveness is caused by the profile of input function and the power function shape.

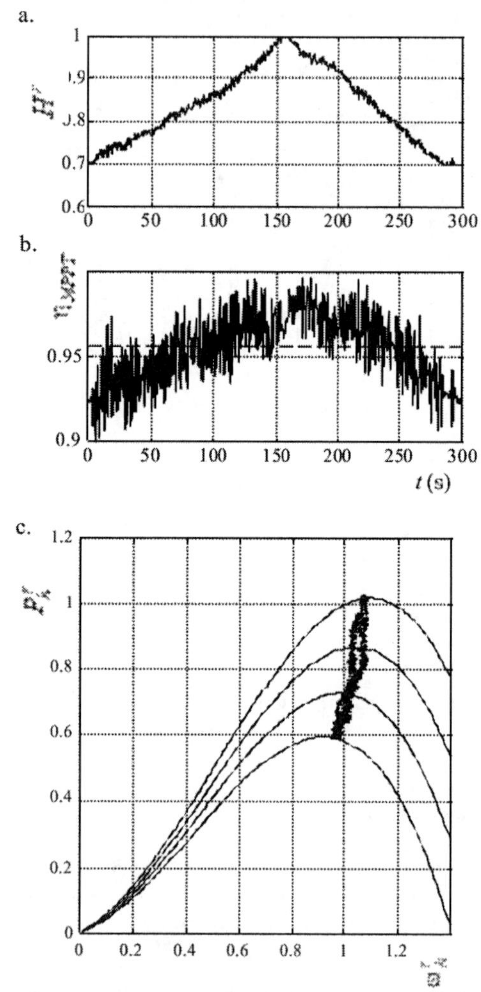

Figure 14. System performance under realistic water head profile: a.) water head, b.) tracking accuracy, c.) power versus angular velocity, in relative values.

4 Summary

This article presents the possibility of using the commercially available PV microinverter in the water and wind applications. The described models of the low power wind turbine and the reservoir small water turbine are based on the torque characteristics in a steady state. The parameters are scaled into relative variables in order to make the performance comparison. The dynamic features are introduced by the laboratory inverter and the machines. The conducted dynamic tests, in the form of stepping up and down the input function, have showed fast systems response comparable in the both cases. The energy production effectiveness of the chosen energy sources with the using of the PV microinverter cannot be easily compared. While this parameter corresponds directly to the power factor for the wind system, the water system requires taking into account the nonlinear efficiency characteristic (hill chart) of the water turbine (Fig. 3a). Therefore the MPPT efficiency parameter was used to compare the system effectiveness. The MPPT tracking accuracy tested under the realistic profiles is similar for both cases: 98% and 96% respectively for the wind and the water systems. These results prove the satisfactory performance of the MPPT method and the PV microinverter in the presented applications.

In conclusion, it may be claimed that the chosen small renewable sources may be controlled by the common indirect MPPT method implemented in the universal microinverter with the satisfactory effectiveness. The designing process of the inverter and the control method for the specific small systems is debatable.

References

1. S. Narendiran, ICCPCT Conf. P., 564-567, (2013)
2. H. Haibing, S. Harb, N. Kutkut, I. Batarseh, Z.J. Shen, IEEE Trans. Power Electron., **28**, 6, 2711-2726, (2013)
3. S.B. Kjaer, J.K. Pedersen, F. Blaabjerg, Trans. Ind. Appl. IEEE, **41**, 5, 1292-1306, (2005)
4. J.D. Bastidas-Rodriguez, E. Franco, G. Petrone, A. Ramos-Paja, G. Spagnuolo, IET Power Electron., **7**, 6, 1396-1413, (2014)
5. B. Subudhi, R. Pradhan, IEEE Trans. Sustainable Energy, **4**, 1, 89-98, (2013)
6. S.M. Raza Kazmi, H. Goto, G. Hai-Jiao, O. Ichinokura, IEEE ECCE Conf. P., 4075-4082, (2010)
7. X. Yuanye, K.H. Ahmed, B.W. Williams, IEEE Trans. Ind. Electron., **60**, 3, 1122-1132, (2013)
8. J. Fraile-Ardanuy, J.R. Wilhelmi, J.J. Fraile-Mora, J.I. Perez, IEEE Trans. Energy Convers., **21**, 2, (2006)
9. D. Borkowski, T. Węgiel, IEEE Trans. Energy Convers., **28**, 2, 452–459, (2013)
10. W. Mazgaj, Z. Szular, T. Węgiel, T. Sobczyk, Electrical Rewiev, **87**, 5, 282-287, (2011)
11. L. Belhadji, S. Bacha, I. Munteanu, A. Rumeau, D. Roye, IEEE Trans. Energy Convers., **28**, 1, 34-43, (2013)
12. Borghetti, M. DiSilvestro, G. Naldi, M. Paolone, M. Alberli, PSCC Conf. P., 447-454, (2008)
13. D. Borkowski, Technical Transactions – Electrical Engineering, **1-E**, 15-24, (2015)
14. S. Heier, *Grid Integration of Wind Energy Conversion Systems*, Hoboken, NJ: Wiley, (2006)
15. S. B. Kjaer, *Design and control of an inverter for photovoltaic applications*, Ph.D. dissertation, Inst. Energy Technol., Aalborg University, Denmark, (2005)
16. Z.M. Dalala, Z.U. Zahid, Y. Wensong, C. Younghoon, L. Jih-Sheng, IEEE Trans. Energy Convers, **28**, 3, 756-767, (2013)
17. T. Kerekes, R. Teodorescu, M. Liserre, R. Mastromauro, A. Dell'Aquila, , IEEE 11th OPTIM Conf. P., 427-432, (2008)
18. T. Andrejašič, M. Jankovec, M. Topič, IET Renew. Power Gen., **5**, 4, 281-286, (2011)

E3S Web of Conferences 10, 00005 (2016) DOI: 10.1051/e3sconf/20161000005
SEED 2016

Using of CFD software for setting the location of water stream micro turbines

Łukasz Borsuk[1,a]

[1]*AGH University of Science and Technology, Faculty of Mining Surveying and Environmental Engineering, Department of Environmental Management and Protection, Poland.*

Abstract. The aim of this work was to estimate the efficiency of CFD software in calculating flow velocity magnitude in natural water streams. These kinds of estimations are essential for setting the locations of water stream micro turbines. These devices can be useful to provide electricity in areas remote from power generating facilities or as backup power supply in case of power grid failure. The analysed water stream has length of 100 m and its average slope was approximately 10%. Water velocity varies in the range from 0.5 m^3*s^{-1} to 5 m^3*s^{-1}. Additionally, the influence of ground roughness on the stream velocity was also an important factor. Results proved to be satisfactory. In the analysed stream, velocities were in a range which allows the proposed micro turbine to be effective. Calculation grid created by CFD software did not have many areas which may raise doubts. Also, the influence of changes in the ground roughness factor was noticeable. Preliminary CFD simulations allow to estimate where in the stream the micro turbine will be most efficient. On the other hand, despite these calculations, profitability and return on the investment still can be questionable.

1 Introduction

The aim of this work was to estimate the efficiency of Computational Fluid Dynamics (CFD) software in calculating flow velocity magnitude in natural water streams. This task is complicated because the river bed is not the only factor that should be taken into consideration. Also ground roughness coefficient and yearly changes in river flow are very important factors in such estimations.

This type of calculation can be used as a support for decisions concerning the placement of water stream micro turbines. Additionally, information about water stream velocity is crucial to estimate the optimal type of turbine for a particular part of the river. Also, as the optimal placement of the turbine determines how much power that turbine can generate, it is possible to approximate potential benefits from the investment.

Micro water stream turbines can have a variety of uses, from providing power to households remote from the power grid, through alternative power supply in case of blackout accidents, up to supplying power in the tourism industry, e.g. on a camping near the river side. Additional aim of this work was to estimate how profitable this type of device can be.

CFD Software is successfully used as a help for designing turbines. But in that kind of usage, the influence of environment is not treated with sufficient attention. In literature [1-3], uneven river bed surface is an overlooked issue. Also turbine models are created with flat ground and steady stream flow conditions in mind. This is a result of problems with creating an accurate Digital Elevation Model of the river bed ground level. But according to the latest literature [4-6], it seems reasonable to think about development in that direction. New sonar devices and advanced methods of laser scanning can be used with success to create a sustainable and trustworthy ground model of the river bed, as exemplified in figure 1.

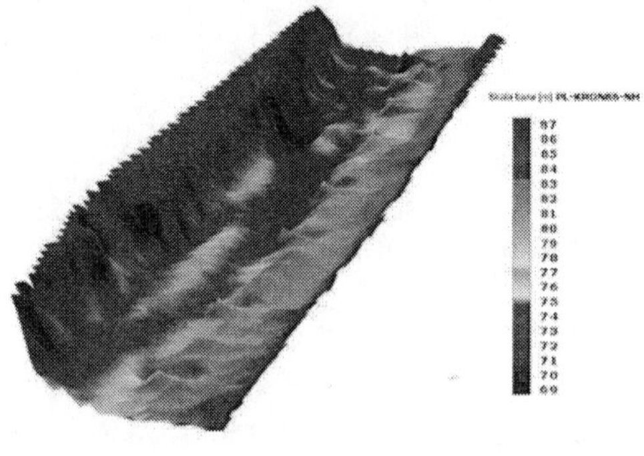

Figure 1. River bed model of flooded area, source: [4]

[a] Corresponding author: borsuk@agh.edu.pl

 DOI: 10.1051/e3sconf/20161000005

Assuming that Digital Elevation Model will be available for reasonable price, it is important to answer the following question: will the CFD model based on this type of data prove a reliable solution to engineering problems?

2 Materials and methods

2.1 River bed ground level

AutoCAD Civil 3D was used to create a river bed model necessary for this analysis. The terrain model is 100 m long and its width varies from 1.5 m to 3 m. River bed was generated from over 5000 points using surface modelling tools. Although those tools are designed to calculate land masses, they also can be applied in this unorthodox way. River bed is irregular: there are areas of shallow water, especially upstream. Additionally, near the river curves, there are also deeper areas of the river. This part of watercourse was not based on any existing example; it is only a hypothetic situation created for the purpose of this study. On the length of 100 m the average slope of the ground level was around 10%. Additionally, to increase realism and to test the software in more difficult conditions, river bed curves three times. This situation and the complete river bed is shown in figure 2.

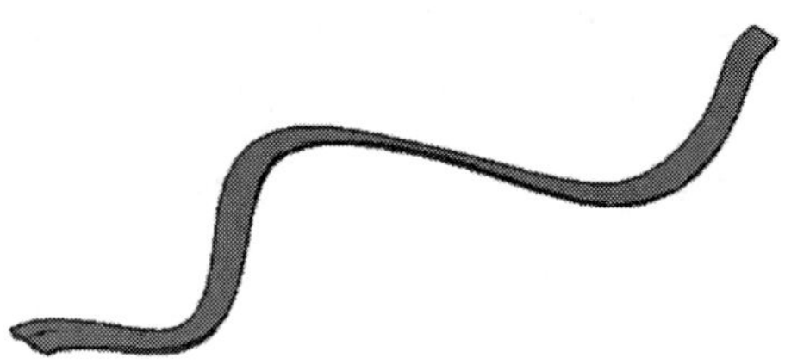

Figure 2. River bed model, source: own elaboration.

2.2 Variable flow velocity

The change of velocity with time is an important parameter which has to be considered to perform a reliable analysis. Calculations were run using three different variants. First of them, according to the existing studies [7], was low-flow which can occur during light rain, so that velocity does not exceed 0.5 m^3*s^{-1}. Second variant is close to average flow rate from multi-year equal to 2 m^3*s^{-1}. Last case concerns the flows that can be measured in months with highest amount of rainfall and it is 5 m^3*s^{-1}. These flow velocities were often measured in Polish rivers from years 1965 to 2015 [7]. An additional assumption is that this watercourse is classified as a small lowland river.

2.3 Ground roughness coefficient

Ground roughness was also taken into consideration. The Manning's n values were established according to classic literature [8]. For that, three variable conditions were created. The first one was assumed for clean river bed without serious disturbances and the roughness was equal to n=0.025. In the second variant, calculations was run for n=0.05, as this value is characteristic for heavily overgrown water streams. The third case was extreme: river had significant amount of water plants and additional roughness was added by fallen trees and rocks. For this condition Manning's coefficient was n=0.150. The reason for presenting this wide variety of ground types was to assess the sensitivity of software to roughness. If the results were similar, it would mean that Autodesk Simulation CFD underestimates the influence of this factor on the calculations.

2.4 Software

According to paragraph 2.1., elevation model of river bed was created mainly by the AutoCAD Civil 3D software. All the calculations inside the model were run in Autodesk Simulation CFD. Both of them were created by the same company so cooperation between them and data exchange does not generate errors. Autodesk Simulation CFD has tools which can be used to implement various boundary conditions such as flow rates, roughness, gravity force etc. Additionally, surface models imported from Autodesk Civil 3D were processed to calculating mesh by automatic scripts. It gives an opportunity to check quality of algorithms used by software in these cases. Unfortunately, one of the decisive defects of the Autodesk Simulation CFD is lack of tools that can be used to perform multiphase calculations. Therefore, influence of atmosphere (for example wind) must be omitted.

3 Results

3.1 Results of analysis for different velocities

Results for the different velocity conditions are shown in figures 3-5. Water volume in which velocity exceeds the level mentioned in the caption was highlighted. Also it is important to remember that Manning's roughness coefficient for figures 3-5 was equal n=0.05. Turbine [9] used to calculate generated power starts working at velocity of 1 $m*s^{-1}$ and it reaches optimal power at 4 $m*s^{-1}$.

According to figure 3, even with the lowest flow rate in this particular river bed it is possible to find places in which a turbine will work properly. Unfortunately, the generated power will be low because flow speed is barely enough to start the turbine. In this case, the amount of electric power generated by this micro turbine can be estimated at10 W.

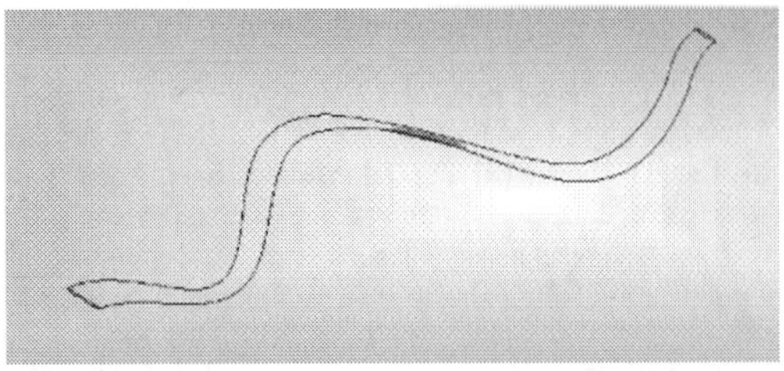

Figure 3. Surface of area in which velocity is higher than 1 m*s^{-1} for flow rate of 0.5 m^3*s^{-1}, source: own elaboration.

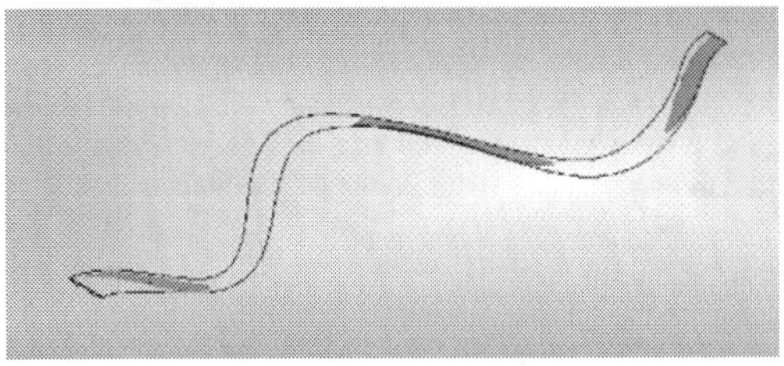

Figure 4. Surface of area in which velocity is higher than 2,5 m*s^{-1} for flow rate of 2 m^3*s^{-1}, source: own elaboration.

In this case, velocities highlighted in figure 4 are high enough to reach 50% of turbine efficiency; it can give approximately 60 W of power. Unfortunately, there is no area in the river in which a turbine can work in optimal way.

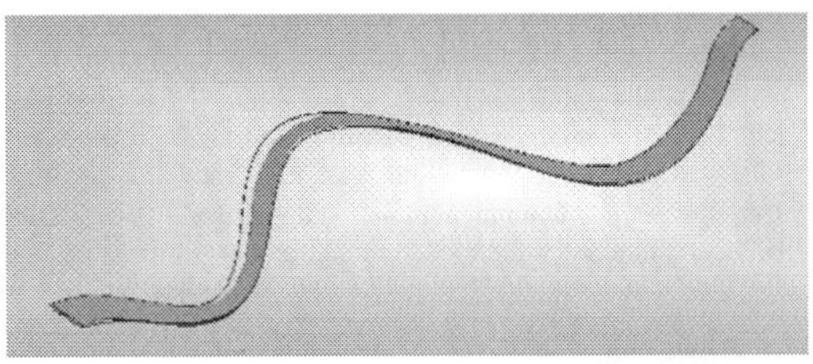

Figure 5. Surface of area in which velocity is higher than 4 m*s^{-1} for flow rate of 5 m^3*s^{-1}, source: own elaboration.

The last case shows that if the water level is high enough and stream velocity is significant, a turbine will have optimal conditions for work almost in any area of the river. Power generated with these variables can approximate to 100 W.

3.2 Analysis of return on the investment

According to the obtained results, return on the investment is unclear. Purchase cost of the considered turbine [9] is near 2200$. Assuming an extremely optimistic scenario in which the average flow is near 2.5 m^3*s^{-1} for the whole year and turbine works without any conservation breaks, the generated power will be approximately 525 kWh. Actual market price of electric power in Poland is at the level of 0,16 $ for 1 kWh. Taking into account the presented data, the investment will fully pay off in about 26 years. And if the scenario is be more realistic, for example there are gaps in worktime or usage of the turbine generates some repair costs, this kind of investment seems to be unprofitable. Additionally, even as an alternative source of power supply it is barely useful. The generated power of 60 W will not be sufficient to supply most of the common electrical devices. In this scenario, it will be necessary to invest in more than one turbine or some kind of batteries which will lead to further increase in the investment cost.

3.3 Quality of automatically generated mesh

Automatic generation of computing mesh allows to optimize the time necessary for running calculations. Unfortunately, it seems that this algorithm does not generate mesh dense enough to show exactly what is happening with the velocities on the border between fluid and ground surface. In figure 6 it is possible to observe that this area is treated without necessary attention. In commercial analysis, density of computing mesh near the ground level should be noticeably increased. In addition to this, size of edge folds shown in figure 6 will significantly decrease.

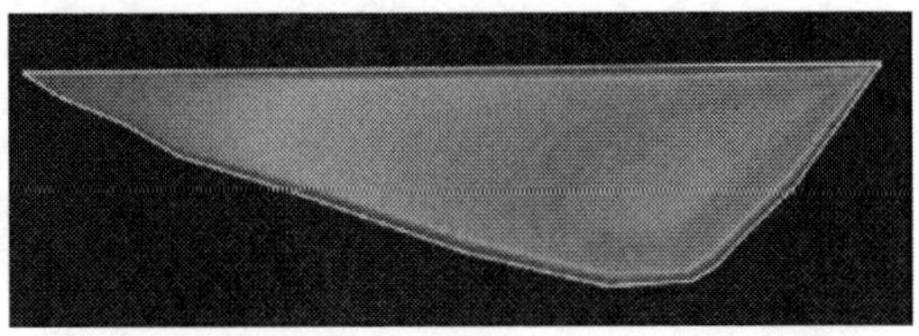

Figure 6. Cross-section example, source: own elaboration.

3.4 Influence of Manning's roughness coefficient on the results

Another element which can influence the quality and reliability of the analysis is the impact of ground roughness on the results. This factor was noticeable in all cases of the analysis. Migration of the highest velocity field in the stream to the middle of the river bed increases significantly with n value. Unfortunately, according to the previous paragraph and to figure 6 it is impossible to track specific changes near the edges of the analysed stream.

4 Discussion

Method shown in the presented analysis seems to be promising. If the geodetic data, like those in figure 1 and literature [4-6], are more available, there will be an opportunity to perform this type of analysis more often. Consequently, the results of this kind of calculations will be improved. Also, it seems that it is possible to perform this type of analysis not only on even surfaces but also on rough terrain which constitutes a development of concepts shown in the literature [1-3]. Results for the different velocities do not raise crucial doubts. But the usage of micro turbines in practice is not as obvious as it seems. On the one hand, it was shown that if the flow magnitude is high enough, it is possible to create optimal performance conditions for water stream micro turbines. On the other hand, this type of devices generates too low power to be considered as a reliable source of electric energy. Also, full return on the investment takes 26 years (in perfect conditions), which may raise doubts about the profitability of the investment.

5 Conclusions

Autodesk Simulation CFD can be used to estimate stream velocity in river bed with complicated ground level. In the simulation influence of flow change and Manning's roughness coefficient is noticeable. Also, use of denser computing mesh and more reliable geodetic data will significantly increase the quality of analysis. However, profit from the usage of water stream micro turbines is unclear. Generated power may raise doubts about profitability of the investment.

References

1. S.C. Tatum , C.H. Frost , M. Allmark, D.M. O'Doherty , A. Mason-Jones , P.W. Prickett , R.I. Grosvenor , C.B. Byrne , T. O'Doherty. *Wave–current interaction effects on tidal stream turbine performance and loading characteristics,* (International Journal of Marine Energy, 2015)
2. M.I. Shahidul, S.S Syed Tarmizi, Abdullah Yassin, Al Khalid Othman, Hushairi Zen, Ting Ching Hung, Lee Man Djun. *Modeling the Energy Extraction from In-stream Water by Multi Stage Blade System of Cross Flow Micro Hydro Turbine,* (Procedia Engineering Volume 105, 2015, Pages 488–494)
3. Taimoor Asim, , Rakesh Mishra, Kuldip Ubbi, Karina Zala. *Computational Fluid Dynamics Based Optimal Design of Vertical Axis Marine Current Turbines.* (Procedia CIRP Volume **11**, 2013, Pages 323–327)
4. Laskowska K., Lesiak M., *Rola pomiarów batymetrycznych w bezpieczeństwie przeciwpowodziowym* (XXIX Seminarium Kół Studenckich Wydziału Inżynierii Lądowej I Geodezji)
5. Łubczonek J., Zaniewicz G., *Modeling and visualization of digital terrain model and sea bottom for port security geoinformatic system,* (Archiwum Fotogrametrii, Kartografii i Teledetekcji Vol. **24**, 2012, s. 187 – 198)
6. Gołuch P., Dombek A., Kapłon J., *Evaluation of data accuracy obtained from bathymetric measurement using fish finder Lawrence LMS-527C DF iGPS,* (Archiwum Fotogrametrii, Kartografii i Teledetekcji, Vol. **21**, 2010, s. 109–118)
7. Kanclerz J., Murat-Błażejewska S., Sojka M., *Zmienność przepływów małej rzeki nizinnej,* (Zeszyty problemowe postępów nauk rolniczych 2008 z. 528: 71-78)
8. Chow V.T., *Open-Channel Hydraulics,* (Mc Graw – Hill Book, New York, 1959)
9. http://www.seamap.com/products/ampair/ampair-uw100/ (Access 17.04.2016)

Cost–effective Polythiophene Counter Electrodes for Dye Sensitized Solar Cells

Merve Celik Tolu[1,a], Buket Bezgin Carbas[1] and Savas Sonmezoglu[2]

[1] *Karamanoglu Mehmetbey University, Faculty of Engineering, Department of Energy Systems, Karaman, Turkey*
[2] *Karamanoglu Mehmetbey University, Faculty of Engineering, Department of Metallurgical and Materials, Karaman, Turkey*

Abstract. Dye sensitized solar cells (DSSCs) are most promising devices among third–generation solar cells because of low cost, easy production, environmental friendliness, and relatively high conversion efficiency. Counter electrode (CE), which is an important component in DSSCs, functions as an electron transfer agent as well as the regenerator of redox couple. Hitherto, various methods and materials were used to prepare different counter electrodes.Among these materials, conducting polymers have been widely investigated and employed in various applications such as sensors, supercapacitors, energy storage devices, DSSCs and others. In this study, Polythiophene (PTh) conducting polymer was successfully synthesized by electrochemical deposition method, and employed as an alternative to expensive platinum (Pt) CE for DSSC. Besides, PTh conducting polymer was electrochemically deposited via cyclic voltammetry method on FTO substrates. The morphology of the PTh film was characterized by SEM and AFM. Finally, the photovoltaic performance of PTh CE based DSSC was compared with PEDOT CE based device. This new concept—along with promising electrocatalytic activity and facile electron transfer—provides a new approach to enhance the photovoltaic performances of Pt–free DSSCs.

1 Introduction

The energy required for production of pure silicon is too high so new photovoltaic devices were developed alternative to Si–based solar cells [1]. In 1991, Oregan and Gratzel were developed dye sensitized solar cell (DSSC) and gain new perspective to commercial photovoltaic devices [2]. DSSCs have become developing technology continuously due to their low cost, environmentally friendly fabrication processes, and acceptable energy conversion efficiencies [3–6]. Standart DSSC consist of working electrode (WE/photoanode), liquid electrolyte, and counter electrode (CE/photocathode) [3].

In general, each component performs a specific task and the CE is one of the most important component in DSSC [3]. Because its task is to reduction of the redox species used as a mediator in regenerating the dye molecules that injected electrons [1,6]. Optimization of cell components are important for cost effective energy production. Over the past few decades, researchers have made consistent progress in replacing components of DSSC with appropriate materials for optimization of device [3].

Until now, Pt was commonly used as CE. Although, Pt provides a high conductivity and a good catalytic activity, because of its rarity and high cost, researches focus on optimization of CE with alternative materials [1,3–6]. These alternatives are carbon derivatives, transition metal compounds, and conductive polymers [7–9]. Electrochemical synthesis of conducting polyheterocycles (polypyrrole, polythiophene (PTh), poly(3,4–ethylenedioxythiophene) (PEDOT) etc.) have been reported in literature and some of these studies are about possible application of these polymers in electronic devices such as solar cells [10,11]. Saito *et al.* reported the preparation of PEDOT on FTO by electrochemical deposition and used as CE in DSSC which is exhibited the efficiency of 3.6% under 100 mW.cm^{-2} [12]. Torabi *et al.* achieved the highest efficiency of 2.76% by employing 3–methyl–thiophene (P3MT), synthesized with electrochemical deposition method, as CE in DSSCs[13]. They examined the improvement of photovoltaic performance as a function of some parameters such as morphology of polymer, duration of polymerization and so on.

In this study, PTh and its derivative PEDOT polymers were successfully synthesized via electrochemical deposition method, and applied as the catalyst material for the counter electrode of a DSSC. Also, their morphological and photovoltaic properties were characterized by AFM as well as SEM and I–V measurements, respectively.

[a] Corresponding author: mervecelik@kmu.edu.tr

 DOI: 10.1051/e3sconf/20161000006

2 Experimental Procedure

2.1 Equipments

The electrochemical measurements were conducted by Ivium compactStat device. The morphology of the conducting polymers was determined using Nanomagnetics Instruments atomic force microscopy (AFM) and JEOL 6390–LV scanning electron microscope (SEM) with an accelerating voltage of 20 kV in the secondary electron image mode. The current-voltage (I-V) curves of DSSCs were recorded using Keithley 4200 SCS characterization system and standard solar irradiation of 300 Wm^{-2} (xenon arc lamp with AM 1.5 filter) as the light source (Solar Light XPS 300 solar spectrum).

2.2 Preparation of Counter Electrodes

Electrodeposition method was used for deposition of PTh and PEDOT conducting polymers on fluorine doped tin oxide (FTO) glass. Three electrode system that consists of FTO glass working electrode (WE), Ag/AgCl reference electrode (RE), and Pt sheet as CE were used. Thiophene monomer was dissolved in the medium of dichloromethane, tetrabutylammonium hexafluoro phosphate while EDOT monomer was dissolved in the medium of acetonitrile, lithium perchlorate, and ionic liquid. The oxidation potential of these monomers was found as 2.3 V for thiophene and 1.4 V for EDOT via cyclic voltammetry (CV) method. By using this data, polymers were deposited on the electrode via chronoamperometry method with 50 $mC.cm^{-2}$ deposition charge capacity. Finally, the samples deposited on FTO were annealed at 60 °C for 30 minute in the furnace.

2.3 Preparation of Photoanodes

The photoanode was prepared according to our previous report [14]. In briefly, the TiO_2 paste (approximately 10 μm thickness) were made from 21–nm TiO_2 nanopowder in ethanol and printed to FTO substrates using the doctor blade technique and annealed at 450 °C for 1h before sensitization. To fabricate the photoanodes, the TiO_2 films were immersed in a dye solution (3.0×10^{-4} M mixture of the ruthenizer 535–bisTBA (N719) in methanol) for 18 h.

2.4 Preparation of Electrolyte

The iodine/iodide (I^-/I_3^-) electrolyte was prepared according to our previous report [15]. To describe the process briefly, 0.01 M iodine, 0.6 M 1–butyl–3–methylimidazolium iodide, 0.1 M 4–tert–butylpridine, and 0.1 M lithium iodide hydrate were prepared in 3–methoxypropionitrile solvent, separately. Then, the prepared solutions were mixed and subjected to magnetic stirrer for 2 h.

2.5 Assembly of DSSCs

We fabricated the device according to the following method: The photoanode was placed face up on a flat surface, and CEs were placed on top of the photoanode. These two opposing glass plates were offset from one another so that the entire photoanode was covered by the CE. The electrolyte solution was placed at the edges of the plates, and the liquid was drawn into the space between the electrodes via capillary action. We used an epoxy adhesive to hold the electrodes together. Thus, we fabricated the PTh and PEDOT CE based DSSCs having an active area of ~1.0 cm^2. The schematic illustration of the DSSC is shown in Figure 1.

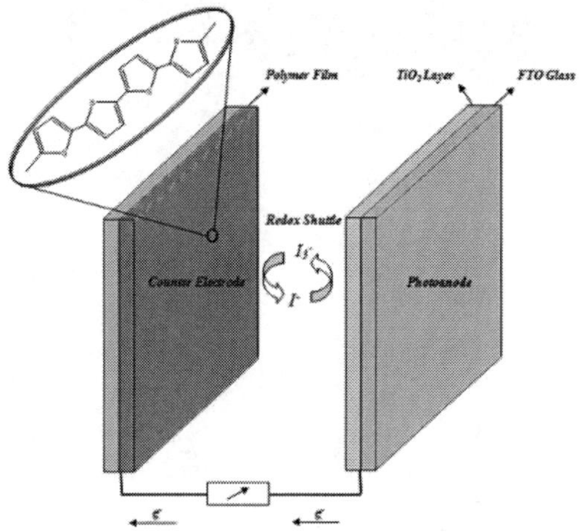

Figure 1. The illustration of DSSC structure with PTh– and PEDOT–based CEs.

3 Results and Discussions

The SEM and AFM images of the PTh and PEDOT polymer deposited films are shown in Figure 2. From the SEM images, it was found that average particle size for PTh and PEDOT conducting polymers is 200 nm and 400 nm, respectively. From the AFM images, the surface roughness of PTh and PEDOT films was found as 40.37 nm and 240.80 nm, respectively. Furthermore, all images indicate homogeneous and uniform distributions of grains.

The photovoltaic performances of the DSSCs based on PTh and PEDOT CEs are shown in Figure 3, and the photovoltaic parameters that control the device performance are tabulated in Table 1. The obtained results indicated that conversion efficiency of PTh CE based DSSC (0.03%) is lower than PEDOT CE based DSSC (4.76%).

Adhesion of PTh on FTO glass is not good as PEDOT. Therefore, the efficiency of PTh is lower than that of PEDOT. The lower conductivity of PTh and more grain boundaries can be additional reasons of lower efficiency. In future, studies on polymerization medium

DOI: 10.1051/e3sconf/20161000006

and conditions will be optimized for good adhesion of PTh on FTO.

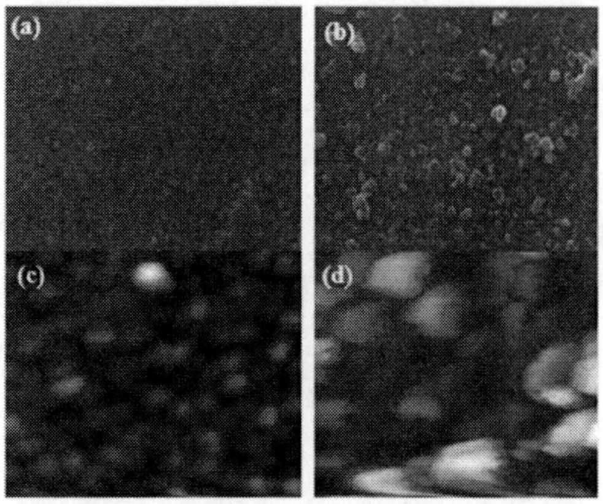

Figure 2. **(a,b)** SEM images and **(c,d)** AFM images of PTh and PEDOT films, respectively.

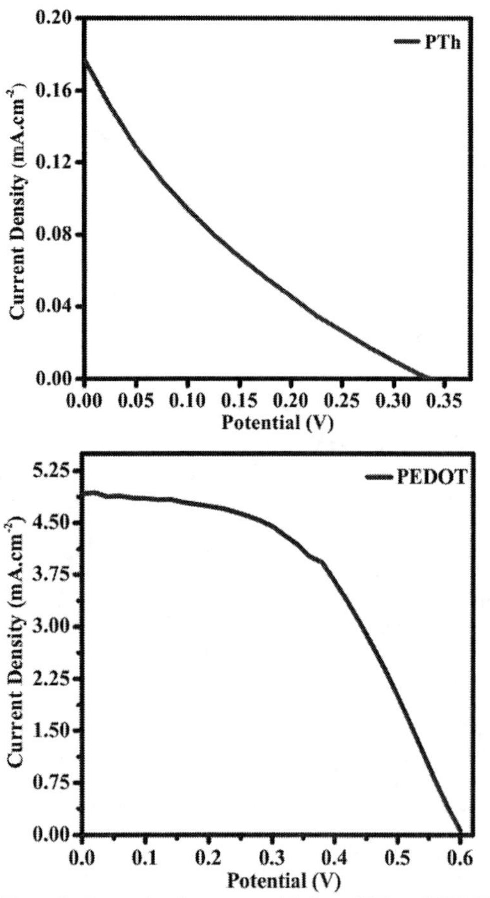

Figure 3. Current density vs. potential plots of PTh and PEDOT CE based DSSCs.

Table 1. The photovoltaic parameters of the conducting polymers CE based DSSCs.

Devices	J_{sc} (mA.cm^{-2})	V_{oc} (V)	FF	η (%)
PTh	0.18	0.33	0.17	0.03
PEDOT	4.96	0.60	0.48	4.76

4 Conclusions

This study focuses on the fabrication of conducting polymer based CEs in DSSCs. These conducting polymers were synthesized via electrochemical deposition method which is an effective way of polymerization. The polymerized PTh and PEDOT conducting polymers with the same charge density were characterized for their electrochemical and morphological properties. The results suggest that PEDOT based electrode can enhance the catalytic activity between polymer–electrolyte interface of a DSSC. As expected, the power conversion efficiency of DSSC was increased by PEDOT based CE. Efficiency of PTh CE based DSSC is lower than PEDOT CE based device due to the weak adhesion, therefore our studies are still going on to overcome this problem. As a result, DSSC performance can increase by using new polymers creating by substitution of polymer.

References

1. W. Wei, H. Wang, Y. H. Hu, Int. J. Energy Res. **38**, 1099 (2014)
2. B. O'regan, M. Gratzel, Nature, **353**, 737 (1991)
3. S. Akın, M. Gülen, S. Sayın, H. Azak, H.B. Yıldız, S. Sönmezoğlu, J. Power Sources, **307**, 796 (2016)
4. F. Özel, A. Sarılmaz, B. İstanbullu, A. Aljabour, M. Kuş, S. Sönmezoğlu, Sci. Rep. **6**, 29207 (2016)
5. S. Sönmezoğlu, C. Akyürek, S. Akın, J. Phys. D: Appl. Phys. **45**, 425101 (2012)
6. S. Sönmezoğlu, V. Eskizeybek, A. Toumiat, A. Avcı, J. Alloys Compd. **586**, 593 (2014)
7. Y. Tang, X. Pan, S. Dai, C. Zhang, H. Tian, Key Eng. Mater. **451**, 63 (2011)
8. S. Thomas, T.G. Deepak, G.S. Anjusree, T.A Arun, S.V. Nair, A.S. Nair, J. Mater. Chem. **2**, 4474 (2009)
9. K. Li, Z. Yu, Y. Luo, D. Li, Q. Meng, J. Mater. Sci. Technol. **23**, 577 (2007)
10. S. Sönmezoğlu, R. Taş, S. Akın, M. Can, Appl. Phys. Lett. **101**, 253301 (2012)
11. C.T. Li, C.P. Lee, M.S. Fan, P.Y. Chen, R. Vittal, K.C. Ho, Nano Energy, **9**, 1 (2014)
12. Y. Saito, W. Kubo, T. Kitamura, Y. Wada, S. Yanagida, J. Photochem. Photobiol. A: Chem. **164**, 153 (2004)
13. N. Torabi, A. Behjat, F. Jafari, Thin Solid Films, **573**, 112 (2014)
14. R. Taş, M. Gülen, M. Can, S. Sönmezoğlu, Synth. Met. **212**, 75 (2015)
15. S. Sönmezoğlu, C. Akyürek, H. Akış, IET Optoelectron. **8**, 270 (2014)

E3S Web of Conferences **10**, 00007 (2016)
SEED 2016
DOI: 10.1051/e3sconf/20161000007

Effect of soot formed in result of coal combustion with two types of solid fuel additives on soot – soil bacteria interaction – biodegradation of soot components by soil bacteria

Barbara Kalicka[1,2], Marek Chyc[3,4], Monika Bojko[1], Monika Olchawa-Pajor[1,3] and Dariusz Latowski[1,a]

[1]*Department of Plant Physiology and Biochemistry, Faculty of Biochemistry, Biophysics and Biotechnology, Jagiellonian University, Gronostajowa 7, 30-387 Kraków, Poland*
[2]*Department of Pharmaceutical Biochemistry, Faculty of Pharmacy, Jagiellonian University Medical College, Medyczna 9, 30-688 Kraków, Poland*
[3]*Department of Environmental Protection, State Higher Vocational School in Tarnów, Mickiewicza 8, 33-100 Tarnów, Poland*
[4]*Silesian Environmental Doctoral Study, Plac Gwarków 1, 40-166 Katowice, Poland*

Abstract. During coal combustion soot formation is unwanted. There are many fuel additives used to reduce soot production. In this study toxicity and biodegradation of three types of soot by two soil bacteria strains i.e. *Pseudomonas aeruginosa* and *Bacillus subtilis* were compared. Two additives were selected to produce soot samples: (i) commonly used and containing significant quantities of NaCl (40-44%) and copper (Cu-additive), (ii) newly developed without NaCl and with triiron tetraoxide instead of copper (Fe-additive). Average number of bacterial colonies was similar when soot produced during combustion without additive and with Fe-additive were used. It was about 50% higher than number of colonies when soot formed with Cu-additive was applied. The soot biodegradation was tested by HS-GC-MS analysis. The highest biodegradation level (about 80%) was obtained for samples with *P. aeruginosa* cultures and Fe-additive soot. It can be concluded that newly developed additive does not increase toxicity of the soot to soil bacteria and will be removed more quickly from the environment.

1 Introduction

Coal is one of the largest sources of energy, especially for electricity generation, but it is also the largest worldwide anthropogenic source of many pollutants emitted to environment during its combustion. The main of these pollutants emitted into the atmosphere are carbon dioxide and soot. They are theorized to be the two-largest causes of global warming [1].

Soot is formed as a by-product of combustion or pyrolysis of carbon-containing materials, such as coal, wood, fuel oil, waste oil, paper, plastics, and household refuse. Although its chemical compositions and properties are highly variable and depend on the type of starting material and the combustion conditions, soot is always a chemical and physical pollutant of air, water and soil. It is known that soot is a powerful absorbent of sunlight and contributes to climate warming and melting of glaciers covered with so called dry deposition (particulates) causing the greenhouse effect [2, 3]. Soot is causing not only a greenhouse effect but also circulatory and respiratory system diseases. Because it is a mixture containing crystal amorphous carbon (black carbon), polyaromatic hydrocarbons (PAHs), persistence organic radicals [4] or heavy metals [5] it can be primarily associated with proven properties such as mutagenicity and carcinogenicity [2]. Therefore, soot reduction constitutes a significant component of air protection programmes as well as the low emission reduction programmers (LERP).

Reduction of soot level can be achieved by using of solid fuel additives. Although additives reduce soot formation during coal combustion their influence on toxicity of soot formed in result of the combustion with these additives has not been enough documented.

The investigation of the soot effect on organisms still do not provide conclusive evidence about the pollution impact on soil microbes [6]. Similarly, the effect of various solid fuels additives on soot biodegradation by soil microbes is still unknown [7].

In this study soot produced as a result of combustion with and without solid fuel additives was tested. Two additives were selected. One was commercially available and commonly used additive containing significant quantities of NaCl and $CuSO_4$ (Cu-additive). The second one was newly developed, NaCl free additive with triiron tetraoxide instead of copper compound (Fe-additive).

The main purpose of this work was to determine how various solid fuels additives modulate interaction soot – soil bacteria including biodegradability of soot by these microbes.

2 Materials and methods

2.1 Procedure of soot formation

Three types of soot were tested. One of them was called unmodified (UM) and no additives were used during its formation. Two other types of soot were produced with solid

[a] Corresponding author: dariuszlatowski@gmail.com

fuel additives. One of them was commercially available and commonly used additive containing significant quantities of NaCl i. e. 40 – 44% and $CuSO_4$ (Cu-additive). The second one was newly developed, NaCl- and $CuSO_4$ free additive with triiron tetraoxide (Fe-additive). All soot samples were obtained from combustion chamber of central heating system boiler (Viadrus Hercules U22, rated power 29.1 kW) by the combustion of 10 kg of hard coal (KWK Piast) without or with additives. The Fe-additive contains triiron tetraoxide, potassium, sodium and ammonium nitrate, sodium percarbonate and anti-caking agent. This invention reduces the emission of particulate matter, CO, CH_4 and unburned volatile organic compounds (VOCs) exhaled with the flue gases into the atmosphere. Fe-additive maintains the surface of the fireside heat exchanger free of soot thereby increasing the thermal efficiency of the heating system. The boiler used for combustion was cleaned after burning to reduce the impact of residues prior preparation. Before burning with the fuel additive the boiler was not cleaned to achieve the appropriate amount of soot in the combustion chamber.

2.2 Soot sterilization procedure

To study soot – bacteria interactions samples of each soot type were grinded carefully in a mortar and sterilized.

Two methods of soot sterilization were tested. One of them was dry sterilization by UV radiation and another one was moist heat sterilization in autoclave. Soot samples (200 mg) were illuminated by UV lamp in laminar chamber for 20 min or sterilized in the autoclave (120°C, 20 min) as a suspension in a growing medium. To control soot sterilization UV- and autoclave- treated samples were placed in aseptic liquid medium LB MIX (Tab. 1) and incubated (see: 2.3. Method of bacteria culturing).

Table 1. Composition of liquid nutrient medium LB MIX used for bacteria culture (pH 7.00).

Composition	Mass [g/L]
Casein peptone	10.00
Yeast extract	5.00
Sodium chloride	10.00

After incubation about 0.2 mL of medium were spread on LB MIX-agar plates, incubated (24 h, 37°C) and checked whether the microorganism colonies were present. Samples were sterile if no colonies of microbes were detected.

2.3 Method of bacteria culturing

For the bacteria multiplication two 100 mL conical flasks containing 50 mL of sterile liquid medium LB MIX (Tab. 1) were inoculated with one of two the most common soil bacteria strains, i.e.: *Pseudomonas aeruginosa* or *Bacillus subtilis*. Then both flasks were incubated for 24 h at 37°C with shaking (200 RPM). Growth of liquid monocultures was monitored by optical density (OD) measurement at 600 nm using UV-Vis spectrophotometer. The measurements were done after 24 hours of incubation . For further experiments proper volume of each culture was taken to obtain the new cultures with the same OD value (0.300) i.e. the same amount of bacteria cells.

2.4 Influence of soot on bacterial growth

The appropriate volumes of 24-hours bacteria cultures were used for inoculation of 50 mL new liquid cultures of *P. aeruginosa* or *B. subtilis*. Each bacteria culture was treated by one of three types of soot (UM, Cu- or Fe- additive) and compared with control (culture without soot). The study was performed for two concentrations (w/v) of soot in the nutrient medium i.e. 0.30% and 6.00% (w/v). Culture conditions were the same as described in paragraph 2.3. After 24 hours appropriate volume of each culture was taken, diluted with the growing medium and seeded on LB MIX - agar Petri plate. Dilution and volume of each culture were estimated so as to obtain on Petri dish number of colonies which were possible to count. All tests were repeated 10 times

2.5 Biodegradation of soot components by soil bacteria

Samples of each type of soot (0.20 g) were weighed out to 100 mL volume conical flask. Then each flask was filled with 50 mL of BSM mineral nutrient medium containing no carbon source (Tab. 2). Obtained suspensions were sterilized.

Table 2. Composition of BSM mineral nutrient medium containing no carbon source (pH 7.00).

Composition	Mass [g/L]
Potassium dihydrogen phosphate	0.38
Dipotassium hydrogen phosphate	0.60
Magnesium sulfate heptahydrate	0.20
Ammonium chloride	1.00
Iron(III) chloride hexahydrate	0.08

After sterilization, soot (UM, Cu-, or Fe- additive) were left as control sample (with no bacteria). Other two containing the same type of soot were inoculated with the same amount of cells one of two bacterial strains: *P. aeruginosa* or *B. subtilis*, respectively. The cultures were kept in darkness to prevent possible photolysis of sample components and incubated for 4 weeks (30°C, RPM 150). The test was repeated 10 times.

After 4 weeks of incubation 15 μL of soot solution with bacteria from each of the flasks were collected, spread on LB MIX-agar plates and incubated (24 h, 37°C) to check whether the microorganisms survived the incubation period. The rest of the solution from the flasks was filtered on a Büchner funnel under reduced pressure with a hard filter (degree: 390, weight: 84 g/m^2). The filters with residues were placed in glass vials, allowed to dry at room temperature for 5 days and then sealed. The samples were analyzed to determinate analytes concentrations by static headspace gas chromatography with mass spectrometry detection (HS-GC-

MS) after about 15 min of vials termostation at 150°C just before the gas phase injection (50 μL). The analysis was performed in a gas chromatograph (Agilent Technologies 6890N Network GC system) coupled with a mass spectrometer 5975C VL MSD quadrupole type with electron beam ionization (70eV Electron Impact, EI). The injector temperature was set at 250°C, the division of the stream of carrier gas (helium) was 10:1, flow through the column at 1 mL/min. and a linear speed of 36 cm/sec. The programmed temperature was used: 60°C for 1 min, then temperature was increasing in 10°C/min steps until it reached its maximum value of 250°C (held for 1 min). The analyses were performed using a capillary column HP-1 MS (Agilent Technologies) with a length of 30 m, a diameter of 0.25 mm, and a thickness of 1 μm film. The transfer line was maintained at 250°C, MS source temperature was 230°C and the quadrupole temperature 150°C, dwell time - 100 ms. The samples were analysed by the sum of four components: phenol (m/z=94, RT=about 7.78 min), naphthalene (m/z=128, RT=10.04 min), phenanthrene (m/z=178, RT=about 17.67) and anthracene (m/z = 178, RT = about 17.79 min) in selected ion monitoring acquisition (SIM). Despite close relations and retention parameters m/z ions for isomers: phenanthrene and anthracene achieved adequate separation of peaks.

2.6 Statistical analysis

The data were analyzed in STATISTICA12. Levene'a test of homogeneity of variances, ANOVA and HSD Tukey multiple comparisons of means were used to determine, which means are significantly different from each other.

3 Results

In presented experiments two methods of sterilization were tested. One of them was UV radiation and another one was sterilization in autoclave. We observe that in all solutions containing soot sterilized in autoclave no organisms were detected on Petri plates with agar LB MIX. However, UV-radiation turned out to be useless for soot sterilization. All Petri plates shown bacteria colonies when medium with soot UV radiation treated were spread on agar plates (Tab. 3).

Average number of bacterial colonies was similar when soot produced during combustion without additive (UM) and with newly developed additive (Fe-additive) were used. It is also worth to notice that average number of the bacterial colonies is almost the same when cultures without soot were compared with samples treated with UM or Fe-additive soot. Additionally, the number of colonies grown in presence of Cu-additive soot was about 50% lower than in the samples without soot or in the samples treated by UM and Fe-additive soot.

When biodegradation of soot components were tested it was observed that the bacteria inoculated into soot suspensions were alive during whole 4-weeks period of experiments.

The HS-GC-MS analysis of the sum of the four selected aromatic components has shown the changes in the chemical composition of soot due to the two bacteria strains used for biodegradation (Fig. 1).

Table 3. Effect of different types of soot on growth of two strains of soil bacteria i.e. *B. subtilis* and *P. aeruginosa*.

Type of soot	**Soot concentration**	**No. of *B. subtilis* colonies**	**No. of *P. aeruginosa* colonies**
Soot without additives	0.3% (w/v)	$10.2^{a} \pm 3.8$	$11.1^{a} \pm 3.0$
	0.6% (w/v)	$9.4^{a} \pm 2.5$	$10.5^{a} \pm 1.7$
Fe-additive soot	0.3% (w/v)	$9.8^{a} \pm 2.5$	$10.9^{a} \pm 3.5$
	0.6% (w/v)	$9.6^{a} \pm 1.7$	$10.1^{a} \pm 3.5$
Cu-additive soot	0.3% (w/v)	$4.1^{b} \pm 2.0$	$3.8^{b} \pm 2.5$
	0.6% (w/v)	$4.2^{b} \pm 1.5$	$4.5^{b} \pm 2.5$
Control - bacteria without soot	0%	$10.6^{a} \pm 3.5$	$10.9^{a} \pm 3.5$

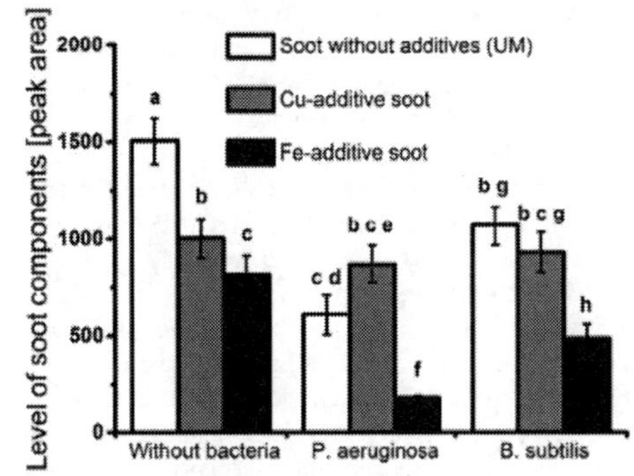

Figure 1. Comparison of the sum of: phenol, naphthalene, phenanthrene and anthracene analysed by (HS-GC-MS) in the biodegradation studies; in three soot types.

Table 4. The degree of biodegradation of aromatic compounds in soot samples.

Type of soot	**Level of biodegradation with *P. aeruginosa* [%]**	**Level of biodegradation with *B. subtilis* [%]**
Soot without additives	59.6	29.2
Fe-additive soot	77.9	40.2
Cu-additive soot	13.2	7.2

After four weeks of biodegradation it could be observed that *P. aeruginosa* is more efficient in degradation of all soot types components than *B. subtilis* (Fig. 1, Tab. 4). Although tested components of all using types of soot could be degraded by bacteria, the best biodegradation was observed for soot formed during combustion with the new-developed additive, i.e. Fe-additive (Fig.1, Tab. 4). The lowest degree of biodegradation was detected for Cu-additive soot (Fig. 1, Tab. 4).

The degree of biodegradation (Tab. 4) was compared to the results obtained from soot without and with bacteria after 4-weeks incubation. The biggest declines in the value of biodegradation degree were recorded for Fe-additive soot (77.9% for *P. aeruginosa* and 40.2% for *B. subtilis* cultures).

4 Discussion

Soot as a pollutant is an important factor affecting biosphere. Because of its toxicity, reducing soot production is one of the primary tasks in modern energy production. One of the most efficient ways to reduce soot formation is the use of fuel additives [8]. On the market there are many additives to the coal as a solid fuel-reducing soot formation. However their influence on toxicity of soot formed during coal combustion with these additives has not been enough documented. Similarly, not enough attention has been also paid to the question what various additives to solid fuels modulate interaction soot – soil microbes including biodegradability of soot by these microbes. In our studies we compared effect of two solid fuel additives on soot toxicity on soil bacteria and how these additives result in soot biodegradability by these bacteria. The first was commercially available and commonly used additive containing significant quantities of NaCl and copper (Cu-additive). The second one was newly developed, NaCl free additive with Fe_3O_4 instead of based on copper compound items (Fe-additive).

To study the interaction soot – bacteria, soot sterilization method is required. Our results have shown that although dry sterilization by UV radiation was ineffective, moist heat sterilization in autoclave can be used successfully.

Effect of three different types of soot on growth of the two most common strains of soil bacteria i.e. *P. aeruginosa* and *B. subtilis* was tested. In the case of both tested bacterial strains, regardless of types and concentrations of soot bacteria growth was observed. Thanks to the applied dilutions, performing the quantitative tests was possible. Obtained results clearly confirm that using the additives increase efficiency of fuel combustion [9]. It is worth to notice that although sum of phenol and three selected polycyclic aromatic hydrocarbons (PAH) was lower in Cu-additive soot, the toxicity of this soot were the highest (Tab. 3).

The large number of primary components of fuels, complex combustion mechanisms and the heterogeneous interactions during soot formation made that soot is a mixture of various chemical components [10]. Our results shown that phenol, naphthalene, phenanthrene and anthracene were not responsible for soot toxicity on soil bacteria (Tab. 3). The highest toxicity of Cu-additive soot arise probably from contents of NaCl and Cu. Both NaCl and Cu are known as factors adversely affecting on the soil microflora [11, 12].

Using of fuel additives in coal combustion has positive influence on an industry, but also might have negative impact to the environment. The basic components of a majority of fuel additives are NaCl and Cu compounds. The use of NaCl has a positive effect on increasing the efficiency of the energy sources combustion, but also is responsible for releases of small quantities of nitrogen oxides (NO_x) and carbon monoxide (CO). On the other hand the compound during its combustion is a source of gaseous chlorine and hydrogen chloride and it is responsible for heating devices corrosion. Those gases contribute to the formation of particularly dangerous organochlorine compounds like dioxin. Copper salts not only efficiently catalyze oxidation of soot, but also may be responsible for a formation of other toxic xenobiotics [13].

It can be concluded that soot obtained from combustion with newly-developed additive with Fe-compound is not only less toxic but also will be more quickly removed from the environment.

Both *Bacillus* and *Pseudomonas* are typical soil bacteria [14]. The highest biodegradation rates were shown for *P. aeruginosa* (Tab. 3). This result can be attributed to a wide spectrum of hydrocarbon degradation, which is characteristic for bacteria of the genus *Pseudomonas* [15].

References

1. T.C. Bond, S.J. Doherty, D.W. Fahey, P.M. Forster, T. Berntsen, B.J. Deangelo, M.G. Flanner, S. Ghan, B. Kärcher, D. Koch, S. Kinne, Y. Kondo, P.K. Quinn, M.C. Sarofim, M.G. Schultz, M. Schulz, C. Venkataraman, H. Zhang, S. Zhang, N. Bellouin, S.K. Guttikunda, P.K. Hopke, M.Z. Jacobson, J.W. Kaiser, Z. Klimont, U. Lohmann, J.P. Schwarz, D. Shindell, T. Storelvmo, S.G. Warren, JGR: Atmospheres **118**, 5380 – 5552 (2013)
2. J. Bachmann, *Black Carbon: A Science/Policy Primer, Pew Center on Global Climate Change* (2009)
3. R.F. Service, Science **319**, 1745-1745 (2008)
4. W. Gehling, B. Dellinger, Environ. Sci. Technol. **47**, 8172-8178 (2013)
5. P. Lu, C. Li, G. Zenga, X. Xiea, Z. Caia, Y. Zhoua, Y. Zhaoa, Q. Zhana, Z. Zenga, J. Hazard. Mater. **199**, 272 – 281 (2012)
6. J. Lehmann, M. C. Rillig, J. Thies, C.A. Masiello, W.C. Hockaday, D. Crowley, Soil Biol. Biochem. **43**, 1812-1836 (2011)
7. Ch. Anyika, Z.A. Majid, Z. Ibrahim, M.P. Zakaria, A. Yahya, Environ. Sci. Pollut. Res. **22**, 3314-3341 (2015)
8. T. Farjon, O. Boinet, *Soot disaggregating combustible agent,* patent USA No. 5882365
9. E. Lipstein, *Combustion catalyst*, patent USA No. 8048234
10. H. Omidvarborna, A. Kumar, D-S. Kim, Recent studies on soot modeling for diesel combustion, RENEW SUST ENERG REV Renew. Sust. Energ. Rev. **48**, 635–647 (2015)
11. Ch.L. Dupont, G. Grass, Ch. Rensing, Metallomics **3**, 1109-1118 (2011)
12. M. Diaz-Raviña, E. Bååth, Appl. Environ. Microbiol. **62**, 2970–2977 (1996)
13. M. Chyc, Research Reports of Central Mining Institute, Mining Environment **1**, 5-17 (2012)
14. D. Hank, N. Saidani, A. Namane, A. Hellal, J. Eng. Sci. Technol. Rev. **3**, 123-127 (2010)
15. M. Kumar, M.P. Singh, D.K. Tuli, Adv. Microbiol. **2**, 26-30 (2012)

DOI: 10.1051/e3sconf/20161000008

The hybrid two stage anticlockwise cycle for ecological energy conversion

Piotr Cyklis[1,a]

[1]*Politechnika Krakowska, Department of Mechanical Engineering, Institute of Power and Process Engineering Poland*

Abstract. The anticlockwise cycle is commonly used for refrigeration, air conditioning and heat pumps applications. The application of refrigerant in the compression cycle is within the temperature limits of the triple point and the critical point. New refrigerants such as 1234yf or 1234ze have many disadvantages, therefore natural refrigerants application is favourable. The carbon dioxide and water can be applied only in the hybrid two stages cycle. The possibilities of this solutions are shown for refrigerating applications, as well some experimental results of the adsorption-compression double stages cycle, powered with solar collectors are shown. As a high temperature cycle the adsorption system is applied. The low temperature cycle is the compression stage with carbon dioxide as a working fluid. This allows to achieve relatively high COP for low temperature cycle and for the whole system.

1 Introduction

The energy conversion, as well as the heat transport from sources of low to high temperatures requires the application of a thermodynamic cycle. The anticlockwise cycle is used for heat transport in refrigeration, air conditioning and heat pumps applications. The most efficient cycle in terms of COP (Coefficient of Performance) is still the compression Linde cycle. The compression cycle requires a refrigerant as a working fluid. Due to the Montreal 1982 and later EU regulations the application of the F-gases as refrigerants is limited. The low ODP (Ozone Depletion Potential), GWP (Global Warming Potential) and TEWI (Total Equivalent Warming Impact) factors substances are needed. The newly developed refrigerants such as R1234yf and others have many disadvantages. This is favourable for natural refrigerants such as carbon dioxide, water and ammonia. However the natural refrigerants have strong temperature application limits. The theoretical application limits for the Linde cycle is within the temperatures of the triple and critical points. Therefore the cooperation of the two energy carriers as for example water and carbon dioxide in the double stage anticlockwise refrigeration cycle is the interesting subject of the investigation and development. In the paper the possibilities of the two stages hybrid ecological refrigerating cycle will be presented. The possibilities of the chosen solutions, as well as some experimental results of the adsorption-compression double stages cycle powered with solar collectors for refrigeration will be shown.

As a high temperature cycle the adsorption system is applied where the waste or renewable energy can be used as the energy source. Using this solution the low COP of the high temperature adsorption cycle is not influencing the total energy consumption. The low temperature cycle is the compression stage with carbon dioxide as a working fluid. This allows to achieve relatively high COP for the low temperature cycle and for the whole system. The system is working at the Laboratory of Thermodynamics and Thermal Machines Measurements. The possibility of the whole year system use (as a heat pump [1], refrigeration and air conditioning setup) makes the economy of the system more economically efficient.

2 Hybrid double stage cycle

To apply the CO_2 compression system for refrigeration there are two possibilities. First is the one stage compression refrigeration system. In this case the cooling temperature for the ambient air cooled heat exchanger is 45-55°C. This temperature exceeds the critical point temperature of the CO_2, so the condensation is not possible. Therefore the gas cooler instead of the condenser is applied. This has important disadvantages such as: big gas cooler dimensions, low system COP, and high pressure in the gas cooler [2]. Second possibility is the application of the double stages system where CO_2 low temperature (LT) stage condenser is cooled by the high temperature (HT) stage evaporator. As the HT stage the other compression cycle (for example R410) may be used, but this requires additional electric power for the HT stage compressor, lowering the total COP of the system. The application of the sorption cycle for the HT stage has many advantages, such as: water as the working fluid, waste

[a] Corresponding author: pcyklis@mech.pk.edu.pl

or renewable heat utilisation, etc. The absorption or the adsorption cycles may be used. One design concept is shown in figure 1. The absorption LiBr-H_2O cycle is applied as the HT stage [3-6]. The advantage of this cycle is the ecology of the working liquid. There are several absorption cycle designs with different COP, shown in figure 2. The disadvantage of the absorption system is that the required desorber temperature is at least 90°C for the one effect system [3]. The half effect cycle allows for lower temperature supply, but the efficiency is lower and design and operation is more complicated.

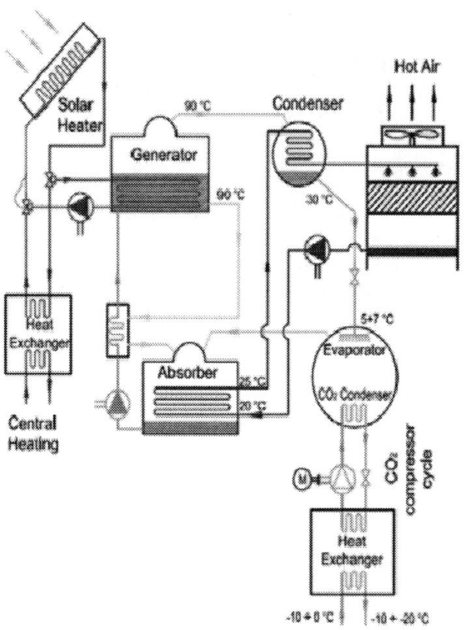

Figure 1. The concept of the ecological hybrid absorption-compression system [4].

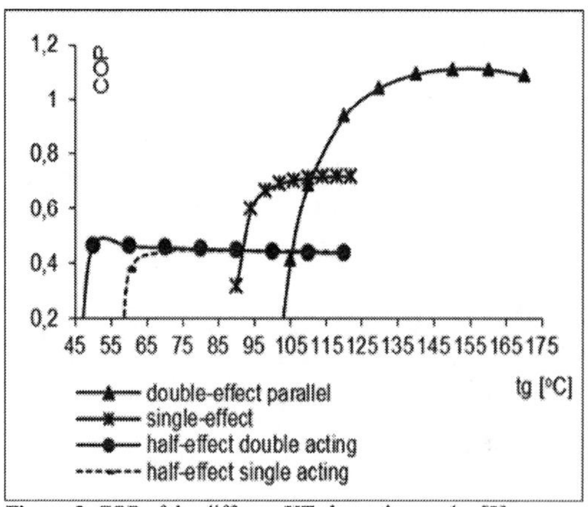

Figure 2. COP of the different HT absorption cycles [3].

If the renewable or waste heat source temperature exceeds 90°C, the refrigeration absorption cycle shown in figure 1 may be applied with high total COP.

In the Mid-European climate conditions the temperature in the solar collectors is usually low, and frequently the waste heat sources operate with lower temperatures. In this case the HT cycle based on adsorption is more applicable because it may operate with the desorption heat source as low as 55°C [7,8].

The concept of the hybrid cascade refrigeration system with the adsorption cycle is presented in figure 3. In figure 4 the COP for 75°C desorption temperature is shown. The COP for the adsorption system depends of the t_m re-cooling temperature. Therefore the water sprayed cooling tower is the best choice for the re-cooling.

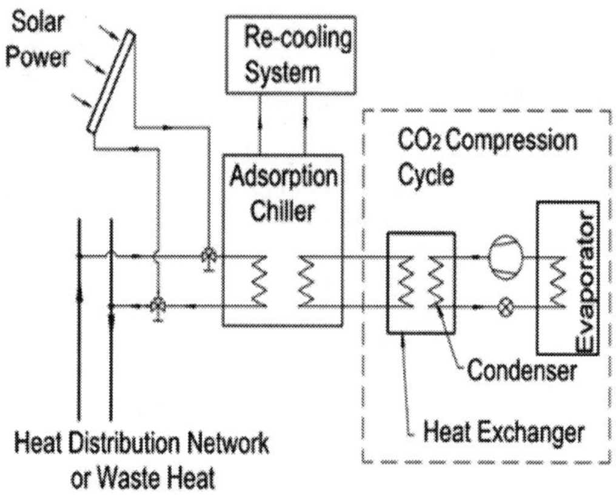

Figure 3. Two stage adsorption compression system for refrigeration [4] [7].

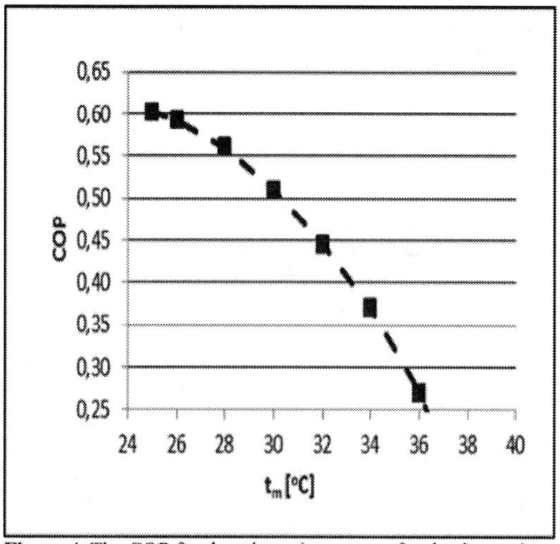

Figure 4. The COP for the adsorption system for the desorption temperature 75°C with the respect to the re-cooling temperature.

The efficiency for the adsorption cycle is shown in figure 4, for 75°C. The COP in the shown case is related to the re-cooling temperature, where the heat from the sorption cycle is removed to the ambient air. The adsorption chiller works periodically, with 2, 4 or 8 adsorption/desorption beds where the water inflow for cooling or heating of the sorption bed is controlled automatically with the on/off valves, since the adsorption bed is made of solid materials such as for example zeolite.

DOI: 10.1051/e3sconf/20161000008

Nevertheless the periodic work of the adsorption chiller does not affect the LT stage CO_2 condensation [7]. This is due to the thermal capacity of the condenser. The results in terms of TEWI for the hybrid sorption compression system is significantly lower than the result of combined or single stage compression system. The energy efficiency is the challenge, since the compression cycles have higher COP than the sorption systems. Therefore the investigation presented here has been worked out for the energy efficiency comparison.

3 Laboratory tests

In the Laboratory of Thermodynamics and Thermal Machines Measurements at the Cracow University of Technology the refrigeration hybrid compression adsorption laboratory stand was built. The main elements are shown in figure 5.

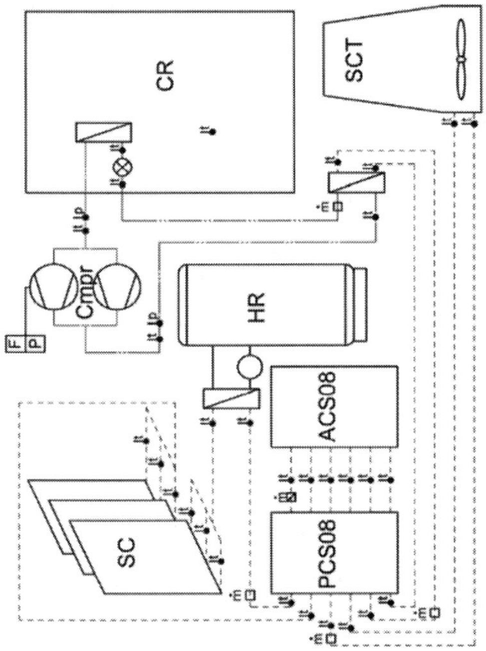

Figure 5. The laboratory stand with hybrid adsorption-compression cycle for refrigeration. SC- solar collectors, PCS08- pumps with controls and valves, ACS08-adsorption system, Cmpr- compressors, SCT- Sprayed Cooling Tower DECSA REF-C 05, HR- heat reservoir, CR- Refrigeration chamber. Measurements: t- temperature, m-flow, p-pressure.

The details of the system are shown in figs 6,7,8. The adsorption chiller made by Sortech ACS08 [9] is used, for the compression cycle two of the DORIN CD300H reciprocating compressors are applied. The solar collectors are tube type with about 16m^2 surface area. During the night time when no sun radiation is available the ACS08 unit is switched off and the CO_2 condenser is cooled directly from the water sprayed cooling tower DECSA-C05. The system was put into operation in 2011/12 and working until now. During this time the adsorption chiller was changed in 2014 for a newer, more efficient unit Ecocool also by Sortech.

Figure 6. The LT refrigeration compression stage with mass flow meter below, and compressor control case.

Figure 7. Pipe connections to the HT sorption refrigeration cycle, with thermometers and flow meters.

The measuring equipment used is as follows:

- glycol flow: turbine flow meters HOFFER HO1/2x1/2 class of accuracy 0.25% with impulse meters KEP BATRTM2AC class 0.01%,
- CO_2 flow: SIEMENS SITRANS F C MASSFLO 2000 (Coriolis type) class 0.1% on liquid side,
- Pressure: Vegabar 17 transducers class 0.5%,
- temperature: Pt100, current loop class 0.3%,
- electric power: LUMEL M30P-LCTM74 meters class 0.5%.
- Data acquisition system is realized with eight 16-bit ICP DAS measuring cards with range ±20mA, 0~20mA, 4~20mA class 0.1%.

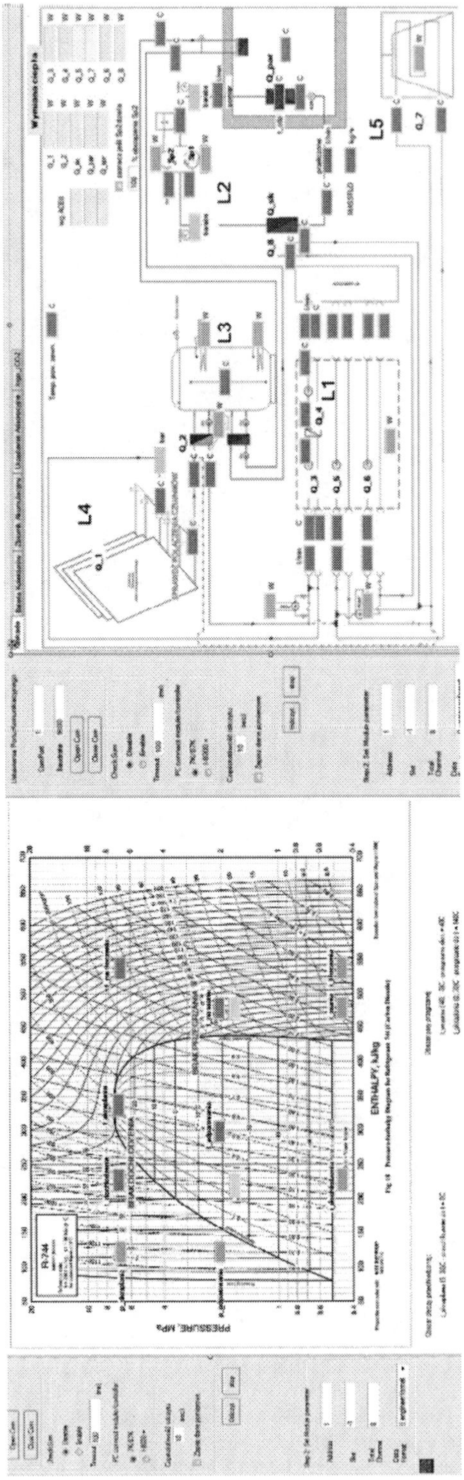

Figure 8. Screen view of the system and its operation (compressor refrigeration cycle shown on the pressure-enthalpy chart.).

- For data acquisition PC software have been prepared with visualization possibilities. The system has been connected via the RS485 port using DCON protocol.

Calculated measurement uncertainty for enthalpy of all liquids better than 4.3%, calculated uncertainty for electric power reading including averaging 1%.

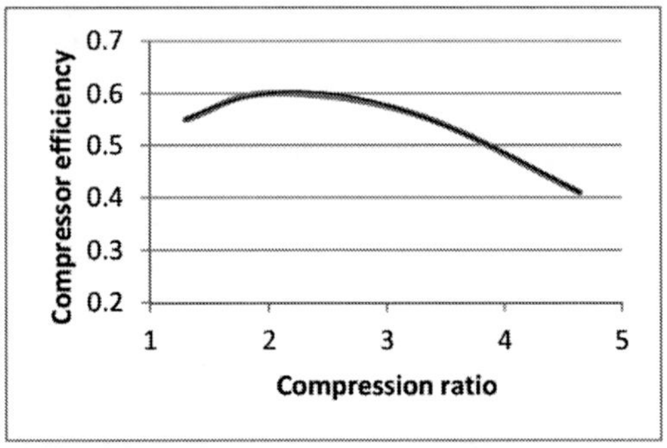

Figure 9. CO_2 compressor efficiency related to the compression ratio.

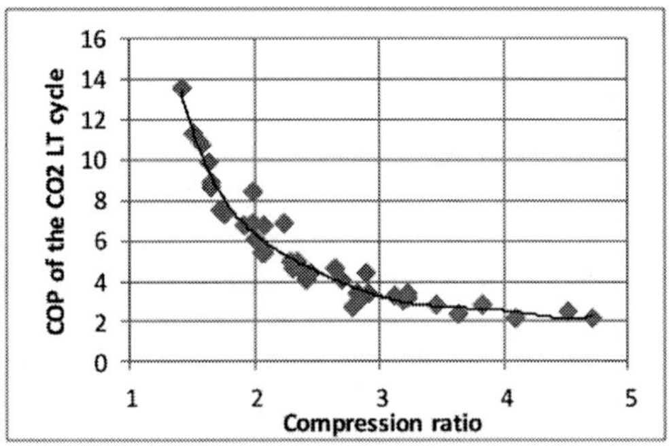

Figure 10. The COP of the LT refrigeration compression cycle.

The compressor efficiency calculated on the basis of the measurement results is shown in fig. 9 with the respect to the compression ratio. The real compressor efficiency has a significant influence on the total system COP (Coefficient of Performance), through the LT cycle efficiency. The COP for the compressor refrigeration system means the ratio of the cooling power related to the compressor power. This relation is shown in figure 10. In case of the full double stage hybrid cycle, the COP is defined as the ratio of the cooling power to the total electric energy supplied for the system. In general this includes also measuring equipment, fans, pumps etc.

The COP calculated on the basis of the experimental results for the presented system is shown in figure 11, compared with two ideal cycles: Carnot cycle and ideal Linde cycle with the compressor efficiency equal to 1. This chart shows the possibility and limits for increasing the cycle efficiency using better compressors and other possibilities for the real system efficiency improvements. Important part of the system is the automatic control algorithm and design which is shown in the other paper.

DOI: 10.1051/e3sconf/20161000008

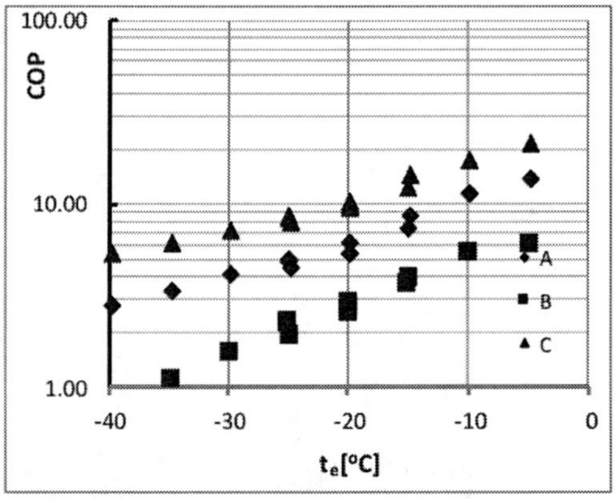

Figure 11. Measured COP of the CO_2 compression cycle comparison for the condenser temperature within 4-7°C. A-calculated Linde ideal cycle, B – real measured cycle, C – Carnot efficiency

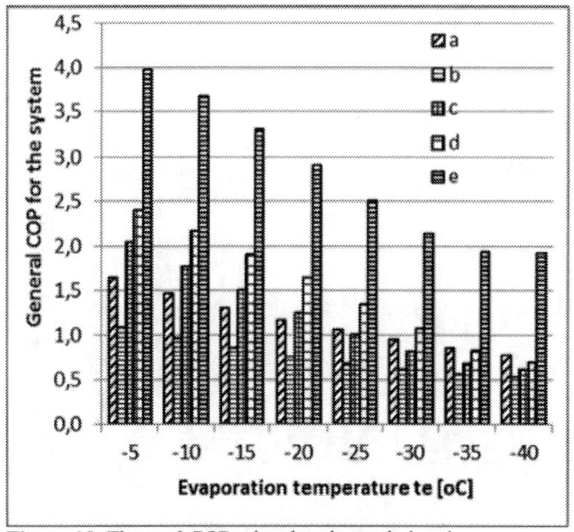

Figure 12. The total COP related to the total electric energy consumption of the system, for five cycles: a-two stages compression R410+CO2, b-transcritical one stage CO2, c-CO2 compression with the condenser cooled by water sprayed cooling tower, d–double stage hybrid adsorption compression CO2 , e-double stage hybrid with the recalculation for ideal compressor.

4 Analysis of the results

For the adsorption systems powered by the solar radiation, the heat container allows to store the solar heat for a few hours only. Therefore during the most part of a night, the CO_2 condenser, has to be cooled directly by the cooling tower. In the night in summer the temperature from the cooling tower may be lower than the ambient temperature if the water sprayed cooling tower is applied instead of commonly used dry re-coolers (dry heat exchangers). The condensation temperature in the compression LT cycle rises then, from optimal 3-5°C up to 15-25°C in Mid-European conditions. In most cases 17-18°C was obtained during the tests in hot June-August period in the nights in Poland.

The COP of the CO_2 compression cycle is related to the condensation temperature, through the compression ratio and the cycle efficiency.

Five cycles have been compared in figure 12 and 13:

a) compression two stage cycle with CO_2 and R410A,

b) transcritical CO_2 one stage compression cycle,

c) hybrid adsorption-compression with cooling directly from sprayed cooling tower (about 18°C in the CO_2 condenser),

d) hybrid adsorption-compression with cooling from adsorption (about 5°C of CO_2 condenser).

e) hybrid adsorption compression as in point d) but with the compression efficiency ideal

The results of the COP calculations for those cycles are shown in figure 12.

Comparison of the two stages systems presented in figs 12 and 13, shows that in most cases the electric energy consumption, which is directly related to the COP, is lower for hybrid system, so in case that the thermal energy is available the hybrid system has all advantages, due to the fact that the TEWI is in each case lower than all other systems.

The TEWI coefficient was calculated for comparison for the same refrigeration power according to the formula:

$$TEWI=GWP*L*n+GWP*m*(1-f)+n*E*z \quad (1)$$

$$E=N*t_r \quad (2)$$

Where:

GWP - Global Warming Potential with relationship to CO_2,
L - emission to the atmosphere (kg/yr),
n - life of the cycle (yr),
m - amount of refrigerant in installation (kg),
f - recovery of the refrigerant (-),
E - energy use of the cycle per year (kWh),
z - CO_2 emission for the electric energy; Poland *z*=0,94 (kgCO_2/kWh),
N - supplied electric power of the cycle (kW),
t_r - yearly number of working hours (h).
(Better TEWI means lower value, better COP means higher value)

Another comparison for the refrigeration system is the hybrid system compared to the one stage refrigerating compressor system with the commonly used R407C refrigerant.

The energy – COP comparison is shown in figure 14, and ecological –TEWI comparison in figure 15. The comparison is shown for the wide range of evaporation temperatures. The practical refrigerating application for the hybrid system is the condensing temperature 5-10°C. For the one stage refrigerating compressor cycle the real condensation temperature with the air cooled condenser is about 40-45°C.

DOI: 10.1051/e3sconf/20161000008

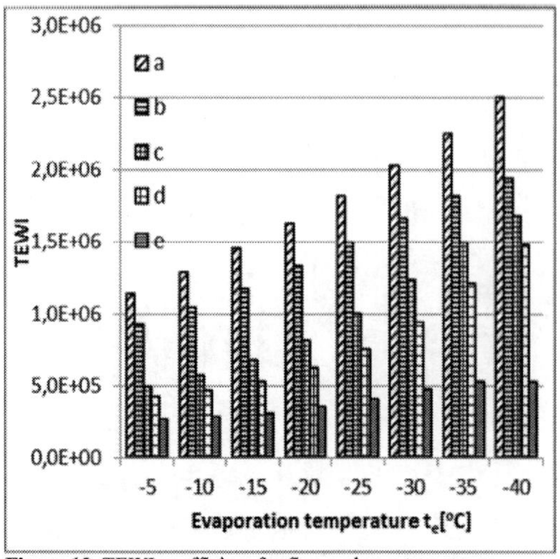

Figure 13. TEWI coefficient for five cycles: a-two stages compression R410+CO_2, b-transcritical one stage CO_2, c-CO_2 compression with the condenser cooled by water sprayed cooling tower, d–double stage hybrid adsorption compression CO_2, e-double stage hybrid with the recalculation for ideal compressor.

As an addition other condensing temperatures are depicted in fig 14 and 15. It can be noticed that the COP of the hybrid system is higher than one stage R407C compression system in each case when the condensing temperature comes from air heat exchanger (45°C) and for ice water cooling (10°C) the carbon dioxide compression cycle is more efficient below evaporation temperature -25°C. In each case TEWI index is lower for the investigated hybrid system.

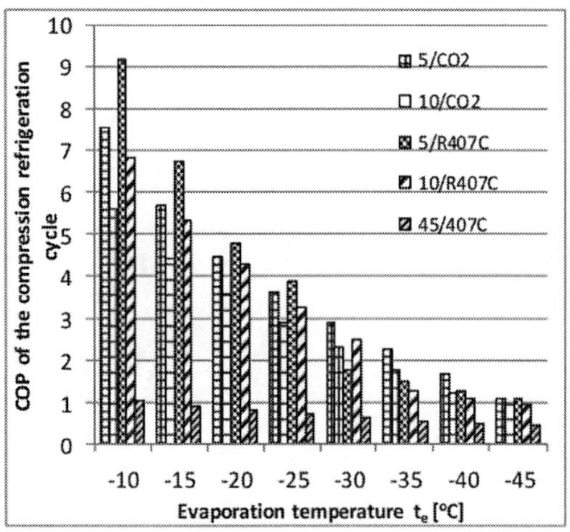

Figure 14. The COP coefficient of the compression refrigeration cycle related to the evaporation temperature for: 5/CO2-CO_2 cycle with condensation temperature 5°C, 10/CO2-for condensation temperature 10°C, 5/R407C,10/R407C,45/R407C R407 compressor refrigeration cycle for evaporation temperature respectively 5,10,45°C.

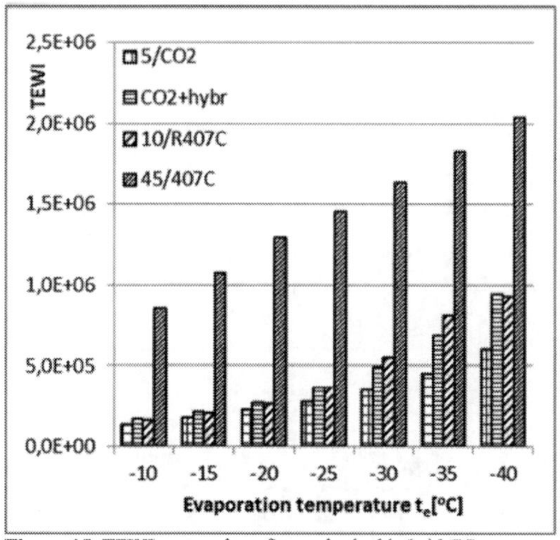

Figure 15. TEWI comparison for ecological hybrid CO_2 and adsorption cycle and one stage refrigeration compression cycle with popular R407C.

5 Conclusions

The sorption refrigeration cycles are still popular due to the possibility of the low temperature waste or renewable heat utilisation. In Asia, Australia and USA adsorption systems are commonly used, while in Europe the absorption systems are more popular. The absorption and adsorption systems are used usually for the air conditioning purposes or ice water production, however the possibilities of the application of the sorption technology in refrigeration is much wider. Due to the lack of the really non-harmful refrigerants and temperature limits for carbon dioxide, the application of the hybrid systems combining CO_2 compression cycle and adsorption HT is welcome. They can be used in refrigeration where the required evaporation temperature is -45°C up to -15°C. Properly designed hybrid refrigeration system assures both ecology and economy benefits. This investigation shows that when needed temperature for the evaporation is lower than -25°C the hybrid system has higher COP than commonly used refrigerants R407C or R410 single or double stage, even when the R407C cycle is ice water cooled. The water sprayed cooling tower allows for system operation when the HT cycle heat source is limited, without exceeding the critical point temperature. This makes it possible to operate the system powered by the solar radiation. In this case during the day the CO_2 condenser is powered by solar collectors and during the night or colder weather the cooling tower is enough to keep the temperature of the evaporator below 20°C. The total COP of the system is in this case lower, but the TEWI coefficient is always better. The flexibility of the cycles combination allows for different solutions such as additional air conditioning when HT adsorption cycle has higher cooling power than the refrigeration cycle. In winter time adsorption cycle may be reversed and used for heating as a heat pump. This makes the system more economically efficient.

Acknowledgements

This project has been sponsored by the Polish National Centre of Research and Development within grant no.: N R06 0002 10 0936/R/T02/2010/10

References

1. A. Grzebielec, *Archives of thermodynamics* **30**, (2009)
2. Y. Hiroshi and Z. Xin-Rong, *International Journal of Refrigeration* **32**, (2009)
3. P. Cyklis and G. Brak, *Transactions of the Institute of Fluid-Flow Machinery* **121**, (2008)
4. P. Cyklis and R. Kantor, *XXIII International Congress of Refrigeration*, proceedings, Prague, Czech Republic, August (2011)
5. Z. Danxing and M. Xuelin, *Energy Conversion and Management* **56**, (2012)
6. H. Wei, S. Liuli, D. Zheng, J. Hongguang, M. Sijun and J. Xuye, *Applied Energy* **106**, (2013)
7. P. Cyklis, *International Journal of Refrigeration,* **48**,(2014)
8. B. Saha, L. El-Sharkawya, A. Chakrabortya, S. Koyama, N. Banker, P. Duttab, M. Prasad I K. Srinivasana, *International Journal of Refrigeration* **29**, (2006)
9. SorTech, „SorTech Adsorption Chiller ACS 08/ACS 15, Design manual, Version 2.2, SorTech, Halle, (2009)

The process of fly ash magnetic separation impact on hydrothermal synthesis of zeolites

Natalia Czuma[1,a], Katarzyna Zarębska[1], Paweł Baran[1], Wojciech Franus[2]

[1]*AGH University of Science and Technology, Poland*
[2]*Lublin University of Technology, Poland*

Abstract. The aim of the experiment was modifying raw fly ash through the demagnetisation process and determining its influence on the efficiency of the synthesis of zeolite materials out of fly ash. A series of experiments have been performed on modified samples and, for the purpose of verification, on non-modified fly ash. No direct correlation has been confirmed in relation to synthesis efficiency as well as the type of obtained zeolite material. The research results indicate that the composition of fly ash determines the type of the zeolitic phases obtained. Based on the analysis of experimental data, it was found that the demagnetisation process allows to obtain additional zeolite phases, while there is no direct impact of the process used on the increase of synthesis reaction efficiency.

1 Introduction

Zeolites are crystalline aluminasilicates, formed by a continuous network of oxygen-sharing AlO_4 and SiO_4. They are characterized by a highly developed inner structure, which is responsible for very good sorption, catalytic and ion-exchange properties. Zeolites have regular and repeatable pores and channels with dimensions 0,4-0,15 nm, which enable, among other things the separation of gas mixtures made out of components of different active molecule diameter [1]. Due to high industrial demand for zeolites, different methods of obtaining synthetic zeolitic materials have been developed. Zeolites synthesized out of fly ash are an attractive alternative for natural zeolites as well as for zeolites synthesized out of pure chemicals [2].

Currently the processes of zeolite synthesis can be divided mainly into: fusion method, hydrothermal synthesis and molten-salt method. In each variant it is possible to use the methods jointly and/or introduce modifications enhancing the synthesis process. Due to its functionality, the most widespread method of obtaining synthetic zeolitic materials out of fly ash is hydrothermal synthesis [3,4]. It is a complex physicochemical process, taking place in liquid phase in alkaline environment. Crystalline and amorphous substances which dissolve and crystallize are involved in the process [1]. The hydrothermal method with the use of fly ash makes it possible to obtain zeolitic materials type X, A and P [3, 4, 5].

According to recent literature on the subject [1,6,7], the positive outcome of the synthesis is determined by the quality of fly ash, i.e. mainly the silica/alumina ratio, the selection of appropriate base solution and its amount, the synthesis conditions (temperature, time, pressure), the proportion of the solution used and the presence of additives.

One of the methods of fly ash modification discussed in literature is magnetic separation. The process is based on the removal of ferromagnetic minerals out of fly ash and its efficiency is determined by the intensity of the magnetic field used [1].

Magnetic separation-demagnetisation has been presented in literature as one of the methods of modifying fly ash before the synthesis, resulting in higher efficiency of the synthesis [8] or in the occurrence of new zeolite phases [9]. The aim of the process is to remove any ballast, which may have a negative influence on the formation of zeolites. Fly ash demagnetization can be performed with the use of the „wet method" [10], without the use of liquid [11] or by a combination of both methods [11]. The use of acids in order to remove the metallic components is also possible [12].

For the purpose of examining the influence of the magnetic separation process, a series of hydrothermal synthesis were performed, aimed at receiving zeolite material out of non-modified fly ash from Polish Power plant (marked with symbol L in the work) as well as the same fly ash subjected to demagnetisation process (marked with symbol LO in the work). Due to its relative simplicity and low cost (no excessive energy penalty), the dry demagnetisation method was used in the experiment.

The synthesis was performed based on seven formulas differing in the concentration of the base solution, time and temperature of synthesis as well as in the use

[a] Corresponding author: nczuma@agh.edu.pl

of synthesis enhancing additives. The aim of the experiment was to test whether the removal of ferromagnetic oxides will have a significant influence on the type of the zeolite material obtained and on the synthesis efficiency.

An additional advantage of the demagnetisation process is a potential possibility of industrial application of the residues from the demagnetisation process [10].

2 Materials

2.1 Non-modified fly ash

The fly ash marked L fly ash was used. The examined fly ash was characterized by Si/Al=2,45. The silica-alumina ratio in natural zeolites is approximately 1-6. From the value of Si/Al in the fly ash used one can infer that the fly ash examined will be appropriate for the zeolite synthesis [6, 13, 14].

The analysis of the XRF fly ash data before the modification made it possible to determine the qualitative and quantitative composition of fly ash: Na_2O-2,3%, MgO-1,769%, Al_2O_3-15,706%, SiO_2-38,449%, P_2O_5-0,743%, SO_3-1,038%, K_2O-2,773%, CaO-3,578%, TiO_2-1,001%, V_2O_5-0,044%, Cr_2O_3-0,045%, MnO-0,168%, Fe_2O_3-10,085%, NiO-0,024%, CuO-0,085%, ZnO-0,987%, Ga_2O_3-0,006%, As_2O_3-0,023%, Rb_2O-0,014%, SrO-0,047%, Y_2O_3-0,004%, ZrO_2-0,03%, Nb_2O_5-0,003%, SnO_2-0,022%, BaO-0,094%, PbO-0,298%.

It was demonstrated that the fly ash contained the highest percentage of silicon and aluminum oxides along with a significant share of iron oxides, which prompted the decision to choose this fly ash for the research.

The mineral composition of a fly ash sample L was determined via powder X-ray diffraction. The XRD measurements (of both substrates and products) were carried out using PANalytical – Empyrean diffractometer, equipped with Cu-Kα (λ=1.5406Å) radiation source, within the 2θ range from 5-90° and in a step size of 0.02 °/min. XRD measurements were performed in temperature 25°C. The diffractogram (Fig. 1) presents the fly ash sample L. It can be observed that the diffraction intensities characteristic for m-mullite, q-quartz, Fe-magnetite and hematite are present.

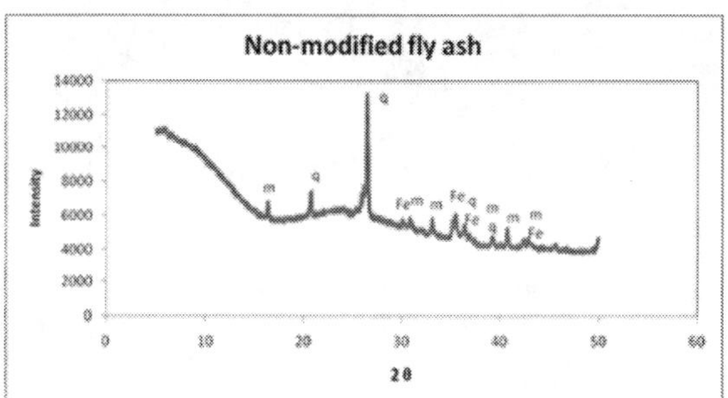

Figure 1. XRD diffractogram of non-modified fly ash

The iron oxides identified in fly ash in the form of magnetite and hematite derive mainly from the decomposition and oxidation during the combustion proces of the iron-bearing compounds from the coal (pyrite, siderite, ankerite, Fe^{2+} -illite –clay mineral) [10].

Apart from magnetic components, the chemical composition of ferromagnetic compounds defined as ferrospheres is silica, alumina, calcium etc. [10]. These forms have a characteristic structure that may be recognised on SEM microphotographs.

In order to verify the presence of ferrospheres and attempt to identify them, SEM microphotographs of non-modified fly ash were taken (Fig. 2). The SEM microphotograph shows spherical forms characteristic of fly ash; additionally a number of amorphous forms can be observed which most probably originate from mullite, quartz and unburned carbon. In the center of the microphotograph, a spherical form with distinctive spots can be seen. It is assumed to be a ferrosphere with a smooth surface with ferromagnetic elements present.

Figure 2. SEM microphotograph of non-modified fly ash

A further microphotograph analysis (Fig. 3) allowed to confirm the presence of spheres with a rough surface in the investigated material. By analogy to the ferrospheres described in literature, it is suspected that this stucture will probably be composed of magnetic elements.

In order to prove this hypothesis, the EDS analysis was performed.

Figure 3. SEM microphotograph of ferrosphere, a fly ash particle

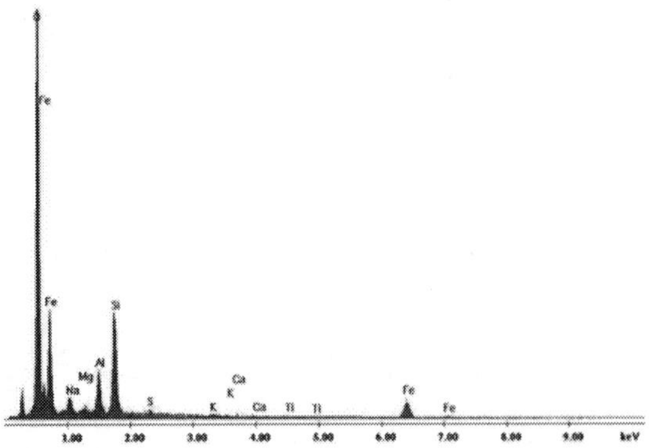

Figure 4. EDS spectrum of particle present in the Fig. 3

The analysis results presented in Fig. 4 reveal a significant iron content, which proves that the rough structure is a ferrosphere, composed mainly of iron, with silicon and aluminium

2.2 Modified fly ash

The process of the demagnetisation of fly ash was performed in a way presented schematically in Fig. 5.

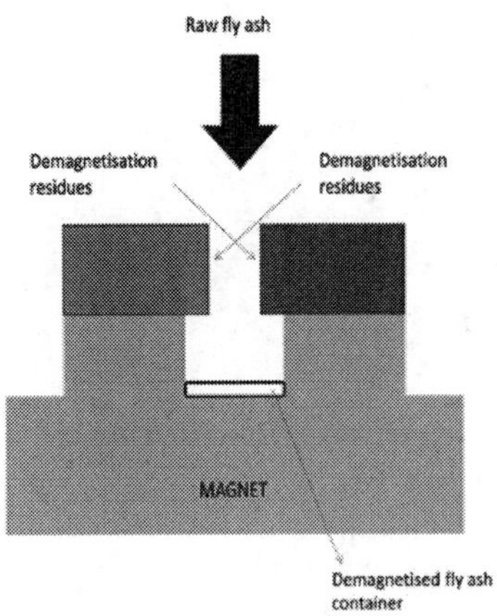

Figure 5. Scheme of the demagnetization process (own elaboration)

The magnetic separation process used in the experiment resulted in obtaining fly ash of the following composition: Na_2O-2,141%, MgO-1,7%, Al_2O_3-15,443%, SiO_2-38,651%, P_2O_5-0,719%, SO_3-1,35%, K_2O-2,773%, CaO-3,298%, TiO_2-0,992%, V_2O_5-0,043%, Cr_2O_3-0,034%, MnO-0,123%, Fe_2O_3-7,104%, NiO-0,019%, CuO-0,081%, ZnO-0,97%, BaO-0,071%.

The diffraction patterns of modified (demagnetised) fly ash have been presented in Fig. 6. XRD data analysis, confirmed the presence of diffraction intensities characteristic for m-mullite and q-quartz.

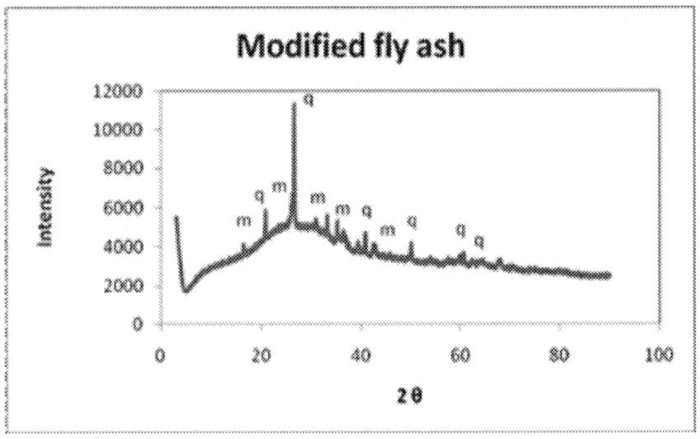

Figure 6. XRD diffractogram of modified fly ash-demagnetized

2.3 Fly ash demagnetisation residues

Additionally, the XRD analysis of residues from the demagnetisation process was performed, SEM microphotographs were taken and the EDS analysis carried out (Fig. 8, 9) . The diffractogram (Fig. 7) shows the peaks characteristic of m-magnetite and h-hematite.

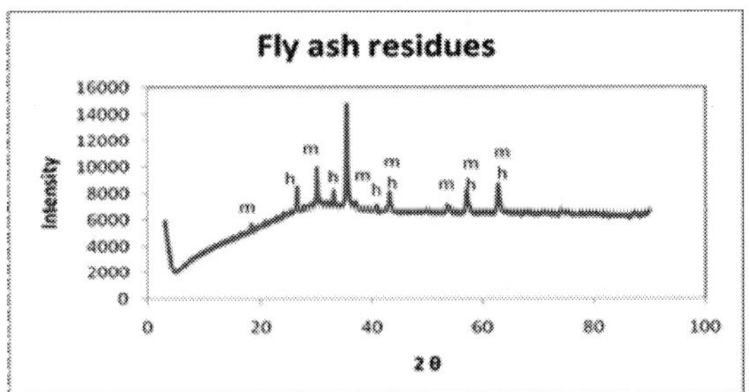

Figure 7. XRD diffractogram of magnetic residues after the demagnetisation of fly ash

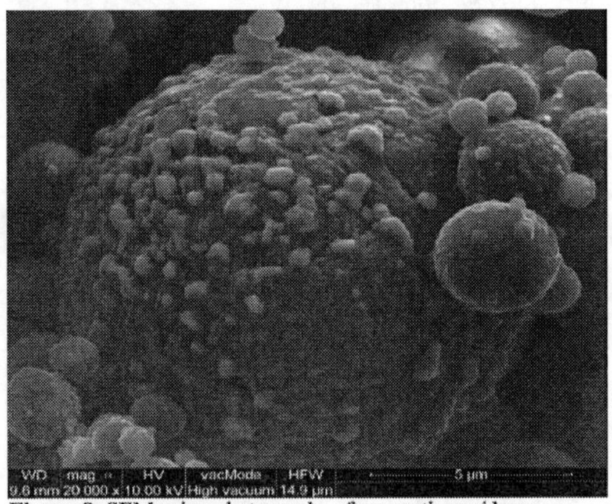

Figure 8. SEM microphotograph of magnetic residues after thedemagnetisation process

SEM-EDS data point taken at the centre of sphere presented in the Fig. 8, showed the predominant iron content in the chemical composition of the investigated point.

DOI: 10.1051/e3sconf/20161000009

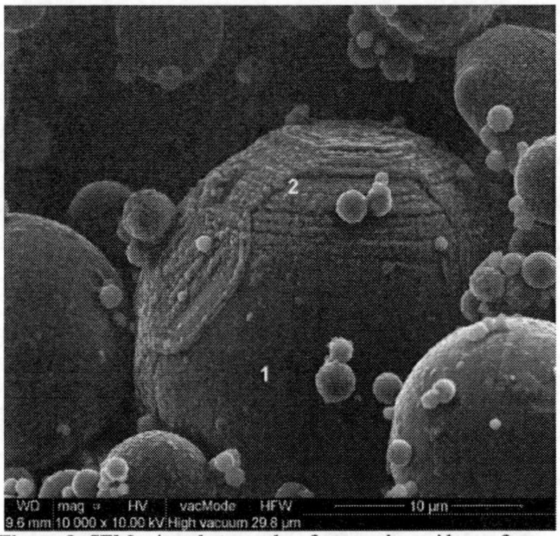

Figure 9. SEM microphotograph of magnetic residues after demagnetization process-microphotograph 2

An EDS analysis in point marked as 1 (Fig. 9) demonstrated the presence of iron and silicon, which proves that the sources of silicon are also being removed in the process of demagnetisation. The rougher surface in point marked as 2 is characterized by significant iron content.

In order to compare the impact of the modification process on the physical properties of fly ash, a helium density analysis was performed. The results have been presented in table 1.

Table 1. Helium density of non-modified and modified fly ash

	Density, g/cm^3
Non-modified fly ash	**2,3211**
Modified fly ash	**2,2333**

The analysis of the helium density showed an expected decrease in the density of the fly ash subjected to modification, which was caused by the removal of higher-density material such as hematite, magnetite.

3 Experimental

For the purpose of the experiment, the hydrothermal synthesis method was adopted. It was performed on non-modified and modified fly ash samples, using 7 different formulas for synthesis, in time ranges up to 24 hours, temperature from the range 80-105^{0}C and base solution concentration of 1-5 mol/dm^3. The main synthesis parameters are presented in the table 2.

As a result of the syntheses, light powders were obtained. The powders were rinsed up to pH~10, dried and analyzed with the XRD diffractometer in order to verify the effectiveness of the zeolite material synthesis and to estimate the influence of the demagnetization process on the zeolite phases obtained.The results of the synthesis performed have been presented in table 2.

Table 2. Comparison of zeolitic products and synthesis parameters obtained in Methods 1-7 from syntheses based on non-modified and modified fly ash

	Non-modified fly ash - L	Modified fly ash - LO	Concentration of base solution used, mol/dm^3	Time, hours	Additives	Temperature, ^0C
Method 1	A,P1	A,P1	2	24	-	80
Method 2	A, X	A, X	3	24	-	80
Method 3	P1, A	X, sodalite	1	24	NaCl	80
Method 4	sodalite	sodalite	5	24	NaCl	105
Method 5	A, chabazite	A, chabazite	3	7	Al	95
Method 6	A, sodalite	A,sodalite	1,5	7	Al	95
Method 7	P1, X	P1, X, A	2	22	-	95

In table 3 the comparison of peak intensities from zeolite phases obtained from samples synthesised out of L and LO fly ash (estimated area under reflections), the aim of which was to compare the efficiency of the zeolite conversion process.

Table 3. Comparison of the efficiency of the zeolitic products obtained in Methods 1-7 from syntheses based on non-modified and modified fly ash

	Non-modified fly ash - L	Modified fly ash - LO
Method 1	higher efficiency	lower efficiency
Method 2	lower efficiency	higher efficiency
Method 3	different products received, no possibility of efficiency comparison	
Method 4	higher efficiency	lower efficiency
Method 5	higher efficiency	lower efficiency
Method 6	lower efficiency	higher efficiency
Method 7	higher efficiency	lower efficiency

4 Results

In the Fig. 10-16 diffraction patterns for each of the syntheses performed have been presented. The results of the syntheses made using non-modified fly ash and the one subjected to modification process were compared in order to identify the differences resulting from obtaining a different type of zeolite material.

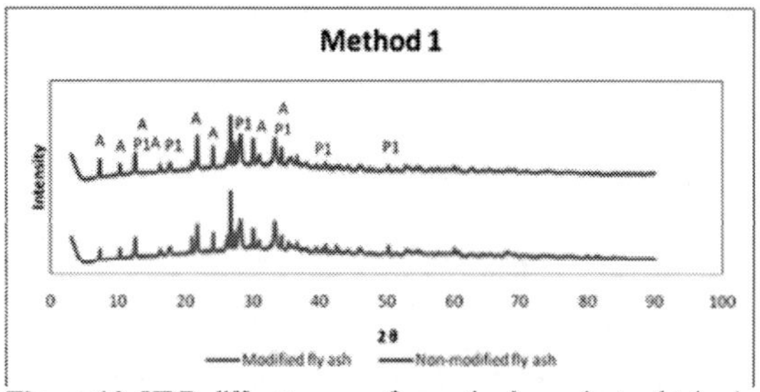

Figure 10. XRD diffractograms for synthesis products obtained using Method 1: red curve- the product of non-modified fly ash synthesis, blue curve- the product of modified fly ash synthesis (A-zeolite A, P1-zeolite P1)

In Method 1, based on a diffractogram analysis (Fig. 10) for the samples synthesised from non-modified and modified fly ash, it can be observed that some of the reflections overlap. For the reflections characteristic of zeolite phase A and P1, the ones from non-modified samples are more pronounced. By contrast, diffraction reflections of quartz are more significant for modified fly ash. Due to the lower conversion of quartz-Si source to the solution, lower diffraction intensities of zeolite phases are present in modified fly ash.

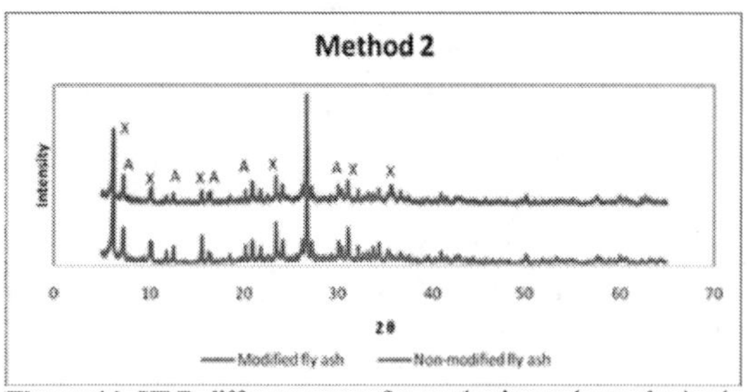

Figure 11. XRD diffractograms for synthesis products obtained from Method 2: red curve- the product of non-modified fly ash synthesis, blue curve- the product of modified fly ash synthesis (A-zeolite A, X-zeolite X)

Using Method 2, on the XRD diffractogram (Fig. 11), the overlaping of reflections were also observed, but there were differences in intensities. It is clearly visible that for a sample synthesised from modified fly ash, reflections of zeolite A and X are higher. The reflections characteristic of quartz are more intense for the sample synthesised from non-modified fly ash.

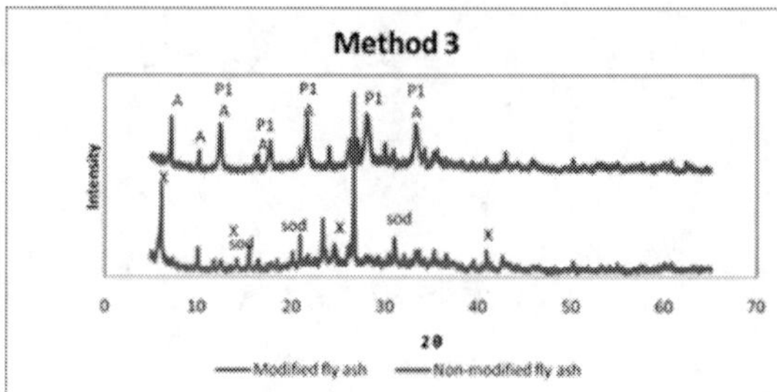

Figure 12. XRD diffractograms for synthesis products obtained from Method 3: red curve- the product of non-modified fly ash synthesis, blue curve-the product of modified fly ash synthesis (A-zeolite A, X-zeolite X, sod-sodlite)

In Method 3, based on the XRD diffractogram analysis (Fig. 12), it was found that reflections do not overlap. This proves the presence of other substances during the synthesis. For the sample synthesized from modified fly ash, the diffraction intensities characteristic of zeolite X of high intensity were present and those of sodalite. In the sample from non-modified fly ash the presence of zeolite A and P1 was confirmed. What is interesting is the presence of intensive peaks from quartz along with relatively high intensity peaks from zeolite X.

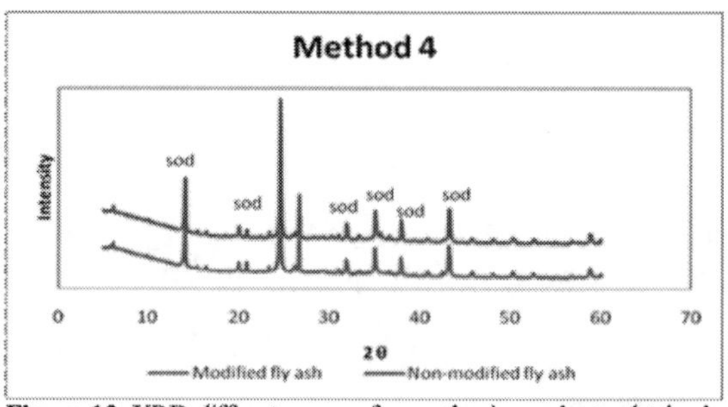

Figure 13. XRD diffractograms for synthesis products obtained from Method 4: red curve- the product of non-modified fly ash synthesis, blue curve-the product of modified fly ash synthesis (sod-sodalite)

Analyzing the synthesis results of Method 4 (Fig. 13), an overlap of diffraction intensities can be observed. The reflections from material synthesized from non-modified fly ash are more pronounced than those from modified fly ash. It is worth noticing that in this synthesis the addition of NaCl was used. It may be assumed that the type of the crystallizing zeolite was determined by this additive.

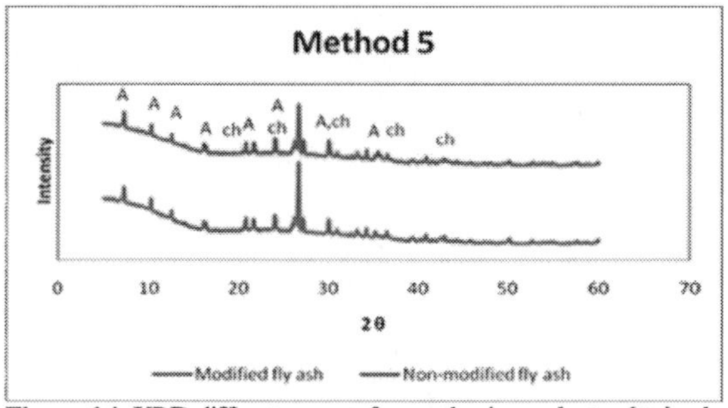

Figure 14. XRD diffractograms for synthesis products obtained from Method 5: red curve- the product of non-modified fly ash synthesis, blue curve-product of modified fly ash synthesis (A-zeolite A, ch-chabazite)

In Method 5, based on XRD diffractogram analysis (Fig. 14) it can be observed that the reflections from the substance synthesized with the use of non-modified fly ash overlap with most reflections from modified fly ash. The only reflections of higher intensity are representative of quartz.

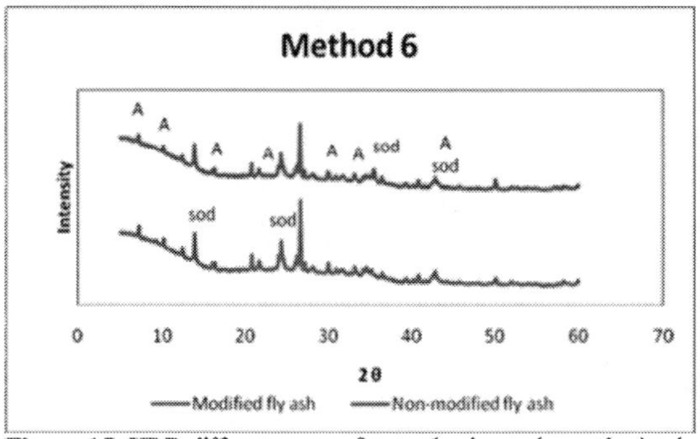

Figure 15. XRD diffractograms for synthesis products obtained from Method 6: red curve- the product of non-modified fly ash synthesis, the blue curve-product of modified fly ash synthesis (A-zeolite A, sod-sodalite)

Looking at the Method 6 XRD diffractograms (Fig. 15), a slight difference in reflections intensities may be observed. The modified fly ash peaks had higher intensity and in particular the zeolite A and quartz peaks are more intense.

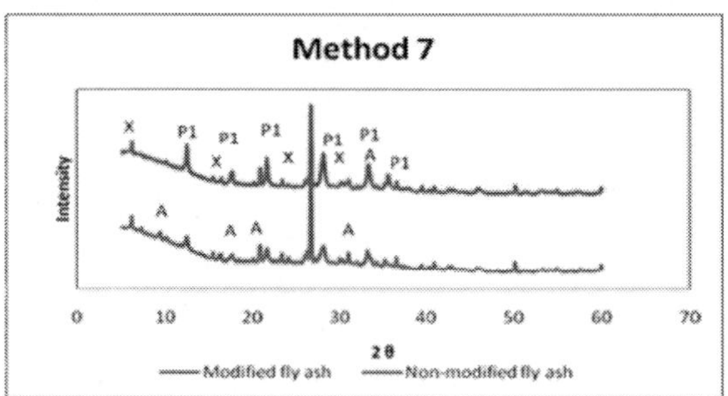

Figure 16. XRD diffractograms for synthesis products obtained from Method 7: red curve- the product of non-modified fly ash synthesis, blue curve-product of modified fly ash synthesis (X-zeolite X, A-zeolite A, P1-zeolite P1)

In Method 7 (Fig. 16), reflection intensities of peaks from zeolite phases are higher for samples synthesized from non-modified fly ash. For a sample from modified fly ash the presence of small reflections of zeolite A was noticed, which was not observed for the non-modified fly ash sample.

5 Conclusions

In theory, the demagnetisation of fly ash involving ballast removal should increase the efficiency of the zeolite synthesis from fly ash.

However, the results obtained in the research reveal that there is no direct correlation when it comes to the influence of the process of the fly ash demagnetisation on the synthesised material. The type of products obtained and the XRD diffraction intensities do not present a consistent pattern.

Only in one case (Method 3) different zeolite phases were identified, depending on the type of fly ash used for the synthesis.

It should be emphasised that in this method the base solution of the lowest concentration of 1 mol/dm^3 was used. In Method 7 the presence of a new zeolite phase was observed when a modified fly ash sample was used.

In the remaining examples the synthesis products differed in the diffraction intensities (the surface area under the most intensive peaks characteristic of a compound), which in approximation provides information about the quantity of the material obtained.

In Methods 5 and 6, it was noted that the addition of aluminum during the synthesis determined the crystallisation of a specific zeolite phase, i.e. zeolite A, due to obtaining a Si/Al ratio supporting the zeolite A synthesis. According to sources [7], the ratio Si/Al, should be in ranges 1-1,5 to favour zeolite A synthesis.

In the analysis of the syntheses results, it was helpful to identify zeolite phases as well as determine reflections intensities, understood as the surface area under the most intensive peaks characteristic of the given phase. Taking this criterion into account, in four samples (Method 1,4,5 and 7) higher reflection intensities from zeolite phases synthesised from non-modified fly ash were identified. In two samples (Method 2 and 6) on the other hand, higher reflection intensities were observed for modified fly ash samples.

In the syntheses performed, the only factor which had an impact on the type of the zeolite phase obtained was the fly ash demagnetisation process.

On the basis of the research conducted, a conclusion has been reached that in most cases the modification of raw fly ash through demagnetisation did not have the expected influence on the zeolite synthesis when it comes to the efficiency of the zeolite phase crystallisation.

It is not possible to draw a definitive conclusion whether the demagnetisation process of fly ash will impact the zeolite crystallisation process positively or negatively, depending on the synthesis parameters. Neither it is possible to determine if the demagnetised fly ash will increase or decrease the efficiency of the synthesis of the zeolitic material. However, since a new zeolite phase was obtained in one case when modified fly ash was used, it may be inferred that it is possible to obtain new zeolite phases providing that favourable parameters are used. The above analysis of the results obtained points to the necessity of conducting a parallel synthesis experiment on fly ash samples derived from a different power station.

Acknowledgements

The paper was financed from the AGH University of Science and Technology Statutory Research Fund no. 11.11.210.244.

References

1. T. Suchecki, Zeolity z popiołów lotnych. Otrzymywanie i aplikacje w inż. Środowiska, Ossolineum (2005)
2. B. Szala, P. Turek, A. Jeleń, T. Bajda, *Inżynieria Środowiska*, **150**, 30 (2013)
3. W. Franus, M. Wdowin, M. Franus, *Environ. Monit. Assess*. **186**, 9, 5721-5729 (2014)
4. M. Wdowin, M. Franus, R. Panek, L. Badura, W. Franus, *Clean Technol. Envir.* **16**, 6, 1217-1223(2014)
5. X. Querol, J.C. Umaña, F. Plana, A. Alastuey, A. López-Soler, A. Medinaceli, A. Valero, M.J. Domingo, E. Garcia-Rojo, *Fuel*, **80**, 857–865 (2001)
6. Wang Yanxin, Guo Yonglong, Yang Zhihua, Cai Hesheng, Xavier Querol, *Sci China Ser D*, **46**, 9 (2003)
7. D. Sztekler, I. Majchrzak-Kucęba, W. Nowak, Monografia pod red. J. Ozonka, M. Pawłowskiej, **1**, 58, 313-320 (2009)
8. R. Sommerville, R. Blissett, N. Rowson, S. Blackburn, *Int J Miner Process*, **124**, 20–25 (2013)
9. A.M. Cardoso, A. Paprocki , L.S. Ferret , C.M.N. Azevedo, M. Pires, *Fuel*, **84**, 1351–1363 (2005)
10. Qing-feng Xue, Sheng-gao Lu, *J Zhejiang Univ-Sc A*, **9**, 11, 1595-1600 (2008)
11. A.G. Anshits, O.M. Sharonova, N.N. Anshits, S.N. Vereshchagin, E.V. Rabchevskii, L.A. Solovjev, World of Coal Ash (WOCA) Conference-May 9-12,2011, in Denver, CO, USA
12. K. Ojha, N.C Pradhan, A. N. Samanta, *B Mater Sci*, **27(6)**, 555-564 (2004)
13. Duangkamol Ruen-ngam, Doungmanee Rungsuk , Ronbanchob Apiratikul , Prasert Pavasant, *Japca J Air Waste Ma*, **59**, 10 (2009)
14. H.A. Ibrahima, A.M. El-Kamasha, M. Hanafyb, N.M. Abdel-Monemb, *Chem Eng J*, **144**, 1, 67–74 (2008)

Analysis of the influence of fusion synthesis parameters on the SO_2 sorption properties of zeolites produced out of fly ash

Natalia Czuma[1,a], Katarzyna Zarębska[1] and Paweł Baran[1]

[1]*AGH University of Science and Technology, Poland*

Abstract. Fly ash zeolites are interesting alternative for zeolies produced out of pure chemicals as well as for natural zeolites. It can be applied as SO_2 sorbents out of flue gases. In the work the experiments were aimed at investigation of SO_2 sorption capacity of samples produced out of fly ash in fusion process. The influence of fusion reaction parameters on the type of obtained zeolites and on the efficiency of synthesis was investigated. The samples containing the same types of zeolites, but with different production yield were testes on SO_2 sorption capacity.

1 Introduction

Fly ashes are solid residues form the process of fossil fuels burning, captured from flue gases by electrostatic precipitator. According to its composition may be used in various industrial areas. The examples of its use may be reclamation of mining excavations (brown coal fly ashes), production of solidified ash mixtures for curing of lands, embankments, dumps, local roads and others, production of non-cement binders, Portland clinker production, as an active cement additive, concrete production, the light aggregates production, road building, use in agriculture and gardening, for the self-solidifying filling in mining [1]. Fly ash production in Poland is at a level of around 4,5 mln tonnes [2]. Poland is a country where the reuse of fly ashes is very efficient (87%) [2]. Due to the need of decreasing the amount of fly ash landfills (especially in the countries of lower degree of fly ash use [3]), searching for new processes for fly ash utilization is needed.

The way allowing for the reduction of fly ash landfill sites along with the production of valuable material is the process of zeolite production with the use of fly ashes. Zeolites can be obtained in the processes of: hydrothermal synthesis, fusion synthesis, molten salt synthesis method [4], by the combination of different methods and for example by utilizing microwave and ultrasound energies [5]. The comparison of literature data of particular methods allowed for the selection of fusion method for the research presented in this work. Adoption of this method was dictated by the reports stating that highly efficient zeolite synthesis process can be expected [5,6].

Burning process of fossil fuels is related to the emissions of particulate matter and gaseous pollutants, including sulphur dioxide (SO_2). A common way to capture SO_2 is to use calcium compounds in the dry and semi-dry methods of flue gas purification. It is necessary to search for alternative ways for sulphur dioxide capture out of flue gases. Zeolites as materials used for soil, sewages as well as gases purification [7], can be used as SO_2 adsorbents. The possibility of fly ash zeolites use as SO_2 sorbents would have a positive ecological outcome due to use of waste material as a substrate.

The series of experiments were performed aimed at selecting the best conditions for the zeolites formation. Based on the outcome of synthesis two samples were selected. In the chosen samples the same types of zeolites were present, but the efficiency of zeolite material formation was different.

In the work the SO_2 sorption capacity of material synthesized out of fly ash with the use of fusion method on selected samples was investigated, as well as regeneration properties of received materials.

2 Material and measurements methods

For the research fly ash from one of Polish heat-power plants was selected. Chemical composition analysis have been preformed with the use of XRF method. XRF analysis was preformed with the use of a Philips spectrometer PW 1404 with an X-ray tube equipped with dual Cr–Au anode with a maximum power of 3 kW as the excitation source. The investigated fly ash was marked with symbol L in this work and presents the following the chemical composition: Na_2O-2,3%, MgO-1,769%, Al_2O_3-15,706%, SiO_2-38,449%, P_2O_5-0,743%, SO_3-1,038, K_2O-2,773%, CaO-3,578%, TiO_2-1,001%, V_2O_5-0,044%, Cr_2O_3-0,045%, MnO-0,168%, Fe_2O_3-10,085, NiO-0,024%, CuO-0,085%, ZnO-0,987%, Ga_2O_3-0,006%, As_2O_3-0,023%, Rb_2O-0,014%, SrO-0,047%, Y_2O_3-0,004%, ZrO_2-0,03%, Nb_2O_5-0,003%,

[a] Corresponding author: nczuma@agh.edu.pl

 DOI: 10.1051/e3sconf/20161000010

SnO_2-0,022%, BaO-0,094%, PbO-0,298%. Due to standard specification [8], fly ash L was classified as siliceous fly ash. In literature more widespread classification based on American standard specification [9], classified fly ash L as F class fly ash. The XRD analysis have been also performed. The results of analysis were presented in the figure 1.

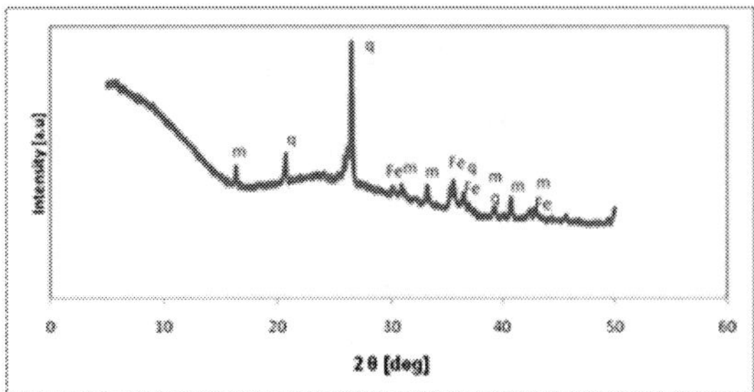

Figure 1. XRD diffractogram of fly ash L (m-mullite, q-quartz, Fe-hemaite/magnetite)

In the experimental part the synthesis of zeolites by the fusion method combined with hydrothermal aging process of solutions was preformed. Synthesis were differentiated in case of the temperature of synthesis process, used fly ash/base ratio and the time during which the aging of solution took place in the evaluated temperature after the fusion process. Received materials were washed to pH~10 and then dried.

The estimated efficiency of zeolite synthesis was calculated with the use of XRD data, based on equation (1) [10]. In the estimated calculations there was no pattern used, the comparison was based on 10 most intensive reflections characteristic for a given zeolite compared with the other synthesized sample- reference sample.

$$WK_{XRD} = \frac{\text{The sum of 10 individual reflections relative intensities of fly ash zeolite}}{\text{The sum of 10 individual reflections relative intensities of commercial zeolite (pattern)}} \times 100 \qquad (1)$$

Performed fusions were differentiated in case of the process temperature: Fusion 550 - 550°C, Fusion 600 - 600°C, Fusion 650 - 650°C, Fusion 700 - 700°C. The time of fusion process was in cases of all examined samples the same - 1 hour.

3 Results of synthesis

Prepared powders were analysed in order to confirm the presence of zeolite materials with the use of XRD method. The comparision of recieved synthesis products was presented in table 1.

Table 1. Comparison of synthesis results

	6-hour hydrothermal aging time		24-hour hydrothermal aging time	
	Fly ash/base ratio	Zeolite type	Fly ash/base ratio	Zeolite type
FUSION 550	2	-	2	trace X
	1	X, trace A	1	X, sodalite
	0,8	X, A	0,8	X, sodalite
FUSION 600	2	-	2	sodalite
	1	X, A	1	X, sodalite
	0,8	X	0,8	sodalite
FUSION 650	2	-	2	X, A
	1	X, A	1	X, sodalite
	0,8	X	0,8	X, sodalite
FUSION 700	2	-	2	-
	1	X, A	1	sodalite
	0,8	X, A	0,8	sodalite

4 Results of synthesis analysis

After the analysis of results it was fund that the elongation of the hydrothermal aging after the fusion process up to 24 hours, in most cases lead to sodalite formation. In part of the samples zeolite X was present, however it was noticed, based on the analysis of area under the most intensive reflections characteristic for X zeolite phases, that along with the appearance of sodalite reflections, the intensity of reflections characteristic for zeolite X is decreasing, what in approximation gives idea of the amount of formatted zeolite. It is believed that in the 24-hour aging the secondary dissolution of silica-alumina sources takes place (also from zeolite phases) and the recrystallization in the form of sodalite is observed [6].

It was discovered that the used fly ash/base ratio influences the formation of zeolite phases. For the purpose of determination of the best conditions for zeolite formation it was assumed that zeolites of high potential application should be obtained with high estimated efficiency of synthesis. Least of all profitable fly ash/base ratio is 2. The use of fly ash/base ratio of 1 and 0,8 in majority of samples resulted in zeolite formation. While comparing the areas under the reflections characteristic for analyzed phase, conclusions can be drawn that the higher reflections intensity, in approximation representing the zeolite formation efficiency, can be observed for fly ash/base ratio 0,8.

Detailed analysis of results allowed to draw conclusions of the temperature influence on the synthesis process. It was found that the most profitable temperature in relation to efficiency of received zeolite materials is 700°C.

5 Sorption properties of selected samples

Due to possibility of application of zeolites synthesized out of fly ash for adsorption of SO_2, the experiments aimed at determination of sorption capacity of selected samples have been performed.

 DOI: 10.1051/e3sconf/20161000010

The experiment was performed in the temperature of 25⁰C. The analyzed material was not additionally prepared for the sorption experiments. For the analysis two samples were selected, with different quantity of present zeolite. In both samples the presence of zeolite X and A was confirmed. The first sample was received in the fusion preformed in the temperature 700⁰C, the ratio of fly ash to base was 0,8, the fusion was combined with 6-hour hydrothermal solution aging. This sample in the work was marked with symbol F700-0,8-6. The second sample was marked with symbol F550-0,8-6 and was received in the fusion process performed in the temperature 550⁰C with the use of fly ash and base ratio of 0,8, combined with 6-hour hydrothermal solution aging. Diffractograms of both samples were presented in figure 2.

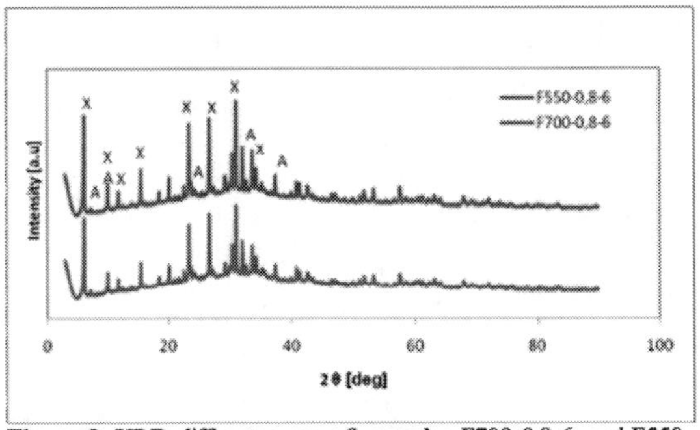

Figure 2. XRD diffractogram of samples F700-0,8-6 and F550-0,8-6 (X-zeolite X, A- zeolite A)

Based on the equation (1) the efficiency of zeolite synthesis was estimated, with the use of commercial zeolites type A and X as reference samples. For sample F700-0,8-6 the efficiency of zeolite X synthesis was approximately 47%, while in the same sample the efficiency of zeolite A synthesis was about 4%. For sample F550-0,8-6 the efficiency of zeolite X synthesis was estimated at 44%, while for zeolite A the value was 7%.

6 Sorption capacity results

The results of sorption capacity experiments performed on selected samples were presented in figures 3 and 4.

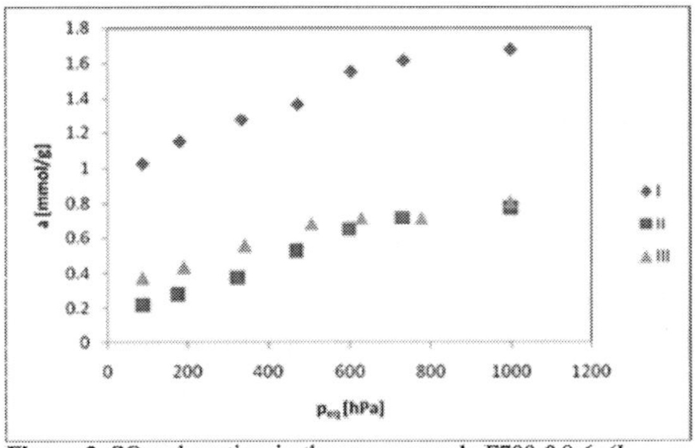

Figure 3. SO_2 adsorption isotherm on sample F700-0,8-6 (I-first sorption cycle, II- second sorption cycle, III-third sorption cycle)

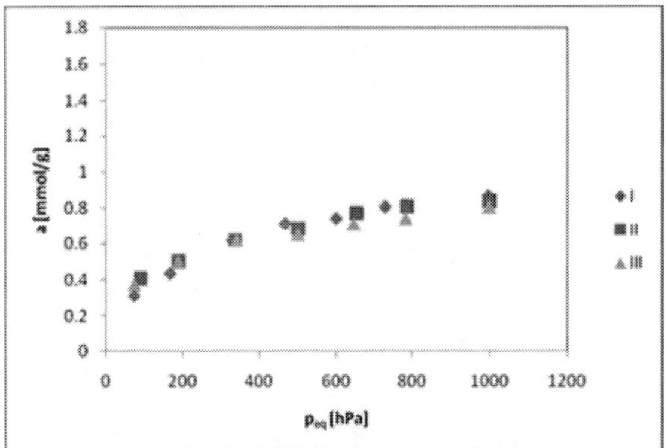

Figure 4. SO_2 adsorpion isotherm on sample F550-0,8-6 (I-first sorption cycle, II- second sorption cycle, III-third sorption cycle)

In table 2 the comparison of SO_2 sorption values were presented.

Table 2. Comparison of SO_2 sorption capacity for investigated samples

Sample name	Cycles of sorption capacity a, mmol/g		
	I	II	III
F700-0.8-6	1.68	0.78	0.81
F550-0.8-6	0.87	0.84	0.81

7 Results of sorption analysis

During the analysis of sorption experiments it was noted that sample F700-0.8-6, characterized by the higher amount of zeolite material synthesized, presented higher SO_2 sorption capacity. It was also observed that after the first sorption-desorption cycle the sorption capacity decreases, but in the following cycles it remained on constant level. Authors noticed similar observations for other samples containing zeolites synthesized out of fly ash [11]. Sorption capacity decrease after the first cycle may be related to permanent connection of sulphur dioxide with the non-reacted fly ash materials. Sample F550-0.8-6 characterized by lower content of zeolite material has presented lower sorption capacity value in relation to SO_2. This trend is consistent with author's predictions, the lower zeolite material amount will result in lower area able to capture SO_2. From the application point of view, sample F550-0.8-6 is highly attractive due to very good regeneration properties of examined material. In the second and third sorption-desorption cycle only slight decrease in the sorption capacity was observed. Authors considered performing additional experiments for sample F550-0.8-6 to be intentional.

8 Conclusions

In the work the investigation of fusion process parameters on the formation of zeolite phases synthesized out of fly ash was presented. The series of synthesis were performed, differentiated in case of fusion temperature, the ratio of amount of fly ash to base used and the hydrothermal aging time after the fusion process.

As a result of performed experiments it was fund that the highest efficiency of zeolites synthesis, estimated as an area under the most intensive reflections for given zeolite, was for the fusion temperature 700^{0}C. This trend was observed for all samples with the same fly ash/base ratio and with the same aging time. The use of fly ash/ base ratio of 0,8 was proven to give the best results, lower efficiency of zeolites formation was noticed for the use of fly ash/base ratio equal 1. With the use of smaller amount of base (ratio 2), in most cases, no zeolite phases were present in the investigated material. The time of hydrothermal aging after the fusion process is believed to have a significant influence on the zeolite formation. It was fund that the time of hydrothermal reaction of 6 hours was more profitable due to formation of zeolite phases of higher application potential. Extended to 24 hours time of hydrothermal aging lead to appearance of sodalite in predominant number of samples. It was stated that formatted zeolite phases in the 6-hour hydrothermal aging, were dissolved and the secondary crystallization took place as a result of which sodalite was formed. Confirmation of this fact are available literature data [6] as well as the fact that in case of presence of different than sodalite zeolite phases (also present in the sample subjected to 6-hour hydrothermal aging), the reflections intensity of this different phase is lower than for analogical sample subjected to 6-hour hydrothermal aging process.

The analysis of sorption properties proved the use possibility of received zeolites for the SO_2 sorption. Selected samples were differentiated in case of the yield of zeolite formation. In case of sample characterized by higher zeolite synthesis efficiency F700-0,8-6, the sorption capacity was 1,68 mmol/g in the first cycle and 0,78 and 0,81 mmol/g in the second and third sorption-desorption cycle accordingly. After the first cycle the decrease in the sorption capacity was observed, while in the second and the third cycle it remained on constant level. For the sample F550-0,8-6 lower sorption capacity was observed 0,87 mmol/g, which is related to the lower amount of zeolite present in the sample. Very good regeneration properties of this sample is highly interesting feature (0,84 and 0,81 mmol/g in the second and third sorption-desorption cycle respectively). Sorption capacity only slightly decreased in the second and the third sorption-desorption cycle, what classified the sample for future research. According to no detailed data in relation to the quantity of zeolite material present in the samples synthesized out of fly ash, the description of efficiency calculation methods and the use of different methodology of the sorption experiments performed in other works aimed at SO_2 sorption analysis of the zeolites synthesized out of fly ash, it is difficult to perform comparative analysis of received values. In general it can be stated that the received in this work values are in the range specified by other authors or are higher [12-14].

Acknowledgements

The authors are grateful to the AGH University of Science and Technology (Project 11.11.210.244) for its financial support of this work.

References

1. http://www.spalanie.pwr.wroc.pl/badania/witryfikacja/popioly.htm (accessed 15.04.2016)
2. Główny Urząd Statystyczny, Ochrona Środowiska 2014, Warszawa 2014
3. Y. A. Alhamed, S. U. Rather, A. H. El-Shazly, S. F. Zaman, M. A. Daous, A. A. Al-Zahrani, Korean J. Chem. Eng. Korean Journal of Chemical Engineering, **32**, 723-730, 2015
4. T. Suchecki, *Zeolity z popiołów lotnych. Otrzymywanie i aplikacje w inż. Środowiska*, (Ossolineum, 2005)
5. S. S. Bukhari, J. Behin, H. Kazemian , S. Rohani , Fuel, **140**, 250–266 (2015)
6. N. Shigemoto, H. Hayashi, K. Miyaura, J. Mater. Sci., **28**, 17, 4781-4786 (1993)
7. M. Franus, M. Wdowin, L. Bandura, W. Franus, Fresen. Environ. Bull., **24**, 3a, 854-866 (2015)
8. BN-79/6722-09
9. ASTM C618-12a
10. I. Majchrzak-Kucęba, *Mikroporowate i mezoporowate materiały z popiołów lotnych* (Częstochowa, Wydawnictwo Politechniki Śląskiej, 2011)
11. Documenta Geonica 2015/1 10th Czech and Polish Conference " Geology of Coal Basins", 13-15.10.2015 Ostrava; 164-169
12. S. Bopaiah, M. W. Grutzeck, Preprints of Papers-American Chemical Society, *Division* of *Fuel Chemistry* **45**, 522-526 (2000)
13. T. T. Suchecki, T. Wałek, M. Banasik, Pol. J. Environ. Stud. **13**, 6, 723-727 (2004)
14. A. Srinivasan, M. W. Grutzeck, Environ. Sci. Technol. **33**, 9, 1464–1469 (1999)

E3S Web of Conferences **10**, 00011 (2016) DOI: 10.1051/e3sconf/20161000011
SEED 2016

Sequestration of carbon dioxide – influence of coal surface chemistry

Agnieszka Orzechowska-Zięba[1,a], Katarzyna Zarębska[1], Paweł Baran[1] and Agnieszka Ćwik[1]

[1]*AGH University of Science and Technology, Faculty of Energy and Fuels, Aleja Mickiewicza 30, 30-059 Cracow, Poland*

Abstract. The physical gas adsorption is a widely used method for the characterisation of the solids porosity. The water steam, primarilydue to its physicochemical properties and ease of use in the experiment has great potential as a sorbate. When applied to coal, water steam allows to determine the quantity of primary adsorption centers as measurement of interaction of molecules adsorbed to the surface of the adsorbent. In order to determine the adsorption capacity and the chemical nature of the coal surface, adsorption / desorption of water vapour to the selected coals was examined at 303K, using a volumetric method. The presence of water in the coal may affect on the sorption properties of other molecules. The analysis of the results show that the coals of low rank and a high content of oxygen functional groups, which are the active sites, showed a greater affinity to absorbing water molecules. Adsorption isotherms were compiled through approximating the Langmuir and BET linear equation to measurement data. Based on the adsorption equation, the amount of adsorption centers have been specified, which can potentially be involved in the adsorption of CO_2 during the injection of gas into the coal seams.

1 Introduction

According to the international agreement, Poland is obliged to reduce the greenhouse gases emissions to the atmosphere, mainly carbon dioxide [1,2]. There are many ways which leads to achieving this condition: reducing fossil fuels burning, improving the energy efficiency of power plants and increasing the share of energy production from renewable sources and nuclear energy. Poland is a state whose national energy is based on fossil fuels and therefore the problem of minimizing the emission of gases into the atmosphere is particularly difficult [3]. Thus, the above mentioned activities may not be sufficient to reduce the global CO_2 emissions.

Among the alternatives, one of the most promising ideas is the use of fossil fuels combine with CCS (Carbon Capture Storage) - capture and storage of CO_2 [4,5]. The idea of storing the carbon dioxide in underground tanks for a long time is termed sequestration of CO_2. Therefore, a lot of scientific research is conducted on the possibility of injection and storage of carbon dioxide in deep geological structures, for example anticlines, depleted oil and gas fields. In addition, one of the indicated possibilities are also depleted coal mines.

To be able to apply an innovative method of waste gas emissions disposing by CO_2 injection in porous geological deposits, required criteria must be fulfilled (environmental impact and long-term storage). Due to the speciality of heterogeneous coal rocks construction, for each case individual analysis of the gas-coal system must be conducted, in the context of permeability and storage capacity impact of other system components (water, inert gas) on the capacity of the storage center [4].

[a] Corresponding author: azieba@agh.edu.pl

2 Experimental

2.1 Characteristic of the samples

The research object was the five samples of coal selected from polish mines. Samples of coal have been obtained in accordance with PN-90/G-04502, and then comminuted in a laboratory conditions to a size of 0.5 - 1.25mm, in order to averaging their physicochemical properties. The choice of sites from which samples were taken was dictated by diverse parameters petrographic, chemical and technology.

The Central Mining Institute (GIG) in Katowice made elemental and technical analysis, according to the requirements and procedures laid down in the Polish standards, regarding classification of coal. Parameters of chemical in and technical analysis are summarized in Table 1 and Table 2. In addition, Table 1 shows special signature of the samples that will be used in the work, instead of the full names of mines because of the clarity and readability of the work.

The content of the basic elements contained in the coal material was examined. Also, the moisture (PN-80/G-04511) and ash (PN-80/G-04512). The oxygen content was calculated as the remainder to 100%, after the moisture and ash content subtraction. Among the selected coals, the smallest content of the elemental carbon in the dry and ash-free state was shown in the coals from the mines: "Bogdanka" (B), "Jankowice" (J) and "Zofiówka" (Z2). The content of the carbon ranged from 75.16 - 89.06%. The difference could also be observed in the ash content (3.17 - 8.92%), moisture (0.68 - 4.36%) as well as volatile matter 19.72 - 36.25%.

Table 1. Chemical (elementary) analysis of the coal samples.

		Coalmine/seam				
		Bogdanka p. 391	Jankowice p. 505	Zofiówka p. 410	Zofiówka p. 404/2	Zofiówka-Borynia p. 502
		Coal sample				
		B	J	Z1	Z2	Z3
Elementary Analysis [%]	C^{daf}	75.16	75.71	84.04	78.62	89.06
	S^{daf}	1.67	0.92	0.78	0.35	0.36
	H^{daf}	4.92	4.41	4.65	4.37	4.14
	N^{daf}	1.70	1.27	1.32	1.15	1.38
	O^{daf}	8.09	10.49	4.35	6.17	1.12
	$(O+N)^{daf}$	9.79	11.76	5.67	7.32	2.50

Table 2. Technical analysis of the coal samples.

Coal sample		B	J	Z1	Z2	Z3
Technical Analysis [%]	W^a	4.36	2.82	1.42	0.60	0.95
	A^a	4.17	4.45	3.51	8.92	3.17
	V^{daf}	36.25	30.93	24.53	27.93	19.72

Explanation: W^a: moisture content, air-dried basis, A^a: ash content, air-dried basis, V^{daf}: volatile matter content, air-dried basis, C^{daf}: content of element C, dry-ash-free basis, H^{daf}: content of element H, dry-ash-free basis, N^{daf}: content of element N, dry-ash-free basis, S^{daf}: content of element S, dry-ash-free basis

As an additional characteristic of the research material petrographic analysis was performed. Microscopic studies were carried out using petrographic microscope POLMI type and microscope OLYMPUS. An important parameter characterizing the coal is its reflectivity and the maceral content. Petrographic analysis results (Table 3) shows that selected coals have the light reflectivity value in the range of 0.68 - 1.3%. The highest index value R_0 is recorded for the sample Z3, and the smallest for the sample B. The samples differ significantly in the vitrinite content. This value is in the range from 54 - 91%.

The highest value of this parameter is shown in the sample Z2 and Z3 has the smallest. Number of macerals from the group liptinite and inertinite and ranged as follows: 0 - 5% and 8 - 45%. It is surprising that the sample Z1 did not contained any egzynite, and Z2 contains only minor liptynite and inertynite content.

The selected group of samples contains low, medium and high-carbon coals (Figure 1) as evidence of the relationship between the vitrinite reflectance (R_0) and the content of volatiles (V^{daf}).

Table 3. Petrographic analysis of the coal samples.

	Samples	B	J	Z1	Z2	Z3
Maceral group [%]	vitrynite	84	74	88	91	54
	liptynite	5	3	0	1	1
	inertynite	11	23	12	8	45
Mineral matter MM [%]		0	2	1	4	2
Reflectivity R_0 [%]		0.68	0.80	1.19	1.01	1.30

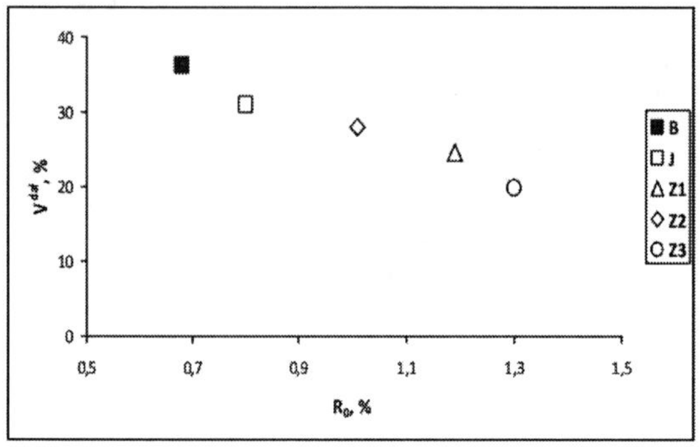

Figure 1. The dependence between the volatile mater content and the vitrinite reflectance of the examinated samples.

The actual density was determined using a camera Micrometrics - Accu Pyc 1330, which is an automatic device for determining the actual density. As pycnometric liquid helium was used. Helium, a non-adsorbant medium, do not react with the tested substance and has great ability to penetrate the porous structure of coals [6].

Table 4. Helium density values for the selected samples

Coal sample	B	J	Z1	Z2	Z3
d_{He}, g/cm^3	1.35	1.37	1.38	1.10	1.32

Comparing the value d_{He} can be seen that carbon B had the largest true density, and the smallest Z.

2.2 Sorbate characteristic

Water molecules are non-linear. The bonds between oxygen and hydrogen are strongly polarized, thus the water has a permanent dipole moment (Table 5) and is strongly polar. Additional advantages of the water as sorbate is the high evaporation heat, thermal stability and non-toxicity [7].

Water used as a sorbate in the process of sorption on the coal is sensitive to the chemical nature of the hard coal, since only the polar sorption occurs [8]. Water vapour is one of the most frequently used sorbate to study the hydrophilicity of hard coal, depending on the metamorphism degree and the fragmentation level of the sample. Water sorption can distinguishe coals very well [9].

DOI: 10.1051/e3sconf/20161000011

2.3 Research methodology

Adsorption isotherms for selected coals were determined using volumetric apparatus - microburette liquid adsorption type [10] at 303K on samples of a grain size 0.5 - 1.25mm. Apparatus with liquid microburretes allows to determine isotherms of sorption/desorption polar and apolar vapours at a pressure range from vacuum to the absolute vapour pressure. The construction of the apparatus and the principle of measurement is described in the Klinik's work [11].

Table 5. Selected parameters of the physical and chemical properties of water.

Mm, [g/mol]	Boiling temperature at atm. pressure, T_{nbp}, [K]	Solidification temperature at atm. pressur, T_{mp},[K]	Density d w 298 [K], [g/cm³] ·10³	Dipole moment / [D]	Vapour pressure at 298K [10⁵ Pa]	Occupied surface area ω, [m²/cząst]·10¹⁰
18.0	373.12	273.15	997	1.82	0.0317	1.05

Before the sorption measurement the samples had to be properly prepared. Therefore, it is necessary to clean the surface of the samples from the previously absorbed gases and vapours, and also removing from the micropores other, absorbed specific molecules. For this purpose, tested material was degassed to static vacuum 10^{-3}Pa. In addition, to get a better cleaning effect on the surface and porous structure of the research material, several rinsing with helium under pressure was performed. Helium atoms are not significantly being adsorb on the adsorbent surface and provide the necessary kinetic energy for the absorbed particles vapours and gases, thereby knock them from the surface of the sorbent [12]. After each rinse, the sample was again degassed with helium. The prepared samples were than subjected to the experiment. During the investigation the optimal time was chosen in order to reach the state of equilibrium (24 hours), after which there was no change in pressure equilibrium.

3 Analysis

Sorption/desorption curves for the investigated systems are presented in figures 2 - 6. The sorption isotherms are shown as a relation between amounts of absorbed water vapour (a, mmol/g) and the relative pressure (p/p_0). The complexity of the physicochemical nature of coal makes the interpretation of the sorption properties of coal in terms of low molecular weight substances as qualitative.

For each coal sample, the adsorption isotherm is the II type, according to the IUPAC classification, which is characteristic for the sorbents containing micro- and mesopores. It indicates that there is the initial adsorption of a monolayer, followed by the multi layers. Creating a monolayer shows isotherms of sorption at the initial relative pressure range.

Depending on the amount of oxygen, the slope of the isotherm is more or less abrupt. For example, for the carbon samples B and J filling of the monolayer takes place at a relative pressure of 0.12 and for the sample Z3 monolayer is already filled at 0.07. This indicates a higher level of Z1 sample coal metamorphism and thus, the presence of a smaller amount of oxygen functional groups which are responsible for the hydrophilic nature of coal and reduced sorption absorbency.

After formation of the adsorption monolayer, successive filling of next layers occurs. Figure 7 corresponds to the passage of the isotherms in a straight line.

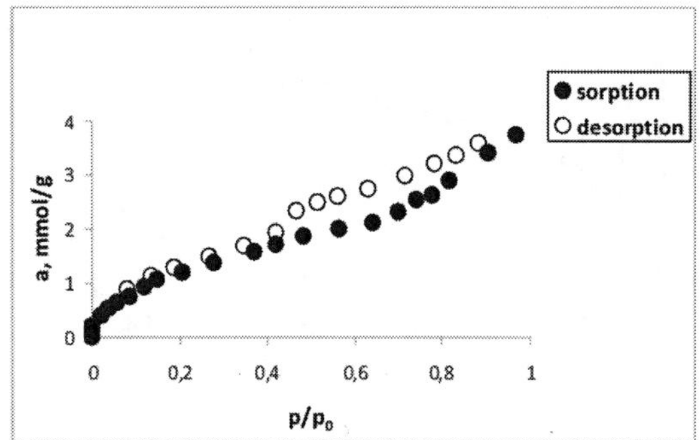

Figure 2. Sorption / desorption isotherm of H_2O vapour on coal B

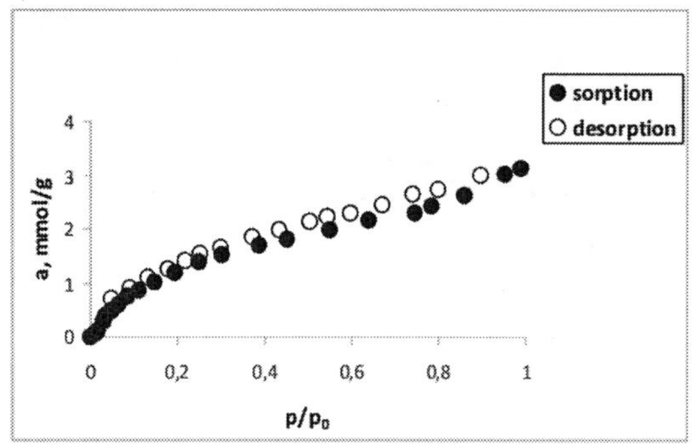

Figure 3. Sorption / desorption isotherm of H_2O vapour on coal J

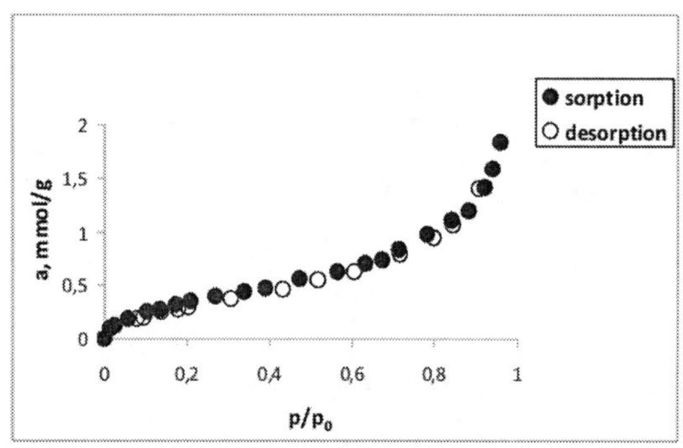

Figure 4. Sorption / desorption isotherm of H_2O vapour on coal Z1

The degree of successive filling of adsorption layers is closely related to the amount of polar groups, which are the sorption centers. Absorbed water molecules become the sorption centers for the next sorbate particles. The process then takes the island nature and clusters are being formed [13]. Isotherms of coals B and J indicates that there is a significantly higher secondary sorption surface area when compare with the samples Z1, Z2, Z3. Since these samples are older higher rank coals, they do not contain in their structure so many oxygen species and hence, they have reduced ability to absorb polar water molecules.

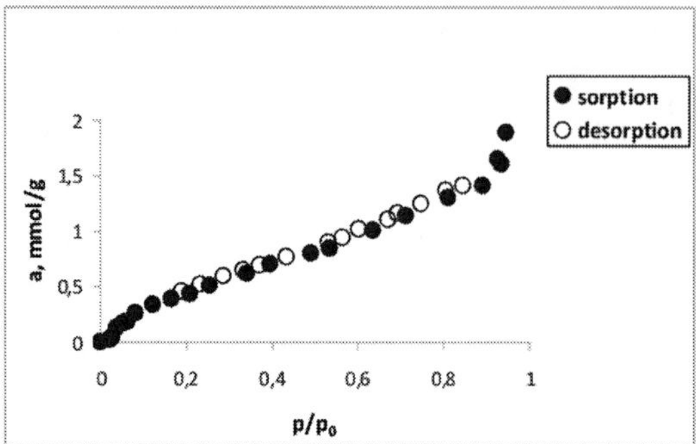

Figure 5. Sorption / desorption isotherm of H_2O vapour on coal Z2

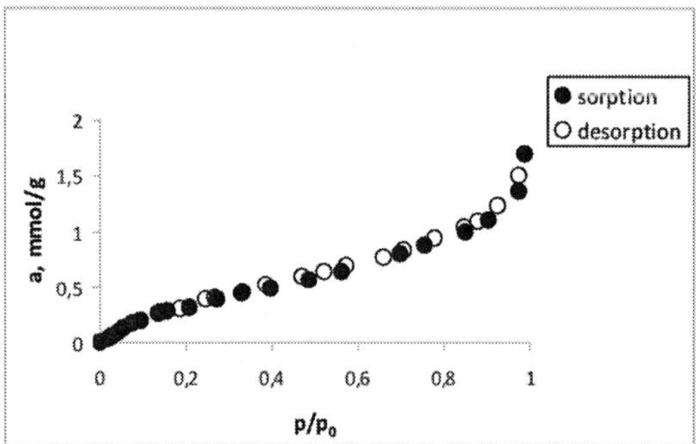

Figure 6. Sorption / desorption isotherm of H_2O vapour on coal Z3

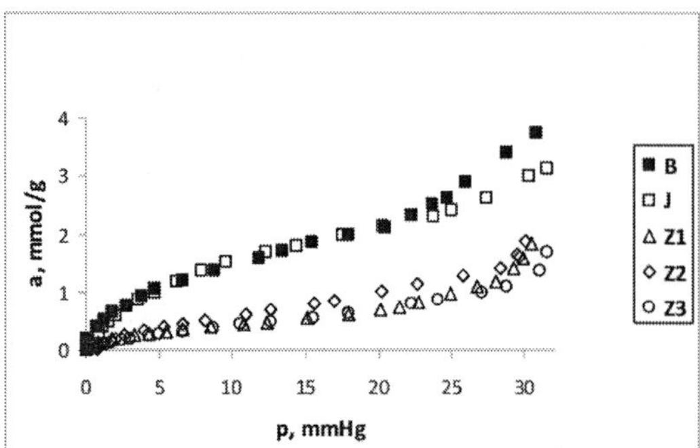

Figure 7. The sorption isotherms of water vapour on selected samples of coal - a comparison of sorption capacity.

Noticeable fact is that the isotherm of carbon Z1, Z2, Z3, in contrast to other coal samples at higher relative pressures tilts upward. This could indicate that there is a phenomenon of coal swelling, or sorbate vapours condense on the inner surface of the pores. However, considering the size of the sorption for the coals Z1, Z2, Z3, significant increase in sorption capacity may be associated with the deposition of sorbate molecules and formation of clusters.

In all cases, on each sorption / desorption curve there is closed hysteresis loop. This fact indicates the presence of capillary condensation. For coals B, J desorption isotherm does not coincide with the sorption isotherm, what indicates a higher degree of chemical sorption when compare with coal samples Z1, Z2, Z3, where desorption and sorption isotherms almost cover each other in the entire range of relative pressures.

Experimental sorption isotherms have been developed by the approximation of a Langmuir and BET linear equation with the measurement data [14]. Knowing that the isotherms of type II compliance experimental data of BET linear equation is in the range of relative pressures between 0.05 to 0.35 and produces a constant linear section is calculated monolayer capacity (a_m), and other constant (C) for each sample of coal. The results of calculation for BET theory are given in Table 6.

The monolayer capacity for each coal varies. As expected, absorption capacity is the smallest for coal Z3, with the highest carbon content and the minimum content of oxygen functional groups, responsible for the coal hydrophilicity.

Table 6. The adsorption parameters according to the BET theory

Coal sample	BET	
	a_m	C
B	1.15	18.48
J	1.30	11.55
Z1	0.36	16.32
Z2	0.51	9.01
Z3	0.35	9.72

There is four times higher oxygen groups presence in coal J, which are the centers of adsorption on the surface of coal. Because of that, to fill the surface with one molecular adsorbent layer there is a need for four more times of water vapour molecules. It is obvious that there is an impact of the oxygen species content on the sorption parameters. Picture 8 shows the relationship between the total content of O and N and the size of the sorption. The coal ability to linearly sorb depends on the presence of oxygen functional groups. The more oxygen species in the form of sorption centers, the greater the coals possibility of sorption.

Vitrinite reflectivity is one of the fundamental parameters determining hard coal rank. Figure 9 shows that the samples Z1 and Z3, with the highest vitrinite reflectance has the lowest water absorption capacity. In contrast, sorption on coals B and J with reflectance order of 0.68 and 0.8 is four times greater than carbon Z1 and Z3.

The content of vitrinite reflects the coal possibility of water vapour absorption. Macerals of vitrinite group are rich in oxygen, as they belong to the youngest macerals. Coals B and J with the content of vitrinite respectively 84% and 74% have the highest absorption capacity for water. On opposite, coal Z3 with vitrinite reflectivity equal 54% has the smallest sorption capability (Figure 10).

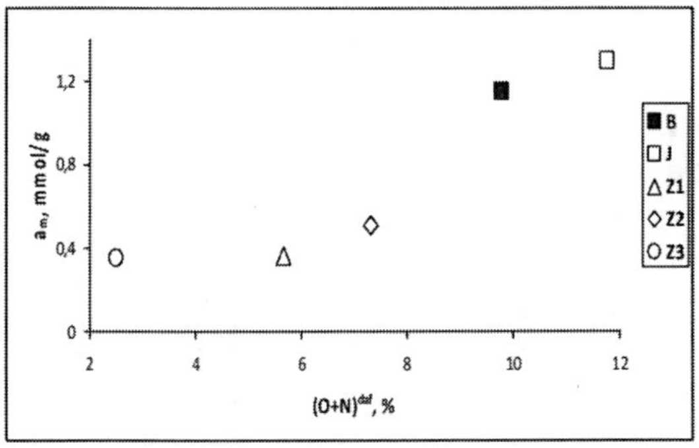

Figure 8. The dependence of the sorption capacity on oxygen and nitrogen content

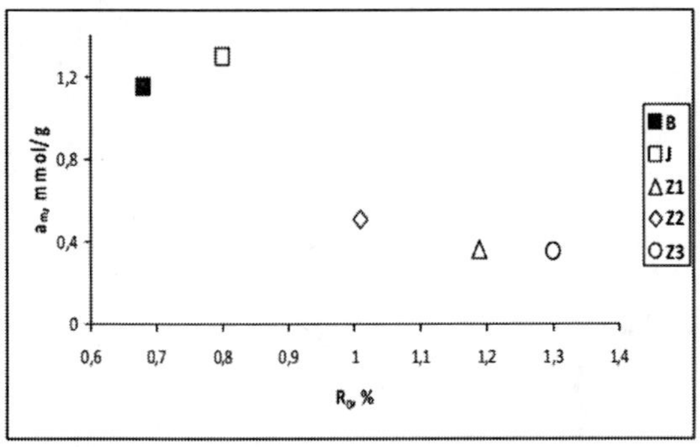

Figure 9. The dependence of sorption capacity on vitrinite reflectivity

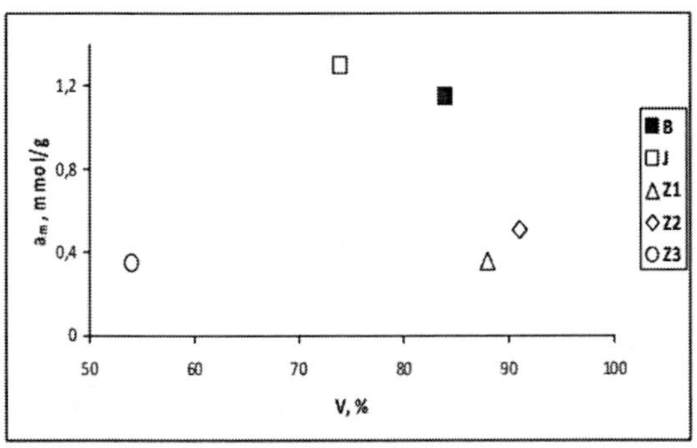

Figure 10. The dependence of the sorption capacity on the vitrinite content

4 Summary and conclusions

The object of this investigation was the six coal samples from Polish coal mines. Examined coals differ in the degree of metamorphism. Tests were performed to determine the sorption capacity with the respect to water vapour. From the CO_2 capture point of view, it appears to be very important because the presence of moisture in coal sometimes causes a significant reduction in the sorption capacity of coal regarding gases (methane and carbon) [15-17]. Therefore, quantitative analysis determining the water vapour sorption on coal allow us to evaluate the sample in terms of chemical nature, as well as for potential storage center.

As expected [8], studies have shown that the coal with the highest degree of metamorphism (Z3 carbon) showed the lowest sorption capacity. Also, since the oxygen groups, which are closely connected with the coal rank, are responsible for the amount of sorption centers, their presence is the most important factor influencing the sorption capacity of coal. They are also responsible for the chemical nature of the carbon surface, which hydrophilic properties are increasing with higher number of these groups.

Carbons B and J showed the highest ability to absorb water vapour, which is closely related to the amount of oxygen species on their surface. Z3 coal, which has the highest content of carbon and the lowest number of reactive oxygen groups, absorbed the least amount of water molecules from the tested samples.

It is therefore necessary to stress once again that in the deliberations on the underground storage of CO_2, the storage capacity in relation to this gas may not be the most important parameter in assessing the suitability of such deposits. It is important to determine the sorption capacity of water vapour on the coal, and even better, to determine the absorbency of CO_2 on coal pre-saturated with H_2O.

Acknowledgments

Financial support for this study was provided by the AGH University of Science and Technology (framework no. 11.11.210.244).

References

1. C.M. White, D.H. Smith, K.L. Jones, A.L. Goodman, S.A. Jakich, R.B. LaCount, S.B. DuBose, E. Ozdemir, B.I. Mors, K.T. Schroeder, Energy & Fuels, **19**, 659–724, (2005)
2. G.X. Wang, P. Massarottoa, V. Rudolpha, International Journal of Coal Geology, **77**, 127–136, (2009)
3. A. Wójcicki, Przegłęd Geologiczny, **57**, 138-143, (2009), in Polish
4. Y. Gensterblum, A. Busch, B.M. Krooss, Fuel, **115**, 581-588, (2014)
5. M. Wdowin, R. Tarkowski, W. Franus, International Journal of Coal Geology 130, 79-88, (2014)
6. Rao M.B. Carbon, 29 (1991) 813-815
7. J. Gancarz, *Woda*, Wydawnictwa Szkolne i Pedagogiczne, Warszawa, (1975), in Polish
8. K. Kreiner, M. Żyła, Górnictwo i Geoinżynieria, AGH, R. **30**, z. 2, 19 – 34, Kraków, (2006) in Polish
9. A. Orzechowska – Zięba, Rozprawa doktorska, AGH, Kraków, (2009) in Polish
10. M. Lasoń, M. Żyła, Chemia Analityczna, **8**, 279 – 286, (1963), in Polish
11. J. Klinik, *Tekstura porowatych ciał stałych,* Ośrodek Edukacji Niestacjonarnej, Kraków, (2000), in Polish

12. S. Saha, B.K. Dharma, S. Kumar, G. Sahu, Y.P. Badhe, S.S. Tambe, B.D. Kulkarni, Fuel, **86**, 1594-1600, (2007)
13. K. Kreiner, M. Żyła, P. Baran, Gospodarka Surowcami Mineralnymi, **23**, zeszyt specjalny nr 3, 167-174, (2007), in Polish
14. J. Ościk, *Adsorpcja*, Powszechne Wydawnictwo Naukowe, Warszawa, (1983), in Polish
15. J.I. Joubert, C.T. Grein, D. Bienstock, Fuel, Vol. **52**, Issue 3, 181-185, (1973)
16. C.R. Clarkson, R.M. Bustin, International Journal of Coal Geology, Vol. **42**, Issue 4, 241-271, (2000).
17. P.J. Crosdale, T.A. Moore, T.E. Mares, International Journal of Coal Geology, Vol. **76**, Issues 1–2, 166-174, (2008)

Kinetics of methane and carbon dioxide sorption and sorption–induced expansion of coal – kinetic equations assessment

Katarzyna Czerw[1,a], Agnieszka Ćwik[1] , Paweł Baran[1] and Katarzyna Zarębska[1]

[1]*AGH University of Science and Technology, Faculty of Energy and Fuels, Aleja Mickiewicza 30, 30-059 Cracow, Poland*

Abstract. The aim of this study was to investigate the ability of kinetic equations to describe the sorption kinetics and expansion rate of solid coal samples. In order to address his issue the sorption kinetics of methane and carbon dioxide on bituminous coals were studied. At the same time, the changes occurring in the sample's overall dimensions, which accompanied sorption processes, were monitored. Experiments were carried out at high pressure by means of the volumetric method on a cubicoid solid samples. Several literature-based modeling approaches and equations are proposed to fit the kinetic curves of gas deposition, as well as the adequate kinetics of coal swelling. First equation represents the traditional approach to interpret experimental data in terms of fast and slow sorption process and consider the combination of two first-order rate functions. The other empirical kinetic equations are: the pseudo-second-order kinetic equation (PSOE), Elovich equation and the stretched exponential equation (SE). Two of the four equations are suitable to describe the kinetics of methane and carbon dioxide sorption and have been successfully used to quantify the observed dilatometric phenomena rates. The stretched exponential equation gave the best fit to the experimental data.

1 Introduction

Interest in sequestration of carbon dioxide and the enhanced coal bed methane recovery (ECBM) process, that is, recovery of methane from carboniferous deposits with simultaneous CO_2 sequestration results primarily from ecological issues, for example to lower the emission of carbon dioxide, but also has prospective economic aspects [1-5]. Investigation and understanding of the relations between the expansion/contraction of hard coal and the amount of gas sorbed in its porous structure in coal–mine gas (mainly CH_4 and CO_2) systems are fundamental issues in carbon dioxide sequestration and recovery of methane from carboniferous formations. Important issues that need to be focused on are coal–gas (CH_4, CO_2) interactions within a coal bed, their analytical and numerical description and the underlying mechanisms [6].

Mine gases are deposited in coal beds in different forms: (1) adsorbed in micropores and on the surface of larger pores; (2) absorbed in the coal molecular structure; (3) as free gas in fissures and larger pores, which becomes significant at higher sorbate pressures; and (4) dissolved in deposit waters [7-8]. Investigation of vapors and gases sorption process on solid hard coal samples has shown the occurrence of sorption-induced swelling of the sorbent. Further research into this phenomenon has led to the conclusion that coal is a biporous, involving transport and sorption process. It is a system in which eleveted pressure of gases causes compression of microporous regions and expansion of macropores. Gas sorption gives an opposite effect of induced expansion of microporous regions and narrowing of transport pores and the *in situ* a decrease of system permeability [9-10]. The majority of research in the field shows that the swelling of coal is approximately proportional to the amount of sorbed gas implicating a linear relation between sorption and expansion [6, 11-16]. Studies that suggest non linear dependence are a minority [9, 17].

Sorption kinetics concerns the time dependence or rate of reactions, usually described in terms of a mathematical relationship between the amount of sorbed substance, equivalent to weight gain or volume of accumulated agent, and time. To interpret and quantify the observed gas uptake rates in coal two models have been used: unipore model, based on the solution to Fick's second law for spherical symmetric flow [18-22] and bidisperse sorption/diffusion gas transport model for coals characterized by two distinct pore systems [18, 20, 23-28]. It is generally confirmed that the amount micropores increase and mesopores decrease with coal rank. As a result, although not generally confirmed, unipore models seem to better represent the sorption kinetics of high rank coals (medium-volatile bituminous coal to anthracite) while bidisperse models better represent low rank coals [23].

In the light of this consideration, in our previous investigation we assumed that it might be possible to use the same approach to determine equations that provide

[a] Corresponding author: kczerw@agh.edu.pl

a mathematical description of the kinetic curves for both sorption and swelling of coal [29]. To obtain the experimental results for further mathematical evaluation, sorption of methane and carbon dioxide on cuboidal hard coal samples was investigated at high pressure. The pressure value of gas in the sample cell at the beginning of experiments was about 4 MPa. At the same time as the sorption kinetics were monitored, changes of the sample's overall dimensions accompanying the sorption processes were measured. Coal samples (high-volatile bituminous C) were obtained from the Brzeszcze-Silesia mine [29]. Experiments were performed using the apparatus and test procedures described by Majewska et al. [17]. Our findings translate into general conclusions:

- for two equations based on the bidisperse gas transport model, the equation proposed by Busch et al. [18] gives better fit than the equation proposed by Siemons et al. [26],
- the pseudo-second-order equation (PSOE) of Marecka and Mianowski [30] was unsuitable for the data
- the Elovich equation fits the results well, because the sorption and sorption-induced expansion were far from equilibrium, however, because the function goes to infinity it should be avoided
- the stretched exponential equation gives the best fit and is in very good agreement with the experimental results [29].

In this study, we evaluated the ability of four sorption kinetic equations to describe sorption kinetics and the rate of expansion/contraction of cuboidal hard coal samples in coal-carbon dioxide and coal-methane systems in evaluated pressure range, as the pressure value of gas in the sample cell at the beginning of experiments was somewhat above 1 MPa. First approach proposed by Busch et al. [18] consider the combination of two first-order rate functions and represent the traditional approach to interpret experimental data in terms of fast and slow sorption process. The other empirical kinetic equations are: the pseudo-second-order kinetic equation (PSOE) proposed by Blanchard et al. [31], in original time-integrated form [32]; Elovich equation developed to describe kinetics of chemisorption [33-34] and stretched exponential equation (SE) also known as the KEKAM equation [35-36].

2 Experimental methods

Primary experiments were performed using the apparatus and test procedures described by Baran et al. [11]. The experimental program uses a purpose-built apparatus, enabling the concurrent measurements of sorption and dilatometric parameters of cuboid-shaped coal samples. Measurements of sorption capacity are taken by the manometric method. The device is placed inside a water thermostat, maintaining a constant temperature of 298 K. Gas of known pressure and volume becomes decompressed and moves from the reference cell to the sample cell containing the sorbent. Knowing the dead volume of the apparatus and the volume of the dosing unit, the amounts of absorbed/adsorbed gas can be computed on the basis of gas laws. Coal sample deformations were measured with a strain gauge engineered at the Strata Mechanics Research Institute of the Polish Academy of Sciences. Linear strains are measured with an electric resistant wire strain gauge incorporating a resistance (bridge) transducer [11].

Table 1. Specification of coal samples: proximate and ultimate analysis.

Proximate analysis				
Coal	W^a [%]	A^a [%]		V^{daf} [%]
Pniówek	0.68	3,78		25.50
Budryk	0.70	1.80		32.60
Ultimate analysis				
Coal	C^{daf} [%]	H^{daf} [%]	N^{daf} [%]	S^{daf} [%]
Pniówek	84.96	4.60	1.70	0.58
Budryk	90.84	5.81	1.45	0.65

W^a: moisture content, air-dried basis, A^a: ash content, air-dried basis, V^{daf}: volatile matter content, air-dried basis, C^{daf}: content of element C, dry-ash-free basis, H^{daf}: content of element H, dry-ash-free basis, N^{daf}: content of element N, dry-ash-free basis, S^{daf}: content of element S, dry-ash-free basis

Table 2. Specification of coal samples: maceral composition.

Coal	Pniówek	Budryk
Telinite [%]	0.6	0.2
Collotelinite [%]	26.0	15.2
Vitrodetrinite [%]	1.0	0
Collodetrinite [%]	23.0	48.3
Corpogelinite [%]	1.0	0
Gelinite [%]	1.0	0.6
Total vitrinite [%]	**52.6**	**64.3**
Sporinite [%]	5.3	3.8
Resinite [%]	0	0.4
Liptodetrinite [%]	3.0	0
Cutinite [%]	0	1.0
Total liptinite [%]	**8.3**	**5.2**
Fusinite [%]	20.8	10.7
Semifusinite [%]	8.0	5.2
Macrinite [%]	1.9	2.0
Micrinite [%]	0	0.2
Funginite [%]	1.6	5.3
Inertodetrinite [%]	6.8	7.1
Total inertinite [%]	**39.1**	**30.5**
Mineral matter [%]	**0**	**0**

For the purpose of measurements, a cuboid samples (18 × 18 × 40 mm) were cut from the basic coal fragments, obtained from the Pniówek mine and Budryk mine.). The results of analysis performed on the basic coal samples are shown in Table 1 and Table 2. Experiments on coal Pniówek followed test procedures described by Baran et al. [11]. Experiment on coal

DOI: 10.1051/e3sconf/20161000012

Budryk were performed using the apparatus and test procedures described by Majewska et al. [17].

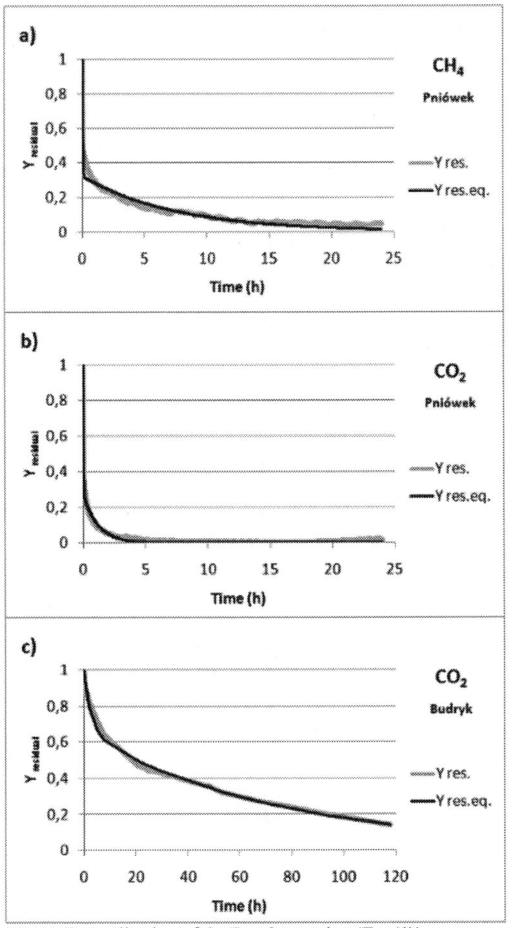

Figure 1. Application of the Busch equation (Eq. (1)) to the kinetics for (a) CH_4 and (b) CO_2 sorption on coal Pniówek and (c) CO_2 sorption on coal Budryk.

On the opposite walls of coal samples, resistance strain gauges were attached, arranged alongside and transverse to the bedding plane. The sample was connected to the strain gauge and placed inside a test ampule. Prior to the main experiment, the sample was degassed, the dead volume of the sample cell was determined using helium, and the system was degassed again to achieve the static vacuum of the order of 10^{-2} Pa. Carbon dioxide or methane was then admitted to the reference cell, and when the constant pressure value was established, the gas was admitted to sample cell. The moment that gas was introduced, the recording of the kinetics of the pressure drop in the measurement system began. The values of the sorption progress (V [NTPcm^3/g]) were then calculated. The data on kinetics of linear strains of the coal sample were stored simultaneously in the internal memory of the strain meter [11]. On the basis of the obtained kinetics of coal-induced linear strains in the perpendicular ($E_\perp$ [‰]) and parallel ($E_{/\!/}$ [‰]) to the bedding plane, the kinetics of volumetric strains (E [‰]) were calculated in accordance with the relation: $E = E_\perp + 2\,E_{/\!/}$.

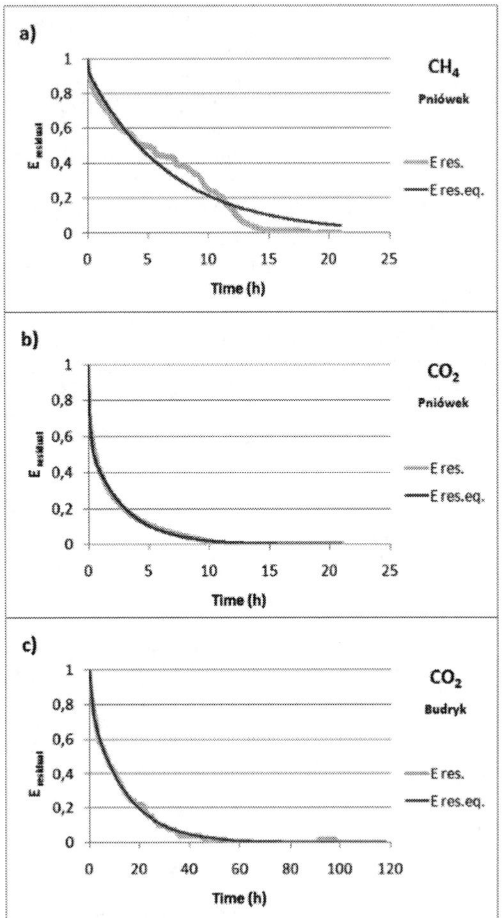

Figure 2. Application of the Busch equation (Eq. (2)) to the kinetics for (a) CH_4 and (b) CO_2 sorption-induced swelling of coal Pniówek and (c) CO_2 sorption-induced swelling of coal Budryk.

3 Results and discussion

3.1 Kinetic equations

The approach of Busch et al. [18] is based on the bidisperse sorption/diffusion model and describes the gas sorption kinetics as a linear combination of two first-order rate functions because the transport is being split to fast and slow process:

$$Y_{residual}(t) = Y'\exp(-k't) + Y''\exp(-k''t) \qquad (1)$$

$Y_{residual}$ (Y_{res}) stands for the relative residual (unoccupied) sorption amount, k' and k'' are different rate constants, Y' and Y'' (with $Y'' = 1 - Y'$) are normalized sorption at equilibrium [18].The expansion kinetics can be described by the equation:

 DOI: 10.1051/e3sconf/20161000012

$$E_{residual}(t) = E'\exp(-k't) + E''\exp(-k''t) \qquad (2)$$

where $E_{residual}$ (E_{res}) is the relative residual expansion, which is the potential relative swelling, k' and k'' are different rate constants, and the parameters E' and E'' represent expansion at equilibrium (with $E'' = 1 - E'$).

The mathematical form of the PSOE was first proposed by Blanchard et al. [31] to describe the kinetics of heavy metal removal by natural zeolites. It is the most commonly used equation to describe sorption kinetic data [32]. The general time-integrated form of the equation is:

$$V(t) = V_{eq} \frac{kV_{eq}t}{1 + kV_{eq}t} \qquad (3)$$

where V_{eq} is the equilibrium sorption amount and k is the rate constants. For the purpose of this analysis, Eq.(3) was transformed to Eq. (4) for the swelling progress:

$$E(t) = E_{eq} \frac{kV_{eq}t}{1 + kV_{eq}t} \qquad (4)$$

where E_{eq} is the equilibrium sorption amount and k is the rate constants.

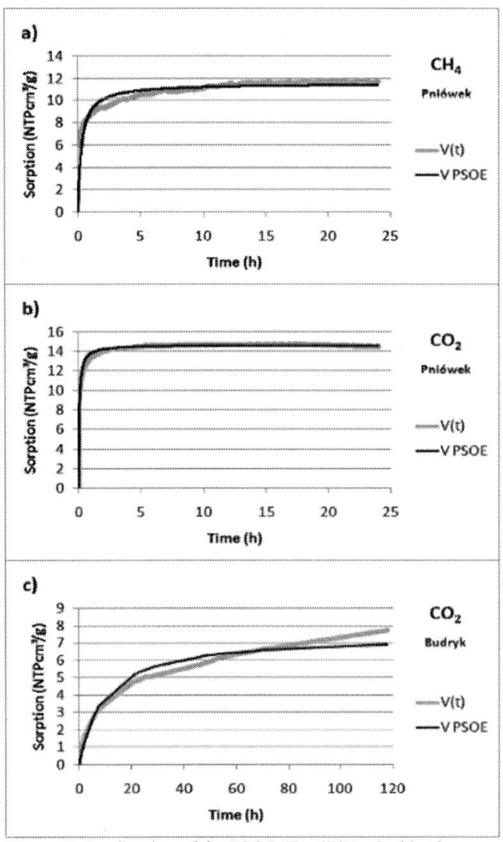

Figure 3. Application of the PSOE (Eq. (3)) to the kinetics for (a) CH_4 and (b) CO_2 sorption on coal Pniówek and (c) CO_2 sorption on coal Budryk.

The Elovich equation [33-34] corresponds to chemisorptions. It is characterized by the two parameters A and B.The sorbed amount has no maximum value it goes to infinity and should be avoided. However, it was used in this analysis in its integrated forms:

$$V(t) = \frac{1}{B}\ln(ABt + 1) \qquad (5)$$

$$E(t) = \frac{1}{B}\ln(ABt + 1) \qquad (6)$$

The stretched exponential (SE) equation was developed as a global kinetic equation by Kolmogorov, Erofeev, Kozeeva, Avrami, and Mampel, and it is called the KEKAM equation [35-36]. It is characterized by parameter A and rate coefficient k. For the purpose of this analysis, the following forms of the equation were used:

$$V(t) = V_{eq}\{1 - \exp[-(kt)^A]\} \qquad (7)$$

$$E(t) = E_{eq}\{1 - \exp[-(kt)^A]\} \qquad (8)$$

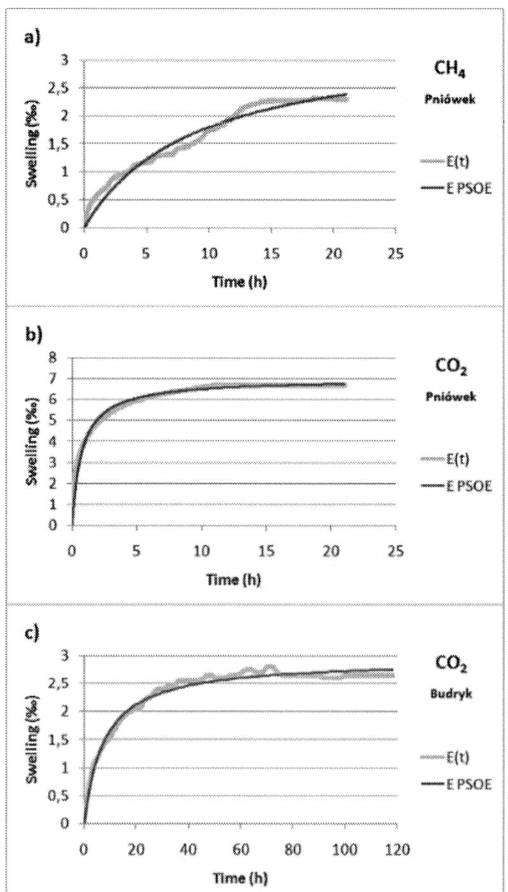

Figure 4. Application of the PSOE (Eq. (4)) to the kinetics for (a) CH_4 and (b) CO_2 sorption-induced swelling of coal Pniówek and (c) CO_2 sorption-induced swelling of coal Budryk.

3.2 Analysis

Figs. 1-7 show experimental kinetic curves of the accumulation of CH_4 and CO_2 in the porous structure of coal and the kinetic curves of coal expansion that accompany the sorption processes as well as plots resulted from the application of the kinetic equations.

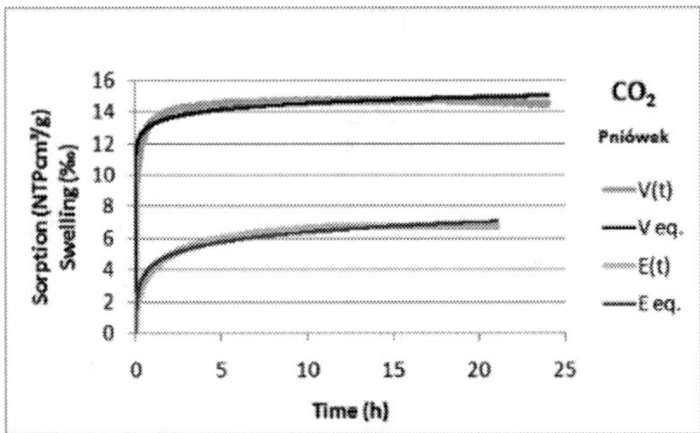

Figure 5. Application of the Elovitch equations to the kinetics for CO_2 sorption (Eq. (5)) and the sorption-induced coal swelling (Eq. (6)) of coal Pniówek.

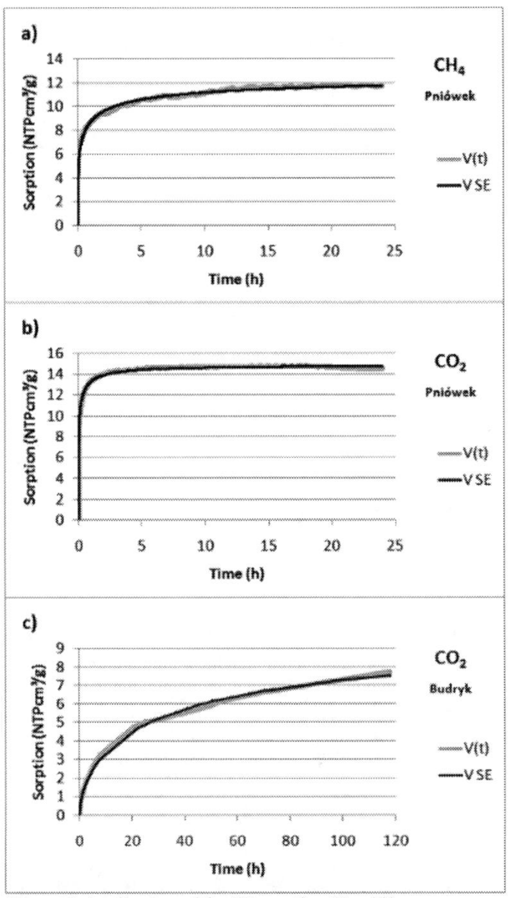

Figure 6. Application of the SE equation (Eq. (7)) to the kinetics for (a) CH_4 and (b) CO_2 sorption on coal Pniówek and (c) CO_2 sorption on coal Budryk.

The fitting results of Busch et al. [18] equation are shown in Figs. 1-2. The residual (unoccupied) relative sorption amount ($Y_{residual}$) (Fig. 1) or the amount of residual expansion ($E_{residual}$) (Fig. 2) is shown on the y axis and the x axis is time [h]. The fitting results of the PSOE, Elovitch equation, and SE are shown in Figs. 3–7, respectively. In these figures, the y axis is the amount of sorbed gas (Sorption [NTPcm^3/g]) and/or the volumetric swelling of coal (Swelling [‰]) for parts c and d. The x axis is time [h].

The first approach using the equation of Busch et al. based on two combined first-order rate functions gives a reasonable fit to the experimental data for CO_2 sorption and swelling for both coals, but for CH_4 the approximation seems to be second best in the case of the sorption and expansion data (Figs. 1-2).

The PSOE shows good agreement with the experimental data obtained for CO_2 sorption and swelling for coal Pniówek (Figs.3b and 4b). Application of the equation to the experimental data of sorption CH_4 on coal Pniówek and CO_2 on coal Budryk is unsuitable (Fig. 3).

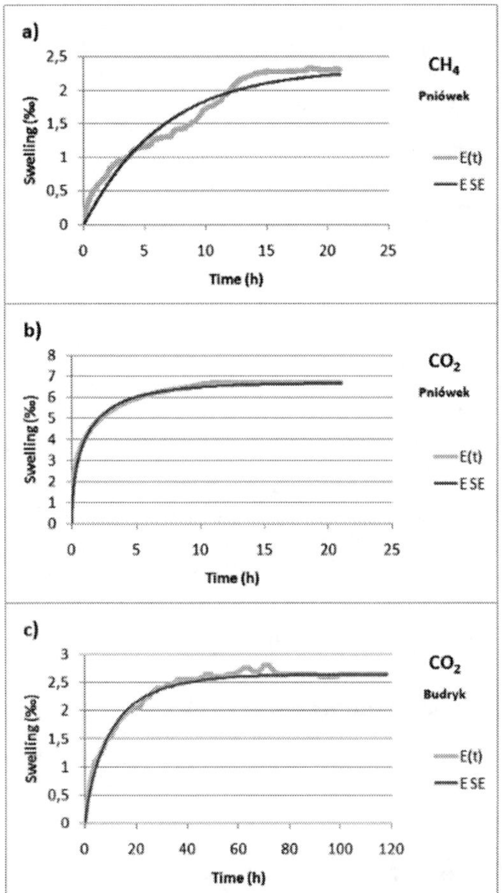

Figure 7. Application of the SE equation (Eq. (8)) to the kinetics for (a) CH_4, (b) CO_2 sorption-induced swelling of coal Pniówek, (c) CO_2 sorption-induced swelling of coal Budryk.

However, it seems to be quite suitable for the corresponding swelling kinetics (Fig. 4). The PSOE underestimates the equilibrium values for lower rate sorption (Fig. 3a,c). For the two main components of mine gas, methane has lower diffusivity, which is why its sorption rate is lower than that of carbon dioxide [18, 20, 24]. The conclusion is that the PSOE might be unsuitable for experiments with solid samples, in contrast to traditional research of sorption on powders and grain fractions, that take longer time to reach equilibrium state.

From Fig. 5, it is clear that the Elovitch equation is unable to accurately describe the sorption kinetic of CO_2 as well as the CO_2-induced swelling behavior of coal Pniówek. The same fitting results were obtained for CH_4 sorption kinetic on coal Pniówek and CH_4–induced swelling behavior for both coals and for that reason they are not included herein. As mentioned, because of the form of the equation the calculated sorbed goes to infinity and definitely should be avoided in cases like these.

Figs. 6-7 shows the results of applying the SE kinetic equation. The fitted results are in excellent agreement with the test results for both sorbats and both coals. This equation is the most promising approach to fit the experimental results of both the sorption processes and expansion phenomena of solid coal samples despite the rate of the process and time required to reach equilibrium state.

4 Conclusions

Four approaches were used to fit experimental sorption and dilatometric data of methane and carbon dioxide sorption to two coals. Equation based on the bidisperse gas transport model, proposed by Busch et al. [18] is very good to interpret and quantify observed gas uptake and expansion rates in coal. The PSOE is suitable for he obtained data, except the sorption kinetics with a lower rate than that obtained for CO_2 sorption on coal Pniówek. The Elovich equation does not fit the results obtained in this study, because the sorption and sorption-induced expansion processes reached close to equilibrium state and the problem of a lack of a maximum sorbed amount value in the equation did occur. Therefore the function should be avoided. The SE equation gives the best fit and is in very good agreement with all experimental results.

Acknowledgments

Financial support for this study was provided by the AGH University of Science and Technology (framework no. 11.11.210.244).

References

1. B. Dutka, M. Kudasik, Z. Pokryszka, N. Skoczylas, J. Topolnicki, M. Wierzbicki, Fuel Processing Technology **106**, 95–101 (2013)
2. G.S. Jodłowski, P. Baran, M. Wójcik, A. Nodzeński, St. Porada, J. Milewska-Duda, Applied Surface Science **253**, 5732–5735 (2007)
3. J.P. Kaszuba, D.R. Janecky, *Carbon Sequestration and its Role in the Global Carbon Cycle*, Geophysical Monograph **183** (2013)
4. M. Wdowin, R. Tarkowski, W. Franus, Arab. J. Sci. Eng. **39**, 295–309 (2014)
5. M. Wdowin, R. Tarkowski, W. Franus Int. J. Coal Geol. **130**, 79–88 (2014)
6. K. Czerw, Int. J. Coal Geol. **85**, 72–77 (2011)
7. M. Mastalerz, H. Gluskoter, J. Rupp, Int. J. Coal Geol. **60**, 43–55 (2004)
8. K. Zarębska, G. Ceglarska-Stefańska, Int. J. Coal Geol. **74**, 167–174 (2008)
9. Z. Pan, L.D. Connell, Int. J. Coal Geol. **69**, 243–252 (2007)
10. H. Seewald, J. Klein, Glückauf – Forschungshefte **47**, 149–156 (1986)
11. P. Baran, K. Zarębska, M. Bukowska, Energy & Fuels **29**, 1899–1904 (2015)
12. R.M. Bustin, X. Cui, L. Chikatamarla, AAPG Bulletin **92**, 15–29 (2008)
13. X. Cui, M.R. Bustin, L. Chikatamarla, J. Geophys. Res. **112**, B10202 (2007)
14. S. Harpalani, G. Chen, Fuel **74**, 1491–1498 (1995)
15. S. Ottinger, R. Pini, G. Storti, M. Mazzotti, Adsorption **14**, 539–556 (2008)
16. J.D. St. George, M.A. Barakat, Int. J. Coal Geol. **69,** 83 – 115 (2001)
17. Z. Majewska, G. Ceglarska–Stefańska, S. Majewski, J. Ziętek, Int. J. Coal Geol. 77, 90–102 (2009)
18. A. Busch, Y. Gensterblum, B.M. Krooss, R. Littke, Int. J. Coal Geol. **60**, 151–168 (2004)
19. D. Charrière, Z. Pokryszka, P. Behra, Int. J. Coal Geol. **81**, 373–380 (2010)
20. C.R. Clarkson, R.M. Bustin, Fuel **78**, 1345–1362 (1999)
21. Z. Pan, L.D. Connell, M. Camilleri, L. Connelly, Fuel **89**, 3207–3217 (2010)
22. J.D.N. Pone, P.M. Halleck, J.P. Mathews, Energy & Fuels **23**, 4688–4695 (2009)
23. Busch, A., Gensterblum, Y., Int. J. Coal Geol. **87**, 49–71 (2011)
24. X. Cui, M.C. Bustin, G. Dipple, Fuel **83**, 293–303 (2004)
25. E, Ruckenstein, S. Vaidyanathan, G.R. Youngquist, Chem. Eng. Science **26**, 1305–1318 (1971)
26. N. Siemons, K.-H.A.A. Wolf, J. Bruining, Int. J. Coal Geol. **72**, 315–324 (2007)
27. J.Q. Shi, S. Durucan, *SPE*, paper 84342 (2003)
28. J.Q. Shi, S. Durucan, Fuel **82**, 1219–1229 (2003)
29. K. Czerw, K. Zarębska, B. Buczek, P. Baran, Adsorption 22, 791–799 (2016)
30. A. Marecka, K. Mianowski, Fuel 77, 1691–1696 (1993)
31. G. Blanchard, M. Maunaye, G. Martin, Water Res. **18**, 1501–1507 (1984)
32. W. Płaziński, J. Dziuba, W. Rudziński, Adsorption **19**, 1055–1064 (2013)

 DOI: 10.1051/e3sconf/20161000012

33. C. Aharoni, Y. Suzin, J. Chem. Soc., Faraday Trans. 1. **78**, 2321–2327 (1982)
34. C. Aharoni, F.C. Tompkins, Trans. Faraday. Soc. **66**, 434–444 (1970)
35. M. Avrami, J. Chem. Phys. **7**, 1103–1112 (1939)
36. F. Brouers, O. Sotolongo-Costa, Phys. A **368**, 165–175 (2006)

Talented Employees in the Field of Brownfields

Marcela Davidová[1,a] and Petra Horváthová[2]

[1] *VŠB-Technical University of Ostrava, Faculty of Mining and Geology, Institute of Environmental Engineering, 17. listopadu 15, 708 33 Ostrava-Poruba, Czech Republic*

[2] *VŠB-Technical University of Ostrava, Faculty of Economics, Department of Management, Sokolská tř. 33, 701 21 Ostrava 1, Czech Republic322-442*

Abstract. The article is aimed at bringing information on one of the important terms for successful redevelopment, recovery and operation of brownfields – efficient people, employees. Not only brownfields, but generally all organizations that want to be competitive and successful, want to outperform their present rivals and considerably increase their added value have to pay attention to talent management. The article is concerned, on the basis of available theoretical information and the existing practical experience, at describing three particular processes of talents management (their identification, development and retention) which are necessary for the successful use of talent. The purpose of the article is providing recommendation how to help operators of brownfields identify and retain highly talented employees, build on their strong points, reward their success, provide them with a chance to make progress, and increase their overall efficiency.

1 Introduction

Surely we can agree with the statement that human resources are becoming, in the present brownfields environment and in the conditions of competition and globalization increasing pressures, for organizations a more and more pivotal factor. The main competitive advantage is not financial means, modern and efficient technology and technique or quality strategy but people, efficient employees. Organizations-investors that want to survive and expand, to outperform their present rivals and considerably increase their added value in such environment have to attract, cultivate and maintain talented employees – they have to pay attention to talent management. A number of talented individuals in the population are quite limited [1].

Precondition for effective talent management in an organization is its introduction and its support by the organization's management, the interconnection of the talent management strategy (which is the basis for talent management working with the organization's business strategy) and also the allocation of a sufficient quantity of resources (and not just financial) for its implementation and use in the organization.

The aim of a talent management strategy is to ensure a pool of highly talented, skilled, committed, and loyal individuals able to contribute to the achievement of present as well as future requirements of the organization.

The objective of the article is providing recommendations to brownfields´ operators how to manage talents that´s mean to help them identify and retain highly talented employees, build on their strong points, reward their success, provide them with a chance to make progress, and increase their overall efficiency.

2 Talent management processes

On the basis of available theoretical information and the existing practical experience we can describe three particular processes of talents management – identification, development and retention of talents. Each of these processes includes quite a number of activities which will be described in detail in the following sections.

2.1 Identification of talents

The first step the organization has to take is to identify key roles. These are such positions in the organization which, if they are not filled, may seriously disturb working of the organization. The organization has to assess whether there are available employees with key competencies who will be needed in the future in dependency with business strategy. As long as it has not sufficient existing resources then it has to find these talents. It may identify them among its present employees, from the internal resources or obtain them from the external resources.

[a] Corresponding author: marcela.davidova@vsb.cz

The whole process of talent identification consists of three basic steps: first – the identification of key competences and creation of a rating scale to specify talents' contributions to the organization, second – performance appraisal and a potential prediction, and the creation of measuring scales for their measurement, third – the creation of a talent-pool.

An important step that the organization has to take in talent management is to identify key competences of the organization, which are competences crucial for the success of every employee and thus of the whole organization.

After that key competences are determined and defined, performance appraisal and a prediction of individuals' potential follow. Performance appraisal is the measurement of actually achieved results within those areas for which the specified individual is responsible, and/or competencies understood as critical for the success of performed work as well as of the entire organization. A measuring scale most often used in organizations is as follows: considerably exceeds expectations (5), exceeds expectations (4), comes up to organization's expectations (3), below expectations (2), deep below expectations (1).

Prediction of a potential is in fact a prognosis by how many levels within the organization an employee can advance on the basis of their past/current performance appraisal, training and development, preferences in their careers, and the current and planned levels of competencies. A scale most often used by organizations is as follows: a high potential (5) → it is possible to promote by two levels or even more, a higher potential (4) → it is possible to promote by maximum two levels, an average potential (3) → it is possible to promote by one level, a limited potential (2) → ready for a lateral redeployment, no potential (1) → it is not possible to promote at the moment [1].

There are a number of methods the organization can use for performance appraisal and a forecast of employees' potential both from the internal and external sources. Talents from own sources are usually sought for through a working system of regular appraisal within which a number of methods are utilized. To the most used methods belong the method of assessment according to set objectives (MBO - Management by Objectives), assessment centre, the analysis of critical events, and a 360° feedback. In order to identify talents from the external sources the tests of fitness, a method of assessment centre and behavioural (competent) talks can be used.

After performance appraisal and a potential prediction, the very group of talented employees is identified – a talent-pool is being shaped. Individuals meeting in advance determined criteria (e.g. performance has to be evaluated as such which exceeds expectations (5) or considerably exceeds expectations (4), or meets expectations, and a potential has to be predicted as high (5), or higher (4), or average (3) and it can be jointly indicated as talents or this group of talented individuals can be furthermore divided into three groups – top talents, talents and potential talents. The decision whether only talents or top talents or talents or all three groups will be included into programs for talents depends on the organization [2].

But this is not the end of the systematic approach to talent and talent management. Further steps follow – development and retention of talent.

2.2 Developmentx of talents

The organization should offer talented individuals the chance to develop their strengths, improve their total performance as well as their particular competences; strengthening their motivation and making their career development possible.

A special development plan should be made for talented employees in close cooperation with their managers. This plan should resemble the integral program for the specified group of talent and should be supplemented by the particular participants' individual needs.

Long-term training programs are not recommended in the current rapidly changing world. Organizations should prefer shorter but well drawn up programs that are relevant to their needs and for which there is a faster return on investment.

A development program may include a combination of various methods, activities or instruments of talent development, but always according to the specific needs of the specific organization. The methods employed are both the methods used for development at a workplace on job performance (on-the-job) and the methods used for development outside a workplace (off-the-job). Greater emphasis is laid on methods used for education at a workplace but, of course, both groups of methods are used [2].

Programs of talent development are very closely connected with carrier management; that is with career planning and succession planning. They provide talent with opportunities to grow in their current job roles and to move forward into roles of a higher level.

Career planning determines the advancement of talent within an organization based on the consideration of the organization's needs. Talented people should go through a certain sequence of experience supplemented by couching and development programs that will help them to be prepared for more demanding job positions in the future. Succession planning should ensure that an organization will have enough talent to satisfy its business needs in the future. Succession planning is about planning key roles, in particular.

By the development of talent, however, work with talent is not finished. It is necessary to achieve the situation in which talent individuals remain in the organization they do not leave it. It is necessary to strive for their retention in the organization.

2.3 Retention of talents

Also activities ensuring retention and stabilization of talent in the organization are inseparable part of talent management. The aim is to ensure that talented employees stay in the organization as its devoted members and in their work engaged people that they do not have a tendency to leave because their departures have usually extraordinary impact on organization's operation, which is irrelevant to their number.

The organization should do its best to be 'an attractive employer', to be a place where people like to work, 'an excellent workplace', and 'an excellent place for work performing'. Among factors contributing for the organization to be an attractive employer there is:

- The offer of interesting and valued work.
- Providing the opportunity to learning and development and advancement in career.
- Respecting a balance between the life at work and a private life by the organization
- The offer of a flexible role to talented employee.
- The offer of quality work conditions and equipment.
- Providing a feeling of recognition and respect.
- The offer of an adequate reward.
- Promotion of organization's responsibility to the society.

Among the three factors that the most influence retention of talented employees in the organization belong, the opportunity of advancement in employees′ career, the offer of interesting and satisfying work, and ensuring opportunities for training and development [1].

The above-mentioned facts create a valuable offer that sound values will be observed in the organization, employees can expect good leadership, freedom and autonomy, highly incentive work, opportunities for professional advancement in employees′ career and adequate remuneration. It is necessary to ensure that talented individuals would be aware of being appreciated and that they have a considerable or even a principal value for the organization [3].

For the purpose of retention of talents the organization should address the following potential activities:

Take an interest in the systems of remuneration that are not competitive, they are not unjust and unbiased. It is necessary to realize that not only financial means are in question; talent should be offered also other forms of rewards – interesting work, relevant way of acting and the like.

Tasks and job opportunities in the organization should be created in such way so that a variety of work, the importance of tasks, autonomy, checking of own work, and feedback could be ensured. Also the opportunity of education, growth and career advancement should be ensured. Some roles can be 'made-to-measure' and thus accommodate concrete individuals' needs.

The organization should do its best in order to ensure loyalty to work not only by creating work tasks and job opportunities, but also by work organization in connection with projects the talent can identify with better than with the organization as a whole.

Building of social bonds should be encouraged in the organization.

Decisions on the choice or professional advancement should be properly made so that individuals' abilities could be duly interconnected with requirements of work which is to be performed by those people.

The organization should try, in connection with preventing talent from leaving, its employees in the period after joining-up the organization will be able to adapt themselves to their work, team, and the organization as best as possible. In the starting phases of their work for the organization they should be offered the appropriate assistance, education and training.

The organization should put into effect activities that will lead to the improvement of balance between a life at work and extra-work life. This includes for example a chance of flexible work regime or flexible ways of employment respecting extra-work needs of talent.

Unfavourable work conditions should be eliminated and excessive occupational stress should be removed or made lower.

The organization should choose, inform and train managers and heads of teams in such way that they are aware of positive contribution by which they can contribute to better retention of talent. However, they have to choose the right way of managing their teams.

There are many possibilities how to retain talent in the organization, not only those that have been mentioned above. The main idea is quite simple – the organization has to pay real attention to its talent. It should talk with them about their work, motivate and reward them, help them accomplish their careers etc.

3 Conclusions

Successful redevelopment, recovery and operation of brownfields make higher and higher demands on organizations. For organizations/investors the main asset to manage those difficult conditions are people. Without a skilled workforce, without employees equipped by required abilities and knowledge, a number of organizations/investors will not be able not only to keep up with competition but they even may not to survive under specified conditions.

One of the possible instruments of human resources successful management is the system of talent management. For organizations a conception of talent management does not bring any special requirements from the view of human resources management as well as of particular personnel activities. It concerns only a careful application of the best principles and approaches that have proved in practice, mainly in the field of acquisition and choice, education and development, remuneration, and socio-cultural and welfare activities for employees.

Talent management can be considered as one of fundamental instruments of human resources management in the organization. But what should be the organization's attitude to the management of its talented employees? It should offer them a real

DOI: 10.1051/e3sconf/20161000013

perspective and pay attention to the strategy of their management. Not to use only formal programs of talent development and concentrate on career management and succession planning but to apply a comprehensive system of talent management that will enable those talented individuals a real development and a deeper identification with the organization. This system then helps organizations identify and retain highly talented employees, build on their strengths, reward their success, provide them with opportunities for advancement and increase their overall efficiency and thus the efficiency of the whole organization.

The article described in detail how it is possible to apply talent management in organizations and also brought individual process steps for a wider application of this comprehensive and systematic approach to work with talented individuals.

Acknowledgement

The research was supported by the European Social Fund within the project CZ.1.07/2.3.00/20.0296 and by the Student Grant Competition of the Faculty of Economics, VŠB-Technical University of Ostrava within the project SP2016/123.

References

1. L.A Berger, D.R Berger. Talent management Handbook: Creating Organizational Excellence by Identifying, Developing & Promoting Your Best People, second ed. McGraw-Hill, New York (2010)
2. J. A. Cannon, R. McGee. Talent Management and Succession Planning, CIPD, London (2007)
3. M. Armstrong, S. Taylor. Armstrong´s Hanbook of Human Resource Management Practice, thirteen ed., Kogan Page, Philadelphia (2014)

Sustainability considerations in the operation of Wastewater Treatment Plant 'Swarzewo'

Alina Dereszewska[1,a] and Stanislaw Cytawa[2]

[1]*Gdynia Maritime University, Department of Chemistry and cargo Commodity Science, Morska Str.81-87, 81-225 Gdynia, Poland*
[2]*Swarzewo' Wastewater Treatment Plant, Wladyslawowska Str. 84, 84-120 Wladyslawowo, Poland*

Abstract. The Wastewater Treatment Plant (WWTP) 'Swarzewo' plays a special role in the protection of coastal waters of the Baltic Sea area and the management of solid wastes in the region. This paper discusses several options implemented in the WWTP area in order to achieve sustainability. The first one was the inclusion of WWTP into municipal waste management plan to increase the biogas production and to reduce volume of organic waste in the region. Nowadays, daily production of about 2000 m^3 of biogas is gained. The energy balance shows a considerable benefit from the co-fermentation of sludge with municipal organic wastes. The next goal was to obtain a favorable economical balance of energy and high level of pollution reduction. The last one was the involvement of local communities in a conscious segregation of waste 'at source'. For the purpose of this paper bio-energy production, nutrient elimination, composting, and research, have been selected as indicators of sustainable development. Furthermore, in this study the methods of nutrient recovery from wastewater were explored. Struvite precipitation and compost production were presented as an example of nutrient elimination with 'end of waste' production. Depending on the struvite precipitation conditions, recovery of 4 Mg of phosphorous and 1,8 Mg of nitrogen is possible to obtain annually.

1 Introduction

Protection of coastal areas from excessive pollution is a challenge for many countries [1,2]. The specificity of each region determines the system of sewage treatment and waste management, so that it could possibly correspond to the behavior principles of responsible, sustainable development. The Baltic Sea Coast is significantly burdened with pollution that is caused by intensive tourism and highly loaded wastewater fish industry. One of the most serious environmental problems of the region is eutrophication caused by excessive loads of nitrogen and phosphorus discharges of the water reservoirs. Nutrient compounds present in wastewater act as fertilizers. Those compounds, when excessive in their quantities in the sea environment, can disturb the ecological balance. The eutrophication devastates the ecosystem, hinders the use of marine resources and affects the recreational activities. The solution of this problem is strictly connected with the sustainability concept.

There is no consensus on the sustainability definition. Several concepts have been used to describe sustainability but a common principle is that sustainability should be seen in terms of the integration of three dimensions: environmental, economic and social [3-6]. Economic sustainability implies the incurrence of costs that should not exceed the benefits. Ecological sustainability refers to support its long-term development of environment through the protection of natural resources and reduction of emissions. Sustainability also has a social aspects. The social dimension is based on the education, which forms public awareness. In fact sustainable development can be achieved only in a community which feels that behind this development is responsible, understands the challenges and accepts the consequences of appropriate actions [4,7].

In relation to the wastewater treatment system, the most important task is to protect aquatic life. Increasing the awareness (including the appropriate decision-makers) of the impact of sewage pollution on the Baltic Sea waters results in the increased emphasis on the consistent sustainable reconstruction of the wastewater treatment plants [2]. The adoption of low-cost wastewater treatment technologies is very important in the implementation of this strategy. It maximizes the efficiency of wastewater treatment, reduces the amount of consumed energy simultaneously (thereby reducing CO_2 emissions) and seeks the minimization of waste generation. Furthermore the awareness of the local community must also be raising largely as well as bear the costs of wastewater treatment plant operating because it is exposed to the side effects of its work (eg. odors) [4,8,9]. In order to find the best way to sustainable region development it should be find a compromiseshould be found between the cost

[a] Corresponding author: a.dereszewska@wpit.am.gdynia.pl

 DOI: 10.1051/e3sconf/20161000014

of waste water treatment and environmental benefit from reduction of the pollution of the Baltic water.

In this paper strategy for the operation of Wastewater Treatment Plant 'Swarzewo' (WWTP) based on the principles of sustainable development has been briefly described. The indicators of sustainability of particular relevance for the work of WWTP and the coastal region of the Baltic Sea were selected. Among them are both measurable indicators, (ie. the cost, energy saving, effluent parameters) and irrational outcomes, associated with the waste segregation propagation, education or scientific research were presented. The paper also presents the examples of applied specific technological solutions.

2 Characteristic of the region and WWTP 'Swarzewo'

The catchment area of the Wastewater Treatment Plant 'Swarzewo' includes land belonging to the Puck District and is the most northerly situated of the treatment plants in Poland. It has a significant impact on the protection of coastal waters of the Baltic Sea. The biological part of this treatment consist of the six SBR reactors (Sequential Batch Reactors) enable to removal of nutrients.

WWTP 'Swarzewo' was modernized in the years 2014-2016 in order to generate less wastes and produce more valuable products (compost and biogas). Among many changes, as a part of modernization digesters and cogeneration have been incorporated into the existing technological sequence, resulting in the production of biogas for electricity and heat. The modernization also included additional solutions for protection against excessive discharge of nutrients into the Baltic Sea. There was the oxygenation chamber for leachate from digestive chambers an additional secondary sedimentation tank, for treated sewage. In the summer months (July and August) sewage treatment plant purifies up to 15 000 m^3/day, while during the remaining period - about 5 000 m^3/day. About 95 percent of waste water is domestic, the rest is received from the fish industry, heavily loaded of fats and proteins.

The specificity of the Puck region is a production of large amount of municipal waste, including those that are susceptible to biological degradation. A significant amount of it is exported as mixed waste to distant landfills up to 60 km. About 45% of waste produced by households is a biodegradable organic waste. If landfilled, then methane, carbon dioxide, hydrogen sulfide and other harmful gases are emitted into the atmosphere. Rains cause the formation of leachate polluting groundwater.

Coordinated cooperation with the Authorities responsible for protecting of the coastline led to the extension of the tasks of WWTP for solid waste utilization enable for fermentation. This solution is beneficial both for the regional units of public utilities, because it reduces transport of waste to distant landfills, and for the treatment plant which can thus increase the production of energy from biogas [9,10]. The condition for the implementation of co-fermentation (fermentation sewage sludge and organic waste) is to educate inhabitants in the segregation of waste 'at source'. It involves detailed segregation conducted in households by the instruction drawn up by the waste customer [11].

3 Indicators of sustainability

Sustainability of wastewater treatment systems can be estimated through analysis of different indicators. Some of them (eg. energy saving, cost analysis, environmental impact) are possible to quantify, others are immeasurable. Many of indicators are listed in literature but sometimes self- defining of them is more effective than achieving a sustainability in region [1,5].

Criteria of conducted in the WWTP sewage-waste management, for the sustainable development of the Baltic Sea region, can be summarized in the following points:

1. Minimizing, within an acceptable cost, impurities in the outflow.
2. The maximum recovery of nutrients and their conversion into a full-fledged product ('end of waste' philosophy).
3. The production of 'green energy' and commitment to self-sufficiency in energy.
4. Processing of waste being harmful to the region.
5. Education in the field of waste segregation and development using research.
6. Reduction of the effects burdensome to the local community (eg. odors).

The work was chosen the analysis of several aspects important for the region to achieve those objectives.

3.1 Nutrient reduction and recovery

Removal of nitrogen and phosphorus from wastewater is one of the priorities of WWTPs, since the presence of nutrients in the outflow contributes to eutrophication of the Baltic Sea. In addition, deposits of phosphate are intensively used as the main source for the production of industrial fertilizers. The amount of these minerals is limited which implies the need for the recycling of phosphorus [7]. To the WWTP flows an average of 296 Mg on nitrogen and 43 Mg of phosphorus per year. The nutrient concentration, allowed by standards in the outflow is 10 g/m^3 for nitrogen and 1 g/m^3 for phosphorus. It is difficult and costly to achieve these parameters. Exceeding these standards by 1%, monthly is discharged additionally 15 kg of nitrogen and 1,5 kg of phosphorus. In summer, this value increases up to three times. Environmental penalties incurred by WWTP 'Swarzewo' for exceeding the limit of nutrients are even PLN 750.000 per year. Average concentration of nutrients in the effluent and the load of phosphorus and nitrogen discharged by treatment plant for the Baltic Sea annually is shown in Table 1. Data for 2016 are estimated on the basis of the results obtained after modernization, in March 2016.

Table 1. The concentration of nitrogen (C_N) and phosphorous (C_P) in the effluent. The amount of nutrients supplied by the WWTP 'Swarzewo' annually to the Baltic Sea (N-load; P-load).

year	C_N [g/m^3]	N-Load [Mg/year]	C_P [g/m^3]	P-Load [Mg/year]
2010	15.8	28.8	0.9	1.6
2011	15.5	28.3	1.0	1.8
2012	10.9	19.9	1.0	1.8
2013	12.8	23.4	1.2	2.2
2014	12.1	22.1	0.9	1.6
2015	17.5	31.9	2.2	4.0
2016	**6.0**	**11.0**	**0.8**	**1.4**

The majority of nitrogen is dissolved in wastewater as ammonium salts. In WWTP it is removed from the wastewater biologically in the nitrification / denitrification processes. The decisive role in its removal is fully controlled by of the concentration of ammonium compounds during the aeration process of activated sludge. Before the modernization, control process of wastewater treatment was carried out, as in most of this type of treatment plants, basis on timing control suited to the aeration of waste water volume. The studies conducted jointly with the Gdynia Maritime University have shown that this control often leads to incomplete removal of nitrogen or to unnecessary aeration [12]. Analysis of the results has led to a changes in the control system. After the modernization, nitrogen transformations are controlled by sensors of ammonia and nitrate nitrogen during process. With this change, the nitrogen concentration in the effluent drop down to 6 g/m^3, below the limit. Thus, the flow of nitrogen into the Baltic Sea can be reduced by up to three times compared to the state before upgrading.

The biological removal of phosphorus from wastewater is much harder to control. A large part of it is removed together with the excess sludge composted and sold as valuable products. An sustainable alternative to remove it is composting of excessive sludge [13]. Taking care of a high standard of composted organic material led to the production of certified compost which can be used as organic fertilizer in agriculture. In this way non-degradable waste from WWTP 'Swarzewo' are minimized to 2%. When the new product appears on the market, produced from waste and complies with required standards criteria with no harmful effects on the environment and people, meets the criteria 'end of waste' philosophy [14]. Phosphorus is in a very large extent absorbed by sludge, but it is only partial remove of the phosphorus from wastewater. Finally, it's removing is performed chemically, using a ferric coagulant.

3.1.1 Nutrient recovery research

During the anaerobic process of sludge digestion is released a significant amount of nitrogen and phosphorus. Therefore, the post-fermentation leachate contains a high concentration of these elements (both nitrogen and phosphorus - up to 300 g/m^3). These leachate are returned to the chambers SBR and again cleaned. Oxidation of such large loads of nitrogen results in significant energy consumption (aeration) and difficults to obtain proper standards of wastewater discharged into the sea. It causes decrease in throughput of WWTP [8,14].

Currently, research on the nutrient elimination from the leachate by precipitation of magnesium ammonium phosphate (struvite) is carried out. This method not only allows the removal of nutrients from wastewater, but also its recovery in the form of a compound with very desirable properties as fertilizer [15,16]. Struvite is precipitated by using a cheap secondary raw material (magnesium salt), thereby reducing the use of ferric coagulant to the process of phosphorus removal (approx. 250 t/year). Using traditional coagulant, significant amounts of iron is accumulated in the compost produced by the WWTP, which is disadvantageous in the process of fertilization. The use of the cheaper waste magnesium salt to struvite precipitation also reduces the costs associated with the purchase of coagulant (annual cost now stands at PLN 50 000).

At this stage the laboratory test were carried out and results are promising. Depending on the precipitation conditions recovery of 90% of phosphorous and 20% of nitrogen from the leachate were obtained. Selected results of the test are presented in Table 2.

Table 2. The percentage of phosphorous ($P\text{-}PO_4$) and nitrogen ($N\text{-}NH_4$) recovery depend on the pH and initial concentration of nutrients in some of leachates investigated.

Initial concentration (g/m^3)		pH	Nutrient recovery [%]	
$P\text{-}PO_4$	$N\text{-}NH_4$		$P\text{-}PO_4$	$N\text{-}NH_4$
120	-----	7.5	30	---------
		8.5	17	
		9.5	72	
105	550	9.3	79.2	21.6
98	110	8.5	69	----
315	412	8.5	74	24
315	412	9.5	77	27
330	504	9.5	90	22.1

Simultaneously, the test with using the model of prototype device and leachate from fermentation chambers was carried on. Depends on crystallization conditions, struvite crystals with different size and form were obtained (Figure 1). Regardless of their size, struvite crystals exhibited good sedimentation properties. This is important from a technological point of view, since it reduces process time and makes it easy to separate the crystals from the leachate.

DOI: 10.1051/e3sconf/20161000014

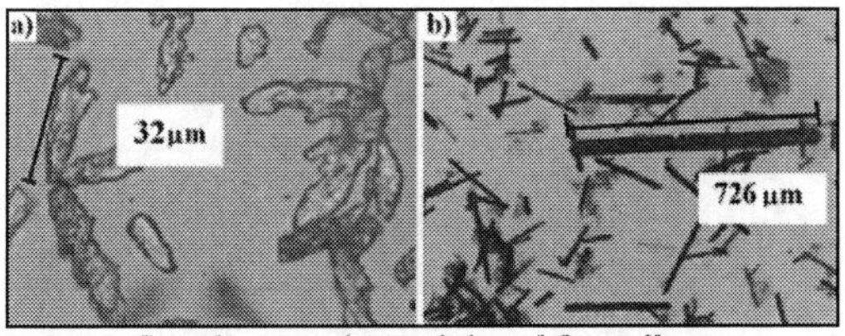

Figure 1. Struvite crystals precipitated from digestate leachate at pH 9.5 a) crystallization time 1h, concentration P-PO_4 315 g/m^3 b) crystallization time 48 h, concentration P-PO_4 153 g/m^3.

Summarizing, the technology based on struvite precipitation offers the following benefits:

- Reduction of phosphorus and nitrogen return loads and in treated wastewater,
- Reduction of energy consumption,
- Reduction of ferric coagulants consumption,
- Prevention of struvite crystallization in pipes,
- Simple and quick struvite precipitation process,
- Obtaining of bio-available and slow-release fertilizer
- Reduction of fossil minerals consumption (eg. apatites).

The expected amount of post-fermentation leachate in the WWTP 'Swarzewo' is 60 m^3 / day. Assuming that the same efficiency as in the preliminary study will be achieved, it is possible to recover 4 Mg of phosphorus and 1.8 Mg of nitrogen per year. The costs and expenses associated with the production of struvite include, as appropriate:

Savings: coagulant consumption, the elimination of aeration; sale of struvite;

Expenditure: a magnesium salt; pH adjustment, depreciation, electricity

This analysis indicated that financial gain for the WWTP will be approximately PLN 40 000, The profit is not significant, however, the most important is environmental effect and the possibility of utilizing the resulting waste.

After detailed study and obtaining the appropriate certification, the precipitated struvite can be a valuable bio-fertilizer. In this case, it will be the next WWTP product which meets the criteria for 'end of waste'.

3.2. Energy

Wastewater treatment is a process demanding high energy consumption [9,10]. WWTP 'Swarzewo' consumes approx. 2 500 MWh of electricity per year. One of the ways to optimize energy consumption in the plant is the implementation of the fermentation process for biogas production. It is possible to co-digest selectively segregated municipal waste with sludge from sewage treatment and some industrial waste (ie. grease from fish industry, agricultural waste). There are two co-generators installed by the treatment plant, which convert the resulting biogas into electricity and heat (1: 1). Part of the biogas is burned directly in boilers. Quantity of biogas obtained in the first six weeks of working chambers, and the quantities of energy (heat and electricity) obtained in this way is shown on the Figure 2.

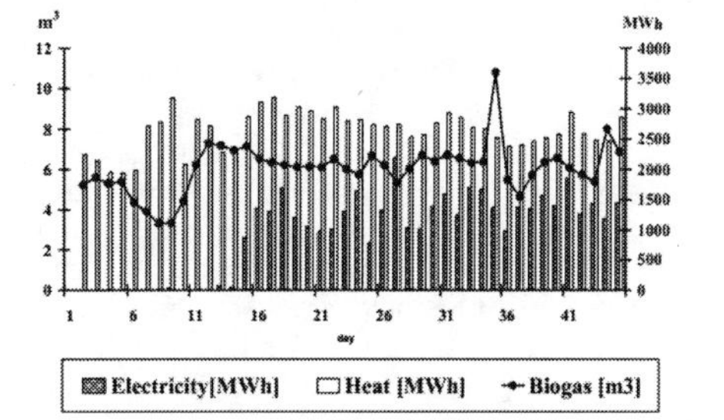

Figure 2. Biogas and energy production in WWTP 'Swarzewo' after modernization (1.03-18.04.2016)

The energy balance of WWTP shows significant benefits from the co-fermentation of sludge with municipal waste. In this way an average of 1925 m^3 of biogas is produced daily, which is converted into 12,5MWh (total - heat and electricity). It is estimated that the plant can become energy self-sufficient while maintaining the same level of fermentation during the whole year and can produce 4560 MWh/year of energy. It should be taken into account, however, after modernization, a number of new devices was installed and energy consumption increase.

The planned target capacity of digestive chambers is 8000 m^3 of biogas / day and is largely dependent on collected wastes to co-digestion. The possibility of co-digestion of waste is also a factor for segregation of garbage in the region. It mobilizes inhabitants generating waste to extract one stream fit for fermentation. Catchment area of treatment plant includes 32 thousand population, each of them produces approximately 200 kg / year of green waste. The waste, after a suitable separation, can be incorporated into co-fermentation or to produce compost. Part of the waste for fermentation passes from the fish industry. These are high-energy waste (grease, protein) troublesome for manufacturers, and favorable for biogas production. The WWTP 'Swrzewo' contract is to acquire, for the purposes of the fermentation, 1500 Mg per year of this type of waste.

The WWTP 'Swarzewo' is capable of receiving 26.000 Mg of organic waste/year, thus it is facilities for the production of energy for own needs and even to sell the surplus. First quarter of the year of operating modernized WWTP showed that proceeds from the selective collection of waste represents only 4% of potential. Acquisition of organic waste is therefore a fixed point in the development plans of WWTP.

Energy balance of wastewater treatment plants can also be improved by reduction of energy consumption in the process of aeration. Wastewater treatment is based on the mineralization of organic compounds by the activated sludge microorganisms. This process requires a sufficient amount of oxygen delivery to micro-organisms and involves a considerable energy consumption. Due to the fact that the oxidation of nitrogen compounds, is the longest mineralization process, this parameter can be used as an indicator of the time required for their purification. As already mentioned the use of nitrogen probe eliminates unnecessary aeration of wastewater and thereby improves energy saving. Pilot studies have shown that using control system of the purification process, based on indications of ammonium and nitrate nitrogen, the consumption of energy for aeration can be reduced about 10-15% [12]. Another important change in the process of wastewater treatment, resulting after the nitrogen sensors using, is a significant prolongation of denitrification phase. The process do not requires aeration, and consumes relatively small amounts of energy, but reduce the capacity of the single reactors by increase of the retention time in the reactor. For this reason two additional reactors were built to increase the capacity of the entire plant.

3.3 Social dimension

3.3.1 Waste segregation and composting

The results of sewage treatment is waste in the form of excess sludge as a source of carbon- organic fertilizer and it is rich of valuable substances. The sludge mixed with selected organic waste can be processed into a valuable compost [2]. Wastewater treatment plant produces compost under the name of "Ulkomp' with certified quality since 2006. The low price is also an encouragement for local communities to include them in the segregation system. The obtaining a compost that meets all the requirements excludes the use of contaminated input materials. Such purity can be provided only by segregation of waste 'at source' at homes.

WWTP runs an educational campaign since 2001. The detailed instructions of segregation organized among the inhabitants of the region were developed. The campaign has enjoyed great popularity among the inhabitants, also contributed to the growth of ecological awareness. After the modernization of WWTP the waste are supplied by authorized individuals and the WWTP was joined to the system of waste management. The area covered by the system of waste segregation involves 104 000 of inhabitants (the average value takes into account the visiting tourists). Basis of the quantity of waste collected it can be concluded that 10% of the population is involved in the segregation system.

The cooperation with the local community also contributed to greater acceptance of WWTP, which belongs to the object to some extent onerous. The modernization included the reduction of odors, which are burdensome to the local community and have negative influence on tourism development in the region. The artificial aeration was applied to composting for accelerate the processes, also a number of filters to eliminate odors were installed.

3.3.2 Education and research

There are several research and development project on wastewater treatment. Pro-ecological profile of company resulting in the opening of the new technologies. An important aspect is the cooperation with the academic community. It is known, that this collaboration have an impact on both: regional innovation and academic research. to regional innovation [17,18]. Exchange of experience between operators of the WWTP 'Swarzewo' and researchers from universities (Technical University of Gdansk and Gdynia Maritime University), results in the search for new solutions, especially designed to protect the Baltic Sea environment. Moreover, there are also students included in the research work This allows them to confront the theoretical knowledge acquired at university with its practical application. Preliminary modeling studies allow estimation of both the benefits and expenses resulting from the possible implementation of new solutions. The discussed earlier pilot studies are the example of the WWTP 'Swarzewo' co-operation with Gdynia Maritime Academy. Those studies are aimed at assessing the possibility of incorporation of the prototype device for struvite precipitation and purification in the technological line of wastewater treatment.

4 Summary

The need to be more efficient with the use of resources and with environment protection has become a important issue in water treatment management. Sustainability issues as the energy balance, carbon dioxide emissions, total chemical usage, the nutrients and other type of contamination present in effluent are now of critical importance in the design and operation WWTPs. It results with a number of changes in wastewater management, including the development of new technologies for recovery of nutrients and energy. On the other hand WWTPs managers have to regard to economic aspects and find the balance between high level of wastewater treatment and costs. The sustainability also has social aspects. Local community acceptance is very significant considering that largely they bear the costs of wastewater treatment and modernization of the plant. Very important is the education of the local community to understand the impact that has on the environment and WWTP and thus the meaning of the costs incurred.

This paper discussed several sustainable considerations implemented in wastewater treatment plant 'Swarzewo' or planned for its future development. Treatment plant was modernized, adapting to the processing of organic waste. The installation is based on obtaining detailed waste-selected 'at source' by the local inhabitants. The co-fermentation of sludge from municipal waste generated benefits as a decrease of energy consumption, and is promising to achieve energy self-sufficiency in WWTP. In this area WWTP seeks to maximize the use of their own waste.

Cooperation of WWTP operators with Academic, and results of research work carried out, point the way to further changes.

References

1. A.N. Bdour, M.R. Hamdi, Z. Tarawneh, Desalination **237** (2009)
2. J. Pubule, A. Blumberga, F. Romagnoli, D. Blumberga, J. Clean. Prod. **88** (2015)
3. S.Ch. Jhansi, S.K. Mishra, J. Sust. Develop. **10**,1 (2013)
4. H. E. Munga, J. R. Mihelcic, J. Environ. Manag. **88** (2008)
5. A. J. Balkema, H. Preisig, R. Otterpohl, F.J. Lambert, Urban Water, **4** (2002)
6. R. Reardon, S. Deslauriers, R. Samstag, L. Baumberger, Florida Wat. Res. J. **1** (2011)
7. U. Wiessmann, I.S. Choi, E. Dombrowski, *Fundamentals of Biological Wastewater Treatment*, (WILEY-VCH Weinheim, Germany, 2007)
8. I. Bodik, M. Kubaska, Environ. Prot. Eng. **39**, 2 (2013)
9. O. Novak, P. Enderle, P. Varbanov, J. Cleaner Prod. **88** (2015)
10. K.Y. Chae, J. Kang Energy Conversion and Management **75** (2013)
11. T.H. Christensen (Ed), *Solid Waste Technology & Management,* (Blackwell Publishing Ltd, Chichester UK, 2011)
12. A. Dereszewska, S. Cytawa, Ekonomia i Zarządzanie **4**, 1 (2012) (in Polish)
13. W. Mo, Q. Zhang , J. Environ. Manag. **127** (2013)
14. H. Saveyn, P. Eder, *IRC Scientific and policy reports Technical Proposal EU* (2014)
15. L.E. de Bashan, Y. Bashan, Water Res. **38** (2004)
16. Y. Ueno, M. Fujii Environ. Technol. **22**, 11 (2001)
17. R. Ponds, F. Oort, K. Frenken, J Econ. Geogr. **1** (2009)
18. I.Ramos-Vielba, M.Fernandez- Esquinas, E. Espinosa de los Monteros, Scientometrics **84** (2010)

Numerical investigation of a straw combustion boiler – Part I: Modelling of the thermo-chemical conversion of straw

Andrea Dernbecher[1,a], Andreas Ortwein[1] and Fouzi Tabet[1]

[1]*DBFZ - Deutsches Biomasseforschungszentrum gGmbH, Torgauer Straße 116, D-04347 Leipzig, Germany*

Abstract. In the framework of a European project, a straw combustion boiler in conjunction with an organic Rankine cycle is developed. One objective of the project is the enhancement of the combustion chamber by numerical methods. A comprehensive simulation of the combustion chamber is prepared, which contains the necessary sub-models for the thermo-chemical conversion of straw and for the homogeneous gas phase reactions. Part I introduces the modelling approach for the thermal decomposition of the biomass inside the fuel bed, whereas part II deals with the simulation of the gas phase reactions in the freeboard.

1 Introduction

1.1. Alternative solid fuels

Solid biofuels are used with increasing tendency in Europe. Already in 2012, wood and solid biofuels covered 47.2 % of the primary energy production by renewables in the European Union [1]. Solid biofuels are advancing, since they provide sustainable and almost CO2 neutral energy. Additionally, biomass can be utilized based on energy demand to compensate fluctuations in energy production from other renewables as wind or sun. However, a large biomass potential is still not yet exploited. Especially energetic utilization of agricultural residues, for example straw, can contribute to the primary energy production. But combustion of alternative solid fuels is associated with high emissions of pollutants. Straw contains significant amounts of inorganic salts, which tend to form aerosols during combustion. Aerosols are altered during their path through combustion chamber and flue pipe and form fine particles by nucleation, agglomeration, coagulation and surface reactions. The fine particles and larger particles of fly ash are entrained and carried out of the combustion chamber or precipitate on surfaces. Due to the low melting point of ash, precipitation can lead to slagging or fouling. Additionally, highly corrosive product gases can be formed from the inorganic salts, for example due to chlorine [2, 3].

1.2. Biomass fuelled cogeneration technologies

Responding to a growing interest in decentralized production of electricity from renewable resources, small and middle-scale technologies for combined heat and power (CHP) are investigated and further developed at present. Biomass fuelled technology plays an important role in this field due to the before mentioned advantages of biogenetic fuels. The organic Rankine cycle (ORC) is of interest for the combination with biomass combustion, as it is capable to use low temperature levels for electricity production. Thermal energy of the flue gas is transferred to an organic working fluid, which powers a tubine. The organic working fluid has a lower boiling point than water and therefore provides higher efficiency at moderate temperatures. This attribute allows a modular operation of ORCs with full or partial load at reasonable efficiency. Additionally, ORCs are robust to corrosive flue gases due to a heat exchanger between combustion chamber an turbine. As ORCs are operated at low temperature and pressure, less personnel is needed to supervise the process [4]. A flow scheme for a typical ORC is shown in figure 1.

An organic Rankine cycle in conjunction with straw combustion is developed in the framework of a European project. A straw combustion boiler with approximately 1 MW power output is investigated and optimized. The boiler is used to gain thermal energy from whole bales of straw and can be operated with an additional ORC to produce electricity. For the optimization of the combustion chamber with regard to efficiency increase and emission reduction, simulations with computational fluid dynamics (CFD) are conducted.

[a] Corresponding author: andrea.dernbecher@dbfz.de

DOI: 10.1051/e3sconf/20161000015

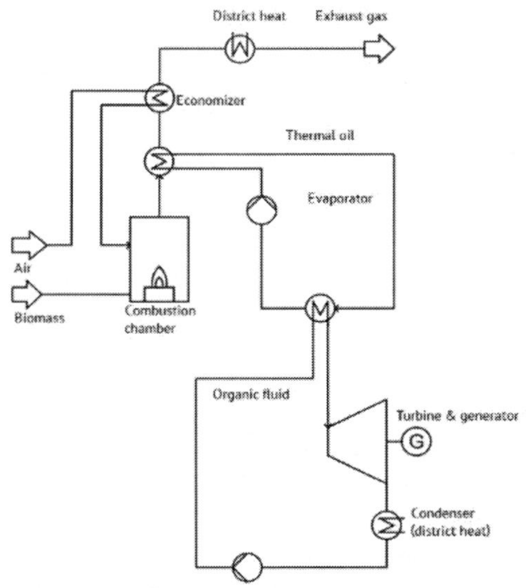

Figure 1. Flow scheme for a typical organic Rankine cycle (ORC) [5]

1.3. CFD-Simulations for optimization of biomass combustion

A comprehensive CFD simulation has to cover all processes inside the combustion chamber. The turbulent reactive flow in the freeboard and the thermo-chemical conversion of biomass in the fuel bed are crucial for the simulation. Additional models can be used to describe other processes, for example pollutant formation [6]. For the numerical investigation of the straw fired boiler, the bed model is particularly important, as whole bales of straw are burned inside the combustion chamber. They take a large space inside the firing chamber and influence the combustion process. Due to the transient character of the batch process, the converting biomass changes the gas flow inside the combustion chamber.

This study is divided into two parts. The first part deals with the development of a suitable bed model for straw combustion. This model contains all processes to describe thermo-chemical conversion of straw. A single particle of biomass is simulated as a test case to enhance the performance of the model and improve the sub-models.

In parallel, project partners from AGH in Krakow are working on the simulation of the homogeneous gas phase reactions. Details and results of their work are introduced in part II.

2 Methods

A three-dimensional bed model, which can be integrated into the computational domain, is needed for the description of the whole bales of straw inside the boiler. Therefore, a bed model based on a porous medium is developed. During development, a single particle model is used as a test case. The single particle model can be used for detailed improvement of the mechanisms and sub-models. Biomass undergoes the phases heating up, drying, pyrolysis, gasification and combustion during thermo-chemical conversion in oxidizing atmosphere. All phases have to be represented in a bed model.

Simulations of this study were conducted with the open source software OpenFoam [7]. The modelling approach is based on the solver biomassGasificationFoam, which was developed for gasification of biomass [8]. The solver was adapted and further developed to describe the combustion of straw. The pyrolysis mechanism was enhanced by experimental determination of kinetic parameters for straw. Furthermore, equations for the heterogeneous combustion of char were added.

The cubic particle (edge length 3.2 mm) is placed inside a box, in which the walls are heated with a constant heating rate. Air enters at the inlet at the bottom of the box and streams upwards, as it can be seen in figure 2.

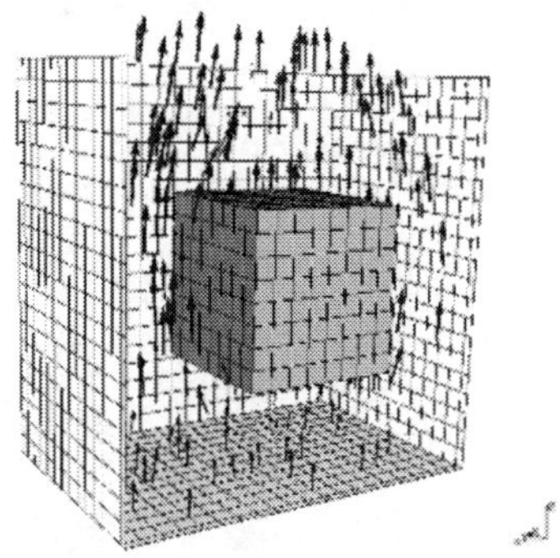

Figure 2. Geometry and mesh of the simulated straw particle [5]

Heating up of the particle is determined by heat transfer at the surface of the particle from gaseous to solid phase and heat transfer inside the particle. The temperature of gaseous and solid phase are not equal, heat transfer between the phases is described by equation (1) [9].

$$Q = h\,S\,(T_G - T_S) \qquad (1)$$

In which Q is the heat flow, h is the heat transfer coefficient, S is the surface area of the pores, T_G is the temperature of the gaseous phase and T_S is the temperature of the solid phase. Thermal conductivity of the solid phase is dependent from the components, which are present in the porous medium and their properties.

The initial components of biomass are *humidity, cellulose, hemicellulose, lignin* and *ash*. Humidity is consumed by drying, which is described by an Arrhenius expression in the numerical model. Therefore, evaporation of water is described by chemical reaction of the form:

$$biomass \rightarrow dry\ biomass + water \qquad (2)$$

Kinetics of this reaction is described by pre-exponential factor, activation energy and reaction order of the according Arrhenius expression (eq. (3)). These parameters were determined in an experiment with thermogravimetric analysis (TGA).

Pyrolysis is represented by three parallel reactions for the decomposition of *cellulose, hemicellulose* and *lignin*. The parameters for the kinetics of all three reactions were determined experimentally by TGA.

$$d\alpha/dt = A \exp(-E_A/RT)(1-\alpha)^n \quad (3)$$

Pre-exponetial factor A, activation energy E_A, reaction order n and initial concentration c for each component are the result of a mathematical fitting of the model equations to the measured mass loss of the biomass. R and T in equation (3) are the universal gas constant and the temperature in Kelvin, respectively. α represents the conversion of the component.

$$cellulose \rightarrow volatiles + char \quad (4)$$

$$hemicellulose \rightarrow volatiles + char \quad (5)$$

$$lignin \rightarrow volatiles + char \quad (6)$$

The kinetic parameters for drying and pyrolysis of the present study are summarized in table 1.

Table 1. Kinetic parameters for drying and pyrolysis determined by TGA.

	humidity	cellulose	hemicellulos	lignin
c (daf)	-	0.5609	0.2258	0.2133
logA / s^{-1}	7.738	12.772	6.172	5.243
E_A / kJ/mol	63.983	168.22	86.064	96.193
n	1.885	1.322	1.071	3.355

Volatiles are composed of CO_2, CO, CH_4, H_2 and H_2O in this study. The composition of product gases is based on an experimental analysis by Yang et al. [10].

Char is formed during pyrolysis and consumed by heterogeneous reactions afterwards. *Char* is assumed to be pure carbon for the heterogeneous reactions. Combustion of *char* is described by equation (7) [11].

$$C + \varphi O_2 \rightarrow 2(1-\varphi)CO + (2\varphi - 1)CO_2 \quad (7)$$

In which φ is the char oxidation parameter. φ depends on the temperature T_S of the solid phase. Heterogeneous gasification of *char* is represented by equation (6) [11].

$$C + CO_2 \rightarrow 2CO \quad (8)$$

Homogeneous gas phase reactions are not activated for the simulations for this study, as they are discussed in depth in part II of this paper as part of the gas phase simulation.

3 Results and Discussion

In Figure 3 the resulting normalized mass of the simulated single straw particle is shown. Simulation has been conducted with a heating rate of 10 K/min.

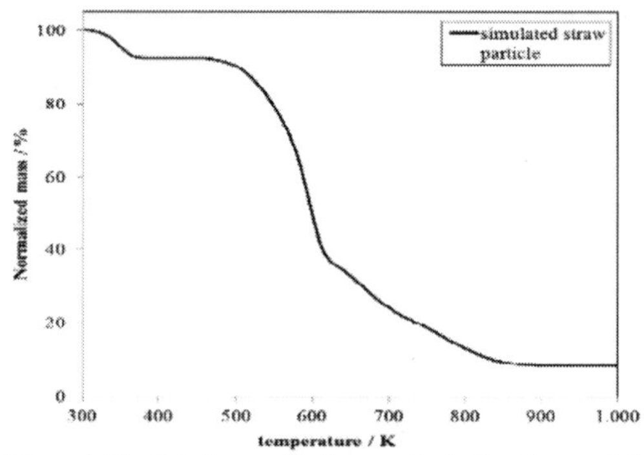

Figure 3. Simulated normalized mass of a singles straw particle in oxidizing atmosphere, heating rate 10 K/min

In the region between 300 K and 400 K, drying occurs. The mass loss due to evaporation of water can clearly be identified in the graph. Starting at 450 K, pyrolysis with decomposition of hemicellulose, cellulose and lignin occurs. Additionally to pyrolysis, heterogeneous gasification and combustion of charstart at approximately 650 K. and increase the mass loss. After thermo-chemical conversion of the biomass particle, ash remains.

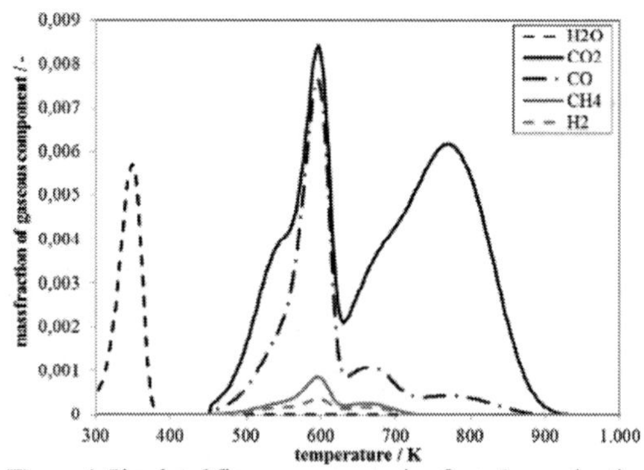

Figure 4. Simulated flue gas concentration from the combustion of a single straw particle

Concentration of the flue gases, which are formed during combustion, is shown in figure 4. The processes of thermo-chemical conversion of biomass can be clearly identified in the curves. *Water vapour* has the highest concentration in the region between 300 K and 400 K, where drying occurs. Pyrolysis generates large amounts of CO and CO_2 between 450 K and 600 K. Combustion of char contributes to the concentration of CO_2 at temperatures above 600 K. In comparison to CO and CO_2, mass fractions of *methane, hydrogen* and *water vapour* are less significant.

As different concentrations of product gases can be assigned to the decomposition of *cellulose, hemicellulose* and *lignin* and as the decomposition of these components depends on temperature, a distribution of product gas release can be observed during heating of the particle. The solver is capable of simulating a thermally thick biomass particle with a temperature gradient inside the particle. The resulting temperature distribution inside the simulated particle of the present study can be seen in figure 5. Due to the small size of the particle and the slow heating rate of 10 K/min, only a slight temperature gradient can be observed. In a scaled up bed model, local temperature differences will be more influential. The proposed modelling approach is capable of describing local differences in temperature and product gas composition.

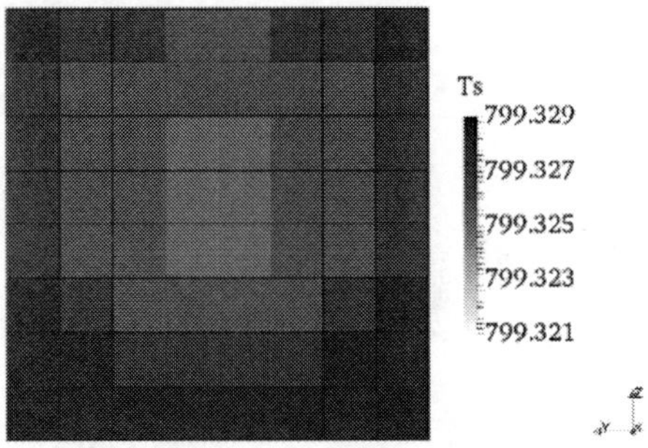

Figure 5. Simulated temperature of the solid phase of the straw particle after 3000 seconds (intersection of the particle)

4 Conclusions

A modelling approach for the thermo-chemical conversion of straw was presented in this study. Drying, pyrolysis and heterogeneous reactions of the thermal decomposition of a straw particle were simulated. The solver biomassGasificationFoam forms a reliable basis for future simulations of a whole biomass bed inside a biomass combustion stove or boiler. The porous medium approach is suitable for the comprehensive simulation with the biomass inside the computational domain. Additionally, the composition of the product gases can be described in dependency of time and space.

5 Outlooks

Further improvement of model with regard to emission prediction, scale-up and accuracy is intended. To conduct comprehensive simulations of a straw firing boiler, the coupling with gas phase simulations, which are introduced in part II, is planned. The bed model can provide boundary and input conditions for the gas phase model. The comprehensive CFD simulation should be used to improve the combustion conditions inside the firing chamber of the straw boiler and adept it to the use in a cogeneration system.

Furthermore, the presented modelling approach can be used in the development of systems, which are used for energy generation by thermo-chemical conversion of biomass. Manufacturers of these systems can use CFD simulations for the development and design of new or for the improvement of existing firing systems with regard to emission reduction and efficiency increase.

Acknowledgements

The work presented in this study has been supported by KIC InnoEnergy SE within the KIC Innovation Project "Construction of cogeneration system with small to medium size biomass boilers" (Reference: 13_2014_IP92_BioORC).

References

1. European Union, Eurostat, Energy, transport and environment indicators, Luxembourg, (2014) doi:http://dx.doi.org/10.2785/56625
2. A. K. Christensen, M. Stenholm, H. Livbjerg, J Aerosol Sci **29**, 4 (1998)
3. M. Jöller, T. Brunner, I. Obernberger, Energ Fuel **19**, 1 (2005)
4. L. Dong, H. Liu, S. Riffat, Appl Therm Eng, **29**, No. 11-12, 2119–2126 (2009)
5. A. Dernbecher, F. Tabet, A. Ortwein, *Proceedings of the European Combustion Meeting*, P5-36, March 30-April 2, 2015, Budapest, Hungary, (2015)
6. F. Tabet, I. Gökalp, Renew Sust Energ Rev **51** 1101–1114 (2015)
7. [online] [access: April 11th, 2016] http://www.openfoam.org/
8. K. Kwiatkowski, P.J. Zuk, M. Dudyński, K. Bajer, J Phys: Conf Ser **530** 12-15 (2014)
9. K. Kwiatkowski, B. Górecki, J. Koroto, W. Gryglas, M. Dudyński, K. Bajer, Numer Heat Tr A-Appl, **64** 216-234 (2013)
10. H. Yang, R. Yan, H. Chen, D. H. Lee, C. Zheng, Fuel **86** 1781-1788 (2007)
11. M.A. Gómez, J. Porteiro, D. Patiño, J.L. Míguez, Fuel **117** 716-732 (2014)

Light clay straw bale solutions in the contemporary housing as an element of sustainable development. Selected issues

Wojciech Drozd [1,b]

[1]*Institute of Building and Transport Management, Tadeusz Kościuszko Cracow University of Technology, Warszawska 24 St., 31-155 Kraków, Poland*

Abstract. The article presents issues related to the solutions with light clay and straw bale in the contemporary housing. Building using straw bale and light clay is simple, eco-friendly and accessible to all. It fits in with the idea of sustainable development, supporting local businesses and giving people the opportunity to integrate in the design and construction of the house. The article presents the thermal analysis for both walls made of straw bale and of light clay. The analysis showed a very good performance. All positive aspects allow treating straw and light clay as a viable alternative to the commonly used technologies for erecting buildings.

1 Low-tech, a natural way of building

Modern material and technological solutions - concrete, steel, glass as well as intelligent systems have become the synonyms of modernity and luxury[1]. Low-tech is on the other extreme of modern construction. Low-tech solutions are based primarily on tradition and local, low-cost raw materials that do not require special treatment, which are readily available, such as soil, clay, straw and sand. Human hands are the main building force, while the use of complicated techniques or expensive expertise is limited. Family, friends, or neighbours may be involved in the construction process, which largely reduces the cost of construction. Low-tech construction skips complex technical solutions and supports natural, generally accessible, raw materials. The construction market, dominated by giant construction companies, does not support the development of such solutions, which are not profitable. This makes it difficult to popularise this building method on a large scale. Despite this, low-tech buildings are becoming more common and perfectly fit into the idea of sustainable development.

Sustainable development means "socio-economic development, where political, economic and social actions are integrated, while maintaining environmental balance and sustainability of basic natural processes, in order to guarantee the ability to meet basic needs of individual communities or citizens of the present generation and future generations[a]". The most important features of a building characterised by sustainable development can be described using four R's: reduce, reuse, recycle and recover[c]. Less material and energy is used to build such a building compared to conventional construction. The materials used should be recycled and should allow for re-use after the end of the life of the building. Natural construction derives the materials from the surrounding environment. Straw, clay, or wood is obtained from local, often family businesses. This supports local development and cultural independence of the region. Natural low-tech construction is friendly to the environment. Production of materials does not require high energy and high temperature processed, does not produce CO_2, so it consumes less energy needed to construct the building than conventional building construction. Transport of materials is only done locally. And the structure itself is completely biodegradable and after the end of its life it does not leave harmful waste, hard to break down by the environment. Natural building allows for carrying out a lot of the work on one's own or with the help of not necessarily highly skilled people. It is available to all, creates local jobs and integrates people. The building material is cheap. All these features are very characteristic for sustainable development [3].

In the context of sustainable development in each of the stages of evaluation: design, construction, operation, points are awarded for the same use of natural materials.

In the selected issue some of the parameters of the envelope of natural materials were compared such as: clay and straw, in accordance with the procedures of evaluation: LEED, BREEAM, DNGB (The world's foremost environmental assessment method and rating system for buildings). In the article the heat transfer

[a] Environmental Protection Act Article 3 Section 50

[c] Jerzy Baryłka, „Environmental requirements in the construction industry as an example of construction law and the law on construction products”

[b] Corresponding author: wdrozd@ztob.pk.edu.pl

coefficient U, which determines the operation of the facility as a result of the impact on the environment, was analyzed.

2 Thermal and moisture properties of building partitions

The PN-EN ISO 6946 [4] standard provides a method for calculating the thermal resistance and heat transfer coefficient of building components that separate the internal and external environment, with the exception of doors, windows, glazed components and those that transfer heat to the ground, as well as components, through which air supply is expected. Heat transfer coefficient is an important parameter that allows for determining the thermal properties of a partition. It is defined for homogeneous-layer building partitions, building partitions with at least one composite layer and for components with layers of varying thickness.

The heat transfer coefficient is expressed by the equation:

$$U = \frac{1}{R_T}, [\frac{W}{m^2 \cdot K}] \quad (1)$$

Where:
R_T - total thermal resistance of a building partition

Total thermal resistance R_T for homogeneous partitions is described by the formula:

$$R_T = R_{si} + R_1 + R_2 + \ldots + R_n + R_{se}, [\frac{m^2 \cdot K}{W}] \quad (2)$$

Where:
R_{si} - heat transfer resistance on the inner surface
$R_1 + R_2 + \ldots + R_n$ - design thermal resistance of each layer
R_{se} - heat transfer resistance on the outer surface

Thermal resistance of a homogeneous layer with thickness *d* is obtained from the equation:

$$R = \frac{d}{\lambda}, [\frac{m^2 \cdot K}{W}] \quad (3)$$

Where:
d - thickness of the material
λ - design thermal conductivity of the material

The PN-EN ISO 13788 [5] standard presents a simplified method of calculating the temperature of the inner surface of the building component, under which, at a known temperature and humidity of indoor air, mould or condensation on surfaces can develop. It also provides methods for assessing the risk of indoor condensation due to water vapour diffusion. The risk of mould growth occurs at relative humidity above 80% for several days. To prevent the formation of mould, it is important to fulfil the condition:

$f_{Rsi} > f_{Rsi,min}$

where:
f_{Rsi} - temperature on the inner surface,
$f_{Rsi,min}$ - minimum permissible temperature on the inner surface, determined on the basis of the internal and external climate.

The calculation is carried out for each month of the year [6]. The internal humidity and acceptable volumetric moisture at saturation are determined. It is also possible to calculate the partial pressure of saturated steam at the surface and assume a critical relative humidity at the surface. Then, the minimum temperature of the analysed area is determined.

To calculate the temperature and humidity parameters, a partition composed of homogeneous layers without a layer of air is used. In order to simplify the calculations it was assumed that the supporting structure made of the timber, if it is constructed on the entire thickness of the wall as a continuous structure, has no substantial effect on the conduction of heat through the wall. This is associated with a large wall insulation thickness in relation to the cross-section of the superstructure columns. Flow of heat takes place in the horizontal direction, which is defined by R_{si}=0.13 and R_{se}=0.04. Internal moisture conditions are adopted for constant internal temperature of 20°C and a constant relative humidity of 50%. When checking the criterion for preventing surface condensation of partitions, regardless of the type of partition, the heat transfer resistance value on the inner surface of the partition of $R_{si}=0.25\ [\frac{m^2 \cdot K}{W}]$ is adopted.

The first step compares the heat transfer coefficients for different wall thickness, made in straw-bale and light clay technology (Figure 1.). The analysis was performed for the same wall thicknesses, calculated together with the clay plaster. In the case of light clay, double-sided plaster with a thickness of 3 cm on each side was used. For straw bales, 7cm of external plaster and 4cm of internal plaster was assumed.

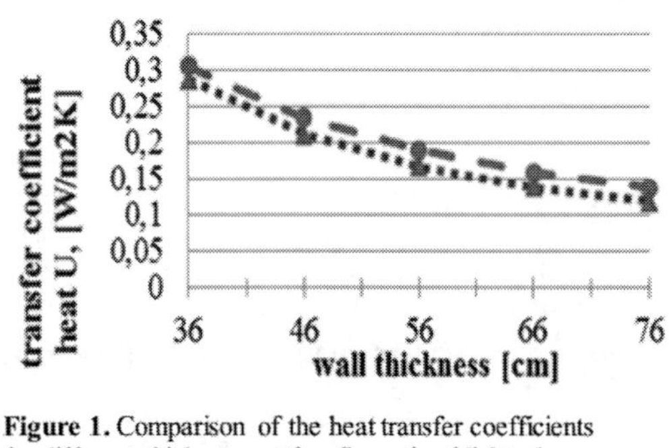

Figure 1. Comparison of the heat transfer coefficients for different thicknesses of walls made of light clay • or pressed straw • (Source: own work)

"Regulation on technical conditions to be met by buildings and their location" gives the limit value of heat transfer coefficient U=0.25 [W/m²K] for the outer wall. In addition to meeting the requirements of heat transfer, it is worth considering the optimal thickness of the walls. Due to the low cost of insulating the walls, it is best to choose the thickness of about 50cm. This will create a well-insulated partition, satisfying, e.g. in the case of straw bales, almost twice the thermal requirements. It will also have an appropriate heat storage mass and greater rigidity. At the same time, the wall thickness of about 50cm will provide the required sound insulation. The use of thicker insulating material is associated with a greater reduction in living space and possible higher costs of the wooden frame. The following section presents a thermal and moisture analysis of a building with wall thickness of 56cm (Table 1, Table 2.).

Construction of a partition:

Table 1. Properties of the light clay wall materials used.

No	The layer	d [m]	λ [W/mK]	μ	R [m²K/W]	Sd [m]
Outside						
1	clay plaster	0,03	0,58	10	0,052	0,3
2	light clay	0,5	0,1	10	5	5
3	clay plaster	0,03	0,58	10	0,052	0,3
Inside						

Table 2. Properties of the straw bale wall materials used.

No	The layer	d [m]	λ [W/mK]	μ	R [m²K/W]	Sd [m]
Outside						
1	clay plaster	0,07	0,58	10	0,121	0,7
2	straw bale	0,45	0,08	1,3	5,625	0,585
3	clay plaster	0,04	0,58	10	0,069	0,4
Inside						

The effective value of the temperature coefficient on the inner surface of the barrier is determined by the value of heat transfer coefficient U element and the heat transfer resistance at the inner surface R_{si}.

For light clay:

Total thermal resistance of a partition: $R_t = 5.273$ [m²K/W]

The heat transfer coefficient of the building partition (without taking into account corrections for leakage and connectors ΔU and allowance for linear bridges ΔUk): U = 0.19 [W/m²K]

The value of temperature coefficient for the partition: $f_{Rsi} = 0.968$

For straw bale:

Total thermal resistance of a partition: $R_t = 5.985$ [m²K/W]

The heat transfer coefficient of the building partition (without taking into account corrections for leakage and connectors ΔU and allowance for linear bridges ΔUk): U = 0.167 [W/m²K]

The value of temperature coefficient for the partition: $f_{Rsi} = 0.972$

Due to the same value for the temperature coefficient of the two partitions, a minimum value of the f_{Rsi} coefficient in individual months is as follows (Table 3.):

Table 3. Value of the minimum f_{Rsi} coefficient.

	Months	fRsi, min
1	January	0,679
2	February	0,659
3	March	0,583
4	April	0,385
5	May	-0,054
6	June	-1,235
7	July	-2,688
8	August	-1,837
9	September	-0,117
10	October	0,359
11	November	0,548
12	December	0,64

The highest value $f_{Rsi,min} = 0.679$ was obtained for the critical month of January. the condition that $f_{Rsi}=0.968 > f_{Rsi,min}=0.679$ is met. Therefore, the analysed building partitions are designed properly in terms of avoiding mould growth.

The calculated value of the dew point temperature is T_s=9.23°C for all months of the year. This indicates that the temperature on the inner surface of the building partition, which fluctuates around 19.5°C, is higher than the dew point temperature plus 1°C. Partitions were therefore designed in accordance with the technical requirements (Regulation of the Minister of Infrastructure of 12 April 2002) on the dew point.

3 Conclusion

Construction using light clay and straw is still not very common. Lack of standardisation of straw as a building material, as well as the old standards for building using clay are a major barrier against the rapid development of this type of natural building in Poland [7, 8, 9]. There is a lack of promotion, modern scientific and research base that could afford to put these materials on the construction market on an industrial scale. On the other hand, rapidly vanishing natural resources, rising energy costs and tightening thermal standards of the European Union may soon help change the attitude of the state and the people to natural building methods. These technologies are needed as they make it possible to reduce CO_2 emissions and reduce environmental pollution [10, 11, 12].

The thermal analysis showed a very good performance for both walls made of straw bale and of light clay.

All these positive aspects, as well as good thermal calculation results allow treating straw and light clay with full responsibility as a viable alternative to the commonly used technologies for erecting buildings.

References

1. Law on construction products of 16 April 2004
2. PN-EN ISO 6946:2008 Building components and building elements
3. BN – 62/6738 – 01, Masses of cement - clay fillers
4. BN – 62/6738 – 02, Construction of clay. The masses of clay

5. BN – 62/8841 – 04, Construction of clay. The walls of rammed clay. Technical conditions for execution and acceptance
6. PN-EN ISO 13788:2003 Thermal and moisture properties of building components and building elements
7. M. Hyła, D. Kupiec-Hyła, *Houses of light clay – Guide*, Published by: Grupa Cohabitat on Feb 20, 2011
8. M. Zatylny, *Formal and legal conditions for natural building in Europe and Poland with particular emphasis on the application of straw bale*, United Nations Development Programme, Warszawa 2011
9. M. Jagielak, *Building with straw bale in Poland against the general characteristics of this technology*, United Nations Development Programme, Warszawa 2011
10. A. Lecyk, *Architecture inherently -low-tech*, Intelligent Building No. **5**/2012
11. B. Backiel – Brzozowska, *Construction of clay and straw - Preliminary assessment of selected aspects of sustainability*, Environmental Engineering, **40**/2014
12. M. Spisakova, D. Maćkova, *The Use Potential of Traditional Building Materials for the Realization of Structures by Modern Methods of Construction*, Journal of Civil Engineering, December 2015

Modernised DC traction substation recuperating energy of braking

Piotr Drozdowski[1,a] and Arkadiusz Duda[2]

[1]*Cracow Univ. of Tech., Institute on Electromech. Energy Conv. Warszawska 24, 31-155 Kraków, Poland*
[2]*Cracow Univ. of Tech., Institute on Electromech. Energy Conv. Warszawska 24, 31-155 Kraków, Poland*

Abstract. The traction substation composed of the 3-phase transformer and the output diode rectifier is still the fundamental device for DC traction supply. In this paper two technical solutions (power electronic converters) allowing for braking energy recuperation to the AC mains are discussed. This has been thought as the modernisation of the existing traction substations allowing for high speed trains regenerative braking and braking of heavy train hollers. Such a solution brings some profits: energy recuperation, mechanical brakes saving and enhanced quality of traction vehicle control. For the proposed converters technical and economical advantages and drawbacks are indicated.

1 Introduction

In the last time the DC traction vehicles are driven with cage induction motors supplied by VS inverters. These are the heavy train hollers and the hollers for trains of high speed. For braking of both the trains a high energy dissipation is demanded. The vehicles are supplied with the traction substations composed of transformer and diode rectifier. The transformer is supplied from high voltage mains giving at the output the low voltage appropriate for the DC output of the rectifier (3 kV DC voltage for railway traction, 0.6 kV DC voltage for urban transportation and additionally 750V and 1.5 kV). Such a supply does not allow for energy recuperation during regenerative braking of the vehicles. The braking energy can be dissipated only for the need of other vehicles or, in case of lack of such vehicles, the dynamic braking with braking resistors can be performed. Electrical braking is usually supported by the mechanical. For some traction vehicles without electric braking the mechanical braking is the only possibility. The braking energy amounts to about 10 to 30 % of the whole vehicle energy depending mainly on the kind of traction, the time-table of trains, the railway line profile and some others [5]. So, the energy recuperation to the AC mains could be considered as profitable according to the amount of this energy, mechanical brakes saving and flexible control of the vehicle drive.

The solution for bidirectional power flow from AC mains to the DC side and vice versa is well known in the power electronics by employing two back-to-back line-frequency converters. One solution assuring such a power flow is the voltage source inverter (VSI) that is in fact the parallel connection of the line frequency converter (feedback diodes) and the 3-phase bridge of full controlled semiconductor switches (GTO, IGCT, IGBT) – Fig. 1. Another proposed bridge composed of SCRs instead of the mentioned turn off power electronic switches must be supplemented with two additional cut-off thyristors (GTO or SCR) – Fig. 3. Both the 3-phase bridges must be controlled with respect to the supplying AC voltages. So, these are in fact the line-frequency converters.

The researches of the modernised traction substations were performed for a low power model to indicate if the real system seem to be promising. The substations were loaded by the drive with the input pulse converter. This could be a vector controlled VSI asynchronous drive or a DC motor drive with the supplying chopper. Both the proposed installations that could supplement the existing traction substations are compared in the conclusions.

2 Analysis of the modernised traction substations

The utility interface converter assuring the power flow during the motor operation and the regenerative braking of the traction drive is presented in Fig. 1. This converter is connected to transformer output terminals *L*1, *L*2, *L*3 (AC side) of the traction substation and supplies, from terminals *a-b* (DC side), the traction drive *DRIVE*. So, the inverter *INV* is connected in parallel to terminals *p-n* of the traction substation rectifier *RECT*. The drive has on its input the pulse converter *PUL* (DC chopper supplying the DC motor or VSI with PWM supplying the AC motor). The inverter *INV* is composed of GTOs, IGCTs or IGBTs with anti-parallel feedback diodes creating switches *S*1, *S*2,..., *S*6. So, the whole system has the same structure as the single voltage source inverter.

[a] Corresponding author: xpiotrd@wp.pl or pdrozdow@pk.edu.pl

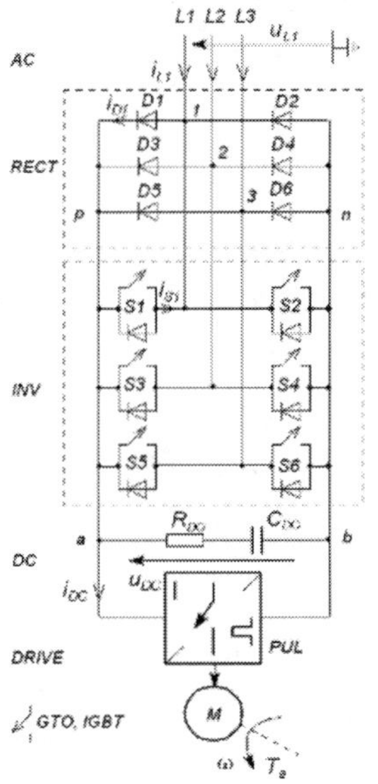

Figure 1. Modernised traction substation allowing for energy recuperation with back-to-back connected rectifier *RECT* and inverter *INV*

During regenerative braking the polarity of u_{DC} remains the same but the direction of i_{DC} is reversed. The power transmitted to the AC side must be as great as possible. Thus, the inverter is controlled with single pulses (without PWM) with so called quasi-square-wave mode. The method for gate signals processing is presented below. Generated gate pulses are synchronised with signals u_1, u_2, u_3 that are proportional to input phase voltages u_{L1}, u_{L2}, u_{L3}. These synchronising signals are usually produced on the outputs of *R-C* low pass filters connected to the output of measurement transformer. The *R*, *C* elements assure the phase angle shift of 30 or 60 degree. Thus the signals u_1, u_2, u_3 are sinusoidal even if the voltages u_{L1}, u_{L2}, u_{L3} are distorted from sinusoids. Assuming the relative amplitude of the synchronising signals equal to 1 the following phase shift signals are generated

$$\left.\begin{aligned} V_1 &= \operatorname{sgn}[u_2 - u^{ref}\cdot\operatorname{sgn}(u_3-u_1)] \\ V_2 &= \operatorname{sgn}[u_3 - u^{ref}\cdot\operatorname{sgn}(u_1-u_2)] \\ V_3 &= \operatorname{sgn}[u_1 - u^{ref}\cdot\operatorname{sgn}(u_2-u_3)] \end{aligned}\right\} \quad (1)$$

The sign function represents an ideal comparator. The triggering signals Q_1, Q_2, Q_3 create six 0-1 gate signals q_1, q_2,..., q_6 of 120 degree time duration, where $q_k = 1$ turns on the GTO.

$$\left.\begin{aligned} Q_1 &= (V_1 - V_2)/2 \\ Q_2 &= (V_2 - V_3)/2 \\ Q_3 &= (V_3 - V_1)/2 \end{aligned}\right\} \quad (2)$$

$$q_1 = (|Q_1| - Q_1)/2\,;\; q_2 = (|Q_1| + Q_1)/2$$

$$q_3 = (|Q_2| - Q_2)/2\,;\; q_4 = (|Q_2| + Q_2)/2 \quad (3)$$

$$q_5 = (|Q_3| - Q_3)/2\,;\; q_6 = (|Q_3| + Q_3)/2$$

For $q_k = 1$ the appropriate switching element of\ the switch *Sk* is turned on and for $q_k = 0$ is blocked ($k = 1, 2,..., 6$). The control signal u^{ref} assures the phase shift of the gate pulses with respect to supply voltages. For $u^{ref} > 0$ the inverter operates as the rectifier, whereas for $u^{ref} < 0$ there is just the inverter operation. Thus, the proper operation of the inverter assuring the minimum extinction angle is for u^{ref} slightly lower than -1 (e.g. $u^{ref} = -1.05$).

The system operation was checked for a single asynchronous drive controlled with field orientation. The illustrating waveforms are presented in Fig. 2 for an interval comprising three states: stable motor operation at regulated constant speed, regenerative braking to the lower speed and stable motor operation at this lower speed. Particularly interesting is the interval at braking.

The waveforms ω (motor speed) and T_e (motor torque) indicate only the drive actual operation. Line current i_{L1} is significantly distorted from the sinusoid, however the line voltage u_{L1} is practically sinusoidal. The DC voltage u_{DC} containing pulsations is constant, whereas the DC current i_{DC} supplying the drive reverses its average value during braking. Currents i_{D1} and i_{S1} of single switching elements from both converters show that the rectifier *RECT* works all time, whereas influence of *INV* appears practically during braking.

ALSTOM and ABB have proposed the similar solutions [Internet]. The ALSTOM system is composed of controlled SCR rectifier and a parallel IGBT inverter, whereas ABB offers the system for metro without the controlled rectifier. Both the systems can operate additionally as active filters. SECHERON proposes the inverter connected between AC MV grid and the DC traction line via additional transformer.

The next proposal for modernising the traction substation is shown in Fig. 3. The proposed converter is based on the 3-phase thyristor-bridge *INV* connected anti-parallel from the one side to the AC mains of the traction rectifier *RECT* and from the other side to the DC output terminals *p*2-*n*2 trough two cut-off SCRs *TP*1, *TP*2 [4]. Terminals *a-b* are connected to those cut-off thyristors. Such a substation is loaded directly with the pulse converter *PUL* supplying the vehicle motors.

Principal operation of the system is illustrated in Fig. 4. For simplicity the system was loaded with *R-L-E* receiver. During the motor operation the vehicle is supplied from the diode rectifier *RECT* via cut-off thyristors *TP*1, *TP*2 (interval 0 – t_1). However, when the vehicle changes to the regenerative braking the DC voltage u_{DC} on terminals *p*2-*n*2 increases with respect to the normal voltage on terminals *p*1-*n*1 (by increasing the internal voltage *E* of the loading) and the cut-off thyristors *TP*1, *TP*2 are naturally blocked (while t_2). Then the thyristor bridge *INV* is turned on to the inverter mode

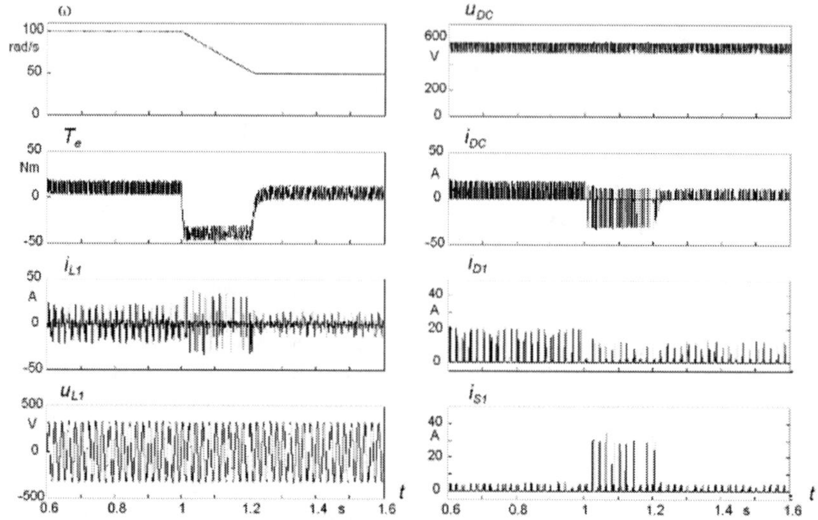

Figure 2. Waveforms for the substation from Fig. 1 operating at traction drive braking

(while t_3). The electric power produced by the vehicle due to braking is transmitted to the AC mains (i_{L1} inverses phase with respect to u_{L1}). The situation lasts up to the while when this DC voltage u_{DC} falls to the normal or a lower level (while t_5). Then, at t_6, the cut-off thyristors *TP*1, *TP*2 are turned on again. These thyristors are indispensable to protect the connection of *RECT* and *INV* against internal short circuits.

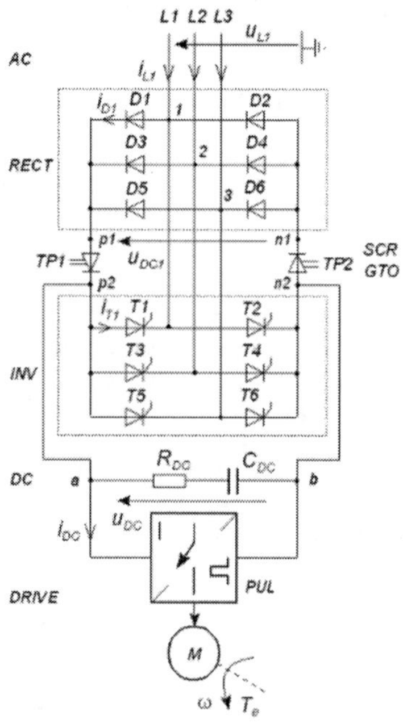

Figure 3. The modernised traction substation with the thyristor bridge *INV* and two cut-off thyristors

The main gate pulses of thyristors are generated on the rising slope of pulses described by (3). Instead of single pulses the train of short pulses could be keyed by q_1, q_2,..., q_6. Instead of SCRs the GTOs could be used.

In such a case the cutting off the both converters can be proceed in any while. That extends the control possibilities with one difference: on the contrary to soft switching of SCRs the switching off GTOs is hard.

From the point of view of the AC mains, this

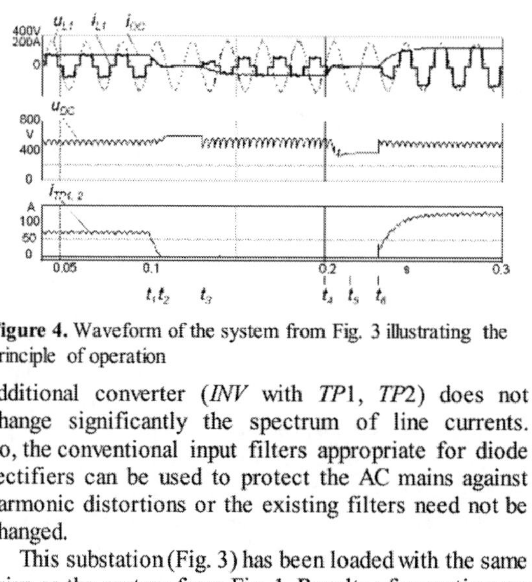

Figure 4. Waveform of the system from Fig. 3 illustrating the principle of operation

additional converter (*INV* with *TP*1, *TP*2) does not change significantly the spectrum of line currents. So, the conventional input filters appropriate for diode rectifiers can be used to protect the AC mains against harmonic distortions or the existing filters need not be changed.

This substation (Fig. 3) has been loaded with the same drive as the system from Fig. 1. Results of operation are shown in Fig. 5. Observing the waveforms it operates similarly as the first substation (Fig. 1), though the line currents are distorted from the sinusoid typically as for the diode rectifier. The first system generates additional distortions caused by simultaneous operation of switches in *RECT*_and *INV*. The main drawback of the second system is the necessity of special control of *TP*1 and *TP*2, whereas the first system is prepared for inverter operation during braking all the time. For *TP*1, *TP*2 control voltages u_{DC1} and u_{DC} must be measured and additionally the direction of current i_{DC}.

In paper [3] another converter assuring energy recuperation during braking was presented. This structure

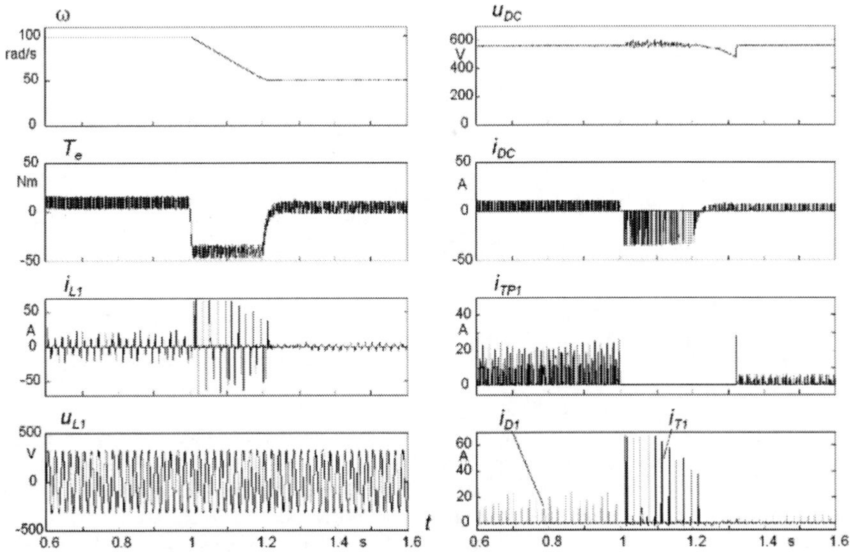

Figure 5. Waveforms for the substation from Fig. 3 operating at traction drive braking

was created connecting terminals *a*-*b* of the loading to terminals *p*1-*n*1 (Fig. 3) resulting the scheme presented in Fig. 6. However this system was not predicted by the authors for modernising the traction substation and was considered only as an extension for the standard PWM inverter drives.

In this circuit *TP*1 and *TP*2 must be a gate full-controlled element (e.g. GTO, IGCT or IGBT) playing the role of cut-off switches protecting *RECT* and *INV* against internal short circuits. These elements were controlled in a quite different way than the cut-off elements in the system from Fig. 3, where *TP*1 and *TP*2 could also be gate turn-off elements. The similar idea of this system for regenerative braking was presented earlier in [1] and [2]. Additionally analogous schemes as in [1, 2, 3] were found in the patents described in the Chinese and therefore (despite the best efforts) we could have not mentioned it in the bibliography.

3 Conclusions

• The modernisation of DC traction substation for energy recuperation is to attach a parallel converter operating as an inverter.

• All the analysed systems of substations operate properly.

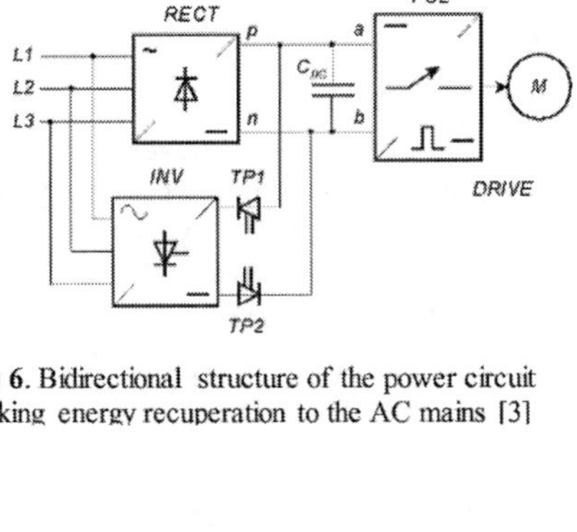

Figure 6. Bidirectional structure of the power circuit for braking energy recuperation to the AC mains [3]

• The simplest for control and the most reliable is the substation presented in Fig. 1.

• The inverters of remaining systems (Figs. 3 and 6) can be made of SCRs. They have soft commutation causing slightly lower current distortions, however sensitive to AC line disturbances that may cause failures.

The amount of recuperated energy depends on actual loading of the traction vehicle. Having the profile of the line and the vehicle parameters it can be easy appreciated answering if the modernisation is economically satisfying. To decide finally about the choice the systems of a real power should be studied in different conditions

References

1. W. Linden, W. Gratzke, H. Gehm, *Schaltungsanordnung mit einem Wechselrichter* (Siemens AG, Germany Patent 2531391, 1975).
2. J.M. Liptak, F.C. Joyner, J.C. Guyeska, *Regenerative Controller for a Voltage Source Inverter Drive* (Textile Ind. Tech. Conf., Charlotte, NC, pp. 10/1-10/7, 1989).
3. J.W. Kolar, H. Ertl, E.C. Zach, V. Blashko, V. Kaura, R. Lukaszewski, *A Novel Concept for Regenerative Braking of PWM-VSI Drives Employing a Loss-Free Braking Resistor* (APEC'99. Conf. Proc. 1997, Twelfth Annual, **1**, 297-305).
4. P. Drozdowski, A. Duda, *Braking energy recuperator* (Polish patent application No. P.412159, 2015-04-30).
5. S.F. Seyed, F. Saman, K.S. Babak, *Eenrgy-Efficient Emplacement of Reversible DC Traction Power Substation in Urban Rail Transport trough Regenerative Energy Recovery*. Int. Journal of Railway Research, July 2015.

DOI: 10.1051/e3sconf/20161000018

The control system of the ecological hybrid two stages refrigerating cycle

Piotr Cyklis[1,a] and Roman Duda[1]

[1]*Poland, Cracow University of Technology, Department of Mechanical Engineering, Institute of Power and Process Engineering*

Abstract. The compression anticlockwise cycle is mostly used for refrigeration. However due to the environmental regulations, the use of classic refrigerants: F-gases is limited by international agreements. Therefore the combined compression-adsorption hybrid cycle with natural liquids: water/carbon dioxide working as the energy carriers is a promising solution. This allows to utilize the solar or waste energy for the refrigeration purpose. In this paper application of the solar collectors as the energy source for the adsorption cycle, coupled with the low temperature (LT) refrigerating carbon dioxide compression cycle is shown. The control of the system is an essential issue to reduce the electric power consumption. The control of the solar heat supply and water sprayed cooling tower, for the adsorption cycle re-cooling, is presented in this paper. The designed control system and algorithm is related to the LT compression cycle, which operates according to the need of cold for the refrigeration chamber. The results of the laboratory investigations of the full system, showing the reduction of the energy consumption and maximum utilization of the solar heat for different control methods are presented.

1 Introduction

Compression and sorption systems are widely used in refrigeration and air conditioning systems for refrigeration and heat pumps cycles [2][3][4][5][6]. Restrictions for "F-gases" limits its application in such systems because of greenhouse and ozone depletion effect. The carbon dioxide cycle is commonly used refrigerant in the low temperature stage (LT) of the two stage compression refrigerating cycle. As a refrigerant in one stage refrigerating compression cycle only the transcritical cycle can be applied due to the low critical temperature of CO_2. This requires high discharge pressure and high volume gas heat exchanger instead of the condenser. Coupling the adsorption cycle with water as working fluid as a high temperature stage (HT) and compression CO_2 cycle, gives the possibility of standard Linde compression refrigeration cycle as LT stage with CO_2 as natural refrigerant. This hybrid double stage cycle achieves higher COP than single stage cycle. Coupling this two systems: adsorption at the high temperature stage and CO_2 compression at the low temperature stage is a new idea [1][2], allowing to utilize low temperature waste or solar heat as energy source for high temperature stage desorption.

The system with this level of complexity, where the required working time is continuous for the whole year operation, demands the optimisation of the automatic control of all subsystems and machines, to allow for automatic unwatched operation.

First requirement for the optimal control is to maintain all important measured parameters of the system within the required limits and react with proper system shutdown in unexpected cases. The system operation outside of the required limits may cause damage of the machines. Safety shutdown of a subsystem or individual machine with warning message for the service is required. The optimization part of the designed algorithms is to reduce the energy waste and minimize the power consumption.

The main purpose of this paper is to show the possibilities of the electric power consumption reduction using optimized control algorithm for chosen subsystems.

2 The test stand

In the Laboratory of Thermodynamics and Measurements of Thermal Machines at the Cracow University of Technology a test stand with hybrid refrigeration adsorption-compression system has been designed and constructed [2]. The HT stage is the adsorption cycle based actually (after upgrading) on the eCoo unit from German manufacturer (SorTech AG). This system is coupled with the tube solar collectors, the heat storage tank and the wet cooling tower for MT (medium temperature) glycol heat rejection system. The solar subsystem L4 contains 17 vacuum tube solar collectors HEWALEX KSR-10 and the heat storage tank of 2000l capacity. The evaporative (water sprayed) cooling tower DECSA REF-C-005, with maximum cooling power about 75kW is used for the adsorption unit cooling. The low temperature compression cycle is equipped with the two CO_2 compressors Dorin CD300H ph3,

[a] Corresponding author: pcyklis@mech.pk.edu.pl

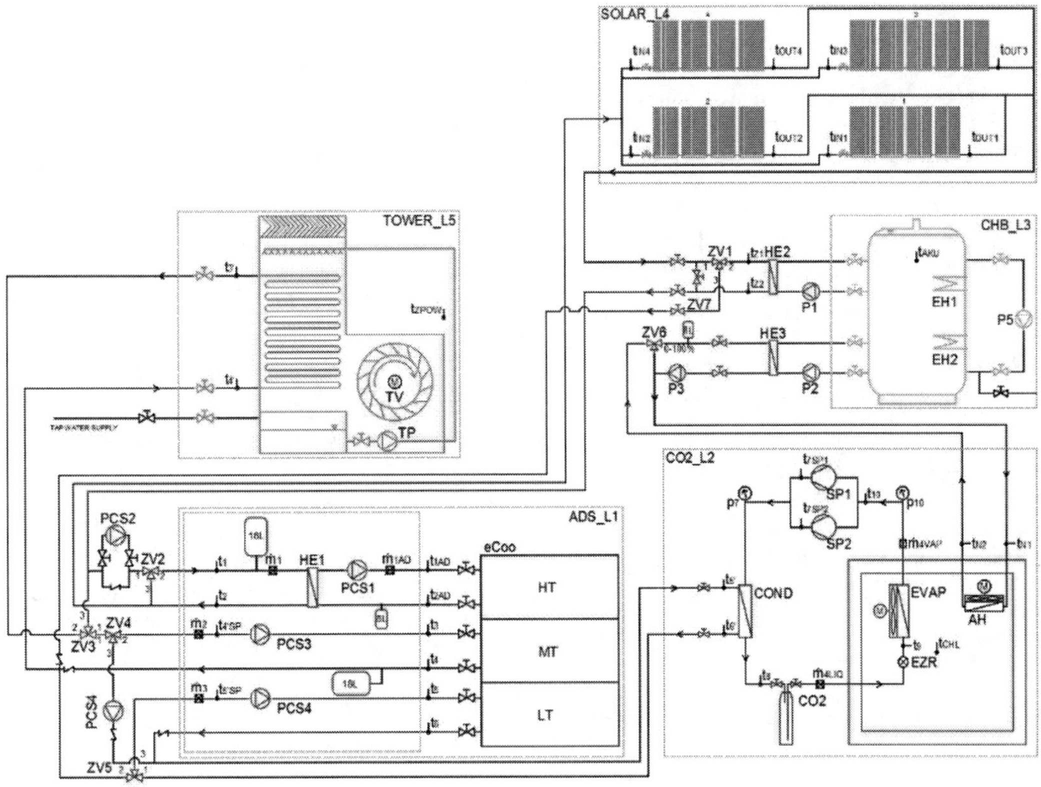

Figure 1. Schematic diagram of a complete hybrid refrigeration system installed in The Laboratory of Thermodynamics and Thermal Machines Measurements

working in the parallel cascade. One compressor is equipped with the frequency inverter ABB ACS355-03-08A8-4. The refrigeration chamber has internal dimensions of 1960x1920x2690 mm and the wall insulation thickness of 200mm.

The chamber evaporator was delivered by Gunter CXGHF 040.2H/17-ENW50.E with the fan VT0398U and the high pressure expansion valve CX4. The lamellar heat exchanger SWEP B16DWHx64/1P-SC-U is applied as a CO_2 condenser.

The ethylene glycol is used as a secondary liquid in the evaporative wet tower, the adsorption unit and solar subsystem. The air heater AH Flowair Leo FB9 was installed in the cooling chamber to simulate the thermal load. The scheme of the system is shown in figure 1.

The system has to work continuously with different load and weather conditions, so the wet cooling tower applied in our laboratory stand has more cooling capacity than required only for adsorption cooling. The adsorption cooling of the CO_2 condenser is possible only in the period when the waste or solar heat is available. The heat container solar charging, allows only for a few hours operation, until the temperature drops below 55°C, so only part time of the night operation of the adsorption cycle work may be powered from the storage tank. Then the cooling tower has to cool down the condenser directly. In case of extremely high air humidity, the electric heater may be used as a backup, which is not desirable due to the energy consumption and total COP decrease. The advantage of the adsorption system is, that it may work with 55-65°C desorption temperature source, depending on the adsorption unit design.

Several operation regimes for each subsystem were designed for system operation:

The subsystem L1 the adsorption unit - has four main regimes defined as:

1. A summer day operation (SDW): when the input temperature from the heat source exceeds 55°C, and the ambient air temperature is higher than 13°C. Then the adsorption subsystem works in the cooling mode. Since the outlet cooling temperature from the adsorption unit is a function of cooling load, it is sufficient to leave the cycle for lowest possible cooling temperature as possible. This self-adjustment function of the adsorption unit allows to achieve lowest possible condensing temperature, reducing total energy consumption. This regime works until two above mentioned temperatures used as control functions will not be exceeded.

2. A summer night operation (SNW): the adsorption unit operation is no longer possible when the heat source temperature decreases below 55°C, then the adsorption

system has to be switched off, and only the wet tower is used to cool down the CO_2 condenser.

3. A heat pump operation (HPW): this option is used when the ambient air temperature decreases below 13°C. Then the adsorption unit may be used separately as a heat pump collecting all waste heats and ambient heat and use the heat collector as HT source, while the MT (medium temperature heat) is used for heating.

4. An additional heater operation (AHW): The temperature of the heat source is below 60°C and the required condensing temperature cannot be obtained by the cooling tower. This may be the case only in special ambient conditions with high relative air humidity and high temperature when no waste or solar heat source is available.

The subsystem L2 – CO_2 compression cycle.

The basic function of the CO_2 compression system is commonly known as compression-refrigeration cycle control, and will not be discussed here in details. However there are two important differences:

1. The cooling requirements for CO_2 condenser may in some cases be limited, and additional cooling power will be required. In this case there are two possibilities: the control signal is released to run AHW program in the L1 subsystem, in the meanwhile reducing the compressor revolution speed.

2. The frequency inverter mounted on one of the cascade controlled compressors, allows for the variable refrigeration load control instead of the on/off control.

The subsystems L3,L4,L5 – heat container, solar collectors, wet cooling tower, have two or triple way valves and variable speed pumps to adjust temperatures and flow directions.

The solar pump PCS2 operational speed before andafter solar energy accumulation is controlled using the temperature in solar collectors as control parameter. Four below given equations are used for the control algorithm for the pump revolution speed. The optimal control of this pomp allows for power consumption decrease and the solar heat maximum utilization. Below given equations are related to:

(1) – the energy balance at the outflow from the solar collectors,

(2) - the energy balance at the inflow to the solar collectors,

(3) - the collectors temperature increase over the ambient temperature,

(4) – the temperature difference between the outflow and inflow to the solar collectors

$$W_1 = Log\left(\frac{t_{OUTkol\_MAX} + 273.15}{t_{OUTkol\_AVG} + 273.15}\right)\cdot C \qquad (1)$$

$$W_2 = Log\left(\frac{t_{INkol\_MAX} + 273.15}{t_{INkol\_AVG} + 273.15}\right)\cdot D \qquad (2)$$

$$W_3 = Log\left(\frac{t_{OUTkol\_MAX} - t_{zpow} + 273.15}{t_{zpow} + 273.15}\right)\cdot E \qquad (3)$$

$$W_4 = Log\left(\frac{t_{OUTkol\_MAX} + 273.15}{t_{INkol\_AVG} + 273.15}\right)\cdot F \qquad (4)$$

$$O_{PCS2} = A\cdot\left(W_1 + W_2 + W_3 + W_4\right) + B \qquad (5)$$

where:

A - the set point of the maximum rotational speed of the pump,

t_{OUTkol_MAX} - the temperature maximum value at the solar collector outflow,

$$t_{OUTkol\_MAX} = \mathrm{MAX}\left(t_{OUT1}, t_{OUT2}, t_{OUT3}, t_{OUT4}\right) \qquad (6)$$

t_{OUTkol_AVG} - the average temperature value at the solar collector outflow,

$$t_{OUTkol\_AVG} = \frac{t_{OUT1} + t_{OUT2} + t_{OUT3} + t_{OUT4}}{4} \qquad (7)$$

the same form of equations are used for calculations of inflow temperatures t_{INkol_MAX} and t_{INkol_AVG}, t_{zpow} is the external ambient temperature,

B=1000, C=5, D=10, E=2,5, F=3,5 are coefficients experimentally adjusted after tests testing.

If the average temperature of the outflow from the solar collectors increases above the temperature in the heat container, the solar energy accumulation starts:

$$t_{OUTkol\_AVG} >= t_{AKU} + hysteresis_{PID} \qquad (8)$$

Then when the t_{OUTkol_AVG} decreases below $t_{AKU} + 0{,}5\cdot hysteresis_{PID}$ the PID controller is switched off and equations (1)-(5) are used for control.

$$t_{OUTkol\_AVG} <= t_{AKU} + 0.5\cdot hysteresis_{PID} \qquad (9)$$

The hysteresis added in the equation 8 and 9 was observed during experiments. In this case the it is value of 8K.

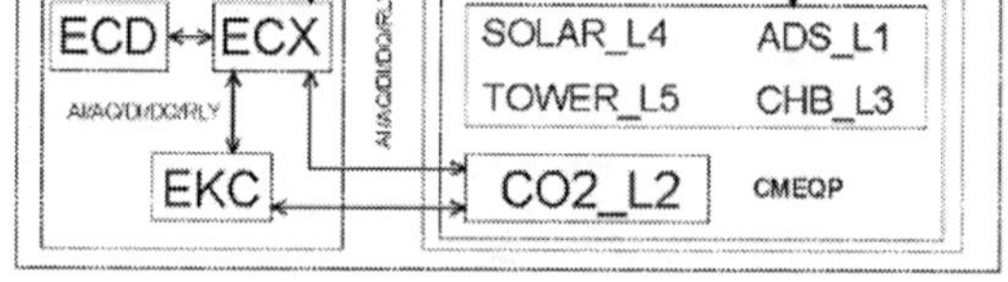

Figure 2. Schematic block diagram of the control system.

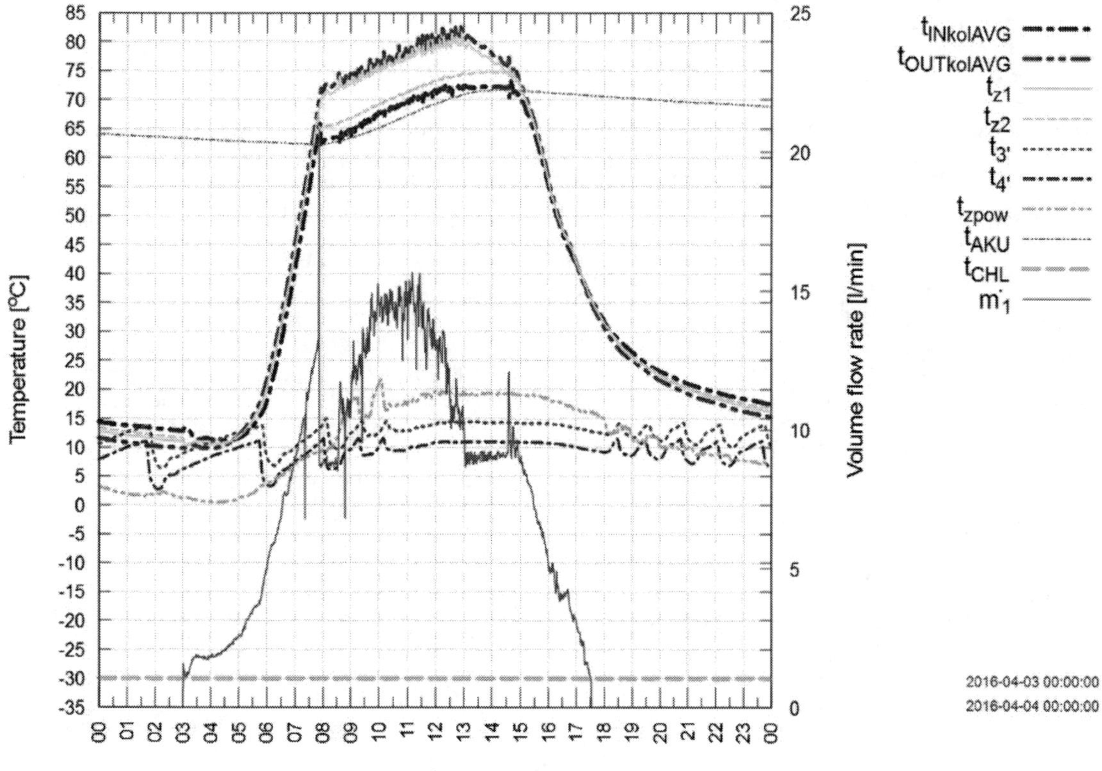

Figure 3. Temperature distributions SOLAR+CO2+TOWER, $t_{INkolAVG}$ - the temperature averaged value at the solar collector inflow, $t_{OUTkolAVG}$ - the temperature averaged value at the solar collector outflow, all other parameters notations are shown in Fig. 1.

When the temperature "t_{AKU} + *hysteresis*$_{PID}$" is reached the P1 pump and PCS2_PID controller is switched on and accumulation starts.
During the night time the pump is switched off.

The heat container subsystem is simple in operation. It has electric heaters EH1 and EH2 with on/off system only for L1 AHW mode. The valves ZV and pumps have to work accordingly to other subsystem modes.

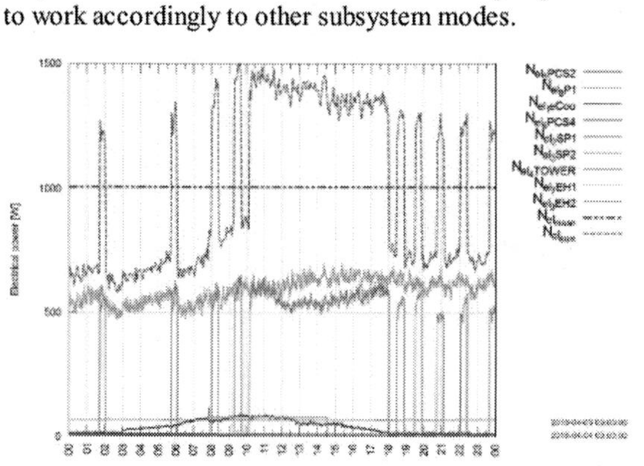

Figure 4. Electrical power consumption of the system SOLAR+CO2+TOWER, Nel_{mean} - power consumption averaged during the operation period, Nel_{sum} – summary power consumption during the operation period, all other parameters notations are shown in fig 1

1. Mode SAC – the solar heat accumulation is controlled using the temperature difference t_{z1}-t_{z2} with the control set point 4K. This set point adjusts the PCS2 pump speed through the PID controller. In the case of the full heat container load (95°C) the mode SAC is changing into the mode SWS.

2. The mode SW (solar operation)– the heat accumulator is fully loaded (95°C), then P1 pump is switched off. If the L1 subsystem is on the SDW mode the glycol is directly heating the adsorption unit. In the case of the too high solar temperature, the subsystem L1 is switched off, or the mode SW change into SWS mode.

3. The SWS mode – the solar waste mode. This is the case when no heat source is needed (L1 is not operating, accumulator is full) and there is a considerable amount of solar radiation and the temperature readings in solar subsystem exceeds the set points (about 100°C) then pump PCS2 has to be switched on and all heat is lost in the cooling tower.

The measurement system operates with the designed in the Laboratory under Visual Studio software „kaskada". Part of this software called kaskada_vi was written using the Lab View. The NIST REFPROP software for refrigerant properties was implemented. The CoolPack and Mathcad software for data calculations were used. The values of the 36 temperatures, 6 flow rates, 5 gauge pressures, 9 electric power meters are recorded each minute. Schematic block of the system with measurement system is shown in figure 2.

An example of the temperature distributions in the system during one operating day is presented in figure 3.
This data is used for calculations and simultaneously send to the main plc SIEMENS S7 1217 C unit.

The SIEMENS S7 unit controls the system programmed using SIEMENS TIA Portal software.

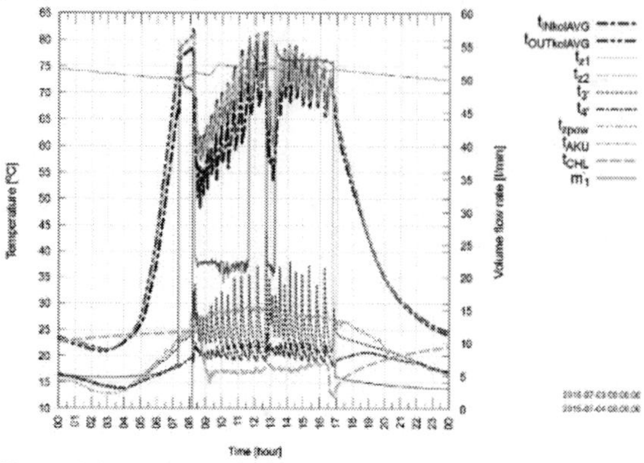

Figure 5. Temperature distributions SOLAR+CO2+ADS, $t_{INkolAVG}$ - the temperature averaged value at the solar collector inflow, $t_{OUTkolAVG}$ - the temperature averaged value at the solar collector outflow, all other parameters notations are shown in fig. 1

3 The control system

In figure 2 the schematic diagram of the control system is presented with the following depicted elements:
CC1 MAIN CASE – measurements of: temperature and pressure, mass and volume flow rates, electric power for electric devices, depicted in figure 1. Also PLC controller for the whole system is included.

The main parts are:
- ICP_DAS 8 pcs. M-7017RC cards, for data acquisition, this as a whole measuring system gives 64 analogue inputs signals of which 52 are used at this development stage.
- PLC SIEMENS S7 CPU 1217C controller with Ethernet/PROFINET interface and CM 1241 RS485
- HMI KTP700 Basic PN operator panel with Ethernet/PROFINET interface for communication.
- EQP subsystems of the refrigeration stand,
- CO2_L2 compression system CO_2,
- SOLAR_L4 solar subsystem,
- CHB_L3 heat container,
- ADS_L1 adsorption unit,
- TOWER_L5 water sprayed cooling tower,
- CMEQP common part of the acting elements: servomotors for valves, circulating pumps,
- ETH_SW network switch LAN 1000/100/10 Mbit/s,

CC2 subsystem CO2 CASE – consists of independent controller EKC with independent safety subsystem, overheating controller for expansion valve ECX, this controller has a build in www page where operating

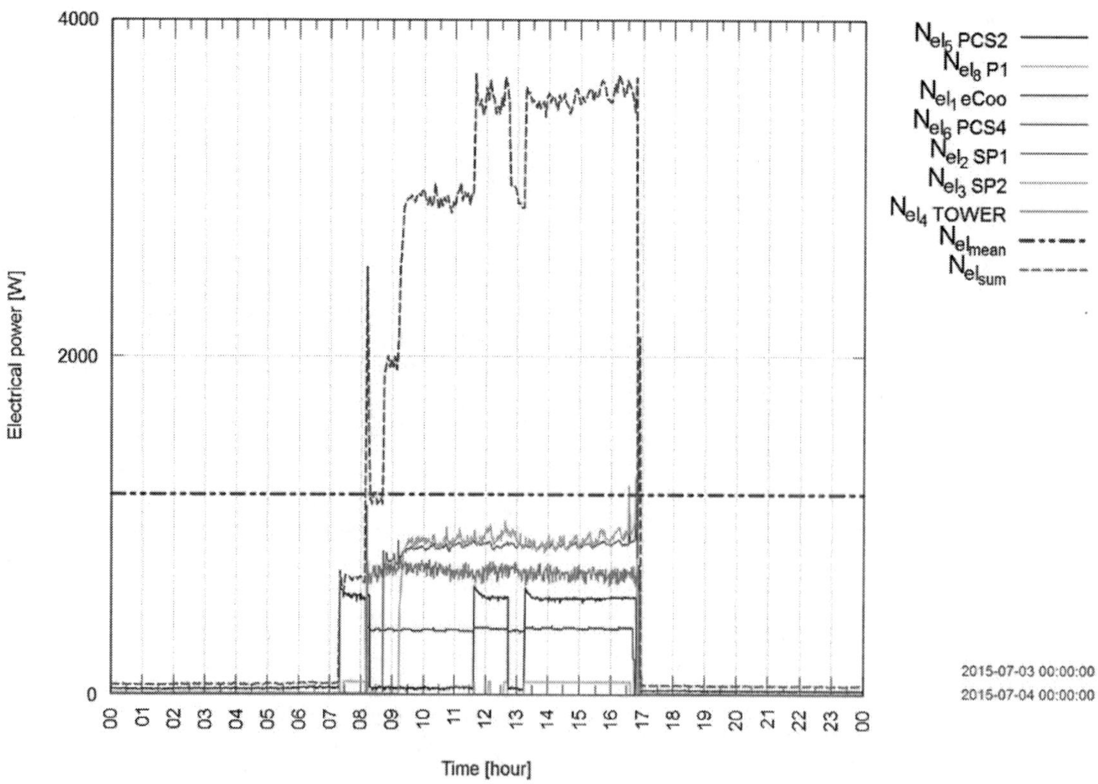

Figure 6. Electrical power consumption of the system SOLAR+CO2+ADS, Nel_{mean} - power consumption averaged during the operation period, Nel_{sum} – summary power consumption during the operation period, all other parameters notations are shown in fig. 1

parameters can be changed remotely using web page browser, ECD controller parameters display.

4 Tests results

The selected day of the system work with the solar energy accumulation and CO_2 condenser cooled directly from evaporative cooling tower is presented in figure 3 (temperatures) and figure 4 (power consumption).
The selected day of the system operation with adsorption unit switched on is presented in figure 5 (temperatures) and fig. 6 (power consumption). The power reduction and savings using the control system is clearly visible as Nel_{mean}.

This value represents electric power consumption averaged during the operation period. If no optimised control would be introduced, the electric power consumption would be the maximum value of Nel_{sum} for all the operation time, which is considerably higher.

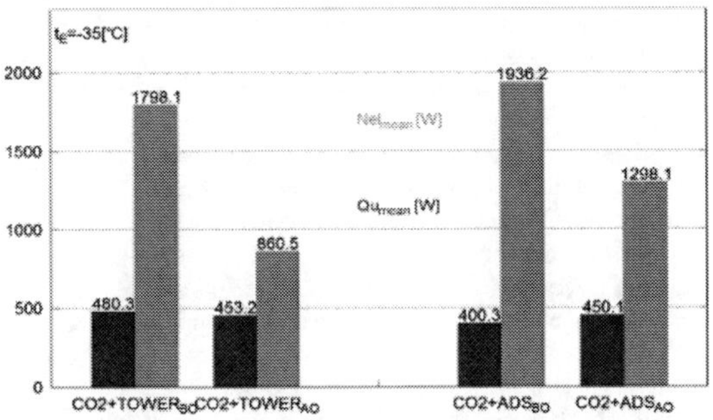

Figure 7. Electric power consumption and heat transferred from CO_2 evaporator EVAP. BO – before optimization; AO – after optimization, t_E – CO_2 evaporation temperature, all other parameters notations are shown in fig. 1

Summary of the results is shown in figure 7. It is worth noticing that in all cases the electric power consumption was several times lower than before the optimisation.

5 Conclusions

The hybrid system worked out in the Laboratory of Thermodynamics and Thermal Machines Measurements combines the ecology of the natural refrigerant with high efficiency of the refrigeration system. To achieve this task the control system with specially designed algorithm was implemented. The control system contains frequency inverters for the pumps and compressor revolution speed. The control parameters are measured every 10 seconds and are recorded for future analysis. Parallel the control system adjusts the frequency inverters to achieve desired flows. The main assumptions for control system the algorithm design were the minimisation of the electric power consumption with the maximisation of the solar heat usage. This requires the frequency inverter controls for pumps, to keep the solar liquid flow on the level which allows for the heat container loading, even in low sun radiation condition. The algorithm is designed for the whole year operation. Selected results are shown in this paper. The uncontrolled system without control optimisation has in some cases twice the power consumption than after the control optimisation algorithm application. This increases the COP of the system, making this system more economically efficient. The results of the control algorithm operation are shown in the paper.

References

1. P. Cyklis, G. Brak, *Trans. IFFM*, **121**, 49-66, (2008)
2. P. Cyklis, *INT J REFRIG*, Vol. **48**, s. 121-131, (2014)
3. P. Cyklis, K. Janisz, *Chem. and Proc. Eng.*, Vol. **36**, No. 3, s. 321-330, (2015)
4. A. Grzebielec, *AoT*, **30**, 189-200, (2009)
5. W. Han, L. Sun, D. Zheng, H. Jin, S. Ma, X. Jing, *Applied Energy*, **106**, 383–390, (2013)
6. Suamir I., Tassou S. *Appl. Therm. Eng.*, **50**, 1487-1495, (2013)

Utilisation of coal for energy production in fuel cells

Magdalena Dudek[1,a], Tadeusz Olkuski[1], Barbara Tora[2], Przemysław Grzywacz[1] and Alicja Rapacz-Kmita[3]

[1] *AGH-University of Science and Technology, Faculty of Fuels and Energy 30-059 Cracow, Av. Mickiewicza 30, Poland*
[2] *AGH-University of Science and Technology, Faculty of Mining and Geoengineering 30-059 Cracow, Av. Mickiewicza 30, Poland*
[3] *AGH-University of Science and Technology, Faculty of Materials Science and Ceramics 30-059 Cracow, Av. Mickiewicza 30, Poland*

Abstract. In this paper a brief characterization of fuel cell technology and its possible application in sustainable energy development was described. Special attention was paid to direct carbon fuel cell technology. The direct carbon fuel cell is an electrochemical device which directly converts the chemical energy of carbonaceous based fuel into electricity without 'flame burning'. The electrical efficiency of a DCFC is indeed very high (in practice exceeding 80%), and the product of conversion consists of almost pure CO_2, eliminating the most expensive step of sequestration: the separation of carbon from flue gases. In this paper the process of electrochemical oxidation of carbon particles on the surface of oxide electrolytes at 8% mol Y_2O_3 in ZrO_2 (8YSZ) as well as cermet anode Ni-8YSZ was analysed. The graphite, carbon black powders were considered as reference solid fuels for coal samples. It was found that the main factors contributing to the electrochemical reactivity of carbon particles is not only the high carbon content in samples but also structural disorder. It was found that structurally disordered carbon-based materials are the most promising solid fuels for direct carbon solid oxide fuel cells. Special impact was placed on the consideration of coal as possible solid fuels for DC-SOFC. Statistical and economic analyses show that in the coming decades, in developing countries such as China, India, and some EU countries, coal-fuelled power plants will maintain their strong position in the power sector due to their reliability and low costs as well as the large reserves of coal and lignite in the world. Coal is mined in politically stable areas, which guarantees its easy and safe purchase and transport. The impact of the physiochemical properties of raw and purified coal on the performance of the DC-SOFC was studied. An analysis of the stability of electrical parameters was performed for a DC-SOFC operating under a load over an extended period of time. The tests indicated that DC-SOFCs fed with de-ashed coal were characterized by stable operation, with a power density greater than 100 mW/cm^2 from a single cell.

1 Introduction

The power sector has been facing serious restructuring in generation technologies in many countries in the world. The economic, cultural, and technological development of societies may be impeded as a result of the growing demand for energy (particularly in rapidly developing countries such as China, India, and the countries of South America and Africa), which cannot be fully satisfied due to the depletion of fossil energy resources. Protection of the environment from pollution from emissions is also an important factor in developing new power technologies [1,2]. Relevant efforts have been aimed at increasing the efficiency of power generation; introducing new energy conversion technologies that impact renewable energy sources; separating and storing CO_2; and developing efficient technologies for the storage, transmission, and distribution of energy (smart grids) [3, 4]. The strategy of sustainable development assumes energy production from diverse sources, not only large-scale power plants but also small power generators operating on the basis of local resources and renewable energy sources. Hydrogen is also expected to play an important role in the energy economy in the future, based on environmentally clean fuel and secondary energy carriers [5,6]. The most important place in hydrogen technologies, as well as in sustainable energy developments overall, is reserved for fuel cells. A fuel cell (FC) is an electrochemical device which directly converts the chemical energy of fuel into electricity and waste heat. A FC can operate as long as fuel (mainly hydrogen) and an oxidant (usually oxygen taken from the air) are supplied. Although the concept of fuel cells was born early in the 19th century their rapid development began in the mid-20th century and was directly connected to the application of hydrogen as fuel[7].

Fuel cells are generally classified into two basic categories according to types of electrolyte used or temperature of operation. In the electrolyte used, five main types of fuel cells are distinguished:

a) Proton exchange membrane fuel cells -PEMFCs, operating temperature 30- 80°C,

b) Alkaline fuel cells AFCs, operating temperature 50-200°C,

[a] Corresponding author: potoczek@agh.edu.pl

c) Phosphoric acid fuel cells - PAFCs, operating temperature ~ 200°C,

d) Molten carbonate fuel cells (MCFCs, operating temperature ~ 650°C),

e) Solid oxide fuel cells (SOFCs, operating temperature 500-1000°C),

Solid oxide fuel cells can be powered by not only gaseous but also liquid and solid fuels from a variety of sources, including hydrogen gas, hydrocarbons, ammonia, ethanol, synthesis gas, carbon monoxide from the Boudouard reaction, and even powdered coal [8].

Carbon-based materials are also used frequently for the construction of fuel cells, batteries, supercapacitors, and devices for analytical chemistry[9]. They can also be applied as fuel for a fuel cell. This can be done in two ways: directly and indirectly. In the first case, solid carbon fuel is introduced directly into the anode chamber (without any preliminary gasification process) of a reactor placed outside the fuel cell [10,11]. Direct carbon fuel cells (DCFC) have a long and uneven history. The first examples were constructed over 120 years ago by Yablochkov (1877) and Bacquerelle (1885). A carbon anode and a molten alkaline electrolyte were applied. The DCFCs operated within a temperature range of 400–600°C. For a long time, it was believed that carbon was not electrooxidised to CO or CO_2 but reacted with OH^-. The 1990s saw renewed interest in direct carbon fuel cell technology. The main reason for this interest in this type of electrochemical device was its several promising advantages compared to other types. Carbon-based solid fuel appears to be an attractive and abundant source of fuel for DCFCs, due to its easy storage, transport, and high level of reactivity at high operating temperatures. DCFCs can be fuelled by coal, lignite, coke, a variety of biomass-origin solid fuels, and waste carbonaceous-based materials originating from various sources. Another advantage of this type of cell is that carbon fuel and the reaction product, CO_2, exist as pure substances which do not mix with each other; their chemical potentials are constant and independent of the degree of consumption (utilisation) of fuel, as well as of its distribution in the cell. This enables the cell's voltage to remain constant (at approximately 1.0 V) from the moment of the fuel's introduction to its total consumption [12].

Hydrogen gaseous fuel cells do not operate in this way. In operation, the fuel cell is consumed gradually, forming a gaseous mixture of the fuel and reaction product (steam). The gradually decreasing concentration of the gaseous fuel (hydrogen) in the mixture flowing into successive parts of the electrode causes voltage differences between different parts of the electrode.

Contrastingly, during operation of a DCFC, CO_2 is produced at the anode. Thus, it isolates carbon fuel contained in the cell from the oxygen in the external atmosphere, preventing it from being oxidised in a direct chemical reaction which would produce no electricity. In this way, no carbon fuel is lost as a result of unproductive reactions in the cell. Furthermore, in contrast to coal combustion in the air, CO_2, during its evolution at the anode, does not mix with the non-combustible components, mainly nitrogen. This leads to a very high concentration of CO_2 in the exhaust gas stream from the cell, greatly facilitating the process of CO_2 sequestration. In a case where solid carbon is used as fuel for a DCFC, the danger of explosion is incomparably lower than for hydrogen-powered fuel cells. Various types of DCFCs have been investigated recently. One of the most popular classifications is based on the kind of electrolyte used in the cell. Carbon fuel can be oxidised in an environment of alkaline, molten carbonate, or solid oxide electrolytes. The last-mentioned system, i.e. a direct carbon fuel with a solid oxide electrolyte (DC-SOFC), is believed to be the most promising due to its uncomplicated construction and maintenance [13,14].

In a DC-SOFC [15,16] carbon can be electrochemically oxidised directly to CO_2 according to reaction (1):

$$C + 2O^{2-} \rightarrow CO_2 + 4e^- \quad (1)$$

as well as in a sequence of reactions (2) and (3):

$$C + CO_2 \rightarrow 2CO \quad (2)$$

$$CO + O^{2-} \rightarrow CO_2 + 2e^- \quad (3)$$

The Boudouard reaction (2), involving CO_2 and C as reactants, is an additional source of the CO consumed in reaction (3).

On the cathode side the reduction of oxygen process took place according to reaction (4)

$$O_2 + 4e^- \rightarrow 2O^{2-} \quad (4)$$

The complex course of the reactions (1–3) may constitute a potential asset for this fuel cell, because it enables a choice of different design options leading to the optimisation of its performance. It should be stressed that DCFCs, just like hydrogen-oxygen SOFCs, can be designed in planar or tubular geometry. In addition, DC-SOFCs can operate using a solid or a fluidised coal bed.A DC-SOFC usually operates within a temperature range of 700–900°C [12,17].

These electrochemical devices are expected to be applied in power generation with a rather modest power range, 10–100W. In the construction of such devices, the minimum electrical parameter obtained from a single cell should be not lower than 150 mA/cm^2, whereas the power should not be higher than 100 mW/cm^2. The durability and stability of the electrical parameter is also a crucial requirement for the construction of an efficient small power system [18].

The physicochemical properties of the applied solid fuels, the construction of the cell, the composition of the electrode and electrolyte materials, and the participation of CO produced by the Boudouard reaction (2) in the overall electrooxidation process all influence the amount of power generated by a DC-SOFC [19,20].

One of the industry's main strategic efforts is the elaboration of a method of utilising coal-based fuels in solid oxide fuel cell technology. Coal is a very abundant fossil fuel around the world. Statistical and economic analyses show that in the coming decades, in developing

DOI: 10.1051/e3sconf/20161000019

countries such as China, India, and some EU countries, coal-fuelled power plants will maintain their strong position in the power sector due to their reliability and low costs and to the existence of large global reserves of coal and lignite. Coal is mined in politically stable areas, which guarantees its safe and easy purchase and transport [21-23]

The goal of this paper is to analyse the performance of solid oxide fuel cells supplied by solid powdered carbon. Special attention has been paid to the modification of the physicochemical properties of raw and purified coal using a cost-efficient method.

2. Experimental

One of the main criteria for the construction of efficient power sources based on direct carbon fuel cell technology is the selection of appropriate solid fuel. In this paper, two kinds of representative solid fuel were chosen for analysis. The first group consisted of carbon black and graphite powders. These materials constituted reference solid fuels for the raw and purified coal being considered as potential fuels for DC-SOFC application.

The electrochemical oxidation of carbon was studied using two types of electrochemical cells built with the support of 200-μm-thick commercial electrolyte discs:

(1) C|8YSZ|LSM–GDC|LSM|O_2

(2) C|NiYSZ|8YSZ|LSM–GDC|LSM|O_2

(3) C|(Ni-Fe-Ti)8Ce15Sm5Pr|NiYSZ|8YSZ|LSM GDC|LSM|O_2

where 8YSZ was an oxide electrolyte made of 8 mol% Y_2O_3 in ZrO_2, $La_{0.8}Sr_{0.2}MnO_3$(LSM) was a cathode material, LSM–GDC was a composite cathode buffer coating made of LSM + 10 mol% Gd_2O_3 in CeO_2 (GDC), and Ni-YSZ was a cermetallic anode material.

In case of cell (3) the new elaborated cermet material consisted of 50 wt. % $Ni_{0.85}$-$Fe_{0.12}$-$Ti_{0.03}$ particles in a matrix of $Ce_{0.8}Sm_{0.15}Pr_{0.05}O_{1.9}$ was applied in DC-SOFC [24]

Schematics of the two types of DC-SOFC used in the experiments are shown in Figures 1 and 2. During the experiment, argon was supplied continuously to the anodic chamber as a shield gas. The geometric area of the active electrolyte surface was ca 1.18 cm^2. The electrochemical measurements were performed within a temperature range of 700–900°C using a potentiostat (PGSTAT 300N) equipped with GPES (CV) and FRA (EIS) modules. The experimental setup and measurement procedure were similar to those described in our previous paper [25] Ar was used as a shield gas in the anode chamber.

In order to determine the electrochemical activity of the carbon and graphite fuel, symmetrical DC-SOFCs were measured for impedance. For this purpose, a special setup was elaborated for testing solid button oxide fuel cells and other types of small FCs [26].

The concept of the electrochemical oxidation of carbon particles on an 8YSZ electrolyte (cell 1) is presented in Figure 1.

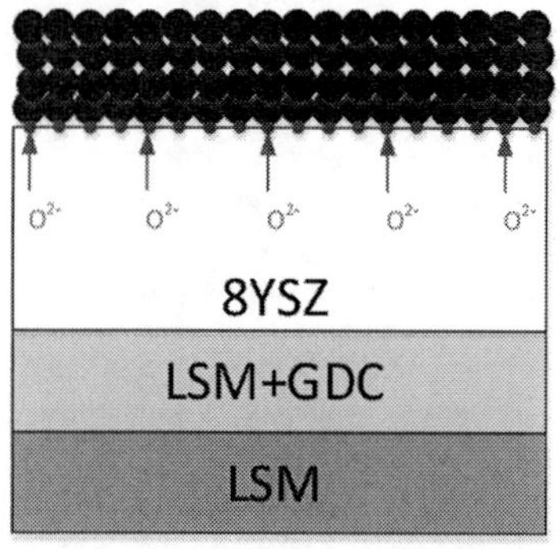

● – electrochemical reaction zone of oxidation of carbon particles on the 8YSZ surface

Figure 1. The concept of the electrochemical oxidation of carbon particles on an 8YSZ electrolyte.

In the case depicted in Figure 1, the electrochemical oxidation of carbon particles mainly occurs according to reaction (1) [27,28]. The current and power generated from DC-SOFC (1) could be directly addressed to direct carbon oxidation according to reaction (1), making it easily possible to determine the activity of solid particles in the anode reaction of the direct electrochemical oxidation of fuel.

The electrochemical oxidation of carbon particles on the Ni-YSZ surface is presented in Figure 2.

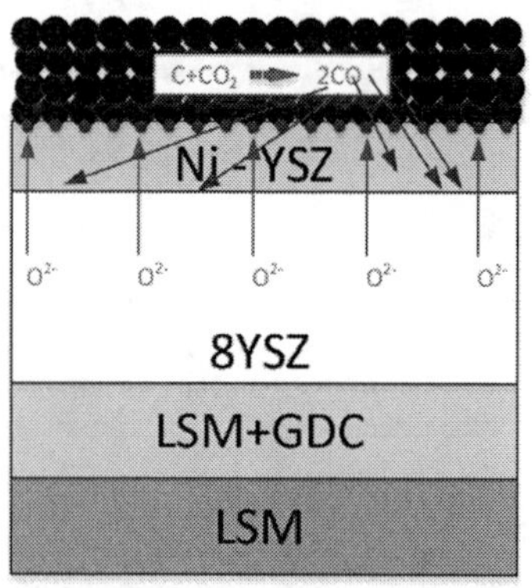

Figure 2. The concept of electrochemical oxidation of carbon particles on a Ni-YSZ anode material

In the case of cell (2) and (3), the electrochemical oxidation of carbon-based fuel could occur according to reaction (1) as well as sequences of reaction (2) and (3).

3.Results

In Table 1, the characteristics of graphite and carbon black powders used as reference carbon fuels for DC-SOFC studies are summarised. The data are limited to carbon content and surface area, which exert a considerable impact on electrochemical oxidation.

Table 1. Characteristics of graphite and carbon black powders tested in the DC-SOFC

Designated in this paper as	Type of carbon fuel	Surface area S(m²/g)	Carbon content (wt. %)
1G	Timrex® Ks 15 (Imerys Graphite and Carbon, Switzerland)	12	98.3
2G	Natural graphite (MS Graphite, India)	11	96.8
2CB	Vulcan® XC72 (Cabot, USA)	198	99.2
3CB	Ensaco 250G (Imerys Graphite and Carbon, Switzerland)	67	98.1
4CB	Carbon black N220 (Konimpex, Poland)	61	98.3

As can be seen, samples of all solid powders possessed similar carbon content.

The typical morphology of the 3CB sample is presented in Figure 3.

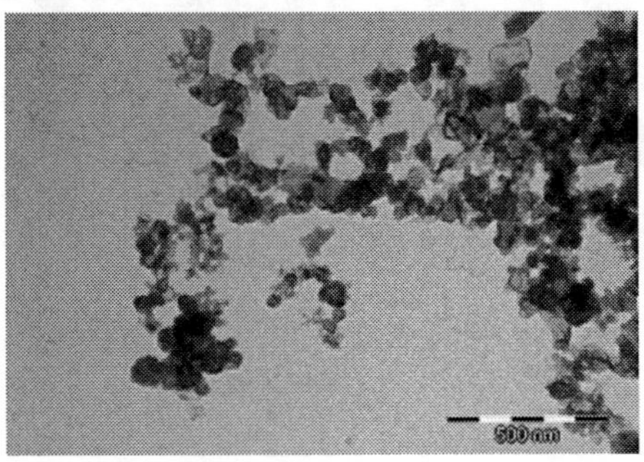

Figure 3. The SEM morphology of carbon black (3CB)

Carbon black (3CB) consisted mainly of isometric particles with dimensions ranging from 20–100nm. Isometric particles were observed in the case of all solid powders used to supply DC-SOFCs (1) and (2) -(3).

This is also an important feature in the electrochemical oxidation of carbon particles on the surface of an 8YSZ electrolyte. In the case of needle carbon particles, due to point contact with the surface of the oxide electrolyte an increase in polarisation resistance, and consequently very low power output, was observed [15,19].

In Figure 4a, b, typical values of voltage (U) vs current density (I) and power (P) vs current density (I) are shown for a DC-SOFC (4) supplied with graphite or carbon black powders. The measurements were recorded for cell (4) at 700–900°C.

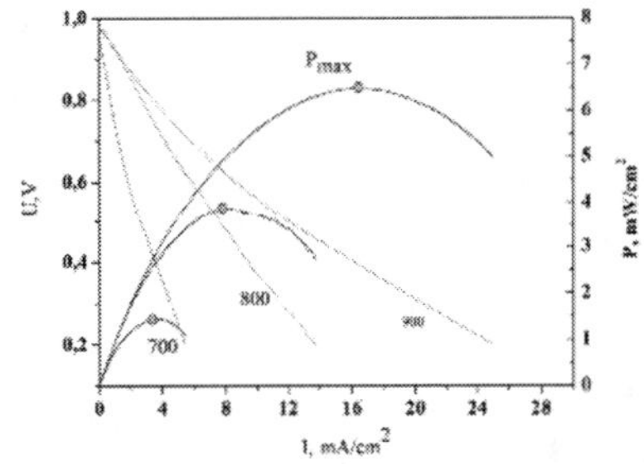

Figure 4a. U-I curves recorded for DC-SOFC (4) supplied by powder 2G

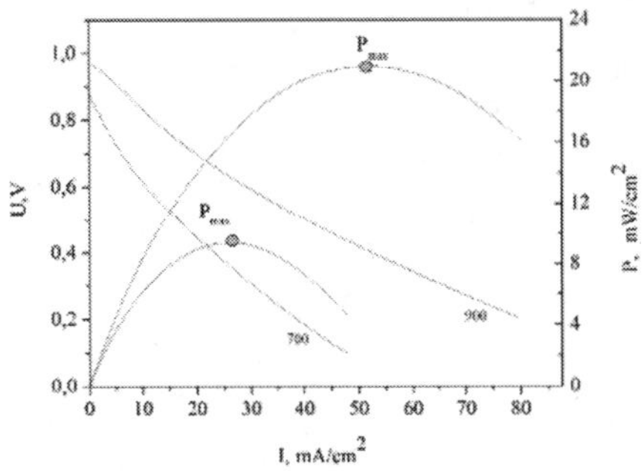

Figure 4b. U-I curves recorded for DC-SOFC (4) supplied by powder 3CB

Based on the performed cyclic voltammetry measurements (family of U-I recorded curves), the characteristic maximum power output P_{max} of cell (1) was calculated. Table 2 presents a comparative summary of the power output (P_{max}) of a DC-FC (1) supplied by graphite and carbon black powders.

Table 2. The maximum power density P_{max} of DC-SOFC type (4) supplied by graphite and carbon black powders.

Designated in this paper as	P_{max} (mW/cm²)	
	700°C	900°C
1G	0.8	4.1
2G	1.3	6.4
2CB	7.1	19.6
3CB	8.4	22.2
4CB	10.8	16.8

As shown in Table 2, the highest power output P_{max} was recorded for cell (4) supplied by carbon black powders. Graphite powders are characterised by very low electrochemical reactivity in a DC-SOFC compared to carbon black powders. These observations are in strong agreement with data presented in previous papers [15,16,19].

These results were confirmed by electrochemical impedance spectroscopy measurements. In Figure 5, the typical a impedance measurement recorded for DC-SOFC (1) supplied with graphite powder (2G) is presented.

 DOI: 10.1051/e3sconf/20161000019

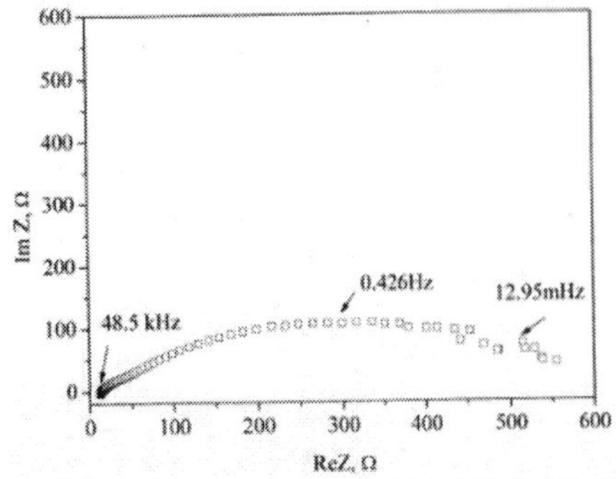

Figure 5. Impedance spectra recorded at 700°C for DC-SOFC (1) fed with graphite powder (2G)

Similar impedance spectra were recorded for DC-SOFC (1) supplied with other investigated solid fuels. The data measured by electrochemical impedance were simulated by the electrical circuit presented in Figure 6.

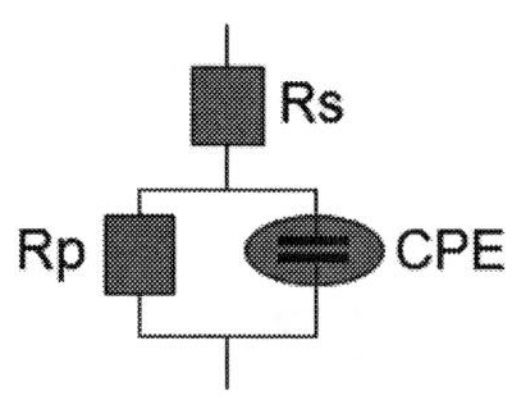

Figure 6. The electrical circuit used to fit measured electrochemical impedance data. R_s-ohmic resistance, R_p-polarization resistance, CPE –constant phase element

The calculated values of R_p for samples 2G and 3B equalled 600 Ω (2G) and 350 Ω (3CB), respectively. According to the measurements of impedance spectroscopy, the DC-SOFC cell powered directly with graphite (2G) powders was characterised by the highest value of R_p resistance, which is directly connected with the slow kinetics of the anodic reaction of electrochemical oxidation of graphite particles as opposed to carbon black particles.

In this study, investigations of 8YSZ electrolyte chemical stability in direct contact with carbon-black-type solid fuel were performed. The experiment was conducted over two months, at 900°C. The sample was kept in an argon gas atmosphere. No structural or microstructural changes were found on either the surface or a cross section of the 8YSZ electrolyte during the performed tests.

Coal, in contrast to high-purity carbon-based materials, is characterised by a number of organic and inorganic contaminants, which may have a significant impact on the process of electrochemical oxidation of carbon particles in a coal-powered DC-SOFC and thus on the values of current and power density produced by the cell [29,30]

There is no data concerning the impact of ash content on the process of electrochemical oxidation of coal on the surface of an 8YSZ electrolyte.

It is difficult to establish the level of ash that should be removed from coal to ensure high power output and durability for a DC-SOFC. In order to perform preliminary selection, certain coals were selected.

Table 3 presents the total contents of carbon, hydrogen, sulphur, ash, and volatile matter in the selected coal samples.

Table 3. Total contents of carbon, hydrogen, sulphur, ash, and volatile matter in selected coal samples, wt %.

Designated in this paper as	C^d	H^d	S^d	ash	Volatile matter
C1	84.5	7.78	0.24	2	30.4
C2	83.1	4.88	0.33	3	29.6
C3	83.2	5.13	0.42	6	31.8
C4	64.7	3.31	0.81	14	24.7
C5AF	81.5	5.30	0.35	2	33.1
C6 BF	77.4	4.96	0.54	8	28.6

C5 AF – coal after the flotation enrichment
C6 AF - coal before the flotation enrichment

In the case of coals the total content of carbon varied between ~65 and ~84 wt%. The total sulphur level was no higher than 0.81 wt%. All of the coal samples were grounded in an attritor mill to achieve a surface area of 10–15 m^2/g. In Figure 7, a representative SEM microphotograph of coal particles (1C), recorded in the course of scanning microscopy observations, is presented.

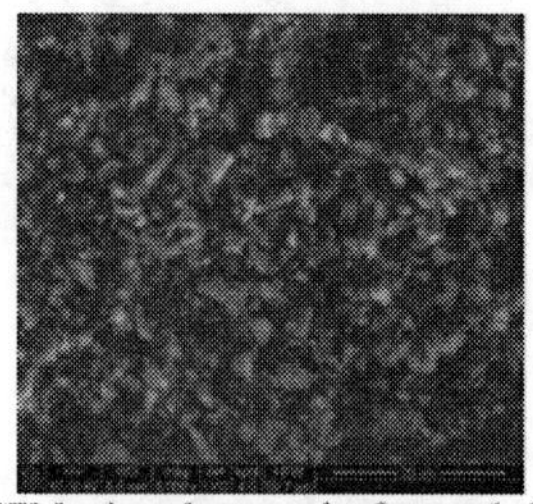

Figure 7. The SEM microphotograph of groundedcoal sample 1C

These coal particles are characterised by particle sizes ranging from about 0.2 to about 2 μm. The grains are isometric, which promotes better adhesion to an 8YSZ surface or to Ni-YSZ anode material.

Typical voltage vs current density (U-I) and power density vs current density dependencies (P-I) (Figure 8) is shown for a DC-SOFC (4) supplied with C3 powder.

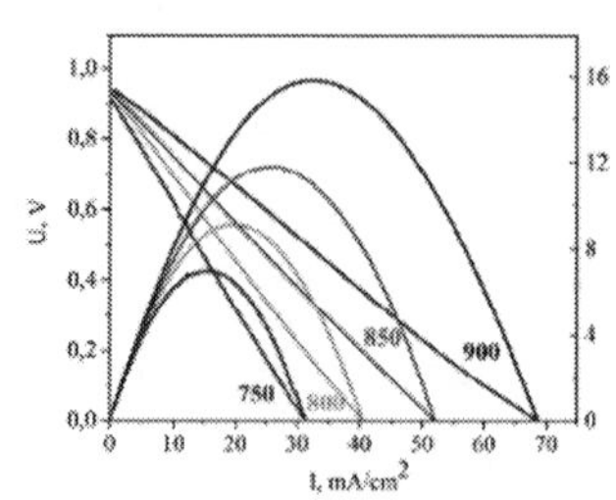

Figure 8. The family of U-I and P-I curves recorded for a DC-SOFC supplied by coal powder 3C

 DOI: 10.1051/e3sconf/20161000019

An increase in current density as well as in power density vs temperature was observed for DC-SOFC (4).

Figure 9 presents the power output (P_{max}) of DC-SOFC (4) vs total mineral matter content in coals used as fuel.

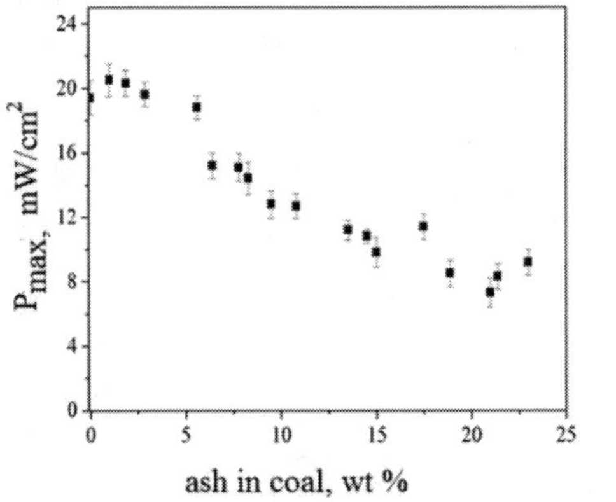

Figure 9. The impact of mineral matter in coals used as fuel on the maximum power acquired from DC-SOFC (4) at 900°C.

On the basis of Figure 9, it can be concluded that the maximum power density (P_{max}) of DC-SOFC cell (4) supplied by coals decreases slightly in the case of a solid fuel containing not more than 5 wt% mineral matter The presence of higher levels of impurities in the solid fuel has already been tied to the decrease of power (P) due to the increase in the resistance of carbonaceous fuel and the consequent total resistance of DC-SOFC cell (4).

In the case of the same DC-SOFC cell supplied with purified/upgraded coal powders, the obtained values of maximum power density (P_{max}) vs temperature (Figure 10) are close to those obtained for a DC-SOFC supplied with carbon black powders.

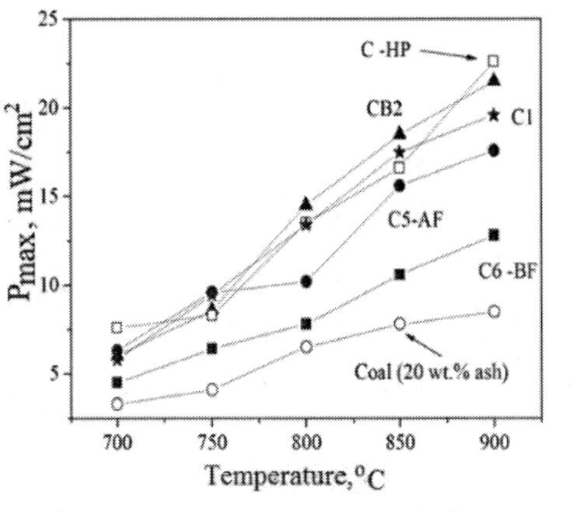

Figure 10. Dependence of the maximum power density of DC-SOFC (4) fed with coal samples vs temperature.
The data for de-ashed coal (Hyper-coal, HP Japan) have also been added [31, 32]

As a result of the analysis of data presented in Figure 10, it can be stated that the highest values of P_{max} were recorded for DC-SOFC (1) supplied with upgraded coal (with ash content no higher than 5 wt.%). The values are close to P_{max} obtained for the same DC-SOFC (1) supplied with Hyper-coal samples [31, 32] or carbon black samples (CB2).

In this graph, a direct comparison of coal before and after the flotation enrichment used as fuel is presented. An increase in P_{max} from 12 to 17 mW/cm^2 was recorded.

The chemical stability of an 8YSZ electrolyte in direct contact with coal samples was also studied. Tests were performed at 900°C for 1month.

No structural or microstructural changes were found on either the surface or a cross section of the 8YSZ electrolyte during contact with purified coals.

On the other hand, following the same experiments performed with coals involving amounts of ash greater than 10 wt.%, some adsorption of impurities on the 8YSZ surface (Figure 11) as well as some mechanical cracks were observed.

Figure 11. An SEM image recorded for 8YSZ samples after the corrosion test.

3.1 Electrochemical oxidation of carbon-based solid fuels on the Ni-8YSZ anode

In the case of cell (2), carbon can be oxidised according to reaction (1) as well as in a sequence of reactions (2) and (3): $C + CO_2 \rightarrow 2CO$ (2) and $CO + O^{2-} \rightarrow CO_2 + 2e^-$ (3) [31]. In this section, the electrochemical performance of DC-SOFC (2) supplied with coal samples will be discussed.

Compared to DC-SOFC (1), the impact of gaseous products on the performance of DC-SOFC (2) is expected to be greater. It is also inevitable that coal samples decompose during DC-SOFC operation and release pyrolysis gases, which may affect the cell's performance and change the anode reaction pathway. Pyrolysis gases oxidise more easily than solid carbon particles on the Ni-YSZ surface. In this case, the electrochemical oxidation of carbon particles is less significant than the gaseous fuel [33].

Figure 12 presents the typical U-I curves recorded for DC-SOFC (5) supplied with purified C1 samples.

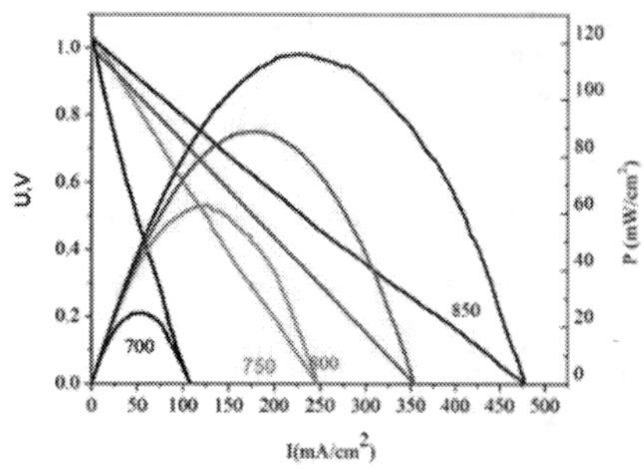

Figure 12. The family of U-I curves recorded for cell (2) supplied with C1samples.

A typical increase in current density as well as power density was observed vs temperature for the DC-SOFC. In the case of the same DC-SOFC cell supplied with purified coal powders, the values of maximum power P_{max} (Table 4) are higher compared to those obtained for a DC-SOFC supplied with carbon black powders, and slightly lower than those for an application of de-ashed coal (Hyper-coal).

Table 4. The maximum power density P_{max} of DC-SOFC type (5) supplied by graphite, carbon and coal samples

Designated in this paper as	P_{max} (mW/cm²)	
	700°C	850°C
1G	8.5	14.6
2G	11.9	20.3
2CB	22.6	64.5
3CB	28.7	70.1
4CB	19.5	56.4
C1	22.5	124.5
C2	24.5	105.6
C-HP	33.6	130.6
C5 AF	25.8	113.4
C6 BF	17.5	80.6

All comparable experiments were performed on the same type of cell (5) in similar conditions. However, low power output was observed in the case of the application of graphite as solid fuel. Based upon these preliminary investigations, upgraded coals seem to be adequate solid fuels for a DC-SOFC. The data existing in literature [30,33,34] indicated that power output P_{max} for DC-SOFC supplied by coals reached values close to 100-120 mW/cm²

3.2 The stability of an Ni-YSZ anode in a DC-SOFC operated with coal as fuel

The most commonly used anode material in a hydrogen-oxygen solid oxide fuel cell is a porous composite material, Ni-YSZ cermet, in which metallic nickel particles are distributed uniformly in a matrix of fully yttria-stabilised zirconia (8YSZ). This cermet is often applied as an anode in a DC-SOFC at this stage of developing technology. The main drawback of such material (especially nickel) seems to be low chemical stability vs carbon. This can lead to an undesirable chemical reaction between nickel and carbon, which in turn causes adverse changes in the microstructure of the cermet, as manifested in grain growth, cracks on the surface of the anode material, clogging of pores, and consequently a gradual degradation of the anode material, as well as a decline in the quality of the DC-SOFC's operation. Furthermore, Ni-YSZ cermet used as a component in DC-SOFCs is characterised by limited resistance to sulphur, which is the main impurity in fossil fuels, as well as to other gaseous hydrocarbon fuels used for supplying DC-SOFCs. The problem of the application of de-ashed coal samples containing different levels of the organic form of sulphur as fuel for DC-SOFCs was presented in a previous paper [35,36].

In this paper, new an anode material 50 wt % $Ni_{0.85}$-$Fe_{0.12}$-$Ti_{0.03}$ particles in a matrix of $Ce_{0.8}Sm_{0.15}Pr_{0.05}O_{1.9}$ [24] with higher corrosion resistance to sulphur attack was proposed and preliminarily tested in DC-SOFC (5). The results are compared to a classic Ni-YSZ anode. The data are presented in Table 5. The coal sample C1 was applied as a solid fuel in DC-SOFC (5).

Table 5. The parameters of the DC-SOFC (5) during operation under voltage load with respect to the initial value

Anode material	P_{max} (mW/cm²) at 800°C	changes in current density [I/cm²] under a load of U=0.5 V after 300 h with reference to the initial value
$Ni_{0.85}$-$Fe_{0.12}$-$Ti_{0.03}$ in a matrix of $Ce_{0.8}Sm_{0.15}Pr_{0.05}O_{1.9}$	114	7%
Ni-YSZ	84	26%

Based on data from the analysis presented in Table 5, it was observed that the newly-elaborated anode materials, i.e. Ni-Fe-Ti particles in a co-doped ceria solid solution, enabled us to obtain higher values of power output (Pmax) compared to classical Ni-YSZ; moreover, a DC-SOFC with this anode exhibited less variation in current density under load. This is probably related to improved chemical stability with regard to sulphur corrosion.

4.Conclusion

In this paper, a brief description of direct carbon solid oxide fuel cell technology was presented. It was found that carbon black powders were characterised by superior electrochemical reactivity on the surface of an 8YSZ electrolyte when compared to graphite powders. Based on the electrochemical investigation, it was found that purified coal samples (with ash content no higher than 6 wt%) exhibited high values of power output, which may have been directly connected to the four-electron reaction and direct electrochemical oxidation of carbon particles to CO_2. Upgraded coal samples also seem to be valuable solid fuels for a DC-SOFC with either Ni-YSZ or the newly-elaborated anode materials

$Ni_{0.85}$-$Fe_{0.12}$-$Ti_{0.03}$ in a matrix of $Ce_{0.8}Sm_{0.15}Pr_{0.05}O_{1.9}$. The tests indicated that DC-SOFCs fed with de-ashed coal were characterised by stable operation, with a power density greater than 100 mW/cm^2 from a single cell. Improvement in the stability of electrical parameters was observed in a DC-SOFC operated with the new anode material compared to traditional Ni-YSZ.

Acknowledgements

The work was carried out under the contract no. 11.11.210.217 (AGH University of Science and Technology, Faculty of Fuels and Energy, Cracow, Poland)

References

1. D. Quiggin, R. Buswell, Energy **98**, 1 (2016)
2. P.D. Lund, Energy **35**, 2 (2010)
3. Ch. Karakosta, Ch. Pappas, V. Marinakis, J. Psarras, Renew. Sust. Energy Rev. **22** (2013)
4. P. Crespo, D. Granado, Z. Pang, S.W. Wallace, Appl. Energ. **170**, 15 (2016)
5. A. Züttel, A. Borgschutle, L. Schlapbach, *Hydrogen as a future energy carrier*, Wiley-VCH (2008)
6. P. Alonso, R. Hewitt, J. Pacheco, Renew. Energ. **89** (2016)
7. R.S. Garcia, D. Weisser, Renew. Energ. **34** (2006)
8. B. Sørensen, Hydrogen and Fuel Cell Edition Academic Press (2011)
9. F.Beguin, E. Frackowiak, Cjarbons for Electrochemical Energy Storage and Conversion Systems (red.), CRC Press, USA (2010)
10. A. Dicks, J. Power Sources **156**, 2 (2006)
11. S. Giddey, A. Kulkarni, C. Munnings, S.P.S. Badwal Energy **68** (2014)
12. S. Giddey, S.P. Badwal, A. Kulkarni, C. Munnings, Prog. Energy Combust. Sci. **38** (2012)
13. S. Giddey, A. Kulkarni, C. Munnings, S.P.S. Badwal, J. Power Sources **284** (2015)
14. W. Hao, X. He, Y. Mi, Appl. Energ. **135** (2014)
15. P. Desclaux, H.C. Schirmer, M. Woiton, E. Stern, M. Rzepka, Int. J. Electrochem. Sci. **8** (2013)
16. M. Dudek, P. Tomczyk, M. Skrzypkiewicz, J. Jewulski, Int. J. Electrochem. Sci. **8** (2013)
17. S. Li, A. Lee, R. Mitchell, T.M. Gür, Solid State Ionics **179** (2008)
18. P.Tomczyk, J. Power Sources 160 (2006)
19. M. Dudek, M. Sitarz, P. Tomczyk, J. Solid State Electrochem. **18** (2014)
20. K. Xu, Ch. Chen, H. Liu, Y.Tian, X. Li, H. Yao Int J Hydrogen Energ **39** (2014)
21. A. Kacprzak, R. Kobyłecki, R. Włodraczyk, Z. Bis, J. Power Sources **255** (2014)
22. V. Sarhosis, A.A. Jaya, H.R. Thomas, Energy **107** (2016)
23. C.F. You, X.C. Xu, Energy **35** (2010)
24. M.Dudek, Polish patent application P.410775
25. M. Dudek, P. Tomczyk, Catal. Today **176** (2011)
26. M. Dudek, P. Tomczyk, Chamber for testing small fuel cells, Utlity Mode PL 68129 Y1 (2015)
27. X. Yu, Y. Shi, H. Wang, N. Cai, C. Li, R.I. Tomov, J. Hanna, B.A. Glowacki, A.F. Ghoniem, J. Power Sources **243** (2013)
28. P. Desclaux, S. Nürnberger, M. Rzepka, U. Stimmin Int J Hydrogen Energ **36** (2011)
29. K. Xu K. Ch. Chen Ch, H. Liu, Y. Tian, Li H.Yao, Int J. Hydrogen Energy **39** (2014)
30. J.Jewulski, M. Skrzypkiewicz, M.Struzik, I. Lubarska-Radziejewska Int J Hydrogen Energ **36** (2011)
31. M. Dudek, P. Tomczyk, R. Socha, M. Hamaguchi, J. Hydrog. Energy **39** (2014)
32. M. Dudek, Solid State Phenomena **1** (2015)
33. A.C. Raddy, S. Giddey, A. Kulkarni, S. Badwal, S. Bhattacharya, Electrochim. Acta **143** (2014)
34. A.C. Rady, S. Giddey, A. Kulkarni, S. P.S. Badwal, S.Bhattacharya, Electrochim. Acta **178** (2015)
35. T. Chun, J.D. Mumford, T. Ramanarayanan, J Electrochem. Soc. **147** (2000)
36. S.P. Jiang, S.H. Chan, J. Mater. Sci. **39** (2004)

 DOI: 10.1051/e3sconf/20161000020

Optimization of Renewable Energy-Based Autonomous Device Operation Using Simulation

Sebastian Ernst[1,a]

[1]*AGH University of Science and Technology, Department of Applied Computer Science, Kraków, Poland*

Abstract. Off-grid devices, such as autonomous lamps, are finding numerous applications in locations with no power grid availability. These devices rely on built-in generators, usually small PV panels or wind turbines, and batteries which allow the energy to be stored and utilized at appropriate time. Dedicated power device simulators are used to establish possible operation paths given uncertain parameters with regard to power generation, storage and consumption. The obtained data series are integrated into a graph search space which can be processed using various planning algorithms. An example of such scenario is given, along with example data obtained during preliminary research.

1 Introduction

Protection of the environment and strong emphasis on sustainability have led to increased use and popularisation of renewable energy sources. However, this has resulted in numerous problems related to management of modern power grids: power generation by renewables largely depends on weather conditions, such as insolation for photovoltaic devices and wind intensity for wind turbines. This makes it difficult to react to changes in demand, leading to insufficient power generation during cloudy or still weather or excess power being generated in times of low demand. In extreme cases, this may lead to energy being sold for prices close to zero.

This discrepancy between energy generation and consumption characteristics is sometimes avoided through the use of power storage (e.g. batteries), which can provide a "buffer" for better use of energy.

A more radical version of this concept is materialised as *off-grid* devices, which are autonomous with regard to power generation and consumption. The primary characteristics of such devices are:

• often misaligned power generation and consumption, apparent e.g. in solar-powered lamps,

• susceptibility to running out of power, especially during periods of limited generation.

The main operating goal of these off-grid devices is to maintain the desired autonomy, i.e. to guarantee operation until the next opportunity to accumulate more energy. However, algorithms used to control the generation and consumption of energy in such devices are simplistic and do not always yield the optimal results. For example, off-grid lamps often use a simple algorithm which switches the light source on at sunset and off at sunrise. No planning takes place and no knowledge regarding e.g. weather forecast is taken into account, which may result in the energy being consumed before more can be generated.

This problem can be avoided by computing a dispatch plan which dims the lamps during predicted periods of cloudy or windless days. Still, the accuracy and reliability of weather prediction is limited, and the dispatch plan should allow the device to adapt its operation to actual conditions. Therefore, it is important to generate a space of all situations with a certain degree of probability and calculate optimal control strategies for each of them.

The proposed solution relies on using a simulator to predict the behaviour of power generators given assumed uncertainty of input parameters (i.e. weather forecast), possible output parameter values (i.e. the light intensity desired by the user and the minimum admissible one) and synthesising them into a search space in the form of a graph.

2 State of the Art and Related Work

Simulation tools for generation, storage and consumption of energy are at the core of the proposed methods. They are described in the following sections.

2.1 Available Simulation Tools

There are numerous simulation tools available in the field of power systems. They can be divided into the following categories:

• *isolated simulators* – software which simulates an elementary device within the power network,

[a] Corresponding author: ernst@agh.edu.pl

e.g. a single PV panel, implemented as dedicated applications or Matlab modules,
• *dedicated simulators* – software, which simulates a certain class of power network devices operating within a system, e.g. a farm of photovoltaic devices; usually additional functionality can be supplied (for instance, to calculate the autonomy of a renewable-powered system under given load),
• *integrated simulators* – software able to simulate power networks integrating different types of elements, including generators, loads and various power flow analyses.

The choice of the appropriate simulation tool must be based on the support for the desired components, as well as the ease of integration in the workflow, especially if the simulator is to be used as part of an automated process in a batch manner. For that reason, the simulation tool of choice for the presented research was the GridLAB-D software package [1]. Developed by the Pacific Northwest National Laboratory and supported by the US Department of Energy, is an advanced, open-source power system simulation tool based on the BSD license. GridLAB-D is a modular system: besides the "central" power flow simulation module, it has dedicated modules for precise simulation of loads (including residential and commercial customers), distributed generation, weather, energy market components, reliability testing (including metrics such as SAIFI or SAIDI), power storage or electric vehicle charging.

The scenario for its utilisation in this research differs from the usual workflow: instead of simulating an entire power grid, it will mostly be used to simulate individual components, such as renewable power source generators or (in the following stages) power storage devices.

2.2 Weather Forecast and Modelling

Weather is the basis for prediction of behaviour of renewable generators such as PV panels or wind turbines. In practice, it usually originates from one of the following sources:
• statistical data, often contained in Typical Meteorological Year [2], especially popular in America,
• weather forecast service APIs, which allow for acquisition of conditions on a particular date, including deviation from typical weather on a given day of year.

Weather forecasting methods themselves are out of scope of this paper. However, it is important to note the possible attributes for weather forecast.

Using GridLAB-D as example, it is possible to distinguish the following attributes for weather [3]:
• outside air temperature,
• humidity,
• recorded wind speed,
• direct and indirect solar input,
• hourly rainfall,
• depth of snow layer.

Each of the attributes is timestamped. Therefore, it is convenient to store weather data in tabular form, often as CSV (comma-separated values) files.

It must be noted that cloudiness is not included in GridLAB-D's weather model, and at the same time, it is crucial to simulate output from PV panels. It may be modelled by adjusting the *solar input* parameter, but that may require an additional and non-trivial step after the weather data is obtained. Another solution is to use the *shading factor* parameter of the `solar` module and use a *player* object to play back the shading value based on the obtained cloud density.

2.3 Simulation of Renewable Power Sources

In order to predict energy generation by renewable sources, two approaches can be used:
• *simulation-based prediction*, based on weather forecast (cf. Section),
• *statistics-based prediction*, based on previously recorded behaviour of renewable sources.

The former, simulation-based approach requires a mathematical model which simulates the behaviour of power generators, as well as data regarding the parameters which affect the generators' behaviour – namely, weather data. The advantage of this approach is that no prior operational data needs to be collected to make it work. However, any discrepancy between the model and the actual device characteristics will reduce the accuracy of prediction.

The latter, statistics-based approach utilises collected operational data to predict how a given generator will operate. It requires that a substantial amount of operational data for a given device type is collected under various conditions. However, it can provide more accurate behaviour if the model is not precise or if the characteristics of the device have shifted with time (i.e., the solar panel became dirty).

2.3.1 Simulation of PV devices

The `solar` GridLAB-D module is responsible for simulation of solar panel behaviour. Although it was initially conceived to assess the impact of solar installations on the thermal performance of buildings [4], several years of development and validation have brought it to the form of a general-purpose solar generation simulator.

The model is parameterised and includes the following attributes [3]:
• panel technology (single/multi-crystal silicon, amorphous silicon, GaAs thin film, etc.),
• tilt model for tilted array calculations,
• method used to adjust the PV array output efficiency for temperatures of the cells,
• coefficients for temperature correction formulae, nominal operating cell temperature,
• calculated internal temperature of the PV module, outside air temperature,
• solar radiation incident upon the solar panel,
• insolation level that the cell is rated for,

- coefficient for the effects of temperature changes on the actual power output,
- coefficient for the effects of temperature changes on the DC terminal voltage,
- maximum operating voltage of the PV module,
- Voc_{max} of the solar module, open circuit voltage (as specified by the PV manufacturer),
- efficiency of power conversion from the solar insolation to DC power,
- surface area of the solar module, soiling of the array (representing dirt on the array) panel derating to account for manufacturing variances,
- actual power delivered to the inverter,
- shading factor for scaling solar power to the array,
- tilt angle of PV array,
- and the facing direction of the PV array.

In addition, a `solar` object needs a weather object to perform simulations (or, for initial prototyping, default weather data can be used).

2.3.2 Simulation of wind turbines

Wind turbine simulation is handled by the `windturb_dg` GridLAB-D model [5].

The model includes the following parameters [3]:

- generator type (synchronous, induction),
- control mode (constant voltage, constant real power, constant real and reactive output),
- turbine model (allowing a selection of vendor-specific and generic models),
- turbine geometry, including height and blade diameter,
- minimum/maximum operating wind speeds,
- maximum/rated coefficients of performance,
- power conversion ratio,
- and various other power-related parameters.

Similar to the `solar` module, `windturb_dg` also needs a weather object.

2.4 Simulation of Power Storage

To simplify the model at this stage of research, power storage is assumed to have linear characteristics and constant capacity. Of course, in reality, the behaviour of storage devices such as batteries is much more complex, and their behaviour depends on various factors, such as ambient temperature. Therefore, in the following stages of research, a separate module will be used to simulate precise behaviour of power storage.

GridLAB-D also contains a module to simulate power storage. Therefore, integration should be easy, as it uses the same input/output schemes and weather data as the aforementioned generator simulator modules.

2.5 Simulation of Power Consumption

The third crucial factor regarding operation of off-grid devices is the consumption of power. The method for its prediction depends on the type of the device and the user's requirements. As stated in the Introduction, there usually are several options to choose from when it comes to controlling the device operation. For instance, a lamp can be operated on various levels of dimming.

Usually, one of the choices will be the *preferred* operation mode, and one will fulfil the *minimum requirements*.

These choices will also depend on other parameters, such as the time of day. There is no need to operate lamps during the day; therefore, given the date and geographic location, a rule will determine the time of sunrise and sunset and set both the *required* and *preferred* operation levels to zero between these hours each day.

Also, for more complex solutions, the *preferred/required* labelling can be extended to an ordered set of priorities, as the users' preferences can depend on multiple parameters. More details on preference modelling can be found e.g. in [6].

2.6 Solving Planning Problems via State-Space Search

The actual algorithms used to determine the optimal dispatch plan for a device in the state space are out of scope of this paper. However, a properly formulated search space should allow for utilisation of a wide range of algorithms, including heuristic search algorithms [7, 8] such as A-star, or multivariant planning tools such as Wavefront. It is also possible to use hybrid algorithms, e.g. as described in [9].

3 Proposed Solution

This section describes the proposed algorithm for generation of a search space based on results of simulation.

3.1 Structure of States

Time is assumed to be discrete. The granulation of time depends on the resolution of the simulator, weather data, dynamics of the device and required control agility. In the presented examples, 15-minute resolution was used.

Let s_i represent a possible state of the system in the form of a tuple:

$$s_i = (t_j, g, h, c)$$

where:

- t_j is the timestamp,
- g [Wh] is the amount of energy generated during the period (t_j, t_{j+1}),
- h [Wh] is the charge of the battery at time t_j,
- c [Wh] is the amount of energy consumed during the period (t_j, t_{j+1}).

DOI: 10.1051/e3sconf/20161000020

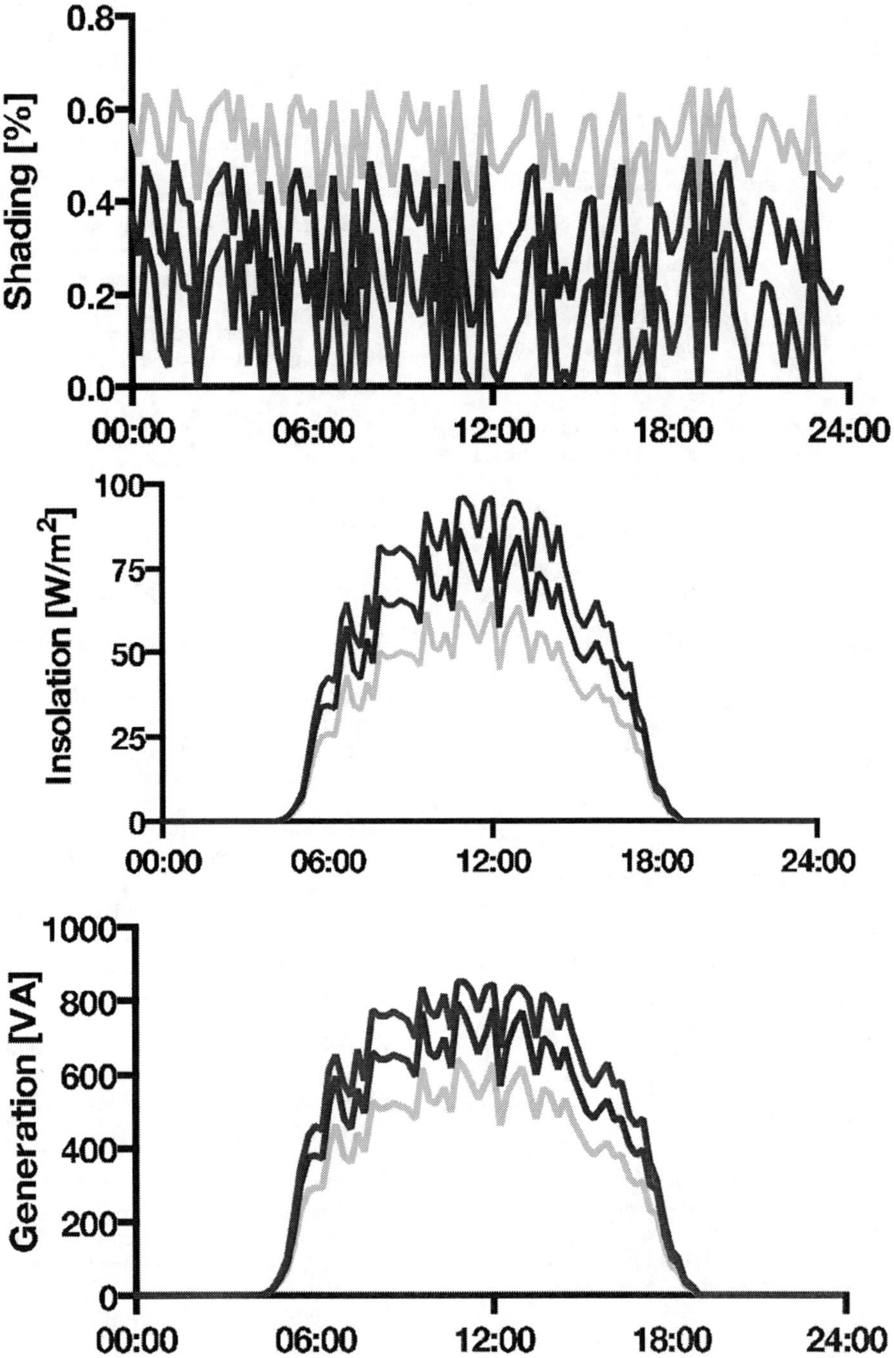

Figure 1. Example results for simulation of PV panel behaviour; values have been calculated for predicted cloudiness as well as lower and upper bounds for v_i=0.3.

3.2 Simulation-based generation of states

The approach used to generate state can be summarised as follows: for any given time t_{j+1}, generate states for all combinations of uncertain parameter values and link them to all preceding states with time t_j in respective simulated data series.

Uncertainty of parameters can have the following reasons:

• for g (energy generation), it results from assumed uncertainty of weather forecast attributes as described below,

• for h (battery charge), no uncertainty is assumed at this time, but it will be introduced in the following stages of research e.g. to compensate for SoH (state of health) variations,

• for c (energy consumption), it results from the possible modes the device can operate in during the period (t_j, t_{j+1}).

Generation of all possible c parameter values is straightforward, as it results directly from device operation rules. However, more attention needs to be given to generating all possible g values.

As stated above, uncertainty of power generation will result mostly from uncertain weather forecast.

Let us assume that the weather forecast for time t consists of n attributes (e.g. temperature, insolation, precipitation):

$W(t) = (w_1, w_2, \ldots, w_n)$.

For each attribute, we establish v_i – the *variation factor*, which defines how much a value can differ from its assumed, predicted value, thus defining the lower and upper bound for the attribute's value.

For example, if we assume that the only uncertain parameter is cloudiness (i.e., v_i=0 for all other parameters), at each time point t_j, there will be three possible levels of power generation.

Actual values for this example have been presented in Figure 1: Figure 1a presents the cloudiness levels obtained from the weather forecast, as well as lower and upper bounds calculated for v_i=0.3; Figure 1b presents the insolation calculated for these shading values, and Figure 1c shows the power generated under these conditions by an example PV device.

In this simple case, at any given point, there are at most three possible levels of power generation, which directly relates to the amount of power accumulated in the battery.

Modelling of possible operation modes of devices is out of scope of this paper, but if we assume that the lamp can be either off, on, or operating at half power, at any point there will also be three distinct possibilities for energy consumption.

As the time interval was set at 15 minutes, there will be 96 states in every 24-hour period. Therefore, the maximum number of states in such scenario will be 864. However, simple heuristics can be used to reduce this number significantly *a priori*:

1. During periods of no power generation (e.g. during the night), there is no uncertainty, as output from PV panels equals zero. However, the reduction will not be as efficient if the device is also equipped with a wind turbine.
2. At no point will all three modes for operation be available; if the lamp is to be off, there is no uncertainty; if the lamp is to be lit (during the night), the option to have it off will probably not be acceptable.

The generated state space will actually be a graph containing all possible states the device can achieve at any time, unless the parameter variations exceed the predicted limits.

Such a graph can be used with any planning algorithm to determine the optimal mode of operation. One solution is to apply a penalty for each disallowed state, e.g. states where power consumption (c) exceeds the amount of power accumulated in the battery (h).

That way, the optimal (lowest-cost) paths in the graph will represent any control schemes leading to acceptable behaviour of the device under consideration.

4 Conclusions and Future Work

The paper presents a novel approach to generating a search space for off-grid device dispatch planning using power device simulators. The advantages of this approach, compared to reactive control or simple stochastic algorithms, is precise representation of the device behaviour and support for uncertainty in any area of operation (e.g. weather, power generation and consumption in different modes).

Most other approaches to predicting the performance of renewable energy sources rely on machine learning, artificial neural networks and similar techniques. This means that they require operational data to have been collected in order to let the algorithms tune their predictions. In applications similar to off-grid lamps, each instance can have different characteristics due to non-uniform component quality and factors such as shading. Simulation-based prediction may improve the accuracy of such a system in the early stages of operation, and can be supplemented by data-based algorithms later on.

Future work involves introduction of more advanced modules for simulating the behaviour of power storage, integration with a rule-based engine designed to run on the embedded controller and field testing. The paper presents a novel approach to generating a search space for off-grid device dispatch planning using dedicated simulators.

References

1. D.P. Chassin, K. Schneider, C. Gerkensmeyer, *GridLAB-D: An open-source power systems modeling and simulation environment*, in *2008 IEEE/PES Transm. Distrib. Conf. Expo.* (2008), pp. 1–5
2. S. Wilcox, W. Marion, *NREL/TP-581-43156, Users Manual for TMY3 Data Sets* (2008)

3. Pacific Northwest National Laboratory, *GridLAB-D 3.0 Documentation* (2015), http://gridlab-d.sourceforge.net/doxygen/3.0/
4. F. Tuffner, JL Hammerstrom, R Singh, *Incorporation of NREL Solar Advisor Model Photovoltaic Capabilities with GridLAB-D*
5. J. Fuller, K. Schneider, *Modeling Wind Turbines in the GridLAB-D Software Environment* (2009), http://www.osti.gov/scitech/biblio/1052120
6. R. Klimek, I. Wojnicki, S. Ernst, *State-Space Reduction through Preference Modeling* (2013)
7. B. Bonet, H. Geffner, *Planning as heuristic search* (2001), http://linkinghub.elsevier.com/retrieve/pii/S0004370201001084
8. B. Bonet, H. Geffner, *Planning as Heuristic Search: New Results*, in *ECP '99 Proc. 5th Eur. Conf. Plan. Recent Adv. AI Plan.* (1999), Vol. 46, pp. 368–369
9. I. Wojnicki, S. Ernst, W. Turek, *A Robust Planning Algorithm for Groups of Entities in Discrete Spaces* (2015), http://www.mdpi.com/1099-4300/17/8/5422/htm

 DOI: 10.1051/e3sconf/20161000021

Solvent effects on extraction of polycyclic aromatic hydrocarbons in ambient aerosol samples

Mira Flasch[1], Magdalena Kistler[2,a], Bernadette Kirchsteiger[1] and Anne Kasper-Giebl[1]

[1]*Institute of Chemical Technologies and Analytics, Vienna University of Technology, Vienna, Austria*

Abstract. Polycyclic Aromatic Hydrocarbons (PAHs) in the ambient particulate matter pose one of the most important issues in the focus of environmental management. The concentration of their representative, Benzo(a)pyrene (BaP), undergoes limitations according to European Union directive. However, a successful control over the pollution levels and their sources is limited by the high uncertainty of analytical and statistical approaches used for their characterization. Here we compare differences in PAH concentrations related to the use of different solvents in the course of ultrasonic extraction of a certified reference material (PM10-like PAH mixture) and filter samples of ambient particulate matter collected in Austria for the CG-MS PAH analysis. Using solvents of increasing polarity: Cyclohexane (0,006), Toluene (0,099), Dichloromethane (0,309), Acetone (0,43) and Acetonitrile (0,460), as well as mixtures of those, filters representing high and low concentrations of particulate matter were investigated. Although some scatter of the obtained concentrations was observed no trend related to the polarity of the solvent became visible. Regarding the reproducibility, which can be expected of PAH analysis no significant difference between the different solvents was determined. This result is valid for all compounds under investigation.

1 Introduction

Polycyclic aromatic hydrocarbons (PAHs) are organic compounds consisting of two or more condensed aromatic rings. They are ubiquitous in the atmosphere and they are burdened with different adverse health effects [2]. Thus they are known to be carcinogenic and mutagenic compounds [3, 4]. Consequently, emissions and ambient concentrations have to be controlled and monitored. Principally PAHs are formed during the incomplete combustion of organic matter, i.e. often by anthropogenic activities based on the combustion of biomass or fossil fuels. The European emission inventory (EMEP) states that emission densities of PAHs in Europe are strongly influenced by domestic heating [5], while earlier the main influence of industries and transportation was outlined [6].

Due to its high toxicity Benzo(a)pyrene (BaP) is most thoroughly studied. It often is used as an indicator for the overall concentration of PAHs in ambient air. A limit value for BaP was defined within the EU. This value equals 1 ng/m^3, to be determined in particulate matter samples with a size cut of 10 μm a.d. (PM10) as an average value for one calendar year [1]. During the last years this limit value was exceeded at a number of stations in Europe [7]. Comprehensive investigations of PAH concentrations cover additional compounds as well. Besides their actual concentrations the toxicity of the other PAHs can be expressed by a 'benz(a)pyrene-equivalent' (BaPE) [8]. The main PAHs required to derive the BaPE can be found predominantly in the particulate phase [9].

To assure the conformity within the monitoring networks the analytical approach for BaP (and other PAHs) in the ambient aerosols is defined by the European Norm EN 15549 [10]. However, this document allows a choice of analytical approaches for the analysis of BaP (GC-MS or HPLC-FLD). This fact results in a relatively wide spectrum of the approved extraction methods and solvents. In this contribution we present the results of an experiment comparing the effects related to the use of different solvents in the course of ultrasonic extraction of filter samples for the CG-MS PAH analysis.

2 Experimental

2.1 Sample preparation

For the comparison two filter samples collected in Austria in winter 2010/11 were used. Ambient particulate matter was loaded onto quartz fibre filters, using High-Volume samplers. Sampling time was 24 hours and the sampled air volume was approximately 720 m^3. To account for different filter loadings one sample was collected in an urban area (i.e. the city of Klagenfurt, Carinthia), representing a highly loaded filter (Filter 1). The second sample (Filter 2) was loaded in a rural region (Arnfels, Styria) and represents conditions typical for lower particulate matter concentrations. Furthermore, a standard

[a] Corresponding author: magdalena.kistler@tuwien.ac.at

 DOI: 10.1051/e3sconf/20161000021

reference material (1649a, Urban Dust) was used for the intercomparison.

To obtain equally loaded sub-samples, the filters were cut into sixteen segments of equal area using a special cutting tool. For each analyses one of these segments was used. The filter segments were placed in glass Petri dishes. For the determination of the extraction recovery the samples were spiked with a deuterated standard (d_{12}-perylene). Afterwards the filter segments were transferred to a glass test tube. For analysis of the reference material approximately 150 mg of the material was directly weighed into the test tube and spiked there. The respective solvents were added to a total volume of 5 mL. Five different solvents in seven different combinations were investigated. The applied solvent combinations are displayed in Table 1.

Table 1. Solvents and solvent mixtures used for the sample extraction

Name	Solvent 1 [mL]	Solvent 2 [mL]	Ratio
DCM	Dichloromethane	-	-
DcH1	Dicloromethane	Cyclohexane	3:2
DcH2	Dicloromethane	Cyclohexane	3:7
cHex	Cyclohexane	-	-
AcH	Acetone	Cyclohexane	1:1
AcN	Acetonitrile	-	-
TocH	Toluene	Cyclohexane	3:3

Subsequently the samples were put in an ultrasonic bath to ensure thorough extraction of the filters. The extraction time was 30 min. The procedure was repeated once and the combined extracts were then spiked with an internal standard (d12-BaP). In a next step the extracts were evaporated to about 200µl in a nitrogen flow. Only the samples extracted with acetonitrile were evaporated to dryness, then dissolved in 3 mL cyclohexane and put in an ultrasonic bath for 30 min before finally reducing the liquid volume to approx. 200 µl. As solids (washed off insoluble particulate matter or fibres of the quartz filters) could be present, especially in the extracts of the samples of the reference material, the extracts were centrifuged and just the clear supernatant was used for analysis.

2.2 Analytical methods

The PAHs determination was performed with a GC-MS method previously used by Kistler et al. [11]. The HP-5890 Gas Chromatograph is equipped with a HP-7683 auto sampler and a split/split-less injector (300°C) operated in a split-less mode (2 min). A deactivated fused silica pre-column (1m x 0.32mm) was used in combination with an analytical capillary column (DB-5 MS- 95% dimethyl-, 5% phenylsiloxane, 30m x 0.25mm ID x 0.25µm film thickness) for the separation. The temperature programme started at 50°C for 2 min, followed by an increase of the temperature to 150°C with a rate of 20°C min^{-1}. Afterward the temperature was increased to 300°C, with a rate of 10°C min^{-1} and the final temperature was held for 20min. The temperature was 230°C for the source, 150°C for the quadrupole and 300°C for the transfer line. Helium 5.0 (Messer) with a flow rate of 1 mL min^{-1} was used as a carrier gas. A quadruple mass spectrometer HP-5973 (70eV) was applied as a detector. PAHs were recorded in the full scan mode and selected ion monitoring (SIM) mode.

Standards were prepared using a mix-standard (PAK-Mix 18, Neochema) containing 18 PAHs to identify and quantify the compounds. Naphthalene, 2-methylnaphthalene, 1-methylnaphthalene, acenaphthene, acenaphthylene, phenanthrene, fluorine, anthracene, benz[a]anthracene (BaA), chrysene (Chr), pyrene (PY), fluoranthene (FLT), benzo[b]fluoranthene (BbF), benzo[k]fluoranthene (BkF), benzo[a]pyrene (BaP), indeno[1,2,3-cd]pyrene (IcdP), benzo[ghi]perylene (BghiP) and dibenz[a,h]anthracene (DahA) were included in the mix-standard. However, PAHs lighter than pyrene were not quantified. Samples and standards were measured twice each time.

3 Results and Discussion

The results of the analysis of filter 1 (highly loaded filter) and filter 2 (light loaded filter) are summarized in Tables 2 and 3, respectively. Loadings of the respective PAHs are given as ng per mm^2 of filter area to highlight differences in filter loadings. Table 4 lists the results obtained for the reference material. Standard deviations given for the single compounds and measurement were calculated based on the variations obtained for the multiple injection (n=5) of the DcH2 extract. Additionally, the mean value, determined for all extractions was calculated and listed in the last column. To assign an overall uncertainty to that average we use variations, more precisely the expanded uncertainties for the single PAHs and concentration levels, determined within an inter-laboratory intercomparison of PAH analyses on quartz filters [12]. Consequently, the variations given for the mean do not reflect standard variations of our measurements, but an estimate for an expanded uncertainty which can be expected for analysis using various methods for the determination of those compound. In case of filter 1 and the standard reference material the overall expanded uncertainties determined for higher concentration levels (F30 – as given in [12]) for the respective compounds were applied, while the overall expanded uncertainty determined for lower concentrations (F10 – as given in [12]) was used for filter 2. Values of the single analyses falling out of the overall expanded uncertainty of the mean are marked in bold print.

Table 2. Results of the determination of PAHs on filter 1 with different solvents including standard deviations of the measurement (calculated as given in the text)

Analyte [ng/mm²]	DCM	DcH1	DcH2	cHex	AcH	AcN	TocH	Mean
PY	30.3±1.4	40.0±1.9	32.7±1.5	26.7±1.2	48.3±2.3	37.6±1.8	31.7±1.5	35.3±9.3
FLT	28.7±1.9	37.0±2.5	30.1±2.0	24.2±1.6	**47.5±3.2**	35.4±2.4	27.5±1.9	32.9±7.4
BaA+Chr	41.0±0.9	56.5±1.2	53.9±1.1	35.1±0.7	47.5±1.0	51.3±1.1	44.4±0.9	47.1±14.9
BbF+BkF	60.0±2.6	77.5±3.4	70.2±3.0	56.7±2.5	58.1±2.5	71.5±3.1	**36.9±1.6**	61.5±19.3
BaP	45.0±6.9	56.5±8.7	49.7±7.6	39.5±6.1	34.2±5.2	54.7±8.4	50.5±7.8	46.6±13.8
IcdP	48.4±5.7	55.0±6.5	53.8±6.3	39.5±4.6	56.3±6.6	53.0±6.2	77.2±9.1	54.7±20.3
BghiP	79.9±15.9	96.3±19.2	88.9±17.7	61.3±12.2	72.3±14.4	95.2±19.0	87.3±17.4	83.0±33.7
DahA	15.6±0.8	16.7±0.9	15.5±0.8	14.2±0.8	12.1±0.6	18.1±1.0	15.7±0.8	15.4±5.1

Table 3. Results of the determination of PAHs on filter 2 with different solvents including standard deviations of the measurement (calculated as given in the text)

Analyte [ng/mm²]	DCM	DcH1	DcH2	cHex	AcH	AcN	TocH	Mean
PY	5.8±0.3	7.3±0.3	7.3±0.3	7.1±0.3	12.1±0.6	6.7±0.3	5.8±0.3	7.4±2.7
FLT	5.8±0.4	6.4±0.4	7.1±0.5	6.0±0.4	**11.5±0.8**	6.2±0.4	5.2±0.4	6.9±2.8
BaA+Chr	2.8±0.1	3.6±0.1	3.5±0.1	3.3±0.1	4.1±0.1	3.5±0.1	2.9±0.1	3.4±1.5
BbF+BkF	4.1±0.2	5.7±0.2	5.8±0.3	4.5±0.2	6.4±0.3	5.5±0.3	6.2±0.3	5.9±1.8
BaP	4.3±0.7	4.3±0.7	4.5±0.7	4.2±0.6	4.7±0.7	4.5±0.7	4.7±0.7	4.4±2.2
IcdP	6.8±0.8	8.4±1.0	7.1±0.8	6.4±0.7	8.4±1.0	8.5±1.0	7.5±0.9	7.6±2.9
BghiP	5.7±1.1	6.5±1.3	6.1±1.2	4.7±0.9	7.3±1.5	8.3±1.7	6.5±1.3	6.4±2.8
DahA	4.1±0.2	5.5±0.3	4.5±0.2	4.2±0.2	5.5±0.3	5.0±0.3	4.5±0.2	4.8±2.7

Table 4. Results of the determination of PAHs within the reference material with different solvents including standard deviations of the measurement (calculated as given in the text)

Analyte [ng/mg]	DCM	DcH1	DcH2	cHex	AcH	AcN	TocH	Mean
PY	6.6± 0.3	4.0±0.2	4.1±0.2	3.4±0.2	6.5±0.3	5.4±0.3	2.5±0.1	5.0±1.3
FLT	6.8±0.5	7.5±0.5	7.7±0.5	**3.8±0.3**	**10.0±0.7**	9.6±0.7	5.6±0.3	7.3±1.6
BaA+Chr	2.6±0.1	2.0±0.1	2.0±0.1	1.4±0.1	3.3±0.1	2.8±0.1	1.4±0.1	2.2±0.7
BbF+BkF	3.8±0.2	3.2±0.1	2.8±0.1	**1.0±0.1**	**5.1±0.2**	3.8±0.2	2.0±0.1	3.1±1.0
BaP	1.0±0.2	0.7±0.1	0.8±0.1	0.4±0.1	1.0±0.2	0.6±0.1	0.5±0.1	0.7±0.2
IcdP	2.1±0.2	2.3±0.3	1.7±0.2	**0.9±0.1**	2.4±0.3	1.7±0.2	1.3±0.2	1.8±0.7
BghiP	5.6±1.1	5.1±1.0	3.4±0.7	**0.9±0.2**	4.9±1.0	2.8±0.6	2.5±0.6	3.5±1.5
DahA	0.9±0.1	0.9±0.1	0.7±0.1	0.4±0.1	0.9±0.1	0.8±0.1	0.6±0.1	0.7±0.3

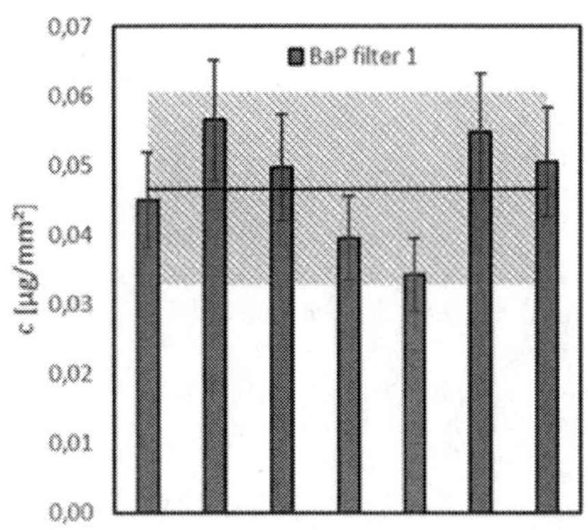

Figure 1. Filter loadings of BaP on filter 1, extracted with various solvents. Error bars reflect the standard deviation calculated by multiple injections of a single extract. The solid line represents the mean of all extractions, while the shaded area reflects the expanded uncertainty of the mean (calculated as given in the text).

Figures 1, 2 and 3 highlight the results exemplarily for BaP. The solid line represents the mean value calculated based on all extracts, while the shaded area describes the expanded uncertainty obtained within an independent inter-laboratory intercomparison [12] and is explained in detail above.

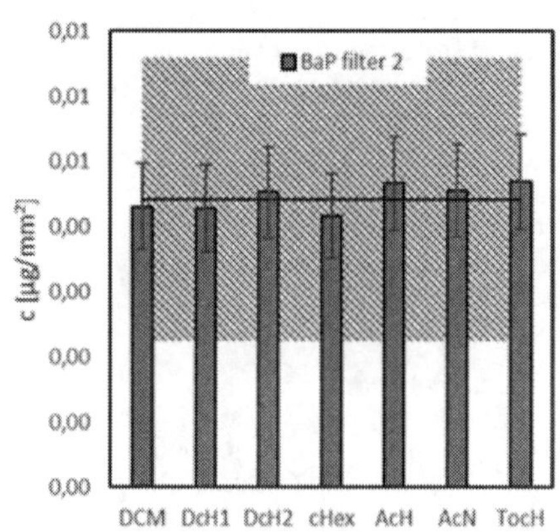

Figure 2. Concentration of BaP on filter 2, extracted with various solvents. Error bars reflect the standard deviation calculated by multiple injections of a single extract. The solid line represents the mean of all extractions, while the shaded area reflects the expanded uncertainty of the mean (calculated as given in the text).

Although some scatter of the BaP concentrations determined with the respective solvents used for extraction is visible for the highly loaded filter and the reference material, all results remain within the overall expanded uncertainty of the mean. Very good agreement of the results, however, was obtained for filter 2. In this case the variations of the filter loadings are within 7.5%, even lower than the variations of multiple injections.

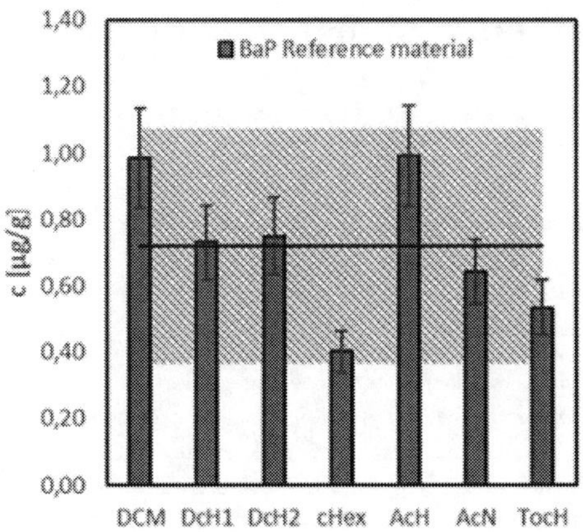

Figure 3. Concentration of BaP in the standard reference material, extracted with various solvents. Error bars reflect the standard deviation calculated by multiple injections of a single extract. The solid line represents the mean of all extractions, while the shaded area reflects the expanded uncertainty of the mean (calculated as given in the text).

The results obtained for the other PAHs are comparable to the findings shown for BaP and filter 1. The excellent agreement obtained for BaP loadings on filter 2 extracted with the respective solvents was not seen for the other compounds. Still some variations of the measurements cannot be explained by the overall expanded uncertainty only. Using TocH as solvent BbF+BkF (filter 1) fall below the expanded uncertainty area. In AcH FLT (filter 1+2, reference substance) and BbF+BkF (reference substance) exceed the expanded uncertainty range. A more general trend was observed when using cHex as a solvent. In this case the underestimation of FLT, BbF+BkF, IcdP and BghiP in the reference material cannot be explained with the overall expanded uncertainty. Analyses of the filter samples do not clearly reflect or disprove this trend. Further investigations are needed to investigate this questions.

Overall the results indicate that the used solvents do not influence the extraction efficiency as much as expected.

Acknowledgement

Filter samples investigated here were collected within the EU project PMInter (www.pminter.eu).

References

1. Directive 2004/107/EC of the European Parliament and of the Council of 15 December 2004 relating to arsenic, cadmium, mercury, nickel and polycyclic aromatic hydrocarbons in ambient air, Official Journal of the European Union, L23/3-16 (2005)
2. T. Rengarajan, P. Rajendran, N. Nandakumar, B. Lokeshkumar, P. Rajendran, I. Nishigaki, Asian Pac. J. of Trop. Biomed. **5**, 182-189 (2015)
3. A.O. Asita, M. Matsui, T. Nohmi, A. Matsuoka, M. Hayashi, M. Ishidate Jr., T. Sofuni, M. Koyano, H. Matsushita, Mutat. Res. Lett. **264**,7-14 (1991)
4. P.H. Danielsen, S. Loft, A. Kocbach, P.E. Schwarze, P. Møller, Mutat. Res.-Genet. Tox. **674**, 116-122 (2009)
5. EEA/LRTRAP, Air pollutant emissions data viewer (LRTAP Convention), Available online: http://www.eea.europa.eu/data-and-maps/data/data-viewers/air-emissions-viewer-lrtap (accessed on April, 2016)
6. Working Group on PAH, *Ambient Air Pollution by Polycyclic Aromatic Hydrocarbons (PAH) Position Paper*, Office for Official Publications of the European Communities, Luxembourg, ISBN 92-89-2057 (2001)
7. R. Gehring, *Polyzyklische aromatische Kohlenwasserstoffe im PM10 an ausgewählten Stationen des NABEL sowie kantonalen Stationen*, measurement report 2010, EMPA report Nr. 203056/17, CH 8600 Dübendorf
8. EPA, *Provisional Guidance for Quantitative Risk Assessment of Polycyclic Aromatic Hydrocarbons*, Office of Research and Development Washington, DC 20460, EPA/600/R-93/089 (1993)
9. K. Hytönen, P. Yli-Pirilä, J. Tissari, A. Gröhn, I. Riipinen, K.E.J. Lehtinen, J. Jokiniemi, Aerosol Sci. Tech. **43**, 442-454
10. EN 15549:2008
11. M. Kistler, *Particulate matter and odour emission factors from small scale biomass combustion units*, PhD, Vienna University of Technology (2012)
12. E. Grandesso, K. Kowaleswski, P.P. Ballesta, *First EC-JRC PAHs inter-laboratory comparison on PM10 quartz filters*, JRC Scientific and Technical Reports (2012)

Underestimation of nuclear fuel burnup – theory, demonstration and solution in numerical models

Paweł Gajda[1], Grzegorz Kępisty[1,a] and Michał Orliński[1]

[1]*AGH University of Science and Technology, Faculty of Energy and Fuels, Department of Nuclear Energy, al. Mickiewicza 30, 30-059 Krakow, Poland*

Abstract. Monte Carlo methodology provides reference statistical solution of neutron transport criticality problems of nuclear systems. Estimated reaction rates can be applied as an input to Bateman equations that govern isotopic evolution of reactor materials. Because statistical solution of Boltzmann equation is computationally expensive, it is in practice applied to time steps of limited length. In this paper we show that simple staircase step model leads to underprediction of numerical fuel burnup (Fissions per Initial Metal Atom - FIMA). Theoretical considerations indicates that this error is inversely proportional to the length of the time step and origins from the variation of heating per source neutron. The bias can be diminished by application of predictor-corrector step model. A set of burnup simulations with various step length and coupling schemes has been performed. SERPENT code version 1.17 has been applied to the model of a typical fuel assembly from Pressurized Water Reactor. In reference case FIMA reaches 6.24% that is equivalent to about 60 GWD/t_{HM} of industrial burnup. The discrepancies up to 1% have been observed depending on time step model and theoretical predictions are consistent with numerical results. Conclusions presented in this paper are important for research and development concerning nuclear fuel cycle also in the context of Gen4 systems.

1 Introduction

Continuous energy Monte Carlo burnup codes such as MCB [1] and SERPENT [2] are precise tools for depletion calculations in nuclear critical and subcritical systems. The biggest advantages of this approach is exact treatment of neutron transport physics in arbitrary geometries. This leads to reliable estimation of reaction rates that are applied in transmutation procedure. Solutions obtained by Monte Carlo burnup codes are considered as reference ones and the methodology is widely applied in fuel cycle studies.

In this paper we present a numerical problem that may exhibit as underestimation of numerical burnup produced by the code. The bias can be observed in all kinds of burnup codes, also deterministic ones. As will be shown, the observable underestimation of nuclear fuel FIMA (Fissions per Initial Metal Atom) can be present when long time steps are applied and the simplest coupling scheme is in use. The problem is independent from the nuclear data applied in neutron transport and numerical treatment of the Bateman equations. Problem is relevant for comparative studies of fuel cycles, benchmark calculations and experimental validations, in which numerical FIMA may be underestimated in comparison to the burnup in GWd/t_{HM} imposed by code user or measured by the operator of industrial facility.

In Section 2 the methodology of burnup calculation is presented. Section 3 shows origins of numerical bias. Model applied in numerical tests is described in Section 4 and the results are analyzed in Section 5. Section 6 summarizes the research and provides conclusions for the good practices in fuel cycle studies.

[a] corresponding author: gkepisty@agh.edu.pl

2 Methodology

The approach applied in large variety of currently existing Monte Carlo burnup codes is explained in detail in the paper [3]. The key aspect of this methodology is a statistical solution of neutron transport Boltzmann equation obtained at each point of declared time mesh. Monte Carlo code run provides physical quantities (so called *tallies*) such as reaction rates of nuclear reactions ($\mathbb{R}$) and heating per source neutron (h) that are corresponding to the critical configuration of nuclear system. Nuclear fuel isotopic composition $\mathbb{N}$ at given point $\bar{r}$ and time t is driven by Bateman Equations presented below:

$$\frac{d\mathbb{N}(\bar{r},t)}{dt} = \mathbb{M}(\mathbb{R},S)\cdot\mathbb{N}(\bar{r},t) \quad (1)$$

Transmutation matrix $\mathbb{M}$ appearing in Bateman equations is a product of reaction rates and neutron source intensity S plus decay matrix $\mathbb{D}$. The formula is presented below:

$$\mathbb{M}(\mathbb{R},S) = \mathbb{R}\cdot S + \mathbb{D} \quad (2)$$

It can be deduced that neutron source intensity describes dynamics of fuel depletion and stands for a normalization factor in transmutation problems. S can be obtained using power normalization equation, which can be formulated in the following way:

$$S = P/h \quad (3)$$

where : P is total fission power constrained in the system model.

The set of Bateman equations can be solved using a variety of numerical methods optimized for this purpose and several of them are discussed in the work by [4]. For example MCB applies Transmutation Trajectory Analysis (TTA) and Chebyshev Rational Approximation Method (CRAM) is implemented in SERPENT. Formal mathematical solution of the Bateman equations is called matrix exponent presented below:

$$\mathbb{N}(\vec{r}, t) = \exp[\mathbb{M}(\mathbb{R}, S)(t - t_0)] \cdot \mathbb{N}_0(\vec{r}) \quad (4)$$

where : $\mathbb{N}_0$ is isotopic vector at the beginning of step. It should be noted that coefficients of transmutation matrix $\mathbb{M}$ are assumed constant and on the other hand $\mathbb{R}$ and S can in general vary significantly over time in realistic nuclear systems. The compromise commonly applied in burnup codes is a limited length of burnup steps, during which the variation of reactor model properties is neglected. The most basic step model is called Euler predictor (aka. staircase), however predictor-corrector methodology becomes more commonly applied in currently developed codes [5]. In the next section it will be shown that if variation of transmutation coefficients is not taken into account, the results suffer from inherent numerical bias.

3 Bias of numerical burnup

FIMA indicator of nuclear fuel burnup can defined as a ratio between fissioned heavy metal atoms (ΔN_{HM}) and their initial number ($N_{HM,0}$) in fresh fuel as shown below:

$$FIMA = \frac{\Delta N_{HM}}{N_{HM,0}} \quad (5)$$

In practice, it is easier to measure the amount energy released in industrial reactor per initial heavy metal mass (m_{HM}) and equivalent formulation of burnup (BU) is often in use:

$$BU = \frac{\int_T P(t)\, dt}{m_{HM}} \quad (6)$$

As long as the energy released per fission in system is approximately constant (roughly 200MeV), it can be assumed that both formulations are physically equivalent.

Most often the users of burnup codes declare the burnup of fuel using its second formulation (6) and the staircase function. It appears that numerical solution of Bateman equations does not automatically take into account that heating per source neutron h decreases over time step. This should normally result as an increase of neutron source intensity S, however in the simplest case only the beginning of step value is applied. The problem is schematically shown in Fig. 1 below.

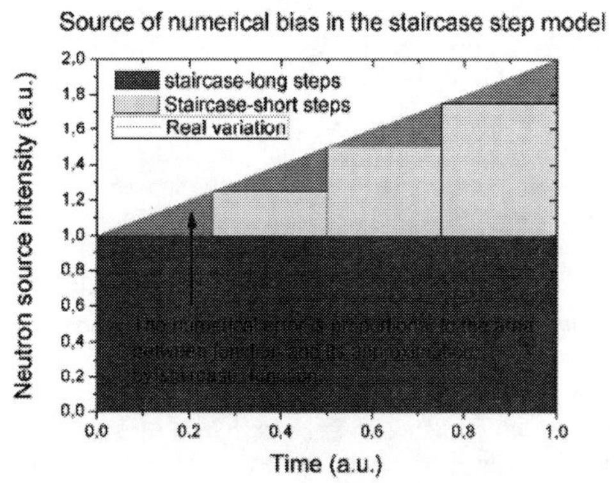

Figure 1. Scheme explaining the source of possible numerical burnup underestimation in depletion simulations.

Neutron source intensity governs the speed of nuclear reactions in model during each step and it can be assumed that the number of fissions is directly proportional to the area under the curve in Fig. 1.

The error of the fuel burnup at each time step is proportional to the area between the real curve and staircase function (green color in Fig. 1). The considered bias of FIMA indicator over irradiation period T is defined by the ratio of integrals:

$$\frac{\Delta FIMA}{FIMA} = \frac{\int_T (S(t) - S(0))\, dt}{\int S(t)\, dt} \quad (7)$$

As will be shown, the variation of S in numerical models can be well represented by linear function, thus the absolute or relative bias can be estimated analytically. The error can be described as the sum of n triangular surfaces representing contributions from each time step:

$$\frac{\Delta FIMA}{FIMA} = \frac{\frac{1}{2}\frac{\Delta S}{\Delta T} \cdot \Delta t \cdot \Delta t \cdot n}{\frac{1}{2}(2S_0 + \Delta S)T} = \frac{\Delta S \cdot \Delta t \cdot \frac{T}{n} n}{(2S_0 + \Delta S)T^2} = \frac{\Delta S}{(2S_0 + \Delta S)} \frac{\Delta t}{T} \quad (8)$$

This prediction has prominent implications for the accuracy of the Euler predictor step model. The underestimation of FIMA is dependent on time grid applied in numerical model. The bias is directly proportional to length of time step. Logarithm can be applied to both sides of Eq. (8) so as to provide the form more useful in comparative analyses:

$$\log(\Delta FIMA) = \log(C) + \log(\Delta t) \quad (9)$$

where: $C = FIMA_{EOL} \cdot \frac{\Delta S/T}{2S_0 + \Delta S}$. Under such formulation the analyzed error should vary linearly with logarithm of applied time step. This model will be verified in the following sections.

More advanced coupling schemes such as predictor-corrector variants (described by [6]) takes into account the variation of neutron source intensity and reaction rates. The simplest solution is application of an average values from the beginning and end of each step as an effective input for transmutation coefficients. In such a way the error originating from variation of S should disappear. Simultaneously this approach requires at least one more

Monte Carlo neutron transport run at each time step, which is a computational drawback.

The bias of FIMA described in this section is only one among several sources of systematic errors possibly present in burnup simulations. More examples of numerical and modeling difficulties and corresponding solutions are presented in the papers by [7] and [8].

4 Test model of PWR assembly

In order to demonstrate the presence of the problem explained in previous section, the model of typical fuel assembly from Pressurized Water Reactor (PWR) has been prepared. It consists of 17x17 lattice of fuel pins among which 25 control rod guide tubes are localized. Uranium dioxide is surrounded by zirconium cladding and light water coolant. The section of this 2-dimensional model is shown in Fig. 2 below.

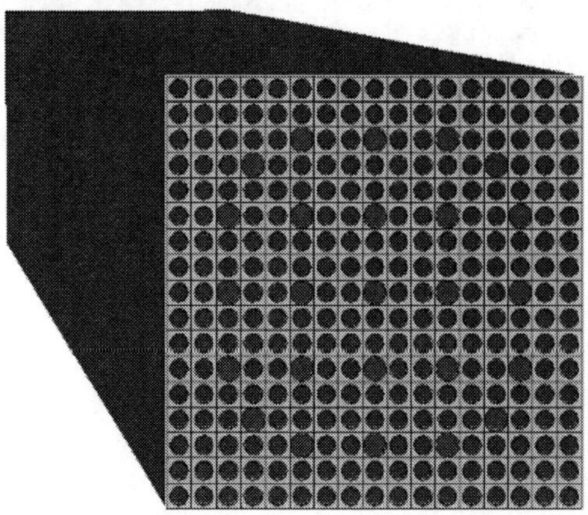

Figure 2. Horizontal section of geometry applied in model. Fuel pins are navy blue, guide tubes are red and coolant/moderator is yellow.

The geometry and compositions applied in this model are presented in detail in the article by [7] and also very similar model has been applied in the study by [6]. Both descriptions are sufficient for reproducing our results and the model is representative for commercially applied PWR reactors. The only modification applied in the current research is the enrichment of uranium, which has been increased up to 5%.

The burnup studies have been performed using the SERPENT code version 1.18, which belongs to family of continuous energy Monte Carlo burnup codes. Bateman equation solver has been set to CRAM method and neutron transport precision has been set to 40 pcm of standard deviation on k_{eff}. Cross-section libraries applied to neutron transport is JEFF3.1 and the rest of nuclear data (decay constants, fission yields, branching ratios) have been provided by the code author [9]. Power in the system has been normalized to $14MW_{th}$ per assembly and various number of time steps during irradiation period of 2000 days has been analyzed. In each case we compared staircase and predictor-corrector time step models.

5 Computational results

In this chapter we present computational results concerning the output FIMA in numerical model. Firstly, the fuel burnup at the End-of-Irradiation for considered step models and lengths of step (between 5 and 2000 days) is shown in Fig. 3 below.

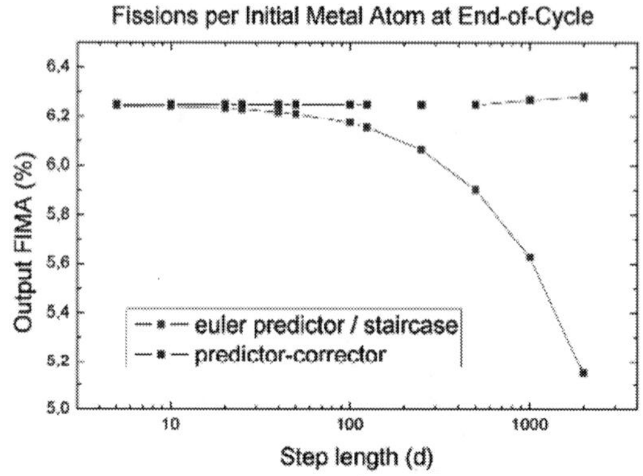

Figure 3. Nuclear fuel FIMA at the End-of-Irradiation (2000 days) for various time discretization and coupling schemes.

It is clearly visible that both the applied step length and the coupling scheme have systematic influence on the output burnup of fuel. We analyze here FIMA indicator only, however its discrepancies indicate differences for all the isotopic concentration separately as well as. For staircase model the bias quickly increases with time step length reaching 1% in the extreme case. On the other case, predictor-corrector model gives roughly same FIMA regardless from step length applied.

Next, the most exact result (predictor-corrector, 5days step) is treated as the reference solution and the absolute discrepancies from its FIMA are studied. The observed differences are presented in Fig. 4.

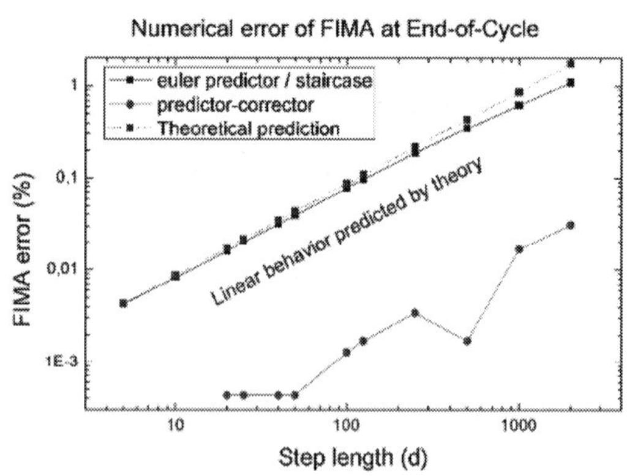

Figure 4. Convergence of the output FIMA values for two considered step models.

The error of results obtained using staircase model varies linearly as predicted in Section 3. In case of predictor-corrector the numerical bias is always over one order of magnitude lower. For smaller lengths of steps we observe saturated behavior of error due to the limited numerical precision of variables in output files.

Finally, the evolution of the neutron source intensity and the heating per source neutron (obtained with identical model with MCB code) is exposed in Fig. 5.

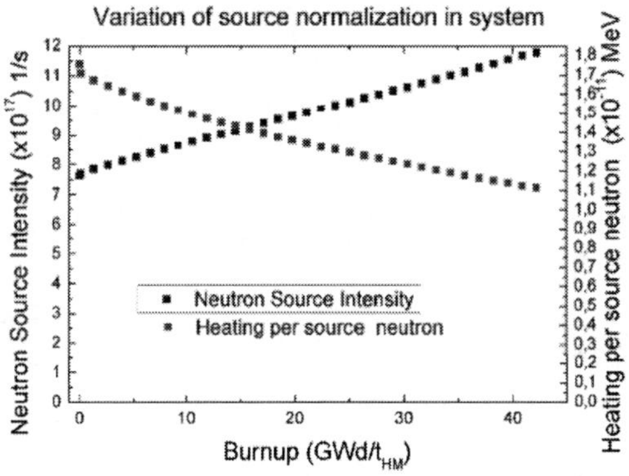

Figure 5. Variation of heating per source neutron during irradiation and resulting change of neutron source intensity.

We deduce from the figure that an assumption about linear evolution of S is reasonable and its variation is non-negligible in practical applications. Obtained linear trend can be applied to compare theoretical prediction (Eq. 8) and the value of error from numerical tests. We performed calculation for the case of 2000-day time step. Analytical approach gives the bias value of -1.7% while the numerical test shows FIMA underestimation of about -1.1%. The values are not exactly equal as the other sources of error exist in the numerical model and can partially compensate each other. In this case we attribute the difference to spectral changes of neutron flux, which influence reaction rates in the system and partly compensate bias caused by the variation of S. Observed bias is meaningful in comparison to the FIMA at the end of cycle (~6.24%).

6 Conclusions and summary

The problem of numerical underestimation of Fissions per Initial Metal Atoms in burnup codes has been studied theoretically and numerically. Analytical considerations indicate that step model assuming beginning-of-step coefficients gives systematic error in predicted fuel burnup. The bias value is directly proportional to the length of time step of numerical model. The problem originates from the variation of heating per source neutron in a system, which modifies transmutation coefficients present in the Bateman equations.

Numerical tests performed with SERPENT code confirmed presence of the problem for the most basic step model. The second applied coupling scheme – predictor-corrector – provided more reliable estimation of fuel burnup for any step length. We conclude that advanced step models eliminate analyzed bias in first approximation. Such solution can be recommended for all upcoming burnup simulations based on computationally expensive Monte Carlo neutron transport method. Usage of predictor-corrector coupling scheme provides more exact results than simple reduction of time step length while decreasing or maintaining overall computational cost.

Analyzed type of error is not specific to Monte Carlo burnup methodology, but can be present also in results of deterministic codes. It belongs to systematic errors present in modeling of nuclear systems and awareness about it inevitable to provide reliable results. Fortunately, this type of bias can be effectively excluded/diminished in the codes currently available for nuclear reactor engineers. The neglect of considered bias can arise following problems:
1) overestimation of energy extracted from the fuel,
2) underestimation of the heavy metal consumption,
3) propagated errors of burnup-dependent temperature coefficients (in case of safety assessments).

The results and analysis are meaningful basically for all types of calculation performed in research and development of nuclear fuel cycle including: validation of codes using experimental measurements, code-to-code comparison, benchmark problems, comparison of various fuel cycles scenarios, nuclear wastes radioactivity and transmutation assessments. In fact, all types of safety assessments account for burnup feedback. Reliable calculations in this field are fundamental for upcoming concepts of Gen-III+ and Gen-IV reactors.

Acknowledgements

This research was supported in part by PL-Grid Infrastructure. Partial financial support of this study through the funds of the Polish Ministry of Science and Higher Education is kindly acknowledged.

References

1. M. Oettingen et al. (2012) Nucl. Eng. Des. 242, 399–412;
2. J. Leppänen et al. (2015) Annals of Nuclear Energy, Volume 82, August 2015, Pages 142-150;
3. A. Isotalo, (2013) Aalto University Publication Series, Aalto University Publication, Helsinky, Finland, ISBN 978-952-60-5397-4;
4. A.E. Isotalo, P.A.Aarnio (2011) Ann. Nucl. Energy 38, 261–268;
5. G. Kępisty, J. Cetnar (2015) Nucl. Eng. Des. 286, 49–59;.
6. D. Kotlyar, E. Shwageraus (2013) Annals of Nuclear Energy, Volume 58, August 2013, Pages 228–237;
7. G. Kępisty, J. Cetnar, P. Stanisz (2016) Annals of Nuclear Energy 94 (2016) 189–207;
8. J. Cetnar (2014) Proceedings of the HTR 2014, Weihai, China, October 27-31, Paper HTR2014-51291;
9. J.Leppänen (2015) http://montecarlo.vtt.fi/download/Serpent_manual.pdf (15.04.2016);

Carbon dioxide emission in hydrogen production technology from coke oven gas with life cycle approach

Piotr Burmistrz[1], Leszek Czepirski[1] and Magdalena Gazda-Grzywacz[1,a]

[1]*AGH University of Science and Technology, Faculty of Energy and Fuels, al. Adama Mickiewicza 30, 30-059 Cracow, Poland*

Abstract. The analysis of Carbon Footprint (CF) for technology of hydrogen production from cleaned coke oven gas was performed. On the basis of real data and simulation calculations of the production process of hydrogen from coke gas, emission indicators of carbon dioxide (CF) were calculated. These indicators are associated with net production of electricity and thermal energy and direct emission of carbon dioxide throughout a whole product life cycle. Product life cycle includes: coal extraction and its transportation to a coking plant, the process of coking coal, purification and reforming of coke oven gas, carbon capture and storage. The values were related to 1 Mg of coking blend and to 1 Mg of the hydrogen produced. The calculation is based on the configuration of hydrogen production from coke oven gas for coking technology available on a commercial scale that uses a technology of coke dry quenching (CDQ). The calculations were made using ChemCAD v.6.0.2 simulator for a steady state of technological process. The analysis of carbon footprint was conducted in accordance with the Life Cycle Assessment (LCA).

1 Introduction

Coal is one of the most important primary energy sources and, in contrast to oil and natural gas, its resources are much larger and located in all geographical regions [1-2]. World coal resources, estimated by EURACOAL [3] account 726 mld tones, while according to WEC - 826 mld tones [4]. Coal is not only used as a fuel to generate heat and electricity, but also as a raw material in pyrolysis process, gasification and for direct hydrogenation of coal.

Nowadays, excluding electric power industry, the pyrolysis process is the most popular way of coal processing. The raw material is heated without the access of oxygen to about 1300 K and the main products are: coke, coke oven gas, tar, benzol and depending on the gas purification technology: ammonium sulfate, sulfur or sulfuric acid. Currently, about 80% of the global coke production is located in Asia, including nearly 70% in China. CRUs forecasts [5] confirm, that a similar proportions in the sector will sustain in the coming years. According to CRU - world coke production will rise and the estimated production in 2017 will reach about 793 million tones [5, 6-8]. Coking plants creates a threat to the environment [9-11]. It is worth noting that coking process gives the possibility of obtaining hydrogen from purified coke oven gas - which contains more than 50 vol.% H_2 [12, 13].

Hydrogen is not present in large quantities and concentrations on Earth [14]. It is mainly obtained from biomass, water and fossil fuels due of their wide availability [15]. Interest in hydrogen as a fuel [16, 17] due to the fact that during reaction with oxygen a significant amount of energy is being released - 143.1 MJ/kg and the only reaction product is water. Hydrogen being used in fuel cells allows direct conversion of chemical energy in the reaction with oxygen into electricity. It is also a base material of many chemical syntheses, in which many substances, both organic and inorganic may be produced [18]. Global hydrogen production comes mainly from fossil fuels processing without CCS (*Carbon Capture and Storage*) technologies: 48% from natural gas, 30% from the refinery off-gases, 18% from coal and the rest comes from electrolisys and biomass [19-24].

Environmental footprints are widely use to assess the impact of technology on the environment. Scientists have developed and implemented a lot of indicators showing the impact of human activities on the environment, for example: energy footprint [25], water footprint [26], nitrogen footprint [27] and carbon footprint [28]. Carbon footprint (CF) is the most objective method of assessing the technology impact on the environment in the aspect of the greenhouse effect [29, 30]. CF is a measure of the total amount of Greenhouse Gas emission (GHG - greenhouse gas), both direct and indirect, throughout the life cycle of a product or technology. The methodology of analysis and calculation of the CF is based on Life Cycle Assessment (LCA), a research into potential aspects of environmental impact in the whole life cycle that is, on each of its stages – from obtaining raw materials,

[a] Corresponding author: magdago@agh.edu.pl

through production, until the utilization and/or recycling of the main product [31-36].

The aim of this study was to analyze the CF of hydrogen production from purified coke oven gas. Calculations were made and an evaluation of indicators of direct and indirect CO_2 emissions resulting from the cycle of obtaining hydrogen were performed, including the following steps: coal mining, transport of coal to coking plants, coking process and hydrogen production from purified gas, and operations of CO_2 capture and sequestration. For this calculations methodology of LCA described in ISO 14040:2009 was used [37].

2 General information, data and assumptions

An analysis of CF for the technology of hydrogen production from coke oven gas derived from coking of coal blend with the characteristics given in Table 1 was performed. CF of the hydrogen production was calculated in accordance with standard ISO 14067:2014 [38] and included both direct and indirect emissions associated with electricity and heat consumption during coal mining, coal transport from mine to coking plant, coal blend preparation, coking process, cooling and cleaning gas and its reforming.

Table 1. Characteristics of coal blend

Mark	As received	Dry basis
C [%]	77.4	84.3
H [%]	4.2	4.6
N [%]	1.2	1.3
S_t [%]	0.5	0.5
O [%]	2.4	2.6
W_t [%]	8.2	-
A [%]	6.1	6.6

2.1 Coal mining processes

Investigated coking coal blend (Tab. 1) is composed of several coking coals obtained from mines with high absolute methane content. Therefore, the assumption was made, that 60% of methane is captured and used to produce electricity and heat, while the rest present in the ventilation air (methane concentration below 2%) is catalytically combusted. To estimate the values of CO_2 emission in the production of coal was assumed that: to produce 1 Mg of coal a mine consumes on average 68.02 kWh of electric energy and 43.98 MJ of heat (by commercially available technology),

- the results of calculations show that in the case of the ventilation air containing 0,5% of methane, it is necessary to provide the 486 kJ heat on 1 m^3 CH_4 for the catalyticall combustion process
- emission of CO_2 from burning methane at the power plant is 1.9643 kg CO_2/m^3 and 1.9951 kg CO_2/m^3 from catalytic combustion,
- the rate of electricity consumption to produce compressed air is 0.0833 kWh/m^3,
- in Polish conditions production of 1 kWh of electricity generates CO_2 emission equal to 0.9124 kg, while producing 1 GJ of heat energy generates 63.4600 kg CO_2 emission.

Based on the above the calculated CF from the combustion of methane is 20.6027 kg CO_2 per 1 Mg of mined coal, while the catalytic combustions generates CF value of 33.2449 kg CO_2/Mg_{coal}. As a result of methane combustion 28.4317 kWh of electricity, 0.0701 GJ of heat and 5.1 m^3 of compressed air per 1 Mg of coal is obtained. This allows to save 0.4248 kWh/Mg. Thus, coal mining under the precise conditions present in the mine consumes net 39.1590 kWh / Mg_{coal} electricity and -0.0261 GJ /Mg heat. A negative value means, that the amount of heat produced by the power plant (at the mine) is greater than mine requires.

2.2 Coal transporting to coking plant

Based on the data obtained from the company transporting coal from the mine to the coking plant a rate of electricity consumption during by rail transportation (0.0789 kWh/Mg·km) was assumed. This generates indirect emission of carbon dioxide associated with the transport of raw materials 0.072 kg CO_2/Mg·km (0.003 kg CO_2/(GJ·km)). It was assumed that coking plant is located 100 km from the mine.

2.3 Coking process

The coking process is carried out in the chambers of the coke oven battery system with a volume of 30 m^3 useful chamber. The battery is heated exclusively by own coke gas. After completion of the coking process, the basic product - coke at a temperature of 1000°C, is pushed out of the chamber, cooled, sorted and transported to customers. Raw coke oven gas is purified by ammonia absorption in the scrubber, where the reduction of H_2S and NH_3 to a level below 0.5 g/m^3 and 0.03 g/m^3 is obtained, respectively. The composition of raw coke oven gas is given in Table 2. The mixture of steam and gas from absorption system is moved to catalytic decomposition of NH_3 and sulfur production by Claus process. In the next step, benzol is absorbed from purified gas by ammonia method. The purified gas is divided into two equal streams, one of them is used for own coking plant requirements and the other is used for the hydrogen production.

Energy supply for preparation and coking coal departments were estimated based on the information on technological solutions in coking plants available on a commercial scale, specifically these using technology of coke dry quenching. During coke dry quenching a large part of the physical enthalpy of the hot coke is recovered and it is used to produce steam and electricity for coking plants own consumption. Summary consumption of energy factors for preparing and coking processes and coal derivatives department for 1 Mg of dry coal are given in Table 3.

Table 2. Composition of raw coke oven gas

Component	m/m%	v/v%
H_2	6.0	41.7
CO	4.6	2.3
CH_4	22.6	19.7
CO_2	3.1	1.0
C_2H_6	5.6	2.6
H_2S	1.4	0.6
NH_3	2.3	1.9
H_2O	36.2	28.0
Benzol	4.3	0.8
Tar	14.0	1.5

Table 3. Usage of energy and coke oven gas for coking plant

Energy carrier	Expenditure	Income	Difference
Electricity [kWh/Mg_{ws}]	74.07	39.63	-34.44
Heat [GJ/Mg_{ws}]	1.57	1.32	-0.25
Coke oven gas [MJ/Mg_{ws}]	2687	6560	+3873

2.4 Hydrogen production

After benzol absorption step, gas stream is being subjected to the reforming technology, which comprises the following process units: coke oven gas reforming system, oxygen production installation, CO conversion system and the hydrolysis of COS, desulphurization system and CO_2 removal system, H_2 separation system and other plant. The calculations for hydrogen production were made using ChemCAD v.6.0.2 simulator for a technological process in a steady state. Calculations were made for the following assumptions:

- cleaned coke oven gas was used in reforming system at 25°C and saturated with water vapor and tar,
- battery heating uses 50% of the coke oven gas mixed with purified residual gas,
- separated in the Selexol system compounds of sulfur are moved to Claus installation, belonging to the technological system of coking plant

Table 4 shows the mass balance for hydrogen production process from purified coke oven gas. Table 5 shows the consumption of electricity according to installation requirements. Table 6 presents efficiency of obtaining hydrogen from purified coke oven gas, which was calculated as the ratio of chemical enthalpy of products and the enthalpy of chemical raw materials, i.e. excess gas and tar.

The separation of CO_2 generated during the hydrogen production from purified coke oven gas was included. As a result carbon dioxide is being produced as a byproduct, in an amount of 3 835 Mg/h (28 555 Mg/year), with a degree of purity of over 99%. It can be a commercial product or a storage object. In the case of sequestration of carbon dioxide separated during hydrogen production from coke oven gas, compression, transportation and congestion in selected places of geological storage are required.

Table 4. Mass balance for hydrogen production plant

Stream name	Expenditure/production				
	[Mg/24 h]	[Mg/year]	Purified coke oven gas [kg/kg gas]	Gas for reformer[kg/kg gas]	[kg/kg H_2]
In (entrence)					
1. Coke oven gas	47	14 706	1.00	1.04	4.30
2. Oxygen	36	11 099	0.75	0.79	3.24
3. Tar	15	4 579	0.31	0.32	1.34
4.Water supplement	44	13 775	0.94	0.97	4.03
5. Steam	40	12 360	0.84	0,87	3.61
Summary	**182**	**56 520**	**3,84**	**4,00**	**16,52**
Out (exit)					
1. Hydrogen	11	3 425	0.23	0.24	1.00
2. CO_2	92	28 555	1.94	2.02	8.34
3. Acid gas	2	663	0.05	0.05	0.19
4. Gas for battery	31	9 553	0.65	0.68	2.79
5. Waste	45	13 939	0.95	0.99	4.07
6. Steam production	1	387	0.03	0.03	0.11
Summary	**182**	**56 523**	**3.84**	**4.00**	**16.52**

Assumptions concerning the conditions for the transport of carbon dioxide and its injection in geological deposits are:

- carbon dioxide will be transported in the liquid phase,
- the pressure at the inlet to a transport system (pipe) is 120 bar, which allows to transport of CO_2 at about 100 - 150 km distance and the congestion it in geological structures without additional compression.

The values of electric energy requirements for CO_2 compression system for transport and storage, and corresponding carbon dioxide emissions are shown in Table 7. Table 8 shows the CF associated with net production of electricity and heat for hydrogen production from purified coke oven gas.

Based on the collected data (Tables 3-5) from process simulator (ChemCAD) production indicators and consumption of electricity and heat throughout the life cycle of the technology of hydrogen production from purified coke oven gas at all process stages were calculated. The values of these indicators are referenced to the unit mass of coking coal and the unit mass of produced hydrogen.

Table 5. Electric energy usage for hydrogen production plant

Department	Installed power [kWe]	Expediture of energy		
		kWe/kg gas	kWe/kg H_2	GWh
1. Oxygen	564	0.29	1.23	4.20
2. Gas reformer (power pumps)	17	0.01	0.04	0.10
3. Gas pressing	147	0.08	0.32	1.10
4. Desulphurization	-	-	-	-
5. Gas cooling system	12	0.01	0.03	0.1
6. Gas compresion	1 177	0.62	2.56	8.80
7. Removal of acid gas - Selexol	213	0.11	0.46	1.60
8. Claus/Scot-installation	-	-	-	-
Summary	**2 130**	**1.12**	**4.64**	**15.9**

Table 6. Process efficiency

Process efficiency	Wd
Gas after reformer	90.24 %
Gas PSA	85.02 %
Hydrogen	71.28%

Table 7. Electric energy requirement for the CO_2 compression step and related emissions.

Parameter	Value
CO_2 stream for sequestration, Mg per year	28 559
CO_2 stream for sequestration, kg CO_2/Mg_{coal}	227.24
CO_2 stream for sequestration, kg CO_2/Mg H_2	8 346.03
Compression electricity power, MWe	0.32
Energy consumption for compressing, kWh/Mg_{coal}	19.13
Energy consumption for compressing, kWh/Mg H_2	702.71
CO_2 emission indicator, kg CO_2/Mg_{coal}	**17.46**
CO_2 emission indicator, kg CO_2/Mg H_2	**641.16**

Based on the above-mentioned indicators, CF associated with the consumption of heat and electricity in the coking process and coke oven gas reforming in the hydrogen production were calculated. Coke oven gas is only one of the products of the coking process therefore, to the individual coking products used to produce hydrogen, carbon dioxide emissions in proportion to their contribution to the total amount of energy in all products of coking has been attributed.

Based on the above-mentioned information about distribution of energy in the coking products and basic balance data of the major process streams (ChemCAD), share of CO_2 emission from the coking process connected to raw materials supplied to reformer have a value of 0.1465 for hydrogen production from purified coke oven gas.

Table 8. CF related to the production of net electricity and heat for the hydrogen production from purified coke oven gas.

H_2 production	CO_2 emission indicators related with net production of electricity	
	Electricity	
	kg CO_2/Mg_{coal}	kg CO_2/Mg $_{H2}$
From purified coke oven gas.	119.36	4 383.84
Including: Coking process, H_2 production	4.23 115.13	155.19 4 228.66
	Heat	
	kg CO_2/Mg_{coal}	kg CO_2/Mg H_2
From purified coke oven gas	14.00	513.79
Including: Coking process, H_2 production	4.10 9.91	149.72 364.10
	Summary	
	kg CO_2/Mg_{coal}	kg CO_2/Mg H_2
From purified coke oven gas.	133.35	4 897.63
Including: Coking process, H_2 production	8.30 125.04	304.91 4 592.72

3 Conclusions

Hydrogen production technology from purified coke oven gas was subjected by CF analysis.

According to the LCA methodology including: coal mining, transport of coal to coking plants by railway, coking process with auxiliary operations, obtaining hydrogen from purified coke oven gas and sequestration of captured CO_2.

Hydrogen yield for considered technology was: 0.1859 Mg H_2 per 1 Mg of coal and 0.0076 Mg H_2 per 1 GJ net calorific value contained in coal. Conversion efficiency of chemical enthalpy is equal to 71.3%.

Carbon Footprint counted for 1 Mg of dry coal was 1 106.3 kg CO_2 and consists of emissions from the production of raw materials (14.0 kg CO_2/Mg_{coal}), transport of coal (1.1 kgCO_2/ Mg_{coal}), net energy consumption (938.9 kg CO_2/ Mg_{coal}), CO_2 indirect emission (22.5 kg CO_2/Mg_{coal}), CO_2 captured for sequestration (1689.8 kg CO_2/Mg_{coal}) and energy consumption for sequestration (129.8 kg CO_2/ Mg_{coal}).

CF counted in reference to 1 GJ of chemical entalphy contained in raw coal is 41.6 kg CO_2/ GJ, it consists of emissions from the production of raw materials (0,5 kg CO_2/ GJ), coal transportation (0.0 kg CO_2/ GJ), net energy consumption (35.3 kg CO_2/ GJ), CO_2 indirect emission (0.8 kg CO_2/ GJ), CO_2 captured for sequestration (63.5 kg CO_2/ GJ) and energy consumption for sequestration (4.9 kg CO_2/ GJ).

Value of CF in reference to 1 Mg of produced hydrogen was 5 463.7 kg CO_2. This contributes to the emissions resulting from the production of raw materials (69.3 kg CO_2/ Mg H_2), coal transportation (5.7 kg CO_2/ Mg H_2), net energy consumption (4637.4 kg CO_2/ Mg H_2), CO_2 indirect emission (110.2 kg CO_2/ Mg H_2), CO_2 captured for sequestration (8 346.0 kg CO_2/ Mg H_2) and energy consumption for sequestration (641.2 kg CO_2/ Mg H_2).

Acknowledgments

Financial support for this study was provided by the AGH University of Science and Technology (framework no. 11.11.210.244)

References

1. Massachusetts Institute of Technology, *MIT 2007: The Future of Coal*
2. H.H. Schobert, C. Song, Fuel **81**, 16 (2002)
3. *Coal industry across Europe 2008* (Eurocoal, Bruksela, 2008)
4. *Survey of Energy Resources. Interim Update* (World Energy Council, Londyn, 2009)
5. www.crugroup.com
6. J. Mielnikiewicz, M. Ściążko, R. Cieślar, Karbo - Wydanie Specjalne, 4-12 (2007)
7. Coke Market Report, Analysis of the Global Coke & Coking Coal Markets (Resource-Net, Brussels, Belgium, 2007)
8. Coke Market Survey (Resource-Net, Brussels, Belgium, 2007)
9. J. L. Wang, X. C. Quan, L. B. Wu, Y. Qian, W. Hegemann, Process Biochemistry **38**, 777–781 (2002)
10. E.E. Chang, H.J. Hsing, P.C. Chiang, M.Y. Chen, J.Y. Shyng, Journal of Hazardous Materials **156** , 560–567(2008)
11. W.H. Zhang, C.H. Wei, C.H. Feng, B. Yan, N. Li, P.A. Peng, J.M. Fu, Science of the Total Environment **432**, 396–403 (2012)
12. A.Karcz, A. Sobolewski, 2009, Karbo **2**, 112-120 (2009)
13. Q. Sun , J. Dong , X. Guo , A. Liu, J. Zhang, Ind. Eng. Chem. Res. 51 (17), 6205–6211 (2012)
14. G. Stiegel, M. Ramezan, International Journal of Coal Geology **65**, 173– 190 (2006)
15. J.A. Turner, Science **305**, 972-974 (2004)
16. B. Behera, R. Balasundaram, K. Gadgil, D. Sharma, Energy Sources **Part A 29**, 761–7 (2007)
17. K.L. Kovács, G. Maroti, G. Rákhely, International Journal of Hydrogen Energy **31**, 1460–8 (2006)
18. T.A. Moore, S. Pearce, International Journal of Coal Geology **65**, 171–172 (2006)
19. IEA Energy Technology Essentials, Hydrogen Production & Distribution, Apr. 05 (2007)
20. T. Wiltowski, K. Mondal, A. Campen, D. Dasgupta, A. Konieczny, Int. J. Hydrogen Energy **33**, 293 (2008)
21. A. J. Minchener, Fuel **84**, 2222–2233 (2005)
22. Y. Wu, J. Wang, S. Wu, S. Huang, J. Gao, Fuel Processing Technology **92**, 523-530 (2011)
23. T. Take, K. Tsurutani, M. Umeda, Journal of Power Sources **164**, 9–16 (2007)
24. L. Shen, Y. Gao, J. Xiao, BIOMASS AND BIOENERGY **32**, 120-127 (2008)
25. M. Wackernagel, W.E. Rees, New Society, (Gabriola Island, British Columbia, 1996)
26. A.Y. Hoekstra, P.Q. Hung, Value of Water Research Report Series 11 (UNESCO-IHE Institute for Water Educa-tion, 2002)
27. A.M. Leach, J.N. Galloway, A. Bleeker, J.W. Erisman, R. Kohn, J. Kitzes, Environ. Dev. **1**, 40–66 (2012)
28. T. Wiedmann, J. Minx, Nova Science Publishers 1, 1-11 (Hauppauge, NY,USA, 2008)
29. J. Kitzes, M. Wackernagel, Ecological indicators **9**, 812–817 (2009)
30. D. Yajie, X. Beicheng, C. Weidong, Environmental Science & Policy Volume **44**, 181–189 (2014)
31. A. Galli, T. Wiedmannb, E. Ercinc, D. Knoblauchd, B. Ewinge, S. Giljumf, Ecological Indicators **16**, 100–112 (2012)
32. G.R. Cranstona, G.P. Hammonda, Ecological Indicators **16**, 91–99 (2012)
33. K. Fang, R. Heijungs, G. R. de Snoo, Ecological Indicators **36**, 508– 518 (2014)
34. D. Burchart-Korol, K. Czaplicka-Kolarz, A. Fugiel, A. Rejman-Burzyńska, Przemysł Chemiczny **94** (9) , 1584-1589 (2015)
35. K. Czaplicka-Kolarz, A. Fugiel, D. Burchart-Korol, Polityka Energetyczna **17** (4), 263-275 (2014)
36. P. Krawczyk, J. Krzemień, D. Burchart-Korol, K. Czaplicka-Kolarz, Przemysł Chemiczny **93** (12) , 2232-2236 (2014)
37. PN-EN ISO 14040:2009
38. ISO/TS 14067:2014

Gasification and pyrolysis of different biomasses in lab scale system: A comparative study

W. Gądek[1a], A. Mlonka-Mędrala[2], M. Prestipino[3], P. Evangelopoulos[4], S. Kalisz[1], W. Yang[4]

[1]*Silesian University of Technology, Institute of Power Engineering and Turbomachinery, 44100 Gliwice Konarskiego 20, Poland*
[2]*AGH University of Science and Technology, Faculty of Energy and Fuels.30059 Krakow Mickiewicza 30, Poland*
[3]*University of Messina, Department of Engineering. C.da Di Dio, 98166, Messina, Italy*
[4]*Royal Institute of Technology, Department of Material Science and Engineering, 100 44 Stockholm Brinellvägen 23, Sweden*

Abstract. Gasification and pyrolysis are very promising technologies for clean energy production especially from low rank fuels. Biomass and wastes with high chlorine, alkali and even heavy metals content are fuels preferential for thermal utilization. However, several problems during combustion in conventional steam boilers occurs e.g. slagging, fouling, chlorine corrosion, boiler efficiency deterioration. New efficient and cost effective technologies are needed, even in small-scale applications. The main objective of this work was to compare the thermochemical behaviour and process parameters effects of different biomass under air gasification and pyrolysis conditions. Three important fuels for European power industry were selected: woody biomass and two residual biomass, such as oat straw and dried citrus wastes. In order to evaluate the possibility to use different feedstocks or to combine and/or integrate them in thermochemical processes, a comparison among typical and untypical feedstocks is needed. Tests performed on small scale fixed bed reactor show the gas yield, its composition and LHV parameter. The results were performed in Royal Institute of Technology (KTH) in Sweden during BRISK program (Biofuels Research Infrastructure for Sharing Knowledge).

1 Introduction

Nowadays renewable sources of energy are of great importance. This is strongly associated both with the climate change and the current policies of EU. According to the "2030 Climate and Energy Framework", which extends the existing EU climate policy, the main goal is to reach at least 40% reduction of greenhouse gasses by 2030 comparing with the values of1990 [1]. Furthermore, another goal set until the 2030 is to increase the share of renewable sources for energy production to 27% [1]. EU regulations can strongly influence the research community for targeting more efficient and more environmental friendly energy production. Biomass has the largest share in the renewables market and great potential for clean energy production [2]. Biomass is considered to be a "carbon neutral" fuel, considering that the emitted CO_2 during its combustion will be reabsorbed during the trees growth, which may be an answer for meeting the EU's future goals and challenges. Even though, a lot of effort has been made for implementing biomass as an alternative source of fuel, the EU has not conclude to an efficient energy production from a specific source of biomass. Therefore, more research is needed in order to accomplish a feasible energy production from biomass.

1.1 Biomass utilisation (Poland, Sweden, Italy)

Poland has a large potential in case of biomass utilization. During the year 2012, the energy produced from biomass represents 82.2% of Poland's total renewable energy production and 9.6% of the national energy production which accounts for 81 TWh [3]. Biomass is mainly utilized by combustion and co-combustion in Mega-plants such as biomass co-combustion Rybnik (PF Pulverised Fuel combustion 1775 MW_{el} – 11.05% biomass usage by weight) and Połaniec biomass power plant (CFB Circulated Fluidized Bed 205 MW_{el}, fuel: mix of forestry biomass, seeds, sunflower husk and straw) [4,5]. The forestry (mainly waste from wood industry), agriculture (straw mixture, organic waste), waste biomass are main source of biomass in Poland [3].

Biomass in Sweden is mainly represented by forestry biomass, agriculture, waste biomass from paper and pulp industry, sawmills which generates huge quantities of bio-wastes [6]. In 2011 total energy production from biomass and wastes reaches 129 TWh which corresponds 11.1% of total national energy production [7]. For consideration deserve Swedish biomass gasification power plants such as: GoBiGas/Gothenburg indirect gasification 32 MW_{th} (the target is to reach 100 MW_{th}) fuel: solid biofuels, forestry wastes; Värö by Götaverken/Värö fluidised bed gasification 28 MW_{th} fuel: dried bark and wood wastes [8].

[a] Corresponding author: waldemar.gadek@polsl.pl

Biomass gasification market in Sweden is growing rapidly and has a significant potential [8].

Italy has large biomass resources mainly straw biomass, olive and grape residues, pruning's and forestry biomass. In 2010 energy production from biomass reaches the 73.3TWh which corresponds to 20.6% of total national energy production [9]. Total thermal power installed in biomass power plant amounts 1,243 MW [10]. Considering the different distribution of forestry and the different climate situation in Italy, the district heat production from solid biomass is mainly concentrated in the northern regions [11]. Furthermore, since residual biomass are wide dispersed in the Italian territory, the national government has encouraged the installation of sub-megawatt plants for combined heat and power production (CHP). For these reasons, especially in the southern Italians regions, residual biomass from agro-food industries have a greatunexploited potential.

1.2 Alternative technologies to combustion

According to climate change the new alternative technologies for energy production are essential. It is associated with reduce of carbon dioxide emission. Combustion of fossil fuels (coal, oil, natural gas) increase the greenhouse effect [12]. The first easiest idea to reduce the emission of CO_2 was to use biomass instead of coal. However, combustion and co-combustion of biomass mainly waste biomass or agro-biomass is associated with biomass feeding, pre-treatment problems, fouling, corrosion, modification of boiler operation and flue gas cleaning section, modification of heat transfer in the boiler [13]. These serious challenges contributed to the research for other technologies of biomass utilisation. Finally the application of pyrolysis and gasification process might be promising alternative to combustion. The research to improve efficiency of gasification and pyrolysis process is essential.

1.3 Pyrolysis and gasification of biomass

Gasification is thermochemical process that enable to convert solids or liquids in a value added gas product (syngas). This process use an oxygenating agent, with an equivalence ratio usually between 0.2-0.4, depending on the process temperature (700-1200°C).

Pyrolysis is a process similar to gasification, but in comparison to other technologies of fuel thermal conversion, pyrolysis occurs at relatively low temperature range (300-600°C) and in the absence of an oxidizing agent.

The advantages of using pyrolysis and gasification systems instead of direct combustion of solid biomass are related to environmental aspects and to the possibility of polygeneration. Indeed, syngas produced from biomass gasification or pyrolysis can be used for combined synthesis of high-value chemicals synthesis and for energy production. Other products: liquid tars and solid char may be used as direct fuels or to make new products like ethanol from tars or fertilizers from char. From the environmental point of view, it has to pointed out that gasification plants produce a lower amount of major air pollutants, like SO_2, NO_x and particulates, than direct combustion [14]. Differences between chemical composition of biomass and coal affects not only gaseous phase but mineral phase as well. Fly ash from biomass combustion contains much more potassium, magnesium and calcium, simultaneously less iron, aluminum and silicon [15]. Biomass combustion may cause severe problems concerning fly ash agglomeration and corrosion of heat exchanging surfaces. Therefore, gasification and pyrolysis are better technologies for thermal conversion of biomass containing high amounts of chlorine and alkali metals.

In addition, gasification is preferable route for generation of electricity from small plants in remote areas, where gasification/internal combustion engine systems are more convenient than combustion/steam engine.

2 Methodology

2.1 Small scale fix bed reactor

Both gasification and pyrolysis experiments has been conducted on a semi batch type fix bed reactor located at the Royal Institute of Technology, Stockholm, Sweden. The reactor has a cylindrical shape with 0.0254 m inner diameter and 1 m total length, while inside the reactor, a fine stainless steel grid has been placed on the middle of the reactor's length. For every experiment, 10 grams of biomass has been placed on that grid in order to make sure that the heated N_2 in case of pyrolysis and the air in case of gasification experiments was passing through the reacted material. High purity nitrogen has been used as inert gas before every experiment in order to purge the whole reactor from the remainings of O_2. The reactor was heated with a 1 kW cylindrical electric furnace to the desired temperature, while a K thermocouple was attached to the outer wall of the reactor, close to the position of the grid (0,5m from the top of the reactor) for controlling the temperature of the furnace.

Pyrolysis experiments

The pyrolysis experiments were conducted for the temperatures of 400°C, 500°C and 600°C for three examined fractions of biomass. During each experiment, the reactor with the grid and the sample was placed inside furnace which initially was at room temperature. A stable flow of 50 ml/min of N_2 was introduced from the top of the reactor for sufficient time before the heating up in order to make sure that there was no oxygen inside the reactor before every experiment. Additionally, the facility was leak tested in order to make sure that no leaks existed on the different connections. During the heating up of the furnace, from the 100°C until the desired temperature, the gas produced plus the additional nitrogen that was added on the system was collected on the first container. When the furnace has reached to the desired temperature, the produced gas was collected on the second gas container. Both containers were initially filled up with water, so during the gas collection, the volume of gas collected was measured according to the water displaced from the containers. Each

DOI: 10.1051/e3sconf/20161000024

sample of biomass was kept at the desired temperature for 15 minutes in total. On the bottom of the reactor, before the gas collecting containers, three empty gas washing bottles was connected and placed on a cooling bath of 0°C for condensing the oils and the gases produced. Finally, the gas composition collected on the containers was analyzed.

Gasification experiments

The gasification experiments were conducted for the temperatures of 700°C and 800°C for the three examined biomass fractions. The reactor, the grid and the sample was again placed on the same position inside the furnace while nitrogen was used to purge the remaining O_2 for sufficient time before the heating up. During the heating up, the nitrogen was switched to air with a stable flow according to the chosen λ for each biomass type. All the products of the gasification process were passing through two gas washing bottles placed on a cooling bath for condensation of the condensable fractions and finally they were passing through a gas washing bottle filled up with isopropanol in order to trap any tar remained. Finally, a portable gas chromatographer was sampling the gas produced every three minutes. The examined cases are summarized on the graph (Figure 5.), an average gas composition for temperature range from around 600 °C, when gas analysis showed first combustible gas components, up to desired temperature - 800 °C.

2.2 Micro GC gas analysis

The gas collected on a gas containers and the online gas composition measurement during the gasification process was analyzed with an Agilent 490 Micro gas chromatographer (GC). The micro-GC is equipped with four channels and each one of them contain different columns and a thermal conductivity detector. The first channel has a Molsieve 5Å column, which is able to detect the permanent gases (H_2, He, O_2, N_2, CH_4, CO), the second channel is equipped with a CP PoraPlotU column which mainly separates and analyze the CO_2, H_2S and C2 hydrocarbons, while the third channel has a CP PoraPlotU column, which can detect and analyze the C3 and C4 hydrocarbons. The final channel has a CP-Sil 5 CB column for detection of higher molecular weight hydrocarbons. Argon was used as a carrier gas on the first two columns while helium was used on the third and fourth column. The method of analysis used for the analysis of the gases was lasted for two minutes on stable temperature conditions while the total sampling time was chosen to be 30 seconds, for more accurate results.

2.3 Materials

Wood

Wood is widely used for energy production in Poland and Sweden. Particularly important is utilization of forestry waste, wood chips, and waste from wood and paper industry. Polish and Swedish furniture sector produces a large quantities of wooden bio-wastes which have to be utilized. For experiments wooden mixed waste from wood industry was chosen. They are characterized by high LHV 18,5 MJ/kg, low ash content 0,31 % compared to other investigated fuels. Particles were sieved in the order to obtain a homogenous mixture (0.7<d<1.4). The biomass properties comes within the scope of literature data [16].

Oat straw

Straw is a mixture of dried stalks and blades after the ripe crop threshing. Straw is mainly a byproduct from a plant production on farmlands. In Poland, despite a wide straw use in agriculture, a significant overproduction is observed, which might be used for power production purposes. It is assumed that due to variable straw resources, it would be possible to use approx. 30-50% of existing straw resources. According to previous assumptions, power industry can take advantage of 1.74 - 2.90 Mtonnes of dry straw pulp per year [17].

In the experiment, oat straw (d<1mm) was selected as an example of agricultural wastes used for power industry purposes. In comparison to other fuels selected for the experiment, oat straw has quite high ash, sulfur and chlorine content (Table 1). Therefore, it is a difficult fuel for direct combustion.

Citrus waste

Agricultural wastes in Italy are of great interest because of the lack of intensive forest presence in some area of the country, if compared with northern countries in Europe. Furthermore, Italy is well known worldwide for the food industry due to the large variety of typical agricultural activities. Citrus pulp and peels are residues from agro-industrial activities that are typical of the Mediterranean area, and it implies a significant cost for juice manufacturer, because of its fermentability and the high residue yield [18]. This waste was selected because to date, it is not enough exploited for energy production and it has a different structure and composition with respect to others well known lignocellulosic residues in the field of biomass gasification. In this work, dried citrus wastes were supplied by a company in Sicily (Italy), where the total amount of citrus wastes for the region is close to 33000 t/year (dry matter) [19].

Table 1. Proximate and elemental analysis biomass samples.

Parameter	Unit	Oat straw	Citrus wastes	Wood
Carbon content, C^{ar}	%	46,39	41	50,70
Hydrogen content, H^{ar}	%	5,83	5,6	6,10
Nitrogen content, N^{ar}	%	0,56	0,8	0,18
Sulfur content, S_t^{ar}	%	0,11	0,1	0,012
Chlorine content, Cl^{ar}	%	0,147	-	-
Ash content, A^{ar}	%	7,6	8,2	0,31

Moisture content, W^{ar}	%	4,83	7	8,47
HHV	kJ/kg	18 170	15 212	20 460
LHV	kJ/kg	16 970	14 082	18 572

3 Results and discussion

Gas pyrolysis yield ($\%_{wt/wt}$) is shown in Figure 1. For different biomasses and different temperatures (400, 500 and 600°C). Wood gas yield increased slightly from 400 to 500°C (3.24%) and it was almost constant from 500 to 600°C but in the same temperature range oat straw showed a marginal increase (3.9%). Citrus peel had the highest gas yield increase as the temperature increases, varying from 14.17 to 30.89% (from 400 to 600°C). The last was also the highest gas yield value among the investigated biomass. This behavior could be explained with the low concentration of lignin in citrus peel [18], resulting in an easier decomposition process at higher temperatures (600°C), when the cellulose decomposition is almost complete [20]. Figure 2. shows pyrolysis gas LHV for the different biomass at different temperatures. It can be noticed that the higher heating value (17.1 MJ/m^3) was obtained for wood at 600°C, which is the biomass with the higher carbon content. Syngas low heating values were also evaluated for gasification test for all the samples at 700 and 800°C, which results are shown in Figure 3. The highest LHV was observed for wood, that was 2.26 MJ/m^3 at 700°C, while for citrus peel and oat straw it was 2.01 and 1.94 MJ/m^3 respectively. Gas composition was also evaluated for both pyrolysis and gasification tests. Examples of gas composition are shown in Figure 4 and Figure 5 for pyrolysis at 600°C and gasification at 800°C, respectively.

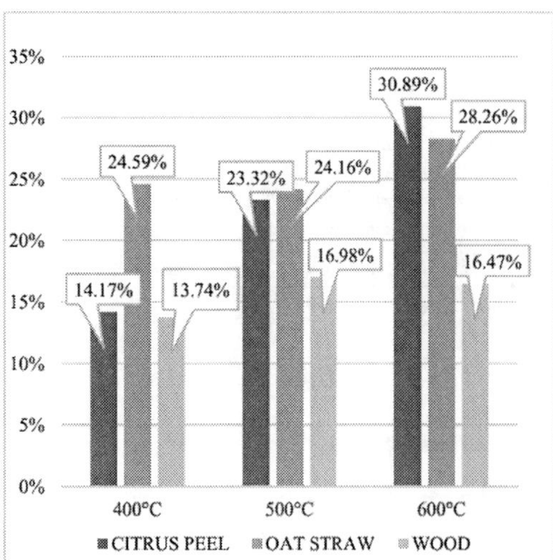

Figure 1. Pyrolysis gas yield ($\%_{wt/wt}$) for citrus peel, oat straw and wood at 400°C, 500°C and 600°C

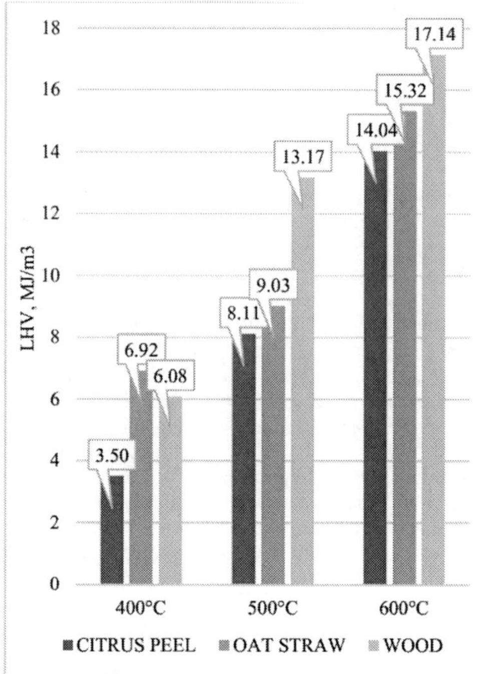

Figure 2. Normalized pyrolysis syngas LHV [MJ/m^3] for citrus peel, oat straw and wood at 400°C, 500°C and 600°C ([N_2], [O_2] = 0)

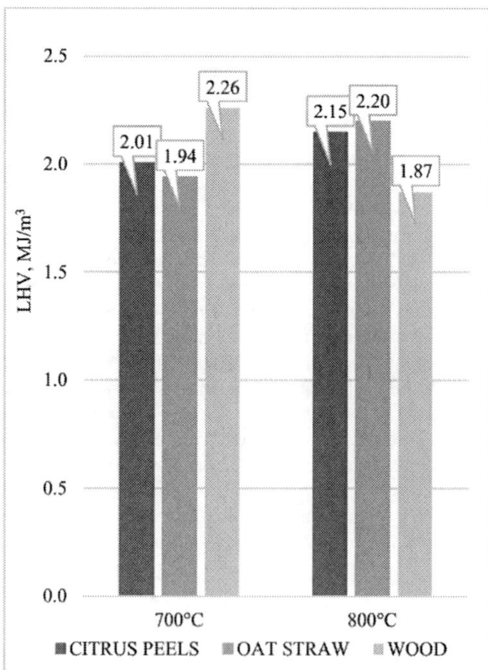

Figure 3. Gasification syngas LHV [MJ/m^3] for citrus peel, oat straw and wood at 700°C and 800°C

 DOI: 10.1051/e3sconf/20161000024

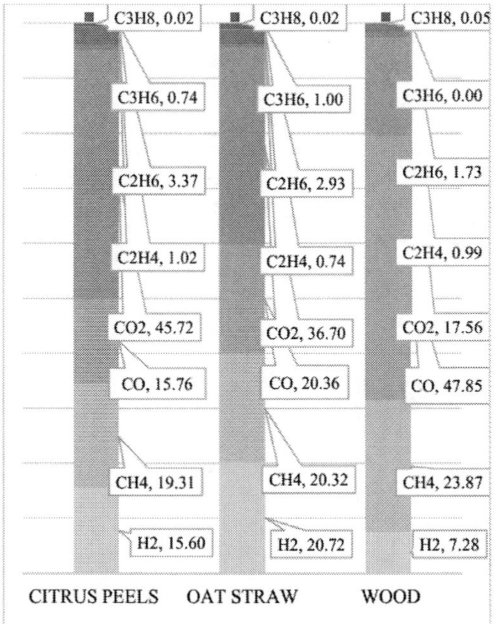

Figure 4. Normalized pyrolysis syngas composition for citrus peel, oat straw and wood at 600°C ($[N_2]$, $[O_2] = 0$)

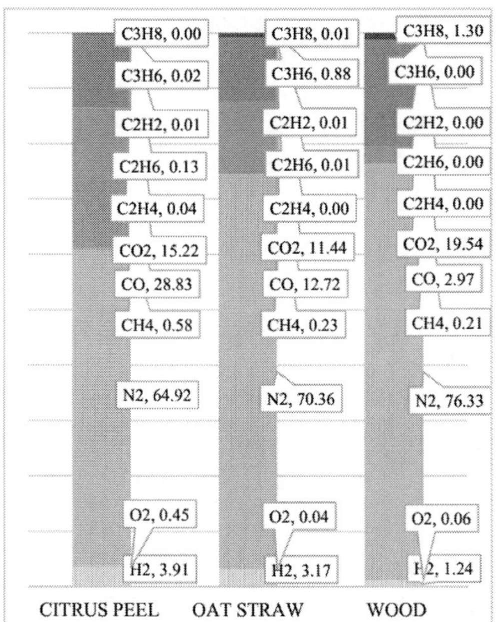

Figure 5. Gasification syngas composition for citrus peel, oat straw and wood at 800°C

4 Conclusions

The behaviour of different biomass from different European regions under air gasification and pyrolysis conditions was studied in order to evaluate the possibility of using alternative local feedstocks that are typical of different continental areas. Results showed that oat straw and citrus peels can be exploited as feedstocks for pyrolysis and gasification conversion, as well as wood. Although citrus peel showed the worst performance during pyrolysis tests in terms of LHV, it had higher gas yield (30.89%) than wood (16.47%) and oat straw (28.26%) at 600°C. Gasification tests showed that wood produced the highest syngas LHV (2.26 MJ/m^3) and oat straw the lowest LHV (1.94 MJ/m^3) at 700°C. It can be concluded that residual biomass, such as oat straw and citrus peel can be considered as alternative to or combined with wood for thermochemical biomass conversion.

References

1. EU Council, EUCO 169/14, (2014)
2. D. Bourguignon, EPRS, Members' Research Service, **PE 568.329** (2015)

A. Piwowar, A. Dzikuć, Renewable and Sustainable Energy Reviews, **54**, 415-420, (2016)

3. EDF Polska S.A. http://www.edfrybnik.pl/art,646,struktura-paliw.html (access 18.04.2016)
4. ENGIE Energia Polska S.A. http://www.gdfsuez-energia.pl/dokument/20130607/30
5. (access 18.04.2016)
6. W. Wei, P. Mellin, W. Yang, C. Wang, A. Hultgren, H. Salman, Report I, Swedish Energy Agency, **35819-1**, (2013) https://www.diva-portal.org/smash/get/diva2:681739/FULLTEXT01.pdf, (access 18.04.2016)
7. OCED/IEA, Energy Policies of IEA Countries – Sweden 2013 Review, (2013) http://www.iea.org/publications/freepublications/publication/Sweden2013_free.pdf (access 18.04.2016)
8. Ridjan, M. B. Vad, D. Connolly, Aalborg University Denmark, (2013) http://vbn.aau.dk/files/123284438/A_review_of_biomass_gasification_technologies_in_Denmark_and_Sweden.pdf (access 18.04.2016)
9. OCED/IEA, Energy Policies of IEA Countries – Italy 2009 Review, (2009) http://www.iea.org/publications/freepublications/publication/italy2009.pdf (access 18.04.2016)
10. GSE, Statistical Report 2011 – Renewable Energy power Plants in Italy, (2011) http://www.gse.it/it/Dati%20e%20Bilanci/GSE_Documenti/ENG/Italy%20RES%20Stastistical%20Report%202011%20WEB%20def%2015-11-2012%20%20tag.pdf (access 18.04.2016)
11. F. Di Mario, G. Braccio, V. Pignatelli, N. Colonna, F. Zimbardi, Quaderni ENEA: Biomasse e Bioenergia. ENEA, Rome, 2011
12. IEA/OCED, CO2 Emission from Fuel Combustion, (2014)
13. S. Kalisz, M. Pronobis, D. Baxter, Energy, **33**, 1770-1778, (2008)
14. P. Basu, *Biomass Gasification, Pyrolysis and Torrefaction - Practical Design and Theory – Second Edition*, (Oxford: Academic Press Elsevier 2013)

15. T. Fulczyk, E. Głowacki, Energetyka, (2010) http://www.cire.pl/pliki/2/ProblemyEksploatacyje_EF_IOS_biomasa.pdf (access 18.04.2016)
16. J. Wandrasz, A. Wandrasz, Paliwa formowane, (Wyd. Seidel Przywecki sp. z o.o., 2006)
17. G. Wiśniewski, K. Michalowska-Knap, A. Oniszk-Poplawska,: Określenie potencjału energetycznego rejonów Polski w zakresie odnawialnych źródeł energii- wnioski dla Regionalnych Programów Operacyjnych na okres programowania 2014-2020, (2011) https://www.funduszeeuropejskie.2007-2013.gov.pl/dzialaniapromocyjne/Documents/raport OZE_druk_korekta_m.pdf (access 18.04.2016)
18. F. R. Marin, C. Soler-Rivas, O. Benavente-Garci, J. Castillo, J. A. Perez-Alvarez; *Food Chem.*, **100**, 2, 736 - 741 (2007)
19. ISPRA, *Studio sull'utilizzo di biomasse combustibili e biomasse rifiuto per la produzione di energia* (Italy, ISPRA, 2010)
20. P. Giudiciani, G. Cardone, R. Ragucci ; Journal of Analytical and Applied Pyrolysis, 100, 213-222 (2013)

Energy audit in feed production plant

Jacek Gembicki[1,a]

[1]*Szkoła Główna Gospodarstwa Wiejskiego w Warszawie Wydział Inżynierii Produkcji*

Abstract. The energy audit in a feed production plan is intended to specify its energy consumption and, if it is too high, propose changes aiming to reduce it. The energy audit may be internal or external, depending on whether it is carried out by trained workers of the plant or by an external company. The internal audit requires inconsiderable financial expenses but in effect does not assure a full picture of the energy consumption in the production plant. Internal audits may show disadvantages of the plant which the management board should pay special attention to. The external audit in turn provides a wider view of the plant and shows things unnoticed or skipped during the internal audit. The stages of the audit have been specified in relevant rules of law. The results of the audit and their implementation do not need to require considerable financial expenses, but may instead lead to substantial savings, thanks to which the production is more profitable and the plant more competitive on the market

What is the Energy Audit and what is its purpose? To answer the question, let's refer to the definition of the Energy Audit, which will also be the basis for scrutinizing the subject. The energy audit is an elaboration that specifies the scope, technical and economic parameters of the project, as well as indicates the optimal solution from the point of view of costs in this respect. The audit also serves as a useful elaboration for the system user (entrepreneur) and a basis for gaining the right to thermo-modernization bonus. It is also an assumption for the building permit design related to the project in question.

The legal basis for the Energy Audit and its completion is included in the following documents:

- Resolution of the Minister of Infrastructure of 17 March 2009 (on a detailed scope and forms of energy audits, templates of audit charters, as well as algorithm of assessing profitability of the thermo-modernization project)
- Act on supporting thermo-modernization and renovations of 21 November 2008
- Polish construction standards
- Resolution of the Minister of Infrastructure of 6 November 2008 (on methods of calculating energy performance of buildings and residential accommodations or part of the building being an independent technical and functional entity, as well as a method of preparing and templates of certificates of their energy performance).

The factors determining energy consumption also have an impact on the energy audit. Certain factors have a considerable influence on the entire plant and it is difficult to change them, whereas others can be easily modified [1].

Much depends on whether a given plant has enough funds to introduce new technologies.

- Productive capacities utilization factor,
- Production profile (range of products),
- Proper selection of performance characteristics of devices and energy carriers,
- Device work system (permanent, periodical),
- Degree of mechanization and automation of processes and production operations,
- Proper selection of engines and installed power structure,
- Machine operation and maintenance system
- Quality of materials,
- Spatial development and cubic volume of rooms,
- Material delivery schedule and production organization,
- Efficiency of energy use inspection and ability to draw proper conclusions.

As far as the energy audit is concerned, it is advisable to make reference to energy certificates and answer the following questions: what is it and what is it for? The questions listed below will be of assistance as well.

What is the energy performance certificate? The document prepared and certified by the authorized specialist that specifies the amount of energy consumed for the purposes of heating, preparing hot water, ventilating and air conditioning the building/accommodation. The energy consumption index (E) is expressed in kWh/(m ^{2}year), and specifies the amount of energy consumed throughout the year per square meter. Depending on its value, the certificate assigns one of seven energy classes (A-G rated) to the building/accommodation.

[a] Corresponding author: gembicki.jacek@gmail.com

Who is responsible for preparing the energy performance certificate? The owner of the building/accommodation is obligated to prepare the certificate. The certificate should be provided to the purchaser or tenant/renter of the building. With regard to new buildings which are handed over, the energy performance certificate is a document required to gain occupancy permit, whereas in the case of selling the building/accommodation, it is necessary to sign the notarial deed.

What are the benefits of preparing the energy performance certificate? The energy class assigned by the certificate specifies the energy quality of the building/accommodation, and therefore its attractiveness in terms of sale or rental. Additionally, the certificate defines the guidelines and recommendations related to potential improvements to be made by the owner in order to reduce the energy demand. Thanks to the certificate, purchasers or renters/tenants are able to estimate power consumption costs related to the use of the building/accommodation and choose the offer that suits their financial situation.

Now we know what the energy audit and energy performance certificates are. Let's see how to carry out a basic energy audit.

1. Preparing checklists.

Before the energy auto-audit is carried out, it is obligatory to collect and organize data related to the production plant. For this purpose, it is necessary to prepare the so-called checklists which specify the list of recommendations, checks, measurements and analyses relevant to plant assessment. We can distinguish 14 groups of factors. Each of them describes detailed recommendations with regard to measurement, data collection and their analysis [2].

2. Preparing a device operation schedule within a day (working day photograph) and analyzing a load duration curve.

We prepare a list of intended uses with regard to devices in the product plant and then develop a method of using these devices in order to eliminate downtimes and ineffective use of machines.

Table 1. 14 groups of factors use in Preparing checklists

Item	Factors	Detailed description
1.	Spatial development, production facilities	Location, location in terms of a part of the world, dimensions, cubic volume, shape, building age, building surface and cubic volume use factor.
2.	Primary energy	Place of storage, consumption, primary source of energy costs, possibility of changing deliveries and energy carrier suppliers.
3.	Electric energy	Power supply and maximum load, transformer's condition, possibility of correcting peak power, possibility of using alternative rates.
4.	Energy conversions	Type, model, age of boilers and burners, quality of burning, watt-hour efficiency; consumption of fuels by each device.
5.	Cool management – cooling devices	Equipment, type, class and age of compressors; operation and use conditions; efficiency of cooling devices.
6.	Energy distribution	Leakage and losses in distribution of hot water, water steam and heated liquids; leakage and losses in the cooled water distribution system.
7.	Room heating	Type of heating system; heating device operation schedule with reference to real use of the facility; adjustment of room temperature to production technology requirements and operation conditions.
8.	Air conditioning and ventilation	Operating conditions of air conditioners and ventilating devices with reference to facility use.
9.	Domestic hot water	Condition and capacity of storage water heaters; feed water parameters.
10.	Heating and cooling processes	Degree of adapting real parameters to project-related conditions.
11.	Lighting	Type of lighting; measurements of luminous flux; conditions, settings, activation of existing lighting fittings.
12.	Compressed air	Installed power, age and degree of wear of air compressors; equipment condition; use and operation conditions.

13.	Waste energy	Possibility of recovering heat from products, post-production waters.
14.	Machine use and operation	Friction wear, lubricating.

3. Establishment of permissible loads (limits) and explanation of reasons for exceeded values.

We set real and planned use of machines, we check the power intended for them, and attempt to optimize their operation to assure effective and economical use.

The devices which are to be switched off at peak hours must be equipped with information boards which specify the shutdown hour.

In the event of exceeding the power limit, the user (entrepreneur) is obligated to cover expenses arising from the rate set out in the agreement with an energy provider.

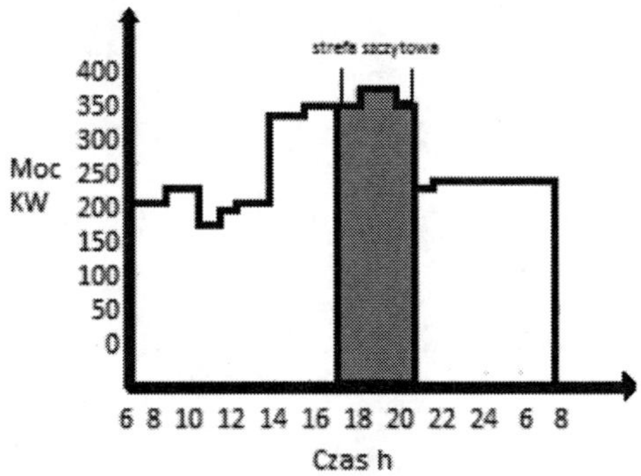

Figure 1. Use of electric energy before improvements

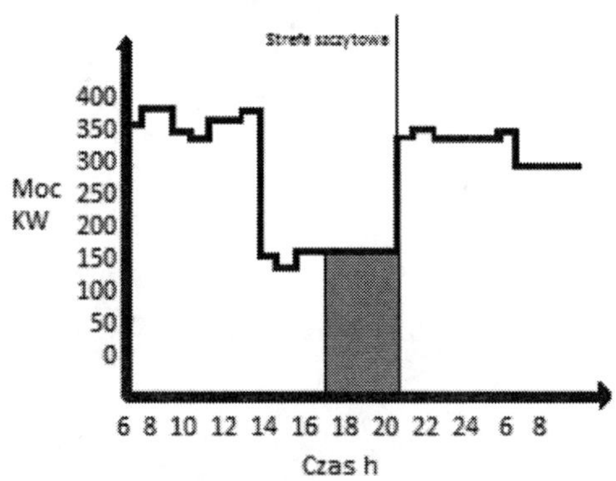

Figure 2. Use of electric energy after improvements

4. A detailed analysis of heat energy receivers that includes water heater or steam boiler efficiency. When carrying out the energy audit, it is obligatory to inspect other energy load points (energy receivers) which may affect the audit of the entire plant.

The example presented below shows what we can achieve by using the energy audit.

When we use technical solutions, we aim to create such conditions in which electric energy is somehow accumulated in the form of work performed. There is a number of usage possibilities in production processes. When the production cycle is complex, then by storing part of the work at one of the stages (where electric energy serves as a source), we achieve the same effect as if the electric energy was stored. [3]

Referring to organization-related projects, we may state that it is possible to modify operation of not only receivers that work for part of the day but also those which operate permanently but are not fully loaded behind peak hours. The quantitative production drop is used by increased machine load and increase in their efficiency during other hours of operation. [4]

For instance, if the production line operates for 3 shifts at the average efficiency of 82%, maximum hour efficiency

A_{hmax} then its day output is

$$A_{dob} = 24 \times 0{,}820 \times A_{hmax} = 19{,}68\ A_{hmax} \quad (1)$$

This output will not decrease if it is suspended for three peak hours, and during other 21 hours will operate with the efficiency of 0,937 A_{hmax}

$$A'_{dob} = (24 - 3) \times 0{,}937\ A_{hmax} = 19{,}68\ A_{hmax} \quad (2)$$

When we stop the line in question even for 4,3 hours during peak hours and resume operation in the remaining period at full load, we will be able to achieve output volume in accordance with the following:

$$A''_{dob} = (24 - 4{,}3) \times A_{hmax} = 19{,}70\ A_{hmax} \quad (3)$$

It sometimes happens that a receiver faces time-related challenges, and it is not only its insufficient efficiency but also a need to maintain continuous work of particular links of the entire production chain. In these conditions a proper solution is the use of equalizing tanks suitable to a given production factor. These tanks should have such capacity that despite shutdown of particular elements of production line devices it is possible to assure continuous work of the entire cycle. It applies to e.g. tanks for disintegrated material which when accumulated allows suspending the work of grinders for a specific period of time. It also applies to liquid material tanks or liquid half-finished product tanks which allow suspending pumps at peak hours. The use of equalizing tanks and containers allows a vast array of adaptations of these solutions, and thus energy savings. [5]

Positive effects of the energy audit:

- Increased efficiency at non-peak hours leads to equalization of load curve in the power system (effect in macro scale),
- Reduced energy consumption of the national economy (macro scale) and decreased CO_2 emission and other atmosphere pollution,
- Reduced impact of energy costs in production expenses

To recapitulate, audits and building energy performance certificates should base on guidelines and the algorithm stipulated in the legal regulations and standards. This is the requirement to assure that the documents are useful and applied when planning and carrying out the investment intended to reduce the energy consumption in the production plant. A reliable preparation of such documents is possible if the contractor cooperates closely with the manager or owner of the production facility. In turn, the certificate is a required document when a given building is to be handed over (commissioned), sold or rented. Such audit may entail new investments in the production plant, yet with time these investments may lead to substantial savings and increased competiveness of the facility.

References

1. T.Ekvall, A. Tillman, Journal of Cleaner Production, **13** (2005)
2. J. Wojdalski, B. Dróżdż, MOTROL, Motoryzacja i Energetyka Rolnictwa, Podstawy analizy energochłonności produkcji zakładów przemysłu rolno-spożywczego, **8A** (2006)
3. W. Franklin, Resources, Conservation and Recycling, Life Cycle Assessment – a remarkable tool in the era of sustainable resource and environmental management, **14** (1995)
4. M. Węgrzyn, Inżynieria Rolnicza, **7**, 82 (2006)
5. D. Muller, F. Marechal, Applied Thermal Engineering, **27** (2007)
6. A. Björklund, Environmental Impact Assessment Review, **32** (2012)
7. J. Witczak, J. Kasprzak, LCA related aspects. **19** (2014)
8. B. Bilska, W. Grzesińska, Projektowanie technologiczne zakładów przemysłu spożywczego. Wybrane zagadnienia, (2011)

 DOI: 10.1051/e3sconf/20161000026

Kinetics examination of pressurised steam gasification of beech wood

Przemysław Grzywacz[1,a], Grzegorz Czerski[1], Tadeusz Dziok[1], Stanisław Porada[1] and Katarzyna Zubek[1]

[1]AGH University of Science and Technology, Faculty of Energy and Fuels, al Mickiewicza 30, 30-059 Cracow, Poland

Abstract. In Poland, among all renewable energy sources, share of biomass in electricity production is the largest. Great potential in biomass utilization involves use of gasification technology. The course of biomass gasification process in steam atmosphere is the subject of numerous scientific studies. The aim of this study was kinetics examination of pressurized beech wood waste gasification, that can be successfully used in industrial gasification. Measurements were carried out on a unique laboratory installation, that allows kinetics examination of solid fuels gasification with steam at wide range of pressures, via using thermovolumetric method. Formation rates of main gaseous products were determined, moreover, composition of post-reaction gas, as well as biomass conversion degree depending on the temperature were specified. Kinetic parameters of gasification were calculated and series of kinetic models were used to their designation ie Isoconvesional method, Integrated Core Model (ICM), Grain Model (GM) and Random Pore Model (RPM).

1 Introduction

In accordance with the requirements of the European Union, in 2020 the share of energy from renewable sources in gross final energy consumption is expected to reach 20% [1], and in Poland, according to a government document 'Energy Policy of Poland until 2030' at least 15% [2]. Over the past years the share of renewable sources in Poland is steadily increasing and in 2013 amounted to 8.7% [3], however it should be noted that this increase is insufficient. Among renewable energy sources the largest share in electricity production accounted for biomass [4].

Biomass for energy purposes can be used in many processes, ie combustion, co-combustion, gasification and pyrolysis, while the share of the first two technologies is dominant. It should be noted however, that using biomass for energy is associated with certain limitations, mainly due to the type of boilers used, properties of coal used [5], as well as the increased risk of chloride corrosion of steam superheater and fire hazard [6]. In addition, biomass requires adequate preparation to ensure the stable quality parameters [7].

The great potential in the use of biomass is linked with application of gasification technology. This applies particularly to low-power units, representing local source of clean energy, but also larger objects dedicated to the production of chemicals and liquid fuels. Currently, compared to other raw material the share of biomass in gasification process is small [8]. Biomass gasification plants, due to the high cost of obtaining a large amount of feedstock, are usually low-power units. According to the data compiled in the World Gasification Database [9], currently there are 35 biomass gasification plants working on an industrial scale, with a total capacity of 787 MW_{th}. In the vast majority the raw materials for gasification are wood and wood waste, much less, e.g. olive waste or rice husk. Biomass is subjected to gasification in order to generate electricity and heat, as well as to produce gaseous and liquid fuels. Technologies used include, among others: Envirotherm Circulating Fluidized Bed Gasification Process, Nexterra, Foster Wheeler Atmospheric Circulating Fluidized Bed Gasifier, Primenergy, Valmet, B&W Volund, ZeroPoint.

The course of the biomass gasification process in atmosphere of steam is the subject of numerous scientific studies. Research are carried out on the use of various types of biomass in the reactors of diverse construction, among others: pine chips in a fluidized bed [10] and fixed bed reactors [11]; bark of pine, Japanese oak [12] and pomace from olive oil production [13] in a fixed bed; oak, maritime pine and eucalyptus [14] as well as wood [15] and wood pellets [16], digestate from the production of ethanol [17], almond shell [18] in the fluidized bed reactors; waste from the processing of sugar cane in the cyclone reactor [19,20]; husks of sunflower, rapeseed, pine cones, cotton waste and pomace oil in thermogravimetric analyzer [21]; animal waste compost in a two-stage fluidized bed reactor [22]. During research influence of temperature, particle size, amount of steam and air fed to the reactor, as well as the type of biomass and catalysts used on the amount and the calorific value of the resulting gas are analyzed. Analyses of conversion degree, thermal efficiency and the amount of unreacted char are also made. Furthermore, the subjects of research are composition

[a] Corresponding author: grzywacz@agh.edu.pl

DOI: 10.1051/e3sconf/20161000026

and properties of the ash and its influence on the biomass gasification process [19].

The key parameter for the gasification of biomass process is temperature, which determines the composition of the resulting gas [10,15,17,20,23] and its calorific value [14]. That is why the influence of temperature on the gasification process was investigated in this paper. Tests were carried out on a unique laboratory installation, which allows measurement of kinetics of solid fuels gasification process in steam atmosphere at wide range of pressures. Waste from the processing of beech wood, which can be successfully used on an industrial scale, were subjected to gasification process. For the tested biomass release rate of the main gas product, composition of resulting gas, as well as biomass conversion degree depending on the temperature were determined. Kinetic parameters of gasification process were also calculated and a series of kinetic models were used for their designation, ie isoconvesional method, Integrated Core Model (ICM), Grain Model (GM) and Random Pore Model (RPM).

2 Material and method

The subject of this study was woody biomass, which characteristics are shown in Table 1. Measurements were carried out at a pressure of 1.5 MPa and at temperatures of 700, 800 and 900 °C. Samples with a grain size of less than 0.2 mm. and a weight of 0.5 grams were used in the measurements. Flow rate of steam was 0.3 g/min, and the inert gas 1.5 l/min.

Table 1. Characteristic of woody biomass.

Parameter	Symbol	Value, %
Proximate analysis		
Moisture	M^a	6.12
Ash	A^a	0.2
Volatile matter	VM^{daf}	83.91
Ultimate analysis		
Carbon	C^{daf}	51.0
Hydrogen	H^{daf}	6.28
Sulfur	S^{daf}	0.01
Oxygen&Nitrogen*	$(O+N)^{daf}$	42.7

*calculated by difference

Measurement of the kinetics gasification process was carried out under isothermal conditions, using the installation which schematic diagram is shown in Figure 1. Installation and methodology used in this research has been successfully applied to examine coal and chars gasification process [24, 25]. The installation consists of several basic systems: a pressure reactor with a heating system (1-2) a system for feeding the reactor with the gasifying agent - steam (3-5), carrier gas - argon (6-7), and biomass, (8) system for collecting and purifying the resulting gas (9-11) and the gas analysis system (12-14). In the resulting gas the content of carbon monoxide and carbon dioxide is continuously controlled. For this purpose the analyzer operating on the principle of infrared radiation adsorption was used. In addition, the content of hydrogen and methane in the resulting gas was analyzed using gas chromatographs. The gas chromatograph used to analyze content of methane was equipped with a flame ionization detector (FID) and the chromatograph for the analysis of hydrogen content with thermal conductivity detector (TCD).

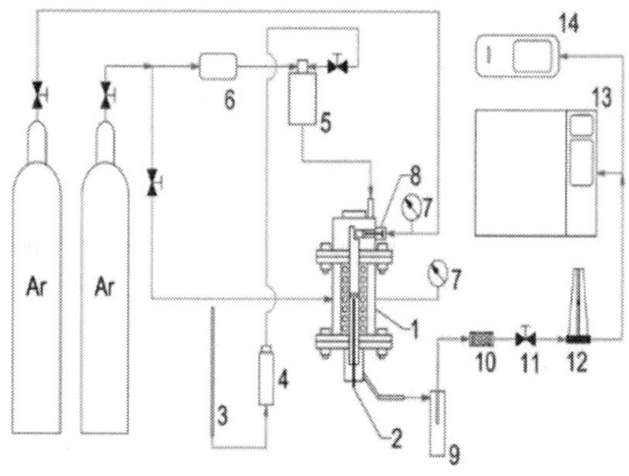

Figure 1. The laboratory equipment for kinetic examinations of biomass gasification process.
1 - reactor, 2 - thermocouple, 3 - water tank, 4 - water pump, 5 - steam generator, 6 - mass flowmeter, 7- pressure gauge, 8 – biomass feeder, 9 - cooler, 10 - filter, 11 - pressure regulator, 12- rotameter, 13 - gas chromatographs, 14 - CO, CO_2 IR analyser.

Based on the measurement of concentrations of carbon monoxide, methane and carbon dioxide in the resulting gas, the changes in formation rates of those gaseous products during gasification as a function of time were calculated, allowing determination of the gas composition and the conversion degree. The yield of each gasification product was calculated by determination of the area under the curve $dV/dt = f(t)$ and carbon conversion degree was evaluated based on the volume of a gaseous product containing this element:

$$X(t) = \frac{(V_{co}(t)+V_{CO2}(t)+V_{CH4}(t))\cdot M_C}{V_{mol}\cdot m\cdot C^{daf}} \cdot 100\ \% \quad (1)$$

where:
$V_{CO}(t)$, $V_{CO2}(t)$, $V_{CH4}(t)$ - volume of released gas component at standard conditions as a function of time, dm^3/g
M_c- molar mass of carbon, g/mol
m – sample mass, g
C^{daf}- dry ash free carbon content, -
V_{mol} - volume of one mole of gas at temperature of 273 K and pressure of 101325 Pa, dm^3/mol

In next step curves of conversion rate determined for different temperatures were used to analyze the kinetics of biomass gasification process.

The effect of temperature and concentration of gaseous reactant on the gasification process rate can be expressed by the equation (2):

$$\frac{dX}{dt} = k(T)G(C)f(X) \quad (2)$$

where:
k(T), and *G(C)* include the effect of temperature and the gas reactant concentration on the process rate, respectively and *f(X)* accounts for physical or chemical

changes occurring with the progress of the reaction. The apparent reaction rate constant *k(T)* and the concentration of the reagent *G(C)* are generally described in the form of an Arrhenius equation:

$$k(T)G(C) = k_0 e^{\frac{-E_a}{RT}} C^n \quad (3)$$

where:
k_0 is pre-exponential factor, E_a is activation energy and n is an order of the reaction taking into account the concentration of the gas reactant.

In order to analyze the kinetics of the gasification process two approaches were used. Firstly, isoconversional method, which allows to determine the activation energy from the slope of *ln(t) = f(1/T)*, was applied. Combining and transforming equations (2) and (3), provides:

$$\int_0^x \frac{dX}{f(X)} = \int_0^t C^n k_0 e^{\frac{-E_a}{RT}} dt \quad (4)$$

which after integration and taking the logarithm gives:

$$\ln(t) = \ln\left(\frac{F(X)}{C^n k_0}\right) + \frac{E_a}{RT} \quad (5)$$

The first term on the right hand side is a function of the reaction progress and the gaseous reactant concentration. If the concentration is constant and the time to reach a certain, founded conversion degree is a function of temperature, then equation (5) allows to determine the activation energy from the slope of *ln(t)=f(1/T)*, even though *f(X)* in equation 2 is unknown. Then, in order to determine the activation energy and pre-exponential factor three models, describing heterogeneous gas-solid reaction, were used ie Integrated core model (ICM) Grain model (GM) and Random Pore Model (RPM).

The first of the models used was the ICM. Rate constant in this model is determined by the selection of the exponent n (representing the order of reaction) in equation (6), to obtain the best fit of experimental data to the model.

$$\frac{dX}{dt} = k_{ICM}(1 - X)^n \quad (6)$$

GM model assumes that the surface of the reaction decreases non-linearly as the reaction proceeds, and RPM model assumes that the reaction surface changes as a result of two competing processes: the increase of the pore structure due to the gasification process and its destruction due to the coalescence of adjacent pores. The GM and RPM models are described by equations (7) and (8), respectively.

$$\frac{dX}{dt} = k_{GM}(1 - X)^{\frac{2}{3}} \quad (7)$$

$$\frac{dX}{dt} = k_{RPM}(1 - X)\sqrt{1 - \Psi \cdot \ln(1 - X)} \quad (8)$$

In RPM model Ψ parameter is associated with the pore structure of the test material and can be calculated from the equation:

$$\Psi = \frac{2}{2\ln(1 - X_{max}) + 1} \quad (9)$$

where X_{max} is the conversion degree at maximum reaction rate. Due to the fact that in gasification process the greatest reaction rate are observed at the beginning, it can be assumed that $X_{max} = 0$ and then Ψ = 2.

3 Results and discussion

The achieved carbon conversion degree of tested biomass, calculated according to the equation (1), varies from about 80% at 700 °C to about 100% at a temperature of 900 °C (Figure 2). Simultaneously, time of gasification process is reduced. During the process unreacted char and tar, which was not analyzed, were also formed.

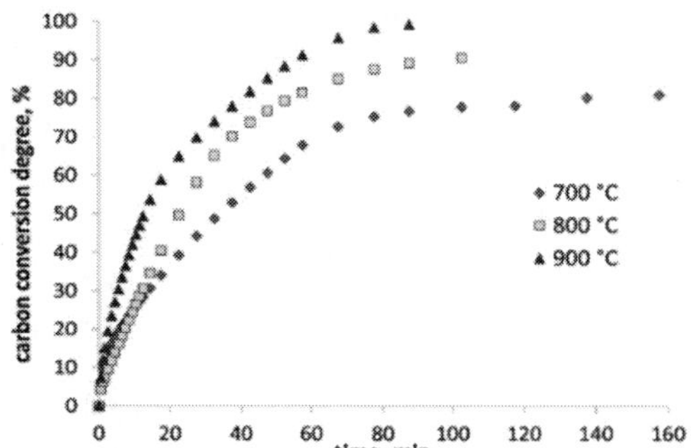

Figure 2. The curves of the carbon conversion degree.

With increasing temperature increases, however, the yield of carbon dioxide at the expense of the most important products ie hydrogen and carbon monoxide (Figure 3), which leads to reduction in calorific value of the fuel. The only component which yield does not change depending on the temperature is methane, that is formed in the first minutes of the process, during pyrolysis stage.

The kinetics analysis of the biomass gasification was started by determining the activation energy using isoconversional model (equation (5)). Calculations of activation energy were made for three conversion degrees, ie X = 0.3, 0.5, and 0.7. Fit of linearized isoconversional model to experimental data of biomass gasification process for conversion degrees 0.3, 0.5, 0.7 is shown in Figure 4, and activation energy, obtained on this basis, amounted to 43 kJ/mol.

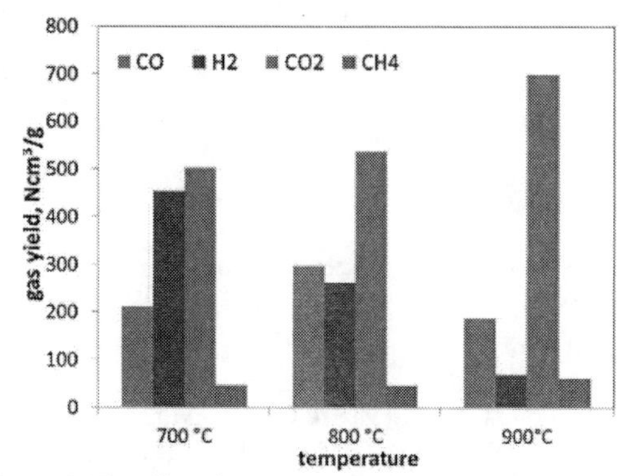

Figure 3. The effect of temperature on the resulting gas composition.

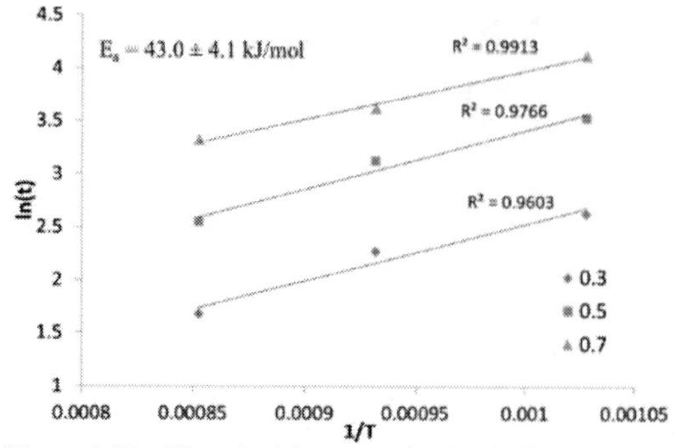

Figure 4. Fit of linearized isoconversional model to experimental data of biomass gasification process for conversion degrees 0.3, 0.5, 0.7.

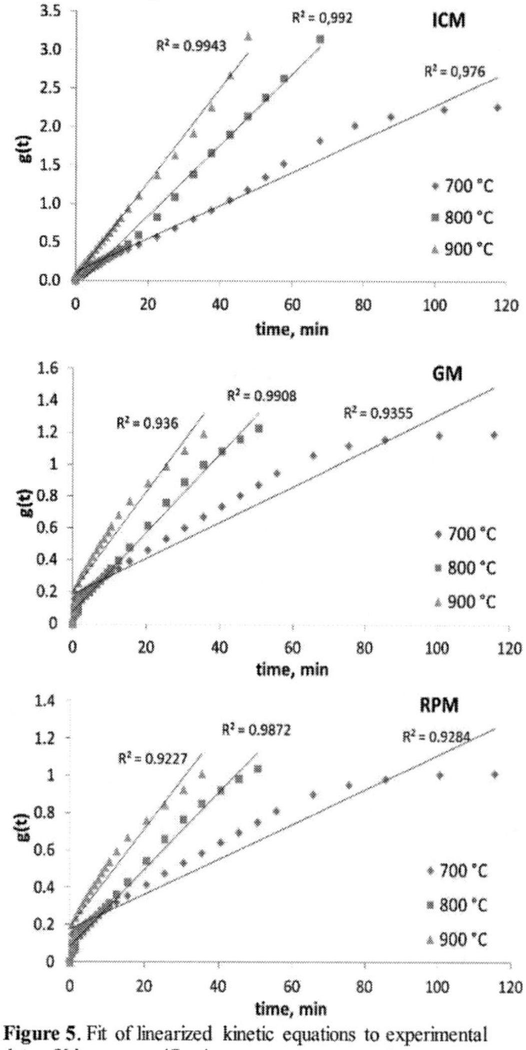

Figure 5. Fit of linearized kinetic equations to experimental data of biomass gasification process.

Due to the compensation effect of pre-exponential factor the knowledge of exclusively activation energy is not sufficient. That is why in the second approach typical models for this type of reaction (ICM, GM, RPM) were applied to determine both activation energy and pre-exponential factor. Fit of the linearized form of these equations to the experimental data are presented in Figure 5. The best fit, measured via coefficient of determination R^2, was obtained for the ICM model, while GM model and RPM did not give such good results. The kinetic parameters, calculated based on determined curves, are summarized in Table 2. The obtained values of the activation energy are similar to each other and amounted to 43.0 - 50 kJ/mol. The values of pre-exponential factor are comparable and were in the range between 7.59-02 and 1.86-01 1/s. The obtained values of activation energy are similar to the literature data, for example according to Haykiri-Acma et al [21] different types of biomass are characterized by diverse values of the activation energy, from 21.4 to 33.8 kJ / mol. When using the ICM model exponent n was also determined. The calculated value was 1.5 and does not correspond to any of the basic kinetic equations, and therefore it should not be considered as a reaction order (especially, that gasification process consist of several reactions) but only as a mathematical constant.

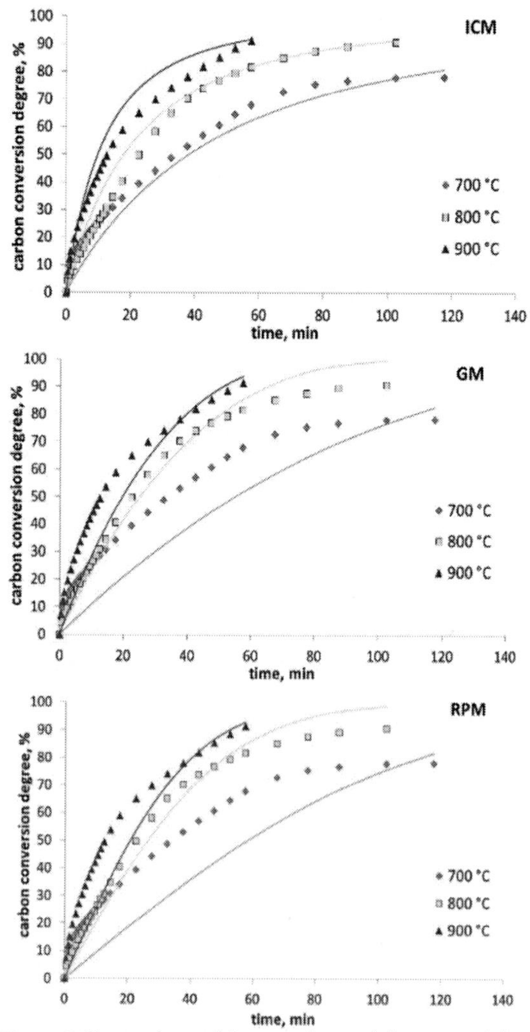

Figure 6. Comparison of the experimental data to model curves of conversion degree of the gasification process.

Table 2. Calculated values of kinetic parameters.

Model free	E_a, kJ/mol		
	43.0		
ICM	E_a, kJ/mol	k_0, 1/s	
	50.0	1.86E-01	
	E_a, kJ/mol		
GM	49.4	9.03E-02	
	E_a, kJ/mol	k_0, 1/s	Ψ
RPM	49.5	7.59E-02	2

The determined kinetic parameters allowed for modeling of conversion degree curves and comparison of models with experimental data (Figure 6). ICM model reflects very well the course of gasification process while GM and RPM models do not coincide satisfactorily with the experimental data, which suggests that they are not suitable to describe steam gasification of woody biomass process.

4 Conclusion

The process of steam gasification of waste wood industry for the production of synthesis gas seems to be a viable alternative to conventional combustion and co-combustion processes. The results obtained for beech wood confirmed the significant influence of temperature on the course and yields of the gasification process. With increasing temperature of gasification process the conversion degree and yield of carbon dioxide increases, while yield of hydrogen and carbon monoxide decreases. Activation energy determined for the steam gasification of beech wood was 43-50 kJ/mol, depending on the model used. The best fit of model calculations to experimental data was obtained for Integrated Core Model.

Acknowledgments

This paper was prepared as a part of the statutory activity of the Faculty of Energy and Fuels at the AGH University of Science and Technology, No. 11.11.210.213

References

1. Directive 2009/28/EC of the European Parliament and of the Council of 23 April 2009 on the promotion of the use of energy from renewable sources
2. Ministerstwo Gospodarki RP, *Polityka energetyczna Polski do 2030 roku* , Warszawa (2009)
3. GUS, *Gospodarka paliwowo-energetyczna w latach 2013 i 2014*, Warszawa (2015)
4. T. Mirowski, T. Surma, Gospod Surowcami Min, **24**, 211-221 (2008)
5. M. Ściążko, K. Kubica, Karbo, **11**, 325-330 (2002)
6. J. Zuwała, Karbo, **4**, 165-172 (2008)
7. T. Dzik, G. Czerski, Gospod Surowcami Min, **30**, 145-160 (2014)
8. Ch. Higman, *State of the Gasification Industry: Worldwide Gasification Database 2014 Update*, Gasification Technologies Conference, Washington, (2014)
9. *World Gasification Database*, Gasification Technology Council (2015)
10. P.M. Lv, Z.H. Xiong, J. Chang, C.Z. Wu, Y. Chen, J.X. Zhu, Bioresour Technol, **95**, 95-101 (2004)
11. S. Luo , B. Xiao, X. Guo, Z. Hu, S. Liu, M. He, Int J Hydrogen Energ, **34**, 1260-1264 (2009)
12. T. Hanaoka, S. Inoue, S. Uno, T. Ogi, T. Minowa, Biomass Bioenerg, **28**, 69-76, (2005)
13. J.F. González, S. Román, D. Bragado, M. Calderón, Fuel Process Technol, **89**, 764-772 (2008)
14. C. Franco, F. Pinto, I. Gulyurtlu, I. Cabrita, Fuel, **82**, 835-842 (2003)
15. Y.-J. Son, S. J. Yoon, Y. K. Kim, J.-G. Lee, Biomass Bioenerg, **35**, 4215-4220, (2011)
16. C. Lucas, D. Szewczyk, W. Blasiak, S. Mochida, Biomass Bioenrg, **27**, 563-575 (2004)
17. A. Kumar, K. Eskridge, D. D. Jones, M. A. Hanna, Bioresour Technol, **100**, 2062-2068 (2009)
18. S. Rapagna, A. Latif, Biomass Bioenerg, **12**, 281-288 (1997)
19. M. Gabra, E. Pettersson, R. Backman, B. Kjellstrom, Biomass Bioenerg, **21**, 351-369 (2001)
20. M. Gabra, E. Pettersson, R. Backman, B. Kjellstrom, Biomass Bioenerg, **21**, 371-380, (2001)
21. H. Haykiri-Acma, S. Yaman, Fuel Process Technol, **88**, 417-425, (2007)
22. X. Xiao, D. Dung Le, L. Li, X. Meng, J. Cao, K. Morishita, T. Takarada, Biomass Bioenerg, **34**, 1505-1512 (2010)
23. T. Song, J. Wu, L. Shen, J. Xiao, Biomass Bioenerg, **36**, 258-267 (2012)
24. S. Porada, G. Czerski, P. Grzywacz, T. Dziok, D. Makowska, Przem Chem, **93**, 2059-2063 (2014)
25. S. Porada, G. Czerski, T. Dziok, P. Grzywacz, D. Makowska, Fuel Process Technol, **130**, 282-291 (2015)
26. G. De Micco, A. Nasjleti, A.E. Bohé, Fuel, **95**, 537-543 (2012)

Estimating the daily solar irradiation on building roofs and facades using Blender Cycles path tracing algorithm

Mateusz Ilba[1,a]

[1]*AGH University of Science and Technology in Krakow, Poland*

Abstract. The paper presents the development of an daily solar irradiation algorithm with application of the free software Blender. Considerable attention was paid to the possibilities of simulation of reflections of direct and diffuse solar radiation. For this purpose, the rendering algorithm "Cycles" was used, based on the principle of bi-directional path tracing – tracing random paths of light beams. The value of global radiation in this study is the sum of four components: direct beam radiation, reflected beam radiation, diffuse radiation and reflected diffuse radiation. The developed algorithm allows calculation of solar irradiation for all elements of the 3D model created in Blender, or imported from an external source. One minute is the highest possible time resolution of the analysis, while the accuracy is dependent on the resolution of textures defined for each element of a 3D object. The analysed data is stored in the form of textures that in the algorithm are converted to the value of solar radiance. The result of the analysis is visualization, which shows the distribution of daily solar irradiation on all defined elements of the 3D model.

1 Introduction

When assessing the possibilities of solar energy gain, the most frequently used are the analyses of two-dimensional space. Analysis of data in urban areas takes into consideration horizontal and sloping surfaces of roofs. This is due to the easy application of solar cells, and a large amount of energy which can be obtained. Looking at big cities, the surface of walls of tall buildings is much bigger than the roof surface. For this reason, the information about the amount of energy received by the building walls is very valuable. From year to year, the possibility of the non-invasive obtaining energy from the walls of buildings is increasing, through the application of transparent photovoltaic cells [1,2]. Identification of the best mounting locations of this type of cells requires knowledge of the distribution of exposure to the solar irradiation on the whole surface of the building.

Performing the solar irradiation analysis for surfaces of a given orientation and inclination is relatively simple. In the three-dimensional environment one must take into account shading of the source of illumination by individual objects of the city infrastructure, which complicates the analysis. Calculations of even greater complexity must be performed if we plan to take into account the effect of sunlight reflections of individual objects. Reflections are of great importance in urban areas with many tall buildings covered with a reflective material (glass, smooth concrete). In this case sun beams (although severely weakened by a small value of reflections) can frequently appear in most unexpected places. This also creates the risk of accumulation of a large amount of solar radiation in one place. Especially dangerous effects can be caused by buildings having the shape which focuses the reflected sun beams. An example of such a threat is the building at 20 Fenchurch Street, "Walkie-Talkie," which led to melting of the plastic body of a Jaguar [3].

In the case of 2D analyses, the analysis is based on the input data in the form of GRID raster. In addition to the x and y data representing the position of the individual elements, we have additional data on the height of the element, stored in the form of pixel values. Analysing the solar irradiation, we can see the result in 3D thanks to the projection, whereas the correct data is only on flat and sloping elements. The increasing gradient of the surface results in inaccuracies, vertical surfaces along their entire length are represented by individual pixels. Raster data analysis can be performed in commercial programmes, e.g. ArcGIS [4] or open source, e.g. GRASS GIS [5]. In both cases, the input data is a raster with geographic mapping.

Analyses in 2D have serious limitations. They cannot be used to perform analysis of solar irradiation of vertical walls and surfaces. Therefore, there is a considerable interest in applications which use full 3D data. The most popular include Ecotect Analysis 2011 Daysim and Diva for Rhiano [6,7]. All the applications use full 3D data in the form of a mesh. After finishing the analysis, we have access to the results in the form of a 3D model

[a] Corresponding author: mateuszilba@wp.pl

visualization. In addition to commercial solutions, there are also original analyses of the solar irradiation, e.g. SOL algorithm to perform the analysis using data acquired with LIDAR [8].

2 Methodology

This article presents the method of calculating the solar irradiation in a dense urban area by means of an algorithm called by the author "Blend-Sun". One of the main principles of the algorithm is considering the influence of reflection and diffusing of light by buildings which can be made of different materials. The whole analyses will be carried out in a full three-dimensional environment.

Modelling the natural and artificial lighting is one of the main factors shaping the image in computer graphics. In rendering (creating the photorealistic image presenting the projection of a 3D model), advanced algorithms can trace millions of individual light beams in order to determine the colour of a particular pixel in the final image. Since the algorithm can simulate the lighting of the virtual world, it can be used to calculate the solar irradiation of individual parts of a 3D model.

The greatest opportunities useful when developing an algorithm of the sunlight exposure can be provided by the open source Blender, with the integrated rendering algorithm called Cycles. It allows creating own scripts using Python 3.4. The algorithm calculating the light distribution is based on the method of Bi-directional path tracing [9], which analyses random paths of light beams. Path tracing is a computer graphics Monte Carlo method of rendering images of three-dimensional scenes such that the global illumination is faithful to reality. With appropriate number of paths (samples), we receive light distribution pattern comparable to that found in the real world.

The selected software has a great advantage over other applications of this type, allowing easy way to obtain information about render lighting of all objects in the virtual world. In Blender there is an option named "bake", used to store data from the rendering engine directly in the texture maps in the form of images, related to the analysed 3D model. The textures related to the object are realized by means of the UV map [10], through which each pixel of the texture has its own stable position on a 3D object. The use of textures as a container for data on the solar radiation also allows any modification of the resolution analysis without dividing the input shape of objects. We can determine that one pixel of the texture reflect any size unit on the model (e.g. meter, centimetre).

In the analysis of solar irradiation it is necessary to use direct and indirect sunlight. This paper will describe how to obtain the data for the two light sources. The reflected light will be modelled by the program algorithm.

Rendering in Blender allows saving the components of the lighting in empty or defined texture maps laid on the objects. Some of these components are "shadow", direct illumination intensity - "diffuse direct" and reflected -"diffuse indirect". After rendering the image is created with the selected data type in a black-and-white scale. The image can be converted into a matrix representing the pixel value in numerical form, which makes it possible to further analyse the data.

In the presented method of obtaining data on the solar irradiation, the individual components of global radiation were divided into four separate parts: direct beam radiation, reflect beam radiation, diffuse radiation and reflect diffuse radiation.

2.1 Materials

For the accurate simulation, on all the objects materials representing the surface properties which they can be produced in the real world. The main materials include light-diffusing (diffuse BSDF (**B**idirectional **S**cattering **D**istribution **F**unction)), mirror-like light-reflecting (Glossy BSDF) and transparent (Transparent BSDF). In the program there is a possibility of defining other materials, but they will be omitted due to the lack of, or a very little effect on the results of the analysis.

Materials such as Glossy and Diffuse affect the formation of light reflections, Transparent material has an impact on the intensity of the resulting shadows. Each material can be combined with another one. It is thus possible to simulate a glass material which reflects light in a specific quantity, but also allows the sun beams to pass through the material with a reduced energy. An example of how different materials behave can be seen in Figure 1. The information on lighting was collected for the horizontal surface, while the vertical surface has a defined material, the angle of incidence of the light beam is 45' for both surfaces. In parts (A) and (B), we can see two different types of light reflections, diffused and mirror-reflected. In parts (C) and (D), we can see two different types of shadows, complete and partial.

The materials allow the control of the albedo values of individual objects. The perfect white material defines the full, one-hundred-percent reflection of light, whereas the perfect black material simulates the full absorption of light and lack of its reflection

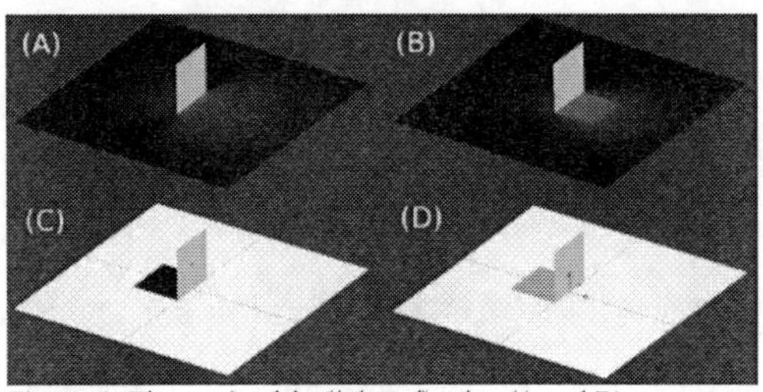

Figure 1. The result of the light reflection (A and B) and shading (C and D).

2.2 Direct and reflect beam radiation

The value of direct beam radiation for a given moment can be calculated with the method described by [11].

The result represents the theoretical value of the energy carried by the direct sun beam, assuming a standard clear sky day at a specific time. To calculate the value of the solar beam radiation, one should take into account the angle of incidence of the beam on the analysed surface, its shading and reflections. These components can be determined using the virtual light source named "Sun", defined in Blender, acting as the natural sun. It sends parallel beams, whose direction is determined by the vector. Knowing the altitude and azimuth of the natural sun, we will automatically have the vector needed to define the correct direction of the beam in the virtual 3D environment. The altitude and azimuth of the sunshine can be calculated using the algorithm specified by [12], along with the correction NOAA atmospheric refraction [13], the input data is the coordinates of the place of the analysis and the date. The algorithm allows calculation of the position of the sun for the set date with a minute accuracy.

The sun vector is needed to determine the angle of incidence of the light beams on the analysed model surface. This angle (in equation used Ai symbol) can be defined as the angle of intersection of the vector of light - "Sun" and the normal vector of the surface (the direction perpendicular to the surface which in Blender is given as the information describing the single surface). With the knowledge of the angle of incidence of sunlight on the surface, we can easily reduce the calculated value of beam radiation (Eqs. 1), dbr denotes direct beam radiation calculated with the method described by [11].

$$direct\ beam\ radiation' = dbr*\cos\ (90°-Ai) \quad (1)$$

In order to estimate the final value of the direct beam radiation we can use the shadow map texture created after rendering. It tells us whether at the given moment the light beam falls on the given element of the model or not. The pixels of the generated texture map in Blender function have the value of 1 in the lack of shadows, and the value of 0 at full shading. All intermediate values indicate the passage of light beams through the semi-transparent surface (e.g. Fig. 1C and 1D). Multiplying the values of the texture pixel (in equation used x symbol) by the reduced value of direct beam radiation we receive the value corrected by the shading factor (Eqs. 2).

$$direct\ beam\ radiation = x*direct\ beam\ radiation' \quad (2)$$

The last part is the analysis of reflections of the direct beam radiation. For this purpose a diffuse indirect texture map generated in Blender function was used. Depending on the location of the adjacent 3D objects and their materials, distribution of light reflections in the form of texture map was obtained (e.g. Fig. 1A and 1B). The pixel values will be from 1 (the maximum intensity reflection) to 0 in the absence of incidence of the reflected light. Several measurements in the application indicate that pixels with a value of 0.6 correspond to the incidence of light of one-hundred-percent intensity (when the beam is perpendicular to the surface). Measurement was made using two parallel planes. One of them pass up the rays of light and was used to retrieve the value of reflection. The second reflected rays completely. Analysing the individual degrees of reflection of light beam "Sun", it was found that the pixel value and the percentage of reflection do not change linearly (e.g. Fig. 2), but they can be approximately described by a third degree polynomial. It is caused by the properties of the Cycles rendering engine. The texture with the pixel values corresponding to the degree of the light reflection can be converted into the reflected solar beam radiation value using the following equation (Eqs. 3), where x is the colour value of the pixel of the image in question, rbr is abbreviation the reflected beam radiation:

$$rbr = dbr * (1.03 * x^3 + 1.98 * x^2 + 0.08*x) \quad (3)$$

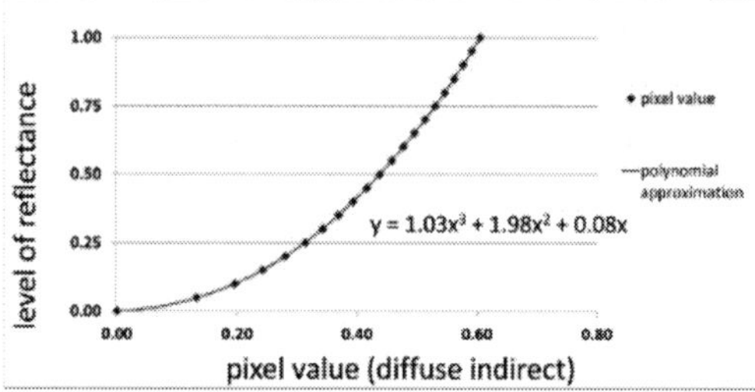

Figure 2. Chart of the reflection of the model light according to the colour of the resulting pixels of the image with the third-degree polynomial approximation

2.3 Isotropic diffuse radiation

Another component of the global solar radiation is the radiation diffused by the atmosphere. For the horizontal unshaded surface, the theoretical value of clear-sky diffuse radiation can be estimated from the relationship described by [14]. The inclusion of shading and different inclination and orientation of the surface can be done with the lighting named "sky" available in Blender. The shades generated by the lighting are soft, the light comes from the virtual sphere surrounding the objects. Shading is stronger when the light source "sky" becomes less visible. As with the direct beam radiation, we can consider not only the direct light but also that reflected.

With direct diffuse radiation we generate the 'bake diffuse direct' texture. Measurements determined that the value of the pixel 0.737 corresponds to 100-percent visibility of the light "sky", while the value of 0 means the total shading. Measurement was made for not shaded plane with light "sky". Multiplying the pixel values by a given value of diffuse radiation, we get the value reduced by the amount of energy that reaches the surface in question (Eqs. 4), dr denotes diffuse radiation calculated with the method described by [13], x are pixel value.

$$diffuse\ radiation = dr * (x / 0.737) \quad (4)$$

The reflection analysis was performed on the diffuse indirect texture map created after rendering in Blender. Depending on the location of objects and their materials we can obtain the image presenting the distribution of reflections. The same as with direct beam radiation, the value of 0.737 corresponds to the incidence of the reflected light with the 100-percent intensity. The value of reflect diffuse radiation is calculated using the above equation (Eqs. 4). In this case, the colour of pixels changes linearly with respect to the value of reflection.

The final value of diffuse radiation is the sum of the direct and reflected diffuse radiation values.

2.4 Global solar radiation

In the presented algorithm, the global solar radiation is composed of four values calculated separately and then added up in order to determine the final value. The algorithm flow chart for one point in time is shown in Figure 3. To analyse the solar irradiation for a certain period of time we must choose the step of the calculation. In the above solution, the step of the calculation can be adopted with the accuracy of one minute, which is also the smallest possible step, and the time range of the analysis.

In the presented solution, theoretical calculation of values for beam radiation and clear-sky diffuse radiation was used. They can be replaced by other values, e.g., derived from long-term observations (meteorological year data).

3 Results

To test the algorithm of the solar irradiation exemplary models representing buildings and a section of terrain were used. They were made by the author. The individual buildings had a material added that simulates the albedo value reflection of 0.35, which roughly corresponds to concrete materials. One model building (number 1 in Figure 5E) was defined with a material imitating glass surfaces with the value of reflectance of 20%. The ground surface was defined with a material with the albedo value of 0.35, corresponding approximately to cobblestones. The location of the analysis was central Warsaw (52.232728N, 21.004202E). The test models can be seen in Figure 5E. Resolution of the analysis was 2 x 2 m.

First tested was the algorithm in the analysis of the total daily solar irradiation for the direct beam radiation value on 1st May without considering reflections. The result of the analysis can be seen on the visualization (e.g. Fig. 4A). For the calculation the time interval was set for 15 minutes. A pixel of the generated textures corresponds to the surface of 2 x 2 m on the model. The maximum total value of direct solar irradiation is 4978 Wh/m2.

Another component is the diffused solar irradiation. The result of the analysis is presented in the Figure 4B. In this case, the maximum total value is 1180 Wh/m2. Visible is the impact of the individual buildings on the level of obstruction, sky dome and thereby reducing the clear sky radiation.

The reflected beam radiation component is a more advanced simulation made on the basis of the Bi-directional path tracing algorithm programmed in Cycles. The result of this analysis can be seen in Figure 4C. The greatest impact on the accumulation of the reflected energy is from the building with the reflective material (glass). The maximum observed total value of energy is about 710 Wh/m2.

The last considered component is the analysis of diffuse radiation reflection (e. g. Fig. 5D). In this case also the biggest influence on the result of the analysis is from the building with the reflective material (glass). The observed value of the reflected energy has the least impact on the result of the analysis, and it does not exceed 191Wh. By adding up the individual values of solar irradiation we receive the complete finish result (e.g. Fig. 5F). The distinct shadows of the buildings reflect the adopted large time interval of 1 h. The maximum daily solar irradiation value observed was 6223 Wh/m2.

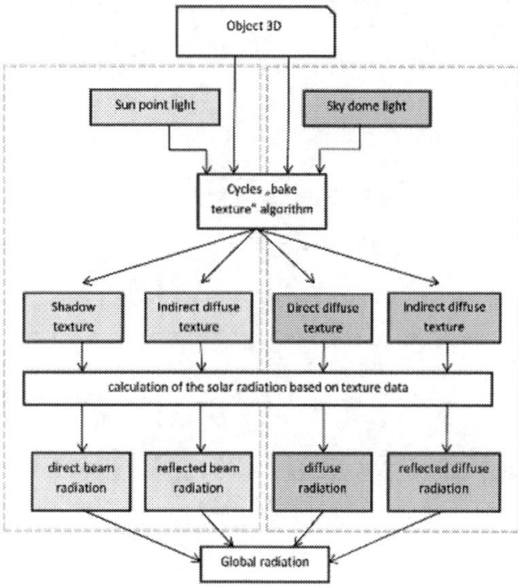

Figure 3. Scheme of the algorithm.

There were additional tests done on three days: 21st June - the day of summer solstice, 20th March - the day of spring equinoxes, and 22nd December - the day of winter solstice (e.g. Fig. 6). The values of solar radiation were analysed separately for each surface of the building. On 21st June, the highest values of solar radiation were measured on roof surfaces, the lowest on the surfaces with a northern exposure. On this day, the reflections of solar radiation are also the highest. On 20th March the highest values of solar radiation were on surfaces with a southern exposure and roof surfaces. On 22nd December the highest values of solar radiation were on surfaces with a southern exposure. In all other exposures they stay at a very low level. Reflections of solar radiation are also very low.

In addition to the solar irradiation test conducted for the regular buildings, the author checked the application of the algorithm for buildings focusing the reflected light. Figure 7 presents the analysis of solar energy reflections on the building 'Walkie-Talkie'. The model building was obtained from [15].

Due to the fact that the actual reflection factor is unknown, in the example a material reflecting the sun beams in 15 percent was used. During the analysis the geographical location of the building was taken into account (51.511103N, 0.083567W); the analysis was made for the longest day in the year, i.e. June 21st . In the absence of data, the adjacent buildings around were not taken into account. The results indicates there are areas which have a higher energy balance than others. In order to eliminate the problem, the building it can be covered with an anti-reflective coating limiting the reflection of sun beams.

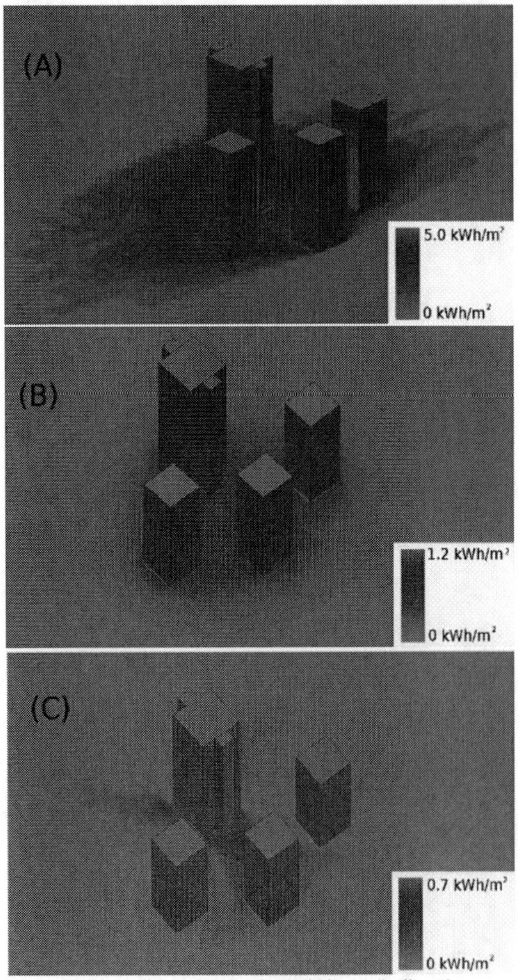

Figure 4. Steps of calculating solar radiation: (A) direct beam radiation component, (B) diffuse radiation component, (C) reflected beam radiation component for one day (May 1st)

No annual analyses of solar irradiation were conducted, as the presented solution is based on the assumption of a clear sky day.

4 Discussion

Looking at the results it can be seen that the analysis of solar irradiation using Blender has considerable potential. Applying different materials we can freely model the analysed area. The particularly unique option is analysis of the reflections of light beams. It allows finding areas with elevated energy balance, which can be used for various purposes, from the design of solar energy generation systems to forecasting risks arising from the accumulation of energy. The evidence can be the result of the simulation for the building 'Walkie-Talkie', which shows the potential risks generated by the building (e.g. Fig. 7).

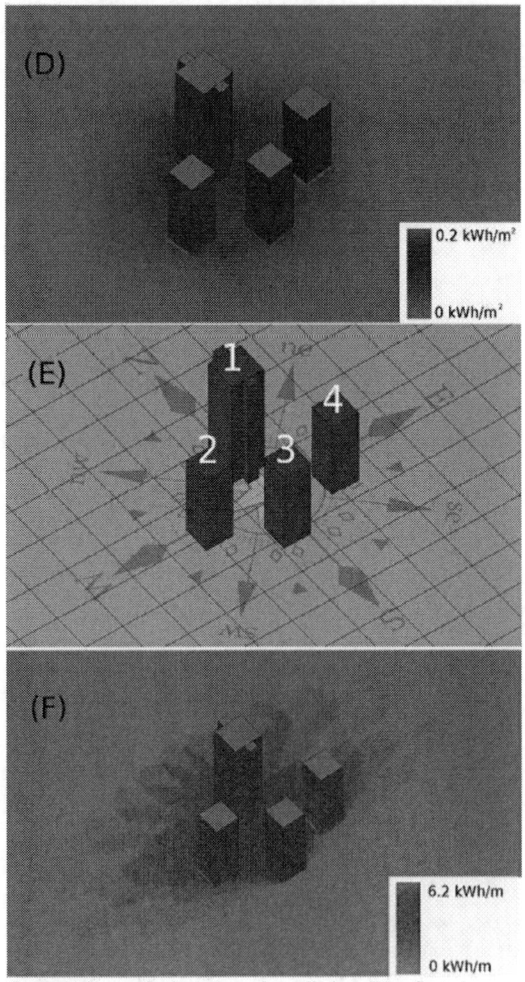

Figure 5. Steps of calculating solar radiation: (D) reflected diffuse radiation component, (E) model with compass rose, (F) the total result of the analysis for one day (May 1st).

DOI: 10.1051/e3sconf/20161000027

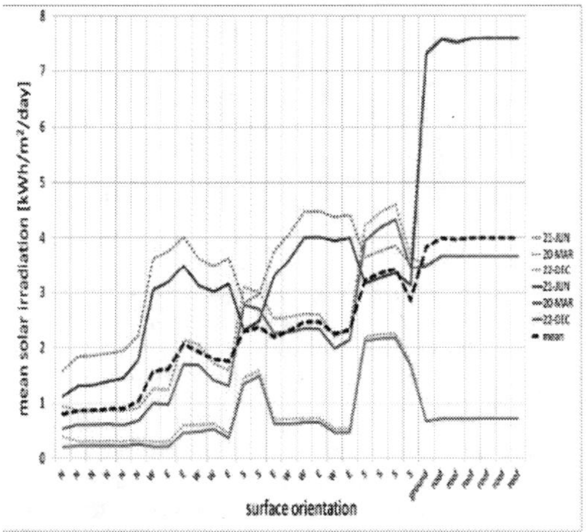

Figure 6. Mean solar irradiation for each building surface for three specific days, with reflection (dotted line), without reflection (solid line) and mean (dashed line)

When reflections are taken into consideration in the analysis, the value of solar irradiation increases on average by 20%, compared to the analysis without taking them into account in case of building facades, but they do not affect the solar radiation on horizontal roofs situated at the same height.

Figure 7. Daily solar irradiation analysis for ground around the building "Walkie-Talkie" on 21st June

In Figure 8 we can see the distribution of the reflected solar radiation calculated with the algorithm on the facades. On the longest day of the year, 21st June, a large amount of reflected radiation can be observed. On the day of the spring equinox, 20th March, reflected beam radiation is visible from the building with a reflective mirror material. On the shortest day of the year a small amount of reflected radiation can be seen.

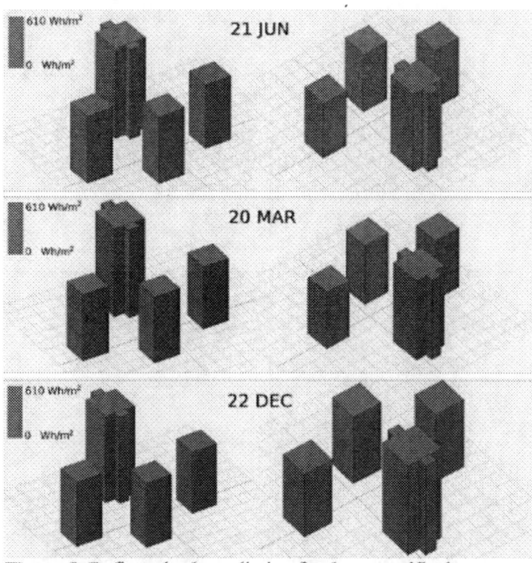

Figure 8. Reflected solar radiation for three specific days

The developed solution also has a downside. One of the weak points is significant lengthening of the analysis duration when analysing reflections. The algorithm Bi-directional path tracing "Cycles" used for the calculation of the reflections and diffusing of light is slow. It can, however, use computers with a number of multi-core processors. In addition, the cycles algorithm is optimized for efficient handling of graphics cards in Nvidia CUDA technology [16], allowing shortening the analysis time with a suitable graphics card several times.

In case of a one-day analysis, its duration depends on: the total number of pixels in textures of the individual elements (analysis resolution), accuracy of the analysis (number of virtual beams), the step of the analysis and the time scope in which the analysis is performed. Table 1 presents the illustrative representation of the efficiency of the calculations on a quad-core i5 3470 processor 3.2 GHz. The analysis time increases linearly with the analysis step, analysis resolution and time scope. The accuracy of the analysis results from the number of random virtual beans (diffuse bounce samples) generated by the algorithm "cycles". With a small number of beams, the reflection analysis will be distorted. By increasing the number of beams we get a better representation of the natural light. In Figure 9 we can see an example of the difference in generating reflections at settings (A) 100, (B) 5,000 beams per pixel of the resulting image. In the case of a small number of beams, we can see artefacts generated during light reflection.

Comparing the presented algorithm with the applications available on the market some similarities can be noticed. Daysim application is based on backward-ray-tracing algorithm. It is also the algorithm used in rendering. However, it does not allow an analysis of mirror reflections.

DOI: 10.1051/e3sconf/20161000027

Table 1. The analysis of time of the solar irradiation for one day at different algorithm settings

Time analysis [minutes]	Resolution (size of pixel in [m])	Time scope of the day [hour]	Step of the analysis [minutes]
16	2x2	8	60
23	2x2	12	60
31	2x2	16	60
92	2x2	12	15
95	1x1	12	60

Diva for Rhino is a complete tool for determining the energy balance of buildings. The algorithm is based on a hybrid deterministic/stochastic (Monte Carlo) ray tracing. It allows the use of materials assigned to analysing surfaces making it possible to conduct analyses taking into account the reflections of light.

Autodesk Ecotec has many features for analysing energy performance of buildings. The author has not found documentation specifying what type of algorithm is used. The surfaces to be analysed are split into smaller equal parts of a predetermined size, or the surface is dashed with points in which the solar radiation is calculated. In the program we have the possibility to define sunlight reflection from the model elements.

In the algorithm "Blend-Sun" presented by the author there are further improvements planned, including adding the ability to load year meteorological data. The next step is to develop an effective calculation of solar irradiation for long time scopes (month, year). Due to the possibility of defining different types of surface textures for the sky dome and operation of the standard HDRI [17] it is planned to more accurately determine the value of diffuse radiation, taking into account circumsolar diffuse and horizon brightening. It is also planned to optimize the script code to work with CityGML data [18], the standard defined for the description and exchange of information on urban infrastructure with the geometry of buildings and descriptive data. In the nearest future it is planned to allow the free access to the script in the form of an addition to the application Blender, together with the necessary detailed documentation.

5 Conclusion

The main aim of the author was to create a flexible application to be used in analysis of the daily solar irradiation in urban areas. The main purposes included providing a simple solution based on open source applications, able to analyse sun exposure, taking into account light reflection and diffusing. Selecting the solution based on application Blender it was possible to create the tool for this purpose.

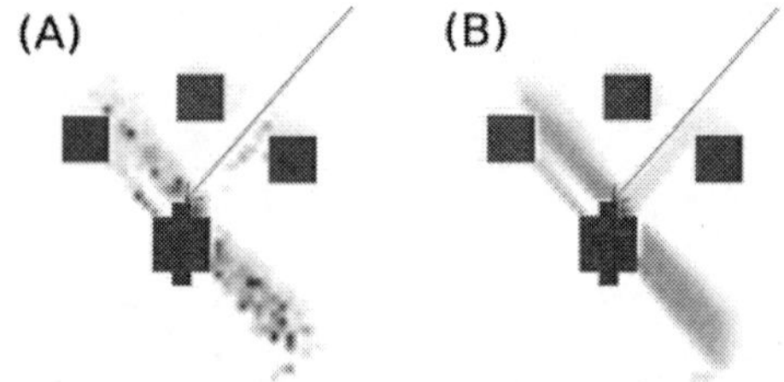

Figure 9. The difference in the application of (A) 100 virtual beam samples per pixel and (B) 5,000 beam samples per pixel of the image. Top view of the test scene. The grey colour shows the reflection of light, the red arrow shows direction of the sunlight

The algorithm "Blend-Sun" has advanced capabilities of conducting analysis of the direct and diffuse radiation reflections. The resolution of the performed analysis is very flexible, and depends on the resolution of the specified textures. The smallest possible time scope of the analysis is one minute. The possibilities of the presented solution can contribute to a greater popularization of advanced solar irradiation analyses by specialists from various areas of science, who will be able to modify the algorithm code to suit their own purposes. These results are useful for the development of solar dissemination policies, architects and urban planning.

Acknowledgement

This paper was supported by Anna Pasek Foundation, scholarship 2015, 7 edition.

References

1. R. R. Lunt, V. Bulovic, Appl. Phys. Lett., **98**, 113305, (2011)
2. Y. Zhao, G. A. Meek, B.G. Levine, R. R. Lunt, Adv. Opt. Mater., **2**, 606-611, (2014)
3. BBC NEWS. *http://www.bbc.com/news/uk-england-london-23930675*, (2013, accessed 12.18.15)
4. Y. Choi, J. Rayl, C. Tammineedi, J. R. Brownson, Sol. Energy, **85**, 2924-2939, (2011)
5. J. Hofierka, J. Kaňuk, Renew. Energ., **34**, 2206-2214, (2009)
6. D. Ibarra, C. F. Reinhart, *Proceedings of Building Simulation 2011* (12th Conference of IBPSA, Sydney, 2011)
7. J. A. Jakubiec, C. F. Reinhart, *Proceedings of Building Simulation 2011* (12th Conference of IBPSA, Sydney, 2011)
8. P. Redweik, C. Catita, M. Brito, Sol. Energy, **97**, 332-341, (2013)
9. E. P. Lafortune, Y. D. Willems, *Proceedings of Third International Conference on Computational Graphics and Visualization Techniques Compugraphics' 93* (ACM, Alvor, 1993)
10. P. S. Heckbert, IEEE Comput. Graph., **6**, 56-67, (1986)
11. H. C. Hottel, Sol. Energy, **18**, 129-134, (1976)

DOI: 10.1051/e3sconf/20161000027

12. J. H. Meeus, *Astronomical algorithms* (Willmann-Bell, Incorporated, 1991)
13. ESRL. Atmospheric Refraction Effects . *http://www.esrl.noaa.gov/gmd/grad/solcalc/calcdetails.html* (2010, accessed 12.18.15)
14. B. Y. Liu, R. C. Jordan, Sol. energy, **4**, 1-19, (1960)
15. 3D Warechouse 20 Fenchurch Street, *https://3dwarehouse.sketchup.com/model.html?id=595d345ff08aed16e724fc90297634ab* (2008, accessed 03.20.15)
16. D. Kirk, *Proceedings of the 6th International Symposium on Memory Management*,(ISMM 2007, Montreal 2007)
17. E. Reinhard, G. Ward, S. Pattanaik, P Debevec, *High Dynamic Range Imaging: Acquisition, Display, and Image-Based Lighting* (Morgan Kaufman Publisher, San Francisco, 2005)
18. G. Gröger, L. Plümer, ISPRS J. Photogramm. Remote Sens., **71**, 12-33 (2012)

 DOI: 10.1051/e3sconf/20161000028

Possibility analysis of combustion of torrefied biomass in 140 t/h PC boiler

Katarzyna Jagodzińska[1,a], Marek Pronobis[1] and Bartłomiej Hernik[1]

[1] *Division of Boilers and Steam Generators, Silesian University of Technology, Konarskiego 20, 44-100 Gliwice, Poland*

Abstract. The study attempts to evaluate the impact of combustion of torrefied willow (Latin: *Salix viminalis)* and palm kernel shell (Latin: *Elaeis guineensis)* on the heat exchange in a 140 t/h PC boiler through an analysis of 6 cases for different boiler loads (60 %, 75 % and 100 %) and a comparison with coal combustion. The analysis is premised on a 0-dimensional model based on the method presented in [15, 16, 17] and long-standing experimental measurements. *Inter alia,* the following results are presented: the temperature distribution of flue gases and the working medium (water/steam) in characteristic points of the boiler as well as heat transfer coefficients for each element thereof. The temperature distribution of both fluids and the heat transfer coefficients are similar for all analysed fuels for each boiler load. However, the flue gas temperature at the outlet is higher in the case of torrefied biomass combustion. Due to that, there is an increase in the stack loss, which involves a decrease in the boiler efficiency. The conclusion is that torrefied biomass combustion is possible in a PC boiler without the need to change the boiler construction. However, it would be less effective than coal combustion.

1 Introduction

Due to the environmental policy of the European Union, there is a steady increase in the use of biomass for energy generation. The utilization of biomass in existing pulverized coal-fired (PC) boilers is related to challenges connected with its properties, such as fibrous and tenacious nature and the high moisture content [1, 2]. Compared to coal, biomass is also characterized by lower energy densities, which makes it necessary to prepare larger amounts of fuel [3]. Due to that, biomass milling and firing in already existing pulverized coal-fired systems is very difficult. Therefore, in order to make it possible, new technologies of biomass pre-treatment need to be developed.

One of such technologies is torrefaction, which is thermal processing of biomass. The process is carried out in inert atmosphere, in nearly atmospheric pressure and in temperatures usually about 200-400 °C so that volatiles, moisture and a considerable amount of acetic acid contained in biomass should be removed [4, 5]. The aim is also disintegration of lignin and cellulose contained in biomass to improve its grindability [6, 7, 8]. Thereby, biomass has a larger energy density (per a mass unit) than raw biomass and a higher lower heating value (LHV). For example, wood torrefaction causes a 30 % mass reduction (converted into gas) and only a 10 % energy reduction at the same time [9].

Torrefied biomass also contains more fixed carbon. Owing to that, it burns longer and the heating conditions in the furnace are more common to coal combustion [10]. Compared to coal, agricultural residues usually contain more chlorine compounds, which is potentially harmful to the boiler heating surfaces due to chlorine-associated corrosion. Torrefaction reduces the chlorine content in biomass up to 50 %. Furthermore, thermal pre-treatment of biomass reduces the microbiological risk associated with the fuel transportation and storage [4].

To conclude, torrefaction makes biomass suitable for grinding in coal mills, but it is not established if biomass can substitute for coal in respect of the heat transfer in existing PC boilers.

The majority of studies on torrefied biomass combustion concern co-firing of biomass with coal or lignite. The 100 % fuel-switching issue was analysed in [11], where results of CFD modelling of torrefied palm kernel shell (PKS) combustion in a 650 t/h PC boiler were presented. Therefore, previous work has focused on the possibilities of using torrefied biomass instead of coal in existing boilers without changing the boiler construction. It reports on fundamental investigations into the combustion of torrefied PKS and willow (W) in an existing 140 t/h PC boiler through an analysis of 6 cases, for three different boiler loads (60, 75 and 100 %), and a comparison with coal (C) combustion.

PKS is residue from the African oil palm (Latin: *Elaeis guineensis*). In Poland it is considered as biomass provided it contains not more than 3 % of non-biomass contamination in the dry ash-free state [12]. Compared to other types of biomass, it is characterized by a higher LHV, a lower content of ash and a smaller amount of potassium and sodium in ash. It also contains less sulphur and chlorine than other types of agro biomass

[a] Corresponding author: katarzyna.jagodzinska@polsl.pl

[13]. Owing to that, it is characterized by a low value of the fouling index (F_u), similar to values for coal, so it has a relatively low potential for causing operational problems in boilers [14].

On the other hand, owing to its availability and resistance to extreme climatic conditions, willow (Latin: *Salix viminalis*) constitutes a popular type of biomass in Poland. Its chemical composition is common to agricultural biomass such as straw or Virginia mallow - it contains more sodium, potassium, chlorine and sulphur than PKS.

The fuels under analysis (PKS and willow) were torrefied in the temperature of about 350-370 °C with a 10 minute residence time. The flue gas from the torrefaction gas combustion was used as the torrefaction medium. The torrefaction degree, defined as the ratio of volatiles released during the process to the initial volatile content of raw biomass (on the dry ash-free basis) has the following value: 32.2 % for PKS and 27.6 % for willow.

The analysis was performed by means of a 0-dimensional model based on the method described in [15, 16, 17, 18] and supported by industrial tests. In the case of thermal calculations of the furnace, the KTI method was used [15, 16, 17], while thermal calculations of the boiler convection part was based on the method presented in [18].

2 Input data

The thermal calculations are based on a 140 t/h PC boiler (Figure 1). The markings of the boiler elements and the temperatures in the boiler characteristic points shown in Figure 1 correspond to those used in the calculations.

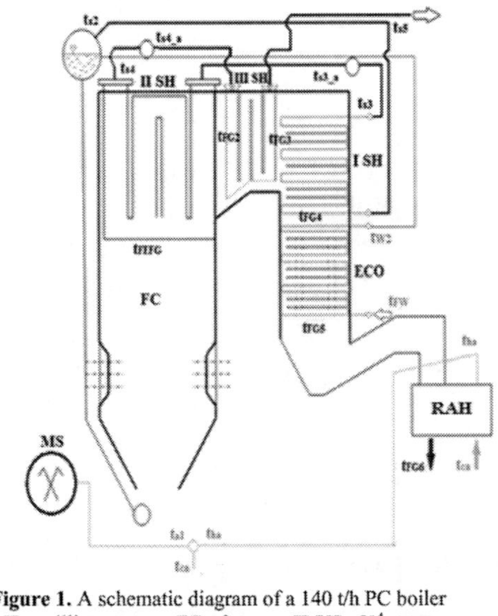

Figure 1. A schematic diagram of a 140 t/h PC boiler MS - milling system; FC - furnace; II SH - 2nd stage superheater; III SH - 3rd stage superheater; I SH - 1st stage superheater; ECO - economizer; RAH - rotary regenerative air preheater

The boiler basic operating parameters are listed in Table 1. Table 2 shows the LHV and the elemental composition of fuels.

Table 1. Basic operating parameters of the boiler

Parameter	Symbol	Value	Unit
Nominal operating capacity of the boiler	$\dot{m}_s$	140	t/h
Nominal temperature of live steam	t_{s_nom}	540	°C
Nominal pressure of live steam	p_{s_nom}	9.8	MPa
Temperature of feed water	t_{FW}	215	°C
Temperature of water in the attemperator	t_{W_a}	160	°C
Unburnt carbon in slag	C_s	7	%
Unburnt carbon in fly ash	C_{fa}	5	%

Table 2. LHV and elemental composition of fuels (as-received state). C - coal; PKS - palm kernel shell; W - willow

Parameter / Symbol		Value / Fuel			Unit
		C	PKS	W	
Lower heating value	LHV	21.5	23.8	18.5	MJ/kg
Carbon content in fuel	C^{ar}	54.2	64.8	51.6	%
Oxygen content in fuel	O^{ar}	6.7	22.3	22.8	%
Hydrogen content in fuel	H^{ar}	3.4	4.3	3.7	%
Nitrogen content in fuel	N^{ar}	1.1	0.9	0.6	%
Sulphur content in fuel	S^{ar}	0.8	0.04	0.03	%
Moisture content in fuel	M^{ar}	10.3	1.1	17.7	%
Ash content in fuel	A^{ar}	23.5	6.5	3.6	%
Chlorine content in fuel	Cl^{ar}	-	0.04	0.01	%

3 Results and discussion

This study attempts to evaluate the impact of the combustion of torrefied PKS and willow on the heat transfer in a 140 t/h PC boiler through an analysis of changes in fuel and boiler load.

Three relative boiler loads were taken into consideration - 60, 75 and 100 % (of the boiler nominal load). The temperatures obtained in the boiler

 DOI: 10.1051/e3sconf/20161000028

characteristic points are listed in Table 3. Table 4 shows values of the stack loss, the boiler efficiency and the fuel flow rate for each case under analysis. Table 5 presents values of the heat transfer coefficient on the flue gas side in each element of the boiler.

A change in fuel involves a change in the fly ash chemical composition and, as a result, a change in the character of the heat transfer in the boiler. Biomass contains more sodium, potassium and chlorine compounds that intensify slagging and fouling of the surfaces of heat exchangers and consequently, decrease their thermal efficiency ψ [19]. Due to that, it was assumed that thermal efficiency of the furnace waterwalls is reduced by 6 % in the case of torrefied biomass combustion. It was further assumed that thermal efficiency of the boiler heating surfaces is reduced by 20 % compared to coal combustion. The values are based on own unpublished research conducted by the staff of the Division of Boilers and Steam Generators of the Silesian University of Technology.

Table 3. Temperatures in the boiler characteristic points. C - coal; PKS - palm kernel shell; W - willow

Parameter / Symbol		**Units**	**Relative boiler load**								
			60 %			**75 %**			**100 %**		
			Value / Fuel								
			C	**PKS**	**W**	**C**	**PKS**	**W**	**C**	**PKS**	**W**
Furnace exit gas temperature	t_{FEFG}	°C	892	956	1025	993	1007	1074	1098	1134	1180
Flue gas temperature at the outlet of the 2nd stage superheater	t_{FG2}	°C	663	703	783	722	743	826	806	859	931
Flue gas temperature at the outlet of the 3rd stage superheater	t_{FG3}	°C	611	648	724	652	682	762	711	741	796
Flue gas temperature at the outlet of the 1st stage superheater	t_{FG4}	°C	457	528	593	472	551	619	509	604	658
Flue gas temperature at the outlet of the economizer	t_{FG5}	°C	310	389	442	311	403	459	331	446	497
Exit flue gas temperature	t_{FG6}	°C	154	203	237	128	199	235	145	229	264
Temperature of feed water	t_{FW}	°C	215	215	215	215	215	215	215	215	215
Temperature of water at the outlet of the economizer	t_{W2}	°C	275	257	278	269	255	275	278	260	281
Temperature of steam at the outlet of the drum	t_{s2}	°C	310	310	310	310	310	310	310	310	310
Temperature of steam at the outlet of the 1st stage superheater	t_{s3}	°C	389	350	378	385	349	376	405	356	381
Temperature of steam at the outlet of the attemperator downstream the 1st stage superheater	t_{s3a}	°C	376	350	361	374	349	367	357	354	345
Temperature of steam at the outlet of the 2nd stage superheater	t_{s4}	°C	509	476	525	502	467	523	486	486	498
Temperature of steam at the outlet of the attemperator downstream the 2nd stage superheater	t_{s4a}	°C	509	476	504	502	467	502	485	482	453
Live steam temperature	t_{s5}	°C	540	499	540	540	492	540	540	540	540

 DOI: 10.1051/e3sconf/20161000028

Table 4. Stack loss, boiler efficiency and fuel flow rate. C - coal; PKS - palm kernel shell; W - willow

Parameter / Symbol		Units	Relative boiler load								
			60 %			75 %			100 %		
			Value / Fuel								
			C	PKS	W	C	PKS	W	C	PKS	W
Stack loss	S_2	%	7.3	8.12	11.9	5.9	7.9	11.8	6.8	9.3	13.4
Boiler efficiency	η	%	90.7	90.6	86.9	92.3	90.8	87.1	91.4	89.5	85.6
Fuel flow rate	$\dot{B}$	kg/s	2.6	2.7	3.8	3.8	3.3	4.7	5.2	4.7	6.4

Table 5. Heat transfer coefficient on the gas side in each element of the boiler. C - coal; PKS - palm kernel shell; W - willow

Parameter / Symbol		Units	Relative boiler load								
			60 %			75 %			100 %		
			Value / Fuel								
			C	PKS	W	C	PKS	W	C	PKS	W
Heat transfer coefficient on the gas side in the 2nd stage superheater	α_{FG_IISH}	$W/m^2{\cdot}K$	117.4	72.8	94	132.6	80.4	106.9	163.7	115.3	129.2
Heat transfer coefficient on the gas side in the 3rd stage superheater	α_{FG_ISH}	$W/m^2{\cdot}K$	70.4	60.8	74.5	76.8	66.7	82	182.4	171.4	195
Heat transfer coefficient on the gas side in the 1st stage superheater	α_{FG_IIISH}	$W/m^2{\cdot}K$	63.5	42.8	53.2	68.3	47	58.5	83.5	58.8	70.5
Heat transfer coefficient on the gas side in the economizer	α_{FG_ECO}	$W/m^2{\cdot}K$	91.1	59.5	73.4	96.2	63.9	79.3	116.7	80.3	96.5
Heat transfer coefficient on the gas side in the rotary regenerative air preheater	α_{FG_RAH}	$W/m^2{\cdot}K$	95.5	56.8	68.5	98.1	59.5	72.2	119.7	73.3	87.4

3.1 Furnace exit gas temperature

Substitution of torrefied biomass for coal results in an increment in the furnace exit gas temperature t_{FEFG} (Figure 2). It is expressed by A.M. Gurvich's formula, which is the basis of the CKTI furnace calculation method.

$$t_{FEFG} = T_a \cdot \frac{Bo^{0,6}}{M{\cdot}a_{fu}^{0,6} + Bo^{0,6}}, K \quad (1)$$

T_a - adiabatic flame temperature, K,
Bo - Boltzmann number, -,
M - temperature field coefficient, -,
a_{fu} - furnace emissivity, -.

Temperature field coefficient M depends on the relative level of burners in the boiler, the fuel properties, such as the state and the volatile fraction content, and the furnace type. None of these parameters change for PKS, willow or coal, so coefficient M could be considered as constant.

However, other parameters change due to the fuel replacement (Figure 3). Adiabatic flame temperature T_a depends on the amount of heat released from fired fuel and heat introduced into the furnace by air [17]. Mostly it depends on heat released from fuel, i.e. there is a direct dependence on the fuel LHV. Compared to coal or willow, PKS is characterized by a higher LHV, so in this case T_a has the highest values.

Boltzmann number Bo value is proportional to the fuel flow rate. Therefore, willow has the lowest

DOI: 10.1051/e3sconf/20161000028

LHV of the fuels under analysis, and it has the highest values of the fuel flow rate and, consequently, of the Boltzmann number.

Furnace emissivity a_{fu} depends on flame emissivity a_{fl}. Flame emissivity depends directly on the flue gas composition, specifically - on the contents of fly ash, moisture and triatomic gases. The fly ash content has a dominant influence on a_{fl}, because of its better radiative properties compared to other components. Torrefied biomass contains less ash than coal (by about 17-20 %). For this reason, flue gases contain less fly ash, deteriorating the radiative properties. Due to that, a smaller amount of heat is absorbed by the furnace waterwalls, and, in consequence, more heat remains in flue gases.

The flue gas temperature at the furnace outlet is affected by any change in the parameters under consideration, but it is the Boltzmann number that has the greatest effect. According to it, t_{FEFG} increases by an average of 20 °C for PKS and 66 °C for willow combustion compared to coal combustion.

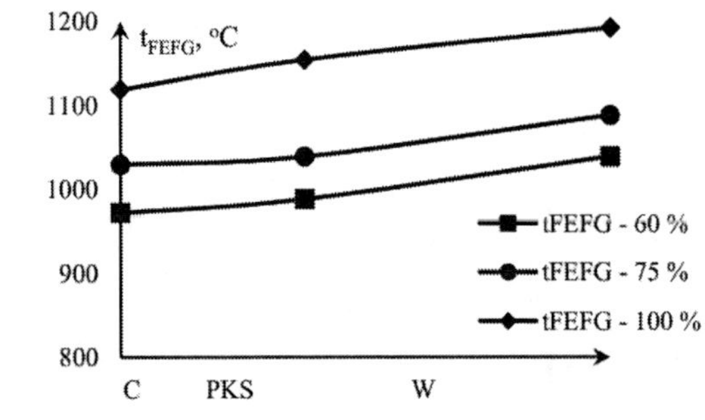

Figure 2 Furnace exit gas temperature as a function of the fuel type and the boiler relative load (60, 75 and 100 %)

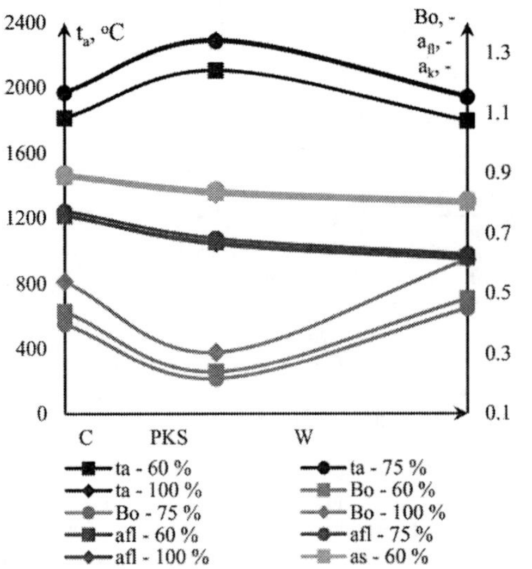

Figure 3 Adiabatic flame temperature, Boltzmann number, flame emissivity and furnace emissivity as a function of the fuel type and the boiler relative load (60, 75 and 100 %)

3.2 Flue gas and water/steam temperatures in the boiler

An increase of furnace exit gas temperature in the case of torrefied biomass combustion results in a rise in other temperatures in the boiler characteristic points. Nevertheless, the nature of the flue gas temperature distribution is similar for each fuel combustion. Due to that, it could be considered that PKS, willow and coal are characterized by similar parameters of the heat exchange, therefore torrefied biomass has the potential to substitute for coal in existing PC boilers.

Figures 4, 5 and 6 show the temperature distribution in the boiler for each relative load and fuel type.

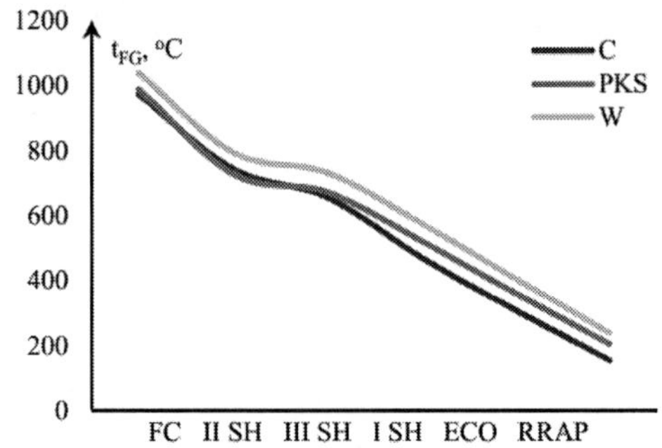

Figure 4. Flue gas temperature distribution in the boiler characteristic points for the boiler relative load of 60 %

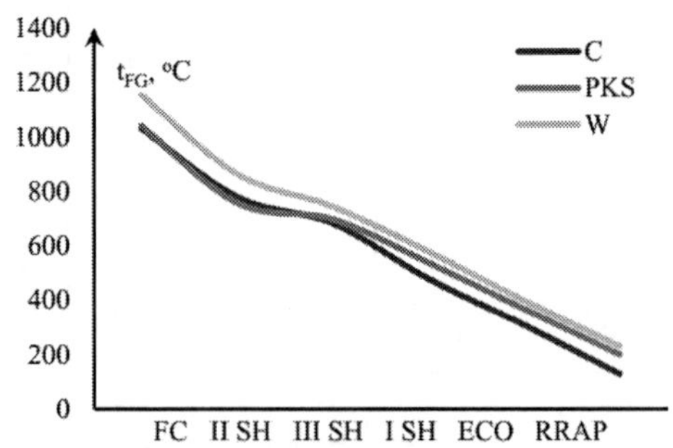

Figure 5. Flue gas temperature distribution in the boiler characteristic points for the boiler relative load of 75 %

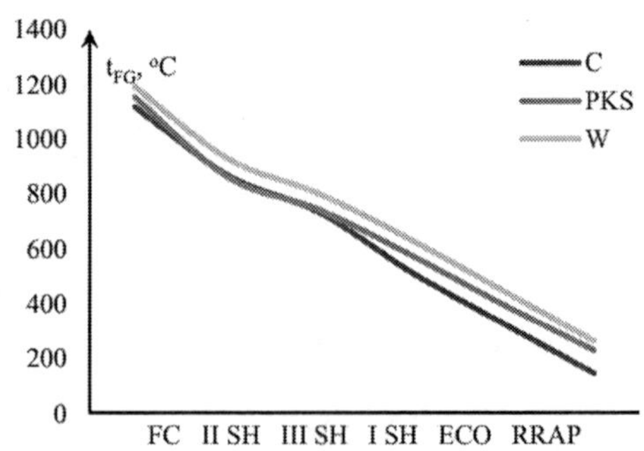

Figure 6. Flue gas temperature distribution in the boiler characteristic points for the boiler relative load of 100 %

There are some differences between the analysed fuels in the case of the working fluid (water/steam) temperature distribution in the boiler. The temperature distribution nature is still similar, but in the case of PKS combustion, the nominal exit steam temperature (540 °C) is not maintained for the boiler relative load included in the range of 60 to 75 % (Figures 7, 8 and 9). This is caused by the PKS high LHV, which results in the low fuel flow rate. Due to that, the amount of flue gases is much lower compared to coal or willow combustion, and this means the flue gas lower velocities. Considering also reduced thermal efficiency of the boiler heating surfaces, the heat transfer is significantly worse in the case of PKS combustion. It might be concluded that in such a case it will be necessary to enlarge the boiler heating surfaces, which would involve the need to change the boiler construction.

However, most types of palm kernel shell are characterized by lower LHV's than the one analysed herein (23.8 MJ/kg) – the values are usually about 19-21 MJ/kg (depending on the temperature of torrefaction) [20, 21], whereas the PKS higher heating value (HHV) is usually about 22.9 MJ/kg [22]. In such a case, the live steam temperature is maintained.

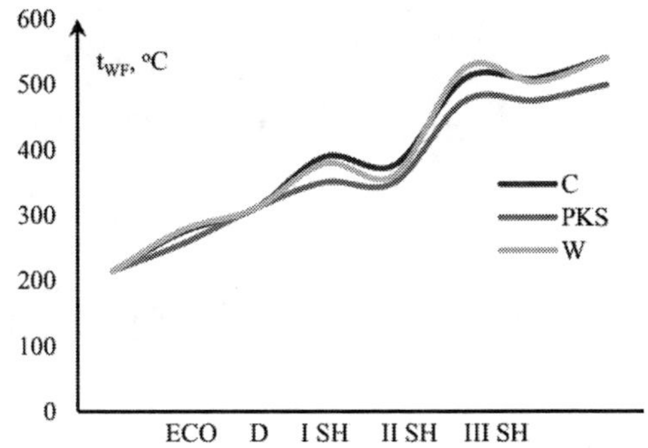

Figure 7. The working fluid temperature distribution in the boiler characteristic points for the boiler relative load of 60 %

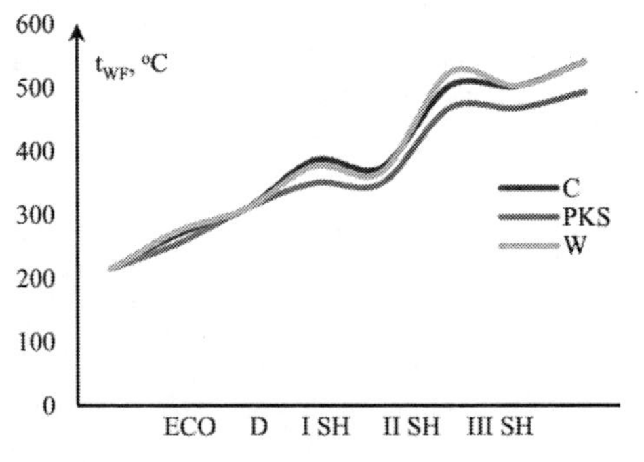

Figure 8. The working fluid temperature distribution in the boiler characteristic points for the boiler relative load of 75 %

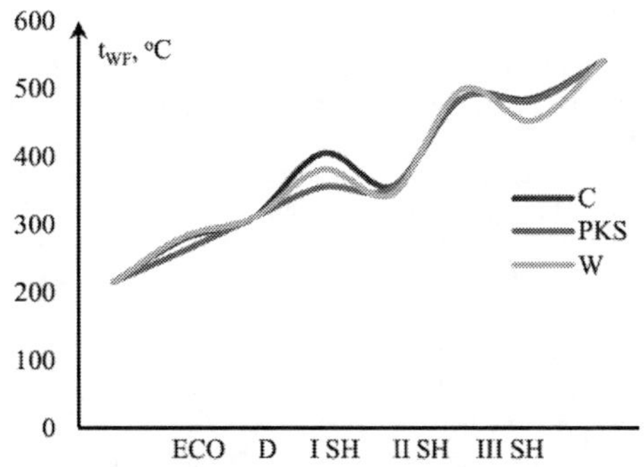

Figure 9. The working fluid temperature distribution in the boiler characteristic points for the boiler relative load of 100 %

3.3 Boiler efficiency

An increase in the furnace exit gas temperature and a reduction in thermal efficiency of the boiler heating surfaces involve a rise in the exit flue gas temperature. This results in an increase in the stack loss (Figure 10) and, in consequence, a decrease in the boiler efficiency which is reduced on average by 1.2 percentage points (pp) for PKS and by about 5 pp for willow combustion compared to coal combustion (Figure 11).

However, taking account of environmental and economic benefits of biomass combustion, these values are not significant enough to discredit it. In this case, consideration should be given to the use of such waste heat, as an additional benefit of the fuel replacement.

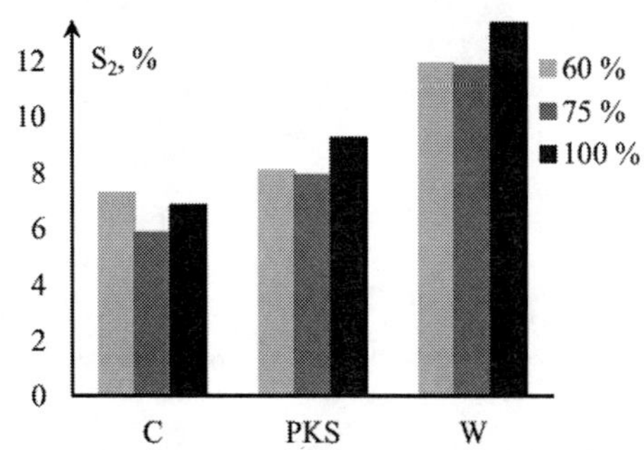

Figure 10. Stack loss as a function of the fuel type and the boiler relative load (60, 75 and 100 %)

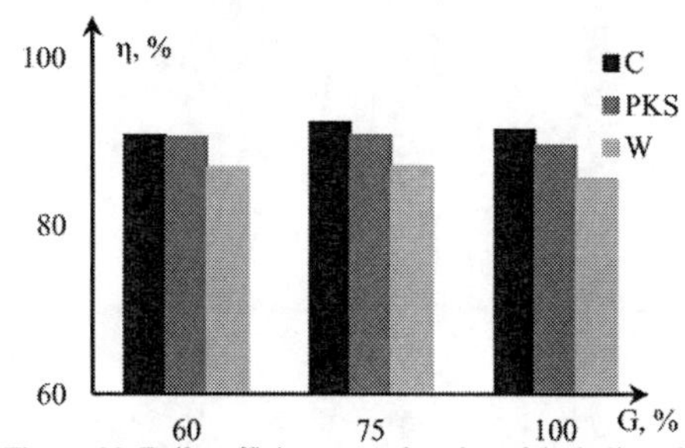

Figure 11. Boiler efficiency as a function of the boiler relative load (G) and the fuel type

3.4 Flue gas heat transfer coefficient

According to the similar temperature distributions of the fluids, the nature of the heat exchange in the boiler elements should be similar for each fuel type. It is well demonstrated by the overall heat transfer coefficient *k*. Figures 12 and 13 show it on the example of the 2nd and the 1st stage superheater. These components were chosen because of the difference between the kind of the heat exchange – the 2nd stage superheater is classified as the radiant and the 1st stage superheater as the convective type of a heat exchanger.

Otherwise, the heat transfer coefficient on the gas side α_{FG} changes slightly according to changes in fuel (Figures 14 and 15). Despite an increase in the mean temperature difference in all heat exchangers of the boiler, the values of α_{FG} are reduced in the case of torrefied biomass combustion. It is particularly noticeable in the case of the 2nd stage superheater. This is due to the deterioration in the flue gas radiative properties and, in consequence, in the gas emissivity. It also results from a reduction in the flue gas velocity, which deteriorates the convective heat exchange. It is mostly the effect of the change in the flue gas temperature and density.

Due to lower live steam temperatures in the case of PKS combustion, the values of α_{FG} are in fact lower for that type of torrefied biomass. If PKS has a lower LHV, its values of α_{FG} would be close to values for willow.

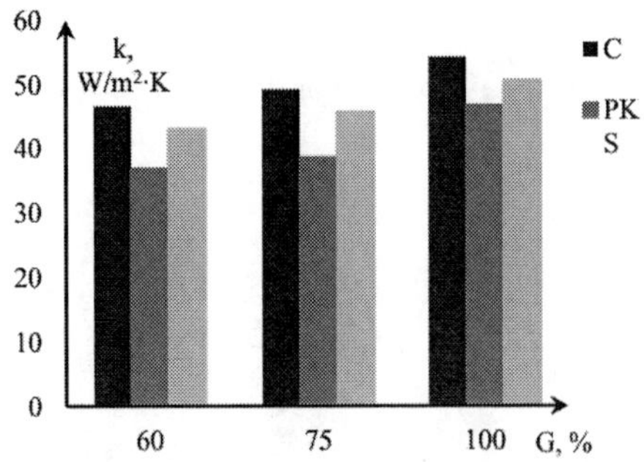

Figure 12. Overall heat transfer coefficient in the 2nd stage superheater as a function of the boiler relative load (G) and the fuel type

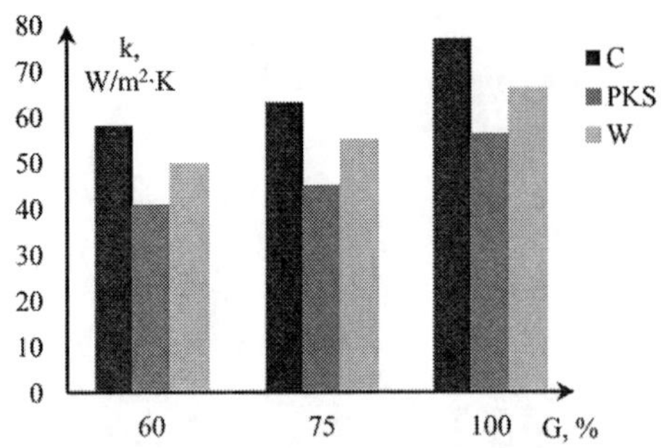

Figure 13. Overall heat transfer coefficient in the 1st stage superheater as a function of the boiler relative load (G) and the fuel type

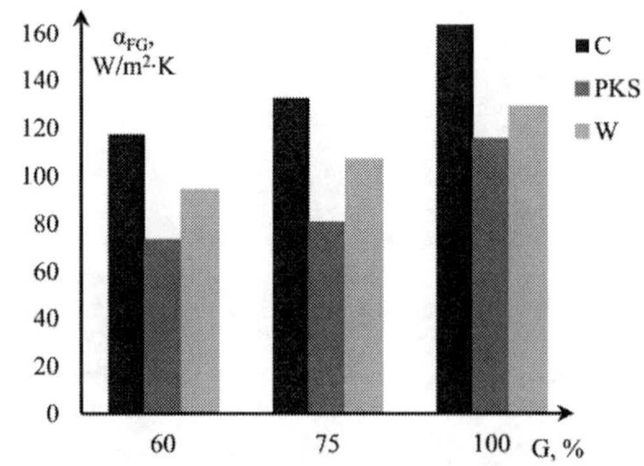

Figure 14. Heat transfer coefficient on the gas side in the 2nd stage superheater as a function of the boiler relative load (G) and the fuel type

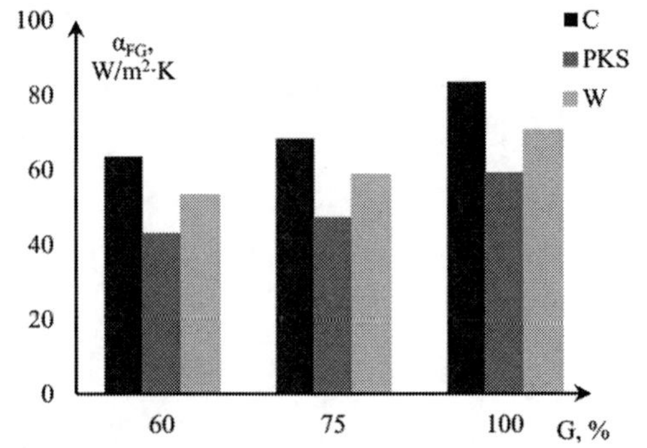

Figure 15. Heat transfer coefficient on the gas side in the 1st stage superheater as a function of the boiler relative load (G) and the fuel type

4 Conclusion

The possibility of substituting torrefied biomass for coal depends on the biomass type, especially on its elemental composition and the LHV. As most types of agro biomass are similar to willow, it might be assumed that most of them could be used instead of coal in existing PC boilers.

In the case of PKS combustion, live steam temperatures for the boiler relative load of 60 and 75 % are not maintained as a result of the fuel high LHV. However, most torrefied palm kernel shells have lower LHV's compared to the type analysed herein, so it could be assumed that in most cases of PKS combustion thetemperatures would be equal to the nominal value.

Combustion of torrefied biomass in a PC boiler causes an increase in the furnace exit gas temperature and a decrease in thermal efficiency of the heating surfaces. In consequence, there is a rise in the exit flue gas temperature, which causes a reduction in the boiler efficiency by about 1.2 pp for PKS and by 5 pp for willow combustion. A further analysis is required to establish whether this would be compensated for by the benefits of biomass combustion.

In addition, it should be noted that PKS shipment from distant regions of the world requires energy derived from fossil fuels. In view of the increasing number of power plants firing imported biomass, this issue has raised concern and become the subject of analyses. It is estimated that emissions from the transport of such biomass constitutes about 15% of total avoided emissions compared to coal combustion [23]. Besides, as mentioned in [24], a reduction in savings of emissions from biomass combustion could reach 15-50 % due to its importation and to using nitrogen fertilisers. Therefore, the regulations concerning classification of PKS and other imported biomass types as a 100% renewable energy source should be corrected by appropriate legislation.

Acknowledgments

The investigations presented in this work were financed within the KIC InnoEnergy innovation Project BioPoGen "Biocoal for Power Generation".

References

1. V. Repellin, A. Govin, M. Rolland, R. Guyonnet, Biomass Bioenerg. **23**, 923-930 (2010)
2. B. Arias, C. Pevida, J. Fermoso, M.G. Plaza, F. Rubiera, J.J. Pis, Fuel Process. Technol. **89**, 169-175 (2008)
3. T. G. Bridgeman, J. M. Jones, I. Shield, P.T. Williams, Fuel **87**, 844-856 (2008)
4. M. Kopczyński, J. Zuwała, *Biomass torrefaction as a way of elimination of technical barriers existing in large-scale co-combustion* (Polityka Energetyczna, Kraków, 2013) [in Polish]
5. M. Kratofil, R. Zarzycki, R. Kobylecki, Z. Bis, *Investigation of biomass torrefaction process* (Zeszyty naukowe Politechniki Rzeszowskiej 291, Rzeszów, 2015) [in Polish]
6. D. T. Van Essendelft, X. Zhou, B.S.-J. Kang, Fuel **105**, 103-111 (2013)
7. M. Phanphanich, S. Mani, Bioresour. Technol. **102** 1246-1253 (2011)
8. L. Kokko, H. Tolvanem, K. Hamalainen, R. Raiko, Biomass Bioenerg. **42**, 219-223 (2012)
9. M. J. C. van der Stelt, H. Gerhauser, J. H. A. Kiel, K. J. Ptasinski, Biomass Bioenerg. **35**, 3748-3762 (2011)
10. B. Batidzirai, A. P. R. Mignot, W. B. Schakel, H.M. Junginger, A. P. C. Faaij, Energy, **62**, 1-19 (2013)
11. J. Li, A. Brzdekiewicz, W. Yang, W. Blasiak, Appl. Energ. **99**, 344-354 (2012)
12. Energy Regulatory Office, *Appendix 1 to Communication 30/2011 on biomass classification for energy purposes* (Warszawa, 2011) [in Polish]
13. B. Jagustyn, I. Patyna, A. Skawińska, Chemik **67, 6**, 552-559 (2013)
14. R. Rajczyk, Magazyn Biomasa, **4**, 9-14 (2015)
15. P. Orłowski, W. Dobrzański, E. Szwarc, *Steam boilers* (WNT, Warszawa, 1979) [in Polish]
16. *Teplowoj rasczet kotelnych agregatov (Normativnyj metod)* (Energia, Moskwa, 1973) [in Russian]
17. P. Basu, C. Kefa, L. Jestin, *Boilers and Burners Design and Theory* (Springer, New York, 2000)
18. M. Pronobis, *Heat transfer in contaminated convection boiler surfaces* (Zeszyty Naukowe Politechniki Śląskiej seria: Energetyka, Gliwice, 1992) [in Polish]
19. M. Pronobis, W. Wojnar, Energ. Convers. Manage. **74**, 461-470 (2013)
20. A. A. Jaafar, M. M. Ahmad, Int. J. Chem. Mol. Nucl. Mater. Metall. E. **12**, 1106-1109 (2011)
21. H. Pawlak-Kruczek, M. Czerep, Nowa Energia **1**, 31-38 (2016)
22. W. Kordylewski, A. Tatarek, Archiwum Spalania **12, 3**, 109-116 (2012)
23. A. Adamkiewicz, W. Zeńczak, Rynek Energii **6**, (2015)
24. S. Korhaliller: *The UK's Biomass Energy Development Path* (International Institute for Environment and Development, London, 2010)

E3S Web of Conferences 10, 00029 (2016) DOI: 10.1051/e3sconf/20161000029
SEED 2016

ICCP cathodic protection of tanks with photovoltaic power supply

Mirosław Janowski[1,a] and Agnieszka Wantuch[2]

[1]*AGH University of Science and Technology, Department of Fossil Fuels, Krakow, Poland*
[2]*AGH University of Science and Technology, Electrical Department, Krakow, Poland*

Abstract. Corrosion is the result of the electrochemical reaction between a metal or composite material usually having conducting current properties. Control of corrosion related defect is a very important problem for structural integrity in ground based structures. Cathodic protection (CP) is a technique to protect metallic structures against corrosion in an aqueous environment, it is employed intense on the steel drains in oil and gas industry, specifically to protect underground tanks and pipelines. CP is commonly applied to a coated structure to provide corrosion control to areas where the coating may be damaged. It may be applied to existing structures to prolong their life. There are two types of cathodic protection systems: sacrificial (galvanic) anode cathodic protection (SACP); the other system is Impressed Current Cathodic Protection (ICCP). Majority of the structures protected employ impressed current system. The main difference between the two is that SACP uses the galvanic anodes which are electrochemically more electronegative than the structure to be protected - the naturally occurring electrochemical potential difference between different metallic elements to provide protection; ICCP uses an external power source (electrical generator with D.C.) with inert anodes, and this system is used for larger structures, or where electrolyte resistivity is high and galvanic anodes cannot economically deliver enough current to provide protection. The essential of CP is based on two parameters, the evolution of the potential and the current of protection. A commonly accepted protection criterion used for steel is a potential value of minus 850 mV. ICCP system consist of anodes connected to a DC power source. As power sources may be used such as solar panels, wind turbines, etc. The object of this study is analysis of the possibilities and operating parameters of ICCP system supplied with photovoltaic solar panels. Photovoltaic generator made up of the following elements: photovoltaic modules of solar cells, a control and regulation system, a storage system..

1 Introduction

Corrosion is the result of the electrochemical reaction between a metal or composite material usually having conducting current properties. When two metals having different potentials are coupled together, current will flow. Variations in the metal may be the result of temperature, stress, metal composition or the presence of impurities. In general, as the soil resistivity becomes lower, ground water chemistry can also impact soil aggressiveness, so the corrosion of a buried metal becomes easier. Corrosion will occur at the point where positive current leaves the metal surface.

There are two basic mechanisms by which metals in electrolytes corrode:

• Electrolytic Corrosion
• Galvanic Corrosion

Electrolytic corrosion is a result of direct current from outside sources entering and then leaving a particular metallic structure by way of the electrolyte. A corrosion cell is a circuit consisting of an anode, a cathode, an electrolyte, and an electrical contact between the anode and cathode. In underground work, this type of corrosion is often referred to as stray current corrosion and results from currents entering the soil from sources of DC.

To live a long and productive existence, propane tanks and metallic piping need protection. The first line of defense against corrosion is coating the steel tank like many of the manufacturers do in the factory. The second line of defense is cathodic protection, which is now well established on a large variety of immersed and buried metallic structures as well as reinforced concrete structures, and provides effective corrosion control [1]. It can extend the life of an underground tank by helping to prevent corrosion and rust. Cathodic protection systems prevent the oxidation process from occurring by creating a current flow from the cathodic protection system to the structure. This system prevent corrosion for many years.

2 Cathodic protection

Control of corrosion related defect is a very important problem for structural integrity in ground based structures. In some cases, cathodic protection (CP) is required by policy or regulation for example in case of cathodic protection for tanks. Problem of protection against corrosion has been described in many publications. Review of numerical methods applied for corrosion protection issue can be found at [2, 3]. Cathodic protection

[a] Corresponding author: janowski@agh.edu.pl

DOI: 10.1051/e3sconf/20161000029

works by preventing the anodic reaction of metal dissolution occurring on the structure under protection. It is a technique to protect metallic structures against corrosion in an aqueous environment, it is employed intense on the steel drains in oil and gas industry, specifically to protect underground tanks and pipelines.

CP is commonly applied to a coated structure to provide corrosion control to areas where the coating may be damaged. It may be applied to existing structures to prolong their life. There are two types of cathodic protection systems:

• Sacrificial anode cathodic protection (SACP);
• Impressed Current Cathodic Protection (ICCP).

Majority of the structures protected employ impressed current system. The main difference between the two is that SACP uses the galvanic anodes which are electrochemically more electronegative than the structure to be protected – the naturally occurring electrochemical potential difference between different metallic elements to provide protection. ICCP uses an external power source (electrical generator with DC) with inert anodes, and this system is used for larger structures, or where electrolyte resistivity is high and galvanic anodes cannot economically deliver enough current to provide protection [4, 5]. The power source must be able to deliver direct current (DC) and examples are transformer rectifier units, solar generating units etc.

The essential of CP is based on two parameters, the evolution of the potential and the current of protection. A commonly accepted protection criterion used for steel is a potential value of minus 850 mV [6].

3 ICCP system

Galvanic protection works well with small tanks usually meant for residential and small commercial applications [7]. A magnesium anode has the power of a 1 volt battery [8]. In many applications, the potential difference between the galvanic anode and the steel structure is not enough to generate sufficient current for cathodic protection to occur. For larger structures, or where electrolyte resistivity is high, ICCP systems (fig. 1) are used, 'cause galvanic anodes cannot economically deliver enough current to provide protection.

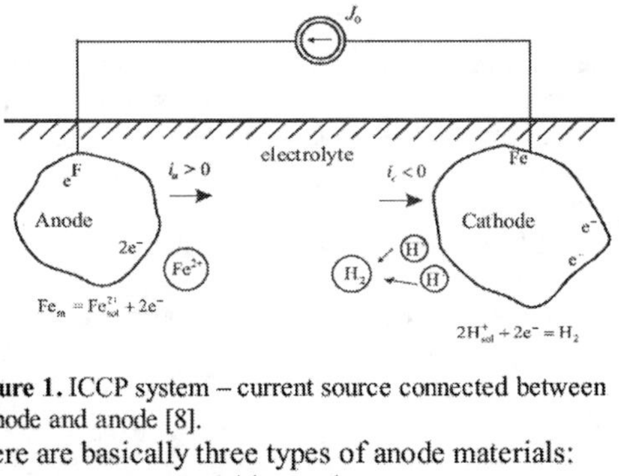

Figure 1. ICCP system – current source connected between cathode and anode [8].

There are basically three types of anode materials:

• Inert or non consumable anodes
• Semi-consumable anodes
• Consumable anodes

ICCP system consist of anodes connected to a DC power source. As power sources may be used such as solar panels, wind turbines, etc. The object of this study is analysis of the possibilities and operating parameters of ICCP system supplied with photovoltaic solar panels. Photovoltaic generator made up of the following elements: photovoltaic modules of solar cells, a control and regulation system, a storage system.

In ICCP systems since the driving voltage is provided by the DC source there is no need for the anode to be more active than the structure to be protected. Typically, the DC output will be in the range 15–100 V and 5–100 A, but 200 V/200 A units are also known.

Anodes for ICCP systems are available in a variety of shapes and sizes. Typical anodes are titanium coated with mixed metal oxide (MMO) or platinum, silicon cast iron, graphite and other materials. The output of the ICCP system should be optimised to provide enough current to provide protection to the protected structure.

One of the most important parameters in the design of cathodic protection systems is the electrical resistivity of the environment. Resistivities encountered for tank environments vary from 1 ohm cm for brackish river water to greater than 500,000 ohm cm in non-porous granite [1]. One of the protection criterion based on field experience, is that a negative potential change of 100 - 300 mV from the free corrosion potential is a good measure of adequate protection. Another condition depending on normal component of current density says that for bare metal in the ground, a current of 11 to 22 mA/m^2 of bare metal-electrolyte surface has been found sufficient, except under extreme or extraordinary conditions [8].

4 Equations and mathematics

Corrosion is a continuous process, practically independent of the operating status of the protected steel structure. When using ICCP systems it is important that their use is continuous and that it requires a constant supply of electricity. In spite of the fact that the power grid failures are negligible, such situations arise due to repairs and random events. A properly chosen photovoltaic system is a kind of an independent source of electricity. A photovoltaic set may be the primary or emergency power supply of the ICCP system. In case of using PV as an emergency source, we should take into account the frequency and duration of potential disruptions in energy supply. Energy must be saved in a storage system which is powered by a low power solar set. In the case of using the solar system as a primary energy source, a PV system should be selected with energy storage system in such a way as to provide one hundred percent coverage of the demand for energy by the ICCP system throughout the entire period of operation. We encounter here the daily and annual variability in the production of energy by the solar PV system. Therefore, the solar PV system should be "oversized" so as to be able to power the ICCP system in the most unfavourable sunlight in the area. This requires the optimization of the photovoltaic panel in terms of its performance, as well as batteries for the capacity ensuring the operation during the periods of unfavourable conditions of sunshine.

The problem can be solved by a simple balance sheet method, taking into account the transmission losses, the energy consumption of the system and the efficiency of the system components including storage.

For the solar PV system as the primary power source of the ICCP we have a formula evaluating the availability of energy E based on the worst sunlight in E_{min}

$$E_{pv} = A_{pv} * \eta_{pv}(T_0) * E_{min} * B \qquad (1)$$

where:

A_{pv} - active surface of photovoltaic cells

$\eta_{pv}(T_o)$ - conversion efficiency of photovoltaic cells for the ambient temperature (T_o) corresponding with E_{min} period

E_{min} - minimum annual daily amount of solar radiation on a surface inclined at an angle optimal for its duration

B - safety factor for solar PV systems from 1.5 to 2.5 (for installation operating in continuous mode)

It is assumed that the extra energy E is stored in the system of accumulating batteries. Its storage efficiency is η_{ak}, and it is sufficient for B•24 hours to cover the power demand of the ICCP system and the energy needs of the solar PV system controller and the transmission losses.

The formula is

$$E_{el} = \eta_k\big((E_{pv} - E_{pw} - E_l) * \eta_{ak} * \frac{T_{ak}}{24} +$$

$$+ (E_{pv} - E_{pw} - E_l) * \frac{T_d}{24}\big) \qquad (2)$$

where:

E_{pv} - the amount of energy produced by photovoltaic cells

η_k - efficiency of a photovoltaic controller

E_{pw} - energy used for the solar PV system needs

E_l - energy losses in transmission lines in the solar PV system

T_d - time of work when the ICCP is directly powered by a PV set

T_{ak} - operating time when using the battery it can be assumed that T_d = time from sunrise to sunset, while T_{ak}=24h - T_d

Battery capacity Qa must be chosen in such a way that the battery capacity can supply the ICCP system for (B•24)/η_{ak} hours.

Table.1. Assumed current levels and energy needs (DC output 48V) .

level	protection current	ICCP set efficiency	ICCP set power needs	ICCP set day energy needs
	mA/m²	%	mW/m²	mWh/m²
I	0,03	80	1,152	27,648
II	0,20	80	7,680	184,32
III	4,00	80	153,6	3686,4
IV	30,00	80	1152,1	27648

The amount of energy defined in this way should correspond to the ICCP electricity needs. Based on the sample data the four levels of operating parameters of the ICCP system of the protected surface have been chosen. All the data in the table refer to one square meter of the protected surface. Table.1.

After determining the power of electricity necessary for the proper functioning of the ICCP system in all sample ranges, the efficiency and performance of devices that generate current flow and the appropriate parameters for individual cases have been calculated. The data refer to 1m² of surface of a metal underground tank or part of tank (bottom). Calculations were done for solar exposure conditions of the city of Krakow. Polycrystalline silicon solar cells with an efficiency of conversion of sunlight into electrical energy is amounting to 16% were taken into account for the analysis. The efficiency of the solar controller was assumed as 90%, the losses on the electrical wiring of PV system as 4% [10]. The optimum angle of inclination of cells which ensures the best possible power generation, based on the insolation and the solar PV cell temperature coefficient is between 58° and 68° (depending on the parameters of the cell) [9]. The results are shown in Table 2.

Table.2. Parameters of PV panels

level	Protection current	PV unit power needs	power using coefficient	The unit area of PV cells	battery capacity
	mA/m²	W/m²	%	m²/m²	mWh/m²
I	0,03	0,0310	12,2%	0,216*10⁻³	81,7
II	0,20	0,220	12,2%	1,440*10⁻³	576,0
III	4,00	4,40	12,2%	28,82*10⁻³	11520,0
IV	30,00	31.00	12,2%	216,1*10⁻³	81600,0

For example; I - for the tank with an area of 700m² and the required protection current density of 0.3 mA/m² (good anti-corrosion coating), use a cell with an active area of 1,512m² and batteries with a capacity of 567 Wh.

II - for the tank or steel structure with an area of ground contact of 700m² and the required protection current density of 4.0 mA/m² (poor anti-corrosion coating), use a cell with an active area of 20,16m² and batteries with a capacity of 8,064 kWh.

5 Conclusions

The above considerations permit on the basis of a second formula to estimate the size of the photovoltaic cells needed for the ICCP grid power supply system protecting the surface of 1m² of the metal underground tank (construction). It is clear that the size of the panels required for a particular case is directly dependent on the solar energy available in a particular area. But for the territory of Poland it can be assumed with high probability that the size calculated for the conditions of Krakow will be sufficient due to the relatively low variability of solar radiation in the whole of the country. However, you should pay special attention to the specific local conditions (frequent dense fog and shade, especially during the winter period), determining the need for a larger field of PV cells and increased battery capacity by increasing the rate of battery work (B) to coefficient 2.5. In a situation when the batteries work at ambient temperature, it is important to take into account the battery temperature coefficient for the calculation of their size.

Selection of PV cells inclination angle should be performed taking into account the temperature coefficient of PV cells used.

The relatively large area of cells per 1m^2 of the protected surface is determined by a very low rate of capacity utilization of the solar ratio which amounts to 12.2%, but unfortunately this is a requirement to guarantee virtually 100% coverage of electricity demand by the ICCP system. Thus, the use of solar systems will be well founded for the ICCP systems of low protection current density, and away from the power grid, where the cost of the power supply connection compensates for the use of a relatively expensive solar power system.

References

1. A. Wantuch, Electrical Review, R. **86**, pp. 167-169 (2010)
2. P. Marcassoli, A. Bonetti, L. Lazzari, and M., Materials and Corrosion, vol. **66**, pp. 619-626 (2015)
3. G. Qiao, B. Guo, and J. Ou, *Fifth International Conference on Instrumentation and Measurement, Computer, Communication and Control* (IMCCC), Qinhuangdao, China (2015)
4. G. Jayapalan, G. Agnihotri, D.M. Deshpande, The Institution of Engineers (India): Series B, vol. **93**, pp. 259-265 (2012)
5. G. Cui, Z.L. Gan, Li, C. Yang, M. Wang, Petroleum Science, vol. **13**, pp. 135-145 (2016)
6. R.A. Gummow, *Examining The Controversy Surrounding The -850 mV CP Criteria*, Pipeline and Gas Journal, vol. **237**, 11 (2010)
7. J. Gopalakrishnan, G. Agnihotri, D. M. Deshpande; The Institution of Engineers, **93** , pp. 259–265 (2013)
8. E. Kurgan, A. Wantuch, Electrical Review, R. **87**, pp. 96-99 (2011)
9. M. Janowski et al., *Odnawialne źródła energii w Małopolsce,* Stowarzyszenie Polska Siec « Energie Cites », Kraków, **ISBN 978-83-924306-1-2** (2007)
10. B. Szymański, *Instalacje fotowoltaiczne*, GLOBEnergia, Kraków, **ISBN 978-83-64339-00-4** (2014)

 DOI: 10.1051/e3sconf/20161000030

Emission of particulate matter during aircraft landing operation

Remigiusz Jasiński[1,a], Jacek Pielecha[1] and Jarosław Markowski[1]

[1]*Poznan University of Technology, Institute of Combustion Engines and Transport, 60-965, Poznan, Poland*

Abstract. The article presents the results of studies conducted in the area adjacent to the airport located in Poznan, to determine the impact of aircraft landing operation on the concentration of particulate matter. Measurements were carried out using Engine Exhaust Particle Sizer Spectrometer 3090, enabling the measurement of the particle size distribution. The research allowed to determine the nature of the emitted particles and adds to the evaluation of impact on air quality. The tests were carried out during four aircraft landing operations. In each of the four cases total concentration of particulates number before landing (reference level) and during the landing of the aircraft was measured. In addition, the size distribution of reference level particles and its change after the landing operation was measured. Based on the conducted research, it was found that landing aircrafts have a significant impact on the concentration of particulate matter in the area adjacent to the airport. A single landing operation causes an substantially increase of the value of particle number concentration.

1 Introduction

Deterioration of atmospheric air quality in particular in urban areas results in various types of efforts to reduce toxic emissions. Currently one of the biggest problems concerning the air in urban areas is the increased concentration of particulate matter [1], resulting in a reduction of visibility in the form of smog and negative effects on human health appearing as diseases of the heart and lungs [2, 3]. Legislative measures taken by the authorities of the European Union are focused on reducing emissions of harmful compounds from automotive vehicles, however, more and more attention in particular scientific community attracts exhaust emissions from aircraft engines.

One of the fundamental issues in the assessment of air quality is the concentration of particulate matter. Particulate matter is a term generally used to describe a type of air pollutant, consisting of different particles mixtures, which differ in size, composition and spot of formation. The main sources of creating this type of pollution are: factories, power plants, incinerators, motor vehicles and many others. The basic division of solid particles due to their aerodynamic diameter, allowed for the determination of two main groups: PM2.5 and PM10 (Particulate Matter) diameter respectively of less than 2.5 μm and 10 μm [4]. The dynamic development of research topics related to particulate matter properties change depending on their size forced an introduction of a more detailed division [5]. It was assumed that the smallest particles (ultrafine particles) have a diameter of less than 0.1 μm and fine particles (nanoparticles) less than 1 μm.

In recent years, more and more attention is paid to the emissions in areas of airports, due to the rapid increase of the air traffic volume and the outlook of further development in order to meet the transport needs for the subsequent years. Most of the research and publications indicate inadequate explanation of the phenomena which are the subject of intensive research because as the literature shows air transport emissions can significantly affect the quality of the air in the vicinity of airports and their surroundings [6].

Aircraft engines is just one of the many sources of pollution in the area of airports. It is widely recognized that aircraft emissions is responsible for most of the pollution in areas of airports, but it is necessary to be aware of other sources particularly dangerous at local scale. Among them we can point the products of wear of tires, brakes and runways. In addition the strong air turbulence generated by aircraft leads to re-suspend these pollutants in the air. Physical and chemical characteristics of particulate matter generated in the tribological process were based only on a few publications [7], which indicates a strong need to develop this issue [8, 9].

Increasing knowledge of the processes in the aircraft engines and the dynamic development of the measuring devices necessitate the creation of new standards and rules for aircraft engines certification. The first emission standards for aircraft engines were introduced in the late 70s of XX century by the International Civil Aviation Organisation (ICAO). The key rules related to reduce the emission of harmful compounds from the aircraft were introduced by ICAO in Volume II of Annex 16 to the Convention on International Civil

[a] Corresponding author: remigiusz.w.jasinski@doctorate.put.poznan.pl

Aviation [10]. In this document, the recommendations relating to the measurement of carbon monoxide, nitrogen oxides, hydrocarbons and smoke were introduced.

These actions led to a significant reduction in direct emissions of harmful compounds from aircraft in the past twenty years. However the benefit may be offset by the increasing air traffic, due to the dynamic growth of air transport. In addition, the regulation introduced by ICAO applies only to the four main pollutants. In order to proper assess of the exhaust emissions more thorough analysis is required, including the analysis of chemical, physical and numerical characteristic of particle matters emitted by aircraft engines. The literature leaves a lot of questions in terms of emission of harmful compounds from aircraft in real traffic conditions, and it results in particular from the use of different methods of measurement, different way of presenting the results and the method of analysis of the research results [11, 12].

2 Methodology

2.1 Purpose of the research

The study was conducted in the vicinity of Poznan airport in western Poland. Location of the airport enforces frequent flights over the city center. Low altitude does not allow dilution of harmful compounds, as it is in the case of upper atmosphere level on a standard cruising altitude. It can be assumed that air transport has a negative impact on air quality over the area of airport (especially in the Poznan agglomeration). Large values of pollutants concentration in the Poznan air confirm reports of institutions established for air quality control. In 2014, it was found more than twice the permissible number of days with exceeded the norm PM10. The bad condition of the air is caused by all kinds of pollution sources, among other aircraft. The aim of the study was to evaluate the impact of landing aircraft on the numerical concentration of particulate matter in the area of Poznan airport.

2.2 Test objects

The objects of the research were cruise passenger aircraft, Boeing 737-500 (Fig. 1) and Boeing 737-800 (Fig. 2) landing at the Poznan-Lawica airport. These aircraft were selected because of their high frequency of landings in Poznan, which allowed for an increase in measurement attempts.

Figure 1. View of passenger plane Boeing 737-500.

Figure 2. View of passenger plane Boeing 737-800.

Selected versions of the Boeing 737 in addition to the differences in the construction of the airframe, differed in applied jet engines. In the case of the Boeing 737-500 was CFM56-3 engine, while the Boeing 737-800 was equipped with CFM56-7B engines. Basic technical parameters of these engines were included in Table 1.

Table 1. Technical parameters of aircraft.

Engine / Parameter	CFM56-3 (Boeing 737-500)	CFM56-7B (Boeing 737-800)
Maximum thrust [kN]	89	95
Bypass ratio	6.0	5.3
Airflow (takeoff conditions) [kg/s]	300	320

2.3 Research method

The measurements were carried out near the airport Poznan-Lawica, approximately 2 km from the aircraft touchdown point. In the day of the measurements there was no rain, and the wind speed did not exceed 12 km/h. Measuring device was set in the axis of approaches passenger aircraft. Measurement of particle number concentration was performed 4 times during the landing of each aircraft.

For measuring the size distribution of particle matters and their total concentration in the air a spectrometer TSI Incorporated – EEPS 3090 (Engine Exhaust Particle Sizer Spectrometer) was used (Fig. 3). It enabled the measurement of a discrete range of particle diameters (from 5.6 nm to 560 nm) on the basis of their differing speeds. The degree of electric mobility of particulate matter is changed exponentially, and measurement of their size is carried out at a frequency of 10 Hz. In order to reduce the impact of dust located at the ground, probe was placed on a stand with a height of 4 m.

Tests were carried out on four aircraft in the order shown in Table 2. In each of the four cases total particle matter concentration before landing (reference test) and during the landing of the aircraft was measured. In addition, the size distribution of PN (particle number) before and after aircraft landing operation was measured.

Figure 3. Mass spectrometer EEPS 3090.

Table 2. List of tested aircraft.

	Date	Aircraft	Engine
1	17.09.2015	Boeing 737-800	CFM56-7B
2	17.09.2015	Boeing 737-800	CFM56-7B
3	17.09.2015	Boeing 737-800	CFM56-7B
4	17.09.2015	Boeing 737-500	CFM56-3

3 Measurement results and their analysis

Parameterization of the emission limits from aviation engines shown in Annex 16 of the ICAO. Convention provides information on the exhaust smoke. In order to expand the information about the quality of exhaust gases from jet engines the studies of size distribution of particulate matters have been carried out. Using the measurement apparatus the diameter distributions of particles were determined.

The first aircraft was Boeing 737-800 equipped with CFM56-7B engine. The total average concentration of particles in the air immediately before the first aircraft landing was $3{,}2 \cdot 10^3$ cm^{-3}. During the landing operation the concentration of PN increased 47-fold to $1{,}5 \cdot 10^5$ cm^{-3} (Fig. 4). The increased particulate concentration was maintained for about 60 seconds.

In the case of the PN size distribution of the first aircraft particles with diameters of 5–15 nm were dominated (Fig. 5). The characteristic value of the diameter of the particles (the most common) obtained distribution was 9 nm. Particles with diameters greater than 20 nm were negligible part of the total concentration of PN.

The average concentration of PN in the air before landing the second plane was slightly different compared to the level before landing the first plane and amounted to $2.8 \cdot 10^3$ cm^{-3} (Fig. 6). The average concentration of particulates during landing operation of the second airplane reached $2.7 \cdot 10^5$ cm^{-3}. A 96-fold increase in the number concentration of particulate matter was found. The Increased concentration was maintained for about 110 seconds.

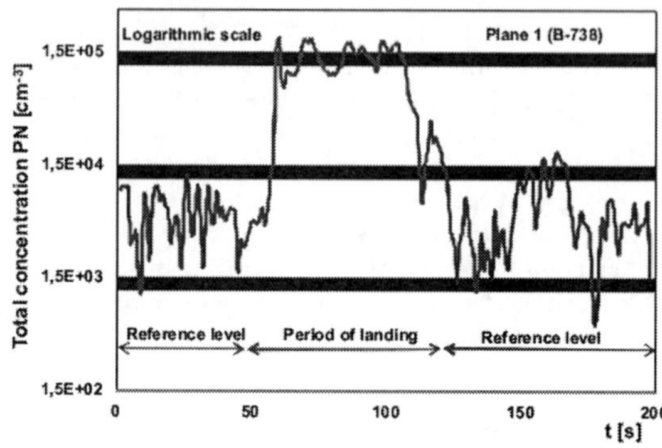

Figure 4. Total concentration of PN obtained for first aircraft.

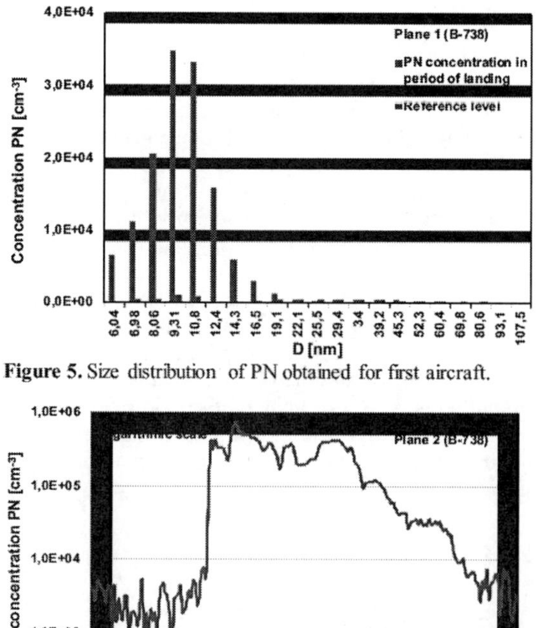

Figure 5. Size distribution of PN obtained for first aircraft.

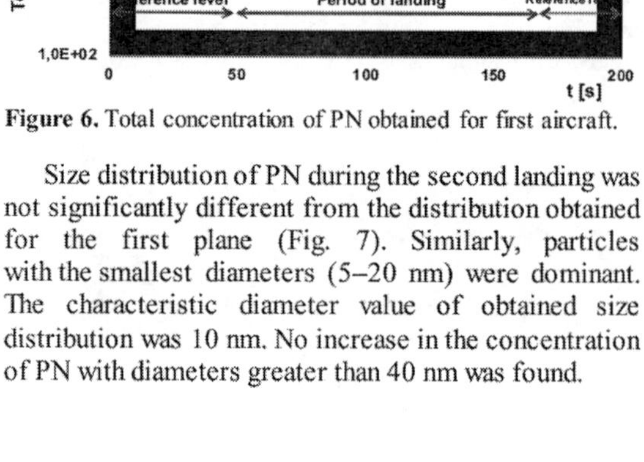

Figure 6. Total concentration of PN obtained for first aircraft.

Size distribution of PN during the second landing was not significantly different from the distribution obtained for the first plane (Fig. 7). Similarly, particles with the smallest diameters (5–20 nm) were dominant. The characteristic diameter value of obtained size distribution was 10 nm. No increase in the concentration of PN with diameters greater than 40 nm was found.

The effect of the third airplane landing was a 20-fold increase in particulate concentration of 3,8 · 10^3 cm^{-3} to 7.9 · 10^4 cm^{-3} (Fig. 8). The increased particulate concentration in the test area was maintained for about 100 s.

Size distribution as for other aircraft equipped with an engine CFM56-7B was characterized by the presence of PN from the range 5–15 nm (Fig. 9). The values of concentration of particles with diameters greater than 20 nm was negligible.

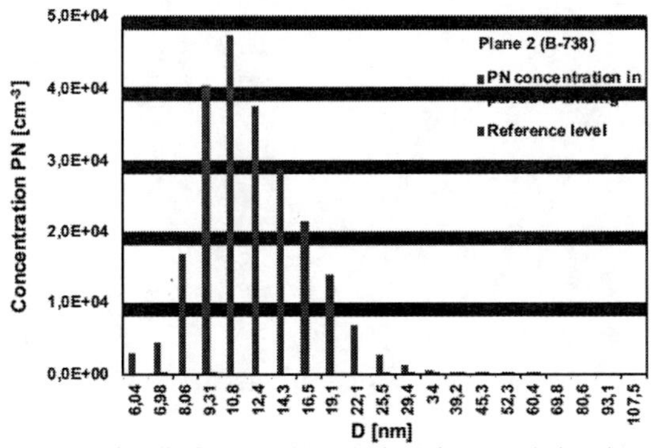

Figure 7. Size distribution of PN obtained for second aircraft.

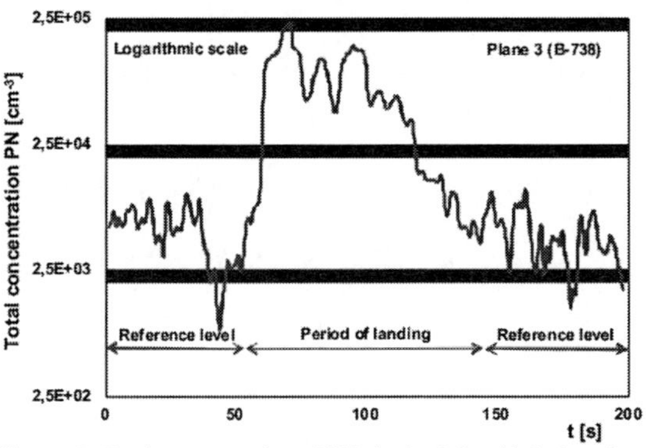

Figure 8. Total concentration of PN obtained for third aircraft.

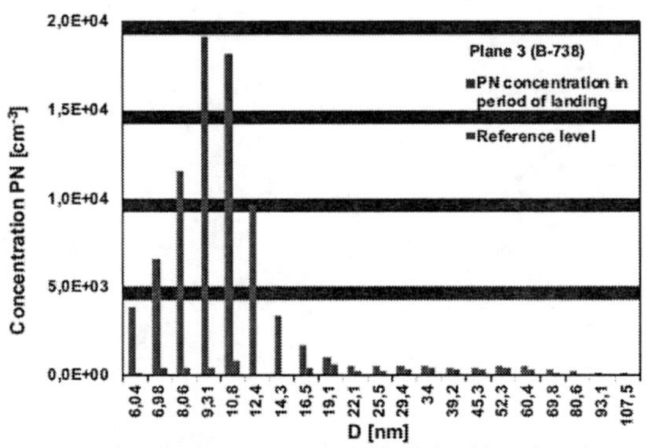

Figure 9. Size distribution of PN obtained for third aircraft.

The result of fourth passenger plane approach was a 18-fold increase in the average particle number concentration to of 6.9 · 10^4 cm^{-3} (Fig. 10). Increased concentration was maintained for a period of about 60 seconds.

Size distribution obtained in result of measurements during the fourth aircraft landing operation was significantly different from the distributions obtained for the other aircraft. In addition to the dominant particle number in the range of 6–10 nm increased concentrations of particulate characterized by diameters of 15–40 nm was observed (Fig. 11). On the basis of obtained size distribution two characteristic values of diameters (10 nm and 22 nm) can be pointed. No significant increase in the concentration of particles of a size greater than 50 nm was found. The differences between the resulting size distribution obtained for the fourth airplane and distributions obtained for other test objects were caused by differences in the design of the fourth aircraft engine. Dimensional distribution of particulates obtained for aircraft equipped with the older model CFM56 engine (fourth object) shows similarity to the characteristic size distribution of particulates contained in the exhaust of motor vehicles.

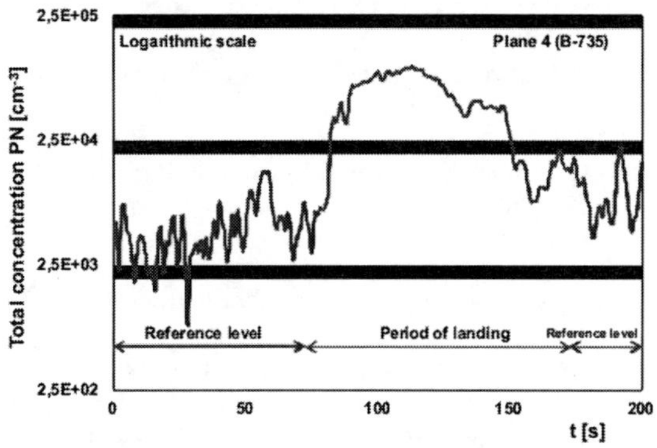

Figure 10. Total concentration of PN obtained for fourth aircraft.

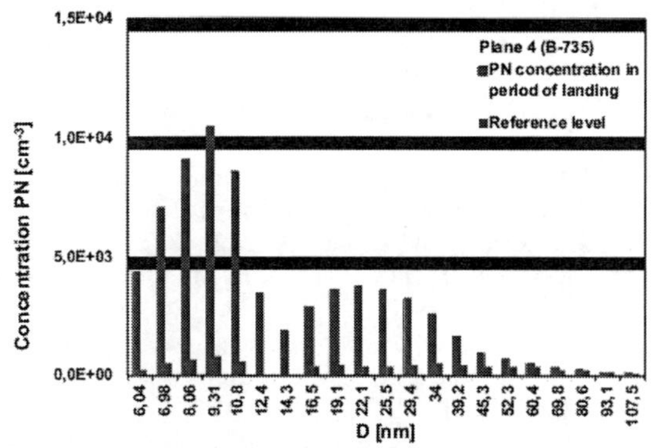

Figure 11. Size distribution of PN obtained for fourth aircraft.

4 Summary

On the basis of the measurements it was found that an aircraft approach causes significant effect on the quality of air in the area adjacent to the airport. The main symptom is a change in the concentration of particulate matter of diameter less than 100 nm. A single landing operation causes an increase of the value of particle number concentration tens of times. The increased number of particulates in the air is maintained for a period of not less than 60 seconds. The resulting particle size distributions indicate that particulate matter emitted by the aircraft engine are characterized by a set of dimensions in the range of 5–40 nm. This fact is significant that these particles are most harmful to human health.

There were significant changes in size distribution of PN depending on the jet engine. The older generation engines emit particles with a distribution similar to the distributions derived from motor vehicles. The fact that the measurements was performed at a distance of about 2 km from the aircraft touchdown point, suggests that areas of the runway, taxiways and aprons can be much more polluted than the area where the tests were carried out. This is particularly important in the context of workers of ground handling service.

References

1. V. Ramanathan, Y. Feng, Atmos. Environ. **43**, 37–50 (2009)
2. M. Masiol, R.M. Harrison, Atmos. Environ. **95**, 409–455 (2014)
3. N.P. Hyslop, Atmos. Environ. **43**, 182–195 (2009)
4. M. Kampa, E. Castanas, Environ. Pollut. **151**, 362–367 (2008)
5. F. Amato, T. Moreno, M. Pandolfi, X. Querol, A. Alastuey, A. Delgado, M. Pedrero, N. Cotsc, J. Environ. Monitor. **12**, 854–862 (2010)
6. D.S. Lee, G. Pitari, V. Grewec, K. Gierens, J.E. Penner, A. Petzold, M.J. Prather, U. Schumann, A. Bais, T. Berntsen, D. Iachetti, L.L. Lim, R. Sausen, Atmos. Environ. **44**, 4678–4734 (2010)
7. M. Bennett, S.M. Christie, A. Graham, B.S. Thomas, V. Vishnyakov, Environ. Sci. Tech. **45**, 3533–3538 (2011)
8. A. Unal, Y. Hu, M.E. Chang, M.T. Odman, A.G. Russel, Atmos. Environ. **39**, 5787–5798 (2005)
9. H. Peace, J. Maughan, B. Owen, D. Raper, Environ. Modell. Softw. **21**, 532–538 (2006)
10. ICAO, Environmental Protection (Annex 16) Aircraft Engine Emission, International Standards and Recommended Practices **2** (2008)
11. M. Mazaheri, G.R. Johnson, L. Morawska, Atmos. Environ. **45**, 3500–3507 (2011)
12. H. Agrawala, A.A. Sawanta, K. Jansena, J.W. Millera, D.R. Cocker, Atmos. Environ. **42**, 4380–4392 (2008)

Methane rich gasification of wood pellets

Magdalena Joka[1,a] and Sławomir Poskrobko[1]

[1]*Białystok University of Technology, Department of Civil and Environmental Engineering, Division of Agri-Food and Forestry Engineering, ul. Wiejska 45A, 15-351 Białystok, Poland*

Abstract. In the work there are shown the results of experimental studies on methane rich gasification of pinewood pellets in Bio-CONOx technology. The experiment was carried out on a laboratory scale gasifier (5 kW), which design features allow producing a high quality gas with a high methane content. In the results there was identified the impact of the quantity of Bio-CONOx on the amount of flammable gas compounds (methane, hydrogen and carbon monoxide) in the synthesis gas and the gas calorific value. The additive was added in 10,20,30 and 50% concentrations to the gasifier chamber. It has been shown that increasing the amount of the additive has a positive effect on the calorific value of the synthesis gas (Fig.1,2). Gas with a high content of methane (and high calorific value) was obtained from gasification of biomass with a 50% addition of Bio-CONOx. There was also examined the proportion of blowing air (gasifying medium) for which the properties of obtained syngas were the best.

1 Introduction

The growing demand for heat and electricity, and the specter of depletion of fossil fuels affects the ongoing development of renewable energy. Biomass provides a number of possibilities for energy production, among others, through its gasification. By the term "gasification" are named biomass thermal processings, which lead to obtain the synthesis gas. During gasification the solids are heated in an atmosphere of a gasification medium which can be air, oxygen, steam or a mixture of the ones above [1]. The process has been known for over 200 years, but recent technological development has increased interest in it. The main reason for this state of affairs is the trend of development, thus popularizing the technology of producing energy in diffuse sources that can meet the energy needs of a small scale for local needs. An example of such cogeneration can be a co-gasifier supplied by solid fuels (eg. coal, biomass or fuel from waste RDF) combinated with a gas turbine systems or even implementing more efficient gas power associations of circulating steam power plants i.e. IGGC [2,3,4]. In the practical application of cogeneration, small power can successfully provide a source of energy for the industrial plant and for small towns and villages. Such technologies require the implementation of high-performance transformations implemented at every stage of the manufacturing energy process. Specific values of the energy conversion chain is attributed to processes for gas production, as already mentioned above, in gasification of solid fuels there is a great flexibility of feed material selection (almost any type of biomass can be subjected to the process). Widely used in power generation gasification technologies are characterized by relatively low efficiency of fuel/gas conversion estimated at approx. 60-65% [5]. The efficiency of the gasifier, defined by the quotient of the calorific value of the fuel to the calorific value of obtained in this gasifier syngas, unambiguously determines its progress in technological development. In the literature, much attention was paid to research the technology of gasification of different solid fuels by picking as an aim high efficiency. A first aspect of differentiating known technologies is the difference of used gasifying agent, ie. in the case of air, heating value of the gas produced fluctuates within 4-7·MJ m^{-3} for gasifying wood chips. Relatively better calorific values are achieved when gasification takes place with the use of oxygen or water vapor wherein it is possible to achieve calorie gas order of 10-18 MJ·m^{-3} [6,7]. The few known domestic installations gasifying solid fuels [8-11] are mainly based on the gasification on the fixed grate or circulating fluidized bed. They allow to obtain a gas from wood chips with a low calorific value of approx. 4.5-5.5 MJ·Nm^{-3} and a methane CH_4 content from 1.3 to 5%. It should be emphasized that the high methane content is a significantly associated priority element of the heating value of the synthesis gas obtained which is shown by the VDI computational algorithm used to examine the calorific value of gaseous fuels:

$$W_o=126\cdot[\%CO]+108\cdot[\%H_2]+359\cdot[\%CH_4]+665\cdot[\%C_nH_m] \qquad (1)$$

Polish gasifier EKOD (containg a fixed grate) allows to obtain a gas having a methane content to 4.4%

[a] Corresponding author: m.joka@pb.edu.pl

for the synthesis gas calorific value of 5.65 $MJ \cdot m^{-3}$, which limits the industrial use of this constitution [9]. Superior parameter, both in the country [12] and abroad, are obtained with the fluidized bed gasifiers, where the contact gasifying agent/fuel factor is increased and which increases the reaction rate and efficiency of the process. Typical levels of methane in a fluidized bed technology are 5-12%, and the calorific value reaches even 14 $MJ \cdot m^{-3}$ [13]. There are also known methods of improving the quality of synthesis gas by the use of catalysts and mineral puffs dosaged directly to the interior of the reaction chamber [14,15,16] in which for example Rapagnano et al. received gas at high hydrogen content of approx. 56% but a very low methane quantity of 2%

Another application of synthesis gas rich in reduction ingredients is attributed to low-emission technologies of combustion in boilers supplied by coal dust i.e. NO_x reduction by the reburning method [17]. An example of the use of such low-carbon technologies in a power unit with a capacity of 600 MWel is the Longannet power station in Scotland, where is achieved a 40-60% reduction in nitrogen oxide emissions [18]. It is worth noting, which indicates the size of the gasifier, that the reburning gas is approx. 20% of the whole chemical energy supplied to the block. Fine reduction properties are achieved during reburnig with the use of natural gas of which, as known, the dominant component is CH_4. For example, Hassan et al. using methane (purity of 99%) reached a 94% reduction of NOx [19]. In syngas there was not obtained a high methane concentration. The methane content is usually instigated at approx. 0.8-2.5%. In contrast, as indicated by the reports in the literature [20], a high reduction efficiency is obtained by a methane-rich reburning gas. Taking into account the importance of methane in the clean combustion of solid fuels in terms of reduction the emissions of NO_x and as well considering the importance of the efficiency of the process, ie. high calorific value of synthesis gas, this paper presents the results of experimental studies of methane-rich gasification of biomass. High content of CH_4, and hence high gas calorific value was obtained by the addition to the biomass a fuel preparation so called Bio-CONOx [17,18].

1.1 Methane creating reactions

Methanation is an exothermic process that aims in the methane production from hydrogen and carbon oxides (Eq.2 and Eq.3):

$$CO + 3H_2 \leftrightarrow CH_4 + H_2O \quad -206\ kJ \cdot m^{-1} \tag{2}$$

$$CO_2 + 4H_2 \leftrightarrow CH_4 + H_2O \quad -164\ kJ \cdot m^{-1} \tag{3}$$

Stoichiometaly, methantion of carbon monoxide releases 206 $kJ \cdot m^{-1}$ of energy (heat) and the reaction of carbon dixoide 164 $kJ \cdot m^{-1}$. The two reactions are influence by pressure and temperature, where high pressure favours the methanation processes. In contrast, high temperature limits the methanation processes [21].

Obtained, in the experiment, low temperatures in the fuel zone (of approx. 400°C) contributed to the achievement of a high methane content in the producted syngas. Mostly known in the literature instalations which require the methane production are controling methane production by the conversion of CO and CO_2 but not taking into anounce the reactions of heavy hydrocarbons cracking. The cracking reactions require the presence of oxygen when hydrocarbons may decompose to inter alia hydrogen (Eq. 4):

$$C_{10}H_8 + 5O_2 \rightarrow 10\,CO + 4H_2 \quad -1\,257\ kJ \cdot m^{-1} \tag{4}$$

As a result of the abow mentioned equilibrium, in the stream of the raw gas from gasification, the hydrogen concentration is very high. Hydrocarbons which did not react with oxygen, may be reacted with hydrogen, which is presented in the reactions of hydrocraking (Eq. 5, Eq. 6) [22]:

$$C_{10}H_8 + 7H_2 \rightarrow C_6H_6 + 4CH_4 \quad -368\ kJ \cdot m^{-1} \tag{5}$$

$$C_{10}H_8 + 16H_2 \rightarrow 10CH_4 \quad -900\ kJ \cdot m^{-1} \tag{6}$$

Reactions 4, 5 and 6 are strongly exothermic and their presents results in a significant increase in the temperature. In the conducted gasification experiment, abow the fuel layer, were observed temperatures of approx. 800°C, which taking into anounce the high purity of the obtained gas, were the result of cracking and hydrocracking reactions.

2 Materials and methods

2.1 Materials

2.1.1 Pellet fuel properties

The used in the experiment biomass were pinewood pellets which were divided to a particle size in the range of 1.25 mm to 3.15 mm. Properties of pellets are given in the table below.

Table 1. Pellet fuel properties

Parameter	Value
Carbon C	50,00 %
Hydrogen H	6,00 %
Oxygen O	43,00 %
Nitrogen N	0,20 %
Humidity	8,84 %
Ash content	<1,50 %
LHV	18,50 $MJ \cdot kg^{-1}$

DOI: 10.1051/e3sconf/20161000031

2.1.2 Bio-CONOx fuel properties

Bio-CONOx additive is a methane-formating waste fuel of natural orgin. It wasused in the gasification process in weight concentrations of 10%, 20%, 30% and 50% of the sample. The table below presents fuel properties and chemical composition of the sorbet.

Table 2. Bio-CONOx fuel properties

Parameter	Value
Carbon C	49,30 %
Hydrogen H	6,96 %
Sulfur S	1,38 %
Chlorine Cl	0,001 %
Oxygen O	35,59 %
Nitrogen N	6,76 %
Humidity	6,60 %
Ash content	6,50 %
LHV	15,70 $MJ \cdot kg^{-1}$

Nitrogen contained in the Bio-CONOx affects the autonomous reduction of acidic nitrogen oxides NO_x and it eliminates the need for special installation injecting urea or insufflating ammonia. Low content of chlorine Cl in the preparation is not conducive to the occurrence of high-temperature corrosion of boiler components. Thus, the addition of Bio-CONOx to solid fuels reduces the mass content of Cl in the fuel mixture. Chlorine Cl is often found in biomass fuels in an amount greater than 0.1%, which drastically increases the risk of corrosion. Bio-CONOx has a relatively high proportion of sulfur S, wherein the preferred mass ratio of sulfur to chlorine (S/Cl=2,2) inhibits the formation of low-melting chloride. The fulfillment of the above ratio conditions results in the formation of a protective layer on the elements of the alkali sulfates utility system, thereby heat exchange surfaces are protected against chloride corrosion. It should be stressed that the preparation, if added to the fuel, strongly influences the formation of reducing zones in the gasification process, but the quality of the gas from the gasification of Bio-CONOx alone no longer accepts such a high performance of the process [1, 18].

2.2 Methods

The fuel was task to the tray (shown in Figure 1) and transported by a screw conveyor into the reaction chamber. Gasification occured in a stream of heated air, blown through the chamber from the bottom and controlled by a inverter and two valves. Airflow was measured using a flowmeter (Figure 2). Due to the fact that the system does not have a burner, start-up takes place when the auto-ignition temperature is reached. After stabilizing the temperature in the boiler, the gas was sampled from the upper part of the reaction chamber. Sampled gas flown through the filtering system and was subjected to composition analysis by the 3100 Syngas Gas Analyser.

Figure 1. Gasifier (1-gasification chamber, 2-fuel feeder)

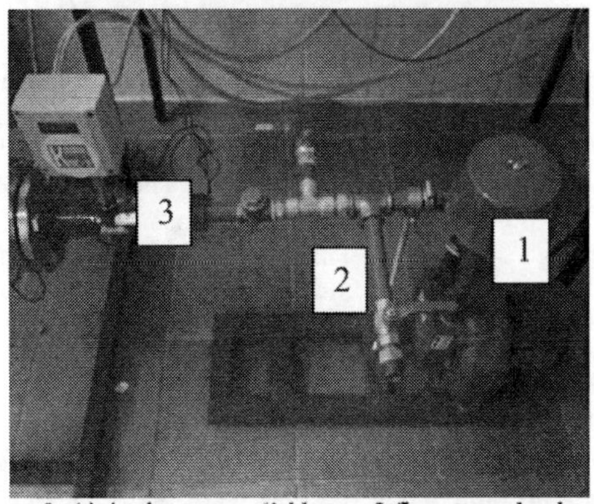

Figure 2. Air intake system (1-blower, 2-flow control valves, 3-flowmeter)

The gasification process was conducted in conditions of a temperature of approx. 720°C and at atmospheric pressure prevailing in the reactor. Efficient fuel flow supplied to the reaction chamber was constant at 0.42 $g \cdot s^{-1}$. Under these conditions, the average ratio of fuel to air was Air/Fuel = 1.38. There have been changes in the value of the gasifying agent flow: 0.52; 0.55; 0.60; 0.65 $g \cdot s^{-1}$ and for it the values of the coefficients Air/Fuel were successively 1.23; 1.31; 1.43; 1.55.

3 Results

The table below shows the values of obtained syngas for different concentrations of Bio-CONOx in the fuel mixture.

Table 3. Calorific value of obtained syngas

Bio-CONOx concentration %	Maximum LHV (MJ·m⁻³)	Avrage LHV (MJ·m⁻³)
10	4,30	3,50
20	5,23	4,43
30	5,57	5,12
50	12,04	8,27

Shown in figures 3-6 results of the conditions in which was obtained the methane rich syngas establish the combustible syngas components in dependence of oxygen content in the syngas for various fuel compositions containing Bio-CONOx preparation. Variable gas composition was obtained by various Bio-CONOx additives to the fuel mixture i.e. 10%, 20%, 30% 50%. The most favorable process conditions, where was obtained methane rich gas, are sought by changing the flow of blowing air into the reaction zone of the gasifier. In each case airflows were the same: 0.50; 0.55; 0.60; 0.65 $g{\cdot}s^{-1}$. The O_2 content in the past-process gas is a real reflection of the reactivity of the used fuel, which was enriched varying additions of Bio-CONOx. Fig. 3 shows the process conditions of wood pellets gasification without the addition of Bio-CONOx. Also, noteworthy is the high, comparing to conventional gasification technology [12] proportion of CH_4 which reached approx. 8%, where in the case of gasification in a fixed bed was achieved only 0,5-2% of CH_4, in a fluidised bed about 4% CH_4. The high share of CH_4 is achieved by tubular gasifier structures with an intense air purge into the fuel layer. The nature of the gasyfing medium flow, through the gasification bed, is determining not typical temperature distribution in the reaction chamber, ie.: the temperature in the fuel layer is 370-420°C, the temperature above the fuel layer - 700-850°C. The temperature in the fuel layer favors the methanation processes of CO and CO_2 [17], and temperature conditions abow the fuel layer are assisting the processes of tar cracking where one of the products is CH_4 [17]. A similar character of the process in terms of temperature distribution, was examined in the case of gasification of pellets and 10%, 20%, 30% 50% of Bio-CONOx. The test results shown in fig. 4,5,6,7 clearly show the increase in the proportion of CH_4 in the syngas, with an increasing share of the Bio-CONOx. At the same time increasing the participation of the additive, in each case, with the same air stream, causes a change in the reactivity of the bed in relation to the gasifying agent.

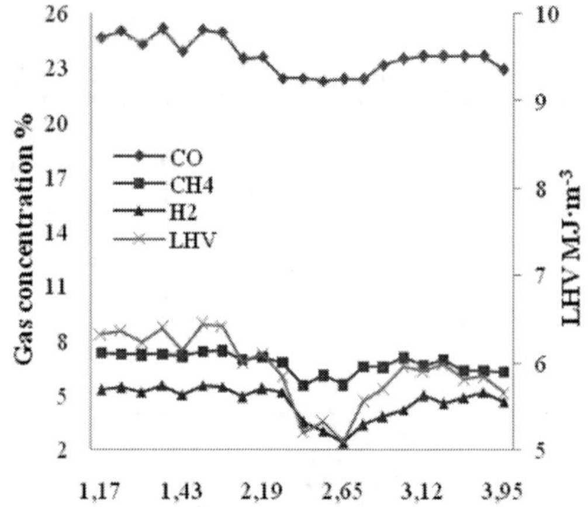

Figure 3. Synthesis gas composition in the amount of oxygen for particulate pellets without any addition

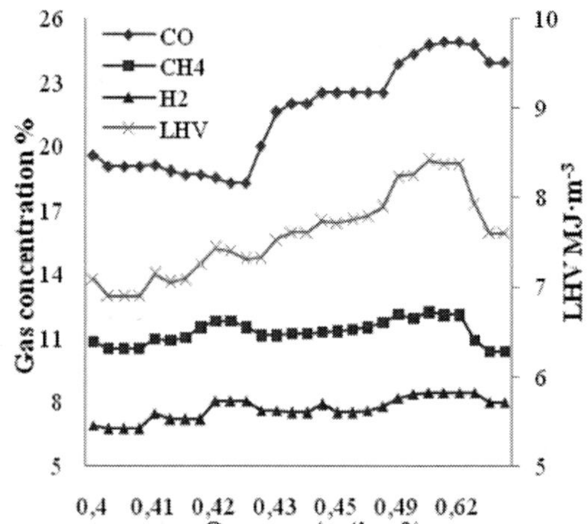

Figure 4. Synthesis gas composition in the amount of oxygen for particulate pellets with a 10% Bio-CONOx addition

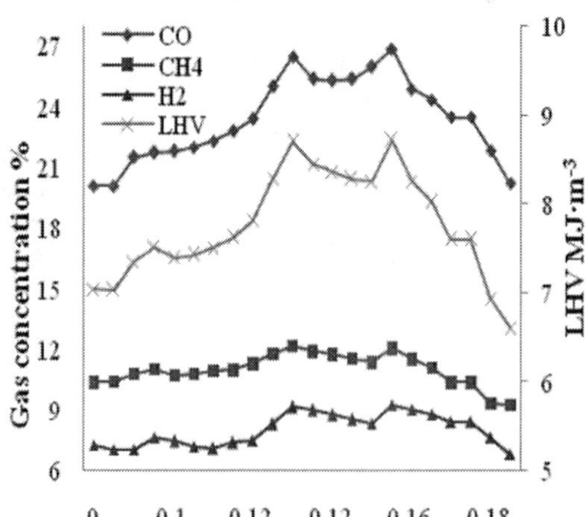

Figure 5. Synthesis gas composition in the amount of oxygen for particulate pellets with a 20% Bio-CONOx addition

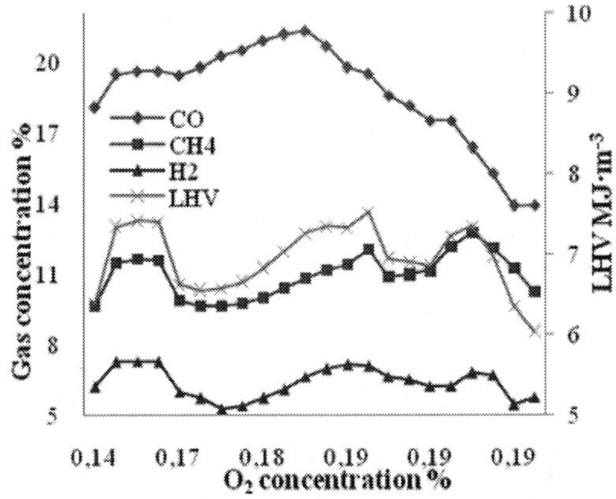

Figure 6. Synthesis gas composition in the amount of oxygen for particulate pellets with a 30% Bio-CONOx addition

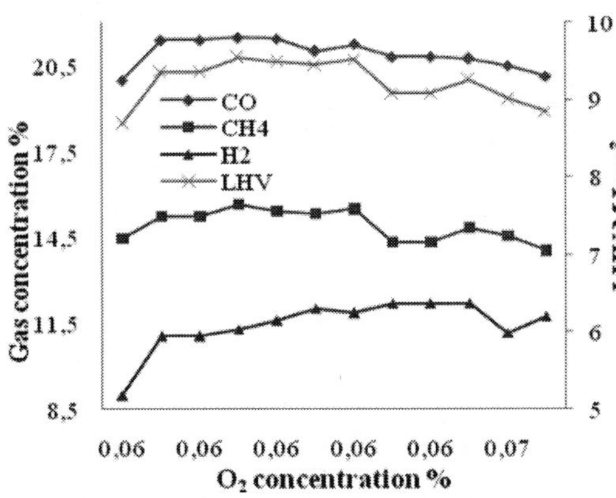

Figure 7. Synthesis gas composition in the amount of oxygen for particulate pellets with a 50% Bio-CONOx addition

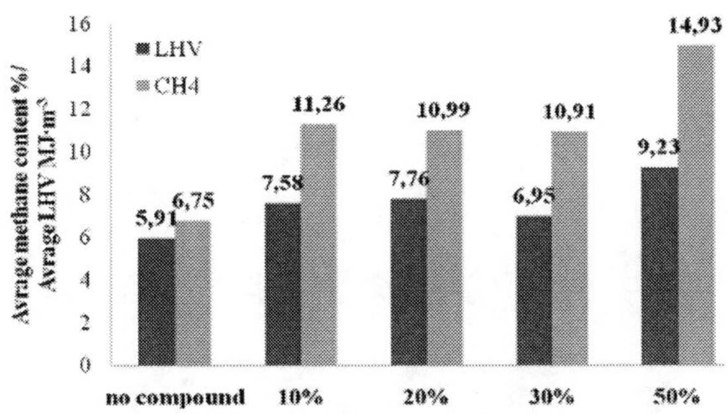

Figure 8. The calorific value of the syngas depending on the content of methane and the amount of Bio-CONOx in the fuel mixture

In the case shown in fig. 3 (without Bio-CONOx) high proportion of O_2 (1.17-3.95%) shows its lower reactivity compared with the other cases (fig. 4,5,6,7), where the part of the preparation increases. Thus, the addition of Bio-CONOx to the pellets resulted in increased reactivity of deposits which affect the activity, among other things, methanation process. The addition of Bio-CONOx in the reaction chamber of the gasifier, easily persisted reducing conditions, which favor the formation of methanogenic components such as CO and H_2. Shares of CO and H_2, in each test case, did not undergo major changes despite increased CH_4 content. This indicates the methanation process occurring ie. the hydrogenation of CO.

In the table abowe and at Fig. 8 are shown the syngas calorific values obtained from the preparation of various concentrations of Bio-CONOx. Calorific value was significantly dependent on the content of CH_4 as shown in fig. 3-8 where the nature of the curve is similar to a curve illustrating the change in the CH_4 content.
The best parameters of synthesis gas (the highest calorific value) was obtained in a sample of 50% addition of Bio-CONOx.

Methane-rich gasification process was carried out on a laboratory stand which cointained an atmospheric fixed bed reactor with a maximum output of 5 kW, using air as a gasifying agent. Gasified biomass was in the form of crushed pinewood pellets with different amounts of Bio-CONOx addition ie. 10,20,30 and 50% to the fuel mixture. Specific temperature lyers were obtained in the atmoshperic conditions process ie. the temperature in the fuel zone was approx. 400°C, where took place redutiong reactions of methanation (hydratation) of CO and CO_2; abow the fuel lyer the temperature reached 800°C. The high temperature conditions favour the reactions of tar craking and hydrocraking where the main gas product is methane. In the research process action was taken at increasing the calorific value of syngas by inreasing the methane content in the gas mixture. The indicated resaerch show that Bio-CONOx, thanks to it's sepcific chemical structure, favours the methane production while gasification processes ie. it increases the reactions rate occurring during gasification i.e. hydrogenation and methanation. The avrage amount of methane occured from gasyfing pinewood pellets was 5,91% and it let to obtain the gas LHV at 6,75 MJ·m^{-3}. Increasing the amount of Bio-CONOx resulted in higher syngas methane content and the highest effectiveness of gasification was observed using a 50% supplement of Bio-CONOx (Fig. 7,8) where avrage CH_4 content was 9,23% and the syngas LHV 14,93 MJ·m^{-3}.

In the frame of the experiment was analyzed the stability of the gasification process. The assessment was made by referring to oxygen kontent in the obtained syngas. Dependence on content of CH_4, H_2, CO, CO_2 from the O2 content show high reactivity of the used fuels. On the resulting high gas performance primarily has an affect high content of calorific methane (due to the tendency of creating reduction zones), by which it can be used in prosumer power generation eg. to drive Capstone microturbines.

Further studies should be performed to optimize the amount of the additive in the fuel while maintaining high efficiency of the gasification process.

Acknowledgements

Studies have been carried out in the framework of the statutory work No. S/WBiIŚ/2/15 and funded by the Polish Ministry of Science and Higher Education

DOI: 10.1051/e3sconf/20161000031

References

1. S. Poskrobko, J. Łach, D. Król, Energetyka i Ekologia (2010).
2. P.N. Pressley, T.N. Aziz a , J.F. DeCarolis, M.A. Barlaz a , F.He, F. Li, A. Damgaard, J. Cl. Pro., 70 (2014).
3. I. Hwang , J. Kobayashi, K. Kawamoto, Waste Manage., 34 (2014).
4. C. Cormos, INT J Hydrogen Energ., 35 (2010).
5. M. Asadullah, Renew. Sust. Energ. Rev., 29(2014).
6. A.V. Bridgwater, Fuel, 74 (1995).
7. G. Schuster, G. Loffler, K. Weigl, H. Hofbauer, Bioresource Technol., 77 (2001).
8. M. J. Stolarski, S. Szczukowski, J. Tworkowski, M.Krzyzaniak, Renew. Energ., 59 (2013)
9. T. Chmielniak, M. Ścieżko, T. Iluk, K. Matuszek Przemysł Chemiczny, 8-9 (2006).
10. M. Ścieżko, A. Sobolewski, P. Billig, Patent PAT.208616. Zgazowarka ze złożem stałym.
11. A.Sobolewski, J.Koowicz, K.Matuszek, T. Iluk, Polityka Energetyczna, 14 (2011).
12. Monograph. Instytut Energetyki (2010).
13. S. Heidenreich, P. U. Foscolo Prog. Energ. Combust., 46 (2015).
14. S. Rapagnà, K. Gallucci, M. Di Marcello, P.U. Foscolo, M. Nacken, S. Heidenreich, M. Matt, Fuel 97 (2012).
15. J.A. Ruiz, M.C.Jua, M.P.Morales, P.Munoz, M.A.Mendıvil, Renew. Sust. Energ. Rev., 18 (2013).
16. P. Chaiprasert, T. Vitidsant, Korean J. Chem. Eng., 26 (2009).
17. D. Król, S. Poskrobko, Gaz reburningowy – wysokometanowe zgazowanie preparatu Bio-CONOx.
18. D. Król, Arch. Gosp. Odpad. Ochr. Środ., 2 (2010).
19. M. I. Hassan, T.W. Wu, K. Saito, Fuel 108 (2013).
20. M.W. Sackett, S. Nakaparksin, D. Dalrymple, Advances in Organic Geochemistry (1966).
21. S.Rönsch, J. Schneider, S.Matthischke, M. Schlüter, M. Götz, J. Lefebvre, P. Prabhakaran, S. Bajohr, Fuel 166 (2016).
22. P. Babiński, G. Łabojko, Polityka energetyczna 15 (2012).

Assessing temporal complementarity of solar, wind and hydrokinetic energy

Jakub Jurasz[1,a], Adam Piasecki[2] and Marcin Wdowikowski[3]

[1]*AGH University, Department of Management Engineering, 30-067, Gramatyka St., Cracow, Poland*
[2]*AGH University, Department of Economics, Finance, and Environmental Management, 30-067, Gramatyka St., Cracow, Poland*
[3]*Institute of Meteorology and Water Management - National Research Institute, 01-673 Warsaw, Podleśna 61 Street, Poland*

Abstract. Renewable energy sources (RES) exhibit various characteristics when it comes to their availability in time and space domain. Some are characterised by significant variability and limited predictability. This makes their integration to the power grid a complicated task. Temporal and spatial complementarity of RES is perceived as one of the possible ways to facilitate the process of integration. This paper investigates the concept of temporal complementarity of solar wind and hydrokinetic energy in case of two sites in Poland. Obtained results indicate existence of some beneficial complementarity on inter-annual and annual time scale. Combination of those three RES in one hybrid system makes power source more reliable.

1 Introduction

In case of some power systems (e.g. Germany [1], California [2], Denmark [3]) RES start to play a very significant role and strongly contribute to the reduction of: greenhouse gasses emissions and dependence on conventional fuels. However their integration to the power grid comes with additional challenges – resulting mainly from RES variability in time and space. Jones [4] addressed those problems and exemplified them based on several case studies – both from practical and theoretical point of view. Issue of RES integration draw much attention in the most recent literature. Highly cited work of Jacobson and Delucchi [5] pointed to seven ways of efficient design and operation of variable renewable energy (VRE) in power systems. To VRE one may classify *inter alia* photovoltaics (PV), wind turbines (WT) and hydrokinetic turbines (HKT). Naturally their variability is different in scale, frequency and predictability. One of the approaches recommended by [5] is to design power systems in such a way that it utilizes energy sources which are complementary one to another.

For the best of the authors' knowledge so far conducted research did not investigate the complementarity of those three energy sources. Several papers addressed wind and solar complementarity in different parts of the world: China [6], Mediterranean Islands [7], Germany [8], Italy [9], Iberian Peninsula [10] and India [11]. Hydropower (using height difference in contrary to hydrokinetic which utilizes flow speed), solar and wind complementarity has been investigated in case of: Brazil [12], Northeast Brazil [13] and one city in Poland (Piła) [14]. Conclusions coming from all above mentioned papers indicate, that wind and solar energy exhibit strong complementarity from the perspective of their annual variability patterns. Usual abundance of solar energy during dry period may result in improving water budged – meaning that water can be stored in reservoirs and do not have to be used for propelling water turbines (this beneficiary aspect of resources complementarity has been pointed by [13]).

This study deals with small scale PV, WT and HKT power sources which can be used by for domestic purposes. Hydrokinetic turbines are usually perceived from the perspective of large-scale technologies which use the energy of waves, tides and ocean currents. However there is a huge potential of using this technology by applying it to the energy of flowing water. Vermaak et al. [15] gave a thorough analysis and review of current technology status and trends of micro-hydrokinetic turbines for rural applications. Authors point to the fact that there are significant gaps in literature when it comes to the technical, economic and environmental aspects of utilizing HKT. Those gaps may be one of the reasons why this technology keeps in the background.

The goal of this study was twofold. First it aimed at analysing the complementarity of mentioned energy sources from the perspective of various time scales. This analysis was divided into two subsections: potential energy and energy generated from power sources based on their individual characteristics. Secondly it investigated the behaviour of PV-WT-HKT hybrid power source from the perspective of hourly energy generation values.

2 Methods and data

Two sites in Poland were selected for the purpose of this study. The first one is a small city Lądek Zdrój (50.34N, 16.88E) which is situated in Southern part of Poland,

[a] Corresponding author: jakubkamiljurasz@gmail.com

Lower Silesia Voivodeship. Lębork a slightly bigger city (54.53E, 17.74E) is the second site and lies in Northern Poland, Pomeranian Voivodeship. Those both sites are characterised by the fact that the Institute of Meteorology and Water Management (IMGW) is making there measurements of basic meteorological and hydrological parameters such as wind speed and flow rate. It is important to note, that the rivers in both sites have different regimes and pattern of annual flow (see Figure 1) additionally both exhibit various annual patterns of wind speed, as shown on Figure 2. When it comes to the solar energy, the values of mean hourly irradiation observed in those two locations are almost the same. Those recorded in Lądek Zdrój are by 5% greater than those from Lębork, respectively 1.62 kWh/m²/day and 1.55 kWh/m²/day.

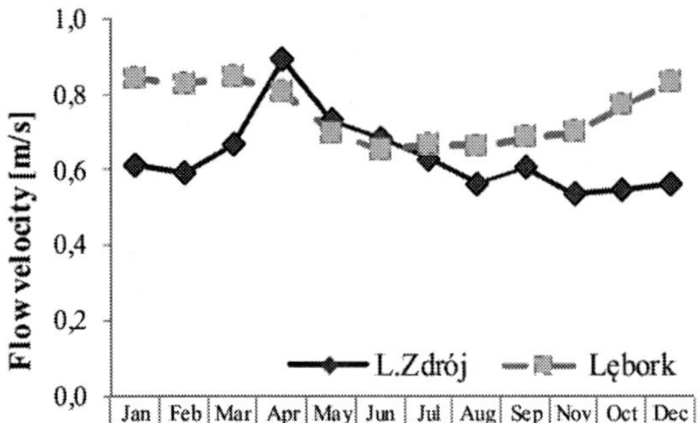

	Jan	Feb	Mar	Apr	May	Jun	Jul	Aug	Sep	Nov	Oct	Dec
L.Zdrój	0,61	0,59	0,67	0,90	0,73	0,69	0,63	0,56	0,61	0,54	0,55	0,56
Lębork	0,85	0,83	0,85	0,81	0,70	0,65	0,67	0,67	0,69	0,70	0,77	0,84

Figure 1. Mean monthly flow velocity in Lądek Zdrój and Lębork. Values were calculated based on 2006-2015 time series of hourly meter readings.

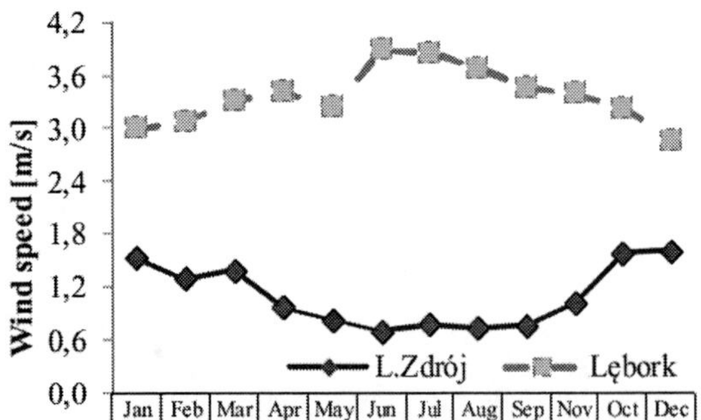

	Jan	Feb	Mar	Apr	May	Jun	Jul	Aug	Sep	Nov	Oct	Dec
L.Zdrój	1,53	1,29	1,39	0,97	0,83	0,70	0,77	0,73	0,76	1,03	1,58	1,60
Lębork	3,02	3,09	3,33	3,42	3,25	3,90	3,86	3,69	3,48	3,40	3,24	2,87

Figure 2. Mean monthly wind speed in Lądek Zdrój and Lębork. Values were calculated based on 2006-2015 time series of hourly meter readings.

In both cases data on flow rate (later recalculated into flow velocity based on method described in following paragraph) and wind speed has been downloaded from IMGW NRI databases. Wind speed is measurements are taken at 10 meters and this height has been assumed to be the height of small wind turbine rotor. Irradiation time series have been obtained from SoDa Solar Radiation Data (http://www.soda-pro.com/) which provides access to a broad range of services and web services related to *inter alia* meteorological data. Time series from SoDa did not have any missing observations, therefore were treated as a benchmark for the two remaining. Both in case of Lądek Zdrój and Lębork some values of flow rate and wind speed were missing. When a single value was missing and its predecessor and successor were available then it was replaced by the mean of those two mentioned. However if missing values represented longer continuous time series subset, then the whole period was removed from further analysis. In result, considering 2006-2015 times series with hourly time step in case of Lądek Zdrój 1034 values records were removed whereas in case of Lębork this number was greater and amounted to 3068. Those values constitute to respectively 1.2% and 3.5% of total number of observations.

2.1 Flow rate and flow velocity

River flow rate (Q) also known as a river discharge, in simple form of hydraulic relationship, is described as an exponential equation – rating curve [20, 21]:

$$Q = \alpha\,(h - e)^{\beta} \tag{1}$$

where: h is the river depth or gauge height of the water surface (cm), e is gauge height of effective zero flow (cm), α, β are the equation parameters depend on the river channel shape. Flow rate is given in cubic meters per second unit (m^3s^{-1}).

Stage-discharge relations for opened channel stations with uniform flow can be governed by the Manning equation, as it applies to the reach of controlling channel downstream from a gauge [20, 21]:

$$Q = n^{-1} A R^{2/3} S^{1/2} \tag{2}$$

where A is cross-section area, in square meters (m^2), R is hydraulic radius, in meters (m), S is friction slope, and n is channel roughness.

In Manning's formula, combination of n, R and S values gives a local river flow velocity computed in a specific time. Stage-discharge, rating curve, stage-velocity or discharge-velocity relationships can be developed using hydrometric field measurements [22]. Comparison between rating curve, stage-velocity plots and approximation of flow rate data series gives flow velocity series. In this paper hourly flow velocity series were approximated for two water gauges Lądek Zdrój and Lębork for 2006-2015 time period.

Because of the natural cycle of the analyzed part of rivers as well as the purpose of this research, the ranges of used flow velocity values were limited to the average conditions of flow rate. In order to do so, flow rate data series in 2006-2015 period were compared to the characteristic value of *average of maximal annual flows* (SWQ), which for Lądek Zdrój and Lębork were 50.5 m^3s^{-1} and 14.2 m^3s^{-1} respectively. Every value from data series higher than SWQ was removed from the string to further computations. Using SWQ flow series threshold ensures that extreme values derived by natural and unpredictable flooding are excluded (have no influence) from the analysis.

2.2 Potentially available energy

The energy accumulated in flowing air masses (E_W) can be calculated based on formula [16]:

$$E_W = 0.5 \cdot A \cdot \rho_{air} \cdot v_{wind}^3 \cdot t \quad (3)$$

where: A is the wind turbine area (m^2), ρ_{air} is the density of air (1.2 kg/m^3), v_{wind} is the current wind speed (m/s) and t is the time.

The energy which can be generated by a perfect hydrokinetic system (E_{HKT}) can be expressed as [17]:

$$E_{HKT} = 0.5 \cdot A \cdot \rho_w \cdot v_{water}^3 \cdot t \quad (4)$$

where: A is the hydrokinetic turbine area (m^2), ρ_{water} is the density of water (1000 kg/m^3), v_{water} is the current water flow speed (m/s) and t is the time.

Potential of solar energy has been directly taken from irradiation time series.

2.3 Power sources characteristics

In order to simulate energy generation from wind and hydrokinetic turbines their power curves were applied (see Figure 3 and 4).

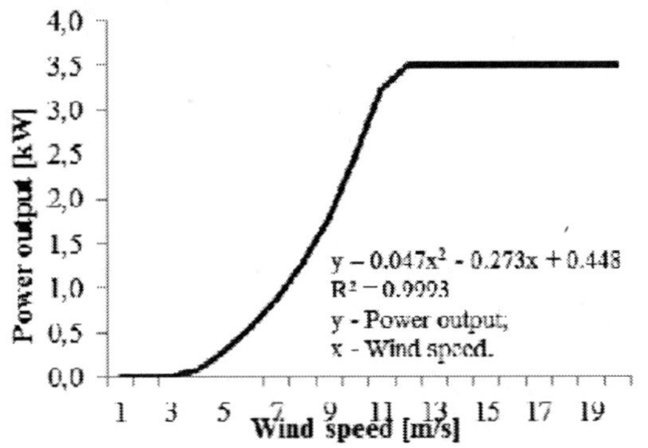

Figure 3. 3.5 kW Sonkyo Windspot HAWT power curve. Second degree polynomial approximates power output for wind speed ranging from 4 m/s to 12 m/s. Data source [18].

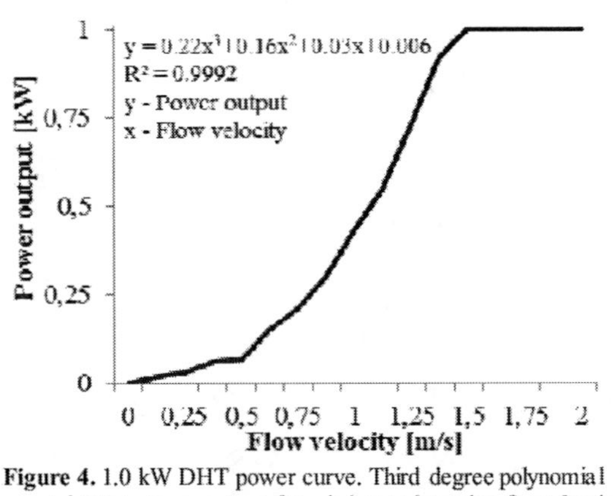

Figure 4. 1.0 kW DHT power curve. Third degree polynomial approximates power output for wind speed ranging from 0 m/s to 1.5 m/s. Data source [19].

Small scale units were selected because of the assumption that they will be used for household purposes. In case of hydrokinetic turbine its energy output has been multiplied by 3.5, because its nominal power is 1 kW and the goal was to make it comparable with wind turbine. The nominal power of PV installation has been assumed to be also 3.5 kW. Energy generation from PV system has been calculated based on formula:

$$E = \frac{G\, P^{PV}}{STC} \eta^{PV} \quad (5)$$

where: E – energy yield [kWh], G – global irradiation [kWh/m^2], P^{PV} – nominal power of PV installation [kW] η^{PV}- overall efficiency [%], STC – standard testing conditions irradiance [kW/m^2].

3 Results and discussion

This section is divided into three parts which describe obtained results. Two first sections deal with complementarity of: potential energy and energy generated based on power sources individual characteristics.

3.1 Complementarity of potential energy

As it has been already said, considered in this study energy sources tend to vary in time and space. Figures 5-8 illustrate this variability over seven consecutive days for both sites in January and June. Please note that the maximal available energy per hour is three times smaller in January than in June. Figures 5-8 additionally indicate that the energy from coming from flowing water has bigger potential than other two energy sources to operate as a base load power unit.

Table 1 summarizes basic statistical parameters of solar, wind and hydrokinetic energy time series over the year 2014. One statistical measure which is not sensitive to the range of time series values and therefore enables comparison among those energy sources is the coefficient of variation (CV) defined as a ratio of the standard deviation to the mean. Values of CV in Table 1 indicate that in case of Lądek Zdrój solar energy is the one with lowest variability, whereas in Lębork it is hydrokinetic energy. In both sites wind energy have the greatest value of CV. Differences in CV values in case of hydrokinetic energy can be attributed to the regimes of those rivers. Biała Łądecka River in Lądek-Zdrój has a mountainous catchment with a relatively rapid outflow. Biała Łądecka catchment is in addition six times smaller than the lowland catchment of Łeba River which flows through Lębork.

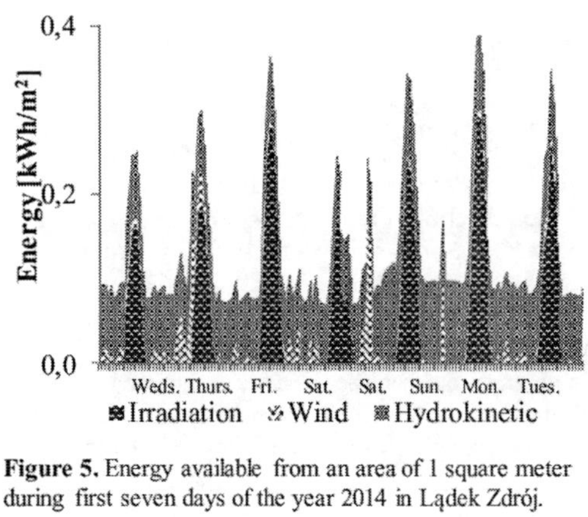

Figure 5. Energy available from an area of 1 square meter during first seven days of the year 2014 in Lądek Zdrój.

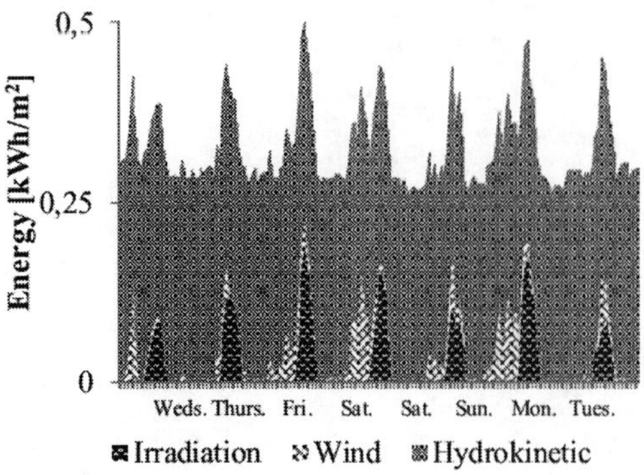

Figure 6. Energy available from an area of 1 square meter during first seven days of the year 2014 in Lębork.

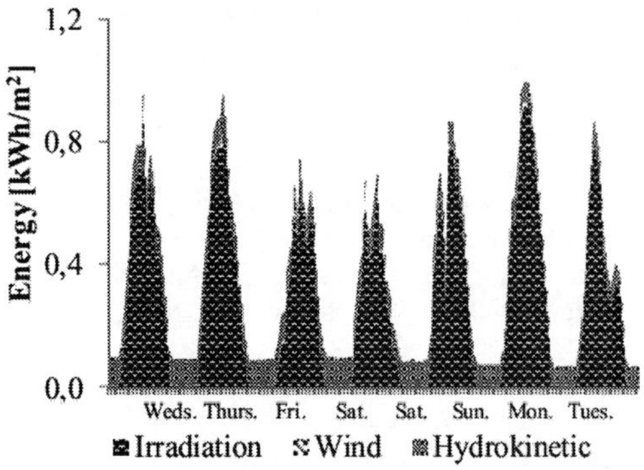

Figure 7. Energy available from an area of 1 square meter over the period 18-24.06.2014 in Lądek Zdrój.

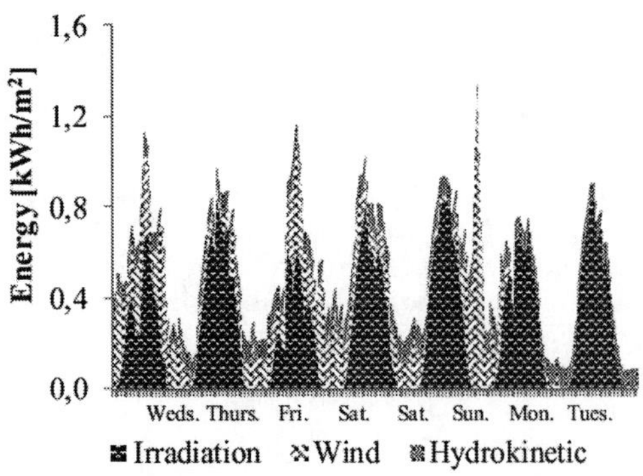

Figure 8. Energy available from an area of 1 square meter over the period 18-24.06.2014 in Lębork.

From the perspective of hybrid power source design and optimisation the potential and temporal complementarity of energy sources play crucial role. Perfect temporal complementarity can be described as a situation when dwindling amount of energy available from one source is instantaneously replaced by energy available from another one. A statistical parameter which is often used to calculate this phenomenon is correlation coefficient (CC) which gives values between +1.0 and -1.0 inclusive. Already mentioned perfect temporal complementarity is observed when the value of CC is equal to -1.0.

Table 1. Statistical parameters of investigated time series in over the year 2014.

	Lądek Zdrój		
	Irradiation	Wind	Hydrokinetic
Mean [kWh]	0.139	0.005	0.130
StD [kWh]	0.215	0.017	0.261
CV [%]	154.2	298.3	200.6
Var [kWh²]	0.046	0.000	0.068
	Lębork		
	Irradiation	Wind	Hydrokinetic
Mean [kWh]	0.136	0.056	0.159
StD [kWh]	0.213	0.125	0.095
CV [%]	156.1	221.2	60.0
Var [kWh²]	0.213	0.125	0.095

key: StD – standard deviation, CV – coefficient of variation, Var – variation.

Tables 2-5 present values of CCs calculated for various time steps. Values above diagonal (filled with 1.0) refer to Lądek Zdrój, whereas those bellow to Lębork. In case of annual sums of potentially available energy (Table 1) only hydrokinetic and solar (here denoted as irradiation) in Lębork exhibited beneficiary negative CC value. It means that during "sunnier" years one should expect lower potential of hydrokinetic energy. Moderate dependence (CC value of 0.478) can be also observed in case of wind energy and irradiation in case of Lądek Zdrój – this means that a year abundant with wind energy may be also have good potential for solar energy.

Table 2. Temporal complementarity based on annual sums

Lądek Zdrój/ Lębork	Irradiation	Wind	Hydrokinetic
Irradiation	1.000	0.478	-0.296
Wind power	-0.316	1.000	-0.200
Hydrokinetic	-0.633	-0.044	1.000

Table 3. Temporal complementarity based on monthly sums

Lądek Zdrój/ Lębork	Irradiation	Wind	Hydrokinetic
Irradiation	1.000	-0.590	0.172
Wind power	0.363	1.000	-0.178
Hydrokinetic	-0.480	-0.211	1.000

Table 4. Temporal complementarity based on daily sums

Lądek Zdrój/ Lębork	Irradiation	Wind	Hydrokinetic
Irradiation	1.000	0.013	0.011
Wind power	-0.228	1.000	-0.018
Hydrokinetic	0.034	-0.027	1.000

Table 5. Temporal complementarity based on hourly values

Lądek Zdrój/ Lębork	Irradiation	Wind	Hydrokinetic
Irradiation	1.000	0.013	0.011
Wind power	0.054	1.000	-0.018
Hydrokinetic	-0.186	-0.043	1.000

Table 3 focuses on monthly sums of individual parameters. Interesting value of CC (from the perspective of complementarity) occurs in case of wind and solar energy (Lądek Zdrój) and solar and hydrokinetic energy

DOI: 10.1051/e3sconf/20161000032

in Lębork. Both exhibit moderately negative values, which implies that they may be to some extent complement one another. Analysis conducted from the perspective of daily sums and hourly values of irradiation, wind and hydrokinetic energy did not reviled any significant correlation between those time series. This indicate that the complementarity on a daily and hourly time scale does not exist. But one must bear in mind, that those energy sources will still exhibit some individual daily patterns of available energy as it is the case of monthly values (see Figures 9-10).

3.2 Complementarity of energy considering power sources characteristics

Potential energy of wind, solar radiation and flowing water differes from the energy which can be derived from generators using them. This is due to the individual characteristics of: PV modules or wind and hydrokinetic turbines. For example wind turbine is characterized typicaly by three values of wind speed. One is know as *cut-in* speed – that is when blades start to rotate, *rated* speed – when turbine generated power at its rated capacity and finally *cut-off* speed when the wind velocity is to fast and blades have to be stoped from safety reasons. Thats the main reason is why not whole potential wind energy can be converted into electricity.

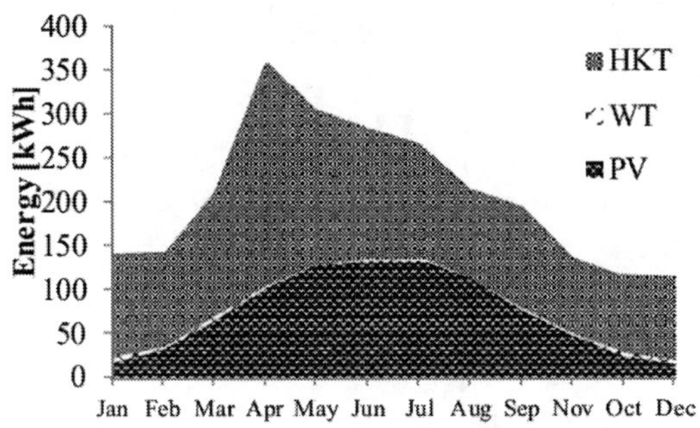

Figure 9. Mean monthly sum of energy generated per kW of installed capacity of given power source in Lądek Zdrój.

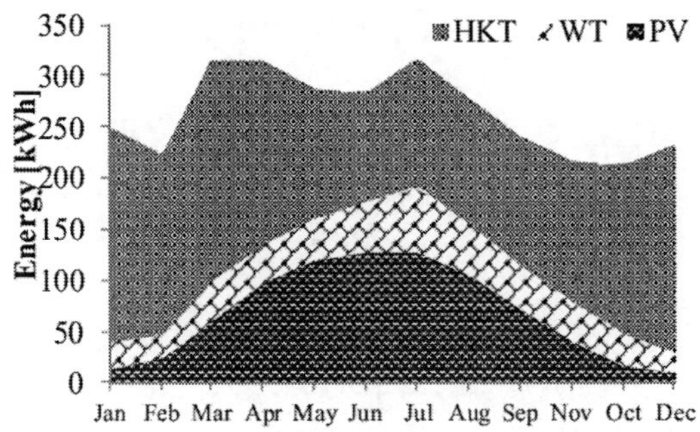

Figure 10. Mean monthly sum of energy generated per kW of installed capacity of given power source in Lębork.

Additionally please take a look at power curves depicted in Figures 1-2 and exponent for speed in formulas 1-2. Basically if wind speed increasy two times, then the energy output may increase eight times. Presented above short characteristic of wind turbines principles of operation are the main reason for conducting the analysis of CC values once again. This time for time series exhibiting energy generated. Figures 9-10 depict mean monthly sums of energy generated from different power sources with rated capacity of 3.5 kW each. A peak observed in Figure 9 in case of HKT can be attributed to thew (Lądek Zdrój is situated in Kłodzka Basin sourounded by Śnieżnik Masiff and Złote Mountains).

Comparison of values presented in Tables 2-5 and Tables 6-9 shows a significant increase in the correlation coefficient between hydrokinetic and solar energy in case of Lądek Zdrój (Table 6) which was calculated based on annual sums. The value of this CC i s greatest of all observed. It indicates that a year with smaller sums of irradiation may be also a year with greater availability of hydrokinetic energy. One more change was observed in case of daily sums in Lębork. The CC value between irradiation and hydrokinetic energy has significantly decreased. It is still rather relatively low, but may indicate that during sunny day energy generation from hydrokinetic turbine may be lower, this may be of course not the case of for example spring when irradiation and temperatures above 0°C might lead to thaw and greater energy yield from HKT.

Table 6. Temporal complementarity based on annual sums

Lądek Zdrój/ Lębork	Irradiation	Wind	Hydrokinetic
Irradiation	1.000	0.194	-0.692
Wind power	-0.289	1.000	-0.049
Hydrokinetic	-0.596	-0.097	1.000

Table 7. Temporal complementarity based on monthly sums

Lądek Zdrój/ Lębork	Irradiation	Wind	Hydrokinetic
Irradiation	1.000	-0.522	0.205
Wind power	0.369	1.000	-0.149
Hydrokinetic	-0.463	-0.218	1.000

Table 8. Temporal complementarity based on daily sums

Lądek Zdrój/ Lębork	Irradiation	Wind	Hydrokinetic
Irradiation	1.000	-0.192	0.076
Wind power	0.139	1.000	-0.021
Hydrokinetic	-0.385	-0.069	1.000

Table 9. Temporal complementarity based on hourly values

Lądek Zdrój/ Lębork	Irradiation	Wind	Hydrokinetic
Irradiation	1.000	-0.030	0.029
Wind power	0.055	1.000	-0.012
Hydrokinetic	-0.190	-0.056	1.000

3.3 Energy generation values

Using hybrid renewable system when energy sources do not perfectly complement one another will entail usage of energy storage device. In case of small residential hybrids usually acid-lead or lithium-ion batteries would be applied. However batteries are costly and their longevity depends on their operation regime. It is not recommeded to deeply discharge them. Combining different power sources, which are to some extent complementary may enable battery bank capacity redcution and prolong its operation. Assuming that load is continuous (there are no periods with zero load) please take a look at Figures 11-18. If battery bank is being charged only from one source of energy (e.g. PV or WT) there will often occure situations when there is energy demand but it all must be covered from battery bank. This is the case especialy for WT in Lądek Zdrój or PV in both locations. The former results from daily availability of solar energy, which for each point on the Earth can be calculated based on clear-sky models.

Figures 17-18 indicate that combining all those energy sources makes energy more available – meaning that it is generated almost all the time. However volume of its generation may be sometimes relatively small.

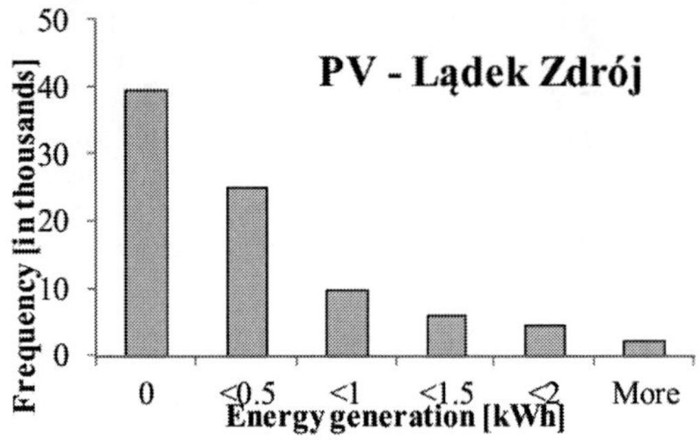

Figure 11. Histogram of energy generation values from PV in Lądek Zdrój covering period 2006-2015.

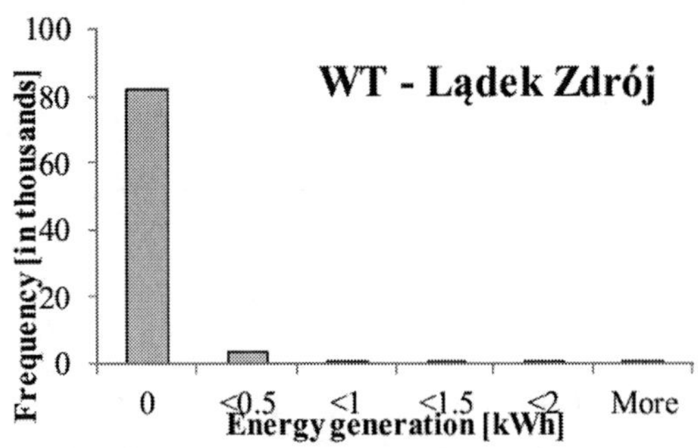

Figure 12. Histogram of energy generation values from WT in Lądek Zdrój covering period 2006-2015.

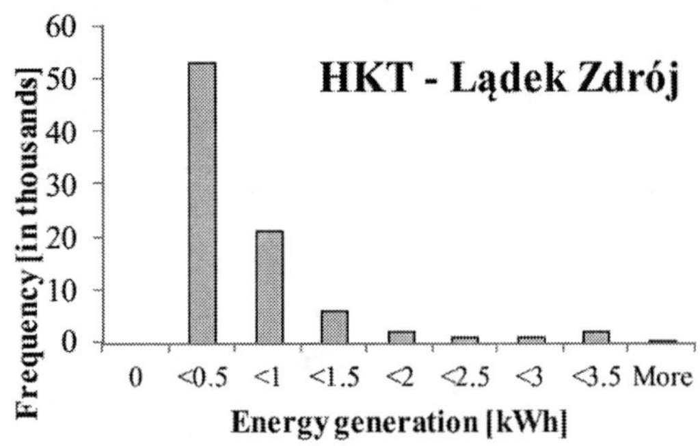

Figure 13. Histogram of energy generation values from HKT in Lądek Zdrój covering period 2006-2015.

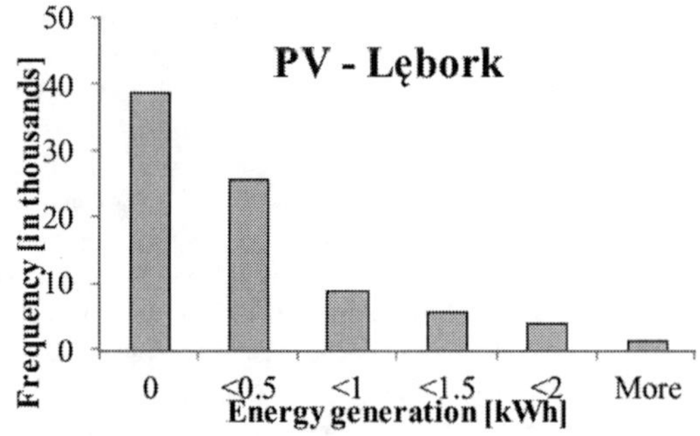

Figure 14. Histogram of energy generation values from PV in Lębork covering period 2006-2015.

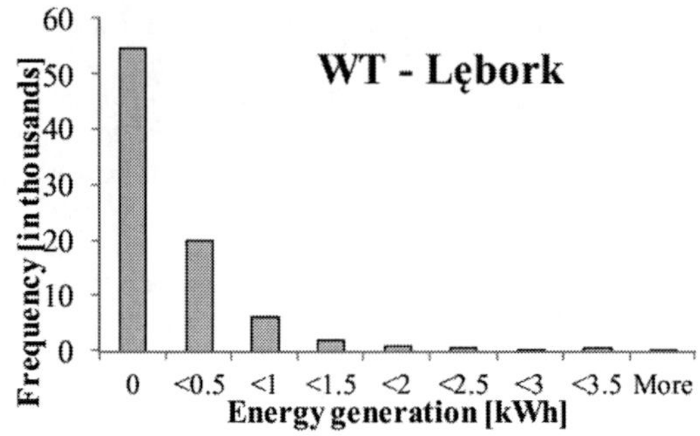

Figure 15. Histogram of energy generation values from PV in Lębork covering period 2006-2015.

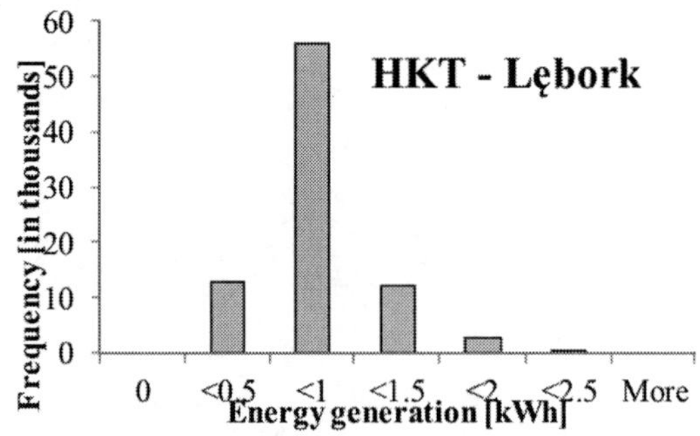

Figure 16. Histogram of energy generation values from HKT in Lądek Zdrój covering period 2006-2015.

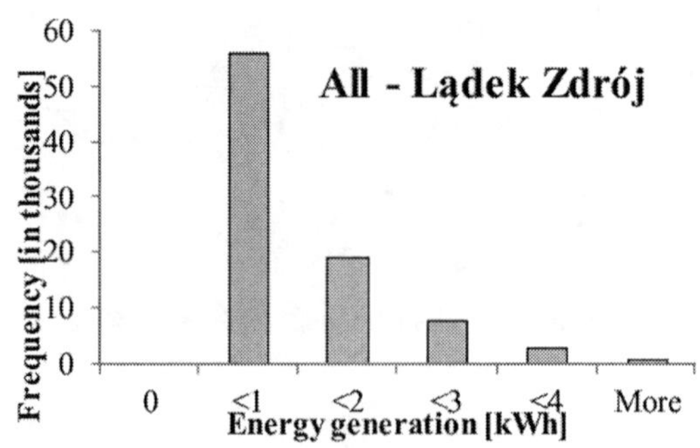

Figure 17. Histogram of energy generation values from PV-WT-HKT (3.5 kW per each source) in Lądek Zdrój covering period 2006-2015.

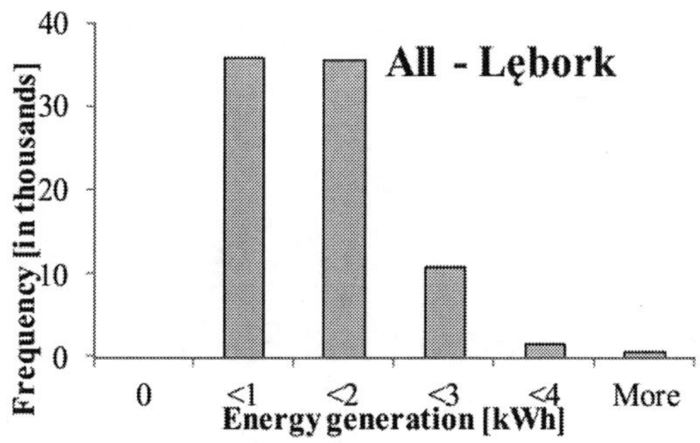

Figure 18. Histogram of energy generation values from PV-WT-HKT (3.5 kW per each source) in Lębork covering period 2006-2015.

4 Conclusions

According to the authors' knowledge the study described in this paper was the first one ever conducted. There are many papers which dealt with energy sources complementarity. However only several studies investigated this concept from the perspective of three types of energy sources. Additionally instead of hydrokinetic so far conducted research concentrated on large-scale hydropower which utilizes the potential energy of falling water masses. Results obtained from this analysis point to a moderate complementarity between solar and wind, solar and hydrokinetic energy on the annual and monthly time scale. Values of complementarity indices are not as spectacular as in referenced papers. This may be due to the fact that this study focused on small scale power sources, and to some extent the potential of wind energy may be distorted. Wind speed values were taken from measurements conducted 10 meters above the ground level, whereas industrial wind turbines are utilizing winds which are blowing 100 meters or more above the ground and are more stable. However obtained results indicate that development of polish energy sector should be to some extent based on RES that complement each other despite the fact that in some situations energy yield from a given source in certain locations may not be economically justified. What is more, in order to overcome problems (namely variability in energy output) encountered in case of power sources utilizing single RES one may use complementary energy sources such as wind and solar – as for example shown in [23].

After conducting this analysis there seems to be several interesting directions for future research. First of all, more sites representing various wind availability patterns and river regimes should be analyzed. Secondly a study should be conducted in order to determine the minimal needed storage capacity considering standard household load profile and various combinations of power sources.

References

1. F. Mormann, D. Reicher, V. M. Hanna, Stanford Environmental Law Journal **35,** 55.M (2016).
2. H. Shaker, Z. Hamidreza, D. Wood, Renew. Sust. Energ. Rev. **58,** 761-774 (2016).
3. P. Frączek, Arch. Min. Sci. **60,** 1, 387-401 (2015).
4. L. E. Jones *Renewable energy integration: practical management of variability, uncertainty, and flexibility in power grids*. Academic Press, (2014).
5. M. A. Delucchi, and M. Z. Jacobson, Energ. Policy, **39,** 3, 1170-1190 (2011).
6. H. H. Chen, H. Y. Kang, & A. H. Lee, Renew. Sust. Energ. Rev., **14,** 1, 413-421 (2010).
7. G. Notton, S. Diaf, L. Stoyanov, Energy Procedia, **6,** 666-677 (2011).
8. S. Killinger, K. Mainzer, R. McKenna, N. Kreifels, W. Fichtner, Energy, **84,** 563-574 (2015).
9. F. Monforti, T. Huld, K. Bódis, L. Vitali, M. D'Isidoro, R. Lacal-Arántegui, Renew. Energ., **63,** 576-586 (2014).
10. S. Jerez, R. M. Trigo, A. Sarsa, R. Lorente-Plazas, D. Pozo-Vázquez, J. P. Montávez, Energy Procedia, **40,** 48-57 (2013).
11. S. Lolla, S. B. Roy, S. Chowdhury, Energy Procedia, **76,** 187-192 (2015).
12. J. Schmidt, R., Cancella, A. O. P. Junior, A. Olímpio, **No. 572014** (2014).
13. P. De Jong, A. S. Sánchez, K. Esquerre, R. A. Kalid, E. A. Torres, Renew. Sust. Energ. Rev., **23,** 526-535 (2013).
14. J. Jurasz, A. Piasecki, Acta Energetica, **2,** 27, (2016).
15. H. J. Vermaak, K. Kanzumba, P. Sandile Koko, Renew. Sust. Energ. Rev., **29,** 625-633 (2014).
16. J. Zimny, *Odnawialne źródła energii w budownictwie niskoenergetycznym*, Kraków-Warszawa, Wydawnictwa Naukowo-Techniczne (2010).
17. R. H. Clark, **33,** John Wiley & Sons, (2007).
18. WindSpot by Sonkyo Energy http://usa.windspot.es/ [accessed 17.04.2016]
19. Darrieus turbine. Alternative hydro solutions. Available from http://www.althydrosolutions.com; 2014 [accessed 18.04.2014]
20. M. Ozga-Zielińska, J. Brzeziński, *Hydrologia stosowana*, PWN, Warszawa, (1997) 58–64
21. WMO-No.1044, *Manual on stream gauging. Volume II Computation of discharge*, (2010)
22. K. Wójcik, M. Wdowikowski, Interdyscyplinarne zagadnienia w inżynierii i ochronie środowiska **6,** 447-465 (2015).
23. J. Jurasz, J. Mikulik, Architecture Civil Engineering Environment, **8,** 4, 5-10 (2015).

Manufacturing technology of AS-SOFC prepared with different commercially available precursors

M. Kawalec[1,a], R. Kluczowski[1] and M. Krauz[1]

[1]*Institute of Power Engineering Ceramic Department CEREL*

Abstract. Fuel cells are devices converting the chemical energy into the electrical energy and heat as result of the electrochemical reaction between gaseous fuel and a gas oxidant in flameless combustion process. Because of omission of thermo-mechanical steps that are present in any traditional energy conversion technology (e.g. gas turbine) fuel cells show increased efficiency in comparison. Compact sizes and modular scalability predestines this technology for distributed energy generation including but not limited to renewable energy sources (e.g. wind, solar). Fuel cells technology also addresses other very important part of distributed renewable energy generation. Because of the unreliable energy production rates and the usual for renewable energy sources mismatch between energy supply and demand, some sort of energy storage is needed to store surplus of produced energy and release it when needed. Reversible fuel cells, that generate hydrogen from available surplus of energy and then generate energy from that stored fuel when needed are cheaper and more ecologically friendly alternative to usually used batteries. This technology is still under development, including research at IEn OC CEREL. In the early development of reversible fuel cells, new types of nickel oxide and porosity forming carbon was evaluated for this task. This work compares the electrical and mechanical parameters of SOFC manufactured with JT Backer NiO and Carbon Polska carbon with cells made from other commercially available materials. Based on evaluated quality, purity, availability and cost, following materials were selected for comparison: Novamet NiO, 99,9 % pure, grain size 1-2 μm and Aldrich carbon with parameters similar to graphite used previously. Preliminary tests show clear changes in the microstructural, mechanical and electrical parameters.

1 Introduction

As part of previous work related to the development of Solid Oxide Fuel Cells [1-9] at IEn OC CEREL, research and development of half-scale manufacturing technology of flat, thin Anode Supported Solid Oxide Fuel Cell (AS-SOFC) with dimensions of 50 x 50 mm and 100 x 100 mm was conducted. In this technology, anode supports are produced by means of highly reproducible, novel method of high-pressure injection molding and the other functional layers are made by screen printing method [7,9]. Previously, main components of the anode supports were nickel oxide obtained from JT Baker and graphite originating from Carbon Polska, with following parameters supplied by respective manufacturer:

• Nickel Oxide NiO
Manufacturer – J. T. Baker
Average grain size– 2 μm
Purity – 99,5 %

• Graphite
Manufacturer– Carbon Polska
Average grain size – 1-2 μm
Purity – min. 99% mas.

In order to maintain production continuity the need arose for search and usage of new raw materials for anode supports - nickel oxide as well as graphite. Based on our market research, taking into account price, quality and availability, nickel oxide type „F" originating from Novamet and graphite from Aldrich were selected, with following parameters supplied by respective manufacturer:

• Nickel Oxide NiO – type F
Manufacturer – Novamet
Average grain size– 1-2 μm
Purity – 99,8 %

• Graphite
Manufacturer– Aldrich
Average grain size < 20 μm
Purity – min. 99 % mas.

Application of new type of raw materials required trial batch of anode supports and complete fuel cells based on said supports as well as mechanical and electrochemical testing in order to compare the influence of this change on the mechanical and electrochemical parameters and fine tune production process.

[a] Corresponding author: kawalec@cerel.pl

DOI: 10.1051/e3sconf/20161000033

2 Experimental

Particle size distributions were determined for both nickel oxide powders using granulometer Mastersizer 2000. Particle size distribution curve and summation particle size curve for both powders are presented in the collective chart (see Figure 1).

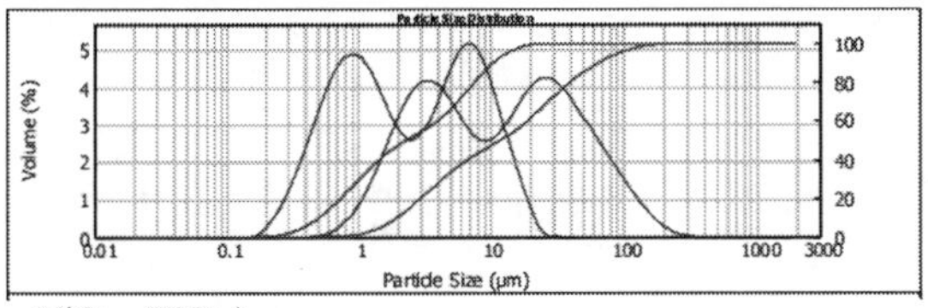

– NiO – JT Baker
– NiO – Novamet typ „F"

Figure 1. Particle size distribution for different nickel oxide powders.

For both powders d_{50} had been determined:

- NiO (JT Baker) - d_{50}=12,117 µm
- NiO „F" (Novamet) - d_{50}=2,340 µm

In both cases, particle size distribution is bimodal and in the case of nickel oxide from Novamet chart is clearly shifted towards smaller grain size. Average grain size of NiO powder originating from JT Baker is bigger in comparison to the number provided by the manufacturer. This discrepancy can be explained by differences in measuring methods. NiO obtained from Novamet corresponded to information from data sheet.

For both types of graphite ignition loss was determined as test of their purity:

SCarbon = 0,85 % mas.

SAldrich = 0,76 % mas.

In both cases, remainder of ignition loss does not exceed 1% mas, which is consistent with provided data.

Anode supports dedicated for comparative tests of two types of nickel oxide and two types of graphite were made by high-pressure injection molding. For this task, previously developed raw anode support composition was used, as follows:

Composition 1:

- NiO (JT Baker) – 66 % mas.
- 8YSZ (TOSOH) – 34 % mas.
- Graphite (Carbon Polska) – 25 % vol. of powder
- thermoplasticizer

Composition 2:

- NiO (Novamet) – 66 % mas.
- 8YSZ (TOSOH) – 34 % mas.
- Grafit (Aldrich) – 25 % vol. of powder
- thermoplasticizer

The prepared mixtures of powders, poreformer and thermoplasticizer were homogenized using an eccentric mixer and a technologic line for pellet production. After homogenization, masses were formed using high-pressure injection molding, obtaining raw anode supports with different kinds of nickel oxide. This process was accomplished by usage of high-pressure injection-molding device BOY XS (Figure 2.) with following base parameters:

- Mold clamping force: 100 kN
- Max. injection pressure: 2298 bar
- Max. injection volume: 6,15 cm3

Figure 2. High-pressure injection molding device BOY XS

In order to manufacture anode supports, high-pressure injection mold was used with internal dimensions of 62x62x1.2 mm. Taking sintering shrinkage into account, resulting supports after sintering have dimension of 50x50x1 mm.

There were no significant differences observed at the stage of homogenization or injection molding of prepared anode support in comparison to previously used nickel oxide. Anode supports (Figure 3) were manufactured into full AS-SOFC, using previously developed, optimized composition of individual layers (Figure 4).

Figure 3. Anode support made by injection moulding method

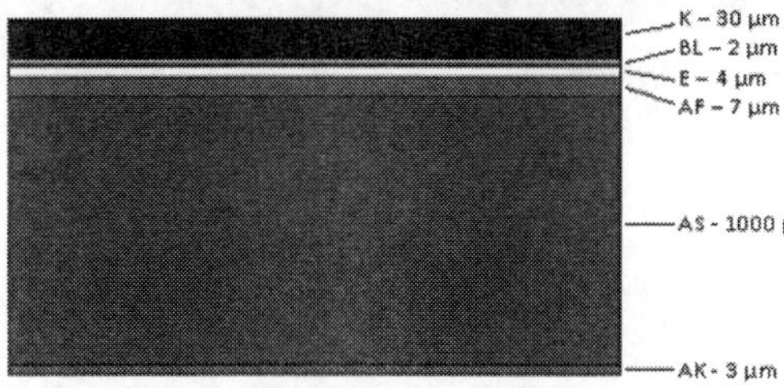

Figure 4. Scheme of solid oxide fuel cell AS-SOFC

- **K** – Cathode layer – $La_{0,6}Sr_{0,4}Fe_{0,8}Co_{0,2}O_{3-\delta}$ (Praxair), thickness 30 µm, porosity 25 vol. %

- **BL** – barrier layer $Gd_{0,1}Ce_{0,9}O_2$ (Praxair), thickness 2 μm
- **E** – electrolyte 8YSZ (TOSOH) thickness 4 μm
- **AF** – anode functional layer – NiO/8YSZ(TOSOH) 50/50% wt., thickness 7 μm,
- **AS** – anode support – NiO/8YSZ(TOSOH) 66/34 % wt., poreformer 25 vol. %, thickness 1,0 mm
- **AK** – anode contact layer – NiO, thickness 3 μm

Manufactured anode supports were debinded and pre-sintered at 1100 °C for 3 h. All of debinding and pre-sintering processes were conducted in high-temperature electric furnace HT 17/64 HDB (Nabertherm). After this process, anode functional layers 8YSZ (TOSOH) + NiO 1:1 and electrolyte layers 8YSZ (TOSOH) were applied using screen printing method on pre-sintered supports. Anode supports based on nickel oxide from Novamet and graphite from Aldrich had significantly lower mechanical strength and cracked during printing process. Anode supports based on previously used nickel oxide (JT Baker) and graphite (Carbon Polska) were not damaged by screen printing process. In order to mitigate this problem, two additional batches of anode supports were sintered, at 1150 °C and 1180 °C, both with retention time of 3h. Pre-sintering temperature of 1150 °C proved to be enough to ensure mechanical strength needed for screen printing process. After deposition of anode functional layer and electrolyte layers, anode supports were air dried at 80 °C and sintered at 1400 °C for 3 h. Then, $Gd_{0,1}Ce_{0,9}O_2$ (Praxair) layer were deposited using screen printing, dried (75 °C) and sintered at 1350 °C for 1h. After sintering, cathode layers of $La_{0,6}Sr_{0,4}Fe_{0,8}Co_{0,2}O_{3-\delta}$ (Praxair) were screen printed, dried (75 °C) and sintered at 1100 °C for 1h. Completed fuel cells with dimensions of 50x50x1 mm (**Figure 5**) were tested in order to determinate physical, mechanical and electrochemical properties.

Figure 5. The photo of a single AS-SOFC

3 Results and discussion

There were no visual differences observed between different types of manufactured fuel cells. No visible defects, deformation or cracks. Open porosity and apparent density were measured using hydrostatic weighting in water, deflection was measured using Coordinate Measuring Device Crysta-Apexc 7106. Bending tests were conducted using tripoint method on five samples with dimensions 8x10x60 mm manufactured by means of high-pressure injection molding of corresponding anode mass, debinded and then sintered at 1400 °C for 3 h. Test results are presented at Table 1.

Table 1. Results of open porosity, apparent density, deflection and bending tests for fuel cells with anode supports based on nickel oxide from JT Backer and Novamet sintered at 1400 °C

Support	Open porosity [%]	Apparent density [g/cm³]	Deflection [mm]	Bending strength [Mpa]
NiO (JTBaker) /8YSZ	26.1	4.79	0.08	131.45
NiO (Novamet) /8YSZ	24.2	4.91	0.11	112.38

Applying new type of NiO (type F, Novamet) and graphite (Aldrich) resulted in decreased porosity and increased apparent density by about 2 % and decreased bending strength by more than 15 % in comparison to anode supports manufactured with previously used raw materials. Microstructure measurements were conducted on samples using SEM microscope HITACHI S-3400N. Below, microstructure of anode supports before reduction (Figure 6,7) were presented.

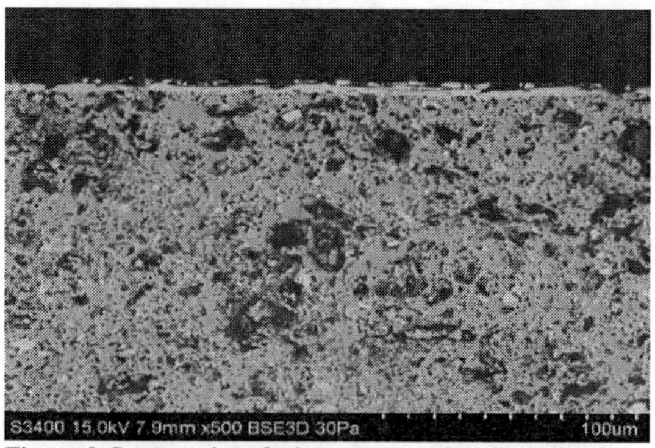

Figure 6. Cross section of microstructure of anode made of composition of 8YSZ (TOSOH)/NiO (JT Baker)

Figure 7. Cross section of microstructure of anode made of composition 8YSZ (TOSOH)/NiO (Novamet)

In case of anode supports made with nickel oxide from JT Baker and graphite from Carbon Polska (Figure 6) pores have spherical shape. In case of anode supports made with nickel oxide from Novamet and graphite from Aldrich (Figure 7) pores are more elongated with sharp ends in horizontal direction. This can be responsible for lowered mechanical and bending strength as well as lowered electrochemical parameters related to gas fuel transport through the anode. Both types of fuel cells were tested on the specialized single cell test rig. Tests were conducted at temperature of 800 ±3 °C, with pure H_2 flow of 1.0 Nl/min acting as fuel and standard air flow of 2.0 Nl/min as cathode feed. Results of electrical testing were presented below as I-V and I-P curves (Figure 8-9).

Measured open circut voltage (OCV) for both types of fuel cells:

• OCV=1,050 V – JT Baker
• OCV=1,058 V – Novamet

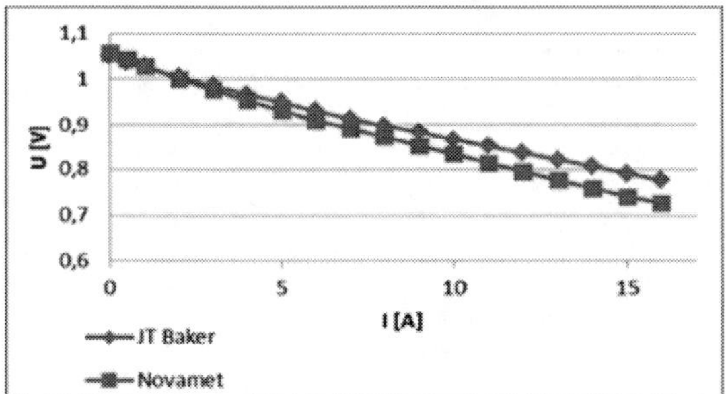

Figure 8. Polarisation curves U-I recorded for SOFC with anode based on nickel from Novamet and JT Backer

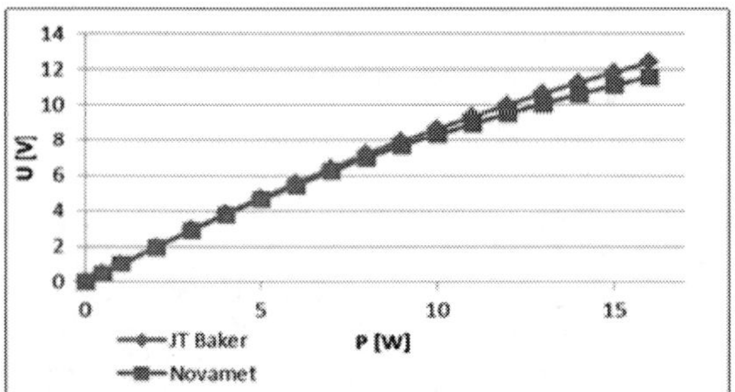

Figure 9. Power-current relation for fuel cells with anode based on nickel from Novamet and JT Backer

As can be gathered from above figures, there are slight differences between the polarization curve of the fuel cell with nickel oxide from Novamet and the one with nickel oxide from JT Backer. Power density with load of 1 A/cm² was as follows:

• 0,776 W/cm2 – JT Baker
• 0,726 W/cm2 – Novamet

In case of fuel cells made with new type of raw material maximum power density decreased by 6 %.

4 Conclusions

Replacement of base anode support raw materials (NiO-Jt Baker, graphite-Carbon Polska) with ones with similar characteristics (NiO-Novamet, graphite-Carbon Polska) results in changes of mechanical strength, which in turn required significant adjustments to the pre-sintering of anode supports. Additionally, changes were observed in the strength and the electrochemical parameters of fully manufactured cells, which will require additional adjustment to the process parameters in order to maintain established properties of produced SOFC. Research has shown that in case of big, planar fuel cell (above 50x50 mm in dimension) change in any of base materials, even for one with the same or similar characteristics can lead to significant changes of the production process and the properties of final product.

Acknowledgement

This work is supported by National Center of Research and Development, as a Sector Program INNOLOT, project EPOCA „Device for power supplying and controlling on-board and ground equipment", um. Nr INNOLOT/I/1/NCBR/2013

References

1. Z. Rak, R. Kluczowski, M. Krauz, Ceramika 459-466, vol. **96** (2006).
2. M. Krauz, K. Krząstek, R. Kluczowski, Conference materials "XXXI International Conference of IMAPS Poland Chapter" 503-506, Krasiczyn 23-26 wrzesień (2007).
3. R. Kluczowski, R. Śliwa, Inżynieria Materiałowa, 925-929, vol. **6** 160, (2007).
4. R. Antunes, T. Golec, M. Miller, M. Krauz, R. Kluczowski, K. Krzastek, Journal of Fuel Cell Science and Technology **7** (2010) 011011-1-6.
5. T. Golec, M. Miller, R. Antunes, J. Jewulski, M. Krauz, R. Kluczowski, K. Krzastek, M. Błesznowski,, Journal of Fuel Cell Science and Technology **7** (2010) 011003 -1-5.
6. R. Tomov, M. Krauz, J. Jewulski, S. Hopkins, R. Kluczowski, D. Głowacka, B. Głowacki, Journal of Power Sources, **195** (2010) 7160-7167
7. R. Kluczowski, M. Krauz, M. Kawalec, J.P. Ouweltjes, Journal of Power Sources **268** (2014) 752-757
8. R. I. Tomov, M. Krauz, A. Tluczek, R. Kluczowski, Venkatesan V. Krishnan, K. Balasubramanian, R.V.Kumar, B. A. Glowacki, Mater Renew Sustain Energy (2015) 4:14
9. Praca zbiorowa pod redakcją naukową T. Golca, R. Kluczowski, M . Krauz, M. Kawalec, rozdz. „*Wytwarzanie i charakterystyka ogniw SOFC*", Wydawnictwo Naukowe Instytutu Technologii Eksploatacji – PIB, Warszawa (2015)

E3S Web of Conferences **10**, 00034 (2016) DOI: 10.1051/e3sconf/20161000034
SEED 2016

Wind Power Plants for Low Rated Wind Speed Regions: Feasibility Analysis and Simulation of a System

Mehmet Numan Kaya[1,a] and Faruk Köse[2]

[1]*Karamanoglu Mehmetbey University, Mechanical Engineering Department, Karaman, Turkey*
[2]*Selcuk University, Mechanical Engineering Department, Konya, Turkey*

Abstract. The use of wind power has become an important and growing part of the electrical energy supply in many countries all over the world. Wind turbines are the main components of wind power stations and every single development on these turbines affects the market. Latest developments in wind turbine technologies caused to decrease in costs, thus, besides high-wind-speed regions, low-wind-speed regions are taken into account as a feasible option to generate electricity. The present work investigates the feasibility of a wind power station located in a low rated wind speed region of Turkey providing a general approach for selection of the suitable wind turbine. Economic analysis was performed by calculating the key financial figures such as net present value (NPV), basic payback period (BPP) and the internal rate of return (IRR) for five various options and hourly mean, monthly and seasonal power productions are simulated for the most feasible option using the wind speed measurements in the region. Results show that it is possible to obtain a reasonable capacity factor in low rated wind speed regions with wind turbines that have larger rotors. Although the average wind speed is low in the examined region, a wind power station is still found to be feasible. According to the simulation results, maximum power is produced during the evening hours on a daily basis and during the summer months on a seasonal basis, ensuring power flow usually when maximum power consumption levels are reached. All the data used in this study are based on long-term measurements.

1 Introduction

Energy has been an essential component of human being since decades to perform various works and the need for energy will never end. Not harming the environment while producing energy is on the front burner of many researchers since people realized that environmental effects has the potential of negatively impacting the daily lives of individuals. The first thing that spring to mind is renewable energy when the topic is sustainable energy production. Today, most of the developed and developing countries are planning to increase energy production from renewables since these are environment friendly, native and unlimited. Wind power is one of the most known renewables and the use of wind to produce energy increases day by day. Global cumulative installed wind power capacity has increased from 17,400 to 432,419 MW from 2000 to 2015 [1]. Studies on wind energy, especially about feasibility of wind power, have effect on the increase. In a feasibility study for İzmir, Turkey, authors stated that the larger the installed capacity, the smaller the generating cost per kWh and the higher IRR of the investment [2]. Another study for Turkey investigated the competition potential of wind power plants and it was concluded that the Marmara, southeast Anatolian and Aegean regions are highly suitable for wind power generation since wind speeds exceed 3 m/s in most of these areas [3]. Shaahid et al. investigated economic feasibility of 75 MW wind power plants on four coastal locations - Al-Wajh, Jeddah, Yanbu and Jizan - of Saudi Arabia and found out that capacity factors and unit costs vary from 12% to 21% and from 0.0423 to 0.0711 US$/kWh, respectively [4]. Khabit et al. studied on Assessment of electricity generation by wind power in nine costal sites in Malaysia specified that average unit cost of the energy produced by a wind power system in Malaysia is 1.6–7.29 USD/kWh and the use of wind power systems as standalone systems is not recommended for the selected sites [5]. Celik pointed out in his study that the cost of wind electricity per kWh can be significantly reduced if the components of wind energy systems were exempted from taxes and subsidies were introduced [6]. Mostafaeipou et al. studied on wind energy feasibility study for city of Shahrbabak in Iran and recommended to install small size wind turbines for electricity supply of public buildings and private houses [7]. Blackler and Iqbal specified that a wind farm project at the Holyrood thermal generation station site is feasible in their study on Pre-feasibility study of wind power generation in Holyrood, Newfoundland [8]. Many other studies can be found in the literature

[a] Corresponding author: mnumankaya@kmu.edu.tr

 DOI: 10.1051/e3sconf/20161000034

on feasibility of wind power plants [9-13], in addition, there are some review studies available for some countries [14-15].

The present study investigates the feasibility of a wind power station located in a low rated wind speed region of Turkey. Most proper wind turbine among various options is selected the for the region considering the low rated wind speed and it is simulated for different time periods. In addition, economic analysis was performed by calculating BPP, NPV and IRR values.

2 Measured wind data

Wind speed measurements have been performed in Cumra, Turkey since 2006 with the help of a wind pole in the region. Five anemometers are installed at various heights on the wind pole and the wind speed data measured at 80 m height in 2013 are used in the present study. Monthly mean wind speeds are shown in Figure 1.

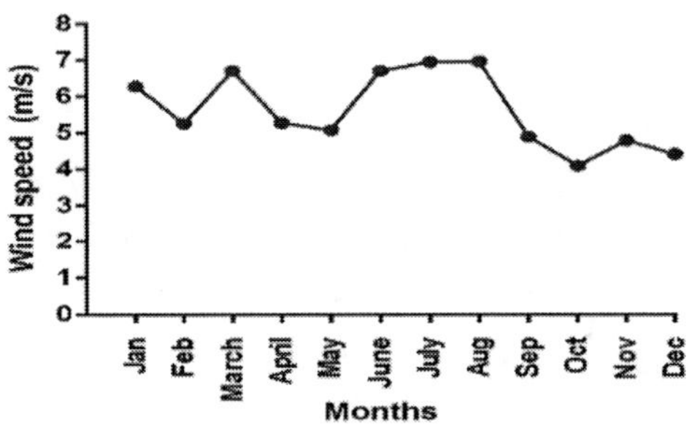

Figure 1. Monthly mean wind speeds

As seen from Fig. 1, mean wind speed is higher in summer months compared to other months and highest mean wind speed is obtained in August. Wind rose plot depicting wind speed frequency and energy potential according to the wind direction is given in Fig. 2.

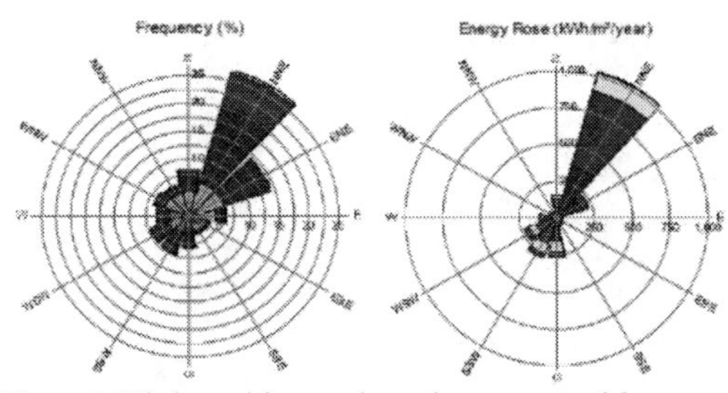

Figure 2. Wind speed frequencies and energy potential

3 Selected wind turbines

Six commercial wind turbines are selected to predict annual energy production and perform economic analysis. Total installed power is assumed to be 6 MW according to the power requirement in the region and wind turbines are selected considering this value. Characteristics of wind turbines are given in Table 1.

Table 1. Characteristics of the selected turbines

Wind turbines	Cut-in WS (m/s)	Cut-out WS (m/s)	Rated Power (kW)	Rotor diameter (m)
WT - 1	3	25	3000	115
WT - 2	3	21	3000	122
WT - 3	3	22	2000	100
WT - 4	3	25	1500	100
WT - 5	3	24	2000	100
WT - 6	3	25	1500	86

4 Calculation Methodology

Capacity factor is the ratio of annual output to potential output, and it is calculated using the Eq. 1 where P_{annual} is the annual energy production (MW) and $P_{potential}$ is the annual potential energy production in full capacity (MW).

$$C_p = P_{annual} / P_{potential} \quad (1)$$

Unit cost is calculated using Eq. 2. In this equation, E_{wt} is the cost of the wind turbine (€) and n is the lifespan (year) of the project.

$$UC = E_{wt} / (P_{annual} \text{ x } n) \quad (2)$$

Basic payback period is the ratio of total expenditures, E_{total} (€) to annual savings (€).

$$BPP = E_{total} / AS \quad (3)$$

Total expenditures include wind turbine cost, total operation and maintenance (O&M) cost in the lifespan period, foundation, transmission line and other costs. It is given in Ref. [16] that the O&M costs are approximately 0.3-0.4 c€/kWh during the first two years and approximately 0.6-0.7 c€/kWh after six years for wind turbines. In our calculations, we considered the value 0.5 c€/kWh for wind turbines. Foundation, transmission line and other costs are taken as 20% of the wind turbine price.

Net present value (NPV) is calculated by discounting all future income and expenditure flows to the present with Eq. 4 [9].

$$NPV=\sum[(B-C)/(1+r)n] \quad (4)$$

Where, B is the benefit, C is the cost, r is the discount rate and n is lifecycle year of the project. In this study, the project lifespan was taken as 25 years for the analysis and the overall annual interest rate (r) is assumed to be 2.5%. Salvage cost was not taken into account which was estimated to be equal to the disassembly cost of the wind power system components at the end of the project lifespan. IRR is the rate, which would make NPV value zero and it can be calculated with Eq. 5, where the parameters are same as the ones of NPV [9].

$$\sum[B/(1+r)n]=\sum[C/(1+r)n] \quad (5)$$

Table 2. Annual electricity productions, capacity factors and unit prices of selected wind turbines

	WT - 1	WT - 2	WT – 3	WT - 4	WT – 5	WT - 6
Rated Power (MW)	3 MW	3 MW	2 MW	1.5 MW	2 MW	1.5 MW
WT Cost* (€)	4,050,000	3,350,000	2,175,000	2,525,00	2,575,000	1,875,000
Number of WT	2	2	3	4	3	4
Foundation, Transmission line and other costs (€)	1,620,000	1,340,000	1,305,000	2,020,000	1,545,000	1,500,000
O&M costs for 25 years (€)	1,769,125	1,557,500	1,776,500	2,135,875	1,816,125	1,642,625
Total Cost(€)	11,489,125	9,597,500	9,606,500	14,255,875	11,086,125	10,642,625
Annual Electricity Production (MWh)	14,153	12,460	14,212	17,087	14,529	13,141
Total generated electricity in 25 years (MWh)	353,825	311,500	355,300	427,175	363,225	328,525
Unit cost of the electricity (€/kWh)	0.0325	0.0308	0.0270	0.0334	0.0305	0.0324
Capacity Factor, Cp (%)	26.9	23.7	27	32.5	27.6	25

*Cost of one wind turbine including its installation cost.

Table 3. Economic feasibility analysis results

Turbine type	**WT - 1**	**WT - 2**	**WT - 3**	**WT – 4**	**WT - 5**	**WT - 6**
Investment cost[1] (€)	9720000	8040000	7830000	12120000	9270000	9000000
Annual energy production (MWh/year)	14153	12460	14212	17087	14529	13141
Annual cost saving (€/year)	1556830	1370600	1563320	1879570	1598190	1445510
Operation and maintenance cost (€/year)	70765	62300	71060	85435	72645	65705
Net annual cost saving[2] (€/year)	1486065	1308300	1492260	1794135	1525545	1379805
Basic payback period (years)	6.54	6.15	5.25	6.76	6.08	6.52
NPV (€)	17,659,820	16,064,611	19,663,960	20,935,819	18,837,215	16,422,046
IRR (%)	14.8	15.9	18.8	14.3	16.1	14.9

[1]Investment cost includes installation and other additional costs except O&M costs.

[2]Net annual cost saving is calculated by subtracting yearly O&M costs from annual cost saving.

5 Results

The power outputs, capacity factors and unit prices of the selected turbines are given in Table 2. As mentioned before, total installed power is assumed to be 6 MW. According to the Table 2, capacity factors and unit prices change between 25 - 32.5 % and 0,027 – 0,0325 €/kWh, respectively. Although WT – 4 has the highest capacity factor, WT – 3 has the best unit price because of its cheaper price. Economic feasibility analysis results are presented in Table 3. All the options have basic payback periods that change between 5,25 and 6,54 years, and the most feasible one is found to be the WT -3.

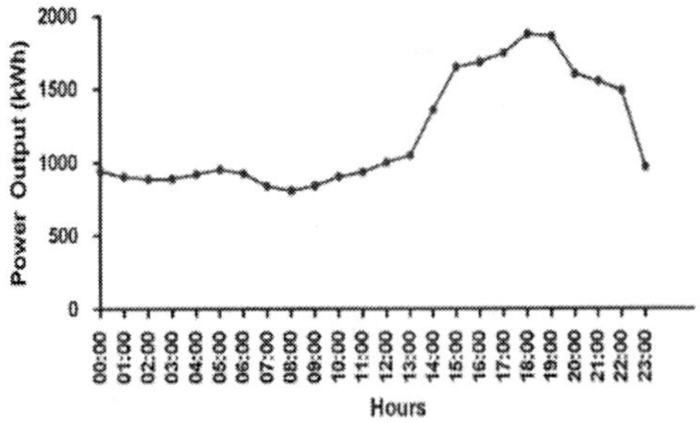

Figure 3. Hourly mean power production

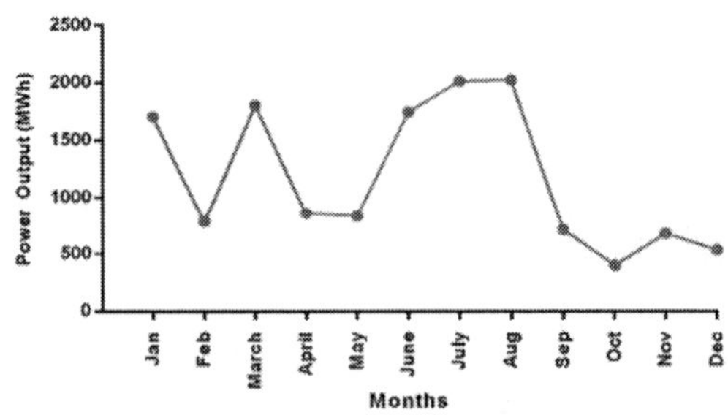

Figure 4. Monthly power production

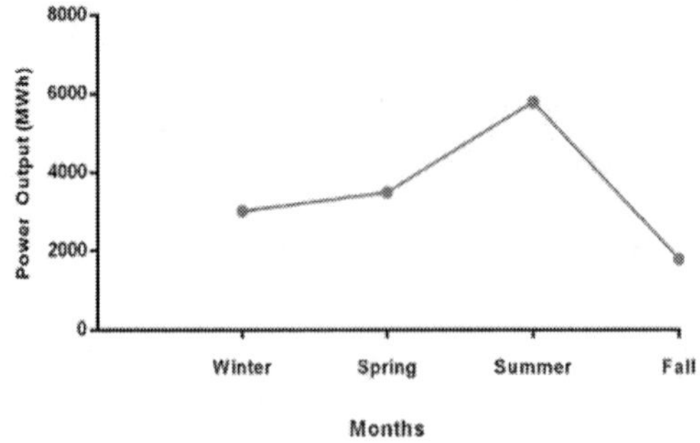

Figure 5. Seasonal power production

Hourly mean, monthly and seasonal power productions are simulated for the most feasible option, the one including three of WT – 3s. Hourly, monthly and seasonal mean power productions are presented in Fig. 3, 4 and 5, respectively. As it is can be seen from the figures, maximum power productions are obtained during evening hours on a daily basis and during summer months on a monthly basis.

6 Conclusion

In the present study, feasibility of a wind power plant in a low rated wind speed region is investigated and it is found out that it can also be feasible to install wind power plants in low rated wind speed regions. A good capacity factor that is about 32 % is obtained for a commercial wind turbine, however, it was not the most feasible option because of the higher price. A reasonable basic payback period, 5.25 years, is obtained in the region where the average wind speed is under 6 m/s at 80 m height. This shows that the more the wind turbine technology enhances, the more feasible wind power plants will become in low rated wind speed regions in the future. Simulation results show that maximum power output from the wind power plant is obtained during the times when maximum energy consumption levels are reached.

Acknowledgement

Authors would like to thank to Alibeyhuyugu Irrigation Cooperation for providing the data.

References

1. GWEC- Global Wind Energy Council, Global Wind Power Statistics (2015), available online at : http://www.gwec.net/wp-content/uploads/vip/GWEC-PRstats-2015_LR_corrected.pdf.
2. B. Ozerdem, S. Ozer, M. Tosun, Feasibility study of wind farms: A case study for Izmir, Turkey J. Wind Eng. Ind. Aerodyn. 94, 10 (2006).
3. A. Demirbas, Competition Potential of Wind Power Plants, Energy Sources 27, 7 (2005).
4. S. M. Shaahid, L.M. Al-Hadhrami, M.K. Rahman, Economic feasibility of development of wind power plants in coastal locations of Saudi Arabia – A review, Renewable Sustainable Energy Rev. 19 (2013).
5. T. Khatib, K. Sopian, M. Z. Ibrahim, Assessment of electricity generation by wind power in nine costal sites in Malaysia, Renewable Sustainable Energy Rev. (2012).
6. A. N. Celik, A Techno-Economic Analysis of Wind Energy in Southern Turkey Int. J. Green Energy 4, 3 (2007).
7. A. Mostafaeipour, A. Sedaghat, A. A. Dehghan-Niri, V. Kalantar, Wind energy feasibility study for city of Shahrbabak in Iran, Renewable Sustainable Energy Rev. 15, 6 (2011).
8. T. Blackler, M.T. Iqbal, Pre-feasibility study of wind power generation in Holyrood, Newfoundland, Renewable Energy 31, 4 (2006).
9. M.S. Genç, M. Gökçek, Evaluation of Wind Characteristics and Energy Potential in Kayseri, Turkey, J. Energy Eng. 135, 2 (2009).
10. F. Kose, M.N. Kaya, Analysis on meeting the electric energy demand of an active plant with a wind-hydro hybrid power station in Konya, Turkey: Konya water treatment plant Renewable Energy 55 (2013).
11. F. Kose, M. H. Aksoy, M. Ozgoren, An assessment of wind energy potential to meet electricity demand and economic feasibility in Konya, Turkey, International Journal of Green Energy 11 (2014).
12. M. R. Nouni, S. C. Mullick, T. C. Kandpal, Techno-economics of small wind electric generator projects for decentralized power supply in India, Energy Policy 35 (2007).
13. F. Kose, M.H. Aksoy, M. Ozgoren, "Economic Feasibility of Wind Energy Potential in Konya, Turkey", International Scientific Conference, UNITECH09, Proceedings (2009) Gabrovo, Bulgaria.
14. S.M. Shaahid, L.M. Al-Hadhrami, M.K. Rahman, Economic feasibility of development of wind power plants in coastal locations of Saudi Arabia – A review, Renewable Sustainable Energy Rev. 19, (2013).
15. R. Majumder, I. Mukherjee, B. Tudu and D. Paul, Review on feasibility of wind energy potential for India, Non Conventional Energy (ICONCE), 1st International Conference on, Kalyani (2014).
16. The European Wind Energy Association (EWEA), Wind energy 2 (2011).

E3S Web of Conferences **10**, 00035 (2016) DOI: 10.1051/e3sconf/20161000035
SEED 2016

Risk assessment of children's exposure to potentially harmful elements (PHE) in selected urban parks of the Silesian agglomeration

Kicińska Alicja[1,a]

[1]*AGH University of Science and Technology, Department of Environmental Protection, Faculty of Geology, Geophysics and Environmental Protection, Mickiewicz av. 30, 30-052 Cracow, Poland*

Abstract. The author determined the total contents of selected elements potentially hazardous for health (PHE): As, Be, Cd, Co, Cu, Ni and Sn in soils, sand of sandboxes and airborne dust collected in three urban parks of the Silesian agglomeration. The upper limit of Cd content (a Polish regulation) was exceeded in the soils of two largest and most frequented parks, the Silesian Park and the Kościuszko Park. The mean Cd contents in soils are 9 and 7 mg/kg, respectively. The metal contents of the sand from sandboxes are generally much lower than those of the soils: Cu 28 times on average, As 13 times, and Cd and Ni around 4 times, while the Co and Sn contents of sand are comparable with those of soils. Airborne dusts are a significant source of metals: they contain Cd (1–20 mg/kg), Co (2–17 mg/kg) and Cu (6–143 mg/kg). The quotients of the health risk indicate a potential health risk caused by As, Cd and Ni for children, particularly those with a low (below 15 kg) body weight. The risk level of 1–4% PTMDI (Provisional Maximum Tolerable Daily Intake) origins from an accidental swallowing of soil.

1 Introduction

Several century-long traditions of bituminous coal mining and metallurgy of Zn-Pb and Fe ores have left their marks not only on the landscape of Upper Silesia, but also on the environment quality and the health of local population [1, 2]. There are many papers dealing with the presence of Zn, Pb and Cd in individual components of the environment (soils, air, surface waters, etc.) and with their impact on living organisms [3, 4, 5, 6]. The reports on the elements generally considered to be less harmful, such as Co, Cu, Ni and Sn, or on those with lower abundance in the nature, for instance As and Cd, are rather rare. It was a reason to undertake the investigations of selected environmental components of the Upper Silesian industrial region, one of the oldest in Poland and currently the largest one. Research included the analysis of contents of some potentially hazardous elements (PHE): As, Be, Cd, Co, Cu, Ni and Sn, which can affect the health of population, particularly of the youngsters.

A prolonged exposure to As may lead to cancerous changes and disorders in the functioning of the nervous, circulatory and respiratory system [7]. Arsenic is commonly considered a significant source of increasing cancer-related diseases [8]. Cadmium is another element strongly toxic to human organism. The children and newborns being put at a risk of contact even with traces of Cd may suffer irreversible changes of their central nervous systems. Cd is also a carcinogenic element, belonging to the group I of such pathogens [7, 8]. In turn, cobalt has got important functions in producing red blood cells, metabolism of proteins and nucleic acids. An excess of Co is a cause of damages of the thyroid, bones, kidneys, and also of the DNA structure which probably results in carcinogenic changes. An excess of copper, the next of the elements considered, although only occasionally occurring, is also hazardous to the human health. Its symptoms include nervousness, irritability and even depressions or psychosis. High build up of copper results in the Wilson disease and is linked to a mental retardation, muscular tremors and an impaired coordination of movement [7]. The nickel-containing compounds that enter the human organism via the respiratory system belong to the 1st group of carcinogens. In excessive amounts, this element damages the structure of nucleic acids and causes tumors of the upper respiratory system. The last of the selected PHE is tin, characterized by a weak toxicity, except for its alkyl compounds that paralyze the nervous system. However, Sn is a metal of low bioavailability [5].

The aim of the investigation carried out in three urban parks of the Silesian urban agglomeration was two-pronged:

1. to determine total contents of the selected PHE in the environment components which children come into contact with on playgrounds. The sample material included adjacent soils and sand of sandboxes, and also urban dusts deposited on recreational equipment and facilities used by children;
2. to establish a degree of health risk of children playing in Silesian parks.

[a] Corresponding author: kicinska@geolog.agh.edu.pl This paper was supported by AGH grant no. 11.11.140.199

2 Study area and methods

2.1 Study area

The investigations were carried out in three parks of the Silesian agglomeration (Fig. 1). Two of them are the most frequented by the Upper Silesia population: the Tadeusz Kościuszko Park in Katowice (no. I in Fig. 1) with an area of 72 ha, founded in 1888, and the Śląski (Silesian) Park in Chorzów (no. III in Figure 1) with an area of 520 ha, founded in 1950. As a comparative area was chosen a small park (no. II in Fig. 1) of a housing development Bażantowo in the Kostuchna district of Katowice. The area of this park is barely 0.08 ha.

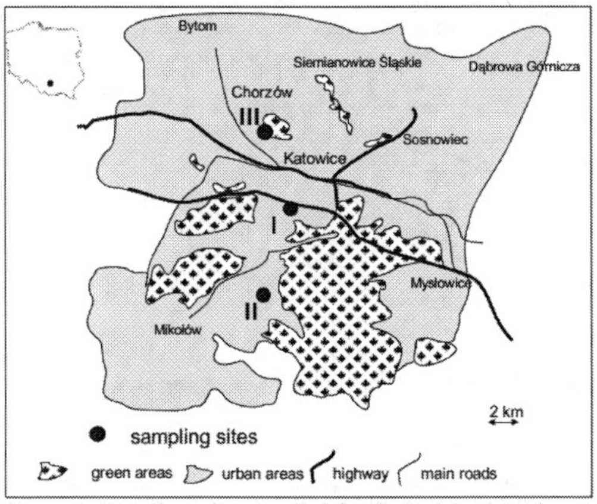

Figure 1. Samplings sites: I – Kościuszko Park in Katowice, II – Park in Kostuchna district of Katowice, III – Silesian Park in Chorzów.

2.2 Sampling and measurement procedures

In each of both large parks three sampling sites within the playgrounds including sandboxes were selected. In each of them the author collected 15 samples of the topsoil (around 1 kg each), 15 samples of sand (around 1 kg each) and 5 samples of urban dusts rinsed with distilled water from the toys and playing facilities (1 dm^3 each). In the small comparative park the list of sample types was the same, but only five samples of each type were taken because of its considerably smaller area (3 and 4 orders of magnitude).

From the topsoil samples (a depth of 0–30 cm) the grain fraction >2 mm and plant fragments were removed, next the material was dried, homogenized and quartered. All kinds of analytical samples were dissolved in a mixture of concentrated acids (HNO_3, $HClO_4$ and HCl) in a microwave oven, at a solid to liquid ratio of 1:20. Then, the concentrations of As, Be, Cd, Co, Cu, Ni and Sn in the solutions were determined with an ICP-MS analyser at a detection limits of $0.2 \cdot 10^{-5}$ mg/dm^3. The active (in H_2O) reaction of soils was determined at a solid to liquid ratio of 1:5.

The health risk was calculated according to the following formula [9]:

$$D = C \cdot \frac{LC}{BW} \cdot \frac{FE}{T} \quad (1)$$

where:
D – dose of metal absorbed (μg/kg body weight/day),
C – concentration of a heavy metal in soil (mg/kg),
LC – length of contact with a soil per unit of time (accepted as 200 mg soil/day),
FE – frequency of exposure (accepted as 2 hours/day),
BW – body weight (accepted as 15 kg for a three-year-old child and 19 kg for a six-year-old child),
T – time of exposure (accepted as 1050 days for three-year-old child or 2100 days for six-year-old child).

3 Results

3.1 PHE in soils

The pH of soils varied from 5.7 to 6.8, classifying them [10] as slightly acid, which results from the prevalence of luvisols and brown soils in the area. The lowest pH values were determined in the park no. II (mean 5.73), in the other two parks the mean pH values are almost equal (6.1–6.2).

The contents of As in soil ranges from 2.3 to 6.1 mg/kg (Tab. 1), with the highest mean (4.1 mg/kg) calculated for the park no. I and a slightly lower mean (3.8 mg/kg) for the park no. III. Distinctly lower is the calculated As mean contents of the soils in park no. II – 2.3 mg/kg. Similar trends have been found in the case of Co, Cu, Ni and Sn (Tab. 1). The highest mean soil contents of these elements occur in the park no. I (12, 25, 23 and 5 mg/kg, respectively), slightly lower in the park no. III (11, 15, 12 and 4 mg/kg, respectively), and distinctly lower in the park no. II (7, 5, 9 and 1 mg/kg, respectively).

Table 1. PHE content in soils from Silesia playground.

Park	n	As	Cd	Co	Cu	Ni	Sn
		[mg/kg]					
I	15	4.1	7	12	25	23	5
II	5	2.3	3	7	5	9	1
III	15	3.8	9	11	15	12	4
min.–max.	35	2.3–6.1	3–12	7–19	5–45	7–40	1–7
av.		3.7	7.4	11	31	16.5	4
SD		*0.07*	*0.16*	*0.12*	*0.14*	*0.3*	*0.02*
accept. values*	**A**	**20**	**1**	**20**	**30**	**35**	**20**
	B	**20**	**4**	**20**	**150**	**100**	**20**

*Due to [11]: A – protected areas, B – other areas, n – number of samples, min. – minimum, max. – maximum, av. – average arithmetic, *SD* – standard deviation.

The contents of all the PHE within the park no. I are the highest in the soils of the sampling site no. 2 (Fig. 2). In the park no. III the differences among the PHE contents of soils are less significant than in the park no. I, being the highest in the sampling site no. 2 (Fig. 3). In both parks these two sites are located relatively far from major town transit systems and the park playgrounds have been set in open areas (without shrubs and higher trees). Their significant contaminations must be, thus, related to the fall of airborne particles containing heavy metals among various pollutants.

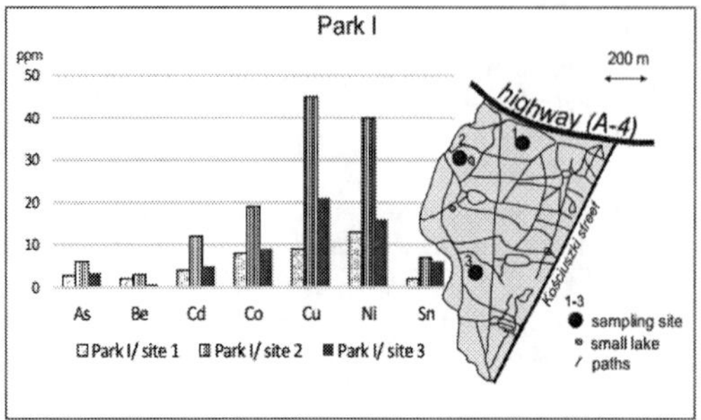

Figure 2. The concentration of PHE in Kościuszko Park in Katowice.

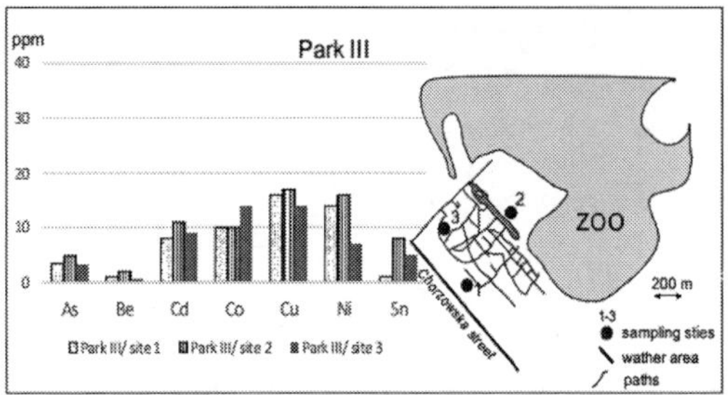

Figure 3. The concentration of PHE in Silesian Park in Chorzów.

The soil metal values of the three parks do not exceed upper permitted values of As, Co and Sn (20 mg/kg for each of them) of the Polish soil quality standards established by the Minister of the Environment [11] for the soils of the protected areas (i.e., the areas A). In the case of Ni and Cu, their limits of 35 and 30 mg/kg respectively, have been exceeded in respect to the type A soils for all the samples collected in the sites no. 2 of the parks no I and no. III. However, the upper limits of the metal contents of the soils occurring in the areas other than the protected and industrial ones (i.e., the areas B) have not been exceeded for Ni and Cu.

The cadmium soil contents of the playgrounds are most burdensome. They exceed generally not only 1 mg/kg, which is the upper border for soils of the protected areas (type A), but also 4 mg/kg, the upper limit of the type B areas (i.e., the areas other than protected and industrial ones). The exception make the soils of the park no. II, whose mean Cd content is 3 mg/kg and classifies them in this respect between the protected and industrial soils.

3.2 PHE in sand

Common habit of young children which is putting dirty hands into the mouth, is a serious hazard of entering significant amounts of toxins into their organisms via the digestive tract. Therefore, the sand of the sandboxes, if metal-contaminated represents the same threat as contaminated, soils if accidentally swallowed by children. The metal contents of sand samples (Tab. 2) are generally lower than the contents of the same PHE determined in the soils. Their means are 13 times the lower in the case of As, 4 times in the case of Cd, and 29 times in the case of Cu. The quantities of the PHE of sand range as follows: As 0.12–0.65, Cd 1–3, Co 5–14, Cu 0.4–4 Ni 4–9 and Sn 1–7 mg/kg.

Table 2. PHE content in sand from Silesia playground.

		As	Cd	Co	Cu	Ni	Sn
Park	n	[mg/kg]					
I	15	0.38	2	10	1.6	7	4.6
II	5	0.18	2	12	0.4	6	5
III	15	0.21	2	8	0.7	4	1.6
min. –max.		0.12–0.65	1–3	5–14	0.4–4	4–9	1–7
av.	35	0.28	2	9.8	1.1	6.8	3.4
SD		*0.03*	*0.04*	*0.1*	*0.02*	*0.1*	*0.01*

n – number of samples, min. – minimum, max. – maximum, av. – average arithmetic, *SD* – standard deviation.

The sand samples of the park no. II are an exception as they contain higher mean quantities of cobalt and tin than those established for soils. Their mean metal quantities of sand and soil are 12 and 7 mg/kg of Co, and 5 and 1 mg/kg of Sn.

3.3 PHE in dusts

The most contaminated soils and sand in the parks no's I and III have been found in their sampling sites numbered 2. Both of them are exposed to a relatively free fall of airborne dusts.

The mean of arsenic and tin quantities in the playing ground dusts are considerably lower than those of soils: calculated for all the samples they are 7 times lower in the case of As and 13 times lower in the case of Sn. The quantities of Cu and Ni are comparable and also those of Cd and Co (Tab. 3).

Table 3. PHE content in dusts from Silesia playground.

		As	Cd	Co	Cu	Ni	Sn
Park	n	[mg/kg]					
I	5	0.44	5	8	19	7	0.37
II	5	0.74	8	10	25	10	0.28
III	5	0.47	7	7	60	34	0.27
min.–max.		0.36–0.74	1–20	2–17	6–143	9–34	0.19–0.41
av.	15	0.5	6.5	8.1	38	15	0.31
SD		*0.07*	*0.15*	*0.12*	*0.81*	*0.33*	*0.03*

n – number of samples, min. – minimum, max. – maximum, av. – average arithmetic, *SD* – standard deviation.

A comparison of the metal contents of dusts to those of sand shows higher quantities of As and Ni (around 2 times), Cd (3 times) and Cu (even 34 times) and lower of Co (slightly) and Sn (even 10 times).
The results indicate a significant contribution of dusts to the contamination of the park soils with Cd, Co, Cu and Ni.

3.4 Risk assessment of the children's health

A health risk of children resulting from the presence of metals in the environment was evaluated by calculating a daily intake of the metals (D) using the formula (1). The following assumptions were made: the daily quantity of accidentally swallowed soil is 200 mg, an average body mass of a three-year-old child is 15 kg and that of a six-year-old child 19 kg, a daily exposure to a toxic element is 2 hours. The daily doses D (Tab. 4) do not exceed in all the parks the permitted values of the Provisional Maximum Tolerable Daily Intake (PTMDI) values recommended by the WHO [12-16]. Cadmium present in soils is a troublesome element, particularly for the youngsters of age below 3 years and for the three-year-old children of the body mass below 15 kg. The doses D calculated for the parks no's III and I are 4% PTMDI, whereas slightly above 1% PTMDI for As and Ni in the case of younger children.

Table 4. Potential intake of PHE by children in the case of incidental swallowing of soil in Silesian playground.

Park	As		Cd		Co	
	15 kg	19 kg	15 kg	19 kg	15 kg	19 kg
	μg/kg bw per day					
I	0.10	0.04	0.19	0.07	0.30	0.12
II	0.06	0.02	0.08	0.03	0.18	0.07
III	0.10	0.04	0.23	0.09	0.28	0.11
PTMDI*	2		0.36		-	
	30	38	5.4	6.84	-	-
Park	Cu		Ni		Sn	
	15 kg	19 kg	15 kg	19 kg	15 kg	19 kg
	μg/kg bw per day					
I	0.63	0.25	0.58	0.23	0.13	0.05
II	0.13	0.05	0.23	0.09	0.03	0.01
III	0.38	0.15	0.30	0.12	0.10	0.04
PTMDI*	500		5		2000	
	7500	9500	75	95	30000	38000

* Provisional Max. Tolerable Daily Intake due to [12-16] " –" lack data

It considered that only parts of the total quantities of the elements are bioavailable and can enter via the digestive tract the chain of metabolic transformations [5]. On the other hand it also has to be considered that the metal absorption of children is higher than that of adults, and the children are additionally sensitive to metal absorption via the respiratory tract and the skin [17]. Besides, the investigations of the regions affected by the mining and metallurgical industries have revealed that the cation exchangeable and easily leachable carbonate-bound metal forms may contribute in the case of cadmium even to 60-70% of its totals [18].

4 Conclusion

The determination of the total quantities of the selected PHE in soils of the playing grounds in three urban parks of the Silesian agglomeration make the basis of the pollution assessment of this environmental segment. The worse is cadmium pollution of two largest and most frequented parks: the Silesian one in Chorzów and the Kościuszko one in Katowice. In both of them the Cd contents of soils exceed the upper limit of the B-type soils set in the Regulation of the Minister of the Environment [11]. Lower than permitted quantities of all the PHE considered characterize the soils of the park no. II, located within a new housing development in Katowice, as well as the As, Cd and Cu contents of playground sand. However, the park toys and recreation facilities of this park are covered with airborne dusts with the highest As, Cd and Co quantities. The highest quantities of As, Co, Cu, Ni and Sn occur in the soils of the park no. I except of the Cd contents being the highest in the park no. III.

In sand of the park sandboxes the metal contents are significantly lower than those of soils, which is due to the inert chemical nature of quartz grains, by far the major component of every sand. The metal contents of the sand originate from airborne dusts, which are carriers of many elements, particularly of Cd, Co and Cu.

The quotients of the health risk indicate to a potential risk of As, Cd and Ni for children, particularly for those with a low body weight, less than 15 kg. The risk level is 1–4% PTMDI and results from an accidental (via dirty hands) swallowing of soil. This finding requires attention because As, Cd and Ni are toxic and included in the group I of carcinogens.

References

1. E. Wcisło, J. Bronder, J. Długosz, M. Korcz, *Ocena środowiskowych zagrożeń mieszkańców dużych miast Polski*, Wyd. Ekon. i Środ., Białystok, **132** (2008)
2. E. Wcisło, *Ocena ryzyka zdrowotnego(..)*, Wyd. Ekon. i Środ., Białystok, **29-63** (2009)
3. I. Norska-Borówka, J. Bursa, J. Rzempołuch, R. Wawryk, Arch. Ochr. Środ., **3-4** (1990)
4. G. Dziubanek, R. Baranowska, K. Oleksiuk, J. Ecol Health. **16, 4** (2012).
5. K. Ljung, A. Oomen, M. Duits, O. Selinus, M. Berglund, J. Environ. Sc. Health Part A, Tox. Hazard. Subst. Environ. Eng. **42, 9**, 1241 (2007).
6. J. Nieć, R. Baranowska, G. Dziubanek, D. Rogala D, J. Ecol. Health. **17, 2** (2013).
7. A. Kabata-Pendias, B. Szteke, *Pierwiastki śladowe w geo- i biosferze*, Puławy (2012)
8. A. Kabata-Pendias, H. Pendias, *Biogeochemia pierwiastków śladowych*, PWN, Warszawa (1999)
9. J. B. Diatta, W. Grzebisz. Polish J. Envi. Stu. **20, 6** (2011)
10. BN-75/9180-03 *Oznaczanie wartości pH.*
11. Roz. Min. Środ. Dz.U. 2002 nr 165, poz. 1359.
12. WHO, Techn. Rep. Ser. 776, Geneva (1989)
13. WHO, Food Add. Ser. No. 17, Geneva (1982)
14. WHO, Guid. for drink. water. qual.2, Geneva (1996)
15. EFSA J. 10 : 2551-2558 (2012)
16. IECFA, FAO/WHO, Rome (2005)
17. US EPA: *Guidelines for Human Health Risk assessment of Chemical Mixtures.* Federal Register (51 FR 34014-34025), Washington (1986).
18. A. Kicińska, Envir. Prot. & Natur. Res. 49 (2011)

Sulfur as a parameter in the suitability assessment of gangue from coal mining for reclamation of opencast excavation, taking into the requirements regarding protection of the soil

Beata Klojzy-Karczmarczyk[1a], Janusz Mazurek[1] and Jacek Mucha[2]

[1] *The Mineral and Energy Economy Research Institute of the Polish Academy of Sciences, Krakow, Poland*
[2] *AGH University of Science and Technology, Krakow, Poland*

Abstract. In the process of mining coal considerable amounts of waste material are generated from which it is possible to produce aggregates which could find application in the reclamation of opencast excavations. One of the basic characteristic parameters which may be used to assess the suitability of aggregates for filling reclaimed opencast excavation is sulfur. Total and leachable forms of sulfur content from gangue samples were analyzed in the study. It has been shown that it is possible to select aggregates which are safe for the environment when the main assessment criterion was the percentage of sulfur in the aggregates and sulfate concentration in leaching tests. The research confirmed the role of sulfur as the most important and characteristic parameter which determines where and in what specific conditions the use of gangue or mining waste may be possible. Studies of sulfur content in the samples should be carried out with sufficient periodicity and in a manner enabling to obtain most reliable results which would refer to the entire batch of material. The statistical evaluation makes it possible to estimate to what extent the obtained results of sulfur content in the samples are close to actual values for the entire mass of the material.

1 Introduction

The process of extraction and processing of coal produces significant amounts of waste material. It is mainly gangue removed during the preparation of the deposit and the waste material generated by the extraction and processing of run-of-mine coal. An important issue is the proper management of waste material including the requirements for the prevention of pollution of soil and water environment. These materials may be treated as minerals accompanying the coal deposits [1]. It is possible and practiced to use them in a variety of economic processes, including production of aggregates for civil engineering, hydro-engineering, road engineering and as a raw material for the production of building materials, or as material for underground backfilling and reclamation of degraded areas [1 - 4]. These materials can be used also in the process of filling open-pit excavations. Filling the post-exploitation excavation with external material is the primary stage of the technical phase of the reclamation process in open-pit mining [5]. Waste materials for which there is no use, become mining waste, classified as group 01 and most often neutralized by storage [6, 7].

Both waste rock, considered as raw material and material deemed mining waste may be used for reclamation of excavations by filling. There is a number of legal conditions in force, application of which allows for safe filling of the excavation pit. It should be noted that there are frequent changes of regulations regarding the possibility of using by-products from coal mining to fill various types of excavations [6, 8 - 11]. In each case it is necessary to conduct a wide analysis of possible use of mining waste from coal production or products based on waste rock for reclamation, which is a complex issue in relation to the soil and water conditions and to the location of the project. The relevant regulations provide limits of physicochemical parameters for mining waste, which - if met - allow for considering the waste to be neutral to the environment. However, in regard to gangue as a product (aggregate), which is a side effect of coal production, and may be used in the process of filling various open-pit mines, no clear legal provisions relating to the determination of their safety for the environment are in force. The material may be considered safe for the environment, if it meets the stringent requirements for inert mining waste and demonstrates no negative impacts on the soil and water environment [11].

The analysis presented in the work, concerns gangue from Janina Mine in Libiąż, which is a side effect of coal production and is used for aggregate production. The conducted studies allow for drawing conclusions regarding the content of sulfur in the analyzed material and whether it is possible in this case to separate particular fractions of the aggregate, which will meet

[a] Corresponding author: beatakk@min-pan.krakow.pl

DOI: 10.1051/e3sconf/20161000036

the requirements for the quality of inert waste in order to ensure safety for the soil and water environment.

2 Limit values of the sulfur content

The produced aggregate, depending on the source, contains varying amounts of mineral impurities, including elements commonly recognized as possibly contaminating the soil and water environment. One of the basic characteristic parameters, which can be used to assess the suitability of aggregates for use as a filling material for reclaimed open-pit mines is sulfur, which is released slowly from the sulfides present in the waste and is introduced to the water in the form of sulfates and can lead, in extreme cases, to pollution of the environment. There is a great number of information in the literature concerning the possibility of oxidation of sulfides and consequently the activation of sulfates from landfills, where the stored waste includes waste from coal mining [12 - 16]. The literature data indicates definitive variation in the sulfur content in the mining waste, also due to the presence of this element in the run-of-mine coal [6, 8, 10, 15 - 19]. In [3] different sulfur contents in the mining waste of Janina Mine are provided, where - depending on the place of its origination - the contents of this element ranges from 0.4 to 3%.

It can be assumed that the perspective material for reclamation includes mining waste or gangue-based materials (aggregates) for which the total sulfur content is less than 1%. Such a limit for inert mining waste may be adopted where the neutralizing potential ratio (defined as the ratio of neutralizing potential to the acidity potential) is greater than 3. Otherwise, one should adopt the limit of sulfur content at 0.1% [20]. The maximum content of sulfide sulfur in the mining waste defined this way allows for classifying it as inert waste or not.

3 Methodology of the sulfur content analysis in the aggregates

To investigate the variability of total sulfur content in the aggregates from the waste material from Janina Mine, 16 primary samples were collected (15 samples with a mass of about 200 kg each and one sample with a mass of about 100 kg) (Fig. 1). Samples were taken for a year at intervals of 2-3 weeks. Sampling was conducted directly from the run-of-mine winning (mining waste rock) before its transport to the dumping ground. Therefore, the analyzed material originated directly from the mine's production.

Raw samples were sieved to 14 fractions in the range of 0 to 200 mm (0-6, 6-8, 8-10, 10-16, 16-20, 20-25, 25-30, 30-35, 35-40, 40-50, 50-60, 60-80, 80-120, 120-200 mm) and above 200 mm. It has been proven that in the aggregate fractions (mining waste) of larger grains diameter, the total sulfur content is low enough that this material can be considered inert for the environment [21]. The key issue in this case was to determine the lower limit of the fraction of rock above which the average sulfur content does not exceed the limit value with risk of error small enough and possible for practical acceptance. Based on the study of the indicator of neutralizing potential, and in accordance with [22], the limit for sulfur content was set at 1%. In order to verify the obtained results, samples called secondary samples were prepared in laboratory conditions through combining all fractions with a grain diameter exceeding 20 mm in the respective percentages. In practice, this has resulted in the removal of the finest fractions from each raw sample. For these samples studies of the content of the total and leachable forms of sulfur compounds were conducted. Leaching was conducted using static method (batch tests) on the basis of 1:10 tests (material: deionized water), according to [23]. Tests for the total sulfur content in the individual fractions of the raw samples were carried out using high-temperature method (Laboratory of Janina Mine). Examination of the total sulfur content in the secondary samples and of the sulfates in the solutions (Tab. 2) were conducted using the ICP OES method (AGH Hydrogeochemical Laboratory) and the spectrophotometric method (MEERI PAS Laboratory).

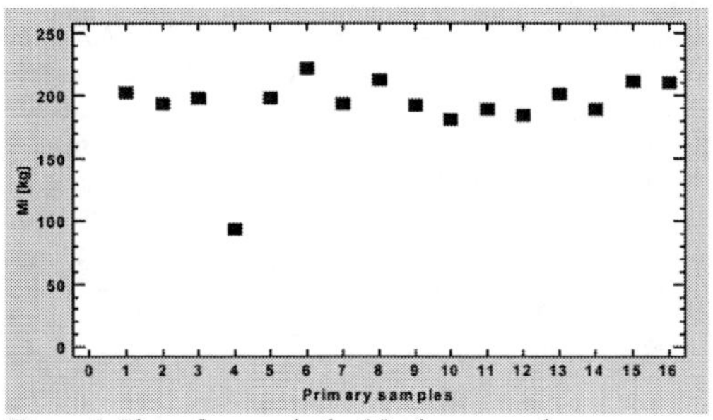

Figure 1. Plots of masses in the 16 primary samples (Mi – masses of the individual samples).

4 Statistical analysis of the total sulfur content in the primary samples

The samples of aggregates with weight of approx. 200 kg each obtained directly from the production were called the raw samples. In these samples the total sulfur content was determined in the individual fractions [21]. As the result of the estimation of the average sulfur content using the weighted average, in which the weight is the mass of the sample associated with the determination of sulfur content for each of the 15 grain fractions separated mechanically, the concept of the primary samples was introduced.

The basis of the statistical analysis constituted of the following:

- determination of total sulfur on dry matter (S_t^d) in a 14 isolated fractions separately for each of the 16 raw samples [21],
- the sulfur content in the primary samples calculated as the weighted averages mass fraction (Fig. 3),
- the weighted averages of the sulfur content in each of the 14 fractions in the combined rock material of the 16 primary samples (Fig. 4).

For all 16 primary samples the cumulative masses of the individual fractions of the collected material are in the range of 100-300 kg with the exception of the finest fraction (0-6 mm), the mass of which is significantly higher and is about 750 kg (Fig. 2).

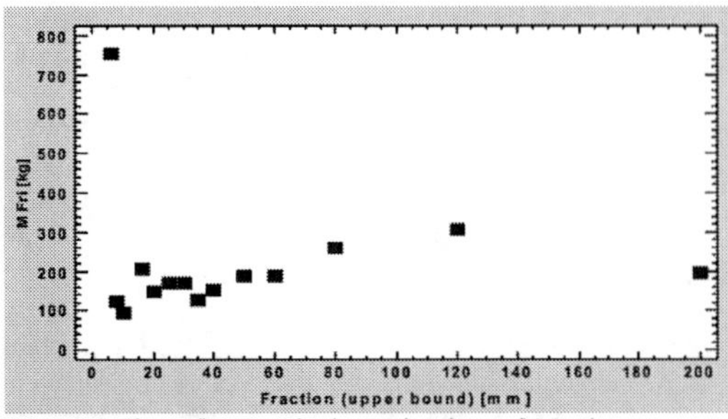

Figure 2. Plots of masses in the 14 fractions of 16 primary samples (M-Fri - masses of the individual fractions).

The sulfur content in the primary samples shows strong variation, from about 0.5% to about 4% (Fig. 3). Even greater variation is exhibited by the sulfur contents in the fractions of the material (cumulated for the 16 primary samples) ranging from 0.4% to over 5% (Fig. 4). It should be noted that there is an evident tendency for reduction of the sulfur content with the increase in the fraction size, which described in a statistically significant (P-value = 0.00) and highly satisfactory manner by the non-linear model of dependence (Fig. 5) with a very high coefficient of determination $R^2 = 97.3\%$.

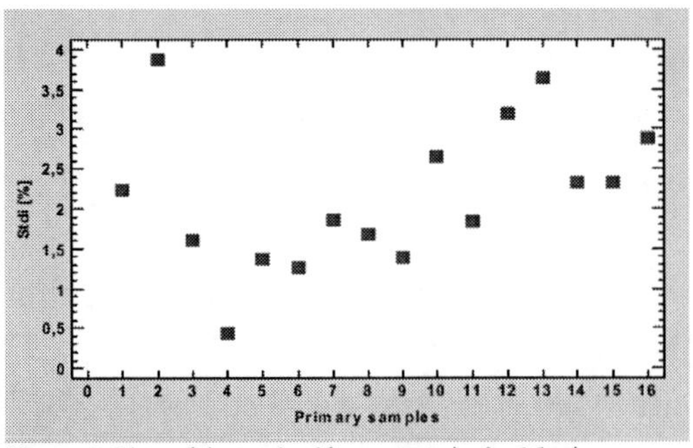

Figure 3. Plots of the total sulfur contents in the 16 primary samples (Stdi – the sulfur content (S_t^d) in the individual samples).

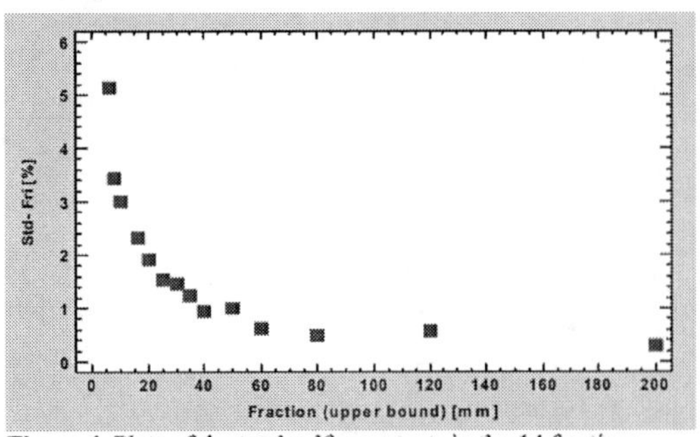

Figure 4. Plots of the total sulfur contents in the 14 fractions of 16 primary samples (Std-Fri - the sulfur content (S_t^d) in the individual fractions).

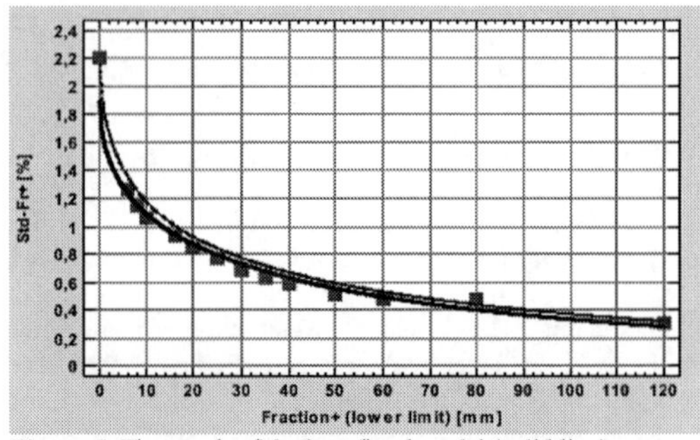

Figure 5. The graph of the best fitted model (solid line) of empirical dependence between the sulfur content and the lower limit of fraction of material (dashed line – upper confidence bound for probability level P=0.99);
$S_{sd}[\%] = e^{0.696-0.174\sqrt{Fr+[mm]}}$, P-value = 0.00, coefficient of determination $R^2 = 97.3\%$,
(Std-Fr+ - the sulfur content (S_t^d) in the cumulative fractions above grain diameter).

The variability of the sulfur content in the primary samples is large, and when measured by the coefficient of variation is at 56%. The Shapiro-Wilk's test does not give grounds to reject the hypothesis of normal distribution of the sulfur content (at the significance level of 0.05). Average sulfur content in the primary samples is 1.79% and goes significantly beyond the reference value of 1% permissible for inert mining waste. Given the increasing trend of the sulfur content with decreasing size of fractions (Fig. 4), it can be expected that after removal of the finer fractions of the material, the sulfur content in the remaining (thicker) portion of the rock material will be less than 1%.

The conducted statistical analysis showed that only after the separation of the fraction finer than 16 mm, the average sulfur content in the 16 primary samples has a value of less than 1%, and in this case is 0.93% (Tab. 1). There is no guarantee, however, that it will stay below the limit (permissible value) of the sulfur content in subsequent batches of the waste rock material because of its relatively high volatility. For this purpose, the sulfur content in the rock material after elimination of subsequent fractions (<20 and <25 mm) was also calculated. The Shapiro-Wilk's test (for the level of significance of 0.05) does not give grounds to reject the hypothesis of normal distribution of the sulfur content in other rock material of the fractions >16, >20 and >25 mm. The variability of the sulfur content is still high and is at about 40%.

To determine the minimum size of the fraction above which the probability of exceeding the average sulfur content will be small and will not exceed the adopted value of 0.01, three statistical techniques were used, including:
- defining the upper limit of the confidence interval for the average sulfur content with the risk of exceeding it being less than 0.01,
- the use of the Student's t-test to verify the null hypothesis that the average sulfur content is equal to 1% against the alternative hypothesis that it is less than 1%,

 DOI: 10.1051/e3sconf/20161000036

- using the upper confidence limit for the non-linear model of dependence binding the sulfur content with the lower limits of the fractions (Fig. 5).

Table 1. Statistics of the total sulfur contents in distinguished fractions of combined 16 primary samples.

Statistics of sulfur content	Fraction >16 [mm]	Fraction >20 [mm]	Fraction >25 [mm]
Arithmetic mean [%]	0.93	0.85	0.78
Coeff. of variation [%]	38	39	43
Minimum/Maximum [%]	0.28/1.44	0.27/1.32	0.23/1.19
Upper confidence bound for S_t^d mean content [%] (probability level P=0.99)	1.16	1.06	0.997
t-test: P-value Null hypothesis: mean S_t^d =1.0[%] Alternative hypothesis: mean S_t^d <1[%]	0.216	0.043	0.009
Upper confidence bound for S_t^d contents [%] predicted from fitted model for P=0.99 (Fig. 5)	1.01	0.93	0.84

The applied approaches give very similar but not identical results (Tab. 1), which is associated with a certain variation of the masses of samples and fractions. It should be noted, however, that - beginning with the fraction >25 mm - all 3 methods agreeably show that the probability of exceeding the reference sulfur content of 1% is less than 0.01. It can therefore be assumed that the boundary of separation of the fraction equal to 25 mm is suitable for "safe" distinguishing of fractions meeting one of the most important requirements for inert waste mining.

5 The sulfur content in the secondary samples

To verify the results of statistical analysis, which allowed for identifying the lower limit of the grain size, after the rejection of which it is possible to obtain material with low total sulfur content, the analysis focused on the secondary samples, in order to determine the value of the basic studied parameter. The secondary samples no. 4 to 16 were formed as a result of the rejection of fines with a diameter of <20 mm. Static batch test of contaminants leaching at 1:10 was conducted to determine the content of individual elements in the leachable form which may potentially be introduced into the aquatic environment (Tab. 2). For samples 1 to 3 the secondary samples were not prepared because of insufficient amount of material in the individual fractions. The obtained results were compared with the permissible parameters for waste water, according to [24] and with the permissible values for surface and groundwater in accordance with the regulations issued under the Water Act [25, 26, 27]. The results were additionally compared with the criteria for waste admitted to inert waste landfills [28], as a complement to the classification of inert mining waste [20], to which the analyzed aggregate relates for the purposes of this work. The results obtained in the contaminants leaching test indicate that the tested samples (after the rejection of fines) will not introduce abnormal amounts of sulfates in relation to qualifying the waste for inert waste landfill. The limit value set out by the above-mentioned regulation [28] is 1000 mg/kg, while the values obtained in the leaching tests were in the range of 258-862 mg/kg on dry matter (based on Tab. 2). On the other hand, with regard to the solutions quality test (water extracts 1:10) the measured sulfates content is within the acceptable limits for wastewater. With respect to the limit values for the quality of surface water intended for human consumption [25], to the classification of groundwater quality [26], as well as to the classification of surface water quality [27], in no case exceeding of water quality was observed.

Table 2. The total content of sulfur and sulfates leachability in the secondary samples from the Janina Mine aggregates.

Secondary samples Fraction >20 [mm]	Neutralizing potential ratio NPR	S total (S_t^d) [%] DM	SO_4^{2-} batch tests 1:10 [mg/dm^3]*
Sample 4	40.00	0.038	52.47
Sample 5	8.00	0.202	49.35
Sample 6	1.16	1.383	71.46
Sample 7	3.08	0.520	35.37
Sample 8	6.15	0.262	31.02
Sample 9	4.32	0.373	57.21
Sample 10	4.21	0.382	86.22
Sample 11	5.33	0.295	26.25
Sample 12	3.64	0.438	31.05
Sample 13	2.42	0.663	25.83
Sample 14	2.81	0.571	57.27
Sample 15	3.20	0.500	46.86
Sample 16	1.95	0.816	84.66

* pH of the solutions of 7.7-8.1
DM – dry matter

In the light of the results of neutralization potential ratio NPR in the secondary samples (Tab. 2) it was concluded that the maximum content of sulfide sulfur in the secondary samples may be required at 1%. This parameter expresses the ability of the tested aggregates to neutralize the acidity of the environment caused by the potential decay of sulfide minerals. For the secondary samples created from the combination of grains with a diameter >20 mm, in most cases this value is greater than 3. Only in individual cases, the value is slightly lower. It should be noted that the average rate of neutralizing potential ration, as calculated for the whole batch of the analyzed secondary samples is 2-fold higher than 3 and is equal to 6.63. In the secondary samples with grain size greater than 20 mm exceedance of the limit value adopted for the total sulfur content is not observed, beyond the sole exception of sample 6, where the sulfur content reaches 1.4%.

To clarify the issue of occurrence of much higher sulfur contents in the fine fractions, an analysis of forms of sulfur occurrence in the surveyed samples was conducted (Tab. 3). Observations for the averaged samples were conducted using scanning electron microscopy (SEM, AGH Laboratory for Phase, Structural and Textural Research). Sulfur compounds identified using this method in all tested samples are found to be pyrite. The form of pyrite occurrence in all tested samples is similar (irregular fragments of most likely larger crystals). Spread in the distribution of the pyrite fragments is relatively small, and the majority of tested fragments has a size of several tens of micrometers. There are also smaller objects, having a size less than 10 micrometers. Only in the analyzed fraction of sample 5 framboid pyrite was observed, with crystal size of about 1 micrometer. No other sulfide or sulfate minerals were found in any of the analyzed samples. For the same analyzed fractions another study of the total sulfur content using high-temperature method was conducted (Chemical Laboratory of the Department of Forest Ecology and Reclamation in the University of Agriculture in Krakow).

Table 3. The form of sulfur occurrence and its content in the selected fractions of the aggregate from the Janina Mine.

The fraction separated in the sample	S total (S_t^a) [%]	Form of sulfur
Fraction 0–6 mm sample 8	4.766	pyrite
Fraction 0–6 mm sample 13	8.709	pyrite
Fraction 8–10 mm sample 14	5.703	pyrite
Fraction 50–60 mm sample 5	0.125	pyrite, framboid pyrite
Fraction 50–60 mm sample 6	2.050	pyrite
Fraction 60–80 mm sample 12	0.335	pyrite

The conducted study showed a high content of pyritic sulfur in the fine fractions 0-6 and 8-10 mm in excess of 5%. The coarser fractions had low total sulfur content, thus confirming the results obtained in the previous step of the work for particular fractions sieved from the raw samples.

6 Summary and conclusions

The study confirmed the role of sulfur, as the most important and characteristic parameter that determines where and under what conditions, the use of waste rock as raw material and/or mining waste is possible. In the process of reclamation of open-pit mining objects with simultaneous filling of the mined pits, each time it is necessary to conduct a broad analysis of the quality of the material and the conditions of the project location. It is important to relate the results of research to the legal requirements applicable to both mining waste as well as to the quality and protection of soil and water environment. The studies of sulfur content in the samples should be carried out in appropriate number and periodicity, allowing for obtaining the best results, that relate to the entire batch of material.

The study analyzed the sulfur in numerous samples of aggregate as both total content and in the leachable form. It has been shown that there is a possibility of selecting aggregate fractions safe for the environment if using the percentage content of sulfur in the aggregates and the concentration of sulfate in the static pollutants leaching tests as the main criteria for assessment, supplemented by testing of the aggregates buffering abilities (study of the neutralizing potential ratio). The conducted studies of total sulfur content in the secondary samples has shown that the best quality parameters for waste rock from Janina Mine are found in the aggregates deprived of the fine fractions. Based on the conducted study, the grain diameter of 20 mm can be considered as the lower limit, above which the materials will have appropriately low sulfur content. The statistical analysis showed similar results but simultaneously displays that even higher safety (with the risk of error <0.01) for the soil and water environment can be achieved using a lower limit of grain diameter of 25 mm.

The applied statistical evaluation of the results of the sulfur content in the individual fractions, allows for estimating the extent to which the results of the primary samples are close to the actual values obtained in the laboratory conditions for the secondary samples. The laboratory studies also indicate the possibility of seeking lower limit of the grain diameter above which the material will have low sulfur content, which will be continued in subsequent works.

Acknowlegements

The authors thank TAURON Wydobycie S.A. in Jaworzno for providing data and enabling the conduct of this paper. The publication was carried out partly in the framework of the statutory works of MEERI PAS.

References

1. K. Galos, J. Szlugaj, Górnictwo Odkrywkowe, **51(2)**, 25-31 (2010)

2. K. Galos, J. Szlugaj, Gospodarka Surowcami Mineralnymi - Mineral Resources Management, **30(4)**, 51-64 (2014)
3. A. Szymkiewicz, A. Fraś, R. Przystaś, Wiadomości Górnicze, **7-8**, 435-441 (2009)
4. J. Wróbel, A. Fraś, T. Pierzchała, R. Przystaś, A. Machnik, J.J. Hycnar, In: *Proceedings of the XXVI-th Conf. „Zagadnienia surowców energetycznych i energii w gospodarce krajowej"*, 113-127 (2012)
5. P. Strzałkowski, U. Kaźmierczak, Mining Science – Mineral Aggregates, **21(1)**, 203-213 (2014)
6. S. Góralczyk Ed., *Gospodarka surowcami odpadowymi z węgla kamiennego* (IMBiGS Warszawa, 2011)
7. K. Bojarska, Z. Bzowski, Górnictwo i Geologia, **7(2)**, 101-113 (2012)
8. S. Góralczyk, I. Baic, Polityka Energetyczna - Energy Policy Journal, **12(2/2)**, 145-157 (2009)
9. I. Baic, B. Witkowska-Kita, Rocznik Ochrona Środowiska (Annual Set The Environment Protection), **13**, 1315-1326 (2011)
10. Z. Korban, Górnictwo i Geologia, **6(1)**, 109-120 (2011)
11. B. Klojzy-Karczmarczyk, J. Mazurek, Zeszyty Naukowe IGSMiE PAN, **90**, 67-78 (2015)
12. I. Twardowska, J. Szczepańska, S. Witczak, *Wpływ odpadów górnictwa węgla kamiennego na środowisko wodne. Ocena zagrożenia, prognozowanie, zapobieganie,* (Prace i Studia 35, 1988)
13. S. Witczak, A. Postawa, In: *Proceedings of the Fourth Intern. Sym. on the Reclamation, Treatment and Utilization of Coal Mining Wastes,* **1**, 37-44 (1993)
14. J. Szczepańska, J. Krawczyk, Gospodarka Surowcami Mineralnymi - Mineral Resources Management, **9(3)**, 529-546 (1993)
15. B. Klojzy-Karczmarczyk, *Zastosowanie odpadów energetycznych w ograniczaniu transportu zanieczyszczeń ze składowisk odpadów górniczych,* (Studia, Rozprawy, Monografie 117, 2003)
16. B. Klojzy-Karczmarczyk, J. Mazurek, Polityka Energetyczna - Energy Policy Journal, **17(4)**, 289–302 (2014)
17. J. Mucha, M. Wasilewska, Gospodarka Surowcami Mineralnymi - Mineral Resources Management, **21(1)**, 5-21 (2005)
18. Z. Kokesz, Gospodarka Surowcami Mineralnymi - Mineral Resources Management, **26(3)**, 95-110 (2010)
19. Cz. Rosik-Dulewska, *Podstawy gospodarki odpadami, 5-th ed.* (PWN, 2012)
20. Dz. U. [Journal of Law] of 2011, No. 175, Item 1048 (www.isap.sejm.gov.pl)
21. B. Klojzy-Karczmarczyk, J. Mazurek, K. Paw, Gospodarka Surowcami Mineralnymi - Mineral Resources Management (to be published, 2016)
22. Polish Norm PN-EN 15875, PKN Warszawa (2011)
23. Polish Norm PN-EN 12457-4, PKN Warszawa (2006)
24. Dz. U. [Journal of Law] of 2014, Item 1800 (www.isap.sejm.gov.pl)
25. Dz. U. [Journal of Law] of 2002, No. 204, Item 1728 (www.isap.sejm.gov.pl)
26. Dz. U. [Journal of Law] of 2016, Item 85 (www.isap.sejm.gov.pl)
27. Dz. U. [Journal of Law] of 2014, Item 1482 (www.isap.sejm.gov.pl)
28. Dz. U. [Journal of Law] of 2015, Item 1277 (www.isap.sejm.gov.pl)

DOI: 10.1051/e3sconf/20161000037

The properties of doped sand-lime products

Ryszard Dachowski[1], Katarzyna Komisarczyk[2,a]

[1] PhD. DSc, Kielce University of Technology, Department of Civil and Environmental Engineering, Al. 1000-Lecia P.P.7 Kielce, Poland
[2] MSc, Kielce University of Technology, Department of Civil and Environmental Engineering, Al. 1000-Lecia P.P.7,25-317 Kielce, Poland

Abstract. Sand-lime products are natural materials consisting of lime, sand and water with the least content of radioactive elements in comparison to other masonry elements. They are characterized by very high compressive strength, high acoustic insulation, good thermal properties, provide a friendly atmosphere and prevent the spread of mold and bacterial flora. In addition they are fully recyclable. White bricks through the porous structure and the occurrence of capillaries have the ability to rising water. The height of capillary action is dependent on the contact angle and the size of existing pores in the material. This property affects the frost resistance and other characteristics of durability of wall materials operated under conditions of intense exposure to moisture. The aim of the study is to determine the impact modifier on the properties of autoclaved sand-lime products. For testing used autoclaved sand-lime brick dimensions 40x40x160 [mm]. The weight of the products consists of 5% lime, 90% sand and 5% lithium water glass (MP=2,6 and MP=7,0). The produced samples were subjected to autoclaving at temperatures of 203°C and pressure of 1.6 MPa in collaboration with the Silicate Production Plant in Ludynia. Three finished sets of samples (standard, modified with lithium silicate 2.6 and 7.0) have been immersed in water to the desired height during certain time. The results show the diversity of the internal structure of the tested products. In particular pore distribution, size and volume.

1 Introduction

The economic development of the XXI century contributes to upsetting the balance in the environment. The way to prevent the negative effects of this development is the concept of sustainable development. It ensures meeting the needs of current generations without compromising the ability of future needs. Construction aimed at satisfying the basic needs of society is one of the key elements of sustainable development. The Directive Council of the European Communities [1] and polish building regulations need to ensure that products intended for use in buildings meet the criteria such as: mechanical resistance and stability, fire safety and utility, noise protection, energy saving and heat protection of buildings. For the construction of houses and flats suitable building materials are used. Their properties should take into account their impact on health, hygiene and environment.

Sand-lime products are natural materials consisting of lime, sand and water, with the smallest content of radioactive elements [2,3]. They are characterized by very high compressive strength, standard products are manufactured in a class of 15-20 MPa. Because of their considerable weight they are characterized by high acoustic insulation. Due to strong alkaline reaction they are resistant to biological corrosion in confined spaces. Thanks to the ability of heat accumulation they have very good thermal properties [4]. Additionally they are completely recyclable. White bricks are therefore materials that meet the requirements for modern construction.

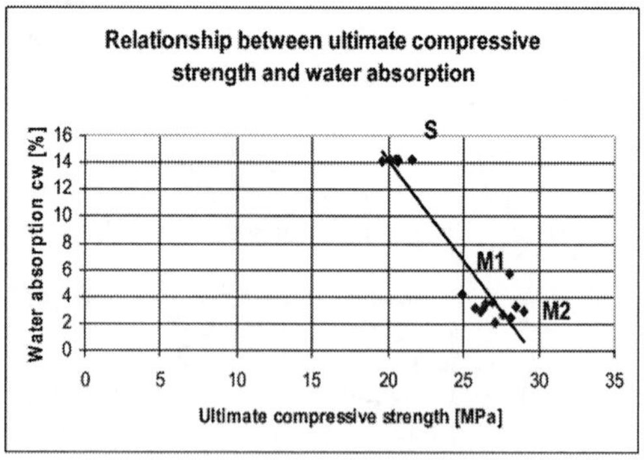

Figure 1 Relationship between ultimate compressive strength of sand-lime products and their water absorption. [4]

Because silicate products are compatible with sustainable development, all modifications, including the selection of additives introduced to the sand-lime mixture must also meet certain criteria. Interference in the basic composition of products through the use of various types of fillers has an impact on the internal structure as a consequence of changing the physical and mechanical properties. Changing parameters such as porosity and tightness will directly

[a] Corresponding author: komisarczyk.kasia@gmail.com

affect the compressive strength, sound insulation and frost resistance, but the ability of the material to resist the penetration of water.

White brick is a material which absorbs about 8% of H2O less than the other wall products for construction of interior and exterior walls of buildings. According to research performed by A. Stępień, use of an basalt additive with a grain size of 2-4 mm leads to a significant improvement in strength of the product and to reduce its absorbency. The compressive strength of the modified product increased to almost 40 MPa, and the absorption capacity was reduced by about 50%. The use of barite of the grain size of 0-2 mm increased the absorption by 40%, the compressive strength increased to a similar level. (Fig.1) The addition of liquid admixture - lithium silicate resulted in doubling of the compressive strength and reduction in water absorption by about 20%. [5,6] However, modification of white brick not always results in the improvement of their parameters. The article present the impact of admixture in the form of lithium water glass with different molar ratio on the properties of final silicate products.

2 Examination of sand-lime products

The main aim of my research is to determine the effect of modification of sand-lime products on their physical properties. Samples have been performed inncollaboration with the Silicate Production Plant in Ludynia following 4 stages of the production process. A mixture of quicklime and quartz sand was placed in reactors for 4-6 h, where under the conditions of elevated temperature by the hydration process of CaO and MgO, and the curing process, the lime slaking process took place. The mass of moisture content of 4-5% (Fig.2) was subjected to hydraulic pressing (approx. 20 MPa) and then transported to an autoclave. During the autoclaving process, with the steam pressure of 16 bar and a small energy distribution at 203◦C, the silicates were cured. After 6 hours of hardening the sample were prepared for the last stage of their formation- cooling.

Figure 2 Compacted samples. Prepered for autoclaving process.

The experiment was conducted on the samples with dimensions of 40 x 40 x 160 mm (Fig. 3) I have prepared three sets of three samples each. To constant weight of the dry ingredients I added a slightly-turbid, lithium water glass having a molar ratio of 2.6 and 7.0. Admixture, which includes lithium hydroxide (1,0-5,0%), colloidal silica- SiO2 (10,0-25,0%) and water (50,0-75,0%), I added in the amount of 5 % by weight of the total mixture.

Table 1 The content of ingrediens in the test samples.

	Sand+ Lime [g]	Water [g]	Admixture [g]
N	1800	90	-
A (7,0)	1800	-	90
B (2,6)	1800	-	90

After 21 days after preparation of the samples, the products were subjected to the study of compressive strength, water absorption due to capillary action and absorption (by complete immersion in water). The samples were marked with letters: N, A and B. The sample marked with the letter N is the traditional sand-lime product, A-modified with water lithium glass with molar ratio of 7.0 and sample B – modified with water lithium glass having a molar ratio of 2.6. (Tab. 1)

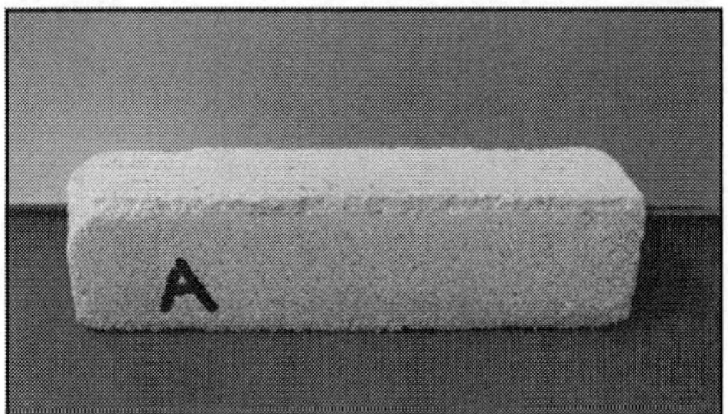

Figure 3 The sample prepared for testing.

3 Research analysis

3.1. Capillary test

The amount of capillary action is inversely proportional to the contact angle (liquid type), and the capillary diameter (pore size). Fig. 4 shows that the addition to the silicate mass lithium water glass having a mole ratio of 2.6 resulted in the greatest increase in the high of capillary action. (Fig. 5) Reduction of the diameter of the pores present in the product was noted. This was confirmed by the results of Mercury Intrusion Porosimetry. The use of porosimeter allows to determine the pore diameter in the range of 0,003- 400 µm. According to the IUPAC classification in this range we have mesopores (2 <d <50 nm) and macropores (d> 50 nm), and the micropores (d <2 nm) were not included in the measurement. In the sample A mesopores constitute 43% of all pores, which is above 20% more in comparison to the normalized sample.

 DOI: 10.1051/e3sconf/20161000037

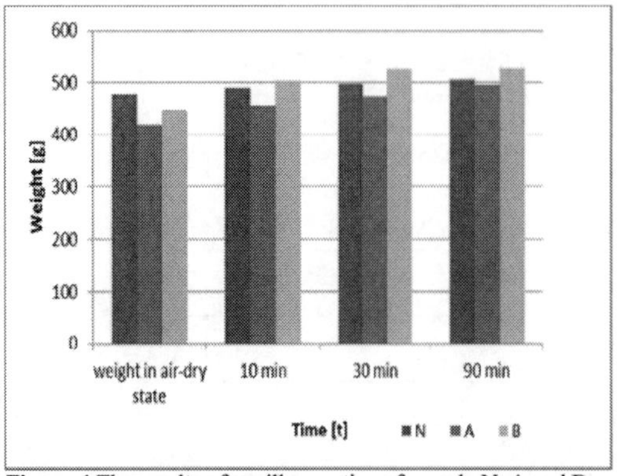

Figure 4 The results of capillary action of sample N, A and B.

Figure 5 Capillary action of samples N and B.

3.2. Compressive strength

The research on the lime-sand products shows that the use of lithium silicate admixture have a negative impact on one of the main mechanical properties - compressive strength. Fig. 6 shows the results of the measurements. In the sample A reduction in the strength to 3.2 MPa was observed. In the sample B - to 5.9 MPa, which is three times less than the strength of the standard sample (17.29 MPa).

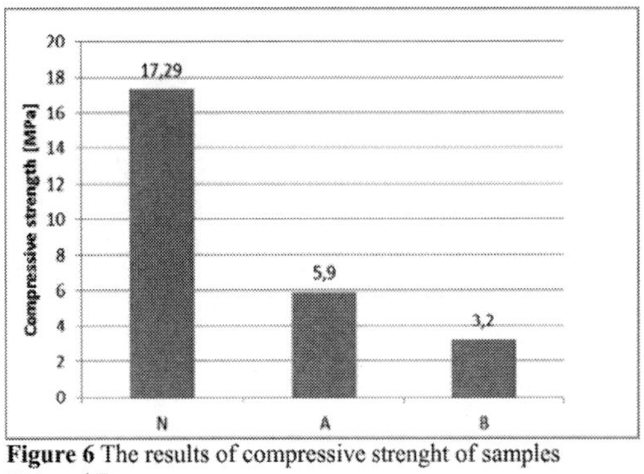

Figure 6 The results of compressive strenght of samples N,A and B.

3.3. Water absorption

Compressive strength is inversely proportional to absorption. The smaller the resistance, the greater the absorption.

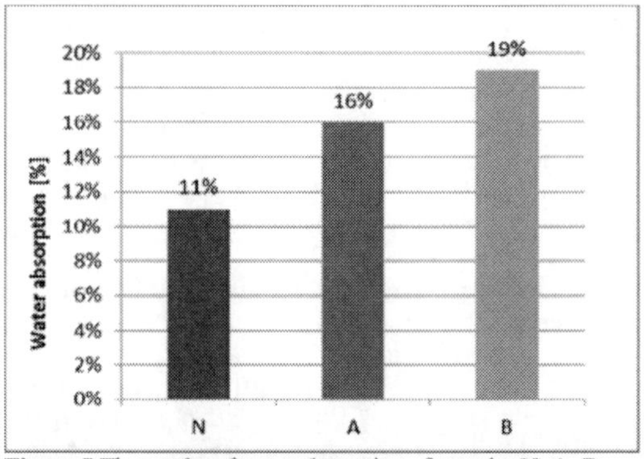

Figure 7 The results of water absorption of samples N, A, B.

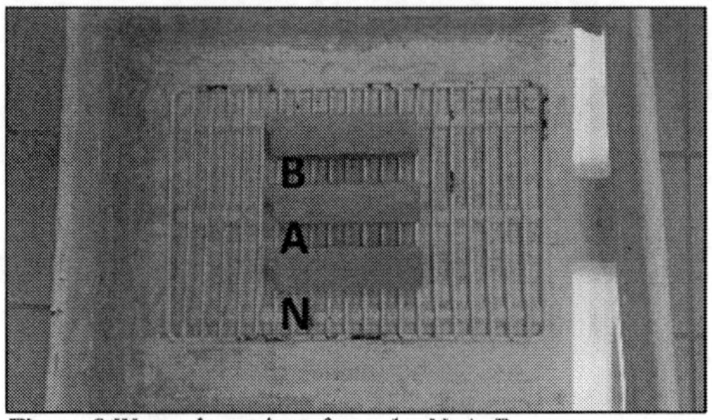

Figure 8 Water absorption of samples N, A, B.

Figure 7 shows that sample A increased the absorbency of a 5%, sample B, in turn, 8% with respect to a standard (11%). The results therefore indicate that the use of lithium water glass admixture of different molar ratio changes the pore structure. (Fig 8) Changing the shape of the pores from the spherical to flattened influences the change in frost resistance.

4 Conclusion

Controlling of dosing of admixtures and additives can change the ability of silicate material to resist the penetration of water.

The choice of additives and admixtures does not always improve the performance of the resulting product. Lithium water glass reduces the strength of the products, increases water absorption and the amount of capillary.

The value of the molar ratio of lithium water glass is inversely proportional to the compressive strength, water absorption and capillary action.

Lithium water glass affects the pore structure. Changes the shape of the pores (pores flattened) and increases the number of pores having a diameter of 2-50 nm.

References

1. Ustawa z 27 marca 2003 r. o planowaniu i zagospodarowaniu przestrzennym.
2. P. Parker Lilly E. World Outlook for All Silica. Brick and Shapes Excluding SemiSilica. *The 2006-*

DOI: 10.1051/e3sconf/20161000037

2011, Insead (Singapore and Fontainebleau, France 2005)
3. PN-EN 772-10. Określenie wilgotności elementów silikatowych i elementów z autoklawizowanego betonu komórkowego.
4. A. Stępień, R. Dachowski Structural modification of sand –lime blocks regarding their physical and mechanical features. *8th PhD Symposium in Kgs. Lyngby.* (Denmark, 2010)
5. A. Stępień, R. Dachowski. The impact of various additives on the microstructure of silicate products. *International Conference on Green Buildings and Sustainable Cities.* (2011)
6. A Stępień, The impact of barium sulfate on the microstructural and mechanical properties of autoclaved silicate products. *The 9th International Conference "Environmental Engineering.* (Vilnius, Lithuania, 2014)

Economic efficiency of coal gasification in Poland in reference to the price of CO_2 emission rights

Michał Kopacz[1,a]

[1]*MEERI PAS, Poland*

Abstract. The article presents the impact of prices of carbon dioxide on the economic efficiency of 14 coal gasification technologies employed for producing electricity, hydrogen and methanol measured with the use of NPV method. All technical, technological and economic assumptions in the assessment have been made for Polish conditions. The impact of CO_2 prices were examined in the range of 30-200 PLN/Mg. The production capacity of the base technology corresponds with the fuel consumption of indicative coal having the calorific value of 20.5 GJ/Mg, used in the amount of 100 Mg/h. On the basis of the conducted research, with respect to all technical and economic assumptions, it can be stated that for the base scale there is a clear impact of prices of CO_2 emission allowances above the 90 PLN/Mg CO_2. Such a level of carbon dioxide prices makes the decision concerning construction of geological sequestration systems (CCS, carbon capture and storage) worthwhile. This applies in particular to the production of electric energy. For the variants focused on hydrogen production there is a dominance of variants with CCS system only at the price exceeding 120 PLN/Mg CO_2, and in the case of methanol such a situation occurs above 150 PLN/Mg CO_2.

1 Introduction

Gasification of coal has a relatively long history in Poland, although it has never been implemented on industrial scale. Globally, increasing interest in that technology has been noted for the production of power (and heat), hydrogen, or methanol, which are raw materials for chemical synthesis, and further processing [1, 2].

Hydrogen is used for the production of ammonia and – further on – chemical fertilizers [3,4]. Ammonia production capacity in Poland is assessed at some 3.0 Tg a year [5]. Methanol, among other things, is used as an intermediate product in MTO processes for production of olefins, gasoline, etc. [6, 7, 8].

At present, some 40 Tg of hydrogen are produced worldwide every year, which amounts to 1% of the world demand for primary energy. Increased interest in hydrogen production is also confirmed in the forecasts of EIA and BP; in future one may expect that the increase of demand for natural gas will be most noticeable, with less steep increase of demand for oil and coal [9, 10, 11]. Production of hydrogen by means of coal gasification appears to be also a technology competitive in terms of costs, feasible on industrial scale [12, 3].

Gasification of coal belongs to the so-called clean technologies, the aim of which is to minimize the negative influence of energy generation and chemical processes upon the environment. In the processes of methanol synthesis, the amount of carbon dioxide emitted to the atmosphere is reduced [13]. CO_2 may also be stored underground, in the processes of geological sequestration, yet the development level of that type of CCS technology is not significant. That development depends on the substantial requirements concerning he place of storage, rock mass, environment, and safety of people on the surface.

2 Statement of reasons for dealing with the topic

Below, there are some arguments, which justify the author's interest in coal gasification in Poland, as well as the assessment of the influence of CO_2 upon the efficiency of that process:

• there is a substantial interest in coal gasification technologies for production of liquid and gas fuels worldwide, particularly in Asia [14],

• documented recoverable resources of hard coal, as of 31.12.2014, amount to some 52 billion Mg, while those of brown coal (lignite) amount to 23.5 billion Mg [15],

• the project of the Polish Ministry of Economy, entitled *Polityka Energetyczna Polski do 2050* (Energy Policy of Poland until 2050) assumes conducting development studies concerning technologies, the application of which will be advantageous over a longer time, such as: gasification of coal, CCS technology, and

[a] Corresponding author: kopacz@meeri.pl

technologies for utilization of captured carbon dioxide [16],

- the European Commission, in its document entitled *Energy 2020* states the aim – reduction of greenhouse gases emission by 80-95% in 2050 [17]. In the *Roadmap 2050* it is proposed to reduce the greenhouse emissions by 80% by the year 2050, in comparison with emission level of 1990 (in the entire European Union), stating at the same time the required reduction of greenhouse gas emissions in subsequent years: 25% in 2020; 40% - 2030; 60% - 2040; 80% - 2050. In order to ensure the achievement of the goals set, the European Union Emissions Trade Scheme (EU ETS) has been envisaged, with the new emission reduction rate of >1.74% a year [18],
- Also in the Industrial Emissions Directive (IED) of 2010 [19], measures have been indicated to prevent environmental pollution and to control emission levels.
- In the provisions of the *Accession Treat* as well as the *Climate and Energy Package* it is additionally assumed to introduce new standards for construction of new power plants in the CCS-ready system, as well as to conduct research for geological sequestration of CO_2.
- In the documents referred to above, it is also indicated that – due to the policy conducted – a significant increase of CO_2, prices can be expected, which can result in diminishing the economic efficiency of power generation and chemical undertakings, which will have high levels of carbon dioxide emissions. Road Map 2050 assumes that the price of CO_2 emissions allowances in 2030 may exceed 25 EUR per 1 Mg of CO_2, whereas in IEA forecasts [20] those prices, depending on the scenario, may increase even to 75 EUR per 1 Mg of CO_2.
- Besides the above arguments, professional literature in Polish does not contain many studies, which would analyse the influence of prices of emissions allowances upon the efficiency of coal gasification technologies in a uniform assessment standard, and for common assumptions as to the scale and type of fuel consumed [21, 22, 23, 24]. The world professional literature contains studies on efficiency of coal gasification, taking into account the influence of prices of CO_2 emissions allowances, conducted among others by: Herzog, Simbeck, Rubin et al. [25, 26, 27]. The results of studies in that respect can also be found on the website of the US Department of Energy [28].

3 Approach to valuation and key assumptions

3.1 Approach to valuation

Income-based valuation has been used to assess the economic efficiency of the analysed coal gasification technologies – discount model based on the forecast of FCFF (free cash flow to firm). It is preferred when assessing technologies for power generation and chemical processing [29,30]. The net present value (NPV) has been calculated, as well as related indicators, such as IRR or NPVR. In the publication, inference has been based solely only NPV.

In the investment outlays, the following have been identified: costs of technological infrastructure of processes, auxiliary infrastructure, construction works, project documentation, or replacement investments. The investment outlays have been calculated at the level of TPC (total plant costs), as well as after consideration of other overheads and provisions (working capital requirements, demand for materials and chemicals in the start-up phase, as well as provisions amounting to 15% of TPC) – as total as spent costs (TASC) [31].

The calculation of operating costs comprised the cost of fuel used as charge – coal – as well as raw materials and other materials, energy and utilities, services, taxes and fees (real property taxes and insurance premiums, cost of CO_2 emission, fees for use of the environment), remuneration and related costs, depreciation, as well as other costs of operation (cost of plant closure, costs of CO_2 monitoring, waste management). Costs related to general administration (administration and support) amounted to 30% of the cost of personnel. On the other hand, the costs of sales have been calculated as TPC overheads in the amount of 0.25%. The costs of taxes and fees have been calculated as 0.7% of TPC, while the costs of insurance as 0.5% of TPC. Costs of repairs, servicing, and overhauls are calculated individually. The costs of fixed asset depreciation have been calculated taking into account the depreciation rates used for balance sweet purposes, adjusted to the useful life of a given technology.

The value of revenues has been calculated by multiplying the quantities of main products (electricity, methanol, and hydrogen), as well as respective intermediate products (heat, sulphur), and their prices, in line with assumptions made in table (Tab. 3).

3.2 Key assumptions

3.2.1 Key technical and technological assumptions

Table 1 shows the mass balance with particular emphasis on the final products and the quantity of CO_2 emitted or stored. These calculations are the underlying assumptions in the assessment of economic efficiency of the analysed technologies.

Table 1. Technical and technological assumptions in the assessment – mass balance

Name of tech.	Coal	CCS	Final product	Raw input (flow rate)	Dried input (flow rate)	Capacity factor	Cal. value of dried coal	Cal. value of raw coal
Technology description				[Mg/h]	[Mg/h]	[%]	[GJ/Mg]	[GJ/Mg]
Shell	H.c.	0	E.en.	85,1	69,0	80%	24,9	20,2
Shell	H.c.	1	E.en.	85,1	69,0	80%	24,9	20,2
Shell	B.c.	0	E.en.	161,6	85,8	80%	20,1	10,6
Shell	B.c.	1	E.en.	161,6	85,8	80%	20,1	10,6
Shell	H.c.	0	H2	95,7	77,7	90%	24,9	20,2

Shell	H.c.	1	H2	95,7	77,7	90%	24,9	20,2
Shell	B.c.	0	H2	181,8	96,5	90%	20,1	10,6
Shell	B.c.	1	H2	181,8	96,5	90%	20,1	10,6
GE	H.c.	0	H2	95,7	73,4	90%	26,4	20,2
GE	H.c.	1	H2	95,7	73,4	90%	26,4	20,2
Shell	H.c.	0	MeOH	95,7	77,7	90%	24,9	20,2
Shell	H.c.	1	MeOH	95,7	77,7	90%	24,9	20,2
Shell	B.c.	0	MeOH	181,8	96,5	90%	20,1	10,6
Shell	B.c.	1	MeOH	181,8	96,5	90%	20,1	10,6

H.c. – hard coal, B.c. – brown (lignite) coal, E.en. – electric energy, H2- hydrogen, MeOH – methanol. Own study.

Table 1a. Technical and technological assumptions in the assessment – mass balance (cont.)

Description	Coal	CCS	El. energy [GJ/h]	MeOH [kg/h]	H2 [kg/h]	CO_2 emission [kg/h]	CO_2 stored [kg/h]
Shell	H.c.	0	736	-	-	164 921	-
Shell	H.c.	1	547	-	-	17 137	147 785
Shell	B.c.	0	737	-	-	171 778	-
Shell	B.c.	1	526	-	-	17 528	154 250
Shell	H.c.	0	-	-	9 306	184 850	-
Shell	H.c.	1	-	-	8 477	20 061	164 789
Shell	B.c.	0	-	-	8 421	192 799	-
Shell	B.c.	1	-	-	7 237	20 643	172 156
GE	H.c.	0	-	-	7 977	182 764	-
GE	H.c.	1	-	-	6 637	18 431	164 333
Shell	H.c.	0	-161	57372	-	105 662	-
Shell	H.c.	1	-254	57372	-	6 465	99 196
Shell	B.c.	0	-177	54214	-	118 024	-
Shell	B.c.	1	-276	54214	-	6 110	111 915

Own study.

The assessment period was 30 years for all the technologies, with the building phase of 3 years. The intensity of the capital expenditures was assumed 20%, 45% and 35% in the third year.

The useful life comprised the construction phase, operation (26 successive periods/cycles), preceded by the commissioning and start-up (half a year) and one year for liquidation. The manufacturing scale, expressed as the amount of energy at input to the system (equivalent of 100 Mg of coal with the calorific value of 21.5 GJ/Mg) was equal for all the technological variants analysed.

Table 2 contains the characteristics of physical and chemical parameters of coal grades that underwent gasification. Grades of similar characteristics may be considered the most frequently occurring in Poland.

Table 2. Elemental composition and energy specifications for hard and brown (lignite) coal

Description	U.m.	Hard coal	Brown coal /lignite
humidity	[%]	21.3	48.8
coal	[%]	54.26	32.00
hydrogen	[%]	3.11	2.48
nitrogen	[%]	0.84	0.39
chlorine	[%]	0.00	0.01
sulphur	[%]	1.10	0.03
ash	[%]	9.30	5.49
oxygen	[%]	10.09	10.80
Calorific value			
High heat value HHV	[GJ/Mg]	21.43	12.39
Low heat value LHV	[GJ/Mg]	20.22	10.65

Own study.

Gasification process in Shell technology is based on entrained-flow gasifier operating at full commercial scale. Gasification reaction takes place at 4.2 MPa and 1430°C to produce syngas. The Shell process uses a dry feed system (hard coal is crashed and dried to 6% of moisture and the lignite to 12%). The product gas from the gasifier is cooled using a syngas recycle quench. Syngas goes through a raw gas cooler, which lowers the gas temperature to a minimum of 230°C (450°F), and contributes to the production of high pressure steam for use in the steam cycle. The solids and liquid slag formed in a water bath are removed through a lock hopper system. Ash carried over with the syngas is removed in a ceramic candle filter. Syngas enters the scrubber for removal of chlorides and entrained particles, principally unconverted carbon, slag and metals. In the acid gas removal process, COS hydrolysis, mercury clean-up about 15 percent of the CO_2 along with the H_2S and COS is removed. The residual CO_2 passes through the SRU, the hydrogenation reactor and is recycled to the AGR. The SRU utilises oxygen instead of air. The Claus plant produces sulphur, of witch recovery exceeds 99%. Clean syngas exiting the sulfinol absorber is reheated, diluted with nitrogen from the ASU, and enters the advanced F-Class gas turbine in the IGCC plant. Then exhaust gas is transported to the HRSG where additional heat is recovered. The steam form the HRSG is delivered to the steam turbine and generator.

In the case of Shell hydrogen production the water-gas-shift reactor provides the optimal syngas composition. Hydrogen is separated with an efficiency of 80% and is purified to the required level of 99.9%. Tail gas is burned in the CT burner to produce heat and vapour for the turbine generator.

However, in the case of methanol production, cleaned syngas is sent to the methanol synthesis unit. Raw methanol is then subjected to purification (99.9% pure). The tail gas is combusted in a flare, or if it is sufficiently rich in combustible components, passes through the boiler and the turbine connected to the generator (fig. 1).

In the case of the General Electric hydrogen production plant a wet feed system is used and gasification process occurs at a temperature of 1320°C and pressure of 5,6 MPa. The oxidant is 99% oxygen separated in the ASU. Raw synthetic gas is then transported to battery filters and cyclones to remove particles, NH_3, SO_2 and other impurities and in the quench water system and scrubbers is cooled to 230°C. The 99,9% of particles, 90% of chlorides and the alkaline compounds are removed so far.

 DOI: 10.1051/e3sconf/20161000038

The water-gas-shift reaction takes place at 5.4 MPa enhancing the amount of

H_2. After WGS conversion syngas is cooled to 35°C and cleaned of mercury, acid components (COS) in the selexol unit. Hydrogen is separated at a level of 80% and purified to 99,9%. The tail gas is combusted in the boiler, and steam produced enters the turbine connected to the generator.

For technologies with the CCS, CO_2 extracted from syngas is compressed and then with the pipeline system is transported assuming geological sequestration.

In both technologies, the exhaust gases having a temperature of 140°C are sent to the chimney (fig 2).

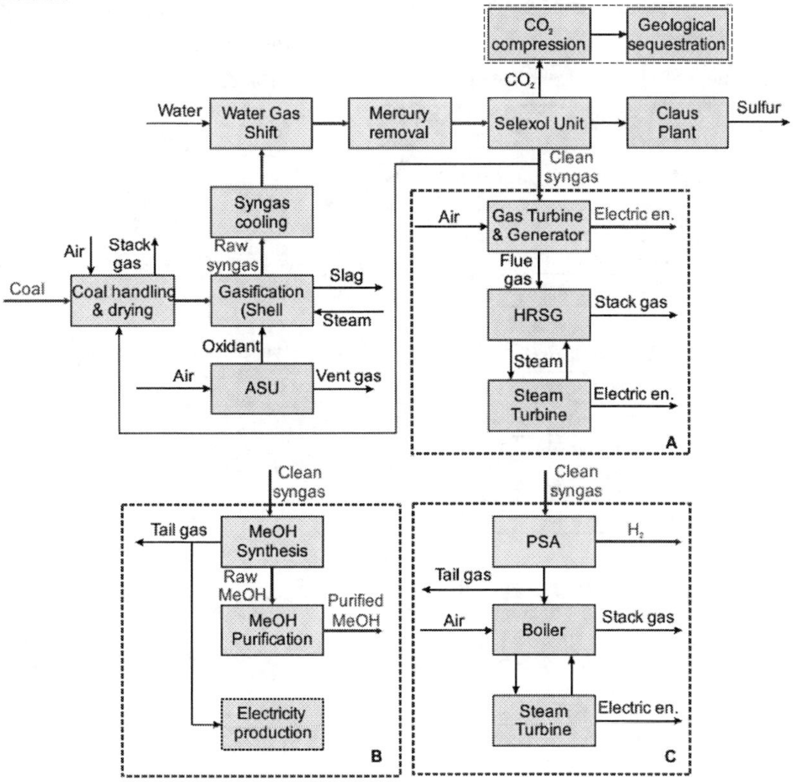

Figure 1. Block diagram of the Shell technology; A – electric energy, B – methanol, C – hydrogen production. Own study.

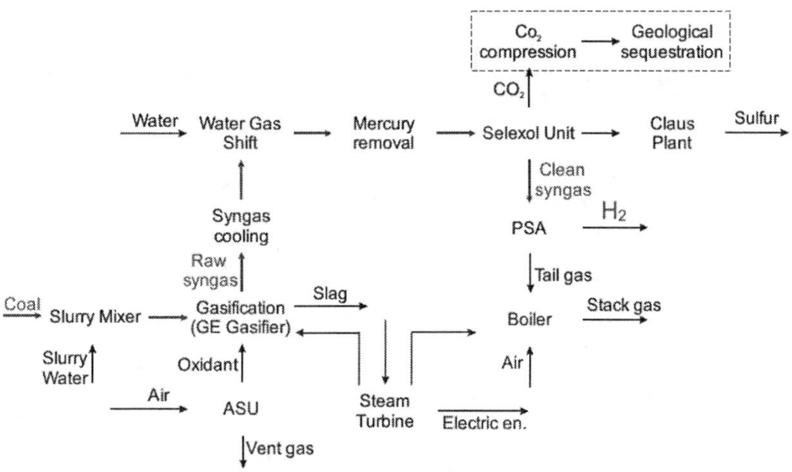

Figure. 2. Block diagram of the General Electric technology; hydrogen production. Own study

3.2.2 Key economic and financial assumptions

Table 3 contains the key macroeconomic assumptions, which serve the purpose of evaluating the analysed technological variants of coal gasification. All the calculations have been made in nominal terms (the detailed forecast comprised 10 consecutive one year periods). The prices of specific end products and of coal have been estimated according to individual growth paths (Tab. 3).

Table 3. Gasification of coal to produce electricity

Description	2015	Assumptions in 2016-2025
Inflation CPI	1.1%	2016: +1.6%; 2017-2025: +2.5%
Price of hard coal	197.0 [PLN/Mg] (9.74 PLN/GJ	+ 0.25% y/y
Price of brown coal	76.6 [PLN/Mg] (7.2 PLN/GJ)	+ 0.3% y/y
Price of methanol	1,450 [PLN/Mg]	+ 1.25% y/y
Price of hydrogen	7,400 [PLN/Mg]	+ 0.75% y/y
Price of electricity (sold)	170.0 [PLN/MWh]	+ 1% y/y
Price of electricity (buy)	290.0 [PLN/MWh]	+ 1% y/y
Price of natural gas	1.11 [Nm^3] (110 PLN/MWh)	+ 0.25% y/y
Price of CO_2 allowance	30.0 [PLN/Mg]	+ 1% y/y
Financing	50% of Debt	
Risk	RADR (risk adjusted discount rate 6-10%)	

Own study.

It has been assumed in the analyses that 50% of the capital expenditures will be financed by a loan, having the interest rate of 5.84% p.a. The discount rate has been assessed individually for each variant analysed by experts, while the results of such assessment are presented in Table 4. Variants with CCS demonstrate higher risk (premium of 2-3%), due to the fact that those technologies are not so advanced in Poland, for the time being.

Table. 4. Risk assessment and discount rates (RADR)

Sp.	Tech.	Coal	CCS	Final product	Base [%]	Risk premium [%]	RADR
1	Shell	H.c.	0	El.en.	4.2%	1.8%	**6.0%**
2	Shell	B.c.	0	El.en.	4.4%	1.8%	**6.2%**
3	Shell	H.c.	1	El.en.	5.3%	3.7%	**9.0%**
4	Shell	B.c.	1	El.en.	5.3%	3.7%	**9.0%**
5	Shell	H.c.	0	H_2	4.2%	1.7%	**5.9%**
6	Shell	B.c.	0	H_2	4.2%	1.7%	**5.9%**
7	Shell	H.c.	1	H_2	5.3%	3.7%	**9.0%**
8	Shell	B.c.	1	H_2	5.3%	3.7%	**9.0%**
9	GE	H.c.	1	H_2	5.3%	3.6%	**8.9%**
10	GE	H.c.	0	H_2	4.2%	1.8%	**6,0%**
11	Shell	H.c.	0	MeOH	4.2%	1.7%	**5.9%**
12	Shell	B.c.	0	MeOH	4.2%	1.7%	**5.9%**
13	Shell	H.c.	1	MeOH	5.3%	3.7%	**9.0%**
14	Shell	B.c.	1	MeOH	5.3%	3.7%	**9.0%**

H.c. – hard coal, B.c. – brown coal (lignite).

Table 5 presents the investment outlays related to construction of installations. Additionally, the expenditures concerning geological sequestration of carbon dioxide have been specified (if applicable). In case of CO_2 storage, its previous compression has been assumed and transport by pipeline along a distance of some 80 km. The investment outlays are in the range of 2.91-3.91 billion PLN. The investment costs concerning CCS module amount from 300 to 430 million PLN, depending on the variant.

Table 5. Capital expenditure for the coal gasification technologies – electricity production [mln PLN]

Description	Shell technologies			
	Hard coal		Brown coal/Lignite	
	w/o CCS	CCS	w/o CCS	CCS
Total Capex	2 990	3 450	3 440	3 910
Capex for CCS	-	390	-	400

Own study.

Table 5a. Capital expenditure for the coal gasification technologies – hydrogen production [mln PLN]

Description	Shell technologies				GE technologies	
	Hard coal		Brown coal/Lignite		Hard coal	
	w/o CCS	CCS	w/o CCS	CCS	w/o CCS	CCS
Total Capex	2 910	3 400	3 510	4 010	2 680	3 160
Capex for CCS	-	420	-	430	-	420

Own study.

Table 5b. Capital expenditure for the coal gasification technologies – methanol production [mln PLN]

Description	Shell technologies			
	Hard coal		Brown coal/Lignite	
	w/o CCS	CCS	w/o CCS	CCS
Total Capex	2 900	3 200	3 240	3 570
Capex for CCS	-	300	-	330

Own study.

Table 6 contains the key assumptions used calculation of operating costs pertaining to the consumption of chemicals and materials, whereas Table 6a, respectively, contains the assumptions for salary cost estimation. Total employment amounted to 50 people (including 24 persons of maintenance personnel and 3 persons of technical inspection).

Table 6. Costs of production – key assumptions

Description	U.m.	Value
Carbon (Mercury Removal)	[PLN/kg]	12.6
Claus catalyst	[PLN/m^3]	23,500
COS catalyst	[PLN/m^3]	13,000
MU & WT chem.	[PLN/kg]	2.1
Slag	[PLN/Mg]	30.0
Spent mercury catalyst	[PLN/kg]	5.0
Sulfinol solution	[PLN/l]	9.5

Description	U.m.	Value
Water	[PLN/m^3]	1.3
Water gas shift catalyst	[PLN/m^3]	90,000.0
Selexol solution	[PLN/l]	33.5

Own study.

Tabela 6a. Labour costs – key assumptions

Description: labour	U.m.	Cost
Operator	[PLN/mth]	5,000
Skilled operator	[PLN/mth]	10,000
Foremen and administration	[PLN/mth]	12,000

Own study.

4 Results of studies

Tables 7, 8, and 9 show the results of assessing the economic efficiency of coal gasification, as well as recommendation of the most favourable investment decisions, as regards the generation of carbon dioxide. Also the assessment of NPV result sensitivity to the change in the price of CO_2 emissions allowances in the range of 30-200 PLN/Mg has been added.

When Table 7 is analysed, one can notice that – for the assumptions made – the technologies of hard coal and lignite gasification to produce electricity are not efficient. At the price of CO_2 emissions allowances below 90 PLN/Mg of carbon dioxide, it is more profitable to emit the gas to the atmosphere. Only when the price exceeds 120 PLN/Mg of CO_2, efficiency advantage measured by means NPV is gained by the variants with CCS. The sensitivity of NPV results in case of gasification for power generation from coal grades analysed was relatively low, which is due to the conditions of the valuation model and low NPV values in the base scenario (CO_2 emissions allowance trading at 30 PLN/Mg).

Table 7. Gasification of coal for power generation

Description	Shell Technology			
	Hard coal		Brown coal/Lignite	
	w/o CCS	CCS	w/o CCS	CCS
CO_2 price [PLN/Mg]	**NPV [FCFF] [mln PLN]**			
30 (ref. Scen)	-2 818	-3 952	-3 014	-4 338
60	-3 325	-3 994	-3 542	-4 380
90	-3 831	-4 036	-4 070	-4 423
120	-4 338	-4 077	-4 597	-4 466
150	-4 844	-4 119	-5 125	-4 508
200	-5 689	-4 189	-6 004	-4 579
CO_2 price [PLN/Mg]	**Decision to invest**			
30 (ref. Scen)	EU ETS	-	EU ETS	-
60				
90				
120	-	CCS	-	CCS
150				
200				
CO_2 price [PLN/Mg]	**NPV sensitivity (to reference scenario)**			
60	18%	1%	18%	1%
90	36%	2%	35%	2%
120	54%	3%	53%	3%
150	72%	4%	70%	4%
200	102%	6%	99%	6%

Own study.

In case of gasification of coal to hydrogen, all the technological variants are also inefficient, taking into consideration the NPV criterion and assessment conditions. It can be stated that generally better results are achieved in case of hard coal gasification using the technology of Shell than that of General Electric (GE). Higher NPV values have been achieved in case of hard coal gasification. The analysis of Table 8 leads to the conclusion that only in case of prices of CO_2 emissions allowances exceeding 120 PLN/Mg, the decisions concerning construction of CCS are justified (in case of lower prices, the purchase of emissions allowances is more advantageous). The net present value is particularly sensitive to changes in the emissions allowances price level, in particular for those variants, for which the economic efficiency measured by means of the NPV indicator is close to 0.

Table 8. Coal gasification to hydrogen

Description	Shell Technology				GE Technology	
	Hard coal		Brown coal/Lignite		Hard coal	
	w/o CCS	CCS	w/o CCS	CCS	w/o CCS	CCS
CO_2 price [PLN/Mg]	**NPV [FCFF] [mln PLN]**					
30 (ref. Scen)	-339	-1 497	-1 040	-2 547	-745	-2 185
60	-687	-1 531	-1 470	-2 597	-1 149	-2 230
90	-1 061	-1 566	-2 046	-2 648	-1 701	-2 275
120	-1 567	-1 601	-2 638	-2 698	-2 262	-2 320
150	-2 135	-1 636	-3 231	-2 748	-2 824	-2 364
200	-3 081	-1 696	-4 218	-2 832	-3 759	-2 439
CO_2 price [PLN/Mg]	**Decision to invest**					
30 (ref. Scen)	EU ETS	-	EU ETS	-	EU ETS	-
60						
90						
120						
150	-	CCS	-	CCS	-	CCS
200						
CO_2 price [PLN/Mg]	**NPV sensitivity (to reference scenario)**					
60	103%	2%	41%	2%	54%	2%
90	213%	5%	97%	4%	128%	4%
120	363%	7%	154%	6%	204%	6%
150	530%	9%	211%	8%	279%	8%
200	809%	13%	306%	11%	405%	12%

Own study.

In case of gasification of selected grades of coal to methanol, the base variant – for the emissions allowances price of about 30 PLN/Mg CO_2, assumed prices of coal and methanol over the next 30 years generates the positive value of NPV at the level of 135 million PLN. In other cases the net present value was negative. Only in case of lignite (brown coal) gasification and prices of emission allowances exceeding 150 PLN/Mg of CO_2 the option of constructing CCS has an efficiency advantage. NPV sensitivity to the change of emission allowances varies depending on the variant. The highest level NPV variability is demonstrated for those variants, for which the base NPV was relatively close to 0. The least sensitivity is demonstrated by the variants with strongly negative NPV (they include CCS systems, and the emission to air does not exceed 10%).

Table 9. Coal gasification to methanol

Description	Shell Technology			
	Hard coal		Brown coal/ Lignite	
	w/o CCS	CCS	w/o CCS	CCS
CO_2 price [PLN/Mg]	**NPV [FCFF] [mln PLN]**			
30 (ref. Scen)	135	-1 072	-261	-1 513
60	-64	-1 082	-483	-1 523
90	-263	-1 093	-708	-1 533
120	-462	-1 103	-938	-1 543
150	-664	-1 113	-1 188	-1 553
200	-1 021	-1 130	-1 716	-1 571
CO_2 price [PLN/Mg]	**Decision to invest**			
30 (ref. Scen)	EU ETS	-	EU ETS	-
60	EU ETS	-	EU ETS	-
90	EU ETS	-	EU ETS	-
120	EU ETS	-	EU ETS	-
150	EU ETS	-	EU ETS	-
200	EU ETS	-	-	CCS
CO_2 price [PLN/Mg]	**NPV sensitivity (to reference scenario)**			
60	148%	1%	85%	1%
90	295%	2%	171%	1%
120	443%	3%	259%	2%
150	593%	4%	354%	3%
200	858%	5%	557%	4%

Own study.

5 Summary and conclusions

The study presents the results of economic efficiency assessment for 14 technological variants of hard and brown coal (lignite) gasification to produce electricity, hydrogen, and methanol. Higher NPV was decisive for the selection of a more advantageous economic decision (purchase of EU ETS or construction of CCS system). In the entire assessment process, also other criteria of costs, values, or risks have been considered, such as: COE – cost of electricity, LCOE, yearly costs of operations, investment outlays (also individual ones), levels of CO_2 emission, dynamic discount rate, RADR (risk-adjusted discount rate), IRR, or NPVR, they have not been presented in this paper, though. The entire results of analyses may be found in [24].

The analysis concerned technological solutions of two producers, namely Shell and General Electric (Texaco).

The analyses conducted allow to state that for the assumed scale of energy consumption analysis in the charge fuel, parameters of selected coal grades, and specified prices of end products of gasification, most of the technological variants failed to obtain efficiency confirmation, using NPV as base. Only one variant of hard coal gasification achieved the NPV value of nearly 135 million PLN. However, this is a case in which the entire fuel is converted into end products, while the internal consumption is covered by external purchase of electricity. In case of gasification for power generation, at allowances prices exceeding 90 PLN/Mg of CO_2, geological sequestration of carbon dioxide seems to be more advantageous. In case of hydrogen production, the justification for decision change occurs at the allowance price of 120 PLN per 1 Mg of CO_2 or higher. In case of methanol, only the allowance price level higher than 150 PLN/Mg of CO_2 may justify decisions to invest in the construction of CCS systems.

The results sensitivity of NPV to change of the price level of CO_2 emissions allowances differs in different variants, with the most significant changes occurring in case of methanol production, similar in case of methanol, and less profound in case of electricity.

In conclusion, one can state that the assumptions of rigorous requirements concerning CO_2 emission limits, which result in significant increases of the emissions allowance price, may greatly influence the assessment of coal gasification efficiency. This applies in particular to the production of electricity and hydrogen. In case of methanol production, a part of CO_2 is bound in end products, which significantly enhances the efficiency and attractiveness of MeOH generation. A significant increase of the price of emissions allowances will also result in substantial reduction of attractiveness of investment in conventional power engineering and heat generation.

References

1. T. Borowiecki, J. Kijeński, J. Machnikowski, M. Sciążko: *Czysta energia, produkty chemiczne i paliwa z węgla – ocena potencjału rozwojowego.* Wyd. ICHPW, Zabrze 2008.
2. M. Taniewski, Przemysł Chemiczny, **(91/4)**, str. 492-498 (2012),
3. K. McHugh, S. Eisele, J. Nestell: *Hydrogen production methods*, MPR Associates, Inc. MPR-WP-0001, (2005).
4. T. Chmielniak, *Badania symulacyjne technologii wytwarzania wodoru w aspekcie emisji CO_2 – wydobycie, transport i przetwórstwo wegla*, Wydawnictwo Politechniki Śląskiej, Gliwice, (2014),
5. W. Lubiewa-Wieleżyński: *Prognozy rozwoju przemysłu chemicznego w Polsce do 2020 roku*, Warszawa, (październik 2014), materiał niepublikowany.
6. UOP LLC, Methanol to olefins, IOCL Conclave, Des Plaines, Illinois, USA, 2014.
7. M. Taniewski, Przemysł Chemiczny, **(92/2)**, str. 140-147, (2013),
8. R. Luque, J. Speight: *Gasification for syntetic fuel production*, Woodland Publishing Series In Energy, **69**, Elsevier Ltd. (2015),

9. BP, British Petroleum: Energy Outlook 2035, http://www.bp.com/content/dam/bp/pdf/energy-economics/energy-outlook-2015/bp-energy-outlook-2035-booklet.pdf, on-line (15.04.2016),
10. J. B. Tennant: Gasification Systems Overview, US Department of Energy, National Energy Technology Laboratory, http://www.netl.doe.gov/File%20Library/Research/Coal/energy%20systems/gasification/doe-gasification-program-overview.pdf, (on-line 15.04.2016)
11. EIA, U.S. Energy Information Administration: Annual Energy Outlook 2015 with projections to 2040.: http://www.eia.gov/forecasts/aeo/, (on-line 15.04.2016)
12. IEA, International Energy Agency: Energy Technology Essentials - Biofuel Production, Jan. 2007, https://www.iea.org/publications/freepublications/publication/essentials2.pdf, (on-line 15.04.2016),
13. M. Ściążko, J. Kijeński red.: *Studium koncepcyjne wybranych technologii, perspektywicznych procesów i produktów konwersji węgla – osiągnięcia i kierunki badawczo-rozwojowe.* Zgazowanie węgla, Tom 1. Wyd. ICHPW, Zabrze 2010,
14. C. Higman, State of the Gasification Industry – the Update Worldwide Gasification Database, *International Pittsburgh Coal Conference*, Beijng, China, (2013),
15. PIG, Państwowy Instytut Geologiczny, Bilans zasobów PLNóż kopalin w Polsce wg. stanu na 31.12.2013, Warszawa, (2014).
16. MG, Projekt Polityki Energetycznej do roku 2050, Ministerstwo Gospodarki RP 2014, www.mg.gov.pl (on-line 15.04.2016),
17. EC, European Commission: Energy 2020. A strategy for competitive, sustainable and secure energy, KOM(2010) 639, (2010).
18. EC, European Commission: Energy Roadmap 2050; KOM (2011) 885/2, (2011),
19. EC, European Commission: EC, Industry Emission Directive. 2010/75/EU, (November 2010),
20. IEA, International Energy Agency, The World Energy Outlook 2014. Paris. France, http://www.worldenergyoutlook.org/weo2014/, (on-line 15.04.2016),
21. M. Ściążko, L. Zapart, K. Dreszer, Polityka Energetyczna, IGSMiE PAN, **9**, special editon, (2006)
22. K. Stańczyk K, M. Bieniecki, Górnictwo i Geoinżynieria, **2/31**, pp. 575-586, (2007)
23. K. Janusz-Szymańska, Rynek Energii, **5/2012**, str. 24-30, (2012),
24. K. Kwaśniewski, M. Kopacz, P. Grzesiak, R. Kapłan, E.J. Sobczyk, *Coal gasification – conditions, effectiveness and perspectives for development*, Wydawnictwa AGH, Kraków, (2015)
25. H. Herzog, Greenhouse Gas Control Technologies, Elsevier Science Ltd., pp. 101-106 (1999).
26. D.R. Simbeck, *CO_2 Mitigation Economics for Existing Coal-Fired Power Plants. Paper read at Pittsburgh Coal Conference*, NSW, Australia. (Dec.4 2001),
27. E.S. Rubin, C. Chen, A.B. Rao, 2007, Energy Policy, **35(9)**, 4444-4454 (2007),
28. DOE NETL: Advanced CO_2 Capture R&D Program; Advanced CO2 Capture R&D Program, http://www.netl.doe.gov/research/coal/carbon-capture/capture-handbook (on-line 15.04.2016),
29. K. K. Humphreys, S. Katell: *Basic cost engineering*. Marcel Dekker Inc, New York, (1981)
30. K. Kwaśniewski, M. Kopacz, P. Grzesiak, R. Kapłan, D. Kryzia, Przemysł Chemiczny 95/1, str. 130-137, (2016),
31. DOE NETL: Quality guidelines for Energy System Studies: Cost Estimation Methodology for NETL Assessments for Power Plant Performance: DOE/NETL-2011/1455, (April 2011),

DOI: 10.1051/e3sconf/20161000039

Coal from the waste disposal site of the Siersza mine (Trzebinia, Poland) and its properties as a possible alternative fuel

Beata Kosa[1,a] and Alicja Kicińska[1]

[1]*AGH University of Science and Technology, Faculty of Geology, Geophysics and Environmental Protection, Mickiewicza 30 av.,30-059 Kraków, Poland*

Abstract. The authors recovered bituminous coal from the dump of the closed down Siersza mine in Trzebinia, determined its properties and evaluated a possibility of applying it as an alternative fuel for the cement industry. The dump material was enriched in a laboratory using the gravity (jigs, shaking tables) and flotation methods of coal cleaning. The gross calorific value of the processing products, measured with a calorimetric bomb, ranges from 13.588 to 22.543 $MJ \cdot kg^{-1}$. The contents of heavy metals, determined with XRF, are as follows: Fe 3.76–5.25 $mg \cdot kg^{-1}$, Mn 0.037–0.132 $mg \cdot kg^{-1}$, Ti 0.325–0.493 $mg \cdot kg^{-1}$ and Zn 0.113–0.329 $mg \cdot kg^{-1}$, while of Hg (determined with ASA) from 0.02689 to 0.21870 $mg \cdot kg^{-1}$. The content of sulphur is 0.7349–1.0484 wt.% and of chlorine Cl 0.131–0.135 wt.%, the net calorific value of 13.446–22.538 $MJ \cdot kg^{-1}$. The results indicate that the laboratory jigging and flotation provide products that meet the parameters of the solid fuels selected by applicable to the needs of cement plants.

1 Introduction

Continuously growing amounts of waste enforce new ways of their neutralization and disposal. One of the methods of the economic management of coal-related waste is using them as alternative fuel in power stations, heat and power generating stations and cement plants [1-7]. However, such a use of processed waste requires an assurance that the material obtained complies with legal requirements. The plant utilizing an alternative fuel must not only know or recognize its properties but also consider whether the process of burning under current technological conditions will not break the environmental standards that set the emission limits of selected elements (compounds) into the environment.

The authors describe the tests that were conducted on samples of hard coal recovered from the dump of the closed down Siersza coal mine in Trzebinia. The aim of the study was to evaluate the possibility of using this material as a potential alternative fuel for cement plants.

2 Material and methods of study

The material tested in the study comes from the dump of the closed down Siersza coal mine in Trzebinia (S Poland). Its sampling grid was irregular due to the presence of stagnant water pools and the lack of clear-cut dump boundaries (Fig. 1). A total of 40 samples were collected with a weight of approx. 2 kg each.

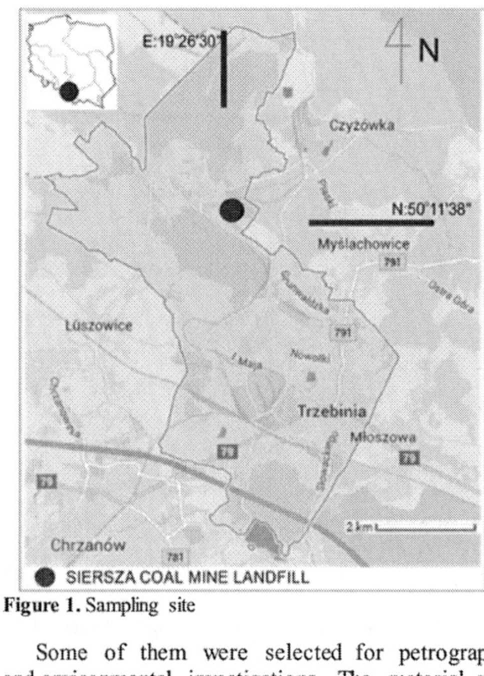

Figure 1. Sampling site

Some of them were selected for petrographic and environmental investigations. The material was mixed, averaged, dried and subjected to mechanical pre-treatment: crushing and screening into two grain fractions, finer (<1 mm) and coarser (1–20 mm) ones.

[a] Corresponding author: beata.kosa@wp.pl

The finer material was processed using two methods: the froth flotation and the gravity concentration on a shaking table, the latter preceded by hydrocyclone desilting. Three methods were used for the coarse material: the gravity concentration either in heavy media or in jigs and the flotation with the earlier crushing the material to the size <0.5 mm. Further analyses were conducted on the samples generated by the methods that provided most promising results of coal waste cleaning, i.e., a high coal recovery and a low ash content of the concentrates. Such conditions are fulfilled only by the group of 1–20 mm sized concentrates:

(1) from the concentration table (Fig. 2), hereafter referred to as the sample CT,
(2) from the jig separator, hereafter referred to as the sample P, and
(3) from the flotation (Fig. 3), hereafter referred to as the sample F.

Figure 2. Concentration table.

Figure 3. Flotation machine.

The total contents of selected metals: Fe, Mn, Ti, Zn, Cd and Tl were assayed using X-ray fluorescence (XRF) and an Olympus X-5000 device. Determinations of sulphur and chlorine were carried out using the Eschka method [8, 9]. Determinations of the heat of combustion and the calorific value were done according to the standard [10]. In the mercury determinations a dedicated atomic absorption spectrometer AMA-254 (ALTEC company) was used. The apparatus is a modification of classic AAS spectrometers and measures mercury contents of solid and liquid samples with a detection limit of few nanograms. The sample is mineralized within the analyzer itself, immediately prior to the detection step. This solution reduces losses of mercury that may be caused in the classic AAS devices.

3 Results

The coal waste represent a diversified assemblage of rocks predominantly sandstones, claystones and coal shales. The coal content was visually evaluated at 1–6%, and the pyrite content at 2–5%.

Environmental studies shown that the accumulated waste have a negative impact on the environment [11]. Some of the samples contain slag of an unknown origin, which reported to the gravity concentrates with the highest coal recovery and increased their ash content [12].

3.1 Chlorine and sulphur contents

The Cl contents of the concentrates are almost the same and range from 0.131 to 0.135 wt.% (Tab. 1). The higher value (0.135 wt.%) corresponds to the F sample (from flotation) and the lower (0.131 wt.%) to the CT sample (from the concentration table). Looking at the recommendations made by Sarna et al. [1] concerning the upper Cl limit (≤0.2 mass %) in the material that could be an alternative fuel, none of the samples exceeds this value.

Table 1. Contents of chlorine and sulphur in coal samples.

Sample	Cl	S
	[wt. %]	
CT	0.131	1.0484
P	0.132	0.7349
F	0.135	1.0335
x	0.133	0.9389
σ	0.002	0.1768
V	*0.016*	*0.1883*
PLiF*	**0.2**	**2.5**

x – arithmetic mean; σ – standard deviation, *V* – coefficient of variability (σ/x), *PLiF – permissible level in fuels, according to [1]

The S content ranges from 0.7349 to 1.0484 wt.% (Tab. 1), being the highest in the CT sample (1.0484 wt.%) and the lowest in the P sample (0.7349 wt.%). According to Sarna et al. [1], the sulphur content of the potential alternative fuel should not exceed 2.5 wt.%, and even the highest sulphur value of the P sample is 2.5 times below this limit.

3.2 Heavy metal and mercury contents

The contents of individual metals are: Fe 3.76–5.25, Mn 0.037–0.132, Ti 0.325–0.493 Zn 0.113–0.329 and Hg 0.02689–0.2187 mg·kg^{-1}. They show a considerable variability, expressed as the coefficient of variation V (the ratio of standard deviation to arithmetic mean). The spread of the values ranges from high Hg (V=0.99) to lesser Zn and Mn (V = 0.62 and 0.54, respectively) and to low Ti and Fe (V = 0.22 and 0.17, respectively).

The highest are contents of Fe (5.255 mg·kg^{-1}), Ti (0.493 mg·kg^{-1}) and Hg (0.219 mg·kg^{-1}) of the coal concentrate from the concentration table (CT). In the flotation concentrate (F) the highest are Mn (0.132 mg·kg^{-1}) and Zn (0.329 mg·kg^{-1}) contents.

The lowest contents of Fe (3.76 mg·kg^{-1}) and Ti (0.325 mg·kg^{-1}) occur in the flotation concentrate (F), whereas those of Mn (0.037 mg·kg^{-1}), Zn (0.113 mg·kg^{-1}) and Hg (0.027 mg·kg^{-1}) the jig concentrate (P). The range of the Hg determinations is from 0.02689 mg·kg^{-1} (P concentrate) to 0.21870 mg·kg^{-1} (CT concentrate). The Cd contents of the P and F concentrates are below the detection limit, while in the CT concentrate the Cd content is 0.021 mg·kg^{-1}. The thallium content in all the concentrates is below its detection limit.

The values cited above do not exceed the upper limits of these metals set by Sarna et al. [1] for alternative fuels.

Table 2. Contents of selected elements in coal samples.

Sample	Fe	Mn	Ti	Zn	Hg
	mg·kg^{-1}				
CT	5.25	0.097	0.493	0.135	0.219
P	4.87	0.037	0.483	0.113	0.027
F	3.76	0.132	0.325	0.329	0.062
x **σ**	4.63 0.78	0.09 0.05	0.43 0.09	0.19 0.12	0.102 0.02
V	*0.17*	*0.54*	*0.22*	*0.62*	*0.99*
PLiF*	-	-	-	-	**10**

x – arithmetic mean; σ – standard deviation, V – coefficient of variability (σ/x), *PLiF – permissible level in fuels, according to [1]

3.3 Heating and calorific value

Calorific value ranges from 13.588 kJ·kg-1 of the CT sample to 22.543 KJ·kg^{-1} for the P sample (Tab. 3). The calorific values show the same trend: from the lowest value of the CT sample (13.446 KJ·kg^{-1}) to the highest value of the P sample (22.538 KJ·kg^{-1}).

Table 3. Heating and calorific values of coal samples.

Sample	Heating value	Calorific value
	MJ·kg^{-1}	
CT	13.59	13.45
P	22.54	22.54
F	20.64	20.61
x **σ**	18.92 4.72	18.87 4.79
V	*0.25*	*0.25*
PLiF*	-	**>14.00**

x – arithmetic mean; σ – standard deviation, V – coefficient of variability (σ/x), *PLiF – permissible level in fuels, according to [1]

The sample obtained on the concentration table does not meet the minimum calorific value criterion of alternative fuels (i.e., >14,000 KJ·kg^{-1}), whereas the two other concentrates significantly exceed it.

3.4 Other parameters

The list of parameters that characterize any type of solid fuels includes also its contents of ash and volatile matter as well as of the humidity (Tab. 4).

The ash content varies from 21 to 26 wt.%. in the concentrates P and F, while it is nearly doubled (55 wt.%) in sample CT. Ash contents >20% are obviously undesired in the heat generation, and may also affect the process of cement production [13, 14,15].

The content of volatiles is the lowest (12 wt.%) in the CT sample, and in the remaining samples P and F is more than twice higher (31 and 28 wt.%, respectively). The moisture contents of the concentrates CT, F and P after their ageing (1 month, ambient temperature) are 13, 15 and 21 wt.%, respectively. High water contents of coal fuels is undesired because they lower calorific values. However, it should be explained why the jig concentrate (P) has the highest calorific value of the three samples despite its relatively high moisture value.

Table 4. Other parameters measured in tested samples of coal.

Sample	Parameters		
	Ash content	Volatile matter content	Wetness
	[%]		
P	21	31	21
CT	55	12	13
F	26	28	15
x **σ**	34 18.35	23.67 10.21	16.33 4.16
V	*0.54*	*0.43*	*0.25*

x – arithmetic mean; σ – standard deviation, V – coefficient of variability (σ/x),

4 Conclusions and discussion

Two samples, i.e., P and F concentrates from jigging and flotation, respectively, meet the parameters required for alternative fuels in cement plants. The third one, obtained from a concentrating table, cannot be utilized in this way. The major reason is probably either its low coal content in the starting material that represented the fine-grained fraction or a low degree of preliminary treatment of the material, which undoubtedly contributed to the quality and ash content of the sample [9, 16].

In addition to the parameters determined in this paper, cement plants themselves check their internal, mainly operational parameters, which concern, e.g., emission standards, an impact of fuel on the cement quality, effects imparted by fuel on the kiln performance.

Comparing the results with the literature data on the quality of alternative fuels [17, 18], the concentrate obtained from jigging falls into the class II of such fuels. The requirements for this class are as follows: calorific value ≥20 MJ·kg^{-1}, Cl content ≤0.6 %wt., Hg content ≤0.03 mg·kg^{-1}. The concentrate obtained from flotation is classified as the class III of alternative fuels. The requirements of this class are: calorific value ≥ 15 MJ·kg^{-1}, Cl content ≤1%, Hg content ≤0.08 mg·kg^{-1}. The concentrate obtained from the concentration table meets the requirement of only the class V of alternative fuels: calorific value ≥3 MJ·kg^{-1}, Cl content ≤3%, Hg content ≤0.5 mg·kg^{-1}.

Summarizing, the following conclusions can be made:

(1) The coal concentrates obtained using laboratory jigging and flotation meet the literature recommendation for alternative fuels of cement plants;
(2) The contents of S and Cl in the concentrates do not exceed the limits for alternative fuels specified in the literature [1];
(3) The Fe contents of the concentrates are much higher (up to about 130 times) than the contents of the remaining metals considered (Mn, Ti, Zn, Cd and Hg;
(4) The highest is the content of Hg (0.2187 mg·kg^{-1}) of the sample obtained on the concentration table;
(5) High heating and calorific value of jigging and flotation concentrates indicate that the two methods are effective in recovering coal from the waste accumulated on the dumps of coal mines;
(6) All concentrates have high ash contents, which may be a problem if using the coal recovered as an alternative fuel in cement plants;
(7) All concentrates have relatively low contents of volatiles;
(8) All concentrates have high moisture contents after one-month ageing which would lower their calorific value if used as alternative fuels;
(9) The concentrates meet the requirements of the alternative fuel of the class II (jig sample P), class III (flotation sample F) and class V (concentrate table sample CT).

Acknowledgments

The research has been funded by AGH University of Science and Technology Statutory Research KOŚ No. 11.11.140.199.

References

1. M. Sarna, E. Mokrzycki, A. Uliasz-Bocheńczyk, Zesz. Nauk. Wydz. Bud. Inż. Środ.Pol. Kosz. Inż. Środ., **21**, 309-316 (2003).
2. M. Czop, Pol. J. Environ. Stud. **4**, 1369-1372 (2014).
3. M. Czop, M. Kajda-Szcześniak, Arch. Environ. Prot. **39**, 71-80, 4 (2013).
4. J. Bieniek, M. Domaradzka, K. Przybysz, W. Woźniakowski, Act. Agr., **17**, 277-288, 2 (2011).
5. A. Sobolewski, R. Wasilewski, K. Dreszer, S. Stelmach, Przem. Chem., **85**, 8-9 (2006).
6. A. Kicińska, B. Kosa, Envi. Prot. & Nat. Res., **27**, 1-6, 2(68) (2016)
7. H. Karcz, M. Kantorek, K. Folga, T. Butmankiewicz, A. Kozakiewicz, D. Maciejak, Arch. Gosp. Odp. I Ochr. Środ., **6**, 23-32 (2007).
8. B. Klojzy-Karczmarczyk, J. Mazurek, En. Pol. J. **13**, 289-302, 2 (2014).
9. PN-ISO 334:1997
10. PN-ISO 587:2000
11. PN-ISO 1928:2002
12. S. Postrzednik, En., **10**, 607-611 (2011).
13. https://works.bepress.com/robert_oleniacz/24/
14. S. Góralczyk, I. Baic, En. Pol. J., **12**, 145-157, 2(2009)
15. J. Duda, W. Kalinowski, Pr. Nauk. Pol. Warsz. Konf., **22**, 343-351, 1 (2002)
16. E. Mokrzycki, A. Uliasz-Bocheńczyk, En. Pol. J., **5**, 53-69, 1(2002)
17. S. Postrzednik, En., **7**, 395-398 (2011).
18. A. Sobolewski, R. Wasilewski, S. Stelmach, En. Pol. J. **10**, 379-389, 2 (2007).

DOI: 10.1051/e3sconf/20161000040

Impact of polyethylene glycol on porosity and microstructure of sand-lime product

Paulina Kostrzewa[1,a] and Ryszard Dachowski[2]

[1]*MSc, Kielce University of Technology, 25-315 Kielce, Poland*
[2]*Dr Eng. Prof. PŚk, Kielce University of Technology, 25-315 Kielce, Poland*

Abstract. Autoclaved sand-lime products are environment-friendly building materials in the form of bricks and blocks. Due to great interest in silicate products, there have been numerous studies and modifications aimed at improving the properties of the material. The article presents the most important characteristics of sand-lime products, both traditional as well as modified with polyethylene glycol. The purpose of described research was to estimate porosity (determine the type, size and pores distribution) and to analyze the microstructure of aforementioned samples. From the practical point of view, these characteristics are particularly important in the construction materials. The following tests have been carried out: bulk density using the hydrostatic method, capillarity, water absorption and mercury porosimetry of samples containing polyethylene glycol. The samples were also subjected to observation of the microstructure under the scanning electron microscope. Preliminary research findings clearly show the positive influence of the used substrate on the basic properties of modified sand-lime product. In comparison to the reference sample, modified products showed a decrease in water absorption and reduced capillarity. The confrontation of the measurements made using a mercury porosimeter with observation of the microstructure, showed differences in the distribution of pores and their size. Besides the additive, the production process itself has a huge impact on the properties of silicates (mixing process, time and temperature of autoclaving).

1 Introduction

Autoclaved sand-lime products (silicates) are manufactured from natural raw materials, so they are classified as healthy and ecological materials [1]. Silicates, which are characterized by high compressive strength and durability, are an excellent acoustic and thermal insulator. In addition, they are completely non-flammable, resistant to frost and growth of micro-organisms [2, 3].

Silica brick is becoming increasingly important as a construction material. It is willingly used for the construction of load-bearing walls, partitions, curtain walls and foundation. Sand-lime blocks are characterized by dimensional accuracy and allow to reduce the risk of improper leveling of each layer [4, 5]. Low water absorption of silicates allows the erection of silicates cladding walls. Most of masonry elements must be quickly plastered, while walls made of sand-lime blocks can be safely left without plaster. They will not be damaged under the influence of atmospheric factors. Polish Standard PN-EN 1996-2: 2010 [6] says that silicates can be used in wet and humid conditions. Silica brick does not emit unpleasant odor, even when saturated in water and its bright color provides a friendly interior of the basement.

The aim of the work is modification of the composition of traditional sand-lime products, which will result in better porosity of the material, which as a consequence will be more resistant to water penetration. The authors wish to maximize these properties of silicate bricks, which are important in terms of the usefulness of this material for making foundation and cladding walls.

The composition of traditional sand-lime products was modified by introducing the additive in the form of polyethylene glycol. Selection of the additive was based on its interesting properties, such as: polyethylene glycol penetrates into the wood structure displacing water; it is resistant to solutions of acids, alkalies and salts, non-toxic and colorless [7]. Polyethylene glycol is used in the pharmaceutical and cosmetic industries as well as in the textile industry, leather industry and wood industry [8].

2 Methodology of research

In order to carry out laboratory tests of modified sand-lime products, cuboidal samples were prepared with dimensions of 40x40x160 mm. The samples were prepared in the Silicate Production Plant in Ludynia (Poland).

[a] Corresponding author: pkostrzewa@tu.kielce.pl

The sand was mixed with quick lime and water in appropriate proportions. Then the mixture was placed in a reactor, where the process of lime slaking took place.

The next step was the addition of the substrate in a percentage presented in the Table 1. After mixing all the ingredients, the mass was placed in a trigeminal steel mold and pressed under a pressure of 20 MPa. Then the mass was placed in an autoclave for 8 hours at a temperature of 203°C and a pressure of 1.6 MPa. Removed from the molds, the samples were stored for 21 days under required conditions of temperature and humidity.

Table 1. Composition of the samples.

Sample symbol	Sand-lime mass [%]	PEG 400 [%]
N	100	0
P1	99	1
P2	98	2
P3	97	3
P4	96	4
P5	95	5

"A" represents a traditional sample and "P1-P5" are modified samples (depending on the amount of the additive). The mass composed of: quick lime (about 7% by weight of the product), sand (about 90%) and water (about 3%). The used additive was polyethylene glycol (PEG 400) having a molecular weight of 400 Da..

2.1 Density

Specific density is a ratio of mass of dry sample to its pore volume. The study was conducted using a helium pycnometer on the reference sample and samples with polyethylene glycol (with the best strength parameters).

2.2 Buk density

Bulk density is a ratio of mass of dry sample to it's the sample to the apparent density (including pores). Because of the irregular shape of the sample, the test was performed using a hydrostatic method.

2.3 Porosity

Porosity determines what part of the volume of the material constitutes the pore volume. It is an important feature, on which characteristics such as strength, frost resistance, thermal and acoustic insulation depend [10].

2.4 Mercury porosimetry

In this method mercury is injected into the pores under pressure. Knowing the amount of mercury used in the test, the pore volume can be estimated [9]. Moreover, when the pressure rises, smaller pores are penetrating and thanks to that determining their size is possible [2]. IUPAC classification of pores due to their size: micropores (dp <2 nm), mesopores (2 ≤ dp ≤ 50 nm), macropores (dp > 50 nm). The minimum diameter that can be measured depends on the maximum pressure obtained during mercury pressing.

2.5 Microstructure

Samples of the sand-lime products were subjected to observation of their microstructure under the scanning electron microscope to determine the distribution and pore geometry.

2.6 Capillarity test

Capillarity is a phenomenon of transporting water upward through the element. The amount of capillary water increases with decreasing pore diameter [9]. Water absorption effects are visible in the form of moisture stains in the lower parts of walls. Capillarity depends on the porosity and pore size. To determine this characteristic of the material, properly prepared samples were placed in water to a depth of 5mm. When placed in water, samples were weighed and measured after 10, 30 and 90 minutes according to the recommendations of the PN-EN 772-11: 2011 [11]. The level of moisture and weight increase of the sample were investigated.

2.7 Absorbability

Absorbability is the mass of water which can be absorbed by a sample, completely immersed in water, at atmospheric pressure. Typically, absorbability of building materials depends on the porosity. This is due to the fact that water cannot get into the closed pores, while in the case of large pores – it only moistens their walls [7].

3 Results and discussion

The study shows that the modified products have slightly lower specific density and bulk density. The volume of pores in the samples constitutes 33.5%, and the tightness (both of the sample with the additive and traditional one) is 0.665%. These studies enable to determine the total volume of the pores. However, they do not provide any information on their shape and size.

Table 2. The test results density and porosity.

Sample symbol	A	P1
Density [g/cm^3]	2,595	2,585
Bulk density [g/cm^3]	1,726	1,719
Tightness [%]	0,665	0,665
Porosity [%]	33,488	33,511

The results of tightness, density and porosity tests are shown in Table 2. Sample “A” is a traditional sand-lime

 DOI: 10.1051/e3sconf/20161000040

product, and P1 is a sample containing 1% of polyethylene glycol (characterized by improved compressive strength).In order to determine the pore size mercury porosimetry was used. Modified sand-lime products is characterized by lower content of mesopores than traditional product, which is about 15%. The number of macropores in the sample P1 is about 85%. Small number of pores with low diameter is beneficial, because the high of capillary rising is inversely proportional to the diameter of the capillary. It means that the modified material absorbs less water compared to the standard sample. Volume pore size distribution is shown in Figure 1.

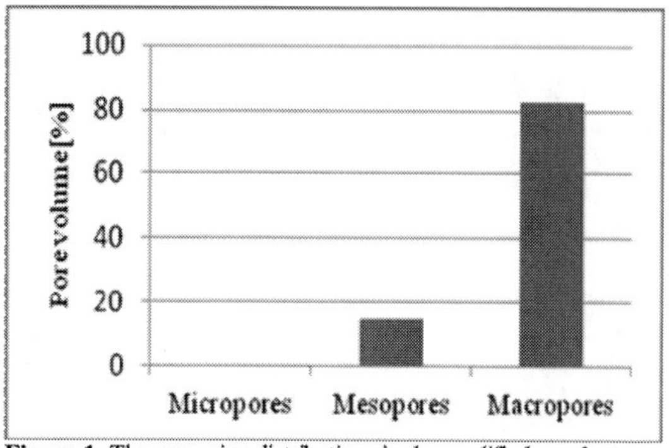

Figure 1. The pore size distribution in the modified product.

SEM observations allowed to point out differences in the distribution, shape and size of pores in tested materials. In a silicate with the addition of polyethylene glycol (Fig. 2) evenly distributed pores with a circular shape are present. Pores in size from 100 microns to 400 microns are visible. Traditional sample (Fig. 3) has smaller pores with an oblong and irregular shape, the size of which is from 50 microns to 350 microns. Microscopy with higher resolution would probably register pores with significantly smaller diameters. Moreover, a disadvantage of this method is a small representative of the sample which is analyzed.

Figure 2. SEM picture of the modified sample.

Figure 3. SEM picture of the traditional sample [12].

Behavior of the material exposed to moisture have been investigated using capillarity (Fig. 4) and absorbency measurements. The results of capillarity (Fig. 5) revealed a significantly lower capillary rising in the samples with the addition of polyethylene glycol. Modified samples absorbed about 60% water less than traditional product.

Figure 4. Examination of capillarity.

The use of a modifier in an amount of 1% or 5% gave a similar result in water adsorption test. With the increase of the substrate in the sample, the absorbency of samples slightly decreased.

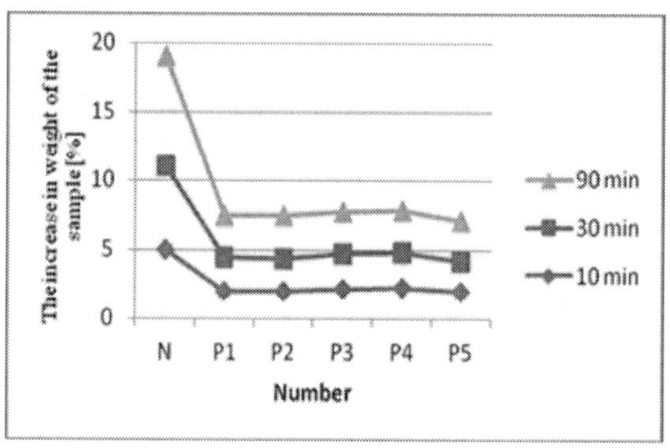

Figure 5. The results of the capillarity test.

Saturation of the material with water depends mainly on the type of the material and type of pores. The study of absorption showed that polyethylene glycol has a positive impact on this feature in the modified products. Decrease in water saturation from 6% to 3% was noted with 4% of polyethylene glycol in comparison to traditional silicate products (Fig. 6).

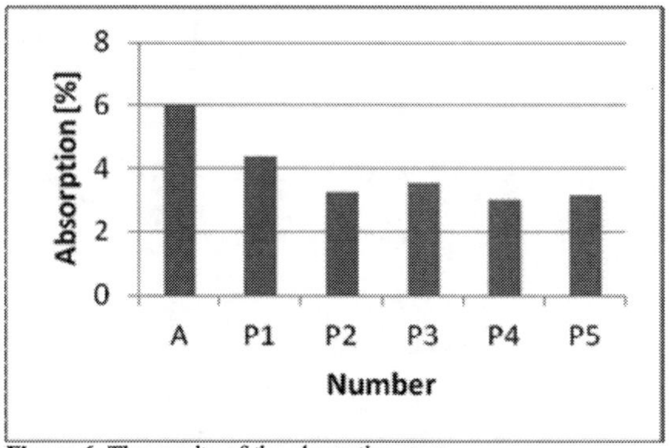

Figure 6. The results of the absorption test

4 Conclusions

1.The addition of polyethylene glycol has a beneficial effect on the properties of the resulting sand-lime products, what depends on the amount of the substrate.
2.Change in the composition of the silicates has improved physical features such as capillarity and water absorption.
3.Studies conducted by using mercury porosimeter showed that the modified silicate products are characterized by reduced number of mesopores and increased number of macropores in comparison to the traditional products.
4.Despite the small differences in the density and porosity of the traditional product and product with the additive, a significant reduction in capillarity and water absorption was observed. This is caused by changes in the pore structure.

References

1. S. Lewowicki, *Zarys technologii materiałów budowlanych* (2000)
2. L. Lichołai, A. Szalacha, *Materiały budowlane i ich badania laboratoryjne* (2005)
3. A. Sieniawska-Kuras, *Tradycyjne i nowoczesne materiały budowlane* (2011)
4. M. Nowek, The 9th International Conference "Environmental engineering, pp 1-6 (2014)
5. R. Dachowski, A. Stepien, Procedia Engineering, **21**, pp. 1173-1178 (2011)
6. PN-EN 1996-2:2010 - Projektowanie konstrukcji murowych -- Część 2: Wymagania projektowe, dobór materiałów i wykonanie murów
7. M. Dziewońska, *Związki wielkocząsteczkowe – polimery* (1995)
8. J. Ciabach, Ochrona Zabytków **36/3-4 (142-143)**, pp. 223-226 (1983)
9. J. Małolepszy, *Materiały budowlane. Podstawy technologii i metody badań* (2008)
10. B. Stefańczyk i inni, *Budownictwo Ogólne*, **1** (2005)
11. PN-EN 772-11 :2011 - Metody badań elementów murowych -- Część 11: Określenie absorpcji wody elementów murowych z betonu kruszywowego, kamienia sztucznego i kamienia naturalnego spowodowanej podciąganiem kapilarnym oraz początkowej absorpcji wody elementów murowych ceramicznych
12. A. Stępień, K. Komisarczyk, Ochrona przed Korozją, **59,** pp.49-51 (2016)

Analysis of briquetting process of sewage sludge with coal to combustion process

Bogdan Kosturkiewicz[1,a] and Andrzej Janewicz[1]

[1] AGH University of Science and Technology, Faculty of Mechanical and Robotics, al. Mickiewicza 30, 30-059 Cracow, Poland

Abstract. Energy recovery from sewage sludge can be achieved by several thermal technologies, but before those processes sewage sludge requires special pretreatment. The paper presents the investigation of the sewage sludge with coal briquettes as a fuel for combustion process. Research is conducted at Department of Manufacturing Systems and Department of Thermal Engineering and Environmental Protection, AGH University of Science and Technology to develop a technology of briquette preparation. The obtained results showed possibility of briquetting of municipal sewage sludge with coal in roll presses, equipped with asymmetric thickening gravity feed system. The following properties were determined for the obtained briquettes: density, drop strength and compressive strength. Based on physical and chemical analysis of prepared briquettes it was confirmed that briquettes have good fuel properties to combustion process. Thermal behaviour of studied sewage sludge and prepared mixture was investigated by thermogravimetric analysis (TG). For the thermo gravimetric analysis (TG) the samples were heated in an alumina crucible from an ambient temperature up to 1000 °C at a constant rates: 10 °C/min, 40 °C/min and 100 °C/min in a 40 ml/min flow of air.

1 Introduction

Article 14 of the Council Directive of 21 May 1991 regarding municipal sewage sludge obliges members of the EU to re-use the waste whenever it is appropriate. According to Statistical Yearbook of the Republic of Poland [1], around 85 thousand tons of dry mass of communal sewage sludge is being thermally converted. According to National plan of operation in range of energy from renewable sources [2], in 2020 over 400 thousand tons (around 57% of the waste) will be converted. In order to achieve such growth, development of new technologies regarding thermal converting waste will be necessary. For several years in Department of Manufacturing Systems, AGH University of Science and Technology, Krakow, research in order to develop new technologies in binding municipal sewage sludge [3-4]. Results confirmed possibility of obtaining briquettes from wastes using roll presses, which are characterised by compressive strength, allowing transporting the raw material and later utilising it in stoker-fired boilers installed in incineration plants. Calorific value of briquetting sewer sludge equalled around 8.7 MJ/kg. Analysis of literature regarding this problem [5] suggests that even though there is no problem in combustion of such sewage sludge in thermal municipal plants (minimal calorific value of wastes allowing autothermal combustion on stoker equals 5800 kJ/kg [6]), calorific value of produced briquette is too low for Polish cement plants. Their technicians claim that calorific value of waste fuel (of which energy is effectively used in process of clinker production) cannot be lower than 12 MJ/kg for solid fuel. Therefore, for example, Polish cement plants demand that minimal average weekly calorific value of sludge should equal 14 MJ/kg and average daily 11,7 MJ/kg. Combustion of sewage sludge in cement industry is effective way for thermal utilization of sewage sludge as well as can reduce the consumption of coal. In order to increase calorific value of alternative fuel, the briquettes of municipal sludge with addition of coal and burnt lime were made. Next, the combustion process with thermal analysis (TA) of obtained briquettes was examined. The advantages of thermogravimetric analysis are its rapid assessment of the fuel value, the temperatures at which combustion starts and ends, and other characteristics, such as the maximum reactivity temperature, amount of ash and total combustion time. Thermal methods, such as Thermogravimetry (TG), Derivative Thermogravimetric Analysis (DTG) and Differential Thermal Analysis (DTA) have been used to study a variety of areas of combustion. Thermal analysis determines a set of methods for studying the selected physical properties of a substance under the influence of temperature or atmosphere [7-10]. As can be observed from the literature, the combustion and pyrolysis processes of sewage sludge from wastewater treatment plants and mixture of sewage sludge and coal are not explained sufficiently,

[a] Corresponding author: kostur@agh.edu.pl

which could be due to the complex composition of sewage sludge and the diversity of the constituents.

The main aim of the study was to determine the possibility of briquetting mixture of municipal sewage sludge, burnt lime and coal. The obtained briquettes can be combustion in cement industry.

2 Experimental

Municipal sewage sludge from Municipal Water and Sewage Company in Poland was used in this study. The municipal sewage sludge from waste water treatment plant contains a lot of organic matter and some heavy metals and pathogenic organisms. The total content of organic matter is 43,7% (Table 1).

Table 1. Chemical analysis of municipal sewage sludge after dewatering process (Magdziarz A. et al.2011).

Denotation	Value
Organic matter, % d.m.	43,7
Nitrogen, % d.m.	2,9
Phosphorus, % d.m.	3,3
Calcium, % d.m.	2,4
Magnesium, % d.m.	0,33
Lead, mg/kg d.m.	75,6
Cadmium, mg/kg d.m.	3,6
Mercury, mg/kg d.m.	0,6
Nickel, mg/kg d.m.	22,8
Zinc, mg/kg d.m.	1183
Copper, mg/kg d.m.	103
Chromium, mg/kg d.m.	36,1

Studied sludge contains a relatively large amount of organic matter, which suggests that it has got adequate fuel properties. Sewage sludge contains, of course, those components that are present in municipal waste, and thus trace metals whose levels may be higher than in the raw waste. Zinc is the most common of the trace elements. There are also copper and lead. It should be noted that in the investigated sludge the amount of harmful heavy metals (Pb, Cd, Hg) is low. It is concluded that in terms of harmful substances per unit of energy, sludge is characterized by its fuel properties suggesting possibility of thermal utilization in industry.

The first step of sewage sludge briquetting process was homogenisation of the material composition. The samples were mixed in a mixer with a preheated water jacket. Taking into account the limitation of heat demand for sludge dewatering and appropriate biostabilization, for better uniting waste burnt lime was added. Mixing time was determined based on the study of amount of material moisture. The moisture value of tested sample was approximately 23.5%. The percentage composition of studied mixture is presented in Table 2.

Table 2. Percentage composition of prepared mixture for briquetting process.

Mixture	Sewage sludge, %	Burnt lime, %	Coal, %
I	45	5	50
II	29.5	3.5	67

Mineralogical characteristics and XRD spectrum of the sample of ash/slag from burning coal particular presented in Figure 1 and Table 3.

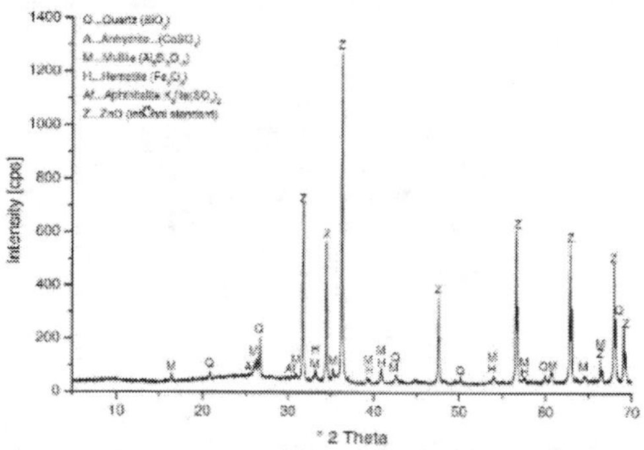

Figure 1. XRD spectrum of the sample of ash/slag from burning coal particular

Table 3. Results of the XRD analysis of the sample "Ash/slag" from burning coal particular

Parameter	Unit	Results
Amorphous	% by mass	77,8
Hematite	% by mass	< 0,5
Mullite	% by mass	17,1
Anhidrite	% by mass	< 0,5
Aphthitalite	% by mass	< 0,5
Quartz	% by mass	4,2

 DOI: 10.1051/e3sconf/20161000041

Prepared sample was fed into the hopper located above the zone of gravitational compaction of roller press LPW 450, which is equipped with forming rings for briquetting process without the parting plane in the shape of a saddle (Fig.2).

Figure 2. Asymmetrical layout compacting system of laboratory roll press LPW 450

It is known that the "saddle" shape of the working surface of the rollers allows high pressure units in the briquetting process. That kind of roller press is adequate for the uniting of difficult materials to briquetting. The amount of fine-grained material was 5 ÷ 7 kg. The sample was united at 0,1 ÷ 0,4 m/s of tangential velocity corresponding to rotational speed 4,25 ÷ 17,0 rpm. The initial value of the interspace between the rolls during briquetting process was d ~ 1,0 mm. The briquettes were obtained in shape presented in Figure 3.

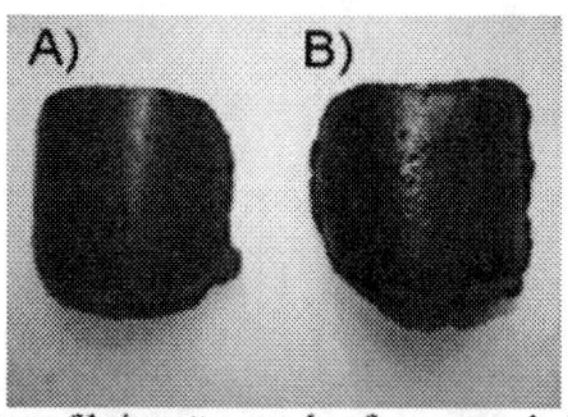

Figure 3. Photo of briquettes made of sewage sludge, lime and coal (A) mixture I, B) mixture II).

After leaving the cavities, the briquettes were stored in a container, from which samples were taken at random (in batches 10 pcs. each) in order to determine parameters characterising their mechanical properties. The following properties were determined for the obtained briquettes: the density (ρ_b), drop strength (K_{168}) and compressive strength (D_{168}).

Considering the loads character the briquettes are endangered to during their reloading, transportation and thermal management, it was decided that indexes that reflect their resistance to destruction at best shall be: compression strength and resistance to drop. Compressive strength tests of briquettes were performed at the measurement stand equipped with ZWICK 1120 press of pressure range 0 ÷ 2000 N. The briquettes were pressed between two parallel planes at rate $v = 0.001$ m/s and the direction of pressure were perpendicular to those planes. The strength tests were performed after 168 h seasoning in order to minimize the influence of atmospheric conditions (humidity, temperature) on the value of obtained results. Drops of briquettes in batches 10 each from height 2 m onto steel plate of thickness 60 mm. This test was performed three times, each time sieving the broken mass at a sieve of mesh 18 x18 mm. Size of sieve was determined to 2/3 of average calculated from two maximum briquette dimensions measured in mutually perpendicular directions.

Thermogravimetric analysis was carried out using a Mettle Toledo TGA/SDTA 851 apparatus. The TGA instrument was calibrated with indium, zinc and aluminium. Its accuracy is equal to 10^{-6} g. For the thermo gravimetric analysis (TG) the samples were placed in an alumina crucible. C.a. 15 mg of sample was heated from an ambient temperature up to 1000 °C at a constant two rates: 10 °C/min, 40 °C/min and 100 °C/min in a 40 ml/min flow of air. The measurements for each sample have to be done under exactly the same condition including e.g. the range of temperatures, atmosphere, heating rate, etc. To determine the most repeatable and precise results. The TG curves represent the instantaneous weight percentage of the tested fuels in relation to the initial weight. The TG curves were used in order to assess the thermal characteristics of studied samples.

3 Results and discussion

Test results were processed statistically according to procedures specified for the estimation carried out by evaluation in points. Based on them authors described the influence of rotational speed on density and strength parameter of briquettes. Right after binding process they presented they showed a plastic material characteristics due to high moisture of samples of approximately 23.5 %. Plastic characteristic prevented determination of raw compressive strength of measured samples.

The obtained briquettes were characterized by a high drop of strength coefficient after the test had been made and after 168 hours had passed. Compressive strength of briquettes was very high (force needed for permanent decohesion exceeded 900 N) (Fig. 4).

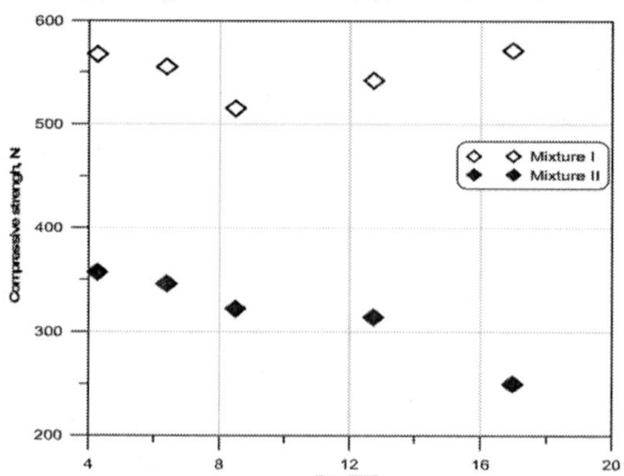

Figure 4. Compressive strength via rotational speed of rollers.

It should be taken into consideration that value of rotational speed of rollers did not cause major changes in briquettes parameters. With increase of rotational speed, strength parameters of briquettes slightly fell. It may suggest that strength parameters are determined by chemical binding rather than physical interactions.

Power consumed during briquetting process N_b was determined by the relation:

$$N_b = 2 \cdot (M_t \cdot n_w)/9.55, \text{ kW} \quad (1)$$

where: M_t - torque moment on a roll press, kNm; n_w - roll press rotational speed, rpm.

Mass performance of roll press equipped with saddle-shaped forming cavity rings W_b was determined by the relation:

$$W_b = 60 \cdot V_b\, i_b\, n_w\, \rho_b, \text{ Mg/h}, \quad (2)$$

where: V_b - briquette volume, m^3; i_b - cavities quantity, i_b= 90; n_w - roll press rotational speed, rpm; ρ_b- briquette density, Mg/m^3.

The density of briquette is one of the most important parameter in such processes. High-density briquette are desirable in terms of transportation, storage and handling. Those technological parameters bring economic benefits. The values of density of studied briquettes are in the ranges 1,30 ÷ 1,41 g/cm^3 for mixture I and 1,28 ÷ 1,35 g/cm^3 for mixture II. Dependence of briquette density via rotational speed of rollers is presented in Fig. 5.

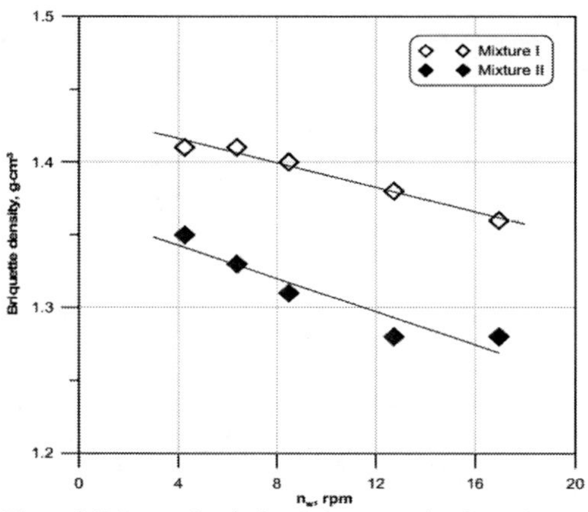

Figure 5. Briquette density in relation to rotational speed of rollers.

Unit energy demand of briquetting process (Z_e) was determined by the equation:

$$Z_b = N_b/W_b, \text{ kWh/Mg}, \quad (3)$$

where: N_b - power consumed during briquetting process, kW; W_b - mass yield of the roll press, Mg/h.

Fig. 6 shows influence of unit energy demand unitary versus rotational speed of rollers during briquetting process.

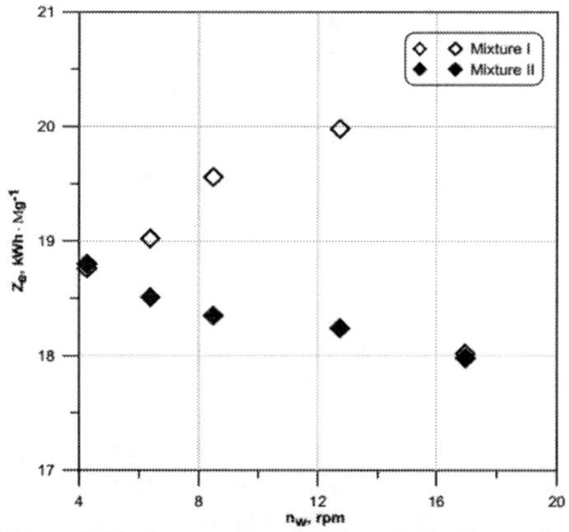

Figure 6. The impact of rotational speed of rolls on unit energy demand during briquetting process.

In this study, the obtained briquettes achieved proper physical parameters such as density, drop and comprehensive strength. The briquettes of sewage sludge and additions in volume 6,5 cm^3 were characterised the following parameters:

- compressive strength (D_{168}) after 168 h in the range 250 ÷ 585 N,
- direct drop strength, K_0 = 98,2 ÷ 99,6 %,
- drop strength after 168 h, K_{168} = 90,2 ÷ 98,2 %,
- briquette density, ρ_b = 1,28 ÷ 1,41 g/cm^3,
- unit energy demand, Z_b = 17,98 ÷ 19,98 kWh/Mg.

The results indicated that the increase of coal amount in mixture and the increase of rotational speed of rollers make loss of briquette strength parameters. The addition of coal to mixture dose not change unit energy demand of briquetting process.

Combustion of sewage sludge especially in the mixture with coal can be attractive option for thermal utilization of sewage sludge. The comparison of combustion process for sewage sludge and mixture I is shown in Fig. 7.

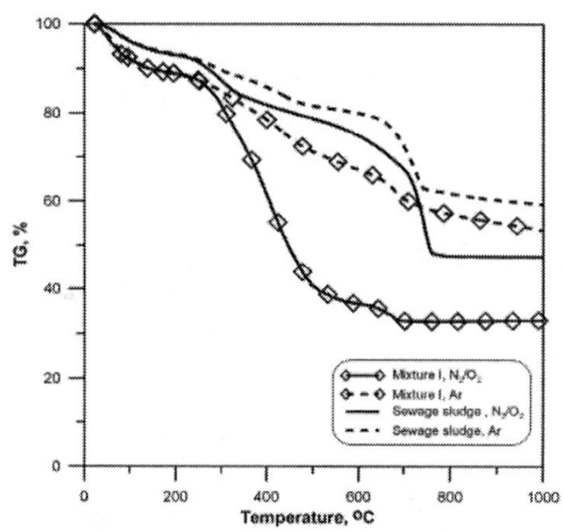

Figure 7. TG curves for mixture I and sewage sludge at 10 °C/min in air.

This figure shows the TG profiles for mixture I and sewage sludge at 10 °C/min heating rate at temperatures up to 1000 °C under air atmosphere. On raising temperature, combustion of studied samples takes place with associated weight loss. The differences between the TG profiles of mixture I and sewage sludge are obvious because of organic and inorganic matter characteristics of the samples. Generally in both cases, they can be divided into three different stages. For the temperature below 300 °C the slight decay of the tested samples weight are due to the drying process and the release of some light volatiles. The weight loss of the studied samples are about 10 %. In the second stage (300 °C ÷ 700 °C) a significant drop in weight of mixture I is observed caused by the thermal decomposition. Once the temperature is higher than 700 °C weight loss is not significant. For sewage sludge the thermal conversion takes place about to 800 °C. Fig. 7 presents the comparison between combustion (N_2/O_2) and pyrolysis (Ar) processes, too. Pyrolysis is the process of thermal decomposition of organic substances in an inert atmosphere. For sewage sludge, pyrolysis process is very complicated because of a wide variety of components. The pyrolysis process of sewage sludge has similar TG curve profile to combustion, and ends before 800 °C. The thermal decomposition is 50 % of total mass at 800 °C for combustion, but for pyrolysis is less than 40 % of total conversion. In the case of mixture I TG curves have not the same profile. The most abundant release of volatile matter during combustion is observed in the temperature range 200 ÷ 500 °C. The weight loss in the this step is about 60 %. At final combustion temperature (1000 °C), the residue is nearly 30 %. The pyrolysis process for mixture I takes place in different temperature ranges and the rate of thermal decomposition is other. The main stage of thermal decomposition occurs to 700 °C. Higher up 700 °C, the small mass loss takes place, too.

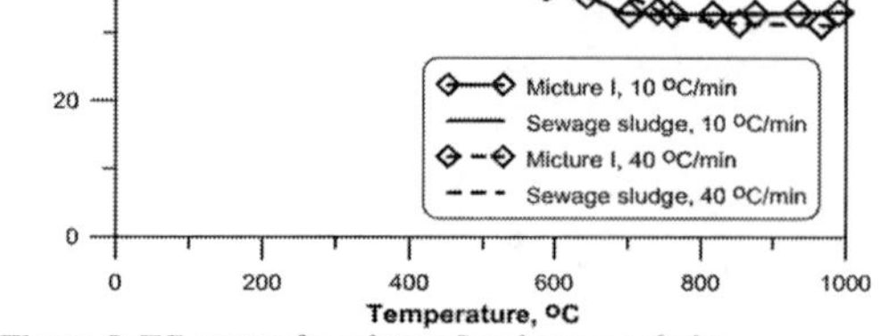

Figure 8. TG curves for mixture I and sewage sludge 10 and 40 and 100 °C/min in air.

It may be conclude that amount of final residue depends on kind of thermal decomposition atmosphere as well as chemical composition of sewage sludge and the addition of coal to blend.

Fig. 8 shows the influence of heating rate (10 °C / min and 40 °C / min) on TG profiles of mixture I and sewage sludge in the air atmosphere.

As we can observe the increase in heating rate does not cause the change in profile curves. The increase of heating rate cause the shift of combustion curve towards higher temperatures. Combustion of sewage sludge and coal briquettes is more efficient than combustion of sewage sludge without and addition, because process can begin in lower temperature range and amount of ash is lower.

Amount of electricity or heat produced by combustion of sewage sludge and coal briquettes E_{SSC} can be calculated by using following formula:

$$E_{SSC} = m_C W_C + m_{SS} W_{SS} - E_Z - E_t - 3{,}6\, m_b Z_e,\ \mathrm{MJ} \quad (4)$$

where: m_C - mass of coal, Mg; W_C – lower heating value of coal, MJ/Mg; m_{SS} - mass of sewage sludge, Mg; W_{SS} - lower heating value of sewage sludge, MJ/Mg; E_Z - electric or heat energy for dewatering process, MJ; E_t - electric energy for preparing of mixture to unite the briquette, MJ; m_b - mass of briquetting material, Mg; Z_e – unit energy demand for briquetting process, kWh/Mg.

4 Conclusion

Accession of Poland to European Union caused numerous changes in waste management. Those changes influenced renewable energy sector [11]. Solutions that helped competitiveness and forced betterment of economical effectiveness (National Plan of Operation in Range of Energy from Renewable Sources, 2010) were introduced. Despite growing significance of ecological methods in municipal sewage sludge in Poland until 2014, huge part of this waste is still being lodged in dumps. In order to counter this process Polish government takes administrative steps which are going to force a discontinuance of this process. One of the main ways to counteract is constant growth of taxes based on environmental harmfulness and dump storage. In addition tax rate depends strongly on moisture of sewage.

The obtained results show possibility of briquetting sewage sludge with waste burnt lime and coal in technological roll press with gravity fed system, equipped with a saddle-shaped cavity forming rings. Usage of this technology can be considered reasonable in wastewater treatment plants localized near agglomerations with at least 100 000 citizens, in other words, where building of special incineration plants is uneconomic. According to National Programme for Municipal Sewage Treatment in 2015 42% of mentioned waste will be produced in those agglomeration in Poland. Sewage sludge briquetting allows for easier transport to the utilization place and addition of waste burnt lime gives opportunity of combustion in cement kilns. This kind of sewage sludge utilization, not only reduces usage of natural resources, but also lowers amount of waste, considered hazardous of the environment in their raw state, thus protects natural environment. Obtained results such as calorific value of sewage sludge and unit energy demand of briquetting process can useful during planning of design municipal power plant.

Acknowledgements

This article has been prepared within the scope of statutory activity, Contract No. 11.11.130.957.

References

1. Statistical Yearbook of the Republic of Poland, Warsaw 2012, 104.
2. National plan of operation in range of energy from renewable sources, Warsaw 2010, 127.
3. B. Kosturkiewicz, B. Tora, Waste Forum, **4**, 385-388, (2010)
4. A. Magdziarz, M. Wilk, B. Kosturkiewicz, Chem. Proc. Eng., **32**, 299-309, (2011)
5. A. Magdziarz, M. Wilk, Energy Conversion Manage., **75**, 425-430, (2013)
6. Resolution no. 217 issued by the Council of Ministers on December 24, 2010, on "National wastes management plan 2014", Official Journal - Monitor Polski, 101, 1183, 2010.
7. L.I. Darvell, J.M. Jones, B. Gudka, X.C. Baxter, A. Saddawi, A. Williams, A. Malmgren, Fuel, **89**, 2881-2890, (2010)
8. A. Skreiberg, O. Skreberg, J. Sandquist, L. Sorum, Fuel, **90**, 2189-2197, (2011)
9. D. Vamvuka, E. Karouki, S. Sfakiotakis, Fuel, **90**, 1120-1127, (2011)
10. A. Magdziarz, S. Werle, Waste Manage. **34**, 174-179, (2014).
11. R. Machnik, B. Karwat, M. Nocuń, J. Niedźwiedzki, Przem. Chem., **94(9)**, 1530-1533, (2015).

Isolation, screening and identification of mercury resistant bacteria from mercury contaminated soil

Anna Kowalczyk[1,4], Magdalena Wilińska[1], Marek Chyc[2,3], Monika Bojko[1] and Dariusz Latowski[1,a]

[1]Department of Plant Physiology and Biochemistry, Faculty of Biochemistry, Biophysics and Biotechnology, Jagiellonian University, Gronostajowa 7, 30-387 Kraków, Poland
[2]Silesian Environmental Doctoral Study, Plac Gwarków 1, 40-166 Katowice, Poland
[3]Department of Environmental Protection, State Higher Vocational School in Tarnów, Mickiewicza 8, 33-100 Tarnów, Poland
[4]Department of Environment Protection, Faculty of Geology, Geophysics and Environment Protection, AGH University of Science and Technology, Mickiewicza 30 30-059 Kraków

Abstract. New bacterial strains resistant to high concentration of mercury were obtained and characterized focusing on their potential application in bioremediation. The biological material was isolated from soil contaminated with mercury. The ability to removal of Hg from the liquid medium and the effect of the various pH and mercury concentrations in the environment on bacterial strains growth kinetics were tested. The selected strains were identified by analysis of the 16S ribosome subunit coding sequences as *Pseudomonas syringae*. The analysis of Hg concentration in liquid medium as effect of microbial metabolism demonstrated that *P. syringae* is able to remove almost entire metal from medium after 120 hours of incubation. Obtained results revealed new ability of the isolated strain *P. syringae*. Analyzed properties of this soil bacteria species able to reduce concentration of Hg or immobilize this metal are promising for industrial wastewater treatment and bioremediation of the soils polluted especially by mercury lamps scrapping, measuring instruments, dry batteries, detonators or burning fuels made from crude oil, which may also contain mercury. Selected bacteria strains provide efficient and relatively low-cost bioremediation of the areas and waters contaminated with Hg.

1 Introduction

Nowadays energy industry brings numerous benefits for humans, but on the other hand it also carries the risk of natural environment pollution. One of the most hazardous pollutants associated with energy industry is mercury. Even though this element is natural component of environment and forms variety of species revealing different chemical features and level of toxicity, the progress of energetic technologies causes, that the concentration of Hg in environment significantly increases [1, 2]. The main anthropogenic source of Hg emission to air is fuel combustion both in power plants and in residential or commercial boilers [3].

Burning of fossil fuels releases to the atmosphere more than 800 t Hg per year and it is thought to be the most important anthropogenic source of this pollutant. Generally, it is thought that annually the global Hg emission due to fossil fuels burning is between 374 t and 1121 t and more than 90% of Hg coming from this kind of fuels is released to the atmosphere as a result of stationary combustion [4]. The content of Hg in the fuel as well as level of Hg removal from exhaust gases formed during fuels burning are important factors effecting on Hg emission while combustion process. Among the fossil fuels main source of Hg emitted to atmosphere is coal but also combustion of another sorts of fuels like oil, wood or even biofuels connected with power or heat production causes Hg emission. It is worth to consider that not only stationary combustion, but also mobile sources can cause increase of Hg content in the atmosphere. Although it was shown that Hg emission during fuels combustion by cars is relatively low if compared to other sources of this pollutant, it should be mentioned that long-range atmospheric transport can be also important factor increasing air pollution by Hg. Another Hg point sources connected with energy production are batteries, fluorescent lamps, wiring devices or electrical switches, which together with various measuring and control instruments cause the rise of the Hg emissions into environment. The intensifying use of energy since the Industrial Revolution has increased the Hg content in the air, water and soil. In 1990 global Hg emission was about 1800 t, in 1995 it was more than 2200 and in 2005 some results shown value close to 3000 t. Additionally, in 2005 it

[a] Corresponding author: dariusz.latowski@uj.edu.pl

was postulated that more than 60% of the global mercury emission were generated by energy production [5]. The level of Hg emission varies greatly dependently on continent or even country [6]. If controls of Hg emission to environment will not be extended worldwide, the rise of the mercury emission is predicted. In the environment, mercury can exist in the form of molten metal, mercury bound in natural minerals, vaporous mercury in the atmosphere (HgO), as well as in the form of ions particularly in water and soil. Most of Hg forms are highly toxic even at low concentrations, particularly for fetuses and children [7, 8]. Even though detection of Hg emission in 2013 was about 1000 t lower than in 2005, mercury still seriously threatens life on our planet and as a strong pollutant can exist in the air more than 350 years. It was also shown that Hg in soils is much more persistent than in fresh or sea water ecosystems or even in biomes [1]. What is worse, mercury can relocate among long distances through different ecosystems and can easily penetrate various components of biosphere. In this way, mercury mainly from water and soil, entering to food chains strongly degrades ecosystems and finally seriously endangers to humans health [9]. Accordingly, it is urgent need for rapid introduction of methods for efficient removal Hg from the environment components or to transform it into its most stable and less toxic forms *in situ* [10, 11]. Among numerous remediation techniques applying to neutralize of Hg contamination in environment [12, 13] one of the most promising method is bioremediation with mercury tolerant microorganisms. In this paper screening, identification and characteristics of six new high mercury concentration resistant *Pseudomonas* strains will be presented.

2 Materials and methods

2.1 Soil sampling

Five soil samples were collected from the top layer of the ground of 10 cm depth on mercury contaminated area of the abandoned electrolysis factory. The area was not remediated after cessation of the factory activity in 2006. The sampling points intervals were regular and about 50 m. The original samples were then homogenized to prepare the representative laboratory sample. The concentration of mercury in the soil sample was determined by cold vapour atomic absorption spectrometry (CVAAS).

2.2 Microorganisms isolation

3 g of representative soil sample was suspended in 75 ml of sterile 0.9% NaCl solution and shaken for 48 h (150 rpm, 20°C). 300 µl of the extract was inoculated to agar plates with Luria Bertani broth medium (LB) (NaCl 10 g/L; peptone 10 g/L; yeast extract 5 g/L) supplemented with $HgCl_2$ to reach the final pure mercury concentration: 0.03; 0.50; 1.00 and 1.50% (w/v) in the first step and 0.001; 0.002; 0.004; 0.008; 0.015; 0.020; 0.025% (w/v) in the second step of the experiment. Medium without Hg was used as a control. The plates were incubated for 48 h, 20°C. The colonies grown on medium with highest concentration of Hg were seeded on streaking plates with LB supplemented with Hg. Single colonies distinguished basing on morphology were inoculated to liquid LB and incubated with shaking (150 rpm, 20°C) 4 days, and then streaking plates were prepared. The procedure was repeated 4 times and resulted in obtaining pure strains. Isolated strains were stored on LB supplemented with 0.01% (w/v) Hg agar plates and liquid cultures were stored with 70% glycerol solution (4:1) under -80°C. Six selected pure strains were stained by Gram staining and analyzed by microscope (Nikon Eclipse TE200 Inverted Fluorescence & Brightfield Microscope).

2.3 Bacteria identification

DNA was extracted from selected six 4-days liquid LB cultures supplemented with 0.01% Hg, marked 12.4, 15.5, 16.2, 16.7, 17.1 and 17.5 according to phenol-chloroform extraction method. The concentration of the extracted material was measured at 260 nm wavelength. The purity of isolated DNA was estimated as a 260 nm/280 nm measurements ratio together with its concentration by a nano-drop spectrophotometer (Implen NanoDrop). DNA was amplified with PCR on thermal cycler (Bio-Rad S1000™ Thermal Cycler) with standard 8F (5'-AGAGTTTGATCCTGGCTCAG-3') and 1492R (3'-GGTTACCTTGTTACGACTT-5') primers under following conditions: 95°C, initial denaturation, 15 min; 95°C subsequent denaturation, 1 min; annealing, 54°C, 1 min; elongation 72°C, 2 min; repeated 30 times; 72°C final elongation, 10 min. The PCR product was subjected to ethidium–bromide agarose gel electrophoresis separation (20 min, 80 V) and purified with SynGen GEL/PCR Mini Kit according to the attached protocol. The isolated, amplified and purified DNA samples of 6 selected strains were sequenced. The data set were aligned using Chromas Lite Version 2.1.1 and BioEdit 7.2.5 and identified basing on V-4 hypervariable region with BLAST database, determining the genus and species by the highest similarity.

2.4 Growth kinetics dependence on Hg concentration

4-days liquid LB without Hg cultures of six selected isolates were inoculated to 15 ml of liquid LB medium supplemented with different concentrations (% w/v) of mercury: 0.002, 0.005, 0.01 and without Hg (control) to obtain initial optical density (OD) at 600 nm between 0.01–0.08. The cultures were incubated with shaking (200 rpm) in 37°C.

The growth kinetics was measured by OD measurements at 600 nm wavelength (OD_{600}) using Metertek SP-830 Spectrophotometer every 2 hours till cultures reached initial stage of death phase after 45 hours on average.

2.5 Growth kinetics dependence on pH

4-days old cultures in liquid LB supplemented with 0.01 (% w/v) Hg were inoculated to 15 ml of LB medium solved in ammonium acetate–acetic acid buffer of pH 3, 5, 6, 7 and 8 supplemented with 0.01% (w/v) of Hg and of pH 3, 5, 6, 7 and 8 without Hg. The initial OD_{600} value was in the range of 0.01–0.02. The cultures were inoculated in triplicate. The OD_{600} measurements were carried on every 4 hours till cultures reached initial stage of death phase, i.e. after 50h on average. The pH influence on growth kinetics was presented as specific growth rate of bacteria culture (μ) calculated according to following equation:

$$\mu=\Delta lnOD_{600}/\Delta t, \quad (1)$$

where t is time (hours).

2.6 Mercury removal by the bacterial isolates

Samples to determine the mercury removal by isolated strains were prepared as follows: the 4-days liquid LB cultures without Hg were inoculated to 20 ml of LB 0.01% (w/v) Hg medium and cultured with shaking (200 rpm) in 37°C. The cultures were prepared in triplicate and the initial OD_{600} values were between 0.01–0.08. 10 ml of cultures were collected for analyses of Hg concentration after 12 and 120 h. The samples were centrifuged (5500 rpm, 20 min, 4°C). Supernatant was collected and stored under -20°C before analysis. An Automated Mercury Analyzer MA-2000 (Nippon Instruments Corporation) was used for total mercury determination in examined samples. The method is based on the thermal release of mercury vapours (Hg^0) from solid or liquid materials and capturing it through the amalgamation process. After the thermal release of atomic mercury vapours from the amalgam, the atomic absorption measurement was performed in the measurement vessel at the wave length of 253.7 nm (measuring the range 0–1000 ng/sample, limit of detection LOD-0.002 ng). The samples were combusted at 850°C in filtered dry air. The main advantage of used method is its ability to process samples without conducting sample mineralization. The correctness of the measurements was verified by means of standard reference material.

3 Results

3.1 Isolation and identification of microorganisms

The analysis of contaminated soil samples, the source of isolated bacteria performed by cold vapour atomic absorption spectrometry demonstrated that the content of mercury is 153±23 mg/kg d.m. (1.5% w/w). The result of this analysis was the basis for selection of mercury concentrations to isolate bacteria possessing the strongest resistance to this pollutant.

At the first stage of the experiment the highest concentration of the mercury in selective medium was similar to the concentration in collected soil and was 1.5% (w/v) and the lowest one was 0.03% (see part 2.2). Despite of bacteria growth on the control plates (not supplemented with Hg) there were no observed any bacterial colonies on Hg containing media. Therefore in the next stage of bacteria isolation the concentrations were reduced to the range 0.001–0.25% (w/v) (see part 2.2). Under provided conditions the bacteria cells grew on every selected concentrations, but after several passages they lost capability of growth when treated with concentrations of Hg higher than 0.015% (w/v).

The result of applied procedures was obtaining of 54 pure bacterial cultures differing with morphology and type of growth on LB-agar medium. Six strains showing the most visible morphological differences and resistance to highest content of mercury in medium were selected to research. Gram staining and microscope analysis demonstrated, that all selected isolates belong to Gram negative bacteria and demonstrate features characteristic for *Pseudomonas* genus. The identification of all isolated strains by V-4 hypervariable region highest similarity comparing to GenBank data determined the genus as Pseudomonas (*P. syringae*, access no NC_005773.3, GenBank), despite the morphological differences observed for particular cultures. Isolates were marked by following numbers: 12.4, 15.5, 16.2, 16.7, 17.1, 17.5.

3.2 Growth kinetics in presence of Hg

The characterized *Pseudomonas* strains cultured in medium without Hg did not demonstrate significant differences in growth kinetics (Fig. 1A). The increase of mercury concentration in the medium effected in differentiation of cells growth, the more significant, the higher concentration of Hg (Fig. 1B-D).

The first difference observed just at the lowest concentration of 0.02% was that the strains 16.7 and 17.5 demonstrate the fastest growth, but probably 16.7 initiates cells division earlier in comparison to 17.5 under given conditions. 3 hours after inoculation there was visible enlargement of 16.7 strain population, whereas all other strains show similar increase from 5th hour of incubation and the more decelerated, the higher concentration of Hg in the medium is. There was also observed, that the kinetics of 12.4, 15.5, 16.2 and 17.1 growth are similar, which is clearly visible especially in cultures supplemented with 0.005%

(w/v) of Hg (Fig. 1C). In case of cultures supplemented with 0.01% (w/v) of Hg the diversity of growth is the most visible and allows to put the strains according to the growth rate in a following order, correlated with the decreasing resistance to mercury: 16.7>>15.5=17.5>16.2=17.1>12.4 (Fig. 1D).

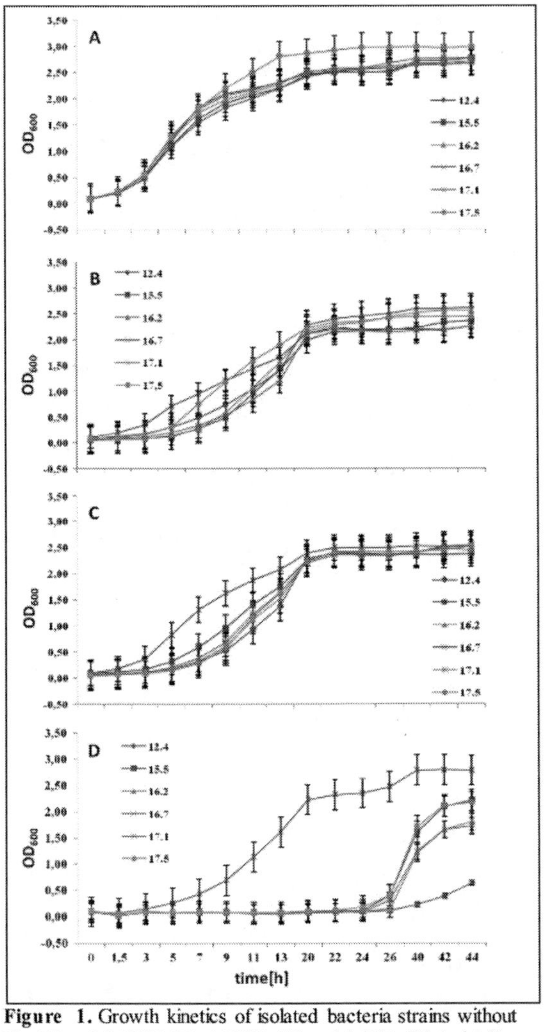

Figure 1. Growth kinetics of isolated bacteria strains without Hg (A) and 0.002% (B), 0.005% (C) and 0.01% (D) (w/v) Hg.

3.3 Mercury removal

The capability of mercury removal by investigated strains was analyzed after 12 and 120 hours of incubation (Tab.1). After 12 h there are no observed differences of mercury content in the medium, whereas after 120 h the removal of Hg was over 99% in case of strains 15.5, 16.7, 17.1 and 16.2. The order of bacterial growth rate is related to the level of mercury removal from medium, which is the higher, the faster growth demonstrate studied *Pseudomonas* strains. Only 17.5 strain growing as well as 15.5 strain has shown the lowest removal of mercury from medium. It was even lower than 12.4 strain characterized by weaker growth of the culture.

Table 1. Mercury removal by isolated strains. Initial concentration of Hg was 0.01% (w/v)

Strain	Time [h]	
	12	120
12.4	0.0016	0.000150
15.5	0.0022	0.000049
16.2	0.0020	0.000089
16.7	0.0018	0.000050
17.1	0.0021	0.000057
17.5	0.0024	0.000200

3.4 Effect of pH on bacteria growth

In pH 3.0 and pH 5.0 no growth of investigated strain cultures either in medium supplemented with mercury or without presence of this element was observed. There were also no significant differences of growth kinetics dependent on pH among selected strains (Tab. 2).

Table 2. The effect of different pH values on specific growth rate (μ) of selected strains treated and untreated with Hg

Strain No	Control (without Hg) [x10^{-3}/h]		
	pH 6.0	pH 7.0	pH 8.0
12.4	10.00±0.61	43.89±0.51	43.83±0.60
15.5	11.00±0.41	43.92± 0.21	43.97±0.45
16.2	10.52±0.47	43.87±0.45	43.85±0.55
16.7	11.06±0.51	43.91±0.54	43.99±0.51
17.1	11.03±0.47	43.85±0.39	43.88±0.45
17.5	10.50±0.51	43.89±0.45	43.85±0.53
Strain No	**With 0.01% (w/v) Hg [x10^{-3}/h]**		
	pH 6.0	**pH 7.0**	**pH 8.0**
12.4	4.20±0.05	19.10±0.50	19.18±0.45
15.5	5.11±0.04	21.78±0.51	22.21±0.53
16.2	4.60±0.06	21.24±0.45	20.99±0.47
16.7	5.11±0.05	24.21±0.47	25.01±0.53
17.1	5.02±0.05	21.45±0.55	21.65±0.39
17.5	4.91±0.03	21.59±0.35	22.12±0.45

The optimum pH values for all strains were pH 7.0 and pH 8.0, in which the bacterial growth was about 80% more efficient than at pH 6.0. In presence of Hg the specific growth rate (μ) of particular strains was almost two times lower than without Hg, but the ratio of μ in case of buffered pH 7.0 and non-buffered LB medium was close to 1. This result indicates, that the growth-limiting factor is the presence of toxic metal, not the pH value. The higher pH values have not been tested, as the Hg compound precipitated in more alkaline conditions.

4 Discussion

Performed experiments resulted with isolation and preliminary characteristics of mercury resistant *Pseudomonas* strains. The source of bacteria was soil contaminated with high concentration of Hg

(1.5% w/w). Under laboratory conditions this concentration occurred to be extremely toxic for selected microorganisms to adapt. During the progress of isolation procedure the highest concentration of Hg permanently tolerated by the most resistant bacteria was 0.015, which was 100 times lower than in soil. Admittedly, at the beginning bacteria growth was observed on media with 0.025% Hg (w/v) but in successive passage no growth of these strains was observed. One of the putative reason for this result is fact that the chemical and biological composition of the soil provide another conditions than laboratory medium. Mineral compounds as well as edaphon may decrease the bioavailability of Hg present in the soil. It is also possible that the provided laboratory conditions of culturing were not optimal for investigated strains. The other explanation is that separation of single strains functioning in soil as a consortium so far, resulted with increase of sensitivity to Hg. It is possible that in natural environment they derive benefits from interrelationships [14, 15], what will be researched in future experiments.

The analysis of growth kinetics of selected strains cultured with different mercury concentrations demonstrated that one strain (16.7) possesses more efficient mechanisms of resistance in comparison to the another isolates (Fig. 1A-D). It is reported several mercury resistance mechanisms such as methylation of mercury [16-18], involving methylation proceeded by bacteria excreting methylcobalamin [17], Hg^{2+} to Hg^0 reduction by reductase encoded by *mer* operon possessed by numerous bacterial strains [19-24], presence of plasmids providing resistance to another heavy metals, correlated with antibiotics resistance [17], precipitation in form of HgS [25] and the synthesis of thiols or another chelates binding the Hg compounds [17, 26]. All of these mechanisms work efficiently and decrease mercury toxicity to the cells.

While analysis of the effect of Hg concentration on the growth kinetics of particular strains it was noted, that even in case of the highest concentration of mercury, i.e. 0.01% (w/v), there are no differences between strains 15.5 and 17.5 as well as 16.2 and 17.1 (Fig. 1D). Basing on that observation it may be concluded, that they are the same strain, but on the other hand when consider the Hg removal from the medium after 120 hours of incubation, in case of 15.5 the content of Hg was more reduced than in case of 17.5, which removed Hg less efficiently even than strain 12.4 presenting the weakest growth. For unmistakable explanation if selected bacteria are the same or different strains further studies on Hg resistance molecular mechanism are needed. One of the possible mechanisms may be mercuric reductase activity. The pH optimum for activity of the enzyme is 8 [27], what corresponds with pH value favourable for bacteria growth, which was observed while incubation of investigated isolates (Tab. 2). However, the optimal pH is the same (pH 8.0) for cultures independently if they are treated or untreated with mercury. Therefore it cannot be concluded that the most efficient cells growth at pH 8.0 is a consequence of mercuric reductase detoxification activity.

The selected strains characterizes strong resistance to high concentrations of mercury and by their activity contribute to Hg removal from the solution (Tab. 1). The results and conclusions of carried experiments are promising for use of *P. syringae* in bioremediation of areas contaminated with mercury. The next stage of research in this field is to investigate all 54 isolated strains and exploring the mechanism of their resistance to mercury, focusing on utility for neutralization the harmful effect of Hg on environment.

References

1. J. Xu, A.G. Bravo, A. Lagerkvist, S. Bertilsson, R. Sjöblomb, J. Kumpiene, Environ. Int. **74** (2015)
2. U.C. Branch, *The global atmospheric mercury assessment: sources, emissions and transport* (Geneva: UNEP-Chemicals, 2008)
3. E.G. Pacyna, J.M. Pacyna, K. Sundseth, J. Munthe, K. Kindbom, S. Wilson, F. Steenhuisen, P. Maxson, Atmos. Environ. **44**, 20 (2010)
4. N. Pirrone, S. Cinnirella, X. Feng, R.B. Finkelman, H.R. Friedli, J. Leaner, R. Mason, A.B. Mukherjee,G.B. Stracher, D.G. Streets, K. Telmer,Atmos. Chem. Phys. **10** (2010)
5. E.G. Pacyna, J.M. Pacyna, F. Steenhuisen, S. Wilson, Atmos. Environ. **40** (2006)
6. P. Holmes Sci. Tota.l Environ. **408**, 2 (2009)
7. P. Nance, J. Patterson, A. Willis, N. Foronda, M. Dourson Regul. Toxicol. Pharmacol. **62**, 3 (2012)
8. M. Gabriel, Environ. Geochem. Hlth. **26**, 3–4 (2004)
9. W.W. Cui, X.L. Wang, J.J. Duan, M . Yang, W. Zi, X.C. Liu, Guizhou. Agric. Sci. **7** (2011)
10. B.V. Tangahu, Sheikh S.R. Abdullah, H. Basri, M. Idris, A.N. M. Mukhlisin, Int. J. Chem. Eng. **2011** (2011)
11. P. Randall, S. Chattopadhyay, J. Hazard Mater. **114**, 1 (2004)
12. R.B. Richter, H. Flachberger, BHM **155**, 12 (2010)
13. O. Rodríguez, I. Padilla, H. Tayibi, A. López-Delgado, J. Environ. Manag. **101** (2012)
14. M. Ledin, Earth Sci. Rev. **51** (2000)
15. A.R. Sprocati, C. Alisi, L. Segre, F. Tasso, M. Galletti, C. Cremisini, Sci. Total Environ. **366**, 2-3 (2006)
16. M. Pepi, C. Gaggi, E. Bernardini, S. Focardi, A. Lobianco, M. Ruta, V. Nicolardi, M. Volterrani, S. Gasperini, G. Trinchera, P. Renzi, M. Gabellini, S. E. Focardi, Int. Biodeter. Biodegr. **65**, 1 (2011)
17. J.B. Robinson, O. H. Tuovinen, Microbiol. Revs. **48**, 2 (1984)
18. N. S. Reyes, M. E. Frischer, P. A. Sobecky, FEMS Microbiol. Ecol. **30** (1999)
19. D.H. Nies Appl Microbiol Biotechnol 51 (1999)

DOI: 10.1051/e3sconf/20161000042

20. T. Barkay, S.M. Miller, A. Summers, FEMS Microbiol. Rev. **27** (2003)
21. T. Barkay, I. Wagner-Döbler, Adv. Appl. Microbiol. **57** (2005)
22. A.O. Summers, J. Bacteriol. **174** (1992)
23. M. Narita, K. Chiba, H. Nishizawa, H. Jshij, C.C. Huang, Z. Kawabata, S. Silver, G. Endo, FEMS Microbiol. Lett. **223** (2003)
24. S. Silver, J.L. Hobman, Microbiol. Monographs. **6** (2007)
25. D.W. Boening, Chemosphere 40 (2000)
26. S. Silver, L.T. Phung, J. Ind. Microbiol. Biotechnol. **32** (2006)
27. P. Giovanella, L. Cabral, F.M. Bento, C. Gianello, F.A.O. Camargo, New Biotechnol. **33**, 1 (2016)

E3S Web of Conferences **10**, 00043 (2016) DOI: 10.1051/e3sconf/20161000043
SEED 2016

Protection of historic buildings against environmental pollution of vibrations

Alicja Kowalska-Koczwara[1,a] and Krzysztof Stypuła[2]

[1] *Cracow University of Technology, Cracow, Warszawska 24 St., 31-155, Poland*
[2] *Cracow University of Technology, Cracow, Warszawska 24 St., 31-155, Poland*

Abstract. Historic buildings in Poland are largely neglected objects that small percentage is preserved in its original form. Unrelenting in the case of historic buildings is a time that brings with it the natural processes of aging of buildings, but also the history of the object which is often marked by military conflicts, fires or even incompetently carried out reconstruction. Nowadays historic buildings are also destroyed by the rapid development of infrastructure and residential construction. This development could lead to changes of water in the soil, make changes in the geological structure or cause exposure of the historic building to the new influences (eg. traffic vibrations), to which building has not been subjected so far. Vibrations are often omitted in environmental issues, although the protection against noise and vibration has its place in the Law on Environmental Protection. This article presents the methodology for the measurement and interpretation of vibration influence on historic buildings and the assessment methods of technical condition of historic building on the example of dynamic measurements made on St. Nicholas Church in Krakow. The importance of well-done crack-by-crack documentation and characterization of damages based on damage index is shown. Difficulties that can be encountered when determining the causes of technical condition of historic buildings are also shown. Based on the example of the Church in the article are also given the possible protection solutions of historic structures from vibrations.

1 Introduction

Bad condition of preservation part of the historic resources in Poland does not require a deeper explanation. The owners of historical buildings, despite of their best intentions often do not have the funds for their renovation. Time for historic buildings plays a key role. This time is closely connected with environmental influences. Environmental influences are often mistakenly identified only with air or/and water pollution. Meanwhile, air pollution [1, 2] and water [3], although it may have a significant impact on historical buildings, is not the only environmental pollution, which should be taken into account in the context of protection of historical monuments. According to art. 3 in Polish Law of Environmental Protection [4] noise and vibration are also pollution against which people and buildings should be protected. In the era of fast development of road and rail infrastructure this type of pollution can have a decisive impact on the durability of historic buildings. Unfortunately this aspect is often unnoticed or neglected.

On the effectiveness of the legal protection of historical buildings the negative influence have a lot of factors. One of them is the absence of policy adopted in the field of legal protection of historical buildings, which significantly impair the action of the administration and conservation. Another aspect of legal protection is absence of engineering standards on protection of this specific types of buildings which are historical buildings. In this matter we can learn from our neighbours across the western border, who in a very transparent way describe the treatment of historic buildings in the event of vibration pollution [5]. German standard [5] apart from the whole group of buildings three categories for which vibration influences should be evaluated:

- commercial buildings, industrial buildings and building of the similar design,
- residential buildings (dwellings) and building of the similar design or occupancy,
- structures that, because of their particular sensitivity to vibration, cannot be classified as the first or second group of buildings, and are of great intrinsic value, for example listed buildings under preservation order.

Historic buildings are then considered acc. [5] in the third group of buildings, which has special conditions much stricter then first two groups. In the draft of the new Polish standard on influence of vibration on structures [6] also can be seen the regulation relating to historic buildings. This regulations says that in the case of historic buildings and buildings with valuable wall paintings, stucco, etc. through lowering the zone boundaries by assigning zone II to zone III effects.

[a] Corresponding author: akowalska@pk.edu.pl

The scope of the performance of expertise of historic buildings is described in [7]. In article [7] seven main point of historic building expertise are described. This points are as follows:

- analysis of history of the building on the background of the history of the village and the neighbouring buildings (archival query, historic photos, archaeological research etc.),
- description of the existing state (plans, cross-sections, constructional inventory, crack by crack documentation),
- "in –situ" measurements depending on the requirements (for ex. research of geotechnical conditions, research of cracks of the structure, examination of the humidity and salinity walls etc.),
- engineering calculations (static and / or dynamic) including verification of the existing structure,
- data analysis that will help in derivation of conclusions about the causes of damage,
- determining the scope of repair work, to remove or reduce unwanted influences,
- specify, where appropriate, specific solutions (including drawings and solutions supported by static calculations).

These seven point should be always considered in the case of historic buildings, regardless of the cause of existing damage. All these points could be found in well-done measurements made on historic buildings no matter whether it is a construction: timber-framed [8], masonry tower [9, 10] or masonry building [10]. In this article detailed measurements and interpretation methodology of expertise of historic building subjected to dynamic influences is presented. Two main situations will be considered: when dynamic source of excitation exists and when it doesn't. The importance of well-done crack-by-crack documentation in according to assessing of the state of cracked structure is also presented and supported by the results of expertise [11].

2 Measurement and interpretation methodology

In the diagnosis and design of buildings, including the influence of vibration can be divided into five cases. Diagnostic and design cases are based on the state in which the building (design, exploitation) and the source of vibration (plan, exploitation) is. In the cases of historic buildings we have got two possible situations: when the source of vibration exists and when it is in design phase, because historic building has already existed. These two cases according to [12] could be named as follows: diagnosis (vibration source exists) and diagnosis with forecast (vibration source is in design phase). These cases require a different approach to the subject. That is why they will be described in separate subsections.

2.1 Methodology in diagnosis situation of historic building

Diagnosis of vibration influence on historic building could be divided into following steps:

- historical analysis,
- crack-by-crack documentation with photos and cracks inventory,
- "in- situ" measurements of the dynamic influence on historic building from the existing vibration source,
- assessment of vibrational influence on historic building based on vibration results (if the historic building is subjected by standard regulation, ex. SWD scales acc. [6])
- assessment of vibrational influence on historic building based on numerical calculations (if the historic building is not subjected by standard regulation for ex. on the basis of the adopted percentage of material effort of construction).
- proposal of technical solutions to reduce vibrations in the case if the measured values (according to assessment criterion) can influence on the accelerated wear or damage to the structure of the historic building.

2.2 Methodology in diagnosis with forecast situation of historic building

Diagnosis with forecast differ from diagnosis in measurement situation because "in-situ" measurement cannot be done because source of vibration does not exist. "In-situ" measurements should then be replaced by so called dynamic background measurements. First of all this measurement will diagnose if existing sources of vibration influence on structure of historic building. Secondly these measurements will serve as an excitation force to validate the FEM model of the historic building.

So diagnosis with forecast will include the following steps:

- historical analysis,
- crack-by-crack documentation with photos and cracks inventory,
- dynamic background measurements,
- assessment if existing sources of vibration influence on the structure,
- creating FEM model based on results from background measurements,
- validated FEM model must then be forced by forecasted excitation taken from the similar situation,
- assessment of vibrational influence on historic building based on numerical calculations (based on standard regulation or based on the basis of the adopted assessment criterion)
- proposal of technical solutions to reduce vibrations

In the case of diagnosis with forecast the FEM model is the most important. Good FEM model should be validated according to modal analysis and to background measurement results.

Assessment criterion should be assumed basing on whether historic building is subjected to standard regulation. For example if it is subjected to SWD scales then according to Polish standard [6] basing on vibrogram received from FEM model, analysis in 1/3 octave band could be made. If the historic building is not subjected to SWD scales then additional criterion should be adopted. It could be for example percentage of material effort derived from Huber-Mises-Hencky criterion.

3 Consideration of cracks in the methodology for assessing the influence of vibration on historic buildings

Historic buildings are often in bad condition and have many scratches and cracks or another damages in their structure. That is why there are some steps in above mentioned methodology which deserve special attention because they can effect on the proper performance of evaluation.

3.1 Crack-by-crack documentation

Examination of buildings scratches should include an inventory of scratches and cracks on internal and external elements of the object and possible observation of their stability.

During the inventory of scratches it is important to determine whether visible damage is only scratch on a plaster or is it already a crack in the structural element. Often to determine this fact it is necessary to remove the plaster, which of course in the case of historic buildings may be impossible. In [7] authors proposed classification of scratches which is as follows:

- scratch - visible element discontinuity of small length and usually to 0.5 mm width across,
- slot - scratch that has width across greater than 0.5 mm,
- crack - inclined slot of a significant length (eg. through the entire length of the wall).

It is noticeable that the above mentioned classification may be useful in assessing the state of scratches, but it should be remembered that in historic buildings determining what is a scratch and what is a slot can be troublesome and often impossible. What is important is to distinguish between scratch and crack because cracks can change static scheme of the structure. Damage which is identified as a crack is usually visible inside and outside of the building wall (ex. fig. 1. and fig. 2).

Assessment of damages in historic buildings is often difficult by the lack of direct access to damage, eg. to the vault in the historic church. It is therefore important experience and curiosity of people performing an inventory of damages in such buildings. In the cases of difficulties in access to damage modern technologies can be useful, for example: digital photogrammetry, 3D

Figure 1. Crack visible on the north elevation of the St. Nicholas Church in Krakow.

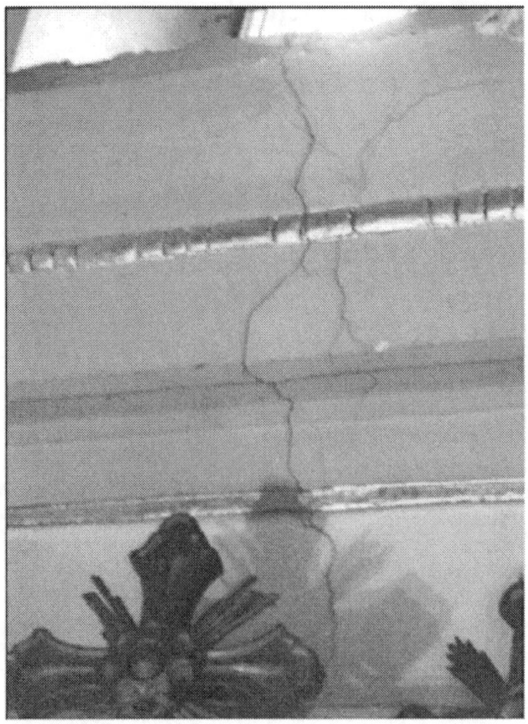

Figure 2. The crack on the wall of the north side nave of the St. Nicholas Church in Krakow.

DOI: 10.1051/e3sconf/20161000043

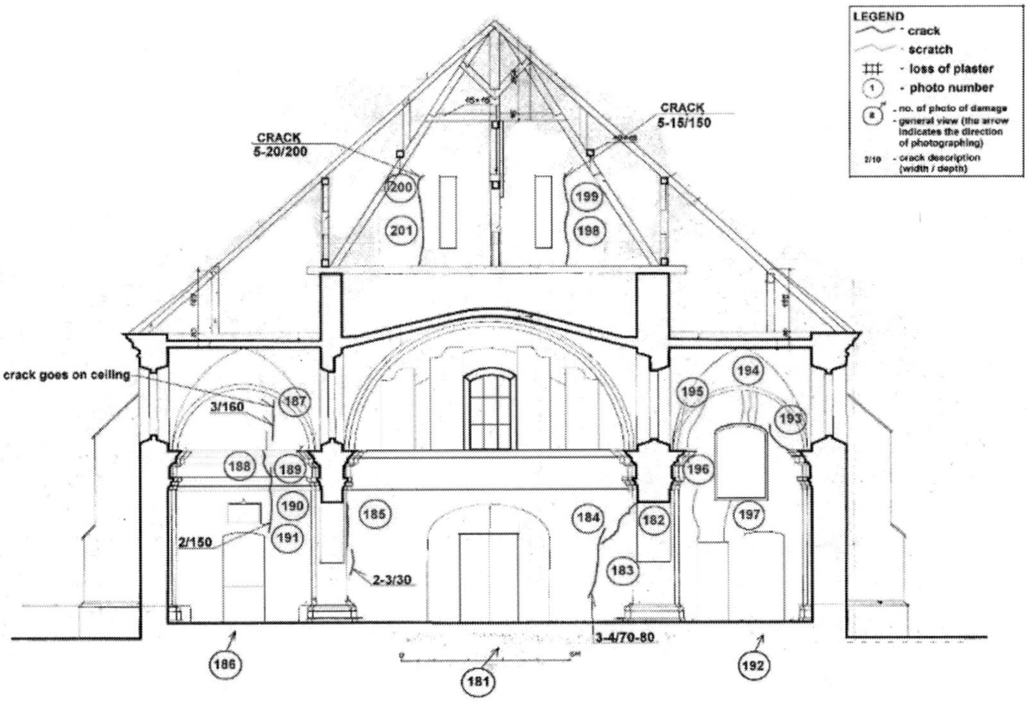

Figure 3. View of damages on the lateral walls of St. Nicholas Church in Krakow.

scanning [13] or increasingly popular thermography [14, 15].

For investors, but also for anyone who will be reading the results of the inventory of damage, it is important that the documentation of the inventory should be readable. That is why it is always good to performed the appropriate inventory drawings illustrating the place of occurrence of damage, containing the type and shape and number of the photo of damage. Inventory drawings of all facades, horizontal plans and important cross-sections including all necessary information were made during performance of expertise [11] (see ex. Fig. 3). On the Fig. 3 location of damages, shape of damage, type of damage (scratch or crack), description of major damages including the width of the crack is visible. Numbers in circles including, by the contrast, the number of photography attached to the expertise with an arrow (if necessary) which describes the direction in which the photos were made.

Measurements of the stability of cracks should be done if there is a suspicion that the observed damages will have been enlarged. Then a pointer crack width could be set up to check this suspicion. Changes in width of cracks may have periodical character (due to transport excitation), could be daily repeated (due to solar radiation) or have long term character (conditioned by the change of seasons). It is therefore important to check width of cracks and observe if changes in width of cracks have periodical or permanent character.

3.2 FEM modelling of cracks

Inventory of damages of a building not only describes the status quo but also has an effect on the formation of computational model of a building. Nowadays, in the era of numerical calculations, there are several FEM programs allowing engineers to learn the behaviour of structures subjected to different types of loads without need for measurements on a real building. But it is important to perform a structural model accurately with regard to correctly modelled structural system, prepared on the basis of existing technical documentation or inventory of building. In the case of a damaged building, it is important to take cracks in structural elements into account.

There are mainly two approaches to this subject. The first one is to reduce stiffness in the area of crack by reduction in Young modulus [16]. Which is a good idea when the behaviour of the building as a whole is considered and when we do not need to analyse the behaviour of individual components. In this approach is good to calculate modal analysis to check accuracy of numerical model with measurements, because cracks can influence on values of natural frequencies (making them lower than for undamaged structure). It is also good to thicken the FEM mesh in the area of cracks because in stress analysis in the wicker elements is a need for more accurate calculation [17].

The second approach in FEM modelling is known as "detailed micro-modelling"[18]. In this approach cracks are modelled as a discontinuum elements. There is then a need for adjastement of contact elements, which always makes calculations difficult and long in time. This type of approach has an advertisement when is a need to know better the behaviur of single structural element [19].

4 Measurement program of historic building

The scope of well-done measurements should include some points. First of all researcher should know the aim of measurements and the way according to standards procedures to achieve this aim.

When vibration influence on building structure is considered is good to know if this building is located in the so called"zone of dynamic influences". These zones have various ranges depending on the type of vibration sources, their propagation path and the dynamic characteristics of the building. Table 1 (acc. [12]) summarizes the zones of dynamic influences of vibrations generated by different sources.

Table 1. Range of zones of dynamic influences.

Source of vibration	Range of the dynamic influence zone
Railway	25-50 m
Tram, road	15-25 m
Shallow underground	40 m
Driving of foundation piles	40-60 m
Soil compaction by vibratory roller	20-60 m
Driving of sheet piling with vibratory hammer	30 m
Driving of sheet piling with shock hammer	20 m

According to Table 1 if historic building is in one of this zones then dynamic influence should be considered and could be the reason of cracks observed during inventory. So the dynamic measurements should be done and they should include following points:

- every measurement made on historic building should be done under conservatory supervision,
- local vision,
- study on historic documentation of the building, which could help to make decision on location of measurements points,
- location of measurements points in the way to take maximum information about structural behaviour under dynamic excitation and do not damage valuable elements of historic structure,
- dynamic signal recording,
- analysis of results according to assessment criterion (SWD scales for building which are supported to these scales, percentage of material effort in other cases).

4.1 Analysis of historical documentation

Study on historic documentation is always a good direction because sometimes already on this stage of expertise the cause of existing damages can be determined. During expertise [11] such recognition was also done. Fig. 4 shows acc. [20] further changes which occurred on the structure of Church of St. Nicholas .

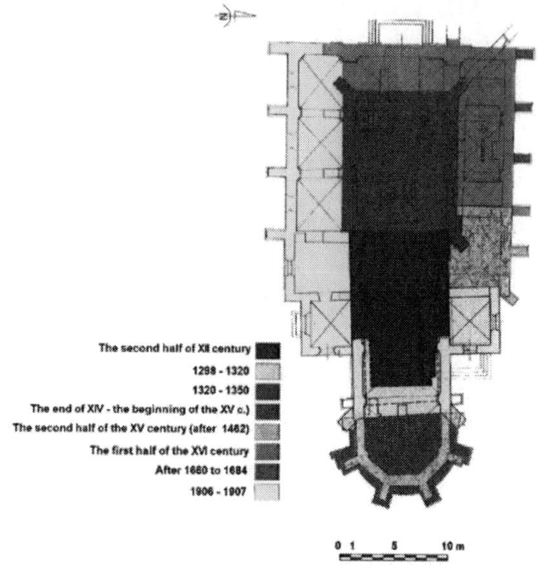

Figure 4. Development phases of the Church of St. Nicholas in Krakow (acc. [20])

On the example of this Church can be seen that historic buildings throughout history are subject to numerous reconstructions and expandability. Of course, such historical interventions can have an impact on existing damage. Historic building should therefore be evaluated in a comprehensive manner, taking into account historical conditions.

4.2 Location of measurement points and measurements realization

Location of measurements points should be selected so that one side does not violate the historically valuable building structure and on the other hand, to receive as much as possible information that could be used for data analysis.

To assess the influence of vibration on the building structure measurement points should always be located on the foundation of a building or on a basement wall in the ground level in a rigid node. It is good to locate another measurement points higher on the structure but in the same cross-section Location of measurement points in one plan in St. Nicholas Church is shown on Fig. 5.

Vibration recorded at the measuring points were excited by trains passages and vehicles passages along Copernicus St. St. Nicholas Church is located in Cracow near two kinematic excitation sources:

- road located m from the building,
- railway located 9 m from the building (distance measured from the railhead).

 DOI: 10.1051/e3sconf/20161000043

They were measured vibration acceleration in three directions: x - horizontal direction perpendicular to excitation from passing trains, y – horizontal direction along the railway line and z - the vertical direction.

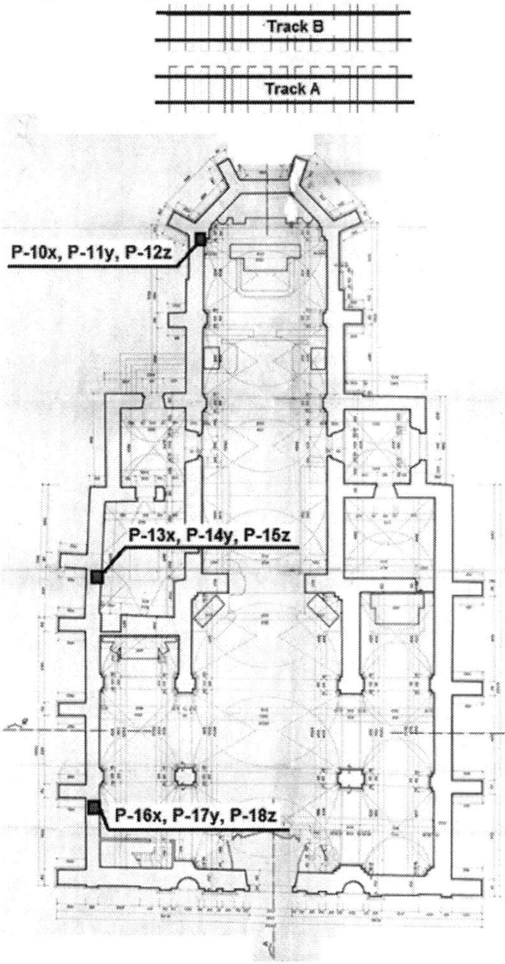

Figure 5. Location of measurement points.

The example of recorded signal in the x direction with one of the highest values of acceleration caused by train passage is shown on Fig. 6.

Measured vibration level of train traffic significantly exceed vibrations level recorded from passing cars on the Copernicus St.

Analysis of measurement first of all was used to validate the numerical model because the structure of St. Nicholas Church is not subjected to SWD scales. When the numerical model was validated then highest signals recorded on foundation of the Church were used as excitation in the numerical model of the structure.

5 Mistakes in the diagnosis of historical buildings

Historic structures are usually not very high masonry buildings that is why the most common mistake is using SWD scale as an assessment criterion to churches which are not subjected to SWD scales. SWD scales could be used for some of historic building such as presbytery or old residential houses but not for churches. Churches mostly has got much high open space without floors which could stiffer the structure in horizontal direction. That is why they cannot be assessed using SWD scales.

The second mistake particularly seen in relation to historic buildings is incorrect positioning of the measuring point. That is why in the case of historic building could be a problem in access to some selected points because of conservatory supervision. According to standard [6] to assess the influence of vibration on building structure recorded signal must be kinematic excitation. Signal recorded in different points then on buildings foundation or on the structural wall in rigid point on the ground level are signals that cannot be used in dynamic assessment.

Assessment of dynamic influence on building structure is based on maximum values of acceleration. However it sill happens that in the diagnosis of vibration influence on buildings RMS values are used.

There are also mistakes in measurement data analysis. The most common mistake of data analysis is the length of analysed signal. This mistake results from unclear provisions in ISO standard [21]. Fortunately in Polish standard duration of vibration is defined simple and clear.

Data analysis should cover this vibrogram section in which the values of acceleration amplitudes are greater than 0.2 of maximum value (see fig. 7).

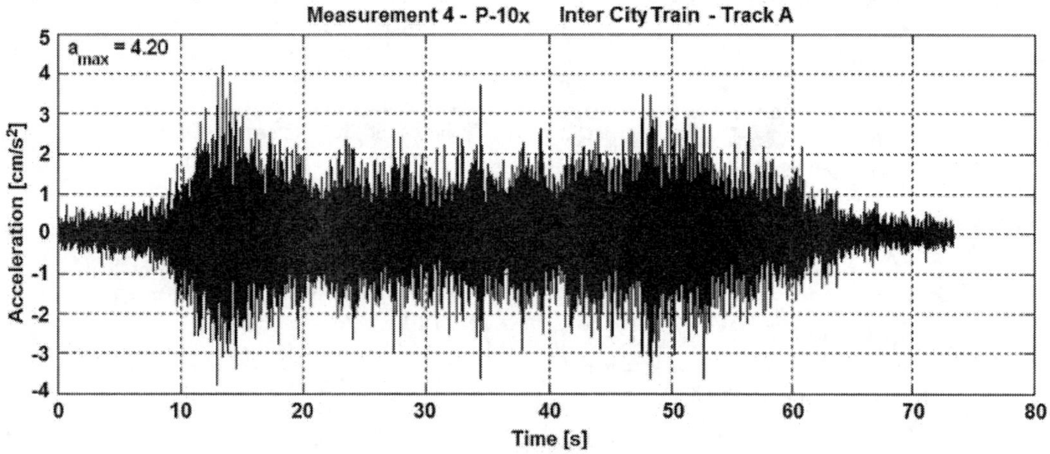

Figure 6. Measured acceleration signal in x direction on the wall on St. Nicholas Church.

 DOI: 10.1051/e3sconf/20161000043

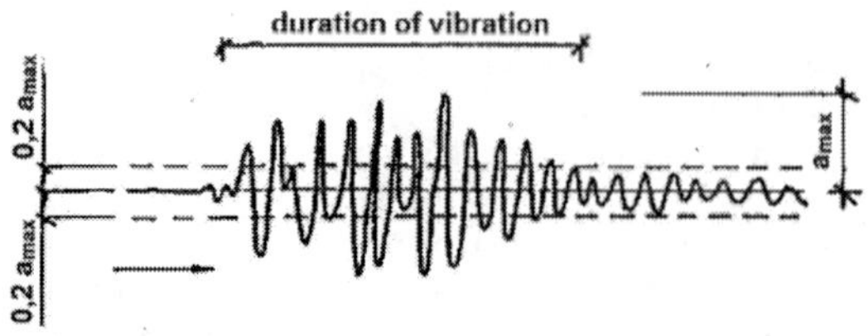

Figure 7. Duration of vibration.

There is a whole bunch of errors associated with numerical calculations. It is worth to specify these most frequent:

- mistakes in geometry of the building and in structural system,
- using triangular elements in dynamical simulations (these elements introduce numerical efforts because of their high stiffness),
- too large size of mesh elements or dismiss of cracks,
- a large number of flattened finite elements,
- wrong materials constants introduced to the model (it is difficult especially in historical structures when we do not know what kind of material was used in some elements)
- dynamic excitation not introduced in mesh node.

6 Protection against vibrations

Diagnosis of the building always ends with proposition of solution the existing problem. With regard to vibration pollution there are three main ways to protect the building from vibrations:

- in the source of vibrations (ex. by using vibration insulation),
- on the way of vibration propagation (ex. by partitions in the soil),
- in the building (ex. by using vibration insulation in the floors)

Considering historical building to introduce vibration insulation in the building could be impossible according to conservatory restrictions.

Using partitions on a way of vibration propagation could be troublesome and expensive. Another thing is good design of partition to fulfil its functions, the partition should not be too close and too far to the building. There is still a matter of location the partition and rights to land on which it would stand.

The best way to protect historical building from vibration influence is to design vibration insulation in the source. Even more the investor or owner of the transport infrastructure is forced under Polish law to eliminate the causes of pollution caused by vibration. In the final recommendations of the expertise [11] such a solution was selected.

7 Conclusion

The specificity of historic buildings may be associated with the lack of direct access to the visible damage or conservatory restrictions. However, this does not relieve the diagnostician to carry out reliable diagnosis of visible damage of historic structure.

Although troublesome to carry out an inventory of damage is an important element in the diagnosis of historic buildings. It provides information of existing status and is helpful in shaping the computational model of the building, with the help of which different loads scenarios can be carried out.

Well-done diagnosis should include all points mentioned in this article, and watch out for mistakes that may arise during its duration.

Each diagnosis should end with giving causing of the phenomenon and should include guidelines for the removal of the cause of damages and proposition of repairs. Especially in the case of historic buildings subjected to dynamic influences this could be the problem. That is why protection in these cases should be focused on eliminating the sources of vibration or reducing dynamic excitation coming from the source. Sometimes if necessary partitions in the soil could be considered.

References

1. F. Corvo, J. Reyes, C. Valdes, F. Villaseñor, O. Cuesta, D. Aguilar, P. Quintana. Water Air Soil Poll **205,** 359–375 (2010).
2. Y. Nuhoglu, E. Oguz, H. Uslu, A. Ozbek, B. Ipekoglu, I. Ocak, I. Hasenekoglu. Sci Total Environ **364,** 272–283 (2006).
3. H. Bravo, R. Soto, R. Sosa, P. Sa´nchez, A.L. Alarco´n, J. Kahl, J. Ruı´z. Environ Pollut **144,** 655-660 (2006).
4. Act of 27 April 2001. Environmental Law (in Polish)
5. DIN 4150-3. Structural vibration. Part 3 : Effects of vibration on structures
6. prPN-B-02170P. Ocena szkodliwości drgań przekazywanych przez podłoże na budynki.
7. J. Banaszak, A. Halicka. Building and Architecture **9** 51-68 (2011).
8. A. Kandemir-Yucel, A. Tavukcuoglu, E.N. Caner-Saltik. Infrared Phys Techn **49,** 243-248 (2007).
9. A. Carpinteri, S. Invernizzi, G. Lacidogna. Eng Struct **27,** 387–395 (2005).
10. S. Ivorraa, F. J. Pallares. Eng Struct **28,** 660–667 (2006).
11. K. Stypuła (head of the team of authors): Specialized constructional expertise with performing vibration measurements, dynamic computer analysis and geotechnical subsoil researches of St. Nicholas Church at St. Copernicus 9 in Krakow. Cracow University of Technology, Institute of Structural Mechanics, November 2009.
12. J. Kawecki, A. Kowalska-Koczwara, K. Stypuła. WASET **59,** 1496-1502 (2011).
13. A.Rzońca. Ph.D. Thesis, Krakow 2008
14. D. Paoletti, D. Ambrosini, S. Sfarra, F. Bisegna. J Cult Herit **14,** 116–121 (2013).
15. L. Binda, A. Saisi, C. Tiraboschi. Constr Build Mater **14,** 199-233 (2000).

16. I. Lubowiecka, J. Armesto, P. Arias, H. Lorenzo. Eng Struct **31,** 2667–2676 (2009).
17. J. F. Unger, S. Eckardt, C. Konke. Comput. Methods Appl. Mech. Eng. **196,** 4087–4100 (2007).
18. P. B. Lourenco. Prog. Struct. Engng Mater. **4,** 301–319 (2002).
19. T. Belytschko, R. Gracie, G. Ventura. Comput. Methods Appl. Mech. Engrg. **196,** 4087–4100 (2007).
20. H. Rojkowska, W.Niewalda. Dokumentacja z badań historyczno – architektonicznych dla Kościoła św. Mikołaja przy ul. Kopernika 9 w Krakowie. Kraków 1995-97r
21. ISO 4866 :2010. Mechanical vibration and shock— Vibration of fixed structures — Guidelines for the measurement of vibrations and evaluation of their effects on structures.

E3S Web of Conferences 10, 00044 (2016) DOI: 10.1051/e3sconf/20161000044
SEED 2016

Green public procurement – legal base and instruments supporting sustainable development in the construction industry in Poland

Renata Kozik[1,a] and Izabela Karasińska - Jaśkowiec[2]

[1]*Cracow University of Technology, Faculty of Civil Engineering ,ul. Warszawska 24, 31-155 Cracow, Poland*
[2]*Cracow University of Technology, Faculty of Civil Engineering ,ul. Warszawska 24, 31-155 Cracow, Poland*

Abstract. In the respect of value, public procurement in the construction industry belongs to one of the largest ones in the domestic market. Therefore, green procurement for construction works should become the center of attention of public authorities in a broad sense, due to its scale and importance for the sustainable development. The authorities and contracting entities who spend public money should have the opportunity to apply such legal articles and instruments that allow them to both optimize public expenditures and consider the environmental factor, such as decreasing carbon emission. To make the idea of sustainable development a reality as European Union's the most vital aim, EU law is implemented in Poland. Local authorities' duty is to appropriately shape their policies and use the vital instrument of sustainable development, namely green public procurement. This paper presents a comparative analysis of legal regulations to illustrate the actual Polish and EU laws concerning the construction industry. Even though the generally applicable law allows to implement the idea of sustainable development efficiently, local self-government units in their regional policies do not report any need for specific solutions, or they do so only occasionally.

1 Legislation in Poland and the European Union

In the existing European Union (EU) law and economy public procurement is the vital instrument influencing the way the markets of individual EU countries operate and, as a result, the European Union as a whole. The quality of law becomes the factor indispensable for the completion of public contracts reliably, taking into consideration the specificity of the ordering institution and the context it works in, which positively influence the execution of its policies and aims.

According to the sustainable development principle, the ecological, economic and social aspects have equal weight and importance. For this reason caring for the natural environment is one of the main aims of countries and international organizations. Green public procurement (GPP), in accordance with the definition proposed by the European Committee, mean the policy in which public entities include criteria and/or environmental requirements to procurement procedures and search for solutions minimizing the negative influence of products/services on the environment, providing for the whole life cycle of products [1, 2]. In this way they influence the development and enhancement of environmental technologies.

1.1 Provisions of the Polish law

The most important law for the Polish legal order is the Constitution of the Republic of Poland [3], which in Article 5 and indirectly in Art. 68 and Art. 74 draws attention to the issue of sustainable development. It indicates the duties of the public authority which include the prevention of the negative effects of environment degradation on health, the pursuit of policy ensuring ecological safety of present-day and future generations, protection of the environment and support of citizens' incentives to protect and improve it.

Public authorities, according to Art. 74 of the Constitution of the Republic of Poland, are obliged to adopt a policy that ensures ecological safety for the present-day community and future generations. Therefore, the inclusion of green records in tender documentation should become a common phenomenon in accordance with Art. 74 paragraph 1 of the Constitution of the RP.

In Poland the issue of public procurement is formally regulated by the Act of 29 January 2004, Public Procurement Law [4] with the implementing acts. Apart from these, other vital acts constructing the public procurement system in Poland include the following:

- Civil Code [5]
- The Law on Public - Private Partnership [6]
- Construction Law [7]
- Act on Concessions for Construction Works or Services [8]
- The Public Finance Act [9]
- The Law on Liability for Violation of Public Finance Discipline [10].

[a] Corresponding author: rkozik@izwbit.pk.edu.pl

GPP development is additionally promoted by numerous documents concerning the development and future of Poland. They include, for instance: „Strategy of Energy Security and Environment - the prospect of 2020"[11], which assumes that green procurement will be one of the instruments supporting environmental improvement and by 2020 they should constitute a half of all orders in Poland; „Efficient State Strategy 2020" [12] stating that spreading the use of GPP will allow for more efficient management of physical and financial resources of public administration; finally „Dynamic Poland" - Strategy for Innovation and Economic Efficiency for the years 2012 - 2020 [13] according to which public procurement should serve the consolidation of ecological behavior, thus enabling the development of products and services characterized by high environmental and social standards.

Polish law is shaped by the law created in the European Union.

1.2 EU regulations in Poland

The reform of the public procurement law at the European level which took place in 2014 involved the introduction of changes facilitating GPP development and the encouragement to apply it. Poland, as a EU member, is thus required to include EU legislation in her domestic legal order. At present, however, the Public Procurement Law stays in accordance with the following: the so-called Classic Directve (2004/18/WE) [14] concerning the coordination of the procedures of granting public contracts for construction works, deliveries and services; the so-called Utilities Directive (2004/17/WE)[15] that coordinate the procurement procedures of entities operating in the water, energy, transport and postal services sectors; the so-called Remedies Directives (2007/66/WE [16], 89/665/EWG [17], 92/13/EWG) [18]; and the so-called Defence Procurement Directive 2009/81/WE [19] that coordinates contracting procedures for construction works, deliveries and services by institutions and entities ordering in the defence and security sectors, changing the 2004/17/WE and 2004/18/WE Directives. The Directives stress the importance of the contract offering the best value for money but also allow to include environmental issues in the subsequent stages of the public procurement procedure:

- at the description of the object of contract stage (Art. 23 Dir. 2004/18/WE),
- at the contractor selection stage (Art. 45, 48 i 50 Dir. 2004/18/WE),
- at the stage of optimum tender selection through the environmental evaluation criteria (Art. 53 Dir. 2004/18/WE),
- at the stage of determining terms and conditions (Art. 26 Dir. 2004/18/WE).

On 28th March 2014, new Directives concerning public procurement were published: Directive of the European Parliament and of the Council 2014/24/UE of 26th February 2014 [20] on public procurement and repealing Directive 2004/18 / EC; Directive of the European Parliament and of the Council 2014/25/UE of 26th February 2014 [21] on procurement by entities operating in the water, energy, transport and postal services, repealing Directive 2004/17/WE; Directive of the European Parliament and of the Council 2014/23/UE [22] of 26th February 2014 on the award of concession contracts. The Directives came into force on the twentieth day following its publication in the Official Journal of the EU. The deadline for the implementation of the directives into the legal systems of EU Member States falls on 18th April 2016.

The noteworthy element of the implementation of EU Directives is the amendment introduced by the Act of 29th August 2014 amending the Public Procurement Law, which came into force on 19th October 2014. It changed, among others, the criteria for bid evaluation introducing the solutions included in Article 67 of the new Classic Directive to Polish law. Yet the new Classic Directive offers greater possibilities for GPP than those only relating to bid evaluation criteria. It allows to exclude the contractors (Art. 24 paragraph 1 point 3-8) who commited a crime against the environment. To prove that a bid fulfils the orderer's requirements, the the orderer may demand, for instance (Art. 25 paragraph 2 PPL), a certificate issued by an independent entity dealing with certifying the compliance of the contractor's activities with European standards of environmental management if the orderers indicate environment management measures which the contractor is to use while executing the construction works or services contract, referring to the environmental management system and audit (EMAS) or environmental management standards based on European or international standards certified by bodies operating under the laws of the European Union, European or international standards concerning certification. The contractor may also submit equivalent certificates issued by entities established in another Member State of the European Economic Area or other documents confirming the use by the contractor of equivalent quality assurance measures and the use equivalent environmental management measures.

The orderer may withdraw from the terms of reference through the Polish , European or international standards if they provide an accurate description of the contract by indicating the functional requirements. These requirements may include the description of the impact on the environment (Art. 30 paragraph 6.).

Non-price criteria (Art. 91 of PPL) concerning the bid evaluation criteria enumerated in the contract notice and/or in the specification of essential terms of the contract (SETC), depending on the public procurement procedure type. On the basis of this provision the contracting authority chooses the best tender basing on the bid evaluation criteria specified in the SETC.

The current legal status allows it to be „the price or price and other criteria relating to the object of the contract, in particular the quality, functionality, technical parameters, environmental, social, innovation, service, term of the contract and operating costs".

The use of sustainable public procurement is promoted by the European Commission and the Council of Ministers of Poland. The main activities for the implementation of GPP have been undertaken by the Public Procurement Office (PPO) and the strategic document is the National Action Plan on sustainable procurement for the years 2013-2016 [23]. Thus the Public Procurement Office undertook activities aiming to inform, train, educate and counsel in order to spread and promote tender procedures with solutions contributing to the sustainable development and innovations in the domain of environment protection.

GPP implementation is also mentioned in the European Commission's Communication of 3rd March 2010: „Europe 2020 strategy for smart, sustainable and inclusive growth"[24]. According to this document public procurement plays a key role as one of the market-based instruments used to achieve a smart, sustainable growth and inclusive growth while ensuring the most efficient use of public funds.

2 The structure and characteristic of the public procurement market

In Poland, as well as in the Member States of the European Union, a significant part of the demand for products, services and works comes from public institutions, including the administrations of various levels.

The multiplicity of EU projects and related funds allocated to spending causes that public procurement is one of the main elements that stimulate economic growth. Yearly public institutions in the EU spend about 2 trillion euros for tenders, which represents 19% of Gross Domestic Product of the European Union [25].

2.1 The structure of the public procurement market

The value of the public procurement market in Poland in the subsequent years is presented in Table 1. In 2011 – 2014 the size of this market significantly exceeded 130 billion PLN annually. A large slowdown in procurement was recorded in 2015, influenced mainly by the change of the threshold from 14,000 to 30,000 euros, from which the PPL provisions should apply. The change has been in force since October 2014, so only in 2015 the effects were fully seen. A large share of this market belongs to construction works contracts, with an average of over 40%.

Table 1. Public procurement market value in Poland, own study based on [26]

Year	The value of order (billions PLN)	The percentage of the value of contracts awarded on construction works [%]
2011	144.1	39.0%
2012	132.7	45.0%
2013	143.2	38.0%
2014	133.2	40.0%
2015	38.4	60.0%

The value of orders executed by public administration units greatly exceeds the value of contracts awarded by other entities obliged to apply the PPL Act, which Table 2 illustrates. Entities other than government administration and local government which are also obligated to apply the PPL such as courts, public schools, public hospitals, etc., were summarized ("Other entities").

Table 2. A percentage of the total value of contracts awarded by entities obliged to obey PPL, own study based on [26]

	Goverment administratio n, central and field	Local government	Other entities
2011	6.69%	52.62%	40.69%
2012	7.21%	48.44%	44.35%
2013	6.00%	50.25%	43.75%
2014	5.65%	51.44%	42.91%
2015	5.76%	49.31%	44.93%

The decrease in the number of orders in 2015 is also a result of the change of the threshold above which public orders are obliged to use the PPL Act. Local government has a significant influence on the shape of this market. Table 3 presents the share of individual contracting in the total number of orders in the years 2011 – 2015.

Table 3. Percentage of individual contracting in the total number of public procurement, own study based on [26]

Year	Total number of contracts awarded	Government administra - tion central and field	Local govern - ment	Other entities
2011	186 232	6.90	40.64	52.46
2012	188 478	6.68	40.71	52.61
2013	210 352	5.92	41.40	52.68
2014	174 886	5.54	42.95	51.51
2015	140 482	5.81	39.52	54.67

Green public procurement can be a powerful tool for innovative approaches and an encouragement for entrepreneurs to develop new products and services focused on the environment. They may significantly influence the execution of local and regional aims within the scope of sustainable development.

The concept of sustainable procurement is inextricably linked to sustainable development. It indicates an integrated approach of ordering institutions to public procurement voluntarily respecting environment protection (the so-called Green Public Procurement) and important social interests.

One of the most vital positions in activities executing the principles of sustainable development belongs to the construction industry. Communication from the Commission of 16th July 2008 „Public procurement for a better environment" [1] indicates 10 sectors with the highest potential for GPP. Construction ranks in the first place among them (including raw materials such as wood, aluminum, steel, concrete and glass; as well as construction products, such as windows, wall and floor coverings, heating and cooling equipment, aspects concerning the maintenance of buildings and end-of-life services, building maintenance and execution of works contracts in place).

Table 4. The percentage of construction works in total orders by type of authorities, own study based on [26]

Year	Goverment administration, central and field	Local government	Other entities
2011	34.55%	38.37%	27.08%
2012	38.26%	37.30%	24.44%
2013	34.47%	36.30%	29.23%
2014	37.2%	42.37%	20.43%
2015	39.13%	43.36%	17.51%

It was estimated that, among others, in this sector the offered opportunities to improve the environment, public spending, the potential impact on the supply side, the example to private or corporate consumers, political sensitivity, the existence of relevant and easy-to-use criteria , the availability of suitable products on the market and economic efficiency are vital. Therefore, a product card was created for Construction, providing a set of recommendations related to public works procurement including the provision of related services (for instance, including cooling, electricity supply, heating and ventilation). All phases of design, construction, operation and decommissioning of office buildings were taken into consideration. For each of these criteria proposals for environmental criteria were given.

One of the EU legislation related to green procurement is Directive 2010/31/UE [27] concerning energy performance of buildings and Directive 2012/27/UE [28] for energy efficiency, which translate into activities rationalizing energy consumption on the side of contracting authorities.

Construction in accordance with the principles of sustainable development is inextricably linked with the development of modern technologies which support the construction of buildings. It seems necessary to remind the BIM technology which concerns mostly information and the management process in the whole investment process. The solutions developed earlier, for instance in the sustainable construction, will efficiently support the public procurement related to construction works. Green public procurement related to construction works focuses on achieving longevity and high quality solutions in materials, construction and design to improve and shape new building objects by, for example, efficient use of energy, water and other natural resources, as well as the reduction of waste, pollution and the degree of environmental degradation.

2.2 Characteristics of the public procurement market

According to the Public Procurement Office's statistics, the evaluation of tenders for public procurement market is mainly based on the price. Price was the only criterion for the evaluation of tenders for the most procurement procedures (91 – 92 %) in 2011-2014 (Table 5). After the amendment of the PPL Act in October 2014 before which price was the only criterion for the evaluation of tenders, the price criterion was allowed only when object of the contract is widely available and has established quality. As a result, in 2015 th percentage of procedures in which price was the only criterion decreased to 12%.

Table 5. A percentage of the total value of all contracts awarded where the price was the only criterion, own study based on [26]

Year	Below EU thresholds, publ. in Public Procurement Bulletin (PPB)	Above EU thresholds publ. in Tenders Electronic Daily (TED)
2011	91.0%	73.0%
2012	92.0%	76.0%
2013	92.0%	85.0%
2014	84.0%	75.0%
2015	12.0%	11.0%

For orderers a procedure with price as the decisive factor is a safe one because it consistent with applicable law and does not cause problems in the event of a control. According to Art. 91 paragraph 2 of PPL, the orderer may use the most advantageous offer using not only the price criterion but also other criteria, such as quality, functionality, technical parameters, use of the best available technologies in terms of environmental impact, operating costs, service and the date of contract completion. Including the environmental criteria is therefore voluntary, except two cases: for orders regarding means of transport and transport services (Directive of the EP and the Council (WE) no. 2009/33/WE [29] of 23th April 2009 on the promotion of clean and energy efficient road transport vehicles) and procurement of office, computer and audiovisual equipment (Regulation of the EP and the Council (WE) no. 106/2008 [30] of 15th January 2008 on a Community energy efficiency labeling program for office equipment). For comparison, in some EU countries the application of green criteria is mandatory for certain groups of products (for instance, in Denmark and Germany) or orderers (for example, Belgium) [31]. In other cases, the contracting authority is not obliged to take into account any criteria related to environmental protection. Therefore, the majority of orderers who currently belong mainly to local authorities (Table 2) consider price as the only selection criterion and choose the cheapest bid (Table 6).

Table 6 The number of procedures involving selection of the bidder based on the lowest price [26]

Year	The total number of contracts awarded	Choosing the cheapest bid (the percentage of proceedings)
2011	186 232	88.0 %
2012	188 478	86.8 %
2013	210 352	86.5 %
2014	174 886	85.7 %
2015	140 482	no data

The remaining procedures with more than one criterion the number of criteria typically amounted to 2-3. In the procedures for construction works the situation was similar. The most frequent criteria, apart from the price one, included guarantee (terms and conditions) and the execution date. In contracts for public works orderers occasionally used such environmental criteria as «energy consumption» or «operating costs».

The most common mode of public procurement in Poland is unlimited tender (more than 90 %).

3 Duties, powers and capabilities of local government units in the area of green public procurement

Public procurement is a legal way to control the spending of public funds by the trustees of these funds, for instance, by local government units. Given the role of public funds in the economy, the public procurement system has become one of the elements of the environmental policy of the Union, including Poland. Therefore, when creating a legal framework for public procurement, a possibility to execute a contract on construction works supporting the sustainable development of the environment and its protection was introduced.

As indicated earlier, the principle of sustainable development in Poland has a constitutional basis, so its character is defined as a political system. Thus, all the entities of public administration should act in accordance with the Constitution. The particular measures for sustainable development that are of a sufficiently high rank include the local development strategies which define the long-term goals and tasks of the municipality and the main directions of development. It is these strategies into which municipalities should include their policies regarding the conduct of public procurement, among others, in the context of their greening.

The holders of public funds should take into account the costs generated in the whole life-cycle of products and use environment friendly solutions. The selection of a bid for works based on green public procurement should refer to the evaluation of the life-cycle costs (that is, the purchase, use and disposal) and not just the lowest purchase price criterion. Commercially available computational tools help to accurately determine life cycles through the use of various parameters, such as energy consumption, recycling of materials and maintenance costs. The fully balanced public procurement should take account of ecological criteria throughout the whole supply chain and product life cycle.

Green public procurement meets the local government administrations' undertaking to solve the problems relating to environment protection by reducing emissions of greenhouse gases and toxins. They are a powerful tool that should be used by public institutions to reduce CO_2 emissions and the implementation of the chosen environmetal objectives.

When the entity is not aware of how to describe the object of the contract to meet the expected criteria for sustainable public procurement, before the relevant procedure begins, under Article 31a-31c of the Public Procurement Law, they may conduct the so-called technical dialogue. As part of the technical dialogue, the entity can obtain advice or information necessary to prepare the description of the subject of the contract, the essential terms of reference or to determine the conditions of the contract.

The ability to perform a technical dialogue allows for a better understanding of the market and a better preparation of the proper tender procedure. This procedure does not require from the orderer to bear any costs that would arise in the case of ordering the description of the object of contract to an external entity [32]. Thus the public finance law is vital from the point of view of economy.

It is estimated that the technical dialogue is an institution still very rarely used by orderers. Unfortunately, as in many cases its application could facilitate the work of the procuring entity, it can not be subject to any claims on the part of the tenderer. The object of the dialogue may involve all aspects of the contract : the strictly technical, organizational or legal ones. The subject of consultations in the course of the dialogue not only can but also should be areas where there is a deficit of knowledge on the part of the contracting authority, and whose identification is vital from the point of view of the proper preparation and execution of the procedure for public procurement [33]. The technical dialogue, though slightly modified, was regulated also in the new European directives concerning public procurement, whose implementation time fell on the 18th April 2016. According to Art. 40 of Directive 2014/24 / EU - "The new classical directive" and Art. 58 of Directive 2014/25/UE - "The new Utilities Directive" before launching a procurement procedure, the contracting entities may conduct market consultations with the aim of preparing orders and informing contractors of their plans and requirements for the contract. What is more, for this purpose the contracting authorities may use the advice of independent experts, authorities or market participants. The only limitation for the consultation process is a prerequisite that the received help and advice would not cause distortion of competition or violation of the principles of non-discrimination and transparency.

3.1. Environmental certification of buildings

An important element of the implementation of the principles of sustainable development into the area of construction is the process of environmental certification of buildings. For the investment and construction market there exist several types of certificates, with the following available in Poland: LEED (Leadership in Energy and Environmental Design) designed by the U.S. Green Building Council (USGBC) and BREEAM (Building Research Establishment Environmental Assessment Method), managed by BRE Global Ltd., being part of the BRE Group and a subsidiary company of the BRE Trust.

LEED considers those parameters for buildings which remain of the greatest importance: energy saving , rational water consumption, carbon emissions reduction, improvement of the quality of indoor environment, resource management and the sensitivity to effects. The building is assessed according to 10 categories, each of which consists of a particular number of sub-categories and translates into a percentage result of the certification.

BREEAM is a method of assessmet of any building type. The building is subject to a point-based evaluation in the following categories: management, health and well-being, energy, transport, water, materials, land use and ecology, pollution and innovations. The number of points in each category gives a result on the basis of which a design or a building (if the evaluation is performed after it was constructed) is assessed. Having been summed up, the points give the total score.

The BREEAM International Version, which refers to Europe, is designed for the certification of office, industrial and commercial buildings.

It is therefore another tool to take austerity measures, which should attended to by the investors administering public funds during the design and planning of the investment.

3.2 Own activity of municipalities in the promotion of sustainable development

Another possible action of support for municipalities is that they can join the initiatives supporting sustainable development which, by exchanging experiences and basing on specific rules of conduct, seek to make public procurement green as comprehensively and efficiently as possible. For instance, *Covenant of Mayors – CoM – The Agreement between the Mayors for Climate and Energy*, is a popular European movement bringing together local authorities interested in increasing energy efficiency and using renewable energy sources in the areas they administer. The aim of the movement formed in 2008 on the initiative of the European Commission is to reduce CO_2 emissions by 40% by 2030 (compared to the 1990 base year). The new integrated Covenant of Mayors for Climate and Energy was launched by the European Commission on 15th October 2015 during a ceremony at the European Parliament in Brussels [34].

4. Conclusions

The implementation of sustainable public procurement strengthens and justifies the administrative actions, and contributes to the development of a positive image of the purchasing entity. Due to the public interest, including the need to improve the environment, it is necessary to include environmental aspects in public prcurement as extensively as possible. The activities of the public administration should relate in particular to supporting the solutions that reduce CO_2 emissions by relating to, for example, energy consumption for heating, cooling, ventilation, hot water and electricity supply, or cold water consumption, which may be attractive to the orderer due to the economic benefits in the short and in the long term.

Municipalities have the opportunity to take into account environmental issues in subsequent stages of the procurement procedures, in particular on the following stage: description of the object of contract (Art. 23 PPL), contractor qualification (Art. 45, 48 i 50 PPL), selection of the best bid by using environmental criteria for tender evaluation (Art. 53 PPL), at the stage of determining contract conditions (Art. 26 PPL), description of the subject of the proceedings (Art. 29–30 PPL), when creating the specification of essential terms of the contract and the description of the criteria that the orderer will use during bid selection, with an explanation of the criteria and the way of tender selection (Art. 36 paragraph 13 PPL), as well as at the stage of selection of the optimum bid on the basis of the criteria specified in the terms of reference (Art. 91 paragraph 1 and 2 PPL) and signing a contract (Art. 94 PPL). On all of the stages of the proceedings, municipality may take into account environmental issues and pursue a policy of green procurement.

Green public procurement is a specialized tool for environmental protection. Even though in Poland environmental criteria are often overlooked in the plans of spending the public funds by local government units, actions promoting green public procurement should continue to be undertaken, as in the most developed European countries. The increased knowledge of the environment certainly contribute to the increased number of environmental investigations. The orderers who wish to undertake green procurement have adequate legal tools. At the same time, including the elements of sustainable development in green public procurement should be done in a conscious and consistent way, since a sporadic action in the application of environmental criteria will not bring any expected effect.

The public procurement for construction works plays a significant role in the budget of municipalities and, as a result, of the country, allowing to influence not only the environmental aspects but also the economy. Properly managed public procurement system implemented by municipalities has a significant impact on the growth of public trust in government institutions and public expenditures.

The current, mediocre level of implementation of solutions for green public procurement clearly does not correspond with the content of the strategic documents and the necessity of pro-environmental public spending. The number of public procedures in which green procurement are implemented is still small but steadily growing [35]. It can be also seen growing interest in this process among scientists [36].

It needs to be noted that local government units in their policies should take into account the aspect of sustainability, not only because of the constitutional provisions and PPL, but also because of the Public Finance Act, in particular the wording of Art. 44 paragraph 3, which specifies the basic principles of public spending. In accordance with the Act discussed here, expenditure should be done in a purposeful and frugal way, on the basis of the following principles: obtaining the best results with particular expenditures and optimal selection of methods and means to achieve the goals planned. The sense of the principle of effectiveness is described as an optimization of methods and measures to achieve the goals planned or an optimization of results assuming specified expenditures. Efficiency focuses on obtaining the best results from the given outlays; therefore, it is related to efficiency. According to the Supreme

Chamber of Control, the saving is the standard indicating "good management of public money" [37]. Thus, it is clear that GPP for construction works is to fulfill these rules. Yet in each case a deep analysis of all variables is needed to make the right decision.

The evaluation of the policy concerning the implementation of environmental criteria to public procurement for construction works by local government units clearly shows that a continuous enhancement of both orderers' and contractors' skills and knowledge is vital. It is also necessary to change the attitude of the controllers from regional accounting chambers and other supervising institutions, who distrustfully approach any contract based on criteria other than price.

List of abbreviations

EMAS - Eco-Management and Audit Scheme
EU – European Union
GPP – Green Public Procurement
PLN – Polish currency
PPB - Public Procurement Bulletin
PPL – Public Procurement Law
PPO – Public Procurement Office
RP – Republic of Poland
SETC - The specification of essential terms of the contract
TED - Tenders Electronic Daily

References

1. Communication from the Commission to the European Parliament, the Council, the European Economic and Social Committee and the Committee of the Regions Public procurement for a better environment {SEC(2008) 2124} {SEC(2008) 2125} {SEC(2008) 2126} https://www.uzp.gov.pl/__data/assets/pdf_file/0022/26518/KOM-2008-400_Zamowienia-publiczne-na-rzecz-poprawy-stanu-srodowiska.pdf (access: 16.04.2016)
2. M. Kalubanga, *Sustainable procurement. Concept, and Practical Implications for the Procurement Process*, IJEM Science, 1 (2012), pp. 01-07
3. *Constitution of the Republic of Poland* of 2 April 1997 (Journal of Laws No 78, item 483)
4. Act of 29 January 2004 - *Public Procurement Law* (Journal of Laws of 2015, item 2164)
5. *Civil Code* od 23 April 1964 r. (Journal of Laws No 16, item 93) text of 17 Februay 2016 r. (Journal of Laws of 2016, item 380)
6. *The Law on Public - Private Partnership Act* of 19 December 2008 (Journal of Laws of 2009, No 19, item 100); text of 7 May 2015 r. (Journal of Laws of 2015 item 696)
7. *Construction Law* of 7 July 1994 (Journal of Laws No 89, item 414) text of 9 February 2016 r. (Journal of Laws of 2016 item 290)
8. *Act on Concessions for Construction Works or Services* of 9 January 2009 (Journal of Laws No 19, item 101) text of 13 January 2015 (Journal of Laws of 2015, item 113)
9. *The Public Finance Act* (Journal of Laws of 2009 No 157, item 1240 text Journal of Laws of 2013 item 885)
10. *The Law on Liability for Violation of Public Finance Discipline* of 17 December 2004 (Journal of Laws of 2005 No 14, item 114) text of 27 September 2012 (Journal of Laws 2013 item 168)
11. *Strategy of Energy Security and Environment* - the prospect of 2020, (Polish Monitor of 2014 item 469)
12. *Efficient State Strategy 2020*, Resolution of the Council of Ministers No 17 of 12 February 2013 (Polish Monitor of 201, item 136)
13. Resolution of the Council of Ministers No 7 of 15 January 2013, *Strategy for Innovation and Economic Efficiency for the years 2012 - 2020*, Polish Monitor of 13 February 2013, item 73 http://monitorpolski.gov.pl/mp/2013/73/1
14. Directive 2004/18/EC of the European Parliament and of the Council of 31 March 2004 on the coordination of procedures for the award of public works contracts, public supply contracts and public service contracts (Official Journal of the European Union L 134 of 30 April 2004);
15. Directive 2004/17/EC of the European Parliament and of the Council of 31 March 2004 coordinating the procurement procedures of entities operating in the water, energy, transport and postal services sectors (Official Journal of the European Union L No. 134 of 30 April 2004) ;
16. Directive 2007/66/EC of the European Parliament and of the Council of 11 December 2007 amending Council Directives 89/665/EEC and 92/13/EEC with regard to improving the effectiveness of review procedures concerning the award of public contracts (Text with EEA relevance) (Official Journal of the European Union L 335 of 20 December 2007 r.)
17. Council Directive 89/665/EEC of 21 December 1989 on the coordination of the laws, regulations and administrative provisions relating to the application of review procedures to the award of public supply and public works contracts Official Journal of the European Union L 395 of 30 December 1989) ;
18. Council Directive 92/13/EEC of 25 February 1992 coordinating the laws, regulations and administrative provisions relating to the application of Community rules on the procurement procedures of entities operating in the water, energy, transport and telecommunications sectors (Official Journal of the European Union L 76 of 23 March 1992)
19. Directive 2009/81/EC of the European Parliament and of the Council of 13 July 2009 on the coordination of procedures for the award of certain works contracts, supply contracts and service contracts by contracting authorities or entities in the fields of defence and security, and amending Directives 2004/17/EC and 2004/18/EC (Text with EEA relevance); Official Journal of the European Union L 216
20. Directive 2014/24/EU of the European Parliament and of the Council of 26 February 2014 on public

procurement and repealing Directive 2004/18/EC Text with EEA relevance (Official Journal of the European Union L 94 of 28 March 2014)
21. Directive 2014/25/EU of the European Parliament and of the Council of 26 February 2014 on procurement by entities operating in the water, energy, transport and postal services sectors and repealing Directive 2004/17/EC Text with EEA relevance, (Official Journal of the European Union L 94 of 28 March 2014)
22. Directive 2014/23/EU of the European Parliament and of the Council of 26 February 2014 on the award of concession contracts Text with EEA relevance
23. *Krajowy Plan Działań w zakresie zrównoważonych zamówień publicznych na lata 2013 - 2016* (Eng: National Action Plan on sustainable procurement for the years 2013-2016 https://www.uzp.gov.pl/__data/assets/pdf_file/0019/26353/KPD_ZZP_2013_2016.pdf(access: 14.04.2016)
24. Communication from The Commission Europe 2020 A strategy for smart, sustainable and inclusive growth;http://ec.europa.eu/eu2020/pdf/1_PL_ACT_part1_v1.pdf (access: 14.04.2016)
25. *Zrównoważone zamówienia w praktyce. Modelowe klauzule społeczne i środowiskowe* (Eng: Sustainable procurement in practice. Model social and environmental clauses) http://www.czp.com.pl/userfiles/files/Zrownowazon.pdf (access: 14.04.2016)
26. Sprawozdania o funkcjonowaniu systemu zamówień publicznych w 2011 - 2015r, (Eng. Reports on the functioning of public procurement system in 2011 – 2015) https://www.uzp.gov.pl
27. Directive 2010/31/EU of the European Parliament and of the Council of 19 May 2010 on the energy performance of buildings (Official Journal of the European Union L 153 z 18.06.2010)
28. Directive 2012/27/EU of the European Parliament and of the Council of 25 October 2012 on energy efficiency, amending Directives 2009/125/EC and 2010/30/EU and repealing Directives 2004/8/EC and 2006/32/EC (Official Journal of the European Union L 315 of 14 October 2012)
29. Directive 2009/33/EC of the European Parliament and of the Council of 23 April 2009 on the promotion of clean and energy-efficient road transport vehicles (Text with EEA relevance) (Official Journal of the European Union L 120 of 15 May 2009)
30. Regulation (EC) No 106/2008 of the European Parliament and of the Council of 15 January 2008 on a Community energy-efficiency labelling programme for office equipment (recast version) (Official Journal of the European Union L 39 z 13.02.2018)
31. Assessment and Comparison of National Green and Sustainable Public Procurement Criteria and Underlying Schemes, Report to the European Commission, 26th November 2010 ec.europa.eu/.../gpp/.../AEA_7%20June%202010.pd (access: 14.04.2016)
32. Doświadczenia z dialogiem technicznym ; wrzesień-październik 2013 (Eng. The experience with technical dialogue);September-October2013); http://zamawiajacy.pl/wpcontent/uploads/Do_wiadczenia-z-dialogiem-technicznym-artyku_-przyk_adowy.pdf; (access: 14.04.2016)
33. P. Wójcik, *Dialog techniczny zbyt rzadko stosowany*, LEX Zamówienia Publiczne (Eng: Technical dialogue used too rarely, LEX Public Procurement http://www.zamowienia-publiczne.lex.pl/czytaj/-/artykul/dialog-techniczny-zbyt-rzadko-stosowany (access: 14.04.2016)
34. Covenant of Mayors – CoM – The Agreement between the Mayors for Climate and Energy http://www.covenantofmayors.eu/index_en.html
35. J. Korytarova, T. Hanak, R., R. Kozik, E. Radziszewska Zielina, *Exploring the contractors' qualification process in public works contracts*, Creative Construction Conference (CCC) Kraków, 21-24.06.2015, PE, **123** (2015), pp. 276-283
36. H Walker, J. Niemczyk., T. Johnsen, R. Spencer, *Sustainable procurement: Past, present and future*, JPSM, **18** (2012) pp. 201–206
37. *Standardy Kontroli Najwyższej Izby Kontroli, publikacja elektroniczna* (Eng: Standards Control of the Supreme Chamber of Control, electronic publication), www.nik.gov.pl/kontrole/standardy-kontroli-nik/ (access: 14.04.2016)

Influence of excitations originating from moving subway onto transferring vibrations from ground to structure

Krzysztof Kozioł [1,a]

[1] *Cracow University of Technology, Cracow, Warszawska 24 St., 31-155, Poland*

Abstract. In the majority of these influences parameters describing them are known. Nevertheless, in case of vibration it is not always simple to determine their parameters, especially when the source is located outside. The paper presents problems associated with the building loads generating by the subway passages. Subway train moving in a tunnel, through the layout of the track, transfers energy to the housing tunnel, then to the ground which then propagates in the ground layers to the building. This causes movement of foundations generating kinematic excitation for the entire facility. Generally way of propagation of vibration is complex. During designing new facility, it is possible - at its future location – to measure vibration level of existing sources on the ground. Soil movements in the vicinity of the foundation of the house are not the same - they are quite often very different from each other. This is related to wave transitions through the border between the different soil layers. Therefore there is a need to determine the transfer function between the vibrations registered on the soil and potential real construction loadings. Such analysis were made for an existing building with six floors loaded with excitations caused by movement of the subway in a nearby tunnel. According to Polish Standard [1] PN-85/B-02170 ("Hazard rating of vibrations transmitted through the ground to the buildings") only vibrations with frequency up to 100Hz in terce octave bands (1/3 octave) were considered. As is apparent from the analysis the global amplitude value of vibrations registered on the ground is several times less than on the foundation of the building. However, in particular one-third octave bands (up to several Hz) it may happen that the acceleration vibration of the foundation is higher than on the surrounding terrain. Increased amplitudes may be important in the effort analysis of structure cross-sections as well as the effects of vibrations on people staying inside. The issue of loading building with vibrations caused by the movement of subway should take into account an analysis of the nature of amplitude-frequency at the interface soil-foundation.

1 Introduction

Every structure is exposed to a variety of loadings having the "static" [2] or "dynamic" [1,3-6] character. Static loads as opposed to dynamic actions are quite well discerned, described and standardized. In case of dynamic loads, in particular those for whom excitation is not directly associated with the object, the route (path), must be considered in which vibrations are affecting the structure (Figure 1) [7-8].

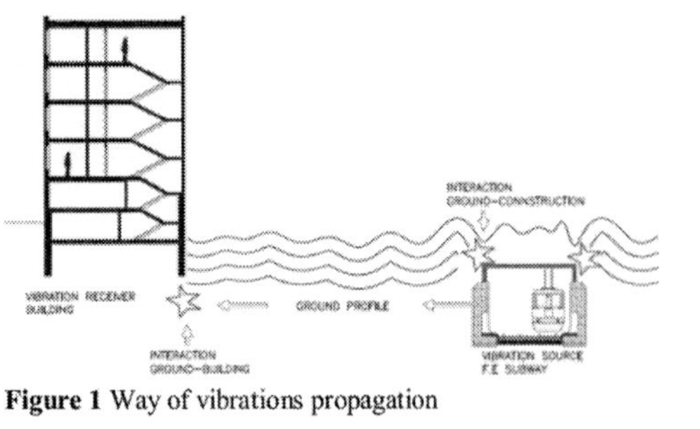

Figure 1 Way of vibrations propagation

During passage, subway transmits energy to the ground through the subgrade layers, from which energy propagates through the soils layer to the structure embedded in it causing movements of foundations, building and then the entire structure. It should be noted that the movement of the foundation and the soil located in the immediate vicinity are not the same, and quite often very different from each other. This issue is related to the wave transition through the various phases. The study presents the relationship between the load registered on the ground and fixed node structure located at its foundation level for building B (Figure 2).

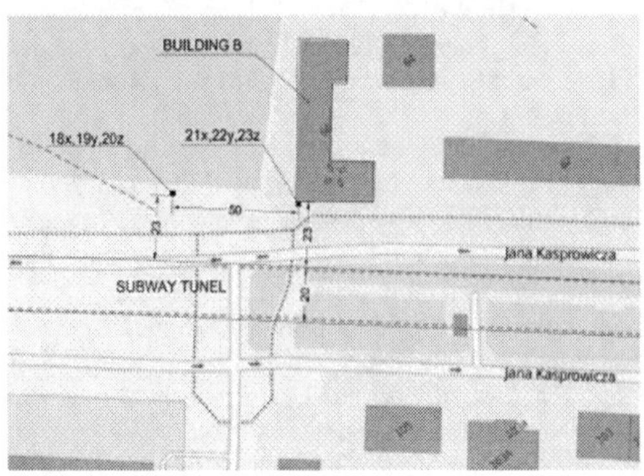

Figure 2 Building location

[a] Corresponding author: kkoziol@pk.edu.pl

 DOI: 10.1051/e3sconf/20161000045

Building B (Figure 3) it is six-storey structure with commercial units on the ground floor and living quarters on the top floor. It is made using traditional technology - brick walls, ceilings constructed on steel beams, roof of wooden structure. The building is located 23m horizontal and 14m vertical far away from the nearest subway track.

Figure 3 Photo of the building

Energy propagated in the ground has a wave character, so in order to avoid a possible interaction between the wave caused by subway and the reflected wave from the building the gauge located on the ground was shifted in relation to the building. In case of building B value of the sensors offset from the object was 50 m [8].

In order to measure acceleration, several gauges (accelerometers) were mounted on the building as well as on the ground. All measurements were carried out by the accredited Laboratory of Structural Strains and Vibrations located on Cracow University of Technology. This laboratory has a PCA accreditations. The study was performed in accordance with the following procedure:

- sensors calibration (PCB 393B12),
- installation of the sensors on the building (Figure 4, 5)
- installation and checking the measuring path,
- trial measurements in order to determine correct apparatus settings,
- performing the measurement - recording vibrations at the measurement points caused by individual events (subway rides and various vehicles) using LMS analyzer Scadas Mobile (16-channel version). Each event was written down in the measurement protocol,
- checking calibration after measurements,
- removal of the measurement path (after completion of all registrations).

Vibration measurements were made simultaneously in all measurement points located onto building.

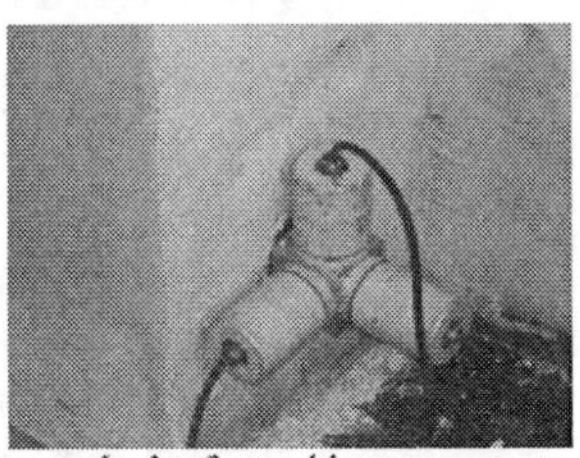

Figure 4 The method of attaching sensors to the structure (aluminum cube)

Gauges were attached to the building (Figure 4) using aluminum cube, rigidly glued to the corner of the building where two mutually perpendicular walls are facing - it allowed to neglect the impact of deformation of structural elements.

In case of sensors located on the ground, there was used special structure (Figure 5) in the form of a steel cross, attached to the ground by means of four ground anchors (bars ϕ16, length 25 cm).

Figure 5 The method of attaching sensors onto ground (steel cross)

All measurements were performed using PCB Piezotronics accelerometers 393B12 type having sensitivity of 10 V / g. Registration was performed using LMS Scadas Mobile analyzer.

2 Analysis

As result of registration, there were obtained waveforms of acceleration vibrations recorded for ground and foundation of the building due to subways. In total, for building there were made 44 registrations in three directions: horizontal "x" - parallel, "y" – perpendicular and "z"- vertical to the subway line.

Exemplary acceleration waveforms in the horizontal "y" direction are visualized in Figs. 6 and 7. As shown in Figure 6 and 7 on the phases boundary (i.e. ground and structure), reduction of extreme values of vibration acceleration amplitude is observed. This paper attempts to identify quantitative (signal amplitude) and qualitative (signal frequency) change in the structure of recorded signal after passing between the phases.

Analysis of relationship between measured peak ground acceleration and acceleration of the building was presented in relation to size of the recorded ground acceleration next to the building. For this purpose, the dimensionless coefficient P was defined as the ratio of acceleration vibration of the building to ground acceleration. This ratio is non-linear (Figure 8), wherein

DOI: 10.1051/e3sconf/20161000045

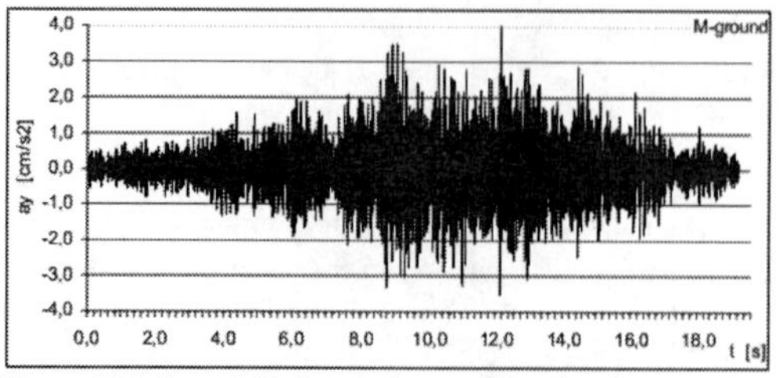

Figure 6 Horizontal (y-direction) ground acceleration

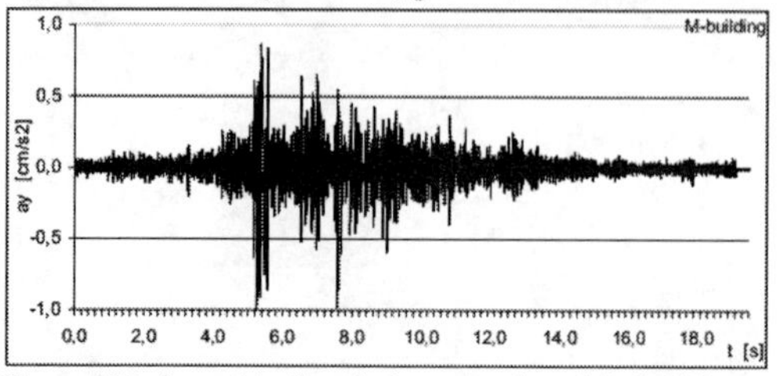

Figure 7 Horizontal (y-direction) acceleration of the building

the determination of fit coefficient of power curve is satisfactorily only for direction z (R^2=0.5547). In the horizontal direction determination of fit coefficient of power curve is unsatisfactory (it has value less than 0.5).

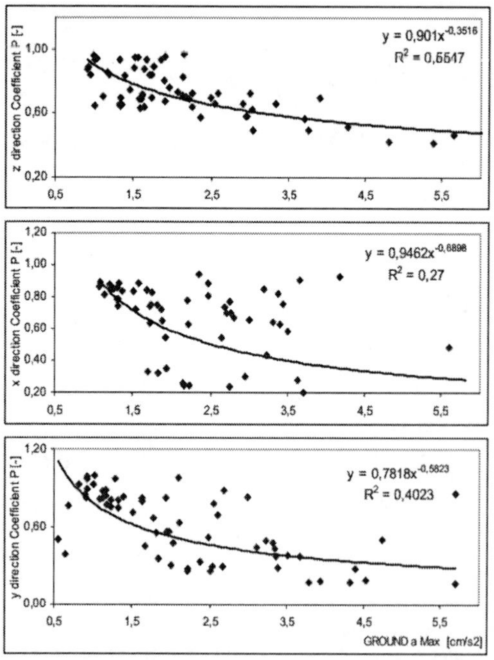

Figure 8. Change of the global maximum acceleration amplitude for the z, x y directions

However, it should be noted, that in each of the analyzed vibration direction growth of the excitation (which representative is the signal recorded on the ground) the P ratio is lower for each of the analyzed direction. So then vibration with higher amplitudes are more reduced.It was decided to separate the response of the building and ground for 1/3 octave band within range of the maximum amplitude in each one-third octave bands [1,5,6]. The value of this ratio (named ZM) on z, x, y directions z, x, y in each one-third octave bands was shown in Figure 9 - maximum change of coefficient ZM and Figure 10 - arithmetic mean average of coefficient ZM.

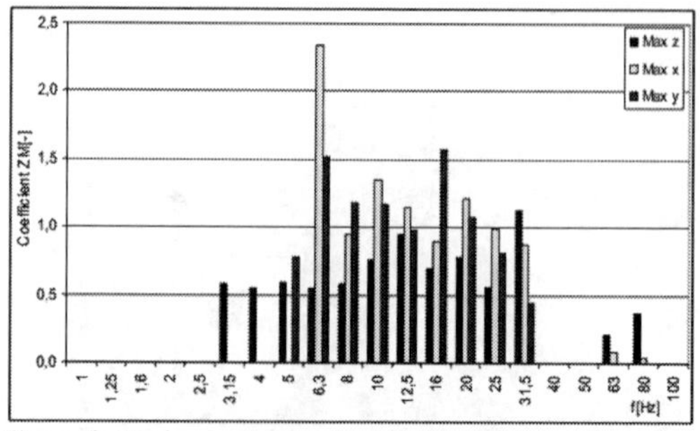

Figure 9. Maximum change of coefficient ZM factor depending on the center frequency band

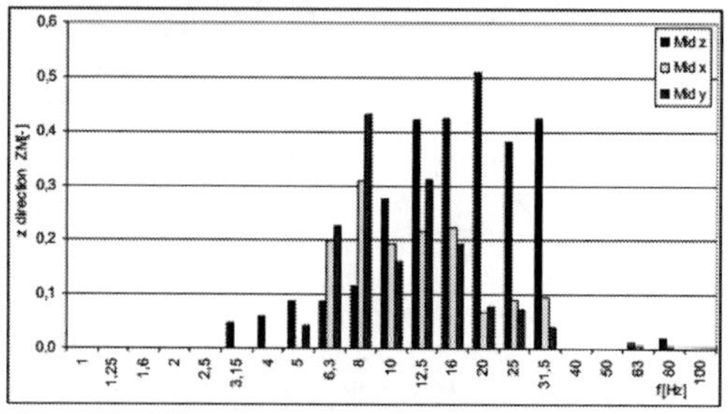

Figure 10. Arithmetic mean average of coefficient ZM factor depending on the center frequency band

3 Conclusions

Variation amount of the acceleration amplitude during transition from ground to building depends on excitation direction (subway location with respect to the object). Value of the global amplitude registered on the foundation of the building is lower than on the ground around it (Figure 8). Additionally it could be noted that the transfer ratio between ground and foundation of the analyzed object has a non-linear characteristics depending on the size of the acceleration amplitude of the ground. Nonlinearity coefficient increases with the amplitude of acceleration. Amplitude-frequency analysis (Fig. 10) show that the mean values of vibration acceleration of the building foundation are lower than the surrounding soil. As shown by the analysis of the test results it may happen that some of the components of acceleration of the foundation of the building (Figure 9) will be higher than the surrounding soil. For the considered building in case of horizontal direction within the range of 1/3 octave center frequencies it is the range from 6.3Hz to 20Hz. In contrast, for the vertical direction at a frequency 31,5Hz it can be seen 10% amplitude greater than in case of the acceleration of the building foundation of the surrounding soil.

Table 1 shows the coefficients of the trend function defined as P (for global acceleration amplitude) and ZM coefficients (assumed as a power function ($y=\alpha * aMax^{\beta}$))

DOI: 10.1051/e3sconf/20161000045

for individual center values of 1/3 octave filters of the maximum amplitudes of the vibration signal depending on the frequency mid-band period. The last line in Table 1 refers to the extreme amplitudes that occurred in the recorded signal – P (Figure 8).

Table.1 Coefficients of the trend function for the transfer function between the ground and the foundation of the building accelerations vibration for the center frequency in the 1/3 octave analysis

f	DIRECTION					
	x		y		z	
	α	β	α	β	α	β
[Hz]	[s*cm/s^2]	[-]	[s*cm/s^2]	[-]	[s*cm/s^2]	[-]
1	0,00	-1,01	0,00	-0,77	0,00	-0,91
1,25	0,00	-0,91	0,00	-1,42	0,00	-1,07
1,6	0,01	-0,62	0,00	-1,44	0,00	-0,70
2	0,12	-0,24	0,01	-0,65	0,00	-1,56
2,5	0,02	-0,50	0,00	-1,01	0,00	-1,14
3,15	0,00	0,40	0,02	-0,48	0,00	-1,03
4	0,00	-0,75	3,55	0,43	0,00	-0,79
5	0,04	-0,34	1,10	0,26	0,01	-0,43
6,3	0,06	-0,32	1,88	0,22	38,42	0,99
8	0,32	-0,16	0,00	1,30	0,67	-0,11
10	0,17	-0,32	1,07	0,04	0,64	-0,11
12,5	0,03	-0,85	0,45	-0,22	0,70	-0,08
16	0,06	-0,83	0,12	-0,61	0,53	-0,19
20	1,19	0,15	6,57	0,64	0,64	-0,19
25	3,25	0,56	2,84	0,50	0,56	-0,17
31,5	0,32	-0,34	0,67	0,06	0,53	-0,27
40	0,18	-0,27	0,17	-0,12	0,83	-0,09
50	0,17	-0,25	0,09	-0,09	0,65	-0,09
63	0,14	0,72	0,08	0,67	0,18	-1,42
80	0,06	0,73	0,05	1,53	0,26	-1,13
100	0,07	-2,00	0,05	-2,32	0,37	-0,79
GLOB	0,95	-0,69	0,78	-0,58	0,91	-0,35

Analysis of the slopes of the trend function P parameter revealed a large scatter, which depends on third-octave band center frequency (Figure 11-13).

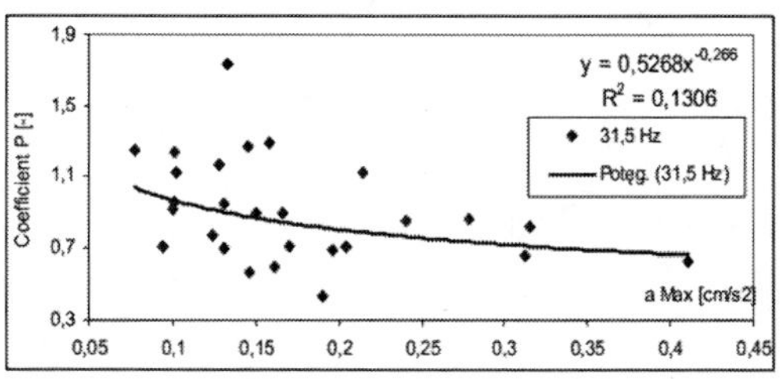

Figure 11 Change of the acceleration amplitude for the 31,6 Hz terce in the z-direction

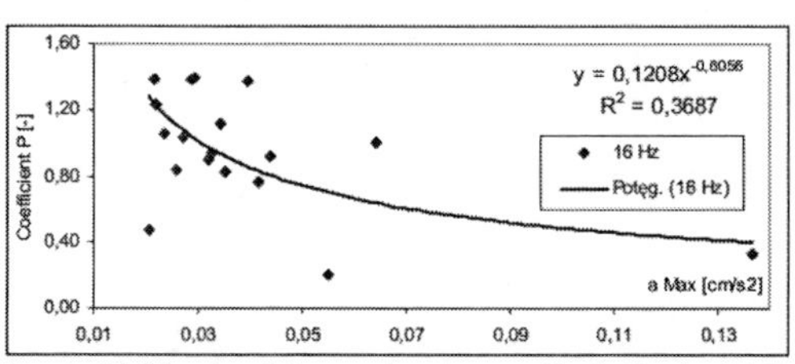

Figure 12 Change of the acceleration amplitude for the 16 Hz terce in the y-direction

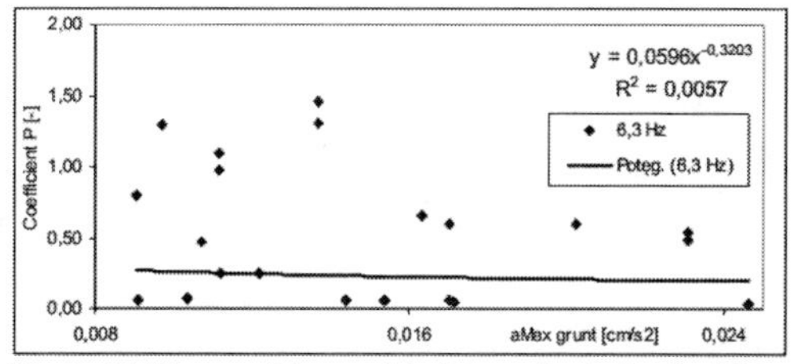

Figure 13 Change of the acceleration amplitude for the 6,3 Hz terce in the x-direction

Due to the non-linear relationship between the size of the vibration acceleration amplitude change at the interface between ground and foundation of the building during determining the kinematic force coming from subway it is necessary to verify it experimentally. The parameter of this verification besides the vibration amplitude should be its frequency and direction of incoming wave - location of the subway in relation to the building.

References

1. PN-85/B-02170 (polish standard), Hazard rating of vibrations transmitted through the ground to the buildings
2. ISO 6897: Guidelines for the evaluation of the response of occupants of fixed structures, especially buildings and off-shore structures, to low frequency horizontal motion (0,063 tol Hz), 1984, International Organization for Standardization
3. ISO 10131:2007, "Bases for design of structures – Serviceability of buildings and walkways against vibrations"
4. BS 6472-1:2008, Guide to evaluation of human exposure to vibration in buildings, Part 1: Vibration sources other than blasting, 2008,
5. DIN 4150-3, Structural vibration, Part 3: Effects of vibration on structures
6. ISO 2631-2, Guide to the evaluation of human exposure to whole body vibration. Part 2- Vibration in buildings, 2003, International Organization for Standardization.
7. Kozioł K., Stypuła K.: Vibration measurements and computer simulations in the design of surface rail vibration isolation. Mat. VIII conf. Modern technologies and management systems for railways. Papers Scientific and Technical Association of Engineers and Technicians of Poland, Branch in Cracow, Zakopane 2-4.12.2009 r.
8. Stypula K., Koziol K.: Effects of changes to the track superstructure on human perception of rail-induced vibrations in a selected building near the rail tunnel in Warsaw – predictions and measurements. Proceedings of the 7th European Conference on Structural Dynamics, EURODYN 2008, 7-9 July 2008 Southampton, s. 58.

E3S Web of Conferences 10, 00046 (2016) DOI: 10.1051/e3sconf/20161000046
SEED 2016

Impact of vibrations on people residing in buildings from different types of traffic loads according to the VDV scale

Krzysztof Kozioł [1,a]

[1] *Cracow University of Technology, Cracow, Warszawska 24 St., 31-155, Poland*

Abstract. Development of communication lines as well as reducing the distance of its location from buildings contributes to the deterioration of human comfort residing in buildings. For that reason it is necessary to determine the size of this nuisance. This paper presents a comparative analysis in order to determine influence of vibrations originating from selected means of transport for people in buildings using ISO 2631-1, British BS 6472-1 and Polish pr PN/B-02171 standards. The analysis focuses mainly on the parameter called dose of vibrations - VDV. In the comparative analysis - as the load acting on the human - vibration acceleration of ceilings caused by crossing subway, truck, tram and train were used. Presented in the paper analysis show, that each types of considered traffic excitations gives similar results for all concerned standards. In case of BS648 standard values of VDV are smaller in vertical direction than for other standards. However, for extortions, in which in the amplitude-frequency characteristics dominate lower natural frequencies, difference between obtained results decreases.

1 Introduction

Comfort of the people staying in buildings is affected by floor deflections as well as its response in time, which can be caused by following factors:

- reaching the occupied space from outside causing the movement of the building foundations - it could be a seismic or so called paraseismic vibrations, transmitted from ground to building or vibrations transmitted through the outside barriers (walls) affecting changes in air pressure (eg. acoustic wave),
- generated in the building in which there is a room dedicated to people accommodation as well as generated by people - for example, walking in the room.

Considering the question of comfort people staying inside the rooms of existing building subjected to dynamic excitations we have to choose assessment criteria and determine the parameters describing them. It is quite obvious that for almost identical dynamic loads, analysis result of vibrations impact on people will be different for standards binding in different countries. However, there is a question about discrepancies between the different dynamic influences (eg movement of cars, trams, subway etc.) which are defined according to the same criteria but using different standards.

2 Historical background

At the beginning of the twentieth century (in 1931) the Reicher's and Meister's scales were published, which are defining relationship between frequency of vibration and movement of the structure in the vertical direction (Fig. 1). These scales were prepared on the basis of the behavior of people subjected to stationary excitations. Therefore, they cannot be used for any kind of structure, in particular for the buildings with high damping, in which response to the generated vibration is non-stationary. Lenzen's [1] investigations concerning reinforced concrete and steel ceilings subjected to impact loads, as well as Murray's [2] studies including implementations for ceiling construction different types of steel; contributed to introduce proposals to reduce the applicability (due to the force type) of modification (due to frequency response of the object) the Reicher 's and Meister's scales.

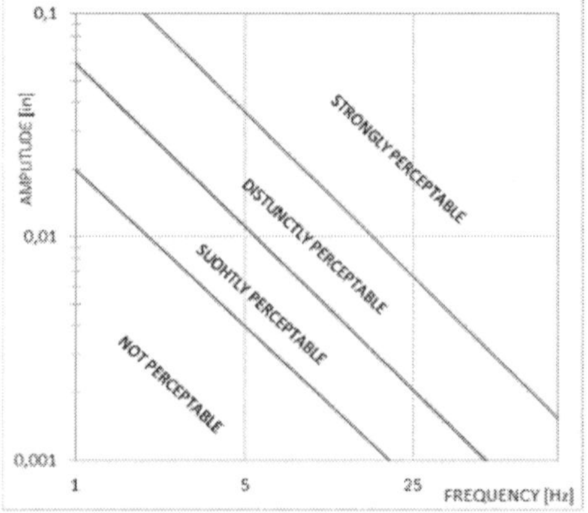

Figure 1 Reicher's and Meister's scale [7]

[a] Corresponding author: kkoziol@pk.edu.pl

DOI: 10.1051/e3sconf/20161000046

In 1974 it was published McCormick work [3] concerning selection of the design criteria for the construction of floors. According to his research, damping is an important parameter determining about admission of ceiling for application in construction. In 1974, Wiss and Parmelee have developed experimental five-level scale (from harmless to harmful) connecting vibrations of the floor and evaluation of its response for human. In 1974 also appeared first edition of ISO standard concerning the human response to vibration of the floor. These guidelines are result of research carried out by Crowley. This standard through the work of Irwin (1978) has been modified and published in 1985 under the ISO 2631name. On its basis guidelines of standards are prepared for many regions, in particular British BS 6481 and Polish pr-PN / B-02171.

3 Assumptions

Standards: ISO 2631-2 [4], BS 6472-1 [5] British and Polish pr PN/B-02171 [6] introduces following factors assessing the impact of vibration on human:

- purpose of the room in building,
- part of the day, in which vibrations occur,
- nature of vibrations and their repeatability,
- direction of vibrations and position of human body while receiving vibrations.

During evaluation of impact of vibrations on people in buildings, it is essential to select proper criteria and related to it assessment methods and parameters, which include:

- acceleration (velocity, displacement) during vibrations (in considered frequency range),
- amplitude-frequency spectrum of acceleration (velocity, displacement), in particular the effective value of vibrations (RMS) in 1/3 octave band,
- vibration dose value (VDV),
- Maximum Transfer Value of the vibrations transition (MTV).

According to ISO 2631, BS 6472-1 British and Polish prPN/B-02171 as a basic evaluation parameter it is assumed analysis of effective response of structure RMS. However, in justified cases, it is recommended to use additional methods: VDV and MTV. These situations include cases, in which the peak factor during analyzed measurements is greater than 6. As the event time one should take time interval in which the beginning and end value reaches 20% of the extreme amplitude of measured parameter (in this case - acceleration).

Peak factor "As" (equation 2) is defined as ratio of the absolute value within extreme values of measured parameters to mean value (RMS-equation 1) of this parameter during measuring time.

$$\mathrm{RMS} := \left(\frac{\int_0^T a_w(t)^2 \, dt}{T} \right)^{0.5} \tag{1}$$

$$A_s := \frac{A_{max}}{\mathrm{RMS}}, \tag{2}$$

where: $a_w(t)$ – is weighted acceleration as a function of time [m/s2];
T – is the duration of measurement [s],
A_{max} - extreme acceleration amplitude

VDV (equation 2) parameter is called Vibration Dose Value and is defined by following relationship:

$$VDV = \left[\int_0^T a_w^4(t) dt \right]^{\frac{1}{4}} \tag{3}$$

It should be noted that for acceleration $a_w(t)$ it is imposed filter, which is visualized in Fig. 2. and table 1

Table 1. Weighting values for accelerations of the vertical (z-axis) and horizontal (x,y-axis) directions in a 1/3-octave band spectrum 1-80Hz

RMS ferguency	ISO 2631, prPN/B02171		BS 6481-1		% average	
	Vertical	Horizontal	Vertical	Horizontal	Vertical	Horizontal
[Hz]	[-]		[-]		[%]	[%]
1,00	0,482	1,011	0,500	1,000	-4%	1%
1,25	0,484	1,008	0,559	1,000	-13%	1%
1,60	0,494	0,968	0,632	1,000	-22%	-3%
2,00	0,531	0,890	0,707	1,000	-25%	-11%
2,50	0,631	0,776	0,791	0,800	-20%	-3%
3,15	0,804	0,642	0,887	0,635	-9%	1%
4,00	0,967	0,512	1,000	0,500	-3%	2%
5,00	1,039	0,409	1,000	0,400	4%	2%
6,30	1,054	0,323	1,000	0,317	5%	2%
8,00	1,036	0,253	1,000	0,250	4%	1%
10,00	0,988	0,212	0,800	0,200	24%	6%
12,50	0,902	0,161	0,640	0,160	41%	1%
16,00	0,768	0,125	0,500	0,125	54%	0%
20,00	0,636	0,100	0,400	0,100	59%	0%
25,00	0,513	0,080	0,320	0,080	60%	0%
31,50	0,405	0,063	0,254	0,063	59%	0%
40,00	0,314	0,049	0,200	0,050	57%	-1%
50,00	0,246	0,039	0,160	0,040	54%	-3%
63,00	0,186	0,030	0,127	0,032	46%	-8%
80,00	0,132	0,021	0,100	0,025	32%	-16%

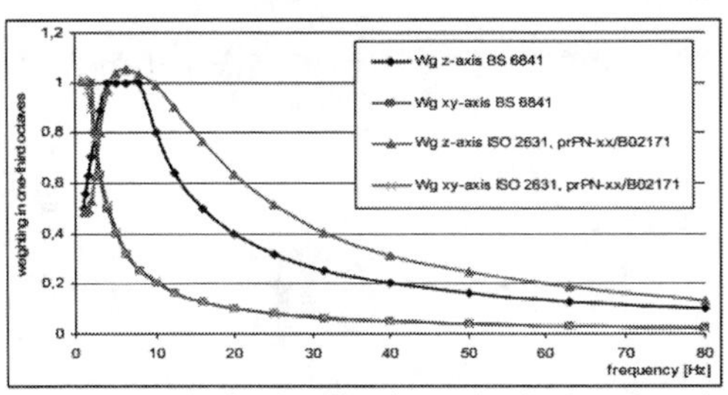

Figure 2 A graph of weight filter for acceleration in horizontal and vertical directions for ISO2631, British BS 6481 and BS Polish pr-PN/B-02171 standard.

As is clear from figure 2 and table 1 weight factors for the ISO 2631 and Polish pr PN-8xx / B-02171 are the same regardless of the direction of vibration. In case of horizontal direction for each of mentioned standards weight coefficients of filters are similar. The aim of the study is to show differences between values of the

DOI: 10.1051/e3sconf/20161000046

same parameter (here VDV) obtained using different standards.

If the vibration exposure is comprised of two or more events "i" which are characterized by VDV_i values, and during day or night occurred N events that characterize the VDV_i, therefore total value of dose vibration VDV could be determined from formula (4):

$$VDV = [\sum(VDV_i)^4]^{0,25} \quad (4)$$

According BS 6472-1 (Table 2) is a low probability of adverse comment or disturbance to building occupants at vibration values below the preferred values. Complaints may be expected if vibration values approach the maximum values. Activities should be designed to meet the preferred values where an area is not already exposed to vibration.

Table 2. Acceptable vibration dose values for intermittent vibrations [$m/s^{1.75}$]- [5]

Location	Daytime (7.00am to 10.00pm)		Night-time (10.00pm to 7.00am)	
	Preferred value	Maximum value	Preferred value	Maximum value
Critical areas1	0.10	0.20	0.10	0.20
Residences	0.20	0.40	0.13	0.26
Offices, schools, educational institutions and places of worship	0.40	0.80	0.40	0.80
Workshops	0.80	1.60	0.80	1.60
Examples include hospital operating theatres and precision laboratories where sensitive operations are occuring. These criteria are only indicative, and there may be a need to assess intermittent values against the continuous or				

4 Analysis

The paper analyses the VDV factor for vibration accelerations recorded on ceilings for real objects. These accelerations are caused by moving cars, trams, subway a and trains (Fig. 3).

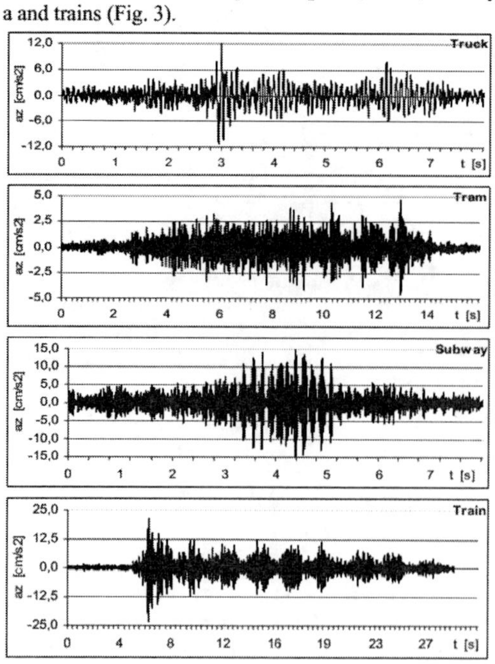

Figure 3 Plot of vertical accelerations of vibrations of the floor caused by the movement of truck, tram, subway and train.

These are vertical components of ceiling vibration acceleration (unweighted) registered on real buildings. Car and subway as well as tram movements are associated with the same building.

As is apparent from the performed amplitude-frequency analysis (Figure 4) the ceiling responses caused by moving truck are dominated by the frequency of 8 to 12 Hz, for which the weight factor of vibration acceleration varies in the range of 1.036 to 0.902 for the ISO standard and from 1.00 to 0.640 for BS standard (Table 1)

Influence of the floor acceleration amplitudes in the other frequency bands is negligible. Considering the fact that in the dominant frequency range of excitation associated with passing car, differences between the ISO and BS standards are lowest - ranging from 4% to 24% (Table 1), discrepancy in the VDV ratio are equal to 27% - Table. 3.

Other structure responses are characterized by vibrations amplitudes components with higher frequencies.

- for tram - dominant component of acceleration amplitudes are focused around a frequency of 20Hz,
- for subway one can observe in floor responses accelerations with frequencies ranging from 40Hz to 80Hz, and the largest accelerations amplitudes have frequencies around 55Hz,
- influence of train is characterized by floor accelerations with frequencies above 30 Hz and dominant frequency is 45Hz.

While above the frequency 16 Hz the ratio of the weighting factor applied to the acceleration proposed in ISO standard

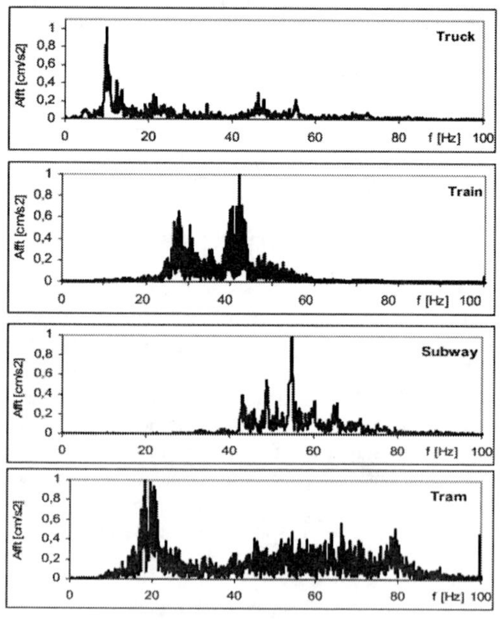

Figure 4. Amplitude-frequency analysis of excitations coming from car, tram, subway and train traffic

to its counterpart applied in the BS standard is more than 50% (with a slight reduction to the 32% at a frequency of 80Hz) afforded the difference between VDV for excitations which generate vibrations with frequencies greater than 16Hz at a similar level (from 52% to 58%) - Table. 3. The peak factors for analyzed acceleration

DOI: 10.1051/e3sconf/20161000046

waveforms (non-weighted) are respectively 12.34 for cars, 6.02 for trams, 15.9 for subway and 22.1 for trains. Applying of the weight factors for accelerations in the present case was the cause of a reduction in the peak factor:

- for ISO standard 28.7% in case of cars, 75.8% for subway, 44.8% for trams and 36.4% for trains,
- for BS standard 46.4% in case of cars, 84.1% for subway, 64.8% for trams and 77% for trains.

Such a large reduction of peak coefficients after weighing stems from the fact of wide spectrum of acceleration and a small difference between the dominant signal amplitudes and other amplitudes. In the case of car movements the dominant amplitude of the ceiling response expressly dominates over the other, and the frequency spectrum is very narrow therefore weighting the signal has a very small impact on change of the event time (signal duration) and the change in amplitude is close to its simple rescaling.

Table 3. Analyzes results

Value	Signal/change	Dimension	Tram			Train		
			ISO 2631 pr PN-/B-02171	BS 6472-1	Difference ISO 2631 ,pr PN-/B-02171/BS	ISO 2631 pr PN-/B-02171	BS 6472-1	Difference ISO 2631 ,pr PN-/B-02171/BS
RMS	a	[cm/s²]	0,63		-	2,70		-
	a_w	[cm/s²]	0,30	0,19	55,8%	0,96	0,61	58,1%
	change	[%]	52,2%	69,3%		64,4%	77,4%	
Peak factor	a	[cm/s²]	6,02		-	21,22		-
	a_w	[cm/s²]	2,58	1,65	56,7%	7,73	4,89	58,1%
	change	[%]	57,2%	72,7%		36,4%	77,0%	
VDV	a	[cm/s^1,75]	2,67		-	10,99		-
	a_w	[cm/s^1,75]	1,39	0,89	56,5%	3,97	2,61	58,2%
	change	[%]	47,9%	66,7%		36,1%	77,2%	
	daytime	[cm/s^1,75]	5,79	3,70	-	11,05	6,99	-
Value	Signal/change	Dimension	Truck			Subway		
			ISO 2631 pr PN-/B-02171	BS 6472-1	Difference ISO 2631 ,pr PN-/B-02171/BS	ISO 2631 pr PN-/B-02171	BS 6472-1	Difference ISO 2631 ,pr PN-/B-02171/BS
RMS	a	[cm/s²]	1,44		-	1,61		-
	a_w	[cm/s²]	1,22	0,96	26,9%	0,37	0,24	51,9%
	change	[%]	15,4%	33,3%		77,0%	84,9%	
Peak factor	a	[cm/s²]	12,34		-	15,90		-
	a_w	[cm/s²]	8,80	6,62	32,9%	3,85	2,52	52,7%
	change	[%]	28,7%	46,4%		75,8%	84,1%	
VDV	a	[cm/s^1,75]	5,28		-	8,50		-
	a_w	[cm/s^1,75]	4,40	3,46	27,2%	1,96	1,29	52,0%
	change	[%]	16,6%	34,4%		77,0%	84,9%	
	daytime	[cm/s^1,75]	18,32	14,40	-	8,14	5,36	-

Table.3 shows daily values of VDV (daytime) coefficients. It was assumed 300 passages of trams, subways and cars as well as 60 train passages. As we can see in each of the analyzed cases there is low probability of occurrence of complaints residents because the rooms in which the measured acceleration have residential destination.

5 Conclusion

Estimation of vibration harmfulness felt by people in buildings using VDV ratio according to ISO2631 and estimated in accordance with standards binding in various countries (eg PN, BS) does not lead to convergent results. The reason for this are the different weighting functions imposed on the measured acceleration (Fig. 2). Due to the non-linear relationship between weighting function imposed on the amplitude and frequency of their prevalence, resulting from the analysis VDV values are a function of the spectral distribution of the recorded acceleration. Due to the fact that the recorded ceilings responses (resulting from car, tram, metro and train passages) acceleration amplitude at frequencies below 5 Hz are negligible (less than 0.5%) and VDV ratios calculated according ISO standard are greater than if they were calculated according BS standard. The smallest differences between the analyzed standards are achieved for the passage of the car. This is due to the smallest differences in the weighting functions applied to the amplitudes occurring in the dominant frequency response of the structure. The differences in the VDV values (for the tram, train and subway) obtained in accordance with ISO and BS are at 52% -58% and are due to differences (from 46% to 60%) between the weighting factors for the acceleration amplitudes of the dominant frequencies present in the signal.

So without any information about the amplitude-frequency structure of response, referencing VDVs obtained in accordance with various standards to the vibrations perceptibility threshold can cause errors.

According to pr PN/B-02171 standard, analysis using the VDV is treated as informational used to assess complaints of inhabitants. Author proposes usage of weighting coefficients for the acceleration in accordance with ISO, and the threshold values compatible with the BS. This means greater than in BS-standard probability of residents complaints due to traffic loads.

References

1. Lenzen, K. H., "Vibration of Steel Joist-Concrete Slab Floors," Engineering Journal, American Institute of Steel Construction, 3rd Qtr., pp. 133-136, 1966.
2. Murray, T. M., "Design to Prevent Floor Vibrations," Engineering Journal, American Institute of Steel Construction, 3rd Qtr., pp. 82-87, 1975.
3. McCormick, M. M., Office Floor Vibration: Design Criteria and Tests, Melbourne Research Laboratories, The Broken Hill Proprietary Company Lmt.f Melbourne, Australia, 1974.
4. ISO 2631-1, Mechanical vibration and shock – ewaluation of human exposure to whole-body vibration Part 1 General requirements 1997 International Organization for Standardization.
5. BS 6472-1:2008, Guide to evaluation of human exposure to vibration in buildings, Part 1: Vibration sources other than blasting, 2008,
6. Pr PN/B-02171, Evaluation of vibrations influence on people in buildings
7. Lara, K.M, Ricadro, C. "Parametric study on the vibration sensitivity of hollow-core slab floors" Eurodyn 2014

 DOI: 10.1051/e3sconf/20161000047

Impact of the vibrations on the environment caused by passages of trains at variable speed

Barbara Kożuch[1,a] and Tadeusz Tatara[1]

[1]*Institute of Structural Mechanics, Cracow University of Technology, Poland*

Abstract. The paper deals with negative environmental impact caused by the passages of different kinds of trains at variable speed. The study is based on the measurement results which took place in Poland in 2013 on the railway line no. 4. The effect of the traction unit – Pendolino (EMU 250) on the vibration climate was analysed. The impact of passages of new trains was compared to currently operated rolling stock. The speed of trains was varying between 40 and 250 km/h. Vibration measurements were conducted by stuff of an accredited Laboratory of Structural Mechanics at Cracow University of Technology (Accreditation No. AB 826). The influence of the indicated vibrations due to passages of the trains on the building in the neighbourhood of the line was investigated. The vibration assessment was done for horizontal components of vibrations according to Polish standard code. Assessment of environmental impact was presented by indicator of perceptibility of vibration through construction (WODB), which refers to the Scales of Dynamic Influences (SDI scales). The limits specified by standards in any of the passages have not been exceeded. The change of speed or rolling stock resulted in a change in the characteristic of the vibration spectrum.

1 Introduction

Although it is not commonly associated with environmental pollution, vibration from roads and railways has a negative impact on the surrounding. Numerous standards and laws (both Polish [1,2] and foreign [3-5]) govern issue of vibration emission. Using world literature in the field of mechanical vibration, among others [6-8], it can be stated that the issue of vibration excited by trains is an important issue of the modern world. The problem is increasing in proportion to the increase in train speed and the creation of new high-speed line. As mentioned in [9], ground vibration limits are exceeded in 44% of assessment (worldwide investigation). Review of over 2300 technical railway vibration papers and holistic vibration prediction (from track to nearby building) we can found in [10].

2 Background information

In November 2013, before the authorization for the use of newly purchased trains, the Polish network line manager organized test during the passage of Pendolino (EMU 250) at speed varying between 40 and 293 km/h. This opportunity was taken to perform the vibration measurement e.g. on the track, ground and building. Electric Multiple Unit was crossing section Psary - Góra Włodowska (approximately 36 km) located within the railway line No. 4 (CMK). The train was following track "A" with closed two tracks for other vehicles during three weekends. During weekdays analogical studies were done for rolling stocks running the CMK section. Similar studies were performed and discussed in paper [11].

The paper presents selected results of measurements of building vibration (horizontal components of vibration: x – perpendicular and y – parallel to the axis of the track), in one of the three measuring polygons. The assessed building was located within 50 m of the track. Passages of the EMU 250, the InterCity and the InterRegio trains induced vibrations. The data from trains runs with a speed of 40 and 250 km/h is a base of comparison.

The accredited Laboratory of Structural Mechanics at Cracow University of Technology (Accreditation No. AB 826) performed in situ measurements.

Analyses were performed based on the knowledge and experience of the researchers and are also available in references [12-17].

3 Measuring polygon and equipment

Vibration measurements were performed with piezoelectric accelerometers' PCB Piezotronics and analyzer SCADAS Mobile's LMS International. The relative standard uncertainty of the maximum signal acceleration must not exceed ± 11.61%, which is the sum of the deviations of nominal sensors and amplifier distortion by the digital recorder and the accuracy and linearity of the amplifier. In the test, schematic arrangement of sensors is shown in Fig. 1. Tested

[a] Corresponding author: kozuchbm@gmail.com

residential building was 50 m from the track A. The tracks and structures were separated by shallow drainage ditch.

Two sensors recording the horizontal vibrations: perpendicular (labelled as P-01x) and parallel (labelled as P-02y) to the x and y axis respectively. Accelerometers were fixed on the building foundation, at ground level. Mounting the sensor on the basis was carried out according to the method featured in [18].

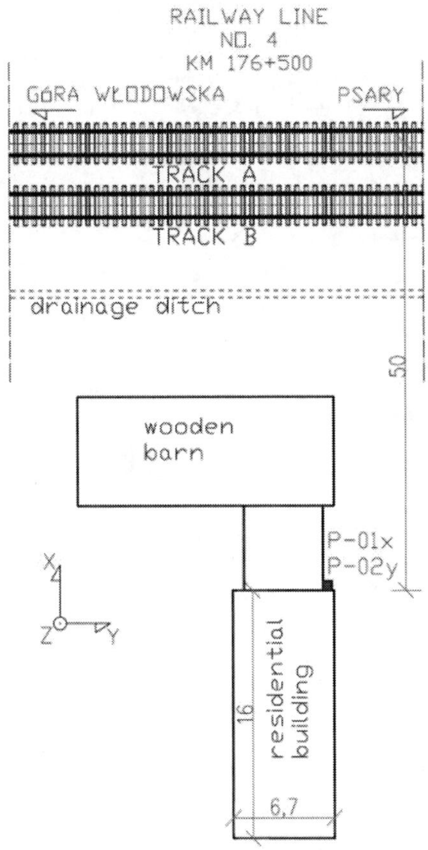

Figure 1. Schematic arrangement of sensors in the measuring profile

4 Analyses

4.1 Analysis of acceleration records

There were analysed all measured time histories and frequency contents of the horizontal accelerations of the residential building. The exemplary time history of the building acceleration recorded at the ground level during the passage of Pendolino with speed 250 km/h is presented in the Fig. 2.

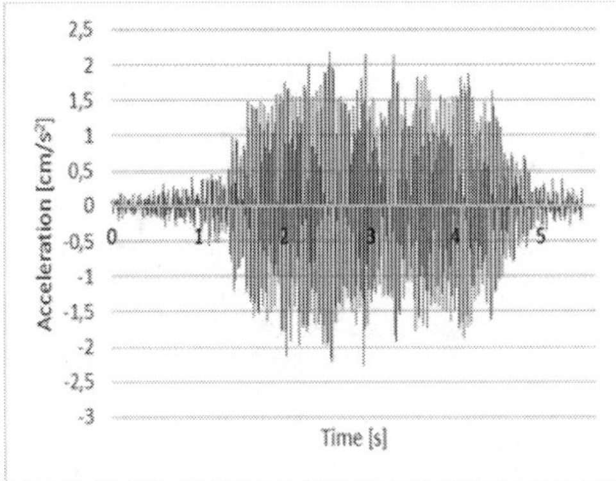

Figure 2. Measured time history of horizontal acceleration of the building at ground level during passage of Pendolino with a speed v=250 km/h

Fig. 3 shows the maximum values obtained from the time history of the acceleration of horizontal components P-01x of vibrations caused by passages of the Pendolino and other trains with speed 40-250 km/h. Similarly, Fig. 4 presents the maximum values of P-02y sensor.

The growth curves of maximum values of vibration acceleration versus velocity of the EMU 250 were fitted and shown in Figures 3 and 4. For each set of data there is the exponential function given. It describes the dependence of the maximum acceleration of train speed. The coefficients of determination R^2 for both functions is above 0.8. The maximum values of vibration acceleration showed significantly difference depending on the type of train (Pendolino or InterRegio). The extremal values of other trains are often higher than Pendolino's once.

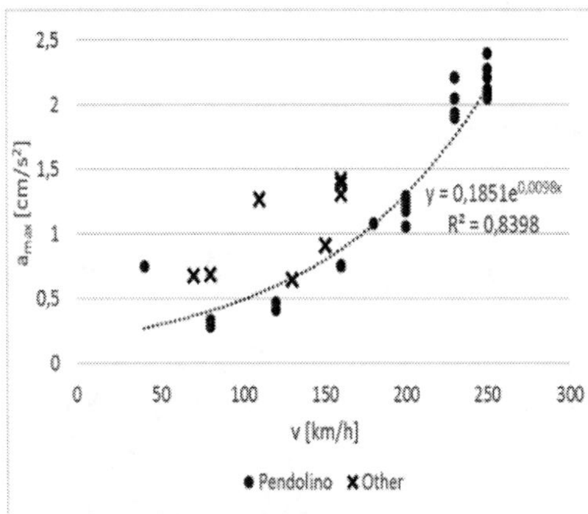

Figure 3. The maximum value of the horizontal component x of vibrations from the recorded accelerations caused by trains at variable speed

DOI: 10.1051/e3sconf/20161000047

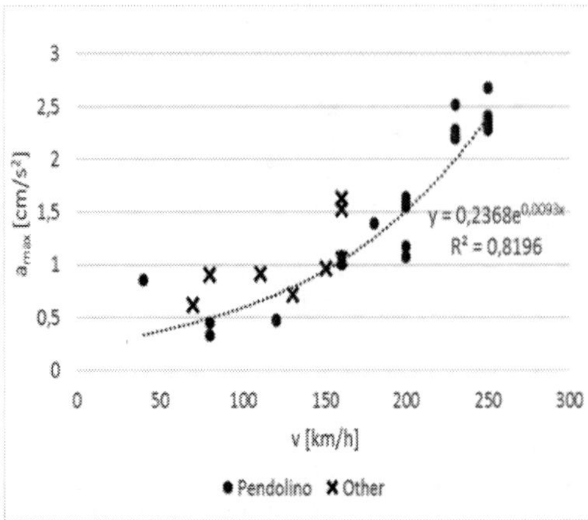

Figure 4. The maximum value of the horizontal component y of vibrations from the recorded accelerations caused by trains at variable speed

4.2 Vibration spectrum analysis – SDI scale

Assessment of vibration harmfulness for the building was done for the horizontal components of vibrations (X and Y) according to Polish standard [19]. The investigated structure was classified to scale SDI-II. The measured time histories of the horizontal acceleration were analysed using one-third octave band frequency filters and distributed in one-third octave bands of the mid-band frequency from 1 to 100 Hz. For example, Fig. 5 summarize the results of the horizontal vibration (sensor P-01x) recorded from Pendolino passages at speeds of v = 160 km/h (the blue columns). The obtained results are compared to the Scales of Dynamic Influences (SDI scales) (the red line). The maximum accelerations up to 6,3 Hz are smaller than for higher frequencies. All acceleration values in 1/3 octave bands are located in I zone of the SDI scale (below red line). It means that vibrations are not harmful for the whole building.

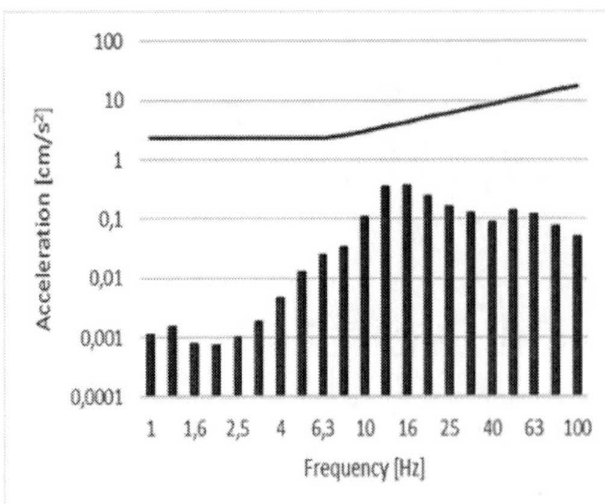

Figure 5. Vibration spectrum in one-third octave frequency bands with SDI scale – the measuring point P-01x – Pendolino trains – track A – speed trains v = 160 km/h

4.3 Vibration spectrum analysis – WODB

Assessment of environmental impact was presented using indicator of perceptibility of vibration through construction (WODB). In our case, indicator WODB is the largest value of the maximum values of the ratio of the vibration acceleration parameters in one-third octave bands to the acceleration corresponding to the lower limit consideration of dynamic influence on buildings due to the SDI scales in the same frequency band. The indicator is expressed as two numbers: dimensionless, designated in the manner mentioned above and middle frequency of one-third octave band in which the WODB is determined. The indicator was established for each passage. The usefulness of this ratio is fact, that the final result becomes independent of the frequency band. The WODB indicator shows directly how many times the lower limit consideration of dynamic influence on buildings has been exceeded [13].

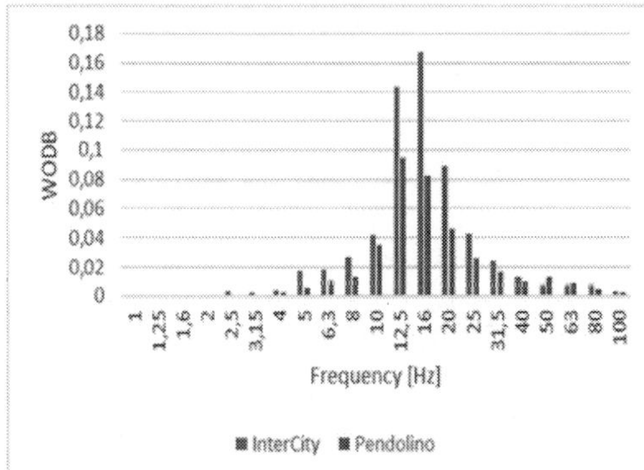

Figure 6. WODB spectrum in one-third octave frequency bands – the measuring point P-01x – Intercity and Pendolino trains – track A – speed trains v = 160 km/h

It was compared two types of rolling stocks: Intercity and Pendolino at speed 160 km/h (Fig. 6. for sensor P-01x and Fig. 7 for sensor P-02y). Both trains excite the greatest values of acceleration in the middle bands (12,5 – 20 Hz), but the values obtained by EMU 250 are lower especially in mentioned bands for horizontal vibration perpendicular to the axis of the track. The vibration parallel to the axis of the track is not as unambiguous. The differences between the values are smaller. The dominant bands extended between 10 – 20 Hz. For the frequency 12,5 Hz the EMU 250 train gets higher indicator. Nevertheless, this proves that using the new trains reduce negative impact of the vibration on the environment.

Fig. 8 and 9 present the impact of vibration on building depends on track which the train follow by. The analysis has been made for the Intercity train at speed v=160 km/h. Despite of the fact that the track B is closer the building, the results are not obvious. The records from P-01x and P-02y sensors have higher values of acceleration parameters for frequencies up to 12,5 Hz for track A.

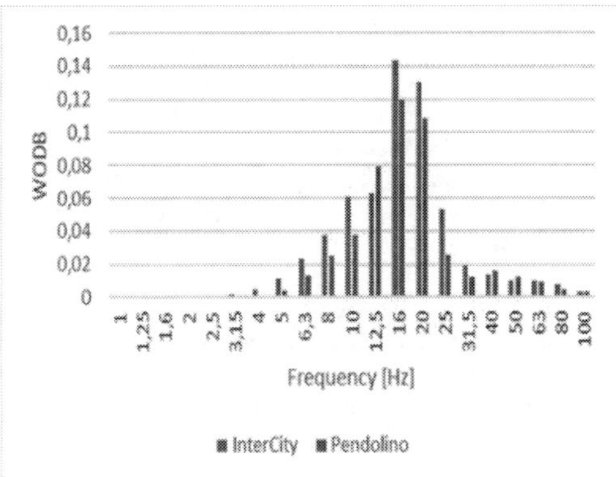

Figure 7. WODB spectrum in one-third octave frequency bands – the measuring point P-02y – Intercity and Pendolino trains – track A – speed trains v = 160 km/h

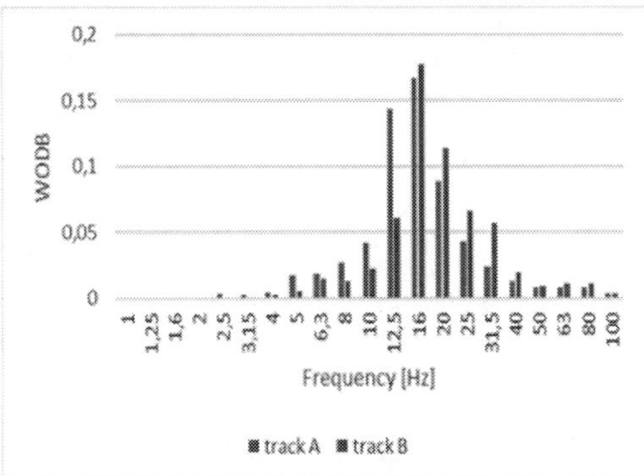

Figure 8. WODB spectrum in one-third octave frequency bands – the measuring point P-01x – Intercity trains – track A – speed trains v = 160 km/h – track A and B

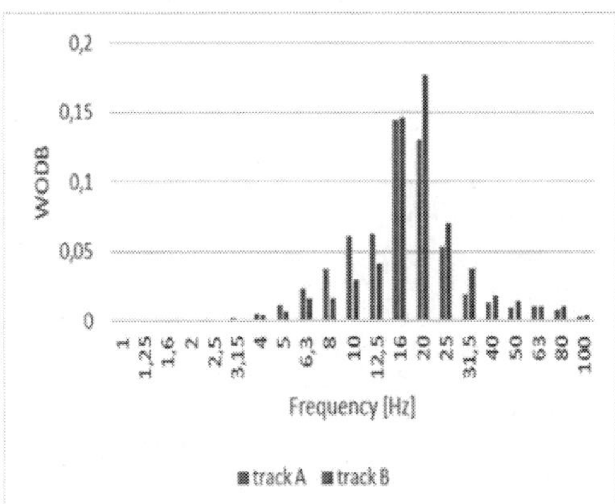

Figure 9. WODB spectrum in one-third octave frequency bands – the measuring point P-02y – Intercity trains – track A – speed trains v = 160 km/h – track A and B

To compare the vibrations perpendicular and parallel to the track axis Fig. 10 presents obtained maximum values of the coefficient WODB from all passage of Pendolino train for accelerometer P-01x and P-02y. The biggest difference reveals in the band 16 Hz, in which vibrations parallel to the axis have stronger influence on building. On the other side, it was compared similarly vibration excited by Intercity train at speed 160 km/h. The received results are less similar. The parameters of component y of vibrations are even two times higher than for component x for the frequency band 12,5 Hz. For frequency band 20 Hz, sensor P-02y measures higher acceleration.

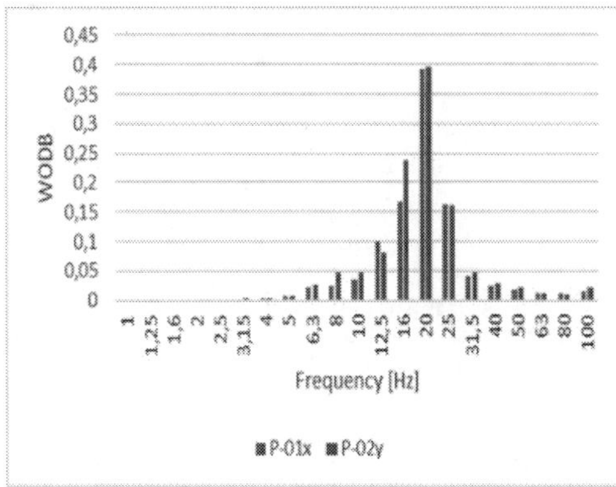

Figure 10. WODB spectrum in one-third octave frequency bands – the measuring point P-01x and P-02y – Pendolino – maximum values

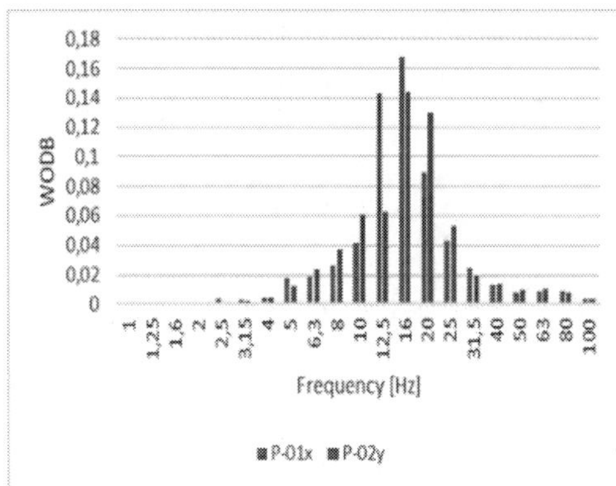

Figure 11. WODB spectrum in one-third octave frequency bands – the measuring point P-01x and P-02y – Intercity train – speed 160 km/h

Finally, passages of Pendolino have been divided into two categories: the passage with speed up to and above 160 km/h. For each of categories it has been choose the worst values of the coefficient WODB. The results obtained for P-01x and P-02y sensors have been compared and presented in Fig. 12 and Fig. 13, respectively. The diagrams show that not all frequencies increase depending on the growth of speed of the train. The large increase in frequency bands is noticed for the values in the bands 16-25 Hz. Other frequencies do not present such huge differences. Even higher speed refers to slightly smaller value of coefficient WODB, e.g. for 10 Hz band. For this reason, if we want to increase the velocity of train, we should focus on reducing the vibration in appropriate frequency bands rather than the entire range.

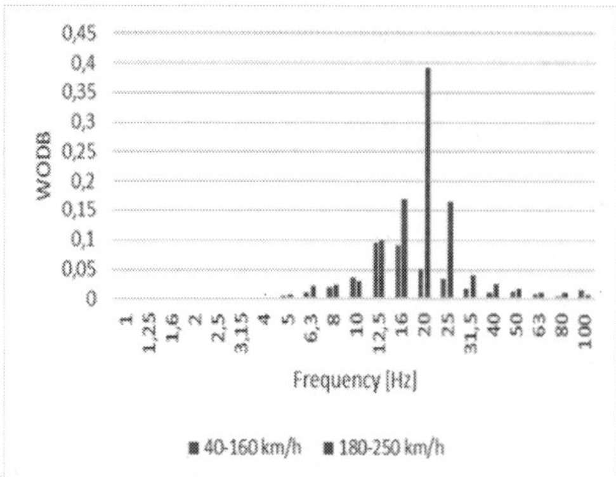

Figure 12. WODB spectrum in one-third octave frequency bands – the measuring point P-01x – Pendolino trains – the maximum values for speed trains v =40-160 km/h and v=180-250 km/h

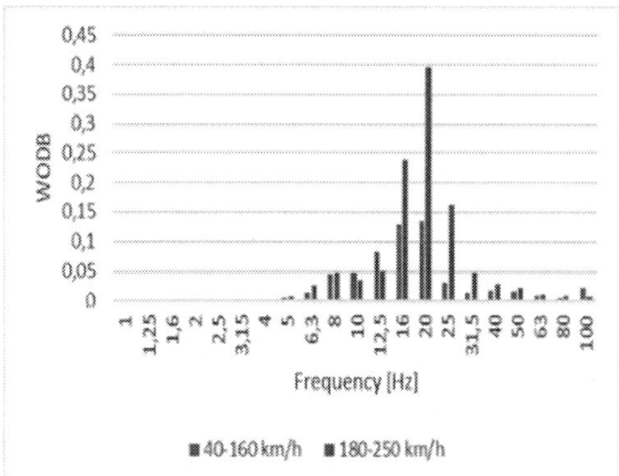

Figure 13. WODB spectrum in one-third octave frequency bands – the measuring point P-02y – Pendolino trains – the maximum values for speed trains v =40-160 km/h and v=180-250 km/h

5 Conclusions

This paper analysed negative environmental impact caused by the passages of trains. The following conclusions were made:

Train vibration is an area that affects environmental condition. Taking into account information from literature it is a global concern. It also applies to well-developed countries, where the railway infrastructure is advanced with the use of new technologies with the European standards. For this reason, transport has a great role in shaping the quality of the environment.

Assessment of environmental impact was presented by indicator of perceptibility of vibration through construction (WODB), which refers to the Scales of Dynamic Influences (SDI scales). The limits specified by standards [19] in any of the passages have not been exceeded. The change of speed or rolling stock resulted in a change in the characteristic of the vibration spectrum.

The limits specified by standards in any of the passages have not been exceeded on examined building. However replacement of old rolling stock trains by the new trains will reduce negative impact of the vibration on the environment.

Propagation of free-field vibrations in ground with their effect on the structural response of buildings near railway lines are complex phenomenon. Therefore it is difficult to assess the impact of vibration on buildings without in situ measurements. E.g. as proven above a train moving on a track further away from the building does not necessarily reach lower values of acceleration for each frequency band.

If we want to develop a network lines and increase the velocity of train to replace a part of road transport (because of air pollution), we should not forget about vibration pollution. In conclusion we must adjust rail system to be safe for all aspects of environment.

Acknowledgements

Investigations of vibrations have been performed in situ by accredited Laboratory of Structural Mechanics at Cracow University of Technology. The authors are grateful to the staff of the laboratory.

References

1. The Act of 27.04.2001 - *Environmental Protection Law* (Journal of Laws of 2001. No. 62, item. 627). (in Polish)
2. The Act of. 3.10.2008 - *Act on Providing Information on the Environment and Environmental Protection, Public Participation in Environmental Protection and on Environmental Impact Assessment* (Journal of Laws of 2008. No. 199, item. 1227). (in Polish)
3. ISO 14000 - *Environmental management*
4. Directive 85/337/EEC on the assessment of the effects of certain public and private projects on the environment.
5. Directive 2001/42/EC of the European Parliament and of the Council of 27 June 2001 on the assessment of the effects of certain plans and programmes on the environment.
6. T. Lawrance, *Noise and vibration from road and rail* (CIRIA, Londyn, 2011)
7. D. Tompson, *Railway noise and vibration, Mechanisms, Modelling and Means of Control* (Elsevier, Oxford, 2009)
8. V. V. Krylov, *Noise and vibration from high-speed trains* (Thomas Telford, Londyn, 2001)
9. Connolly D. P. et al., *The growth of railway ground vibration problem – A review*, Sci Total Environ, p.1-7 (Elsevier, 2015)
10. D. Connollya, G. Kouroussisb, O. Laghrouchea, C. Hoc and M. Forded, *Benchmarking railway vibrations – Track, vehicle, ground and building effects*, Construction and Building Materials, no. **92**, p. 64-81, (Elsevier, 2015)
11. L. S. G. Degrande, *Free field vibrations during the passage of a Thalys high-speed train at variable speed*, Journal of Sound and Vibration, no. **247** (1), p. 131-144, (Academic Press, 2001)

DOI: 10.1051/e3sconf/20161000047

12. R. Ciesielski, E. Maciąg, *Road vibrations and their impact on buildings* (Communication and Connectivity Publishing, Warsaw, 1990) (in Polish)
13. J. Kawecki, P. Stecz, K. Stypuła, *On the nesessity of use of simulation calculations of the vibration isolation in the tram track,* Technical Transactions,vol. Series : Civil Engineering, no. **3-B**, p. 163-173 (2011) (in Polish)
14. K. Stypuła, *New investments and environmental protection against vibration*, Isolations **10** (Medium Group, 2008) (in Polish)
15. R. Ciesielski, A. Kwiecień, K. Stypuła, *Vibration propagation in layers closed to the surface of the soil – Experimental investigation in situ* (Cracow University of Technology, Cracow, 1999) (in Polish)
16. T. Tatara, K. Stypuła, *Analysis of the results of measurements of ground vibration and some elements of the surface during the test of Pendolino train and other trains on the line CMK in agreed locations in the track No. 1, section Psary - Góra Włodowska,* (Cracow University of Technology, Cracow, 2013) (in Polish)
17. K. Stypuła, *Effect of vibration of communication on buildings and people staying in them*, p. 120-131, Building Materials, no. **3** (SIGMA-NOT, 2009)(in Polish)
18. E. Maciąg, J. Chełmecki and T. Tatara, *Investigation of ground and low-rise buildings vibrations caused by city traffic, Engineering and Construction*, pp. 135-140, **5** (PZIiTB, 2005) (in Polish)
19. PN-85/B-02170:1985, *Evaluation of the harmfulness of building vibrations due to ground motion* (in Polish).

Environmental degradability of polycaprolactone under natural conditions

Katarzyna Krasowska[a], Aleksandra Heimowska and Magda Morawska

Gdynia Maritime University, Department of Chemistry and Industrial Commodity Science, Morska 81-87 str., Gdynia, Poland

Abstract. The aim of this work was an estimation of susceptibility of biodegradable poly(ε-caprolactone) (PCL) to environmental degradation in different natural environments. The commercial poly(ε-caprolactone) film, the trade name "CAPA 680", was degraded in the compost, pond, open and harbour area of the Baltic Sea. Characteristic parameters of all natural environments were monitored during the incubation of polymer samples and their influence on degradation of PCL was discussed. Susceptibility of PCL to degradation in natural environments was evaluated based on changes of weight, crystallinity and polymer surface morphology. The rate of environmental degradation of PCL depended on the incubation place, environmental conditions and decreased in order: compost>harbour area of the Baltic Sea>open area of the Baltic Sea>pond.

1 Introduction

In recent decades world consumption of polymers has increased exponentially. Polymers are used in many areas, especially in the packaging, agriculture, medicine etc. In the process of consuming products humans generate plastic waste, which are responsible for the problem of environmental pollution. Nowadays, plastics which are not incinerated, landfilled or recycled are found in the nature and they polluting for example seas, rivers, oceans, lakes, beaches, forests etc. Very often the natural environment must serve as a waste repository, either by absorbing or recycling them into useful or at least harmless substances. When the waste products exceed the environment's ability to absorb them, the result is water, soil and air pollution.

The increase in volume of synthetic non-degradable polymers, particularly in the form of one-trip packaging, presented a potential threat to the natural environment. The plastic packaging litter is not only the aesthetically undesirable but it causes the possibility of risk for people, animals, birds on the land and in the water.

Recently the increasing attention is paid to attempt rational replacement commonly used synthetic polymers by biodegradable polymers. The use of biodegradable polymers in consumer products should not lead to the generation of toxic or otherwise environmentally unacceptable chemicals in the natural environment.

Biodegradable polymers are susceptible to biological degradation, resulting in the disintegration and mineralization under the action of living organisms. The initial breakdown of a polymer, which is the first step of biological degradation process can result from physical and biological forces, which can cause mechanical damage such as cracking of polymeric materials and fragmentation. Most polymers are too large to pass through cellular membranes, so they must be depolymerized to smaller molecules before they can be adsorbed and degraded within microbial cells. The monomers, dimers and oligomers of a polymer's repeating units are much easily degraded and mineralized, because they can be assimilated through the cellular membrane and then further degraded by cellular enzymes. Under oxygen conditions, aerobic microorganisms are mostly responsible for degradation of polymer. Biomass, carbon dioxide and water are the final products of deterioration. As opposite to this, under anoxic conditions, anaerobic microorganisms play the main role in polymer destruction. The primary products are methane, water and biomass [1, 2].

In the case of environmental degradation under natural conditions very often synergistic action various factors including temperature, pH, humidity, aeration, sunlight, macroorganisms, microorganisms and enzymes leads to degradation of biodegradable polymers.

Biodegradable polymers have been found to degrade more rapidly if a combination of microbes is used rather than one specific microbe. Thus the presence of large varieties of microbes in the natural environment support faster biodegradation [3, 4].

An example of a biodegradable polymer which undergoes microbial degradation, both in contact with living organisms, and in terms of the natural environment is poly(ε-caprolactone), which can degrade in several biotic environments, including sea, river and lake water, sewage sludge, farm soil and compost [5-12].

According to the literature degradation of poly(ε-caprolactone) in a living environment can result

[a] Corresponding author: k.krasowska@wpit.am.gdynia.pl

of simple chemical hydrolysis of ester bonds or from enzymatic attack or both [13].
Chemical hydrolysis can be resulted in decreasing of molecular weight. Enzymatic degradation takes place only on the polymer surface, and therefore is associated usually with erosion process and may be quantitatively characterized by weight loss and fragmentation of the sample. At the result the polymer is disintegrated in the natural environment. Obviously, disintegradability does not indicate on total sample degradation. The small particles of the sample are constantly degraded and bioassimilated by microorganisms present in the environment, until the samples would be completely converted to carbon dioxide and water.
The aim of this work was an estimation of susceptibility of poly(ε-caprolactone) (PCL) to environmental degradation in different natural environments. The results of PCL degradation in compost, pond, open and harbour area of the Baltic Sea are compared.

2 Experimental

2.1 Material

The commercial poly(ε-caprolactone) film (PCL, M_w= 80000) with the trade name "CAPA 680" used in this work was kindly supplied by Solvay (England).

2.2 Environments

The incubation of PCL samples took place in the several environments under natural weather depending conditions, such as: the compost, pond, open and harbour area of the Baltic Sea. For comparison degradation of PCL samples was also performed under the laboratory conditions in a liquid medium containing sea water with the sodium azide (NaN_3) and in a liquid medium containing pond water with NaN_3.

2.2.1 The compost pile

The compost pile was prepared under natural conditions of municipal waste treatment plant in Gdynia. It consisted of the dehydrated sewage sludge, burnt lime and straw. Burnt lime (0.45kgCaO/1kg dry mass of compost) was added to ravage phatogenic bacterium and eggs parasites, to deacidificate sewage sludge and to convert sludge to compost. The straw was added to maintain the higher temperature of the compost pile and to loosen the structure of the compost pile. The compost pile prepared under natural conditions was not adequately aerated, so it was expected that a combination of conditions from aerobic at the upper part of pile, microaerophilic in the middle part and facultative anaerobic at the bottom of the pile could occur for microorganisms growth [2].
The Figure 1 represents cross-section of compost pile in natural environment.

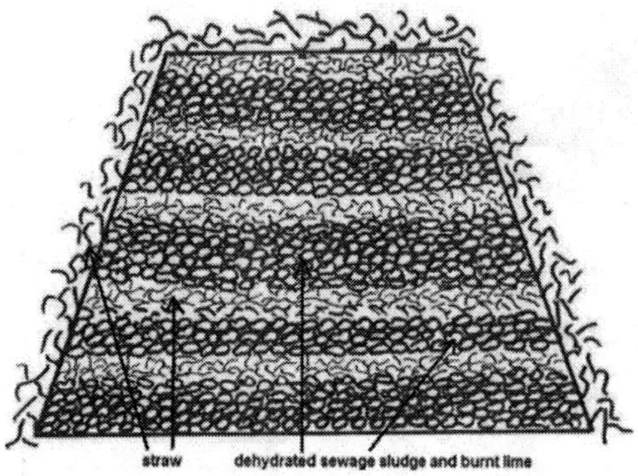

Figure 1. The cross-section of compost pile prepared in natural environment

The PCL samples were put into the special perforated basket and buried of 1 m in depth of the compost pile. The perforated structure of basket allowed for free access of macro, micro-organisms and enzymes existing in this environment to degraded material.

The characteristic parameters of the compost such as: temperature, pH, moisture content, activity of dehydrogenases were measured during environmental degradation process of PCL samples and are shown in Table 1 [9].

Table 1. Characteristic parameters of compost

Months	Parameters			
	temperature [°C]	pH	moisture content [%]	Activity of dehydrogenases [mol mg^{-1} d.m.]
June	17.6	7.8	42.4	0.0197
July	18.2	5.8	45.8	0.0568
August	19.2	6.3	43.5	0.0485

2.2.2 The Baltic Sea water

The incubation of PCL samples took place in two places of the Baltic Sea under natural conditions:
- in the open area of the Baltic Sea in Łeba, on the west side of the western breakwater, in the waters of the East current which allow for constant flow,
- in the harbour area of the Baltic Sea in Gdynia Harbour on the Norwegian Pier near the ship of the Polish Ship Salvage Company.

The PCL samples were located in a special basket made of perforated galvanized steel, which was suspended from a rope at 2 m depth under the surface of the sea. The perforated baskets structure allowed for free movement of seawater and the access of micro-organisms and enzymes dissolved in the water to degraded material. The places of environmental degradation of PCL in the open and harbour area of the Baltic Sea are shown on Figure 2.

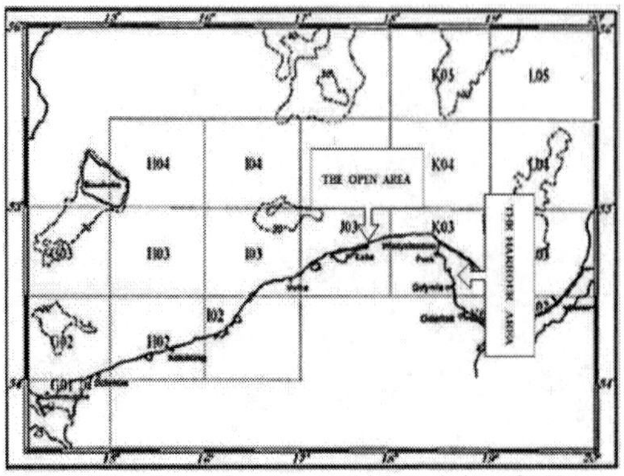

Figure 2. The places of environmental degradation of PCL in the Baltic Sea

The temperature, pH, oxygen and salt content of the Baltic Sea were monitored during environmental degradation process of PCL samples. The characteristic parameters of the sea water in the open and harbour area according to the State Environmental Monitoring, Inspectorate of Environmental Protection are shown in Table 2 and 3 [9, 14].

Table 2. Characteristic parameters of water in open area of the Baltic Sea

Months	**Parameters**			
	temperature [°C]	pH	oxygen content [cm³/dm³]	salt content [ppt]
July	17.8	8.3	6.68	7.50
August	18.6	8.2	6.64	7.53
September	18.2	8.1	6.40	7.46
October	13.7	8.4	6.38	7.75

Table 3. Characteristic parameters of water in harbour area of the Baltic Sea

Months	**Parameters**			
	temperature [°C]	pH	Oxygen content [cm³/dm³]	Salt content [ppt]
June	17.6	8.5	7.55	5.36
July	20.3	8.2	7.57	5.58
August	19.3	8.9	7.42	6.01

2.2.3 The pond

The environmental degradation of PCL samples took place in Rumia's ponds in a special perforated basket at two meters depth under the water surface. The scheme of the pond construction is shown on Figure 3. The Zagórska Struga River is divided into two parts. There are four ponds between that river parts. The basket with polymer samples were located in the one of them.

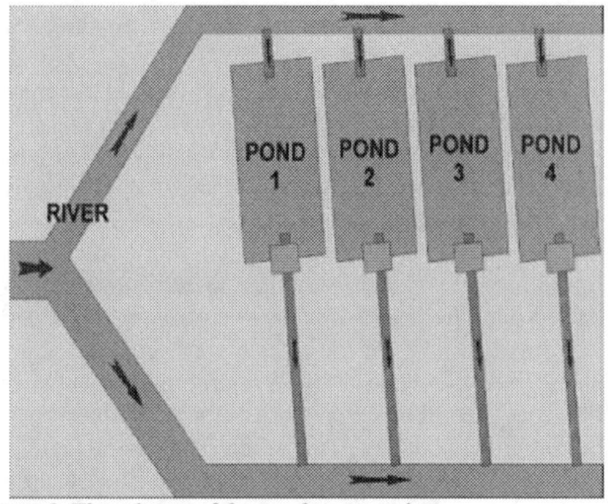

Figure 3. The scheme of the pond construction

The characteristic parameters of the pond such as: temperature, pH were measured during environmental degradation process of PCL samples and are shown in Table 4 [15].

Table 4. Characteristic parameters of water pond under natural conditions

Months	**Parameters**	
	temperature [°C]	pH
June	17.8	9.0
July	19.3	8.6
August	18.5	8.4
September	12.3	8.5
October	10.0	7.7

2.2.4 The laboratory tests

The degradation of PCL samples was also performed in the laboratory in a liquid medium containing sea water with sodium azide or pond water with sodium azide (NaN_3). The sodium azide (0.195gNaN_3/1000ml) [16] was added to the natural water (derived from a pond or sea) to exclude the activity of micro-organisms and to evaluate the resistance of the polymer to hydrolysis. The PCL samples were located in the glass aquarium equipped with an aeration pump. The characteristic parameters of liquid media under laboratory conditions are shown in Table 5.

Table 5. Characteristic parameters of sea water and water pond under laboratory conditions

Months	**sea water with NaN_3**		**pond water with NaN_3**	
	temperature [°C]	pH	temperature [°C]	pH
June	23	8.0	21.7	8.6
July	22	8.1	23.4	6.8
August	23	8.0	21.9	7.7
September	20	8.1	19.5	8.2
October	19	8.0	19.5	7.6

2.3 Measurements

2.3.1 Characterization of the natural environments

The characteristic parameters of the compost, pond and aqueous laboratory environments were measured during degradation process of PCL samples.

2.3.1.1 The humidity of compost

The moisture content of the compost was determined by drying at 105°C until constant weight was obtained.

2.3.1.2 The pH of compost, pond and aqueous laboratory environments

The pH of the compost was determined with a Teleko N 5172 pH-meter.

2.3.1.3 The biochemical activity of compost

To estimate the biochemical activity of microorganisms in sludge, the activity of the dehydrogenases was measured by a spectrophotometric method using triphenyltetrazolium chloride (TTC). The method is based on the dehydrogenation of glucose added to the compost with a subsequent transfer of hydrogen to the colourless biologically active compound of TTC, which undergoes a reduction to triphenyloformazan (TF). The intensity of red colour compound TF was measured using a Specol colorimeter at 490 nm [2, 9].

2.3.2 Investigations of PCL samples

After incubation time the samples were taken out from the environment, washed with distilled water and dried at room temperature until constant weight. The environmental degradability of PCL samples was investigated by changes of weight, surface morphology and crystallinity of polymer after incubation in the each environment.

2.3.2.1 The changes in polymer surface

The morphology of the PCL surfaces was investigated using the optical microscope ALPHAPHOT-2YS2-H Nikon linked to the photo camera Casio QV-2900UX. The view of PCL samples surface before and after degradation was compared. PCL samples were observed with and without polarizer. The pictures were taken before and after incubation in natural and laboratory environment.

2.3.2.2. The changes of weight

The dried samples of PCL were weighed on an analytical electronic balance (Gibertini E 42s, repeatability 0.1 mg). The weight of clean and dried samples of PCL after incubation in the natural and laboratory environment was compared with those before incubation. The percentage weight changes of PCL samples were calculated according to the following equation:

$$x = \frac{m - m_1}{m} \cdot 100 \quad (1)$$

where:
x – changes of weight [%],
m – weight of PCL sample before incubation [g]
m_1 – weight of PCL sample after incubation [g]

2.3.2.3 The changes in crystallinity

The changes of crystallinity PCL samples were also performed based on the differential scanning calorimetry using analyser Setaram Labsys TG-DTA/DSC. The heating scans at the rate of 10°C /min in the temperature range 20–200°C in nitrogen atmosphere, in nitrogen flow 20 cm^3/min were recorded. Based on determined melting enthalpy of PCL samples the percent of crystallinity was calculated [17]. According to the following equation:

$$x = \frac{\Delta H}{\Delta H_{100\%\ PCL}} \cdot 100 \quad (2)$$

where:
x – crystallinity [%],
ΔH - melting enthalpy of PCL sample [J/g]
$\Delta H_{100\%\ PCL}$ – 139.5 [J/g], melting enthalpy of 100% crystalline PCL [18].

3 Results and discussion

3.1 The characteristic of environments

The characteristic of natural environments is determined by its chemical composition and biological properties. Both physico-chemical and biological properties of the environment influence on its suitability for various purposes. Some of them are particularly useful for assessing the quality of environment and occurring in the degradation processes.

Characteristic parameters of all natural environments were monitored during the incubation of polymer samples (Tables 1-4) and their influence on degradation of PCL is discussed.

The temperature of each natural environment was depended on the weather conditions and had been fluctuating a lot during experiment. Only the temperature of natural environments during summer months was more preferable for enzymatic degradation (20-60°C) [19].

Looking at the parameters presented in the Tables 1-4 we can state, that the average temperature in the compost was about 18°C, in the open area of the Baltic Sea – 17°C, in the harbour area – 19°C and in the pond about 16°C.

Temperature fluctuation occurred in the open water of the Baltic Sea. It was connected with the water exchange in this natural environment. The period of water exchange in the open water of the Baltic Sea

is less than 7 days, while in the closed water takes to 52 days [20].
Temperature fluctuations in the pond were related to the small size the water reservoir, which facilitates the heating and cooling of the water pond.
However the average temperature in all natural environments (compost, sea water, pond) was lower as that preferred for enzymatic degradation (20-60°C) [19].

Analyzing characteristic parameters presented in the Tables 1-4 we can also state that there were significant differences in the pH values of all natural environments. The average pH in the compost was slightly acid (pH 6.6), but in the open area value of pH of the Baltic Sea was 8.1, in the harbour area 8.5 and in the pond 8.4. In all liquid natural environments pH was above the upper limit of pH (5-8) recommended in the biodegradation process [19].
The best value of pH for growth bacteria in freshwater is 6-8 [21]. In order to maintain a proper mechanism of water self-purification and a favorable climate for all inhabitants of the pond - the pH should be between 7.4 and 8.4. Water from the pond with a high alkalinity is able to neutralize acidic rainwater supply and the high acidity may be indicative of conditions in which toxic components are activated [22]. The water in the pond must not contain toxic components, especially since in the pond are bred fishes (trouts).

The water has the capacity to dissolve gases. This solubility decreases with increasing temperature and salinity. Pond water is better than sea water. The amount of oxygen depends on the ability of water to the self-cleaning due to the mineralization of organic substances. Mineralization is done with the help of micro-organisms, especially aerobic bacteria. The average concentration of oxygen in the clear river waters in temperate climates is about 5-7mg/dm^3[21].

The rather low temperatures (below 20°C) and slightly acid pH (~6) of compost under natural weather depending conditions (Table 1) caused that psychrotrophic acidophilic microorganisms (fungi) could play the main role in the degradation process. It is known that the most favorable conditions for fungal growth are at a pH below 6 and a high oxygen supply [21].
The activity of dehydrogenases depends on the degree growth of microorganism populations, which are producing enzymes involved in degradation process. During the degradation time of PCL samples the activity of dehydrogenases had been changing and depending on both biotic and abiotic conditions in this environment. The weather, as well as respiration of microorganisms have an influence on fluctuation of moisture content in the compost. With decreasing of moisture content lower absolute value of the activity of dehydrogenases was observed (Table 1).

The Baltic Sea is both a very interesting and complicated natural environment for degradation process because microorganisms, animals, salt, sunlight, fluctuation of water, rain etc. all play a part in degradation in nature.

The relatively low temperature and alkalinity of sea water (Table 2 and 3) could have an influence on the activity of psychrotrophic bacteria, which are able to adapt to so changing conditions.
At the beginning of the summer period of incubation of PCL samples in sea water in the natural environments (June in the harbour area and July in the open area) we could observe the higher oxygen content, than at the end of incubation time. This result is connected with the production of organic matter by micro-organisms, which consumed oxygen contained in the water to degradation processes [2,14].
Looking at the parameters presented in the Table 2 and 3 we can see significant differences in the salinity of both natural environments.
Salt content of the Baltic Sea water is changed naturally and depends on tributary of rivers or heavy rains [20].

Considering the characteristic abiotic parameters of all natural environments: the compost, sea water and pond presented in Tables 1-4 and the different microbial communities (fungi in compost, bacteria in sea water and algae, heterotrophic and epilithic bacteria in pond), we could expect the different rate of environmental degradation of PCL samples.

Incubation of PCL samples in the laboratory in a liquid medium containing sea water with NaN_3 and pond water with NaN_3 (Table 5) was performed in a stable temperature (about 21°C in medium containing sea water and the same in pond water) and under near neutral environment conditions (pH about 7.8-8.0).

3.2 The evaluation of polymer changes during environmental degradation

Susceptibility of PCL to environmental degradation is evaluated based on weight changes [%] of degraded polymer samples. The results of the weight changes of PCL samples after incubation in all natural environments and in the laboratory in a liquid medium containing sea water with NaN_3 or pond water with NaN_3 are presented in Table 6.

Table 6. Weight loss [%] of PCL samples after degradation in natural and laboratory environments

Enviro - nment	Degradation time [weeks]					
	2	4	5-6	10	12	15
compost	5.6	86.3	disintegration after 5 weeks			
open area of the Baltic Sea	2.2	4.0	8.4	14.9	29.2	disintegration after 12 weeks
harbour area of the Baltic Sea	3.5	20.7	33.8	disintegration after 6 weeks		
pond	-	2.4	3.1	4.0	-	5.6
sea water with NaN_3	1.2	1.4	1.5	1.5	1.5	1.5
pond water with NaN_3	-	1.6	1.9	2.8	-	2.7

[-] not measured at that time

The obtained results testify that PCL degrades in all investigated natural environments such as compost, the Baltic Sea water and pond. It confirms susceptibility of PCL to biological degradation. Generally, the results presented in Table 6 reflect well, that the environmental degradability of PCL depends on the kind of natural environment – abiotic and biotic parameters.

Disintegration of PCL samples is observed after 5 weeks incubation in compost, 6 weeks in the harbour area of the Baltic Sea and 12 weeks in open area (Table 6). Disintegration, means fragmentation and loss of visibility of samples in natural environment. However, biological degradation has been still continuing. According to the literature polymer samples have been degraded by macro and microorganisms and/or enzymes living in natural environment until they have been converted to carbon dioxide and water [1, 3].

At the beginning of incubation of PCL samples in all natural environments (2 weeks) the small changes of weight (~2-5%) are observed because of similar low temperature conditions (~17°C). The slightly higher changes of weight after incubation of PCL in compost could be caused by different living microbial communities in this environment.

Considering the parameters of compost pile under natural weather depending conditions (temperature and pH) we could state that the psychrotrophic acidofilic microorganisms (fungi) are responsible for the level of degradability of PCL. In contrary to this, there are conditions in the Baltic Sea water, where the low temperature and the alkaline pH are favourable for the development of aerobic psychrotrophic bacteria.

In the next weeks of incubation in the compost and the Baltic Sea water (4-6 weeks) the weight changes of PCL are more visible.

The activity of dehydrogenases in compost influences the variety of microorganisms which are producing enzymes involved in biodegradation process. The parameters presented in Tables 1 show that after 4 weeks of incubation of PCL (July-August) the moisture content and activity of dehydrogenases are higher than in the beginning of incubation. Finally after 5 weeks of environmental degradation PCL samples in compost disintegration is observed.

The higher weight losses of PCL in the harbour than in open area of the Baltic Sea could be caused by different temperature conditions (Table 2 and 3). The temperature of the Baltic Sea water in harbour area was more preferable for enzymatic degradation (20-60°C) [19]. It also suggest, that in the harbour water are probably more favourable conditions to degradation of PCL samples, than those in the open area water. As a result of the higher temperature more visible weight changes and quicker disintegration of PCL samples in the harbour area are observed.

PCL samples have required more time to be destroyed in the pond. After 15 weeks of incubation in natural pond, the weight loss of PCL was only 5.6%. This is most likely related to the high variety of microbial life in the pond (bacteria, protozoa, actinomyces, fungi, algae and fishes), where (at the first) the micro and macroorganisms prefer to consume natural organic compounds present in the environment.

During incubation of PCL in the laboratory medium containing sea water with NaN_3 or in pond water with NaN_3 the weight changes are insignificant even though the temperature was higher, than in natural environment. This was due to an absence of microorganisms in laboratory tests.

The very small changes of weight might be explained by very slow nonenzymatic hydrolytic ester cleavage, what is complied with literature reports that the hydrolytic degradation of PCL is slow because of its hydrophobic nature [23].

One reason of slightly higher weight losses of PCL after incubation in pond water with NaN_3 is the lack of mineral salts which facilitates the diffusion of water within the polymer [24]. Salt contents in laboratory sea water probably have an influence on slow chemical hydrolysis of PCL during incubation in this environment.

The environmental degradation of PCL samples is also evaluated on the basis of changes of surface morphology. After incubation in natural environments the PCL samples were not homogeusly destroyed over the whole polymer surface and there were different images depending on the place the picture was taken.

Photomicrographs of the surface of PCL samples after incubation in all natural environments are presented on Figure 4.

The surface of PCL before incubation when observed under optical microscope is grained and rough (Figure 4a). The microscopic observations after incubation in all natural environments have shown vulnerability of PCL samples to the microbiological attack and deterioration of the surface is observed. The changes of PCL surface during the environmental degradation lead to the erosion (damage) in the form of the observed cracks and holes (Figure 4 b, c, g, e).

Similarly to the changes of weight of the PCL the microscopic observations of surface show that, the rate of changes is faster in compost than in natural aqueous environments (sea and pond water).

The stronger effect of environmental degradation of PCL in compost, than in sea water testify that the psychrotrophic acidophilic microorganisms (fungi) could play the main role in the environmental degradation process.

Noticed faster environmental degradation of PCL in the harbour area of the Baltic Sea is probably because of more favourable temperature conditions for the biodegradation process and lower salinity, which caused the growth of different microorganisms able to degrade PCL samples.

The process of biodegradation in freshwater is carried out with the participation of microorganisms, including bacteria, protozoa, actinomyces and algae [25].

Abiotic parameters of the pond as well as various microbial life in the pond contribute to the slow degradation of PCL (Figure 4 h, i).

The results of changes in crystallinity of PCL samples after incubation in all natural environments are shown on Figure 5.

The changes of PCL crystallinity during incubation in natural environments lead to conclusion that the environmental degradation of PCL occurs in two stages. The first stage consists in the degradation of amorphous phase and as a result an increase in crystallinity of polymer occurs. The second stage starts when most of the amorphous regions are degraded and subsequently the crystalline phase is degraded. The DSC analysis of PCL samples after degradation in all natural environments revealed the differences in their crystallinity (Figure 5). Only during incubation of PCL in the compost the evident of two-step process of degradation is observed. The higher

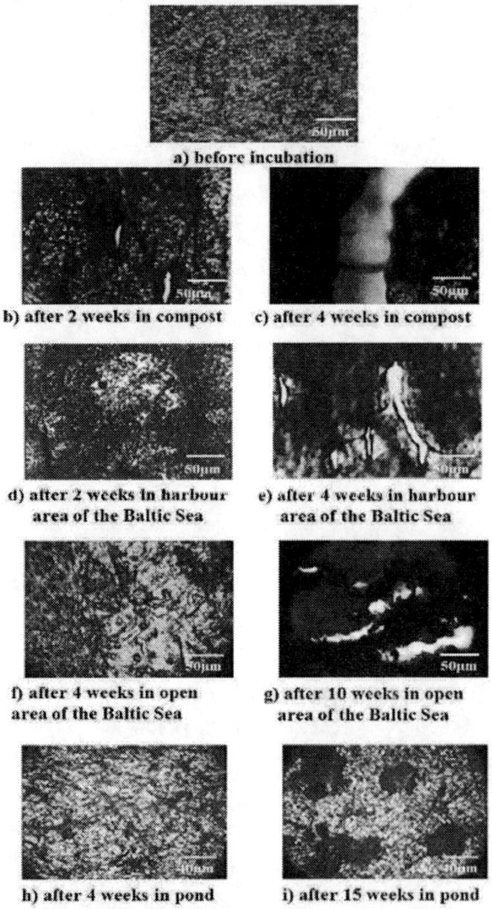

Figure 4. Photomicrographs of the surface of PCL before and after environmental degradation

crystallinity of PCL after 2 weeks of incubation suggests that at first the amorphous phase is degraded, which is accompanied by an increase of crystallinity and next the crystalline phase is degraded (4 weeks).

4 Conclusions

The environmental degradation studies confirm the vulnerability of poly(ε-caprolactone) to the attack of living organisms. The obtained results reveal that the rate of environmental degradation of PCL is dependent on the incubation place, environmental conditions and decreased in order: compost>harbour area of the Baltic Sea>open area of the Baltic Sea>pond.

The characteristic abiotic parameters (temperature, pH, salinity, humidity and oxygenation) of all natural environments and the different microbial communities (psychrotrophic acidofilic microorganisms in compost, psychrotrophic bacteria in sea water and algae, heterotrophic and epilithic bacteria in pond) cause the different rate of environmental degradation of PCL samples.

Disintegration of PCL in the compost is observed faster (after 5 weeks) than in the aqueous environments, because of more favourable biotic and abiotic conditions.

The significant weight loss and erosion of the PCL surface after environmental degradation are the result of enzymatic hydrolysis of ester bonds, which play the key role in the process under natural conditions.

The microscopic changes of PCL surface and crystallinity after incubation in natural environments suggest that the enzymatic degradation of PCL is a two-step process. The first the amorphous phase is degraded, which is accompanied by an increase of crystallinity and next the crystalline phase is degraded.

Insignificant weight losses of the PCL samples incubated in laboratory liquid media may be explained by slow chemical hydrolysis of ester bonds.

The results of research indicate that the degradation of PCL in natural environments such as compost, sea water and pond is the result primarily of enzymatic hydrolysis, which plays a predominant role in the degradation process.

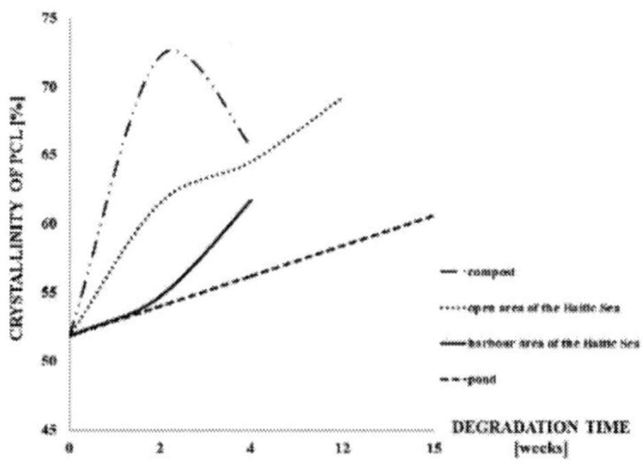

Figure 5. The changes in crystallinity of PCL after environmental degradation

References

1. M. Falkiewicz-Dulik, K. Janda, G. Wypych, *Handbook of Biodegradation, Biodeterioration and Biostabilization* (ChemTec Publishing, Toronto 2010)
2. K. Krasowska, A. Heimowska, M. Rutkowska, [in] *Thermoplastic Elastomers Synthesis and Applications,* (INTECH, Croatia 2015)
3. N. Karak, *Vegetable Oil-based Polymers, Properties, Processing and Applications* (Woodhead Publishing Limited, Oxford-Cambridge-Philadelphia-New Delhi, 2012)
4. K. M. Dean, P. Sangwan, C. Way, M. A. Nikolic, [in] *Characterization Techniques for Polymer*

Nanocomposites (Wiley-VCH Verlag GmbH & Co. K.GaA, 2012)
5. K. Fukushima, C. Abbate, D. Tabuani, M. Gennari, P. Rizzarelli, G. Camino, Mater Sci Eng **C30**(4), 566 (2010)
6. H. Tsuji, K. Suzuyoshi, Polym Degrad Stabil **75**(2), 347 (2002)
7. H. Tsuji, K. Suzuyoshi, Polym Degrad Stabil **75**(2), 357 (2002)
8. D. S. Rosa, D. R. Lopes, M. R. Calil Polym Test **24**(6),756 (2005)
9. K. Krasowska, A. Heimowska, M. Rutkowska, Polimery **51**(1), 21 (2006)
10. M. Rutkowska, K. Krasowska, A. Heimowska, I. Steinka, Polimery **47**(4), 262 (2002)
11. M. Rutkowska, A. Heimowska, K. Krasowska, I. Steinka, H. Janik, J. Haponiuk, S. Karrlsson, Pol J Environ Stud, **11**(4), 413 (2002)
12. J. Kotek, D. Kubies, J. Baldrian, J. Kovarova, Eur Polym J, **47**(12), 2197 (2011)
13. S. Li, M. Vert, *Degradable Polymers* (Chapman and Hall, Londyn, 1995)
14. K. Krasowska, K. Suchocka, M. Rutkowska, Pol J Com Sci, **1**(38), 93 (2014)
15. A. Heimowska, M. Morawska, A. Bocho-Janiszewska, Pol J Com Sci, **4**(45), 19 (2015)
16. A.C. Albertson, C. Berenstedt, S. Karlsson, J Appl Polym Sci, **51**, 1097 (1994)
17. D. Haynes, N. Abayasinghe, G. Harrison, K. Burg, J. Smith, Biomacromolecules **8**, 1131 (2007)
18. Q. Guo, G. Groenineks, Polymer, **42**, 8647 (2001)
19. R. W. Lenz, [in] *Advances in Polymer Science* (Springer-Velay, Berlin, 1993)
20. W. Krzymiński, *Wstępna ocena stanu środowiska wód morskich polskiej strefy Morza Bałtyckiego* (GIOŚ, IMGW-PIB, Gdynia 2013)
21. B. Kołzwan, W. Adamiak, *Podstawy mikrobiologii w ochronie środowiska*, (Politechnika Wrocławska, Wrocław 2005)
22. W. Chełmicki, *Woda. Zasoby, degradacja, ochrona*, (PWN, Warszawa 2013)
23. M. Avella, M. E. Errico, P. Laurienzo, E. Martuscelli, M. Raimo, R. Rimedio, Polymer **41**(10), 3875 (2000)
24. M. Deroine, A. Le Duigou, Y-M. Corre, P-Y. Le Gac, P. Davies, G. Cesar, S. Bruzaud, Polym Degrad Stabil **108**, 319 (2014)
25. Z. Podbielskowski, H. Tomaszewicz, *Zarys hydrobotaniki*, (PWN, Warszawa 1996)

The role of the silica fly ash in sustainable waste management

Anna Król[1,a]

[1]Opole University of Technology, Department of Environmental Engineering, 5 Mikołajczyka Street, 45-271 Opole, Poland

Abstract. Mineral construction materials play an important role in waste management. They are often used for solidification of hazardous mineral waste.These processes involve common cement or binders that apart from cement contain other mineral additive, which indeed are post-process by-products from other industries. Among approved mineral additives there is silica fly ash from hard coal combustion in conventional power plants. Such fly ash used during solidification allow formation of stable and durable matrices of high immobilisation potential for heavy metals.The paper presents analysis of effect of addition of fly ash on solidification process of galvanic sewage sludge in comparison to matrices prepared using Portland cement.

1 Introduction

The concept of sustainable development is an assumption of continuous economic and social growth in harmony with the natural environment. This is a pursuit of such an economic model that would provide progress of mankind without destroying supporting it natural systems. This means using natural resources, while respecting them. A sustainable development is not contrary to the progress, but rather a call for keeping equilibrium and moderation in order to reconcile key areas for mankind: ecology, economy and society [1].

Also application of by-products from other industries in production of clinker, cement or concrete is in line with the sustainable development concept. These products can be successfully replaced with natural minerals or clinker in cement production. This way it is possible to significantly reduce output of natural raw materials and improves profitability of the process, without compromising the quality [1].

Also utilization of alternative fuels in the production of Portland cement is consistent with the sustainable development concept. The most of waste with proper minimum energy value can be used for production of alternative fuels. The use of alternative fuels in cement industry has positive impact on clinker production economics and improves environmental effects of fuel combustion (such as reduction in atmospheric emissions of gases and dust). Cement kilns are one of the few plants that use waste-based fuels as a component of conventional fuel. Combustion of alternative fuels combined with fossil fuels is both effective and environmentally safe solution that form a part of energy waste recovery [1-3]. The consumption of alternative fuels in the cement industry in Poland in 2012 was 1180.5 thousands Mg.

Moreover, replacing clinker with other active mineral additives starts to play more and more important role in cement production. In the global cement production, this process is justified by both economy and environment protection, but also by technological reasons. There is a tendency to obtain cements of such properties that would allow obtaining durable concrete, and the application of proper mineral additives can help in that [4].

Cement mineral additives are mainly combustion by-products from power industry and waste generated by iron and steel industry, i.e. silica fly ash (denoted in cement name as V) and granulated blast-furnace slag (denoted as S) [5-7]. Such materials exhibit high pozzolanic and/or hydraulic activity. An important position in mineral cement additives is gypsum used in amount of 4 to 6% cement weight as a binding time regulator. For that purpose mainly gypsum from flue desulfurization of power boilers are used (often called reagypsum).

In the recent years, the interests of cement industry involved also lime fly ash (W) from lignite combustion [8].

According to the data of the Polish Cement Association [9] in 2012, the Polish cement industry consumed over 4,200 thousand Mg secondary raw materials for cement, including over 1,800 thousand Mg of fly ash.

This shows that activity of the cement industry is associated with some environmental benefits. Production of clinker and cement often involves use of by-products from other industries, which helps to remove waste from the environment, while saving reserves of resources or fossil fuels.

Additionally, technology development leads to use of mineral binders for neutralization of hazardous waste in a solidification/stabilization process (s/s). This method

[a] Corresponding author: a.krol@po.opole.pl

has been developed in the mid-20th century for neutralization of radioactive waste. In time, it has been adapted for utilization of other waste, especially of industrial and hazardous, as well as sludge. Waste solidification process allows to modify waste physical and chemical properties, as well as reduce solubility and leaching of hazardous compounds [10,11]. Common cements (compliant with standard [6] requirements in mortars and concretes), and mineral binders of different composition based in particular on mineral additives, are used in order to permanently solidify waste.

The paper focuses on use of silica fly ash in production of binders allowing successful solidification of hazardous industrial waste.

2 Role of silica fly ash in production of mineral binders

Silica fly ash from coal combustion in power industry are relatively well recognized combustion by-products. They are widely used in cement and concrete processing as [8]:
- mineral additives for production of different cement grades (e.g. CEM II, CEM IV/A,B, CEM V/A,B),
- active mineral additive and micro-aggregate for normal and self-compacting concrete,
- one of components for production of high performance concrete,
- component of cellular and road concrete.

Silica fly ash is a material obtained by electrostatic or mechanical precipitation of dust particles from flue gases from conventional furnaces fired with coal dust at temp. approx. +1450°C. Their chemical composition is usually similar to composition of roasted carboniferous slate, which is a main incombustible component of coal. The main components of discussed ash are SiO_2 and Al_2O_3, CaO, SO_3, MgO, Na_2O and K_2O from the degradation of silty materials, pyrite and calcite. A component mainly determining properties of pozzolanic fly ash is unstable silica-aluminum-potassium glass produced as a result of separation of inorganic matter during coal combustion and its subsequent melt and solidification in form of fine, spherical particles (Fig. 1) [8]. Crystalline phases that occur most often in fly ash from coal are quartz and mullite, as well as haemetite and magnetite [8].

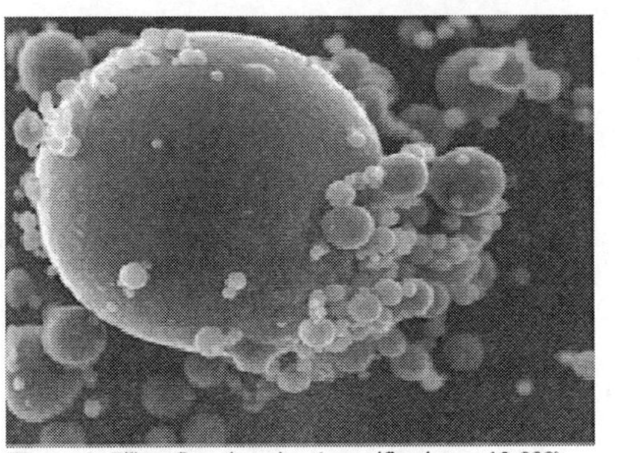

Figure 1. Silica fly ash grains (magnification x 10 000).

A beneficial impact of fly ash on mineral composites consists mainly of [12]:
- increasing grout tightness, due to change of microstructure of set binder matrices,
- formation of additional hydration products, i.e. hydrated calcium silicates (C-S-H phase) and hydrate calcium aluminates filling the pores,
- sealing structures by not hydrated fly ash particles,
- advantageous strength levels in long-term ageing of materials with addition of fly ash.

The listed effects on one hand result from ash acting as aggregate that tightens the grout structure, and on the other hand are related to pozzolate reaction of ash in the grout. Pozzolate reaction is related to formation of additional amount of hydrated calcium silicates that seal grout pores. Comparison of total porosity of ash cement grout with Portland cement grout shows that they are similar, but pore size is smaller in ash cement grout. This is due to the fact that incorporation of fly ash to the grout causes formation of larger amount of gel pores and filling capillary pores with new hydration products. This in turn, reduces permeability which hinders migration of liquid containing among other heavy metals [13, 14]. This also confirms positive effect of silica fly ash on binding heavy metals.

3 Advantages of fly ash application for solidification of hazardous waste

Taking into the account the fact that the aforementioned advantages of silica fly ash regarding formation of permanent structures of mineral composites, it may be said that this is a material applicable for production of matrices for solidification of hazardous industrial waste. Solidification with cement materials allows modification of waste physical properties. This way waste material of reduced (usually low) liquid phase content that is mainly responsible for transport of environmentally hazardous substances.

Waste solidification effectiveness in cement matrix is characterized by two properties: matrix durability and heavy metal leaching [15 - 17]. Unfortunately, these two properties are characterised through many others parameters, which makes unambiguous assessment of composite very difficult. Indeed, only matrix durability depends on its: strength, absorbability, water permeation depth, porosity. And as we know, concretes used for solidification of hazardous waste stored in natural environment or used as aggregate or granulated product in engineering construction can be exposed to many different, unfavourable environmental conditions [18].
The reason for using fly ash from power industry for production of composites solidifying hazardous waste is obtaining matrices of increased durability. Such ash is also to support immobilization of heavy metals in the composite.

4 Material and methods

The conducted tests involved design and preparation of concretes using the following cements:
- Portland cement CEM I 32.5R (denoted hereinafter CEM I);
- Fly Ash Portland cement CEM II/B-V 32.5 R-HSR, in which silica fly ash content was approx. 30% (denoted hereinafter CEM II/B-V).

Contents of heavy metals in cements and in the fly ash used in the studies are presented in Table 1.

A composition of concrete mix was the following: cement 300.0 kg/m^3; sand - 685.2 kg/m^3; gravel 2÷8 mm - 600.4 kg/m^3; gravel 8÷16 mm - 628.6 kg/m^3; water - 180.0 kg/m^3; water/cement (w/c) ratio in all mixes was equal to 0.6. The mixes prepared in such a manner was used as reference samples. Moreover, samples were prepared with 10% of cement replaced with hazardous waste (GO) – galvanic sewage sludge. Contents of heavy metals in waste is presented in Table 2.

Obtained concrete mixes were formed into cubes of dimensions 10x10x10 cm. After 24h, the cubes were subjected to leaching tests or physicochemical tests. Water extracts from concrete samples or waste samples were produced acc. to PN-EN 12457-4 [19]. According to the procedure described therein, the particulate material to grain size < 10 mm, undergoes shaking with water for 24 hours, while keeping liquid solid (L/S) ratio equal to 10. Heavy metal content in eluates was determined using atom absorption spectrometry.

Pore size distribution was measured using mercury porosimeter PoreMaster 60. The measurements were conducted in range 3.5 nm to 10 micrometers.

Compressive strength of concrete was determined according to the standard PN-EN 12390-3 [20].

Table 1. Content of selected heavy metals in cement and fly ash.

Heavy metal	Content [mg/kg]		
	CEM I	CEM II/B-V	Fly ash
Cr	54	52	97
Zn	316	262	142
Cd	< 1	< 1	11
Pb	24	31	34
Co	7	11	32
Ni	18	26	41
Mn	288	323	482
V	34	63	176
Cu	60	54	71
As	6	11	140
Hg	< 0.08	< 0.08	7
Tl	< 5	<5	160

5 Results

The waste selected for tests cannot be classified as neutral waste. In eluates, significant exceeding of content of heavy metals such as chrome, nickel and cadmium (Table 3) was observed in comparison to allowable metal concentrations in water extracts from waste stored on neutral waste landfills [21].

Table 2. Content of heavy metals in dry matter of galvanic sewage sludge (GO)

Heavy metal	Content [mg/kg]
Cr	270100.7
Zn	1441.1
Cd	1133.5
Pb	545.8
Co	163.7
Ni	12789.6
Mn	140.2
V	185.6
Cu	38443.5
As	20.5
Hg	31.8
Tl	0.234

Hazardous waste often before their disposal undergo solidification. This allows to reduce heavy metal leaching to water or soil.The results obtained for immobilisation of heavy metals from galvanic sewage sludge in matrices made of Portland cement and cement with fly ash addition were determined in two test environments. Some samples were exposed for one year to natural environment, while others were in constant contact with distilled water for the same period

Table 3. Content of heavy metals in water extract from galvanic sewage sludge (GO)

Heavy metal	Concentration in water extract [mg/kg]	Allowable value acc. to [21] for waste water extracts [mg/kg]	
		neutral waste	non-hazardous and non-neutral waste
Cr	3.04	0.5	10.0
Zn	0.44	4.0	50.0
Cd	0.71	0.04	1.0
Pb	0.017	0.5	10.0
Co	0.64	-	-
Ni	8.86	0.4	10.0
Mn	6.54	-	-
V	0.030	-	-
Cu	0.48	2.0	50.0
As	0.015	0.5	2.0
Hg	0.0053	0.01	0.2
Tl	0.0027	-	-

 DOI: 10.1051/e3sconf/20161000049

Table 4. Leaching of selected heavy metals from hazardous waste (GO) solidifying concrete.

Concrete sample exposure conditions*	Sample designation	Heavy metal content [mg/dm³] in water eluate					
		Cr	Zn	Ni	Cu	Cd	Pb
Distilled water	CEM I+GO	0.054	0.021	0.002	0.020	0.0001	0.005
	CEM II/B-V+GO	0.098	0.101	0.004	0.069	0.0001	0.023
Natural environment	CEM I+GO	0.186	0.001	0.020	0.006	0.001	0.044
	CEM II/B-V+GO	0.500	0.001	0.013	0.007	<0.001	0.041

*Samples were exposed for a year at various test conditions.

The obtained concentration of selected heavy metals are presented in Table 4.

The measured leaching is low, while the used concrete mixes allow to significantly reduce leachability of heavy metals. It shall be noted that fly ash Portland cement in comparison with Portland cement increases immobilisation degree of heavy metals (e.g. Ni, Cd, Pb, when matrices are exposed to natural environment).

Decreasing heavy metal leachability may be a result of changes occurring at the level of grout microstructure. Addition of fly ash increases total porosity of samples prepared from CEM II/B-V compared to CEM I, however it also allows reduction of pore sizes. Reduction of average pore size occurs also with the addition of only waste to the concrete mix. Pore size decreases also with time. This is a result of pores being filled with subsequent hydration products. Distribution medians presented in Table 5 indicate that average pore size in the tested grout
is in the range corresponding to gel pores, which is important in terms of heavy metal leaching. Gel pores are closed pores, which inhibits liquid transport outside the matrix. The tests results on matrix porosity are presented in Table 5.

The matrices used for neutralization of hazardous waste must have high durability under conditions of the During the studies, concrete samples with addition of by natural environment. This property is determined mainly compressive strength of such composites hazardous waste were measured for their strength. The tests were conducted both for samples stored in distilled water, and for samples exposed to atmospheric environment for one year.

The strength test results presented in Table 6 show that cement with addition of fly ash (CEM II/B-V) not only maintains similar strength as Portland cement when the samples are exposed to natural environment, but when samples are constantly exposed to water for one year, it shows greater strength. This shows high durability potential of such matrices and obtaining better dependability of solidification process stability under various exposure conditions.

6 Conclusions

Hazardous waste management poses a challenge for all communities, thus becoming a very important element of global system for environmental care. The system requires that used methods and processes are in line with foundations of the sustainable development. Therefore, waste management shall be conducted in such a manner to protect natural environment, and if possible, use also

Table 5. Open porosity (P) and pore size distribution in grout with and without hazardous waste (GO), after 1 and 6 months of hydration.

Hydration period	Grout using:	P [%]	Median value [μm]	Modal values
1 month	CEM I	10.32	0.105	0.20 – 0.01 – 0.006 -<0.004
	CEM II/B-V	11.24	0.074	0.26 - 0.013 - 0.01 - 0.004
	CEM I+GO	13.61	0.072	0.55 - 0.05 - 0.01 - 0.004
	CEM II/B-V+GO	15.46	0.055	0.52 – 0.05 – 0.01 -<0.005
6 months	CEM I	14.61	0.094	0.52 - 0.09 - 0.02
	CEM II/B-V	15.86	0.057	0.54 - 0.06 - 0.006
	CEM I+GO	20.47	0.048	0.54 -0.015
	CEM II/B-V+GO	24.45	0.042	0.58 – 0.04

Table 6. Compressive strength of samples stored for one year under different conditions.

Concrete designation	Storage conditions	Compressive strength [MPa]
CEM I+GO	Natural environment	40.3
	Distilled water	40.6
CEMII/B-V+GO	Natural environment	38.3
	Distilled water	50.4

Waste disposal method that is solidification/stabilisation may be one such example. For neutralization of hazardous waste common cements are used, but also combustion by-products such as silica fly ash from coal combustion in the power industry.

Use of fly ash may result in reduction of consumption of cements, which production is expensive and energy-consuming.

Study results presented in the paper show that use of fly ash in solidification of galvanising sludge is a correct, or even a desired, action. This way matrices of increased content of gel pores are obtained, where release of pore liquid to the natural environment is reduced. Thus reducing transport of liquid containing heavy metals from wastes. This leads to increasing immobilization degree of heavy metals in solidifying composites containing fly ash. Additional advantage is increase in strength of matrices containing fly ash in comparison to Portland cement-based matrices (CEM I).

Hence, fly ash gives hope and opportunity to conduct waste solidification process in a stable manner while observing environmental protection principles. The application of silica fly ash in this process is therefore in line with the concept of sustainable waste management.

References

1. A. Pawłowski, L. Pawłowski: Monographs of the Committee of Environmental Engineering PAS, vol. 25, Lublin (2004)
2. M. Wzorek, M. Kozioł, W. Ścierski, J Air Waste Manage, **60**, 12, 1487-1493 (2010)
3. M. Wzorek, T. Baran, M. Ostrowski, Cement Wapno Beton, **18**, 4, 207-215 (2013)
4. R. Roskovic´, D. Bjegovic´, Cement Concrete Res., **35**, 974-978 (2005)
5. PN-EN 15167-2:2006 *Ground granulated blast furnace slag for use in concrete, mortar and grout. Definitions, specifications and conformity criteria*
6. PN-EN 197-1:2012 *Cement. Composition, specifications and conformity criteria for common cements*
7. PN-EN 450-2:2006 *Fly ash for concrete. Conformity evaluation*
8. Z. Giergiczny: Monograph 325. Cracow University of Technology. Cracow 2006.
9. Bulletin of the Polish Cement Association Version of 2016.
10. A. Król, Environ. Prot. Eng. **38**, 4, 29-40 (2012) [11] A. Król, Przem Chem., **86**, 10, 971-973 (2007)
11. M.L. Berndt, Constr Build Mater, **23**, 2606–2613 (2009)
12. X. Fu, Z. Wang, W. Tao, Ch. Yang, W. Hou, Y. Dong, X. Wu, Cement Concrete Res, **32**, 1153-1159 (2002)
13. J.D. Bapat, Adv Cem Res, **13**, 4, 139-155 (2001)
14. A. Król, D. Jagoda, Cement Wapno Beton, **17**, 2, 90 – 101 (2012)
15. B. Batchelor, Waste Manage, **26**, 689-698, (2006)
16. R. Malviya, R. Chaudhary, J Hazard Mater, **B137**, 267–276 (2006)
17. J.R. Fitch, C.R. Cheeseman, J Hazard Mater, **A101**, 239–255 (2003)
18. PN-EN 12457-4:2006 *Characterisation of waste - Leaching - Compliance test for leaching of granular waste materials and sludges - Part 4: One stage batch test at a liquid to solid ratio of 10 l/kg for materials with particle size below 10 mm (without or with size reduction)*
19. PN-EN 12390-3:2011 *Testing hardened concrete. Compressive strength of test specimen.*
20. Regulation of the Ministry of Economy and Labour of 7 September 2005 on the criteria and procedures allowing wastes to be stored at different types of landfills. Journal Of Laws, No. 186/2005, item 1553

Directions of development of research methods in the assessment of leaching of heavy metals from mineral waste

Anna Król[1,a] and Kamila Mizerna[1]

[1]*Opole University of Technology, Department of Environmental Engineering, 5 Mikołajczyka Street, 45-271 Opole, Poland*

Abstract. There are many test methods to assess the level of the release of heavy metals into the environment from mineral waste materials. Leaching methods can be different depending on the leaching time periods, leaching dynamics, sample preparation method or the pH of the elution medium. In Poland, little attention is paid to the research on the relationship between the leaching of particular heavy metals from mineral wastes and changes in environmental conditions, including the pH of the environment. Tests being carried out abroad have started to pay great attention to the pH-dependent impact of the environment and the liquid being in contact with the material on the degree of leaching contaminants from wastes. The solubility of all metals depends on the value of the pH. Authors of the paper will try to prove that Polish methods of waste characterization is incomplete and inconsistent with opinions prevailing in the global literature. The procedure described in the Polish standards are insufficient to determine the actual level of leaching of heavy metals having regard to the impact of multiple external conditions on the level of leaching of heavy metals. Paper will present a directions of development of research methods in the assessment of leaching of heavy metals from mineral waste.

1 Introduction

Improper management of industrial wastes can impact all components of the environment. Mineral wastes from mining, power and metallurgy industry are characterized by varying properties and the level of contamination caused by heavy metals, depending on the source and technology being used. The knowledge of the content and chemical properties of wastes determines the way in which they are managed. The most desirable direction is the recycling by entering the waste for the second time into technological cycle or for its re-use by other industrial sectors. However, not all waste can be utilized in such a way or such action can be economically unviable. Therefore, wastes are being subjected to temporal deposition in heap or in landfills, although it is an ultimate and least preferred direction in the waste management [1].

Wastes placed in the natural environment are exposed to a variety of physical, biological and chemical factors having an impact on the level of the release of heavy metals. Climatic conditions occurring in a landfills or in a heaps, especially the amount of precipitation have a significant impact on the volume of water, which is infiltrating through the mass of waste being deposited in landfills. Water is an agent facilitating implementation of a variety of chemical processes. It aids in the dissolution and leaching of contaminants from wastes, which in turn can lead to contamination of soil and water environment. The main factors, which influence the process of leaching of heavy metals are: fragmentation and shape of the waste, the ambient temperature, the ratio of liquid to solid (L/S ratio), redox potential and contact time with rainwater [2, 3]. One of the key parameters governing the leaching process is also the pH level of the environment in which the waste is located [4]. Significant decrease in the pH of the eluent can lead to the situation that the leaching process can be accompanied by solubilisation of wastes, which can widely promote changes in the levels of the release of heavy metals from great amounts of wastes.

Research of heavy metals leaching under the impact of environmental factors is possible by using various leaching tests. Performing leaching tests by taking into account the impact of atmospheric conditions or simulation of such conditions in the laboratory is one of the objectives for a comprehensive and correct assessment of the level of leaching heavy metals from various materials having contact with the natural environment [5]. Worldwide there are applied different leaching test methods. These methods can be divided according to the leaching time periods, leaching dynamics, sample preparation method or the pH of the elution medium. Subject of this paper was undertaken due to the fact that in the legislation being in force in Poland, the explicit standard regarding waste characterization concerning the leaching of heavy metals is PN-EN 12457:2006 [6]. However, the procedure described in the standard is insufficient to determine

[a] Corresponding author: a.krol@po.opole.pl

the actual level of leaching of heavy metals having regard to the impact of multiple external conditions on the release of heavy metals.

2 The current state of knowledge

In the last two decades, research on the assessment of leaching behaviour of particular heavy metals from various types of materials (including construction mineral materials, granular materials and wastes) were undertaken. H. A. van der Sloot [5, 7-9] has contributed to expand the knowledge about factors and processes determining the leaching of heavy metals mainly from mineral construction materials in different exposure conditions. Due to the fact that wastes, similar as mineral materials, after being places in the natural environment are exposed to contact with its various components (soils, surface water, air), these factors can also affect the leaching of contaminants from such kind of materials.

In worldwide research have started to pay great attention to the pH – dependent impact of the environment and the liquid being in contact with the material on the level of leaching contaminants from wastes [3, 4, 8, 9-13, 14-16]. The solubility of all metals depends on the value of the pH [17]. The common view being described in the literature is that cations of (Ni, Cu, Zn, Cd, Pb, Al, Fe) are characterized by the highest rate of the release in an acidic environment and anions of (Cr, Mo, As, Se, Sb) in the basic environment (Fig. 1). However, the tendencies of changes in the leaching of heavy metals as a function of the pH can vary depending on the type and level of contamination of the test material and its physico-chemical characteristics.

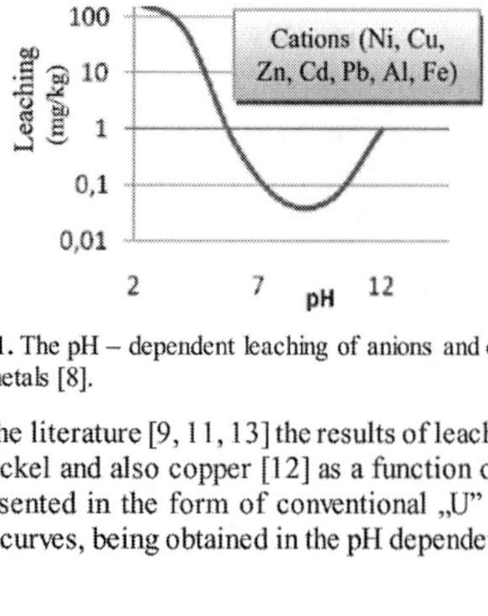

Figure 1. The pH – dependent leaching of anions and cations of heavy metals [8].

In the literature [9, 11, 13] the results of leaching zinc, lead, nickel and also copper [12] as a function of the pH are presented in the form of conventional „U" or „V" – shaped curves, being obtained in the pH dependence tests [18, 19]. With these methods consisting on constant adding for 48 hours an acid or alkali in order to maintain the desired pH value, there are observed trends in the behaviour of leaching contaminants depending on changes in the pH of the environment in which the waste is placed. Leaching tests of heavy metals from various waste materials conducted by van der Sloot [9] and Vitková [13] show that the highest concentrations of zinc, nickel and lead are being obtained in an extremely acidic environment and the lowest ones at pH ranging from 7 to 10, depending on the test material. In the basic medium, another increase in concentrations of the above mentioned heavy metals is being observed. For cooper, the leaching curve does not always adopts the shape of „U" letter. In the research by van der Sloot [9] there are observed downward as well as upward trends in the level of leaching of this element as the eluent's pH increases. This is caused by various capacities to neutralize acid or alkali by the test material. Tests performed by van der Sloot [9] also indicate the leaching tends of chromium, the concentrations of which are the highest in a strongly alkaline environment.

In Poland, little attention is paid to the research on the relationship between the leaching of particular heavy metals from mineral wastes and changes in environmental conditions. Therefore, the problem of the impact of environmental factors on the leachability of heavy metals from wastes was undertaken by the authors of the paper. Authors have a results of preliminary research on the release of heavy metals from industrial wastes [20-22], which indicate the need for drawing attention in the leaching process on factors such as: particle size of the waste and the pH of the eluent (elution medium). Król [23-25] has also carried out the research on the impact of temperature change and acting of chemically aggressive fluids on the level of the release of heavy metals from mineral composites containing solidified hazardous wastes. The research unambiguously indicates the meaning of changing conditions of conducting the process on the level of heavy metal release, and thus the fact of how important is the carrying out the leaching tests using various methods.

3 Tests for leaching of heavy metals from mineral wastes

The need to develop complex methods for waste characterization and assessment of the release of heavy metals was observed in the 90s of the twentieth century [15]. Leaching tests have proven to be an indispensable tools to evaluate the behaviour of release of hazardous substances to the environment from different types of materials under natural conditions and under the impact of external factors. The European Committee for Standardization (CEN) has been a pioneer in the creation of a unified draft of the new leaching tests partly based on the already used procedures. The Dutch Standards Institute (NEN) has also contributed to development of leaching methods. Research of heavy metals leaching from mineral wastes using several different tests allows us to characterize the waste in a broader range and evaluate the behaviour of the heavy metals release in different environmental conditions.

Methods of heavy metals leaching can be divided according to:

- leaching time periods:
- long-term – test examples: tank test [26], column test [27]
- short-term – test examples: batch test [6], maximum availability leaching test [28];
- leaching dynamics:
- static – the quantity of the sample solution is specified and joined by a fixed period of time [16]; product of this connection is the leached substance (eluate); test examples: tank test [25], pH dependence tests [18, 19], batch test [6]
- dynamic – the substance resulting in leaching constantly flows through the sample; test example: column test [27];
- sample preparation method – the sample can be: monolithic, cut off from a monolith and fragmented; the monolithic samples mean integrated samples. In the tank test [26], monolithic samples are used that are placed for a specified period of time in containers with liquid (demineralized water). While, in the batch test [6] a fragmented sample with grain size of less than 4 mm or 10 mm and with a mass of 100 g is being poured with an appropriate amount of water (the ratio of the liquid phase to the solid phase is L/S=10 or 2) and agitated for 24 hours. The eluent is here in contact with a large surface area of the particulate waste material. Column tests in which the material is fragmented to a grain size of <4 mm is packed in a column are utilizing the percolation process for leaching heavy metals from waste materials [27];
- the pH of the elution medium – neutral, acidic or alkaline. In the tank test [26] is used a liquid (demineralized water) having a neutral pH, which covers the monolithic sample. While, in the procedure pH dependence tests [18, 19], the test sample is subjected to the impact of the liquid having various pH levels (the pH from 4 to 12).

When selecting the research methods for assessment the level of leaching, the form in which the test material is deposited is a very important issue. This relates to various processes being accompanied by the heavy metal release from the waste materials. For waste in monolithic forms this phenomenon is an effect of diffusion or dissolution processes imposed by the leaching from the surface [26]. Conditions for performing the research should consider various factors that influence the behaviour of particular heavy metals. The ongoing structural changes caused by external factors (temperature changes, changes in the pH, water contact) can lead to an increase in level of heavy metals being released from materials into the natural environment [5]. The processes as oxidation, carbonation and other corrosive effects of aggressive media entail the need of understanding the impact of particular factors on the phenomenon of leaching heavy metals, considering the entire "life cycle" of the test material. Therefore, the research study on the impact of various factors on the release of heavy metals from wastes will be performed by using a number of leaching tests or their modifications. The different methods of heavy metals leaching from various waste materials are presented in Table 1.

The pH dependence tests are used in order to assess the impact of a wide range of pH on the leaching process of heavy metals release. Tests consist in maintaining

Table 1. Comparison of leaching tests.

Leaching test		**Test category**	**Type of waste material**	**Sample particle size**	**Dry mass or volume of the sample**	**pH of the eluent**	**L/S ratio**	**Procedure duration**	**Mixing/ shaking**
Batch test [6]	Part 1	static test	all waste materials (granular)	< 4 mm	175 ± 5 g	5÷7,5	2 dm³/kg	24 h	yes
	Part 2				90 ± 5 g		10 dm³/kg	24 h	
	Part 3				175 ± 5 g		2 and 8 dm³/kg	6 h + 18 h	
	Part 4			< 10 mm	90 ± 5 g		10 dm³/kg	24 h	
pH dependence tests [18,19]		static test	all waste materials (granular)	< 1 mm	15, 30 or 60 g ± 10%	target 4÷12	10 dm³/kg	48 h	yes
Maximum availability leaching test [28]		static test	all waste materials (granular)	< 0.125 mm	16 ± 2 g	target 7 and 4	50 dm³/kg	6 h	yes
Tank test [26]		static test	all waste materials (moulded or monolithic)	one sample dimension of at least 40 mm	not applicable	5÷7,5	not applicable	64 days	no
Dynamic monolithic leaching test [29]		dynamic test	all waste materials (moulded or monolithic)	one sample dimension of at least 40 mm	not applicable	5÷7,5	not applicable	36 days	no
Column test [27]		dynamic test	all waste materials (granular)	< 4 or < 10 mm	0,6 or 2,4 dm³ after compaction	approx. 7	10 dm³/kg	approx. 30 days	no

constant pH level of the eluent for a predetermined period of time (with the target for 48 hours), and thereby, to assess the release of inorganic components from the waste in an extremely acidic and basic conditions. The test involves reaching the equilibrium condition between the test material and the eluent. This condition is reached when pre-determined (computed) amount of acid or base (standard EN 14429:2015 [19]) is being added to the water, in order to achieve and maintain the desired pH value at constant level or the value is continuously monitored and controlled by adding automatically an acid or a base (standard EN 14997:2015 [18]). The standards apply to all types of waste materials (with the grain size of < 1 mm). The test will cover eight selected values of pH ranging from 4 to 12 (the range can be expanded if necessary). The pH values are being measured after 4, 44 hours and after 48 hours from the time of pouring the sample with eluent. The capacity of waste to neutralize acid (ANC) or base (BNC) can be determined in the discussed tests. This property is obtained based on the amount of acid or base being added to achieve the final predetermined pH level [18, 19].

The maximum availability leaching test allows researchers to assess the leaching behaviour of contaminants by using liquid with the pH level of 7 and 4. Based on the results, the capability to neutralize acid by wastes at a given pH can be also determined. The test is applied to determine leaching of inorganic components, which can be leached from fine-grained waste materials exposed to extreme weather conditions or being after the dissolution of the material or in the case of loss of the capacity for neutralization of acids. The availability test is a two-step analysis performed from the moment of reaching by the eluate the pH at the level of 7 in the first stage and at the level of 4 during the second stage. At least 95% of dry weight of the sample should be of particles with the size smaller than 125 μm. The aqueous extract is obtained from 16g of such fragmented material. The material is mixed with water for a total of 6 hours by using a stirrer in the ratio of L/S = 50 dm^3/kg. By using the pH–meter, the pH level of the eluate should be measured after one and ten minutes from the moment of agitation. This is used for classification of the test material (in terms of acid or alkaline reagent). Based on this classification, the concentration of nitric acid being added to lower the pH of the eluate is being determined. Extracts obtained during the two stages were mixed together and analysed according to the content of selected impurities [28].

Assessment of processes accompanying the leaching of heavy metals from waste materials and the impact of time during which the liquid is in contact with the sample on the leachability will be carried out in the tank test or dynamic monolithic leaching test (DMLT). Procedure of both tests is similar. The difference is due to the duration of the tests. DMLT lasts 36 days and tank test lasts 64 days. Waste in a monolithic form (minimum size of one sample dimension = 40 mm) is subjected to leaching in a closed tank. The sample with known surface area and volume is poured with distilled, so the volume of the liquid will be 2÷5 times greater than the volume of the sample (in the tank test) or 8 times greater than the surface of the sample (in DMLT). The distance of all surface areas of the sample from walls of the tank should be at least 2-centimeter long. Eluates are collected in the 8 test periods. After each collection of the eluate, the liquid is being exchanged in its entirety and then a sample is fixed inside again. Subsequently, after the eluates filtration, the pH level is measured and the concentration of heavy metals is determined for each eluate fraction (8 fractions). For each of the analysed heavy metal, following can be determined during the research tests: the leachability in the eluate fraction, the total leaching per unit surface area and emerging leaching processes (diffusion, elution from surface area, depletion, dissolution, delayed diffusion or dissolution) [26, 29].

By performing column tests it is possible to assess the leaching of heavy metals in certain percolation conditions that are dynamic conditions. Wastes can contain unstable substances, which are decomposed under the influence of weather conditions, and this causes the release of additional pollution loads. Therefore, dynamic tests simulate real conditions of the leaching of contaminants in the landfill aeration zone. Column test also allows researchers to perform the observations on how the change of the liquid to solid (L/S) ratio influences the obtained concentrations of particular components. The test material is being packed in a column of a suitable size. On the applied size of the column (with diameter of 5 cm or 10 cm) decides the waste particle size. The column packed with waste is being weighed and then the material is saturated. The saturated material is left for three days to stabilize the system. Then, the distilled water is provided to the column, which flow upwards the column with a determined flow rate until reach the eluate with predetermined liquid to solid ratio (L/S = 0.1; 0.2; 0.5; 1.0; 2.0; 5.0; 10.0 dm^3/kg). In this manner, seven eluates are derived, which are after filtration subject to the analysis on the content of heavy metals [27].

The batch test is relatively quick and often used procedure. The scheme of batch test procedure was presented in Figure 2. Test should be carried out according to the part 1 of standard [6]: PN-EN 12457-1:2006 (single batch test at L/S = 2 dm^3/kg), part 2: PN-EN 12457-2:2006 (single batch test at L/S=10 dm^3/kg), part 3: PN-EN 12457-3:2006 (two stage batch test at L/S = 2 and 8 dm^3/kg) or part 4: PN-EN 12457-4:2006 (single batch test at L/S=10 dm^3/kg). Particulate material with relevant grain size of < 4 mm or < 10 mm is subject of agitation with water for 24 hours. Shaker's rotations should be set on in such a way to prevent deposition of solid particulates. After the shaking process ends, the containers should be left for about 15 minutes, so the process of sedimentation of solid particles take place. Subsequently, the eluate is being filtered through membrane filter and analysed. In the legislation being in force in Poland, the explicit standard regarding waste characterization concerning the leaching of heavy metals is PN-EN 12457:2006 [6]. The requirement for the waste to be deposited in a designated type of a landfill (inert, hazardous or non-hazardous) is fulfilling relevant criteria on the leaching of contaminants. However, the batch test doesn't include the conditions and their changes occurring during waste deposition in the environment (heaps, landfills). On the basis of own preliminary tests [22] in accordance with the

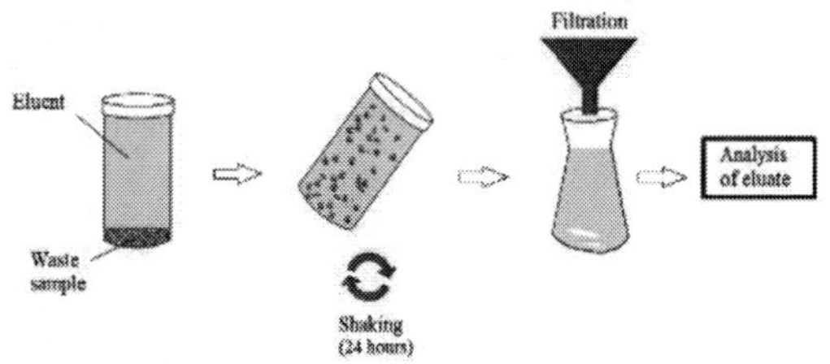

Figure 2. Scheme of batch test according to PN-EN 12457:2006 [6].

abovementioned standard, it was noted that if a little change decreasing or increasing the pH level of the eluent influences the release of varying amounts of heavy metals. Therefore, a change in the pH environment, for example as a result of acting rainwaters or leaching the waste in the deposited mass, can cause an increase in the level of the release of contaminants from wastes. This is important when classifying the waste or when determining the capacity of its deposition in landfills of a given type. In Poland, wastes are classified based on leachability criteria set out in the Regulation of the Ministry of Economy of 16 July 2015 on the acceptance of waste for landfill [30]. Due to various impacts affecting the waste being deposited in a landfill for inert wastes, the limit values for the leaching of heavy metals determined for such type of landfills can be exceeded. Then, it can turn out that the waste should be deposited in hazardous waste landfills.

4 Conclusions

Development of different research methods in assessment of heavy metals leaching from mineral waste was an important step towards the development of waste studies. Research of release of heavy metals from waste by several leaching methods allow us to assess the leaching behaviour in different waste exposure conditions. It is also possible the proper determination and prediction of heavy metals leaching taking into account the impact of many factors and processes (the pH of environment, amount of rainwater, landfilling duration). Complex testing methods of leaching of heavy metals from mineral waste make it possible to assess the short- and long-term effects of leaching. Authors in the paper sought to indicate the relevant methods and tests allowing a comprehensive assessment of mineral industrial wastes in terms of the release of heavy metals. Both the structure, type and method of potential waste management decide on the method to assessment the process of leaching of contaminants. Using a various tests can assess how the waste is hazardous and which risks to the environment. This allows us to choose the appropriate direction of development of waste. It therefore seems that the adopted in Poland leaching methods and thereby the aggregation of the research results connected with databases on waste categorization into particular groups (neutral, hazardous) provides incomplete image on the state of risk, which is posed by waste deposited in the environment in terms of the release of heavy metals.

References

1. Polish Act on Waste of 14 December 2012 on Poland's Journal of Laws dated January 8, 2013, item 21,http://isap.sejm.gov.pl/DetailsServlet?id=WDU20130000021
2. A. Król, Arch. Civ. Eng. Environ. **4**, 3, 71-76 (2011), http://acee-journal.pl/1,7,20,Issues.html
3. H.A van der Sloot, A. van Zomeren, Mine Water Environ. **31**, 92-103 (2012), doi: 10.1007/s10230-012-0182-8
4. M. Vitková, V. Ettler, O. Šebek, M. Mihaljevič, T. Grygar, J. Rohovec, J. Hazard. Mater. 167, 427-433 (2009), doi: 10.1016/j.jhazmat.2008.12.136
5. H.A van der Sloot, A. van Zomeren, J.C.L Meeussen., P. Seignette, R. Bleijerveld, J. Hazard. Mater. **141**, 2, 354–369 (2007), DOI: 10.1016/j.jhazmat.2006.05.106
6. PN-EN 12457:2006. Characterisation of waste. Leaching. Compliance test for leaching of granular waste materials and sludges. Part 1-4
7. H.A van der Sloot, J. Dijkstra, Development of horizontally standardized leaching tests for construction materials: a material based or release based approach? Report ECN-C-04-060 (2004)
8. H.A van der Sloot, J. Dijkstra, P. Seignette, O. Hjelmar, G. Spanka, Characterisation leaching tests as basis of reference for quality control and decisions on acceptability of alternative materials in construction, Energy Research Centre of the Netherlands, Wascon Conference, Lyon (2009)
9. H.A van der Sloot, Horizontal standardization and harmonization of leaching test methods for waste, secondary raw materials, construction materials and (contaminated) soil, *Proceedings of the Wascon Conference*, San Sebastian (2003), https://www.ecn.nl/docs/society/leaching/wascon_005.pdf
10. M.J. Quina, J.C.M. Bordado, R.M. Quinta-Ferreira, The influence of pH on the leaching behaviour of inorganic components from municipal solid waste APC residues, Waste Manage. **29**, 2483-2493 (2009), doi: 10.1016/j.wasman.2009.05.012
11. D. Centioli, R.N.J. Comans, S. Gaudino, C. Galas, M. Belli, Ann. Ist. Super. Sanita **44**, 3, 252-257 (2008), http://www.iss.it/binary/publ/cont/252%20-%20ANN_08_36%20Centioli.1224497272.pdf
12. V. Cappuynus, R. Swennen, J. Hazard. Mater. **158**, 1, 185-195 (2008), doi: 10.1016/j.jhazmat.2008.01.058
13. M. Vitková, V. Ettler, O. Šebek, M. Mihaljevič, Mineral. Mag. **72**, 1, 521–524 (2008), doi: 10.1180/minmag.2008.072.1.521

14. J. Yao, W.B. Li, Q. Kong, F. Xia, D.S. Shen, Fuel, **93**, 99-104 (2012), doi: 10.1016/j.fuel.2011.11.026
15. J.L.T Hage, E. Mulder, Waste Manage. **24**, 2, 165–172 (2004), doi: 10.1016/S0956-053X(03)00129-6
16. J. Kalembkiewicz, E. Sitarz-Palczak, J. Ecol. Eng. **16**, 1, 67-80 (2015), doi: 10.12911/22998993/589
17. N. Rothe, K.O Gundermann, F. Jentsch, Zentralbl. Bakteriol. Mikrobiol. Hyg. B. 187, 2, 112-124 (1988)
18. EN 14997:2015 Characterization of waste. Leaching behaviour test. Influence of pH on leaching with continuous pH control.
19. EN 14429:2015 Characterization of waste. Leaching behaviour test. Influence of pH on leaching with initial acid/base addition.
20. K. Mizerna, A. Król, Inż. Ekolog. **43**, 1-6 (2015) (in Polish), doi: 10.12912/23920629/58898
21. A. Król, K. Mizerna, Chemik, **69**, 10, 670-673 (2015), http://www.chemikinternational.com/wp-content/uploads/2015/10/10_15-ro%CC%81l_GB.pdf
22. K. Mizerna, A. Król, Appl. Mech. Mat. **797**, 408-414 (2015), doi:10.4028/www.scientific.net/AMM.797
23. B. Augustyniok, D. Jagoda, A. Król, R. Roszczyk-Walczak, Release of heavy metals from concrete as a result of impact of high temperature, *Proceedings of ECOpole*, **1**, 1/2, 81-84 (2007) (in Polish). http://tchie.uni.opole.pl/ecoproc07/Augustyniok_07.pdf
24. A. Król, Z. Giergiczny, J. Hazard. Mater. **160**, 2-3, 247-255 (2008), doi: 10.1016/j.jhazmat.2008.03.007
25. A. Król, Environ. Prot. Eng. **38**, 4, 29-40 (2012), doi: 10.5277/EPE120403
26. EA NEN 7375:2004 Leaching characteristics of moulded or monolithic building and waste materials. Determination of leaching of inorganic components with the diffusion test. The Tank Test
27. CEN/TS 14405:2004 Characterization of waste. Leaching behaviour tests. Up-flow percolation test
28. EA NEN 7371:2004 Leaching characteristics of granular building and waste materials. The determination of the availability of inorganic components for leaching. The maximum availability leaching test
29. EN 15863:2015 Characterization of waste. Leaching behaviour test for basic characterization. Dynamic monolithic leaching test with periodic leachant renewal, under fixed test conditions
30. Regulation of the Ministry of Economy of 16 July 2015 on the acceptance of waste for landfill, Poland's Journal of Laws dated September 1, 2015, item 1277

E3S Web of Conferences **10**, 00051 (2016) DOI: 10.1051/e3sconf/20161000051
SEED 2016

CSR, environment-friendly investments and innovations - the three elements necessary to build a modern and strong coal mining company?

Katarzyna Hałasik[2] and Joanna Kulczycka[1,a]

[1] *Mineral and Energy Economy Research Institute, Polish Academy of Sciences, Krakow, Poland*
[2]*AGH-UST, AGH University of Science and Technology, Faculty of Management, Krakow, Poland*

Abstract. Currently, strongly increasing globalization forces companies around the world to search new ways to build competitive advantage and achieving goals. Society having wider access to information and products, began to pay attention not only to the attractiveness of the goods, but also to the credibility of the environmental and social of functioning companies. This makes, that the concept of corporate social responsibility and the green revolution, which are strongly associated with innovations, in last years become widely discussed in the literature, as well as popular in the activities of enterprises. This article is an attempt to answer the question, whether the combination of these three elements is a guarantee for the creations of a positive image, and thus the value of a modern and strong company. The research literature aims to identify the current approach to these three factors and their characteristics. In contrast, the analysis of Polish hard coal mining sector indicates the role of innovation, environmental protection and CSR in the industry - whether the actions in this area are present and to what extent. In addition, it will be presented to outline the current state of Polish hard coal mining sector.

1 Introduction

Exponential technological progress that took place during recent years has caused a series of far-reaching changes in the world. The first immediate result was undoubtedly created a number of innovative products and services. Further consequences are now the changes on many levels: the social, organizational or cultural, related to the process of globalization and faster flow of information, capital and goods.

As a result of these changes, we began to pay much attention to the ethical, environmental or social matters within the organization. Consequently, companies that pay little attention to zero emissions, pollution prevention, ecolabelling and other environmentally related matters will sooner or later loose their customers [1]. The growing popularity of corporate social responsibility (CSR) and sustainable development also confirms changes, that have occurred in the perception of the main objectives and values which should guide the functioning of companies. In recent years in Poland the entrepreneurs' interest in the concept of sustainable development and CSR has been increased - the analysis of reports from studies enables to note positive changes in a way of thinking and perception of CSR by the Polish entrepreneurs and managers, though it is still far to satisfactory results [2].

This article aims to illustrate the current situation of Polish hard coal mining sector, in the context of the implementation of the assumptions arising from the concept of CSR, environmental protection and innovation. The aim is to assess achieved benefits and costs for these reasons. This will enable to present an answer to the question, whether these factors are the basis for building a competitive advantage and a source of a number of other benefits for companies in the sector which interferes with the natural environment.

2 Main principles coming of CSR for modern enterprises

The growing popularity of CSR makes companies begin to function under social pressure. It is expected, that their actions do not only meet the requirements of ethical and environmental within the law, but at the same time meet the expectations of different social groups. As indicated by Jenkins (2009) [3], CSR relates to the activities of businesses, particularly in terms of their contribution to achieving economic, social and environmental sustainability. To remain competitive, companies need to be able to adapt to these new demands from the market and society in which they operate.

CSR, despite its popularity, is still variously defined and understood. In addition, in the literature you can meet different positions regarding the significance of CSR, which is making even more controversies surrounding

[a] Corresponding author: kulczycka@meeri.pl

the theory. In the 1990s, but also presently there are views that the whole concept is not relevant for business (Freeman & Liedtke, 1991) [4], by those who see the importance, but say it is a bad business idea (Friedman, 1962) [5] for a wide range of different authors, who believe that CSR is of strategic importance for the business [6]. Therefore, while investigating the CSR we have to take into account the fact that different countries have different priorities and values that shape the way business operates [7], which also has a decisive influence the knowledge and practice issues of responsible business.

CSR derives from business ethics, and the growth of interest showed mostly large companies in order to overcome the anti-globalization and anti-capitalist protests in the 90s of the 20th century [8]. Informing about the actions taken and their results is still voluntary, however on 12.06.2014 for large companies (meeting the criteria for the employment over 500 people and balance sheet total over EUR 20 million, or income over EUR 40 million) came into force the EU directive on disclosure of publication non-financial information and information on the diversity of some large companies and groups. Poland and other EU countries are obliged to implement the provisions of the directive into national law by 12.06.2016. However, recognition of CSR is current also among small and medium-sized enterprises (SMEs), which began to practice CSR activities.

Among the many definitions, as an example you can quote those created by international institutions, such as:
- the definition given by the European Commission, according to which corporate social responsibility is a concept whereby companies integrate social and environmental concerns in their business operations and in their interaction with their stakeholders on a voluntary basis [9],
- the definition in accordance with ISO 26000, which says that social responsibility means inclusion by business of social and environmental concerns in the commercial (economic) activities and taking responsibility for the impact of their activities on society and the environment [10],

or in the national context, such as the definition given by the Ministry of Economy in Poland, according to which:
- the basic premise of CSR is responsible and ethical conduct of business in relation to social groups, which interacts with the greatest possible respect for the natural environment. Building a socially responsible business model based on the concept of sustainable development, stressing the need to take into account in business, in addition to the economic dimension, the social and environmental factors in order to meet the needs of not only the present but also for future generations [11]. Indicated definitions clearly show obligations of enterprises, arising from the implementation of CSR principles. However, the important issue need to be borne in mind is that, a properly implemented CSR strategy also leads to raise a number of benefits and intangible. Most frequently reported are:
- increased interest of investors,
- extending the consumers and stakeholders loyalty,
- improvement of relations with the community and local government,
- increased competitiveness,
- improvement of the organizational culture of the company,
- shaping a positive image of the company among employees,
- obtainment and retention the best employees [12].

Benefits obtained from CSR depend primarily on the scope and implementation of CSR activities, their level of implementation and fitting to the strategy of the entire enterprise, the correct understanding of the theory and tools which were used. In this range, a number of international institution formulate standards for reporting, for example GRI. Guidelines of the Global Reporting Initiative (GRI), include a benchmark reporting corporate responsibility and sustainable development for companies around the world [13]. As indicated by Levy, Sznejwald Brown and de Jong (2010) [14] standards GRI has become embedded in the operational routines and practices of hundreds of large companies in multiple countries. It has gained widespread legitimacy, as demonstrated not just by corporate compliance, but also by the attainment of official recognition by governmental agencies and multilateral organizations such as the UN Global Compact [14]. The principal aim of GRI is that corporate social responsibility has become a reliable and routine reporting tool, such as financial reporting [15]. The principles promoted by GRI are a number of clear guidelines and rules regarding the reporting of companies on the basis of sustainable development and CSR, formulated in such a way that they can be used in the most of enterprise. The first guidelines G1 were issued in 2000, while in 2011 five supplements were published: for companies from the energy industry, financial services, food processing, mining and metals, and non-governmental organizations, which include sets of additional indicators reporting for companies [15]. The latest and current guidelines are G4.

3 Implementation of CSR as a straight path to innovation and environmental protection

CSR is not the only, popular in recent years, way to find new sources of competitive advantage and other intangible benefits for businesses around the world. With the development of technology, many companies began to invest in innovative projects, often promoting environmental protection. Innovations are strongly stimulated not only by new technologies, the development of technical skills and consumer demands, but also by strong and global the importance of CSR issues. Companies in order to be successful and innovative nowadays, must consider the social and environmental impact of their operational processes, stimulate the creativity of their workforce, and cooperate with their suppliers, customers, and other business partners in designing and developing new and innovative products and services [7]. Admittedly, as indicated by Bocquet and Mothe (2013) [16], on an academic ground, the relationship between CSR and technological innovation has still not been evidenced. In literature definitions of CSR does not relate directly to the issue of innovation. On the other hand, the strategic management and dynamic evolutionary approaches offer an interesting angle for reconciling CSR and technological innovation [16].

To investigate, whether CSR can become a source of innovation, and hence a number of intangible benefits for the company, the relationship between CSR and innovation should be considered. CSR has opened the way for many innovative products and processes, in particular ecology and prosocial. Gallego-Alvarez, Manuel Prado-Lorenzo Garcia-Sanchez (2011), studying the relationship between innovation and CSR proposed to formulate the following hypotheses:
- there is a bidirectional relationship between CSR practices and investment in innovation,
- CSR practices positively influence the volume of investment in innovation,
- investments in innovation positively influence CSR practices [17].

Thus, there is a positive correlation between investments in innovation and CSR. This is confirmed by research done by MacGregor and Fontrodona (2008) [18] on SMEs had either a proactive or a reactive approach to CSR and innovation. It has been shown that:
- the most successful companies were proactive in their approach and also, interestingly, had the highest level of CSR implementation,
- proactive CSR is easier than proactive innovation,
- CSR adoption can be a starting point for proactive innovation and also for companies that want to improve their position in their industry but are generally risk-averse [7].

Important combination of CSR and innovation are the eco-innovation. The Organisation for Economic Co-operation and Development (OECD) clearly defined, what is meant by eco-innovation: Innovation that reflects the concept's explicit emphasis on a reduction of environmental impact, whether such an effect is intended or not. It is not limited to innovation in products, processes, marketing methods and organisational methods, but also includes innovation in social and institutional structures [19]. As indicated by Rexhepi, Kurtishi and Bexheti (2013) [7], eco-innovation is thus seen as an overarching concept which:
- provides direction and vision for pursuing the overall societal changes needed to achieve sustainable development,
- leads to progress in eco-efficiency, and eco-efficiency is a management philosophy to guide and measure companies and other actors' development in environmental performance.

Such an approach to eco-activity, shows that it is very strongly associated with CSR - is a kind of tool to realize CSR' assumptions related to the environment - it is intended to create value by reducing the negative impact on the environment. There's a definition of CSR which says: CSR is a concept, in which the organization functioning in the environment, voluntarily takes into account its expectations and the responsibility to care for it [20]. Surroundings, in this context, means all stakeholder groups, in other words entities that are under the influence of the organization or have an influence on it. Keep in mind that according to the concept of CSR, the natural environment is also a stakeholder. Therefore, there cannot be activity in the field of responsible business without an activity to protect the environment.

Despite the strong correlation between CSR, eco-innovation and environmental protection, as well as the growing awareness of ecological and social in Poland, the specificity of the Polish market imposes many barriers to these issues. According to Zuzek (2015), the most common include:
- economic factors (eco-investitions tie in with considerable financial expenditures: equipment, procedures, transaction costs)
- a complex organization (eco-innovation involves various disciplines and a large number of actors from different areas of research)
- the need for well understanding and vision of future needs (relative lack of clarity environmental issues and information problems occurring in the process of eco-innovation) [21].

Poland among EU countries is seen as a country with one of the lowest levels of implementation of eco-innovation, and this situation has not been changed for several years (Figure 1). The Eco-Innovation index based on indicators in five areas: eco-innovation inputs, eco-innovation activities, eco-innovation outputs, resource efficiency outcomes and socio-economic outcomes [22]. This confirms that in Poland there are many barriers (mostly economic barriers) to the implementation of eco-innovative solutions. Responsible field of business without an activity to protect the environment.

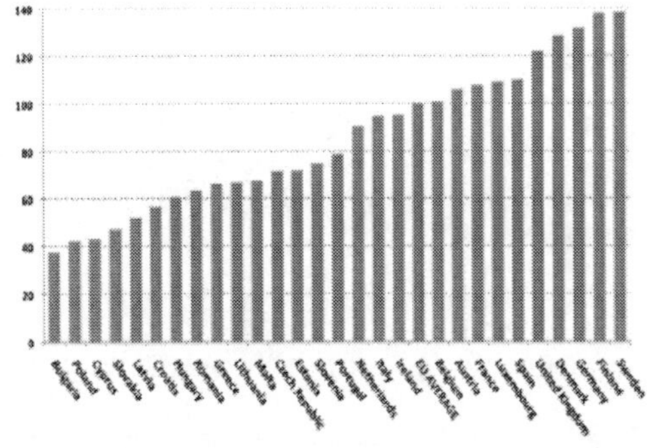

Figure 1. The level of implementation of eco-innovation in the EU – April 2014 [23]

Increasingly common practice of CSR, as well as the availability of funds for pro-environmental investment give hope, that in the near future this unfavorable trend will change and there will be new eco-innovative solutions, which are increasingly visible in the mining industry. This could be proven by the activity of KGHM Polska Miedź, which spends on environmental investments in Poland about 180 million PLN [24] every year or by an example

DOI: 10.1051/e3sconf/20161000051

of eco-innovation developed in the Central Mining Institute in Katowice (GIG), in collaboration with the Mine Limestone "Morawica" SA, in the form of a mixture of limestone, on the basis of limestone from "Morawica" Mine and waste materials from energy which is applicable in various mining techniques [25].

4 The methodology and the current situation of the Polish hard coal mining sector

4.1 Methodology of research

As shown Kulczycka and Wirth (2010) [26], a lot of mining companies in Poland take a number of CSR activities, environmental protection, promotion of local employment, investment in new projects, increase spending on R&D. But usually, there is a lack of publicly available information in this area, and mining companies which begin investments are often treated by the local community as a threat. It is recommended to create coordinated CSR activities, in the whole mining industry and implement standards for reporting in this industry, containing the joint declarations and missions [26].

This applies particularly to the hard coal mining industry, which on grounds of a significant impact on the environment, but also of great importance for regional development should clearly inform about the strategic actions affecting local communities. Therefore, the aim is to analyze, on the basis of publicly available information and reports, the activities of companies Polish hard coal mining sector companies in terms of CSR, innovation, eco-innovation and environmental protection. These activities will be confronted with the financial results of the investigated companies.

4.2 Polish hard coal sector

Currently, Polish hard coal sector includes eight companies, which deal with hard coal mining:

1. Kompania Węglowa S.A. (100% owned by the State Treasury) - 15 mines,
2. Katowicka Grupa Kapitałowa (100% owned by the State Treasury) - it consists of Katowicki Holding Węglowy S.A., includes 4 mines and KWK Kazimierz Juliusz, includes 1 mine, which 100% of shares has KHW S.A.,
3. Jastrzębska Spółka Węglowa S.A. (listed on the Warsaw Stock Exchange, the majority shareholder is the State Treasury) - 5 mines
4. Południowy Koncern Węglowy S.A. - 1 mine, which 50.45% shares has a Kompania Węglowa S.A. and 49.55% Południowy Koncern Energetyczny S.A.
5. Lubelski Węgiel „Bogdanka" S.A. (listed on the Warsaw Stock Exchange, in which the State Treasury holds no shares)
6. Przedsiębiorstwo Górnicze Silesia (private company, which is joined ownership by Czech group Energetický a Průmyslový Holding and employees of PG Silesia)
7. Siltech Sp. z o.o. (100% owned by a private owner)
8. Spółka Restrukturyzacji Kopalń S.A. (leads the liquidation of mines and utilizes non-productive assets of closed mines) [27].

It should be emphasized that the both mining of hard coal and lignite in Poland are very essential. These are the basic raw materials for the production of electricity in the country (88% of the energy came from coal in 2012 [28]). Unfortunately, most of the technology which is used is outdated and unprofitable, that is why the issue of investment and eco-innovations are so important for the energy security of the country.

Another factor demonstrating the validity of the hard coal mining industry in Poland are social and environmental factors. The deployment of mines is regionally focused on the areas with mining culture and traditions. The social, environmental and economic consequences of closing Polish mines would be significant - mass layoffs, reclamation of mining areas (removal of equipment, cleaning the area, etc.) would absorb enormous financial expenditure [29]. Therefore, new investments, social acceptance and meeting the environmental standards for the sector of hard coal mining in Poland are extremely important, in the context of the functioning of the whole country.

5 The analysis of selected entities in the mining sector in Poland in terms of innovation, CSR and environmental protection

The analysis was carried out in two stages for the analyzed companies: first for CSR activities and second for investments in innovation and contribution to the protection of the environment, which will allow for comparison of companies surveyed in a given range. Then it will be compared the indicator of the return on sales for 2014, to compare the activity of the analyzed range to companies' financial position. The criterion for selection of companies for the analysis was the access to the information needed for analysis.

In Table 1 are summarized the CSR activities for the Kompania Węglowa S.A., Katowicka Grupa Kapitałowa and the Lubelski Węgiel "Bogdanka" S.A. [30, 31, 32]. Only three of the eight entities provide relatively detailed and valuable information on CSR, innovation and environmental protection. Jastrzębska Spółka Węglowa S.A. provides information in this field, but these are very general and ideological information, while the other four entities do not present such kind of data. Only two companies, i.e. Katowicka Grupa Kapitałowa and Lubelski Węgiel „Bogdanka" S.A. publish CSR reports, with indicators of GRI (*Global Reporting Initiative*), only recently. This means, that both companies follow the international standard of reporting. Also, only those companies have a code of ethics, which converted the most important essential standards and values of companies.

 DOI: 10.1051/e3sconf/20161000051

Table 1. CSR activities of analyzed companies [own work]

Company name	CSR Report	The main areas of CSR activities
Kompania Węglowa S.A.	no	1. Social dialogue - establishing and regular contacts with the trade unions operating in Kompania Węglowa S.A. and the Employers' Association of Hard Mining Coal - keeping records of social dialogue in Kompania Węglowa S.A. and correspondence in this range with the trade unions and the National Labour Inspectorate - keeping records and correspondence related to collective disputes, including mediation - coordinating the work of thematic groups in Kompania Węglowa S.A. in the range of occupational safety, wages, employment restructuring and other cooperation with social partners - cooperation with the Voivodeship Commission for Social Dialogue 2. Modern, organized, creative and efficient work system - having a specialized and well-qualified staff - integrating employees in order to the achievement of the defined objectives - investing in the development of professional and personal employees and the local environment - motivating staff to achieve high results of work - compliance with standards of safety management system and occupational health standards PN-N-18001:2004 - the strong association with the region, creating new jobs and contributing to the social security of the inhabitants of Upper Silesia, co-cultivate its culture and traditions of mining 3. The Integrated Management System - implementation of the Integrated Management System in accordance with the requirements of ISO 9001:2008, ISO 14001:2004 and PN-N-18001:2004 - using Integrated Management System Policy, Quality Management System, Environmental Management System, Occupational Safety and Health Management System
Katowicka Grupa Kapitałowa	yes (including GRI indicators for 2013, 2015)	1. Code of Ethics - contains the rules of ethics, fair competition and high standards of corporate governance, a set of values 2. Spokesman for Ethics - counseling and care for the observance of ethical standards - monitoring the observance of ethical principles - the promotion of ethical behavior - participation in the activities of communication and education for the benefit of the company's corporate culture - the prevention of social conflicts arising from the breach of the Code of Ethics 3. The most important value of CSR (described in detail in the report) - the promotion of cultural values - improving the quality of life of the inhabitants of the district of Katowice - raising awareness and knowledge of the natural sources of energy in the society - supporting the education of mining in the region - minimizing the negative impact on the environment - engaging in charity - supporting public benefit organizations - the possibility of employee development through projects aimed at raising the competence - fair employee appraisal system - encouraging employees to participate in programs that can raise the standard of living of the local community - achieving a policy of sustainable development - combining efforts to economic success with care for the environment and society

[a] Corresponding author: kulczycka@meeri.pl

Lubelski Węgiel „Bogdanka" S.A.	yes (including GRI indicators, for 2012-13, 2014)	1. Code of Ethics - a set of norms and values at each position and in every cell of the organization and any body corporate 2. Emphasis on human capital - testing the organizational culture and the internal communication system - the provision of adequate staff competencies through a system of training and schooling - striving to develop a formalized appraisal system, reward and promotion - Incentive Systems for Employees: bonuses and commendations 3. Ensuring the safety and promote the development of local communities - social dialogue - counteracting the effects of the social consequences of mining damage - activities for the local community - education and development in the mining industry 4. Integrated Management System - compatible with PN-EN ISO 9001:2009, ISO 14001:2005, PN-N-18001:2004 - Occupational Safety and Health Management System is compatible with British BS OHSAS 18001:2007

The activities of the analyzed entities in the field of CSR partially overlap, but as you can see they put the emphasis on different aspects of the activity in this area and given information to the general, which means that to a certain extent otherwise perceive corporate social responsibility and this confirms that Polish companies have relatively little experience and knowledge of the activities of CSR. In contrast, compliance with GRI guidelines in reporting shows that these companies follow global trends.

All the analyzed entities conduct extensive activities, mainly in the field of environmental protection and then innovation. This is due to the specific nature of their business, which permanently changes the landscape and significantly affect the environment. The scope of activities of environmental protection is broader than innovation, the reason are high costs and risks associated with the implementation of innovations. However, regulations and environmental standards, both national and international, pressure coal mining industry to reduce the negative impact on the natural environment, which indirectly is also reflected in investments and innovations.

Available information in this range is quite general and imprecise. Especially in the case of the Lubelski Węgiel „Bogdanka" S.A., which calls itself "innovation leader" and informs that has one of the most modern mine in Poland and also in the world. Unfortunately, the report does not set out specific innovative solutions, it is only given information that innovations are used and yield positive effects.

To summarize the activities of analyzed companies in the range of CSR, innovation and environmental protection must be noted that the activities in this area are certainly present. Lubelski Węgiel „Bogdanka" S.A. and Katowicka Grupa Węglowa S.A. look the best in this collation (both companies have a code of ethics and publish social reports, in contrast to Kompania Węglowej S.A.), what does not mean that the activity and information from the tested range are fully satisfactory. The indicator of the return on sales (ROS) for 2014 can confirm that - for Lubelski Węgiel „Bogdanka" S.A. amounts 13.53, for the Katowicka Grupa Węglowa S.A.13.79, while for Kompania Węglowa S.A. -1.29. The higher value of this rate means the better financial condition of the company (the company generates higher profits). Of course, the rate of ROS is influenced by many complex aspects, mainly economic, influence the rate of ROS but keep in mind that a properly executed strategy CSR can bring a lot of intangible benefits, which can then be translated into tangible benefits. In contrast, innovation (including also those related to environmental protection) can have a direct impact on the achievement of economic and financial advantage. As indicated by Bek-Gaik and Rymkiewicz (2015) [33], it is difficult to properly and accurately determine the impact of corporate social responsibility on the financial measures of achievements of the organization, and what is more, it is still not specified variable, which in a measurable way would describe the social responsibility of business - but without doubt the subject of the impact of corporate social responsibility on the financial achievements is extremely important and needs for research in this area.

Table 2 shows the activities in the range of innovation and environmental protection of analyzed companies [30, 31, 32].

6 Summary

The economic entities function in specific conditions - economic, social, environmental, legal and organizational - which does not exempt them from the observance of international norms of behavior [11]. In recent years, CSR has become a kind of standard or required action by the environment. Due to the fact that CSR activities should be implemented at three levels: economic, social and environmental, they are strongly associated with the innovation, eco-innovation and environmental protection. Especially sectors, which heavily affect the environment and local communities. Hard coal mining is an example of such operation. Additionally, the specificity of the Polish market - the results of research conducted by the Institute for Private Enterprise and Democracy, and the Responsible Business Forum confirmed that Polish companies are increasingly using in their business CSR tools [34], however, it is still a new concept for many Polish firms. The analysis showed that Polish company of hard coal mining industry still do not treat CSR as a constant part of their management strategy, in which they could present innovative solutions and activities of environmental protection.

The analysis confirmed that the issues of social and environmental represent a new challenge for Polish companies of tested sector - some of them do not provide information in this area to the public information, so their activity is unknown in this field, while the other companies provide information but doing it in (more or less) a general way. Undoubtedly, entities that follow global guidelines in the field of CSR reporting, the introduction of codes of ethics or integrated management systems achieve the greatest benefits. Therefore, it is clear that the activity in these three areas: social responsibility, innovation and environmental protection brings measurable tangible and intangible effects, and could be a source of competitive advantage.

Table 2. The activities in the field of innovation and environmental protection of analyzed companies [own work]

Company name	The main areas of innovation activities	The main areas of environmental protection activities
Kompania Węglowa S.A.	1. Cooperation with research and development units, local governments and economic entities, co-create research-and-development projects 2. Develop technology that will make coal fuel environmentally friendly 3. Current analysis of developments in the fuel and energy sector in the country and in the world with special emphasis on EU policies in these areas 4. Any employee may submit a project inventive, idea or technology improvement design	1. Compliance with the requirements in range of European Union law by: - design of mining operations with regard to limiting the negative impact on the environment, - reducing the amount of waste going to dump by using them for reclamation and any engineering works - reducing the negative impact of mining of salt water to surface water 2. Implementation of an Environmental Management System 3. Waste management in the most favorable forms 4. Properly planned and carried out reclamation and also exploitation post-mining land, what may lead to give them a new natural values, often more valuable than the original
Katowicka Grupa Kapitałowa	Implementation of investments for the years 2013 - 2020: 1. The introduction of innovative methods of methane drainage decks 2. Heat recovery from the cooling system group 3. The construction of a fossil-fuel power station for low calorie fuel 4. Implementation of the monitoring system of work underground in the mining 5. The method of pre-methane drainage and gasification of coal 6. Implementation of the control system powered roof support, adapted for use in an unmanned longwall machinery 7. Implementation of the identification system for monitoring employees, vehicles, machinery, equipment and materials	1. Environmental Management System (PN-EN ISO 14001:2005) 2. Good environmental practices - minimizing the amount of waste mining - reducing the amount of waste mining together with the continuation of possible maximum waste management on the land and in underground mines - recovery of foreign waste (waste from power plant) in underground mines - the introduction of selection of underground water - identification of characteristics of methane found in coal mines KHW SA - intended use of methane to generate electricity and heat
Lubelski Węgiel „Bogdanka" S.A.	1. Cooperation with research and development units 2. The leader of innovation and efficiency in the domestic mining industry - the information from the integrated report for 2014 3. Developing innovative technical solution entitled "Frame to dump output carriage" (pol. *„Rama do wywrotu wozów urobkowych"*)	1. Integrated Management System for Quality & Environment compatible with PN-EN ISO 9001:2009, ISO 14001:2005, PN-N-18001:2004 2. The Book of the Integrated Management System, which regulate environmental management by: - identification and assessment of environmental aspects (PZ/S/04/01) - monitoring of key characteristics of the environmental aspects (PZ/S/04/02) - identification of legal and other requirements and assessment of conformity (PZ/S/04/03) - developing environmental programs (PZ/S/04/04) 3. The report on environmental impact 4. Waste management 5. Protecting the earth's surface

References

1. Brakman, S., *Nations and firms in the global economy: An Introduction to International Economics and Business*. (Cambridge university press, 2006).
2. A. Kuzior, B. Knosala, Zesz. Nauk. PŚl., Org. Zarz. **81**, 119-129 (2015)
3. H. Jenkins, Business Ethics: A European Review **18**, 21-36 (2009)
4. R. E. Freeman, J. Liedtka, Business Horizons, **34**(4), 92-98 (1991)
5. M. Friedman, *Capitalism and freedom*, (Chicago: University of Chicago Press, 1962)
6. J.J. Asongu, Journal of Business and Public Policy **1(3)**, 1-21 (2007)
7. G. Rexhepi, S. Kurtishi, G. Bexheti, Procedia - Social and Behavioral Sciences **75**, 532-541 (2013)
8. van Oppen, C.,Brugman, L. Organizational Capabilities as the Key to Sustainable Innovation. Accesable at: http://www.squarewise.com/Capabilities.pdf
9. A. Czubała, Kons. i Rozw. **1**, 58-67 (2011)
10. Kromer,B. Stud. Ekonom., **180**, 131-140. (2014).
11. P. Dulewicz, A. Perzyńska, K. Witkowski, Logistyka **4**, 3890-4023 (2014)
12. D. Szymanek, Zrównoważony Rozwój - Debiut Naukowy **1**, 97-103 (2011)
13. Mossakowska T., Wasilewski M., Przegląd Górniczy Nr 10 (1067) **66**, 47-58 (2011)
14. Levy, D. L., Brown, H. S., De Jong, M. Bus. & Soc, **49(1**), 88-115 (2010)
15. A. Paszkiewicz, A. Szadziewska, Zeszyty Naukowe Uniwersytetu Szczecińskiego Nr 668 Finanse, Rynki Finansowe, Ubezpieczenia **41**, 627-644 (2011)
16. R. Bocquet, C. Mothe, Revue Sciences de Gestion **80**, 101-119 (2011)
17. I. Gallego-Alvarez, J. Manuel Prado-Lorenzo, I.M. Garcıa-Sanchez, Management Decision **49**, 1709-1727 (2011)
18. van Tonder, E., Roberts-Lombard, M. A. Journ. of Bus. & Econ. Res., **11**(12), 503 (2013)
19. Machiba T. *Sustainable Manufacturing and Eco-Innovation: Framework, Practice and Measurement - Synthesis Report* (Paris: OECD, 2009)
20. K. Kazojć, Ekonomia i Środowisko **3(50)**, 122-129 (2014)
21. D. Zuzek, Rocz. Nauk. Stow. Ekon. Rol. i Agrob. **17**(1), 240-244 (2015)
22. http://database.eco-innovation.eu/:
23. http://www.eco-innovation.eu/
24. http://kghm.com/pl/eko-innowator
25. https://www.nfosigw.gov.pl/
26. Kulczycka, J., Wirth, H.. Zeszyty Naukowe IGSMiE PAN **79**, 147-157 (2010)
27. https://weglokoks.com.pl/organizacja wegla.pdf
28. *Polski Sektor Energetyczny Raport*, (Polska Agencja Informacji i Inwestycji S.A. 2012)
29. M. Białas, *Materiały XXV Konferencji z cyklu Zagadnienia surowców energetycznych i energii w gospodarce krajowej*, 17-29 (2011)
30. https://www.kwsa.pl/
31. http://www.khw.pl/#
32. http://www.lw.com.pl/
33. B. Bek Gaik, B. Rymkiewicz, Zeszyty Naukowe Uniwersytetu Szczecińskiego Nr 854 Finanse, Rynki Finansowe, Ubezpieczenia **73**, 151-166 (2015)
34. M. Ratajczak, Zeszyty Naukowe SGGW w Warszawie - Problemy Rolnictwa Światowego, **10(25)**, 75-82 (2010)

The effects of storing heat excess in a stone battery in a horticultural premise

Sławomir Kurpaska, Jakub Sikora, Krzysztof Mudryk, Hubert Latała and Jarosław Knaga

Faculty of Production and Power Engineering, University of Agriculture in Krakow, Poland

Abstract. This work presents the results of research connected with the flow of air through a stone battery. The research was carried out in a foil tunnel with a four-section stone battery. The battery deposit consisted of porphyry stone of 30 – 63mm granulation. The process of inlet from the inside of a tunnel to a battery (manager using an algorithm) enabled the unbundling of heat storing cycle and the cycle of delivering the stored heat to the inside of the tunnel. The amount of heat and exchanged mass was described (in relation to the unitary surface and a particular cycle). Correlational equations were formed for the found values and they enable the calculation of a unitary heat stream and the masses of stone battery exchanged cycles.

1 Introduction

The concern for environment and the fear of attrition of fossil fuels sources boost the search for technical resolutions enabling heat saving. One of the premises in agricultural production are greenhouses and foil tunnels at which, in average conditions, demand for heat is around 2100 MJ/m^2 (greenhouses) and 900 MJ/m^2 (foil tunnels). These premises are characterised by high transparency and are prone to solar radiation. As a consequence, a conversion of solar radiation into the increase of heat occurs. The excess of heat is the cause of too high temperature inside the premises what results in an abnormal development of plants. There are some recommended values of air temperature which, depending on the type of plant, are located within 14 to 25 °C [1]. Too high temperature favours the intensification of gas exchange causing the advantage of breathing process over photosynthesis. Too low temperature causes irregular development of above-ground parts of the plants. Abnormal temperature influences the speed, amount and quality of the crops of plants. Hence, ventilation of the premise is necessary. Therefore, the demand for heat increases. Storing the excess of heat in Energy batteries is an alternative solution leading to increased ventilation. The issue of storing excess heat in premises under covers was the subject of analyses and the developed systems concerned the issues of storing heat in bodies in which thermal capacity, phase changes (PCM) and chemical and photochemical changes were used. The results of experimental research of the theoretical analysis connected with heat storing in batteries filled with solid can be found in the works: [2- 5], and in liquid batteries in works [6,7].

Analysing the research data, it is possible to note that the amount of stored and allocated heat to warm the premise depends not only on the temperature and the amount of inlet but also on the heat capacity of the deposit of the battery as well as the used overcame heat loss while storing. The research results in which the heat was stored in bodies undergoing phase changes (melting- coagulation, crystallization-dissolving, sublimation- resublimation, evaporation-condensation, boiling- condensation) and crystal structures' changes that go Ahead with obtaining and giving heat can be found in the works: [8- 11]. The obtained results show that this method can be used successfully as means to lower the usage of heat in heated premises. Independent studies (carried out for areas of plural climate conditions) show thermal effects of liquids in ponds (together with temperature gradient) and the heat's complicity in an overall balance of the heated greenhouse. Kurt et al. [12] analysed the plural composition of liquid in a solar pond. They worked out a mathematical model describing temperature stratifications in the pond and verified the results in laboratory conditions and actual scale. Ozturk and Bascetincelik [13] have calculated the support foil tunnel heating with the heat stored in soil to annual savings in the heat usage showing the usefulness of this system in the examined climate conditions.

The presented overview unambiguously shows that the issue of storing heat excess from the inside of horticultural premise in batteries was the subject of analyses. The considered processes, however, concerned different climate conditions and premises with plural (constructive and geometrically) type of batteries. Therefore, the aim is to define the effects occurred while storing heat in stone battery through which the air flew in a controlled manner.

2 Materials and method

The research was done in a battery situated in a foil tunnel of 144 m^2. The battery in the tunnel has 4 sections of 1.7x11m filled with stone deposit (augen stone of 30 – 63 mm granulation) on depth 0.7 m. The outline of the stand is presented in fig.1.

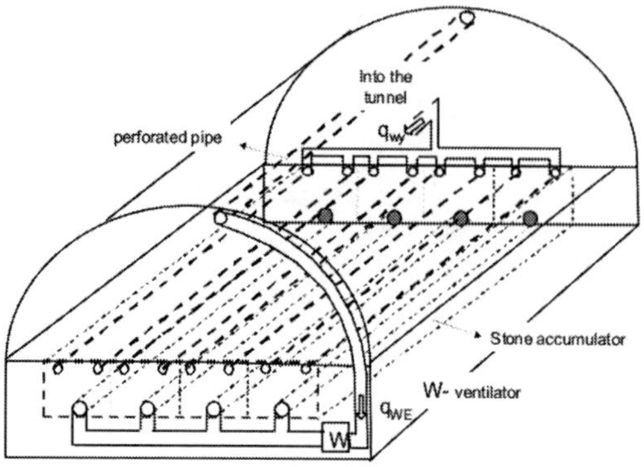

Figure 1. The outline of test stand

The following values were measured while carrying out research experiments:

- parameters of the surrounding climate: solar radiation volume- using LP PYRA 02AV pyranometer, wind speed- MAX 40H gauge, relative humidity and temperature of the air,
- parameters of the air flowing through a ventilator: speed(Mini Air64 gauge) and air temperature
  and humidity,
- parameters of the air flowing out of the battery: temperature and humidity,
- parameters inside the premise: air temperature and humidity.

The air temperature was measured using PT1000 gauges. Relative air humidity measurement was done using integrated HD 4917TC gauges. Managing the battery work (number of sections, battery work cycles) was realised thanks to an algorithm, where the managing signal was based on temperature difference: between the air being intaken by a ventilator and the air flowing out of the battery (charging cycle) and between the inside of the tunnel and the air flowing from the battery (discharging cycle). The stream of battery inlet was realize by means of radial ventilator cooperating with an inverter.

The amount of heat, in a differentia time $d\tau$ stored in a battery describes the following relation:

$$dQ_{ak} = \frac{\pi \cdot d^2}{4} \cdot V_{pow} \cdot \rho_{pow} \cdot \int_{\tau_1}^{\tau_1+\tau} (i_{WE} - i_{WY}) d\tau \quad (1)$$

The change in steam density in the air pumped through the battery as in the equation:

$$dm_{H2O} = \frac{\pi \cdot d^2}{4} \cdot V_{pow} \cdot \int_{\tau_1}^{\tau_1+\tau} (f_{WE} - f_{WY}) d\tau \quad (2)$$

where: ρ_{pow}- air density, kg/m^3; d- measurement wire diameter, m; i_{WE}, i_{WY}- enthalpy of inlet (i_{WE}) and outgoing (i_{WY}) from a battery, kJ/kg; f_{WE}, f_{WY}- irrespective humidity of inlet (f_{WE}) and outgoing (f_{WY}) from a battery, kg/m^3.

To generalise the results, an average measurement of parameters' values (W_{avg}) was done which determine the changes in the amount of stored heat and the changes in the amountof water in the battery. These values were calculated using their temporary values:

$$W_{avg} = \frac{1}{\tau} \cdot \int_{\tau_1}^{\tau_1+\tau} w(\tau) d\tau \quad (3)$$

All parameters necessary for determining those dependencies were determined for each cycle using the standard psychometric dependencies. The speed of inlet was measured in the wire of an internal diameter of d=0,3m.

3 Results and discussion

460 battery charging cycles and 410 battery discharging cycles were analysed. The time of charging cycles was located within 626 to 53637s; time of discharging was between 557 to 259241s. The rest of the parameters were as follows:

a) Charging cycle: battery surface (F_{ak}), $18.7 \leq F_{ak} \leq 74.8$m^2; temperature of the surrounding (t_{ot}), $-6.2 \leq t_{ot} \leq 37.1$°C; solar radiation volume (R_{sl}), $5 \leq R_{sl} \leq 890$W/m^2; wind speed (V_{wiatr}), $0.1 \leq V_{wiatr} \leq 3.47$m/s; battery initial temperature (t_{ak0}), $2 \leq t_{ak0} \leq 37.7$°C; battery final temperature (t_{ak1}), $6.7 \leq t_{ak1} \leq 39.7$°C; air initial temperature (twE), $11.1 \leq t_{pow} \leq 49.5$ °C; air initial humidity (ϕ_{WE}), $11 \leq \phi_{WE} \leq 75$%; air final temperature (twY), $6.4 \leq t_{pow} \leq 33.07$ °C; air final humidity (ϕ_{WY}), $29.4 \leq \phi_{WE} \leq 94.4$%; inlet speed ($V_{pow}$) $1.34 \leq V_{pow} \leq 4.72$m/s.

b) Discharging cycle: battery surface (F_{ak}), $18.7 \leq F_{ak} \leq 74.8$m^2; temperature of the surrounding (t_{ot}), $-11.1 \leq t_{ot} \leq 18.7$°C; wind speed ($V_{wiatr}$), $0.1 \leq V_{wiatr} \leq 2.79$m/s; battery final temperature (t_{ak1}), $4.1 \leq t_{ak0} \leq 37.2$°C; air initial temperature (twE), $-4.55 \leq t_{pow} \leq 28.58$ °C; air initial humidity (ϕ_{WE}), $45.3 \leq \phi_{WE} \leq 93.2$%; air final temperature (twY), $5.9 \leq t_{pow} \leq 33.71$ °C; air final humidity (ϕ_{WY}), $31.7 \leq \phi_{WE} \leq 91.5$%.

The range of changes of the inlet speed was in the same range as in battery charging cycles. Figure 2 shows the influence of 2 independent variables (the speed of inlet, inlet enthalpy) on the amount of stored heat in a unitary surface of a battery while charging cycle.

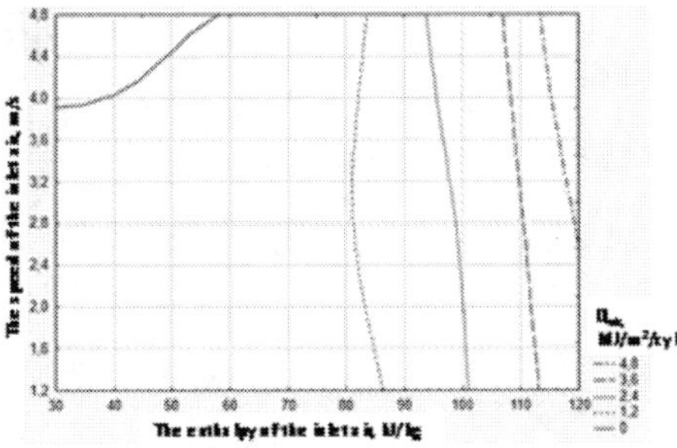

Figure 2 The influence of speed and enthalpy of inlet on the unitary amount of stored heat in a battery.

Figure 3 shows the influence of the same variables on the change of a mass of water absorber in the examined battery.

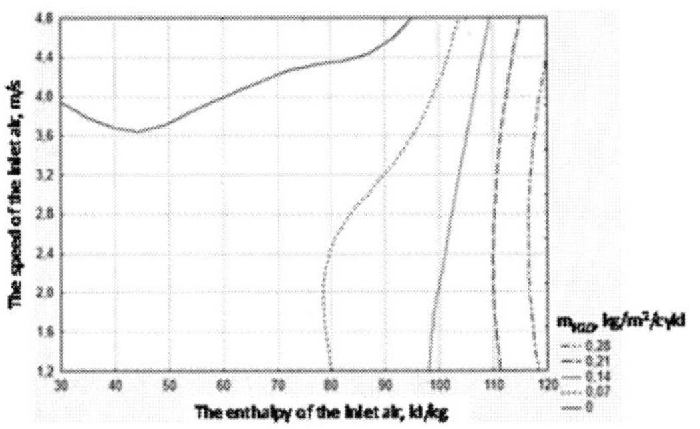

Figure 3. The influence of speed and inlet enthalpy on the change of water absorber in the battery while charging cycle.

As we can see, together with the increase of inlet enthalpy there is an increase of both the amount of heat and the mass of water in the battery. The increase in the inlet speed is not so clear, causing minor increase in the amount of stored heat (with higher enthalpy), in other cases- causing a negative effect both on the amount of stored heat and the amount of absorbed water. The entire range of changes in the amount of stored heat is within -3.9 to 4.7 MJ/m^2/cycle. In the case of the change of absorbed mass of water it is -0.19 to 0.92 kg/m^2/cycle.

Considering the time of particular cycles, the results were calculated into a unitary heat stream (q) and mass (m). The correlational equation mentions the relation (exponent model chosen on the basis of the highest value of resolving coefficient. This dependency was determined as nonlinear estimation using the quasi-Newton's method while a preserved coefficient of 0.001) takes the form of:

a) Heat stream (W/m^2):

$$q_{lad} = -53.02 \cdot t_{ak0}^{0.62} - 24.57 \cdot V_{pow}^{-1.9} + 29.7 \cdot i_{WE}^{0.64};$$

R^2=0.72

b) Mass stream (g/m^2/min):

$$m_{lad} = -0.079 \cdot t_{WE} + 0.061 \cdot \phi_{WE} - 0.169 \cdot V_{pow}$$

R^2=0.58

Figures 4 and 5 show the course of the same variables while charging cycles.

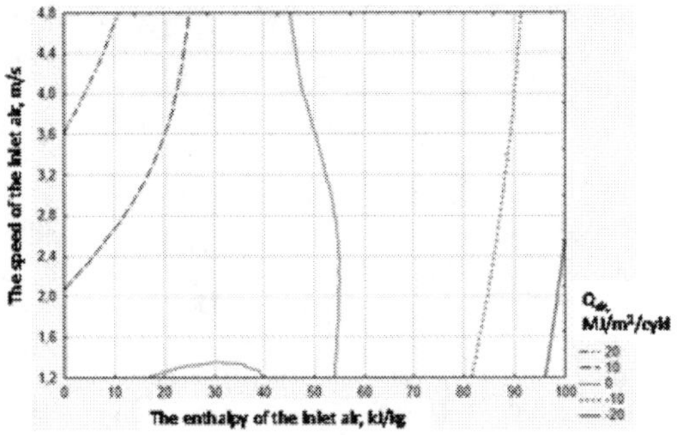

Figure 4. The influence of enthalpy and speed of inlet on unitary amount of absorbed heat from battery in its discharging cycle.

It can be noticed that the increase of enthalpy creates a negative effect on the amount of heat delivered from the battery to the inside of a foil tunnel. The increase in speed causes the increase of the amount of removed heat.

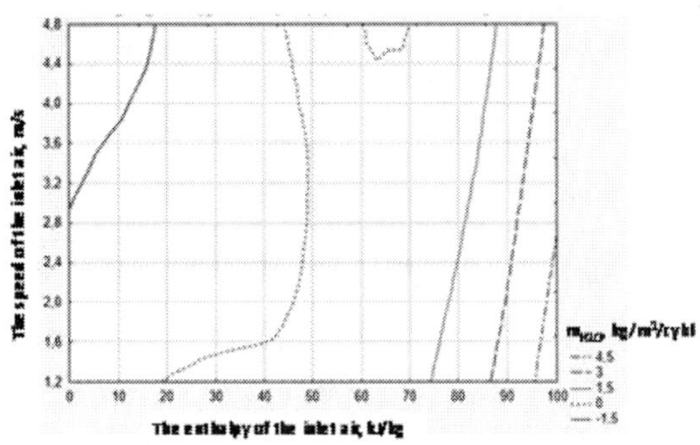

Figure 5. The influence of enthalpy and the speed on inlet on the amount of water mass removed from the battery in its discharging cycle.

The entire range of changes in the amount of heat delivered from the battery is within -21.5 to 28.1 MJ/m^2/cycle. In the case of the mass of absorbed water, it is -6.17 to 4.96 kg/m^2/cycle. Analysing these changes, we may state that the increase of inlet enthalpy causes an increase in the intensity of exchange of mass from the examined battery. This trend is reverse at the increase of speed. The observed negative values (both for the heat and the mass of water removed from the battery) shows that, for the existing conditions of carried out experiments, there is heat storing (fig. 4) and humidifying of the air coming out of the deposit. Alike before, correlational equations take the form of:

a) Heat stream (W/m^2):

$$q_{rozl} = 21.18 \cdot V_{wiatr} + 15.83 \cdot t_{ak1} + 6.36 \cdot t_{ot} - 24.63 \cdot t_{WE} - 0.19 \cdot \phi_{WE} - 2.68 \cdot V_{pow}$$

R^2=0.68

b) Mass stream (g/m^2/min):

$$m_{mat} = -0.101 \cdot t_{akt} + 0.184 \cdot t_{wz} - 0.0041 \cdot \phi_{wz} - 0.0394 \cdot V_{pow}$$

$R^2=0.61$

The equations can be used in the range of given initial variables. Analysing the obtained results, we can unambiguously say that apart from the discussed issues connected with heat while charging process, there is a desiccation of air. Whereas during the discharging process, the air is humidified. The observed phenomena are a consequence of the inflow of humid air which causes vaporization or condensation of steam if it has contact with a surface of humid battery deposit. As a result, there is drainage (if condensation occurs) or humidifying (if vaporization occurs) of air coming from the deposit. The issues connected with heating (storing and heat recovering) are the result of both the occurring difference of temperatures between the flowing air and the temperature in deposit as well as the occurring steam thermodynamic transformation.

4 Conclusions

1 In the examined range of variables, the range of changes in the amount of stored heat while the charging process is within -3.9 to 4.7, whereas while its discharging process, it is within-21.5 to 28.1 MJ/m²/cycle.

2 The range of changes in the amount of exchanged mass while air flowing through stone battery is within -0.19 to 0.92 kg/m²/cycle (battery charging), and while battery discharging, it is -6.17 to 4.96 kg/m²/cycle.

3 The correlational equations enabling the calculation of unitary heat stream and mass stream while analyses cycles of battery work can be used in the range of experiment factors change and consist of easily measured parameters connected with the flowing air, temperature of the deposit and the parameters of the surrounding air.

Acknowledgements

This research was financed by the Ministry of Science and Higher Education of the Republic of Poland

References

1. C.A. Campiotti, G. Morosinotto, G. Puglisi, C. Schettini, G. Vox, G.. Agriculture and Agricultural Science Procedia **8**, (2016), 664 – 669
2. Bouhdjar, M. Belhamel, F. E. Belkhiri, A. Boulbina. Renewable Energy, **9**, 1-4 (1996), 724- 728
3. Kurklu, S. Bilgin, B. Ozkan. Renewable Energy, **28**, 5, (2003), 683- 697
4. W. Chen, W. Liu. Solar Energy, **76**, 5 (2004), 623-633
5. S. Kurpaska, H. Latała. Renewable Energy, **35**, 12 (2009), 2656- 2665
6. S. Kurpaska, H. Latała, D. Baran, P. Konopacki, R. Hołownicki. Agricultural Engineering, **3**, 155 (2015), 47-57
7. D. Lindenberger, T. Bruckner, H. Groscurth, R. Kümmel. Energy, **25**, 7 (2000), 591 -608
8. T. Ayhan, O. Comakli, K. Kaygusuz. Energy Conversion and Management **33**, 3 (1992), 165- 173,
9. A.A. Badran, B. A. Jubran, E. M. Qasem, M. A. Hamdan. Applied Energy **58**, 1 (1997), 57-72
10. H.E.S. Fath. Energy Conversion and Management, **31**, 2 (1991), 149- 155
11. S. Zhu, J. Deltour, S. Wang. Aquacultural Engineering **18**, 3 (1998), 201-217
12. H. Kurt, M. Ozkaymak, K. Binark. Applied Energy **83**, 4 (2006), 324-342
13. H. H. Ozturk, A. Bascetincelik. Biosystems Engineering **86**, 2 (2003), 231-245

Thermal effects of the stone battery depending on the bed volumes

Hubert Latala[1,a], Slawomir Kurpaska, Jakub Sikora, Krzysztof Mudryk and Jarosław Knaga

[1]*Faculty of Production and Power Engineering, University of Agriculture in Krakow, Poland*

Abstract. The search restrictions energy consumption in horticultural production there are in progress continually. They concern on the one hand reduce heat loss and, second, better use of available energy. Greenhouse efect in the greenhouse forced ventilation process especially in the surplus heat. The paper presents the manner of management of excess heat in the stone battery. Its structure included four segments with a volume of 13.1 m^3 each. Segments of the battery were fed a stream of air in an independent manner. This enabled the work of individual segments or a few at a time. The battery can work in two stages: charging and discharging. Selecting the stage took place automatically according to the developed algorithm. This decision was dependent on the parameters of air flowing through the bed and the thermal state of the battery. During the process of charging and discharging monitored changes in air enthalpy at the inlet and outlet of the battery, and the temperature and relative humidity of air in the bed of battery. For different bed volumes increases its temperature, heat storage capacity and the ability to absorb water vapoure was determined.

1 Introduction

Environmental control in horticultural crops plays an important role. Possibility of adjusting the relevant parameters for plant growth conducive to intensification of production and achieving a good quality crop. To achieve this purpose, a number of technical solutions are applied. These include, inter alia, various types of coverage, modern heating systems and systems of accumulating heat.

The heat production under cover plays a key role especially in times of its deficiency in total production costs can reach up to 70%. One of the solutions to reduce these expenditures is the development of excess heat in the plastic tunnel caused by the greenhouse effect. Normally, excess heat is removed by ventilation. It is the heat flux lost forever. Reasonable, therefore, it is to use a system that accumulated excess heat. However, the differences in access of energy and heat demand affect the dispatch ability and performance [1].

In principle, the heat can be stored using the specific heat, heat of fusion and solidification. It depends on the type of materials and, more specifically whether during heating/cooling undergo a phase transition. The ability of the heat storage material in terms of the specific heat is highly dependent on the heat capacity. In addition to this criterion must be considered also good thermal conductivity and cost material [2].

In literature different types of media are used to heat accumulation. Using only the specific heat of these materials should be characterized by low cost of utilisable heat, high heat transfer coefficient in the process of charging and discharging the battery, ease of use and be friendly to the environment [3-6].

Storage of excess heat in the battery water for energy needs in the greenhouse or plastic tunnel is not very effective. Only such a solution is useful for the preparation of process water [7]. A better solution to this is to use a base stone. Ease of obtaining heat and supplying it to the battery on one side and on the other hand, heat distribution for the rows on the existing cultivation gutters provide a simple solution. Presented in articles [8-12] results indicate that the storage of surplus heat brings tangible results.

Due to varying environmental conditions available flow of surplus heat possible to the development needs of different size battery. Similarly, the use of its potential to meet the needs of heat resulting from the impact of climate outside. This is evidenced by searching multiple authors presented above. Therefore, in operation seeks to provide determination of the increase in temperature, the heat accumulation capacity and the absorption of water vapour in the battery depending on its volume.

2 Materials and methods

The experiment was conducted in a plastic tunnel, which was a four-segment battery stone. It was built below the level of cultivation area. All the walls and interior partitions were isolated 10 cm insulating material. Each segment contains 12.23 m^3 of stone (porphyry) of grain size of 30 - 63 mm. Jobs individual segments in both process of charging and discharging was supervised by the measurement system. This system also archived all the

[a] Corresponding author: rtlatala@cyf-kr.edu.pl

DOI: 10.1051/e3sconf/20161000053

parameters monitored battery working conditions and the data associated with its surroundings.

The all segments of battery was supplied by stream of hot air during charging process. The stream of warm air from the upper layers of plastic tunnel was powered segments of stone battery. During this process was observed in the bed of battery changes of temperature, the quantity of accumulated heat and moisture content (Fig.1.). At the output from the battery to be monitored the same parameters as was the case at the entrance. Comparison of the state of the air inlet and outlet was the basis for the calculation of heat that has accumulated by battery.

The stone battery worked in two cycles. The first concerned the charge process, and the second discharge process. In the process of charging the system supplied to the battery the stream of warm air from the upper part of the tunnel film during the day. In the discharging process the system supplied to the battery the stream of cooled air came from the lower part of the tunnel film during night. In both cycles of the air flow coming out of the battery was directed to the interior of the object.

During the experiment the parameters of work of battery and the parameters of air flow were monitored. During the experiment monitored parameters regarding battery mining and parameters of air flow.

In the study were monitored and recorded the following data:

- Temperature of stone battery,
- The operating status of individual segments,
- Temperature and relative humidity over heat screens and inside the tunnel,
- Temperature and relative humidity on the output of the battery,
- The mass of air flowed through the battery was calculated on the basis of measuring the speed of air stream.

These values were used to determine the amount of heat and mass of the steam during simultaneous operation of I, II, III or IV of the battery segments. In the charging process was specified amount of accumulated heat and the mass of water vapour which was absorbed by each segment of the bed according to formulas 1 and 2. Similarly, in the discharge process the characteristics were defined above. To differentiate the processes of charging and discharging the heat accumulated in the accumulator is positive, and the heat comes from battery (discharge process) and put into the tunnel is negative. Similarly, the sign associated with the mass of water vapour - is positive absorption (dehumidify the air passing through the bed) of water vapour in the bed. While a negative sign indicates moistened air leaving the battery (evaporation of water accumulated in the deposit).

The quantity of heat accumulated in the battery (dQ_{batt}) was defined as the change in enthalpy of air at the inlet and outlet of the battery as result the flowing air stream in the time interval (dτ).

$$dQ_{batt} = m_{air} \cdot \int_{\tau_1}^{\tau_1+\tau} (i_{in} - i_{out}) d\tau \qquad (1)$$

The quantity of water vapour (dm_{water}) stored in the stone accumulator was calculated as the difference of concentration of water vapour at the inlet and outlet of the battery as result the flowing air stream in the time interval (dτ).

$$dm_{water} = V_{air} \cdot \int_{\tau_1}^{\tau_1+\tau} (x_{in} - x_{out}) d\tau \qquad (2)$$

where: m_{air} - air mass, kg; i_{in}, i_{out} - enthalpy of inlet (i_{in}) and outlet (i_{out}) from a battery, kJ/kg; V_{air} - air volume, m^3; x - concentration of water vapour (kg$_{vapour}$/m^3) inlet (x_{in}) and outlet (x_{out}).

Enthalpy and concentration of water vapour of the air stream input and output have been calculated according to the psychometric formulas for the process of charging and discharging battery.

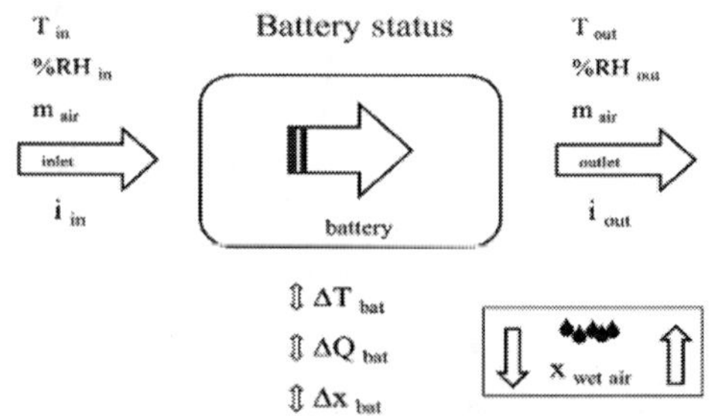

Figure 1. Scheme of process for charge and discharge the stone battery.

3 Results and discussion

For the analysis was selected cycles of charging and discharging the stone battery operating in same time for four configurations: I - one segment, II - two segments, III - three segments, IV - four segments.

This meant monitoring battery for four different volumes. The research took place in the conditions of a real object which was the plastic tunnel.

Figures 2 and 3 show the comparison of the time of charging and discharging for individual segments of the battery. About twice longer the discharge time of segments was resulted of a lower temperature differential inputs and outputs compared to the charging process. Energy accumulated in the battery (Fig.2.) was calculated based on the difference in enthalpy of the air at the entry and exit of individual segments of the battery. This energy was also dependent on the mass of air flowing through the stone bed during the charging cycle.

Discharge cycle of individual segments of the battery (Fig.3.) shows the amount of useful energy. Heat is supplied to the interior of the tunnel. Negative values shown in the graph was used to distinguish the stored energy from useful energy.

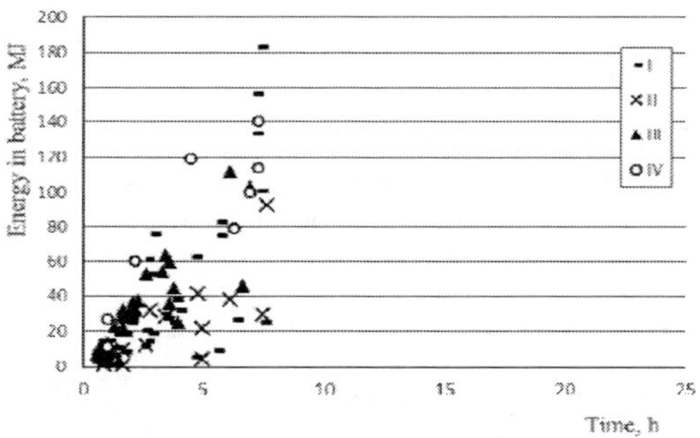

Figure 2. Storage energy in the battery depending on charging time for various bed volume.

The amount of stored energy in the battery during charging cycle also affects the temperature of the bed in each segment (Fig.4.). In the course of the experiment the temperature of a single segment increased from 0.1 to 19.2 ^{0}C with an average value of stored energy at 30.8 MJ. For the two segments change in temperature of the bed contains from 0.1 to 6.8 ^{0}C. For the three and the four segments of these changes ranged from 0 to 3.7 ^{0}C and 0.2 to 2.9 ^{0}C respectively.

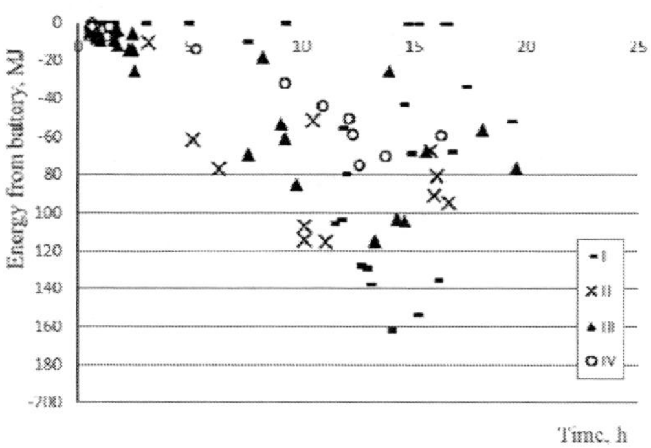

Figure 3. Useful energy of the battery depending on discharging time for various bed volume.

Simultaneous increase an amount of concurrent segments, and thus the volume of the bed caused heating of the battery (segments) to a lower temperature. The average amount per unit of energy stored in the individual segments of the battery for the above temperature ranges were as follows: II-segments - 19.4 MJ, III-segments - 30.7 MJ, IV-segments - 81.4 MJ.

The process of charging the different segments of stone battery was conducted under the conditions which were the following: battery volume (V_{ak}), $12.3 \leq V_{ak} \leq 48.9$ m^3; air temperature of the inlet (t_{NC}), $10.1 \leq t_{NC} \leq 38.2$ ^{0}C; air relative humidity (ϕ_{NC}), $10.7 \leq \phi_{NC} \leq 67.2\%$; mass of air flowing through the bed (m_{air}), $0.1 \leq m_{air} \leq 1658.7$ kg, cycle time (τ) $0.5 \leq \tau \leq 7.6$ hours.

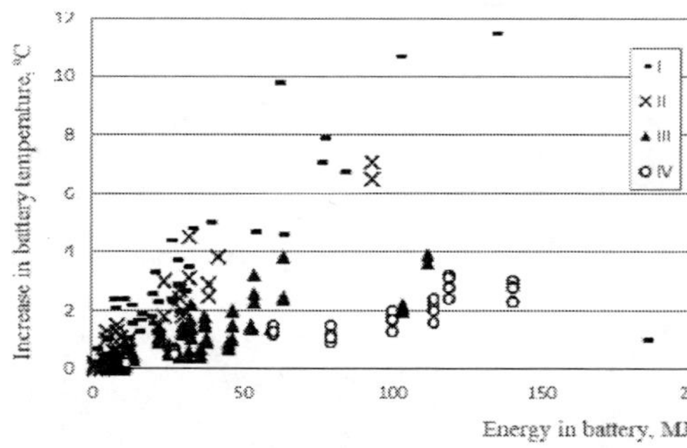

Figure 4. Increase the temperature of the battery depending on stored energy for various bed volume.

Figure 5 shows the dependence of the energy stored in the battery at different temperatures in the bed, depending on the number of segments working simultaneously.

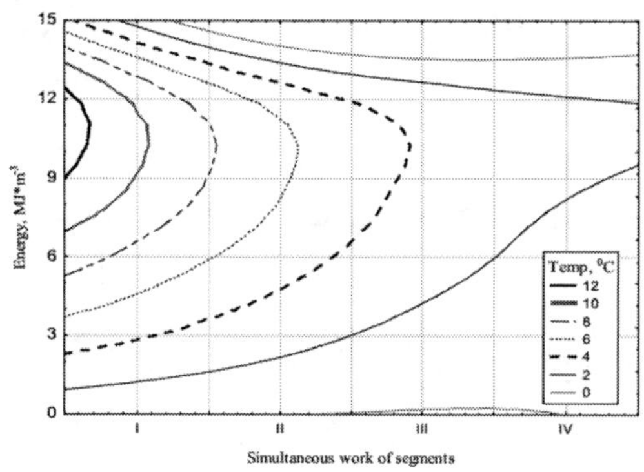

Figure 5. Accumulated energy in the battery depending on its temperature for various bed volume.

Next figure (6) describes how the absorption of water in the battery depending on the temperature and the amount of concurrent segments. Positive values indicate the absorption of condensation of water vapour in the bed and also the drying air leaving the battery. This is particularly noticeable during the charging cycle if the battery temperature rises above 2 ^{0}C Negative values concern to the process of evaporation of water from the battery. Then the air at the output of the battery becomes moist.

The correlation equations for the flow of heat and mass were determined on the basis of the monitored operating parameters of the individual segments in the course of the experiment . The values of these streams are dependent on the parameter input and output, taking into account their interdependence. The significance of the parameters determining the value stream was set at a significance level of less than 0.05 as linear estimation using the quasi-Newton's method.

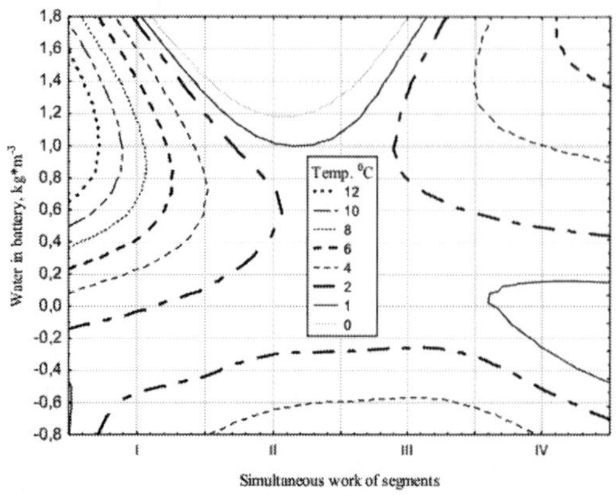

Figure 6. Absorbed water in the battery depending on its temperature for various bed volume - charging process.

Charging process of stone battery (**one segment**):
Heat accumulated in the battery (MJ/m³):

$$q_{LOAD} = 0.563 \cdot t_{batt} + 0.051 \cdot t_{NC} - 0.038 \cdot \phi_{NC} + 0.83 \cdot m_{vapour}$$

R^2=0.68

Mass of water vapour absorbed in the battery (g/m³):

$$m_{Lwater} = 126.861 \cdot q_{LOAD} - 50.548 \cdot t_{batt} - 5.762 \cdot t_{NC} + 3.271 \cdot \phi_{NC}$$

R^2=0.82

Charging process of stone battery (**two segments**):
Heat accumulated in the battery (MJ/m³):

$$q_{LOAD} = 0.006 \cdot m_{air} + 0.256 \cdot t_{NC} - 0.06 \cdot \phi_{NC} - 9.662$$

R^2=0.96

Mass of water vapour absorbed in the battery (g/m³):

$$m_{Lwater} = -1.238 \cdot m_{air} - 20.428 \cdot t_{NC} + 26.567 \cdot \phi_{NC}$$

R^2=0.80

Charging process of stone battery (**three segments**):
Heat accumulated in the battery (MJ/m³):

$$q_{LOAD} = 0.006 \cdot m_{air} + 0.256 \cdot t_{NC} + 0.06 \cdot \phi_{NC} - 9.662$$

R^2=0.96

Mass of water vapour absorbed in the battery (g/m³):

$$m_{Lwater} = -159.872 \cdot t_{batt} + 0.539 \cdot m_{air} - 7.757 \cdot t_{NC} + 10.581 \cdot \phi_{NC}$$

R^2=0.43

Charging process of stone battery (**four segments**):
Heat accumulated in the battery (MJ/m³):

$$q_{LOAD} = 3.251 \cdot t_{batt} - 0.148 \cdot t_{NC} - 0.176 \cdot \phi_{wew}$$

R^2=0.96

Mass of water vapour absorbed in the battery (g/m³):

$$m_{Lwater} = 79{,}826 \cdot q_{LOAD} - 23.915 \cdot t_{NC} + 31.245 \cdot \phi_{wew} +$$

R^2=0.89

The process of charging the individual segments of the battery showed two phenomena inside. The first is the rise of temperature of the battery with the supplied energy. The carrier is a stream of warm air from inside the tunnel. The second phenomenon is more complex because it relates to the transformation of water vapour in to the battery. A stream of warm air, which supplies the battery also provides a certain amount of water vapour. This pair in contact with a cool bed may condense. This means that absorption of water vapour by the accumulator. Analysing the accumulation of water vapour in the segments turned out that the segments I, II and III followed by condensation and evaporation of water. The extent of these changes were in the range from -741.5 to 1,877.4 g/m³ bed. Only the work of four segments characterized by only a process of condensing level of 42.9 to 754.2 g/m³.

Lowering the temperature of the individual segments of the battery and the amount of useful energy obtained by a discharge process were marked as the negative values (Fig. 7.). The average amount of energy supplied from the individual segments of battery was as follows: I-segment - 48.2 MJ, II-segments - 31.7 MJ, III-segments - 9.8 MJ, IV-segments - 28.1 MJ. Delivery the heat to the inside tunnel lowered the temperature of the individual segments. The average values of decreasing temperature were respectively: I-segment - 2.6 °C, II-segments - 0.9 °C, III-segments - 0.1 °C, IV-segments - 0.5 °C.

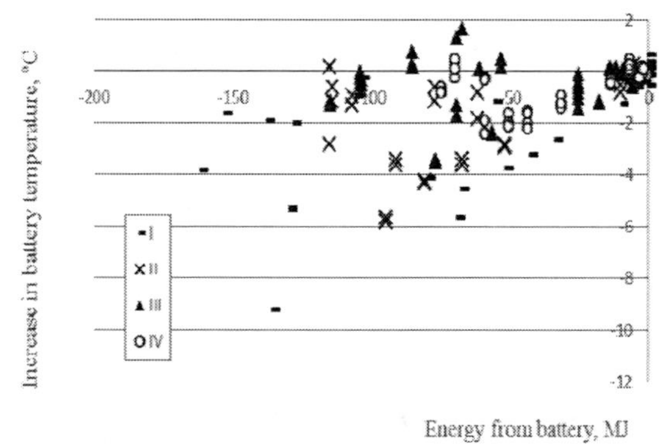

Figure 7. Decrease the temperature of the battery depending on downloaded energy for various bed volume.

Discharging cycle: battery volume (V_{ak}), 12.3≤ V_{ak} ≤48.9 m³; air temperature of the inlet (t_{wew}), -3.6 ≤ t_{wew} ≤ 21.2 °C; air relative humidity (ϕ_{wew}), 42.0≤ϕ_{wew} ≤ 90.2%; mass of air flowing through the bed (m_{air}), 0.1 ≤ m_{air} ≤ 625.1 kg, cycle time (τ) 0.5≤ τ≤19.6 hours.

Figure 8 shows the dependence of the energy supplied from the battery at different temperatures in the bed, depending on the number of segments working simultaneously. For the analysis of the available energy from the battery in the process of discharge (useful energy) assumed negative values. This assumption is made to quickly identify the processes of charging and discharging of the battery.

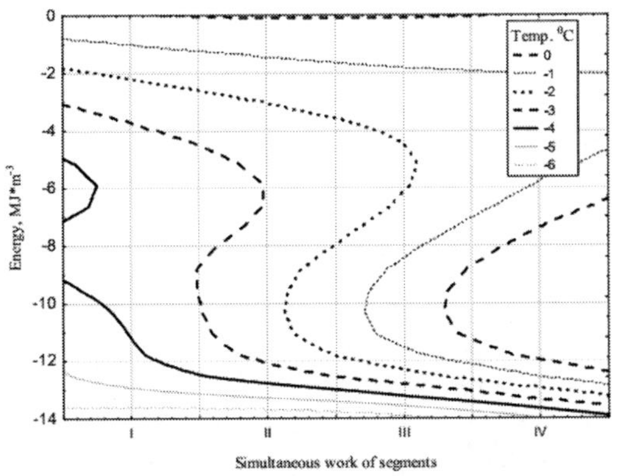

Figure 8. Downloaded energy from the battery depending on its temperature for various bed volume.

The absorption of water in the battery depending on the temperature and the amount of concurrent segments in the process of discharge (Fig. 9) has a negative value. That is, in this cycle, the water contained in the bed evaporates and humidified air leaving the battery. This is particularly evident during the operation of a single segment, for cooling the bed which contributes to a smaller evaporation of water. The work of two or three segments had little effect on humidification air heated from the battery in comparison to the work of one or four segments. The lack of air humidification in the process of discharging the battery is vital for plants grown in plastic tunnel. This is important because of the need to remove excess water formed in the process of plant growth.

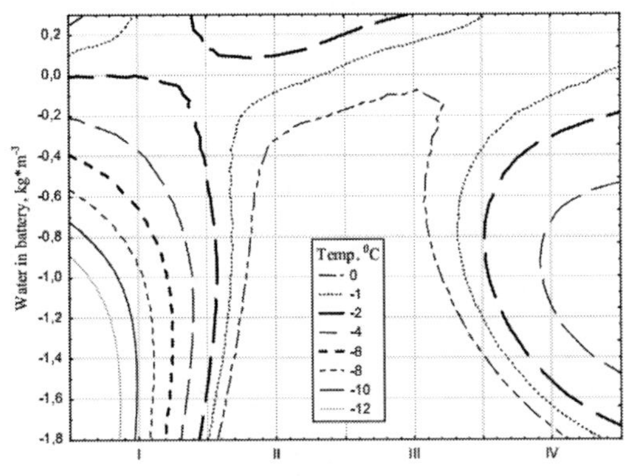

Figure 9. Absorbed water in the battery depending on its temperature for various bed volume - discharging process.

For the process of discharging the correlation equations for the flow of heat and mass were determined on the same way as for process of charging.

Discharging process of stone battery (**one segment**):
Downloaded heat from the battery (MJ/m³):

$$q_{unload} = 0{,}314 \cdot t_{batt} - 0{,}02 \cdot m_{air} + 0.106 \cdot t_{wew}$$

R^2=0.66

Mass of water vapour absorbed in the battery (g/m³):

$$m_{Uwater} = 10.02 \cdot m_{air} + 115.25 \cdot t_{wew} + 58.49 \cdot \phi_{wew} - 4608.33$$

R^2=0.87

Discharging process of stone battery (**two segments**):
Downloaded heat from the battery (MJ/m³):

$$q_{unload} = -0.016 \cdot t_{wew} - 0.023 \cdot m_{air} - 0.002 \cdot \phi_{wew} + 0.002 \cdot m_{vapour} + 0.359$$

R^2=0.99

Mass of water vapour absorbed in the battery (g/m³):

$$m_{Uwater} = 406.67 \cdot q_{unload} + 9.427 \cdot m_{air} + 6.433 \cdot t_{wew} + 0.851 \cdot \phi_{wew} + -144.246$$

R^2=0,99

Discharging process of stone battery (**three segments**):
Downloaded heat from the battery (MJ/m³):

$$q_{unload} = 0.067 \cdot t_{batt} - 0{,}023 \cdot m_{air}$$

R^2=0.99

Mass of water vapour absorbed in the battery (g/m³):

$$m_{Uwater} = 410.392 \cdot q_{unload} + 9.623 \cdot m_{air} + 3.792 \cdot t_{wew} - 65.202$$

R^2=0.99

Discharging process of stone battery (**four segments**):
Downloaded heat from the battery (MJ/m³):

$$q_{unload} = -0.023 \cdot m_{air} + 0.004 \cdot m_{vapour}$$

R^2=0.99

Mass of water vapour absorbed in the battery (g/m³):

$$m_{Uwater} = 9.875 \cdot m_{air} + 5.021 \cdot t_{wew} - 1.183 \cdot \phi_{wew} + 421.347 \cdot q_{unload}$$

R^2=0,99

The process of discharging the individual segments of the battery showed that the most useful energy referred to the unit volume of deposits can be obtained with the single segment. The average value of this energy was 3.9 MJ/m³. The energy obtained from the segment - double and quadruple was at the average level of 2.6 MJ/m³ and 2.3 MJ/m³ respectively. The smallest useful energy was in the triple segment. It was about 0.8 MJ/m³.

In the process of discharging the most important role played the water vapour during operation of a single segment. On average, the amount of water which vapour exhausting out of the battery was at 133.8 g/m³ bed with a maximum value of 1939.2 g/m³. This meant that during the operation of this segment of the air coming out was mainly humidified. Better results achieved for the other segments. During their work followed by absorption of water vapour. The amount of vapour absorbed by these segments were contained on average from 10.0 to 59.0 g/m³. Operating variables segments II, III and IV also cause evaporation of the water. `For single discharge cycles the amount of water vapour absorbed by the air stream didn't exceed, respectively for segments: II - 89.9 g/m³, III - 66.1 g/m³ and IV - 51.2 g/m³.

4 Conclusions

1. In the process of charging the highest average amount of accumulated heat (6.6 MJ/m^3) was recorded for the four operating segments simultaneously. While the lowest (1.6 MJ/m^3) for the two segments.
2. Only in the process of loading the four segments at the same time there were no evaporation of water from the reservoir. In other cases they occurred in both of evaporation and condensation.
3. The biggest average gain of useful heat was obtained in the process of discharging a single segment (3.9 MJ/m^3) and the lowest value 0.8 MJ/m^{3}for triple segment.
4. The dried air from the battery bed in the process of discharge is easier to achieve using more segments of the battery compare to the one.

Acknowledgements

This research was financed by the Ministry of Science and Higher Education of the Republic of Poland

References

1. U. Hermann, D.W. Kearney, J. of Sol. E. Eng. **124**, 124-132 (2002).
2. A. Gil, M. Medrano, I. Martorell, A. Lázaro, P. Dolado, B. Zalba, L.F., Ren. and Sust. E. Rev. **14**, 31-55 (2010).
3. A.I. Fernandez, M. Martínez, M. Segarr, I. Martorell, L.F. Cabeza, Sol. En. Mat. & Sol. Cells **94**, (2010) 1723-1729.
4. D. Sragovich, Solar Energy **43**, (1989) 7-16.
5. D.E. Laing, W.-D. Steinmann, R. Tamme, C. Richter, Sol. En. **80**, (2006) 1283-1289.
6. M. Geyer, Proposal to the 5th Framework Program of theEuropean Union, Spain. http://www.p2pays.org/ref/18/17991.pdf.
7. S. Kurpaska, H. Latała, D. Baran, P. Konopacki, R. Hołownicki. Agr. Eng.. **3**, 47-57 (2015).
8. P. Konopacki, R. Sabat, R. Hołownicki, S. Kurpaska, H. Latała. Inżynieria Rolnicza **147**, 129-138 (2013).
9. S. Kurpaska, H. Latała, M. Sporysz, B. Łapczyńska-Kordon, J. Knaga, P. Konopacki, R. Hołownicki. J. of Env,. Sci. and Eng. **3**, 351-356 (2014).
10. S. Kurpaska, H. Latała, P. Konopacki, R. Hołownicki. Inżynieria Rolnicza. **151**, 71-83 (2014).
11. S. Kurpaska, H. Latała, M. Sporysz, J. Sikora, K. Mudryk, P. Konopacki, R. Hołownicki. J. of Env. Sci. and Eng. **A4**, 154-160 (2015).
12. S. Kurpaska, H. Latała, R. Hołownicki, P. Konopacki, J Nowak. Inżynieria Rolnicza. **145**, 179-189 (2013).

 DOI: 10.1051/e3sconf/20161000054

Analysis of sewage sludge ashes from air and oxy-fuel combustion in a circulating fluidized-bed

Aneta Magdziarz[a1], Monika Kosowska-Golachowska[2], Agnieszka Kijo-Kleczkowska[2], Katarzyna Środa[2], Krzysztof Wolski[2], Damian Richter[2] and Tomasz Musiał[2]

[1]AGH University of Science and Technology, Mickiewicza 30 Av., 30-056 Krakow, Poland
[2]Czestochowa University of Technology, Institute of Thermal Machinery, Armii Krajowej 21,42-201 Czestochowa, Poland

Abstract. The ashes from sewage sludge combustion in air versus O_2/CO_2 atmospheres with oxygen concentrations in the range of 21-40% vol. at temperature of 850°C in a 12 kW bench-scale CFB combustor were characterised. The chemical and phase composition of ashes were studied by XRF and XRD. The morphology of studied ashes were examined by SEM method. The slagging and fouling indices were calculated to study the deposition tendencies of ash. The thermal behaviour of ashes was studied by TG-DSC techniques, focusing on the mass loss and thermic effects with the increasing of temperature up to 1200°C.

1 Introduction

Recently, increasing sewage sludge production and disposal, as well as the properties of sewage sludge, have been currently affecting the environment, which has resulted in legislation changes in Poland. Based on the Economy Minister Regulation of 16 July 2015 regarding the criteria and procedures for releasing wastes for landfilling, the thermal disposal of sewage sludge is important due to its gross calorific value, which is greater than 6 MJ/kg, and the problems that result from its use and application. Consequently, an increasingly restrictive legislation that began on 1 January 2016 was introduced for sewage sludge storage in Poland. Sewage sludge incineration is an attractive option because it minimizes odour, significantly reduces the volume of the starting material and thermally destroys the organic and toxic components of the off pads. Currently, as many as 11 plants use sewage sludge as fuel in Poland; thus, this technology must be further developed in Poland while considering the benefits of co-combustion with other fuels.

There are several thermal technologies for utilising municipal sewage sludge to obtain useful forms of energy, such as pyrolysis, gasification, combustion, and co-combustion processes [1,2]. The combustion processes of sewage sludge and its combustion in relation to coal and biomass fuels and co-combustion are presented in [3-6].

Circulating fluidized bed (CFB) combustion has been considered as a common technology for waste sewage sludge due to the fuel-like characteristics of sewage sludge composed of lots of organic components. The capacity of commercial fluidized bed combustion plants ranged from 100 to 200 tons of waste sewage sludge per day. Those plants have utilized excess air for combustion, and have generated lots of carbon dioxide [7]. Oxy-fuel combustion is one of the leading technologies considered for capturing CO_2 from power plants with CCS. This technology can significantly reduce emissions of NO_x and improve the thermal efficiency of the combustion process by reducing the flue gas volume. In the oxy-fuel combustion, the fuel is burnt in a mixture of pure oxygen and recycled flue gas. Because N_2 is eliminated from the oxidizing gas, the flue gas that leaves the combustion chamber is highly enriched in CO_2 which means that the combustion process takes place in an O_2/CO_2 environment [8].

Sewage sludge is characterized by high moisture and ash contents, a low calorific value and the content of significant concentrations of substances that are environmentally unfriendly [9]. It should be noted that sewage sludge contains microorganisms and harmful substances including heavy metals, poorly biodegradable organic compounds, bacteria, viruses, pharmaceuticals, hormones and dioxins. The combustion of sewage sludge can cause operating problems because of the high moisture content, which is a significant and important issue regarding combustion and a large amount of ash. Additionally, gaseous pollutants (CO, C_xH_y, NO_x, SO_x, PAHs, dioxins and furans) and particular matters are generated [10].

However, despite the fact that technology concerning sewage sludge combustion has been developing well, new problems have appeared. They concern the amount and properties of ash. Burning sewage sludge generates a large amount of ash containing varied mineral compounds (even more than 30 %). The main elements

[a] Corresponding author: amagdzia@agh.edu.pl

in the sewage sludge ash are silicon, phosphorus, calcium, magnesium, sulphur, potassium, chlorine and sodium. The amount and the chemical composition of ash can cause some risks during thermal conversion processes. The most important of them are slagging, bed agglomeration, fouling, and corrosion of the combustion devices, which degrade their performance and severely damage the firing equipment. The intensity of these phenomena depends on the chemical character of the ash, boiler design and operating conditions [11, 12]. The most aggressive compounds of ash are alkali metals associated with sulphur and chlorine, which can form an agglomerated bed particles deposited on the metal surface limiting heat transfer and accelerating the corrosion process. Different correlations and methodologies have been used for predicting the slagging and fouling tendencies of solid fuel ashes. There are some empirical indices for predicting the slagging and fouling tendencies of ash correlating the chemical composition of ash. Most of them have been used for coal as well as biomass ashes [13-15] but Wang [16] used it for sewage sludge ashes.

The major objective of the paper is to present detailed characteristics of ashes obtained from air and oxy-fuel combustion in a circulating fluidized-bed. The aim of this paper was to study the influence of the combustion atmosphere on the chemical composition of the ash.

2 Experimental

2.1.1 CFB combustor

Air and oxy-fuel combustion tests were conducted in a 12-kW bench-scale CFB combustor shown schematically in Figure 1. The facility consists of a combustion chamber (1), a cyclone (2), a downcomer (3) and a loop seal (4). The electrically-heated rectangular combustion chamber (riser), 680×75×35 mm, is the main component of the unit. Silica sand (particles smaller than 400 μm) to a mass of 0.3 kg constituted the inert bed. The gases that make up gas mixtures are supplied from cylinders (12) to a mixer (17) and then transferred via a preheater (8) directly into the combustion chamber. The flow rates of gases are controlled by valves (16) and measured by rotameters (15). During combustion tests, the superficial gas velocity was kept at a constant level of about 5 m/s. The temperature was held at 850°C by means of a microprocessor controller (11). S-type thermocouples (T1–T3) measured the temperature at three different levels inside the combustion chamber with an accuracy of ±2°C.

2.1.2 Experimental procedures

Once the fluidized bed had reached 850°C, a 2 g sewage sludge sample was introduced to the riser and kept there for a period of time that corresponded to the time required for the complete combustion. The experiments were performed in air (base case) and mixtures of O_2/CO_2 with oxygen concentrations in a range from 21% to 40% vol.

Sewage sludge samples in the form of spherical granules from a sewage treatment plant serving large urban-industrial agglomeration in Poland were used in this study. The proximate and ultimate analyses of the tested sewage sludge are presented in Table 1.

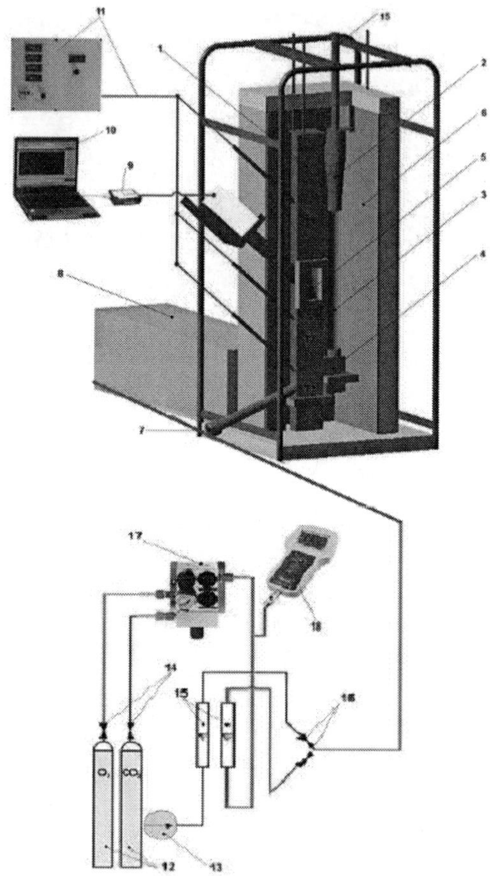

Figure 1. Schematic diagram of the experimental apparatus [8] 1-combustion chamber, 2-cyclone, 3-downcomer, 4-loop seal, 5-biomass particle, 6-insulation, 7-drain valve, 8-preheater, 9-card, 10-computer, 11-temperature control system, 12-gas cylinders, 13-air compressor, 14-pressure regulators, 15-rotameters, 16-valves, 17-mixer, 18-gas analyser, 19-ventilation duct, T1–T3-S-type thermocouples

For this study, four sewage sludge ashes were obtained from air and oxy-fuel combustion experiments. The samples were denoted as follows:

- Ash_1 – air (base case),
- Ash_2 – 21%O_2/79%CO_2,
- Ash_3 – 30%O_2/70%CO_2,
- Ash_4 – 40%O_2/60%CO_2.

The most important parameter determining the character of the ash is the chemical composition. The X-ray fluorescence (XRF, Riau ZSX Primus II) has been used to determine the main chemical composition of sewage sludge ashes (expressed

DOI: 10.1051/e3sconf/20161000054

in the form of oxides in Table 2). There were two kinds of oxides then the chemical properties were taken into account. The basic oxides are Na_2O, K_2O, MgO, CaO and Fe_2O_3, and the acid oxides are SiO_2, Al_2O_3 and TiO_2.

The phase analysis of the sewage sludge ash was examined by X-ray diffraction (XRD) using PANanalytical EMPYREAN DY 1061 with Cu Kα radiation in Bragg-Brentano geometry. The operating conditions of the XRD were 40 KV and 40 mA Cu Kα(λ=1.54 Å) radiation and step-scanned in the range 2θ = 10-90°, step 0.06°, time 10s. For phase identification, a data base PDF-4+ product of ICDD was used.

A scanning Electron Microscopy with Energy Dispersive Spectroscopy (SEM-EDX) was applied to perform microanalysis of the ashes. The FEI Quanta 3D FEG Dual Beam microscope apparatus was used. Particles of the studied ashes were mounted on metal stubs by double sided carbon adhesive discs. The samples were observed in a high vacuum mode by secondary electrons (SEM) using the accelerating voltage - 1 kV.

TG–DSC technique (Thermogravimetry and Differential Scanning Calorimetry) was conducted using Netzsch STA 449 F3 Jupiter. The samples were heated in alumina crucibles from ambient temperature to 1200°C at a constant rate of 10°C/min and at a 40 ml/min flow of air.

Table 1. Proximate and ultimate analyses of the tested sewage sludge (dry basis) [1]

Proximate analysis	
Moisture, wt.%	4.9
Volatiles matter, wt.%	51.4
Ash, wt.%	36.4
Fixed carbon, wt.%	45.6
Higher heating value, MJ/kg	12.57
Ultimate analysis	
Carbon, wt.%	30.8
Hydrogen, wt.%	3.9
Nitrogen, wt.%	4.3
Sulphur, wt.%	1.4
Oxygen (by difference), wt.%	18.3

The presented proximate and ultimate analyses of the studied sewage sludge responded as typical sewage sludge. The volatile matter of 51.4 wt. % corresponds to a higher heating value of 45.6 MJ/kg. The sludge contains very high amount of ash, 36.4%. The carbon content of 30.8% is an average value for this kind of renewable fuels. The moisture content is low, at 4.9%, because the sample followed the drying process. Normally, raw sewage sludge has a moisture content of approximately 90%. The H, N and S contents are comparable to other studied sewage sludge samples [2, 6]. The presented sewage sludge has "good fuel properties" and can be successfully used as a fuel for the combustion process.

3 Results and discussion

The content of the major inorganic elements in the investigated ashes in the form of oxides is given in Table 2. The main compounds of the ashes are SiO_2, Fe_2O_3, P_2O_5, CaO, Al_2O_3, MgO, TiO_2, K_2O, Na_2O and SO_3. Sewage sludge contains high amounts of phosphorus, calcium and iron as well as silicon.

Table 2. Ash composition expressed as oxides, wt. % (XRF method)

	Ash_1	Ash_2	Ash_3	Ash_4
SiO_2	26.83	28.05	27.48	28.29
Al_2O_3	7.13	6.95	7.01	6.89
Fe_2O_3	20.63	20.03	19.95	20.21
TiO_2	1.05	0.94	0.99	0.93
CaO	18.12	17.55	17.88	17.56
Na_2O	0.39	0.38	0.39	0.36
K_2O	0.83	0.79	0.80	0.81
MgO	2.46	2.35	2.44	2.32
Cr_2O_3	0.18	0.17	0.17	0.19
CuO	0.12	0.12	0.11	0.11
ZnO	0.75	0.76	0.77	0.77
P_2O_5	18.82	18.31	18.58	18.27
SO_3	2.11	2.96	2.80	2.63
Cl	0.01	0.03	0.02	0.02
R_B	42.43	41.10	41.46	41.26
$R_{B/A}$	1.21	1.14	1.17	1.14
$R_{B/A+P}$	1.75	1.65	1.69	1.65
S_R	39.43	41.26	40.56	41.37
R_S	1.69	1.60	1.64	1.60
F_u	1.48	1.33	1.39	1.33
Fe_2O_3/CaO	1.14	1.14	1.12	1.15

 DOI: 10.1051/e3sconf/20161000054

The concentrations of K and Na are low. The ashes were obtained in different combustion conditions (atmospheres), but the chemical composition and concentration of elements for all of the studied ashes are quite similar.

Minor differences have been observed between ashes obtained under air and oxy-combustion atmospheres. It suggests that combustion conditions do not significantly influence the chemical composition of ash.

The amount of ash is a serious problem in the combustion process of sewage sludge because of the high risk of slagging, bed agglomeration, fouling, and corrosion in the combustion devices. The presence of ash deposit material reduces the heat transfer and combustion efficiency, and damages combustion chambers when large particles break off. There are some empirical indices correlating the functional temperature and standardized chemical composition of ashes, thus making it possible to predict the ash behaviour and deposition tendencies [13-16] (R_B – sum of percentage of basic components, $R_{B/A}$ – ratio of basic to acidic components in ash, $R_{B/A+P}$ – ratio of basic to acidic components with considering phosphorus in ash, S_R – viscosity index, R_S – slagging index where S^d is the percentage of S in dry fuel, F_u – fouling index):

$$R_B = (Fe_2O_3 + CaO + MgO + Na_2O + K_2O) \quad (1)$$

$$R_{B/A} = (Fe_2O_3 + CaO + MgO + Na_2O + K_2O)/(SiO_2 + Al_2O_3 + TiO_2) \quad (2)$$

$$R_{B/A+P} = (Fe_2O_3 + CaO + MgO + Na_2O + K_2O + P_2O_5)/(SiO_2 + Al_2O_3 + TiO_2) \quad (3)$$

$$S_R = 100\, SiO_2/[SiO_2 + eq(Fe_2O_3) + CaO + MgO] \quad (4)$$

$$R_S = R_{B/A}\, S^d \quad (5)$$

$$F_u = R_{B/A}\, (Na_2O + K_2O) \quad (6)$$

The slagging tendency depends on R_S value and if $R_S < 0.6$ – low slagging inclination $R_S = 0.6 \div 2.0$ – medium, $R_S = 2.0 \div 2.6$ – high, and $R_S > 2.6$ extremely high.

The fouling index shows a tendency to the sintering of deposits and if $F_u \leq 0.6$ – low fouling inclination, $0.6 < F_u \leq 40$ – high, and $F_u > 40$ extremely high.

Calculations of slagging and fouling indices by the formulas (1-6) are shown in Table 2. As can be seen from the base-to-acid ratio, the fouling tendency of all of the ashes is medium. The chemical composition of the mineral part of sewage sludge indicates its slagging tendency. The sludge ash includes a high amount of iron and phosphorus, which leads to a decrease of the melting temperate of ash [13, 14]. The values of F_u and R_S indicate that the studied ashes of not have a strong tendency towards the deposition process. It should be emphasized that sewage sludge ash contains a very low amount of sodium and potassium, which are the most aggressive elements of ash. The chlorine content is not high, too. Therefore, it can be assumed that during the combustion of sewage sludge it should not be found a lot of operating problems concerning the deposition of ash as well as corrosion risk should not be found, because of the low amount of potassium, sodium and chlorine in sewage sludge ash.

The crystalline mineral species in ash samples were identified by XRD and X-ray patterns are shown in Figure 2. A slight difference can be observed from the XRD patterns. It should be noted that only the main crystalline phases were observed by XRD in all of the samples. Since there were too little amounts of amorphous materials and crystalline phases, it was not possible to identify them. For all of the ashes, the main identified mineral phases are Fe_2O_3, SiO_2 and $Ca_{2,589}Mg_{0,411}(PO_4)_2$. The main crystalline phases are dominated by P, Si, Ca and Mg contents in ash, naturally occurring in sewage sludge. Silicates were not detected, although they would likely be generated in the reaction between Al_2O_3, SiO_2, and CaO. Silicon exists as quartz SiO_2. Iron mineral compound is present only as hematite, Fe_2O_3. Iron oxide can react with quartz to form eutectics having low meting point temperatures, but they are from the amorphous phase and non-detective for XRD [16]. The most interesting is the phosphate of calcium and magnesium as a low-solubility mineral named whitlockite. The presence of whitlockite mineral in sewage sludge ashes was confirmed by [17,18]. Although the sewage sludge was oxy-combustion and the concentration of CO_2 was higher than air combustion the carbonates are not detected. XRD analysis confirms the absence of phases which can cause corrosion e.g. the potassium or sodium phases.

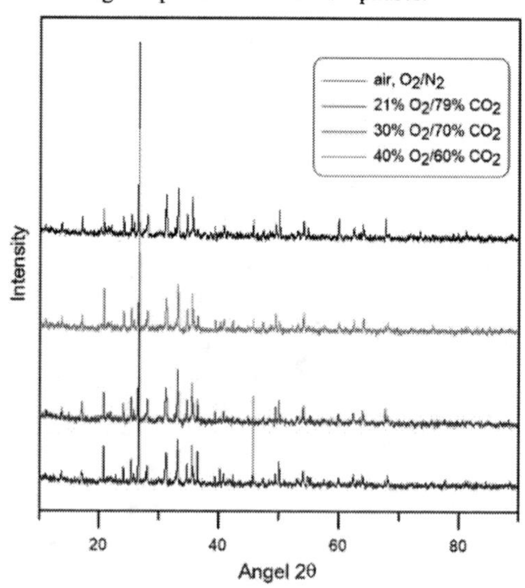

Figure 2. X-ray diffraction patterns of sewage sludge ashes obtained under different conditions.

Figure 3 presents the TG and DSC curves during heating the studied ashes under air atmosphere. TG corresponds to mass change and DSC reflects heat flow. The TGA analysis was conducted to study the thermal behaviour of sewage sludge ash. The slight differences

DOI: 10.1051/e3sconf/20161000054

in thermal behaviour have been observed for ash obtained under air atmosphere and oxy-combustion conditions. For Ash_1 a little mass gain from 100 to 900°C was observed. This could take place during the oxidation of an element of ash. Comparing the TG for A_2 to A_3 ashes, the influence of oxygen addition was not observed. For all of the ashes the thermal behaviour up to 1200°C can be divided into three steps. The first step concerns moisture release (up to 100°C), which is absorbed on the ash surface during storage and preparation for investigations. The second step, between the temperature range of 120 – 900°C is without mass loss, and has a wide exothermic peak. The third stage, above 900°C, corresponds to the decomposition of other compounds, probably due to phosphates and sulphates. The mass loss was not significant and was less than 4 %. It suggests the beginning of the melting process, but defined endothermic effects were not observed.

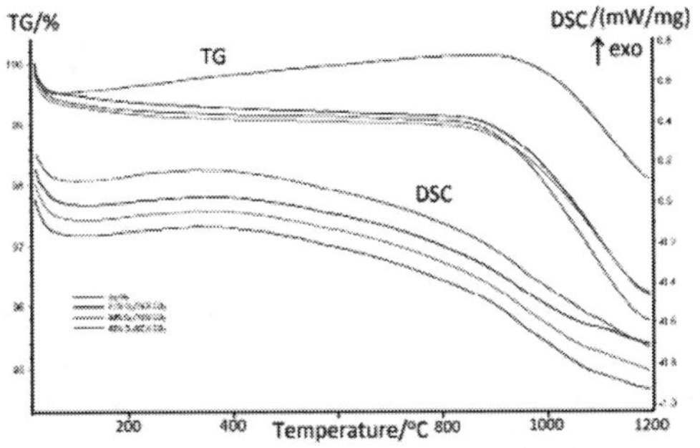

Figure 3. TG and DSC curves of sewage sludge ashes during the thermal process under air atmosphere.

Figure 4. SEM images of sewage sludge ashes Ash_1.

The SEM method makes it possible to investigate the particle morphology and surface structure of ashes. ashes. Figures 4 and 5 present the morphology of the studied ashes. The particles of ash are different size and irregular. There are solid particles generated with smaller ones. Darker particles with white layers can also be identified. The particles of sewage sludge ashes seem to be angular with a sharp edge and a porous surface.

Figure 5. SEM images of sewage sludge ashes Ash_3.

3 Conclusions

This study presents a comprehensive analysis of sewage sludge ashes using analytical techniques such as XRF, XRD, SEM and TGA. The study was focused on getting information concerning whether the atmosphere of the combustion process influences the chemical composition of the ash. The obtained results indicate that there is only a very slight impact. It is difficult to confirm evident influence. Thus, the results show that sewage sludge can be successfully combusted in oxy-fuel technology. Generally, the ashes are dominant in the inorganic matter: P, Ca, Si and Fe. It was observed that the main mineral phases came from phosphorus and silicon and iron-based minerals. XRD analysis had confirmed the presence of the Fe_2O_3, SiO_2 $Ca_{2,589}Mg_{0,411}(PO_4)_2$ in a crystalline forms. The tendency of fouling and slagging calculated by chemical composition of the ashes in not very strong. Finally, the oxy-CFB combustion technology can be effectively used for sewage sludge thermal conversion processes.

Acknowledgments

The research leading to these results has received funding from the Polish-Norwegian Research Programme operated by the National Centre for Research and Development under the Norwegian Financial Mechanism 2009-2014 in the frame of Project Contract No Pol_Nor/208189/105/2015.

References

1. P. Manara, A. Zabaniotou, Renewable Sustainable Energy Rev. **16**, 1081-1087 (2012)
2. A.I. Calvo, L.A.C. Tarelho, E.R. Teixeira, C. Alves, T. Nines, M. Duarte, E. Coz, D. Custodio, A. Castro, B. Artinano, R. Fraile, Fuel Process. Technol. **114**, 58-68 (2013)

3. A. Kijo-Kleczkowska, K. Środa, M. Kosowska-Golachowska, T. Musiał, K. Wolski, Waste Manage. **46**, 459-471 (2015)
4. A. Kijo-Kleczkowska, K. Środa, M. Kosowska-Golachowska, T. Musiał, K. Wolski, Fuel **170**, 141–160 (2016).
5. A. Kijo-Kleczkowska, K. Środa, M. Kosowska-Golachowska, T. Musiał, K. Wolski, Waste Manage. (to be published) http://dx.doi.org/10.1016/j.wasman.2016.04.021
6. A. Magdziarz, S. Werle, Waste Manage. **34**, 174–179 (2014),
7. H.N. Jang, J.H. Kim, S.K. Back, J.H. Sung, H.M. Yoo, H.S.Choi, Y.C. Seo, Fuel **170** 92–99 (2016)
8. M. Kosowska-Golachowska, A. Luckos, K. Klos, T. Musial, Proceedings of the 21st International Conference on Fluidized Bed Combustion, 611–618 (2012)
9. G. Garcia, J. Aruazo, A. Gonzalo, J.L. Sanchez, J. Abrego, Chem. Eng. J. **222** 345–352 (2013)
10. V.K. Tyagi, S.L. Lo, Sust. Energ. Rev. **25** 708–728 (2013)
11. A. Magdziarz, A.K. Dalai, J.A. Koziński, Fuel **176** 135-145 (2016)
12. L. Weidong, L. Ming, L. Weifeng, L. Haifeng, Fuel **89** 1566–1572 (2010)
13. M. Pronobis, Biomass Bioenergy **28** 375–383 (2005).
14. M. Pronobis, Fuel **85** 474–480(2006)
15. D. Vamvuka, D. Zografos, G. Alevizos, Bioresour. Technol. **99**, 3534-3544 (2008)
16. L. Wang, G. Skjevrak, J.E. Hustad, M.G. Gronli, Fuel Process. Technol. **96** 88–97 (2012)
17. M.B. Folgueras, M. Alonso, J.R. Folgueras, Fuel Process. Technol. **131**, 348–355 (2015)
18. M. Atienza-Martinez, G. Gea, J. Arauzo, Sascha R.A. Kersten, A. Maarten J. Kootstra, Biomass Bioenergy, **65**, 42–50 (2014)

E3S Web of Conferences **10**, 00055 (2016) DOI: 10.1051/e3sconf/20161000055
SEED 2016

Local Balancing System from the Business Model Canvas Perspective

Bożena Ewa Matusiak[1,a]

[1]*University of Łódź, Faculty of Management, Department of Computer Science, Matejki 22/26, 90237 Łódź, Poland*

Abstract. The paper presents an overall view of the business model (BM) for the e-balance system for: balancing energy production and consumption in energy efficient, smart neighbourhoods (the e-balance project, FP7-SMARTCITIES-2013) along with its functionalities, based upon the Osterwalder's canvas methodology. Additionally, this is the second, after two years of work, more incisive evaluation of the BM from the user's and demo site's perspective (Bronsbergen, the Netherlands). The aim of this paper is to present results and assess the above mentioned BM in the face its commercialisation and applicability to Europe.

1. Introduction

Smart grids, dispersed energy resources (DER) development, active demand response management (ADRM) and energy flexibility (EF) on the power market – these are the main elements which play a critical role in understanding of the energy policy ambitions in the future European power system [1, 2]. Demand response [3, 4], a need for more energy flexibility and better use of them in local destinations (smart neighbourhoods) trigger many innovative projects, platforms/applications, as well as new business model approaches and propositions [5].

The e-balance project proposes a new system on the energy market with a large number of prosumers who will not only consume the energy, but mainly produce and save it depending on the market and price /incentives conditions [6, 7]. The available energy capacity (i.e. energy flexibility) for demand response can be executed by dynamic management and a special Triana algorithm [8]. The new, most important role in this system will be played by market-makers, referred to as aggregators. So far the aggregator is not a clearly identifiable electricity market actor. Such a role and areas of its responsibilities are still in development [9, 10].

The e-balance system/platform proposes one of the solutions to involving the aggregators into the bottom-up processes of the energy flexibility aggregation (in comparison with [13]).

This paper presents the current state of the e-balance system and its business model's picture [6] using the Osterwalder's canvas after two years of project work.

1.1. Tool and method used

Within the last decade, the Osterwalder's business model canvas approach [11] has become a very popular tool and method of thinking not only in the internet start-ups, but also in the energy market and smart grid projects. In the area of innovative ideas, it is hard to find a more adequate tool for the first BM assessment, systematic recognition and answering the resultant main questions: what the system will be, to whom it will apply and where the added values are. Investors are not eager to trust in an overall description of new systems or technology without a methodological approach or even a detailed business plan.

Many complementary analyses and first step studies might be used in order to find the effective methods of building the reliable systematic BM framework. These methods include: the analysis of market competition / suppliers / manufacturers and users, SWOT analysis, business or use cases descriptions of needs, KPI's definition for adequate BM validation [6], deep insight analysis after a feedback from the users, followed by the first assessment of risk analysis and a business plan.

The most innovative BMs in several European projects such as: Sustainable Places, CityOpt, Besos, Optimus, Ambassador, Indicate etc. are considered to have used mainly the Osterwalder's tool for structured BM specification [12].

The earlier mentioned BM Osterwalder's canvas consists of several blocks. Each block needs additional analysis and studies to be answered, including: Customers' segmentation, Value proposition, Dissemination channels, Customers' relations, Revenue / financial, Key resources, Key activities, Key partners and Cost structure description.

[a] Corresponding author: bmatusiak@pocztawz.uni.lodz.pl

 DOI: 10.1051/e3sconf/20161000055

2. Business Model

The next part of this paper presents the prepared e-balance BM description according to the above mentioned nine blocks with an overall technical description of the system.

2.1. Key activities' description

The end user (mainly prosumers with the smart meter, smart devices, and the dedicated GUI at home) is the first and the most important actor in the e-balance system. The Customer Management Unit (CMU) is an additional device provided by the e-balance system to the user/prosumer, which allows them to interact with the system in a technical sense [14]. It will function as the energy manager, data hub and (tele) communication portal (See Fig. 1.)

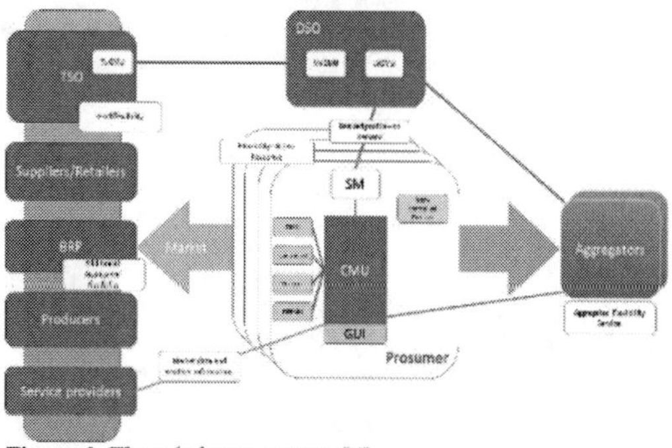

Figure 1. The e-balance system. [6]

The CMU will be connected to the Smart Meter (SM) for the local data analysis purposes. It may be connected to the local LAN via a direct connection or via WLAN. This connection will be used to acquire additional information for data analysis. After the initial, physical connectivity, other devices that need to be directly connected can be attached (DMUs for many smart devices (IEDs) and sensors). The end user may also possess the devices that are not connected to the CMU. Such devices will not be operable by steering signals from the system, but their demand will be estimated by the system. By using the GUI, the end user can set the initial system parameters (home scenario/ strategy). If the DER (micro generation) was connected as a set of PV panels, the typical characteristics can be delivered to the system.

The end-user of the e-balance system will have a CMU used to also predict (using information from the weather forecast agency) [14], monitor and control the energy usage and/or production. The CMU will gather all the desired energy usage of Intelligent Energy Devices (IEDs) and dispersed generation and saving (DER installations/batteries) in the shape of power profiles.

Subsequently, through the aggregators' activities, the users' power profiles are combined with a prediction of the energy usage of all the other devices connected to the local installation within the smart area. The resulting aggregated profile will be communicated to the DSO. This will be repeated every 15 minutes; sending out the predicted aggregated power profile for the next 24 hours and sent to the DSO by the aggregator.

The role of aggregators is to aggregate and to incentivise end users to react to the DSO /Triana algorithm signals. The DSO may respond with a general request to reduce or increase the energy exchange with the grid in either direction (usage or feed-in) for each of the 15-minute intervals. Using incentives (technical as a steering signals), the aggregator is supporting the DSO with an information exchanging process and the calculation of benefits (financial) for each customer involved in this process. The aggregator's CMU will negotiate with the smart devices and with the user's CMU to determine to what extent it is possible to comply with the general DSO request [14]. A measure determined by the CMU (on the aggregator's side) will indicate how well it can assist the DSO. This single number will be delivered back to the DSO, which will then again inform the aggregator's whether or not the newly negotiated power profile should follow the originally sent profile.

The DSO can provide and maintain the e-balance main infrastructure in the user's home, but this is not at all obvious. The CMU's, DMU's and SM can be provided by other market actors. The DSO will facilitate the controlled balancing of energy, provide information and access to the energy market, as well as assist all stakeholders.

The aggregator (that can also be the ESCO company) supports and aggregates all financial processes and data in cooperation with retailers , which will prepare the final settlements. The most important mechanism on how to settle the costs and revenues will be hidden in the price and incentives mechanisms proposed under a win-win strategy for all sides of the contract [6]. It is indeed interesting how to go about preparing and matching a system of tariffs for each of the groups of aggregated users. This is separated area of next researches in this project.

In the low voltage grid, the most significant component of the e-balance infrastructure will be the Low Voltage Grid Management Unit (LVGMU). The LVGMU will request every CMU of the connected aggregators or end users' CMUs to send out a power profile with a 15-minute interval (covering a total time span of 24 hours). It is likely that not every connection to the low voltage grid will be equipped with a CMU. Hence, the LVGMU will have to make a prediction of the energy exchange at the particular points of connection. The predictions and communicated CMU power profiles are gathered to form a "neighbourhood profile". This profile represents the unadjusted, expected power usage for the following 24 hours (in this sense it corresponds with Day Ahead Market and Intraday Market).

The newly formed neighbourhood profile will be compared to a pre-programmed "desired profile". At the beginning, this profile will represent a default setting and might be, for instance, a completely flat profile. The absolute value of this profile is the subject

DOI: 10.1051/e3sconf/20161000055

of the e-balance system research. The desired profile will be influenced by the local grid limitations and requests from other management units. During or after the negotiation phase, the LVGMU will perform a technical "rationality check" by using the supplied profile as an input to the electrical load flow. This is a calculation that will show whether grid components are overloaded and if the grid voltages remain within the grid code limits.

The result of the comparison will be a power profile, indicating at which times there will be an energy shortage or surplus and how large it will be. The CMUs will respond with a single number specifying how well they are capable of assisting in solving the shortage or consuming the surplus. The LVGMU will rank these responses and select the CMUs that together can reduce the difference profile in the most efficient way.

All remaining difference profiles are communicated by a group of LVGMUs, upwards to the Medium Voltage Grid Management Unit (MVGMU). The MVGMU will go through a similar process as the one described above.

Any remaining differences after these negotiations will be communicated upwards to the Top Level Grid Management Unit (TLGMU). This TLGMU will provide balance amongst the MVGMUs, large centralised power plants, and the grid connections to the neighbouring regions (interconnectors) [14].

It is important to recognise that, on a conceptual level, the centralised power plants will be reduced to an imbalance functionality and the transport grid will only be used when Distributed Energy Resources (DER) are insufficient to deliver the desired power to a given region. In practice, considering the currently low amounts of DER and low flexibility of loads and generators in most regions, the centralised power plants will be retained to have a peak and base load functionality.

It should be noted that in the e-balance system, the aggregator works on behalf of the aggregated customers and can offer flexibility for their customers towards balance responsible parties (BRPs) as an additional service.

The BRP might want the end users (by aggregators) to reduce energy consumption, or increase production in order to compensate for the lack of production elsewhere, thus enforcing better energy availability elsewhere (on the Intra Day Market). In such a case, the user's/aggregator's CMU will not aid the DSO by complying with the LVGMU's request, but instead will additionally comply to the BRP's request. As such, the BRP and DSO will both provide incentives to influence the behaviour of the customers, but these incentives are not necessarily the same.

The settlement and invoicing will be prepared according to the measures taken from the smart meters and will correspond to the contracts made with the aggregators and retailers/suppliers. The aggregators together with suppliers offer the additional services as a "demand response flexibility" on the market and are also responsible for the system of prices and incentives [6, 7].—This is the most important mechanism for achieving a satisfactory level of cooperation with end users.

In this sense, the key activity is to provide the energy flexibility on the market. The four actors mainly involved in this process are: the end user, the aggregators of end-users, the DSOs and the retailers.

2.2. Customers' segmentation

The e-balance platform is dedicated to the prosumers – as the main users of the system and to those who are involved and will have any interactions with the dynamic energy management within the whole, single energy system.

However, prosumers are not the only ones who can have a contract to use the e-balance system. Depending on the features, retail customers, RES/DER owners, commercial users, customers who are interested in energy efficiency and decreasing costs, city municipalities or real estate owners, can also be the main consumers of our product. They can focus on saving money and reaching the high energy efficiency and security of the energy supply (by using the new energy system and smart appliances or sensors, home batteries, smart cars). They are willing to provide feedback regarding the service and to adapt an e-balance system into their premises [15, 16].

They have their own smart devices, as well as the contract for the direct devices control with the e-balance direct provider (which can be the aggregator party).

2.3. Value proposition

The deeper value proposition analysis for all stakeholders was prepared during project work and is visualized in the Fig. 2 , described in detail in [6].

Stakeholder/market	Benefits from the e-balance system as a part of the DSM
Prosumers	Savings on the energy bill
	Other non-monetary benefits
	Increase awareness in participation in local balancing
	Satisfaction of being integral with environment and with their neighbours
	Detailed market data and detail data about his own load/production
Retailer	Lower exposure to high price/demand
	Accurate prognosis
	New tariffs structure possible
Aggregator	Detailed customer's data
	Optimization and power aggregation
	Balancing in neighbourhood
TSO/DSO	Power flexibility and power balancing in local area
	Lower number of congestions
	Lower number of outages
	Better Fault, Isolation, Restoration management (FIR)
	Lower losses and slower aging of components due to flattened power profiles
Supplementary producers	Reduced peak load generation
	Less need for new power plants and peak units.
	Less need for capacity reserve
	Higher share of RES is possible
Market (in general)	Lower price volatility
	Increased demand/supply flexibility
	Lower risk, and costs of imbalance
Environment	Increased integration of RES made possible
	Less fossil import, reduction CO_2 emissions,
	Energy efficiency effect.

Figure 2. Benefits for market roles and the environment when e-balance is utilised.

A user wants to receive lower bills for the energy consumption and to be satisfied about being ecological and efficient in her/his local environment. The neighbourhoods will become more cooperative and aware in terms of their energy saving.

The DSO has a more stable and optimal power flow in part of the grid, with a minimum cost of its maintenance, losses and faults' decrease. The DSO also has quality benefits from grid monitoring and maintenance service offered in the e-balance system.

For other partners who have a greater demand for their products and services, sometimes this can even mean a new service or business for them. There is a possibility of increasing local activities on the basic, neighbourhood level.

Third parties - aggregator/retailers or ESCOs - benefit from increasing the energy efficiency in the area of their activity.

A Smart City has a modern system for the energy optimisation which will develop in a scalable/fractal way with benefits for the environment of future generations.

The main value of the e-balance system is to provide the flexibility in the energy exchange with the grid. At request, the amount of power drawn from the grid can be reduced or increased (both in amplitude and direction i.e. withdrawal or feed-in). It is solely the users' strategies that determine the amount and shape of the flexibility provided.

2.4. Dissemination channels

Website and internet apps, mobile apps, hot-lines, calls with experts, as well as in-home display(GUI) will be the main channels for consumer/prosumer communication and their feedback signals from the users [17, 18].

- GUI (in home display with dedicated apps) meaning all types of devices providing graphics visualisation of the energy consumption and production data on screen, smartphone apps etc.
- Website with access to the e-balance system and modules.
- Machine to machine communication for operations and technical system realisations (through the dispersed services/modules of the e-balance system).
- Internet infrastructure with wireless communication (all telecommunication stack).
- Informative correspondence. It refers to the information regarding all important data given to the users (e-books or newsletters, on line and real time information, on line bills).

2.5. Customer relations

All relationships between partners will be based on understandable contracts, cooperation and trust. The owner such as the DSO and the aggregator/ the retailer will be the main providers of services (the energy balancing and grid resilience). They can have ~~a~~ cooperation contracts for balancing service performance, where the aggregator/the retailer will work on behalf of the DSO in the area of the e-balance functionalities. The part of the system such as the CMU tools can be distributed by retailers or technical providers. The retailers can use the CMU installations for additional auxiliary services preparation as well. It means that the retailers can have their own system of price incentives, different from the aggregators' incentives.

As a main mechanism for the e-balance system, a set of financial mechanisms will be delivered along with help/guidance on every level. Furthermore, some gamification methods, non-monetary incentives and feedback systems are also considered, especially at the beginning (during the project's time and testing). The systems of tariffs, as well as incentives will be chosen for each type of users' activity in the e-balance system. The right choice will depend on the in-lab testing, the detailed analysis of the previous load profile and the type of strategy savings. Overall, finding the win-win strategy of rewards for each participant of the system was the most crucial goal. Although, such analysis has been done [6], the basic conflict of interests was also identified among the roles of the aggregator, retailer and the DSO.

2.6. Revenue

In the first stage of the e-balance system (for MVP), the revenue model is thought to be a supplemental payment for given values as a transaction fee. The end user will also pay for additional apps only when using them. Moreover, the licence fee is also re-considered (yearly or monthly). The application for the e-balance system and service maintenance will have a fixed monthly charge (added to the bill) [6]. The in-home display will be provided as a consumer leasing system of payment. It can be proposed to the system owner to have the additional revenues on the adverts of products recommended as compatible with the system on the website. A multiple flow schedule, in which the monthly fee, extra rewards for customers, long-term contracts and system of incentives, is considered.

The decision about the final recommendation will be made after the in-lab testing.

2.7. Key resources

Key resources were considered on every level of the e-balance system/apps, as well as within the technical grid infrastructure (on LV/MV levels). The detailed list of appliances/resources was prepared in [19]. A technical grid specification was listed in the e-balance deliverables: [14, 20]. This is on the assumption that the smart meter and the Internet access had been installed in the user's house earlier. Instead of developing a DMU on our own, the main focus was put on standard solutions which improved the system scalability and applicability. The same concerned the communication technologies to be used within the Home Area Network (HAN) that connect the DMUs to their corresponding CMU. The CMU is equipped with several communication modules or gateways supporting interoperability.

In terms of human recourses, it was assumed that a professional staff, equipped with necessary skills corresponding to the BM needs, will be involved.

 DOI: 10.1051/e3sconf/20161000055

In such a case, the engineers will operate and maintain the e-balance resources and web applications, whereas the marketing and sales staff will co-create the best vision of the services and the added value for the customers.

Apart from all of the above mentioned items, the financial resources are the most important. Financing should be available considering this business model, which was developed regarding increasing grid flexibility. Naturally, the first investments funds will be provided by DSOs and other cooperative partners, as a joint venture capital. The accepted model of reimbursement should guarantee a profit and return on the investment.

2.8. Key partners

During the project work, ten different stakeholders were selected and described as the most important actors along the value chain (Fig 2) [6]. Their current and future role has been proposed and described. The DSO is a main coordinator and a market facilitator for the e-balance platform. The relations among the end users- aggregators-DSO and retailers are crucial for balancing and the financial processes flow.

2.9. Cost structure

Today, the actual investment costs in Bronsbergen can be assessed only in regards to the physical resources. This had been incorporated into the planning of the project. The actual, future costs for small installation such as the Bronsbergen one, will only be known when the demo results are out.

Table 1. The basic investment cost of the e-balance system infrastructure in Bronsbergen demo site - from DoW

Alliander	Total cost: 30 000 euro
Other partners (LW; IHP; UTWE)	Total cost: 7000 euro
Grand Total:	37 000 euro

At this moment, it is established that over 50 users (home users) together with two LVGMU and one MVGMU will be deployed [21]. Smart devices like dish washers or washing machines will be installed in users' homes for free. On the integration level, the APIs of the Whirlpool devices will be tested. The cost and revenue analysis will be prepared and estimated after obtaining the results from the testing. Table 1 presents the basic investment costs planned for the project. As yet, there are no set calculated final costs for its implementation.

3. Results and difficulties

The e-balance system will be tested at the demonstrator site in Bronsbergen (The Netherlands).

Up until now, in cooperation with the Increase project, around 50 houses/users have expressed their intention (level of intent) of participating in the test. The deployment of e-balance system is on its way. The first project participants will receive modern equipment from the Whirlpool company free of charge.

During planning of the e-balance business model, several obstacles have been found [12]. The list of the most important ones is as follows (a lesson learnt during the e-balance and Sustainable Places):

- Main barriers for the implementation/testing of new business models are regulatory constraints.
- A competitive market is the key to the introduction of new business models. No competitive market environment exists as yet; problems with changing regulations, limited possibilities for new data-driven business models.
- Low incentives for existing market actors to encourage them to adopt new business services, as currently any striving for more efficiency does not get rewarded under current legislation.
- It is very important to identify, from the very beginning, the stakeholders (current and potential) involved in each aspect of the proposed solution (project), in order to appoint the system owners and the revenue models scenarios.
- Identification of future energy markets, future actors and their roles is difficult in the unstable EU market.
- Energy companies (DSO/TSO) need to engage more with customers/users. Feedback and engagement with end users is very important to introduce new business models.
- In DR strategies, it is difficult to identify the revenue flow.
- Limited information available for the definition and quantification of the costs and revenues streams.
- A successful replicability of the business models depends on the ability to retrieve data from the sites to be analysed.
- A win-win strategy based on the market rules should be guaranteed in order to avoid being rejected by some stakeholders. Any proposed revenue model should compensate less income with additional savings or higher prices with high-value services.

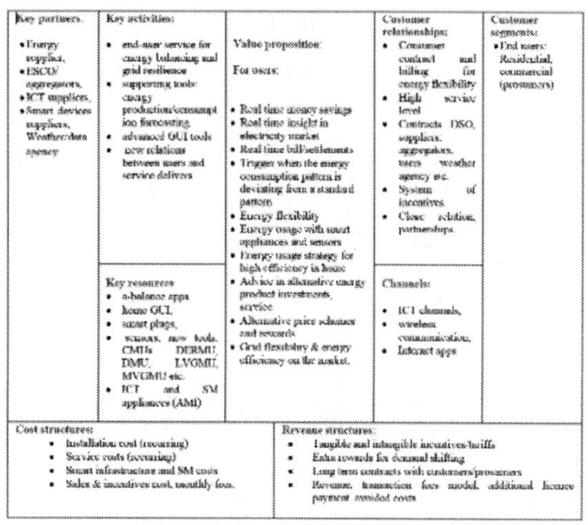

Figure. 3 The Osterwalder's canvas prepared for e-balance BM

These constraints demonstrate how difficult it is to build a credible business model/BM canvas (Fig.3) for innovative solutions, even in countries where the level of development of a new energy market is much higher than in Poland. The present business model is still being developed and improved accordingly.

Acknowledgment

This work has been partially funded by the e-balance project (Project Number: 609132), under the European Commission's 7th Framework Programme (FP7-SMARTCITIES-2013) and the grant of the Ministry of Science and Higher Education: 3009/7.PR/13/2014/2; (PL). http://www.e-balance-project.eu.

All project's participants: IHP, LessWire: Germany, CEMOSA, UMA: Spain, Alliander, UTWENTE: The Netherlands; ULODZ, OPI: Poland; INOV, EFACEC, EDP: Portugal. All pictures have been prepared by author of this paper.

References

1. European Parliament and Council Directive 2009/72 / EC of 13 July 2009. Concerning common rules for the internal market in electricity.
2. European Parliament and Council Directive 2012/27 / EU of 25 October 2012. Concerning energy efficiency.
3. *The future of the electric grid*, An interdisciplinary MIT study 2011, https://mitei.mit.edu/system/files/Electric_Grid_Full_Report.pdf [online: 04.2016]
4. J. Torriti, M. G. Hassan, M. Leach, Energy **35**, 4 (2010), http://epubs.surrey.ac.uk/749580/1/DR%20experience%20in%20Europe.pdf [online: 04.2016]
5. B.E. Matusiak, et.al.: *Energy Management Using the Business Model Approach; European Energy Market*, EEM15, Lisbon (Portugal), May 19-22, 2015; doi: 10.1109/EEM.2015.7216612
6. B.E. Matusiak et.al.; Deliverable D2.3 "*Market Assessment and Business Models*; The e-balance project proceedings; www.e-balance-project.eu; 2015
7. M. Muratori, G. Rizzoni, IEEE Trans. on Pow. Syst. **31**, 2 (2016)
8. V. Bakker, *Triana: a Control Strategy for Smart Grid*, PhD thesis on University of Twente, 2012
9. G. Kotsis, et.al.: *Demand Aggregator Flexibility Forecast: Price Incentives Sensivity Assessment;* EEM15, European Energy Market International Conference 2015 Lisbon 19-22 May
10. E. Koliou, et.al.: *Complexity of demand response integration in European electricity markets*, EEM15, European Energy Market International Conference 2015 Lisbon 19-22 May
11. A. Osterwalder, Y. Pigneur, *Business Model Generation*, Wiley 2012
12. Symposium and proceedings SP15, S. Capato, *Business models dimension of EU projects*, Savona, Sept. 16-18, 2015 Italy, http://sustainable-places.eu
13. USEF 2013, https://ec.europa.eu/energy/sites/ener/files/documents/xpert_group3_summary.pdf [online: 03.2016]
14. http://www.e-balance-project.eu/publication_deliverables.html D6.1 [online: 03.2016]
15. S. Hill, *How European Utilities can engage customers at a key point to boost NPS, loyalty & sales*; Presentation Opower company on the V Smart Communication & Technology Forum 11.06.2015 Warszawa
16. B.E. Matusiak, et.al.: *How to Balance the Energy Production and Consumption in Energy Efficient Smart Neighbourhoods*, MedPower 2014, Athens, ISBN: 978-1-78561-146-9.
17. http://www.e-balance-project.eu/publication_deliverables.html D4.1 [online: 03.2016]
18. http://www.e-balance-project.eu/publication_deliverables.html D4.3 [online: 03.2016]
19. http://www.e-balance-project.eu/publication_deliverables.html D3.1 [online: 03.2016]
20. http://www.e-balance-project.eu/publication_deliverables.html D3.2 [online: 03.2016]
21. http://www.e-balance-project.eu/publication_deliverables.html D5.2 [online: 03.2016]

Application of multi-criteria methods to compare different solutions of supplying buildings in electricity from photovoltaic systems

Barbara Mendecka[1] and Alicja Stoltmann[2,a]

[1] *Industrial Engineering Department, Florence University, Italy*
[2]*Faculty of Electrical and Control Engineering, Gdańsk University of Technology, Poland*

Abstract. Nowadays, the technologies of electricity generation in distributed systems are usually associated with Renewable Energy Sources (RES). The choice of the construction site depends mainly on the availability of the power system. However, energy planning, especially in case of RES, is a complex process involving multiple and often conflicting objectives. The complexity of the selection of the electricity system is typically addressed with the use of multi-criteria tools, involving all of the considered criteria and also different methods of their aggregation. The result is a final ranking of the available alternatives. This paper describes the application of a multi-criteria decision tool for the comparative analysis of the use of alternative options of the PV technology for electricity production. Four decision variants are considered, including the different construction of solar farms (static and movable structure) and different types of configuration of individual installation (off and on-grid). The construction of each new sources of electricity generation, including PV, is the multi-threaded and multi-dimensional decision problem. The criteria used in the analysis combine economic, environmental and social issues. The first of the considered criterion is the Net Present Value (NPV) which determines the economic viability of the project. The second criterion, thermo-ecological cost (TEC), connecting energy and environmental issues. Finally, the Land Use (LU) is considered as a social criterion. As aggregation function, the Weighted Sum Method (WSM) is used. The sensitivity analysis of the criteria weights was performed with the use of a novel method involving Monte Carlo simulation and a method of data reconciliation.

1 Introduction

One of the main consequences to the accession to the European Union (EU) by member countries is the transformation process in all sectors of economy, as well as in the energy one.

Concerning the latter, one of the essential points is the reduction of emissions of greenhouse gasses. This aspect leads to the needs of developing renewable technologies and reducing the amount of existing non-renewable energy sources. Renewable Energy Sources (RES) are those sources of energy that are naturally regenerated over a short time scale and they are, for this reason, considered inexhaustible. RES include energy from wind, water, sun, geothermal heat, wave, tidal and biofuels.

Deployment of energy from RES in the EU member states is regulated by documents and normative acts of the EU, mainly [1]. Specifically, as far as Poland concerns, such regulations are contained in the legislation through the Environmental protection Low Act [2] and the Energy Law Act [3]. In the framework of environmental obligations, the EU has set quantitative targets for 2020 time horizon (3x20% energy package). Such targets contain: a reduction in greenhouse gas emissions by 20% compared to 1990; a reduction of energy consumption by 20% compared to projections for 2020; an increasing of the share of renewable energy to 20% of the total energy consumption with, in particular, an increasing of the use of renewable energy sources in transport to 10%.

Recently, a report by the Intergovernmental Panel for Climate Change pointed out the risk of an increasing of global temperature up to 3-5 °C, if the CO_2 concentration will keep increase, which would lead to dramatic consequences for the earth [4]. Energy production by means of RES may represents a possible solution, which may lead to a reduction of the environmental impact of the energy production sector Polish Energy Policy establishes an increase of the share of energy from RES up to 7% of the total energy production within the 2030 [5].

The Polish power system of the southern part of the country, which is particularly developed, is a centralized system, characterized by low substantial losses in electric energy transfer (such losses are proportional to the transfer distance). In contrast, due to the low number of large power generation sources in the northern Poland, the power system of such area is non-centralized. Because of this, the northern Poland (especially the north-east region of Poland) is the most suitable place for investment in distributed generation [6]. Distributed generation, due to the nature of production

[a] Corresponding author: alicja.stoltmann@pg.gda.pl

DOI: 10.1051/e3sconf/20161000056

may be a good solution to ensure energy supply to objects who have no access to the grid [7]. Currently, in EU, one of the most developed system from RES is the photovoltaic one (PV), as well as the thermal and concentrated solar power technologies. It is estimated that, in UE 178 GW_p (W_p –power measured in Standard Test Conditions) in the EU energy comes from PV installations. Among the members of UE Germany is the country which is characterized by the highest PV energy production rate in (474.1 W_p per capita). It follows Italy with 303.5 W_p per capita. In Poland, the production rate from PV technology is 0.6 W_p per capita (with a total production in PV installations of 39.2 MW_p) which data demonstrates that the photovoltaic market in this country is developing [8]. The Institute of Renewable Energy predicts that in the next two years the price of PV installations will drop significantly, causing an increasing interest in this way of energy production. Additionally, the latest legal regulations, introducing the concept of prosumer (defined as a person who consumes and produces a media) [8] assisted by subsidy programs [9], will contribute to the development of small PV systems that can be used by individual users.

The electricity demand can be cover by the production of energy both in a residential installation as well as in photovoltaic power plants. Such solutions have a common drawback, which is the cost of installation per 1 kW of installed capacity, being this the reason why PV installations still require to be subsided. On the other hand, advantages of PV installations concern the reduction of transmission losses and the fact that the technology is environmentally friendly.

The Polish legislation, classify energy production from PV installations as a distributed energy sources, that results in low-power sources connected to the distribution network. Distributed energy sources include sources with installed capacity not exceeding 50 MW. The PV technology is characterized by a direct processing of the primary energy into electrical energy. PV installations can be of '*on-grid*' as well as 'off grid' type, depending whether the installation is connected to the grid [10].

In case of '*on-grid*' system, voltage parameters do not depend on the source state. This because a PV installation is affected by voltage drops due to the impedance of the connection line between the source and the grid. When working off-grid, PV installation influences the parameters of the network [11].

Moreover, PV installations do not work continuously. Therefore, it is necessary to store the exceeding energy in order to provide it when necessary or to deliver it to, higher power hybrid systems that are those systems consisting of renewable and non-renewable energy sources. Few example of such systems are the wind and PV technology [12], and the Diesel engine [13]. In the recent literature, much attention is given to the design and the modeling of PV installations [12], [14], [15]. Selection of PV installations and its location is a multi-criteria decision problem. A great amount of works focusing on PV construction projects and in decision making processes are available in literature. Among these, the Multi-Criteria Decision Making methods (MCDM) is one of the most interesting and used. In [16], [17] the Preference Ranking Organization METHod for Enrichment Evaluations (PROMETHEE) model is used to choose the most suitable energy policy among many possibilities which include the employment of small and large size PV installations. In order to choose the most sustainable electricity production technology, the multi-criteria Technique for Order Preference by Similarity to Ideal Solutions (TOPSIS) was presented in [18]. Another popular multi-criteria method is the Analytic Hierarchy Process (AHP), which was applied in [19] to help a Spanish company in the decision making process concerning the investment for a particular solar power plant project. In [20] the Weighted Sum Method (WSM) for the quantification and evaluation of the potentials of available PV was presented. The use of fuzzy (linguistic) quantifiers for specified parameters in multi-criteria evaluation procedure, which is based on GIS-based analysis, was presented in [21] to assess the land suitability for large PV farms. The combination of GIS and MCDM methods for side selection studies are widely used, for example in [22], to determine suitable site selection for solar farms by using GIS and AHP in the study area, and in [23] using GIS and Analytic Hierarchy Process (AHP) method to weight the criteria in order to evaluate potential sites to locate a solar plant.

It has been observed that, in polish solar exposure conditions, both on-grid and off-grid PV individual installations are not economically profitable. In contrast, it is estimated that PV farms may be suitable to cover the electricity demand of whole towns.

The aim of the present study is to analyze and compare the different options of employing PV technology to electricity supply in the rural areas of the northern Poland. The assessment is made using a multi-criteria decision tool in order to compare the specified variants of the electricity generation including different types of PV farm construction (with static and movable structure) and different configuration of individual installations (off-grid and on-grid). The criteria used in the analysis combine economic, environmental and social issues. The first of the considered criterion is the Net Present Value (NPV) which determines the economic viability of the project. The second criterion, thermo-ecological cost (TEC), connecting energy and environmental issues. Finally, the Land Use (LU) is considered as a social criterion. As aggregation function, the Weighted Sum Method (WSM) is used. The sensitivity analysis of the criteria weights is performed with the use of a novel method involving Monte Carlo simulation and a method of data reconciliation.

2 Material and methods

Multi-criteria decision analysis (MCDA), which engages economic, environmental and social issues through participatory and analytical tools, is widely adopted for energy technologies selection. Moreover, multi-objective methods provide decision makers with an opportunity to negotiate and explore the different

DOI: 10.1051/e3sconf/20161000056

decision options. Complex interactions among multiple objectives (goals) have insisted to integrate multi-criteria assessment techniques in the framing of appropriate energy management directions.
MCDA is also a form of integrated sustainability evaluation and can be defined by a General Index of Sustainability [24]. In the present study, to perform such evaluation, the generic multi-criteria approach base on the Weighted Sum Method (WSM) is proposed.
Three different criteria in terms of economic, environmental and social impacts are also defined.
For the economic assessment, the Net Present Value (NPV, also known as Net Present Worth, NPW [25]) which is defined as the sum of the present values of incoming and outgoing cash flows over a period of time, is proposed. In the profitability analysis of investments such criterion represents a standard method to appraise long-term energy projects. Used for capital budgeting, it measures the excess or shortfall of cash flows, in present value terms, once financing charges are met. Also, it is often used to assess the early stage of decision making process [25], [26]. The NPV indicator is given by the following formula:

$$NPV(i,N)=\sum_{t=0}^{N}\frac{R_t}{(1+t)^t} \quad (1)$$

where:
t is the time of cash flow, i is the discount rate and R_t represents the net cash flow.

A positive value for the NPV indicates that the projected earnings generated by a project or investment overcome the anticipated costs. Thus, if the NPV is positive, it means that the investment is profitable and the project may be accepted. On the contrary, a negative value for the NPV indicates a net loss for the investment and the project should be rejected.
To consider environmental implication of electricity produced in a PV systems, the impact assessment is carried out using thermo-ecological cost analysis in whole life-cycle (TEC-LC). According to J. Szargut [27], [28] the TEC-LC indicator is defined as the cumulative consumption of non-renewable exergy connected with the manufacturing of a particular product. The TEC-LC includes, in addition, the consumption resulting from the necessity of compensation of environmental losses caused by rejection of harmful substances in the natural environment. The detailed description of the TEC-LC method, including the set of equations, can be found in [29], [30] and for this reason is not reported here. In previous works, TEC referring to RES was implemented, for example, to optimize solar collector used for production of hot water [31], as well as for evaluation of biofuel conversion processes [32] or electricity produced in micro- wind turbines [33]. The total thermo-ecological cost can be presented as follows:

$$TEC\ LC_{total}=TEC\cdot(DR+IF+E) \quad (2)$$

Where:
DR – direct resources
IF – instant flows
E – emissions

The social criterion, the Land Use (LU) is defined as an external urban area required for the PV plant construction.
The selected functional unit for the comparison was 1 kWh of produced electricity, assuming a service life time equal to 25 years.
As mentioned before, the multi-criteria approach applied in this work employs the Weighted Sum Method as an aggregated objective function.
Since, the considered criteria are characterized by different units, the normalization procedure is required. Normalized value of the j-th criterion is calculated using the following equation:

$$\overline{C}_j=\frac{\Delta C_j}{C_{j,\max}-C_{j,\min}} \quad (3)$$

$$\text{where}\begin{cases}\Delta C_j=C_j-C_{j,\min} & \text{for benefit criteria}\\ \Delta C_j=C_{j,\max}-C_j & \text{for cost criteria}\end{cases}$$

After criteria normalization, the aggregated objective function is defined as follows:

$$\overline{C}=\sum_{i=1}^{n}W^T\overline{C}_j \quad (4)$$

where:

$$\mathbf{W}=\left\{w\in R\ :w\geq 0\ \sum_{j=1}^{M}w_j=1\right\} \quad (5)$$

w is a vector of weights of the decision criteria.

Finally, a ranking of the analyzed solutions is determined. An optimal solution is a variant with the maximal value of the aggregated objective function (eq.4).
Since their subjective features, the criteria weights are the uncertain parameters in the decision model. Thus, the uncertainty analysis of the final results has to be performed. For this purpose, the novel method of minimizing the uncertainty of criterion weight functions using Monte Carlo simulation and method of data reconciliation is applied. Fully description of this method is presented in [34].

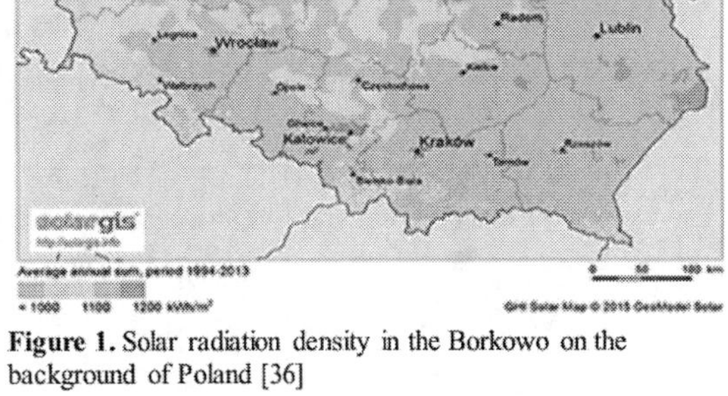

Figure 1. Solar radiation density in the Borkowo on the background of Poland [36]

 DOI: 10.1051/e3sconf/20161000056

Table 1. Monthly sunshine hours

Month	I	II	III	IV	V	VI	VII	VIII	IX	X	XI	XII	Σ
Sunshine hours [h]	239,6	264,8	360,4	417,2	492,5	506,34	503,52	445,88	363,9	305,7	238,4	220,9	4359

3 Results

The comparative analysis of the alternative solutions of the electricity generation from PV systems is performed for the case study Borkowo. Borkowo is a town located close to the city of Gdansk, in the Pomorskie Voivodship. This particular region is characterized by an average insolation of 1000 kWh/m^2 per month. About the 80% of the total annual insolation occurs during six months of the year, from April till September. Moreover, during the summer months, the sunshine hours are about 16 h per day while in winter – 8 h per day [35].

Table 2. Case study assumptions

	Area, m^2	Electricity demand, MWh/a
Single-family buildings	5408	1372
Multi-family buildings	978	360
Individual public utility buildings	946	1076
School and kindergarten	2630	329

As it is shown in Table 1, the total hours of sunshine (length of day) are 4359 h per year, with about 500 h per month during five spring-summer months. Therefore, this area is worth of attention in case of building PV installation. Furthermore, in Pomorskie Voivodship, in the Ustronie Morskie community, there is currently a 1MWp PV farm located on former waste dump. The plant supplies electricity to community buildings and allows to save 80-90% of the electricity costs of the communityThe electricity demand in the selected case study concern the consumption in the single and multi-family buildings, two individual public utility buildings, a trade and a service area, a school and a kindergarten. In the present analysis, it is assumed that the energy demand for individual public utility buildings is entirely covered from the power grid. Also, the energy produced from PV installations is assumed to not cover the domestic hot water heating. The estimated electricity demand for the present case is presented in the Table 2.

In order to cover the electricity demand, four different solutions for the electricity supplying by a PV installations are proposed. In the first variant (*Variant I*) the electricity is delivered from a PV farm mounted on a fixed structure. In the second scenario (*Variant II*), the construction of the solar farm is located on a mobile servo system, which allows to obtain a higher amount of electricity production than the fixed structure. In the next alternative solution (*Variant III*), it is assumed that electricity is delivered from the individual off-grid PV installations. In this case, the recipient uses energy derived from a PV installation at the same time of the production or, in case, it uses energy that has been previously accumulated in batteries during a lack of demand. In absence of available energy, this needs to be acquired from the power grid. The last option *(Variant IV)* considers that the electricity is produced by individuals on-grid PV installations. This allows to sell electricity to the power grid when the PV installation produces energy, whenever the demand is fulfilled. The characteristic of the analyzed solutions is presented in the Table 3. The results of energy analysis are shown in the Table 4.

Table 4. Energy generation in the analyzed PV options

	Type	PV output, MWh/a	Deficit, MWh/a
Variant I	PV farm static	15 708	0
Variant II	PV farm movable	21 069	0
Variant III	Individual off-grid	5 906	123
Variant IV	Individual on-grid	5 906	213

Table 3. Characteristics of the analyzed PV alternatives

	Description	PV area, m^2	Type of PV	Number of inverters	Type of inverter	Number of batteries	Type of battery
Variant I	PV farm static	11 048	Mono-Si ZXP6-60-240	369	Sunny Boy 3300	-	-
Variant II	PV farm movable	11 048	Mono-Si ZXP6-60-240	369	Sunny Boy 3300	-	-
Variant III	Individual off-grid	11 241	Mono-Si ZXP6-60-240	1153	Sunny Boy 1200	1776	Victron Energy GEL 12V/220Ah
Variant IV	Individual on-grid	11 241	Mono-Si ZXP6-60-240	1068	Sunny Boy 1200	-	-

 DOI: 10.1051/e3sconf/20161000056

3.1 Economic analysis

The NPV is calculated on the basis of the current price data. In order to predict the electricity price during the next 25 years, an average evolution rate of the electricity price is calculated based on the past 25 years (2%). The base value of electricity price for the first year of the analysis is assumed to be equal to 445 zł/MWh for the Variants I and II [37] and 750 zł/MWh for the Variant IV [38]. The total investment costs are shown in the Table 5. The contribution of particular system elements in the total investment cost is presented in Fig. 2.

Table 5. Total investment cost

	Type	Total investment cost, zł
Variant I	PV farm static	7 397 290
Variant II	PV farm movable	8 642 096
Variant III	Individual off-grid	11 697 352
Variant IV	Individual on-grid	9 670 265

Source: primary data

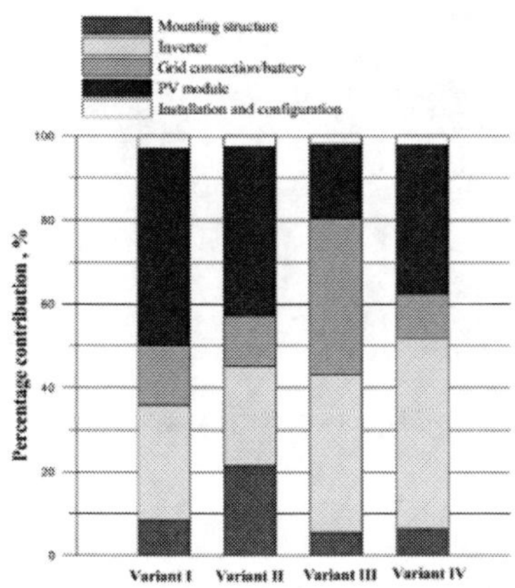

Figure 2. Contribution of the system components in the total investments cost.

In case of scenario I and II, the highest contribution to the investment cost is related to PV modules. However, for the scenario III, the grid connection and battery costs are the most significant, while for scenario IV, the inverter cost is the highest. The costs of installation and configuration give the lowest contribution for all of the analyzed cases.

The NPV in terms of electricity produced within 25 years is depicted in Fig. 3. The NPV is the benefit criterion. Thus, the maximal value refers to the optimal solution.

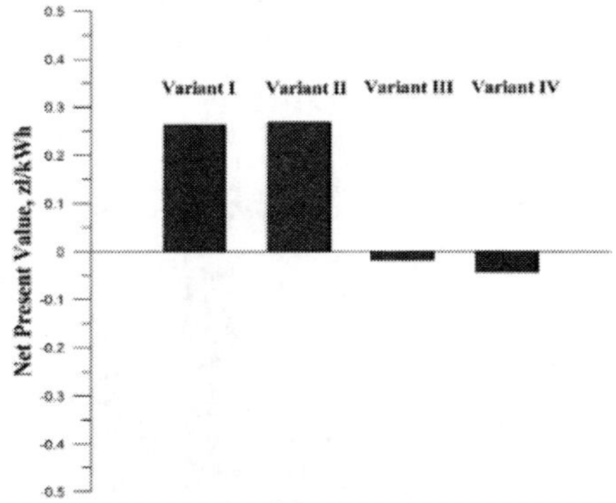

Figure 3. Net Present Value in terms of electricity produced within 25 years

Comparing the analyzed alternative solutions, the choice of variant II with movable PV farm seems to be the most economically justified. In this case, the highest positive effect is due to the highest rate of electricity produced and sold to the grid. The worst scenario from the economic point of view is the employing of the individuals *'on-grid'* PV.

3.2 Environmental analysis

The thermo-ecological cost include the construction and operational phases and is calculated using eq.2. The data on processes and materials are collected on the basis of technical specification of the devices as well as other system elements. The data on the cumulative exergy demand due to material consumption are retrieved from the Ecoinvent database and [29], while the exergetic costs of compensation for harmful substances are assumed on the basis of [30] (Table.6). The total thermo-ecological cost of the analyzed systems is presented in the Table. 7

Table 6. Thermo-ecological cost of harmful substances

Substance	External cost, EUR/kg	Cost of compensation, kJ/kg
SO_2	8.24	37.8
NOx	4.07	18.7
PM	3.99	18.3

Table 7. The total thermo-ecological cost of the analyzed systems

	TEC, construction phase, MJ_{ex}	TEC, operational phase, MJ_{ex}	Total TEC, MJ_{ex}
Variant I	78 386 152	0	78 386 152
Variant II	87 776 761	0	87 776 761
Variant III	52 517 935	1 599	52 519 534
Variant IV	41 367 957	2 766	41 370 724

The results are expressed in terms of unit of electricity produced by the PV panels within the whole life cycle (25 years) and are shown in Fig. 4. The TEC-LC is the cost criterion. Thus, the minimal value refers to the optimal solution. In Fig. 5, the influence of the different system components on the total value of TEC is presented.

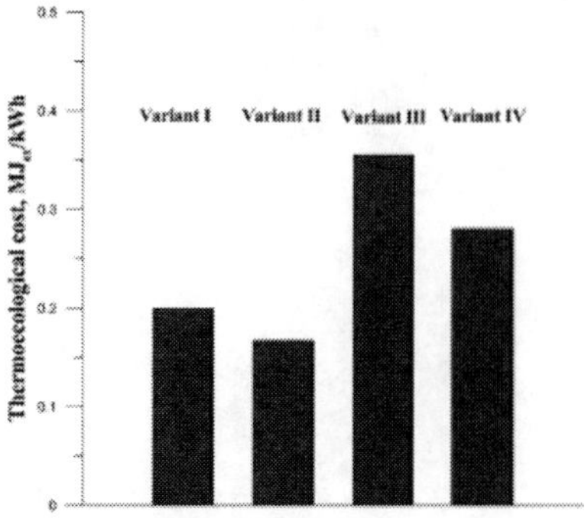

Figure 4. Thermo-ecological cost in terms of electricity produced within 25 years

Comparing the TEC-LC results, the same variant (variant II – PV movable farm) as in economic analysis, is indicated as the most preferable.

Analyzing the component contributions, it can be also observed that the most influencing component for the variants I and II is the mounting structure. This is due to high concrete and steel consumption, which are characterized by the high exergy consumption. On the contrary, for individual systems, the PV panels production contribute the most in total energy consumption.

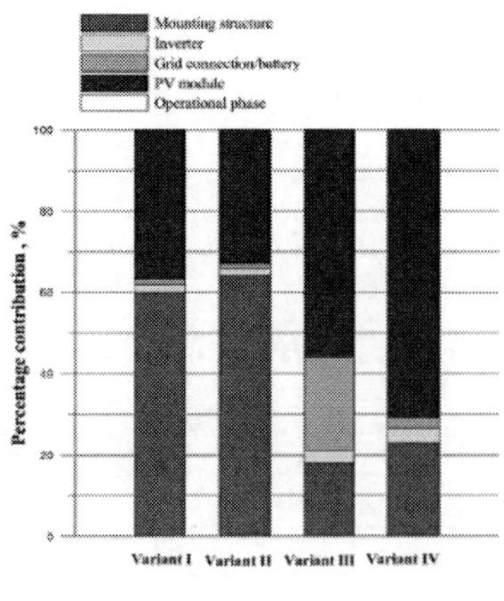

Figure 5. Contribution of the system components in the total thermo-ecological cost

3.3 Social analysis

The social criterion is determined by any additional external urban area required for the PV plant construction. Such factor is often the major constraint in the energy technology choice.

For the PV farm plants (variant I and II), the Land Use is estimated assuming the area of a single PV panel and the distance between the panels. Thus, for static installation (variant I), the total Land Use is equal to 2.97 ha, while for dynamic structure (variant II), which in general requires more space, is equal to 3.78 ha.

In the options employing PV panels for individual use which are installed on the rooftop of the buildings, any external urban area is not required. Thus, for those cases, the LU equal to 0. Similarly to TEC-LC, the Land Use is a cost criterion.

In Fig. 6, the results, in terms of electricity produced, are presented. Comparing the ranking for this criterion with the previous ones, an opposite result is obtained.

3.4 Multi-criteria analysis

Since the economic, environmental and social analysis give a different conclusion, the multi-criteria approach is in this case justified. Firstly, the ranking of the alternative solutions is evaluated assuming deterministic weights that are equal to each other. The normalized values of the criteria as well as the value of the aggregated objective function (eq.3 and 4) are calculated using the data reported in previous sections. The results are shown in Fig. 7

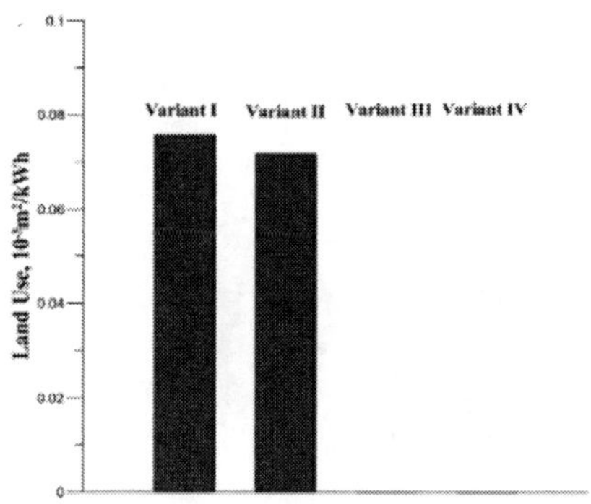

Figure 6. Land Use in terms of electricity produced within 25 years

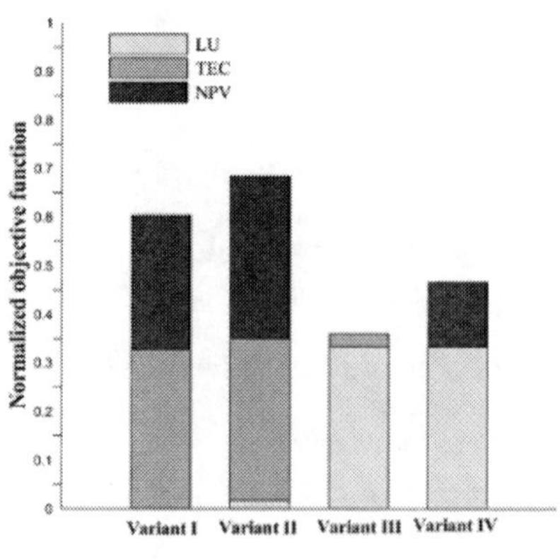

Figure 7. Final ranking of the analyzed scenarios with the determined weights

Considering the deterministic assumption regarding the criteria weights, the variant II is characterized by the maximal value of the aggregated objective function. Nevertheless, the criteria weights are subjective parameter in the model. Variability of this parameters

DOI: 10.1051/e3sconf/20161000056

has an influence on the final results. Thus, in order to evaluate the uncertainty of the resulting value of the aggregated objective function, a Monte Carlo simulation with a data reconciliation method is conducted. The random values of criteria weights are selected in 10000 runs. After that, the forecast distribution, for the final result, is obtained. The Monte Carlo simulation is performed assuming the highest weight variability described by a uniform distribution between 0 and 1 for all of the considered parameters. Then, in order to fulfill the constrain condition (eq. 5) the weight vector is recalculated in accordance to rules of the reconciliation calculus.

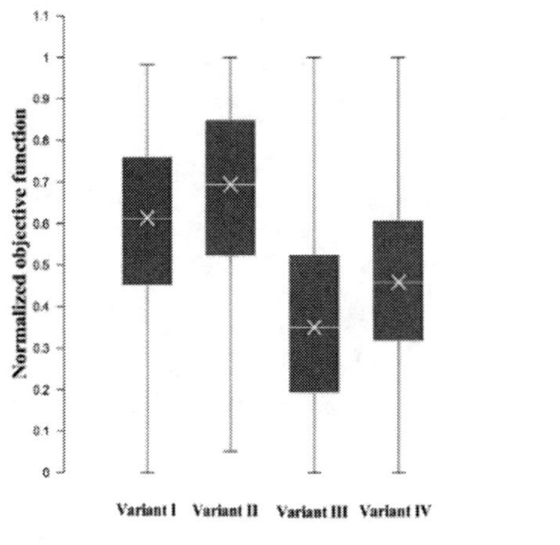

Figure 8. Final ranking of the analyzed scenarios with the uncertain weights

In Fig. 8, the obtained variabilities of the final results are presented. Such variabilities are presented for selected parameters of the probability distribution (min, max, 1st, 2nd and 3rd quartiles). Assuming the marginal weights composition e.g. (0, 0, 1), the results correspond to those presented in Fig.3, 5 and 6.

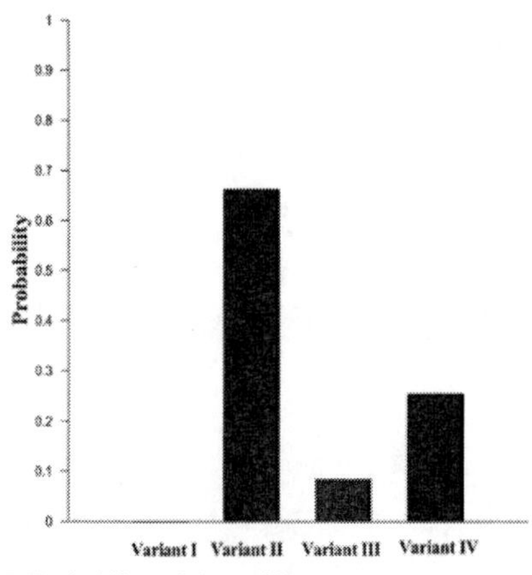

Figure 9. Probability of the ranking score

The results are characterized by high variability. The coefficients of variation for the analyzed scenarios are: 26%, 32%, 63% and 44% for the variant I, II, III, and IV, respectively. In order to determine the most suitable option among all of the analyzed ones, the probability of obtaining the maximal value of the aggregated objective function is evaluated. As it is shown in Fig. 9, variant II is indicated as the preferable solution with the most frequency. Moreover, independently on the criteria weights, the variant I never achieves the maximal value of the objective function. This is because such solution is not preferable, even if the marginal weight vector is assumed.

4 Conclusions

The purpose of this study was to analyze the economic, environmental and social impacts due to different alternative solutions of PV use. Four decision variants, including the different construction of solar farms (static and movable structure) and different types of configuration of individual installation (off and on-grid) were compared between themselves. The assessment was carried out using the multi-criteria analysis tool. The first of the considered criterion was the Net Present Value (NPV) which determines the economic viability of the project. The second criterion, thermo-ecological cost (TEC), connecting energy and environmental issues, was evaluated. Finally, the Land Use (LU) was considered as a social criterion. As aggregated objective function, the Weighted Sum Method (WSM) was used. The sensitivity analysis of the criteria weights was also performed using a novel method involving Monte Carlo simulation and a method of data reconciliation.
Results indicate that, for economic and environmental criteria, the solutions considering the PV solar farms are more preferable than individual PV installations. In case of Land Use, the ranking indicates that the individual PV are more sustainable.
In accordance with multi-criteria ranking, and including high uncertainty of criteria weights, the solution based on the movable PV farm results to be the most effective.

References

1. 'Directive 2009/28/EC of the European Parliament and of the council of 23 April 2009 on the promotion of the use of energy from renewable sources and amending and subsequently repealing Directives 2001/77/EC and 2003/30/EC', Official Journal of the European Union.
2. 'Environmental protection Low Act (in Polish)', Dz.U. 2001 Nr 62 poz. 627.
3. 'Energy Law Act (in Polish)', Dz.U. 1997 Nr 54 poz. 348, vol. 2006, no. 54, 2014.
4. A. Tartaglia, G. Fracastoro, C. Mele, and M. L. Ruggiero, 'The role of renewables in the energy crisis', E3S Web of Conferences, **2**, (2014)

5. National Energy Policy 2030 (in Polish)', Załącznik do uchwały nr 202/2009 Rady Ministrów z dnia 10 listopada 2009 r., **85**, 9 (2009)
6. A. Chmielewski, R. Gumiński, J. Mączak, S. Radkowski, and P. Szulim, Ren. and Sust. En. Rev., **60** (2016)
7. J. C. Rojas-Zerpa and J. M. Yusta, Ren. and Sust En. Rev. **52**, (2015)
8. 'Instytut Energetyki Odnawialnej'. [Online]. Available: http://ieo.pl/pl/projekty/mikrosieci-i-instalacje-fotowoltaiczne. [Accessed: 02-Apr-2016].
9. 'Renewable Energy Sources Act (in Polish)', Dz. U. z 2015 r. poz. 478, 2365.
10. S. Mandelli, J. Barbieri, R. Mereu, and E. Colombo, Ren. and Sust. En. Rev. **58** (2016)
11. C. P. Vineetha and C. A. Babu, 'Economic Analysis of Off Grid and On Grid Hybrid Power System', IEEE International Conference on Circuit, Power and Computing Technologies, (2014)
12. G. Merei, C. Berger, and D. U. Sauer, 'Optimization of an off-grid hybrid PV-Wind-Diesel system with different battery technologies using genetic algorithm', Solar En. **97** (2013)
13. Y. Li, D. Zhao, and Y. Llu, 'Optimal Design and Sensitive Analysis of Distributed Generation System with Renewable Energy Sources', IEEE China International Conference on Electricity Distribution, (2014)
14. R. Rawat, S. C. Kaushik, and R. Lamba, Ren. and Sust. En. Rev. **57** (2016)
15. M. Obi and R. Bass, Ren. and Sust. En. Rev. **58** (2016)
16. T. Tsoutsos, M. Drandaki, N. Frantzeskaki, E. Iosifidis, and I. Kiosses, En. Pol. **37**, 5 (2009)
17. M. Troldborg, S. Heslop, and R. L. Hough, Ren. and Sust. En. Rev. **39** (2014)
18. D. Streimikiene, T. Balezentis, I. Krisciukaitien, and A. Balezentis, Ren. and Sust. En. Rev. **16**, 5 (2012)
19. P. Aragonés-Beltrán, F. Chaparro-Gonzalez, J.-P. Pastor-Ferrando, and A. Pla-Rubio, Energy **66** (2014)
20. K. van Alphen, W. G. J. H. M. van Sark, and M. P. Hekkert, Ren. and Sust. En. Rev. **11**, 8 (2007)
21. Y. Charabi and A. Gastli, Renewable Energy **36**, 9 (2011)
22. M. Uyan, Ren. and Sust. En. Rev. **28** (2013)
23. J. M. Sánchez-Lozano, J. Teruel-Solano, P. L. Soto-Elvira, and M. Socorro García-Cascales Ren. and Sust. En. Rev. **24** (2013)
24. N. H. Afgan and M. G. Carvalho, Energy **27**, 9 (2002)
25. J.-J. Wang, Y.-Y. Jing, C.-F. Zhang, and J.-H. Zhao, Ren. and Sust. En. Rev. **13**, 9 (2009)
26. J. C. Mourmouris and C. Potolias, En. Pol. **52** (2013)
27. J. Szargut, Exergy Method: Technical and Ecological Applications. (WIT Press, 2005)
28. J. Szargut, A. Ziębik, and W. Stanek, Energy Convers Manag, **43** (2002)
29. W. Stanek, Metodyka oceny skutków ekologicznych w procesach cieplnych za pomocą analizy egzergetycznej. (Wydawnictwo Politechniki Śląskiej, 2009)
30. Czarnowska, 'Thermoecological cost of products with emphasis on external environmental costs. Doctoral thesis.', Silesian University of Technology, (2014)
31. Szargut and W. Stanek, Energy, **32** (2007)
32. Piekarczyk, L. Czarnowska, K. Ptasiński, and W. Stanek, Energy, **62** (2013)
33. L. Lombardi, E. Carnvale, B. Mendecka, G. Santoni, and W. Stanek, 'Environmental impacts of electricity production of micro wind turbines with vertical axis', in Proceedings of 29th international conference on Efficiency, Cost, Optimisation, Simulation and Environmental Impact of Energy Systems (ECOS) (2016)
34. B. Mendecka and J. Kozioł, Arch. of Thermodyn. **36**, 1 (2015)
35. B. Igliński, A. Iglińska, M. Cichosz, W. Kujawski, and R. Buczkowski, Ren. and Sust. En. Rev., **58** (2016)
36. 'SolarGIS © 2016 GeoModel Solar, Solar Radiation Map: Global Horizontal Irradiation Map of Poland', 2016. [Online]. Available: http://solargis.info/doc/free-solar-radiation-maps-GHI#P.
37. 'Order of the Polish Minister of Economy of 13 November 2015 establishing the reference price of electric energy from renewable energy sources in 2016 (in Polish)' Dz. U. z 2015 r. poz. 2063
38. 'Renewable Energy Sources Act (In Polish)' Dz. U. z 2015 r. poz. 478, 2365.

Production and import of energy - Poland against EU countries

Sylwia Lorenc[1] and Marcin Migza[1,a]

[1] *AGH University of Science and Technology, Department of Economy and Management in Industry, al. Mickiewicza 30, 30-059 Krakow, Poland*

Abstract. The purpose of this paper is to present the Polish energy market compared to the countries of the 'old' (UE-15) and the 'new' (UE-28) European Union. Today, the EU consists of 28 states of different size, population, economy, and energy structure. Without presenting the political-economic context, a comparison between energy market in Poland to those in the other 28 EU countries would not produce a reliable picture. The existence of these discrepancies was a motivating factor in the authors' efforts to account for the Polish energy market as compared to the 'old' and 'new' EU states. The data were presented and analysed according to the different kinds of energy sources.

1 Introduction

Changes occurring in the European economy require application of pioneering solutions in the energy policy scope. This necessity caused by unfavorable effects occurring last years in the world economy including a growing environment pollution, energy power failures and an abrupt decrease of the power industry raw materials prices.

Poland as a member country of the European Union is actively participating in creating the common energy policy, but due to unfavorable tendencies occurring in the sector and the necessity of fulfilling the ecological obligations is facing a significant challenge. From one side there is a growing demand for energy and strong dependence on external supplies of natural gas and crude oil, while from the second side there are obligations concerning the environment protection within the energy packet „3x20" [1]. Within the strategy „Europe 2020" assumptions of the European Commission oblige Poland to reduce greenhouse gases emission by 20% to the year 2020 (as compared with the year 1990), to increase the energy efficiency by 20% and to increase the fraction of renewable energy sources by 20% in the general energy production. The climatic policy of the EU caused that the Polish coal-based economy increased fractions of alternative and low-emission energy sources [1].

Pursuing a uniform energy policy, at the present configuration of the Union, without in-depth market analysis in individual countries, seems impossible. High diversification of the European energy markets causes that the detailed specification of the energy situation in each member country is necessary. The energy sector structure in the member countries is determined by the EU directives, which demand the energy markets liberalization. The aim of such policy is striving to the energy safety improvement, decrease of prices and facilitation of international cooperation by means of forming the competitive electric energy market. These elements should finally contribute to the reduction of the environment pollution. On account of this, alternative and low-emission energy sources started to play the key role in the energy policy of the EU.

One of the most important threats influencing the energy policy of the Union is a high dependence of the European Union on external supplies of energy carriers. In the year 2014, 53.4% of gross inland energy consumption in the EU was imported [2]. This value stabilized in last years in the whole EU and in Poland even slightly decreased, which is shown in Table 1. However, it should be remembered, that the growing rate of the Polish import of energy carriers during last 18 years was much higher, which is presented in Figure 1.

Table 1. Energy dependence of EU in 2010-2014 [2].

	2010	2011	2012	2013	2014
EU-28	52,6%	54,0%	53,3%	53,1%	53,4%
Poland	31,3%	33,4%	30,6%	25,6%	28,6%

In order to decrease the EU energy dependence, it is necessary to conduct the energy policy not only directed towards limitations of consumption, but first of all promoting the energy production in the EU countries.

The current social-economic situation inside the Community is not uniform. The present day Union consists of 28 countries of various sizes, populations, power industry structures and economic conditions. Significant differences in economic potentials, between the 'old' and 'new' European Union countries, can be still observed, including the energy sector.

[a] Corresponding author: migza@agh.edu.pl

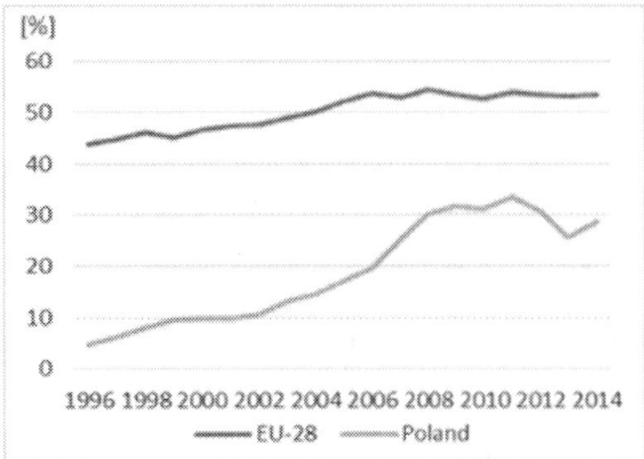

Figure 1. Energy dependence of EU in 1996-2014 [2].

2 Main dividing axle of the EU – accession date

The efficiently operating energy production sector is one of the key factors of increasing the economic potential of developing countries as well as of highly developed countries of a significant industrialization degree.

One of the most important measures of the economic growth is the Gross Domestic Product (GDP) per capita. Values of this index for individual EU countries calculated on the basis of Eurostat data are presented in Table 2.

Table 2. GDP per capita in EU 2010-2014 [3,4].

	2010	2011	2012	2013	2014
EU-28	25394	26096	26618	26694	27486
EU-15	29582	30315	30935	31051	31842
EU-N13	9612	10103	10231	10426	10729
Poland	9509	9988	10227	10373	10808

The values of the GDP per capita index indicates that countries of the so called 'new Union' (ten countries which entered the Union in the year 2004 : Cyprus, Czech Republic, Hungary, Latvia, Lithuania, Estonia, Malta, Poland, Slovenia, Slovakia, two in 2007: Bulgaria, Rumania and Croatia in 2013) still do not have the economic potential allowing to compete with economics of the 'old Union' (fifteen countries which created the Union or entered into it before the year 2000 : Austria, Belgium, Dania, Finland, France, Greece, Spain, Holland, Ireland, Luxemburg, Germany, Portugal, Sweden, Great Britain and Italy). The GDP per capita in Poland nearly ideally reflects the average value of this index for the EU-N13 countries. This dependence is presented in Figure 2.

The Pearson correlation coefficient between values of the index in Poland and the average for 13 'new' EU countries, calculated for years 2007-2014 equals more than 0.988, while the correlation level between Poland and the average of the whole Union equals only 0.837. This shows that countries of the 'new' Union are a good comparing background for Poland for further analyses in the energy production sector.

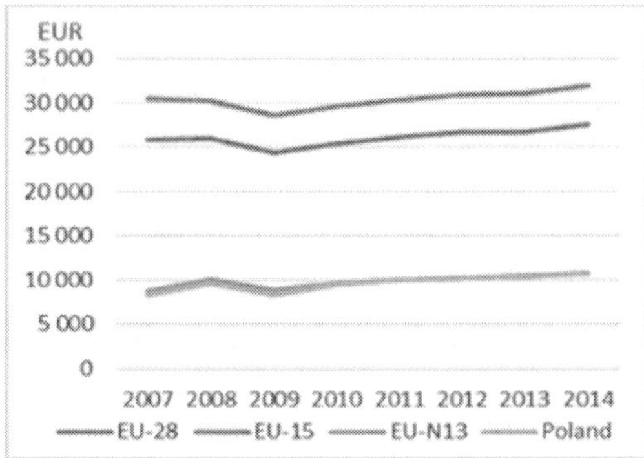

Figure 2. GDP per capita in EU, in 2007-2014 [3,4].

3 Energy production in the European Union

The main problem with which the EU is wrestling, is warranting the energy supply safety. One of the reasons of the high dependence on the import is a decrease of the energy production in the EU. The data are presented in Table 3.

Table 3. Primary production of energy (1 000 tonnes of oil equivalent) [5].

	2010	2011	2012	2013	2014
EU-28	835704	802122	795148	790398	771682
EU-15	665026	628534	619796	619362	603734
EU-N13	170678	173589	175351	171036	167948
Poland	66740	67681	70971	70571	66867

In the year 2014 the total production of the primary energy from all sources in the UE-28 equaled less than 772 million tonnes of the oil equivalent. This constitutes the continuation of the European decreasing tendency, which could be observed during last years. Among the 'new' thirteen countries this tendency is significantly weaker. Poland does not share these decreasing tendencies and in the energy production per capita is significantly above the average of EU-28, EU-15 and EU-N13 (Table 4 and Figure 3).

Table 4. Primary production of energy per capita in EU (tonnes of oil equivalent) [4,5].

	2010	2011	2012	2013	2014
EU-28	1,6588	1,5882	1,5758	1,5573	1,5196
EU-15	1,6703	1,5729	1,5518	1,5443	1,498
EU-N13	1,6153	1,6463	1,6665	1,6289	1,603
Poland	1,7543	1,7781	1,8646	1,8552	1,7591

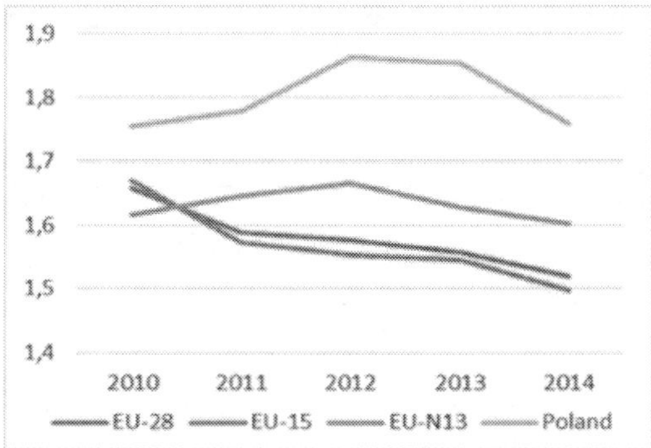

Figure 3. Primary production of energy per capita in EU (tonnes of oil equivalent) [4,5].

The fact that the Energy sector is also subjected to the division into the 'old' and 'new' Union is the best seen on the example of the energy production method. The participation of individual energy carriers in the general energy production in the EU in 2014, is shown in Table 5.

Table 5. Share of different energy sources in energy production in the EU [5].

	2014			
Energy source	**EU-28**	**EU-15**	**EU-N13**	**Poland**
Nuclear heat	29,3%	33,4%	14,7%	0,0%
Renewable	25,4%	27,2%	18,8%	12,0%
Natural gas	15,3%	16,9%	9,4%	5,6%
Crude oil + NGL	8,8%	10,1%	4,0%	1,4%
Solid fuels	19,4%	10,2%	52,2%	80,2%
Others	1,9%	2,2%	0,8%	0,8%
SUM	100,0%	100,0%	100,0%	100,0%

A high participation of energy obtained from solid fuels is characteristic for the 'new' Union countries. In this respect Poland with a participation above 80% stands out even at the EU-N13 background.

In order to characterize better the individual energy carriers and to show actual trends, the main groups of energy sources are characterized separately.

3.1 Nuclear heat

The primary energy produced in EU-28 in the year 2014 originated from various sources, out of which the highest fraction constituted the nuclear energy (29.3 % of the total production). Nuclear fuel was of a especially high meaning in the 'old' Union countries (33.4%). In years 2012-2014 an insignificant decrease of the atomic power industry in EU-28 countries could be observed. Poland is one of 14 member countries, which do not use the atomic energy. In consideration of a negative reception of the nuclear energy all attempts of building nuclear blocks were ended with a fiasco.

Table 6. Primary production of energy from nuclear (1 000 tonnes of oil equivalent) [5].

	2010	2011	2012	2013	2014
EU-28	236563	234007	227719	226285	226132
EU-15	213004	209739	203261	202211	201438
EU-N13	23558	24268	24458	24074	24694
Poland	0	0	0	0	0

Table 7. Share of nuclear energy in energy production in EU [5].

	2010	2011	2012	2013	2014
EU-28	28,3%	29,2%	28,6%	28,6%	29,3%
EU-15	32,0%	33,4%	32,8%	32,6%	33,4%
EU-N13	13,8%	14,0%	13,9%	14,1%	14,7%
Poland	0,0%	0,0%	0,0%	0,0%	0,0%

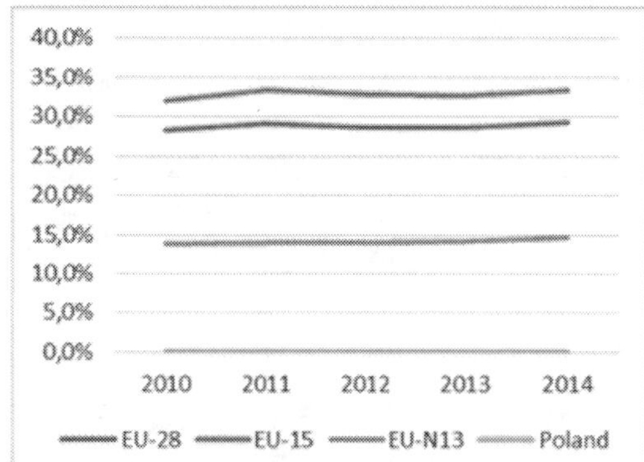

Figure 4. Share of nuclear energy in energy production in EU [5].

The main factor influencing weakening of the public interest in obtaining energy from nuclear power plants are potential effects of nuclear failures. Each member country can independently decide whether it wants to develop nuclear energy plants. The European Commission remains officially neutral with regard to the nuclear energy development. In October 2007 the European Parliament promoted the nuclear heat and power plants development as a necessary option for the European Union, emphasizing that each country has the right to decide on the nuclear energy introduction, but the effects of its decision will influence electric energy prices in the whole Union [6].

Table 8. Primary production of energy from renewable in EU (1 000 tonnes of oil equivalent) [5].

	2010	2011	2012	2013	2014
EU-28	167890	164267	180577	192838	195846
EU-15	140119	136885	150916	161682	164204
EU-N13	27771	27381	29660	31156	31642
Poland	6845,5	7439,4	8465,6	8511,5	8054,4

In order to realize the strategic purpose of the energy related EU policy, it means the diversification of energy sources, the development of the production technology of rational costs, especially nuclear energy with zero emission of gases and dusts harmful for the environment and with a low sensibility to increased uranium prices, is necessary [1].

From the point of view of the Polish natural resources, utilization of the nuclear energy is not necessary. Current fundamental oppositions of the social party cause that the vision of the nuclear power plants development in the nearest years is not much real.

3.2 Renewable

Due to the necessity of implementing the new, energy related policy an increased meaning of renewable energy sources has been observed (Table 8). One fourth of the total production of the primary energy in UE-28 originates from this source (25.4%) (Table 9). An increase of the primary energy from the renewable sources was

Table 9. Share of renewable energy in energy production in EU [5].

	2010	2011	2012	2013	2014
EU-28	20,1%	20,5%	22,7%	24,4%	25,4%
EU-15	21,1%	21,8%	24,3%	26,1%	27,2%
EU-N13	16,3%	15,8%	16,9%	18,2%	18,8%
Poland	10,3%	11,0%	11,9%	12,1%	12,0%

higher than in case of all other energy kinds, and stayed at the similar level for the largest part of the period: 2010 to 2014. Poland, at the background of the remaining EU countries, is still behind. Only 12% of the total energy production originates from alternative sources. A stable increase of the meaning of the renewable energy sources in individual member countries instills optimism. This dependence is presented in Figure 5.

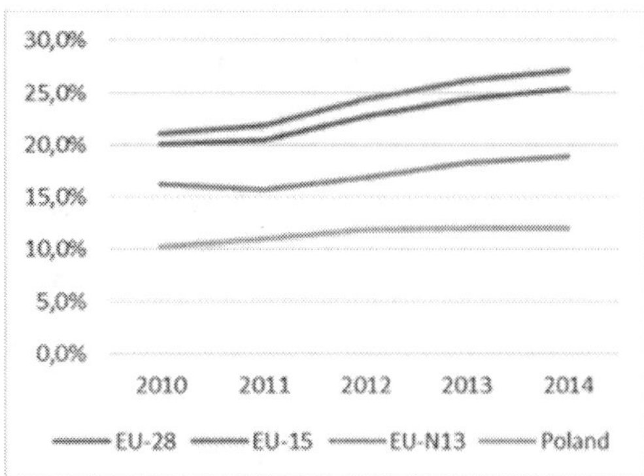

Figure 5. Share of renewable energy in energy production in EU [5].

The European Union is the leader in the range of the renewable sources technologies, into which the wind, solar, hydroelectric, oceanic, geothermal, biomass and bio-fuels energy can be numbered, constitute the alternative for fossil fuels [7]. The increase of utilizing these sources ensures the local energy supply safety, decreases the dependence on uncertain conditions in the fossil fuels markets and influences a reduction of contaminations emitted to the environment. Poland has an access to large and various resources of renewable energy sources. However, there are infrastructural limitations, which significantly hamper utilizing this potential [1].

Table 10. Primary production of energy from natural gas (1 000 tonnes of oil equivalent) [5].

	2010	2011	2012	2013	2014
EU-28	159676	141619	133054	131823	117986
EU-15	142560	124335	116410	115801	102155
EU-N13	17116	17284	16644	16022	15831
Poland	3693	3849,8	3906,8	3823,1	3726,3

3.3 Natural gas

The production level of the primary energy of natural gas was decreased by app. 8% among EU-28 and EU-15 countries, in the period: 2010-2014. The smallest fluctuations occurred in the 'new' Union countries, including Poland (Table 8 and Table 9). The natural gas fraction in the total energy production is maintained at the stable level oscillating round 5%. This dependence is presented in Table 10, Table 11 and Figure 6.

Table 11. Share of natural gas in energy production in EU [5].

	2010	2011	2012	2013	2014
EU-28	19,1%	17,7%	16,7%	16,7%	15,3%
EU-15	21,4%	19,8%	18,8%	18,7%	16,9%
EU-N13	10,0%	10,0%	9,5%	9,4%	9,4%
Poland	5,5%	5,7%	5,5%	5,4%	5,6%

The energy production of natural gas constitutes 1/6 of the total production in the EU-28 countries, 1/11 in the 'new' Union, while only 1/20 in Poland.

The energy production of natural gas is less harmful for the environment than other energy sources: the CO_2 emission from gas burning is up to 30% lower than in case of crude oil and up to 60% lower than in case of coal. Emissions of other chemical substances such as mercury, sulphur and nitrogen dioxide are also significantly reduced [8].

The power industry safety of the EU in case of natural gas is threatened by shrinking resources of this raw material in the world and by strong dependence on the import, which probably is the cause of the decrease of the natural gas consumption in Europe.

The Polish natural gas market has a high potential. The energy policy within this space is based on the diversification of gas supply sources, which positively influences the energy safety in Poland. The creation of technological and physical possibilities of the natural gas import from the western direction indicates favorable perspectives for this energy sector.

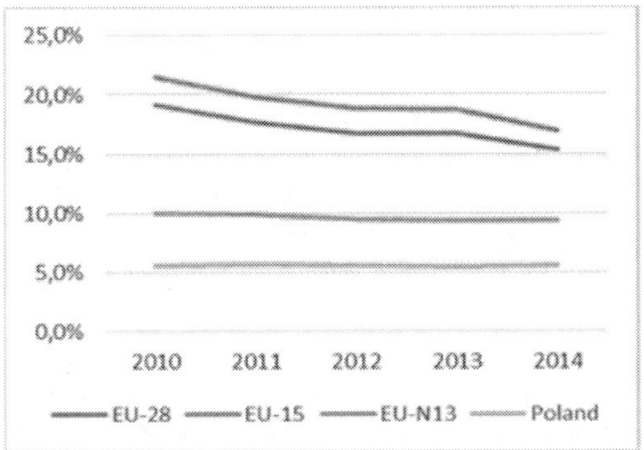

Figure 6. Share of natural gas in energy production in EU [5].

3.4 Crude oil - NGL

The energy production of crude oil and natural liquid gas decreases year by year (Figure 7); the highest declines occurred in EU-28 and EU-15 countries (Table 12). The gas and crude oil utilization in the 'new' Union countries and in Poland is at the stabilized level: 'new' Union (app. 4%), Poland (app. 1%) of the total primary energy production (Table 13).

Table 12. Primary production of energy from crude oil and NGL (1 000 tonnes of oil equivalent) [5].

	2010	2011	2012	2013	2014
EU-28	94162	81966	74007	69595	68004
EU-15	86867	74978	67422	62680	61231
EU-N13	7294,5	6988,4	6584,4	6915,5	6773,3
Poland	681	603,4	663,7	949,2	932,2

The dependence of the EU on the energy import, especially crude oil and - in the last period - also natural gas is the reason of anxieties related to the supplies safety. In a similar fashion as in case of natural gas the drop in utilisation of crude oil and natural liquid gas is probably caused by the realisation of the energy related, Union policy. A small fraction of these energy sources in the Polish energy balance causes that there is not a direct threat of losing the energy stability. However, on the other side, even small supplies cause threatening of the monopolistic price policy of the suppliers. In order to minimize the risk, similarly as in the natural gas case, it is necessary to look for various directions of supplies and building storage accommodations for liquid fuels to ensure the supply continuity in an eventual crisis situation.

Table 13. Share of crude oil and NGL in energy production in EU [5].

	2010	2011	2012	2013	2014
EU-28	11,3%	10,2%	9,3%	8,8%	8,8%
EU-15	13,1%	11,9%	10,9%	10,1%	10,1%
EU-N13	4,3%	4,0%	3,8%	4,0%	4,0%
Poland	1,0%	0,9%	0,9%	1,3%	1,4%

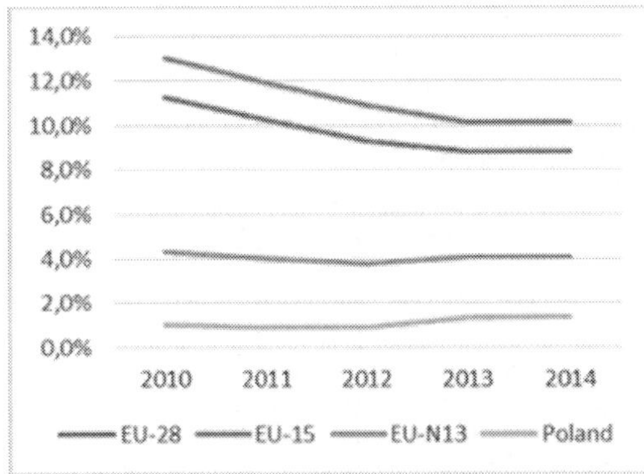

Figure 7. Share of crude oil and NGL in energy production in EU [5].

3.5 Solid fuels

According to the Eurostat data the fraction of solid fuels (hard coal, brown coal) decreases year by year (Table 14). In the year 2014 only 19,4% of energy consumed in EU-28 originated from solid fuels (Table 15). A declining trend can be seen also in the EU-15 and 'new' Union countries. Hard coal is first of all localized in Poland (the largest resources), Czech and to a lower degree in Spain, Hungary, Great Britain and Germany. The brown coal resources spread from Germany to Greece. Solid fuels constitute the largest fraction in the energy production in Poland, in the year 2014 they equaled above 80%. Poland, as the country having one of the largest hard coal resources, is able to be independent of the external supplies of solid fuels. Fractions of solid carriers in the total obtaining the primary energy in individual comparative groups is highly diversified (Figure 8). The EU recommendations oblige to the realization of the low-coal policy, especially to reduce the greenhouse gases and to increase the renewable energy sources utilization. During the last years the influence of these directives and the climatic package '3x20' on the participation of solid fuels in the total energy production, can be seen.

Fractions of solid carriers in the total obtaining the primary energy in individual comparative groups is highly diversified (Figure 8). The EU recommendations oblige to the realization of the low-coal policy, especially to reduce the greenhouse gases and to increase the renewable energy sources utilization. During the last years the influence of these directives and the climatic package '3x20' on the participation of solid fuels in the total energy production, can be seen.

Table 14. Primary production of energy from solid fuels (1 000 tonnes of oil equivalent) [5].

	2010	2011	2012	2013	2014
EU-28	164005	166608	166056	155826	149335
EU-15	70011	70024	69132	64156	61701
EU-N13	93994	96584	96924	91670	87634
Poland	55077	55298	57507	56835	53639

For Poland as the country, having at its disposal significant hard coal resources the low-coal economy is unacceptable. Economic, social and environmental advantages determined by the sustainable development are possible without the necessity of resigning from the only one perspective and competitive energy source. The Tables of natural solid fuels resources in Poland indicate that Poland is totally independent of external supplies [9].The main fuel raw material in Poland is hard coal. Despite the self-sufficiency, Poland imports significant amounts of this raw material. This results from lower coal prices in other markets and better qualitative parameters of imported coal. The solution of this problem seems to be building new, more efficient power units, implementing modern processing technologies and restructuring of mines, all these at costs decreasing and joining the mining with the power industry.

On account of the fact that solid fuels still constitute an important element of the energy balance of the EU, the European Commission calls for introducing ecological technologies of the coal burning. The application and development of these technologies becomes the real perspective. Necessary financial investments in their development can result in the future in the breakthrough of the conventional energy production.

Table 15. Share of solid fuels in energy production in EU [5].

	2010	2011	2012	2013	2014
EU-28	19,6%	20,8%	20,9%	19,7%	19,4%
EU-15	10,5%	11,1%	11,2%	10,4%	10,2%
EU-N13	55,1%	55,6%	55,3%	53,6%	52,2%
Poland	82,5%	81,7%	81,0%	80,5%	80,2%

4 Conclusions

A dynamic development of the European economy causes that the demand for the electric energy increases day by day. To satisfy the increasing demand for the electric current the diversification of the primary energy carriers and directions of their supplies is necessary (the so-called Energy Power Mix).

Decreasing production of the primary energy with using hard coal, brown coal, crude oil, natural gas and in the last period also the nuclear energy causes the situation in which meeting the EU demands is dependent in more than 50% on the primary energy import.

There are still considerable differences between countries of 'old' and 'new' EU regarding economic capabilities, which can also be observed in the energy sector.

Polish energy sector should be compared not only to the EU average, but above all with countries that have similar history and potential. However, even against countries of the 'new' EU Poland is still too much reliance on solid fuels (coal), and in far too little on renewable sources. Moreover, Poland should reduce the speed of growth of energy dependence, not imitating the old EU countries in this aspect.

Next to the problem of growing imports, the attention should be drawn to other aspects of creating the energy policy. There is a pressure of the EU to implement innovatory technologies in order to limit environmental hazards connected with coal, crude oil and natural gas. But there are also threats of applying pure technologies e.g. nuclear. These aspects cause that managing uniform energy policy in all European Union countries is quite a challenge.

The improvement of the energy efficiency and limitation of growing needs for fuels and energy, at maintaining the sustainable development regulations can become the recipe for the energy safety increase.

The energy crisis is for the Union a real threat and therefore it should develop more loyal mechanisms based on harmonizing interests of all member countries. The main task - in front of the European Union - is the development of common and efficient mechanisms of the energy policy, which will help member countries to manage with an eventual crisis related to energy supplies.

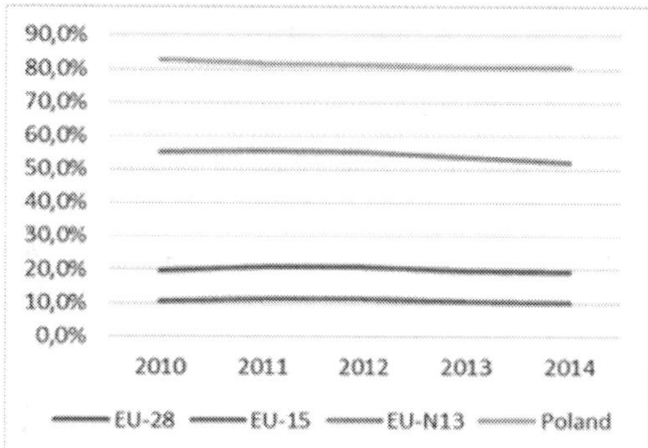

Figure 8. Share of solid fuels in energy production in EU [5].

References

1. http://www.cecgr.com/EnergyPolicy.pdf
2. http://ec.europa.eu/EnergyDependence
3. http://ec.europa.eu/eurostat/database
4. http://ec.europa.eu/eurostat/Demography
5. http://ec.europa.eu/eurostat/EnergyProduction
6. http://www.europarl.europa.eu/ConvEnergy
7. www.europarl.europa.eu
8. http://www.europagazniekonwencjonalny.pl/
9. http://ppg.ibngr.pl/

Environmental management reporting in Polish mining companies in the context of World CSR trends

Paweł Bogacz[1] and Marcin Migza[1,a]

[1]*AGH University of Science and Technology, Department of Economy and Management in Industry, al. Mickiewicza 30, 30-059 Krakow, Poland*

Abstract. CSR is very fast evolving management concept in the whole world. Therefore environmental management development is one of the most important strategic directions for modern enterprises. It is particularly relevant for mining enterprises, because of their impact on the natural environment. The authors aimed to verify if this trend is reflected in polish mining companies actions. Mining companies are obligated to non-financial reporting. In the paper the level of Global Reporting Initiative (GRI) usage in mining companies was considered. GRI is now the most frequently used method of non-financial reporting in the world..

1 Introduction

All mining activities, both underground and surface mining have a significant, most often a negative impact on the surrounding environment. The changes may affect lithosphere, atmosphere, hydrosphere, pedosphere and biosphere [1].

Environmental is one of the most important sphere of CSR reporting. It is particularly relevant for mining companies, because their impact on the natural environment is in principle significant.

In December 2014 Directive on disclosure of non-financial and diversity information by some big companies came into force. Member states should transpose it into national laws within two years. Therefore the first company reports should be published in 2018, regarding the financial year 2017-2018.

The measures established by the Directive will enhance the transparency and accountability of roughly 6000 businesses in the EU. These so-called 'public interest entities' with more than 500 employees will be:

• Obligated to report on environmental, social and employee-related, human rights, anti-corruption and bribery matters;

• Obligated to present their business model, results and hazards of the policies on the above topics, and the diversity policy in management and supervisory bodies;

• Persuaded to use recognized frameworks such as GRI's Sustainability Reporting Guidelines, the United Nations Global Compact (UNGC), the UN Guiding Principles on Business and Human Rights, OECD Guidelines, International Organization for Standardization (ISO) 26000 and the International Labour Organization (ILO) Tripartite Declaration.

Regarding to this law, mining companies in EU are obligated to non-financial, CSR reporting.

The aim of this article is to verify level of environmental management reporting in Polish mining companies. It focuses particularly on the GRI reporting standards. The observations and conclusions were made from quantitative and qualitative perspective.

2 Importance of environmental sphere in CSR reporting for mining companies

CSR is the way in which companies integrate social, environmental and economic matters into their values. It is reflected in companies' culture, strategy, operations and allows to establish better practices within the firm as well as to improve society [2].

According to European Commission CSR is the responsibility of companies for their influence on society. CSR should be company led. Public authorities can play a supporting role.

Companies can follow social responsibility by:

• compliance with the law;

• including social, environmental, ethical, consumer, and human rights matters into the way they run business [3].A sustainability report presents the economic, environmental and social impacts caused by actions of a company. It also shows the organization's values and governance model, and demonstrates how its strategy contributes to a sustainable economy.

Sustainability reporting can be also called as triple bottom line reporting or corporate social responsibility (CSR) reporting. It is also an intrinsic element of integrated reporting; a more recent development that combines the analysis of financial and non-financial performance [4].

[a] Corresponding author: migza@agh.edu.pl

According to International Council on Mining and Metals survay, environmental concern is one of the most important challenges. Figure 1 presents the most serious issues for the international mining industry, in table 1. Importance of different kind of areas for the European mining industry was presented.

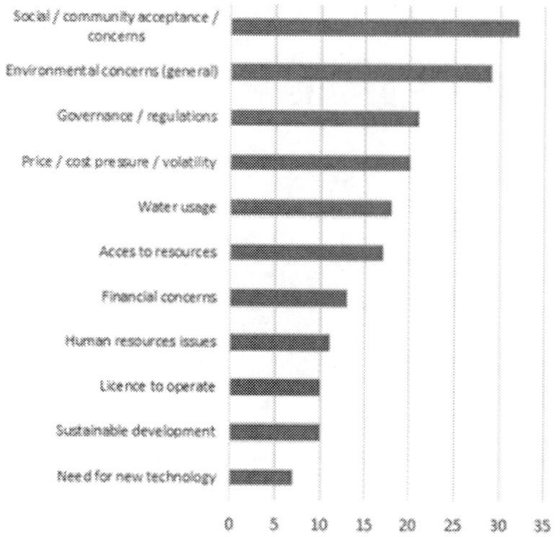

Figure 1. The most serious issues that the international mining industry will face in the next five years [5].

Table 1. Importance of different kind of areas for the European mining industry [5].

	Europe
Reducing impacts and making improvements to the environment where possible	91
Improving community relations	82
Improving its approach to health and safety	53
Improving the transparency of paid taxes and royalties to governments	63
Providing economic benefits and jobs	55

The environmental aspect of sustainability regards an organization's influence on living and non-living natural systems, including ecosystems, land, air, and water. Environmental Indicators cover performance related to inputs (e.g., material, energy, water) and outputs (e.g., emissions, effluents, waste). Additionally, they cover performance related to biodiversity, environmental compliance, and other important information e.g. environmental expenditure and the impacts of products and services [6].

3 GRI Environmental Performance Indicators for Mining and Metals Sector

The GRI (Global Reporting Initiative) guidlines are the most common way of CSR and sustainability reporting. Sustainability reports are released by companies and organizations of all types, sizes and sectors, from every corner of the world.

Thousands of companies across all sectors have published reports that reference GRI's Sustainability Reporting Guidelines. Public authorities and non-profits are also big reporters. GRI's Sustainability Disclosure Database features all known GRI-based reports [4].

The Global Reporting Initiative (GRI) encourages companies to use the sustainability reporting to become more sustainable. It aims to make sustainability reporting standard practice by sharing free Sustainability Reporting Guidelines. GRI, a not-for-profit, networkbased organization, engages many professionals and organizations of various backgrounds [7].

GRI in cooperation with International Council on Mining and Metals (ICMM) have prepared sustainability reporting guidance for the mining and metals sector. This Supplement has the sector-specific sustainability reporting guidance integrated in the GRI G3 Guidelines, the most widely-used guidelines for reporting on economic, environmental, and social performance. The Supplement covers biodiversity, indigenous rights, labor, community, artisanal and small-scale mining, resettlement, closure planning and materials stewardship [8].

GRI Environmental Performance Indicators for Mining and Metals Sector are presented in table 2.

Table 2. GRI Environmental Performance Indicators for Mining and Metals Sector [6]

Aspect	Imp.	No.	Description
Materials	Core	EN1	Materials used by weight or volume.
	Core	EN2	Percentage of materials used that are recycled input materials.
Energy	Core	EN3	Direct energy consumption by primary energy source.
	Core	EN4	Indirect energy consumption by primary source.
	Add	EN5	Energy saved due to conservation and efficiency improvements.
	Add	EN6	Initiatives to provide energy-efficientor renewable energy based productsand services, and reductions in energyrequirements as a result of these initiatives.
	Add	EN7	Initiatives to reduce indirect energy consumption and reductions achieved.
Water	Core	EN8	Total water withdrawal by source.
	Add	EN9	Water sources significantly affected bywithdrawal of water.
	Add	EN1 0	Percentage and total volume of waterrecycled and reused.
Biodiversity	Core	EN1 1	Location and size of land owned, leased, managed in, or adjacent to, protected areas and areas of high biodiversity value outside protected areas.
	Core	EN1 2	Description of significant impacts of activities, products, and services on biodiversity in protected areas and areas of high biodiversity value outside protected areas.

 DOI: 10.1051/e3sconf/20161000058

	Core	MM1	Amount of land (owned or leased, and managed for production activities or extractive use) disturbed or rehabilitated.
	Add	EN13	Habitats protected or restored.
	Add	EN14	Strategies, current actions, and future plans for managing impacts on biodiversity.
	Core	MM2	The number and percentage of total sites identified as requiring biodiversity management plans according to stated criteria, and the number (percentage) of those sites with plans in place.
	Add	EN15	Number of IUCN Red List species and national conservation list species with habitats in areas affected by operations, by level of extinction risk.
Emissions , Effluents , and Waste	Core	EN16	Total direct and indirect greenhouse gas emissions by weight.
	Core	EN17	Other relevant indirect greenhouse gas emissions by weight.
	Add	EN18	Initiatives to reduce greenhouse gas emissions and reductions achieved.
	Core	EN19	Emissions of ozone-depleting substances by weight.
	Core	EN20	NO, SO, and other significant air emissions by type and weight.
	Core	EN21	Total water discharge by quality and destination.
	Core	EN22	Total weight of waste by type and disposal method.
	Core	MM3	Total amounts of overburden, rock, tailings, and sludges and their associated risks.
	Core	EN23	Total number and volume of significant spills.
	Add	EN24	Weight of transported, imported, exported, or treated waste deemed hazardous under the terms of the Basel Convention Annex I, II, III, and VIII, and percentage of transported waste shipped internationally.
	Add	EN25	Identity, size, protected status, and biodiversity value of water bodies and related habitats significantly affected by the reporting organization's discharges of water and runoff.
Products and Services	Core	EN26	Initiatives to mitigate environmental impacts of products and services, and extent of impact mitigation.
	Core	EN27	Percentage of products sold and their packaging materials that are reclaimed by category.
Compliance	Core	EN28	Monetary value of significant fines and total number of non-monetary sanctions for noncompliance with environmental laws and regulations.
Transport	Add	EN29	Significant environmental impacts of transporting products and other goods and materials used for the organization's operations, and transporting members of the workforce.
Overall	Add	EN30	Total environmental protection expenditures and investments by type.

4 Environmental management reporting in polish mining companies

The most important Polish mining companies were subject to non-financial reporting. Not all of them were able to implement this. Table 3 presents the most important Polish mining companies and their level of CSR reporting.

Table 3. Sustainability reporting in polish mining companies in 31.12.2015

Company	Sustainability reporting	GRI Guidelines
LW Bogdanka S.A.	+	+
Famur S.A.	-	-
JSW S.A.	+	-
KGHM Polska Miedź S.A.	+	+
KHW S.A.	+	+
Kopex S.A.	+	+
KW S.A.	-	-
PG Silesia	+	-
PGE S.A.	+	+
Tauron Polska Energia S.A.	+	+
Węglokoks S.A.	-	-

The analysis of Table 3 shows that among the 11 most important Polish mining companies, only 3 have not introduced any sustainability reporting. Companies that have not yet implemented the non-financial reporting are Famur SA, KW SA and Węglokoks SA.

No CSR reporting may be especially surprising in the case of KW SA, which is the largest mining company in Europe, employing over 50 000 employees. Mining company with such a powerful impact on the environment and the local community should be particularly committed to a sustainable development.

Among the companies that have implemented CSR reporting, the vast majority (6 of 8) has adopted GRI guidelines. This trend is reflected in the whole world where GRI is the most widely used standard.

Table 4 shows in detail the level of reporting the specific areas within environmental reporting of CSR in Polish mining enterprises, using GRI standard. Five areas of CSR reporting, which are described below, are completely missed by all Polish mining companies: EN2, MM2, EN24, EN27, EN29.

DOI: 10.1051/e3sconf/20161000058

There is only one sphere, which appears in the sustainable reports of all the companies presented in Table 4. This is an EN8 area - Total water withdrawal by source. Among the Polish mining companies declaring reporting environmental sphere according to GRI guidelines KGHM Polska Miedź S.A. copes definitely the best. From 33 detailed areas 25 are reported by this company, two more: EN14 (Strategies, current actions, and future plans for managing impacts on biodiversity) and EN21 (Total water discharge by quality and destination) are reported in an incomplete or inadequate manner.

The areas completely ignored in reports of KGHM are also missed in all other mining companies. These are: EN2, MM2, EN24, EN27, EN29. The exception is EN19 (Emissions of ozone-depleting substances by weight), not reported by KGHM, but which can be found in the report of KHW S.A.

Table 4. Reporting of GRI Environmental Performance Indicators for Mining and Metals Sector in polish mining companies.

Aspect	Imp.	No.	KGHM Polska Miedź S.A.	KHW S.A.	Tauron Polska Energia	LW Bogdanka	PGE S.A.	Kopex S.A.
Materials	Core	EN1	+	-	+	+	+	-
	Core	EN2	-	-	-	-	-	-
Energy	Core	EN3	+	-	+	+	+	+
	Core	EN4	+	+	-	+	-	-
	Add	EN5	+	+	+	+	+	-
	Add	EN6	+	-	-	+	-	-
	Add	EN7	+	-	-	-	-	-
Water	Core	EN8	+	+	+	+	+	+
	Add	EN9	+	-	-	+	-	-
	Add	EN10	+	+	+	+	+	-
Biodiversity	Core	EN11	+	+	-	+	-	-
	Core	EN12	+	-	+	+	+	-
	Core	MM1	+	-	-	-	-	-
	Add	EN13	+	+	-	+	-	-
	Add	EN14	+/-	-	+	+	+	-
	Core	MM2	-	-	-	-	-	-
	Add	EN15	+	-	-	+	+	+
Emissions, Effluents and Waste	Core	EN16	+	+	+	+	+	-
	Core	EN17	+	-	-	-	-	-
	Add	EN18	+	+	+	-	+	-
	Core	EN19	-	+	-	-	-	-
	Core	EN20	+	+	+	+	+	-
	Core	EN21	+/-	+	+	+	+	-
	Core	EN22	+	+	+	+	+	-
	Core	MM3	+	-	-	+	-	-
	Core	EN23	+	-	+	+	+	+
	Add	EN24	-	-	-	-	-	-
	Add	EN25	+	+	+	-	+	-
Products and Services	Core	EN26	+	+	-	+	-	-
	Core	EN27	-	-	-	-	-	-
Compliance	Core	EN28	+	+	+	+	+	-
Transport	Add	EN29	-	-	-	-	-	-
Overall	Add	EN30	+	+	+	-	+	-

KHW S.A. reports less than a half (16 of 33) of areas identified in the GRI guidelines. A similar level of reporting is observed in Tauron Polska Energia (16 areas) and PGE S.A. (17 areas).

These three companies are reporting mostly the same, or similar areas, which suggests that they used common patterns in the process of creating the reports

Reporting at the level of approximately 50% of the guidelines is far too poor result. But reports of these companies already provide an image of the business activity impact on the environmental sphere.

Slightly wider range is contained in LW Bogdanka SA's report, in which references to 21 of 33 environmental areas indicated in the GRI guidelines can be found.

Definitely the worst was the outcome of Kopex S.A., which report covered only 4 areas: EN3, EN8, EN15 and EN23. At this level it is difficult to speak about CSR reporting or especially about the use of the GRI guidelines, which has been declared by the company. The report is too narrow to draw conclusions based on it regarding the impact of business on the environment.

5 Summary

In conjunction with introduction of the duty of non-financial reporting in EU, the most important Polish mining companies have been forced to sustainable reporting. EU authorities left some flexibility in choosing the method of reporting, but the most commonly used standard is the GRI. Similar situation is in the metal and mining industry, for which GRI together with ICMM prepared special guidelines. Also in Poland, among mining companies that have already implemented the CSR reporting, majority declares the use of this standard.

Due to intense impact of mining companies on the environment, the environmental area is the most important one of non-financial reporting for these companies. The authors analyzed in detail the reports of Polish mining companies declaring environmental reporting according to the GRI guidelines. The only company that in a satisfactory manner fulfills the recommendations of this standard is KGHM Polska Miedź S.A. Other audited companies meet approximately 50% of the requirements in their reports. Definitely the worst is the result of Kopex S.A., which report contains only 4 elements of the GRI guidelines.

Generally, Polish mining companies follow the same path as world companies, but still a considerable part of them have not implemented the non-financial reporting. Those that already fulfill this requirement, still do not do this with reasonable accuracy and commitment. The optimistic phenomenon is the rapid increase in the importance of CSR in Polish mining industry. With the appropriate amount of effort Polish mining companies have a chance to catch up with the world leaders, as exemplified by KGHM Polska Miedź S. A.

References

1. Lapcik, V., Lapcikova, M., J. of the Polish Min. Eng. Soc. **12**, 1-10, (2011).
2. http://www.iisd.org/pdf/2007/csr_guide.pdf
3. http://ec.europa.eu/corporate-social-responsibility/
4. https://www.globalreporting.org/default.aspx
5. http://www.icmm.com/document/8615
6. Sustainability Reporting Guidelines & Mining and Metals Sector Supplement. Version 3.0/MMSS Final Version, ©2000-2010 GRI
7. GRI G4 Guidelines and ISO 26000:2010 How to use the GRI G4 Guidelines and ISO 26000 in conjunction
8. www.globalreporting.org

DOI: 10.1051/e3sconf/20161000059

How might residential PV change the energy demand curve in Poland

Jakub Jurasz[1,a], Magdalena Krzywda[1] and Jerzy Mikulik[1]

[1]*AGH University, Department of Management Engineering, 30-067, Gramatyka St., Cracow, Poland*

Abstract. Photovoltaics (PV) in terms of installed capacity play a minor role in the portfolio of renewable energy sources (RES) in Poland. However current market tendencies indicate that residential PV installations are gaining on popularity and may in future significantly contribute to covering national energy demand. This study investigates the potential impact of numerous residential PV installations on the shape and statistical properties of the polish energy demand curve. Analysis employed statistical data on mean household energy consumption in different districts, typical energy demand patterns and hourly values of irradiation for the year 2012. Obtained results indicate that there is a possibility to integrate in total as much as 300 000 residential PV installations (0.9 GW) from which generated energy will be utilized by households within given district. Further analysis has shown that to some extent increasing number of residential PV decreases the value of energy demand coefficient of variation.

1 Introduction

Great diversity of factors shapes the volume of energy demand and its patterns. But it is the advent of RES and particularly variable renewable energy (VRE) that lead to a situation when not only demand is varying in time and space but also energy supply. Therefore the integration of non-dispatchable power sources such as PV and wind turbines creates as serious challenge for power grid operators. Jones [1] highlights several possible ways to effectively overcome major difficulties associated with VRE. One of them is to use energy sources which complement each other in time and space domain – this is for example the case of wind and solar energy [2, 3].

It is clear that wind and solar energy exhibit diurnal and annual patterns of their availability. Solar ones can be calculated based on clear-sky models [4], but they are usually disturbed by clouds, fog and particulate matter. This makes forecasting energy yield from PV installations an arduous task but according to literature not an impossible one [5]. The results presented in [6] indicate that PV energy generation patterns positively correlate with polish power load curve during the first hours of each day – just after sunrise. This indicates that this power source may beneficially contribute to covering energy demand.

The aim of this research is investigate the impact of rather small (3kW$_{peak}$) residential PV installations on energy demand curve in Poland. The main task was to evaluate the changes in energy demand statistical parameters and the occurrence of energy surpluses for scenarios considering different number of households owning PV installation.

[a] Corresponding author: jakubkamiljurasz@gmail.com

2 Method and data

Poland is divided into 16 voivodeships, these are further divided into districts (380 in total) and these in turn are divided into communes (about 2 500). Central Statistical Office CSO (http://stat.gov.pl/) provides data on mean annual household energy consumption on a district level. For the purpose of this study only households in the countryside were selected, therefore districts with city rights were excluded from further investigation. This leaves a total of 314 districts for further analysis. Figure 1 shows location of those districts – points indicate the centre of their surface.

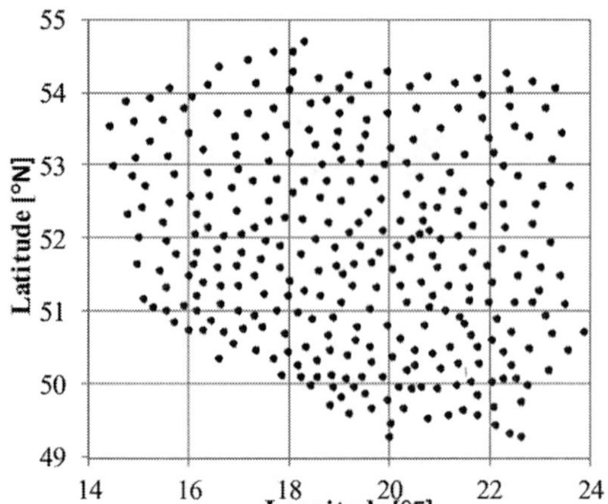

Figure 1. Sites for which irradiation time series where collected, and simultaneously districts surface centres.

Energy consumption in households comes from various appliances, electronics, lighting, water heating, air conditioning and space heating. Demand patterns depend on number of inhabitants, life style, external temperature and time of the year. In Poland, several studies were conducted in order to estimate typical household energy demand patterns [7, 8, 9]. This research bases on their results, by using those patterns. In case of each district the typical daily energy demand pattern (see for example Figure 2) has been recalculated based on annual demand pattern (see Figure 3) and specific for each district sum of energy consumed per year (see Figure 4).

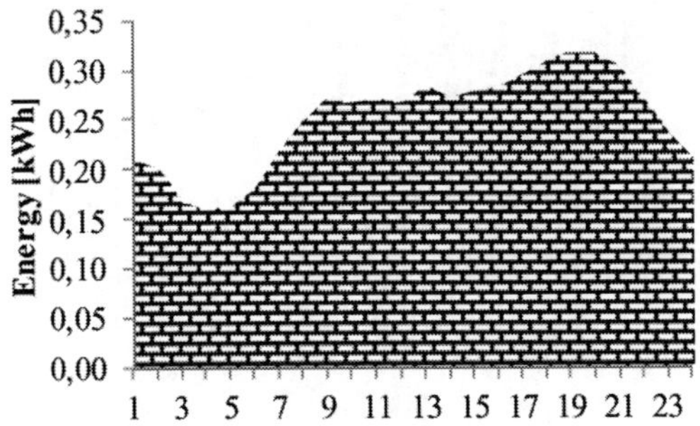

Figure 2. Mean hourly energy demand for a household in Bolesławicki district in June.

Additionally for each district data on the number of households were collected – this information was then used to estimate the number of possible PV systems which can be installed in a given district. In case of one district (Tczewski) which exhibited abnormally high energy consumption (over 8 MWh/annum per household) data from the year 2011 was used, in all remaining cases from the year 2012.

In order to calculate the energy yield from PV installations, hourly values of irradiation for the year 2012 were obtained from Solar Radiation Data SoDa (http://www.soda-pro.com/), which has both good spatial and temporal resolution. In total 314 irradiation time series for different points (which distribution is shown in Figure 1) were gathered. Energy yield from PV installation in all cases has been calculated based on formula (1):

$$E = \frac{G\, P^{PV}}{STC}\eta^{PV} \qquad (1)$$

where: E – energy yield [kWh], G – global irradiation [kWh*m⁻²], P^{PV} – nominal power of PV installation [kW] η^{PV}- overall efficiency [%],STC – standard testing conditions irradiance [kW* m⁻²].

The performance ratio here denoted as overall efficiency has been assumed to be 80% and the nominal power to be 3 kW.

Each district is characterised by mean annual energy consumption (see Figure 4), number of households and its irradiation time series. Investigated districts have in total over 4 million single-family households and their mean annual energy consumption amounts to 2.35 MWh. Energy derived from PV installations is assumed to be utilized on-site, within given district. If energy generation exceeds total household/district demand, then this excess energy is perceived as a surplus.

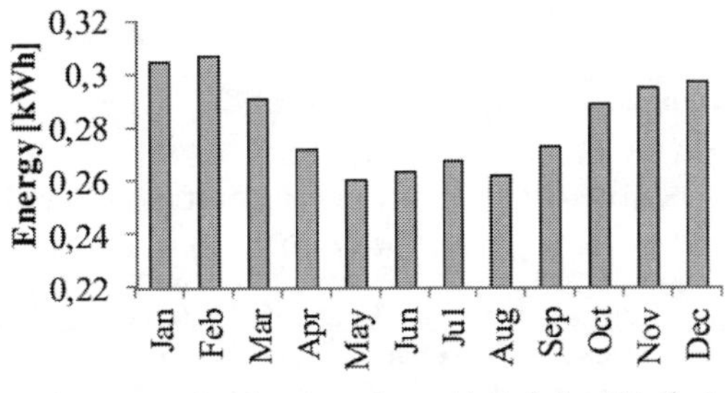

Figure 3. Mean hourly energy demand in Bolesławicki district.

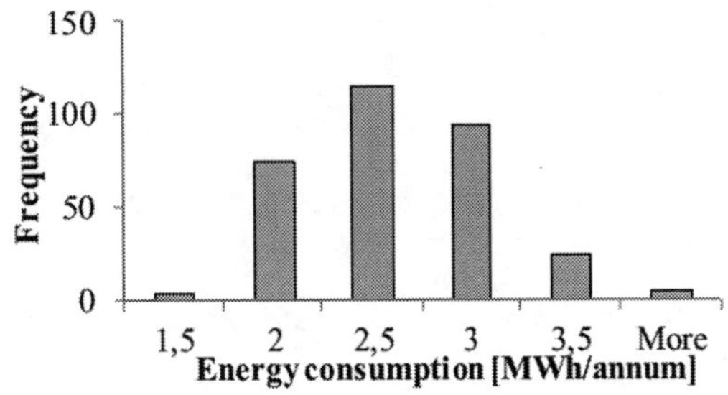

Figure 4. Mean annual energy demand per household in polish districts over the year 2012.

From the perspective of the method proposed in this study it is also important to note that supporting development of small and distributed PV installations entails a number of benefits. According to [10] distributed generation leads to a decreased transmission losses, improves power quality and system reliability. Additionally spatial distribution of PV installation enables reduction in energy output fluctuations [11,12]. Therefore on may expect that on a national or a district level an ensemble of small PV installations will generate much smoother power output than one single PV farm.

3 Results and discussion

Based on formula (1) and collected irradiation time series the average energy yield per kW of installed PV capacity was 0.952 MWh. The maximal energy output was 1.01 MWh/kW whereas the minimal was 0.877 MWh/kW. According to the CSO considered households have consumed almost 11.4 TWh of electrical energy, and this was 7% of the total electrical energy used in Poland over the year 2012.

As previously mentioned it is a known phenomenon that solar irradiation values vary and exhibit both annual and diurnal patterns. The same can be observed in case of household's energy demand, however those exhibit also one more pattern, namely a weekly one. Solar energy in an adequate form is not available for PV systems overnight, therefore as a rule it is not possible to cover whole energy demand. Although this may be achieved by coupling PV with batteries and another complementary energy source, as it was for example presented for a polish household in [13].

Summative analysis was conducted for aggregated hourly values of energy yield from PV and households energy demand. Over the whole study, energy demand from households remained constant.

3.1 Impact on households demand

First of all, results on the potential impact of residential PV on households energy demand is presented. As stated earlier different levels of households owning a PV installation were investigated. Figure 5 illustrates how they might change residential energy demand in July. Due to the diurnal patterns of solar energy, this is usually not available during late afternoon. Unfortunately late afternoon and evening is the period when usually greatest energy demand is observed. In those hours energy generated from PV installations would have much greater value. Please note how residential PV systems reduce energy demand in July (Figure 5) and how they eliminate the midday demand peak. Exceeding number of 40% households having PV system leads on average to situation when some part of energy generated cannot be used by households and must be transferred via power grid to other consumers.

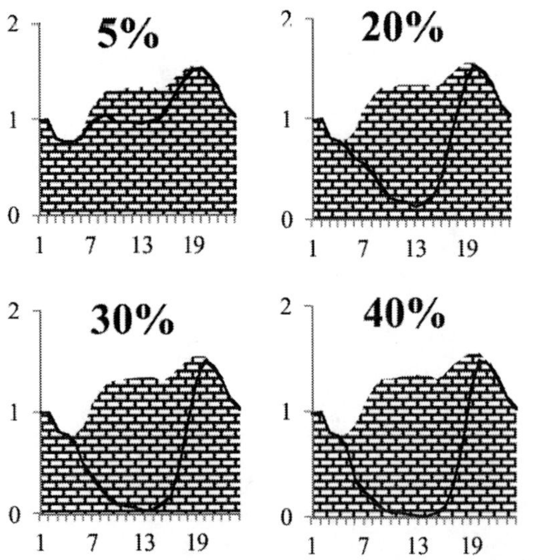

Figure 5. Average residential PV impact on aggregated households load in July. Key: OX axis - hour of the day, OY axis - energy demand in GWh, brick pattern – normal load, continuous line – load reduced by residential PV generation. Percentages indicate households with 3 kWp PV installation.

In general, over individual months PV installations have different capability of reducing average hourly energy demand. Those demand and generation patterns were investigated considering 30% households owning PV system. This translates into almost 1.47 million PV installations with nominal capacity of 4.3 GW. Analysis based on the mean hourly values indicates that such number of PV systems enables reduction in maximal hourly energy demand from April to August by accordingly: 50; 23; 85; 68 and 19 MWh. In this 30% scenario the mean hourly values of energy demand over December 2012 were reduced by 7% and as much as 50% in May. Detailed values of this reduction are presented in Figure 6. This observation is important mainly from one reason; current tendencies of load curve in Poland [14] indicate that the ratio of energy demand in winter to that in summer is decreasing. In other words, increasing demand in summer may to some extent be covered from residential and commercial PV installations.

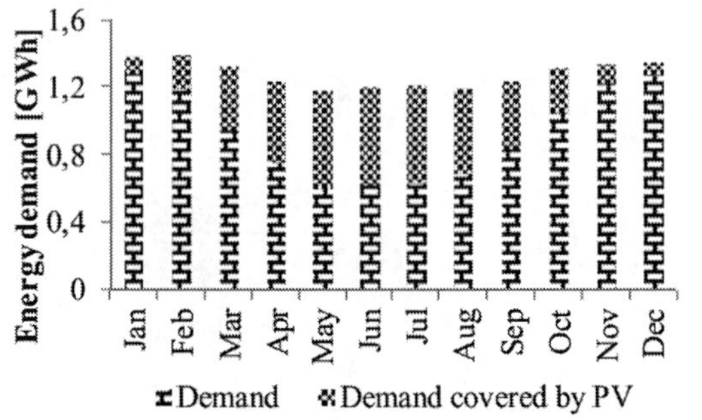

Figure 6. PV systems impact on mean hourly demand in individual months.

3.2 National energy demand curve change

As previously mentioned households located on rural areas constitute to roughly 7% of total energy demand in Poland. However considering various numbers of residential PV installations the impact of national energy demand curve becomes visible. Dotted line in Figure 7 represents values of energy demand coefficient of variation (CV) for each month. The CV values have been calculated as a ratio of the sample standard deviation to its mean. According to the data from the year 2012 the highest value of CV (exceeding 20%) can be observed in December, over remaining months CV ranges from 13.5% to 15.8%.

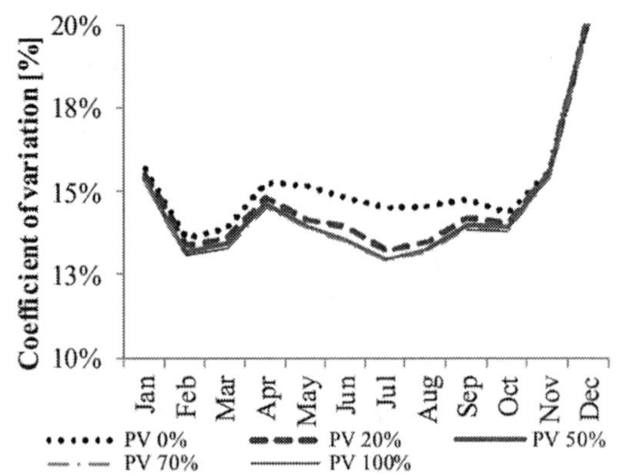

Figure 7. National power load mean monthly CV considering different share of residential PV, energy surpluses are discarded.

The CV values presented in Figure 7 were calculated based on an assumption that occurring energy surpluses do not have impact on the national energy balance. This shows that equipping even all households with PV

 DOI: 10.1051/e3sconf/20161000059

systems will reduce the CV value of energy demand curve. In short, variable energy source used to cover household's energy demand does not introduce additional disturbance to the energy market. However discarding energy surpluses is not correct from scientific and economic point of view. Figure 8 illustrates how changes CV when energy surpluses from PV installations are reducing national energy demand. Now it is clear that introducing almost 4.8 million of 3 kW PV installations (unbroken line 100%) would cause a significant increase in energy demand curve variability. From April to June CV values exceed 20%, and in case of almost all months (with exception to December – February) CV are greater than in baseline scenario (PV 0%). Interestingly in case of 20% and 50% scenarios the impact on national energy demand is positive. See Figure 9 to compare those values from scenarios presented in Figure 7 and 8.

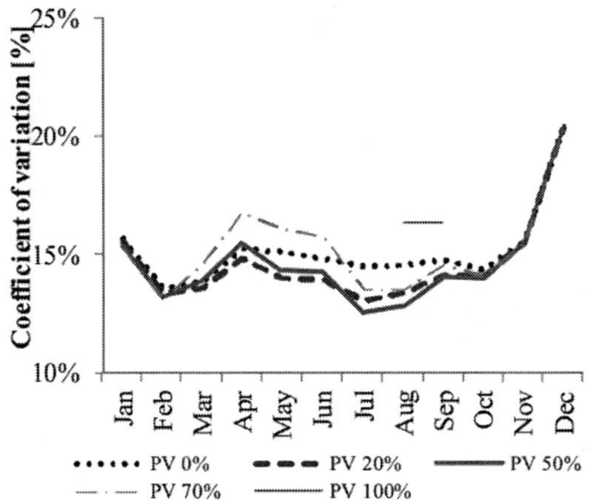

Figure 8. National power load mean monthly CV considering different share of residential PV and occurring energy surpluses

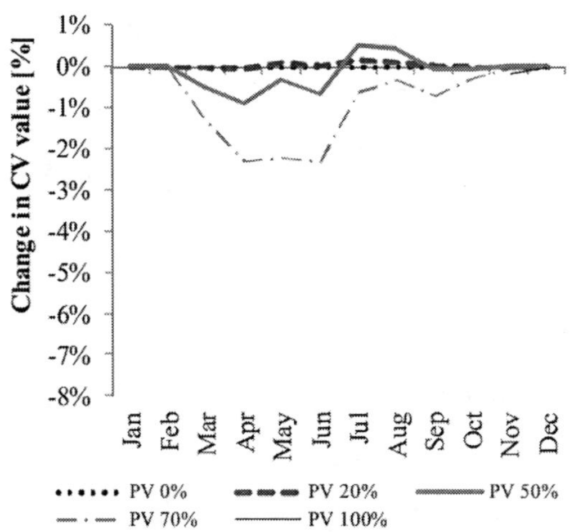

Figure 9. Differences between individual values of CV for two cases presented in Figures 7 and 8.

3.3 Energy surplus from residential PV

Energy surpluses mentioned in previous section demand more thorough analysis. An increasing number of residential PV installations undoubtedly will lead to a situation when some excess energy will be generated. This is mainly due to the energy demand patterns which make it impossible to utilize the whole energy generated from VRE. In the worst case scenario the energy may be available when it is not needed. Take for example summer, when all households' residents are running their everyday errands and energy demand is at its lowest. That is why nowadays energy storage technologies in case of VRE draw so much attention. They are perceived as a viable solution to the mismatch between energy demand and its availability.

Chart presented in Figure 10 illustrates how with increasing number of PV installations change the ratio of energy surplus to energy generated. Note that up to 10% of households owning PV system (which translates into almost 0.5 million PV installations of total nominal capacity equal to 1.4 GW) energy surpluses on a national scale does not occur. For number of households having PV installation ranging from 10% to 40% the increase of energy surplus to energy generation is much faster. This phenomenon was also observed in [6] however that study investigated the impact of PV installations on a national level, not from the perspective of household's energy demand pattern. Those results are coherent with observations made in previous section. Increasing number of PV installations, which installed capacity exceeds maximal energy demand in given area in some conditions (high irradiation) will lead to energy surpluses. Those surpluses (if not forecasted) will be perceived by power system operator as a disturbance. Additionally, according to the current market regulations so called *green energy* has priority before energy generated from conventional fuels – which in case of Poland is dominated by hard and brown coal [15]. This results in a situation when conventional power plants to some extent are forced to ramp up and down to adjust to the changing energy generation from VRE. This comes with additional costs and negative impact on power plant efficiency [16].

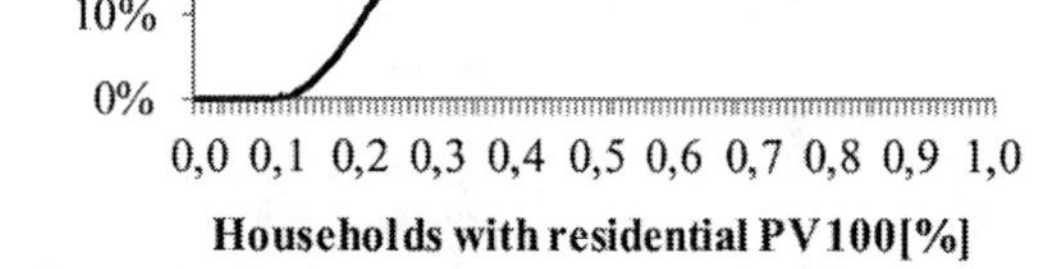

Figure 10. Ratio of energy surplus to energy generation considering various number of households with PV systems.

Considering the fact that over typical year about 80% [17] of solar energy available in Poland can be used only from April to October, and minimal values on energy

DOI: 10.1051/e3sconf/20161000059

demand in households fall on this period there is no wonder that biggest energy surpluses from residential PV will occur over those months. Based on 30% scenario and 2012 irradiation values, those surpluses in May can constitute to over 50% of energy generated (see Figure 11). From April to August over 30% of the generated energy is perceived as a surplus, in June and July it is even higher and exceeds 40%. Increasing demand on air-conditioning may in future be a viable option to utilize this excessive energy [18]. It is important to note that air-conditioning is commonly used in office buildings which are rapidly developing in Poland. This may be a very important market niche for so called building integrated or building applied photovoltaics (BIPV/BAPV).

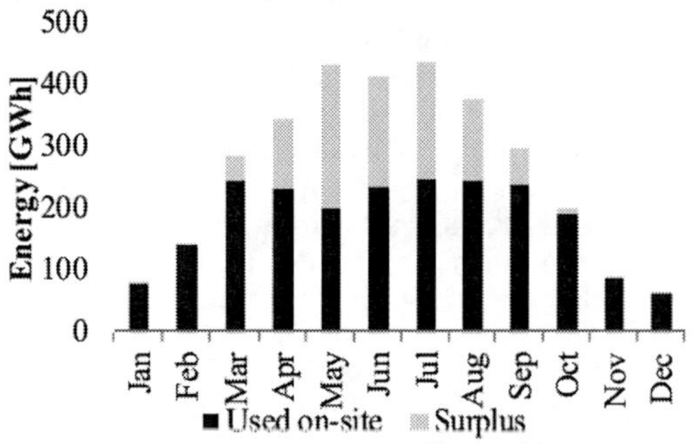

Figure 11. Energy consumed by households and energy surplus in case of 30% households with installed PV systems.

Interesting phenomenon can be observed when analysis of energy surpluses is conducted on the district level of its generation in consumption. According to the data presented in Figure 12 almost 90% districts have energy surpluses smaller than 5 GWh but clearly there are some districts where those surpluses are much greater than in others.

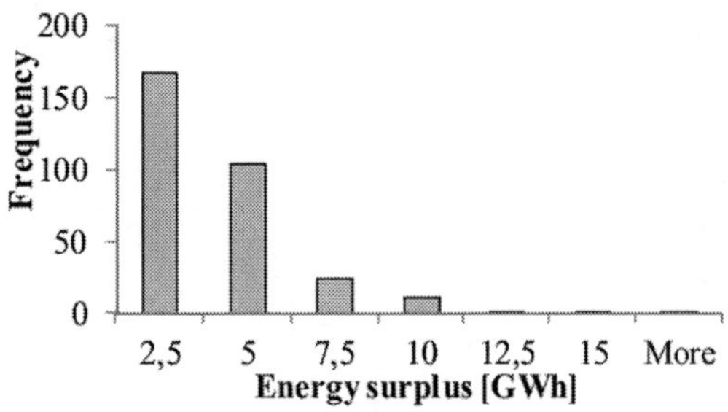

Figure 12. Histogram of annual sum of energy surpluses in different districts, for 30% PV scenario.

Scatter plots presented in Figures 13-15 show the relation between energy surpluses and three factors which may potential influence those, namely: mean energy generation, mean energy demand and the number of households. All charts were obtained for 30% PV scenario. According to the presented values of the coefficient of determination (R^2) only number of households exhibits positive correlation with energy surpluses. Interestingly two outliers on Figure 14 (marked by broken line) are examples of districts with highest mean energy demand, however the occurrence of energy surpluses in those cases is clearly driven by significantly greater number of households.

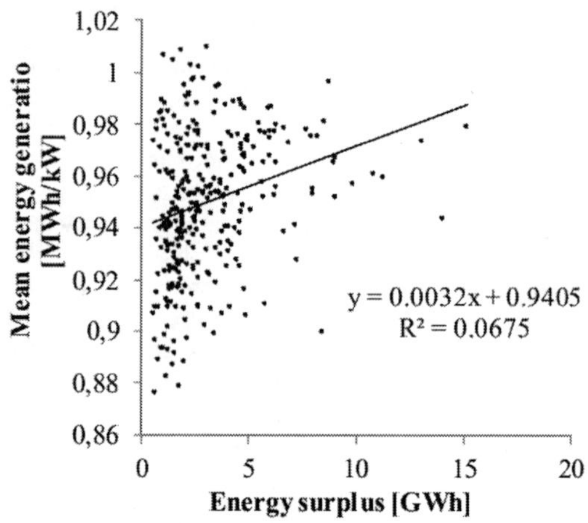

Figure 13. Relationship between mean energy generation and energy surpluses.

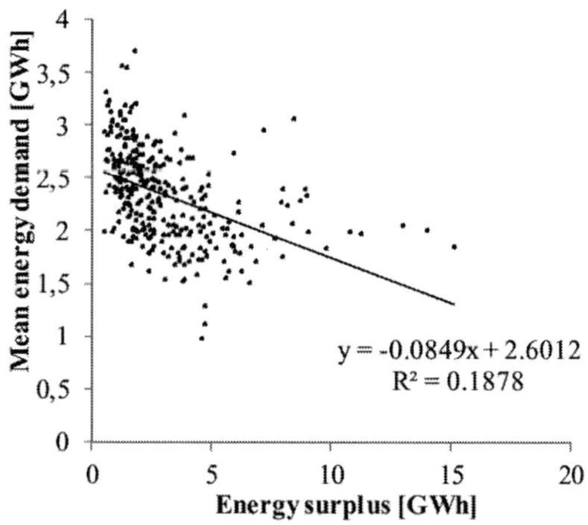

Figure 14. Relationship between mean energy demand and energy surpluses.

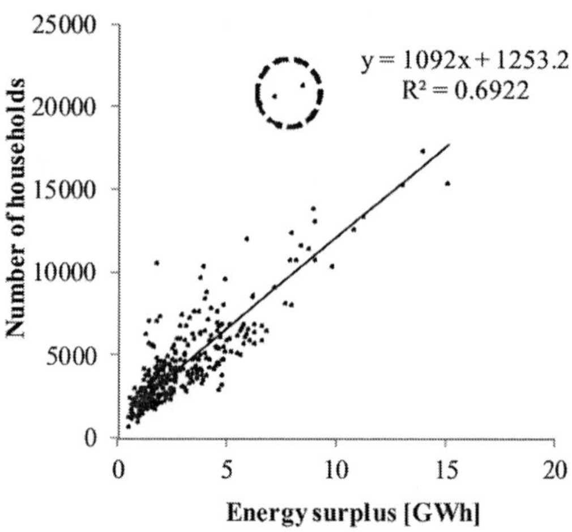

Figure 15. Relationship between number of households and energy surpluses.

4 Conclusions

Conducted analysis and obtained results indicate that PV systems have huge potential to cover a significant part of households energy demand. It has been also found that daily energy generation PV patterns match well with typical energy demand in early daily hours. In case of residential energy demand, photovoltaics are capable of reducing mean hourly demand but due to the nature of solar radiation they cannot contribute to covering daily peak demand. Further analysis on the impact of PV on the energy demand curve in Poland has shown that: integrating as much as 4.3 to 5 GW of PV may still lead to a decrease in energy demand coefficient of variation, and distributing 1.4 GW of PV among 10% households of each district will no entail an occurrence of energy surpluses. Additionally, according to the recently conducted analysis [19] in Poland the temperature impact on energy demand in getting on importance. Meaning that not only cold days lead to a greater energy demand but also sunny and hot ones. This may result from an increasing interest in air-conditioning (AC) devices. They become irreplaceable as daily temperatures are becoming higher and higher and one has to ensure adequate working conditions in offices. What is more, the residential AC market in Poland is developing. Therefore due to the positive correlation between irradiation values, temperature and air-conditioning demand the future role of PV in providing extra power when AC is operating seems to be obvious.

References

1. L. E. Jones, *Renewable energy integration: practical management of variability, uncertainty, and flexibility in power grids*. Academic Press, (2014)
2. Ch. E. Hoicka, I. H. Rowlands. Ren. E. **36.1**, 97-107 (2011).
3. J. Jurasz, A. Piasecki, Acta Energetica, **2** (2016).
4. R. E. Bird, and R. L. Hulstrom. *Simplified clear sky model for direct and diffuse insolation on horizontal surfaces*. No. SERI/TR-642-761. Solar Energy Research Inst., Golden, CO (USA), 1981.
5. Vaz, A. G. R., Elsinga, B., van Sark, W. G. J. H. M., & Brito, M. C. Ren. E. ***85***, 631-641 (2016).
6. J. Jurasz, J. Mikulik, *Investigating Theoretical PV Energy Generation Patterns with Their Relation to the Power Load Curve in Poland.* International Journal of Photoenergy, vol. 2016, Article ID 3789840, 7 pages, (2016). doi:10.1155/2016/3789840
7. R. Frąckowiak, T. Gałan. *Electric Load in the G12 Tariff for Consumers Assigned to the PTPiREE C-profile*. Acta Energetica (2014).
8. R. Frąckowiak, T. Gałan. Przegląd Elektrotechniczny **85,** 156-159 (2009).
9. R. Frąckowiak, T.Gałan. Rynek Energii **4** (2012).
10. P. P. Barker, and R. W. De Mello. *Determining the impact of distributed generation on power systems. I. Radial distribution systems.* Power Engineering Society Summer Meeting, (2000). *IEEE.* Vol. **3**. IEEE, 2000.
11. E., Boudewijn, and W. van Sark.. Prog. in Photov.: Res. and App. **23.10,** 1390-1397 (2015).
12. J. Jurasz, J. Mikulik. Rynek Instalacyjny, **11**, 35-37 (2015).
13. J. Jurasz, J. Mikulik. Architecture, Civil Engineering, Environment, **4**,:5-10 (2015).
14. J. Jurasz, J. Mikulik, Przegląd Elektrotechniczy **92,** 152-154 (2016):.
15. H. L. Gabryś, Energetyka **03-04** (2012).
16. K. Van den Bergh K, and E. Delarue. Energy Conversion and Management **97,** 70-77 (2015):.
17. Wacławek M, Rodziewicz T. *Ogniwa słoneczne. Wpływ środowiska naturalnego na ich pracę. (Solar cells. Influence of environment on their performance). II edition.* Warszawa: WNT; (2015).
18. Davis, L. W., & Gertler, P. J. Proc. of the Nat. Acad. of Sci., **112**(19), 5962-5967 (2015)
19. J. Jurasz, J. Mikulik and A. Piasecki. Przegląd Elektrotechniczny **97** (2016): *In Press Corrected Proof*

DOI: 10.1051/e3sconf/20161000060

Chloride corrosion in biomass-fired boilers – Fe-O-Cl system thermodynamic analysis

Robert Kaczmarczyk[1,a] and Agata Mlonka-Mędrala[1]

[1] *AGH University of Science and Technology, Faculty of Energy and Fuels, 30 Mickiewicza St., 30059 Krakow, Poland*

Abstract. The most common and easiest alternative technologies for conventional fossil fuel combustion are biomass combustion and co-combustion. However, high-chlorine fuels (Cl^{ar}>0,2%) like: biomass, waste and high chlorine coals generate the risk of intensified corrosion process and a limited steel mechanical strength is observed.

The paper presents a thermodynamic analysis of chloride-induced corrosion in the Fe-O-Cl system. The ranges of the metallic, oxide and chloride phase stability are determined within the temperature range T = 750-1000 K. Based on the parametric equations the equilibrium concentration of gaseous phase determined by Deacon reaction are presented. The effect of H_2O concentration in the gaseous phase on high-temperature corrosion process and gaseous NaCl influence on $NaFeO_2$ formation in the passive oxide scale layer ($FeO/Fe_3O_4/Fe_2O_3$) are discussed as well. The results are correlated with available in the literature laboratory experimental data and industrial corrosion process observations. Presented thermodynamic analysis is compared with assumptions of "active oxidation" model. The results may be used for experimental research prediction and a corrosion prevention in the industry.

1 Introduction

In the last update of LCP BREF document, a significant change in emission limits, especially for fossil fuel combusting units was introduced [1]. Therefore, modernization of flue gas treatment systems in existing power plants will be necessary and the energy production costs will increase. The Polish 2020 renewables target assumes 15% share of renewable energy in gross final energy consumption. [2] Biomass combustion and co-combustion are one of the most developed RES technologies in Poland. A major problem associated with biomass combustion is a limited material strength and corrosion resistance of boiler heat exchanging surfaces at higher temperatures. To reduce negative effects, the steam temperature in biomass fired boilers is decreased to 725-825 K. [3] However, it also reduces the efficiency of the unit and as a consequence increases the energy production cost. The highest risk of the intensified corrosion process is observed during combustion of fuels with chlorine content greater than 0,2% - Cl^{ar}> 0.2% (high-chlorine coal, biomass, waste). [4] Chlorine compounds in flue gas, fly ash and eutectic mixtures agglomerated on the heat exchanging surfaces has been reported as the main sources of high-temperature chloride-induced corrosion [3, 5]. The kinetics of the corrosion process is determined by: flue gas and fly ash deposits composition (especially the chlorine content), corrosive species concentration in a tube boundary layer and temperature gradient defined by a combustion process parameters [5]. An average gas composition for selected fuels is presented in Tab. 1.

Table 1. Approximate flue gas composition in coal, biomass and waste firing units [3]

GAS \ FUEL	COAL	BIOMASS	WASTES
O_2 [%]	~4-5	~5-10	~5-11
CO_2 [%]	~12	~8-15	~8-14
H_2O [%]	~4-16	~10-20	~10-20
SO_2 [ppm]	~400-1200	~0-70	~0-150
HCl [ppm]	~10-50	~25-1000	~250-1300
KCl +NaCl [ppm]	-	~5-50	<120

[a] Corresponding author: robertk@agh.edu.pl

2 Chloride-induced corrosion process of steel - an "active oxidation" model interpretation

Laboratory experiments and field measurements indicate an acceleration of steel degradation process by gaseous HCl in flue gas. The analysis of gaseous chlorine compounds shows that chlorine (Cl_2) is a much more corrosive agent then hydrochloric acid (HCl). Although, chlorine in biomass and waste is released mainly as HCl and (K,Na)Cl (Tab. 1).

Chloride-induced corrosion rate of iron and selected steel grades in Ar/He-O_2-HCl atmosphere was examined in [6, 7, 8, 9, 10]. The oxidation of iron exposed to alkali and heavy metal chlorides and sulfates mixtures was discussed in [11]. The corrosion process in a solid-liquid phase with a presence of low-melting eutectic mixtures in the deposits was also investigated in [12, 13, 14, 15]. Some fundamental conclusions come from experimental results. A passive oxide scale composed of iron oxides is formed during boiler operation. The rate of oxide scale formation in Fe-O system is rather slow and it depends on steel grade. The oxygen concentration in a gas phase does not affect the kinetics of the process [7, 16]. Direct reaction of iron (Fe) and hydrochloric acid (HCl) occurs only in elevated temperatures, in reducing atmosphere with high HCl concentration [17]. However, experimental results shows that high temperature corrosion proceeds already in 5% O_2/500-3000 ppm HCl atmosphere and $FeCl_2$ is formed according to an active oxidation model mechanism presented in Fig. 1. [6-11] Fundamental assumptions of an active oxidation model [18, 19] include:

1. Diffusion of gaseous Cl_2/HCl through pores and cracks of a metal oxide scale to the scale/metal interface where iron chloride is formed:

$$Fe + Cl_2 \leftrightarrow FeCl_2 \quad (1)$$

2. Volatility of $FeCl_2(s) \rightarrow FeCl_2(g)$ in the metal wall boundary layer and further iron chloride diffusion through porous and loose oxide scale layer back to flue gas, where chloride vapors oxidation takes place:

$$3FeCl_2 + 2O_2 \leftrightarrow Fe_3O_4 + 3Cl_2 \quad (2)$$

$$2FeCl_2 + \frac{3}{2}O_2 \leftrightarrow Fe_2O_3 + 2Cl_2 \quad (3)$$

The diffusion of Cl_2 through the Fe_3O_4/Fe_2O_3 oxide scale, formation of $FeCl_2(g)$, which penetrates the scale backward to flue gas and final oxidation (reactions (2), (3)) form a closed cycle of reactions. Such repeatability causes a severe oxide scale destruction and steel protective properties against chloride corrosion are lost. Exothermic Deacon oxidation reaction of hydrochloric acid catalyzed by metal oxides (Weldon process) is a one of the main sources of chlorine in gas phase:

$$2HCl + \frac{1}{2}O_2 \leftrightarrow Cl_2 + H_2O \quad (4)$$

During combustion process of biomass gaseous alkali metal chlorides are released. The chlorides in gas phase can react with oxide scale and cause its further destruction and additional chlorine is formed:

$$2NaCl + Fe_2O_3 + \frac{1}{2}O_2 \leftrightarrow Na_2Fe_2O_4 + Cl_2 \quad (5)$$

A reaction cycle according to proposed model is presented below.

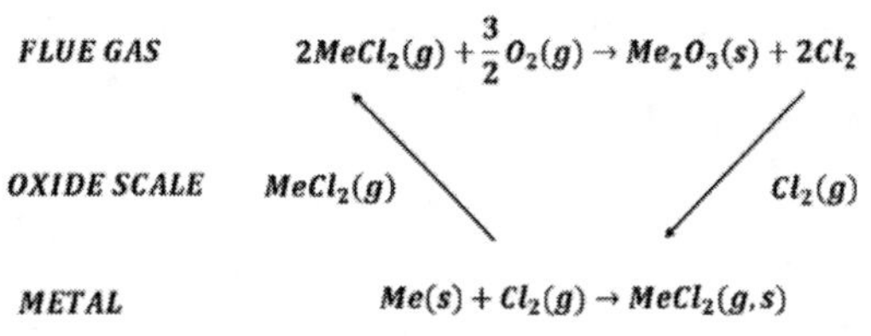

Figure 1. Cl-induced corrosion cycle described in [3, 18]

A steel oxidation process accelerated by chlorine presence in flue gas called an active oxidation model is widely described in literature. However, it determines the chloride-induced corrosion process only to Fe-Cl-O system and limits the diffusion through oxide scale layer to Cl_2 and $FeCl_2(g)$ only. Selectivity of oxide scale layer Fe_3O_4/Fe_2O_3 in diffusion process of gas phase components between flue gas and metal surface and neglect of other gas phase components (e.g. H_2O) influence on corrosion process arouses some controversy among scientists.

3 Chloride-induced corrosion process of iron - a thermodynamic analysis

A characteristic of Cl-induced corrosion process of iron determined by [O_2-Cl_2-H_2O-NaCl] gas phase was accomplished based on thermodynamic fundamental principles, thermodynamic properties of pure substances taking part in the reaction and properties of solutions formed in a gas phase. For a random chemical reaction, the Gibbs free energy change ΔG^0_T was determined according to relation below:

$$\Delta G_T^0 = \sum_{i=1}^{m_1} k_{B_i} \cdot \mu_{B_i}^0 - \sum_{i=1}^{m_2} k_{A_i} \cdot \mu_{A_i}^0 \quad (6)$$

Where: A_i, $i = 1, \dots m_1$ and B_i, $i = 1, \dots m_2$ are reactants and products respectively, k_{Ai} and k_{Bi} are stoichiometric coefficients for reactants and products in the reaction, $\mu_{A_i}^0$ and $\mu_{B_i}^0$ represents standard chemical potential of pure substances for reactants and products. In the calculations was suggested a relation: $\Delta G_T^0 = A + BT + CTln(T)$. The function shape comes from thermodynamic properties of pure components (C_p). The relation between Gibbs free energy change and thermodynamic equilibrium constant K was used as well:

$$\Delta G_T^0 = -RT \ln K \quad (7)$$

where: K represents a limit the process is tending to and is determined by reagents activities a_i.

$$K = \frac{\prod_{i=1}^{s} a_{B_i}^{k_{B_i}}}{\prod_{i=1}^{s} a_{A_i}^{k_{A_i}}} \quad (8)$$

Activity of components in solid or liquid phase, in a pure state is equal to unity. Although for gases, the activities of chemical reaction components might be determined by partial pressures expressed as the components mole fractions $P_i=x_iP_{tot}$ (P_{tot} – total pressure). In isothermal-isobaric conditions, the reaction gas phase composition changes linearly and a conversion from initial to final (equilibrium) state can be described by a vector parallel to the linear function [20, 21, 22, 23]:

$$x_i = x_i^0 + \tau \frac{k_i - x_i^0 \sum_{i=1}^{s} k_i}{\sqrt{\sum_{i=1}^{s} (k_i - x_i^0 \sum_{i=1}^{s} k_i)^2}} \quad (9)$$

where: x_i, x_i^0 – equilibrium and initial molar fraction of „i" component, $\tau \in R$ – straight line parameter, k_i– „i" reagent stoichiometric coefficient. Positive stoichiometric coefficients (k_i) are defined for products, negative for reactants and zero for reagents doesn't taking part in the reaction. The direction cosines are time independent and they are functions of initial composition and stoichiometric coefficients only. In reaction ($\sum_{i=1}^{s} k_i \neq 0$) the phase composition change proceed along straight lines intersecting at a so-called "characteristic point" ($\bar{x}_i$). In case of reaction ($\sum_{i=1}^{s} k_i = 0$) the direction cosine depends on stoichiometric coefficients only. Initial composition does not affect the direction cosine and a change of reagents concentration proceed along straight lines parallel to each other. The presented computational process analysis was carried out based on thermodynamic data [24, 25] and a numerical computing environment - MATLAB.

Table 2. Equilibrium composition of gaseous phase in Deacon reaction (4) defined by parametric equations.

$\sum_{i=1}^{5} k_i$	i	k_i	$\bar{x}_i = \frac{k_i}{\sum_{i=1}^{5} k_i}$	$x_i = x_i^0 + (\bar{x}_i - x_i^0)\tau$	T=800 K			
					x_i^0	x_i	x_i^0	x_i
-0.5	HCl	-2	4	$x_{HCl} = x_{HCl}^0 + (4 - x_{HCl}^0)\tau$	$3 \cdot 10^{-3}$	$1.311 \cdot 10^{-3}$	$3 \cdot 10^{-3}$	$2.979 \cdot 10^{-3}$
	O_2	-0.5	1	$x_{O_2} = x_{O_2}^0 + (1 - x_{O_2}^0)\tau$	$5 \cdot 10^{-2}$	$4.960 \cdot 10^{-2}$	$5 \cdot 10^{-2}$	$4.999 \cdot 10^{-2}$
	Cl_2	1	-2	$x_{Cl_2} = x_{Cl_2}^0 + (-2 - x_{Cl_2}^0)\tau$	0	$8.449 \cdot 10^{-4}$	0	$1.056 \cdot 10^{-5}$
	H_2O	1	-2	$x_{H_2O} = x_{H_2O}^0 + (-2 - x_{H_2O}^0)\tau$	0	$8.449 \cdot 10^{-4}$	0.35	0.35
	N_2	0	0	$x_{N_2} = x_{N_2}^0 - x_{N_2}^0\tau$	0.947	0.9474	0.597	0.597

In a temperature range T=750-1000 K the phase stability areas of iron compounds in [Fe-O-Cl] system were determined based on the thermodynamic equilibrium condition of chemical reactions. At given temperature, phase stabilities of oxides (FeO, Fe_3O_4, Fe_2O_3), chlorides ($FeCl_2$) and pure iron (Fe) depends on Cl_2 and O_2 concentration in gas phase. A phase stability diagram for [Fe-O-Cl] system at T=800K is presented in Fig. 2. Protection of iron against chloride and oxide corrosion at T= 800K is possible in low partial pressures of chlorine and oxygen $P_{Cl_2} = 10^{-16}$ [atm.], $P_{O_2} = 10^{-27}$ [atm.]. At partial pressure of oxygen $P_{O_2} = 10^{-8}$ [atm.] and chlorine $P_{Cl_2} = 10^{-3}$ [atm.], formation of chloride phase in contact with pure hematite ($FeCl_2/Fe_2O_3$) might be observed. In a full-scale biomass fired units and in laboratory experiments carried out in conditions similar to biomass combustion, the oxygen content in flue gas is around 5% and HCl is 500-3000 ppm. It is not enough to form an iron chloride phase and an iron(III) oxide - Fe_2O_3 is the stable phase in these conditions. The oxide corrosion is a main process in these conditions. Oxygen (O_2) and chlorine (Cl_2) are the main reagents in the gas phase. Introduction of gaseous NaCl with oxygen to the environment creates a new [Fe-O-Cl-NaCl] system. When oxygen is present in the atmosphere, Fe_2O_3 reacts with NaCl and $2NaFeO_2$ with Cl_2(g) are formed according to reaction (5). Assuming activity of solid phase components equal to unity ($a_{2NaFeO_2} = a_{Fe_2O_3} = 1$), the change of gas phase composition for P_{NaCl}=100 ppm in an oxide corrosion area is shown in Fig. 2. Sodium oxide (Na_2O), a product of NaCl oxidation is an active component of $2NaFeO_2$ formation. At T=800K, calculated activity of sodium oxide in the reaction with hematite:

$$Na_2O + Fe_2O_3 \leftrightarrow 2NaFeO_2 \quad (10)$$

equals to $a_{Na_2O} = 10^{-10}$ and corresponds to the results in [26]. When NaCl is present in the atmosphere, $2NaFeO_2$ is formed on a hematite surface. Such reaction can be destructive for passive oxide layer (Fe_2O_3). During biomass or waste combustion, low-melting eutectic mixtures are formed. They may deposit on a surface of iron oxide layer and $2NaFeO_2$ can dissolve in the liquid phase of the deposits. It led to an activity decrease ($a_{2NaFeO_2} < 1$) and as a consequence an iron oxide corrosion process is intensified. Proposed mechanism of high-temperature corrosion of iron in [O-Cl] atmosphere seems to be correct, however it includes gaseous NaCl reactions on boundary layer of [NaCl-O] – [$Fe_2O_3/2NaFeO_2$] within oxide corrosion range only. In analyzed temperature range and gas phase composition (O_2 - 5%, HCl <1300 ppm), presented above mechanism of chloride corrosion has no thermodynamic basis. According to reaction (4) HCl introduced with oxygen atmosphere determines formation of water vapor in the system. The actual composition of the gas phase was defined by a set of parametric equations. [20-23]. The calculation results presented in Tab. 2 were obtained for initial concentrations of oxygen, chlorine and nitrogen expressed in terms of molar fractions ($x^0_{O_2} = 5 \cdot 10^{-2}$, $x^0_{HCl} = 3 \cdot 10^{-3}$, $x^0_{N_2} = 0.947$). Total pressure of the system was assumed to be 1 atm. (P_{tot}=1 atm). The initial parameters correspond to conditions of the experiment described in [8]. Calculations for the Deacon reaction (4) show a formation of steam and chlorine in equilibrium concentrations $x_{H_2O} = x_{Cl_2} = 8.5 \cdot 10^{-4}$. At constant steam concentration P_{H_2O}=const., equilibrium concentrations Cl_2=f(O_2) tend to gaseous chloride corrosion with a decrease of moisture in the flue gas. According to an active oxidation model, it may cause an activation of a corrosive agent – Cl_2. This conclusion coming from the thermodynamic calculations was confirmed by results of laboratory experiments [12].

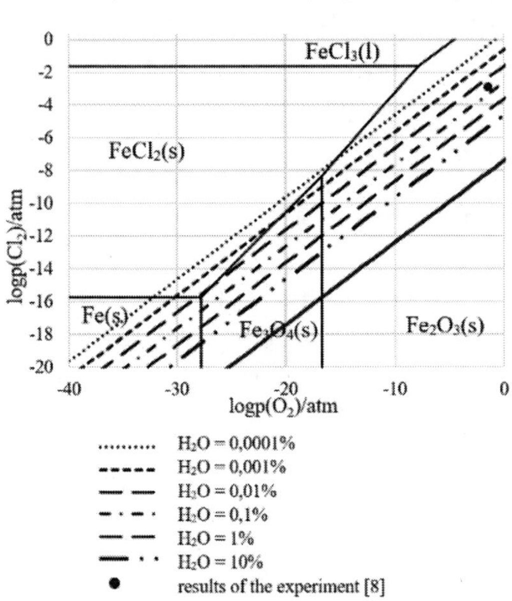

Figure 2. Phase stability diagram for Fe-O-Cl at T = 800K; HCl=1300ppm

For pure iron, nickel and low-chromium alloys (16Mo3) it is confirmed experimentally that corrosion rate decrease with steam concentration increase in flue gas. However, when additional deposits rich in chlorides (39$PbCl_2$–50KCl–11$ZnCl_2$ wt.%, T=673K) are introduced to the system, anticorrosive properties of iron and chromium alloys decrease. The mechanism of the process is changed and the influence of water vapor on chloride-induced corrosion is rather insignificant. In this case steam accelerates the oxide corrosion rate. According to reactions (4) and (5), an increase of steam content in flue gas and dissolution of $2NaFeO_2$ in low-melting ash components cause a decrease of hematite activity:

$$a_{Fe_2O_3} = \frac{K_4}{K_5}\left(\frac{P_{HCl}}{P_{NaCl}}\right)^2 \frac{a_{2NaFeO_2}}{P_{H_2O}} \qquad (11)$$

At T=800 K, for a gas atmosphere 1000 ppm-HCl, 100 ppm-NaCl, 10 %-H_2O just above an oxide phase Fe_2O_3/$2NaFeO_2$ an activity decrease $a_{Fe_2O_3} < 1$ in molten ash deposits appears already at $a_{2NaFeO_2} > 10^{-3}$. Steam concentration in flue gas P_{H_2O}=45 atm. will prevent a dissolution of $2NaFeO_2$ in the deposits. Similar to the behavior of NaCl(g), components of molten eutectic mixture KCl – $ZnCl_2$ may react with hematite according to relations:

$$2KCl + Fe_2O_3 + \frac{1}{2}O_2 \leftrightarrow 2KFeO_2 + Cl_2 \qquad (12)$$

$$ZnCl_2 + Fe_2O_3 + \frac{1}{2}O_2 \leftrightarrow Fe_2ZnO_4 + Cl_2 \qquad (13)$$

In this case, the corrosion rate of iron increases ($a_{Fe_2O_3} < 1$) when in deposit molten phase activities are equal to $a_{Fe_2ZnO_4} > 10^{-7}$; $a_{2KFeO_2} > 10^{-27}$ at the gas phase saturation point (1000 ppm HCl, 10 % H_2O). It seems that the corrosion process of iron in atmosphere rich in chlorine compounds and steam has a two-way mechanism. The increase of the water vapor concentration and the presence of the low-melting components in the deposits accelerate the corrosion rate in the oxide area. Although, a decrease of steam partial pressure in gas phase activates chlorine toward gaseous chloride corrosion area. Moisture content in the flue gas depends on oxidizing/reducing combustion conditions according to reaction:

$$CO + H_2O \leftrightarrow CO_2 + H_2 \qquad (14)$$

And is inversely proportional to CO concentration in the gas phase:

$$P_{H_2O} = \frac{P_{CO_2}}{P_{CO}}\frac{P_{H_2}}{K_{13}} \qquad (15)$$

In reducing conditions (O_2 < 1%, CO > 2%) steam concentration is decreasing. Low steam concentration promotes gaseous chloride corrosion of iron. Equilibrium concentrations for reaction (14) determined based on parametric equations method are presented in Fig. 3.

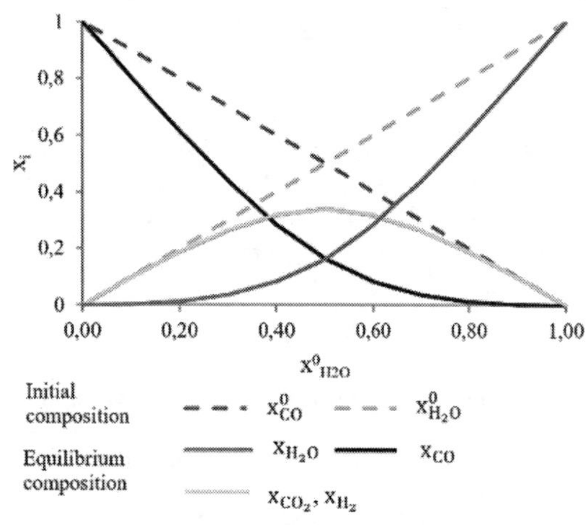

Figure 3. Equilibrium concentrations x_i gaseous phase components of the reaction: $CO + H_2O \leftrightarrow CO_2 + H_2$; T=800K, according to parametric equations model.

4 Conclusions

Several conclusions come from the presented analysis of high-temperature corrosion of iron in gas atmosphere rich in chlorine compounds. In the temperature range T=750-1000K and gas phase (O_2 - 5%, Cl_2 <1300 ppm), chloride corrosion process mechanism doesn't have any thermodynamic basis. Introducing of HCl with oxygen to the system determines steam formation. At constant water vapor concentration P_{H2O}=const., equilibrium concentrations Cl_2=f(O_2) tend to gaseous chloride corrosion with a decrease of moisture in the gas phase. According to an active oxidation model it may cause activation of the corrosive agent – Cl_2. However, when deposits rich in chlorides ($PbCl_2$, KCl, $ZnCl_2$) are introduced to the system, the influence of steam on chloride corrosion process is small. In this case steam accelerates the rate of oxide corrosion. It seems that the corrosion process of iron in atmosphere rich in chlorine compounds and steam has a two-way mechanism. The increase of the water vapor content and the presence of a low-melting ash components in the system accelerates the corrosion rate in the oxide area. Although, a decrease of H_2O partial pressure in the flue gas activates Cl_2 towards gaseous chloride corrosion area. In reducing conditions steam concentration in the system is decreasing. Low steam concentrations promotes gaseous chloride corrosion. Presented conclusions can be used for combustion processes optimization as well as in interpretation of steel chloride-induced corrosion experiments.

Acknowledgments

The present work was supported by the Polish Ministry of Science (Grant AGH No. 11.11.210.312).

References

1. *Final meeting of the technical working group (TWG), for the review of the BAT reference document for large combustion plants (LCP BREF), final meeting, background paper (BP)* (Seville, 2015)
2. European Commission, *Renewable energy progress report* (Brussels, 2015)
3. P.Viklund, *Superheater corrosion in biomass and waste fired boilers. Characterization, causes and prevention of chlorine-induced corrosion* (Sweden, Division of Surface and Corrosion Science, Department of Chemistry, School of Chemical Science and Engineering, 2013)
4. M. Pronobis, Modernizacja kotłów energetycznych (Wydawnictwo Naukowo-Techniczne, Warszawa 2002)
5. T. Hardy, W, Kordylewski, K. Mościcki, *Arch. Spal.*, **9**, 181-195 (2009)
6. A. Zahs, M. Spiegel, *Mat. Sci. Eng.*, **50**, 561-578 (1999)
7. M. Spiegel, Beeinflussung der Korrosionsgeschwindigkeit von den Verhaltnissen im rohrwandnahen Bereich bei der HT-Chlor-Korrosion. Vortag VDIWissensforum „Belage und Korrosion", 14.06.2005 in *Dampferzeugerkorrosion*, M. Born (Freiberg, SAXONIA Standortentwicklungs-und-Verwaltungsgesellschaft mbH, 2005)
8. D. Bramhoff u.a.: Einfluss von HCl und Cl_2 auf die Hochtemperaturkorrosion des 2,25Cr1Mo-Stahls in Atmospharen mit hohen Sauerstoffdrucken. Werkstoffe und Korrosion in *Dampferzeugerkorrosion*, M. Born (Freiberg, SAXONIA Standortentwicklungs-und-Verwaltungsgesellschaft mbH, 2005)
9. K. Zhang, Y. Niu, C. Zeng, W. Wu, J. *Mater. Sci. Technol.*, **20**, 2, 213-216 (2004)
10. M.A. Uusitalo, P.M.J. Vuoristo, T.A. Mantyla, *Mater. Sci. Eng.*, **346**, 168-177 (2003)
11. M. Spiegel,, *Mater. Corro.*, **50**, 373-393 (1999)
12. M. Sanchez Pasten, M. Spiegel, *Mater. Corro.*, **57**, 2, 192-192 (2006)
13. B. J. Skrifvars, M. Westén-Karlsson, M. Hupa, K. Salmenoja, *Corro. Sci.*, **52**, 1011–1019 (2010)
14. B. J. Skrifvars, R. Backman, M. Hupa, K. Salmenoja, E. Vakkilainen, *Corro. Sci.*, **50**, 1274–1282 (2008)
15. Y. S. Li, Y. Niu, W.T. Wu, *Mater. Sci. Eng.*, **A345**, 64, 64-71 (2003)
16. J.R. Nicholls, R. Newton, N. J. Simms, J.F. Norton, *Mater. High Temp.*, **20**, 2, 93-108 (2003)
17. K. Salmenoja, M. Hupa, R. Backman, Laboratory Studies on the Influence of Gaseous HCl on Fireside Corrosion of Superheaters in *Impact of Mineral Impurities in Solid Fuel Combustion*, edited by R. P. Gupta, T. F. Wall, L. Baxter (Springer US, New York, 1999)
18. H. J. Grabke, E. Reese, M., *Corro. Sci.*, **37**, 7, 1023-1043 (1995)
19. M.A. Uusitalo, P.M.J. Vouoristo, T.A. Mäntylä, *Corro. Sci.*, **46**, 1311-1331 (2004)
20. W. Ptak, M. Sukiennik, *Bulletin de l'Academie Polonaise des Sciences. Serie des sciences techniques*, **17**, 21-25 (1969)
21. W. Ptak, M. Sukienni, R. Olesinski, R. Kaczmarczyk, *Arch. Metall.*, **32**, 355-362 (1987)
22. R. Kaczmarczyk, S. Gurgul, *Arch. Metall. Mater.*, **59**, 4, 1379-1383 (2014)
23. R. Kaczmarczyk, S. Gurgul S, *Arch. Metall. Mater.*, **59**, 1, 145-148 (2014)
24. O. Knacke, O. Kubaschewski, K. Hesselmann, *Thermochemical Properties of Inorganic Substances* (Berlin, Springer, 1991)
25. Outokumpu HSC Chemistry, *Chemical Reaction and Equilibrium Software*, (Outokumpu Research Oy, Finland, 2002) ISBN 952-9507-08-9.
26. N. Masuko, *Mater.Tran.*, **45**, 8, 2489 - 2495 (2004)

E3S Web of Conferences 10, 00061 (2016) DOI: 10.1051/e3sconf/20161000061
SEED 2016

Modelling changes in the energy efficiency of buildings using neural networks on the example of Zielona Góra

Andrzej Łączak[1], Maria Mrówczyńska[2] , Anna Bazan – Krzywoszańska[2] and Marta Skiba[2,a]

[1] *State School of Higher Professional Education in Sulechów, Institute of Law and Tourism, 66-100 Sulechów, Armii Krajowej 51, Poland*
[2] *University of Zielona Góra, Faculty of Civil Engineering, Architecture and Environmental Engineering, 65-417 Zielona Góra, Licealna 9, Poland*

Abstract. The objective of this article is to find a way to pursue optimum spatial policy on the local level to meet the assumptions of the energy policy of the European Union. One of the possible ways of developing energy efficient civil engineering is varied town policy and programmes supporting energy efficient buildings. And the second is the use of renewable energy sources as a factor improving the energy safety of built areas and reducing the emission of greenhouse gases. And the third is the optimization of expenditure on these goals in towns. Although our current research and estimations based on it are limited to a medium-sized town in the west of Poland, the observations included in this article may be important for other regions that are interested in reducing energy consumption in buildings, residential areas and towns. Taking into account the geographical context, it is especially important for these regions of Europe that are obtaining financial aid from the European Union in the perspective for the years 2014-2020.

1 Introduction

The participation of cities in the global consumption of energy is constantly increasing and there is no indication that this trend will change. Increasing population, increasing demand for construction services and raising the level of comfort, with the increase in time spent in buildings, provides the be upward trend in energy demand. Reducing the amount of energy needed to ensure and maintain thermal comfort is a common goal that all developed societies, driven both by rising energy costs and environmental concerns, such as reducing carbon dioxide emissions. The increase in the level of heat and power consumption may not, however, be reduced below the thresholds set out in the strategies, and based only on insulation or ventilation control. There is also a need for energy from natural and renewable sources [1].

Reducing consumption of thermal, electrical and gas energy and an increase in the efficiency of devices in buildings is the main objective of the energy policy at regional, national and international levels today. This goal is also reflected in the local policy, setting out the scope for investments. The popularity of smart ways to design and construct more energy-efficient buildings, as well as increase their comfort has been growing in construction and architecture [2,3]. Numerous case studies are known in the literature, that describe and analyze individual projects to increase energy efficiency, depending on the environmental conditions, the characteristics of the building and supporting technology used to reduce energy consumption, energy costs and CO [2] emissions in buildings, also through the active technologies, such as photovoltaic panels [4,5], solar panels [6], wind turbines [3.7] or heat pump systems [8]. Important economic benefits were described due to the application of the simultaneous production of electricity, heating and cooling in cogeneration [9].

The economic justification of the future of energy-efficient construction without the possibility of obtaining funds from the European Union (also in the new perspective 2014-2020) is very difficult to achieve [10]. There is a need for public intervention in the forms of colour energy certificates, subsidies, tax exemptions, promotions and good practices to encourage a more efficient use of energy in households. Communal and city municipal authorities are the main bodies responsible for monitoring processes of reducing energy demand.

One of the possible ways of development of energy-efficient construction is the diversity of urban policy and the launch of programmes to promote energy-efficient construction. The second one is the promotion of good practices and examples, for instance by the pressure on low-energy construction of municipal and local authorities buildings. Third, is the use of renewable energy sources as means to improve safety and energy balance. And the fourth one is the optimization of spending the financial resources intended for these purposes in a city [10].

[a] Corresponding author: M.Skiba@aiu.uz.zgora.pl

Presented modelling, you made on the basis of the GIS data for the city[1] of medium size in the western part of Poland. The observations and conclusions presented in this article may be relevant for regions that are interested in reducing the energy consumption of buildings, housing estates and towns. Taking the geographical context into account, it is especially important for those regions which benefit from financial support of the European Union over the 2014-2020 perspective. The process of reaching the values of the energy balance, provided for by the European Commission, will not be quick if these values are referred to only by the insulation parameters of particular elements of wall barriers, increase in devices ' efficiency and energy characteristics of an apartment or a building without specifying the requirements of improving energy efficiency in the scale of the housing estate , the district and the city.

Making actions aimed at improving the efficiency and reducing the demand for heat and electricity dependant only from the economic benefits, the influences on the economic situation of families, which is related to the notion of energy poverty. Energy poverty depends primarily on family income and energy prices (Household Budget Survey [2], GUS 2014). Anti-energy poverty strategy requires actions to be taken in order to reduce energy prices and its availability (rationing) and taking actions aiming to intensive use of the infrastructure, which is contributed by the concentration of people [11]. These objectives can be achieved, inter alia, by conducting a responsible urban policy requiring control-direct action and economic incentive.

2 Energy efficiency requirements

Poland, like the countries that joined the European Union in the 21st century, faces the need to adapt the development policy to the requirements of the Union in terms of reducing the production of greenhouse gases, mainly by reducing the operational demand for heat, electricity, gaseous fuels [12-14]. In accordance with article 9 par. 3 item (b) of the directive 2010/31/EU, a National Energy Efficiency Action Plan for Poland 2014 [15] was prepared, which includes, inter alia, the intermediate targets to improve the energy performance of new buildings for the year 2015, with a view to the preparation of implementing the assumption that until 31 December 2020 all newly constructed buildings are near-zero energy buildings, and after 31 December 2018 new buildings occupied and public buildings are near-zero energy consumption buildings.

On the basis of article 4 of the 2012/27/EU directive Supporting Investments for Buildings Modernisation [16] document was prepared in the Ministry of Infrastructure and the Development. It is an annex to the National Energy Efficiency Action Plan for Poland 2014, which was prepared due to the obligation of reporting to the European Commission on the implementation of the directive 2012/27/EU, as well as on the basis of the obligation imposed on the Minister of Economy in article 6 par.1 of the Energy Efficiency Act of 15 April 2011 [17,18]. This document contains, inter alia, a description of the planned energy efficiency improvements specifying the actions taken to improve energy efficiency in various sectors of the economy, necessary to implement the national target for economical energy management for 2016, as well as a review of the final energy savings to be reached in 2020.

It seems, however, that if communes and cities do not cooperate, and the effects of the public consultation do not become part of the arrangements of local zoning plans [19], a mechanism to support small and micro-installations will not be an incentive for the development of renewable energy sources (RES). One of the reasons is that the regulations concerning the energy efficiency of buildings are constructed in such a way that meeting the requirements in force from 1 January 2021 and thus achievement of the required values of Primary Energy Factor using conventional solutions (including the use of better thermal insulation wall barriers and recovery and improving the efficiency of central heating and hot utility water systems) is associated with significant investments limitations of a building features or functionality, to which we are accustomed [20]. This is a deliberate measure, which is meant to encourage investors to use alternative sources of energy, including renewable energy, heat pumps and cogeneration heat and power in buildings. The effectiveness of this measure will depend on the cost of the applied solutions.

2.1 Support for energy efficiency

Supporting energy efficiency in public buildings and residential sector has become one of the tasks carried out under the EU funds for 2014-2020 in Poland. These tasks are implemented in the Infrastructure and Environment Operational Programme, and 16 regional programmes. In Zielona Góra there is a possibility to use two programs of the Regional Operational Programme - Lubuskie 2020 and National Infrastructure and Environment Operational Programme. Three following actions supporting energy efficiency in buildings are planned:
-supporting energy efficiency in public buildings,
-supporting energy efficiency in the residential sector,
-creating a nationwide system of advisory support on energy efficiency and RES for the public, housing and business sectors .

The implementation of these activities is intended to contribute to the deep comprehensive energy retrofit, which may include:
- thermal insulation of the building
- exchange of equipment objects on energy-efficient equipment, among others, replacement of windows, exterior doors, the modernization of the internal heating and preparing hot utility water facilities, and common lighting parts to energy-efficient.

License no DR-G.6642.1059.2016.AW_0862_P of 14.03.20E16

[2] http/stat.gov.pl

- reconstruction of heating systems and mechanical ventilation systems,
- installation of renewable energy sources in the energy-efficient modernized buildings,
- introduction of energy management systems.

The scope of the project must lead to a reduction in final energy consumption by at least 25%. Total funding for this purpose within the Operational Programme for Infrastructure and Environment is 421 278 811 EUR [21].

The second program available in Zielona Góra is the Lubuskie Regional Operational Programme - Lubuskie 2020, priority third - low carbon Economy (Table 1). 108 005 260 euro is provided for the implementation of the entire priority. Under this priority, investments in renewable energy, energy efficiency, low-emission limit in cities and cogeneration are provided.

Table 1. The value of the funding in priority third Regional Operational Programme-Lubuskie2020 for 2014-2020 [EUR]. Source: Own studies based on the Regional Operational Programme-Lubuskie 2020.

The title of the action	The value of the grant in EUR
Renewable energy sources	19 601 052
Energy efficiency	39 202 104
Low emission reduction in cities	36 134 736
Cogeneration	13 067 368

The maximum level of granting in all programs is 85% of the eligible expenditure of the project.

2.1.1 Public buildings

In the current financial perspective of the EU public buildings have gained in importance. The legal basis under a new perspective, was being prepared from 2010. Article 9 par.2 of the directive 2010/31/EU specifies that Member States "shall furthermore, following the leading example of the public sector, develop policies and take measures such as the setting of targets in order to stimulate the transformation of buildings that are refurbished into nearly zero-energy buildings, and inform the Commission thereof in their national plans" [13].

Article. 4 of the directive of the European Parliament and of the Council of 2012/27/EU of 25 October 2012 on energy efficiency, amending Directives 2009/125/EC and 2010/30/EU and repealing Directives 2004/8/EC and 2006/32/EC (Official Journal of the EU L 315 of 14.11.2012, p. 1, as amended), hereinafter referred to as "the directive 2012/27/EU" imposes the obligation to establish a long-term strategy for mobilising investment in the renovation of residential and commercial buildings, both public and private (Table 2). This strategy includes:

- an overview of the national building stock based, as appropriate, on statistical sampling;
- identification of cost-effective approaches to renovations relevant to the building type and climatic zone;
- policies and measures to stimulate cost-effective deep renovation of buildings;
- including staged deep renovations;
- a forward-looking perspective to guide investment decisions of individuals, the construction industry and financial institutions;
- an evidence-based estimate of expected energy savings and wider benefits"[14].

In accordance with § 3, point 6. Regulation of the Minister of Infrastructure on the technical specifications which should correspond to the buildings and their position of 17 July 2015. (Journal of Laws of 2015 item 1422) [22], a public building - shall mean "the building designed for the needs of public administration, justice, culture, worship, education, higher education, science, upbringing, health care, social care, banking, trade, catering, services, including postal services or telecommunications, tourism, sport, serving passengers in transport by rail, road, air, sea or inland waterway, and another building designed to perform similar functions; public building shall also mean an office or amenity building".

Table 2. Intermediate objectives to improve the energy performance of new buildings. Source: Own studies based on the Regulation of the Minister of Infrastructure on the technical specifications which should correspond to the buildings and their position of 12 April 2002. (Journal of Laws No. 75, item. 690).

near-zero energy consumption buildings	
after 31 December 2018	after 31 December 2020
new buildings occupied and owned by municipalities	all new buildings

2.2 Funding and support

Thermal renovation of buildings is generally highly profitable, but requires incurring significant initial costs, therefore, many building owners may not accomplish the thermal renovation without a financial support of the state. The State support system for buildings owners has been specified in The Act of supporting thermomodernisation and renovation of 21 November 2008 (Journal of Laws 2008, No. 223 item 1459) [23]; The system aims to facilitate funding thermal renovation in order to reduce energy consumption and costs of heating the buildings and preparing hot utility water. The act introduces new granting rules for thermal renovation purposes, and the helping system is introduced to encourage a certain group of renovations.

In the field of energy efficiency and cogeneration on the territory of the town of Zielona Góra a total of 10 projects were completed. The projects were implemented within the two programs - Infrastructure and Environment Operational Programme, and Lubuskie Regional Operational Programme (Table 3). In the field of renewable energy only one project was completed (Table 4).

3 Method and object of research

The research focused on the buildings owned by the City Hall (municipalities) within the city of Zielona Góra, which is located in the western part of Poland. In order to determine the energy potential, meant as the ability to make effective use of local resources, the city was

Table 3. Projects completed in the field of energy efficiency in Zielona Góra within the energy efficiency in the financial perspective 2007-2013 in EUR (1EUR=4,36 PLN for day 22.07.2016 r. [29]). Source: Own studies based on the KSI-SIMIK (2007-2013) (National Information Technology System- Information Technology System for Monitoring and Financial Control of the Structural Funds and the Cohesion Fund).

Title of the project	The Operational Programme	Total value	Eligible Expenditure	Funding	EU funding	Beneficiary's name
Modernization and reconstruction of heat supply system in Zielona Góra	The Infrastructure and Environment Operational Programme	1 2581 135	9 776 965	8 310 420	8 310 420	Elektrociepłownia "Zielona Góra" S.A.
Low carbon economy in the functional area of the provincial city of Zielona Góra	The Infrastructure and Environment Operational Programme	18 804	18 804	15 372	15 372	Zielona Góra city
Thermal renovation of the main building of the C.Norwid Provincial and Municipal Public Library in Zielona Góra	The Lubuskie Regional Operational Programme	2 149 567	2 018 633	1 378 171	1 378 171	C. Norwid Provincial and Municipal Public Library in Zielona Góra
Thermal renovation of the Police Headquarters building in Zielona Góra	The Lubuskie Regional Operational Programme	414 571	414 571	352 385	352 385	Provincial Police Headquarters in Gorzow Wielkopolski
Thermal renovation of the Roman Catholic Parish of the Holy Spirit in Zielona Góra buildings including the exchange of a heat system for an eco-friendly hybrid heating system based on RES	The Lubuskie Regional Operational Programme	773 228	773 228	657 243	657 243	Roman Catholic Parish of the Holy Spirit in Zielona Góra
"Thermal renovation of a Tennis Hall and the Facility Building "	The Lubuskie Regional Operational Programme	818 079	544 456	381 119	381 119	Lubuskie Voivodeship represented by the Provincial Sports and Recreation Centre in Drzonkow
Thermal renovation of educational buildings in Zielona Góra	The Lubuskie Regional Operational Programme	964 642	964 642	819 945	819 945	Zielona Góra city
Thermal renovation of the Provincial Labour Office building in Zielona Góra	The Lubuskie Regional Operational Programme	173 689	126 080	107 168	107 168	Lubuskie Voivodeship represented by the Provincial Labour Office in Zielona Góra
Reconstruction of the boiler-room in State Music School building in ul. Chrobrego 26, Zielona Góra, from coal to gas supplied.	The Lubuskie Regional Operational Programme	33 082	25 509	21 683	21 683	The State Music School of 1st and 2nd level
Thermal renovation of the Public Health Care Centre of the Ministry of the Interior buildings in ul. Wazów 42, Zielona Góra	The Lubuskie Regional Operational Programme	778 208	522 193	443 864	443 864	Independent Public Health Care Centre of the Ministry of the Interior in Zielona Góra
Total		18 705 005	15 185 081	12 487 370	12 487 370	

DOI: 10.1051/e3sconf/20161000061

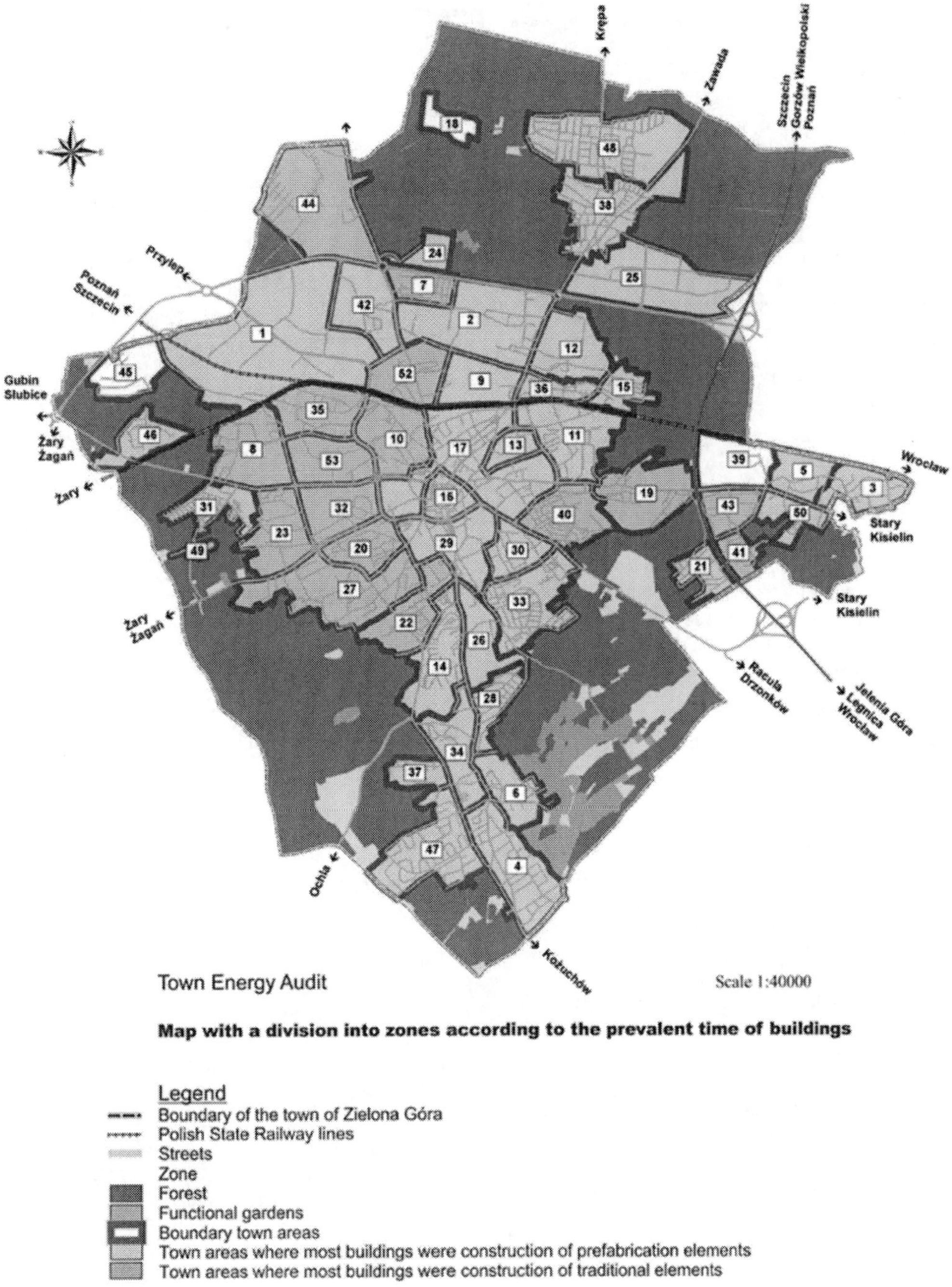

Figure 1. The division of research area. Source: [20].

Table 4. RES projects realised in Zielona Góra under the EU funds in the 2007-2013 in EUR. Source: Own studies based on the KSI-SIMIK (2007-2013) (National Information Technology System- Information Technology System for Monitoring and Financial Control of the Structural Funds and the Cohesion Fund).

Title of the project	The Operational Programme	Total value	Eligible Expenditure	Funding	EU Funding	Beneficiary's name
The modernization of the heating systemwith the use of solar panels in the "L" building of the Provincial Hospital SP ZOZ in Zielona Góra	The Lubuskie Regional Operational Programme	139 115	136 406	113 871	113 871	Karol Marcinkowski Provincial Hospital Independent Public Health Care Centre in Zielona Góra

divided into 50 districts, differentiated in terms of functions, construction technologies, construction date, and buildings condition (fig. 1).

3.1 Research area - Zielona Góra

The study examined two areas with the highest and lowest index of development with public buildings. Public buildings are 55.6% (quarter 17) and 20.2% (quarter 32) of buildings in designated areas for which factors affecting energy consumption and estimated energy potential were examined. These factors included: technology and construction year(most related to the used building materials), dominant functions and the location of the district, associated with a source for obtaining heat and energy (solid fuels, gas, cogeneration). In this way, the data was obtained, which was used to determine the potential savings arising from achieving the standard 95 kWh/m²/year, and the standard 70 kWh/m²/year in public buildings.

3.1.1 Sigmoid neutral networks

Neural networks are modelling technique, which allows the representation of extremely complex functions. As a convenient non-linear tool they are used in solving problems in such unrelated areas as: finance, medicine, engineering, geology, physics, and many others. Neural networks work well wherever a linear approximation is not applicable to describe the phenomena, because models created with the use of neural networks easily reflect non-linear relations. Sigmoid artificial neural network represent a class of mathematical algorithms which have a strong theoretical basics and usability in practice. They are systems in the form of interconnected neurons, where computing power allows to carry out mapping of the input space to the output [24]. This mapping is implemented by a process of training the network. Network training is carried out by forcing it to such a reaction, which meets the relevant requirements with respect to the input signal. Scheme of mapping the input vector

$$\mathbf{x} \in R^N \quad (1)$$

in the output vector

$$\mathbf{y}(\mathbf{x}) \in R^M \quad (2)$$

is shown in Figure 2 [25]. Symbols used in the formulas (1) and (2), and in Figure 2 are defined as follows: **x** - input vector; **y** - output vector; *RN*, *RM* - *N* and *M*-dimensional set of real numbers; *v* -the hidden layer weights, *w*- the output layer weights.

In the presented work sigmoid multilayer neural network trained by back propagation error technique were used to model the potential savings resulting from the need to reduce energy consumption in existing public buildings to 95 kWh/m²/year until 2017, and to 70 kWh/m²/ year until 2021 [22].

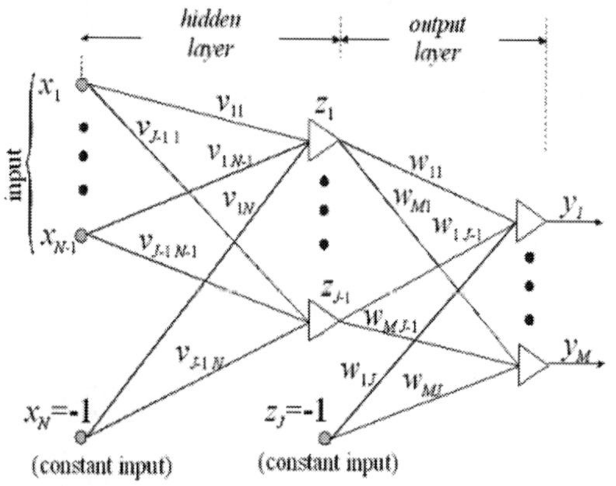

Figure 2. The mapping of the input vector **X** in the output vector **y**.

Training and testing the network was conducted on two areas of the city of Zielona Góra, in which the largest (quarterly 17) and the smallest (quarter 32) number of public buildings was located (table 5). Neural network used in the work to the approximation of potential savings within the analysed quarters, was a two-layer network of bipolar activation function (hyperbolic tangent). The input vector were the data for the location of the public buildings in space (coordinates *x* and *y* of the centre of gravity of the object), the surface of the building and the technology of the implementation (coded numerical form). The output vector ware the values of potential savings arising from the improvement of the buildings to the standards specified in table 5, estimated on the basis of energy audits carried out in the area of Zielona Góra. Training set consisted of 65% of the data in particular quarters, and test set consisted of 35% of the data. Due to the limited number of data, a validation set that could be used in the process of the ongoing training, was not created. Graphic illustration of approximated obtained using a network of potential savings is featured on the figures contained in table 5.

4 Results and discussion

Research results presented in the work made it possible to identify potential in savings and the differences arising from the plot ratio of development with public buildings in various parts of the city of Zielona Góra. This knowledge allows us to adapt the policy of the city development to optimize the benefits of improving public buildings to standards (95 and 70 kWh/(m 2/year)) set in the technical conditions and legislative documents. Energy requirements of cities are complex and multi-faceted. As a consequence of the State policy adopted, cities should improve the current system of power supply and energy consumption and implement new solutions in a coordinated manner, thus optimizing synergies between all the possible energy savings solutions. Growing demand for electricity and heat and an increase in its prices, as well as the use of renewable sources, is an important energy challenge, which would be better dealt with in its complexity and not separately as it is often the case [4].

Table 5. Numerical characteristics of analysis performed.

Quarter	No. 32	No. 17
Total developed area	25300 m^2	311200 m^2
Area of development by public buildings	5100 m^2	173000 m^2
The potential savings a result of achieving the standard 95 kWh/m^2/year in public building	1,230,000 kWh/m^2/year	8,905,200 kWh/m^2/year
The potential savings as a result of achieving the standard 70 kWh/m^2/year in public buildings	1,480,000 kWh/m^2/year	10,715,200 kWh/m^2/year

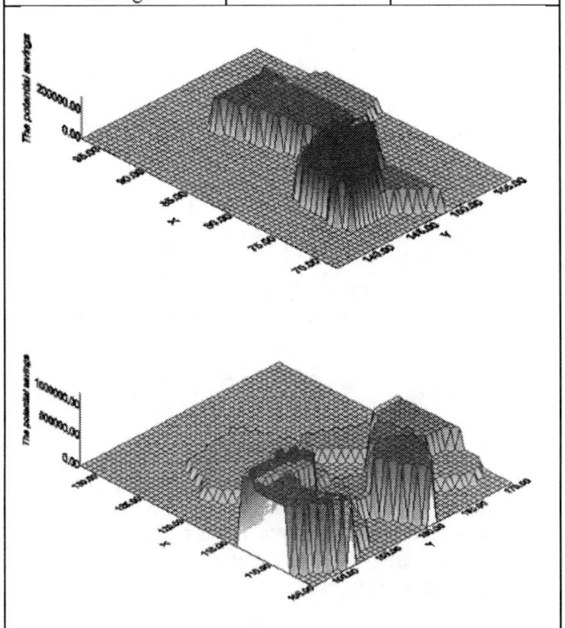

Distribution of potential savings resulting from energy-friendly activities for public buildings in certain year quarters

Fabrri and Tronchin (2015) described another challenge for the construction real estate industries, which is formed as a question: is there a link between the buildings with low and high demand for energy and the quality of their use? Quality, which, in case of buildings which are representative for municipal authorities and the city should not be dependent on the standards defining the height of rooms, transposition and technical solutions. EN 15251 standard specifies the internal input parameters for the design environment and assess the energy performance of buildings for indoor air quality, thermal comfort, lighting and acoustics [26].

The use of renewable energy sources is increasing especially in Mediterranean climate, it works as a strategy to reduce energy consumption in buildings. The reduction of the energy demand of the building is based on the use of construction elements, which are able to absorb passively, distract or collect energy [27].

In 2011 and 2012, in four cities: Shanghai, Tianjin, Shenzhen, Chongqing - a pilot programme of increasing energy efficiency in modernized commercial buildings was implemented [28]. On the basis of the actual research and expert interviews, a comparative analysis of motivation used in local politics was conducted. The results indicated that, in order to maximize efficiency and reduce energy consumption changes in administrative policy and support should be applied in the form of fiscal subsidies to existing commercial and public buildings, such as schools. Taking into account local economic conditions and fiscal capacity, the actual energy savings and a policy of incentives in order to modernize the economy was proposed. It is specified in the technical solutions that power management by: optimization of working time, changes in human behavior, lighting system and modernization of an air conditioning system can be done without a subsidy [28].

5 Summary

Public buildings in every city are the municipalities representative buildings, courts, swimming pools, schools, libraries, museums, and more. They often decide on the reception and perception of the city. Buildings of important social role. In the urban tissue, they make up one third of the volume. City policy should support changes to increase the energy efficiency of these buildings. Currently established technical criteria (energy characteristics) do not comply with the existing buildings. Newly built investment will fail to reach the criteria without the use of devices based on obtaining energy from RES. Energy policy of a city should be more diverse. It would be advisable to run programs to promote energy-efficient construction, so as to recommend potential investors beneficial solutions along with a possibility of financial support.

Artificial neural networks allow to obtain beneficial results in practical applications such as: approximation, interpolation, recognition and classification of patterns, compression and many more. It is worth noticing that modelling potential savings which result from the improved energy efficiency, can be successfully applied with artificial neural networks, which due to their non-linear nature represent an advanced modelling technique capable of mapping complex relations. In this article, the use of neural networks, which as input vector used the coordinates (x, y) of the centre of gravity, the area of a public building and its building technology, has allowed modelling of the deployment of energy savings potential on two quarters of the city of Zielona Góra. At the same time, it is worth indicating that the used approach will be available in the future to model the deployment of potential savings for the whole city.

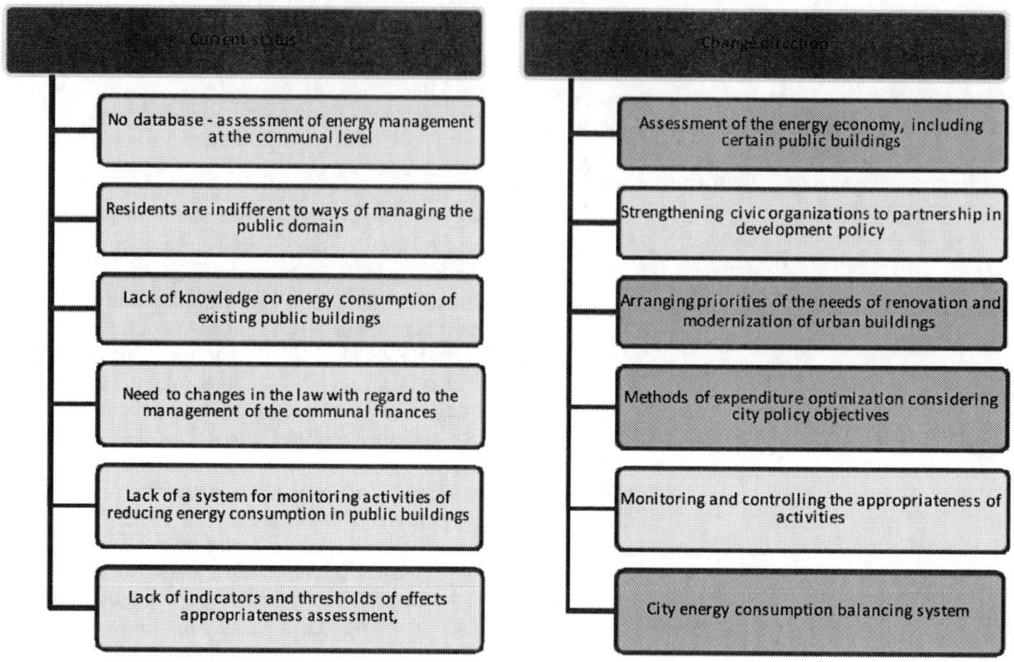

Figure 3. The possible practical actions supporting city management for increasing energy efficiency in public buildings. Own studies

References

1. M. Saffari, A. de Gracia, S. Ushak, L. Cabeza, Energ Buildings **112,** 159–172 (2016)
2. K. Fabbri, Energy **89**, 244-258 (2015)
3. M. Santamouris, Sol Energy **128**, 61-94 (2016)
4. C. Calvillo, A. Sánchez-Miralles, J. Villar, Renew Sust Energ Rev **55**, 273–287 (2016)
5. A. Stefanović, M. Bojić, D. Grodić, Energ Buildings **76**, 57-63 (2014)
6. D. Teodorescu, A. Vartires,Energ Procedia **85**, 530-538 (2016)
7. A. Tummala, R. Velamoti, D. Sinha, V. Indraja, V. Krishna, Renew Sust Energ Rev **56**, 1351 -1371 (2016)
8. F. Scarpa, L. Tagliafico, Appl Therm Eng **100**, 820-828 (2016)
9. A. Chmielewski, R. Gumiński, J. Mączak, S. Radkowski, P. Szulim, Renew Sust Energ Rev **60**, 930-952 (2016)
10. A. Bazan-Krzywoszańska, M. Mrówczyńska, A. Łączak, M. Skiba, Energ Buildings **118**, 170–180 (2016)
11. L. Bettencourt, J. Lobo, D. Helbing, C. Kuhnert, G. West, P Nat Acad Sci of the USA **104** (17), 7301–7306, (2007)
12. Directive 2009/29/EC of the European Parliament and of the Council of 23 April 2009 amending Directive 2003/87/EC *so as to improve and extend the greenhouse gas emission allowance trading scheme of the Community*
13. Directive 2010/31/EU of the European Parliament and of the Council of 19 May 2010 *on the energy performance of buildings*
14. Directive 2012/27/EU of The European Parliament And Of The Council of 25 October 2012 *on energy efficiency*, amending Directives 2009/125/EC and 2010/30/EU and repealing Directives 2004/8/EC and 2006/32/EC
15. Ministerstwo Gospodarki, *Krajowy Plan Działań dotyczący efektywności energetycznej dla Polski 2014*, Warszawa (2014)
16. Ministerstwo Infrastruktury i Rozwoju, *Wspieranie Inwestycji w Modernizację Budynków*, Warszawa (2014)
17. Ustawa z dnia 15 kwietnia 2011 r. *o efektywności energetycznej* (Dz. U. z 2015r. poz. 2167 t.j. z późn. zm.)
18. Ustawa z dnia 10 kwietnia 1997 r. *Prawo energetyczne* (Dz. U. z 2012r. poz. 1059 t.j. z późn. zm.)
19. Ustawa z dnia 27 marca 2003 r. o planowaniu i zagospodarowaniu przestrzennym (Dz.U. z 2015 poz. 199 z późn. zm.)
20. *Energetyczny Audyt Miejski dla miasta Zielona Góra*, Zadanie badawcze nr 1: Analysis of the possibilities and socio-economic impact of raise in energy efficiency in the construction industry.

Umowa Nr SP/B/1/91454/10, NCBiR, Zielona Góra (2011)
21. Program Operacyjny Infrastruktura i Środowisko 2014-2020, Ministerstwo Infrastruktury i Rozwoju, Warszawa (2015)
22. Obwieszczenie Ministra Infrastruktury z dnia 17 lipca 2015r. w sprawie ogłoszenia jednolitego tekstu rozporządzenia Ministra Infrastruktury *w sprawie warunków technicznych, jakim powinny odpowiadać budynki i ich usytuowanie* (Dz. U. z 2015 r. poz. 1422)
23. Ustawa z 21 listopada 2008 r. *o wspieraniu termomodernizacji i remontów* (Dz. U. z 2014 r. poz. 712)
24. Ch. M. Bishop, *Pattern Recognition and Machine Learning*, Springer, New York (2006)
25. M. Mrówczyńska, *Studium nad doborem metod inteligencji numerycznej do rozwiązywania problemów z geodezji inżynieryjnej*, Oficyna Wydawnicza Uniwersytetu Zielonogórskiego, Zielona Góra (2015)
26. K. Fabbri, L. Tronchin, Energ Procedia **78**, 2778–2783 (2015)
27. J. Tenorio, J. Sánchez-Ramos, A. Ruiz-Pardo, S. Álvarez , L. Cabeza, Energy **8**, 8630-8649 (2015)
28. J. Hou, Y. Liu, Y. Wu, N. Zhou, W. Feng, Energ Policy **88**, 204–215 (2016)
29. http://www.nbp.pl/transfer.aspx [date of access: 23.07.2016]
30. http/stat.gov.pl [date of access: 23.07.2016]
31. License no DR-G.6642.1059.2016.AW_0862_P of 14.03.20E16

Performance calculations of thermoelectric module p-type leg composed of $(Bi_2Te_3)_x(Sb_2Te_3)_{1-x}$

Michał Musiał[1,2,a], Marcin Borcuch[1,2] and Krzysztof Wojciechowski[2]

[1]*AGH University of Science and Technology, Faculty of Energy and Fuels, Department of Thermal and Fluid Flow Machines, Poland*
[2]*AGH University of Science and Technology, Thermoelectric Research Laboratory, Poland*

Abstract. The paper presents the results of numerical and simple analytical performance investigations of p-type leg of thermoelectric module (TEM). Calculations of $(Bi_2Te_3)_x(Sb_2Te_3)_{1-x}$ p-type leg for different chemical compositions (x=0.16,0.20,0.24,0.26) have been carried out to estimate the power output and conversion efficiency. Study has been performed for constant and temperature dependent thermoelectric parameters: Seebeck coefficient α, electrical resistivity ρ and thermal conductivity k. Results of modelling for constant material parameters fit very well with the analytical solution, pointing that numerical tools can be useful in development of thermoelectric modules and generators.

Nomenclature

A Cross section area of p-type leg $[mm^2]$
I Current generated in the circuit $[A]$
k Thermal conductivity $[Wm^{-1}K^{-1}]$
L Length of p-type leg $[mm]$
P_L Power generated on the load $[W]$
$\dot{Q}_{abs}$ Heat absorbed at the hot junction $[W]$
$\dot{Q}_C$ Heat flux released at the cold junction $[W]$
$\dot{Q}_H$ Heat flux absorbed at the hot junction $[W]$
R Internal resistance of p-type leg $[\Omega]$
R_L Load resistance $[\Omega]$
T Temperature $[K]$
T_C Temperature at the cold junction $[K]$
T_H Temperature at the hot junction $[K]$
V_L Voltage on the load $[V]$
V_{OC} Open circuit voltage $[V]$
zT Thermoelectric figure of merit $[-]$
α Seebeck coefficient $[\mu VK^1]$
η_{mat} Conversion efficiency $[\%]$
ρ Electrical resistivity $[\Omega\, mm]$

1 Introduction

Thermoelectricity is a mature technology applied in large variety of fields and robustly growing year after year. As a result of Seebeck effect thermoelectric generators (TEGs) are able to directly convert heat into electricity using thermoelectric modules. Since 1960s thermoelectric generators have been used to power space probes [1] (e.g. Voyager 1 and 2) and Mars rovers (Spirit and Opportunity in 2004, Curiosity in 2011). Due to undeniable advantages such as no moving parts, no working fluid, compact design, quiet operation and no pollution generation the vast majority of these generators are still working till today. As a result of harsh EU environmental politics the issue of waste heat utilization became popular research topic worldwide, providing solutions how to improve the energy generation efficiency and limit the environment degradation problem. Solutions available on the market are expensive, complicated, requiring constant maintenance and occupying large amount of space. Thermoelectric generators since last decades have been noticed in the field of waste heat utilization. As the TEG efficiency do not satisfy the requirements to compete with other technologies, the efficiency improvements have to be done. In general TEG efficiency depends on the efficiency of thermoelectric modules and effectiveness of implemented heat exchanger [2]. Nowadays a lot of research has been carried out in the area of thermoelectric materials to develop more efficient modules to be applied in thermoelectric generators [3]. Commercially available modules with efficiency around 5% relies on thermoelectric materials developed in 1960s [4]. State-of-the-art

[a] Corresponding author: michal@agh.edu.pl

solutions provide materials to construct thermoelectric modules with efficiency reaching even 15% [5]. Developed in Thermoelectric Research Laboratory segmented thermoelectric modules can operate with in higher temperatures than commercial ones with efficiency exceeding 9% [6]. Thermoelectric materials efficiency is described using dimensionless figure of merit zT which directly corresponds with heat into electricity conversion efficiency.

$$zT = \frac{\alpha^2}{\rho \cdot k} \cdot T \qquad (1)$$

Recent research provides materials with zT values over 1 for many materials, such as: Bi_2Te_3 alloys, PbTe, skutterudites, SiGe etc. The highest figure of merit 2,6 has been shown for SnSe single crystals [7].

It has been proven that thermoelectric module leg (thermocouple) length ratio for maximum efficiency and power output is not the same. According to results presented in [8] maximum output power and conversion efficiency decrease with increase of the thermocouple length. Conventional thermoelectric module generates losses due to having ceramic insulator plates. Integration of TEM into heat exchangers was presented and discussed [9]. Steady-state and transient models have been prepared in order to simulate different operating conditions. Spacer-inserted thermoelectric device (SITED) has been proposed by Hoon Kim [10]. The main idea is to improve power generation and reduce thermoelectric material usage by putting low thermal conductivity material spacers. As a result of this modification electrical resistance increases, while thermal conductivity of thermoelectric modules decreases. Conclusions show the $/W ratio can be reduced by 10% sustaining 60-70% of maximum power output in contrast to maximum power output for a given zT. The importance of selecting appropriate soldering alloys was discussed in [11]. It is essential to provide stress relaxation by deformation of the solder to avoid the breakage of the legs of thermoelectric materials. The main aim of the paper is to present results of analytical and numerical investigations of simplified thermoelectric module consisting only of p-type leg. This results are first stage in further research, leading to develop numerical model of the entire thermoelectric module, taking into account both p-type and n-type materials, ceramics, connectors and thermal pastes.

2 Simplified analytical model

Scheme of analysed p-type leg of thermoelectric module is shown on Figure 1. Cross section area is A=72 mm² and length is L=12 mm. Heat is absorbed ($\dot{Q}_H$) at the hot junction and liberated at the cold junction ($\dot{Q}_C$). As a result of Seebeck effect, due to the temperature gradient voltage is generated. Load resistance is connected to the p-type leg and the electrical current appears in the circuit. Simple analytical performance calculations have been carried out only for constant thermoelectric properties. For temperature dependent values it is impossible to solve the equations analytically.

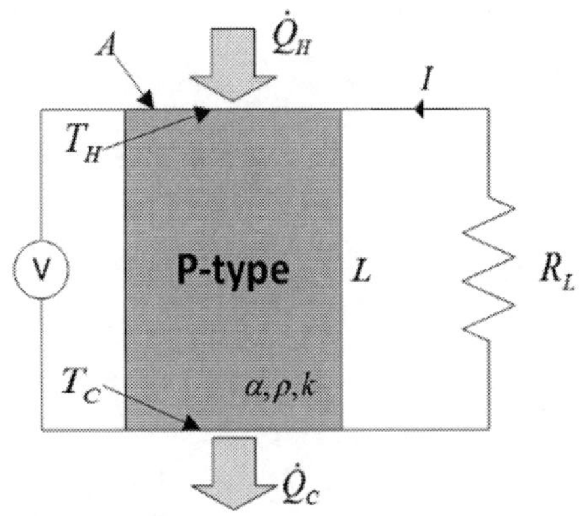

Figure 1. Scheme of p-type leg of thermoelectric module.

P-type thermocouple is made of $(Bi_2Te_3)_x(Sb_2Te_3)_{1-x}$, material presented and discussed in [12] with different chemical compositions (x=0.26, 0.24, 0.20, 0.16). Thermoelectric parameters for independent case are presented in Table 1.

Table 1. Temperature independent thermoelectric properties of p-type leg [12].

x	α [μV·K⁻¹]	ρ [Ω·mm]	k [W·m⁻¹K⁻¹]
0.26	225	0,0168	1,697
0.24	223	0,0147	1,767
0.20	219	0,0134	1,883
0.16	204	0,0101	1,870

Parameters values for temperature dependent case are shown on figures below.

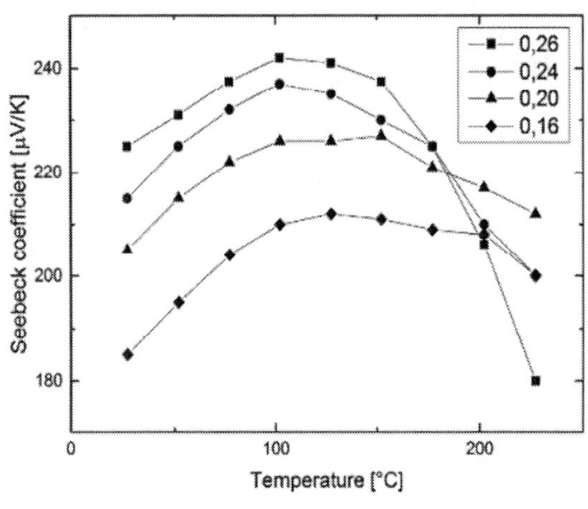

Figure 2. Temperature dependent Seebeck coefficient for different chemical compositions of p-type leg material [12].

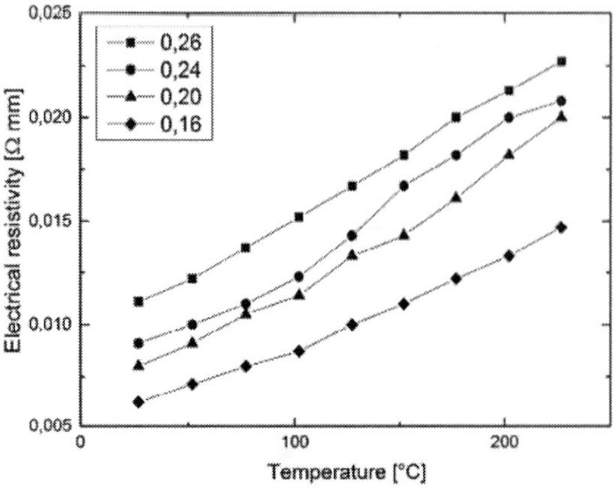

Figure 3. Temperature dependent electrical resistivity for different chemical compositions of p-type leg material [12].

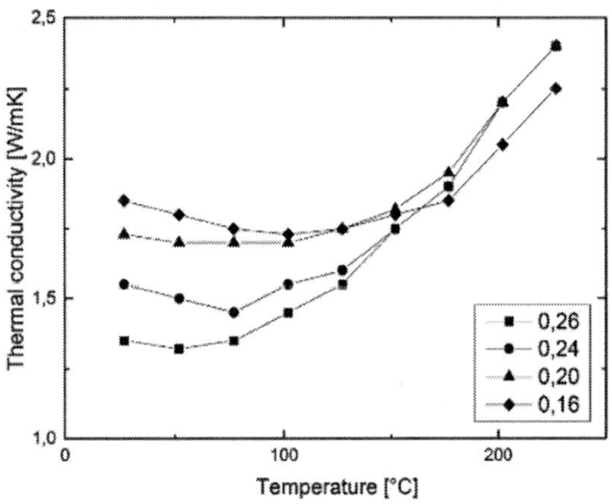

Figure 4. Temperature dependent thermal conductivity for different chemical compositions of p-type leg material [12].

Analysis has been done for hot side temperatures in the range from 27°C to 227°C with the step of 20°C, cold side temperature remained at the level of 20°C. As a result of Seebeck effect, due to temperature difference between hot and cold side voltage is generated. Seebeck coefficient defines amount of electrical energy generated with temperature difference of 1K. Voltage generated
by p-type leg can be calculated as follows:

$$V_{OC} = \alpha \cdot (T_H - T_C) \quad [V] \tag{2}$$

Internal resistance of the leg can be calculated using electrical resistivity and dimensions using equation presented below:

$$R = \frac{\rho \cdot L}{A} \quad [\Omega] \tag{3}$$

External load resistance R_L was assumed to be equal to internal resistance of p-type leg, which allows to calculate performance parameters for maximum power conditions. Electrical scheme of p-type leg with load resistance connected is presented on figure 4.

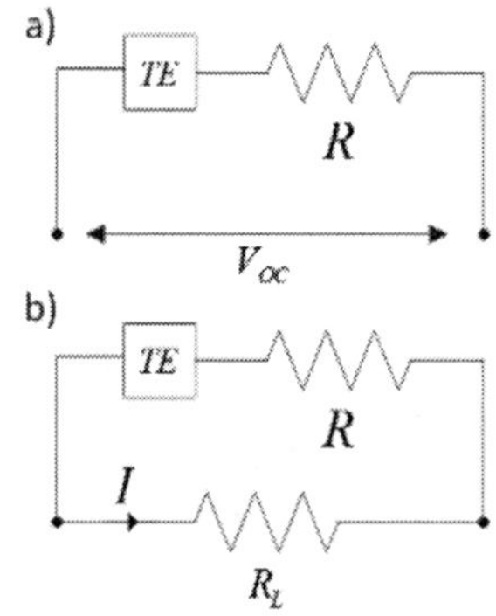

Figure 5. Electrical scheme of thermoelectric p-type leg with load resistance.

Figure 5a presents open circuit voltage, while scheme on Figure 5b represents p-type leg connected with load resistance. Current in the circuit is calculated as follows:

$$I = \frac{V_{OC}}{R + R_L} \quad [A] \tag{4}$$

Internal resistances for different x compositions of p-type leg material are presented in Table 2.

Table 2. Temperature independent thermoelectric properties of p-type leg [12].

x	R [Ω]
0.26	0,00280
0.24	0,00245
0.20	0,00224
0.16	0,00169

Voltage across the load and maximum power generated on the load can be obtained by:

$$V_L = I \cdot R_L \quad [V] \tag{5}$$

$$P_L = I^2 \cdot R_L \quad [W] \tag{6}$$

Heat absorbed by the material at the hot junction is calculated using:

$$Q_{abs} = \alpha \cdot T_H \cdot I - \frac{1}{2} \cdot I^2 \cdot R + \frac{k \cdot A}{L} \cdot (T_H - T_C) \quad [W] \tag{7}$$

Conversion efficiency, defined as the ratio of power generated by the module and heat absorbed is as follows:

$$\eta_{mat} = \frac{P_L}{Q_{abs}} \cdot 100\% \quad [\%] \tag{8}$$

This material efficiency do not take into account ceramics, connectors and copper base resistances. It is assumed that the efficiency of entire thermoelectric module is around one fourth of material efficiency.

3 Numerical model

Steady-state numerical model has been prepared in order to simulate the thermoelectric effect in p-type leg of thermoelectric module. Fine hexahedral mesh presented on Figure 6, composed of 12000 elements with orthogonal quality equals to 1 has been developed to decrease calculation time.

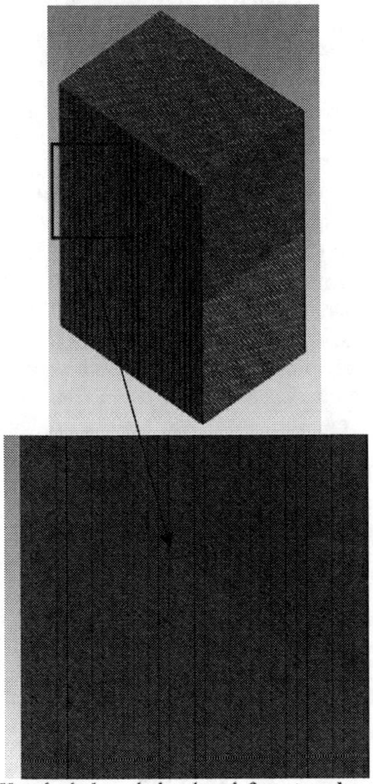

Figure 6. Hexahedral mesh developed for p-type leg of thermoelectric module.

Heat flux and electric current flows from hot side to cold side of the p-type leg. Simulation takes into account only heat transport due to Fourier's law and voltage generation as a result of Seebeck effect. Load resistance is included to assess the power generated.Thermal and electrical boundary conditions for the model have been defined as follows:

- Hot side temperature 27°C – 227°C (with the step of 20°C);
- Cold side temperature remained constant: 20°C;
- Thermal insulation condition has been defined to exclude heat transfer between p-type leg and surrounding;
- Voltage at the hot side of p-type leg has been set as 0V;
- Load resistance for constant α,ρ,k has the same value as internal resistance of p-type leg – simulations for maximum power generated on the load;
- Load resistance for temperature-dependent α,ρ,k has the value equals to temperature-averaged internal resistance of p-type leg.

4 Results

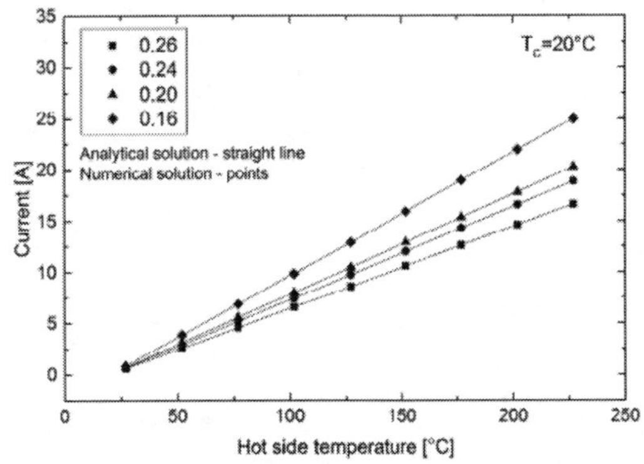

Figure 7. Results of analytical and numerical calculations of current in the function of hot side temperature for different chemical compositions and constant α, ρ, k.

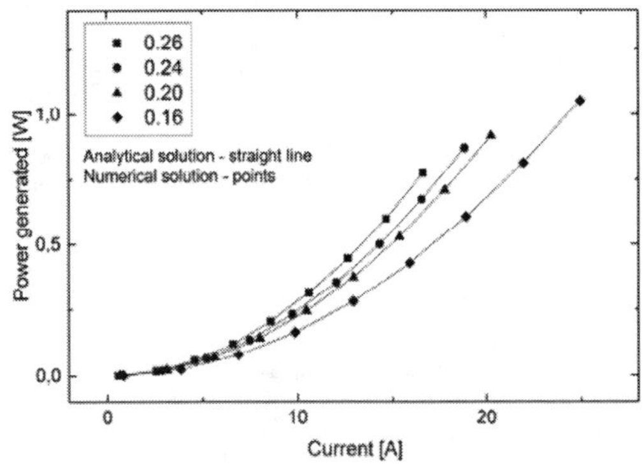

Figure 8. Results of analytical and numerical calculations of generated power in the function of current for different chemical compositions and constant α, ρ, k.

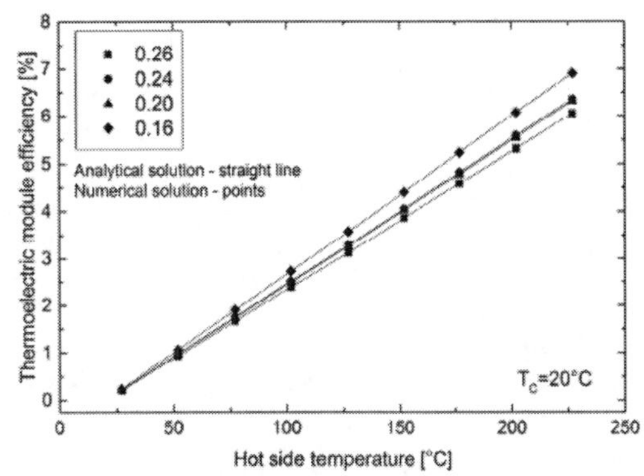

Figure 9. Results of analytical and numerical calculations of conversion efficiency in the function of hot side temperature for different chemical compositions and constant α, ρ, k.

Figures 7-9 presents the results of analytical calculations compared with numerical simulations. Maximum performance parameters are noticed for x = 0.16. This material has the highest conversion efficiency 6,9% for T_H = 227°C.

 DOI: 10.1051/e3sconf/20161000062

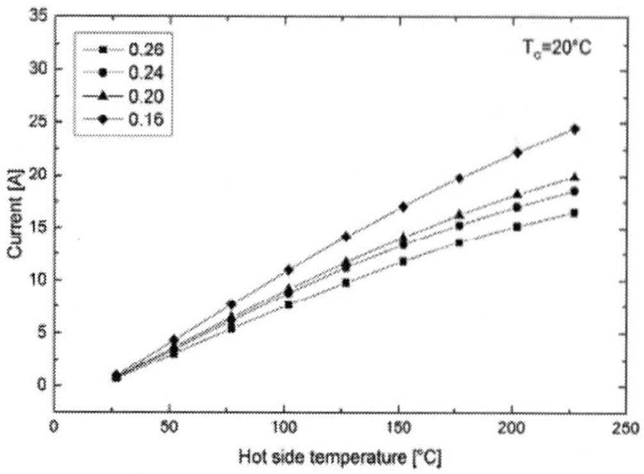

Figure 10. Results of numerical calculations of current in the function of hot side temperature for different chemical compositions and temperature dependent α, ρ, k.

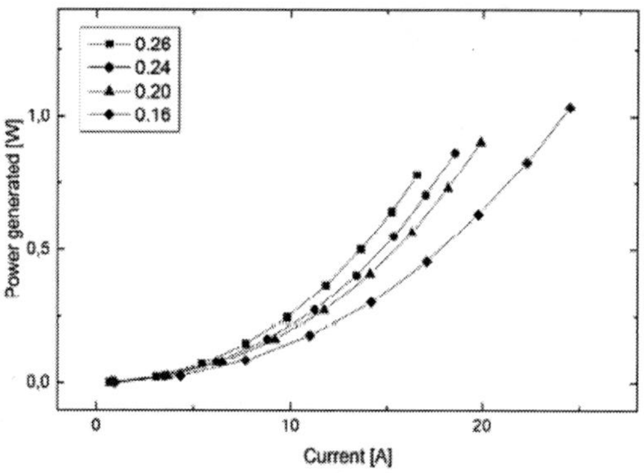

Figure 11. Results of numerical calculations of generated power in the function of current for different chemical compositions and temperature dependent α, ρ, k.

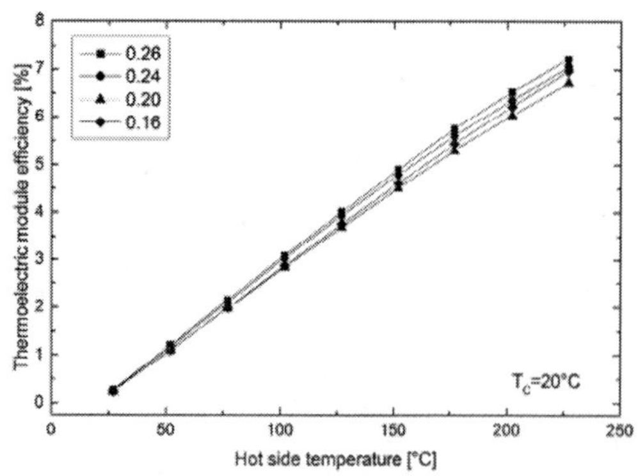

Figure 12. Results of numerical calculations of conversion efficiency in the function of hot side temperature for different chemical compositions and temperature dependent α, ρ, k.

Results of calculations and simulations for temperature dependent thermoelectric properties are shown of Figures 10-12. In contrast to constant α, ρ, k case performance parameters are not linear functions of temperature. For this case maximum efficiency exceeds 7,32% in T_H = 227°C for x=0.26.

5 Conclusions

The paper presents the results of performance calculations for p-type leg of thermoelectric module composed of $(Bi_2Te_3)_x(Sb_2Te_3)_{1-x}$ p-type leg for different chemical compositions (x=0.16, 0.20, 0.24, 0.26). Simplified analytical model concerning only thermoelectric material has been developed to estimate current, power generated and conversion efficiency of p-type leg. Two cases have been considered, the first with constant Seebeck coefficient, electrical resistivity and thermal conductivity, while second for parameters varied with temperature. Due to non-linear problems that can occur when temperature dependent thermoelectric parameters are taken into consideration the analytical calculations have been done only for the case with constant α, ρ and k. Numerical calculations have been carried out for both cases. Numerical model results have been confirmed with simple algebraic calculations. Current generated in the circuit is clearly the highest for x=0.16. For temperature independent part relation between current and hot side temperature is linear, while for temperature dependent case is not. The most power is generated on the load for x=0.16 in both cases. Conversion efficiency for temperature independent case has the highest value for different material than for temperature dependent case. It is also noticeable that for temperature dependent case the relation between efficiency and hot side temperature in not linear.
Maximum efficiency exceeded:

- 6,90% for T_H = 227°C, T_C = 20°C, R_L = 0,00169Ω for x = 0.16 (constant thermoelectric parameters)
- 7,32% for T_H = 227°C, T_C = 20°C, R_L = 0,00169Ω for x = 0.26 (temperature dependent thermoelectric parameters)

Nowadays, the numerical tools play an important role in development processes worldwide. The share of thermoelectric generators in the field of waste heat utilization increasing constantly. According to report [13] the market of thermoelectric generators will reach $875 million by 2023. Recent cutting edge solutions from the field of thermoelectric materials will contribute to the growth of more efficient and cheaper thermoelectric generators. Next steps in the research will cover:

- Investigations of performance parameters for thermocouple composed of p-type leg and n-type leg;
- Developing thermocouple model with soldering, contact layers, ceramics.

Acknowledgments

This work was supported by the funds of National Science Center granted on the basis of the decision No. DEC-2013/09/B/ST8/02043 and AGH grant No. 15.11.210.326

DOI: 10.1051/e3sconf/20161000062

References

1. H. Wei, Z. Gan, Z. Xingxing, J. Jie, L. Guiqiang, Z. Xudong, Appl. Energy **143**, 1-25 (2015),
2. G. Schierning, R. Chaves, R. Schmechal, B. Balke, G. Rogl, P. Rogl, Transl. Mater. Res. **2** (2015),
3. D. Aswal, R. Basu, A. Singh, Energy Convers. Manage. **114**, 50-67 (2016)
4. Z. Tian, S. Lee, G. Chen, Cond-mat.mtrl-sci (2014),
5. T. Caillat, J. Fleurial, G. Snyder, A. Borshchevsky, online: http://thermoelectrics.caltech.edu/SegmICT01.pdf (2001)
6. K. Wojciechowski, J. Merkisz, P. Fuć, J. Tomankiewicz, R. Zybała, J. Leszczyński, P. Lijewski, P. Nieroda, CE **154**, 60-71 (2013)
7. L. Weishu, J. Qing, K. Hee Seok, R. Zhifeng, Acta Mater. **87**, 357-376 (2015)
8.T. Hua, J. Na, J. Qi, S. Xiuxiu, S. Gequn, L. Xingyu, Energy Procedia **75**, 590-596 (2015)
9. D. Crane, J. Electron. Mater. **40**, 561-569 (2010)
10. H. Kim, W. Kim, Appl. Energy **139**, 205-211 (2015)
11. M. Picard, S. Turenne, D. Vasilevskiy, R. Masut, J. Electron. Mater. **42**, 2343-2349 (2013)
12. J. Jun, C. Lidong, B. Shengqiang, Y. Qin, W. Qun, J. Cryst. Growth **277**, 258-263 (2005)
13. H. Zervos, *Thermoelectric Energy Harvesting 2016-2025*, 2016
14. H. Lee, *Thermal Design*, 2010
15. X. Zheng, C. Liu, Y. Yan, Q. Wang, Renew. Sustainable Energy Rev. **32**, 486-503 (2014)
16. E. Antonova, D. Looman, IEEE, 215-218 (2005)
17. S. Yongming, Z. Zhixiang, D. Yuan, Z. Wei, C. Xin, Z. Yongsheng, Energy Convers. Manage. **101**, 713-720 (2015)
18. N. Espinosa, M. Lazard, L. Aixala, H. Scherrer, J. Electron. Mater. **39**, 1446-1455 (2010)

Investigation of the efficiency of a high temperature heat storage charging

Artur Nemś[1,a], Agata Hołowczak[1] and Magdalena Nemś[1]

[1]*Wroclaw University of Technology, Faculty of Mechanical and Power Engineering, Poland*

Abstract. The article presents the results of the test of a heat storage filled with chamotte brick. The air parameters were used as the basis for determining the heat stream represented by a working medium, which was then compared with the stream accumulated in the material filling the deposit. Additionally, heat streams were shown for individual segments of the deposit. The efficiency of heat accumulation was determined for each of them. They showed that the efficiency of accumulation process was highest in the deposit's first segments and the lowest in the last segments, where the air was the coldest. The dynamics of the efficiency's shift was different for each of the observed segments. In the area closest to the inlet of hot air, the dynamics of the process of heat accumulation decreased the fastest and after one hour only a slight increase of efficiency was visible in the function of time. In the accumulation material placed closest to the outlet, the process of absorbing heat was very fast, starting from the 20th minute before the tests were completed. The smallest changes in the process' character, throughout the whole time of testing, were recorded in the middle layer.

1 Introduction

Accumulation of heat is a very important issue, especially when there is a periodically available source of waste heat. Sensible heat storage materials are rocks, minerals, water, oil and different kinds of composite materials. Their alternative is accumulation in phase-change materials or materials that use chemical reactions. Apart from the ability to work in a broad range of temperatures, sensible materials have many advantages, such as: durability and low requirements as for the process' conditions, as well as for the heat storage's construction itself. Working media in such systems are usually gases, most often air.

There are studies devoted to accumulation in sensible heat storage materials. The following have been considered: the influence of shape on heat transfer [1], pressure drops [2], and optimisation of the process of charging and discharging [3, 4]. The most popular practical application is using them in connection with solar installations. Low-temperature rock storages are used in warm climate countries, where in combination with flat solar collectors they level the temperature in residential buildings [5]. Also natural rock caves are used as storages, being seasonal heat accumulators for whole housing estates, and they are charged during summer [6]. High-temperature installations are used in advanced systems, usually with concentrating solar collectors. They can be found in solar power plants as well as in heating systems still in the test phase [7, 8]. Also, a combination with high-temperature waste heat from technological processes and with air fireplaces using biomass is possible.

Further part of the article presents experimental tests of the efficiency of charging a heat storage filled with chamotte bricks. The installation is at the research stage and its purpose is cooperation with air system such as solar concentrating collector or the aforementioned biomass fireplace.

2 Set-up

The set-up shown in Fig. 1 is made of cuboidal sheet metal housings, having between them insulator in the form of mineral wool, 15-centimetre thick on vertical walls and 30-centimetre thick on horizontal walls. Internal housing is filled with accumulating material. The installation works in the following way: a ventilator sucks in ambient air, which flows from the inlet channel through the gas-meter, to the storage. Then, the air is heated until it reaches the desired temperature. Inside the deposit, the air transfers the heat to the material filling the storage and the deposit is charged. Cold working medium is taken out through the outlet channel. The system works in an open cycle.

Fig. 2 shows the measuring set-up during conducted tests. Air inlet at the bottom of the smaller housing is placed under metal mesh, on which chamotte plates were put. Such a solutions allows for an even charging of the deposit by distributing the air over the whole surface of the filling.

[a] Corresponding author: artur.nems@pwr.edu.pl

 DOI: 10.1051/e3sconf/20161000063

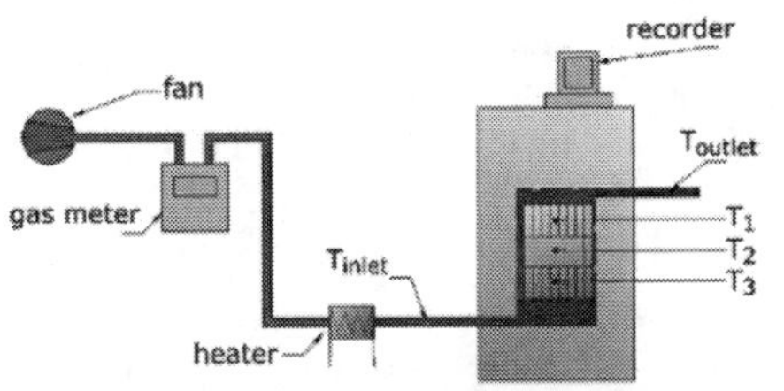

Figure 1. Measuring set-up marked with temperature taking points and with measuring equipment.

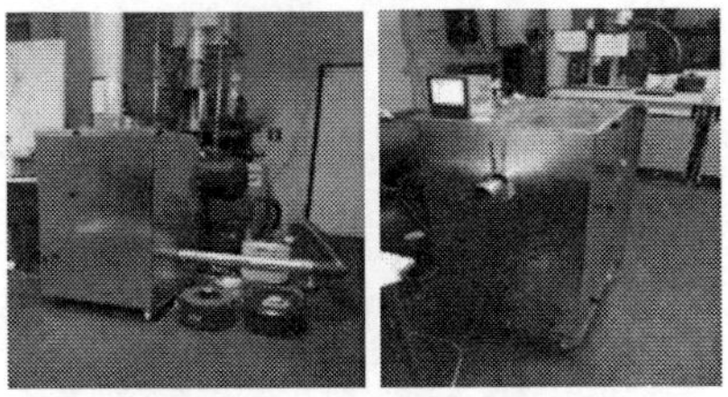

Figure 2. Research set-up. Air inlet together with gas-meter, autotransformer and ventilator visible on the right, outlet and measurements recorder on the left.

The deposit was filled with KNAP chamotte bricks with specific heat of 0.84 kJ/kg·K and single brick mass of 1.74 kg. The accumulating material was laid alternately in three layers, as shown on Fig. 3 and 4.

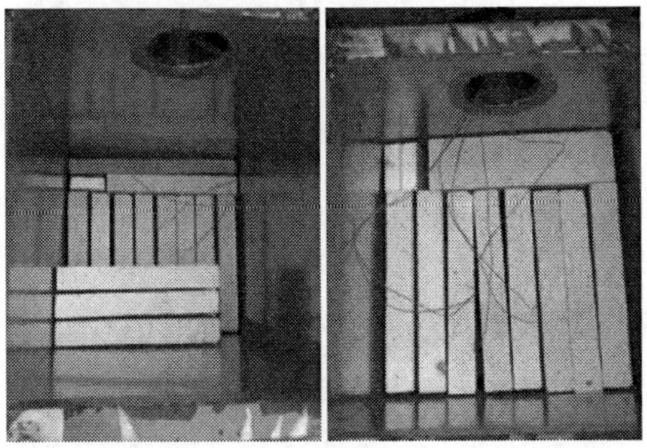

Figure 3. Layout of material filling the storage.

Fig. 4 shows the layout of plates in the storage in each of the three segments. The number of bricks is different and it is 9, 7 and 9 pieces on the lower, middle and upper level respectively. Also, empty spaces without filling material are visible. They result from the specific size of the bricks, making it impossible to fill the storage accurately. Those places were filled with different material, visible on Fig. 3, characterised by low heat capacity. Such a solution allows for an even distribution of air in the storage and low thermal conductivity coefficient allows to omit those elements in further calculations due to their insignificant influence on the process.

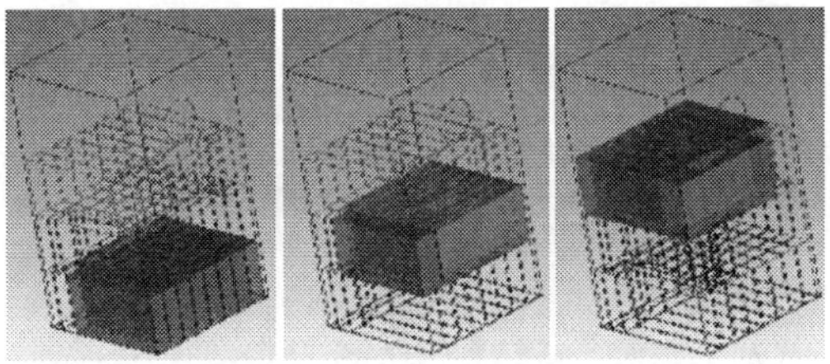

Figure 4. Chamotte brick layout in the storage on the lower, middle and upper level.

The deposit's purpose is accumulation of thermal energy supplied, for instance, by a solar collector. In order to determine the amount of heat possible to be accumulated, the measurements of temperature and volume of air were conducted in laboratory conditions. The volume was measured with ACD-G16C gas-meter, initially every 5, 10, and then every 30 minutes. Thermocouple readings were recorded with a 24-channel KD7 screen recorder. The main aim during the research was the determination of the storage's charging characteristics. Thermocouples distributed according to the scheme shown in Fig. 1 allowed for the recording of temperature, at one-minute intervals, of the air at the inlet and outlet, as well as of the accumulating material, in each of the 3 layers. The study of the charging process lasted for 2 hours and 44 minutes.

In the examined storage, the following assumptions were accepted in order to describe the heat transfer:

- stability of the physical properties of the accumulating material (bricks) and the heating medium (air),
- stability of the air temperature at the inlet of the storing system in the whole section,
- stability of the air speed in the deposit's cross-section and its invariability along its length,
- not taking into account the heat transfer along the air flow direction [9].

3 Analysis of the results

The following values were obtained as the result of conducted tests:

- ambient temperature: t_{amb}=18°C,
- volume of the flowing air: $\dot{V}_{avg} = 0.00433 \frac{m^3}{s}$,
- temperatures in different areas of the storage were put together in Table 1.

The calculations allowed to determine the characteristics of storage charging, actual and maximum amount of accumulated heat and the process' efficiency. It enabled the analysis of high-temperature heat accumulation in the deposit filled with chamotte bricks.

Fig. 5 shows the rise of the storage's temperature during charging in individual segments: lower (T1), middle (T2) and upper (T3). The quickest temperature rise is observed on the lowest lever of the storage, right above the air inlet. It results from the biggest difference between the temperature of air and the deposit. The slowest rise of temperature is observed in the upper part of the storage, reached by air of lower temperature. The air flowing through the storage cools down, transferring heat to the material. When analysing the diagram, one can notice that 80 minutes after the beginning of the experiment the increase ratio of temperatures becomes constant, which is shown by means of parallel straight lines on the diagram.

Table 1. Partial results of temperature measurements in different areas of the storage.

τ, min	t_{inlet}, °C	t_{outlet}, °C	t_1, °C	t_2, °C	t_3, °C
1	61.3	18.8	17.9	18.3	18.5
10	155.1	26.1	20.7	24.3	19.1
20	159.3	29.3	29.3	30.5	20.0
30	159.8	33.1	39.1	36.8	21.3
40	160.2	36.6	48.7	42.8	23.0
50	160.8	39.9	57.6	49.1	25.3
60	160.9	42.9	65.9	55.1	27.9
80	161.0	47.9	80.0	66.2	30.9
100	161.2	55.3	91.5	76.1	34.3
120	161.3	62.3	100.8	85.0	37.7
140	161.9	68.9	108.2	93.1	41.5
160	161.0	75.3	114.4	100.1	45.3
164	160.9	76.1	115.2	101.1	49.3

Temperature measured in different parts of the storage enables the determination of the stream of accumulated heat. The most crucial value is the amount of actually accumulated heat, resulting from the material's (chamotte bricks') accumulating capacity. The calculations below allowed to determine the amount of accumulated energy, the actual and maximum stream of heat. This lets us determine the speed of storage's charging and the process' efficiency.

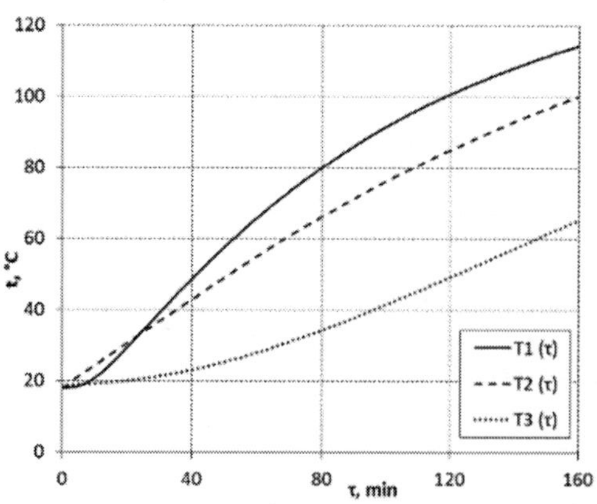

Figure 5. Distribution of temperature in individual segments of the storage in the function of time.

The amount of accumulated energy is described with equation (1).

$$Q_{zm} = m_z \cdot c_{p_z} \cdot \Delta t_{avg} = \\ = m_z \cdot c_{p_z} \cdot \left(t_{avg2} - t_{avg0}\right), kJ \tag{1}$$

where:

- m_z – total mass of chamotte bricks, kg,
- c_{p_z} – specific heat of the bricks equal to 0.84 $\frac{\text{kJ}}{\text{kg}\cdot\text{K}}$,
- t_{avg2} – average temperature of the storage at the moment of measurement, °C,
- t_{avg0} – average temperature of the storage at the beginning of measurements, °C.

The average temperature is the temperature averaged in individual layers of the storage (2).

$$t_{avg} = \frac{t_1+t_2+t_3}{3}, °C \tag{2}$$

Total mass of the storage m_z is:

$$m_z = m_{z1} \cdot n, kg \tag{3}$$

where:

- m_{z1} – single brick mass equal to 1.74 kg,
- n – number of bricks.

The increase of the accumulated air in the storage is shown in Fig. 6.

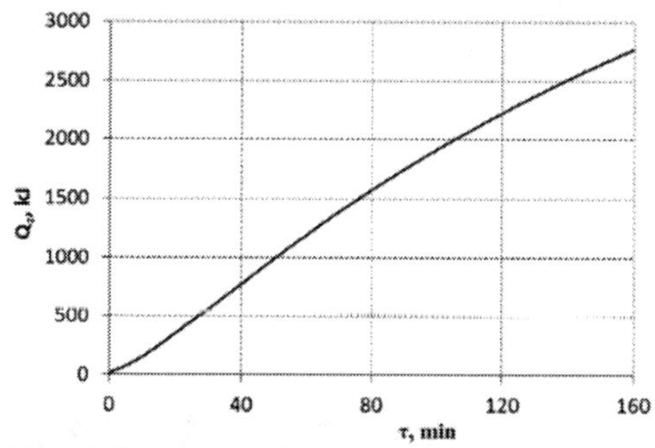

Figure 6. The amount of thermal energy accumulated in the storage.

On the basis of the obtained results, one can see an almost linear relationship between the amount of accumulated thermal energy and the time of the process' duration. The temperature keeps rising during the experiment, which directly influences the increase of accumulated heat.

Maximum heat stream possible to be accumulated in the storage is equal to the stream of heat carried by heated air. The air flowing through the storage cools down and gives its energy to chamotte bricks. The amount that the gas can transfer was determined on the basis of equation 4. Treating air as ideal gas, the following were assumed [10]:

- density: $\rho_{air} = 1.2 \frac{kg}{m^3}$,
- specific heat: $c_{p_{air}} = 1.005 \frac{kJ}{kg\cdot K}$.

$$\dot{Q}_{max} = \dot{m}_{air} \cdot c_{p_{air}} \cdot \Delta t = \\ = \dot{V}_{air} \cdot \rho_{air} \cdot c_{p_{air}} \cdot (t_{inlet} - t_{outlet}), kJ \tag{4}$$

where:

- t_{inlet} – temperature of air at the inlet, °C,
- t_{outlet} – temperature of air at the outlet, °C.

Fig. 7 shows calculation results of maximum heat stream possible to be transferred by flowing air.

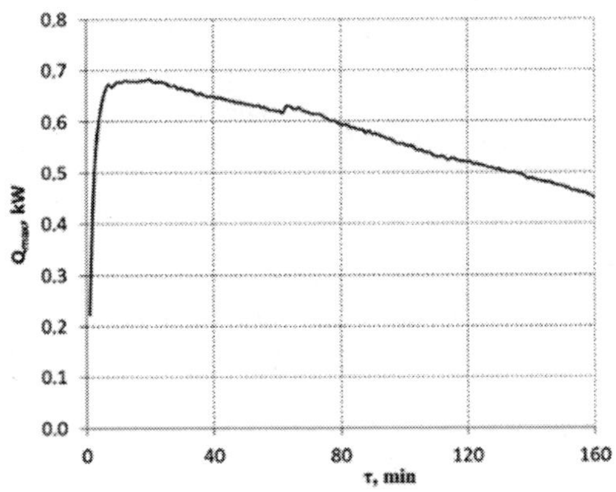

Figure 7. Characteristics of the maximum heat possible to be accumulated in the function of time.

When analysing the results, one can notice a low maximum value of the heat stream in the first minutes of the measurements. The reason behind this may be the storage's temperature equal to ambient temperature. Unheated set-up (cool metal housings) absorbed large amounts of heat from the flowing air at the beginning. However, as the storage is further charged, the value of maximum heat stream decreases. It's main reason is the growing heat loss (from the material to the surroundings) as chamotte bricks are being heated. It results in a lower drop of the temperature of air flowing through the storage. It leads to a decreasing difference of temperatures between the inlet air (relatively constant value) and the outlet air (rising with the duration of the experiment), which directly influences the decrease of the value of maximum heat possible to be accumulated. The stream of accumulated air was determined on the basis of equation 5 and it is shown in Fig. 8.

$$\dot{Q}_{zm} = \frac{m_z \cdot c_{p_z} \cdot \Delta t_{avg}}{\tau} = \frac{m_z \cdot c_{p_z} \cdot (t_{avg2} - t_{avg0})}{\tau}, kW \quad (5)$$

where:
- τ – duration of the experiment, s.

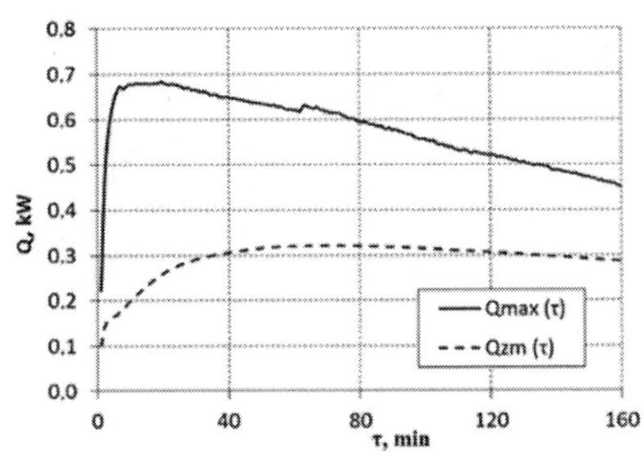

Figure 8. Characteristics of the maximum and actual heat in the function of time.

Fig. 8 shows the increase of the stream of actually accumulated heat in the function of time. It is caused by growing temperature of the storage, and by increasing temperature difference determined on the basis of equation 5. After 73 minutes of the measurements, value $\dot{Q}_{zm}$ slightly decreases and the process of storage charging is slower from that moment on. It results from the decreasing relationship between temperature difference and duration of the test, which in turn can be caused by greater losses to the surroundings. When comparing both characteristics, it can be seen that the value of maximum heat stream decreases with the increase of the value of actual stream. The diagram allows us to draw the conclusion that the difference between these values is levelled with time. The rising temperature of the storage with a stable temperature of inlet air cause an increase of heat losses from chamotte bricks to the surroundings, which directly leads to the rise of outlet temperature.

In order to picture the process of the storage's charging, the stream of accumulated heat in each of the deposit's segments was calculated, according to equation 5. In this case, the mass of filling material was calculated on the basis of equation 3, only the number of plates is different:
- on level 1 (lower): n=9,
- on level 2 (middle): n=7,
- on level 3 (upper): n=9.

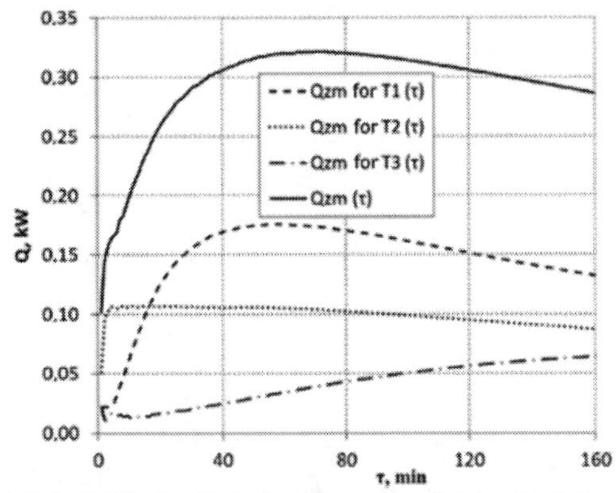

Figure 9. Relationship between the stream of accumulated air and the time for individual segments of the storage.

Calculation results are shown in Fig. 9. The characteristics for the lower part of the storage is analogous to the diagram of the total stream of accumulated heat (Fig. 7). Bot characteristics show a quick increasing tendency during the first hour of the process. After approx. 60 min. value $\dot{Q}_{zm}$ for the lowest level of the storage starts to decrease. In the middle segment, value $\dot{Q}_{zm}$ remains almost on an even level. The lowest value of the stream of accumulated heat is observed in the upper segment, but differently from other layers, there is a growing tendency there. It is worth stressing that values $\dot{Q}_{zm}$ for individual layers become more similar to one another with time. Thus, it can be concluded that the longer the process of storage charging, the more level is the distribution of accumulated heat in its layers.

The obtained characteristics result from temperature characteristics of the deposit – the higher the temperature of the deposit, the higher the value $\dot{Q}_{zm}$. The sum of three characteristics of individual segments equals the total value of actual heat stream.

Knowing the maximum and actual stream of accumulated air, we can determine the process' efficiency (equation 6) being the relationship between actually accumulated heat and the maximum heat determined on the basis of equation 4. The characteristics of the efficiency of storage charging is shown in Fig. 10.

$$\varepsilon = \frac{\dot{Q}_{zm}}{\dot{Q}_{max}} \tag{6}$$

The efficiency of the process of charging increases with the time of the experiment's duration. On the basis of the diagram's shape it is possible to conclude that the increase of efficiency is logarithmic. It means that with time the charging process will stabilise, which results from the gradual levelling of the values of maximum and accumulated heat stream. Final efficiency in the conducted experiment is on the level of over 60%. The curve shows a clear relationship between the efficiency value and the time of the process. Theoretically, in infinitely long time, this coefficient would be 100%.

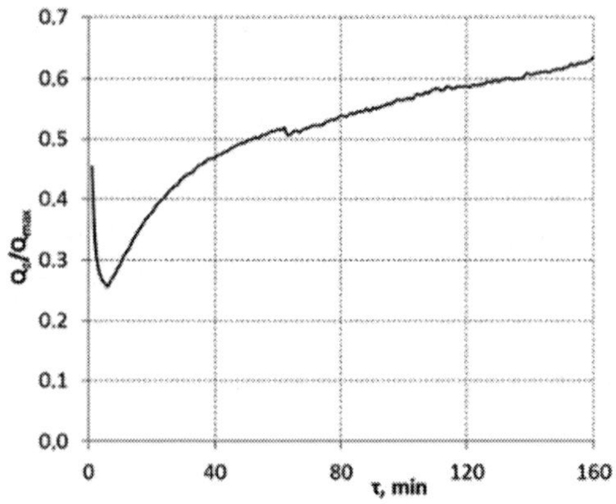

Figure 10. Diagram of the efficiency of charging process in time.

The efficiency for individual segments was determined on the basis of equation 6. The value of actually accumulated energy $\dot{Q}_{zmi}$ was assumed as the accumulated heat and the value of maximum heat stream for the whole storage was assumed as $\dot{Q}_{max}$. The results are shown in Table 2.

Table 2. Partial results of partial calculations of the process' charging efficiency.

τ, min	$\dot{Q}_{zm}$, kJ	$\dot{Q}_{max}$, kW	$\dot{Q}_{zm}$, kW	ε, -	$\dot{Q}_{zm1}$, kW	$\dot{Q}_{zm2}$, kW	$\dot{Q}_{zm3}$, kW	ε_1, -	ε_2, -	ε_3, -
10	119	0.678	0.199	0.294	0.064	0.107	0.013	0.094	0.158	0.019
20	311	0.683	0.259	0.379	0.126	0.107	0.016	0.185	0.156	0.024
30	523	0.666	0.290	0.436	0.156	0.107	0.020	0.234	0.161	0.031
40	733	0.649	0.306	0.470	0.169	0.106	0.025	0.261	0.163	0.038
50	946	0.635	0.315	0.497	0.175	0.106	0.030	0.275	0.167	0.047
60	1152	0.620	0.320	0.516	0.176	0.105	0.034	0.283	0.170	0.055
70	1348	0.618	0.321	0.520	0.174	0.104	0.039	0.281	0.169	0.063
80	1537	0.594	0.320	0.539	0.170	0.103	0.043	0.287	0.173	0.073
90	1714	0.576	0.317	0.551	0.166	0.101	0.047	0.289	0.175	0.081
100	1885	0.556	0.314	0.565	0.162	0.099	0.050	0.290	0.178	0.091
110	2046	0.532	0.310	0.582	0.157	0.097	0.053	0.294	0.182	0.100
120	2202	0.520	0.306	0.588	0.152	0.095	0.056	0.292	0.183	0.108
130	2347	0.507	0.301	0.594	0.146	0.093	0.059	0.289	0.184	0.116
140	2488	0.489	0.296	0.606	0.142	0.091	0.061	0.290	0.187	0.124
150	2620	0.473	0.291	0.615	0.137	0.089	0.063	0.289	0.189	0.132
160	2745	0.450	0.286	0.635	0.132	0.087	0.064	0.294	0.194	0.142
164	2781	0.446	0.283	0.634	0.130	0.086	0.064	0.292	0.194	0.143

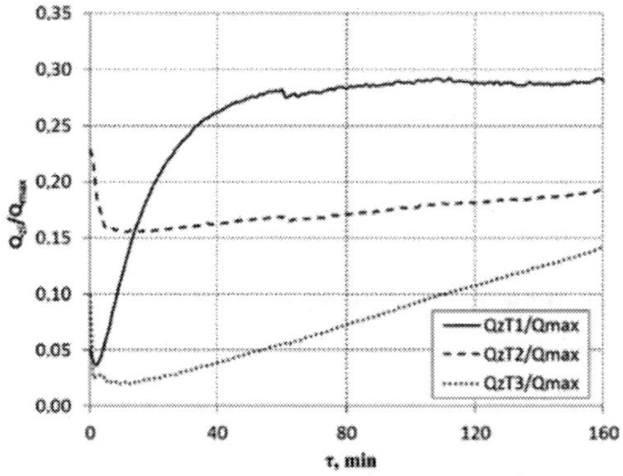

Figure 11. Diagram of the efficiency of the storage's charging in individual segments in the function of time.

As can be concluded on the basis of Fig. 9 and 11, both in the case of the characteristics of accumulated heat and of the charging efficiency, the highest increase is observed for the lowest segment of the deposit. The increase of efficiency for this layer shows a logarithmic character reaching the determined value on the level of less than 30% as early as after one hour. In the middle segment the efficiency slightly rises, because it increases by less than 5% during the whole experiment. A relatively high linear increase can be observed for the storage's upper segment. The main reason behind it is a slight increase of accumulated heat accompanied by the decrease of maximum heat stream as the process continued.

4 Conclusions

The increase of temperature in the storage influences the rising value of the stream of actually accumulated heat. Analysing the amount of energy accumulated in individual segments of the storage it can be concluded that the longer the process of storage charging, the more level is the distribution of accumulated heat in its layers. The first - lowest level of the storage absorbs the greatest amount of energy.

The efficiency of the process of charging increases with the time of the experiment's duration. It signifies an increase of the heat absorbed by accumulating material from the air. Final efficiency in the conducted experiment is on the level of approx. 60%. In individual segments the highest efficiency is observed for the storage's lowest segment.

The research presented can be treated as an introduction to the analysis of different materials filling the deposit and their influence on the amount of accumulated heat. In the actual installation, the amount of accumulated energy will be directly related to the potential of the heat source supplying the storage. However, it seems that the ability to adjust work parameters at the laboratory set-up are sufficient to simulate actual conditions and can be a major input in the further research of the process of heat accumulation.

References

1. R. Singh, R.P. Saini, J.S. Saini, Sol. Energy **80,** (2006)
2. K. Sagara, N. Nakahara, Sol. Energy **47**, 3 (1991)
3. N.G. Barton, Appl. Therm. Eng. **55,** (2013)
4. A. Mawire, M. McPherson, R.R.J. van den Heetkamp, S.J.P. Mlatho, Appl. Energy **86,** (2009)
5. I.A. Abbud, G.O.G. Lof, D.C.Hittle, Sol. Energy, **54,** 2 (1995)
6. J. Xu, R.Z. Wang , Y. Li, Sol. Energy **103,** (2014)
7. H.B. Madessa, T. Veslum, J. Løvseth, O.J. Nydal, 30th ISES Biennial Solar World Congress 2011, SWC 20113 (2011)
8. M. Nemś, J. Kasperski, Energy Procedia **57,** (2014)
9. R. Domański, *Magazynowanie energii cieplnej* (PWN, Warszawa, 1990)
10. E. Kalinowski, *Przekazywanie ciepła i wymienniki* (Oficyna Wydawnicza Politechniki Wrocławskiej, Wrocław, 1995)

Linear concentrating collector as an air heater in the heating system of building in Polish climatic conditions

Magdalena Nemś[1,a], Agnieszka Manikowska[1] and Artur Nemś[1]

[1]*Wroclaw University of Technology, Faculty of Mechanical and Power Engineering, Poland*

Abstract. The article presents the analysis of the performance of a concentrating collector in the heating system of a residential building. Air was used as the working fluid. The heating requirements of the building were determined for each day of the year. The amount of direct irradiation reaching the absorber's surface on all the days of the year was determined with the use of hourly meteorological data for Wroclaw, shared by the Ministry of Infrastructure and Growth. It was assumed that the collector is equipped with a tracking system working in one axis. Calculations and comparisons were made for the amount of solar irradiation for three values of the receiver's inclination angle: β_1=60°, β_2=90° and β_3=30°. Statistical method was used in order to determine the optimum inclination of the mirror and the amount of flowing air. This method involves creating a plan of experiment with three levels of changeability for two input factors. In the last stage, the amount of heat obtained from the installation during all the days of the year was analysed. The gains were juxtaposed on the diagram with the building's heat demand. The analysis has shown that the heat requirements can be met only partially.

1 Introduction

The solar energy market in Poland has been dominated by flat solar collectors. Concentrating collectors are quite rare, and their manufacturers usually use water as the working medium. Selection of medium includes choosing appropriate construction parameters of the collector; the degree of concentration of mirror cannot be too high so that the final temperature of water obtained was not too high. Concentrators are used for purposes of heating of buildings, unlike countries with high insolation; in solar power plants, to run technological processes such as utilization, or production of process heat. The use of air as the working medium is not that popular. In literature on the subject there is research on flat solar collectors that heat the air. The main subject of this research is heat exchange intensification through changing the absorber's design. Its profile is changed: fins [1, 2], wire mesh [3] or other elements such as internal beds made of scrap metal are added. However, this does not change the low temperature range on the basis of which flat collectors operate. There are few manuscripts and papers devoted to concentrating air collectors. Research conducted by scientists from Norway, who analysed different types of absorbers for solar concentrating parabolic dish [4], can be considered significant, whereas Turkish researchers studied the cone-shaped collector, which was ultimately used as a solar cooker [5].

Climatic conditions in Poland make solar collectors used in heating on a seasonal basis, mainly for generation of hot water in houses. It is associated with low insolation and winter temperatures which freeze the working medium. An important advantage of concentrating collectors is the possibility of obtaining high temperatures of the medium, even in the winter period. Further in this article the concept of a heating system in a single-family building based on a large-size, linearly focusing air collector is described.

2 The heating system concept

The main element of the analysed heating system is a focusing tube collector that heats the air. The hot working medium from the collector is distributed around a building through tubes (Fig. 1). As a backup system that supplements heat shortage a biomass fireplace that heats the air and uses the same heat distribution system around the house was used. The selection of the collector's working medium was not accidental. Despite a low thermal capacity, the air has many advantages. These include: universal accessibility, safety in case of installation leakage, and a simple structure, which translates into a low price of structural elements of the system. Another advantage is that the air from the collector can be directly used in the heat distribution installation in the building.

Calculations were made for the heating system dedicated for a single-family building with an area of 82 m^2. As the geographic location the city of Wroclaw, located in the western part of Poland, was chosen.

[a] Corresponding author: magdalena.nems@pwr.edu.pl

The heat demand for the house calculated according to the methodology of the Ministry was 45 kWh/m^2·year.

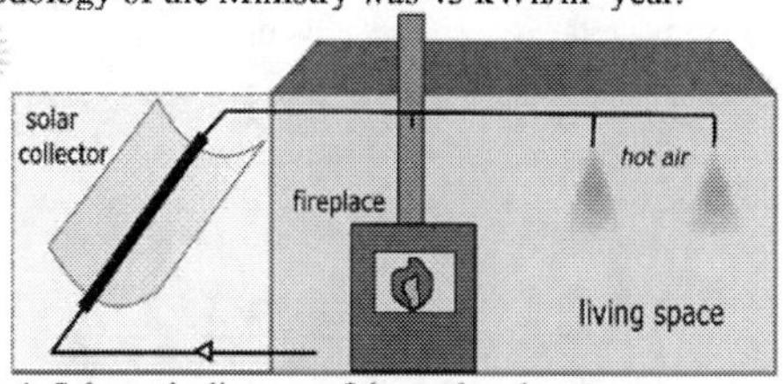

Figure 1. Schematic diagram of the analysed system.

To be able to conduct a detailed analysis of operation of the solar heating system the daily demand for heat in the building was calculated. For this purpose, a definition of heating degree-days was adopted [6]. The number of heating degree-days was calculated using the following equation (1) [6]:

$$Sd = \sum_{i=1}^{12} \sum_{i=1}^{31} (t_i - t_e(d)) \quad (1)$$

where:
t_i - base temperature, i.e. the adopted ambient temperature for which in the building there is thermal comfort,
$t_e(d)$ - average daily temperature on a given day [7].

It was assumed that a given day is part of the heating period if the average daily temperature on a given day is lower than the base temperature. For the needs of calculations it was assumed that the base temperature for Wroclaw was 15°C, and an additional relation was introduced:

$$\begin{cases} if\ t_e(d) \geq 15°C\ then\ Sd(d) = 0 \\ if\ t_e(d) < 15°C\ then\ Sd(d) = (t_i - t_e(d)) \end{cases} \quad (2)$$

On the basis of the relation (2) the daily number of degree-days was calculated for each day of the year for Wroclaw. This allowed calculating the daily demand for heat from the equation (3) [6]:

$$Q_{hd} = Q_h \cdot \frac{(t_i - t_e(d))}{Sd} \quad (3)$$

where:
Q_h - annual heat demand.

The daily demand for heat is shown in Fig. 2.
Additionally, it was assumed that the heating system should be activated when for 3 days in a row the temperature is below 15°C; in this case, the heating season lasts from mid-September to mid-May. In the remaining months the building does not have to be heated.

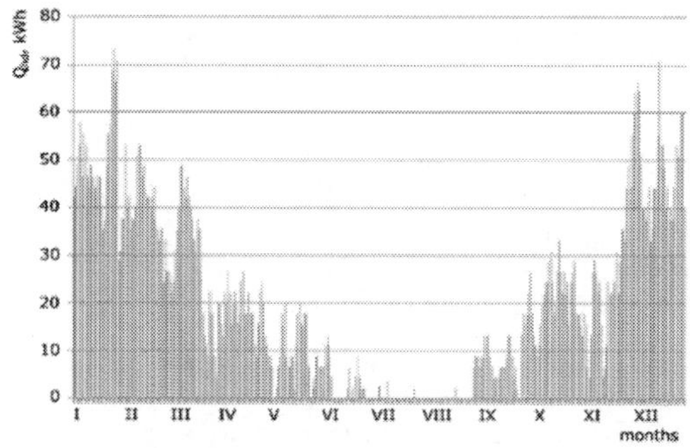

Figure 2. Daily heat demand.

3 Collector as an air heater

To examine the annual work of the collector and its efficiency its model was required. It was taken from [8]. The basic values used in the calculations are shown in Fig. 3. The copper tube in which there is the medium (air) was surrounded by a glass shield. In the space between the tube and the shield there was vacuum.

It was assumed that a 7-metre absorber has a selective coating - black chrome. The transparent shield is made of glass with a low content of Fe_2O_3 and with a thickness of 3 mm. The adopted width of the collector's mirror was 3 m, and the internal diameter of the absorber was $DW = 0.096$ m, the outer diameter was $D_z = 0.1$ m, and the diameter of the absorber's shield was $D_c = 0.14$ m. The heat obtained from the collector was calculated on the basis of an equation by Hottel-Whillier-Bliss (4) [8]:

$$Q_u^* = A_f F_R \left(S - \frac{A_p}{A_f} U_g (T_{w1} - T_a) \right) \quad (4)$$

where:
A_f - non-shaded area of the aperture,
F_R - heat removal coefficient from the receiver,
S - solar radiation absorbed by the receiver,
A_p - surface of the receiver,
U_g - losses through transparent cover,

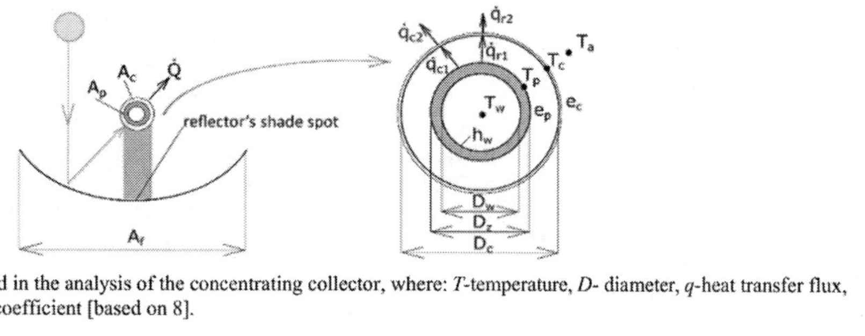

Figure 3. The symbols adopted in the analysis of the concentrating collector, where: *T*-temperature, *D*- diameter, *q*-heat transfer flux, *e*- emissivity, *h*- heat transfer coefficient [based on 8].

T_{wI} - inlet temperature of the medium (it was assumed that the air is taken from the room and T_{wI} = 20°C was adopted),
T_a – ambient temperature.

Collector efficiency expressed by equation 5 [8]:

$$\eta_m = \frac{Q_u^*}{A_{f\,gross} \cdot G_b} \qquad (5)$$

where:
$A_{f\,gross}$ - total area of the aperture,
G_b - density of direct radiation.

The formula for the efficiency of the model collector was determined using the iterative method. The formula included its size, solar radiation and power output.

3.1 Analysis of solar radiation

To a large extent the efficiency of a solar collector depends on solar radiation falling on the collector. In the case of concentrating collectors it is direct radiation. For this reason, the concentrating collector must have a follow-up system, which will direct the optical system throughout the day towards the sun. The analysed air heater had one axis of rotation. Additional three angles of inclination of the collector towards the horizontal plane were analysed: 30, 60 and 90 degrees.

The amount of radiation that reached the plane of the collector directly can be calculated using the following relation (6) [8].

$$G_b = G_{on} \cdot cos\theta_{\beta}, \frac{\mathrm{W}}{\mathrm{m}^2} \qquad (6)$$

where:
G_{on} - radiation flux density measured on a normal plane directed towards radiation,
θ_β - angle of incidence of radiation on the receiver's plane measured between the direction of direct radiation, and the normal plane directed towards the receiver.

To be able to make calculations of density of radiation a new angle of incidence of radiation on a plane of a moving concentrating collector was determined. For this purpose, relation (7) was used which allows calculating angle θ_β by taking into account the so-called optimum angle Ψ, for which the concentrating collector rotates optimally towards the current direction of sunlight [9].

$$\cos^2\theta_\beta = cos\psi[sin\beta sin\varphi\, cos(\alpha-\gamma) + cos\beta cos\varphi] + \\ + sin\psi sin\varphi\, sin(\alpha-\gamma) \qquad (7)$$

where:
ψ - optimum angle of rotation of the concentrating collector around its diagonal axis,
β - inclination of the receiver towards the horizon (three variants of inclination of the collector were assumed: $\beta_1 = 30°$, $\beta_2 = 60°$, $\beta_3 = 90°$),
φ - zenith angle, the angle of incidence of solar radiation on a horizontal plane,
α - solar azimuth,
γ - receiver's azimuth, deviation from the local meridian measured relative to the southern direction; γ = 0 was adopted.

The optimum angle of rotation of the collector around the diagonal axis is calculated from equation (8) [9]. It falls within the range of -90° < ψ <+90°.

$$tan\psi_{opt} = \frac{sin\varphi\, \sin(\alpha-\gamma)}{sin\beta sin\varphi \cos(\alpha-\gamma) + cos\beta cos\varphi} \qquad (8)$$

On the basis of relation (6) - (8) the amount of direct radiation that reaches the collector at any hour of the day throughout the year was calculated for the three variants of inclination. The maximum values of radiation received for angle 30° amounted to 550 W/m², for 60° - 490 W/m², and for 90° - 320 W/m². The obtained amounts were compared throughout the year and for further analysis the angle of inclination of β = 30° was assumed (Fig. 4).

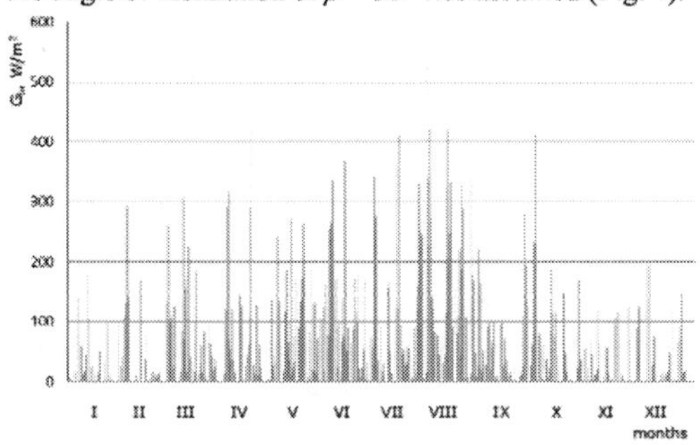

Figure 4. Amount of direct radiation that reaches the collector for the angle of inclination of β = 30°.

3.2 The experiment planning method

A key issue before conducting the analysis of annual work of the conceptual system was to determine a dependence for the operating efficiency of the collector.

The analysis of the collector's work shows that variable parameters that affect the efficiency of the system include the density of direct radiation and the value of a given air flow stream. When determining the effect of more than one factor on the result of a given process, special plans are used to simplify the courses of research and development of results. To determine the equation, the procedure of conduct of a static experiment planning was used for two input factors at three levels of variation, in short PS/DK 3^2 [10]. The use of the PS/DK 3^n plan guarantees accurate analysis of the process, even if it had been unknown, and allows obtaining a mathematical model of an examined issue in the form of a second degree polynomial [10].

In the selected programme the input factors take on value on three levels of variation [10]:

- higher, marked as +1.
- medium, also called basic or zero, marked as 0,
- lower, marked as -1.

DOI: 10.1051/e3sconf/20161000064

The equation that is obtained in the three-level planning has the form of a second degree polynomial in the form of equation (9) [10]:

$$y = b_0 + \sum b_k x_k + \sum b_{kk} x_k^2 + \sum b_{kj} x_k x_j \qquad (9)$$

where:
x_k, x_j,... - input factors,
b_0, b_k,... - regression coefficients,
y - result factor.

For the tested model the three-level plan was with two input factors - direct radiation and flow stream - was adopted. Levels of variation, which are shown in Table 1, were selected for the given factors. The values were selected on the basis of collected meteorological data and predicted possibilities of changes in the flowing air stream.

Table 1. Parameters of input factors.

Item	Parameter	G_b (W/m^2)	m^* (kg/s)
1	Zero level (0)	450	0.03
2	Upper level (+)	800	0.05
3	Lower level (-)	100	0.01
4	Range of changes	350	0.02
5	Codes of variables	$x_1 = \left(\frac{G_b - 450}{350}\right)$	$x_2 = \left(\frac{m^* - 0.03}{0.02}\right)$
6	Result factor	$y=\eta_{sys}$	

The efficiency of the collector was calculated from equation (9) after calculating regression coefficients [10], and pre-coded factors (Table 2).

Table 2. Values of regression coefficients.

b_0	0.5956
b_{11}	-0.0057
b_{22}	-0.0129
b_{12}	0.0002
b_1	0.0074
b_2	0.0198

Table 3 shows a list of input factors and efficiency values calculated for these factors from model η_m of the collector, and efficiency values calculated from experiment planning η_{sys}. The obtained results were similar. The efficiency calculated using the experiment method has almost the same values as the one calculated from the model. Differences are observed only in hundredths of a percent therefore the calculations made were considered correct.

On the basis of the above calculations a universal formula for efficiency of the modelled collector, dependent on direct radiation and the air flow stream in the absorber, was obtained. This formula has the following form (10):

$$\eta = 0.5956 + 0.0074 \cdot \frac{G_b - 450}{350} + \\ +0.0198 \cdot \frac{m^* - 0.03}{0.02} - 0.0057 \cdot \left(\frac{G_b - 450}{350}\right)^2 + \\ -0.0129 \cdot \left(\frac{m^* - 0.03}{0.02}\right)^2 + \\ +0.0002 \cdot \frac{G_b - 450}{350} \cdot \frac{m^* - 0.03}{0.02} \qquad (10)$$

Table 3. A comparison of the efficiency of the collector from model η_m and the efficiency calculated from experiment planning η_{sys}.

Item	G_b,W/m^2	m^*,kg/s	η_m, %	η_{sys}, %
1	800	0.05	60.42	60.43
2	800	0.03	59.73	59.72
3	800	0.01	56.43	56.43
4	450	0.05	60.25	60.24
5	450	0.03	59.57	59.56
6	450	0.01	56.27	56.29
7	100	0.05	58.91	58.91
8	100	0.03	58.24	58.25
9	100	0.01	55.02	55.01

The obtained equation was substituted to the calculated data. The changing efficiency of the collector for radiation G_b (β_1) at different values of m^* was analysed. Three values of the flow stream were compared: $m_1 = 0.05$ kg/s, m_2 = 0.03 kg/s, and m_3 = 0.01 kg/s. The results were presented in the form of a graph shown in Fig. 5. Fig. 6 shows the changing efficiency of the collector with three different values of solar radiation: G_{b1} = 100 W/m^2, G_{b2} = 450 W/m^2, G_{b3} = 800 W/m^2.

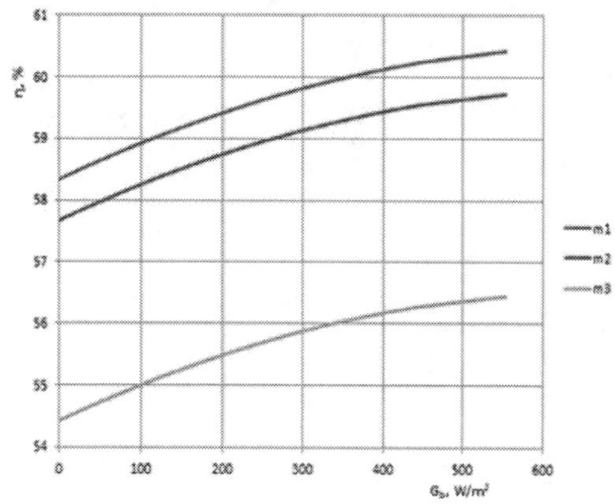

Figure 5. Collector efficiency change chart depending on available solar radiation at different values of the flow stream.

Charts presented in Fig. 5 and 6 show that the efficiency of the collector is highest for flow stream m_1 = 0.05 kg/s, and solar radiation G_{b3} = 800 W/m^2.

 DOI: 10.1051/e3sconf/20161000064

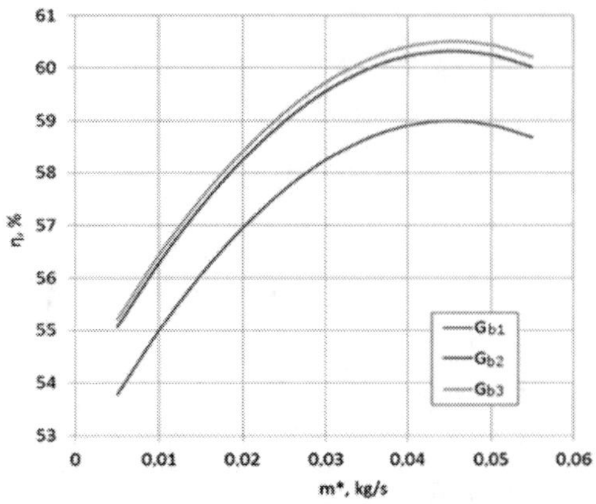

Figure 6. Collector efficiency change chart depending on the flow stream at different values of solar radiation G_b.

4 Analysis of annual operation of the system

After determining the general formula for the efficiency of the concentrating collector depending on the flow stream of the medium and the amount of solar radiation, it was possible to determine the changeability of the amount of power output of the collector per year. Using formula (5) the thermal power of the analysed collector was calculated for various combinations of inclinations of the device and the flow stream of the medium.

After making the calculations it can be seen that the greatest power output of the analysed concentrating collector is observed for angle β = 30° and flow stream m = 0.05 kg/s. The maximum power output of the collector was Q = 168 kWh at the end of July. During the heating season the heating power reaches 90 kWh. For the variant with angle β = 60° the maximum thermal power of the collector does not exceed 160 kWh, and for variant β = 90° - 100 kWh. In the case of the analysed variants, thermal power of the collector changes by a few kWh only.

On the basis of Figure 7, the degree to which the solar collector meets the heat demand of the building can be assessed. The highest heat demand of a single-family house are in the months from September to May. Whereas the collector has the most power in the summer months from May to September, when the greatest amount of sunlight is available. In the months from March to October the designed solar collector is enough to heat the building however in the remaining months of the heating season and additional heating source is required. In the months from November to February heat gains due to the short day are insufficient. Due to the fact that the working medium of the concentrating collector is air and it is used to heat the building, a fireplace with a hot air distribution function was selected as the additional heating source.

5 Summary and conclusions

The most popular methods of heating in Poland are traditional solutions such as gas furnaces, stoves, solid fuel furnaces or urban heating. Heat distribution in rooms is carried out with the use of the water system. In recent times, it has been recognized that air heating is increasingly used for heating houses. Hot air distribution fireplaces, but also air solar collectors, are used for this purpose. The latter are dominated by flat collectors. Concentrating collectors are not that common, especially in Poland.

This manuscript analysed the work of a concentrating solar collector. Heat exchange calculations were made to determine efficiency, and the amount of solar radiation reaching the plane of the collector was determined. Thanks to the experiment planning method the general formula for the efficiency of the collector, which depends on the flow stream of the working medium and the amount of solar radiation reaching the collector, was worked out. The changing efficiency of the collector for three different variants of inclination and three variants of flow of the medium was analysed. It has been shown that the best solution, in which the efficiency of the system is the highest, is the variant with the angle of inclination of β=30° and the flow stream of m = 0.05 kg/s. For this solution the collector reaches the greatest thermal power. The maximum amount of energy exceeds 160 kWh, whereas in the least efficient variant (β=90° and m = 0.05 kg/s) the amount of energy gained from the collector is less than 100 kWh.

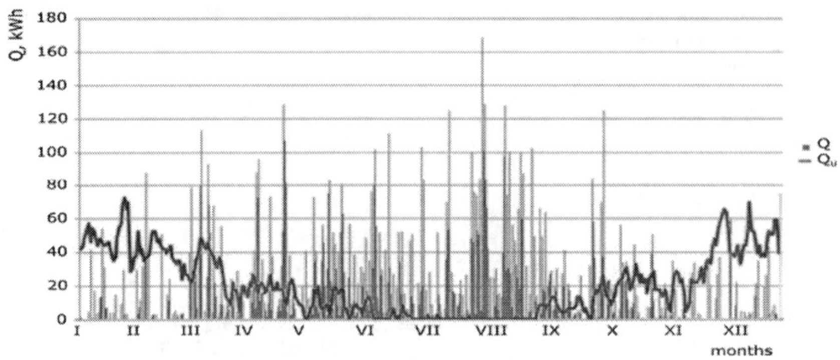

Figure 7. Heat demand in building Q_u and power output of collector Q.

The last step in the analysis was to compare heat demand in the building with the amount of heat provided by the collector for the best of the considered variants. The designed collector covers the thermal needs of the building completely in the months from March to October. In the remaining months the heat from the collector is not enough to heat the entire building therefore an auxiliary heating source, which is a biomass furnace, was chosen. In both cases heat is delivered to the heated rooms through insulated ducts. Both systems can constitute an integral whole. Then, the building will be provided thermal comfort in every day of the year, and renewable energy sources will be used to heat it. It is also possible to use a heat accumulator which could accumulate excess heat and supplement heat deficiency when the temperature in the building drops below thermal comfort.

References

1. J. Muñoz, A. Abánades, Appl. Energy **88,** (2011)
2. J. Kasperski, M. Nemś, Appl. Therm. Eng. **58,** (2013)
3. P. Verma, L. Varshney, Sustain. Energy Technol. Assess., **10** (2015)
4. H.B. Madessa, T. Veslum, J. Løvseth, O.J. Nydal, 30th ISES Biennial Solar World Congress 2011, SWC 20113, (2011)
5. I.T. Togrul, D. Pehlivan, Appl. Therm. Eng. **25,** (2005)
6. J. Dopke, *The calculation of monthly number of heating degree* (in Polish): www.ogrzewnictwo.pl, accessed : 22.04.2016
7. Meteorological data for Wroclaw: http://air.wroclaw.pios.gov.pl/, accessed : 22.04.2016
8. Z. Pluta, *Theoretical basis of photothermal conversion of solar energy* (in Polish), (Publishing house of Warsaw University of Technology, Warsaw, 2006)
9. J. Kasperski, M. Lewkowicz, S. Pietrowicz, (in Polish), District Heating, Heating, Ventilation **14,** 6/7 (2007)
10. M. Korzyński, Methodology of the experiment (in Polish), (Scientific-Technical Publisher, Warsaw, 2006)

Application Of Relevance Maps Method To Evaluate The Suitability Of Coal Samples For Fluidal Gasification Process

Dariusz Jamroz[1], Tomasz Niedoba[2,a] and Agnieszka Surowiak[2]

[1] AGH University of Science and Technology, Faculty of Faculty of Electrical Engineering, Automatics, Computer Science and Biomedical Engineering, Department of Applied Computer Science, al. Mickiewicza 30, 30-059 Kraków, Poland
[2]AGH University of Science and Technology, Faculty of Mining and Geoengineering, Department of Environmental Engineering and Mineral Processing, al. Mickiewicza 30, 30-059 Kraków, Poland

Abstract. The coal gasification process is one of the technologies which gain more and more attention among technologists dealing with processing and utilization of coal. In case of ground gasification, one of such technologies is fluidized bed gasification. For such gasification the guidelines were elaborated within the scientific project of NCBiR no. 23.23.100.8498/R34 Authors chose main of these guidelines, concerning the required levels of certain coal features. In purpose of investigating coal for its suitability to gasification, samples were collected from two Polish mines: ZG Janina and ZG Wieczorek and processed properly.
The methods being used to visualization of multidimensional data through transformation of multidimensional space into two-dimensional space allow to present multidimensional data on computer screen. Among such methods, relevance maps method can be found which was used in this paper to present and analyze set of seven-dimensional data describing coal samples originating from both mines. It was decided to check whether this method of visualization of multidimensional data allows to divide the samples space into subspaces of various usefulness to the process of fluidal gasification or not. The method enables the visualisation of the optimal subspace containing the set requirements concerning the properties of coals intended for this process.

1 Introduction

In the paper, coal samples coming from two coal mines - Janina Mining Plant (ZG Janina) and Wieczorek Coal Mine (ZG Wieczorek) were analysed. The above mentioned samples were taken in order to evaluate their suitability for on-surface gasification in fluidised bed. The properties of coals directed to gasification must comply with the limits [3, 5-8, 32-33, 40-41], wherein it must be noted that they are linked with each other. The evaluation of coal suitability for gasification should be therefore conducted multidimensionally with the use of multidimensional distributions of properties and their statistics [4, 15-24, 27, 30-31, 35-38, 43-53]. It is a natural thing that the analysis of multidimensional properties and statistics begin with the analysis of particle density distribution and particle size distribution of coals and then is extended on the basis of further coal properties, especially the contents of components and reactions to the processes of its processing. The analysis of coal in terms of distribution of the so-called class-fraction is the initial information on the coal capability of developing the surface area and concentration of flammable and volatile parts and ash. While multidimensional methods of visualisation allow for the combined interpretation of all measured properties in tested terms. Recently, with development of information science, multidimensional statistical analyzes gained significant meaning. It is worthy to mention here multidimensional visualization methods which can be treated as modern analysis tools. Among such types of methods the observational tunnels method can be found], parallel axes method, Kohonen maps, relevance maps, Principal Component Method, multidimensional scaling or autoassociative neural networks. The sources where is possible to find additional information about the applied methodology and similar methods are [1-2, 12-14, 25-26, 28-29, 34, 39, 42, 54].

2 Data

In total, from both coal mines, 99 samples (50 from ZG Janina and 49 from ZG Wieczorek were obtained containing such parameters as: total sulfur content, hydrogen content, nitrogen content, chlorine content, total carbon content, heat of combustion and ash content. The card of technological suitability of coal was used additionally [5-6], in which key, relevant and additional parameters, having effect on the gasification process, were described. On the card's basis coal samples, which are subjected to the gasification process in an effective way, were identified. Conditions used are: calorific value [kJ/kg]>18000, ash contents<25%, chlorine contents<0.1%, total sulfur contents<2%, elementary carbon contents>60%, 3.5% ≤ hydrogen contents ≤ 5.5%, nitrogen contents<2%. On the basis of the presented conditions, among the analysed 99 samples only 18 samples were identified as those which can be subject to gasification in an efficient way. Among those 18 samples, 17 came from ZG Janina and only one sample came from ZG Wieczorek.

[a] Corresponding author: tniedoba@agh.edu.pl

Table 1. Elemental analysis of coal in layer II after beneficiation in the jig - ZG Janina

Fraction d [mm]	Total sulphur content S_t^a [%]	Hydrogen content H^a [%]	Nitrogen content N^a [%]	Chlorine content Cl^a [%]	Total carbon content C^a [%]	Heat of combustion Q_s^a [kJ/kg]	Ash content A^a [%]
< 1.00	1.09	3.75	1.06	0.14	62.70	24 727	9.20
1.00 – 2.00	0.88	3.99	1.05	0.15	66.10	26 074	5.10
2.00 - 3.15	0.86	4.01	1.08	0.12	66.30	26 116	4.70
3.15 – 5.00	0.88	3.88	1.03	0.09	65.10	25 785	4.80
5.00 - 6.30	0.95	4.21	1.09	0.08	66.70	26 489	4.80
6.30- 8.00	0.87	4.33	1.15	0.08	66.00	26 380	4.80
8.00 – 10.00	1.01	4.2	1.15	0.09	66.90	26 486	5.00
10.00 - 12.50	0.83	4.17	1.09	0.09	67.60	26 633	4.70
12.50 – 16.00	0.84	4.31	1.13	0.07	68.40	26 771	4.90
16.00 – 18.00	0.69	4.45	1.11	0.08	68.00	27 267	4.70

In order to investigate the mineral beneficiation capability intended for the process of gasification in fluidised bed - bituminous coals from ZG Janina (coal of 31.2 type) and ZG Wieczorek (coal of 32 type) - each of them was subjected to the beneficiation process in the laboratory ring jig (10 rings, coal in the class of 0-18mm). After the completion of the separation process, material was divided into 5 layers (with 2 rings) and each of them was sieved on sieves into 10 particle fractions, establishing yields of layers and fractions. Then, products obtained in such a way – particle fractions, after the separation of analytical samples, were subjected to chemical elemental and technical analysis of coal in order to characterise features influencing gasification processes. Table 1 presents an example of the obtained data for layer II from ZG Janina.

3 Relevance maps

3.1 Description of the method

Relevance maps method on plane serving for data visualization is based on placing special points called relevance points which represent individual features of considered object [22, 29, 44]. For each feature (coordinate) the relevance point representing this feature is assigned. That means that by seven-dimensional data set 7 such points are placed on plane which represent individual coordinates. The distribution of the points representing presented multidimensional data shows relations between these data and features. The more i^{th} feature is present in certain object (what means that i^{th} coordinate is higher) the closest point representing certain object according to relevance point representing i^{th} feature (coordinate) should be. In this way each relevance point representing certain feature divides plane on areas more or less dependent on i^{th} feature (more or less distanced from relevance point representing i^{th} feature).

3.2 Algorithm

Set of initial data compounds of elements described by n features. It can be then treated as set of n-dimensional vectors. Let mark i^{th} initial data vector as $x_i=(x_{i,1}, x_{i,2}, \ldots x_{i,n})$. Algorithm serving to realize visualization by means of relevance map consists several steps, which are:

- Scaling of initial data. Individual features, represented by individual data dimensions are scaled in the way assuring their fitness to the same specific range. It was decided to scale the individual coordinates (features) of data set vectors to range (0, 1).
- Randomization of initial location of points images in 2-dimensional space. Let assume that image of jth point x_j, what means the point related to it in 2-dimensional space, will be marked as $p_j=(u_j, v_j)$. In previous point each of data vectors coordinates was scaled to range (0, 1). That is why the initial coordinates u_j and v_j were drawn also from range (0, 1) by plate probabilistic distribution applied.
- Randomization of initial location of relevance points in 2-dimensional space. For n-dimensional data n such points are created. These points will be marked as $w_i=(y_i, z_i)$. Initial coordinates y_i and z_i are drawn also from range (0, 1) by means of application of plate probabilistic distribution.

Next steps are realized for each initial data vector:

- For certain j^{th} vector of data x_j the distance of its image p_j from each relevance point w_i is calculated by means of Euclidean metrics:

$$D_{ij} = \sqrt{(y_i - u_j)^2 + (z_i - v_j)^2} \qquad (1)$$

- The location of image p_j is changed taking into consideration distance from each relevance point w_i. It is changed in the way assuring the distance D_{ij} from relevance point w_i

to be relevant to i^{ith} coordinate of data vector x_j. The goal is to achieve such image p_j which by high value of ith coordinate of point x_j (close to 1) is possibly close to relevance point w_i. For small value of ith coordinate of point x_j its image p_j is far from relevance point w_i:

$$\tilde{u}_j = u_j + 0.01\frac{(u_j - y_i)(1 - x_{j,i} - D_{ij})}{D_{ij}} \quad (2)$$

$$\tilde{v}_j = v_j + 0.01\frac{(v_j - z_i)(1 - x_{j,i} - D_{ij})}{D_{ij}} \quad (3)$$

where $\tilde{u}$ and $\tilde{v}$ are values after change. Equations (2) and (3) cause that point p_j is closer to point w_i when $D_{ij}>$ 1-$x_{j,i}$ and farther from w_i when $D_{ij}<$ 1-$x_{j,i}$. The dislocation of point p_j occurs according to straight line crossing through points p_j and w_i. Constant value 0.01 means the speed of points moving in proper directions.

Next steps are realized for each relevance point:

- For given i^{th} relevance point its distance D_{ij} is calculated from each image p_j of data vector by means of equation (1).
- The location of relevance point w_i is changed taking into consideration distance from each image p_j of data vector x_j. It is changed in the way assuring distance D_{ij} from relevance point w_i to be relevant to i^{th} coordinate of data vector x_j. Similarly like in step 5 the goal is to obtain possibly closest location of image p_j to relevance point w_i by high value of ith coordinate of point x_j (close to 1). For small value of ith coordinate of point x_j its image p_j is far from relevance point w_i:

$$\tilde{y}_i = y_i + 0.01\frac{(y_i - u_j)(1 - x_{j,i} - D_{ij})}{D_{ij}} \quad (4)$$

$$\tilde{z}_i = z_i + 0.01\frac{(z_i - v_j)(1 - x_{j,i} - D_{ij})}{D_{ij}} \quad (5)$$

where $\tilde{y}$ and $\tilde{z}$ are values after change. Equations (4) and (5) cause that relevance point w_i is closer to point p_j when $D_{ij}>$ 1-$x_{j,i}$ and farther from it when $D_{ij}<$ 1-$x_{j,i}$. Disslocation of relevance point w_i occur according to straight line crossing through points p_j and w_i. The constant value 0.01 means speed of points moving in proper directions.

The whole procedure covered in steps 4-7 is repeated ITER times (where ITER means the parameter accepted in certain moment).

In this way the image of multidimensional initial data points in two-dimensional space were obtained. It is sufficient then to present image of each vector on computer screen. It is realized by drawing symbol representing fraction to which data vector x_i is related in place of coordinates (u_j, v_j). Additionally, symbols representing relevance points in in screen location (y_i, z_i) can be drawn. In this way the image of multidimensional points is created on computer screen.

4 Experimental results

In purpose of visualization of seven-dimensional data describing coal samples the computer program was constructed which was based on assumptions presented in previous part of the paper. It was written in C++ language by means of Microsoft Visual Studio software. The obtained results were presented on Figs. 1-4. These views show how seven-dimensional data are transformed by means of relevance maps to just two dimensions. The algorithm of visualization by means of relevance map works in the way ensuring that despite significant reduction of dimensions number the distances between point representing certain data vector and relevance points depend possibly the most on coordinates of this data vector. In such way the important features of seven-dimensional data can be seen on computer screen.

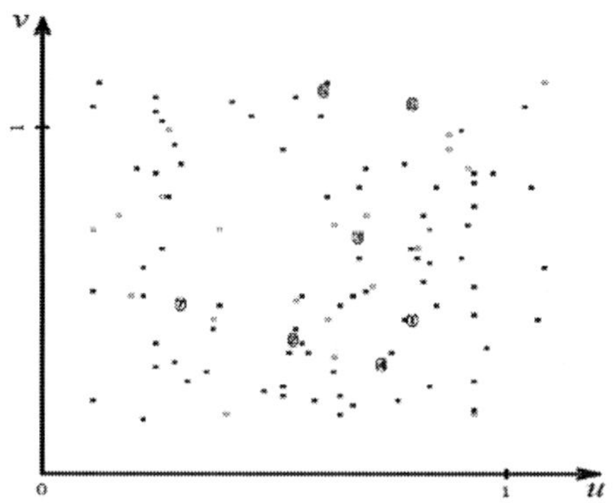

Figure 1. Initial position, in which each image of point representing seven-dimensional coal samples and each relevance point are in random positions. By (■) the images of point representing coal samples less suitable to fluidal gasification were marked, (o) – samples well suitable to fluidal gasification. Digit of the value i means relevance point representing i^{th} coordinate.

It was decided to check whether the method of relevance maps of data visualization allows to divide samples space into areas of various suitability for on-surface fluidal gasification process. Figures 1-4 present views of the points representing seven-dimensional data vectors describing coal samples obtained from ZG Janina and ZG Wieczorek. It is visible on Figs. 1-3 how the grouping of points representing discussed data according to the division on samples of well suitable and less suitable coals occur with the growth of ITER parameter. It can be seen that point being images of the data representing the same coal fractions start to gather in separated subareas and start to group. The clearness of the space division grows with the growth of ITER parameter what means better fitting of the distance D_{ij} between points images and relevance points in two-dimensional space and related coordinates of seven-dimensional data vector x_j. Figure 1 shows the initial position in which each image of entry data point

DOI: 10.1051/e3sconf/20161000065

and each relevance point is located in random position on the screen. Figure 2 presents that even by small value of the ITER parameter being equal to 100, the images of points representing the same fractions start to group. Figure 3 shows the most clear result which was obtained for the discussed data according to the division of samples into well and less suitable coals to gasification process. It occurred by the value of ITER parameter being equal to 2025. It can be visible on this screen that images of points representing samples of well and less suitable coals to gasification are located in separated subareas and gather in groups. Also, it can be seen that these gatherings on the whole picture can be easily separated. That means that through application of the relevance maps method is possible to divide samples space on the areas of various suitability for fluidal gasification process. Then, this method allows to qualify other, unknown samples of coals into well or less suitable for gasification ones through the visualization process. It is important mainly because samples of well suitable coal gather inside of the seven-dimensional cuboid – what is a simplification. It occurs directly from the fact that the accepted conditions determining affiliation to this group ("Card of technological suitability of coal") are simple inequalities by which means is easily to verify such affiliation. In reality however, it can occur that the affiliation area can have much more complicated shape. In such case, on the basis of larger amount of samples which would be empirically affiliated to the well suitable for gasification ones, the method of relevance maps allows to try to get the division of space into areas representing samples of coals well and less suitable to gasification process. In this way, it can occur that the obtained mapping will more precisely image the reality. That is why the previous statement that relevance maps method allows to divide the samples space into areas of various suitability to fluidal gasification process becomes particularly important.

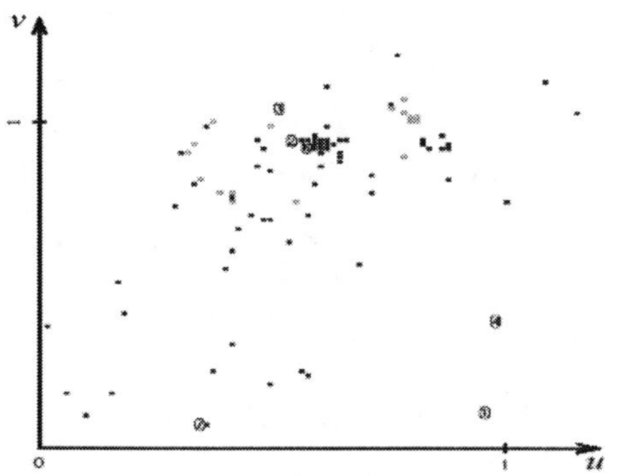

Figure 2. View of seven-dimensional data representing coal samples of various suitability to fluidal gasification process by the value of ITER parameter being equal to 100. Starting of grouping is visible. By (■) the images of point representing coal samples less suitable to fluidal gasification were marked, (o) – samples well suitable to fluidal gasification. Digit of the value *i* means relevance point representing i^{th} coordinate.

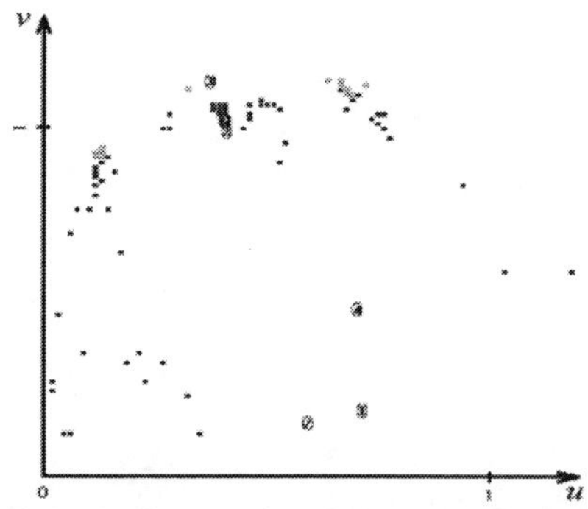

Figure 3. The most clear obtained view of seven-dimensional data representing coal samples of various suitability to fluidal gasification process by the value of ITER parameter being equal to 2025. By (■) the images of point representing coal samples less suitable to fluidal gasification were marked, (o) – samples well suitable to fluidal gasification. Digit of the value *i* means relevance point representing i^{th} coordinate.

Figure 4 presents the view of the discussed data according to completely new sort division – on samples of coals originating from ZG Janina and ZG Wieczorek. It can be seen clearly that the images of the points representing samples of coals from different mines are located in separated subareas and form gatherings. Furthermore, these gatherings can be easily separated on the picture. On the basis of this Figure it can be stated that the method of relevance maps allows to divide space on areas being characteristic for various mines. Thanks to this fact is possible to apply it to analyze new, unknown samples and through their visualization to qualify them according to their origin to the appropriate group.

It should be noticed that the algorithm of relevance maps method does not use the information of affiliation of points representing data to certain fractions. In this situation this how images of points representing certain fraction will be grouped depends only on some data properties noticed by the algorithm. That is why Figs 3 and 4 differ between themselves only by affiliation of individual points to various fractions. It occurs from the fact that both Figures were created as the result of work of relevance maps algorithm on precisely the same seven-dimensional data and by the same value of ITER parameter being equal to 2025 (with omitted information of point affiliation to individual fractions). That is why the location of points on Figs 3 and 4 is identical – only the assignment of individual points to appropriate fractions is different.

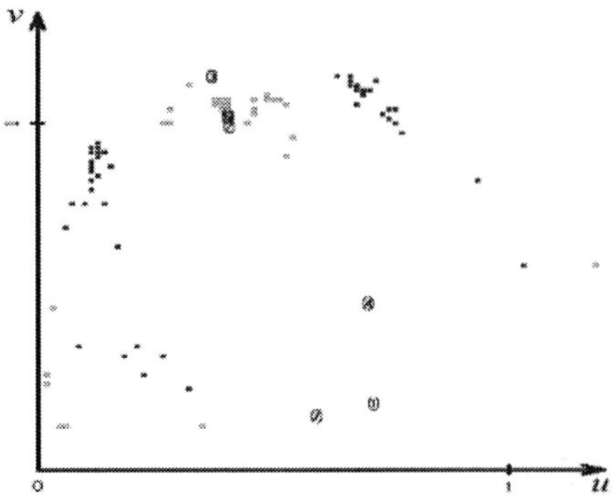

Figure 4. View of seven-dimensional data with division into extraction points by the value of ITER parameter being equal to 2025. By (■) the images of points representing coal samples obtained in ZG Janina were marked, by (o) – samples of coal obtained in ZG Wieczorek. Digit of the value *i* means relevance point representing i^{th} coordinate.

5 Conclusions

The conducted experiments based on visualization of seven-dimensional data by means of relevance maps methods allow to conclude that:

- Multidimensional visualization by means of relevance maps allow to state that the images of points representing samples of coal well or less suitable coals are located in separated subareas and form gatherings which can be easily separated. The relevance maps method allow then to divide the samples space on areas of various suitability to fluidal gasification process. It is possible then to qualify coals to groups of well suitable or less suitable for gasification.
- As a result of visualization applying relevance maps method is possible to state that images of points representing coal samples from ZG Janina and ZG Wieczorek are located in separated subareas and form gatherings which can be easily separated. It is possible then to divide samples space into areas representing various mines and to identify origin of the coals (ZG Janina or ZG Wieczorek) by means of relevance maps.
- The algorithm of visualization by means of relevance maps does not use the information about the affiliation of points representing data to certain fractions. In such situation this how images of points representing certain fraction will be grouped depends only on some data properties noticed by the algorithm. This is done independently on their affiliation to individual fractions.
- The same division of samples space conducted by means of relevance maps method groups points representing analyzed data according to their origin (point of extraction – ZG Janina or ZG Wieczorek) and also their suitability for fluidal gasification process.
- On the basis of "Card of technological suitability of coal", among analyzed 99 samples of coals only 18 of them were marked as the ones which gasify efficiently. Among these 18 samples, 17 originated from ZG Janina and only one from ZG Wieczorek.
- Clearness of the results grows with more precise fitting of the distance D_{ij} between points images and their relevance points in two-dimensional space and appropriate coordinated of seven-dimensional data vector x_j (with the growth of the value of ITER parameter).
- Clearness of obtained results depends highly on accepted parameters.
- A problem occurring by such visualization is necessity of selection of the parameters in purpose of getting view which clearly presents information searched by the researcher. It should be mention that during conducted experiments the views were obtained by the value of ITER parameter from 1 to 2000. The experiments were conducted multiple times by renewed generating of random initial values. Sometimes it lead to obtain more clear results. Also, various speeds of learning were assumed before the one presented in equations 3 and 4 was accepted. The presented results are the clearest ones from the obtained ones.

Acknowledgement

The paper is an effect of the statutory project no. 11.11.100.276.

References

1. A. Glowacz, Measurement Science Review **15**(4), 167-175 (2015)
2. A. Inselberg, *Parallel Coordinates: VISUAL Multidimensional Geometry and its Applications* (Springer, 2009)
3. A. Kosminski, D.P. Ross, J.B. Agnew, Fuel Processing Technology **87**(11), 943-952 (2006)
4. A. Nowak, A. Surowiak, Archives of Mining Sciences **58**(4), 1209–1220 (2013)
5. A. Sobolewski, T. Chmielniak, T. Topolnicka, G. Swieca, Polish Mining Review **2**, 174-183 (2012) [in Polish]
6. A. Sobolewski, T. Chmielniak, T. Topolnicka, N. Gieza, Karbo **1**, 28-38 (2013) [in Polish]
7. A. Strugala A., G. Czerski, Chemical Review **91**(11), 2181-2185 (2012) [in Polish]
8. A. Strugala, K. Czalicka-Kolarz, M. Sciazko, Energy Policy Journal **14**(2), 375-390 (2011)
9. A. Surowiak, M. Brozek, Physicochemical Problems of Mineral Processing **52**(1), 228–243 (2016)
10. A. Surowiak, M. Brozek, Archives of Mining Sciences **59**(2), 553–562 (2014)
11. A. Surowiak, M. Brozek, Archives of Mining Sciences **59**(1), 269–282 (2014)
12. A.K. Jain, J. Mao, In: *Proc. IEEE Internat. Joint Conf. On Neural Networks* (Baltimore, MD, **3**, 335-340, 1992)
13. C. Aldrich, Pattern Recognition Letters **19**(8), 749-764 (1998)

14. D. Asimov, SIAM Journal of Scientific and Statistical Computing **6**, 128-143 (1985)
15. D. Foszcz, M. Duchnowska, T. Niedoba, T. Tumidajski, Physicochemical Problems of Mineral Processing **52**(1), 98–111 (2016)
16. D. Jamroz, Archives of Mining Sciences **59**(2), 413-425 (2014a)
17. D. Jamroz, In: Gruca A., Czachórski T., Kozielski S. (eds.), *Man-Machine, Interactions 3* (AISC, Switzerland, Springer International Publishing, **242**, 443-450, 2014b)
18. D. Jamroz, Physicochemical Problems of Mineral Processing **50**(2), 719-734 (2014c)
19. D. Jamroz, In: Cyran K., Kozielski S., Peters J., Stanczyk U., Wakulicz-Deja A. (eds.), *Man-Machine, Interactions*, (AISC, Heidelberg, Springer-Verlag, **59**, 445–450, 2009)
20. D. Jamroz, T. Niedoba, Archives of Mining Sciences **60**, 1, 39–50 (2015)
21. D. Jamroz, T. Niedoba, Physicochemical Problems of Mineral Processing **50**(1), 185-202 (2014)
22. D. Jamroz, T. Niedoba, Physicochemical Problems of Mineral Processing **51**(2), 769–784 (2015)
23. D. Saramak, Physicochemical Problems of Mineral Processing **49**(1), 495–512 (2013)
24. D. Saramak, Archives of Mining Sciences, **56**, 3, 517-526 (2011)
25. G.J. Lyman, Coal Preparation **13**, 179-195 (1993)
26. H. Hotelling, Journal of Educational Psychology **24**, 417-441 and 498-520 (1933)
27. H.A.M. Ahmed, J. Drzymala, Physicochemical Problems of Mineral Processing, **39**, 129-139 (2005)
28. I.T. Jolliffe, *Principal Component Analysis* (Series: Springer Series in Statistics, Springer, NY, 2002)
29. J. Assa, D. Cohen-Or, T. Milo, Visual Computer **15(5)**, 217-234. (1999)
30. J. Drzymala, *Basics of minerallurgy* (Oficyna Wydawnicza Politechniki Wrocławskiej, 2009) (in Polish)
31. J. Drzymala, *Mineral processing: foundations of theory and practice of minerallurgy* (Oficyna Wydawnicza Politechniki Wrocławskiej, Wroclaw, 2007)
32. J. Marciniak-Kowalska et al. Project report NCBiR, *Elaboration of technology of coal gasification for highly efficient fuel and energy production*, report from part of the project *Investigations of coal beneficiation by means of mechanical processing*, (not published work, 2012-13) [in Polish]
33. J. Marciniak-Kowalska, T. Niedoba, A. Surowiak, T. Tumidajski, Archives of Mining Sciences **59**(3), 677–690 (2014)
34. J.B. Kruskal, Psychometrika **29**, 1–27 (1964)
35. M. Brozek, A. Surowiak, A. Jarosinski, Archives of Metallurgy and Materials **60**(3), 1737-1744 (2015)
36. M. Brozek, A. Surowiak, Archives of Mining Sciences **55**, 21-40 (2010)
37. M. Brozek, A. Surowiak, Physicochemical Problems of Mineral Processing **41**, 397-413 (2007)
38. M. Brozek, A. Surowiak, Physicochemical Problems of Mineral Processing **39**, 199-210 (2005)
39. R. Snopkowski, A. Napieraj, Archives of Mining Sciences **57**(1), 121-138 (2012)
40. T. Borowiecki, J. Kijenski, J. Mochnikowski, M. Sciazko M. (ed.), *Pure energy, chemical products and coal made fuels – evaluation of development potential* (IChPW, Zabrze, 2008) [in Polish]
41. T. Chmielniak, G. Tomaszewicz, Karbo **3**, 191-201 (2012) [in Polish]
42. T. Kohonen, *Self Organization and Associative Memory* (Springer-Verlag, 1989)
43. T. Niedoba, A. Surowiak, Proceedings of the XXVI International Mineral Processing Congress, New Delhi **1**, 3844-3854 (2012)
44. T. Niedoba, Archives of Mining Sciences **60**(1), 93–106 (2015)
45. T. Niedoba, D. Jamroz, Archives of Mining Sciences **58**(4), 1317-1333 (2013)
46. T. Niedoba, *Multidimensional characteristics of random variables in description of grained materials and their separation processes,* (Wydawnictwo Instytutu Gospodarki Surowcami Mineralnymi i Energią PAN, Krakow, 2013a) [in Polish].
47. T. Niedoba, AGH Journal of Mining and Geoengineering **33**(4), 235-244 (2009) [in Polish].
48. T. Niedoba, Physicochemical Problems of Mineral Processing **50**(2), 575-589 (2014)
49. T. Niedoba, Physicochemical Problems of Mineral Processing **49**(1), 175-188 (2013b)
50. T. Niedoba, in *Proceedings of the XIV Balkan Mineral Processing Congress* (Tuzla, Bosnia and Herzegovina, 57-59, 2011)
51. T. Olejnik, A. Surowiak, T. Gawenda, T. Niedoba, T. Tumidajski, AGH Journal of Mining and Geoengineering **34**(4/1), 207-216 (2010) [in Polish]
52. T. Tumidajski, D. Saramak, *Methods and models of mathematical statistics in mineral processing,* (Wydawnictwo AGH, Krakow, 2009) [in Polish]
53. T. Tumidajski, *Stochastic analysis of grained materials properties and their separation processes*, (Wydawnictwo AGH, Krakow, 1997) [in Polish]
54. W.S. Cleveland, R. McGill, Journal of the American Statistical Association **79**, 807-822 (1984)

DOI: 10.1051/e3sconf/20161000066

Performance of sand-lime products made with plastic waste

Milena Nowek[1,a]

[1]*Kielce University of Technology, Al. 1000-lecia PP 7, 25-314 Kielce, Poland*

Abstract. The paper describes the studies on the sand-lime (silicate) masonry units modified with recycled plastics in various forms: regranulate, regrind and powder. The following materials were examined: high impact polystyrene (HIPS) and acrylonitrile butadiene styrene (ABS). The results of the functional properties tests, such as compressive strength, softening behavior, bulk density and water absorption are presented in the article. The microstructure of the products was analyzed using SEM and XRD methods.Obtained results show that the properties of modified product largely depend on the type, form and amount of used polymer. The highest compressive strength was achieved with 15% of HIPS regranulate in the product (by weight). ABS does not improve the strength of the sample, however, it does block the capillary action in the sand-lime product. The lowest softening coefficient was obtained in the sample modified with HIPS regranulate. The examined polymers contributed to decrease in bulk density of the samples as well as lowered their water absorption. The samples with pulverized polymer have the worst properties. All the results are compared with those of the traditional sand-lime bricks and sand-lime product modified with other additives. The analysis indicate that silicate masonry units with post-production and post-consumption plastic waste can possess interesting functional properties what brings a new potential possibility to dispose of still growing number of plastic waste.

1 Introduction

Sand-lime products (silicates) are environmentally friendly construction materials. Traditional silicates contain in their composition only sand, lime and water and their characteristic properties are obtained thanks to hydrothermal treatment during autoclaving process.

In recent years, much attention has been devoted to research on modified sand-lime products containing among other waste materials, such as waste molding sands, post reclamation dusts, hematite tailings, fly ashes, bottom ashes, copper tailings, foam glass as well as recycled plastics [1-8].

The overall purpose of the study was to improve certain characteristics of silicates and at the same time dispose of plastic waste.

It is known from the literature [8] that regranulate of high impact polystyrene has very beneficial effect on compressive strength of silicates.

Any kind of acting during processing of plastic waste, for example cutting or grinding, can cause changes in the material properties. Thanks to comprehensive research, changes in molecular structure after multiple processing of both pure and polluted polymers are well known. Multiple of processing, as well as heat treatment during a particular operation, result in degradation of macromolecules of the polymer. Nevertheless, research has shown that there are grounds for wider than previously use of polyolefins recyclates for manufacturing of various products without reducing their quality requirements [9]. Plastic waste can be blended with virgin plastic in order to reduce cost without sacrificing their properties. The problem related to plastics wastes has increased tremendously since the use of plastics increased in most industrial, commercial and residential applications. Plastic waste causes severe environmental problems when incinerated or open burned on roadsides or in illegal dumpsites [10].

Mechanical recycling of plastics involves cleaning, sorting, cutting, shredding, agglomeration, pelletizing and finally reprocessing by injection molding, blowing or extrusion according to the required products [10]. High value of the recycled material justifies the incurred expenditures. The condition of proper material recycling is to obtain recyclates with good, repeatable and defined properties. Problems with the quality and repeatability of the properties of recyclates can be solved if the process of their recovery will be treated as a process of production of new materials, shaped in terms of their most efficient uses [11].

[a] Corresponding author: mnowek@tu.kielce.pl

2 Methodology

In order to carry out laboratory tests of modified sand-lime products, the author has prepared cuboidal samples with dimensions of 40x40x160mm. The samples were made in Silicates Production Plant in Ludynia. The sand for production of sand-lime products has to meet certain requirements concerning the minimum SiO_2 content of 80%, chemical composition and particle size. The sand used in the technological process has the grain size of 0-2 mm. Grain size distribution of sand influences the mechanical properties of silicate materials. The sand is mixed with quicklime and water in appropriate proportions. The sand constitutes approximately 90% of the product weight, lime is approx. 7% and water - 3%. In steel reactors, the process of lime slaking in silicate mass takes place. When exposed to water, quicklime is converted into slaked lime. As a result of elevated temperature derived from the reaction of lime slaking and alkaline environment the surface of sand grains loses its crystalline structure. In this way, the grains of sand get the proper form for further transformation into silicate in the later stages of the production process [12].

In the next stage, the author introduced to the sand-lime mass 15% of polymer (per weight). The polymer was in the form of: high impact polystyrene (HIPS) regranulate, regrind and powder, as well as copolimer acrylonitrile-butadiene-styrene (ABS) regranulate.

HIPS is a lightweight material with high stiffness, which is impact and crack resistant. Polystyrene HIPS contains caoutchouc bound physically or chemically. Accordingly to the share of the caoutchouc, physical and mechanical properties of the material are different. Products made of polystyrene HIPS are used among others to make equipment cases, elements of lighting fixtures and in the food industry to manufacture of packaging.

ABS is a material with high hardness and impact resistance which has good insulating properties and is characterized by a high softening temperature. ABS is used in the production of machinery and equipment housing, interior elements of vehicles, various types of containers or elements of furniture [13].

The amount and type of used in the study polymers resulted from the previous research [8], which has shown a significant improvement in the compressive strength of sand-lime products containing 20% of HIPS regranulate and 30% of its regrind.

HIPS regranulate in the form of slightly flattened cylinders with the maximum dimension of 4 mm was used. HIPS regrind was added as irregular blades with various shapes and dimensions. HIPS had been previously pulverized to the grain size of 630 μm. ABS was used in the form of irregular cylinders with dimensions up to 3 mm.

Next, all the components were precisely mixed with a small amount of additional water and then placed in steel molds. The specimens were compressed under the pressure of 20 MPa and cured in industrial autoclaves under the following conditions: saturated water steam temperature equal to 203°C and pressure equal to 1.6 MPa, total time of curing - 8 hours. Table 1 presents the qualitative and quantitative composition of the individual samples.

Table 1. Composition of the samples.

Sample No.		Form of polymer	Share of polymer in the sample weight [%]
1	x	x	x
2	HIPS	regranulate	15
3	HIPS	regrind	15
4	HIPS	powder	15
5	ABS	regranulate	15

3 Results and discussions

All the tests have been carried out on 6 specimens. Compressive strength test of dry samples indicated, that the value of this parameter strongly depends on the type of used polymer and its form. In the case of HIPS regranulate the value was the highest and equal to 32.04 MPa, which is higher by almost 55% than the strength of the reference sample The other polymers has decrease the compressive strength of the silicate. The lowest strength equal to 2.84 MPa was obtained for the samples with HIPS powder. HIPS powder constitutes a large part of the specimen (high volume fraction) and physically blocks the reactions between lime and silica. The pulverization of the regranulate might have a major impact on the deterioration of the properties of the material. Loose structure and many relatively large pores were observed. During hydrothermal curing all the polymers have partially melted down and to some extent filled the spaces between the other ingredients.

Previous studies performed by the author [8] indicated, that the addition of 10 and 20% of HIPS regranulate improves the compressive strength of such modified sand-lime products by respectively 43 and 163%. The addition of HIPS powder in an amount of 10% resulted in decrease in the compressive strength by 70%. The findings of the study on the use of mixed polypropylene and polyethylene regrinds are also interesting. PP and PE also improved the compressive strength of sand-lime products, however, their impact was slightly lower than the impact of high impact polystyrene.

There are known descriptions of the research on the use of HIPS regranulate as an additive for the autoclaved aerated concrete [14]. At 10% of the regranulate in the sample weight, the compressive strength of the products slightly rises, but further addition of HIPS results in reduced strength. HIPS recyclate has been also tested as the additive in cement mortar [15]. Decilne in the compressive strength by 12%, 22% and 49% was obtained for mortar containing 10%, 20% and 50% HIPS, respectively. Polymer-modified mortars with recycled ABS powder have been also investigated [16]. The compressive modulus increased at 8% and 15% but decreased at 25% of ABS in the mortar. The addition of ABS powder increases the porosity of the material.

Softening-coefficient determines the destructive effect of water on building materials. The coefficient was calculated by dividing the compressive strength of a saturated sample by the compressive strength of a dry sample. For the sample containing HIPS regranulate the softening-coefficient has the highest value, while the lowest coefficient was noted for the sample with HIPS regrind. In this case, water absorption due to capillary action was the highest. The absorption of the other samples was lower than the absorption of the reference sample. Extremely low water absorption, equal to 1.59% was obtained for the sample containing ABS regranulate. The central part of the cross-section of this sample remained dry after 3 days of full immersion in water.

Added plastics in varying degrees influenced the decrease in the bulk density of of the sand-lime products. The lowest bulk density was achieved with the addition of HIPS powder. Table 2 presents the results of the compressive strength, softening-coefficient, water absorption and bulk density of the samples.

Table 2. Tests results.

Sample No.	Compressive strength [MPa]	Softening coefficient	Bulk density [kg/m^3]	Water absorption [%]
1	**20.7**	**0.87**	**1944**	**13.5**
2	32.0	0.96	1530	12.5
3	17.0	0.56	1417	15.7
4	2.8	0.61	1089	11.9
5	13.7	0.71	1629	1.59

The basic components of sand-lime products are hydrated calcium silicates, which bind together the grains of sand. Besides them in traditional sand-lime products also some residues of unreacted sand and lime as well as water and air in pores are present. The amount of particular phases has an impact on the mechanical strength of resulting products and their other properties. Stabilized compounds with crystalline structure and dimensions up to several tens of microns are obtained thanks to the hydrothermal treatment of the components of $CaO-SiO_2-H_2O$ system at the temperature exceeding 100°C [17]. Many years of research show that in the autoclaved sand-lime products mostly the following hydrated calcium silicates can be found: C-S-H (I), C-S-H (II), 11Å tobermorite, xonotlite and C_2SH (A) [3].

The phase composition of the samples was characterized by XRD. X-ray diffractometry revealed the presence of quartz, calcium silicate and tobermorite 11 A in each specimen. XRD analysis of samples containing plastic waste indicated also presence of calcite in small amount. In addition, a large amount of an amorphous substance (C-S-H) in the modified products, which does not give sharp peaks, was present. The measurements were made using a monochromatic radiation of a wavelength corresponding to the emission line K of copper in the angular range of 5-90°, scale 2.

The samples were also subjected to the SEM observation. The C-S-H phase is a configuration of hydrated calcium silicates with a different level of structural arrangement: $CaO-SiO_2-H_2O$, which emerges as a result of the reaction between calcium silicates and water. The tobermorite is the most ideal form of the C-S-H phase. It is a hydrated calcium silicate characterized by an arranged structure and specified inter-plane spaces [18,19].

Figure 1 shows a grain of sand surrounded by the C-S-H phase, which then transforms into a higher and more perfect form, i.e. tobermorite. The tobermorite phase in a traditional silicate product appears in the form of blades with flat endings. Figure 2 is a diphractogram presenting the elemental structure of traditional silicate product [18]. A small amount of carbon comes from the preparation of the samples for SEM.

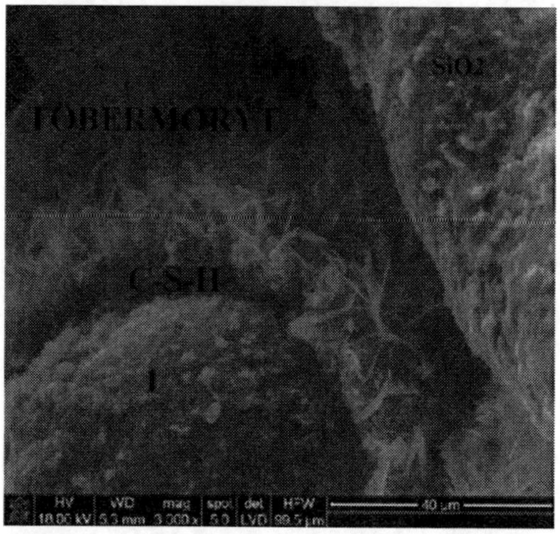

Figure 1. Microstructure of traditional silicate product [18].

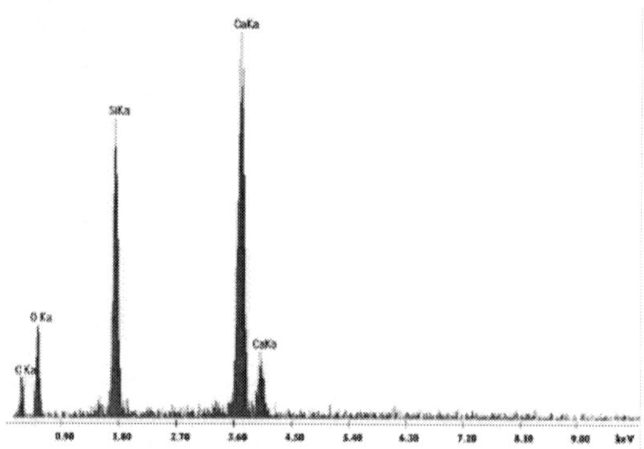

Figure 2. EDS spectrum of traditional silicate product – in point 1 [18].

Observation of modified products revealed diverse forms of the synthesis products. In the samples with HIPS powder fibrous tobermorite, lathlike tobermorite and amorphous C-S-H phase broken by the polymer

 DOI: 10.1051/e3sconf/20161000066

(Figure 3) were found almost exclusively. High compressive strength of silicate product containing HIPS regranulate is a result of good contact between the polymer and sand-lime mass ingredients. Well developed plate-like tobermorite and poorly crystallized C-S-H phase were found. In the sample with ABS regranulate hydrated calcium silicates with more and less ordered structure were found. Fibrous C-S-H phase was the predominant product.

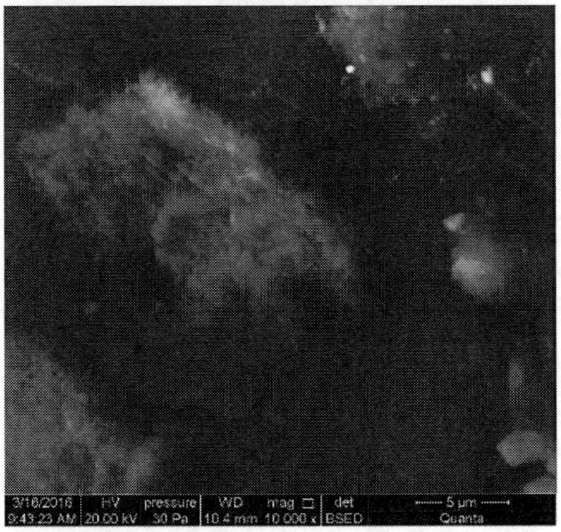

Figure 3. Microstructure of sample with HIPS powder – C-S-H phase broken by polymer.

4 Conclusions

Introducing to the sand-lime mass properly selected recycled plastics can improve the compressive strength of the resulting products or significantly reduce water absorption.

Additives in the form of recycled plastics have a considerable influence on the characteristics of the modified sand-lime products, which depends on the type of polymer, form of recyclate and its quantity in the sample. The biggest impact on the compressive strength increase has HIPS regranulate, while ABS regranulate significantly reduces capilary action in the products. Obtained results show that the properties of modified product largely depend on the type, form and amount of used polymer. The highest compressive strength was achieved with 15% of HIPS regranulate in the product (by weight). ABS does not improve the strength of the sample, however, it does block the capillary action in the sand-lime product. The lowest softening coefficient was obtained in the sample modified with HIPS regranulate. The examined polymers contributed to decrease in bulk density of the samples as well as lowered their water absorption. The samples with pulverized polymer have the worst properties.

References

1. Z. Pytel, Archives of Foundry Engineering, **10** Special Issue 2(2010)
2. Y. Zhao, Y. Zhang, T. Chen, Y. Chen, S. Bao, Const. and Build. Mat. **28**, 450-455 (2012)
3. S. Wolfke, Technologia wyrobów wapienno-piaskowych. Arady, Warszawa (1986)
4. Z. Pytel, Ceramic Mat., 60, **4** (2008)
5. Y. Fang, Y. Gu, Q. Kang, Q. Wen, P. Dai, Const. and Build. Mat. **25**, 867-872 (2011)
6. R. Dachowski, I. Jasińska, Archives of the Institute of Civil Engineering, **13** (2012)
7. I. Jasińska, R. Dachowski, Praca zbiorowa pod redakcją Joanny Bzówki pt. Badania doświadczalne i teoretyczne w budownictwie, monografia (2012)
8. M. Nowek, Tech. Transactions. Czasopismo Techniczne, 1-B (2014)
9. praca zbiorowa pod redakcją prof. Dr hab. Andrzeja K. Błędzkiego, Recykling materiałów polimerowych, Wydawnictwa Naukowo-Techniczne, Warszawa (1997)
10. S. El-Haggar, Sust. Indust. Design and Waste Managem. (2007)
11. J. Kijeński, A. K. Błędzki, R. Jeziórska, Odzysk i recycling mat. Polimer., PWN (2011)
12. J. Jasiczak, Budownictwo Ogólne Tom 1 Materiały i wyroby budowlane, rozdział 7, "Wyroby z zaczynów, zapraw i betonów", Arkady, Warszawa (2005)
13. J. Pielichowski, A. Puszyński, Chemia polimerów. Wydawnictwo Naukowo-Techniczne TEZA, Kraków (2004)
14. R. Dachowski, S. Kapała, Materiały Budowlane **12** (2015), p. 18-20
15. R. Wang, C. Meyer, Cement & Concrete Composites 34 (2012) p. 975-981
16. A. Palos, N.A. D'Souza, C.T. Snively, R.F. Reidy, Cement and Concrete Research **31** (2001), p. 1003-1007
17. E. Szymański, Technologia materiałów budowlanych. Wydawnictwo Politechniki Białostockiej, Białystok (2003)
18. R. Dachowski, A. Stępień, Procedia Engineering **21** (2011) p.1173-1178
19. W. Kurdowski, Chemia cementu i betonu, Wydawnictwo Naukowe PWN, Warszawa (2010)

Numerical design of the Seed-Blanket Unit for the thorium nuclear fuel cycle

Mikołaj Oettingen[1] and Katarzyna Skolik[1]

[1]*AGH University of Science and Technology, al. Mickiewicza 30, 30-059 Krakow, Poland*

Abstract. In the paper we present the Monte Carlo modelling by the means of the Monte Carlo Continuous Energy Burn-up Code of the 17x17 Pressurized Water Reactor fuel assembly designed according to the Radkowsky Thorium Fuel concept. The design incorporates the UO_2 seed fuel located in the centre and $(Th,U)O_2$ blanket fuel located in the peripheries of fuel assembly. The high power seed region supplies neutrons for the low power blanket region and thus induces breeding of fissile ^{233}U from fertile ^{232}Th. The both regions are physically separated and thus this approach is also known as either the heterogonous approach or the Seed-Blanket Unit. In the numerical analysis we consider the time evolutions of infinite neutron multiplication factor, axial/radial power density profile, ^{233}U, $^{235}U$and ^{232}Th.

1 Introduction

The study focuses on the numerical modelling of the advanced nuclear system containing thorium fuel. Thorium occurs in nature in a single form of ^{232}Th and is about three times more abundant than uranium. It is a fertile isotope which captures a neutron and transmutes to the fissile ^{233}U. For this reason thorium fuel has to be combined with another material able to provide sufficient neutron flux, such as ^{233}U, ^{235}U or ^{239}Pu. After irradiation of thorium, the share of power generated by produced ^{233}U is increasing. The process in which ^{232}Th is transformed into ^{233}U is presented below.

$$n + {}^{232}_{90}Th \rightarrow {}^{233}_{90}Th \xrightarrow{\beta^-(22\,min)} {}^{233}_{91}Pa \xrightarrow{\beta^-(27\,d)} {}^{233}_{92}U$$

Introduction of thorium to the general use in nuclear power plants (NPP) would increase fuel diversification and safety of supply. It would also vastly prolong the sufficiency of estimated conventional nuclear fuel, which is about 100 years for today [1]. The advantages of thorium fuel cycle together with its drawbacks comparing to traditional uranium cycle are presented below [2].

Advantages:

•Good chemical and physical properties: higher melting point, higher thermal conductivity and chemical stability, lower coefficient of thermal expansion, less gas fission products released [3, 4],
•Less minor actinides and plutonium in spent fuel,
•Proliferation resistance: irradiated fuel contaminated with easily detectable ^{232}U,
•Possibility of breeding fuel in thermal reactors,
•Thorium is more abundant, safer and more effective in mining.

Disadvantages

•Neutron source required to produce fissile ^{233}U from ^{232}Th,
•Difficult and costly fuel fabrication for potential close fuel cycle because of high γ radiation from ^{232}U,
•Relatively long time of ^{232}Th breeding to ^{233}U due to ^{233}Pa half-life of 27 days,
•Likelihood of using ^{233}U for nuclear weapon,
•Need of many costly researches and licensing processes before starting industrial use.

2 Research programs

There was a significant interest in thorium fuel cycle in the first years of nuclear energy, 1950-1970. The experimental reactors were built in the USA, Germany and UK. Most of them were High-Temperature Reactors (HTR), either with prismatic or pebble-bed fuel configuration, namely Peach Bottom and Fort St Vrain in the USA, Dragon in the UK, Arbeitsgemeinschaft Versuchsreaktor (AVR) and Thorium High-Temperature Reactor (THTR) in Germany [5].

In 1980s the support for nuclear power declined dramatically after the events in the Three Mile Island and Chernobyl. Prices of uranium decreased and the enthusiasm for alternative fuel cycles disappeared. Only in recent years some countries have

DOI: 10.1051/e3sconf/20161000067

become interested again in researches on thorium, some of them are Canada, China, India, Norway and the USA [6].

The most ambitious program was launched in India which has very limited resources of uranium and probably the biggest deposits of thorium in the world. The program is divided in three steps. In the first stage Pressurized Heavy Water Reactors (PHWR) fuelled by natural uranium will be used, where the plutonium is routinely produced. It can be than burned in the second stage with use of fast neutron reactors. Uranium and thorium present on the blanket around the core will produce further plutonium together with ^{233}U. In the last step the Advanced Heavy Water Reactors (AHWR) will burn Th/Pu fuel and continuously breed ^{233}U. An alternative for the third stage is a Molten Salt Breeder Reactor (MSR). Until now about 1 tone of thorium oxide fuel has been irradiated experimentally in the PHWR reactors. The prototype 500 MWe fast breeder reactor is expected to be operating in 2016 at Kalpakkam and six more fast reactors have been announced for construction. Design of the first 300 MWe AHWR is now completed and the construction will start in 2016/17 with operation about 2022.

3 Seed-Blanket Unit

For the usage of thorium in Pressurized Water Reactors (PWR) two main fuel designs are considered: homogeneous and heterogeneous. The first approach assumes using the mixture of thorium and uranium oxides within each fuel rod while the second one considers placing thorium and uranium in separated regions, called the blanket and the seed, respectively. The heterogeneous design is far better for the effective conversion of ^{232}Th into fissile ^{233}U. This approach once again considers two different concepts: Seed-Blanket Unit (SBU) and the whole assembly seed and blanket (WASB) [7]. SBU is also known as Radkowsky Thorium Fuel (RTF), named after Alvin Radkowsky, American nuclear physicist and a big proponent of using thorium as a fuel in NPPs.

The SBU represents the fuel assembly (FA) that consists of uranium "seed" in the center and thorium "blanket" around it as it is presented in Fig 1. Therefore all the assemblies in the reactor core are composed identically. Such design considers open fuel cycle, which means that ^{233}U is continuously produced and burned in the blanket and the fuel is not recycled afterwards. The blanket region is initially composed of ThO_2 and UO_2, where uranium oxide represents about 10% by volume. The seed can be either in metallic (U/Zr) or oxide (UO_2) form. However the second one was already widely used in nuclear power industry and has slightly better economic and neutronic performance than the first one. As such is was accepted as a part of seed in thorium-fuelled PWRs and also chosen for present study. Both in the blanket and in the seed enrichment of uranium is about 20%.

4 Simulations

The numerical simulations of the SBU were performed by the means of the Monte Carlo Continuous Energy Burnup Code (MCB) developed at the Department of Nuclear Energy, Faculty of Energy and Fuels of the AGH University of Science and Technology, Krakow, Poland [8]. The MCB code couples well-known, general-purpose Monte Carlo N-Particle transport code (MCNP) with the state of the art Transmutation Trajectory Analysis code (TTA). The MCNP subroutines are used for the Monte Carlo neutron transport simulation while TTA subroutines for the calculations of isotopic changes due to nuclear transmutation and radioactive decay.

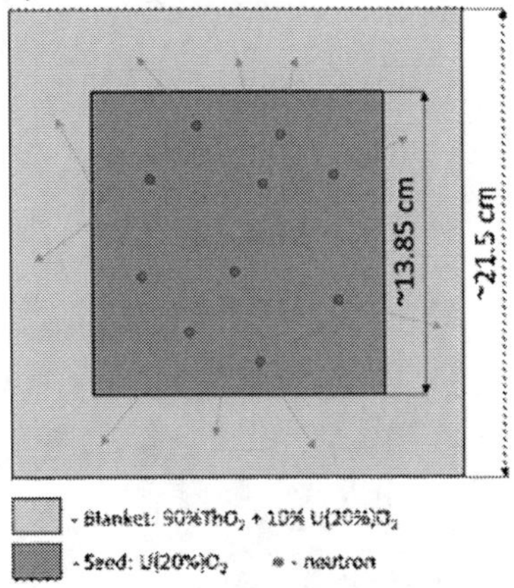

Figure 1. Fuel assembly in SBU design.

The code is implemented on the supercomputer Prometheus of the Academic Computer Center CYFRONET AGH, which allows fast and efficient Monte Carlo simulations. The PROMETHEUS is equipped in 55728 Intel CPUs, which gives the theoretical peak performance of 2399 TFlops. The cluster is ranked 39th in the TOP500 list of world's fastest supercomputers (November 2015) and is the fastest polish supercomputer dedicated for the scientific calculations.

The calculations were executed using two Hewllet Packard XL730f Gen9 nodes with 24 CPUs per node, of the type Intel Xeon E5-2680v3 with base frequency 2.5 GH. The system eigenvalue (neutron multiplication factor) was calculated using the *kcode* mode of MCNP. The well-validated JEFF3.1 nuclear data libraries containing the vast set of cross section data were used for simulations of neutron transport while standard MCB libraries containing radioactive properties of isotopes were used for burnup simulations. The total number of simulated neutron histories equals 7 million including 2 millions of inactive histories for fissile system convergence. The total execution time for one run equals about 5 h and gives the numerical precision below 30 pcm for neutron multiplication factor, which is sufficient for reliable simulations. The simulation was performed for the reactor cycle of one year with twelve burnup steps of one month.

DOI: 10.1051/e3sconf/20161000067

5 Geometry

Fig. 2 presents the horizontal cross section of the applied numerical model representing 17x17 fuel assembly. The transformation of the engineering geometry to the numerical geometry was based on the FA design dedicated for the Evolutionary Power Reactor (EPR) designed by French company AREVA. The main parameters of the fuel assembly are presented in Table 1. It is worth mentioning that the active height of the assembly was enlarged from 3.66 to 4.20 meters, what is strictly related to the larger power output of EPR (1.65 GW_{el}) comparing with PWR of III generation like e.g. 4-loop Westinghouse reactor (1.1 GW_{el}). Hence, the system was normalized to the average thermal power of single EPR fuel assembly of 18.67 MW_{th}.

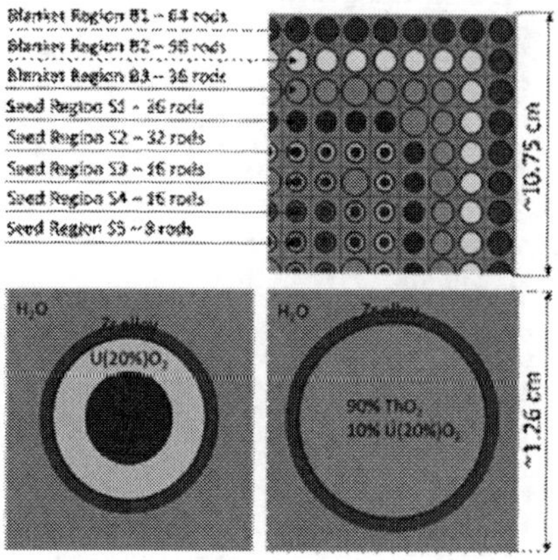

Figure 2. Numerical model of investigated fuel assembly.

According to the Radkowsky Thorium Fuel concept the seed region is surrounded by the thorium blanket region. The fuel rods were grouped into eight radial and seven axial fuel regions, which gives in total 56 independent fuel zones. The length of one axial segment equals 60 cm. This type of division was applied for detailed reconstruction of isotopic changes in nuclear fuel across whole fuel assembly. The seed region comprises five radial regions with angular fuel rods containing enriched uranium fuel, while blanket region three radial regions with compact fuel rods containing fertile thorium fuel with some uranium. The multiple layers of neutron reflector containing water and structural materials were placed below and over active column. The reflective boundary conditions surround the FA.

6 Results

The numerical simulation with the MCB code provides a number of output files containing many sets of numerical data. In this study we mainly focus on the general parameters of the fuel assembly like infinite neutron multiplication factor, power distribution and evolution of fissile ^{235}U, ^{233}U and fertile ^{232}Th.

Fig. 3 shows the evolution of the infinite neutron multiplication factor. The initial large drop in K_{inf} is caused by the formation of absorbing fission products especially ^{135}Xe. Starting from about 30 days of irradiation K_{inf} drops in linear manner, which is strictly associated with the depletion of ^{235}U over reactor cycle. The investigated fuel assembly is strongly supercritical even at the end of cycle (EOC), which is caused by the high enrichment of the initial uranium fuel.

Table 1. Parameters of Seed-Blanket-Unit fuel assembly [9].

Assembly	
Rod Pitch	1.26 cm
Active lenght	4.20 m
Coolant	H_2O
Nr of rods	264
Layout	17x17
Clading type	Zircoalloy
Clad thickness	0.57 mm
He gap thickness	0.08 mm
Blanket	
Fuel Type	90% wt % ThO_2 + 10 wt % UO_2
Enrichemnt	20%
Density	9.4 g/cm^3, 94% of TD
Mass	~418.4 kg
Pellet radius	4.65 mm
Rod radius	0.53 cm
Seed	
Fuel Type	UO_2
Enrichemnt	20%
Density	10.3 g/cm^3, 94% of TD
Mass	~146.5 kg
Pellet outer radius	3.85 mm
Pellet inner radius	2.20 mm
Rod radius	5.30 mm

Fig. 4 shows the net increase/decrease of the ^{235}U, ^{233}U and ^{232}Th over twelve 30 days-long time steps. The net decrease in uranium ^{235}U falls from step to step because the ongoing consumption due to fission. Moreover its consumption is compensated by fission of newly bred ^{233}U and ^{239}Pu. The net decrease of ^{232}Th due conversion to ^{233}U via ^{233}Pa shows average constant value of 245 grams for all considered time steps. It means that ^{232}Th conversion just merely depends on changes of isotopic concentration in the fuel assembly and it can reach some upper limit, providing constant assembly power. The total conversion of ^{232}Th over reactor cycle equals about 3 kg while the depletion of ^{235}U about 9 kg.

Fig. 5 depicts the power profile distribution at the middle of cycle (MOC), which is necessary for the estimation of the hot spots in the fuel assembly for further thermal-hydraulic calculations. The power is mainly released in the inner seed region containing high enriched uranium fuel. The highest power density was determined in the fourth seed radial region at the interface with the blanket thorium region.

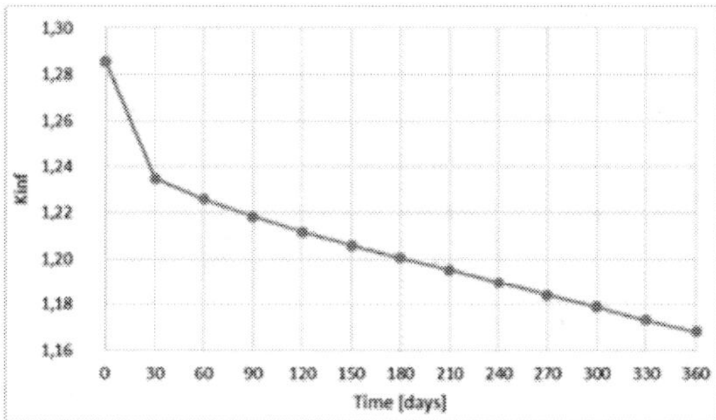

Figure 3. Evolution of infinite neutron multiplication factor.

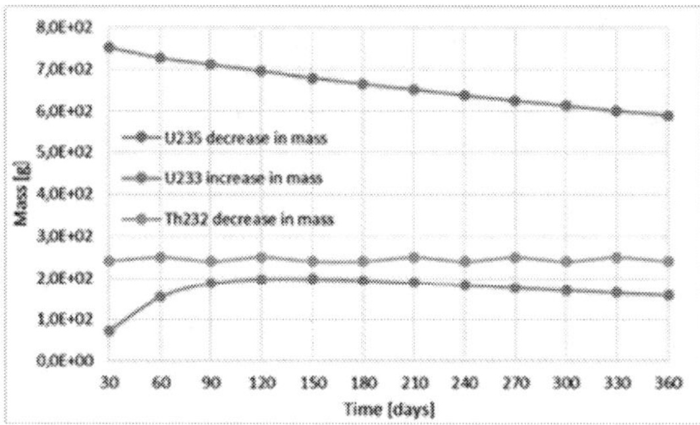

Figure 4. Evolution of ^{233}U, ^{235}U and ^{232}Th.

Fig. 6 shows the detailed axial power evolution for the beginning of cycle (BOC), MOC and EOC for these regions. Moreover, the power profiles undergo flattening due to the enhanced fuel burnup in the central fuel zones.

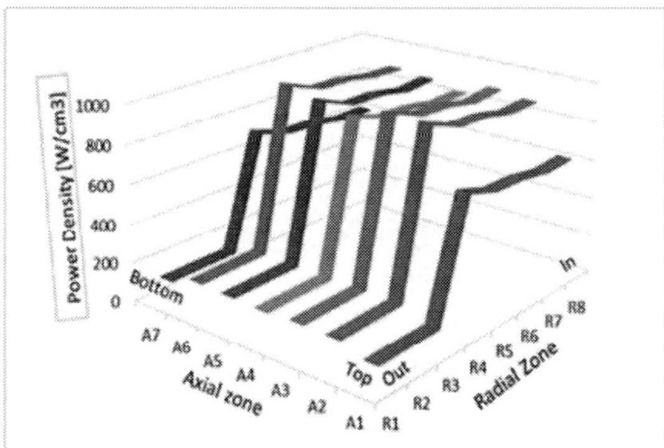

Figure 5. Power density distribution at MOC.

7 Conclusions

In the paper we have presented the Monte Carlo modelling of the EPR 17x17 fuel assembly designed for the thorium utilisation. The advanced numerical tools were used for the creation of the simulation setup for the high multiscale modelling with high performance computer Prometheus. The study shows that:

- The MCB code is the advanced numerical tool, which allows detailed modelling of the nuclear systems working on Th-U fuel,
- The obtained numerical results provide the reliable basis for subsequent research on Th-U fuel foreseen for the PWR reactors,
- The study proves the breeding capability of the fissile ^{233}U from the fertile ^{232}Th in existing and future nuclear reactors cooled and moderated by light water – see Fig. 4.

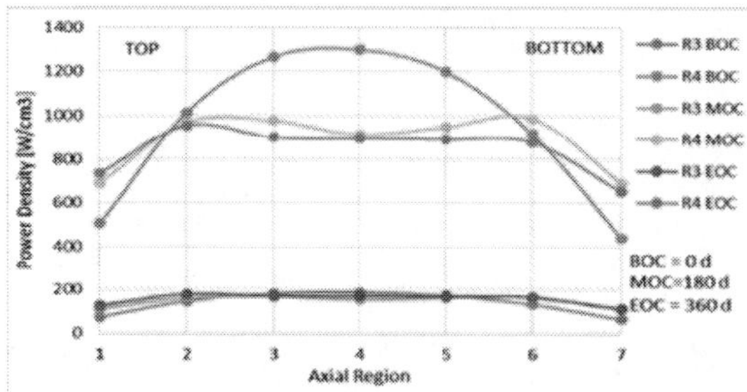

Figure 6. Axial power profile for hottest seed and blanket axial regions.

The future research on the Th-U fuel cycle modelling with MCB should focus on the estimation of safety parameters at the level of the whole reactor core. This, in turn demands development of the more advanced, high resolution, numerical models to determine the characteristics of the fissile system with the high numerical precision. In addition, the comparative analysis of actinide concentrations between numerical simulation and experimental assay would help to validate designed numerical setup.

Acknowledgements

The research was supported in part by PL Grid Infrastructure available at Academic Computer Centre CYFRONET AGH. In addition, partial financial support of this study through the funds of the Polish Ministry of Science and Higher Education is kindly acknowledged.

References

1. D.E. Sefrontein, E.J. Mulder, *Thorium-based fuel cycles: Reassessment of fuel economics and proliferation risk.* Nucl Eng Des **271,** 106-113 (2014)
2. K.I. Björk et al., *Commercial thorium fuel manufacture and irradiation: Testing (Th,Pu)O₂ and (Th,U)O₂ in the "Seven-Thirty" program*, Ann Nucl Energy **75**, 79-86 (2015)
3. L.P. Tucker, A. Alajo, S. Usman, *Thorium-based mixed oxide fuel in a pressurized water reactor: A beginning of life feasibility analysis with MCNP*, Ann Nucl Energy **76**, 323-334 (2015)
4. S.S. Drera, K.I. Björk, J.F. Kelly, *Thorium fuel production and results from beginning of life irradiation*, Prog Nucl Energ **72**, 5-10 (2014)
5. D. Greneche, M. Chhor, *Development of the thorium fuel cycle*, Woodhead Publishing Limited (2012)
6. B. Graves, A. Wong, K. Mousavi, Ch. Canter, A. Kumar, *Techno-economic assessment of thorium power in Canada*, Ann Nucl Energy **85**, 481-487 (2015)

7. IAEA, *Thorium fuel cycle – Potential benefits and challenges*, Vienna, (2005)
8. G.Kępisty, J. Cetnar, P. Stanisz, *Parametric studies of the PWR fuel assembly modeling with Monte-Carlo method,* Ann Nucl Energy **94**, 189–207 (2016)
9. S.F. Ashley et al., *Fuel cycle modelling of open cycle thorium-fuelled nuclear energy systems,* Ann Nucl Energy **69**, 314–330 (2014)

DOI: 10.1051/e3sconf/20161000068

The role of phase change materials for the sustainable energy

Marta Kuta[1,a], Dominika Matuszewska[1] and Tadeusz Michał Wójcik[1]

[1]*AGH University of Science and Technology, Faculty of Energy and Fuels, Department of Thermal and Fluid Flow Machines, al. A. Mickiewicza 30, 30-059 Krakow*

Abstract. Unceasing global economic development leads to continuous increase of energy demand. Considering the limited conventional resources of energy as well as impact on the environment associated with its use, it is important to focus on the rational management of energy resources and on supporting the development of new technologies related to both conventional and renewable energy resources. In a number of cases the use of phase change materials (PCMs) turns out to be a reasonable solution. This paper contains a summary of well-studied and known, previously used solutions based on phase change materials as well as novel possibilities, which are under development. It has been decided to investigate this topic due to the wide range of highly effective solutions. The review is focused on selected applications of PCMs for technologies which are designed to improve energy efficiency and on PCMs used in technologies based on renewable energy sources.

1 Introduction

With the dynamic economic development and consequently incessant increase of the demand for heat and electricity and also increasing requirements on life comfort, we need to face the necessity of rational management of available resources. Among the possibilities of working on this challenge we can list reducing the gap between the demand and supply of energy. For that purpose thermal energy storage (TES) appears as a good solution. Latent heat storage is one of the most efficient way of storing thermal energy. The use of phase change materials is based on the use of latent heat and it is characterised by advantages as [1]: high storage density (as compared to sensible heat) and low difference between temperatures of heat storage and release. In that case PCMs have been used in latent heat thermal storage systems for solar power engineering and heat pumps [1, 2]. Another and also one of the first applications with the use of PCMs for thermal storage is building application [3]. It is also important to improve existing and develop new technologies to support the use of renewable energy resources. Phase change materials can be used successfully in this area. The aim of this paper is to review applications of phase change materials for technologies related for sustainable energy.

2 Phase change materials

Applications of phase change materials for new technologies are valuable. First of all, because of its ability to absorb the heat during the transition and store it for specific and distinctive amount of time, according to the type of material. Fig.1 shows how phase change materials work based on latent heat.

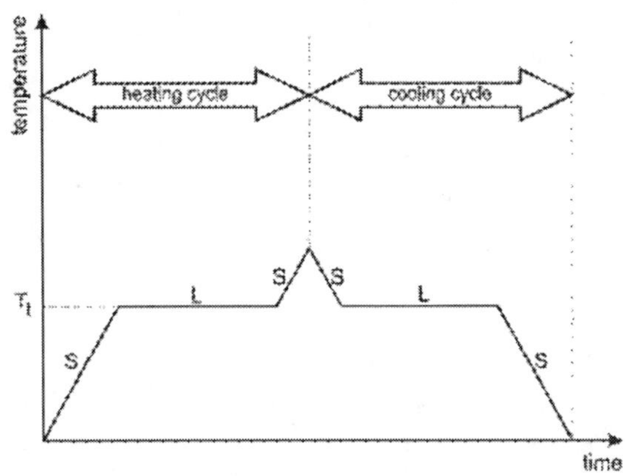

Figure 1. Schematic heat curve for phase change material; S-sensible heat, L-latent heat, T_t – temperature of phase transition.

There are several conditions that determine the possibility and profitability of using the PCMs. The first essential aspect is the temperature range of phase transition which need to be adequate to the application. Nearly always it is important that the material is characterized by possibly high latent heat of transition and specific heat and thermal conductivity. Next important features are: no degradation after many operation cycles, stability of properties at subsequent cycles and minimal subcooling, minimal volume change and gas generation on transition. It is also required that PCM is non-toxic and non-corrosive, non-flammable, non-explosive and easy to recycle, environment friendly. From the economic point of view it is important that price is reasonably low enabling the use of PCM for wider group of applications.

[a] Corresponding author: marta.kuta@agh.edu.pl

There are several criteria used to group PCMs. Fig. 2 and Fig. 3 show how phase change materials can be grouped according to the type of transition (Fig. 2) and to the type of materials (Fig. 3). As it can be seen in table 1 PCMs can be identified as: solid-liquid, liquid-solid, liquid-gas, gas-liquid, solid-solid. The best known and most frequently used are solid-liquid PCMs. It is related to high availability and good thermophysical properties of S-L PCMs. Phase change materials from the solid-solid group are not that widely used because of typically smaller storage capability, however lately S-S PCMs have become interesting because of several advantages such as: no volume change, no leakage or gas generation during the transition.

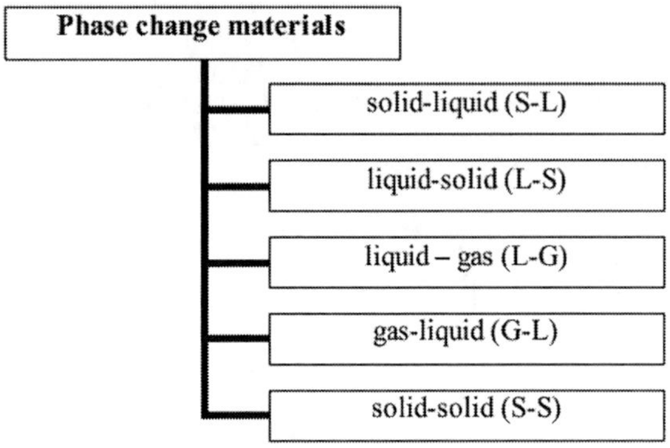

Figure 2. Phase change materials classification – type of transition.

As it has been described before the best known are S-L PCMs. Most researches refer to their properties which are similar within every group. In general, PCMs with solid-liquid phase transformations can be classified into: organic, inorganic or eutectics. Following figure (Fig. 3) shows more detailed classification.

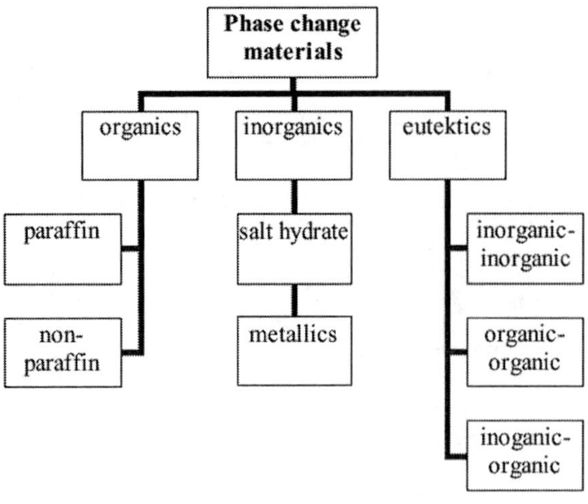

Figure 3. Phase change materials classification.

There is available wide range of phase change materials characterized by different properties. It creates the possibility to use them for various applications.

3 Phase change materials for building applications

3.1 PCMs integrated with buildings envelopes

A building envelope can be described [4] as a part of the building which separates its indoor and outdoor environments. Buildings envelopes plays an important role in maintaining proper internal conditions, especially important are components such as walls, roof, foundation, thermal insulation, external shading elements. Among the various studies on improvement of buildings envelopes properties we can find that many of them are related to application of phase change materials. Some examples of research conducted in this area are presented in this section, paying attention to walls, windows and shutters, roofs and floors.

3.1.1 Walls

There is several solutions considering phase change materials integration into the walls [5, 6]:
- direct incorporation and immersion,
- shape-stabilized PCM composite,
- micro-encapsulation of PCMs,
- macro-encapsulation of PCMs.

Many researchers conduct their researches on those methods to minimize defects and some disadvantages associated with their use.

Phase change materials can be incorporated into the concrete directly. Many researches have been conducted to test this possibility. Lee et al. [7] compared, with good results, the thermal storage performance of ordinary concrete blocks with two different blocks, modified by addition of PCMs - a commercial Butyl Stearate (Emerest 2326), and the other commercial Parafin (Unicere 55). Barreneche et al. [8] have also investigated the possibility of PCMs buildings applications. Additionally authors have compared difference in effect between incorporating PCMs into different types of building materials. They observed how beneficial is addition of various amount of PCMs into both - concrete and gypsium. However, the results were not conclusive. Many researches in this area ended with conclusion that that direct incorporation of PCMs may involve leakage after many work cycles. It can affect mechanical properties of the concrete and its durability [6, 9].

Use of shape-stable composites enables to avoid the leakage [10, 11]. They can be made on the basis of various kinds of components such us [12]: polyethylene, styrene and butadiene. Shape-stable Composites are generally characterized by [8, 13, 14]:

- large specific heat,
- high enough thermal conductivity,
- ability to keep the shape of PCM stabilised over the phase change process,
- good performance of multiple thermal cycles over a long period.

To avoid the leakage, phase change materials can also be incorporated into the wall with the use of micro- or macroencapsulation [15, 16].

 DOI: 10.1051/e3sconf/20161000068

Macroencapsulation involves placing the material in the sealed container. The package can have various shapes such as spheres, tubes, panels or similar [8, 17]. Macroencapsulated materials can be used as a part of the wall or independently. Many possible solutions for macroencapsulation addressed to different types of PCMs are available. Fukahori et al. [18] designed and built capsule with very good corrosive resistance and cycling performance for metallic PCMs. Alam et al. [19] presents technique to encapsulate PCMs with melting point in the range of 120-350°C.

Microencapsulation consists of enclosed a phase change material in a microscopic spherical or rod-shaped or irregular shape capsules with diameter in the range of 1µm to 1mm [9, 13]. There are several physical and chemical methods for the production of microcapsules such us [20, 21]: pan coating, spray drying, coacervation, sol-gel, interfacial polymerization and many others.

Selection of the best suited methods for incorporation of PCMs into the wall is dictated by practical and economic reasons.

3.1.2 Windows and window shutter

Xiang et al. [22] conducted the research and numerical investigation on window-based cooling unit filled with phase change materials. Fig. 4 presents the plan of the office with cooling unit at night (upper picture) and during the day (lower picture).

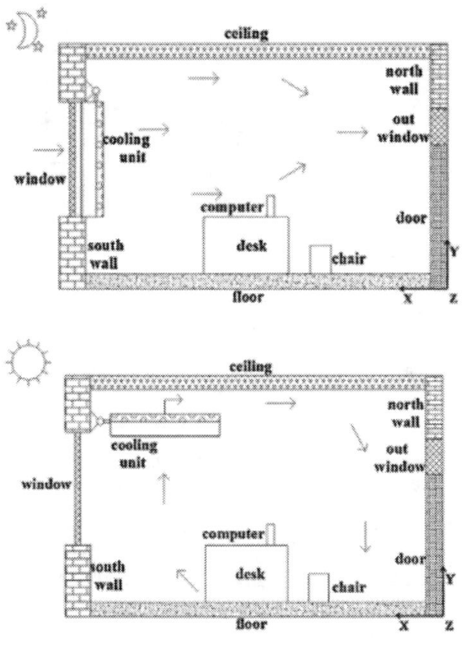

Figure 4. Operation mode: location of the cooling unit in the night (upper) and during the day (lower) [22].

Cooling unit store the coolness over the night by natural ventilation and releases it over day when temperature increase. The authors draw conclusions that proposed system enables to save the energy and improve indoor thermal comfort.

Silvaa et al. [23] were experimentally testing the performance of window shutter with phase change materials. They compared two window shutter-with PCMs and without PCMs in two compartments which were submitted to similar conditions. Fig. 5 shows investigated window shutters and blade filled with PCM.

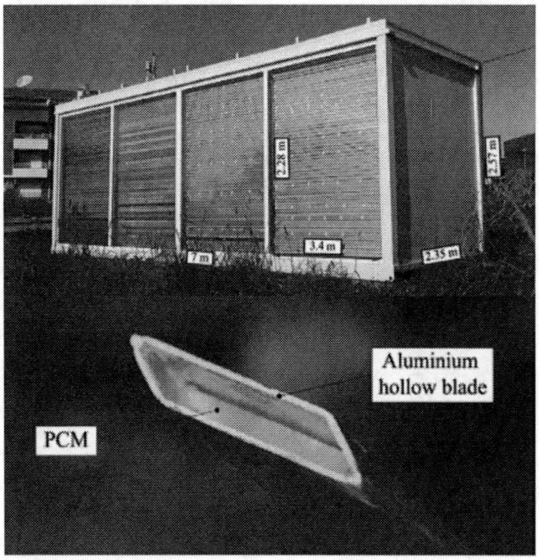

Figure 5. Investigated window shutters (upper) and window blade filled with PCM (lower) [23].

The authors commented that the results reveal potential for using phase change materials for the thermal regulation of indoor spaces as well as improving the energy efficiency of buildings. The authors pointed out in their conclusion how important it is to properly chose the material and optimize all aspects influencing indoor and outdoor conditions.

3.1.3 Roofs

Pisello et al. [24] focused in their study on phase change materials for roofing application. They were working on development and prototyping of an innovative cooling polyurethane based membrane with PCMs.

The authors concluded that the combination of cooling materials and thermal storage effect gives promising results as passive technique for improvement of building thermal-energy efficiency.

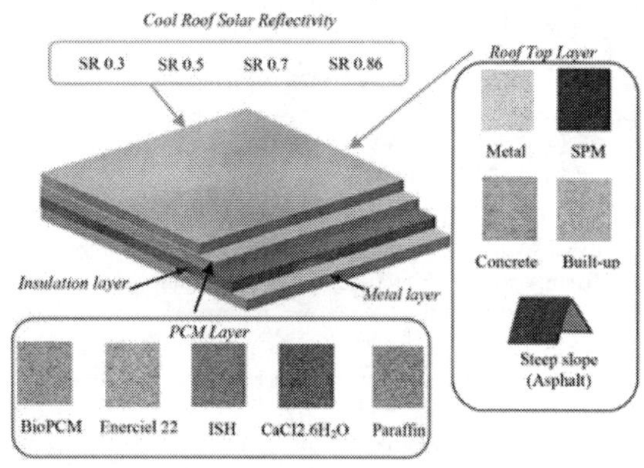

Figure 6 Graphical explanation of the various types of materials that were incorporated throughout thermal simulation study[25].

DOI: 10.1051/e3sconf/20161000068

Roman et al. [25] conducted the comparison of two technologies. Phase change material has been compared to the cooling materials. Simulations were conducted for the range of climate zones across the United States. Various roof types and various PCM types were evaluated. Figure 6 shows graphical description of the various types of materials that were incorporated throughout simulation study.

The authors concluded that technologies with the use of PCMs are the future of energy-saving. Results shows that the PCM roof type maximum through roof heat gains flux is 54% lower than the cool roof and maximum sensible heat flux is 40% lower than the cool roof technology.

3.1.4 Floors

Xia and Zhang [26] proposed in their paper new double-layer radiant floor system with organic phase change material. The new system can work either in summer and winter and the system does not need any additional refrigeration system.

Experiment has been conducted for two experimental boxes (Fig. 7) to compare the performance of different floor structures of the double-layer radiant floor system with phase change material. The environment of both boxes has been controlled.

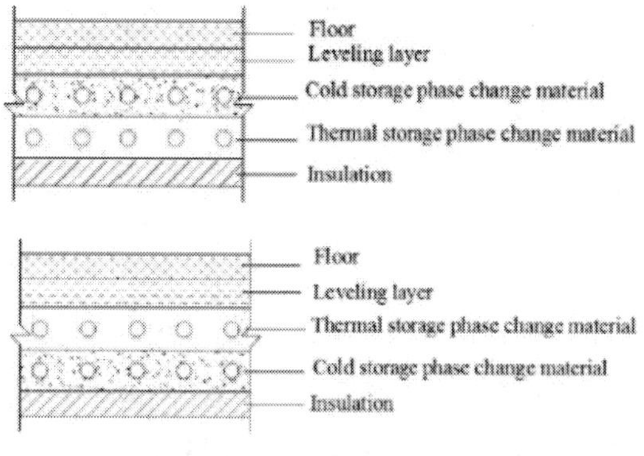

Figure 7. Two experimental floor structures [26].

Floor system proposed by Xia and Zhang can store thermal or cold energy in the off-peak period and use it in the peak period. Due to the opportunity to resign from additional refrigeration system the cost of the system will decrease.

4 Phase change materials for solar energy

4.1 Photovoltaic panels

Ciulaa et al. [27] explains the reason that photovoltaic panel (PV) performance is usually lower than expected. Performance of PV is defined for nominal conditions – for the assumption that maximum electric power is supplied by the panel when it receives an insolation of 1 kW/m^2 and the temperature of the cell is 25°C. Under the operating conditions solar intensity and the temperature of the panel have variable values. Many investigations are carried out to improve PV performance efficiency.

Atkin and Farid [28] conducted research and modelling to evaluate if the use of phase change material infused graphite with an external finned heat sink can be effectively used for PV thermal regulation. They compared four different thermal regulation techniques. The authors commented on results that the combined system of PCM and heat sink shows the potential of use in case of improvement of thermal conductivity.

Huang [29] was testing by numerical simulations how combination of two different PCMs (with different phase transition temperature) will impact thermal regulation effect of the system and length of the thermal regulation time in PVs. The results showed that the system with two types of PCMs can bring the PV temperature closer to its nominal value of 25°C and improve solar to electrical conversion.

4.2 Solar thermal storage systems

Phase change materials due to theirs large thermal energy storage and isothermal behaviour during phase transition are widely investigated as a possibility for solar thermal energy storage medium.

Mahfuz et al. [30] conducted experimental study to test water heating system based on shell and tube thermal energy storage system with paraffin wax to analyse its performance for solar water heating application.

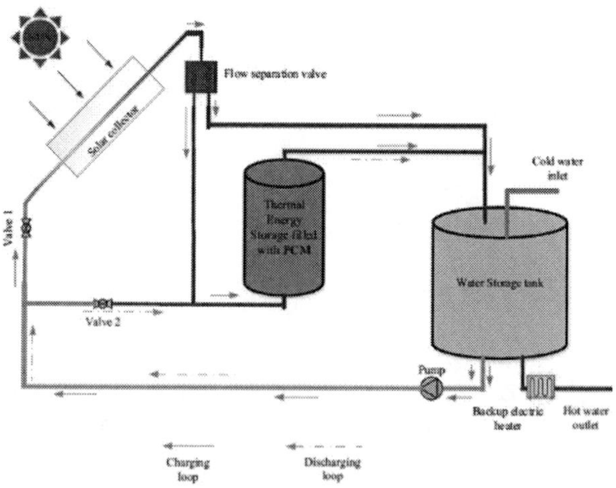

Figure 8. Solar heating water system with the termal energy storage, investigated by Mahfuz et al. [30].

Based on obtained results authors suggested some recommendations for future work related to: optimization of pipe diameter, testing different types of PCMs, testing different heat transfer fluids alternative for water, increasing of heat transfer coefficient.

Padovan and Manzan [31] investigated if application of PCMs in the thermal energy storage system in solar domestic hot water system could have good influence on system performance in the form of energy savings or reduction of TES system size. Fig. 9 shows system used for investigation and Fig. 10 shows possible positions of PCM modules.

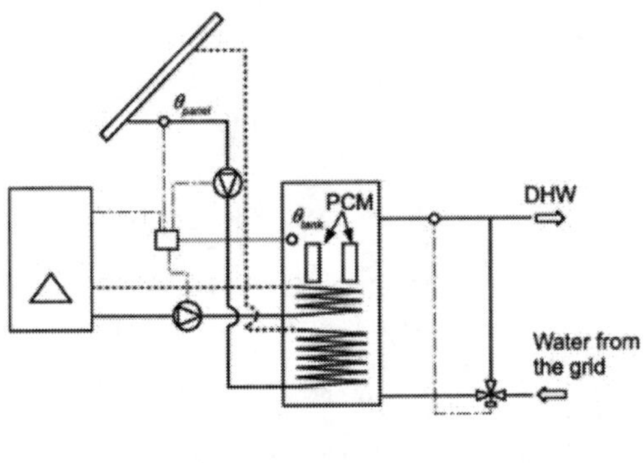

Figure 9. System investigated by Padovan and Manzan [31].

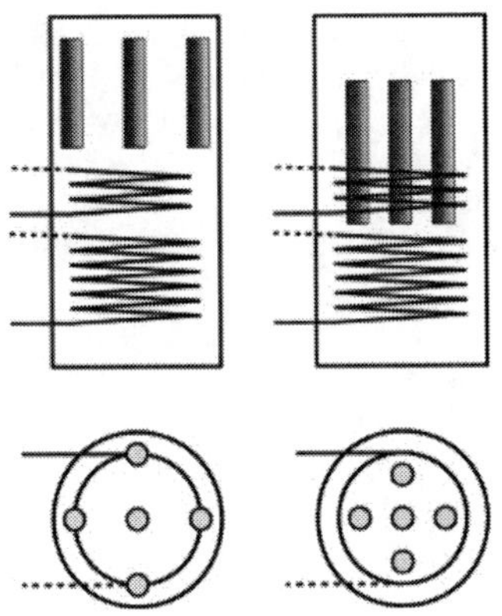

Figure 10. Possible positions of PCM modules in the PCM thermal storage investigated by Padovan and Manzan [31].

The authors concluded that addition of PCMs did not have essential impact on the results. They suggested that application of different PCMs instead of paraffin could be beneficial. It is also strongly possible that this method of application is not the best solution and would be more beneficial if system was redesigned.

5 Phase change materials for heat pumps

The use of heat pumps for buildings cooling and heating is assessed as efficient technology [32, 33]. Additionally it may be even more productive if heat pump cooperates with thermal energy storage system. Selected examples of researches in this area are presented in this section.

Agyenim and Hewitt [34] studied RT58 PCM as a part of a wider study focused on investigation proper phase change material to take advantage of off-peak electricity tariff. Based on obtained results, the authors identified quadratic relationship between heat transfer coefficient and the inlet heat transfer fluid temperature (within temperature range 62-77 °C). They also concluded that it is possible to reduce storage size by 30%, but it is required to heat transfer techniques to charge and discharge heat form the storage.

Zhu et al. [35] carried out numerical simulation and analysis on the system of ground source heat pump integrated with PCM storage tank to study optimal control method for this system. They conducted economic analysis and analysis of energy performance for different conditions - under different cooling storage ratios. They presented an example of optimal usage of the combined system. The scheme of the combined system is presented on the Fig.11.

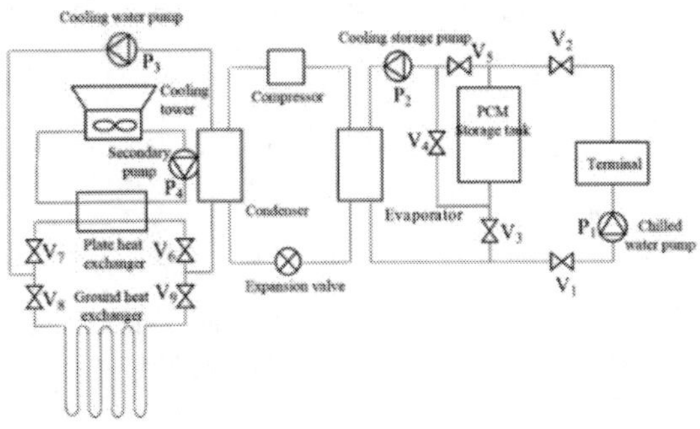

Figure 11. Scheme of combined system of ground source heat pump integrated with PCM [35].

The authors concluded that use of proposed system in comparison to ground source heat pump without PCM storage is characterised by better reliability of the operation performance and reduction of energy consumption and operation cost. Comparison of different cooling storage ratios showed that this factor affects the performance and economy. Optimal cooling storage ratio depends on building type, location and system utilization mode.

6 Summary

This paper reviews, based on literature, applications of phase change materials for technologies related for sustainable energy.

Application of phase change materials for technologies designed to improve energy efficiency and for technologies based on renewable energy sources have been identified by number of researchers as interesting solutions with broad potential.

Described studies are only some examples of researches conducted with the use of PCMs. Phase change materials show

Phase change materials applied in discussed technologies need to meet many criteria to be used with expected effect. The criteria that should be considered are among the others: suitable phase change temperature range, value of latent heat of fusion and thermal conductivity and compatibility with constructional materials.

Based on discussed study results, it can be identified that reasonably used phase change materials and carefully designed systems can have a large impact on improvement of technologies and thereby on energy savings.

Acknowledgments

This work was supported by AGH – University of Science and Technology (Project 11.11.210.216).

References

1. A. Sharma, V.V. Tyagi, C.R. Chen, D. Buddhi, Renewable and Sustainable Energy Reviews **13**, 318–345 (2009)
2. M. M. Farid, A. M. Khudhair, S. A. K. Razack, S. Al-Hallaj, Energy Conversion and Management **45**, 1597–1615 (2004)
3. B. Zalba, J. Marin, L.F. Cabeza, H. Mehling, Applied Thermal Engineering **23**, 251–283 (2003)
4. S.B. Sadineni, S. Madala, R. F. Boehm, Renewable and Sustainable Energy Reviews **15**, 3617–3631 (2011)
5. S.A. Memon, H.Z. Cui, H. Zhang, F. Xing, Applied Energy **139**, 43–55 (2015)
6. F. Kuznik, D. David, K. Johannes, J.-J. Roux, Renewable and Sustainable Energy Reviews **15**, 379–391 (2011)
7. T. Lee, D.W. Hawes, D. Banu, D. Feldman, Solar Energy Materials & Solar Cells **62**, 217-237 (2000)
8. C. Barreneche, M. E. Navarro, A. I. Fernández, L. F. Cabeza, Applied Energy **109**, 428–432 (2013)
9. H. Li, H. Chen, X. Li , J. G. Sanjayan, Applied Energy **135**, 225–233 (2014)
10. L. Wang, D. Meng, Applied Energy **87** (2010) 2660–2665
11. Z.Zhang, G. Shi, S. Wang, X. Fang, X. Liu, Renewable Energy **50**, 670-675 (2013)
12. S. A. Memon, Renewable and Sustainable Energy Reviews **31**, 870–906 (2014)
13. D. Zhou, C.Y. Zhao, Y. Tian, Applied Energy **92** 593–605 (2012)
14. M. Xiao, B. Feng, K. Gong, Energy Conversion and Management, **43**, 103-108 (2002)
15. A. M. Khudhair, M. M. Farid, Energy Conversion and Management **45**, 263–275 (2004)
16. L. F. Cabeza, C. Castellon, M. Nogues, M. Medrano, R. Leppers, O. Zubillaga, Energy and Buildings **39**, 113–119 (2007)
17. Y. Zhang, G. Zhou, K. Lin, Q. Zhang, H. Di, Building and Environment **42**, 2197–2209 (2007)
18. R. Fukahori, T. Nomura, Ch. Zhu, N. Sheng, N. Okinaka, T. Akiyama, Applied Energy **170**, 324–328 (2016)
19. T. E. Alam, J. S. Dhau, D. Y. Goswami, E. Stefanakos, Applied Energy **154**, 92–101 (2015)
20. A. Jamekhorshid, S. M. Sadrameli, M. Farid, Renewable and Sustainable Energy Reviews **31**, 531–542 (2014)
21. Y. Konuklu, M. Ostryc, H. O. Paksoyd, P. Charvat, Energy and Buildings **106**, 134–155 (2015) [25]
22. Y. Xiang, G. Zhou, Energy and Buildings **108**, 267–278 (2015)
23. T. Silvaa, R. Vicentea, F. Rodriguesa, A. Samagaio, C. Cardosoa, Energy and Buildings **88**, 110–121 (2015)
24. A. L. Pisello, E. Fortunatib, S. Mattiolib, L. F. Cabezac, C. Barrenechec, J.M. Kennyb, F. Cotana, Energy and Buildings **112**, 40–48 (2016)
25. K. K. Roman, T. O'Brien, J. B. Alvey, O. Woo, Energy **96**, 103-117 (2016)
26. Y. Xia, X.-S. Zhang, Applied Thermal Engineering **96**, 600–606 (2016)
27. G. Ciulla, V. Lo Brano, M. Cellura, V. Franzitta, D. Milone, Energy Procedia **30**, 198 – 206 (2012)
28. P. Atkin, M. M. Farid, Solar Energy **114**, 217–228 (2015)
29. M. J. Huang, Solar Energy Materials & Solar Cells **95**, 957–963 (2011)
30. M.H. Mahfuz, M.R. Anisur, M.A. Kibria, R. Saidur, I.H.S.C. Metselaar, International Communications in Heat and Mass Transfer **57**, 132–139 (2014)
31. R. Padovan, M. Manzan, Solar Energy **103**, 563–573 (2014)
32. P. Moreno, C. Solé , A. Castell, L. F. Cabeza, Renewable and Sustainable Energy Reviews **39**, 1–13 (2014)
33. M. Iten, S. Liu, A. Shukla, Renewable and Sustainable Energy Reviews **61**, 175–186 (2016)
34. F. Agyenim, N. Hewitt, Energy and Buildings **42**, 1552–1560 (2010)
35. N. Zhu, P. Hu, Y. Lei, Z. Jiang, F. Lei, Applied Thermal Engineering **87**, 615-623 (2015)

The impact of TTIP agreement on the European Union-United States coal trade potential

Tadeusz Olkuski[1a], Andrzej Sikora[2], Piotr Mateusz Sikora[2] and Adam Szurlej [3]

[1] *AGH University of Science and Technology; Department of Sustainable Energy Development; A. Mickiewicza Av. 30, 30-059 Krakow, Poland*

[2]*The Energy Studies Institute, Sniadeckich St. 17, 00-654 Warszawa;*

[34]*AGH University of Science and Technology, Faculty of Drilling, Oil and Gas;*

Abstract. The main aim of the paper is to assess the impact of currently negotiated TTIP agreement (Transatlantic Trade and Investment Partnership) on the use of hard coal in the EU and the US. Hard coal is the most important fuel in global electricity generation. This also applies to the United States, a leading manufacturer and exporter of this energy source. The US coal is exported to the EU market. The article presents the estimated exports of hard coal from the US to the EU. Due to the fact that price has a major impact on the size of exports, the paper presents the estimated prices, including freight costs, of power coal for the analyzed scenarios. According to one scenario, the US and European prices will be equalized (including freight costs) by 2020, while from 2025 on the comparative advantage and competitiveness of the US hard coal will decrease. Taking into account the fact that the export of coal from the United States is free from customs duties, the acceptance of TIPP should not affect the currently existing trade between the two continents and the amount of exported coal. Nevertheless, the question of hard coal economy cannot be separated from other sectors of the energy market, which can be significantly affected by the future agreement.

1 Introduction

The main aim of the *Transatlantic Trade and Investment Partnership* (TTIP) is to create a free trade area, aimed at simplifying and increasing mutual investments, the flow of goods or services. The aforementioned area would allow removing customs barriers and unfavorable regulations having a direct impact on the movement of goods. Removing or reducing customs duties could have a positive impact on the reduction of prices of goods in the United States and the European Union, at the same time forcing the improvement of quality and efficiency, while the exchange of goods and increased sales would become a driving force behind the creation of new jobs. The harmonization of existing regulations and detailed coordination would become the basis for accelerating economic growth in the European Union and the United States. (In the recent years, after the 2008–2009 financial crisis, the GDP growth rate in the US was higher compared to the EU). However, when it comes to coal, the situation is quite specific. The scale of closure in the EU coal mining industry deepens the scale of dependence on external suppliers of energy resources. Similarly as in the EU, the share of coal in electricity generation structure in the US has decreased in recent years. However, the decline in the importance of coal in the US can be explained primarily as a result of the so-called shale revolution, as opposed to the implementation of the EU energy policy objectives, that is, among others, the growing use of renewable energy sources and reducing greenhouse gas emissions [1, 2]. Poland is against further closures in the coal mining industry, since about 85% of Polish electricity is produced from both hard coal and lignite. Coal is and, according to the Energy Policy of Poland, will remain a guarantee of the energy security of Poland until 2030 or even 2050 [3 - 7]. However, the future of coal depends on the global trends in the international arena, which are currently unfavorable or this raw material.

2 The situation in the EU and the situation in the US

Analyzing the available data on hard coal and comparing the IEA baseline scenario [8] estimations of both coal imports from the European Union and exports from the United States, an interesting correlation between these scenarios can be observed (Fig. 1). The EU low-carbon energy policy and criticism of coal, both hard coal and lignite, as an energy source, will cause a gradual shift from the production of coal, not taking into account the constant demand for this raw material [9]. However, increasing prices of emission allowances should result

[a]Corresponding author, email: olkuski@agh.edu.pl

in a gradual decrease of coal imports. According to the IEA, the estimated coal exports from the US by 2020 will amount to 42 Mtoe for coking coal and 14 Mtoe for power coal. When it comes to the European market, coking coal exports will be gradually decreasing, reaching the level of 28 Mtoe in 2030 and 21 Mtoe by 2040. Meanwhile, power coal exports will increase to the level of 18 Mtoe in 2030 and 21 Mtoe by 2040. IHS [10] confirms the aforementioned assumptions in own scenario. (Fig. 2). However, when compared to the IEA estimations, IHS estimates more active EU decarbonisation policy, and therefore assumes lower amounts of the imported US coal.

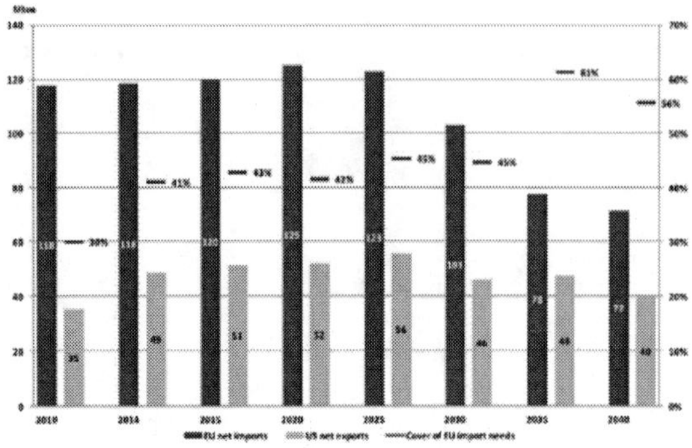

Figure. 1. Net exports of coal from the United States vs. net imports of coal to the European Union according to the IEA baseline scenario [11]

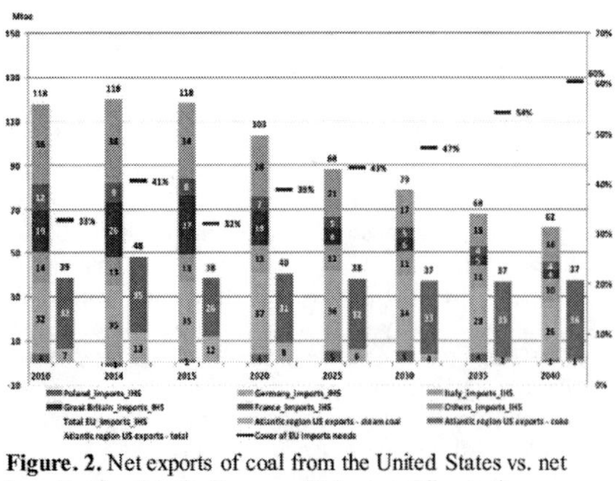

Figure. 2. Net exports of coal from the United States vs. net imports of coal to the European Union according to the estimations of the IHS energy [11]

Taking into account the current condition of the coal sector in Poland and the situation in the global coal market, a general oversupply of this raw material and, above all, incredibly low price of $50/tonne (ARA Cal-16 contract), a scenario developed by the Polish National Energy Conservation Agency (Polish: Krajowa Agencja Poszanowania Energii, KAPE) - still used as a basis for the Polish Energy Policy until 2050 - seems to be unlikely. According to KAPE, the total coal exports will amount to 8-9 Mtoe. The analysis of historical data allows to conclude that coking coal imports and exports (by Polish coking plants and coal mines, respectively) used to be at the same level (about 1.5-2.0 Mtoe). According to the abovementioned estimation, this leaves a surplus of about 6.5-7 Mtoe for power coal (Fig. 3).

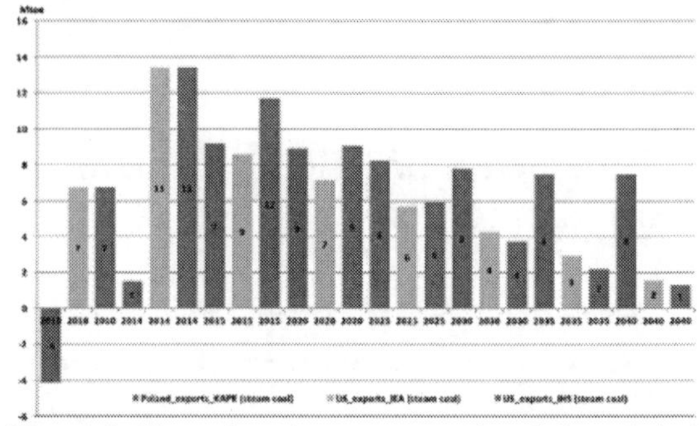

Figure 3. The estimated net exports of coal from Poland according to the Polish National Energy Conservation Agency against the estimated exports of US power coal in the Atlantic market [11]

3 Coal prices and additional costs

The most important and the most cost-effective is the Asia-Pacific coal market, more specifically India, Korea and China, currently facing economic stagnation. The North American market is characterized by a fairly large local and relatively cheap production (low price of coal) and - for some time - an oversupply. The European market is a systematically decreasing, mainly due to climate policy and the decarbonisation of the energy industry [12]. The estimated coal prices (excluding coking coal) for the different scenarios are summarized in Table 1. The IEA confirms the presented distribution of prices. It is clearly visible that the price of US coal (taking into account the freight cost) makes this raw material a very attractive product for European countries and other countries interested in importing coal from the United States (Fig. 4). Coking coal prices depend on the policies of the metallurgical companies. The decrease in the production of steel will automatically lead to lower demand for coking coal [13]. According to Goldman Sachs, coking coal prices can rise only if the market supply will decrease by 30 million tones. Power coal prices estimated by IHS Energy are different than those assumed in the IEA scenario. IHS assumes that European prices are lower while US prices higher. By 2020, the US and European prices (including freight costs) will be equalized, while from 2025 on the comparative advantage and competitiveness of the US hard coal will decrease (Fig. 5). It is worth mentioning that the KAPE estimation is comparable to the IHS scenario

Table 1. The estimated coal prices for the different scenarios with the projected freight costs from the United States to Europe [11]

in USD (2014) per GJ (NCV)	2010	2013	2014	2015	2020	2025	2030	2035	2040
IHS USA spot price	3.32	2.84	2.58	2.84	3.43	3.41	3.43	3.38	3.21
IHS Europe (ARA) spot price	3.92	3.30	3.00	3.28	3.91	3.98	4.06	4.07	3.94
IHS SE Asia spot price	4.17	3.38	2.79	3.34	4.00	3.93	3.92	3.83	3.58
IHS Freight cost East Coast USA - ARA (Western Europe)	0.59	0.48	0.46	0.47	0.52	0.61	0.68	0.74	0.78
The IEA USA base price (spot)		2.47			2.89	2.98	3.10	3.19	3.32
The IEA Europe base price (spot)		3.66			4.25	4.46	4.59	4.68	4.76
The IEA Asia (China) base price (spot)		4.46			4.68	4.76	4.89	5.02	5.10
The IEA OECD average (import price) baseline scenario		3.66			4.59		4.76		4.55
The IEA OECD average (import price)_ current policy scenario		3.66			4.55		4.97		5.27
The IEA OECD average (import price)_ ecological scenario		3.66			3.74		3.32		3.27
KAPE spot price Poland	4.05			3.67	3.94		3.94		3.94
The estimated average variable cash cost of the exported coal according to the IEA * (FOB)_baseline scenario		2.79			3.18				3.98

* The estimated variable cash cost of the exported coal includes mining costs, coal processing costs, cost of land transport to the port, port fees (storage, loading, etc.) and taxes (royalties, etc.). Combined with freight costs (transport, insurance, etc.) they make up the final price in the target market - the spot price or import price.

4 The scenario without changes resulting from TTIP

The above analysis of the situation of the coal sector and the estimated coal prices show a possible potential for export of US raw materials to the European Union. According to the latest data from May 2015 [10], the production of coal in the Appalachian coal basin and the Illinois basin amounts to 187 Mtoe per year and should decrease to 70 Mtoe by 2040. The presented IEA and IHS data, as well as the results of own studies, allow to conclude that coal exports to the EU can amount to up to 7.1-9.1 Mtoe (11.9-15.1 million tonnes of raw material with an average calorific value of 6.000 kcal / kg) in 2020, while around 1.3-1.5 Mtoe (2.2-2.6 million tonnes) by 2040 (Fig. 6).

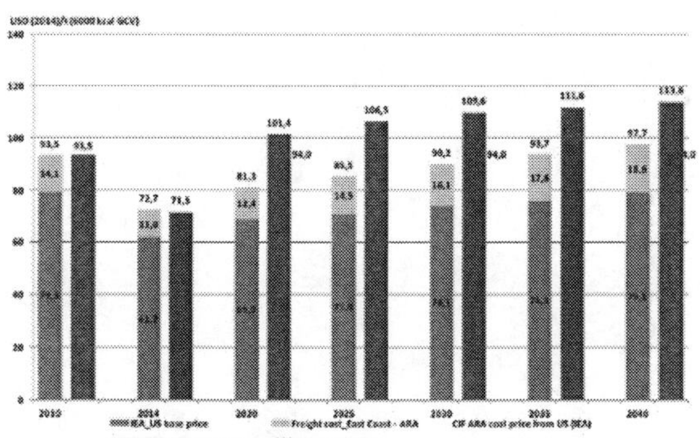

Figure. 4. The prices of power coal imported from the United States compared to baseline prices for the European market according to the estimations of the International Energy Agency [11]

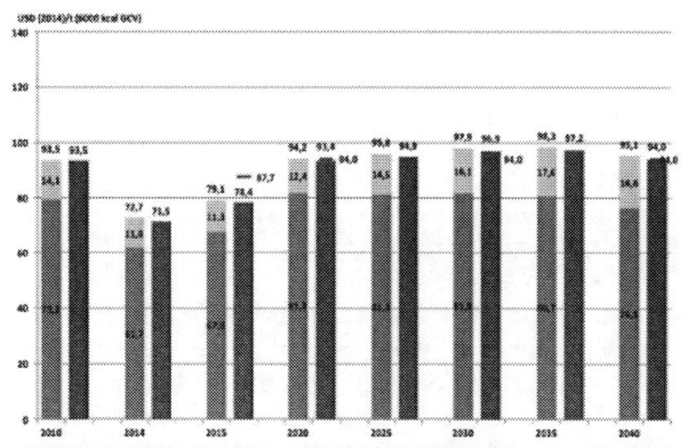

Figure. 5. The prices of power coal imported from the United States compared to spot prices for the European market according to IHS Energy [11]

According to IHS estimates, the production of coking coal in Northern and Central Appalachians is expected to continue at the level of 55-56 Mtoe. As a result, it can be assumed that this raw material will dominate exports from the US to the EU. Taking into account the estimates of the Energy Studies Institute in Warsaw (Polish: Instytut Studiów Energetycznych , ISE) , the maximum potential volume of exports can reach:

DOI: 10.1051/e3sconf/20161000069

- 31-39 Mtoe (46.6 – 58.9 million tonnes of raw material with an average calorific value of 6710 kcal/kg) w 2020,
- 26-33 Mtoe (39.1-49.7 million tonnes) by 2030,
- 20-36 Mtoe (29.3-53.6 million tonnes) by 2040.

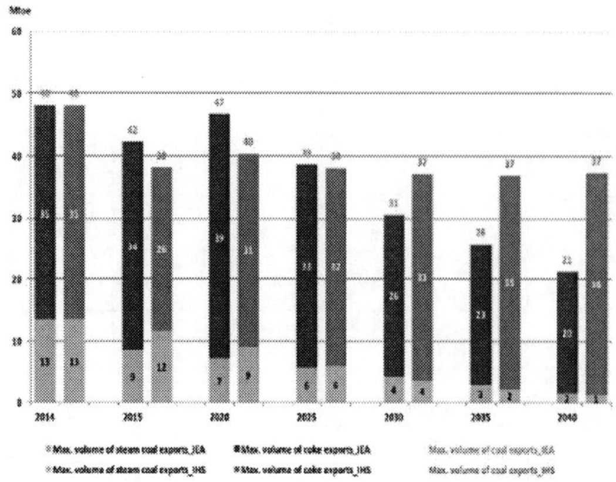

Figure. 6. The maximum potential volume of coal exports from the US to the EU

The ranges are relatively broad because they are directly linked to the rate of depletion of European deposits of coking coal. According to IHS scenario, the EU imports will be significantly increasing in direct proportion to the depletion of coal reserves.

5 Conclusions

The European Union has a long tradition of importing coal. Around ten years ago, Poland - a long-time exporter of this raw material - has become a net importer of coal. Such a situation will take place over the next years, while further intensification of imports can be expected. When analyzing the import of coal from the US to Poland, it should be noted that the coal will be imported through the ports of Gdansk and Świnoujscie, from where the raw material must be transported to the end user. This will cause additional costs, which can make the import of coal uncompetitive when compared to the domestic coal. Transport costs have a significant impact on the final price of coal [14]. Overall, prices in the domestic market depend on the volume of imports, while the lack of legal and logistical barriers facilitates this practice [15, 16].

To sum up, coal trade between the United States and the European Union has been taking place for a long time. The export of coal from the United States is not limited by non-tariff barriers to trade and special restrictions imposed by US law. What is more, the coal imported into the European Union is free from customs duties. Therefore, acceptance or rejection of the TIPP should not affect the currently existing trade between the two continents and the amount of exported coal.

Acknowledgements

The paper is an effect of the statutory project no. 11.11.210.217.

References

1. P. Janusz, M. Kaliski, A. Szurlej. Min. Res. Manag. **31**(3),. 5–24 (2015) (in Polish)
2. J. Paska, T. Surma. Energy Policy Journal, **16(**4), (2013) (in Polish)
3. L. Gawlik, (ed.), *Coal for the Polish energy sector in the perspective of 2,050 years - scenario analyzes.* p. 299 (2013) (in Polish)
4. *The Energy Policy of Poland until 2030*. Ministry of Economy (2009)
5. R. Szczerbowski. En. Pol. J. 16(4), 35-47 (2013) (in Polish)
6. M. Kaliski, A. Szurlej, L. Gawlik, *Perspectives of hard coal sector development in Poland until 2050.* [In:] Współczesne oblicza i dylematy restrukturyzacji. Oprac. i red. nauk. A. Jaki, M. Kowalik, p. 355-365 (2015) (in Polish)
7. *The Energy Policy of Poland until 2050 – Project.* Ministry of Economy (2015) (in Polish)
8. International Energy Agency, *World Energy Outlook 2014* (2015)
9. U. Lorenz. En. Pol. J. **18**(4), 5-18 (2015) (in Polish)
10. IHS Energy *"US supply and demand balance sheet – Monthly"*, May 2015
11. The report from the studies conducted in July and August 2015 by the The Energy Studies Institute (Polish: Instytut Studiów Energetycznych, ISE) (previously unpublished) (in Polish)
12. T. Olkuski. En. Pol. J. **18**(3), 87-98, (2015) (in Polish)
13. U. Ozga-Blaschke. Przeg. Górn. **5**, 26-31 (2014) (in Polish)
14. K. Stala-Szlugaj. En. Pol. J. **17**(4), 65-76 (2014) (in Polish)
15. Z. Grudziński. Przeg. Górn. **5**, 9–16 (2014) (in Polish)
16. Z. Grudziński. En. Pol. J. **17**(4), 37-50 (2014) (in Polish)

E3S Web of Conferences **10**, 00070 (2016) DOI: 10.1051/e3sconf/20161000070
SEED 2016

Municipal Waste-to-Energy plants in Poland – current projects

Maciej Cyranka[1,a], Michał Jurczyk[1]and Tadeusz Pająk[1]

[1]*Department of Power Engineering and Environmental Protection, AGH University of Science and Technology, Krakow, Poland*

Abstract. In 2016, six new municipal waste combustion plants will be operating in Poland. These projects, located in: Białystok, Bydgoszcz, Konin, Kraków, Poznań and Szczecin, will influence waste management in Poland as well as change energy security of the country. So far in Poland only one Waste-to-Energy (WtE) plant, i.e. in the capital city of Warsaw, has been operating. Unfortunately, this is not enough, taking into account Polish economic development in the last years. Polish accession to the European Union (EU) significantly influenced the development of the industry which resulted in increased production of waste while the management system was not prepared to handle it in the proper way so there was a big amount of waste. The main way to deal with waste in Poland was, and still is, landfilling which can be in many ways dangerous for the environment. The EU law have put pressure on Polish Waste Management Systems and results are visible in currently realized Waste-to-Energy projects.

This work analysis of the aforementioned 6 projects from economic (project value, funding from the European Union) and technological (used technology, waste throughout, availability etc.) point of view together with their impact on the entire waste management system and energy security. What is more, history of similar facilities in Poland is presented, what shows how many projects were prepared, how many fell through and time frame of finally realised projects (start of construction and the planned completion).

Conclusions show why in the current situation development of Polish WtE infrastructure is right, i.e. operation of aforementioned plants that will ensure benefits associated with energy production, reduction of landfilling and informing public opinion regarding modern waste management models. Additionally, the article draws attention on the high responsibility that will be put on WtE plants operators and that experience gained during WtE implementation can be used to improve even further for future Polish Waste Management Systems.

1 Introduction

The amount of generated municipal waste is increasing, while European Union (EU) environmental requirements are becoming stricter, forcing the Polish government apparatus to implement reliable and effective waste management programmes. Poland through ratifying the EU Accession Treaty in 2003 oblige itself to pursue a number of commitments in the field of municipal waste management, from which most important are [1]:

- Reduction of municipal waste landfilling;
- Increase of waste recycling levels;
- Reuse of certain waste fractions.

Unfortunately, Polish waste management essentially relies on the waste landfilling, which is a highly unfavorable and undesirable process in terms of suitable development and resource efficiency. In 2014 approx. 10 330 thousand tons of municipal waste were collected in Poland, of which approx. 5 835 thousand tons (56.5% of waste collected in total) were disposed by landfilling or thermal processing without energy recovery [2].

Without any doubt chemical energy contained in municipal waste, instead of wasting on landfills, should be used to produce usable forms of energy. EU impose duties in that matter by Directive 2008/98/WE [3], which states that all Member States shall take appropriate measures to establish an integrated, modern and proportioned network of facilities for the municipal waste recovery or recycling. Directive 2008/98/WE [3] sets the conceptfor EU Member States waste legislation and policy which they shall apply as a priority order, which is shown symbolically on Figure 1.

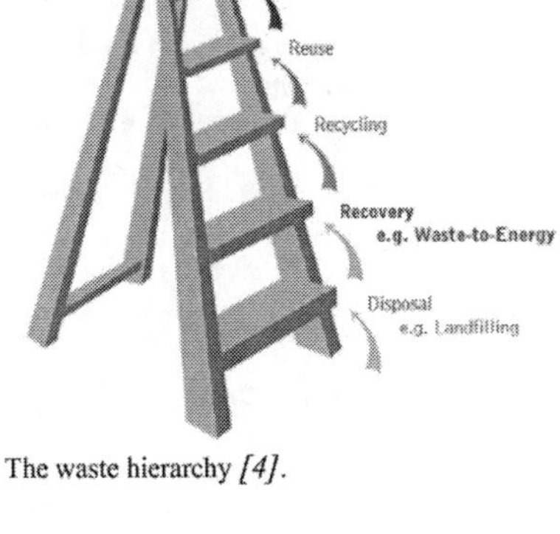

Figure 1. The waste hierarchy *[4]*.

[a]Corresponding author: cyranka@agh.edu.pl

 DOI: 10.1051/e3sconf/20161000070

According to Figure 1 the first step of municipal waste treatment should be re-use or recycling and next energy recovery in Waste-to-Energy (WtE) plants. However, it can be assumed that the development of energy recovery infrastructure contributes also to the development of recycling. Some examples are confirming that communities supplying municipal waste to WtE facilities generally have higher than the average recycling rates [5] [6]. In countries with advanced waste treatment systems (like Germany or Sweden) the WtE infrastructure is an integrated part of the waste management system. Therefore, WtE may be regarded as the natural and necessary step of countries, like Poland, going from landfilling towards higher levels in the waste hierarchy [7].

Energy recovery from municipal waste is carried out primarily by combustion processes, which are conducted in special industrial objects called WtE plants (or MSWI – Municipal Solid Waste Incinerators). Combustion of fossil and biofuels provides more than 90% of the world's energy [8], therefore it is not surprising that this type of thermochemical conversion process is mostly used for energy production from municipal waste. The technical basis of energy recovery from municipal waste combustion, is to use generated thermal energy embedded in flue gases to produce steam at high temperature and pressure. Then, such steam is used to drive a turbine generator and/or for heating purposes. Maximizing the use of energy contained in every fuel, that is also in municipal waste, is currently recommended by international law energy economy model.

Designs of present WtE plants are developed with consideration of numerous criteria related not only with plant energy efficiency but also with impact on the environment, operational costs and available optimization possibilities. Modern WtE facilities that are equipped in advanced systems for waste processing, environmental control, and materials recovery, maintain both high energy efficiency and pollutants emission below EU and regional limits. Moreover, modern plants are able to increase recycling rates and yield residuals (i.e. fly ash, bottom ash, slag, and flue gas cleaning residues) with properties suitable for beneficial reuse, like for example as an aggregate material [5]. However, residuals conditioning can be expensive and because of that comparison analysis of environmental and economic aspects is needed.

The issues of energy, environment, and fuel resources are of great current interest and they have strong and important intersections, with each other and with the field of combustion [8]. Therefore, municipal waste combustion is highly related with aforementioned areas and it turns out that for each of them it brings a number of benefits. Those benefits are described later in this article, mainly on basis of Polish economy example.

2 Present and Past Polish WtE Structure

By the end of 2015 in Poland only one WtE plant has operated, i.e. "ZUSOK" plant in Warsaw, which was built in 2001. Warsaw is the capital and largest city in Poland (1.74 million people [9]) and one of the fastest growing urban centers in the country. In 2014 around 520,7 thousand Mg of municipal waste was collected in this city [10], while the average amount of waste treated in "ZUSOK" is equal to approx. 40 thousand Mg/year [11], what gives 7,7% of waste collected in the Polish capital. Therefore, the current performance of only one WtE plant has a minimum importance in terms of the whole country. Meanwhile, it is predicted that Poland should take into account the needs of even 20-35 new facilities of such type [12]. While in 2011 the total number of WtE plants in the 18 European countries was 455, and in the United States 86 [13].

Newly built WtE plants or those which are currently in the initial phase of operation, i.e. Białystok, Bydgoszcz, Konin, Krakow, Poznań and Szczecin (see Figure 2), due to their scale and the potential impact on the environment, are today one of the most important Polish investments in the environmental protection sector. At the same time the number of planned new WtE investments reaches up to 30 plants, but their estimated total capacity (i.e. approximately 3.8-3.9 million Mg/year) is rather still too small to achieve the planned 2020 reduction in the weight of landfilled biodegradable waste (i.e. 65% weight reduction compared to the year 1995) [14].

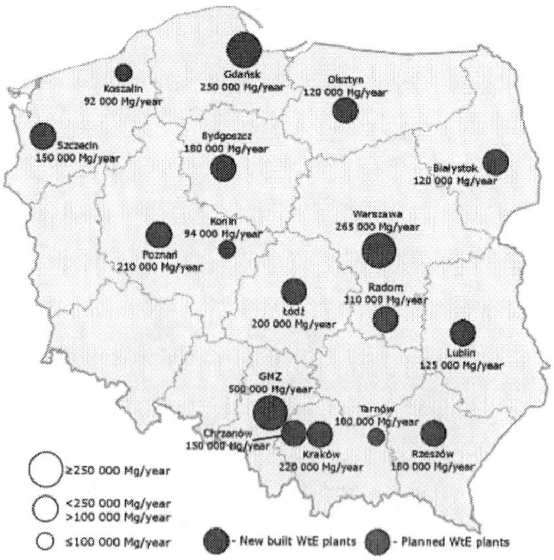

Figure 2. Polish WtE plants projects.

All WtE plants shown in the Figure 2 are intended for medium and large cities, what is an appropriate decision for location of this type of facilities. It is so because municipal waste generation rates are lower in rural areas since, on average, residents are usually poorer, purchase fewer goods, and have higher levels of reuse and recycling [15]. In addition, location of WtE plants in the big cities extends the capabilities of the energy distribution (especially generated heat) and makes it easier to obtain appropriate quality and quantity of municipal waste.

Nearly half of municipal waste in Poland (45.2% in 2008) is generated in big cities with more than 50 thousand inhabitants [16]. Composition of waste originating from such areas is shown in Table 1.

Table 1. Municipal waste composition in Polish big cities.

Place of waste generation	Big cities generally	City of Kraków
Year	2008	2003
Source	[16]	[17]
Type of waste		
Paper and cardboard	19,1%	10,2%
Multi-material waste	2,5%	3,1%
Kitchen and garden waste	34,2%	40,6%
F ine fraction < 10mm	4,2%	8,5%
Textiles	2,3%	2,7%
Wood	0,2%	0,9%
Glass	10,0%	10,1%
Metals	2,7%	1,8%
Plastics	15,2%	12,1%
Mineral waste	3,2%	5,6%
Hazardous waste	0,7%	0,2%
Other categories	5,8%	4,1%
Total municipal waste	5,47 millions Mg	239.1 kg/citizen/year

Research results regarding municipal waste composition and quantities in Polish big cities recently indicate great variability. This is so mainly because of various regimes with different areas and periods of sample collection [17]. However, on the basis of data from Tab 1. we can assume that WtE plants operating in large Polish cities will be supplied with appropriate amount and quantity of waste, since two key requirements for justified WtE use are [18]:

- Supply of municipal waste combustible fraction larger than 100,000 Mg/year.
- The lower calorific value (LCV) must be, on average, at least 7000 kJ/kg and never fall below 6 MJ/kg.

All plants depicted on Fig. 2 will have capacity, larger or comparable, to 100 000 Mg per year. Moreover these plants will most likely process wastes with LCV above 5800 kJ/kg, which is an adopted condition for autothermic municipal waste combustion (i.e. without firing additional fuel) [19]. Municipal waste rich in plastics and paper, like in cities of Kraków (8800 kJ/kg) and Poznań (8400 kJ/kg), can easily have LCV larger than 7000 kJ/kg. What is more, in exceptions, municipal waste mechanical sorting can be used to raise waste average calorific value before energy recovery [15]. Therefore combustible fractions of municipal waste from Polish big cities are of high interest for the WtE industry both in terms of amount and fuel properties.

Present deficiency of the WtE infrastructure in Poland is striking because even before World War II such plants were combusting Polish municipal waste. The first facility of this type was the plant built in Warsaw in 1912 which was working till World War II. The second was the plant in Poznań which was built in 1927 and it was the most modern facility of this type in Central Europe [18]. However, exploitation of this plant was also interrupted by the outbreak of World War II, but after war it was rebuilt, and from 1955 to 1957 plant burned waste again.

During 1906s Western European countries and other high income countries had started to build WtE plants on mass scales. In the late 1970s political measures towards even stricter reduction of allowable emissions of dioxins and other important pollutants, such as heavy metals, chlorine, sulfur dioxide, carbon monoxide and particulate matter have been taken. As a result of new emission standards WtE industry, worldwide, it has been transformed from a past major emitter of pollutants to modern clean power generators, which are subject to the most stringent emission standards from almost all branches of industry [19].

While behind western boundary WtE industry was still under development, at the same time the issue of municipal waste energy recovery was completely neglected in Poland and in second half of 20th century only one WtE plant was built (i.e. ZUSOK). After the political transformation in 1989 the implementation of new Polish WtE projects plants was still postponed by subsequent government cadences, despite the obvious political and economic benefits of waste energy recovery. For example, WtE plant for Cracow has been considered since 1992 [20] and the indicative list of WtE projects in the Operational Programme Infrastructure and Environment (OPIE) for the years 2007 to 2013, takes into account 12 new WtE plants to build in Poland during the 2007-2014 period [21].

The explanation of such delays in the development of Polish WtE industry may be high investment costs which, are much higher for WtE than for any other waste disposal method [15]. For aforementioned 12 WtE plants included in OPIE, total investment costs were estimated on 6,067 billion PLN [21].Such high costs make WtE plants one of the most expensive investments in Polish environment protection policy. However, EU financial funding in the form of grants is helping in the implementation of this type of projects. EU financial support for 12 plants from OPIE was estimated as 61% of total investment costs [21]. Share of EU funding together with the most important dates in the implementation of the 6 most advanced new Polish WtE projects are presented in the Table 2.

EU financial support together with the fact that waste management is a highly regulated industry can be key to the explanation of why WtE is nowadays being exploited more intensively in Poland. Additionally, EU energy targets establish a sustained market pull toward renewable energy, including energy recovered from waste. Those two aspects imply that there exist additional factors which may affect how waste management practices will evolve and ultimately influence the Polish incentive to invest in WtE [7].

DOI: 10.1051/e3sconf/20161000070

Table 2. Summary of ongoing WtE projects together with costs and key dates.

	Throughput [Mg/year]	*Contract value [million PLN]*	*EU funding [million PLN]*	*Environmental decision*	*Date of construction commencement*	*Date of obtaining work permission*
BYDGOSZCZ	180 000	436	262	23.11.2010	30.09.2013	28.08.2015
KRAKÓW	220000	797	372	14.01.2011	06.11.2013	29.09.2015
BIAŁYSTOK	120 000	410	197	30.05.2011	09.12.2013	05.05.2016
SZCZECIN	150 000	666	245	21.06.2011	16.12.2013	18.12.2015
KONIN	94 000	364	150	03.08.2012	03.11.2013	14.12.2015
POZNAŃ	210000	925	330	17.10.2013	30.04.2013	06.05.2016

However, EU on the one hand provides a reliable and relevant, financial and technical support, but on the other hand through its law, is necessary to use in new WtE plants advanced, efficient and expensive exhaust flue gas cleaning systems. Division of costs on example of Konin WtE plant for is shown on Figure 3.

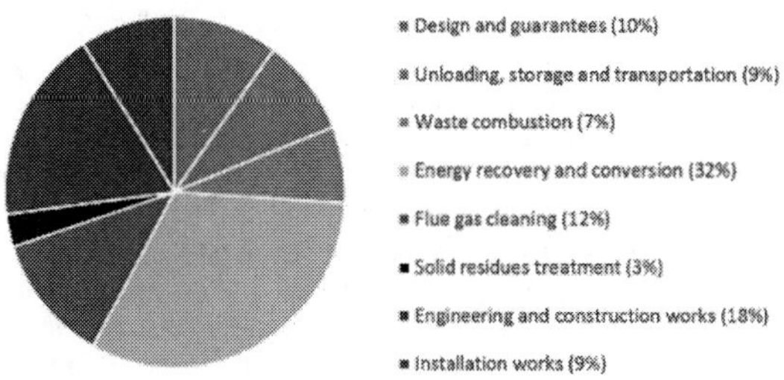

Figure 3. Division of WtE plant investment costs on particular segments *[22]*.

Despite the financial, issues another reason of Polish WtE implementation delays, can be opposition of local communities against construction of waste recovery facilities near residential areas (NIMBY syndrome). However according to data from the study [16] up to 73% inhabitants of Polish cities would agree to build a WtE plant in their neighborhood, therefore positive trend regarding WtE perception among citizens may be observed.

3 Current Polish WtE Projects

In order to minimise environmental impacts and ensure safe operation, all of Polish WtE plants have been prepared on basis of contemporary standards regarding their design, construction and operation. Reference information for WtE plants are included in BREF documents [23], which are a reliable source of information and recommendations about clean, effective and secure design for municipal waste energy recovery. Such design is based on proven techniques and technologies but also on the way in which such facilities should be operated. Nowadays, combustion is the leading process of energy recovery from waste, what usually involves burning waste on the moving grate.

Alternative thermal treatment technology for treating municipal wastes (e.g., fluidized bed, pyrolysis, and gasification), mainly due to the municipal waste physical and chemical properties, have not proven in practice to have as high energy efficiencies as the classical grate-based WtE plants [19]. Table , which presents a summary of technical parameters for most recent Polish WtE plants, shows that the dominant type of used combustion technology is moving grate technology. Moving grate is the leading solution in WtE facilities, mainly because of its simplicity, reliability and energy efficiency [24] (for example, approx. 85% of market share in France and Germany [13]).

All of 6 abovementioned plants have assumed availability equal to approx. 8000 hours per year, what is standard requirement for the modern WtE plants. The operation hours are one of the most important factors for the plant owner, because it has dominant influence on yearly plant income [25]. This fact cause that investors tend to choose well-proven technology, what additionally explains why only grate technology is used to burn Polish municipal waste. What is more all of the examples presented in Table uses state-of-the-art technologies in flue gas cleaning installation what ensures very insignificant environmental impact.

Polish environment protection infrastructure will improve and expand significantly when all of the WtE plants from Table 2 will work with nominal performance in terms of waste disposal and power production. Operation of those medium-scale plants will increase national energy security and will broaden implementation of suitable development rules. The total planned capacity of abovementioned six WtE plants is approximately equal to one million tons per year, which in comparison to the total amount of municipal waste generated in Poland (i.e. 11 330,4 thousand tons in 2014 [2]) shows significant increase in recovery methods in Polish municipal waste management, what is presented in Figure 4.

[a]Corresponding author: cyranka@agh.edu.pl

Table 3. Summary of ongoing WtE projects - technical data.

WtE Plant	*Contractor*	*Technology*	*Flue gas cleaning*	*Availability [h]*	*LCV [kJ/kg]*	*Line throughput [Mg/h]*	*Net Energy Efficiency*	*R1*	*Throughput [Mg/year]*	*Gross electrical energy [MWhe/year]*	*Gross heat energy [MWhth/year]*
BYDGOSZCZ	ASTALDI S.p.A., Termomeccanica Ecologia	Grate furnace integrated with boiler	Wet + semi-dry / SNCR	7800	8 500	2 x 11,5	0,64	0,89	180 000	71 760	216 060
KRAKÓW	Posco Engineering & Construction		Semi-dry / SNCR	8 100	8 800	2 x 14,1	0,60	0,80	220 000	67 860	273 000
BIAŁYSTOK	Budimex S.A., Keppel Seghers		Semi-dry / SNCR	8 050	7 500	1 x 15,5	0,58	0,79	120 000	31 258	118 200
SZCZECIN	MOSTOSTAL S.A, RAFAKO S.A Hitachi PE		Wet /SNCR	7 500	10 500	2 x 10,0	0,50	0,75	150 000	71 439	159 984
KONIN	Integral Eng., ERBUD S.A., Introl S.A.		Semi-dry / SNCR	7 800	8 500	1 x 12,05	0,75	0,94	94 000	26 602	148 200
POZNAŃ	SITA ZE, Hitachi ZI, Hochtief		Semi-dry / SNCR	7 800	8 400	2 x 13,5	0,66	0,89	240 000	78 000	312 000

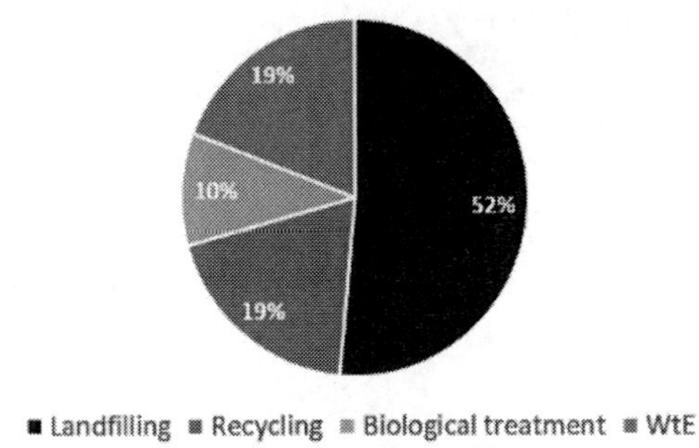

Figure 4. Prognosis of waste management in 2016 (based on *[2]*).

Unfortunately, in the case of energy production, potential impact of the described six WtE plants is not so significant, i.e. those plants can cover only approx. 0.2% of Polish electricity consumption (161,1 TWh_e [26]), and approx. 2.0 % of the heat demand (275,1 PJ_{th} [26]). However, if we consider energy production prognosis in the criteria of renewable energy sources it can be concluded that waste energy recovery will become an important part in Polish "green" energy market, especially in terms of heat production. This is confirmed by results presented onFigure 5 and Figure 6.

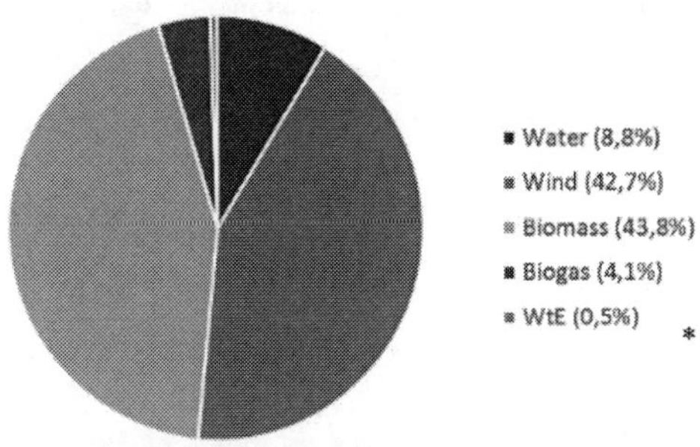

Figure 5. Prognosis of electricity generation from renewable sources in 2016 (based on *[27]*).

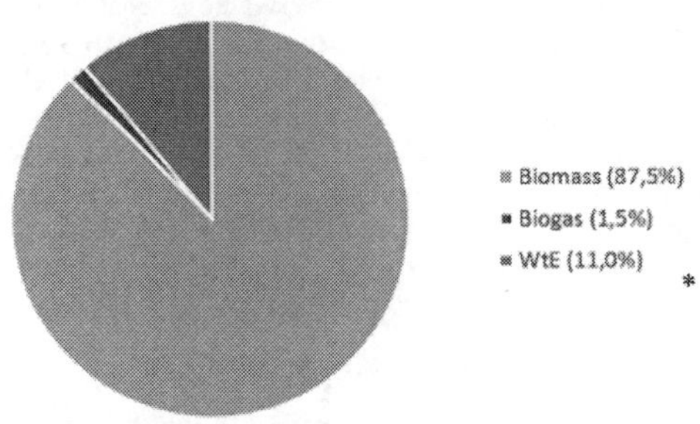

Figure 6. Prognosis of heat generation from renewable sources in 2016 (based on *[27]*).

*- Based on recognition of 50% energy produced in Waste-to-Energy as renewable

[a]Corresponding author: cyranka@agh.edu.pl

 DOI: 10.1051/e3sconf/20161000070

Every Polish WtE plant design (especially designs for six plants described here) assume operation in cogeneration, i.e. synergic production of both electricity and thermal energy. This is because it turns out that the most effective work of plant combusting municipal waste, and therefore plant that is usually located in densely populated areas, is a combination of electricity generation with production of heat which meet the needs of nearby external users, i.e. district heating network or industrial plants [28]. This is caused by fact that in WtE plants far more economically justified is to generate a certain amount of thermal energy than the same amount of electricity. It is so because of basic thermodynamic laws and fact that parameters of steam produced in WtE boiler are significantly reduced due to the highly corrosive nature of the flue gases.

The standard WtE grate-based combustion technology is using steam parameters of 40 bar/400 °C. Typically, these plants exports 500 kWh of electricity per ton of waste [6]. Average efficiency of energy production in European WtE plant is equal to approx. 26.1% in case of producing only electricity, 77.2% in case of producing only heat and 52.1% in case of combined heat and power production [29]. Producing both heat and electricity causes that energy efficiencies of all the mentioned six Polish WtE plants is well above the standard 26.1% (for only electricity production). Energy efficiency is a key parameter for reducing the overall cost as well as the environmental impact of waste energy recovery facilities and it is therefore an important direction of WtE research and development [19].

It can be expected th0at minimum standards for energy efficiency will be raised over time, what has already been implemented by the EU which requires that all new WtE plants must have a sufficiently high energy efficiency indicator, i.e. so called R1 factor. R1 factor is performance indicator introduced by European Union Directive 2008/98/WE [3], indicating the viability and rationality of waste energy recovery process and is used to determine the efficiency of waste chemical energy conversion potential.

According to EU law, R1 value obligatory for WtE plants conducting energy recovery process (R1) must be equal to or greater than 0.65. If R1 value is smaller than 0.65, then WtE plant conduct waste disposal (D10) which is not considered as waste energy recovery. Most likely, when nominal performance parameters will be met, all Polish WtE projects listed in Table will fulfill above requirements during operation. This is mostly by using cogeneration in those plants, without which it would be difficult to meet required R1 factor value.

4 Key Challenges & Opportunities

Ongoing operation of Polish WtE plants will be associated with a number of challenges and responsibilities that must be fulfilled in order to ensure full utilization of such plants.

The most important tasks for operators will include achievement of assumed availability and efficiency of the facility and all its segments. Of course WtE plant operation should be supported by appropriate optimization so that the beneficial effects of energy recovery will be maximized while ecological threats will be maintained on the minimal level. Uncertainties regarding waste feed (e.g. separate collections systems performance and effectiveness) and energy generation (e.g. heat demand and electricity price) are crucial parameters for evaluation and optimization of WtE projects. Unfortunately, those uncertainties make that new WtE facility planning is a very complex task [30].

Even more important than commercial use, step of WtE plant life, is beginning phase of exploitation, i.e. test runs, start-up and commissioning from contractor. All eventual malfunctions and mistakes in design and montage should be revealed latest on this phase. In addition, under the supervision of contractor, plant should be checked for meeting required technical parameters. Another important aspect of preparing WtE plant for regular operation is appropriate selection and training of the crew, which will take care about safe, clean and efficient waste energy recovery during plant operation.

When plant and operators are prepared to run municipal waste energy recovery then it is necessary to ensure incoming waste stream of sufficient quantity and quality. Mass and calorific value of waste must be adequate to boiler design and waste management regime inside plant, which cannot operate in suitable manner when there is too much or not enough of wastes or their proprieties hinder operation of plant subsystems like waste storage and transportation.

WtE plants can be a useful sink, not only for Polish municipal waste, but also for alternative fuels produced from such waste. Current problem is the lack of sufficient infrastructure for disposal of combustible fraction coming from the mechanical-biological waste treatment installations (MBT). Today only available recipient of fuels produced from wastes (RDF/SRF) are cement producing facilities which are able to dispose less than 20% of combustible fraction possible to produce in Poland [31].

Alternative fuels may be input in coal-fired boilers and experience indicates that those fuels can contribute up to 30% of total energy input into such boilers [5]. Using coal for energy generation is in Poland important part of energy/fuel strategy, i.e. coal has approx. 54% share in total energy production and import [26]. However typical cofiring rate for solid fuel is 3% or less, because of metal particles and glass fines which may pose combustion problems, high chloride concentrations which may induce boiler tube corrosion and high levels of ash production which may create residuals management problems [5]. Hence use of municipal waste in the Polish conventional energy industry is rather avoided and seen as unnecessary evasion of new WtE plants development.

In accordance with the suitable development principles energetic usage of coal, as every natural resource, should decrease. Recently built WtE plants replaces an estimated 0.4 tons of coal for electricity generation [32]. Replacement of coal and other fossil fuels, with regionally produced municipal waste without a doubt will increase Polish energy safety.

Energy recovery from municipal waste is capable of displacing large amounts of fossil generation and even additional positive ecological effect occur when raw materials are recovered from the incoming waste and/or from combustion residuals [5]. Therefore, Polish WtE plants operators should optimize not only the energy recovery, but also maximize recovery of recyclable materials from combustion residues. Main goals for optimizing WtE plant in terms of solid residues management are [33]:

- Production of highly inert bottom ashes by adequate combustion control;
- Reduction of fly ashes amount by adequate primary air control;
- Achievement of excellent gas phase and fly ash burnout;
- Rendering the boiler and filter ashes inert and safe for disposal or even utilization;
- Management and safe disposal of the residues from the chemical gas cleaning systems.

Additionally, to technological and organisational challenges that lay before Polish WtE plant operators administrative challenges should be not forgotten. The large costs of this type facilities can attract unwanted political influence, so the local administration should ensure that the technology selection process is as transparent as possible, with all costs over the life of the facility considered [34]. What is more Polish waste managers, during implementation of WtE projects, should consider issues connected with [1]:

- Inadequate ownership of waste streams (Polish Waste Management System is during transformation process);
- Inadequate financial incentives (taxes, fees);
- Poor public awareness / participation;
- Inadequate number of collection points;
- Unreliable data regarding waste;
- Poor administrative capacity for adopting EU funds;
- Lack of public-private partnerships.

No less important than aspects enlisted above is the role of the WtE plant as facility providing society with information about modern waste recovery systems. WtE plants should present themselves to the public opinion as utilities bringing citizens a lot of tangible benefits improving environment in which they are living. Additional advantage, from citizen's point of view, is fact that due to development of the WtE sector there will be a significant increase in direct and indirect jobs aimed on operation and maintenance of such infrastructure (in 2011 only 50 persons were employed in polish WtE industry [35]).

Social information role of WtE plant can be realized mainly by: information campaigns, "open days" visits and especially by full transparency in the operation. Public education in terms of waste management is desirable, because waste managers need to select from among several challenging technology and better decisions are made when the public is part of the decision making process and supports waste management program [34].

WtE should be developed and located in waste management structures taking into account the principles of suitable development and waste hierarchy (Figure 1). Although, there is a risk that European, and therefore Polish, waste policies and the dynamic of the waste system, will become stagnated in recycling/WtE, and even landfill oriented strategies, assigning a lower weight to top of the waste hierarchy [36]. This is particularly relevant in relation to idea of circular economy which contribute to "closing the loop" of product lifecycles through greater recycling and re-use and bring benefits for both environment and economy [37].

EU waste policy pushes for an economy where targets for prevention, reuse, recycling, recovery and landfill should move towards a circular economy, where waste that cannot be eliminated are turned into a resource. Realization of circular economy policy postulates will have great impacts on WtE facilities by limiting energy recovery to non-recyclable materials. This constraint may reduce the future development field of municipal waste energy recovery [7]. Implementation of such plans in Polish economy should affect decisions regarding the implementation of new plants and optimization of plants that are already operated.

5 Conclusion

Due to the low level of any forms of municipal waste recycling or recovery in Polish economy, the development of WtE infrastructure is highly desirable and should be immediately implemented in order to efficiently use energy from Polish municipal waste. Development of Polish WtE industry is right direction and it should be continued at all levels of the national administration.

Taking into account six new WtE plants, with a total capacity of 1 million tons, is a significant success and progress in the fields of municipal waste management and renewable energy sources in Poland. Additionally, experience gained during the entire cycle of projects realization should be used to improve even further implementation of waste recovery and recycling methods for Polish waste management system.

Energy recovered from waste without doubt is a very valuable kind of energy for the national grid, better for the environment than the energy gained from fossil fuels. Moreover, the operation of WtE plants not only provide a regional and renewable source of energy but is also improving Polish economy, especially in the environment protection criteria. Most important advantages resulting from operation of currently operating Polish WtE projects are listed below:

- Source of energy, some of which may be considered as derived from renewable energy sources;
- Reduction of fossil fuels use and increase of national energy security;
- Waste diversion from landfills and thus the environmental load reduction;
- Ability to conduct additional raw materials recovery (especially metals) from combusted waste;
- Adding experience and foundations for further implementation of suitable development rules in Polish waste management;

- Public information role, explaining position and role of WtE in modern and sustainable waste management system.

However, in order to ensure tasks enlisted above, considerable responsibility will be placed on WtE plants operators. Such operators will have to implement appropriate organizational, maintenance and also optimizing measures for safe, efficient and stable WtE plant exploitation.

If six new Polish WtE plants will confirm their technical and operational indicators than national environment protection will be significantly improved, especially in cities and regions where WtE plants will be located. Operation of those plants, apart from the benefits associated with energy production, will also adjust Polish economy to contemporary standards and trends in the EU as well bring closer assumptions of circular economy. What is more, emergence of modern Polish WtE plants is also a great opportunity to change the mindset of local communities and change the public perception of municipal waste energy recovery in Poland.

References

1. A. Tonderski, ISWA (2013)
2. Główny Urząd Statystyczny (2015)
3. P. Aragonés-Beltrán, F. Chaparro-González, J.-P. Pastor-Ferrando, A. Pla-Rubio, Energy **66,** 222-238. (2014)
4. CEWEP, (2009
5. Electric Power Research Institute, Palo Alto, (2011).
6. J. J.E. Martin, R. Koralewska, A. Wohlleben, Waste Manage., **37**, 147 (2015)
7. Coolsweep the 7th Framework Programme (2013
8. R. F. Sawyer, P. Combust. I. 32, 45 (2009)
9. GUS, Ludność. Stan i struktura ludności oraz ruch naturalny w przekroju terytorialnym. Data for 30 VI 2015 r. (2015)
10. Mazowiecki Ośrodek Badań Regionalnych (2015)
11. R. Oleniacz, Geo. Environ. Eng. **4**, 25 (2014)
12. G. Wielgosiński, Międzynarodowa konferencja - Zakłady termicznego przekształcania odpadów jako regionalne instalacje (2013)
13. International Solid Waste Association (2012)
14. G. Wielgosiński, Nowa Ener. **37**, 17 (2014)
15. World Bank, What a waste: A global review of solid waste management (2012).
16. Deloitte, Fortum and 4P Research Mix, Gospodarka odpadami w Polsce - Wyzwania w świetle wymogów unijnych i zmian legislacyjnych, opinie społeczne i perspektywy (2011)
17. E. Den Boer, A. Jędrczak, Z. Kowalski, J. Kulczycka R. Szpadt, Waste Manage. **30**, 369 (2010)
18. B. Kamuk, ISWA (2013)
19. L. Sieja, T. Pająk, Termiczne przekształcanie odpadów, jako niezbędny element systemu gospodarki odpadami komunalnymi dla dużych aglomeracji miejskich – podstawowe uwarunkowania, (2005)
20. Wikipedia, *Spalarnia śmieci na Wilczaku w Poznaniu* (2016)
21. M. Van Berlo, *Renew. Energ.* 1532 (2013)
22. A. Generowicz, J. Pauli Wilga, J. Sacharczuk, *Aktualizacja wyboru lokalizacji zakładu termicznego przekształcania odpadów na podstawie analizy wielokryterialnej*, Krakowski Holding Komunalny S.A. (2009)
23. T. Pająk, *Ocena dobiegających końca budów ZTPOK* (2015)
24. MZGOK w Koninie, *Budowa Zakładu Termicznego Unieszkodliwiania Odpadów Komunalnych w Koninie* (2014)
25. IPPC, *BREF Reference Document on the Best Available Techniques for Waste Incineration (2006)*
26. M. Cyranka, M. Jurczyk, *Agric. Eng.*, **20**, 23 (2016)
27. O. Hedegaard Madsen, *Proceedings of the 15th Annual North American Waste to Energy Conference* (2007)
28. GUS, *Energy - 2015* (2015)
29. GUS, Energia ze źródeł odnawialnych w 2014 r., (2015)
30. M. Cyranka, M. Jurczyk, *Energia i Paliwa* (2015)
31. D. Reimann, Energy Report III (2012)
32. R. Somplák, T. Ferdan, M. Pavlas, P. Popela, Appl. Therm.l Eng. **61**, 106 (2013)
33. A. Majcherczyk, Instalacja Termicznego Przekształcania Odpadów jako alternatywa dla energetyki przemysłowej (2015)
34. N. J. Themelis, C. Mussche, Energy And Economic Value Of Municipal Solid Waste (Msw), Including Non-Recycled Plastics (Nrp), Currently Landfilled In The Fifty States (2014)
35. J. Vehlow, H. Seifert, IEA Bioenergy (2012)
36. D. Hoornweg, P. Bhada-Tata, Waste Wise (2014)
37. EurObser'ER, Renewable Municipal Waste Barometer (2014)
38. M. Mazzanti, R. Zoboli, Resour. Conserv. Recy. **52**, 1221 (2008)
39. Environment Directorate-General of the European Commission, Circular Economy Strategy (2016)

 DOI: 10.1051/e3sconf/20161000071

Waste immobilization and environmental sustainability in glass-ceramics glazes development

M. Reben[1], M. Kosmal[2], N. Pałczyńska[1,a] and P. Pichniarczyk[2]

[1]*AGH University of Science and Technology, Faculty of Materials Science and Ceramics, al. Mickiewicza 30, 30-059 Cracow, Poland*
[2]*Institute of Ceramics and Building Materials, Division of Glass in Cracow, Lipowa 3, 30-702 Cracow, Poland*

Abstract. The current work investigated the formulation of glazes using recycled CRT (Cathode Ray Tube) waste glass powder mixed with basalt, kaolin and other additives. New glaze compositions can be crystallized in order to improve the mechanical and chemical properties of fast-fired glazed tiles. These glazes are characterized by high resistance to abrasion.

1 Introduction

Glazes can be defined as a stable glassy coatings applied to ceramic earthenware and formerly obtained by cooling oxides or minerals applied and melted on the surfaces of ceramic objects. The large increase in demand for tiles with improved and advanced properties, such as high resistance to abrasion, higher hardness, lower closed porosity and improved chemical resistance is observed [1]. A glass-ceramic glaze must provide precursor frit which is technically and commercially compatible with the fabrication conditions normally used in industrial production, where glazes should be obtained by fast firing process. Densification is achieved by sintering in the presence of a viscous flow at temperatures slightly higher than the glass transition temperature (Tg), resulting in a compact layer nearly free of porosity. During fast firing process sufficient crystallization occurs. It resulted in the appearance of crystalline phase in glaze obtained [2]. Moreover the phase composition and structure development under the heat treatment are very important, and fulfill strict requirements in the linear expansion coefficient of the glaze. The greater the difference in the thermal expansion of the ceramic support and the glaze, the faster defects will appear [3]. Glaze suspensions are composed of different types of raw materials such as: plastics, non-plastics, and additives. Plastic materials include clays e.g. kaolin and bentonite, whereas the non-plastic materials consist of oxides, pigments, feldspars and frits. The additives are used to optimize the glazing process are dispersing and deflocculanting agents, binders and defoaming agents [4]. The potential of replacing natural components in ceramics glaze production with wastes is well documented in literature [5]. Replacing fluxes with cullet has been reported in Youssef and Tarvornpanich [6].

P Ketboonruang studied the potential use of Municipal solid waste (MSW), soda lime silica cullet and borax in a low firing temperature glazes production. This allowed them to avoid strong fluxing such as a very toxic lead oxide [7]. This underscore the importance of recycling cullet which will not only turn waste to wealth but will also reduce the stress and cost of seeking fluxes for ceramic glazes. Glass is found to be in different types and properties, moreover there are some types of glass cullet which is more of a nuisance to the environment [8]. Thus for the purpose of this research, exploring the composition and effect of Cathode Ray Tube (CRT) cullet on the glaze properties will significantly reduce the environmental impact of the ceramic glaze production process. CRTs are composed of three different parts: the viewing section known as the screen or panel, made of silica barium-strontium glass; the funnel glass and the neck glass, both of which contain lead oxide which absorbed X-ray radiation [9,10,11]. Various different compositions of glasses used for CRT create obstacles to the increased recovery the glass [12]. Moreover the specific glass composition forbid their recycling in the glass industry for the production of containers, domestic glassware and glass fibers, if crushed and mixed together it cannot be even recycled as cullet, for TV glass, on the other hand there are no manufactures of CRT glass in Europe anymore [13]. The use of CRT cullet as a replacement of float glass and feldspar in glazes production will be a very promising initiative as it will strongly contribute to sustainable development of ceramics industries. Our studies will enhance the use of wastes and also sustain the environmental benefits of the society.

[a] Corresponding author: natalia.palczynska@wp.pl.

DOI: 10.1051/e3sconf/20161000071

2 Experimental

The chemical compositions of CRT glass, float glass, basalt, borosilicate glass and feldspar are presented in table 1. Chemical composition of raw materials used in this experiment was analyzed by X-ray Fluorescence (XRF) with use of spectrometer WDXRF Axios mAX , PANalytical. The thermal linear expansion coefficient was measured on the blocks prepared from the powdery glazes and frits just as shown in table 2. The dilatation of the materials was measured using dilatometer Bahr Thermo Analyse DIL 801 L-Germany with a heating rate 5 K min^{-1}. The kind of crystalline phases which were formed via the thermal treatment of selected glazes at the temperature 1130°C, were examined by XRD method with use of Philips X'Pert X-ray diffractometer with CuKα radiation source. The morphologies of crystalline phases developed under glazes firing were examined by scanning electron microscope (SEM, Nova Nano SEM 200, FEI Company). SEM tests were performed with an attachment for chemical analysis of specimens in micro-areas with energy dispersive x-ray spectroscopy (EDAX). The observations were carried out in low vacuum conditions in the secondary electron mode. The samples were covered with a carbon layer. The abrasion resistance was determined according to the EN-ISO 1545-7 standard (PEI method). Abrasion test was based on a visual assessment of changes in the glazed surface, and wherein the rotational movement moving the shearing load of 100 to 12000 times in comparison with the surface of the tiles.

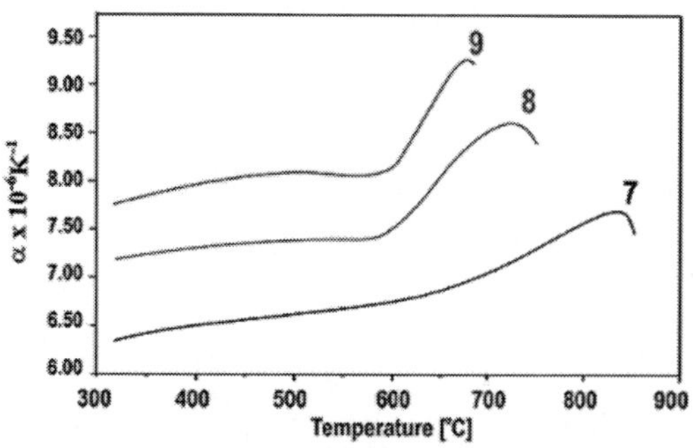

Figure 1. TMA expansion curves of studied glazes 7-9.

Raw materials for the production of glazes were shredded and mixed, then milled in a porcelain ball mill. The raw materials were wet milled for 40 minutes and the milling time was chosen experimentally to obtain expected sieve residue of less than 0.1%. The resulting mixture was applied to a ceramic tile by hand and heated at a rate of 25°C/min up to the firing temperature 1130°C. The entire firing cycle lasted 52 minutes. Tiles after firing were cooled with the furnace. Enamel surfaces were examined visually in terms of surface defects, surface smoothness and gloss. For the determination the basic physical and chemical properties of selected glazes the water absorption and abrasion resistance were tested. Determination of water absorption was performed according to the method described in the standard [PN-EN ISO 10545-3: 1999 Ceramic tiles. Determination of water absorption, porosity and density of the total p.5.1. Water saturation. Cooking method]. After cleaning and drying, the samples were weighed. Distilled water was used to impregnate the samples. Impregnation of samples was carried out by boiling the liquid within 2 hours. After the test, the samples were weighed again.

3 Results and discussion

CRT glass cullet was used in order to formulate in laboratory scale compositions of ceramic glazes as substitute to frit and feldspar in 1-9 glazes series designed and as a partial substitut of these two components in case of 10-14 glazes. As is reported in [14] CRT panel glass is a typical silicate glass containing barium and strontium and shows a tendency to not crystallize with a low melting nature, confirmed by low softening and transition temperatures (respectively Ts=559°C, Tg=490°C) and high thermal expansion coefficient (α=10.3·10^{-6}/°C). The high value of the thermal expansion coefficient of CRT glass cullet does not allow to use this kind of cullet itself for glazes production. The glaze compositions 1-6 prepared after firing showed crazing defect and spot resistance; however the presence of kaolin in glazes 3-6 decreases the intensity of the defect but did not avoid it. Therefore a re-modulation of the raw materials was necessary to obtain a better agreement between glaze and ceramic body dilatometric coefficient. For the laboratory tests different formulations using CRT glass, kaolin and basalt ranging in 20–40, 5 and 10–15 wt.%, respectively, were prepared. It permitted to obtain a glaze No 7 characterized by the by a lower thermal expansion coefficient (6.39·10^{-6}/°C^{-1}) (Fig.1).

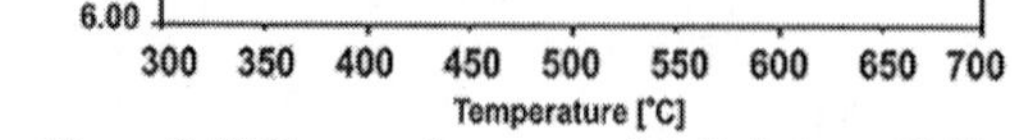

Figure 2. TMA expansion curves of studied glazes 10-14.

The glaze 7 prepared contains 50% of vitreous fraction, substituted only by CRT glass. The glaze appeared to be matt, thus to obtain glossy glaze the second group of glaze compositions was designed (10-14). The glaze 14 prepared contains 85 wt. % of vitreous fraction, substituted by 40 wt % of CRT glass, 30 wt. % of M4 borosilicate glass and 15 wt. % of float glass. The thermal expansion coefficient of glaze 14 was reduced by the addition of 15 wt. % of kaolin and 30 wt. % of M4 borosilicate glass (Fig.2).

Table 1. Chemical composition of gazes components [wt.%].

Component	CRT glass cullet	Feldspar	Float glass cullet	Glass M_4	Basalt
SiO_2	61,82	68,30	71,70	79,00	44,46
CaO	-	0,90	13,00	-	10,53
MgO	0,013	-	-	-	10,88
Al_2O_3	2,0305	18,80	1,30	-	14,95
Fe_2O_3	0,065	0,10	-	-	11,16
Na_2O	7,95	5,80	12,80	-	4,38
K_2O	7,35	6,40	1,20	2,50	1,45
BaO	7,273	-	-		-
B_2O_3	-	-	-	18,50	-

Table 2. The compositions of glazes designed.

	The compositions of designed glazes [wt%]						
Symbol of glazes	CRT glass cullet	Glass M_4	Feldspar	Float glass cullet	Kaolin	Basalt	Firing parameters and appearance
1	100	-	-	-	-	-	1130°C, 1h Matt glazes
2	50	50	-	-	-	-	
3	80	-	-	-	20	-	
4	70	-	-	-	30	-	
5	60	-	-	-	40	-	
6	50	-	-	-	50	-	
7	50	-	-	-	40	10	
8	50	-	-	-	35	15	
9	50	-	-	-	30	20	
10	35	-	20	30	5	10	1130°C, 1h Glazes with gloss
11	35	-	25	35	5	-	
12	15	-	35	45	5	-	
13	10	5	35	45	5	-	
14	40	30	-	15	15	-	

Borosilicate glasses are known for having very low coefficients of thermal expansion (~$5 \cdot 10^{-6}$/°C at 20°C), making them resistant to thermal shock, more so than any other common glass) [15]. By reducing the value of α, B_2O_3 contributes to the density structure of the glazes, making them more chemically resistant (Fig. 2). The tendency to crystallization depend on, among other factors, the chemical composition of the glaze [16]. The glaze 7 crystallize easily because contains basalt. The crystallization is enhanced by the significant amount of the iron oxides and of the oxides of bivalent metals present. As the molar ratio SiO_2:RO is increased from 2-2.5 (glazes 7-9) in the presents of the alkalis the crystallization capability of the glaze decreases. The kind of the crystallizing phases was confirmed by the XRD technique. The XRD pattern of the glaze 7

DOI: 10.1051/e3sconf/20161000071

fired at 1130°C revealed the presence of a three crystalline phase: $KAlSi_3O_8$ - 67% -Sanidine (JCPDS No.99-101-1680),Al_2SiO_8 - 22%-Sillimanite (JCPDS No.99-100-0721) and $BaAl_2Si_2O_8$-11%-Paracelsian (JCPDS No.99-100-0994). The intensity of lines corresponding to $KAlSi_3O_8$ and Al_2SiO_8 phases are narrow indicating the existence of well-developed crystals (Fig.3).

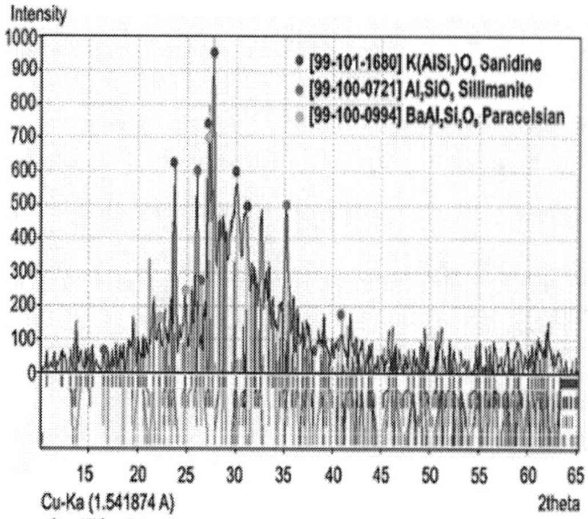

Figure 3. XRD pattern of glaze no. 7 after firing at 1130°C, 1h.

The appearance of not significant amount of $BaAl_2Si_2O_8$, can be explain by its an extenisve polimorphism [17]. Under XRD the crystalline phases in glaze 14 were not detected, enables us to conclude that glaze most probably being amorphous (Fig.4).

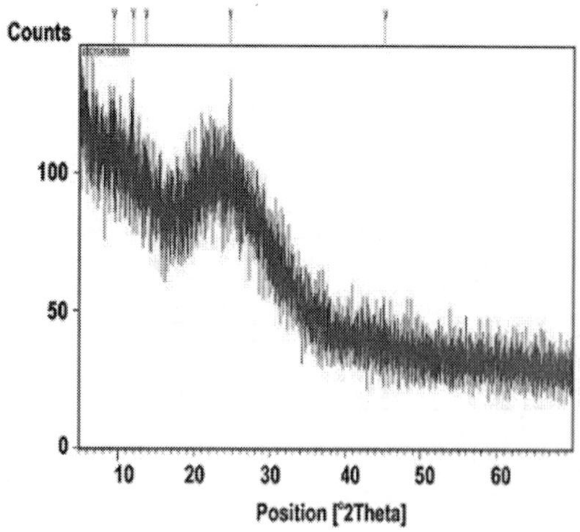

Figure 4. XRD pattern glaze no. 14 after firing at 1130°C, 1h.

Morphology of glaze 7 surface fired at 1130°C analyzed by SEM shows smooth surface with very fine crystals on the surface as shown in Figure 5. Si, Al and K were the major elements analyzed by EDS. That consistent with the results of phase analysis by XRD that shows crystalline phases: $KAlSi_3O_8$, Al_2SiO_8 and not significant amount of $BaAl_2Si_2O_8$ in glaze 7 (Fig.3). Glaze 14 fired at 1130°C shows glassy surface (Fig. 6),

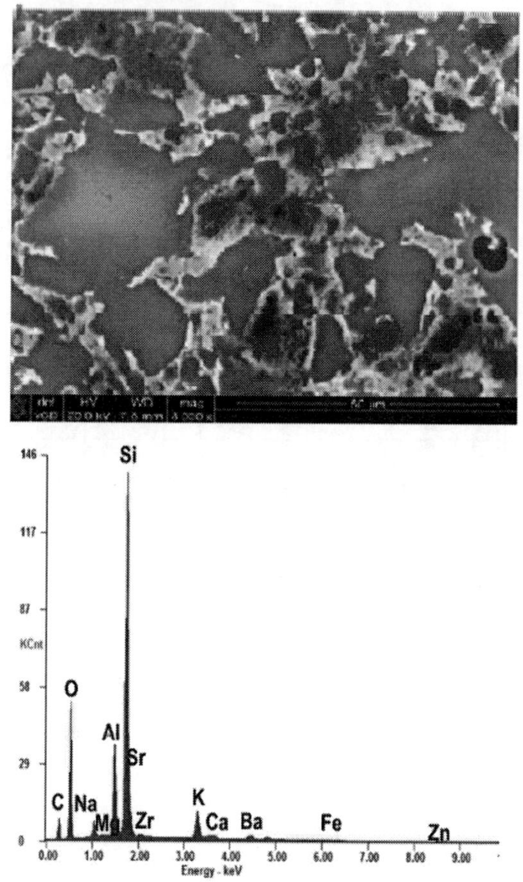

Figure 5. SEM photograph and EDS analysis of glaze 7 (the analysis performed in point 7).

SEM analysis did not show any crystalline phases. EDS revealed more Ca and Na content due to float glass cullet in the glaze composition. The resistance of the glaze tile to mechanical damage resulting from the impact of the abrasive factor was tested. The glazes were divided into abrasion resistance grades (PEI) with a specified number of rotations, of the device that rubs the surface of the glazed tiles after which the permanent traces of abrasion become visible. Based on the abrasion test the glaze 7 is assigned to the class 3 (number of rotations 1500) and glaze 14 (number of rotations 12000) to the class 4. Glaze 3 can be apllied on the tiles for flooring in domestic housing with considerable traffic and normal footware, where small abrasive material may ocassionally be carried out. Glaze 14 can be subjected for more frequent pedestrian traffic. The results of water absorption are presented in Table.3. Water absorption (N) was calculated by the following formula (1):

$$N = \frac{m_2 - m_1}{m_1} \cdot 100\% \qquad (1)$$

where:

m_1 - mass of the dried sample [g]

m_2 - mass of the sample saturated with liquid [g]

 DOI: 10.1051/e3sconf/20161000071

Table 3. Water absorption

Type of sample	Number tiles	Water absorption (%)	Water absorption average (%)
Glaze 7	1/073	5,9	6,0
	2/073	6,1	
	3/073	6,0	
Glaze 14	1/013	4,9	4,9
	2/013	4,9	
	3/013	5,0	

4 Conclusion

The aim of the study was to support, by an environmental point of view, the recycling solution in the ceramic field successfully experimented for the CRT panel glass. In the ceramic industry, CRT cullet used as a raw material in glaze for low-fired products were found possible for CRT glass recycling. Glazes for low firing area with firing temperature of 1130° C utilised up to 50 wt % CRT panel glass. The identification of this limit percentage is related to the importance of avoiding defects like cracking, bubbling. Moreover the attention was paid toward the thermal expansion coefficient that tends to be elevated. The re-modulation of the raw materials in formulating the glaze composition becomes fundamental to obtain acceptable values of this parameter. The matt and amorphous glazes obtained have very similar to originals not containing CRT glass. Based on the surface abrasion tests it was observed that the introduction of CRT panel glass into the formulation meets the requirements of the standard product. However, research conducted have been the starting point for the study with the environmental impact of technologies for the production of ceramic glazes and will contribute to the rational development of the waste and re-introduce it to the market as a full value secondary raw material.

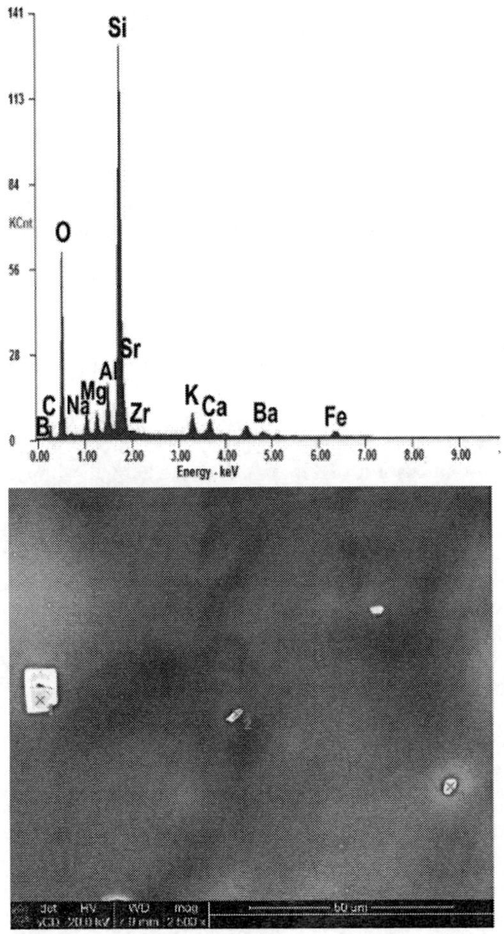

Figure 6. SEM photograph and EDS analysis of glaze 14 (the analysis performed in point 2).

References

1. M.G. Rasteiro, T. Gassman, R. Santos, E. Antunes, Ceram. Int. **33**, 345–354, (2007)
2. M. Llusar, G. Monros, C.M. Rodrigues, J.A. Labrinch, Ceram. Int. **31,** 181–188, (2005)
3. R. Casasola, J. Ma Rinco´n, M. Romero, J. Mater. Sci. **47**, 553–582, (2012)
4. W. Weiand, R. Merz, K. Bonk, Tile & Brick **1(8),** 14–17, (1992)
5. T. Tarvornpanich, University of Sheffield (2006)
6. N.F. Youssef, M.F. Abadir, M.A.O. J. Eur., **18**, 1721-1727, (1998)
7. P. Ketboonruang, S. Jinawat, D.P. Kashima, T. Wasanapiarnpong, P. Sujaridworakun, W. Buggakuptav, N. Traipol, S. Jiemsirilers, , IOP Conf. Series: Materials Science and Engineering, **18**(22), (2011)
8. L.A.R. Siikamaki, Rec. and Reuse of waste Mater., 743–750, (2003)
9. Andreola, L. Barbieri, E. Karamanova, I. Lancellotti, M. Pelino, Ceram. Int., **34,** 1289–1295, (2008)
10. F. Andreola, L. Barbieri, F. Bondioli, I. Lancellotti, P. Miselli, A.M. Ferrari, Int. J. Appl. Ceram. Tec., **7**, 909–917, (2010)
11. F.O. Méar, P.G. Yot, A.V. Kolobov, M. Ribes, M.F. Guimon, D. Gonbeau, J. Non-Cryst. Solids., **353,** 4640–4646, (2007)
12. M. Reben, J. Wasylak, **240**, 159-166, 2013
13. P.Walczak, J. Małolepszy, M. Reben, K. Rzepa, J. Pro. Eng., **108**, 453-458, 2015
14. M. Reben, M. Kosmal, M. Ziąbka, P. Pichniarczyk, I. Grelowska, J. Non-Cryst. Solids, **425**, 118–123, (2015)
15. Toluwalope, Oluwatuase, Int. J Adv. Tech., **6**, 1-3, (2015)
16. Cannillo, T. Manfredini, M. Montorsi, C. Siligardi, A. Sola, Mater. Eng., **14**, 207-228, (2003)
17. F. Yen-Pei, C. Chin-Chung, L. Cheng-Hsiung, Tsung-Shune; Ceram. Int. **30**, 41–45, (2004)

DOI: 10.1051/e3sconf/20161000072

Effect of grinding process on the level of leachability of the contaminants from the fly ashes from combustion of biomass

Aleksandra Pawluk[1a]

[1]*AGH University of Science and Technology, Faculty of Mining and Geoengineering , Cracow, Poland*

Abstract. The most commonly used renewable energy source in Polish energy production companies is solid biomass which is used both as a separate fuel or as a component co-incinerated together with (mostly) hard coal. During its incineration the biomass generates by-products with diverse and variable physicochemical properties. The most of the waste from production of electricity and/or heat are fly ashes. The fly ashes from combustion of biomass are a particular kind of waste distinguished by high level of leaching of contaminants and variable chemical composition. The by-products from combustion can only be used in production when their physicochemical properties meet specific parameters. This article presents results of research on leachability of chemical pollutants from dry ashes of combustion or co-combustion of biomass. The study of the impact of grinding of ashes on the level of their leachability was also conducted. Ashes from combustion of biomass as well as their mixture with fluidized ash from combustion of charcoal were submitted to grinding process. Afterwards level of leachability was measured on them to determine prospect of utility in mining technologies.

1 Introduction

Solid biomass is the third biggest (after solar and wind energy) renewable energy source in the world [8, 3]. It is the most commonly used renewable energy source in Polish energy production. Biomass is used both as a separate fuel as well as a component co-incinerated together with (mostly) hard coal.

Likewise combustion of hard coal and lignite, combustion of biomass generates solid by-products, mainly ashes. As Poland already employs special boilers for combustion of biomass as a single fuel source, a "new generation of by-products" emerges – the fly ashes from combustion of biomass [7].

The by-products from combustion of biomass characterize with diverse and variable physicochemical properties. The most of the waste from production of electricity and/or heat are fly ashes. The fly ashes from combustion of biomass (10 01 03) are a particular kind of waste distinguished by high level of leachability of contaminants and variable chemical composition. The level of leachability of contaminants determines the use of fly ashes in the industry [12].

Fly ashes from combustion of conventional energy sources are commonly used in many branches of industry. Generally in production of building materials and mining technologies. Ashes from combustion of biomass are also utilized in the industry, for example production of building materials.

The side-products from incineration can only be used in production when their physicochemical properties meet specific parameters. One of the important indicators is level of leaching of contaminants. Unexceeding of required levels of leaching of chemical substances can decide on way of utilization.

Taking the above into consideration it is only reasonable to analyze the leachability of side-products from biomass combustion in different kinds of boilers and the possibilities of limitation which could allow to determine potential usage of these kinds of ashes.

This article presents results of research on leachability of chemical contaminants from dry ashes of combustion or co-combustion of biomass. The study of the impact of grinding on the level of leachability was also conducted. Fly ashes from combustion of biomass as well as their mixture with fluidized ash from combustion of hard coal were submitted to grinding process. Afterwards level of leachability was measured on them to determine initial prospect of their utilization in mining technologies. The results of the research on leachability of ashes were compared to PN-G-11011 Mining Norm - Materials For Backfilling And Caulking Of Cavings - Requirements And Tests because fly ashes are commonly used as suspended matter in underground mining [6].

2 Materials used in the research

For this research I used fly ashes from combustion of solid biomass in power production company which were included by the decree of the Ministry of the Environment from 9th of December 2014 for Catalog of Wastes (Dz.U. 2014, poz. 1923), to group 10

[a] grzywa@agh.edu.pl

– the by-products from thermal processes, subgroup 10 01 – by-products from power companies and other enterprises of production of energy by combustion of fuels (excluding group 19).It was also two kinds of by-products from combustion in fluidized boilers and co-combustion in pulverized-fuel boilers.

The research was conducted on two kinds of fly ashes:

a) From combustion of biomass (10 01 03):
- in fluidized boiler with atmospheric circulation (Fig.1a) – SW
- in fluidized boiler with circulating fluidized bed (Fig.1b) – K9,
- in fluidized boiler with bubbling fluidized bed (Fig.1c) – T.

b) From co-incinerated biomass with hard coal (10 01 17):
- In pulverized-fuel boiler (Fig. 1d) – 2MP1.

In the research I also used the ash with code 10 01 82 from combustion of hard coal in fluidized boilers – PF (Fig. 1e), with a low level of leachability of contaminants. This ash was grinded with dry ashes from combustion and co-combustion of biomass in ratio 1:1.

Figure 1. The ashes used in the research

Comparing the fly ashes from combustion of biomass with the ashes from combustion of hard coal, it appears that the qualitative composition of chemical elements of both ashes is very similar. The differences turn up in quantity of particular elements and the compounds in the composition of ashes. In the fly ashes from combustion and co-combustion of biomass the levels of compounds like CaO, MgO, Na_2O, K_2O, P_2O_5 are significantly higher whereas levels of SiO_2, Al_2O_3, TiO_2 are lower compared to ashes from combustion of hard coal [4, 12].

Fly ashes from combustion of biomass characterize by high level of leachability of ions of potassium, chromium, sulfates and chlorides in a scale greatly higher than the levels of ashes from combustion of hard coal in conventional and fluidized boilers. The high level of leachability of contaminants from fly ashes from combustion of biomass depends mostly on the type of fuel, so it can be easily limited by mixing them with ashes from combustion of coal in conventional and fluidized boilers [9].

High content of calcium, alkalis, phosphorus as well as variable and often high content of chlorine can cause intensive corrosion and increase in presence of aggressive sediments in the boiler during the combustion [1,5].

Table 1. The chemical composition of the ashes used in the research, in % [10]

Compound	Type of ash				
	SW	K9	T	2MP1	PF
SiO_2	28,86	62,43	34,06	52,79	51,20
Al_2O_3	2,17	6,21	4,57	22,80	11,27
Fe_2O_3	1,84	2,53	2,90	6,25	5,63
CaO	23,05	8,63	21,13	5,17	4,25
MgO	6,82	3,42	5,64	2,85	3,03
Cl	1,82	0,45	1,19	0,01	b.d.
SO_3	7,12	3,08	9,97	0,41	0,76
Na_2O	0,40	0,46	0,61	1,18	0,90
K_2O	18,11	3,73	13,65	3,06	1,59
Losses of calcination	4,47	4,33	6,71	4,65	0,11

All examined ashes had a high content of SiO_2. Fly ashes from combustion of biomass showed high levels of CaO, K_2O and CaO. It is worth mentioning that the fly ashes had notable levels of SO_3 while in ashes from co-combustion this level was insignificant. On the other hand, in the composition of ashes from co-combustion with hard coal the high content of Al_2O_3 and Fe_2O_3 was observed. And Tests because fly ashes are commonly used as suspended matter in underground mining [6].

3 Analysis of the leachability of contaminants

The research was conducted according to the Norm PN-G- 11011: 1998 Mining - Materials For Backfilling And Caulking Of Cavings - Requirements And Tests.

Subjected for the analysis where the fly ashes from combustion and co-combustion of biomass (SW, K9, T, 2MP1) (Tab. 2), the same ashes after the process of grinding and their mixtures with ash from combustion of hard coal 10 01 82 (PF) in ration 1:1 (Tab. 3).

The results of the research on leachability of ashes were compared with the requirements of the Norm PN-G-11011 Mining Norm - Materials For Backfilling And Caulking Of Cavings - Requirements And Tests (Tab. 4).

DOI: 10.1051/e3sconf/20161000072

Analysed samples of each type of fly ash showed similar pH reaction, between 11,92 and 12,79.

The fly ashes from both combustion and co-combustion characterized by significant leachability of chromium, cadmium, lead, copper, and mercury in comparison to PF ashes (Tab. 2 and 3).

The leachability of sulfate and chloride ions from the fly ashes from combustion and co-combustion of biomass was also a couple or a dozen times higher than in case of the ashes of combustion of hard coal.

Table 2. The leachability of chemical contaminants from the researched ashes, mg/dm³ [9-10]

Leachability	Dry ash				
	SW	K9	T	2MP1	PF
pH	12,79	12,25	12,46	11,96	11,92
Cl^-	1449,02	445,64	658,34	150,32	109,20
K	9446,09	604,31	3160,99	223,19	28,21
Na	82,91	25,36	67,36	26,00	32,62
Cd	0,13	0,34	0,07	0,45	0,01
Cu	164,24	446,99	201,31	44,16	0,01
Cr	1,63	1,16	1,42	1,55	0,02
Hg	0,37	0,14	0,04	1,00	0,02
SO_4^{2-}	5745,81	1436,94	4779,65	790,51	670,80

Table 3. The leachability of chemical contaminants from the researched ashes after grinding separately, mg/dm³

Leachability	Grinded ashes			
	SW	K9	T	2MP1
Cl^-	1329,3750	428,3542	658,0875	138,2550
K	7677,4288	525,1561	3160,9101	199,5052
Na	71,2992	23,2680	60,3095	25,6060
Cd	0,1200	0,1390	0,0660	0,3070
Cu	132,1490	287,6190	156,3770	37,7210
Cr	1,2630	0,8770	0,4870	0,8710
Hg	0,5150	<	0,0400	0,0880
SO_4^{2-}	3391,0427	1293,5499	4906,7672	767,7367

The high leachability of contaminants like potassium, chlorides and sulphates confirms the results of researches conducted by other authors [2,9-11].

Achieved values of leachability of fly ashes from combustion of biomass greatly exceeded the requirements of the Norm PN-G-11011 Mining Norm - Materials For Backfilling And Caulking Of Cavings - Requirements And Test. The process of grinding together with PF ashes caused the decrease in leachability to below 1000 mg/dm³ only in case of chlorides.

The conducted research reviled very high leachability of ions of potassium from fly ashes from combustion of biomass (9446,0938 mg/dm³), co-combustion (223,1981 mg/dm³) and combustion of hard coal in fluidized boilers (10 01 82) (28,21 mg/dm³).

Table 4. The leachability of chemical contaminants from the researched ashes after grinding with ash PF (10 01 82), mg/dm³

Leachability	Ashes grinded with ash PF (10 01 82)			
	SW	K9	T	2MP1
Cl^-	635,1458	194,9750	464,3950	56,7200
K	3028,0372	240,1898	2278,4060	125,8354
Na	43,1195	17,6513	47,8768	26,2035
Cd	0,1280	0,3070	0,0760	0,4240
Cu	70,4970	163,9590	115,3860	41,0440
Cr	0,5940	0,9520	0,5740	1,1440
Hg	0,2680	0,0950	0,0400	0,0310
SO_4^{2-}	3135,1305	601,5172	2345,9992	287,1812

Table 5. The levels of leachability permitted by the Norm PN-G-1101, mg/dm³

Type of chemical contaminant	Permitted levels of leachability by Norm PN-G-11011
Elements	
zink (Zn)	2,0
copper (Cu)	0,5
lead (Pb)	0,5
arsenic (As)	0,2
mercury (Hg)	0,02
cadmium (Cd)	0,1
chromium (Cr)	0,5
chlorides (Cl–)	1000,0
sulphates (SO42–)	500,0

Grinding of the fly ashes and the mixture of ashes SW, K9, T and 2MP1 with PF ash allowed the decrease of leachability of potassium, sodium, copper, chromium, sulphates and chlorides but not to a degree that would be compliant with the Norm (Tab. 5). The process of grinding of fly ashes alone didn't give big impact on the level of leachability, more satisfactory results of reduction of leachability were shown during grinding of mixture of ashes from combustion and co-combustion of biomass.

4 Conclusion

The examined fly ashes from combustion and co-combustion of biomass differ from ashes from combustion of hard coal as to the chemical composition and leachability.

The contaminants for which the leachability was significant was potassium, sodium, chlorides and sulphates. The levels of leachability for these contaminants greatly exceed the values of ashes from combustion of hard coal.

The process of grinding of fly ashes and grinding them together with ashes from combustion of hard coal with the low level of leachability decreased the amount of leaching components. But, despite the reduction, values of leachability for most of the contaminants highly exceed the requirements of the Norm to PN-G-11011 Mining Norm - Materials For Backfilling And Caulking Of Cavings - Requirements And Tests.

The research were conducted as a part of deans grant 15.11.100.912.

References

1. T. Hardy, W. Kordylewski, K. Mościcki, *Zagrożenie korozją chlorkową w wyniku spalania i współspalania biomasy w kotłach*, (Instytut Techniki Cieplnej i Mechaniki Płynów Politechniki Wrocławskiej 2010)
2. A. Jaworek, T. Czech, A. Sobczyk, A. Krupa, J. of Electrost. **71**, 165–175 (2013)
3. M. Kosior- Kazberuk, *Nowe dodatki mineralne do betonu, Budownictwo i inżynieria środowiska*, (ISSN: 2081-3279 , 2011)
4. J. Małolepszy, E. Tkaczewska, *Wpływ popiołów lotnych ze współspalania węgla kamiennego i biomasy na proces hydratacji i właściwości cementu*, (Materiały konferencji Dni Betonu, Wisła 2006)
5. W. Mokrosz, *Migracja związków chloru z biomasy w procesie spalania oraz ich wpływu na procesy korozyjne i eksploatacyjne kotłów*, (MOKROSZ Sp. z o.o, Politechnika Śląska 2014)
6. PN G-11011:1998 Mining – *Materials for Backfilling and Caulking of Cavings – Requirements and Tests*
7. J. Poluszyńska, Możliwości zastosowania popiołów ze spalania biomasy w gospodarowaniu osadami ściekowymi, Prace Instytutu Ceramiki i Materiałów Budowlanych 13 (2013)
8. Raport GUS, Energia ze źródeł odnawialnych, Warszawa (2014)
9. A. Uliasz-Bocheńczyk, A. Pawluk, A. Sierka, Gosp. Sur. Min. – Min. Res. Manag. **31**(3), 145–156 (2015)
10. A. Uliasz-Bocheńczyk, A. Pawluk, Min. Res. Manag. (2016) [to be published]
11. S. Vassilev, D. Baxter, L. Andersen, C. Vassileva, , Fuel **105**, 40-76 (2013a)
12. S. Vassilev, D. Baxter, L. Andersen, C. Vassileva, , Fuel **105**, 19-39 (2013b)

Analysis of Passenger Car Emission Factors in RDE Tests

Jacek Pielecha[1,a], Jerzy Merkisz[1], Jarosław Markowski[1] and Remigiusz Jasiński[1]

[1]*Poznan University of Technology, Institute of Combustion Engines and Transport, 60-965 Poznan, Poland*

Abstract. The article presents a study on emission measurements in passenger cars in tests conducted under real traffic conditions – Real Driving Emissions using a Portable Emission Measurement System type of equipment. A special feature of the outlined RDE tests is that they were performed in Polish road conditions, and thus their parameters may differ from their counterparts adopted in most European Union countries. Based on the findings vehicle emission conformity factors were developed, characterized as the fractional increase (or decrease) of traffic emissions during the homologation test or under normal operation conditions in relation to the emission limit standards (for chosen emission class) of the vehicle. Conducted research and the calculated conformity factors allowed for the environmental impact assessment of the vehicles of various emission classes, while also allowing early actions to restrict the emissions of selected components in passenger vehicles. The methods and measures used can also be applied to other types of vehicles (e.g. heavy duty or off-road vehicles) or vehicles powered by other fuels.

1 Introduction

A fundamental factor leading to technology development in all branches of industry is the need to reduce its negative impact on natural environment. Transport is one of the most dynamically changing fields of economy, particularly due to the ever-changing exhaust emission levels. Increasing emphasis is put on the measurements of exhaust emissions, from combustion engines of machines and vehicles especially under actual operating conditions.

These measurements much better reflect the environmental situation than the tests only simulating actual conditions of operation used thus far or stationary tests. These measurements became possible due to a significant advancement in the area of measurement techniques in recent years. Passenger vehicles are subject to intermittent inspections at inspection stations, however the inspections are carried out for a limited (much lower than in real traffic) operating range of the engine. The assessment of the environmental performance of the vehicles may also be carried out on a chassis dynamometer. It is quite opposite for heavy-duty vehicles, in which case, in order to assess the in-service emissions, the engine must be removed from the vehicle for testing. Statistical research on a chassis dynamometer may sometimes be performed on vehicles of different applications, while the tests used in the methodology designed to assess the emissions are not at all different. The proposal for tests utilizing portable measurement systems is a universal solution as it can be used for vehicles of different applications in which the same engines were fitted.

The growing number of vehicles in the world and the resultant natural environment pollution leads to a growth in the emission related requirements. Current level of technology advancement in all the fields of industry including transport results in a growth of the requirements for the emission measurement devices. In order for these requirements to be fulfilled to comply with the ever-changing regulations a concentration of industry became necessary in this matter. The exhaust emission tests are a sophisticated process. Modern analyzers of the emissions require special laboratory conditions and the homologation procedures comprise tests on chassis and engine dynamometers that do not entirely reflect the emissions under real operating conditions. The latest results of on-road tests show that in relation to some exhaust components the emission is higher by several hundred percent for both the gaseous compounds and particulate matter. Hence, we can see a trend attempting at a formalization and enforcement of the on-road emission testing.

Recent research on pollutant emissions under real traffic conditions performed using mobile devices [1, 2] reflects the state of ecological vehicles very well. Most attention is paid to the possibility of using such research to calibrate engines [3], in such a way so as to limit emissions not only during the emission test [4], but also over the whole engine operating range [5, 6]. Comparative laboratory testing conducted on chassis dynamometers [7] has revealed the compliance

[a] Corresponding author: jacek.pielecha@put.poznan.pl

of vehicles with gasoline and Diesel engines with exhaust emissions standards, while testing carried out under real driving conditions has shown that vehicles with Diesel engines significantly exceed NO_x emissions limits [8-10]. Attention is drawn to the importance of particulate exhaust emissions, mainly in the form of nanoparticles emitted from internal combustion engines (gasoline engines with direct fuel injection, as well as Diesels equipped with a particulate filter (DPF)) [11]. The results of such studies are currently not presented independently, but have been confirmed by publications which encompass investigations spanning several years [12] and comprehensive summaries of vehicles tested under real traffic conditions [13].

2 Polish automotive market analysis

The number of motor vehicles in Poland, particularly passenger vehicles (4/5 of total vehicles), has grown significantly in recent years. In 2015 there were 20.1 million passenger vehicles with the European (European Union) average of 9.3 million per country (Fig. 1) [14].

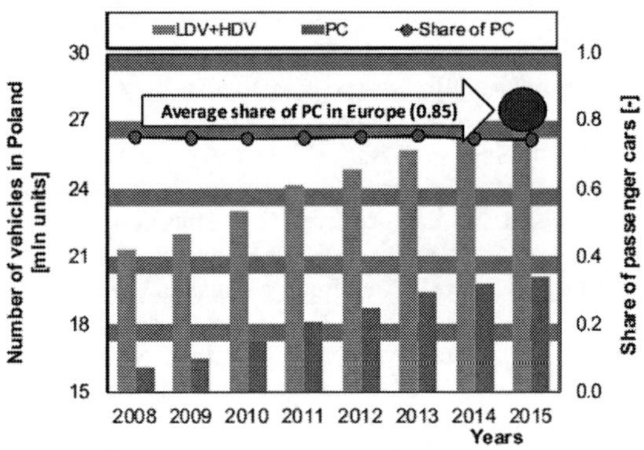

Figure 1. Motor vehicles in Poland [14, 15].

In Poland in 2015, 800 thousand vehicles were registered for the first time including 720 thousand passenger vehicles and 94 thousand heavy-duty trucks and road tractor units (Fig. 2). The increase in the vehicle fleet was mainly caused by private import. In 2013 there were 599 passenger vehicles per 1000 citizens (average for European countries amounts to 564).

Mass import of used vehicles adversely influences the age structure of the vehicles. Ever since the accession of Poland to the European Union, the imported vehicles are mostly older than 4 years of age (92% in 2014), which is why the vehicle fleet in Poland is growing old. In 2014 there were 9.5% of 5 year old or younger vehicles – 0.7% less than in the previous year (Fig. 3) [14]. In 2003, prior to the accession of Poland to the European Union, when old vehicles import limitations applied there were over 18% of vehicles of 5 years old or younger. The share of vehicles from 6 to 15 years old in 2013 was 19.2% (Euro 2 and Euro 3) and 16 –20 year old vehicles – 21.1% (Euro 1) and in both cases this share was over 4% greater than in 2012. There were approx. 7 times more vehicles exceeding 20 years of age in 2012 than vehicles of the age of up to 2 years (Fig. 4).

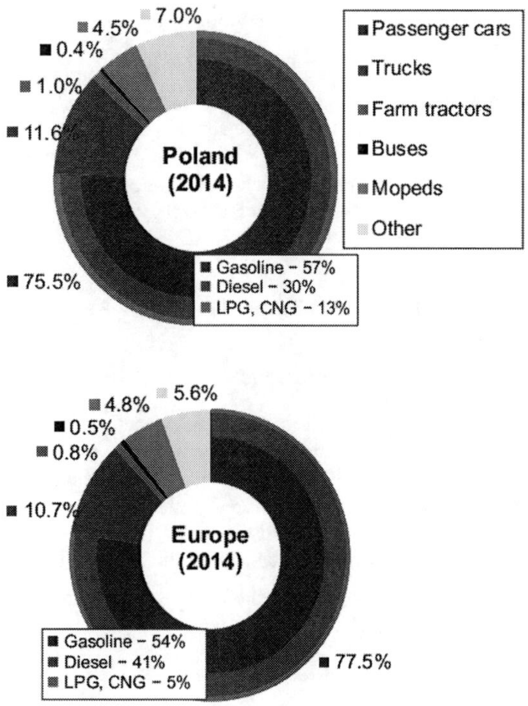

Figure 2. Vehicle structure in Poland and Europe in 2014 (based on [14, 15]).

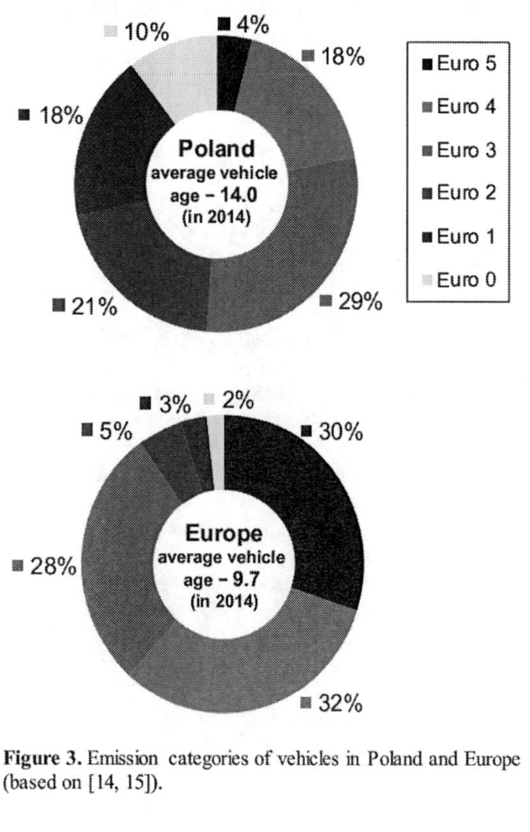

Figure 3. Emission categories of vehicles in Poland and Europe (based on [14, 15]).

In Europe the share of vehicles of up to 5 years constitutes approx. 30% (Euro 5) and at the same time the share of vehicles in the range of 5–10 years (Euro 4) is not much higher. There are not many vehicles older than 10 years – approx. 37% (Euro 3 – Euro 0). The consequence of maintaining an old vehicle fleet (more than half of the vehicles comply to the Euro 2 and Euro 3 exhaust emission standards) is borne not only by the vehicle owners but also the society. Old and deteriorated vehicles are less safe and their environmental performance is poor – they generate much more pollution than new vehicles.

3 Research methodology

3.1 Real Driving Emissions

A variety of conditions that have an impact on the level of emissions occur under real driving conditions. These variables cause results that are not identical to the results obtained during emission tests performed on a chassis test bench, where the conditions are fully repeatable. The relatively worse repeatability of the results obtained is primarily affected by:

- environmental conditions (variable temperature, pressure, humidity, wind rate, rain, snow, etc.),
- the quality of the road surface,
- traffic congestion,
- the driver's behavior: aggressive, neutral, eco-driving.

The authors propose the introduction of emission correction coefficients (as initially presented in a publication from 2009 [16]), being the multiple of the increase (or decrease) of emissions under real driving conditions in relation to the type approval test emissions results. Such coefficients (*CF* – conformity factor) for regulated emissions are defined as follows:

$$CF_j = \frac{E_{\mathrm{RDE},\,j}}{E_{\mathrm{norm},\,j}} \qquad (1)$$

where:

j - the harmful compound for which the conformity factor was specified,

$E_{\mathrm{RDE},j}$ - road emission obtained under real driving conditions ([mg/km]),

$E_{\mathrm{norm},j}$ - value of emission limit in the applicable emission standard ([mg/km]).

The conformity factor can usually be calculated as the value relating to the entire test, which is the ratio of emission of the pollutant in the road test performed under real driving conditions to the normative value. The conditions under which the road tests (Fig. 4) were carried out were characterized by the following properties (as defined in EU legislation on RDE testing):

- 33% of the total time performed under urban conditions, in which the vehicle's speed does not exceed 50 km/h (with cold start),
- 33% of the total time performed under extra-urban driving, in which vehicle's speed is in the range 50-90 km/h,
- 33% of the total time performed in a highway driving in which the vehicle's speed is in the range 90-130 km/h (not to exceed 140 km/h).

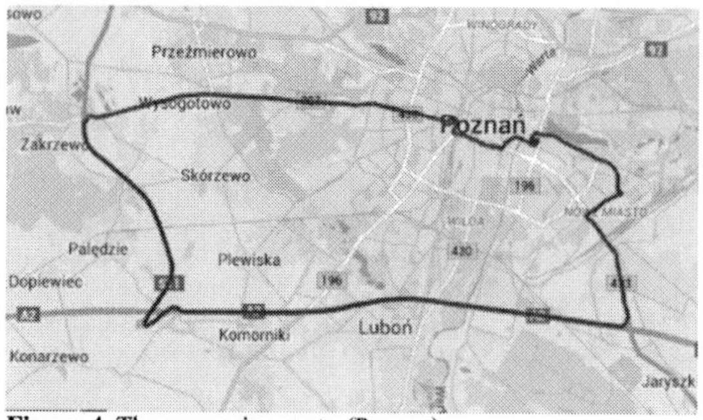

Figure 4. The measuring route (Poznan).

3.2 Tested vehicles

The results of emission testing presented here are the fruit of a 9-year program connected with the study of cars under real traffic conditions. Tests were made on a group of over 150 cars with differing technical parameters (Fig. 5). The study involved vehicles powered by both Gasoline and Diesel engines of various displacements (from 0.7 L to 3.0 L) and fitted with different emission aftertreatment systems. Data presented in this paper relate to road tests performed on different cars with various mileages using various fuel supply systems (various layouts for both gasoline and Diesel engines). The results are divided into categories by emission standard (Euro 4, Euro 5 and Euro 6), while results for vehicles fulfilling lower, earlier emission standards are not included. The road emissions results acquired were ranked from lowest to highest and only extreme values were presented (results obtained under conditions which did not fulfill the requirements relating to on-road testing described in later sections of this paper were rejected).

Figure 5. Vehicles used for tests (examples).

3.3 Research apparatus

The testing apparatus is presented in Fig. 6a. A portable Semtech DS analyser was used for the measurement of exhaust emissions from vehicles (Fig. 6b). It allowed measurements of carbon monoxide, carbon dioxide, hydrocarbons, non-methane hydrocarbons and nitrogen oxides exhaust emissions. A portable AVL M.O.V.E PM (Micro Soot Sensor) analyser was used to measure particle mass (Fig. 7a), while a PMP-compliant AVL condensation particle counter was used to measure particle number emissions (Fig. 7b). The emissions measurement equipment had a maximum mass of 116 kg (Gas PEMS – 25 kg; PM PEMS – 45 kg, PN Counter – 25 kg), together with an additional power supply (generator) – 21 kg.

a) b)

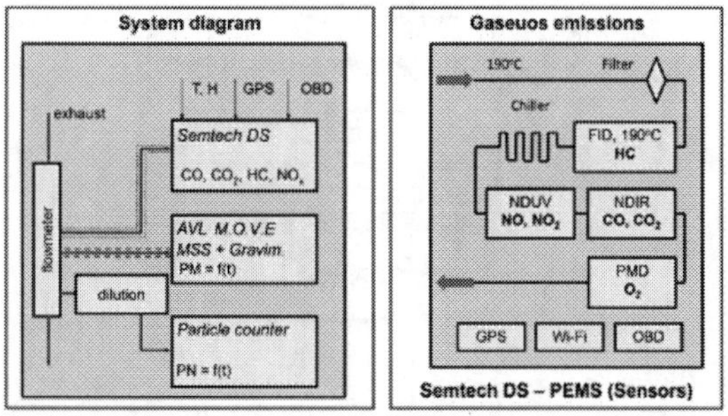

Figure 6. The measuring systems used for testing under real traffic conditions.

a) b)

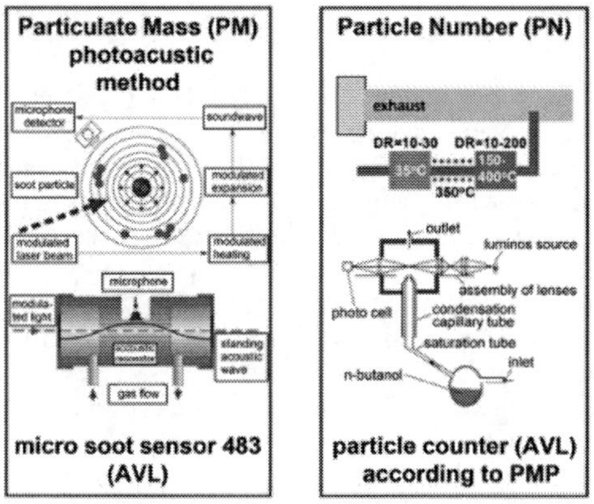

Figure 7. Measurement system used for testing particle emissions: a) analyser for measurement of the particle mass concentration – AVL MOVE (micro soot sensor + gravimetric), b) particle counter (AVL) for measurement of particle number.

4 Results and analysis of road tests

Values of emission correction for pollutants obtained in road tests were as follows:

For cars powered with Gasoline engines (Fig. 8):

- for those meeting the Euro 4 standard (vehicles with multi-point fuel injection and displacement from 0.7 L to 3.0 L were tested): the correction factor for on-road emission of carbon monoxide CF_{CO} was between 0.05 and 0.1 with a median value of 0.08; the conformity factor median value for hydrocarbons CF_{THC} was 0.04 and the conformity factor for nitrogen oxides CF_{NOx} was between 0.08 and 0.2 (with a median of 0.15). The conformity factor for particle mass did not exceed a value of 0.15, with a median value of 0.07 (the 5 mg/km emission limit was adopted as an allowed value for particle mass emission – the same as for the Euro 5 standard);

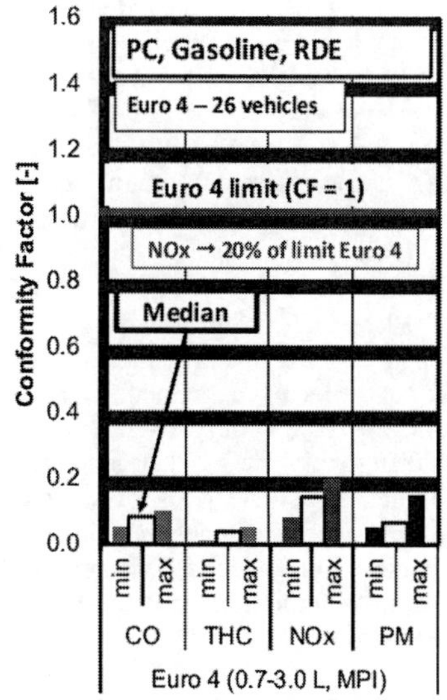

Figure 8. Values of passenger car emission conformity factors (*CF*), for vehicles powered by gasoline engines meeting emission standards of Euro 4 (data acquired from 26 vehicles).

- for those meeting the Euro 5 standard (vehicles with direct injection and displacement from 0.7 L to 2.5 L were tested; Fig 9): the conformity factor of carbon monoxide CF_{CO} was between 0.03 and 0.08 with a median value of 0.06; the median value of the CF_{THC} was 0.05 (minimum and maximum values were 0.03 and 0.08, respectively), conformity factor for nitrogen oxides CF_{NOx} were between 0.2 and 0.8 (with a median value of 0.41); the conformity factor for particle mass did not exceed 0.5 (with a median value of 0.15); for vehicles other than those fulfilling the Euro 5 standard the limit of 5 mg/km was applied;

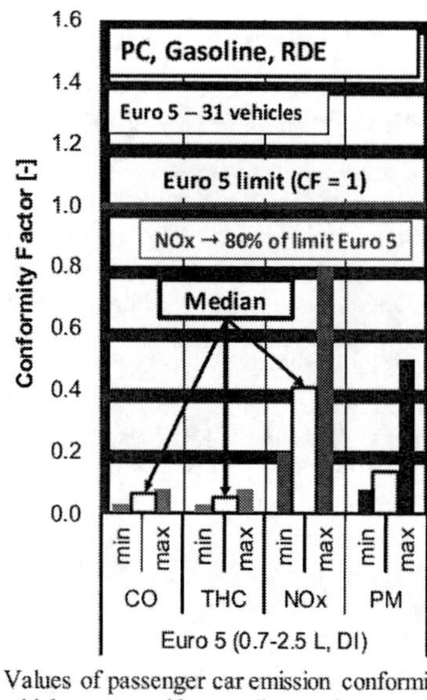

Figure 9. Values of passenger car emission conformity factors (*CF*), for vehicles powered by gasoline engines meeting emission standards of Euro 5 (data acquired from 31 vehicles).

- those meeting the Euro 6 standard (Fig. 10, vehicles with direct fuel injection and engine displacement from 1.0 L to 2.5 L were tested – without gasoline particulate filters): the median value of the conformity factor for on-road emission of carbon monoxide CF_{CO} was 0.09 (with extreme values of 0.02 and 0.11); the median value of the conformity factor of hydrocarbons CF_{THC} was 0.17 (with border values of 0.04 and 0.02); the conformity factor for nitrogen oxides CF_{NOx} was between 0.12 and 0.7 (with a median value of 0.46); the conformity factor of particulate matter emission CF_{PM} did not exceed 0.6 (a median value of 0.42); the conformity factor of particle number behaved somewhat differently – the median value was 1.06 (with extreme values of 0.2 and 1.5); for vehicles powered with Euro 6c Gasoline engines the emission value for particle number was exceeded by more than 50% of the value of the limit (i.e. the highest observed value of CF_{PN} was 1.5).

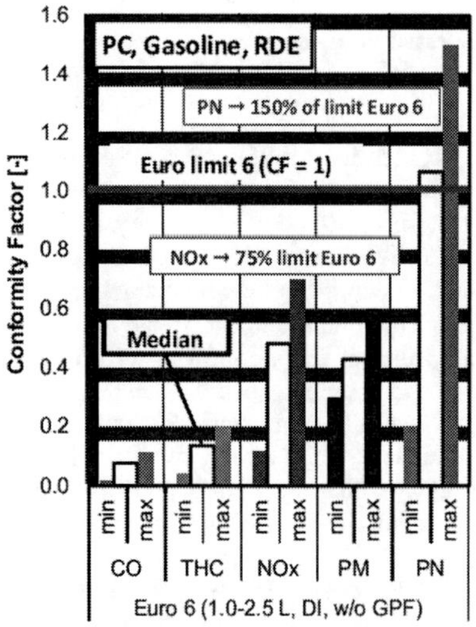

Figure 10. Values of passenger car emission conformity factors (*CF*), for vehicles powered by gasoline engines meeting emission standards of Euro 6 (data acquired from 12 vehicles).

For cars powered by Diesel engines (a deviation from the legislation is used here – THC and NO_x emissions results are reported separately here, although EU legislation specifies a limit in terms of THC+NO_x for vehicles with Diesel engines; the THC limit is equal to (THC+NO_x) – NO_x):

- those meeting the Euro 4 standard (vehicles with displacements between 1.3 L and 2.2 L w/o Diesel particulate filter, Fig. 11): the conformity factor for carbon monoxide CF_{CO} was between 0.0 and 0.05 with a median value of 0.04; the median value of hydrocarbon emission CF_{THC} was 0.4; the conformity factor for nitrogen oxides CF_{NOx} was between 0.8 and 1.4 (maximum exceed over the standard of 40% was observed and a median value of 1.0 was noted). The conformity factor for particulate matter emission did not exceed a value of 1.3 with a median value of 0.7 (so an excess of 30% was observed compared to the emissions standard);

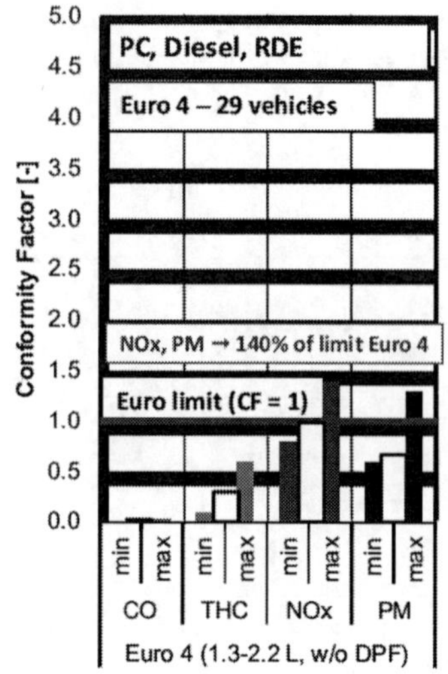

Figure 11. Values of passenger car emission conformity factors (*CF*), for vehicles powered by Diesel engines meeting the Euro 4 emission standards (data acquired from a total of 29 vehicles).

- those meeting the Euro 5 standard (vehicles with displacements ranging from 1.3 L to 2.4 L with Diesel particle filter, Fig. 12); the conformity factor for on-road emission of carbon monoxide CF_{CO} was between 0.01 and 0.03 with a median value of 0.02; the median value of the conformity factor for hydrocarbons CF_{THC} was 0.5 (with minimum and maximum values of 0.3 and 0.7); the conformity factor for nitrogen oxides CF_{NOx} was between 1.4 and 3.7 (the median value was 3.2 – three times the amount allowed in the Euro 5 standard), the conformity factor for particle mass did not exceed a value of 0.2 (with a median value of 0.15); PN emissions were not measured for those vehicles. For this emission class, the greatest excess observed was for nitrogen oxides – exceeding three times the permitted value;

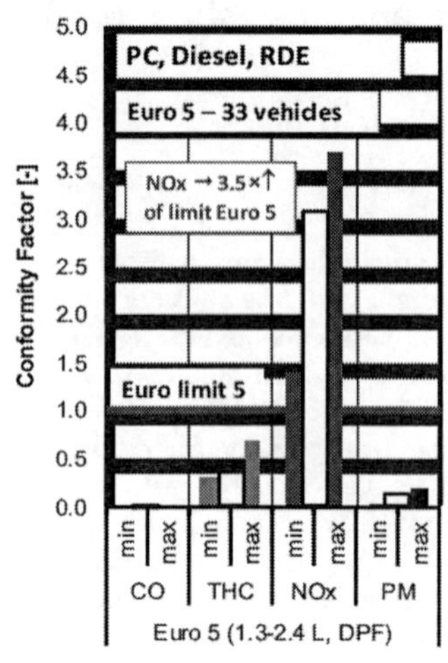

Figure 12. Values of passenger car emission conformity factors (*CF*), for vehicles powered by Diesel engines meeting the Euro 5 emission standards (data acquired from a total of 33 vehicles).

- those meeting the Euro 6 standard (vehicles with displacement ranging from 1.3 L to 3.0 L with Diesel particulate filter, Fig. 13): the median value for the on-road emission conformity factor of carbon monoxide CF_{CO} was 0.2 (with extreme values of 0.02 and 0.3); the median value of hydrocarbon emission factor CF_{THC} was 0.5 (minimum and maximum values of 0.3 and 0.7); the conformity factor for emission of nitrogen oxides CF_{NOx} was between 1.1 and 4.8 (with the median value being 3.4); the conformity factor for particle mass CF_{PM} did not exceed a value of 0.3 (the median value being 0.26); the value for particle number followed a similar trend with a median value of 0.71 (with border values of 0.3 and 0.9); for Diesel engines in the Euro 6 emission class, high excess emissions of nitrogen oxides were observed (more than 3.5 times the limit – with a 4.8 maximum excess value).

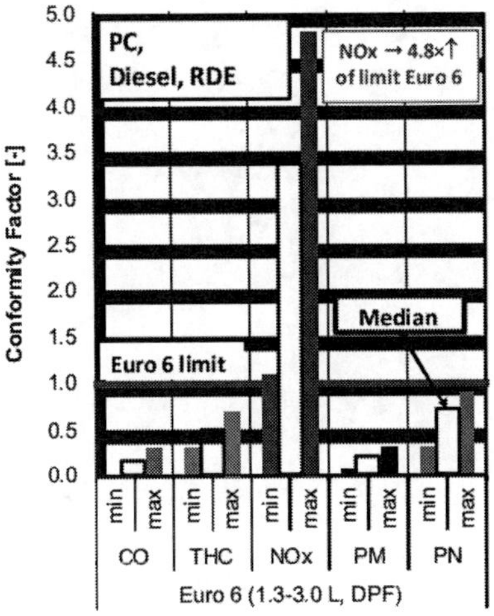

Figure 13. Values of passenger car emission conformity factors (*CF*), for vehicles powered by Diesel engines meeting the Euro 6 emission standards (data acquired from a total of 18 vehicles).

Conclusions

The experimental work and associated numerical analyses presented in this paper allow the following conclusions to be drawn.

Conformity factors for gasoline engines in the Euro 4 and 5 emission classes in road tests (depending on vehicle weight and the level of the engine and aftertreatment technology applied) are CF_j = 0.8 for all emission components. For gasoline direct injection vehicles meeting the Euro 6 standard, the conformity factor values obtained for particle number exceeded the values obtained over type approval test cycles. The average value of the conformity factor for emission of particle number in road tests was CF_{PN} = 1.06 (with a maximum value of 1.5).

The values of the conformity factor for vehicles powered by Diesel engines were found to be different: it has been shown that the values of the correction factors for emission of carbon monoxide and hydrocarbons are lower than 1 (i.e. the emissions limits are met), but for on-road emission of nitrogen oxides values between CF_{NOx} = 0.7–4.8 were observed (only extreme values are shown here). The same goes for particle mass and number (values of the conformity factor were as follows: CF_{PM} = 0.2–1.3 and CF_{PN} = 0.3–0.9).

In European future Portable Emission Measurement Systems will be necessary to measure exhaust emissions in real traffic conditions. The main issue of RDE is nitrogen oxides emissions but European Commission is also interested in particles measurement especially on gasoline direct injection. Wide variety of boundary conditions in RDE test must be considered. These influencing parameters have to be taken into account when evaluating RDE data and are addressed by route criteria, data evaluation tools and conformity factors. In 2016 the RDE Monitoring phase shall start for new vehicle types. Current RDE draft introduces the 'not-to exceed' regulatory concept which will be implemented starting in 2017. The „not-to-exceed" values are simply the Euro 6 limit multiplied by a conformity factor. Temporary conformity factor for nitrogen oxides emissions was suggested at the level of 2.1. Considering measurement uncertainty of PEMS the final conformity factor was suggested on the level of 1.0 plus a margin of 0.5.

References

1. T. Bougher, I. Khalek, S. Trevitz, M. Akard, Verification of a gaseous portable emissions measurement system with a laboratory system using the Code of Federal Regulations Part 1065. SAE Paper Series 2010-01-1069 (2010)
2. M. Weiss, P. Bonnel, R. Hummel, U. Manfredi, R. Colombo, G. Lanappe, Analyzing On-road Emissions of Light-Duty Vehicles with Portable Emission Measurement Systems. Publications Office of the European Union, Technical Report EUR 24697 EN (2013)
3. P. Bielaczyc, A. Szczotka, A. Swiatek, J. Woodburn, A comparison of ammonia emission factors from Light-Duty Vehicles operating on gasoline, liquefied petroleum gas and compressed natural gas. *SAE Int. J. Fuels Lubr.* **5**, 2 (2012)
4. J. Merkisz, J. Pielecha, R. Jasiński, Real driving emissions testing of vehicles powered by compressed natural gas. SAE Paper Series 2015-01-2022 (2015).
5. J. Merkisz, J. Pielecha, P. Bielaczyc, J. Woodburn, Analysis of emission factors in RDE tests as well as in NEDC and WLTC chassis dynamometer tests. SAE Paper Series 2016-01-0980 (2016)
6. J. Merkisz, J. Pielecha, S. Radzimirski, *New trends in emission control in the European Union* (Springer Tracts on Transportation and Traffic, New York, 2014)
7. G. Fontaras, V. Franco, P. Dilara, G. Martini, U. Manfredi, Development and review of Euro 5 passenger car emission factors based on experimental results over various driving cycles. *Science of the Total Environment* **468-469** (2014)

8. G. Kadijk, P. van Mensch, J. Spreen, Detailed investigations and real-world emission performance of Euro 6 Diesel passenger cars. TNO Report, R10702 (2015)
9. M. Weiss, P. Bonnel, R. Hummel, N. Steininger, A complementary emissions test for Light-Duty Vehicles: assessing the technical feasibility of candidate procedures (No. EUR 25572 EN). JRC Scientific and Policy Reports, European Commission **1**, 56 (2013)
10. L. Yang, V. Franco, P. Mock, R. Kolke, S. Zhang, Y. Wu, J. German, Experimental assessment of NOx emissions from 73 Euro 6 Diesel passenger cars. *Environmental Science & Technology* **49**, 24 (2015)
11. C.L. Myung, S. Park, Exhaust nanoparticle emissions from internal combustion engines: a review. *International Journal of Automotive Technology* **13**, 1 (2012)
12. Y. Chen, J. Borken-Kleefeld, Real-driving emissions from cars and light commercial vehicles – results from 13 years remote sensing at Zurich. *Atmospheric Environment*, **88** (2014)
13. V. Franco, F. Posada Sánchez, J. German, P. Mock, White paper: real-world exhaust emissions from modern diesel cars. Part 2: detailed results. The International Council on Clean Transportation (2014)
14. Automotive Industry, Report 2015, Polish Automotive Industry Association (2016)
15. European Vehicle Market Statistics Pocketbook 2013. International Council on Clean Transportation Europe (2014)
16. J. Merkisz, J. Pielecha, W. Gis, Gasoline and LPG vehicle emission factors in a road test. SAE Technical Paper Series 2009-01-0937 (2009)

 DOI: 10.1051/e3sconf/20161000074

Tantalum recycling from waste of electrical and electronic equipment

Andrzej Piotrowicz[1,a] and Stanisław Pietrzyk[1]

[1]*Department of Physical Chemistry and Metallurgy of Non-Ferrous Metals, AGH University of Science and Technology, Cracow, Poland*

Abstract. The possibility of tantalum recycling from waste of electrical and electronic equipment was investigated. Study was carried out using basic physical and chemical methods, ie. mechanical separation via crushing, leaching of silver layer in diluted HNO_3, grinding and oxidation of anodes and thermic reduction with metallic reducing agent. A recovery rate of anodes was determined at 96%, and recycling efficiency of tantalum to pure form was determined more than 50%. Also was made mass balance.

1 Introduction

The necessity of miniaturization of electrical and electronic equipment led to the search for the capacitors on the performance of providing relatively high electrical capacity and stability under working conditions. Metallic tantalum is used for the manufacture of capacitors, because from this metal can be produced a dielectric layer with a high electric capacity.

Typical tantalum capacitor (Figure 1), otherwise Ta and Ta_2O_5, also consists of other factors, such as: MnO_2 (solid electrolyte), Ag (high chemical resistance layer), graphite (layer between MnO_2 and Ag), Sn and Ni (components of solder and wires), epoxy resin (protective layer against mechanical damage). Not only the form of tantalum (pure metal and oxide), but also other above-mentioned components, determines the process of recycling of tantalum capacitors.

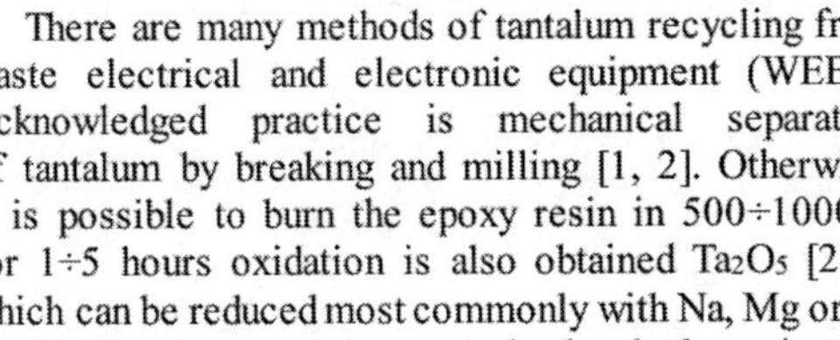

Figure 1. Tantalum capacitors (left), broken capacitor (right): tantalum anode (A), epoxy resin (B), silver foil (C), wires (D)

There are many methods of tantalum recycling from waste electrical and electronic equipment (WEEE). Acknowledged practice is mechanical separation of tantalum by breaking and milling [1, 2]. Otherwise, it is possible to burn the epoxy resin in 500÷1000^{0}C for 1÷5 hours oxidation is also obtained Ta_2O_5 [2-4], which can be reduced most commonly with Na, Mg or Ca [4, 5]. Tantalum capacitors can be leached e.g. in aqua regia, concentrated HCl, NaOH or KOH [1]. The leaching capacitors by using iron chlorides ($FeCl_x$) tantalum chloride is obtained, which is then reduced with Mg [2]. Other methods, such: exfoliation of printed circuit boards containing capacitors [6], tantalum recovery by Cu-Fe-C alloy doped Fe_2O_3 [3], the electrochemical reduction of dissolved tantalum in molten salt by dysprosium ions [5], are reported.

Typically, the tantalum recycling from WEEE (mainly form printed circuit boards) is a multistep process, and the final product may be impure with other elements; therefore, it has still not been development a simple and effective recycling process.

The aim of this study was to develop a process of tantalum recycling from WEEE. Special attention was paid to the use of diluted solutions, low temperature and safe reagents.

2 Experimental

Development process of tantalum recycling from WEEE consists the following steps:

1. coarse breaking the capacitors and grinding the tantalum anodes,
2. leaching anodes in diluted HNO_3,
3. grinding the anodes,
4. oxidation,
5. pressing the oxidised anodes with reducing agents,
6. thermic reduction.

The material used for this study was tantalum capacitors to though-hole mounting (Figure 1), weighting approx. 210 g.

Leaching solution was 8M HNO_3 (700 ml). Reducing agents were: Al, Mg and Si (all technical grade; 1.344, 1.049, 1.815 g, respectively).

2.1 Coarse breaking and grinding

Capacitors were broken (Figure 1), anodes were mechanically separated from wires and resin and grinded. Due to the generation of harmful combustion products of epoxy resin [2-4] was not used from burning process.

[a] Corresponding author: andpio@agh.edu.pl

DOI: 10.1051/e3sconf/20161000074

2.2 Leaching

Leaching was primarily aimed at get rid of Ag from the anodes:

$$4Ag_{foil} + 6HNO_3 \rightarrow 4AgNO_{3dissolved} + \uparrow NO + \uparrow NO_2 + 3H_2O \quad (1)$$

Efficient leaching of Ag requires concentration of HNO_3 above 3M: the best in the range 8÷10M. In addition, the epoxy resin pieces float on the solution surface - they were mechanically removed.

Leaching was lasted 12 h.

2.3 Grinding the anodes

Deprived of Ag and most of resin, anodes were grinding. The purpose of grinding was to allow uniform for future oxidation.

2.4 Oxidation

After oxidation was obtained anodes to Ta_2O_5 "concentrate" [1-3]:

$$2Ta_{anodes} + 5O_2 \rightarrow Ta_2O_{5oxidised\,anodes} \quad (2)$$

In addition, the residual epoxy resin reacts with oxygen to form Si and additionally are formed combustion products (CO_2 and H_2O) [1], and graphite layer - to CO_2.

Oxidation was carried in furnace at 900°C for 1h.

2.5 Pressing the oxidised anodes with reducing agents

Reduction conditions must be such that the material containing Ta_2O_5 (oxidised anodes) had a minimal physical contact with bottom and wall of crucible and other impurity sources [4]. For this purpose, the oxidised anodes (per 12 g) together with respectively reducing agents were pressed to moulded pieces.

It has been applied 20% excess of reducing agents relative to the mass resulting from the stoichiometry.

Pressing made in conditions 160÷240 MPa.

2.6 Thermic reduction

Reducing agents were Al, Mg and Si (all reduce of Ta_2O_5 in process conditions, have high affinity to oxygen and are safe to use). Reductions are as follows:

$$3Ta_2O_{5oxidised\,anodes} + 10Al \rightarrow 6Ta + 5Al_2O_3 \quad (3)$$

$$2Ta_2O_{5oxidised\,anodes} + 5Si \rightarrow 4Ta + 5SiO_2 \quad (4)$$

$$Ta_2O_{5oxidised\,anodes} + 5Mg \rightarrow 2Ta + 5MgO \quad (5)$$

The reduction with Al and Mg were carried at 650°C, and with Si at 1000°C for 9 h duration under Ar atmosphere.

3 Results

3.1 Mass balance

Table 1 shows the mass balance of materials starting from breaking the anodes to thermic reduction.

Tantalum anodes losses were resulted of the layered construction and structure of capacitors. Pressing the oxidised anodes with reducing agents is associated with material loss of 8%.

Table 1. Mass balance in steps: coarse breaking and grinding (1), leaching (2), grinding (3), oxidation (4), pressing (5) and thermic reduction (6) (masses in g)

steps	materials/wastes				
	main stream			wastes / losses / others	
	capacitors 209.554			waste 99.530	
1	anodes 109.485			waste 0.420	
2	leached anodes 106.045				
3	grinded anodes 104.745			anodes not oxidised 8.913	losses 1.300
4	anodes to oxidation 95.832				
	oxidised anodes 106.668			in next step was used 36.000 of oxidised anodes	losses 0.082
5	mixture oxidised anodes &:				
	Al 13.344	Si 13.049	Mg 13.815		
	moulded piece with:			losses 0.089, 0.294, 1.044 (moulded pieces with Al, Si and Mg respectively)	
	Al 13.245	Si 12.755	Mg 12.771		
6	moulded pieces after reduction with:			losses 0.314, 0.247, 0.397 (moulded pieces with Al, Si and Mg respectively)	
	Al 12.931	Si 12.508	Mg 12.374		

3.2. Recovery rate of anodes

Based on data from Table 1, was estimated recovery rate of anodes:

$$\eta_1 = m_{grinded\,anodes} / m_{anodes} \cdot 100\% = 104.745 / 109.485 \cdot 100\% = 95,7\% \approx 96\% \quad (6)$$

where: $m_{grinded\,anodes}$ - mass of grinded anodes, and m_{anodes} - mass of obtained (recovered) anodes (all masses in g).

3.3 Analysis of the chemical composition

The capacitors consists mainly of Ta and Mn, less Ag and Sn (Figure 2A). The confirmation of leaching Ag, Sn and also Mn in HNO_3 is showed in Figure 2. These impurities are passed to solution through chemical and mechanical ways. In oxidised anodes are: Ta_2O_5, manganese tantalum oxide and $Mn(Ta_2O_6)$ (Figure 3), which are oxidation products. Oxidation is resulted in increased concentration of Ta (Figure 2 and Table 2). After thermic reduction, only in case of moulded piece with Mg was obtained pure Ta (Figure 4). The reduction was incomplete, because beside Ta is also formed a complex compound $(Mn_{0.6}Mg_{1.4})Mg_2(Ta_2O_9)$. Mg was also in MnO form (Figure 4). The main impurities are Mn and Mg (Table 3).

Table 2. Comparison of metallic elements contents (% mas.): leached, oxidised and reduced anodes - XRF MINIPAL analysis (without pattern analysis)

	Mn	Ag	Sn	Ta	Mg
leached anodes	16.5	0.4	1.3	81.8	-
oxidised anodes	4.8	0.2	0.2	94.8	-
reduced moulded piece with Mg	4.2	0.2	0.2	85.6	9.7

DOI: 10.1051/e3sconf/20161000074

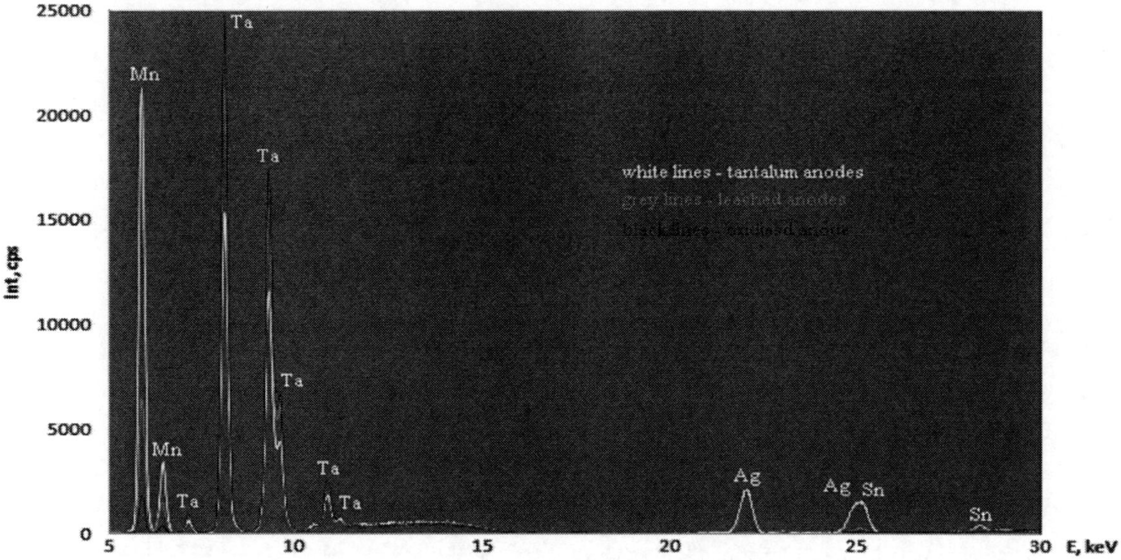

Figure 2. X-ray fluorescence spectroscopy analysis: tantalum anodes (white), leached anodes (grey), oxidised anodes (black)

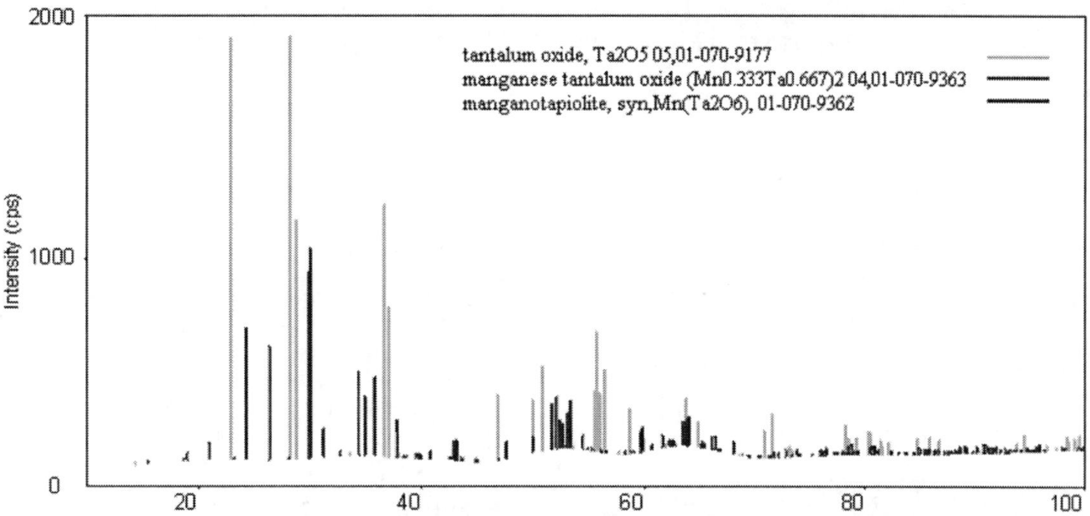

Figure 3. X-ray diffraction analysis of oxidised anodes

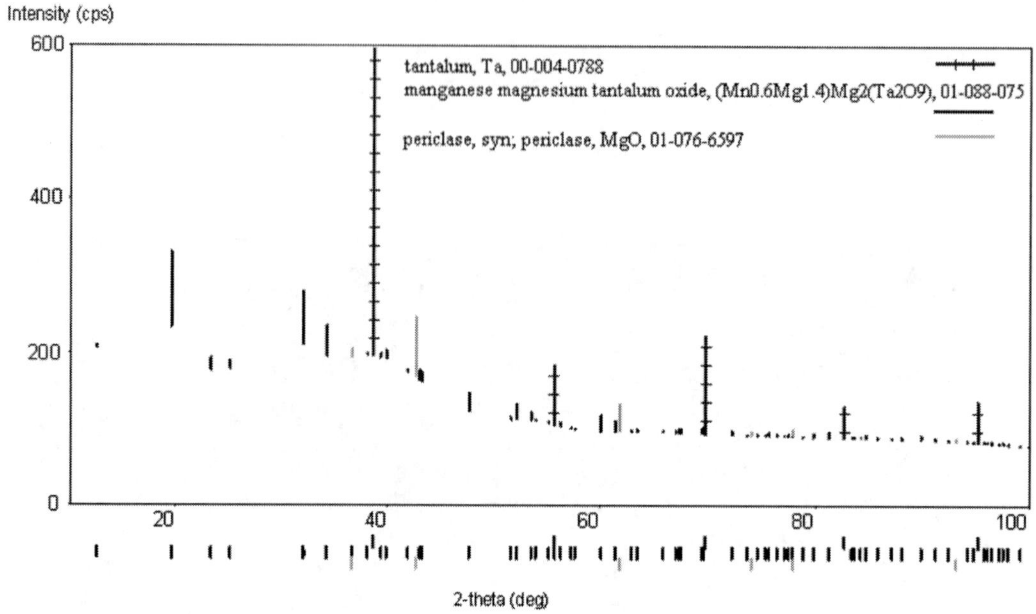

Figure 4. X-ray diffraction analysis of the moulded piece of oxidised anodes with Mg after thermic reduction

3.4 Recycling efficiency of tantalum

Because analysis are without pattern, they do not exclude oxygen in oxidised anodes and reduced moulded pieces. The calculations resulted that in 12 g of oxidised anodes is 1.788 g oxygen and moulded piece with Mg has 1.229 g oxygen. Considering these values, real percentage content Ta in oxidised anodes is equal 81.5% ($\%Ta_{oxidisedanodes}$) and in moulded piece with Mg - 77.1% ($\%Ta_{mp}$), of which 38.6% is occur in bounded form, and 61.4% in pure Ta form ($\%Ta_{pure,mp}$). Mass of Ta, which is contained in oxidised anodes in then:

$$m_{Ta,oxidised\ anodes} = m_{oxidised\ anodes} \cdot \%Ta_{oxidised\ anodes} = 12.000 \cdot 85.1\% = 10.212\ [g] \quad (7)$$

Obtained pure Ta in moulded piece with Mg is then:

$$m_{Ta,mp} = m_{mp} \cdot \%Ta_{mp} \cdot \%Ta_{pure,mp} = 12.374 \cdot 77.1\% \cdot 61.4\% = 5.858\ [g] \quad (8)$$

Recycling efficiency of tantalum can be calculated:

$$\eta_2 = m_{Ta,mp} / m_{Ta,oxidised\ anodes} \cdot 100\% = 5.858 / 10.212 \cdot 100\% = 57.4\% \approx 57\% \quad (9)$$

4 Discussion and conclusions

The presented method of tantalum recycling from WEEE consists of segregation of tantalum anodes and obtaining pure Ta.

Mechanical treatment by breaking the epoxy resin is an alternative to burning. This operation is effective and easy to implement.

This study has shown the ability and efficiency of leaching of Ag layer counting tantalum anodes by using 8M HNO_3.

Recovery rate of tantalum anodes is at 96%. Such a high rate achieved while maintaining the simplicity of this process.

Impurity, which has failed to get rid in a satisfactory degree, is Mn. It should be investigate the possibility of get rid of MnO_2 layer, for example by several leaching in various and fresh solutions. At each stage of process, remained a part of Mn in material. It proves, than MnO_2 has high resistance to diluted by HNO_3.

Only Mg partially reduced Ta_2O_5 in moulded piece to pure Ta. In this case, recycling efficiency of Ta is estimated at 57%. This value is relatively low, but instead a part of Mg to reduce Ta_2O_5, is formed mixture of Mn and Mg oxides. Therefore, it should be investigate the thermodynamic reasons of this phenomenon.

Oxidised anodes or reduced moulded pieces can be considered as a high content concentrate of Ta (81.5 and 77.1% respectively). However, if the goal is to receive a pure Ta, reduced moulded pieces requires further processing, as it contains reduction by-products, namely oxides of reducing agents. Reduced moulded piece may be dissolved in e.g. HCl, CH_3COOH or diluted aqua regia, which in turn would give pure Ta. Further studies are required.

References

1. M. Yoshida, K. Matsuzaki, T. Aoki et al. US 20130014611 A1 patent (2013)
2. K. Mineta, T.H. Okabe, Shigen to Sozai, **121** (2005)
3. R. Kikuchi, T. Yamamoto, M. Nakamoto. Env. & Nat. Res. **4**(1), (2014)
4. T.H. Okabe, N. Sato, Y. Mitsuda, Mat. Trans., **44**, 12, (2003)
5. T.H. Okabe, B. Yuan, Mat. Trans., **48**, 10, (2007)
6. O. Tatsuya, K. Mikio, Nat. Inds. Of Adv. Ind. & Tech., pub. nr: 214352, (2012)

Response of international shipping to the current environmental challenges

Marzenna Popek[1,a]

[1]*Gdynia Maritime University, Department of Chemistry and Industrial Commodity, Gdynia, Poland*

Abstract. The international shipping industry directly facilitates the growth of the world trade, economic development and improvement of the global living standards. As the world economy and population continue to expand, the volume of maritime trade is expected to increase significantly as well. Maritime transport already contributes significantly to the three pillars of sustainable development – social, environmental and economic. Maritime transportation is generally considered environmental friendly when compared to other transportation means, especially if the energy efficiency is measured per tonne transported/per mile. Nevertheless, emissions from the growing maritime transport sector represent a significant and growing source of air pollution. One of future goals in international shipping is to reduce CO_2 emissions. The International Maritime Organization is the United agency responsible for the protection of the environment from the impact of maritime transport. All parts of shipping industry are examining a number of ways to reduce CO_2 emissions, which are primarily linked to reducing fuel consumption. The paper identifies the main areas that should be addressed if maritime sustainable development is to be achieved. It presents analysis of activities focused on environmentally friendly solutions as a form of IMO support for realization of the strategy of sustainable development.

1 Air pollution from ships

The most important sector of the maritime industry is shipping. The international shipping industry is responsible for carriage of about 90% of world trade and is vital to the global economy. Without intercontinental shipping, the bulk trade of raw materials, fuel, affordable food and goods would simply not be possible. Low cost and efficiency of maritime transport has enabled the major shift towards industrial production in Asia.

The World trade is carried across the world's oceans by 90000 marine vessels. Maritime transport has been and still remains to be a very important catalyst of economic development and economic progress of all costal states. The water transport in the EU's directly employs about 200 thousand people, of which approximately 80% work in shipping and 20% in inland waterway transport. Globally, the sea transport industry directly employs about 1, 5 million people (the crew).

In the past three decades, volume of sea transportation grew by about 3,1% per year. Analysis of the volume of freight in maritime transport indicates that it is strictly dependent on the state of the global economy [1]. Global reduction in the gross world product and decrease in the trade in goods drastically affected the value of maritime transport. Positive macroeconomic phenomena have a direct impact on the state of maritime transport and structure. Demand for maritime transport services and seaborne trade volumes continues to be shaped by global economic growth and the need to carry merchandise trade.

An important factor determining the state of the maritime industry is the structure of maritime transport.

Greenhouse gases absorb and emit radiation into the atmosphere, which then influences the average temperature of the earth. Carbon dioxide is the most important greenhouse gas emitted by humans, but several other gases contribute to climate change, too.

Maritime transportation is generally considered environmental friendly when compared to other transportation means, especially if the energy efficiency is measured per tonne transported/per mile. Nevertheless, emissions from the growing maritime transport sector represent a significant and growing source of air pollution.

The fuel consumption od shipping is dominated by three types of ships: oil tankers, container ships and bulk cariers. There are the three most significant sectors of shipping industry from a CO_2 perspective.

For the year 2012, total shipping emissions were approximately 938 million ton. International shipping emitted 769 million tonnes of CO_2 in 2012, that is about 2,2% of the total global CO_2 emissions. By contrast, in 2007, before the global economic downturn, international shipping is estimated to have emitted 885 million tonnes of CO_2, that is 2,8% of the total global CO_2 emissions [2,3].

[a] Corresponding author: m.popek@wpit.am.gdynia.pl

DOI: 10.1051/e3sconf/20161000075

The contribution of global carbon dioxide emission from various sources is shown in Fig.1.

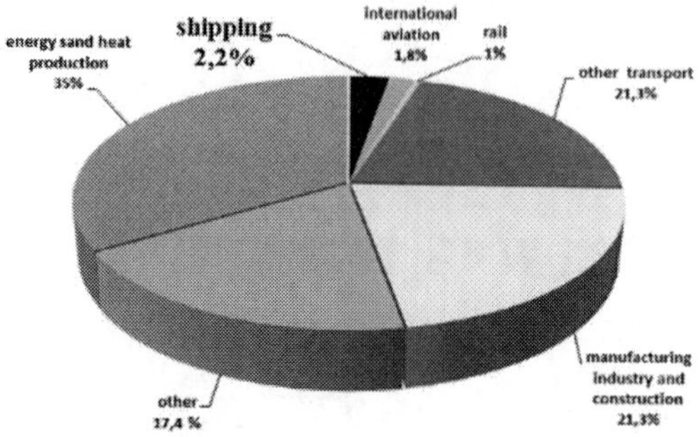

Figure 1. Shipping contribution to global CO_2 emissions
Source : [4]

Exhaust gases are the primary source of emissions from ships. Carbon dioxide is the most important GHG emitted by ships, both in terms of quantity and of global warming potential, other GHG emissions from ships are less important. CO_2 emissions depend on the type of fuel, so the amount of carbon per unit of energy produced.

If international voyages were treated as a country, it would be the sixth largest producer of greenhouse gases. Only the United States, China, Russia, India and Japan emit more carbon dioxide than the world fleet. However, international shipping is the most energy efficient mode of transport and only a modest contributor to the overall emissions of CO_2. During transport of one tonne of freight, the largest emissions of carbon dioxide come from air transport, and the minimum during operation of the largest ships [5].

Nevertheless, a global approach is to improve its energy efficiency and effective emissions control as sea transport will continue growing apace with the world trade.

Forecast scenarios for the medium term suggest that international shipping carbon emissions could increase by 50–250 % by 2050, depending on the expected long term growth of the world economy, population and global energy demand [1].

In addition to carbon dioxide, vessels release other materials such as SOx, NO_x, ODS (ozone depleting substances), VOC (volatile organic compounds) and shipboard incineration, which also contributes to the environmental pollution, but both are less important in terms of quantity and potential of their emissions enhancing global warming.

Another product of fuel combustion is black carbon – small very dark particles, which might warm the atmosphere about one third as much as CO_2 [6]. Because black carbon absorbs much more light than it reflects, it warms the atmosphere through its interaction with sunlight. Black carbon is emitted in a variety of combustion processes and influences the Earth's climate. It has effects on clouds processes, solar radiation, and alters the melting of snow and ice cover.

Marine Environment Protection Committee approved the Bond et al definition of 'Black Carbon' for international shipping as follows:

'Black Carbon is a distinct type of carbonaceous material, formed only in flames during combustion of carbon-based fuels. It is distinguishable from other forms of carbon and carbon compounds contained in atmospheric aerosol because it has a unique combination of physical properties [7].

Although industrial activities are significant sources of black carbon, shipping emissions still represent minor contributions to overall emissions on the global scale. Emission rate of black carbon depends on the combustion process which is influenced by: fuel composition, flame temperature, type of engine, installation upgrades, and fuel and air mix ratio during combustion.

2 Maritime sustainable development

The continuous increase in the number of vessels and their exploitation result in reduced natural resources, in particular crude oil. Pollution from the combustion of fossil fuels contributes to the destruction of the ecosystem and threatens human health.

Shipping is subject to the first ever global and legally binding CO_2 regulations for an entire economic or industrial sector.

Maritime transport already contributes significantly to the three pillars of sustainable development-social, environmental and economic.

Fundamental assumption of the sustainable marine transport may be formulated as follows:

Sustainable marine transport is a transportation system which provides secure, efficient and reliable transport of goods; support global safety and environmental standards, ensures appropriate education and training of crews.

A sustainable marine transport requires coordination at international level. International Maritime Organization (IMO) is a specialized agency of the United Nations, which developing and adopting global standard for the safety, efficiency, security of ships and protection of the environment from shipping operations. Its main role is to create a regulatory framework for the shipping industry. Through IMO, shipping industry works to ensure a continued contribution towards a green economy and sustainable growth. Its achieved through: implementation of global standards, increase of energy efficiency, promotion of new technology for safety, support in education and training, improved maritime security, enhancement of maritime traffic management and improvement of maritime infrastructure.

IMO's original mandate was principally concerned with maritime safety. The Organization assumed responsibility for pollution issues and subsequently has, over many years, adopted a wide range of measures to prevent and control pollution caused by ships and to mitigate the effects of any damage that may occur as a result of maritime operations and accidents.

Of the 51 treaty instruments for the regulation of international shipping IMO has adopted so far. The most important of them concerning the safety of life at sea and the protection of environment. They apply to 99% of the world's merchant fleet.

DOI: 10.1051/e3sconf/20161000075

In 1973 IMO established its Marine Environment Protection Committee (MEPC). MEPC gives extensive consideration to control of GHG emission from ships that contributes to the overall air quality problems encountered by population in many areas, and also affects the environment.

The original focus of its work was the prevention of marine pollution by oil, resulting in the adoption of the first ever comprehensive antipollution convention, the International Convention for the Prevention of Pollution from Ships (MARPOL) in 1973. MARPOL Convention is focused primarily on determining and developing a legal framework for the protection of marine environment and reduction of sea damages.

This has changed over the last few decades to include a much wider range of measures to prevent marine pollution, and the original MARPOL Convention was amended many times to also include requirements addressing pollution from chemicals, other harmful substances, garbage and sewage.

In 1997 the new Annex VI (Prevention of Air Pollution from Ships) of MARPOL Convention was adopted. The regulation seeks to minimize emission of greenhouse gases (GHG) and other relevant substances from ships which contributes to global air pollution.

3 Reduction of emission CO_2 and other relevant gases

Emissions of carbon dioxide derived from ocean vessels, are currently unregulated - the United Nations Framework Convention on Climate Change (UNFCCC), concerning the general obligation of governments to reduce greenhouse gas emissions cannot be applied.

According to the Kyoto Protocol, CO_2 emissions from international shipping cannot be attributed to any particular national economy due to its global nature and complex operations. A ship may be registered in one country while the beneficial owner of the ship may be located in another. Furthermore, the cargo carried by the ship will be of economic benefit to a number of different nations.

Kyoto Protocol principle "*Common but Differentiated Responsibility*" (CBDR) cannot be practically used for shipping without the danger of the significant „*carbon leakage."*

Most shipping companies have the freedom to decide to register their ships with the "*flag state*". Only about 35% of world merchant fleet is registered in Kyoto Annex I countries. The application of CBDR would cause a reduction in the efficiency of maritime transport and the smooth flow of world trade.

Multilateral action is the most appropriate way to address emissions from the maritime transport sector. The IMO principle "*no more favourable treatment*" ensures that requirements adopted for shipping will be applied equally throughout the world, providing maximum environmental improvement. Failure to develop a global CO_2 reduction regime for international shipping may greatly reduce the ability of the shipping as a whole to reduce its emissions.

Global shipping industry is of the opinion that it is possible for shipping to reduce CO_2 emission through a combination of technological developments, building of new and bigger ships, use of new sources of fuel, introduction of the Energy Efficiency Design Index and Ship Energy Efficiency Management Plan (fig. 2.).

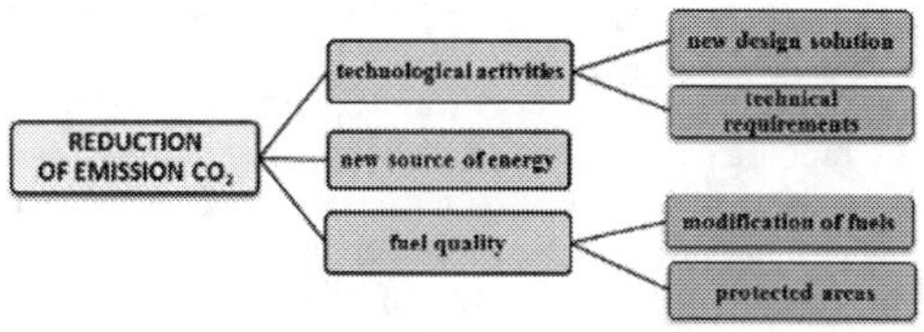

Figure 2. Measures to reduction CO_2

Based on the approved programs and legislative changes the forecast reduction in CO_2 emissions to 2050 is expected. The technical and operational measures will not be sufficient to satisfactorily reduce the amount of GHG emissions from international shipping. Therefore, Market-Based Measures have also been considered

3.1. Technological activities

Gases The various parts of shipping industry are examined a number of ways to reduce CO_2 emissions, which are primarily linked to the reduction of fuel consumption, because bunker cost represents an increasingly significant proportion of ship's operational expenses.

In July 2011 IMO Members agreed to a package of technical requirements for reduction of CO_2 emissions which entered into the force in January 2013. The new rules have been concluded in Annex VI of MARPOL Convention. The amendments include: a system of Energy Efficiency Design Indexing for new ships (EEDI) and Ship Efficiency Management Plan (SEEMP).

The EEDI system calculates vessel's energy efficiency based on ship's emissions, its capacity and its speed. EEDI system was developed for the largest and most energy consuming sectors of the world merchant fleet, because 70% of emissions come from oil tankers and gas carriers, bulk carriers, general cargo ships and container ships.

The base parameters (reductions) taken into account for calculation of EEDI are: lowering of ship speed, use of higher efficiency, derating of engines, use of LPG or LNG, optimisation of the hull and propeller, and coating.

One of the major parameters is the new design of the ship and its propeller.

Ship's lifetime very often exceeds thirty years, and the operating and business environment may change significantly in the course of the time. Better planning at the design stage may lead to a higher potential for reduction of emissions. Optimization of the hull, which is applied to new ship design, is focused on reducing resistance and wetted surface area.

Although the shipping industry is already energy efficient, additional improvements to hull, engine and propeller design are expected to produce further reduction in fuel consumption.

The environmental effect of EEDI is limited, since the requirements apply to new ships, where "*new ship*" means a vessel: the construction of which started in 2013 or was put into service after 1 July 2015.

SEEMP determines the way which allows companies to monitor and improve performance with regard to various factors that contribute to CO_2 emissions. These include: optimising engine power, speed management, use of certain type of fuel, hull management, voyage planning and weather routeing.

Energy efficiency can be improved by using large ships, because it tends to reduce energy consumption. Reduction in speed will increase efficiency, but can be expensive, since they directly affect the number of the ships (more ships will be needed). Traffic management and control system play a role in reducing time in port through more efficient cargo handling.

Optimization and improvement of operational efficiency requires cooperation of several parties. Voyage optimization is the optimization of ship operation, which includes: selection of routes with respect to weather; just –in-time arrival, avoiding unnecessary ballast and trim optimization.

By avoiding unnecessary consumption of energy and parallel operation of electrical generators; optimization of steam plant, fuel clarifier and operation on board; and detection of leaking steam it is possible to reduce energy consumption on board.

Based on the approved programs and legislative changes, the forecast predicts reduction in CO_2 emissions to 2050, due to the use of technological innovations and changes in the operational process is presented in Table 1. The following changes in CO_2 emission are expected: by changes in the design of 10 to 50%, and changes in the operation of 10 to 50%. In total, it is estimated that these changes may result in a reduction of greenhouse gas emissions from 25% to 75%.

Table 1. Assessment of potential reduction of CO_2 emission from shipping

	Saving of CO_2/tonne-mille
DESIGN (new ships):	
Concept, speed, capability	2÷50%
Hull and superstructure	2÷20%
Power and propulsion system	5÷15%
Low-carbon fuel	5÷15%
Renewable energy	1÷10%
Exhaust gas CO_2 reduction	0%
OPERATION (all ships)	
Flat management, logistics	5÷50%
Voyage optimization	1÷10%
Energy management	1÷10%

Source: [8]

3.2 Fuel quality

IMO recognises that some areas need additional protection and MARPOL Convention defines certain areas as "*Special Areas*" (SA). There are areas characterised by heavy maritime traffic and low water exchange caused by the land-locked nature. For a technical reason relating to their oceanographical and ecological conditions, the adoption of special mandatory methods for the prevention of sea pollution is required. These special areas are provided with a higher level of protection than other areas on the sea. The special areas established under Annex V are: the Mediterranean Sea, the Baltic Sea Area, the Black Sea Area, the Gulfs area, the North Sea, the Wider Caribbean Region and Antarctic Area.

In addition, IMO identified "*Particularly Sensitive Sea Areas*" (PSSA) as the areas that need special protection because of their significance for recognized ecological, economic, social or scientific reasons and which may be exposed to damage by maritime activities. The criteria for identification of particularly sensitive sea areas and for the designation of special areas are not mutually exclusive. In many cases PSSA may be identified within a SA and vice versa.

The IMO Guidelines include criteria to allow areas to be designated a PSSA such as: ecological criteria; unique or rare system; diversity of system or vulnerability to degradation by natural events or human activities; social, cultural and economic criteria; significance of the area for tourism; scientific and educational criteria.

When an area is approved as a PSSA, specific measures are used to control the maritime activities in that areas, such as routeing measures and strict application.

Under the MARPOL Convention, the rules, which seek to reduce NO_X and SO_X were adopted and emission control areas were established

SOx controls apply to all fuel combustion equipment and devices on-board ships. The SOx emission control areas (ECA) are: Baltic Sea, North Sea, the North American Area and United States Caribbean Sea Areas. The regulation limits the maximum sulphur content of the fuel used and has two stringency levels: one stringency level that holds in SO_x emission control areas and, another less stringent level for outside ECAs as global requirement (Table 2).

Ships trading in designated emission control areas has to use fuel oil on board with a sulphur content of no more than 0,1% from 1 January 2015.

Table 2. IMO sulphur requirements

Outside ECA (global requirement)	Inside ECA
4,5 % prior to 1 January 2012	1,5 % prior to 1 July 2010
3,5 % on and after 1 January 2012	1,0 % on and after 1 July 2010
0,5 % on and after 1 January 2020	0,1 % on and after 1 January 2015

Source: [9]

DOI: 10.1051/e3sconf/20161000075

As a consequence of the new regulation, ships have started to use low sulphur fuel oil. An increasing number of ships uses as a fuel gas or fuel with low flashpoint. The severe limits concerning the sulphur content of marine fuels will definitely increase of cost of shipping. The derivative effect of the changes could result in a limitation of the competitiveness of the Polish and European economy. Implementation of the regulation will push up the freight rates. Higher ships cost will cause redefinition of cargo flow.
Annex VI of MARPOL Convention introduces a stepwise approach to the reduction of emission NO_x. Regulation of MARPOL sets NOx emission limits for installed marine diesel engines. The IMO emissions standards are commonly referred to as Tier I, II and III. The Tier I standards were defined in the 1993 version of Annex VI, while Tier II and III standards were introduced Annex VI amendments adopted in 2008.
Currently, no specific stringency levels hold for NOx emission control areas, but ships constructed on or after 1 January 2016 have to comply with specific standards when operated in the North American ECA or the United States Caribbean Sea ECA (Table 3).

Table 3. IMO NO_x limits

Tier	Geographical scope	Ships construction data	Total weighted cycle emission limit (g/kWh) n=engine's rated speed (rpm)		
			n<130	n= 130-1,999	n>2000
I	Global	1 January 2000	17,0	$45*n^{-0,2}$	9,8
II	Global	1 January 2011	14,4	$44*n^{-0,23}$	7,7
III	North America and Unites States Caribbean Sea	1 January 2016	3,4	$9*n^{-0,2}$	2,0

Source: [10]

The control of engine NO_X emission is achieved through the survey and certification requirements leading to the issue of an Engine International Air Pollution Preventing Certificate. Different levels of control apply based on the ship construction data and engine's rated speed.

3.3 New source of energy

The need to reduce CO_2 emissions of the shipping sector in order to deliver maritime contributions for the protection of our environment and to prevent climate change has raised the interest in exploring a number of alternative fuel sources.
CO_2 emissions depend on the type of fuel, so the amount of carbon per unit of energy produced. The ability to choose the fuel in the future depends on many factors such as availability, price, durability and regulations and in particular the need to comply with the MARPOL Convention.

Currently, the possibility of using the following energy sources is examined: renewable energy sources such as wind, solar power, second generation biofuels; nuclear propulsion, fuel cells, coal and natural gas.
Renewable energy sources such as wind and solar power, in connection with variable intensity of occurrence, can be used for auxiliary equipment, such as lighting on board ships. Wind and soar conditions differ between regions, so that power is more attractive in certain regions and routes than in others. However, wind and solar power do not provide practically enough power to operate the main engines of vessels (huge physical size of the ship and hence the required power).
First generation biofuels are produced from sugar, starch, vegetable oil, or animal fats. Depending on source, there are certain technical issues, such as stability during storage, acidity, plugging of filters, formation of waxes which may require adapting the engine and selecting the fuels [11].
The new design and construction of engines allow them to operate on some low quality crude vegetable oils, which are considered as the cheapest biofuel available.
The use of liquid biofuel to replace fossil oil in engines carries clear global environmental benefits. Combustion of biofuels results in reduction of greenhouse gases emission and other pollutants, while at the same time allowing for appropriate disposal of waste from biological oils of commercial and industrial origin.
Other arguments in favour of biofuels is the potential for local and regional development, promotion of social and economic cohesion.
First generation biofuels might be considered as possible alternative although there is, considerable public debate about the environmental costs (social effects) of the wider use of such fuels.
Biofuels referred to as second-generation biofuels, produced from residual non-food crops, are considered more sustainable.
Nuclear propulsion for merchant ships is technically possible. Installing nuclear reactor is not taken into account for international shipping due to environmental, political and security reasons.
Fuel cells may be a possibility for new ships in the very long term, although they are currently too limited in range to offer a viable solution.
Coal propulsion is not considered attractive, due to need to remove sulphur oxide emissions, the low thermal efficiency and disposal of the combusted coal residues and ash. Coal could be used as a fuel for steam turbine arrangement.
It is possible to liquid fuels from coal, but emissions of CO_2 related to their production are higher than those of petroleum fuel [11].
Liquid natural gas (LNG) produces lower emissions, and could be solution for some short sea trades if supply infrastructure can be developed. In 2015 IMO prepared and developed the text of the International Code of safety for ships using gases or other low flashpoint fuels (IGF Code). The goal of this Code is to provide safe and environmentally-friendly design, construction and operation of ships and in particular their installations of systems for propulsion machinery, auxiliary power

generation machinery and/or other purpose machinery using gas or low-flashpoint fuel as fuel. The main reason for building vessels fuelled with gas is their environmentally friendly profile both under normal operations and in case of an accident.

Using gas instead of diesel as a fuel is considered beneficial for the environment because of the reduced release of GHG when combusting LNG instead of diesel due to the higher hydrogen to carbon ratio compared to diesel or heavy fuel. The release of NO_X and particles are reduced significantly (82÷84 % and 67 %, respectively). LNG-propelled ships are particularly attractive in emission control areas.

One of the main inconvenience for use of LNG as a fuel for ships is to find sufficient space for the storage of the fuel. In consequence, the potential for reduction of emissions of CO_2 through the use of LNG is mainly relevant for new ships.

To better address the needs of marine customers, a system of small-scale and medium-scale terminals with feeder ships bringing LNG from the import terminal to these terminals and bunker ships, etc., must be established. At present, large LNG import terminals exist in the United Kingdom, the Netherlands, and Belgium and additional terminals are likely to be established by 2020 in France, Finland, Germany, Poland, the United Kingdom, and the Baltic countries. These terminals primarily serve to feed into the natural gas grid and cannot be directly used for maritime purposes. An LNG bunkering infrastructure consisting of fixed terminals, bunker ships, tank trucks, etc., should encompass differences in composition as well as capacities.

LNG containers delivered and loaded on the ships to be used as fuel tanks may become an important introductory solution and a complement in a growing maritime LNG market.

Estimates of future fuel prices contain a large number of uncertainties; however, is important if LNG is to be a competitive fuel for shipping.

4 Conclusion

Shipping is an essential component of any programme for further sustainable economic growth. It is powerful sector of the global economy making a major contribution to global trade and prosperity and has a relatively negative impact on the environment.

The promotion of sustainable shipping and sustainable maritime development is one of the major priorities of IMO.

The IMO regulations that seek to minimise emissions from ships have a significant beneficial impact on the atmospheric environment and on human health, particularly for coastal communities.

The analysis allows to assume that burning of fossil fuels will remains the primary source of energy for the foreseeable future, and the most significant means of reducing CO_2 emission will be achieved by improvements in efficiency across the entire transport chain.

The energy-saving correlated with reduction of emissions CO_2 is difficult to assess, as this depends on how efficiently the vessel will be operated.

Achievement of objectives requires also investment in training and motivating the crew, and in monitoring consumption.

The use of biofuels for reducing emissions of CO_2 from shipping is limited not only by technological issues but also by cost. The biofuels are, currently, more expensive than petroleum fuels.

LNG is a clean fuel and its price is significantly lower than that of distillate fuels which is creating an economic motivation to use it.

Maritime transport shows significant advantages in carbon emissions when compared to road and air freight. Shifting the mode from track to ship may increase emission from ships, but will have an overall beneficial impact on the emissions from the system for movement of goods as whole.

References

1. Review of maritime transport, (UNCTAD, Genewa, 2015)
2. T.A. Boden, R.J. Adres, G. Marland, *Global, Regional and National Fossil-Fuel CO2 Emissions* (CDIAC, Oak Ridge, USA, 2013)
3. G.P. Peters, R.M. Andrew, T. Boden, J.G. Candell, P. Ciais, C. Le Quere, G. Marland, M. R. Raupach, C. Wilson, Nat Clim Chang. **3,** 4-6 (2013)
4. Third IMO Greenhouse Gas Study 2014 (IMO, London, 2014)
5. M. Swahn, *Measuring transport energy use and emissions* (NTM, Sweden, 2009)
6. M.Z., Jacobson, Geophys Res Lett. **27,** 217-220 (2000)
7. T.C. Bond, H. Sun, Environ Sci Technol. **39,** 5921-5926 (2005)
8. O. Buhaugh, Second IMO Greenhouse Gas Study 2009 (IMO, London, 2009)
9. Sulphur Oxides (SO_x)-Regulation 14 (IMO, London, 2014)
10. Nitrogen Oxides (NO_x)-Regulation 13 (IMO, London, 2014)
11. O.P.R. van Vliet, A.P.C. Faaji, Energ. Convers Manage. **50,** 855-876 (2009)
12. D. Vallentin, Energ Policy **36,** 2030-2042 (2008)

DOI: 10.1051/e3sconf/20161000076

Polish experience in the area of management of the waste generated during the exploration for hydrocarbons from unconventional accumulations

Justyna Pyssa[1,a]

[1]*AGH University of Science and Technology, Faculty of Energy and Fuel, Al. Mickiewicza 30, 30-059 Cracow, Poland*

Abstract. The work presents the status of works over the gas exploration from unconventional accumulations in Poland as of the 4th of January 2016. Both the qualitative and quantitative characteristics of waste generated during the exploration for hydrocarbons from unconventional accumulations (data for the years 2010-2014) are presented in the paper. Due to the high content of water and the consistency of waste as well as its changeable and hard-to-predict chemical composition, the waste is difficult to manage. Possible manners of the management of drilling waste connected with the exploration and exploitation of hydrocarbons from unconventional accumulations in Poland and logistical challenges related to it are also discussed in the paper.

1 Introduction

The first borehole in Poland aimed at the exploration for natural gas from shales was drilled in the year 2010. The drilling waste is generated at all stages related to the exploration and exploitation of hydrocarbons from unconventional accumulations.

The natural gas from shale accumulations is released through drilling holes reaching the depth of several thousand meters. In order to cause the flow of gas into the borehole, operations of hydraulic fracturing in a number from several to over a dozen are executed in horizontal sections. The result is the generation of a considerable amount of waste. Managing that waste poses a logistical, technological as well as environmental problem.

To the main sources of pollution residing in the drilling waste one can classify chemical agents used for drilling fluid composition, biocides, oil-derived substances, corrosion inhibitors, and reservoir fluids in a form of brine and crude oil.

2 Area of exploration in Poland

The exploration for and release of unconventional accumulations of natural gas, contained predominantly in oil-bearing black argillaceous and mudstone shale rocks originating from Silurian and Devonian as well as trapped in sandstones of the Rotliegend, is performed mostly through horizontal bores. The reason behind is obtaining a network of artificial crevices resulting from hydraulic fracturing over a large area of the deposit [1]. Till the 1st of February 2016 72 boreholes for the exploration were drilled in Poland. 25 operations of fracturing were performed in the boreholes, in that 13 treatments were taken in vertical boreholes and 12 in crooked-vertical ones [2]. The table 1 presents the breakdown of finished shale gas exploratory boreholes along with the type of performed special treatments.

Table 1. The breakdown of finished shale gas exploratory boreholes along with the type of performed special treatments [2].

Type of treatment performed in borehole	Number of vertical boreholes	Number of crooked-vertical boreholes	Total
Fracturing	13	12	25
Micro-fracturing/ Diagnostic Fracture Injection Test (DFIT)	4	0	4
No fracturing	37	6	43
Total	54	18	72

Locations of Polish deposits range from the middle-east Baltic coast through central Poland to the Lubelskie region. Four boreholes were additionally drilled in the south-west part of Poland (in the Silesian Lowland and the Opole region). The figure 1 shows areas of exploration for shale gas (the condition as of the 4th of January 2016). Green points mark the boreholes which

[a] Corresponding author: jpyssa@agh.edu.pl

DOI: 10.1051/e3sconf/20161000076

are finished. The red colour indicates areas where concessions for the exploration for shale gas are in force and the pink one marks places where requests for concessions for the exploration of natural gas have already been submitted.

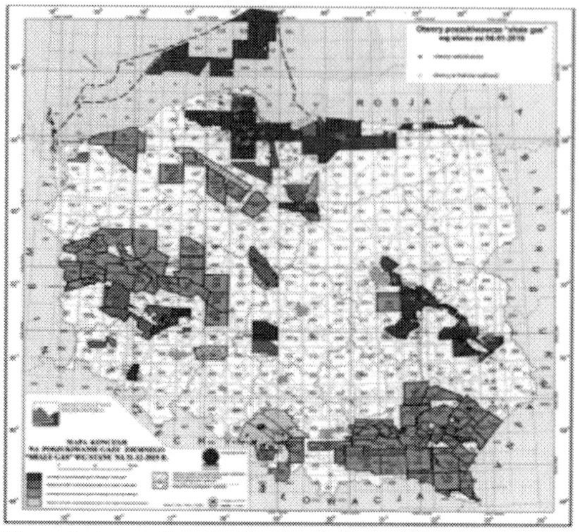

Figure 1. Areas of exploration for shale gas [3].

3 Applied drilling fluids

At the time of drilling a borehole, two or three types of water-dispersing drilling fluids are most often used, depending on the depth and drilled layers. For vertical sections and borehole-completion – these are bentonite and polymeric-and-potassium drilling fluids the properties of which can limit the hydration of rocks and they can also be characterized with inhibitive properties [1, 4]. A typical fracturing fluid contains 95 to 99 percent of volume of water, 3 to 4.5 percent of proppants – used for propping emerging crevices, and 0.5 to 2 percent of chemical additives [5–8]. The composition of such additives may vary depending on the applied technology of extraction and the characteristics of the deposit [5, 8]. The detailed composition of fracturing fluid used for particular boreholes in Poland is placed on the website of The Polish Exploration and Production Industry Organization [9]. The table 2 presents chemical components which are present in fracturing liquids along with their short characteristics.

4 Waste generated during exploitation

As a result of works connected with the exploration and recognition of deposits of natural gas from unconventional accumulations 2 sorts of waste are generated [4, 12, 13]:

1. Waste connected with drilling works – so-called extractive and drilling waste. Here one can classify used drilling fluids, flow-back fluids, reservoir fluids and solid waste produced during the purification of flow-back fluid (the sludge from separators and reservoirs, flow-back proppant etc.);
2. Waste resulting from the functioning of drilling rig (plastics, packages, filters, sorbents, used oil and grease, communal waste, etc.).

Table 2. Common chemical additives from hydraulic fracturing [6, 7, 10, 11].

Additive tape	Example compounds	Purpose	Concentration [%]
Acid	Hydrochloric acid	Clean out the wellbore, dissolve minerals, and initiate cracks in rock	0.123
Friction reducer	Polyacrylamide, petroleum distillate	Minimize friction between the fluid and the pipe	0.088 – 0.125
Corro-sion inhibitor	Isopropanol, acetaldehyde	Prevent corrosion of pipe by diluted acid	0.002
Iron control	Citric acid, thioglycolic acid	Prevent precipitation of metal oxides	0.004
Biocide	Glutaraldehyde, 2,2-dibromo-3-nitrolapropiona mide (DBNPA)	Bacterial control	0.001
Gelling agent	Guar/xantham gum or hydroxyethyl cellulose	Thicken water to suspend the sand	0.056
Cross-linker	Borate salts	Maximize fluid viscosity at high temperatures	0.007
Breaker	Ammonium persulfate, magnesium peroxide	Promote breakdown of gel polymers	0.01
Oxygen scaven-ger	Ammonium bisulfite	Remove oxygen from fluid to reduce pipe corrosion	0.005
pH adjust-ment	Potassium or sodium hydroxide or carbonate	Maintain effectiveness of other compounds (such as crosslinker)	0.011
Proppant	Silica quartz sand	Keep fractures open	3 – 4.5
Scale inhibitor	Ethylene glycol	Reduce deposition on pipes	0.043
Surfac-tant	Ethanol, isopropyl alcohol, 2-butaxyenthanol	Decrease surface tension to allow water recovery	0.085

DOI: 10.1051/e3sconf/20161000076

4.1 Flow-back Fluid

The main mass of waste produced during the process of fracturing is the flow-back fluid which contains small volumes of proppant. The proppant comprises the main part of the dry matter in wastes. After the processes of fracturing are finished, a part of the fracturing liquid returns to the surface (flow-back liquid). The volume of liquids which goes back to the surface depends on local geological conditions. According to the data published in an article [10], in Pennsylvania this volume is in a range from 9 to 53 percent with the average at 10 percent. In China in Sichuan Basin between 15 percent to 80 percent [14]. In Poland in the Pomeranian Basin this value is from 9 to 22 percent, while in the Lublin basin it is from 20.6 to 45 percent [13].

4.2 Used drilling fluids

The used drilling fluid along with extracted output becomes the drilling waste when the drill of the borehole has been finished. This waste can be characterized with considerable variation in terms of chemical composition as well as potential harmfulness to the environment. Each drilling fluid has an increased pH value, increased amounts of undissolved and dissolved solid substances (*Total dissolves solids* – TDS) [8, 14], surfactants, chloride and sulphate ions and dissolved organic carbon (*Total organic carbon* – TOC). It also includes small amounts of petroleum-derived hydrocarbons [14, 15], heavy metals (such as: arsenic, cadmium, barium, chromium, copper, mercury, nickel, lead, antimony, selenium, zinc, tin, cobalt) [8, 11, 14] and also trace amounts of radioactive substances which come from shale formations [1, 8]. The table 3 presents chemical composition of the used drilling fluids.

Table 3. Chemical composition of used drilling fluids [author's own research based on [13].

Elemental	Pomeranian Basin	Lublin Basin
	mg/dm^3	
B	$5.82 \cdot 10^{-3}$ – 0.49	$9.04 \cdot 10^{-2}$ – 0.6
Ba	$1.28 \cdot 10^{-3}$ – 15.91	0.14 – 59.5
Ca	0.23 – 199.56	2.64 – 36.97
Cs	$1.61 \cdot 10^{-3}$ – 20.79	0.11 – 54.57
K	3.28 – 86.65	1.67 – 13.16
Na	0.84 – 601.65	16.72 – 304.63
Se	$4.19 \cdot 10^{-2}$ – 40.58	—
Sr	$8.8 \cdot 10^{-4}$ – 17.3	0.56 – 23.45
Ag	$1.13 \cdot 10^{-2}$ – $3.04 \cdot 10^{-2}$	—
Al	$1.52 \cdot 10^{-2}$ – $3.67 \cdot 10^{-2}$	$3 \cdot 10^{-3}$ – $3.06 \cdot 10^{-2}$
As	$5.52 \cdot 10^{-3}$ – 1.1	—
Cd	$7.7 \cdot 10^{-3}$ – $1.2 \cdot 10^{-2}$	$1.94 \cdot 10^{-4}$ – $3.64 \cdot 10^{-3}$
S	1.29 – 120.36	—
Li	—	$5.66 \cdot 10^{-5}$ – 0.6
Mg	—	0.93 – 3.39
Co	—	$6.16 \cdot 10^{-4}$ – $3.01 \cdot 10^{-3}$
Fe	—	$5.27 \cdot 10^{-3}$ – 1.34
Ti	—	$5.97 \cdot 10^{-5}$ – $3.66 \cdot 10^{-4}$
U	—	$1.69 \cdot 10^{-3}$ – $1.93 \cdot 10^{-2}$
Zn	—	$6.79 \cdot 10^{-4}$ – $2.02 \cdot 10^{-2}$

The chemical composition of post-operation liquids depends not only on the composition of the fracturing fluid (which is selected individually for a given borehole), but also on the rock mass into which it is injected. It can be characterized with significant variability in respect of both the quality and quantity. Differences exist both in case of different areas of Poland (in which boreholes were drilled) and also among different portions of the liquid within the same borehole.

Table 4. Physicochemical characteristics of drilling waste solid phase [13].

Reaction	pH	7.49 – 9.65
Water content	[%]	4.9 – 52.1
Loss on ignition at 850°C	[%]	6.5 – 18.1
Yield point	[%]	12.08 – 28.52
Liquid limit	[%]	18.95 – 54.69
Plasticity index	[%]	5.74 – 26.17
Filtration coefficient	[%]	$2.91 \cdot 10^{-8}$ – $6.24 \cdot 10^{-6}$

The purification of drilling fluids takes place on the area of drilling rig with the use of vibrating, linear and deck sieves, sand separators, gravel sludge, centrifugal separators, hydro-cyclones, flocculation stations. Those treatments enable the reuse of the drilling fluid both for drilling another borehole and for applying as the warp for making a new drilling fluid.

Table 5. Chemical composition of drilling waste solid phase [13].

Elemental	Pomeranian Basin	Lublin Basin
	mg/kg dry weight	
Al	18 370.1 – 48 112.7	25 824.7 – 38 418.3
Fe	20 811.5 – 66 476.4	33 342.7 – 37 149.8
Ca	20 209.2 – 62 542.9	42 527.4 – 47 479.2
Mg	8 432.4 – 29 968.9	17 449.8 – 22 697.6
Na	1 948.4 – 4 4987	2 783.6 – 3 067.1
K	6 794.4 – 22 247.8	9 801.1 – 16 762.7
S	17.3 – 46 180	23 830.2 – 214 204
P	473.4 – 1 191.9	851.6 – 1 051.1
Mn	277.4 – 1 1447.8	362.6 – 411.4
Ba	454.7 – 2 2836.3	877.2 – 2 516.2

4.3 Drill cuttings

Drill cuttings (output) are crumbled fragments of rocks existing in the geological profile of the hole (clays, sands, loam, sandstones, shales, limestones, marlstones, claystones and dolomites) which are contaminated with remains of the drilling fluid, dissolved salts and heavy metals and – in trace quantities – radioactive elements (such as uranium, thorium and their decay products – radium and radon) [1, 4, 13]. The table 4 contains data regarding physicochemical properties of drilling waste solid phase (water content, pH, plastic and liquid limits and others). The table 5 presents the chemical composition, while the table 6 shows the content of organic elements of the waste generated during the exploration for shale gas in the Pomerania and Lublin basin. Doing that research is necessary due to requirements listed in the regulation of the Minister for the Environment as of the 20th of June 2013 [16] concerning the characteristics of the drilling waste. The outcome of the research determines the choice of technology for further management of the waste. As it arises from the data placed in the table 5, drill cuttings coming from the Pomeranian Basin contain considerable amounts of the following elements: Al, Fe, Ca, Mg, K and Na, whereas drill cuttings from the Lublin Basin additionally have very high content of sulphur.

Table 6. Contents of organic compounds in the tested solid phase samples of drilling wastes from Pomeranian and Lublin Basins [13].

Component	Pomeranian Basin	Lublin Basin
	mg/ kg dry weight	
Hydrocarbons	213.79 – 1 616.14	541.03 – 631.72
Gasoline	3.25 – 210.53	8.84 – 181.85
Mineral oil	99.8 – 1 541.38	359.18 – 622.94
Total organic carbon (TOC)	3 058 – 34 241	39 325 –40 650
Dissolved organic carbon (DOC)	1 113 – 7 736	7 548 – 10 190
Phenolic index (phenols)	< 0.5 – 0.8	1.4 – 7
Chemical oxygen demand (COD) (dichromate method)	7 950 – 89 032	108 400 –110 229
Surfactants (anionic)	9.0 – 64.6	34.30

While analyzing the data placed in the table 6 one should pay attention to the high COD, TOC and DOC indicators occurring in the drill cuttings coming from the Lublin Basin. The drill cuttings can also be characterized with lower content of hydrocarbons in comparison to drill cuttings from the Pomeranian Basin.

5 Amount of waste produced during exploitation

In line with regulation of the Minister for the Environment as of the 29th of December 2014 concerning the catalogue of waste (Journal of Laws 2014, item 1923) [17], drilling fluids and waste generated as a result of the exploration for and exploitation of natural gas from unconventional accumulations were classified as:

01 05 05* – oil-containing drilling muds and wastes,

01 05 06* – drilling muds and other drilling wastes containing hazardous substances,

01 05 07 – barite-containing drilling muds and wastes other than those mentioned in 01 05 05 and 01 05 06,

01 05 08 – chloride-containing drilling muds and wastes other than those mentioned in 01 05 05 and 01 05 06.

The amount of waste generated as the outcome of the exploration for hydrocarbons from unconventional accumulations in the area of Poland in the years 2010-2014 is shown in picture 2.

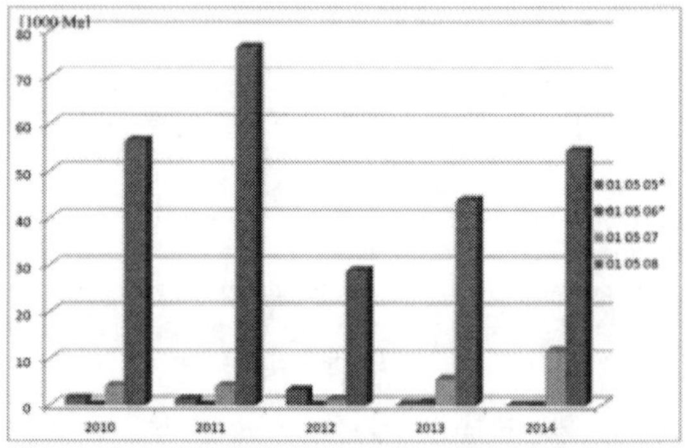

Figure 2. Amount (thousand Mg) of produced drilling waste in the years 2010- 2014 in the area of Poland [author's own research based on [18].

Over the last 5 years a total of 292 700 Mg of waste from the group 01 05 was generated during the exploration for hydrocarbons from unconventional accumulations. The largest part was the waste from the group 01 05 08 (chloride-containing drilling muds and wastes other than those mentioned in 01 05 05 and 01 05 06) which comprised 88.62 of all produced waste. The wastes from the group 01 05 07 (barite-containing drilling muds and wastes other than those mentioned in 01 05 05 and 01 05 06) made up for 9.09% of all waste generated in the analyzed period of time. The hazardous waste from the groups – 01 05 05* (oil-containing drilling muds and wastes) and 01 05 06* (drilling muds and other drilling wastes containing hazardous substances) – comprised respectively 2.12% and 0.17% of all produced waste. The table 7 presents ways of proceeding with the drilling waste.

6 Directions of management

Manner of management of the drilling waste resulting from the exploration for natural gas unconventional accumulations depends on physicochemical properties of the waste. The main feature which makes it harder to manage the drilling waste to a considerable extent is the high content of water: from 5 to 53 percent in case of drill cuttings, from 70 to 90 percent in case of used drilling fluid and up to 99.9 in case of flow-back fluid. Another factor is the consistency of the waste- fluid, semi-fluid and solid. The changeable and hard-to-predict chemical composition of the waste is also an obstacle. In accordance with the article 4.3. of the Act on drilling wastes (Journal of Laws 2008 No 865 item 138) [19] *the owner of the drilling waste is obliged in the very first turn to recover it and, if it is not possible due to technological reasons or not justified on economic grounds, to neutralize it in line with the requirements of the environment protection or the program of drilling waste management, considering the best available techniques.*

From the data placed in the table 7 it follows that the majority of waste from the group 01 05 generated in Poland undergoes the process of recovery: from 71.69% in 2011 to 96.38% in the year 2012.

Table 7. Amount of waste produced in the years 2010 – 2014 in division by subgroups and along with way of managing [author's own research based on [18].

	2010	2011	2012	2013	2014
01 05 05*					
Produced [thousand Mg]	1.5	1.2	3.3	0.2	—
Recovered [thousand Mg]	—	—	3.3	0.2	—
Neutralized [thousand Mg]	1.5	1.2	—	—	—
01 05 06*					
Produced [thousand Mg]	—	—	—	0.5	—
Thermal treatment [thousand Mg]	—	—	—	0.5	—
01 05 07					
Produced [thousand Mg]	4.1	4.1	1.2	5.6	11.6
Recovered [thousand Mg]	3.3	2.6	—	4.8	—
Landfill [thousand Mg]	0.8	1.5	1.2	0.8	1.0[1]
01 05 08					
Produced [thousand Mg]	56.4	76.3	28.7	43.8	54.2
Recovered [thousand Mg]	47.6	55.9	28.7	39.6	—
Landfill [thousand Mg]	8.8	20.4	—	4.2	7.8[1]

[1] The data for the year 2014 relate to the waste treated by the producer on their own and cannot be compared to the preceding years due to changes in the legislation.

The common practice connected with the flow-back liquid management is its utilization in the next fracturing. However, it loses gradually its properties due to the increasing salinity and must finally be neutralized. The purification of technological liquid most often consists in removing suspensions, gases and liquid hydrocarbons. Dehydration of liquid drilling waste (used drilling fluids) is performed in transportable chamber filter press in the process of filtration. The process is supported by coagulation. The most frequently used coagulant is the polyaluminium chloride (its overall chemical formula is $Al_n(OH)_mCl_{3n-m}$). The process of phase separation through filtration in the chamber filter press enables obtaining material which can be characterized with the content of dry matter in a range of 40 to 60 percent and is suitable in this form for use in land reclamation purposes [1].

In the USA the common method is the injection of liquids into reservoirs after exploited conventional accumulations of hydrocarbons or reservoirs of underground saline deposits [8, 20]. In Poland there is only one landfill for waste other than hazardous and neutral. It is situated in the natural gas deposit in Borzecin. In accordance with the concession, waste from the groups 01 05 07 and 01 05 08 can be injected into the rock mass [21].

The hazardous waste from the group 01 05 06* generated in Poland have all undergone the process of thermal treatment. The costs of process are high and vary from 3 215 to 6 440 PLN/Mg [22].

In case of waste containing hydrocarbons of the type 01 05 05* the acceptable methods are incineration and thermal conversion as well as bioremediation (composting, bioreactors, vermiculture and landfarming) [4]. The costs are as follows: up to 4 576 PLN/Mg for waste incineration, and up to 400 PLN/Mg for bioremediation [22].

One of methods of recycling of drilling waste is solidification (immobilisation) of output. It is most commonly used in case of betonite drilling fluids [1]. The method consists in adding binding materials in a form of Portland cement, gypsum, limestone, fly ashes, blast furnace slag and hydraulic and puzzolans binder [23]. It causes hydraulic binding of soluble compounds (chlorides, heavy metals, petroleum-derived substances). This also increases physicochemical stability and mechanical resistance of the obtained material [1, 23]. The materials which are the result of the process are then utilized as additions to road surfaces, base layers of roads. They are also used for earthworks and as an addition to building materials [13]. The costs of immobilization depend on applied materials and vary from 108 to 1000 PLN/Mg [22, 23].

7 Summary

Drilling wastes are a basic byproduct arising as a result of the exploration for shale gas. These are mainly used drilling fluids. The waste is of mineral and organic type. Used drilling fluids are colloidal arrangements comprised of fine-dispersing solids (organic and non-organic), polymers and liquids. Both the composition and properties of the waste may change in a wide range depending on geological and technological conditions of the bore and type of applied drilling liquids.

The main purpose of proper management of waste resulting from the process of fracturing should be minimizing its mass. This can be achieved through the effective separation of solid parts from process water. It allows not only to reduce the mass of waste which ends up in the environment, but also to reuse technological fluid therefore protecting water resources and energy needed for their obtention.

The work is financed within the frames of statutory research no 11.11.210.213 at the Faculty of Energy and Fuels.

References

1. T. Steliga, M. Uliasz, Nafta-Gaz **5** ,273–283 (2012) www.archiwum.inig.pl/INST/nafta-gaz/.../Nafta-Gaz-2012-05-01.pdf
2. http://infolupki.pgi.gov.pl/sites/default/files/czytelnia_pliki/1/02-16-zabiegi.pdf
3. http://infolupki.pgi.gov.pl/sites/default/files/czytelnia_pliki/1/02-16_otwory.pdf

4. J. Macuda, A. Starzycka, Zeszyty Naukowe IGSMiE PAN **85**, 259–271 (2013) https://www.min-pan.krakow.pl/.../ZN85/21-24-macuda-starzycka.pdf
5. W.T. Stringfellow, J. K. Domen, M. K. Camarillo, W. Sandelin, S. Borglin, J. of Haz. Mat. **275,** 37–54, (2014) http://dx.doi.org/10.1016/j.jhazmat.2014.04.040
6. D. Pakuła, Med. Pr. **66(1)**, 99–117 (2015) http://dx.doi.org/10.13075/mp.5893.00147
7. M. Zoveidavianpoor, A. Gharibi, J. of Nat. Gas Sci. and Eng. **24,** 197–209, (2015) http://dx.doi.org/10.1016/j.jngse.2015.03.024
8. L. Torres, O.P. Yadav, E. Khana, Sci. of the Ttl. Env. **539**, 478–493, (2016) http://dx.doi.org/10.1016/j.scitotenv.2015.09.030
9. http://www.opppw.pl/en/fracturing_fluid_composition/23
10. R. D. Vidic, S. L. Brantley, J. M. Vandenbossche, D. Yoxtheimer, J.D. Abad, Science, **340,** 826–836, (2013) http://dx.doi.org/10.1126/science.1235009
11. S. L. Brantley, D. Yoxtheimer, S. Arjmand, P. Grieve, R. Vidic, J. Pollak, G. T. Llewellyn, J. Abad, C. Simon, Int. J. of Coal Geo. **126,** 140–156 (2014) http://dx.doi.org/10.1016/j.coal.2013.12.017
12. PGI-NRI, *Environmental impact of hydraulic fracturing treatment performed on Łubień LE-2H well. Final Report* (Warsaw, November 2011). (http://www.pgi.gov.pl/pl/instytut-geologiczny-informacje-prasowe/4091-raport-z-lebienia-materiaprasowe.html).
13. *The environmental and shale gas exploration. Results of studies on the soil-water environment, ambient air, acoustic climate, process fluids and wastes.*(Directorate General for Environment Protection, Warsaw 2015)
14. D. Zhang, T. Yang, Petrol. Explor. Develop. **42(6),** 876–883, (2015) http://dx.doi.org/10.1016/S1876-3804(15)30085-9
15. W. Orem, C. Tabu, M. Varonka, H. Lerch, A. Bates, M. Engle, L. Crosby, J. McIntosh, Int. J. of Coal Geo. **126,** 20–31 (2014) doi:10.1016/j.coal.2014.01.003
16. Regulation of the Minister for the Environment as of the 20th June 2013 r. concerning the characteristics of drilling waste (Journal of Laws 2013, item 759).
17. Regulation of the Minister for the Environment as of the 9th of December 2014 concerning the catalogue of waste (Journal of Laws 2014, item 1923)
18. Central Statistical Office of Poland, 2016 http://swaid.stat.gov.pl/StanOchronaSrodowiska_dashboards/Raporty_predefiniowane/RAP_DBD_SROD_6.aspx
19. The Act of 10 July 2008 On mining waste (J. of Laws 2008, item 138, No. 865), as amended (J. of Laws 2013 item 1136 and 2014 item 1101).
20. A. Klimkiewicz, K. Korczak, Research Reports Mining and Env. **3,** 43–54, (2012)
21. D. Polak, Min. Res. Manag. **24(3),** 275–284 (2008) https://www.min-pan.krakow.pl/Wydawnictwa/GSM2433/polak.pdf
22. J. Pyssa, Min. Re. Manag. **24(1),** 113–125 (2008) www.min-pan.krakow.pl/Wydawnictwa/GSM2411/pyssa.pdf
23. J. Pyssa, Przemysł chemiczny **89(7),** 927–934 (2010)

 DOI: 10.1051/e3sconf/20161000077

The influence of changes in the structure of electric power generation on the management of energetic waste in Poland

Justyna Pyssa[1,a]

[1] *AGH University of Science and Technology, Faculty of Energy and Fuels, Al. Mickiewicza 30, 30-059 Kraków, Poland*

Abstract. In the paper the changes in structure of production of electric power and heat in Poland has been analyzed. Both the quantity and quality of incineration byproducts is determined by the way of generation of energy and used raw materials. It also depends on the production volume. The possibility of economic utilization of energetic waste from various production sources requires detailed knowledge of its physicochemical composition, properties and vulnerability for further processing in various technologies. Economic directions of using byproducts from incineration is also discussed in detail.

1 Introduction

Hard coal, crude oil and natural gas are the basic carriers of the primary energy. The long-term forecast for Poland appears to show that the position of coal will be steadfast and remain at a high level for at least next several decades. The use of coal as the primary energy is connected with definite ecological effects. In the process of incineration of solid fuels, regardless of the applied technology, dusts, carbon monoxides, sulfur monoxides and nitrogen oxides are emitted into atmosphere. The generation of electrical energy and heat is intrinsically connected with producing significant amounts of waste – regular products of incineration intercepted in a form of fly ash in electrostatic precipitators and bottom ash (also called slag) shed from the deposit. Byproducts of incineration comprise valuable raw materials for many industry branches. Continuous technological changes which occur in the power industry significantly affect their variety which in turn means their physical and chemical parameters. The biggest number of studies regards the use of fly ashes from the hard coal which are generated in conventional pulverized-fuel boilers. Cinder from fluidized bad boilers and the brown coal as well as ashes from co-incineration of the hard coal and biomass are continuously examined and new solutions based on their use are sought.

2 Industry of electric power production in Poland

Polish sector of electric power generation relies mainly on the hard and brown coal – 80,89 percent (figure 1). The share of renewable energy sources is also significant and makes 12,15 percent – it is predominantly the wind energy. Other sources of energy are the natural gas with a share of 5,63 percent and the crude oil with a 1,36 percent stake. By analyzing changes in the production of electric power in the years 2008 – 2014 (figure 2) the largest increase was noted in technologies basing on the renewable energy – from 5,4 Mtoe in 2008 up to 8 Mtoe in 2014. By contrast, the share of energy generated from the hard and brown coal decreased substantially – from 60,5 Mtoe in 2008 to 53,6 Mtoe in 2014.

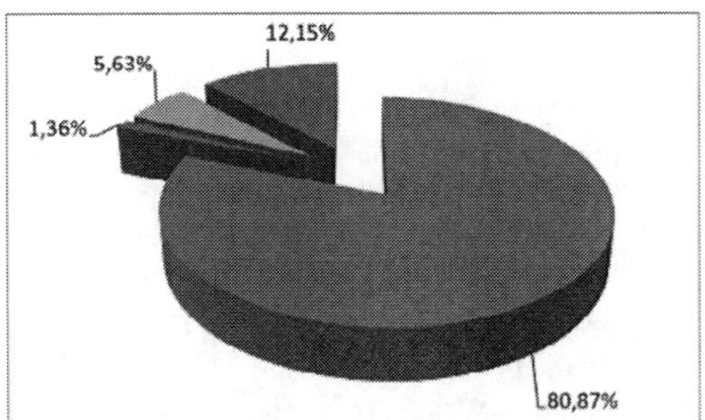

Figure 1. Structure of electric power generation in Poland (in percent) in 2014 (author's own research based on [1]).

Among thermal power plants, 15 power stations use the hard coal for the process of electrical energy production which comprises 65,2 percent of all power plants in that group. The brown coal is incinerated in 6 power stations, while two power plants use the biomass (table 1). A numerous group is created by combined heat and power stations – in a number of 85, and commercial combined heat and power plants – 68.

[a] Corresponding author: jpyssa@agh.edu.pl

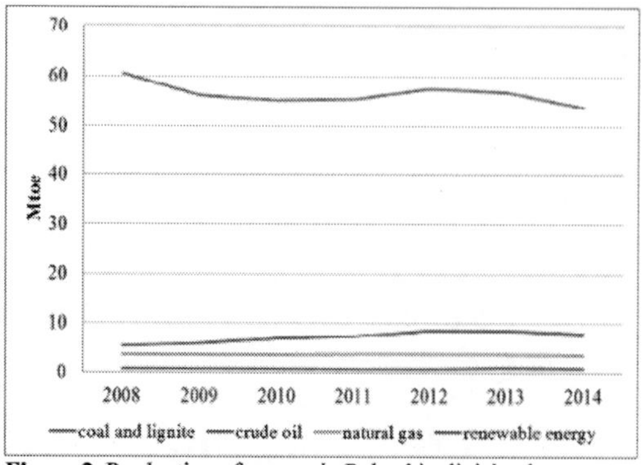

Figure 2. Production of energy in Poland in division by generation sectors (Mtoe) for the years 2008-2014 (author's own research based on [1]).

3 Places of production

The following table 2 and figure 3 present quantities (in thousand Mg) of incineration by-products (according to the Waste Catalogue [3]) generated in particular types of power plants and combined heat and power stations in Poland in the year 2012.

Table 1. Sector of generation of electric power in Poland [2].

Group name	**Number**
Thermal power plants including:	23
— operating on hard coal	15
— operating on brown coal	6
— operating on biomass	2
Combined heat and power stations	85
Commercial combined heat and power stations	68

In combined heat and power stations operating on gas three types of incineration by-products – 10 01 01, 10 01 02 and 10 01 03 – were produced in 2012 (their amount was 17,8 thousand Mg). In power plants and combined heat and power stations operating on biomass 36,6 thousand Mg of incineration by-products from groups 10 01 01, 10 01 03, 10 01 24 and 10 01 82 were generated. Power plants operating on brown coal produced five types of waste: 10 01 02, 10 01 05, 10 01 07, 10 01 80 and 10 01 82 (11 275,9 thousand Mg). Power plants and combined heat and power stations operating on hard coal generate all sorts of waste from the analyzed groups.

4 Physicochemical properties of incineration by-products

Dynamic growth in quantities of production of incineration by-products is connected with the increase in demand for electric power which in Poland is mostly made of the hard and brown coal.

Table 2. Quantities and types of incineration by-products (thousand Mg) generated in particular types of power plants in 2012 (author's own research based on [4].

Code of waste[1]	Power plants operating on brown coal	Power plants and combined heat and power stations operating on hard coal	Combined heat and power stations operating on gas	Power plants and combined heat and power stations operating on biomass
10 01 01	—	1 715,0	10,3	0,6
10 01 02	1 067,0	3 950,6	7,4	—
10 01 03	—	5,6	0,1	21,3
10 01 05	257,8	472,1	—	—
10 01 07	218,1	—	—	—
10 01 17		652,1		
10 01 24	—	548,8	—	10,9
10 01 80	7 480,3	8 386,2	—	—
10 01 81	—	1,9	—	—
10 01 82	2 252,7	1 892,3	—	3,8

where:
— means that the waste was not produced,
[1] – numbers of codes in accordance with the Regulation of the Minister for the Environment as of the 9th of December 2014 concerning the catalogue of waste (Journal of Laws 2014, item 1923) [3]
10 01 01 – Bottom ash, slag and boiler dust (excluding boiler dust mentioned in 10 01 04)
10 01 02 – Coal fly ash
10 01 03 – Fly ash from peat and untreated wood
10 01 05 – Calcium-based reaction waste from flue gas desulphurization in solid form
10 01 07 – Calcium-based reaction waste from flue gas desulphurization in sludge form
10 01 17 – Fly ash from co-incineration other than those mentioned in 10 01 16
10 01 24 – Sands from fluidized beds (excluding 10 01 82)
10 01 80 – Mixture of ash and slug from wet disposal of furnance waste
10 01 81 – Microspheres from fly ashes
10 01 82 – Mixtures of fly ashes and solid waste from calcium methods of desulphurization of waste gas (dry and semi-dry methods of desulphurization of flue gases as well as incineration in fluidized bed).

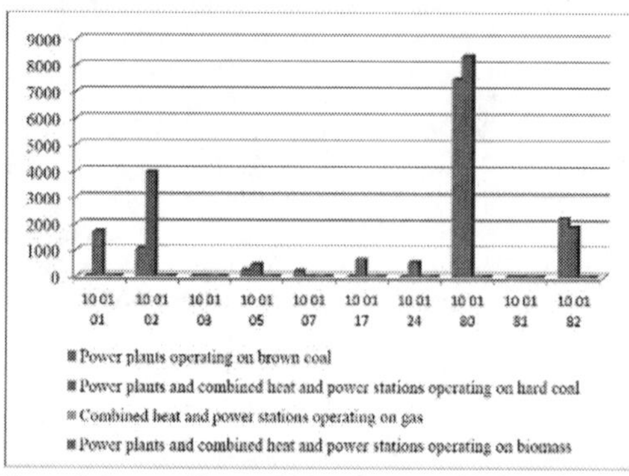

Figure 3. Amount (thousand Mg) of generated incineration by-products UPS in particular types of power plants in 2012 (author's own research based on [4]).

Depending on which type of coal is incinerated – hard or brown – the composition of fly ashes shows different puzzolan properties. Fly ashes obtained from brown coal-fired boilers are of high calcium content amounting to 40 percent of mass and present much better puzzolan properties than fly ashes coming from the combustion of hard coal [5]. Ashes with low content of calcium consist mainly of aluminosilicate glaze. In those ashes – from crystallized phases – sillimanite, mullite and quartz crystallize during cooling. And ferric oxides occur in a form of hematite and magnetite. It is believed that the high content of crystallized phases in the ash of that type decreases its puzzolan activities [6]. In all types of fly ashes the basic chemical elements are oxides of silicon, aluminum, ferric, calcium, magnesium, potassium, sodium and unburned coal. The components of glassy phase are predominantly SiO_2 and Al_2O_3 [5, 7, 8]. The chemical composition of fly ashes with high and low content of calcium is placed in the table 3.

Table 3. Variability of chemical composition of fly ashes with low and high calcium content [9].

Compo-nent	Fly ash with low calcium content	Fly ash with high calcium content
	[% mass]	
SiO_2	34 – 60	25 – 40
Al_2O_3	17 – 31	8 – 17
Fe_2O_3	2 – 25	5 – 10
CaO	0,5 – 5	10 – 38
MgO	1 – 5	1 – 3

In all sorts of fly ashes the essential chemical components are oxides: SiO_2, Al_2O_3 and CaO (95-99%). In ashes there are additionally side components in a form of: MgO, Na_2O, K_2O, SO_3, P_2O_5, TiO_2 (0,5-3,5%) and unburned coal (ignition losses) [6, 10]. The content of coal in ashes is dependent on [11, 12]:

— particle size distribution of incinerated coal dust– the finer it is ground, the lower is the content of unburned coal,

— temperature and combustion time – the higher is the temperature and the longer is the combustion time, the lower are the losses on ignition,

— construction of furnace – bigger remains of coal usually occur in ashes obtained from fluidised boilers.

Ashes can contain trace quantities of such elements as: Mn, B, Ba, Cu, Sr, Ni, Cr, Zn, Cd, Co, Mo, V, Se, Pb, As and others (0,1-0,3%) as well as radioactive elements ^{40}K, ^{226}Ra and ^{232}Th [7, 8, 11-14]. Radioactive elements do not exceed concentration which would prevent the economic use of incineration by-products [5, 14].

Over the past several years new type of energetic waste emerged in Poland – fly ashes from the biomass combustion. The chemical composition of those ashes depends on the type of incinerated material. They usually contain more alkaline compounds than ashes from the hard and brown coal combustion. Biomass can also be marked with lower calorific value and higher humidity which has influence on reducing the combustion temperature and results in lower content of glassy phase and increased content of unburned coal in fly ashes. Those ashes contain also lesser amounts of SiO_2 and Al_2O_3. By contrast, the amount of CaO, K_2O and P_2O_3 increases decidedly. They can be most often marked by the lack of puzzolan properties and hydraulic activity [15].

5 Directions of management of incineration by-products

Incineration by-products are utilized in many branches of economy: in construction industry, ceramics, mining industry, land reclamation and agriculture. Directions of use of incineration by-products in Poland are placed in the table 4.

Table 4. Directions of use of incineration by-products in Poland

Branch	Utilization
Mining industry	Conducting fire prevention (insulation of fire fields) Mechanical reinforcement of rock mass Backfilling of mine workings Use in sealing works thanks to the high porosity of binder
Construction industry	Additive to cement (sulphated component carrier) as well as puzzolan additive (increases tensile strength and resistance against chemical corrosion) Raw material for manufacturing artificial aggregates Production of cell concrete (aerated and foam concrete)
Road and engineering works	Applied in building landfills for waste Used in ground stabilization in road works Applied in flood-related sealing of sites Material for building embankments
Land reclamation	Depending on the chemical composition (lack of heavy metals) – used for reclamation of excavation of clay raw materials and others
Neutralization of sewage and sludge	Thanks to high alkalinity are suitable for neutralization of pH value of sewage High activity of calcium oxide included in them allows to neutralize industrial sewage
Agriculture	Used for soil de-acidification due to alkaline character
Others	Secondary sorbents (because of free lime in their composition)

5.1 Use of incineration by-products in construction industry

Fly ashes are applied to a large extent in manufacturing of cement because of their puzzolan properties and amorphous silicic acid (silica) contained in the glassy part of ash [5, 8]. Most commonly they are used as active additives. Ashes are minced together with the Portland clinker through which, as a result of chemical reactions, functional features of adhesive alter leading to the occurrence of puzzolan cement [16]. The content of ash is between 5 to 14 percent of the total mass of raw material. The addition of ash influences not only the increase in cement mass, but first of all improves its frost resistance, resistance to aggressive chemicals and diminishes concrete shrinkage [17].

Ashes are widely used as raw materials for the production of clinker Portland cement. Fly ashes from the hard coal combustion replace a so-called 'low' material (that is such which calculated on the calcinated substance has less than 67 percent of CaO) in the raw set and is a carrier of Al_2O_3, Fe_2O_3 and SiO_2. The application of fly ashes as a low material allows to reduce the temperature of sintering by 200 to 300°C. It causes the decrease in the use of technological fuel sometimes by 30 percent. Fly ashes are also widely used as a corrective addition. And so they can be utilized as silicon-bearing, iron-carrying and clay-bearing addition depending on particular oxides in the batch [17, 18].

One of the main ways of use of fly ashes is the manufacturing of concrete. Employing ashes causes the reduction in water demand and improves the plasticity of mixture. Ash and microfibre additives have a significant effect on lowering chemical corrosivity through material and structural protection of concrete composites. In the manufacture of cell concrete they are used as aggregates and partially as binders through which production costs are diminished [16].

Ashes are utilized in the production of building ceramics (solid bricks, hollow bricks, checker bricks, breezeblocks), fired ceramics, clinker products, sewer and sanitary ceramics. In the building ceramics industry fly ashes can be used as the basic raw material and as a weakening additive [8, 18]. The application of ashes as the primary raw material takes place in the so-called ash- based technology of manufacture of building ceramics where it replaces the natural loamy material and makes for 60 to 80 percent of brick mass.

5.2 Road and engineering works

The economic situation in the construction market – especially in road works – provoked enormous demand for natural aggregates to which the aggregates produced from industrial waste became an alternative [19].

The manufacture of aggregates from fly ashes can be conducted with use of non-firing method (aggregates are obtained directly from ashes or by granulation of mixture of ashes and mineral binder). In case of firing method the material is granulated first and then fired.

Fly ashes are also used for land valorization. They drain, enhance carrying capacity of soil and make hydraulic or puzzolan factor which allows to bind land and achieve compressive strength and frost resistance [20, 21].

Siliceous and aluminous ashes from fluidized boilers are applied in road building for stabilization and land and base improvement for constructions and creating embankments [5, 22]. Thanks to the high content of calcium, they are very suitable for the production of road binders which allow to stabilize lands and reduce costs of binders while obtaining the same stabilization parameters which are achieved by using clean cement or calcium [18]. Ashes belonging to siliceous ones can be characterized with low content of CaO (0,5-4,7%) and SO_3 (0,3-2,9%) and are used as the substitute for cement in stabilizing substructure of roads with low car traffic and also for creating filtration and insulating layers and building of bodies of levees [5, 17, 22].

Due to the presence of radioactive isotopes (potassium K-40, radium Ra-226 and thorium Th-228) in natural ashes, the examination of samples is conducted on the basis of criteria included in the regulation of the Council of Ministers as of the 2nd of January 2007 (Journal of Laws no 4, item 29) [23]. According to the criteria, incineration by-products fulfill requirements for industrial waste applied in ground construction works built in built-up areas or in areas intended for building in local spatial development plan and for leveling such areas [5].

5.3 Use of incineration by-products in mining technologies

Incineration by-products are also widely used in mining technologies. They are utilized for construction and filling purposes depending on their physicochemical properties [24-27]. Ashes delivered to coal mines are a raw material base for the manufacture of ash-and-water mixtures in mining backfilling installations [5, 18, 26-28]. The mixtures are transported through pipes from the surface to underground mine workings and used mainly within the frames of fire and methane prevention in backfilling and sealing of mine voids [25, 29].

To the main uses of incineration by-products in mining technologies belong [5, 25-27, 30]:

— fire and methane prevention (through sealing and backfilling of gobs which prevents the occurrence of endogenous fires and mitigates the danger of releasing methane and other gases from gobs

— insulation of fire fields and closed mine workings through manufacture of insulating backfilling cork

— backfilling of post-mining voids and mechanical enforcement of rock mass which has impact on reducing the size of the surface deformation

— simplification of ventilation systems in mines through manufacture of insulating and explosion proof cork.

5.4 Use in agriculture

In agriculture there is predominant use of fly ashes from hard coal [14, 31]. They are used as calcium and magnesium fertilizers for soil fertilization [7, 8, 16, 31-34]. They improve physicochemical properties, increase the capacity of absorbent complex, enhances the absorbency to water, and reduce the density and alkalize acidic soils [7, 14, 17, 32, 34, 35].

6 Summary

The state of technological knowledge, which we currently possess, allows for a very wide range of the use of waste (10 01) which comes from power plants. The characteristics one should pay special attention to is the fact that the waste successfully replaces such natural raw materials as gravel and sand the extraction of which causes degradation to the natural environment.

Adding such components not only influences positively the properties of concrete, but also limits the use of natural resources in an easy way.

The future demand for incineration by-products in Poland will be determined to a large extent by changes occurring in the domestic sector of electric power generation. Taking into consideration the current age structure of production units (more than 60 percent of the total power of boilers is installed in equipment which has been working for over 30 years) and stringent environmental standards arising from European directives, one should expect the execution of investments in new powers in the coming years. It is also necessary from the point of view of security of electric energy supply. Possessing considerable, as for European conditions, resources of hard and brown coal may be an argument for investing in new powers based on those energy carriers. Parallel to fulfilling commitments resulting from the climate and energy package, further growth of power in renewable energy sources will be essential.

Summarizing, incineration by-products are a precious raw material for many industry branches. Continuous technological changes taking place in energy economics significantly influence their diversity and in consequence physicochemical parameters – example in here are fly ashes from coal with admixture of co-incinerated biomass. Taking into account the adopted aim of diversification of domestic fuel and energetic structures in existing country's energy policy, one can state that the supply of typical products of hard and brown coal combustion (ashes and slag) will slightly decrease over the course of the next ten years.

Acknowledgments

The work is financed within the frames of statutory research no 11.11.210.213 at the Faculty of Energy and Fuels.

References

1. Eurostat, *Energy, transport and environment indicators* (2015) http://ec.europa.eu/eurostat/en/web/products-statistical-books/-/KS-DK-15-001
2. Agencja Rynku Energii, *Statystyka elektroenergetyki polskiej* (2013)
3. Regulation of the Minister for the Environment as of the 9th of December 2014 concerning the catalogue of waste (Journal of Laws 2014, item 1923)
4. Agencja Rynku Energii. *Emitor 2012. Emisja Zanieczyszczeń Środowiska w Elektrowniach i Elektrociepłowniach Zawodowych* (2013)
5. J. Pyssa, J. Rokita, Przegląd Górniczy **68(12)**, 132–138 (2012) http://yadda.icm.edu.pl/yadda/element/bwmeta1.element.baztech-e36b801d-f368-442d-8733-1763ec5f83bf
6. J. Filipiak, Annual Set The Environment Protection **13,** 1043-1054 (2011) http://ros.edu.pl/images/roczniki/2013/pp_2013_078.pdf
7. N.L. Ukwattage, P.G. Ranjith, M. Bouazza, Fuel **109,** 400–408 (2013) doi:10.1016/j.fuel.2013.02.016
8. Z.T. Yao, X.S. Ji, P.K. Sarker, J.H. Tang, L.Q. Ge, M.S. Xia, Y.Q. Xi, Earth-Science Rev. **141**, 105-121 (2015) http://dx.doi.org/10.1016/j.earscirev.2014.11.016
9. S. Grzeszczyk, G. Lipowski, *Popioły lotne i ich wpływ na reologię i hydratację cementów.* (Wydawnictwo Politechniki Opolskiej. Opole, 2002)
10. P. Iwanek, I. Jelonek, Z. Mirkowski, Min. Res. Manag. **24(4),** 91-104 (2008) https://min-pan.krakow.pl/Wydawnictwa/GSM2444/06-iwanek-jelonek-mirkowski.pdf
11. A. Gonet, S. Stryczek, R. Wojciechowski, Drilling Oil and Gas **22(2),** 551-558 (2005) http://journals.bg.agh.edu.pl/WIERTNICTWO/2005-02/W_2005_2_04.pdf
12. S.M. Shaheen, P.S. Hooda, Ch.D. Tsadilas, J. Environ. Manage **145,** 249-267 (2014) doi: 10.1016/j.jenvman.2014.07.005.
13. D.K. Szponder, K. Trybalski, Górnictwo i Geoinżynieria **33(4),** 287-298 (2009) http://yadda.icm.edu.pl/baztech/element/bwmeta1.element.baztech-article-AGHM-0008-0032/c/Szponder.pdf
14. Z. Ciećko, A. C. Żołnowski, M. Madej, G. Wasiak, J. Lisowski, E. Rolka, Polish J. Soli Sci. **XLVIII(1)**, 1-12 (2015) DOI: 10.17951/pjss/2015.48.1.1
15. M. Gawlicki, J. Małolepszy, Nowoczesna Gospodarka Odpadami, **1-2(8)**, 14-20 (2015) http://www.e-gospodarkaodpadami.pl/rynek/wykorzystanie-odpadow-przemyslowych/
16. R.S. Blissett, N.A. Rowson, Fuel **97,** 1-23 (2012) http://dx.doi.org/10.1016/j.fuel.2012.03.024
17. D.K. Szponder. *Badania wybranych właściwości popiołów lotnych z zastosowaniem analizy obrazu. Rozprawa doktorska* (AGH, Kraków 2012) winntbg.bg.agh.edu.pl/rozprawy2/10519/full10519.pdf
18. M. Ahmaruzzaman, Progress in Energy and Combustion Science. **36**, 327-363 (2010) http://www.sciencedirect.com/science/article/pii/S0360128509000604
19. J. Pyssa, Przegląd górniczy, **66(5)**, 38-44 (2010) http://yadda.icm.edu.pl/baztech/element/bwmeta1.element.baztech-article-BGPK-2859-1183
20. G. Rolka, E. Ślęzak, Prace Instytutu Ceramiki i Materiałów Budowlanych. **9**, 148-155. (2012) http://yadda.icm.edu.pl/yadda/element/bwmeta1.element.baztech-article-BTB6-0004-0044
21. A. Gruchot, T. Zydroń, Annual Set The Environment Protection. 15, 1719-1737 (2013) http://ros.edu.pl/images/roczniki/2013/pp_2013_115.pdf
22. J. Pyssa, Mineral Resources Management. **21(3)**, 83-93 (2005) http://www.meeri.eu/Wydawnictwa/GSM213/pyssa.pdf

23. Regulation of the Council of Ministers as of the 2nd of January 2007 (Journal of Laws no 4, item 29)
24. F. Plewa, M. Popczyk, P. Piontek, Energy Policy Journal **12(2)**, 485- 495 (2009). http://www.min-pan.krakow.pl/se/pelne_teksty23/k23z_pe/k23_plewa_popczyk_piontek_z.pdf
25. Zając A., Madaj M., Energetyka **3**, 143-146 (2010) http://yadda.icm.edu.pl/yadda/element/bwmeta1.element.baztech-article-BPS1-0040-0044
26. F. Plewa, M. Popczyk, A. Zając, P. Pierzyna, Energy Policy Journal **15(3)**, 147-157 (2012) http://www.min-pan.krakow.pl/se/pelne_teksty26/k26pe3/k26pe3_plewa_inni_1_z.pdf
27. F. Plewa, M. Popczyk, P. Pierzyna, Energy Policy Journal **16(4)**, 257-270 (2013) http://yadda.icm.edu.pl/baztech/element/bwmeta1.element.baztech-dc62e6fd-04c9-4341-8b55-27d6b510800c/c/plewa.pdf
28. Z. Piotrowski, *Utilization of fine-grained waste in underground coal mining* (Archives of Mining Sciences, Kraków 2011)
29. J. Palarski, Przegląd górniczy **65(10)**, 45-51 (2009) http://yadda.icm.edu.pl/yadda/element/bwmeta1.element.baztech-article-BGPK-2717-0582
30. Z. Piotrowski, Mineral Resources Management **24(4/1)**, 113-121 (2008) http://www.meeri.pl/Wydawnictwa/GSM2441/piotrowski.pdf
31. E.J. Bielińska, S. Baran, S. Stankowski, Inż. Roln. **6**(115), 7-15, 2009. http://ir.ptir.org/artykuly/pl/115/IR(115)_2525_pl.pdf
32. V.C. Pandey, N. Singh, Agriculture, Ecosystems and Environment. **136**, 16 – 27 (2010) http://www.sciencedirect.com/science/article/pii/S0167880909003442
33. L.C. Ram, R.E. Masto, Journal of Environmental Management **91**, 603 – 617 (2010) DOI: 10.1016/j.jenvman.2009.10.004
34. H.A. Mupambwa, E. Dube, P.N.S. Mnkeni, S. Afr. J. Sci. 111**(7-8)**, 1-6 (2015) http://dx.doi.org/10.17159/sajs.2015/20140103
35. D.K. Szponder, K. Trybalski TEKA Kom. Mot. Energ. Roln. – OL PAN **11**, 373–385 (2011) http://yadda.icm.edu.pl/yadda/element/bwmeta1.element.agro-0a9c223c-7437-4dfb-90db-cfb17ae46a89/c/373_teka_11_2011.pdf

Multicriteria analysis in selecting the optimal variant of solar system

Aleksandra Radziejowska[1,a] and Krzysztof Zima[2]

[1]*Department of Geomechanics, Civil Engineering and Geotechnics, Faculty of Mining and Geoengineering, AGH University Of Science And Technology in Cracow, Poland*
[2]*Institute of Construction and Transportation Engineering & Management, Faculty of Civil Engineering, Cracow University Of Technology, Poland*

Abstract. Alternative energy sources are becoming more serious competition to traditional ways of generating energy. It becomes real integration of eco-energy with ecology, as well as the innovative technologies with low-energy construction. Apart from the cost an important issue are technical parameters of the equipment, durability, ease of installation, etc. The investor therefore is facing with the problem of decision-making to choose the best solution from the point of view of many criteria. In the article, the authors present the proposal to apply the methods of multi-criteria analysis to select the most beneficial variant of the solar system solutions. In this purpose will be use among other method: multivariate analysis of Saaty's AHP, the taxonomic method of weighting factors and, belonging to a group of methods using outranking relations, the Promethee II method. Proposed comparative analysis can be used as a method for decision support during the selection of the most beneficial technological solution of solar installation and to evaluate operational efficiency existing buildings which will have implemented new systems.

1 Introduction

Nowadays, both in residential and commercial buildings are increasingly being used solutions for the application of environmentally friendly energy sources. Application of solar installations as one of the energy sources brings measurable effects of fulfilling the basic principles of sustainable development among which should be mentioned the environmental, economic and social aspects.

Heating water with solar energy is one of the most effective ways of energy conversion of solar radiation. That is why, the most frequently offered solar collectors are used to heat hot water. Such usage of solar installations is the most widely used in countries with high insolation, where the use of solar energy provides measurable benefits for most of the year. However, over time these installations began to arouse greater interest also in countries with temperate climates. However, countries with less favorable climatic conditions were required some modifications hitherto applied solutions, which resulted, among other, of the following occurring the negative temperatures in these regions [1]. Improvements in solar installations were required cold hardiness materials and additional support conventional heat sources.

Currently on the manufacturers of solar collectors market appears a variety of solutions, which are used to convert solar energy into heat. Solar energy that gets directly to the device is converted by the heat transfer medium, which may be a liquid (eg. glycol, water) or gas (eg. air) into heat energy. The analysis can cover flat solar systems, some of which are a gas-filled, liquid-filled or two-phase and flat-vacuum solution, vacuum-tube or concentrating systems.

More often, due to the additional initial costs and long payback period for the building object, selection of the most beneficial source of renewable energy is becoming problem of the private and the public investor. Apart from the cost an important issue are technical parameters of the equipment, durability, ease of installation, etc. The investor therefore is facing with the problem of decision-making to choose the best solution from the point of view of many criteria. In view of the above, authors present the proposal to apply the methods of multi-criteria analysis to select the most beneficial variant of the solar system solutions.

2 Short literature review

The problem of choosing the best solution from the point of view of many criteria is frequently described in the literature. The article focuses on the multi-criteria analysis of the best solution involving the installation of solar collector systems. Huang et al. [2] tried to select an optimum solar ejector cooling system. Authors selected three commercially available solar collectors for comparison and discussed best solution. It seems that in order to select the best solutions it should be supported by mathematical methods.

[a] Corresponding author: aradziej@agh.edu.pl

Aragonés-Beltrán et al. [3] used an AHP (Analytic Hierarchy Process) and the ANP (Analytic Network Process) methods to help the managing board of an Spanish solar power investment company. The system help to decide whether to accept or reject a project according to a set and establishing a priority order among the projects.

J.R. San Cristóbal in [4] shows a Compromise Ranking method (known as VIKOR method), using multi-criteria ranking index based on the measure of "closeness" to the "ideal" solution for renewable energy project in Spain.

The importance of the problem related with choosing panel systems has been addressed, among others, in [5]. Choosing an optimal solution of a solar collector installation for a single-family building is done using the CBR method based on a database containing the costs of 128 projects involving the installation of solar collectors. The example uses 10 evaluation criteria adopted solutions.

The problem of designing related to technical characteristics has been described among others in [6], [7]. In this article authors analyze the differences between the value of the contract and the offers submitted in 2013 in public procurement for a solar collectors.

3 Decision–making methods

For many years, researchers are developing methods of support, which are designed to facilitate resolving the situation ambiguous in various fields of human activity. About the decision-making process in the tender wrote among others in [8], [9]. The search for methods of solving multicriteria problems first appeared in the late nineteenth and early twentieth century. Since then, there was create a number of studies and methods of searching for the optimal solution (Table 1).

Table 1. Application areas of multi-criteria methods [4]

Method	Application area
Multi - objective decision making	RE planning and RE economic planning
Decision support systems	RE planning
Analytical Hierarchy Process	RE planning and wind farm projects
PROMETHEE	Geothermal projects, hydro-site-selection
ELECTRE	RE planning
Multi – attribute utility theory	Solar energy projects and RE planning
Fuzzy programming	Wind site selection and Solar System

The aim of the multi-criteria analysis is the choice of the optimal variant taking into account different criteria to be a major impact on the implementation and functioning of the solution. Multi-criteria analysis mainly focuses on discrete problems, which include issues of countable number of criteria and / or decisions. Today there are many proposals for analysis and synthesis of discrete multicriteria problems, which allow to solve difficult and complex issues [10].

The authors of this paper present the application of three selected methods, each of them belongs to a different group, which differ, among others, approach and way of finding the optimal solution.

The first of these, and also the oldest is classified as taxonomic methods, it consists on the construction of scalar, whose numerical value is a synthetic evaluation index 5 [11]. Aggregation of assessment can be carried out after coding the criteria values, usually expressed in various, not comparable with each other units. Both the choice of the synthetic index and coding method can slightly affect on result of the assessment. Therefore, the authors propose to use two methods of coding, in order to verify the results (Table 2).

Table 2. Coding method and synthetic evaluation index used in analysis.

Coding	Stimulant	Destimulant	Synthetic Evaluation Index
Neuman – Morgenstern	$z_{ij} = \frac{x_{ij} - x_{j\min}}{x_{j\max} - x_{j\min}}$	$z_{ij} = \frac{x_{j\max} - x_{ij}}{x_{j\max} - x_{j\min}}$	Adjusted index of summation $J_i = \sum_{j=1}^{m} (z_{ij} \cdot v_j)$
normalization	$z_{ij} = \frac{x_{ij}}{x_{j\max}}$	$z_{ij} = \frac{x'_{ij}}{x'_{j\max}}; \quad x'_{ij} = \frac{1}{x_{ij}}$	

x_{ij} - maximum value of j-th criterion after maximization
x_{jmax} - maximum value of j-th partial criterion
x_{jmin} - minimum value of j-th partial criterion
v_j –weight of j-th criterion

Second method, that authors propose to apply is Analytic Hierarchy Process (AHP) method, which has been developed by Thomas L. Saaty [12] Users of the AHP method first decompose their decision problem into a hierarchy of more easily comprehended sub-problems and then the decision makers systematically evaluate various elements by comparing them to each other two at a time, with respect to their impact on an element above them in the hierarchy. More than forty years AHP method is the reason for many disputes among scientific elite, because a large subjectivity of the process of finding the optimal solution. However, over the last years, AHP method becomes very popular, and despite the controversy it induces, so far, no one has managed to undermine the correctness of its operation.

The final method that was used to analyzed the problem of multi-criteria is PROMETHEE II (Preference Ranking Organization Method for Enrichment Evaluations)) belonging to the group of method based on identifying relationships outranking. PROMETHEE method uses information about the extent to which (on average) the alternative is preferable over the others (positive flow of preferences), and the extent to which the rest of the alternatives is preferable than given (negative flow of preferences).

4 The set of analyzed variants and criteria

Taking into account a number of assessment criteria analysis using the mathematical methods may include various solar collectors: flat-plate, some of which are gas, liquid-based or two-phase collectors and vacuum, vacuum-tube or focusing solar collectors. In order to assess the suitability of different multi-criteria methods to choose the best solution was analyzed three selected collectors: flat-plate collector Logasol SKN 4.0, vacuum collector Depsol DS 20 and vacuum collector HEVELIUS SCM-20 58/1800.

The collector Logasol SKN 4.0 is an universal flat-plate collector designed primarily for the needs of a few people family with the possibility of installation on a sloping roof, flat, facade and in the roof. Solar collectors Logasol SKN 4.0 are 2,25m^2 net aperture area, and are the ideal solution dedicated to heat utility water and central heating support. The collector includes aluminum absorber covered by a highly selective coating of black chrome applied by the vacuum method, which increases the durability of the collector.

Vacuum collector Depsol DS 20 designed for solar installation with 150-200 liter utility water heater for 2-4 people. Depsol DS 20 is a vacuum collector mounted on sloped roof, flat roof or directly on a vertical wall of the building. Vacuum collector Depsol DS 20 is a combination of the best features of vacuum collectors and flat plate collectors.

The solar collector Depsol DS is similar to the flat panel divided into a plurality of tubular elements. Each vacuum tube is a separate heating unit which is a flat absorber direct flow locked in vacuum. Depsol DS collectors are used to heat the hot utility water and support the central heating. In contrast to the collector Logasol SKN 4.0, Depsol has a copper absorber, coated with a layer with Tinox titanium oxide.

HEVELIUS SCM-20 collector is primarily intended for applications in large investment properties (due to the relatively large absorber area), but also works in Solar sets for several persons households. The collector is designed for use on horizontal surfaces and pitched roofs. It is built from double-walled vacuum glass tubes, made of tempered glass borosilicate.

It has a surface absorption gradually selective Al-N / Al. The distinctive feature of the HEVELIUS SCM-20 collectors is to use the Heat-Pipe technology. It consists in heating the lower end of the tube which is placed within the collector. This causes a phenomenon of evaporation of the liquid by taking heat as steam moves to the upper part of the tube. There subsequently condenses on the cold walls, releasing heat. The efficiency of heat transfer by convection combined with the evaporation is much more efficient than the thermal conduction through the solid matter.

The set of criteria that have been included in conducted multi-criteria analysis and values for the selected variants are presented in Table 3.

Table 3. A set of criteria considered in the multi-criteria evaluation.

Lp	Variant/ Method	A - Logasol SKN 4.0*	B - Depsol DS 20*	C - HEVELIUS SCM-20 58/1800*
c1	Net Aperture Area	2,25 m^2	2, m^2	2,6 m^2
c2	Fluid Content	1,35 l	2,64 l	1,5 l
c3	Dry Weight	41 kg	45 kg	49 kg
c4	Net price	2199,00 zł	2999,00 zł	2977,00 zł
c5	Optical efficiency	77%	78%	69,7%
c6	Absorber Material	aluminum	copper	Al.-N/Al.
c7	A1 index	3,87 W/(m^2 K)	1,936 W/(m^2 K)	1,696 W/(m^2 K)
c8	A2 index	0,012 W/(m^{2}K^2)	0,006 W/(m^{2}K^2)	0,0099 W/(m^{2}K^2)
c9	Warranty	5 years	10 years	5 years
c10	Maximum stagnation temperature	199^{0}C	191,9^{0}C	225,4 ^{0}C
c11	Versatility (grading 1-5)	3	3	4

* data from the manufacturer and dealers.

5 Results and Discussion

A comprehensive review of MCDM (Multi-Criteria Decision Making) methods was undertaken to compare variants. The analyzed methods include weighted sum/product method, analytical hierarchy process (AHP) and preference ranking organization method for enrichment evaluation (PROMETHEE). Below, the authors present partial calculations for each of them. First step in taxonomic methods is coding values which is shown in Table 4.

Table 4. Values coded by using the method of normalization / Neumann - Morgenstern

Criteria	Logasol SKN 4.0		Depsol DS 20		HEVELIUS SCM-20 58/1800	
c1	0,87 /	0,42	0,77 /	0,00	1,00 /	1,00
c2	0,51 /	0,00	1,00 /	1,00	0,57 /	0,12
c3	1,20 /	1,00	1,09 /	0,50	1,00 /	0,00
c4	1,36 /	1,00	1,00 /	1,00	1,01 /	0,03
c5	0,99 /	0,88	1,00 /	1,00	0,89 /	0,00
c6	0,67 /	0,50	0,33 /	0,00	1,00 /	1,00
c7	1,00 /	0,00	2,00 /	0,11	2,28 /	1,00
c8	1,00 /	0,00	2,00 /	0,00	1,21 /	0,35
c9	0,50 /	0,00	1,00 /	1,00	0,50 /	0,00
c10	0,88 /	0,21	0,85 /	0,00	1,00 /	1,00
c11	0,75 /	0,00	0,75 /	0,00	1,00 /	1,00

Coded values were multiplied by the weight vector, then calculated according to the formula showed in Table 2 for adjusted index of summation (Table 5). Weights were based on expert judgment.

Table 5. The adjusted index of summation (IS) for two coding methods

	Normalization			**Neumanna – Morgensterna**		
	A	B	C	A	B	C
c1	0,05	0,05	0,06	0,02	0	0,06
c2	0,04	0,08	0,04	0	0,08	0,01
c3	0,04	0,03	0,03	0,03	0,02	0
c4	0,23	0,17	0,17	0,17	0,17	0,01
c5	0,13	0,13	0,12	0,11	0,13	0
c6	0,05	0,03	0,08	0,04	0	0,08
c7	0,11	0,23	0,26	0	0,01	0,11
c8	0,11	0,23	0,14	0	0	0,04
c9	0,04	0,08	0,04	0	0,08	0
c10	0,07	0,07	0,08	0,02	0	0,08
c11	0,05	0,05	0,07	0	0	0,07
IS	**0,926**	**1,136**	**1,083**	**0,391**	**0,482**	**0,455**

Second method used in proposal analysis was AHP. Decomposition of discussed problem into a hierarchy is shown in Figure 1.

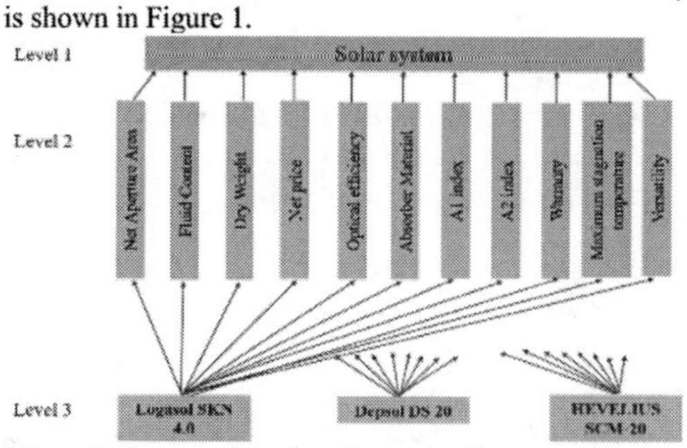

Figure 1. Decomposition for solar system (Source: own).

Calculations for the second level is shown in Table 6.

Table 6. Calculation of eigenvectors for second level in AHP method

No	**Criterion**	Vector wi
c1	Net Aperture Area	0,032
c2	Fluid Content	0,042
c3	Dry Weight	0,016
c4	Net price	0,264
c5	Optical efficiency	0,157
c6	Absorber Material	0,049
c7	A1 index	0,141
c8	A2 index	0,144
c9	Warranty	0,058
c10	Maximum stagnation temperature	0,062
c11	Versatility	0,035
	λmax	**12,12**
	C.I.	**0,11**
	C.R.	**0,07**

And the third level (Table 7).

Table 7. Calculation of eigenvectors for second level in AHP method

Lp	**Logasol SKN 4.0**	**Depsol DS 20**	**HEVELIUS SCM-20 58/1800**	**λ_{max}**	**C.I.**	**C.R.**
c1	0,27	0,12	0,61	**3,10**	**0,05**	**0,09**
c2	0,18	0,68	0,20	**3,04**	**0,02**	**0,03**
c3	0,54	0,30	0,16	**3,01**	**0,01**	**0,01**
c4	0,63	0,17	0,19	**3,01**	**0,01**	**0,01**
c5	0,43	0,43	0,14	**3,00**	**0,00**	**0,00**
c6	0,30	0,16	0,54	**3,01**	**0,01**	**0,01**
c7	0,07	0,35	0,58	**3,04**	**0,02**	**0,04**
c8	0,08	0,34	0,58	**3,04**	**0,02**	**0,03**
c9	0,17	0,67	0,17	**3,00**	**0,00**	**0,00**
c10	0,24	0,14	0,62	**3,03**	**0,01**	**0,02**
c11	0,20	0,20	0,60	**3,00**	**0,00**	**0,00**

Finally we obtain the overall priority (Table 8).

Table 8. Overall priority for AHP method

No	**Variant**	
A	Logasol SKN 4.0	**0,328**
B	Depsol DS 20	**0,311**
C	HEVELIUS SCM-20 58/1800	**0,364**

In the PROMETHEE the intermediate results are presented in Tables 9 and 10.

Table 9. Substitute criterion for partial criteria

	Pj[δi(a, b)]	**Pj[δi(b, a)]**	**Pj[δi(a, c)]**	**Pj[δi(c, a)]**	**Pj[δi(b, c)]**	**Pj[δi(c, b)]**
c1	1,00	0,00	0,00	1,00	0,00	1,00
c2	0,00	1,00	0,00	1,00	1,00	0,00
c3	1,00	0,00	1,00	0,00	1,00	0,00
c4	1,00	0,00	1,00	0,00	0,00	1,00
c5	0,00	0,00	0,45	0,00	0,52	0,00
c6	0,05	0,00	0,00	0,00	0,00	0,5
c7	0,00	1,00	0,00	1,00	0,00	1,00
c8	0,00	1,00	0,00	1,00	1,00	0,00
c9	0,00	1,00	0,00	0,00	1,00	0,00
c10	1,00	0,00	0,00	1,00	0,00	1,00
c11	0,00	0,00	0,00	0,50	0,00	0,50

Subsequently for each pair of decision variants x and y is calculated aggregated indexes of preferences:

$$\Pi(x,y)=\sum_{j=1}^{k} w_j P_j(x,y) \tag{1}$$

$$\Pi(y,x)=\sum_{j=1}^{k} w_j P_j(y,x) \tag{2}$$

Next, for each variant of the decision-making are calculated positive and negative flow preferences (1) and (2).

$$\Phi^{+}(x) = \frac{1}{n-1}\sum_{y \in A}\Pi(x, y) \quad (3)$$

$$\Phi^{-}(x) = \frac{1}{n-1}\sum_{y \in A}\Pi(y, x) \quad (4)$$

The final stage is to calculate net flows for each of the considered variants by using equations (3) and (4) (Table 10).

Table 10. Positive and negative flow preferences and net flows for each variants

Variant	Logasol SKN 4.0	Depsol DS 20	HEVELIUS SCM-20 58/1800
Positive flow	0,29996	0,3783	0,47174
Negative flow	0,43371	0,4037	0,31255
NET FLOW	**-0,1337**	**-0,025**	**0,1592**

Taking into account the results of the encoding using the normalization method and Neuman – Morgenstern method, authors received Variant B as the best. Then the Variant C and Variant A. In a case of normalization, variants are scattered uniformly and in the case of Neuman - Morgenstern method Variant B advantage is significant and variants A and C have similar results (Fig. 2).

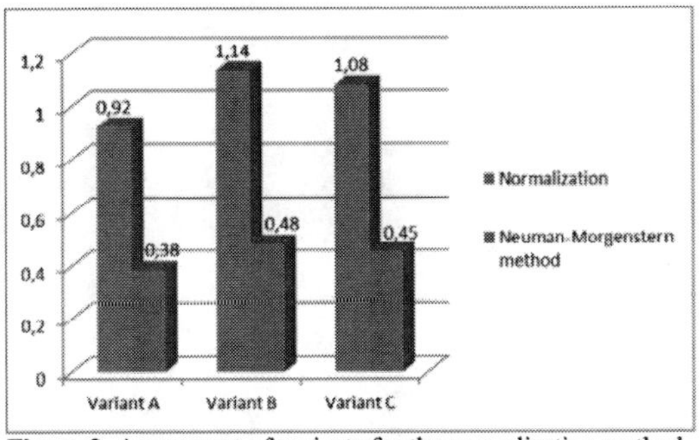

Figure 2. Assessment of variants for the normalization method and Neumann-Morgenstern method (Source: own).

In turn, the AHP method gives quite a different results. The best solution here is variant C, then variant A, and the worst result gives a variant B (Fig. 3).

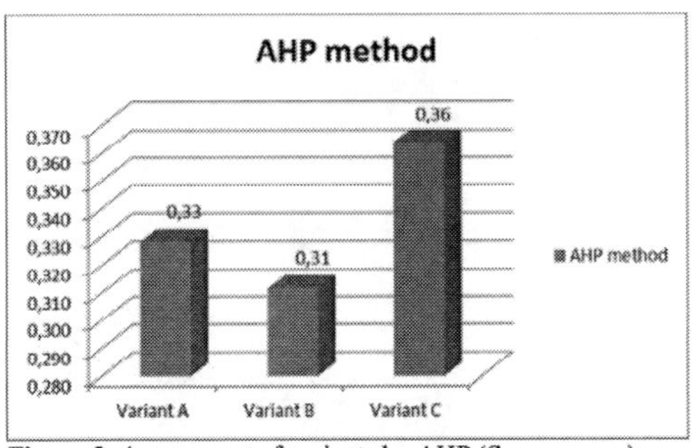

Figure 3. Assessment of variants by AHP (Source: own).

PROMETHEE II method generates results even different. Like the AHP method indicates the variant C as the best solution, but in second place situates variant B, and only as a last variant A (Fig. 4).

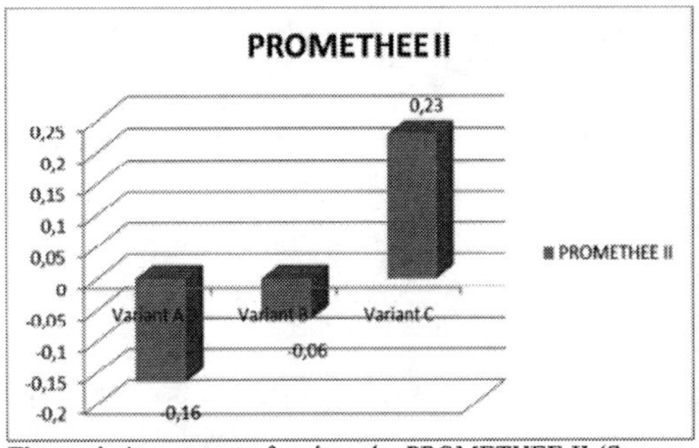

Figure 4. Assessment of variants by PROMETHEE II (Source: own).

6 Conclusions

Sustainability of buildings is currently investigated by a number of evaluation systems that reflect their design, technical, technological and material solutions [13]. In general, a measure of evaluation of alternative solutions is the level of performance criteria specified by the decision maker. The validity of the criteria, their selection, but also the choice of assessment methods substantially affect on the result of the analysis. Experts are expected to assessments consistent with the construction knowledge, reliable, objective and taking into account the specific of the decision situation. It is difficult to clearly define individual preferences, system of value and motives of an expert [14].

The advantage of taxonomic methods is the intuitiveness of use and the ability to give validity to each parameter, which reflects well the actual reality between the different criteria. As a disadvantage most often mentions the need for encryption of individual data, which often, depending on the choice of encoding method may influence the final result.

The advantages of AHP method are simplicity of making choices by comparison, and no need to use complicated mathematical calculations. Critical voices refer mainly to the lack of theoretical bases of the hierarchy construction, as a result, for exact decision situations it is possible to obtain different solutions. The second disadvantage which is most often mentioned is the high sensitivity to small changes and the impact on the final result even neutral criterion.

Application of PROMETHEE II allows for more realistic analysis of preferences, because this technique - using the threshold of equivalence and the threshold preferences - take into account the fact that the real problems of decision-making accompain by an uncertainty, imprecision, indefiniteness and variability for both data as well as evaluations and preferences of participants in the decision making process. On the other hand, the subjective choice of preference function can significantly affect the final result.

The ranking procedure which represents the PROMETHEE II method takes into account uncertainty and imprecision of all the problems accompanying the decision that much more can reflect the reality of assumptions than the method consisting in calculating the weighted average. The best possible result in the PROMETHEE II method was repeated also in the calculation carried out by AHP which confirms the validity indication of Variant 3 as the most favorable.

References

1. K.Neupauer, J. Magiera, Efficiency analysis of different types of solar collectors, Tech. Trans., Issue **4**, (2009)
2. B. J. Huang, V. A. Petrenko, I. YA. Samofatov and N. A. Shchetinina, Solar Energy **71**(4), 269–274.
3. P. Aragonés-Beltrán, F. Chaparro-González, J.-P. Pastor-Ferrando, A. Pla-Rubio, Energy **66**, 222-238 (2014)
4. J.R. San Cristóbal, Ren. En. **36**, 498-502 (2011)
5. K. Zima, Case-Based Reasoning in Employer's Estimates", International Multidisciplinary Scientifif Geoconference, 14th GeoConference on Ecology, Economics, Education And Legislation Conference Proceedings, **III**, 17-26 June 2014, Bulgaria, 75-82.
6. Cinca M., Bancea O., & Adam M. Study of energy saving in buildings for improved comfort conditions, 12th International Multidisciplinary Scientific GeoConference, SGEM 2012 Conference Proceedings, June 17-23, 2012, Vol. 5, 793–800;
7. Sedlakova A., Vojtus J., & Burdova E. K. Design of consumer model and environmental assessment of used building materials, 12th International Multidisciplinary Scientific GeoConference, SGEM 2012 Conference Proceedings, June 17-23, 2012, **5**, 235 – 242.
8. A. Leśniak, Bid assessment with the use of fuzzy sets theory, Proceedings Of The International Conference Of Numerical Analysis And Applied Mathematics 2014 (ICNAAM-2014), AIP Conference Proceedings **1648**, Article Number: UNSP 600006, 2015
9. A. Leśniak, E. Plebankiewicz, J. of Manag. in Eng. **31**(2), 2013. Article Number: 04014032
10. K. Duzinkiewicz, Struktury i algorytmy wspomagania decyzji 2012/2013. Wieloatrybutowe problemy decyzyjne – metody rozwiązywania, Katedra Inżynierii Systemów Sterowania, Politechnika Gdańska [online][12.02.2016] http://slideplayer.pl/slide/59738/.
11. J. Szwabowski, J. Deszcz, Metody wielokryterialnej analizy porównawczej : podstawy teoretyczne i przykłady zastosowań w budownictwie, (Gliwice : Wydaw. Politechniki Śląskiej, 2001)
12. T. L. Saaty, Fundamentals of Decision Making and Priority Theory. (RWS Publications, Pittsburgh, Pennsylvania, 2001)
13. M. Kozlovska, R. Baskova, M. Krajnak, A. Tazikova, Sustainability of construction from perspective of applications technologies for green buildings, 4th International Multidisciplinary Scientific GeoConference SGEM Conference Proceedings, June 19-25, 2014, **6(2)**, 323-330 (2014)
14. M. Książek, Ekspercki system oceny rozwiązań w budownictwie, logitrans - VII Konferencja Naukowo-Techniczna Logistyka, Systemy Transportowe, Bezpieczeństwo w Transporcie, 1135-1145.

E3S Web of Conferences **10**, 00079 (2016) DOI: 10.1051/e3sconf/20161000079
SEED 2016

Analysis Of The Profitability of Investment In Renewable Energy Sources On The Example of A Semi-Detached House

Elżbieta Radziszewska-Zielina[1a] Renata Rumin[2]

[1] *Institute of Management in Construction and Transport, Cracow University of Technology, Poland*
[2]*Cracow University of Technology, Poland*

Abstract. The paper presents an analysis of the profitability of investment in a selection of systems based on renewable energy sources on the example of a semi-detached house with a floor area of nearly 150 m^2. The analysis was conducted in three variants:1) the application of a biomass boiler, solar collectors and photovoltaic panels; 2) a heat pump, solar collectors and photovoltaic panels; 3) a heat pump, solar collectors and a wind turbine. Next, the proposed solutions were compared with a traditional heating system featuring a gas-powered boiler and obtaining electricity through the power distribution grid. The investment and operational costs were calculated after the selection of the appropriate equipment. The example under analysis indicates that variant 3 entailed the highest investment costs (the purchase and fitting of a heat pump as well as the drilling required to install underground pumps). Traditional technical solutions were proven to produce the highest costs. The investment profitability analysis was conducted using both the simple and discounted methods of evaluating profitability. It was established that variant 3 yielded the shortest discounted payback period - less than 13 years. It is estimated that the investment will reach the break-even point after this time. As a summary, it must be noted that the profitability of a given investment depends on the adopted technical solutions. Due to high initials costs, investors often lack interest in the purchase and installation of renewable energy generation systems; however, the contribution of the investors can be lowered with the use of subsidies and price reductions regarding the installation of renewable energy generation systems. We can observe a decrease in the price of the most popular and efficient renewable energy systems.

1 Introduction

Only 10% of Poland's energy is generated using renewable energy sources. The production of energy is still based on fossil fuels, mostly coal and lignite. This leads to a gradual increase in air pollution, something that is particularly felt in urbanised areas. The large emission of dust particles, sulphur and nitrous oxide negatively impacts the health of humans and leads to the degradation of buildings and architectural elements. The use of fossil fuels is also economically tied to the problem of finite resources and the rising difficulty in accessing their deposits. It is estimated that the global crude oil and natural gas deposits are going to be depleted in around 50 years, while those of coal in around 150 years. This directly impacts the costs of energy. The factors presented above, along with an increase in social awareness, are leading to a heightened interest in alternative energy sources [1, 2, 3, 6].

Non-conventional energy sources are used in many fields of everyday life. One can observe an increase in interest in energy efficient building construction, as well as in systems that utilise alternative energy sources for use in residential architecture. Public opinion polls indicate that 45% of citizens prefer the use of renewable energy, mostly in the form of solar or wind power. It is estimated that the number of households that have invested in micro-installations is going to reach around 2500 thousand [4, 5]. Thanks to the rising popularity of alternative energy systems, we can observe a rise in the demand for modern technologies, which is going to result in the lowering of the costs of installing them in the future. Profitability is one of the most important factors in construction projects, with 50% of respondents claiming that the payback period should not exceed 7 years [4, 5]. A large influence on the aforementioned rise in popularity is attributed to changes in legislation that force the adoption of different approaches to designing buildings and the use of alternative energy sources. Pollution emission and energy demand standards are becoming more and more strict, like the ones regarding heat transfer coefficient values [7, 8].

The paper illustrates an analysis of the profitability of a selection of systems that are based on renewable energy sources on the example of a semi-detached single family house.

[a] Corresponding author: eradzisz@izwbit.pk.edu.pl

2 Description of the analysed building

The analysis was conducted for an existing semi-detached single family house, fit for use by two four-person families. The design of the house was developed in 2010. The building is composed of two residential sections, with each of them having a net floor area of 147,42 m^2. The total volume of the structure amounts to 1141,40 m^3. The building is symmetrical in terms of both form and function. It is located in the town of Nowy Sącz in the Małopolskie Voivodship. The house is located in the III climate zone (PN-EN 12831). The building has been determined to be partially covered. It is a single storey house with a usable attic, covered by a gabled roof with a 30 degree incline. The house faces the south. The foundations are composed of both continuous and point footing with a height of 40 cm and a width of 60 to 70 cm. They are placed on a 10 cm layer of lean concrete. The building is insulated against water by the means of a waterproofing compound. The floor slabs are placed on a 15 cm thick layer of sand, with the slabs themselves being 15 cm thick. The slab on grade is waterproofed by the means of two layers of heated bitumen membranes connected by the means of an anionic asphalt emulsion, with thermal insulation provided by 10 cm thick styrofoam plates. The walls of the foundations are made of concrete and are 25 cm thick. The part of the foundations located below ground is waterproofed by a layer of bituminous coating with thermal insulation provided by an 8 cm thick layer of styrofoam that is suitable for use underground. The house has two-layer external masonry walls. The load-bearing layer is composed of 25 cm thick, traditionally bound ceramic blocks. The external layer of thermal insulation is composed of 12 cm thick styrofoam plates. The total thickness of its external walls is equal to 37 cm. The heat transfer coefficient for these walls equals U=0,29 W/m^2K. There is a wooden entry door and a garage door in the northern facade, possessing a heat transfer coefficient of U=1,5W/m^2K, while the heat transfer coefficient of the windows equals U=1,7 W/m^2K. The building in question is covered by a gabled timber purlin and collar-beam roof. The roof cladding is composed of concrete tiles. The roof is thermally insulated using mineral wool with a thickness of 20 cm. The heat transfer coefficient for the roof is equal to 0,18 W/m^2K.

3 The analysed variants

Variant I: natural gas-powered boiler, electricity drawn from the power grid.

The first variant to be analysed is features the use of a natural gas-powered boiler that is used for heating and providing warm water, with electrical power being drawn from the power grid. This variant is going to be used for comparison.

The building already contains a natural gas-powered condensing boiler with the appropriate infrastructure, so it does not qualify as an investment cost. The key element of this variant are the operation costs. These include the costs associated with the transfer of gas fuel, drawing electricity from the power grid and the maintenance costs of the boiler. According to art. 62 of the Construction Law Act [7], the owner of a building is to perform an assessment of the technical condition of its gas installation. Appropriate calculations have been made, yielding the yearly cost of heating the building and its water at 3 993,47 PLN, with the cost of electricity amounting to 1 891,26 PLN, which sums up to a cost of 5 942,86 PLN

Total operational cost: **5 942,86 PLN.**

Variant II: biomass-powered boiler, solar thermal collectors and photovoltaic panels.

The biomass-powered boiler is going to heat the building and provide warm water, with the solar collectors providing the ability to reheat it, while photovoltaic panels are going to be used for energy generation. The Polish market is currently replete with biomass-powered boilers. The most popular ones include those that can use wood or pellets for fuel. The analysis was performed for a pellet-fuelled boiler and a flat set of solar power collectors. Such a set is suitable for a building with 4 inhabitants.

Another piece of equipment utilising the energy of the sun are photovoltaic panels. The analysed case features a photovoltaic power generator with a capacity for generating 3,0 kW. The existing heating system of the building is assumed to be in good condition, without the need to replace it or any of its parts.

Conditions set for the calculations: 50% of the energy generated by the photovoltaic panels is going to be used by the inhabitants, while the other 50% is going to be sold to the power company. The estimated production of power by photovoltaic panels amounts to 3000 kWh.

Total investment costs: **46 669,46 PLN**

Total yearly operational costs: **2 725,74 PLN**

The authors performed an analysis of the profitability of using a biomass-powered boiler to provide heating to a single family house, solar collectors to provide warm water and photovoltaic panels to produce electricity.

The discounted payback period is the time required for the investment costs to be covered by the benefits generated by the investment, while taking into account the possible changes to the value of money. The shorter the payback period the more profitable the investment.

The simple payback period in this case amounts to 14,5 years, while the discounted payback period is 19,5 years. The investment featuring this configuration is thus unprofitable.

Variant III: heat pump, solar thermal collectors and photovoltaic panels.

The purpose of the heat pump is going to be to provide heating for the building as well as to produce warm water, which is going to be reheated using solar thermal collectors, while photovoltaic panels are going to be used to produce electricity.

The analysis is based on a heat pump with the following parameters: the pump operates within the range of 5,8 do 17,2 kW, has a COP coefficient of 4, and its size equals 1049 x 600 x 845 mm. It weights 119 kg and produces <45 dB of noise. It can work in unison with an additional source of heat. The analysis estimates that 40% of the energy needed to provide warm water is going to be obtained from the heat pump.

The operational costs need to be increased to take into account the value of mandatory maintenance.

Total investment costs: **73 059,85 PLN**

Total yearly operational costs: **2 144,49 PLN**

The simple payback period in this case is 19,2 years, while the discounted payback period (with a discount rate of 3%) is 29,7 years. The investment featuring this configuration is thus unprofitable.

Variant IV: heat pump, solar thermal collectors and a wind turbine.

The heat pump is going to be used to provide heating for the building and to produce warm water, while solar thermal collectors are going to be used to reheat it, with the wind turbine used to produce electric power.

The interest in compact wind turbine generators in Poland is continuously rising. For the purposes of this analysis, a turbine with a power generation capacity of 3,0 kW was used.

The turbine is based on a patented structure that features a five-winged fan with a symmetrical, aerodynamic shape. This shape allows it to produce a maximum power output equal to the full capacity of its rotor even at low wind speeds. The effectiveness of this model can exceed that of three-fanned models by up to 40%.

The investment costs are relatively high, which is why it is important to take into account the operational costs when determining the profitability of the investment.

Total investment costs: **79 854,73 PLN**

Total operational costs: **+1 752,27 PLN**

The simple payback period in this case amounts to 10,4 years, while the discounted payback period (with a discount rate of 3%) is 12,7 years. The investment in question has been determined to be profitable using this configuration.

4 Conclusions derived from the analysis

The solutions that utilise unconventional energy generation solutions are currently quite expensive when compared against wages, with investment costs considered quite high. However, we can observe a steady decline in the price of equipment that uses alternative energy sources. There currently exist numerous opportunities to obtain appropriate subsidies, which make the overall investment costs of purchasing and installing such equipment lower, leading to an increased popularity of this type of solution. Fig. 1 shows a comparative plot illustrating the investment costs of each of the variants that were analysed.

As we can see from the plot presented above (fig. 1) the cheapest solution appears to be variant II, which utilises a biomass-powered boiler and photovoltaic panels. The most expensive of the variants is variant IV, utilising a heat pump and a wind turbine. The cost of purchasing and installing a biomass-powered boiler is rather low. The structure of the boiler, which uses heat generated from the burning of pellets, is not complicated. However, it needs to be pointed out that operating it is rather time-consuming. It requires manual operation and is not maintenance-free, in contrast to a heat pump, for instance.

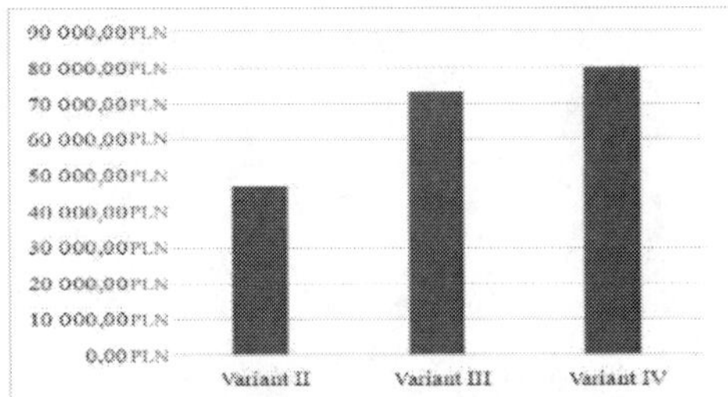

Figure 1. The investment costs for each of the variants

The difference in the amount of investment costs for the variant in which the photovoltaic panels were replaced with a wind turbine is not that large. However, there is a marked increase in the energy that is being produced. Maintenance costs are the most important factor in the comparison between the variants. The analysis presented above features costs that were calculated for a period of one year, which amounted to the purchase of fuel (natural gas, pellets), the purchase and selling of electrical energy (the amount of energy that was used was increased in the case of equipment which made use of renewable energy sources) and the costs of yearly maintenance. Fig. 2 illustrates the operational costs for each variant.

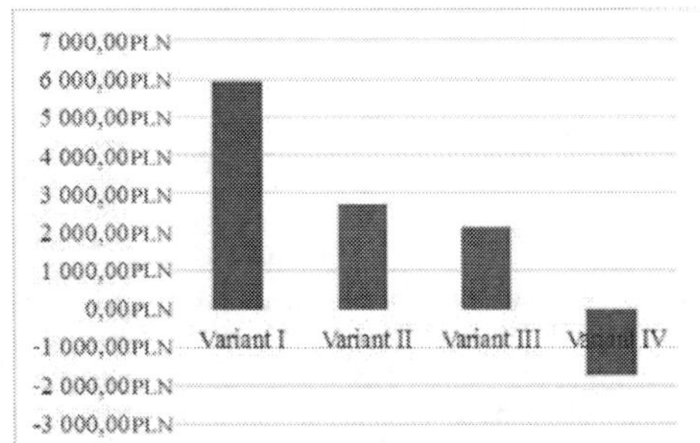

Figure 2. The operational costs for each of the variants

Figure 2 illustrates that the use of natural gas to heat a building and produce warm water is over twice as expensive as using renewable energy sources. The yearly cost of using a boiler consuming natural gas purchased from the gas distribution network amounts to 5 942,86 PLN. The lower yearly operational costs of variant II in comparison to the current equipment installed in the building are based on the fact that the costs of fuel are lower, as well as due to the generation and selling of excess energy back to the power company, in accordance with the Renewable Energy Sources Act. Variant IV, thanks to an increase in the amount of electrical power that it produces thanks to a wind turbine

 DOI: 10.1051/e3sconf/20161000079

rather than photovoltaic panels, allows a homeowner to obtain a profit from this system.

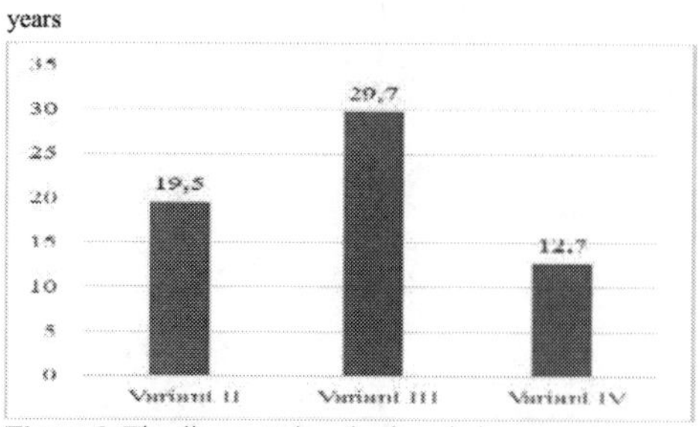

Figure 3. The discounted payback period

The simple payback periods for variants II and IV show that their use can be profitable, as the period for these variants is lower than 15 years. Investing in variant III is unprofitable. The costs of a heat pump are too great for the investment to become profitable in the case of the single family house under analysis.

The discounted payback period (fig. 3) indicates that the most profitable variant is variant IV, as the period is less than 13 years. This is the time it takes for the investment to reach its break-even point. The most profitable investment is variant IV (fig. 4).

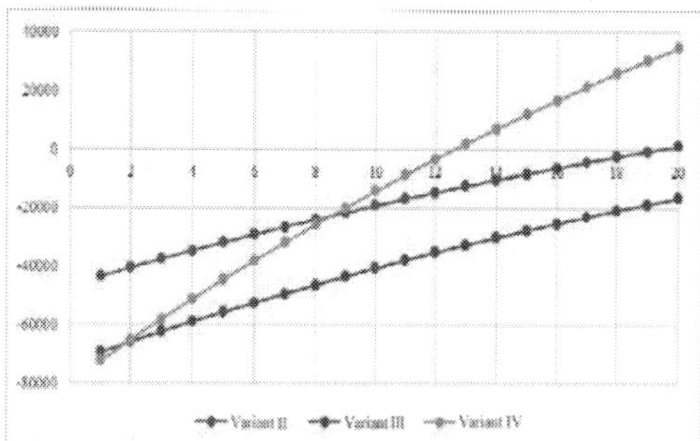

Figure 4. NPV for 20-year period for each of the variants

The investment profitability analysis of investing in renewable energy sources for a single family house has shown that investing in unconventional energy sources can be profitable. However, a potential investor needs to compare the various systems and pick the optimal one for a given structure.

5 Conclusion

The analysis of three variants of systems based on renewable energy sources used for heating and producing warm water and comparing them to an existing system in a semi-detached single family house has shown that such equipment can be profitable, provided its choice optimal. Selecting equipment with different parameters from those that the authors have picked for the analysis may produce different results.

Unconventional energy sources in Poland are relatively expensive when wage-wise, with investment costs being comparatively high. However, we can observe a steady decline in price levels of such equipment. There currently exists a possibility of obtaining subsidies that can make the overall investment costs needed to purchase and install such systems much lower, making their use more and more popular. The operational costs of renewable energy systems are much lower than those of traditional systems. Due to the fact that alternative energy systems mostly make use of free energy, there are no fears related to uncertainty related to an unstable market situation regarding the rising prices of fossil fuels. The investment costs of the systems that use fossil fuels are much lower than those of renewable energy sources. However, their operational costs can be considered high. The profitability of an investment depends on the variant under analysis. Investors are often uninterested in the purchase of such systems. However, thanks to numerous subsidies and price reductions that are obtainable for the purposes of installing renewable energy generation systems, the costs to investors can be substantially reduced. We can also observe a reduction in the price of the most popular and the most effective systems making use of renewable energy sources, as well as a marked increase in the demand for such systems.

References

1. A. Hilarowicz, J. Kozioł (ed.), *Odnawialne źródła energii- badania oddziaływań społecznych*, (Politechnika Śląska, Gliwice 2013)
2. F. Krawiec, *Odnawialne źródła enegii w świetle globalnego kryzysu energetycznego. Wybrane problemy*, (Difin, Warsaw 2010)
3. W. Lewandowski, *Proekologiczne odnawialne źródła energii*, (WNT, Warsaw 2012)
4. „Polacy o źródłach energii odnawialnej. Wyniki badania opinii publicznej. Wycinek z „Krajowego Planu Rozwoju Mikroinstalacji Odnawialnych Źródeł Energii do 2020 roku". Instytut Energetyki Odnawialnej, Związek Pracodawców Forum Energetyki Odnawialnej, Warsaw (2013) http://ieo.pl/pl/raporty/71--24/file(15.12.2015)
5. Krajowy Plan Rozwoju Mikroinstalacji Odnawialnych Źródeł Energii do 2020 roku - Synteza, Instytut Energetyki Odnawialnej, (2013) http://demokracjaenergetyczna.pl/uploads/images/pdf/ekspertyzy_Krajowy_Plan_Rozwoju_Mikroinstalacji_OZE.pdf(15.12.2015)
6. Wnioski z analiz prognostycznych na potrzeby Polityki energetycznej Polski do 2050 roku, Warszawa (2014) http://www.cire.pl/pliki/1/mg_pep_2050_wnioski_z_analiz_prognostycznych_20140811.pdf(15.12.2015)
7. Ustawa z dnia 7 lipca 1994 Prawo Budowlane, (Dz.U. nr 89 poz. 414)
8. Ustawa z dnia 20 lutego 2015 r. o odnawialnych źródłach energii (Dz.U. poz.478)

DOI: 10.1051/e3sconf/20161000080

Bonds convertible to raw materials in the context of bonds convertible to shares and ordinary bonds

Robert Ranosz[1,a]

[1]*AGH University, Department of Economics and Management in Industry, Al. Mickiewicza 30 30-059 Kraków, Poland*

Abstract. This article presents an analysis of bonds convertible to raw materials in the context of ordinary bonds and bonds convertible to shares. The author of the publication decided to explore what kind of benefits an enterprise issuing bonds and a bondholder would receive as exemplified by a hypothetical investment, using various (aforementioned) sources of foreign capital. The article is composed of four parts. The introduction hereto outlines some general assumptions and the purpose of the study. The second part of the article includes the assumptions for the aforementioned comparative analysis. It was suggested that since such variables as the price of raw materials, shares and operating costs of a company are difficult to determine, they should be distributed so that on the basis thereof the final NPV may be established by applying the Monte Carlo method. The third part of the study focuses on the examination (comparison) of the bonds convertible to raw materials in the context of ordinary bonds and bonds convertible to shares. The result of such an analysis is the NPV histogram for particular sources of financing, a probability distribution defining the probability of occurring individual NPV values. The last part constitutes a summary of the entire study.

1 Introduction

The financing of mining investments is one of the most important financial problems faced by the Polish mining industry. Among the tools for financing investments are bonds, which have been used more frequently in the last few years by both Polish and foreign mining companies. Their success results mainly from their flexibility and the relatively low cost of the capital. Due to their high popularity the corporate bond market is developing, and hence new options may be embedded. The additional options lower the cost of the said bonds, in contrast with other bonds without such options. The most popular options embedded in bonds include a possibility of changing (converting) debts to shares. The author of this publication deliberates on bonds convertible to raw materials. It is assumed that this solution retains the features of every convertible bond, and thus lowers the interest on the debt security compared to a regular corporate bond [1,2]. The purpose of this article is to compare the bonds convertible to raw materials with ordinary bonds and bonds convertible to shares in the context of benefits derived by both the issuer and bondholder. As illustrated by the calculation example, it was assumed that a mining company shall be the bond issuer and an electric power company the bondholder.

2 Initial assumptions for the analysis

The efficiency of investment is significant not only for the investor, but also the company using the capital. The investment is more profitable if the return generated thereon is higher. One of the most frequently used methods for estimating the efficiency of the investment is the NPV (Net Present Value) based on the discounted cash flows. The NPV is influenced by many factors, such as: the forecast sales price, forecast level of costs, structure and cost of the capital, and source of the capital. This article focuses on the last of the above-mentioned factors. Due to the nature of this publication, it was decided to analyse only corporate bonds. Coupon bonds, convertible bonds and bonds convertible to raw materials were taken into consideration for the purpose of this analysis. To obtain the most reliable results possible, the Monte Carlo method was applied. Therefore, the result of the analysis shall be the expected NPV (E(NPV)).

Due to the fact that the market price of raw material, price of share, and level of operating costs may be subject to changes in the period of the analysis, and that the length of such a period is difficult to predict, the analysis was performed using the Monte Carlo method. To conduct the analysis correctly, it is crucial to first determine its input parameters (table 1). In the performed study, the distribution of random factors was also assumed

[a] Corresponding author: rranosz@agh.edu.pl

DOI: 10.1051/e3sconf/20161000080

(table 2). In the case of the price of raw material and price of share, it was assumed that the probability distribution is the regular distribution, where such parameters as the expected value and standard deviation [3] shall be established, whereas in the case of the operating costs, it was assumed that it is the triangular distribution, where the minimum, maximum and most probable values [4] need to be determined. Furthermore, it was decided that the bondholder shall convert debt into raw material at the time when the market price of the raw material becomes profitable, and hence at the time when the market price of the raw material exceeds the conversion price increased by the amount of the lost profits in the form of the interest. The bondholder shall always convert half of the debt. A similar assumption was made regarding bonds convertible to shares. However, in such a case, the amount of the lost profits was not included. It was also decided that the bondholder shall not sell the acquired shares and shall derive profits from the dividend due there to [5,6].

Table 1. Initial assumptions for the calculation example

Variable	Unit	Value
Forecast period	[years]	7
Nominal value of a bond	[PLN]	100
Number of issued bonds	[pcs]	20,000
Conversion factor*		0.5
Initial price per share	[PLN]	230
Initial market price per raw material	[PLN/Mg]	230
Initial operating cost	[PLN/Mg]	170
Interest rates on bonds	[%]	5
Income tax	[%]	19
Own (capital) contribution	[ths. PLN]	200,000
Raised capital	[ths. PLN]	200,000
WACC	[%]	8.5

*bonds convertible to raw materials and shares
Source: author's own study

Table 2. Parameters defining the distribution of the adopted variables

Variable	Type of distribution	Parameters defining the distribution		
		Expected value		Standard deviation
Change of the price per raw material	Normal	26%		41%
		Expected value		Standard deviation
Change of the price per share	Normal	81%		94%
		Minimum	Most probable	Maximum
Change of operating costs	triangular	-21.2%	16.1%	90.7%

Source: author's own study

On the basis of the adopted assumptions (table 1), a comparison between individual sources of financing was made (bonds convertible to raw materials, ordinary bonds, and bonds convertible to shares) in the context of benefits obtained by the parties to a contract (the bondholder and the mining company). On the basis of the calculation example, it was also decided that in the case of financing investments through bonds convertible to raw materials, the bondholder (a power plant) shall sign a contract with the mining company for the delivery of raw materials, and thus, in the event that the debt is converted into a raw material, the bondholder shall - in exchange for reducing the debt - receive a part of the said delivery (conversion factor* the amount of the converted bonds) "for free".

3 Analysis of efficiency of investments using various types of bonds

Within the framework of the analysis performed using the Monte Carlo method, 1170 drawings were made in accordance with the assumptions described in the previous chapter.

In the first step, the benefits that the mining company may obtain were analysed. The result of this analysis is the NPV histogram for particular types of bonds (figure 1) and expected values (table 3). To better illustrate the differences between the examined sources of financing, the probability distribution was also provided to show the possibility of reaching the NPV within the individual divisions (figure 2).

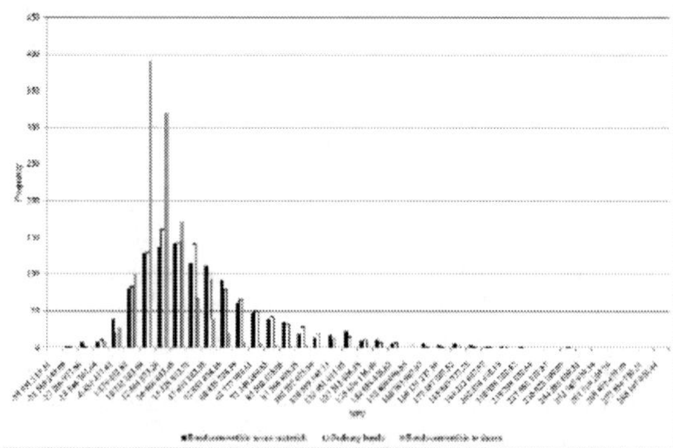

Figure 1. NPV histogram for particular types of bonds (for the bond issuer)
Source: author's own study

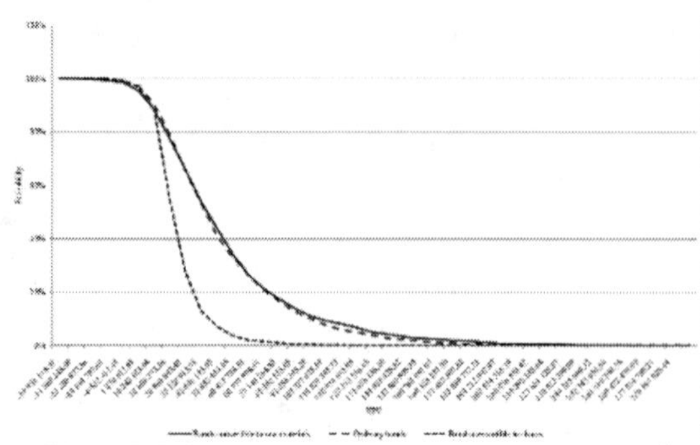

Figure 2. Probability distribution for the NPV depending on the source of financing (for the bond issuer)
Source: author's own study

DOI: 10.1051/e3sconf/20161000080

Table 3. Expected value of cash flows (for the bond issuer) for particular sources of foreign capital

Type of bond	Unit	E(NPV)
Ordinary bonds	[ths. PLN]	41,343,675.99
Bonds convertible to raw materials	[ths. PLN]	43,312,380.24
Bonds convertible to shares	[ths. PLN]	18,289,894.72

Source: author's own study

On the basis of the analysis, it is evident that from the point of view of the bond issuer the most beneficial source of financing for the adopted assumptions shall be the bonds convertible to raw materials used to finance investments. Financing through ordinary bonds is also a good solution, whereas the use of the bonds convertible to shares is definitely the least advantageous. According to figure 2, the probability that the investment value exceeds PLN 43 million in the case of such bonds is nearly 0%, whereas in the case of the ordinary bonds and bonds convertible to raw materials it is almost 40%. The result obtained for the last of the above-mentioned sources of financing mining investments is the least beneficial, mostly due to the fact that the bonds held by the bondholder are converted into the equity of the company, and thus the weighted average cost of the capital of the investment undertaken is increased. Since the bondholder becomes a shareholder in the mining company, it is also entitled to a dividend, i.e. a share in profits, and thus the mining company shall be obliged to allocate some part of its profits for the payment of such a dividend to the bond acquirer.

The issuer of debt securities is not the only entity deciding about the choice of the foreign capital. The decision is also to be made by the bondholder. The aim of such foreign capital shall be to mitigate the risk and maximise the profit. Depending on the bond type, the profits shall be at different levels. In the case of bonds convertible to raw materials, the bondholder shall derive profits, apart from the received interest, with the possibility of a debt-to-raw material conversion at a fixed conversion price, which allows the bondholder to purchase the raw material at a price lower than the settlement price effective at a given time. In the case of bonds convertible to shares, apart from the interest at the time of the debt-to-share conversion, the investor shall also derive profits from the dividend due thereto. In the case of ordinary bonds, the bondholder shall derive profits from the interest. As in the case of the bond issuer, the particular sources of financing have been compared. The result of this analysis is the NPV histogram for particular types of bonds (figure 3) and expected values (table 4). To better illustrate the differences between the examined sources of financing, the probability distribution was also provided to show the possibility of reaching the NPV within the individual divisions (figure 4).

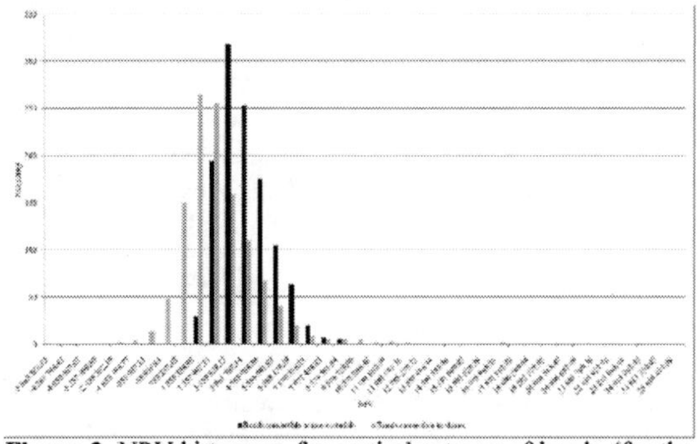

Figure 3. NPV histogram for particular types of bonds (for the bondholder)
Source: author's own study

The histogram presented in figure 3 does not include ordinary bonds due to the fact that in this case the bondholder shall benefit from the amount of the interest obtained for the investment period, which is not subject to change.

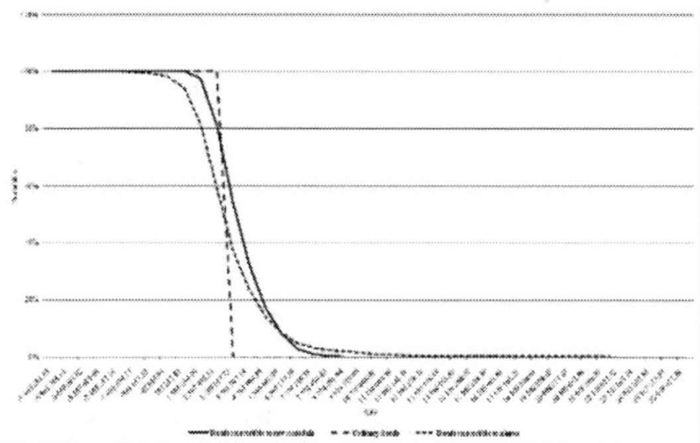

Figure 4. Probability distribution for NPV depending on the source of financing (for the bondholder)
Source: author's own study

Table 4. Expected value of cash flows (for the bondholder) for particular sources of foreign capital

Type of bond	Unit	E(NPV)
Ordinary bonds	[ths. PLN]	2,357,492.31
Bonds convertible to raw materials	[ths. PLN]	3,910,348.08
Bonds convertible to shares	[ths. PLN]	2,626,284.44

Source: author's own study

The results of the analysis explicitly show that from the point of view of the bondholder bonds convertible to raw materials are the most beneficial type of bond. The second choice would be bonds convertible to shares. Ordinary bonds are the least beneficial solution. In such a case, the bondholder derives profits from the interest on the loan capital only. At this point, it should be stressed that ordinary bonds do not provide the investor with any additional option in connection with the above (among those listed); ordinary bonds pose the highest risk to the investor.

 DOI: 10.1051/e3sconf/20161000080

4 Summary

This article was aimed at comparing three sources of capital in the form of bonds convertible to raw materials, bonds convertible to shares, and ordinary corporate bonds. On the basis of the obtained results, bonds convertible to raw materials are the most beneficial type of bonds for the mining company from the point of view of the expected value of the discounted cash flows received based on the Monte Carlo method, whereas bonds convertible to shares are the least beneficial solution. The situation changes when the investment is considered from the point of view of the bondholder. In this case, bonds convertible to raw materials are the most advantageous solution, then bonds convertible to shares, and finally ordinary bonds as the worst solution.

Bibliography

1. F.J. Fabozzi, *Rynki obligacji analiza i strategie.* (Wydawnictwo Fiansnowe WIG-PRESS, Warsaw, 2000)
2. E.F. Brigham & J.F.Houston, *Podstawy zarządzania finansami 1.* (Polskie Wydawnictwo Ekonomiczne Warsaw, 2005)
3. W.Dębski, *Rachunek efektywności inwestycji.* (Oficyna a Wolters Kluwer business. Kraków, 2008)
4. P.Kosowski, *Metody oceny efektywności ekonomicznej podziemnego magazynu gazu.* (Rozprawa doktorska Akademia Górniczo-Hutnicza im. Stanisława Staszica w Krakowie Wydział Wiertnictwa Nafty i Gazu. Kraków, 2008)
5. W.Dębski, *Rynek finansowy i jego mechanizmy.* (Wydawnictwo Naukowe PWN, Warsaw, 2010).
6. A.Rutkowski, *Zarządzanie finansami.* (PWE, Warsaw, 2007)

Proposal for recording bonds convertible to raw materials

Robert Ranosz[1,1]

[1]*AGH University, Department of Economics and Management in Industry, Al. Mickiewicza 30 30-059 Kraków, Poland*

Abstract: this article is devoted to a proposal for recording bonds convertible to raw materials. In view of the fact that the proposed method of recording is a conceptual model, the author of the publication intends to draw further considerations in that direction. The article consists of four parts. The introduction hereto outlines the objectives of this study. The second part of the publication presents the methodology, on the basis of which (according to the author) it shall be possible to record the bonds convertible to raw materials. The third part of the article presents a hypothetical issue of the said debt instrument with an option on the part of the issuer. The last part of the work presents the conclusions of the study. The article ends with a bibliography, based on which this article has been written.

1 Introduction

Over the last few years the Polish mining industry has been subject to constant change, mainly aimed at improving its economic and financial situation [1]. The improvement of this situation also entails the necessity to acquire cheapest capital for financing the business activities of mining companies. Mining companies have increased their share in equities in all sources of financing, and intend to benefit from the more flexible forms of debt, such as long-term corporate bonds [2]. The continuously growing bond market creates new opportunities for investors. Undoubtedly, the possibility of having additional options embedded in bonds contributes to the reduction of the significant cost of capital for mining companies [3]. Bonds convertible to raw materials are new financial solutions proposed by the author of this publication to finance mining investments. Apart from the elements related to the valuation of such debt instruments and presenting a possibility of its application, including benefits for both parties of a contract, the method of its recording also constitutes an important element. This article presents the proposal of recording bonds convertible to raw materials for the issuer as well. From the accounting point of view, bonds convertible to raw materials and shares are relatively similar. The main distinctive feature is the time of debt conversion. In the case of conversion into shares, the investor becomes the owner of a part of the company, whereas the issuer increases its equity. In case of bonds convertible to raw materials, the above-described situation does not take place. At the time of conversion, some part of the issuer's debt is eliminated in exchange for the provided raw material (for which the issuer shall not receive any funds). The author of the publication suggests treating such reduction of debt by the bondholder in exchange for the delivered raw material as debt redemption, which shall increase other operating revenues. Since the revenue is not of a physical nature (no funds are credited to the bank account of a mine), it is essential to make relevant adjustments in the financial flows. Therefore, the objective of this article is to propose a model of recording bonds convertible to raw materials.

2 Methodology

As mentioned in the introduction here to, the methodologies of recording bonds convertible to shares and raw materials shall be similar, yet not identical. In this publication the author does not compare the records of these two different debt instruments with an embedded option, but focuses on the type of book entry for the subject matter hereof. The basic differences between the records of these two different debt instruments are recognised initially and mean that in the case of bonds convertible to shares, we deal with the capital element, whereas in the case of bonds convertible to raw materials, we deal with the value of a derivative instrument.

In compliance with the regulation of the Minister of Finance of 12 December 2001 on detailed principles of recognition, valuation methods, the scope of disclosure, and the way of presenting financial instruments, § 13 sec. 3, first, it is necessary to estimate a fair value of the derivative instrument, and then, on the basis thereof, to estimate the value of the obligation element [4]. The value of the derivative instrument may be estimated by one of three available methods. They include the continuous method applying the Black-Scholes model,

[1] Corresponding author: rranosz@agh.edu.pl

and discreet methods, such as bi- and trinominal models. In the further part of the publication, the value was defined as WR_o. The WR_o value constitutes a discount on the bond (obligation). It is connected with distinguishing the obligation item; in such a case, the bond is considered the financial instrument issued with the discount, which in the initial recognition is equal to WR_o. After establishing WR_o, it is possible to determine the obligation element of the bond convertible to raw materials in compliance with correlation (1).

$$E_z = (N * I) - WR_o \quad (1)$$

Where:
E_z – obligation element
N – nominal value of the bond
I – number of issued bonds

The next step, in accordance with the amortised cost method, is to define the value of the Internal Return Rate (*IRR*). The value is established by correlation (2).

$$E_z = \sum_{i=1}^{n} \frac{O_i}{(1+IRR)^i} + \frac{N * I}{(1+IRR)^n} \quad (2)$$

Where:
O_i – interest (coupon) on the nominal value of the bond in the period "i"

The interest rate is calculated in accordance with formula (3) in the event when the conversion is performed in a given year.

$$O_i = N * r * \frac{y}{12} * I_z \quad (3)$$

Where:
r – value of the bond's coupon (interest on the bond)
y – month in which the conversion was performed
I_z – number of bonds converted into raw materials

The interest is charged as of the beginning of the year and considered interest-bearing liabilities. At the time of the conversion it is necessary to write off some part of the interest payable towards the converted bonds (in accordance with formula (3)). The interest has been accrued, but shall not be paid.

After determining the initial parameters, within the subsequent years of tenor of bonds convertible to raw materials, it is essential to adjust the obligation element by the *IRR* value and interest, following correlation (4):

$$SE_{z(i)} = (E_{z(i-1)} * (1+IRR)) - O_i \quad (4)$$

Where:
$SE_{z(i)}$ – adjusted obligation element
$E_{z(i-1)}$ – obligation element from the previous period (i-1)

Formula (4) assumes that the obligation element is adjusted in the first year after the issue of the bond. In the following years, $E_{z(i-1)}$ shall be equal to $SE_{z(i-1)}$.

It should also be mentioned that at the time of the conversion, the *IRR* value must be adjusted depending on the month in which the conversion is performed: i.e. $IRR*(y/12)$.

In accordance with the assumptions for convertible bonds, the financial cost is defined by correlation (5). The value has been created based on [5,6].

$$KF_i = SE_{z(i)} - E_{z(i-1)} + O_i \quad (5)$$

Where:
KF_i – financial costs in the period "i"

Similarly to the previous case, in the subsequent years $E_{z(i-1)}$ shall be equal to $SE_{z(i-1)}$. The difference in the financial cost in a given period and the interest accrued for the same period shall be the value decreasing the discount on the bond (obligation). The value is established by correlation (6).

$$ZO_i = KF_i - O_i \quad (6)$$

Where:
ZO_i – value decreasing the discount on the bond (obligation)

Table 1. Initial assumptions for the issue of shares convertible to raw materials.

Number of years for which the bond convertible to raw materials is issued.	
Nominal value of the bond	PLN 1,000
Number of issued bonds	2,000
Conversion factor	20
Conversion price	PLN 50
Market price of a raw material at the time of conversion	PLN 65
Nominal value of the issue	PLN 2,000,000
Coupon (interest) of the bond	6%
Standard deviation of the base instrument	25%
Unit cost of excavation of Mg of the material	PLN 35

At this point it should be mentioned that the ZO_i value shall be calculated at the time of the conversion in accordance with correlation (7). Therefore, the value decreasing the discount on the bond shall be defined by the SZO_i value and shall constitute an undepreciated discount on the bond subject to conversion as for the date thereof.

$$SZO_i = \frac{I_z}{I} \sum_{i=1}^{n} ZO_i \quad (7)$$

It is suggested that at the time of the debt to raw material conversion, instead of creating a supplementary capital the same as with bonds convertible to shares, the financial revenues should be created the same way as the supplementary capital. The value of the financial revenues should be established in accordance with formula (8). Created based on [5,6].

$$PF = WOZ + \left(WR_o \frac{I_z}{I}\right) - \left(I_z * N\right) \quad (8)$$

Where:
PF – financial revenues
WOZ – value of bonds converted to raw materials

The WOZ value is estimated based on formula (9):

$$WOZ = \frac{I_z}{I} * E_{z(i)} \left(1 + IRR \frac{y}{12}\right) \quad (9)$$

Similarly to the previous case, in the subsequent years $E_{z(i)}$ shall be equal to $SE_{z(i)}$. At the time of the debt to raw material conversion, no share capital is established as in the case of bonds convertible to shares. It is suggested that instead of share capital in the case of bonds convertible to raw materials, the financial revenues should be created in the event when the value of the converted debt is higher than the cost of the delivered raw materials or the financial costs in the opposite case (the value of the supplied inventory is higher than the value of the converted debt).

3 Example of recording bonds convertible to raw materials

On the basis of the methodology presented in chapter 2 hereof, a model way of recording bonds convertible to raw materials for the issuer has been presented. To perform correct calculations and recordings, it is essential to first assume the initial values defining a given issue. The initial conditions for the issue of bonds convertible to raw materials are presented in table 1. In the calculation example it was assumed that no transactional costs occurred.

On the basis of the above-mentioned correlations, the value $IRR = 37.57\%$ has been estimated in accordance with the amortised cost method. Following the discreet model (trinominal), the market value of the options embedded in the bond convertible to raw materials has been estimated. The value is PLN 792,629.73. On the basis of the nominal value, the value of the obligation element in the amount of PLN 1,207,370.27 (PLN 2,000,000 - PLN 792,629.73) has been obtained. In this case, the recording is in line with table 2.

Table 2. Initial recognition of the issue of bonds convertible to raw materials.

Name of account	Dt [PLN]	Ct [PLN]
Current account	2,000,000	
Bond obligations (nominal value)		2,000,000
Obligations - bonds (discounts)	792,629.73	
Revaluation reserve		792,629.73

After the first year of tenor of the bond convertible to raw materials, the bondholder decided not to convert debt into raw material. The recording at the end of the first year is in table 3.

Table 3. Recognition of bonds on the recording accounts after the first year of tenor of the bond convertible to raw material.

Name of account	Dt [PLN]	Ct [PLN]
Current account		120,000
Financial costs	453,637.22	
Short-term liabilities (interest)	120,000.00	120,000.00
Obligations - bonds (discounts)		333,637.22

Table 4. Calculation of the interest and financial costs for the period up to the conversion time

Name of account	Dt [PLN]	Ct {PLN]
Financial costs	434,244.38	
Short-term liabilities (interest)		90,000.00
Obligations - bonds (discounts)		344,244.38

After 9 months, in the second year of the bonds' tenor, the investor decided to convert debt into raw material. The number of bonds has been established at the level of 1,000 bonds. First, the interest and financial costs for the period of up to 9 months were calculated. The recording of this event is in table 4.

Table 5. Conversion of debt into raw material, including delivery of raw material to the recipient.

Name of account	Dt [PLN]	Ct [PLN]
Bond obligations (nominal value)	1,000,000.00	
Financial revenues		300,000.00
Revaluation reserve	396,314.86	
Financial revenues		383,940.80
Short-term liabilities (interest)	45,000.00	
Obligations - bonds (discounts)		57,374.06
Inventory		700,000

The next step was the debt to raw material conversion. In this part, the delivery of the raw material to the recipient (investor) was assumed. The recording of the conversion of the bonds into raw materials is in table 5.

Table 6. Calculation of the interest and financial costs for the period after the conversion time

Name of account	Dt [PLN]	Ct [PLN]
Financial costs	72,374.06	
Short-term liabilities (interest)		15,000.00
Obligations - bonds (discounts)	57,374.06	

In the next step the interest and financial cost have been calculated until the end of the year (after conversion). The recording entry of this event is in table 6.

DOI: 10.1051/e3sconf/20161000081

In the last step the investor redeemed the bonds. The event is presented in table 7.

Table 7. Redemption of bonds

Name of account	Dt [PLN]	Ct [PLN]
Bond obligations (nominal value)	1,000,000	
Short-term liabilities (interest)	60,000	
Current account		1,060,000

In the last step the remaining part of the revaluation reserve was reclassified to the issuer's financial revenues. The activity is presented in table 8.

Table 8. Redemption of bonds

Name of account	Dt [PLN]	Ct [PLN]
Revaluation reserve	396,314.86	
Financial revenues		396,314.86

4 Summary

The objective of this article is to propose the recording of bonds convertible to raw materials. The assumed goal has been accomplished in this study. The proposal of recording the subject matter hereof is similar to the recording of bonds convertible to shares. However, the events are not identical. In the case of bonds convertible to shares, the investor - at the time of the conversion - becomes a shareholder in the company that issued the debt instrument, whereas in the case of bonds convertible to raw materials, in exchange for the debt redemption, the bondholder receives the raw material at the conversion price, which is usually lower at the time of the conversion than the market price of the raw material.

Bibliography

1. R. Magda, S. Głodzik, J. Jasiewicz, T. Woźny, Arch. of Min. Sci. **54,** 1 (2009)
2. M. Sierpińska, P. Bąk, Arch. of Min. **57,** 4 (2012)
3. R. Ranosz, Arch. of Min. **59,** 1 (2014)
4. RFI - *Regulation of the Minister of Finance of 12 December 2001 on detailed principles of recognition, valuation methods, the scope of disclosure and the way of presenting financial instruments*
5. G.K. Świderska, *Sprawozdania finansowe bez tajemnic.* (Difin Warsaw 2003 p. 8-13)
6. M. Frendzel, *Rachunkowość instrumentów finansowych w świetle regulacji krajowych I międzynarodowych.* (Stowarzyszenie księgowych w Polsce Warsaw 2010)

E3S Web of Conferences 10, 00082 (2016) DOI: 10.1051/e3sconf/20161000082
SEED 2016

Building ceramics with improved thermal insulation parameters

Karol Rzepa[1,a], Wojciech Wons[1] and Manuela Reben[1]

[1]AGH University of Science and Technology

Abstract. One of the most important performance characteristics of masonry units is their high thermal insulation. There are many different ways to improve this parameter, however the most popular methods in case of ceramic masonry units are: addition of pore-creating raw materials and application of proper hole pattern. This study was an attempt to improve thermal insulation of ceramics by applying thermal insulation additives. Perlite dust created as a subgrain from expansion of perlite rock was used. Perlite subgrain is not very popular among consumers, that's why it's subjected to granulation to obtain coarse grain. The authors presented concept of direct application of perlite dust for the production of building ceramics with improved thermal insulation. Fineness of this additive is asset for molding of ceramic materials from plastic masses. Based on the results it was found that about 70% perlite by volume can be added to obtain material with a coefficient of heat conductivity of 0,37 W/mK. Higher content of this additive in ceramic mass causes deterioration of its rheological properties. Mass loses its plasticity, it tears up and formed green bodies are susceptible to deformation. During sintering perlite takes an active part in compaction process. Higher sintering dynamics is caused by: high content of alkali oxides in perlite and glass nature of perlite. Alkali oxides generate creation of liquid phase which intensifies mass compaction processes. Active role of perlite in sintering process causes good connection of its grains with clay groundwork which is important factor for mechanical parameters of ceramic materials. It was also noted that addition of perlite above 40% by volume of mass effectively neutralized negative effect of efflorescence in ceramic materials.

1 Introduction

The importance of renewable source based products for industrial applications becomes extremely visible in recent years with increasing emphasis on the environmental issues, waste disposal, and depleting non-renewable resources. Through innovative ideas in designing the fired building ceramics with dusty expanded perlite which can compete or even surpass with the existing materials on a cost-performance basis with the added advantage of eco-friendliness and thermo insulating properties. One of the most important challenges of future of building ceramics materials is the improvement of their thermo insulating properties, which leads to the reduction of energy consumptions of buildings in all their life phases, from construction to demolition [1]. Constant, or even increasing popularity of this trend shows not only economic justification associated with a decrease in demand for thermal energy, but also has a legal nature. Since 2009 in Poland there is an obligation to perform an energetic audits for new or modernized buildings [2]. Introduced law caused a natural increase in interest of building materials with a good thermal insulation parameters, including ceramics. In the field of thermal insulating of building ceramics there are several ways to improve thermal insulating properties:

1. Use of the additive for porosity increase like: sawdust, cellulose pulp, Styrofoam balls. During firing process these components are burned, leaving pores which are centres of very high thermal insulation [3, 4, 5].
2. Use of the lowest firing temperature for sintering without liquid phase. Liquid phase facilitates sintering process but it also causes flooding of pores which are important to provide thermal insulation of ceramic product.
3. Use of the cavity system in the ceramic products which provide longer heat transfer through ceramic material. This assumption results from fact that ceramics is a material with inferior thermal insulation than the air contained in hollows [6, 7].
4. Use of the thermal insulating additives in ceramic masses. These additives are characterized by very good thermal insulation properties that is why their addition to masses or products improves this parameter [8].
5. Partial or total elimination of grouting, the binder made of mortar between masonry components. Binder is so called thermal bridge, through which heat rapidly flows inside wall. Vertical grouting is completely eliminated by usage of tongue and groove joints while horizontal grouting can be limited by grinding of surface and applying thin binder.

In this research an attempt to improve thermal insulation of building ceramics was made by using expanded perlite as a thermal insulating additive.

Perlite is effusive rock, basically mineral obsidian which can be commonly found in region of Mediterranean Basin. It was formed during underwater volcanic eruption, therefore it is a vitreous substance contains 2-5 % water [9]. Natural perlite has a specific density between 2200 kg/m3

[a] Corresponding author: krzepa@agh.edu.pl

and 2300 kg/m3 [10], and it's not thermal insulating material but it can be transformed into one through the process of expansion. During heating process in the temperature of 850-1150°C, water closed inside glassy phase undergoes rapid expansion [11]. This phenomenon is accompanied by micro-explosions and a dozen times increase in perlite's volume. As a result of the expansion process oval, irregular and glassy grains of cellular microstructure are created. Specific density of expanded perlite varies between 80 kg/m^3 to 240 kg/m^3 [12], also thermal conductivity λ is 0,045-0,059 W/m·K. This material is resistant to chemical and biological corrosion and shows good sound absorption [13]. These and other features of expanded perlite causes its increasing popularity. Perlite also found application in the products of building ceramics [14]. It is used as a light-weight aggregate in concrete and plasters, also as good insulating material [15]. Moulders are often produced through pressing of semi-dry masses or dry masses involving perlite. Proper performance parameters are obtained by autoclaving or firing. Sample parameters of these materials are bulk density of 0,45-0,55 g/cm^3 and minimum compressive strength of 0,7 MPa.

In this paper an attempt to produce a perlite brick using plastic, wet method of forming has been made. This method is common in the production of fired building ceramics, because it allows the production of moulders with complex shapes, including products with a high proportion of cavities. The choice of this method requires the use of dusty expanded perlite. On the one hand it was economically justified, because perlite of such granulation is directly used in a small extent, and also is a low cost material [16]. Moreover, it does not require grinding and can be directly added to the plastic mass without substantial deterioration of operating parameters.

2 Experimental

A series of ceramic materials were prepared with different proportion of perlite in the raw material mixtures. Selected functional parameters of ceramic materials obtained were determined. In addition to perlite in the mixtures, the raw clay was used. Clay and perlite are characterized by the significantly different bulk density ($\rho_{perlite} = (87 \pm 2)$ kg/m3, ρ_{clay} = (1243 ± 22) kg/m3), thus the mixture compositions are given in volumetric and mass ratios (Table1).

Table 1. Symbols and compositions of clay-perlite masses.

Symbol of mass	Volumetric ratio [%]		Mass ratio [%]	
	perlite	lias clay	perlite	lias clay
0	0	100	0	100
P2	20	80	1,7	98,3
P4	40	60	4,5	95,5
P6	60	40	9,5	90,5
P7	70	30	14,0	86,0
P8	80	20	21,9	78,1
P9	90	10	38,6	61,4

Ceramic materials were prepared in the following manner: after weighing the ingredients were dry-homogenized for 30 minutes in a tumbler mixers. Created mixtures were mixed with water until plastic mass was obtained. After 24 hour of homogenization different moulders from masses were formed as follow: 50x50x50 mm cubes, 15x35x50 mm bricks and 80 mm diameter rollers (for thermal conductivity test). Prepared moulders were initially dried under laboratory conditions first at the temperature of 20°C and 60% air humidity, then the moulders were dried in laboratory dryer up to the temperature of 105°C. For dried moldings the necessary measurements to calculate the mixing water, the density after drying and drying shrinkage were made. After these measurements all moulders were fired at a maximum temperature of 950 °C. Designs of the parameters before and after firing are included in [17]. All measurements were performed on 6 samples. Rollers prepared for the thermal conductivity test were grinded on the one side and all pollutants produced during the grinding process were removed with compressed air. In this study the non-stationary method was used - thermal probe [18].

In order to clarify the influence of the perlite on sintering process of ceramic masses measurement in a high-temperature microscope was performed, as well as differential thermal analysis DTA combined with thermogravimetry TG, XRD phase analysis and SEM microstructure study. These tests were made only for the masses and materials marked as „P7" and „0".

High-temperature microscopy consists of recording the surface contour of samples and their shape changes during heating. Based on the test results, characteristic temperatures were determined:
- the beginning of sintering temperature (first shrinkage connected with sintering),
- the maximum of sintering temperature (maximum sintering shrinkage with no deformation effects),
- the temperature of softening point (first sample deformation, e.g. rounded corners).

The measurements were performed with use of high-temperature microscope Hesse Instruments. Cylindrical samples with dimensions of d=h=3 mm were prepared in special manual press. The measurement was carried out in air with a constant heating rate of 10 °C/min to the maximum temperature of 1300°C.

XRD (X-Ray Diffraction) analysis was used to determine the phase composition of materials (diffractometer PANalytical equipped with copper anode). Measurements were carried out with the following performance parameters: range from 10 to 70 2Θ, step size 0,01, time per step 255,905 s.

Complementary research method used for additional analysis of phase composition was differential thermal analysis combined with thermogravimetry. DTA and TG measurements were performed in a STA 449F3 Jupiter Netzsch apparatus for starting materials and P7 ceramic mass. The test was conducted in an atmosphere of synthetic air with a flow of 40 ml/min and the heating rate of 15°C/min. Clay, perlite and ceramic mass samples were milled to a grain size below 0.06 mm and weighted in range from 85 to 95 mg.

DOI: 10.1051/e3sconf/20161000082

The morphology of samples tested was studied with use of scanning electron microscope SEM equipped with EDS microanalyzer. Fractured samples were sputtered with carbon before the measurements, which were performed in a high vacuum.

3 Results and discussion

3.1 Parameters of ceramic bulks

Averages of measured parameters with standard errors (confidence level of 0,95) of 6 samples for each mass before firing are shown in the table 2.

Table 2. Parameters of clay-perlite masses.

Symbol of mass	Mixing water [%]	Drying shrinkage [%]	Bulk density [kg/m³]
0	20,0 ± 0,1	9,3 ± 0,3	2,04 ± 0,01
P2	19,5 ± 0,1	8,1 ± 0,3	1,94 ± 0,02
P4	20,9 ± 0,1	7,4 ± 0,3	1,78 ± 0,01
P6	25,3 ± 0,2	7,7 ± 0,3	1,53 ± 0,01
P7	29,5 ± 0,1	7,8 ± 0,3	1,34 ± 0,01
P8	35,0 ± 0,1	7,5 ± 0,3	1,09 ± 0,01
P9	50,5 ± 0,2	5,1 ± 0,3	0,66 ± 0,01

During making and forming of ceramic masses it was noted that mixes with the highest amount of perlite dust (80 and 90%) had poor formability. Moreover, these masses showed poor plasticity and tore easily, also after forming process moulders loose water and are ductile, which favours their deformation. Due to the high porosity of the perlite, the additive increases mixing water of masses. What is important, perlite does not increase the drying shrinkage, moreover it decreases this parameter. It means that perlite does not increase sensitivity to drying of ceramic masses.

3.2 Parameters of ceramic sintered

Parameters of ceramic sintered obtained are shown in the tables 3-5. The firing shrinkage, compressive strength, bulk density and thermal conductivity dependence of perlite proportion in ceramic mixes are shown in figures 1, 2 and 3.

Table 3. Firing shrinkage, absorbability and open porosity clay-perlite ceramic materials.

Symbol	Firing shrinkage [%]	Absorba -bility [%]	Open porosity [%]
0	1,2 ± 0,3	11,8 ± 0,2	23,0 ± 0,2
P2	1,7 ± 0,2	12,2 ± 0,2	22,9 ± 0,2
P4	1,8 ± 0,3	13,6 ± 0,3	23,5 ± 0,3
P6	2,2 ± 0,2	16,6 ± 0,2	25,0 ± 0,2
P7	2,3 ± 0,3	19,8 ± 0,4	27,2 ± 0,3
P8	3,0 ± 0,2	25,3 ± 0,3	29,5 ± 0,3
P9	5,3 ± 0,4	49,4 ± 1,5	38,4 ± 0,7

Table 4. Parameters of clay-perlite ceramic materials.

Symbol	Bulk density [kg/m3]	Comp. strength [MPa]	Thermal conductivity [W/mK]
0	1,93 ± 0,01	36,9 ± 3,5	0,59 ± 0,03
P2	1,85 ± 0,02	32,8 ± 2,0	0,53 ± 0,02
P4	1,69 ± 0,01	26,6 ± 1,8	0,48 ± 0,01
P6	1,49 ± 0,01	18,4 ± 1,4	0,40 ± 0,01
P7	1,35 ± 0,02	16,2 ± 1,5	0,37 ± 0,02
P8	1,14 ± 0,01	13,3 ± 1,5	0,29 ± 0,01
P9	0,76 ± 0,01	8,7 ± 0,7	0,18 ± 0,01

Table 5. Presence efflorescence and freeze thaw resistance for clay perlite ceramic materials.

Symbol	efflorescence	freeze-thaw resistance 20 cycles (-15/+20°C)
0	+	+
P2	+	+
P4	+	+
P6	-	+
P7	-	+
P8	-	+
P9	-	+

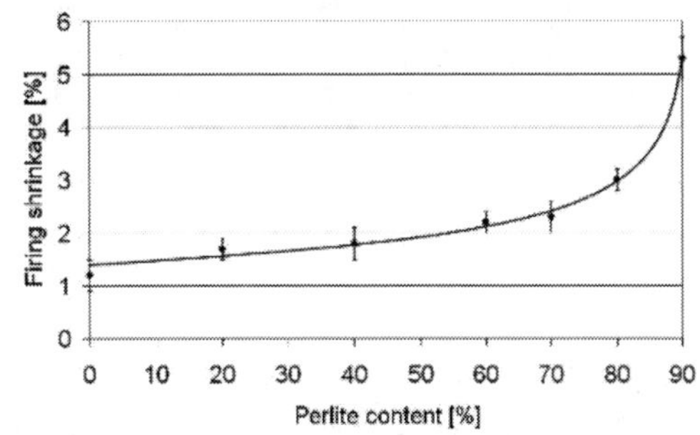

Figure 1. Firing shrinkage of clay-perlite ceramic materials.

Addition of perlite to ceramic materials significantly affects their parameters: increases open porosity and absorbability, reduces the compressive strength, bulk density and thermal conductivity. The greatest impact on compressive strength and absorbability occurs for a high proportion of perlite in ceramic masses, more than fourfold increase in absorbability and more than fourfold decrease in a compressive strength for sample P9. In addition, some similarities in the course of the bulk density and thermal conductivity curves as a function of perlite addition can be noted. Sintered with pearlite successfully passed freeze-thaw resistance research. Essentially, perlite increases the sintering shrinkage (Figure 1). This leads to the idea that when firing, perlite does not behave like an inert filler but actively participates in the process of sintering.

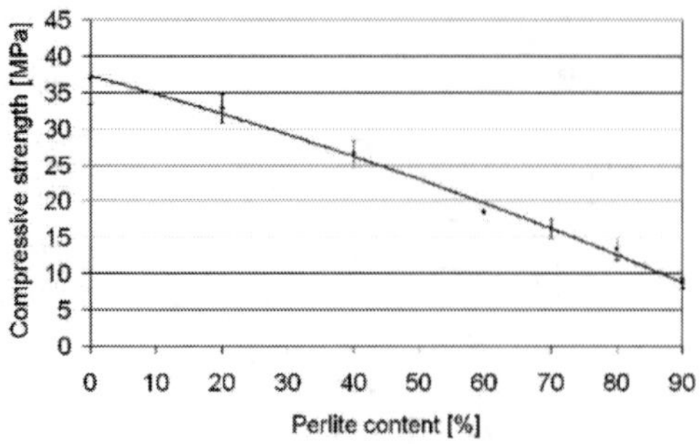

Figure 2. Compressive strength of clay-perlite ceramic materials.

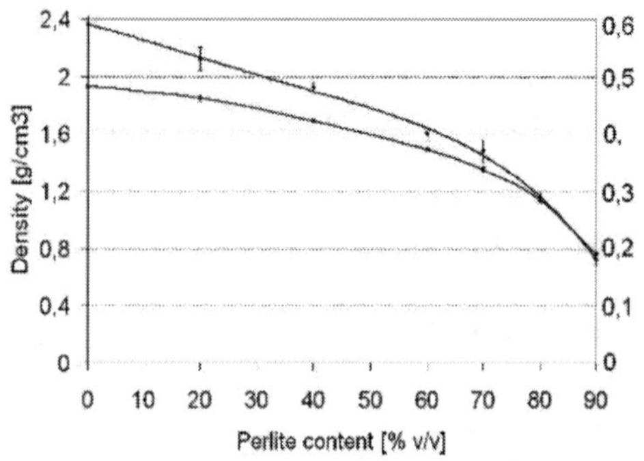

Figure 3. Bulk density and thermal conductivity of clay-perlite ceramic materials.

3.3 High-temperature microscope

Last conclusion in 3.2 chapter is confirmed by the changes in the surface contour, registered during samples heating. Images from high-temperature microscope as well as changes in the surface contour are presented in figure 4.

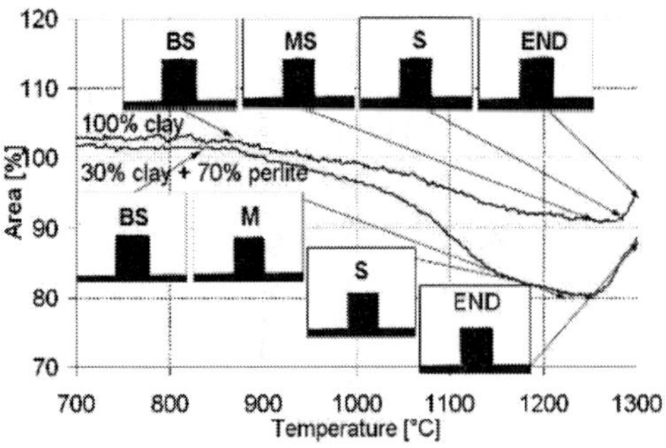

Figure 4. High-temperature microscope results for "0" and P7 samples.

Addition of perlite to loam caused slight decrease in the occurrence of characteristic temperatures, of about 30-40 °C. It did not result in a narrow interval of sintering , that is the difference between the temperature of the end and the beginning of the sintering. Most importantly, the addition of perlite enhances the process of mass thickening during sintering, which is manifested by the larger changes in the surface contour of the samples. The possible reason for these differences is the melting action of perlite related to the high content of alkali oxides in this material.

3.4 Phase composition

In the figure 5 XRD patterns of P7 an "0" samples are shown.Quartz is the main phase component which is present in the sintered clay sample. The presence of hematite and dehydroxylized clay minerals from mica group is not excluded. Samples containing perlite shows the presence of plagioclases. Plagioclases are probably crystallized from glass contained in perlite. The intensity of the peaks of the phases coming from clay decreases proportionally to the mass fraction of perlite.

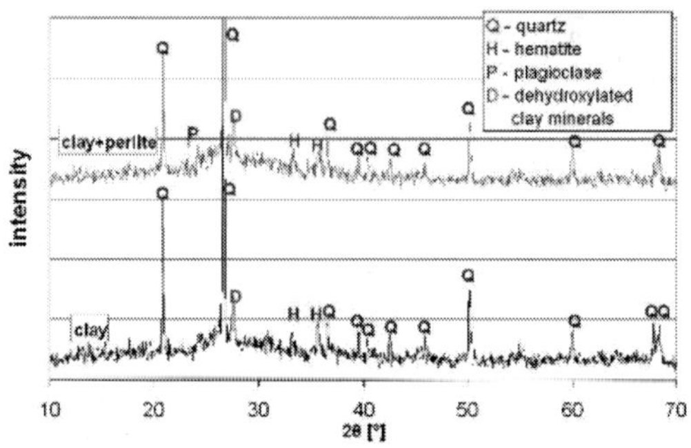

Figure 5. Comparison of XRD patterns of clay made sample and sample with perlite addition.

3.5 DTA/TG analysis

The results of DTA/TG measurements for starting materials and P7 ceramic mass are shown in the figures 6 and 7. In the figure 6 approximate base curves were drawn alongside DTA curves.

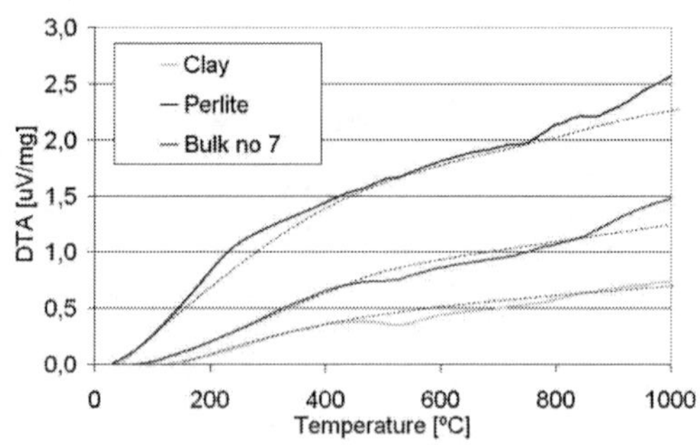

Figure 6. Differential thermal analysis for perlite, clay and P7 ceramic mass.

During heating process minerals which are present in clay are subjected to endothermic, dehydroxylation reaction at temperature range of 400-800°C. This effect is accompanied by weight loss. A small exothermic effect connected with the beginning of sintering process can be observed above 900°C. This effect isn't accompanied by weight loss.For perlite heated in the temperature range of 750-850°C characteristic, exothermic effect connected

DOI: 10.1051/e3sconf/20161000082

to glass transition can be noted, which is accompanied by subsequent, intensive exothermic effect connected to glass crystallisation and sintering. These two thermal effects aren't accompanied by weight loss. P7 mass curve is an approximate component of two curves.

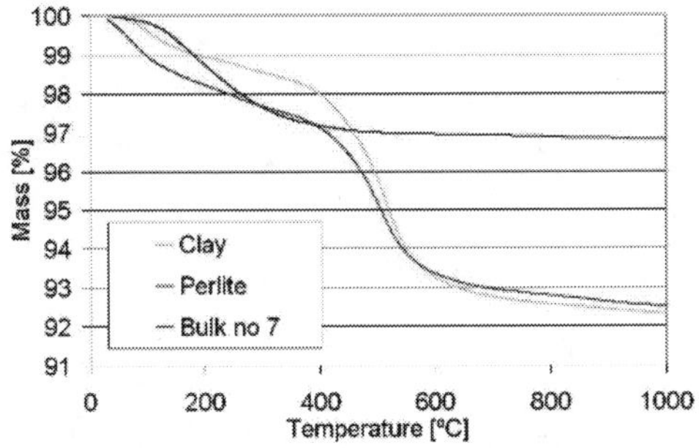

Figure 7. Weight change during heating for perlite, clay and P7 ceramic mass.

3.6 Microstructure - scanning microscope

In figures 8 and 9 microstructure images of ceramic samples P7 and "0" are shown using the same magnification.

Figure 8. Microstructure of building ceramics with perlite – P7 sample

In P7 sample characteristic cellular perlite grains are clearly visible in point 1 and 2 in the figure 8. They are dispersed in a matrix formed of grains of clay, which is shown in point 3 in the figure 8. Perlite grains have clear but thin grain boundary and make an impression of a well-embedded in the clay matrix.

This is confirmed by the fact that, during the creation of the fractured samples, cracking occurred through perlite grains, but not on their grain boundary.

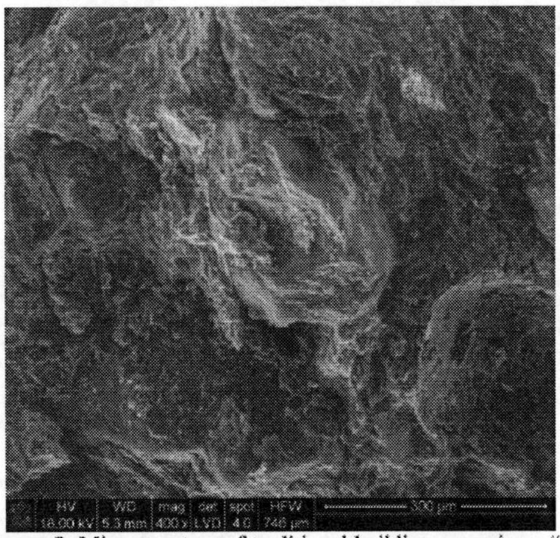

Figure 9. Microstructure of traditional building ceramics – "0" sample.

4 Summary

The results of this research indicated that dusty expanded perlite can be an interesting addition to the ceramic masses for building ceramics with improved thermal insulation. Due to its fine grains perlite can be used in a plastic ceramic masses. At the stage of forming perlite acts as opener for plastic mass and due to rheological parameters of mass, perlite under laboratory conditions can be used up to 70% by volume fraction. The main objective of this additive is to lower the thermal conductivity of ceramic material. About 70% perlite replacement by volume fraction resulted in reducing thermal conductivity of ceramic mass from about 0,59 W/m·K to 0,37 W/m·K. Using a suitable cavity system can further improve the thermal insulation of building ceramics. Addition of perlite to ceramic mixes causes alteration of other operational parameters. It increases absorbability and open porosity, lowers bulk density and compressive strength. While lowering of bulk density is positive aspect, changes of other parameters is unfavourable, especially lower compressive strength. It can also be noted that the use of common pore-making additives such as sawdust and cellulose pulp aggravates strength of materials even more effectively.

Perlite takes an active role in sintering process, making its grains very well build into clay matrix, giving durable composite. It is probable that porosity of materials supplied by perlite reduces their strength by a small amount.

Addition of perlite to ceramic masses has other important advantages. This research showed that perlite as supplement reduces the formation of efflorescence. Moreover, perlite is the sound-absorbing material, so it can be expected that materials made from perlite shows improved sound-absorbing qualities.

References

1. J. Nyers, L. Kajtar, S. Tomić, A. Nyers, Energy Build. **86**, 268 (2015)
2. Regulation of the Polish Infrastructure Minister of 6 November 2008
3. V. Ducman, T. Kopar, Mater. Tehnol. **41**(6), 289 (2007)
4. S. Krebs & H. Mortel, TBI, **15**(1), 12 (1999)
5. M. Sutcu & S. Akkurt, Ceram. Int. **35**, 2625 (2009)
6. J. Kocí, J. Madera, R. Cerný, Energy **83**, 749 (2015)
7. V. Kocí, Z. Bazantová, R. Cerny, Energy Build. **76**, 211 (2014)
8. F. Singer, S.S. Singer, *Industrial Ceramics*, (Springer-Science + Business Media B.V 1963),
9. W.V. Liu, D.B. Apel, V.S. Bindiganavile, Cement Concrete Comp. **53**, 44 (2014)
10. A.G. Celik, A.M. Kilic, G.O. Cakal, Physicochem. Probl. Miner. Process. **49**(2), 689 (2013)
11. A. Sarı, A. Karaipekli, C. Alkan, Chem. Eng. J. **155**, 899 (2009)
12. S. Celik, R. Family, M. P. Menguc, J. Build. Eng. (to be published)
13. X. Yanjun, L. Rui, S. Haiping, J. Ran, Z. Jiayu, TOMSJ **9**, 39 (2015)
14. Ch. Sadik, A. Albizane, I. Amrani, , J. Mater. Environ. Sci. **4**(6), 981 (2013)
15. N. Burriesci, C. Arcoraci, P.L. Antonucci, G. Polizzotti, Mat. Letters **3**(3), 103 (1985)
16. Ł. Kotwica, W. Pichór, W. Nocuń-Wczelik, J. Therm. Anal. Calorim. **123**(1), 607 (2016)
17. E. Brylska, P. Murzyn, J. Stolecki, *Ceramiczne Materiały Budowlane* (Wyd. AGH, Kraków, 2014)
18. J. Małolepszy, M. Gawlicki, W. Pichór, E. Brylska, W. Brylicki, A. Łagosz, W. Nocuń-Wczelik, M. Petri, Z. Pytel, W. Roszczynialski, J. Stolecki, G. Malata, M. Reben, *Podstawy Technologii Materiałów Budowlanych i Metody Badań* (Wyd. AGH, Kraków, 2013)

DOI: 10.1051/e3sconf/20161000083

Biomass fuel based on wastes from the paper industry

Stanisław Budzyń[1] and Barbara Tora[1,a]

[1]*AGH University of Science and Technology, Kraków, Poland*

Abstract. Wastes from paper industry are mostly combustible. It is possible to recycle them with energy recovery. These wastes have a high moisture content (up to 60%) and thus a small calorific value. An alternative to waste incineration is the production of solid recovered fuel. The benefits are: easy adjustment of the physical and chemical properties of the fuel (via the change of proportions of ingredients), low moisture and high calorific value. The study involved the following types of cellulose wastes: - *Belmer* - the rejects from recovered paper, *Krofta* - deinking sludge, sludge - wastewater treatment sludge, bark - the rejects from virgin pulps. The results of investigations of waste produced in one of the biggest Polish paper mill - are shown. Following aspects were investigated: energy properties, content of carbon, hydrogen, sulfur, chlorine and nitrogen, chemical composition of ash. Authors proposed two formulas of the biomass fuel. The properties of the fuel such as the content of carbon, hydrogen, sulfur, chlorine or nitrogen, the chemical composition of the ash were investigated. Due to the fact that the combustion of the biomass fuel is preferred in view of law regulations (zero CO_2 emission, green certificates) the content of biodegradable fraction was examined. It has been shown that the fuel is a biomass one. Fuel from waste can be a substitute for approx. 25% of primary fuel (coal) used by the paper mill.

1 Introduction

Wastes from paper industry are mostly combustible. It is possible to recycle them with energy recovery. These wastes have a high moisture content (up to 60%) and thus a small calorific value. An alternative to waste incineration is the production of solid recovered fuel [1]. The benefits are: easy adjustment of the physical and chemical properties of the fuel (via the change of proportions of ingredients), low moisture and high calorific value. The research was carried out in the company International Paper Kwidzyn (Kwidzyn Pulp & Paper Mill). Following privatization in 1992 the mill at Kwidzyn has gone through a vast modernization program and now occupies a leading position among European and Polish companies [2]. The facility produces office and offset papers. IP also makes Presso® newsprint from deinked wastepaper, while their coated paperboards Arktika and Alaska are widely used in the cosmetics, pharmaceutical and food industry for many household brand names across Europe.

The pulp and paper industry is expected to be a major user of SRF techniques. The impact of this source of wood can best be demonstrated through pulping studies and this procedure has been employed by a number of investigators.

Over the past two decades, increasing environmental awareness has resulted in a much stricter policy with respect to the handling and processing of household and industrial waste. One of the targets of this European Waste Directive is to decrease the amount of waste that ends up on a landfill. Another result of the increased environmental awareness is the attempt to decrease the burning of fossil fuels with the aim to reduce CO_2 emission, and, as a consequence, considerable research effort is directed towards the possibilities of converting waste into solid fuels that are suitable for the replacement of fossil fuels. A solid fuel is a solid material that can be used as fuel to produce energy, such as for example wood, coal or peat.

Several methods for the conversion of waste into solid fuels are known in the art and described in for example US Patent 5302254 and WO 2007/145507 [3]. In US 5302254 a plant for treating industrial and/or urban waste is described, including a stage for drying of the waste, followed by a stage for effecting thermolysis of the dried waste and a stage for effecting recovery of the solids and gases resulting from the thermolysis. Thermolysis of the waste is effected in a reactor by indirect heat exchange with combustion gases. Drying gases for drying of the waste are subsequently treated to remove vapour materials and then recycled to a gas generator for generating the drying gases. Additionally, the plant includes a stage for effecting dechlorination of the solids resulting from the thermolysis by washing the solids with an aqueous liquid, as well as a stage of separating the washed solids and the resulting wash liquid. WO 2007/145507 describes a method for the preparation, by torrefaction, of a solid fuel wherein a starting composition is heated indirectly, and a solid fuel to be obtained by such method. Furthermore, a method for the removal of one or more

[a] Corresponding author: tora@agh.edu.pl

DOI: 10.1051/e3sconf/20161000083

metals from, or the reduction of the "total chlorine content", the sulphur content and the trace element content of a solid fuel thus obtained is disclosed. Also, a solid fuel which can be obtained by the carrying out of such a method and the uses of such solid fuels are described [4].

2 Materials and method

The study includes three parts:

- Study the properties of waste from paper mill,
- Technology for the production of secondary fuel (Solid Recovery Fuel)
- Determine the properties of the fuel.

2.1 Determination of the properties of waste

The study involved the following types of cellulose wastes [5]:

1. *Belmer* – waste code 030305 - the rejects from recovered paper are impurities and consist of lumps of fibers, staples and metals from ring binders, sand, glass and plastics and paper constituents as fillers, sizing agents and other chemicals.
2. *Krofta* – waste code 030310 - deinking sludge. This residue contains mainly short fibers or fines, coatings, fillers, ink particles (a potential source of heavy metals), extractive substances and deinking additives. It is normally reused in other industries (e.g. cement, ceramics), or is incinerated, even though it has a poor heating value)
3. Sludge - waste code 030311 - wastewater treatment sludge. It comes from two sources: primary sludge and biological sludge generated in the second clarifier. These sludges are generally blended together, a polymer added and dewatered together to a 25–40% dry solid content.
4. Bark – waste code 030301 - the rejects from virgin pulps. It consists of sand, bark and wood residues from wood handling, which are undesirable for papermaking.

The results of investigations of waste produced in one of the biggest Polish Paper Mill - International Paper Kwidzyn are shown. Following aspects were investigated: energy properties, content of carbon, hydrogen, sulfur, chlorine and nitrogen, chemical composition of ash.

2.2 Investigation of energetic properties of wastes

The investigations of wastes properties were conducted for their energetic application [6]. The preparation of samples of wastes for research and their analyses were done according to the existing norms concerning analysis of solid fuels. In wastes such parameters were determined: moisture, ash contents, volatile parts contents as well combustion heat and calorific value. Furthermore, coal, hydrogen, sulfur and chlorine contents were determined too. The results are shown in Tables 1 to 5.

Table 1. Waste balance

no	Waste code	quantity [Mg/year]	yield [%]	moisture [%]
1	03 03 05	33 000	11	45
2	03 03 10	24 000	8	52
3	03 03 11	52 000	17	61
4	03 03 01	200 000	64	51

Table 2. Energetic properties of wastes

	W^r_t	V^r	A^r	S^r_t	Q^r_i	Q^d_i
waste code	[%]	[%]	[%]	[%]	[kJ/kg]	[kJ/kg]
03 03 05	45,0	16,1	26,3	0,03	2 107	5 831
03 03 10	52,0	28,1	13,3	0,03	4 095	11 174
03 03 11	61,0	20,1	15,4	0,29	2 420	10 022
03 03 01	51,0	41,3	0,2	0,01	8 500	19 894

Where:
W – moisture content,
V – volatile matter,
A – ash content,
S_t – sulphur content,
Q^a_s – calorific value,
Q^a_i – net calorific value

Table 3. Elements analysis of the waste

	C^a	H^a	Cl^a	S^a_t	N^a
waste	[%]	[%]	[%]	[%]	[%]
03 03 05	23,2	2,26	0,03	0,04	1,42
03 03 10	34,2	4,72	0,03	0,04	1,34
03 03 11	30,3	4,21	0,08	0,73	4,35
03 03 01	53,1	7,06	0,01	0,01	0,08

Table 4. Metals content in ash from waste incineration

	Cr	Cd	Pb	Co	Ti	As	Hg	Sn	Mn	Cu
waste	[mg/kg]	[mg/kg]	[mg/kg]	[mg/kg]	[mg/kg]	[mg/kg]	[mg/kg]	[mg/kg]	[mg/kg]	[mg/kg]
03 03 05	37,22	0,01	46,73	5,79	518,61	23,49	1,32	3,32	188,58	280,81
03 03 10	61,71	0,01	76,92	34,88	847,00	2,15	0,45	3,60	235,23	99,80
03 03 11	48,16	8,12	51,08	7,76	710,31	10,36	1,63	7,49	1431,32	106,06
03 03 01	262,26	64,57	130,47	27,82	472,61	57,25	12,78	72,29		926,57

Table 5. Analysis of ash from the waste

	Na_2O	K_2O	ZnO	Fe_2O_3	SiO_2	CaO	MgO	Al_2O_3	P_2O_5
waste	[%]	[%]	[%]	[%]	[%]	[%]	[%]	[%]	[%]
03 03 05	0,17	0,21	0,02	0,42	26,40	60,81	1,26	9,16	0,44
03 03 10	0,20	0,32	0,05	1,08	43,20	42,23	1,15	10,69	0,60
03 03 11	0,31	0,35	0,26	5,99	39,20	37,16	0,93	3,47	8,80
03 03 01	0,35	4,81	0,66	2,45	22,00	48,24	10,97	1,25	8,72

In ash obtained from combustion of wastes samples the basic metal oxides and heavy metals were determined (Table 4, 5). All of wastes indicated high moisture value achieving in case of sewage sludge the value of 70%. The least beneficial waste to energetic application are slimes from discoloration of wastepaper which have the calorific value being equal to 5.8 MJ/kg in dry conditions. These slimes are 11% of total mass of wastes directed to combustion with recovery of energy what slightly lower the calorific value of the fuels produced from these wastes. The biggest calorific value have wooden wastes (above 19 MJ/tons in dry condition) and they are 64% of total mass of wastes. Table 3 presents the sulfur and chlorine contents in wastes in dry condition. Apart from sewage sludge, which contain 0.73% of sulfur and 0.08% of chlorine and are 17% of total mass of wastes, the rest of the wastes characterize with low contents of sulfur (below 0.04%) and chlorine (below 0.03%).

Table 1 shows the amount of wastes being produced in IP Kwidzyn plant yearly and maximum moisture which these wastes have in initial state. From the data it occurs that about 300 000 Mg of fuel can be produced yearly. The moisture in such fuel is higher than 50% what significantly lowers its calorific value which with this content of moisture is equal to about 7 MJ/kg. For combustion of wastes in such form the delivery of additional energy is required. That is why it is important to elaborate the method of their dewatering what would allow to increase their calorific value and their combustion with recovery of energy.

2.3 Technology of solid recovery fuel production from wastes

The best solution of using wastes from celluloid mass and papers which can be combusted with recovery of energy is their complex application to production of secondary fuel [7]. Omitting the issue of biomass contents in researched wastes, according to the current law, the plant IP Kwidzyn can combust these wastes on the area of the plant without applying regulations concerning thermal transformation of wastes. These regulations are not used for installations which transform wastes thermally if these wastes are fibrous, floral from celluloid mass and paper production and if they are combusted in location of production and the produced energy is recovered (Act of Wastes from 14 December 2012, change Bullettin of Acts and Decrees, 2015. 122, article 163). It is also in accordance with the recommendations contained in reference document BAT for paper industry, which states that the best available method of wastes application, which were created during paper production is their combustion with recovery of energy in the location of their origin.

The studies have been carried out in a specially installed at International Paper pilot installation (Scheme of the installation is shown at Figure 1).

Pilot installation uses BELLMER KUFFERATH Machinery. Bellmer Kufferath offers a range of products able to reach the objective targets [2].

Installation for the production of solid recovered fuel consisted of two main stages

- vacuum filter AKSE® F DISC THICKENER,
- filter press AKUPRESS® CX-M.

AKSE® F disc thickeners dewater fibrous materials from a dry content of 0.5-3% to 5-8%. AKSE® F can be used in various processes either as individual unit or as pre-thickener, e.g. upstream of screw presses. In this case, they help to ensure both the running and the economic efficiency of the screw presses at low solid contents.

AKUPRESS® CX-M presses dewater fibrous materials from a dry content of 3% to 30-55% and meet all requirements made on fiber dewatering in a very particular way:

- maximum and adjustable dry contents,
- high and adjustable dry contents.

2.4 Dewatering pilot plant

Scheme of the pilot plant is shown at Figure 1. The waste-dewatering pilot plant consists primarily of the following components [8]:

- disk thickener AKSE S 175/01 with flocculation reactor FR 1 (Serial No. 2000-3190),
- screw press AKUPRESS BX 250 LL (Serial No. 2000-0149),
- metering system for polymeric flocculation media,
- conveyor system for pressed matter,
- container with control cabinet and process control system WinCC.

Detailed scheme of the pilot installation is shown at Figure 2 and 3.

For transportation, the screw conveyor can be screwed into the base frame as shown in the Figure 3, and can be unscrewed for use according to the diagram (Figure 1).

The system is completely assembled, cased and wired internally. It is delivered on a low loader. The system must be lifted down from the low loader (for the duration of the trials). The total weight without the low loader is 10 tons.

Prior to commissioning, the following external connections must be attached:

- sludge inlet (min. 0.5 bar, DN 50),
- filtrate outlet (DN 100),
- spray water/diluting water for dissolving the flocculation media (freshwater, 4-6 bar, 2"),
- compressed air (min. 5 bar, ½") – only where available at no cost, otherwise supply from own compressor,
- electricity (400 V, 50 Hz, connected load 30 kW, via terminal strip).

Two methods of drying were elaborated in purpose of their application to production of secondary fuel destined to be combusted within the plant.

In first method [6], the technological sewage sludge from mechanical treatment and mechanical and biological treatment as well slimes from discoloration of wastepaper were dewatered by means of the expeller. The attempts of dewatering were conducted on the area of IP Kwidzyn plant by means of installation BELLMER KUFFERATH Machinery. Slimes and sludge were dewatered in two stages, firstly in vacuum filter in purpose of simplifying their dewatering, and then in expeller. The flocculants were added to the wastes in purpose of simplifying their dewatering. The dewatered wastes were directed to screw mixer in which the remaining components of secondary fuel can be added (wooden wastes, biomass of agricultural and forest origin).

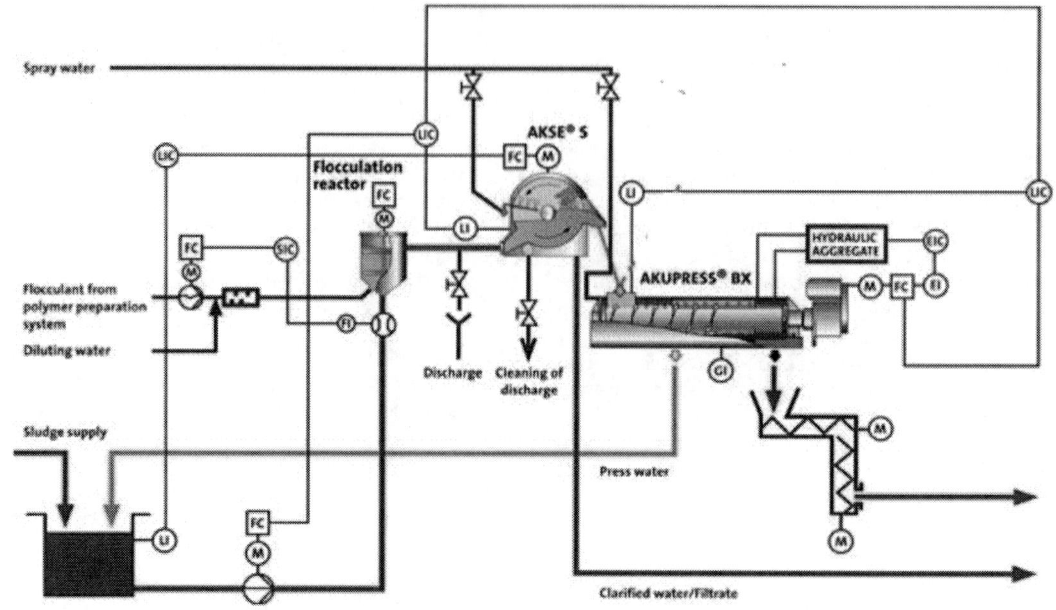

Figure 1. Block scheme of the solid recovered fuel production.

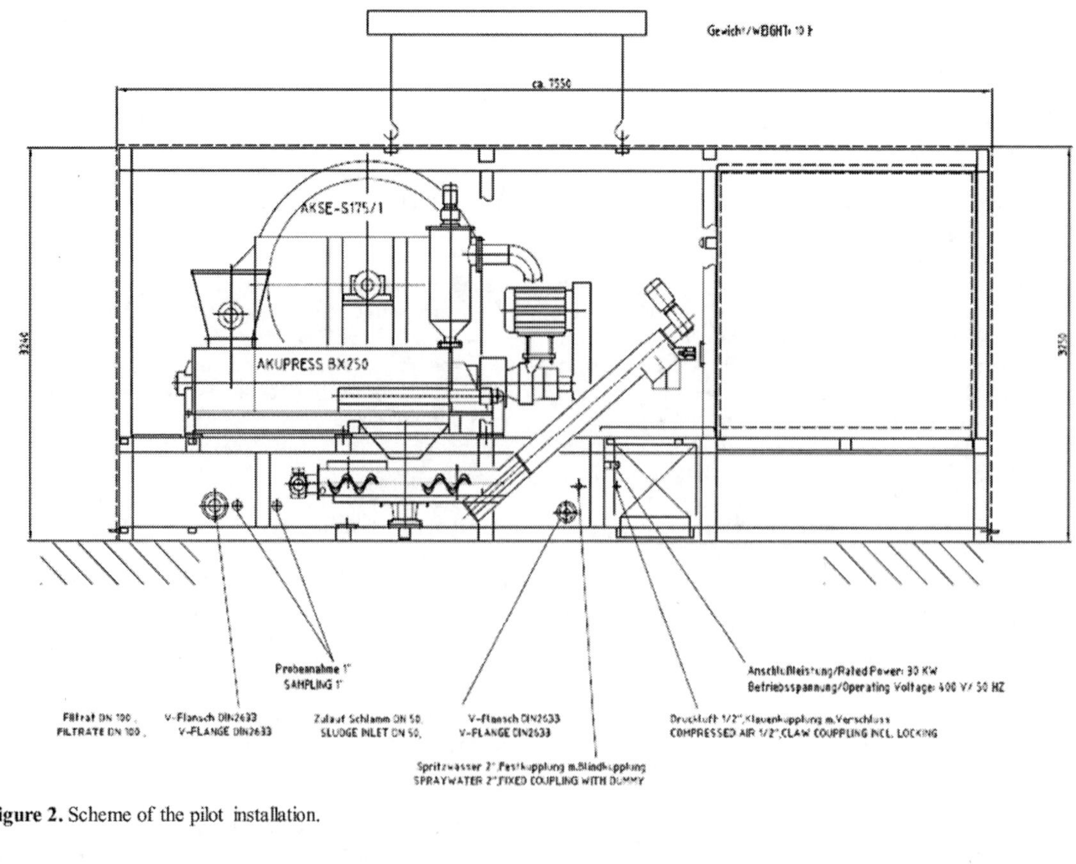

Figure 2. Scheme of the pilot installation.

 DOI: 10.1051/e3sconf/20161000083

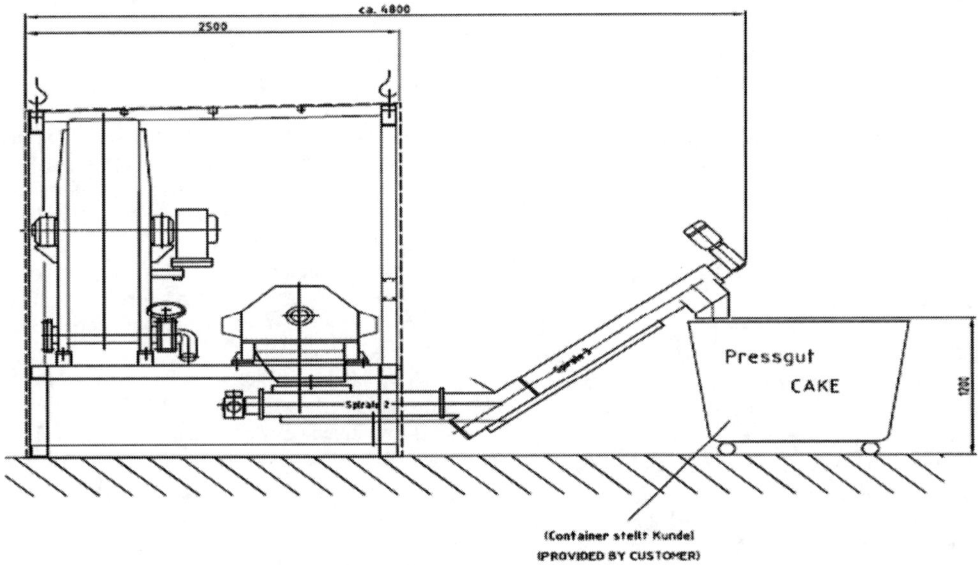

Figure 3. Scheme of the pilot installation – transport by conveyor.

2.5 Analysis of fuel properties

From the dewatered wastes two sorts of secondary fuel were produced: fuel P1, in which total using application of sludge and slimes was taken into consideration (wastes 03 03 05, 03 03 10, 03 03 11) and additionally with 20% share of wooden wastes (bark, silvers, scobs – waste 03 03 01); fuel P2 for which production only fibrous wastes were used (03 03 10) and sewage sludge from sewage treatment plant (03 03 11). The basic energetic properties of elaborated fuels were presented in Tables 6-9, the coal, hydrogen and nitrogen contents in fuel were presented as well the contents of basic oxides and heavy metals in ash obtained from fuel combustion.

The second method of wastes dewatering from celluloid mass and paper production in IP Kwidzyn plant is based on their drying in drying belt in which the waste heat is used. In this technology the wastes are firstly mixed in mixer before drying.

In the following tables (Table 6-9) the results of Solid Recovery Fuels obtained during the pilot investigations are shown.

Table 6. Properties of SRF

SRF	W^r_t	V^r	A^r	S^r_t	Q^r_i
	[%]	[%]	[%]	[%]	[kJ/kg]
P1	6,5	47,1	30,7	0,21	9 759
P2	7,5	47,2	27,2	0,2	10 923

Table 7. Elements analysis of SRF

SRF	C^a	H^a	Cl^a	S^a_t	N^a
	[%]	[%]	[%]	[%]	[%]
P1	30,5	3,36	0,04	0,21	1,79
P2	32,2	3,95	0,04	0,20	1,42

Table 8. Metals content in ash

SRF	Cr	Cd	Pb	Co	Ti	As	Hg	Sn	Mn	V	Cu
	[mg/kg]	[mg/kg]	[mg/kg]	[mg/kg]	[mg/kg]	[mg/kg]	[mg/kg]	[mg/kg]	[mg/kg]	[mg/kg]	[mg/kg]
P1	36,21	2,43	42,28	5,17	227,84	4,40	0,11	4,53	448,43	56,10	222,28
P2	34,12	2,63	45,83	5,43	243,58	3,14	0,10	4,27	514,29	61,40	269,28

Table 9. Analysis of ash from SRS incinerating

SRF	Na_2O	K_2O	ZnO	Fe_2O_3	SiO_2	CaO	MgO	Al_2O_3	P_2O_5
	[%]	[%]	[%]	[%]	[%]	[%]	[%]	[%]	[%]
P1	0,29	0,25	0,11	2,53	33,40	46,38	0,82	6,95	3,42
P2	0,28	0,30	0,11	2,76	33,50	48,86	0,86	7,2	3,73

2.6 Measurement of the biomass fraction of SRF

The biomass fraction of SRF has a monetary value under multiple greenhouse gas protocols, such as the European Union Emissions Trading Scheme and the Renewable Obligation Certificate program in the United Kingdom [8]. Biomass is considered to be carbon-neutral since the CO_2 liberated from the combustion of biomass is recycled in plants. The combusted biomass fraction of SRF is used by stationary combustion operators to reduce their overall reported CO_2 emissions. Several methods have been developed by the European CEN 343 working group to determine the biomass fraction of RDF/SRF. The initial two methods developed (CEN/TS 15440) were the manual sorting method and the selective dissolution method. Since each method suffered from limitations in properly characterizing the biomass fraction, an alternative method was developed using the principles of radiocarbon dating.

A technical review (CEN/TR 15591:2007) outlining the carbon-14 method was published in 2007, and a technical standard of the carbon dating method (CEN/TS 15747:2008) was published in 2008.

In the United States, there is already an equivalent carbon -14 method under the standard method ASTM D6866. Although carbon-14 dating can determine with excellent precision the biomass fraction of RDF/SRF, it cannot determine directly the biomass calorific value. Determining the calorific value is important for green certificate programs such as the Renewable Obligation Certificate program in the United Kingdom. These programs award certificates based on the energy produced from biomass.

Several research papers, including the one commissioned by the Renewable Energy Association in the UK, have been published that demonstrate how the carbon-14 result can be used to calculate the biomass calorific value. Table 10 shows the results of biomass content in fuels from paper wastes.

Table 10. Biomass content in SRF P1 and P2

SRF	Biomass content [%]
P1	67,13
P2	87,34

3 Summary

Eleven million tons of waste are produced yearly by the European pulp and paper industry, of which 70% originates from the production of deinked recycled paper.

Due to legislation [9, 10] and increased taxes, landfills are quickly being eliminated as a final destination for wastes in Europe, and incineration with energy recovery is becoming the main waste recovery method.

Taking into consideration the data contained in Tables which concern calorific values of individual components and fuel composition considering amount of wastes yearly and their complex application to production of secondary fuel, it is possible to obtain fuel of calorific value from 14-16 MJ/kg.

As a result of conducted investigations of energetic properties of wastes produced from paper production in IP Kwidzyn the technology of secondary fuel production was elaborated, which was made on the basis on slime from discoloration of wastepaper (03 03 05), fibrous wastes, slimes from fibers, fillers and covers originating from mechanical separation (03 03 10), sewage sludge from sewage treatment plant (03 03 11) and bark (03 03 01). It allows to apply these wastes totally to energetic purposes on the area of the plant.

The supplement of the fuel components can be biomass of agricultural or forest origin which can be added in purpose of increasing energetic properties of the fuel.

The pulp and paper manufacturing sector is energy and raw materials intensive, with high capital costs and long investment cycles. The industry has an excellent track record in resource efficiency and innovation. Thanks to its knowledge of wood fibre, the pulp and paper industry is at the forefront of developing innovative products alongside more traditional products. It is a pioneer in making the EU low-carbon bioeconomy an industrial reality.

The manufacturing of Solid Recovered Fuels (SRF) for instance – which can be burned for energy instead of coal, oil and gas – represents an area of great potential. The environmental benefits that the International Paper can achieve from this type of Waste to Energy (WtE) process are numerous, and acknowledgement of these benefits is becoming ever-more widespread. But owners and operators of SRF plants cannot neglect the fact that, whilst supporting the polish;l waste agenda, they are also creating a product to sell to customers such as power stations or cement kilns. These SRF customers need a precise and refined particle size, even more-so than RDF which in comparison tends to be a rougher shred with a less specific fraction requirement. The waste processing systems that recyclers invest in to create SRF should therefore be built and operated in the same way as any true production environment, if both the alternative fuel producer and end user are to reap the greatest possible return on investment. This *white paper* is designed to offer specialist insight and much-needed guidance for the best practice design and ongoing management of an efficient SRF plant, presented in a five-point plan. Drawing upon the principles of lean manufacturing this paper examines the ways to structure the SRF production system before considering the importance of input material, equipment and operators when creating a high quality fuel for the end user. Challenges do of course face SRF manufacturers but this advice-led resource explains that, when looking at the plant as a production system not just a recycling process, benefits such as maximum availability, uptime, throughput and return on investment can be achieved.

The dewatering of fibrous materials is a fundamental process in the paper and pulp production, but also in all other manufacturing processes using or treating fiber suspensions.

The principal objectives pursued in fiber dewatering may include:

DOI: 10.1051/e3sconf/20161000083

- separation of water cycles by reducing the dissolved matters in the pressed material ('hydraulic seal');

- the higher the dryness of the thickened stock the higher the efficiency will be:

- increase of the dry content to a value required for downstream processes, e.g. dispersion systems,

- reduction in weight and volume, e.g. for storage and transport.

Due to the large volumes of waste generated, the high moisture content of the waste and the changing waste composition as a result of process conditions, recovery methods are usually expensive and their environmental impact is still uncertain.

Other options such as pyrolysis, gasification, land spreading, composting and reuse as building material are being applied, although research is still needed for optimization of the processes.

References

1. http://www.spp.pl/the_polish_paper_review.php (2016)
2. http://www.internationalpaper.com/ (2016)
3. M.C Monte., E Fuente., A., Blanco, C. Negro.. Waste Management From Pulp And Paper Production in the European Union. Chemical Engineering Department. Complutense University of Madrid. (2014).
4. G. Thompson, J. Swain , M. Kay , C.F. Forster., Bioresource Technology, **77**, 3, 275–286 (2001)
5. S. Budzyń., W. Andrusikiewicz., V. Cablik., B. Tora, W. Gradoń., Journal of the Polish Mineral Engineering Society, **16**, 2, 137–141 (2015)
6. B. Tora, S. Budzyń, M. Krzykowski, W. Gradoń, W. Żmuda, Process for the preparation of secondary biomass solid fuel from waste paper industry, Patent PL 402910, Biuletyn Urzędu Patentowego, ISSN 0137-8015, **16**, (2011)
7. Integrated Pollution Prevention and Control (IPPC), Reference Document on Best Available Techniques in the Pulp and Paper Industry. European Commission. December. (2001).
8. www.bellmer-kufferath.de (2016)
9. Ustawa z dnia 14 grudnia 2012 r. o odpadach, Dz. U. z 2013 r. poz. 21, (2012)
10. Ustawa z dnia 27 kwietnia 2001 r. Prawo ochrony środowiska, Dz.U. 2001 nr 62 poz. 627. (2011)

Electromagnetic therapeutic coils design to reduce energy loss

Przemyslaw Syrek[1], Mikolaj Skowron[1], Szczepan Moskwa[1], Wojciech Kraszewski[1] and Antoni Ciesla[1]

[1]*AGH University of Science and Technology, al. Mickiewicza 30, 30-059 Krakow, Poland*

Abstract. The article introduces the problem of power loss reduction in applicators used in magnetotherapy. To generate magnetic field whose distribution is optimal and to reduce the power loss, the authors establish a set of parameters to evaluate the model of device. Results make it possible to infer that the real power input necessary to operate the magnetic field generator properly may vary significantly depending on construction and localization. The issues raised in this paper should be treated as a basis for further discussion on the construction of applicators used, e.g., in Transcranial Magnetic Stimulation.

1 Introduction

Biological effects of low-frequency electromagnetic field on living organisms have been investigated for few decades. There is wide range of publications that show beneficial influence of EMF, i.e. accelerated tissue regeneration, reduced bone fracture healing time, pain relief, and the anti-inflammatory effect [1,2].

In the extremely low frequency band, both components of electromagnetic field and their influence the human tissues may be considered separately. Both components are used in a wide range of medical problems. Remotely supervised therapies are developed in many branches of medicine, e.g. related to the brain stimulation [3]. Since the fractures and general orthopedic injuries contribute to increase in sick absence from work by several percent, this aspect requires the use of available technical means. Therapy conducted remotely has many advantages: travel time related to rehabilitation is reduced to the barest minimum, whereas the duration of the session is in principle unlimited. In this context, the energy saving problem arises. While searching patent databases, one may find numerous claims concerning magnetic field control systems and power supply methods.

The present article presents the problem of energy loss reduction in order to extend the continuous operation of the portable magnetic field applicator. The energy loss is results from four causes: the energy source itself, the converter and the impulse generator, the anatomical structures absorbing the energy, and the final cause is within the applicator's winding. The first two are beyond the scope of our consideration. Human tissues and their electric conductivity combined with current range induced by devices produce a negligible energy level, especially in the extremely low frequency band. Even the energy absorbed by human tissue, which is an inherent consequence of hyperthermia, could be omitted, if we take into account tissue parameters like heat capacity, density, and the volume under treatment [4].

[a] Corresponding author: syrekp@agh.edu.pl

Therefore, a variety of stimulators coil shapes are considered in order to focus or ensure the uniformity of electric or magnetic field [5]. The coil winding design should take into account energy efficiency as well. The results, which were obtained using evolutionary strategy, are presented for one of many objective functions that may be adopted, which depend on requirements imposed on field distribution.

2 Outline of the problem

In order to avoid the multi-criteria optimization, the problem of the device construction cost is omitted. The applicator winding is to generate a time-varying magnetic field. Depending on the therapeutic requirements, only the same magnetic component of the EMF or indirect utilization of magnetic field is required for generation of an electric field, and thus the eddy currents in the tissues of the body under treatment [6].

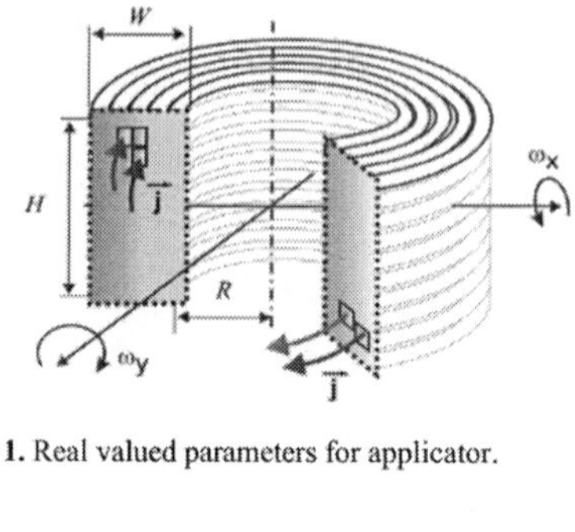

Figure 1. Real valued parameters for applicator.

2.1 Magnetic field generator

The selected aspects of energy loss reduction are considered. The first problem is to specify the relation between resistance and reactance of applicator winding. In the low frequency band, self-inductance, and hence the reactance, is a hundredth part of resistance. Therefore, the dominant is real power compared to the reactive power, despite the fact that the reactive power may be relatively easily and cheaply compensated.

The magnetic field distribution, especially its absolute value, depends on a number of factors: the shape of the winding and its construction (or the modeling of curve representing the wire) in general, wire length, cross sectional area, current supplied to the device. While the winding resistance increases linearly with its length, the energy loss grows as well. In turn, in the linear model, the magnetic field at each point in space is proportional to the current, while the power loss is proportional to the square of the current. Here occurs the main problem concerning the construction of the applicator, which is related to the compromise between the applicator winding and amperage. It follows that, in order to reduce losses, the increase in the length of the applicator is the right direction (successive layers and turns of the winding, fig.1), whereas the current should be reduced. Unfortunately, this approach has its own limitation. Adding another layers means that they are farther away from the target points of its influence, and the decrease in the magnetic field can be assumed to be equal to the square (or almost square) of the distance.

The magnetic induction is calculated using the line integral:

$$\mathbf{B} = (\mu_0/4\pi) \int_V (\mathbf{j}_S \times \mathbf{r})/|\mathbf{r}|^3 \, dV \qquad (1)$$

where $\mathbf{j}_S$ represents current density, $\mathbf{r}$ is the vector pointing from the target point to dV – infinitesimal portion of the coil.

2.2 Applicator parameters

In [7], authors propound the method of placing turns within the coil to concentrate the magnetic and electric field with the minimal energy cost.

Since the magnetic field diminishes with nearly the square of distance, arbitrary specification of the location of applicator relative to the body seems to be inadequate and may significantly disrupt the results of power loss. Thus, the set of parameters for position of the applicator is not to be neglected. In the case of solenoid applicator, five parameters are enough to provide any position [8]. Thus, the best location will be ensured in the context of adopted cost evaluation. We do not distinguish between individual turns of the applicator coil, but replace them with the current density in the cross-section. This method makes it possible to use continuous optimization and real-valued parameters for dimensions and position of the applicator. The parameter j_{RMS} is the RMS of absolute value of current density vector in the wire.

Eight real valued variables describe the vector that represents the applicator:

$$\mathbf{x} = [\, R\ W\ H\ \omega_X\ \omega_Y\ x\ y\ z \,]^T \qquad (2)$$

These parameters are used to describe each individual in the populations used in the evolutionary optimization process. The first three represent the dimensions of the device, next two represent applicator rotation around the axes of the coordinate system and last three correspond to its position in any direction. Table 1 introduces limits for each parameter. The search space is the hypercube in 8 dimensions.

Table 1. Limits for applicator parameters.

Variable	Lower bound	Upper bound
R	0.005 [m]	0.080 [m]
W	0.010 [m]	0.050 [m]
H	0.010 [m]	0.016 [m]
ω_X	0 [rad]	2π [rad]
ω_Y	0 [rad]	π [rad]
x	–0.2 [m]	0.5 [m]
y	–0.2 [m]	0.5 [m]
z	0 [m]	1.0 [m]

2.3 Applicator evaluation

Since the ferromagnetic materials are excluded, and a constant electric conductivity of copper wires is assumed, the model is linear. Due to linearity, the current density j may be omitted in the set of control parameters. During the cost evaluation, the magnetic (or electric) field may be obtained at the arbitrary level of j_{RMS} conventionally equal to 1 A/mm^2. Then, the current density should be rescaled for the sake of the field at the required level:

$$j_{RMS} = (B_{REQUIRED}/B_{OBTAINED}) \cdot 1\ [\mathrm{A/mm^2}] \qquad (3)$$

Whereas the power loss:

$$P_{TOTAL} = (j_{RMS}^2/\gamma_{CU}) \cdot \mathrm{V}\ [\mathrm{W}] \qquad (4)$$

where: γ_{CU} –electric conductivity of copper, V – volume of conducting elements of an applicator.

The quantity (4) is the measure of cost assigned to each individual in the evolutionary strategy for optimization of the title device.

3 Optimization process

Some aspects of optimization process i.e. real-valued parameters of the device, search space boundaries and cost measure definition were presented in the above sections.

To determine the location of target points and to remove and hence "to hollow" the search space, the numerical model of human was used (fig. 2). This is the Virtual Family Model [9], which represents the patient. Exactly, Ella, a 26-years old female model (1.66 m height, 57.3 kg weight) is such a virtual patient. The available models are made of 0.5, 1, 2 and 5 mm voxels. The iso2mesh free library was used to generate 3D meshes on the basis of the voxel [10], and the mesh was developed using generate tetrahedral mesh generation software – TetGen [11].

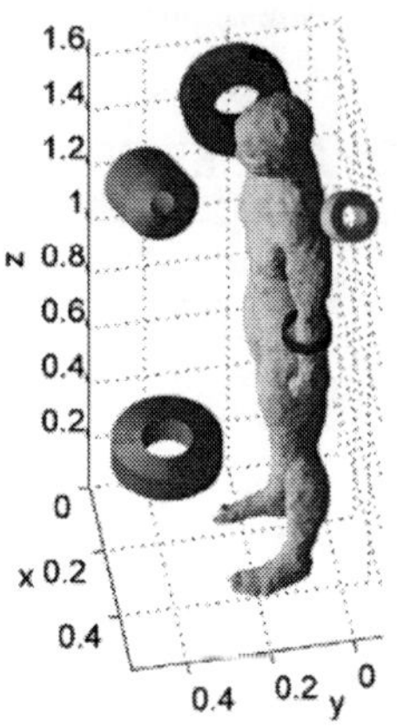

Figure 2. An example set of applicators to evaluate, and the surface for collision detection.

3.1 Violation factor

The position of the applicator is constrained, which results from collision between human body and the device. The solution beyond hypercube as the feasible region, is simple to transform and put at least at the boundary. In the event of a collision with body model, the case becomes more complicated. Both violation factor assignment and restoration of solution point on the border between the patient model and search space are of high computational complexity even when referring to the EMF problems and their calculation costs. Therefore, evolutionary strategies involving the real-valued penalty due to violation degree are omitted. Instead of tolerating a group of individuals slightly in excess of the feasible solution area (and use the exterior penalty factor) to take advantage of an attractive part of their genotype, we propose the elimination of such solutions. It involves the avoidance of unnecessary cost function evaluations.

3.2 Evolutionary strategy

A description of the applicator using real variables allows for a wide variety of evolutionary strategies.As it can be implied from the no free lunch theorem [12], every problem should be matched a strategy that best copes with it. Especially in the case of field problems, where an assessment is usually computationally complex.

Overview and tests of possible strategies pointed to the necessity of elimination. The strategies to be eliminated are the ones that involve a degree of violation, i.e. precise collision detection. Due to the computational complexity of such an approach, these concepts are beyond the area of interest. Imperial Competitive Algorithm is found out to be ineffective when up to five variables may affect finding infeasible chromosome.

The other strategy, Differential Evolution, has a great potential when it comes to seeking the position of the applicator, even if we include the above mentioned problem of removing solutions involving a collision with a body, while the 3 variables responsible for the construction of the applicator and the narrow range of their values makes this method inadequate.

The next aspect is the stop criterion. In [13], $N \cdot 10^5$ candidate solutions evaluation is suggested, where N is the problem dimension. Both the tests of strategies (in terms of population size, reproduction and mutation ratio) and the results of optimization presented below, were conducted through 20000 cost functions evaluations. Instead, each result was obtained by 30 repetitions to control the repeatability of results. The strategy that has eventually been adopted used the population of 150 individuals picked using elitist selection, and the genotype has been enhanced by parameter of mutation strength that evolves with the population [14] to strengthen exploitive abilities of population.

4 Results and discussions

Figure 3 presents the best individual (the best solution to the problem) and the device modeled by it optimized on the basis of the above mentioned for tibia bone piece, 60 mm in length. Applicator induces magnetic field of at least 1 mT within the bone fragment, the real power needed is in this case equal to 0.48 mW with the coil length of 1198 meters.

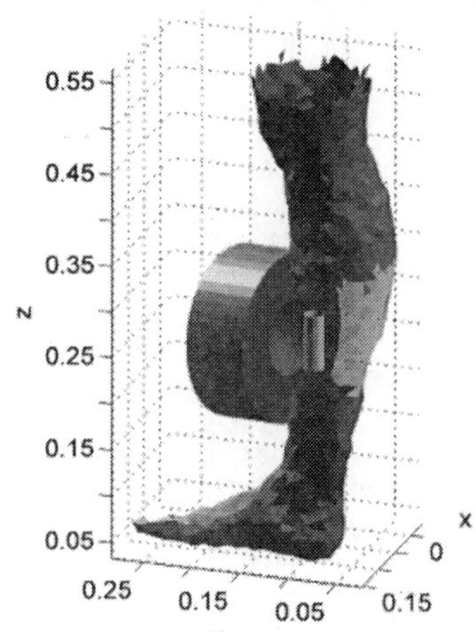

Figure 3. Applicator for therapy of a tibia bone fragment.

Results of optimization are presented in Table 2 and Fig. 4. Applicators and their optimal parameters were obtained for a wide range of magnetic induction that was required within part of tibia bone. A graph on a logarithmic scale shows that power and current increases significantly with the magnetic field specified. The coil length is almost constant. The *W*-parameter, which corresponds to turns of winding, reaches the upper bound of its limitations (parameters marked with * in the tab.2, to stand out feasible solution border) in each case. Otherwise, the *H*-parameter (the possible number of coil layers) increases with the magnetic induction needed. The very important observation on this parameter is that the optimal solution optimal does not converge to its upper bound, what let decrease current supplied to the coil. This conclusion results from the manner the magnetic field diminishes within the target space when the next layers are added. This aspect the fair possibilities to reduce energy consumption by this category of devices.

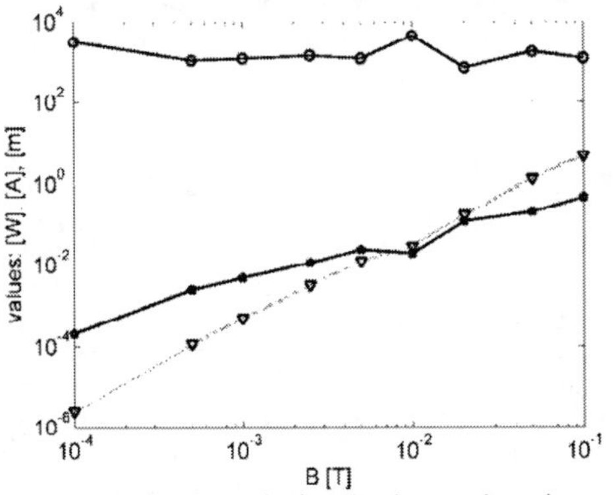

Figure 4. Results of optimization aimed at a variety of magnetic induction values. Plots for: real power (triangles), current for 1 mm² wire (stars), coil length (circles).

5 Conclusions

As a result of the simulations, it was confirmed that not all parameters of the applicator model must be at the upper boundary of the solution space (table 2). On the contrary, further enhancement of certain parameters would result in a faster power loss growth than the component of the magnetic field, which results from the distance of the successive turns of the device from the target points – space to focus the field.

An important observation is that larger applicators may also be used for mobile application. In this case it is limited by the possible discomfort of the patient, but from an energy point of view, it is an attractive construction. Further considerations also need to take the cost of construction that includes the wire and the applicator housing costs into account. The problems presented here should be treated as a basis for further discussion on the construction of applicators used, e.g., in Transcranial Magnetic Stimulation, where it is desirable to achieve the highest value of components of the electric and magnetic field in certain subspaces of the brain, and the technical solutions currently used in this process involve pulse generation using the superconductivity. These solutions are at the limit of modern technology.

Table 2. Results of optimization for real power loss (feasible solution upper bound marked with *).

B [mT]	*P* [mW]	*l* [m]	*R* [m]	*W* [m]	*H* [m]
0.5	0.117	1081	0.028	0.049*	0.063
1	0.486	1198	0.028	0.049*	0.071
5	11.84	1203	0.029	0.049*	0.070
50	13570	1771	0.035	0.050*	0.092

References

1. K. Sieron-Stoltny, L. Teister, G. Cieslar, D. Sieron, Z. Sliwinski, M. Kucharzewski, A. Sieron, Biomed Res. Intern., 896019 (2015)
2. A. Fathi Kazerooni, M. Rabbani, M.R. Yazdchi, IFMBE Proceedings **35**, 458–462.(2011),
3. L.E. Charvet, M. Kasschau, A. Datta, H. Knotkova, M. C. Stevens, A. Alonzo, C. Loo, K.R. Krull, M. Bikson, Frontiers in Systems Neuroscience, **9** (2015), 26, 1-13.
4. E. Kurgan, P. Gas, Przegl. Elektrotech. (Elect. Rev. **87(12b),** 103-106 (2011).
5. D. Zhi-De, S.H. Lisanby, A.V. Peterchev, Brain Stimulation **6,** 1–13 (2013).
6. A. Krawczyk, A. Miaskowski, E. Lada-Tondyra, Y. Ishihara, Przegl. Elektrotech. (Elect. Rev.) **86(12),** 72-74 (2010)
7. L. Darabant, M. Plesa, D.D. Micu, D. Stet, R. Ciupa, A.Darabant, IEEE Transact. on Magn. **3**, 1690-1693 (2009),.
8. A. Ciesla, P. Syrek, Przegl. Elektrotech. (Elect. Rev.) **88**(12b), 124-127 (2012).
9. A. Christ, W. Kainz, E.G. Hahn, K. Honegger, M. Zefferer, E. Neufeld, W. Rascher, R. Janka, W. Bautz, J. Chen, B. Kiefer, P. Schmitt, H.P. Hollenbach, J.X. Shen, M. Oberle, D. Szczerba, A. Kam, J.W. Guag, N. Kuster, Phys. in Med. and Bio. **55** (2), N23-N38 (2010).
10. Q. Fang, D. Boas, Proceedings of IEEE Intern. Symp. on Biomed. Imag., 1142-1145 (2009).
11. Si Hang, ACM Trans. on Math. Soft. **41** (2), (2015).
12. D.H. Wolpert, W.G. Macready, IEEE TransEvolComp **1**,67-82 (1997).
13. K. Opara, J. Arabas, J. of Telec. and Inf. Techn., **4** 73-80 (2011),.
14. H.G. Beyer, H.P. Schwefel, Nat. Comp. **1**, 3–52 (2002).

The possibility of separation of electronic waste by means of their electrical properties

Mikolaj Skowron[1,a], Agnieszka Surowiak[1], Przemyslaw Syrek[1], Antoni Ciesla[1], Wojciech Kraszewski[1], and Damian Lachendrowicz[1]

[1]*AGH University of Science and Technology, al. Mickiewicza 30, 30-059 Krakow, Poland*

Abstract. Electric field has a lot of applications in technology. One of them is electrodynamic separation: electric field influences selectively granular solids of different moments or charges. A mathematical model of the separation process in high voltage drum separator is presented in the paper. Particles are charged both by induction and corona phenomena: next, they are separated by effects of the field forces. Some computational and experimental results are given and analyses.

1 Introduction

Each year, in the European Union around 8 million tons of waste of electrical and electronic equipment (the annual growth rate of formation of these wastes is 3-5% all produced), while worldwide, there are annually around 20-50 million tons of waste. One of the objectives of rational resource consumption and waste is the best use of the primary components of natural resource and recovery largely processed elements from waste products. The processing of various types of waste for recovery of useful components can be used by separation methods in mineral processing. Then the separation of useful chemical elements of the waste is carried out for separation of well-defined characteristics. This attribute, which is called the argument of the chapter, could be the difference in the densities of particle components to be separated, settling velocity of particles, wettability (hydrophobic and hydrophilic), magnetic susceptibility, electrical conductivity and other [1 – 8].

The electromagnetic field is used for different of technological processes (separation of magnetic and electrodynamic [9, 10]), medicine (magnetic therapy, diagnostics) [11, 12]. Electrodynamic separation is the process of separating two or more components with different physical properties, the forces of the electric field. One group of separation processes is represented by the conductance methods. These methods are based on different electric conductance of separated particles, which lead a various time constants of their charging. The applications of electrodynamic separation technologies range from recovery of metallic and insulating materials from industrial wastes [13 – 16], mineral beneficiation [17 – 19] and metals recovery from electronic scarp [20 – 22].

The paper presents how high voltage drum separator works. The analysis of electric field distribution in separator working space was done. Additionally, distribution of forces acting on particles of dielectric and conductive properties was examined, trajectories of charged particles movement were presented as well. The laboratory investigations of electronic wastes were performed, and in such a process the particle of size 0 – 0.5 mm was obtained. It was observed that as a result of selective charging of particles they separate according to surface ability to electrifying what allows to obtain selective separation of components being so-called electronic wastes. The application of electrodynamic drum separator allows to separate such chemical elements as Ti, Cu, Fe, Pb, Sn from plastics occurring in electronic wastes.

The motion of dielectric particles in electric field is described by the formula:

$$\boldsymbol{F}_{el} = \boldsymbol{F}_d + \boldsymbol{F}_e = \mu\nabla\boldsymbol{E} + Q\boldsymbol{E} \qquad (1)$$

where: $\boldsymbol{F}_{el}$ is the electric field force, $\boldsymbol{F}_d$ denotes its dielectrophoresis component, $\boldsymbol{F}_e$ stands for its electrophoresis component, $\boldsymbol{\mu}$ is the electric moment of the particle, $\boldsymbol{E}$ represents the electric field, and Q denotes the electric charge of the particle.

The involved forces of and principal methods of charging dielectric particles are schematically shown in Fig. 1.

[a] Corresponding author: mskowron@agh.edu.pl

DOI: 10.1051/e3sconf/20161000085

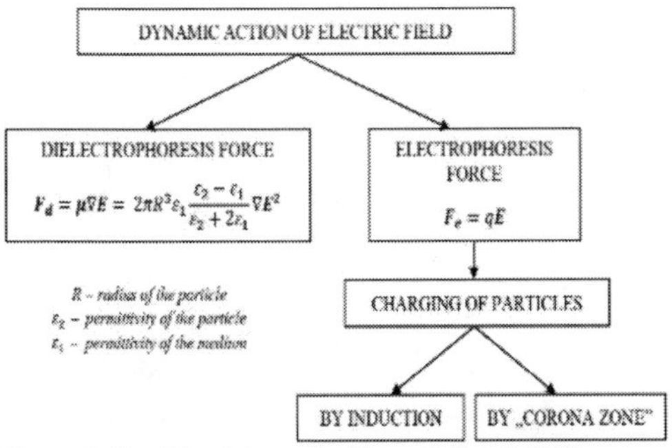

Figure 1. Considered dynamic effects of electric field.

The prevailing structure in mineral processing (beneficiation) technology is the „corona" drum separator invented by Huff [23] (Fig. 2).

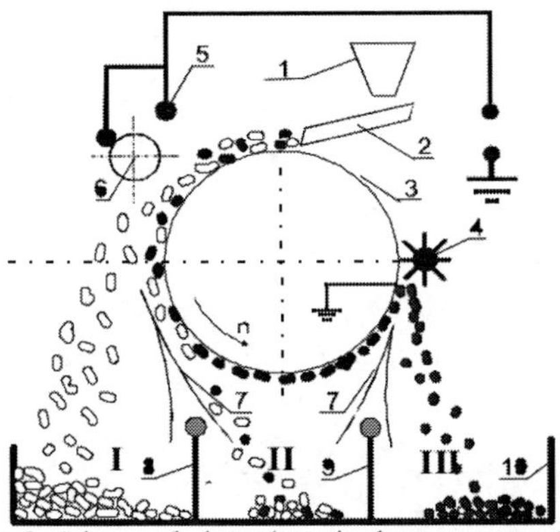

Figure 2. Scheme of electrodynamic drum separator (Huff's separator):
1- feed container, 2-feed hopper, 3-drum, 4-brush, 5-corona electrode, 6-deflecting electrode, 7-separation barriers, 8, 9, 10-bins for separation products.

2 Mathematical model of electric field dynamics in the drum separator

The mixture of grains falls out of the hopper ($\boldsymbol{E} = 0$) into direct field ($\boldsymbol{E}$ = const.) generated between two cylindrical electrodes (one of them rotates). The particles in contact with the rotating electrode obtain an electric charge by induction. The electrical scheme for the „particle – drum electrode" system is presented in Fig. 3. The dependence of the acquired charge on time is defined by the formula [24]:

$$Q(t) = U\cdot\frac{C_pC_z}{C_p+C_z} + U\cdot\frac{C_p^2}{C_p+C_z}\left[1-\exp\left(\frac{-t}{R_z(C_p+C_z)}\right) - Q_0\cdot\exp\left(\frac{-t}{R_z(C_p+C_z)}\right)\right], \tag{2}$$

where: Q_0 is the charge of the particle acquired in corona zone given by the expression

$$Q_0 = \left(1+2\frac{\varepsilon_2-\varepsilon_1}{\varepsilon_2+\varepsilon_1}\right)\varepsilon_0 E\pi r^2 \tag{3}$$

here ε_2 is the relative permittivity of the particle, ε_1 denotes the relative permittivity of the ambient medium and r stands for the radius of the particle. The other parameters C_p, C_z and R_z follow from Fig. 3.

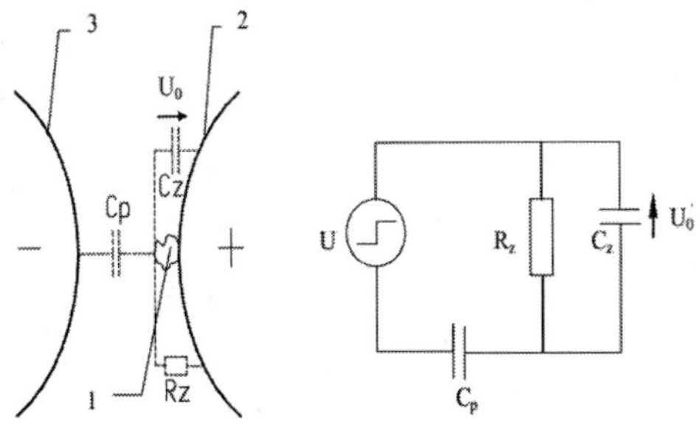

Figure 3. Particle on rotating drum in an electrodynamic drum separator (the sketch and the scheme):
1 – particle, 2 – rotating drum, 3 – counter electrode

The examples of dependences of acquired charges on time are shown in Fig. 4.

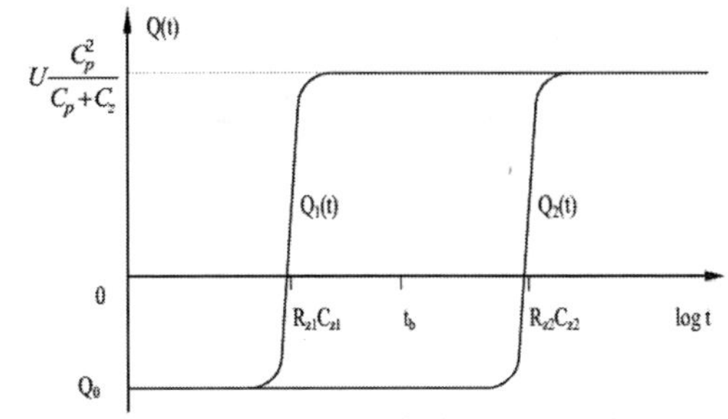

Figure 4. Particle charges vs. time in drum separator for $U_0 \neq 0$.

The analysis leads to the conclusion that the separation takes place when the contact time t_b between the mixture and drum is longer than the time constant of the „conductor" grain, but shorter than the time constant for the „nonconductor" particle (particle of the „conductor", e.g. the component „1" – z_1; particle of the „nonconductor", e.g. the component „2" – z_2). In other words, the separation takes place if

$$R_{z1}C_{z1} \ll t_b \ll R_{z2}C_{z2} \tag{4}$$

A particle, charged according to formula (2), is passing through electric field between two cylindrical electrodes (Fig. 2). The equation for the forces influencing a particle has the form:

$$m\frac{dv}{dt} = \Sigma\boldsymbol{F} \tag{5}$$

where: m is the mass of the particle, v denotes its velocity, and $\Sigma\boldsymbol{F}$ stands for the sum of forces acting on the particle.

Both particles on the surface of the rotating drum and particles passing through the electric field are affected by forces of the following modules:

- *electric field force* $F_e = QE$ (6)

- *„image" force* $F_{im} = \dfrac{Q^2}{4\pi\varepsilon_0\left(2R_2^2\right)}$ (7)

- *centrifugal force* $F_c = m\omega^2 R_2$ (8)

- *gravitational force* $F_g = mg$ (9)

- *aerodynamic resistance* $F_a = 6\pi\mu_d r v$ (10)

where R_2 is the radius of the rotating drum, ω denotes the angular velocity of rotation, g stands for the acceleration by gravity, and μ_d is the coefficient of the dynamic viscosity of the ambient medium.

The scheme of forces influencing a particle during the contact with the rotating drum as well as during the following fall is presented in Fig. 5.

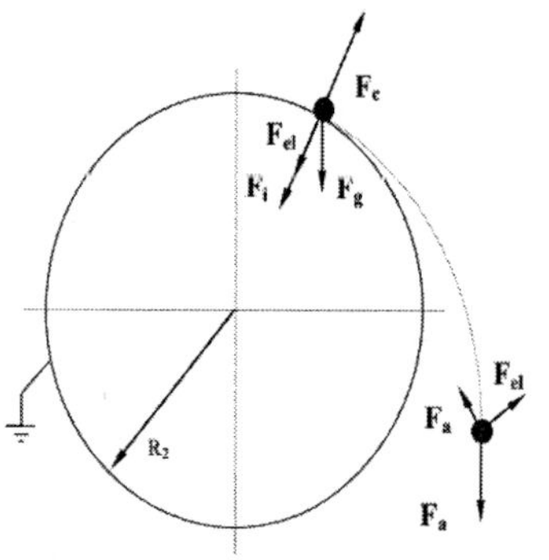

Figure 5. Scheme of forces influencing particle in drum separator.

The trajectories of particles in the separators were calculated on the basis of the field distribution (formula (5) as well as the forces influencing the particles (formulae (6)–(10)). For calculations, the software package „Matlab" was used. The results of the calculations (several trajectories) for various dynamic conditions are presented in Fig. 6.

a)

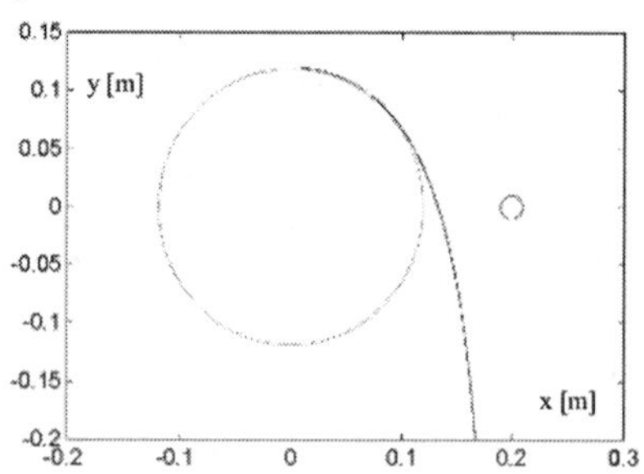

b)

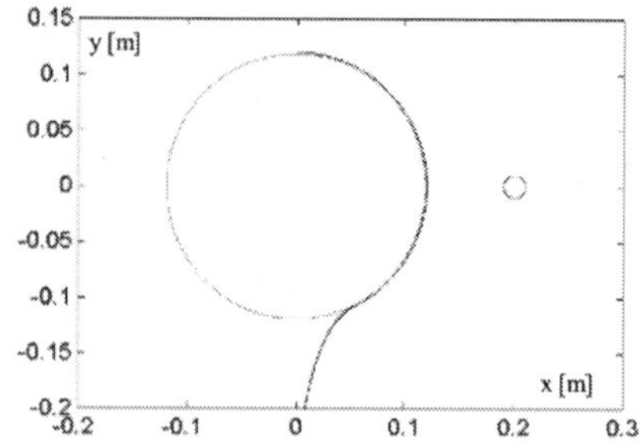

c)

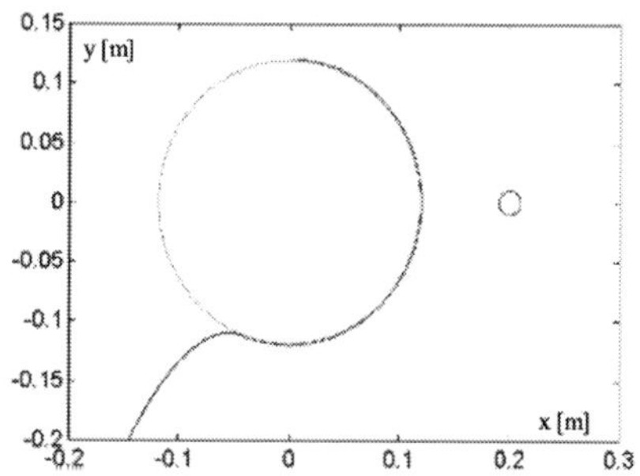

Figure 6. Exemplary particle trajectories (computer simulation): a) R_z = 1E6 [Ω], b) R_z = 4E6 [Ω], c) R_z = 7E6 [Ω], U = 20 [kV]; ω = 8 [rads^{-1}]

3 Experimental investigation

3.1 Research methodology

The high-voltage electrodynamic drum separator consists of two functional parts, namely of the charging part and of the electrophoretic separation part Fig.7.

The key problem of the separation – selective charging of the mixture components – can be based on various physical phenomena. From the practical point of view, these phenomena are electrostatic induction and occurrence of the corona charge. The prevalent construction in mineral processing (beneficiation) technology is the corona drum separator invented by Huff [24]. A scheme of this separator is presented in Fig. 2.

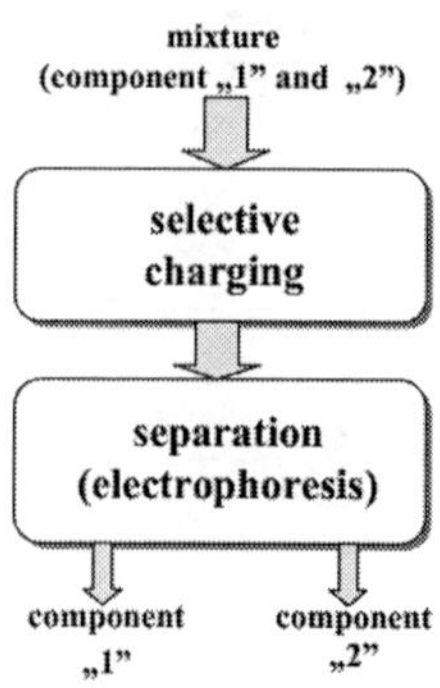

Figure 7. Electrodynamic separation based on the electrophoresis phenomenon

A mixture is poured out of the hopper and passes through the corona zone (the particles assume to have negative charges). Next, the mixture falls on the rotating drum. The particles of small resistivity (consequently, of small relaxation times) lose their charges rapidly. Subsequently, they take up some charges of opposite sign and are repelled by the field forces into the bin I. However, particles of large relaxation time (of great resistances) lose little of no charge: they are „pinned" to the rotating drum by means of the „image" forces. Eventually, the particles are removed with the brush into the bin III. The intermediate product falls into the bin II and may be returned to the feed hopper. A theoretical model of the process was presented by the authors [25].

The device presented above may be analyses as a control object – carrying into effect electro separation as a technological process (Fig. 8). The state of the process is defined by state variables X(*t*). The object can be controlled by a number of input values, called control values U(*t*). Other input values, so called disturbances Z(*t*), are undesirable for the control process.

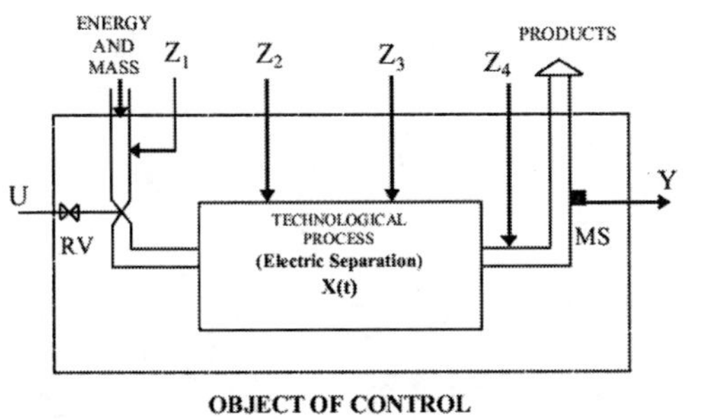

Figure 8. Technological process (electric separation) as an object of control: X, U, Y- vectors of variables of state, control and outlet (respectively), $Z_1 \ldots Z_4$ vectors of disturbances, RV – regulation valves, MS - measurement sensors

The key problem of the separation – selective charging of the mixture components – can be based on various physical phenomena. From the practical viewpoint, these phenomena are electrostatic induction and occurrence of the corona charge. The prevalent construction in mineral processing (beneficiation) technology is the corona drum separator invented by Huff. A scheme of this separator is presented in Fig. 2.

A mixture is poured out of the hopper and passes through the corona zone (the particles assume to have negative charges). Next, the mixture falls on the rotating drum. The particles of small resistivity (consequently, of small relaxation times) lose their charges rapidly. Subsequently, they take up some charges of opposite sign and are repelled by the field forces into the bin I. However, particles of large relaxation time (of great resistances) lose little of no charge: they are „pinned" to the rotating drum by means of the „image" forces. Eventually, the particles are removed with the brush into the bin III. The intermediate product falls into the bin II and may be returned to the feed hopper. A theoretical model of the process was presented by the authors [25].

The device presented above may be analyzed as a control object – carrying into effect electro separation as a technological process (Fig. 8). The state of the process is defined by state variables X(*t*). The object can be controlled by a number of input values, called control values U(*t*). Other input values, so called disturbances Z(*t*), are undesirable for the control process.

The most important input values for the separation process are the voltage V, geometry of the electrodes (their distance D), rotational speed of the drum ω, angle of inclination of the corona electrode α, and concentration of the tested component in the feed β. The disturbances are, for example, the temperature T, pressure p or moisture δ. The state variables are the particle charge Q, electric field $\boldsymbol{E}$, and granulation of the feed d. The separation process described above results in a product of flow off γ and tested component concentration λ, which are the quantities suitable for the evaluation of quality of the separation process.

3.2. Research station

Electrodynamic separation was performed using a laboratory drum separator (Fig. 9). On the basis of the available literature [3], separation and numerical calculations and series of simulations, the separator is selected with the following parameters: the voltage of the electrodes 11 kV and the rotation of the drum 60 rev / min.

Figure 9. Photo electrodynamic drum separator

3.3 Research of separation of electronic wastes

The material directed to separation was electronic wastes from computer mainboards. Firstly, they wer comminuted in knife crusher and the material of granulation <0.5 mm was sieved. Such prepared sample was separated in electrodynamic drum separator by keeping constant temperature and humidity conditions. The products of separation were collected in containers numbered from 1 till 10. The material from each container was weighted and mass shares of separation products were calculated. In each of products collected in individual containers the contents of the following elements were marked: Ti, Fe, Cu, Sn, Pb, Bi. They were presented in Table 1.

The distributions of chosen elements (Ti, Cu, Fu, Pb, Sn) in separated material which gathered in containers numbered from 1 till 10 after separation conducted in electrodynamic drum separator were presented on Figs. 10-15.

Table 1. Contents of individual elements in products of electronic wastes separation

Nr of conteiner	Mass [%]	Ti [%]	Fe [%]	Cu [%]	Sn [%]	Pb [%]	Bi [%]
1	5.4	0.7	7.2	73.2	12.3	13.2	0.0
2	5.4	0.8	8.2	66.8	10.7	11.1	0.0
3	4.3	0.8	10.3	57.5	9.3	9.9	0.1
4	7.3	1.1	11.2	36.1	5.1	5.5	1.1
5	6.5	1.5	8.8	28.6	3.7	4.5	1.0
6	7.0	1.6	4.9	20.7	2.1	2.6	1.2
7	4.8	1.2	7.3	13.8	2.6	2.9	1.3
8	3.8	2.2	6.3	9.7	1.4	2.0	1.7
9	10.8	1.7	5.2	3.3	0.6	1.0	1.8
10	44.9	0.0	4.7	2.9	1.1	1.2	1.0

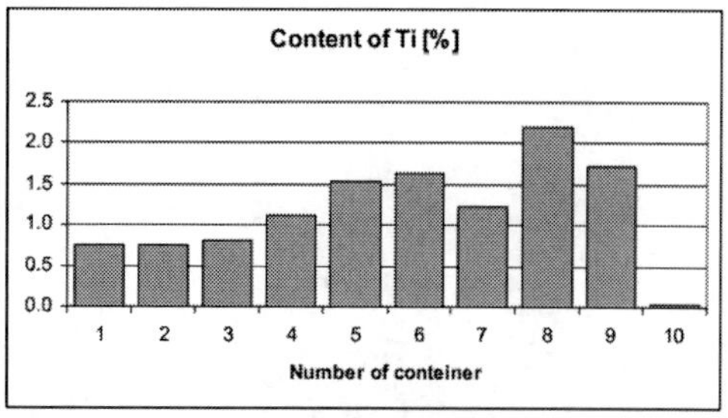

Figure 10. Distribution of Ti contents in separation products

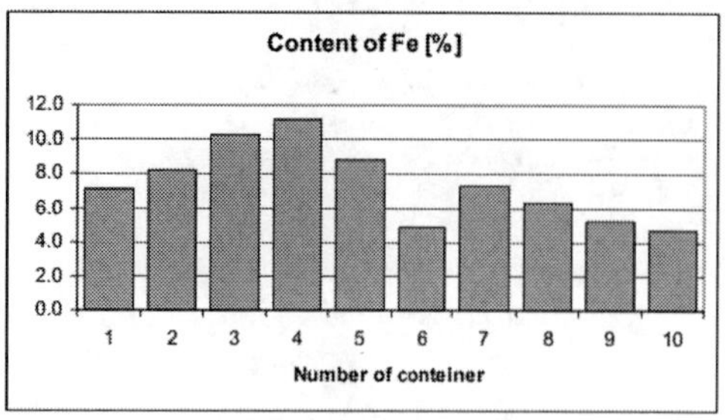

Figure 11. Distribution of Fe contents in separation products

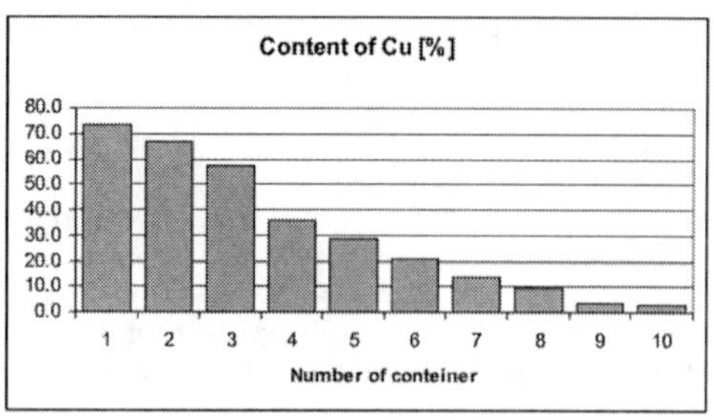

Figure 12. Distribution of Cu contents in separation products

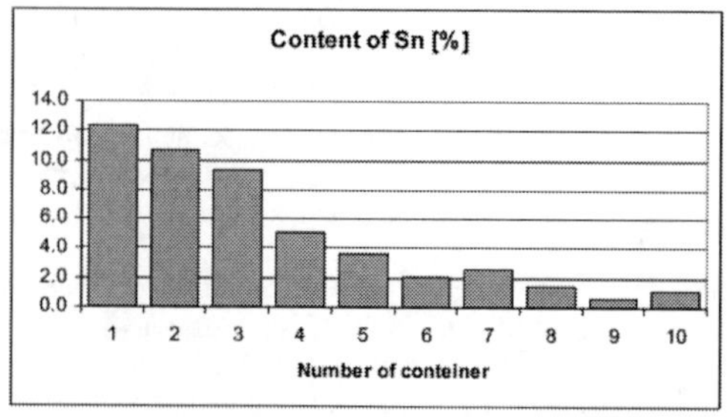

Figure 13. Distribution of Sn contents in separation products

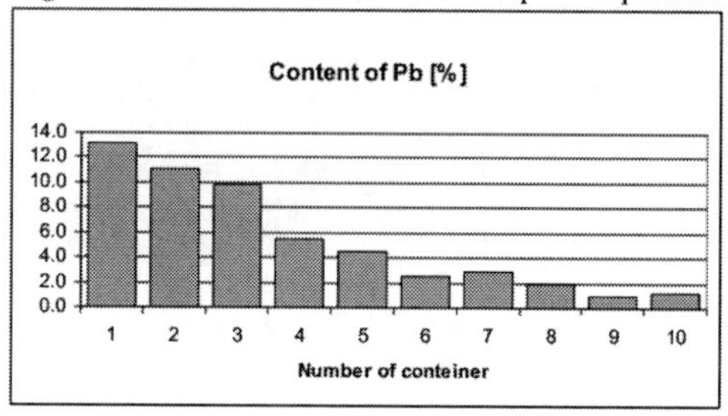

Figure 14. Distribution of Pb contents in separation products

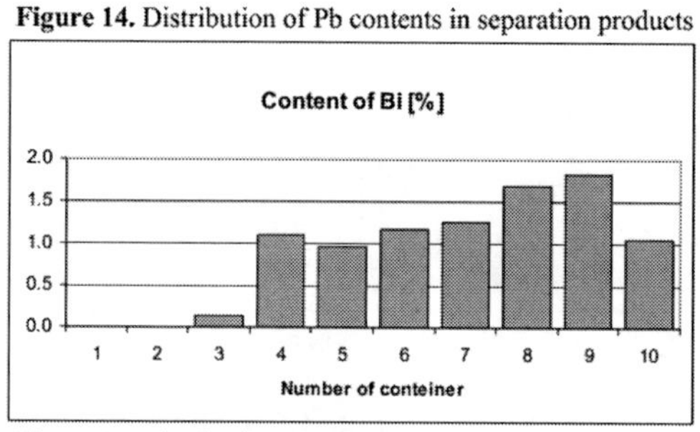

Figure 15. Distribution of Bi contents in separation products

Analysis of the results shows that the titanium content is concentrated in containers numbered from 4 till 9 and it can be seen that quite high levels of Cu and Sn occur in containers with lower numbers (1-3). Distribution of iron contents in containers is rather irregular. The maximum Fe content can be obtained from containers of numbers 3 and 5 while Bi concentrates in containers of numbers 4 till 10. Such a distribution of the elements contents is associated with the particle size distribution of the material, which influences on degree of the elements liberation as well possibility of the surface electrification.

4 Summary

Both theoretical analyses and experiments confirmed the usefulness of electrodynamic drum separators to separation of mixed granular solids. The required condition is that separated materials must characterize with different resistivity. The experiments proved that behavior of particles (treated as control objects) is mainly related to the electric field and rotary speed of the drum. Any unambiguous interpretation of the results is difficult because a lot of mechanical and electrical factors act simultaneously.

Analysis of the results leads to the following conclusions and observations:
- during electrodynamic separation the conductive components can be separated from the non-conductive electronic waste. In addition, the decomposition of some elements can be observed during the process. It can be concluded that the separation of Ti components from the rest is successful. Products containing such elements as Cu, Pb and Sn should be separated by means of another method in purpose of separating these elements. Various metals can be recovered in processes of metallurgy or hydrometallurgy. Products containing iron should be directed to purification by means of magnetic separation.
- advantage of electrostatic separation is that it is conducted in dry conditions. Therefore, there are no additional processes of thickening and dewatering, what eliminates the necessity of additional devices.

References

1. Brozek M., Surowiak A., Physicochem. Prob. of Min. Proc. **41**, 397-413 (2007).
2. Brożek M., Surowiak A., Jarosiński A., Arch.of Metal. and Mat., **60**(3), 1737–1744 (2015).
3. Ciesla A., Kraszewski W., Skowron M., Surowiak A., Syrek P., Min. Res. Manag. **31(1)**, 155 – 174 (2016).
4. Niedoba T. Physicochemical Problems of Mineral Processing, **49(1)**, 175-188, (2013).
5. Surowiak A., Mineral Resources Management, **V. 30**, Iss. 1, pp. 105-122, (2014).
6. Surowiak A., Brożek, M.,. Arch. of Minin. Sci., **59**(1), 269-282 (2014a).
7. Surowiak A., Brożek M., Arch. of Minin. Sci., **59(2),** 553-562 (2014b).
8. Surowiak A., Brożek M., Physicochem. Prob. of Min. Proc. **52(1),** 228–243 (2016).
9..Ciesla A., Przegl. Elektrotech., **87(7)**,: 99-103, (2011).
10. Ciesla A., Przegl. Elektrotech., **88**(12B), 50-53, (2012).
11. Ciesla A., Kraszewski W., Skowron M., Syrek P., Przegl. Elektrotech, **87,** 79-82 (2011).
12. Ciesla, A., Syrek, P., **88(12B),** 124-127 (2012).
13. Lawver E., Dyrenforth W.P., *Electrostatics and its Applications*, (Wiley, New York, 1973).
14. Chong-Qing Wang, Hui Wang, You-Nian Liu, Waste Manag., **35**, 42–47 (2015).
15. Kachi, M., Dascalescu, L., Herous, L., Nemamcha, M., Industry Applications, IEEE Trans., **49(2),** 691 – 698 (2013).
16. Wu Jiang, Li Jia, Xu Zhen-Ming, J. Hazard. Mater. **161**, 257-262 (2009).
17. Bittner J. D., Flynn K. P., Hrach F. J. Intern. Min. Proc. Cong., Chile, (2014).
18. Cui J., Forssberg E., J. of Hazar. Mat. **99(3)**, 243–263 (2003).
19. Lockhart N.C., Powder Techn., **40(1-3)**, 17–42, (1984).
20. Birloaga I., De Michelis I., Ferella F., Buzatu M., Veglio F., Waste Manag., **33**, 935-941 (2013).
21 Veit H.M., Diehl T.R., Salami A.P., Rodrigues J.S., Bernardes A.M., Tenorio J.A.S., Waste Manag., **25**, 67-74, (2005).
22. Li J., Lu H., Liu S., Xu Z., J. of Hazard. Mat. **153,** 269-275 (2008).
23. Ralston O. C., *Electrostatic Separation of Mixed Granular Solids*. (Amsterdam - New York, 1961).
24. Szczerbiński M., J. of Electrostat, 175 – 186, (1983).
25. Cieśla A., Proceedings ISH'99, London, (1999).

DOI: 10.1051/e3sconf/20161000086

Energy balance in a passive solar building. An attempt at economic assessment

Wiktoria Sobczyk[1,a] and Katarzyna Sapa[2]

[1]Ph.D. Eng., Professor AGH, Faculty of Mining & Geoengineering, AGH University of Science & Technology, Al. Mickiewicza 30, 30-059 Krakow, Poland
[2]MSc. Eng., Construction and Design Services Andrzej Bracha, ul. Wschodnia 13/14, 28-200 Staszów, Poland

Abstract. The paper emphasizes possibilities for substantially reducing energy consumption with modern eco-friendly buildings. Passive building construction is a sector of the construction industry that has extremely low demand for the energy for heating house interiors. A passive house requires a small amount of energy to provide thermal comfort, but it requires proper systems (HRU, heat pumps, solar collectors) to accomplish that effect. The modification proposal presented in the paper has dramatically reduced the demand for heat capacity of the building. Unfortunately the passive standard has not been reached, but thermomodernisation would allow for a significant reduction of heating costs. The demand per 1 m2 of heated surface area in a traditional building was 41.9 W/m2, while after the modification – only 15.01 W/m2. The tested building, if located in a warmer climate, with an average outside temperature of ≥ -10°C would certainly perform its function. The use of modern technology ensures high energy savings. Such solutions, however, are not the cheapest on the market, and therefore are less available. The construction of a passive solar buildings is a "green" investment with tangible environmental results. By ensuring the thermomodernisation of a traditional building we can enjoy real environmental and economic benefits.

1 Solar energy conversion methods

Solar energy is an energy source specifically for use by beings living on Earth. The sun produces vast amounts of energy in a form of radiation. Only a small portion of this radiation reaches Earth, yet it is still 15 times more than is demanded. This source has one disadvantage though – it produces low-density energy which makes its conversion problematic [1].

The intensity of solar radiation varies across different regions. Poland has been divided into 11 sun power regions based on their usability for solar energy production. Radiation is the sum of the total radiation exposure and the greatest number of hours of sunshine. The best conditions in this respect can be found in the coastal region (northern Poland) and Podlasie-Lublin region (eastern Poland). The least favourable solar conditions can be found in the regions of Upper and Lower Silesia – this is due to the local severity of anthropogenic air pollution (Fig. 1).

The total radiation per year in Poland is 3600 MJ/m^2 (±10%). Utilisation of radiation energy is effective only for 6 months: from April to September (when 80% of the annual solar insulation is observed). This is caused by the absence of technologies to store solar energy [2]. There are many ways of making use of solar energy reaching the Earth surface. However, three basic methods of solar energy conversion are practicable:

a. thermal energy – photothermal conversion
b. electrical energy – photovoltaic conversion
c. energy related to chemical processes – photochemical conversion.

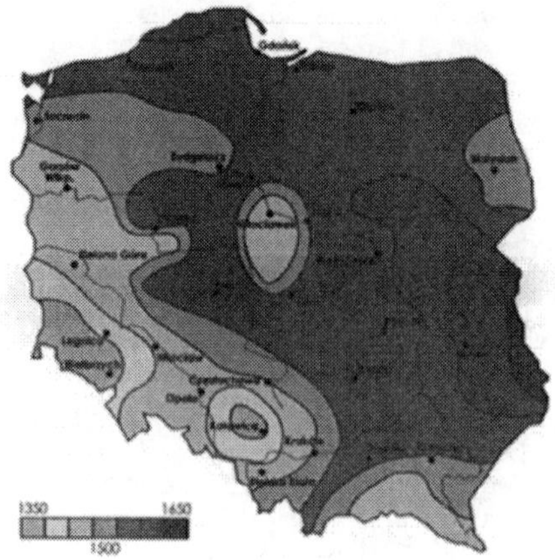

Figure 1. Number of hours of yearly solar insolation for Poland (Wnuk 2007)

<u>Thermal energy – photothermal conversion</u>

Production of hot water is the simplest, most available and efficient method of using solar energy. Efficiency

[a] Corresponding author: sobczyk@agh.edu.pl

 DOI: 10.1051/e3sconf/20161000086

of energy conversion in solar collectors designed to convert solar energy into thermal energy may amount to as much as 90%, while the structure and operation of such devices are relatively uncomplicated. With all these factors taken into account, favourable climate conditions are still required; therefore the expected results will not be achieved everywhere. Solar collectors enable photothermal conversion of solar radiation in order to heat spaces and tap water, to produce electrical energy and to conduct chemical processes. Solar collectors require a carrier to transport and transfer heat. The temperature of the working substance at the device outlet determines the design of the collector, which can be of low-, medium- and high-temperature type. Collectors can also be classified based on the type of working substance, namely liquid and vacuum. The basic element of a collector is the absorber, i.e. a plate coated with a thin layer of black nickel, copper and other material.

The absorber boasts a high coefficient of solar radiation absorption and low coefficient of thermal radiation emission. The flow-through part of the collector is insulated with a material that has a low thermal conductivity coefficient (mineral wool, polyurethane foam). Transparent covers – such as glass panes or transparent insulation – are used for sealing the surface of the collector. Such elements minimize heat loss from the surface of the absorber by convection.

Electrical energy – photovoltaic conversion

The most convenient way to obtain energy is, undeniably, generating it in an electrical form. There are also methods of obtaining electricity from the energy carried by solar radiation – e.g. photovoltaics produce direct current that is usually converted into alternating current. However, it is only profitable to build and install the required devices in highly-insolated areas, where the sun shines for 1800-2000 hours per year on average.

Figure 2. Use of alternative sources of energy: photovoltaic cells and 0.5 kW household wind turbine. "EKOMOTEL na Wierzynka". Wieliczka near Krakow (photo by Ł. Steg)

A photovoltaic cell is a semiconductor which converts solar energy into electricity. A typical photovoltaic cell is a plate made of crystalline or polycrystalline silicon where a barrier for electric potential is formed. The sunlight that reaches a photovoltaic cell generates pairs of electric energy mediums (electrons and positive holes). Under the influence of the electric field, they are scattered, resulting in a potential difference, i.e. PV voltage. When a connection with the receiver of electricity is made, the flow of electric current commences. The thickness of the plates is in the range of 200-400 micrometres. A single cell produces 1-2 watts of power. To provide higher voltages or currents photovoltaic cells are connected in series or parallel forming photovoltaic modules (Figure 2).

Passive solar systems can use some elements of the building (e.g. glazing) to accumulate heat. No intermediate medium is used for heat transport and transfer. Sun rays in the form of visible and infra-red waves which penetrate into the rooms, being accumulated in walls, floors and ceilings. The elements heat up and emit thermal radiation, producing the greenhouse effect [3].

Passive heating methods are effective only for buildings with a low unit energy demand for heating purposes. They may also be an auxiliary source of heat in conventional buildings. Energy related to chemical processes – photochemical conversion

This type of solar energy conversion is mainly used in agriculture (photosynthesis). Solar energy is transformed into usable energy by recuperation of the energy from organic matter by way of combustion or through other decomposition process. Approximately 3% of solar energy can be extracted, but the process itself is complicated and costly.

2 Energy-efficient building and passive building – definitions

Energy-efficient buildings (EeB) use intelligent technologies which allow for high heat comfort and low energy consumption. In this way, operating costs decrease. The basis for any EeB is maintaining air tightness. Building materials play a crucial role, such as insulating materials and heat absorbing windows. EeBs have low energy consumption: E <60 kWh/m^2 per year.

The main advantage of an EeB is not losing heat through outside walls. This kind of building uses the heat produced by a variety of home appliances and utilises natural light as well as the ventilation system for re-use of the heat contained in the air which is removed to the outside. An EeB must meet certain conditions:

- Its location should take into account land relief, sunshine and wind directions;
- It should have a compact design and small windows on the north side;
- It should be equipped with a layer of thermally insulating material;
- Windows and doors should be very tight and well insulated thermally;
- Ventilation should be automatically controlled and feature a heat recovery solution.

A very important design requirement for passive houses is the choice of a suitable building site, providing passive solar heat gains. The correct orientation of the house versus cardinal points allows for optimum use of the properties of glazing (absorption of sunlight). Installing the largest number of windows facing south and west brings greatest benefits [4].

In the case of EeBs, just as with passive housing, modern technologies are used to reduce energy consumption. The most frequent are solar collectors, heat pumps and heat recovery units (HRU).

3 Materials and methods

Research and analyses were carried out based on analysing reports and regulatory documents (EU directives, Polish legislation, strategies, strategic documents of the Ministry of Economy and the Ministry of Environment in Poland).

The following scientific methods were used in the study: modelling and design method, analytical method (calculation method). Heat balance was calculated using Herz OZC, version 3.0 © computer software 1994-2007 P.W.

4 Modelling architectural solutions for passive buildings

Architectural and construction design combines many essential components and solutions necessary for solar energy utilisation: "solar" roof, materials used (passive doors and windows, tight connections, proper insulation), passive systems, extended glazing on south-eastern façades, adequate layout of rooms, as well as geometry and orientation of the building. Unfortunately, even today, in times of energy crisis, not everyone considers using cheap solar energy worthwhile (Figure 3).

Figure 3. Improperly designed new building. a. south-western façade without glazing, b. northern façade – fully glazed, Kraków, Zarzecze street (photo by: W. Sobczyk)

In passive houses hoods are often used to control solar radiation reaching rooms. In the summer, hoods prevent radiation from getting in, while in the winter, when the position of the sun is lower, the rays easily reach rooms. This leads to a reduction in energy consumption by 7% in traditional buildings, while in passive buildings – by up to 30%. The thermal balance can be improved by installing glass porches accumulating heat, shutters or energy-saving roller shutters.

Figure 4. Model of a passive building (by K. Figiel)

The shape of the building should be simple, compact and limiting heat loss through outside walls (Figure 4). Recesses of end walls increase interior sun exposure, so this kind of reinforcement is advantageous when designing passive buildings.

An important element is the layout of rooms. Spaces which require the lowest temperature of all the rooms (staircases, utility rooms) should be facing north, while bathrooms should be located in the middle of the building. Spaces which require intensive lighting should be facing south [5].

Moreover, the proper design of greenery surrounding the building is a very helpful element. The southern façade of the building decorated with deciduous trees increases the efficiency of the energy balance. In the summer, sunlight is partially blocked by tree leaves, thus protecting the building from overheating, while in the winter trees drop their leaves and allow a sufficient amount of energy to reach the building. On the north, coniferous trees should be planted that would protect the building from cold winds.

4.1 Design of a passive building

Buildings that were put on the real estate market after 2009 are required to have an energy certificate that determines their energy efficiency. The basis for quality control is the air-tightness test mentioned above and thermal images.

Figure 5. Southern façade (a) and western façade (b) of the designed building (dev. by the authors)

4.1.1 Design of the solar system

The solar system comprises the Viessmann flat plate collector which ensures high operating efficiency at an affordable price of investment (Fig. 6). The exceptionally high operating efficiency, safety of operation and durability are confirmed by quality tests compliant with the European and Polish standard PN-EN 12975 and Solar Keymark Certificate (www.viessmann.pl). The surface area of Vitosol 200-F collector is 2.3 m^2, and the device itself is easily installable. This particular collector was selected as it can cooperate with compact devices, such as condensation boiler Vitodens 343-F and heat pump Vitocal 343-G.

The heat transfer coefficient U in traditional buildings is 0.242 $W/(m^2{\cdot}K)$. In our modified design, a 16 cm-thick mineral wool insert was replaced with 20 cm-thick ECOTHERM plates whose heat transfer coefficient λ is 0.023 W/m·K. This offers 70% better insulation than in the case of ordinary insulating materials.

The design incorporated, internal walls in the garage and service/utility room insulated with ECOTHERM wall plates which had been previously insulated with vapour-permeable membrane.

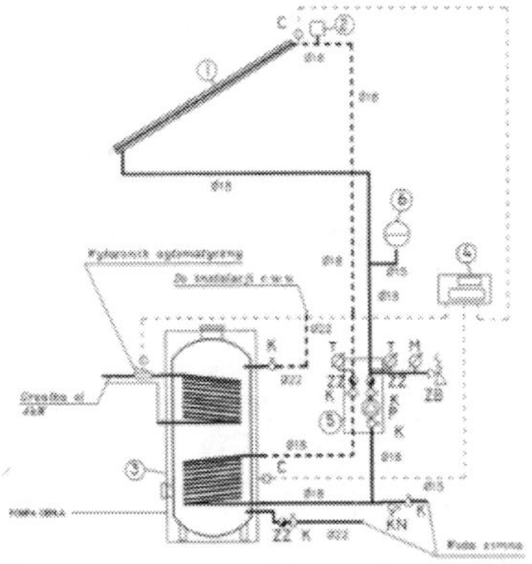

Figure 6. Diagram of the solar system (dev. by the authors) 1. Flat plate collector 2x2.32; 2. Vent; 3. Tray V=3001; 4. Differential controller; 5. Pump group; 6. Diaphragm pressure expansion vessel V= 121; P. Circulation pump; K. Nozzle; ZZ. Return valve; ZB. Safety valve; C. Temperature sensor; T. Thermometer; M. Manometer; KN. Tap with a hose nipple.

We also applied the passive effect by adding a small glazed porch to the southern façade to accumulate heat (Figure 5b). This structure is one of the methods of retaining heat energy and gradually transferring it to the building interior. A part of the wall opposite the windows in the living room will be covered with a heat accumulating material (clinker) [6].

4.1.2 Installation of the HRU

To ensure trouble-free and economical operation, we decided to install the AERIS 350 LUXE VV HRU (Figure 7). It is a top-class counterflow heat exchanger, which offer a 95% efficiency. The built-in filters guarantee clean air, while the polypropylene insulation inside the machine ensures silent operation. By installing an auxiliary heater, the supply air temperature can be controlled. Another advantage of the AERIS HRU is the possibility to connect the controller to the rapid ventilation in the bathroom, kitchen and toilet. The HRU can be programmed according to weekly or hourly options/modes. It can cooperate with the ARTIC cooling system [7]. Before the HRU can be installed, the route of the main cable should be determined, the volume of air should be correctly calculated, the diameter and insulation of ducts should be carefully selected, while the pressure should be equalized. An important element is to ensure an adequate level of humidity in the rooms, especially during the heating season. Moisture and CO_2 sensors that can be connected to the HRU are well worth investing in. In addition to that, all components of the system should be carefully selected: ducts , diffusers, silencers, etc. The HRU must have a low power heater to support heat recovery in the periods when the outside air is very cold.

The design also takes the distribution of air supply and exhaust devices into consideration. The former will be installed in least polluted spaces, i.e. bedrooms, living rooms and guest rooms, while the latter – in the kitchen, bathrooms, hall, garage, i.e. spaces where air can be contaminated or too humid. The air flow rate should not exceed 0.2 m/s. The air supply devices should be spaced away from the places of human occupancy (e.g. over beds, armchairs or desks) and their number should match the cubic volume and throughput of the spaces.

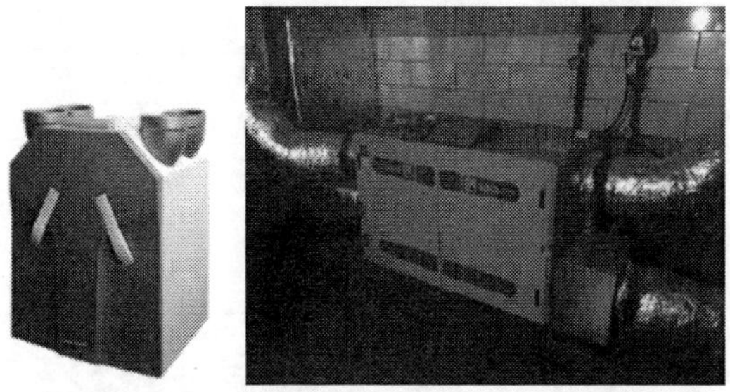

Figure 7. Model of the AERIS 350 LUXE VV recuperator and the same HRU installed in a house (photo by: M. Jarząbek)

Condensation of the water vapour contained in the cooled air takes place in the exchanger; it is then directed to the vents, and then discharged into drains.

4.1.3 Domestic hot water (DHW)

The DHW system is the primary heating system in passive building construction. Heating of rooms takes places simultaneously with the preparation of DHW and is linked to it. The heating system integrated with solar collectors and a heat pump with a ground heat exchanger is a source of hot water. The heat pump will be used for heating the spaces equipped in underfloor heating solutions. In the summer, the solar collector will meet the heat demand required for preparation of DHW. In the winter, the collectors will be supported by the heat pump and an energy-efficient electric heater. In the periods of increased cloudiness in the summer, it is assumed that an electric heater will be used just like in the winter.

In the designed building we have planned to install the Vitocal 343–G heat pump. Its bottom source of heat is a horizontal ground heat exchanger. It is a high capacity, energy-efficient electronic pump of the brine-to-water type, powered with direct current. It has the user-friendly Vitotronic controller and is capable of quiet operation [8]. The heat pump is integrated with floor heating in rooms while it supports the HDW heating system (in the summer it is mainly supplied by the solar collector).

5 Results

A standard for passive houses is achieving a coefficient of heat transfer through the building envelope ≤ 0.15 W/m²·K. In addition to that, the demand for energy for heating should be ≤ 15 kWh/m²·year, and the maximum power demand for heating ≤ 10 W/m². In the calculations we have taken material parameters into consideration in order that a low heat transfer coefficient could be achieved. Data was entered into the software to calculate the energy balance of the building. Having introduced designated construction and insulating materials, we obtained the required building envelope coefficients so that an airtight layer could be obtained. The partitions which ensure creating such a layer have the coefficients as shown in Table 1.

Table 1. Coefficients of heat transfer through the building envelope in passive and traditional buildings (dev. by the authors)

Symbol	Description of the building envelope	Passive building U [W/m²·K]	Traditional building U [W/m²·K]
DZ	external door	0.8	2.0
O	window	0.7	1.5
OD	roof window	0.7	1.5
PD1	floor 1st zone	0.069	0.294 (parquet) 0.451 (terracotta)
PD2	floor 2nd zone	0.069	0.418 (parquet) 0.159 (terracotta)
SZ	external wall	0.137	
SZO	insulated external wall	0.088	
DB	balcony door		1.5
P	external wall		0.506
D	roof		0.242

where: U – heat transfer coefficient

5.1 Calculation of the energy balance

According to the calculations, it was possible to achieve the effect of an airtight building envelope, particular attention was paid to minimizing thermal bridges. A different situation occurred with the maximum demand for heat, which should be no more than 10 W/m². According to software calculations, this value equals as much as 27.3 W/m² (symbol Q_f). However, this amount was calculated without taking into account the recuperation of heat from ventilation (as much as 70.9%). With the efficiency of the HRU we have selected, the coefficient is calculated according to the formula:

$$Q_c = Q_o - (Q_{vent.} \times 0.95)\ [W] \qquad (1)$$

where:

Q_c – designed demand for heat capacity

Q_o – designed demand for heat capacity incl. heat loss through ventilation (designed demand for heat capacity: Q_o=6244 W)

$Q_{vent.}$ – demand for heat for ventilation (designer demand for heat capacity $Q_{vent.}$=2954 W).

Formula:

$$6244\ W - (2954\ W \cdot \times 0.95) = 6244\ W - 2806.3$$

$$W = 3437.7\ W$$

$$3437.7\ W : 229\ m^2 = 15.01\ W/m^2$$

Demand for capacity per 1 m² of heated surface area is 15.01 W.

6 Discussion

The result of 15.01 W/m² resulting from the calculation of demand per 1 m² of heated surface area is satisfactory due to the value characteristic of this traditional building, namely 41.9 W/m². However, it still fails to meet the standard of passive building: ≤10 W/m².

Table 2. Comparison of energy balance of a traditional and passive building (dev. by the authors)

	Traditional residential building	Passive residential building
Location: Oględów		
Designer: Sapa Katarzyna, M.Sc. Eng.		
Date of calculation: 6 January 2015, 5:08 pm		
Locality: Oględów		
Climate zone: 3 Outdoor temp. [°C]: -20		
Heated surface area [m²]: 229 Heated cubic volume [m³]: 540		
Designed demand for heat capacity Qo [W}	**9581**	**6244**
Demand for heat capacity for ventilation Qvent [W]	**2954**	**2954**
Additional heat gains in rooms Qzc [W]	**0**	**0**
Demand per m² of heated surface area Qf [W/m²]	**41.9**	**27.3**
Demand per m³ of heated cubic volume Qv [W/m³]	**17.7**	**11.6**
Annual demand for heat for heating purposes Qh [GJ/year]	**87.10**	**65.17**
Qh [kWh/year]	**24195**	**18103**
Coefficient of seasonal demand for heat EA [MJ/m²·year]	**381.0**	**285.1**
EA [kWh/m²·×year]: 105.8	**105.8**	**79.2**
Coefficient of seasonal demand for heat EV [MJ/m³·×year]: 161.3	**161.3**	**120.7**
EV [kWh/m³·×year]: 44.8	**44.8**	**33.5**
		Herz OZC version 3.0 (c) 1994-2007 P.W.

The coefficient of seasonal demand for heat (in kWh/m²·year) it also too high:

(15.01 W/m² × 1000) : 365 days = 41 kWh/m²·× year.

This result takes into account heat recovery from ventilation enabled by the installed HRU.

The designed heat load of a traditional building Q_c is 9581 W, so the coefficient of seasonal demand for heat E_A is 114.6 kWh/m²·×year:

9581 W : 229 m² = 41.8 W/m²

(41.8 W/m² × 1000) : 365 days = 114.6 kWh/m²·×·year.

This result meets the requirements of the Polish Standard, under which the coefficient for buildings should be EA=120÷180 kWh/m²·×year.

The energy balance of the passive building is as follows:

- transmission heat loss + ventilation heat loss = heat loss
- heat gain from solar energy + gains from internal heat sources = free heat
- free heat • ability to use = heat gains
- heat loss - heat gains = heat demand

The demand for primary energy comprises of meeting all of the energy needs required by the house (heating, HDW preparation, ventilation, air conditioning, lighting, cooking). Energy-efficient appliances (dishwashers, washing machines, dryers) are therefore also required to reduce demand to a minimum [9].

Table 3. Demand for electricity and primary energy for basic household chores [10]

Demand	Demand for primary energy	Demand for electricity
	kWh/year	kWh/year
dishwasher	475	160
dishwasher – hot water	288	
washing machine	307	103
washing machine – hot water	184	
clothes dryer with exhaust ducts	1845	621
energy use by laundry evaporation when drying	244	
cooling / freezing	1236	416
cooking	1384	468
lighting	668	225
electronic devices, appliances		350
total of auxiliary energy		425
Total		2767

Buildings offered on the real estate market since 2009 must have an energy certificate, specifying the energy consumption of a given structure. The basis for the quality control of passive buildings are infrared images and the aforementioned airtight test. The heat loss in passive buildings is much lower than in traditional buildings, assuring indoor thermal comfort. In traditional buildings more than 50% of the energy is spent on heating, while in passive buildings only 20% of energy is required for that purpose.

7 Environmental education of the society – a chance for environmental protection

The environmental education of society in the field of utilising alternative solutions in building thermoregulation should therefore be the top priority for state environmental and energy policy. Raising awareness of the benefits of the implementing non-conventional solutions in residential buildings may bring about huge environmental and social gains. Thoughtless discharge of heated air into the atmosphere through building chimneys is a needless wastefulness [11-15].

So far several thousand passive houses have been built in Europe. Inhabitants of these buildings confirm that the costs of energy have significantly dropped, while the comfort indoors remains high. Thermomodernisation undertakings are a great opportunity for environmental protection, development of the labour market and cities in the future.

In terms of energy use, Poland cannot demonstrate required efficiency. This means that the depletion of natural resources, fossil fuels and environmental pollution are gradually weakening the national economy [16-20]. All the buildings in Poland completed in the past years must undergo thermomodernisation and the construction of new passive buildings should commence. The cost of building a passive house is slightly higher than the cost of a traditional building (an increase by 15% of the investment outlays), however, passive constructions require no financial expenditure for the heating system. Therefore additional costs are compensated for fairly quickly. Furthermore, the construction of a passive house does not require testing new building materials as well-known materials are used, such as are also used in traditional construction. Erecting passive buildings is based on technologies which have already been tried and tested. This means that passive technologies may be utilised in a wide range of applications.

Educating the society about the possibilities offered by the modern construction industry is necessary for passive buildings to become more popular. Investors usually concentrate on standard solutions – cheap and easy to implement, without making the effort to find out what else is available. The designer's task is to present all solutions and to spread knowledge about innovations so that we can enjoy living in healthy, energy-efficient, green houses.

8 Conclusions and policy implications

Today's modern construction industry offers increasingly airtight and warm houses and it pays much attention to the selection of materials which will ensure the lowest energy losses. Warm and airtight windows, modern building materials, minimizing the number of thermal bridges and lowering the heat transfer coefficient due to the excellent insulating materials and their proper installation are all elements significantly affecting the energy demand of a building. Since 2009 all new buildings need to obtain certificates to determine their energy efficiency.

Passive building construction is a sector of the construction industry that has extremely low demand for the energy for heating house interiors. A passive house requires a small amount of energy to provide thermal comfort, but it requires proper systems (HRU, heat pumps, solar collectors) to accomplish that effect.

Modification of a traditional building into a passive building is quite expensive and requires a large amount of work (capital-intensive investment). It is much easier to build a passive house from scratch, which reduces costs and allows for much greater freedom in terms of design. In both cases, subsidies can be obtained for installations that use alternative energy sources.

The modification proposal presented in the paper has dramatically reduced the demand for heat capacity of the building. Unfortunately the passive standard has not been reached, but thermomodernisation would allow for a significant reduction of heating costs. The demand per 1 m^2 of heated surface area in a traditional building was 41.9 W/m^2, while after the modification – only 15.01 W/m^2. The tested building, if located in a warmer climate, with an average outside temperature of $\geq$ -10° C would certainly perform its function. The use of modern technology ensures high energy savings. Such solutions, however, are not the cheapest on the market, and therefore are less available.

Building a passive solar house is an eco-friendly investment and the resultant environmental effects are tangible. Significant environmental and economic benefits can be obtained through thermomodernisation of a traditional building [1]. Furthermore, passive building construction contributes to lower consumption of fossil fuels, the resources of which are shrinking along with the increasing demand for energy.

In Poland today, the degree of electricity and thermal energy utilisation is very high. The interest in eco-friendly investments has been growing with every year. The supporters of energy-efficient buildings, consuming much less energy than traditional buildings, have also been growing in number.

Protection of the natural environment is a standard component of energy policies of any country in the world. The economic balance demonstrates that the application of solar technologies in the construction industry has got a bright future. The global population is growing which entails rising demand for energy.

Many countries, including Poland, are now faced with an energy crisis. The global resources of fossil fuels will soon be depleted. The situation is further aggravated by the fact that fossil fuel deposits are often located in areas subject to strict protection, such as the Natura 2000 network.

In the light of this, searching for alternative energy sources has become crucial to ensuring energy security. Rational utilisation of solar power solves the issue of increased energy demand. It should be noted that the energy of the sun is not only widely available and inexpensive, but also it does involve noxious gas or dust emissions.

Environmental education of the society in the use of alternative solutions of building thermoregulation is a key to environmental and power policies of the state [21]. Raising awareness among the society of the advantages of applying unconventional solutions in residential buildings may bring great environmental and social gains.

The primary goals of EU climate and energy policy are improving energy efficiency and increasing the share of renewable energy in the total energy balance of the European Union. Poland's energy policy is based on the strategy of balancing the safety of energy materials supply, effectiveness of economic processes and environmental protection.

Acknowledgements

Paper developed as part of statutory work no. 11.11.100.482

References

1. A. Ostrowska, W. Sobczyk, M. Pawul. Ocena efektów ekonomicznych i ekologicznych wykorzystania energii słonecznej na przykładzie domu jednorodzinnego. Annual Set The Environment Protection **15** (2013)
2. R. Wnuk. *Instalacje w domu pasywnym i energooszczędnym*. Przewodnik Budowlany (2007)
3. www.passivesolarenergy.info
4. R. Piotrowski, P. Dominiak. *Budowa domu pasywnego krok po kroku*. Przewodnik Budowlany **12** (2008)
5. K. Ochociński. *Budownictwo pasywne*. Biuletyn Wielkopolskiej Okręgowej Izby Inżynierów Budownictwa **1**, 18 (2008)
6. A. Kaczkowska. *Dom pasywny*. Wydawnictwo i Handel Książkami „KaBe" Krosno (2009)
7. www.rekuperatory.pl
8. www.viessmann.pl
9. J. Paska. *Wytwarzanie rozproszone energii elektrycznej*. Oficyna Wydawnicza Politechniki Warszawskiej, Warszawa 2010.
10. W. Feist, U. Münzenberg, J. Thumulla, B. Schulze. *Podstawy budownictwa pasywnego*. Wyd. Polski Instytut Budownictwa Pasywnego, Gdańsk (2008)
11. A. Malada, W. Sobczyk. Uprawa roślin energetycznych jako forma aktywizacji środowisk wiejskich. Zeszyty Naukowe Katedry Inżynierii Procesowej Uniwersytetu Opolskiego, **2** (2005)
12. W. Sobczyk. Uprawa wierzby energetycznej w województwie małopolskim. W: Problemy współczesnej techniki w aspekcie inżynierii i edukacji. Wyd. Akademii Pedagogicznej (2005)
13. W. Sobczyk. Plonowanie wierzby wiciowej - w świetle badań. Polityka Energetyczna **10**, 2 (2007)
14. W. Sobczyk. Wykorzystanie alternatywnych źródeł energii w Zawoi Przysłopie (Małopolska). Folia Scientiarum Universitatis Technicae Resoviensis **252,** Budownictwo i inżynieria środowiska, 47 (2008)
15. W. Sobczyk. Evaluation of harvest of energetic basket willow. TEKA Komisji Motoryzacji i Energetyki Rolnictwa PAN XI (2011)
16. B. Klojzy-Karczmarczyk, A. Karczmarczyk. Systemy grzewcze z pompą ciepła jako element realizacji programów ochrony środowiska. Polityka Energetyczna, **3** (2005)
17. B. Klojzy-Karczmarczyk, A. Karczmarczyk. Zanieczyszczone wody podziemne jako niskotemperaturowe źródło ciepła w systemach grzewczych. Polityka Energetyczna **1**, 9 (2006)
18. B. Klojzy-Karczmarczyk, A. Karczmarczyk, J. Mazurek. Możliwości wykorzystania wód kopalnianych jako dolnego źródła dla pompy ciepła na przykładzie kopalni Trzebionka. Polityka Energetyczna **9** (2006)
19. T. Olkuski, K. Stala-Szlugaj. Pierwiastki promieniotwórcze w węglu oraz w produktach odpadowych powstających podczas jego spalania. Rocznik Ochrona Środowiska Annual Set of Environmental Protection **11**, 2 (2009)

20. L. Styszko, D. Fijałkowska, M. Sztyma. Wpływ warunków pozyskania biomasy na odrastanie pędów wierzby energetycznej w czteroletnim cyklu. Rocznik Ochrona Środowiska Annual Set of Environment Protection **12** (2010)
21. W. Sobczyk., A. Kowalska. The techniques of producing energy from biomass. TEKA Commission of Motorization and Energetics in Agiculture PAN, **XII** (2012)

E3S Web of Conferences **10**, 00087 (2016) DOI: 10.1051/e3sconf/20161000087
SEED 2016

A study of selected aspects of the operation of thermoelectric generator incorporated in a biomass-fired stove

Krzysztof Sornek[1,a]

[1]*AGH University, Faculty of Energy and Fuels, Mickiewicza Av. 30, 30-059 Krakow, Poland*

Abstract. High demands in the field of energy efficiency and clean combustion make it necessary to looking for the new developments in the field of stoves, fireplaces and stove-fireplaces with accumulation. An interesting idea is to use the thermoelectric modules, which receive a heat from flue gas and convert it to the electricity. Electricity generated in this way may be used to power combustion optimizers and other components. This paper shows results of studied carried out to determine the possibility of combined heat and power generation using the stove-fireplace with accumulation. Thermoelectric generator with maximum hot side temperature at a level of 150°C was placed on the surface of the exchanger. Cooling down was realized using the dedicated water exchanger as well as the heat sink without and with an air fan. The experimental results allowed to define the effect of the different cooling systems on the output TEG voltage. Moreover, dependence of the current-voltage characteristics and generated power from the temperature was obtained.

1 Introduction

Thermoelectric generators (TEGs), which are a subject of many worldwide conducted studies, are well suited to recover energy from the low grade waste heat. Among the advantages of TEGs, it can be mentioned e.g. direct transfer heat to electrical energy, light weight, silent operation and not mechanical vibration. With current efficiencies at a level of 5-10%, the heat rejected from the TEG may go back to room heating [1,2].

From a number of carried out investigations, there are several studies, where thermoelectric generators operated with different kinds of stoves. In one of them, the prototypical thermoelectric generator was equipped with two Hi-Z TE modules cooled by an air fan and fitted to a wood stove. This unit was installed in Scandinavian house and produced up to 10 W in the cold mornings, and 4-7 W during the day (when house became warmer). The generated power was sufficient to operate the cooling air fan and charge four 6 V batteries (connected in series/parallel to maintain a 12 V output) using DC/DC converter [3]. Another studies devoted to continuous electric power generation using waste heat from the wood stoves were presented in [4]. TE modules were fitted to the side of a domestic woodstove and were cooling by natural convection using heat sink. The maximum steady state matched load power was achieved at a level of 4.2 W per module. The more advanced power generating system produced up to 9.5 W (the maximum power of electricity available for the end users was around 7.6 W) has been studied in [5]. It was a complete system from the heat source up to the end users. In addition to energy performance, an environmental efficiency of the cook stove equipped with TEGs were considered in [6]. Since obtained a maximum level of 4.5 W from the TEG and less than 1 W was needed to power the blower to provide a clean combustion and reduce indoor air pollution.

One of the moist important aspect in the air cooling systems is electricity demand for power the fans. This problem was considered in studies presented in [7], where the cold side of the TEG was connected to heat pipe heat sink, normally used in CPU cooling. The typical fan supplied with this heat sink operated at 12 V and consumed up to 2.2 W of electricity. To improve the energy performance, the low power DC motor was used instead. It could run the fan from 0.3 V and consumed up to 0.5 W with voltage not exceeded 5 V. Since the fan started from very low TEG voltage, the necessity to use an additional battery was eliminated.

The example of usage forced water cooling system instead of the air cooling systems, is presented in [8]. In this case the electric heater was simulating heat source and heating TE modules. The maximum power generating in modules was obtained at the level of 10 W. It should be noted, that continuous operation of a pump certainly decreased the available output power. The advanced, multifunction device capable to produce a considerable amount of electricity and hot water, was designed by [9]. They used together 21 TE modules divided into seven groups of three, each connected electrically in series. The hot sides of modules were connected to a 50 mm-long aluminium pieces which was attached to the stove body. In this way the upper temperature limited by the TEGs was not exceeded. TEGs were cooling via a water-filled

[a] Corresponding author: krzysztof.sornek@agh.edu.pl

aluminium block, through which the water flow rate could be adjusted. The average electricity generation was at a level of 7.9 W for each module when charging with 9 kg of firewood per hour.

Taking into account above mentioned studies we can conclude, that from technical and an economic point of view, there is huge potential to development of TEG technology. It would constitute an option of sustainable electricity generation that could be developed in the own regional economies where it is used [9] showed, that simple installation and operation for the own users, would give TE modules some comparative advantages respect to e.g. conventional PV solar arrays. Its low operating costs, feature shared with PV systems, becomes thermoelectric generators an attractive option for low-income rural houses. As TEG are destined both to the high and low income houses, the cost is also an important issue. Many studies, including studies performed by [10, 11] have shown that the price per watt may be really close or equal to the price of photovoltaic systems.

Although the worldwide studies consider many configurations of TEGs with different types of stoves and other heat appliances, there are no studies devoted directly to the stove-fireplaces with accumulation (which combine fireplaces and traditional accumulative stoves). Such studies were performed at AGH UST in Krakow and presented in the further part of this paper.

2 Experimental rig

Stove-fireplaces with accumulation (SFA) combine fireplaces and traditional accumulative stoves. Heat produced during wood combustion is stored in the accumulative heat exchanger and dissipated up to 12 hours after the fire has died out. As a result, the thermal efficiency of SFA is high and close to 90% [12,13]. The idea of such device was shown in Fig. 1.

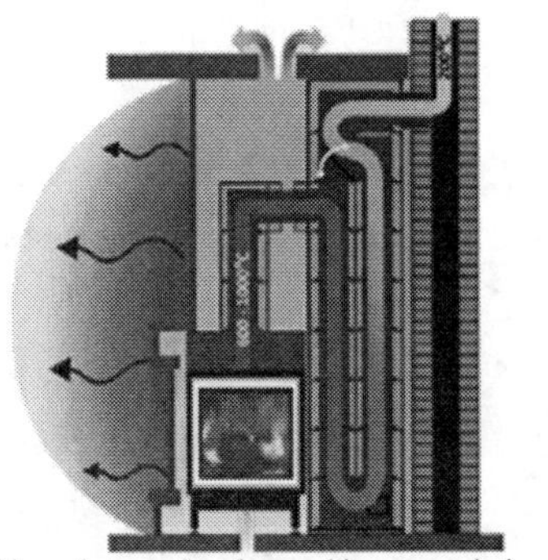

Figure 1. The idea of stove-fireplace with accumulation.

Studies were carried out on the test rig equipped with a stove-fireplace with accumulation composed with a 550 kg furnace and a 1050 kg accumulation exchanger located next to the furnace. Based on a number of previous works, heat exchanger dedicated for TEG was located at the beginning of the accumulative exchanger. The illustration of the tested system is shown in Fig. 2

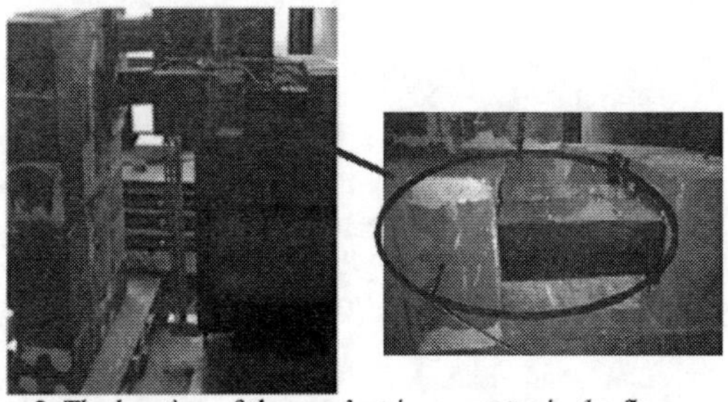

Figure 2. The location of thermoelectric generator in the flue gas channel.

The structure of the heat exchanger dedicated for TEG was made in the form of the rectangular channel with the dimensions of 30 x 16 x 11.5 cm (length x width x height). The exchanger's surface is insulated by a 5 cm layer of mineral wool. Inside the exchanger, a special reducer and radiator were used. Reducer allowed to increase the gas flow in the boundary layer and radiator allowed to intensify the heat exchange between exhaust gas and hot side of TE module (only one module was fitted to the radiator during the tests) [14]. The structure of the heat exchanger with TE module is present in Fig. 3.

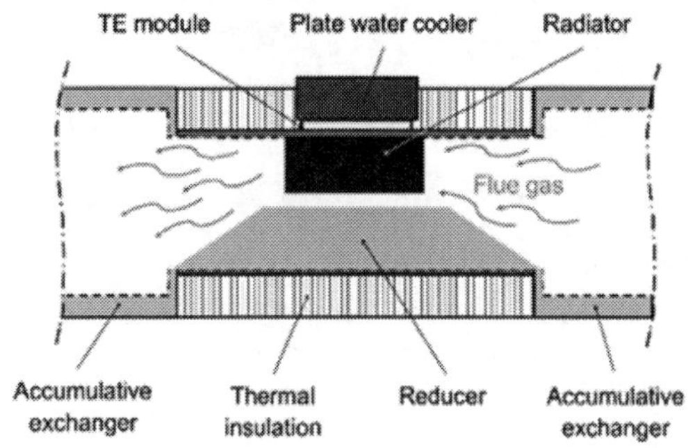

Figure 3. The construction of thermoelectric generator.

Thermoelectric module used in the carried out studies was a Bi_2Te_3 module with a dimensions of 40 x 40 x 3.2 mm and maximum differential temperature at a level of 150°C. It contained 199 thermocouples. Ceramic plates were made from Al_2O_3, and their thermal conductivity was about 15 W/mK.

Hot side of TE module (during described studies only one module was tested) was heating by a flue gas flown through the heat exchanger. Flow and temperature of flue gas was controlled by changing the amount of the air blown to the furnace area. Cooling the cold size of TE module was realized using the dedicated water exchanger and the heat pipe heat sink with or without a fan.

The operation of the stove-fireplace with accumulation was controlled using advanced control and measurement system based on the WAGO PLC controller combined with the set of measurement elements and actuators:

- thermocouple and resistance temperature sensors placed inside the furnace, heat exchanger dedicated for TEG, accumulative mass and chimney;
- resistance temperature sensors placed on the external side of accumulative mass and heat exchanger dedicated for TEG;

- thermoanemometers with a range of 0-2 m/s to measure the flow rate of the air blown into the furnace area;
- flue gas analyser uses the electrochemical method of O_2 concentration measurement and the NDIR method for measuring the concentrations of CO, CO_2, NO, NO_2 and SO_2;
- servo-mechanisms controlled via analogue signals 0-10 V to control the air throttles positions;
- electronic load.

The monitoring and process (measurement) data acquisition system has been developed in the CoDeSys environment on the PC computer. It was possible to observe the real time variations of the measured values, archive measurement data (in the internal memory of the PLC controller or in the PC) and control the SFA operation. Measurements were recorded in the 1s time interval. The location of control and measurement elements is shown in Fig. 4 (only elements used during the described tests were marked).

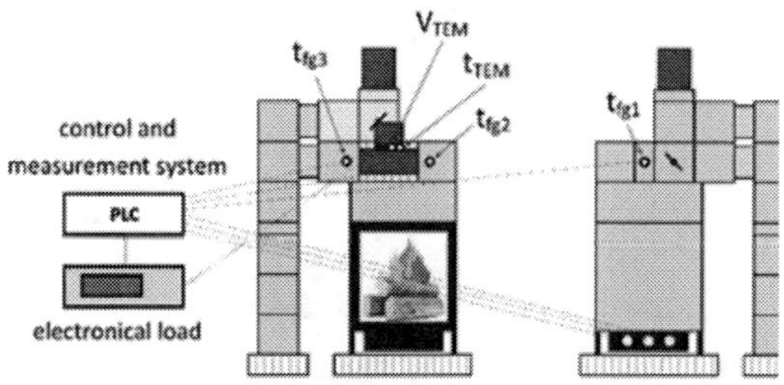

Figure 4. Scheme of the experimental rig.

The investigations were divided into two parts. At the beginning different cooling methods and their impact on the TE module performance have been tested. First, it was used only heat pipe heat sink, in the second approach heat pipe heat sink with an air fan, and finally the water heat exchanger (temperature of the cold water was equal to 22°C). Further experiments were aimed to determine the specific characteristics of the tested TE module. In each of measurement series 10 kg of beech wood briquette has been burned. The shape of briquette was rectangular and its caloric value was in the range of 14 – 17 MJ/kg.

The first tested configuration of TEG contained a heat pipe heat sink typically using in the PC processors cooling (see Fig. 5). The alumina heat sink had four heat pipes with cross-section of 6 mm, which received heat directly from the TE module surface. Dimensions of the whole heat sink was 125 x 52 x 152 mm (length x width x height).

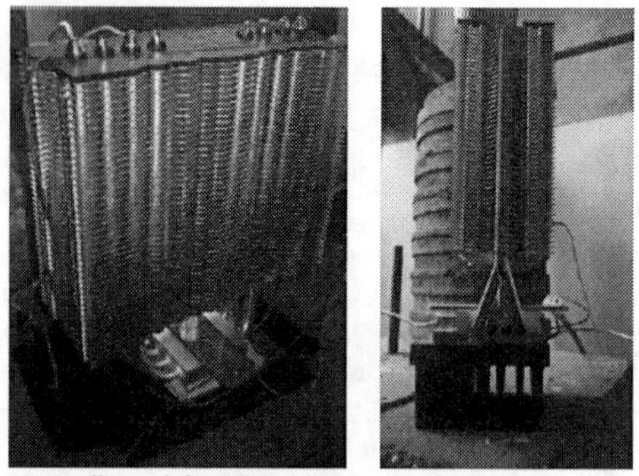

Figure 5. Cooling of the TEG using the heat sink.

In the second approach, in addition to the heat sink, an air fan with dimensions of 120 x 120 x 25 mm was used (see Fig. 6). The operation voltage of the fan was 12 V and current was 0.12 A – what means, that maximum taken power was less than 1.5 W. The fan provided the air flow at a level of 1.7 m^3/h, when the number of rotates per minute was 1400 +/-10%.

Figure 6. Cooling of the TEG using the heat sink with air fan.

The third configuration of TEG was equipped with the dedicated alumina water heat exchanger (see Fig 7). The size of exchanger was suited for the size of TE module's ceramic plate. Water was flown via steel pipes connected with exchanger via inlet and outlet localized at the wall of exchanger.

Figure 7. Cooling of the TEG using the water heat exchanger.

DOI: 10.1051/e3sconf/20161000087

3 Results and discussion

3.1 The impact of the cooling method on the TE module performance

The results of conducted studies show, that the use of proper cooling method has a great impact on the performance of TEG operation. Due to higher heat capacity of water (in comparison to air), water cooling is generally more efficient than air cooling. It can be observed in Fig. 8, where variations of voltage in each of studies cases were presented.

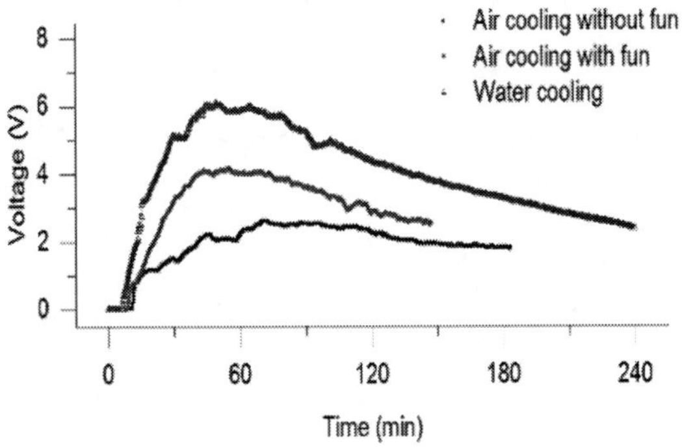

Figure 8. Histogram of TE module voltage (depending on the cooling method).

As we can see in Fig. 8, there is a huge difference between the output voltage obtained in the three considered series. Definitely the most efficient cooling method is water cooling, which allowed to obtain the maximum voltage value above 6 V (there is three times more than using air cooling without fan). On the other hand, during performed tests, temperature of water was constant over time and equal to 22°C. The increase of its temperature in the cooler was less than 4 K (with the flow rate of 5 l/min). In practical solutions it is hard to provide so good conditions. The use of water cooling may be connected with the use of the return water from hot water or heating system.

However, also in the case of using heat sink with air fan, there is necessary to take into account the power taken by the fan. It is about 1.44 W in the studied situation (the operation voltage of the fan is 12 V and current is 0.12 A). The solution is to use one air fan for few TE modules and modulate its rotation speed (depending on the hot side temperature of TE modules). It will allow to minimize the share of power taken by air fan.

In addition to data presented in Fig. 8, in Tab. 1 were shown the maximum and the average voltage values obtained during each of studies cases. The average value was calculated for the range from 10 to 150 minute of combustion process.

Presented above results are in general true for the studied cooling methods. Moreover, to eliminate the possible impact of the variations of combustion process on the output voltage from TE module, it can be introduced a comparative voltage and temperature coefficient (CVTC), calculated as follows:

$$CVTC = \frac{\sum_i^n (V_{TE,i}/t_{ex,i})}{n} \tag{1}$$

where:
$V_{TE,i}$ – output voltage of TE module, V
$t_{ex1,i}$ – temperature of the flue gas measured on the outlet from the furnace, °C
n – a number of the measurements

The values of CVTC calculated for the range from 10 to 150 minute of combustion process, were presented in Tab. 2.

Table 1. The maximum and average voltage values obtained from tested TE module.

Cooling method / Parameter	The maximum voltage	The average voltage
	[V]	[V]
Air cooling (heat sink without fan)	2.64	2.07
Air cooling (heat sink with fan)	4.17	3.21
Water cooling	6.10	4.82

Table 2. The values of comparative voltage and temperature coefficient.

Cooling method / Parameter	CVTC value
	[-]
Air cooling (heat sink without fan)	0.0064
Air cooling (heat sink with fan)	0.0089
Water cooling	0.0144

Data included in Tab. 1 confirms, that the most efficient cooling is connected with the use of water heating system. The relative differences between calculated values are quite lower than in the case of direct voltage values, but still significant. The variations of output voltage of TE module to temperature of the flue gas measured on the outlet from the furnace ratio was presented in Fig. 9.

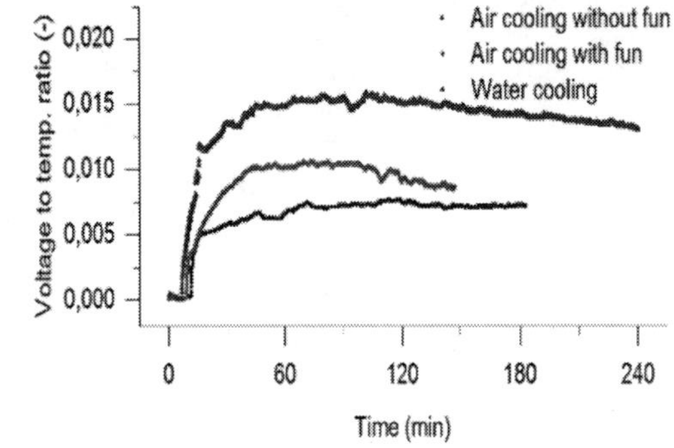

Figure 9. Histogram of the TE output voltage to temperature of the flue gas (measured at the outlet from the furnace ratio).

As we can see in Fig. 9, the courses of presented curves are more stable in time in comparison to output voltage

 DOI: 10.1051/e3sconf/20161000087

curves presented in Fig. 8. This situation confirms once again a direct correlation between the temperature of the cold side of TE module and its performance.

3.2 Specific characteristics of the tested TE module

The operation parameters of the tested thermoelectric module (voltage, current and power), depends not only on the temperature of its cold side. First of all, there are connected with the variations of the hot side temperature. During the studies carried out in the second part of this investigation, the effect of the hot side temperature variations on the current-voltage characteristic and the output power has been determined. The same amount of the fuel was burned (10 kg of briquette). Cooling of the cold side of TE module was realized using water cooling system (flow rate was the same as before, that is 5 l/min).

The current-voltage characteristics for the three chosen values of TE module hot side temperature (110, 130 and 150 degrees Celsius), were presented in Fig. 10.

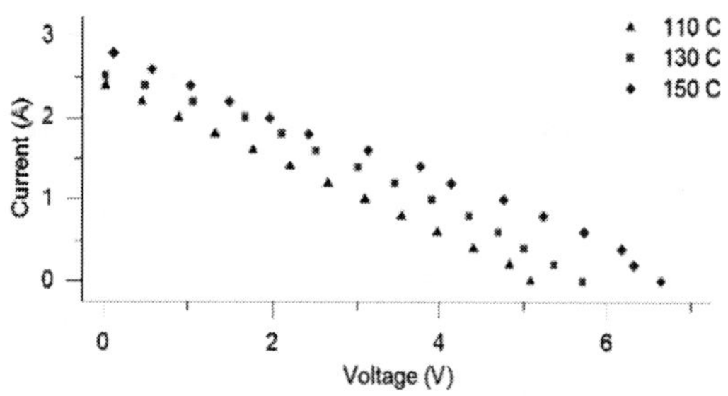

Figure 10. The current-voltage characteristic of the TE module.

The variations of the voltage and current presented in the Fig. 10 depends on the current phase of the combustion process. A great impact is connected also with the velocity of temperature variations of the flue gas and the variations of the flue gas flow (which depends e.g. on the opening degree of the air throttles). As a results the courses of presented curves are non-linear and irregular.

The huge dependence of the power generated in TE module on the temperature was shown also in Fig. 11. This figure illustrates the variations of the power in the function of the voltage and temperature.

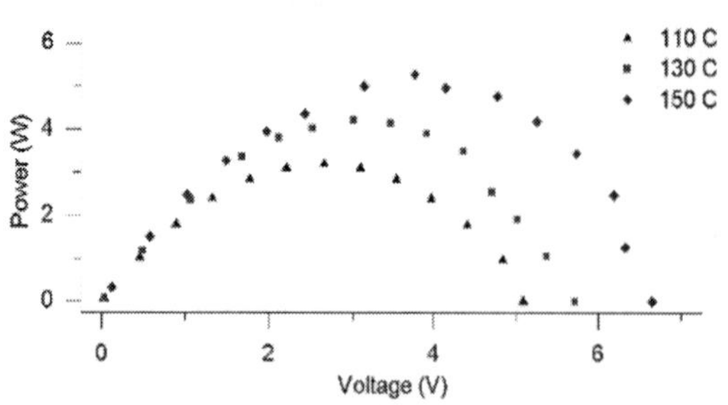

Figure 11. Power generated in the TE module.

As we can see in Fig. 11, the almost literature courses of the power curves in the voltage function were obtained – this illustrates the very good operation of the TE module. Power generated in the tested TE module exceeded 5 W (when temperature of the hot side was at a level of 150 degrees Celsius). In the other cases power was significantly lower – the maximum value measured for the temperature of 110 degrees Celsius was two times lower. However, the voltage obtained under laboratory conditions was about 10.5 V.

The maximum power points (which were marked in the Fig. 11) were also presented in Tab. 3.

Table 3. The maximum power points depending on the temperature.

Temperature of the hot side / Parameter	Voltage	The maximum power
	[V]	[W]
110 degrees Celsius	2.65	3.18
130 degrees Celsius	3.00	4.21
150 degrees Celsius	3.76	5.26

Presented values are related to the use of the only one TE module. Connecting in series/parallel more units, it is possible to generate more power with higher voltage and current. Produced power, using DC/DC converter, may be used to charge the battery. Looking further, there is possible to provide self-sufficient operation of the stove-fireplace with accumulation and also use of the generated electricity to power some of the home appliances (radio, lights, hot water circuit pump etc.).

4 Conclusions

The results of conducted study confirm the possibility of power generation using TEG connected to the stove-fireplace with accumulation.

Comparing different cooling methods, it was concluded, that the water cooling system is significantly more efficient that the air cooling system. The maximum measured output voltage of studied TE module was 6.1 V (in the case when the hot side of TE module was heating by the flue gas, and the cold side was cooling by the water). It is good results if we take into account the fact, that theoretically available output voltage of tested module is about 10.5 V (this value is possible to obtain in laboratory conditions). Also a comparative voltage and temperature coefficient CVTC (introduced to eliminate the possible impact of the combustion process variations on the output voltage from TE module) confirmed a direct correlation between the temperature of the cold side and module's performance. On the other hand, water cooling system may be efficiently used in the stove-fireplaces, which are used to heating water in the hot water circuit or central heating system.

The operation of TE module is strongly dependent not only on the cooling parameters, but also on the actual parameters of the combustion process (in particular flue gas temperature and flow). The variations in the current-voltage characteristics and power generated in TE module were considered as a function of voltage and temperature. As we observed, the maximum power was achieved at a level of 5.26 W, when the hot side temperature was 150°C, voltage – 3.76 V and current – 1.4 A. Connecting together more modules self-sufficient operation of the tested stove-fireplace with accumulation may be provided. On the other hand, some further developments in the construction of the thermoelectric generator are necessary.

Besides technical aspects, also economic reasons are really important and should be taken into account. The expected costs of the power generating system is lower than 30% of the price of tested stove-fireplace with accumulation. On the other hand, the increasing demands in the fields of using renewable energy sources, improvement of energy efficiency and reduce exhaust gas emission to the atmosphere, make such solutions more and more preferred.

Acknowledgements

The work has been completed as part of the statutory activities of the Faculty of Energy and Fuels at the AGH University "Studies concerning the conditions of sustainable energy development" in cooperation with Foundation Institute for Sustainable Energy.
Supervisor: prof. Mariusz Filipowicz

References

1. D.M. Rowe, Proceedings of the institution of electrical engineers, 24 (1978)
2. M. Hodes, IEEE Transactions on Components and Packaging Technologies **28,** 12 (2005)
3. Killander, J.C. Bass, Fifteenth international conference on thermoelectric proceedings ICT '96, 4 (1996)
4. R.Y. Nuwayhid, A. Shihadeh, N. Ghaddar, Energ Convers Menage **46**, 13 (2005)
5. D. Champier, J.P. Bédécarrats, T. Kousksou, M. Rivaletto, F. Strub, P. Pignolet, Energ **36**, 9 (2011)
6. P. Raman, N.K. Ram, R. Gupta, Energ **69**, 13 (2014)
7. S.M. O'Shaughnessy, M.J. Deasy, J.V. Doyle, A.J. Robinson, Appl Energ **156**, 11 (2015)
8. G.F. Rinalde, L.G. Juanico, E. Taglialavore, S. Gortari, M.G. Molina, Int J Hydrogen Energ **35**, 5 (2010)
9. A.M. Goudarzi, P. Mazandarani, R. Panahi, H. Behsaz, A. Rezania, L.A. Rosendahl, J Electron Mater 42, 7 (2013)
10. D. Champier, J.P. Bedecarrats, M. Rivaletto, M. Strub, Energ **35**, 8 (2010)
11. D. Mastbergen, B. Wilson, ETHOS International Stove Research Conference (2005)
12. K. Sornek, M. Filipowicz, L. Kurcz, M. Szubel, M. Rosół, K. Rzepka, J. Ręka, Engineering and Protection of Environment **18**, 16 (2015)
13. K. Sornek, M. Filipowicz, K. Rzepka, Study of clean combustion of wood in a stove-fireplace with accumulation, Journal of the Energy Institute (2016)
14. K. Sornek, M. Filipowicz, K. Rzepka, The development of a thermoelectric power generator dedicated to stove-fireplaces with heat accumulation systems. Energy Convers Manage (2016)

Agricultural utilisation of municipal sludge in willow plantation

Feliks Stachowicz[1,a], Tomasz Trzepieciński[1], Marta Wójcik[1], Adam Masłoń[2], Witold Niemiec[3] and Adam Piech[3]

[1]*Rzeszow University of Technology, Department of Materials Forming and Processing, Al. Powst. Warszawy 8, 35-959 Rzeszów, Poland*
[2]*Rzeszow University of Technology, Department of Chemistry and Environmental Engineering, Al. Powst. Warszawy 6, 35-959 Rzeszów, Poland*
[3]*Rzeszow University of Technology, Department of Water Purification and Protection, Al. Powst. Warszawy 6, 35-959 Rzeszów, Poland*

Abstract. This paper presents the results of analyses of the monitoring selected energetic willow plantations (30 ha) located in Podkarpackie Province (Poland), which have been fertilised using a prototype trailer for land-applying municipal sludge. The fertilisation of the willow plantation is undertaken using this prototype for the subsurface application of municipal sewage sludge into soil. The presented prototype is a component of technology of energetic plant production that has been developed at the Rzeszow University of Technology. The results of soil investigations allow to us conclude that the deposition of sewage sludge using the presented prototype of the agricultural trailer is conducted in accordance with legal regulations and does not cause significant perturbation of the natural trophic chain.

1 Introduction

Recycling sewage sludge to agricultural land to benefit from essential plant nutrients and organic matter it contains, is regarded as an environmentally sustainable approach to managing the sludge generated from sewage-treatment plants [1]. Sewage sludge containing more than 54.2% organic matter can be treated as "organic fertilisers" [2].

During 2010 in Poland, the total amount of produced sewage sludge was 413–450 thousand Mg [3]. It is worth highlighting that the amount of sludge generated in industrial treatment plants decreases from year to year, while the amount of municipal sewage sludge is increasing [4]. The dominant amount of sewage sludge is still stored (about 35%), then about 29% is used for reclamation of land; only 2.6% is designed for fertilisation of plants that are used to produce compost [5]. The utilisation of sewage sludge for fertilisation of agricultural land is the most economic option for inland treatment plants, and is also widely practised in Poland [4, 6].

The use of organic waste in agriculture can improve the soil's productive capacity, and physical and chemical characteristics [7-9]. However, the main problems of an excessive application of sewage sludge are plant toxicity due to accumulation of heavy metals in soils [10] but also an increase in its salt content [11].

To grow properly perennial energy plants (with different structures and different root systems), a sufficient amount of nutrients is required, which for a long period during the establishment of plantation cannot be delivered to the soil in sufficient quantities [12, 13]. Willow shoots are characterised by small concentrations of nitrogen and heavy metals, and using it is less harmful compared to coal combustion [14, 15]. Energy plant plantations are mainly fertilised with mineral fertilisers using well-known specialised machines called fertiliser spreaders. The cultivation of energy plants such as e.g. *Salix viminalis* on fields fertilised with sewage sludge is becoming more and more popular in Poland [16]. Willow plants may be used to extract heavy metals from contaminated soils due to their tolerance to specific heavy metals and adaptation to soil [2]. Salix clones with high-metal accumulating capacity in the shoots may be suitable for phytoextraction of moderately contaminated sites [2, 17]. Willows can thus be used to remove from soil many heavy metals, especially cadmium [18, 19]. The average content of phosphorus in sewage sludge is about 3% of dry mass and is also characterised by small amounts of potassium and calcium [20]. Gondek and Filipek-Mazur [21] claimed that sewage sludge particularly affects the increase in the content of calcium in plants, but this was not confirmed in [5]. According to Sieciechowicz et al. [16], it has not been confirmed that the fertilisation of plantations with sewage sludge exerts a significant impact on the accumulation of lead in the *Salix viminalis* biomass. The use of phosphorus by plants in the first year totals 20% and increases in succeeding years [22].

The application of granulated sludge from a sewage-treatment plant in a willow plantation resulted in biomass yields as high as 45 Mg/ha/year, compared to 15–20 t/ha/year without sludge fertilisation [23]. Lazdina et al. [17] found that, in the first season, the dry mass of control plantations without sewage was 0.2–0.6 Mg/ha,

[a] Corresponding author: stafel@prz.edu.pl

 DOI: 10.1051/e3sconf/20161000088

but in the second season after fertilisation with sewage sludge of 14 Mg/ha (700 kg of nitrogen per ha), the dry mass of shoots reached 4.6–5.5 Mg/ha.

In this paper, a prototype of a trailer for land-applying municipal sludge is presented. Furthermore, the results of analyses of the monitoring of selected energetic willow plantations (30 ha) located in the Podkarpackie Province (Poland), which have been fertilised using sewage sludge, are presented.

2 Agricultural use of sewage sludge

The agronomical value of municipal sewage sludge depends on the content of pathogen microorganisms, heavy metals and certain organic compounds [24]. Whilst the pathogens can be removed by treatment to obtain a well-conditioned product, the organic compounds have to be monitored both with respect to their content and to their possible transfer to the soil and living beings [25]. The sustainability of sewage sludge agricultural disposal requires a value judgment based on knowledge and evaluation of the level of pollution of both sewage sludge and soil [24]. Sewage sludge contains significant amounts of nitrogen and phosphorus, and has value as an effective organic fertiliser [26, 27]. Sewage sludge application to soil also conserves organic matter.

The use of sewage sludge management in agriculture or for reclamation is characterized by the lowest costs ranging from 300–400 PLN per 1 Mg of sludge. In contrast, combustion and co-combustion of sewage sludge are more expensive, even fivefold (Table 1) [28].

Table 1. Predicted costs of sewage sludge management using different methods.

Ways of sewage sludge utilization	Disposal cost of 1 Mg sewage sludge [PLN]
agricultural utilization	300
composting	600
combustion	1500
co-combustion	1750

Agricultural use of sludge is regulated specifically by:

- The Council Directive 86/278/EEC of 12 June 1986 on the Protection of the Environment, and in particular of the soil, when sewage sludge in used in agriculture,
- The Council Directive 91/676/EEC of 12 December 1991 concerning the protection of waters against pollution caused by nitrates from agricultural sources,
- Act on waste materials, of 14 December 2012 (J. of Laws, 2013, item 21),
- Regulation of Minister of Environment on communal sludge of 6 February 2015 (J. of Laws, 2015, item 257), which unified environmental standards for recycling sludge on farmland across the EU.

Sewage sludge may be considered as an important biological resource for sustainable agriculture. The major plant nutrients as nitrogen, phosphorus and potassium in municipal sewage sludges are not substantially removed during sludge processing; therefore the nutrients could improve the soil's nutritional status after sludge application [29]. Sludge also contains essential plant micronutrients such as boron, copper, manganese and zinc. Crops can accumulate toxic elements from sludge-amended soils; where heavily contaminated sludges and excessive rates of application are used, plants may accumulate concentrations that are phytotoxic [30]. In the case of application of sewage sludge in energetic plant plantations, the accumulation of heavy metals does not create health responses to humans, because they are not a part of the human trophic chain. The investigations of Smith and Vasiloudis [31] and Trinidade et al. [32] have shown that the yield of many plant species increases following the application of sewage sludge, which was connected to the increased supply of nutrients as a result of sludge application. After sludge application, the kind of a clone of energetic plant has an impact on the soil quality. The application of sewage sludge causes acidification of soil and increasing bioavailability of metals. The soil quality after sludge application depends on the kind of energetic plants due to different accumulation of heavy metals in tissues. Furthermore, the sewage sludge influences an increase of the calorific value of biomass, especially in the case of miscanthus and reed canary grass.

The analysis conducted by Labrecque et al. [33] showed that willow reacts in a linear increase of yield, to increase doses of sludge, expressed as assimilable nitrogen in kg per ha. A sample fertilized with sludge at the dose equivalent of solid 200 kg of nitrogen per ha, in comparison with a control, obtained following increments: 70% of height of the plants, 72% of the diameter, 415% by weight of shoots, 200% of root mass, 5–40% by weight of the leaves.

Taking into consideration the report of the EEA (European Environment Agency) [34], the total energy of potential of biomass produced in Poland will be 33 megatons of oil equivalent (MtOE) of which 73% will come from biomass crops in year 2020. An increase to 39 MtOE, with 77% share of the crop is expected by 2030. Ericcson et al. [35] estimates needs for area biomass crops on about 2.3 million ha, assuming national energy needs in the field of biomass at the level of 370 PJ per year at an average yield of 9 Mg per ha and calorific value 18 GJ per Mg. Without proper fertilization, including sewage sludge, these assumptions will be difficult to achieve.

3 Methods of sludge application

Land application has been the most common way to dispose of the biosolids, i.e. sewage sludge [36]. The most appropriate application method for agricultural land depends on the physical characteristics of the sludge and the soil, and the species of crops grown. Generally, land application of sewage sludge is carried out by subsurface application, vehicular surface spreading, or irrigation application. The former method of sludge application is performed by disking methods, plough furrowing, or subsurface injection. Liquid or solid fertilisers can also be injected below the soil surface tank trucks fitted with flotation tires and injection shanks. Such equipment minimises odour problems and reduces ammonia volatilisation by immediate mixing of soil and biosolids. Subsurface injection can be used on slopes of up to 15% and will minimise runoff from all soils. Soil incorporation of liquid sludge can be effected by plough-furrow cover and subsurface injection. Vehicular surface spreading is accomplished usually by tank truck spreading. Surface application with its incorporation is normally limited to soils with less than a 7% slope.

Finally, irrigation application is also undertaken by flood irrigation or spray irrigation. Irrigation systems generally should not be used to apply sewage sludge to row crops during the growing season. However, a light application to the stubble of a forage crop following harvest is acceptable. Off-site management of sewage sludge involves loading, hauling and land-applying sewage sludge at a location some distance from the treatment plant. The primary advantage of this procedure is the ability to make use of remote tracts of land that may be scattered some distance from the treatment plant.

Tanks mounted on powered chassis that are equipped with large, low-pressure tires for traversing soft ground are used most commonly for land-applying sewage sludge. The resulting rough surface may be unacceptable for subsequent land use such as grazing or haying.

4 Trailer for land-applying sewage sludge

Fertilisation of perennial plantations using already known devices does not guarantee optimal performance of this operation, i.e. the sewage should be applied at the proper depth and must be covered by a soil layer. These requirements are particularly important in relation to the Act on Waste Materials, of 14 December 2012 (J. of Laws, 2013, item. 21) and Regulation of Minister of Environment on communal sludge of 6 February 2015, and must use fertiliser properly prepared sludge from municipal sewage-treatment plants.

To meet the mentioned requirements Niemiec [37] developed a trailer for subsurface application of sewage sludge in plantations of energetic willow (Fig. 1). The machine is a part of fleet of agricultural machines for the establishment of plantations and the processing of woody biomass [12, 13]. The efficiency of enrichment tract plantation depends on the proper dosage of fertilisers and fertilisation depth. Furthermore, the sewage should be immediately incorporated directly into soil and should never be seen on the surface of the field. Application of organic fertilisers with loose consistency into the soil to the desired depth is the main advantage of the trailer. The sewage sludge is supplied to the vicinity of the plant root without its damaging, which facilitates the absorption of nutrients by plants. The device (Fig. 1) is mounted on the standard currently used manure fertiliser as a separate module, without the need for significant customisation. According to the concept, it is assumed that the device is mounted to a manure spreader and has its own hydraulic system for setting its working position in relation to the ground. The device is equipped with three sections of fertiliser application with a spacing of 0.75 m, which corresponds to a typical system of willow planting.

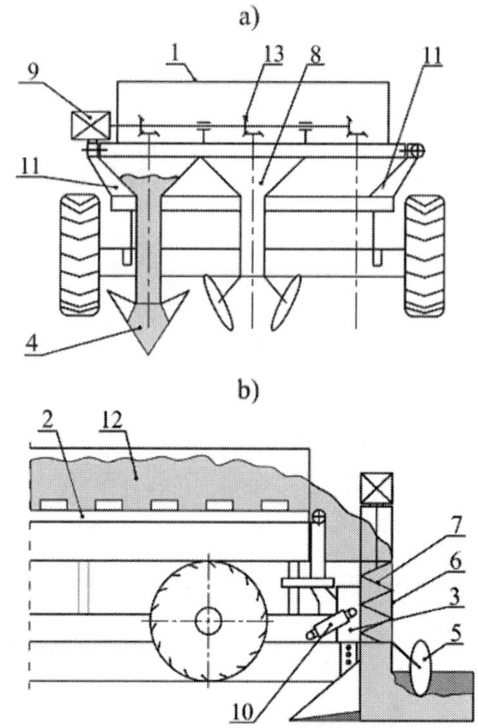

Figure 1. Schematic of the back (a) and side (b) view of the conceptual trailer for subsurface application of loose organic and mineral fertilisers [37]: 1 - trailer, 2 - base frame, 3 - beam, 4 - disk coulter, 5 - ridging plough, 6 - cylindrical housing of screw conveyor, 7 – screw conveyor, 8 - loading chamber, 9 - hydraulic engine, 10 - hydraulic cylinder, 11- arm, 12 - fertiliser, 13 - bevel gear.

The prototype of the trailer (Fig. 2) for subsurface application of sewage sludge into soil was developed at Rzeszow University of Technology in cooperation with "R&D Centre Inventor Sp. z o.o." and "SIPMA S.A", both in Lublin. The device is attached to the trailer. The disc coulters are fastened to the support frame. The disk ridging ploughs are placed behind the disc coulters. Cylindrical housings with the screw conveyors working inside are coupled to the frame of the device. In the upper part of the screw conveyors, the bevel gears driven via a reducer by the hydraulic motor are mounted. The hydraulic system of the trailer is supplied from the tractor. Each of the three screw conveyors alternatively may be individually equipped with a hydraulic motor powered from the tractor's hydraulic system. The location of the disc coulters relative to the frame is established discretely by a mechanical blockage. However, the location of the disc coulters in relation to the frame can be set continuously by the hydraulic system. The arms coupled to a hydraulic cylinder are used to set the machine in the transport position. The depth of the grooves for buckfilling the fertiliser in furrows is set by changing the position of the disc coulters and ridging ploughs in relation to the ground. During the machine operation, the weight of the manure in the transport chamber is moved toward the application area using the floor conveyor. The fertiliser is transported using

 DOI: 10.1051/e3sconf/20161000088

the screw conveyors to the grooves in the soil. Filled with sewage sludge, the grooves are then buckfilled by the ridging ploughs (Fig. 3).

Figure 2. Prototype of device for subsurface application of loose organic and mineral fertilisers manufactured by SIPMA S.A.: 1 - trailer, 2 - base frame, 3 - beam, 4 - arms, 5 – loading chamber, 6 - hydraulic cylinder, 7 – disc coulter, 8 – flap, 9 - screw conveyor, 10 – cylindrical housing of screw conveyor, 11 – ridging plough, 12 – hydraulic engine.

Field tests were carried out on an experimental plantation of energetic plants (Fig. 4), in cooperation with the sewage-treatment plant in Świlcza (Poland). Municipal sewage sludge from the hygienisation line was transported directly on a trailer. The tests of the prototype of the trailer confirmed the correct functional features of the system.

Figure 3. Application of sewage sludge into the soil subsurface.

Subsurface application of sewage sludge with the use of the device for injection dosage of mineral and organic fertilizers is the ideal solution for small and medium wastewater treatment plants. Owing to the high prices of thermal sludge utilization methods, the aforementioned technique is the most economically justified. The sewage sludge can be applied into soil in spring to assure plants nutrients necessary to their growth. Generally, using developed trailer we can apply the sewage sludge through the all year, excluding winter months due to soil freezing.

Figure 4. View of the willow plantation after subsurface application of sewage sludge.

The dosage of properly treated sewage sludge by subsurface application into soil could be an alternative method to the popular in recent years sludge hygienisation with lime and the production of granular fertilizer. The liming of sewage sludge requires the modernization of treatment plants and equipment installation, the cost of which can reach even 500,000 PLN. The process of hygienisation and agglomeration demands a dose of hydrated lime ranging from 0.15–0.25 kg per kg of sewage sludge d.m. which could result in subsequent sterilization and soil degradation. Assuming the price of highly reactive lime ranging from 290–350 PLN per Mg of lime, additional costs of approximately 30,000–100,000 PLN per year are generated for middle-sized sewage-treatment plant.

5 Effect of sewage sludge on soil

Sewage sludge from waste-water treatment plant in Świlcza is applied into the soil using the prototype of trailer (Fig. 2). The experimantal plantation (Fig. 5) is located about 2 kilometers from waste-water treatment plant. The current total willow plantation area is at most 30 ha. Different species of energetic plants are grown on the plantation. However, in this paper the results of soil monitoring before and after sewage sludge application are presented for energetic willow. The potential of biosolids for contaminating soils with heavy metals has caused great concern about their application in agricultural practices. Heavy metals most commonly found in sewage sludges are Cd, Cr, Cu, Ni, Pb and Zn, and the metal concentrations are governed by the type of process employed during the sewage treatment.

Sewage sludge was characterized by high volatility in terms of quality (dry matter, nitrogen, phosphorus) and heavy metal content (Tables 2 and 3). Sewage sludge meets sanitary requirements and legal requirements and can be used for agricultural purposes on medium and/or heavy lands. Monitoring the quality of the soil was carried out in the plantation of energetic willow (1-9 in Fig. 5). The samples of soil were collected according to the standard PN-R-04031:1997. To determine the basic parameters to establish initial and final concentrations the samples were taken from the same place.

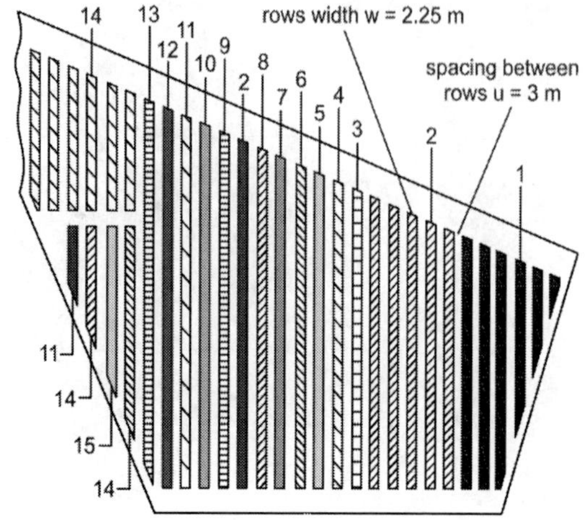

Figure 5. The view of the species arrangement in experimental plantation of energetic plants: 1-9 – genetypes of energetic willow: RF4, RM1, RZ47, RZ52, FR, M, RFM3, RFM2, RZ52, respectively; 10 - mix of willow genotypes, 11 - robinia and mix of willow genotypes, 12 - robinia, 13 - mix of willow genotypes and *Rosa multiflora*, 14 - mix of willow genotypes, 15 – paulownia.

Table 2. Characteristic of sewage sludge used in investigations.

Parameter	Concentration in sewage sludge [% of dry mass]						
	DM	OM	TP	TN	N-NH_4	Ca	Mg
Average	31.7	39.2	0.92	2.64	0.11	22.0	0.28
Minimum	16.4	19.4	0.22	1.0	0.10	11.9	0.15
Maximum	61.2	49.7	1.55	3.76	0.12	33.6	0.35
CV	0.80	0.44	0.62	0.49	0.10	0.44	0.34
SD	25.53	17.16	0.57	1.30	0.01	9.66	0.09
15th percentile	16.76	28.13	0.46	1.54	0.1	14.0	0.20
85th percentile	48.12	49.34	1.38	3.69	0.11	30.1	0.35

DM - dry mass, OM-organic matter, TP - total phosphorus, TN -total nitrogen, N-NH_4 - ammonium nitrogen, CV- coefficient of variation, SD - standard deviation

Monitoring the quality of the soil in plantation of energetic willow showed no negative effect of the application of sewage sludge on the soil quality (Table 4).

Although Cu, Ni, Pb and Zn concentrations increased with the addition of sewage sludge, the concentration of heavy metals are within the limit values for soils of the urban areas defined in the Regulation of the Minister of Environment dated 09 September 2001 on Soil Quality Standards and Quality Standards of Soil (J. of Laws, 2002, No 165, item 1359). Proper agricultural practices ensure the adequate development of energy crops fertilized with sewage sludge. Heavy metals introduced into the soil with sewage sludge as a result of good mixing and distribution on the surface of the whole area undergo bioaccumulation in energy plants.

Table 3. Heavy metals concentration in sewage sludge used in investigations.

	Concentration in sewage sludge [mg per kg of dry mass]						
	Cd	Cu	Ni	Pb	Zn	Hg	Cr
Average	2.5	77.6	6.1	3.67	358.9	0.149	8.03
Minimum	0.3	19.3	2.6	2.5	70.7	0.005	3.37
Maximum	8.8	122.0	9.7	11.2	604.0	0.27	11.5
CV	1.65	0.56	0.47	0.49	0.62	0.80	0.44
SD	4.2	43.32	2.88	3.67	221.9	0.119	3.5
15th percentile	0.38	44.55	4.18	4.68	191.0	0.048	5.25
85th percentile	5.11	109.2	8.12	10.12	522.2	0.248	10.7

Table 4. Characteristic of agricultural soil before and after subsurface application of sewage sludge.

Element	Average±SD accumulation of heavy metals [mg per kg of dry mass]		The limit values in the soil
	Before application of sewage sludge	After application of sewage sludge	
Cd	not tested	0.82±0.212	2.0
Cu	17.83±0.27	21.47±9.26	50.0
Ni	21.41±0.33	29.93±12.28	35.0
Pb	48.48±0.47	31.53±10.93	60.0
Zn	48.65±0.44	58.67±15.66	120.0
Hg	not tested	0.117±0.040	1.2
Cr	78.17±1.108	29.50±9.06	75.0

6 Summary and conclusions

The expanding acreage of energetic plants threatens traditional crops intended for food production. One of many ways to increase the acreages of energetic plants is the management of increasing areas of idle land, wasteland, or areas hitherto unused for agriculture. This is related to the potential use of sewage sludge for fertilisation.

Municipal sewage sludges from biological treatment processes require stabilisation and hygienisation. Prepared in this way, the sewage sludge used as soil fertiliser is free from any health risks. The presented innovative device for subsurface application of organic fertilisers ensures the compliance with both legal and technological requirements of agricultural use of municipal sewage sludge, and the requirements of the Act of 6 September 2001 on Road Transport (J. of Laws, 2001, No 125, item 1371, as amended).

The experimental investigations of the influence of sewage sludge on the quality of the soil were carried out in experimental plantation of energetic plants located in Świlcza. Monitoring the quality of the soil showed no singificant increase of the heavy metals concentration after subsurface application of sewage sludge in experimental plantation of energetic willow. The results of the present study indicate that the addition of sewage sludge does not imply environmental risks, offering a disposal of sewage sludge from sewage-treatment plant.

References

1. S.R. Smith, Philos. T. R. Soc. A, **367**, 4005 (2009)
2. A. Jama, W. Nowak, Pol. J. Agron., **9**, 3 (2012)
3. K.L. Perttu, J. Sustain. For., **5**, 32 (1993)
4. J. Wieczorek, K. Frączek, J. Ecol. Eng., **14**, 36 (2013)
5. A. Jama-Rodzeńska, J. Bocianowski, W. Nowak, D. Ciszek, K. Nowosad, Ecol. Eng., **87**, 212 (2015)
6. S. Werle, Ecol. Chem. Eng. S. **22**, 459 (2015)
7. R. Vaca, J. Lugo, R. Martínez, M.V. Esteller, H. Zavaleta, Rev. Int. Contam. Ambie., **27**, 303 (2011).
8. E. Wołejko, U. Wydro, A. Butarewicz, T. Łoboda, Env. Prot. Eng., **39**, 67 (2013)
9. M.A. Özyazici, Eurasian J. Soil Sci., **2**, 35 (2013).
10. S.P. McGrath, F.J. Zhao, S.J. Dunham, A.R. Crosland, K. Coleman, J. Env. Quality, **29**, 875 (2000).
11. X. Hao, C. Chang, Agr. Ecosyst. Env. **94**, 89 (2003)
12. T. Trzepieciński, F. Stachowicz, W. Niemiec, L. Kępa, M. Dziurka, Croat. J. For. Eng., **37**, 185 (2016)
13. W. Niemiec, F. Stachowicz, T. Trzepieciński, J. Ecol. Eng., **14**, 12 (2013)
14. E. Klang-Westin, J. Errikson, Plant Soil., **249**, 127 (2003)
15. J. Bieniek, A. Żołnierz-Rusinek, Agric. Eng., **4**, 111 (2008)
16. A. Sieciechowicz, Z. Sadecka, S. Myszograj, M. Suchowska-Kisielewicz, Civ. Env. Eng. Report., **12**, 69 (2014)
17. D. Lazdina, A. Lazdinš, Z. Karinš, V. Kāposts, J. Env. Eng. Landscape Manag., **15**, 105 (2007)
18. P. Vervaeke, F. Navez, J. Martin, F.M.G Tack, M.G. Verloo, N. Lust, Workshop: Phytoremediation of Toxic Metals, June 12-15, Stockholm, Sweden (2003).
19. M. Labrecque, T.I. Teodorescu, S. Daigle, Biomass Bioenerg., **12**, 409 (1997)
20. C. Rosik-Dulewska, S. Oleszek-Kudlak, Postępy Nauk Rolniczych, **5**, 47 (2002)
21. K. Gondek, B. Filipek-Mazur, Acta Agrophys., **8**, 83 (2006)
22. A. Jama, W. Nowak, Pol. J. Agron., **9**, 3 (2012)
23. M. Labrecque, T.I. Teodorescu, For. Ecol. Manag., **150**, 223 (2001)
24. C.B. Rizzardini, D. Goi, Sustainability, **6**, 2424 (2014)
25. P. Soler-Rovira, J. Soler-Soler, J. Soler-Rovira, A. Polo, Fert. Res., **43**, 173 (1995)
26. C.D. Tsadilas, T. Masti, N. Barbayiannis, D. Dimoyiannis, Commun. Soil Sci. Plant Anal., **26**, 2603 (1995)
27. C.F. Tester, Soil Sci. Soc. Amer. J., **54**, 827 (1990)
28. A. Henclik, J. Kulczycka, K. Gorazda, Z. Wzorek, Eng. Prot. Env., **17**, 185 (2014).
29. K. Usman, S. Khan, Said Ghulam, M.U. Khan, N. Khan, M.A. Khan, S.K. Khalil, Amer. J. Plant Sci., **3**, 1708 (2012)
30. R.M. Sterritt, J.N. Lester, Sci. Total Env., **16**, 55 (1990)
31. R. Smith, H. Vasiloudis, Water S. A., **17**, 19 (1991)
32. H. Trinidade, J. Coutinho, S. Jarvis, N. Moreira, Eur. J. Agron., **15**, 281 (2001)
33. M. Labrecque, T.I. Teodorescu, S. Daigle, Plant Soil, **171**, 303 (1995).
34. B. Holm, K. Heinsoo, Biomass Bioenerg., **57**, 126 (2013).
35. K. Ericsson, H. Rosenqvist, E. Ganko, M. Pisarek, L. Nilsson, Biomass Bioenerg., **30**, 16 (2006).
36. C. Schaefer, Restor. Reclam. Rev., **5**, 1 (1999)
37. W. Niemiec, Patent PL 214031 (2008)

DOI: 10.1051/e3sconf/20161000089

Energy efficiency and steam coal transport over long distances

Katarzyna Stala-Szlugaj[1a] and Zbigniew Grudzinski[1]

[1] *The Mineral and Energy Economy Research Institute of the Polish Academy of Sciences, Krakow, Poland*

Abstract Coal is one of the most important energy sources in the world. Its main consumers are the energy sector (with a 37-46% share in the years 1990-2014; 1.12-2.34 billion toe) and industry (24-27%; 0.78-1.38 billion toe). Diversified distribution of coal deposits in the world in relation to its consumers means that it has to be transported over often very long distances. The global coal trade is dominated by maritime transport (90-94% in the years 2004-2014), and the share of transport by land is relatively smaller. The aim of the article was to calculate the index describing what part of the energy contained in the coal transported on a particular route is consumed by the train carrying it. Due to the dominant position of the Russian Federation in imports of coal to Poland, it was assumed that coal will be imported by rail from Kuzbass (Russia's largest coal basin). As a result of the calculation, it was found that the rate of energy consumption for transport of imported coal will be somewhere in the range of 9.22-15.26%. In the case of deliveries of hard coal from Polish producers to the power plants the calculated rate changes within the range of 0.55-0.58%.

1 Introduction

Coal, following liquid fuels and natural gas, is the third most important energy carrier used in the world. According to the data for 2014 [1] its share in the global consumption was 30% (4.01 billion toe). Although in comparison to 1990, its share increased by only 3 percentage points, the total world consumption of coal increased by 1.65 billion toe (an increase of 74%). It is anticipated [1] that in 2035 global coal consumption will reach 4.27 billion toe, and its share will drop to 25%.

The main consumer of coal is the energy sector with 37-46% share in the years 1990-2014 (1.12-2.34 billion toe). It is followed by the industry sector (24-27%; 0.78-1.38 billion toe) and other sectors (26-38%; 1.23-1.30 billion toe). According to forecasts [1] in 2035, the use of coal as feedstock for energy transformations will reach 2.49 billion toe and - in comparison with 2014 - will increase by 274 million toe (an increase of 20%). The main contributors to the decrease in coal consumption in the world will be the economy rebalance and the decline in GDP growth in China, as well as the tightening of environmental policy.

Diversified distribution of coal deposits in the world in relation to its consumers means that it has to be transported over often very long distances. The global coal trade is dominated by maritime transport (Fig. 1). In the years 2004-2014, this way accounted for 90-94% of international trade, while the remainder was transported by land.

International markets for coal are usually divided into two regions: Pacific and Atlantic. The Pacific market's major exporters of coal include: Australia, Indonesia and China, and the main consumers are Japan, South Korea and Taiwan. In the Atlantic market coal reaches Europe by sea primarily from: South Africa, Colombia, Indonesia, Venezuela and the United States. Poland and the Russian Federation are also important suppliers of coal to the European market.

In the years 2004-2014 the total volume of global trade by sea varied between 73 and 92 million tons of coal (Fig. 1), and its share decreased from 10% (in 2004) to 7% (2014).

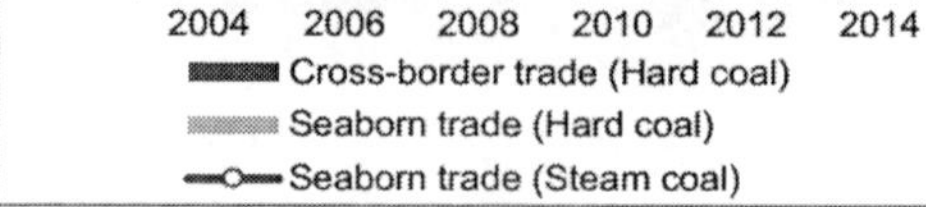

Figure 1. World hard coal trade, 2004-2014 [based on 2]

Due to the high concentration of producers and consumers, coal trade by land usually takes place between the countries situated on the Eurasian continent.

The international market is dominated by sales of steam coal (Fig. 1). In the years 2004-2014, its share in the global trade by sea varied from 74 to 77%.

[a] Corresponding author: kszlugaj@min-pan.krakow.pl

In quantitative terms, turnover increased from 505 (in 2004) to 878 million tonnes (2014).

In the case of Poland, relatively close proximity of the main exporters of coal (the Russian Federation and the Czech Republic) makes the rail transport the most common way of coal imports.

Due to the dominant position of the Russian Federation in coal exports to the Polish market, the objective of this article is to determine the share of energy associated with the transport of coal from the Russian Federation to Poland, in relation to the energy contained in the imported coal.

2 Import of steam coal to Poland

For many years, Poland was one of the leading exporters of hard coal in the world. In the years 1966-1978 the Polish participation in the international coal market averaged as much as 19%. In recent years, Poland was ranked in 9th-10th place on the global scale (with a share of 1-2%), while exports amounted to about 7-11 million tonnes. Mainly steam coal is exported from the country (Fig. 2). In the years 1990-2015 it constituted from 60 to 86% of the exported tonnage.

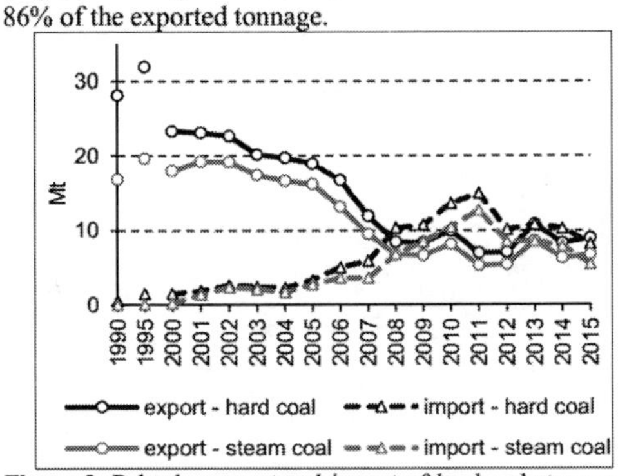

Figure 2. Poland – export and import of hard and steam coal [based on 3, 4]

Before the Polish accession to the European Union, import of hard coal was low. By 1995, it amounted to only a little more than 1 million tons (Fig. 2). Only coking coal, which is used for producing coke, was brought to the country. The increase in coal imports in the 90s of the 20th century (up to approx. 2-4 million tons) was associated with the economic and political transformation in the country. The process of restructuring the hard coal mining industry began at that time [i.e. 5, 6, 7, 8, 9]. A number of legal acts [i.e. 10, 11, 12, 13] were passed to protect Poland against excessive imports of coal and quotas on imports of coal from the Russian Federation and the Czech Republic were introduced [i.e. 14, 15, 16, 17, 18, 19].

With the Polish access to the EU in 2004, the acts protecting Poland against the increased imports of coal lost their legal force [20]. The protection from imports now applied to the whole EU market.

In order to achieve economic efficiency, the Polish mining industry underwent subsequent stages of restructuring. Efforts of the recovery programs together with simultaneous shortages of necessary investments in development works led to a situation where there was a shortage of domestic coal in the market [7, 21]. In 2006-2007 there was an increase in steam coal consumption in the power industry. As a consequence of the underestimation of the needs of the energy sector (lack of relevant mining contracts for delivery in 2007), a threat of a shortage of steam coal occurred. At the expense of export contracts and a reduction in the supply of coal to non-contractual customers, the mining industry complied with its obligations.

However, the domestic customers were left with the impression of "uncertainty" of supplies from domestic producers. As a result, both big and individual consumers of coal began searching for supplies of this material among foreign producers. This situation was used by small coal importers, who have found a source of supply, especially among the CIS countries (The Commonwealth of Independent States). At the same time the profitability of coal imports to Poland was reinforced by the weak position of the US dollar against majority of national currencies.

The consequences of increased imports manifested in 2008 (see Fig. 2), in which Poland for the first time in history became a net importer. In comparison to the previous year, imports of hard coal increased by as much as 75%.

The greatest amount of hard coal (15 million tonnes), was imported to Poland in 2011, of which 85% was steam coal. According to [22], that year the ratio of Poland's dependence on imports of hard coal was more than 10%.

The main exporters of steam coal to the Polish market are the Russian Federation and the Czech Republic (Figure 3). In the years 2004-2014 imports of stem coal ranged from 1.7 to 12.7 million tonnes. Out of that, the two mentioned countries provided (respectively): 73-91% and 1-17% of the material imported to Poland.

The geographical location of the exporters in relation to Poland makes rail transport most convenient. In recent years, this means of transport is responsible for about 70% of the raw material imported to Poland [23, 24]. Coal imported from CIS countries usually crossed the Polish border at railway crossing with the Russian Federation (Braniewo) and Belarus (Kuznica Bialostocka, Terespol). From the Czech Republic it entered Poland through the crossings in Chalupki and Zebrzydowice.

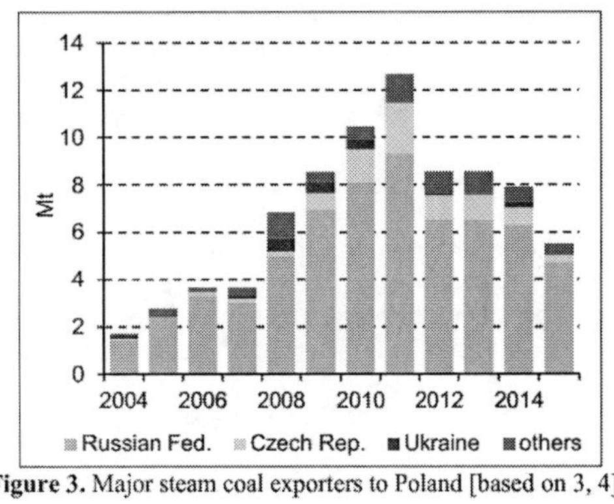

Figure 3. Major steam coal exporters to Poland [based on 3, 4]

 DOI: 10.1051/e3sconf/20161000089

In the case of the largest supplier to the Polish market - the Russian Federation - the raw material is mainly imported from the largest coal basin called Kuzbass. The average distance for the train from Kuzbass to Poland is about five thousand kilometers (Fig. 4). In order to reach the crossings in the north-eastern part of Poland, the train must pass through the territory of Belarus and / or Lithuania.

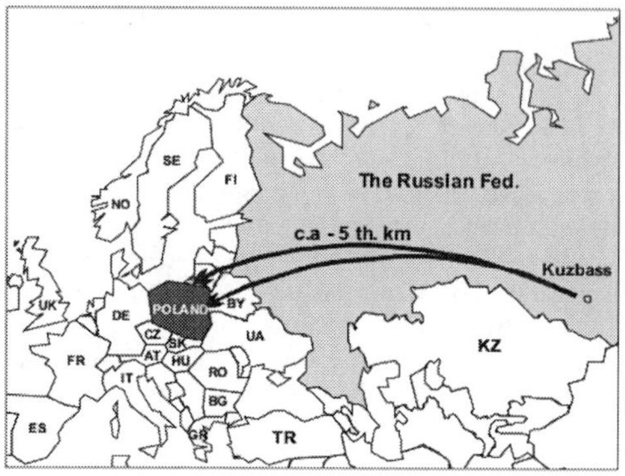

Figure 4. Estimated hard coal transport route from Kuzbass to Poland [own work]

3 Railway Transportation of coal

One of the types of land transport used for transportation of coal is a railway. The advantage of this transport is that it can carry large mass of load.

With the development of railways in the world, a number of railway lines with different track gauge were built. Track gauge is the distance between the inner rails, measured 14 mm below the upper surface of the rail's head [25].

In general, the division associated with the spacing of the rails allows for distinguishing the following types of railway lines [25]:

- standard gauge lines - are characterized by the most popular track gauge of 1435 mm (this is called normal gauge of the track, which is adopted in most European countries),
- broad gauge lines - these are all tracks, whose distance between the rails is higher than normal, i.e. over 1435 mm. Increasing the width of the track was initiated in Tsarist Russia which, in order to obstruct the entry to its territory by rail, began to build tracks with gauge of 1520 mm. Among other European countries with the broad gauge lines of a different spacing one may mention Spain and Ireland (1676 mm).
- narrow-gauge lines - the spacing between the rails is less than the normal gauge. The most common gauge of these lines is 1000 mm. Similarly as in the case of broad-gauge lines, there is a great diversity of gauge and their place of operation. There are about 120 different track widths of narrow gauge in the world.

As shown in chapter 2, the largest exporter of coal to the Polish market is the Russian Federation. The geographic location of this exporter in relation to Poland allows for the use of rail transport.

Both in Poland and in the Russian Federation hard coal is the most important cargo transported by rail. In Poland in the period of 2004-2014 railway was used for transporting from 152.3 million tonnes (2004) to 84.3 million tons (in 2009). Transportation of hard coal in 2014 amounted to 94.5 million tons, and in comparison with 2004, decreased by 59.8 million tonnes.

The decrease in domestic coal production in Poland was reflected in the reduction of the share of hard coal in the total mass of cargo transported by rail. In the years 2004-2014, this share has fallen from 55% (in 2005) to 38% (in 2007) (Fig. 5). A significant increase in carriage in 2008 (by 7 percentage points compared y / y) resulted from the aforementioned increase in coal imports.

In the case of the Russian Federation, both the mass of transported hard coal and its share in the total transportation increased. For the last eleven years, the railway transport of hard coal has increased from 273.5 (in 2004) to 315.4 million tonnes (2014). Its share in the overall tonnage increased from 22% (in the years 2004-2006) to 26% in 2014 (see Fig. 4).

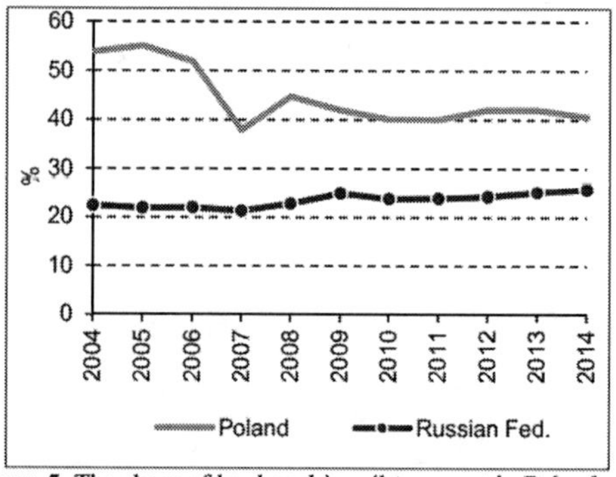

Figure 5. The share of hard coal in rail transport in Poland and in the Russian Federation [own work based on 26, 27]

In the Russian Federation, for the transport of dry bulk goods (coal, ore, wood and such types of goods that do not pollute the atmosphere) so-called coal-wagons are used [Russian: *poluwagony*]. Among them 4, 6 and 8-axis wagons are distinguished. The bearing capacity for these types of wagons varies within a wide range - from 62 tons to 125 tonnes [28, 29].

The length of the train and its weight is dependent on the parameters of the railway lines and the length of station tracks. In the case of trains carrying coal, the longest trainsets can currently be found, among others, in China - 3.9 km [30], South Africa, the USA and Australia - more than 2 km [31, 32, 33]. In the Russian Federation the average trainset length is 806m [34], and in Poland - 600m.

4 Calculations

The aim of the article was to calculate the index describing what part of the energy contained in the coal transported on a particular route is consumed by the train carrying it.

It was assumed that steam coal will be transported by rail from Kuzbass to Poland. The entire cargo of coal would be used to produce electricity, and the train will return without load. Therefore, this indicator will take into account the work of the train on the route back and forth.

In order to accomplish this task one needs to know how much energy is consumed by trains transporting coal from the Russian Federation to Poland. Because the detailed specification of the train's route is unknown, this article is based on empirical indicators calculated on the basis of data from the actual freight trains from the paper [35].

The authors are aware that in order to calculate the amount of electricity consumed by a freight train, one should base on complicated formulas presented in many publications [i.e. 36, 37, 38, 39]. They must, among others, take into account the driving force of the locomotive (eg. Chwieduk's formula), the resistance to the movement of the train (eg. Frank's, CNTK's, Davis' formulas), the braking forces, accelerating forces (eg. Wichter's, Chwieduk's formula), and many other factors.

Input data to calculate the electricity consumption for coal transportation by rail is presented in Table 1. This is the actual data on how to select long-distance transport in Polish conditions. This data must include information on the consumption of electricity on the route, the train weight and the amount of distance traveled. It should refer to both directions of transport (destination / return).

Table 1. The data to calculations [35]

Sample of route	**Out/ Return**	**Av. weight**	**Av. elect. consump.**
	km	t	Wh/tkm
Tarnowskie G.- Gdynia	474	2297 1235	11.30 17.73
Warszawa - Poznań	290	1963 1282	10.90 13.75
Kraków - Medyka	237	2635 1433	8.92 12.42
Warszawa O. - Łazy	271	2381 1305	7.10 15.40

Table 1 shows that transport of cargo to the customer is associated with energy expenditure comprised within 7.10-11.30 Wh / tkm. While the return of the empty train is in the order of 12.42-17.73 Wh / tkm.

The calculation results of the empirical indicator of electricity consumption for coal transportation by rail is presented in Table 2. Column 2 shows the calculated energy consumption for routes in Table 1 in the Polish conditions, column 3 - calculated indicator for routes from Table 1 (Polish conditions). The next step was to recalculate the indicator for routes with a distance of 100 km (column 4). Due to the fact that the average transport distance from Kuzbass to Poland is approx. five thousand km, also the indicator for that distance in the Polish conditions was calculated (column 5).

The last line of Table 2 presents the calculation of the rate of electricity consumption for coal transportation by rail in Russian conditions. Data for calculations is reproduced from [34]. The paper [34] presents the results of studies on the consumption of electricity for freight trains weighing about 4.7 thousand ton over a test distance.

In the absence of information on energy consumption in trains without a load, the proportions of mass and energy, such as the Polish conditions, were adopted.

As shown in Table 2, the rate of electricity consumption for steam coal imports (col. 5) comprises in the range 9.22-14.45%. It has been calculated for the transport of steam coal with a calorific value of 22 MJ / kg and power plant efficiency of 38%.

Table 2. Rate of electricity consumption for hard coal transportation by rail – calculations

Route	**Route: Out + Return**	**Rate of electricity consumption for hard coal transportation, %**		
	MWh/t km	Route: Out + Return	100 km	5000 km
1	2	3	4	5
Poland				
Tarnowskie G.- Gdynia	22.7	0.92	0.19	9.71
Warszawa - Poznań	11.3	0.72	0.25	12.35
Kraków - Medyka	9.8	0.35	0.15	7.40
Warszawa O. - Łazy	10.0	0.40	0.15	7.41
Average:			0.18	9.22
The Russian Federation				
Kuzbass – Polish rail border crossing	33.8	–	0.29	14.45

In the next step of the analysis, empirical indicator of electricity consumption for the transport of coal intended to produce 1 MWh depending on the efficiency of the power plant was calculated. The calculation results are presented in Table 3. Hard coal-fired power plants in Poland have different efficiency and burn coal with different quality parameters [8]. It was therefore assumed that in the plants with different efficiency levels coal with a calorific value of 22 MJ / kg will be burned. The adopted value is the average calorific value of coal delivered to the Polish energy sector.

Table 3 shows the change in the indicator of electricity consumption for coal transportation from the Russian Federation to Poland (distance 5000 km), according to the Polish (column 3) and Russian (column 4) data. For comparison, the last column presents the calculated rate of transport of coal from Polish mines to the Polish power plants over a distance of 300 km. This is the average distance of transport of steam coal in Poland from the producer to the power plant.

Table 3 Rate of electricity consumption for rail transport to produce 1 MWh depending on the efficiency of power plants, steam coal Q=22 MJ/kg

The efficiency of hard coal-fired power plant	**The quantity of hard coal to produce 1 MWh**	**Rail transport**		
		The Russian Fed., 5000 km		Poland, 300 km
		According to Polish data	According to Russian data	According to Polish data
1	2	3	4	5
%	Mt	%		
32	0.511	10.93	17.14	0.66
34	0.481	10.29	16.13	0.62
36	0.455	9.73	15.26	0.58
38	0.431	9.22	14.45	0.55
40	0.409	8.75	13.72	0.52
42	0.390	8.34	13.08	0.50
44	0.372	7.96	12.48	0.48
46	0.356	7.61	11.94	0.46

Comparing columns 4 and 5 with each other, one can see how significant is the energy expenditure associated with the import of coal from the Russian Federation to Poland. For a typical average power plant efficiency of 36-38%, the energy consumption of coal imported from the Russian Federation will be in the area of 9.22-15.26%. However, in the case of deliveries of steam coal from Polish mines it will be within the range 0.55-0.58%. This comparison shows how much of a waste of energy we have to deal with when it comes to the imports of coal. In the transport of steam coal from Polish producers to the power plants it is almost negligible.

Taking into account the requirements of the current climate package, producing 1 MWh from hard coal will emit 1.8 tonnes of CO_2. In the light of current requirements to reduce CO_2 emissions in the European Union, this factor should also be taken into account.

5 Summary

Coal is one of the most important energy sources in the world. Its share of the global energy consumption in 2014 was 30% (4.01 billion toe). According to forecasts for 2035, the use of coal will increase by 274 million toe, and its share in the energy balance of the world will fall.

The main objective of the article was to calculate the index describing what part of the energy contained in the coal transported on a particular route is consumed by the train carrying it.

On the basis of analysis of steam coal imports from the years 2004-2014, it was assumed that coal will be imported to Poland by rail from Russia's Kuznetsk Basin.

In order to estimate energy expenditure, information about energy consumption by freight trains on the Russian market was used. On the basis of this information, the amount of energy required for the transport of steam coal in relation to the energy contained in the transported fuel and burned in the power plants with specific technical parameters, was determined.

As a result of the calculation, it was found that the rate of energy consumption for transport of imported coal will be in the range of 9.22-15.26%. In the case of deliveries of steam coal from Polish producers to the power plants the calculated rate changes within the range of 0.55-0.58%. It can be assumed that the parameter of energy consumption in the transport of steam coal within Poland is negligible because of its low value.

Large level of this ratio (a large share of the energy used for transport in relation to the energy contained in the transported fuel) negatively affects the assessment of the transport of fuels from great distances.

Acknowledgments

Publication prepared within the statutory research of The Mineral and Energy Economy Research Institute of the Polish Academy of Sciences.

References

1. BP Energy Outlook 2035 (BP, 2016, available at: http://www.bp.com/energyoutlook; 12-04-2016)
2. Annual Reports, Facts and trends (Verein der Kohleimporteure, Hamburg, Editions of years 2006-2015, available at: http://www.kohlenimporteure.de/home.html; 12-04-2016)
3. Coal Information (IEA, Paris, Editions of years 2004-2015)
4. IDA JSC, *Podstawowe informacje o rynku oraz sektorze węgla kamiennego* (IDA JSC Branch Office in Katowice within the framework of the "Programme of statistical surveys" - a statistical survey "Mining of hard coal and lignite" conducted by the Ministry of Economy, Katowice, Editions of years 2012-2016)
5. W. Blaschke, L. Gawlik and U. Lorenz, In: *Mat. XIV Konf. „Aktualia i perspektywy gospodarki surowcami mineralnymi"*, Sympozja i Konferencje 63, 15-27 (2004)
6. J. Szlązak, *Restrukturyzacja górnictwa węgla kamiennego w Polsce w latach 1990–2002* (Biblioteka SEP, Kraków, 2004)
7. U. Lorenz, *Gospodarka węglem kamiennym energetycznym* (IGSMiE PAN, Kraków, 2010)
8. Z. Grudziński, *Metody oceny konkurencyjności krajowego węgla kamiennego do produkcji energii*

elektrycznej, (Studia, Rozprawy, Monografie 180, 2012)
9. T. Olkuski, *Analiza produkcji węgla kamiennego i jego wykorzystanie w wytwarzaniu energii elektrycznej*, (Studia, Rozprawy, Monografie 174, 2012)
10. Dz. U. [Journal of Law] of 1997, No 157, Item 1027 (www.isap.sejm.gov.pl)
11. Dz. U. [Journal of Law] of 2001, No 47, Item 477, as amended (www.isap.sejm.gov.pl)
12. Dz. U. [Journal of Law] of 2001, No 123, Item 1352, as amended (www.isap.sejm.gov.pl)
13. Dz. U. [Journal of Law] of 2002, No 125, Item 1063 (www.isap.sejm.gov.pl)
14. M.P. [Official Journal of the Republic of Poland - Monitor Polski] of 1999, No 3, Item 14 (www.isap.sejm.gov.pl)
15. Dz. U. [Journal of Law] of 2001, No 156, Item 826 (www.isap.sejm.gov.pl)
16. Dz. U. [Journal of Law] of 2001, No 156, Item 1825 (www.isap.sejm.gov.pl)
17. Dz. U. [Journal of Law] of 1999, No 83, Item 933 (www.isap.sejm.gov.pl)
18. Dz. U. [Journal of Law] of 2001, No 16, Item 180 (www.isap.sejm.gov.pl)
19. Dz. U. [Journal of Law] of 2000, No 105, Item 1116 (www.isap.sejm.gov.pl)
20. Dz. U. [Journal of Law] of 2004, No 90, Item 959 (www.isap.sejm.gov.pl)
21. U. Lorenz, Przegląd Górniczy, **7–8**, 314–321 (2011)
22. T. Olkuski, Gospodarka Surowcami Mineralnymi - Mineral Resources Management **29(3)**, 115-130 (2013)
23. K. Stala-Szlugaj, *Polish import of steam coal from the East (CIS) in the year 1990-2011, (*Studia, Rozprawy, Monografie 179, 2012)
24. K. Stala-Szlugaj, Przegląd Górniczy (to be published, 2016)
25. Taylor Z. *Rozwój i regres sieci kolejowej w Polsce*, (IGiPZ PAN, Monografie 7, 2007)
26. Transport – activity results (Central Statistical Office of Poland, Warszawa, Editiond of years 2005-2015)
27. Socjalno-ekonomiczeskoje polorhzenije Rossii (in Russian) (Russian Federation – Federal State Statistics Service, Editions of years 2004-2016, available at: http://www.gks.ru/wps/wcm/connect/rosstat_main/rosstat/ru/statistics/publications/catalog/doc_1140086922125; 12-04-2016)
28. http://cargo.rzd.ru/static/public/ru?STRUCTURE_ID=5140&layer_id=3290&refererLayerId=3290&id=2178 (Available at: 12-04-2016)
29. http://railsite.ru/poluvagon.php (Available at: 12-04-2016)
30. http://www.csrgc.com.cn/g981/s2780/t248208.aspx (Available at: 12-04-2016)
31. https://www.ihha.net/sites/default/files/heavy%20haul%20operations%20in%20south%20africa%20wc%20kuys%20110616_f.pdf (Available at: 12-04-2016)
32. http://www.lwv-bf.org/Coal_Train_Fact_Check.pdf (Available at: 12-04-2016)
33. http://www.minerals.org.au/resources/coal/exports/facts_and_figures (Available at: 12-04-2016)
34. W.P. Mogila, *Massa, dlina i skorost dwirhzenija gruzowych pojezdow (in Russian)* (Publ. DWGUPS, Chabarowsk, 2nd ed., revised and enlarged, 2013)
35. W. Wasilewicz, Transport Szynowy, **9**, (1974)
36. J. Podoski, J. Kacprzak and J. Mysłek, *Zasady trakcji elektrycznej (in Polish)* (WKŁ, Warszawa, 1980)
37. A. Rojek, *Tabor i trakcja kolejowa (in Polish)* (PKP, Warszawa, 2010)
38. A. Szeląg and L. Mierzejewski, *Ground transportation systems* (The Wiley Encyclopedia of Electrical and Electronic Engineering, New York, Supplement, 2000)
39. R. Burak-Romanowski, and K. Wozniak, Czasopismo Techniczne – Technical Transactions, **108**, 13, 13-29 (2011)

Technical and Economic Analysis of a Hybrid Generation System of Wind Turbines, Photovoltaic Modules and a Fuel Cell

Radosław Szczerbowski[1,a] and Bartosz Ceran[1]

[1]*Poznan University of Technology, Poznań, Poland*

Abstract. The paper presents the results of the analysis of the economic and manufacturing system consisting of wind turbines, photovoltaic modules, polymer membrane fuel cell and the electrolyzer. The system supplies the customer profile at the assumed wind and solar conditions. Energy analysis was conducted on the basis of the balance equations produced and received electric power. To assess the economic efficiency of investments adopted the following economic indicators: NPV, IRR, MIRR, MNPV, DPP. The authors describe the limits of investment costs intended for the construction, which use hybrid power generation system (HPGS) is viable.

1 Introduction

The rapid development of renewable energy sources and distributed generation makes the electrification system will be changed to the form in which the road transit of electricity will be the shortest possible. A very promising group of distributed energy sources are sources that use renewable energy resources, especially the Sun and the wind. The primary disadvantage of these sources is a strong dependence of the amount of energy produced from the current weather conditions. As a result, forecasting the production of electricity is very difficult.

Way of efficient use of renewable and distributed energy resources the original hybrid power generation systems are, or combinations of connections renewable and conventional sources (small gas turbines, generators, powered by engines, fuel cells, etc.) and/or energy storage (flywheel, batteries, electrolyzers with hydrogen tanks, supercapacitors, etc.). Of these energy storage technology shows potential PEMFC type fuel cell in conjunction with electrolyzers with hydrogen tanks. This system is characterized by a high density of energy stored volume larger than a 500 Wh/kg. The integration of the different technologies in electricity generation in such systems allows the fuller satisfaction of energy needs. In addition, the cooperation of solar power plants and wind farms with storage systems definitely improve their work [1,2]

Storage systems are able to store surplus generated by power plants, energy and power energy stored receivers when the plant is not able to cover the needs of the recipient. This allows to significantly improvethe availability of photovoltaic power plants and wind farms. In addition, the construction of hybrid manufacturing systems that use renewable and distributed energy resources, will help reduce the cost of electricity. This is the reason that these sources are located close to the energy consumers.

Hybrid power generation systems are also of great importance for the protection of the environment. Used in these systems of renewable generation does not emit pollutants.

The concept of a hybrid system is defined, inter alia, in the regulation of the Minister of economy of 18 October 2012 in the document filed, that the hybrid is a unit of generation which produces electricity, or electricity and heat. In the process of generating electricity or heat energy carriers are produced separately in renewable energy sources. There is also the possibility of using the auxiliary fuel to produce electricity or heat.

It seems reasonable to hybrid power generation systems used in areas distant from the traditional sources of energy, or in the case when there are difficulties with fuel to conventional generating units. Hybrid power generation systems most often are made of two or more sources. Build them that way so that you can offset the advantages and disadvantages of individual sources. This fact makes these systems belong to the very expensive due to the need for the use of renewable energy sources and energy storage systems [3].

2 Hybrid generation system

The excess energy produced from renewable sources is used for electrolysis process (decomposition of water into hydrogen and oxygen). Hydrogen is compressed and stored, and then used in a fuel cell to produce electricity, at a time when it is needed by customer. The use of electrolysis process of energy produced

[a] Corresponding author: radoslaw.szczerbowski@put.poznan.pl

by renewable means that operation of the hybrid system is emission-free. The purchase of an additional backup purposes, hydrogen may be charged a small emissions associated with its production, depending on the technology used [4,5].

Hybrid power generation system flowsheet analysis is shown in Figure 1.

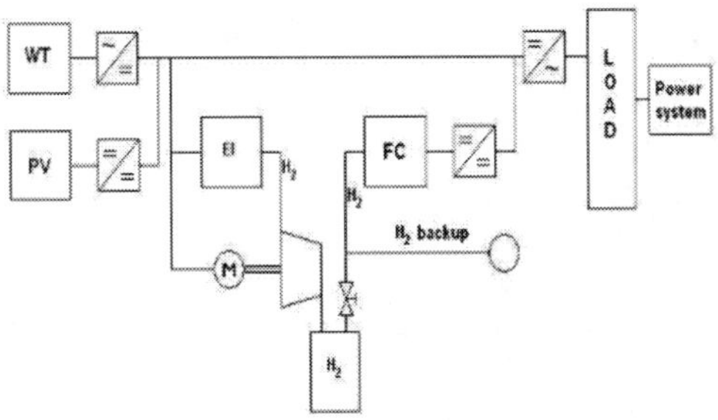

Figure 1. Flowchart-hybrid power generation system (own study)

To analyze the work of the hybrid power generation system consisting of wind turbines, photovoltaic modules, polymer membrane fuel cell and the electrolyzer assumes the customer profile of municipal (Fig. 2) with a maximum power 56 kW during the winter season, 48 kW in the spring-autumn season and 40 kW during the summer and the annual demand for electricity in the amount of 286.9 MWh.

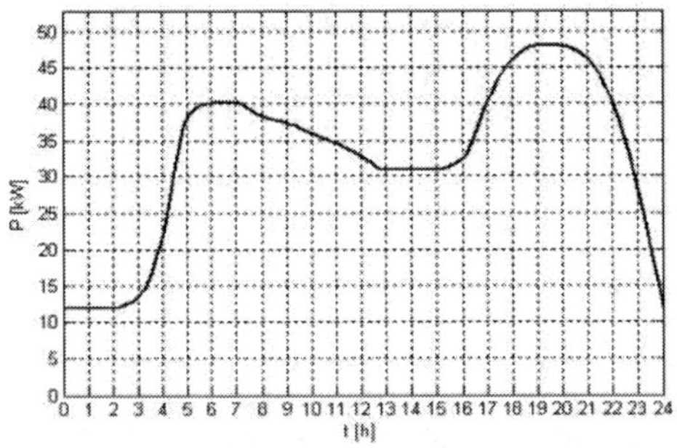

Figure 2. Energy demand profile recipient of municipal - winter season (own study)

In figures 3 and 4 shows the annual timetables sunlight and wind speed technical-economic analysis be adopted.

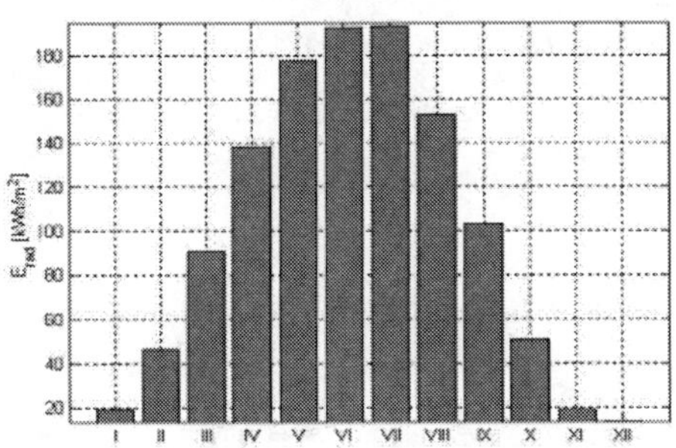

Figure 3. The annual distribution of sunlight (own study)

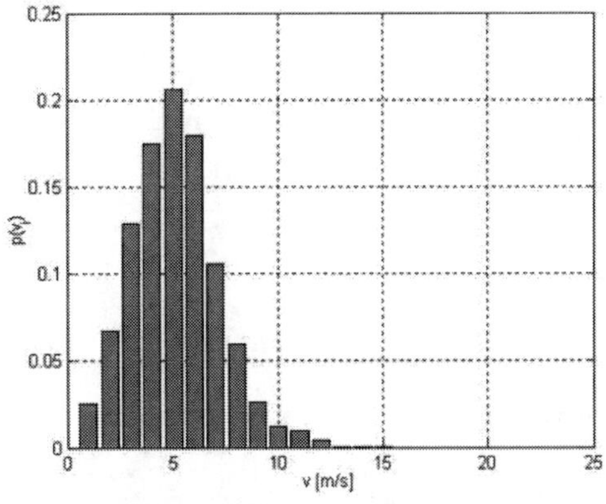

Figure 4. The annual distribution of wind speeds (own study)

2.1 Selection of the composition of the generating units of hybrid power generation system

With a view to the selection of the composition of the generating units to meet the needs of the customer, the following equation can be used [6]:

$$\Delta P = P_{av\_gen} - P_{av\_load} \rightarrow \min \tag{1}$$

$$P_{av\_gen} = n_{cf\_PV} \cdot P_{PV} + n_{cf\_WT} \cdot P_{WT} \tag{2}$$

$$n_{cf\_PV} = \frac{P_{av\_PV}}{P_{PV}} \tag{3}$$

$$n_{cf\_WT} = \frac{P_{av\_WT}}{P_{WT}} \tag{4}$$

where: P_{av_gen} – mean value of power generated by renewable sources, P_{av_load} – average input power by the recipient, n_{cf} – a capacity factor of each renewable energy sources, P_{PV} – rated power of photovoltaic installation, P_{WT} – rated power of wind turbine.

The value under the average for the period T is designated with the dependencies:

$$P_{av\_PV} = \frac{A_{PV}}{T_a} \tag{5}$$

$$A_{PV} = W_k \cdot W_w \cdot \int_0^T P_{av\_PV} dt \tag{6}$$

where: A_{PV} – electricity produced by photovoltaic system at time T, W_k – a correction factor that takes into account the angle of installation referral to the South and the angle of deviation of the modules from the level, W_w - the coefficient of performance.

For the determination of the amount of electricity produced by photovoltaic modules uses the characteristics of the P_{PV} = f (E) developed on the basis of external characteristics family U = f(I) (Fig. 5). Photovoltaic modules system is equipped with an MPPT maximum power point tracking ($P_{PV} = P_{MPP}$).

DOI: 10.1051/e3sconf/20161000090

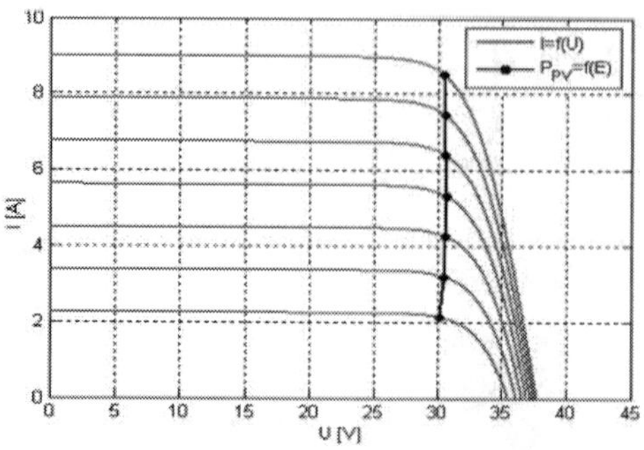

Figure 5. External characteristics of PV module family (own study)

In order to determine the average power produced by wind turbines was operational characteristics P = f (v) and the distribution of wind speeds (eq. 7) presented in the form of a continuous function f(v) using the Weibull distribution [7].

$$f(v) = \frac{k}{\lambda}\left(\frac{v}{\lambda}\right)^{k-1} e^{-(v/\lambda)^k} \tag{7}$$

where: v - wind speed [m/s], k - shape parameter (k > 0), λ - the scale parameter (λ > 1),

$$P_{av\_WT} = \int_0^{\infty} P(v) f(v) dv \tag{8}$$

A graphical interpretation of the determination of the average power is shown in Figure 6.

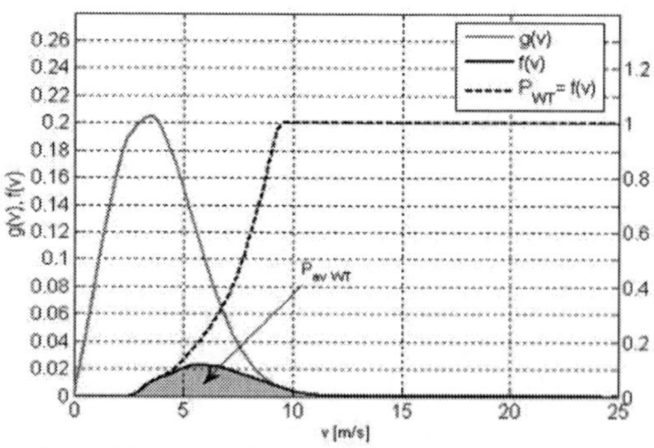

Figure 6. Determination of annual average wind power capacity (own study)

For the customer profile to be analyzed on the basis of equations (1-8) 3 proposed wind turbines with a capacity of 50 kW, photovoltaic installation consisting of 330 modules with a capacity of 250 W each.

2.2 The energy performance of energy storage

Energy storage has been selected based on the following assumptions [6]:

- fuel cell is able to meet the instantaneous maximum demand:

$$P_{FC} = P_{\max\_load} \tag{9}$$

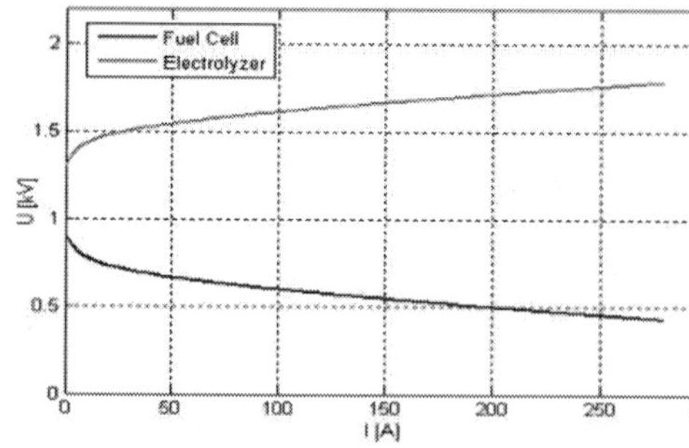

Figure 7. The characteristics of U = f (I) fuel cells and the cell (own study)

- electrolyzer is able to accept temporary power generated by photovoltaic system and wind sources:

$$P_{El} = k \cdot (P_{\max\_WT} + P_{\max\_PV} - P_{\min\_load}) \tag{10}$$

where: P_{EL} – rated power of electrolyzer, P_{max_WT} – the maximum instantaneous wind turbines, P_{max_PV} – instantaneous maximum power photovoltaic installation, P_{min_load} –minimum instantaneous customer demand, k – coefficient simultaneity of maximum power generation by photovoltaic system and wind sources.

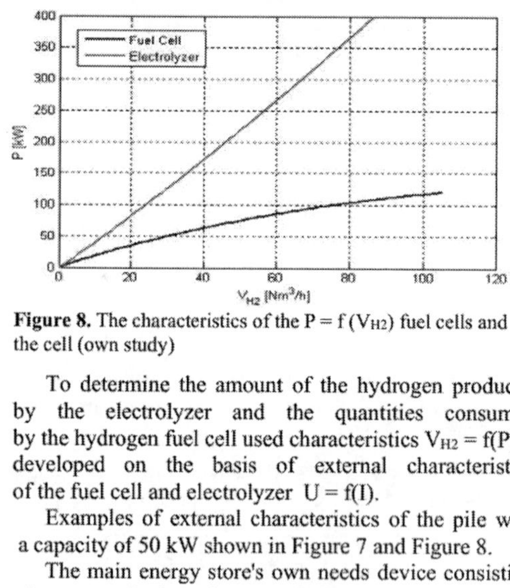

Figure 8. The characteristics of the P = f (V_{H2}) fuel cells and the cell (own study)

To determine the amount of the hydrogen produced by the electrolyzer and the quantities consumed by the hydrogen fuel cell used characteristics V_{H2} = f(P) developed on the basis of external characteristics of the fuel cell and electrolyzer U = f(I).

Examples of external characteristics of the pile with a capacity of 50 kW shown in Figure 7 and Figure 8.

The main energy store's own needs device consisting of a set of electrolyzer - fuel cell is compressor. The power needed for its supply is described by the formula:

$$P_{comp} = m_{H_2} \frac{\kappa}{\kappa - 1} p_0 v_0 [(\frac{p_1}{p_0})^{\frac{\kappa-1}{\kappa}} - 1] \frac{1}{\eta_i} \tag{11}$$

where: m_{H2} [kg/s] – the mass flow of hydrogen, κ – the exponent of the emagram (κ = 1.41 for hydrogen), p_0 [MPa] – hydrogen pressure before compression, v_0 [m³/kg] – an appropriate volume of hydrogen,

p_l [MPa] – hydrogen pressure after compression, η_i – compressor internal efficiency

The relationship required work the compressor pressure function is shown in Figure 9.

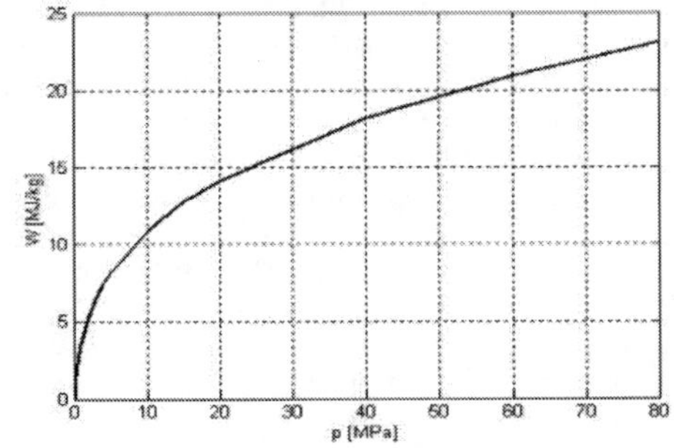

Figure 9. The work of the compressor as a function of the compression pressure (own study)

In the present case, the proposed fuel cell stack with a nominal power of 50 kW, and the electrolyzer with a capacity of 200 kW.

3 Energy analysis of hybrid power generation system

Energy analysis was conducted on the basis of the balance equations produced and received electricity power from hybrid power generation system. The balance of power in the system, depends on the ratio of power generated by RES sources to the demand of the recipient. In the case where the demand is less than the power generated balance sheet equation takes the form [1,6]:

$$P_{load} = P_{PV} + P_{WT} - P_{El} - P_{comp} \quad (12)$$

If the demand exceeds the power produced by sources the receiver power describes the formula:

$$P_{load} = P_{PV} + P_{WT} + P_{FC} \quad (13)$$

In Figure 9 illustrates parsed the energy profile of the recipient and the production of electricity by RES sources on the selected date.

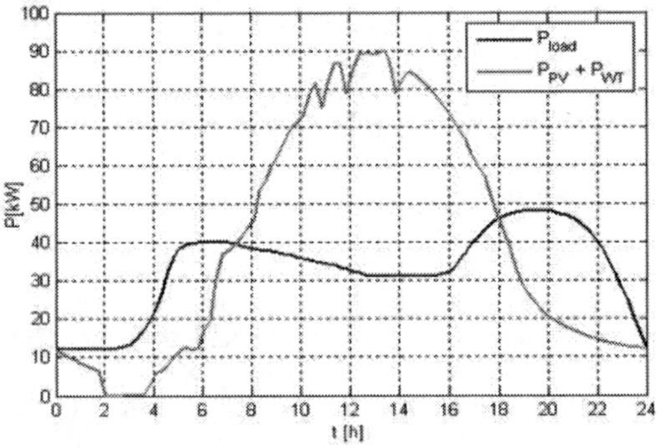

Figure 10. The energy profile of the recipient and the power produced by wind and photovoltaic on the selected day (own study)

In a substantial part of the day (in the present case), the weather conditions do not allow the manufacturing of such quantity of energy that equates demand of the customer. In h. 0-7 and 18-24 missing energy provides a stack of fuel cells. In h. 7-18 working electrolyzer energy in the form of surplus storage of hydrogen (Fig. 10).

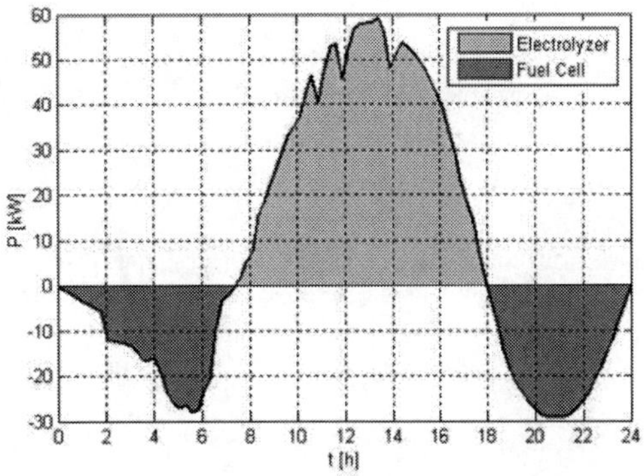

Figure 11. Fuel cells work period and the cell (own study)

Hydrogen is supplied in the first place from the electrolyzer. In the case of the total consumption of hydrogen in the tank, the fuel cell is powered by the backup. When carrying out calculations for the given profiles of sunlight and wind conditions on an annual basis the following results were obtained:

- the volume of hydrogen produced in the process of electrolysis 50 481 Nm^3,
- annual consumption of hydrogen fuel cell stack 124 560 Nm^3 hydrogen,
- with the "backup" 74 079 Nm^3.

The results obtained on the basis of an analysis of energy efficiency form the basis to conduct the economic analysis with that of the hybrid system.

4 Economic analysis of hybrid power generation system

An assessment of the economic efficiency of investment in the energy sector is based on the following economic indicators:

- Net Present Value – NPV:

$$NPV = \sum_{k=1}^{N} PV_{d,k} \times CF_k - CF_0 \quad (14)$$

where: CF_k – positive net cash flows, CF_0 – investment costs, $PV_{d,k}$ – the k-factor of discounting at the discount rate (d) k-this period

- Modified Net Present Value – MNPV:

$$MNPV = PV_{d,N} \sum_{k=1}^{N} FV_{r,N-k} \times CF_k - CF_0 \quad (15)$$

where: $FV_{r,N-k}$ – coefficient of capitalizing on the reinvestment rate (r) and N–k during this period, N - the total duration of the investment project.

- Profitability Index – PI:

$$PI = \frac{\sum_{k=1}^{N} PV_{d,k} \times CF_k}{CF_0} \quad (16)$$

- Internal Rate of Return – IRR:

$$\sum_{k=1}^{N} PV_{IRR,k} \times CF_k = CF_0 \quad (17)$$

where: $PV_{IRR,k}$ – discount factor at a discount rate equal to the internal rate of return (IRR) and the k-th period,

- Modified Internal Rate of Return – MIRR:

$$MIRR = \sqrt[N]{\frac{\sum_{k=1}^{N} FV_{r,N-k} \times CF_k}{CF_0}} - 1 \quad (18)$$

where: $FV_{r,\ N-k}$ – coefficient of capitalizing on the reinvestment rate (r) and N-k during this period, N – the total duration of the investment project,

- Discount Payback Period – DPP:

$$\sum_{k=1}^{n=DPP} PV_{d,k} \times CF_k = CF_0 \quad (19)$$

For the purpose of the economic analysis with that of the adopted investment listed in Table 1. Variant I contains the current value of the investment of the hybrid power generation system devices. Variant II contains the expected investment of equipment per year by 2050 [8- 19].

Table 1. Investment costs

Unit	Variant I [thousand PLN]	Variant II [thousand PLN]
Wind turbine 50 kW	**300**	**250**
PV installation 82,5 kW	**500**	**360**
Fuel cell 50 kW	**780**	**316**
Energy storage (electrolyzer 130 kW, compressor, the hydrogen tank)	**516**	**117**
The cost of hydrogen produced in the process of fuel reforming	**63**	**38**
The cost of hydrogen produced in the process of coal gasification	**100**	**62**
The cost of hydrogen produced in the process of biomass gasification	**140**	**80**

Life of fuel cell, and the electrolyzer is 10 years, remaining service life assumes for 20 years. As the profit is assumed the avoided costs to purchase electricity from electricity system after the unit price 0.56 PLN/kWh a year zero and 0.85 PLN/kWh in the twentieth year of operation for the variant I, for variant II 1 PLN/kWh a year zero and 1.3 PLN/kWh in the twentieth year. Economic calculation assumes that additional hydrogen (backup) is produced in the steam reforming process. The analysis does not include the financial gains associated with receiving certificates for the production green energy. According to the authors of power engineering is a branch of industry which should be profitable without support systems. The value of the rate of return was adopted at 8%, while the value of the reinvestment rate at 10%.The results of the economic analysis are in table 2.

Table 2. The results of the economic analysis

% of investments costs	NPV	MNPV	PI	IRR	MIRR	DPP
Variant I						
100	- 2398	-	-	-	-	-
50	- 580	-	-	-	-	-
33	25	299	1.02	8.3	9.3	19
25	328	602	1.39	12.7	10.9	12.5
Variant II						
-	797	1383	1.5	14.2	11.4	10

Performed economic analysis allows for the formulation of the following conclusions:

- power supply energy customer with established profiles: customer energy, wind and sunlight by the hybrid power generation system is an uneconomic (NPV < 0). This is due to the high investment costs and low capacity values of the coefficients factor,
- in the event of a reduction of 50% of the investment project is still unprofitable (NPV <0),
- the construction and exploitation of the hybrid power generation system generates losses by reducing the total investment by 67%. In this case, the system virtually does not generate profits, because discounted payback time DPP is 19 years,
- construction and maintenance hybrid power generation system would be profitable if investment costs were reduced by 75%, in this case the discounted payback time DPP is 12.5 years,
- in the case of variant II (reduction of investment in relation to variant I about 45% of the unit and increase electricity prices to value of 1 PLN/kWh) construction and operation of the hybrid power generation system is a viable project and time of return, the DPP is 10 years.

At today's electricity prices, investment costs related to the construction of the hybrid power generation system and without certificates for green energy, there is no economic justification. Analyzed hybrid system may be used in situations where the investment is a matter of priority, for example: military applications or pilot installations for research purposes.

References

1. B. Ceran, K. Sroka, Acta Energetica, **2/23,** 36–42 (2015)
2. R. Szczerbowski, B. Ceran, Polityka Energetyczna, **16**, 193 – 205 (2013)
3. J. Paska, M. Kłos, Prz Elektrotech, **8**, 93, (2010)
4. K. Balachander, S. Kuppusamy, P. Vijayakumar, *International Conference on Emerging Trends in Science, Engineering and Technology*, 462-466, (2012)
5. B. Ceran, P.A. Bernstein, Poznan University of Technology Academic Journals, **79**, 157-163 (2014)
6. C. Wang, M.H. Nehrir, *IEEE Transactions*, 957-964, (2008)
7. Z. Lubośny, *Farmy wiatrowe w systemie elektroenergetycznym* (WNT, 2009)
8. P. Bronski, J. Creyts, M. Crowdis, S. Doig, J. Glassmire, L. Guccione, P. Lilienthal, J. Mandel, B. Rader, D. Seif, H. Tocco, H. Touati, *The Economics of Load Defection. How Grid-Connected Solar-Plusbattery Systems Will Compete With Traditional Electric Service, Why It Matters, And Possible Paths Forward* (Rocky Mountain Institute, 2015)
9. D. R. Peterson, P. Zelenay, *ElectroHyPEM Workshop* (2014)
10. The Fuel Cell and Hydrogen, Annual Review, 4th Energy Wave (2015)
11. T. Riis, E. F. Hagen, P. J. S. Vie, Ø. Ulleberg, (IEA, 2006)
12. Technology Roadmap Hydrogen and Fuel Cells, (IEA, 2015)
13. P. E. Dodds, W. McDowall, UKSHEC Working Paper, **6** (2012)
14. T. Smolinka, M. Thomassen, A. Oyarce, F. Marchal, (JU FCH project in the Seventh Framework Programme, 2016)
15. Renewable Power Generation Costs in 2014, (IEA, 2015)
16. E-storage: Shifting from cost to value wind and solar applications, (WEC, 2016)
17. O. H. Mohammed, Y. Amirat, M. Benbouzid, A. Elbast, IEEE ICGE, 119-123 (2014) <hal-01023490>
18. A. A. Elbaset, Journal of Electrical Systems, **7-2,** 270-286 (2011)
19. Technology Roadmap, Solar Photovoltaic Energy, (IEA, 2014)

Polycyclic aromatic hydrocarbons and their nitrated derivatives associated with PM10 from Kraków city during heating season

Katarzyna Styszko[1,a], Katarzyna Szramowiat[1], Magdalena Kistler[2], Anne Kasper Giebl[2], Sylwia Socha[1], Egon Erwin Rosenberg[2] and Janusz Gołaś[1]

[1]*AGH University of Science and Technology, Faculty of Energy and Fuels, Al. Mickiewicza 30, 30-059 Kraków, Poland*
[2]*Institute of Chemical Technologies and Analytics, Vienna University of Technology, Getreidemarkt 9/164, 1060 Vienna, Austria*

Abstract. Polycyclic Aromatic Hydrocarbons (PAHs), their nitro-derivatives (NPAHs) and hundreds of other organic compounds are present in ambient air in gas and particulate form. PAHs and NPAHs originate from diesel and gasoline exhaust emission and other combustion sources. NPAHs are also formed through the nitration of parent PAHs in the atmosphere. Concentrations of PAHs and NPAHs in the particulate matter fraction PM10 collected in the centre of Kraków (27.01.2014 – 17.02.2014) were investigated. The thirteen PAHs and four NPAHs: fluorene, phenanthrene, anthracene, fluoranthene, pyrene, benzo[a]anthracene, benzo[a]pyrene, chrysene, benzo[b]fluoranthene, benzo[k]fluoranthene, indeno[1,2,3-c,d]pyrene, benzo[g,h,i]perylene, dibenz[a,h]anthracene, 2-nitrofluorene, 9-nitroanthracene, 3-nitrofluoranthene and 1-nitropyrene were extracted from particulate matter and analysed applying the GC/MS technique. Depending on the compounds the relative recoveries ranged from 72 to 94%. The concentrations of PM10 in the study period ranged between 23.5 and 153.8 $\mu g \cdot m^{-3}$. The average concentrations of PAHs and NPAHs ranged from 26.6 to 276.4 $ng \cdot m^{-3}$ and from 0.6 to 9.1 $ng \cdot m^{-3}$, respectively. The highest concentrations were observed for benzo[a]pyrene, benzo[a]anthracene, pyrene and fluoranthene. The average concentration of benzo[a]pyrene (BaP), which is a marker for the particle-bound atmospheric PAHs, was 9.5 $ng \cdot m^{-3}$. The concentrations of 3-nitrofluoranthene and 1-nitropyrene were below the quantification limits of the method (< MQL).

1 Introduction

Polycyclic aromatic hydrocarbons (PAHs) and their nitrated derivatives (NPAHs) are common organic compounds occurring in the atmosphere both in vapour phase (low molecular weight compounds) and in particulate-bound form (high molecular weight) [1, 2]. This group of compounds consists of numerous organic compounds with fused benzene rings. Due to their high toxicity, mutagenic and carcinogenic properties PAHs and NPAHs are of great concern to scientists [3-5]. These chemicals are generated as by-products of the incomplete combustion of fuels like coal, wood, gas and oil. Beside that the mobile sources and open burning affect the increase of PAHs concentration in ambient air [6, 7]. At urban and industrial areas polycyclic aromatic hydrocarbons are almost entirely generated within the anthropogenic activities [8]. NPAHs are derived during the two-step reaction including: (a) the addition of an OH or NO_2 radical at position 1 of the parent PAH, followed by (b) a further NO_2 attack (onto position 2 and loss of water or nitric acid, leading to nitration in the 2-position [9]. This reaction can only occur in the gas phase and is rapidly followed by adsorption of NPAH products onto suspended particles [10].

The determination of environmental concentrations of polycyclic aromatic hydrocarbons and their nitro-substitutes seems to be significant especially in areas where increasing atmospheric pollution is reported. Kraków agglomeration, due to the high traffic intensity and the presence of still developing industrial sites, is highly affected by emission of PAHs and NPAHs. The average concentration of benzo[a]pyrene between 2008 and 2013 in Malopolska Voivodship was 8.4 $ng \cdot m^{-3}$ [11], whereas the limit value of the BaP was established on the level of 1 $ng \cdot m^{-3}$ [12]. This situation encouraged the authors of this paper to investigate the amounts of PAHs and their nitrated derivatives in PM10 samples collected during the heating season in Kraków.

2 Materials and methods

2.1 Materials

The molecular formulae, physical and chemical properties of the compounds are summarized in Table 1. The parent polycyclic aromatic hydrocarbons (purity in brackets): acenaphtene (99.9 %), acenaphtylene (99.7 %), anthracene (99.4 %), benzo[a]anthracene (98.9 %), benzo[a]pyrene (99.2 %), benzo[b]fluoranthene (99.9 %), benzo[ghi]perylene (99.6 %), benzo[k]fluoranthene (99.5 %), chrysene (99.9 %), dibenzo[a,h]anthracene (99.9 %),

[a] Corresponding author: styszko@agh.edu.pl

phenanthrene (99.1 %), fluoranthene (99.7 %), fluorene (99.6 %), indeno[1,2,3-cd]pyrene (99.9 %), naphthalene (99.4 %), pyrene (96.6 %) used in this study were supplied by Sigma-Aldrich. 1-bromopentadecane (BrPD, >97%) and the deuterium-labelled benzo[a]pyrene-d12 (d12-BaP, 98%) used as an internal and a surrogate standard, respectively, were acquired from Sigma-Aldrich. Nitro-substituted polycyclic aromatic hydrocarbons (NPAHs), 2-nitrofluorene (98%), 9-nitroanthracene (97%), 3-nitrofluoranthene (90%) and 1-nitropyrene (99%) were supplied by Sigma-Aldrich. Cyclohexane (99.8%) was purchased from Merck.

Table 1. Characteristics of analytes.

Compound	**Structure**	**Molar mass** [g·mol^{-1}]	**Boiling point** [°C]	**Solubility in water** [µg·L^{-1}]
Parent polycyclic aromatic hydrocarbons, PAHs				
naphthalene		128	218	31 000.0
acenaphthene	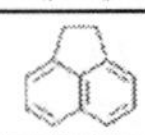	154	279	3 900.0
acenaphthylene	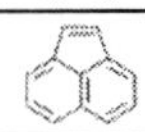	152	265	16 100.0
fluorene	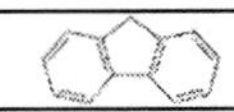	166	295	1 890.0
anthracene	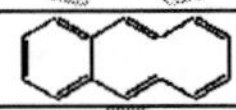	178	340	43.4
phenanthrene		178	340	1 150.0
fluoranthene	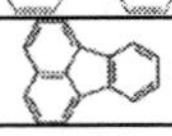	202	384	260.0
pyrene		202	404	135.0
chrysene	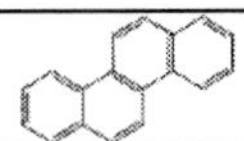	228	448	2.0
benzo[a]anthracene		228	438	-
benzo[a]pyrene	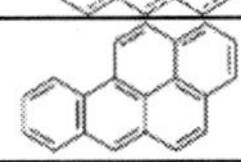	228	438	9.4
benzo[b]fluoranthene	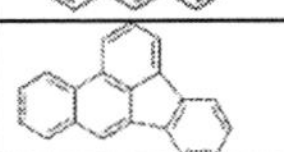	252	481	1.5
benzo[k]fluoranthene	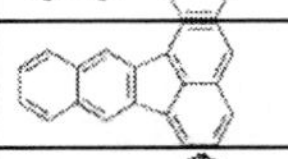	252	480	0.006
dibenzo[a,h]anthracene	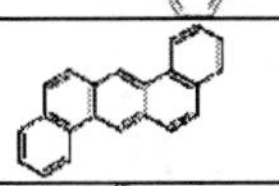	278	524	02.5
benzo[ghi]perylene	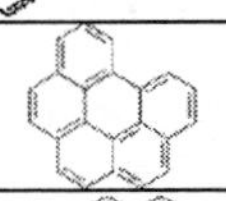	276	550	0.3
indeno[1,2,3-cd]pyrene	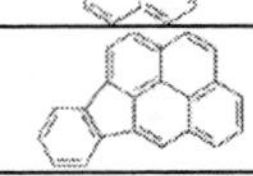	276	536	0.6
Nitro-substituted polycyclic aromatic hydrocarbons, NPAHs				
2-nitrofluorene		211	374	less than 1 000.0
9-nitroanthracene	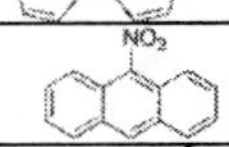	223	275	less than 100.0
3-nitrofluoranthene	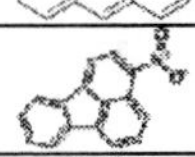	247	445	Insoluble
1-nitropyrene	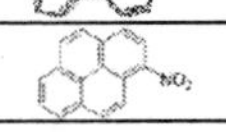	247	472	11.8

2.2 Sampling

The sampling site presented in Figure 1 was located in the centre of Kraków (50°04'01.91" N; 19°54'45.60" E), the second largest city of Poland (near 1 million inhabitants) which faces serious problems of air pollution. Kraków, located in Southern Poland, is an agglomeration with a high intensity of transport, a power plant, a steel metallurgic factory and some small and medium-size industrial factories. During the heating season the, so called, low-stack emission (mainly domestic heating units (DHUs) using hard coal) significantly contributed to the emission sources balance. Sampling site is a built-up area and represents a typical urban location characterized by high traffic intensity. The measuring location is settled near the green areas as well.

The PM10 fractions were collected during the heating season from January 27th to February 17th, 2014 on quartz fiber filters (Pallflex, 47 mm) using low volume samplers. Air flow was set up on 2.3 $m^3 \cdot h^{-1}$. In total, the batch of 14 samples of PM10 fraction was collected. All quartz fiber filters had been preheated for 6 h at 550 ±8 °C and then maintained at temperature 20±1 °C and relative humidity 50 ±5 % for at least 24 h prior to weighting and sampling. After sampling, all filters were conditioned for 48h, weighted in microbalance (OHAUS Discovery DV215CD) up to accuracy of 0.01 mg and then stored in a freezer at -20 °C until analysis. The masses of the filters before and after sampling were obtained as the average of three measurements. PM10 mass was computed as a difference of mean masses of filters after and before sampling, respectively.

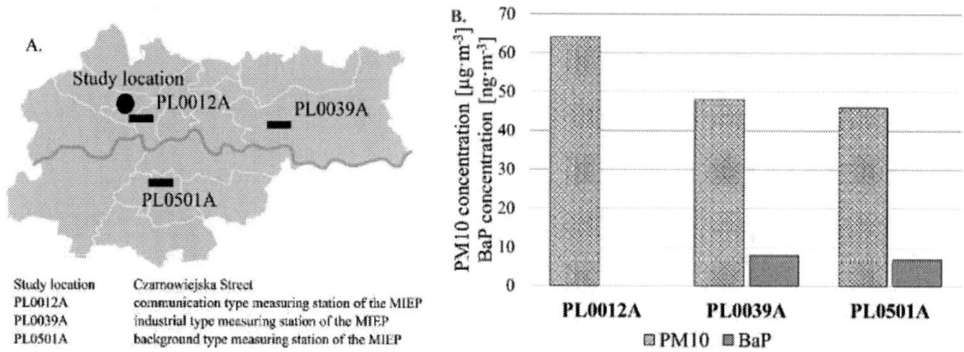

Figure 1. A. Map of the Kraków agglomeration with marked study location and monitoring station of the Malopolska Inspectorate of Environmental Protection; B. PM10 and BaP annual concentrations measured at the station of the Malopolska Inspectorate of Environmental Protection in 2014.

Meteorological data during sampling period were collected from the online platform (http://meteo.ftj.agh.edu.pl/) where the results of the meteorological measurements are directly transferred from the Vaisala WXT520 automatic meteorological station. The station was placed on the roof of the building where the samples were collected. The details on the meteorological conditions during the study period from Jan 27th to Feb 17th, 2014 are summarized in Table 2.

Table 2. Meteorological conditions (diurnal mean values) during the study period in Kraków.

Date	Wind velocity [$m \cdot s^{-1}$]	Temperature [°C]	Relative humidity [%]	Precipitation [mm]	PM10 [$\mu g \cdot m^{-3}$]
27.01.2014	2.0	-8.1	73	0.01	31.7
28.01.2014	2.9	-7.7	72	0	25.9
29.01.2014	3.1	-6.8	71	0	23.5
30.01.2014	3.0	-7.4	69	0	33.0
03.02.2014	1.0	-0.1	66	0	136.6
04.02.2014	0.9	-0.9	60	0	119.2
05.02.2014	1.1	2.7	52	0	149.6
06.02.2014	1.2	2.3	70	0	52.9
07.02.2014	1.0	3.9	65	0	47.4
10.02.2014	2.0	4.8	65	0.83	113.0
11.02.2014	1.3	6.1	69	1.27	35.6
12.02.2014	1.5	3.2	79	3.16	49.2
13.02.2014	1.0	1.2	82	0	79.9
14.02.2014	1.7	3.2	71	0.18	51.8
17.02.2014	1.4	5.8	60	0	97.2
Mean	**1.7**	**-0.2**	**68**	**0.34**	**69.8**
Ranges	**0.9 – 3.1**	**(-8.1 – 6.1)**	**(52 – 81)**	**(0.00 – 3.16)**	**(23.5 – 149.6)**

[a] Corresponding author: styszko@agh.edu.pl

2.3 Chemical analysis of carbon species

The thermo-optical method (Sunset Lab OC/EC Aerosol Analyser, Sunset Lab. Inc.) was applied for the determination of the organic (OC) and elemental (EC) carbon concentrations in PM samples using the *Eusaar2* temperature protocol [13]. Inorganic ions concentration was determined by suppressed isocratic ion chromatography with conductometric detection (Thermo Scientific). For the purpose of the analysis of ion concentrations and the amount of carbonaceous components in PM10, the circular aliquots (10 mm Ø) were cut out from the filters using a steel puncher and dissolved in high purity water (Milli-Q) or methan sulfonic acid (MSA). The details of the analytical procedures are presented in the previous work [14].

2.4 Analysis of PAHs and NPAHs

The concentration of PAHs in particulate matter was determined using gas chromatography coupled with mass spectrometry (Shimadzu QP2010 Ultra GC/MS system equipped with an AOC-5000 autosampler (both Shimadzu, Kyoto)). Concentrations were determined for a total number of 16 PAHs and 4 NPAHs (Table 1).

Table 3. Chromatographic and mass spectrometric characterization of analytes: RT – retention time, mass of characteristic ions, calibration range, correlation coefficients (R^2).

Compound	Retention time [min]	Ions m/z	Range of concentration [$\mu g \cdot mL^{-1}$]	R^2	MQL [$ng \cdot m^{-3}$]
Fluorene	11,870	166, 139	1-7	0.9983	0.18
Phenanthrene	13,057	178, 152	0.5-3.5	0.9988	0.07
Anthracene	13,112	178, 152		0.9989	0.07
Fluoranthene	14,532	202, 101	1-7	0.9984	0.11
Pyrene	14,820	202, 101	0.5-3.5	0.999	0.05
Benzo[a]anthracene	16,758	228, 114		0.9992	0.04
Chrysene	16,848	228, 114		0.9991	0.05
Benzo[b]fluoranthene	19,687	252, 126	1-7	0.9989	0.11
Benzo[k]fluoranthene	19,748	252, 126	0.5-3.5	0.9993	0.04
Benzo[a]pyrene	20,582	252, 126		0.9993	0.04
Indeno[1,2,3-cd]pyrene	24,087	276, 138		0.9988	0.05
Dibenzo[a,h]anthracene	24,188	278, 139	1-7	0.9991	0.09
Benzo[g,h,i]perylene	25,108	276, 138		0.9994	0.07
2-nitrofluorene	14,732	165, 211	1-10	0.9947	0.09
9-nitroanthracene	14,912	223, 193		0.9942	0.11
3-nitrofluoranthene	17,510	247, 217		0.9904	0.11
1-nitropyrene	18,057	247, 201		0.9917	0.09
1-bromopentadecan	13,248	43, 57	1-10	0.9947	-
Benzo[a]pyrene-d_{12}	20,520	264, 132		0.9942	-

Two circular aliquots (10 mm Ø) per filter were punched. Aliquots of two up to four filters were compiled to obtain a mass of sample above 1 mg. The PAHs and NPAHs concentrations were computed according to the PM mass ratio of each sample. The filter aliquots were placed in a glass tube (10 mL) and spiked with 50 μL (10 $ng \cdot \mu L^{-1}$) solution of d12-benzo[a]pyrene (d12-BaP). After 20 min, the analytes were ultrasonically extracted from the sample twice with 5 mL of cyclohexane for 20 minutes. The combined extracts were spiked with 50 μL (10 $ng \cdot \mu L^{-1}$) solution of bromopentadecane (BrPD). The volume of the extract was reduced to 250 μL using a gentle stream of nitrogen in a multiport condenser at 35 °C for about 35 min. The concentrate was successively transferred into 2 mL chromatographic vial equipped with 300 μL insert for analysis by GC/MS.

The PAHs (200 $\mu g \cdot mL^{-1}$) and NPAHs (50 $\mu g \cdot mL^{-1}$) stock solutions were kept in a fridge at 4 °C. The stock solutions served for preparation of calibrations standards with the concentration range from 0.5 to 10 $\mu g \cdot mL^{-1}$ (see Table 3). Standard solutions were diluted in cyclohexane with an addition of 200 $ng \cdot mL^{-1}$ of the internal standard.

The chromatographic separation of target analytes was performed with a DB5MS capillary column (30 m x 0.25 mm i.d. x 0.25 μm film thickness) with 1.2 $mL \cdot min^{-1}$ helium (quality 99.999+%) as a carrier gas. The oven temperature program was as follows: isothermal hold at 70 °C for 4 min, temperature ramp of 20 $°C \cdot min^{-1}$ to 280 °C and then 20 $°C \cdot min^{-1}$ to 300 °C, isothermal hold at 300 °C for 5 min. Injection was performed with a split/splitless injector at a temperature of 320 °C. Splitless time was set to 1 min and injection volume to 2 μL. The GC/MS interface temperature and ion source were kept at 250 °C and 200 °C, respectively.

Detection was performed with a quadrupole mass spectrometer operated in the positive electron impact mode at 70 eV. The mass spectra were obtained in the total ion current (TIC) mode over a mass range (m/z) of 50 – 500 amu and in the single ion monitoring (SIM) mode. Identification of the target compounds in the analysed samples was based on the retention time, quantitative ion and confirmation ion. Quantitative analysis was based on the peak areas of the quantitative ions recorded in the SIM mode. Each GC analysis was repeated at least two times. The retention times, characteristic ions of all analytes and validation parameters of the method are shown in Table 3.

The proposed method was validated using a calibration with standard solutions. Filter blanks were also extracted and analysed in the same way as field samples. Target compounds were not detectable in the blanks. The method quantification limit (MQL) was evaluated on the base of analyses on blank samples, with signal–to-noise ratio of 10:1. The concentrations of the compounds in the ambient air were corrected using the recovery percentages. Recovery efficiencies were determined by evaluating the surrogate standard in the samples. Recovery of surrogate standard was 94 %.

3 Results and discussion

The gravimetrically measured PM10 concentration during the study period equalled 69.8±42.7 $\mu g \cdot m^{-3}$ on average, varying from 23.5 to 149.6 $\mu g \cdot m^{-3}$. The daily maximum value for PM10 in ambient air, set up within the European Commission directive 2008/50/CE (50 $\mu g \cdot m^{-3}$), was exceeded nine times during the 15-days study period. The highest concentrations of PM10 were reported on Feb 3rd (136.6 $\mu g \cdot m^{-3}$), Feb 4th (153.8 $\mu g \cdot m^{-3}$) and Feb 6th (112.9 $\mu g \cdot m^{-3}$). In this period the diurnal mean ambient temperature varied between -0.9 and 2.7 °C, wind velocity between 0.87 and 1.18 $m \cdot s^{-1}$, relative humidity between 52 and 70 %. The lowest concentration of PM10 was noted at the beginning of the study period and equalled 23.5 $\mu g \cdot m^{-3}$. In this period the ambient temperature decreased down to -8.1 °C and the wind was blowing with the velocity of 3.1 $m \cdot s^{-1}$ on Jan 29th. PAHs and NPAHs exhibited similar patterns as PM10. The highest concentrations of examined PAHs and NPAHs were reported on Feb 3rd, 4th and 6th when the sum of analytes concentrations was equal to 386.4 $ng \cdot m^{-3}$, 428.1 $ng \cdot m^{-3}$, 408.8 $ng \cdot m^{-3}$, respectively for PAHs and 26.3 $ng \cdot m^{-3}$, 18.2 $ng \cdot m^{-3}$, 27.5 $ng \cdot m^{-3}$, respectively for NPAHs (Figure 2). Low temperature conditions are favourable for the condensation of the organic compounds and the sorption of the surface of particulate matter. However, higher wind velocities, as it was noted at the beginning of February in Kraków, caused the removal of particulates-bound organic compounds resulting in lower concentrations of PAHs and NPAHs in later part of February.

The presence of two, among four, nitro-compounds was detected in measured PM10 samples. Table 4 presents the mean concentrations of analysed PAHs and NPAHs in ambient air during the study period in Kraków.

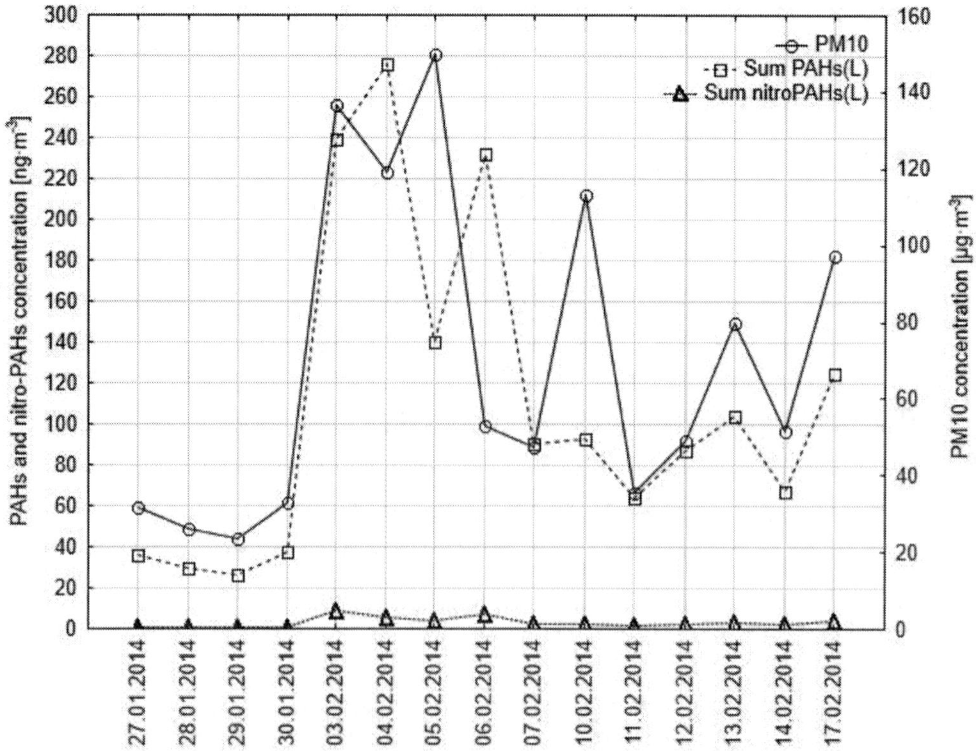

Figure 2. Variation of the sum of PAHs (open squares) concentration, NPAHs concentration (open triangles) and PM10 concentration (open circles) during the study period.

Fluoranthene, pyrene, benzo[a]anthracene and chrysene exhibited the highest concentrations in the ambient air among all analysed PAHs ranging between 4.7 and 46.1 ng·m^{-3} for fluoranthene, between 4.1 and 40.5 ng·m^{-3} for pyrene, between 3.0 and 33.8 ng·m^{-3} for benzo[a]anthracene, between 3.1 and 31.5 ng·m^{-3} for chrysene. In turn, the lowest amounts were reported for flourene and phenanthrene. This may be affected either by the specific emission sources (mostly traffic) from which the PAHs are derived, stable atmospheric conditions which favour the particle concentrations [15], PAHs tendency to sorb onto the surface of particulates at lower temperatures [8], photochemical degradation of PAHs due to ozone and NO_x reactions [16], and lower ability to evaporate of PAHs with higher values of molecular weight of PAHs. Moreover, PAHs which have least than 5 rings occur both in gas and particle bound phase. Results presented in the paper refer to the particle-bound phase what is the reason of low concentrations of fluorine, phenanthrene or anthracene. High molecular weight PAHs are more often associated with atmospheric particulate matter than low molecular weight PAHs which are vapour-phase components [2]. 2-nitrofluorene exhibited the highest mean concentration (2.2±1.7 ng·m^{-3}) among all examined nitrated derivatives of PAHs ranging between 0.5 and 5.6 ng·m^{-3}. The concentration of 9-nitroanthracene was equal to 0.9 ng·m^{-3}, on average. The obtained concentrations of 2-nitrofluorene and 9-nitroanthracene in Kraków were reported on the same level of magnitude as in another industrial cities in Poland: in Katowice and Sosnowiec [17]. In Katowice the mean concentrations were equal to 1.7 and 2.6 ng·m^{-3}, and in Sosnowiec – 1.8 and 2.4 ng·m^{-3}, respectively. In Kraków, during the heating season PAHs and NPAHs, especially chrysene, benzo[a]pyrene or 9-nitroanthracene, are mainly emitted during the combustion of coal, wastes or biomass in domestic stoves [18, 19]. Traffic-related pollutions are another factor which influence the increase of these organic compounds amount. The samples of PM10 were collected near the streets with high intensity of transport what could have an impact on the augmentation of PAHs like: benzo[a]pyrene, benzo[b]fluoranthene, benzo[k]fluoranthene, benzo[ghi]perylene and fluoranthene. Emission of exhausts from diesel engines is considered as a main source of 2-nitrofluorene [20-22].

Table 4. Mean concentrations (ng·m^{-3}) of PAHs and NPAHs bound to PM10 fraction during the study period.

Compound	Mean concentration [ng·m^{-3}]	Compound	Mean concentration [ng·m^{-3}]
Fluorene	0.2±0.4	Benzo[k]fluoranthene	9.3±6.7
Phenanthrene	5.7±4.4	Benzo[a]pyrene	9.5±7.0
Anthracene	1.5±1.2	Indeno[1,2,3cd]pyrene	7.0±5.0
Fluoranthene	17.2±12.8	Dibenzo[a,h]anthracene	1.2±0.9
Pyrene	15.4±11.3	Benzo[g,h,i]perylene	6.0±4.3
Benzo[a]anthracene	14.1±10.3	2-nitrofluorene	2.2±1.6
Chrysene	13.5±9.8	9-nitroanthracene	0.9±0.9
Benzo[b]fluoranthene	9.5±6.8		

The daily concentration limit value of the main atmospheric pollution indicator. BaP, (1 ng·m^{-3}) was exceeded during the entire 15-days study period. Its average concentration was equal to 9.5±7.0 ng·m^{-3}, and ranged between 2.0 ng·m^{-3} on Jan 29th and 23.1 ng·m^{-3} on Feb 4th. The highest concentration of BaP was observed at the beginning of February when in period Feb 3rd – 6th the mean concentration of BaP equalled to 19.2 ng·m^{-3}. In Figure 1 the annual for 2014 concentrations of BaP for two measuring stations belonging to the Malopolska Inspectorate of Environment Protection are showed. BaP amount was equal to 8 ng·m^{-3} and 7 ng·m^{-3} at the industrial type and background type stations in Kraków, respectively. These concentrations are at the similar level as the BaP amount measured by authors of this study, however this paper refers to the short 15-days study period. While comparing the mean concentration of other PAHs it may be noticed that in all cases the annual limit value for BaP was exceeded.

4 Conclusions

Experiments on the assessment of the level of ambient air pollution by airborne particulates-bounded polycyclic aromatic hydrocarbons and NPAHs showed the high concentration of these compounds during the study period. The measurements proved the presence of all examined PAHs and two, among the investigated four, NPAHs. The lowest concentration was reported for fluorene, the highest – for fluoranthene. Among analysed NPAHs, 2-fluorene demonstrated the highest amounts. As the samples of PM were collected in Kraków near the areas of high traffic intensity, the obtained concentrations of 2-nitrofluorene were significantly higher than in Katowice or Sosnowiec, which are mainly polluted by emission from the large-scale coal industry. The, so called, low-stack emission, traffic-related pollution and industrial activities are considered to be main emitters of PAHs and NPAHs in Kraków. Additionally, favourable meteorological conditions, specially low temperature, low wind velocity and low relative humidity force the accumulation of organic compounds and sorption on the particulate matter surface.

The identification of emission sources of air pollutants is very important for the assessment and management of air quality of a determined region. This creates possibilities for further work of the authors.

Acknowledgements

This work was partially supported by AGH University Grant no 11.11.210.244 and by the Rector's Grant for the

AGH Student Association RedoX. The authors acknowledge the financial support of OeaD and of the Ministry of Science and Higher Education (Poland) in the frame of project WTZ (Wissenschaftlich–Technische Zusammenarbeit), No. PL09/2015.

References

1. D.M. Agudelo-Castañeda, E.C. Teixeira, Atmos. Environ. **96** (2014).
2. X. Bi, G. Sheng, P.a. Peng, Y. Chen, Z. Zhang, J. Fu, Atmos. Environ. **37**,2 (2003).
3. K.-H. Kim, S.A. Jahan, E. Kabir, R.J.C. Brown, Environ. Int. **60** (2013).
4. A.J. Grosovsky, J.C. Sasaki, J. Arey, D.A. Eastmond, K.K. Parks, R. Atkinson, Res. Rep. Health Eff. Inst.,84 (1999).
5. H.I. Abdel-Shafy, M.S.M. Mansour, Egyptian Journal of Petroleum **25**,1 (2016).
6. M. Li, S.R. McDow, D.J. Tollerud, M.A. Mazurek, Atmos. Environ. **40**,13 (2006).
7. E.C. Teixeira, D.M. Agudelo-Castañeda, C.D.P. Mattiuzi, Sci. Total Environ. **538** (2015).
8. K. Ravindra, L. Bencs, E. Wauters, J. de Hoog, F. Deutsch, E. Roekens, N. Bleux, P. Berghmans, R. Van Grieken, Atmos. Environ. **40**,4 (2006).
9. J.N. Pitts, J.A. Sweetman, B. Zielinska, A.M. Winer, R. Atkinson, Atmospheric Environment (1967) **19**,10 (1985).
10. R. Atkinson, J. Arey, B. Zielinska, S.M. Aschmann, Int. J. Chem. Kinet. **22**,9 (1990).
11. http://monitoring.Kraków.pios.gov.pl/. [accessed: May 15th 2016].
12. Directive 2004/107/EC of the European Parliament and of the Council of 15 December 2004 relating to arsenic, cadmium, mercury, nickel and polycyclic aromatic hydrocarbons in ambient air
13. F. Cavalli, M. Viana, K.E. Yttri, J. Genberg, J.P. Putaud, Atmos. Meas. Tech. **3**,1 (2010).
14. K. Styszko, K. Szramowiat, M. Kistler, A. Kasper-Giebl, L. Samek, L. Furman, J. Pacyna, J. Gołaś, Comptes Rendus Chimie **18**,10 (2015).
15. P. Di Filippo, C. Riccardi, D. Pomata, F. Buiarelli, Atmos. Environ. **44**,23 (2010).
16. R. Chirico, P. Spezzano, D. Cataldi, Polycyclic Aromat. Compd. **27**,5 (2007).
17. M. Zaciera, J. Kurek, L. Dzwonek, B. Feist, A. Jędrzejczak, Environment protection Engineering **38**,1 (2012).
18. G. Shen, S. Tao, S. Wei, Y. Zhang, R. Wang, B. Wang, W. Li, H. Shen, Y. Huang, Y. Chen, H. Chen, Y. Yang, W. Wang, X. Wang, W. Liu, S.L.M. Simonich, Environ. Sci. Technol. **46**,15 (2012).
19. G. Shen, S. Tao, Y. Chen, Y. Zhang, S. Wei, M. Xue, B. Wang, R. Wang, Y. Lv, W. Li, H. Shen, Y. Huang, H. Chen, Environ. Sci. Technol. **47**,24 (2013).
20. J. Ringuet, E. Leoz-Garziandia, H. Budzinski, E. Villenave, A. Albinet, Atmos. Chem. Phys. **12**,18 (2012).
21. W. Wang, N. Jariyasopit, J. Schrlau, Y. Jia, S.H.U. Tao, T.-W. Yu, R.H. Dashwood, W.E.I. Zhang, X. Wang, S.L.M. Simonich, Environ. Sci. Technol. **45**,16 (2011).
22. I. Zwirner-Baier, H.G. Neumann, Mutat Res **441**,1 (1999).

Carbonaceous species in atmospheric aerosols from the Krakow area (Malopolska District): carbonaceous species dry deposition analysis

Katarzyna Szramowiat[1], Katarzyna Styszko[1,a], Magdalena Kistler[2], Anne Kasper-Giebl[2] and Janusz Gołaś[1]

[1]*AGH University of Science and Technology, Faculty of Energy and Fuels, Department of Coal Chemistry and Environmental Sciences, Av. Mickiewicza 30, 30-059 Krakow*
[2]*Institute of Chemical Technologies and Analytics, Vienna University of Technology, Getreidemarkt 9//164, Vienna, A-1060, Austria*

Abstract. Organic and elemental carbon content in PM10 was studied at three sites in Malopolska District representing the city centre (Krakow), rural/residential (Bialka) and residential/industrial environments (Krakow). The PM10 samples were collected during the winter time study. The highest concentrations of carbonaceous species were observed in Skawina (36.9 μg·m^{-3} of OC and 9.6 μg·m^{-3} of EC). The lowest OC and EC concentrations were reported in Krakow (15.2 μg·m^{-3} and 3.9 μg·m^{-3}, respectively. The highest concentration of carbonaceous species and the highest wind velocities in Skawina influenced the highest values of the dry deposition fluxes. Correlations between OC, EC and chemical constituents and meteorological parameters suggest that a) Krakow was influenced by local emission sources and temperature inversion occurrence; b) Bialka was under the influence of local emission sources and long-range transport of particles; c) Skawina was impacted by local emission sources.

1 Introduction

In recent years, scientists have taken a greater interest in atmospheric aerosols. This is especially due to the adverse impact of particulate matter on human health and due to the climate changes caused by aerosols. Unlike the gases present in the atmosphere, aerosols have a complex structure, and are composed of a wide range of both organic and inorganic chemical compounds. Aerodynamic diameter sizes range from several nanometres to several micrometres. The chemical composition is determined depending on the emission sources and meteorological conditions in the monitored region [1]. Through parameters such as solar radiation, temperature and relative humidity, air pollutants undergo several transformations and chemical reactions. Secondary pollutants, like ammonia sulphates and nitrates as well as organic acids, which are the constituents of the particulate matter, are the products of gas to particle conversion. These chemicals are the result of reactions with gas precursors like SO_2, NO_x, NH_3 or volatile organic compounds [2]. Coarse fraction particles in aerodynamic diameter from 2.5 to 10 μm, are constituted mostly of mineral components, like aluminium silicates, and other metals and non-metals [3]. The carbonaceous species (organic and elemental carbon) concentration contributes to 30 – 40 % of the total mass of the coarse fraction. Elemental carbon (EC) is a primary pollutant, directly generated during the incomplete processes of fuel combustion, mainly diesel oil [4,5]. Thus, EC is frequently considered as a tracer for pollutants emitted from transportation [6,7]. Organic carbon (OC) is either, as in the case of EC, directly emitted during combustion processes (creating primary organic aerosols), or produced during the chemical reactions with gas precursors in the atmosphere creating secondary organic aerosols [8,9].

Due to the increasing threat of high concentrations of particulate matter, the European Commission set up directives regulating the limits of aerosols in the atmosphere. Thus, in accordance with the 2008/50/EC directive, the daily and annual limits for PM10 are equal to 50 μg/m^3 and 40 μg/m^3, respectively. Krakow, like several other cities from the Malopolska District, is highly threatened by the accumulation of air pollution mainly because of the topography of the city and the presence of local emission sources. Krakow is located in a basin and is surrounded by high buildings, which block the natural ventilation of the city.

The aim of the study was to measure the PM10 concentration in the air at three different study locations of the Krakow agglomeration and to determine the contribution of elemental and organic carbon into the PM10 mass. The carbonaceous species dry deposition fluxes were also computed for the purpose of the study.

2 Experimental part

2.1 Sampling

Samples collection of the PM10 fraction was performed using a low volume sampler equipped with a 1 kW pump, a flow meter and a PM10 impactor by Digitel®. The air flow was regulated to the constant value of 1 m^3·h^{-1}. Particulate matter samples were collected on quartz fiber

[a] Corresponding author: styszko@agh.edu.pl

filters by Pallflex® with a diameter of 47 mm. Prior to sampling, filters were pre-heated at 550 °C over 5 h and, afterwards, conditioned at a relative humidity of 50±5 % at 20±2 °C over 24h. Sampling was carried out in accordance with regulations PN-EN 1234:1. In Krakow on weekdays the filters were changed in 24±2 h intervals, while during weekends (Friday, Saturday and Sunday) the filters remained in samplers for around 70 hours. In Skawina and Bialka, filters were changed every 24 hours. Sampled filters were conditioned for 24 hours to achieve a constant humidity prior to weighting. The mass of the particulate matter was obtained as an average of three following weightings of the filter. The OHAUS Discovery DV215CD balance with a accuracy of ±0,01 mg was used for weighting. Detailed data on measurement locations are presented in Table 1.

PM10 samples were collected at three different measurement locations in the Malopolska District: in Krakow, Bialka and Skawina, representing three different types of the environment: city centre, rural/residential area and industrial/residential area, respectively. Krakow is the second-largest city in Poland by population and is inhabited by ca. 760 thousand inhabitants. It is located in the north part of the Malopolska District. Krakow is inconveniently located in a basin, thus the natural ventilation of the city is blocked by the surrounding high buildings. As a result, pollution accumulates over the whole city. In Krakow, samples of PM10 were collected in the Krowodrza District, ca. 1 km from the main street characterized by a high intensity of traffic. Air aspirators were placed on the roof of a three-story building, at the height of ca. 10 m.

Skawina is a medium-sized city, inhabited by ca. 25 000 inhabitants. The city belongs to the industrial part of the Krakow agglomeration and is located ca. 15 km from the Krakow city centre. In Skawina, samplers were placed on the roof of a one-family house, at the height of ca. 5 m, in the residential/industrial part of the city.

Bialka, a small village with 2 000 inhabitants, is located ca. 60 km from Krakow, in the south-east part of the Malopolska District. Bialka is a typical rural/residential area, where almost 90% of households are heated by coal, frequently of a low quality, or by wood (in the authors' own estimation). Samplers were placed in a garden at the height of 2 m, near the main street.

Table 1. Thorough data on study areas.

	Krakow	**Bialka**	**Skawina**
Study period	05 – 27.02.2013	02 – 16.03.2013	28.12.2013 – 10.01.2014
Geographic coordinates	50°04'01.91"N 19°54'45.60" E	49°42'35.80"N 19°41'10.32" E	49°59'00.61"N 19°50'07.61" E
Location type	City centre	Rural/residential area	Industrial/residential area
Filter type applied	Tissuquartz Filters PallFlex®, φ 47 mm	Tissuquartz Filters PallFlex®, φ 47 mm	Tissuquartz Filters PallFlex®, φ 47 mm
Filter changes intervals	24 h (Monday – Thursday) 72 h (Friday - Sunday)	24 h	24 h

2.2 Determination of PM10 chemical composition

The concentration of organic and elemental carbon was determined using the thermal-optical method for OC/EC analysis (OC/EC analyzer by Sunset Laboratory Inc.), developed for atmospheric samples [6]. The *Quartz.par* protocol and laser transmittance methods were applied during measurements. 10 mm circle punches of quartz fiber filters with dust were analyzed without pre-treatment of samples prior to analysis. The concentration of TC, OC and EC was automatically calculated on the base of the flow of the calibration gas (He+CH_4, Air Liquid, Austria). The detection limits of TC, OC and EC were equal to 0.32 $\mu g \cdot m^{-3}$, 0.32 $\mu g \cdot m^{-3}$, and 0.01 $\mu g \cdot m^{-3}$, respectively. The repeatability of results was regularly controlled on the basis of the determination of carbonaceous species in the solution of sucrose containing 200 µg of carbon in 20 µl of solution. The relative standard deviation was equal to 3 %.

Within the presented study the concentration of the following inorganic ions: Na^+, K^+, Mg^{2+}, Ca^{2+}, NH_4^+, NO_3^-, Cl^-, SO_4^{2-} was determined using the method of isocratic ion chromatography. Two 10 mm circle punches were extracted in 2 ml of deionized water (Milli-Qplus 185, Millipore, 18.2 MΩ) for anions concentration determination and in 3 ml of 12 nM methanesulfonic acid (MSA) for cations concentration determination. The extraction was carried out for 20 min in an ultrasonication bath. The DX-3000 (Thermo Scientific) ion chromatograph equipped with ion-exchange column (anions: Ion Pac AS17A, mobile phase: 1.8 mM Na_2CO_3 + 1.7 mM $NaHCO_3$; cations: Ion Pac CS12A, mobile phase: 12 mM MSA) was used in the study. After electrochemical suppression of eluent, analyte concentration was determined using the conductivity detector. Calibration of the method was based on the determination of the concentration of analytes in previously prepared external standard solutions (Merck). Detection limits (obtained multiplying the standard deviation calculated for blank solutions by a constant factor of 3.0) were equal to: 0.1 $\mu g \cdot m^{-3}$ for Na^+ and Cl^- and 0.01 $\mu g \cdot m^{-3}$ for the rest of the ions.

2.3 Meteorological data

The meteorological data (air temperature, wind speed, wind direction, humidity and precipitation volume) of atmospheric conditions in Krakow were downloaded from the online platform (http://meteo.ftj.agh.edu.pl/) where the results of the meteorological measurements are directly transferred from the Vaisala WXT520 automatic meteorological station. The station was placed on the roof of the building where the samples were collected. The meteorological data in Bialka and Skawina were downloaded from the online platforms www.meteoprog.pl and www.freemeteo.com. Table 2 includes the meteorological data recorded during the study periods. In order to compute the backward trajectory of air masses, the HYSPLIT (Hybrid Single-Particle Lagrangian Integrated Trajectory) model was used: (http://ready.arl.noaa.gov/HYSPLIT_traj.php [10]). The backward trajectories for the heating season sampling periods for Krakow, Bialka and Skawina were calculated basing on the following parameters: height of computation – 20, 500 and 1500 m a.s.l.; trajectory duration – 72 h.

Table 2. Summary of chemical composition of aerosols collected in Krakow, Bialka and Skawina and the meteorological parameters for study periods.

Parameter	Krakow	Bialka	Skawina
PM10 [$\mu g \cdot m^{-3}$]	60.3 (36.5 – 128.4)	87.4 (60.4 – 111.6)	62.8 (30.1 – 117.1)
EC [$\mu g \cdot m^{-3}$]	3.9 (1.4 – 11.7)	4.1 (1.5 – 9.9)	9.6 (5.2 – 15.7)
OC [$\mu g \cdot m^{-3}$]	15.2 (7.3 – 35.0)	20.8 (7.4 – 31.8)	36.9 (13.2 – 74.8)
Cl^- [$\mu g \cdot m^{-3}$]	1.5 (0.4 – 5.1)	0.9 (0.1 – 2.2)	3.7 (< LOD – 9.2)
NO_3^- [$\mu g \cdot m^{-3}$]	3.6 (0.8 – 6.5)	2.2 (1.3 – 4.3)	2.3 (< LOD – 4.2)
SO_4^{2-} [$\mu g \cdot m^{-3}$]	6.3 (1.4 – 12.6)	7.6 (3.3 – 16.0)	3.1 (0.02 – 6.5)
Na^+ [$\mu g \cdot m^{-3}$]	0.3 (0.1 – 0.5)	0.3 (0.1 – 0.9)	0.4 (0.2 – 0.7)
NH_4^+ [$\mu g \cdot m^{-3}$]	3.8 (1.4 – 7.3)	3.4 (1.8 – 6.2)	3.5 (2.0 – 5.8)
K^+ [$\mu g \cdot m^{-3}$]	0.2 (0.1 – 0.5)	0.4 (0.1 – 0.8)	0.6 (0.3 – 1.3)
Mg^{2+} [$\mu g \cdot m^{-3}$]	0.01 (< LOD – 0.05)	0.01 (< LOD – 0.06)	0.03 (0.01 – 0.10)
Ca^{2+} [$\mu g \cdot m^{-3}$]	0.1 (0.02 – 0.5)	0.05 (< LOD – 0.6)	0.2 (0.1 – 0.3)
EC/PM10 [-]	0.06 (0.03 – 0.1)	0.05 (0.02 – 0.1)	0.2 (0.1 – 0.5)
OC/EC [-]	4.6 (2.2 – 7.8)	6.4 (2.8 – 10.9)	4.0 (0.8 – 6.7)
Temperature [°C]	0.4 (-2.8 – 4.5)	-0.3 (-6.4 – 5.9)	2.3 (-5.0 – 7.0)
Relative humidity [%]	76 (68 – 84)	84 (56 – 99)	94 (80 – 100)
Wind velocity [$m \cdot s^{-1}$]	1.7 (0.7 – 2.5)	1.4 (0 – 3.1)	5.3 (1.0 – 14.0)

3 Results and discussion

3.1 PM10 concentrations at measuring locations

Average concentrations of chemical constituents of PM10 obtained within analytical measurements of particulate matter samples from Krakow, Bialka and Skawina are summarized in Table 2. The daily accepted value of PM10 (50 $\mu g \cdot m^{-3}$) was exceeded several times during all study periods in Krakow (8 exceedances) and Skawina (7 exceedances), whereas in Bialka extremely high concentrations of PM10 were reported for the whole study period resulting in the average value of PM10 equal to 87.4 $\mu g \cdot m^{-3}$ and ranged from 60.4 to 111.6 $\mu g \cdot m^{-3}$. The mean concentration of PM10 in Krakow was accounted for 60.3 $\mu g \cdot m^{-3}$ ranging between 36.5 and 128.4 $\mu g \cdot m^{-3}$. Occurrence of PM10 particles in Skawina was reported on the average level of 62.8 $\mu g \cdot m^{-3}$ ranging between 30.1 and 117.1 $\mu g \cdot m^{-3}$ (Table 2).

Figure 1 presents the variations of the PM10 concentration measured for Krakow, Bialka and Skawina during the study periods against the relative humidity and ambient temperature. In these plots the episodes of high concentration of PM10 (> 100 $\mu g \cdot m^{-3}$) may be noticeable: on Feb 11th, 18th and 27th in Krakow, on Mar 3rd, 10th, 13th and 16th in Bialka, on Dec 29th and 30th and Jan 5th, 6th and 9th in Skawina. For each sampling period the aerosol prehistory was profiled by conducting 72-hours air masses backward trajectories at 24 h intervals at 20, 500 and 1500 m a.s.l (Figure 2). These backward trajectories enabled to identify the source of origin of aerosols during the above mentioned episodes of PM10 high concentrations. PM10 concentration in Krakow significantly increased on Feb 11th and Feb 18th and reached the maximum values equal to 100,0 and 128,4 $\mu g \cdot m^{-3}$, respectively. These values were reported when Krakow was under the influence of air masses moving from the southeast direction. As the local wind direction on these days was similar to the air mass advection (Figure 1), the origin of PM10 is influence by local sources. The backward trajectories computed for Feb 11th and Feb 18th showed that air masses moved very low, below the 500 m a.s.l., what together with low humidity (76 %), temperature (0.4 °C) and wind speed (1.7 $m \cdot s^{-1}$) may point to occurrence of the temperature inversion. For most of the study period in Krakow the air masses trajectories moved on the height lower than 500 m a.s.l. with exemptions of Feb 8th, 13th and 24th. Krakow was then under the influence of west (Feb 8th and 13th) and southeast (Feb 24th) air masses advection (Figure 2). These masses carried the middle-polluted by aerosols ambient air resulting in the PM10 concentrations on these days equal to: 44.3, 64.9 and 42.6 $\mu g \cdot m^{-3}$. Similar relations were reported for Skawina when on Dec 31st 2013, Jan 4th and 7th 2014 the PM10 concentration was accounted for 43.9, 54.8 and 70.4 $\mu g \cdot m^{-3}$. On these day Skawina was under the influence of long-range air masses which moved at the height of 1000 m a.s.l. Therefore, the long-range transport of aerosols should be considered, beyond the local sources, as the second source of particulate matter in Krakow and Skawina. Especially that in Skawina the wind speed obtained up to 14 $m \cdot s^{-1}$. On the other hand in Krakow the highest concentrations of nitrates and sulfates (3.6 $\mu g \cdot m^{-3}$ and 6.3 $\mu g \cdot m^{-3}$, respectively) were reported. This is especially seen in Krakow where the highest concentrations of nitrates and sulfates (3.6 $\mu g \cdot m^{-3}$ and 6.3 $\mu g \cdot m^{-3}$, respectively) were reported. Strong correlations between ions (NO_3^-/SO_4^{2-} = 0.67 and NO_3^-/NH_4^+ = 0.81, Table 3), additionally proves the origination of aerosols from the secondary sources. In Bialka the augmented concentrations of PM10 which occurred actually during the whole study period were probably the result of prevailing meteorological conditions. Low temperature foster the accumulation and condensation of particles suspended in the atmosphere. The wind velocity in Bialka was as low as in Krakow. As a consequence the pollutants could not be well dispersed.

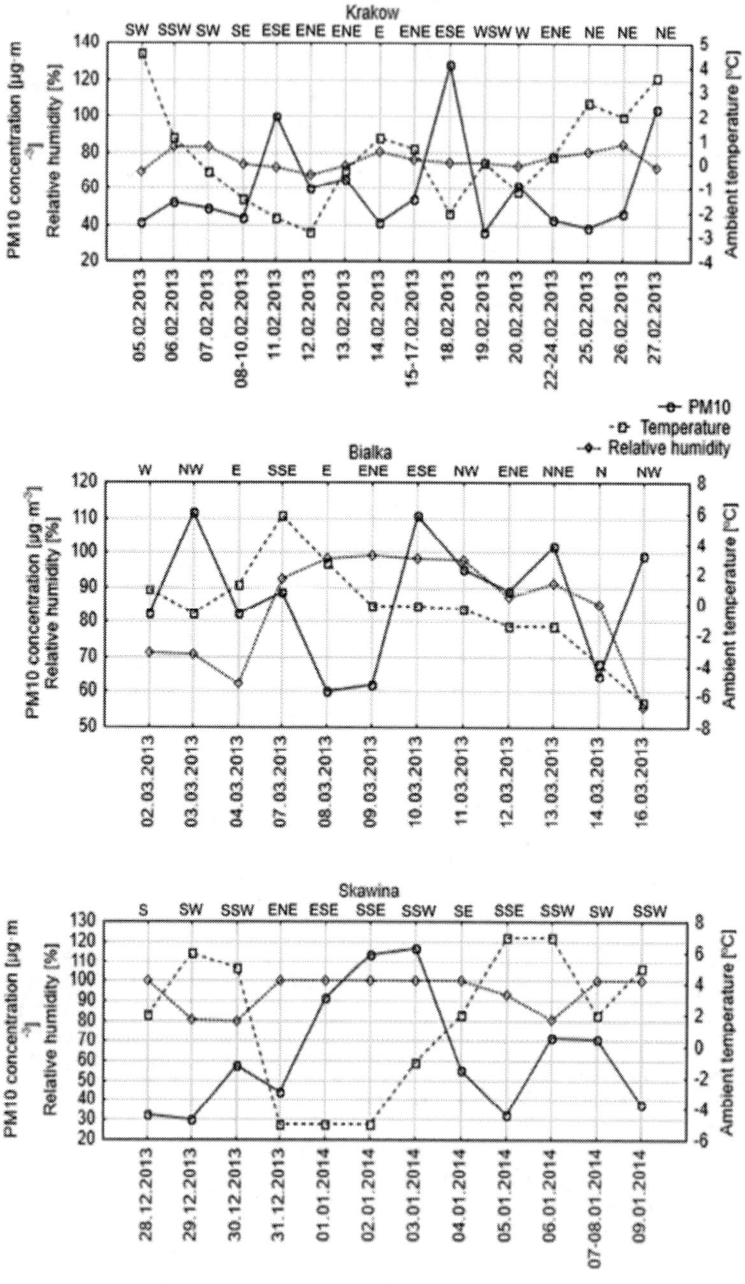

Figure 1. Variation of PM10 concentration at measuring stations during the study periods vs. meteorological conditions.

3.2 Organic and elemental carbon in PM10

The contribution of the total carbonaceous species (TC) into the PM10 mass was accounted to 32 % in Krakow, 28 % in Bialka and 74 % in Skawina reaching the mean values of the TC concentration equal to 19.1±10.6, 24.9±10.6 and 46.5±20.4 $\mu g \cdot m^{-3}$, respectively. The highest contribution of carbonaceous species was observed in Skawina. However, this was affected by the presence of a specific emission source: in Skawina, samplers were placed near a factory which usually uses organic oils for manufacturing processes. The evaporation of these high temperature organic oils resulted in the increased contribution of carbonaceous species. According to the Didyk et al, 2000 study, the concentration of the TC concentration in highly polluted urbanized areas is suspected to be at the level of 85 μg·m^{-3}, as it was investigated in Santiago (Chile) [11]. Significantly lower values were noted in the United Kingdom: 6.4±4.6 μg·m^{-3} at the Bristol Road Site, 5.6±3.2 μg·m^{-3} at the Birmingham city center site and 4.6±3.5 μg·m^{-3} at the Churchill Pumping Station site. However, the TC fraction contributed in 24 % at the Bristol Road Site, in 23 % at the Birmingham city center site, and in 24 % at the Churchill Pumping Station site [12]. Gdynia, in turns, is characterized by 5.5 μg·m^{-3} of TC fraction and 29 % of its contribution into the PM10 mass [8]. Unlike i.e. in South Pole, where the TC concentration was accounted for 0,0015 μg·m^{-3} [13]. Krakow, Bialka and Skawina are more closed to the TC concentrations reported in Santiago. Moreover, the TC concentrations reported in three measuring locations were greater by six orders of magnitude than this one measured in South Pole.

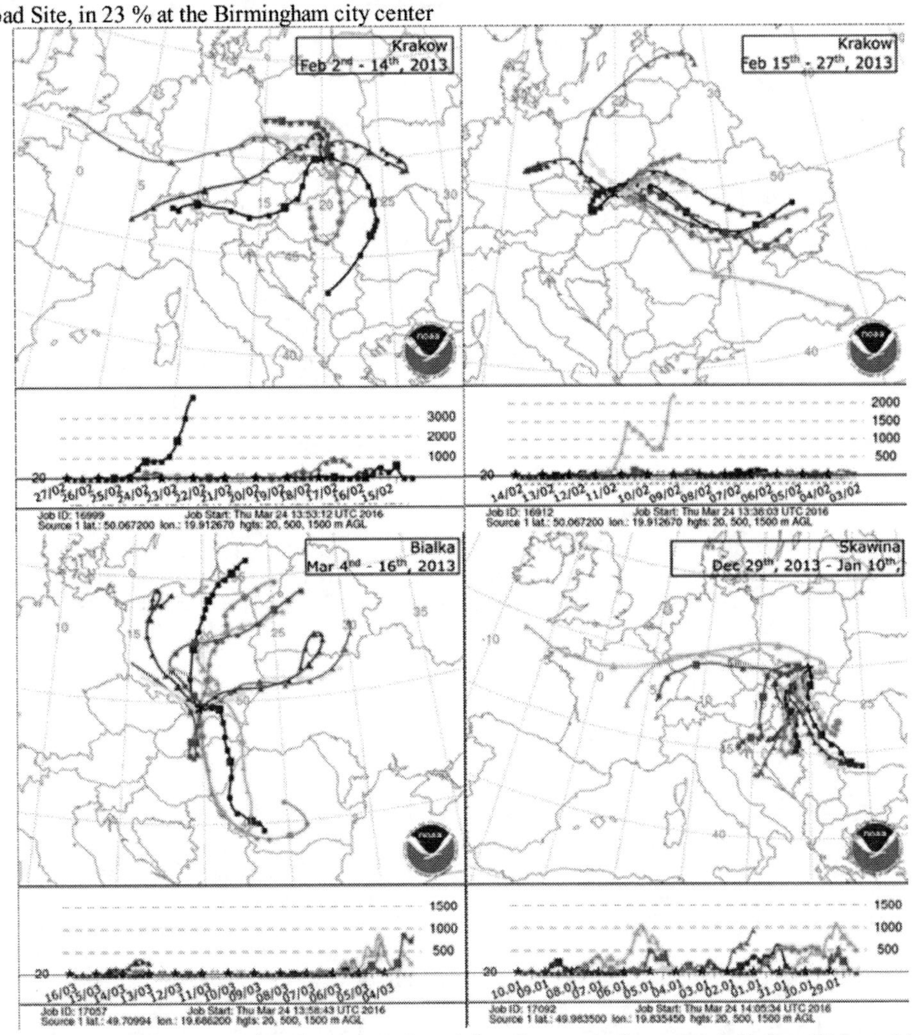

Figure 2. Backward trajectories computed for the whole study periods at three measuring stations in Krakow, Bialka and Skawina.

The mean concentration of elemental carbon was noted on the level of 3.9±2.8 μg·m^{-3} in Krakow, 4.1±3.1 μg·m^{-3} in Bialka and 9.6±3.3 μg·m^{-3} in Skawina (Table 2), contributing averagely in 6 %, 5 % and 18 %, respectively, into the PM10 mass fraction. The EC-to-TC ratio, which performs the origin of carbonaceous species from anthropogenic sources as the EC is a primary pollutant [14], was accounted for 0.19 in Krakow, 0.15 in Bialka and 0.23 in Skawina. The higher the ratio is, the stronger the environment is impacted by the local emission sources. The maximum value of the EC/TC was observed on Feb 2nd in Krakow (0.32), on Mar 4th in Bialka (0.26) and Jan 5th in Skawina (0.54). These days were under the influence of the air masses moving on low heights (~100 m a.s.l.) (Figure 2). The directions of air masses were similar to wind directions reported for measuring locations (Figure 1) what points out to the local sources as incomplete combustion of either diesel oil combustion (Krakow, Skawina) or fossil fuels combustion (Skawina, Bialka). In turns, the minimum values of the EC/TC was observed on Feb 13th in Krakow (0.11), on Mar 13th in Bialka (0.08) and on Jan 7th-8th in Skawina (0.13). On these days, the air masses moved on the high altitudes (above 1000 m a.s.l) carrying the EC absorbed on the particulates surface. Because EC has a residence time of 6 days it could be transported hundreds or thousands kilometers from its source of origin [15].

Table 3. Spearman's rang correlation coefficients computed for carbonaceous species and measured constituents of the PM10 ($p < 0.05$).

Parameter	Krakow	Bialka	Skawina
OC – PM10	0.74	-	0.98
OC – EC	0.76	0.89	-
OC – Cl^-	0.68	-	0.59
OC – NO_3^-	0.52	-	-
OC – SO_4^{2-}	-	-	-
OC – Na^+	0.85	-	0.89
OC – NH_4^+	0.53	-	0.83
OC – K^+	0.63	0.94	0.68
OC – Mg^{2+}	-	-	-
OC – Ca^{2+}	-	-	-
EC – PM10	-	-	-
EC – Cl^-	0.78	0.66	-
EC – NO_3^-	-	-	-
EC – SO_4^{2-}	-	-	-
EC – Na^+	0.72	0.73	0.59
EC – NH_4^+	-	-	-
EC – K^+	-	0.89	-
EC – Mg^{2+}	0.62	0.60	-
EC – Ca^{2+}	-	0.64	-
I_{OC} – relative humidity	-0.65	(-0.53)	(-0.45)
I_{EC} – relative humidity	(-0.29)	(-0.56)	-0.73
I_{OC} – wind speed	-	0.57	0.86
I_{EC} – wind speed	-	0.72	0.84

While transporting the particulates over the lands, the organic compounds could be transformed undergoing photochemical processes or gas-to-particles conversion [16] resulting in the increased contribution of organic carbon into the total carbon fraction: 81 % in Krakow, 92 % in Bialka and 87 % in Skawina on Feb 2nd, Mar 13th and Jan 7th-8th, respectively. The mean concentration of the organic carbon was equal to 15.2±8.0 $\mu g \cdot m^{-3}$ in Krakow, 20.8±8.3 $\mu g \cdot m^{-3}$ in Bialka and 36.9±19.1 $\mu g \cdot m^{-3}$ in Skawina ranging between 7.3 and 35.0 $\mu g \cdot m^{-3}$, 7.4 and 31.8 $\mu g \cdot m^{-3}$, 13.2 and 74.8 $\mu g \cdot m^{-3}$, respectively (Table 2). Significant correlation between organic carbon and PM10 was observed for Krakow (0.74) and Skawina (0.98) pointing out that OC originated from the same source as PM10 (Table 3). Therefore, in Skawina the highest concentrations of Cl^- and K^+ (3.7 $\mu g \cdot m^{-3}$ and 0.6 $\mu g \cdot m^{-3}$, respectively, Table 2) suggested that, beyond the manufacturing processes near the study location, aerosols have an origin in coal and biomass combustion. In Bialka the occurrence of aerosols was probably influenced by the presence of several emission sources: a. biomass combustion as the Spearman's rang correlation coefficient for OC/K^+ and EC/K^+ reached the high values of 0.94 and 0.89, respectively; potassium ions are mainly emitted during the biomass combustion [17]; b. coal combustion as EC was correlated with Cl^- [18] in 66 % and with Na^+ [19] in 73 % . In Krakow, besides the coal combustion, the ambient pollution was affected during the study period by photochemical processes leading to creation of secondary aerosols. This is confirmed by significant correlation between OC and NO_3^- (0.52) and NH_4^+ (0.53). Nitrate ions are created during the secondary reactions with the gas precursor NO_x in the atmosphere [2]. Moreover, nitrates are emitted from Diesel engine exhausts and during combustion processes, mainly for houses-heating purposes. In winter time the phenomenon is especially noticeable as nitrate ions exhibit a stronger association to ammonium ions in urban areas than in rural regions [20]. In Krakow and Bialka the significant correlation between organic and elemental carbon was reported suggesting the common source of origin. The existence of different sources and processes generating various proportions of carbon is reflected by highly variable OC/EC ratios [7,21]. In this study, the mean OC-to-EC ratio was accounted for 4.6 (2.2 – 7.8) in Krakow, 6.4 (2.8 – 10.9) in Bialka and 4.0 (0.8 – 6.7) in Skawina. Accordingly with considerations presented by Sillanpaa et al. 2005, a low OC/EC ratio can be associated with fresh traffic aerosol (2.2 and 0.8 for light-duty gasoline and heavy-duty diesel vehicles, respectively), whereas residential heating (wood combustion 4.15 and natural gas home appliances 12.7), forest fires (14.5) and dust from paved roads (13.1) show remarkably higher ratios [22]. Obtained OC/EC ratios for coarse particles in winter time points out that on measurement stations carbonaceous constituents had a complex origin [23].

3.3 The flux of carbonaceous species dry deposition

Atmospheric aerosols emitted to the atmosphere undergo various processes like transboundary transport, degradation or deposition on the ground. The deposition of particles is a very important process for atmosphere self-cleaning, which, in turn, influences aquatic environment pollution levels [24,25]. The average OC and EC dry deposition flux was accounted for 2.0±0.7 and 0.5±0.3 mg·m^{-2}·d^{-1}, respectively in Krakow, 2.6±1.9 and 0.6±0.5 mg·m^{-2}·d^{-1}, respectively in Bialka, 13.3±10.6 and 3.7±2.9 mg·m^{-2}·d^{-1}, respectively in Skawina. The values of the dry deposition fluxes of carbonaceous species were calculated according to the formulas presented in the previous paper of Styszko et al, 2015 [26]. The values of EC fluxes were lower by one order of magnitude than OC fluxes at all measuring locations. The trend of OC deposition flux reflects the trend of EC deposition flux. Similarly, maximum and minimum values of the dry deposition flux reflect the variations in PM10 concentration. Skawina demonstrates the highest values of carbonaceous species dry deposition fluxes and is higher by one order of magnitude than Krakow and Bialka. In Skawina the widest range of OC and EC fluxes was observed which ranged from 2.3 to 31.6 mg·m^{-2}·d^{-1} and from 0.8 to 8.3 mg·m^{-2}·d^{-1}, respectively. It was undoubtedly caused by the highest contribution of carbonaceous species into the PM10 mass fraction and the highest values of wind speed reported for this study period (Table 2).

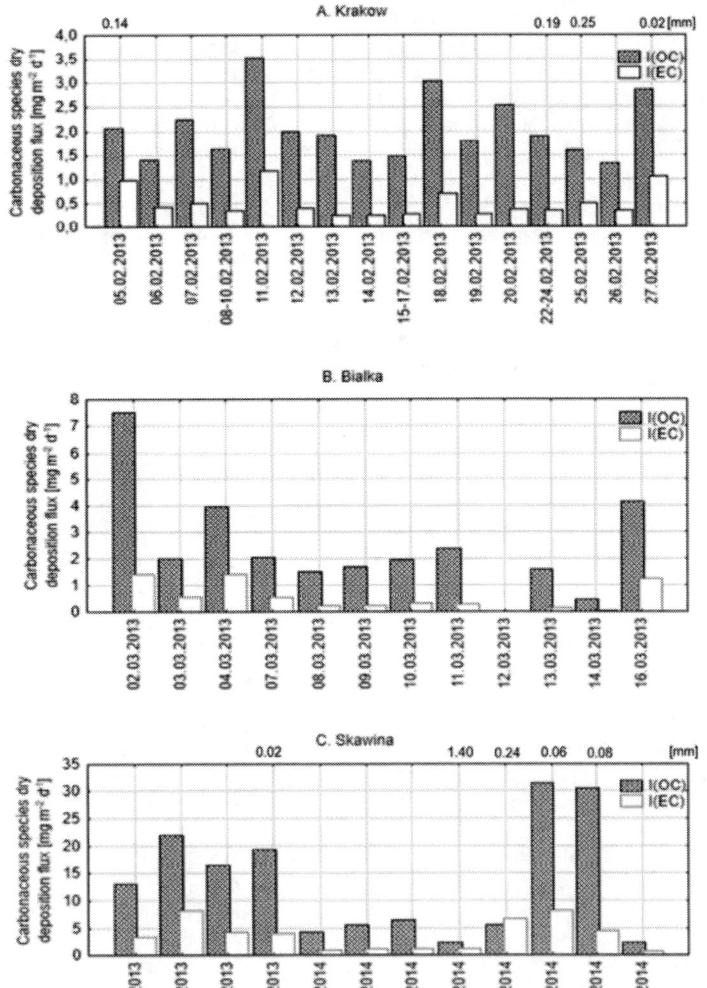

Figure 3. Daily variations of the OC and EC dry deposition flux [mg·m^{-2}·d^{-1}] with inclusion of rainy days [mm].

Figure 3 presents the daily variations of the OC and EC dry deposition fluxes in the measuring locations with the inclusion of the incineration amount. In Bialka no incineration was reported. After each rainy day in Krakow and Skawina, i.e. Feb 22nd – 25th or Jan 4th – 8th, the OC and EC dry deposition fluxes decreased. On these days when rain occurred, the particles were removed through the wet deposition. This is confirmed by the Spearman's rang correlation coefficients computed for OC and EC deposition fluxes and relative humidity (Table 3). The strongest correlation was observed in Krakow (I_{OC}) and Skawina (I_{EC}). The higher the relative humidity is, the lower contribution of the dry deposition into the particulates removal processes. Białka and Skawina additionally performed the relation between fluxes and wind velocity: the increase of the wind speed forces the increase of the dry deposition flux. In Bialka two significant increases of the carbonaceous species dry deposition fluxes were reported: on Mar 2nd and Mar 16th. On these days, Bialka was under the influence of the west and northwest air masses moving on the low altitudes. Additionally, low relative humidity and ambient temperature below 0 °C caused the increase of the concentration of carbonaceous species. Augmented values of OC-to-EC ratios (5.3 and 3.4, respectively) refer to the origin of organic carbon from photochemical processes. In Skawina, at the beginning of the study period (Dec 28th to 31st) the increased values of the Ioc were obtained. In this period the ambient temperature was below 0 °C what was favorable for aggregation of organic compounds and their absorption onto the particulates surface. At the end of the study period (Jan 6th to 8th) Skawina was under the influence of long-range air masses moving from southwest direction. While transporting the organic species underwent many processes resulting in creation of secondary organic aerosols.

4 Conclusions

The mean concentrations of elemental particulate carbon ranged from 3.9 $\mu g \cdot m^{-3}$ at the city center to 9.6 $\mu g \cdot m^{-3}$ at the residential/industrial site. The mean concentrations of organic particulate carbon ranged from 15.2 $\mu g \cdot m^{-3}$ at the city center site and 36.9 $\mu g \cdot m^{-3}$ at the residential/industrial site. The highest correlation between OC and PM10 was observed at the residential/industrial area suggesting the origin of both from the same emission source. The strong correlation between EC and potassium ions and between EC and sodium and chloride ions proves the origination of carbonaceous species from biomass and coal combustion, respectively. This was especially visible in rural site. City center site was additionally influenced by long-range air masses carrying secondary organic aerosols. It was the consequence of emission from Diesel engines and accumulation of pollution during the temperature inversion occurrence. Deposition is the significant process in self-cleaning of the atmosphere. Low humidity, the lack of incinerations and high wind velocities favored the removal of particles through the dry deposition process.

Acknowledgements

This work was financed by AGH University Grant no 11.11.210.244. The authors acknowledge the financial support of OeaD and of the Ministry of Science and Higher Education (Poland) in the frame of project WTZ (Wissenschaftlich–Technische Zusammenarbeit), No. PL09/2015.

References

1. X. Querol, A. Alastuey, C.R. Ruiz, B. Artiñano, H.C. Hansson, R.M. Harrison, E. Buringh, H.M. ten Brink, M. Lutz, P. Bruckmann, P. Straehl, J. Schneider, Atmos. Environ. **38**,38 (2004).
2. B.H. Baek, V.P. Aneja, Q. Tong, Environ Pollut **129**,1 (2004).
3. G. Peng, H. Puxbaum, H. Bauer, N. Jankowski, Y. Shi, Journal of Environmental Sciences **22**,4 (2010).
4. G.-L. Shi, Y.-Z. Tian, Y.-F. Zhang, W.-Y. Ye, X. Li, X.-X. Tie, Y.-C. Feng, T. Zhu, Atmos. Environ. **45**,32 (2011).
5. C. Hung-Lung, H. Yao-Sheng, Atmos. Environ. **43**,26 (2009).
6. M.E. Birch, R.A. Cary, Analyst **121**,9 (1996).
7. F. Mirante, P. Salvador, C. Pio, C. Alves, B. Artiñano, A. Caseiro, M.A. Revuelta, Atmospheric Research **138**,0 (2014).
8. A. Lewandowska, L. Falkowska, D. Murawiec, D. Pryputniewicz, D. Burska, M. Bełdowska, Sci. Total Environ. **408**,20 (2010).
9. H. Puxbaum, B. Gomiscek, M. Kalina, H. Bauer, A. Salam, S. Stopper, O. Preining, H. Hauck, Atmos. Environ. **38**,24 (2004).
10. R.R. Draxler, G.D. Rolph, NOAA Air Resources Laboratory, College Park, MD.
11. B.M. Didyk, B.R.T. Simoneit, L. Alvaro Pezoa, M. Luis Riveros, A. Anselmo Flores, Atmos. Environ. **34**,8 (2000).
12. J. Yin, R.M. Harrison, Atmos. Environ. **42**,5 (2008).
13. A.D.A. Hansen, B.A. Bodhaine, E.G. Dutton, R.C. Schnell, Geophys. Res. Lett. **15**,11 (1988).
14. A.M. Jones, R.M. Harrison, Atmos. Environ. **39**,37 (2005).
15. L. Husain, V.A. Dutkiewicz, A.J. Khan, B.M. Ghauri, Atmos. Environ. **41**,32 (2007).
16. J.H. Seinfeld, S.N. Pandis, *Atmospheric chemistry and physics: from air pollution to climate change*, (John Wiley & Sons, 1986).
17. C. Schmidl, I.L. Marr, A. Caseiro, P. Kotianová, A. Berner, H. Bauer, A. Kasper-Giebl, H. Puxbaum, Atmos. Environ. **42**,1 (2008).
18. A. McCulloch, M.L. Aucott, C.M. Benkovitz, T.E. Graedel, G. Kleiman, P.M. Midgley, Y.-F. Li, Journal of Geophysical Research: Atmospheres **104**,D7 (1999).
19. S.S. Kim, Y.S. Kang, H.D. Lee, J.K. Kim, S.C. Hong, Journal of Industrial and Engineering Chemistry **18**,6 (2012).
20. W. Zhao, P.K. Hopke, L. Zhou, Atmos. Environ. **41**,9 (2007).
21. K. Styszko, M. Kistler, K. Szramowiat, M. Nowak, A. Kasper-Giebl,(to be published).
22. M. Sillanpää, A. Frey, R. Hillamo, A.S. Pennanen, R.O. Salonen, Atmos. Chem. Phys. **5**,11 (2005).
23. M. Viana, X. Chi, W. Maenhaut, X. Querol, A. Alastuey, P. Mikuška, Z. Večeřa, Atmos. Environ. **40**,12 (2006).
24. S.A. Slinn, W.G.N. Slinn, Atmos. Environ. **14**,9 (1980).
25. M.L. Wesely, B.B. Hicks, Atmos. Environ. **34**,12–14 (2000).
26. K. Styszko, K. Szramowiat, M. Kistler, A. Kasper-Giebl, L. Samek, L. Furman, J.M. Pacyna, J. Gołaś, Comptes Rendus Chimie 2015).

DOI: 10.1051/e3sconf/20161000093

Off-grid photovoltaic systems as a solution for the ambient pollution avoidance and Iraq's rural areas electrification

Qusay Hassan[1,2], Marek Jaszczur[1], Muzher Mohamed[2], Katarzyna Styszko[3], Katarzyna Szramowiat[3], Janusz Gołaś[3]

[1]*Department of Fundamental Research in Energy Engineering, AGH University of Science and Technology, Poland*
[2]*Mechanical Engineering Department, University of Diyala, Iraq*
[3]*Department of Coal Chemistry and Environmental Sciences, AGH University of Science and Technology, Poland*

Abstract. The growing demand for local pollution free renewable energy sources in the world considered the solar energy as a one of the important renewable energy technologies and becomes dominant to the renewable energy source being exploited. The paper presents a unique approach to study and analysis off-grid photovoltaic (PV) system in order to provide the required energy for a one fold household in Diyala State, Iraq as well as the CO_2 emissions and life cycle (LCC) and economical aspect was considered. The results of the study by depending on site metrological data shows that the unit of electrical cost of (1 kWh) about ($0.51) and this value is not expensive compared to the current unit cost of electricity and becomes effect and encouraged the use the PV system to electrify the rare sites. The average of CO_2 emissions get avoidance about 1840 kg/year compared with conventional fuels used in Iraq to generate electricity.

1 Introduction

The global challenge to get green energy from reliable and cost-effective services remains one of the major challenges facing the world in this century. The application of green energy solutions significantly impact on the air qualities as well. More than 85% of mercury emissions originate from coal combustion for power generation and industrial processes [1]. The last investigations have shown that concentrations of most air pollutants in Baghdad City have shown a downward trend in recent years, but they are generally in many instances worse than natural ambient air; thus, all pollutants (except CO) fluctuated between high and below limits certified by Iraqi and international standards [2]. Although grid extension still remains the preferred mode of rural electrification [3], an extension of the central electricity grid to geographically remote and sparsely populated rural areas can either be financially unviable or practically infeasible. The efforts in using renewable energies have been often focused on single technologies. Reliance on a single technology generally results in an over-sizing of the system, thereby increasing the initial costs. Recently, the PV energy systems received more interest because of like these systems can overcome the intermittent nature of renewable energy sources, the over-sizing issue and enhance reliability of supply. In the developed countries, often advanced fuel systems such as hydrogen are considered. Examples of such studies include the following Khan and Iqbal [3] who investigated the feasibility of a hybrid system with hydrogen as an energy carrier in Newfoundland, Canada; Barsoum and Vacent [5]; Karakoulidis et al. [6], Giatrakos et al. [7] and Türkay and Telli [8].

Givler and Lilienthal [9] conducted a case study of Sri Lanka, where they identified when a PV/diesel hybrid becomes cost effective compared to a stand-alone small solar home systems (50 W PV with battery).

Solar energy is one of the important renewable energy technologies and photovoltaic system installation has played a big role in renewable energy because PV systems are pollution free, economical, highly reliable for long-time operation and secure energy source. The major obstruction of PV technology is its high capital costs compared to conventional energy sources. In isolated regions and because of the scarcity of means, it is necessary to optimize the solar off grid PV system in order to minimize the costs and to make the PV systems competitive with the other forms of renewable energies [10]. Figure 1 shows the off grid PV system for household [11].

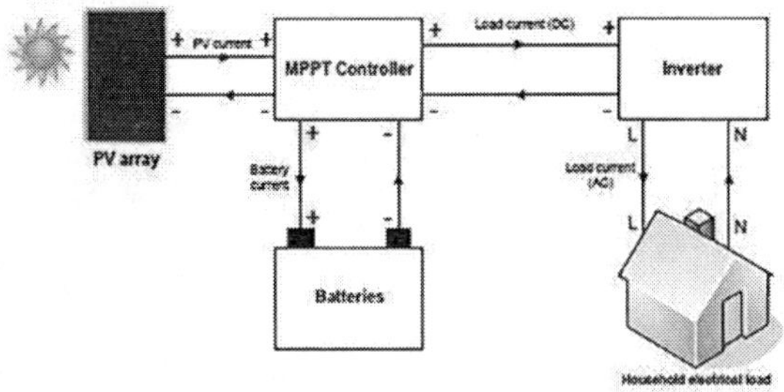

Figure 1. Off-grid PV system.

marek.jaszczur@agh.edu.pl

Iraq is located in very high world's solar radiation and has an excellent solar availability. The annual average total solar radiation over Iraq ranges from about 2490 kWh/m^2/year on the eastern territories to more than 2880 kWh/m^2/year in the western territories (deserts). About 90% of the Iraqi territory have an average total radiation greater than 2680 kWh/m^2/year [4]. The aim of this study is to provide a renewable energy system from the available resources, PV at a given site that can meet the electricity demand. The selected off-grid remote home in rural village Budjah, a small village in the Muqdadiyah district in the Iraqi state of Diyala. The site details are listed in Table 1. The area around the village is partially hilly with flat plains constituting the rest.

Table 1. Details of the selected site.

Particulars	Details
Village	Budjah
District	Muqdadiyah
State	Diyala
Country	Iraq
Latitude	33°58'N
Longitude	44°56'E
Max. Temp. (during summer)	47 °C
Min. Temp. (during winter)	6 °C

2 Household Daily Load Profile

The residential unit is simple and not require heavies of quantities of electrical power used for lighting and small electrical appliances. The Load profile was proposed considering the general hourly based load usage. At midnight hours, the power consumption for the residential unit comes down where only basic electrical appliances are consuming power. The load demand rises up during morning hours when everybody gets-up. Throughout the noon hours the load demand levels are minimum as most of the family members are outside. Again, during the evening hours when all the family members are present, the power consumption rises as everyone switches on various entertainment appliances.

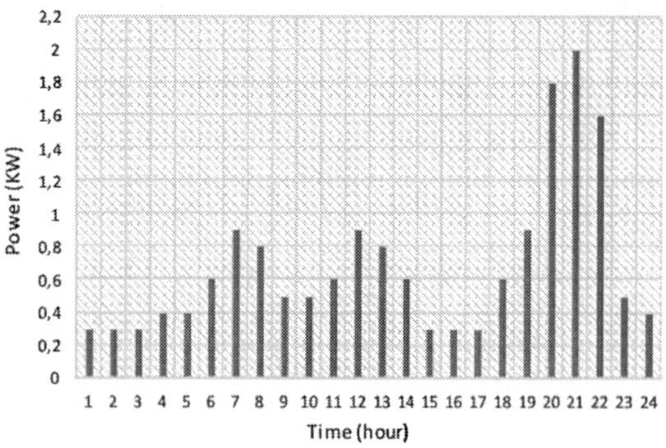

Figure 2. Household load profile.

According to the load profile shown in Figure 2 the average of hourly power consumption is 0.54 kW, minimum of hourly power consumption is 0.3 kW and the maximum of hourly power consumption = 2 kW, the load requirement considered should be maximum hourly load consumption. The average energy consumption of electrical appliances is assumed 13.16 kWh/day i.e. 395 kWh/month. Thus the proposed solar off-grid PV system should be overestimated about 14 kWh/day.

3 Site Meteorological Data

Iraq is among the countries with remarkable potential in solar energy. The off-grid PV system of the interest area that located in the remote village in the Muqdadiyah district at Diyala state. The site meteorological data showed in Table 1 and the solar resource for that site as shown in Figure 3.

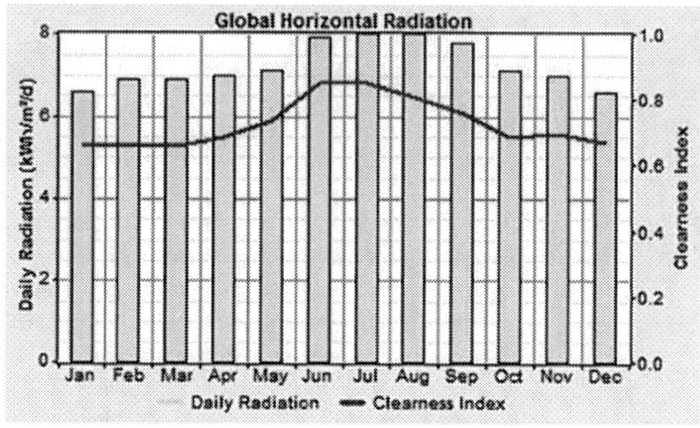

Figure 3. Solar energy profile at the selected site

4 Design and Modeling of the PV System

The main components of the off-grid PV system are namely PV array, battery, controller, inverter and load. The solar PV system is simulated such that the PV module charge the battery through the controller and battery also provides the power to the load when the solar radiation is insufficient. DC/AC inverter provides AC electricity to the required residential AC loads.

4.1 Solar Energy Input Data

The calculation from Figure 3 clarify the annual average of the solar radiation 7.2 kWh/m^2/day and the average clearness index was found to be 0.71 The highest solar radiation was estimated at 7.9 kWh/m^2/day in June while the lowest was 6.6 kWh/m^2/day in December. The (G_{av}) selected in this project to be 7.1 kWh/m^2/day.

4.2 Average Daily Load Demand

The average daily household load demand (EL) was 13.16 kWh/day, 0.54 kW/h showed in Figure 2, and the daily total demand was but the modeled solar off-grid PV system should be overestimated to the value 14 kWh/day, 0.58 kW/h.

4.3 PV Array size required

The size of the PV array, that required to use in this project, can be calculated as the following equation [12]:

$$PV_{size} = \frac{EL}{G_{av} * \mu_{pv} * TCF * \mu_{out}} \quad (1)$$

where,

G_{av} Is the average of solar energy input/day.

TCF Temperature correction factor

μ_{pv} PV module efficiency

μ_{out} Battery efficiency (μ_b) * inverter efficiency (μ_{in}).

Assuming the cell temperature reach to 58^{0}C in the area field, the temperature correction facture (TCF) will be 0.9 and assume the PV module efficiency $\mu_{pv} = 11\%$

$$\mu_{out} = \text{battery efficiency} * \text{inverter efficiency} \; 0.8*0.9 = 0.72$$

by using equation (1) calculated area required for PV modules is 23.4 m^2

$$PV Peak power = = PV_{size} * PSI * \mu_{pv} = 3042 \text{ W}_p \quad (2)$$

The selected model is (Generic, a-Si:H, triple junction) with the following specification at standard test conditions STC:

- Peak Power: 140 W$_p$
- Peak power current: 3.87 A
- Peak power voltage: 36.2 V
- Short-circuit current (Isc) 5.10 A
- Open-circuit voltage (Voc) 46.2 V

The array with 22 modules is required to supply the energy for the residential house. The configuration of series and parallel of the resulted PV array can be adjusted according to the required DC bus voltage and current, respectively. If the DC bus voltage is chosen to be 48 V, then 2 modules will be connected in series and 11 strings will be connected in parallel.

4.4 Model of the Battery

The battery model used in the PV system based on a lead-acid battery PSpice model [14]. The battery model has to work in two operation modes: charge and discharge. The battery is in charge mode when current into the battery is positive, and discharge mode when the current is negative. The battery storage capacity was calculated by using the following relation [13, 14, 15]:

$$\text{Storage capacity} = \frac{Nc * EL}{DoD * \mu_{out}} \quad (3)$$

DoD Battery Maximum permissible depth discharging.

Nc Largest number of continuous cloudy days of selected site.

The largest number of continuous cloudy days *Nc* in the selected site is about 3 days. Thus, for the maximum depth of discharge for the battery *DoD* of 0.8, and battery efficiency 80%, the storage capacity using equation (3) the storage capacity becomes 35156 Wh. Since the selected DC bus voltage is 24 V, then the required ampere-hours of the battery = 65625/24 ≈ 2800 Ah. If a single battery of 24 V with a capacity of 350 Ah is considered, then 2 batteries are required to connect in series and 4 batteries connected in parallel to give 8 batteries for the overall system.

4.5 Inverter Model Required

The required inverter must be able to the maximum expected power of AC load requirements it can be chosen as 15% higher than the rated of the power of the total AC load. Therefore the rated inverter power becomes 1500 W. The specifications of the selected inverter will be 1500 W, 20 V$_{DC}$, 220 V$_{AC}$, and 50 Hz.

4.6 Battery Charge Controller Model

The number of battery charge controller required for the off-grid PV system has to able to carrying short circuit current of PV array it can be calculated as the following [17]:

$$\text{Total maximum power of PV} = P\max * Npv \quad (4)$$

$$\text{Controller Maximum power} = Vo * Ic \quad (5)$$

Number of controller required =

$$\frac{Total\ maximum\ power\ of PV}{Controller\ Max\ Power} \quad (6)$$

where:

Ic Maximum current the controller which can handle from the PV system to the battery bank.

Npv Total number of PV modules required to meet the residential load.

Vo Battery voltage.

5 Analysis the life Cycle Cost

This part computes the estimated values of the life cycle cost (LCC) for off-grid PV system. The LCC of an item consists of the total costs of owning and operating an item over its lifetime, expressed in today's money [16-19]. The costs of an off-grid PV system include acquisition costs, operating costs, maintenance costs, and replacement costs.

Table 2. Cost data of the PV system components.

Item	Cost
PV	$5/W$_p$
Battery	$1.5 /Ah
Charger	$5.878/A
Inverter	$0.8 / W
Installation	10% of PV cost
M&O/ year	2% of PV cost

The LCC of the PV system includes the sum of all the present worths (PWs) of the costs of the PV modules, batteries, battery charger, inverter, the cost of the installation and the operation and the maintenance (O&M) cost of the system. The details of the used cost data for all items are shown in Table 2 [15, 20].

The lifetime *N* of all the PV system items is considered to be 20 years, except that of the battery which is considered to be 5 years. Therefore, there are an extra 3 groups of batteries (each of 8 batteries) should be purchased, after 5 years, 10 years, and 15 years. The inflation rate *i* of 3% has been assumed and a discount or interest rate *d* of 9%. The PWs of all the items are calculated as follows [16, 17]:

- PV array coast C_{pv} = 5*22*140 =$15400
- Initial coast of batters C_b = 1.5 *2800 = $4200

The PW of the first extra group of batteries (purchased after N = 5 years) C_{b1pw} is calculated, to be $3164.5, from:

$$C_{b1pw} = C_b \left(\frac{1+i}{i+d}\right)^N \quad (7)$$

The PW of the second extra group of batteries (purchased after N = 10 years) C_{b2PW} and the third extra group (purchased after N = 15 years) C_{b3PW} are calculated, using an equation (7) and the values are $2384.3 and $1796.4 respectively. Table 3 shows that the LCC cost of controller, inverter and installation.

Table 3. LCC cost with installation of PV system components.

Details	Value
Controller cost	$293.9
Inverter cost	$1200
Installation cost	$1540

The PW of the maintenance cost C_{MPW} is $3583.4, and was calculated using the maintenance cost per year (*M/yr*) and the lifetime of the system (*N* = 20 years), from [18]:

$$C_{mpw} = \frac{M}{yr}\left(\frac{1+i}{1+d}\right)\left[\frac{1-\left(\frac{1+i}{1+d}\right)^{N}}{1-\left(\frac{1+i}{1+d}\right)}\right] \quad (8)$$

The life cycle of the PV system can be calculated as follows:

$$LCC = C_{PV} + C_b + C_{b1PW} + C_{b2PW} + C_{b3PW} + C_{controller} + C_{inv} + C_{inst} + C_{MPW} \quad (9)$$

so the life cycle cost of PV system is $32194.

The annualized LCC (ALCC) of the PV system in terms of the present day dollars can be calculated, to be $2614.8/year, using equation (10) [16, 17]:

$$ALCC = LCC\left[\frac{1-\left(\frac{1+i}{1+d}\right)}{1-\left(\frac{1+i}{1+d}\right)^{N}}\right] \quad (10)$$

Finally, the electrical unit cost (cost of 1 kWh) was calculated, to be $0.51/kWh with the following equation:

$$\text{Unit electrical cost} = A\frac{ALCC}{356EL} \quad (11)$$

Therefore, in remote sites that are so far from the electric grid, the PV suppliers are encouraged to sell the electricity of their PV systems at a price not lower than $0.51 /kWh to earn a profit. It is to be noted that although this price is low compared to the current unit cost of electricity in Iraq and this price will drop if the future initial cost of the PV modules drops also. Meanwhile if the future unit cost of electricity in Iraq increases due to the rapid increase in the conventional fuel prices, the PV power generation will be promising in the future typical home electrification due to its expected future lower unit electricity cost compared to the conventional utility grid.

6 CO_2 emissions avoidance

Iraq relies on dams and natural gas for electricity generation and has 9 major stations spread over its territory, the generation of 1 kW/h of electricity by using natural gas produce 0.465 kg of CO_2 while to generate the same value by using PV produce 0.105 kg of CO_2 [21], the natural gas is classified as a second largest producer for CO_2 after the coals that used in power stations. According to present study the generating of 14 kWh/day by using natural gas produce 6.1 kg/day and at the same time it produce 1.3 kg/day by using a PV system get avoidance in CO_2 emissions about 1840 kg/year compared with conventional fuels used in Iraq to generate electricity.

7 Conclusion

The electrification of rural and remote places worldwide is very important especially in the developing countries as Iraq. Solar photovoltaic technology considered one of the promising renewable energy resources due to their high reliability and safety. At the same time, it represents a vital and economic alternative to the conventional energy generators. This study presents a complete design and the life cycle cost analysis for household in rural site of Diyala State, Iraq is carried out using off-grid PV system. The results of the study shows that electrifying a remote household using PV systems is beneficial and suitable for long-term investments at a price $0.51/kWh it is to be noted that this price is lower compared to the current unit cost of electricity in Iraq, especially if the initial prices of the PV systems are decreased and their efficiencies in the future will increased. The average of CO_2 emissions get avoidance is about 1840 kg/year for the 3042W peak power of the photovoltaic system, compared with conventional fuels used in Iraq to generate electricity.

References

1. K. Styszko, K. Szramowiat, M. Kistler, A. Kasper-Giebl, L. Samek, L. Furman, J. Pacyna, J. Gołaś, C. R. Chim. **18** (2015)
2. A. Rabee, Environ. Monit. Assess. **187**, 1 (2015)
3. S.C. Bhattacharyya, Energ. Sust. Dev. **16**, 3 (2012)
4. M. Khan, M. Iqbal, Renew. Energ. **30,** 6 (2005)
5. N. Barsoum, N. Vacent, *Balancing cost, operating and performance in integrated hydrogen hybrid energy system* in proceedings of the first Asia International conference on modelling and simulation (2007)
6. K. Karakoulidis, K. Mavridis, D.V. Bandekas, P. Adoniadis, C. Potolias, N. Vordos, Renew. Energ. **36** (2011)
7. G. Glatrakos, T.D. Tsoutsos, P.G. Mouchtaropoulos, G. Stavrakakis, Renew. Energ. **34** (2009)
8. B.E. Turkey, A.Y. Telli, Renew. Energ. **36** (2011)
9. T. Givler, T. Lilienthal, Technical Report NREL/TP-710-36774, available from http://www.osti.gov/bridge (2005)
10. J. Abdulateefa, K. Sopiana, W. Kaderb, B. Baisb, R. Sirwana, B. Bakhtyara, O. Saadatiana , Int. J. Sust. Green Energ. **4**, 3-1 (2014)

DOI: 10.1051/e3sconf/20161000093

11. PVDI Solar Energy International, *Photovoltaic Design and Installation Manual*, New Society Publishers (2007)
12. T.M.I. Alamsyah, K. Sopian, A. Shahrir, *Techno-economics Analysis of a Photovoltaic System to Provide Electricity for a Household in Malaysia, Proceedings of the International Symposium on Renewable Energy* (Kuala Lumpur, Malaysia, Environment Protection & Energy Solution for Sustainable Development, 2003)
13. M.M. Mahmoud, I.H. Ibrik, Renew. Sust. Energ. Rev. **10** (2006)
14. M.R. Rivera, *Small wind / photovoltaic hybrid renewable energy system optimization. Master's thesis, electrical engineering* (University of Puerto Rico, 2008).
15. PVDI 2007: Solar Energy International, (Photovoltaic Design and Installation Manual), New Society Publishers.
16. Estimating PV system size and cost. Seco fact sheet No. 24 www.InfinitePower.org.
17. The German Solar Energy Society. 2005. *Planning control apparatus and method and power generating system using them.* Patent US5, 654,883.
18. R. Messenger, J. Ventre, *Photovoltaic Systems Engineering*, CRC Press LLC: Boca Raton, Florida, USA, 2000.
19. A.N. Celik, Renew. Energy. **32** (2007)
20. C.W. Ajan, S.S. Ahmed, H.B. Ahmed, F. Taha, A.A.B.M. Zin, Sol. Energ. **74** (2003)
21. A.N. Celik, Renew. Sust. Energ. Rev. **10** (2006)
22. J. L. R. Proops, P.W. Gay, S. Speck, T. Schröder, Energ. Pol. **24,** 3 (1996)

E3S Web of Conferences **10**, 00094 (2016) DOI: 10.1051/e3sconf/20161000094
SEED 2016

Modelling of the energy use of ventilation air from the mines

Karol Sztekler[1,a], Maciej Komorowski[1] and Dawid Kot[3]

[1]AGH University, Faculty of Energy and Fuels, Mickiewicza Av. 30, 30-059 Krakow, Poland
[3]Fluor S.A, Prymasa Stefana Wyszyńskiego 11, 44-101 Gliwice, Poland

Abstract. In this paper an attempt is a comprehensive presentation of the problem of methane emissions from ventilation air from the mines to the atmosphere and possible solutions are suggested. Methane is released during excavation work in coal mines is a big problem in terms of safety and environmental protection. At the same time, it is a further potential source of energy that has been not yet used on a larger scale [1]. In order to properly identify ventilation air methane as an energy source will be presented its properties and possible complications due to the nature of this type fuel. A very important part will also determine the methane potential from ventilation air in Poland [2]. The work provides an overview of available technology for methane utilization from ventilation air as a fuel. Important from the point of view of the rest of the work is presentation of catalytic flow-reversal reactor technology created in Poland. Main objective of the topic is the development and modeling of heat recovery system cooperating with the IUMK-1000 installation [3]. The premise is the electricity production and also heat for central heating when it is technically feasible and justified in economic terms. Different concepts of heat recovery are modeled by the IPSE-Pro software. Then an analysis of the efficiency and performance of the proposed system is carried out. Based on the results, an individual concepts cycle works were analyzed and assessed and the best solutions possible for implementation on an industrial scale is selected.

1 Introduction

Methane released during excavation work in the coal mines is a big problem in terms of occupational safety and environmental protection. At the same time it is a further potential source of energy that has not been used yet on a larger scale. Emission of methane from coal seams into the atmosphere is harmful to the environment. It is a greenhouse gas whose greenhouse effect potential according to the Third Report (2007 year) of IPCC (Intergovernmental Panel on Climate Change) is tens of times greater than the potential of carbon dioxide. GWP unit (called Global Warming Potential) determines the ability of the IR absorption of the Earth by the individual compounds, wherein the reference point is CO_2 [13]. The most importance (share of total CBM) has methane included in the process of demethylation (firedamp), and ventilation air methane (VAM - Ventilation Air Methane). Currently in Poland firedamp is largely used fuel, mainly in the cogeneration installations with gas piston engine located in the immediate vicinity to the mine. This fuel is a mixture of methane (from 22% to 95% concentrations), nitrogen, oxygen, carbon dioxide [4], and is classified as a low or medium quality fuel. In the case of ventilation air methane energy value is much lower. According to the Decree of the Minister of Economy [5], the methane concentration in the ventilation shafts is 0.75%. According to data from the State Mining Authority absolute methane-bearing, that is the sum of the emitted methane from coal seams per year, remained stable (approx. 830 million m^3 CH_4) in recent years. Relative methane-bearing (the amount of methane per coal tonne) is also not subject to the higher fluctuations and is approx. 11 m^3 CH_4/tonne.

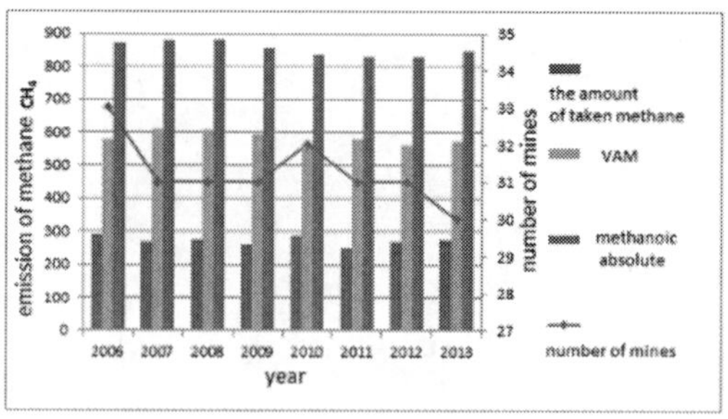

Figure 1. CH4 emission from coal mine [6]

Ventilation air methane in Poland is about 70 % of the total methane emission from coal seams corresponding to 600 million m^3 of methane emitted together with the removable ventilation air. In the world, excluding pilot plant and a small number of commercial installations, it is not used in any way and is the only source of harmful emissions.

[a] corresponding author: sztekler@agh.edu.pl

 DOI: 10.1051/e3sconf/20161000094

2 Disposal methods of ventilation air methane

In the last few years methods of methane disposal from ventilation air were developed. These technologies can be divided into two main groups [7]. In the first group VAM is used as combustion air, and can be regarded as a secondary fuel. This approach does not require significant changes to existing energy systems, but such use of ventilation air is limited by the distance of the ventilation shaft of the system and the required air flow. Devices that can be powered by ventilation air, as an additional fuel, include [7]:

- Internal combustion engines,
- Gas turbines,
- Power and industrial boilers

The air handling unit may be used as the main fuel. This group includes:

- Thermal flow reversal reactors,
- Catalytic flow reversal reactors,
- Gas turbines - turbines and microturbines of higher power (over 0.5 MW)
- Arrangement of the rotary kiln and a gas turbine.

It is also possible to use concentrators of methane to receive higher methane concentration in the gas, and then burning it for example in a gas turbine.

2.1 Thermal flow-reversal reactors

Thermal flow reversal reactor (Thermal Flow Reversal Reactor - TFRR) is a device, which is followed by cyclical processes of spontaneous combustion gas with a low concentration of methane. An example of such device is a VOCSIDIZER reactor units of Megtec Systems [14]. The main part of the TFRR is a heat exchanger in the form of a bed made of a ceramic material, together with the electrical system used to start-up. Start-up system consists in the ventilation air heating to a temperature of at least 1000 °C, allowing oxidation of methane. Ventilation air of a suitable methane concentration (0.3 - 1%) is injected into the reactor in one direction. Under the influence of the temperature, the methane oxidizes and flue gases radiate heat to the heat exchanger. At the same time the opposite side of the heat exchanger is heated and after the temperature reaches the desired one, the direction of gas flow is reversed. VAM fresh stream flowing through the bed, is heated up and in the central part begins to oxidize. The process takes place in a cyclic mood. TFRR reactor by itself can not directly use the energy content of methane in the ventilation air. Energy can be recovered from the hot gases discharged from the reactor. It is therefore necessary to apply heat exchanger, which makes it possible to recover energy from the flue gases. The reactor VOCSIDIZER at a VAM flow of 800 thousand m^3/h of a 1% methane concentration is enough to get 72 MW of heat that can be converted to 18 MW electrical energy.

2.2 Catalytic flow-reversal reactors

The principle of the catalytic reactor in flow-reversal operation (CFRR) is similar to that used in TFRR. CFRR differs by using a catalyst which allows methane oxidation at lower temperatures (up to 350 °C).

This technology has been proposed by Canadian company CANMET. Pilot plant, as well as installations in the semi-industrial and industrial scale were built [8]. For a proper functioning of the device is required a minimum concentration of methane in the ventilation air, so low concentrations required mixing of VAM with other methane source.

2.3 Gas Turbine

There are several types of gas turbines allow for utilization of ventilation air. The first of these are the microturbines working on a mixture of air-methane at CH_4 concentration less than 1%. This solution was developed by Ingersoll Rand Energy Systems (now FlexEnergy in collaboration with Capstone Turbines). Ventilation air is compressed in the compressor then VAM goes to a recovery unit, which heats the flue gas and goes into the combustion chamber where it is possible to add the fuel needed to achieve the minimum operating parameters of the gas turbine [9]. Turbine MT250 has a nominal power of 250 kW and electrical efficiency of 30%. The problem, as in most of the described devices is the need to supplie the minimum concentration of methane at about 1%. The next type of turbines are microturbines for a gas with a low concentration with the catalytic combustion. Production of such turbines with a capacity greater than 30 kW deals with Capstone Turbine Corporation and FlexEnergy [4]. Another possibility is the use of carbureted gas turbines (CGT) and carbureted catalytic gas turbine. As in the previous turbine a minimum concentration of methane in the ventilation air is required at a level of about 1%.

2.4 The arrangement of the rotary kiln and a gas turbine

The arrangement of the rotary kiln and a gas turbine can utilize VAM as a main fuel. Depending on the characteristics of the ventilation air stream suitable low-quality coal amount is supplied into a rotary kiln. Coal waste and the ventilation air are moving in the same direction in the rotary kiln. Coal is combusted on the grate grid and heating the mixture of air and methane. At a temperature of about 1000 °C oxidation of methane occurs. The hot flue gases in the rotary kiln are directed to the heat exchanger where it heats the compressed air to the temperature of 900 °C. Thus the heated air drives the gas turbine coupled to an electric generator. The first system was a pilot installation with a capacity of 1.2 MW located in Queensland, Australia [7.8]. A country where most completed research installation and development work related to the potential use of VAM is Australia (Appin, Australia 2001, reactor VOCSIDIZER). Such initiatives are in Australia strongly supported by state institutions and this is a very important factor for such activity [7]. Extensive experience in the utilization of the ventilation air is also United Kingdom (UK Thoresby 1994, VOCSIDIZER reactor), which formed one of the first plant with a TFRR reactor. In addition to these countries attention should be paid also to the United States of America and China. In the USA (Windsor, USA, 2007) a wide range of investments related to the reduction of methane emissions into the atmosphere is realized, as in Australia. China is the country which has enormous potential in terms of both utilization of firedamp derived from methane drainage and ventilation air. In China (eg. Mine Duerping belonging to the Shanxi Coking Coal Group, 2013, Mine Da Tong in China, 2010) so far many installations where the fuel is VAM are created, and in addition a continuous increase in interest in these types of technologies has been observed. In Poland, as in other countries, actions to reduce methane emissions from coal mines into the air were also taken into action. Most of the investments in this area involve the use of gas piston engines. The fuel used in the power systems of this type is gas from coal mines demethanization. An innovative project, aimed at creation of technology that uses ventilation air exhaust shaft of coal mines as a fuel, is "Environmentally friendly utilization technology of methane from coal mines." The project was developed within the Innovative Economy Operational Programme and co-founded by the European Union. The main objective of the project was to develop technologies for the use of ventilation air with low concentration of methane to produce electricity and heat. The project developed a system with catalytic converter whose operating principle is described above.

The first stage of the project was to investigate the possibility of methane oxidation at concentration below of 1% in the presence of a catalyst. For this purpose, the laboratory IUMK-1 installation (Installation Methane Utilization of Mines) with thermal capacity of 1kW$_{th}$ was built, which helped to determine the course of chemical processes and energy balance for the catalytic oxidation of methane [15]. In the IUMK-1 installation methane oxidation is at temperatures ranging from 350 to 600 °C. The minimum concentration of methane in the ventilation air is 0.4%. The efficiency of the chemical oxidation of methane, that is, the degree of methane conversion from the fuel, reached 97%. The next stage of the project is the design and maintain a pilot scale installation with a thermal capacity up to 100 kW. At this stage, the consortium cooperated with the potential recipient of the technology, which is Jastrzębska the Coal Company. It was agreed the exact location of the installation is in the Jas-Mos Coal Mine. In the case of IUMK-100 installation methane oxidation starts at a temperature of 170 °C. For the temperature of 380 °C methane oxidation efficiency is over of 90% [10].

On the basis of data obtained from the operation of IUMK-100 pilot plant, model of the whole IUMK-1000 system for 1MW$_{th}$ with different variants of the energy use of flue gases was modeled by IPSEpro software of Sim-tech company.

3 Systems of VAM energy use from IUMK-1000 installation

Main objective of the topic is to develop a heat recovery system cooperating with the IUMK-1000 installation (figure 2). The premise is the production of electricity and also heat for central heating when it is technically feasible and justified in economic terms.

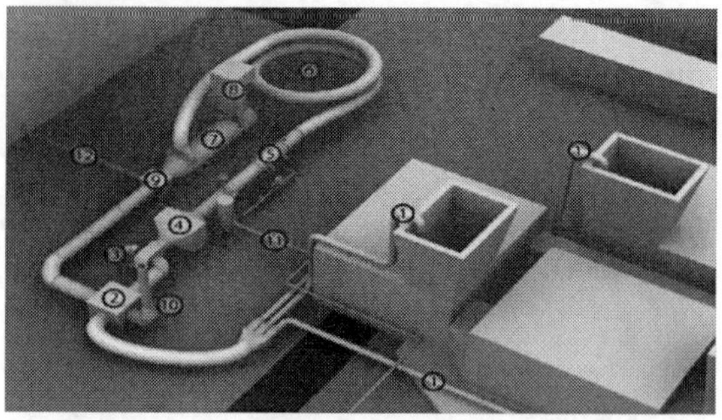

Figure 2. Scheme of installation for methane utilization UMK-1000 : 1. Intake Air VAM, 2. Air dryer, 3. Fan, 4. Air Filter,5. Mixing of gases, 6. Buffer gas, 7. Gas-gas heat exchanger, 8. Catalytic reactor, 9. Gas-water heat exchanger, 10. Outlet gases, 11. Pipeline of methane, 12. Hot Water [4]

In the case of VAM disposal installations developed by the Methane Utilization of Mines Consortium heat source will be gases from the flow-reversal reactor at a temperature of 160 °C (or 500 °C after the system modification).

To develop the model following assumptions were adopted:

Table 1. Parametrs IUMK-1000 [4]

Parametr	Value	-
Stream mass VAM	12.94	kg/s
Pressure VAM	1.15	bar
VAM temperature at the outlet of GAS-GAS exchanger	380	°C
Flue gas temperature at the outlet of the catalytic reactor	500	°C
Useful thermal capacity	1000	kW
Methan concentration	0.7	%

Based on the assumptions of a Table 2 and IUMK-100 operation data base (Figure 3) system was developed by the Ipse Pro software.

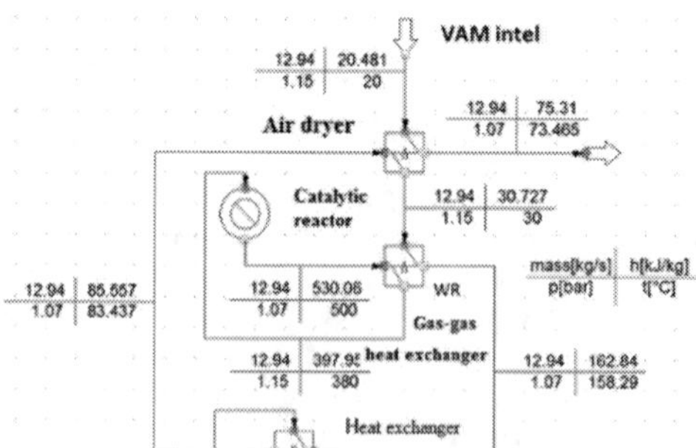

Figure 3. IUMK-100 scheme in IPSEpro software

Applicable technologies using the heat source at the low temperature:

- Steam cycle (Clausius-Rankine cycle) - classic steam cycle power plant, in which water is the thermodynamic factor (water vapor).

It is justified to use steam cycle when the heat source temperature is at least at about 380 °C [11].

- ORC (Organic Rankine Cycle) - The principle of this cycle is the same as in the Clausius-Rankine cycle, excluding the thermodynamic factor, wherein the medium is low - boiling.
- Kalina cycle - is the cycle designed by Alexander Kalina in 1970. It is also a modification of the Clausius-Rankine cycle. In this case, the working medium is a mixture of water and ammonia [17].
- TFC systems (Trilateral Flash Cycle) - is a relatively new solution that enables the use of a dry or wet steam or hot water in the steam screw engine. The essential difference compared to the other cycles is to replace the steam evaporation of water, by a depressurization connected to the vaporising [18]
- Absorption Technologies and heat pumps - the use of absorption chillers for the purposes of air conditioning or heat pump equipment for the purposes of air conditioning and heating [16]
- The internal combustion engine - the use of an external heat source to drive a Stirling engine or a steam engine.
- Thermoelectric devices - are devices that use thermoelectric effect in semiconductor materials. In the present case it is possible to use the Seebeck effect, that is the formation of electromotive force in a cycle made of two metals or semiconductors, which connectors are located in the centers at different temperatures [12].

The rest of the work submitted for consideration will be the steam cycle and ORC cycle. This involves assumptions about the system designing and the limitations resulting from the use of other technologies.

The basic criterion is the ability to produce electricity and heat in combination with acceptable installation performance and effort. Moreover, the advantage of selected technologies should be their current use in real-life conditions and the possibility of basing on a certain experience of their operation.

3.1 Analysis of the results

The various concepts of heat recovery will be presented, and the results of the tests will be described and interpreted. Then an analysis of the efficiency and performance of the proposed system will be carried out. On their basis the best solutions will be selected. There were 2 variant simulations including ORC application and 1 steam cycle variant as a system for electricity production. For the case 1 (Fig. 4a, 4b) and the ORC cycle with R123 and R245fa as working medium.

In ORC unit for a cooling of condeneser water with parametrs 12/55 °C is used but in steam cycle for the cooling of heat exchanger (in this case heat exchanger serves as a condernser) water on parametrs 12/55 °C is used. In the second case (Fig. 4c, 4d) of the ORC cycle, system is provided with an internal regeneration. Assumptions adopted as in case I, in addition to the system evaporator, turbine, condenser and pump also a regenerative heat exchanger is included.

Option 4e steam system realization requires IUMK-1000 installation modification. This involves a low (approx. 101 °C) temperature of the flue gas leaving the gas-gas heat exchanger.

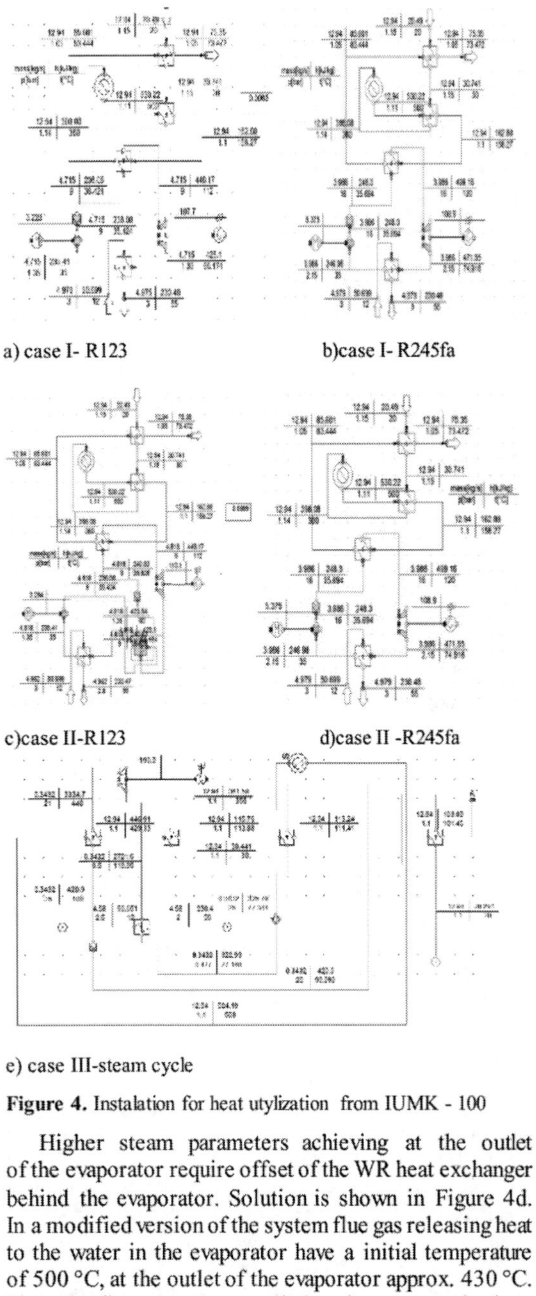

a) case I- R123 b)case I- R245fa

c)case II-R123 d)case II -R245fa

e) case III-steam cycle

Figure 4. Instalation for heat utylization from IUMK - 100

Higher steam parameters achieving at the outlet of the evaporator require offset of the WR heat exchanger behind the evaporator. Solution is shown in Figure 4d. In a modified version of the system flue gas releasing heat to the water in the evaporator have a initial temperature of 500 °C, at the outlet of the evaporator approx. 430 °C. Then the flue gases are supplied to the regenerative heat exchanger where heat the VAM to the desired temperature. In this way it is possible to obtain satisfactory performance of the steam and maintain autothermal operation. Steam system was modeled as ORC cycle, for one parameters cases of the cooling water 12/55 °C. The following parameters were applied for the vapor leaving the evaporator: steam temperature – t_1 = 440 °C, vapor pressure – p_2 = 60 bar. The increased vapor pressure significantly improves the efficiency of the system. Cycle can be implemented at a pressure of 100 bar, but due to the scale of the investment, it is unlikely that the use of steam at such pressure. Fot the parameters selection also results from the WestVAMP system operating in Australia (T = 450 °C, p = 58 bar) were taken into consideration.

Simulations were conducted for determining the suitability of individual cycles and their variants for heat recovery in IUMK-1000 system.

Table 2. Summary of the calculations results.

	Case	Factor	Regeneratio n	Cooling water parmaetrs [°C]	η_{netto} [%]	N_{el_netto} [kW]
1	Ia	R123	–	12/55	9.68	96.82
2	Ib	R245fa	–	12/55	9.56	95.64
3	IIc	R123	+	12/55	9.89	98.93
4	IId	R245fa	+	12/55	10.16	106.35
5	IIIe	water	–	12/55	18.83	188.26

The results of the thermodynamic cycles simulations lead to the conclusion that the most preferred one is the steam cycle with efficiency reaching 18.83 %. However, it requires substantial modification of the IUMK-1000 installation.

Cogeneration systems are able to achieve efficiency in the range from about 9.56% to 10.16%. The hot air stream at a temperature of 500 °C is much better and easier to convert the heat source than the air stream at a temperature of 160 °C. Temperature of heat source determines the value of a exergy stream. Temperature increase causes an increase of exergy stream and at the same useful work results in decreased exergy efficiency.

4 Summary

Management of ventilation air methane from coal mines is a serious problem of coal mining. It should be considered both in terms of environmental protection and the economy. Methane leaving the shafts of mines contributes to adverse changes in the Earth's climate. Equally important it is the under-utilization of a very large number of potential fuel, which is VAM. Existing technologies allow for methane disposal from ventilation air are at an early stage of development and are facing many problems. The main barrier is in any case need to maintain the minimum required concentration of methane in a ventilation air stream. Often this condition can be met only by a VAM mix with available gas mining. Catalytic flow-reversal reactors are the most common methane utilization technology. VOCSIDIZER reactors work in many installations around the world and further investments are planned with their use. Despite the wide use of this technology it is still in the phase of development. Use of heat from the IUMK-1000 reactor is a basic problem of this work. Among the technologies that use low-temperature heat sources to produce electricity Rankine cycle with thermodynamic factors: water and R123 and R245fa organic mediums was choosen. Models of steam and ORC cycles using heat from the IUMK-1000 reactor were created by IPSEpro sotfware. The analysis shows that the most effective solution is to modify the installation and use of steam cycle (efficiency of 18.83%). The efficiency of ORC cycle ranges from 9.5 to 10%. Commisioning of the IUMK-1000 reactor plant and its performance over a longer time may allow for proper evaluation and assumptions revision for the heat recovery installation. The IUMK-1000 reactor can be a good solution for the problem of VAM in Polish coal mines.

Acknowledgements

The article was funded from government money Faculty of Energy and Fuels number 11.11.210.216

References

1. E. Mokrzycki, A. Uliasz-Bocheńczyk, Polityka energetyczna **9**, 1 (2006)
2. N. A. Madlool, R. Saidur, M.S. Hossain, N.A. Rahim, Renew. Sust. Energ. Rev. **15**, (2011)
3. A. Atmaca, R. Yumrutas, Appl. Therm. Eng. **66**, (2014)
4. S. Nawrat, Uczelniane Wydawnictwa Naukowo–Dydaktyczne, Kraków (2009)
5. Rozporządzenie Ministra Gospodarki z dnia 28 czerwca 2002 r. w sprawie bezpieczeństwa i higieny pracy, prowadzenia ruchu oraz specjalistycznego zabezpieczenia przeciwpożarowego w podziemnych zakładach górniczych (Dz.U. 2002 nr 139 poz. 1169)
6. Wyższy Urząd Górnictwa, *Stan bezpieczeństwa i higieny pracy w górnictwie w 2012 roku*. Katowice (2013).
7. *Assesment of the Worldwide Market Potential for Oxidizing Coal Mine Ventilation Air Methane*. United States Environmental Protection Agency (2003)
8. *Ventilation Air Methane (VAM) Utilization Technologies. United States Environmental Protection Agency* (2010).
9. MT250 Series Microturbine. 250kW Continuous Onsite Electrical Power with Integrated Heat Recovery 2015..
10. B. Stasińska, *Katalityczne utlenianie metanu z powietrza wentylacyjnego kopalń*. Towarzystwo Wydawnictw Naukowych Libropolis, Lublin (2014).
11. J. Duda, *Dlaczego ORC jest najlepszym rozwiązaniem do wykorzystania energii odpadowej w cementowni*. Prace Instytutu Ceramini i Materiałów Budowlanych, nr 9, s. 32–43 (2012).
12. L.E Bell, Science, **321** (2008)
13. Forster, *Climate Change 2007: The Physical Science Basis. Contribution of Working Group I to the Fourth Assessment Report of the Intergovernmental Panel on Climate Change*. Cambridge University Press, Cambridge, United Kingdom and New York, NY, USA, s. 210 – 216 (2007).
14. VOCSIDIZER Regenarative thermal VOC oxidation, Materiały reklamowe producenta, MEGTEC Systems (2015).
15. S. Nawrat, S. Napieraj, AGH Journal of Mining and Geoengineering, **36**, 3, Kraków (2013).
16. C.J. Ross, *An overview of industrial waste heat recovery technologies for moderate temperatures less than 1000 °F*. U.S. Department of Energy (2013).

17. B. Saadatfar, R. Fakhrai, T. Fransson, The Journal of Macro Trends in Energy and Sustainability, **2**, 1 (2014)
18. I.K. Smith, N. Stosic, C. Aldis, *Trilateral flash cycle system. A high efficiency power plant for liquid resources.* Materiały konferencji "World Geothermal Congress" (1995).

Utilization of waste heat from rotary kiln for burning clinker in the cement plant

Karol Sztekler[1,a], Maciej Komorowski[2], Milena Tarnowska[3] and Łukasz Posak[4]

[1,a] *AGH University, Faculty of Energy and Fuels, Mickiewicza Av. 30, 30-059 Krakow, Poland*
[2] *AGH University, Faculty of Energy and Fuels, Mickiewicza Av. 30, 30-059 Krakow, Poland*
[3] *AGH University, Faculty of Energy and Fuels, Mickiewicza Av. 30, 30-059 Krakow, Poland*
[4] *Cemex Polska SP.zo.o, ul. Mstowska 10, 42-240 Rudniki*

Abstract. Cement subsector next to the glass industry is counted among one of the most energy-intensive industries, which absorbs approx. 12-15% of the total energy consumed by the industry. In the paper various methods of energy consumption reduction of in the cement industry are discussed. Cement production carries a very large emissions of greenhouse gases, where CO_2 emissions on a global scale with the industry than approx. 5%. Great opportunity in CO_2 emissions reduction in addition to the recovery of waste heat is also alternative fuels co-firing in cement kilns [1], [2]. In the cement sector interest in fitting-usable waste energy is growing in order to achieve high rates of savings and hence the financial benefits, as well as the environment ones [3]. In the process of cement production is lost irretrievably lot of energy and reduction of these losses on a global scale gives a visible saving of consumed fuel. The aim of this study is to investigate the possibility of waste heat use in Rudniki Cement Plant near to Czestochowa. After analyzing of all waste heat sources will be analyzed the heat emitted by radiation from the surface of the rotary kiln at the relevant facility. On the basis of thermal-flow calculations the most favorable radiative heat exchanger will be designed. The calculations based on available measurements provided by the cement plant, a thermal power of the heat exchanger, the heat exchange surface, the geometry of the heat exchanger, and other important parameters will be established. In addition the preliminary calculations of hydraulic losses and set directions for further work will be carried out. Direct benefits observed with the introduction of the broader heat recovery technology, is a significant increase in energy efficiency of the industrial process, which is reflected in the reduction of energy consumption and costs. Indirectly it leads to a reduction of pollution and energy consumption.

1 Introduction

For several years there has been a trend towards the energy efficiency increase of processes and CO_2 emission reduction in accordance with the European Energy and Climate Package 3x20. By reusing the recovered heat is observed: reduction of operating costs of devices (up to approx. 30%), their energy efficiency increase, primary energy savings (fossil fuels), as well as reduction of the pollutants emission (mainly Greenhouse gases). In Poland, the level of utilization of the waste heat potential is very low, mainly due to economic issues (long payback period). Therefore, it should seek to study the subject and take the heat recovery in various industries. Cement subsector next to the glass industry is counted among one of the most energy-intensive industries, which absorbs approx. 12-15% of the total energy consumed by the industry. The cement industry carries a very large greenhouse gases emissions, where CO_2 emissions on a global scale from this industry extend approx. 5%. About half of the emissions from the calcination process occurring in rotary kilns and a half from combustion processes. In the cement sector is growing interest in fitting-usable waste energy in order to achieve high rates of savings and hence the financial benefits, as well as environmental. In the process of cement production is lost irretrievably a lot of energy and reduction of these losses on a global scale, although a fraction of a percent give a visible savings of burned fuel. In this work will be sought the best deal modernization related to the recovery of waste heat that can be made in the Cement Plant Rudniki near Czestochowa.

Waste heat recovery depends largely on the temperature and the rate of heat transfer, the availability of the day and the year. Potential and the most commonly used sources of heat recovery in cement plants are flue gases, radiation from the outer surface of the rotary kiln, the heat carried by the hot clinker leaving the furnace and cooling air for fired material. Direct benefits observed with the introduction of the broader heat recovery technology is a significant increase in energy efficiency of the industrial process, which is reflected in the reduction of energy consumption and costs. Indirectly leads to a reduction of pollution or to the consumption reduction of auxiliary energy [4].

[a] corresponding author: sztekler@agh.edu.pl

 DOI: 10.1051/e3sconf/20161000095

2 Methods for waste heat utilization in cement industry

There are several main heat losses generated by the cement kiln installations, which can be efficiently used. These include, in particular: the flue gases from the furnace; hot air from the clinker coolers; radiation from the surface of the rotary kiln. Figure 1 illustrates the heat loss carried by the cement kiln working in the dry method. The greatest loss is the flue gas enthalpy equals of approx. 20% excess air enthalpy derived from the clinker coolers exceeding 12%, and heat loss from the surface of approx. 10%.

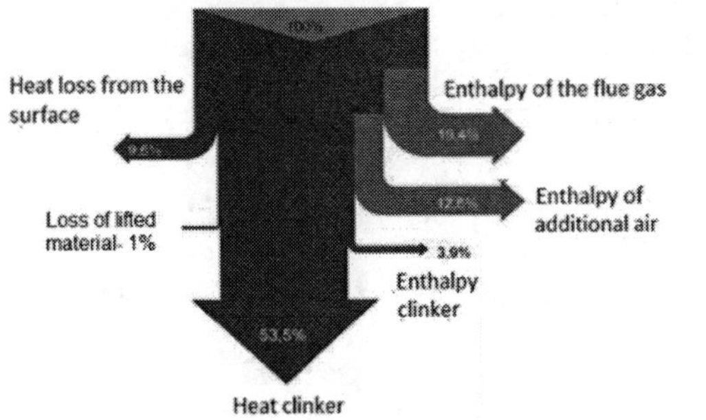

Figure 1. Heat balance of cement kiln [4]

The critical factors that affect the heat loses amount of the rotary kiln are the temperature distribution along the furnace length, thermal stability, thermal conductivity of the refractory within the furnace, environmental conditions, the thermal properties of the outer insulating layer, the rotational speed of the furnace etc. The recovered energy can be used mainly for the electricity generation or hot water preparation [5]. The total yield of the furnace depends on the mechanical efficiency, by appropriately made refrectory within the furnace and the patency of the pipe [6]. The use of waste heat is associated with a visible improvement in thermal efficiency, which translates into reduced fuel consumption, or lower operating costs.

2.1 The use of hot air from the clinker coolers

Rotary kilns are equipped with coolers into which the heated clinker falls out of the furnace to cooling to a temperature of approx. 100 °C. Clinker cooling air heats up and is most often used as primary and secondary air for the combustion of pulverized coal in kiln burner. Furthermore, it can be a factor in drying coal and mills feedstocks. The use of hot air from the clinker cooler as primary air in the combustion of pulverized coal can save a certain amount of heat to warm the air and makes the process faster and there is a higher temperature of fuel combustion [7].

2.2 The heat recovery from the rotary kiln

Hot surface of the kiln, which can reach temperatures of over 400 °C is another significant source of heat loss. Convection losses and radiation may constitute approx. 8-15% of the energy input. Area of the furnace requires frequent observation by the operator to observe the local combustion of the surface, to avoid significant loss of refractory material inside the furnace. The secondary layer may be made of stainless steel, because it has a relatively low emissivity [5].

However, a more interesting use of the external heat exchanger, which are arranged around the furnace and also fulfill the task of the secondary coating but can recover greater amounts of heat.

2.3 The use of heat of flue gases and clinker cooling air

The most profitable in cement plant is the use of waste heat from the excess air from the clinker cooling and flue gas exiting the furnace (about 30%). Effective heat recovery is possible by the use of cogeneration systems in which the same manufacturing process produces two products: electricity and clinker.

The flue gases from rotary kilns reach different temperatures depending on the number grades air heater. Four-stage heater supplies gas temperatures of 300-380 ° C and 5 ÷ 6-stage 200-300 °C. The hot air discharged from the clinker coolers can reach 360 ° C [8]. Both streams of air and flue gases can be directed through the WHRSG steam generator (Waste Heat Recovery Steam Generator, Fig. 2). Waste energy in the cycle causes the generation of vapor, which is then used to drive a steam turbine and electricity production. The generated electricity compensates the part of the purchased energy from the power system [5]. The first implementation of the steam cycle in the cement industry took place in the 70s of the last century. Hot gases flow out from rotary boiler can by used to dry coal in cement plant. Currently this technics utilization heat from flue gases is implemented in many installation for instant in Rudniki Cement Plant.

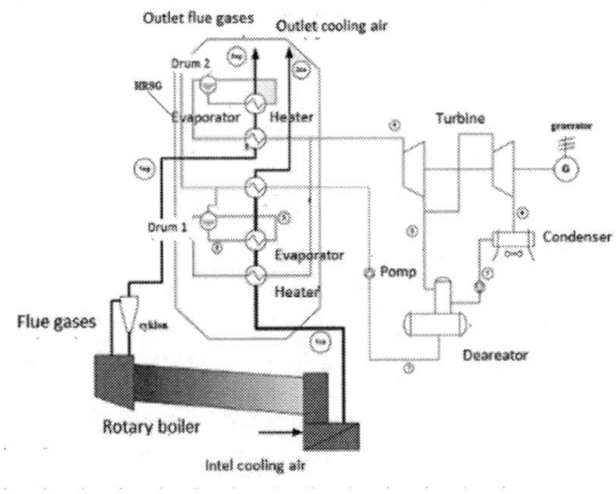

Figure 2. The waste heat recovery system with the Clausiusa-Rankine water - steam cycle [8]

There is also the possibility of electricity generation in a cement plant in other cogeneration cycles: an Organic Rankine cycle (ORC), Kalina cycle, or in the piston steam engine. Very preferred because of its low installation cost and increased efficiency is the use of the TFC system (Trilateral Flasch Cycle) in screw engines where there is no water evaporation, but the expansion with evaporation [11].

Organic Clausius-Rankine cycle (ORC) is often used to recover heat from the energy-intensive production processes in industries: cement, iron, glass industry, the oil and gas industry. ORC is the most effective technology for the recovery of heat and electricity, if it uses a heat source temperatures of 200-400 °C. However, ORC cycle efficiency is limited, and typical examples reach up to approx. 20%.

Figure 3 presents the example of ORC cycle in which there is water flow at high pressure of 30 bar. The first choice of working medium for ORC system is isopentane, because of the highest achievable efficiency of the cycle.

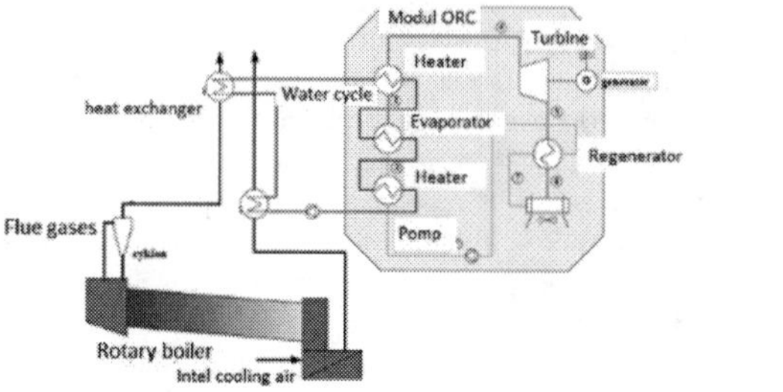

Figure 3. The heat recovery system using the ORC [8]

CR water Steam cycle is more efficient than the ORC when the flue gas temperature is greater than 310 °C. For cement plants that use old technology and flue gas temperature is high, preferred system is the use of waste heat recovery system with CR water steam cycle. But for modern cement plants with higher efficiency and lower temperatures there is advisable to use ORC system [8]. In addition, ORC cycle is more flexible and disposable, can operate at a partial load, and has a wide range of stable operation, while the water steam cycle needs a stable and more uniform operation conditions [9].

The association of rotary kiln with the electricity generation system is beneficial in terms of economic, environmental, and energy for cement plants. Cogeneration among others: facilitates uninterrupted operation of the furnace, apparently increases fuel efficiency, reduces the consumption of refractory materials and allows for higher clinker quality.

According to [9] in the EU-27, there were 259 cement plants, 389 rotary kilns with a total capacity of 247.8 Mt/year. It was estimated that 241 cement plants is possible to use ORC systems whose production potential could reach approx. 576 MW. However, this work is focused on the possibility of heat recovering from the cement shell of the rotary cement kiln which generates more than 10% of the heat losses.

3 Disposal of waste heat in the Rudniki Cement Plant

The possibilities of waste heat use from the cement industry are very large and provide a significant economic and environmental savings, associated with emissions reduction. In this paper, an analysis of the use of waste heat emitted by the radiation cement kiln in the Rudniki Cement Plant near to Czestochowa is presented. Calculations of two variants of the heat exchangers characterized by different dimensions, power and the degree of coverage of the heating needs of the plant were performed. Then construction solution of the radiation heat exchanger is designed and directions for further work was suggested. Currently in the cement plant are 2 utilization ways of waste heat emitted by the installation of the rotary kiln. Produced clinker is cooled in a grate cooler and for cooling process is use cold air. The heated air by the hot clinker is used as secondary air into the furnace, the heat losses are smaller and the heating process of the supplied material is faster. In 2003 the hot gases pipeline received from the furnace which are used for drying of the raw material fed for the raw meal preparation in two ball mills with capacities of 115 t/h and 40 t/h. In 2014 t another second pipeline of flue gases supplied to the coal department was built, where they are used for coal drying in a ball mill. The process is very safe, due to the reduced content of O_2 in the gases (4-8%). Both formed pipes are derived from the dusts chamber. There is no heat recovery from the rotary kilns.

3.1 The concept of heat recovery from the rotary kiln

Rotary kilns by radiation, emit substantial amounts of heat which can be usefully used, eg. for domestic hot water heating used to meet the needs of the plant. The heat to the ambient air through the rotary kiln takes place by radiation and convection. However, in the zone of highest temperature is dominated by heat transfer through radiation (approx. 4 times higher heat transfer coefficient by radiation than convective one) [10].

The heat exchanger must be placed in the zone of the furnace, where the highest temperature, at an appropriate distance from the kiln shell. Taking into account the operational and technological considerations, heat exchanger must not obstruct the operation, repair work or maintenance, so it is expected to placement in suitable distance from the kiln.

Heat exchanger distance from the furnace should be as small as possible, but the above reasons, considering the application of the mechanism governing this distance. Most preferred would be to place the heat exchanger above the kiln, but for the implementing reasons heat exchanger will be located under the furnace equipped with a frame on wheels that can be easily moved. In addition, it can't be in the area of the furnace throat, wherein the diameter changes from 3.45 m to 3,75m, in areas where supports are provided, and where run frequently used road transport. Figure 4 shows the selected location of the radiation heat exchanger placement at a rotary kiln no. 4.

Figure 4. Rotary boiler number 4

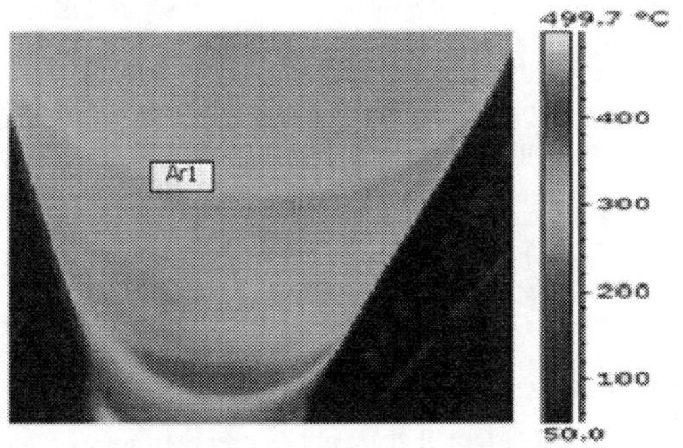

Figure 5. Distribution of kiln shell temperature in section of 16-28 m

Thermal imaging was used for calculations to measure the temperature distribution on the mantle of the furnace (Fig. 5) in a section of 16-28 m initial length of the furnace, where there was on a selected range of Ar1 a time maximum temperature of 348.7 °C.

Two variants of heat exchangers with different lengths will be taken into consideration. For comparison

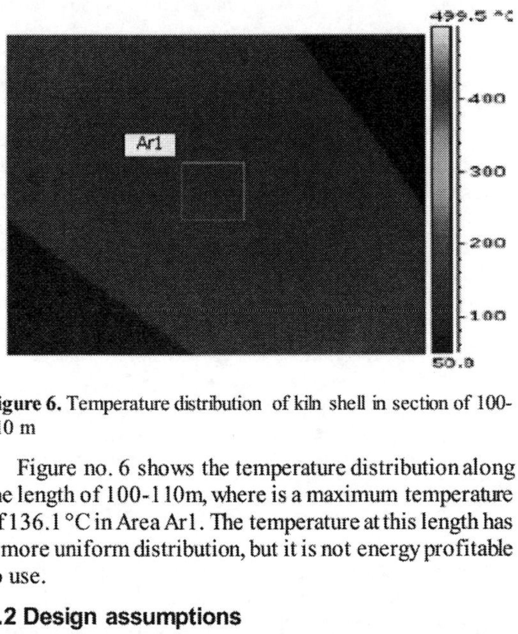

Figure 6. Temperature distribution of kiln shell in section of 100-110 m

Figure no. 6 shows the temperature distribution along the length of 100-110m, where is a maximum temperature of 136.1 °C in Area Ar1. The temperature at this length has a more uniform distribution, but it is not energy profitable to use.

3.2 Design assumptions

It was assumed that the heat exchanger will be made in the form of single-screen seamless steel pipes. The screen will be made in the form of cylinder surface covering 1/6 of the circumference of the rotary kiln, which is the solid angle of 60° at a minimum distance of 1 m from the kiln.

The calculation of the configuration factor and heat flux absorbed by the wall of the heat exchanger of the rotary kiln, the thermal power consumed by the heat exchanger, water velocity in the exchanger tubes, hydraulic losses, heat flux acquired by the exchanger tubes of the rotary kiln were made on the basis of literature references [12,13,14]. Heat exchanger will be equipped with a mechanism for increasing its distance from the kiln shell.

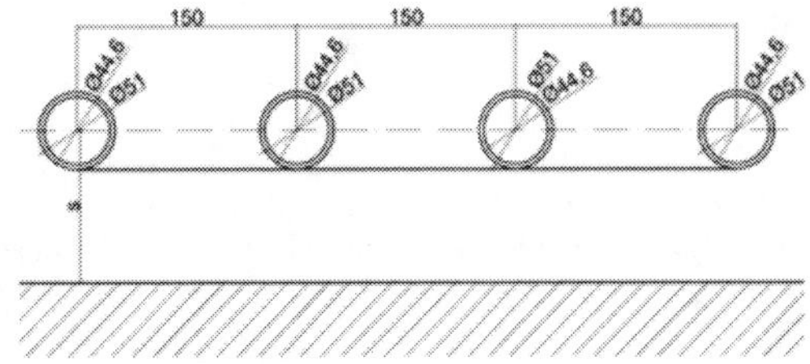

Figure 7. Founded screen size with irradiated pipes: s- exchanger distance from the rotary kiln

Seamless steel boiler pipes were adopted with dimensions:

$\frac{d_z}{d_w} = \frac{51\,mm}{44{,}6\,mm}$, which will be connected by elbows and welded to the screen made of steel. The distance between the pipes was assumed at 0.15 m (Fig.6). The water will flow through the tubes receiving the heat from radiation, the outlet temperature will equal of 65 °C and will be used to satisfy the own needs of cement plant.

Other assumptions for the calculations:

• averaged temperature of kiln shell in section 16-28m;

• averaged temperature of kiln shell in section 16-24m,

• exchanger body length l_1=12m, l_2=8m,

• the maximum daily consumption of hot water in the cement plant in 2014 increased by approx. 30% - 10m^3, which is the average volume flow of 0.12 l/s.

The calculations were performed for the temperature of the heat exchanger wall pipe t_2 from 37.6 at 0.1 °C (values exceeding the value of the reference temperature t_{fm}) for comparing the obtained values of the intensity of the heat flux emitted by radiation through the rotary kiln and absorbed (acquired) through the heat exchanger to choose the right temperature.

Designed heat exchanger coil is recuperator, or apparatus in which two factors: air and water are separated from each other by the pipe wall through which the heat transfer. It is a device in which the temperature field is established and the heat exchange process is continuous.

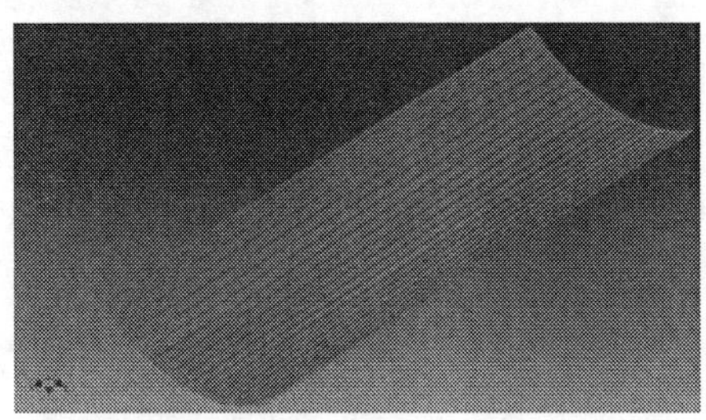

Figure 8. Designed radiation heat exchanger

For the calculations carried out for a heat exchanger minimum distance from the heat exchanger furnace s = 1 m is designed (Fig. 7, 8) wherein: the length of the heat exchanger $l \approx 8$m, the maximum thermal power $\dot{Q}$ ≈65kW; the speed of flowing water v = 0.182 ÷ 0.023 m/s; the water exit temperature from the heat exchanger t_{wy}=65°C; volume flow $\dot{V}$=0,285÷0,036 dm^3/s; size of hydraulic losses Δh_c=0,204÷0,006 m; heat exchanger pipe diameters: d_z = 51 mm, d_w = 44,6mm; the length of the coil $L \approx 161$m; the amount of steel tubes $n = 20$; the number of 90° elbows n_k=38; steel screen size 8,4m x 3,19m; Screen thickness of steel plate: δ = 6mm; the total weight of the heat exchanger > 1.34 t.

Figure 8 shows a heat exchanger visualization, and Fig. 9 an exemplary installation in which the heat exchanger can be operated.

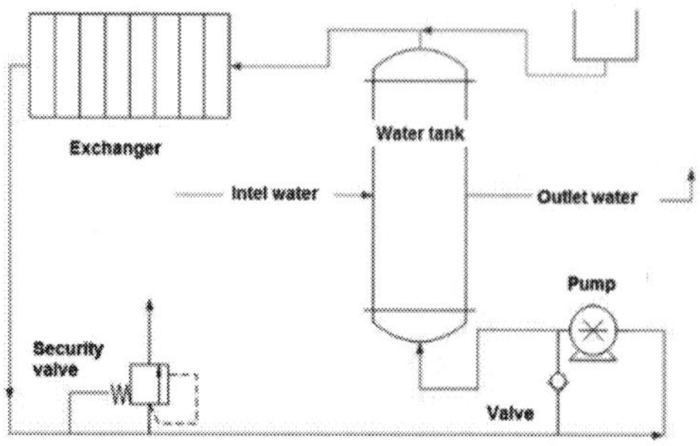

Figure 9. Domestic hot water installation diagram using the designed coil heat exchanger.

It is necessary to equipe an installation in domestic hot water tank of high-capacity, eg. 3000 ÷ 5000 l. Additionally, a buffer tank can be used to the accumulation of excess recovered heat eg. using PCM phase change materials. It is proposed that the exchanger is mounted on the movable platform with dimensions that ensure the stability of the structure. The platform can move on 6 steel rolls, which are movable on steel rails (due to the need to move apart during the overhaul of the furnace heat exchanger, the heat exchanger and in other situations where the heat exchanger is impeded of the kiln). Implementation of the exchanger distance control to the rotary kiln, depending on the demand of domestic hot water may be carried out by a hydraulic system.

4 Conclusions

Thanks to designed heat exchanger with a maximum capacity of approx. 65 kW cement plant will reach a high efficiency of burned fuel. The plant will be able to afford to completely disable the boiler room in summer (May-September), while in winter they will be aided working exchanger. Due to the relatively low density of the heat flux emitted by the kiln shell designed heat exchanger has a fairly large dimensions.

A very important element is to provide a heat exchanger in a continuous temperature measurement of the water input and output, which will give an indication to the approximation, or the heat exchanger distance from the kiln shell depending on the current needs of the plant for the domestic hot water. It is also possible to build several heat exchangers connected together in parallel and, thus, would be a greater amount of heated water, which would be justified when provided would be the continuous reception. The heat exchangers can also be arranged on the sides and top of the rotary kiln so would also be used with a convection heat from it, but this would lead to high costs and possible difficulties during the operation of the kiln. The way to the use of excessive waste heat is possible to prepare a central domestic hot water for eg. a nearby housing estate. However, this requires significant financial investment, eg. from the local government and the residents.

In summary, designed variant of the heat exchanger is a simple and relatively inexpensive investment for easy installation in a cement plant. The rotary kiln does not require the exclusion of the time of installation. In addition, heat has no influence on the quality of the final material produced in the plant and the whole technological process.

Ackowledgements

The article was funded from government money Faculty of Energy and Fuels number 11.11.210.216

References

1. E. Mokrzycki, A. Uliasz-Bocheńczyk, Polityka Energetyczna, **9**, 1 (2006)
2. N.A. Madlool, R. Saidur, M.S. Hossain, N.A. Rahim, Renew. Sust. Energ. Rev. **15**, (2011)
3. A. Atmaca, R. Yumrutas, App. Therm. Eng. **66**, (2014)
4. J. Duda, *Dlaczego ORC jest najlepszym rozwiązaniem do wykorzystania energii odpadowej w cementowni?* Prace Instytutu Ceramiki i Materiałów Budowlanych, Wydawnictwo Instytut Śląski Sp. z o.o, s.32-43 (2012)
5. T. Engin, V. Ari, Energ. Convers. Manage. **46**, (2005)
6. Projekt decyzji BC-10, *Zalecenia techniczne dotyczące bezpiecznego środowiskowo współprzetwarzania niebezpiecznych odpadów w piecach cementowych.* Kartagena, Kolumbia, Konferencja Stron Konwencji Bazylejskiej (2011).
7. S. Żurakowki, S. Hojarczyk, *Piece obrotowe-projektowanie i konstrukcja*. Warszawa, WNT (1969).
8. S. Karellas, A. Leontaritis, G. Panousis, E. Kakaras, Energy **58**, (2013)
9. F. Campana, M. Bianchi, L. Branchini, A. De Pascale, A. Peretto, M. Baresi, Energ. Convers. Manage. **76**, (2013)
10. M. Poniewski, T. Orzechowski, Cz. Cesarz, R. Chajewski, K. Gołaszewski, M. Lubowicz, *Utylizacja ciepła odpadowego cz. II, Zlec. nr 1839/13*, Kielce, Politechnika Świętokrzyska, Samodzielny Zakład Termodynamiki i Mechaniki Płynów (1993).
11. J. Duda, *Dlaczego ORC jest najlepszym rozwiązaniem do wykorzystania energii odpadowej w cementowni?*. Prace Instytutu Ceramiki i Materiałów Budowlanych, Wydawnictwo Instytut Śląski Sp. z o.o., s.32-43 (2012).
12. J. P. Holman, *Heat Transfer*. 10th Edition, Nowy York, McGraw-Hill (2010).
13. T. Hobler, *Ruch ciepła i wymienniki*. Warszawa, Państwowe Wydawnictwa Techniczne (1959).
14. U. Lubczyńska, *Hydraulika stosowana*. Kielce, Politechnika Świętokrzyska (1996).

E3S Web of Conferences **10**, 00096 (2016) DOI: 10.1051/e3sconf/20161000096
SEED 2016

A novel proposal of cooperation between a cage induction generator and the three-phase grid

Zbigniew Szular[1] and Witold Mazgaj

Cracow University of Technology, Institute of Electromechanical Energy Conversion, Kraków, Poland

Abstract. In the paper, a new concept of the cooperation between a cage induction generator and the three-phase grid is presented. The induction generator is excited by a capacitor battery connected to the stator winding terminals. In order to ensure a stable operation, the battery capacity should be continuously adjusted. Three appropriate selected capacitors are connected to the stator winding terminals by means of two bi-directional controlled switches. The generator output voltage is stabilized by a suitable control of these switches. The stator winding terminals are also connected to the diode bridge rectifier. Energy in the DC bus is converted to the parameters of the three-phase grid (400 V, 50 Hz) by means of a three-phase voltage source inverter. Examples of numerical calculation results of the proposed energy generation system are presented in the end part of the paper.

1 Introduction

Two basic types of electrical generators are applied in wind plants and small water plants. Induction generators belong to the first group; the second one includes synchronous generators, first of all, the permanent magnet synchronous generators [1, 2]. Existing systems of energy generation should fulfil, among other, the following conditions [3]:

- generators can operate at a wide range of the turbine velocity; it refers especially to the wind turbines,
- energy generation systems should have a high efficiency of energy conversion,
- higher harmonics of the generator currents should be relatively low; otherwise a certain variable component of the generator torque occurs,
- energy generation system should be inexpensive, reliable and simple in operation.

Energy generation systems in wind and small water plants are often built on the basis of a squirrel cage induction generator (SCI generator) or a double fed induction generator (DFI generator). In the first systems, the SCI generators have been connected to the three-phase grid directly or by means of a soft-starter (Fig. 1). The soft-starter has been used to reduce an inrush current which occurs during connection of the generator to the grid; usually the soft-starter is by-passed in normal operation conditions [4 - 9]. In this case the velocity of the SCI generators can vary within a narrow range; a generator slip is equal to 1 - 2 %. Due to this fact it is not possible to completely use wind energy or water stream energy in a wide range of the generator velocity changes. Mentioned drawbacks of the SCI generator operation do not occur when the generator is connected to the grid by means of a back-to-back converter

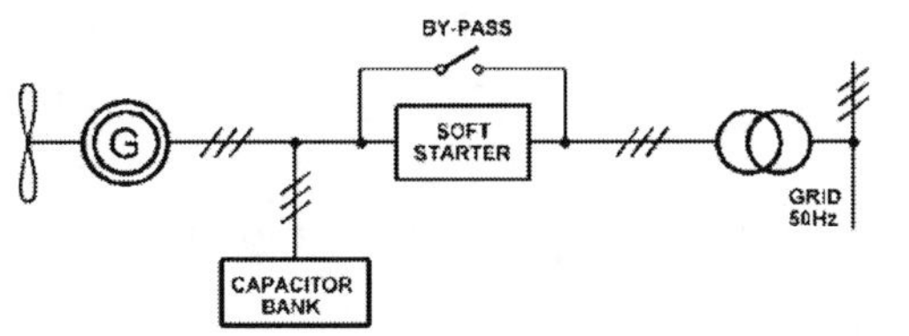

Figure 1. Energy generation system with the SCI generator directly connected to the three-phase grid.

(AC/DC/AC converter) [1 - 4, 8 - 15]. This energy conversion system is shown in Figure 2. With the use of this system, the wind or the water stream energy can be converted to the grid in a wide range of the generator velocity. Reactive power which is necessary to produce a magnetic flux in the SCI generator is provided from a capacitor in the DC bus by means of a PWM rectifier. Vector control methods are usually implemented in these systems in order to keep the DC voltage at an assumed level [11, 13].

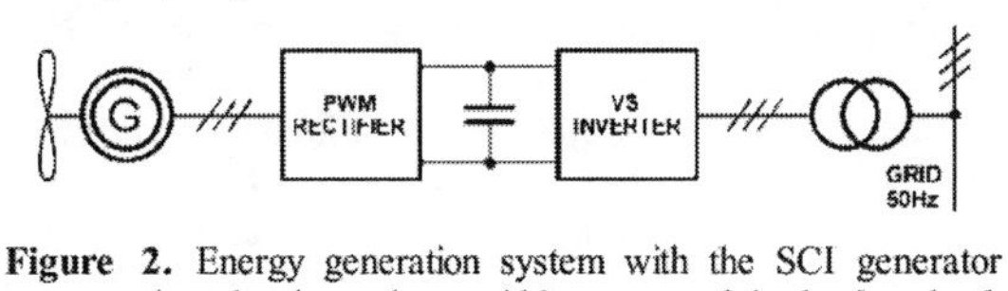

Figure 2. Energy generation system with the SCI generator connected to the three-phase grid by means of the back-to-back converter.

Quite frequently in energy generation systems the DFI generators are used. In this case the stator of the given DFI generator is directly connected to the three-phase grid, whereas the rotor is connected to this grid by means of the back-to-back converter [1, 4, 6, 8, 9]. In these systems the generator velocity can change in the range ± 30 % with respect to the nominal velocity [1].

[1]Corresponding author: aszs@poczta.fm

The presence of slip rings and brushes requires relatively frequent maintenance and it can cause failures in these generators.

The synchronous generators, especially generators excited by permanent magnets, are almost always connected to the three-phase grid via the back-to-back converters [2, 4, 9, 16-20]. These energy conversion systems can operate even at low wind or water stream speed.

2 Topology of a novel energy generation system with the SCI generator

Considering various energy generation systems, the squirrel cage induction generators seem to be an advantageous solution in wind and small water plants, because they have simple construction, low costs and low requirements for the operation and maintenance. The complete use of the mentioned properties requires an application of the back-to-back converters. However, it results in significant complication of the control method of this converter. The advantageous properties of the SCI generators have encouraged us to develop a simpler energy generation system with the use of this generator.

It is worth underlining that a voltage source inverter (VS inverter) and an additional transformer has to be applied on the grid side because the content of higher harmonics in the current flowing from the VS inverter to the three-phase grid should be limited to the level specified by the relevant regulations. However, the requirements for the harmonic content in the generator currents are not critical. Therefore, instead of the PWM rectifier we propose to apply a simpler system whose topology is shown in Figure 3.

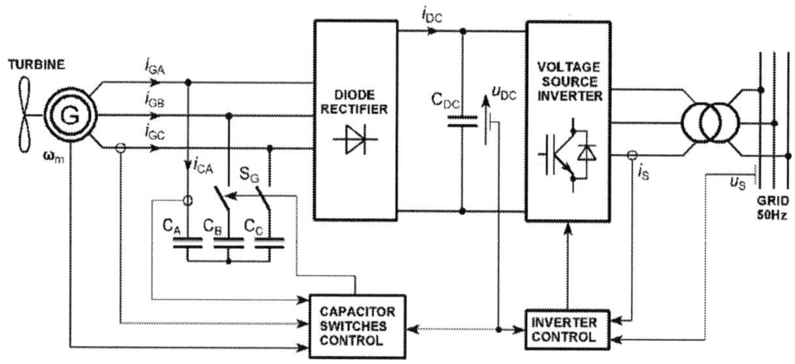

Figure 3. Topology of the proposed energy generation system: G – SCI generator, S_G – full-controlled bidirectional switches.

The stator terminals of the given SCI generator are connected to the diode bridge rectifier. Additionally, these terminals are connected to the three capacitors C_A, C_B, C_C via two full-controlled bidirectional switches S_G; the capacitors are connected in a star configuration. These switches can be implemented as the two series connected Insulated Gate Bipolar Transistors or as only one transistor of this type which is included in the transverse branch of the Graetz rectifier. It is necessary to stress that both switches are turned on and turned off simultaneously by a common control signal. The diode rectifier is connected to the three-phase VS inverter which is coupled to the three-phase grid by means of a transformer, similarly as in existing energy conversion systems that use the back-to-back converter.

Some doubts may appear due to the occurrence of capacitors on both sides of the diode rectifier. However, during the normal operation of the proposed system, the voltage on the DC bus capacitor is higher than the resultant phase-to-phase voltage of the capacitor battery. Otherwise, a high current impulse could occur after the turn-on of the switches causing their damage. Due to the switching frequency of a few several kHz, the diode rectifier should be constructed on the basis of fast diodes. Requirements for the VS inverter, cooperating with the three-phase grid, are the same as in existing systems of the energy conversion.

3 Principles of the system operation and control strategy

The basic assumption of the proposed system operation and control method is the possibility of the energy generation in a wide range of the wind or water stream speed. It was assumed that the generator velocity usually can vary from 50% to 150% with respect to its nominal angular velocity.

Conversion of the wind or water energy by the SCI generator has to be preceded by its self-excitation. Before beginning of this process the capacitors C_A, C_B, C_C have to be pre-charged to the voltage of several dozen volts. This can be done using a simple electronic circuit and accumulator batteries, which should be installed in each energy generation system. During the self-excitation process the switches S_G are still turned on. This process can take up to a few seconds, but this is insignificant in relation to the time period of the energy generation. The self-excitation depends not only on the initial voltage on the capacitor battery. The generator parameters and wind or water stream speed also influence this process significantly. The capacity of the battery capacitors is determined for the assumed minimum value of the generator angular velocity. It is known that the lower the generator velocity is, the higher should be the battery capacity. The required capacity is determined with taking into account the nonlinear magnetization characteristic of the SCI generator. It should be noted that the capacitor in the DC bus is charged to a voltage which results from the connection between the VS inverter and the three-phase grid.

When the generator output voltage reaches the assumed value during the self-excitation process then the switches S_G are turned off, and the capacitor battery is disconnected from the generator. From this moment the generator currents flow through diode rectifier to the DC bus and electrical energy is transferred from the SCI generator to the three-phase grid by means of the VS inverter (Fig. 4a). At the same time, the magnetizing current of the SCI generator begins to slowly decrease, thereby the generator magnetic flux and the voltage on its terminals also decrease. When the generator magnetic flux drops to the minimum value, which depends on the current generator velocity, then the switches S_G are turned on again and the capacitor battery are connected to the generator terminals. Thus, the generator magnetic flux and the output voltage increase; at the same time the energy provided to the grid via the VS inverter causes a certain decreasing of the voltage on the capacitor C_{DC} in the DC bus (Fig. 4b).

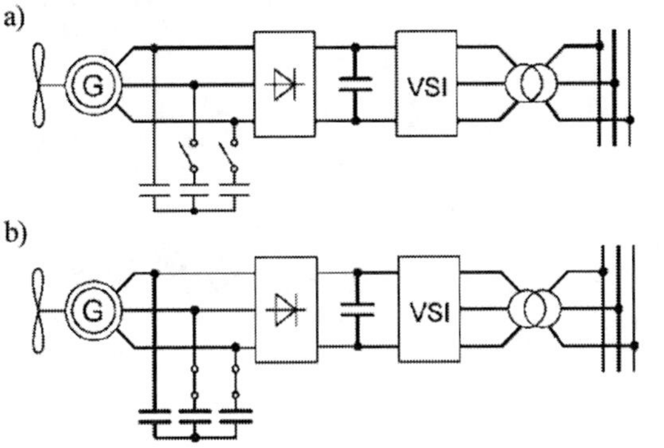

Figure 4. Simplified schematics of the proposed energy generation system: a) switches S_G are turned on, b) switches S_G are turned off.

In practice, the switching frequency is in the range from several hundred Hz to a dozen kHz. Ensuring higher voltage of the capacitor C_{DC} with respect to the highest voltage of the capacitor battery constitutes the basis for the system correct operation. In the time period when the switches S_G are turned off, the voltages of the battery capacitors have a constant value. Due to inductances of the stator windings, the generator currents cannot change its value abruptly. Hence, these currents flowing to the diode rectifier are different from zero; almost always three diodes conduct, unlike the typical operation of such rectifiers. Due to the fact, that the sum of the battery currents is always equal to zero, the current flowing to the DC bus does not change its value during the switching process. It is worth underlining that the capacitor in the DC bus can be discharged by the input current of the VS inverter. As a result, the voltage in the DC bus can be lesser than the maximum voltage of the capacitor battery; it could lead to an unacceptable state of the system operation after the turn-on of the switches S_G (danger of a high impulse current between the capacitor battery and the capacitor in the DC bus). To avoid this risk, the input current of the VS inverter has to be additionally dependent on the DC voltage.

The capacity of the capacitors C_A, C_B, C_C should be determined due to the possibility of the generator self-excitation. The largest capacity is required at the lowest assumed angular velocity at which the generator can still operate correctly. It is known, that the higher the angular velocity is, the smaller can be the battery capacity. The required capacity is determined by the intersection of the nonlinear characteristic of the voltage induced in the generator with a linear characteristic representing the dependence of the capacitor voltage on its current (Fig. 5). It should be emphasized that the normal operation of the SCI generator takes place below the intersection point of these characteristics. Assuming that the voltage in the DC bus can vary within a certain range, the scope of the normal generator operation is shown in Figure 5 as the highlighted area.

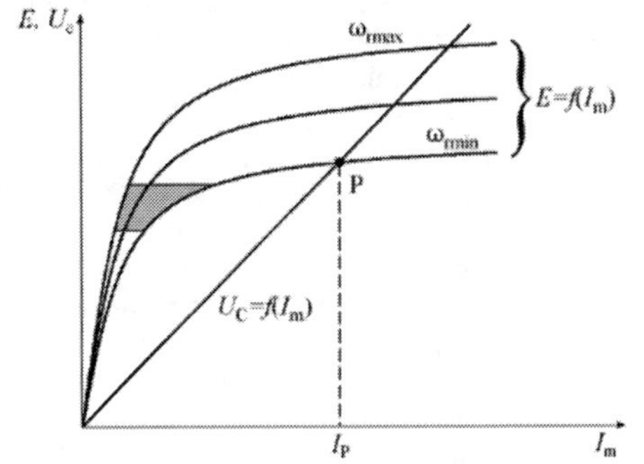

Figure 5. Dependences of the voltage E induced in the generator and the voltage U_C of the capacitor battery on the magnetizing current I_m: P – point of the generator stable operation after its self-excitation.

The self-excitation process should take place in an unloaded state. So, we can assume with a certain approximation that the magnetizing and capacitor currents are equal to each other. On the basis of this condition, we can write the following relationship for the intersection point of the above mentioned characteristics (point P in Figure 5):

$$\psi(I_m)\,\omega_{rmin} = \frac{1}{2\pi f_G C} I_P \tag{1}$$

where: $\psi(I_m)$ – magnetic flux, ω_{rmin} – assumed minimum value of the generator angular velocity, f_G – frequency of the generator current, C – required capacity, I_P – magnetizing current of the SCI generator for the point P, Hence, the capacity should fulfil the condition:

$$C \geq \frac{I_m}{2\pi f_G\, \psi(I_m)\,\omega_{rmin}} \tag{2}$$

The capacitor in the DC bus is selected due to the VS inverter operation. However, voltage fluctuations on this capacitor depend also on the operation of the generator and capacitor battery. The largest increase in DC voltage occurs when the capacitor battery is permanently disconnected from the generator. In this case the almost total magnetic energy associated with the leakage inductances of the generator stator is transferred to the DC capacitor. Hence, it can be written:

$$\frac{3}{4}L_{s\sigma}I_{Gmax}^2 = \frac{1}{2}C_{DC}(U_{DCmax}^2 - U_{DCmin}^2) \quad (3)$$

where: $L_{s\sigma}$ – leakage inductance of the generator stator, I_{Gmax} – maximum amplitude of the generator current, U_{Dcmax}, U_{Dcmin} – admissible maximum and minimum values of the DC voltage, respectively.
So, the capacity should be equal to:

$$C_{DC} = \frac{\frac{3}{2}L_{\sigma s}I_{Gmax}^2}{U_{DCmax}^2 - U_{DCmin}^2} \quad (4)$$

It is necessary but insufficient condition referring to the DC capacitor because its final selection depends on the assumed operation parameters of the VS inverter.

The main principle of the control method is to maintain a constant voltage in the DC bus; in practice, this voltage can vary within a relatively narrow range. The control system generates a control signal, as the PI controller output signal, which is compared with the saw-tooth signal of a frequency of several kHz (Fig. 6). When the voltage induced in the generator should increase (it requires an increase in the magnetic flux), then the value of this control signal has to be appropriately higher. As a result, the time interval in which the capacitor battery is connected to the generator is longer. In time interval t_1 - t_2 the switches S_G are turned on and the capacitor battery is connected to the generator, while in the interval t_2-t_3 these switches are turned off.

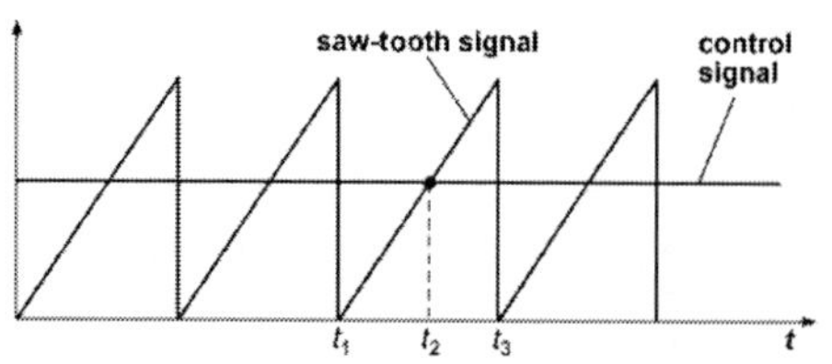

Figure 6. Principle of the capacitor battery switching.

4 Numerical calculations of the proposed system operation

A monoharmonic model of the squirrel cage induction machine was used in numerical analysis of the proposed system operation [21]. It was assumed that the magnetization characteristic is nonlinear but unambiguous. Numerical tests were carried out for the generator with the nominal parameters: power – 22 kW, phase-to-phase voltage – 400 V, angular velocity – 1470 rpm, current – 41 A. For this generator, parameters of its equivalent circuit were following: resistivity of the stator windings – 0.241 Ω, leakage inductance – 1.864 mH, magnetizing inductance – 27.8 mH.

The process of the self-excitation is shown in Figure 7 (the angular velocity is equal to 1500 rpm). The capacity required to the generator self-excitation is equal to 0.32 mF. This process does not last to the point P in Figure 5 because the capacitor battery is disconnected when the generator magnetic flux reaches the required value resulting from the assumed DC voltage and current angular velocity. From this moment the normal operation of the proposed energy generation system begins.

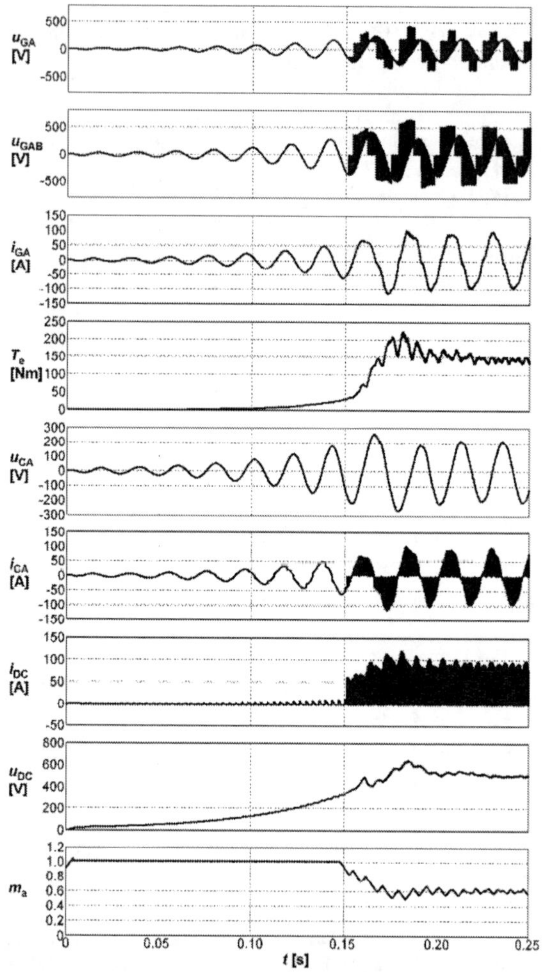

Figure 7. Waveforms of voltages, currents and the generator torque in the proposed energy generation system during the self-excitation process: U_{DC} = 500 V, ω_r = 1500 rpm, C_G = 1.0 mF, C_{DC} = 0.5 mF, f_s = 3 kHz; u_{GA}, u_{GAB} – phase and phase-to-phase generator voltages, respectively, i_{GA} – generator current, T_e – generator torque, u_{CA}, i_{CA} – voltage and current of the capacitor battery, respectively, i_{DC}, u_{DC} – current and voltage in the DC bus, respectively, m_a – ratio of turn-on time of the switches S_G to the switching period.

Figure 8 presents waveforms of chosen voltage and currents and changes of the generator torque during normal operation with the following conditions: U_{DC} = 500 V, ω_r = 1500 rpm, C_G = 1.0 mF, C_{DC} = 0.5 mF, f_s = 3 kHz. In turn, Figure 9 presents analogous waveforms for the minimum value of the generator angular velocity which was assumed as 750 rpm. In this case the number of switching processes is two times higher than before. It should be noted that the smaller the generator velocity is, the higher is the content of higher harmonics in the generator current. Consequently, the variable component

 DOI: 10.1051/e3sconf/20161000096

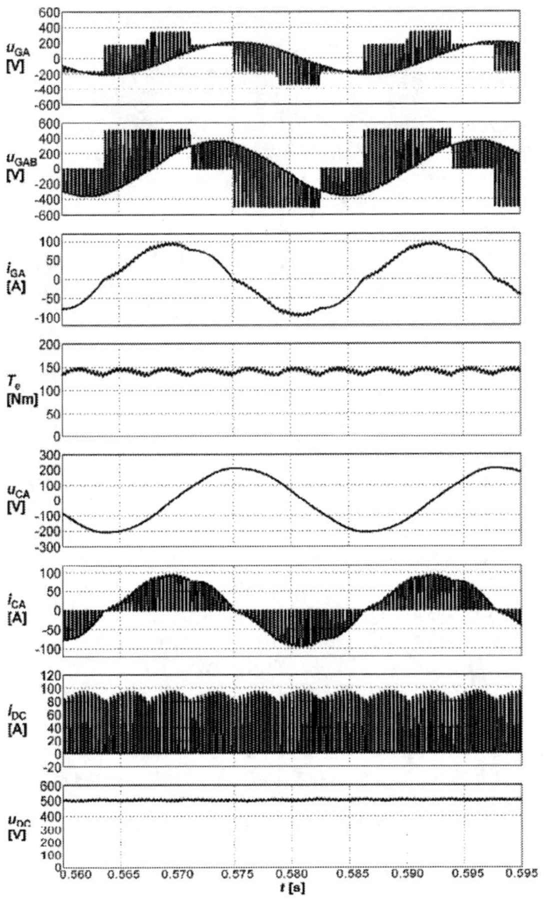

Figure 8. Waveforms in the proposed energy generation system for $\omega_r = 1500$ rpm; descriptions of waveforms as in Fig. 7.

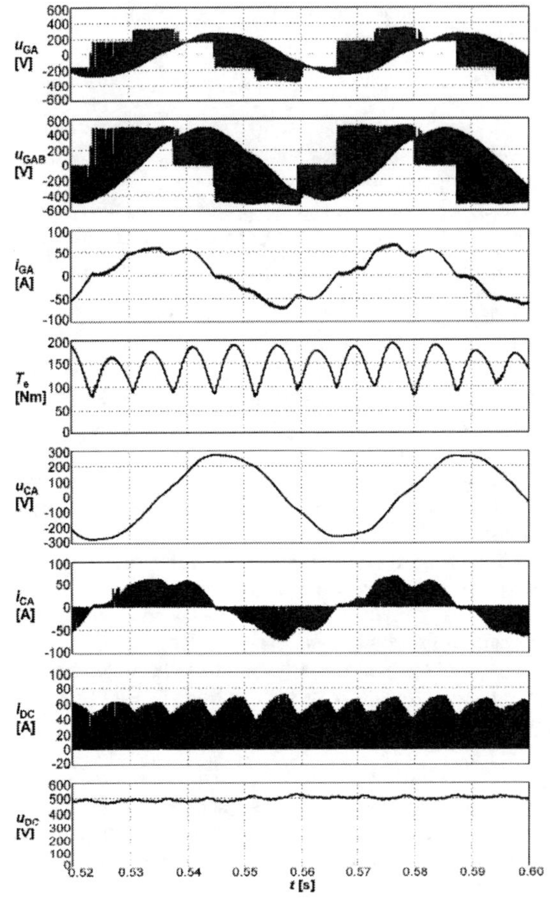

Figure 9. Waveforms of voltages, currents and the generator torque in the proposed energy generation system: $\omega_r = 750$ rpm; descriptions of waveforms as in Fig. 7.

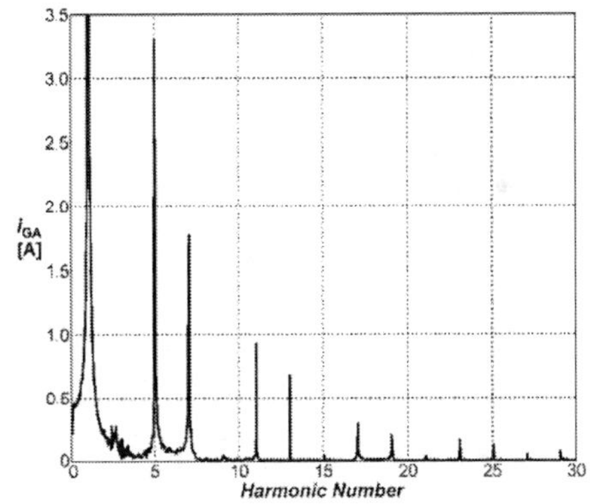

Figure 10. Harmonic spectrum of the generator current presented in Figure 8; amplitude of the first harmonic is equal to 60 A.

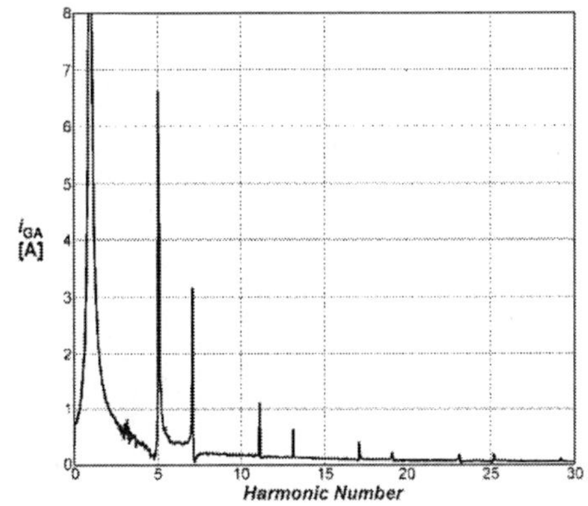

Figure 11. Harmonic spectrum of the generator current presented in Figure 9; amplitude of the first harmonic is equal to 40 A.

of the generator torque is higher; it negatively influences the torque transfer system. Figures 10 and 11 show harmonic spectra of the generator currents presented in Figure 8 and Figure 9 respectively. The fifth and seventh harmonics are significant higher harmonics in both cases. However, in the first case (1500 rpm) the amplitudes of these harmonics are respectively around 10% and 5% with respect to the amplitude of the basic harmonic; in the second case (750 rpm) these amplitudes are higher than previously and they are approximately 16% and 8%. This leads to an increase in the variable component of the generator torque, which is undesirable.

The content of higher harmonics in the generator current depends on its load; analytical description of this dependence is currently a subject of further studies, which allow us to take into account different parameters of the induction generator.

5 Conclusions

In this paper the novel concept of the energy generation system with the use of the squirrel cage induction generator is presented. Numerical calculation results allow us to state that the proposed energy generation system operates correctly for the assumed conditions, especially for different values of the generator angular velocity.

This system can be treated as an alternative to existing solutions which use the cage induction generators and the back-to-back converter. This proposal has several advantages in comparison to existing systems. The control method is simpler because the switches are turned on or turned off at the same time by one common signal. Moreover, the capacitor battery can be switched on by means of only two Insulated Gate Bipolar Transistors.

The proposed energy generation system can operate at various generator angular velocities. It is significant in the cases when the speed of the wind or the water stream varies within a wide range. In order to convert the total energy provided by the wind or the water stream into the electrical energy, the mechanical characteristics of the given turbine should be taken into the control method.

It should be notified that the validation of the correctness of the proposed energy generation system requires comprehensive research with the use of a laboratory stand. This system is especially dedicated to small wind and water plants. It is necessary to stress that the proposed method of energy generation can also be used in autonomous operation systems (island systems). However, in this case the control method of the output voltage source inverter should be modified due to the lack of the external grid.

References

1. M. Cheng, Y. Zhu., Energy Conv. and Manag., **88**, 332–347 (2014)
2. F. Blaabjerg, Z. Chen, R. Teodorescu, F. Iov, *5th International Conference Power Electronics and Motion Control,* **1**, 1-11 (2006)
3. A. Sikorski, A. Kuźma, Bulletin of the Polish Ac of Sc., Tech. Sc., **57**, No. 4, 317-322 (2009)
4. F. Blaabjerg, F. Iov, R. Teodorescu, Z. Chen, *12th International Conference Power Electronics and Motion Control EPE-PEMC*, 1-17 (2006)
5. A.R. Tiwari, A.J. Shewale, A.R. Gagangras, N.M. Lokhande, Multidisciplinary Jour. of Research in Eng. and Techn., **1**, Iss. 2, 129-135 (2014)
6. M. G. Simões, S. Chakraborty, R. Wood, Third Quart. IEEE Power Electr. Soc. News., **19**, 19-23 (2006)
7. G. Quinonez-Varela , A. Cruden, IET Gener. Transm. Distrib., **2**, (2), 301–309 (2008)
8. F. Blaabjerg, Z. Chen, S.B. Kjaer, IEEE Trans. on Power Electr., **19**, No. 5, 1184- 1194 (2004)
9. J.A. Baroudi, V. Dinavahi, A.M. Knight, Ren. Energy, **32**, 2369–2385 (2007)
10. J.G. Trapp, F.A. Farret, F.T. Fernandes, L.C. Corrêa, *10th International Conference on Industry Applications* (*INDUSCON*),1-8 (2012)
11. M. Molinas, B. Naess, W. Gullvik, T. Undeland, *European Conference on Power Electronics and Applications EPE*, 1-10 (2005)
12. A.B. Lehru, D. Chirag, K. Vibhakar, Int. Jour. of Eng. Research and Develop., **April**, 6-13 (2015)
13. R. Cárdenas, R. Peña, IEEE Trans. on Energy Conv., **19**, No. 1, 196-205 (2004)
14. R. Peña, R. Cardenas, R. Blasco, G. Asher, J. Clare, *27th Annual Conference of the IEEE Industrial Electronics Society* (*IECON01*), 1376-1381 (2001)
15. E.J. Bueno, S. Cóbreces, F.J. Rodríguez, Á. Hernández, F. Espinosa, IEEE Trans. on Energy Conv., **23**, No. 3, 932-945 (2008)
16. A. Binder, T. Schneider, *11th European Conference on Power Electronics and Applications*, P10 (2005)
17. A.M. El-Refaie, T.M. Jahns, **27**, Is. 5, 967-984 (2008)
18. W. Mazgaj, Z. Szular, T. Węgiel, T. Sobczyk, Przeg. Elektrot. **5**, 282-287 (2011)
19. T.J. Sobczyk, W. Mazgaj, Z. Szular, T. Węgiel, Arch. of Elec. Eng., **60**, No 2, 159-168 (2011)
20. D. Borkowski, T. Węgiel, IEEE Trans. on Energy Conv., **28**, No 2, 452-459 (2013)
21. Z. Szular, Techn. Trans., Elec. Eng., **2-E**, 149-158 (2015)

Rule-based Mamdani-type fuzzy modelling of thermal performance of fin-tube evaporator under frost conditions

Dilek Nur Ozen[1], Kemal Altinisik[2], Kevser Dincer[2], Ali Ates[2], Ahmet Ali Sertkaya[3] and Muhammed Emin Tolu[4, a]

[1]*Necmettin Erbakan University, Faculty of Engineering and Architecture, Department of Mechanical Engineering, Meram, Konya, Turkey*
[2]*Selcuk University, Faculty of Engineering, Department of Mechanical Engineering, Selcuklu, Konya, Turkey*
[3]*Necmettin Erbakan University, Faculty of Seydisehir Ahmet Cengiz Engineering, Department of Mechanical Engineering,42360, Seydisehir, Konya, Turkey*
[4]*Karamanoglu Mehmetbey University, Faculty of Engineering, Department of Mechanical Engineering, Karaman, Turkey*

Abstract. Frost formation brings about insulating effects over the surface of a heat exchanger and thereby deteriorating total heat transfer of the heat exchanger. In this study, a fin-tube evaporator is modeled by making use of Rule-based Mamdani-Type Fuzzy (RBMTF) logic where total heat transfer, air inlet temperature of 2 °C to 7 °C and four different fluid speed groups (ua1=1; 1.44; 1.88 m s-1, ua2=2.32; 2.76 m s-1, ua3=3.2; 3.64 m s-1, ua4=4.08; 4.52; 4.96 m s-1) for the evaporator were taken into consideration. In the developed RBMTF system, outlet parameter UA was determined using inlet parameters Ta and ua. The RBMTF was trained and tested by using MATLAB® fuzzy logic toolbox. R^2 (%) for the training data and test data were found to be 99.91%. With this study, it has been shown that RBMTF model can be reliably used in determination of a total heat transfer of a fin-tube evaporator.

1 Introduction

An evaporator is a device operating by the vapor compression refrigerating cycle principle and is used for cooling purposes (like cooling the cabin of a refrigerator). If the temperature of the evaporator surface in a refrigerator is below the freezing temperature of the moisture within the air, then frosting forms on the evaporator surface. If the frost keeps staying below the freezing point, then frost will form a layer on the surface. The formed layer acts as an insulator by increasing thermal resistance and thereby leading to dropping the thermal performance of the evaporator [1]. In the existing literature, there are many numerical and experimental studies on the performance of finned tube heat exchangers operating under frosting conditions.

Fuzzy logic is a mathematical discipline that we use every day and helps us to reach the structure in which we interpret our own behaviors. Fuzzy expert system is an expert system that uses a collection of fuzzy rules, instead of Boolean logic, to reason about data [2]. Over the last few years, there have been many investigations on the application of fuzzy logic. Some are briefly mentioned below. Tasdemir et al. [3] studied artificial neural network (ANN) and fuzzy expert system (FES) and their comparison for prediction of performance and emission parameters on a gasoline engine. Tosun and Dincer [4] carried out a study on modelling of a thermal insulation system based on the coldest temperature conditions by using artificial neural networks to determine the performance of building for wall types in Turkey They noted that in this study the ANN approach has been applied accurately to model for the thickness of thermal insulation performance system on the coldest temperature basis for wall types that are mostly used for buildings in Turkey. Tosun et al. [5] investigated Rule-based Mamdani-type fuzzy modeling (RBMTF) of thermal performance of multi-layer precast concrete panels used in residential buildings in Turkey. They reported that prediction of RBMTF modeling approach can be successfully used for the modeling of thermal performance of multi-layer precast concrete panels used in residential buildings in Turkey.

In this study, the performance of a finned tube evaporator under frost forming conditions was modeled by making use of RBMTF technique. Input parameters taken were air entrance temperature and inlet speed while the output parameter was the total heat transfer of the evaporator. When the experimental data were compared with the data obtained from the RBMTF technique, it was found that the two are in correlation with each other. The UA values at different air operating conditions which were not studied experimentally were estimated with the RBMTF technique.

2 Experimental setup

The schematic diagram of the experimental system that shows evaporators' total heat transfer in a no-frost refrigerator operating under frosting conditions is given in Figure 1. The experimental setup consists of an air tunnel, refrigerating system and a data logger. The outside diameter of the test evaporator pipe inserted into the air

[a] Corresponding author: metolu@kmu.edu.tr

tunnel is 8 mm, its thickness is 0.8 mm and the fin thickness is 0.12 mm. during the tests, the air was drawn by a fan from an aperture and circulated around the channel. In order to regulate the air operating conditions within the air tunnel, a demoisturizer, a cooler, a heater and a humidifier were used. The air amount sent to the evaporator was adjusted with a damper. Airstream straightener was used to maintain homogeneous air dispersion to the evaporator. During the experiments, air inlet and outlet temperatures and the relative humidity were measured with a temperature-humidity measuring device while the air speed was measured with the anemometer.

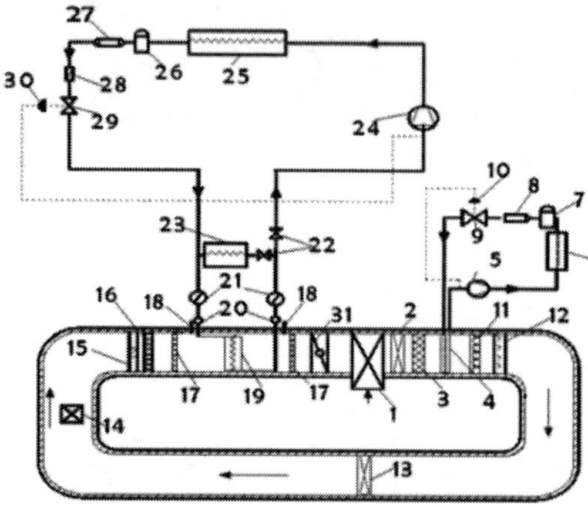

Item No:	Device Name	Item No:	Device Name
1	Vent	17	Temperature-humidity meter
2	Fan	18	Pressure different meter
3	Humidity Absorber	19	Test evaporator
4	Evaporator	20	Thermal couple
5	Compressor	21	Thermocouple
6	Condenser	22	Solenoid valve
7	Gas Tank	23	Temporary evaporator
8	Drier	24	Compressor
9	Expansion valve	25	Condenser
10	Valve regulator	26	Gas Tank
11	Heater	27	Drier
12	Moisturizer	28	Flow meter
13	Fan	29	Expansion valve
14	Vent	30	Valve regulator
15	Damper	31	Anemometer.
16	Flow straightener		

Figure 1. Schematic presentation of experimental system

2.1 Air enthalpy at the evaporator inlet and outlet

The enthalpy of the moisture entering the evaporator is found by using Eq. (1) below:

$$i_a = c_{pa}T_a + w_a(2501.3 + 1.82T_a) \quad (1)$$

Here, c_{pa}, T_a and w_a represent specific heat capacity of air, temperature and specific humidity of the air respectively.

2.2 Heat transfer for air side

The air temperature on the medium where the channel exists is higher than that of the air within the channel. For this reason, there is a heat gain on the test zone of the channel. Air transfer on the air side is found with Eq. (2).

$$q_a = \dot{m}_a(i_{ai} - i_{ao}) - (UA)_{channel}(T_e - T_s) \quad (2)$$

Where, m_a, i_{ai}, i_{ao} and $(UA)_{channel}$ are respectively the inlet air mass flow, inlet air enthalpy, outlet air enthalpy and overall heat transfer coefficient of channel. T_e and T_s are the environment and surface temperatures of the channel respectively.

2.3 Heat transfer for refrigerant side

The heat transferred from the refrigerant is calculated from the following equation.

$$q_r = \dot{m}_r(i_{ro} - i_{ri}) \quad (3)$$

Here $\dot{m}_r$, i_{ri} and i_{ro} respectively, represent the mass flow rate of the refrigerant, evaporator inlet and outlet enthalpies.

2.4 Evaporator's overall heat transfer coefficient

Eq. (4) below, was obtained by making use of the arithmetic mean of the evaporator's heat capacity (q_m), refrigerant's heat capacity (q_r) and the heat capacity on the air

$$q_m = \frac{q_a + q_r}{2} \quad (4)$$

The overall heat transfer coefficient of evaporator (UA), is found with the following equation:

$$UA = \frac{q_m}{\Delta T_m} \quad (5)$$

Here, ΔT_m is logarithmic mean temperature difference and is presented in Eq. (6) below:

$$\Delta T_m = \frac{(T_{ai} - T_{ro}) - (T_{ao} - T_{ri})}{ln\left(\frac{(T_{ai} - T_{ro})}{(T_{ao} - T_{ri})}\right)} \quad (6)$$

where, T_{ai} and T_{ao} are the inlet and outlet temperatures of air to and from the heat exchanger respectively. Here T_{ria} and T_{ro} are the inlet and outlet temperatures of refrigerant to and from the heat exchanger respectively.

3 Developed fuzzy expert system for the performance of finned tube evaporator under frosting conditions

Fuzzy logic is a superset of Boolean-conventional logic that has been expanded to handle the concept of partial truth and truth values between "completely true" and "completely false". Fuzzy theory should be seen as a methodology to generalize any specific theory from crisp to continuous. Fuzzy modeling opens the possibility for straightforward translation of statements in natural language (verbal formulation) of the observed problem into a fuzzy system. Its functioning is based on mathematical tools [6, 7]. There are two types of fuzzy inference systems in the toolbox: Mamdani-type and Sugeno-type. These two types of inference systems vary somewhat in the way outputs are determined. Fuzzy

Table 1. Fuzzy sets of input and output variables.

Membership name	T_a (°C)			u_a(ms^{-1})		UA(W °C^{-1})	
Very Low	L1	T_{a1}	1.17-2.83	u_{a1}	0.34-1.66	UA_1	15.97-23.61
Low	L2	T_{a2}	2-3.66	u_{a2}	1-2.32	UA_2	19.79-27.43
Negative Medium	L3	T_{a3}	2.83-4.49	u_{a3}	1.66-2.98	UA_3	23.61-31.25
Medium	L4	T_{a4}	3.66-5.32	u_{a4}	2.32-3.64	UA_4	27.43-35.07
Positive Medium	L5	T_{a5}	4.49-6.15	u_{a5}	2.98-4.3	UA_5	31.25-38.89
High	L6	T_{a6}	5.32-6.98	u_{a6}	3.64-4.96	UA_6	35.07-42.71
Very High	L7	T_{a7}	6.15-7.81	u_{a7}	4.3-5.62	UA_7	38.89-46.53

Table 2. Actual data (rule, training)

No	T_a (°C)	u_a (ms^{-1})	Rule, Rule #	No	T_a (°C)	u_a (ms^{-1})	Rule, Rule #
1	2	ua1	x, (1)	6	2	ua4	x, (6)
31	3	ua1	x, (11)	36	3	ua4	x, (16)
61	4	ua1	x, (21)	66	4	ua4	x, (26)
71	4.5	ua1	x, (31)	76	4.5	ua4	x, (36)
91	5	ua1	x, (41)	96	5	ua4	x, (46)
121	6	ua1	x, (51)	126	6	ua4	x, (56)
151	7	ua1	x, (61)	156	7	ua4	x, (66)
2	2	ua1	x, (2)	7	2	ua5	x, (7)
32	3	ua1	x, (12)	37	3	ua5	x, (17)
62	4	ua1	x, (22)	67	4	ua5	x, (27)
72	4.5	ua1	x, (32)	77	4.5	ua5	x, (37)
92	5	ua1	x, (42)	97	5	ua5	x, (47)
122	6	ua1	x, (52)	127	6	ua5	x, (57)
152	7	ua1	x, (62)	157	7	ua5	x, (67)
3	2	ua2	x, (3)	8	2	ua6	x, (8)
33	3	ua2	x, (13)	38	3	ua6	x, (18)
63	4	ua2	x, (23)	68	4	ua6	x, (28)
73	4.5	ua2	x, (33)	78	4.5	ua6	x, (38)
93	5	ua2	x, (43)	98	5	ua6	x, (48)
123	6	ua2	x, (53)	128	6	ua6	x, (58)
153	7	ua2	x, (63)	158	7	ua6	x, (68)
4	2	ua3	x, (4)	9	2	ua6	x, (9)
34	3	ua3	x, (14)	39	3	ua6	x, (19)
64	4	ua3	x, (24)	69	4	ua6	x, (29)
74	4.5	ua3	x, (34)	79	4.5	ua6	x, (39)
94	5	ua3	x, (44)	99	5	ua6	x, (49)
124	6	ua3	x, (54)	129	6	ua6	x, (59)
154	7	ua3	x, (64)	159	7	ua6	x, (69)
5	2	ua4	x, (5)	10	2	ua7	x, (10)
35	3	ua4	x, (15)	40	3	ua7	x, (20)
65	4	ua4	x, (25)	70	4	ua7	x, (30)
75	4.5	ua4	x, (35)	80	4.5	ua7	x, (40)
95	5	ua4	x, (45)	100	5	ua7	x, (50)
125	6	ua4	x, (55)	130	6	ua7	x, (60)
155	7	ua4	x, (65)	160	7	ua7	x, (70)

Table 3. Actual data (test)

No	T_a(°C)	u_a (ms^{-1})	No	T_a(°C)	u_a (ms^{-1})
11	2.5	1	46	3.5	3.2
41	3.5	1	106	5.5	3.2
101	5.5	1	136	6.5	3.2
131	6.5	1	17	2.5	3.64
12	2.5	1.44	47	3.5	3.64
42	3.5	1.44	107	5.5	3.64
102	5.5	1.44	137	6.5	3.64
132	6.5	1.44	18	2.5	4.08
13	2.5	1.88	48	3.5	4.08
43	3.5	1.88	108	5.5	4.08
103	5.5	1.88	138	6.5	4.08
133	6.5	1.88	19	2.5	4.52
14	2.5	2.32	49	3.5	4.52
44	3.5	2.32	109	5.5	4.52
104	5.5	2.32	139	6.5	4.52
134	6.5	2.32	20	2.5	4.96
15	2.5	2.76	50	3.5	4.96
45	3.5	2.76	110	5.5	4.96
105	5.5	2.76	140	6.5	4.96
16	2.5	3.2			

Table 4. Predicted values

No	T_a(°C)	u_a (ms^{-1})	No	T_a(°C)	u_a (ms^{-1})
21	2.75	1	26	2.75	3.2
51	3.75	1	56	3.75	3.2
81	4.75	1	86	4.75	3.2
111	5.75	1	116	5.75	3.2
22	2.75	1.44	27	2.75	3.64
52	3.75	1.44	57	3.75	3.64
82	4.75	1.44	87	4.75	3.64
112	5.75	1.44	117	5.75	3.64
23	2.75	1.88	28	2.75	4.08
53	3.75	1.88	58	3.75	4.08
83	4.75	1.88	88	4.75	4.08
113	5.75	1.88	118	5.75	4.08
24	2.75	2.32	29	2.75	4.52
54	3.75	2.32	59	3.75	4.52
84	4.75	2.32	89	4.75	4.52
114	5.75	2.32	119	5.75	4.52
25	2.75	2.76	30	2.75	4.96
55	3.75	2.76	60	3.75	4.96
85	4.75	2.76	120	5.75	4.96
115	5.75	2.76			

inference systems have been successfully applied in fields such as automatic control, data classification, decision analysis, expert systems, and computer vision. Because of its multidisciplinary nature, fuzzy inference systems are associated with a number of names, such as fuzzy-rule-based systems, fuzzy expert systems, fuzzy modeling, fuzzy associative memory, fuzzy logic controllers, and simply (and ambiguously) fuzzy systems [8].

In this study, thermal performance of a finned tube heat exchanger under frost forming operating conditions was modeled with the RBMTF technique. With this aim, the data obtained from the experimental study were used. This stimulus model is constructed into rule-based Mamdani-type fuzzy modeling, using input parameters as T_a, u_a and output parameter as UA, described by RBMTF if-then rules (Figure 2). RBMTF has been designed using the MATLAB R2010b fuzzy logic toolbox. Classification of the variables into a specified number of subsets depends on the nature of the problem. The inputs T_a and u_a together with the output UA were categorized into 7 subsets and their triangle membership functions were determined. The membership functions are presented in Figures 3 and 4. The membership functions were developed for the study, representing Very Low (L1), Low (L2), Negative Medium (L3), Medium (L4), Positive Medium (L5), High (L6), and Very High (L7) linguistic classes (Table 1).

In this study, the data set included 110 data (training+test data). 70 of them were chosen as training data (Table 2), whereas 40 of them were chosen for the test data (Table 3). Values of prediction consist of 50 sets (Table 4).

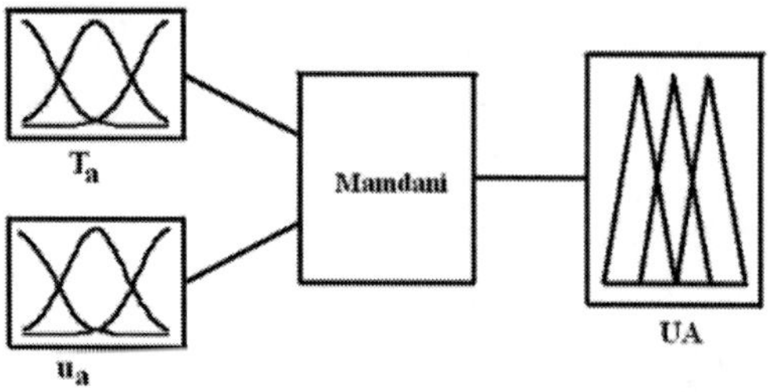

Figure 2. Designed fuzzy modeling structure the performance of a finned tube evaporator operating under snowing conditions

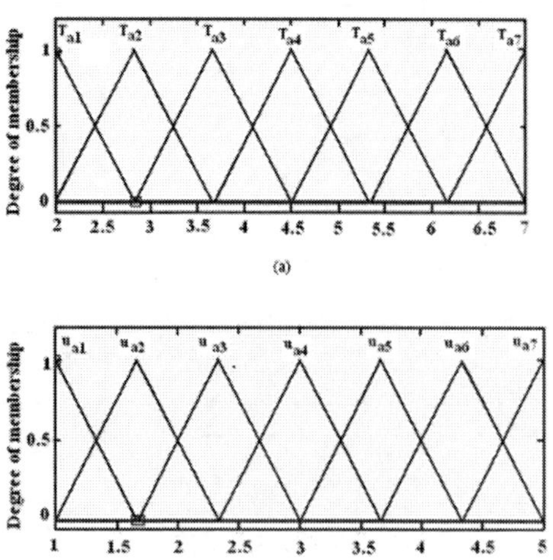

Figure 3. Fuzzy membership functions for two input variables: (a) T_a fuzzy sets graphics; (b) u_a fuzzy sets graphics.

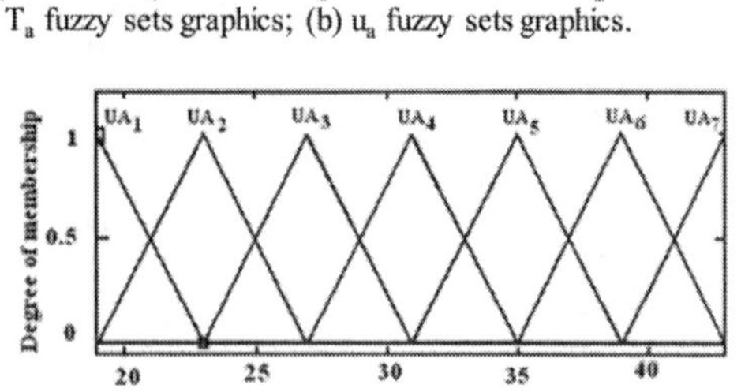

Figure 4. Fuzzy membership functions for one output variable: UA fuzzy set graphic.

Fuzzy membership functions in analytical form are expressed in Eqs. (7) -(10) for UA.

$$UA(x) = \begin{cases} x; & 19.79 \le x \le 42.71 \\ 0; & otherwise \end{cases} \quad (7)$$

$$\mu_{UA1}(x) = \begin{cases} 0; & \text{otherwise} \\ \dfrac{23.61 - x}{3.82} & \text{if } 19.79 \le x \le 23.61 \end{cases} \quad (8)$$

$$\mu_{UA4}(x) = \begin{cases} 0; & x \le 27.43 \\ \dfrac{x - 27.43}{3.82} & \text{if } 27.43 < x \le 31.25 \\ \dfrac{35.07 - x}{3.82} & \text{if } 31.25 < x \le 35.07 \\ 0; & x \ge 35.07 \end{cases} \quad (9)$$

$$\mu_{UA7}(x) = \begin{cases} 0; & \text{otherwise} \\ \dfrac{42.71 - x}{3.82} & \text{if } 38.89 < x \le 42.71 \end{cases} \quad (10)$$

Where $\mu_{UA1}(x)$ is the membership function for UA_1= 15.97-23.61 W °C^{-1}, $\mu_{UA4}(x)$ is the membership function for UA_4= 27.43-35.07 W °C^{-1}, $\mu_{UA7}(x)$ represents the membership function for UA_7=38.89-46.53 W °C^{-1}. Where, UA_1, UA_4 and UA_7 variables are fuzzificated as linguistic variables. An advanced RBMTF model was used for unperformed actual data. T_a and u_a values were predicted because modeling of the necessary UA was made.

4 Results and discussion

If there the surface temperature of evaporators in refrigerating systems, is lower than the freezing point of the water vapor in the air, then a frost film forms on the evaporator surfaces, first; and in case the frost surface temperature continues to be lower than the freezing point, then the frost accumulates over the surface and acts as an insulating agent. Since this situation will increase thermal resistance, the amount of heat absorbed by the refrigerant will decrease, the inter-fin distances will narrow due to the frost accumulation and ultimately, energy consumption will increase [1]. In this study, RBMTF technique has been used to model the performance of a finned tube evaporator at the transient regime for 4 situations and with different u_a values. The findings of this study are presented below:

- Situation for u_a=1; 1.44; 1.88 m s^{-1}: T_a=2-7°C. Number of fuzzy rules=21 (Table 5). Number of RBMTF tests=12. Number of RBMTF prediction=15. UA_{max}=42.7 W°C^{-1} (actual) ; UA_{min}=30.1 W°C^{-1} (actual); UA_{max}=41.5 W° C^{-1} (test); UA_{min}=30.9 W°C^{-1} (test); UA_{max}=39 W°C^{-1} (prediction); UA_{min}=29.6 W°C^{-1} (prediction) (Figure 5).
- Situation for u_a=2.32 ; 2.76 m s-1 : T_a=2-7oC. Number of fuzzy rules=14 (Table 6). Number of RBMTF

tests= 8. Number of RBMTF prediction=10. UA_{max}=38.7 W°C^{-1} (actual) ; UA_{min}=27.2 W°C^{-1} (actual); UA_{max}=37.5 WoC-1 (test); UA_{min}=28 W°C^{-1} (test); UA_{max}=35.5 W°C^{-1} (prediction); UA_{min}=28.43 W°C^{-1} (prediction)(Figure 6).

- Situation for u_a=3.2 ; 3.64 m s-1. Ta=2-7oC. Number of fuzzy rules=14 (Table 7). Number of RBMTF tests= 8. Number of RBMTF prediction=10. UA_{max}=36.1 W°C^{-1} (actual) ; UA_{min}=24.2 W°C^{-1} (actual) ; UA_{max}= 34.9 W°C^{-1} (test) ; UA_{min}= 25.1 W°C^{-1} (test); UA_{max}=34.2 W°C^{-1} (prediction); UA_{min}=24.3 W°C^{-1} (prediction) (Figure 7).
- Situation for u_a=4.08 ; 4.52; 4.96 m s-1. T_a=2-7oC. Number of fuzzy rules=21 (Table 8). Number of RBMTF tests= 12. Number of RBMTF prediction=15. UA_{max}=33.4 W°C^{-1} (actual) ; UA_{min}=19.8 W°C^{-1} (actual) ; UAmax= 32.2 W°C^{-1} (test) ; UA_{min}= 20.7 W°C^{-1} (test) ; UA_{max}=31.6 W°C^{-1} (prediction); UAmin=21.9 WoC-1 (prediction) (Figure 8).

Table 5. Rules of RBMTF for UA
(situation for u_a=1; 1.44; 1.88 m s^{-1})

Rule #	*If*	T_a	*and*	u_a	*then*	*UA*
1	*If*	T_a is T_{a1}	*and*	u_a is u_{a1}	*then*	*UA is* UA_7
11	*If*	T_a is T_{a2}	*and*	u_a is u_{a1}	*then*	*UA is* UA_6
21	*If*	T_a is T_{a3}	*and*	u_a is u_{a1}	*then*	*UA is* UA_6
31	*If*	T_a is T_{a4}	*and*	u_a is u_{a1}	*then*	*UA is* UA_6
41	*If*	T_a is T_{a5}	*and*	u_a is u_{a1}	*then*	*UA is* UA_5
51	*If*	T_a is T_{a6}	*and*	u_a is u_{a1}	*then*	*UA is* UA_5
61	*If*	T_a is T_{a7}	*and*	u_a is u_{a1}	*then*	*UA is* UA_5
2	*If*	T_a is T_{a1}	*and*	u_a is u_{a1}	*then*	*UA is* UA_7
12	*If*	T_a is T_{a2}	*and*	u_a is u_{a1}	*then*	*UA is* UA_6
22	*If*	T_a is T_{a3}	*and*	u_a is u_{a1}	*then*	*UA is* UA_5
32	*If*	T_a is T_{a4}	*and*	u_a is u_{a1}	*then*	*UA is* UA_5
42	*If*	T_a is T_{a5}	*and*	u_a is u_{a1}	*then*	*UA is* UA_5
52	*If*	T_a is T_{a6}	*and*	u_a is u_{a1}	*then*	*UA is* UA_5
62	*If*	T_a is T_{a7}	*and*	u_a is u_{a1}	*then*	*UA is* UA_4
3	*If*	T_a is T_{a1}	*and*	u_a is u_{a2}	*then*	*UA is* UA_6
13	*If*	T_a is T_{a2}	*and*	u_a is u_{a2}	*then*	*UA is* UA_6
23	*If*	T_a is T_{a3}	*and*	u_a is u_{a2}	*then*	*UA is* UA_5
33	*If*	T_a is T_{a4}	*and*	u_a is u_{a2}	*then*	*UA is* UA_5
43	*If*	T_a is T_{a5}	*and*	u_a is u_{a2}	*then*	*UA is* UA_5
53	*If*	T_a is T_{a6}	*and*	u_a is u_{a2}	*then*	*UA is* UA_4
63	*If*	T_a is T_{a7}	*and*	u_a is u_{a2}	*then*	*UA is* UA_4

Table 6. Rules of RBMTF for UA
(situation for u_a=2.32; 2.76 m s^{-1})

Rule #	*If*	T_a	*and*	u_a	*then*	*UA*
4	*If*	T_a is T_{a1}	*and*	u_a is u_{a3}	*then*	*UA is* UA_6
14	*If*	T_a is T_{a2}	*and*	u_a is u_{a3}	*then*	*UA is* UA_5
24	*If*	T_a is T_{a3}	*and*	u_a is u_{a3}	*then*	*UA is* UA_5
34	*If*	T_a is T_{a4}	*and*	u_a is u_{a3}	*then*	*UA is* UA_5
44	*If*	T_a is T_{a5}	*and*	u_a is u_{a3}	*then*	*UA is* UA_4
54	*If*	T_a is T_{a6}	*and*	u_a is u_{a3}	*then*	*UA is* UA_4
64	*If*	T_a is T_{a7}	*and*	u_a is u_{a3}	*then*	*UA is* UA_3
5	*If*	T_a is T_{a1}	*and*	u_a is u_{a4}	*then*	*UA is* UA_6
15	*If*	T_a is T_{a2}	*and*	u_a is u_{a4}	*then*	*UA is* UA_5
25	*If*	T_a is T_{a3}	*and*	u_a is u_{a4}	*then*	*UA is* UA_4
35	*If*	T_a is T_{a4}	*and*	u_a is u_{a4}	*then*	*UA is* UA_4
45	*If*	T_a is T_{a5}	*and*	u_a is u_{a4}	*then*	*UA is* UA_4
55	*If*	T_a is T_{a6}	*and*	u_a is u_{a4}	*then*	*UA is* UA_3
65	*If*	T_a is T_{a7}	*and*	u_a is u_{a4}	*then*	*UA is* UA_3

Table 7. Rules of RBMTF for UA
(situation for u_a=3.2; 3.64 m s^{-1})

Rule #	*If*	T_a	*and*	u_a	*then*	*UA*
6	*If*	T_a is T_{a1}	*and*	u_a is u_{a4}	*then*	*UA is* UA_5
16	*If*	T_a is T_{a2}	*and*	u_a is u_{a4}	*then*	*UA is* UA_5
26	*If*	T_a is T_{a3}	*and*	u_a is u_{a4}	*then*	*UA is* UA_4
36	*If*	T_a is T_{a4}	*and*	u_a is u_{a4}	*then*	*UA is* UA_4
46	*If*	T_a is T_{a5}	*and*	u_a is u_{a4}	*then*	*UA is* UA_3
56	*If*	T_a is T_{a6}	*and*	u_a is u_{a4}	*then*	*UA is* UA_3
66	*If*	T_a is T_{a7}	*and*	u_a is u_{a4}	*then*	*UA is* UA_3
7	*If*	T_a is T_{a1}	*and*	u_a is u_{a5}	*then*	*UA is* UA_5
17	*If*	T_a is T_{a2}	*and*	u_a is u_{a5}	*then*	*UA is* UA_4
27	*If*	T_a is T_{a3}	*and*	u_a is u_{a5}	*then*	*UA is* UA_4
37	*If*	T_a is T_{a4}	*and*	u_a is u_{a5}	*then*	*UA is* UA_4
47	*If*	T_a is T_{a5}	*and*	u_a is u_{a5}	*then*	*UA is* UA_3
57	*If*	T_a is T_{a6}	*and*	u_a is u_{a5}	*then*	*UA is* UA_3
67	*If*	T_a is T_{a7}	*and*	u_a is u_{a5}	*then*	*UA is* UA_2

Table 8. Rules of RBMTF for UA
(situation for u_a=4.08; 4.52 ; 4.96 m s^{-1})

Rule #	*If*	T_a	*and*	u_a	*then*	*UA*
8	*If*	T_a is T_{a1}	*and*	u_a is u_{a6}	*then*	*UA is* UA_5
18	*If*	T_a is T_{a2}	*and*	u_a is u_{a6}	*then*	*UA is* UA_4
28	*If*	T_a is T_{a3}	*and*	u_a is u_{a6}	*then*	*UA is* UA_3
38	*If*	T_a is T_{a4}	*and*	u_a is u_{a6}	*then*	*UA is* UA_3
48	*If*	T_a is T_{a5}	*and*	u_a is u_{a6}	*then*	*UA is* UA_3
58	*If*	T_a is T_{a6}	*and*	u_a is u_{a6}	*then*	*UA is* UA_2
68	*If*	T_a is T_{a7}	*and*	u_a is u_{a6}	*then*	*UA is* UA_2
9	*If*	T_a is T_{a1}	*and*	u_a is u_{a6}	*then*	*UA is* UA_4
19	*If*	T_a is T_{a2}	*and*	u_a is u_{a6}	*then*	*UA is* UA_4
29	*If*	T_a is T_{a3}	*and*	u_a is u_{a6}	*then*	*UA is* UA_3
39	*If*	T_a is T_{a4}	*and*	u_a is u_{a6}	*then*	*UA is* UA_3
49	*If*	T_a is T_{a5}	*and*	u_a is u_{a6}	*then*	*UA is* UA_3
59	*If*	T_a is T_{a6}	*and*	u_a is u_{a6}	*then*	*UA is* UA_2
69	*If*	T_a is T_{a7}	*and*	u_a is u_{a6}	*then*	*UA is* UA_2
10	*If*	T_a is T_{a1}	*and*	u_a is u_{a7}	*then*	*UA is* UA_4
20	*If*	T_a is T_{a2}	*and*	u_a is u_{a7}	*then*	*UA is* UA_3
30	*If*	T_a is T_{a3}	*and*	u_a is u_{a7}	*then*	*UA is* UA_3
40	*If*	T_a is T_{a4}	*and*	u_a is u_{a7}	*then*	*UA is* UA_2
50	*If*	T_a is T_{a5}	*and*	u_a is u_{a7}	*then*	*UA is* UA_2
60	*If*	T_a is T_{a6}	*and*	u_a is u_{a7}	*then*	*UA is* UA_2
70	*If*	T_a is T_{a7}	*and*	u_a is u_{a7}	*then*	*UA is* UA_1

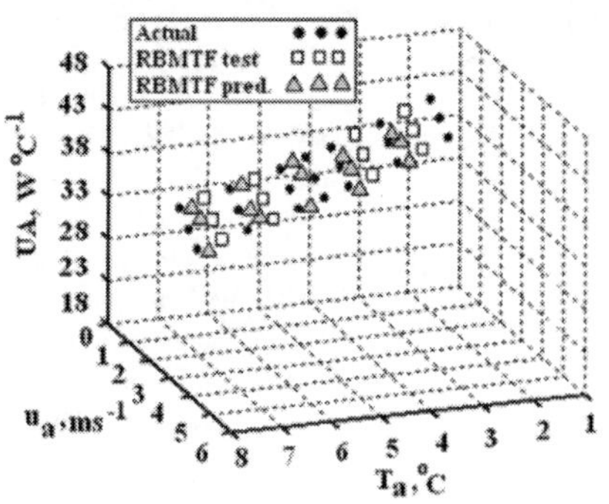

Figure 5. Comparison of actual UA with the UA data obtained from fuzzy technique for u_a=1 ms^{-1}, u_a=1.44 ms^{-1}, u_a=1.88 ms^{-1}

DOI: 10.1051/e3sconf/20161000097

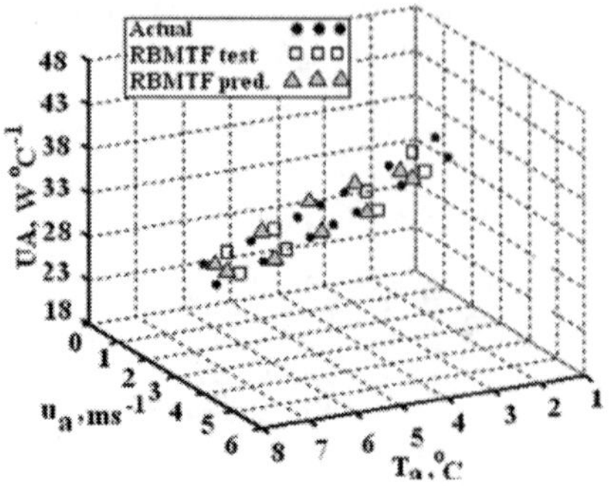

Figure 6. Comparison of actual UA with the UA data obtained from fuzzy technique for u_a =2.32 ms^{-1}, u_a =2.76 ms^{-1}

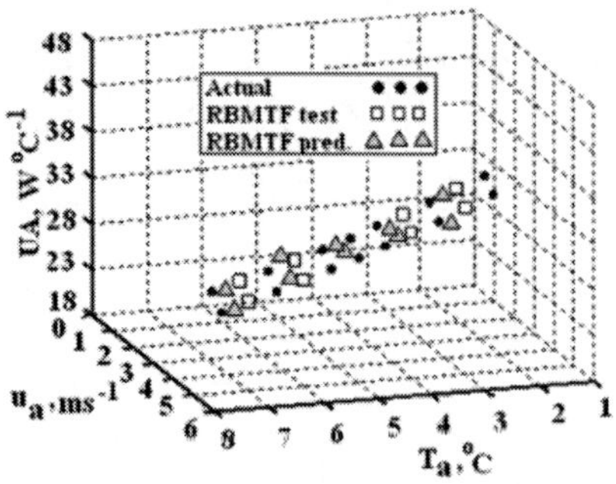

Figure 7. Comparison of actual UA with the UA data obtained from fuzzy technique for u_a =3.2 ms^{-1}, u_a =3.64 ms^{-1}

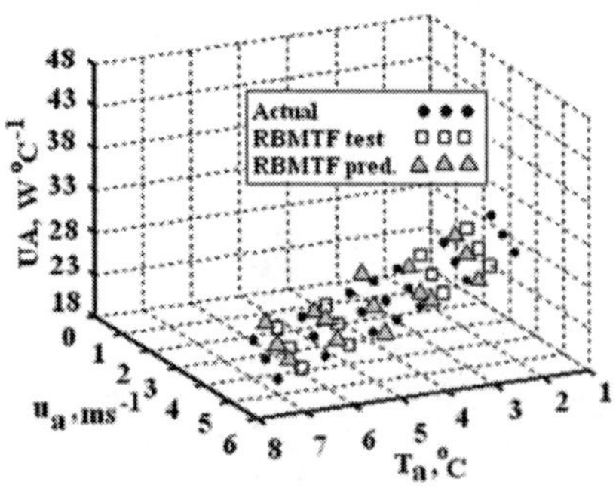

Figure 8. Comparison of actual UA with the UA data obtained from fuzzy technique for u_a =4.08 ms^{-1}, u_a =4.96 ms^{-1}

When Figures 5-8 are evaluated together, it is found that, at minimum air inlet speed and minimum temperature (u_a=1 ms^{-1} and T_a=2°C), UA is a maximum (UA_{max}=42.7 W°C^{-1}), whereas at maximum air inlet speed and maximum temperature (u_a=4.96 m s-1 and T_a=7°C), UA becomes a minimum (UA_{min}=19.8 W°C^{-1}). The increase in the air inlet speed has led to an increase in the amount of moisture in the air. The increase in the amount of water vapor has led ted to the increasing frost layer and hence causing the drop of the total heat transfer in the system.

The error during the learning session is called the root-mean-squared (RMS) value and is defined as follows [9]:

$$RMS = \left((1/p) \sum_j \left| t_j - o_j \right|^2 \right)^{1/2} \tag{11}$$

In addition, the absolute fraction of variance (R^2) and mean absolute percentage error (MAPE) are defined as follows, respectively [9]:

$$R^2 = 1 - \left(\frac{\sum_j (t_j - o_j)^2}{\sum_j (o_j)^2} \right) \tag{12}$$

where t is target value, o is output value, and p is pattern [9].

$$MAPE = \frac{o - t}{o} x100 \tag{13}$$

The statistical values such as RMS, R^2, MAPE, maximum error and minimum error given in Tables 9 and 10 for training and test values. R^2 for the training data is 99.91 % and R^2 for the test data is 99.91% (Figure 9). When Tables 7-8 and Figure 9 are studied, it is found that actual values and the values from fuzzy technique are very close to each other.

Table 9. The statistical error values for UA for training

RMS	R^2 (%)	MAPE (%)	Min error (%)	Max error (%)
0.95	99.91	0.073	0.03	7.03

Table 10. The statistical error values for UA for test

RMS	R^2 (%)	MAPE (%)	Min error (%)	Max error (%)
0.93	99.91	0.78	0.06	8.93

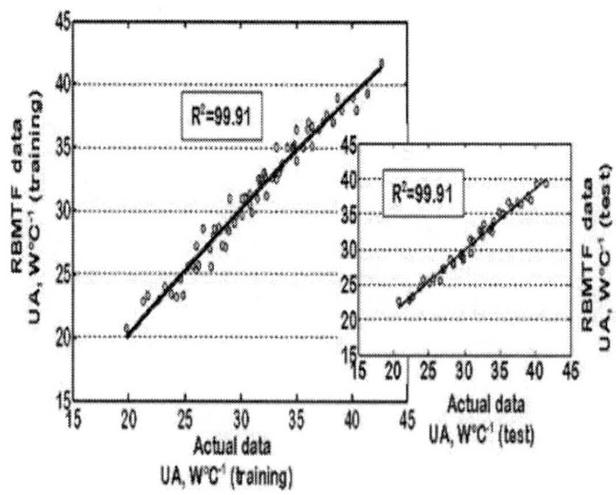

Figure 9. Comparison of the actual and RBMTF results

5 Conclusion

Modeling of performance of a finned-tube type evaporator was conducted in this study by making use of the Rule-Based Mamdani Type Fuzzy logic technique. The evaporator was made to operate under frost forming conditions and the conclusions drawn in this paper are summarized as follows:

- The RBMTF was trained and tested by means of the MATLAB software on a personal computer.
- Input parameters (T_a, u_a) and output parameter UA, were described by RBMTF if-then rules.

- 110 experimental data sets were used, out of which 70 were used in the training step (Table 2) and 40 were for the testing session (Table 3).
- The UA values that were not considered with the experimental study were estimated by using the RBMTF (Figs. 5-8 and Table 4).
- The decrease in inlet air temperature has caused only a slight increase in the frost layer and that's why as the inlet air temperature falls the total heat transfer has increased. (Figs.5-8).
- The amount of water vapor in the air has decreased as a result of the decreasing air inlet speed. This has led to an increasing total heat transfer as the air inlet speed fell. (Figs.5-8).

The actual data and RBMTF results show that RBMTF can be successfully used for the modeling of performance of finned tube evaporator under frost conditions. Future studies may use the data obtained in this study and involve other techniques like artificial neural networks (ANN) and genetic algorithm (GA) under different operating conditions to calculate an optimum performance for total heat transfer of a system.

References

1. D.N. Ozen, Selcuk University Ph.D Thesis, Turkey, (2011)
2. S. H. Liao, Expert Syst. Appl., **28**, 93-103, (2005)
3. S. Tasdemir, I. Saritas, M. Ciniviz, N. Allahverdi, Expert Syst. Appl., **38**, 13912-13923 (2011)
4. M. Tosun, K. Dincer, Int. J. Refrig, **34**, 362-373. (2011)
5. M. Tosun, K. Dincer, S. Baskaya, Expert Syst. Appl., **38**, 5553-5560 (2011)
6. K. Dincer, S. Tasdemir, S. Baskaya, I. Ucgul, B. Z. Uysal, Numer. Heat Transfer, Part B **54**, 499-517 (2008)
7. M. T. Lah, B. Zupancic, A. Krainer, Build. Environ., **40**, 1626-1637, (2005)
8. MathWorks,http://www.mathworks.com/help/toolbox/fuzzy/fp351dup8.html (2012)
9. Sözen, E. Arcaklioğlu, T. Menlik, M. Özalp, Expert Syst. Appl., **36**, 4346-4356 (2009)

Inkjet printing and inkjet infiltration of functional coatings for SOFCs fabrication

Rumen I. Tomov[1], Ryan Duncan[1], Mariusz. Krauz[2], R. Vasant Kumar[1] and Bartek A. Glowacki[1,3,4]

[1]*Department of Materials Science and Metallurgy, University of Cambridge, United Kingdom*
[2] *Institute of Power Engineering - Ceramic Department CEREL, Poland*
[3] *Bernal Institute, Department of Physics and Energy, University of Limerick, Plassey, Ireland*
[4] *Institute of Power Engineering, Warsaw, Poland*

Abstract. Inkjet printing fabrication and modification of electrodes and electrolytes of SOFCs were studied. Electromagnetic print-heads were utilized to reproducibly dispense droplets of inks at rates of several kHz on demand. Printing parameters including pressure, nozzle opening time and drop spreading were studied in order to optimize the inks jetting and delivery. Scanning electron microscopy revealed highly conformal ~ 6-10 µm thick dense electrolyte layers routinely produced on cermet and metal porous supports. Open circuit voltages ranging from 0.95 to 1.01 V, and a maximum power density of ~180 $mW.cm^{-2}$ were measured at 750 °C on Ni-8YSZ/YSZ/LSM single cell 50x50 mm in size. The effect of anode and cathode microstructures on the electrochemical performance was investigated. Two - step fabrication of the electrodes using inkjet printing infiltration was implemented. In the first step the porous electrode scaffold was created printing suspension composite inks. During the second step inkjet printing infiltration was utilized for controllable loading of active elements and a formation of nano-grid decorations on the scaffolds radically reducing the activation polarization losses of both electrodes. Symmetrical cells of both types were characterized by impedance spectroscopy in order to reveal the relation between the microstructure and the electrochemical performance.

1 Introduction

Environmental and economic concerns regarding future use of fossil fuels for energy production have been driving forces behind considerably renewed interest in fuel cell technologies. Solid Oxide Fuel Cells (SOFCs) can facilitate a direct electrochemical conversion of the energy stored in the fuel into electricity and heat without the efficiency limitations inherent to heat engines governed by the Carnot cycle and without polluting emissions. SOFCs high electrical efficiencies (~ 60%) can substantially exceed those typical for coal-fired power plants (~35%). Fuel cells can be scaled across a wide range of sizes - from micro-systems with outputs as small as few W to facilities operating in MW range. Depending on the design, SOFCs can operate at different temperatures within the region of 500-1000 °C [1, 2]. The state-of-the-art commercial SOFCs are based on a combination of cermet anodes (e.g. Ni/YSZ), ion-conducting ceramic electrolyte materials (yttria-stabilized zirconia (e.g. 8YSZ)) and perovskite-based composite cathodes (e.g. $La_{1-x}Sr_xMnO_{3-\delta}$/8YSZ). The above combination offers chemical and thermal stability in oxidizing and reducing atmospheres and good ionic conductivity over a wide range of conditions [3, 4]. Ni-YSZ anodes are preferred due to the exemplary catalytical properties of Ni for hydrogen oxidation as well as their sufficient electrical conductivity, mechanical strength and good compatibility [5]. However operating SOFCs at temperatures of 800-1000 °C leads to some limitations in their design and operation – e.g. necessity for utilization of expensive corrosive-resistant interconnects - and can be detrimental to the durability of the cell causing Ni catalyst degradation through coarsening and poisoning. Currently a shift towards intermediate temperatures (<800 °C) is considered essential for lowering the production and operational costs of SOFCs. The advantages of a reduced-temperature operation include better system compactness and wider materials choice (e.g. stainless steel porous supports providing good electrical conduction, high mechanical strength, favourable thermal distribution due to the high thermal conduction and rapid start-up times) [6]. Lowering the working temperature however results in reduction of the overall electrochemical performance due to increased Ohmic losses and electrodes' polarization losses both related to thermally activated reactions. There are several development strategies for intermediate temperature SOFC. One of them is based on lowering the Ohmic resistance by implementing thinner electrolyte and substituting 8YSZ with higher electronic conductivity material like doped ceria (Gd:CeO_2 – (GDC)). The other is perusing increase cathodes and anodes catalytic activities through nano-engineering of the porous electrodes' surfaces. The performance of the state-of-the-art commercial cathodes based on $La_{1-x}Sr_xMnO_{3-\delta}$ (LSM) is hindered by insufficient oxygen reduction reaction (ORR) activity and low oxygen ion conductivity at intermediate working temperatures. In recent years a significant progress has been made towards reducing the polarization losses and enhancing electrochemical activity via infiltration of active precursors into the electrode scaffolds [7-10]. Transitional metal oxide infiltration of LSM based cathodes is expected to increase substantially the density of ORR sites and lower the polarization related losses.

This study reports on the application of Inkjet Printing technique (IJP) for the fabrication and manipulation of SOFC components – thin electrolytes and engineered porous electrodes. The advantages of IJP were implemented in realization of different strategies for the optimisation of intermediate temperature SOFCs performance – fabrication of thin electrolytes, compositional variations of the anodes and infiltration of porous composite cathode scaffolds. The IJP is simple and cost-effective non-contact "wet" technique for fabrication of patterns and coatings onto variety of surfaces. It allows utilization of very thin fragile porous support (ceramic or metal) and is insensitive to a certain degree of substrate surface irregularity. IJP can reproducibly dispense droplets in the range of pico to nano-L volumes at high rates (kHz). Drop-on-demand (*DoD*) inkjet printing offers excellent thickness and uniformity control and introduces the possibility of printing 2D and 3D patterns. Inkjet systems have wide scale of application - from experimentation platforms working with customized inks to mass manufacturing systems that can print rapidly and competitively on industrial scale. The technology is cost effective and environmentally friendly through waste minimization of expensive precursors. The influence of the major printing parameters and the required optimization steps were explored for both suspension and solution inks. The production of anodes and electrolyte coatings with a modified Domino print head was reported previously by Tomov et al. [9, 10] using suspension inks. Wang et al. [11] deposited GDC electrolytes on NiO-8YSZ cermet anodes using sol–gel-based solutions. Sukeshini et al. [12] employed a DMP-2831 printer for the deposition of 8YSZ electrolyte layers and LSM-YSZ and LSM cathode layers onto NiO–8YSZ supports, reporting maximum power density of 450 mW/cm^2 at 850 °C in hydrogen.

2 Experimental

2.1 Inks preparation

Electrolyte and electrode depositions as well as catalytic precursor (Co-) infiltrations were performed by an electromagnetically (EM) driven print head with 100 μm ruby nozzle orifice and X-Y planar positioning system. EM technology was chosen because it offers simplicity and reliability of use, as well as wider range of ink/suspension compatibility. The preparation of stable suspension and sol-gel inks is of critical importance for achieving repeatable jetting without clogging the nozzles. For the suspension inks commercial NiO, 8YSZ and GDC powders were mixed with alpha-Terpineol and binders, and ball milled with 3YSZ beads in 3YSZ bowls in a planetary mill. The mass load of the ceramic powders was limited by the rheological working window of the nozzles, which defined the regime of stable repeatable jetting. Hence the viscosity of the suspension inks had to be adjusted levels by adding lower viscosity solvents – Methanol (MeOH) or 1-Propanol (PrOH). The cobalt precursor solution ink (0.75M total metal concentration) was prepared by dissolving $Co(NO_3)_2x6H_2O$ in EtOH and adding citric acid as a complexing agent. Jetting of all types of inks was optimized by drop visualization procedure with the aim to produce no satellite drops at practical Weber and Reynolds numbers. The inks were filtered through 3 μm glass micro-fibre filters before being loaded into the nozzle compartments. The nozzles were observed to execute reproducible drop on demand tasks without clogging the internal fluidic pathways of the assembly.

2.2 Cells fabrications

Anode supported cells (50x50 mm) were fabricated by printing electrolyte suspension ink onto pre-fired tape-casted NiO-3YSZ anodes. After sintering at 1400 °C LSM cathodes were screen printing and fired at 1150 °C. Uniform electrolyte printing procedures involved repeated printing of hexagonal lattices with different lattice constants at optimized jetting parameters.

Anode symmetrical cells were fabricated by printing a 10 mm x 10 mm squares of NiO-8YSZ-GDC composite inks (prepared in different 8YSZ:GDC ratios) on each side of commercial circular dense YSZ substrates with a diameter of 20 mm and thickness of 200 μm. After sintering at 1150 °C in air samples were reduced and tested under an atmosphere of Ar/4% H_2 at a flow rate of 150 mL/min. Cathode symmetrical cells were fabricated by inkjet printing of LSM/GDC composite cathodes on each side of circular dense 8YSZ substrates and sintering at 1100 °C in air for 2 hours. The $Co(NO_3)_2$ inkjet printing infiltrations were performed at variable substrate temperature (20 to 40 °C) in order to investigate the influence of the viscosity variation of the ink on the Co-oxide distribution. Each drop (~40 nL as, determined by drop visualisation) contained 0.69 μg of $Co(NO_3)_2x6H_2O$ and each layer of infiltration (resulting from the infiltration of 64 drops) contained 24 μg of Co_3O_4 nanoparticles.

2.3 Characterization

Electrochemical impedance spectra (EIS) of the symmetrical cells were measured with the electrochemical interface and a frequency response analyzer (Solartron 1260) under the open-circuit voltage (OCV) condition. Symmetrical cells featuring each of the three anode compositions (Ni/8YSZ, Ni/GDC, Ni/25:75 GDC:8YSZ) as well as Co-modified LSM/GDC cathode symmetrical cells were heated at 10 °C/min ramping rate to temperatures from 600 to 800 °C, and tested under Ar/4% H_2 or ambient air in a frequency range of 1 MHz to 10 mHz with AC amplitude of 10 mV. The Nyquist plots were constructed and fitted to the semicircles by employing least squares analysis and the Newton-Raphson method via Excel Solver. Silver mesh was painted to the electrodes in order to ensure conductivity along the surface of the electrodes. I-V characteristics of the fuel cells were tested in a specially developed ceramic enclosure equipped with gold current collectors. The tests were performed at 700, 750 and 800 °C using humidified hydrogen as the fuel (at a constant flow rate of 200 ml/min) and ambient air as the oxidant.

3 Results and discussions

The successful application of IJP required three different stages of optimization:

- Inks optimization
- Jetting optimization
- Drops spreading optimization

3.1 Inks optimization

Particle size distributions analyses of the suspension inks produced via ball milling showed typically multi-modal particles distributions. Figure 1 presents the distribution of NiO and GDC inks milled for various time durations. The inks milled for 4 hours showed substantial broad tail. Extending the milling time to 10 hours led to sharper bi- and tri modal distributions with d(0.5) = 0.129 µm for GDC ink and d(0.5)= 0.443 µm for NiO. Similar procedure was used for 8YSZ and LSM inks. Using 3 µm glass fibre filter during the loading of inks into the nozzle chamber ensured stable jetting operation without clogging events and without retention of any significant ink fraction.

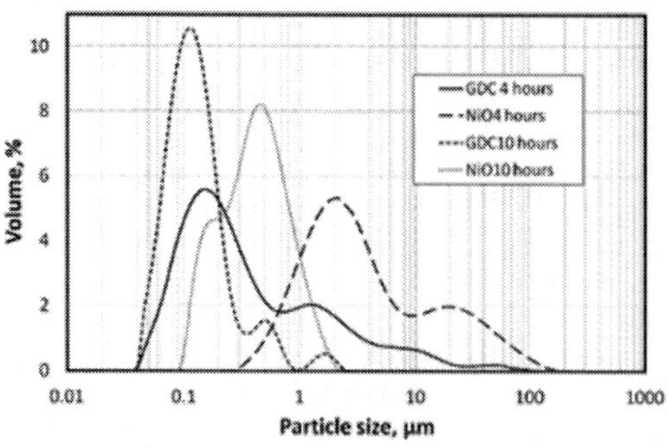

Figure 1. Particle size distribution of GDC and NiO ink diluted with methanol.

The viscosity of two separate NiO and GDC inks as well as NiO-GDC composite inks prepared by dilution in methanol in proportion 1:1 by volume was measured with a rotational viscometer (see Figure 2). The value of the viscosity at γ=30 s^{-1} was 0.0067 Pa s for GDC, 0.0072 Pa s for NiO and 0.0075 Pa s for NiO-GDC mixture. Mixing NiO and GDC into a single anode ink lead to an increase of viscosity due to a partial agglomeration during mixing procedure. The viscosity dynamics was found to follow similar behaviour for all suspension inks over the measured shear rate range. The viscosity of 0.75M sol $Co(NO_3)_2$ ink was tested at different temperatures and lower shear rates (see Figure 3). The viscosity range of 0.005 to 0.025 Pa s was established within 5 to 40 °C temperature variation. The choice of the infiltration temperature was restricted by the empirical rheological nozzle limit of 0.010 Pa s.

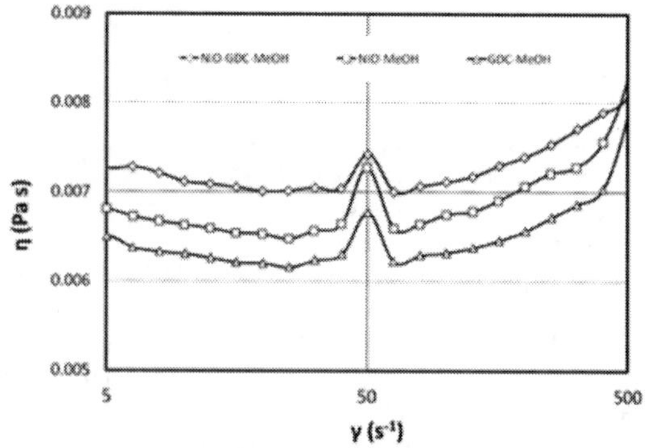

Figure 2. Dynamic viscosity measurement of NiO, GDC and composite NiO-GDC inks.

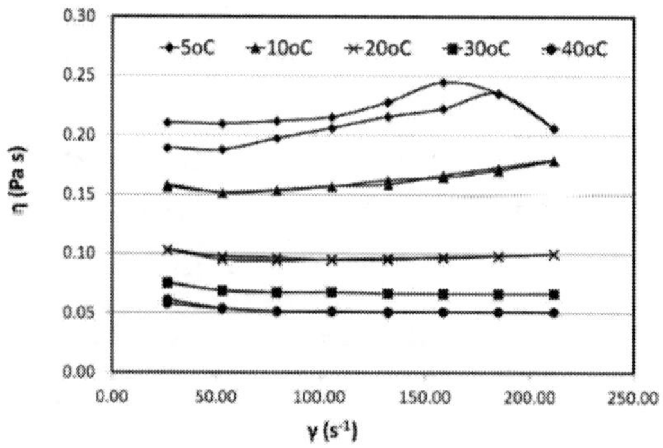

Figure 3. Dynamic viscosity measurement of $Co(NO_3)_2$ - EtOH ink at different temperatures.

3.2 Jetting optimization

The tailored jetting should ideally result in a single drop formation without delayed satellite drops. This is essential for achieving desired thickness control of the coating. The use of drop visualisation system allowed rapid examination of the rheological suitability of all ink. Precise control of the drop volume, velocity and satellite drop formation was achieved by optimization of the printing parameters. Such optimisation for the electromagnetic print heads was based on the variation of carrier/solvent ratio, opening time and nozzle

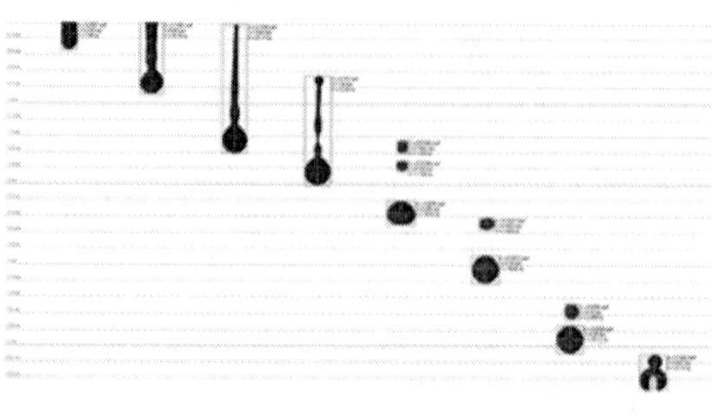

Figure 4. Drop visualisation jetting images of 10 wt% NiO-GDC suspension ink under a pressure of 350 mbar and 250 µs opening time.

pressure. The initially released drop broke into a main drop and one or more smaller drops after it was detached from the nozzle. For optimised printing parameters the smaller drops soon merged with the main drop and formed back to a single drop ensuring that for suitable conditions no satellite drops would be deposited on the substrate. Figure 4 illustrates optimized drop formation behaviour of 10 wt% NiO-GDC suspension ink under a pressure of 350 mbar and 250 μs opening time.

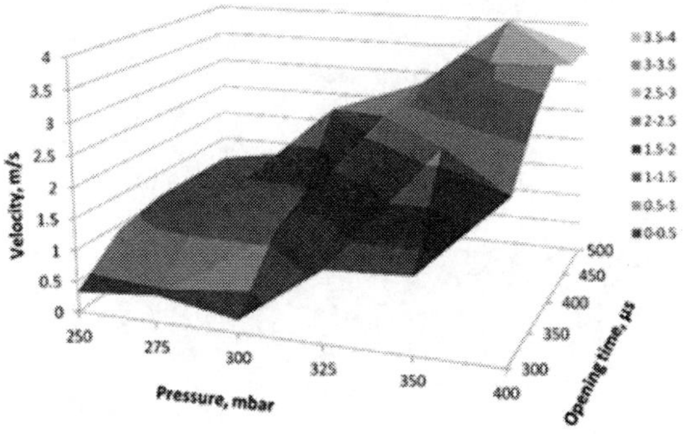

(a)

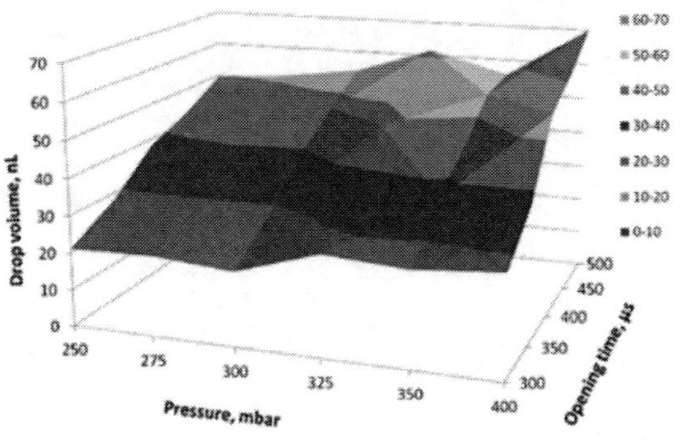

(b)

Figure 5. (a) Velocity data from the visualisation system showing jetting of 10 wt% NiO-GDC suspension ink; (b) Calculated drop volumes versus pressure (250-400 mbar) for tested opening times (300-500 μs).

Figure 5 demonstrates the achievable range of drop volumes and velocities within the practical limits of opening times and pressures. It is clear that the drop volumes depended strongly on the opening time while the velocities were strongly influenced by the nozzle positive pressure.

3.3 Spreading optimization

The optimised velocities and drop volumes presented a problem for the deposition of thin coatings on supports with substantial porosity and wider pore size distribution. The standard printing procedure resulted in a penetration of the ink within the scaffold making deposition of free standing coating impossible. This issue was resolved by implementing inkjet printing on heated supports. The elevated temperature of the substrate surface led to a partial evaporation of the solvent effectively increasing the viscosity and the surface tension of the landing drops. The retention of the ink onto the porous surface allowed for building of free-standing membranes. On the other hand depositions at elevated temperatures can cause an immediate evaporation of the solvent at the periphery of the drop relic where the thickness of the deposit is relatively thin, effecting a net flux of ink radially outward resulting in so called "coffee ring effect". Such mechanism is detrimental, in that a film composed of multiple drop relics will be non-uniform in thickness.

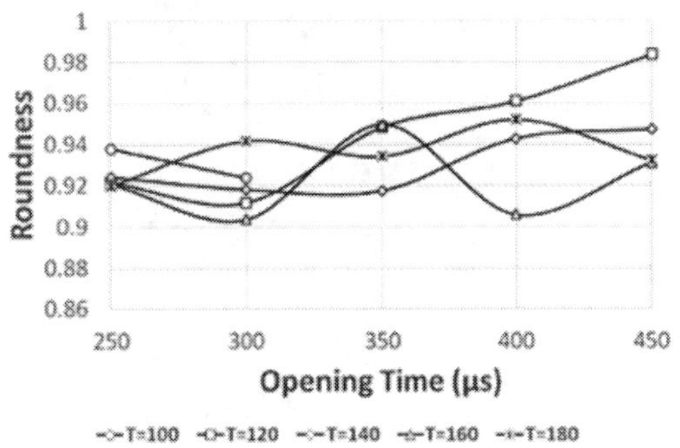

(a)

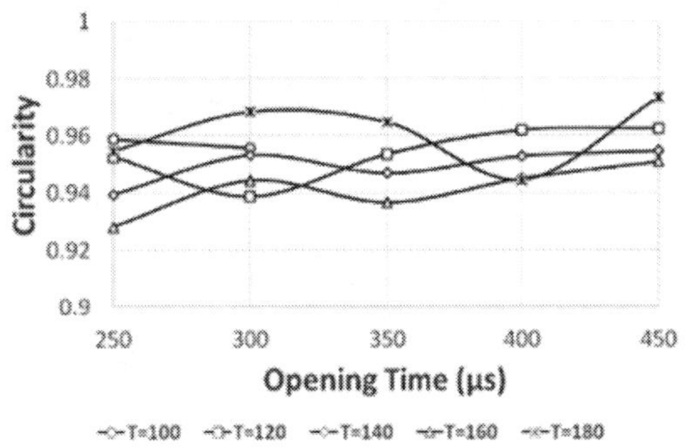

(b)

Figure 6. (a) Circularity and (b) roundness of the printed relics vs. opening times.

The problem was counteracted by the implementation of high vapour pressure solvent as ink carrier (alpha-Terpineol). The evaporation of such solvent at the contact line created a surface tension gradient increasing towards the cooler drop centre and inducing a balancing Marangoni flow. An optimization of drops formation and spreading for the composite NiO-GDC-MeOH ink was performed for the pressure of 350 mbar and opening time (OT) range of 250-450 μs and substrate temperature (T) range 100-180 °C. Rasters of single drops were printed on heated dense 8YSZ electrolyte substrates and the geometric features of the drops were analyzed with ImageJ software. The values of interest were diameters, circularities ($C = 4\pi A/p^2$), and degrees of roundness ($R=4A/\pi a^2$) where A was droplet area, p was droplet perimeter, and a was the major axis of an ellipse fit to the drop relic. Circularity and roundness varied over 0 to1, higher values being an indicator of a shape that is close in character to a perfect circle. Figure 6 (a) and (b) summarizes the image analysis data for drop relics formed at different substrate temperature. The visual representation of selected drop relics relevant to this optical analysis can be seen

DOI: 10.1051/e3sconf/20161000098

in Figure 7. The general preference for small relic diameter with high circularity and roundness numbers led to the choice of one sets of parameters resulting also in the desired absence of "coffee ring effect" - temperature of 120°C, and opening time of 400 μs.

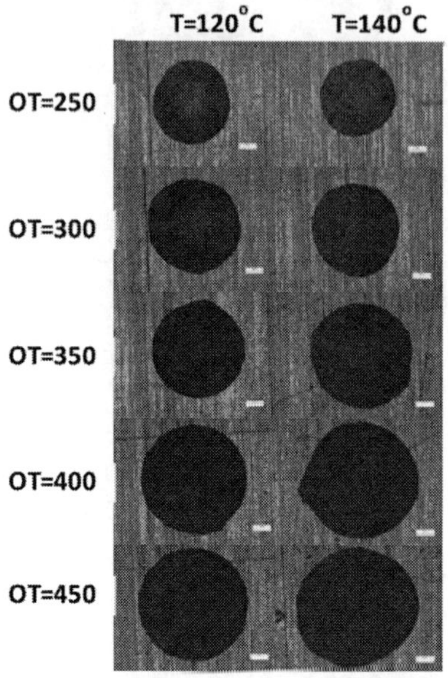

Figure 7. Images of a single drops relics for different opening times and substrate temperature. The white bar on all images represents 200 μm.

3.4 Inkjet printing of thin electrolytes

After choosing the optimum jetting and spreading parameters for 8YSZ inks, complete electrolyte coatings were printed on the 57 mm x 57 mm square NiO-YSZ cermet supports using hexagonal arrays of droplets in reciprocating raster pattern. Figure 8 (a) and (b) shows SEM micrographs of the surface and the cross-section of 8YSZ electrolyte membrane prepared by inkjet printing. 12 coatings of the 5 wt% suspension were printed and sintered at 1400 C. It is important to note that the resulting YSZ membrane, as thin as ~5.7-5.9 μm, appeared to be highly dense and no open porosity was observed. The film had good interfacial cohesion with the support, without any cracks or delamination.

The I-V tests were performed at 700, 750 and 800 °C using humidified hydrogen as the fuel (with different H_2:N_2 mixture ratios at constant flow rate of 200 ml/min) and ambient air as the oxidant. The OCV value of 1.077 V measured at 800 °C confirmed the gas-tightness of the membrane and was consistent with the cross-section appearance. The I-V tests showed maximum power densities of ~ 220 mW/cm^2 for pure humidified hydrogen stream. Diluting the fuel with N_2 as shown in the inset of Figure 8b led to a substantial reduction of the maximum output power without change in the OCV values. The latter suggested the existence of concentration losses due to non-optimized porous structure of the anode support.

Similar printing procedure was employed for deposition composite GDC/NiO functional layer and GDC electrolyte coating on porous metal supports. Stainless steel porous supports were prepared by milling and pressing a mixture of 430L stainless steel powder and organic binder into circular pellets. As-prepared supports were sintered in a tube vacuum furnace at 1150 °C and cooled to room temperature at a ramp rate of 5 °C/min. Such sintering procedure gave sufficient mechanical strength to the supports for the subsequent handling. The density and the open porosity of the pellets were evaluated by Archimedes' method which revealed an open porosity of ~ 40 %. The inspection of the SS430L supports sintered at 1150 °C reveals small degree of surface waviness and relative mechanical fragility due to incomplete sintering. Hence the use of traditional ceramic coating techniques as screen printing was considered challenging. As IJP technique is insensitive to such degree of non-uniformity it was employed for the functional coatings development.

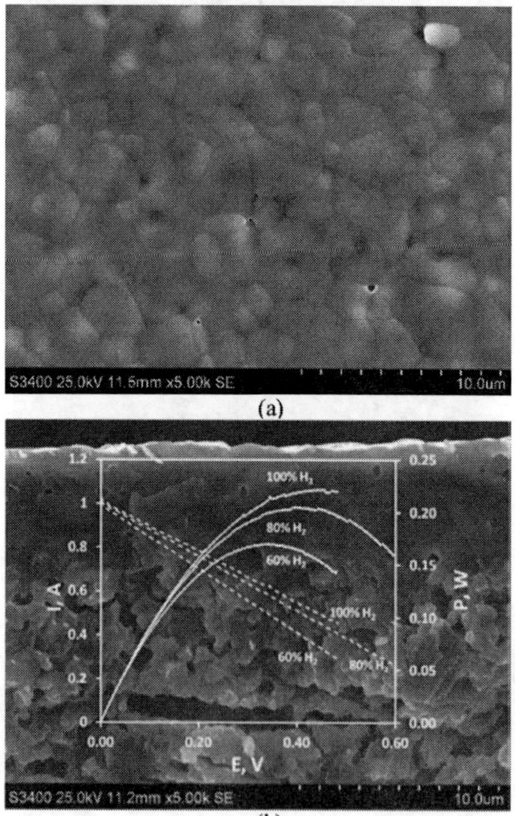

Figure 8. SEM micrographs of the (a) surface and (b) half-cell cross-section sintered at 1400 °C. The inset shows the I-V results of the cell at 800 °C and different fuel dilution levels.

Figure 9 represents cross sectional images of GDC/NiO-GDC/SS430L fractured half-cell fabricated with IJP under the optimized conditions. Electrolyte and anode coatings less than 15 μm thick were deposited as free standing coatings. The sintering temperature of 1350 °C was found sufficient for full densification of the electrolyte. The effect of surface non-uniformity planarization was evident. Some infiltration of the anode ink into the stainless steel support could be detected in the area adjacent to the anode-support interface (see the inset of Figure 9).

Figure 9. Cross sectional SEM image of GDC/Ni-GDC half-cell architecture inkjet printed on 430L stainless steel support and sintered in vacuum at 1350 °C. The inset shows Ni-GDC ink residue infiltrated into an area adjacent to the anode interface.

3.5 Anode symmetrical cells with variable composition

The results of EIS studies of anode symmetrical cells tested with Ar/4% H_2 can be seen in Figure 10a-c. The Nyquist plots obtained at different temperature exhibited suppressed semicircles with varying cross sectional points with X- axis at the low frequency end.

As expected, the polarization resistance (*Rp*) observed correlated with increasing the GDC content in the composite anode for temperatures above 700°C (see Figure 10d). The increase in temperature led to ASR values having weighted contribution from both ion conductive components (GDC and YSZ). This is consistent with GDC's having higher ionic conductivity than that of YSZ at these temperatures hence effectively promoting the charge transfer mechanism in the anode area.

On the other hand an increase of the polarization activation energy (Ea-p) proportional to the GDC composition was observed most likely related to the enhanced electronic conductivity of GDC at higher temperatures (Ea-p = 38.52, 66.55 and 83.58 kJ/mol for YSZ, YSZ+GDC and GDC anodes respectively). Furthermore, as-measured Ohmic resistivity seems to be dominated by GDC component even at volume ratios of only 25 vol% while the activation energy (Ea-o) was found to be almost constant at different GDC levels.

3.6 Co_3O_4 nano decorated LSM/GDC composite cathode

The area-specific resistances (ASRs) of Co- infiltrated and non-infiltrated LSM/GDC cathode symmetric cells were measured at 600°C and 800°C (EIS data shown in Fig 11a and b). The polarization resistance reduction due to Co_3O_4 nano-particles cathode surface decoration was more pronounced at lower temperatures.

A reduction of the activation polarization losses of almost one order of magnitude was observed as a direct result of Co-ink infiltration. The differences in the measured

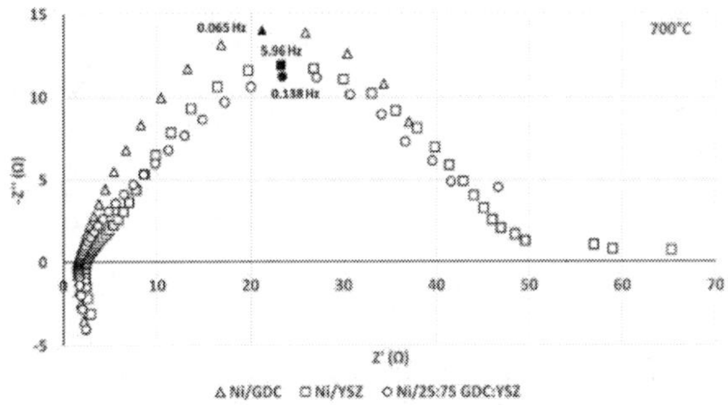

(a)

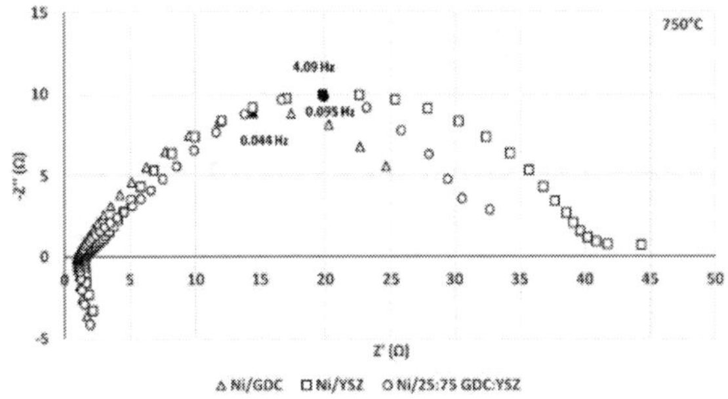

(b)

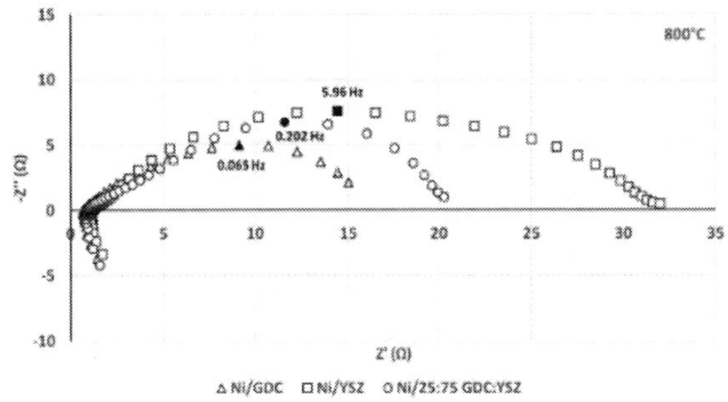

(c)

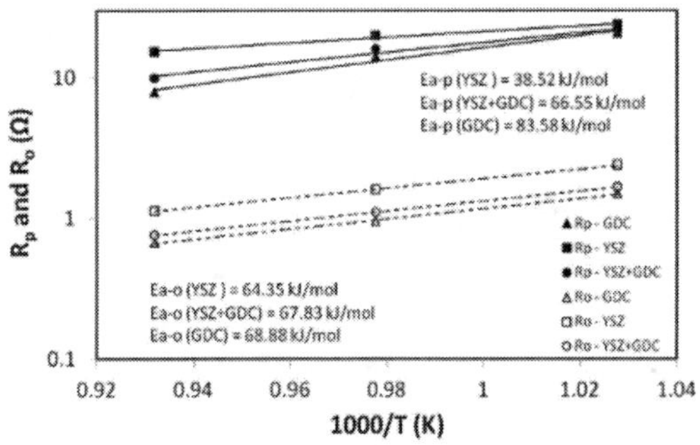

(d)

Figure 10. EIS data of symmetrical cells with anodes of various compositions measured in Ar/4% H_2 at temperatures between 700 and 800°C.

ASR values suggested that at 600 °C the sample infiltrated at elevated temperature (40 °C) had lower polarization losses compared to the one infiltrated at room temperature. One could speculate that this was a result of a better coverage of the cathode scaffold with Co_3O_4 nanoparticles. Such effect was expected due to the lowered viscosity and improved wetting properties of the ethanol-based Co nitrate ink at higher temperatures. At 800 °C the Nyquist plot revealed a second semicircle resolved at the low frequency end of the spectra for the sample infiltrated at 40 °C. This suggested the appearance of concentration losses associated with higher mass load of the infiltrate partially blocking the porous structure of the backbone (see Figure 11b and Figure 12a). The size of Co_3O_4 nanoparticles was estimated to be between 100 and 200 nm. EDS line scans as seen in Fig. 12b and 12c confirmed the superior more uniform depth distribution of Co in the case of the cathode infiltrated at elevated temperature.

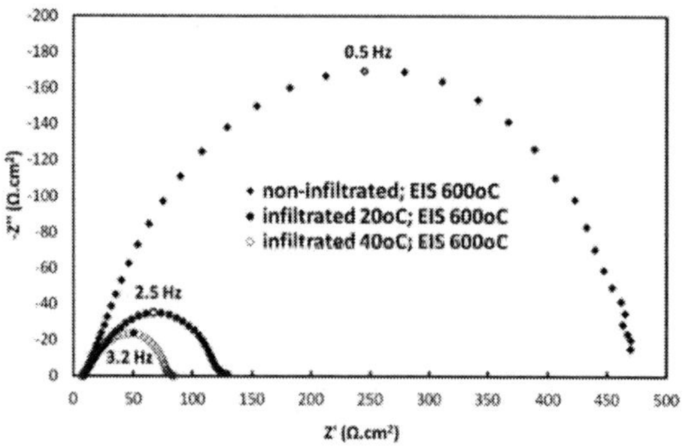

(a)

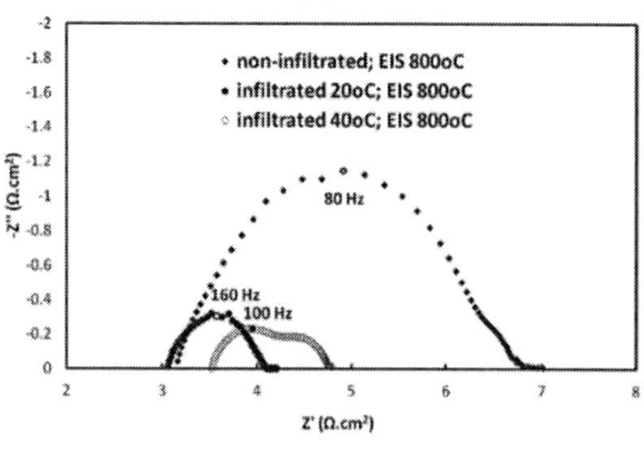

(b)

Figure 11. EIS spectra of LSM/GDC symmetrical cathode cells non-infiltrated and infiltrated at 20 °C and 40 °C substrate temperatures. Data was measured for two different temperatures – 600 °C (a) and 800 °C (b).

4 Conclusions

The results of this study prove the feasibility of IJP technology for the fabrication and manipulation of SOFC functional coatings. The precision of jetting and ink delivery allowed us to print reproducibly 8YSZ free standing membranes as thin as 6μm onto porous uneven surfaces. Anodes optimization was demonstrated via compositional (YSZ:GDC) variation facilitated by the flexibility of composite ink mixing procedure in the inkjet printing processing. The employment of inkjet infiltration of Co nitrate ink into the LSM/GDC composite backbone produced Co_3O_4 nano-decoration of the cathode internal surface which led to substantial performance improvement achieved with precise control of the catalytic element depth distribution and minimum wastage of expensive precursors.

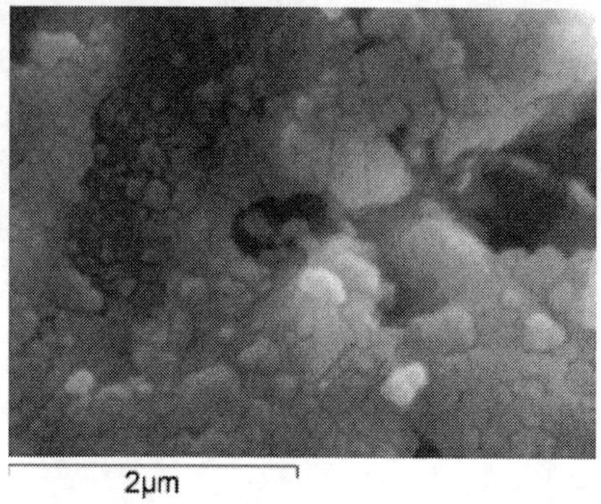

(a)

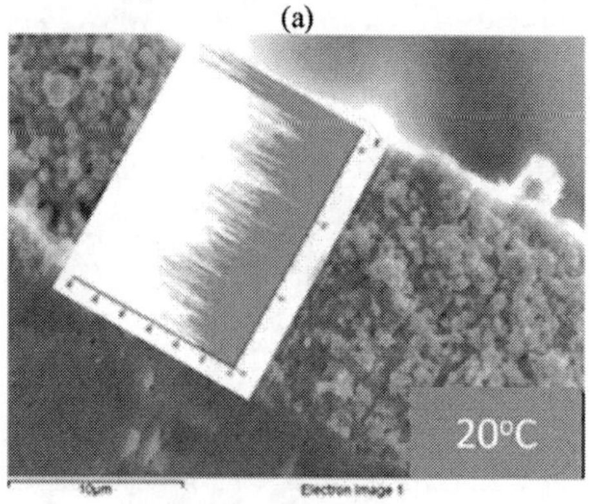

(b)

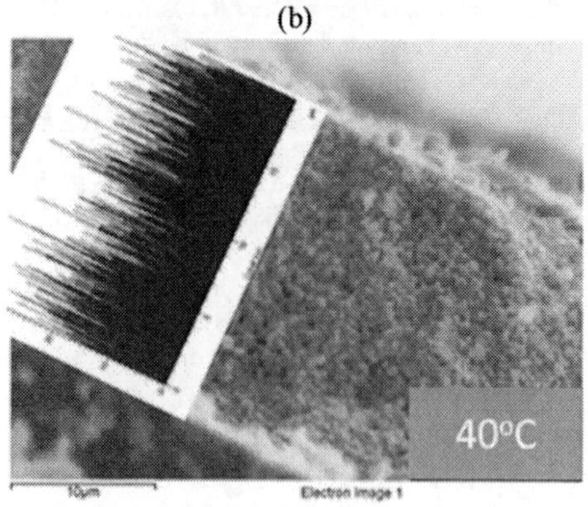

(c)

Figure 12. SEM cross-section micrograph (a) and EDS line scans of Co depth distribution for the samples infiltrated at 20°C (b) and 40°C (c).

Acknowledgements

The authors wish to acknowledge the EU 7th Framework Programme grant agreement 286100 (SENERES), as well as EPSRC grants - "Performance Optimisation of Intermediate Temperature - Solid Oxide Fuel Cells

(IT - SOFCs) by Inkjet Printing on Porous Metal Substrates (*JETCELL*)" and "Tailoring of micro structural evolution in impregnated SOFC electrodes" for the financial support.

References

1. S. Singhal, Solid State Ionics **135** (1–4), 305-313 (2000)
2. Z.P. Shao, S.M Haile, J. Ahn, P.D. Ronney, Z.L. Zhan, S.A. Barnett, Nature **435**, 795–798 (2005)
3. Weber, A., Ivers-Tiffee, E., Journal of Power Sources **127**, 273–283 (2004)
4. I.R. Gibson, G.P. Dransfield, and J.T.S. Irvine, J. Eur. Ceram. Soc. **18**, 661-667 (1998)
5. S.P. Jiang, S.H. Chan, S.H., J. Mater. Sci. **39**, 4405-4439 (2004)
6. N.Q. Mihn, T. Takahashi, *Science and Technology of Ceramic Fuel Cells* (Elsevier Ltd., Amsterdam 1995).
7. D. Ding, X. Li, S.Y. Lai, K. Gerdes and M. Liu, Energy Environ. Sci. **7**, 552-575 (2014)
8. J.A. Kilner and M. Burriel, Annu. Rev. Mater. Res. **44**, 365–393 (2014).
9. A. Atkinson , S. Barnett , R. J. Gorte , J. T. S. Irvine , A. J. McEvoy , M. Mogensen , S. C. Singhal, J. Vohs, Nature Materials **3**, 17- 27 (2004)
10. S.C. Singhal and K. Kendall ed., *High temperature solid oxide fuel cells: fundamentals, design and applications* (Elsevier Ltd, Amsterdam 2003).
9. R.I. Tomov, M Krauz, J. Jewulski, S.C. Hopkins, J.R. Kluczowski, D.M. Glowacka, B.A. Glowacki, Journal of Power Sources **195**, 7160–7167 (2010)
10. Ch. Wang, S.C. Hopkins, R.I. Tomov, R.V. Kumar, B.A. Glowacki, Journal of the European Ceramic Society **32**, 2317–2324 (2012)
11. Ch. Wang, R.I. Tomov, R.V. Kumar, B.A. Glowacki, Journal of Materials Science **46**, 6889–6896 (2011)
13. A.M. Sukeshini, R. Cummins, T.L. Reitz, R.M. Miller, J. Am. Ceram. Soc. **92** (12), 2913–2919 (2009)

 DOI: 10.1051/e3sconf/20161000099

Impact of separated bottom ashes on the parameters of concrete mix and hardened concrete

Daniel Wałach[1,a]

[1]*AGH University of Science and Technology Sciences, Al. Mickiewicza 30, 30-059 Cracow, Poland*

Abstract. Polish energy industry is based primarily on the combustion of coal in conventional boilers and increasingly in fluidised bed boilers. As a result of combustion, by-products are formed, such as fly and bottom ashes. In the case of fly ashes, a number of methods of utilising them have already been developed, e.g. for the production of cement, concrete and in other industries. With regard to bottom ashes, whose properties differ significantly from those of fly ashes, methods are still being researched for the possibility of their industrial application. Similarly to fly ashes, it seems reasonable to move in the direction of bottom ashes being used in the wider construction industry, including for the production of concrete.

This paper analyses the impact of the addition of bottom ashes obtained from hard coal combustion in conventional and fluidised bed boilers on the properties of fresh and hardened concrete. A concrete mix composition was developed by an experimental method, which was then modified with the use of bottom ashes. The impact of the substitution of cement and aggregates with bottom ash on the concrete properties was examined. For all the obtained series of concretes, tests were performed for the consistency of the fresh concrete using the concrete slump test, the compressive strength and tensile strength of the concrete after 3, 7 and 28 days of maturing and their absorption. The experiments have shown significant declines in the strength parameters of the concretes being analysed in the case of the substitution of cement with separated bottom ash. However, substituting relevant aggregate fractions with separated bottom ash resulted in an increase in both the compressive strength and the tensile strength in the analysed concretes.

1 Introduction

As a result of coal combustion, residues in the form of ash, soot and slag are produced. Ash is solid particles transported by gases and can be divided into two types: bottom ash and fly ash, with the latter being captured on electrostatic filters and bag filters. Ash consists primarily of oxides: silicon, aluminium, iron, calcium, magnesium, sodium, potassium, titanium and manganese oxides [1]. Their characteristics and quantity depends on both the properties of coal being burned, i.e. coal granulation, its caking properties, moisture content, and the combustion process technique and parameters, including the type of furnace, combustion temperature, thermal intensity of the combustion chamber, the air and flue gas flow parameters. It should be emphasised that the properties of ashes from conventional pulverised coal boilers [1] differ significantly from those of the ashes from fluidised bed boilers, which are characterized by e.g. the lack or negligible amount of glaze and unburned carbonaceous substance.

Polish energy sector is based primarily on coal and the signs are that this situation will not change any time soon [2]. Thus it seems important to develop comprehensive programmes for the management of combustion by-products produced during the production of energy in Poland. In the case of fly ashes, a number of methods of utilising them have already been developed, e.g. for the production of cement, concrete and in other industries [3]. With regard to bottom ashes, methods are being researched for the possibility of their industrial application. Similarly to fly ashes, it seems reasonable to move in the direction of bottom ashes being used in the wider construction industry. Currently, in the literature can be found some ideas for its application, e.g. in concrete technology in order to improve workability [4], durability [4,5] or sound absorption [6]. Also, as an aggregate in lightweight fly ash geopolymer concrete [7], as well as to improve the workability of high-strength mortar [8]. Nevertheless, because of the significant need to develop methods for utilization of bottom ashes, research work has been conducted to asses the impact of the addition of bottom ashes obtained from hard coal combustion in conventional and fluidised bed boilers on the properties of concrete mix and hardened concrete, the results of which are presented in this paper.

[a] Corresponding author: walach@agh.edu.pl

 DOI: 10.1051/e3sconf/20161000099

2 Characteristics of the used bottom ashes

The tests were conducted on the energy waste resulting from the combustion of coal in a fluidised bed boiler (PDF) and pulverised coal boiler (PDP), which were separated using a sieve with 0.63 mm mesh size. Particle size composition curves for the used ashes are shown in figure 1.

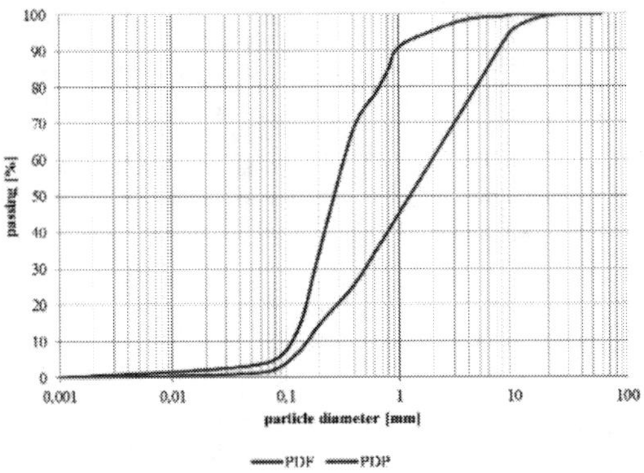

Figure 1. Particle size composition curves of bottom ashes.

The bottom ashes were characterised by a particle diameters of up to 10 mm for the bottom ash from the fluidised bed boiler (PDF) and 25 mm for the bottom ash from the pulverised coal boiler (PDP). The content of fractions below 0.63 mm in the bottom ashes was:
- for PDF: 78.3%
- for PDP: 35.0%

The significant differences in the particle size composition of the bottom ashes resulted from the coal combustion technology used in the boilers and the method of collecting this waste from the bottom of the furnace. The PDF ash contained mainly dust and sand fractions (74.2% and 21.1% respectively) and the bottom ash from the pulverised coal boiler was characterised by a larger particle diameters and contained dust fraction (29.8%), sand fraction (30.9%) and gravel fraction (39.3%).

The average density of the fluidised bed bottom ash (PDF) was 2.68 g/cm³. The bottom ash from the pulverised coal boiler was characterised by lower density of 2.27 g/cm³.

Figures 2 and 3 shows the content of the main chemical ingredients in the analysed ashes by particle size.

The chemical composition of the PDF ash did not differ from the other bottom ashes produced in fluidised bed boilers. A characteristic feature of the tested material was high CaO and SO_3 content, resulting from the process of flue gas desulphurisation. This is associated, in particular, with the addition of calcium sorbent into the boiler, the penetration of desulphurisation products into bottom ash and the residues of unreacted calcium sorbent. A high SiO_2 content in the bottom ashes results from the presence of quartz sand used as a fluidised bed material.

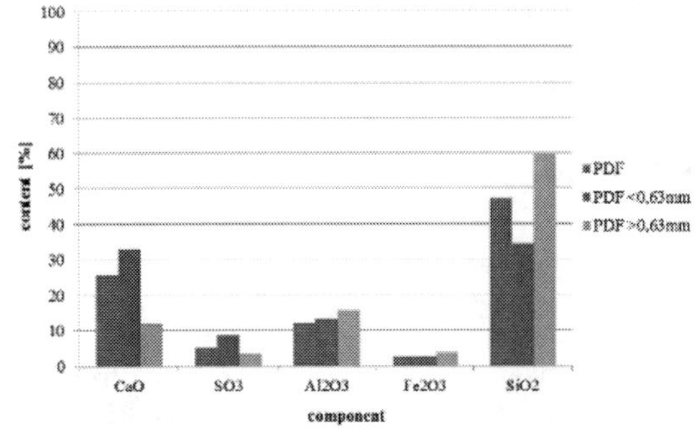

Figure 2. Content of the main ingredients in the PDF bottom ash by particle size.

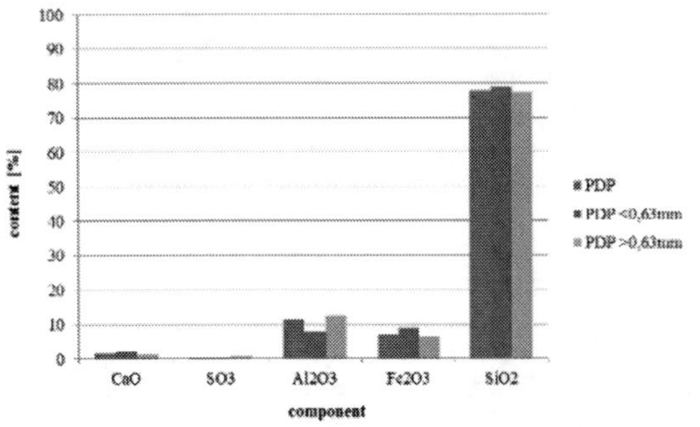

Figure 3. Content of the main ingredients in the PDP bottom ash by particle size.

The PDF ash fractionation caused the diversification of content of individual ingredients within a given fraction. The fraction below 0.63 mm contains considerably more CaO and SO_3 than the fraction with the particle size above 0.63 mm. In turn, the fraction above 0.63 mm contains greater quantities of Al_2O_3 and SiO_2. This situation – i.e. the low content of SO_3 (3,3%), the high content of silicon dioxide and oxides of iron and aluminium – can be very beneficial for the application of the PDF fraction >0.63 mm as an ingredient of concrete and also for meeting the requirements of building standards or obtaining appropriate technical approvals.

Loss on ignition of the PDF bottom ash was 4.4%. It was lower (3.6%) for a finer fraction and 5.3% for the fraction >0.63 mm. All these values meet the requirements of PN-EN 450-1: 2012: Fly ash for concrete [9], in which the maximum loss on ignition of ashes are set at 9%.

The main ingredients of the PDP ash was silica (77.7%), alumina (11.3%) and iron oxide (6.9%). Ash fractionation on a 0.63 mm sieve did not have a significant impact on the content of SiO_2 in the obtained fractions, but there were differences particularly for the content of Al_2O_3 (the fraction below 0.63 mm contained 7.8% and the fraction above 0.63 mm contained 12.3%). Quite clear differences can be seen in the organic matter content. The loss on ignition for the fraction below 0.63 mm was 2.7%, while for the fraction above 0.63 mm, it was as high as 11.5%. Such high loss on ignition may disqualify this fraction for the application as an additive in construction concretes because of the requirements set out for materials used in the manufacture of construction concretes.

3 Test plan

The aim of the tests was to determine the impact of the type and content of the two bottom ashes in cement-ash binder and aggregate particle size distribution on the properties of hardened concretes and concrete mixes. The bottom ashes resulted from coal combustion in the fluidised bed boiler (PDF) and pulverised coal boiler (PDP). Two waste fractions were used with a particle size above and below 0.63 mm. The addition of individual ash fractions was treated as a substitute for some of the cement in the binder or a substitute for the aggregate in the concrete mix.

For the preparation of various concrete mixes, CEM I 42.5R cement was used, whose characteristics are summarised in table 1. The cement was adequately protected from external influences throughout the period of its use.

Table 1. CEM I 42.5R cement properties

Properties	Value
Apparent density [Mg/m^3]	1.82
Setting start time [min]	570
Setting end time [min]	645
Compressive strength [MPa] - after 2 days - after 28 days	 14.99 46.60

Natural river aggregates in the form of sand and gravel were used to produce the concretes. The aggregates that were used to produce the aggregate mix for the tested concretes were sand with a nominal particle size of 0/2 mm and gravel with a nominal particle size of 4/16 mm. The entire aggregate composition used to make the analysed concretes was formulated based on own research on previously produced concrete mixes and the properties of separated fractions of PDF and PDP ashes (PDF>0,63, PDF<0,63, PDP>0,63, PDP<0,63) and non-separated ashes. As a result of the performed analyses and laboratory tests, it was decided to formulate the aggregate mix for concretes using the aggregate mixing mass ratios presented in table 2.

The composition of the B1 reference concrete was determined on the basis of the adopted amount of cement *c* and ratio *w/c*, which was established at 0.4. In addition, the amount of admixtures was determined experimentally in order to obtain a mix with certain rheological properties.

In the case of other mixes (B2–B8), an assumption was adopted regarding the constant *w/c* ratio, as for the reference mix. Therefore, in order to obtain adequate workability of the mixes with the addition of individual fractions of bottom ashes, it was necessary to use admixtures. It should be noted that the superplasticiser was dosed into the individual mixes in amounts that enabled to work them by laying them in moulds and thickening them in an appropriate manner.

The composition of the reference mix was shown in table 3.

Table 2. The mass ratio of the aggregates used in the concrete.

Mix	Aggregate	Mass ratio
B1- reference mix	0/2 mm sand 4/16 mm gravel	32% 68%
B2 – 15% PDF>0.63 mass of sand and gravel	0/2 mm sand 4/16 mm gravel PDF>0.63	21% 64% 15%
B3 – 30% PDF>0.63 mass of sand and gravel	0/2 mm sand 4/16 mm gravel PDF>0.63	10% 60% 30%
B4 – 15% PDF<0.63 mass of cement	0/2 mm sand 4/16 mm gravel PDF<0.63	32% 68% 15% m of cement
B5 – 15% PDF (non-separated) mass of sand and gravel	0/2 mm sand 4/16 mm gravel PDF (non-separated)	27% 58% 15%
B6 – 15% PDP>0.63 mass of sand and gravel	0/2 mm sand 4/16 mm gravel PDP>0.63	21% 64% 15%
B7 – 30% PDP>0.63 mass of sand and gravel	0/2 mm sand 4/16 mm gravel PDP>0.63	10% 60% 30%
B8 – 15% PDP<0.63 mass of cement	0/2 mm sand 4/16 mm gravel PDP<0.63	32% 68% 15% m of cement

Table 3. Composition of the reference concrete mix.

Ingredients	kg/m^3	dm^3/m^3
CEM I 42.5R Cement	350	-
Water	-	140
Gravel	1387.5	-
Sand	653.1	-
Plasticiser	-	2.7
Superplasticiser	-	-

4 Testing the properties of concrete mixes

For each of the developed mixes, tests of consistency were performed using the concrete slump method, as per PN-EN 12350-2:2001 Testing fresh concrete. Part 2: Slump tests [10]. The properties of the concrete mixes were assessed each time by making a batch of concrete from which specimens were formed. The obtained results of the consistency tests are graphically presented in figure 4.

Based on the results of the tests of the impact of the applied bottom ashes on the rheological properties of fresh concrete mixes, it can be said that the impact is significant.

In the case of B1, B4 and B8 concrete mixes, mix consistency was obtained that corresponded to S4 class. For other mixtures, S1 class was obtained. It should be noted that in the case of concrete mixtures B2, B3, B5, B6 and B7 in order to obtain their workability it was necessary to use a superplasticiser. The used

 DOI: 10.1051/e3sconf/20161000099

superplasticiser amounts in individual concrete mixes are shown in figure 5.

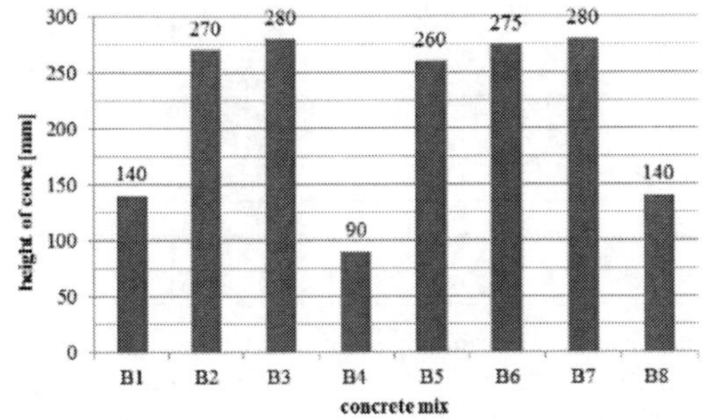

Figure 4. The height of the test cones for individual concrete mixtures.

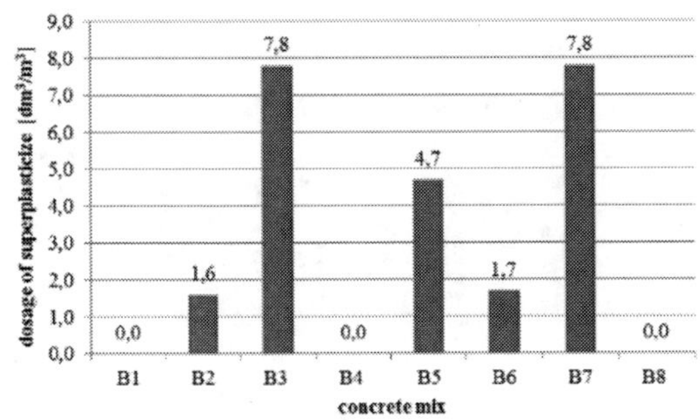

Figure 5. Usage of superplasticiser in individual concrete mixes.

It can be concluded that the use of separated bottom ashes with fractions >0.63 mm as a partial substitute for aggregate (B2, B3, B6 and B7) significantly reduces the wetness of the mixes. In order to obtain their adequate workability, it is necessary to use specialised admixtures.

Application of non-separated PDF fluidised bed bottom ashes (B5) as a substitute for part of the aggregate also resulted in a significant reduction of their workability.

In the case of the application of separated bottom ashes with fraction <0.63 mm as a partial substitute of cement (B4 and B8), one obtains a greater mix wetness, which is comparable to the reference mix (B1).

5 Hardened concrete tests

All the specimens from the analysed concretes were made as part of one-time mixing. 100 mm block specimens were formed in metal moulds. They were compacted on a vibrating table in accordance with PN-EN 12390-2:2001 Testing hardened concrete. Part 2: Making and curing specimens for strength tests [11]. Each time, specimen preparation times from one batch was approx. 30 minutes from the end of mixing. Specimens remained covered with a plastic film until demoulding, i.e. for a period of 28 hours. After demoulding, samples were placed over water in a closed container that maintained a constant temperature, where they remained until the tests were carried out after 3, 7 and 28 days of curing.

The testing of the hardened concretes included determining the following properties:
- the development of uniaxial compressive strength f_c after 3, 7 and 28 days in accordance with PN-EN 12390-3:2002 Testing hardened concrete. Part 3: Compressive strength of the test specimens [12];
- the development of splitting tensile strength f_t after 3, 7 and 28 days in accordance with PN-EN 12390-6:2001 Testing hardened concrete. Part 6: Tensile splitting strength of test specimens [13];
- brittleness characteristics (f_c/f_t relation);
- apparent density after 3, 7 and 28 days in accordance with PN-EN 12390-7:2001 Testing hardened concrete. Part 7: The density of concrete [14];
- water absorption after 28 days in accordance with PN-B-06250:1988 normal concrete [15].

5.1 Compressive strength

Figure 6 shows the results of the compressive strength tests after 28 days of curing of individual concretes and the obtained average strength values are summarised in table 4.

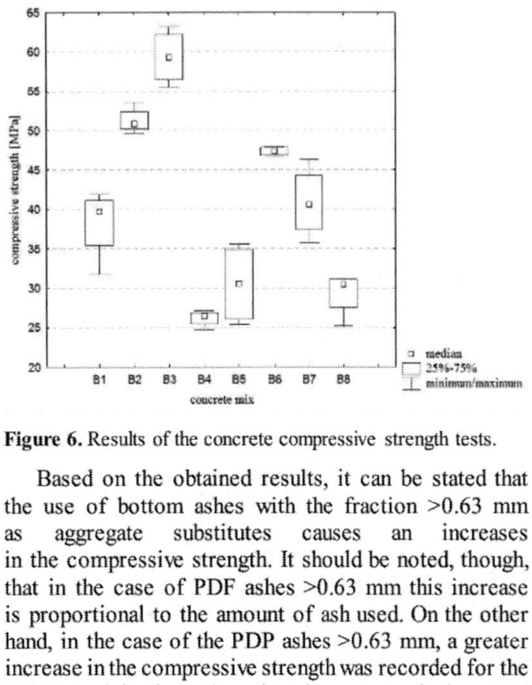

Figure 6. Results of the concrete compressive strength tests.

Based on the obtained results, it can be stated that the use of bottom ashes with the fraction >0.63 mm as aggregate substitutes causes an increases in the compressive strength. It should be noted, though, that in the case of PDF ashes >0.63 mm this increase is proportional to the amount of ash used. On the other hand, in the case of the PDP ashes >0.63 mm, a greater increase in the compressive strength was recorded for the 15% participation than in the case of the 30% participation.

Table 4. Results of the concrete compressive strength tests.

Concrete	Compressive strength f_c [MPa]		
	after 3 days	after 7 days	after 28 days
B1	24.74	27.65	38.30
B2	37.41	38.80	51.25
B3	42.64	54.74	59.33
B4	10.25	16.93	26.20
B5	26.08	28.42	30.51
B6	24.57	33.21	47.33
B7	33.75	37.12	40.86
B8	17.14	26.67	29.31

DOI: 10.1051/e3sconf/20161000099

The highest increase in the compressive strength after 28 days of curing compared to the reference concrete were obtained for concrete B3 (an increase by 54.91%). In concrete B2, this increase was 33.81%. In the case of concretes, where the PDP ash >0.63 mm was used instead of the aggregate, the following increases in the compressive strength were recorded - 23.58% for concrete B6 and 6.68% for concrete B7. Other concretes, i.e. B4, B5 and B8, registered decreases in the compressive strength of up to 31.59%.

The graphs in figure 7 show the increase in the compressive strength of individual concretes. In all the concretes, the strength development after 28 days is quite similar and its intensity is greater than or comparable to that exhibited by the reference concrete.

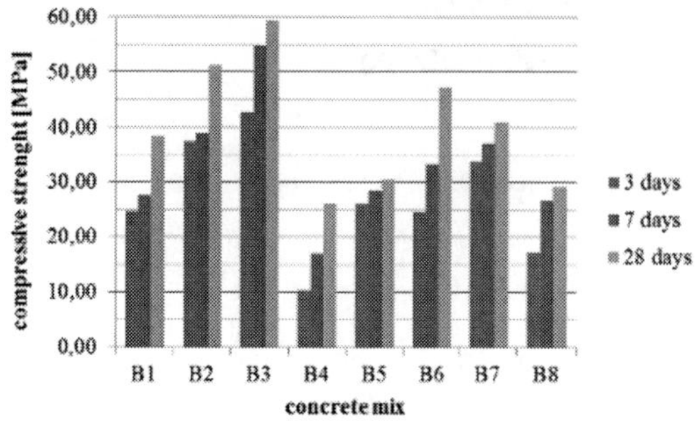

Figure 7. The compressive strength gain of each concrete.

5.2 Tensile Strength

Figure 8 shows the results of the tensile strength tests after 28 days of curing of individual concretes and the obtained average strength values are summarised in table 5.

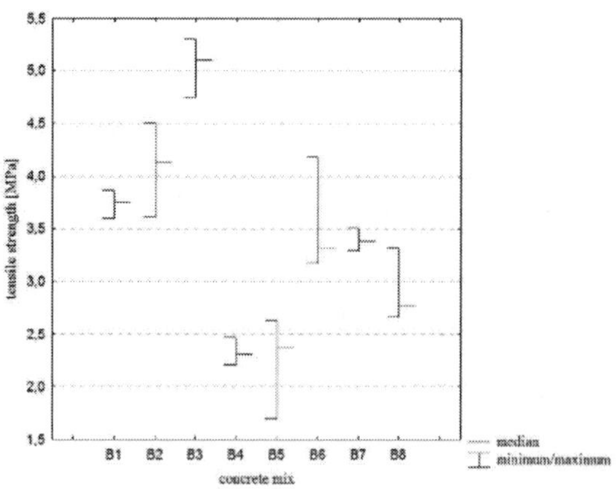

Figure 8. Concrete tensile test results.

Based on the results, it can be stated that the use of bottom ashes with the fraction >0.63 mm as aggregate substitutes causes an increases in the tensile strength.

The highest increase in the tensile strength after 28 days of curing compared to the reference concrete were obtained for concrete B3 (an increase by 35.03%). In the case of concrete B2, this increase was 9.36%. Other concretes recorded decreases in the tensile strength of up to 40.37%, as was the case for concrete B5.

Table 5. Concrete tensile test results.

Concrete	Tensile strength f_t [MPa]		
	after 3 days	after 7 days	after 28 days
B1	2.16	3.19	3.74
B2	3.20	3.69	4.09
B3	4.77	5.02	5.05
B4	1.00	2.11	2.33
B5	2.05	2.20	2.23
B6	2.47	2.83	3.55
B7	2.67	2.75	3.39
B8	2.05	2.75	2.92

The graphs in figure 9 show the increase in the tensile strength of individual concretes. In all the concretes, the strength development after 28 days is quite similar and its intensity is greater than or comparable to that exhibited by the reference concrete.

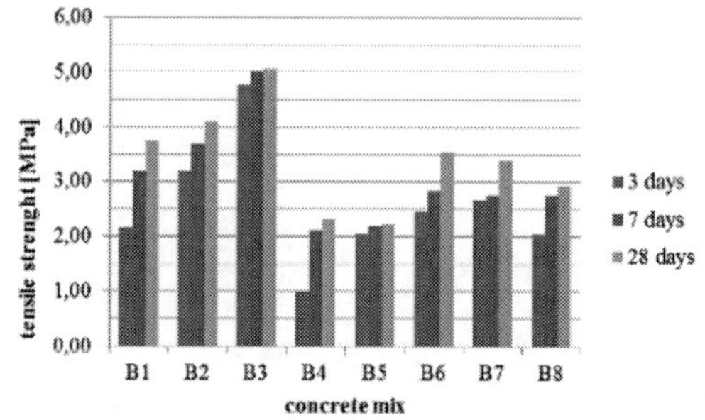

Figure 9. Tensile strength gain of each concrete.

The ratio of the tensile splitting strength f_t to compressive strength f_c that characterised all the test concretes 28 days old or less is in the range from approx. 0.07 to approx. 0.12. The average value of f_t/f_c characterising all the analysed concretes was 0.092. This value tends to decrease slightly with the duration of concrete curing and the increase in the compressive strength. The test results do not indicate that the type and content of bottom ashes have a significant impact on the relationship between the two types of strength in the scope being analysed.

5.3 Apparent density

Figure 10 shows the results of the apparent density tests after 28 days of curing of individual concretes and the obtained average values are summarised in table 6.

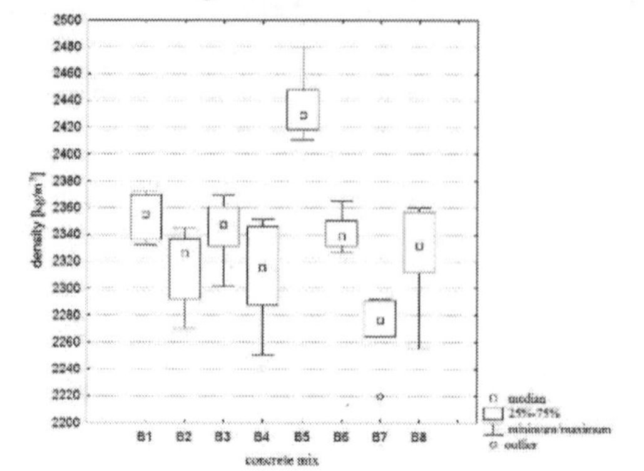

Figure 10. The concrete apparent density test results.

 DOI: 10.1051/e3sconf/20161000099

Based on the obtained results, it can be seen that the apparent density of each concrete depends slightly on the duration of curing. The greatest density was obtained by concretes B5, where 15% non-separated PDF ash was used instead of sand and gravel. For these concretes, average apparent density was 2434.5 kg/m^3. Compared to other concretes, the higher density of concretes B5 results mainly from the difference in the density of the PDF ash itself and the sand and the used gravel. The slightly lower density of concretes B7 (2271.80 kg/m^3) compared to other cements can be explained in a similar way. It results from a lower density of the PDP fractions >0.63 mm compared to that of sand and gravel. The average density values for the remaining concretes ranged from 2312.50 to 2344.00 kg/m^3, with the density of the reference concrete being 2351.60 kg/m^3.

Table 6. Test results for the apparent density of concretes.

Concrete	Density [kg/m^3]		
	after 3 days	after 7 days	after 28 days
B1	2342.1	2345.6	2351.6
B2	2284.1	2307.6	2318.1
B3	2303.9	2302.3	2344.0
B4	2288.0	2302.5	2312.5
B5	2418.1	2431.5	2434.5
B6	2302.1	2276.1	2341.5
B7	2273.3	2261.6	2271.8
B8	2331.6	2332.6	2325.9

5.4 Absorption

Water absorption by weight of all the analysed concretes was in the range from 5.0% to approx. 7.7%. The increase in the absorption compared to the reference concrete was exhibited by concretes B4 and B8, i.e. those where 15% of cement was replaced by bottom ashes <0.63 mm. For other concretes, there were no significant differences in the absorption compared to the reference concrete. Only in the case of B3 concrete, a decrease in the absorption to 4.97% was recorded, which confirms the general principle that the absorption of concretes with the addition of fluidised bed ashes is almost always slightly lower than that of the comparative concretes made of cement only.

The obtained results of water absorption of each concrete is shown in figure 11.

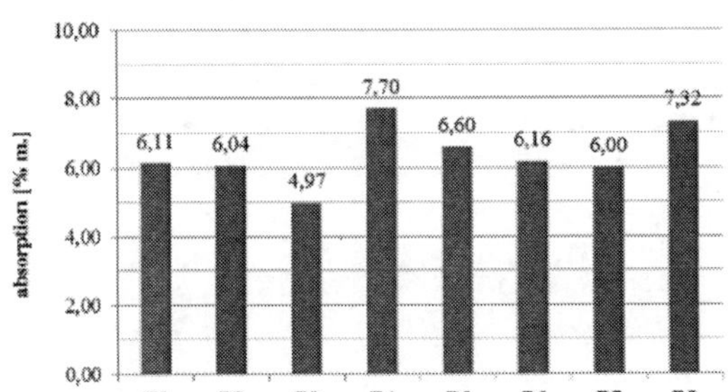

Figure 11. Average values of absorption of the concrete being analysed.

6 Conclusions

Preparation of the concrete mixes with a partial substitution of cement and aggregate with bottom ashes requires special attention, in particular if the aggregate is replaced in large quantities by ash >0.63 mm. This substitution substantially affects the workability of the mixes and therefore it is very important to carefully select the type and amount of plasticiser, superplasticiser and/or the air entraining admixture. A little smaller problem occurs when substituting some cement with ashes <0.63 mm. It should be emphasised that the reduced workability of the mixes can significantly reduce their potential industrial usability and cause negligible economic benefits due to the need to apply substantial amounts of plasticizer and superplasticiser.

If you attempt to use concrete mixes with a partial substitution of cement and aggregate with bottom ashes, it seems necessary to carry out experimental tests to determine the durability of this type of concrete. It is now one of the basic criteria for the selection of materials for the production of building structures.

The PDF ash fractionation caused the diversification of content of individual ingredients within a given fraction. The fraction below 0.63 mm contained considerably more CaO and SO_3 than the fraction with the particle size above 0.63 mm. In turn, the fraction above 0.63 mm contained greater quantities of Al_2O_3 and SiO_2. This situation – i.e. the low content of SO_3 (3,3%), the high content of silicon dioxide and oxides of iron and aluminium – can be very beneficial for the application of the PDF fraction >0.63 mm as an ingredient of concrete.

Loss on ignition of the PDF bottom ash was 4.4%. This result meets the requirements of PN-EN 450-1: 2012: Fly ash for concrete [16], in which the maximum loss on ignition of ashes is set at 9%

It should be stressed that the tests have shown a positive impact of bottom ashes with the fraction >0.63 mm on the strength parameters of the concretes. The highest increase in the compressive strength after 28 days of curing compared to the reference concrete were obtained for concrete B3 (an increase by 54.91%). In the case of B2 concrete, this increase was 33.81%. In the case of concretes, where the PDP ash >0.63 mm was used instead of the aggregate, the following increases in the compressive strength were recorded - 23.58% for concrete B6 and 6.68% for concrete B7. For the tensile strength, the increases were not as high.

The compressive strength and tensile strength development is quite similar and its intensity is greater than or comparable to that exhibited by the reference concrete.

The tests showed no significant effect of the used bottom ashes on absorption, which ranges from approx. 5.0% to approx. 7.7% for all the analysed concretes .

Analysing the results, it can be stated that it will be difficult to meet the criteria set out for construction concretes due to the chemical and phase composition of the analysed bottom ashes if they were to be used. Currently, the good use of this material may be the production of concrete members, whose durability and service life is limited. Such members include e.g.:

- concrete road slabs,
- concrete fence members,
- working platforms to ensure safe working with, e.g pile drivers and cranes,
- concrete facing for mining supports.

In the case of the attempts to use bottom ashes in the production of concrete members, it seems necessary to perform additional testing whose scope should include:

- assessment of the impact of bottom ashes on the durability of standard mortars and concretes in a chloride environment,
- assessment of the impact of bottom ashes on the durability of standard mortars and concretes in a sulphate environment,
- assessment of the resistance of concrete with bottom ash to the permeation of chlorides,
- assessment of the resistance of concrete with bottom ash to the aggression of frost and de-icing agents,
- assessment of the impact on bottom ash on the carbonation of concrete,
- assessment of the impact of bottom ash on the corrosion of steel reinforcement in concrete.

It should be emphasised that these tests are lengthy. However, they can determine the possibility and limitations of using different fractions of bottom ash as an ingredient substituting a part of Portland cement or aggregates in concrete.

References

1. P. Iwanek, I. Jelonek, Z. Mirkowski, Mineral Resources Management, **24**, 91 (2008)
2. A. Tajduś, P. Czaja, Z. Kasztelewicz, AGH J. of Mining and Geoengineering, **3**, 343 (2011)
3. Z. T. Yao, X. S. Ji, P. K. Sarker, J. H. Tang, L. Q. Ge, M. S. Xia, Y. Q. Xi, Earth-Science Reviews, **141**, 105 (2015)
4. M. Singh, R. Siddique, J. Clean. Prod., **112**, 620, (2016)
5. M. H. Wan Ibrahim, A. F. Hamzah, N. Jamaluddin, P. J. Ramadhansyah, A. M. Fadzil, Procedia - Soc. Behav. Sci., **195**, 2280 (2015)
6. C. Arenas, C. Leiva, L. F. Vilches, H. Cifuentes, M. Rodríguez-galán, *Constr. Build. Mater.*, **95**, 585 (2015)
7. A. Wongsa, Y. Zaetang, V. Sata, P. Chindaprasirt, *Constr. Build. Mater.*, **111**, 637 (2016)
8. H. Kim, *Constr. Build. Mater.*, **91**, 57 (2015)
9. PN-EN 450-1: 2012. Fly ash for concrete.
10. PN-EN 12350-2:2001. Testing fresh concrete. Part 2: Slump tests.
11. PN-EN 12390-2:2001. Testing hardened concrete. Part 2: Making and curing specimens for strength tests.
12. PN-EN 12390-3:2002. Testing hardened concrete. Part 3: Compressive strength of the test specimens.
13. PN-EN 12390-6:2001. Testing hardened concrete. Part 6: Tensile splitting strength of test specimens.
14. PN-EN 12390-7:2001. Testing hardened concrete. Part 7: The density of concrete.
15. PN-B-06250:1988 Normal concrete.
16. PN-EN 450-1: 2012. Fly ash for concrete.

 DOI: 10.1051/e3sconf/20161000100

Efficiency analysis of an energy conversion system for a variable speed small hydropower plant

Tomasz Wegiel [1,a], Dariusz Borkowski[1] and Damian Liszka[2]

[1]*Institute of Electromechanical Energy Conversion, Cracow University of Technology, Krakow, Poland*
[2]*Research and Development Centre "Rader-Bis", Paczkow, Poland*

Abstract. Present-day Small Hydropower Plants (SHPs) have a large development potential because of the increasing interest in renewable resources and distributed energy generation. However, the variable hydrological conditions that are found in the run-of-the-river projects require operations over a wide range of water flow and head variations. Special control methods and system topologies are needed to maintain the high efficiency of energy conversion systems. The synchronous generator excited by permanent magnets (PMSG) characterizes higher and more stable efficiency characteristic, in relation to the generated power, then mostly used in SHPs asynchronous machine. This paper investigates the efficiency of an example SHP solution. The researched system is based on an innovative generation unit (hydro-set) containing a propeller turbine integrated with the PMSG. In order to obtain the high efficiency in a wide range of water flow the variable speed operation method has been used. This solution requires a Power Electronic Unit in the energy conversion system to match the load and control the power flow from the generator to the grid. Efficiency analysis concerns all elements of the energy conversion system. Presented results comes from the real SHP of 150kW nominal power.

1 Introduction

Small Hydropower Plants (SHPs) are the objects of a relatively small capacity that not exceed several MW. The investment effectiveness of the SHP depends mainly on the construction costs as well as on hydrological conditions of the specific location [1]. In general SHP are the low head "run-of-the-river" systems that are characterized by the low storage capacity of water. Due to the significant variation of the river hydrological conditions throughout the year SHP operation in the wide range of water flow is desired [2]. This condition may be fulfilled by the using of a double control of rotor blades and guide vanes, what is implemented in a Kaplan turbine. However, this mechanically complex solution, widely used in large systems, is expensive and economically not justified in SHPs. The energy conversion system of small and micro power plants bases on the simple turbine construction with a single control of the guide vanes, such as a Francis turbine and a propeller turbine. However, their efficiency depends considerably on the hydrological conditions (water head H and flow Q) during the constant speed operation. The effective range of the energy conversion system under the hydrological fluctuations can be widen by implementing the variable speed operation of the turbine [3]. This technique may be obtained by using the power electronic converter (PEC) that match the electrical energy taken from the generator to the power system requirements [4]. The important function of the PEC is the regulation of the turbine load torque by an electrical power flow control. This allows to control the turbine speed in the wide range from 40% up to 140% of the nominal speed. The synchronous generator excited by the permanents magnets (PMSG) is especially attractive in that type of application. Its main advantage comparing to the widely used induction machines is the high and more stable efficiency characteristic in relation to the generated power and the rotational speed [5-6]. Moreover, the construction of the synchronous generator allows to achieve a higher pair pole number what can eliminate the mechanical gear.

The hydropower plant costs are dominated by the construction process such as: power house building, inlet and outlet channel, etc. [1]. These costs may be significantly reduced by the using of water turbine integrated with generator (called hydro-set) [6]. This solution eliminates the necessity of extraction of the turbine shaft outside the turbine chamber. Furthermore, the overall dimensions of the hydro-set are reduced and the assembly operation is simplified and fasten.

The innovative solution of the SHP containing two integrated hydro-sets of 150kW total power working with the variable speed is described in [6]. The current paper concentrates on the efficiency analysis of the specific elements of the energy conversion system. The hydro-set with the nominal parameters: P_{NG}= 75 kW, U_{NG} = 500 V,

[a] Corresponding author: tomasz.wegiel@pk.edu.pl

n_N =300 rpm, Q_N =3 m³/s, H_N =3 m consisting of the permanent magnet synchronous generator, the propeller turbine and the guide vanes, cooperates with the separate PEC of AC/DC/AC type. The main target of this research is to highlight the advantages of the innovative solution comparing to the standard one commonly used in SHPs as well as indicate some negative sides.

2 Energy conversion system

The main elements of the energy conversion system of the innovative SHP solution are presented in Fig. 1.The elements that have the main influence on the total efficiency are the turbine integrated with generator and the power electronic converter. The control system decides about the energy conversion effectiveness under the actual hydrological conditions [6-8].

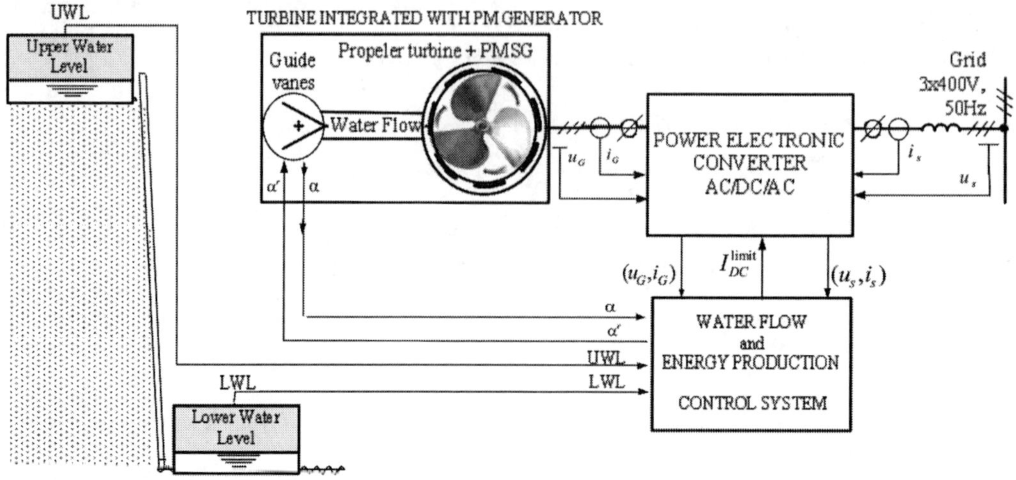

Figure 1. Energy conversion system structure.

2.1 Hydro-set structure

The hydro-set structure is based on the tubular construction (Fig. 2). The guide vanes controls the water flowing through the turbine and functioning also as a turbine closer by blocking the water in a turbine off-state. The generator stator has a classical 3-phase structure, whereas a magnetic circuit has to be designed according to the specific turbine dimensions (Fig. 3). The turbine parameters are calculated taking into account the nominal water head and flow of the specific power plant localisation. Additionally, the stator windings due to the electrical insulation have to be specially protected against water by the dedicated sealing compound. The turbine blades are situated in the generator rotor and fixed to the ring with permanent magnets (Fig 3).

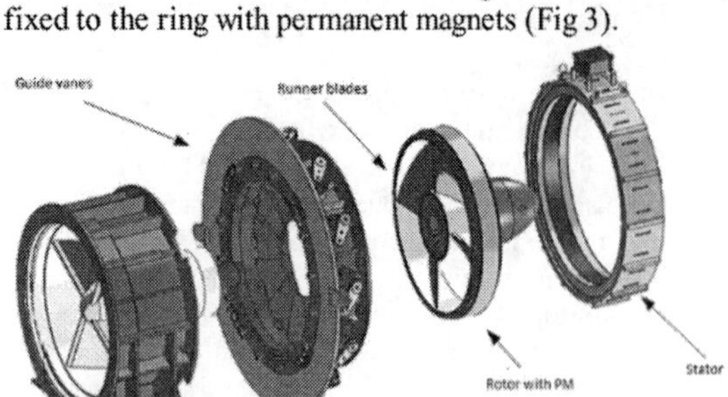

Figure 2. Components of the integrated hydro-set.

The ring replaces the shaft and transfers the mechanical torque from the turbine. Water flowing through the gap between the generator stator and rotor improves the cooling process of the stator windings.

Figure 3. Components of the real hydro-set of 75 kW power: a.) stator, b.) rotor with turbine blades, c.) assembled hydro-set.

2.2 Power electronic converter

The power electronic converter (PEC) of AC/DC/AC type (Fig. 4, Fig. 5) performs two main functions in the presented energy conversion system. Firstly, it match the variable parameters of the electrical energy of synchronous generator with permanent magnets to the constant values of voltage and frequency required by the power system. The second task of the PEC is to control the power flow according to the actual hydrological conditions. Furthermore, this element

 DOI: 10.1051/e3sconf/20161000100

maintains the proper electrical energy quality by minimising higher order harmonics of the PEC current from the grid and generator side. An interesting function is the regulation of a power factor and reactive power compensation of a local system [9]. The analogous PEC solutions are already known in the wind turbine systems. The most popular converter structure is the AC/DC/AC type [4]. The presented SHP uses the PEC with the DPC-SVM algorithm (Virtual Flux—Direct Power Control with SVM modulator). The principle of an operation and the influence of given PEC parameters on the hydro-set control process were described in [6].

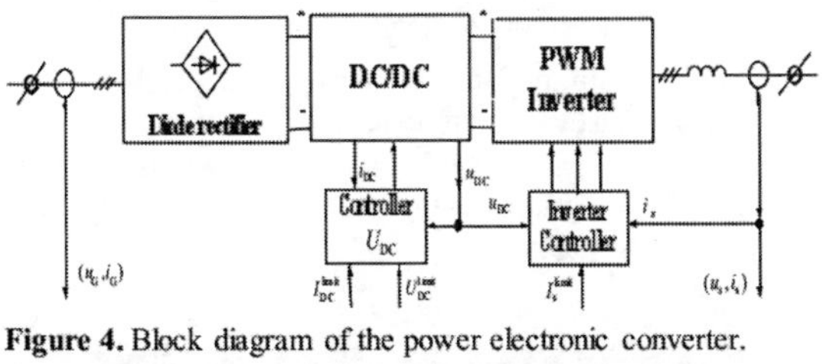

Figure 4. Block diagram of the power electronic converter.

Figure 5. The view of the real power electronic converter of 90kW power.

2.3 Operation principles

The basic characteristic that defines the turbine features and control possibilities is the power as a function of the rotational speed for a different angle of the guide vanes (Fig. 6).

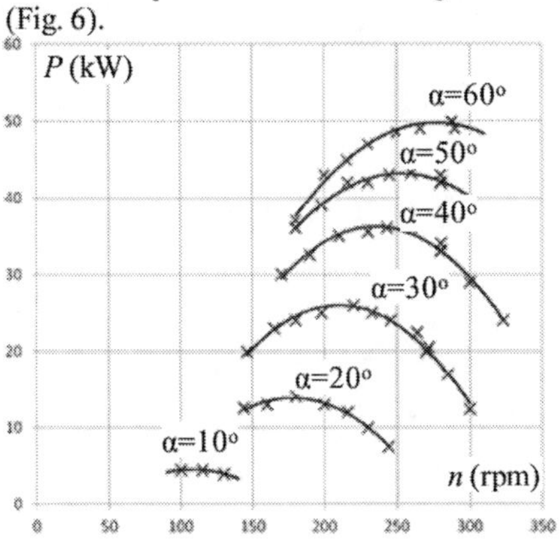

Figure 6. Water turbine characteristics of power versus angular velocity for different angle of the guide vanes and H=3.5m.

The main function of the turbine guide vanes in the "run-of-the-river" SHP is to control the water flow in order to keep the upper water on the constant level. The guide vane angle α influences the rotational speed by changing the turbine driving torque. In the presented solution the maximal power is achieved for the angle equals 60 degrees.

The second set of the control parameters influences the generator load current through the PEC. These are the limits of: the DC voltage U_{DC}^{limit}, DC current I_{DC}^{limit} and AC current I_s^{limit} [6]. They allow to control the turbine load torque because the stator current of the synchronous generator is proportional to its torque.

The turbine can operate in any point of the area covered by the characteristics presented in Fig. 6 by the appropriate settings of the control parameters. The operation point selection is the crucial task of the control system. It has a significant influence on the system operation correctness, its efficiency and utilisation of the water potential.

3 Efficiency analysis of the energy conversion system components

3.1 PM synchronous generator

The synchronous generator excited by the permanent magnets has a higher efficiency in the wide range of the output power comparing to the induction generator. The efficiency curves in the power domain of the investigated synchronous generator and two example induction generators are presented in Fig. 7.

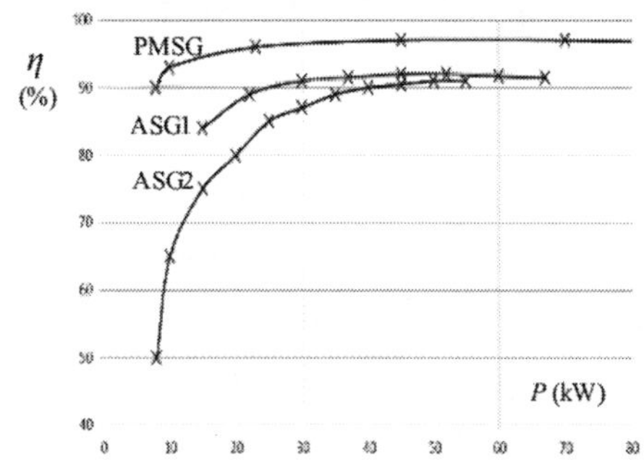

Figure 7. Comparison of the efficiency functions of the PM synchronous generator (PMSG) and two example induction generators (ASG1, ASG2) [10, 11].

It is clearly visible that the efficiency curve of the synchronous generator is higher and more stable in the wide range of the load variation than example induction generators. Similar situation takes place considering the rotational speed influence on the machine efficiency. The nominal efficiency given by the manufacturer equals η_{maxGEN}=97%.

3.2 Propeller turbine

The total hydro-set efficiency depends mostly on the turbine efficiency, which is influenced by the rotational speed and water flow values significantly. This relation is defined by the turbine universal characteristic, known also as a hill chart. This figure (Fig. 8) presents the efficiency contour lines on the speed – water flow plane. The highest possible efficiency may be obtained by the changing the turbine speed according to the water flow variations (thick solid line). The relative increase of efficiency comparing to the constant speed operation (dotted line) is especially significant in the range of low water flow value. This relation is presented in Fig. 9. The nominal efficiency of the propeller turbine given by the manufacturer equals $\eta_{maxTURB} = 80\%$.

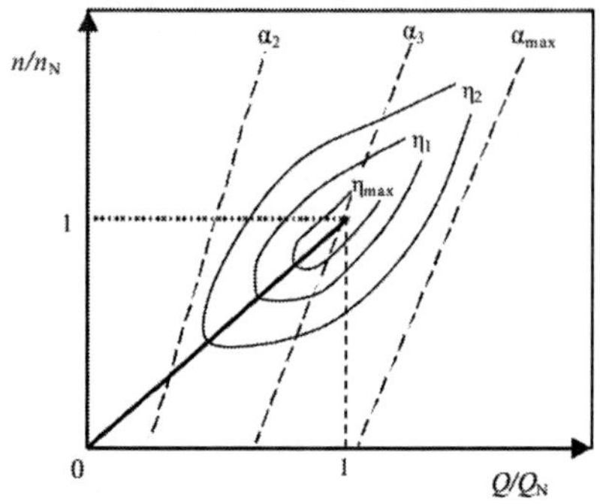

Figure 8. Universal characteristic of the propeller turbine (n – rotational speed, Q – water flow).

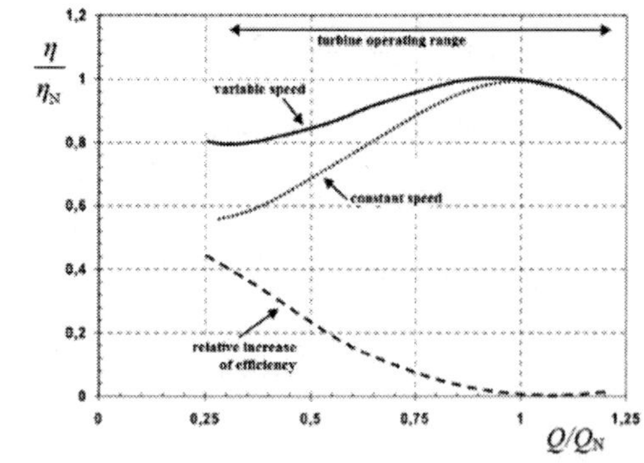

Figure 9. Efficiency curves for a different mode of turbine operation [10].

The second hydrological parameter that influences the turbine efficiency is the water head defined as a difference between the upper and the lower water levels. This relation concerning axial flow turbines is widely described in [12]. It is said, that the variable speed operation provides almost a constant turbine efficiency, independently of the head variations.

The features described above indicate that the variable speed operation improves the energy conversion effectiveness of the propeller turbine especially under the changeable hydrological conditions.

3.3 Additional mechanical losses

The integration of propeller turbine with generator results in the presence of water in the gap between the stator and rotor. This creates the additional power losses of the hydro-set causes by the fluid friction between the rotating elements. These losses can be significant similarly to the submersible motors where they are estimated at about 3-12% of the nominal power [13]. The theoretical analysis of this phenomena may be carried out basing on the two cylindrical model of the hydro-set. The calculations are performed using empirical formulas due to the phenomena complexity. The most likely used formula in that type of calculations is presented below.

$$\Delta P = \lambda \cdot l \cdot D^4 \cdot n^3 \cdot 10^{-1}\ [\mathrm{W}] \tag{1}$$

where: λ – hydraulic resistance coefficient,
l – axial length of the gap [m],
D – rotor diameter [m],
n – rotational speed [rpm].

The hydraulic resistance coefficient depends on many factors, e.g. surface roughness, fluid temperature, pattern of the fluid flow. Its value can be determined using the Moody diagram basing on the relative roughness coefficient given by:

$$\varepsilon = \frac{e}{\delta} \tag{2}$$

where: e – surface roughness [mm],
δ – gap height [mm],

and the Reynolds number, that characterizes the pattern of the fluid flow:

$$\mathrm{Re} = \frac{\omega \cdot r \cdot \delta}{v} \tag{3}$$

where: ω – angular speed $\frac{n \cdot \pi}{30}\left[\frac{\mathrm{rad}}{\mathrm{s}}\right]$,
r – rotor radius $D/2$ [m],
v – kinematic viscosity coefficient of water in temp 15°C ($v = 1.13 \cdot 10^{-6} [\mathrm{m^2/s}]$)

Introducing the real hydro-set dimensions: $l = 0.1$ m, $D = 0.9$ m, $\delta = 0.005$ m and the surface roughness on the 0.001 mm level the value of the mechanical losses for different speed can be calculated (Table 1).

Table 1. Mechanical losses in the hydro-set gap

n [rpm]	100	150	200	300
Re	20850	31275	41700	62550
λ	0.03	0.029	0.028	0.027
ΔP [W]	196	642	1470	4780

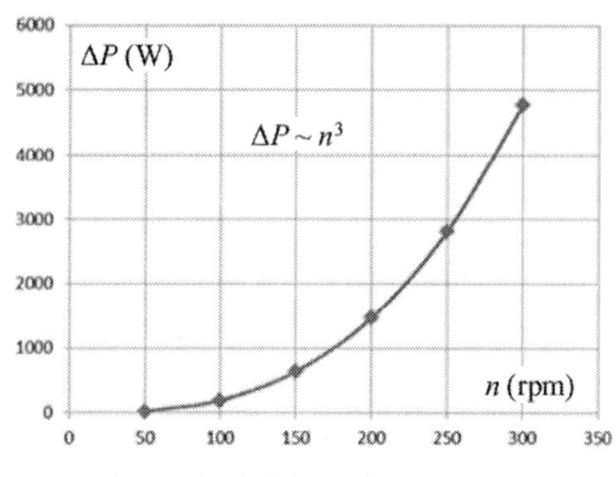

Figure 10. The mechanical losses in the hydro-set gap as a function of the turbine speed.

The changes of Reynolds number caused by the angular speed variations does not influence the hydraulic resistance coefficient significantly. This observation leads to the conclusion, that the mechanical power losses in the gap is proportional to the cube of the rotational speed (Fig. 10).

As presented in Fig. 10, these losses can achieve a significant value and are estimated at about 6.4% of the nominal hydro-set power. It should be emphasized that this calculation concerns the ideal situation, it means that there is no static or dynamic eccentricity. Furthermore, the water contamination which can increase these losses is neglected.

3.4 Power electronic converter

The example time domain waveforms of voltages and currents of the generator and the PEC output are shown in Fig. 11.

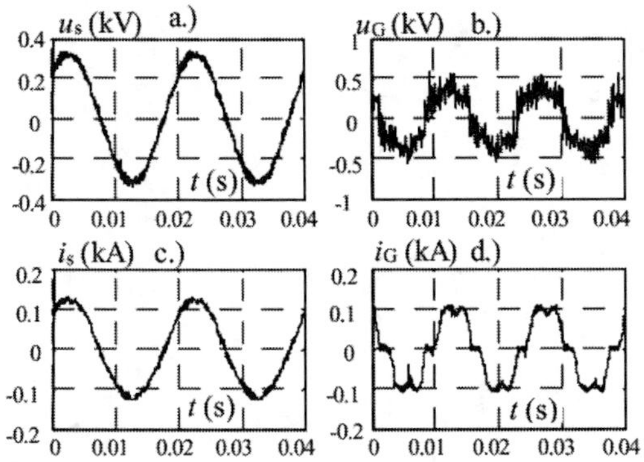

Figure 11. Example time domain waveforms of : a.) grid voltage, b.) generator voltage, c.) grid current, d.) generator current; for parameters: n=298 [rpm], P=54.9 [kW].

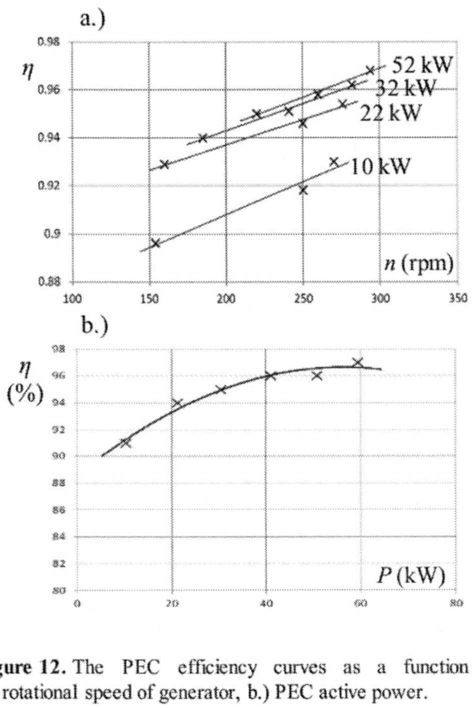

Figure 12. The PEC efficiency curves as a function of: a.) rotational speed of generator, b.) PEC active power.

The time domain signals of the grid are distorted by a low harmonic content and there is no phase shift between voltage and current. The generator signals contain significant harmonic distortion clearly visible in the waveforms. The generator current waveform is typical for systems fed by the diode bridge rectifier.

Presented signals of voltage and current can be used to calculate the input and output active power of the PEC and to define its efficiency.

The efficiency curves presented in Fig. 12 depend on the rotational speed. In order to keep the constant active power at the lower generator speed the current of the generator and PEC has to be increased. This causes the higher switching losses and conduction losses of the PEC. Higher losses at lower speed decreases the PEC efficiency (Fig. 12a).

Figure 12b presents the curve of the average PEC efficiency as a function of the active power. The efficiency values for a power higher than 40 kW exceed 96% (η_{maxPEU} = 97%).

4 Operating aspects

The compact construction of the hydro-set facilitates its assembly on the power plant without an excessive building structure. This construction creates also some operation problems with an electrical insulation caused by the water presence in the gap between rotor and stator of the generator (Fig. 13). The isolation of the inside surface of stator has to be made from the non-conducting material due to edgy currents. The commercially available technologies allow to prepare the insulation that is waterproof. However the main problem concerns the abrasion resistance of such an insulation. The river is contaminated by the mud and sand that abrasive the surface of the stator and rotor. After about two years of the operation the hydro-set surface is damaged and the water leaks into the stator. The humidity migrates into the stator windings what may cause the breakdown of the electrical isolation (Fig. 13).

Figure 13. Insulation fault of the generator stator.

5 Conclusions

This paper presents the efficiency analysis of the energy conversion system of the SHP that contains the innovative construction of the propeller turbine integrated with the permanent magnet synchronous generator. The maximal efficiency value (Table 2) is comparable with the standard solutions used in SHPs.

Table 2. The efficiency values of the individual components of the energy conversion system for chosen two operating points and the nominal hydrological parameters.

Energy conversion system components	η_1 n_1=150 rpm P_1=10kW	η_{max} n_2=300 rpm P_2=60kW
Propeller turbine*	0.55	0.74
Permanent magnet synchronous generator	0.9	0.97
Power electronic converter	0.91	0.97
Total efficiency	**0.45**	**0.70**

* - with gap losses included

The variable speed operation applied in this solution improves the efficiency under changeable hydrological conditions especially at low output power. Furthermore, the operation range is widen significantly (from 30% up to 120% of the nominal turbine discharge) what increase the yearly energy production. The simplification of the mechanical construction of the hydro-set by eliminating the shaft, gear transmission and control system of the turbine blades decreases the investment costs. The main disadvantage concerns the electrical insulation of the generator. Therefore, some further improvements need to be investigated. The dedicated sealing techniques that eliminate water from the gap have to be applied. The alternative solution to the presented one is the bulb unit where the standard generator is in the waterproof bulb submerged in the flow [14].

References

1. G.A. Aggidis, E. Luchinskaya, R. Rothschild, D.C. Howard, Renew Energ, **35** (2010)
2. A. Wijesinghe, Loi Lei Lai, DRPT Conf. P., China (2011)
3. J. Fraile-Ardanuy, J.R. Wilhelmi, J.J. Fraile-Mora, J.I. P´erez, IEEE T Energy Conversion, **21** (2006)
4. S. Nababan, E. Muljadi, F. Blaabjerg, PEDG Conf. P., Denmark (2012)
5. I. Boldea, *Variable speed generators*, CRS Press, Taylor & Francis Group (2006)
6. D. Borkowski, T. Węgiel, IEEE T Energy Conversion, Vol. 28, **2** (2013)
7. W. Mazgaj, Z. Szular, T. Węgiel, T. Sobczyk, Przegląd Elektrotechniczny, **5** (2011)
8. T. J. Sobczyk, W. Mazgaj, Z. Szular, T. Węgiel, Archives of Electrical Engineering, vol. 60, **2** (2011)
9. D. Borkowski, EPE Conf. P., Czech Republic (2015)
10. W. Deprez, A. Dexters, C. Bastiaensen, F. Parasiliti, R. Belmans, EEMODS Conf. P, China, (2007)
11. B. Köhler, Fortschritt-Berichte VDI, Series 21 (Elektrotechnik), **256**, Düsseldorf (1998)
12. D. Roberts, J. Bard, D. Shenhai, Hydropower and Dams, **2** (1998)
13. G. Sčastlivyj, W. Sjemak, G. Fedorenko, *Pogružnyje asinchronnyje elektrodvigateli*, Energoatomizdat, Moskva (1983)
14. P. J. Donalek, Power and Energy Society General Meeting, Pittsburgh (2008).

Application of low-temperature thermal treatment for the reduction of mercury content in coal

Michał Wichliński[1,a], Rafał Kobyłecki[1] and Zbigniew Bis[1]

[1]*Energy Engineering Department, Faculty of Infrastructure and Environment, Czestochowa University of Technology, Poland*

Abstract. The article describes research carried out on four coals samples (raw coals A and B and enriched coals A and B). The fuels were subjected to low-temperature thermal treatment in order to remove as much mercury as possible while still maintaining high energy quality of the fuels. The lab-scale tests were performed in a fluidized bed column and for chosen coal fraction 500-1000μm. Air and a mixture of carbon dioxide (16%) and nitrogen (the rest) were used as the fluidizing medium that as intended to have similar composition as the synthetic flue gas. The coals were tested for five temperatures of 200, 250, 300, 350 and 400°C.The results indicated that it was possible to remove more than 80% of mercury from coal A (raw and enriched), in CO_2/N_2 atmosphere at the temperature of 400°C in the case of coal B, both raw and enriched, the amount of removed mercury at 400°C was roughly 70%. At lower temperature, 350°C, the amount of removed mercury was 70% and 60% for coals A and B, respectively. Apart from the above test the investigation were also carried out in order to determine the loss of volatiles during the thermal treatment of the coals. The results indicated that at 400°C roughly 20% of the volatiles were evolved from the fuel particles. The loss of elemental carbon and fuel higher heating value (HHV) after the treatment were 10% and 8%, respectively.

1 Introduction

The reduction of pollutant emission to the atmosphere is now one of the most important challenges for power plants. Apart from getting rid of compounds such as SO_2, or NO_x, several attempts are carried out nowadays to minimize the emission of other compounds such as CO_2 and mercury. The latter one is emitted to the atmosphere mainly from the combustion of coal and its contribution to the total emission is estimated at roughly 40% of the world mercury emission [1]. The contribution of power plant emission is particularly significant in countries like Poland where more than 63 million tons of lignite and over 73 million tons of hard coal have been mined recently [2] and where over 80% of electricity is produced in hard coal or lignite fired facilities. In 2014, Poland produced approximately 159 TWh of electricity of which more than 74 TWh where generated from hard coal and over 51 TWh from lignite. The recently reports indicate that mercury content in coal in the world varies significant and, in some extreme cases, could reach up to 1000 ng/g [3]. The Polish coals were also examined and several authors report that the average mercury content in Polish hard coals combusted for power generation is roughly 80 ng/g [4-7]. Even though Hg content in Polish is relatively low. Poland is one of the biggest emitters of mercury in Europe with the emission of more than 10 tons/year due to large quantity of coal burned. Głodek and Pacyna [8] report that the emission of mercury from a single Polish power plant may be even 50 kg Hg/year. There are two main groups of methods to reduce mercury emissions to the atmosphere. The first group called 'postcombustion methods' depend on the removal of mercury from the flue gas after the combustion process and the most common sorbent used for this purpose is activated carbon impregnated with sulfur or iodine. The second group of methods called 'precombustion methods' is focused on mercury removal from the fuel before its combustion in power plants. Those methods are briefly described in the publication of Wichliński and others [9]. Particular focus is put to a low temperature thermal treatment of fuel (LT3). The LT3 technology involves the exposition of the fuel to temperature high enough to remove majority of mercury, without significant degradation of coal as a fuel. Except for the removal of mercury, an advantage of this method is also the removal of fuel moisture particularly advantageous in the case of high moisture fuels. The effectiveness of this method depends mainly on the type of mercury compounds in coal (mainly: HgO, HgO_2, $Hg_3(SO_4)O_2$, Hg_2SO_4, $HgCl_2$ as presented by Uruski and others [11]). Except for HgO the decomposition temperatures of the above compounds are below the ignition temperature of coal. Accordingly mercury can be released from the fuel before the ignition of coal. Preliminary results indicate [11- 16] more than 90% of mercury may be removed at temperatures roughly 400°C. Despite such temperature, the reduction of volatile matter and higher

[a] Corresponding author: michal.wichlinski@gmail.com

heat value may be controlled and maintained at low level (Zarzycki and Wichliński [17]). The main goal of this paper is to provide some information on the LT3 technology and corresponding results.

2 Research methodology

2.1 Coal data

For the present studies four coal samples were selected. All of them were hard coals, form Polish mines. Two of them were raw fuels (coals 'A' and 'B'), and the other two were enriched coals (coals 'A' and 'B').The results of the coal analysis are shown in Table 1.

Table 1. Some main parameters of the coal samples used in the experiments (all values are given as 'air dry')

	Unit	Coal A raw	Coal A enriched	Coal B raw	Coal B enriched
Transient moisture	%	9,3	5,3	10,3	8,1
Hygroscopic moisture	%	3,4	3,4	3,5	4,4
Ash	%	28,1	12,6	32,6	22,3
Volatile matter	%	24,0	29,4	23,8	27,1
Fixed carbon	%	44,5	54,6	40,1	46,1
HHV	kJ/kg	22600	28100	20300	23800
LHV	kJ/kg	21700	27000	19500	22800
C	%	54,1	67,4	49,4	56,3
Hg	ng/g	76	75	146	124

2.2 Experimental facility

The experiments of low-temperature thermal treatment (LT3) were conducted on a fluidized bed reactor. Diagram of a fluidized bed reactor was shown in Fig. 1. The bed solids were composed of quartz sand (particle size 250-500μm) and 10 g of ceramic balls (dp=1,6 mm). Depending on the type of experiments, a fluidizing medium was air, or a mixture of nitrogen (84%) and carbon dioxide (16%).

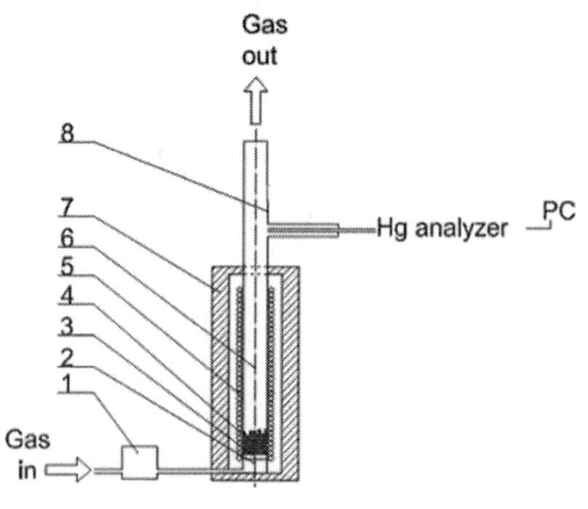

Figure 1. Sketch of the fluidized bed column: 1—mass/gas flow controller, 2—gas distributor,3—grate, 4—fluidized bed, 5—electric heater, 6—quartz tube, 7—thermal insulation, 8—measurement zone.

Regardless of the type of fluidization medium, maintained the same flow of gas volume of 10 dm³/min. The temperature in the fluidized bed column was controlled and adjusted using a thermocouple and adjusted by an electronic controller. For each test 0.5 g of coal of particle size between 500 and 1000μm was introduced into the bed. After the introduction of the sample into the reactor the mercury compounds contained in the coal decomposed. The evolved gases were mixed with fluidizing gas and the Hg content in gas mixture was analysed on-line by the LUMEX RA-915+ spectrometer connected to the system properly prepared test point. The output signals are connected to a computer, which allowed for registration of the measured values. The details on the mercury spectrometer are published elsewhere [18].

The degree of mercury release from coal sample was calculated by integrating the area under the curve of instantaneous mercury concentrations. The calculated value was then substituted into formula (1), to calculate the amount of mercury released during thermal treatment of coal at various temperatures. Eq.(1) allows to determine the mercury concentration in the solid sample from the integration of the area under the mercury concentration curve:

$$S_{Hg} = \frac{144 Fp_{Hg} \dot{V}_g}{m_{pr}} \quad (1)$$

where: S_{Hg} is the mercury content in the sample [ng/g], Fp_{Hg} is the area under the mercury emission curve, [(ng/m³ₙ)s], 144 is a constant factor, m_{pr} is the mass of the sample [g], and $\dot{V}_g$ is the gas flow rate [m^3_n/s].

After the bed was cooled down the remaining coal sample was then removed and combusted in a pyrolysis snap-RP-91c connected to the spectrometer RA-915+. The sum of mercury released during the thermal treatment of coal in the fluidized bed at a given temperature and the mercury released during the combustion in the pyrolytic snap at 850 °C was assumed as the total mercury content in the coal sample. The degree of mercury release during thermal treatment of coal was then calculated as a mass ratio of mercury released during coal thermal treatment at a given temperature to the total mercury in the sample.

3 Results and discussion

3.1 The release of mercury from coal

The research were carried out in air and CO_2/N_2 mixtures, and for a selected fuel fraction of 500-1000μm. The results of mercury release were shown in figure 2. For As can be seen coal A (raw) even at 250°C it was possible to remove roughly 30% of mercury in air while, for coal A (enriched) roughly 35% of Hg was removed. At a higher temperature (300°C) was removed nearly 70% mercury from coal A (raw) already, while only 55% from coal A (enriched). Further increase at temperature 350°C to brought about that 70% of Hg was evolved from both types of coal A (raw and enriched). For the highest temperature (400°C) the amount of released mercury increased to 80% for both coal A (raw and enriched). The replacement of air with CO_2/N_2 mixture favored the release of mercury and over 65% of Hg was evolved from raw coal A already at 250°C. At 300°C over 80% and 75% of Hg was released from raw and enriched coal A, respectively.

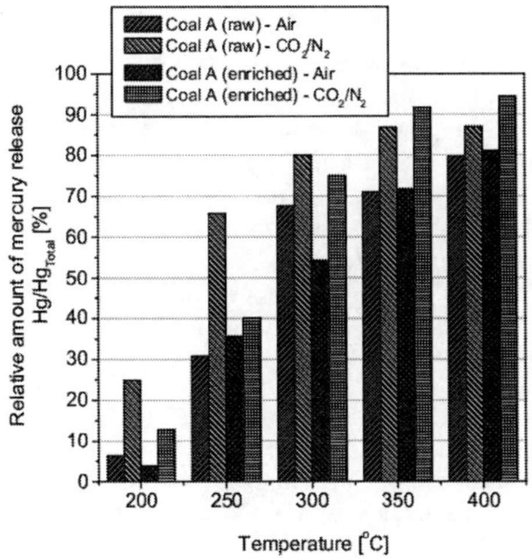

Figure 2. Relative amount of mercury release in selected temperature for coal A

The corresponding values for thermal treatment at 350°C are 90% (enriched coal A) and 85% (raw coal A). In the highest temperature (400°C) the amount of mercury released was similar and yielded at roughly 85% (coal A - raw) and 95% (coal A - enriched).

In the case of coal B at 250°C (fig. 3) and air atmosphere the amount of mercury released was just 15% for both types (raw and enriched) coals while at 300°C over 55% and over 40% of Hg was reduced from raw and enriched coal B respectively. For the temperature of 350°C, the corresponding values were 65% (coal B - raw) and 55% (coal B - enriched). The change of atmosphere form air to CO_2/N_2 mixture has also positively affected the process of mercury removal as can be concluded from fig. 3.

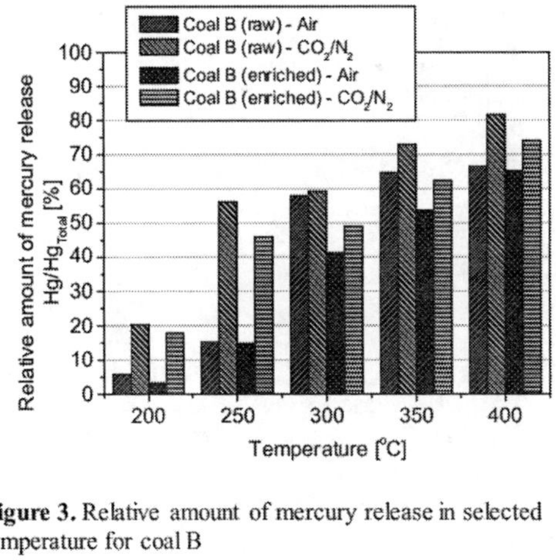

Figure 3. Relative amount of mercury release in selected temperature for coal B

The highest increase was noted, as in the case of coal A for a temperature of 250°C, was roughly 55% coal B (raw) and 45% of coal B (enriched). For other temperatures also noted an increase in the amount of mercury released, but won`t as large as at 250°C. For the temperature of 350°C the amount of released mercury was roughly 75% (coal B - raw) and 65% (coal B - enriched). For the highest temperature (400°C) the amount of mercury released was more than 80% (coal B - raw) and roughly 75% (coal B - enriched).

3.2 The loss of volatile matter during LT3

Except from research of amount mercury released it was also conducted research caused degradation of the fuel subjected to LT3. We investigated the loss of volatile matter (VM) from the various coals as well as change of higher heat value (HHV) and the content of the carbon element in coals. The relative amount of volatile matter loss of were calculated from the difference in the volatile matter content in coal before LT3 and after treatment. In the case of coal A (raw) (fig. 4) at temperatures up to 350°C the amount of VM loss didn`t exceed 13%, regardless of the used atmosphere. Only at a temperature of 400°C the amount of VM loss of were about 20% (for air atmosphere), and about 15% in the atmosphere of CO_2/N_2. The coal A (enriched) amount of VM loss were higher roughly 13% for a temperature of 250°C and atmospheric air, about 15% in temperature of 350°C and about 20% in 400°C. Change of atmosphere for coal A (enriched) caused a reduction in the amount of VM loss for temperatures 250-300°C, and a slight increase for the top two temperatures 350 and 400°C.

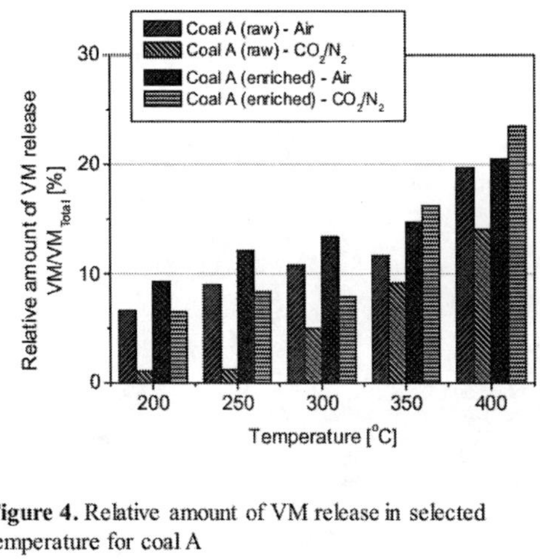

Figure 4. Relative amount of VM release in selected temperature for coal A

For coal B (raw and enriched) (fig. 5) the quantity of VM loss were about 10-12% for both types of coal B (raw and enriched) at temperatures 200-300°C. Higher amount of VM were released until at 350°C and it were roughly 15% for both types of coal B (raw and enriched). In the highest temperature of 400°C the amount of VM released were 18-22% for both types of coal B (raw and enriched).

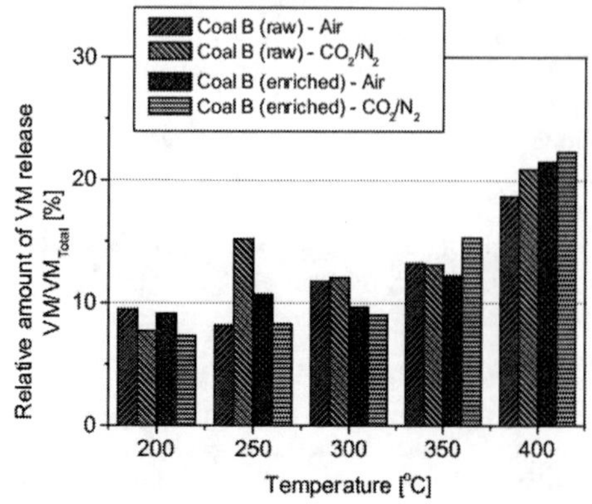

Figure 5. Relative amount of VM release in selected temperature for coal B

3.3 The loss of higher heat value during LT3

In the case of higher heat value (HHV) change for coal A (raw) (fig. 6) in air atmosphere, it was about 7% at 300°C and about 9% at 350°C and roughly 12% for 400°C. Changing the atmosphere for a mixture of CO_2/N_2 significantly reduced the loss of HHV to a level of 8% at 400°C.

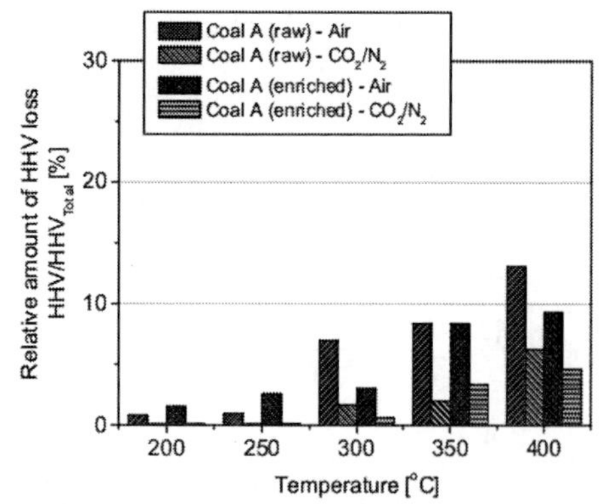

Figure 6. Relative amount of HHV loss in selected temperature for coal A

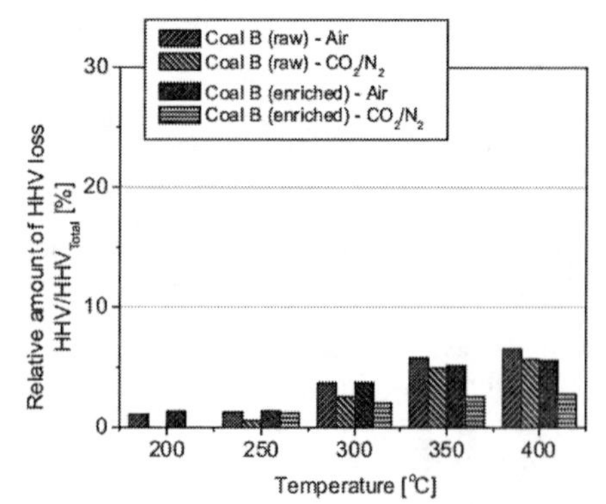

Figure 7. Relative amount of HHV loss in selected temperature for coal B

Also, in the case of coal A (enriched) atmosphere substitutions reduced the loss expressed by HHV from 8% at a temperature of 350°C, up to about 4%, at a temperature of 400°C from about 9% to about 5%.

In the case of coal B (raw and enriched) (fig. 7) change of atmosphere wasn't caused such significant changes as in the case coal A. For the highest temperatures of 350 and 400°C, it was roughly 8%, regardless of the atmosphere and the type of coal.

3.4 The loss of carbon during LT3

The figures 8-9 shown the change in the content of the carbon element as a result of LT3 processing of individual coals. Changes in the content of the element C in the coals A and B (raw and enriched) were very similar. For the temperature of 250°C in the case of coal A loss about 5% of the element C (for both raw and enriched), and in the case of coal B this loss was approximately 3%. At temperatures 350-400°C amount of element C loss were about 9% in the case of coal A, and about 8% in the case of coal B. In both coals could be seen that in enriched coals samples amount of element C loss were less than that in raw coal. Also, in the case of enriched coals changed of atmosphere from air to mixture of CO_2/N_2 decreased the amount of element C loss.

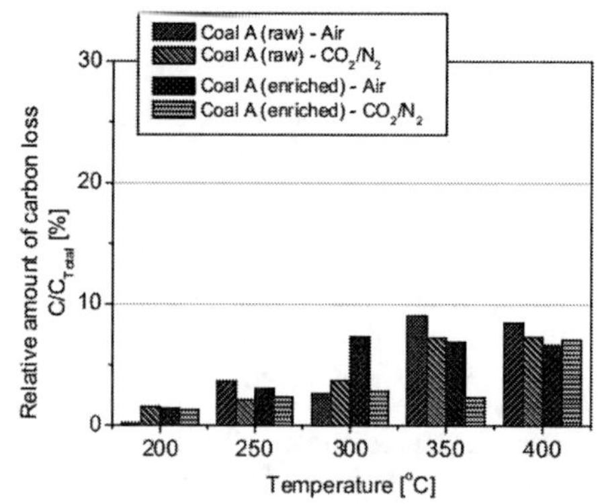

Figure 8. Relative amount of carbon loss in selected temperature for coal A

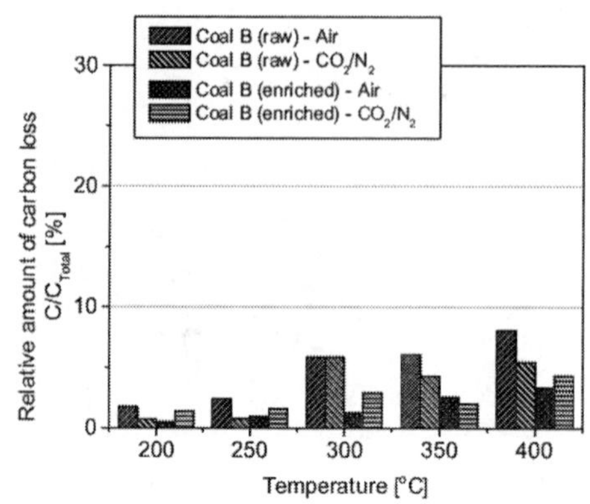

Figure 9. Relative amount of carbon loss in selected temperature for coal B

4 Summary

The results of the research indicate that low-temperature thermal treatment of coals can remove significant amounts of mercury from coal. In the case of coal A maximum degree of mercury release was 95% (enriched coal, atmosphere of CO_2/N_2) and for coal B - 83% (raw coal, atmosphere of CO_2/N_2). The amount of mercury release from the raw coals were in most cases bigger than that for the enriched coals. This is probably due to the fact that enriched coals generally contain less mercury than the raw coals, caused by the release of some mercury in the fuel enrichment processes. A greater amount of mercury released was especially visible in the case of temperatures below 300°C. The amount of mercury released in the atmosphere of CO_2/N_2 was greater than in air, which was particularly evident at temperatures below 300°C for both coals A and B. The amount of volatiles released during LT3 process may seem high, , but as mentioned, it is possible to such lead a process that after the removal of mercury from process gases from LT3 VM could be recycled to the combustion chamber for combustion. This will limit the losses caused by this process, and at the same time improve the combustion of partially degassed coal. LT3 impact on the HHV of coals was approximately 10% for the temperature 400°C - in the case of coal A in air atmosphere, changing atmosphere for a mixture of CO_2/N_2 caused a loss of roughly 6% of the HHV. In the case of coal B changed of atmosphere also caused a reduction in HHV loss to a level of about 4% - in the case of enriched coal, and about 6% for raw coal. The effect of changing atmosphere on carbon loss was also observed. In the case of the raw coal A and B, there was a reduction, but in the case of enriched coals A and B the change of atmosphere caused minor increase in carbon loss.

Acknowledgements

This article is part of the project Programme for Applied Research: Develop a database of mercury in national coals, technical guidance of its further reduction with defining benchmarks for mercury national emission factors - Hg Base (PBS2/A2/14/2013).

References

1. United Nations Environmental Programme, Global Mercury Assessment, UNEP Chemicals, Geneva, 2013
2. Z. Grudziński, 2015; Fakty: Węgiel – Energetyka w Polsce, http://min-pan.krakow.pl/zaklady/zrynek/cf_we, [online: 20.04.2016]
3. Ya. E. Yudovich, M.P. Ketris, Int. J. Coal. Geol. **62** (2005)
4. K. Wojnar, J. Wisz, Mercury In Polish Power Sector, Energetics **4** (2006) (in Polish)
5. M. Wichliński, R. Kobyłecki, Z. Bis, Arch. Environ. Prot. **39,** 2 (2013)
6. B. Klojzy-Karczmarczyk, J. Mazurek, Gospodarka Surowcami Mineralnymi, **29,** 4 (2013)
7. I. Pyka, K. Wierzchowski, Arch. Min. Sci., **55**, 2 (2010)
8. A. Glodek, J.M. Pacyna, Atmos. Environ. **43** (2009)
9. M. Wichliński, R. Kobyłecki, Z. Bis, Polityka Energetyczna **15**, 4 (2012) (in polish)
10. M. Wichliński, R. Kobyłecki, Z. Bis, Polityka Energetyczna **14**, 2 (2011) (in polish)
11. Ł. Uruski, J. Górecki, M. Macherzyński, T. Dziok, J. Gołaś, Fuel Process. Technol. **140** (2015)
12. T. Dziok, A. Strugała, A. Rozwadowski, A. Okońska, Przemysł Chemiczny **93**, 12 (2014)
13. M. Wichliński, R. Kobyłecki, Z. Bis, Fuel Process. Technol. **119** (2014)
14. A. Iwashita, S. Tanamachi, T. Nakajima, H. Takanashi, A. Ohki, Fuel **83** (2004)
15. A. Ohki, K. Sagayama, S. Tanamachi, A. Iwashita, T. Nakajima, H. Takanashi, Powder Technol. **180** (2008)
16. F. Guffy, A. Bland, Fuel Process. Technol. **85** (2004)
17. R. Zarzycki, M. Wichliński, Polityka Energetyczna **17**, 4 (2014) (in polish)
18. S. Sholupov, S. Pogarev, V. Ryzhov, N. Mashyanov, A. Stroganov, Fuel Process. Technol. **85** (2004)

Renewable energy sources (RES): alternative possibilities, which could be implemented in Poland

Anna Wiktor-Sułkowska[1,a]

[1]*AGH University of Science and Technology*

Abstract. The Kyoto Protocol and the EU's Renewable Energy Directive [1] are the World and the EU's legal regulations, which put pressure on most of the countries to extensive use of renewable energy sources (RES), in exchange for a reduction of energy from conventional sources. The main goal of the legal regulations is to reduce greenhouse gas emission. In European Union, from 1990 till 2014, share of the renewable energy sources has increased about 173% [2]. The data show that RES share in energy mix is increasing, and still has to increase. Therefore, all European's countries are obligated to build new RES installations and to improve the efficiency of existing installations. That is the reason of that paper. In this article, on the base of own field research carried out in Germany and in Poland, author presents two alternative methods of energy obtaining, in which former mining excavations are used. In first method the excavations are adapted as the elements of the hydroelectric power stations and in second method they are used as a source of geothermal energy. In the end of the paper, author presents the proposals for the implementation of similar solutions in Poland.

1 Introduction

The directive 2009/28/EC of the European Parliament and Council says "The control of European energy consumption and the increased use of energy from renewable sources, together with energy savings and increased energy efficiency, constitute important parts of the package of measures needed to reduce greenhouse gas emissions and comply with the Kyoto Protocol to the United Nations Framework Convention on Climate Change, and with further Community and international greenhouse gas emission reduction commitments beyond 2012. Those factors also have an important part to play in promoting the security of energy supply, promoting technological development and innovation and providing opportunities for employment and regional development, especially in rural and isolated areas" [1].

It means that, the general policy of the European Union is to put pressure on Member States to extensive use of renewable energy sources (RES), in exchange for a reduction of energy from conventional sources. Figure 1 below shows share of energy sources in European Union, in years 1990-2014. It confirms that implementation of the policy is in advanced phase. In years 1990-2014 share of solid fuels in energy production has decreased about 60%, and share of RES has increased about 173%.

Table 1 below presents share of renewable sources of energy in European Union in 1990-2014. Among RES, the most important source is biomass and renewable wastes (over 64% in 2014). Hydropower is the other main contributor to the renewable energy mix (16% in 2014).

Data in Table 1 show that RES share is increasing and still has to increase. Therefore, all European's countries are obligated to build new RES installations and to improve the efficiency of existing installations. That is the reason of that paper.

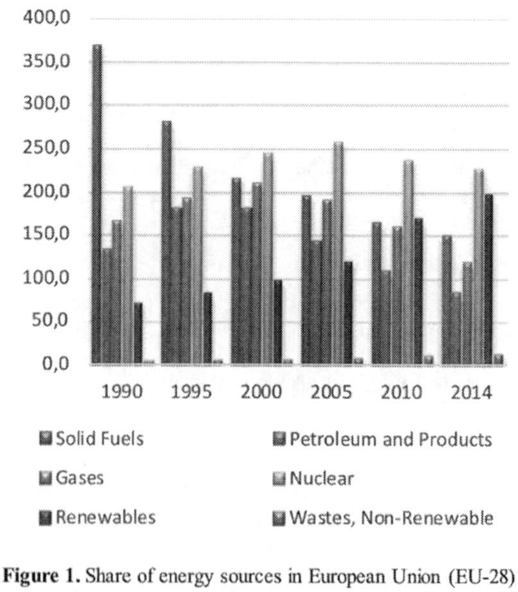

Figure 1. Share of energy sources in European Union (EU-28) in years 1990-2014 [Mtoe] (Source: Own elaboration based on [2])

In this article, author will present two alternative methods of energy obtaining, in which former mining excavations are used. In first method the excavations

[a] Corresponding author: wiktor@agh.edu.pl

 DOI: 10.1051/e3sconf/20161000102

are adapted as the elements of a hydroelectric power station and in second method they are used as sources of geothermal energy. In the end of the paper author will present the proposals for the implementation of similar solutions in Poland.

Table 1. Gross Inland Consumption of RES in European Union (EU-28) in 1990-2014 [Mtoe] (Source: [2])

Renewables	**1990**	**1995**	**2000**	**2005**	**2010**	**2014**
Hydro	25,0	28,5	30,7	26,9	32,4	32,2
Wind	0,1	0,4	1,9	6,1	12,8	21,8
Solar	0,1	0,3	0,4	0,8	3,7	12,0
Biomass and Renewable Wastes	43,5	51,7	60,6	81,9	119,2	129,0
Geothermal	3,2	3,4	4,6	5,3	5,5	6,2

2 Adaptation of former underground mine excavations as hydroelectric power stations

The former mine excavations can be adapted as hydroelectric power stations. In the article, author presents two types of hydroelectric power stations located in mines undergrounds. The first type it is flow power station of special type. That type has been well known in Germany since 19th century, but nowadays it impossible to find working installation like that. The second type is pumped storage plant, where energy can be accumulated. That type of hydroelectric power station is well known in surface space, but not in mines' excavations.

2.1 Flow power station of special type

These type of hydroelectric power station can be applied only in former mines, which will let for that. In that case, the mines have to have dewatering adits. Then water can naturally flow from surface reservoir through shaft, excavations, power generators, dewatering adit to the river. In that place the example of „Drei Brüder shaft" from Freiberg (Brand Ebisdorf, Sachsen, Germany) can be cited.

During the centuries Freiberg was well known from silver mining, but in the beginning of 20th century exploitation of silver became less profitable, what resulted in local economic crisis. In 1899, the local authorities of Freiberg decided to change the local market profile from mining activity to energy production. The market transformation was supposed to help in industrialization process of the Brand Ebisdorf region.

In the beginning, they considered whether to produce energy from coal or using water. If they had produced energy from coal, then they would have had to import it specially. While, water has been brought to Freiberg form Erzgebirge's mines dewatering for over 500 years (for needs of that, 100 km of channels and 15 water reservoirs was built). Therefore, the local authorities have decided to use the potential of water by building the first in the world hydroelectric power station, was supposed to be located in mine's excavations.

In the "Drei Brüder shaft", at a depth of 272 meters, 4 power generators were installed. Officially, on 24th of December 1914, first underground hydroelectric power station has started to produce the electricity. In 1922, the underground power plant was expanded by "Constantin Schacht," where 2 power generators were installed at a depth of 124 m. The mine's excavations in total were able to hold 1.5 million m^3 of water (Figure 2). The Freiberg's power plant then could reach power up to 5.6 MW.

Since 1953, energy production has taken place only in the "Drei Brüder shaft" (Figure 3). In 1972, the power station has been shut [3].

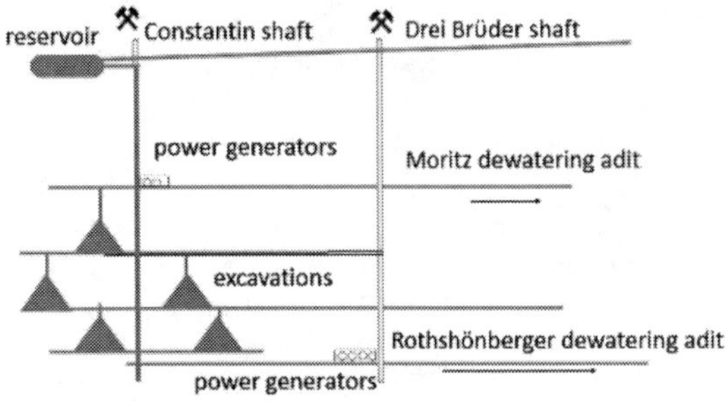

Figure 2. „Drei Brüder shaft" power station in 1922-1953
Source: Own elaboration based on [4]

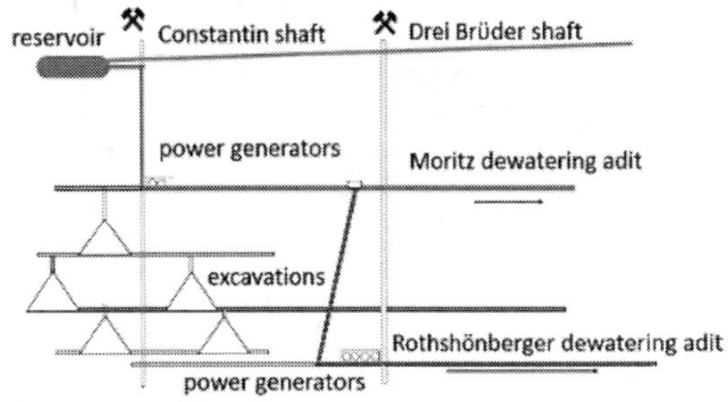

Figure 3. „Drei Brüder shaft" power station in 1922-1953
Source: Own elaboration based on [4]

In the beginning of 20th century power produced the power station could cover whole local industrial and urban demand for electricity. Nowadays, in Freiberg area demand for the electricity is about 40 MW. To cover that demand 90% of energy comes from a state network, and 10% comes from the local wind and solar farms. Simply it can be noticed, if the underground power station produced electricity, then local suppliers all together would cover more than 20% of local demand [3].

Figure 4. "Kaiser Wilhelm" shaft in Clausthal (Source: Own archive).

Second example of flow power station of special type comes from Rammelsberg's mine (Goslar, Upper Hartz). In that mine special system of artificial ponds, channels, tunnels and underground drains has been built. Due to the system power of the water could be used in mining and metallurgical processes. In 1930, when mining activity has finished, the existing water system has been converted into a hydroelectric power plant generating electricity for the city. Power installed in the "Kaiser Wilhelm" shaft (Figure 4) could reached a maximum power of 4.5 MW, and the power installed in the "Ottiliae" shaft reached a power of 1.5 MW. The hydroelectric power plants in this region have been working till early 80's of 20th century.

2.2 Pumped storage plant in former mines excavations

These type of hydroelectric power stations is well known in surface space and is used for energy accumulation. Nowadays, demand for pumped storage plants is increasing, because of increasing need for renewable energy storing. Currently in Germany, two independent groups of scientists are investigating the possibility of adaptation of the mines excavations, as pumped storage, which could store the energy produced from RES (mainly wind and solar sources).

TU Clausthal conducted pilot research in ore mines "Pöhla" and "Grund". Initially, it has been estimated that approx. 104 German's ore mines could be adapted as a pumped storage plants. Each of them would have a capacity of 100 MW, and could generate an average power of 400 MWh, what means that all together could even generate about 40 GWh.

Figures below (Figure 5 and Figure 6) show the proposals of "Pöhla" and "Grund" mines excavations adaptation as the pumped storages plants. Both proposals are similar, it means that in those two examples both reservoirs (upper and lower) are underground. The difference is in the capacity of reservoirs and in the height between them. In the "Pöhla" mine, upper and lower reservoirs will have a capacity of 280 thousand. m^3, and the difference in height between the reservoirs will be 600 m. In the "Grund" mine reservoirs will have a capacity of 240 thousand. m^3, while the difference in height will be about 700 m [5].

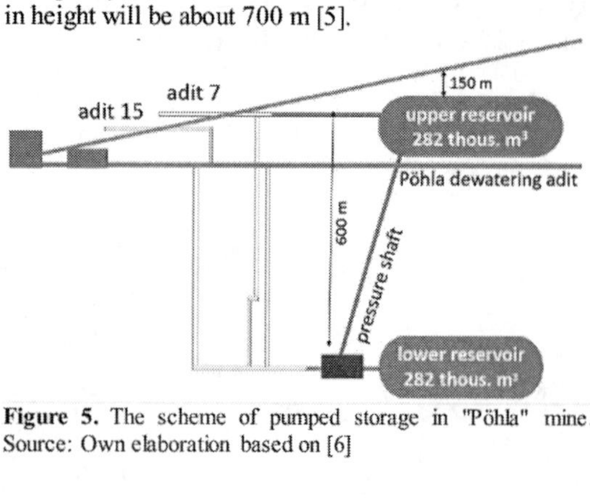

Figure 5. The scheme of pumped storage in "Pöhla" mine. Source: Own elaboration based on [6]

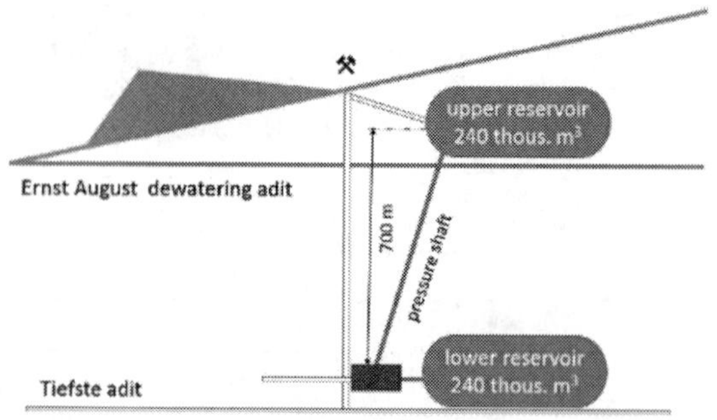

Figure 6. The scheme of pumped storage in "Grund" mine. Source: Own elaboration based on [6]

The total cost of the Bad Grund project is about 180 millions EUR, 60% of it that is cost of mine excavations adaptation [7].

University of Duisburg-Essen conduct the research in Ruhr area. In that case they examine the possibility of coal mines adaptation, as a pumped storage (Fig. 7). The results of the research will be known in 2018. Design assumptions predict location of the lower reservoir at a depth of 581-607m and capacity of 600 thousand m^3. The upper reservoir will be built in the surface. Expected power capacity is about 215 MW, and expected electricity production is about 880 MWh [8].

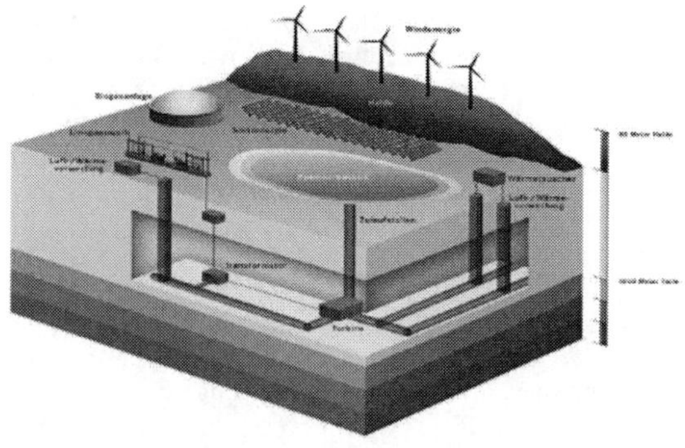

Figure 7. The scheme of pumped storage plant with surface and underground reservoirs. Source: [9]

3 Mine excavations as a source of geothermal energy

Mine water can be used as a medium in a geothermal heat pump installations, which can both, heat and cool the buildings.

The first geothermal heating system was installed in Essen, in 1982. Then hot water from "Henrich" mine heated the nursing house.

In Germany the geothermal heat pump installations are known very well. Only in Saxony is about 10 of them [10]. In the Freiberg area finds 4 geothermal heat pump installations, in Zwickau 1, in Bad Schlema 1, in Lugau/Olesnitz 1, in Marinienberg 1 and in Ehrenfriedersdorf 2 installations.

Table below shows selected examples of mine water use in geothermal heat pump installations.

DOI: 10.1051/e3sconf/20161000102

Table 2. Examples of installation, where the thermal energy of mine water is used (Source: Own elaboration based on [11-13])

Place/Facility	Deepth [m]	T_{hot} [°C]	T_{cold} [°C]	Water flow rate [m³/min]	Heating power [kW]
Reiche Zeche Freiberg/Garmany	228	18	14	2,4	670
Zeche Robert Müser Bochum/Germany	570	20	-	3,07	1 600
Schacht 302 Marienberg/ Germany	107	12,4	7,4	2	690
Zinngrube Ehrenfriedersdorf/ Germany	110	10	4	0,36	approx. 100
Gewerbegebiet Reden/Germany	900	30	------- -	-----------	450
Zeche Henrich Essen/Germany	335-480	22	------- --	0,33	354
Saturn Czeladź/Poland	194	12	------- --	15,10	106,8

In Freiberg, first geothermal heating system was installed in 1990. Then water from "Reiche Zeche" shaft heated the foreman's building. Currently, "Reiche Zeche's" water (approx. 18°C) is used, among others, in heating systems of: city hospital, buildings located near to "Reiche Zeche" shaft, one new building of TU Bergakademie and the "Freudenstein" castle. While, water from the same mine, but gathered in other place, in "Rothschönberger" adit (approx. 14°C), is used in cooling systems [14-15].

In Ruhr area mine water has to be pumped up from the former mine "Robert Müser" for safety reason. That water has temperature of 50°C. In Bochum has been running the project, where "Robert Müser's" mine water is used in heating installations of: primary school, secondary school with a canteen and swimming pool and the main fire station. It has been calculated that due to these investments annually the buildings saves 1,200 MWh, and thus reduces the CO_2 emissions by 245 tons [16].

The geothermal heat pump systems, with mine water as a medium, finds also in Canada, the Netherlands, Czech Republic, Scotland and in Poland. However, any of that countries do not have as much the geothermal heat pump systems as Germany has.

The first, and still only one, polish geothermal heat installation has been working since 2012. Water from "Saturn" mine is used in heating and cooling system of building, which is located next to the shaft.

Table 3. Economical effect of polish geothermal heating installation (Source: [12])

heating system	cost [thous. PLN/year]
Electric	80,3
heating pumps	13,5
yearly savings	66,8

The heating pump installation have costed 352 thousand. PLN. It has been counted that annual savings, w arise from the use of heat pumps can reach 66,8 thousand PLN, so the investment should be recouped after 5-6 years (Table 3).

4 Proposals for the implementation of similar solutions in Poland

If European Union continues to put pressure on Member States regarding to increasing RES share in energy production, then will increase the demand for this energy storage. Already, it is worth to consider the possibility of the former mine workings adaptation as the pumped storages. Especially, taking into account the fact, that in Poland, before implementation of restructuring program, it operated more than 70 coal mines, currently more than 30 still is in operation. All of polish coal mines have approx. 300 km of excavations, what gives huge energy storage potential.

Mine Dewatering Institute (CZOK) has to dewater some of former mines of Upper Silesia region from safety reasons. Yearly it has to pump more than 70 m³ of mine water. The temperature of the water is between 10 to 28°C (vide App. 1). These water could be used in heating and cooling installations. Unfortunately, instead of this, mine water is discharged into surface watercourses, as a waste water. CZOK has to pay environmental fee, which is approx. 8 million PLN / year. In author's opinion, this indicates a lack of economic thought, and a great waste of potential, which carries a mine water.

5 Summary

In this article, on the base of own field research carried out in Germany and in Poland, author has presented two alternative methods of energy obtaining, where former mining excavations are used. In first method the excavations were adapted as the elements of the hydroelectric power stations (flow power station of special type and pumped storage plant) and in second method they were used as a source of geothermal energy.

In the article two the examples has been shown: from Freiberg ("Drei Brüder" shaft) and from Goslar ("Rammelsberg" mine), where were installed flow power stations of special type. That type of hydroelectric power station has been well known in Germany since 19th century, but nowadays it impossible to find working installation like that. These type of hydroelectric power station can be implemented only in specific former mines. In that case, the mines have to have dewatering adits, then water can naturally flow from surface reservoir through shaft, excavations, power generators, dewatering adit to the river. Nowadays, it is hard to find mine like that, because dewatering adits has been turned into water pumped systems.

In next part of paper were shown the examples of "Pöhla" and "Grund" (ore mines), and the example from the Ruhr area (coal mines), where mines excavations were supposed to be adapted as reservoirs of pumped storage plants. Poland has huge energy storage potential, where coal mines excavation could be adapted as reservoirs.

In further part of the article was also indicated that mine water can be used as a medium in a geothermal heat pump installations (both heating and cooling). In these part few examples from Germany and one from Poland were presented. The geothermal heat pump systems finds also in Canada, the Netherlands, Czech Republic, and in Scotland. However, any of that countries do not have as much the geothermal heat pump systems as Germany has. Actually in Poland is only one installation like that. Poland has huge thermal potential of mine water (app.1), but it is totally untapped.

The world and EU's policy can contribute to the development of alternative adaptation of former mine excavations.

Due to adaptation of former mines excavations as RES Poland could save carbon dioxide emission units.

Acknowledgements

Source of financing: university grant no. 11.11.100.693, task no. 4.

References

1. European Commission, *Directive 2009/28/ec of the european parliament and of the council on the promotion of the use of energy from renewable sources and amending and subsequently repealing Directives 2001/77/EC and 2003/30/EC,* (European Commission, 2009)
2. European Commission, *EU Energy in figures. Statistical Pocketbook 2016*, (Publications Office of the European Union, Luxembourg, 2015)
3. M. Schonfeld, J. Leisner, *Power Station in Drei Bruder Schacht,* (Interview, 15th of August 2012)
4. Supaplex,[Online],Available:https://de.wikipedia.org/wiki/Drei-Br%C3%BCder-Schacht (2016)
5. P. Beck, M. Schmidt, *Pumpispeicherwerken in ehemaligen Erzgruben, Konception einer Pilotanlage im Harz*, (Energie-Forschungszentrum Niedersachsen, Zollverein, 2011)
6. P. Beck, M. Schmidt, *Windenergiespeicherung durch Nachnutzung stillgelegter Bergwerke*, (Goslar: Universitätsbibliothek Clausthal, 2011)
7. ndr.de, *Konnen Pumpischer Atomkraftwerke ersetzen?* [Online].Available:http://www.ndr.de/regional/niedersachsen/harz/harz267.html (11th of May 2011)
8. A. Niemman, *Sachstand zur Machbarkeitsstudie einer Nachfolgenutzung der Anlagen des Steinkohlebergbaus im Ruhrgebiet als Pumpspeicherwerke*, (Universitat Dissburg, Essen, 2014)
9. Universitat Disburg Essen, *Unterflur-Pumpspeicherkraftwerke*, [Online]. Available: http://www.unidue.de/imperia/md/images/wasserbau/upw_600.jpg (7th of April 2016)
10. P. Horler, *Der Bergbau in Sachsen*, (Freiberg: ReproMedia GmbH, 2012)
11. T. Grab, Storch T. and others, *Geothermieanlage zur Grubenwassernutzung für Heizung (200 - 670 kW) und Kühlung (155 - 500 kW)*, (Der Geothermiekongress. 17-19 November 2010, Karlsruhe, 2010)
12. CZOK, *Vademecum CZOK*, (Czeladź, 2012)
13. EnEff:Stadt, *Wärme aus Grubenwasser in Bochum-Werne*, [Online]. Available: http://www.eneff-stadt.info/de/waerme-und-kaeltenetze/projekt/details/waerme-aus-grubenwasser-in-bochum-werne/, (16th of Martz 2014)
14. K. Grunt, *Reiche Zeche und Alte Elizabeth*, (Interview, 15th of July 2012)
15. C. Drebenstedt, *Mining in Germany* (Interview, 1st of August 2012)
16. Ch. Geimer, *Thermal heat from pit water," FIZ-Karlsuhe*, (Leibnitz Institute for Information Infrastructure, Eggenstein-Leopoldshafen, 2013)
17. A. Niemann, E. Perau, K. Schreiber and others, *Chancen und Risiken untertägiger Pumpspeicherwerke in Steinkohlebergwerken im Ruhrrevier* (Zeitschrift Wasserwirtschaft 1/2, pp. 66-69, 2014)
18. A. Jasińska, A. Czaplik, *Mine Dewatering Institute in Czeladź*, (Intervew, 23th of January 2013)

 DOI: 10.1051/e3sconf/20161000102

Appendix 1. Upper Silesia's former coal mines water (Source: Own elaboration based on [12])

Pump station	Pumping system	Deep of pumping [m ppt]	Range of pumping [m.a.s.l.]	Water category	Total amount of pumped water in 2012 [thous. m^3]	Amount of pumped water [m^3/min]	Water temperature [^{0}C]
"Saturn"	deep	194	+83,0 / +86,0	IV	7 935	15,10	12,0
"Grodziec"		274	+110,0/+115,0	IV	1 359	2,40	13,3
"Sosnowiec"		227	-------------------	III	3 329	13,00*	13,4
"Paryż"		213	+80,0/+85,0	III	94	9,50	12,9
"Porąbka-Klimontów"		482	-203,0/-200,0	IV	2 864	5,40	17,0
"Niwka-Modrzejów"		411	-175,0/-160,0	V	5 873	11,10	17,5
"Katowice"		483	-204,5/-198,5	V	2 800	5,30	18,0
"Kleofas"		642	-365,0/-358,0	V	3 246	6,20	21,4
"Gliwice"		498	-270,0/-266,0	V	2 902	5,50	21,3
"Pstrowski"	stationary	575	-297,0	V	8 386	5,86	18,8
		825	-559,0			10,04	25,8
„Szombierki"	stationary	630	-370,4	V	2 399	3,91	24,5
		790	-502,8			0,65	27,6
„Powstańców Śl."	stationary	500	-213,0	V	1 421	0,62	21
		650	-356,0			1,04	24,2
		760	-467,5			1,08	23,8
„Dębieńsko"	stationary	112	+136,0	III V	1 727 3 207	1,61	10,7
		202	+46,8			0,74	12,2
		310	-60,92			0,44	15,4
		410	-161,2			3,89	15,8
		600	-350,0			2,28	21,3
		690	-440,2			0,31	26,7
		781	-531,25			0,04	--------------
Total inflow						9,43	
"Siemianowice" – stationary	stationary	206	-27,0	IV	15 648	8,12	14,0
"Siemianowice" – pomp. "Bańgów" p.206 m.		321	-38,0			3,10	17,0
"Siemianowice" – pomp. "Chorzów"		321	-23,5			4,98	17,0
		630	-338,0			6,90	20,0
Total inflow						23,10	
"Jan Kanty"	stationary	270	+11,9	III	12 654	24,00	11,4

DOI: 10.1051/e3sconf/20161000103

A novel method of sewage sludge pre-treatment - HTC

Małgorzata Wilk[1,a]

[1]*Faculty of Metals Engineering and Industrial Computer Science, AGH University of Science and Technology, Al. Mickiewicza 30, 30-059 Krakow, Poland*

Abstract. The aim of the paper is to present a relatively new technology – hydrothermal carbonization (HTC) of municipal sewage sludge. The HTC process was conducted in a stainless steel, Zipperclave Stirred Reactor, with a volume of 1000 ml, equipped with a MagneDrive Agitator. The control panel provides a programme which regulates the heater temperature and mixer speed. The main parameters of the process were temperature (~200°C), pressure (~1.5 MPa) and residence time (4, 7, 10, and 12 h). In order to understand the process, the physical, chemical, thermal, and structural characteristics of the solid product, hydrochar, was investigated. Therefore, the ultimate and proximate analyses, and HHV for raw material and obtained hydrochar are presented. The majority of carbon, of the initial present carbon, remained within hydrochar. TA analysis was used to detect the initial and final temperature for the combustion of raw material and hydrochar. This technique is very useful for predicting the combustion characteristics of carbonaceous material. When considering the use of hydrochar as a solid fuel, more energy can be derived from hydrochar than from e.g. incineration of waste, and its carbon emission should be less significant. Therefore, HTC can be an environmentally beneficial technique for the combustion process.

1 Introduction

In recent years, the rapid population growth, urbanization and industrialization have caused a significant increase in the amount of sewage sludge. Therefore, the development of new utilization methods of sewage sludge is an urgent problem which needs to be solved [1-3]. The most common disposal of sewage sludge is landfilling, farming applications and incineration. Sewage sludge can be treated as a potential energy resource despite containing a high level of pollutants. Nowadays, European and Polish legislation concerning sewage sludge treatment has become very restrictive in imposing the rapid development of thermal methods of sewage sludge utilisation (Figure 1). There are several thermal technologies for utilising municipal sewage sludge to obtain useful forms of energy through the correct pretreatment. One promising option is the hydrothermal carbonization process, which is an effective way to densify the energy content of sewage sludge without the prior drying process.

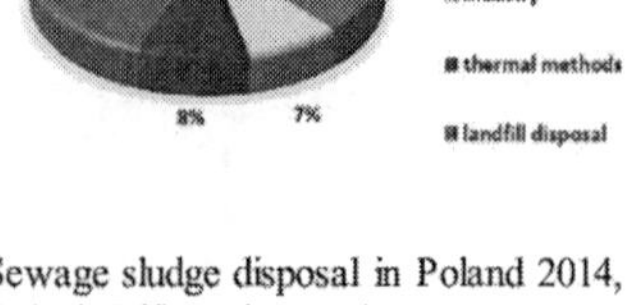

Figure 1 . Sewage sludge disposal in Poland 2014, Central Statistical Office of Poland.

[a] Corresponding author: mwilk@agh.edu.pl

Hydrothermal carbonization (HTC) is a thermochemical process which involves the application of heat and pressure to convert raw material in the presence of water into carbonaceous biofuel. The HTC process results in three types of products: gases, aqueous chemicals, and a solid hydrochar. The gas product is about 1-3% of the raw material, consisting mainly of CO_2. The aqueous extractive compounds can easily be filtered from the reaction solution, avoiding complicated drying schemes and costly isolation procedures. They are primarily inorganic salts, sugars, and organic acids. Hydrochar is a homogenized, carbon rich, and energy-dense solid fuel, biologically sterilized due to thermal treatment. The mechanisms of hydrothermal carbonization are associated with a series of hydrolysis, condensation, decarboxylation, and dehydration reactions. Reaction temperatures are applied within a range of 180-230°C and the various pressures are maintained above the saturation pressure to ensure the liquid state of water.

The HTC method has usually been used for a wide range of biomass feedstocks [4-11], where the effect of temperature, residence (carbonization) time, and carbonization yields have been studied. In the case of sewage sludge, the hydrothermal carbonization process was not so widely investigated, but studies were largely focused on the properties of solid product, energy recovery rate, hydrochar yields and its combustion characteristic, and improved dewaterability of sludge. The effects of hydrothermal carbonization were evaluated by varying the temperature and time [12-19]. In fact, regarding Polish

DOI: 10.1051/e3sconf/20161000103

conditions, there are no published results concerning the HTC of sewage sludge so far, that is why this technology can be named a novel pre-treatment method.

The major objectives of this research are to present the laboratory tests for HTC method of municipal sewage sludge pre-treatment. Therefore, the characterisation of the sewage sludge before and after the hydrothermal carbonization process is presented and the chemical changes after HTC pre-treatment are discussed. The TG/DTG/DSC results are also given to analyse the combustion behaviour and thermal characteristic of hydrochars.

2 Materials and methods

2.1 Materials

Dewatered sewage sludge was collected from the municipal wastewater plant in Gliwice, Poland. Sewage sludge with a moisture content of 85% was stored at 4°C and later used in HTC experiments. Part of the material was dried at room temperature for several days, resulting in 10% of moisture. It was then ground into a fine powder and kept in a sealed plastic container. An image of the sewage sludge is presented in Figure 2. Later, both the proximate and ultimate analyses of the sewage sludge were performed (Table 1).

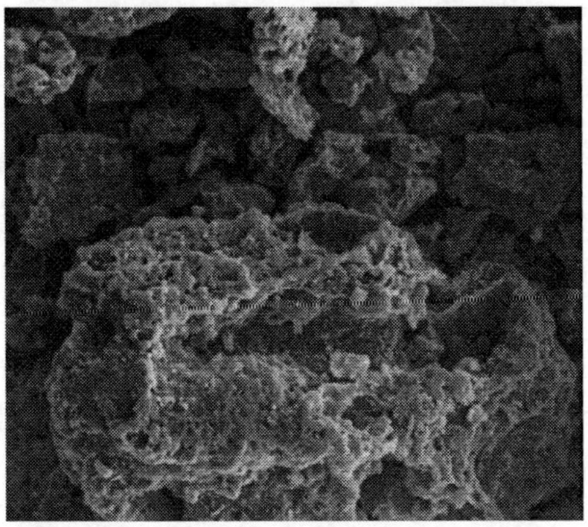

Figure 2. An image of the sewage sludge.

2.2 HTC experiments

HTC of the sewage sludge was carried out in a stainless steel, Zipperclave Stirred Reactor, with a volume of 1000 ml, and equipped with a MagneDrive Agitator (Parker Autoclave Engineers). The Zipperclave is equipped with a low pressure spring closure, which is designed to reduce the time required to open and close the pressure vessel. The reactor is designed to work up to a maximum working pressure of 15.1 MPa and a temperature of 232°C. The magnet zone is capable of cooling down during the operating time, because a temperature in excess of 149°C could cause permanent damage, resulting in lost torque capability. Therefore, a stainless steel cooling jacket is provided between the drive connection and magnet zone to maintain the temperature in the magnet zone at 149°C. The control panel provides a programme which regulates the heater temperature and mixer speed. The main seal and safety pressure release is made of teflon, an elastomer material. A schematic of the HTC apparatus is shown in Figure 3.

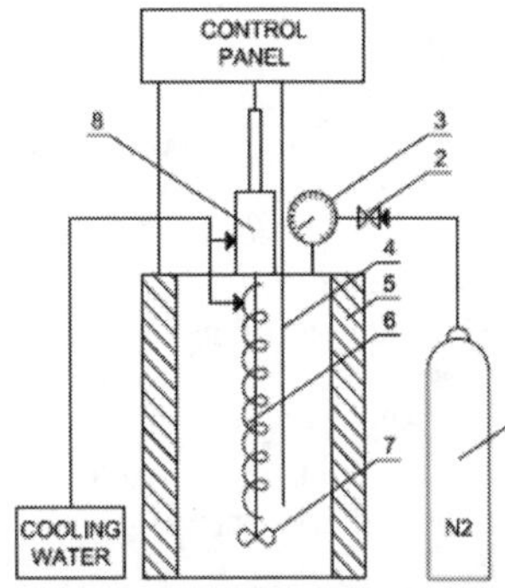

1-nitrogen cylinder, 2-valve, 3-pressure gauge, 4-thermocouple, 5-electrical furnace, 6-cooling coil, 7-stirrer, 8-cooling jacket.

Figure 3. The HTC apparatus.

In this study, 200 g of dewatered SS was dispersed in 300 ml of de-ionized water and placed into the vessel. The mixer was stirred continuously at 150 rpm to ensure the uniformity of reaction temperature in the samples. Then, nitrogen with a purity of 99.99% was supplied from the cylinder to the reactor in order to create an oxygen free atmosphere. The reactor was sealed and heated up to 200°C in an electric furnace. When this temperature was reached, four carbonizations times 4, 7, 10, and 12 h were maintained. The pressure stabilized at 1.5 MPa. At the end of the carbonization time, the heater was turned off and the reactor was taken from the furnace and cooled down by water via a cooling coil. When the pressure and temperature decreased to room temperature, the reactor was opened to collect the remaining products. The mixture of solid and liquid products was firstly extracted from the vessel and placed in a graduated cylinder to observe the sedimentation time. Then, the solid product was separated from the liquid phase by using a vacuum filtration apparatus through microfiltration filter paper and oven dried at 105°C for one day. The liquid phase and hydrochars were weighed and stored in sealed containers until further analysis. Additionally, the liquid phase, so-called HTC water, was analysed by liquid chromatography method. The pH of all studied samples were also measured. The schematic of the hydrothermal carbonization and parameters of the process are summarized in the diagram (Figure 4).

Hydrochars were named as HTC_4, HTC_7, HTC_10, and HTC_12 according to carbonization times used during the process.

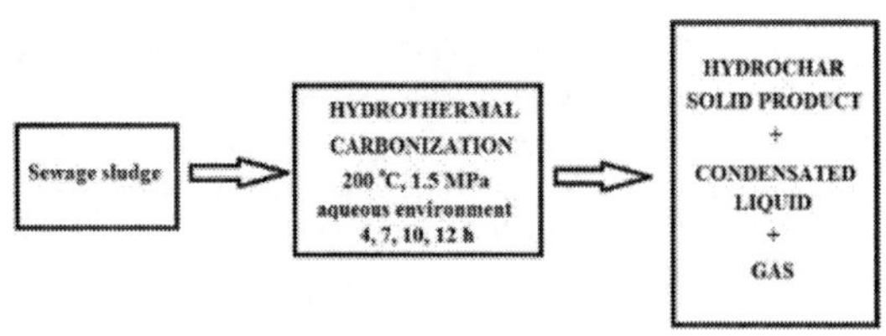

Figure 4. The HTC parameters.

2.3 Analytical methods

The proximate analysis (moisture, ash, volatile matter, and fixed carbon) of the sewage sludge and hydrochars were determined according to European Standards. The moisture content was determined by maintaining 1 g of the samples in the drying system at 105°C to a constant mass according to EN 15934:2012. The ash content was determined by burning a 1 g sample of all studied solid recovered fuels in a muffle furnace at 600 +/- 10°C for 6 h according to PN-EN 14775:2010. The volatile matter was determined by maintaining the 1 g of mass samples at 900 +/- 10°C for 7 min using the EN 15402:2011 method. The fixed carbon content was determined by the difference:

$$FC = 100 - Ash - VM - M, \% \quad (1)$$

where: FC (%), Ash (%), VM (%), and M (%), respectively, indicate the mass percentages of fixed carbon, ash, volatile matter and moisture.

The ultimate analysis (carbon, hydrogen and nitrogen content) were measured by Elemental Analyser Truespec CHN Leco. The LECO CHN628 is a combustion elemental C, H, and N instrument that utilizes only pure oxygen in a furnace, ensuring complete combustion and superior recovery of the elements of interest. An Analyser CHS580 ELTRA was used to determine the sulphur content. The analyser is designed to determine both carbon and sulphur content in a wide variety of organic materials as well as some inorganic materials by combustion with non-dispersive infrared detection.

The higher heating value of the samples was calculated according to the Channiwala and Parikh correlation based on elemental composition:

$$HHV = 0.349 \cdot C + 1.1783 \cdot H + 0.10055 \cdot S - 0.1034 \cdot O - 0.0015 \cdot N - 0.0211 \cdot A, MJ/kg \quad (2)$$

In order to determine the combustion process of sewage sludge and hydrochar thermogravimetric analysis was conducted by the use of Netzsch STA 449 F3 Jupiter. The advantages of thermogravimetric analysis include its rapid assessment of the fuel value, the temperatures at which combustion starts and ends, and other characteristics such as maximum reactivity temperature, ash amount and total combustion time. Therefore, this technique is very useful for describing the combustion processes of fuels of different origin [20,21]. The hydrothermal conversion of sewage sludge and hydrochars were tested under a temperature regime in air atmospheres. The samples were heated in alumina crucibles from an ambient temperature up to 1200°C at a constant rate of 10°C/min and at a 40 ml/min flow of air. The results were presented in the form of TG, DTG, and DSC curves, respectively. Thermogravimetry (TG) presents the weight loss of studied samples in contrast to the initial mass under an increasing temperature. Differential thermogravimetry (DTG) is based on the rate of weight loss. The peak temperature on DTG is the point where the rate of weight loss is at its maximum. Differential scanning calorimetry (DSC) additionally allows for the determination of thermal effects (endothermic and exothermic). The TG, DTG and DSC curves were used in order to assess the thermal characteristics of the studied samples.

3 Results

The proximate and ultimate analyses, fuel and atomic ratio, and HHV (higher heating values) of dried sewage sludge and hydrochars are presented in Table 1.

Table 1. Properties of sewage sludge and hydrochars.

	SS	HTC_4	HTC_7	HTC_10	HTC_12
FC,%	1.07	8.48	8.1	6.29	6.59
VM,%	55.11	39.27	38.00	40.17	39.34
Ash,%	34.29	49.47	52.8	52.84	53.08
M,%	10	2.78	1.1	0.7	1.00
FC/VM	0.02	0.22	0.21	0.16	0.17
C,%	32.1	32.1	32.2	33.8	33.0
H,%	5.26	3.93	3.68	3.8	3.72
N,%	4.7	2.7	2.57	2.8	2.56
S,%	1.56	1.29	1.15	1.33	1.00
O[a],%	23.64	10.51	7.60	5.43	6.64
H/C	1.97	1.47	1.37	1.35	1.35
O/C	0.52	0.25	0.18	0.12	0.15
HHV, MJ/kg	14.42	13.83	13.79	14.73	14.19

[a] - calculated by difference

By analysing the properties of hydrochars, the most evident changes are observed based on the fixed carbon. The fixed carbon, resulting from hydrothermal carbonization ranges from 8.48% to 6.59%, compared with that of the dried sewage sludge - 1.07. The highest FC value are obtained for 4 h of carbonization time, then with an increase of reaction time the FC decreases, but still has c.a. seven times higher values. High FC is obtained, because the volatile matter (VM) of the chemical dehydration and decarboxylation reactions. The VM content for sludge is 55.11%, whereas for hydrochars it seems to be at a similar level, approximately 39.2%, and the carbonization time does not affect it. Furthermore, the fuel ratio (FC/VM) increases from 0.02 to 0.22 for 4h of time used and decreases to 0.17 with increasing HTC time. Additionally, the ash content increases from 34.29% to 53.08, following carbonization due to excess loss of VM caused by HTC reactions, and rises as time is increased. The ash content is significantly higher compared to the initial SS. Sewage sludge ash deposition is a significant problem in the energy industry sector. The formation and transformation of sewage sludge ash stimulates the corrosion processes by slagging and fouling [22]. The hydrothermal carbonization enhances the ash content of hydrochar when the reaction time is increased. Therefore, it suggests that in this case 4 h of carbonization time is enough to obtain good thermal properties of hydrochar.

HTC brings about changes to the elemental compositions of dried sewage sludge caused by different carbonization times. Carbon content slightly increases, but a decrease in hydrogen, nitrogen, and sulphur has been observed when reaction time has been increased, except for carbonization at 10 h. The sulphur content significantly decreases from 1.56 to 1.0, whereas the nitrogen content decreases from 4.7 to 2.56 (Figure 5). Therefore, the observed 35% reduction of sulphur and 45% in the case of nitrogen content caused by the HTC process, positively eliminates the potential risk of generating harmful pollutants such as NO_x and SO_x. During HTC process volatile-N converted to the gas and liquid. It is confirmed by the liquid chromatography detecting a significant rank of heterocyclic compounds contains -N. Finally, the chemical composition of the investigated hydrochars confirms their relevant properties for the combustion process.

The HHVs are calculated taking into account the elemental composition and are low, around 14±15 MJ/kg, compared to lignite. The HTC process has a low impact on the hydrochar HHV, and for HTC_10 h the HHV has slightly increased to 14.73 MJ/kg.

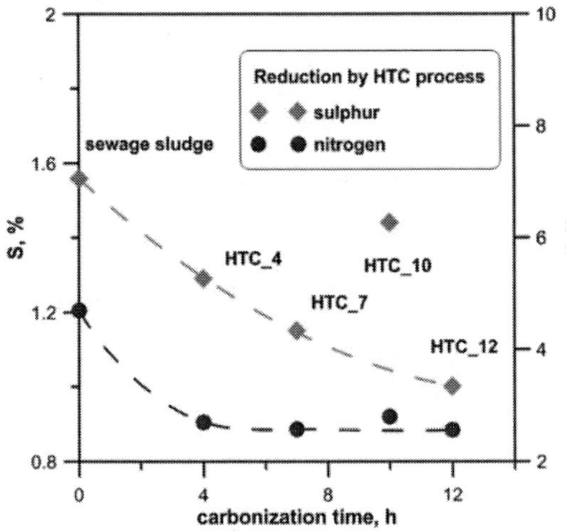

Figure 5. The effect of carbonization time on sulphur and nitrogen reduction under HTC process.

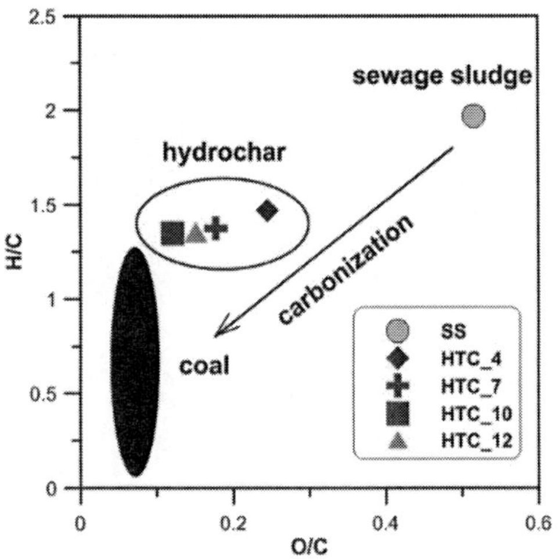

Figure 6. Van Krevelen diagram of sewage sludge and hydrochars.

In order to investigate the difference in the atomic composition of SS and hydrochars, H/C and O/C atomic ratios are depicted in the van Krevelen diagram (Figure 6). As can be clearly seen hydrothermal carbonization led to a decrease in H/C and O/C atomic ratios for all studied char samples. For comparison, the H/C and O/C rations for coal are also plotted. As shown in Figure 2 the atomic ratio of H/C and O/C reduces at a different degree after carbonization, from 1.92 and 0.52 to 1.35 and 0.15, respectively. In this regard, the reduced ratios confirms the significant effect of the HTC method in enhancing the fuel properties of hydrochars. Therefore, HTC can produce a renewable fuel with properties more like coal.

In order to study the thermal behaviour of raw and HTC sewage sludge during combustion, thermal analysis was used. The thermogravimetric studies of sewage sludge and hydrochar are presented in TG, DTG, and DSC curves. The various hydrochar samples obtained under 200°C and 12 h of HTC process were chosen as an example of HTC products. Combustion profiles of the samples are plotted in Figure 7 and 8.

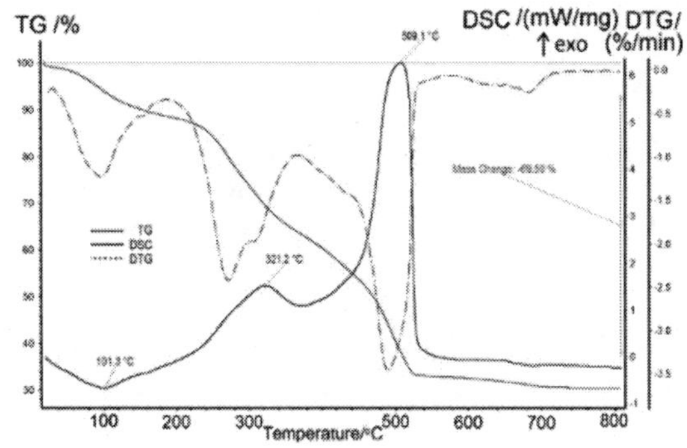

Figure 7. TG, DTG and DSC profiles for sewage sludge.

The sewage sludge combustion profiles differ for both studied samples. The sewage sludge combustion process takes place in three stages: the initial stages of decomposition, main decomposition and final decomposition. The first stage, corresponding to the moisture content, takes place up to c.a. 200°C. The mass loss is 10%. The main decomposition during the combustion process is in the 200-550°C temperature range. The second stage shows the presence of two main effects. Two maximum weight loss rates (DTG curves) occur at 270°C and 490°C, therefore two exothermic effects are observed (321°C and 509°C) connected with oxidation of organic matter. Mass loss after the main stage is 66%. Above 550°C there is no significant mass change. It can be assumed that above 700°C the combustion process is finished and the ash content is c.a. 30%. Based on TGA results the hydrothermal carbonization process causes evident changes in the thermal decomposition of sewage sludge. It was confirmed that the HTC product has insignificant moisture content, the peak connected with moisture loss does not appear. In the main stage (240-500°C) four exothermic effects are observed with the main effects at temperatures of 262°C and 358°C. For the HTC product the combustion process concludes at a lower temperature than raw sewage sludge. However, the final mass change is not observed above 720°C.

Despite that fact, combustion of hydrochar takes place within a narrower range of temperatures as the ash content is evidently higher, which can be an operating problem during HTC product combustion. The characteristic parameters obtained from TG, DTG, and DSC profiles of studied fuels are summarized in Table 2.

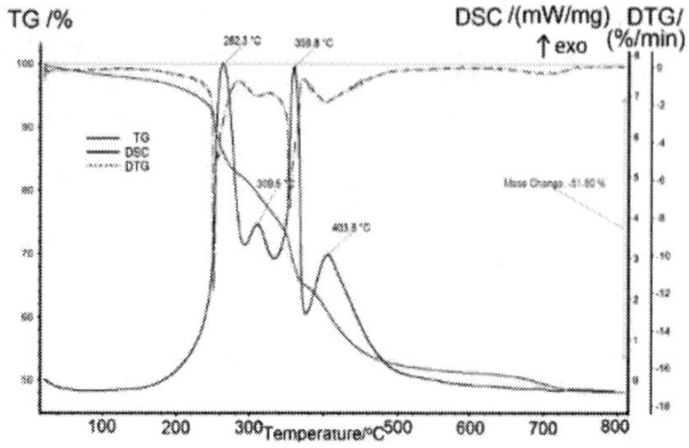

Figure 8. TG, DTG and DSC profiles for hydrochar HTC_12.

Additionally, the pH value of liquid phase which was generated from HTC under 4, 7, 10, and 12 h of carbonisation was measured after the process and was at a similar level, approximately 7.

Table 2. Characteristic parameters obtained from DTG profiles of studied fuels during the combustion process. T_m, temperature of the maximum weight loss rate; DTG_{max}; m, maximum weight loss rate; m_A, solid residue.

T_m, °C	DTG_{max}, %/min	m, %	m_A, %
		SS	
101	1.25	5.7	
321	1.8	38	
490	3.4	57	
800	-	70	30
		HTC_12	
250	11.8	8	
350	7.5	23	
800	-	52	48

4 Conclusions

According to the hydrothermal carbonization investigation of sewage sludge, the process can be named as a promising technology which enhances the thermal properties of raw material. The results confirm that the properties of hydrochars moved from sewage sludge towards coal (higher fixed carbon content and fuel ratio, improved chemical composition). Taking into account the impact of carbonization time on the samples it can be concluded that 4 h of carbonization time is quite enough to obtain good thermal properties of hydrochar. Furthermore, the thermogravimetric analysis confirms the different thermal behaviour of hydrochars for the combustion process. Finally, the HTC process can be used as a novel and alternative method to convert wet sewage sludge into dry solid biofuel.

Acknowledgement

The work was supported by grant AGH no. 11.11.110.423. The Author is grateful to EKOPROD Ltd. the proprietor of the HTC apparatus used in the presented study.

References

1. A. Magdziarz, S. Werle., Waste Manag., **34**, 174 (2014)
2. M. Wilk, A. Magdziarz, I. Kalemba, Energy, **87**, 259 (2015)
3. B. Kosturkiewicz, A. Magdziarz, M. Wilk, Rynek Energii, **5**, 111 (2011)
4. A. Funke, F. Ziegler, Bioresour. Technol., **102**, 7595 (2011)
5. S. Roman, J. M. V. Nabais, C. Laginhas, B. Leesma, J. F. Gonzales, Fuel Proc. Technol., **103**, 78 (2012)
6. J. Stemann, A. Putschew, F. Ziegler, Bioresour. Technol., **143**, 139 (2013)
7. M. Pala, I. C. Kantarli, H. B. Buyukisik, J. Yanik, Bioresour. Technol., **161**, 255 (2014)
8. M. T. Reza, J. Mumme, A. Ebert, Biomass Conv. Bioref. **5**, 425 (2015)
9. G. K. Parshetti, S. K. Hoekman, R. Balasubramanian, Bioresour. Technol., **135**, 683 (2013)
10. I. Oliveira, D. Blohse, H.-G. Ramke, Bioresour. Technol., **142**, 138 (2013)
11. S. Kang, X. Li, J. Fan, J. Chang, Ind. Eng. Chem. Res., **51**, 9023 (2012)
12. C. Peng, Y. Zhai, Y. Zhu, B. Xu, T. W., C. Li, G. Zeng, Fuel, **176**, 110 (2016)
13. C. He, J. Zhao, Y. Yang, J.-Y. Wang, Bioresour. Technol., **211**, 486 (2016)
14. K. Fakkaew, T. Koottatep, C. Polprasert, Bioresour. Technol., **192**, 328 (2015)
15. E. Danso-Boateng, R.G. Holdich, G. Shama, A.D. Wheatley, M. Sohail, S. J. Martin, App. Energy, **111**, 351 (2013)
16. D. Kim, K. Lee, K. Y. Park, Fuel, **130**, 120 (2014)
17. P. Zhao, Y. Shen, S. Ge, K. Yoshikawa, Energy Conv. Manag., **78**, 815 (2014)
18. C. He, A. G., J. Y. Wang, App. Energy, **111**, 257 (2013)
19. J. Zhang, Q. Lin, X. Zhao, J. Inter. Agric., **13**, 3, 471 (2014)
20. A. Magdziarz, M. Wilk, J. Therm. Anal. Calorim., **114**, 2, 519 (2013)
21. A. Magdziarz, M. Wilk, Energy Conv. Manag., **75**, 425 (2013)
22. A. Magdziarz, A. K. Dalai, J. A. Koziński, Fuel, **176**, 135 (2016).

An overview of adsorptive processes in refrigeration systems

Eliza Wolak[1] and Stefan Kraszewski[1]

[1]*AGH University of Science and Technology, Faculty of Energy and Fuels, 30-059 Cracow, Poland*

Abstract. Economic reasons and quest for new solutions based on recovering the energy have provoked an increase of interest in the adsorption technology in the refrigeration industry. The confirmation can be the fact that number of published research is on rise. Adsorption appliances may turn out to be an alternative to compression-type coolers. They use ecological chemical agents instead of substances which are aggressive and harmful to the environment. For regeneration of adsorptive refrigeration systems one can use cheap energy in a form of: industrial waste heat, energy of solar radiation and cheap electric power. The paper presents principles of operation as well as advantages and disadvantages of adsorptive refrigeration systems. Basing on literature the most frequently used adsorbent – adsorbate systems – which are employed in refrigeration industry – have been characterized. A review of construction solutions of systems on both laboratory and industrial scale has been made.

1 Introduction

Economic development entails a rise in the demand for electric power [1]. Considerable amounts of energy are used for work of refrigeration appliances which are present in nearly every household [2]. It is assumed that they consume around 45 percent of electric power drawn by households and public facilities - which comprises at the same time 15 percent of world energy consumption. The statistics are risen by countries located in warmer climates. In times of energy crisis the quest for and recovery of thermal energy are particularly justified in case of high levels of sunlight exposure and limited access to electrical energy [3].

Refrigeration appliances can be classified into refrigerators, freezers and air conditioners. These are usually compression-type appliances which are used on a wide scale because of their high performance. Big disadvantage of such appliances is the presence of substances harmful to the environment – mainly chlorofluorocarbons used as the refrigerant [4].

Adsorptive systems may turn out to be an alternative to compression-type refrigerators. Instead of substances harmful to the environment, they operate on non-toxic chemical agents [5]. In the process of recovery of adsorptive cooling systems one can use cheap energy in a form of: waste heat, energy of solar radiation and electric power. The waste heat occurs where one has to do with combustion, especially in combined heat and power stations, steelworks and heat engines with internal combustion. In an internal combustion engine, the emerging heat (50 to 60 percent of energy coming from fuel burning) is removed outside as the waste, while it could be used to power adsorption cooling devices on boats, rail or wheel vehicles [4, 5]. The shift towards a cooling adsorptive appliance from a conventional one may lead to savings in electricity.

The advantages of adsorptive refrigerating equipment are: simple design, easy handling and control of work. Lack of moving parts is connected with quiet work, lower failure rate and absence of vibrations during the work of appliance. Moreover, refrigerators of that type do not occur problems with corrosion and crystallization of refrigerant due to the applied substances [3, 6].

In spite of many advantages, those appliances have some drawbacks such as intermittent work, large size and low coefficient of performance. Yet - because of their specificity - adsorptive systems can be successfully implemented in all places where the levels of sunlight exposure are high or there is waste heat.

2 Principles of operation of adsorptive refrigerating appliances

The basis for the action of refrigerators is a cyclical phase transition gas – liquid of the coolant. When it is in the liquid state, it collects the thermal energy e.g. from a cold chamber (which gives the cooling effect) which is then used for the change of state. Next it returns the absorbed heat in the process of condensation when it changes its state to liquid again [7, 8].

One can adjust the adsorption process through such parameters as pressure and temperature. If we combine the process of endothermic desorption with the exothermic adsorption into closed thermodynamic cycle, then there is a possibility of usage of adsorptive systems for cooling. The heat emitted in the adsorption process is usually higher than the heat of vaporization (condensation) of adsorbate. After the contact of adsorbent with liquid adorbate, transport of mass will occur in the gas phase because the adsorption shows a bigger driving force than the condensation in liquid-vapour setup. During such process the temperature of liquid phase decreases, while the adsorbent's temperature increases. This effect is utilized in air-conditioning and cooling systems.

The operation of adsorptive refrigerators consists in continuous cycles of adsorption and desorption of the refrigerant from the adsorbent. The refrigerant (adsorbate) condenses and evaporate exchanging the energy in a form of heat with the environment [9].

2.1 Construction of adsorptive refrigerator

Adsorptive refrigerating appliances consist of three main elements. The first one of them is a container/adsorber containing a bed of adsorbent. It acts as a generator allowing the refrigerant to circulate through processes of adsorption and desorption occurring cyclically in a consecutive way. Thanks to heating and cooling of the adsorber the execution of both processes is possible. In compressor systems it is the compressor which plays the role of adsorber [10, 11].

Another element is a condenser. On desorption vapours of the refrigerant diffuse to the condenser where occurs the emission of heat which accompanies the condensation process. Vapours of the coolant change the state of matter to liquid.

The liquefied refrigerant flows then to an evaporator. Its role is opposite to the condenser. At adequately selected pressure the liquid evaporates (boils) absorbing the heat from the environment and emerging vapours are adsorbed in the adsorbent bed. The evaporator is usually placed in the cold chamber as it is the element which generates the cooling effect [12].

2.2 Efficiency of adsorptive refrigeration system

For comparison and assessment of performance of refrigerating appliances two coefficients are used: coefficient of performance (COP) and specific cooling power (SCP).

The COP is the quantity of achieved cooling effect by a refrigeration machine per unit of heat delivered. It is described by the following general formula [13]:

$$COP = \frac{Q_p}{Q_o + Q_d} \tag{1}$$

where: Q_p – the evaporation heat; Q_o – heat used for bed heating; Q_d – heat of desorption.

The coefficient of SCP [W/kg] indicates what cooling capacity is obtained from a unit of mass of adsorbent. It is connected with the adsorption capacity of the adsorbent deposit. The higher its value is, the more efficient is the use of applied sorbent. The SCP is expressed by the formula [10]:

$$SCP \cong \frac{L \cdot \Delta x}{t_c} \tag{2}$$

where: L – latent heat of evaporation, Δx – adsorptive capacity of cycle; t_c – time of adsorptive cycle.

3 Adsorbent - adsorbate working pairs for adsorptive cooling systems

The selection of proper working pair of adsorbent-adsorbate type is the crucial issue during designing of refrigerating appliance as the performance and work of the refrigerator depend on that. The choice of appropriate substances depends first of all on the intended use and work condition of the device [10].

3.1 Adsorbates applied in adsorptive refrigerating systems

The adsorbate, attending as the refrigerant, should fulfil certain criteria so that its properties turns into the maximum possible performance of cooling system.

Good adsorbate must be characterized with the following features: have low temperature of evaporation, high latent heat of evaporation, small size of particles easily adsorbed and desorbed from adsorbent, show thermal stability in the range of operating temperature, possess low pressure of saturation and be environmentally-friendly, non-toxic, non-flammable, non-corrosive [14].

In reality, however, substances that would meet all the above mentioned criteria do not exist. In adsorptive refrigerating systems it is most frequently the ammonia, water and methanol or ethanol which is applied as the adsorbate. Their most important properties are presented in the table 1 [10].

Table 1. The physical properties of the most commonly used adsorbents in the adsorption refrigeration systems [10].

Refrigerants	Normal boiling point [°C]	Density [kg/m³]	Latent heat of vaporizati [kJ/kg]	Latent heat of vaporizati [MJ/m³]
Ammonia	-34	681	1368	932
Water	100	958	2258	2163
Methanol	65	791	1102	872
Ethanol	79	789	842	665

3.2 Adsorbents in adsorptive refrigerating systems

The performance of adsorbents is largely dependent on their surface properties such as: specific surface area, type and number of pores, size of granules, size of powder, size of grains or packets. Adsorbents used in systems of adsorptive cooling should demonstrate the ability to adsorb large quantities of the refrigerant at low temperatures and the desorption of maximum possible amount of the coolant at increased temperature. They should be marked by: high latent adsorption heat (in comparison to emitted adsorption heat), high coefficient of thermal conductivity, the consistency of characteristics over time and reusability, being non-toxic and non-corrosive, being cheap and easily accessible.

Good porous materials, which are most frequently used in systems of adsorptive refrigeration, are: carbonic adsorbents (activated carbon, activated carbon non-woven fabrics, carbon molecular sieves, monoliths [15]), silica gels and zeolite.

3.3 Characteristics of adsorptive system: adsorbent – adsorbate

In literature one can find numerous research in which various systems of adsorbent – adsorbate type are proposed that fulfil requirements of work in adsorptive cooling systems. The most frequently described and applied working pair are: activated carbon-methanol, activated carbon- ammonia, silica gel - water and zoelite - water.

The activated carbon-methanol is often utilized for the production of ice and air-conditioning systems because of high adsorption capacity (up to 0,45 g/g_{carbon}), high latent adsorption heat (ca. 1229 kJ/kg·°C at -30°C) and the ability to work in a temperature below zero. Low values of adsorption heat (1800–2000 kJ/kg) and relatively low temperature of desorption translate into a high COP value for those systems. In spite of many advantages, the pair has several drawbacks and namely: activated carbon catalyses a decomposition reaction of methanol into formaldehyde (HCHO) and dimethyl ether (CH_3OCH_3) at temperatures higher than 150 °C and thus the temperature of desorption should not exceed 120°C. Methanol is a toxicant and this is why it should be handled with precautions taken. The system operating on the pair methanol-activated carbon requires the use of vacuum which makes the construction of appliance somehow more difficult [5,9,10].

Activated carbon and ammonia is another widely used pair. It can be characterized with similar adsorption heat as in case of methanol adsorption. Working in pair with ammonia requires increased pressure which is necessary due to the low boiling point for that substance. Employing high pressure (around 16 bar) results in improving the processes of exchange of mass and shortening the time of refrigeration cycle and is also connected with an increase of SCP indicator. Another benefit is the possibility of using higher temperatures – in a range of 200°C – in the desorption process. Disadvantage of the system is the corrosivity of NH_3 especially in case of use of copper materials. In case of any leakiness an immediate leakage of ammonia into the environment happens. As it is known the ammonia is an irritant with unpleasant smell. One should also mention about a lower adsorption capacity of that pair comparing to activated carbon- methanol systems [5, 10]. The latter ones are used for the manufacture of ice and as refrigerating appliances.

Table 2. Properties comparison of chosen working pairs [18].

Working pair	Operating temperature [°C]	Working pressure [bar]	Adsorption heat [kJ/kg]	Specific cooling capacity	Application
Silica gel – water	Below 90	0.01 – 0.3 (Vacuum conditions)	1000 – 1500	High	Air conditioning
Activated carbon – methanol	Below 120	0.01 – 0.35 (Vacuum conditions)	1800 – 2000	Normal	Ice making
Activated carbon – ammonia	Up to 150	3 – 10.4	2000 – 2700	Higher than silica gel- water	Refrigeration/ Ice making
Zeolite – water	Up to 200	3.4 – 8.5	3300 – 4200	Normal	Air conditioning
Zeolite – ammonia	150 – 200	3.5 – 7	4000 – 6000	High	Refrigeration/ Ice making

The silica gel - water system is marked by slightly higher adsorption heat (about 2500 kJ/kg) than it is in case of carbon and methanol. A big advantage of that pair is low desorption temperature beginning already at 55°C. However, it cannot exceed 120°C as the silica gel loses its sorption properties and this is why one uses in practice desorption temperatures in a range of 75 – 90°C. The disadvantage of the system is low sorption capacity of silica gel (0,2 $kg_{water}/kg_{adsorbent}$) and the range of temperature of water evaporation which cannot be lower than 0°C. This system is successfully used in air-conditioning, cold storage appliances [16] and solar energy-powered systems because of its characteristics [5].

Zeolite – water pair shows the highest adsorption heat among the above-mentioned working pairs which amounts to 3300 – 4200 kJ/kg. The system is stable at high temperatures and enables the occurrence of desorption at temperatures exceeding 200°C. The isotherm of adsorption in that working pair is virtually flat and because of that it can work with similar performance in a wide range of temperatures. High adsorption heat causes a lower value of COP coefficient than in previous cases in low and middle temperature ranges. That situation changes in case when the temperature of heat source for desorption exceeds 200°C. Then the system works with constant effectiveness, while all other pairs reach their maximum of temperature or are not capable to work at all. The drawback for the system is, as in the previous case, the low-end range of water evaporation as well as limited rate of mass exchange arising from the need of using lowered pressure. For the above reasons, that pair can be successfully used in air-conditioning systems with the high-temperature source of waste heat and heat pumps [17]. The table 2 presents properties of the selected adsorbent – adsorbate systems.

4 Review of design solutions of adsorptive refrigerating systems

The type of source of delivered heat can be one of criteria for classification of adsorptive cooling appliances. In literature one can find many various construction solutions. Yet, in every case one can distinguish three basic elements of the system: generator – condenser – evaporator.

4.1 Solar energy-powered systems

The solar energy is often utilized as in adsorptive refrigerating systems because of the possibility of their use in areas of limited access to electricity. In the figure 1 a prototype of refrigerator designed for storing food in countries having a Mediterranean climate is presented.

DOI: 10.1051/e3sconf/20161000104

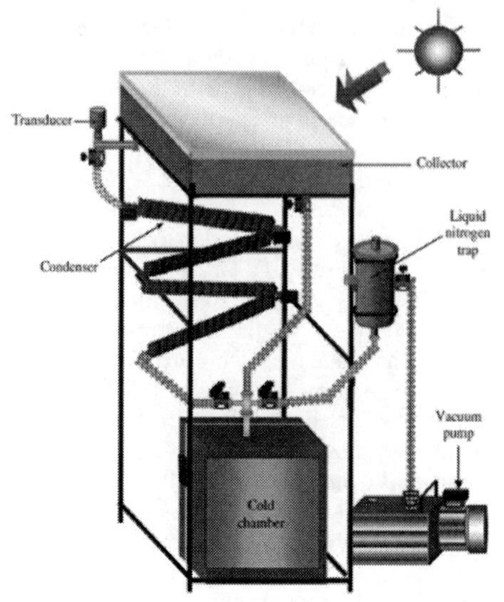

Figure 1. Scheme of solar adsorption refrigerator [19].

The system consists of an adsorber which is the sun collector at the same time. It has a form of chest filled with 14.5kg of activate carbon. The box is separated inside with copper ribs for better transport of heat in the bed. The copper ribbed pipe placed under the collector plays the role of condenser in which the refrigerant (methanol) change its state matter as well as flows into the evaporator, also made of copper, which is simultaneously a collecting container for the methanol. That part is situated in a thermally-insulated cold chamber. There are also a vacuum pump together with a trap with liquid nitrogen linked with the system. The pump enables lowering the pressure in the system without losses of methanol, pressure transmitter and thermocouple for process control.

The refrigerator works periodically – during the day high sunlight exposure causes desorption of the coolant, while at night the re-adsorption of methanol is possible as well as achieving the cooling effect when the temperature falls. The capacity of work depends on weather conditions and because of that the COP varies from 0.02 to 0.07 and the output is low temperature obtained in the cold chamber – from + 6.2 to – 11.1 °C [19].

Another example of use of the solar energy in adsorptive refrigeration is a solar refrigerator designed for the production of ice (figure 2). The working pair in this case is also activated carbon and methanol. The adsorber is made of stainless steel and accommodates 24 kg of activated carbon. It is distinguished by a condenser cooled with water thanks to which the methanol passes directly to the evaporator in the liquid phase. The evaporator is situated in a thermally-insulated water tank [20].

The COP coefficient of performance for this system varies from 0.083 to 0.127 depending on weather conditions, while the quantity of ice produced within one cycle is from 30 to 35 kg.

The drawback of adsorptive refrigerators powered by solar energy is undoubtedly very long time of cycle dependant on daily insolation phases (recovery of bed occurs by day and adsorption at night) and work productivity which also depends on weather conditions.

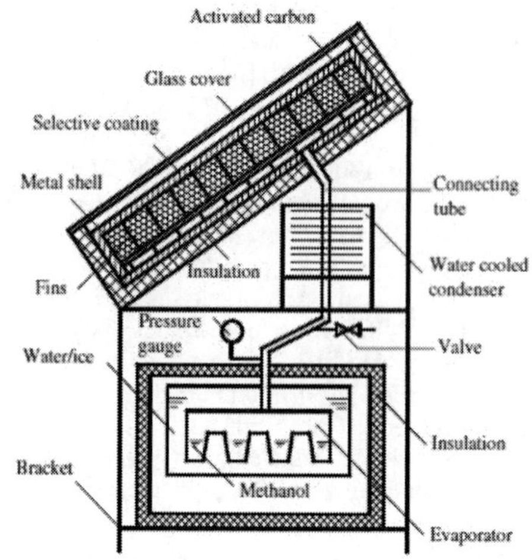

Figure 2. Scheme of adsorption device for ice production [20].

4.2 Systems powered by waste heat of exhaust fumes

The next group of cooling appliances are systems using the waste heat of internal combustion engines. The temperature of exhaust fumes and internal combustion engine can be utilized for air-conditioning of compartments in vehicles which may bring the reduction of fuel consumption.

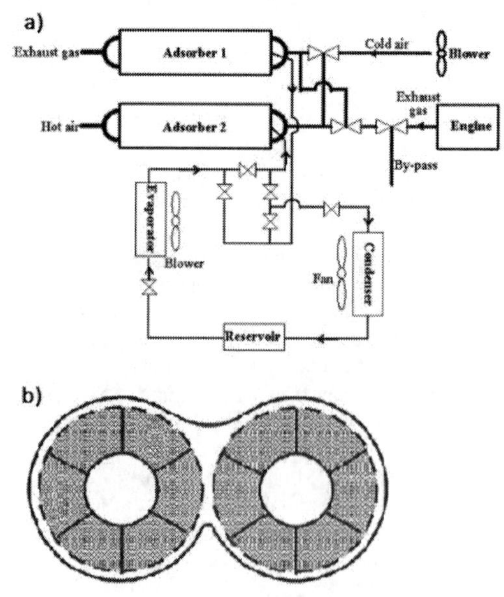

Figure 3. a) Scheme of adsorption cooling system working on car engine exhaust gases waste heat; b) A cross section of applied adsorber [21].

A sample of such solution is a prototype of car air-conditioner on laboratory scale operating on the activated carbon-methanol system which is presented in the figure 3. It is made of two adsorbers working in opposite phases of the cycle which ensures an almost continuous cooling effect. Each of them consists of two carbon beds (mass is 08, kg for each). They have a form of concentrically placed pipes among which the sorbent material is located (external pipe is made of steel grid). Hot exhaust fumes coming from the internal combustion engine or cool air from the surroundings flow through internal pipes (150 - 200 °C) (figure 3b). The whole is locked in airtight case made of stainless steel which enables the transport of refrigerant from or to remaining part of the system. In the bed there are steel fins for improving the exchange of heat. The condenser is a finned aluminum pipe which is cooled by air from the fan. The role of evaporator in the system is to chill the air pressed to it by the fan. The air is then used for cooling the interior of compartment in vehicle [21].

Another system making use of waste heat of exhaust fumes is an air-conditioner cooling the interior of cockpit in a locomotive which is made on a scale allowing to use it practically (figure 4). In that case the working pair is zeolite 13X and water. In this system there is one generator with a mass of bed of 140 kg which is heated by hot exhaust fumes and cooled by ambient air depending on the phase cycle. In the evaporator there is a separate external water cycle which is chilled during the adsorption phase. Cooled water is transported by a pipeline and pump to the cockpit where it collects heat from the air present over there. This air-conditioner is able to chill the air on the cockpit of locomotive from 33 to 25 °C. Its cooling capacity is 4.1 kW and the COP value is 0.25 [22].

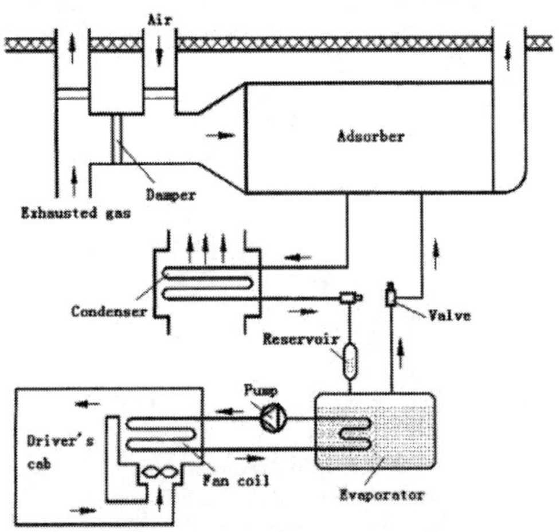

Figure 4. Adsorption air conditioner working on locomotive engine exhaust gases waste heat [22].

Adsorptive refrigerating appliances using the waste heat from internal combustion engines appear to be a promising solution which could replace widely used air-conditioning systems in vehicles. There are also some disadvantages related to them such as the necessity of using a set of generators in order to ensure continuous cooling and requirements connected with appropriately long working time of the engine for the purpose of assuring proper working conditions for adsorptive appliance.

5 Commercially produced adsorptive refrigerators

Significant part of adsorptive refrigerating systems described in the literature is a prototypical appliance usually made on a small laboratory scale and theoretical models constructed on the basis of experimental data with the aid of computer programs. There are also appliances created on a bigger scale serving till today as air-conditioners for facilities, but they are still a single instance. The development of technology of adsorptive refrigeration is primarily limited by high costs of the design of equipment and still low efficiency of such appliances.

In the market there are several companies which deal with the manufacture of adsorptive cooling systems. Their products are air-conditioners operating on the working pair of silica gel- water and zeolite – water. The main producers offering appliances of that type are: German SorTech and InvenSor, Japanese Mayekawa and American HIJC and Eco – Max.

The figure 5 presents an adsorptive aggregate eCoo 2.0 by SorTech company along with its plan. The system is made of two absorbers working in opposite cycles. The working pair is silica gel and water and the cooling capacity is up to 16 kW. It also chills the water which is utilized later on in air-conditioning in a temperature range from 8 to 21°C. For the proper operation of the system a source of heat is necessary – it is water at a temperature from 50 to 95°C. For that purpose the water can be heated by an optional source of waste heat, with the aid of sun collectors or with use of both variants in a manner of cogeneration [23].

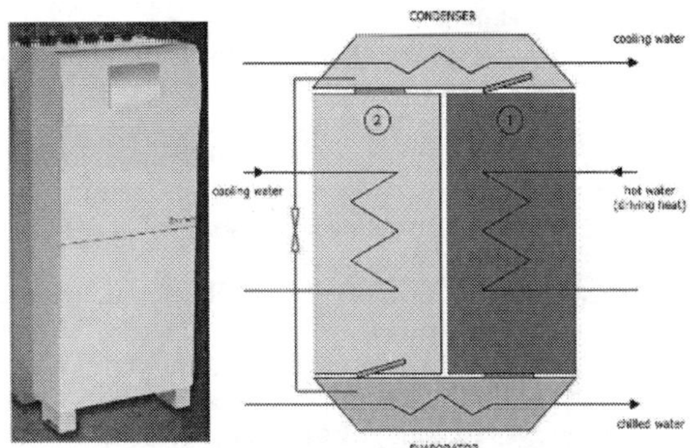

Figure 5. Adsorption cooling unit eCoo 2.0 produced by SorTech company. Photograph and scheme of the system (1 – adsorber in desorption phase, 2 – adsorber in adsorption phase) [23].

They only differ by values of COP, the cooling capacity and sizes. Commercial adsorptive refrigerating apparatus is often installed in associated cooling systems in which several different types of heat source are used. In the year 2000 in a hospital in Kammenz (Germany) for example an associated Cogeneration system for Cooling, Heat and Power (CCHP) was installed for the very first time in which the heat collector is powered by the waste heat from a fuel cell and solar energy. It delivers both heat and cold to hospital spaces

 DOI: 10.1051/e3sconf/20161000104

In combination with adsorptive refrigerating system. The aggregate's cooling capacity is 105 kW.

In Japan Macom company has been dealing with the production of adsorptive refrigerator with a working pair of silica gel and water since 2003. In Japan, too, Tokai Optical Co. company from Nagoya introduced an associated adsorptive system (CCHP) with the use of waste heat from the diesel engine in 2003.

6. Directions of improvement and development of adsorptive refrigerators

In spite of many advantages, adsorptive refrigerating appliances have also some disadvantages. The main drawbacks are a relatively low COP effectiveness, cyclicity, big sizes and weight of systems. As far as the intermittent operation can be easily changed by the use of two or more generators in the system, then the increase in COP is combined with the necessity of improving the heat exchange in the generator bed which is still quite problematic. As it is known, adsorbents are marked by low heat conductivity coefficient. Adsorbents undergo various modifications in order to intensify the exchange of heat in the adsorbent's bed. According to the literature, there are many ways as for example: placing metal partitions in the bed [19, 21], mixing the adsorbent with metal filings and formulation of composites like e.g. composite of zeolite and aluminum in a form of foam surrounding particles of adsorbent [24]. Adsorbents undergo modifications with the purpose of improving sorption properties and packing in the bed through processing. That processing consists in the abrasion of external layers of adsorbent grains and the change of nature of surface through oxidation [25].

The COP coefficient can also be increased by the improvement of energy use – delivered and emitted during the process. Due to that reason, advanced adsorption cycles for systems consisting of at least two generators were worked out. Their purpose is more effective energy usage, improvement of mass and heat transport in the adsorbent deposit.

One can distinguish the following advanced adsorption cycles: heat recover cycle, mass recover [18], thermal wave cycle, convective thermal wave cycle, cascade cycles [5, 10]. The cycles are still being investigated and developed in order to obtain the best possible effects and cooling capacities. One also attempts to combine the above described cycles with each other. This, however, results in increase of complexity of systems and makes them sophisticated and that, it turn, has an effect in failure rate of adsorptive refrigerating appliances. Despite that the advanced adsorption cycles are a promising direction for development for this branch of practical use of the desorption phenomenon.

The main direction of development of adsorptive cooling technologies are works over increasing the process efficiency, intensification of the heat exchange in the bed and lowering the costs of system production. In order to achieve that numerous actions are taken on challenges of development of generator/adsorber design, improvement of use and recovery of heat and modification and quest for more and more effective adsorbents.

As one can see from variety and number of research on that topic it is still current and interesting issue both from the scientific and practical point of view.

Acknowledgements

The work was carried out within the research to keep the research potential AGH (11.11.210.244).

References

1. W. Tkaczyk, A. Kozieł, H. Mikołajuk, G. U. S., (2014)
2. M. Turski, R.Sekret, Civil and Envirnomental eng. **1** (2010)
3. M. Gwadera, K. Kupiec, Czas. techn., z. **8**, 108, Wyd. P. K. (2011)
4. M. Hamdy, A. A. Askalany, K. Harby, N. Kora, Ren. and Sust. En. Rev., E. **51** (2015)
5. D.C. Wang, Y.H. Li, D. Li, Y.Z. Xia, J.P. Zhang, Ren. and Sust. En. Rev., E. **14** (2010)
6. C.Y. Tso, K.C. Chan, C. Y.H. Chao, C.L. Wu, Inter. Jour. of H. and M. Trans., E. **85** (2015)
7. E. Klimowska, B. Buczek, Inż. I Och. Środ., **7**, 3-4, (2004)
8. K. M. Gutowski, *Chłodnictwo i klimatyzacja*, W. N. T. (2003)
9. A. A. Askalany, M. Salem, I.M. Ismail, A. Hamza, H. Ali, M.G. Morsy, Ren. and Sust. En. Rev., E. **16** (2012)
10. R.Wang, L. Wang, J. Wu, *Adsorption refrigeration technology*, J. W. & S., (2014)
11. S. Gajczak, *Absorpcyjne urządzenia chłodnicze*, P. W. T., (1958)
12. A. Grzebielec, A. Rusowicz, Chłodn., t. XXXIX, nr **1-2**, (2004)
13. E.E. Anyanwu, Energy Convers. Mgmt. **45** (2004)
14. M.A. Alghoul, M.Y. Sulaiman, B.Z. Azmi, M.A. Wahab, App. Ther. Eng., E. **27** (2007)
15. B. Buczek, E. Wolak, Mat. Sci.-Pol, **30**, 2 (2012)
16. M. Szyc, W. Nowak, Chem. and Proc. Eng., **35**, 1 (2014)
17. E.E. Anyanwu, Ener. Conv. and Mgmt., 44 (2003)
18. P. Goyal, P. Baredar, A.Mittal, A. R. Siddiqui, Ren. and Sust. En. Rev., E. **53** (2016)
19. F. Lemmini, A. Errougani, Ren. En., E. **32** (2007)
20. H. L. Luo, Y.J. Dai, R. Z. Wang, R. Tang, M. Li, En. Conv. and Man., E. **46** (2005)
21. H. R. Ramji, S. L. Leo, M. O. Abdullah, App. En., E. **113** (2014)
22. Y. Z. Lu, R. Z. Wang, M. Zhang, S. Jiangzhou, En. Conv. and Man., P. **44** (2003)
23. http://www.sortech.de/en/
24. P. Hu, J.-J. Yao, Z.-S. Chen, En. Conv. and Man., E. **50** (2009)
25. B. Buczek, E. Klimowska, E. Vogt: Adsorption **11**, 769, (2005)

High-temperature nuclear reactor power plant cycle for hydrogen and electricity production – numerical analysis

Michał Dudek[1,a], Marek Jaszczur[1], Katarzyna Skolik[2], Mateusz Malicki[2] and Ludwik Pieńkowski[2]

[1]*AGH University of Science and Technology, Faculty of Energy and Fuels, Department of Fundamental Research in Energy Engineering, Krakow, Poland*
[2]*AGH University of Science and Technology, Faculty of Energy and Fuels, Department of Nuclear Energy, Krakow, Poland*

Abstract. High temperature gas-cooled nuclear reactor (called HTR or HTGR) for both electricity generation and hydrogen production is analysed. The HTR reactor because of the relatively high temperature of coolant could be combined with a steam or gas turbine, as well as with the system for heat delivery for high-temperature hydrogen production. However, the current development of HTR's allows us to consider achievable working temperature up to 750°C. Due to this fact, industrial-scale hydrogen production using copper-chlorine (Cu-Cl) thermochemical cycle is considered and compared with high-temperature electrolysis. Presented calculations show and confirm the potential of HTR's as a future solution for hydrogen production without CO_2 emission. Furthermore, integration of a high-temperature nuclear reactor with a combined cycle for electricity and hydrogen production may reach very high efficiency and could possibly lead to a significant decrease of hydrogen production costs.

1 Introduction

Energy consumption is one of the key indicators showing the development stages of countries. The growing demand for electricity and contemporary development of nuclear power technology, computer technology, and materials science allow today design and implement new solutions for nuclear energy, energy security as well as an energy conversion system, to provide a lower unit cost of energy conversion and new possibilities in energy and fuel production [1]. A new generation of high-temperature nuclear reactors (HTR) is one of the most innovative concept among current advanced nuclear reactor technologies [2,3]. A new program for future nuclear energy systems Generation IV has been created in an effort to provide next-generation technologies that will compete in all markets with the most cost-effective technologies expected to be available over the next three decades [4]. Gas-cooled reactor systems have several fundamental characteristic features that distinguish them from other types of reactors and provide significant operational advantages. In particular, the fuel is in the form of small ceramic-coated particles capable of operating at very high temperatures, the moderator is solid graphite and the coolant is helium or carbon dioxide inert for neutrons [4]. VHTR are designed with average coolant outlet temperatures above 900°C and operational fuel temperatures above 1250°C. These designs provide the potential for increased energy conversion efficiency and for high-temperature processes heat applications, such as coal gasification or thermochemical hydrogen production. In contrary to the fossil fuels, hydrogen is cleaner and less harmful to the environment. By using this fuel we avoid all kinds of pollution, greenhouse gases, and acid rains. Several methods of production of this gas are known. The most common methods are steam reforming of natural gas, high-temperature electrolysis, direct thermal, thermochemical, electrochemical and photoelectrochemical decomposition of hydrocarbons. The most promising thermochemical method for hydrogen production are sulphur-iodine and copper–chlorine cycles [4]. The efficiency of high-temperature I-S cycle is about 52%, while the medium-temperature Cu-Cl cycle efficiency is about 47%.

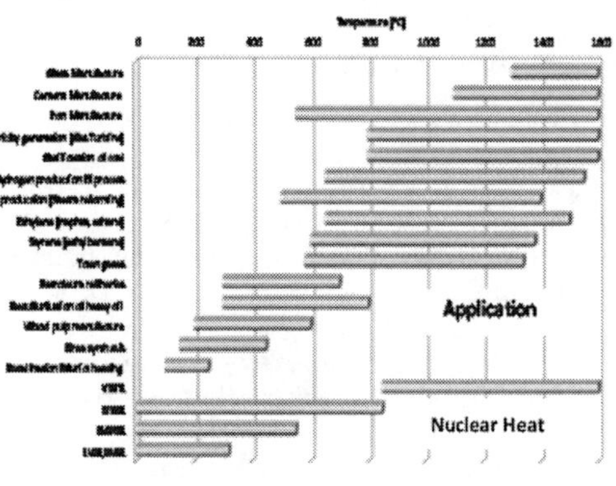

Figure 1. Operational temperatures in various industry application vs operational temperatures in various nuclear reactor types [4].

[a] Corresponding author: michald@agh.edu.pl

DOI: 10.1051/e3sconf/20161000105

2 HTR development status

Although the HTR's are a very promising concept for using high outlet temperature for industrial purposes, there is only one demonstration site in the world where such reactors are currently in construction. This place is Shidao Bay NPP in China and the reactors are two 250 MW_t HTR-PM (High-Temperature Reactors-Pebble-bed Module), together able to generate power of 210 MW_e. The construction has begun in 2012 and the operation is predicted for 2017. The outlet temperature of about 750°C is in the lower range of heat needed for efficient hydrogen production [5]. Several projects were implemented and investigated by means of test reactors in the UK with the DRAGON reactor, the US with the PeachBottom reactor and in Germany with the pebble bed AVR reactor.

The HTR's designed to supply higher temperatures, up to 1000°C are called VHTR's but they are still at the research stage with no industrial use. One project launched by the United States Department of Energy (DOE) is a Next Generation Nuclear Power Plant (NGNP) which assumes designing a VHTR similar to Areva's SC-HTGR (Steam-cycle high-temperature gas-cooled reactor) with the aim of using up to 30% of a process heat to produce hydrogen. The prototype of NGNP reactor is likely to be deployed by approximately 2030 [6,7].

Another program dedicated to VHTRs' development was submitted by Korean Atomic Energy Research Institute (KAERI). In cooperation with China and Japan, they plan to start operation of their first VHTR, coupled with hydrogen production in 2020 [8].

With the outlet temperature of 900-1000°C supplied by a Very High-Temperature Reactor, the hydrogen is to be produced on efficient processes of the sulfur-iodine cycle or the electrolysis. However, with such high temperatures the problem of safe and reliable heat exchangers occurs [9]. The aim of the present work is to analyse the possibility of hydrogen production with the lower temperatures and using the nuclear systems that will be introduced to the industry in the near future. That is why the process of the Cu-Cl cycle was chosen which has a temperature requirement of about 500°C while maintaining a reasonable efficiency over 40%.

3 Gas Turbine Combined Cycle

The operation of a gas turbine combined cycle (GTCC) employs a gas turbine and heat recovery steam generator (HRSG) that uses exhaust gas from a turbine in order to produce high-quality steam, which is then supplied to a steam turbine [2,10]. The main constraint in the operation of the GTCC power plants is HRSG which is located directly after gas turbine where changes in temperature and pressure of the exhaust gases may cause significant thermal and mechanical stresses [10]. The most common type of HRSG for combined cycle power plant contains three sections of heat exchanger modules – for high pressure (HP), intermediate pressure (IP) and low-pressure (LP) steam. Additionally, when such plant is operated in a common load-following operation mode it can lead to a large thermal stress in practical application and can eventual damage some components of the system. Operating conditions of the steam turbine are in the presenting system directly connected to the gas turbine and heat recovery steam system. The situation has become even more complex since the additional heat exchanger is installed after or before the gas turbine in order to extract heat from a high-temperature thermochemical process using I-S or Cu-Cl cycle. Determining the proper thermodynamic and mechanical conditions for all components to ensure efficient work of the cycle has become a very complex problem [2,10].

The exhaust gases from the classical gas turbine in combined cycle can actually reach 600°C or more while exhaust gases from high temperature combined cycle can achieve 830°C what significantly increases thermal and mechanical stresses in the first section of Heat Recovery Steam Generator and therefore generate additionally outlays on materials structure [11]. HT-GTCC may produce electricity, heat, and hydrogen and allow as to optimize investment and maintenance cost. An example of classical gas turbine combined cycle is referred to figure 2.

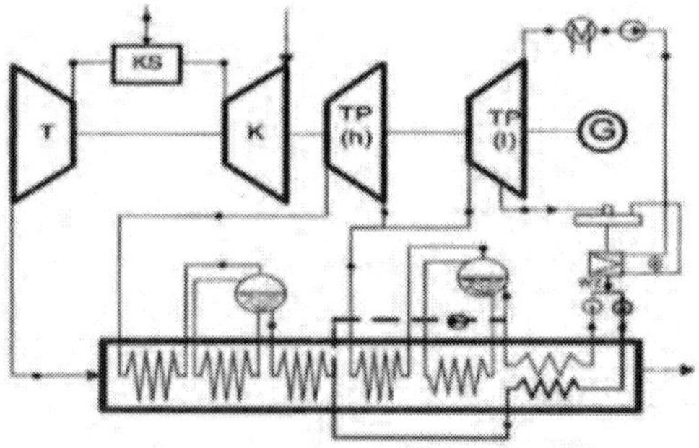

Figure 2. Gas Turbine Combined Cycle with HP and LP Heat Recovery Steam Generator system

The main thermodynamic parameters of high temperature combined projects developed in the word are described at table 1.

4 Power system

High-temperature gas turbine combined cycle (HT-GTCC) analysed in this paper which employs nuclear reactor is presented in figure 3. It consists of three subsidiary circuits. In primary loop, the helium directly flows through the reactor and by the intermediate heat exchanger it heats the gas to 700°C–950°C. The gas is then moved into the gas turbine and HRSG system in the secondary circuit. The steam is produced in the last cycle which realizes an independent Rankine cycle. The first loop of presented system is equipped with High-Temperature Nuclear Reactor (HTR), Intermediate Heat Exchanger (IHX) and Blower. The second cycle is equipped with High-Temperature Gas Turbine (GT), Medium Temperature Heat Exchanger, Heat Recovery Steam Generation system and Main Helium Compressor. The independent Rankine cycle is realized by three-stage Steam Turbine (ST), Condenser, Pump, Economizer, Evaporator, and Superheater.

Table 1. Main thermodynamic parameters of high-temperature gas turbine combined cycle concepts [12].

Parameter	HTR Gas Turbine Plant Concepts			
	GTMHR	**GTHTR300**	**ANTARES**	**VHTR NGTCC**
Power [MWth]	600	600	600	350
Thermodynamic cycle	Inter-Cooled Recuperated Brayton Cycle	Recuperated Brayton Cycle	Combined Cycle	Combined Cycle
Power conversion working fluid	He	He	He/N_2	He
Reactor inlet/outlet temperature [°C]	491/850	587/850	355/850	400/950
Turbine inlet temperature[°C]	850	850	800	950
Reactor gas pressure [MPa]	7.1	7.0	5.5	7.1
Compression ratio	2.86	2	2	1.94
Plant Net Power [MWe]	286	274	280 (80 GT, 200 ST)	180 (50 GT, 130 ST)
Thermal efficiency [%]	47.6	45.6	47.0	51.5
Turbine blade cooling	Uncooled	Uncooled	Uncooled	First two stages cooled

In the literature several types of Cu–Cl cycles are presented [13]. Three different variations of the Cu-Cl cycle are currently under investigation: 3-step, 4-step and 5-step cycles. The copper-chlorine thermochemical cycle uses a series of intermediate copper and chlorine compounds. The chemical reactions form a closed internal loop that recycles all chemicals in a continuous basis, without emitting any greenhouse gases or waste. The thermochemical copper - chlorine cycle, realized in four steps is carried out in a lower temperature range and requires heat and electricity supply. It is recommended to employ when the working fluid is at low temperature but still at least 500°C (the efficiency of hydrogen production is 47% at 800°C for HTR [13], see figure 4 for reference). The main chemical reactions are as follows:

Hydrolysis at 370 – 400°C
$2CuCl_{2(s)} + H_2O_{(g)} \rightarrow Cu_2OCl_{2(s)} + 2HCl_{(g)}$

Copper oxychloride decomposition at 500–550°C
$Cu_2OCl_{2(s)} \rightarrow 2CuCl_{(s)} + \frac{1}{2}O_{2(g)}$

Hydrogen production at 25–100°C
$2HCl_{(aq)} + 2CuCl_{(aq)} \rightarrow H_{2(g)} + 2CuCl_{2(aq)}$

Drying at 80–100°C
$2CuCl_{2(aq)} \rightarrow H_2O_{(g)} + 2CuCl_{2(s)}$

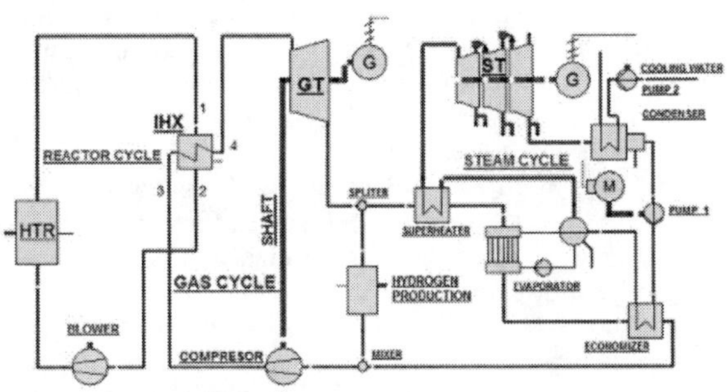

Figure 3. High temperature combined cycle for hydrogen and electricity production.

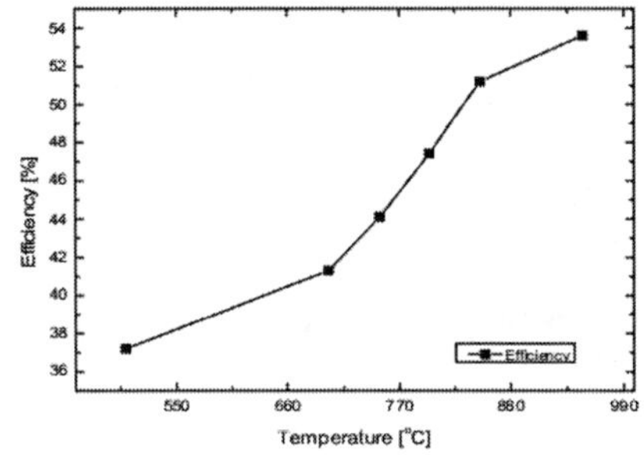

Figure 4. The thermal efficiency of copper-chlorine (Cu-Cl) cycle.

5 Numerical Model

HT-GTCC power plant is a main topic of present research for hydrogen and electricity production. General data and main thermodynamic assumptions are described in Table 2.

The main mathematical model of the high-temperature hybrid combined cycle is presented below.

Thermal reactor (HTR) power was calculated as follow:

$$\dot{Q}_r = \dot{m}_{1He}\, c_{pHe} \left(T_{out} - T_{in}\right) \tag{1}$$

where: $\dot{m}_{1He}$ - helium mass flow rate [kg/s], c_{pHe} - helium specific heat [kJ/kgK], T_{out} - reactor outlet temperature [K], T_{in} - reactor inlet temperature [K].

The heat flux supplied in primary helium circuit:

$$\dot{Q}_s = \dot{m}_{1He}\left(h_1 - h_2\right) \tag{2}$$

where: h_1 - helium specific enthalpy at the inlet to the heat exchanger hot side [kJ/kg], h_2 - helium specific enthalpy at the outlet from the heat exchanger hot side.

The heat flux received in the second circuit:

$$\dot{Q}_R = \dot{m}_{2He}\left(h_4 - h_3\right) \tag{3}$$

where: $\dot{m}_{2He}$ - helium mass flow rate [kg/s], h_4 - enthalpy at the outlet from the heat exchanger cold side [kJ/kg], h_3 - enthalpy at the inlet to the heat exchanger cold side [kJ/kg].

Table 2. General model data and thermodynamic assumptions for HT-GTCC.

Main Parameter	Value	Unit
Reactor Cycle		
Power	600	MWth
Pressure	6	MPa
Coolant	Helium	--
Outlet Temperature	750-1000	°C
Helium Mass Flow	140,2 -541,1	Kg/s
Gas Cycle/Steam Cycle		
Pressure	6	MPa
Gas Turbine Inlet Temperature	750-1000	C
Gas Turbine –Isentropic Efficiency	0,9	-
Gas Turbine – Mechanical Efficiency	0,99	-
Expansion Ratio	1-0.2	-
Compression Ratio	1-5	-
Generator Electrical Efficiency	0,9856	-
Steam Turbine – Isentropic Efficiency	0,88	-
Steam Turbine – Mechanical Efficiency	0,998	-
Motor Efficiency – Electrical/Mechanical	0.85/0.998	-
Compressor Mechanical/Isentropic Efficiency	0.99/0.9	-

High Temperature heat exchanger **(IHX)** enthalpy balance:

$$\dot{m}_{1He}c_{pHe}\left[T_1 - T_0\right] + \dot{m}_{2He}c_{pHe}\left(T_3 - T_0\right) = \\ = \dot{m}_{1He}c_{pHe}\left(T_2 - T_0\right) + \dot{m}_{2He}c_{pHe}\left(T_4 - T_0\right) \quad (4)$$

where: m_{1He}/m_{2He} - helium mass flow at hot/cold side [kg/s], c_{pHe} - helium specific heat [kJ/kgK], T_1/T_2 – inlet/outlet temperature to/from heat exchanger - hot side [K], T_3/T_4 - inlet/outlet temperature to/from heat exchanger - cold side [K], T_0 - reference temperature [K].

Gas turbine electrical power was calculated from the following equation:

$$P_{GT} = \eta_{gen}\dot{m}_{2He}(\eta_{iGT}\eta_{mech}w_{GT} - w_C / \eta_{mech}\eta_{iC}) \quad (5)$$

where: η_{gen} - electrical efficiency, η_{iGT} - gas turbine isentropic efficiency, η_{iC} - compressor isentropic efficiency, η_{mech} - mechanical efficiency, w_{GT} - gas turbine work [kJ], w_C - compressor work [kJ].

Steam turbine electrical power was calculated as follows:

$$P_{ST} = \eta_{gen}\eta_{mech}\dot{m}_{p1}\left(\sum_{WHPST}^{LPST}\eta_i\eta_{mech}w_i\right) - \dot{m}_w\, w_{P_1} / \eta_{ip}\eta_{mech} \quad (6)$$

where: η_{gen} - generator efficiency, η_{mech} - mechanical efficiency η_i - isentropic steam turbine efficiency, η_{ip} - isentropic pump efficiency, m_{p1} - steam mass flow rate [kg/s], w_i - high/medium/low pressure steam turbine work [kJ/kg], w_{p1} - water pump work [kJ/kg], $\dot{m}_w$ - water mass flow [kg/s].

For cases when hydrogen production has been taken into account the following formula for cycle efficiency was used.

$$\eta_c = \frac{P_{GT} + P_{ST} - W_B - W_{P2} + \eta_{CuCl}\Delta H_{H2}}{Q_r} \quad (7)$$

where: P_{GT}/P_{ST} - gas turbine/steam turbine electrical power [MW_e], W_B - blower work [MW], W_{P2} - water cooling pump work [MW], η_{CuCl} - thermochemical copper chlorine efficiency, ΔH_{H2} - enthalpy H_2 [kJ], $\dot{Q}_r$ - is the nuclear reactor thermal power. All calculations have been performed using Ebsilon Professional software which was delivered by (Steag).

6 Results

Results of numerical calculation for high-temperature hybrid combined cycle without hydrogen production are presented in figure 5. The first graph shows gas turbine power P_{GT} vs compression ratio and different inlet temperature into the gas turbine. This graph shows that gas turbine in high temperature combined cycle coupled with nuclear reactor effectively works when optimal compression ratio varies from 1.4 to 2.5 at different nuclear reactor temperature. However, from a thermodynamic point of view not all cycle conditions the as possible to realize, due to a significant power decrease in the steam system (see figure 6). The system will work effectively only when the compression ratio is between 1.2 and 2 for the lowest nuclear reactor outlet temperature and for higher compression ratio if higher reactor outlet temperature is available. From a thermodynamic point of view, the optimal value of expansion ratio for effective work varies from 1.5 to 3.5.

It can be seen from figure 6 and 7 that it is possible to achieve more than 50% of thermal efficiency when reactor outlet temperature achieves 1000°C. The highest system efficiency has been obtained for compression ratio 1.5 up to 2.5.

Figure 8 shows inlet temperature available for the steam turbine. These temperature calculations are based on the assumption of a constant temperature difference between inlet and outlet temperature delivered to the HRSG system equal 15°C. The graph shows that it is possible to achieve more than 40% of thermal efficiency when the sluice of heat is equal 15% at the lowest of available nuclear reactor outlet temperature. The achievable efficiency is higher with higher reactor outlet temperature.

Figure 9 shows the numerical calculation for high-temperature hybrid combined cycle with hydrogen production at a thermochemical cooper-chlorine cycle

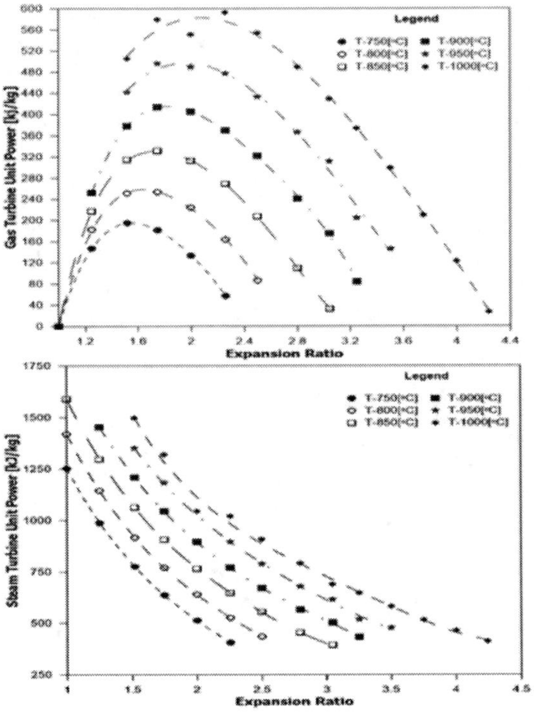

Figure 5. Gas turbine and steam turbine electrical power vs different outlet temperature from a nuclear reactor.

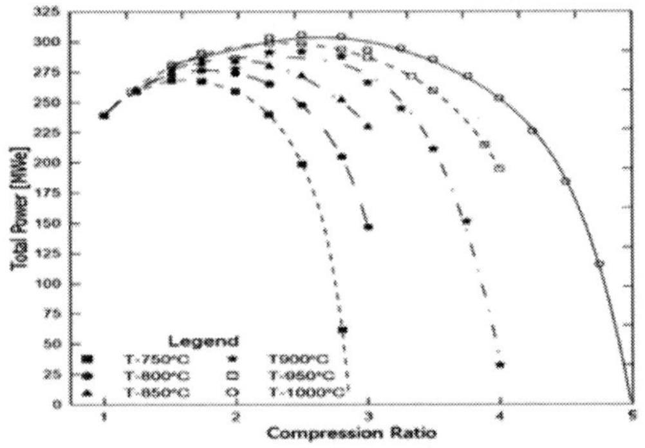

Figure 6. High-temperature hybrid combined cycle - total power vs different nuclear reactor outlet temperature and different compression ratio.

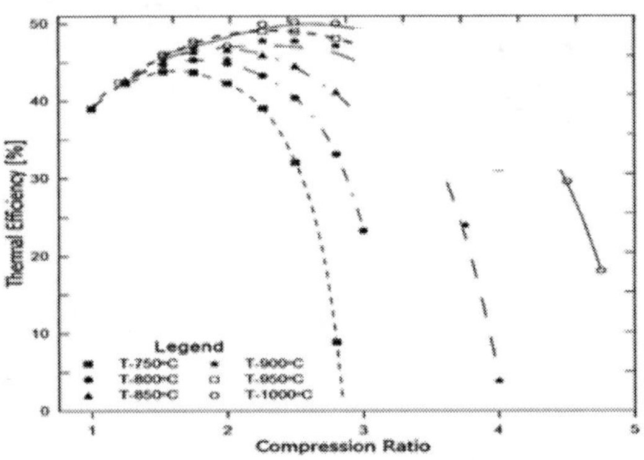

Figure 7. High-temperature hybrid combined cycle – thermal efficiency vs different nuclear reactor outlet temperature, compression ratio.

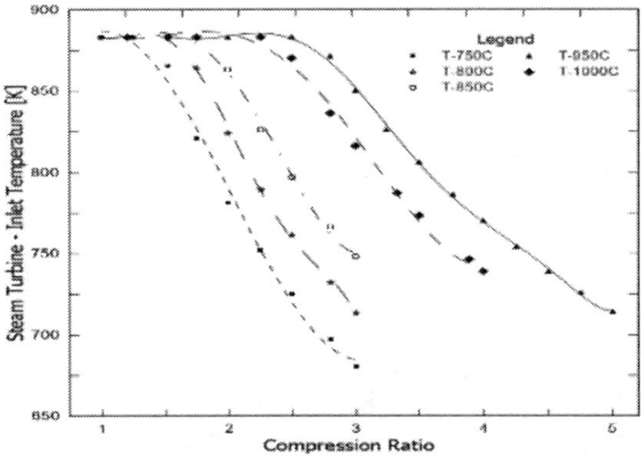

Figure 8. Steam turbine inlet temperature vs different nuclear reactor outlet temperature and compression ratio.

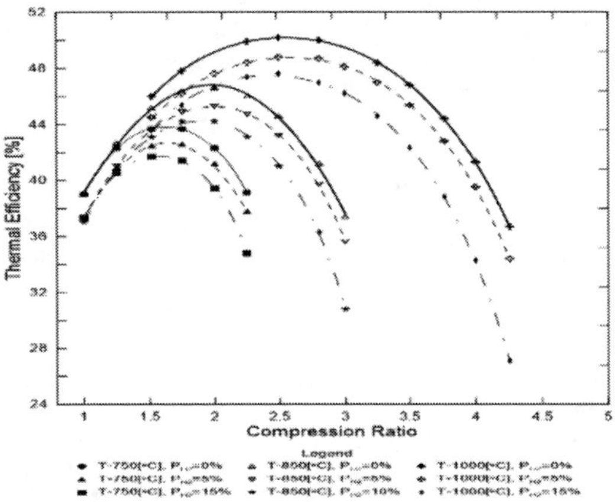

Figure 9. High-temperature hybrid combined cycle – thermal efficiency vs different nuclear reactor outlet temperature, compression ratio.

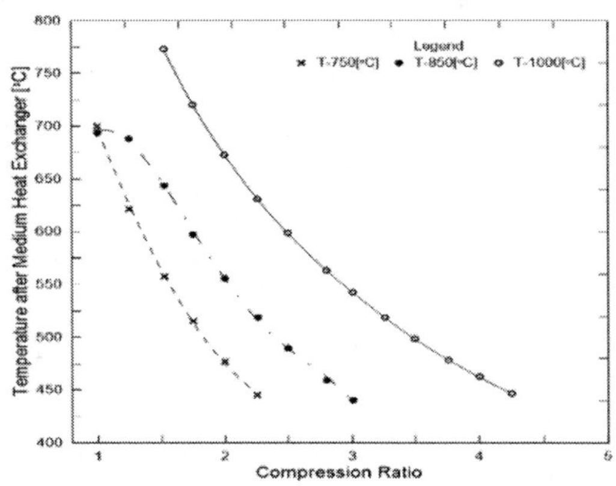

Figure 10. Temperature available after medium heat exchanger to realize hydrogen production process vs different nuclear reactor temperature and different compression ratio.

for thermal efficiency vs compression ratio and different nuclear reactor outlet.

Figure 10 shows temperature distribution vs different expansion ratio for medium temperature heat exchanger for hydrogen production. A prerequisite for the realization of the thermochemical cooper-chlorine process is keeping the temperature at a level needed to realize Copper oxychloride decomposition reaction at 500–550°C,

$Cu_2OCl_{2(s)} \rightarrow 2CuCl_{(s)} + ½O_{2(g)}$. The analysis shows that it is possible to achieve required temperature when the compression ratio varies from 1.2 to 2 for the lowest available nuclear reactor outlet temperature 750°C.

Conclusions

High or very high-temperature nuclear reactors can be effectively used to produce electricity and hydrogen with the Cu-Cl thermochemical cycle. The power conversion system has been analysed with and without hydrogen production in order to maximize the cycle efficiency. From a thermodynamic point of view, the most appropriate value of the compression ratio for high-temperature gas turbine coupled with high-temperature nuclear reactor should be maintained between 1.2 and 1.8 at the lowest nuclear reactor outlet temperature 750°C. When the outlet temperature from a nuclear reactor is increased up to 1000°C it is possible to work with higher compression ratio what gives the opportunity to increase gas turbine power, steam turbine power, and thermal efficiency. The analysis shows that high-temperature nuclear reactor allows effective realization of coupled electricity generation and thermochemical hydrogen production process. What is more the analysis show that it is possible to achieve a high value of thermal efficiency with and without hydrogen production. When the high temperature combined cycle works without hydrogen production it is possible to achieve 44% at the lowest temperature and much higher when reactor temperature is increased.

The temperatures are not only parameter to improve the thermal efficiency of high-temperature hybrid combined cycle. Another most important parameters are isentropic and mechanical efficiencies of the turbines and the compressor which has significant influence. High-temperature reactors used for electricity and hydrogen production have significant potential to improve efficiency by raising the reactor outlet temperature or steam temperature. Safety and compact unit size, low costs and high temperature may offer a lot of benefits for various applications beyond electricity generation (e.g., district heating). The high-efficiency energy conversion, hydrogen production with a thermochemical process that requires large amounts of heat can be accommodated with HTR or VHTR.

Acknowledgments

The present work was supported by the Polish Ministry of Science (Grant AGH No. 11.11.210.312). Participation in SEED conference was supported by Grant AGH No. 11.11.210.200.

References

1. S.J. Bae, J. Lee, Y. Ahn, J.I. Lee, Ann Nucl Energy **75**, 11-19 (2015).
2. U. Guven, G. Velidi. Proceedings of International Congress on Advances in Nuclear Power Plants, ICAPP (2011).
3. B.J. Marsden, S.L. Fok, G. Hall, International Conference on Global Environment and Advanced Nuclear Power Plants (2003).
4. *Generation IV Roadmap: R&D Scope Report for Gas-Cooled Reactor Systems*, Generation IV International Forum (2002).
5. Z. Zhang et al., *An Engineering and Technological Innovation*. Engineering **2**, 112-118 (2016).
6. *Next Generation Nuclear Plant Research and Development Program Plan*, Idaho National Laboratory, (2005).
7. *International Prismatic Block HTGR Commercial Deployment Meeting Summary in Washington DC*, March 8-10 (2016).
8. S.K. Mylavarapu et al., Appl Thermal Eng **65**, 605-614 (2014).
9. Y. Shin et al., Appl Energy 173, 460-469 (2016).
10. M.M. Rahman, K.T. Ibrahim, A.N. Abdalla, Int Journal of Phys Sci **6(14)**, 3539-3550 (2011).
11. J. Mougin, Comp Hydr Energy, 225-253 (2015).
12. C.F. McDonald, Appl Thermal Eng **73**, 82-103 (2014).
13. Z.L. Wang, G.F. Naterer, K.S. Gabriel, R. Gravelsins, V.N. Daggupati, International Journal of Hydrogen Energy **35**, 4820–4830 (2010).

Dragon bridge - the world largest dragon-shaped (ARCH) steel bridge as element of smart city

Luong Minh Chinh[1], Anna Adamczak[2], Aleksandra Krampikowska[2], Grzegorz Świt[2,a]

[1]*Thuy Loi University, Civil Engineering Faculty, Vietnam*
[2]*Kielce University of Technology, Faculty of Civil Engineering and Architecture, Poland*

Abstract. Dragon Bridge - The world's largest dragon-shaped steel bridge, with an installation cost of $85 million USD, features 6 lanes for two separate directions, 666 meters of undulating steel in the shape of a dragon in the Ly Dynasty, the symbol of prosperity in Vietnamese culture. This unique and beautifully lit bridge, which also breathes fire and sprays water. It's the purposeful integration of the lighting hardware articulates the dragon's form, and the fire-breathing dragon head. This project transcends the notion of monumental bridge with dynamic colour-changing lighting, creating an iconic sculpture in the skyline that is both reverent and whimsical. The signature feature of the bridge was the massive undulating support structure resembling a dragon flying over the river. The dragon is prominent in Vietnamese culture as a symbol of power and nobility. Dragon Bridge stands out as a model of innovation. It has received worldwide attention in the design community and from the global media for its unique arch support system. Dragon Bridge serves as an example of how aesthetic quality of a design can serve cultural, economic and functional purposes. The article presents design solutions of the object and the evaluation of the technical condition before putting the facility into service.

1 Introduction

To realise its full economic and tourism potential, Da Nang (the port city in south central Vietnam) needed a new bridge across the Han River to its developing eastern sectors, famed beach resorts and the route to the historic town of Hoi An. Charged with initiating a bridge plan, Da Nang sought a unique structure that would become an attraction unto itself. Da Nang is a transportation hub and important commercial and educational centre and Da Nang wanted a bridge that would complement and enhance its surroundings (Figure 1).

Figure 1. Picture of Dragon Bridge on the panorama of Da Nang city.

In 2005, Da Nang sponsored an international competition to find a bridge design fitting for Da Nang's imagination. Six entrants presented their concepts to the committee in a public meeting (Japan Bridge & Structure Institute, Inc., Nippon Engineering Consultants Co., Ltd, The Louis Berger Group (LBG) Inc., Egis Bceom International, Kraus & Nguyen Architekten Ingenieure). And the team of Louis Berger and Ammann & Whitney won the competition with a design that caught the Da Nang's imagination (Figure 2). It also provided safe and seamless road access to the city centre on the west end with a low deck that did not block scenic views of the river and city. The signature feature of the bridge was the massive undulating support structure resembling a dragon flying over the river. The dragon is prominent in Vietnamese culture as a symbol of power and nobility.

Figure 2. Final design of Dragon Bridge by Louis Berger and Ammann & Whitney.

For this project Louis Berger and Ammann & Whitney, a Louis Berger company, won a Diamond Award for engineering excellence in the category of structural systems from the American Council of Engineering Companies (ACEC) of New York for the Dragon Bridge in Da Nang, Vietnam. The Diamond Award is the highest honour given for projects outside the state of New York.

Although the dragon shape was a factor in winning the design competition, other matters of a more practical

[a] Corresponding author: gswiter@gmail.com

nature also were important. One key consideration was that two prominent structures- the city's television and cultural museum- are located quite close to the bridge's western end. To avoid obstructing views of these edifices, the design team the navigation channel eastward so that the western end of the bridge would touch down at ground level at the water's edge. In this way the bridge designers achieved the minimum required vertical clearance over the navigation channel without having to construct a large embankment or a high-level approach structure on the bridge's western end.

Opened in March 2013, the 666-meter-long, multi-arch structure features [1]:

- Six 3.75-meter-wide vehicle lanes.
- Two 2.5-meter-wide sidewalks.
- 2,500 LED lights that illuminate the structure at night.
- The capability to "breathe" fire or water to mark special occasions (Figure 3,4)

Figure 3. Breathing fire of Dragon Bridge.

Figure 4. Breathing water of Dragon Bridge.

Dragon Bridge lights up the more than 900 LED luminaires. Great, golden dragon looks quite impressive, especially at night (Figure 5). Every year come to see it approx. 3 million tourists from all over the world. Additionally, it has improved the economic situation of the whole region, because tourism translates into increased revenue.

The intelligent LED lighting solution painted the iconic dragon with a myriad of colours that can change in 5 different colours, customizable for any occasion, festival or holiday. The eye-catching design of this lighting solution clearly positions the Dragon Bridge as a gateway for Vietnam and the rest of the world to the city of Da Nang. Lighting solution brought to life the design of the dragon that seems to leap out of the Han River through a stunning display of colours, with lighting that is both environmentally sustainable and energy efficient [2].

Figure 5. Dragon Bridge at night.

2 The solutions

The main bridge is an arch structure consisting of the 5 major spans, the pylon from P0 to P5 (64 + 128 + 200 + 128 + 72)m = 592m (Figure 6) [1]. Concrete extreme spans (64 m) on both sides have varying heights of 1,775m - 3.5 m. Centre span (200 m) is a steel structure with a fixed height of 3.5m. While the indirectly span of length 128 m (2 and 4) is a steel structure with a constant height of 3.5m. In places above supports and pylons there is prestressing concrete structure with a fixed height of 3.5m. On both sides there are flyovers leading to the main bridge. They are consisting of prestressed concrete beams with a length of 24m and 26 m [1].

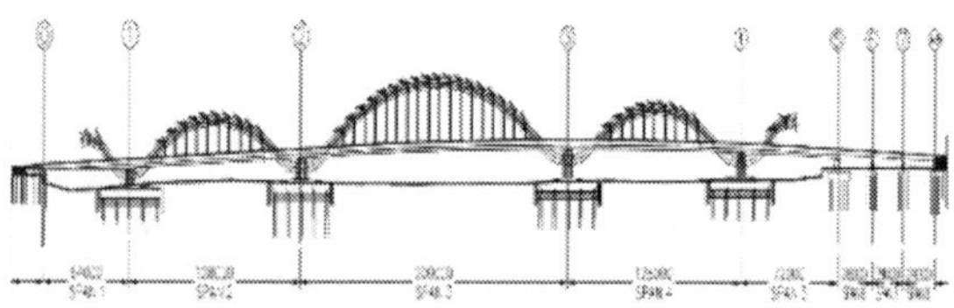

Figure 6. General plan of Dragon Bridge [1].

2.1 Main girders

Steel spans (Figure 7) of three consecutive spans between pylons 1 and 4 have a width of 14m and are supported on the concrete pylons and on a special lower arches which form in this way a framework support for the upper element. The main building spans are suspended by hangers to the main upper steel arcs. To enlarge the width of the roadway designed steel support (so-called. wings) on both sides of the main beams of steel with a width of 10.5 m. However, in the reinforced concrete beams further expanded width of 0.5m - 1.25m on both sides to create the changing width of the bridge along the sidewalks, which is an interesting element of design and aesthetic of bridge. Concrete spans (Figure 8) are supported on pylons by bearing neoprene (rubber-steel) which are also provide support for the upper arches and steel girders. Designed additional transverse prestressing strands of the bridge deck with the "wings" which increases the stiffness of the bridge and also increases the capacity of the bridge particularly in terms of negative moments bending bridge deck.

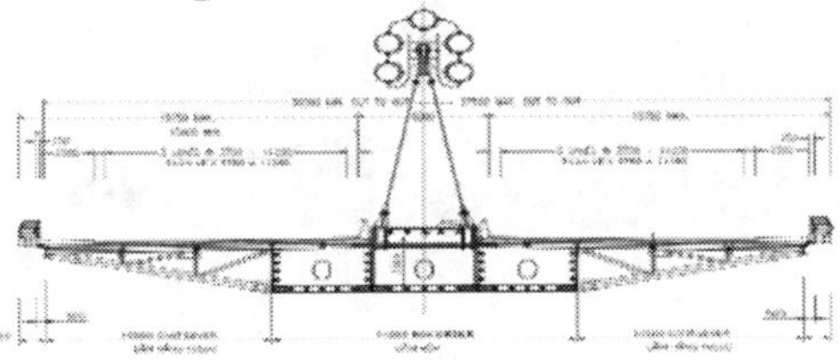

Figure 7. Cross section of steel girder.

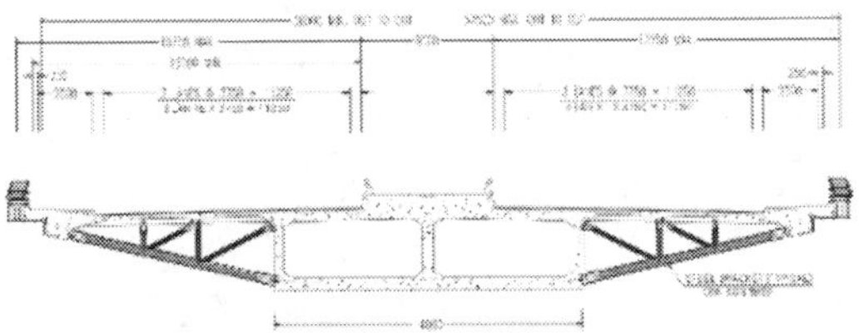

Figure 8. Cross section of concrete girder.

2.2 The bridge deck

The bridge deck is a hybrid structure, consisting of posttensioned concrete sections above the piers and steel box girders section in the suspended portions of the spans (Figure 9 and Figure 10). The two types of sections will be posttensioned together to form a continuous structure. The superstructure is 36m wide, accommodating ample shoulders and sidewalks in addition to three lanes of highway traffic in each direction. The arches have only one rib, but for aesthetic reasons that rib is made up of five steel tubes. Steel plates and connect the tubes, which measure 1.2m in diameter. The arch rib penetrates the bridge's superstructure in the centre of the roadway in the space provided by a 6 m wide median.

Figure 9. Prepare steel girder to transport.

Figure 10. Cable anchor at steel girder.

2.2.1 Joints between concrete and steel elements of span

In normal structures between the spans there is dilatation, but the Dragon Bridge has a rigid connection. It creates continuous and rigid manner hybrid span consisting of concrete and steel parts alternately. To create the connection to the cross-section designed 64 strand tendons between the structural elements (Figure 11 and Figure 12).

Figure 11. Cable tension in joints between sections.

Figure 12. Prestress cable between steel and concrete section.

2.3 Steel arches

The main arches are the steel structures of the five pipes with a diameter of 1200mm and a thickness of 19mm each. To the centre of the arch pulled a concrete to increase the stiffness of it. Steel pipes are fixed to each other by means of a specially designed steel fastener for every 8 m. In the areas of contact between the steel pipe and the connecting element is a layer of neoprene rubber to cushion the influence of spans transferred to arches by steel hangers (Figure 13). They are an added plus, ie. reduce the bending moment in arches, which has been tested in the calculation. Each steel hanger consists of three steel bar having a diameter of 65mm type Hi-Am with a high capacity to fatigue. On both sides of the hanger there are special heads for attachment to the arches and to spans with a diameter of 350 mm. On the one side they are hung between tube connectors and on the other side are attached to spans by screws.

DOI: 10.1051/e3sconf/20161000106

Figure 13. Steel hangers.

To provide high load capacity of arches, they were made by methods:

- CNC plasma cutting/Flame cutting (Figure 14)
- High frequency bending and welding.

Figure 14. CNC plasma cutting/Flame cutting - High frequency bending and welding.

To ensure the best protection for hangers against external influences and corrosion chosen the following solutions:

• Galvanizing and further applying a protective coating on the steel cables
• Hangers are made with external additional coating PE

The ratio of arch height to the length of the normal arch bridges are the values 1/4 ÷ 1/6. For Dragon Bridge this ratio is 25/200 = 1/8 through the use of a continuous bonding between the elements of the girders, which creates a continuous span. In addition, three steel arches are the main load-bearing element of the bridge with the framework support so be able to use the arch of small height (Figure 15). The ratio of height to length of the girder in the bridge is 1/57 which much closer to the upper limit specified in the standard (normal, this ratio is between 1/30 and 1/60). Due to the above mentioned solution was able to firstly reduce the height of the bridge arches to fit in the designated areas on both sides of the river, but still could maintain adequate clearance for maritime traffic under the bridge. Simultaneously raise the aesthetic value of the bridge.

Figure 15. Construction of steel arch [foto by Ehrin Macksey/Noi Pictures].

Concrete elements of girders were poured on site, later installed steel element after reaching the proper stiffness of concrete.

2.4 Pylons of bridge

The concrete pylons are connected with the lower arches (ie. the belly of the bridge) and are poured on site. Pylons 1, 2, 3 and 4 have a width of 7.5m which allows the structure to move lateral forces and large bending moments. To reduce the dead weight of the bridge part of the pylon are empty and have a partition in the middle, but the upper part of the pylon connecting with the lower arches concrete structure is compact.

The upper part of the bridge is relatively low, because a large height of girders enforces low height of pylons. Pylon of main bridge consists of an upper part and a lower portion at the base of the pylon, which is an element connecting pylon with piles drilled with a large diameter and a depth of 45m. Upper proper grade of pylon's base is lower than the lowest recorded water level of 1m. Loads from the top of the bridge are transmitted through the pylons to the foundation by means of neoprene bearings and by a rigid connection between the upper steel and lower concrete arches, which are able to bear not only the imposed loads and also loads due to external factors (such as wind, temperature, shrinkage of concrete and a seismic event).

To lessen the influence of the collision of vessels plying the river to the lower parts of the bridge, designed to separate the steel columns filled with sand and gravel in front of the pylons on both sides of the opposite directions of traffic. They also represent the prevent erosion elements of the bottom of the river bed at the base of the foundation of the bridge.

3 Assessment of structure state

Before putting the facility into service on March 29th, 2013. Da Nang Transport Department commissioned to independent design office to evaluate the technical condition of the object. Was found a lot of scratches with a width of 0.1 to 0.2mm mostly in places of contact technology at the base of arches and the main girders [3-5]. Perpendicular cracks are also present on the walls of the front of eastern abutment of the bridge (Figure 16). During the study engineers found that cracks are "normal" and accepted ranges of code, mainly there are cracks caused by concrete shrinkage [6,7]. But advocated monitor and regularly check the development of cracks in places prone, especially in connecting elements between the steel girder and concrete, in the middle of the spans, in concrete arch to be possible to prevent unexpected symptom [7,8].

In addition to the above, the independent engineers claimed that:

• During the tests the bridge operates in the elastic, there is no abnormal deviations.

• The values of stress and stiffness of the object are accepted in intervals of code.

• Displacement peak of arch and its amplitude of the own vibrations are consistent with theoretical calculations.

• The bridge is able to bear the load utility provided in the standards.

Figure 16. Cracks appeared in the abutment after construction.

The final verdict is that the cracks caused by concrete shrinkage and temperature fluctuations during construction. During the study there were no changes in the created scratches and they do not influence the normal operation of the facility.

Figure 17. Injection epoxy glue to fill the cracks.

The object is still within the warranty period and the contractor of the bridge is intended to eliminate of scratches, monitor and supervise of development of the resulting cracks and the emergence of a new scratches. Simultaneously prevent the influence of refrigerants external on quality components and concrete construction of bridge. Adopted the following treatments:

• injecting epoxy adhesive Sikadur 572 high strength and low viscosity to visible scratches. Application of epoxy mortar Sikadur 731 in places of small cracks (Figure 17)

• cracks occurring in concrete technology breaks at the base of arch and the abutment in the main spans will also be secured in a similar manner by means of epoxy adhesive Sikadur 572 and outer finishing mortar Sikadur 731.

• the contractor must forge outer layer of concrete and pour additional concrete layer with a capacity of 30MPa outside, near the joints to create a sloping prevent collecting water at the base of the abutment.

• in places of soaking water, workers must forge outer layer of concrete soaked with rainwater and apply a layer of special protective of Lemax 7WP (1 liter / 5m^2).

References

1. Project documentation of bridge
2. http://www.ledinside.com/lighting/2013/11/dragon_bridge_in_da_nang_vietnam
3. B. Goszczyńska, G. Świt, W. Trąmpczyński, A. Krampikowska, *Application of the acoustic emission (AE) method to bridge testing and diagnostics comparison of procedures* (IEEE Proceeding Prognostics and System Health Management, 1-10, Bejiging (2012)
4. P. Olaszek, G. Świt, J. Casas, *Proof load testing supported by acoustic emission. An example of application,* Bridge Maintenance, Safety, Management and Life-Cycle Optimization: Proceedings of the Fifth International IABMAS Conference, 1-133, Philadelphia, USA, 11-15 July (2010)
5. G. Świt, *Diagnostics of prestressed concrete structures by means of acoustic emission* (Reliability, Maintainability and Safety, 2009. ICRMS 2009. 8th International Conference on IEEE, 958-963 (2009)
6. G. Świt, A. Krampikowska, L. Minh Chinh, A. Adamczak, *NhatTan Bridge – the biggest cable-stayed bridge in Vietnam*, Procedia Engineering (to be published)
7. G. Świt, *Predicting failure processes for bridge–type structures made of prestressed concrete beams using the acoustic emission method*, Publishing Kielce University of Technology, 1-178 (2011)
8. B. Goszczyńska, G. Świt, W. Trąmpczyński, Bulletin of the Polish Academy of Sciences Technical Sciences, **63**, 1 (2015)
9. B. Goszczyńska, G. Świt, W. Trąmpczyński, Bulletin of the Polish Academy of Sciences Technical Sciences, **61**, 1 (2013)

 DOI: 10.1051/e3sconf/20161000107

Unconscious aspects of sustainability in modern and postmodern religious architecture in Poland

Joanna Gil-Mastalerczyk[1,a]

[1]*Kielce University of Technology, Faculty of Civil Engineering and Architecture, Al. Tysiąclecia Państwa Polskiego 7 , 25-314 Kielce, Poland*

Abstract. The emergence in recent decades of the idea of sustainability and sustainable development significantly affected the innovative approach to architectural design and urban planning. The article presents contemporary examples of the Polish works of sacred architecture and urban planning, which prove the unconscious use and incorporation of the concept of the architectural values of the environment, including climate, solar energy and other natural elements such as landform, structure or green water. The analysis of the contemporary examples of architecture, realized in the second half of the twentieth century, confirms that such actions are taken. The search for a friendly urban space combines the comfort of living with the comfort of the natural environment. The contemporary objects and sacral complexes testify to integrated planning with nature. The system of architectural and urban planning objectives is clearly reflected in the functioning of thecity and how to organize the use of space, including an attempt to rebuild a sustainable relationship between the infrastructure and the natural environment. When designing the contemporary forms, sacred has become very important in addition to the historical past, the identity of individual urban structures, at the same time, maintaining the high quality of the environment. Projecting of harmony with the local climate and the use of natural conditions in urban planning.

1 Introduction

The contemporary architecture and urban planning impose environmental measures. Sustainable architecture has three main areas of conditions: environmental, social and economic. The Ecological Urbanism introduces principles of ecology to the urban planning and emphasise the social inclusion and sensitize to the environment. On the other hand the idea of design in the vein of ecological urban planning is based on the interaction of the human measures and natural processes of the city (air, earth, water and life, ecosystems), ie. the adjustment of the city to the natural processes [1].

Harmonious relationships between created environment and the nature, tender different scales and forms, serve to human comfort as well as settle the unique quality of each architectural style and urban composition. Mentioned relationships enrich the urban space with a new values, improving its health and aesthetic benefits as well as its attractiveness [2].

In the context of implementing the concept of sustainability and sustainable development, the intensity of the integration between the urban fabric and the environment may be one of the most important criteria for assessing the quality of the contemporary urban space.

Relationships between the architecture and the nature can be seen in various aspects and scales. The rating depends on the integration of the architectural and urban planning plank with the environment and the landscape. It also depends on analysis of the undertaken actions on a small scale, such as the inclusion of various forms of nature in the architectural development [2].

The arrival of the idea of sustainability and sustainable development in recent decades significantly affected the innovative approach to architectural design and urban planning. These ideas have an impact on the new view into the modern city forming.

The achievements in the field of applied technology - in terms of its compliance with the natural environment have been changed.

This article presents examples of creating harmonious relations linking the created environment with the nature. It also raise a question of selected issues related to sustainable architectural and urban planning sacral ensemble designed - in Malopolska (Poland).

The unity of architecture and environment - maintaining a balance between architecture and natural ambient or shaped by a human - has an positive impact on the quality of the surrounding environment. Relations with the surrounding architecture are defined by the functional, spatial and esthetic interassociations. functional, spatial and aesthetic.

It`s regarding the relationship between architectural form and the shaping of the ambient space [3].

[a] Corresponding author: joanna.gil@onet.pl

2 The aspect of sustainability and sustainable development in the urban planning shaping – exampled on modern sacral architecture

In 1987, in the report "Our Common Future" (known as the Brundtland) prepared by the World Commission on Environment and Development (UN) the concept of sustainable development was defined It was found that the rules of sustainable development should be implemented by all countries. Only then it will be possible to meet the needs of present and future generations [4].

Sustainable architecture is one that is realized in accordance with the principles of sustainable development. The strategy of sustainable development encompasses three dimensions - environmental protection (ecology), social inclusion and economic growth as well as seven key challenges in the area of economic, ecological and social:

- *Limiting climate change and promoting clean energy,*
- *Ensuring that transport systems meet the requirements of environmental protection,*
- *Promoting sustainable patterns of production and consumption,*
- *better management and prevention of excessive exploitation of natural resources,*
- *promoting high-quality public health,*
- *creation of a society based on social inclusion, and ensuring high quality of life,*
- *active promotion of sustainable development and to ensure the compliance of the EU in this aspect* [5].

Analysis of examples modern and post modern Polish sacred architecture (80s and 90s of the twentieth century), reveals that the realization of these features (admittedly unconscious) formation sustainable environmental future. The creators of the buildings did not take any attention to achieve integrated governance - social, economic, environmental and spatial planning [6]. However, contemporary religious assumptions (implemented in Poland in the second half of the twentieth century), testify to the use of urban values of the environment and integrate into the architectural and urban planning concept. There are various forms of wildlife, such as climate, solar energy and other natural elements like - topography, the structure of greenery and water which are betaken.

Sacral architecture has become one of the most recognizable elements of the landscape of Southern Poland (Region of Małopolska). Natural lay of the land aids the exposal of number of the modern realizations, which have predominant architectural features of the local, artistic and cultural importance.

In the process of architectural and urban planning design the landscape becomes an important part of the context. The urban context loom large. It eventuates from elicitation" the place identity" and "it is an element of inspiration for creative investment" [3]. Instead of that, building`s forms often have the regional connotations.

Contemporary sacral assumptions are featured by the attempt to rebuilt the sustainable relationship between built and natural environment. Areas of specific cultural and natural values (due to geological, geographical and historical conditions) can have a significant impact on the contemporary forms and spatial layout creation of groups and single sacral objects.

2.1 Sanctuary of St. Joseph the Guardian of Family in Kielce (Sanktuarium Świętego Józefa Opiekuna Rodziny) – complement of the urban and public space, public-oriented, combining recreational and cult functions

The post-conciliar form of Sanctuary of St. Joseph the Guardian of Family in Kielce (1975-1995, designer W. Pieńkowski), in plan, refers to the rich geological past of the region of Kielce. The fifth elevation of monumental solid of the church recales its shape imprinted shell in the rock. The monumental concrete solid of the church dominate in a special way in the natural landscape of the city and is a clear sign of this place (Fig. 1).

The Sanctuary constitute the important place in the city. It is surrounded by housing and leafy territory it provides quietude, safety and proper coexistence between nature and urban structures to the public.

Native limestone sourced from the quarries of Kielce Region played the key role in the way of shaping the solid of the Sanctuary[a]. Sourcing a limestone is associated with centuries-old history of the exploitation of primary commodities of the rocks. They are offering stone coming from different geological ages extracted from still active mines.

The use of the Pińczów limestone deposits can be observed in the outside architecture of the temple - as a building material and also used for making stone details (Fig. 2-3).

Figure 1. The Sanctuary of St. Joseph the Guardian of Family in Kielce, general view of the sacral solid (photo by J.Gil-Mastalerczyk, 2014).

The whole mold is topped with a roof, with the slope referring to the neighboring hills.

The natural surroundings of the church, including the local conditions - a significant difference in levels, a huge number of leafy territory has influenced on the final shape of the building. They had an effect on the

[a] The Quarry - area Chęcin, Słopca, Sukowa, Tumlin, Pińczów [7].

functional and spatial arrangement of routes. Several entrances using the natural environment and the different decline lead to split-level interior (Fig. 2).

Figure 2. Sanctuary of St. Joseph the Guardian of Family in Kielce, views of solid, blocks, sequence of paths (photo by J.Gil-Mastalerczyk, 2014).

Figure 3. Sanctuary of St. Joseph the Guardian of Family in Kielce, the architecture of the exterior walls – Pińczów limestone of stones' detail (photo by J.Gil-Mastalerczyk, 2014).

The facades of the church designed in the form of undulating walls are inspired by the natural lay of the land. Also the interior organization of the temple, just as the architecture of sacral group of buildings, clearly refers to the character of the natural landscape. The advantages of natural environment are used by the spatial organization and the architectural detail. The glazing windows - in the shape of colored stained-glass windows with its composition and divisions, recall the branches of trees filtering to the interior of the solid the colorful light. The external light falling down through the huge glazing is creating the metaphorical air of the inside.

Figure 4. Sanctuary of St. Joseph the Guardian of Family in Kielce, view of the interior - natural light coming from the top to the inside of the temple (photo by J.Gil-Mastalerczyk, 2014).

The natural light has become a factor in shaping the image and specific mood of the object (Fig. 4).

The use of large glazing has a beneficial effect on the level of natural lighting and the aspect of saving energy. Glazing used in external walls play an important role in the development of energy-efficient architecture. Appropriate use of glass also affects the improvement of the energy balance of the building.

Interconnection of materials, their texture and color as same as relationship with natural environment has become an important factor in shaping the solid of the church and its perception. Regional materials, advantages of natural environment and landscape context played the key role in shaping the modern form, despite of using reinforced concrete construction.

The use of local stone materials (limestone, granite) with a long shelf life has an economic dimension. The availability of local raw materials - in economic terms - reduces construction costs and provides speedy transport.

The solid was integrated into the natural environment and was bed in relationship with high and low leafy neighborhood (bushes and well-kept lawns).

Regional materials, used by the creators, are featured by the creative adaptation to contemporary forms while respecting the identity of the place. The architecture of the church is the result of a dialogue of modernity with tradition of the region and the surrounding nature.
It creates a welcoming space for the community.

In summary, the whole founding of the architectural and urban-planning of St. Joseph the Guardian of Family in Kielce is a representative example of contemporary ideas, which are exposing the value of regional traditions. They have a positive influence on the architecture of sacral objects, their dialogue with the environment and the way of functioning in the surrounding area. The authors of the project offered modern architecture which is a continuation of the modernistic tradition and which is harmoniously united with the natural landscape of the region of Kielce.

The building in Kielce represents contemporary Polish sacral architecture with high aesthetic values – for example the way of using advantages of the surrounding landscape. The church is a part of it.

It should be emphasized that features such as integration with the landscape, take into account the cultural context, the design of large green areas, the implementation of the social - characterized by today's investment of balance. The compound creates the composition of the public space attractive in terms of natural surrounding. The building is important for the status, image and urban-planning composition of the city.

In general, it should be noted that the Church has the characteristics of building a sustainable environment. Building creates a safe and healthy environment for the community staying. Rewarding users from staying in it and a pleasure to contact the outside for passing. The architecture of the church reflects the social and cultural aspirations of society. The building does not have a negative impact on environmental pollution (no hazardous waste). In shaping the architecture of the building it becomes visible aspect of the use of solar energy - provided by your desired orientation of rooms and glass partitions in relation to the world.

In the 80s and 90s of the twentieth century, when the building was designed, did not focused on the problem of energy efficiency. The introduction of renewable energy sources or activities preferring ecological transport.

The solutions reduce water consumption, saving raw materials are not taken into account, innovation and quality of design solutions, at this time (according to the principles of sustainable development).

The creators of the object represented a lot of attention to the integration of the church with the environment and related to the comfort of use. The architecture of the church, particular attention was paid to regional priorities (the use of local materials - stone, tradition of building) and the ecological values of the area.

At present, according to the standards, with an emphasis on issues of sustainable development (LEED- Leadership in Energy and Environmental Design, USA; GBC - Green Building Challenge, Europe, Japan, Canada, USA, BREEAM-Building Research Establishment Environmental Assessment Method, United Kingdom; BEPAC-Building Environmental Performance Assessment Criteria, Canada) the buildings should be considered in the full cycle of operation (life cycle). Important become all stages, such as stage design, implementation, operation building and its utilization [5].

2.2 The architectonic sacral complex of St. Peter`s the Apostle in Wadowice - a harmonious fusion of the cult places with the socio-recreational function and whit the landscape

The architectonic sacral complex of St. Peter`s the Apostle in Wadowice (1986-1991, designer Gyurkovich J., E. Węcławowicz-Gyurkovich, P. Szafer), is an example of organized urban space using the values of the natural landscape, while becoming a part of it. The project is a part of a typical natural conditions of the location, a context of the existing buildings and the context of the existing landscape.

It should be noted that the designers have taken action (though unconscious) for sustainable development at the regional level. At present, sustainable development is presented in relation to three (economy, society, environment) or four (in addition spatial order) dimensions and relationships between them. The elements of the natural environment can be replaced besides the resources (soil, water, vegetation, animals) values of intangible and difficult to quantify, such as landscape or space (relevant and important from the point of view of economics).

The sacral complex has been surrounded by modern housing estates from two sides. From the side of the city is in immediate vicinity of the old cemetery, set on the background of beskid Maly slope`s, which is towering above. *The amorphous space of "modern" housing state provoke the need to create clearly defined space, referring to the historical traditions of monasters` compexes which are separating sacrum and profanum, from the urban principles that form building lines and urban interiors - streets, public squares and parks* [8].

The sacral complex of the buildings has an axial composition. The church of St. Peter`s the Apostle (on the traditional Latin cross plan, with a longitudinal layout) was located in the center of it and was surrounded by a "wall" of didactic, social and residential buildings, referring with its composition to the complexes of historical monasteries. *The "wal`sl" line is continuated on the south side by a double row of trees* [9] (Fig. 5).

The gate - in the form of a triumphal arch (which has not been done) reffering to the form of western elevation and the dome of the Vatican Basilica - was designed in the axis of the complex [10].

The whole complex also includes the freestanding towering tower, the chapel of Christ the King, the building with rooms to learn (converted to residential[b]) and also situated behind the church and using the natural

[b] Author's interview with E. Węcławowicz-Gyurkovich, Krakow, 11 March 2013.

lay of the land – The Way of the Cross, also known as "The Wadowice`s Golgotha"[c] (Fig. 6).

Figure 5. The sacral complex of buildings of St. Peter`s the Apostle in Wadowice, the solid view of the church from the green hills (photo by J.Gil-Mastalerczyk, 2013).

The architecture of sloping roofs or small scale of detail are a continuation of the local, regional traditions. The composition of the processional way with symbolic gates refers to the architecture of the old streets of Krakow, and the internal and external colonnade repeat the atmosphere of the Vatican [11] (Fig. 7).

The presented characteristics indicate on application of the principles of sustainable development (respect for the terrain and the landscape context, integration with the landscape, use of local building traditions, the implementation needs of the community and enhancing social cohesion, the use of safe materials, proper orientation of the building, windows and rooms - impact on energy savings).

The organization of the postmodern church`s interior uses natural advantages of the environment. The natural light inside the temple has particular importance. It is inextricably linked with the imagery of the sanctuary.

The natural light is an important factor in forming the image of the building - not just architectural image. The effect of natural light plays a key role in creating and reading the interior of the temple. Conscious creation of light, as one of the of the natural environment parameters, brings structure`s harmony, underlines the space and perfect form`s proportions.

In summary the sacral complex of buildings of St. Peter`s the Apostle in Wadowice, is an example of a new architectonic creation, while respecting the quality of traditional architecture and landscape context.

The spatial and architectonic solutions of the sacral complex of St. Peter`s the Apostle come from the interpretation of the overall conditions of the location. The compositional structure of the whole architectonic complex set *"a strong form"* of amorphous space of the surrounding housing estate [8].

Figure 6. The sacral complex of St. Peter`s the Apostle in Wadowice, the view of "Wadowice`s Golgotha" (photo by J.Gil-Mastalerczyk, 2013).

Figure 7. The sacral complex of St. Peter`s the Apostle in Wadowice, the walking paths, "processional way" (photo by J.Gil-Mastalerczyk, 2013).

The designers have offered a modern architecture, open to the natural environment and harmonically weld with the landscape [11]. These are important elements in the process of sustainable development. According to A. Baranowski balancing relationship "built environment - natural environment" is one of the primary aims of sustainable architectural design [6].

All buildings, located on a meadow close to an old river basin of the Skawa evoke in perception from different sides aesthetic impressions marked with the sacral imagery and the cohesiveness of the composition with surrounding environment [12].

It was uniquely integrated with undulating land blurring the institutional boundary between the sacrum and the profanum. The solid form a contemporary interpretation of processed simple geometric forms creating a distinctive atmosphere in the existing urban fabric and also facilitating finding in the structure of the city an important public space, which combines the social and recreational functions with buildings of religious worship. Owing to that, the complex does double duty as the parochial center, votive places of worship and it provides live comfort and convenient contact with the nature.

The complex has good composition in the ground. The church in a creative way compose a new place of

[c] The whole sacral complex has a deep symbolic dimension. According to the authors of the project: (...) *Processed Wadowice`s, Cracow`s, and especially Rome`s architecture quotes, included in this complex are (...) to symbolize the Way of the Great Pole to the Holy See* [9].

DOI: 10.1051/e3sconf/20161000107

symbolic shape subserving to both liturgical and secular purposes.

Figure 8. The sacral complex of buildings of St. Peter`s the Apostle in Wadowice, the view of the interior (photo by J.Gil-Mastalerczyk, 2013).

The whole complex significantly influenced on changing the layout of visual city elements and has become a modern symbol of the sequence of the city. It is an example of changing the existing architectonic patterns and creating other formal solutions in contemporary sacral architecture. The place fully integrated with the natural terrain has been created to provide users` comfort with the comfort of the natural environment. The sacral buildings have become a part of the space in which they exist. Through its construction material and its size they dialogue with the surrounding streets and buildings. A wide context of the environment, landscape and larger fragments of the nature are part of the construction project.

In conclusion it should be noted that the assumption sacred in Wadowice was formed, taking into account (though unconscious) sustainable three fundamental aspects of development: social, environmental and economic.

Social aspect: extensive social participation - in the process of use; increasing social cohesion through the implementation of social space, new public spaces (character sacred and secular) have the ability to produce human coexistence – in a diverse open space and architecture; formation new meeting space - with different activities; development of safe, open structure of the city through a smooth transition between public space and private, attractive external program stimulates a rich social life; the use of safe materials (respect for the user).

Ecological aspect: the introduction of diverse habitats in the landscape of the city – development of various forms and green areas allow to exist rich flora and fauna in the city center; integration with the landscape and taking into account the cultural context; increase the area of biologically active; in the area of the new structure is planned pedestrian and bicycle traffic (environmentally friendly); building structure with traditional materials (non-toxic);

Economic aspect: the introduction of functional diversity and spatial; Buildings (secular-residential, educational, religious) and recreational areas were built with the aim of perspective functioning; use of materials and technical solutions with a long durability period.

Unfortunately, it is does not apply to new methods and effects, which should be of high quality technical solutions. The lack of new technologies 3Rs (reduce, reuse, recycle), energy savings and use of alternative energy sources and minimizing pollution.

3 Conclusions

In the present time (in the threat environment and economic crises) the role of architecture and construction in the creation of sustainable development is very important. Architecture and construction represent largest sector of the economy in the economic aspect and in terms of the flow of raw materials. In building invested the largest amount of financial capital and mineral resources was invested. One of the main aim in the construction industry becomes improve the quality of life, which should be permanent in nature, should involve current and future generations and should not cause degradation of the natural environment. In contrast, the activity aimed to meet the needs should take into account environmental protection as a condition for proper management.

According to the so called new economy *environmentally sustainable development aims to ensure that all people living today and future generations sufficiently high standards of environmental, economic and socio-cultural limits of the natural strength of the Earth, applying the principle of intra-generational and intergenerational justice* [13].

It should be emphasized that the concept of sustainable development is assumed to maintain a balance between the general dimensions (social, economic, environmental). None of them should be treated as a parent. However, in practice, it seems difficult to obtain.

Professor G. Sznajder-Skalska claims that among scientists and practitioners (in the field of architecture and urban planning) there is a consensus as to the general features characterizing investments balanced. *These are investments in which the following principles:*

- *Respect for the area* (taking into account the cultural context, the use of local materials and traditional building (...), integration with the landscape, increasing the area of biologically active (...)),

- *Respect for the user* (realization of the needs of each individual, high quality architecture and space (...)),
- *Reduce, reuse and recycling (Rule 3 r),* covering a group of actions that result should be saving and re-use of area, the volume and materials (...),
- *Energy efficiency* (use of efficient insulation, proper orientation of the building and facilities, minimizing heat loss and passive solar energy gain (...)),
- *The use of alternative energy sources* (use of high-quality technical solutions in order to obtain solar, wind, geothermal sources (...)) [14].

In the presented examples (Wadowice, Kielce) many of features of sustainable investment may be foud. Although adopted (unconsciously) solutions in the 80s and 90s were not seen in terms of design and sustainable development.

The arrival of the idea of sustainability and sustainable development in recent decades significantly affected on the innovative approach to architectonic design and urban planning. The analyzed examples characterized by spatial and ecological variety of solutions, as well as a harmonious relationship with the landscape and the immediate surroundings.

They are distinguished by the high quality of architecture and the systems of composition, friendly attitude towards the private and the public space, a strong relationship with the nature. It creates all together integrated totality – harmoniously integrated architecture in environment creating a harmonious relationship with nature and built-on area.

Diverse and complementing functions harmoniously coexisting in the urban structure provide environmentally sustainable development of urban area. *Opening the architecture on the natural environment combines activities in the economy, ecology and society. (...) It implements the ideas of forming the economical city, among others through cost-effective management of the terrain and resources of existing buildings (...), and natural resources: water, raw materials, materials (...)* [15].

The expressive continuation of the landscape and traditional sacral architecture of region Malopolska is presented in buildings` architecture. The landscape context and confrontation with the environment of analyzed buildings had a significant impact on making the constructions` decisions by the designers. The architectural forms have been harmoniously integrated into the natural landscape at the same time not exerting a strong contrast on the present buildings.

The architectonic approaches stem from the landscape`s features, especially its scale. They were developed as a result of creative transformation process and interpretation of the local forms.

In the contemporary solutions regional elements of architecture, such as a silhouette, slope roofs regard to the climate conditions – were used.

In summary, all presented contemporary buildings and sacral complexes attest about an integrated planning with the nature. The system of architectonic and urban planning projects clearly reflected in the way of functioning of the city and in the way of organizing and using the space, including an attempt to rebuild a sustainable relationship between built environment with the natural environment.

The concurrent maintaining of a high quality of natural environment instead of the past and particular urban structures` identity has became very important in designing contemporary sacral forms. Design stays in harmony with the local climate and the use of natural conditions in urban planning.

It should be noticed that the presented buildings of contemporary sacral architecture attest the high sensitivity of the designers on the context. The shaping environment related to religious building, is to be performed with respect for public space natural and cultural landscape with the use of contemporary solutions, functional and spatial, technological, and material creates the optimal conditions for the society. The analysis of contemporary examples of architecture, done in the second half of the twentieth century, confirms that such actions are taken. They express a guest for a friendly urban space, which combines the comfort of living with the comfort of the natural environment. The buildings are created in harmony with the natural environment and respect the conditions of the environment. This applies to relationships of coming into being buildings in the urban fabric with both the natural landscape and the urban space, at which a new quality of the architectonic and urban planning and architecture being in line with the modern times.

References

1. A. Spirn, *Ecological Urbanism: A Farmework for the Design of Resilient Cities* (2012).
2. S. Wehle-Strzelecka, *Współczesne miaston jako środowisko zycia człowieka zintegrowane z przyrodą/ The modern city as a human environment integrated with nature,* Architektura 2/2015, Wyd. Politechnika Świętokrzyska w Kielcach, 5 (Kielce, 2015).
3. W. Seruga, *Jednośc architektury z otoczeniem/ The Unity of Architecture and its Surroundings*, Czasopismo Techniczne Architektura/ Technical transactions Architecture, 3-A/2011, Z. 12, rok 108, Issue 12, Year 107, p. 133-145 (Wyd. Politechnika Krakowska, Kraków, 2011).
4. World Commission on Environment and Development. Our cammon future, the rapport of the World Commission on Environment and Development, (Oxford University Press, New York, 1987).
5. L. Kamionka, *Standardy architektury zrównoważonej jako istotny czynnik miasta oszczędnego na przykładzie wybranych programów certyfikacyjnych* [w:] Czasopismo Techniczne Architektura/ Technical Transactions Architecture, Issue 14, 6-A /2010 Year 107, p. 27-38 (Wyd. Politechnika Krakowska, Kraków, 2010).
6. A. Baranowski, *Projektowanie zrównoważone w architekturze,* p. 44, 89 (Wyd. Politechnika Gdańska, Gdańsk 1998).

7. J. Jędrychowski, *Kieleckie kamienie w architekturze* [w:] Renowacje i zabytki, No. 4 (52), p. 34-40 (2014).
8. J. Gyurkovich, *Znaczenie form charakterystycznych dla kształtowania i percepcji przestrzeni, Wybrane zagadnienia kompozycji w architekturze i urbanistyce*, Seria Architektura, Monografia 258, p. 109-110 (Wyd. Politechnika Krakowska, Kraków, 1999).
9. E. Węcławowicz-Gyurkovich, *Współczesna polska architektura sakralna w twórczości krakowskich architektów. Wybrane zagadnienia*, [w:] Kwartalnik Architektury i Urbanistyki, tom XLI, zeszyt 3-4, Polska Akademia Nauk Komitet Architektury i Urbanistyki, p. 327 (Wyd. Naukowe PWN, Warszawa, 1996).
10. E. Węcławowicz-Gyurkovich, *Postmodernizm w architekturze*, [w:] Encyklopedia Kultury Polskiej XX wieku, od Awangardy do Postmodernizmu, p. 472 (Instytut Kultury, Warszawa, 1996).
11. J. Gil-Mastalerczyk, *O przemianach architektury sakralnej Archidiecezji Krakowskiej w latach 1945-2000,* doctoral dissertation under the direction of Dr. hab. Eng. architect Andrzej Białkiewicz, prof. PK, (Kraków, 2013).
12. T. P. Szafer, *Sacrum Wadowickie,* [w:] Udział pracowników Politechniki Krakowskiej w życiu Kościoła katolickiego za pontyfikatu Jana Pawła II, Flaga K., Paluch M., Pawlicki B.M. (red.), p. 157-162 (Kraków, 1998).
13. H. Rogall, *Ekonomia zrównoważonego rozwoju. Teoria i praktyka,* p. 44 (Wyd. Zysk i S-ka, Poznań 2010).
14. G. Sznajder-Skalska, *Projektowanie zrównoważone – zbliżenie do realizacji* [w:] Czasopismo Techniczne Architektura"/ Technical Transactions Architecture, Issue 3-A/2007, p. 257-265 (Wyd. Politechnika Krakowska, Kraków 2007).
15. S. Wehle-Strzelecka, *Close-to nature housing – austrian examples/ Architektura mieszkaniowa otwarta na przyrodę – przykłady austriackie* [w:] Czasopismo Techniczne Architektura/ Technical Transactions Architecture, Issue 3-A (3) Year 2014 (111)/ Z. 3-A (3) Rok 2014 (111), p. 147-160 (Wyd. Politechnika Krakowska, Kraków, 2014).

 DOI: 10.1051/e3sconf/20161000108

Low-emission economy plans as a chance for sustainable development of municipalities in Poland

Kinga Świtalska[1,a]

[1]*Poznan University of Economics, Faculty of Commodity Science, Department of Commodity Science and Ecology of Industrial Products, Al. Niepodległości 10, 61-875 Poznań, Poland*

Abstract. Adequate low-emission policy should activate society to take action to achieve the benefits of improving the environment. According to the National Programme for the Development of Low-emission Economy, low-emission transformation is a direction for changes of Polish economic model, which could solve the problem of natural resources depletion. The document, that determines the direction of economic transformation, is the Low-emission Economy Plan (LeEP). It is a derivative of the commitments it has made by Poland in the framework of international agreements to reduce emissions. The results of the study will be presented, which include the analysis of the plans of low emission economy for municipalities from Wielkopolska region, which has been developed in order to meet the requirements of the Ministry of Economy and will be used to develop model guidelines on the preparation of such documents. The municipalities have potential to improve energy efficiency and use of resources of local energy sources, as well as the opportunities associated with the management of public and private transport, especially in urban areas. Low-emission economy plan has to be a document that will allow local governments to establish the right direction to a low-emission economy transformation.

1 Introduction

Climate change, rational management of environment and its efficient use are global problems, considered at all levels of management - regional, national and international. The undertaken actions of reducing the harmful substances emissions and stimulating energy efficiency increase, have specially dedicated documents specifying the action in this direction at all levels of coordination. Low-emission policies implemented at the local level should activate citizens to take action to achieve the benefits of improving the environment. On the 10th of January 2007 the European Commission presented energy and climate package, which has set targets for the member states of European Union. The so-called "3x20" package indicates a reduction in greenhouse gas emissions by 20% to 2020 (compared to the base year 1990), increasing the share of renewable energy in final energy consumption to 20% (including a 10% share of biofuels in transport) and increasing energy efficiency by 20% [1, 2].

2 Low emission and low-emission economy

The society is becoming increasingly aware of environmental protection need and to prevent its further contamination. The two definitions of "low emission" and "low-emission economy" became very popular nowadays, but they are often confused, especially in terms of emissions reduction [3, 4, 5, 6].

The concept of "low emission" is usually associated with inefficient combustion, accompanied by the release of large amounts of dust and gases. The term "low-emission economy" is usually linked to concepts such as energy efficient buildings, thermo-modernization, energy efficient transportation and progress in emissions reduction resulting from implementation of the BAT technology to convert fossil fuels to electricity or heat. Implementation of low-emission policy in the municipality has to favourably affect on functioning and actions of community [3, 4, 5, 6].

In majority of Polish municipalities "low emission" is associated with emissions of particulates and noxious gases coming mainly from individual heating systems based on coal derivatives in which coal combustion often takes place in an inefficient manner. A large number of low heights stacks emitters causes great nuisance for population. Contamination accumulates around the site of emission, usually in areas with dense housing development. This process has the highest density during a heating season [1].

3 Sources of international law

Climate change and continuous growth of air pollution has forced the world's leading economies to take actions that are intended to improve the state of the atmosphere and the reduction of greenhouse gas concentration.

The first international convention regarding problems of environment protection and human impact on its

[a] Corresponding author: kinga.switalska@ue.poznan.pl

 DOI: 10.1051/e3sconf/20161000108

transformation was organized in 1972 Stockholm Conference. The second conference of the environment, and the first, which shown an increase of greenhouse gases in the atmosphere as one of the causes of these changes was conference organized in Rio de Janeiro in 1992. The Framework Convention on Climate Change, adopted within the framework of this meeting, imposed an obligation to stabilize greenhouse gases concentrations in the atmosphere at not threatening levels. Then the Framework Convention has been ratified by the European Community by decision 94/69/EC in 15th in December 1993. This decision came into force a year later. During another conference in 1997 in Kyoto detailed arrangements were taken by the United Nations on the elimination of greenhouse gases from the atmosphere. The Kyoto Protocol introduced greenhouse gas emission limits for individual countries.

The European Union conducts a relatively strict environment and climate policy. One of the main objectives set out in the Functioning of the European Union Treaty in article no. 191 is the protection and improvement of the environment, as well as problem-solving nature of the global and regional, especially related to climate change. In the "Europe 2020" document quantitative targets which should be achieved by 2020 has been established.

In 2014 at the European Union climate summit further quantitative targets were presented with the perspective of 2030. These are: reducing carbon dioxide emissions by 40% to 2030 (compared to 1990) and increasing the share of renewable energy sources in the energy balance by 27%. There are also a lot of directives aimed at controlling and regulating the provisions related to the protection of air, climate change and energy saving [2, 7].

4 Sources of national law

Polish law contains a lot of acts governing energy planning and management of pollution produced in the process of energy production. General issues are regulated by the Environmental Protection Law act of the 27th of April 2001. The Act contains general rules and guidelines for air quality, among other things, requiring the maintenance of substances in the air below the permissible levels.

Another source of law is the Act of 15th of April 2011 Energy Efficiency. It was implementation of Directive 2006/32/EC on energy end-use efficiency and energy services into Polish law. The main objective of the Act is the cost of energy, indicating that for 2016 final energy savings will be not less than 9%. It also determines the tasks of the public sector in energy efficiency and introduces a system of energy efficiency certificates.

The Act of 20th of February 2015 Renewable Energy Sources defines a number of objectives, such as increased energy security, effective and rational use of renewable energy sources. It identifies international obligations in the field related to renewable energy sources and their participation in the country's energy balance.

A several national strategic documents related to the subject of low-emission were also created. The main of them is the National Programme for the Low-Emission Economy Development (NPRGN) adopted on the 4th of August 2015. The main objective is the development of low-emission economy while ensuring sustainable development of the country. The Programme has been prepared to transform current economy to a low-emission economy. It is mainly to create optimal model of modern material- and energy-saving economy which is oriented towards innovation and competitive on the European and global market by 2050. One of the measurable effects of this transformation will achieve reducing greenhouse gas emissions and other harmful substances.

The Programme should enable Poland active role in the process of determining the European and global targets to reduce emissions of greenhouse gases and other substances, as well as achieving a comparable level of energy efficiency per unit of GDP to developed countries in Europe. The essence of the program is to achieve economic, social and environmental impacts of actions. Its establishment and implementation is necessary because of the reduction commitments specified in the Kyoto Protocol and the European Union climate-energy package.

The NPRGN target are entrepreneurs of all economy sectors, local authorities, as well as non-governmental organizations and the business environment. Its implementation will facilitate adaptation to the requirements of all low-emission economy industries. The program provides for specific objectives, such as:

- low-emission energy, which addresses such issues as: network modernization and expansion of the power grid, the development of renewable energy, dissemination of alternative ways of generating energy;
- improving the efficiency of materials management (including waste), which focuses on optimizing the use of raw materials and the development of low-emission waste management;
- sustainable production in industry, construction and agriculture with the dissemination of low-emission production technology and improve energy standards in existing buildings (thermal efficiency improvement);
- low-emission transformation in the distribution and mobility, including the development of intermodal transport, support for public and non-motorized transport and introduction of low-emission fuels;
- promotion of sustainable consumption patterns with particular emphasis on educational campaigns on this subject [8, 9, 10].

5 Low-emission Economy Plan (LeEP)

Low-emission Economy Plan is document which determines the direction of the economy transformation. According to the financial perspective for 2014-2020, municipalities which will develop and implement a low-emission economy plans will be able to apply for external sources of financing for tasks related to air protection, energy efficiency and renewable energy [2]. The Low-emission Economy Plan is not formally required by law. It is a result of commitment that Poland has made in conjunction with its international agreements to reduce emissions [10, 11].

Low-emission Economy Plan is a document that use the information about the size of energy consumption and carbon dioxide emissions in the municipality, to achieve the objectives of increasing energy efficiency, reducing greenhouse gas emissions and increasing the share of renewable energy in overall energy consumption. Baseline emission inventory is the first step of the LeEP development. This will help to determine problem areas in the municipality and choose the appropriate action to prevent the emissions. Creating action plan for a low-emission economy requires a constant vision of the municipality development, which includes investment and non-investment activities, spatial planning, green public procurement, education and information. LeEP cannot be treated as a finished document. It changes over time, requires at least once a year to supplement the database and continuous analysis of the municipality operations and development. LeEP must be monitored and if necessary, updated.

Low-emission Economy Plan suggests action in the field of energy management, aimed at reducing emissions and optimum use of available resources. A result of planned actions is to reduce emissions into atmosphere, including carbon dioxide, sulphur compounds, nitrogen oxides and particulates. In the long term LeEP involves action to promote energy efficient products and services, and affecting the change in attitudes of local energy consumers initiatives [12].

The general obligation to reduce CO_2 emissions must be translated into concrete actions and measures, together with the associated reduction in CO_2 emissions by 2020. LeEP is referred to as "equivalent document" in relation to the Sustainable Energy Action Plan (SEAP) by local governments for national and regional regulations on financial support for investments to improve energy efficiency.

LeEP key objectives are:

- inventory of the greenhouse gas emissions sources located within the municipality,
- thermo-modernization of public and private buildings,
- reduction of energy consumption for heating in buildings,
- reduction of energy consumption in public buildings,
- connecting new customers to the gas network leading to progressive gasification of municipality,
- replacement of heat sources for new low-emission,
- increase the share of energy from renewable sources in the overall energy balance,
- construction of the community platform structure for information exchange to improve energy efficiency and installations of renewable energy sources.

The scheme of Low-emission Economy Plan includes:

- summary;
- overall strategy, describing the strategic goals, current state, identify problem areas, organizational and financial aspects (organizational structure, human resources, stakeholders, budget, sources of financing, financial resources for monitoring and evaluation);
- the results of the carbon emissions base inventory;
- actions / tasks and measures planned for the entire period covered by the plan, including long-term strategy, objectives and commitments; short / medium term actions and targets (description, entities responsible for the implementation, schedule, costs, indicators) [7].

6 Purpose of research and preliminary results

The aim of the launched studies is to determine the effectiveness of energy planning with usage of the new tools - low-emission economy plans - in selected municipalities of Wielkopolska region. The analysis will be made using Sectional Evaluation Plans Method (SEPM). SEPM is an original method proposed to evaluate the implementation of municipal waste management plans. For the purpose of this studies Method will be modified in the range of indicators used in the development and implementation of low-emission economy plans. Method is rapid, measurable and representative of municipal LeEPs. On the one hand it has to give the ability to self-esteem municipalities, on the other hand facilitate the control authorities work. The method will also allow to compare the effects of energy planning in the individual municipalities. The most of municipalities in Wielkopolska region have developed LeEP (several documents are just being completed or updated). Still there are 30 municipalities that do not have this document (about 15%). Some of municipalities has received funding from the National Fund for Environmental Protection and Water Management or Regional Fund for Environmental Protection and Water Management in Poznań to develop that document. Currently, existing low-emission economy plans for municipalities from the Wielkopolska region are reviewed for compliance with the requirements of the Ministry. Research will develop future guidelines for the preparation of such documents, as well as demonstrate the necessary of additional components and allows for more efficient energy planning at the level of local government [13].

The previous analysis gives conclusion that low-emission economy plan has to be a document showing the direction of development of the municipality in terms of sustainable energy planning, in fact the content value and the municipalities involvement are unsatisfactory. Documents are drawn up by an external contractor, and the responsible for implementing the tasks personnel do not have the proper knowledge and preparation. It is necessary to engage municipal power engineer, which will be responsible for the assumptions implementation made in the LeEP. There are no adequate and uniform guidelines, the methodology used for estimating CO_2 emissions is not the same and correct in all documents.

As part of the doctoral dissertation arise guidelines for developing a standard LeEP and effective methods of implementation tasks planned in the document in the field of low-emission economy in the municipality.

7 Summary

The municipalities have huge potential to improve energy efficiency and usage of local energy sources, as well as opportunities associated with the management of public and private transport, especially in urban areas. Local governments are the basic units in the implementation of energy policy and climate and energy package guidelines, as well as the objectives of the National Programme for the Low-emission Economy Development. Low-emission economy plan has to be a document that will allow local governments to choose the right direction in transformation towards a low-emission economy terms [12].

Low-emission policies, implemented at the local level, should generate strong incentives to encourage citizens to take action to achieve the benefits of improving the environment. They are connected with reduction of emission nuisance accompanying the combustion of fuels and help the local community to achieve a higher level of awareness in subject of air protection and effective response to observed side effects.

One of the elements of this policy is to effectively obtain financing and overcome barriers related to limitations of information access on the opportunities for reducing low emissions. LeEP have to help in sustainable development of the municipality, also through implementation of the concept "the exemplary role of the public sector" for the coming years [14].

References

1. K. Brzeziński, M. Bukowski, *Niskoemisyjne dylematy. Jak ograniczyć emisję gazów cieplarnianych i co to oznacza dla polskiej gospodarki?* (Instytut Badań Strukturalnych, 2011)
2. A. Hinc, Studia BAS, **1**, 29, 109–136 (2012)
3. K. Świtalska, Logistyka odzysku, **4**, 13, 36-38 (2014)
4. K.. Żmijewski, (Forum Inicjatyw Rozwojowych)
5. Misja Emisja (http://misja-emisja.pl, accesed: 27.03.2016)
6. Niska Emisja (http://nowamisja-niskaemisja.pl, accesed: 05.04.2016)
7. Urząd Publikacji Unii Europejskiej, *Jak opracować plan działań na rzecz zrównoważonej energii (SEAP)?* (2010)
8. Ministerstwo Środowiska (http://www.mos.gov.pl, accesed: 20.02.2016)
9. Ministerstwo Gospodarki, Ministerstwo Środowiska, *Założenia Narodowego Programu Rozwoju Gospodarki Niskoemisyjnej* (2011)
10. B. Poskrobko, *Gospodarowanie energią na poziomie lokalnym. Podręcznik dla gmin* (Wyższa Szkoła Ekonomiczna w Białymstoku 2011)
11. Sz. Firląg, W. Terlikowski, A. Węglarz, *Nowa misja – niższa emisja. Gospodarka niskoemisyjna w gminach* (Krajowe Stowarzyszenie Inicjatyw 2014)
12. K. Świtalska, *Plany gospodarki niskoemisyjnej jako narzędzia wspomagania zrównoważonego rozwoju energetycznego* (Energia i Paliwa 2015)
13. T. Alankiewicz, *Skuteczność funkcjonowania gospodarki odpadami na przykładzie jednostek samorządowych województwa wielkopolskiego* (Uniwersytet Ekonomiczny w Poznaniu 2009)
14. S. Tokarski, J. Janikowski, *Wreszcie konkurencyjna i niskoemisyjna gospodarka?* (Polska Energia **6**/2013)

E3S Web of Conferences 10, 00109 (2016)
SEED 2016
DOI: 10.1051/e3sconf/20161000109

Thermogravimetric analysis in the study of solid fuels

Robert Wróblewski[1] and Bartosz Ceran[1]

[1]Institute of Electrical Power Engineering, Poznan University of Technology, Poznan, Poland

Abstract. Depletion of fossil energy is the reason for the exploration of the possibility of the use of renewable energy resources. In the article describes a method of thermo-gravimetric analysis and concept, design and execution of the instrument to thermogravimetric measurements placed in the Laboratory of Fuel and Energy Conversion of Institute of Electrical Power Engineering of Poznan University of Technology. The further part of the paper contains the results of the tests carried out on two types of solid fuels (pellets from sawdust and energy willow wood chips) in the form of a thermogravimetric curves. This analysis is to determine the level of the pyrolysis process temperature and degree of conversion of solid fuels into fuel gas. These studies are conducted by looking at opportunities to improve the energy efficiency of the gasification process of biomass.

1 Introduction

Electricity and thermal energy are two of the primary energy carriers needed in municipal economy. Traditionally these energy carriers are made from fossil fuels. These fuels as non-renewable are going to exhaust. At the moment, the emphasis is placed on increasing the use of renewable energy resources. In addition to wind and solar energy in Poland we have a significant amount of available biomass. This fuel can partly replace fossil fuels. Technology characterized by high fuel use efficiency are cogeneration systems. Using biomass as fuel in them we get high-technology based on renewable energy sources. One possible solution is to use a gas-powered internal combustion engine cogeneration module from the gasification of biomass. Still new concepts are being developed of installation of gasification of biomass [1]. High-efficiency gas-producing energy solutions are still being looked for with a low content of tar. The gasification process is endothermic and requires the delivery of thermal energy. The source of this energy is the partial combustion of the fuel. Successive stages of gasification of fuel are drying, pyrolysis, partial combustion and gasification of carbon residue. For the purpose of drying and pyrolysis waste heat can be used which will increase the energy efficiency of the conversion process of solid fuel into gas [1, 2]. The drying process takes place at a little more than 100 ° C. In order to determine the level of the temperature necessary for the process of the partial thermal decomposition, thermogravimetric research of fuels can be used. They will allow to specify for which temperatures and to what extend solid fuel will undergo degassing.

In the case of biomass, which has a high content of volatile substances, pyrolysis process allows to convert, in considerable part, to transform this solid fuel into gas fuel.

2. Thermogravimetrics

Thermogravimetrics is one of the methods of thermal analysis. It is a method of measuring the change in mass of the analyzed sample of the substance, resulting from the impact of the temperature value on this sample in controlled atmosphere (oxidizing - oxygen, air or neutral - nitrogen, argon). Mass variation is due to the physical and chemical changes occurring as a result of the environmental impact on a sample of the appropriate temperature and atmosphere. Thermogravimetric analysis is indicated as TGA. Due to the versatility of the thermogravimetric analysis and the ability to use it for a variety of materials, it is used in many fields of science and technology. [3,4]

In the case of studies of fuels in the neutral atmosphere (simulating the process of pyrolysis) mass variation is due to the moisture evaporation and the thermal decomposition of a solid substance as a result of which gases and condensing compounds arise. TGA analysis, by which thermogravimetric curves are obtained, let us estimate the degree of degassing of sample of function of changes of temperature process and the course of process over time.

Unfortunately, commercially available apparatus for thermogravimetric analysis is expensive. Purchase of a thermobalance is the cost of PLN 200 000. An even greater cost is associated with the analyzer of gas composition resulting from the thermal decomposition-spectrometer. Due to the high price of devices an attempt was undertaken to build a thermogravimetric snap to a precision scale for the purpose of equipment of the Laboratory of Fuels and Energy Conversion of Institute

[a] Corresponding author: robert.wroblewski@put.poznan.pl

of Electrical Power Engineering, Poznan University of Technology.

The project aims to minimize the construction costs of the system. The premise is to allow the stand to perform a thermogravimetric analysis of fuels for laboratory and classroom purposes. The further part of this research paper presents the concept, design and construction of the device to thermogravimetric measurements and the results of tests carried out on two types of solid fuels (pellets and energy willow wood chips).

3. Description of the test stand

A typical commercial thermobalance is characterized by the following parameters described in table 1

Table 1. The selected technical data for examples of the commercial thermobalances [9,10]

type TG	Maximum temperature	Range of weigh	Resolution of weigh	Volume of crucible
	°C	mg	mg	ml
TG 209 F1 Libra	1100	2000	0,0001	0,35
TG 209 F3 Tarsus	1000	2000	0,0001	0,35
TGA 2 XP1	1100	1000	0,001	0,02-0,9
TGA 2 XP1U	1100	1000	0,0001	0,02-0,9
TGA 2 XP5	1100	5000	0,001	0,02-0,9
TGA 2 XP5U	1100	5000	0,0001	0,02-0,9

Weight in commercial thermogravimeters is characterized by a high resolution which is one of the reasons for the high price of the device. In addition, low maximum mass of a sample and especially the low volume of typical evaporating dishes may cause difficulty in placing a sample of the fuel in the raw form (e.g. wood in the form of wood chips) to such attempts wood chips should be additionally crushed.

Laboratory of the Institute of Electric Power is equipped in precise scale of range of 320 g and a resolution of 1 mg and analytical scale AS 60.3 of range 60 g and a resolution of 0, 1 mg. In order to carry out the analysis of thermogravimetric fuels the concept to build a thermogravimetric snap-in to already owned scales. A schema of designed measuring system is shown in Figure 1.

Measurement system, whose schema is shown in Figure 1 is situated in the Laboratory of Fuel and Energy Conversion of Institute of Electrical Power Engineering of Poznan University of Technology.

The various elements of the system are: 1-the nitrogen bottle, 2-pressure reducer for nitrogen, 3-rotameter with a valve to regulate the flow of gas, 4-thermogravimeter body, 5-the crucible with the sample on it, 6-analytical balance, 7-control system with integrated power switch and two temperature regulators type PID, T1, T2-temperature sensors-thermocouples type K.

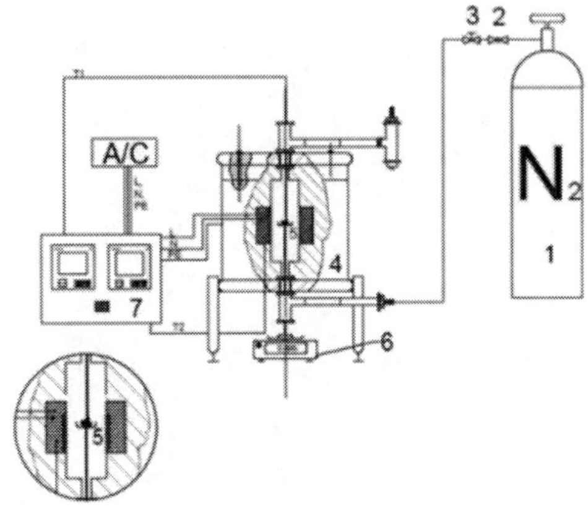

Figure 1. The schema of thermobalance.

In addition to the previously described scale, the key element of thermobalance is the heating chamber. The basis for the construction of the heating chamber is ceramic muffle 0.4-MFO - interior size Ø70x120mm with heating element 1100W (230V) (Figure 2).

Figure 2. Ceramic muffle

Nominal temperature of the heating muffle is 1100 °C (maximum 1200 °C). The supporting structure of heating muffle is made from ceramic mat with a thickness of 12 mm soaked with IZO 12 hardener. The whole has been additionally sealed with a rock wool and covered with galvanized sheet (Figure 3b). For temperature measurement K-type thermocouples were used. Thermocouple T1 is placed directly above the test sample. T2 thermocouple is placed in the ceramic material covering heating element.

An evaporating dish to a scale is made of bottom - board stainless steel with a diameter of 5/4 and 6/4 inch and stainless steel pipe 6 and 8 mm (Figure 3a).At the bottom of the snap-in a bushing was made to the cantilever of the crucible and delivery of nitrogen to the heating chamber. Nitrogen is used to displace oxygen from the heating chamber while simulating the process of pyrolysis. In the upper part of the casing of the snap-in, a bushing for sheathed thermocouple and process gas outlet were made. Conduits are made of brass hydraulic components 1/2 inch. To control temperature the PID type controller is used.

DOI: 10.1051/e3sconf/20161000109

a) b)

Figure 3. Thermobalance evaporating dish (a) and thermobalance(b).

4. Test results

After placing the sample in the crucible of the heating chamber it is washed out with inert gas (N_2) in order to remove the oxygen. The presence of oxygen in the air would cause that, instead of the pyrolysis process, combustion process would occur. Then the process of heating of the sample begins to the assumed temperature controlled by PID-type controller. In the course of the study the change of the mass of the sample is recorded and the temperature value.

The subject of the study were samples of fuel in the form of sawdust pellets from the sawmill and wood chips from energy willow. The results of the technical analysis of these fuels are presented in table 2. These are the average values of the three measurements. Thermobalance tests were carried out by heating the weighed sample of fuel to the required temperature by keeping on this temperature for some period of time. The following thresholds of temperature were adopted: 300 °C, 350 °C, 400 °C and 500 °C. In addition, the study was carried out with a multistep raising of the temperature of the same sample of fuel. In the first step the temperature was raised to 200 °C and then gradually by another 50 °C. In the course of the research the change of temperature and the mass of the sample were recorded. The results of the research in the form of thermogravimetric curves are shown in figures 4-14.

Both tested fuels are characterized by a small ash content. Energy willow has a slightly higher moisture content and lower heating value.

Table 2. Characteristics of selected parameters pellets and willow chips

	Ash	Moisture	Calorific value
	[%]	[%]	[kJ/kg]
pellet	0.85	6.3	18347
willow	0.75	8.5	17476

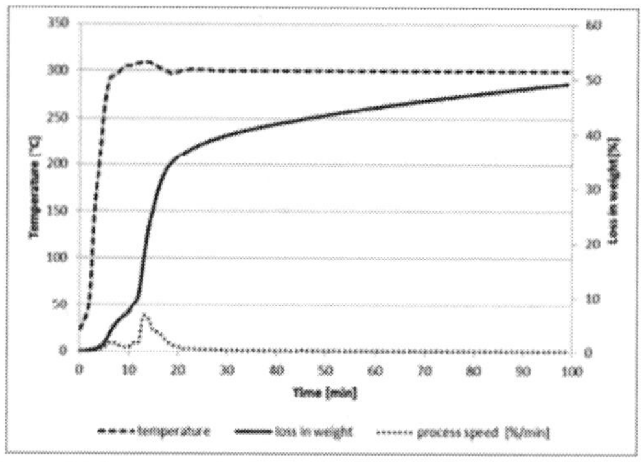

Figure 4. Thermogravimetric curves for wood pellets in real time for temperature 300°C.

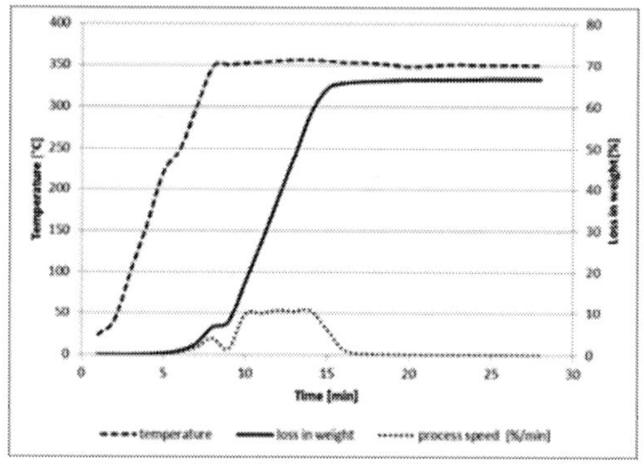

Figure 5. Thermogravimetric curves for wood pellets in real time for temperature 350°C

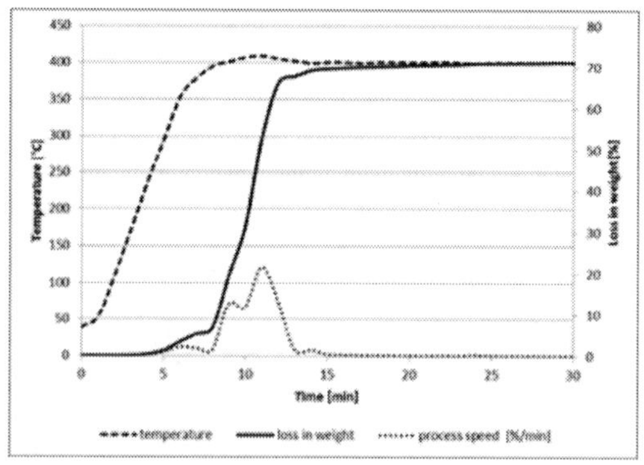

Figure 6. Thermogravimetric curves for wood pellets in real time for temperature 400°C.

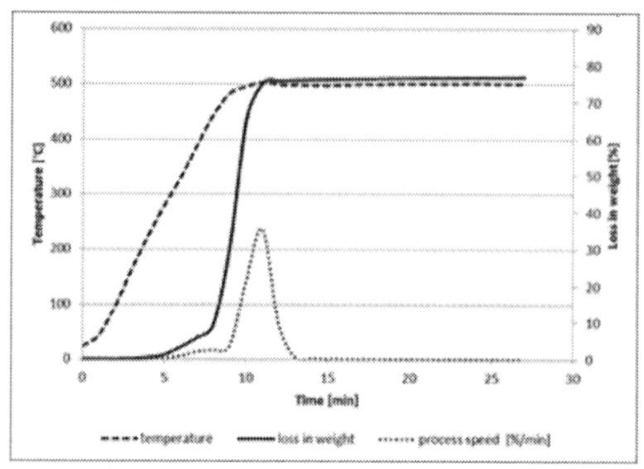

Figure 7. Thermogravimetric curves for wood pellets in real time for temperature 500°C.

DOI: 10.1051/e3sconf/20161000109

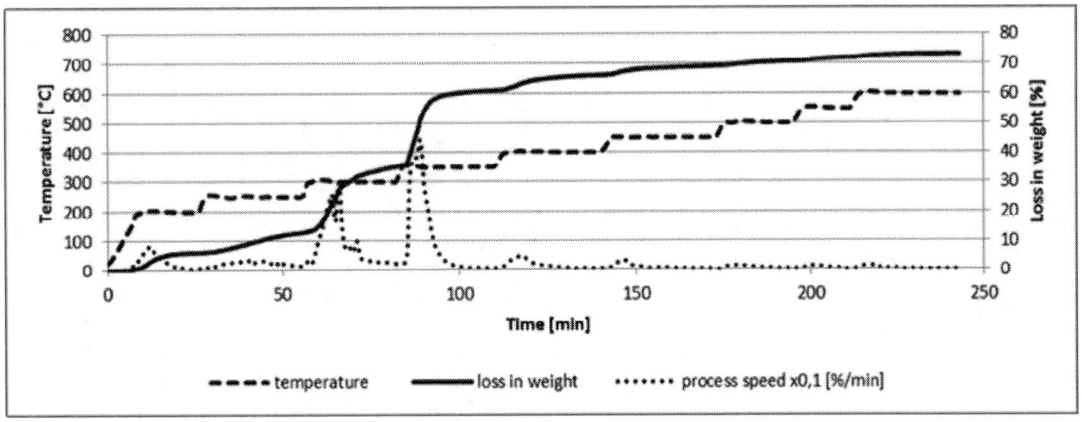

Figure 8. Thermogravimetric curves for wood pellets in real time for a multi-step heating

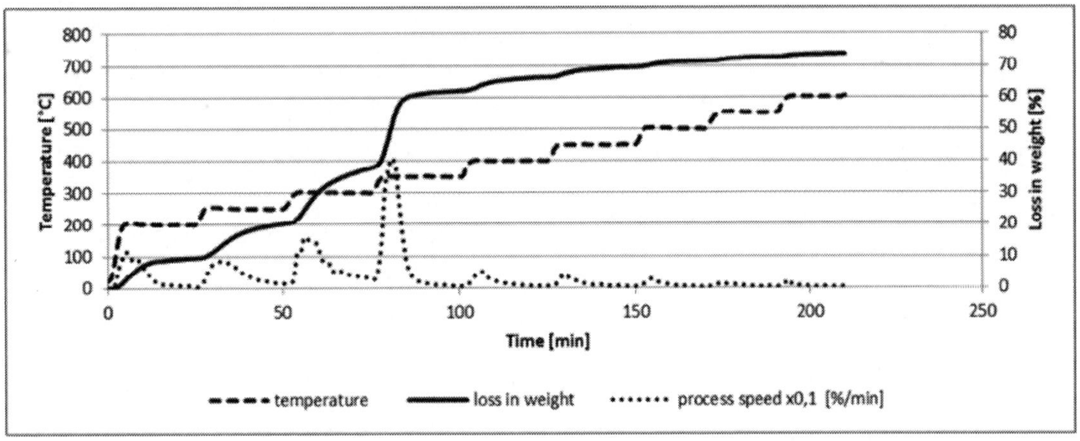

Figure 9. Thermogravimetric curves for willow wood chips in real time for a multi-step heating

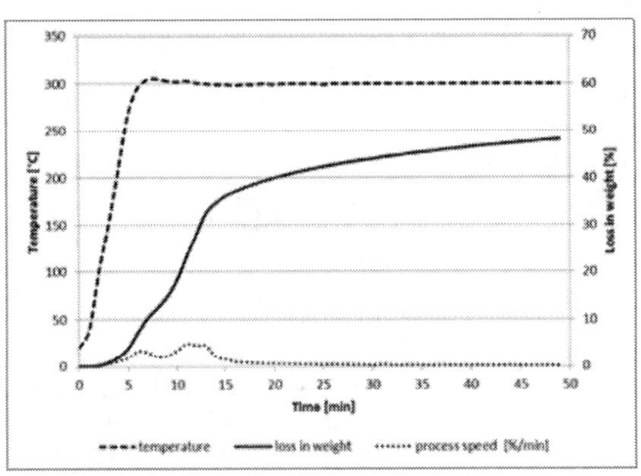

Figure 10. Thermogravimetric curves for Willow wood chips in real time 300°C.

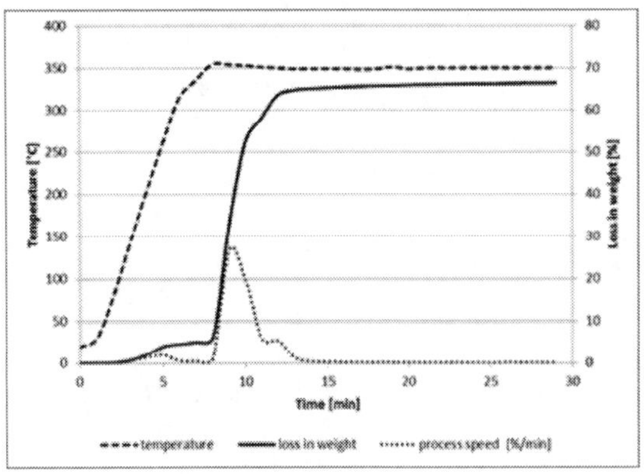

Figure 11. Thermogravimetric curves for Willow wood chips in real time 350°C.

DOI: 10.1051/e3sconf/20161000109

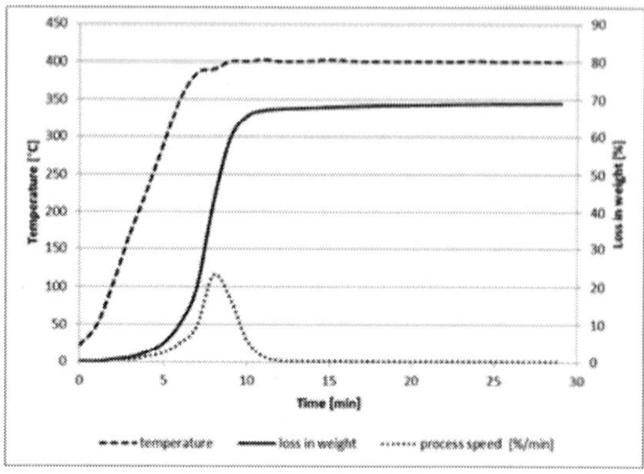

Figure 12. Thermogravimetric curves for Willow wood chips in real time 400°C.

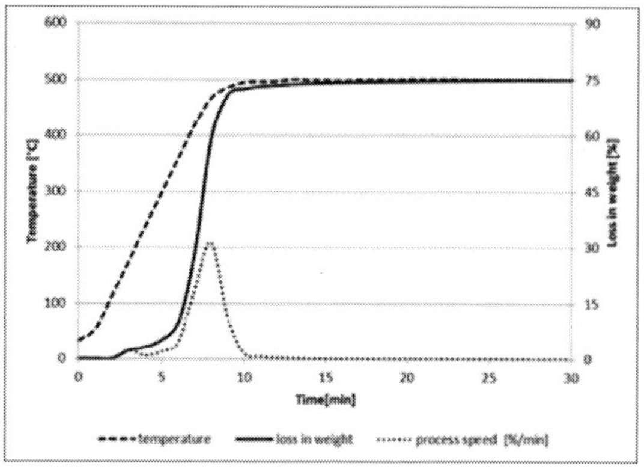

Figure 13. Thermogravimetric curves for Willow wood chips in real time 500°C.

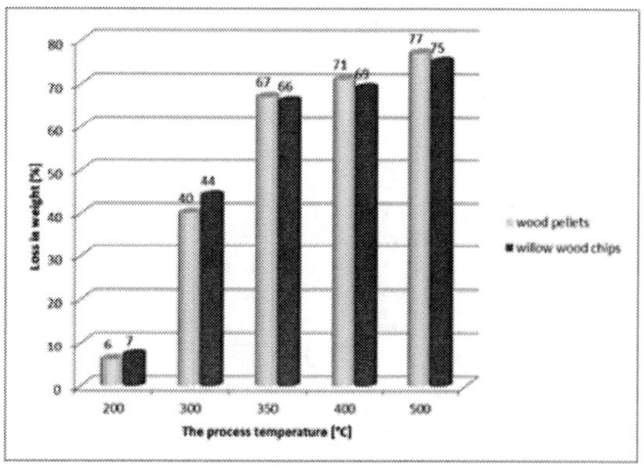

Figure 14. The dependency of the conversion degree of tested fuels from the process temperature

5 The analysis of the obtained results

As is evident from the thermogravimetric curves (fig. 4) the heating chamber has reached the preset temperature after about 5 min. After about 10 minutes the process of heating up the sample and its drying ended. For the next 10 min lasted an intense pyrolysis to achieve 35% weight loss. Then the process of thermal decomposition was crawling with definitely less intensity. The next charts (Figures 5-7) show that increasing temperature of the process causes minor lengthening of the waiting time to achieve the preset temperature while reducing time when intensive pyrolysis occurs. In these cases, the process of further slow loss of mass is far less intense or virtually unnoticeable. Figure 8 shows the thermogravimetric curves for the gradual raising of temperature by 50 °C starting from 200 °C. In this case, we can notice at which temperature ranges the pyrolysis process is the most intense. The greatest weight losses were reported for 250-350 °C temperature range. Similarly, in a study came out with the energy Willow shavings (Figure 9). Here also the biggest weight loss that is the greatest intensity of pyrolysis has been reported for temperatures in the range of 250-350 °C. Similarly, as in the case of the pellet process, looks pyrolysis of wood chips at a temperature of 300 °C (Figure 10). In this case, the process of intense pyrolysis lasted for a shorter time. After a period of intensive weight loss pyrolysis continued but definitely less intense. In other cases (Figure 11-13) increase in temperature causes a shortening of the time of thermal decomposition. Pyrolysis of willow wood chips carried out in a little shorter time than for the corresponding temperature the pyrolysis of wood pellets. Figure 14 shows the dependence of the degree of conversion of solid fuel into gas depending on the temperature of the process. In case of raising the temperature of fuel to 300 °C about 40 % of the fuel conversion of solid fuel into gas was obtained. Similarly for temperatures of 350 °C 66 % conversion is obtained. For subsequent temperature thresholds, these growths are smaller. For 400 °C 70 % is obtained and up to 500 °C approx. 77% degassing of fuel.

6 Summary

Constructed apparatus allows to study thermal decomposition of solid fuels and obtained thermogravimetric curves are similar to those that can be performed on commercial thermobalances [5-9]. The advantage of the device is definitely lower price than of commercial solutions (approx. 100 times). Preferred option seems to be also a larger size of a chamber and evaporating dish allowing the analysis of larger samples of fuel. Commercial solutions have a cooling system that allows for relatively fast exit of the heating chamber and enabling the implementation of the next measurement. Therefore the cooling system would be a welcome improvement in the next version of this type of device. A welcome improvement would be to use temperature regulator with the possibility of registration data and control from your PC. Despite these imperfections it can be considered that building of such a device meets the goal of thermogravimetric testing of fuels.

On the basis of an analysis of the results of thermogravimetric research of sawdust pellets and energy willow wood chips, it can be concluded that for the tested biofuels the most intensive process took place in temperature range around 300-400 °C. As a result of pyrolysis for this temperature range for tested biomass, a high conversion level of solid fuel into gas was obtained. This allows to draw the conclusion, at least in significant part, that the conversion of fuel towards providing the necessary heat for that endothermic process coming from hot waste gases is possible. Usually the heat of this type of gas is used for heating purposes. The use of them in wood gas generator would let reduce the amount of heat produced in the combustion of the fuel components. In small wood gas generator systems for combustion air is provided. This causes the appearance of ballast in the form of nitrogen in producer gas. Partial combustion and nitrogen from the air reduce the energy value of arising gas fuel. Application of waste heat to the process of initial pyrolysis will help to limit the amount of heat generated in combustion process. This will limit the amount of air needed. As a result, it will allow to increase the efficiency of the gasification process of solid fuel and improvement of the quality of this gas. To confirm this thesis, however, you should carry out more trials of this type as well as broaden the scope of tested biofuels.

References

1. H. A. M. Knoef, Handbook Biomass Gasification Second Edition, BTG (2012)
2. R. Wróblewski, Koncepcja małego układu kogeneracyjnego zintegrowanego ze zgazowaniem biomasy, Polityka Energetyczna, **17**, 159-170 (2014)
3. M. Szumera, Charakterystyka wybranych metod termicznych. Cz. 1, LAB Laboratoria, Aparatura, Badania, 6, 28-34 (2012).
4. D. Schultze, Termiczna analiza różnicowa, PWN (1974)
5. H. Liu, K. Han, M. Liu, S. Niu, C. Lu, Energ. Source Part A, **37**, 22 (2015)
6. C. Deyong, L. Shaohua, J. Jia, Z. Rui; Thermogravimetric Analysis of Pyrolysis Characteristics of Biomass; World Automation Congress (WAC), 2012
7. W. Jin, K.a Singh, J. Zondlo, Agriculture, **3**, 1, 12-32 (2013)
8. M. Carrier, A. Loppinet-Serani, D. Denux, J.-M. Lasnier, F. Ham-Pichavant, F.Cansell, C. Aymonier, Biomass and Bioenergy, **35** (2011)
9. L. Gašparovič, Z. Koreňová, Ľ. Jelemenskýl; Kinetic study of wood chips decomposition by TGA; 36th International Conference of SSCHE May 25–29, 2009,
10. http://www.mt.com
11. http://www.netzsch-thermal-analysis.com

Evaluation of the possibility of forecasting the displacements of the Solina Dam based on observation of the feeler gauges

Aleksandra Bąk[1]

[1]*AGH University of Science and Technology, Faculty od Mining Surveying and Environmental Engineering, Cracow.*

Abstract. These studies are the continuation of previously conducted researches based on displacements of the Solina Dam. To support the monitoring of the dam, the EDF model were used. This model determines the dependence between displacements observed on the elements of the dam and the level of the upper water, cyclicality of temperatures and the impact of time. The use of model, in both the imaging of displacements of the Solina Dam and forecasts made for the benchmarks located on the crest of the dam, gave satisfactory results, therefore further analysis were undertaken. Displacement forecasting was made based on data from feeler gauges located at the pillars, retaining walls and four galleries. Analysis based on 4 different forecasting periods enabled for the conclusion that to obtain reliable forecasts, measurement data based on 10 years of operation with frequency of 1 per month is required. The use of such a comprehensive database allowed for the achievement of full compliance between modelled values and observed displacements within the established confidence intervals.

1 Introduction

Unsatisfactory condition of hydraulic structures, as dams, is confirmed by various institutions, including, among others, the Chief Inspectorate of Construction Supervision [1]. Said technical condition of the facilities, is the result of the influence of various factors on the dam. Undoubtedly the safety of dams is affected by rheological and meteorological factors, excessive filtration, and errors during the design and execution stage, and difficulties to predict extreme natural phenomena [2]. Growing experience of designers has enabled to minimize the risk of errors at the design stage, as well as eliminating the shortcomings in the execution phase. Currently, the improper operation of the dams is the biggest threat to their safety. An important aspect, from the point of view of the safety of dams, is also the impact of natural factors, which could exacerbate the small displacements and deformations.

In order to improve the safety of the dam there is a need to take actions to monitor the behaviour of this type of facilities. The currently used monitoring and control system is based i. a. on the conduct of regular geodetic measurements of displacements of the elements of the facility, the collection and interpretation of data. One of the possibilities to improve the safety of hydro-technical facilities could be proper interpretation of measurement data that enable to draw conclusions about the technical condition of the dams. The interpretation of the current state of the dams is supported by numerous methods [3]. These methods are focused on a variety of factors, ranging from extreme phenomena, to natural, cyclical factors, such as seasonal variability, the influence of time or temperature cyclicality.

The article - as a continuation of previously conducted researches [4] - focuses on assessing the impact of natural, cyclical factors on the possibility to reflect changes observed on the structures during normal operation of the facility. Similarly, to previous researches, this analysis was based on model developed by Électricité de France. This is a tool created to interpret the behaviour of arch dams. This model focuses on assessing the impact of rheological changes, temperature cyclicality, and the assessment of impact of the hydrological factor. This article focuses on modelling measured data from the feeler gauges located on the Solina Dam during normal operation of the facility. Analysis confirmed utility of the EDF model in reflecting the displacements observed on the feeler gauges, therefore the predictions based on explanatory factors were undertaken.

2 Methods and materials

As it was mentioned, presented analysis, is based on the model proposed by Électricité de France.

2.1 The EDF model

The EDF model is the first model created by Électricité de France to assess the state of arch dams. This model describes the dependence between natural cyclical factors and displacements observed on various part of the structure. Previously conducted analysis confirmed that the method developed by Électricité de France is a valuable tool to interpret the behaviour of concrete dams, as the Solina Dam. The EDF model (1) takes into

[1] corresponding author: abak@agh.edu.pl

account the impact of three key factors: hydrostatic effect, the value of the temperature and the impact of time on elements of the dam. The influence of the rheological and seasonal factors is determined as a date function (t), while the variability caused by the hydrological factors is presented using the upper water level (2).

$$M(z,t) = T(t) + S(t) + H(z) \quad (1)$$

M(z, t) – the value of the displacement;
T(t) – rheological impact;
S(t) – seasonal effect;
H(z) – the effect of hydrological impact.

$$M(z,t) = A + B \cdot e^{-Ct} + D \cdot \sin(\frac{2\pi t}{365} + E) + F \cdot \sin\left(\frac{4\pi t}{365} + G\right) + Hz + Iz^2 \quad (2)$$

A, B, C, D, E, F, G, H, I – coefficients specific to the measuring point;
t – time;
z – water level.

Proposed method performs adequately and reflects the behaviour of dams; however, numerous scientists have called into question its usefulness, as a reliable tool for the analysis of the hydraulic measurements. These conclusions are dictated by the inclusion of phenomena of the delayed piezometer response on the changes of hydrological factor [5]. Another limitation of the model, as a result of its simple structure, is associated with seasonal impact. The influence of seasonality, according the EDF model is presented as a date function. This leads to the inclusion of fixed conditions of the dam on a specific day of the year, but despite this limitation, the results were considered to be very good [7].

Despite numerous limitations the model is burdened with, it was decided to continue research on the displacement of the Solina Dam. Bearing in mind these simplifications and approximations, it was decided to continue the assessment of the possibility of forecasting the displacements of the measurement points based on the data from feeler gauges located on the Solina Dam.

2.2 The Solina Dam

Analysed structure is the tallest hydro-technical object in Poland. It measures 81.8 meters in height, forming a lake whose water surface reaches over 2 100 hectares, while the capacity of the created reservoir is estimated at 474 million cubic meters. The maximum depth of the created lake is 60 m. Construction of the Solina dam has allowed to launch in 1968, at the Solina Hydroelectric Power Plant. It is a pumped storage power plant equipped with four Francis type turbine sets with installed power output of 200 MW, which ensures an annual production in order of 112 GWh.

This analysis is focused on displacements observed using feeler gauges. The Solina Dam is equipped with feeler gauges located at 4 sections of gallery (Fig. 1), the retaining walls and pillars (Fig. 2). Feeler gauges are observed with frequency of 4 measurements per month.

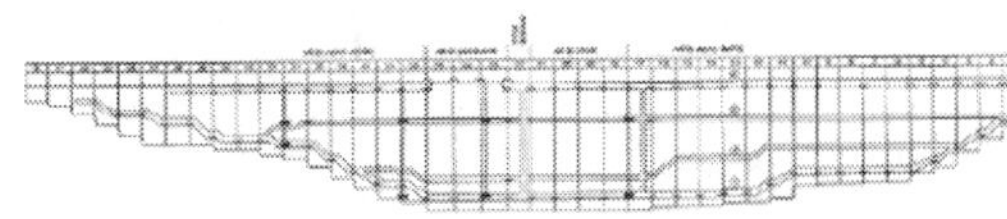

Figure 1. Feeler gauges' deployment in the body of the Solina Dam

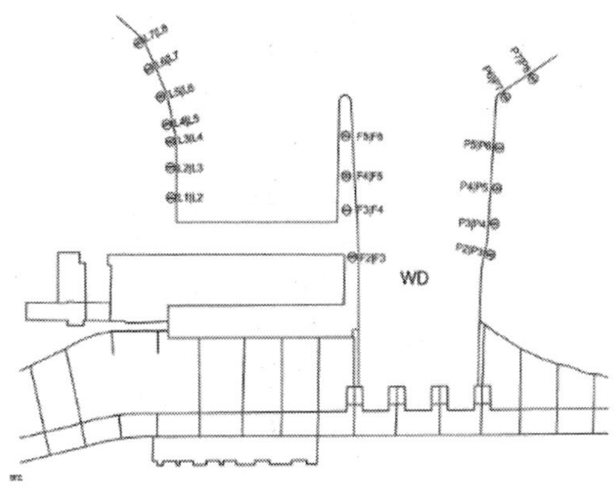

Figure 2. Feeler gauges' deployment in the retaining walls and pillars.

2.3 Methods

Analysis indicated a reliable reflection of the behaviour of the dam using said model. This allowed for attempt of the possibility of forecast of the displacements based on data obtained from monitoring of feeler gauges placed on the Solina Dam. Analysis was based on measured data obtained from the years: 1992 - 2011. Displacements were analysed based on data from measurement equipment located in 4 sections of the dam gallery, pillars and retaining walls. Analysis was performed using 4 different sequences of measurement data: 10, 12, 15 and 18 years.

Presented modeling was conducted using MATLAB program. Mentioned model was implemented to MATLAB programme. The entered data (time and level of upper water) made it possible to generate specific coefficients of the model for individual measurement points. As the results of the analysis there were obtained the graphs that allow to compare actual and predicted values of displacements.

Analysis indicated a reliable reflection of the behaviour of the dam using said model. This conclusion allowed for forecasting, which was then evaluated for compliance of the forecast with the absolute deviation test [8]. The commonly used test shows that 99.7% of all observations is placed in the $[\mu-3\sigma,\mu+3\sigma]$ interval. If the measurement differs by more than 3 standard deviations from the expected value, it can be concluded that it is disturbed due to incorrect definition of the correlated values, or incorrect assessment of the measurement uncertainty.

3 Results

In line with the assumptions of the analysis, the forecast of displacements based on observations of feeler gauges of the Solina Dam was made.

As opposed to the modeling of the displacement based on measurement data from the crest of the Solina Dam,

 DOI: 10.1051/e3sconf/20161000110

there is a possibility to forecast changes observed on the feeler gauges based on 10 years of operation. It is caused by the frequency of measurements. Measurements of the crest of the Solina Dam are performed with frequency of 2 per year, while feeler gauges were measured once per month. As extensive measurement database allowed to determine coefficients of the model, that ensure compliance in 87% of analyzed cases. Based on measurement data from 10 years of exploitation of the dam the best forecasts were performed in case of the feeler gauge located in section 30 – 31, at the gallery no. 4 (Fig. 3). In this case, the average error was 0,0409 mm.

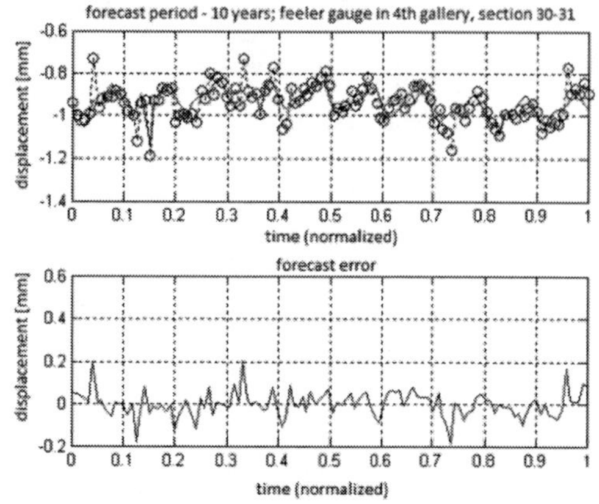

Figure 3. Approximation of displacements observed in the feeler gauge, section 30-31, gallery no. 4. Forecast period 10 years.

The model, that was generated based on the 10-years of operation of the dam, worse illustrated the displacement observed on the pillars and retaining walls of the Solina Dam, e.g. feeler gauge no. L 5-6, located on left retaining wall (Fig. 4). In this case an average error was about 0,5 mm.

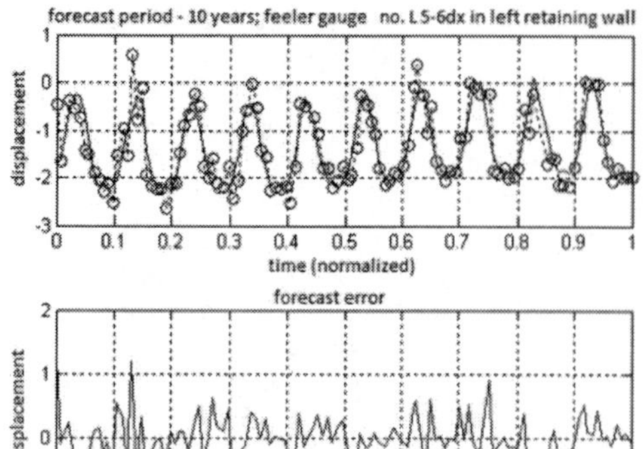

Figure 4. Approximation of displacements observed in the feeler gauge no. L5-6, retaining wall. Forecast period 10 years.

Increasing the forecasting period by 2 years allowed to create the model, that ensure compliance in 90% of analyzed measurement points Modeling based on 12 years of exploitation of the dam allowed for better reflection of the direction of the displacements. Unfortunately the forecasts were burned with higher error than based on 10 years of exploitation. Similarly, to the previous forecast period, the model better reflects the displacements observed on galleries of the dam, less well changes observed on the pillars and retaining walls were illustrated. Based on forecasting period of 12 years' feeler gauge in section 25-26, gallery 4 (Fig. 5) was the best reflected. An average error was 0,047 mm.

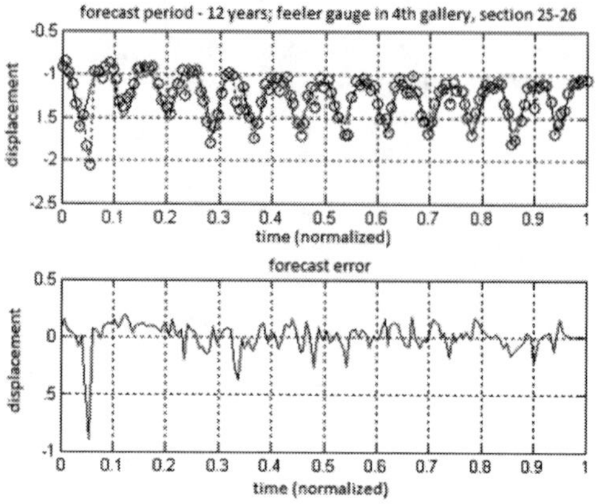

Figure 5. Approximation of displacements observed on the feeler gauge, section 25-26, gallery no. 4. Forecast period 12 years.

In this analysis the worst reflection was observed on the retaining wall no. M 6-7 (Fig. 6), where an average error was about 3 mm.

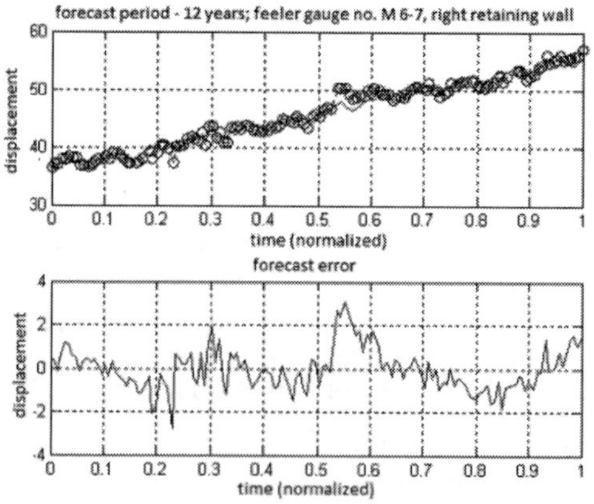

Figure 6. Approximation of displacements observed in the feeler gauge, retaining wall M 6-7. Forecast period 12 years.

Modelling based on the forecast period of 15 years overlaps to a large extent with the modelling based on 12 years of operation. It was best adopted to the displacement observed on the feeler gauges of the galleries, and the worst reflection of the model was again observed in the case of pillars. Increasing the forecasting period up to 15 years allowed for compartmental forecast properly reflecting the actual displacement observed on the elements of the dam. When assuming such comprehensive database, the point forecasts are very close to the actual observations of movements in subsequent measurement periods. Models that were created based on 15 years of exploitation correspond to the observations in 85% of the feeler gauges. This is caused by incomplete measurement database of the feeler gauges located on the retaining walls. Created models illustrate changes in 66 % of measurement points. Incomplete observations made it impossible to forecast displacement in the next measurement step. An example can be retaining wall no. M 7-8 (Fig. 7), where an average displacements were about 30 mm, and an average error reached 15mm.

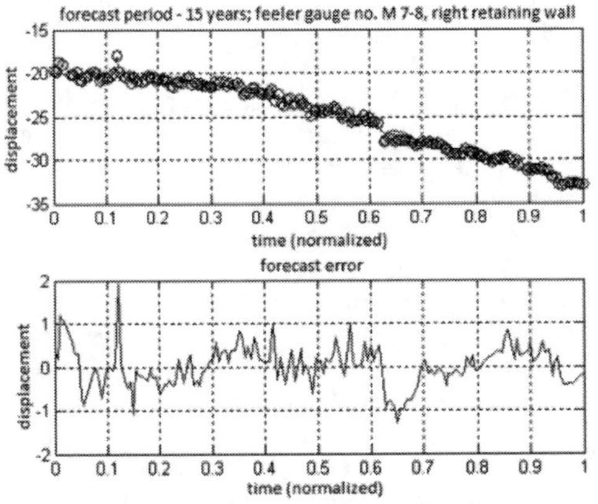

Figure 7. Approximation of displacements observed in the feeler gauge, retaining wall M 7-8. Forecast period 15+1 years.

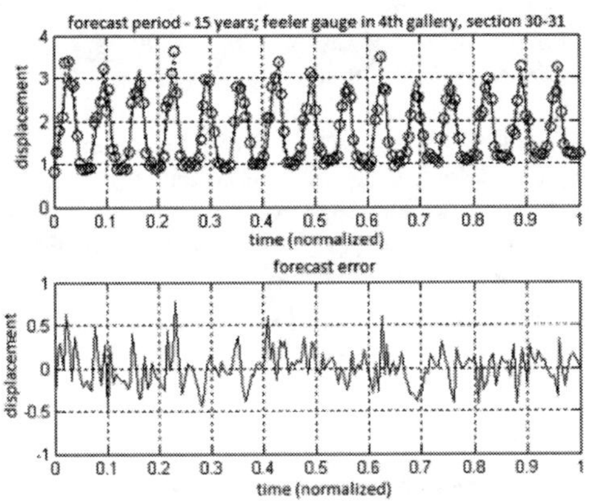

Figure 8. Approximation of displacements observed in the feeler gauge, section 30-31, gallery no. 4. Forecast period 15 years.

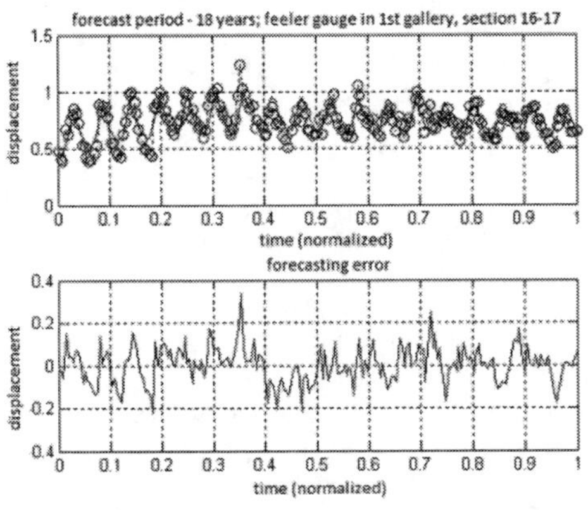

Figure 9. Approximation of displacements observed in the feeler gauge, section 16-17, gallery no. 1. Forecast period 18 years.

The best reflection based on 15 years of exploitation of the dam was observed on feeler gauge located in gallery 4 in section 4 (Fig. 8). An average error was 0,056 mm.

Modelling based on the longest of the proposed forecasting periods, i.e. 18 years reaffirms that best reflected by the EDF model were the displacements of feeler gauge located in the section 16-17, gallery no. 1 (Fig. 9). In this case an average error was 0,084 mm.

The models created based on 18 years of operation of the dam the worst reflected displacements observed on the feeler gauge no. M 6-7, located on the right retaining wall (Fig. 10). Similarly to previous analysis it was caused by incomplete measurement database. Some of those feeler gauges did not correctly reflects the direction of changes.

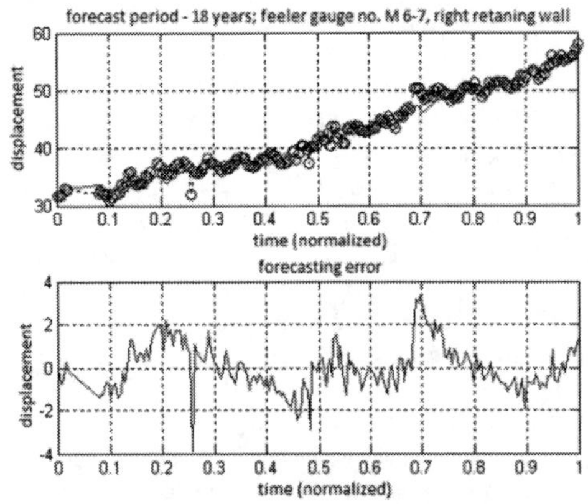

Figure 10. Approximation of displacements observed in the feeler gauge no. M 6-7, retailig wall. Forecast period 18+1 years.

Obtained results confirm that the EDF model, although originally designed for arch dams well, reflects the behavior of concrete, gravity dams. This report - just like the previous one - also allows us to draw the conclusion that the model can also be a useful tool in terms of forecasting of future displacements. In order to assess the possibility of the forecast, 4 different forecasting periods were used. However unlike to the previous analysis of forecasting – crest of the Solina Dam – it was possible to predict future displacements based 10 years of exploitation of the dam. This situation is caused by large sequence of measurement data. In case of the crest of the dam, measurements were performed in frequency of 2 per year and feeler gauges were measured with frequency of 1/month. The displacement forecast results based on 12 years of operation of the facility have prompted similar conclusions. Database of 12 years of operation allowed to improve not only coefficients of the model, but also forecasts. Further analysis corresponding to 15 and 18 years of operation of the dam, did not allowed to created models that improved the direct of displacements. It was also caused by incomplete measurement data. Some of the feeler gauges located on the retaining walls made it impossible to forecast future displacements. However, based on complete measurements data sequences – all of analysed galleries – forecasts were in line with observed displacements.

4 Discussion

Model, used to analyse displacements of the elements of the Solina Dam, is not without its faults and limitations. Despite these limitations, the flood risk associated with exploitation of hydro technical facilities, as dams, caused that further steps towards monitoring and control were undertaken. The monitoring of dams, and their control play a vital role in ensuring the safety of such facilities.

The article makes assessment of the possibility to expand activities in the field of data interpretation of the displacements of elements of the Solina Dam based on the EDF model. Ultimately, the model has been created having arch dams in mind, but it is also used for other types of structures. The EDF model has several limitations. First of all - it is widely believed that statistical models seeking dependence between the causes and quantitative effects have some weaknesses. Quoted analyses show a significant limitation of the model, one of them is the delay of the response of the structure to the measurements of the explanatory factors. The authors, due to the tendency of analysed dam in a form of excessive filtering, propose the delay of analysis on the basis of the method based on Darcy's law and Richard's equation. This addition takes into account the use of a linear, dynamic analysis system of the share of factors that do not cause aging of dams on changing of the pore pressure. Additionally, as a result of analysis, the authors ruled out the possibility of applying the model for the forecasts due to the mentioned difference in pore pressures, and the lack of taking into account the effects of precipitation. A similar statement was drawn by the authors of the article "Delayed response analysis of dam monitoring data" [6], who support the lack of forecast opportunity based on the EDF model. The main cause is also considered to be the delay phenomenon due to distracting behaviours, such as filtering or viscoelasticity. As a solution they propose the inclusion of an impulse response, other than the aging of materials in the statistical analysis.

Despite these limitations, the EDF model is considered to be stable and capable of providing reliable results. Assuming all restrictions, the model is burdened with, an attempt was made to predict the dam displacement. Previous analysis in terms on prediction displacement based on measured data from the crest of the Solina Dam was correct, thus further modeling was undertaken.

5 Summary

The measures taken constitute the continuation of analysis of displacement based on observation feeler gauges located on the elements of the Solina Dam. Forecast based on displacement data of measurement points was continued based on the EDF model. It shows correlation between displacements observed on the elements of the dam, and cyclical natural factors. It is estimated that the greatest impact on the behaviour of the dam are temperature factors, seasonality, and the influence of time. Being fully aware of the limitations, the model is burdened with, an attempt was made to put a forecast of displacements as a function of time.

Opposite to previous analysis of the displacements observed on the crest of the Solina Dam, forecasting of changes of feeler gauges based on 10 years of measurements were possible. Increase of the length of forecasting period across all of analyzed feeler gauges were worst reflected. It was caused by incomplete measurement data of the feeler gauges located on the retaining walls. In the other feeler gauges that were analyzed, increase of the length of measurements database improved the forecasts of previous displacements.

The overall objective of the analysis was to evaluate the possibility of forecasting changes observed on the elements of the Solina Dam based on data obtained from feeler gauges. The undertaken analysis confronts the problems of the unsatisfactory conditions of dams and the necessity to minimize the threat posed by their operation. There are plans to achieve it through actions aimed at improving the safety of hydraulic structures by expanding activities in terms of interpretation of measurement data, and forecasts of changes observed in the dams. Forecasts of displacements of the feeler gauges of the dam are only the beginning for further study. Subsequently, there are plans to forecast the displacements studied in other parts of the structure. Based on differing displacement models and forecasts of individual elements, a holistic model will be created that will enable to treat the facility in a comprehensive manner and thus will allow to assess the state of the facility across the dam.

Implementation of the presented analysis will allow for early detection of disturbing phenomena, destructive processes or discontinuities in the displacements of dams, significant evolutions, irreversibility and other disorders. It will also allow for assessment of the current technical state of dams, which - to a large extend - will streamline the decision-making process and indicate the necessary renovation and repair actions or the necessity to take measures of emergency protection, thus allowing those responsible to make expert judgement, technical and economic analysis.

References

1. K. Parylak, W. Wolski *Oceny stanu technicznego budowli piętrzących w Polsce w świetle standardów unijnych.* XXIII konferencja naukowo – techniczna, Szczecin – Międzyzdroje (2007)
2. Z. Kledyński *Monitoring i diagnostyka budowli hydrotechnicznych,* cz.1. Nowoczesne Budownictwo Inżynieryjne, s. 54 – 61 (2011)
3. K. Fiedler *Awarie i katastrofy zapór – zagrożenia, ich przyczyny i skutki oraz działania zapobiegawcze.* IMGW (2007)
4. A. Bąk *Analiza przydatności modelu EDF dla celów monitoringu odkształceń elementów zapory Solina.* IV International Conference of Young Scientists „Nauka dla Środowiska Przyrodniczego" Poznań, (2015)
5. Bonelli S., Royet P. *Delayed response analysis of dam monitoring data.* ICOLD European Symposium on Dams in a European Context, GEIRANGER, (2001)
6. S. Bonelli, R. Tourment, H. Felix *Analysis of earthdam monitoring data.* Selected problems of water engineering, p. 133 – 150. Kraków (2003)
7. L. Chouinard, R. Larivière, P. Côté, W. Zhao *Analysis of Irreversible displacements in multiple arch concrete dam using principal component analysis.*

Joint International Conference on Computing and Decision Making in Civil and Building Engineering, Montréal, Canada (2006)

8. W. Łuczek *Podstawy statystycznego opracowania pomiarów*. University of Technology in Gliwice, Gliwice (1995)

Magnetic nanofluid properties as the heat transfer enhancement agent

Aleksandra Roszko[1,a] and Elzbieta Fornalik-Wajs[1]

[1]*AGH University of Science and Technology, Department of Fundamental Research in Energy Engineering, 30 Mickiewicz Ave., 30-059 Krakow, Poland*

Abstract. The main purpose of this paper was to investigate an influence of various parameters on the heat transfer processes with strong magnetic field utilization. Two positions of experimental enclosure in magnetic environment, two methods of preparation and three different concentrations of nanoparticles (0.0112, 0.056 and 0.112 vol.%) were taken into account together with the magnetic field strength. Analysed nanofluids consisted of distilled water (diamagnetic) and Cu/CuO particles (paramagnetic) of 40-60 nm size. The nanofluids components had different magnetic properties what caused complex interaction of forces' system. The heat transfer data and fluid flow structure demonstrated the influence of magnetic field on the convective phenomena. The most visible consequence of magnetic field application was the heat transfer enhancement and flow reorganization under applied conditions.

1 Introduction

Regarding the forthcoming thermo-ecological requirements there are a lot of issues devoted to improvement of existing technologies, increasing the efficiencies, emissions reduction, looking for a new devices and systems. In industrial applications increasing efficiency is very often identify with higher heat rate and flux transfer. Therefore one of the investigated ways to achieve it is connected with new fluids and their properties. Nanofluids become one of the analyzed options. The reason of their popularity comes from the properties and possible applications, e.g. heating of the buildings, solar absorption, energy storage, industrial cooling, friction reduction, nanodrug delivery, transportation, heat transfer intensification, etc [1]. Their properties differ from the common fluids, therefore they can replace them in existing processes or be used in new areas. There are many research devoted to the heat transfer processes intensification [2]-[5]. The first report on nanofluid successful preparation was in 1995 [6]. Since then, interest in such fluids is constantly growing.

Going toward the higher heat fluxes transfer the nanofluids appeared as the promising working fluids due to changes in the base fluid physical properties. Authors' goal was to check if the magnetic field can enhance heat transfer by low concentration nanofluids. The thesis that the nanofluid properties are the main parameters features and taking part in these processes was verified.

In this paper the experimental analysis of heat transfer and flow structure of diamagnetic, low concentration Cu/CuO nanofluids under the strong magnetic field are presented. Two positions of experimental enclosure in magnetic environment, two methods of preparation and three different concentrations of nanoparticles (0.0112, 0.056 and 0.112 vol.%) were taken into account together with the magnetic field strength. The fluctuations of temperature field were analyzed by Fast Fourier Transform and spectral analysis what led to the understanding of flow behavior under applied conditions. Through the results of mentioned analysis the magnetic field influence on the convective phenomena was valuated.

2 Experimental analysis

A schematic view of experimental set up and enclosure are presented in Figures 1 and 2. Experimental apparatus consisted of an cubical enclosure, a heater control system, a thermostating bath and a data acquisition system connected to a personal computer. A heating element was located under the bottom copper wall and was constructed with a nichrome wire. It was connected to a power supply monitored with the multimeters.

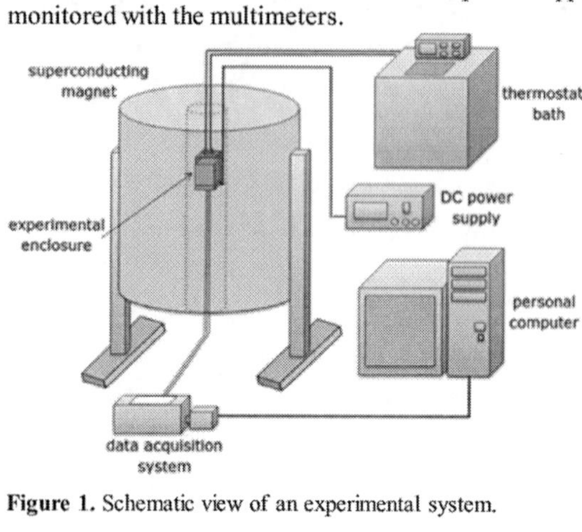

Figure 1. Schematic view of an experimental system.

[a] Corresponding author: roszko@agh.edu.pl

The cooling chamber being a part of enclosure was placed above the upper copper wall. Water flowing through the chamber assured the temperature maintained by the thermostating bath. The test vessel interior was a cube of 0.032 m size with heated bottom and cooled top horizontal walls. In one of the Plexiglas side walls six T-type thermocouples were placed to control temperature in the vicinity of that wall. The other six K-type thermocouples were located within the heated and cooled copper plates. The thermocouples' signals were recorded during whole experiment and enabled the post-processing analysis. The enclosure was filled with working fluid and positioned in the superconducting magnet test section. The cube locations were chosen for the sake of maximal value of magnetic induction square gradient (grad$\mathbf{B}^2_{max}$) occurrence, where magnitude of the magnetic force has the greatest influence. The working fluids were nanofluids with distilled water, as a base fluid, containing Cu/CuO nanoparticles of 40-60 nm diameter, prepared by two-step method. The properties of all examined fluids are presented in Table 1.

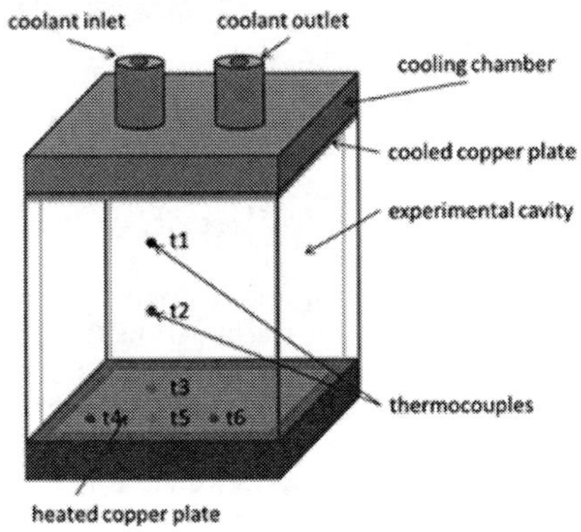

Figure 2. Experimental enclosure

The particular stages of experiment differed from each other in some details due to analysis of various parameters. However the main procedure remained the same. The heat transfer was evaluated through the Nusselt and thermo-magnetic Rayleigh numbers calculation. Definition of the Nusselt number applied in the studies took form:

$$\mathrm{Nu} = Q_{\mathrm{conv}} \cdot \left(Q_{\mathrm{cond}}\right)^{-1} \quad (1)$$

where Q_{conv} is the convective heat rate and Q_{cond} is the conducted one. The thermo-magnetic Rayleigh number was define as follows:

$$\mathrm{Ra}_{\mathrm{TM}} = \mathrm{Ra}_{\mathrm{T}}[1 - \gamma B_z \left(B_{\mathrm{max}}\right)^{-1} (\partial B_z / \partial z)\left(B_{\mathrm{max}}\right)^{-1}], \quad (2)$$

where $\mathrm{Ra_T}$ is thermal Rayleigh number $\mathrm{Ra_T} = g\beta\rho^2 c_p \left(\mu k\right)^{-1} d^3 \Delta T$; g is the gravitational acceleration; β is the thermal expansion coefficient; ρ is the density; c_p is the specific heat; μ is the dynamic viscosity; k is the thermal conductivity; d is the characteristic dimension; ΔT is the temperature difference. The magnetization parameter γ was described by equation:

$$\gamma = \chi_m B^2_{\mathrm{max}} \left(\mu_m g d\right)^{-1} \quad (3)$$

where: χ_m is the mass magnetic susceptibility; B_{max} is the magnetic induction in the centre of coil; B_z is the magnetic induction in the position of enclosure centre; μ_m is the vacuum magnetic permeability.

The first stage of whole measurement procedure was to determine heat losses dependence on temperature difference. It was assumed in the basis of [7] and verified experimentally that heat losses did not depend on phenomena occurring in the enclosure. Therefore, the formulas obtained in this stage were applied in all analysed cases. The heat losses were determined during the conduction state, which was established in the experimental cube turned upside-down (heated wall was at the top, while cooled one at the bottom). After stabilization of fluid and temperature stratification the measurements for various temperature differences were done. Detailed information and calculation scheme were presented in [8]. Following equations were obtained:

$$Q_{\mathrm{loss\_Cu0.0112}} = 0.0621 \cdot \Delta T \quad (4)$$

$$Q_{\mathrm{loss\_Cu0.056}} = 0.0636 \cdot \Delta T \quad (5)$$

$$Q_{\mathrm{loss\_Cu0.112}} = 0.0640 \cdot \Delta T \quad (6)$$

The next stage of experimental procedure was connected with Rayleigh-Benard convection and magnetic influence on it. The temperature difference was set when magnetic field was off. Temperature of cooled wall was maintained at 18°C, what was equal to ambient temperature inside the magnet test section. Temperature of heated bottom wall depended on supplied heater power. The temperature difference was kept constant during the experiment. Therefore when heat transfer processes in the enclosure were changing

Table 1. Thermo-physical properties of all working nanofluids (at 293 K), * values measured

Property	Symbol	Unit	Cu0.0112		Cu0.056	Cu0.112
Thermal conductivity	k_{nf}	[W·(mK)$^{-1}$]	0.59862		0.59942	0.60042
Density	ρ_{nf}	[kg·m^{-3}]	999		1003	1007
Specific heat	c_{pnf}	[J·(kg·K)$^{-1}$]	4181		4180	4177
Thermal expans. coef.	β_{nf}	[K^{-1}]	$20.58 \cdot 10^{-5}$		$20.57 \cdot 10^{-5}$	$20.56 \cdot 10^{-5}$
Dynamic viscosity	μ_{nf}	[kg·(m·s)$^{-1}$]	$100.39 \cdot 10^{-5}$		$100.28 \cdot 10^{-5}$	$100.14 \cdot 10^{-5}$
Electrical conductivity	σ_{nf}	[S·m^{-1}]	$5.502 \cdot 10^{-6}$		$5.509 \cdot 10^{-6}$	$5.518 \cdot 10^{-6}$
Mass magn. susceptibility	χ_m	[m^3·kg^{-1}]	$-14.5 \cdot 10^{-9}$*	$-13.4 \cdot 10^{-9}$*	$-12.4 \cdot 10^{-9}$*	$-12.4 \cdot 10^{-9}$*
Nanofluids process mixing	-	-	mechanically	ultrasonically		

the supplied heater power was adjusted to obtain required temperature difference. After system stabilization the convection measurements were performed for the magnetic induction of 0, 2, 4, 6, 8 and 9T. When one step was completed the magnetic field induction was changed to next value and then it required about 1-2 hours to stabilize the system, once again.

The recorded thermocouples signals were a basis in the Fast Fourier Transform (FFT) and spectral analysis. These methods were chosen due to the nanofluids opaqueness and failure of optical ones.

To understand the influence of magnetic field on the convective phenomena occurring in the cubical enclosure the following parameters were studied: location of the enlosure in magnetic environment, the type of mixing (mechanical and ultrasonic agitation) and the nanoparticles concentration at various magnetic induction values.

3 Influence of maximal gradient location

This phase of experiment concerned position of the experimental enclosure in the magnet test section. The working fluid was nanofluid with distilled water, as a base fluid, containing Cu/CuO nanoparticles of 40-60 nm diameter, prepared by two-step method with mechanical mixing utilization for 8 hours. The resulting concentration of nanoparticles was 0.0112 vol.%. The temperature difference was about 3 K. The cube locations were chosen due to maximal value of gradient of magnetic induction square (grad$\mathbf{B}^2_{max}$) occurrence, where magnitude of the magnetic buoyancy force had the biggest influence. Theoretical considerations of forces acting in the systems led to that two locations of the cubical enclosure in magnet test section schematically shown in Figure 3. The scale of the arrows does not represent the real relation between the gravitational and magnetic buoyancy forces' magnitudes. They only indicated existed differences.

The forces acting on fluid in the system were defined as:

- the gravitational buoyancy force acting on diamagnetic fluid:

$$\mathbf{F}_g = -\mathbf{g}\rho\beta(T - T_0)\,, \tag{7}$$

- the magnetic buoyancy force acting on diamagnetic fluid:

$$\mathbf{F}_m = -\chi_m \rho\beta(T - T_0)(2\mu_m)^{-1}\nabla\mathbf{B}^2_{max}\,, \tag{8}$$

- the magnetic force acting on paramagnetic particles:

$$\mathbf{f}_{m\_particle} = -\chi_m \rho(T - T_0)(T_0 2\mu_m)^{-1}\nabla\mathbf{B}^2_{max}\,, \tag{9}$$

- and the resultant force of the buoyant and gravity forces acting on paramagnetic particles:

$$\mathbf{F}_b = \rho_w \mathbf{g}V + \rho_{CuO}\mathbf{g}V \tag{10}$$

where: T_0 is the reference temperature equal to the arithmetical average of cooled and heated walls' temperature, χ is the volume magnetic susceptibility, V is the volume of particle.

In the position I the maximal value of grad$\mathbf{B}^2_{max}$ was situated in the middle of experimental enclosure and is represented as an ellipse in Figure 3. The magnetic

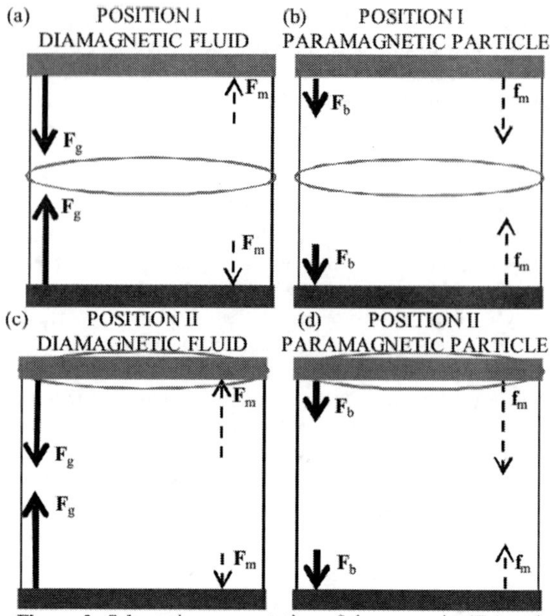

Figure 3. Schematic representation of the magnetic ($\mathbf{F}_m$) and gravitational ($\mathbf{F}_g$) buoyancy forces appearing in the system at position I acting on (a) diamagnetics, (b) paramagnetics and position II acting on (c) diamagnetics, (d) paramagnetics

and gravitational buoyancy forces acting on diamagnetics (water) are in opposite directions (Figure 3 (a)). The magnitudes of magnetization forces were similar due to the same distance between them and grad$\mathbf{B}^2_{max}$ position. Simultaneously, the magnetic buoyancy force acting on paramagnetic particles Cu/CuO were directed toward the maximum location (see Figure 3 (b)). The force magnitudes for paramagnetics were equal to each other but higher than for diamagnetics of about 10^3 times, however the particle concentration was low.

In the position II the maximal value of grad$\mathbf{B}^2_{max}$ was located on the upper (cooled) wall. Directions of acting forces were the same as in position I, although the magnitude of magnetic bouyancy forces acting in upper part of cube were greater than the ones in lower part of enclosure. It was caused by the distance from the maximal value of grad$\mathbf{B}^2_{max}$, the higher distance led to reduction of the magnetic buoyancy force value.

Summarizing the magnetic buoyancy forces acting on the diamagnetics caused convection attenuation while acting on the paramagnetics caused its enhancement. This complex reciprocal interaction of the forces came from the different magnetic properties of nanofluid components.

Dependence of the Nusselt number on thermo-magnetic Rayleigh number for two positions of the enclosure in the magnet test section is presented in Figure 4 (a). In both cases enhancement of heat transfer processes with increasing magnetic induction was of about 26% and about 13% in the first and the second enclosure positions, respectively. The theoretical considerations concerning the system of forces in both positions indicated that the attenuation of convection should be observed due to magnetic properties of water (which is diamagnetic).

DOI: 10.1051/e3sconf/20161000111

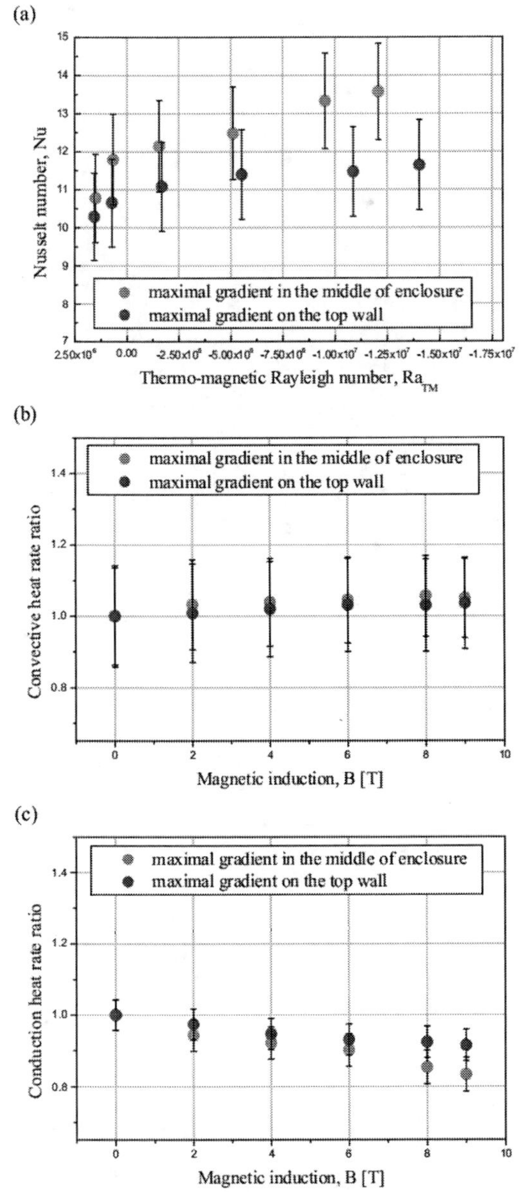

Figure 4. Heat transfer characteristics for various position of experimental enclosure in the magnetic field (a) The Nusselt number versus thermo-magnetic Rayleigh number; (b) convective and (c) conduction heat rate ratios versus magnetic induction

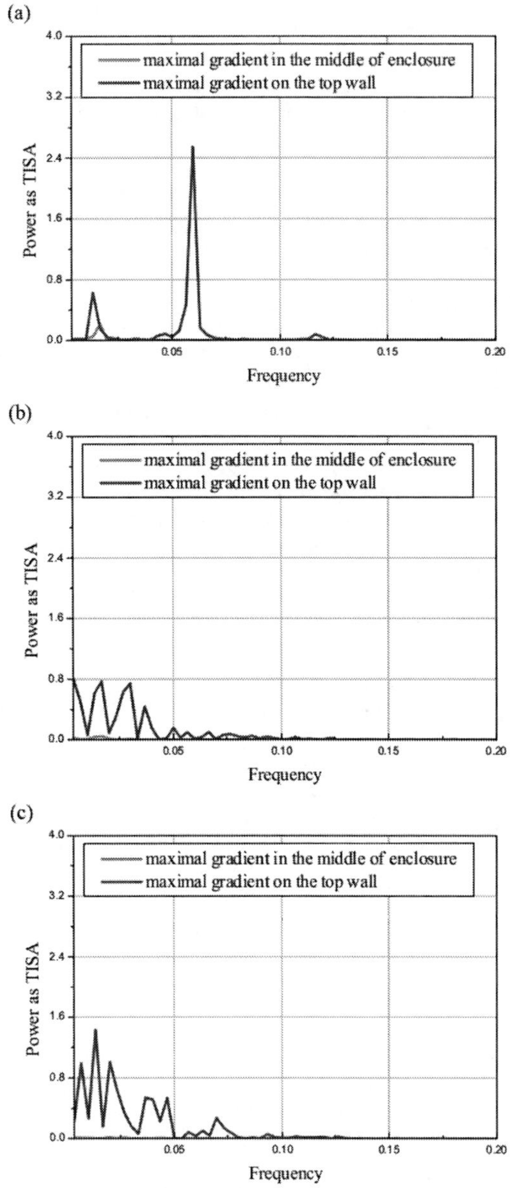

Figure 5. The power spectrum versus frequency for t6 thermocouple signal at (a) 0T (natural convection) (b) 4T and (c) 9T of magnetic induction for different maximal gradient location $\Delta T = 3$ K and $Ra_T = 1.5 \cdot 10^6$

However the results revealed that addition of paramagnetic particles caused change in the heat transfer nature. Possible explanation of this situation could be that the paramagnetic Cu/CuO nanoparticles acted as mixing agents what caused heat transfer enhancement. What may be concluded that nanofluid should not be treated as one-phase fluid. Moreover, it should be noted that convection changed more in position I, when the magnetic buoyancy forces magnitudes were similar to each other. To understand in detailed the heat transfer processes occurring in the enclosure the convective and conduction heat rates ratios versus magnetic field strength were presented in Figure 4 (b) and (c). The heat rate ratios were calculated as the ratio of actual heat rate to the reference heat rate obtained without magnetic field influence (at natural convection state). The convective heat rates ratios slightly increased with increase of magnetic induction for both experimental enclosure locations. Whereas, the conduction heat rates ratios decreased with magnetic field strength increase for both examined positions. It suggested that conduction was reduced at the expense of convection. The example of one thermocouple signal spectral analysis is presented in Figure 5. Thermocouple denoted as t6 in Figure 2 was placed inside the experimental enclosure, close to the bottom on the right side. The thermal power spectrum versus the frequency at different maximal gradient location for Cu0.0112 at and Ra_T = $1.5 \cdot 10^6$ and magnetic induction of 0T (natural convection state), 4 T and 9 T are shown in Figure 5 (a), (b) and (c), respectively. It could be seen that convection for case of maximal gradient located in the middle of enclosure was almost suppressed. Its slight appearance could be found at 9 T (Figure 5 (c)). In Figure 5 (b) the convection was present when maximal gradient was on the top wall, but 9 T of magnetic induction enhanced it.

4 Influence of preparation method

Two-step method of nanofluid preparation involves nanoparticles addition to the base fluid and mixing. Even though, it looks like a simple step, it causes a lot of problems. Therefore there are many publications about proper nanofluids preparation with satisfactorily dispersed particles [9]–[11]. Furthermore, there are no general instruction to prepare all kinds of nanofluids [12]. However the National Institute of Standards and Technology (NIST) and the Center for the Environmental Implications of Nanotechnology (CEINT) published a protocol with some guidelines [13] related to dispersion from powdered nanomaterials with ultrasonic disruption.

The aim of this stage investigation was to verify if the preparation method had an influence on the nanofluid behaviour in the magnetic field. The working fluids were prepared by two different methods, the first one was mechanical agitation for 8 hours, while the second one was ultrasonic homogenization for 30 minutes (with a few pauses preventing fluid overheating) in accordance with [13]. The temperature differences during the thermo-magnetic convection were about 3 K.

The results of Nusselt number dependence on thermo-magnetic Rayleigh number for two kinds of mixing processes (mechanical and ultrasonic agitation) during preparation of Cu0.0112 nanofluid were presented in Figure 6 (a). For mechanically mixed fluid the higher values of Nusselt number could be found. It came from the agglomerates, which were not broken by the mechanical mixing. Therefore, the magnetic buoyancy force had bigger influence on them and in the consequence on the heat transfer processes. To get deeper understanding the convective and conduction heat rates ratios were also calculated and are presented in Figure 6 (b) and (c). The convective heat rates ratios were almost constant for both kinds of preparation methods, however the conduction heat rates ratios slightly increased for the nanofluid being prepared by ultrasonic mixing. At the same time, the conduction heat rates ratio decreased for mechanically mixed nanofluid in higher magnetic fields. The particles uniformly dispersed caused increase in the conducted heat rate. Therefore it could be written that convection was dominated phenomena.

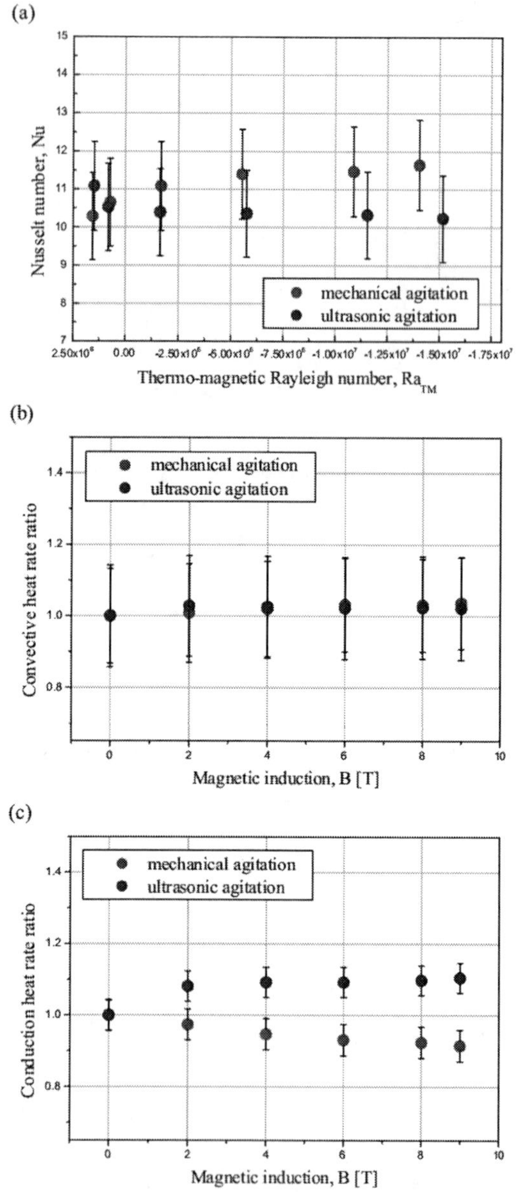

Figure 6. Heat transfer characteristics for various nanofluid preparation methods (a) Nusselt number ratio versus thermo-magnetic Rayleigh number; (b) convective and (c) conduction heat rate ratios versus magnetic induction

The thermal power spectrum versus the frequency at different mixing process for Cu0.0112 and $Ra_T = 1.5 \cdot 10^6$ and magnetic induction of 0T (natural convection state), 4 T and 9 T are shown in Figure 7 (a), (b) and (c), respectively. In Figure 7 (b) two large peaks with low frequency for ultrasonic mixing and a few small ones for mechanical agitation at 4 T were visible, while peaks with small amplitude at 9 T (Figure 7 (c)) were noticed. It could pointed out on big swirls and their reorganization to smaller vortices occurring in the experimental enclosure.

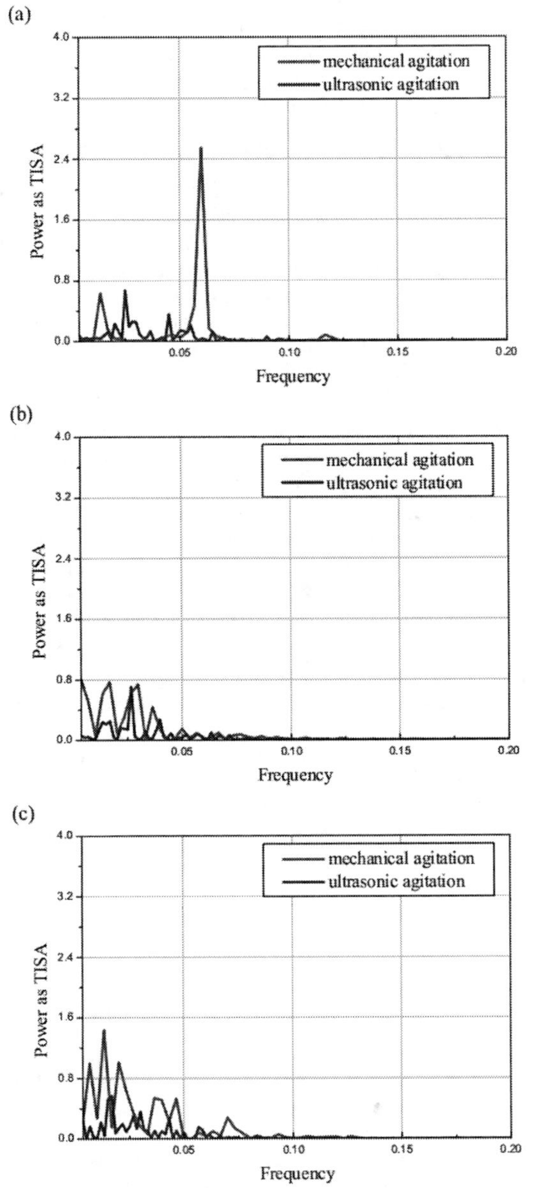

Figure 7. The power spectrum versus frequency for t6 thermocouple signal at (a) 0T (natural convection) (b) 4T and (c) 9T of magnetic induction for the various prepartion method (mechanical and ultrasonic mixing) at $\Delta T = 3$ K and $Ra_T = 1.5 \cdot 10^6$

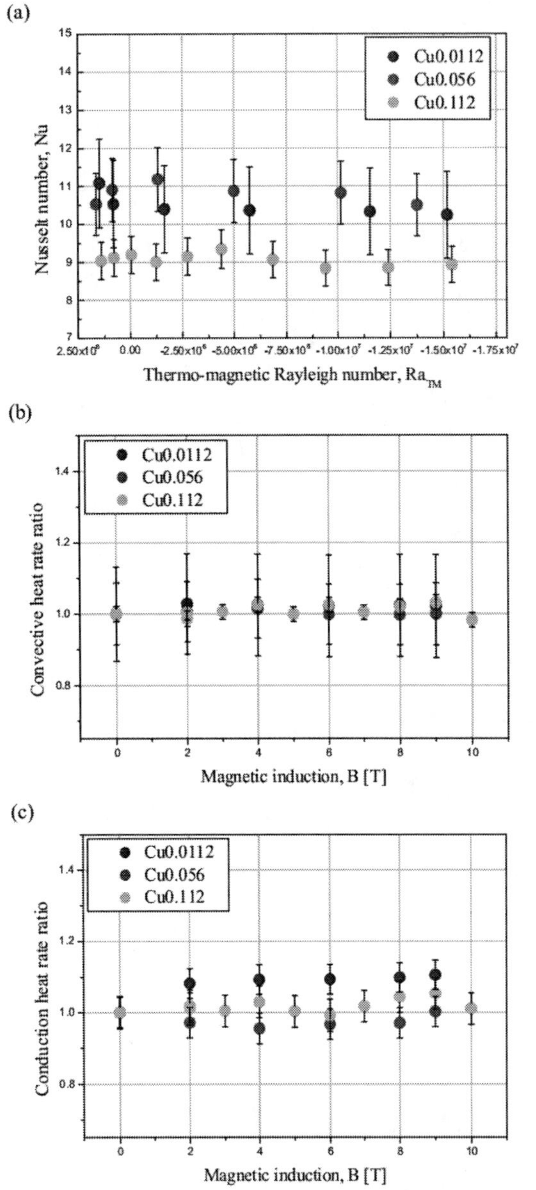

Figure 8. Heat transfer characteristics for various concentrations of nanoparticles (a) Nusselt number versus thermo-magnetic Rayleigh number; (b) convective and (c) conduction heat rate ratios versus magnetic induction

5 Influence of particles concentration

Low concentration nanofluids were examined due to relatively low cost of materials and assumption about electrical non-conductivity. The fluid electrical conductivity level for which it is considered as non-conductive was specified in [14] and equal to 21 S/m. The electrical conductivity of Cu0.112 was 5.518·10-6 S/m (Table 1) [15]. The nanofluids were prepared by two-step method with ultrasonic mixing process and their concentrations were 0.0112, 0.056 and 0.112 vol.%, which corresponds to abbreviations: Cu0.0112, Cu0.056 and Cu0.112. All properties of working fluids were calculated on the basis of formulas presented in [15] except the magnetic susceptibility which was measured (see Table 1).

The Nusselt number versus thermo-magnetic Rayleigh number for temperature difference of about 3 K for three nanoparticles concentrations at the various magnetic induction values is shown in Figure 8 (a). The Nusselt number decreased with increase of the magnetic field. After an initial increased of the Nusselt number value, it decreased at higher thermo-magnetic Rayleigh number for all showed cases. The convective heat rate ratio for various magnetic inductions is presented in Figure 8 (b).

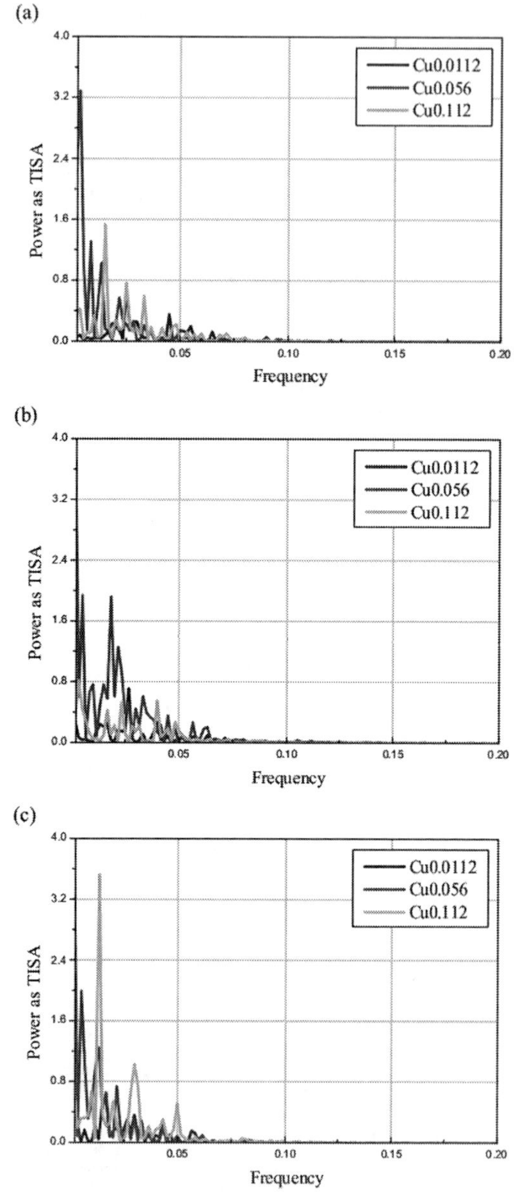

Figure 9. The power spectrum versus frequency for t6 thermocouple signal at (a) 0T (natural convection) (b) 4T and (c) 9T of magnetic induction for the various concentration of copper nanoparticles (ultrasonic mixing process) at $\Delta T = 3$ K and $Ra_T = 1.5 \cdot 10^6$

The ratios remained almost the same for all presented nanofluids concentrations. The conduction heat rate ratio for various magnetic field strength is shown Figure 8 (c). The ratios values are also similar to each other and there is not clear tendency.

Whereas, the thermal power spectra versus the frequencies for various copper nanoparticles concentration at magnetic induction of 0T (natural convection state), 4 T and 9 T are shown in Figure 9 (a), (b) and (c), respectively. For each fluid the peaks changed their frequency and amplitudes. The most important conclusion from the FFT and spectral analysis results is that increasing magnetic induction reorganized flow structure and also energy transported by them, what may have influence on the heat transfer.

6 Discussion and results summary

In this paper the experimental analysis of thermo-magnetic convection of weakly-magnetic, low concentration Cu/CuO nanofluids was presented. The few aspects were taken into account: influence of maximal gradient location, influence of mixing process during nanofluid preparation, influence of Cu/CuO nanoparticles concentration and influence of magnetic induction values. The goal of the paper was accomplished by proven that the magnetic field can enhance heat transfer by low concentration nanofluids. Moreover, it was verified that the nanofluid properties are the main parameters features and taking significant part in these processes.

Interaction of the gravitational and magnetic buoyancy forces, caused by different magnetic properties of the base fluid and nanoparticles, was complex. The theoretical analysis of system forces indicated that there should be convection attenuation observed under applied conditions. However the magnetic buoyancy force acting on the paramagnetic particles led to the heat transfer enhancement. The particles themselves acted as "mixer", which intensified the convection. Therefore, the nanofluid flow cannot be simplified as singe-phase flow in the presented system and under applied conditions.

The results confirmed that the nanofluid preparation is crucial to proper experimental analysis and is strongly related to the nanoparticles size and properties. Moreover, among all of studied parameters the preparation method had the greatest influence on the obtained heat transfer results.

Thermo-magnetic convection of nanofluids with different Cu/CuO nanoparticles concentration showed no clear trend. Therefore, the unequivocal interpretation of this part of presented results cannot be given. This aspect should be thoroughly examined and it will be the next step of the research.

The magnetic field application changed the flow structure what could be observed due to FFT and spectral analysis utilization. The most important conclusion from this research part is that increasing magnetic field reorganized flow structure and therefore, energy transported by the fluid. The experimental analysis will be continued and complemented with the numerical analysis to obtain the description of analysed aspects.

Acknowledgment

The present research was supported by the Polish National Science Centre (Project No. 12/07/B/ST8/03109). The participation in the conference was partially supported by the AGH UST Vice-Rector for Student Affairs, Professor Anna Siwik.

References

1. W. Yu, H. Xie, J. Nanomater. **2012**, 1 (2012)

2. M. T. H. Mosavian, S. Z. Heris, S. Gh. Etemad, M. N. Esfahany, J. Nanoparticle Res. **12**, 7, 2611 (2010)
3. Y. Xuan, Q. Li, Int. J. heat fluid flow **21**, 58 (2000)
4. K. Khanafer, K. Vafai, M. Lightstone, Int. J. Heat Mass Transf. **46**, 9, 3639 (2003)
5. D. Wen, G. Lin, S. Vafaei, K. Zhang, Particuology **7**, 2, 141 (2009)
6. S. U. S. Choi, J. A. Eastman, ASME International Mechanical Engineering Congress & Exposition, (2014)
7. H.Ozoe, S.W.Churchill, AIChE Symp. Ser. Heat Transf. **69**, 126 (1973)
8. A. Roszko, E. Fornalik-Wajs, J. Donizak, J. Wajs, A. Kraszewska, L. Pleskacz, MATEC Web Conf. **6**, 18, 1 (2014)
9. W. Williams, I. Bang, E. Forrest, W. Hu, J. Buonogiorno, Proc. NSTI **2**, 408 (2007)
10. M. Drzazga, Inż. Ap. Chem. **5**, 213 (2012)
11. S. Mukherjee, S. Paria, IOSR J. Mech. Civ. Eng. **9**, 2, 63 (2013)
12. Md. E. Kabir, M. C. Saha, S. Jeelani, Mater. Sci. Eng. A **459**, 1–2, 111 (2007)
13. J. S. Taurozzi, V. A. Hackley, M. R. Wiesner, NIST Spec. Publ. 1200-2 (2012)
14. A. Y. Jianwei Qi, Nobuko I. Wakayama, Int. J. Heat Mass Transf. **44**, 16, 3043 (2001)
15. Y. Xuan, W. Roetzel, Int. J. Heat Mass Transf. **43**, 3701 (2000)

E3S Web of Conferences 10, 00112 (2016) DOI: 10.1051/e3sconf/20161000112
SEED 2016

Green infrastructure practices – strategies how to sustain life in metropolitan areas

Anna Zaręba[1,a], Alicja Krzemińska[1], Krzysztof Widawski[1] and Piotr Oleśniewicz[2]

[1]*University of Wrocław, pl. Uniwersytecki 1, 50-137 Wrocław, Poland*
[2]*University School of Physical Education in Wroclaw, Banacha 11, 51-617 Wrocław, Poland*

Abstract. Green Infrastructure is 'an interconnected network of green space that conserves natural ecosystem values and functions and provides associated benefits to human populations' [1, p. 12]. It contributes to long term strategic development goals for data and information management. The research paper was introduced with the aim to protect green areas in urban environment by improving biodiversity, leisure and recreation, tidal and fluvial flood risk management, grey water treatment and quality of life. Planning and design of green system should provide appropriate responses to the distinctive local circumstances. Green spaces encourage social interactions, form new places to practise sport and contribute to public health and fitness, as well as have indirect benefits for a range of environmental services. The research hypothesis is that Green Infrastructure is a framework for conservation and development and we need to design Green Infrastructure systems strategically to connect across urban, suburban, rural and wilderness landscapes and incorporate green space elements and functions at the state, regional, community, neighbourhood and site scales [2]. Analysis and design of emerging Green Infrastructure system take a holistic view which links related physical, environmental, economic, social and cultural aspects of local communities.

1 Introduction

Intensive urbanization hampers the development and quality of municipal green areas, new terrain is covered by newly erected structures, pollution increases, forest area is reduced, erosion intensifies, biodiversity decreases due to fragmentation of habitats. The ecological functioning of green areas is largely dependent on the size of plant clusters, immunity, ecological relations, durability and use. As important as their role may be, urban ecosystems are, nevertheless, increasingly at risk of being devastated by an ever-expanding city. Planning solutions within the framework of Green Infrastructure are of extraordinary importance in urban environment as it is inhabited by over 60% of the EU population. Green Infrastructure in cities may have a beneficial effect on general health in the form of clean air, better quality water and wider accessibility of recreational grounds.

2 Objectives of the research

The main goals of the paper are: to identify opportunities for the restoration and enhancement of naturally functioning systems in already developed areas, to provide mechanism to identify green areas used for multiple functions: including ecological, recreational, cultural, aesthetic and other uses. The task of the paper is to present models of the Green Infrastructure strategy construction using various examples of metropolitan areas as case studies. The basic idea is to make comparison between Green Infrastructure practices and to create the review of scientific analysis of the GI typology and assessment which in further practical research could be used to develop urban planning models flexible enough to accommodate new development, but at the same time constant in protecting valuable terrain for recreation and the environment. The idea is to introduce the Green Infrastructure examples, which could be applied in different spatial settings.

The question is how to change municipal industrial and residential areas to fulfil different expectations of the inhabitants and optimize the use of accessible land for multiple functions related to polycentric city-regions – recreation, water management, health, biodiversity, renewable energy, sustainable transport, etc. The idea of the paper is to introduce models of ecological integrity which will help in regeneration of the environment and reclamation of derelict landscapes in cities. This paper takes a new approach to framing the role of green spaces within a multifunctional Green Infrastructure, and to developing a scientific analysis which promote spatial integration, sustainable transport, health and accessibility. Traditional land conservation and green infrastructure planning both focus on environmental restoration and preservation, but Green Infrastructure also concentrates on the pace, shape, and location of development and its relationship to important natural resources and amenities.

[a] Corresponding author: anna.zareba@uwr.edu.pl

 DOI: 10.1051/e3sconf/20161000112

3 From Urban Ecological Network to Green Infrastructure

Urban Ecological Networks concept introduced at the end of 20th century to connect green network and green structure concepts can be defined as: protecting or enhancing natural resources and systems while linking the urban area for recreation use [3]. The idea gives possibility to integrate of urban areas into a structural network as legally recognised instrument of planning in different scales (from regional to city and habitat). The principles of the concept were used in developing of Pan European Network for Central and Eastern Europe as well as in the "Green Hand" Copenhagen idea. Design features combine: ecological, recreation, cultural and aesthetes functions. The Ecological Network, was introduced with the aim to maintain the integrity of environmental processes by promotion of sustainable use of natural resources in order to reduce the impacts of human activities on biodiversity and/or to increase the biodiversity value of man-aged landscapes [4,2]. This system include vital ecological components: core areas (hubs), corridors, mix land used buffer zones (transitional areas) and sustainable-use areas. In the USA planning the Urban Ecological concept influenced the beginning of "greenways" idea, which corresponds to question how to create the large scale combination of green and blue ways connecting city, suburbs, exurbs and hinterland. Greenways, in the concept, are networks of land including linear elements that are planned for multiple purposes as: ecological, recreation, cultural, aesthetic, or other [5]. Greenways are multi-objective corridors, which have a key meaning in maintaining biodiversity, urban flood damage reduction, enhancing water quality and wildlife protection. The modern US greenways combine various landscape ecology features (examples: corridors, patches, matrix). Multi-function Rose Kennedy Urban Greenway, opened to the public in 2008, joins the park and public spaces in downtown of Boston.

Green infrastructure is a relatively new term in urban planning. It was used for the first time in 1994 report on land conservation strategies by the Florida Greenways Commission. Introduced initially as a new way of looking at nature conservation, with intent to elevate the value and functions of natural lands to the same level of importance as roads, constitute "grey infrastructure", assumed the proportion of the most important tool of sustainable development. In spatial planning across the EU the term Green Infrastructure (GI) corresponds to planning and management of green networks of multi-functional open spaces in a way that allows to support biodiversity and quality of life of inhabitants, particularly in derelict lands. Its design and management should also respect and enhance the character and distinctiveness of an area with regard to habitats and landscape types. Natural England defines Green Infrastructure as: "a strategically planned and delivered network comprising the broadest range of high quality green spaces and other environmental features. It should be designed and managed as a multifunctional resource capable of delivering those ecological services and quality of life benefits required by the communities it serves and needed to underpin sustainability…" [6, 7]. The GI combines: parks and gardens, natural and semi-natural urban greenspaces - including woodlands, wetlands, open and running water, green and blue corridors, outdoor sport facilities, amenity greenspaces, allotments, community gardens, urban farms, green roofs and walls. In the US spatial planning apart from the above mentioned comprehensive view about the significance of green areas, according the guidelines of the United States Environmental Protection Agency, Green Infrastructure strategies should first of all draw attention to problems of reduction or divert water from the sewer system though infiltration, re-usage and evapotranspiration. Green infrastructure refers to "systems and practices that use or mimic natural processes to infiltrate, evapotranspiration (the return of water to the atmosphere either through evaporation or by plants), or reuse storm water or runoff on the site where it is generated [7,8].

Green Infrastructure is a combination of interconnected hubs of nature lands. Hubs, including reserves, managed native landscapes, working lands, regional parks and preserves, community parks and natural areas differ in shapes and sizes. The continuity of that green network is possible thanks to system of linkages: landscape linkages, conservation corridors, greenways, green corridors, ecobelts [9]. Green Infrastructure technics include storm water infrastructure, tree planting, ecological restoration, green building and green roofs. Storm water infrastructure allows to direct runoff to areas that retain and infiltrate water through bioswales (natural drainage ways implemented to receive and absorb runoff), wetlands, rain gardens (vegetation planted in natural or artificial depressions) and permeable surfaces). Planting of trees as the basic technic of GI can be implemented in different scales: from single parcel to neighbourhood, boulevards, ecological corridors and regional parks. GI application in ecological restoration include among other things: renaturation of river valleys, replanting of native vegetation and removal of invasive species.

Green Infrastructure delivers a broad range of benefits to environment which can be grouped in three categories: environmental, economic and social. This multiscale benefits are particularly visible in highly damaged and limited urban environment. GI benefits include:

- reduced storm water runoff volumes – through usage of natural retention and absorption capabilities of vegetation
- enhance groundwater recharge - through implementation of natural infiltration of GI technologies
- improving environment quality - storm water pollutant reduction and absorption pollutants from the air through leaf uptake and contact removal
- reduced sewer overflow events – through usage of natural retention and absorption capabilities of vegetation allows to reduce runoff volumes and delay storm water discharges
- mitigation of Urban Heat Island Effect - implementation of trees, green roofs, green walls, permeable pavements can help in decreasing temperature in city centres and reduce demand for energy for air conditioning and heating

• increase biodiversity
• additional recreational space - new parks, urban forests, wetlands, greenways, boulevards and other forms of Green Infrastructure create accessible recreational space for city inhabitants
• improved human health and wellbeing – GI has positive impact on local communities
• increased land values in surrounding residential areas [7].

4 Multi-function of green infrastructure in regeneration of metropolitan areas – case studies

Theory of patchiness in landscape ecology allows to understand dynamics in urban ecosystem expressed in the flow of energy, matter and species across the countryside. It gives the tools to explain spatial heterogeneity and multi-scale of Green Infrastructure seen as a bond between different forms of urban continuum: city, inner suburbs, suburbs, exurbs and hinterland divided further into land use types, neighbourhoods, blocs, etc. [10]. In other words Green Infrastructure as a system consists of parts (ex. trees, soil, infrastructure etc.) that are organized in patterns (green core areas, green corridors) and performs various functions (ex. climate mitigation, recreation, environmental protection etc.). Green Infrastructure is subjected to hierarchy rules: 'it incorporates multiple subsystems (e.g., hydrology, vegetation, and movement) and in turn is a subsystem within a larger system (e.g., region, city, or neighbourhood), where it interacts with other systems (e.g., transportation, economy, and governance)' [8].The contemporary Green Infrastructure concepts put emphasis on multi-purpose usage of green areas. Metropolitan regions that effectively incorporate greenspace and Green Infrastructure into their urban fabric use natural areas not only for maintaining and enhancing biodiversity and habitat conservation but also in innovative storm water management, recreation, sustainable food production and delivery systems.

4.1 The Emerald Necklace Expanded Vision Plan (2014)

In the 1930 the Los Angeles Chamber of Commerce concerned at the growing air pollution, overpopulation and the lack of inner city green areas introduced the repair plan for the future - The Emerald Necklace, completed by the famous urban park pioneer Frederick Law Olmsted, Sr. The plan, rooted in the "emerald necklace" vision for Boston. designed in the late 1800's wasn't adopted and during last decades the urbanization sprawled dramatically. In 2005 the vision plan was significantly extended to connect 17-mile loop of parks and greenways along the Río Hondo and San Gabriel Rivers in eastern Los Angeles, inhabited by nearly 500,000 residents. The guidelines for the Plan included: restoration of the Los Angeles and San Gabriel rivers and tributaries for recreational open space, conservation of native habitat, and put emphasis on education. After three years the necessity of unification of heavy overpopulated area of Southern California assumed the proportion of the most important planning goal and The Emerald Necklace Expanded Vision Plan to connect 1,500 acres of parks and open spaces along an greenways around the Río Hondo, San Gabriel, and the lower Los Angeles Rivers was initiated. The planning target is to link Los Angeles County with interconnected multi-purpose network of parks, rivers and land [11].

4.2 Trinity River Corridor Project

The Trinity River Corridor Project of converting the 2,300-acre floodway of the Trinity River near downtown Dallas to a sustainable Green Infrastructure joins flood control, environment, recreation, transportation, and economic development actions. The project key concept apart from regeneration of the downtown and increasing the level of green amenity areas, is to modify existing levees to achieve sustainable flood protection by introducing recreational lakes (built on the place of extraction of materials needed for construction of new roads), marshlands, wetlands which help to maintain water cleansing, aquatic habitat, and recreational value. Green corridors are perfect examples to illustrate a rule of interconnectivity of watercourses linking rural, suburban, and urban areas within a region, introduced for multi-function providing multiple benefits for communities. The scale of actions could vary from typical urban investments, less managed landscape in suburbs to riparian woodland in rural area [12].

4.3 The Regional Plan of Territorial Planning in the Metropolitan Area of Lisbon (2010)

The plan was introduced with the idea to incorporate existing and new green urban areas into regional spatial planning. The Metropolitan Ecological Network (*Rede Ecológica Metropolitana*, REM) was established to coordinate a network of green areas between cities and with the urban system The objectives of the Metropolitan Ecological Network (*Rede Ecológica Metropolitana*, REM) are to maintain and enhance ecological connectivity. The plan is orientated not only to ecological goals as: revitalization of degraded urban areas, renaturalisation of watercourses, prevention and reduction of natural risks and protection of water resources, soil and landscape in urban environment, but also to society benefits as: tourism, recreation and leisure. Green areas in REM were organized on three hierarchical levels according to their contribution to the ecological structure of the metropolitan area (*Rede Primária, Rede Secundária e Rede Complementar*) [13].

5 Green Infrastructure practices – how to build healthier environment and sustainable communities

Green Infrastructure solutions can be applied at different scales (regional, local, subarea, district, neighborhood and site) and in different contexts. At the regional level implementation of Green Infrastructure guidelines can help in urban revitalization and landscape preservation (e.g. in maintaining and enhancing watersheds, ecological zones), at the scale of local government in maintaining the continuity of green areas in urban and suburban zones (e.g. through greenways), at the district and neighborhood in delivering of new amenity space in master planning and urban design. The benefits of implementation of GI solutions at the site scale concern green storm water infrastructure, green roofs, green walls, etc.

5.1 Data Inventory

The data include information about: geographical environment (e.g. division of administrative units, water pattern, slopes, settlement pattern, transport network), nature base (e.g. biotopes, designated areas, natural vegetation areas), land base (e.g. soil type, land quality, soil erosion sites, land cover) and water resources.

5.2 Defining the Green Infrastructure Network

A Green Infrastructure network map illustrates long-term planning pattern of GI areas and shows their relation to other components of the community's spatial guidelines. The map is the result of data inventory with distribution of existing and potential Green Infrastructure. The guidelines how to create the GI Network are included in Chapter: Smart Growth & Smart Conservation in the State of Maryland of Benedict and McMahon book [9] describing a comprehensive approach to designing the GI, which was used in Maryland and Florida in 1997. According to Benedict and McMahon, natural ecosystem values and functions are primary criteria to identify network components. The methodology uses the largest, highest-quality areas of natural habitat and the most important ecological links, which bond together GI patches. In Maryland's Green Infrastructure Assessment, the minimum size of defined hubs is 250 acres and links 1,100 feet in width. The Maryland GIA used GIS methods to identify high priority lands, for example: the series of watershed maps was created as the result of analysis of data layers e.g.: natural vegetation types, stream systems and topographic features. The methodology prioritize conservation targets and preservation of the most important hubs of natural areas introduced for multiple ecological benefits (e.g., water quality protection, biodiversity, conservation and restoration of natural areas) [14,8]. The hubs and corridors were marked through consideration of factors: land cover, wetlands, sensitive species, roads, terrestrial and aquatic conditions, floodplains, soils and development pressure. After the identification of hubs and corridors GIS methods were used to assess the network's components for vulnerability to development. Further analysis included consideration of ecological value of components. The ultimate comprehensive map showed composite ranking of components according to ecological value, vulnerability to growth and current degree of protection. The task of ranking was to recommend the areas that should be at the top of the state's acquisition list [14].

Above mentioned approach to Green Infrastructure can easily be applied in the area of intact natural resource land. In urbanized, highly fragmented environment looking only for the largest areas of virgin nature seems irrelevant. Investigation on planning models should be based on GI classification : agricultural land, allotment, community garden or urban farm, cemetery, churchyard or burial ground, derelict land, general amenity space, grassland, heathland, green roof, institutional grounds, orchard, outdoor sports facility, park or public garden, private domestic garden, street trees, water body, water course, wetland, woodland, green corridors, (not GI) etc. and attributing green areas to functions such as: habitat for wildlife, corridor for wildlife, soil stabilization, water supply, wind shelter, recreation, learning, etc. Each GI element can perform multiple functions, for example: green travel route, aesthetic, shading from sun, noise absorption, habitat for wildlife, corridor for wildlife, soil stabilization, water supply, wind shelter learning. Using the smallest possible units of functionality allows a distinction to be made between GI elements, but also provides a platform for grouping functions (eg. into benefits) at a later date. In the next step of GI strategy guideline needs for particular functions are assessed which allows to introduce the Intervention Plan. This method has been developed by the North West Green Infrastructure Unit for mapping Green Infrastructure. North West Green Infrastructure Guide consists of five steps: Partnership and Priorities, Data Audit& Resource Mapping, Functionality Assessment, Needs Assessment and Intervention Plan [15].

Allen [16] established a straightforward set of GI project typologies that are organized into three categories: 1) Income Generating (urban forest, commercial harvesting, urban agriculture, food production in re-used buildings, alternative energy, 2) Compliance/Regulatory (stream/riparian restoration, storm water park phytoremediation (i.e. plantings to Sunflower planting for toxics absorption mitigate contamination without excavation) impervious surface reduction, and 3) Community Benefit (local recreation facilities, recreational trail linkages, non-production garden, community garden).

5.3 Plan Goals and Policies

GI plans include goals which define the vision that a community would like to achieve, objectives how to achieve the planned out goals, policies to guide decision making and implementation strategies and actions. Comprehensive plans involve Green Infrastructure solutions in community systems (community facilities and services, etc.), for example green boulevards, belts of trees along rivers can incorporate gray infrastructure: pedestrian, bicycle and car roads into native plantings and natural storm water management features [8]. Policy priorities identified by Natural Economy North West, a partnership between Natural England and the North West Development Agency represent 11 benefits of Green Infrastructure 'interventions' grouped in 3 categories – economic (economic growth and employment), environmental (protect and enhance cultural heritage, protect and enhance the landscape, geodiversity and natural environment, biodiversity conservation and enhancement, climate change mitigation and adaptation and social (promoting sustainable transport and reducing the need to travel by car, community cohesion and lifelong learning; volunteering, healthy communities; health and wellbeing) [17].

5.4 Implementation and Monitoring

Plans include an implementation program that establishes action priorities, timelines, and indicators of progress. Green infrastructure implementation tools are grouped in four main categories: land dedication, regulations, incentives, and physical investments [8].

Conclusions

Natural environmental qualities are a significant value in spatial management. In many metropolitan areas, green space is rapidly disappearing. Preservation of urban greenery is the underlying principle of metropolitan regions eco-development. Identification and evaluation of green areas is one of the rationales behind spatial management. Absence of information on the potential and risks connected with expanding human activity, which directly affect environmental qualities, will rule out the possibility of appropriate and systematic management of these areas. The analysis of green spaces identification, typology and assessment should be combined to form different models of Green Infrastructure. The main purpose of comparing different concepts is to show the sphere of activity of the Green Infrastructure in different conditions. The GI models contribute to long term strategic development goals for data and information management. Analysis and design of emerging Green Infrastructure system take a holistic view which links related physical, environmental, economic, social and cultural aspects of local communities. Planning and design of green system should provide appropriate responses to the distinctive local circumstances.

References

1. M. Benedict, E. McMahon, *Green Infrastructure: Linking Landscapes and Communities.* Washington, D.C.: Island Press (2006)
2. A. Zaręba, Problems of Sustainable Development, **9**, 149-156 (2014)
3. C.W. Thompson, Landscape and Urban Planning **60** 59–72 (2002)
4. G. Bennet, *Integrating Biodiversity Conservation and Sustainable Use: Lessons Learned From Ecological Networks*, Gland/Cambridge, IUCN (2004)
5. J. Ahern, Landscape and Urban Planning, Special Greenways Issue, **33**, 131-155 (1995)
6. www.naturalengland.org.uk (accessed April 26, 2016)
7. https://www.epa.gov/green-infrastructure/benefits-green-infrastructure (accessed May 10, 2016)
8. D. C. Rouse, I. F. Bunster-Ossa, *Green Infrastructure: A Landscape Approach*, American Planning Association, Planning Advisory Service Report Number 571, American Planning Association, (2013)
9. M. Benedict, E. McMahon., *Green Infrastructure: Smart Conservation for the 21st Century*, (2001), http://www.sprawlwatc h.org/greeninfrastructure.pdf
10. C.Wayne et al, Ecological Applications, **10**, 685-688 (2000)
11. http://www.conservationfund.org/projects/an-emerald-necklace-for-los-angeles (accessed April 26, 2016)
12. http://www.trinityrivercorridor.com (accessed April 26, 2016)
13. Regional Plan on Spatial Planning of the Metropolitan Area of Lisbon (*Plano Regional de Ordenamento do Território da Área Metropolitana de Lisboa*, PROT AML) (2010) http://protaml.inescporto.pt/
14. Maryland's Green Infrastructure Assessment and GreenPrint Program 2004, http://www.conservationfund.org/images/programs/files/Marylands_Green_Infrastructure_Assessment_and_Greenprint_Program.pdf(accessed April 26, 2016)
15. http://www.greeninfrastructurenw.co.uk/resources/GIguide.pdf (accessed May 10, 2016)
16. W. L. Allen, Journal of Conservation Planning **10**, 43 - 51 (2014)
17. http://enviroeconomynorthwest.com/download/environmental_quality/projects/A%20Future%20for%20the%20Natural%20Economy.pdf (accessed April 26, 2016)

DOI: 10.1051/e3sconf/20161000113

Corrosive components of nutshells and their chars

Mateusz Karczewski[1,a], Marta Marczak[1], Piotr Burmistrz[1] and Dorota Makowska[1]

[1]*AGH University of Science and Technology, Faculty of Energy and Fuels, al. Mickiewicza 30, 31–464 Krakow, Poland*

Abstract. Biomass combustion stands among various technologies pointed at fossil fuels consumption decrease. Biomass can be found in very diversified sources spread more evenly across the globe, can be burned with use of traditional combustion solutions and is more CO_2 neutral in combustion than their fossil fuel counterparts. On the other hand biomass has several problems with composition that despite its potential diversity. Problem of excess moisture can be already solved by material selection or by preliminary pyrolysis. The main problem concerns however biomass ash composition. Biomass ashes are more prone to have higher quantities of potentially corrosive components than their coal counterparts. The example of such constituents are alkali metals, sulphur and chlorine. Ash basic composition is also important due to various ash properties like its melting temperature and slagging or fouling tendencies. To address the problem, several indices for fast properties prediction and earlier problem identification can be appointed. This work concentrates on ash quality evaluation for potentially attractive biomass fuel from nutshell materials and their corresponding char obtained by pyrolysis in 300, 450 and 550 °C. Pistachio and hazelnut shells with their chars will be analysed for corrosive compounds and their potential influence on combustion process.

1 Introduction

Biomass solid fuels became increasingly popular source of renewable energy in recent years. The relative low investment cost for implementation and possibility to use seemingly unwanted side products of different processes related to agricultural, industrial or municipal activity were one of the main drives for such trend. There are also significant advantages over conventional coal fuels like lower sulphur and nitrogen content, similar qualitative composition for coal and reduction in waste storage scale [1-5].

On the other hand, there are many drawbacks connected with its use. Most of potential biomass fuels possess large quantities of water in its content, an unfavorable constituent that lowers calorific potential of given fuel and increases its volume needed to substitute conventional solid fuels, effectively increasing the transportation cost [6]. One of possible solution is selection of low-moist biomass materials like nutshells [7]. Among other possible operations, pyrolysis as pretreatment method seems to be most promising. Treating fuel with high temperature in lack of oxidant leads to water evaporation and to release of volatiles due to temperature induced breaks in fuel structure. Removal of both constituents results in noticeable increase in carbon content and fuel quality.

Although pyrolysis allows for better fitting of biomass fuels for existing coal combustion technologies, the biomass inorganic constituents are largely unaffected by its influence. Various studies on the topic of biomass utilization agree that biomass ash pose much bigger of a challenge in process handling. Due to very diversified ash composition between various biomass types and often different quantities than that found in coal ashes. [8,9].

Chlorine is one of the examples of unfavorable ash constituent. In biomass that is generally scarce in sulphur many alkali metals are present in a form of alkali chlorides. Alkali chlorides can act as a catalysts in a process of iron oxidation in the structure of pipes [8]. If there is also sulphur present in ratio at least twice as low as chlorine concentration the corrosion of equipment might even worsen due to complex hydrochloric acid-based corrosion reactions [10,11]. Abundance of alkali metals is also important in this process, as they are needed for creation of salts with sulphur and chlorine, thus increasing their deposits in the system and form various eutectic with corroded steel e.g. KCl-$NaCl$-$FeCl_2$, further developing the damage of its structure [12] and is also suspected for catalysts deactivation [13]. Apart from corrosion caused by certain ash constituents, the combustion can be influenced by general ash properties. Biomass ash is susceptible for slagging, agglomeration and fouling[14,15] as a result of its lowered ash-fusion temperatures due to presence of elements like Ca, Cl, K, Mg, Na, P, Al, Fe, Ti, Si and S in certain ranges [16]. Several indices were proposed to predict biomass ash properties among which

[a] Corresponding author: mateusz.karczewski@agh.edu.pl

Nomenclature

A	ash yield	SP	slagging possibility
a	analytical basis	ST	ash softening temperature
AI	alkali index	ppm	parts per million
BSEI	biomass slagging evaluation index	PS	pistachio shells
daf	dry, ash free basis	PSC	pistachio shell chars
db	dry basis	$R_{b/a}$	base-to-acid ratio
HS	hazelnut shell	UHS	untreated pistachio shells
HSC	hazelnut shell char	UPS	untreated hazelnut shells
Y	Yield	V	volatile matter
M	moisture	%	percent by weight

Cl and S ratios [8,17] e.g. Cl/S ratio [18,19], Biomass slagging evaluation index [20] and Alkali index [16].

More on index calculations can be read elsewhere [21,22]. Given the importance of biomass ash composition, a complex analysis needs to ensure biomass fuel quality. In this work the ash composition of raw hazelnut and pistachio nutshells and their chars were evaluated in order to ensure their potential as a biomass material for combustion and co-combustion processes. The materials were analyzed for corrosion-related constituents and various indices will be used to state the potential behavior of their ashes during the process.

2 Experimental

2.1 Material preparation

Base material was obtained from food industry residue waste materials from pistachio and hazelnut distribution. The shells were air dried for up to 48h and ground in ring roller mill (EKOLAB LAB-09-200). The resulting powders were the sifted for fractions between 0.5 and 2.0 mm in diameter. In order to obtain chars of respective nutshell, the materials were placed in ceramic containers with lids and put into a muffle furnace (CZYLOK FCF 7 SM/pl) where they were heated to 300, 450 or 550 °C for 30 minutes. Raw nutshell or their respective chars were then characterized.

2.2 Proximate and ultimate Analysis

Nutshells and their chars elemental composition was conducted with Eltra CHS-580 analyser according to polish standard PN-G-04571:1998 and PN-G-04584. The measurement provided information on carbon, hydrogen and sulphur content.

Proximate analysis comprised of moisture, volatile matter ash content and lower heating value was also performed in order to obtain data for indices calculations. Moisture, volatile matter and ash analysis was performed with thermogravimetric analyzer (Eltra Thermostep). The lower calorific value measurement was performed with KL-11 Mikado calorimeter according to polish standard PN-EN 14918:2010.

2.3 Ash constituents determination

For determination of main ash components content Atomic Absorption Spectrometry (Hitachi High Technologies Z-2000) was used. Measurement required conversion of the material into liquid medium.

Table 1 Summarisation of most important fuel constituents and parameters for pistachio and hazelnut shells and their chars obtained from technical analysis, elemental analysis and chlorine spectrophotometric. Some measurement errors was not added due to their neglibility

Sample	C^a [%]	H^a [%]	M^a [%]	A^a [%]	V^a [%]	Cl^{db} [%]	S^a [ppm]	Cl/S [-]	Y^a [%]	LHV^{daf} [MJ/kg]
PS (Raw)	47.4±0.30	6.9±0.30	6.1±0.05	1.1±0.02	76.9±0.01	0.5±0.03	61±5	86.8	-	16.8
PSC (300°C)	59.3±0.30	5.6±0.30	0.5±0.05	1.1±0.02	62.8±0.02	0.8±0.03	73±5	109.1	63.8	21.9
PSC (450°C)	82.5±0.30	3.4±0.30	0.2±0.05	1.6±0.02	18.2±0.02	0.8±0.03	63±5	121.1	26.1	31.0
PSC (550°C)	88.8±0.30	2.6±0.30	0.1±0.05	3.0±0.02	8.2±0.02	0.8±0.03	44±5	179.3	22.4	34.4
HS(Raw)	50.1±0.30	6.9±0.30	8.0±0.05	0.3±0.02	75.8±0.02	0.1±0.03	92±5	3.7	-	16.1
HSC (300°C)	61.2±0.30	5.6±0.30	1.3±0.05	2.3±0.02	61.2±0.02	0.1±0.03	112±5	9.0	63.6	16.9
HSC (450°C)	81.8±0.30	3.4±0.30	0.8±0.05	3.3±0.02	25.7±0.02	0.1±0.03	114±5	8.7	31.7	31.1
HSC (550°C)	88.6±0.30	2.8±0.30	0.2±0.05	3.0±0.02	8.1±0.02	0.1±0.03	39±5	23.7	23.2	34.3

DOI: 10.1051/e3sconf/20161000113

To achieve this, 0.2 g of every sample was moved into separate Teflon tube and treated with mixture of 6 ml HNO_3 (69%, POCH, for spectral analysis) and 2 ml of HF (J.T. Baker, analytical solution up to 60%), closed and placed into microwave digestor (Speedwave Four Berghof) for 2.5 h, in order to react all solid mineral matter. Sample vessels were then cooled, opened and treated with 20ml of saturated boric acid solution and placed in digester for another 40 minutes. Volume of resulting samples was then adjusted to 50ml with deionized water with 50µl addition of cesium solution (100mg/l). Concentrations of Na, Ca, Mg, P, Ti, Si, Fe, Al, K, were then measured on AAS flame atomizer. Phosphorus content was determined by X-ray fluorescence
using EDXRF spectrometer (Panalytical Epsilon 3^{XLE}).

Chlorine content have been evaluated as chlorine anions in water solution with use of direct reading spectrophotometer (DR/2000 HACH). Chlorine present in sample was transferred into liquid solution by burning 0.5 g under 0.5 g layer of chlorine-absorbing Eschka mixture (60÷72% MgO and 30÷36 % Na_2CO_3, POCH) and boiling remaining residue in deionized water with addition of mercuric thiocyanate (99.99%, POCH) and ferric ion solution (POCH), according PN-ISO 587:2000 standard.

3 Results

3.1 The influence of pyrolysis temperature on material properties and content of sulphur and chlorine

General outlook on nutshells properties after thermal treatment shows increase of carbon content followed by decrease of other constituents like moisture and volatile matter, typically for pyrolytic processes. Beside major enhancement of lower heating values with increase of pyrolysis temperature, such loss of carbonaceous compounds is followed by visible loss of sample mass, from around 33-36% on lowest treatment temperature up to around 77% in highest, as pictured by data on char yield (Table 1).

Although the overall quality of fuels is improved that way, some more temperature resistant biomass compounds also increase their mass contribution. One of most important of them is chlorine, which increases its content from 0.52% to around 0.8% in case of pistachio shells and from 0.03% up to 0.1% in case of hazelnut shells, as seen in Table 1 The rapid gain in chlorine content between samples treated in 300 °C and raw sample followed by stable Cl concentration in rest of chars for both materials might suggest some process of weaker bonded chlorine loss, balanced with mass decrease as it declined during higher pyrolysis temperatures. It is important to note that literature data

Table 2 Chemical composition of ash in pistachio and hazelnut shells and their corresponding chars after recalculation for oxides, shown as weight content in percents [%], parts per million [ppm] or mixed in specified order [ppm/%]. The result assumes that detected elements are present in ash in their most stable oxide, presented by weight percent. Literature comparison is marked with asteriks [26,27].

Sample	MgO [%]	Na_2O [%]	K_2O [%]	Fe_2O_3 [%]	SiO_2 [%]	Al_2O_3 [%]	CaO [%]	P_2O_5 [%]	TiO_2 [ppm/%]
UPS (Raw)*	3.3	4.6	18.7	36.3	8.4	2.2	10.3	12.1	0.2
PS (Raw)	3.8±0.20	14.8±0.08	21.7±0.90	28.3±0.42	8.7±0.08	2.1±0.06	12.7±0.28	12.79±0.02	647±1.50
PSC (300°C)	3.9±0.21	12.8±0.08	22.9±0.80	28.5±0.36	8.7±0.06	2.3±0.14	12.8±0.34	10.45±0.01	645±1.50
PSC (450°C)	3.8±0.30	11.0±0.08	20.8±1.25	28.8±0.55	9.0±0.42	2.1±0.09	12.7±0.25	8.47±0.02	645±1.50
PSC (550°C)	3.5±0.30	8.3±0.04	22.0±0.50	28.9±0.73	8.9±0.4	2.1±0.09	12.8±0.19	6.82±0.01	646±1.50
UHS (Raw)*	7.9	1.3	30.4	3.8	33.7	3.1	15.4	3.2	0.1
HSC (Raw)	3.3±0.40	1.6±0.03	34.6±0.32	1.4±0.28	37.9±0.08	3.0±0.1	13.7±0.19	5.98±0.02	0.2±0.02
HSC (300°C)	3.7±0.04	1.4±0.02	34.4±0.54	1.5±0.4	37.7±0.22	3.1±0.07	13.9±0.24	3.31±0.03	0.2±0.02
HSC (450°C)	3.5±0.10	0.9±0.02	34.7±0.25	1.4±0.13	36.9±0.35	2.9±0.31	13.7±0.27	2.05±0.03	0.2±0.02
HSC (550°C)	3.7±0.07	0.6±0.02	33.4±1.47	1.4±0.24	36.7±0.26	3.0±0.14	13.6±0.34	2.00±0.02	0.2±0.02

on pistachio shows lower concentrations of chlorine of up to 0.21%. The difference for hazelnut is much smaller as literature claims it to be around 0.2%, which in comparison with 0.03% might origin from measurement precision. Such differences might result from salt addition to nutshells during their processing. Sulphur content is one of many nutshell virtues as it is present in their structure in amounts as small as 61ppm for pistachio and 92 ppm in hazelnut. What is more, its content, although increasing with pyrolysis temperature up to 450°C treatment temperature, finally decrease with temperatures beyond that mark, up to 44ppm and 39ppm, respectively. Literature data gives much higher sulphur concentrations of around 1,6% and 0.44% for pistachio and nutshell respectively. The reason between such differences can only be explained by growth environment. In order to examine those subsequent processes in respect for their influence on corrosion potential, a ratio of chlorine to sulphur was Cl and S contents. The results shown that slagging intensity would be much lower if the nutshells used by literature data were used rather than pistachio shells post those that have been post-processed.

3.2 Influence of ash constituents on corrosive properties

Ashes are the most dangerous solid fuel counterpart due to potential presence of vast range of either metallic of non-metallic origin. The summary of their share in ash composition was presented in Table 2. Data in the table was prepared with assumption that most of ash compounds are present in their most stable oxide forms. Pistachio and hazelnut being in the same category of biomass products, have large similarities in their ash composition. Both nutshells are abundant in potassium and calcium, elements typical for biomass fuels. High concentration of phosphorous and iron with relatively low amounts of aluminium and magnesium are normal

Table 3 Summarisation of calculation results of ash quality indices for pistachio and hazelnut shells and their corresponding chars, with additional explanation of results. The indices present in the table are starting from the left : Cl and S Ratio, Biomass Slagging Evaluation Index, Alkali Index, Base-to-acid Ratio, Slagging intensity based on Cl and S ratio, Slagging possibility based on AI and BSEI and Softening Temperature based on Base-to-acid Ratio. Samples marked with asteriks were calculated basing on literature data [26,27].

Sample	Cl Ratio [-]	S Ratio [-]	BSEI [-]	AI [kg Na_2O + K_2O/ GJ]	$R_{b/a}$ [-]	Slagging	SP	ST [K]
UPS (Raw)*	0.14	1.40	1.87	0.14	6.72	Slight	Low	1000-1200
PS (Raw)	3.05	2.89	1.29	0.23	7.27	Serious	Possible	1000-1200
PSC (300°C)	3.33	3.19	1.43	0.43	7.47	Serious	Certain	1000-1200
PSC (450°C)	2.93	3.34	1.63	0.48	6.94	Serious	Certain	1000-1200
PSC (550°C)	2.83	3.21	1.67	0.49	6.96	Serious	Certain	1000-1200
UHS (Raw)*	0.09	0.45	0.8	0.19	1.59	Slight	Possible	1000-1200
HSC (Raw)	0.98	2.70	0.43	0.35	1.38	Serious/ Severe	Certain	<1000
HSC (300°C)	0.87	2.09	0.48	0.56	1.34	Serious/ Severe	Certain	<1000
HSC (450°C)	0.92	2.24	0.49	0.53	1.39	Serious/ Severe	Certain	<1000
HSC (550°C)	0.79	1.94	0.52	0.50	1.26	Severe/ Slight	Certain	<1000

defined. This parameter seems to be highly exceeded benchmark ratio of 2 for all pistachio materials, suggesting that their potential to dissociate steel is strong. On the other hand the amount of chlorine versus sulphur in hazelnuts is also high but the result is smaller by around tenfold with their pistachio counterparts, suggesting weaken chlorine/sulphur alkali corrosion [10,11] for hazelnuts than for pistachio shells. Cl/S ratio was also calculated for literature data, due to varying for pistachio nutshells, according to literature [26]. Its high sodium content of around 14% that is not confirmed in other works also hints along with high chlorine content to be an effect of salt addition during pistachio processing. Hazelnut has more biomass-proper composition with more corrosion safe compounds like potassium, aluminium and calcium oxides dominating with low presence of potentially corrosive sodium and iron. Its composition is also in agreement with literature data [27].

The important conclusion ash content data is general stability of compounds concentration is ashes for every obtained chars. The only decreases are shown by more stable compounds such as sodium, potassium and phosphorous oxides. Stability of their presence ensures stability of ash properties as they are dependent only on ash composition instead of presence in fuel.

In order to fully evaluate process several indicators were introduced in order to efficiently examine the quality of examined biomass fuel. One of important ash properties is its susceptibility to form alkali-based glass deposits (slag) on the surface of pipes and various combustion equipment. Such ability depends largely on presence of chlorine and sulphur based salts and can be characterized by Cl (1) and S-ratio indexes (2):

$$Cl\ ratio = \frac{\%(Cl + K_2O + Na_2O)}{\%(SiO_2 + Al_2O_3)} \quad (1)$$

$$S\ ratio = \frac{\%(S + K_2O + Na_2O)}{\%(SiO_2 + Al_2O_3)} \quad (2)$$

In addition, alkali index was also used as it addresses higher proximity of biomass chars to coal than their raw counterparts and also relates to fuel volume. Alkali index (AI) can be calculated with use of fuel lower heating value LHV:

$$AI = \frac{K_2O + Na_2O}{LHV} \quad (3)$$

Where LHV is a lower heating value of given material in GJ/kg. To evaluate temperature ranges in which ash softens, biomass slagging evaluation index (BSEI) was calculated:

$$BSEI = \frac{\%(MgO + Al_2O_3 + Fe_2O_3)}{\%(MgO + P_2O_5)} \quad (4)$$

Finally, changes in fouling tendency with pyrolysis temperature of nutshell chars was evaluated with base to acid ratio:

$$R_{b/a} = \frac{\%(MgO + Al_2O_3 + Fe_2O_3)}{\%(SiO_2 + Al_2O_3 + TiO_2)} \quad (5)$$

Results of calculations can be found in Table 3. According to Cl and S ratio, as every material received S ration above 1.9 and in most cases Cl ratio above 1.9, both pistachio and hazelnut shells combustion can lead to ash slagging despite the examined biomass pyrolysis pre-treatment. Although high slagging potential for both biomass, hazelnut shells have lower slagging risk. Alkali index confirms however that in both cases presence of slagging during the process is certain. On the other hand receiving BSEI value lower than 7 hazelnut seems to have ash softening temperature (AST) set below 1000 °C setting some combustion technologies for hazelnut on much more strict level, than in case of pistachio that according to BSEI have AST between 1000 and 1200 °C. Hazelnuts have base-to-acid ratio much lower that pistachio which suggest lower presence of fouling with use of this material. Additionally fouling seems to decrease with increase of pre-treatment pyrolysis temperature in both cases.

4 Conclusions

Hazelnut and pistachio shells were pyrolyzed in three different temperatures and resulting chars were evaluated in respect for their corrosion potential in combustion processes. Elemental and spectrophotometric analysis has shown very low sulphur concentrations of under 100 ppm and minimal chlorine concentrations.

Despite relatively low char yield for both nutshells, sulphur and chlorine concentration haven't increased dramatically due to their partial release during pyrolysis. Sulphur and chlorine ratio analysis shown that both pistachio and hazelnut shells have considerable potential for alkali chlorine/sulphite corrosion with pistachio shells possibly weaker intensity in case of hazelnut due to relatively lower Cl/S values.

Further analysis aimed at ash composition revealed typical potassium, calcium and aluminium rich ashes in case of hazelnut and additionally large presence of sodium and iron and phosphorous in case of pistachio shells. Estimation of ash properties with several ash quality evaluation indices like BSEI, Cl and S ratio, Alkali index and base-to-acid ratio pointed that both materials are highly susceptible to cause slagging during combustion with higher fouling potential for pistachio shells but lower ash softening temperature for hazelnut (<1000 in comparison with 1000-1200 for pistachio). As a result both processed pistachio and hazelnut shells chars and their raw counterparts nevertheless of pyrolysis temperature are unsuitable for a fuel because of potential equipment maintenance need. It is generally beneficial properties like high lower heating value and low moisture content suggest however that they could be suitable as additive to other solid fuels. In case of untreated nutshells reported by literature the corrosive influence would noticeable weaker and would possibly allow for its use as standalone fuel material.

Acknowledgement

This paper was prepared thanks to funding from Faculty of Energy and Fuels at the AGH University of Science and Technology as part for PhD Students Grants 2016, No. 15.11.210.341.

References

1. R. Saidur, G. BoroumandJazi, S. Mekhilef, H.A. Mohammed, Renew. Sust. Ener. Rev. **16**, 2 (2012).
2. P. McKendry, Biores. Tech. **83,** 1 (2002).
3. J. A Ruiz, M. C. Juárez, M. P. Morales, P. Muñoz, M. A. Mendívil, Renew. Sust. Ener. Rev. **18**, 174-183 (2004).
4. S. G. Sahu, N. Chakraborty, P. Sarkar, Renew. Sust. Ener. Rev. **39**, 575-586 (2014).
5. D. A. Tillman, (2000). Biom. Bioener. **19**, 6 (2000).
6. S. V. Vassilev, C. G. Vassileva, V. S. Vassilev, Fuel **158**, 330-350 (2015).

DOI: 10.1051/e3sconf/20161000113

7. M. Marczak, M. Karczewski, D. Makowska, P. Burmistrz, Agric. Eng. 3, 159 (to be published)
8. J.L Easterly., M. Burnham, Biom. Bioener. **10**, 79-92 (1996).
9. P. McKendry, Biores. Tech., **83**, 1 (2002).
10. M. G., Kim, M. G. Kanatzidis, A. Facchetti, T. J. Marks, Nat. Mater. **10**, 5 (2011).
11. M. Metzger, G. Polakow, Ind. Inf. IEEE Trans. **7**, 4 (2011).
12. R. T. Symonds, D. Y. Lu, V. Manovic, E. J. Anthony, Ind. Eng. Chem. Res. **51**, 21 (2012).
13. J. E. Dengler, A. Doroodian, B. Rieger, J. Organomet. Chem. **696**, 24 (2011).
14. J. Beck, R. Müller, J. Brandenstein, B. Matscheko, J. Matschke, S. Unterberger, K.R.G. Hein, Fuel **84**, 1911–1919 (2005).
15. S. V. Vassilev, D. Baxter, C. G. Vassileva, Fuel **112**, 391-449 (2013).
16. X. Liu, X. Gao, W. Wang, L. Zheng, Y. Zhou, Y. Sun, Renew. Ener. **44**, 463-468 (2012).
17. S. V. Vassilev, D Baxter, L. K. Andersen, C.G. Vassileva, Fuel **89**, 5, (2010).
18. D.C. Dayton, B.M. Jenkins, S.Q. Turn, R.R. Bakker, R.B.Williams, D. Belle-Oudry, L.M. Hill, Ener. Fuel, **13**, 860 (1999).
19. K.O. Davidsson, L. E. Åmand, B. M. Steenari, Elled, A. L., Eskilsson, D., Leckner, B. Chem. Eng. Sci. **63**, 21 (2008).
20. Y. Niu, Y. Zhu, H. Tan,, S. Hui, Z. Jing, W. Xu, Fuel Proc.Tech. **128**, 499-508 (2014).
21. H. Oleschko, M. Müller, Ener. Fuel. **21**, 6 (2007).
22. M. Díaz-Ramírez, C. Boman, F. Sebastián, J. Royo, S. Xiong, D. Boström, Ener. Fuel. **26**, 6. (2012).
23. Y.,Hu, S. Cheng, P Sun, J. Xie, H. Zhang, Therm. Pow. Gen., **40**, 8-12 (2011).
24. D. Vamvuka, D. Zografos, G. Alevizos, Biores. Tech., **99**, 9 (2008).
25. Y. Niu, H. S. Tan, E. Hui, Prog. Ener. Comb. Sci, **52**, 1-61 (2016).
26. A. Demirbas, Prog. Ener. Comb. Sci. **30**, 2 (2004).
27. T.R. Miles, J.T.R. Miles, L.L. Baxter, R.W. Bryers, B.M. Jenkins, L.L. Oden, *Alkali deposits found in biomass power plants. A preliminary investigation of their extent and nature* (NREL Laboratory, Golden, USA, 1995).

Low energy building in MHM and PHE systems as an example of an ecological building solution

Mateusz Hodurek[1,a]

[1]*PhD Student, AGH University of Science and Technology, Faculty of Mining and Geoengineering, Department of Geomechanics, Civil Engineering and Geotechnics, hodurek.mateusz@gmail.com*

Abstract. The aim of this paper is to examine the technology of MHM and PHE systems used in the single family housing project of the architectural company Natkaniec Olechnicki Architekci. The project consists of the ten one-storey houses which will be constructed in the village Chyby, located 20 kilometers from Poznań (Poland). The unusual form of the buildings required a close collaboration between an architect, structural engineer and the contractor. Buildings fulfill low energy building condition.

1 The idea

The land on which the building will be situated is located in close neighbourhood of Kierskie lake near Poznań. The investor and architect decided to create there 10 single-family houses.

Figure 1. Settlement with 10 single-family houses [1].

The buildings have simple form with symmetric gabled roof. Configuration of the roofs was dictated by solar panels and gaining provided from solar heating. The project assumes that the buildings will not only be low energy but also due to photovoltaic installation energy autonomy energetically safety. The project was prepared by Natkaniec Olechnicki Architekci and the structural design developed Pracownia Inżynierska Czesław Hodurek. The unusual form of the buildings required a close collaboration between the architect, the structural engineer and the contractor.

[a] Corresponding author: hodurek.mateusz@gmail.com

Figure 2. Visualisation of the building Z [1].

Figure 3. Visualisation of the building W [1].

2 Construction and applied materials

2.1 Description of designed construction

The analysed buildings are supported on foundation slab 20cm thick with peripheral beam 20x94cm. Due to complicated ground conditions indirect foundations on CFA columns are applied. Load bearing structure is made in MHM system (walls) and PHE system (slabs). Walls on ground floor are located on larch board 310 mm width and 30 mm thick. The top level of foundation slab is 12 cm below ground, consequently water insulation is raised 50 cm above the ground. The MHM wall is connected with the foundation by steel connectors presented below (see figure 4).

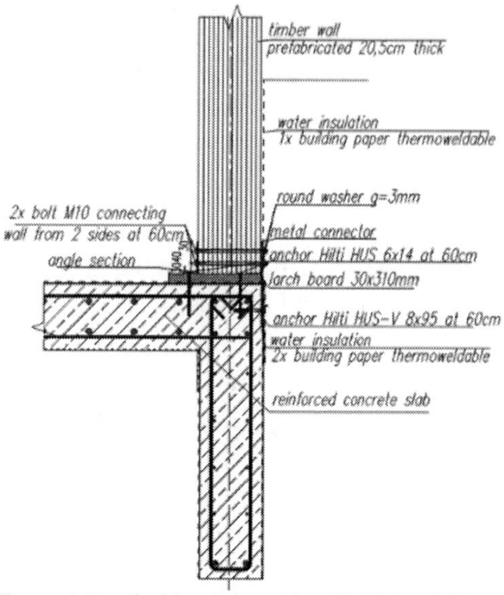

Figure 4. Detail of foundation slab and MHM wall [2].

In order to fulfil ultimate and serviceability limit states the design frame from C-section C200 was proposed. Additionally, the hangers from I-beam HEA 200 or two C-section UPE 200 are recommended. The roof formwork is a queen post. The proposed rafters are Steico joist SJ60/360. Steico joist with 40 cm spacing were also applied to create subfloor where all installations are located.

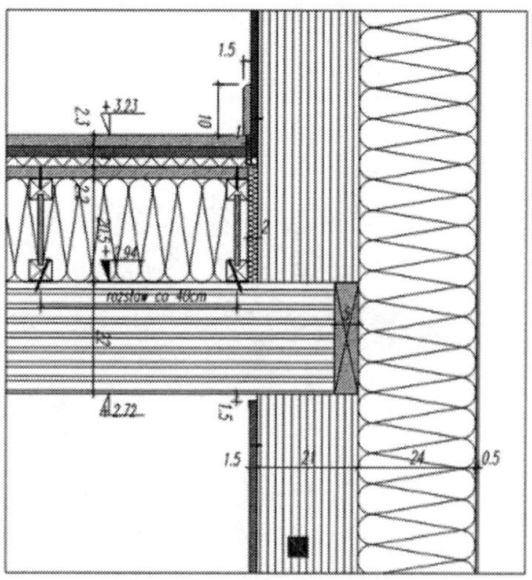

Figure 5. Detail of subfloor [1].

2.2 Timber walls in MHM system

Massive-Holz-Mauer is a high-tech construction material. The system uses 3 dimensional CAD program which creates model and stages for separate elements such as: knee walls, lintels, gablets and walls with openings on doors and windows. All drills in the elements for installations such as water, heating and electricity have to be introduced into the element by computer-controlled tools. Fully automated production process allowed achieving accuracy level of +- 2mm.

MHM system is made from coniferous wood panels sorted according to width. Timber is technically dried up to 15% +-3% with 23mm thickness. These wood panels are brought together into a continuous strand by a finger jointing station which is integrated in the plant. The lamella is profiled on one side to create a finely-structured uniform bottom layer. To achieve requested dimensions individual layers of board are stacked one on top of the other, compressed and joined together by aluminium groove pins layer by layer (according to ETA approval ETA-13/0801). Final product ranks between thickness of 10 to 24 cm, width up to 120cm and the length from 4m up to 12m. The only waste from production occurs from planning, which should not exceed 5%. According to fire standards analyzed material is classified as F90B and one storey house should fulfil F30B fire retardancy standard.

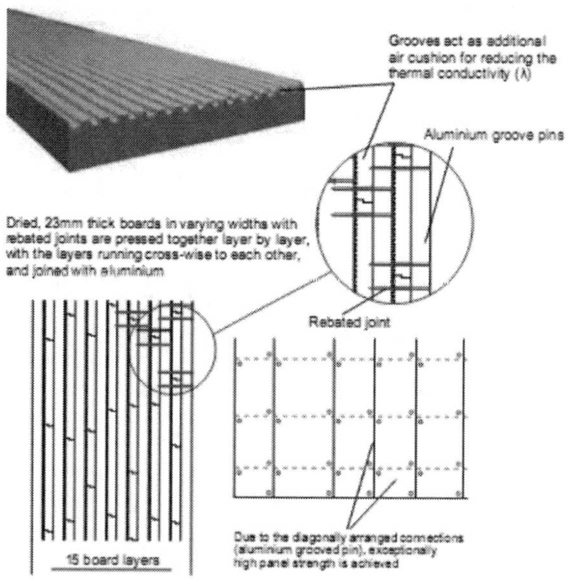

Figure 6. Cross-section by MHM wall [8].

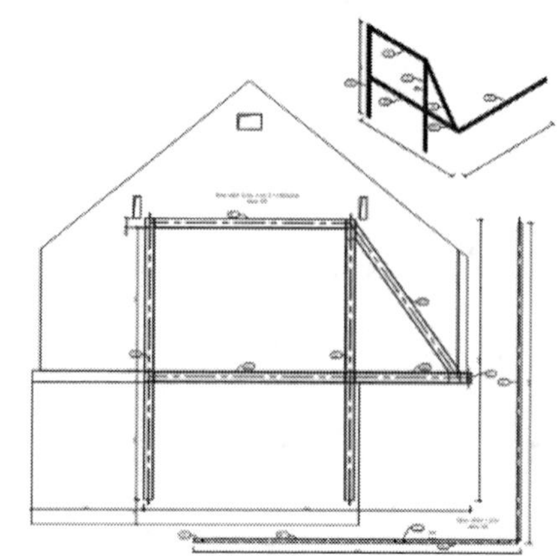

Figure 7. Steel structure in building Z – front view, top view and 3D view [2].

2.3 Timber slabs in PHE system

Profiled timber elements (PHE) complete the MHM wall system. Both systems together create complete structural construction made with solid timber. Technical data for PHE system is the same as MHM wall element.

2.4 Steel structure

To fulfill ultimate and serviceability limit states in buildings W and Z steel structure is used. On the other hand in building Z design frame from C-section C200 is proposed to minimize cantilever deflection.

In order to reduce span between supporting hangers from two C-section UPE 240 and columns from square pipe 100x100x5 were recommended.

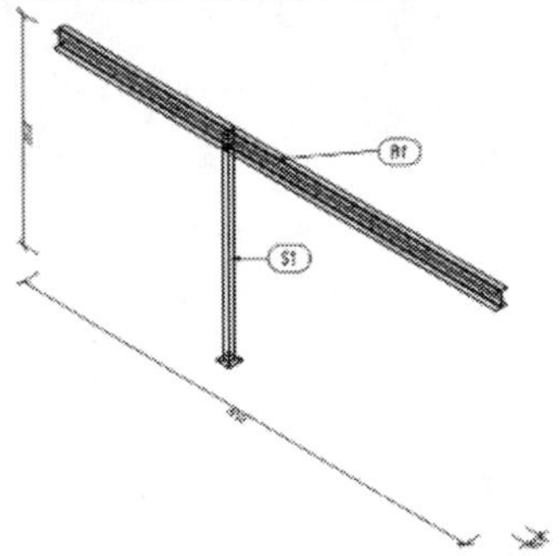

Figure 8. Steel hanger in building Z – 3D view [2].

The hanger presented above (see figure 8) bears load from MHM system 20.5 cm thick. Hanger supports loadings on column and walls. The figure 9 below shows solution of mounting MHM wall on 2 UPE 240 C-section.

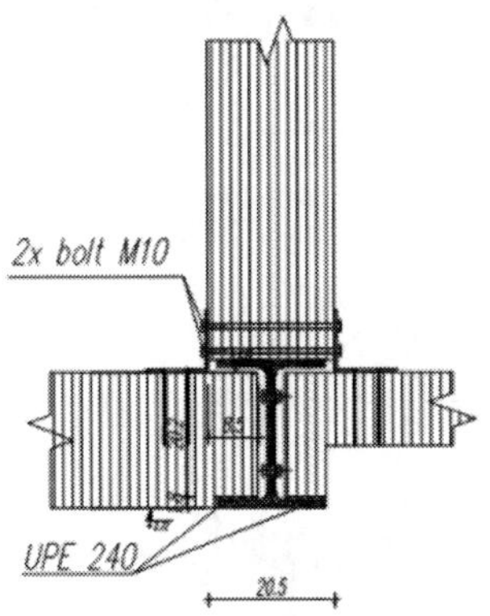

Figure 9. Steel hanger in building Z – cross section [2].

In the building W lintel from two UPE 240 C – section supported on MHM wall was designed.

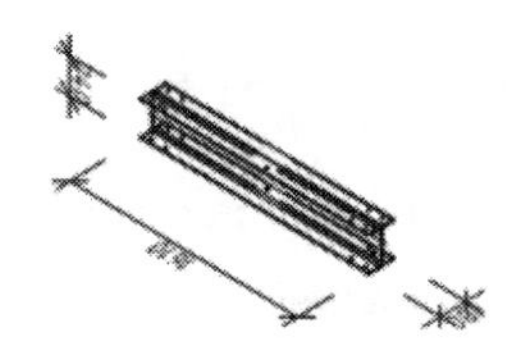

Figure 10. Lintel in building W made from 2 UPE 240 profiles [2].

In building W part of the first floor is overhanged. To connect slab with beam or wall "plate" steel elements were recommended.

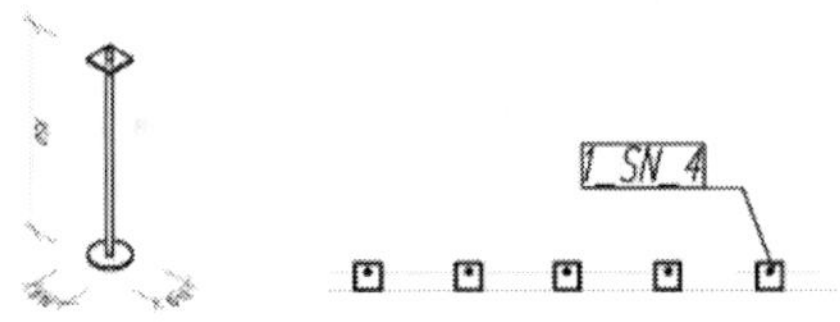

Figure 11 and 12. "Plate" – steel element connecting PHE slab with beam or wall – 3D view and location in 1_SN_4 [2].

In the building W introduce frames made with I-beam HEA 200 and square pipe 120x120x5 as a column were recommended. This solution shortened span between supporting and fulfilled serviceability and ultimate limit states.

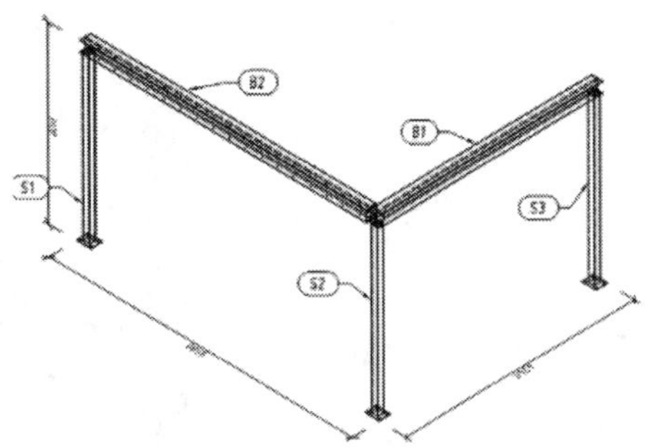

Figure 13. Frame in building W. Columns are made from square pipes 120x120x5 and beams are made from HEA 200 [2].

2.5 Steico components

Steico insulation products are characterized as environment friendly. The wood fibre insulation is formed from raw timber. The production process of Steico insulation includes heating wood chips through the defibration process, drying fibres and forming mats. The final product fulfils EN 13171 standards.

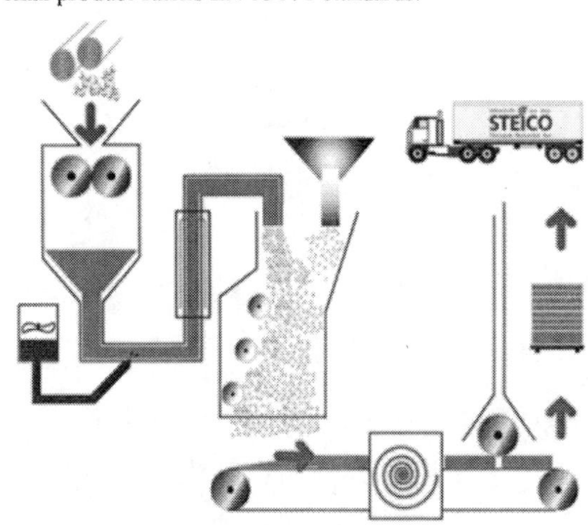

Figure 14. Production process of insulating mat [7].

The applied rafters and the joist Steico product is an I-web and consists of flanges made with timber according to the norm EN 385:2001. The web consist of hard fibreboard in accordance with EN 622-2:2004. The connection between the web and the flange is made by gluing the web into a groove in the centre of the wide face of the flange [7].

Figure 15. Joist with solid timber flanges - cross section [7].

2.6 Noise insulation

The sound insulation is important for human comfort. The project has to include solution of connecting elements and insulation openings.

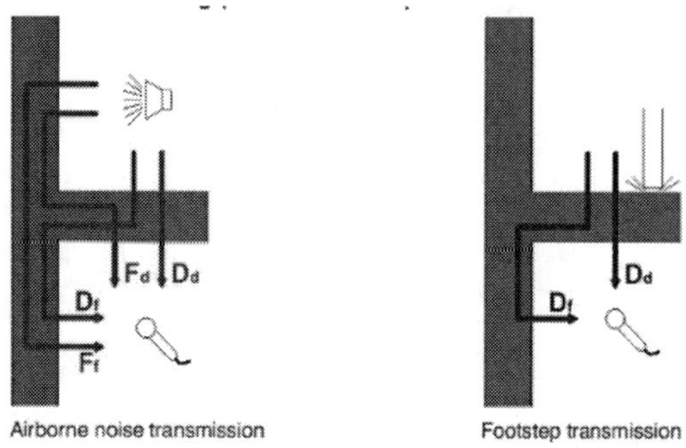

Figure 16. Noise transmission [8].

To prevent noise two-layer wall was recommended. This kind of wall consists of plasterboard from two side filling with mineral wool which reduces noise transmission.

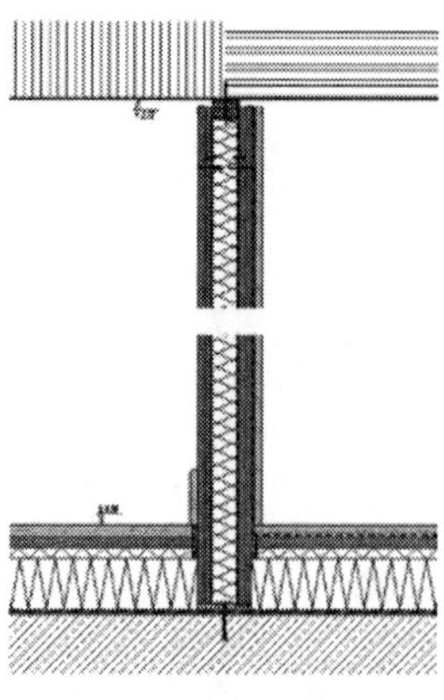

Figure 17. Cross section through the partition wall [1].

Depending on the type of construction walls and ceilings the project may require additional solutions to reduce noise transmission. The figures 18 and 19 below show when additional bearings are required. If the facing layers on walls and ceilings are applied no bearings are required. On the other hand if the facing layers are not applied we have to apply bearings. When extra ceiling layers are applied footstep transmission is reduced.

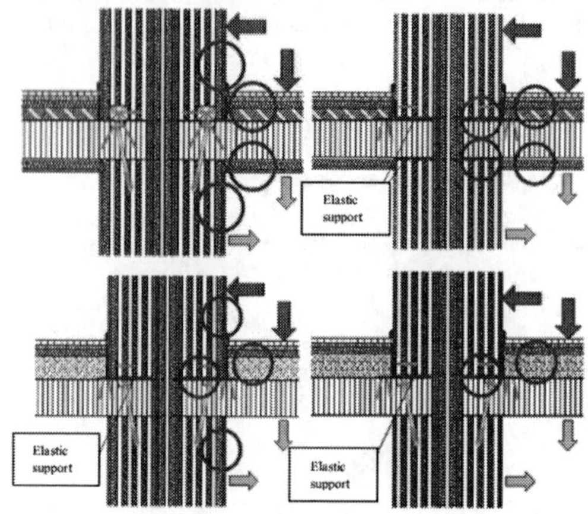

Figure 18. Noise transmission. Location of elastic supports to reduce noise transmission [8].

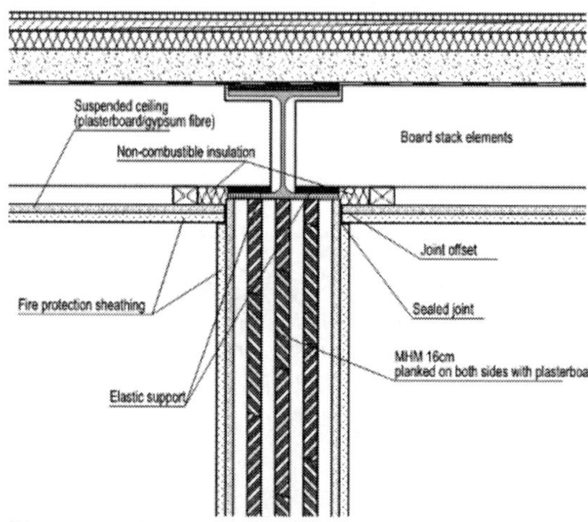

Figure 19. Noise transmission. Location of elastic supports to reduce noise transmission [8].

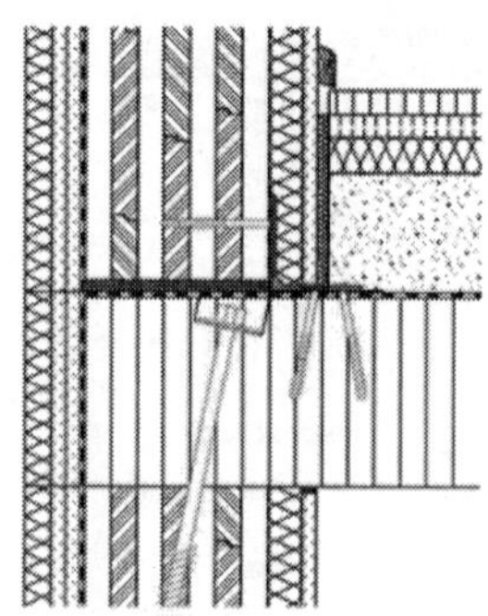

Figure 20. Elastic support - separating construction elements to prevent noise transmission [8].

3 Low energy building solutions

3.1 Low energy building criteria

The low energy building has to fulfil certain criteria. According to NFOŚiGW (National Fund for Environmental Protection and Water Management) fulfilling standard NF40 gives subsidy for construction or to buying house. NF40 means that energy demand for heating and ventilation does not exceed 40 kWh/m²·year.

Low energy houses are required to achieve a whole building air change rate of no more than 1.0 ac/hr under forced pressurisation and depressurisation testing at 50Pa minimum according to PN-EN 13829 standard. Such a high tightness of building reduces heating in winter and creates optimal thermal comfort for occupants.

Proving compliance on site blower door testing by certified testers is used.

Ventilation in the low energy building should be mechanical with a heat-recovery system. Efficiency of the heat-recovery system is more than 85%. It is ensured by rotary wheel or fixed play exchanger. Energy consumption of the ventilator is less or equal $0.45 W/m^3 \cdot hr$.

3.2 Thermal insulation – materials and barriers

MHM walls have very low coefficient of thermal conductivity λ = 0.094 W/(m·K) according to MFPA Leipzig GmbH or λ = 0.11 W/(m·K) according to Polish Standards. The external walls are composed with MHM 20.5cm thick and Steico protect dry L240. For Steico protect dry L240 declared thermal conductivity equals to 0.041 W/(m·K). Heat transmission coefficient of external wall barrier equals U_c = 0.11 [W/m²·K] and fulfils standard of $U_c \leq 0.12$.

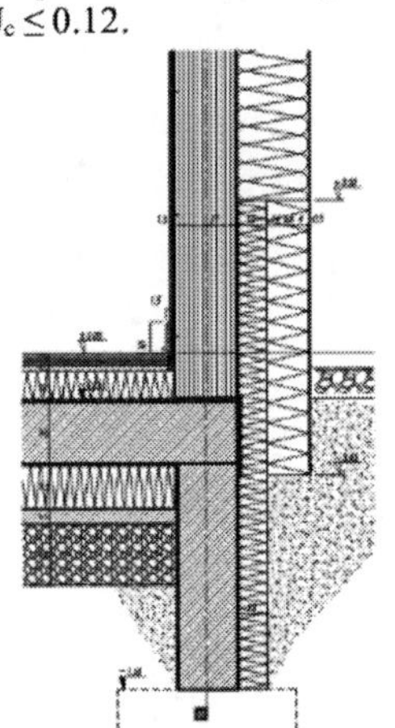

Figure 21. Architecture detail of foundation and wall barrier – cross section [1].

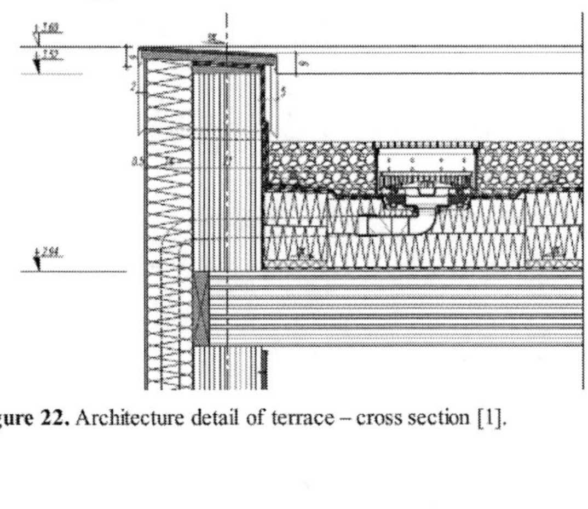

Figure 22. Architecture detail of terrace – cross section [1].

For slab supported on ground expanded polystyrene (EPS) as insulation was applied. Heat transmission coefficient of this barrier equals to U_c = 0.11 [W/m²·K] and fulfils criteria $U_c \leq 0.12$.

Heat transmission coefficient of terrace equals to U_c = 0.11 [W/m²·K]. To insulate terrace expanded polystyrene (EPS) was applied. This solution fulfils these requirements.

To insulate roof between Steico joist rafters 36 cm wood fibre insulation is applied. The coefficient of heat transmission equals to U_c = 0.10 [W/m²·K] was achieved and conditions were fulfilled.

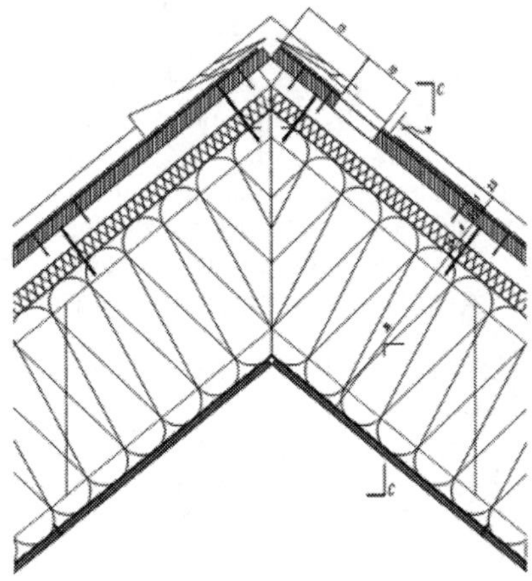

Figure 23. Architecture detail of ridge – cross section [1].

3.3 Woodwork

The triple glazed insulated windows with gas filled spaces and low emissivity (low-E) coatings fulfilling criteria U < 0.9 [W/m²·K] were applied. The tight timber construction allowed to minimize losses caused by the ventilation.

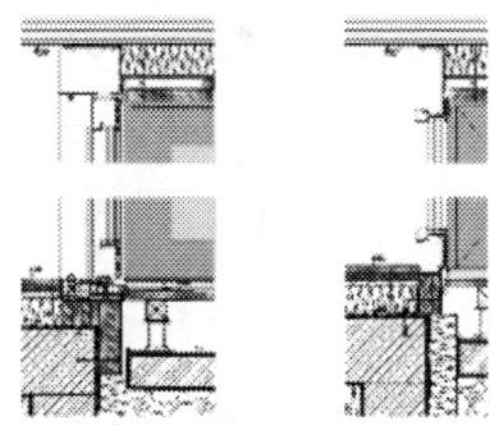

Figure 24. Architecture detail of woodwork construction – cross section [1].

3.4 Installations

The mechanical ventilation was applied in the building. The efficiency of heat recovery is minimum 85%. The ventilation includes recuperator with counter-flow heat exchanger. For heating and cooling the building air-source heat pump was recommended.

In order to reduce use of electricity to about 80% designed photovoltaic installation at every building with power ~10kW were applied. The installation is connected to energy network to balance electricity production and demand. Photovoltaic installation reduces CO_2 emission and provides energy safety and autonomy.

According to WT 2014 the analyzed building fulfils EP criteria with EP equalling to 24 kWh/(m²·year), where standards for low energy building is less or equal 40 kWh/(m^2·year). In energetical characteristic checked criteria on thermal insulation of external barriers, condition of windows area and condensation.

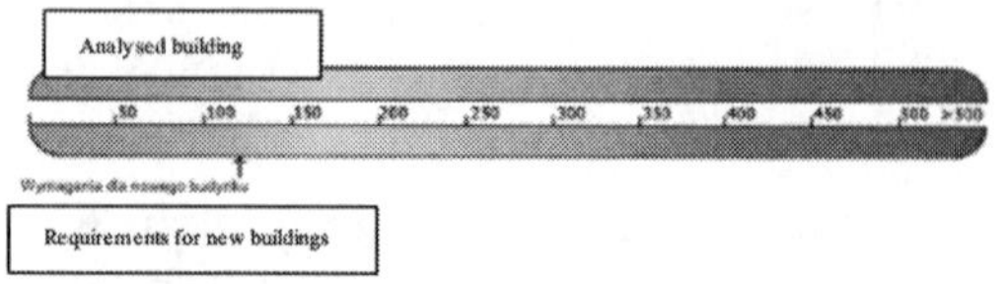

Figure 25. Heating and energy demand ratio per year qualify analysed building as a low energy [9].

4 Sustainable and green building

Due to natural character of the elements no hazardous or any chemical substances are used. Natural construction of the ceiling and the walls positively affects permeability of air humidity due to phenomenon of construction breathing. The solid construction gives the building considerable heat storage larger than timber skeletal building filled with insulation. The buildings have high retention capacity. Consequently, during night time when temperature is lower than during day time buildings keep the temperature and the buildings do not need to be heated. Comparing brick with solid timber one can say that time in which a square metre of wall of bricks cools down by 1°K, is up to four times shorter than a timber.

Due to the fact, that all elements used in the buildings are prefabricated, the construction is very fast. It may takes approximately 3 days and is shorter than in case of standard buildings.

Figure 26. Prefabricated wall elements [8].

During construction of the building, the amount of waste is reduced to minimum. Waste from timber elements can be reused to create smaller elements or as a biomass fuel. The examination performed by German scientist proved that the elements in PHE and MHM systems shield off high-frequency waves effectively due to their large mass. The structure creates a zone which neutralizes signals from mobile phones.

5 Summary

In the investment process related to the implementation of MHE and PHE systems close collaboration between the architect, the structural engineer and the contractor was important. Many problems were solved during the coordination meetings. The designers created an interesting project which fulfils low energy criteria. The applied materials and the construction system make the investment sustainable and environment friendly. The pre-planned solutions in distribution installations eliminated later changes and allow to reduce waste. The analyzed investment is a significant example of modern technology in erecting buildings. It will allow an introduction of a synergy between the economical and environmental aspects, allowing for affordable housing that is cost effective and comfortable for living.

References

1. P. Natkaniec, P. Bieroński, M. Faber,Natkaniec Olechnicki Architekci, Architectural design, Cracow, 2015
2. M. Hodurek, Cz. Hodurek, Pracownia Inżynierska Czesław Hodurek, Structural design, 2015
3. European Technical Approval ETA-13/0799
4. PN-EN 12831 Heating systems in buildings
5. PN-EN 1995-1-1: Design of timber structures. Part 1.1: General rules and rules for buildings
6. [online] www.massivholzmauer.de, 14 May 2016
7. [online] www.steico.com, 14 May 2016
8. [online] www.dreamarthome.pl, 14 May 2016
9. P. Kliński, G. Piskorz, Termo Studio s.c., Energy consumption characteristic, Poznan, 2015

 DOI: 10.1051/e3sconf/20161000115

Solid Oxide Fuel Cells coupled with a biomass gasification unit

Marek Skrzypkiewicz[1,a], Michał Wierzbicki[1] and Michał Stępień[1]

[1]*Institute of Power Engineering, Thermal Processes Department, Ul. Augustówka 36, 02-981 Warsaw, Poland*

Abstract. A possibility of fuelling a solid oxide fuel cell stack (SOFC) with biomass fuels can be realized by coupling a SOFC system with a self-standing gasification unit. Such a solution enables multi-fuel operation, elasticity of the system as well as the increase of the efficiency of small-scale biomass-to-electricity conversion units. A system of this type, consisting of biomass gasification unit, gas purification unit, SOFC stack, anode off-gas afterburner and peripherals was constructed and operated successfully. During the process, biomass fuel (wood chips) was gasified with air as gasification agent. The gasifier was capable of converting up to 30 kW of fuel to syngas with efficiencies up to 75%. Syngas leaving the gasification unit is delivered to a medium temperature adsorber for sulphur compounds removal. Steam is added to the purified fuel to maintain steam to carbon ratio higher than 2. The syngas then is passed to a SOFC stack through a fuel preheater. In such a configuration it was possible to operate a commercial 1.3 kW stack within its working regime. Conducted tests confirmed successful operation of a SOFC stack fuelled by biomass-sourced syngas.

1 Introduction

Biomass gasification coupled with solid oxide fuel cells (highly efficient power generation technology) gives the opportunity to increase the amount of electricity generated from biomass in overall power generation. The use of renewable fuels such as biomass is justified from the environmental point of view as well as to ensure the energy resources availability. The disaster at the Fukushima nuclear power plant in Japan has resulted in reduction plans for nuclear power in several countries. On the other hand, the increase of power generation from renewable energy sources (wind, solar) has already faced some barriers associated with the balance of electricity demands and supply which cause needs for large amounts of energy storage. This problem has not been solved yet, despite of extensive research. Production of electricity from biomass, due to the continuity of the generation process is not subjected to these constraints.

Fuel cell technology can be used as a stand-alone system or integrated with complementary technology for electricity generation (organic Rankine cycle - ORC, gas turbine (GT) or recuperative micro gas turbine - MGT). Because of the typical for fuel cell separation of the exhaust gases (on the anode side residual fuel gas, on the cathode side depleted air), it is possible to use the residual fuel for the additional electric power generation (e.g. in a gas turbine) [1]. Alternatively, through the burning of residual fuel, it is possible to generate additional heat for the purpose of:

a) thermal integration,
b) production of domestic hot water,
c) additional electricity generation (ORC, Stirling engine).

Direct use of conventional technologies (internal combustion engine) and Stirling engine for the generation of electricity from biomass has been considered in small scale CHP systems [2]. These solutions have relatively small electrical efficiency [3]. For example, a Danish company DK Sterling offers a 140kW system operating on wood chips with electrical efficiency of 17% [4]. However, the Stirling engine technology is promising in the case of availability of waste heat in the system.

The integrated power generation system based on gasification reactor and solid oxide fuel cell is considered to be one of the promising technologies [5]. Generation of fuel for SOFC can be performed in gasification reactor with a use of number of gasifying agents: a mixture of steam and air, mixtures of oxygen, carbon dioxide and water vapour, as well as pure carbon dioxide (Boudouard reaction).

The results presented in this paper concentrate mainly on generation of electricity in SOFC stack fed with fuel gas produced in a fixed bed, counter - current biomass gasification reactor with a use of air as a primary gasifying agent.

Micro-CHP systems based on solid oxide fuel cells technology offer the highest fuel-to-electricity conversion efficiency (up to 60%), when compared to systems based on internal combustion engines (ICE), and Stirling engines (both ca. 20%) [6, 7]. They are dedicated for continuous operation and the excess of electrical energy can be supplied to the grid.

[a] Corresponding author: marek.skrzypkiewicz@ien.com.pl

An important aspect for the implementation of the micro-CHP/SOFC systems fed by fuel from biomass gasification is the cleaning process of the fuel stream. It is necessary to limit the content of pollutants which may cause performance drop or anode degradation to the level accepted by fuel cells.

Research in the field of purification of gasification products led to the development and implementation of effective method of low temperature gas cleaning for SOFC.

2 SOFC systems fed with syngas from biomass gasification – state of the art

Thorough analysis of the biomass gasification reactor (air as gasifying agent) coupled with a 1-2 kW SOFC system has recently been carried out. Analysis takes into account the moisture content in the biomass feedstock [8]. Appropriate integration of the systems enables increase of the electrical efficiency from 24.7% to 37.7%. The increase of the electrical efficiency is directly related to the fuel utilization factor in the fuel cell module. Optimisation analysis of the system fuelled by wet biomass and using air as a gasifying agent indicates that the ratio of water to biomass (STBR) of 1.5 leads to the highest efficiency with a minimum amount of fuel cells [10]. The SOFC fuel utilization at 75% leads to the highest system efficiency. Additionally, partial anode gas recirculation has been analysed. The optimum level of residual fuel recirculation from the anode side has been optimised at 60% [9]. A method of gas cleaning process has an important effect on the efficiency of integrated gasification-SOFC system. Fuel purification is necessary due to the purity requirements of the solid oxide fuel cells. Comparison of two possible scenarios of syngas treatment leads to the conclusion that the high-temperature cleaning allows higher thermal efficiency compared to the low temperature cleaning [10]. Due to the higher cost of high-temperature cleaning, the decisive economic element is the income from the sale of the thermal energy generated in the system.

Carbon deposition in SOFC stack is an important problem for development of SOFC based systems, fuelled by syngas from biomass gasification. It is a result of the thermodynamics and kinetics processes on the anode surface in the presence of syngas. The processes of carbon deposition were the subject of a number of experimental and thermodynamic analyses [1].

Carbon deposition on the surface of the anode side of SOFC decreases with increase of water vapour content in the fuel. Increasing the content of water vapour in the feed gas can be realized by:

a) increase of the steam addition (gasifying agent) during the gasification of biomass,
b) additional steam addition to the stream of fuel gas before the SOFC stack.

Carbon deposition on the anode of SOFC is also influenced by the current load. The process has smaller effect when the current load increases. There is a limit value of the current above which the carbon deposition does not occur. This is an effect of direct carbon oxidation on the anode surface by oxygen ions O^{2-}, which activity increases with current load [11]. However, the deposition of carbon may occur on other elements of the stack and block the access for the reactant gases to the surface of the anode.

High tar content in the fuel is a significant factor affecting and intensifying carbon deposition processes. Tars influence on the cell degradation is often determined by experimental test with the use of representative composition of the tar (naphthalene, benzene, and toluene). However, in reality, degradation is much lower in the presence of real tars then for example in the presence of the toluene [12].

3 Experimental investigations of SOFC stack

The tests of a SOFC stack coupled with biomass gasification reactor (Figure 1) have been conducted in laboratory of the Institute of Power Engineering (IEn). The test stand is divided in several sections:

a) fuel supply and conditioning (peristaltic pump, gas flow controllers, electric gas heaters, steam generator, heat exchangers),
b) the air supply (air flow controller, electric heater),
c) fuel cell section (SOFC stack, impedance spectrometer FuelCon, current and voltage connections),
d) residual fuel utilization (fuel afterburner, cooler, a set of gas analysers, ventilation),
e) data acquisition and control system.

The section of fuel supply and conditioning was responsible for preparation of fuel for testing purposes. Fuel was pre-heated to a temperature of 300°C and humidified by the addition of steam from the steam generator. The humidified fuel is then delivered to the primary heater, which provided adequate inlet temperature of the fuel before the fuel cell stack (approx. 800°C). In the first phase of testing the reference gases were supplied from gas cylinders with a use of mass flow controllers.

Next, when the fuel was delivered from biomass gasification reactor with a use of peristaltic pump, the flows of reference gases (from gas cylinders) through the mass flow controllers were gradually reduced to zero.

The air supply section besides main goal of providing the oxidant to the electrochemical reaction in fuel cells allows the temperature control inside the stack by controlling the flow rate and air inlet temperature. Air flows were regulated in such a way to ensure the oxidant utilization below 30%.

The electrochemical tests of SOFC stack integrated with gasification reactor were conducted with a use of the impedance spectrometer FuelCon with electronic load. Measuring apparatus allowed control of current load and voltage readout in the several sections of the fuel cell stack. In the first stage of the tests the analysis of fuel cell performance at different current loads was conducted. Next during long term stability tests the current was kept at constant value of 18A.

The analysis of SOFC stack performance during operation on fuel from biomass gasification reactor included the following tests:

a) Performance monitoring (at different current loads) of the fuel cell stack before and after the long term test at reference fuel (H_2, N_2) and the simulated fuel (H_2, CO, CO_2, CH_4),
b) Performance monitoring (at different current loads) of the fuel cell stack on the fuel delivered from biomass gasification reactor (H_2, CO, CO_2, CH_4) before and after the long term test,
c) Long term stabilization test (at constant current) of the fuel cell stack on the fuel delivered from biomass gasification reactor (H_2, CO, CO_2, CH_4).

Figure 1. Experimental setup for SOFC stack testing fuelled by syngas coming from biomass gasification process: 1) Gas analysers, 2) Control box, 3) Mass flow controllers, 4) Steam generator, 5) Ventilation system, 6) Air heater, 7) Afterburner, 8) SOFC stack, 9) Fuel heater, 10) Fuel pump, 11) Electronic load.

Experimental studies of the stack consisting of 60 electrolyte supported cells have been performed based on the acquisition of current – voltage curves for fuel mixture with different chemical composition. The air flow was variable and dependent on the stack temperature.

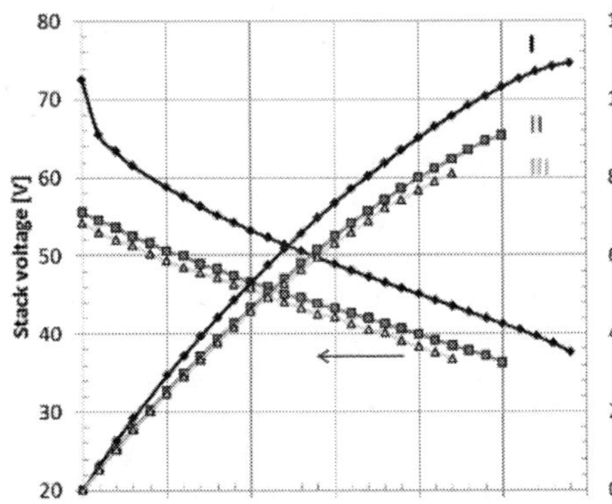

Figure 2. Current – voltage and power curves at stack temperature 850°C before long term test for three considered fuels

First stage of experimental campaign just before long term test using syngas from gasifier was devoted to current – voltage analysis for dry fuel reference mixture H_2/N_2 containing 40% of hydrogen (Fuel I) at flow that corresponds to 75% of fuel utilization at stack operating temperature 850°C. The electrical power was equal 1079 W for electric current of 29 A and the stack voltage equal 37 V.

Next experimental investigation has been performed using synthetic syngas (Fuel II) with chemical composition of 19% CO, 10% CO_2, 3% CH_4, 13% H_2, and 54% N_2. The stack performance obtained in this case is lower than in the previous case and the electric power reached 895W at stack current 25A and voltage of 36V.

Another current – voltage curve was performed for flow of real syngas obtained from the gasification of biomass (Fuel III). Chemical composition of syngas mixture delivered from gasification reactor (gasification agent: air, fuel: wood chips)was measured as 18.24% CO, 12.35% CO_2, 2.6% CH_4, and 12.35% H_2. The electrical power of the stack for Fuel III was 788W at electric current 22 A and stack voltage of 36 V. The results of all performed measurements for selected fuels are shown in Fig. 2.

Table 1. Power, current and voltage values for considered fuels at stack temperature 850°C before long term test.

Fuel	Stack Power [W]	Stack Current [A]	Stack Voltage [V]
Fuel I	1079	29	37
Fuel II	895	25	36
Fuel III	788	22	36

From obtained experimental results it can be concluded that the amount of hydrogen in the fuel stream has significant effects on the performance of the SOFC stack. With the increase of hydrogen content in the fuel, the performance of the stack increases. The highest power values of the stack have been obtained for a dry reference fuel mixture containing 40% H_2. However, in case of fuel from the gasification of biomass, the performance was lower by ca. 290 W.

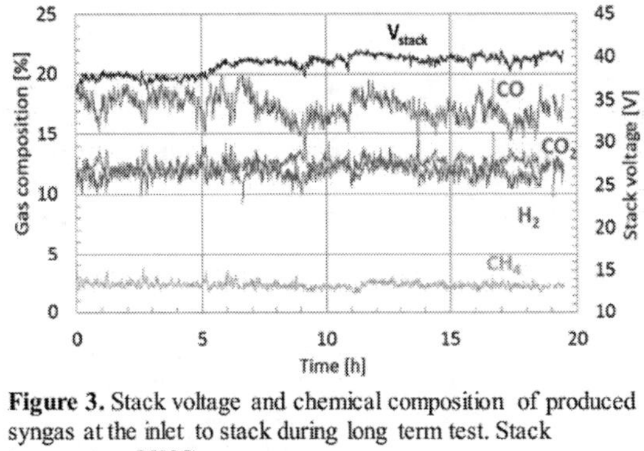

Figure 3. Stack voltage and chemical composition of produced syngas at the inlet to stack during long term test. Stack temperature 850°C.

Another series of tests has been carried out and was related to monitoring voltage of the stack under a constant current load. Applied electric current was equal 18A. During this experiment, the composition of syngas from biomass gasification was also checked

DOI: 10.1051/e3sconf/20161000115

and recorded. The resulting stack voltage variation as a function of time and the composition of syngas from gasification is presented in Fig. 3.

It can be observed that the voltage of the stack during the long term operation of the stack (about 20 hours) under constant current load seems to be stable and their fluctuations were estimated within range of 4 V. Thus, it can be concluded that tightness of the tested system was good and the chemical composition of fuel mixture delivered from biomass gasification was stable.

In turn, Figure 4 shows the exhaust gas from the stack during realization of a long term test under constant current load. It can be clearly seen that concentration of hydrogen and carbon monoxide is much lower than for the composition of the inlet fuel mixture. This is due to consumption of these components as the result of electrochemical reaction. In addition the higher concentration of carbon dioxide can be observed which is caused by formation of carbon dioxide during "water shift" process.

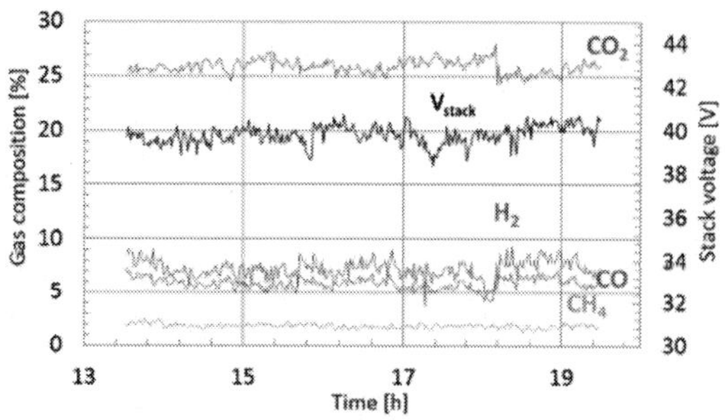

Figure 4. Stack voltage and chemical composition of produced syngas at the inlet to stack after long term test. Stack temperature 850°C.

The next stage of the experimental work was related to perform and analysis of polarization curves after the long term test carried out using syngas coming from gasification process. Four current - voltage characteristics has been recorded for different fuel chemical compositions. Similarly to the first stage of experimental investigation, the highest stack electric power was obtained for the dry reference fuel mixture H_2/N_2 (Fuel I) at current 27A and stack voltage equal 40V. The output power of the stack was about 1095W. It should be noted that the resulting stack performance is very similar to those which have been obtained just before start of long term test. This demonstrates the good quality of the tested system including purity of the fuel without presence of any degradation processes [13].

Next current-voltage curves for reference fuel mixture H_2/N_2 humidified with water vapour (Fuel IV) has been recorded. For such a fuel composition the stack performance characterized by electric power was equal 955 W at electric current 26 A and stack voltage of 37 V. Another current – voltage characteristics was performed for synthetic syngas (Fuel II) and syngas (Fuel III) obtained directly from the gasification of biomass. The results of performance of the stack were comparable to each other and were equal to 854 W and 832 W respectively.

During measurements of these current – voltage curves the stack temperature was maintained at the level of about 850°C. The above described polarization curves were presented in Fig. 5.

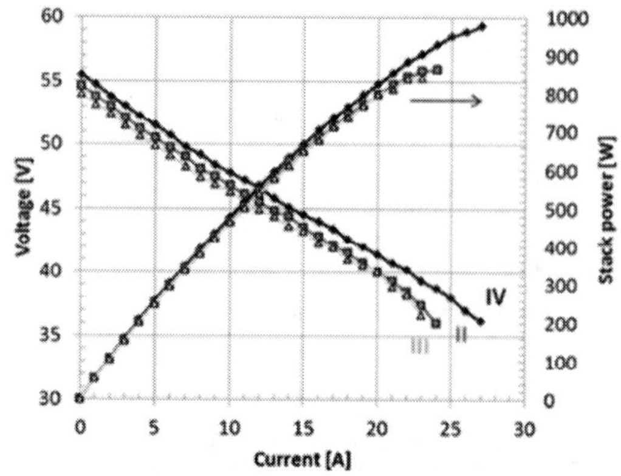

Figure 5. Current – voltage and power curves at stack temperature 850°C after long term test for considered fuels

All more relevant results of the whole experiment has been collected and presented in Table 2.

Table 2. Power, current and voltage values for considered fuels at stack temperature 850°C after long term test.

Fuel	Stack Power [W]	Stack Current [A]	Stack Voltage [V]
Fuel I	1095	27	40
Fuel II	854	24	36
Fuel III	832	22	38
Fuel IV	955	26	37

4 Summary

Possibility of fuelling the SOFC fuel cell stack with fuel coming from biomass gasification process makes the electricity generation system based on gasification reactor one of the promising technologies. Such solutions can be competitive in distributed power generation systems (less than 1 MW). However, firstly it is necessary to prepare gaseous fuel in the gasification reactor and also some amount of steam in order to prevent carbon deposition. In addition, very important issue is the effective gas cleaning system for pollutants removal that cause decreasing of performance or degradation of the material components of the stack.

The performed study of fuel cell stack allowed for a detailed analysis of operating conditions concerning fuelling with syngas coming from biomass gasification reactor.

The current-voltage characteristics and power curves of SOFC stack fuelled with the reference fuel, synthetic syngas and the real syngas coming from gasification of biomass has been performed. These characteristics were recorded both before and after the long term test. Analysis of the results shows that the highest stack power of 1.1 kW has been reached for the reference fuel. However, in the case of synthetic syngas and the syngas obtained from the gasification process electric power of the stack is lower than for reference fuel about 250 W. It was observed that operating parameters on synthetic

DOI: 10.1051/e3sconf/20161000115

syngas are very close to parameters for real syngas obtained directly from biomass gasification process. It shows good tolerance of cells in the SOFC stack to a small amount of impurities in the fuel.

Long term tests were carried out for stack fuelled with syngas obtained from the gasification process under a constant current load. Operation of SOFC stack was stable during whole test. Operating voltage of the stack within 20 hours was nearly constant and equal about 39 V.

Performed experimental studies confirmed the possibility of fuelling SOFC based systems with purified syngas, together with maintaining high operation performance of the fuel cell stack. As a result, the experimental data has been obtained, which have been used to verify the design and modelling calculations [14, 15]. In addition, the experimental results will be used for further work aimed to developing technology of cogeneration systems, including micro - CHP systems based on SOFC stacks [16]. The cleaned fuel coming from biomass gasification process is considered as one of the most promising fuels for similar systems for combined production of electricity and heat.

Acknowledgements

The work was done within the National key project of the Polish Ministry of Science and Higher Education (POIG.01.01.02-00-016/08).

References

1. C. Athanasiou, F. Coutelieris, E. Vakouftsi, V. Skoulou, E. Antonakou, G. Marnellos, A. Zabaniotou, Int. J. Hydrogen Energ. **32**, 337-342 (2007)
2. Y. Huang, D.R. McIlveen-Wright, S. Rezvani, M.J. Huang, Y.D. Wang, A.P. Roskilly, N.J. Hewitt, Appl. Energ. **112**, 518-525 (2013)
3. L. Dong, H. Liu, S. Riffat, Appl. Therm. Eng. **29**, 2119–2126 (2009)
4. Gasification in Stirling Engine Applications, Stirling DK marketing materials, October 2010.
5. C.O. Colpan, F. Hamdullahpur, I. Dincer, Y. Yoo, Int. J. Hydrogen Energ. **35**, 5001 – 5009 (2010)
6. J. Jewulski, J. Kupecki, M. Błesznowski, Instal **1**, 11-15 (2014)
7. E. Podesser, Renew. Energ. **16**, 1049–1052 (1999)
8. G. Campitelli, S. Cordiner, M. Gautam, A. Mariani, V. Mulone, Int. J. Hydrogen Energ. **38**, 320-327 (2013)
9. S. Wongchanapai, H. Iwai, M. Saito, H. Yoshida, J. Power Sources **216**, 314-322 (2012)
10. A.O. Omosun, A. Bauen, N.P. Brandon, C.S. Adjiman, D. Hart, J. Power Sources **131**, 96–106 (2004)
11. I.V. Yentekakis, J. Power Sources **160**, 422–425 (2006)
12. E. Lorente, M. Millan, N.P. Brandon, Int. J. Hydrogen Energ. **37**, 7271-7278 (2012)
13. M. Błesznowski, J. Jewulski, A. Zieleniak, Centr. Eur. J. Chem. **11**, 960-967 (2013)
14. J. Kupecki, K. Motylinski, M. Ferraro, F. Sergi, N. Zanon, Journal of Power Technologies **96**, 63-71 (2016)
15. J. Kupecki, J. Milewski, A. Szczesniak, R. Bernat, K. Motylinski, Int. J. Hydrogen Energ. **40**, 15834–15844 (2015)
16. S. Sieniutycz, M. Błesznowski, A. Zieleniak, J. Jewulski, Int. J. Heat Mass Tran. **15-16**, 3984–3994 (2012)

 DOI: 10.1051/e3sconf/20161000116

Carbon as a fuel for efficient electricity generation in carbon solid oxide fuel cells

Marek Skrzypkiewicz[1,a] and Magdalena Dudek[2]

[1]*Institute of Power Engineering, Thermal Processes Department, Ul. Augustówka 36, 02-981 Warsaw, Poland*
[2]*AGH University of Science and Technology, Faculty of Energy and Fuels, Al. A. Mickiewicza 30, 30-059 Cracow, Poland*

Abstract. In this paper, the impact of the physicochemical properties of carbonaceous solid fuels on the performance of a direct carbon solid oxide fuel cell (DC-SOFC) was investigated. High-purity synthetic carbon powders such as carbon black N-220 and Carbo Medicinalis FP5 were chosen for analytical and electrochemical investigations in a DC-SOFC. The research focussed on choosing an optimised, cost-effective, high-purity carbon powder which could be applied as a solid reference fuel for all tests performed on a single DC-SOFC cell as well as on DC-SOFC stack constructions. Most of the electrochemical investigations described in this paper were performed using square DC-SOFCs with dimensions of 5 × 5 cm. The relationship between structure, physicochemical properties, and electrochemical reactivity in a DC-SOFC was analysed.

1 Introduction

The economic, cultural, and technological development of present-day societies (particularly rapidly developing countries such as China, India, and the countries of South America and Africa) may be impeded as a result of the growing demand for energy, which cannot be fully satisfied due to the depletion of fossil energy resources. For this reason, many countries have been carrying out intensive research and development work on the use of renewable energy sources (RES), as well as on improving the efficiency of the conversion of fossil fuels in the commercial and distributed power industries, with particular emphasis on protection of the natural environment. In the nearest future, for technological and economic reasons, it will not be possible to replace conventional power plants with new generators on a large scale using renewable energy sources [1, 2].

Therefore, technologies referred to as ‘intermediate’ are being introduced and developed in many countries, constituting a bridge between conventional and renewable energy technologies. New energy technologies enable more efficient use of fossil resources of energy, often through the co-generation of electricity and heat, which also results in a significant reduction in emissions of greenhouse gas and other harmful substances into the environment. A great deal of hope is placed in fuel cells, which are marked by very high rates of conversion efficiency of the chemical energy of fuels into electricity and by their potential for use in distributed systems. So far, the most commonly used fuel for fuel cells is hydrogen [3, 4].

Currently, the most rapid technological development can be observed in proton exchange membrane fuel cells (PEMFCs) and solid oxide fuel cells (SOFCs). There is a growing interest in utilising different alternative fuels in fuel cell technology. Many authors have presented performance results for different types of fuel cells operated with ethanol, methanol, formic acid, biogas, syngas, or methane ammonia [5, 6].

Carbon-based fuels can also be used to supply fuel cells. This can be done both directly and indirectly. In the first case, coal or carbonaceous-based solid fuels can be gasified, and the resulting gas, usually rich in CO, can be used as a reagent in fuel cells. In some types of fuel cells, the direct conversion of solid carbon-based fuels is also possible. Direct carbon fuel cells (DCFCs) are powered by solid fuel. Powdered carbonaceous-based solid fuels are fed directly to the anode, where gaseous CO_2 is formed during an electrochemical oxidation reaction [7, 8].

DCFCs are of great importance for the global economy, as they require large deposits of hard coal and lignite. In DCFCs, the chemical energy of coals can be directly converted into electricity with very high efficiency (over 70%). The sole product of this conversion is pure CO_2; thus the expensive process of separating CO_2 from flue gases can be avoided [9]. These potentially attractive characteristics of DCFCs make them attractive to a broad spectrum of power sector investors.

Carbon particles can be directly oxidised on the surface of an oxide electrolyte and cermet and oxide-based anode materials in SOFCs according

[a] Corresponding author: marek.skrzypkiewicz@ien.com.pl

DOI: 10.1051/e3sconf/20161000116

to reaction (1). Cells of this type fed directly with solid carbonaceous fuel seem to be a very attractive option in direct carbon fuel cell technology. As early as in 1988, Nakagawa and Ishida [10] described the mechanism of carbon oxidation as a pair of anodic electrochemical half-reactions (2–3).

$$C + 2O^{2-} \rightarrow CO_2 + 4e^- \quad (1)$$
$$C + O^{2-} \rightarrow CO + 2e^- \quad (2)$$
$$CO + O^{2-} \rightarrow CO_2 + 2e^- \quad (3)$$

The additional source of carbon monoxide for reaction (3) is the chemical Boudouard reaction (4):

$$C + CO_2 \rightarrow 2CO \quad (4)$$

Direct carbon fuel cells with solid oxide electrolytes (DC-SOFCs) seem to constitute the most promising option in DCFC technology. DC-SOFCs operate at high temperatures of 700–900°C. The increased temperature improves the kinetics of the electrochemical oxidation of carbon particles in the anode process. A variety of SOFC components (electrolyte, anode, cathode and catalysts) can be considered for design and construction of a DC-SOFC stack. Engineers also have a range of construction options: they can use planar and tubular as well as fluidised-bed designs [11, 24].

However, the introduction of DC-SOFC technology into the mass market depends on improvements in carbon conversion rate, electrical efficiency, and fuel utilisation attainable in actual conditions. The DC-SOFCs most frequently investigated in R&D institutions are typically built using planar geometry on the basis of a small button-size cell (diameter: ca. 2 cm) or a larger, square MEA with dimensions of 5 × 5 cm [12, 13]. Performance data for DC-SOFCs with large dimensions (above 10 cm^2) are rather scarce. A DC-SOFC stack is generally constructed from MEAs with larger dimensions (5 × 5 or 10 × 10 cm). The problem of scaling up an SOFC system from small cells (active area of electrode materials ~1–2 cm^2) to ~16–80 cm^2 is a difficult one. A single SOFC with larger dimensions enables the acquisition of higher values of current and power than can be obtained from a small button solid oxide fuel cell. Tests of planar DC-SOFCs require more complicated current collectors and electrical connections to avoid electrical losses during measurements. Sealing the electrochemical chamber is also more complicated. Following successful tests of such single MEAs, the construction of a short stack, which usually combines 2–4 single cells, can be planned.

This research focusses on choosing an optimised cost-effective high-purity carbon powder which can be applied as a reference solid fuel for all tests performed on a single SOFC with larger dimensions.

2 Experimental

In order to carry out a comparative interpretation of the performance data, two different types of carbon fuels were investigated in parallel. A commercial carbon black N-220 (Konimpex, Poland), referred to as N-220, and a charcoal-type activated carbon Carbo Medicinalis FP5 (Gryfskand, Poland), referred to as CM, were analysed. Both carbon-based solid fuels are typical representatives of high-purity synthetic carbon and are readily available and commonly applied in different areas of industry. Raman spectra of carbon particles were collected using a FTS 6000 Bio-Rad Spectrometer with a Raman section (Nd: YAG Spectra Physics T10, 1064-nm laser). The spectra were collected after 10,000 scans at a resolution of 4 cm^{-1}. The specific surface area was determined by multipoint nitrogen adsorption at -196°C (Quantachrome PoreMaster). The contents of elements such as carbon (C), nitrogen (N), sulphur (S), and hydrogen (H) were determined in all of the investigated samples using an elemental analyzer TruSpec Micro CHNS Leco. The proximate analyses were done with a calorimeter AC 500 Leco, laboratory drier and a muffle oven. The chemical reactivity of the carbon-based powders vs CO_2 gas was determined using the thermogravimetric method within a temperature range of 25–850°C. The temperature was increased at a rate of 5°C/min up to 850°C. The mass losses of carbon powders vs CO_2 gas at a constant rate were also analysed at 850°C for 300 min.

2.1 Experimental setup

A planar, electrolyte-supported SOFC cell, 5 × 5 cm, with the structure Ni-YSZ|Ni-GDC|8YSZ|LSM-GDC|LSM and a 150 μm thick YSZ electrolyte layer was provided by Fuel Cell Materials. The cathode and anode surface areas were equal 16 cm^2. Gold current collectors were applied in a four-probe configuration. Carbon black N-220 and charcoal-type activated carbon CM were introduced into the anode chamber for both single-batch test runs, respectively. The experimental DC-SOFC setup used was similar to the one used in our group's previous paper [11], as schematically presented in Fig. 1.

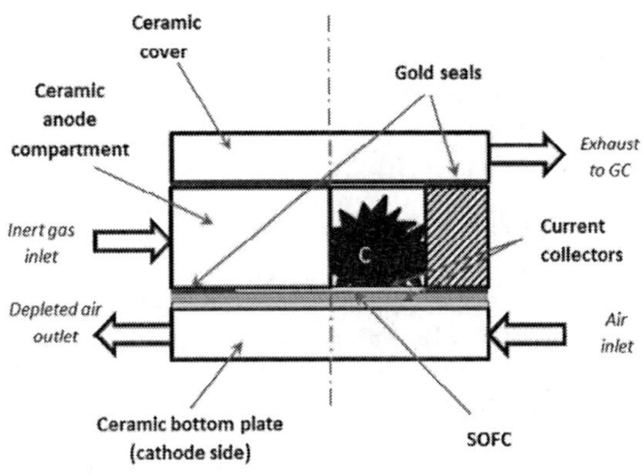

Figure 1. Schematic diagram of the DC-SOFC's housing configuration.

The DC-SOFC current-voltage characteristics were acquired at a temperature of 850°C using a Chroma 6310 electronic load with a Keithley 2000 multimeter. The anodic exhaust gas was analysed with a Varian CP-4900 gas chromatograph (GC) for determining concentrations of CO and CO_2 during the experiment.

DOI: 10.1051/e3sconf/20161000116

2.2 Experimental procedure used for the DC-SOFC test

Before testing each cell with the carbonaceous fuel, both cells' anodes were reduced under hydrogen flow in another test runs in a standard SOFC ceramic setup with gold current collectors. After cooling, each cell was placed in a DC-SOFC housing configuration and filled with carbon. In both experiments, ca. 25 ml of carbon powders was applied. This volume corresponds to 9.25 g of the carbon black N-220 and ca. 5g of charcoal CM. The system was heated to 850°C with a constant flow of 5% H_2 + 95% Ar gas mixture through the carbon bed compartment and air on the cathode side. After the operational temperature had been reached, the flow of hydrogen gas to the anode chamber was stopped and the cell was purged with Ar (190 Nml/min) for 1 hour. Then the argon flow was stopped and the cell operated without any gas supply; however, a minor leak of nitrogen from the inert-gas shield into the cell compartment was detected. The exhaust from the anode compartment remained open at all times and the gas chromatograph was connected to the beginning of the exhaust pipe with a T-junction. After preliminary I-V measurements, the cell was loaded with a current of 0.65 A (40.5 mA/cm^2). CO and CO_2 composition in the anode exhaust was measured by GC with 3–10 min intervals.

In the experiment using CM as fuel, no GC measurements were made. Instead, following the initial electrochemical characterisation of the DC-SOFC, the influence of CO_2 flow on electrochemical performance was experimentally verified. For this purpose, the cell was subsequently subjected to a flow of CO_2 of 50, 25, 5, 2, and 0 Nml/min through the anode chamber. The changes in flow were initialised under stable OCV conditions, and a 5-min stabilisation was performed in OCV conditions after each change, allowing the OCV to reach a plateau. Next, an I-V measurement was performed and the subsequent CO_2 flow investigated. This procedure was repeated five times to obtain comparable measurements for all 5 flows of CO_2 through the anode chamber.

3 Results and discussion

3.1 Analyses of solid fuels

Selected physicochemical properties of carbon black N-220 powder and charcoal were the subject of previous investigations. The impact of their physicochemical properties on direct carbon button solid oxide fuel cells were described in papers [14, 15].

Within the scope of the performed research, solid powders of carbon black N-220 and charcoal (CM) were subjected to analytic investigation aimed at identifying the comparative physicochemical properties crucial for application of these solid fuels in a large-scale DC-SOFC. First, the properties of solid carbon-based fuels were determined using proximate and ultimate analysis. The concentrations of major elements in the solid fuels as well as proximate analysis results are listed in Table 1.

Table 1. Proximate and ultimate analysis of applied carbonaceous fuels.

Parameter	Unit	N-220	CM
Water content	%	1.1	5.7
Ash	%	0.4	0.6
Volatile matter	%	1.42	2.54
HHV	kJ/kg	32,408	30,735
LHV	kJ/kg	32,290	30,463
C_t^a	%	90.6	90.3
H_t^a	%	0.4	0.6
N_t^a	%	0.45	0.47
S_t^a	%	0.96	0.03

The data presented in Table 1 clearly indicates that both chosen solid fuels were characterised by similar amounts of carbon, hydrogen, and nitrogen. A difference occurred in the case of sulphur. The carbon black N-220 exhibited a considerably higher content of sulphur than did Carbo Medicinalis. The chemical composition and content of ash and volatile matter have a considerable impact on DC-SOFC performance; thus, this parameter should be monitored before each electrochemical test of a DC-SOFC.

Raman spectroscopy was used for studies aimed at showing the detail of atomic disorder of carbon in these samples. Raman spectra recorded for N-220 and CM samples are presented in Fig. 2a and b.

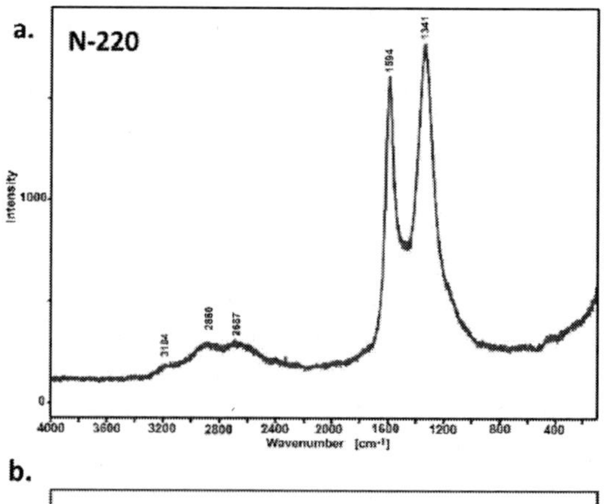

Figure 2. Raman spectra recorded for carbon black N-220 (a) and Carbo Medicinalis (b).

Analysis of the Raman spectra revealed a distinctively visible G band (ca 1594 and 1603 cm^{-1}), which confirms the sp^2 carbon hybrid structure of both samples.

The presence of a D band at ca 1341 and 1353 cm^{-1} indicated the presence of a large number of structural defects in the investigated carbon-based samples. The ratio of the intensities of the G and D bands can be attributed to the degree of disordered carbon structure in both samples.

This kind of Raman spectra for carbon black was previously interpreted by Sadezky et al. [16] as a highly disordered graphite structure. The first-order spectrum of the soot generally exhibited two broad and strongly overlapping peaks with maxima at ~1341–1353 cm^{-1} and ~1594cm^{-1}–1603cm^{-1}. Both of these investigated samples represent the short-range graphite or turbostratic structure of the sample [17, 18].

The chemical reactivity of carbon powder with respect to CO_2 may be another crucial factor with a considerable effect on DC-SOFC power output in the temperature range 700–850°C.

One feature preferred for gasification through the Boudouard reaction (4) is high reactivity of carbon with CO_2, i.e. the ability to reduce CO_2 to CO, according to reaction (4) [19]. The presence of CO as an additional reagent for anodic oxidation of carbonaceous fuel should lead to an increase in the current density as well as in the power acquired from the DC-SOFC [20]. It was found that, in a DC-SOFC, a convection mechanism can occur inside the carbon bed, enabling gaseous CO to be effectively transferred to the triple phase boundary areas inside a porous cermet SOFC anode [21].

The reactivity of carbon black N-220 and activated charcoal CM was evaluated using thermogravimetric analysis. Fig. 3 depicts the variation in carbon mass during a reaction with flowing CO_2 with a temperature increase from 25 to 850°C as well as at a constant temperature of 850°C.

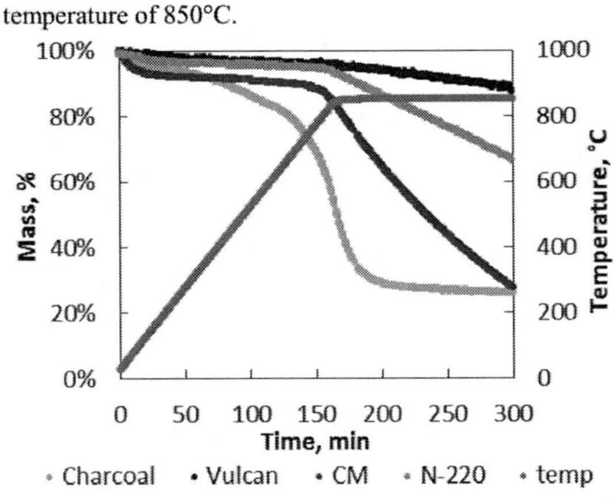

Figure 3. TGA analysis for investigated (N-220, CM) and reference (Vulcan, charcoal) carbon samples in CO_2. Heating rate: 5°C/min.

As can be seen, the thermogravimetric analysis showed a continuous decrease in total carbon mass weight in the course of heating from 25 to 850°C, as well as at a constant temperature of 850°C. The analysed dependence of mass losses vs temperature or time indicated a difference in the chemical reactivity of the investigated solid fuels, i.e. both of the analysed carbon black powders, CB-Vulcan and CB-N220, were characterised by rather poor reactivity vs CO_2.

The charcoal-based samples (charcoal from Sigma Aldrich and Carbo Medicinalis) exhibited greater reactivity vs CO_2 than the carbon black samples. Therefore, application of charcoal as fuel is expected to enable higher performance levels in DC-SOFCs, caused by more intensive gasification processes inside the anodic compartment of the cell.

3.2 Electrochemical performance of a DC-SOFC with a Ni-YSZ anode

The polarisation curves (U-I) measured at 850°C for DC-SOFCs fuelled by charcoal and by carbon black N-220 are presented in Fig. 4. As can be seen, the open circuit voltage for a DC-SOFC supplied by the investigated solid fuels reached values of 1.006 V, close to the theoretical OCV values calculated for reaction (1) (OCV_{cal} = 1.026 V). For the measurements, the following methodology was employed: a 16-cm^2 fuel cell was loaded with a stepwise increasing current, with each step followed by 1 min of voltage stabilisation. The comparative analysis of U-I and P-I curves recorded for an identical DC-SOFC setup supplied with carbon black N-220 or CM indicated different electrochemical outputs.

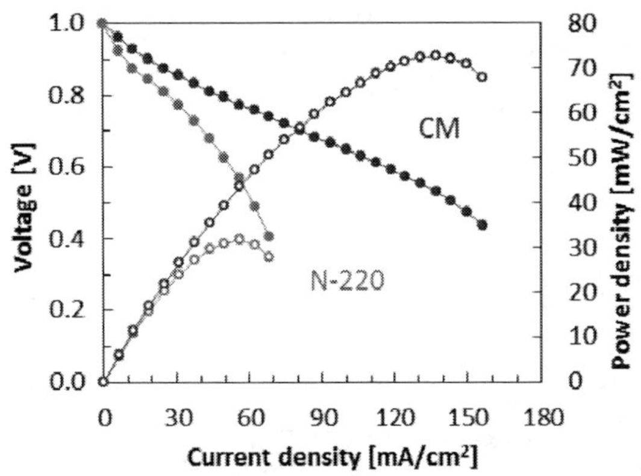

Figure 4. Polarisation and power density curves, cell temperature 850°C, anode flow: none, cathode flow: 200 Nml/min of air.

Analysis of the polarisation curve for the N-220 fuel reveals concentration losses already present in a current density regime of 30–40 mA/cm^2. This is a consequence of the relationship between reactions 1–4. The concentration losses occur because of the low level of reactivity of the applied carbon to the Boudouard gasification reaction (4). This fact was confirmed by gas chromatography measurements performed under load. The ratio of CO to CO_2 in the outlet stream lied in range 0.89-0.65, being far from the equilibrium state of the CO-CO_2-C system, which in 1 atm and 850°C presents CO/CO_2 ratio of 13 [22].Therefore, under higher current densities, the CO flux from reaction (4) occurring in the carbon bed to the anode surface is significantly

 DOI: 10.1051/e3sconf/20161000116

diluted by the CO_2 generated on the anode in reactions 1 and 3. Therefore no higher current densities can be achieved. In the case of a DC-SOFC supplied by Carbo Medicinalis solid fuel, a significantly higher level of performance was observed.

In the case of utilisation of carbon black N-220 as a solid fuel in DC-SOFC, the values of power output and current density were two times lower in comparison to the same cell fed with charcoal CM. The area specific resistance (ASR) calculated according to the linear part of the polarisation curve at ca 0.8 V equalled 5.34 Ωcm^2 and 2.93 Ωcm^2 for N-220 and CM, respectively. The various reasons for this behaviour should be analysed.

The electrochemical process of carbon particle oxidation occurred on the YSZ electrolyte according to reaction (1). It had previously been found that electrochemical oxidation of carbon black particles enabled the acquisition of P_{max} ranging from 18 to 25 mW/cm^2 at 850°C [23], depending on the presence and chemical composition of chemical impurities on the surface of the carbon particles, surface area and ratio of structurally highly-ordered graphite phase to the highly disordered phase of soot. In the case of a DC-SOFC (Ni-YSZ) with dimensions of 1.38 and 16 cm^2, a similar P_{max} was achieved, ranging from 30 to 50mW/cm² [15, 23].

This finding clearly indicates that the process of electrochemical oxidation occurred mainly in reaction (1). A different situation was observed in the case of a DC-SOFC supplied with Carbo Medicinalis. It was also found that the electrochemical oxidation of carbon particles on a YSZ surface enabled the acquisition of about 20–25 mW/cm^2 at 850°C. An increase in the surface area of Carbo Medicinalis powders from 600 to 1200m^2/g had practically no influence on DC-SOFC performance, with values ranging from 23 to 28 mW/cm^2. In the case of a DC-SOFC with a Ni-YSZ anode, the impact of CO from the Boudouard reaction as an additional reagent was considerable. The previous data recorded for a DC-SOFC with dimensions of 1.38 and 16 cm^2 were comparable; in that case, P_{max} reached values from 50 to 70 mW/cm^2 [15]. Fig. 5 presents the variations of a DC-SOFC with Ni-YSZ anode vs the specific surface area of the applied charcoal solid fuels. This measurements were performed for the direct button solid oxide fuel cells (surface active area ~1.38cm^2).

As can be seen, with an increase in the surface area of a charcoal solid fuel applied in a DC-SOFC, the power output increases considerably. Higher values of P_{max} were obtained from a DC-SOFC with the application of CO_2 as a shielding gas. This finding clearly indicated that the presence of CO as a possible product of the reaction between carbon and CO_2 led to an increase in the power output of the DC-SOFC.

In the case of a DC-SOFC supplied with charcoal, the electrochemical oxidation proceeded due to reaction (1) and a combination of possible reactions (2)–(4).

Fig. 6a presents the impact of the volumetric flow of CO_2 to the anode chamber on the power output of a DC-SOFC with a 16 cm^2 surface area, supplied with Carbo Medicinalis powder.

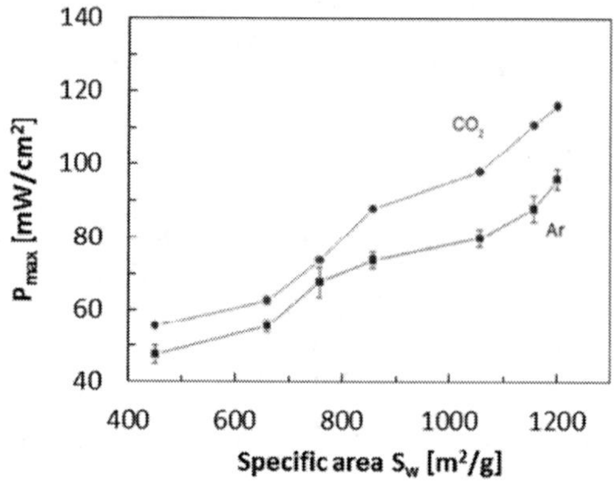

Figure 5. Variation of P_{max} vs specific surface area S_w of applied charcoal as a solid fuel.

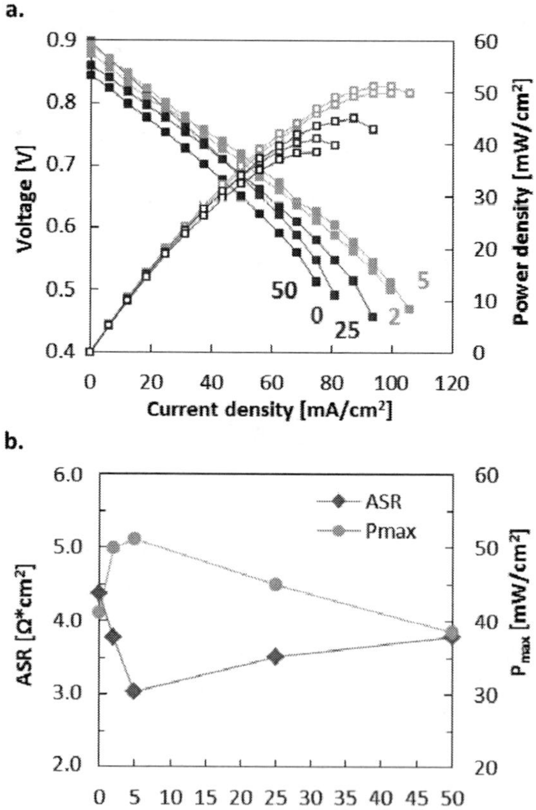

Figure 6. Polarisation and power density curves (**a**) and ASR and P_{max} (**b**) for 0, 2, 5, 25 and 50 Nml/min of CO_2 flow through the anode chamber, cell temperature: 850°C, cathode flow: 200 Nml/min of air.

Analysis of the family of voltage (U)-current (I) as well as power (P)-current (I) curves (Fig. 6a) indicated that the power output P_{max} increases slightly with the introduction of CO_2 into the anode chamber up to 25 Nml/min.

An additional increase in the flow of CO_2 to the anode of the DC-SOFC caused a decrease in power output P_{max}. In Fig. 6b, introduction of a minimal flow of CO_2 to the anode chamber enabled significant improvement in the ASR and maximum power of the cell. The optimum (in terms of DC-SOFC performance) CO_2 flow was experimentally determined to be ca. 5 Nml/min. This finding may be directly connected with adsorption of CO_2 on the surface of carbon particles and the establishment of equilibrium between carbon-CO_2 and the rate of CO synthesis as a gaseous product of reaction (4). These results are consistent with Gür's [24] proposed 'CO shuttle mechanism' to explain the mechanism of carbon oxidation under CO_2 or inert gas atmospheres in several types of direct carbon fuel cells. Gür suggested that this reaction involved oxygen on the carbon surface desorbing as CO into the anode and subsequently undergoing electrochemical oxidation to CO_2 at the TPB. Furthermore, at temperatures above 800°C, the Boudouard reaction is thermodynamically favourable. Within this temperature regime, the surface oxygen desorbs CO_2 formed at the anode from the carbon particles on the anode surface.

4 Conclusions

In this paper we analysed physicochemical high-purity synthetic carbonaceous materials which are crucial for application as solid fuels in DC-SOFC technology. Based on this analysis, charcoal-based carbonaceous materials seem to be appropriate reference fuels. The power output of a DC-SOFC supplied by solid fuels of the Carbo Medicinalis type is high compared to the same DC-SOFC supplied by carbon black N-220. One of the main reasons for this was the high chemical reactivity of carbon particles in a CO_2 gas atmosphere, producing CO, which acts as an additional reagent during cell operation.

Acknowledgements

The present work was carried out under the contract no. 11.11.210.217 (AGH University of Science and Technology, Faculty of Fuels and Energy, Cracow, Poland) and under the statutory program no. CPC/11/STAT/16 (Institute of Power Engineering, Warsaw, Poland) realised under the support of the Ministry of Science and Higher Education (Poland).

References

1. H.A. Linde, A.A.M. Sayigh, Renew. Energ. **16**, 869–871 (1999)
2. H. Zhang, L. Li, D. Zhou, P. Zhou, Renew. Energ. **63**, 330–336 (2014)
3. N. Sammes, *Fuel Cell Technology, Reaching Towards Commercialization*, (Springer, 2006)
4. J. Garche, *Encyclopedia of electrochemical power sources* (Elsevier, 2009)
5. Y. Chen, C. Su, T. Zheng, Z. Shao, J. Power Sources **220**, 147–152 (2012)
6. F. Ishak, I. Dincer, C. Zamfirescu, J. Power Sources **212**, 73–78 (2012)
7. S. Zecevic, E. M. Patton, P. Parhami, Carbon **42**, 1983–1993 (2004)
8. D. Cao, Y. Sun, G. Wang, J. Power Sources **167**, 250–257 (2007)
9. K. Hemmes, J.F. Cooper, J.R. Selman, Int. J. Hydrogen Energy **9**, 8503–8513 (2013)
10. N. Nakagawa, M. Ishida, Ind. Eng. Chem. Res. **27 (7)**, 1181–1185 (1988)
11. M. Skrzypkiewicz, J. Jewulski, I. Lubarska-Radziejewska, Int. J. Hydrogen Energy **40**, 13090–13098 (2015)
12. S. Giddey, S.P.S. Badwal, A. Kulkarni, C. Munnings, Prog. Energy. Combust. **38**, 360–399 (2012)
13. A. Elleuch, K. Halouani, Y. Li, J. Power Sources **28**, 350–361 (2015)
14. M. Dudek, P. Tomczyk, K.L. Juda, R. Tomov, B.A. Glowacki, S. Batty, P. Risby, R. Socha. Int. J. Electrochem. Sci. **7**, 6704–6721 (2012)
15. M. Dudek, P. Tomczyk, R. Socha, M. Skrzypkiewicz, J. Jewulski. Int. J. Electrochem. Sci. **8**, 3229–3253 (2013)
16. A. Sadezky, H. Muckenhuber, H. Grothe, R. Niessner, U. Pöschl, Carbon **43**, 1731 (2005)
17. F. Tuinstra, J. L. Koenig, J. Chem Phys **53**, 1126 (1970)
18. G. A. Zickler, B. Smarsly, N. Gierlinger, H. Peterlik, O. Paris, Carbon **44**, 3229 (2006)
19. C. Li, Y. Shi, N. Cai, J. Power Sources **195**, 4660–4666 (2010)
20. T.M. Gür, M. Homel, A.V. Virkar, J. Power Sources **195**, 1085–1090 (2010)
21. R. Antunes, M. Skrzypkiewicz, Int. J. Hydrogen Energy **40**, 4357–4369 (2015)
22. T. Reed, *Free Energy of Binary Compounds* (MIT Press 1971)
23. M. Dudek, M. Sitarz, P. Tomczyk, J. Solid State Electrochem. 18 (11), 3023-3032 (2014)
24. T. Gür, *J. Electrochem. Soc.* **175(5)**, B571 (2010)

DOI: 10.1051/e3sconf/20161000117

Investigations of the effectiveness of lead disposal from hard coal through the cleaning process

Dorota Makowska[1,a], Andrzej Strugała[1], Faustyna Wierońska[1] and Adam Włodek[2]

[1]AGH University of Science and Technology, Faculty of Energy and Fuels, Al. A. Mickiewicza 30, 30-059 Krakow, Poland
[2]AGH University of Science and Technology, Faculty of Geology, Geophysics and Environmental Protection, Al. A. Mickiewicza 30, 30-059 Krakow, Poland

Abstract. Lead, belonging to the category of toxic heavy metals, poses a major threat to the environment and the health and life of living organisms. Anthropogenic sources of lead emissions to the atmosphere include coal combustion, pyrolysis and gasification processes. This metal enters the atmosphere through the fly ashes and gases, causing chemical imbalance in the ecosystem. One way of reducing the emissions of heavy metals into the atmosphere due to the thermochemical processing of coal involves removing impurities from the fuel prior to its use, e.g. through the coal cleaning processes. The removal of heavy metals from coal through these processes is possible due to the fact that these elements are mostly related to the mineral matter, which is separated from the organic matter. This paper presents the results of investigations of the removal efficiency of lead from coals through the coal cleaning processes in industrial plants. Additionally, the distribution of lead among individual products of these processes was identified. Raw coals, coal concentrates, intermediate products and waste from five plants of mechanical processing of coking coal were analyzed. The determination of the lead content was performed with the use of Flame Atomic Absorption Spectrometry (FAAS). Moreover, for all the samples, a proximate analysis and ultimate analysis as well as the determination of the content of main elements of mineral matter were conducted. In order to confirm the mode of occurrence of lead in coal mineral matter, an elemental analysis was conducted with the use of an electron microprobe. The results of the investigations showed that with a significant reduction of mineral matter in the coal cleaning processes, the lead content is decreased, as a result of both, the application of the gravity and flotation methods. The analyzed metal accumulates primarily in the waste and/or intermediate products of coal cleaning. The lead content in the obtained coal concentrates does not exceed 25 mg/kg. The microprobe examinations confirmed the presence of lead in sulphides, mainly in the form of galena.

1 Introduction

Lead belongs to the group of heavy metals (metals with density greater than 4.5 g/cm^3) with an extremely harmful effect. It can enter the body through inhalation, ingestion (water, food), and through the skin, and cause immune function disorders, renal disfunctions, or neurotoxic and reproductive disorders [1]. This metal exhibits a toxic effect and is unnecessary for the proper functioning of living organisms. Therefore, Pb poses a major threat to the natural environment as well as health and life [2-3].

One of the major anthropogenic sources of lead emission to the environment is constituted by the thermochemical coal processes (mostly combustion). In Poland, nearly the entire amount of electricity and heat is obtained from hard and brown coal combustion [4]. Therefore, the industrial and nonindustrial combustion processes cause the release of 466 tons of lead to the atmosphere per year, representing 83% of total lead emission resulting from human activities [5].

The large share of the combustion processes in lead emission can be attributed to the fact that coal is usually enriched with heavy metals, including lead [6]. A number of geochemical research works have shown that Polish coals, i.e. not only those from the Upper Silesian Coal Basin (USCB) and of the Lublin Coal Basin (LCB) but also those from the Lower Silesian Coal Basin LSCB (no longer exploited) contain a higher amount of lead compared to the world average content of this element in hard coal. The average content of Pb in the coals of the USCB was 19 ÷ 28 mg/kg (ppm) [7-12]. Differentiation of the lead distribution in each layer of this basin and the impact of coal organic and inorganic matter on the Pb content indicates the great dynamic processes of accumulation and dispersion of this element [13]. Solutions rich in lead, flowing through the bed at every stage of its creation, were the main factor affecting the lead content in organic and inorganic coal matter. Lead is generally associated with coal mineral matter. It occurs mainly in sulphides (PbS - galena, PbSe - clausthalite) [14] and can be associated with pyrite (FeS_2) [15]. It may also be present in alumino-silicates

[a] Corresponding author: makowska@agh.edu.pl

and carbonates [16-18]. Studies [19-20] on the coals from the USCB have linked lead mainly with the mineral matter. Coals from the Chełm deposits of the LCB are characterized by a similar lead content (27 mg / kg) to coals from the USCB [21]. These coals also have shown a higher affinity of lead to the mineral matter than to the organic matter of coal. The coking coals from the LSCB contain a higher amount of Pb with the average value of 32.7 mg/kg [22].

The coal combustion processes result in the enrichment of solid by-products with heavy metals. Lead passes primarily to fly ash and flue dust [23]. The enrichment factor for these products, i.e. the ratio of the Pb content in the product to its contents in the combusted coal, is 3.78 and 21.11, respectively. The fly ash from the combustion of the coals from the USCB accumulates lead mainly in particles with a grain size of less than 100 µm [24]. For the ash grains <10 µm, the lead content was 265 ppm (mg/kg). These results are confirmed by other researchers [25], and support the view [26] about an increase in the content of elements in the series of As-Se-Sb-Zn-Pb-Cd-Hg with a decrease in the particle diameter of the solid by-products of coal combustion. The enrichment of these products depends on the ionization potential and the vaporization temperature of the elements [27-29]. The highest content of Pb was found in a group of fly ash particles with a size less than 20 µm comprising mainly magnetite. The lead content in the solid by-products of combustion depends primarily on its mode of occurrence in coal (connection with mineral or organic matter) [30-31]. Combustion conditions are of secondary importance.

The smallest particles of fly ash (<10 µm) are not fully captured from the flue gas and get into the atmosphere in the form of suspended particles (TSP). The study [32] demonstrated the effect of combustion on the emission of lead into the atmosphere through dust. The concentration of Pb in the TSP increased 1.5 times during the heating season.

In the pyrolysis processes, lead also has a tendency to pass to the gas phase at temperatures above 700 °C [33]. However, about 50% of lead remains in the solid phase (char or coke) up to 1000 °C. In addition, pyrolysis changes the mode of lead occurrence in solid residues. This was confirmed by other studies [34]. It has been shown that the lead sulphides, which were considerably quantitatively higher than other forms of lead bonding, were reduced in char relative to the raw coal, contrary to the other forms of Pb (carbonate, organic, alumino-silicate and pyrite).

Currently Poland's policy is to reduce the emissions of heavy metals, i.e. Hg, Pb and Cd into the atmosphere [35]. Reducing the emission of these metals from coal combustion, pyrolysis and gasification is possible to achieve in two ways: (1) by removing pollutants from fossil fuels before they are used in industrial processes, or (2) by the purification of exhaust and flue gases [36-38]. The first group of methods includes the coal cleaning processes. Among them gravity methods (enrichment in the heavy media and jigs) and flotation methods can be distinguished [39]. These processes involve the removal of mineral matter from raw coal (feed), resulting in a high- quality coal concentrate. Therefore, the removal of lead from coal by its cleaning is possible due to lead association with coal mineral matter, mainly in sulphide form.

Studies [40-41] on the removal of heavy metals, (including lead) from coal by the coal cleaning process in a dense medium on a laboratory scale were carried out. These studies confirmed lead affinity to coal mineral matter and the possibility of its removal from the raw coal through this process. The preliminary study [42] on the possibility of reducing the content of selected heavy metals (lead, zinc, copper, chromium, manganese, and nickel) from hard coals by their enrichment was also conducted by the authors of this paper. Tests of samples from two coking coal preparation plants have shown that the content of these metals can be effectively reduced by up to about 80%. In the case of lead, the coal cleaning in dense media resulted in a reduction of its content up to 56%; enrichment in jigs removed 44% of Pb, and in the case of flotation - 30%. The lead content in the coal concentrates did not exceed 16 mg / kg.

The research carried out within the framework of this study was conducted to determine the mode of lead occurrence and to confirm the results obtained in earlier studies on the possibility of lead removal from coal by its enrichment in industrial coal preparation plants. It was possible by way of expanding the scope of research.

2 Experimental

2.1 Examined material

The examined material consisted of samples from the coal preparation plants of five Polish coking coal mines. The study focused on both, samples of raw coal (coal output) as well as on individual products of the coal cleaning processes. Each of the processing plants conducts the enrichment of coal in a different way. In view of the above, a summary of the examined mines and a description of the coal cleaning processes are shown in Table 1.

Table 1. Description of examined coal cleaning processes.

Mine	Sections of coal cleaning process		
	Dense medium washer (grain class)	**Jig washery (grain class)**	**Flotation (grain class)**
A	Yes (20÷150 mm) two-product dense media bath DISA	Yes (0÷20 mm) two-product water pulsation jigs	Yes (<0.5 mm from dewatering of jig wash product) Flotation machines IZ
B	No	Yes (0÷80 mm) three-product water pulsation jigs	
C	Yes (20÷150 mm) three-product dense media bath DISA	Yes (0÷20 mm) three-product water pulsation jigs	
D	Yes (20÷200 mm) three-product dense media bath DISA	Yes (0÷20 mm) three-product water pulsation jigs	
E	Yes (20÷120 mm) two-product dense media bath DISA	Yes (0÷20 mm) three-product water pulsation jigs	

A total of 56 samples were analysed, including: raw coals, coal concentrates, intermediate products, waste and a dehydrated sediment from flotation. For the purposes of the electron microprobe analysis, samples of waste and intermediate products, which were characterized by a higher content of lead, were selected.

2.2 Analytical methods

For all samples, the determinations of the content of ash, total sulphur and pyritic sulphur in accordance with ISO standards were performed. For raw coals and coal concentrates, the calorific value was determined. The analysis of the lead content was performed using Atomic Absorption Spectrometry with flame atomization. The Hitachi Z-2000 spectrometer with the Zeeman background correction effect was applied. To determine the lead content all the samples were incinerated at 600 °C and subjected to microwave digestion in a Berghoff SpeedWave4 microwave system using concentrated nitric acid (V). The quantification limit of lead determination is equal to 0.055 mg/dm^3 (for prepared solutions) and the relative standard deviation is equal to 10%.

The microscope analysis and the phase chemical composition analysis of selected waste and intermediate products were conducted with the use of the Electron Probe Microanalyzer (EPMA) Jeol JXA-8230, located at the Laboratory of Critical Elements KGHM – AGH. The analysis of the chemical composition enabled identification of mineral phase and determination of accompanying trace elements. The polymorphic type of pyrite was determined based on visual assessment of grain shape. The detection limit of lead determination by this method equals 195 mg/kg (ppm). Samples for the microprobe analysis were taken in the form of sample cut.

3 Results and discussion

3.1. Mode of lead occurrence in examined coals

Based on the performed analyses, the correlations between the lead content and the ash, total sulphur and pyritic sulphur content in the examined samples were examined (Fig. 1-3). The aim of this analysis was to determine the association of lead with coal mineral matter and sulphides.

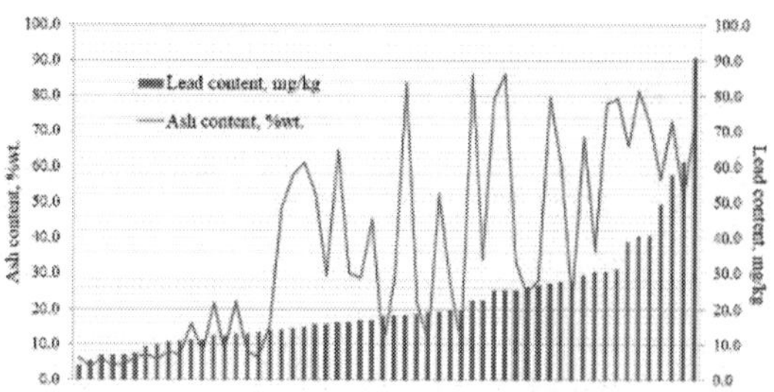

Figure 1. Correlation between lead and ash content in examined samples.

The results indicate a weak correlation between the lead content and the ash content for the examined samples (Fig. 1). However, a general increase in the content of lead can be observed along with an increase in the ash content. There is no visible relationship between the lead content and the total sulphur content for the analysed samples (Fig. 2). Nevertheless, a correlation between the lead and pyritic sulphur content (Fig. 3) is evident, which indicates the probably connection of lead with the sulphide form.

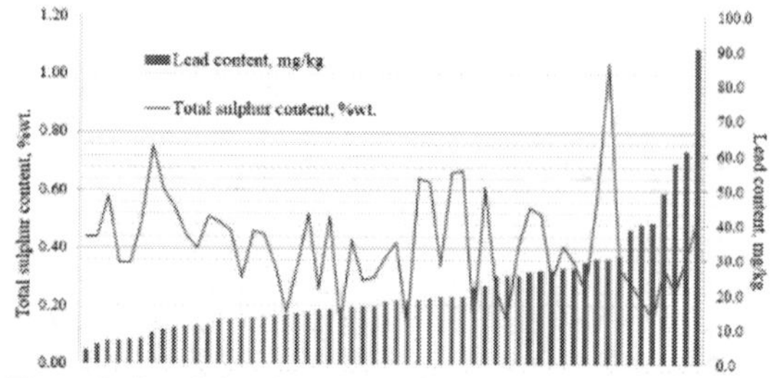

Figure 2. Correlation between lead and total sulphur content in examined samples.

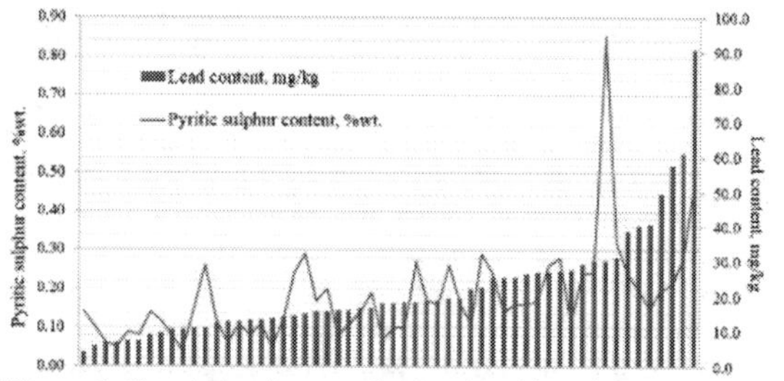

Figure 3. Correlation between lead and pyritic sulphur content in examined samples.

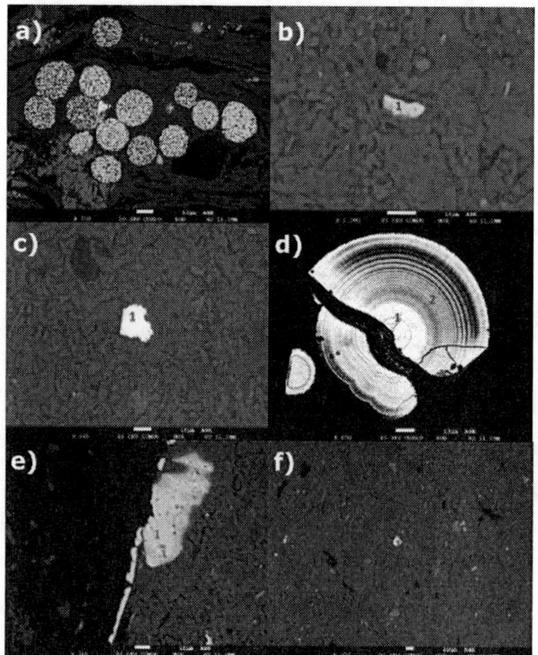

Figure 4. Results of lead content determination in sulphides: a) framboidal pyrite with chalcopyrite inclusion, b) chalcopyrite, c) galena, d) marcasite, e) pyrite, f) pyrite.

The relation between the analysed metal and the sulphide form is confirmed by the analysis of the chemical composition using a microprobe. Fig. 4 shows microscopic pictures with marked measuring points. Table 2 summarizes the results of the analysis

DOI: 10.1051/e3sconf/20161000117

of the chemical composition. The results show that, apart from lead sulphide (galena), lead can be found in relatively large amounts in other sulphides. Inclusions or individual grains of regular pyrites and concentrations of framboidal pyrite contain from 0.116 to 0.227 %wt. of Pb. Also chalcopyrite contains about 0.1 %wt of lead. A high content of this metal (0.2 ÷ 0.5 wt.%) was determined in marcasite. Additionally, carbonates, iron oxides, and alumino-silicates were analysed in respect of the lead content, but the results did not exceed the detection limit.

The results confirm the presence of lead in sulphuric form in the analysed coal and thus, the coal cleaning process aiming at removing this toxic metal becomes justified. Sulphides are characterised by a high density (chalcopyrite 3.2 g/cm^3; pyrite 4.95 ÷ 5.10 g/cm^3; galena 7.2 ÷ 7.6 g/cm^3) and they are usually located in the exterior mineral matter of coal, which may be removed through the coal cleaning processes.

Table 2. Results of analysis of sulphides.

No.	Content, % wt.				Mineral
	S	Cu	Fe	Pb	
Fig. 4a -1	32.313	30.060	27.840	0.110	chalcopyrite
Fig. 4a - 2	52.752	0.010	44.934	0.116	framboidal pyrite
Fig. 4b - 1	30.150	29.186	24.341	0.098	chalcopyrite
Fig. 4c - 1	11.514	0.043	0.069	81.224	galene
Fig. 4d - 1	51.056	0.062	42.620	0.248	marcasite
Fig. 4d - 2	50.763	0.110	41.212	0.537	marcasite
Fig. 4e - 1	40.747	0.038	37.938	0.122	pyrite
Fig. 4f - 1	51.438	0.019	43.938	0.227	pyrite

3.2 Lead distribution in products of coal cleaning processes

The results of the lead content determination in the analysed samples of coal and enrichment products are summarized in Table 3. The average lead content in the tested raw coals is 22.9 mg/kg, which is close to the average value of the Pb content in coals from USCB and more than twice the value of the world average Pb content in hard coal, which is 9.0 ± 0.7 [6]. The clark value (global average value) of lead for sedimentary rock is 12 mg/kg; thus the enrichment factor EF [43] for the analyzed raw coals is 1.9. The ashes from these coals are characterized by the enrichment factor of 3.3, which constitutes a little lower value than the CAI (Coal Affinity Index), which in the case of lead is 3.9 [6] and represents the ratio of the global average lead content in the coals to the clark value of lead in sedimentary rocks.

Coal concentrates from dense medium washery contain the lowest amount of the analysed metal (7 mg/kg), while the highest concentration was determined for coal concentrates from flotation (14.3 mg/kg).

In order to better illustrate the distribution of lead among the different coal cleaning products, results are shown in the graphs (Fig. 5-7). The lead content in the feed for the enrichment in the dense media washers is at the same level for all the mines (14.3 ÷ 16.2 mg/kg). Coal concentrates contain a significantly lower amount of this element, i.e. from 4.0 to 11.1 mg/kg. Lead accumulates mainly in waste (28.8 mg/kg) and, to a smaller extent, in intermediate products (17.6 mg/kg).

Table 3. Concentration of lead in individual products of coal cleaning processes.

Type of coal cleaning	Material	Lead content, mg/kg					
		Mine A	Mine B	Mine C	Mine D	Mine E	Average
Raw coals before classification*		13.0	16.9	19.2	15.7	49.5	22.9±15.1
Dense medium washery	Raw coals	14.3	-	15.0	16.2	14.6	15.0±0.8
	Coal concentrates	7.1	-	4.0	5.6	11.1	7.0±3.0
	Intermediate products	-	-	22.6	12.6	-	17.6±7.1
	Waste	18.4	-	30.6	40.6	25.4	28.8±9.3
Jig washer	Raw coals	16.7	16.9	15.8	16.4	25.4	18.2±4.0
	Coal concentrates	6.8	9.0	6.7	7.2	12.8	8.5±2.6
	Intermediate products	-	18.5	19.5	27.6	27.0	23.2±4.8
	Waste	22.2	25.4	31.3	27.3	40.4	29.3±7.0
Flotation	Raw coals	27.7	30.4	26.4	18.2	11.0	22.7±8.0
	Coal concentrates	18.0	19.5	9.7	10.6	13.5	14.3±4.4
	Waste	61.2	29.4	57.7	38.8	91.0	55.6±23.8
	Dehydrated sediments	13.8	-	-	10.8	13.3	12.6±1.6

*) raw coal before grain segregation into two sorts processed in dense media bath or in jig washers.

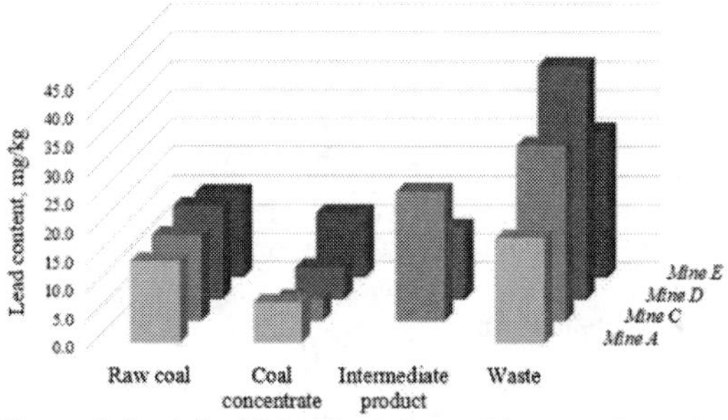

Figure 5. Lead distribution in products of dense media washers.

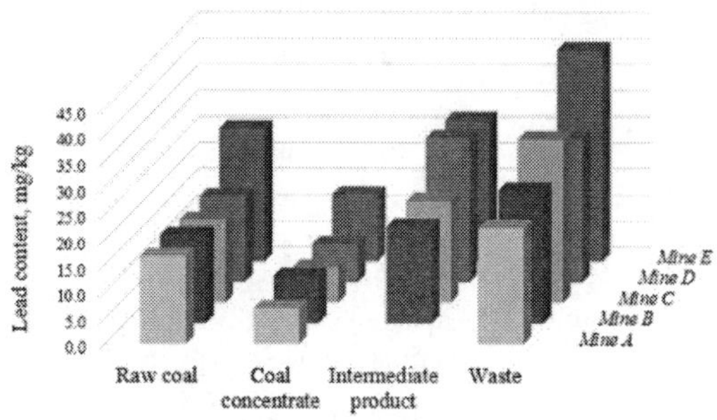

Figure 6. Lead distribution in jig washers.

The coal intended for cleaning in the jigs has an average of 18.2 mg Pb per kg of the tested material;

 DOI: 10.1051/e3sconf/20161000117

in the case of coal concentrates the Pb content do not exceed 12.8 mg/kg. The Pb contents in waste and intermediate products are similar to each other and are equal to 29.3 mg/kg and 23.2 mg/kg, respectively.

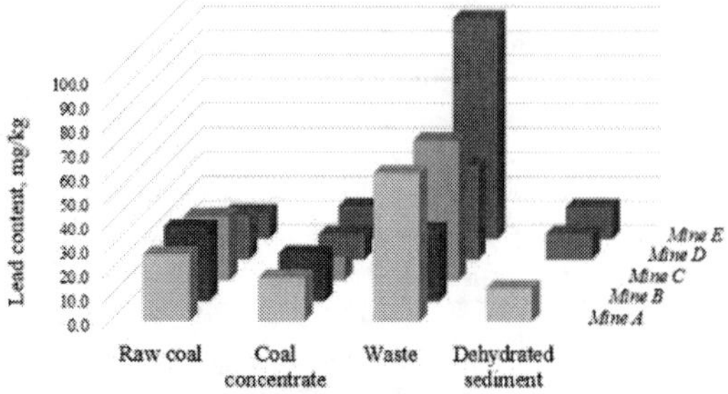

Figure 7. Lead distribution in flotation.

The feed for flotation contains the biggest amount of lead compared to the feeds for other sections of the coal cleaning process (22.7 mg/kg), which affects its high content in the waste, even up to 91.0 mg/kg. The dehydrated sediments contain a similar amount of lead to the coal concentrates (12.6 mg/kg).

3.3 Effect of lead reduction in coal through coal cleaning process

The effectiveness of the reduction of the lead content in coal was determined by comparing the content of the analysed element in the coal before and after the cleaning process, according to the formula (1) [42]:

$$\eta_{Pb1} = \frac{c_{raw\_coal} - c_{coal\_concentrate}}{c_{raw\_coal}} \cdot 100 \qquad (1)$$

where η_{Pb1} represents the reduction ratio of the lead content in coal, %; c_{raw_coal} - the Pb content in the feed, mg/kg, and $c_{coal\_concentrate}$ - the Pb content in the coal concentrate, mg/kg.

The effectiveness of the reduction of the content of the investigated metal in coal was calculated by means of the ratio which takes into account the calorific value of the feed and the coal concentrate according to the formula (2) [42]:

$$\eta_{Pb2} = \frac{\frac{c_{raw\_coal}}{Q^a_{i\_raw\_coal}} - \frac{c_{coal\_concentrate}}{Q^a_{i\_coal\_concentrate}}}{\frac{c_{raw\_coal}}{Q^a_{i\_raw\_coal}}} \cdot 100 \qquad (2)$$

where η_{Pb2} represents the Pb reduction ratio, %; $Q^a_{i\_raw\_coal}$ - the calorific value of the feed, kJ/kg (in analytical state); $Q^a_{i\_coal\_concantrate}$ - the calorific value of the coal concentrate, kJ/kg (in analytical state).

The results of these calculations for each of the coal cleaning process sections are shown in Table 4. The highest ratio of lead content reduction in the examined coals was recorded for the gravity methods, for both dense medium washers and jigs. This type of coal cleaning eliminates up to 53% of the lead content in coal for dense medium washers and 54% for jig washers. The flotation also enables a high degree of lead removal from coal, except for the case of mine E, for which the result indicates the enrichment of the concentrate with lead. Therefore, excluding the mine E, the average value of the degree of lead content reduction in coal is 44%.

Table 4. Lead content reduction index for examined coal cleaning processes.

Factor	Type of coal cleaning	Mine					
		A	B	C	D	E	Average
η_{Pb1} [%]	Dense medium washery	50	-	73	66	24	53±22
	Jig washer	59	47	58	56	50	54±5
	Flotation	35	36	63	41	-22	31±32
η_{Pb2} [%]	Dense medium washery	75	-	92	90	59	79±15
	Jig washer	71	71	70	69	65	69±3
	Flotation	46	58	71	56	-6	45±30

The effect of reducing the Pb content in coal is even more clearly visible when its content is compared to the calorific value of raw coal and the coal concentrate obtained in an individual section of the coal cleaning process (η_{Pb2}). This illustrates the real environmental effect of the process for coal utilization (combustion, gasification and pyrolysis). The lead content in coal per unit of its calorific value (Fig. 8) was significantly reduced as a result of the coal cleaning process to not more than 0.35 mg/MJ in the coal concentrates from dense medium washery, 0.40 mg/MJ in the coal concentrates from jigs and 0.68 mg/MJ in the coal concentrates from flotation. The reduction ratio of the lead content η_{Pb2} for cleaning in the dense medium washers is relatively diverse and ranged from 59 ÷ 92 %.

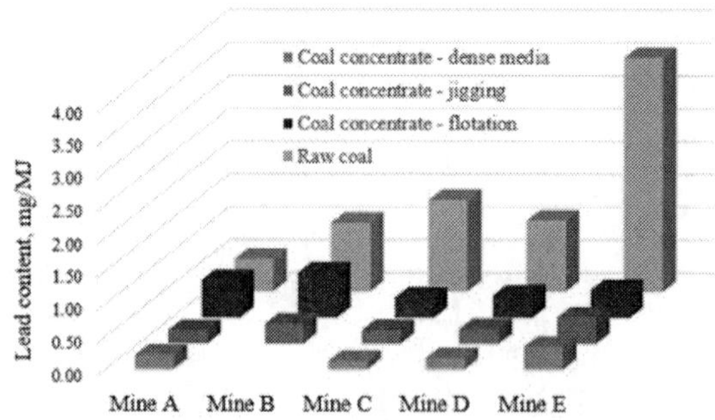

Figure 8. Lead content in raw coals and coal concentrates per unit of calorific value.

The average value of this ratio is 79%, which indicates a very high efficiency of removal of Pb from coal before its industrial use. For cleaning in jig washers, the η_{Pb2} ratios for individual mines reach similar values, yielding an average rate of 69%. Flotation is characterized by a ratio of 45%, and, rejecting the result obtained for the mine E, the η_{Pb2} value would be 58%.

4 Conclusions

The study on the possibility of removing lead from coal through its cleaning process in industrial plants allows for the formulation of the following conclusions:

• Lead in the examined samples of coking coal from the Upper Silesian Coal Basin is associated with coal mineral matter, mainly in the form of galena (PbS) as well as pyrite and marcasite inclusions.
• The industrial processes of coal cleaning successfully remove at least 30% of lead from coal.
• The greatest efficiency of the removal of lead from coal is achieved through cleaning in dense media and jig washers which reduce 53-54% of lead.
• The received coal concentrates had an average content of lead equal to 9.3 mg/kg; a lead content per unit of their calorific value was equal to 0.48 mg/MJ (on average).

Acknowledgements

This paper was prepared as a part of the statutory activity of the Faculty of Energy and Fuels at the AGH University of Science and Technology, No. 11.11.210.213

References

1. G.F. Nordberg, B.A. Fowler, M. Nordberg, L. Friberg, *Handbook on the toxicology of metals* (Elsevier 2007, 3rd edition)
2. A. Kabata-Pendias, H. Pendias, *Biogeochemia pierwiastków śladowych* (PWN Warszawa 1999)
3. D. C. Adriano, *Trace elements in terrestrial environments. Biogeochemistry, bioavailability, and risks of metals* (Springer-Verlag, New York 2001)
4. *Energy Statistics in 2013 and 2014 - Statistical Information and Elaborations*, Central Statistical Office (Warsaw 2015)
5. *Environment 2015 – Statistical Information and Elaborations*, Central Statistical Office (Warsaw 2015)
6. M.P. Ketris, Ya.E. Yudovich, Int J Coal Geol **78**, 135-148 (2009)
7. H. Parzentny, A. Różkowska, Przegląd Geologiczny **2**, 52-56 (1990)
8. H. Parzentny, Przegląd Górniczy **6**, 33-36 (1991)
9. H. Parzentny, Przegląd Górniczy **2**, 25-29 (1992)
10. H. Parzentny, A. Różkowska, L. Róg, Geol Q **43** (3), 365-374 (1999)
11. H. Parzentny, L. Lewińska-Preis, Chem Erde-Geochem **66**, 227-247 (2006)
12. H. Parzentny, Energetyka **443**, 169-171 (1991)
13. H. Parzentny, Geol Q **38**, 1, 43-58 (1994)
14. J. Hower, J.D. Robertson, Int J Coal Geol **53**, 219-225 (2003)
15. S.F. Diehl, M.B. Goldhaber, J.R. Hatch, Int J Coal Geol **59**, 193-208 (2004)
16. F. Goodarzi, Fuel **81**, 1199-1213 (2002)
17. R.B. Finkelman, Fuel Process Technol **39**, 21-34 (1994)
18. J. Wang, A. Sharma, A. Tomita, Energ Fuel **17**, 29-37 (2003)
19. H. Parzentny, A. Różkowska, Przegląd Geologiczny **11**, 656-659 (1992)
20. H. Parzentny, Przegląd Górniczy **46**, 3 (798), 16-19 (1990)
21. M. Marczak, H. Parzentny, Przegląd Geologiczny **12**, 680-683 (1985)
22. H. Parzentny, Górnictwo i Geologia **2**, 4, 69-81 (2007)
23. J.N. Pacyna, Pr. Nauk. Inst. Inż. Ochr. Środ. Pol. Wrocł. 47, seria Monoografie **17** (1980)
24. H. Parzentny, L. Róg, Górnictwo i Geologia **2**, 3, 81-91 (2007)
25. J. Stanisz, A. Pasoń-Konieczyńska, J. Konieczyński, , Arch Environ Prot **26**, 7-20 (2000)
26. J. Kuhl, Przegląd Górniczy **2** (1980)
27. H. Parzentny, L. Róg, Przegląd Górniczy **7-8**, 52-60 (2001)
28. J.A. Ratafia-Brown, Fuel Process Technol **39**, 139-157 (1994)
29. R. Yan, D. Gauthier, G. Flamant, Fuel **80**, 2217-2226 (2001)
30. S.V. Vassilev, G.M. Eskanazy, C.G. Vassileva, Fuel Process Technol **72**, 103-129 (2001)
31. M. Xu, R. Yan, C. Zheng, Y. Oiao, J. Han, C. Sheng, Fuel Process Technol **85**, 215-237 (2004)
32. Y. Liu, C. Liu, Y. Shen, F. Tian, H. Liu, F. Shi, Pol J Environ Stud **25**, 1, 205-212 (2016)
33. R. Guo, J. Yang, D. Liu, Z. Liu, J Anal Appl Pyrolysis **70**, 555-562 (2003)
34. S. Scaccia, R. Mecozzi, Microchem J **100**, 48-54 (2012)
35. *Krajowa Strategia Ograniczania emisji metali ciężkich*, Ministerstwo Środowiska (Warszawa 2002)
36. A. Strugała, J. Gołaś, *Mercury as a coal combustion pollutant* (Oficyna Drukarska – Jacek Chmielewski, Kraków 2014)
37. K. Jóźwik, M.I. Szynkowska, P. Kula, P. Niedzielski, R. Pietrasik, J. Góralski, Przem Chem **93**, 8, 1339 (2014)
38. C. Zhu, H. Tian, K. Cheng, K. Liu, K. Wang, S. Hua, J. Gao, J. Zhou, J Clean Prod, 1-9 (2015)
39. W. Blaschke, *Przeróbka węgla kamiennego. Wzbogacanie grawitacyjne,* Instytut Gospodarki Surowcami Mineralnymi i Energią PAN (Kraków 2009)
40. E. Zajusz-Zuber, J. Konieczyński, Arch Environ Prot **40**, 1, 115 (2014)
41. H. Aleksa, F. Dyduch, K. Wierzchowski, Górnictwo i Geoinżynieria **31**, 3, 35 (2007)
42. D. Makowska, K. Bytnar, T. Dziok, T. Rozwadowska, Przem Chem **93**, 12, 2048-205 (2014)
43. A. Strugała, D. Makowska, K. Bytnar, T. Rozwadowska, Polityka Energetyczna - Energy Policy Journal **17**, 4, 77-88 (2014)

E3S Web of Conferences **10**, 00118 (2016) DOI: 10.1051/e3sconf/20161000118
SEED 2016

Influence of FIMA burnup on actinides concentrations in PWR reactors

Mikołaj Oettingen[1] and Katarzyna Skolik[1]

[1]*AGH University of Science and Technology, al. Mickiewicza 30, 30-059 Krakow, Poland*

Abstract. In the paper we present the study on the dependence of actinides concentrations in the spent nuclear fuel on FIMA burnup. The concentrations of uranium, plutonium, americium and curium isotopes obtained in numerical simulation are compared with the result of the post irradiation assay of two spent fuel samples. The samples were cut from the fuel rod irradiated during two reactor cycles in the Japanese Ohi-2 Pressurized Water Reactor. The performed comparative analysis assesses the reliability of the developed numerical set-up, especially in terms of the system normalization to the measured FIMA burnup. The numerical simulations were preformed using the burnup and radiation transport mode of the Monte Carlo Continuous Energy Burnup Code – MCB, developed at the Department of Nuclear Energy, Faculty of Energy and Fuels of AGH University of Science and Technology.

1 Introduction

The aim of the study is a comparative analysis between measured and calculated actinide concentration in function of FIMA burnup to assess its influence on the *C/E* ratios (calculated-to-experimental), which define agreement between experimental measurements and numerical calculations. The change in the nuclear fuel composition caused by the series of nuclear transmutations and radioactive decays is called fuel burnup. The burnup in FIMA units (Fission per Initial Metal Atom) describes the net change of heavy metal in the specified piece of a nuclear fuel at the specified irradiation time. The measured burnup of the irradiated nuclear fuel is the crucial parameter for the numerical reconstruction of the irradiation experiment e.g. in the terms of validation of numerical models, libraries and codes. In this way, using measured experimental data and the data obtained in the numerical simulation the relationship between the real world and the computational model could be defined. However, this kind of study must be based on some reference measurements obtained in dedicated experiments.

The Ohi-2 irradiation experiment was chosen for the study because the reported experiment specification was sufficient for the reliable numerical reconstruction using the MCB code. The reference data applied in the analysis are mainly based on the two scientific publications: the paper of T. Adachi et al. [1] and the paper of K. Suyama et al. [2]. The papers include the comprehensive knowledge about performed post irradiation examination of the spent fuel samples, irradiation history and fuel assembly design. Thus, the papers provide the reliable and robust source of information for numerical reconstruction.

2 PWR reactors

Pressurized Water Reactor (PWR) is the most popular and best known type of nuclear reactor in the world. Currently (May 2016) there are about 285 PWRs in commercial operation which represent approximately 65% of the entire nuclear fleet – see Fig. 1 [3]. They use common and cheap light water as a coolant and a moderator, employ uranium fuel with enrichment no bigger than 5% and consist of two separated circuits what helps to minimize potential radioactive releases to the environment. The pressure of about 15 MPa in primary circuit allows to maintain liquid state of the water that reaches temperature above 300°C.

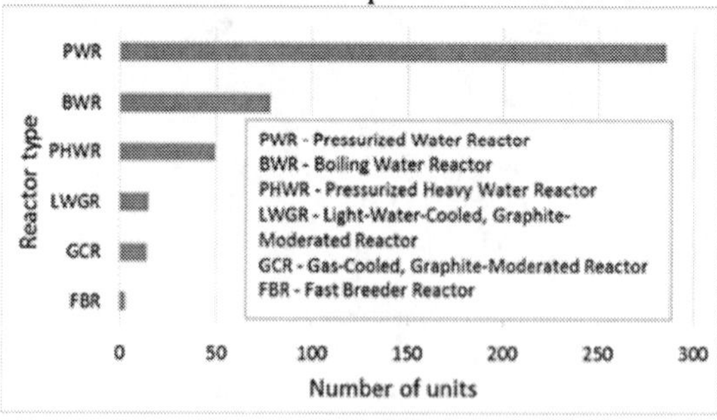

Figure 1. Nuclear reactors by type [3].

Since the first full-scale commercial PWR was connected to the grid in 1957 (Shippingport Nuclear Power Plant (NPP), USA), the design was constantly being improved in order to achieve bigger power output, longer lifetime of the reactor and better safety features. PWRs that are in operation nowadays belong mostly to generation II of nuclear reactors [4]. Specific models originate from the USA (Westinghouse, Babcock&Wilcox and General Electric designs), France

DOI: 10.1051/e3sconf/20161000118

(Framatome /now Areva/), Russia (VVER reactors by OKD Gidropress), China and South Korea. Nearly 50% operating NPPs in the world are based on US Westinghouse technologies. The electrical power output from one reactor is 1000 MW_e on average and predicted lifetime of a plant is approximately 60 years. Evolution of PWR design still continues and now generation II+, III and III+ reactors are developed and constructed, mostly in the countries mentioned above [5].

The newest PWR designs in the USA are AP1000, APWR and APWR+. Westinghouse AP1000 is the first generation III+ reactor with the approval from U.S. Nuclear Regulatory Commission. The first unit will start its operation in 2017 in China and four more are now under construction in the USA. APWR and APWR+ are developed by Mitsubishi Heavy Industry. The second one is planned to have increased power output (up to 1700 MW_e) and full Mixed-Oxide Fuel (MOX) fuel ability.

European Pressurized/Evolutionary Power Reactor (EPR), with the output electrical power of 1650 MW_e was developed mainly by Framatome (Areva) in France and Siemens in Germany. Four units are under construction, one each in France and Finland and two in China. However all of them suffer from long-term delays because of complexity and expensiveness of the project.

Chinese models, CPR1000 (gen. II+) and ACPR1000 (gen. III) are the first reactors created in this country intended for the international market. By now they are only operating or being constructed in China – about 25 in total.

The latest version of the Russian VVER is called VVER-1200, offered for domestic and export use. A lot of reactors of this type are planned in Bangladesh, China, Hungary, Indonesia and Vietnam and are being constructed in Russia and Finland [5].

A common features of all new PWR designs are increased power, extended lifetime of a plant and lower risk of failure. The bigger output power is intended to improve economic performance according to the effect of scale (the profits gained during operation time exceeds the additional investments during construction). The increased level of safety is achieved by introduction of passive safety systems [6], installation of redundant devices and systems and, in many cases, presence of so-called core catcher.

3 Irradiation

The actinide concentrations were obtained in the destructive assay of two uranium fuel samples irradiated in the 17x17 fuel assembly at the Japanese Ohi-2 PWR of the Kansai Electric Power Company Inc. The fuel samples indexed as 89G08 and 89G010 were cut from the UO_2 (3.2 wt.% ^{235}U) fuel pin named F4 – see Fig 2. The considered leading 17x17 fuel sub-assembly was irradiated during two reactor cycles from Jul. 1984 to Feb. 1987. The first cycle lasted 410 full power days, while the second one 427 full power days. The operational outage between cycles lasted 104 days. The leading fuel assembly was discharged at the average burnup of 31.5 GWd/t.

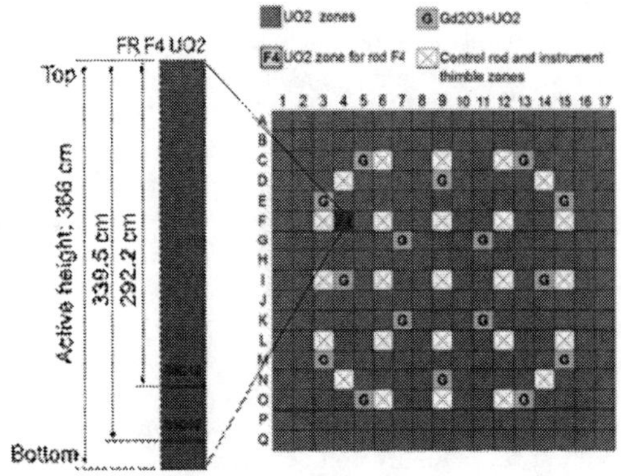

Figure 2. Location of the investigated fuel samples in the fuel rod and assembly.

4 Experimental

The post irradiation examination of the spent fuel samples was conducted at the Japan Atomic Energy Research Institute in 1987, just after removal of the fuel assembly from the reactor core. The non-destructive testing included: gamma scanning of fuel rods, mass spectrometry of the gaseous fission products, visual examination, crude analysis, profilometry, fuel stack elongation measurement and X-ray radiography. Fig. 3 shows the flowchart of the destructive assay, which follows to the final measurements of the actinide concentrations and burnups. The measured actinide concentrations are necessary to quantify sample burnup, which is the crucial parameter for the burnup calculations using the MCB code. The burnups in FIMA units were calculated using measured concentrations of ^{148}Nd, U, Np, Pu, Am, Cm and effective fission yields using ^{148}Nd method.

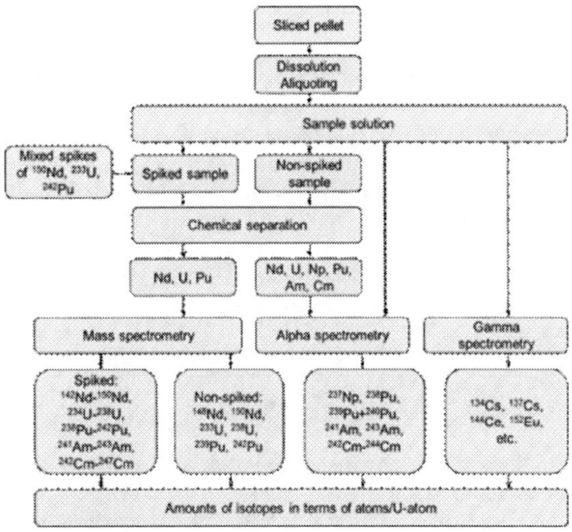

Figure 3. Flowchart of destructive assay.

5 Numerical

Usually, in the neutron transport and burnup calculations the numerical reconstruction focuses only on the most important components of the engineering geometry influencing neutronic parameters, like neutron fluxes

DOI: 10.1051/e3sconf/20161000118

and spectra. Therefore, the engineering geometry does not need to be exactly transferred to the computational geometry of the numerical model.

Thus, the fuel in fuel rods was not divided into particular fuel pellets. However, it was divided into six radial and fourteen axial burnup zones. Hence, the whole fuel in the assembly was represented by the 84 independent zones – two of them correspond to the investigated fuel samples. This kind of division was purposefully chosen to reflect the physics of the fuel sample with high accuracy.

All layers above the fuel column constitute the top reflector while all layers below fuel column constitute the bottom reflector. The modelling of the reactor core at the level of the fuel assembly demands introduction of the reflective boundary conditions on the vertical edges of the numerical model, which was done in all presented calculations.

The critical or subcritical fissile system reconstructed using the MCB code has to be normalized to the total thermal power released during operation time [7, 8]. The simulations of the 17x17 fuel assembly were normalized to the total thermal power using the Q_{fiss} values implemented in the JEFF3.1 cross section libraries [9]. Providing the average assembly discharge burnup B, 31.5 GWd/t, duration of the reactor cycles t, 837 days and initial content of heavy metal in the fuel m_{init}, 451.76 kg, the Eq. (1) can be used to calculate absolute thermal power P, which equals 16.94 MW_{th}. The initial mass of heavy metal was derived using the reference fuel assembly geometry and fuel composition. The average system power was applied to perform initial MCB simulations.

$$B = \frac{P \cdot t}{m_{init}} \qquad (1)$$

The final mass of heavy metal in the investigated fuel zones – m_{final} was obtained in the initial simulations. Then, FIMA burnup in fuel zones representing fuel samples could be calculated using Eq. (2). To adjust the initial FIMA from simulation to the reference FIMA from experimental measurements the initial calculations were rerun applying the same system thermal power (16.94 MW_{th}) but changing irradiation time. This procedure allows the direct adjustment of the calculated FIMA to the reference FIMA.

$$FIMA = \left(1 - \frac{m_{final}}{m_{init}}\right) \cdot 100 \ [\%] \qquad (2)$$

In the next step the sensitivity analysis of the actinide concentration to the changes in FIMA burnup was performed. For this reason the irradiation time was modified to increases or decrees FIMA burnup up to 2%. This procedure can be applied because the FIMA behaves in linear manner, which helps to predict targeted FIMA after assumed irradiation time.

The calculations for all time steps were performed in the *kcode* mode with the initial neutron multiplication factor set to 1.0. The number of inactive generations was set to 40 and the number of active generations to 100. In each generation 50 000 neutrons were simulated, which gives $7x10^6$ particle histories.

5.1 Sample 89G08

Table 1 presents the *C/E* ratios for the lower sample 89G08 with the reference measured FIMA burnup of 3.129. The green fields show the acceptable agreement in the range of 5%, the orange fields medium agreement in the range of 10% while the read field poor agreement above 10%, comparing with the measured concentrations. Among isotopes of uranium and plutonium the best agreement was achieved either for reference burnup or for burnup increased by 1%. The results for isotopes of americium and curium show unpredictable behaviour comparing to the experimental measurement. However, general trends in concentration changes are proper i.e. *C/E* ratios of all not-fissile isotopes increase with FIMA burnup, which means their accumulation in the fuel assembly.

Table 1. Calculated-to-Experimental ratios for sample 89G08.

C/E	FIMA				
	-2%	-1%	0%	1%	2%
U232	0.78	0.80	0.82	0.85	0.87
U234	1.02	1.01	1.00	1.00	0.99
U235	1.08	1.07	1.05	1.03	1.02
U236	0.98	0.99	0.99	1.00	1.00
U238	1.00	1.00	1.00	1.00	1.00
Np237	1.00	1.02	1.03	1.04	1.06
Pu236	1.25	1.28	1.32	1.35	1.38
Pu238	0.87	0.90	0.92	0.94	0.97
Pu239+Np239	1.01	1.01	1.01	1.01	1.01
Pu240	0.99	1.00	1.01	1.01	1.02
Pu241	0.97	0.98	0.99	1.01	1.02
Pu242	0.94	0.96	0.99	1.01	1.04
Am241	0.86	0.87	0.89	0.89	0.91
Am242m	0.97	0.99	1.01	1.02	1.04
Am243	0.99	1.03	1.06	1.10	1.14
Cm242	0.76	0.78	0.80	0.83	0.85
Cm243	0.63	0.65	0.67	0.69	0.72
Cm244	0.93	0.98	1.02	1.08	1.13
Cm245	1.00	1.06	1.13	1.21	1.28
Cm246	0.79	0.85	0.91	0.99	1.07
Cm247	0.65	0.71	0.76	0.84	0.93

5.2 Sample 89G10

Table 2 presents the *C/E* ratios for the upper sample 89G10 with the reference measured FIMA burnup of 3.980. The burnup for this samples is higher because of increases of absolute neutron flux value towards centre of fuel assembly. The *C/E* ratios for almost all actinides show much better agreement with experimental measurements comparing to the sample 89G08. The reason for this could be attributed to the better reconstruction of irradiation conditions towards centre of fuel assembly because the spectral effects related to the influence of bottom reflector partly disappear. The best results were achieved for the reference FIMA or FIMA increased by 1% similarly to sample 89G08. However, this sample shows large improvements for ^{232}U, ^{241}Am, and curium isotopes except ^{246}Cm.

Table 2. Calculated-to-Experimental ratios for sample 89G08.

C/E	FIMA				
	-2%	-1%	0%	1%	2%
U232	0.98	1.01	1.04	1.06	1.09
U234	1.01	1.00	0.99	0.98	0.98
U235	1.09	1.07	1.05	1.03	1.01
U236	0.98	0.99	0.99	0.99	1.00
U238	1.00	1.00	1.00	1.00	1.00
Np237	1.05	1.07	1.08	1.09	1.10
Pu236	1.15	1.17	1.20	1.23	1.26
Pu238	0.90	0.92	0.95	0.97	0.99
Pu239+Np239	1.00	1.00	1.00	1.00	1.00
Pu240	0.96	0.97	0.98	0.99	1.00
Pu241	0.98	0.98	0.99	0.99	1.00
Pu242	0.93	0.95	0.97	0.99	1.01
Am241	0.90	0.91	0.92	0.92	0.93
Am242m	1.19	1.20	1.22	1.22	1.23
Am243	0.96	0.99	1.02	1.05	1.08
Cm242	0.91	0.92	0.94	0.96	0.98
Cm243	0.85	0.87	0.89	0.91	0.93
Cm244	0.92	0.95	1.00	1.04	1.08
Cm245	0.95	1.00	1.04	1.12	1.19
Cm246	0.77	0.82	0.88	0.94	1.01
Cm247	0.88	0.95	1.03	1.12	1.21

6 Conclusions

The results of the study provide a technical basis for the sensitivity study on actinide concentration in function of FIMA burnup. The results show that MCB code reflects the physics of 17x17 fuel assembly with high accuracy. The general performances associated burnup calculations show good consistence with theoretical predictions and the final actinide concentrations obtained in burnup calculations present rather good agreement with experimental measurements.

However, some inconsistence in concentrations of ^{232}U, ^{239}Pu, ^{241}Am, ^{242m}Am and Cm isotopes were observed. The large deviations in *C/E* ratios may either indicate to some errors in experimental measurement or in the numerical set-up. The applied sensitivity analysis via change in FIMA burnup can lead to improvements for some particular isotopes but it deviates the final results for other isotopes. The improvement in *C/E* ratios could be achieved rather by increasing than decreasing FIMA burnup. In addition, it is worth mentioning, that the better coherency was achieved for isotopes of uranium and plutonium because the transmission chains following formation of these isotopes are quite simple unlike for americium and curium isotopes.

Acknowledgements

The research was supported in part by PL Grid Infrastructure available at Academic Computer Centre CYFRONET AGH. In addition, partial financial support of this study through the funds of the Polish Ministry of Science and Higher Education is kindly acknowledged.

References

1. T. Adachi, Y. Nakahara, N. Kohno, K. Gunji, T. Suzuki, T. Sonobe, M. Onuki, K. Kato, E. Tachikawa, S. Inoue, M. Takayasu, M. Yoshikuni, S. Kobayashi, Z. Teruya, T. Kawamoto, M. Kawamura, J. Nucl. Sci. Technol. **31**, 10 (1994)
2. K. Suyama, M. Murazaki, K. Ohkubo, Y. Nakaharac, G. Uchiyama, Ann. Nucl. Energy **38**, 930-941 (2011)
3. The Power Reactor Information System, International Atomic Energy Agency, [online], [access: 13-05-2016], Available in World Wide Web: www.iaea.org/pris/
4. S. M. Goldberg, R. Rosner, *Nuclear Reactors: Generation to Generation*. American Academy of Arts & Sciences (2011)
5. World Nuclear Association, [online], [access: 13-05-2016], Available in World Wide Web: www.world-nuclear.org
6. S. Er-bing, C. Fang, C. Wang, G. Xia, C. Zhao, Ann. Nucl. Energy **85**, (2015)
7. J. Cetnar, G. Domańska, P. Gajda, J. Janczyszyn, Nukleonika **59**, (2014)
8. J. Cetnar, *General Solution of Bateman Equations for Nuclear Transmutations*, Ann. Nucl. Energy **33**, (2006)
9. G. Kępisty, J. Cetnar, P. Stanisz, Ann. Nucl. Energy **94**, (2016)

 DOI: 10.1051/e3sconf/20161000119

Nonlinear modeling of activated sludge process using the Hammerstein-Wiener structure

Paweł Frącz[1,a]

[1]*Opole University of Technology, ul. Prószkowska 76, 45-758 Opole, Poland*

Abstract. The paper regards to physical model of the Activated Sludge Process, which is a part of the wastewater treatment. The aim of the study was to describe nitrogen transformation process and the demand of chemical fractions, involved in the ASP process. Moreover, the non-linear relationship between the flow of wastewater and the consumed electrical energy, used by the blowers, was determined. Such analyses are important from the economical and environmental point of view. Assuming that the total power does not change the blower is charging during a year an energy amount of approx. 613 MW. This illustrates in particular the scale of the demand for energy consumption in the biological aeration unit. The aim is to minimize the energy consumption through first building a model of ASP and then through optimization of the overall process by modifying chosen parameter in numerical simulations. In this paper example measurement and analysis results of nitrite and ammonium nitrogen concentrations in the aeration reactor and the active power consumed by blowers for the aeration process were presented. Further the ASP modeling procedure, which uses the Hammerstein-Wiener structure and example verification results were presented. Based on the achieved results it was stated that the developed set of methodologies may be used to improve and expand the overriding control system for system for wastewater treatment plant.

1 Introduction

1.1. The Activated Sludge Process

The Activated Sludge Process (ASP) is a method used for biological wastewater treatment and is widely used in sewage treatment plants, for example [1]. Basic ASP process flow diagram is shown in Figure 1. The wastewater is brought into the biological reactor. The activation of wastewater process is based on providing oxygen to increase the number of microorganisms, responsible for contaminants reduction in the aeration tank. The activated sludge is then formed; it consists of flocks, mixes with new incoming wastewater and finally takes form of slurry. The flocks contain bacteria but also ciliates, flagellates, protozoa, nematodes, rotifers, larvae of insects, arachnids and mushrooms. Typical flock size varies from 50 μm to 100 μm. The sludge from the biological reactor is then transported to secondary settling tank (also called clarifier), in which the activated sludge is separated. A considerable part of the sludge is returned back to the reaction chamber (biological reactor) and the excessive part of the sludge is removed. The advantage of this method is its significant performance of harmful chemicals reduction. The main disadvantage is connected with high costs of electrical energy, which supply is needed for to assure a suitable amount of oxygen bowing into the aeration tank. Bacteria's growth requires a continuous supply of energy. The heterotrophic bacteria require oxygen and organic carbon as a substrate. The autotrophic bacteria grow in oxygen environment using inorganic carbon substrate.

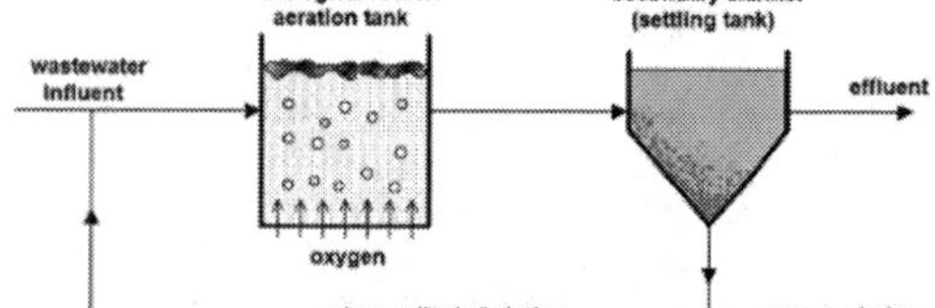

Figure 1. The basic principle of Activated Sludge Process.

1.2. Biological removal of nitrogen compounds

Nitrogen is present in the wastewater in various forms, e.g. free form of ammonia, nitrates, nitrites, and different kinds of organic compounds. The total content of nitrogen is a sum of its particular components. Nitrogen is essential for biological growth, but in certain forms it is toxic and therefore, its reduction is desirable. Ammonia is toxic for aquatic organisms, among others, for fish. Nitrate presence causes excessive demand for oxygen. Nitrogen, as a nutrient, causes excessive growth of aquatic plants.

There are two basic nitrogen transformation processes in the wastewater treatment plants: the nitrification and the denitrification. Both processes result in the removal of nitrogen from wastewater. Nitrification is the oxidation of ammonia and ammonium salts into nitrites and nitrates. This process is driven by nitrifying bacteria. Autotrophic nitrification takes place in the presence of oxygen at a concentration of at least 2 mg O_2/dm^3. Nitrification

[a] Corresponding author: p.fracz@po.opole.pl

DOI: 10.1051/e3sconf/20161000119

reaction consists of two phases. In the first stage bacteria oxidize ammonium to nitrite:

$$NH_4^+ + 1.5O_2 \rightarrow NO_2^- + H_2O + 2H^+ \qquad (1)$$

In the second stage nitrite ions are oxidized to nitrate by bacteria:

$$NO_2^- + 0.5O_2 \rightarrow NO_3^- \qquad (2)$$

The condition of autotrophic nitrification occurrence is, beside the oxygen presence, the presence of coal, as an energy source for autotrophic bacteria (e.g. carbon dioxide) and the appropriate pH value (around 7.5).

Denitrification is a reduction of nitrite to nitrogen. This process takes place under anaerobic conditions. Oxygen is present in the nitrite. Bacteria use it instead of dissolved oxygen. A simplified description of the denitrification reaction is as follows:

$$2NO_3^- + 2H^+ \rightarrow N_2 + H_2O + 2.5O_2 \qquad (3)$$

1.3. Modeling of the Activated Sludge Process

The practical aim of our work is to develop a set of methodologies that will be used to improve and expand the precedent control system for wastewater treatment plant. The overriding control system provides measurement data to the subordinating control systems of the plant and replaces in this way human service for sewage treatment plants. Conducted works aim to determine the nature of changes, dependencies, trends and to make other observations of the wastewater parameter values, which characterize the treatment process. The study uses empirically acquired and digitally processed data in order to identify develop and verify a non-linear simulation model of the wastewater treatment plant. The hypothesis relates to confirmation the possibility of use of a physical model of wastewater treatment plant, based on Hammerstein-Wiener structure, to optimize the treatment process. According to the latest global trends, mathematical modeling has become an integral part of the design and operation of wastewater treatment systems, particularly those using the ASP [2]. Simulation of the activated sludge systems (experiment carried out on the model) proved to be an extremely useful tool for operators, designers and consultants, as well as for the scientific community. The use of mathematical models allows one for to examine in a short time and with low financial outlay of many technological solutions, and for simulation of events from outside the range of typical real system conditions.

The most common physical model is the ASM1 (Activated Sludge Model No.1), which was the first bio kinetic model of activated sludge. Implementations of the ASM1 in several popular simulation programs were compared in an exhaustive manner [3]. Examples of modified, supplemented or improved models of ASM1, known respectively as ASM2, ASM2d and ASM3 are described in detail in [1]. In addition to the ASM family models, in which the mass balance is based on the COD, there are also models that use the mass balance of the BOD, for example the ASAL models [4]. Models describing in detail the metabolism of organisms involved in the pollutants decomposition, which include, among others, TUDP model are described in [5-6]. The activated sludge system model is composed not only from the activated sludge (biochemical model, e.g. ASM), but also it includes models of other processes, in particular: he sedimentation model (settler model) and the mass transfer model (hydraulic model, aeration system model). Most of the simulation models, described in the literature, were carried out for the biological purification stage and thus the system consisted of the activated sludge reactor and the secondary sedimentation/settling tank [7]. The newest trends in modeling of wastewater treatment processes, however, seek to include also other system parts/processes, instead of only the specified area, in the multiphysical model. In such as a manner the whole wastewater treatment plant constitutes the object under study and the designed model includes additionally the sewage, sewage treatment plant and the receiver [8]. A factor impeding modeling of the technological sequence of the entire wastewater treatment plant lies in the fact that not all models of wastewater or sludge treatment processes have a common set of variables. For example one may consider a combination of the ASM1 model with a settler model. Such a hybrid model, however, involves the concentration of slurries from the settler as state variable, but this value is not present in the activate sludge process. This problem might be solved by using a complex variable, corresponding to the concentration calculated from the respective ASM1 variables, which is, however, not a state variable. Another example of biochemical transformations model, which uses a different set of variables, is the ASM model and the methane fermentation - ADM1 (Anaerobic Digestion Model No.1) model.

The most important part prior modeling is collection of measurement data, which allows the model to be accurately calibrated and validated for all conditions. The modern modeling approaches and algorithms are relatively easy to understand for engineering application purposes. They enable to model the wastewater treatment facilities and to calibrate them, which provide then accurate predictions of the system behavior under different operational conditions. Nowadays in every facility there are specialized measurement and SCADA devices installed for registering and collection of hydraulic and technological parameters. Given sufficient data, a modeler can accurately model the activated-sludge process using different models.

Developments in the field of modeling of sludge treatment processes, including both dynamic and steady-state conditions, are well presented in [9]. Based on a literature review Author enhances the importance of computer modeling of activated-sludge process during design of new and optimization of existing wastewater plants. Readers who are interested in other interesting research results dealing with modeling of activated sludge tanks are referred [10-14].

In this paper chosen analysis result of nitrate and ammonium nitrogen and blower active power values, measured in experiments, are presented. Furthermore, the ASP modeling procedure and results of this model verification are described.

E3S Web of Conferences 10, 00119 (2016) DOI: 10.1051/e3sconf/20161000119
SEED 2016

2 Experimental results analysis

2.1. Analysis results of nitrate nitrogen concentration

Nitrate nitrogen NO_3-N is next to nitrite nitrogen NO_2-N the main product of decomposition of ammonium NH_4-N. This value was registered during measurements by using an ion-selective probe from Hach-Lange. The measuring sensor was placed inside the oxygen reactor. Under normal operational conditions, the gathered data should indicate significantly lower values in the effluent as compared to these in the influent of the reactor.

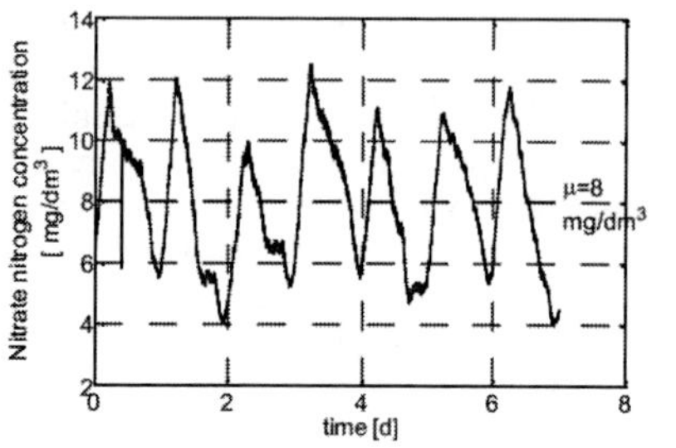

Figure 2. Time course of the nitrate nitrogen NO3-N concentration.

In Figure 2 example time course of the nitrate nitrogen measured within a week is presented. A periodical pattern can be recognized: the values are generally lower at night and higher at mornings. The nitrate nitrogen concentration varies in the range from 4 to 12.5 mg/dm^3. The average value μ equals 8 mg/dm^3.

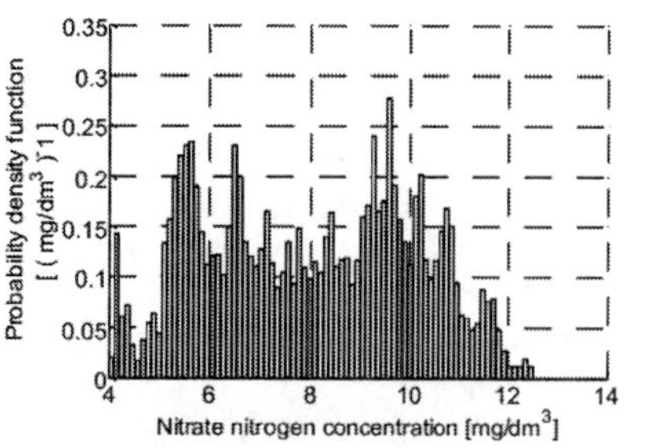

Figure 3. Probability density function calculated over the nitrate nitrogen concentration gathered during the measurement campaign.

In Figure 3 probability density function calculated over the nitrate nitrogen concentration gathered during the measurement campaign is presented. The probability density function depicts no uniform mode. Variability in the entire interval is relatively equal.

In Figure 4 power spectrum density calculated over the nitrate nitrogen concentration gathered during the measurement campaign is presented. Power spectrum depicts several periodicity components of the sludge what is indicated by picks in the spectrum. The most significant value is the fundamental frequency component equal to 1*10e-5, the conversion of 24 hours.

In Figure 5 the auto covariance function calculated over the nitrate nitrogen concentration gathered during the measurement campaign is presented. It indicates a stochastic component and significant own variances for time delays, which are multiples of 24h day.

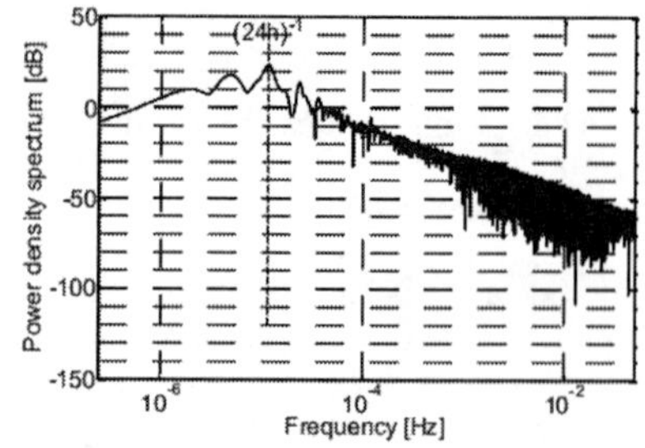

Figure 4. Power spectrum density calculated over the nitrate nitrogen concentration gathered during the measurement campaign.

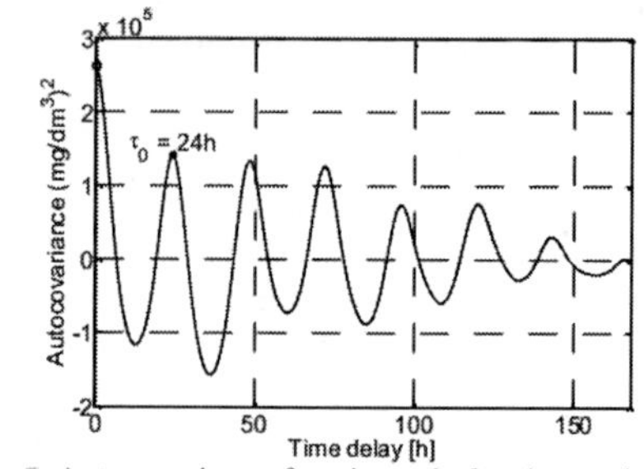

Figure. 5. Autocovariance function calculated over the nitrate nitrogen concentration gathered during the measurement campaign.

Correspondent result was obtained by examining variation of the frequency structures over time, what is illustrated in Figure 6. The scale graph depicts no significant harmonics in the gathered data.

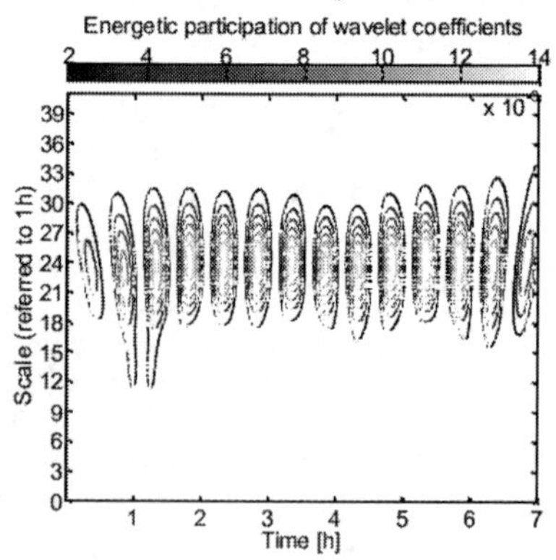

Figure 6. Scalegraph calculated over the nitrate nitrogen concentration gathered during the measurement campaign.

2.2. Analysis results of ammonia nitrogen concentration

The concentration of ammonium nitrogen NH_4-N was measured using two probes: the Hach-Lange and the Endress-Hauser sensors. Both sensors were placed inside the oxygen reactor. In Figure 7 time course of the ammonium nitrogen concentration measured with two different probes is presented. A periodical pattern can be recognized. But in contrast to the nitrate concentration, ammonium nitrogen indicates higher values in the afternoons and at nights and lower at mornings. Values measured by the Hach-Lange probe were slightly lower as compared with data registered by the Endress-Hauser probe. The ammonia nitrogen concentration varies in the range from 0.8 to 19.5 mg/dm³. The average values μ=8 mg/dm³ for the Hach-Lange and μ=9 mg/dm³ for the Endress-Hauser probes. The standard deviation equals 4.09 for the Hach-Lange probe, and it is approx. 16% higher for the Endress-Hauser probe. In further studies the Hach-Lange probe was applied.

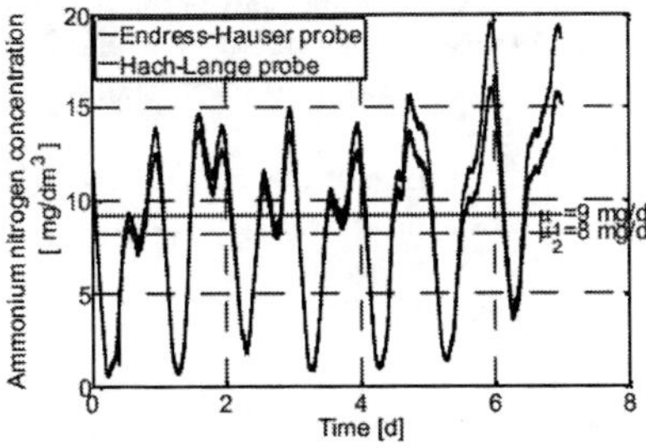

Figure. 7. Time course of the ammonium nitrogen NH_4-N concentration measured with two different probes.

In Figure 8 probability density function calculated over the ammonium nitrogen concentration gathered during the measurement campaign by the Hach-Lange probe is presented. The highest probability occurred for the concentration equal to about 14 mg/dm³. The distribution has a fuzzy character and values are within the range from 0 to 16 mg/dm³.

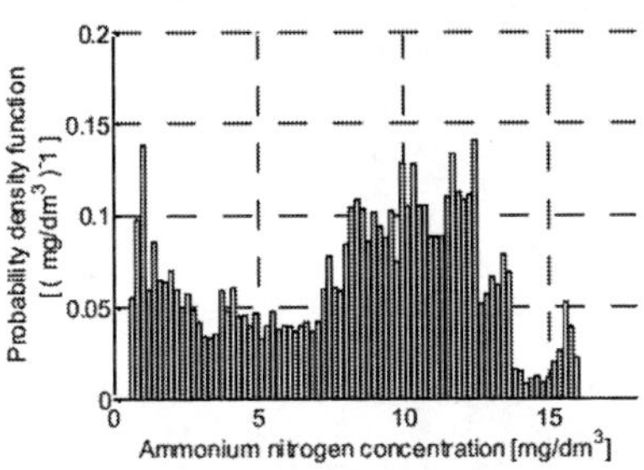

Figure 8. Probability density function calculated over the ammonium nitrogen concentration gathered during the measurement campaign by the Hach-Lange probe.

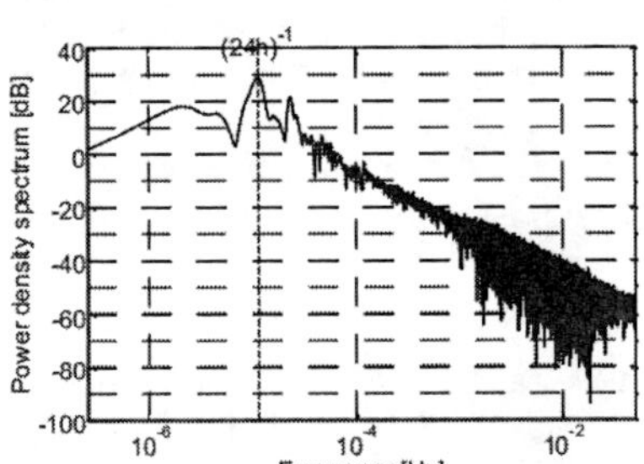

Figure 9. Power spectrum density calculated over the ammonium nitrogen concentration gathered during the measurement campaign by the Hach-Lange probe.

In Figure 9 power spectrum density calculated over the ammonium nitrogen concentration gathered during the measurement campaign by the Hach-Lange probe is presented. The spectrum depicts periodicity of the sludge what is indicated by a pick near the 1*10e-5 frequency value. Certain importance is give also to the second harmonic component.

In Figure 10 auto covariance function calculated over the ammonium nitrogen concentration gathered during the measurement campaign by the Hach-Lange probe. Here a high value of the covariance of time delay, which is equal to 24 h, and for the subsequent multiples, is depicted.

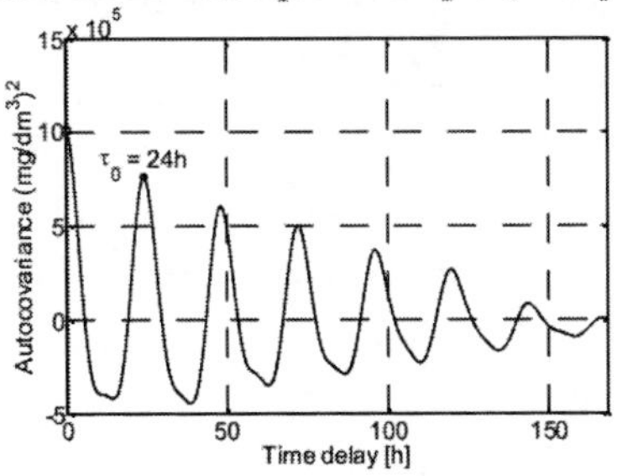

Figure 10. Autocovariance function calculated over the ammonium nitrogen concentration gathered during the measurement campaign by the Hach-Lange probe.

In Figure 11 scale graph calculated over the ammonium nitrogen concentration gathered during the measurement campaign by the Hach-Lange probe is presented. One can recognize similar periodic components as compared to the nitrate nitrogen concentration. For the entire duration of measurement the 24h daily component does not change its energy. Wavelet coefficients for the scale corresponding to the 1/(12h) frequency occur sporadically.

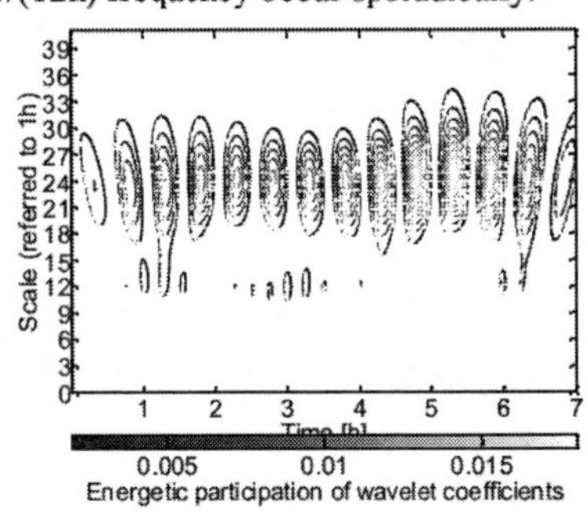

Figure 11. Scalegraph calculated over the ammonium nitrogen concentration.

2.3 Analysis results of the blower active power

The next parameter analyzed was the active power applied for three different blowers that are installed in the reactor. The time courses for the various blowers are shown in Figure 12. The total value of blowers active power is presented in Figure 13. The average value of active power μ was equal to approx. 70 kW.

Additional the probability density function was calculated over the active power values (Figure 14), which was approximated by using normal distribution function achieving very high fit (Pearson determination coefficient R^2=0,89).

Calculated standard deviation value depicts that during the entire measurement campaign the blowers have had relatively small active power, which was equal to 70 kW.

The momentary power reaches the maximum value of 120 kW.

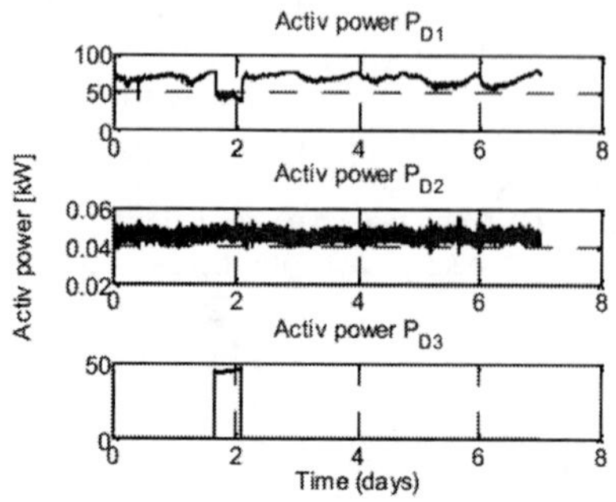

Figure 12. Time course of the active power applied for particular blowers.

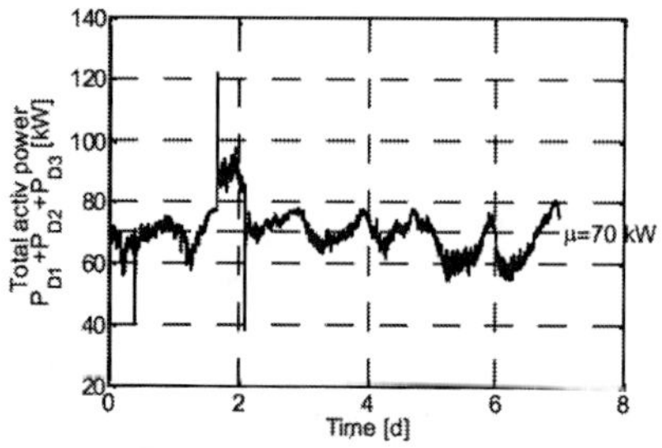

Figure 13. Time course of the total blower active power.

Assuming that the total power does not change during the year blowers charge an energy amount of approx. 613 MW. This illustrates in particular the scale of the demand for energy consumption in the biological aeration unit.

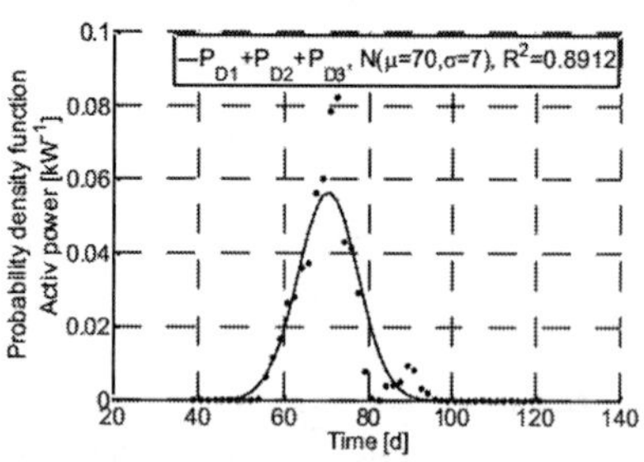

Figure 14. Probability density function calculated over the active power values. The solid line corresponds to approximation estimated by means of normal distribution.

3 Non-linear modeling of chosen processes involved in the ASP

The study was carried out to identify certain processes related to waste water treatment. The purpose of these studies is an alternative description of the treatment of nitrogen and chemical fractions of demand. Moreover, the aim of research is to determine the non-linear relation between the flow of sewage and electric energy consumed. The process is treated as a Single Input Single Output (SISO). The identification was carried out using Hammerstein-Wiener model. The study was conducted using MATLAB Toolbox System Identification Toolbox. In the studies two sets of data were applied: one for identification and one for validation. Weekly measurement data were used for the modeling process. The available waveforms were divided in two half: one half - 3.5 day, was used for identification of the model, as the learning sequence (LS); the remaining 3.5 days was used for testing purposes, as a test sequence (TS). All data sets applied in the study consisted of 336 measuring points, i.e. 84 hours with four samples per hour. In a series of simulations an optimal nonlinearity on the input and output of the model were determined. Also an appropriate structure of the linear dynamic member was selected. For the estimation of non-linear members a piecewise-linear model was applied. Three different criteria for evaluation of the model quality were used:

1. C1: Best fit for learning sequence FIT(LS),
2. C2: Best fit for test sequence FIT(TS),
3. C3: Best fit for the sum over the learning and test sequences FIT(LS) + FIT(TS).

3.1. Non-linear process identification of nitrate nitrogen S_{NO} production in the aerobic reactor

In the first step the model for nitrate nitrogen (S_{NO}) was identified and verified. The input of the SISO model was the outflow of sewage. The output variable was the course of S_{NO}. The sampling period for both variables was 15 min. Both variables were changing over time. A genetic algorithm was used for to determine five parameter values, which were as follows:

- number of zeros (plus 1) – P1,
- number of poles – P2,
- input delay – P3,
- number of piecewise-linear estimators of the input non-linearity (plus 1) – P4,
- number of piecewise-linear estimators of the output non-linearity (plus 1) – P5.

For model goodness indication, the following equation was applied:

$$FIT = 100 \cdot \left(1 - \frac{norm(Y-C)}{norm(C-\bar{C})}\right) [\%] \quad (4)$$

Where: Y is the model response, C is the measured data. This value is given in percent, therefore the higher the result, the better the fit.

The summary results of studies, which aimed determination of the best structure for the different criteria analyzed, are listed in Table 1. All the structures were found to have the same number of zeros P1=3, the same number poles P2=7 in the dynamic part and the same delay, which was equal to one step P3=1. Differences were calculated for the number of piecewise-linear estimators of the input and output nonlinearities.

Table 1. Comparison of the best fit Hammerstein-Wiener structures found for the three considered criteria; output – the nitrate nitrogen S_{NO} concentration, input – sludge flow.

Crit.	P1	P2	P3	P4	P5	FIT(LS) [%]	FIT(TS) [%]
C1	3	7	1	4	4	87.55	68.27
C2	3	7	1	3	2	78.33	76.19
C3	3	7	1	1	3	83.48	74.80

When used as criterion the best fit of learning sequence (no. 1), the optimal structure consisted of 5 piecewise-linear (P4x=4+1)in both nonlinearity estimators.

An example time course of nitrate nitrogen model identified using the learning sequence is shown in Figure 15. The model relatively properly reproduces the measured nitrate nitrogen values adjusting to changes in the following days. Negligible mismatches occur only at the hourly/minutely variability level of the data.

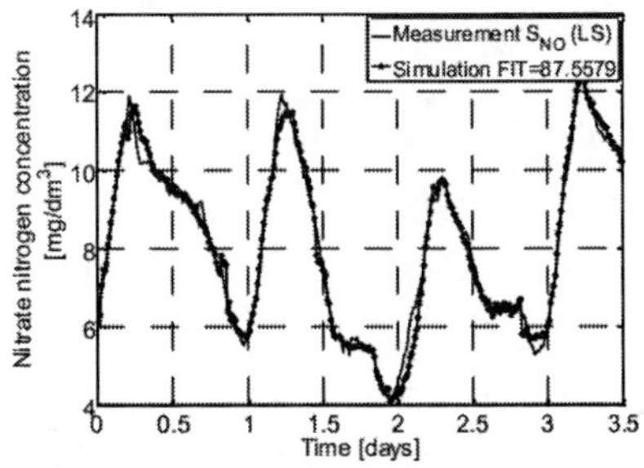

Figure 15. Timing charts of the Hammerstein-Wiener model in response to nitrate nitrogen data contained in the LS (first 84 hours); Identification according to the criterion C1.

The residuals time course of the Hammerstein-Wiener model in response to learning sequence, when identification considered the criterion C1, is presented in Figure 16. The values are in the range from -0.8 to 1mg/dm³.The stochastic nature of the residuals within time course, confirms that the model has been selected properly.

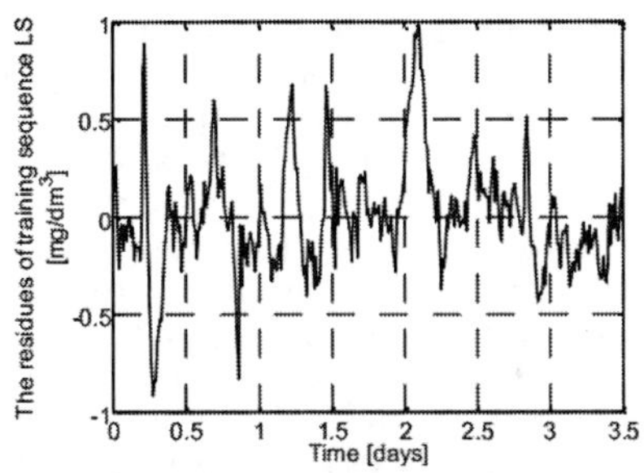

Figure 16. The residuals of the Hammerstein-Wiener model in response to LS; Identification according to the criterion C1.

3.2. Non-linear process identification of ammonium nitrogen NH4-N production in the aerobic reactor

We then consider the process, in which the wastewater outflowwas given as the input of the SISO system, and the discharged ammonium nitrogen concentration (NH_4-N) was assumed as output. The sampling period for both datasets was equal to 15 min. Just as for the nitrate nitrogen identification, during these studies it were used two datasets: one for the identification and second for the validation. Also the same genetic algorithm, the same number of parameters of the objective function, the same goodness indicator and the same criteria were applied. The summary of the results obtained are listed in Table 2.

The optimal model structure, for which the best matching value was calculated, occurred for identification with the LS (criterion C1). It had a significant number of zeros P1=7 and poles P2=7. It had respectively 2 (P4) and 6 (P5) points of inflection of the input and output non-linearity estimators. It was found that criteria C2 and C3 meet the same structure with two zeros and three poles in the dynamic part. The static nonlinearity input estimator had 6 piecewise-linear (P5=5+1), while the output nonlinearity estimator included four piecewise-linear.

Table 2. Comparison of the best fit Hammerstein-Wiener structures found for the three considered criteria; output – the ammonium nitrogen concentration N-NH_4, input – sludge flow.

Crit.	P1	P2	P3	P4	P5	FIT(LS) [%]	FIT(TS) [%]
C1	7	7	1	2	6	85.46	55.69
C2	2	3	1	5	3	81.18	62.17
C3	2	3	1	5	3	81.18	62.17

An example time course of ammonium nitrogen model identified using the LS is shown in Figure 17. The residuals (Figure 18) are in the range of from -2.7 to 1.5 mg/dm³. Stochastic nature of the residuals time course confirms that the model has been selected properly.

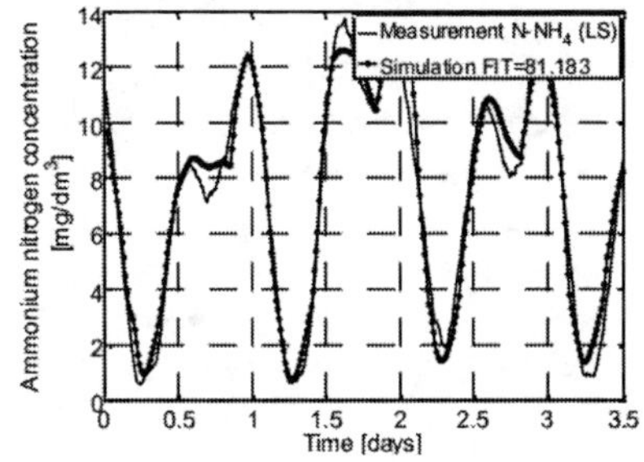

Figure 17. Time course of the Hammerstein-Wiener model in response to ammonium nitrogen data contained in the LS (first 84 hours); Identification according to the criterion C3.

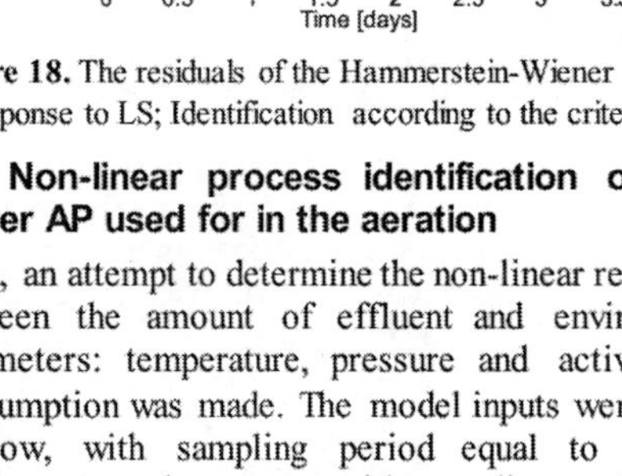

Figure 18. The residuals of the Hammerstein-Wiener model in response to LS; Identification according to the criterion C3.

3.3. Non-linear process identification of active power AP used for in the aeration

Then, an attempt to determine the non-linear relationship between the amount of effluent and environmental parameters: temperature, pressure and active power consumption was made. The model inputs were: sludge outflow, with sampling period equal to 15 min, temperature and pressure, with sampling period equal to 60 min. The model output was the total active power, which constitutes the sum of active power values measured on three blower devices mounted for aeration in the biological reactor. All datasets were interpolated in order to obtain a uniform sampling rate (1/4 h -15 min). As in the previous studies, two sets of data, 84 hours - measuring period, were used for the model identification and validation. A genetic algorithm was used for to determine 11 variables: P1 for sewage outflow: P1S, P1 for Temperature: P1T, P1 for Pressure: P1P, P2 for sewage outflow: P2S, P2 for Temperature: P2T, P2 for Pressure: P2P, P3 for sewage outflow: P3S, P3 for Temperature: P3T, P3 for Pressure: P3P, P4 and P5.

For estimation of model goodness eq. (4) was applied. Three different criteria were used for evaluation of the model quality, when checking for the best fit within the learning, testing and the total sum of this both values. The final results of studies aiming determination of the best structure for the different criteria are listed in Table 3. The best fitted structure was meet for the learning sequence (criterion C1).

Table 3. Comparison of the best fit Hammerstein-Wiener structures found for the three considered criteria; output –the active power consumed by the blower.

Crit.	P1	P2	P3	P4	P5	FIT(LS) [%]	FIT(TS) [%]
C1	3	2	1	2	1	12	1
C2	3	2	1	2	1	12	2
C3	3	1	1	2	2	12	3

An example time course of active power model identified using the test sequence is shown in Figure 19. The model relatively correctly provides the active power values when based on the given input data. The residuals (Figure 20) are in the range of from -8 to 10 kW. They indicate a single error in the measurement data around the 50th hour of measurement.

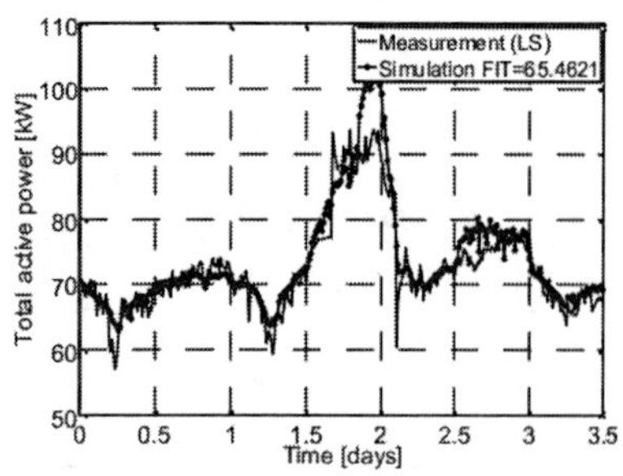

Figure 19. Timing charts of the Hammerstein-Wiener model in response to active power data contained in the LS (first 84 hours); Identification according to the criterion C2.

Summing up the results it is to conclude that the developed ASP process model using the Hammerstein-Wiener structure allows conducting simulation studies, which concern determination of the amount of energy consumption. The electrical energy is used by the blower in the aerobic reactor of a wastewater treatment plant for aeration. Presented results confirm the efficacy of the developed methodology for application to carry out simulation studies aimed at optimization of the ASP process.

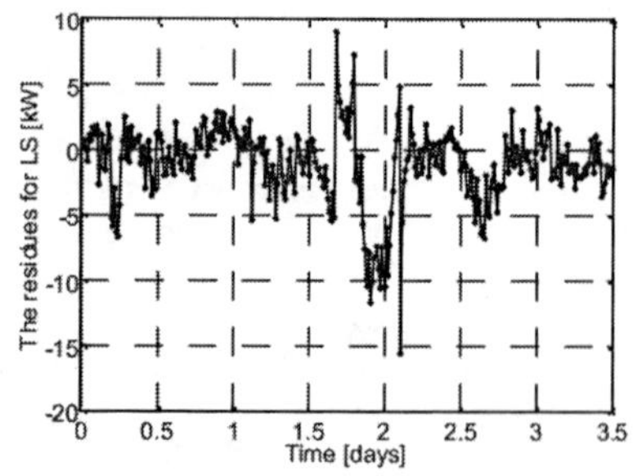

Figure 20. The residuals of the Hammerstein-Wiener model in response to LS; Identification according to the criterion C2.

4 Conclusions

Modeling results of three selected processes occurring in the aeration reactor of wastewater treatment plant are presented in this paper. The Hammerstein-Wiener structure was applied for identification of the process of nitrate nitrogen S_{NO} production, the process of ammonium nitrogen N-NH_4production and the process of active power supply AP during aeration. For each of the models identified three criteria were applied:

C1: Best fit for the learning sequence FIT(LS),

C2: Best fit for the test sequence FIT(TS),

C3: Best fit for the sum of the learning and test sequences FIT (LS)+FIT(TS).

Analysis of the results allowed for to define the following conclusions:

1. All the models received reproduce relatively correctly the actual measurements. The best fit for the S_{NO} model was found using the criterion C1 for the LS, which was equal to 87.55%. The best fit for the N-NH_4 model was found for the criterion C1 for the LS, which was equal to 85.46%. The best fit for the AP model was found by using the criterion C1 within the LS, and which equaled 87.55%. The best fit for the TS was obtained for criterion C2, for all identified models.
2. Residuals were differently for each of the models. For criterion C1, residuals were stochastic and were not correlated with the time courses, by which it can be concluded that the chosen models are appropriate. For other criteria residual values ranged from -4 to 5 mg/dm^3 for both nitrogen models and from -15 to +12 kW for the active power model.
3. The achieved results confirmed the hypothesis that it is possible to use a physical model of the ASP based on Hammerstein-Wiener structure for to optimize the wastewater treatment process.

References

1. G. Bitton, *Wastewater Microbiology*, 3rd edition, Chapter 8. Activated Sludge Process
2. M. Henze, W. Gujer, T. Mino, M. van Loosdrecht, *IWA Scientific and Technical Report No. 9,* IWA Publishing. London, UK (2000)
3. J. B. Copp, Office for Official Publications of the European Community, Luxembourg. ISBN 92-894-1658-0
4. A. J. Stokes, J. R. West, C. F. Forster, W. J. Davis, *Water Res.*, **34** (4), 1296-1306 (2000)

5. E. Murnleitner, T. Kuba, M. C. van Loosdrecht, J. J. Heijnen, *Biotechnol. Bioeng.* **54**, 434-450 (1997)
6. H. M. van Veldhuizen, M. C. van Loosdrecht, J. J. Heijnen, *Water Res.*, **33**, 3459-3468 (1999)
7. J. Makinia, M. Świnarski, E. Dobiegala, *Wat. Sci. Techn.*, **45** (6), 209-218 (2002)
8. W. Gujer, *Wat. Sci. Tech.*, **53** (3), 111-119 (2006)
9. M. Smith, J. Dudley, *Wat. & Envir. J.*, **12** (5), 346-356 (1998)
10. Z. Lin, S. Lu, *Wat & Envir. J.*, **25**, 573-587 (2011)
11. S. Plessis, R. Tzoneva, *Water SA*, **38** (2), 287-306 (2012)
12. Z. Sadecka, A. Jędrczak, E. Płuciennik-Koropczuk, S. Myszograj and M. Suchowska-Kisielewicz, *Chem. Biochem. Eng. Q.*, **27** (2), 185-195 (2013)
13. C. Fall, E. Millán-Lagunas, K. M. Bâ, I. Gallego-Alarcón, D. García-Pulido, C. Díaz-Delgado, C. Solís-Morelos, *J. Environ. Manage*, ***113***, *71-77 (2012)*
14. H. Boursier, F. Béline., E. Paul, *Bioresource Technol.*, **96** (3), 351-358 (2002)

Various aspects of magnetic field influence on forced convection

Lukasz Pleskacz[a] and Elzbieta Fornalik-Wajs

AGH University of Science and Technology

Abstract. Flows in the channels of various geometry can be found everywhere in industrial or daily life applications. They are used to deliver media to certain locations or they are the place where heat may be exchanged. For Authors both points of view are interesting. The enhancement methods for heat transfer during the forced convection are demanded due to a technological development and tendency to miniaturization. At the same time it is also worth to find mechanisms that would help to avoid negative effects like pressure losses or sedimentation in the channel flows. This paper shows and discuss various aspects of magnetic field influence on forced convection. A mathematical model consisted of the mass, momentum and energy conservation equations. In the momentum conservation equation magnetic force term was included. In order to calculate this magnetic force Biot-Savart's law was utilized. Numerical analysis was performed with the usage of commonly applied software. However, user-defined functions were implemented. The results revealed that both temperature and velocity fields were influenced by the strong magnetic field.

1 Introduction

The magnetic force presence had been discovered relatively early by Faraday in 1847. Then the progress in research stopped for almost one hundred years. The next person interested in the magnetic field utilization was Pauling, who developed the oxygen analyzer in 1946 [1].

Nevertheless, the real revolution was brought in 1986 by Bednorz and Muller [2]. They discovered high-temperature superconductivity and were granted with Nobel prize for this achievement. This discovery contributed to the quicker construction of super-conducting magnets, that nowadays can generate magnetic fields up to 26 T.

Super-conducting magnets made it possible to conduct research on strong magnetic field influence on weakly magnetic substances, especially on the ground of fluid mechanics and heat transfer analysis. Braithwaite et al. [3] were the first to confirm experimentally that the strong magnetic field can enhance heat transfer during natural convection. They also proposed mathematical model that described this phenomenon. Soon after that many interesting phenomena regarding the magnetic field were reported. For example: combustion promotion [4] or magnetic breath support [5]. The most famous paper from that time is [6], which presented the levitation of the droplet of water. After that it was found out that the still air can be moved by the magnetic field (so called magneto-thermal wind blower) [7].

Simultaneously to the experimental investigation, the numerical approaches were also undertaken. Tagawa et al. presented set of papers connected to numerical model of natural convection in the strong magnetic field for paramagnetic [8] and diamagnetic [9] fluids. As the basis they took Bai et al. [10] analysis. Then the investigations were expanded on another geometries like cylindrical [11], cylindrical annuli [12] and rectangular one [13].

Nowadays the research directions have branched significantly. The emphasis is placed, for example, on the transition regime between laminar and turbulent convection [14], nanofluids behaviour in strong magnetic field [15] or magnetic field impact on forced convection. The last issue was investigated at first by Ueno and Iwasaka in 1994 [16]. Their paper revealed that the flow of diamagnetic fluid can be completely stopped by the magnetic field. In [17] Ozoe showed numerical investigation of Graetz problem under the influence of magnetic field. Authors continued this work and presented papers concerning flow structure for various boundary conditions.

This paper raises a subject of forced convection subjected to the strong magnetic field in a very common geometry - circular channel with an elbow, this geometry which can be found in numerous industrial applications such as power engineering, motorization or food industry. For this type of pipe junction a typical problem is sedimentation near the longer wall of the elbow. However, magnetic force might appear to be helpful in order to get rid of these sediments.

Authors verified in [18] the influence of magnetic field on the flow in wide range of Reynolds number. They have found that such influence can be important for low Reynolds number flows, high temperature differences and high magnetic induction values.

[a] Corresponding author: pleskacz@agh.edu.pl

The flows of low Reynolds number occur widely in the industrial applications. The fluids like tars, oils or honeys are taken into account. Many of these substances are paramagnetics, which are a group of substances weakly attracted by the magnetic field. Their magnetic susceptibility is described by the Curie's law.

Therefore presented scope of research concentrated on low Reynolds number flow (Re = 0.84) of paramagnetic fluid through the elbow.

2 Mathematical formulation

The mass, momentum and energy conservation equations are a core of mathematical model used in the paper. They were supplemented by the Biot-Savart's law, which was utilized to calculate magnetic induction and magnetic force distribution.

The continuity equation was formulated with the following assumptions: flow was stationary, without sources, it was incompressible, laminar and three-dimensional. The final form could be presented as:

$$\nabla \cdot \mathbf{u} = 0 , \tag{1}$$

where **u** is the velocity (m/s).

The momentum equation, taking into account above assumptions and adding to them that magnetic and gravitational forces were external body forces, could be written as:

$$\rho(\mathbf{u} \cdot \nabla \mathbf{u}) = -\nabla p + \mu \nabla^2 \mathbf{u} + \mathbf{F}_b , \tag{2}$$

where: ρ is the density (kg/m^3), p is the pressure (Pa), μ is the dynamic viscosity (Pa·s), $\mathbf{F}_b = \mathbf{F}_g + \mathbf{F}_{mag}$ represents the body forces (N/m^3), $\mathbf{F}_g$ is the gravitational force (N/m^3), $\mathbf{F}_{mag}$ is the magnetic force (N/m^3).

In order to calculate the magnetic induction distribution Biot-Savart's law was used in the following form:

$$\mathbf{B} = \frac{\mu_m i}{4\pi} \oint_c \frac{d\,\mathbf{s} \times \mathbf{r}}{|r|^3} , \tag{3}$$

where: **B** is the magnetic induction (T), μ_m is the magnetic permeability (H/m), i is the electrical current magnitude (A), $d\,\mathbf{s}$ is the infinitely small segment of the coil (m), **r** is the position vector (m).

The magnetic force influencing the paramagnetic fluid flow could be described as follows:

T_w is the heated wall temperature (K), T_f is the inlet

$$\mathbf{F}_{mag} = -\left(1 + \frac{1}{T_0 \beta}\right) \frac{\chi_m \rho \beta (T - T_0)}{2\mu_m} \nabla \mathbf{B}^2 , \tag{4}$$

where: $T_0 = (T_w - T_f)/2$ is the reference temperature (K), fluid temperature (K), T is the local fluid temperature (K), β is the thermal expansion coefficient (K^{-1}), χ_m is the mass magnetic susceptibility (m^3/kg).

With the following assumptions: flow was steady, without external heat sources, the energy conservation equation is presented in the form:

$$\mathbf{u} \cdot \nabla T = \frac{\lambda}{\rho c_p} \nabla T^2 , \tag{5}$$

where: λ is the thermal conductivity (W/(m·K)), c_p is the specific heat (J/(kg·K)).

2.1 Dimensionless parameters

In this paper Prandtl number and Reynolds number were used. They represent the most important features of the system geometry, flow and fluid characteristics. Due to the non-dimensional parameters it is possible to compare various systems. Definitions of important for present research parameters are as follow:

$$\mathrm{Re} = \frac{\rho u_{avg} d}{\mu} , \tag{6}$$

$$\mathrm{Pr} = \frac{c_p \mu}{\lambda} , \tag{7}$$

where: u_{avg} is the average inlet velocity (m/s), d is the pipe diameter (m).

3 Problem specification and numerical approach

The analysis was carried out for three-dimensional circular duct (pipe) of 0.01 (m) diameter with an elbow. The pipe starts with the inlet section of length equal to 0.1 (m), which passed to the elbow section of external radius of 0.015 (m). The outlet section's length was 0.035 (m). The studied geometry is presented in Figure 1.

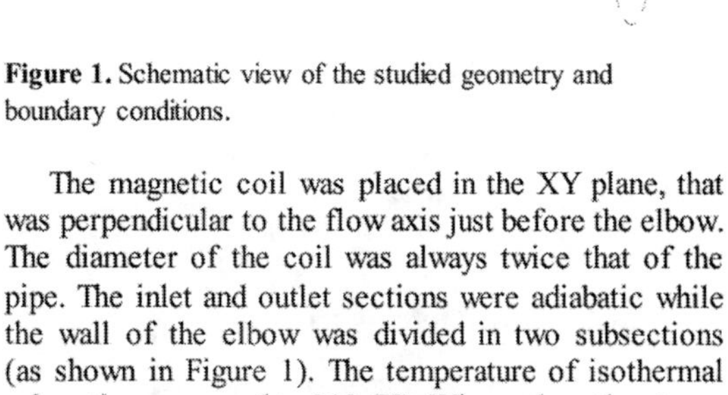

Figure 1. Schematic view of the studied geometry and boundary conditions.

The magnetic coil was placed in the XY plane, that was perpendicular to the flow axis just before the elbow. The diameter of the coil was always twice that of the pipe. The inlet and outlet sections were adiabatic while the wall of the elbow was divided in two subsections (as shown in Figure 1). The temperature of isothermal subsection was equal to 310 (K). When subsection 1 was treated as isothermal the subsection 2 was adiabatic. It was the case called variant I. When the subsection 2 was isothermal and subsection 1 adiabatic the variant was called II. The inlet velocity profile was assumed to be parabolic, while the constant value of pressure (equal to environment pressure p_0 = 101325 (Pa)) was set at the outlet.

The generated grid was unstructured and consisted of 186800 wedge-shaped elements. The validation of the grid was performed iteratively with the usage of maximum velocity value (known from analytical solution). Gambit 2.3 was chosen as a geometry-creator as well as meshing utility. The computations were performed with Ansys Fluent 14.5. However, special user-defined functions (UDF) were implemented in order to determine parabolic inlet velocity profile, calculate the magnetic induction and magnetic force distribution and to introduce the last force to the momentum conservation equation.

The chosen discretization scheme for momentum and energy equations was first order upwind scheme. As it comes to pressure interpolation, the standard scheme was utilized. If the difficulties with the convergence occurred, the standard interpolation scheme was replaced with body-force weighted scheme.

The residuals for continuity and momentum equations were set at level of 10^{-6}, while for energy equation at 10^{-8}.

The reference fluid used to perform computations possesses strong paramagnetic properties. It is a solution of gadolinium nitrate hexahydrate (C_{mass} = 80%) - ($Gd(NO_3)_3 \cdot 6H_2O$) with the molar concentration of gadolinium of C_{gado} = 0,8 (mol/kg). It was characterised by Prandtl number equal to Pr = 584. Table 1 presents the properties of the working fluid used in computations.

Table 1. Thermo-physical and magnetic properties of the reference working fluid at 298 K [1].

Property	Symbol	Unit	Value
Density	ρ	kg/m³	1463
Dynamic viscosity	μ	Pa·s	$8.689 \cdot 10^{-2}$
Thermal expansion coefficient	β	K^{-1}	$0.52 \cdot 10^{-3}$
Volumetric magnetic susceptibility	χ	-	$3.38 \cdot 10^{-4}$
Magnetic permeability	μ_m	H/m	$4\pi \cdot 10^{-7}$
Thermal conductivity	λ	W/(mK)	0.397

The Reynolds number of each studied flow was equal to Re = 0.84, while the Prandtl number varied from Pr = 40 to Pr = 584. The magnetic properties for all analyzed fluids were kept the same. The magnetic induction in the center of the coil was always 10 (T).

4 Results and Discussion

Figure 2 presents the distribution of velocity (Figure (a)) and temperature (Figure 2 (b)) fields and served as a reference case for the further investigations. The presented distributions were obtained for the fluid with Prandtl number equal to Pr = 584. Very thin thermal boundary layer could be seen in Figure 2 (b).

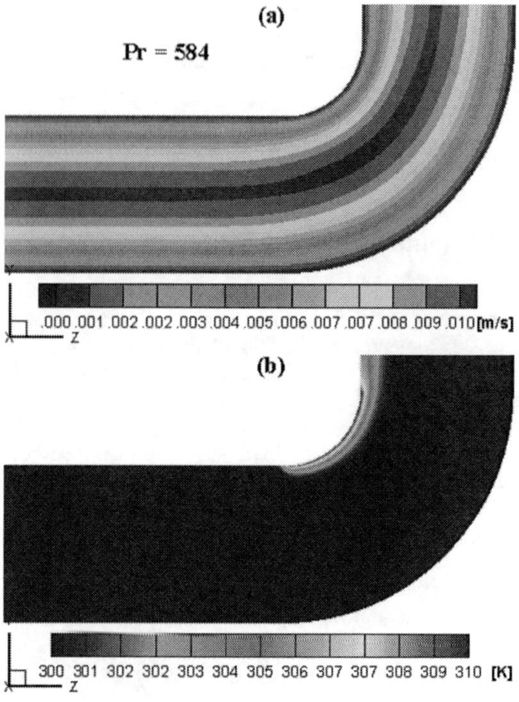

Figure 2. (a) Velocity and (b) temperature distributions for the case of flow without magnetic field.

Figures 3 and 4 present flow characteristics for variants I and II and Pr = 584 under the influence of strong magnetic field. In both cases similar trends could be observed. The fluid stream was attracted by the magnetic field in the vicinity of heated wall and accelerated near it, while in the vicinity of the cold wall deceleration zone appeared (Figures 3 (a) and 4 (a)). However, only in variant II the flow velocity rose above the maximum initial velocity ($u_{max} = 2 \cdot u_{avg}$). In variant I only the local acceleration could be distinguished. The temperature distribution weakly responded to velocity distribution. Different situation could be found in other cases of variant I for lower values of Prandtl number (Figures 5 (a) and 6 (a)), where velocity and temperature fields influenced each other. The key parameter to understand the fluid behaviour is magnetic force distribution (Figures 3 (c) and (d), and 4 (c) and (d)). The magnetic force changed its direction streamwisely after crossing the fluid reference temperature T_0 (in accordance with Eq. (4)) by the fluid. This change caused the flows acceleration near the heated wall.

Analyzing Figure 5 (variant I), it could be observed that deceleration zone near the cold wall reduced its size as the Prandtl number increased. The acceleration value was diminished with the growth of this parameter. It appeared that that the magnetic field and boundary conditions for variant I were inneffective in reducing negative effects, because they might cause sedimentation

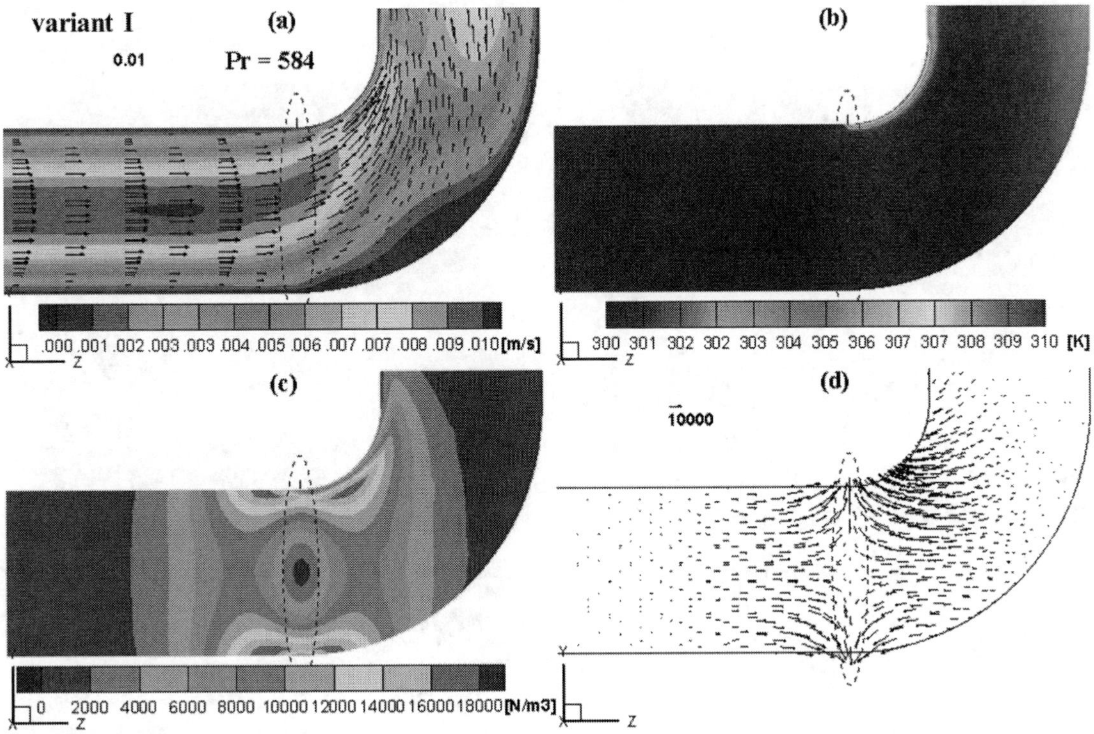

Figure 3. Flow characteristics for variant I: (a) velocity vectors and contours, (b) temperature contours, (c) magnetic force contours, (d) magnetic force vectors.

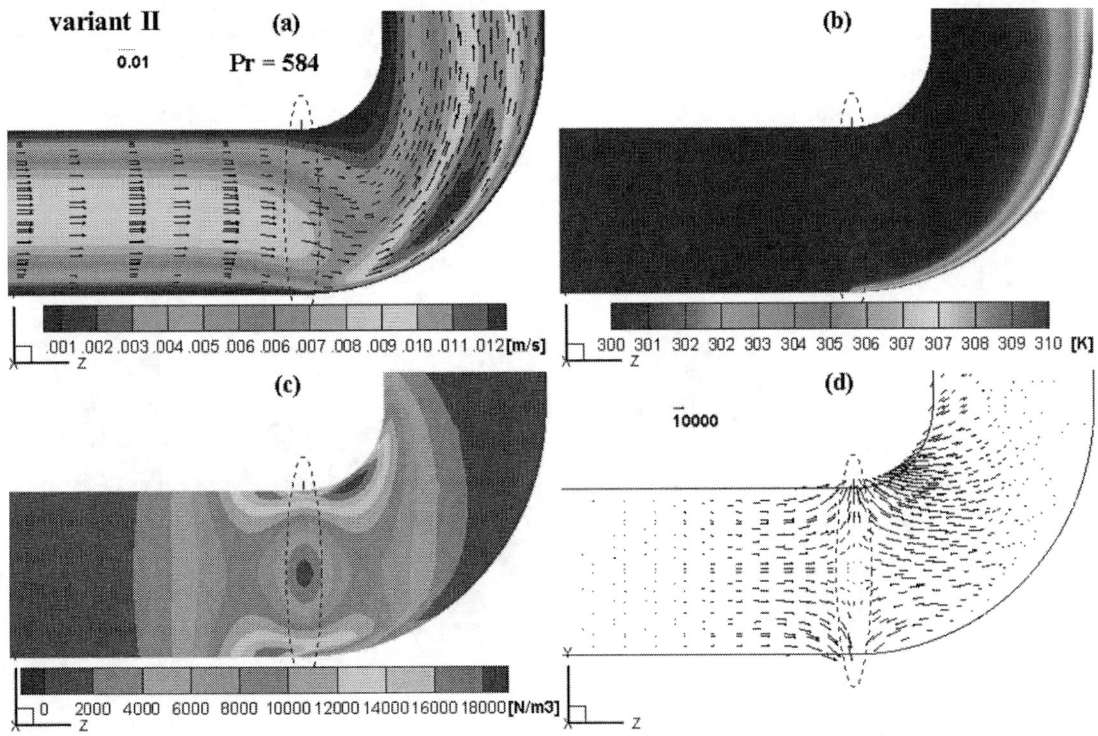

Figure 4. Flow characteristics for variant II: (a) velocity vectors and contours, (b) temperature contours, (c) magnetic force contours, (d) magnetic force vectors.

DOI: 10.1051/e3sconf/20161000120

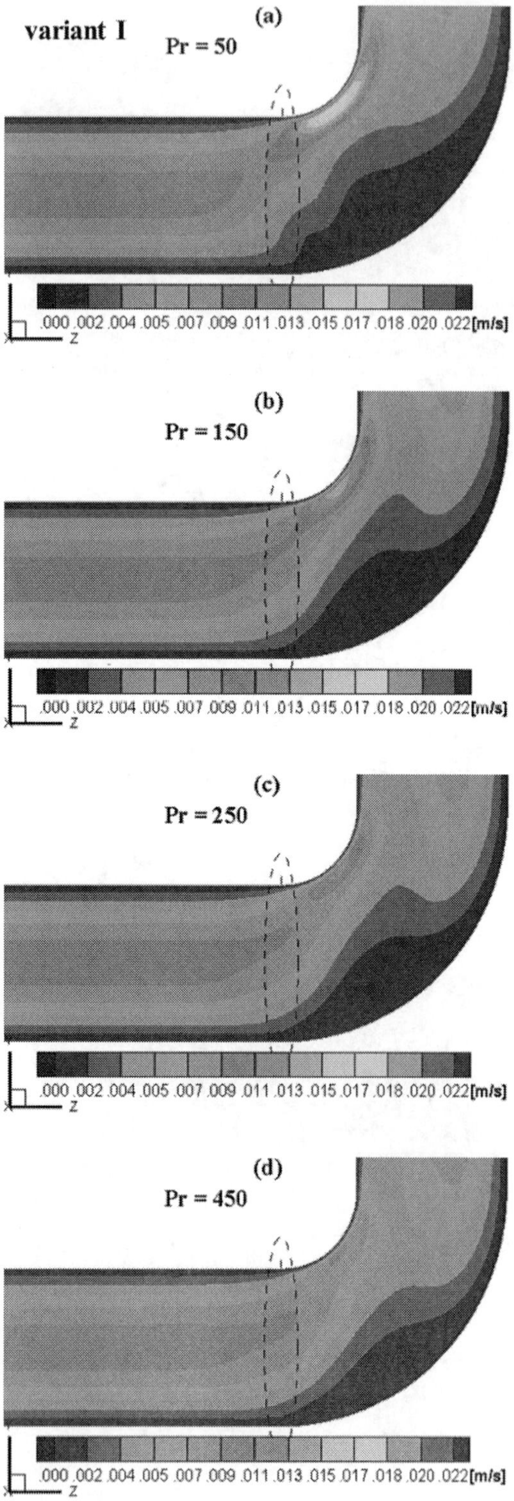

Figure 5. Velocity contours for various Prandtl number: (a) Pr = 50, (b) Pr = 150, (c) Pr = 250, (d) Pr = 450.

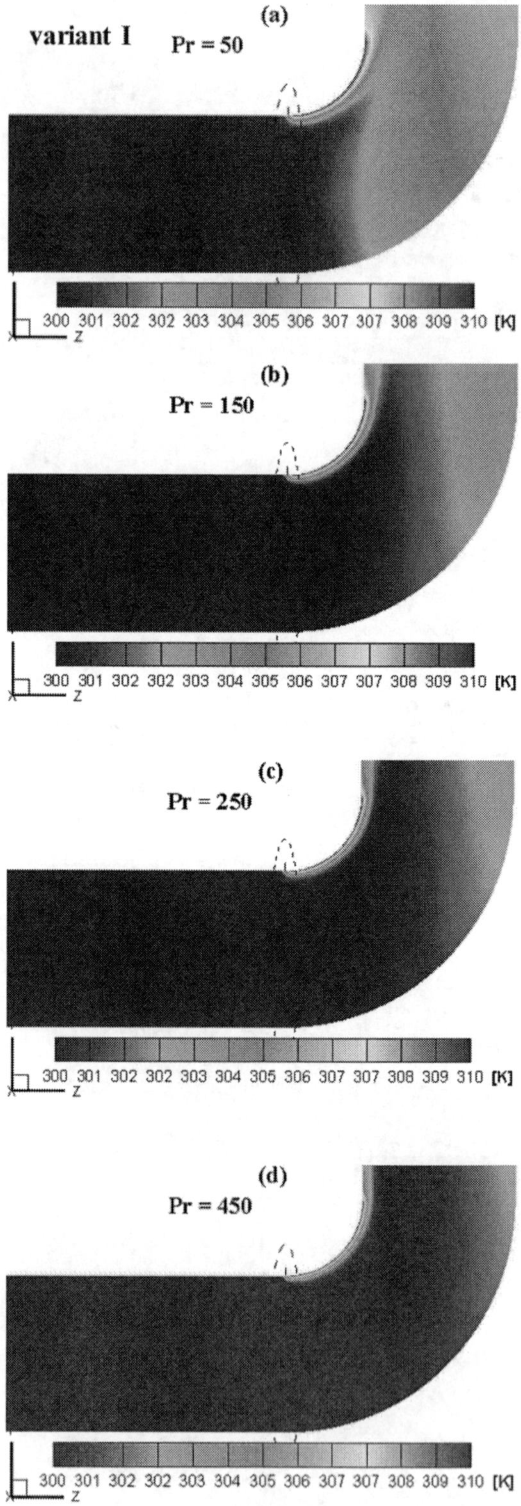

Figure 6. Temperature contours for various Prandtl number: (a) Pr = 50, (b) Pr = 150, (c) Pr = 250, (d) Pr = 450.

DOI: 10.1051/e3sconf/20161000120

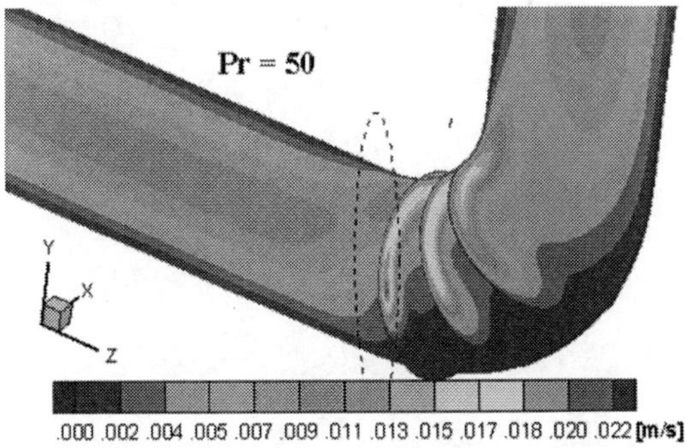

Figure 7. Velocity distribution on various cross-sections for variant I.

near the cold wall. On the other hand, variant II seems to be promising and requires further investigation.

What is worth mentioning is a fact that the obtained structures are fully three-dimensional. It could be confirmed with the glance at Figure 7, where the highest velocity region is placed beyond the axial cross-section of the flow.

Figure 8 presents the distribution of dimensionless velocity ($u_d = u / u_{avg}$) versus Prandtl number. The strongest accelerating effect of magnetic field can be observed for lower values of Prandtl number.

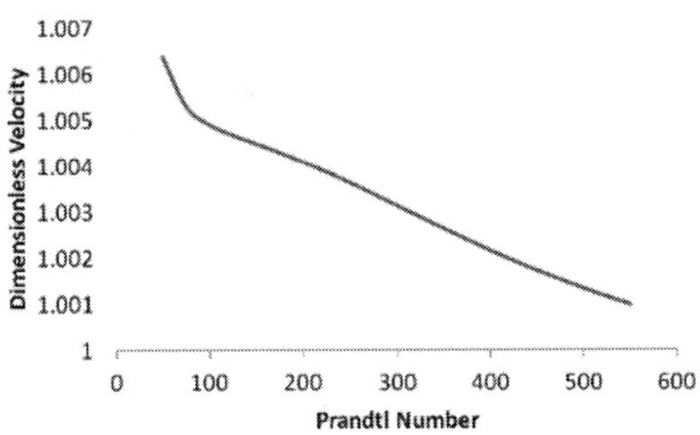

Figure 8. Distribution of dimensionless velocity versus Prandtl number under the influence of strong magnetic field.

5 Conclusions

The question whether the magnetic field enhances or reduces the forced convection cannot be answered in an explicit way. The effect of magnetic field strongly depends on the applied configuration and boundary conditions.

It can be stated that the presence of magnetic field accelerates the fluid in the heating area of the flow, but also suppresses the flow in the cold area. From this point of view, the more promising approach is to heat the wall no. 2 (as in Figure 4) what can prevent the sedimentation and accelerate the fluid near the higher curvature of elbow.

Another option is to incline the magnetic coil and to find an optimal configuration. This topic will be undertaken in the future papers.

Acknowledgements

This work was supported by the Academic Computer Centre CYFRONET AGH, under award no. MNiSW/IBM-BC-HS21/AGH/060/2013.

References

1. T. Bednarz, *Numerical and experimental analyses of convection of paramagnetic fluid in a cubic enclosure* (PhD Thesis, Kyushu University, 2004)
2. J. G. Bednorz, K. A. Muller, Z. Phys. B. Con. Mat. **64,** 189 (1986)
3. D. Braithwaite, E. Beaugnon, R. Tournier, Nature **354,** 134 (1991)
4. N. I. Wakayama, M. Sugie, Physica B: Condens. Matter **216,** 403 (1996)
5. N. I. Wakayama, M. Wakayama, Jpn. J. Appl. Phys. **39,** 262 (2000)
6. Y. Ikezoe, N. Hirota, J. Nakagawa, K. Kitazawa, Nature **393,** 749 (1998)
7. H. Uetake, J. Nakagawa, N. Horota, K. Kitazawa, J. Appl. Phys. **85,** 536 (1999)
8. T. Tagawa, R. Shigemitsu, H. Ozoe, Int. J. Heat Mass Tran. **45,** 267 (2002)
9. T. Tagawa, A. Ujihara, H. Ozoe, Int. J. Heat Mass Tran. **46,** 4097 (2003)
10. B. Bai, A. Yabe, J. Qi, N. I. Wakayama, AIAA J. **37,** 1538 (1999)
11. E. Fornalik-Wajs, P. Filar, J. Wajs, A. Roszko, L. Pleskacz, J. Phys. Conf. Ser. **530,** 012041 (2014)
12. W. Wrobel, E. Fornalik-Wajs, J. S. Szmyd J. Phys. Conf. Ser. **395,** 012124 (2012)
13. A. Kraszewska A *An analysis of thermo-magnetic convection of paramagnetic fluid in rectangular enclosure Conference: Konferencja „Innowacyjne pomysły młodych naukowców: Nauka – Startup – Przemysł"* Book of Abstracts 47 (2015)
14. S. Kenjeres, W. Wrobel, L. Pyrda, E. Fornalik-Wajs, J. S. Szmyd, Flow Turbul. Combust. **92,** *371* (2014)
15. A. Roszko, E. Fornalik-Wajs, Trans. IFFM **128,** 29 (2015)
16. S. Ueno, M. Iwasaka, J. Appl. Phys.**75,** 1.356686 (1994)
17. H. Ozoe *Magnetic Convection* (Imperial College Press, 2005)
18. L. Pleskacz, E. Fornalik-Wajs, J. Phys. Conf. Ser. **530,** 012062 (2014)

E3S Web of Conferences **10**, 00121 (2016) DOI: 10.1051/e3sconf/20161000121
SEED 2016

Germanium content in Polish hard coals

Dorota Makowska[1,a], Faustyna Wierońska[1], Andrzej Strugała[1] and Katarzyna Kosowska[1]

[1]*AGH University of Science and Technology, Faculty of Energy and Fuels, al. A. Mickiewicza 30, 30-059 Krakow, Poland*

Abstract. Due to the policy of the European Union, it is necessary to search for new sources of scarce raw materials. One of these materials is germanium, listed as a critical element. This semi-metal is widely used in the electronics industry, for example in the production of semiconductors, fibre optics and solar cells. Coal and fly ash from its combustion and gasification for a long time have been considered as a potential source of many critical elements, particularly germanium. The paper presents the results of germanium content determination in the Polish hard coal. 23 coal samples of various coal ranks were analysed. The samples were collected from 15 mines of the Upper Silesian Coal Basin and from one mine of the Lublin Coal Basin. The determination of germanium content was performed with the use of Atomic Absorption Spectrometry with Electrothermal Atomization (GFAAS). The investigation showed that germanium content in the analysed samples was at least twice lower than the average content of this element in the hard coals analysed so far and was in the range of 0.08 ÷ 1.28 mg/kg. Moreover, the content of Ge in the ashes from the studied coals does not exceed 15 mg/kg, which is lower than the average value of Ge content in the coal ashes. The highest content of this element characterizes coals of the Lublin Coal Basin and young coals type 31 from the Vistula region. The results indicate a low utility of the analysed coal ashes as a source of the recovery of germanium. On the basis of the analyses, the lack of the relationship between the content of the element and the ash content in the tested coals was noted. For coals of the Upper Silesian Coal Basin, the relationship between the content of germanium in the ashes and the depth of the seam was observed.

1 Introduction

In 2008 the European Commission initiated a discussion on the rare raw materials [1] resulting from China's economic policy. The first report of the working group "Raw Materials Supply Group" of the European Commission "Enterprise and Industry" was announced in 2010 [2], and the next in 2014 [3]. These reports set out the situation of raw materials in Europe and a list of elements and minerals necessary for the further development of European countries was compiled. The list includes germanium, as a critical element, which is characterized by limited resources and a lack of alternatives; at the same time it is important for the development of the world economy, especially for the electronics industry, telecommunications and automotive industries. Most of the demand for critical elements (including germanium) in the European Union is satisfied by imports from countries such as eg. China, USA, South Africa, Brazil. For this reason, it becomes necessary to search for new sources of these metals. A number of studies in the literature discussed the issue of critical raw materials for national mineral resources [4-5] and the current resource market in Poland is monitored by Mineral and Energy Economy Research Institute, Polish Academy of Sciences [6-7].

Coal and coal fly ash for a long time have been considered as a potential source of trace elements, especially germanium [8-15]. Also, Polish coals could be seen as a potential raw material for the recovery of germanium. In particular, coals from the Lublin Coal Basin were reported to contain 50 ÷ 1720 mg of Ge per ton of their ash [15]. In hard coals from the Upper Silesian Coal Basin content of this element was determined in the range of 10.5 ÷ 52 mg/Mg [16-18]. During the 60's and 70's of the 20th century, J. Kuhl's research group carried out detailed investigations of the germanium occurrence in coals from the Upper Silesian Coal Basin [19-20].

Research [21] has shown that germanium is present in solid fuels regardless of its coal rank. It can be found in both the peat and lignite or in coal. In recent years, many researchers have tried to determine whether germanium is closely associated with coal mineral matter, or perhaps with any particular form of coal organic matter. Scientists have not established a clear relationship between the content of germanium in coal and the coal rank. However, it was found that there are tendencies that the content of germanium decreased with the age of coal. Studies [19] have shown that coals contain germanium partially in coal macerals and in the free, inorganic mineral matter. It was also shown that the germanium content in coals is dependent on the geological conditions of the seam and the composition of the mineral matter. Higher concentration of germanium was recorded in coals containing more calcium and magnesium carbonates,

[a] Corresponding author: makowska@agh.edu.pl

as well as iron compounds. In the free mineral matter Ge is associated primarily with the pyrite and the occurrence of this form can represent up to 40% of the total content of this element in coal. The highest concentration of germanium was reported in young coals type 31 in the proximity of zinc and lead deposits.

An interesting issue is how germanium has been stored in coal seams. Based on years of observation [21], no connection has been found with the storage of Ge during the growth of the plant, from which coal was formed. On the other hand, a number of other researchers believe that the accumulation of Ge by Carboniferous plants played a major role in the storage of this component in coal [19]. It is believed that the main factor influencing the amount of germanium in coal seams was Ge sorption from aqueous solutions during peat formation. It is presumed that the chemical sorption by gel humic compounds had a special significance. Therefore, the differences in the content of germanium in coals are the consequence of different conditions of the peat formation process. The thickness of the coal seam is also significant. An important role in the accumulation of germanium in coal is played by sorption from water circulating in the seam during the process of diagenesis and metamorphism, as shown by increased content of Ge in the zone near the roof and the floor of the seam.

Due to the current policy of the European Union on raw materials, the authors of this paper have researched the presence of certain critical elements (including germanium) in the waste from coal cleaning processes [22] and in fly ashes from coal combustion [23]. Aguścik [24] has made an attempt to estimate the resource of germanium in Polish coal deposits.

A number of methods for the recovery of germanium from fly ash have been used [25-27]. Currently in Spain intensive research on the recovery of germanium from the ashes produced in the coal gasification process carried out in the industrial IGCC Puertollano is being conducted [28-31].

Accordingly, coal fly ashes might be expected to be rich in germanium. The present work aimed at determining whether the Polish coals may indeed be a potential source of this precious metal. Due to the fact that germanium is recovered from coal ash obtained in industrial combustion or gasification processes, it is important to analyse the content of this element in the coals for these processes. Therefore, the content of germanium was analysed in averaged commercial coal delivered currently to Polish power plants. This was in fact the basic assumption of the preliminary analysis of the legitimacy intention of germanium recovery from the ashes, which was conducted based on the results of the research presented in the article.

2 Experimental

2.1 Examined material

The research material consisted of 23 samples of bituminous coal with different coal ranks (from steam coal to coking coal), of which 21 came from 15 mines of the Upper Silesian Coal Basin and 2 from the Lublin Coal Basin mine. The examined samples represent averaged commercial coals delivered to Polish power plants and they were taken and prepared in accordance with the standard PN-G-04502: 2014-11. All samples are listed with their specification in Table 1. For all of the samples, the determination of ash content, the content of the main ash components (SiO_2, Al_2O_3, CaO, Fe_2O_3, MgO) and the determination of germanium were performed.

Table 1. Examined samples specifications.

Sample ID	Type of coal		Seam group	Coal Basin
BG-1	steam coal	32,2	300	Lublin Coal Basin
BG-2	steam coal	32,2	300	
BO-1	coking coal	35,1	400	Upper Silesian Coal Basin
ZO-1	coking coal	35,1	400	
JM-1	coking coal	35,2	500	
KR-1	coking coal	34,1	300	
PN-1	coking coal	35,1	300/400	
BU-1	coking coal	34,2	400	
SB-1	steam coal	31,1	200/300	
JN-1	steam coal	31,1	100	
ST-1	steam coal	32,2	400/500	
WR-1	steam coal	32,1	500	
WJ-1	steam coal	32,2	400/500	
BB-1	steam coal	32,1	500	
ZM-1	steam coal	31,2	200	
ZM-2	steam coal	31,2	200	
ZM-3	steam coal	31,2	200	
ZM-4	steam coal	31,2	200	
BŚ-1	steam coal	32,1	400	
BŚ-2	steam coal	32,1	400	
HL-1	steam coal	32,2	400/500	
HL-2	steam coal	32,2	400/500	
HL-3	steam coal	32,2	400/500	

2.2 Analytical methods

Coal samples were prepared in accordance with Polish standard PN-G-04502:1990. For all the samples, the determination of ash content in analytical condition was carried out in accordance with the standard PN-80 / G-04512 by slow ashing. The analysis of the main components of ash was conducted with the use of Flame Atomic Absorption Spectrometry and the determination of germanium content was conducted with the use of Graphite Furnace Atomic Absorption Spectrometry with the application of a tandem spectrometer Hitachi Z-2000 with Zeeman background correction effect. As a source of radiation, hollow cathode lamps were used.

In order to perform the analyses by AAS, ash samples were subjected to complete digestion in a Berghoff SpeedWave4 microwave system. About 100 mg of ash was digested with spectrally pure nitric acid (V) and hydrofluoric acid. As the complexing reagent supersaturated boric acid was used.

The determination of germanium content by GFAAS was conducted with the following parameters: analytical spectral line 265.2 µm; gas Ar flow 200 cm^3/min (atomization – 30 cm^3/min); slit width 0.4 nm; modifier Pd-Mg and Na solution. The quantification limit of this method is equal to 0.36 mg/kg and the uncertainty of the measurement results reaches the level of 7.0% resp.

3 Results and discussion

Results of the determination of the ash content and the content of germanium are summarized in Table 2. The ash content of the examined samples fluctuates between 2.7 and 56.8 %wt. The germanium content in the analysed samples of coal ash is lower than the world average content of germanium for coal fly ashes, which is 18 ± 1 mg/kg [32], while calculating the germanium content in coal samples, the maximum value obtained is 1.28 mg/kg, which is almost twice lower than the world average germanium content in hard coals (amounting to 2.4 ± 0.2 mg/kg) and even lower than the average content of this element in sedimentary rocks (1.4 mg/kg) or lithosphere (1.7 mg/kg). The results are also lower than those given in the cited literature, which may be associated with the specificity of the analysed coal samples, namely samples of averaged commercial coals currently used in combustion processes in power plants.

Table 2. Germanium content in examined samples of Polish coals.

Sample	Ash content A^a [%wt]	Ge content [mg/kg]	
		in ash	in coal
BG-1	4,3	10,63	0,46
BG-2	8,7	14,71	1,28
BO-1	56,8	1,02	0,58
ZO-1	52,7	1,04	0,55
JM-1	22,2	0,95	0,21
KR-1	46,0	2,26	1,04
PN-1	53,4	1,62	0,87
BU-1	36,9	2,40	0,89
SB-1	6,5	4,10	0,27
JN-1	8,7	10,98	0,96
ST-1	20,3	4,50	0,91
WR-1	2,7	3,23	0,09
WJ-1	16,7	0,72	0,12
BB-1	9,5	2,84	0,27
ZM-1	7,1	7,88	0,56
ZM-2	6,4	3,11	0,20
ZM-3	6,7	4,15	0,28
ZM-4	21,4	5,13	1,10
BŚ-1	17,5	1,83	0,32
BŚ-2	29,9	3,56	1,06
HL-1	13,8	0,56	0,08
HL-2	29,2	1,59	0,46
HL-3	7,7	<LOQ	-

In order to verify the association of germanium with mineral matter, the relationship between the content of this element (in both coal samples and its ashes) and ash content was examined (Fig. 1). It can be concluded that there is no correlation between the content of germanium in the examined coals and ash content. Similarly, the correlation between the content of Ge in ash and ash content is relatively low. It can be observed, however, that coals with low ash content have the highest concentration of Ge.

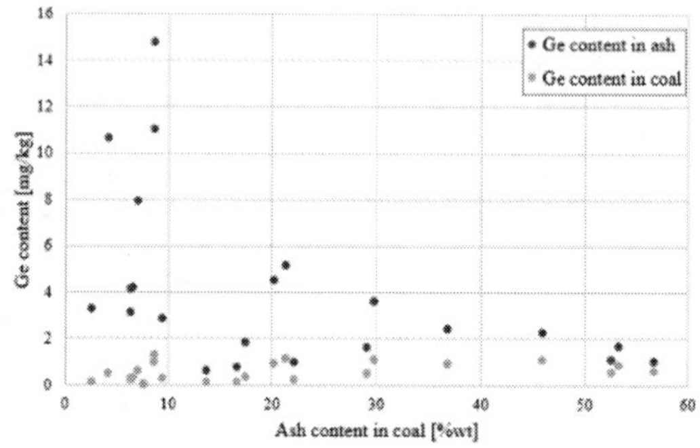

Figure 1. Relationship between germanium content in examined coals and their ashes, and ash content in coal.

The dependence of the seams depth on germanium content was also examined (Fig. 2). The presented graph indicates the presence of the correlation between the content of the tested element and the depth of the seam for coals from the Upper Silesian Coal Basin; the deeper the seam is located, the smaller the content of germanium becomes. However, due to the amount and type of the analysed samples shown in Fig. 2 generalized conclusions cannot be drawn on the correlation between the group of coal seam and the content of germanium. It would be possible based on the examination of a large amount of geological samples. The presented graph is merely aimed at showing the results only within the framework of this work.

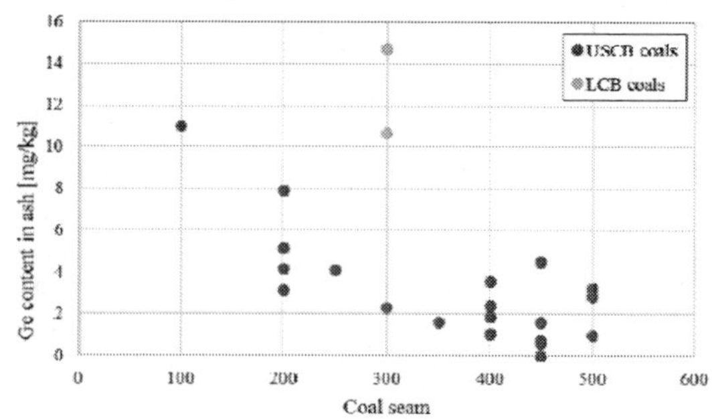

Figure 2. Germanium content in examined coal ashes in relation to seam group of analysed coals.

A similar correlation can be observed in the case of the relationship between the germanium content in ash and the type of coal (Fig. 3), namely, the higher coal rank of the examined samples is, the lower the germanium content becomes.

Therefore, these results confirm the tendency of the content of this element to decrease along with the age of coal.

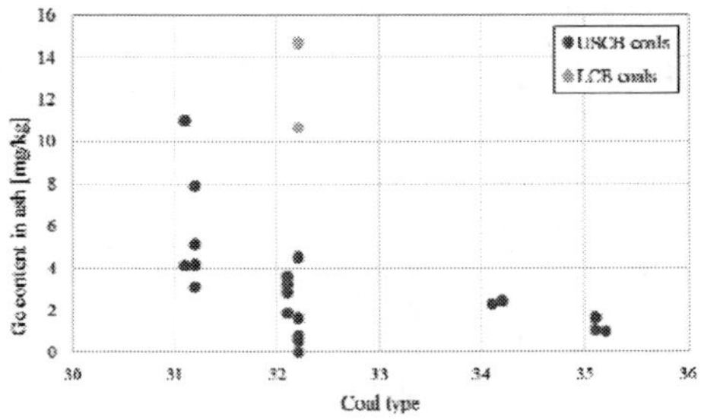

Figure 3. Relationship between germanium content in examined coal ashes and coal type.

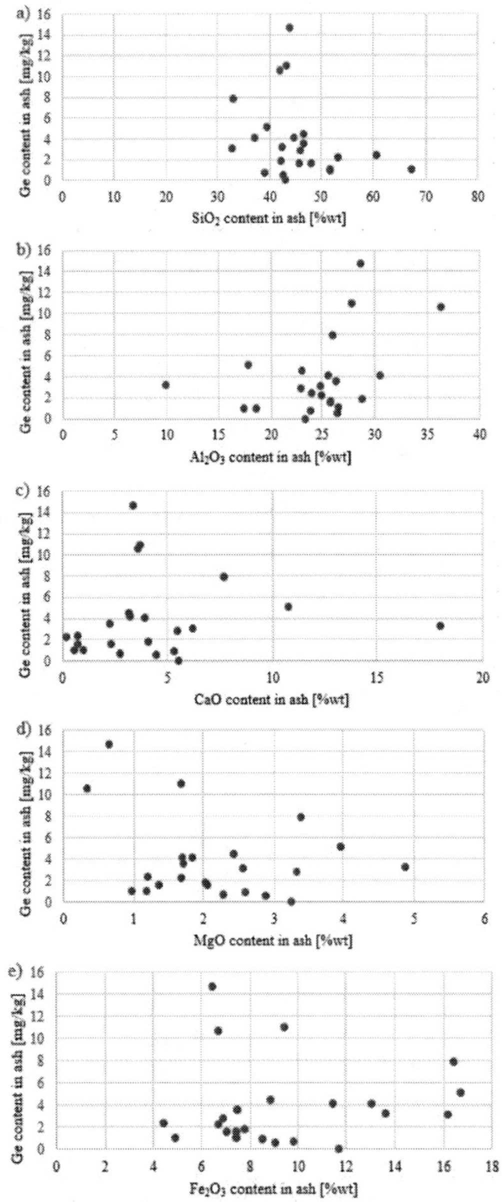

Figure 4. Relationship between germanium content in examined coal ashes and content of main ash compounds: a) silicon; b) aluminium; c) calcium; d) magnesium and e) iron.

Based on the analysis of ash composition of the examined samples, the impact of the individual components of the mineral matter on the accumulation of germanium in coal was also examined. The correlation between the Ge content and the content of SiO_2, Al_2O_3, CaO, Fe_2O_3, MgO is shown in Fig. 4.

The presented graphs show no relationship between the germanium content in the examined coal ashes and the content of each of the main ash components. As a result, the findings of the earlier research [19] have not been confirmed.

To determine whether the ashes from the examined Polish coals can be considered as a potential source of germanium, the Enrichment Factor *EF* was calculated for each ash sample from the equation below (1) and plotted in Fig. 5.

$$EF = \frac{A_e}{A_c} \qquad (1)$$

Where A_e is the concentration of Ge in the fly ash sample (mg/kg) and A_c – the average Ge content in sedimentary rocks (mg/kg). The graph also presents the Coal Affinity Index *CAI* [32]. This index defines the ratio of the world average content of germanium in coal ashes to its average concentration in sedimentary rocks.

The Enrichment Factor for most of the analysed samples is greater than 1.0, but no value exceeded *CAI*. The highest values of *EF* were observed for coal samples from the Lublin Coal Basin and the coal samples from the Vistula region (JN-1 and ZM-1).

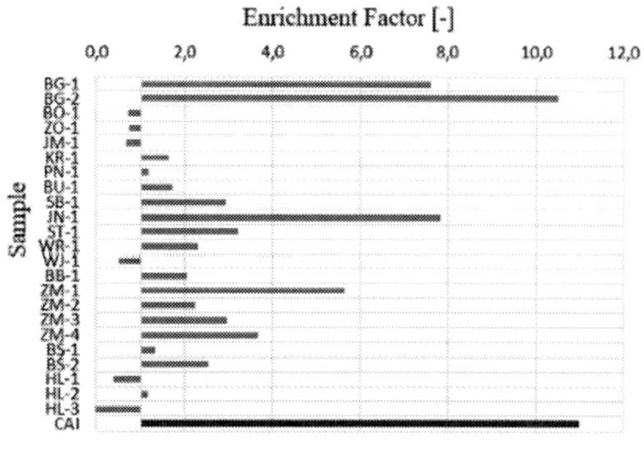

Figure 5. Germanium enrichment factor for examined coal samples.

4 Conclusions

The aim of the study was to update the information on the content of germanium in Polish coals and to evaluate ashes from these coals as a potential source of germanium. Based on extensive investigations, the following conclusions can be drawn:

• germanium content in the examined Polish coals does not exceed the average content of this element in the earth's crust;

• no relationship between the Ge content in the samples and ash content was noticed, although the highest content of this element in the ash was determined in coals with the lowest ash content;

• no association was found between the content of germanium in the ash and the content of each of the main elements of mineral matter, i.e. Si, Al, Ca, Mg, Fe;

• germanium content in the ashes of the analyzed samples is lower than the world average germanium content in the coal ashes, which indicates the usefulness of the examined coal ashes as a raw material for the recovery of germanium, however, these results cannot exclude the possibility of recovery of this element in the future, when other coal seams are operated.

• the biggest Enrichment Factors have been noted for the samples of coal from the Lublin Coal Basin and the coals from Vistula region.

Acknowledgements

This paper was prepared as a part of the statutory activity of the Faculty of Energy and Fuels at the AGH University of Science and Technology, No. 11.11.210.213

References

1. The Raw Materials Initiative – Meeting our critical needs for growth and jobs in Europe, Commission of the European Communities, Brussels 2008
2. Critical Raw Materials for The EU. Report of the Ad-hoc Working Group on defining critical raw materials, Raw Materials Supply Group, Brussels 2010
3. Report on Critical Raw Materials for The EU. Report of the Ad-hoc Working Group on defining critical raw materials, Brussels 2014.
4. B. Radwanek-Bąk, Mineral Resources Management **1**, 5-19, (2011)
5. T.J. Smakowski, Bulletin of the Mineral and Energy Economy Research Institute of the Polish Academy of Science **81**, 59-68 (2011)
6. T.J. Smakowski, T. Galos, E. Lewicka [red.], *Bilans gospodarki surowcami mineralnymi Polski i świata 2013* (IGSMiE PAN, Warsaw 2015)
7. A. Burkowicz, K. Galos, K. Guzik, et al., *Minerals Yearbook of Poland 2012* (IGSMiE, Warsaw 2013).
8. J. Całus-Moszko, B. Białecka, Mineral Resources Management **29** (1), 67-80 (2013)
9. G. Eskenazy, Fuel and Energy Abstracts 37, 3, 170 (1995)
10. M. Marczak, *Geneza i prawidłowość występowania pierwiastków śladowych w węglach złoża Chełm w Lubelskim Zagłębiu Węglowym* (Wydawnictwo Uniwersytetu Śląskiego, Katowice 1985)
11. M. Mastalerz, A. Drobniak, Int. J. Coal Geol. **94**, 302-313 (2012)
12. H.R. Parzentny, *Wpływ nieorganicznej substancji mineralnej na zawartość niektórych pierwiastków śladowych w węglu Górnośląskiego Zagłębia Węglowego* (Wydawnictwo Uniwersytetu Śląskiego, Katowice 1995)
13. V.V. Seredin, R.B. Finkelman, Int. J. Coal Geol. **76**, 253-289 (2008)
14. Ya.E. Yudovich, Int. J. Coal Geol. 53, 223-232 (2003)
15. M. Plewa, *Pierwiastki śladowe w węglu Lubelskiego Zagłębia Węglowego, II Seminarium: Metodyka rozpoznawania i dokumentowania złóż kopalin stałych* (Wydawnictwo AGH, Kraków 1990).
16. S. Cebulak, A. Różkowska, Geological Quarterly **27** (1983)
17. A. Różkowska, Geological Exploration Technology no 2 (1984)
18. A. Idzikowski, Arch. Miner. **23**, no 2, 271 (1959)
19. J. Winnicki, Prace GIG **18**, Komunikat 354, 159-195 (1964)
20. J. Winnicki, Praca Naukowa Instytutu Chemii Nieorganicznej i Metalurgii Pierwiastków Rzadkich Politechniki Wrocławskiej no **18** (1975)
21. K. Akerman, *German* (WNT, Warsaw 1967)

22. A. Strugała, D. Makowska, K. Bytnar, et al., Energy Policy Journal **17** (4), 77-88 (2014)
23. D. Makowska, F. Wierońska, GeoScience Engineering **62**, no 1, 31-36 (2016)
24. J. Aguścik, Bulletin of the Mineral and Energy Research Institute of the Polish Academy of Sciences **88**, 8-13 (2014)
25. R.S. Blissett, N.A. Rowson, Fuel **97**, 1-23 (2012)
26. V.G. Nosach, et al., Fuel and Energy Abstracts **38** (6), 431 (1997)
27. L. Zhang, Z. Xu. Z., Journal of Hazardous Materials, **312**, 28-36 (2016)
28. F. Arroyo, F.A. Torralvo, C. Fernandez-Pereira, Minerals Engineering **24**, 35-41 (2011)
29. F. Arroyo, et al., Fuel Processing Technology **124**, 222-227 (2014)
30. O. Font, X. Querol, A. Lopez-Soler, et al., Fuel **84**, 1384-1392 (2005)
31. A.S. Meawad, D. Bojinova, Y.G. Pelovski, Waste Management **30**, 2548-2559 (2010)
32. M.P. Ketris, Ya.E. Yudovich, Int. J. Coal Geol. **78**, 135-148 (2009)

 DOI: 10.1051/e3sconf/20161000122

Intensification of heat transfer with an application of strong magnetic gradients

Anna Kraszewska[1,a], Lukasz Pyrda[1] and Janusz Donizak[1]

[1]*AGH University of Science and Technology, Department of Fundamental Research in Energy Engineering, Faculty of Energy and Fuels, Krakow, Poland*

Abstract. Experimental analysis of a strong magnetic field influence on heat transfer by paramagnetic fluid was conducted. A rectangular enclosure, filled with working fluid, was placed inside superconducting magnet's working section in Rayleigh-Bénard configuration and three temperature differences between heated and cooled walls ΔT= 3, 5, 11 [°C] were applied. On the basis of performed measurements heat transfer analysis in the form of Nusselt number, conduction and convection heat fluxes calculations was conducted. Obtained results demonstrated that Nusselt number strongly depends on the temperature difference between thermally active walls as well as on the magnetic induction applied to the system. An application of external high magnetic field from 0 [T] to 10 [T] caused an increase of the heat transfer rate – about 250% for ΔT = 3 [°C] and over 340% for ΔT = 5, 11 and 20 [°C] and can be successfully implemented to heat transfer intensification for paramagnetic fluids.

1 Introduction

Buoyancy-driven natural convection control in closed system is crucial in many applications, such as: heat exchangers, chemical reactors, space industry or non-gravitational state research.

All surrounding substances may be divided into three basic types, depending on magnetic properties: ferromagnetic, diamagnetic and paramagnetic. While ferromagnetics are strongly attracted towards magnetic field (i.e. iron volume susceptibility = $2\cdot10^5$), paramagnetics are poorly attracted by magnetic field source (i. e. oxygene volume susceptibility = $3\cdot10^{-8}$), and diamagnetics are weakly repelled by it (i.e. water volume susceptibility = $-9\cdot10^{-6}$).

To influence a fluid flow with paramagnetic or diamagnetic properties, a high magnetic field gradient should be applied. Then, depending on mutual configuration of magnetic and gravitational forces, different results could be achieved.

First published results with a paramagnetic fluid convection enhancement and suppression was presented by Braithwaite [1]. Tagawa [2] developed a model equation for thermo-magnetic convection. Bednarz [3] studied thermo-magnetic convection in configuration with one side wall heated and the opposite one cooled. Pyrda et al. investigated heat transfer enhancement in cubical enclosure [4] and developed non-dimensional analysis of paramagnetic fluid in Rayleigh-Bénard configuration [5]. Kenjeres et al. investigated oscillatory states in thermal convection of a paramagnetic fluid in a cubical enclosure subjected to a magnetic field gradient [6] and turbulence pockets in thermal convection of paramagnetic fluid subjected to strong magnetic field gradients [7]. His group performed numerical and experimental study of Rayleigh–Bénard–Kelvin Convection [8].

In present paper authors will focus on heat transfer enhancement with utilization of thermo-magnetic convection of paramagnetic fluid.

2 Experimental setup

2.1 Experimental apparatus

Apparatus used to performed experimental analysis of thermo-magnetic convection is shown in Figure 1. It consisted of an experimental enclosure placed in the bore of a superconducting magnet (HF10-100VHT-B, *Sumito Heavy Industries*, Ltd. Japan), thermostating bath with constant temperature flow, a heater control system, and a data acquisition system connected to a personal computer. The experimental enclosure of dimensions 0.032 x 0.032 [m] in base and 0.016 [m] in height is presented in Figure 2.

Experimental enclosure, heated from one horizontal wall and cooled from the opposite one, consisted of following elements: cooling chamber filled with cold water (18[°C]) pumped by thermo-stated bath washing over a top copper plate, cavity made from plexiglass and heating chamber, where bottom copper plate was heated by nichrome wire connected to a DC power supply to maintain constant temperature (21,23,29,38 [°C]). Six T-type thermocouples, inserted into small holes in bottom and top copper plates constantly measured temperature of thermally active walls. Another five T-type thermocouples, inserted into small holes in front wall of

[a] Corresponding author: anna.kraszewska@agh.edu.pl

DOI: 10.1051/e3sconf/20161000122

plexiglass enclosure, measured temperature of working fluid during experiments.

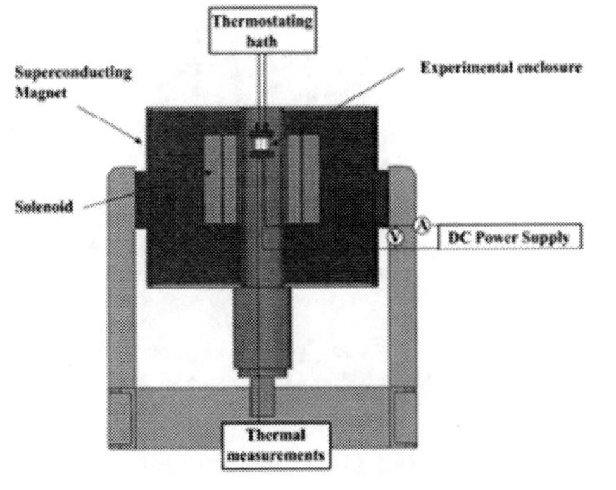

Figure 1. Schematic view of experimental apparatus.

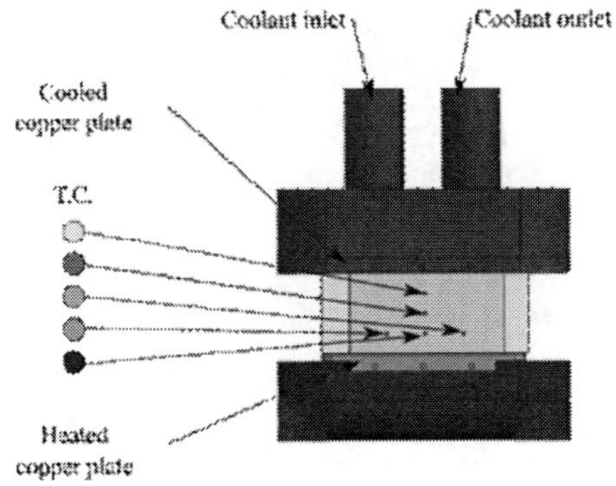

Figure 2. Schematic view of experimental enclosure with marked positions of thermocouples.

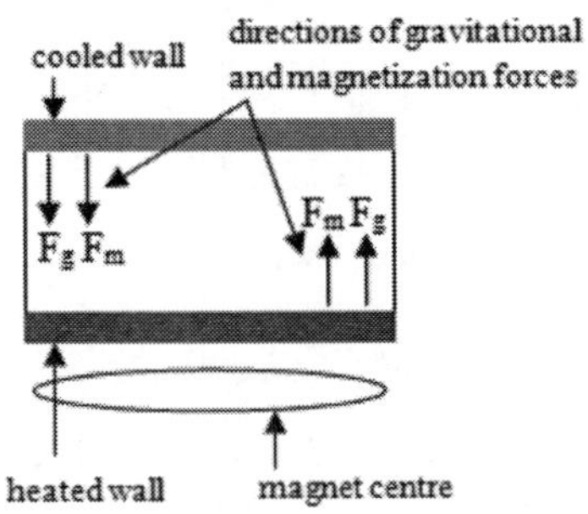

Figure 3. Directions of gravitational and magnetization forces.

Measuring vessel was placed in the upper half of the superconducting magnet, where gravitational and magnetic forces acted in the same direction, causing enhancement of convective flow, as shown in Figure 3. This position is correlated to the location of a maximal value of grad $\mathbf{b_0}$, where $\mathbf{b_0}$ stand for magnetic induction, and it was 0.095 [m] from the magnet's top.

2.2 Working fluid

As a working fluid, 50% volume glycerol aqueous solution with an addition of 0.8 mol/(kg of solution) gadolinium nitrate hexahydrate $Gd(NO_3)_3{\cdot}6H_2O$ to make it paramagnetic, was chosen. Properties of the working fluid are listed in Table 1.

Table 1. Properties of the working fluid.

Property	Sym.	Value	Unit
Density	ρ	1411	[kg/m^3]
Kinematic viscosity	ν	$9.25{\cdot}10^{-6}$	[m^2/s]
Dynamic viscosity	μ	$1.30{\cdot}10^{-2}$	[kg/m·s]
Heat capacity	c_p	$2.92{\cdot}10^{3}$	[J/kg·K]
Thermal diffusivity	α	$9.13{\cdot}10^{-8}$	[m^2/s]
Thermal expansion coefficient	β	$1.21{\cdot}10^{-5}$	[1/K]
Thermal conductivity	λ	0.376	[W/m·K]
Mass magnetic susceptibility	χ_m	$2.39{\cdot}10^{-7}$	[m^3/kg]

2.3 Experimental procedure

Experimental procedure consisted of two steps: an analysis of a conduction state and analysis of a thermo-magnetic convection. In the first step, the enclosure filled with water was placed in the magnet in a position, where the bottom horizontal wall was cooled, and the opposite one was heated (reverse Rayleigh-Bénard configuration). Temperature on a bottom wall was set to constant temperature of 18 [°C], the same as an ambient temperature in the magnet working section, and temperature on the top wall was sequentially: 21, 23, 26, 29, 32, 34 and 38 [°C]. After linear temperature distribution was achieved, heating power was measured. Assuming one-dimensional conductive heat flow, the heat flux can be calculated from Fourier's law of conduction, and therefore the difference between directly measured heat flux and the heat flux calculated from Fourier's law is the heat loss from the experimental enclosure.

The next step was connected with analysis of a thermo-magnetic convection. Experimental enclosure, filled with working fluid, was rotated 180 degrees to Rayleigh-Bénard configuration, with bottom wall heated and the opposite one cooled. The power supply was set to obtain chosen temperature difference between thermally active walls and the setup was left to obtain a stable state. Then temperature, electrical current and voltage were recorded and magnetic field was applied to the system by stages of 1 [T].

3 Heat transfer analysis

Temperature signals recorded during experimental analysis allowed investigation on a heat transfer rate, which was established by calculation of dimensionless parameter called Nusselt number. This criterion, speaking of a heat transfer in analysed system, can be written as follows:

$$\mathrm{Nu} = \frac{Q_{\mathrm{net\_conv}}}{Q_{\mathrm{net\_cond}}} \tag{1}$$

The net convection ($Q_{\mathrm{net\_conv}}$) and net conduction ($Q_{\mathrm{net\_cond}}$) heat fluxes were estimated accrording to method proposed in [9]:

 DOI: 10.1051/e3sconf/20161000122

$$Q_{\text{net\_cond}} = Q_{\text{cond}} - Q_{\text{loss}} \tag{2}$$

$$Q_{\text{net\_conv}} = Q_{\text{conv}} - Q_{\text{loss}} \tag{3}$$

Assuming that the heat loss depends only on the temperature of the heated wall, conduction measurements were made and heat losses were estimated from :

$$Q_{\text{loss}} = Q_{\text{cond}} - Q_{\text{Fourier's\_law}} \tag{4}$$

where:

$$Q_{\text{Fourier's\_law}} = \frac{a^2 \lambda \Delta T}{d} \tag{5}$$

d - enclosure height – 0.016 [m]; a – enclosure width 0.032 [m]; λ – thermal conductivity of the fluid [W/m·K]; ΔT – temperature difference between heated and cooled walls [K].

The heat flux was calculated for the conduction area of a base of the measuring enclosure 0.032 [m] x 0.032 [m]. The calculated heat losses were approximated linearly:

$$Q_{loss} = 0.0907 \cdot \Delta T \tag{6}$$

Heat losses dependency on temperature differences between thermally active wall is presented in Figure 4.

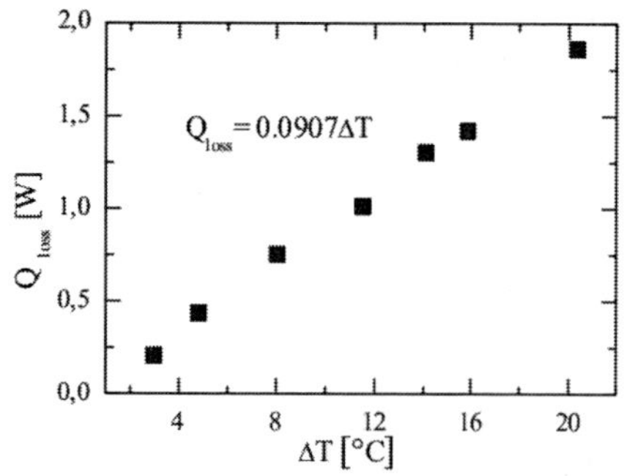

Figure 4. Heat losses in the measurement vessel versus temperature differences.

By applying equations (2),(4) and (5) to (1) expression for Nusselt number can be expressed as follows:

$$\text{Nu} = \frac{Q_{\text{conv}} - Q_{\text{loss}}}{\frac{a^2 \lambda \Delta T}{d}} \tag{7}$$

The results of heat transfer analysis are presented in Figure 5 - 7.

4 Results and discussion

Figure 6 presents results of heat transfer analysis as a Nusselt number calculation versus magnetic induction in the center of the magnet. Nusselt number significantly escalates with an increase of magnetic induction. For natural convection cases, without magnetic induction applied to the system, Nusselt number values start from 3.46, 2.95, 3.62 and 4.51 for ΔT=3, 5, 11 and 20 [°C] respectively.

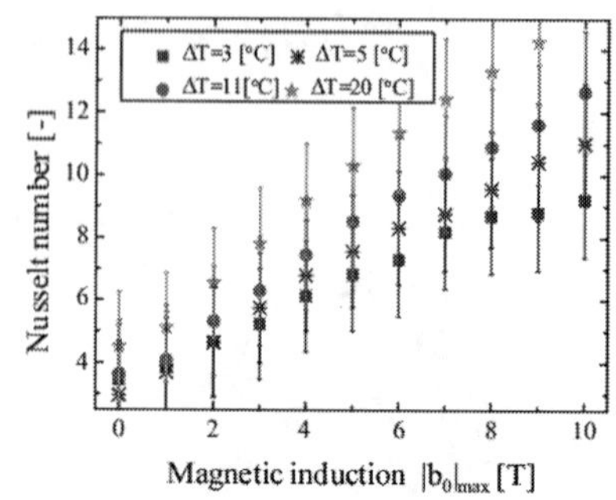

Figure 5. Nusselt number versus magnetic induction in the centre of the magnet $|b_0|_{max}$.

Such fluctuations of Nusselt number values are probably related to the fact that the working section of the magnet could not be tightly insulated and ambient temperature of the laboratory (measurements were performed in winter months and temperature in the room was about 15.6-17.3[°C]) had a significant impact on heat transfer in natural convection. In [4] Authors report Nusselt number values for natural convection in a cubical enclosure (of dimensions 0.032x0.032x0.32 [m]) to be ~7.5 for ΔT=5 [°C] and ~10 for ΔT=11 [°C], therefore a higher experimental geometry is characterized by greater heat transfer. An application of magnetic induction to the system and increasing its value to $|b_0|_{max}$= 10 [T] causes heat transfer growth by 260% for ΔT=3 [°C] and over 340% for ΔT=5, 11 and 20 [°C] cases, which is similar to increasement reported in [4].

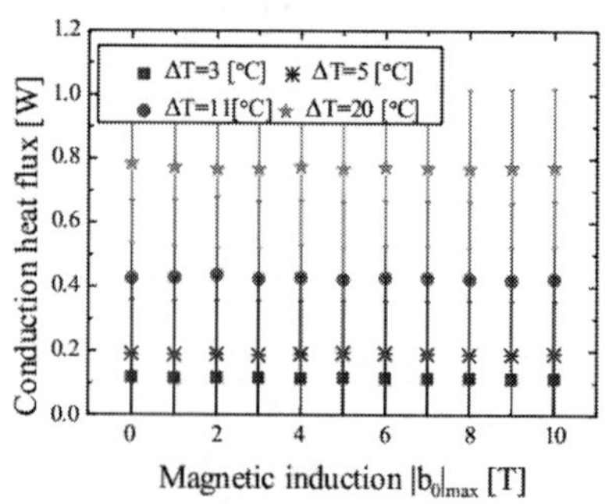

Figure 6. The conduction heat flux versus the magnetic induction in the centre of the magnet.

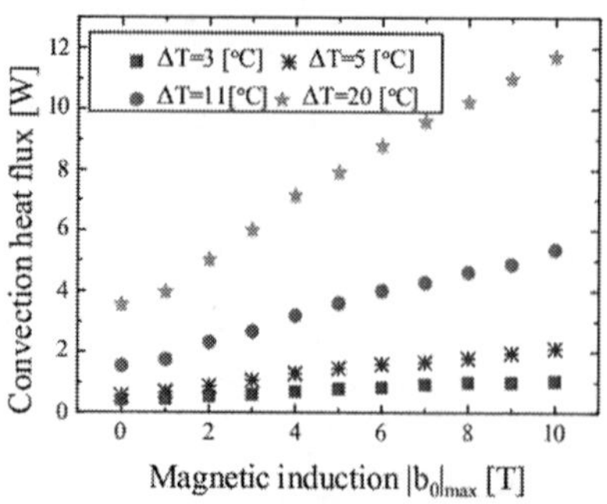

Figure 7. The convection heat flux versus the magnetic induction in the centre of the magnet.

According to equation (1) Nusselt number is a ratio of a convective heat flux to a conduction heat flux in the analysed system. Figures 6 and 7 presents results of conduction and convection fluxes calculations as a function of magnetic induction.

The conduction heat flux is the higher the grater the temperature difference and does not change significantly with magnetic induction increase. For $\Delta T = 3$ [°C] Q_{cond} value oscillates around 0.11 [W] while for $\Delta T = 20$ [°C] it is about seven times bigger and approximately equals 0.77 [W]. For all cases convective heat fluxes are significantly higher than Q_{cond}. For $|b_0|_{max} = 0$ [T] Q_{conv} are 0.41, 0.56, 1.54 and 3.54 [W] for $\Delta T = 3, 5, 11$ and 20 [°C] respectively. Application of external magnetic field results in a major increase in convective heat flux. For $\Delta T = 3$ [°C] and maximal magnetic induction in the center of the magnet convective heat flux rises over about 250% in comparison to natural convection case. For higher temperature differences this increase is even bigger and equals over 330%.

5 Summary

In this paper experimental analysis of thermo-magnetic convection of paramagnetic fluid was conducted. Four temperature differences between heated bottom wall and the opposite one cooled were applied and an influence of magnetic field induction on experimental fluid from 0 [T] to 10 [T] was tested. Performed analysis of signals obtained from thermocouples placed in the experimental enclosure enabled calculation of heat transfer indicator in the form of a Nusselt number and conduction and convection heat fluxes. Obtained results led to the following conclusions: (a) the Nusselt number strongly depends on temperature difference between thermally active walls, as well as on magnetic induction in the system; (b) an application of external magnetic field from 0 [T] to 10 [T] causes an increase of the heat transfer rate – about 250% for $\Delta T = 3$ [°C] and over 340% for $\Delta T = 5$, 11 and 20 [°C]; (c) conduction heat flux increase with higher temperature differences, but does not depend on magnetic induction; (d) convective heat flux strongly depends on magnetic induction in the center of the magnet and the increase in its rate with magnetic induction up to 10 [T] is over 250% for every temperature difference; (e) external magnetic field can be successfully implemented to heat transfer intensification for paramagnetic fluids.

Acknowledgments

The present research was supported by the National Science Centre (Project No. 12/07/B/ST8/03109).
The participation in the conference was partially supported by the AGH UST Vice-Rector for Student Affairs, Professor Anna Siwik.

References

1. D. Braithwaite,E. Beaugnon, R. Tournier, Nature **354** 134–6 (1991)
2. T. Tagawa, R. Shigemitsu, H. Ozoe, Int. J. Heat Mass Transf. **45** 267–77 (2002)
3. T. Bednarz, E. Fornalik, T. Tagawa, H. Ozoe, J.S. Szmyd, Int. J. Therm. Sci. **44** 933–43 (2005)
4. L. Pyrda, S. Kenjeres, E. Fornalik-Wajs, J.S. Szmyd J. Phys. Conf. Ser. **395** 012125 (2012)
5. L. Pyrda, J. Phys. Conf. Ser. **530** 012060 (2014)
6. S. Kenjereš, L. Pyrda, W. Wrobel, E. Fornalik-Wajs J.S. Szmyd, Phys. Rev. E **85** 1–8 (2012)
7. S. Kenjereš, W. Wrobel, L. Pyrda, E. Fornalik-Wajs, J.S. Szmyd, J. Phys. Conf. Ser. **318** 072028 (2011)
8. S. Kenjereš, L. Pyrda, E. Fornalik-Wajs, J.S. Szmyd, Flow, Turbul. Combust. **92** 371–93 (2013)
9. H. Ozoe, S.W. Churchill, AIChE Symposium Series, **131**, 69, 126-133 (1973)

Validation of VHTRC calculation benchmark of critical experiment using the MCB code

Przemysław Stanisz[1,a], Mateusz Malicki[1] and Mariusz Kopeć[1]

[1]*AGH University of Science and Technology, al. A. Mickiewicza 30, 30-059 Kraków, Poland*

Abstract. The calculation benchmark problem Very High Temperature Reactor Critical (VHTR) a pin-in-block type core critical assembly has been investigated with the Monte Carlo Burnup (MCB) code in order to validate the latest version of Nuclear Data Library based on ENDF format. Executed benchmark has been made on the basis of VHTR benchmark available from the International Handbook of Evaluated Reactor Physics Benchmark Experiments. This benchmark is useful for verifying the discrepancies in k_{eff} values between various libraries and experimental values. This allows to improve accuracy of the neutron transport calculations that may help in designing the high performance commercial VHTRs. Almost all safety parameters depend on the accuracy of neutron transport calculation results that, in turn depend on the accuracy of nuclear data libraries. Thus, evaluation of the libraries applicability to VHTR modelling is one of the important subjects. We compared the numerical experiment results with experimental measurements using two versions of available nuclear data (ENDF-B-VII.1 and JEFF-3.2) prepared for required temperatures. Calculations have been performed with the MCB code which allows to obtain very precise representation of complex VHTR geometry, including the double heterogeneity of a fuel element. In this paper, together with impact of nuclear data, we discuss also the impact of different lattice modelling inside the fuel pins. The discrepancies of k_{eff} have been successfully observed and show good agreement with each other and with the experimental data within the 1 σ range of the experimental uncertainty. Because some propagated discrepancies observed, we proposed appropriate corrections in experimental constants which can improve the reactivity coefficient dependency. Obtained results confirm the accuracy of the new Nuclear Data Libraries.

1 Introduction

The objective of the high-temperature gas-cooled reactor benchmark VHTR [1] is to determine the temperature coefficient of reactivity by measuring the temperature dependence of reactivity for the HTGR type reactors. The temperature in large commercial core can change largely between a cold shut-down state and a hot normal operation state. Thereby large change in temperature will impose significant change in reactivity. Proper estimation of the temperature reactivity coefficient is very important as it influences the designed shut-down margin and has significant consequences in economics. Moreover, presented benchmark considers the case with low-enriched UO_2 that fits in with IAEA strategy to use as a fuel low-enriched uranium (LEU) rather than high-enriched uranium (HEU).

2 Description of experiment

The VHTRC (Very High Temperature Reactor Critical Assembly) assembly is precisely described in „International Handbook of Reactor Physics Experiments Benchmark" [2] distributed by the NEA data Bank. Originally the experiment was located at the JAEA Tokai Research and Development Centre in Japan [3]. The core was designed as thermal, graphite moderated assembly. Each graphite rod is filled with approximately 20000 randomly distributed fuel capsules of the BISO-type The capsule consists of the centrally positioned fuel sphere (2 % and 4 % enriched UO_2 kernels, respectively) covered by two layers of pyrolytic carbon. Scheme of experimental system is shown on Figure 1. The core prism itself is made of two 2.4 m height hexagonal assemblies folded in 2.4 m long core. One assembly is always in fixed position and the second one is movable. Each assembly consists of smaller hexagonal blocks, except the side blocks of trapezoidal shape.

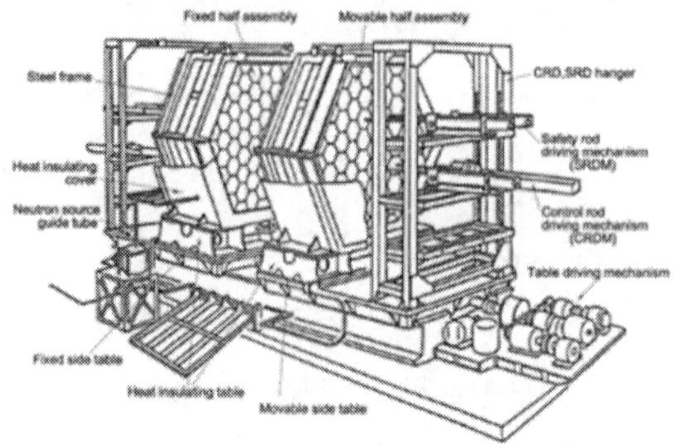

Figure 1. Overhead view of VHTRC

[a] Corresponding author: pstanisz@agh.edu.pl

The original experiment was designed in three different load patterns: HP, HC-1, HC-2. However, present work is focused only on the HP case shown in Figure 2. Core was made mostly with fuel rods called B-4 and one B-2. Difference between them is mainly weight enrichment, respectively 4 % and 2 %, but there are also small differences in size. Proposed benchmark based on experiment considers comparison of criticality in 5 temperature points. First criticality measurement was performed in 25.5°C. In the course of experiments the core was evenly wormed up by electrical heaters until almost 200°C. Further reactivity values were measured by pulsed neutron method for specific temperatures: 71.2, 100.9, 150.5 and 199.6°C.

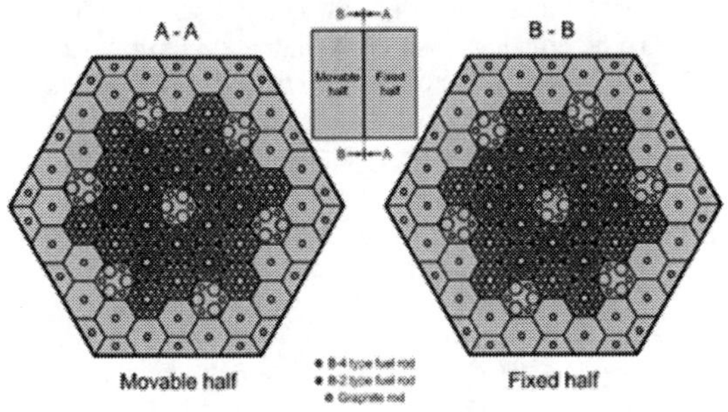

Figure 2. Core Loading Pattern of HP Core

3 Description of the calculation model

In order to simulate the benchmark conditions, a numerical representation of VHTR experiment was prepared for the MCB (ver. 5) Monte Carlo code used for calculations[4][5]. Two recent sets of nuclear data: American ENDF-B-VII.1 and European JEFF-3.2 cross section libraries were used. They represent the best available knowledge on the neutron interactions with matter and nuclide properties. It guaranteed the most precise results produced by the MCB code. Performed calculations consider three cases of successive problems that could have impact on obtained results: type of placement of the BISO fuel; temperature cross section applications and recalculation of variables used in presented benchmark.

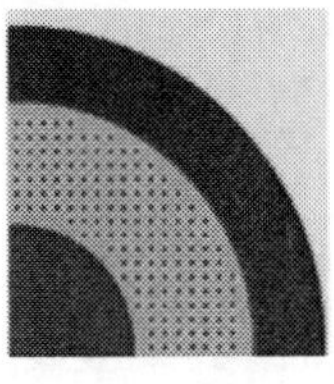

Figure 3. Horizontal sections quarter of or the fuel block with 4% enrichment, simple cubic.

MCB uses the same geometrical model as the well-known MCNP code. In the current form direct random placement of fuel kernels is impossible for large number of geometrical elements, however we show in the paper [6] that for models with high packing ratio (as in the considered problem) using random distribution of places could be easy and satisfactory changed using the URAN option. Because of that, for our investigation, two approaches have been introduced in order to understand analysed results:

– Cubic lattice without BISO particle touching the rod surface (*simple cubic*) – presented in Figure 3,

– Cubic lattice with additional 3D stochastic universe option enabled (*cubic URAN*) – described by Brown [7].

In case of the second mentioned problem we were faced with temperature interpolation. Typical cross sections are provided for standard temperatures: 300, 400, 500, 600, 700, 800, 900, 1000, 1200 and 1800 K. On the other hand, the experimental values were determined for very precise temperatures: 298.65, 344.35, 374.05, 423.65 and 472.75 K. Due to this fact we had to do some preparation. Generally there are few main approaches widely used for this:

- direct processing of nuclear data from evaluated nuclear data in the ENDF format;
- manipulation of cross sections library with Doppler broadening of the resolved resonances data and interpolate unresolved resonance probability tables to the new temperatures using higher and lower temperatures;
- calculation for the given library temperatures and interpolation between obtained results;
- mixing two materials of higher and lower standard temperatures and averaging the obtained results (linear or square root interpolation).

Only the first method assumes direct processing of data from the Evaluated Nuclear Data Files and converts them into suitable tables. The other three methods are based on using the ACE data files by MCNP and MCB. For the presented calculation we use the most convenient 2nd method for neutron interaction using the *makxsf* code delivered with MCNP6. Special treatment with 1st method was carried out only for cross section of the thermal neutrons scattering. In general, thermal cross sections are important in highly thermalized systems, where neutron scattering cross section values are additionally affected by the scattering from waves that are able to interfere with each other in a solid, liquid, or gas. Similarly may contribute the elastic and inelastic scattering of neutrons in materials. Generally additional effects could occur as neutrons give energy to material excitation or gain energy from it. These effects change reaction rates and modify the energy and angle distribution of scattered neutrons. Application of thermal scattering law is represented by two variables α – the momentum transfer and β – the energy transfer and is described by function $S(\alpha,\beta)$. This function is prepared within NJOY code [8]. The authors used the LEAPR module, based on General Atomics physics models, to obtain new ENDF files for thermal libraries. Once the data files were complete, they were used into ACE format to recompute the thermal scattering files for used temperatures.

4 Results

Application of broadening method performed by the *makxsf* program and use of the LEAPR module for thermal libraries in order to receive the $S(\alpha,\beta)$ thermal scattering cross section fulfil usefulness into benchmark problem, which results in good agreement concerning the temperature interpolation. Results obtained by the authors with generated cross sections – called AGH results - are compared with that of data sets available in the ACE format (NEA sets). All results are plotted in Fig. 4. Interpolated trend curve obtained from the JEFF-3.2 cross sections well overlap with curve from new calculations or produced points. Results are in good agreement. We assume that using the same methodology successful results will be obtained for the ENDF-B-VII.1 cross sections. In the same plot results from the benchmark are presented. Some discrepancy with the experiment is observed at higher temperatures resulting into discrepancies of thermal coefficient of reactivity. As a first approximation, we investigate this problem with simple parametric study.

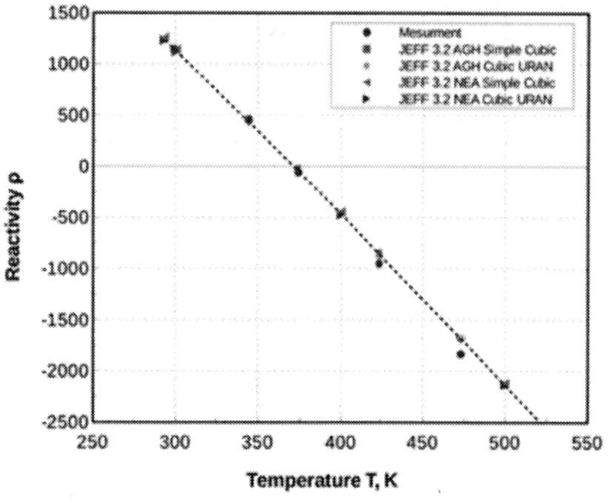

Figure 4. Reactivity change due to assembly temperature rise

In the following step, several parameters have been taken into account in order to see their impact on the thermal coefficient. These are successively: second set of cross section (ENDF-B-VII.1), random placement of the fuel particle through launching the URAN option in the simulation and the impact of distances between BISO particles. In the last approach we increase and decrease the distance by 2% with the same packing factor. Packing factor of the fuel kernels in considered geometry is very high. However, it is good to see how this parameter will change the results. All results are presented in Figure 5 where temperature dependent differences are compared as a function of reactivity (pcm – range) differences between benchmark results and our calculations.

Experimental uncertainty in the benchmark problem is around 300 pcm and our calculation has 6 pcm. The = discrepancy between two the farthest cases presented in Figure 5 has roughly 200 pcm. Thereby, all results show good agreement with the benchmark data. Numerical results present, on average, a 26 pcm decrease in URAN option compared to the Simple Cubic geometry. Differences between JEFF and ENDF nuclear data library, are not higher than 150 pcm. Interestingly, change (in the case of JEFF tables) of local pitch of fuel kernel by ± 2% results in bias of 100 pcm. This difference shows that geometry parameters are very viable for different variation of lattice even for high packing factor. The disparity of these results has the same order of magnitude as the differences between JEFF and ENDF data. Therefore, it is almost impossible to indicate which cross section set leads to better agreement with the benchmark problem. Similar problem was studied in our previous work [6] and we found its importance for burnup simulation. Decisively, none of used parametric improvements change the impact on the temperature coefficient. All trend lines have the same inclination. We propose a change of some parameter in order to improve thermal reactivity coefficient, comparing to original paper [4]. Authors of this paper noticed that a few constants (as beta effective or correction values), which are used for recalculation of the reactivity measurement, have been taken from calculations performed with the diffusion code CITATION [9]. To be consistent with only one methodology, in presented analysis, it will be more accurate to use only the Monte Carlo method to obtain all constants. We decided to simulate the experiment by using more suitable routine SOURCE available in MCNP/MCB. Next, Sjöstrand corrected method with spatial effects correction [10] was chosen to calculate area ratio obtained from MCB and recalculate measurement with appropriate correction factor. In original paper [4] β_{eff} was calculated by the CITATION code only for the lowest temperature and was used for all temperatures. It was decided in this work to replace β_{eff} by the values obtained for particular temperatures by the Monte Carlo calculated ones.

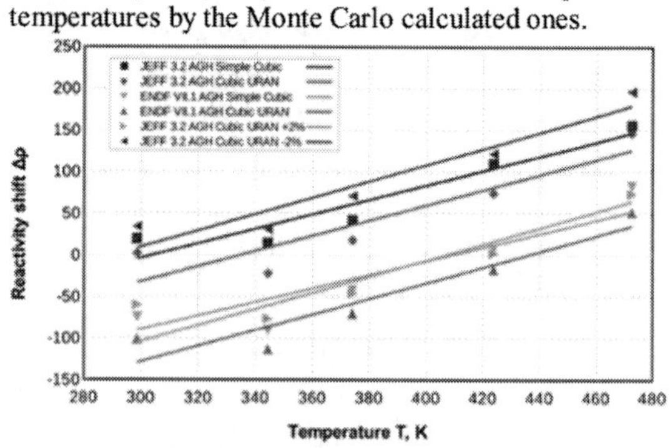

Figure 5. Difference of reactivity in the VHTRC model with various approaches (relative to the benchmark results)

In the original paper [4] areas obtained from BF_4 detectors is reported for four temperatures. The Sjöstrand method is strongly affected by spatial effects and corrections of the spatial effects should be done for each detector position. In our calculations we used correction factor (Eq. 1) obtained by the same Monte Carlo method, assuming relation between the expected value of measured reactivity in given detector position and the simulated global reactivity.

$$f_{cor} = \frac{\rho_{global,MCNP}}{\rho_{local,MCNP}} = \frac{\rho_{global,MCNP}}{A_{p,MCNP}/A_{d,MCNP}} \quad (1)$$

Assumption of the same relation in the experiment provides the corrected value of reactivity that is given by Eq. (2):

$$\rho_{cor} = f_{cor}\,\rho_{measur} \quad (2)$$

The effective delayed neutron fractions β_{eff} in VHTRC model is calculated by the MCB5 code using prompt transport method (Eqn. 3). This method needs two calculations: first, the normal run (considers prompt and delayed neutrons), and second, only with prompt neutrons:

$$\beta_{eff} \approx 1 - \frac{\chi_p \upsilon_p}{\chi \upsilon} \approx 1 - \frac{k_p}{k} \quad (3)$$

Differently than in original paper [4], we recalculated β_{eff} for each temperature point. The results of the measurement after applying different methods of correction are listed in Table 1.

Table 1. Benchmark and our values of β_{eff}

Temp. [K]	Original β_{eff}	MCB β_{eff}
298.65	0.007287	0.00710(7)
344.35	0.007287	0.00709(7)
374.05	0.007287	0.00711(7)
423.65	0.007287	0.00712(7)
472.75	0.007287	0.00703(7)

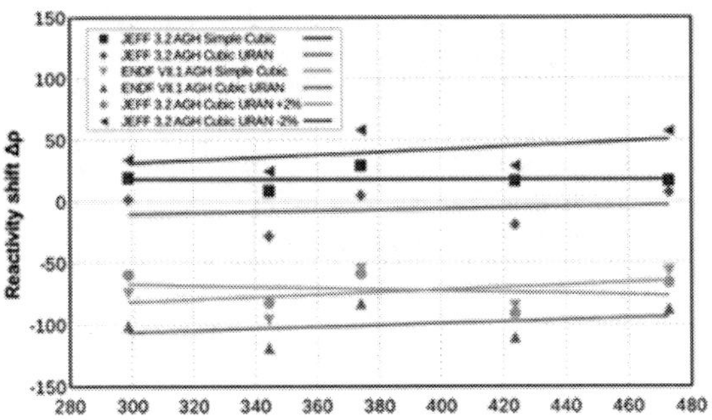

Figure 6. Difference of reactivity in the VHTRC model with various approaches (relative to the updated benchmark results)

5 Conclusions

In order to understand temperature effect on reactivity in thermal reactors, extensive benchmarking tests were undertaken to probe the quality of the new data library. As shown here, these studies have considered results of calculations based on the Monte Carlo method compared to results of the critical benchmark measurements. In general, in our calculations we used the most recent available nuclear data in all cases where the results stay within standard uncertainty. However a constant proportional difference in reactivity was observed through increasing the temperature. The performed parametric studies give explanation behaviour on reactivity. It show that in the investigated VHTR model the impact from geometry parameters, regarding fuel arrangement, introduce the same order of uncertainty as the use of different cross section libraries. However, this parametric studies did not improve the temperature reactivity coefficient dependency. It turns out that after recalculation of computable constant values used in the benchmark, we were able to show improvement in simulated experiment results eliminating temperature dependency. Our results suggest taking into account examination of the used computing parameter used in the benchmark measurement and recalculation final reactivity values.

Acknowledgments

This work was supported by the funds of the Polish Ministry of Science and Higher Education. This research was supported in part by PL-Grid Infrastructure.

References

1. NEA, "Temperature effect on reactivity in VHTRC-1 core," VHTRC-GCR-EXP-001, CRIT-COEF, NEA/NSC/DOC(2006)2, 2006.
2. NEA, *International Reactor Physics Experiment Evaluation (IRPhE) Project*, https://www.oecd-nea.org/science/irphe/irphe-handbook/handbook.html
3. T. Yamane, H. Yasuda, F. Akino, Y. Kaneko J. Nucl. Sci. Technol. **27**, 2 (1990)
4. J. Cetnar, *User Manual for MCB professional*
5. J. Cetnar, M. Kopeć, M. Oettingen, *Assessment of Pu and MA utilization in deep burn Prismatic HTR by Monte Carlo Method – MCB* - Project PUMA; http://www.kej.agh.edu.pl/publikacje/puma_kej1.pdf
6. G. Kępisty, P. Stanisz, J. Cetnar, Ann. Nucl. Energy. **92** (2016)
7. F.B. Brown, *Stochastic Geometry and HTGR Modeling with MCNP5*, Conference Paper. The Monte Carlo Method: Versatility Unbounded In A Dynamic Computing World
8. E. MacFarlane, *NJOY99 – code system for producing pointwise and multigroup neutron and photon cross-sections from ENDF/B data*, Los Alamos Nat. Laboratory, Los Alamos, NM, Rep. RSIC PSR-480, 2000.
9. T.B. Fowler, D.R. Vondy, G.W. Cunningham, *Nuclear Reactor Core Analysis Code CITATION ORNL-TM-2496*, Revision 2 (July 1971)
10. P. Gajda, J. Janczyszyn, W. Pohorecki, Nukleonika **58**, 2 (2013)

E3S Web of Conferences 10, 00124 (2016) DOI: 10.1051/e3sconf/20161000124
SEED 2016

Approach to novel design of CO_2 based centrifugal compressor

Tomasz Kura[1,a] and Elżbieta Fornalik-Wajs[1]

[1]*AGH University of Science and Technology, Department of Fundamental Research in Energy Engineering, al. Mickiewicza 30, 30-059 Cracow, Poland*

Abstract. Even though turbomachinery design issues have been investigated almost since the beginning of engineering, its optimization process is still important. With the development of refrigeration devices and ORC based distributed generation facilities, a need for efficient and low-energy compressors and turbines became even more demanding. Such machines working with typical fluid, like air, are well described, but there is a room regarding the fluids like CO_2, vapour of organic fluids, etc. The main objective of present studies is to propose a numerical model of the centrifugal compressor, with CO_2 as the working fluid. Such unit may be a part of refrigeration cycle. Commonly, the scroll or piston compressors are used in such cases, however some discussed disadvantages show that the novel designs should be looked for. Properly designed centrifugal compressors can have higher efficiency than the presently used. Three dimensional analyses of proposed geometries were conducted – using a model including heat, mass and momentum conservation laws as well as ideal gas law. Verification of the proposed mesh and results was performed in the basis of values obtained using theoretical and empirical equations. With about 700 000 control volumes in the validated model, error of the results was no higher than 5%, with only about 1% in regards to the thermal parameters. Two design proposals were analysed, with performance maps as the main comparison factor. Apart from performance characteristics, the pressure and velocity fields were presented, showing the process of flow structure optimization. The main goal was to reduce negative effects of pressure and velocity gradients on the performance. Proposed precursory design might be a good starting point for further development of compressors. The results of numerical analysis were promising and shows the possibility of proposed design usage in practical applications, however to obtain deep understanding of the problem, the experimental analysis should be also done.

1 Introduction

One of the most important issues of modern world is to develop sustainable energy systems, that would work in spite of circumstances outside the system. Heat pumps, solar collectors, photovoltaic systems – they all are parts of such trials. More and more advanced usage of low-temperature heat sources became one of the breakthroughs that led to development of distributed energy generation. However it became clear for scientists and engineers, that water as the working medium, present in almost all big-scale energy systems, would not be suitable for many proposed solutions. Commonly, when it comes to small generation or refrigeration issues, there can be just one or only a few fluids, that suit particular design of machine. Best way is to construct operating maps for analysed devices, to achieve full image of the problem and the best solution [1]. Scroll or piston compressors are being used in such cases more often, but due to some slight drawbacks, such as worse steering sensibility or need to be oiled [2], researchers looked for the possibility of centrifugal compressors usage.

Not only type of machine, but also working fluid were a matter of investigation. For proper selection of the working fluid thermo-physical properties are considered, but chemical stability in desired temperature range or safety should also be taken into account. These issues are important from the application point of view, because small-scale systems can be installed directly in households [3]. In ORC systems, organic fluids are used, namely isobutene, n-pentane aromatic hydrocarbons or similar. All of them have properties, that make them more efficient in low temperature ranges than water, especially in the superheating area of phase diagrams. These properties play major role in the design of expanders [4], but since ORC, heat pump and similar systems consist also of compressors, proper desing of them is also very important.

In the paper, analysis of radial compressor's rotor, with various design proposals, is presented. Carbon dioxide was chosen as the working fluid. Couple of articles or dissertations on the implementation of that fluid in turbomachinery were published in previous years, for example [5, 6]. Monje in [5] presented comprehensive analysis of particular model of CO_2 compressor designed for supercritical Brayton cycle. Only numerical results were included based on one case. For purposes of that work, Monje prepared literature review – and stated, basing on it, the necessity of further exploration, as CO_2 behaviour in turbomachinery is promising. Therefore there

[a] Corresponding author: kura@agh.edu.pl

is still a need for further compressor development and it becomes an aim of this paper.

2 Mathematical model

Analysed flow was three-dimensional, stationary and compressible. The conservation laws of mass, momentum and energy coupled with ideal gas law were applied in the model. Continuity equation was considered in following form [7]:

$$\frac{\partial}{\partial x_j}(\rho U_j) = 0, \tag{1}$$

where ρ is the fluid density, kg/m^3, U is its velocity, m/s. Momentum equation was represented by [7]:

$$\frac{\partial}{\partial x_j}\left(\rho U_i U_j\right) = -\frac{\partial p}{\partial x_i} + \frac{\partial}{\partial x_j}\left(\tau_{ij} - \rho\overline{u_i u_j}\right) + S_{Cor} + S_{cfg} \tag{2}$$

where p is the pressure, Pa, τ is the molecular stress tensor, S_{Cor} is the source term accounting for the effect of Coriolis force and S_{cfg} is the centrifugal force source. Energy equation was applied in the form [7]:

$$\frac{\partial}{\partial x_j}\left(\rho i_{tot} U_j\right) = \frac{\partial}{\partial x_j}\left(\lambda\frac{\partial T}{\partial x_j} - \rho\overline{u_j i}\right), \tag{3}$$

where T is the temperature, K, λ is the fluid thermal conductivity, W/(m·K) i_{tot} is the specific total enthalpy, J/kg i is the specific static enthalpy, J/kg and k is the turbulence kinetic energy, J/kg. ANSYS 14.5 software, particularly CFX and Turbogrid modules, were chosen to achieve results. CFX uses coupled solver to conduct the simulation, which main advantage is the computing speed, but significant memory resources are needed, because all equations are being determined at the same time. High resolution upwind advection scheme was applied, second order, to avoid numerical diffusion and therefore to get more certain results. Also, SST k-ω turbulence model was chosen, due to some advantages [7] over other two-equations models. Comparing this model with typical k-ε or k-ω models, SST k-ω, being the combination of both, gives more reliable results. It is suggested for simulation of high pressure gradients or flow separation areas [7] – those phenomena occur in turbomachinery flows, with high rotational speeds. Moreover SST k-ω model allows obtaining feasible near-wall results. Research was based on the comparison of rotor's blade performance when compressing air and CO_2. Therefore, two working fluids were used in the model. The air had density value of 1.225 kg/m^3, the thermal conductivity – 0.0242 W/m·K, the viscosity value of 1.79E-05 kg/(m·s) and the specific heat – 1006.4 J/(kg·K) in standard conditions, while CO_2 had density of 1.7878 kg/m^3, the thermal conductivity – 0.0145 W/m·K, the viscosity value of 1.37E-05 kg/(m·s) and the specific heat – 840.37 J/(kg·K) [7].

Performance maps were constructed for varying rotational speed of the rotor. Speeds were chosen in the basis of calculation procedure presented in [8], to avoid off-design conditions and non-reliable results. Rotation occurred around the axis shown in Figure 1, together with an indicated direction. Initial temperature in the whole passage as well as the blade was 288 K. Also inlet pressure remained constant and was equal to 1 bar. For outlet boundary condition, "supersonic" type was chosen, as the proper one for simulations with possible compressibility effect [7]. As all analysed flows had the Mach number at the inlet above 0.3, that effect was expected. That boundary condition demand no other parameters to be defined.

All implemented inlet boundary conditions are presented in Table 1.

Table 1. Boundary conditions at the inlet.

Rotational speed, 10^3 rpm	26	28	30	32	34	40
Temperature, K	288					
Pressure, bar	1					

3 Studied cases and applied solutions

Two different blades designs were proposed. Both of them based on the data from literature [8] but modified to fit the needs of presented approach.

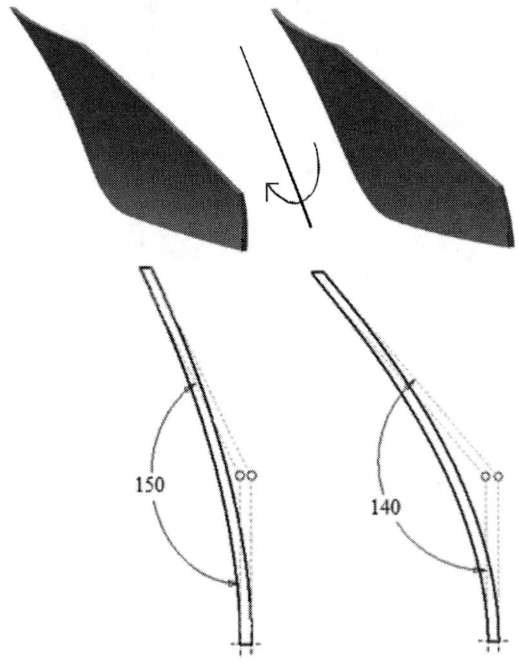

Figure 1. Blade 1 (B1) – up left, and blade 2 (B2) – up right. Visible angles of curvature and schematic view of blades shape, B1 – down left, B2 – down right.

One passage per simulation was discretized since the software is capable to interpolate the results for whole device. It led to reduction of computer resources necessary for analysis. Figure 1 presents schematic view of studied blades. Rotors consisted of 15 blades per machine. Both blades had the same length, around 10 cm, and height, varying from 5 to 2 cm. The only difference between two blades was the angle of curvature along the blade height. For blade 1 (B1) it was 150° and for blade 2 (B2) 140°. Mentioned angles for blades are also presented in Figure 1. Ansys software is based on finite elements method. Meshing and Turbogrid modules were used to generate mesh. Hexahedral elements were generated in the majority of the model. Their number was high enough to maintain elements shape close to cube in areas of high curvature, which is important for gaining proper results and avoiding artificial gradients of calculated parameters. In Table 2, number of nodes and elements for both types of blades is listed. Figure 2 shows generated mesh, for blade B1. Refined mesh near the blade can be noticed, as shown in Figure 3. Setting an approximate Reynolds number value to be 0.5E07, y+ values did not exceed 2.

Table 2. Number of nodes and finite elements.

Blade	B1	B2
Nodes	739 860	699 825
Elements	694 235	693 676

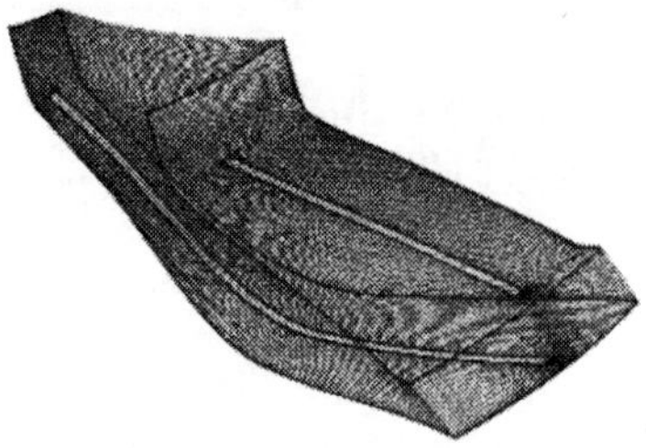

Figure 2. Mesh for the passage with B1 blade. Direction of rotation.

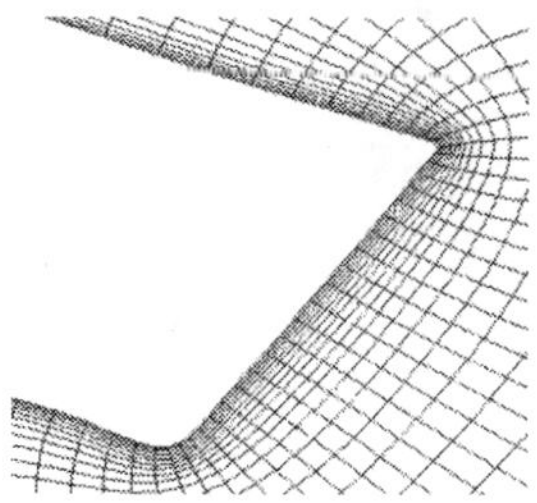

Figure 3. Near-blade mesh refinement.

Verification of mesh was conducted through the comparison with theoretically obtained results, according to methods suggested in [8]. In Table 3, conformity of results is presented. Values in the second column are the differences between theory and numerical analysis with the proposed mesh. Coherence was higher than 96%, therefore the mesh was considered fine enough to continue the studies. Increase in the number of elements did not change the conformity significantly, therefore around 700 000 elements was kept for the whole analysis. Static and total pressure, used for verification purposes, came directly from Bernoulli's equation. Total value is the sum of its dynamic and static part. Temperature was evaluated basing on the same manner but also with use of equation of state.

Table 3. Verification of the proposed mesh.

Variable	Coherence, %
Static pressure	95.8
Total pressure	94.4
Static temperature	99.6
Total temperature	99.1
Density	96.6

4 Results and discussion

The main objective of presented studies was to propose numerical model of CO_2 compressor, which would be competitive to other available devices. The compression ratio (CR) can be an indicating factor for comparison between particular units since it is one of the main parameters describing device performance. It is defined as:

$$CR=\frac{p_{t,out}}{p_{t,in}}, \quad (4)$$

where $p_{t,in}$ is the total pressure at the inlet and $p_{t,out}$ is the total pressure at the outlet. Both pressures were calculated using area integral. Total pressure is a sum of static and dynamic pressures and needs to be calculated in the cases with high speeds of the flow, where dynamic parameters have significant values.

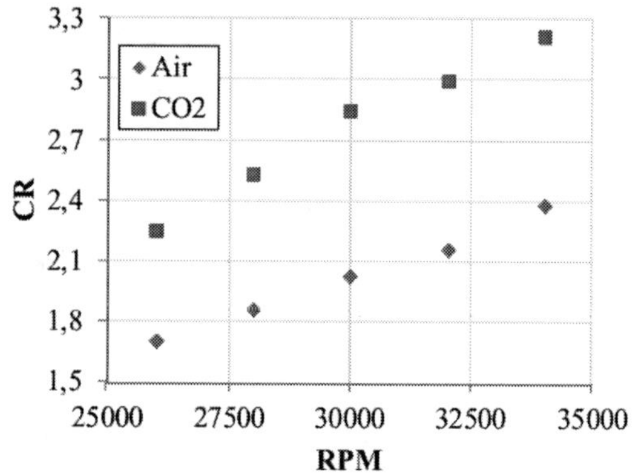

Figure 4. Compression ratio vs rotational speed, B1 design.

The comparison of CR versus rotational speed for two fluids and blades B1 and B2 is presented in Figures 4 and 5, respectively. In both blade cases CR was higher for CO_2 of about 25 to 30%. Moreover the difference was increasing with an increase in rotational speed. Higher values of CR could be noticed for B1 design in almost whole studied range, except the highest analysed rotational speed for which B2 could generate the highest value of CR.

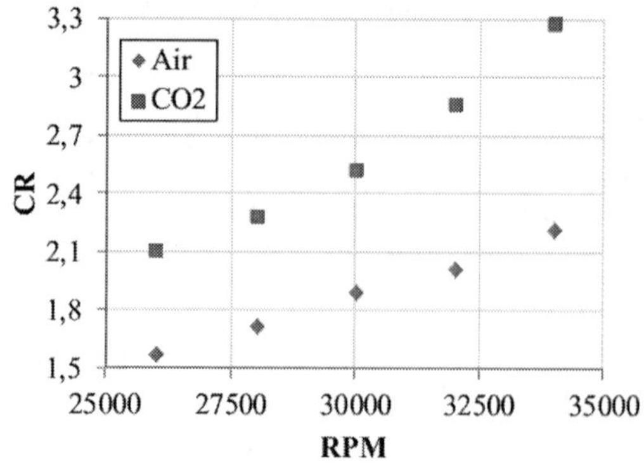

Figure 5. Compression ratio vs. rotational speed, B2 design.

As previously mentioned, compressibility effect was expected. To confirm that, Mach number, defined as the ratio of local fluid velocity to the local speed of sound, was calculated in the passages. Figure 6 presents its values for the case with 30 000 rpm and B1 design. Mach number was higher than 0.3 in almost whole passage, what confirmed the usage of supersonic boundary conditions, mentioned in previous section with mathematical model.

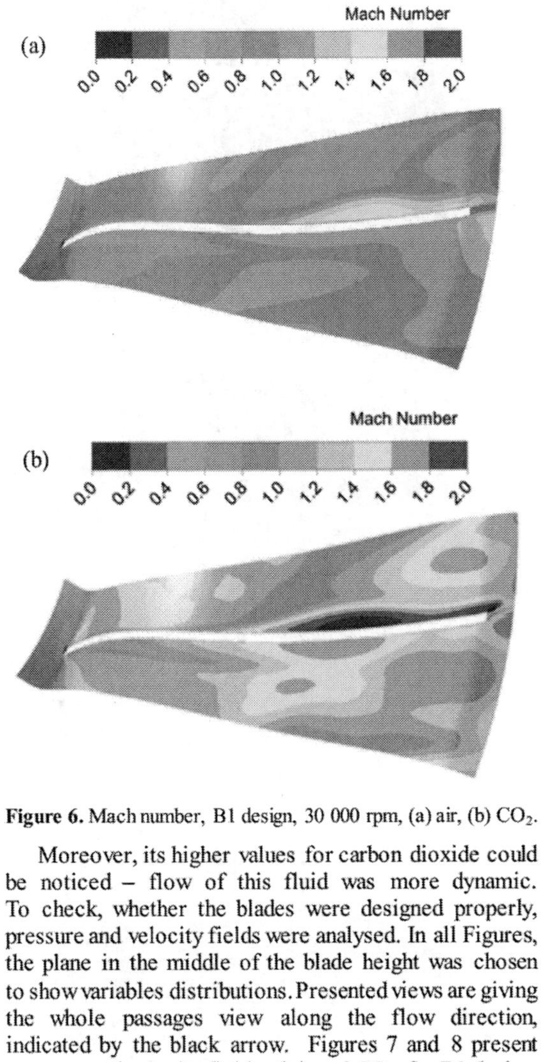

Figure 6. Mach number, B1 design, 30 000 rpm, (a) air, (b) CO_2.

Moreover, its higher values for carbon dioxide could be noticed – flow of this fluid was more dynamic. To check, whether the blades were designed properly, pressure and velocity fields were analysed. In all Figures, the plane in the middle of the blade height was chosen to show variables distributions. Presented views are giving the whole passages view along the flow direction, indicated by the black arrow. Figures 7 and 8 present pressure and velocity fields of air and CO_2, for B1 design, at 30000 rpm. The lowest pressure was observed for CO_2, even though the flow of both fluids was highly irregular. Higher velocity could be noticed for CO_2 case. Figures 9 and 10 present the same variables, but for B2 design, and only for CO_2 fluid, for comparison purposes. Analysing B1 and B2 designs (Figures 7, 8, 9, 10), the biggest influence of the shape on the pressure could be observed. Red arrow in Figure 9 indicates low pressure area in B2 blade's passage, that may be one of the reasons of lower CR values for that design. In Figures 11 and 12, CO_2 pressure and velocity fields for B1 and B2 case and 34000 rpm are presented. With visible irregularity of pressure distribution, it can be stated, that due to it, CR value for B1 is lower than in the case of B2. It can suggest, that B2 design is more suitable for higher rotational speed.

Table 4 presents isentropic efficiencies calculated for all simulations, which are counted as ratio of actual work in the machine to the corresponding work in reversible process (see Equation 5). Even though values for air were higher, it did not correspond with CR values. Explanation of this situation is connected with power needed to compress fluids.

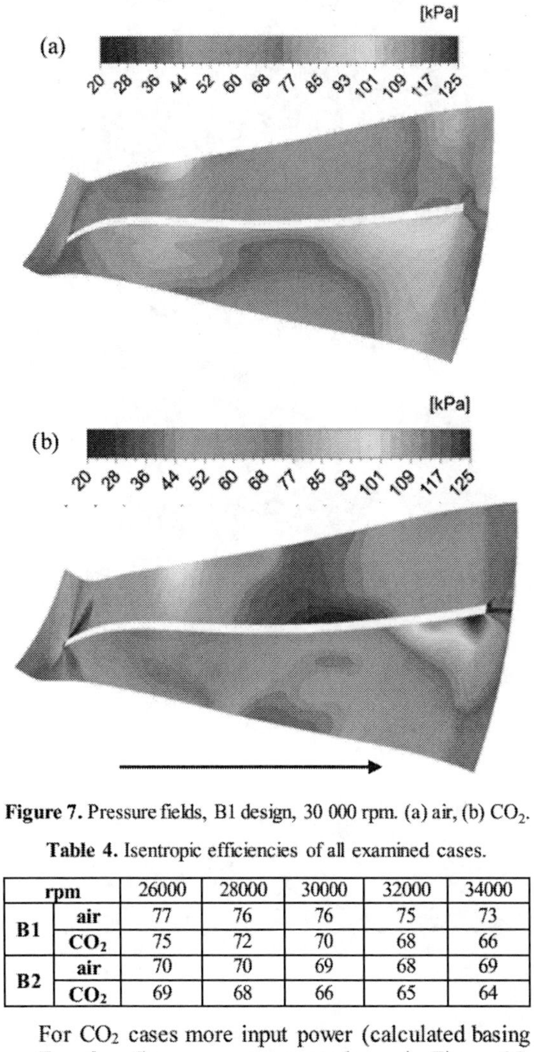

Figure 7. Pressure fields, B1 design, 30 000 rpm. (a) air, (b) CO_2.

Table 4. Isentropic efficiencies of all examined cases.

rpm		26000	28000	30000	32000	34000
B1	**air**	77	76	76	75	73
	CO_2	75	72	70	68	66
B2	**air**	70	70	69	68	69
	CO_2	69	68	66	65	64

For CO_2 cases more input power (calculated basing on Equation 6) was necessary, as shown in Figure 13, mostly due to its higher density. That impacted the values of isentropic efficiency, which is defined as:

$$\eta = \frac{H_{2s}\text{-}H_1}{H_2\text{-}H_1}, \tag{5}$$

where H is the total enthalpy, J. Indexes mean: 1 – the inlet, 2 – the outlet, s – the total enthalpy value in isentropic process.

Input power, W, was defined for steady process, as:

$$IP = H_2\text{-}H_1, \tag{6}$$

Work is connected with fluids enthalpy changes more than pressure, what explains the discrepancy between CR and efficiency changes.

DOI: 10.1051/e3sconf/20161000124

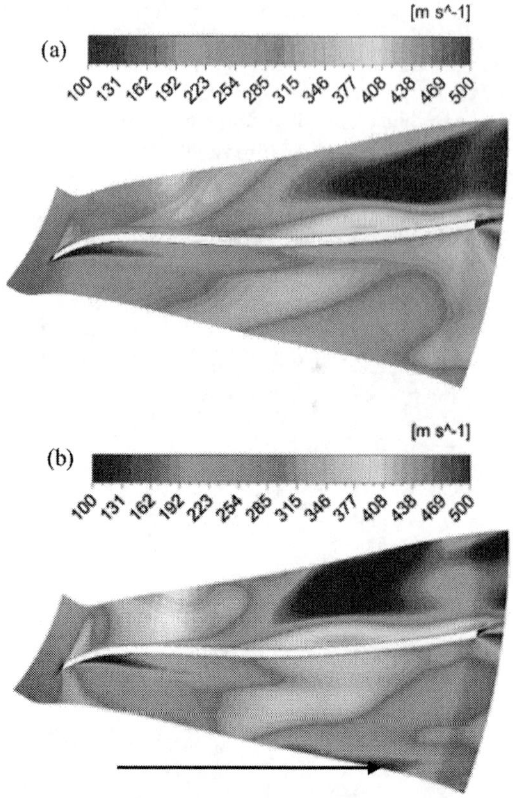

Figure 8. Velocity fields, B1 design, 30 000 rpm, (a) air, (b) CO_2.

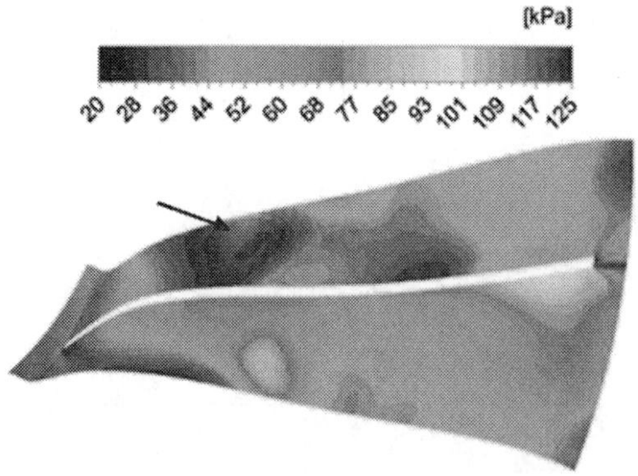

Figure 9. Pressure field, B2 design, 30 000 rpm, CO_2.

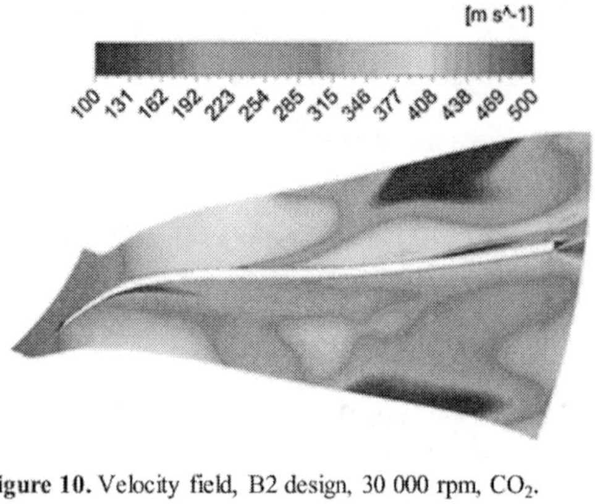

Figure 10. Velocity field, B2 design, 30 000 rpm, CO_2.

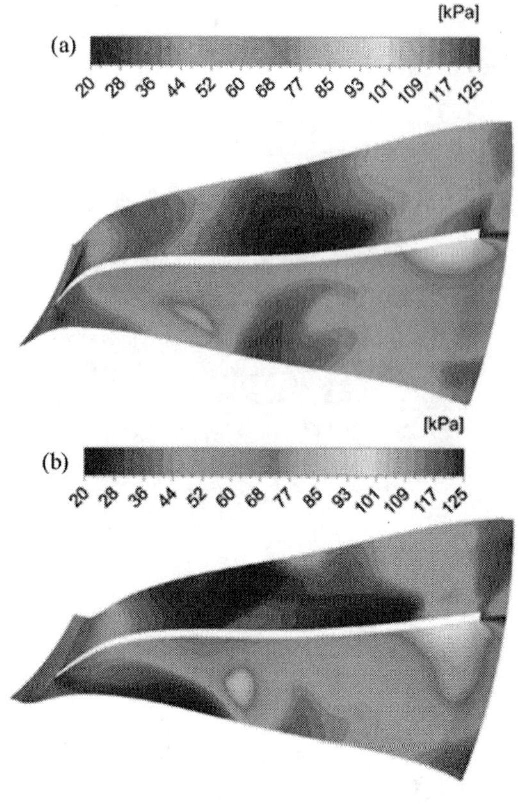

Figure 11. Pressure fields, 34 000 rpm, CO_2, (a) B1, (b) B2.

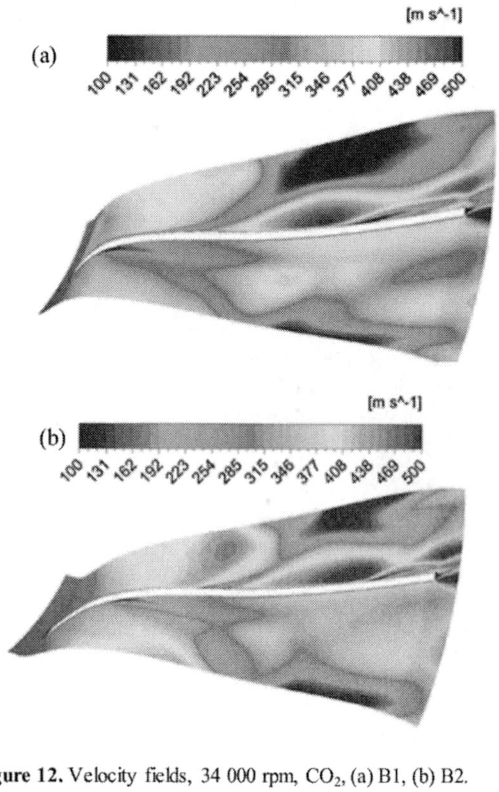

Figure 12. Velocity fields, 34 000 rpm, CO_2, (a) B1, (b) B2.

DOI: 10.1051/e3sconf/20161000124

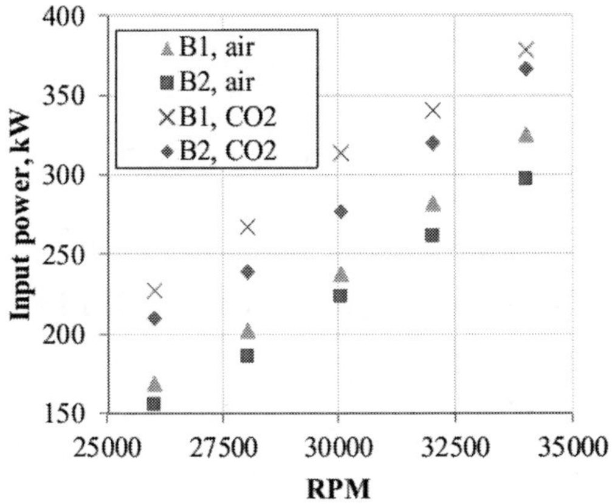

Figure 13. Input power comparison.

5 Summary

Radial compressor's blade designs were investigated, with various rotational speeds, to check the difference in the flow behaviour between two fluids – CO_2 and air. As CO_2 is thought to be one of the possible fluids to be used in ORC or refrigeration systems, its behaviour should be precisely investigated. For proposed compressors, CO_2 turned out to give more effective pressure output in whole examined range.

In overall, B1 design turned out to be better, with higher CR, which was the main factor for comparison. The only disagreement with that relation occurred for CO_2 compression at 34000 rpm – possible answer for that situation is, that the required power input for blade B2 is lower than for B1. Although it has to be investigated, it did not impacted main conclusion and was left for future research.

Further examination needs to be undertaken to develop adequate model of CO_2 compressors. The experimental data would be necessary for the proper image. However it may be concluded, that presented results construct the basis for further development as some important issues characterizing the flow have been described and analysed.

Acknowledgements

The present work was financially supported by the Polish Ministry of Science and Higher Education (Grant AGH No. 15.11.210.344) and by AGH UST Vice-Rector for Student Affairs, Professor Anna Siwik. Present work was also supported by Academic Computer Centre CYFRONET AGH, under award no. MNiSW/IBM_BC_HS21/AGH/021/2015.

References

1. S. Quoilini, S. Declaye, A. Legros, L. Guillaume, V. Lemort, Int. Compressor Eng. Conf. at Purdue, **2012**, 1546 (2012)
2. T. Turunen-Saaresti, P. Roytta, J. Honkatukia, J. Backman, Int. J. of Refrigeration, **33**, 1152 (2010)
3. H. Chen, D. Yogi Goswami, E. K. Stefanakos, Ren. and Sust. En. Rev, **14**, 3059 (2010)
4. L. Moroz, C. Kuo, O. Guriev, Y. Li, B. Frolov, Proceedings of the ASME Turbo Expo 2013: Turbine Technical Conference and Exposition, **5A** (2013)
5. B. Monje, *Design of supercritical carbon dioxide centrifugal compressors* (PhD thesis, Universidad de Sevilla, 2014)
6. B. Monje, D. Sanchez, R. Chacartegui, Proceedings of the ASME Turbo Expo, **3**, 405 (2012)
7. *ANSYS® CFX-Solver Theory Guide 2013* (ANSYS Inc., Cecil Township, 2013)
8. A. Witkowski, *Sprezarki wirnikowe Teoria, konstrukcja, eksploatacja* (wyd. Politechniki Slaskiej, Gliwice, 2013)

Sustainable design in the contemporary architecture of tall buildings

Joanna Gil-Mastalerczyk[1,a]

[1]*Kielce University of Technology, Faculty of Civil Engineering and Architecture, Al. Tysiąclecia Państwa Polskiego 7 , 25-314 Kielce, Poland*

Abstract. The tall buildings in Singapore - one of the main financial centers of the world - reflect the metropolitan character of the city. These buildings emphasize the importance of the center in the urban landscape. They determine and create a new landscape of the city. Objects, mainly about the function of commercial and prestigious, distinguished by the natural environment-friendly image. Through financial expenditures, they excel in the implementation of innovative solutions in the field ecology. The applied design, facade systems and functional systems are subject to the use of renewable energy sources. Promoting the idea the development of the city "inside". Ecological aspect is the durability of the structure. Modern tall buildings have completely changed the image of the spatial part of the city-state of Singapore and contributed to enhance the individual character and highlight the specifics of the layout and position of the center (combining functions of business, commercial, recreational and residential).

1 Introduction

The twenty-first century is characterized by significant transformations of civilization, social, economic, political, information and rapid development of any innovative technology. According to the professionals - in the field of architecture and urban planning – *also century unknown new urbanism from nature, the protection of cultural values, the introduction of new forms of residence into a sustainable housing environment. (...) the era of man-friendly functional and spatial transformations within urban structures (...) different perspective on issues related to the evolution of the city and a variety of new forms of residence to meet all the expectations of man as a result of new current challenges, eg. due to significant climate change, protection of the natural environment (...)*[1].

The architecture and the construction are the greatest economic sector in economic aspects and in terms of the flow of raw materials. In the sector of the construction a huge amount of the financial capital is being invested as well as natural. Therefore the role of architecture and the construction in creating the sustainable development is very important.

The assumption of sustainable development should take into account three main priorities with each other in balance: ecology (environmental aspect), economy, society and its culture.

In relation to the architecture - sustainable development - the first time presented in 1998 in GÄVLE and published in Agenda 21 - (document entitled "Sustainable buildings") [2].

The contemporary architecture newest high buildings is the answer and present the result of creative searches, offered modern technology, and sometimes experimental solutions - enabling use on a large scale renewable energy sources.

Designed buildings are not only visually stunning but also built with high performance and sustainable functionality in mind. Innovative building design and responsible construction practices minimize the immediate impact associated with the extraction, manufacturing, and transportation of building materials, and provide the foundation for conservation of natural resources during the years of operation. It is important, to embed sustainable practices in the early stages of design and throughout construction.

The author uses example of modern architecture high buildings in Singapore, presents new trends in the architectural form (sustainable architecture) - in relation to the environment - in harmony with nature and the surrounding landscape.

The modern building construction Singapore completes the network - designed a grand scale - road connections. On a world scale a road infrastructure is singling itself out - bridges and walking-bicycle footbridges. The public transport support 4 metro lines - classified in the world to the most automated. Singapore City authorities attach great importance to the development of green areas. Tropical and humid climate favors the vegetation lush vegetation in the many parks that permeate the urban areas. Under the Governmental program, transforming Singapore into a garden city in 2012 created the world famous gardens Gardens by the

[a] Corresponding author: joanna.gil@onet.pl

Bay ". It became a place of social integration and the area of highest recreational activities

2 Sustainable design in the architecture of tall buildings

The sustainable architecture is one that is carried out in accordance with the principles of sustainable development. The strategy of sustainable development has three dimensions - environmental protection, social inclusion and economic growth.

Sustainable architecture should take into account the whole range of natural and civilization. The process of architectural design should respect the general rules of sustainable design. These include social, cultural and aesthetic. The rules of sustainable design include energy efficiency, renewable energy and the 3R rule - saving and reuse area, volume, materials, energy, water and waste. It should also be noted that the idea of sustainable development into account aspects related to the architectural design. In particular - is associated with a respect for the place - its natural conditionality and the local building tradition. It is also associated with a respect for the user - including aspects of health, social contacts and a sense of security [3].

Architecture designed buildings should be primarily "a response to the means employed in relation to the challenges of the powerful forces of nature" [4].

These measures should be a component of many factors, including but not limited investor's expectations and vision of the designer. Natural conditions impose different possibilities and solutions bioclimatic. It should also take into account an important humanistic context. The use ecological solutions with local materials is justified by rules of energy efficiency and respect for the place [3].

Generally, it must be found that the next idea of sustainable development is rational and context-sensitive approach to architectural design.

2.1 Modern solutions - for example Singapore

City-state Singapore is located in Southeast Asia, near the southern tip of the Malay Peninsula. It is a small island in the Strait of Malacca (an area of about 650 km^2 and a coastline of about 138 km), surrounded by a team of 56 smaller islands (the area around the archipelago is about 697 km^2) [5].

A result of work carried out on the acquisition of land from the sea area of Singapore is constantly increasing. The original island was overgrown with vegetation equatorial - due to its location in over 137 km north of the equator, the climate zone humid, tropical climate and rainforests.

Currently, the city-state Singapore is one of the most developed and richest countries in Southeast Asia and the world[a]. It is inhabited by over 5 million people. By the end of 2019 years is expected to exceed 6.5 million people.

Lack of space has become an investment for the city one of the main barriers to development. A significant part of the investment is localized under the surface - deep in the ground.

Like most big cities, life in Singapore takes place partly underground. Under the ground are located metro, multi-level shopping centers and highways.

The new resort set on the coast, in the south-eastern part of Singapore, was a complex of luxury high building. It combines the features of business, commercial, recreational and residential buildings. It is now one of the largest financial centers, commercial and banking world.

The complex newest buildings of the city is made up of high buildings and multi-storey buildings, towers commercial, service, residential and underground parking. The towers are equipped with numerous green vertical gardens. In addition to high usability, they represent about ecological character of this part of the city.

2.1.1 CapitaGreen, Singapore, designer Toyo Ito & Associates (Figure 1-4)

"This office tower presents fresh opportunities to develop a city centre which is at the cutting edge of sustainable design. It is a green building both physically and environmentally (...)" - Mr Tan Chuan-Jin – Minister of State for National Development and Manpower [8]

CapitaGreen is an iconic landmark on the skyline of Singapore with its striking sculptural petalled funnel on the crown of the building and its impressive façade of greenery. Standing tall at 245 metres, building offers approximately 702,000 square feet of premium Grade A office space - level 3 to 36.

The development will feature specially commissioned art installations by internationally acclaimed artists on level 1 and exceptional facilities such as an exclusive club on level 38, and a sky forest and restaurant on level 40. Located in the heart of Singapore's Central Business District, CapitaGreen have close access to the station Raffles Place and Telok Ayer MRT stations [9].

CapitaGreen is an ecological building with its environmentally-friendly features and a green 'living' façade (Figure 1-2).

With the inclusion of numerous state-of-the-art energy-saving features, the completed office tower is designed to achieve the Green Mark Platinum award, the highest accolade in recognising environmentally-friendly buildings from the Building & Construction Authority of Singapore [10].

---

[a]Singapore flesh out state model of development based on a highly centralized, hierarchical institutions. In the city-state created one of most favorable environments for the development of competitive companies. Specific feature of the model in Singapore has become the realization of selective industrial policy, aimed at promoting strategic sectors of the future, the acquisition of foreign capital, expertise and technology. As a result of increased participation of Singapore in global trade, especially exports of services, goods, high technology and rapid economic growth.
In addition, improving efficiency and stimulating innovation translates simultaneously increasing the attractiveness of a location for foreign direct investment in the area of high technology. These changes additionally enhance national long-term competitiveness by increasing the attractiveness of the location for inflows of high-tech foreign direct investment [6].

 DOI: 10.1051/e3sconf/20161000125

Figure 1. CapitaGreen (2012-2014), Architect Toyo Ito & Associates Architects. Innovatively shaped architecture of the buildings with the introduction of greenery and trees in the entire layout, Singapur (photo by autor, 2016).

Figure 2. CapitaGreen (2012-2014), Architect Toyo Ito & Associates Architects. Innovatively shaped architecture of the buildings with the introduction of greenery and trees in the entire layout, Singapur (photo by autor, 2016).

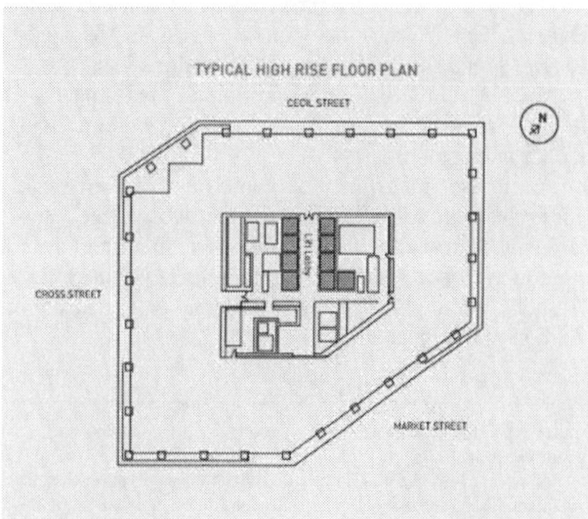

Figure 3. CapitaGreen (2012-2014), Architect Toyo Ito & Associates Architects. Typical high rise floor plan (source: [9]).

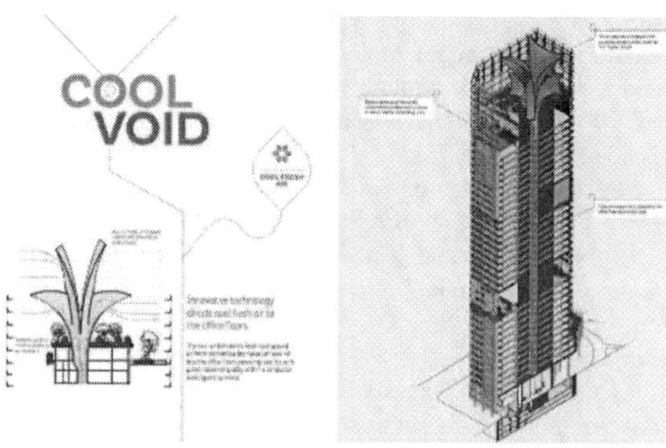

Figure 4. CapitaGreen (2012-2014), Architect Toyo Ito & Associates Architects. Scheme air flow throughout the building, effectively regulating the building's temperature (source : [8]).

This is articulated through the use of extensive sky gardens on three levels and Sky Forest on the roof, which forms part of a unique air circulatory system that Arup designed [10] (Figure 4).

Installed with a wind scoop, the air circulatory system, much like that of a living plant's, draws cool air from the sky forest where tropical trees are planted. The air is then delivered through the 'Inner Tube', an empty central core, to each floor. Overall energy consumption is significantly lowered as a result [10].

A cool void at the centre core of the property permits air flow throughout the building, effectively regulating the building's temperature [8].

CapitaGreen houses a large expanse of greenery. Apart from the sky forest on level 40, there are also sky terraces on levels 5, 14 and 26, and greenery on every floor, allowing tenants greater proximity to and intimacy with nature [9].

To irrigate the building's plants, it will harness innovative water-saving techniques such as rainwater harvesting.[8]

Air-conditioning for the office floors is provided via innovative technology which takes in fresh cool natural air from the crown of CapitaGreen at lower energy consumption, providing tenants with better indoor air quality [9] (Figure 4).

Each floor features generous 3.2m floor-to-ceiling heights to provide uninterrupted views of Marina Bay and the rest of the CBD. A double glazed facade system reduces solar heat gain, while an accompanying green façade behind it is designed to lower surface temperature, diffuse sunlight and provide beautiful dappled natural light into the building. The result is a well-lit and cool indoor environment that tenants and visitors can be in [10].

In meeting CapitaGreen's high design standards, Arup's team of vertical transport specialists designed a premium grade, energy-efficient lift service that integrates a strategic selection of destination control service and regenerative drive technologies. These greatly benefit the building's crowd management as tenants and visitors are brought to their destinations in the most time and energy-efficient way possible [10].

2.1.2 Marina Bay Sands, Singapore, designer Moshe Safdie (Figure 5-7)

"Marina Bay Sands is really more than a building project, it is a microcosm of a city rooted in Singapore's culture, climate, and contemporary life" – architect Moshe Safdie [11].

The complex of commercial and prestigious tall buildings is located at the "Marina Bay Sands". The building complex is characterized by an innovative functional and spatial large urban complex. An individually shaped architecture (with vertical layout) distinguishes the each on.

The object Marina Bay Sands was designed by world-renowned architect - Moshe Safdie for the Las Vegas Sands Corporation (cooperation: James Carpenter, Antony Gormley, Ned Kahn Sol Lewitt and Chongbin Zheng). Building has 929.000 square meter (10 million square-foot)) [b] [11].

The building Marina Bay Sands was designed in a way that uses the principles of sustainable development. The objective of sustainable development has become the growth prosperity and the harmonious arrangement of the relationship between man and nature (Figure 5-7).

In the architecture are used a different system energy efficiency of the building. The environmental impact of building operation lifetime is much higher than the construction industry. Implementation of responsible operating practices reduces the consumption of natural resources and as a result improves the comfort of the guests and team members.

It is important to embed sustainable practices in the early stages of design and throughout construction.

During the construction and renovation the unique design features of the environment, the optimization of mechanical systems, select sustainable materials and recycling construction waste were used.

Building Marina Bay Sands was designed to introduce abundant natural daylight to illuminate indoor areas, reducing the need for additional lighting. Specially designed convex and concave roofs reduce heat gain into the podium interior. Sun shading across the glass façade in the form of balconies, vertical fins, and canopies were installed as well [12].

The basis for action has become a quest for continuous improvement, regularly examine and test new products to stay at the cutting edge of technological innovation in the industry.

In the building regenerative drives on elevators, which require 40% less energy than comparable non-regenerative elevators were used.

---

[b]*This new urban place integrates the waterfront promenade, a 74,000 square meter (800,000 square-foot) multi-level retail arcade, and the iconic Museum of ArtScience on the promontory. Located along the network of public paths are also two theaters with a combined 4,000 seats, a casino, a 9,000 square meter (96,000 square-foot) convention and exhibition center, and a hydraulically adjustable public event piazza of 5,000 square meters (54,000 square feet). Combining indoor and outdoor spaces and providing a platform for a wide array of activities, this vibrant, 21st-century cardo maximus, or grand arcade, also connects to the subway and other transportation. A series of layered gardens provide ample green space throughout Marina Bay Sands, extending the tropical garden landscape from Marina City Park towards the Bayfront. The landscape network reinforces urban connections with the resort's surroundings and every level of the district has green space that is accessible to the public. Generous pedestrian streets open to tropical plantings and water views. Half of the roofs of the hotel, convention center, shopping mall, and casino complex are planted with trees and gardens. Three 55-storey hotel towers anchor the district and are connected at the top by the 1 hectare (2.5-acre) SkyPark. An engineering marvel 200 meters (656 feet) above the sea, the SkyPark spans from tower to tower and cantilevers 65 meters (213 feet) beyond. It accommodates a public observatory, gardens, a 151 meter-long (495 foot-long) swimming pool, restaurants, and jogging paths and offers sweeping panoramic views, a formidable resource in a dense city like Singapore. Shielded from the winds and lavishly planted with hundreds of trees, the SkyPark celebrates the notion of the Garden City that has been the underpinning of Singapore's urban design strategy* [11].

DOI: 10.1051/e3sconf/20161000125

Figure 5. Marina Bay Sands (2009-2011), architect Moshe Safdie (photo by autor, 2016).

Figure 6. Scheme strategy (Marina Bay Sands). The Sands Eco360° strategy consists of four pillars: Green Buildings, Environmentally Responsible Operations, Sustainable Meetings and Stakeholder Engagement (source: [13]).

Figure 7. Marina Bay Sands (2009-2011), architect Moshe Safdie, the view from the 'Gardens by the Bay' (photo by autor, 2016).

The escalators at the mall retail and convention center are designed to stop when it is not in use and start when it detects incoming people [12]

In addition energy, water and resource conservation projects, is an extensive waste management strategy, including food waste diversion, to help to improve waste reduction, diversion and total environmental impact. In convention centre there are over 250 rooms. A special air conditioning system works in a building - air conditioning is turned on only during meetings [12].

Global sustainability programme, Sands ECO360°, is the ethos that guides the company in responsible operations, reducing our carbon footprint and overall environmental impact. Sands ECO360° integrates the company's best practices, technologies and methodologies in the area of sustainability across all properties. It consists of four priorities: Green Building, Environmentally Responsible Operations, Green Meetings and Sustainability Education and Outreach [12] (Figure 6).

The building is constructed with high performance and sustainable functionality in mind. Innovative building design and responsible construction practices minimize the immediate impact associated with the extraction, manufacturing, and transportation of building materials, and provide the foundation for conservation of natural resources during the years of operation [12].

The attractive location allows to make custom solutions functional and spatial and urban infrastructure.

In spite of dense and steadily expanding development (Marina Bay Sands Singapore), development of the area are interesting and attractive solutions. There are places for rest and recreation equipment. There are walking paths, communication a coastal, marina, cycle routes, waterways.

Intensive building Marina Bay Sands contrasts with the lush by greenery squares, parks, botanical gardens. Located near the 'Marina Boulevard' world-famous gardens 'Gardens by the Bay' are the site of frequent cycling and walking. Above the gardens dominates the Singaporean group super-trees (Figure 7). These are extraordinary, multi-functional structures - referring to a form of real trees. Their height is 25 to 50 m. Super trees are vertical space breeding many species of plants. It provide communities the visiting shadow in their environment and are a kind of environmental engines, producing electricity (photovoltaic). Also collect rain water (irrigation system and fountains) and are part of the cooling systems.

The whole assumption is distinguished by appropriate shaping of composite architecture. The developed spatial concept, high-density building, was located over the water. The area is a kind a edge of the city and in contact with water - rich natural area and the coast.

The convenient location allows to make custom solutions functional and spatial and urban infrastructure. Despite the dense and steadily expanding development, development of the area are interesting and attractive solutions.

Intensive building contrasts with the lush greenery of numerous squares, parks, botanical gardens. Located near the 'Marina Boulevard' gardens 'Gardens by the Bay' area of over 100 ha are the site of frequent cycling and walking.

A method of forming architecture addressing a client's needs, their lifestyle and climatic conditions. Style features surround of the buildings correspond to the priorities of sunshine, in terms of the location of the mutual, proper shading and ensure high environmental values of work and life.

All objects distinguishes the natural environment-friendly image. Through a considerable financial effort, excel in the implementation of innovative solutions in the field of ecology. Applied design, facade systems and functional systems are subject to the use of renewable energy sources. Promoting the idea of development of the city "inside". It reduce the absorption of natural areas. Ecological aspect is the durability of the structure.

Forms of skyscrapers (with different bases) form polyhedra arranged next to one another. Standing close together, are a group of towers emphasizing reach metropolitan area - the coastal part of the city. Dynamic and light forms - on the shoulders tapering - form the structure of a vertical city. It is possible here to find the idea of the sustainable development. Contemporary tall buildings have completely changed the image of surround this part of town. Contributed to enhance the individual character and highlight the specific nature of the facility and its location. Emphasize the importance of the center in the urban landscape and define and create a new landscape of the city.

3 Conclusions

The overall conclusion is that sustainable, energy-efficient architecture of modern tall buildings must be reasonable. Climatic factors and increasing requirements for energy efficiency in buildings will be determined by formal solutions and functional space.

Presented solutions design tall buildings are characterized by innovative and experimental development of buildings and urban social space - in relation to nature. The natural landscape, the assumptions on water and recreation areas.

Original architectural forms are distinguished by extraordinary visual effects and aesthetic - in symbiosis with nature and landscape. This is characterized by the search for new ideas and visions to shape the environment to live and work.

The search for new forms of architectural related to the environment. As a result, design solutions in innovative shaped vertical structure of urban development (commercial and prestigious), will contribute to improvement of the work environment, house a living environment and the quality of architectural space.

Generally it should be noted that the presented solution is an example of sustainable design in the architectural and urban planning and architectural scales. Urban context has played a significant role in shaping the architectural because "there is no architecture without urban planning and urban design without the architecture".

Design and realization of architecture objects promotes the idea of "saving the city." The aspect "savings" in the consumption of energy, water, natural resources, reduce waste, and preference for green transport.

In summary that contemporary architecture and sustainable building (for example, tall buildings of Singapore):

- create a safe and healthy environment to live and work,
- does not pollute the environment,
- generates low maintenance,
- gives users the satisfaction staying in it and a pleasure to contact passers and the public,
- the modern building represents a technical level corresponding to the needs of society and civilization, reflects the social and cultural aspirations of society,
- responds to the needs of the market supply and demand.

This is confirmed by numerous research programs (created in the nineties the twentieth century.), Focusing on issues of sustainable development. Their objective is to test systems for assessing and improving the quality of buildings in terms of environmental impact (These are e.g.: LEED (Leadership in Energy and Environmental Design, USA), GBC (Green Building Challenge, Europe, Japan, Canada, USA), BREEAM (Building Research Establishment Environmental Assessment Method, United Kingdom, BEPAC (Building Environmental Performance Assessment Criteria, Canada)).

Sustainable architecture designed and implemented in accordance with the standards of sustainable development has a positive impact on the environment - helping to reduce consumption of natural resources and reduce environmental degradation (environmental benefits). Sustainable architecture has a beneficial effect on the health and safety of users - contributes to improving the comfort, health and safety of users. It contributes to improve the quality of life (social benefits). It has an impact on increasing the earned benefits and profits (economic benefits).

References

1. W. Seruga, *Nowe formy zamieszkania w zrównoważonym środowisku mieszkaniowym/ New forms of residence in a sustainable housing environment* [w:] Środowisko mieszkaniowe/ Housing Environment, W. Seruga (red.), 12/2013, pp. 114-133 (Wyd. Politechnika Krakowska, Kraków, 2013).
2. Agenda 21 on Sustainable Construction, CIB Report Publication 237 – ISBN 90-6363-015-8, Rotterdam.
3. B. Ostrowska, *Zrównoważony dom jutra a lokalna tradycja/ Tomorrow's Sustainable House and Local Tradition* [w:] Środowisko mieszkaniowe/ Housing Environment, W. Seruga (red.), 12/2013, pp. 34-41 (Wyd. Politechnika Krakowska, Kraków, 2013).
4. A. Cząstka, *Architektura a natura. Problem mimesis w architekturze,* p. 163 (Wyd. Politechnika Krakowska, Kraków, 2007).
5. K. Kubiak, *Twierdza, której nie było* (Wyd. Książka i Wiedza, 2010).
6. M. Żmuda, *Źródła przewagi konkurencyjnej małej gospodarki otwartej w dobie powiązań globalnych – Singapur w latach 2007-2012/ Sources of Competitive Advantage of a Small Open Economy in the Era of Global Interconnections – Singapore in the Years 2007-2012,* Zeszyty Naukowe Uniwersytet Ekonomiczny w Krakowie, 9(945), pp. 60-72 (Kraków, 2015).
7. W. Seruga, *Jedność architektury z otoczeniem/ The Unity of Architecture and its Surroundings*, Czasopismo Techniczne Architektura/ Technical transactions Architecture, 3-A/2011, Z. 12, rok 108, Issue 12, Year 107, p. 133-145 (Wyd. Politechnika Krakowska, Kraków, 2011).
8. http://capitagreensingapore.com, (July 2016).
9. http://www.cct.com.sg/system/misc/Portfolio_CapitaGreen.pdf, (July 2016).
10. http://www.arup.com/projects/capitagreen (July 2016).
11. http://www.msafdie.com/file/1698 (July 2016).
12. https://www.marinabaysands.com (July 2016).
13. http://www.marinabaysands.com/content/dam/singapore/marinabaysands/master/main/home/company-information/environmental-sustainability/eco_mbs-green-facts.pdf, (July 2016).

Modelling of the work processes high-pressure pump of common rail diesel injection system

Katarzyna Botwinska[1,a], Remigiusz Mruk [2] and Łukasz Krawiec[3]

[1,2,3] *Warsaw University Of Life Sciences, Faculty of Production Engineering, Department of Production Organization and Engineering, Nowoursynowska 164, Warsaw, Poland.*

Abstract. Common rail injection systems are becoming a more widely used solution in the fuel systems of modern diesel engines. The main component and the characteristic feature of the system is rail injection of the fuel under high pressure, which is passed to the injector and further to the combustion chamber. An important element in this process is the high-pressure pump, continuing adequate pressure in the rail injection system. Common rail (CR) systems are being modified in order to optimise their work and virtual simulations are a useful tool in order to analyze the correctness of operation of the system while varying the parameters and settings, without any negative impact on the real object. In one particular study, a computer simulation of the pump high-pressure CR system was made in MatLab environment, based on the actual dimensions of the object – a one-cylinder diesel engine, the Farymann Diesel 18W. The resulting model consists of two parts – the first is responsible for simulating the operation of the high-pressure pump, and the second responsible for simulation of the remaining elements of the CR system. The results of this simulation produced waveforms of the following parameters: fluid flow from the manifold to the injector [m^3/s], liquid flow from the manifold to the atmosphere [m^3/s], and manifold pressure [Pa]. The simulation results allow for a positive verification of the model and the resulting system could become a useful element of simulation of the entire position and control algorithm.

1 Introduction

Nowadays, diesel engines are becoming more and more popular; modern diesel drives combine both driving dynamics and economy, which for many vehicle users is a key advantage. It is also the dominant power unit in the heavy vehicles sector, vehicle fleets, and machinery and equipment. It is also possible to achieve stricter emission reduction targets in terms of CO_2 into the air [1] with these engines. To optimise the operation of these engines and their emissions requires constant modifications and design improvements. Experimenting with real objects is usually very costly, time-consuming, and can result in damage or total destruction of the object. Therefore, nowadays computer simulations of the work of individual components as well as whole appliances are becoming increasingly popular. This allows you to estimate the functioning of the device at the given parameters without the risk of damage, enabling you to choose the best working configuration and implement it directly into the system, omitting the options that do not meet the requirements or expectations [2]. Because of the adjustable parameters of the injection system, and the resulting higher efficiency, power output, and lower noise and emission of the engine, CR injection systems are more commonly found in diesel engines. Their structure is based on the injection common rail - a rail wherein fuel is supplied by high-pressure pump from a fuel tank. The pump provides the appropriate parameters of fuel pressure, which travels to the injectors via the rail and further to the combustion chamber. The injection rail keeps the fuel pressure constant until the injection. Due to the high operating pressure, together with a short injector opening time (milliseconds), the whole structure requires enormous precision and proper control. The cost of experiments on such sensitive systems is high, limiting the possibility of system failure [2,3].

The present study used simulation models to analyze the work of the high-pressure pump as an important component of the CR system, which provided the operating parameters of the system and is a useful tool in the process of designing and planning new systems.

1.1 The aims of the study

The aims of the study were:

1. To create a simulation of the processes occurring in the working fluid in a high pressure system in an engine with the common rail fuel system, based on a test stand to study the combustion process at the Department of Production Engineering at Warsaw University of Life Sciences (made and used by development project no. R10 037 03 Topic: "Application rapeseed crude oil as a fuel for diesel engines of tractors and agricultural vehicles");

[a] Corresponding author: katarzyna_botwinska@sggw.pl

2. To create a simulation which will be useful as software for a control as well as collecting research data to study the combustion process.

1.2 Scope of the study

The scope of the study included:

• Description of the overall construction of the Common Rail with special attention to the conditions of liquid flow in the elements of the fuel feed pump, high-pressure lines, and manifold and the processes of mechanical work elements;

• The mathematical description of the flow of liquid through the above-indicated part of the high pressure system;

• The construction of the simulation model (simulation is not in real time) and analysis of the work of selected parts of the high pressure system and correct operation of control algorithms.

2 Materials & Methods

The simulation model was made in the Matlab Simulink environment. To build the model, the parameters of the real object - Farymann Diesel 18 W engine – were used. This is a single cylinder, four-stroke diesel engine used most often as drive for simple machines, e.g. Power generators and pumps. Below is a real object with the main elements:

Figure 1. View of the engine - the real object with the main elements [(A) – Engine, (B) – Synchronous generator (brake), (1) – the exhaust pipe , (2) – bellows expansion joint, (3) – lock gas sampling (analyser chemical composition) (4) – sluice of thermocouple, (5) – silencers]

The key element for the simulation model part of the engine was the common rail injection system. Relevant for the simulation were:

• Common Rail system with Bosch CR / V4 / 10 -12S injector - a popular storage tank used in the CR systems of four-cylinder car engines. This rail allows connection of 4 injectors, however, due to the use of a single-cylinder, only one injector (also Bosch) was included;

• High-pressure pump Bosch model CR / CP1H3 – this provides the appropriate pressure up to a maximum of 150 [MPa]. Due to the reliability of the measurement results, the pump is supplied from an external source, controlled by an inverter.

The system with the main components is presented on Figure 2.

Figure 2. A view of Common Rail injection system in object [(1)– storage tank, (2) – CR injector electronically controlled].

The scheme of the high-pressure pump is presented in Figure 3.

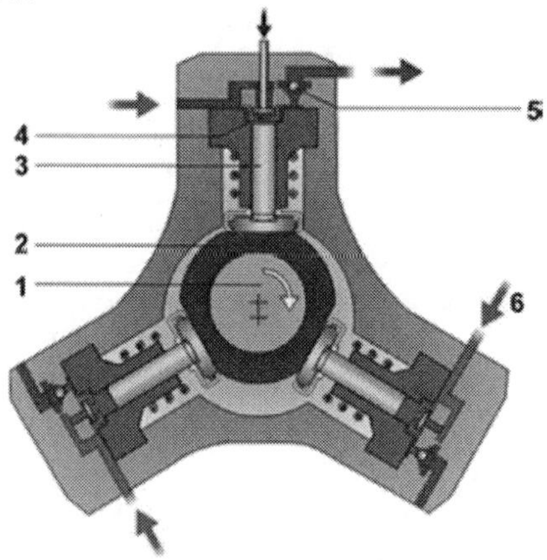

Figure 3. Diagram of the high pressure pump CP1 type [1) Drive shaft, 2) eccentric cam, 3) pressuring section, 4) inlet valve, 5) outlet valve, 6) fuel inlet], (source:http://www.wtryskiwacz.com/jakie-mamy-pompy-wysokiego-cisnienia-Delphi-a-jakie-Bosch-naprawa-regeneracja.html).

To build the model the following assumptions were adopted [4,5]:

• Fuel is a compressible liquid with elastic modulus E and is subject to the Hooke law;

• The elastic deformations of the injection pipe caused by changes in fuel pressure were ignored;

• The flow of fuel in the injection pipe is treated as a one-dimensional movement;

• The flow of fuel in the injection pipe is isothermal;

• The impact of friction is taken into account as are the movements of the elements of the injection system due to the effect of inertia, and damping forces of the springs and the fuel pressure of the system;

• Nominal pressure in the manifold is $135 \cdot 10^{6}$ [Pa]

• Pump flow is equal to $2 \cdot 10^{-2}$ [m^3/s];

• The diameter of the high-pressure line is equal to $2 \cdot 10^{-3}$ [m];

• The length of pipe from the pump to the collector is $450 \cdot 10^{-3}$ [m] ;

 DOI: 10.1051/e3sconf/20161000126

• The length of pipe from the manifold to the channel inlet valve is $1 \cdot 10^{-3}$ [m];
• The length of pipe from the collector to the injector is equal to $300 \cdot 10^{-3}$ [m];
• The density of the fuel is constant and is equal to 0,8247 =$0{,}8247 \cdot 10^{3}$ [kg/m³];
• Atmospheric pressure (p_{atm}) is constant and is equal to $1013{,}25 \cdot 10^{2}$ [Pa];
• Kinematic viscosity of diesel oil was assumed to be constant (independent of temperature) at $3 \cdot 10^{-6}$ [m²/s];
• Radius of the ball on the valve head is equal to $5 \cdot 10^{-3}$ [m];
• The mass of the valve head is equal to $5 \cdot 10^{-3}$ [kg];
• The spring constant is equal to 2500000 [N/m];
• The volume of the manifold is equal to $29 \cdot 10^{-6}$ [m³];
• The modulus of elasticity of the liquid (Ep) are taken as a constant value independent of the temperature and equal to $1441 \cdot 10^{6}$ [Pa].

Next, block diagrams [6,7,8,9,10] were constructed. To reflect the real operation of the system, a detailed model of the CR high-pressure pump was taken into account which consisted of parts listed below as Models to describe the movement of the piston pumping section. Below, the calculation diagram for the pumping section of the high-pressure CR pump is shown.

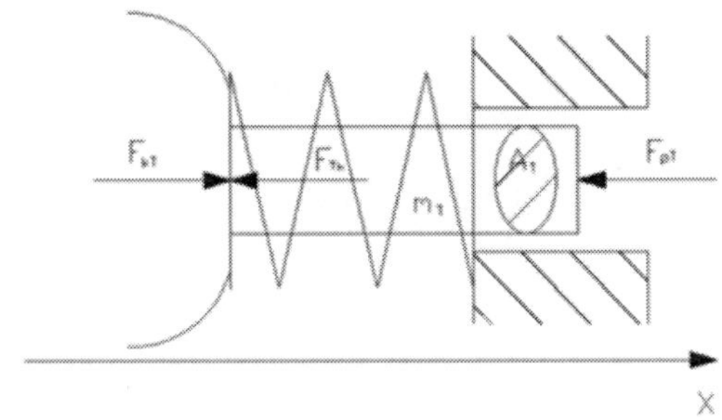

Figure 4. The calculation diagram of pumping section of the high pressure CR pump.

The movement of the piston pumping section was described by the mathematical model as follows:

$$X_k = (\cos \omega t)e + r \tag{1}$$

$$m_t \frac{d^2 x_t}{dt^2} = F_{kt} - F_{pt} - C\frac{dx_t}{dt} - F_{st} \tag{2}$$

$$F_{pt} = P_p A_t \tag{3}$$

$$F_{kt} = \begin{cases} 0: X_t - X_k \geq 0 \\ (X_t - X_k)K_t: X_t - X_k < 0 \end{cases} \tag{4}$$

where:
m_t – the mass of the piston,
F_{kt}- force on the piston from the cam,
F_{pt}-force of the fluid pressure,
F_{st}-spring force,
C – loss factors.

The force acting on the cam is calculated based on the instantaneous position of the piston and cam height with regard to the elasticity of the material. Then, using the Matlab Simulink environment, the simulation model was built to describe the movement of the piston pumping section, which can be seen below.

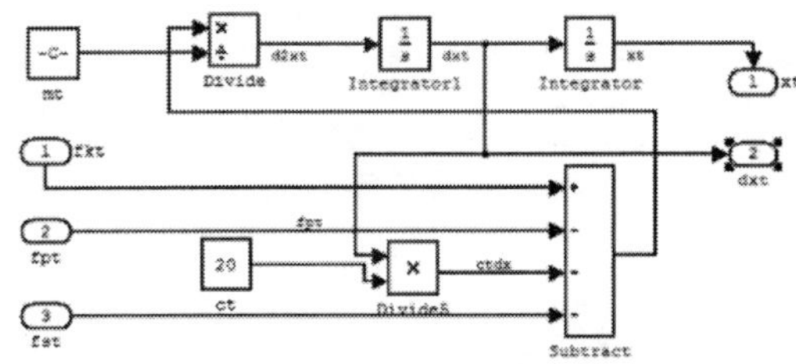

Figure 5. The simulation model to describe the movement of the piston pumping section.

Next, models of the high-pressure chamber of the CR pump were built. The calculation diagram for the high-pressure chamber of the CR pump which was used is shown below.

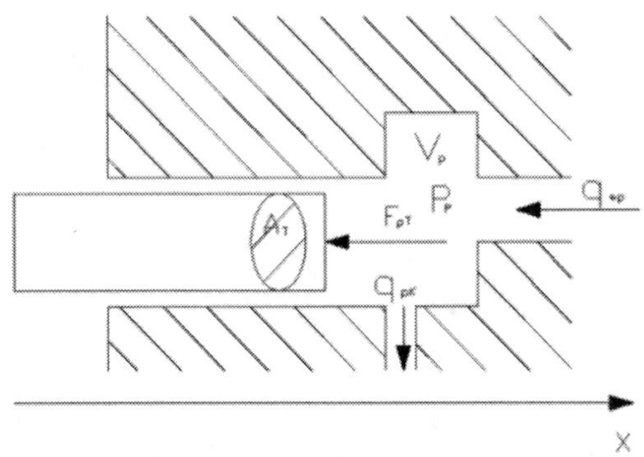

Figure 6. The calculation diagram of high pressure chamber of the CR pump.

The high-pressure chamber of the CR pump was described by the mathematical model as follows:

$$\frac{V_p}{E_p}\frac{dp_p}{dt} = q_{wp} - q_{pk} + A_t\frac{dx_t}{dt} \tag{5}$$

where:
V_p – volume of the chamber,
p_p- chamber pressure,
q- liquid streams,
X_t- the position of the piston,
A_t – cross-sectional area of the piston.

Using the Matlab Simulink environment, the simulation model for the high-pressure chamber of the CR pump was built which can be seen below.

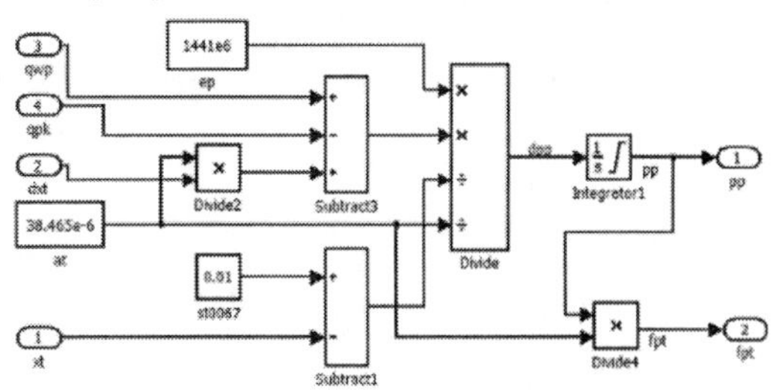

Figure 7. Simulation model of the high pressure chamber of the CR pump.

Another section was the mathematical model of the ball valve, the calculation diagram is shown below.

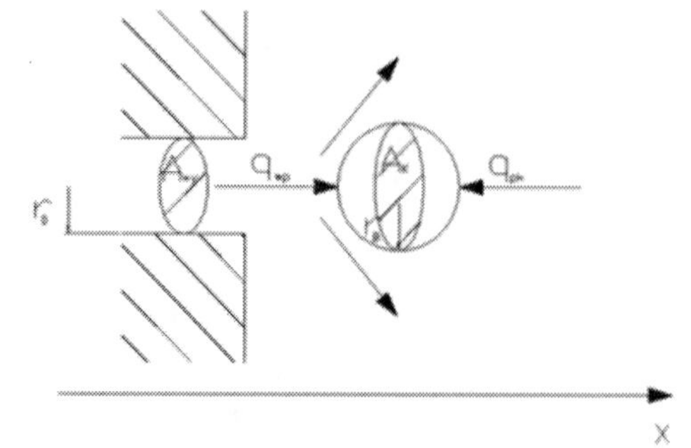

Figure 8. The calculation diagram of ball valve.

The ball valve was described by the mathematical model as follows:

$$m_k\frac{d^2X_k}{dt^2} = F_{kwy} - F_{wek} + F_{0k1} - F_{0k2} - C\frac{dx_4}{dt} \quad (6)$$

$$F_{kwy} = A_{kul}P_{kom} \quad (7)$$

$$F_{wykul} = A_{kul}P_{wy} \quad (8)$$

$$F_{0kul1} = \begin{cases} 0: X_k \geq 0 \\ -K_kX_k: X_k < 0 \end{cases} \quad (9)$$

$$F_{0kul} = \begin{cases} O: X_k + P_k \leq 5 \\ K_{kul}(X_k - P_k): X_k + P_k > 5 \end{cases} \quad (10)$$

where:
m_k – ball mass,
F- forces acting on the ball,
C- loss factor,
X_k- the position of the ball.

Then, using the Matlab Simulink environment, the simulation model of the ball valve was built, which can be seen in Figure 9.

Mathematical models of the plate valve were also built and the calculation diagram of the plate valve is shown in Figure 10.

The final part of this section is a comprehensive model of the high-pressure pump. Below, the complete mathematical model to describe the movement of the piston pumping section is shown.

$$m_z\frac{d^2X_z}{dt^2} = F_{0zp1} - F_{0zp2} + F_{cpz} + F_{wez} - F_{pz} \quad (11)$$

$$F_c = -C_{pz}\frac{dx_{pz}}{dt} \quad (12)$$

$$F_{wezp} - F_{pzp} = \left(P_{we} - P_p\right)A_p \quad (13)$$

$$K_{0zp1} = \frac{E\left(A_{zp} - A_{we}\right)}{l_{zp}} \quad (14)$$

$$F_{0p1} = \begin{cases} 0: X_{zp} \geq 0 \\ -K_{zp}X_{zp} - C_{0zp}\dfrac{dx_{zp}}{dt}: X_p < 0 \end{cases} \quad (15)$$

$$F_{0p2}\begin{cases} 0: X_p - 1 \leq 0 \\ X_p - 1 > 0: \left(X_p - 1\right)K_{0zp2} \end{cases} \quad (16)$$

$$q_{PK} = C_D\pi D_{pz}X_p\sqrt{\frac{2\left(P_{we} - P_{wy}\right)}{\rho}}sign\left(p_{we} - p_{wy}\right) \quad (17)$$

where:
m_z –weight of the valve plate,
F- forces acting on the plate,
C- loss factor,
X_k- position of the plate.

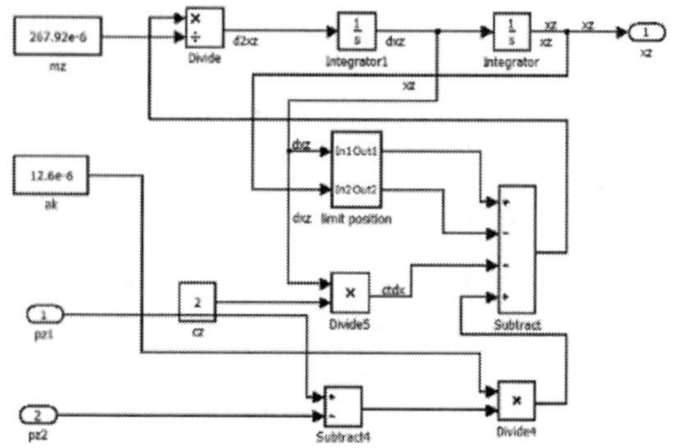

Figure 9. Simulation model of the ball valve.

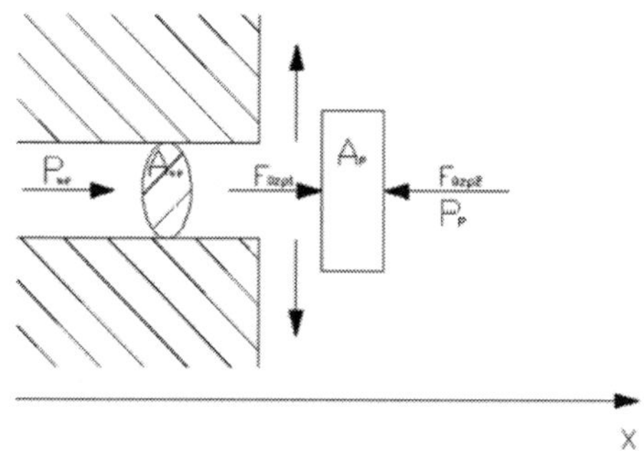

Figure 10. The calculation diagram of plate valve.

Another large section in providing the real operation of the system was construction of a simulation model for the other elements of the CR (simplified models). For this purpose, a calculation scheme is shown below.

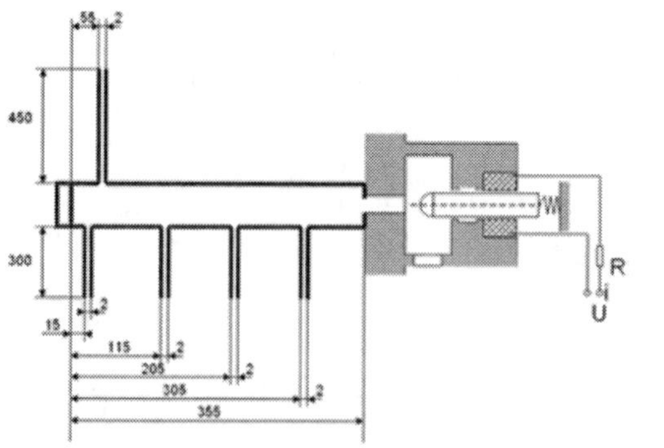

Figure 11. The calculation diagram of part of the high pressure CR system.

This section consisted of:

1. The equation of continuity in the manifold, as shown below:

$$0 = u_{zpompa} \cdot A_{zpompa} \cdot \sqrt{\frac{2}{\rho}} \cdot \sqrt{p_{str}} - u_{zatm} \cdot A_{zatm} \cdot \sqrt{\frac{2}{\rho}} \cdot \sqrt{p_z - p_{atm} - p_{str}} - u_{z1} \cdot A_{z1} \cdot \sqrt{\frac{2}{\rho}} \cdot \sqrt{p_z - p_{atm} - p_{str}} - \frac{V_z}{E_p} \cdot \frac{dp_z}{dt} \quad (18)$$

where:
A – surface for the identified cables,
u_{zpompa} - flow loss factor between the pump and the manifold,
p_z – pressure in accumulation pipe,
p_{pompa} – pressure in the fuel supply channel to the chamber,
p_{atm} – output pressure in control chamber,
μ - flow losses factor.
and the part of simulation model responsible for liquid continuity equation in manifold.

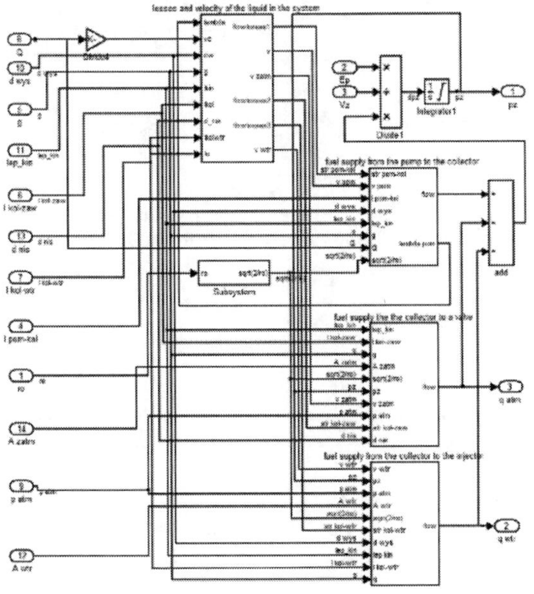

Figure 12. Scheme corresponding to liquid continuity equation in manifold.

2. The flow loss factor between the pump and the manifold. The hydraulic flow loss factor between the pump and the manifold is described by the equation (19) and modelled in Figure 13.

$$u_{zpompa} = \lambda \frac{l \cdot v^2}{d \cdot 2 \cdot g} + 0{,}5 \tag{19}$$

where:
λ – linear loss factor of turbulent flows,
l – length of the pipe from the pump to the manifold,
v – average velocity of fluid flow,
d – the diameter of the cable leading from the pump into the manifold,
g – acceleration of gravity.

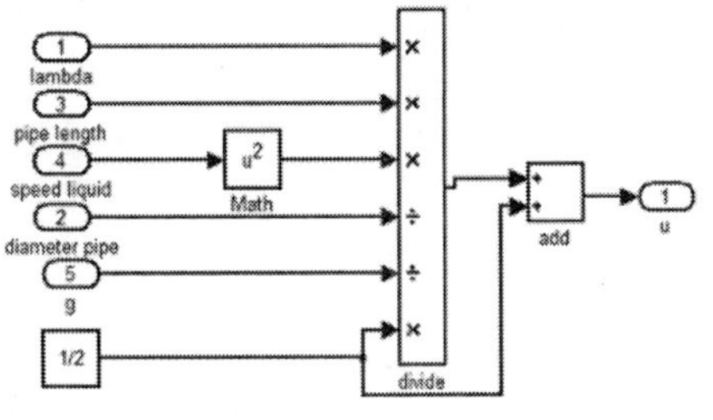

Figure 13. The flow loss factor.

3. The part of simulation model responsible for fuel supply to the injector.

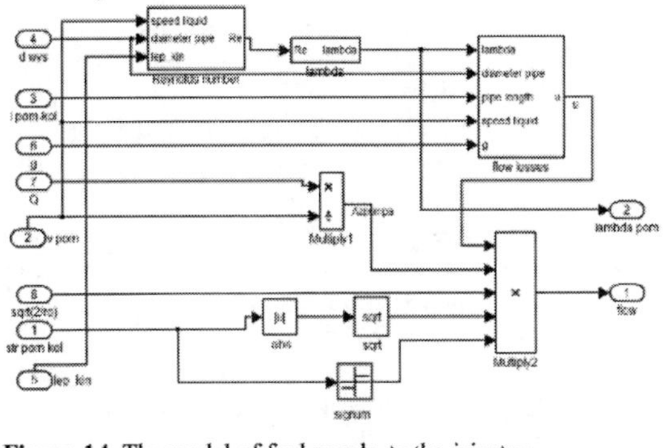

Figure 14. The model of fuel supply to the injector.

4. The mathematical equation in the model is the equation of valve head movement of the pressure regulating valve in the manifold.

$$m\frac{d^2w}{dt} = -u\frac{dw}{dt} - k \cdot w + B \cdot i \tag{20}$$

where:
m – movable part mass,
υ – kinematic viscosity factor,
w – displacement of the movable part,
k – spring constant,
B·i – force produced by the acting magnetic field on, current flowing through the coil.

5. The part of the model responsible for temporary opening of the valve, depending on the parameters prevailing in the system.

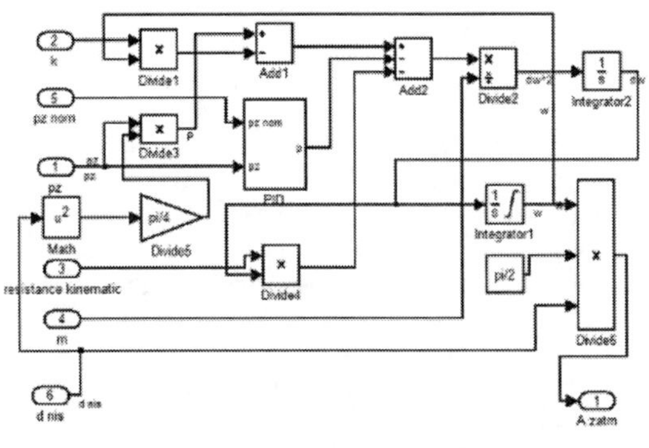

Figure 15. Model describing temporary opening the valve, depending on the parameters prevailing in the system.

6. The scheme for the PID regulator manifold pressure (moved to the control software) shown below.

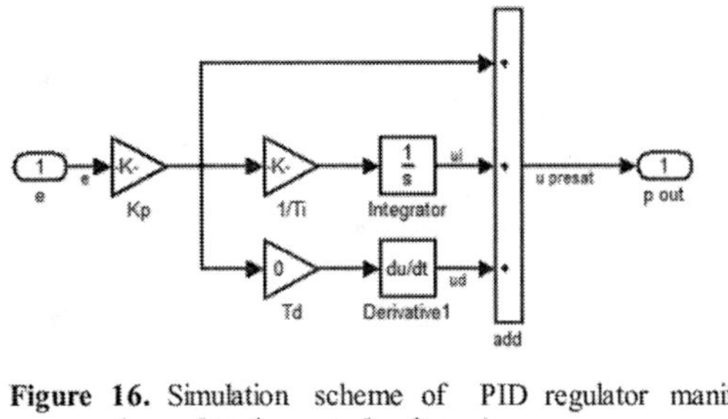

Figure 16. Simulation scheme of PID regulator manifold pressure (moved to the control software).

3 Results

Having finished the simulation model, a computer simulation for the given parameters was conducted. The following presents the results of simulations in terms of waveforms: liquid flow from the manifold to the injector [m^3/s], liquid flow from the manifold to the atmosphere [m^3/s], and pressure in the manifold [Pa] for set parameters and a period of time to 0.7 [s].

- The pressure waveforms in the manifold at a pressure of 100 mPa at a given time from 0 to 0,07 [s] (initial state) – Figure 17.

In the initial phase it can be seen that the fluid flow from the manifold to the injector takes the form of regular peaks every 0,01 [s], starts at 0 to 0,001 [m^3]. The liquid flow from the collector to the atmosphere at the beginning changes rapidly in a range from 0,0013 to 0,0023 [m^3], then over a period of 0,01 [s] reaches a value of about 0,003 [m^3], then decreases to a value of 0,00255 [m3] and stabilizes. Manifold pressure in the initial phase (from 0 to 0,02 [s]) varies in the range from 0,00114 to 0,0096 [Pa], and then stabilizes in and around 0,001 [Pa].

- The pressure waveforms in the manifold at a pressure of 100 [MPa] and the opening of the injector (1%) over a period of 0,04 to 0,07 [s] (Figure 18).

During the opening of the injector (1%) in a range from 0,04 to 0,07 [s], a slight increase in the flow time of liquid from the manifold to the injector 0 - 0,001 [m^3] can be observed. The liquid flow from the collector into the atmosphere during 0,01 [s] initially falls rapidly, then stabilizes, increases, and after that is recurrent in time intervals. Pressure in the manifold is more variable. In each interval (0,01 [s]), pressure initially drops sharply (the minimum value for a single pitch 0,0094 [Pa]) and then increases (the maximum value for a single stroke 0,01015 [Pa])The pressure waveforms in manifold at a predetermined pressure of 100 [MPa] and the opening of the injector (5%) over a period of from 0,04 o 0,07 [s].

- The pressure waveforms in the manifold at a predetermined pressure of 100 [MPa] and the opening of the injector (5%) over a period of 0,04 to 0,07 (Figure 19).

During the opening of the injector (5%), elongation of fluid flow in time can be observed as well as increase of flows over 0,001 [m^3]. For fluid flow from the manifold to the atmosphere, stabilizing in the value in the time interval of 0.01 [s] can be observed. The pressure in the manifold assumes a greater range of variation in comparison with the opening of the injector (1%).

- The pressure waveforms in the manifold at a predetermined pressure of 100 [MPa] and the opening of the injector (10%) over a period of 0,04 to 0,07 [s] (Figure 20).

During the opening of the injector (10%), further elongation of the liquid flow over time can be observed, together with an increase in the range of values for liquid flow from the manifold to the atmosphere from 0,00256 to 0,0028 [m^3], and increase in the range of changes of pressure in the manifold.

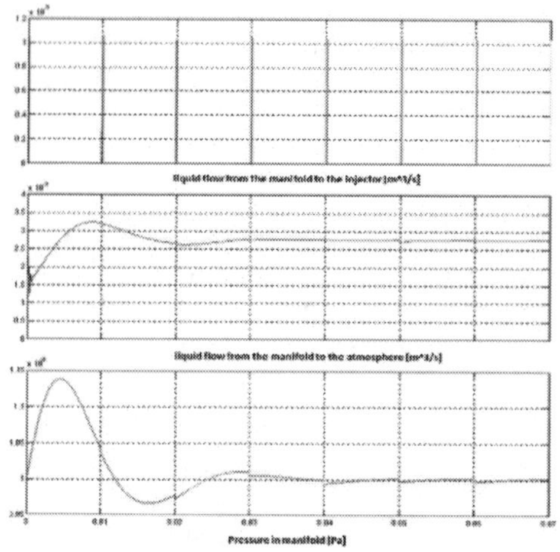

Figure 17. The pressure waveforms in the manifold at a pressure of 100 mPa at a given time from 0 to 0,07 [s] (initial state).

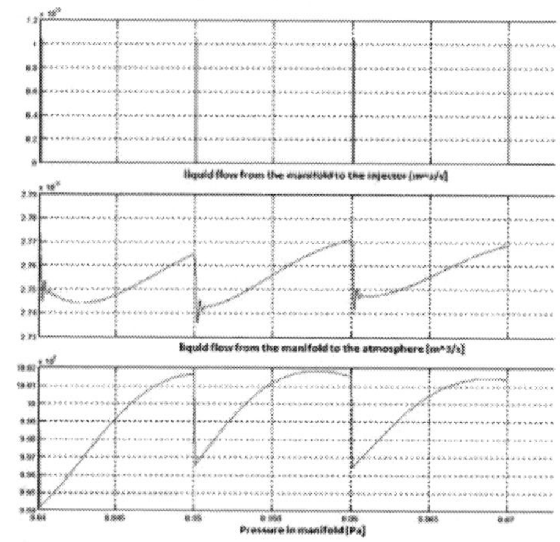

Figure 18. The pressure waveforms in the manifold at a pressure of 100 [MPa] and the opening of the injector (1%) over a period of 0,04 to 0,07 [s].

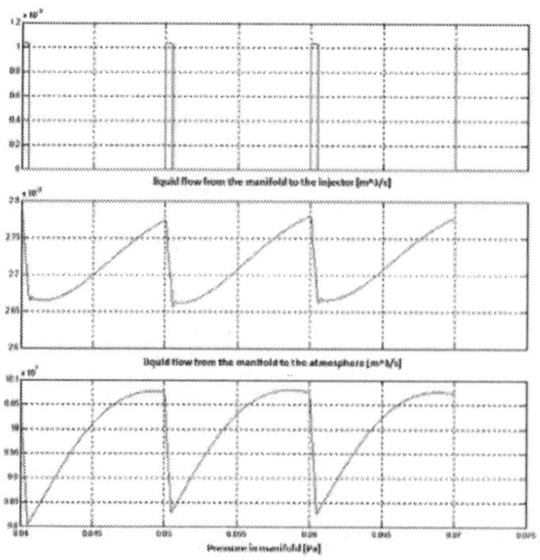

Figure 19. The pressure waveforms in the manifold at a predetermined pressure of 100 [MPa] and the opening of the injector (5%) over a period of 0,04 to 0,07 [s].

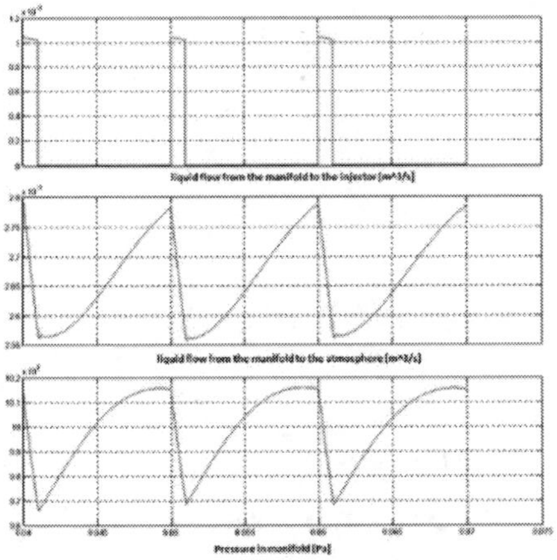

Figure 20. The pressure waveforms in the manifold at a predetermined pressure of 100 [MPa] and the opening of the injector (10%) over a period of 0,04 to 0,07 [s].

4 Conclusions

On the basis of the constructed model, simulations conducted, and the results obtained, the following conclusions can be made:

- The basic aim of this study was to build a dependence which automatically delineating change the manifold pressure in the common rail system during engine operation on the basis of certain parameters.
- The authors have shown and described the simulation model and then checked for correctness on the basis of simulations.
- The results of simulation allow for verification of correct construction of the structure diagrams - Full verification of the functionality of the model will be carried out on a real object in future studies.
- Correct results indicate that the high-pressure control system, built in the simulation, would operate in a stable manner.

References

1. Press materials Robert Bosch GmbH, *The future of the diesel engine - will be electrified?,* http://mojafirma.infor.pl/moto/wiadomosci/rynek/324193,Przyszlosc-silnika-Diesla-zostanie-zelektryfikowany.html.
2. A. Szuliborski, *Controlling compression ignition engines information Technical BOSCH* (Publishing Transport and Communications, Warsaw 2004)
3. H. GüntheR, *Common Rail injection systems in practice workshop: construction, testing, diagnostics* (Publisher of Communications and Communications, Warsaw 2010)
4. N.H. Chung, B. G. Oh, and M. H. Sunwoo, *Modelling and injection rate estimation of common-rail injectors for direct-injection diesel engines* (Journal of Automobile Engineering 222.6 (2008): 1089-1001)
5. 5. C. Gautier, O. Sename, L. Dugard, G.Meissonnier, *Modelling Of A Diesel Engine Common Rail Injection System* (IFAC 16th Word Congress, Prague 2005)
6. B. Mrozek, Z. Mrozek, *Matlab Simulink 2.x 5.x User's Guide* (PLJ Publisher, Warsaw 1998)
7. T.A. Kumar, *MATLAB and Simulink for Engineers* (Oxford 2011)
8. D. Xue, Y.Q. Chen, *System Simulation Techniques with MATLAB and Simulink*, (Wiley Publisher, 2013)
9. M. Gołębiewski, M. Walkowski, *Common Rail systems in the combustion engines and selected phenomena occuring in the fuel pipes of high pressure during the fuel injection* (Scientific Journals 2009, 17(89) pp. 38–43)
10. T. GancarczyK, T. Knefel, *Modelling Analyses of Common Rail High Pressure Pump* (Mechanik, 2013, R. 86, nr 2CD.)

Rapid prototyping methods for the manufacture of fuel cells

Piotr Dudek[1,a], Andrzej Raźniak[2] and Bartłomiej Lis[2]

[1]*AGH University of Science and Technology, Faculty of Mechanical Engineering and Robotics, Al. A. Mickiewicza 30, 30-059 Kraków, Poland*
[2]*AGH University of Science and Technology, Faculty of Energy and Fuels, Al. A. Mickiewicza 30, 30-059 Kraków, Poland*

Abstract. The term rapid prototyping (RP) is widely used to describe technologies which create physical prototypes directly from digital data. The first methods for rapid prototyping became available in the late 1980s and were used to produce models and prototype parts. Three-dimensional (3D) rapid prototyping methods are widely employed in various areas of manufacturing, research, and education. Typical applications include engineering, architecture, and medicine. The present level of technical and commercial demand requires the development of faster and cheaper methodologies for the design and execution of structures. Various techniques based on CAD platforms and rapid prototyping are available. Parts can be produced in a variety of materials: polymers, metal, paper, ceramics, and composites.
The RP method also enables the production of prototypes useful for analysing the characteristics of a complex system (e.g. interference between dynamic parts, geometric evaluation, quality, and reliability). RP technologies can produce very complex parts which are impossible or difficult to produce by traditional methods.
This paper presents typical applications of rapid prototyping technology for the manufacture of mechanical parts for fuel cells, such as housing parts or bipolar plates which supply reactants (hydrogen to anodes and oxygen to cathodes), conduct electrons from one cell to the next, remove waste heat from cells, and provide mechanical support for cells in a stack. Conventional graphite or composite materials can be replaced by lighter metallic materials characterised by vastly superior manufacturability and cost effectiveness, greater mechanical strength, increased durability and resistance to shock and vibration, and zero permeability.
The potential for the application of this method for the manufacture of metallic bipolar plates (BPP) for use in proton exchange membrane fuel cells (PEMFCs) is presented and discussed. Special attention is paid to the fabrication of light elements for the construction of PEMFC stacks designed for mobile applications such as aviation technology and unmanned aerial vehicles (UAVs).

1 Introduction

A traditional generator converts fuel to electricity by using machinery to convert thermal energy from combustion of the fuel into first mechanical and then electrical energy. A fuel cell, on the other hand, is a clean, quiet electrochemical device that converts hydrogen into electricity with no emissions (so-called 'zero emissions'). Today, while fuel cell technology is still considered pre-commercial, it offers significant economic and environmental benefits and is already being introduced in a number of applications. One of the critical steps in the commercialisation of fuel cells will be reducing the cost of the technology [1-3].

Rapid prototyping (RP) technologies can automatically manufacture near-net-shape parts with complicated geometry from computer-aided design (CAD) data. The 3D part is built up by means of powder consolidation, solidification of fluids, or gluing sheets in layers (the 'additive' or 'generative' process). For this reason, these techniques are often referred to as solid freeform fabrication or layered manufacturing [4,5].

For instance, in contrast to traditional 'subtractive' processes (e.g., drilling, milling, grinding), 'additive' RP methods enable the fabrication of products with complex geometries such as undercuts, internal cavities, or overhangs, as well as the generation of parts with oriented internal pore structures which cannot be manufactured using other approaches.

Due to this technology, it is possible to significantly shorten fabrication times and costs. Technologies such as selective laser sintering (SLS), direct metal laser sintering (DMLS), stereolithography (SLA), or fused deposition modelling (FDM) can be used for the fabrication of fuel cell prototypes, particularly for planar configuration.

It must be stated that RP technologies will never completely replace other manufacturing techniques, especially in large production batches, in which mass production is more economical. For short production runs, however, the cost of using RP is much lower, since no tooling is required. This technology is also ideal

[a] Corresponding author: pdudek@agh.edu.pl

for the production of custom parts to meet the exact specifications of the user [6].

Rapid prototyping is beginning to change the way companies design and build products. The introduction of non-polymeric materials such as ceramics and composites enable RP users to produce functional, strong, and durable parts. The use of ceramics, metals, and composite materials greatly expands the range of products that can be made by means of rapid manufacturing.

Fuel cells are devices known for the most efficient electrochemical conversion of fuel into electrical energy. There are several papers showing the potential for using RP methods to fabricate elements for fuel cell with complicated shapes or housing characterised by reduced weight and improved heat dissipation designed for mobile applications such as aviation technology and unmanned aerial vehicles (UAVs) [1, 7-11].

2 Definition and classification of rapid manufacturing

All RP techniques employ the same basic process, as shown in Figure 1:

- Creating a design model and converting the CAD model to an STL format;
- Slicing the STL file into thin cross-sectional layers;
- Construction of a model, layer upon layer;
- Cleaning and finishing the model.

Creating a CAD model and converting it to STL

The first step is to create the object to be built using CAD software. This process is identical for all RP construction techniques. CAD/STL parts can also be prepared using reverse engineering techniques using 3D scanners or CT/MRI machines and appropriate software. It must be mentioned that following conversion to STL (stereolithography) format, some objects lose a certain degree of precision. STL is the standard file type used by most additive manufacturing systems. An STL file is a triangular representation of three-dimensional surface geometry. The surface is tessellated or broken down logically into a series of small facets (triangles). Each triangle is described by three points representing the vertices (corners) of the triangle and a direction perpendicular to it. These data are used by a slicing algorithm to determine the cross sections of the 3D shape to be built by the rapid prototyping machine.

Slicing the STL file into thin cross-sectional layers

In this step, a slicing program processes the STL file in order to create slices of the part to be printed. Several programs are available, most of which are incorporated into printer control software, enabling the user to adjust the size, location and orientation of the model. The orientation is important for several reasons. First, the properties of prototypes vary with the direction of the coordinate. For example, prototypes are usually less accurate and weaker in the z direction (vertical) than in the x-y (horizontal) plane. Furthermore, the orientation of the part partially determines the period of time required to build the model. Placing the shorter dimension in the z direction reduces the number of layers needed, thereby reducing construction time. The program can also generate an auxiliary structure to support the model during the building process, which is necessary for some technologies.

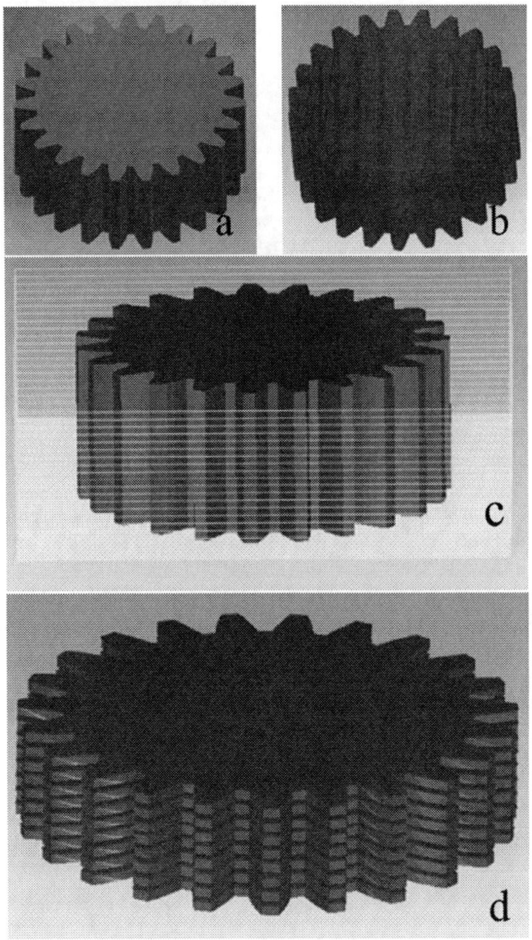

Figure 1. Schema of RP methods (a – a CAD model, b – the model is converted to STL triangles, c – the model is sliced, d – the model is built layer by layer).

Building the layered model

The next step is the physical construction of the part. Using one of several techniques (described in the next section), an RP machine builds one layer at a time from a polymer, paper, metal, plastic powder, or other starting material. Most machines are fairly autonomous, requiring little human intervention.

Cleaning and finishing the model

The final step is post-processing work, which involves removal the prototype from the machine, separation of some supports, cleaning and removal of unused powder, and additional hardening of the final parts. Some technologies require special procedures for the proper manufacture of a part. For example, photosensitive materials need to be cured completely before use, while other technologies require cleaning and surface treatments such as sanding.

There are various ways to classify rapid prototyping technologies; one is based on the form of the starting material (Figure 2). In this chapter, some of the most important technologies are presented.

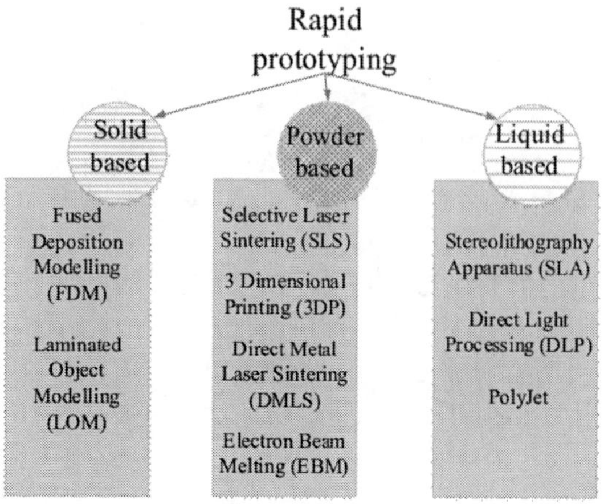

Figure. 2. Schema of RP methods.

2.1 Stereolithography

This process, patented as a means of rapid prototyping in 1986 by Charles Hull, co-founder of 3D Systems, Inc., initiated the rapid prototyping revolution. In this process, a uniquely designed 3D printing machine called a stereolithograph apparatus (SLA) converts liquid plastic into solid objects.

SLAs have four main parts: a tank filled with liquid photopolymer, a perforated platform that is lowered into the tank, an ultraviolet laser, and a computer controlling the platform and the laser.

As shown in Figure 3, the model is built on a platform located under the surface in a liquid photopolymer. A low-power laser traces the first layer, solidifying the first cross section of the model.

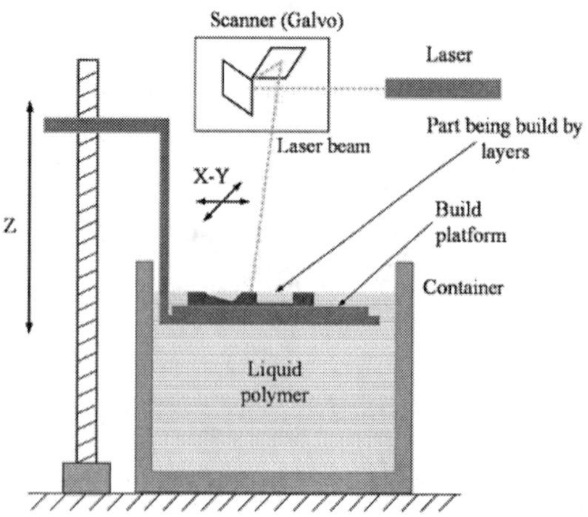

Figure 3. The stereolithography (SLA) process.

Once the initial layer has hardened, the platform is lowered, exposing a new surface layer of liquid polymer. The laser traces the next cross section of the object being printed, which instantly bonds to the hardened section beneath it. This process is repeated again and again until the entire object has been formed and is fully submerged in the tank. The platform is then raised to expose a three-dimensional object. After being rinsed with a liquid solvent to free it of excess resin, the object is baked in an ultraviolet oven to further cure the plastic; subsequently, any supports are broken away.

Currently available photocurable materials include acrylic and epoxy resins . Researchers are investigating the possibility of manufacturing parts reinforced with carbon fibres. Additionally, ceramic powders can be added to UV curable resins so as to form a so-called 'green' part, and then thermally treated to eliminate the resin and sinter the part. Objects made using stereolithography are generally characterised by smooth surfaces, but the quality of the object depends on that of the SLA machine used to print it.

2.2 Fused deposition modelling

Fused deposition modelling (FDM) is one of the best-known methods of rapid prototyping. A thermoplastic filament is unwound from a coil and material is supplied to an extrusion nozzle, assembled to a mechanical head which can be moved over a table according to the required slice geometry, which deposits a thin bead of extruded plastic to form each succeeding layer. The nozzle is heated to the melting point of the plastic and includes a mechanism which enables the flow of melted plastic to be stopped or continued. The result of lamination of the solidified material to the preceding layer is a plastic 3D model built up one layer at a time (Figure 4).

FDM printers use two kinds of materials: modelling material which constitutes the finished object, and support material which acts as a scaffolding to support the object as it is being printed. Once the part is completed, the support columns are removed and the surface is finished.

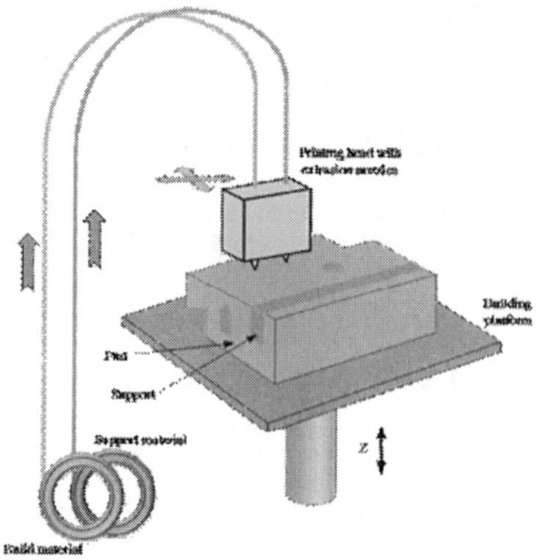

Figure 4. The FDM process.

2.3 Selective laser sintering

The SLS process was developed and patented in the 1980s by Carl Deckard, then an undergraduate

 DOI: 10.1051/e3sconf/20161000127

student at the University of Texas, and his mechanical engineering professor, Joe Beaman. Objects printed using SLS are made with powder materials, most commonly plastics such as nylon, which are dispersed in a thin layer on top of the building platform inside an SLS machine. A computer-controlled laser sends pulses down to the platform, tracing a cross section of the object onto the powder (Figure 5).

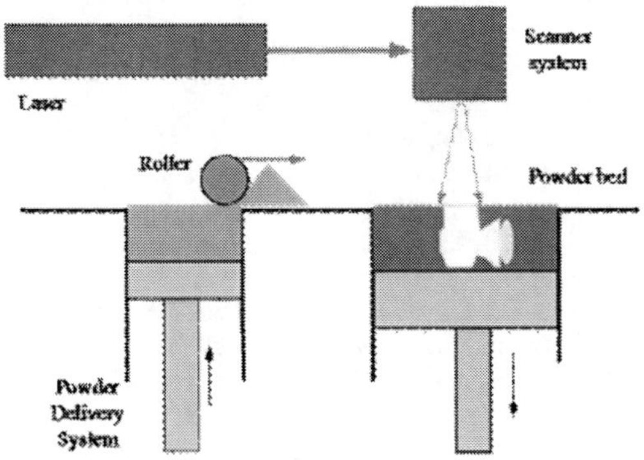

Figure 5. The selective laser sintering process.

The platform is lowered to the height of the next layer and powder is reapplied. The process continues until the part is complete. Excess powder in each layer helps support the part during construction. The process chamber is heated to prevent part warping and filled with nitrogen to reduce powder aging. When the object is fully formed, it is left to cool in the machine before being removed.

The materials available for SLS technology include polyamide, polyethylene, elastomers, ceramics, and metals, using polymers as additives. This technique is still in development for ceramic materials, due to difficulties arising from sintering; however, structures can be fully densified via sintering in the liquid phase.

2.4 Direct metal laser sintering/selective laser melting

Like SLS, this technology employs a laser as a way to consolidate powder, but uses metal powder as source material.

Preparation for a build requires support, even though parts are sintered in a bed of powder (as in SLS). The rapid melting of each metal layer can create residual stresses in the part during DMLS. Supports are required to hold parts in place in order to prevent any warping that might occur from these stresses as each layer cools.

DMLS uses a laser system that draws onto a surface of atomised metal powder. Using a focussed laser beam, this technology melts metal powder locally to fuse it into solid parts. Upon the completion of each layer, a blade adds a fresh layer of powder and repeats the process until the final metal part is formed (Figure 6).

Once the build is finished, it is necessary to remove the support structures and carry out any other post-processing work required; for example, surfaces which were in contact with support structures must be cleaned, and any bead blasting and deburring needed to finish the parts must be carried out.

In SLM, the material is fully melted, whereas in DMLS, the material is heated just enough for the molecules to fuse. Both technologies are capable of producing parts of over 98% density. Leftover powder can be reused repeatedly, so waste levels are low.

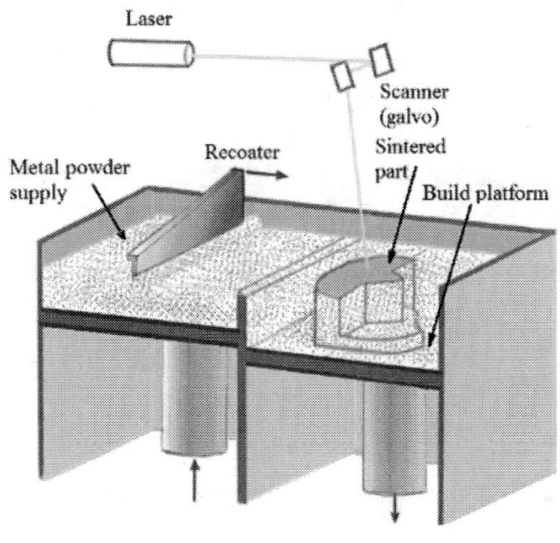

Figure 6. The DMLS process.

Compared to SLS, this technology has different requirements for the design of parts, and also special requirements for positioning parts on the platform.

3 Application of 3D printing for manufacturing some components of PEMFC fuel cells

At present, proton exchange membrane fuel cell (PEMFC) technology shows a relatively high level of development. Various energy power systems based on PEMFC technology are already available in the marketplace. They can be used as power generators with a wide range of electric power from 5 We to 200 kWe, from stationary and military to automotive and aviation applications, for example: (i) the small (5 We, cooled by heat spreaders) and portable Minipak mini charger from the Horizon company, for charging USB-compatible portable devices [12]; (ii) Panasonic 750-We domestic CHP plants with PEMFCs (hydrogen from reformed gas) supplying electricity and hot water [13]; (iii) Toyota Mirai, the first mass-produced car to run on hydrogen, using a TFCS (Toyota Fuel Cell System) with a capacity of 114 kW, powered by hydrogen derived from composite hydrogen tanks at a pressure of 700 bar [14].

A proton exchange membrane fuel cell (PEMFC) is an electrochemical device in which hydrogen and oxygen react electrochemically, resulting in the generation of electricity (up to 55 % based on the high heating value of hydrogen), heat, and water [3].

The membrane, which acts as the electrolyte, is compressed between two porous, electrically conducting electrodes, typically made out of carbon cloth or carbon fibre paper. At the interphase between the porous electrode and the polymer membrane is a layer

 DOI: 10.1051/e3sconf/20161000127

with catalyst particles, typically carbon-supported platinum [2, 3].

One of the challenges involved in commercialisation is finding a new method of fabrication which would enable the production of reproducible components at reduced costs [2].

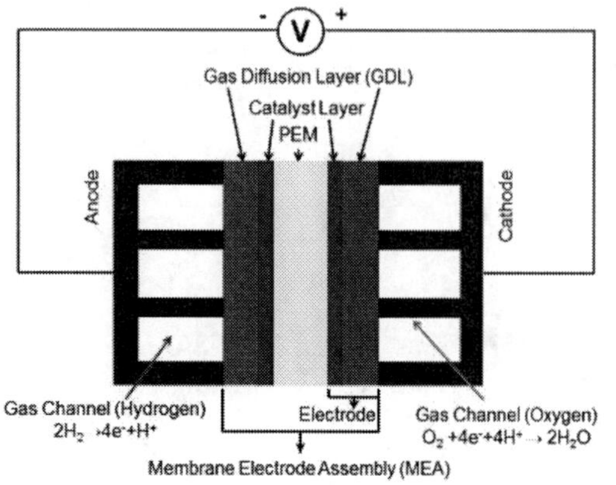

Figure. 7. The idea behind the construction of a single PEMFC fuel cell.

Some examples of the application of injection printing technology to the fabrication of Nafion polymer electrolytes exist. This technique can substitute for commonly-used membrane foil. A Nafion® dispersion is deposited directly onto the catalyst layers of anode and cathode gas diffusion electrodes. The two electrodes are then pressed together with their membrane layers facing each other. For the manufacture of PEMFC components it is necessary to precisely deposit a very thin layer of platinum, needed for the oxidation and reduction reactions, with a high level of utilisation efficiency. This can be critical in terms of taking this technology to the masses. Another process is screen printing, but this is done by hand and necessitates compromises in terms of uniformity and time efficiency. The process is greatly improved by using the 3D printing process to deposit the layer of platinum. The inkjet printing method is 4 to 5 times faster than screen printing [8, 11].

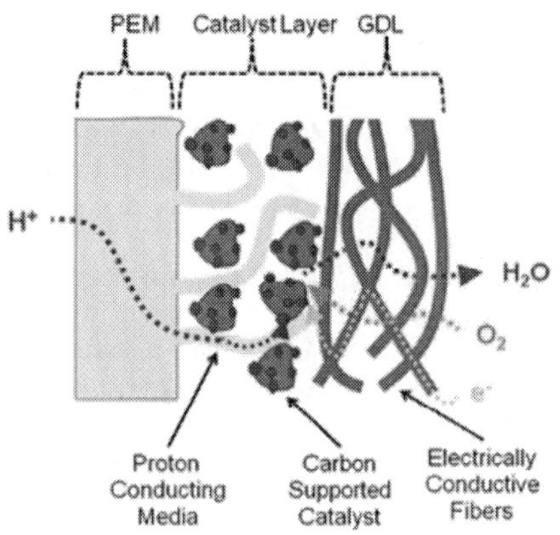

Figure 8. The idea behind MEA construction and operation.

The main parameters that determine the application of a single membrane in a PEMFC designed stack are power output, stability of electrical parameters under load, and water management. Water is necessary in order to hydrate the membrane, increasing its electrical conductivity. However, in excess, water can flood the pores of the GDL (gas diffusion layer) and hinder the reactants from reaching the active centres of the catalyst layer. Thus, an efficient balance between generated and removed water is necessary to obtain good performance from a PEMFC. Adequate hydration in the PEM and GDL enables the attainment of higher current values, increasing the power provided by the PEMFC [9].

The distribution of gaseous reagents is also a crucial parameter determining the power output of a single-membrane PEMFC stack. Special architecture of the gas diffusion layer can lead to enhanced performance by a PEMFC. It was found that the performance of a single MEA was enhanced by the utilisation of microchannels (0.05–1 mm) to improve gas routing. Tests performed on distributed water in actual fuel cell microchannels microfabricated using the stereolithography technique seem to confirm these statements. Side channels or slots perpendicular to the main microchannel are used in the microfabricated samples, whereas the stereolithography samples incorporate a sandwiched GDL for distributed water injection into the main channel [15, 16]

A PEMFC stack consists of a number of single MEA cells stacked so that the cathode of one cell is electrically connected to the anode of the adjacent cell. The key aspects of fuel cell stack design are: uniform distribution of reactants to cells and within each cell; maintenance of required temperature in each cell; a minimum of resistive losses (choice of materials, configuration, uniform contact pressure). A special role is played by bipolar plates, which are the backbone of a PEMFC stack. They act as current collectors in the fuel cell environment and provide conduits for the reactant gases. Bipolar plates are usually fabricated from graphite and metal. In special application in unmanaged aerial vehicles (UAV) technology, the mass of the entire system is very important. The capacity for hydrogen storage is also critical for UAV application, as flight duration depends on it. Graphite bipolar plates are typically used in PEMFC stacks designed for stationary applications, whereas in the case of PEMFC applications in UAVs and transports, mass seems to be one of the crucial parameters. Some data in the literature indicate that the application of metallic bipolar plates enables not only a reduction in mass but also relative savings in hydrogen consumption. Comparative results using bipolar plates made from graphite and metal indicated that hydrogen consumption decreased by at least of 18 % due to higher electrical and thermal conductivity [10, 17].

4 Application of 3D printing for manufacturing air inlet channel

The design and execution of a PEMFC stack as a basic unit to power an electric motor in an unmanned aerial vehicle requires knowledge of certain electrical

 DOI: 10.1051/e3sconf/20161000127

characteristics: the voltage (U)-current (I), and power (P)-current(I) curves produced by the DC power sources. Another requirement is adjustment of the operating parameters of the PEMFC stack to the scope of permissible values of current and voltage consumed by the electric motor. In the case of aeronautical applications, it is also crucial not to exceed the permissible mass of a PEMFC stack with a hydrogen fuel tank.

Even though products dedicated to unmanned aerial vehicle applications are already available on the fuel cell market, it is not always possible to use a commercial product due to the specific design solutions needed for UAV construction. At AGH University of Science and Technology, Faculty of Fuels and Energy, research was undertaken to adapt a commercial 300-W PEMFC stack (Horizon Singapore) to an electrical engine in order to supply a UAV. It should be emphasised that this route has also been taken by most of the leading R&D centres dealing with applications of fuel cell technology in aerial vehicles [18, 19]. Our team planned each step of design changes aimed at reducing the weight of the generator by: i) removing the metal casing and replacing it with a structure made of 3D-printed plastics; ii) removing an electronic circuit used to power fans and to humidify and purify the PEMFC stack; iii) eliminating fans and replacing them with air jets to cool the stack. Figure 9 shows examples of 3D-printed models of supply channels for the air jets aimed at reducing the number of fans, which not only increase the weight of the system but also draw power from the fuel cell stack.

Figure 9. The examples of 3D printed models.

Figure 10 presents a photo of an experimental setup designed for investigations of a 300-W PEMFC stack, including an air jack fabricated by means of FDM technology. The main parts of the experimental setup comprised an electrical engine with a propeller, an electronic load, and a model of the fuselage manufactured from styrofoam.

The main function of this model was to reflect the actual fuselage constructed by the UAV producer. In the course of the experimental investigations, aimed at determining the suitability of the 3D-printed air jack, the electric motor was supplied from an external DC source. The air flow, driven directly into the tight channel, passed through the PEMFC fuel cell stack and exited the fuselage. The speed of the air flow directed to the inlet of the fuselage was 24 km/h. Under these conditions, the characteristics of the voltage (U)-current (I) and power (P)-current (I) curves were determined.

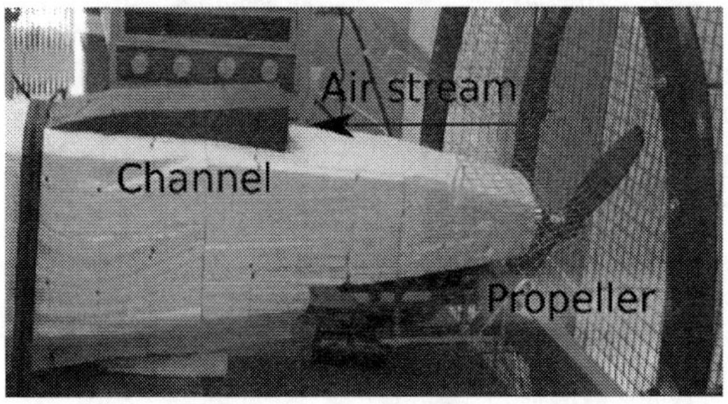

Figure 10. Photo of an experimental setup designed for investigations of a 300-W PEMFC stack with air inlet channel.

Concurrent temperature measurements were taken at points T_1 and T_2 using thermocouples placed at different locations in the fuselage. Thermocouple 1 (T_1) was placed inside the front part of the stack; the second thermocouple (T_2) was placed in the middle of the PEMFC stack, which was, in turn, placed in the fuselage. To avoid any unexpected temperature increases inside the PEMFC stack during operation, the fuel cell was gradually loaded (first part of Fig. 11). It was established that the maximum temperature of the stack would not be allowed to rise above 55°C. In Figure 11, variations in power (P) and temperature T1, T2 of the PEMFC stack vs time was analysed. As can be seen, the PEMFC stack attained about 200 W of electrical power. No significant difference was observed between T1 and T2. At 2300 s, the electronic load was switched off. In the second part, the PEMFC stack was fast-loaded to reach 200 W. In this part of the experiment as well, the temperature inside the PEMFC reached values no higher than 55°C. Based on these preliminary results, it seems important to carry out further research on the analysis of the parameters of air flow and the construction of the channel.

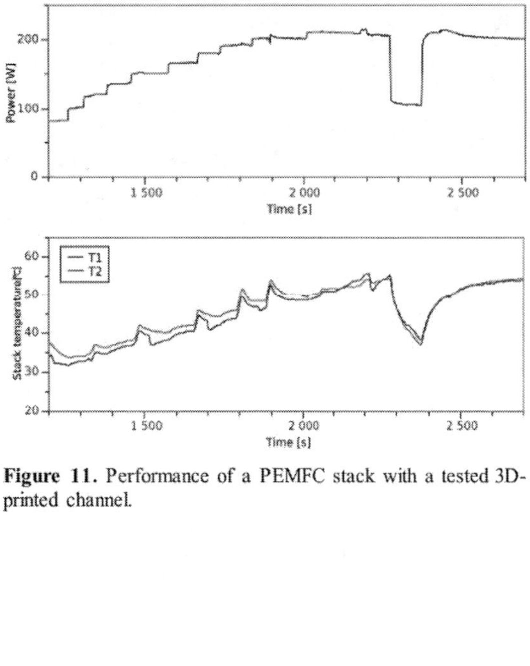

Figure 11. Performance of a PEMFC stack with a tested 3D-printed channel.

The concept of reverse engineering is defined as follows by Bradley and Currie [4]: 'the process of duplicating an existing part, subassembly, or product without drawings, documentation, or a computer model is known as reverse engineering. Reverse engineering is also defined as the process of obtaining a geometric CAD model from 3D points acquired by scanning/digitising existing parts/products'.

Sometimes it is necessary for some reason to change a commercial product to meet the requirements of specific solutions. In this case it was impossible to order parts or a solution that fit our requirements; therefore it was necessary to modify the product.

Due to these requirements, research was performed at the Department of Manufacturing Systems at AGH to develop lightweight housing elements for PEMFCs. These elements were required to be mechanically strong, as rigid as the original, and characterised by good thermal conductivity for appropriate heat dissipation.

In order to develop these elements, a measurement arm with an integrated scanning head was used. A scanned 3D housing element was processed using reverse engineering programs to create CAD elements, which were then subjected to additional processes in order to reduce their weight (Figure 12). Weight reduction in housing parts without a commensurate reduction in strength is very important for the use of fuel cells in aerospace applications. Additive Manufacturing (AM) affords the flexibility to create complex part geometries that are difficult to build using traditional manufacturing. It can build parts that incorporate design features such as internal cavities, pipes, and lattice structures that help to reduce the weight of parts without compromising their mechanical performance. Furthermore, AM machines produce less scrap than traditional machines, a critical attribute when using expensive aerospace materials; thus this process enables the use of materials such as titanium at an acceptable cost.

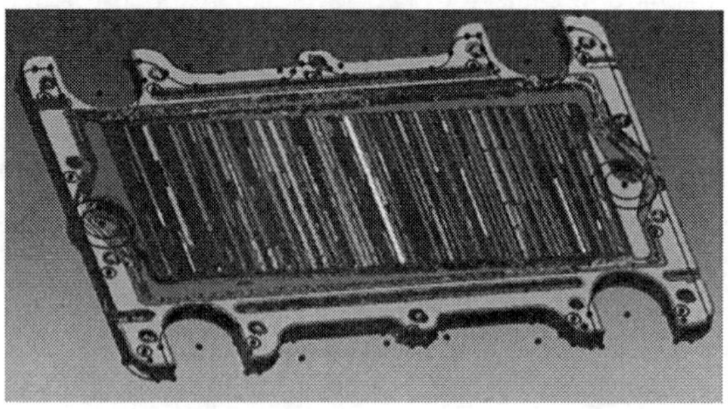

Figure 12. Scanned housing element of a PEMFC stack in a reverse engineering application.

After the necessary research, new PEMFC housing elements were created. In the next stage, CAD parts were converted to STL format and printed on an EOS Formiga P 100 SLS (stereolithography) machine using the polyamide powder PA2200 (also known as PA12). Printed parts (original project based on scanned elements and developed new) were compared in order to find differences in stress and strength between the old and new elements. Cooling channels were integrated into the housing, reducing the risk of damage to the PEMFC due to a coolant leak.

Conclusions

Rapid prototyping is characterized as a low-cost manufacturing process with high flexibility for the production of functional prototypes and customized products. Its working principle is based on the imposition of layers forming a three-dimensional design of varying complexity, using various techniques. There is an increasing interest in the polymer fuel cell technology for constructing new reliable power sources for stationary, transport, avionic, and military applications. One of main goals of the PEMFC technology is lowering manufacturing costs for continuous production. The conventional fuel cell manufacturing processes requires long time and high costs, especially in the stages of product development and optimization. The techniques of Selective Laser Sintering (SLS), Laminated Object Manufacturing (LOM), and Three Dimensional Printing (3DP) may be considered as promising alternatives for the production of fuel cell prototypes. SLS, despite the high initial cost of the equipment, can conform and sinter the components simultaneously. LOM uses tapes, produced by tape casting, to obtain laminates. 3DP already has the lowest cost of acquisition of equipment and manufacturing of the prototype. Compared to other methods, RP can produce very complex parts which are impossible or difficult to manufacture by traditional methods. PEMFC tackles niche applications in military and aviation, which require the special construction of PEMFC stacks as they need to fit to the entire electrical system. In case of UAVs, the weight of the power source is also a critical parameter. The replacement of fans with a 3D-printed channel allows for mass reduction, and limits electric power consumption for PEMFC stack cooling. Cooling possibilities of air taken from stream located behind the propeller are proportional to fuel cell load by the main electric motor. Further studies will concentrate on the design of the channel shape and structure, which will ensure the effective cooling the elements in the PEMFC stack.

Acknowledgements

The work was carried out under the project no. PBS3/A6/24/2015 The National Centre for Research and Development (2015-2018)

References

1. T.H. Bradley, B.A. Moffitt, D. Mavris, D.E. Parekh, *Applications – Transportation Aviation: fuel cells in: Reference Module In Chemistry, Molecular Sciences And Chemical Engineering Encyclopedia Of Electrochemical Power Sources*, (Elsevier 2009)
2. Y. Wang, K.S. Chen, J. Mishler, S.C. Cho, X.C. Adroher, Applied Energy, **88** (2011)
3. D. Stolten, *Hydrogen and Fuel Cells: Fundamentals, Technologies and Applications* (John Wiley & Sons, 2010)

 DOI: 10.1051/e3sconf/20161000127

4. C. Bradley, B. Currie, Computer-Aided Design & Applications **2**, 5 (2005)
5. J. P. Kruth, CIRP Annals—Manufacturing Technology **40**, 2 (1991)
6. I. Gibson, D.W. Rosen, B. Stucker, *Additive manufacturing technologies: rapid prototyping to direct digital manufacturing*, (New York, 2010)
7. O. Z. Sharaf, M. F. Orhan, Renewable and Sustainable Energy Reviews **32** (2014)
8. M. Klingele, M. Breitwieser, R. Zengerleab, S.Thieleac, J. Mater. Chem. A **3** (2015)
9. J.A. Salva, A. Iranzoa, F. Rosab, E. Tapia Energy **101** (2016)
10. B.D. Gould, J.A. Rodgers, M. Schuette, K. Bethune, S. Louis, R. Rocheleau, K. Swider-Lyons, J. Solid State Sci. Technol. **4** (2015)
11. Chen-Yu Chen, Yun-Che Wen, Wei-Hsiang Lai, Ming-Chang Chou, Biing-Jyh Weng, Ching-Yuan Hsieh and Chien-Chih Kung, J. Fuel Cell Sci. Technol. **8** (2010)
12. http://www.horizonfuelcell.com/#!minipak/c156u (2016-05-07)
13. http://www.hvacr.pl/panasonic-opracowal-domowe-ogniwo-paliwowe-ene-farm-209 (2016-05-07)
14. https://www.toyota.pl/new-cars/mirai (2016-05-06)
15. https://gcep.stanford.edu/pdfs/SI3U6jOMPAIgwkaiBD_77Q/goodson_ers06.pdf (12.05.2016)
16. C. H. Hidrovo, T.A. Kramer, E.N. Wang, S. Vigneron, J.E. Steinbrenner, Jae-Mo Koo, Fu-Min Wang, D.W. Fogg, R.D. Flynn, Eon Soo Lee, C.H. Cheng, T.W. Kenny, J.K. Eaton, K.E. Goodson, Heat Transfer Engineering, **27** (2006)
17. B. Lyons, M. Batalov, P. Mohanty, S. Das, Rapid Prototyping of PEM Fuel Cell Bi-polar Plates using 3D Printing and Thermal Spray Deposition, Proc. SFF Symposium, Austin, TX (2005)
18. J.M. Krawczyk, A.M. Mazur, T. Sasin, A.W. Stokłosa, Transactions of the Institute of Aviation **4** (2014)
19. O. Gonzalez-Espasandin, T.J. Leo, E. Navarro-Arevalo, The Scientific World Journal **2014,** 497642 (2014)

Selected aspects of the design and diagnostics of solid oxide fuel cells

Bartłomiej Lis[1,a]

[1]*Faculty of Energy and Fuels, AGH University of Science and Technology in Cracow, Av. A. Mickiewicza 30, Poland*

Abstract. An increased growth in demand for energy accompanied by efforts to limit its negative impact on the environment is forcing society to seek new, more efficient energy sources. Fuel cells are one of the most promising solutions among the widely developed new generation of electrical generators. Fuel cells directly convert chemical fuel into electricity. Water and waste heat are by-products of fuel cell operation. Solid oxide fuel cells (SOFCs) have proven to be one of the most interesting solutions among the five types of technologically advanced fuel cells, for their ability to operate at temperatures above 800°C. Furthermore, SOFCs are characterized by other advantages in comparison to PEMFCs, including: (1) no need to use expensive catalysts (e.g. platinum, the price of which is high, and its resources limited), (2) the possibility of direct, internal conversion of hydrocarbon fuels, (3) lower sensitivity to contaminants (in particular, hydrogen fuel containing CO, which is useful for SOFCs instead of the platinum catalyst poisoners), and (4) the possibility of using waste heat in a gas turbine, or for heating or other industrial purposes. The paper discusses selected issues regarding the construction and characteristics of planar solid oxide fuel cells. Selected results of the following electrochemical investigations: $Ba_{0.9}Ca_{0.1}Ce_{0.9}Y_{0.1}O_3$-based proton electrolyte as possible components of SOFCs operating at intermediate temperature range (500-700°C).

1 Introduction

A fuel cell (FC) is a type of a galvanic cell, along with primary cells, electrochemical accumulator batteries, and flow-through accumulators. As in any galvanic cells, what occurs in an FC is a direct (single-stage) conversion of chemical fuel energy into electricity (and waste heat). A characteristic feature of FCs is that the electroactive substances, i.e. the oxidiser and the fuel, are delivered to them from outside, and the reaction product is removed from the fuel cell [1-7].

Currently, the most rapid technological development can be observed for proton exchange membrane fuel cells (PEMFCs) and solid oxide fuel cells (SOFCs). Small units with fuel cells (with power ranging from several up to a dozen watts, mainly PEMFCs) have already appeared on the commercial market; they can be used, for example, to power portable electronics or mobile phones. Slightly larger stacks of fuel cells (approximately 250–1000 W) can serve as power sources for electric motors for unmanned aerial vehicles (drones). Even larger fuel cells (1–200 kW) are being built for application in cars and small planes. Stationary units, with power in megawatts [8–13], have been commercialised for phosphoric acid fuel cells (PAFCs) and molten carbonate fuel cells (MCFCs), whereas the development of PEMFCs and SOFCs remains consistently at the level of advanced prototype units.

Among five types of technologically advanced fuel cells special attention should be given to solid oxide fuel cells (SOFCs). Solid oxide fuel cells which currently utilise ZrO_2-Y_2O_3 as solid electrolytes operate at temperatures of 850–1000°C and require the use of components (i.e. electrodes and interconnectors) made from materials characterised by high thermal and chemical stability as well as high corrosion resistance in an oxidation-reduction environment. This vastly increases the cost of producing fuel cells, due to the high cost of suitable materials, and limits their use [14].

The reduction of operation temperature to 550–700°C seems to be most important strategic goal of SOFC technology. There are two main approaches to solve this problem. The application of some oxygen ionic conductors as well as proton conducting ceramics, which exhibited higher ionic conductivity than 8YSZ at intermediate temperature range seem to be most often considered solution. Ceria-based solid electrolytes still seem to be the most attractive oxide conducting ceramic membranes at intermediate temperature ranges (600-800°C) [15-17]. The utilization of ceramic proton conducting membrane instead to oxygen ionic conductor as an electrolyte in SOFCs can lead to simplification of the SOFC stacks construction. In SOFCs with a ceramic proton conducting membrane, the water evolves on the cathode, avoiding the fuel dilatation on the anode side, as well as reduction of cell voltage.

At present, high protonic conduction is reported for several perovskite-type oxides, such as doped $BaCeO_3$, $SrCeO_3$, $BaZrO_3$, and $SrZrO_3$. Proton conduction can be

[a] Corresponding author: blis@agh.edu.pl

 DOI: 10.1051/e3sconf/20161000128

thermally activated at comparatively low temperatures (400–700°C) due to the low activation energy (0.3–0.6 eV) required by this process [18,19].

Currently, research and development for SOFCs are conducted in order to improve the efficiency of conversion of chemical energy of fuels into electrical energy in these devices. These efforts are also aimed at improving the reliability of fuel cells, prolonging their life, and reducing the costs of single cells and their stacks. The results of this work are modifications in the already produced SOFC stack systems, and the construction of a new type of a SOFC [20].

The latter requires confirmation of the effectiveness of its performance. Due to the cost of SOFC components, as well as the cost production of SOFCs, electrochemical tests are usually performed for SOFCs of small sizes. For this reason, the most often applied systems are button solid oxide fuel cells. The basic element of this measurement station is a button solid oxide fuel cell with electrodes with a maximum geometric surface area of 2 cm^2. Standardising procedures can be easily introduced for this kind of SOFC construction, as they will take into account, among other factors, the size of the active surface of the electrode material, the impact of the ohmic polarisation, and the area of the surface changes in the electrode material, or the electrolyte after long-term testing for all single SOFC performances.

The aim of this study is to determine the possibility of designing a SOFC operating in the temperature range of 600–800°C using a $BaCeO_3$-based ceramic proton conductor.

2 Experimental

A solid-state proton-conducting electrolyte ($Ba_{0.9}Ca_{0.1}Ce_{0.9}Y_{0.1}O_{3-d}$, 10CBCY) was fabricated via the sol-gel method. Stoichiometric amounts of $BaCO_3$, $CaCO_3$, and Y_2O_3 were dissolved in nitric acid and mixed with a $Ce(NO_3)_2$ solution. In a separate beaker, EDTA was dissolved in an ammonia solution diluted with distilled water. The two solutions were mixed and heated to 80°C. The ratio of total metal cations to EDTA was maintained at 1:1.1. As a result of continuous heating and stirring, excess water was evaporated and a viscous gel was formed. The gel was placed in a furnace and heated to 900°C. White ash was collected and milled in an agate mortar in the presence of isopropanol. Subsequently, the powder was heated to 1000°C for 2 h.

The powder was grounded in an agate mortar and uniaxially pressed in a 28mm pressing die under 9 tonnes of force. Then the sample was sintered at 1500°C for 2 h. The sintered sample was cut into the shape of a circle with a diameter of 20 mm. Both sides of the samples were grounded and polished. Platinum paste electrodes were screen-printed on both sides of the cylindrical cell. In order to sinter the electrodes with the cell surfaces, the sample was heat-treated in air at 815°C for 15 min. The anode and cathode surfaces of the commercial cell were screen-printed with a platinum honeycomb network current collector (Fig.1a-b). The prepared powder and sintered electrolyte were examined using the X-ray diffraction method.

Solid oxide button fuel cells (proton- and oxide-conducting) were tested using an elaborated setup (Fig.1c-d). Single SOFCs cells were placed inside a ceramic socket and pressed with another alumina tube in order to obtain a gastight configuration, gold gaskets were placed on both sides of the cell.

Hydrogen was supplied from a bench-scale PEM electrolyser using a liquid-gas water separator. In the case of the solid oxide button cell, 20% oxygen in an argon mixture was fed to the cathode. A Solartron electrochemical interface (SI1287) with a frequency response analyser (FRA 1255B) was used for electrochemical measurement. Impedance spectroscopy measurements were also performed to identify the sources of electrochemical losses during SOFC performance.

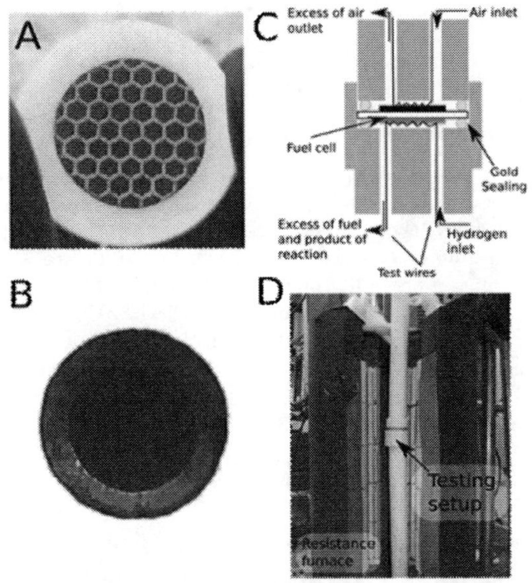

Figure 1. Photographies of tested electrolyte –supported solid oxide fuel cells. A) The cathode side of the solid button oxide fuel cell, B) The 10CBCY electrolyte with printed Pt electrodes. C) The scheme of electrochemical testing setup, D) The electrochemical testing setup inside opened furnace.

3 Results and discussions

The details of structural and physicochemical investigations of $Ba_{1-x}Ca_xCe_{0.9}Y_{0.1}O_{3-d}$-based solid solutions were presented in the previous paper [21]. Analysis of the electrical conductivity of both series of investigated materials showed that the highest ionic conductivity, in air and wet 5–10 % H_2 in Ar, was attained by compositions of x = 0.02–0.05 $(Ba_{1-x}Ca_x)(M_{0.9}Y_{0.1})O_3$, M=Zr, Ce.

Figure 2 presents the XRD diffraction pattern obtained for $BaCe_{0.9}Y_{0.1}O_{3-d}$ (BCY) as well as $Ba_{0.9}Ca_{0.1}Ce_{0.9}Y_{0.1}O_{3-d}$ (10CBCY) sintered samples. The introduction of 10 % calcium cations in place of barium in the solid solution $Ba_{0.9}Ca_{0.1}Ce_{0.9}Y_{0.1}O_{3-d}$ left the perovskite structure unchanged. Button solid oxide fuel cells incorporating electrode material with an active surface area of about 1.3–2 cm^2 are very often used in solid oxide fuel cell technology as standardised ceramic elements for preliminary electrochemical tests. Ceramic shaped parts in these sizes are very often offered by well-known fuel cell manufacturers as a reference for tests of single SOFCs [22,23].

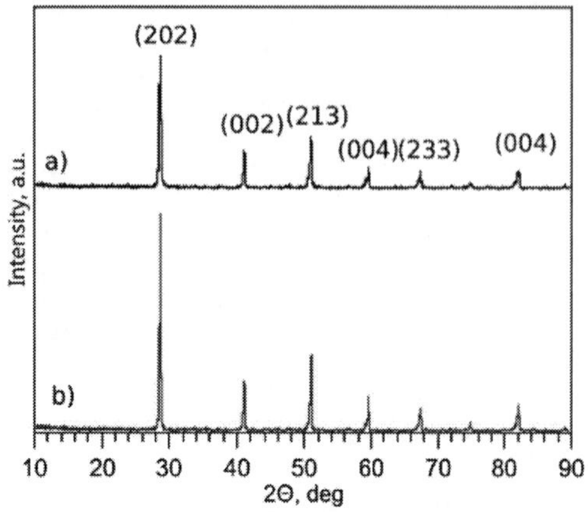

Figure 2. The X-ray diffraction pattern recorded for BCY (a) and 10CBCY (b) sintered samples.

In this study efforts were aimed at the manufacture of gastight $Ba_{0.9}Ca_{0.1}Ce_{0.9}Y_{0.1}O_{3-d}$ sintered samples with diameters about 2 cm and the performance tests of hydrogen-oxygen solid oxide fuel cells. Figure 3a-b presents a photo of commercial 8 %mol Y_2O_3 in ZrO_2 8YSZ ceramic foil (Fuel Cell Material, USA) used as a reference electrolytic material and a newly manufactured ceramic 10CBCY proton-conducting electrolyte. Based on this comparison, plane-parallel ceramic discs based on 10CBCY were obtained.

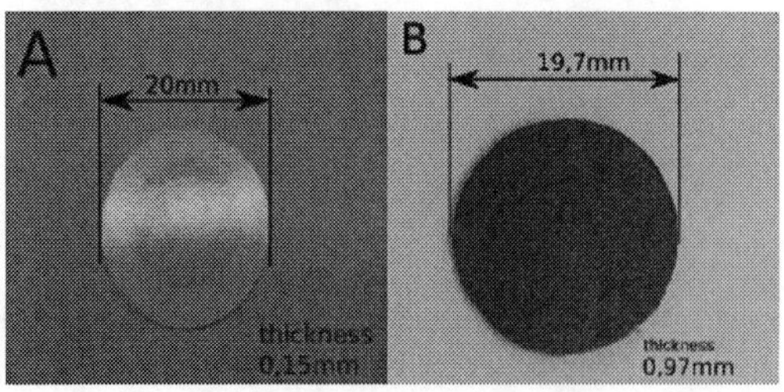

Figure 3. a) Photo of commercial 8YSZ ceramic electrolyte foil (supplier Fuel Cell Material, USA)b) Photo of a newly manufactured ceramic 10CBCY proton conducting electrolyte

The microscopic investigations performed on 10CBCY, using confocal, atomic force microscopy as well as scanning electron microscopy (Figure 4 a, b), showed no open porosity in the sintered sample.

The image of surface of the elaborated cell was compared with surface of commercial electrolyte presented on figure 4 (c,d). No significant differences in roughness was observed.

Electrochemical tests of a single SOFC were performed for reference solid button oxide fuel cells (Fuel Cell Materials, USA). The aim of these investigations was the verification of correct performance and the reproducibility measurements.

The characteristic U-I and P-I curves was presented in Figure 5.

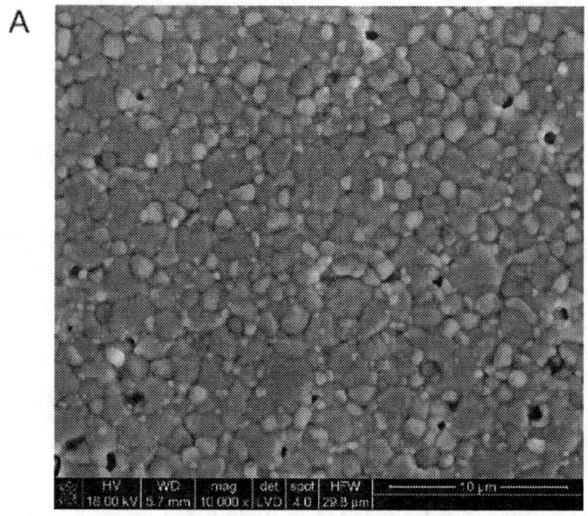

Figure 4a. Scanning electron microscopy image of the microstructure of a 10CBCY sample sintered at 1500°C for 2h.

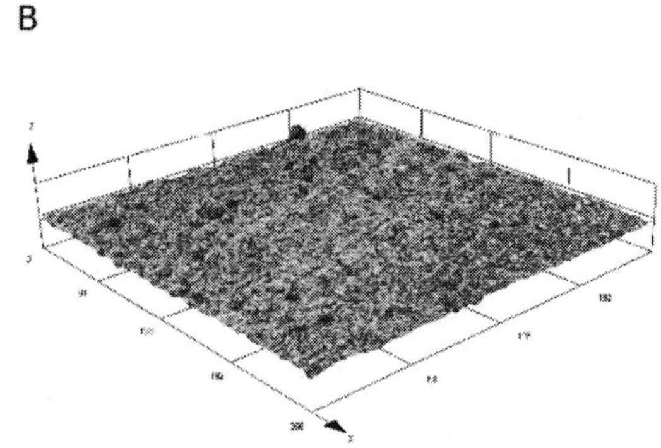

Figure 4b. Image of the surface of a 10CBCY sample sintered at 1500°C for 2h. The observations were carried out under optical confocal microscopy.

The data obtained for a commercial E-SOFC (Fuel Cell Material) was compared with the parameter declaration provided by the manufacturer [22].

As can be seen, current density (I) and power density (P) increase with rising temperature. The obtained power outputs P_{max} are very close to data provided by the supplier of the components of this SOFC. Figure 6 presents the family of voltage (U)-current (I) and of power (P)-current (I) curves recorded for an H_2,Pt|10CBCY|Pt,O_2 fuel cell over a range of 550–750°C. An increase was reported in the current density (I) as well as power density (P) in an SOFC involving a 10CBCY solid electrolyte. These results were compared with data in the literature. Matsumoto et al. [12] investigated an IT-SOFC in which a $BaCe_{0.9}Y_{0.1}O_{3-d}$ electrolyte with a thickness of 0.5 mm was tested. In the case of H_2,Ni-BCY|BCY|LSCF,O_2, he obtained a power output of 60 mW/cm^2 at 700°C. The main difference between the lower power output (15 mW/cm^2) obtained for a SOFC involving 10CBCY as an electrolyte and the SOFC investigated by Matsumoto can be directly attributed to the difference in the thickness

 DOI: 10.1051/e3sconf/20161000128

of the electrolyte and the different cathode and anode materials applied in the cited paper. The results obtained by the author in this study appear to be valuable for further electrochemical investigations.

C

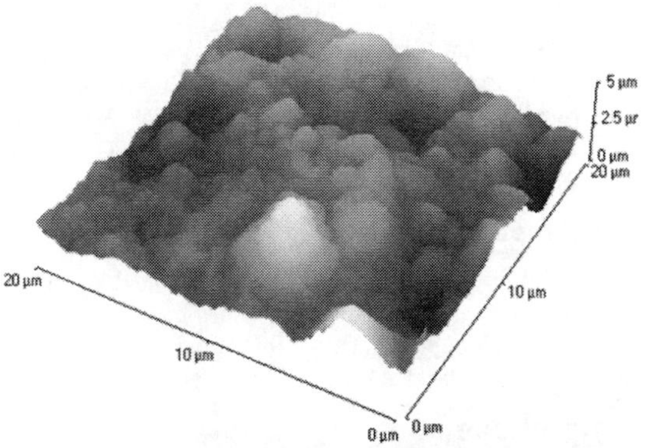

D

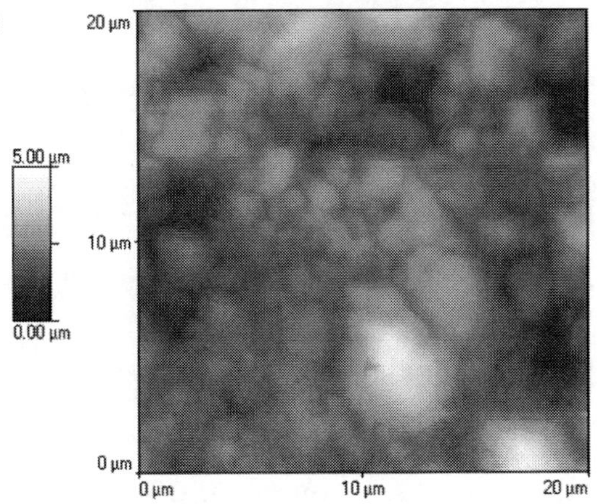

Figure 4.c,d Image of the surface of commercial 8YSZ ceramic electrolyte foil (supplier Fuel Cell Material, USA) obtained by atomic force microscope (AFM).

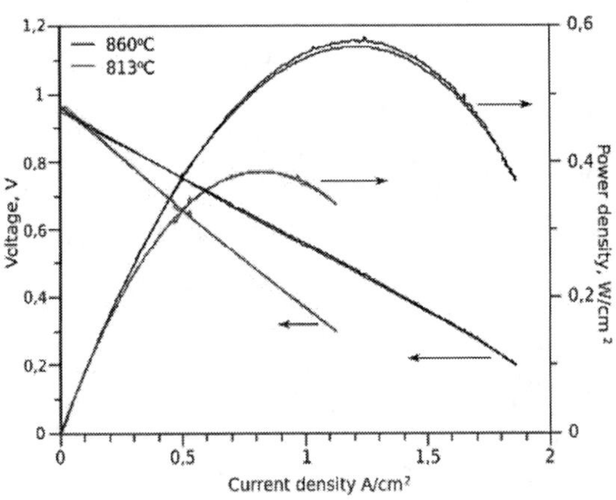

Figure. 5 The U-I curves recorded for an E-SOFC (Fuel Cell Material, SOFCs)

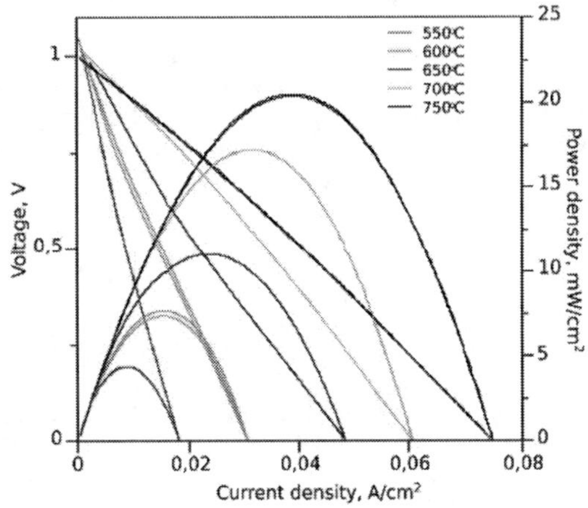

Figure 6. Voltage - Current density and Power density characteristics recorded for an H_2,Pt|10CBCY|Pt,O_2 fuel cell over a temperature range of 550–750°C

The electrochemical impedance spectroscopy (Figure 7) method was used to determine the impact of ohmic resistance (R_s) and (R_p) on the performance of the tested SOFC with a 10CBCY electrolyte. The data are collected in Table 1.

Table 1. The values of ohmic resistance (Rs) and polarisation resistance (Rp) determined for a H_2,Pt|10CBCY|Pt,O_2 fuel cell within a temperature range of 600–650°C

Temperature [°C]	R_s [Ω]	R_p [Ω]
600	14.9	18.1
650	11.1	5.9

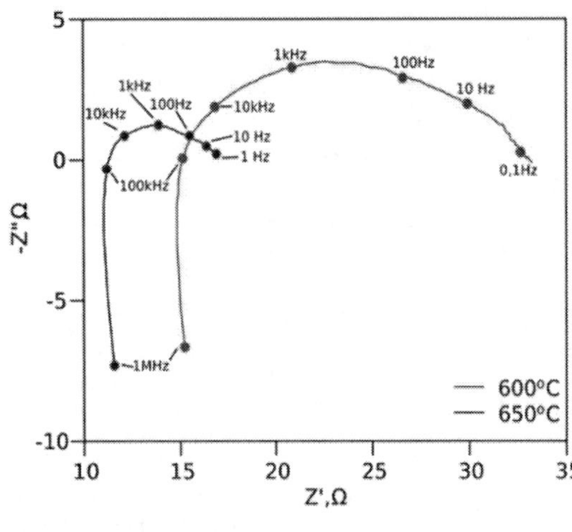

Figure 7. Electrochemical impedance spectra recorded for a H_2,Pt|10CBCY|Pt,O_2 fuel cell at various temperatures.

Based on data presented in Table 1, it was found that the values of Rs and Rp decrease with increasing temperatures of SOFC operation. This phenomenon is typical for SOFCs which are constructed correctly. The values of both R_s and R_p obtained by EIS method, are in good agreement with data obtained from U-I measurements.

Electrochemical impedance spectroscopy measurements were also performed in an oxygen-rich and a hydrogen-rich gas atmosphere. The recorded impedance plots in the Z''-Z coordination are presented in Figure 8a - b. The values of R_p and R_s are collected in Table 2.

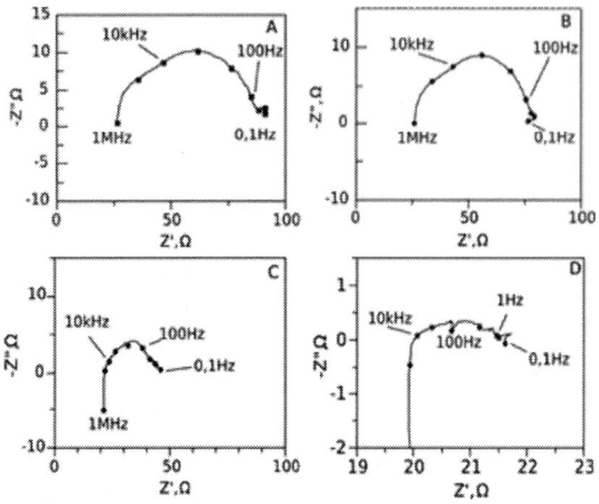

Figure 8. Electrochemical impedance spectroscopy of a symmetrical Pt|10CBCY|Pt cell in various gas atmospheres at 550°C: a) air, b) humidified oxygen (20%) with argon, c) fuel cell conditions; d) humidified hydrogen.

Table 2. Values of Rs and Rp recorded for Pt|BC10CY|Pt in different gas atmospheres

Gas atmospheres	R_s[Ω]	R_p[Ω]
H_2 (humidified)	19.9	1.7
Fuel cell condition, H_2 (humidified) + 20% O_2 in Ar (humidified)	2.3	24.8
20% O_2 in Ar (humidified)	26.0	26.0
Air	52.0	65.0

Analysis of data collected in Table 2 indicated that values of R_s and R_p decrease with increasing H_2 content in gas atmospheres. The highest values of ohmic resistance Rs and polarisation resistance Rp were found in an oxygen-rich atmosphere. In a highly rich oxygen gas atmosphere, $BaCe_{0.9}Y_{0.1}O_3$-based electrolytes exhibited rather mixed electronic conductivity. This caused an increase in values of polarisation resistance, especially in the cathode reduction process. On the other hand, in the case of a hydrogen-rich gas atmosphere, the BCY-based electrolytes exhibited practically pure proton conductivity. These results clearly indicated that cathode polarisation has a considerable impact on the performance of a SOFC with a BCY electrolyte.

Conclusions

$Ba_{0.9}Ca_{0.1}Ce_{0.9}Y_{0.1}O_{3-d}$ (10CBCY) proton-conducting disc-shaped ceramic electrolytes were obtained. Based upon confocal and scanning microscopy observation, no cracks or other surface defects were found for sintered samples of 10CBCY; only low closed-porosity content was observed in the case of 10CBCY sintered samples. The obtained 10CBCY proton-conducting ceramic was applied as an electrolyte for IT-SOFC operating within a temperature range of 550–750°C. An increase was observed in the performance of H_2,Pt|10CBCY|Pt,O_2 in this range. The results of these initial electrochemical tests appear to justify further investigations of an SOFC utilising 10CBCYwith reduced thickness.

Acknowledgements

I would like to thank Dr Magdalena Dudek for useful discussions and advices during experiments and manuscript preparation. This work was supported by the contract no. 15.11.201.345.

References

1. *Fuel Cell Handbook* (Seventh edition) by EG&G Technical Services, Inc Morgantown (2004)
2. W. Lewandowski Proekologiczne odnawialne źródła energii, Red; Warszawa WNT 2013, ISBN 978-83-7926-041-6
3. N.H. Behling *Fuel cells: Current Technology Challenges and Future Research Needs,* Elsewier Science Technology (2013)
4. A.B. Stambouli, E. Traversa Renew. Sust. Energ. Rev. **6** 433-455 (2002)
5. Z. Gao, L.V. Mogni, E.C. Miller, J.G. Railsback, S.A. Barnett, Energy Environ. Sci. **9** 1602-1644 (2016)
6. O.Z. Sharaf, M. F. Orhan, Renew. Sust. Energ. Rev. **32** 810-853 (2014)
7. P. E. Dodds, I. Staffell, A.D. Hawkes, F. Li, P. Grunewald, W. McDowall, P. Ekins, Int. J. Hydrogen Energy **40** 2065-2083 (2015)
8. P. Tomczyk, J. Power Sources **160** 858-862. (2006)
9. R.O.Stroman, M.W. Schuette, K. Swider-Lyons, J.A. Rodgers, D.J. Edwards, Int. J. Hydrogen Energy **39** 11279-11290 (2014)
10. A. Alaswad, A. Baroutaji, H. Achour, J. Carton A. Makky, A.G. Olabi, Int. J Hydrogen Energy (to be published) (2016)
11. D.J.L. Brett, A. Atkinson, N.P. Brandon, S.J. Skinner, Chem. Soc. Rev. **37** 1568-1578 (2008)
12. H. Matsumoto, I. Nomura, S. Okada, T. Ishihara, Solid State Ionics **179** 1486-1489 (2008)
13. A. Buonomano, F. Calise, M.D. Accadia, A. Palombo, M. Vicidomini, App. Ener. **156** 32-85 (2005)
14. S. Mekhilef, R. Saidur, A. Safari Renew. Sust. Energ. Rev. **16** 981-989 (2012)
15. Y. Xiong, K. Yamaji, T. Horita, N. Sakai, H. Yokokawa J. Electroch. Soc. **151** A407-A412 (2004)
16. J.G. Lee, J.H. Park, Y.G Shul, Nat. Commun. **5** 4045 (2004)
17. M. Dudek, Journal of the European ceramic society **28** 965-971 (2008)

DOI: 10.1051/e3sconf/20161000128

18. D.A. Medvedev, J.G. Lyagaeva, E.V. Gorbova, A.K. Demin, P.Tsakaras, Prog. Mater. Sci. **75** 38-79 (2016)
19. E. Fabbri, D. Pergolesi, E. Traversa, Chem. Soc. Rev. 39 4355-4369 (2010)
20. B. Timurkutluk, C. Timurkutluk, M. D. Mat, Y. Kaplan Renew. Sust. Energ. Rev **56** 1101-1121 (2016)
21. M. Dudek, B. Lis. A. Rapacz-Kmita, M. Gajek, A. Raźniak, E. Drożdż, Mater. Sci-Poland vol. **34** 1 (2016)
22. https://fuelcellmaterials.com/
23. http://www.kerafol.com/en/sofc/

 DOI: 10.1051/e3sconf/20161000129

Sustainability reporting in the energy sector

Barbara Kowal[1,a] and Arkadiusz Kustra[2]

[1]*AGH University, Department of Economics and Management in Industry, Al. Mickiewicza 30 30-059 Kraków, Poland*
[2]*AGH University, Department of Economics and Management in Industry, Al. Mickiewicza 30 30-059 Kraków, Poland*

Abstract. Development of the concepts of sustainable development and corporate social responsibility has a great impact on reporting in companies. The increase of their importance has resulted in a need to create a reporting system that would provide information on not only the methods but also the results of implementation of those concepts in companies. Globally, there are many organizations that promote and support companies in the area of integrated reporting. The most popular standard for reporting non-financial data that is used by a number of companies worldwide is the Global Reporting Initiative (GRI) Guidelines. The main objective of the GRI is to support the development of sustainable economy in which companies take responsibility for the economic, social, and environmental consequences of their operations, manage that responsibility, and report all their actions. An example of a sector where the concept of sustainable development and its transparent reporting has an impact on the formation of values is the energy sector, which creates value for stakeholders and, together with the financial sector, has the greatest impact on national economies.

1 Introduction

The problem of strategic management based on the concepts of sustainable development and social responsibility focuses on building value for stakeholders. The value is created in appropriate assets of the company that determine its tangible and intangible capital. The forms of disclosure of information about capital are frequently financial and non-financial, which demonstrates the need for company reporting in the form of reports prepared for stakeholders.

It must be noted that the need for financial information is fulfilled by reporting which is the product of financial accounting systems. In the light of the concept of sustainable development, another important issue is reporting on intangible areas related to the company's non-financial assets. In this regard, the solutions that are adopted, although they are not obligatory, focus on sustainability reports prepared based on the guidelines of the Global Reporting Initiative (GRI).

Sustainability reports are a source of information about the effectiveness of companies in non-financial areas that are related to economic, environmental, and social efficiency. The data disclosed in the aforementioned areas fill the information gap that is due to the limitations of standard financial reporting, which focuses on tangible materials owned by companies and the resulting financial aspects. At the same time, the information gap is the cause of underestimation of the sustainable value for the stakeholders. Thus, closing this gap and proper estimation of the value for the stakeholders must be based on both financial reporting and sustainability reporting, which together constitute integrated reporting.

The concept of integrated reporting is relatively new and, as it develops dynamically, constitutes an important source of information, not only for the shareholders. At the same time, it meets the requirement of full transparency, in particular in sectors and companies whose operations depend on and are determined by a large group of stakeholders.

An example is the energy sector whose functioning often constitutes a strategic activity in a specific economy. The present article presents solutions related to the concept of reporting of sustainable development of energy sector entities and attempts to integrate this reporting with data from financial reports.

2 Importance of the energy sector to the economy

The energy sector, next to the financial sector, determines the strength of national economies and the possibility of their growth. The impact of the sector on the economy should be considered in a balanced approach, taking into account financial, fiscal, social, welfare, and environmental aspects.

In Poland, the share of the energy sector in the gross domestic product (GDP) was about 8% in 2014, 3.8% out of which has a direct impact. At the same time, the energy sector has a significant impact on the functioning of other industries, such as mining, commerce, processing, construction, and transport. The energy sector, together with its associated sector, employs over 600,000 persons or 3.3% of all employees

[a] Corresponding author: bkowal@agh.edu.pl

DOI: 10.1051/e3sconf/20161000129

in the economy; as a result, the social aspects of its operations is significant [7].

The profitability of Polish companies is above average, even though revenues of companies in energy sectors are decreasing due to lower prices. The financial results of energy sectors translate into the taxes they pay. Of the 100 largest companies that pay the highest corporate income tax in 2013, about 20 are companies from the energy sector, which have paid in total about 2.3 billion zlotys in taxes [8].

The largest final consumers of energy are households (approx. 31%), transport (27%), and industry - mostly construction (23%) and services (13%). The energy intensity of the Polish economy is gradually decreasing as a result of more efficient generation sources and lowed demand by individual consumers and businesses. Despite the drop of energy consumption in Poland, the demand reported by the economy is much higher than in countries of the European Union [7].

According to government forecasts, by 2030, despite the increase of the GDP, there may be a "zero energy" increase of demand for energy.

According to a PwC report published in 2016, the biggest challenges to the Polish energy sector include [6].

- maintaining profitability of production taking into account a mix of sources;
- innovativeness associated with business objectives;
- concentration on customers and offering them suitable and uncomplicated services;
- further integration on the operational level that ensures competitive advantage; and
- digitalization of decision-making processes.

3 Sustainable development of the energy sector in the light of value creation

Due to their special and important role in the economy of the country and their strategic importance, companies in the energy sector must focus not only on building their value but also on increasing that value. Consequently, their supreme objective is to generate value for all groups of stakeholders, which include, among others, shareholders, owners, members, the management board, customers and employees, as well as collaborating companies, financial institutions and investors (Figure 1).

Due to the large number of stakeholders, efforts to meet their expectations are not easy but they determine the priorities in operations. According to the study on awareness of Corporate Social Responsibility in Poland, stakeholders identify three main issues on which companies from the energy and fuel sector should concentrate [3]:

- care for the national environment (81%);
- care for the safe labor of employees (81%);
- ensuring secure supply of energy (80%).

All such activities must be subordinated to the supreme objective and, at the same time, must meet the criteria of sustainable development and social responsibility. Value creation by companies from the energy sector should be performed in a sustainable manner.

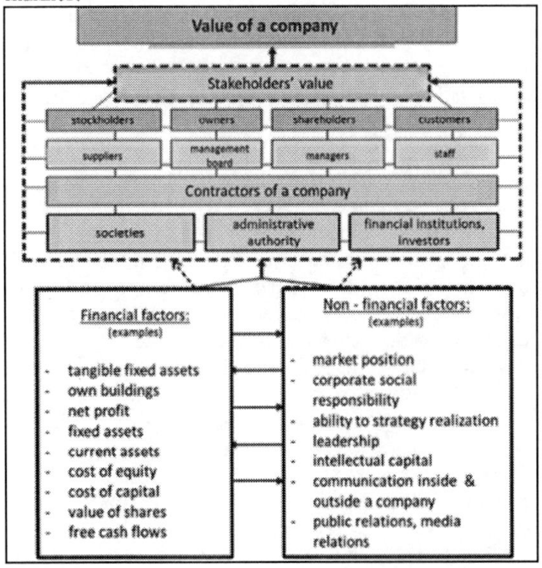

Figure 1. Elements of company's value creation (Source: author's own study)

This means combining tangible and intangible aspects (factors) on the one hand and financial, social, and environmental aspects on the other hand. Taking into account the aforementioned factors that influence generation of value of such companies leads to creation of a durable, sustainable value for the company, its shareholders, and the society at the same time.

4 Integrated reporting

The increase of the importance of the notion of sustainable development, or the concept of corporate social responsibility which has grown on its basis, has resulted in a need to create a reporting system that will provide information on the methods and results of implementation of those concepts in companies. The result of inclusion of environmental and social aspects in reporting on company operations is integrated reporting, described as expanded business reporting [1]. An integrated report is a document in which the picture of the operations and the achievements of an entity (both financial and non-financial) is presented in accordance with the concept of environmental and social effectiveness, and corporate governance.

Globally, there are many organizations that not just promote but support companies in the area of integrated reporting; these include International Standards Organization, Sustainability Accounting Standards Board, Social Accountability International, and UL Environment. However, the following two organizations have gained the best reputation in this area: Global Reporting Initiative and International Integrated Reporting Committee (IIRC). The objective of the GRI is to ensure the best possible combination and continuous improvement of the guidelines for preparation of sustainability reports and an integrated reporting framework. The IIRC, on the other hand, strives to provide tips concerning links between financial reporting and sustainability reporting. The latter organization is also committed to encourage companies to use the guidelines created by the Global Reporting Initiative when writing their reports [2]. Both organizations and their recommended guidelines are mutually supplementary.

5 GRI standard guidelines

The most popular standard for reporting non-financial data is the GRI Guidelines. For many years worldwide, the Guidelines have become the standard for reporting on sustainable development and corporate social responsibility. The Guidelines are a voluntary tool that can be used by companies to prepare integrated reports that take into account the non-financial aspects of their operations, which are so important to the creation of the value of the companies. The Guidelines include principles and indicators that companies can use to measure and report their results in the economic, environmental, and social areas. The first version of the Guidelines, G1, was published in 2000 and the version that is valid now is the fourth version of the Guidelines, G4, published in 2013 [4]. The latest version makes it possible to create reports using quantifiable and comparable data. It expands the scope of social reporting in the area of involvement of stakeholders in the process of preparation of the report and emphasizes the role of the value chain.

Companies that use the GRI Guidelines in their reporting process become more attractive to their stakeholders, thus contributing to a more sustainable business practice. Moreover, the Guidelines quite flexible, which enables their use by various businesses, regardless of their size, sector, or location.

6 Example

Due to the unique nature of its operations and its impact on the natural environment, the energy sector should use integrated reporting. Examples of energy sector companies that present reports from their operations in an integrated manner are Iberdrola and Vattenfall.

Iberdrola is a Spanish energy sector group which has been present in the Polish renewable energy market since 2005. It focuses on designing, building, and operating wind farms [9]. Vattenfall, on the other hand, is a European energy sector corporation owned by the Swedish government. It develops and offers to its customers non-standard products based on the gas, electricity, renewable energy sources, and emission allowances markets. It also conducts trade on exchanges and OTC markets in such countries as Czech Republic, Hungary, and Slovakia [10].

Both companies actively participated in the study performed by Tim Steinweg and Joseph Wilde-Ramsing of the Centre for Research on Multinational Corporations (SOMO) as a part of the project Use of the Global Reporting Initiative in Sustainability Reporting by European Electricity Companies [5]. They reviewed sustainability reporting in 20 European energy sector companies. Their study consisted in checking:

- whether companies use the GRI Guidelines and, if they do, on what level is their application and whether their reports are internally verified;
- whether companies use the Electricity Utilities Sector Supplement (EUSS) and, if they do, they were asked to state 10 indicators;
- and in pointing at gaps in the reports of the studied companies and trying to find their causes.

Table 1. Short information on the reporting by electric utility companies (Source: author's own study on the basis of [5])

	Published a CSR	Use of GRI	Use of the EUSS
Yes	19	18	13
No	1	2	7

The aforementioned companies were among the best, i.e. in the 95% of companies that publish social responsibility reports, in the 90% that use the GRI Guidelines in their reporting, and in the 60% that use the EUSS. Also, along with four other companies, their reports had applications on the highest level, which indicates broad use of GRI Guidelines and internal verification of their reports.

In the review of selected 10 indicators, Iberdrola and Vattenfall also received good scores as they reported either fully or partly on each specific indicator.

In the second part of the study, companies that were found to have any discrepancies received a draft with tips and a number of questions concerning the discrepancies that were found. The companies were allowed two weeks to answer the questions and to make corrections to the draft. Only seven companies, or 36%, participated in this part of the study. Iberdrola and Vattenfall did participate, along with such companies as Dong, Fortum, RWE, Statkraft, and Verbund. Cooperation of those companies contributed to preparation of certain recommendations for various groups of stakeholders aimed to improve corporate sustainability, in particular to increase the accuracy of reporting and the credibility of the system used.

7 Summary

In the recent years, the great impact of the concepts of sustainable development and corporate social responsibility on company reporting, also in the energy sector, has been observed. As a result, there has been pressure on improvement of guidelines that cover presentation of information from the ESG area. The cohesive framework created by the Global Reporting Initiative and sustainable reporting are used worldwide. As the above examples show, European energy sector companies know the GRI Guidelines and use them widely. The scope of the information that is disclosed covers intangible aspects of creation of value and effectively supplements the tangible data coming from the traditional financial reporting.

 DOI: 10.1051/e3sconf/20161000129

Bibliography

1. J. Krasodomska, Sprawozdawczość zintegrowana – aspekty teoretyczne i praktyczne [Integrated reporting - theoretical and practical aspects], in: M. Remlein, ed., Rachunkowość odpowiedzialności społecznej – teoria i praktyka [Accounting of social responsibility - theory and practice].
2. M. Mazurowska, *Studia Oeconomica Posnaniensia*, **3**, 1 (2015)
3. Report: Challenges to sustainable development of the energy sector, http://www.pwc.pl/pl_PL/pl/publikacje/PwC_Odpowiedzialna_energia.pdf, accessed on 23 April 2015
4. E. Różańska, *Studia Oeconomica Posnaniensia*, **3** (2015).
5. T. Steinweg, J. Wilde-Ramsing, *Use of the Global Reporting Initiative (GRI) in Sustainability Reporting by European Electricity Companies*, Amsterdam, December 2012.
6. Report: 7 temptations of the Polish energy sector, PwC 2016.
7. Report: Directions 2014 - the energy sector DNB Bank, Deloitte 2014.
8. http://www.sii.org.pl/7595/aktualnosci/newsroom/najwieksi-platnicy-cit-w-2014-roku.html
9. http://www.phig.pl/pl/firma/iberdrola-renewables-polska-sp-z-o-o/
10. https://corporate.vattenfall.com/about-vattenfall/contact-us/vattenfall-poland/

 DOI: 10.1051/e3sconf/20161000130

The use of CeO_2-Co_3O_4 oxides as a catalyst for the reduction of N_2O emission

Maria Rajska[1,a], Bartosz Wójtowicz[1], Łukasz Klita[1], Łukasz Zych[1] and Rafał Żybała[2]

[1]*AGH University of Science and Technology, Faculty of Materials Science and Ceramic, Al. Mickiewicza 30, 30-059 Krakow, Poland*
[2]*Faculty of Materials Science and Engineering, Warsaw University of Technology, Wołoska 141, 02-507 Warsaw, Poland*

Abstract. The morphological characterization of a series of cobalt-cerium oxide composites prepared by the deposition of CeO_2 onto Co_3O_4 powder with a molar ratio of cerium oxide to Co_3O_4 in the range of 0 to 1 was performed. The powders were also impregnated with a solution of K_2CO_3 to obtain the theoretical content of potassium atoms 2at·nm^{-2}. To investigate the effect of adding specific amount of CeO_2 on the catalytic activity, the X-ray diffraction, SEM-EDX, laser particle size distribution and BET surface area measurements were used. The catalysts were tested through the low-temperature decomposition of nitrous oxide in the temperature range of 50°C to 700°C. The addition of CeO_2 and K always moved the temperature of a complete N_2O conversion towards lower temperatures (480°C-540°C to 340°C-420°C). The best catalytic properties were shown by the samples in which the ratio of cerium oxide to cobalt oxide ranged from 0.4 to 0.7.

1 Introduction

N_2O emission to the atmosphere is considerably less intense than in the case of other greenhouse gases, but the GWP (Global Warming Potential) of nitrous oxide is 310 times higher than for carbon dioxide [1]. It is estimated that the current contribution of N_2O oxide to the total greenhouse effect is approximately 6.5 vol.% [2]. Nitrous oxide has the ability to accumulate infrared radiation and indirectly destroy the ozone layer through the chemical reactions of substances formed from N_2O. In the upper stratosphere, compounds derived from nitrous oxide, reacting readily with ozone, are formed, thus destroying the natural protective layer of the Earth [3].

The reduction of N_2O can be carried out in two different temperature ranges, which resulted in selecting catalysts suitable for the pre-defined reaction conditions. In terms of higher temperatures, i.e. these ranging from 800°C to 900°C, for the reduction of N_2O systems characterized by good thermal stability, including spinels, perovskites and mayenite [4-9], are used. For the lower temperature range between 250°C and 450°C, zeolites, known as oxide catalysts with a high specific surface area and a small amount of precious metals or transition metal oxides are applied [10-13]. It was found that cordierite could be a very good catalyst support for other catalytically active phases in deN_2O reaction [14]. This kind of catalysts, based on spinel K/Zn-Co_3O_4 are successfully checked in pilot plant conditions, with maintaining thermal stability, structure, morphology and high catalytic activity [15].

Lowering of the temperature of the N_2O reduction reaction is economically advantageous, at the same time meaning other chemical composition of the catalyst than in the case of high-temperature reduction. At present, catalysts with different promoters used for the low-temperature reduction of N_2O have to act selectively in a catalytic reduction process [1]. The catalysts based on a Co-Ce oxide composite with different promotors were prepared. The alkali metals (Na, K, Rb, Cs) showed a better catalytic activity in the case of the low-temperature reduction of N_2O than the alkaline earth metals (Mg, Ca, Sr, Ba). The promotion effect followed the Li<Na<K<Rb<Cs order for the alkali metals, and the Mg<Ce<Sr<Ba order for the alkaline earth metals. The alkali metals may act as electronic promoters and thus could be improved the redox ability Co^{2+} [1]. Grzybek and co-workers analysed a series of Co_3O_4-CeO_2 catalysts prepared by the incipient wetness impregnation of CeO_2. The Co-Ce catalyst (1wt.% Co_3O_4-1wt.%CeO_2) exhibits the highest apparent reaction rate per mass of Co_3O_4. The authors proposed an explaination that the observed phenomenon entailing the high catalytic activity of the Co_3O_4-CeO_2 catalyst with 1wt.% content results from the dispersion effect. The catalytic activity of the Co-Ce composites was intensified as a consequence of facilitating the electron donation process, which improved the reducibility of Co^{3+} to Co^{2+} [16]. It was found that the catalysts with a high efficiency during the low-temperature N_2O reduction reaction contained rhodium supported on zirconia, praseodymium-doped ceria or mixed systems incorporating titanium oxide and cerium oxide, additionally coated with rhodium

[a] Corresponding author: rajska@agh.edu.pl

[17-19]. However, research into the cobalt-based catalysts and promoted alkali metals: sodium and potassium shows that it is possible to achieve complete conversion at a temperature below 450°C [20,21]. Such catalysts are very active in this reaction and do not contain precious metals, which consequently further reduces their cost [22]. The presence of potassium ions significantly improves the catalytic properties of cobalt-containing catalyst systems or manganese spinels [23,24]. Potassium-promoted cobalt spinel catalysts obtained from the K_2CO_3 precursor have both high stability and low work function [23].

In the present work, we report on the catalytic activity of a series of K-promoted Co-Ce oxide composites with the molar ratio of CeO_2/Co_3O_4 oxides ranging from 0 to 1 respectively. We try to find the optimum range of molar participation of two oxides forming a composite material.

2 Experimental

A series of Co-Ce composite oxide catalysts was prepared by the deposition of CeO_2 onto Co_3O_4 powder (99%, Merc), as shown in Table 1. Samples were obtained through the precipitation of the precursor of CeO_2 ($Ce(NO_3)_2 \cdot 6H_2O$) on the Co_3O_4 powder at pH = 10±1 using a 3M ammonia solution [25,26]. The resultant precipitate was aged for two hours at room temperature and then several times eluted from the excess of hydroxide ions to obtain the near-neutral pH. The slurry prepared in this manner was dried for 12 hours at 40°C and calcined for 3 hours at 500°C [27]. The resultant powders have a different molar content of CeO_2 (from 0.0 to 1.0 mol) and a constant content of Co_3O_4 (1 mol).

The impregnation of the catalysts was made using the dry impregnation method with $2K_2CO_3 \cdot 3H_2O$ (to obtain 2 atoms $K \cdot nm^{-2}$) in an amount sufficient to fill the pores of the support [28]. Before performing the impregnation of each sample the sorption capacity was measured twice by titrating the carrier in distilled water.

Table 1. The sample symbols.

Sample	Before impregnation	After impregnation
Co_3O_4	Co	Co_K
Co_3O_4-0.1CeO_2	Co0.1Ce	Co0.1Ce_K
Co_3O_4-0.2CeO_2	Co0.2Ce	Co0.2Ce_K
Co_3O_4-0.3CeO_2	Co0.3Ce	Co0.3Ce_K
Co_3O_4-0.4CeO_2	Co0.4Ce	Co0.4Ce_K
Co_3O_4-0.5CeO_2	Co0.5Ce	Co0.5Ce_K
Co_3O_4-0.6CeO_2	Co0.6Ce	Co0.6Ce_K
Co_3O_4-0.7CeO_2	Co0.7Ce	Co0.7Ce_K
Co_3O_4-0.8CeO_2	Co0.8Ce	Co0.8Ce_K
Co_3O_4-0.9CeO_2	Co0.9Ce	Co0.9Ce_K
Co_3O_4-1.0CeO_2	Co1.0Ce	Co1.0Ce_K

2.1 Methods

The morphology of the particles was investigated using scanning electron microscopy with energy dispersive spectroscopy (SEM-EDX) using the NOVA NANO SEM 200 scanning electron microscope from FEJ EUROPE. The crystalline size of the powders was evaluated based on X-ray diffraction measurements (XRD) performed using the Philips Analytical X'PERT PW-3710 equipped with an adapter for a low angular geometry GID. The specific surface area of the samples was determined by the BET method using nitrogen adsorption-desorption isotherms with ASAP 2010 Micrometrics. The particle size distribution (DLS) was characterized by a laser method employing the Malvern MasterSizer 2000.

2.2 Catalytic measurements

A fixed-bed, quartz flow reactor connected to a quadrupole mass spectrometer (QMS, Thermostar QSD 300, BALZERS) was used in the performed catalytic tests. Before proper catalytic measurements, for each sample (0.3 g) a standardization procedure, consisting of keeping the sample in a helium atmosphere at a gas flow of 100 cm^3min^{-1} for 30 min until reaching 500°C, was applied. After this treatment, the reactor was cooled down to room temperature and the used gas was changed to the target mixture.

The catalytic tests were carried out at a gas flow of 100 cm^3min^{-1}, in the temperature range of 50°C to 700°C, with a gas mixture containing 1000 ppm N_2O in He. For a more complete characterization, all catalytic measurements were performed three times for each sample. To characterize the catalytic activity in a more precise manner, the temperature at which the conversion of N_2O to O_2 and N_2 was 50% (T_{50}) and 90% (T_{90}), respectively, was determined.

3 Results

3.1 XRD

Figure 1 presents the XRD spectra of all samples before impregnation. The identified reflections are from the cobalt (II, III) oxide phase (ICDD 04-006-9070) and the cerium dioxide phase (ICDD 03-065-5923).

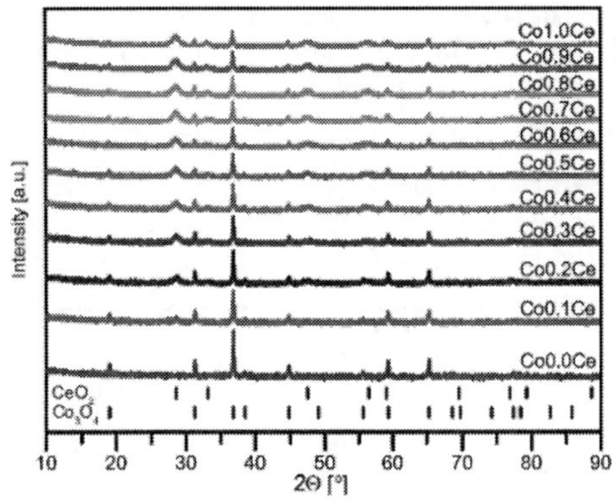

Figure 1. The XRD patterns of Co_3O_4-CeO_2 catalysts before impregnation.

In the XRD spectra, no other phase containing Co and/or Ce was observed. Similarly, in Figure 2 one can see reflections from all phases after the impregnation process. After impregnation, reflections were not detected from any of the compounds containing potassium. The CeO_2 crystallite size, regardless of its content in the sample, was approximately the same in the case of all samples, reaching about 27 nm. The diffraction lines of Co_3O_4 were changed after the addition of CeO_2. They were broadened.

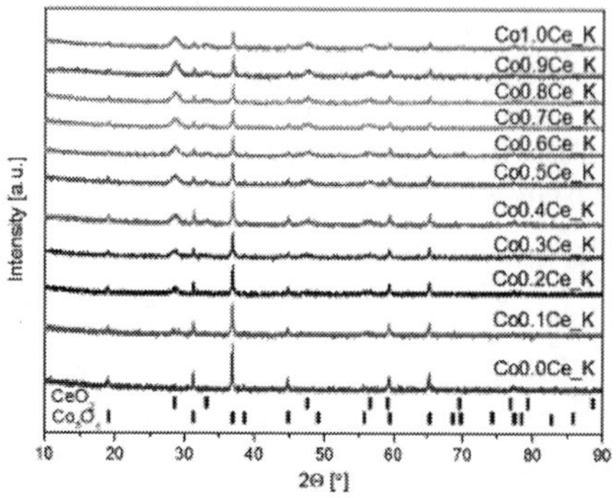

Figure 2. The XRD patterns of Co_3O_4-CeO_2 catalysts after impregnation.

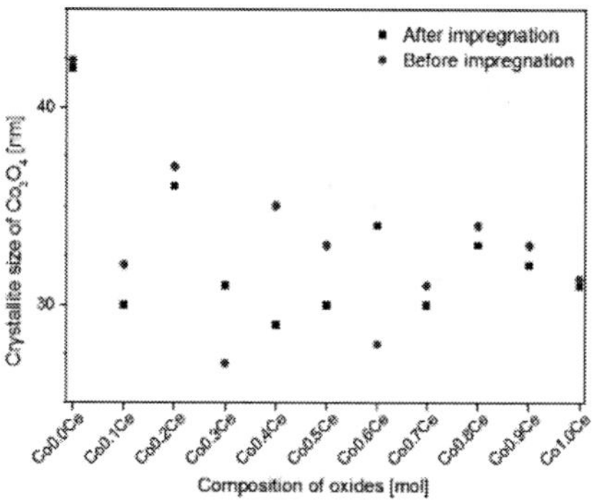

Figure 3. The crystallite size for Co_3O_4.

It was calculated that the crystalline size of the pure cobalt (II, III) oxide phase was 42 nm and after having added CeO_2 it was reduced to about 32 nm. No clear trend for a systematic decrease or increase in the size of the crystallite of cobalt oxide (II, III) was shown with increasing the content of CeO_2.

3.2 BET

The specific surface area of catalysts was enlarged from 8.67 m^2g^{-1} to 49.61 m^2g^{-1} before impregnation and from 8.21 m^2g^{-1} to 43.46 m^2g^{-1} after impregnation respectively.

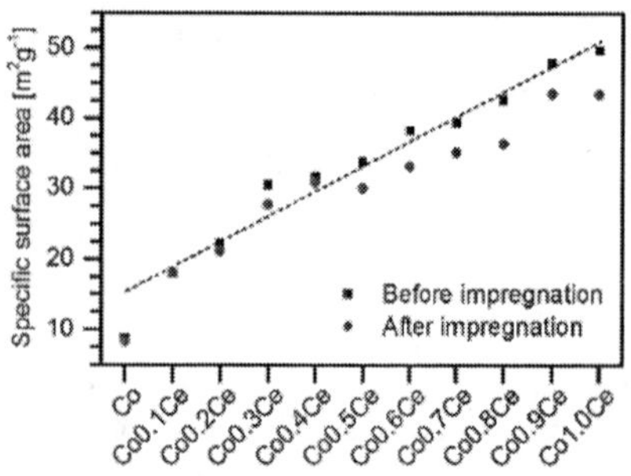

Figure 4. The specific surface area of catalysts.

Figure 4 presents the specific surface area of samples when increasing the CeO_2 content. When compared to cobalt oxide, a systematic development of the specific surface area (as indicated by the black line) was observed with increasing the content of cerium oxide. After impregnation, the specific surface area was reduced for each sample because of filling the pores with potassium.

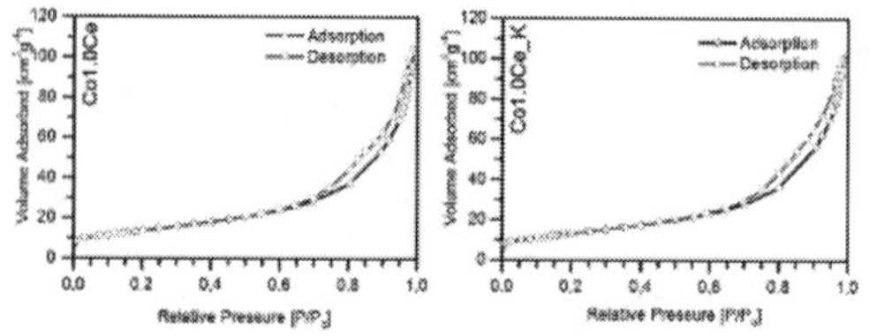

Figure 5. The exemplary isotherm adsorption-desorption curves.

Figure 5 is a collection of graphs showing desorption and adsorption for samples Co1.0Ce and Co1.0Ce_K. It can be seen that as a result of impregnation the size of surface pores was similar and there was an almost unnoticeable change in the specific surface area. For the remaining samples, the hysteresis curve of adsorption-desorption resembled the results presented in Figure 5. The average adsorption pore diameter before impregnation was calculated at 13.2 nm and after impregnation at 13.0 nm. The hysteresis curve, according to the de Boer classification, indicates the type of pores which have an elongated, tubular shape. At the same time, the hysteresis loop of adsorption and desorption isotherms represents the case in which adsorption involves the formation of a multilayer adsorbate film according to the principles of a classical BET theory.

3.3 Particle size distribution

The particle size distribution with respect to the quantity and volume of the selected samples, before and after impregnation, is presented in Figure 6. The mean particle size determined on the basis of number of particles in the case of sample Co0.0Ce is 64.5 nm both before and after impregnation with potassium ions. For this sample, before impregnation, was determining particle size on the basis of volume distribution and there are three peaks when it comes to the maximum particle size, i.e. 120 nm, 10 μm and 180 microns. The presence of such substantial grains may indicate the formation of large agglomerates, which was confirmed by the SEM examination. A similar dependence was found in the case of the same sample after impregnation (Co0.0Ce_K), where the two marked peaks were observed for the particle size of about 130 nm and 90 microns, which is also connected with the formation of large agglomerates of powder.

Figure 6 shows the volume of grains as a percentage relative to the particle size for another sample, Co0.5Ce. The largest amount of particles (in terms of volume) is observed for the particle size range of approx. 590 nm to approx. 420 microns for the sample before and after impregnation. However, the grains larger than 600 nm are actually agglomerates which was confirmed by quantitative analysis and morphological SEM observations. The quantitative percentage of the samples

DOI: 10.1051/e3sconf/20161000130

showed that for the majority of particles the size was in the 35-275 nm range. Such a marked difference between the analysis of the particle size distribution showed as a volume percentage and the quantitative percentage was noticed due to the presence of a few but large agglomerates in powder (e.g. over 10 microns). The particle size distribution of the powder with the highest content of CeO_2 (Co1.0Ce) does not differ from the previously described ones. A cumulative curve based on a number of particles for samples Co1.0Ce and Co1.0Ce_K shows that the average grain size in both cases was about 56 nm. For samples Co0.0Ce and Co0.0Ce_K it equalled 65 nm and 56 nm and for samples Co1.0Ce and Co1.0Ce_K it reached 56 nm and 64 nm respectively.

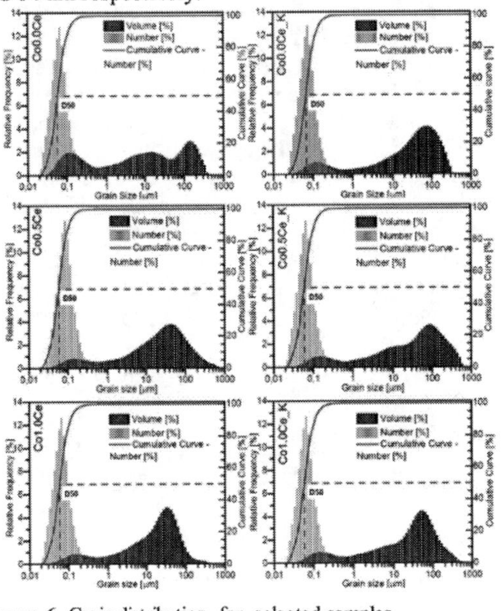

Figure 6. Grain distribution for selected samples.

The weighted average particle size was calculated and collected in Figure 7 based on the data obtained when analysing the particle size distribution for a number of particles.

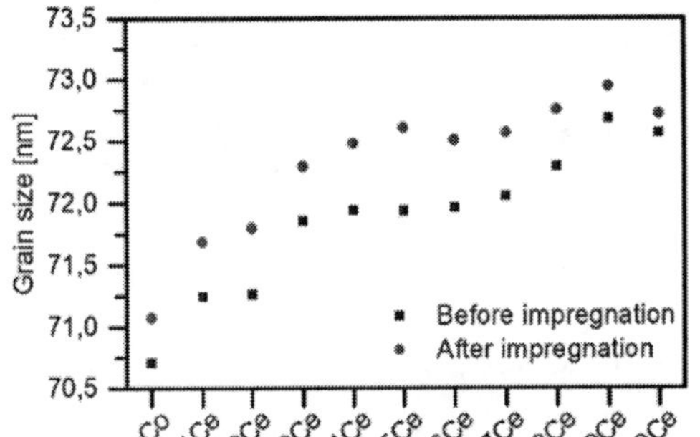

Figure 7. The weighted average particle size for all samples.

It was observed that the higher the content of CeO_2, the larger the weighted average grain size that was calculated in relation to the measurements of the number grains. These results coincide with the information obtained from the BET analysis.

3.4 SEM

The selected catalysts are shown in the micrographs collected in Figure 8. All grains have irregular shapes and form large agglomerates, measuring about a few microns. The shape of these agglomerates is the same for all samples, irrespective of whether the amount of CeO_2 is low (0.1 mol CeO_2) or high (1 mol CeO_2). The chemical analysis in terms of chemical composition showed no significant deviation from the assumed chemical composition, as shown in Table 2 and Table 3.

The average size of grains determined based on the images was about 0.12 µm. After impregnation, the shape and size of agglomerates did not change significantly. These observations are consistent with the data obtained from the analysis of the particle size distribution.

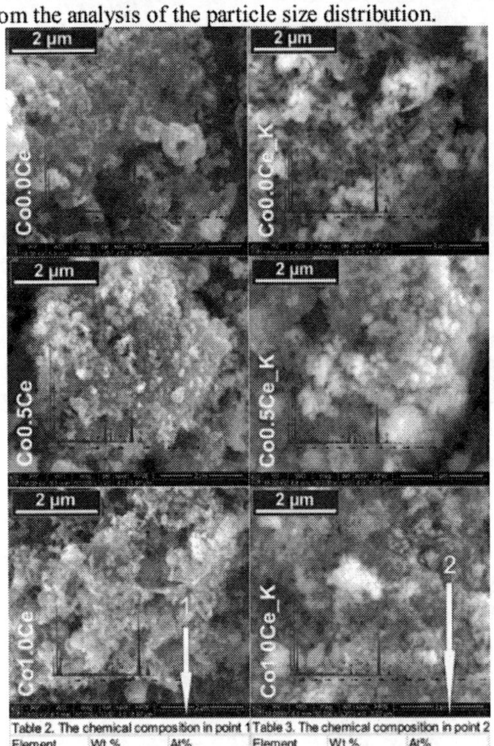

Table 2. The chemical composition in point 1

Element	Wt %	At%
O	17.27	45.48
Ce	11.17	3.36
Co	71.56	51.16

Table 3. The chemical composition in point 2

Element	Wt %	At%
O	30.37	63.87
K	0.75	0.64
Ce	11.6	2.79
Co	57.29	32.71

Figure 8. SEM images of catalysts.

3.5 Catalytic tests

The Co-Ce catalysts were active in N_2O decomposition, which is shown in Figure 9 and Figure 10. The N_2O conversion was negligible up to about 350°C for all samples. It was observed that the addition of cerium oxide, regardless of its quantity, always improves the catalytic properties of pure Co_3O_4. In addition, ceria oxide can stabilize metal oxide supports and modify the catalytic activity by changing the redox properties of metal oxides [22].

The temperature at which the system reaches the N_2O decomposition of 50% is on average lower than about

DOI: 10.1051/e3sconf/20161000130

30°C for samples containing CeO_2. The same tendency is observed for the temperature determining the conversion of 90%. T_{90} is lower for the samples containing CeO_2 and, at the same time, 60-80 Celsius degrees higher than T_{50}, as shown in Figure 11.

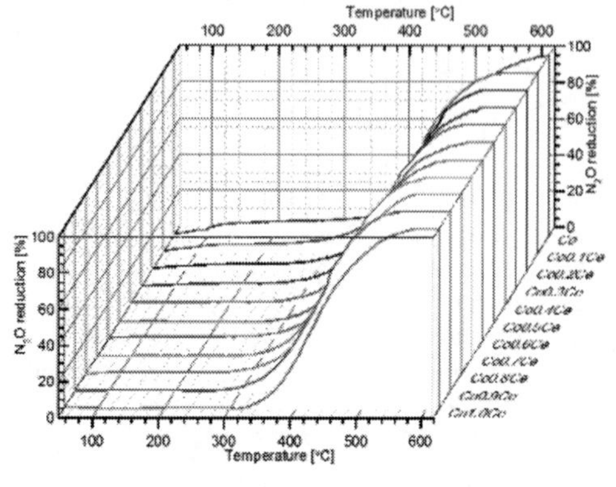

Figure 9. The N_2O conversion before impregnation.

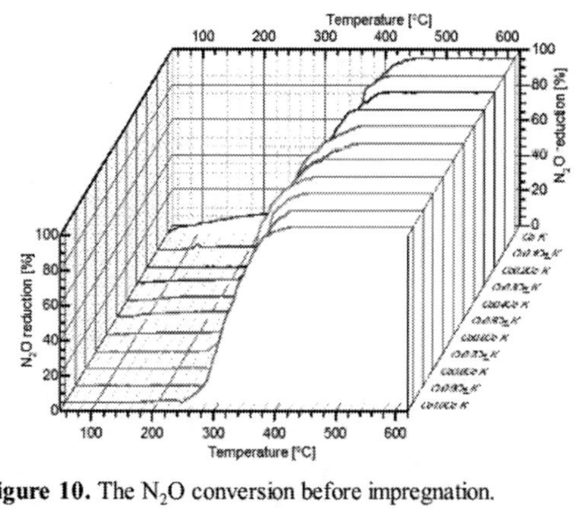

Figure 10. The N_2O conversion before impregnation.

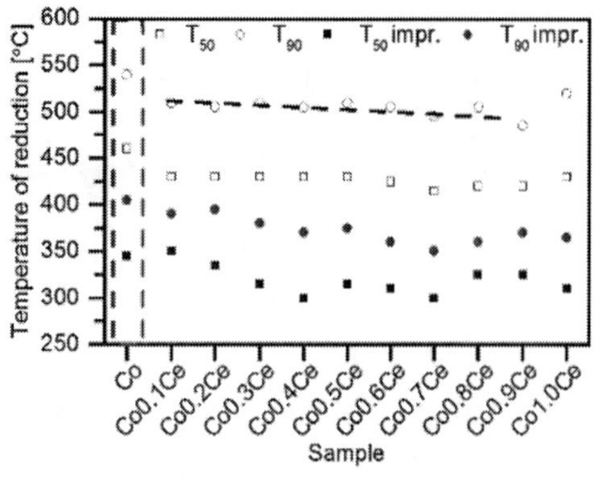

Figure 11. Temperatures for N_2O conversion of 50% and 90%.

After impregnation, a significant improvement in the catalytic properties of all analyzed samples was demonstrated. The temperatures at which complete conversion was achieved were shifted towards lower values, which is presented in Figure 10. The catalytic activity was improved to the biggest extent for the samples in the case of which the CeO_2/Co_3O_4 molar ratio was in the range of 0.4 to 0.7. Lowering of the temperature at which the system reached the N_2O conversion of 50% was observed from approximately 130 to 210°C. The sample showing the best catalytic properties was Co0.4Ce_K.

4. Summary

A series of Co-Ce oxides composites was characterized by the XRD, SEM-EDX, DLS and BET techniques. All analysed catalysts were tested for the low-temperature N_2O decomposition.

The results suggested that the addition of cerium oxide causes an increase in the specific surface and an improvement in the catalytic activity. The specific surface area of the catalysts obtained after impregnation was reduced by filling the pores with potassium. X-ray examination carried out for a series of catalysts obtained by the precipitation of CeO_2 on Co_3O_4 showed the presence of phases from CeO_2 and Co_3O_4 oxides only. The crystallite size calculated using the Scherrer equation showed that for an impregnation process the size reduction of the crystallite derived from the Co_3O_4 phase occurred in the case of all samples. The specific surface area was enlarged when increasing the CeO_2 content for all samples. It was also found that the micropores in powders have an elongated cylindrical shape. The addition of CeO_2 considerably lowered the temperature at which the system reached the conversion of 90%.To sum up, impregnation exerts the most significant impact on the improvement of the catalytic properties.

Acknowledgment

M. Rajska acknowledges the Statutory Project for Science (Grant for Young Scientists no.15.11.160.763) 2015 at the Department of Inorganic Chemistry, Faculty of Materials Science and Ceramics AGH UST.

This work has been partially financed as a research project no. DEC-2014/12/S/ST8/00582 from the resources assigned for science by National Science Centre (NCN, Poland) and partially financed by the National Centre for Research and Development (NCBR, Poland) under the project no. PBS3/A5/49/2015.

References

1. L. Xue, H. He, Ch. Liu, Ch. Zhang, B. Zhang, Environ. Sci. Technol. **43**, 890 (2009)
2. E.A. Davidson Environ. Res. Lett. **7**, 024005 (2012)
3. E.A. Davidson, D. Kanter Environ. Res. Lett. **9**, 105012 (2014)
4. U. Chellam, Z.P. Xu, H.C. Zeng, Chem. Mater. **12**, 650 (2000)
5. J. Pérez-Ramiréz, F. Kapteijn, Appl. Catal. B **47**, 177 (2004)
6. Y. Wu, C. Dujardin, C. Lancelot, J.P. Dacquin, V.I. Parvulescu, M. Cabié, C.R. Henry, T. Neisius, P. Granger J. Catal. **328**, 236 (2015)
7. P. Granger, P. Esteves, S. Kieger, L. Navascues, G. Leclercq Appl. Catal. B **62**, 236 (2006)
8. J. Petryk, E. Kołakowska, Appl. Catal. B, **24**, 121 (2000)

9. E. Iwanek, K. Krawczyk, J. Petryk, J.W. Sobczak, Z. Kaszkur, Appl. Catal. B, **106**, 416 (2011)
10. E. Pachatouridou, E. Papista, A. Delimitis, M.A. Vasiliades, A.M. Efstathiou, M.D. Amiridis, O.S. Alexeev, D. Bloom, G.E. Marnellos, M. Konsolakis, E. Iliopoulou, Appl. Catal.,B **187**, 259 (2016)
11. I.V. Yentekakis, G. Goula, P. Panagiotopoulou, S. Kampouri, M.J. Taylor, G. Kyriakou, R.M. Lambert Appl. Catal., B **192**, 357 (2016)
12. L. Obalová, M. Reli, J. Lang, V. Matějka, J. Kukutschová, K. Kočí, Catal. Today **209**, 170 (2013)
13. D. Pietrogiacomi, M.C. Campa, L.R. Carbone, S. Tuti, M. Occhiuzzi, Appl. Catal., B **187**, 218 (2016)
14. G. Grzybek, P. Stelmachowski, P. Indyka, M. Inger, M. Wilk, A. Kotarba, Z. Sojka, Catal. Today **257**, 93 (2015)
15. M. Inger, M. Wilk, M. Saramok, G. Grzybek, A. Grodzka, P. Stelmachowski, W. Makowski, A. Kotarba, Z. Sojka, Ind. Eng. Chem. Res. **53 (25)**, 10335 (2014)
16. G. Grzybek, P. Stelmachowski, S. Gudyka, P. Indyka, Z. Sojka, N. Guillén-Hurtado, V. Rico-Pérez, A. Bueno-López, A. Kotarba, Appl. Catal. B **180**, 622 (2016)
17. A. Bueno-López, I. Such-Basanez, C. Salinas-Martinez de Lecea, J. Catal. **244**, 102 (2006)
18. S. Suarez, M. Yates, F.J. Gil Liambias, J.A. Martin, P. Avila, J. Blanco, Stud. Sur. Sci. Catal. **143**, 111 (2000)
19. S. Imamura, J. Tadani, Y. Saito, Y. Okamoto, H. Jindai, C. Kaito, Appl. Catal. A **201**, 121 (2000)
20. W. Piskorz, F. Zasada, P. Stelmachowski, A. Kotarba, Z. Sojka, Catal. Today **137**, 418 (2008)
21. L. Xue, Ch. Zhang, H. He, Y. Teraoka, Catal. Today **126**, 449 (2007)
22. L. Xue, Ch. Zhang, H. He, Y. Teraoka, Appl. Catal. B **75**, 167 (2007)
23. G. Maniak, P. Stelmachowski, A. Kotarba, Z. Sojka, V. Rico-Pérez, A. Bueno-López, Appl. Catal., B **136-137**, 302 (2013)
24. P. Legutko, W. Kaspera, T. Jakubek, P. Stelmachowski, A.Kotarba, Topic Catal., **56**, 745 (2013)
25. G. Zhang, Y. Guo, L. Lu, Z. Zhang, L. Wang, L. Zhou, J. Shao, Mat. Lett., **125**, 162 (2014)
26. L. Li, L. Song, H. Wang, Ch. Chen, Y. She, Y. Zhan, X. Lin, Q. Zheng, Int. J. Hydrogem Energy **36**, 8839 (2011)
27. M. Mazaheri, S.A. Hassanzadeh-Tabrizi, M. Aminzare, S.K. Sadrnezhaad, Ceramic Materials, **62**, 529 (2010)
28. F. Zasada, P. Stelmachowski, G. Maniak, J.-F. Paul, A. Kotarba, Z. Sojka, Catal. Lett. **127**, 126 (2009)

Effect of Support Structure in Au/Al_2O_3-TiO_2 Catalysts in Low-Temperature CO Oxidation

Maria Rajska[1,a], Piotr Długosz[2] and Rafał Zybała[3]

[1]*AGH University of Science and Technology, Faculty of Materials Science and Ceramic, Al. Mickiewicza 30, 30-059 Krakow, Poland*
[2]*Foundry Research Institute, Department of Non-Ferrous Metal Alloys, Zakopiańska 73, 30-418 Krakow, Poland*
[3]*Faculty of Materials Science and Engineering, Warsaw University of Technology, Wołoska 141, 02-507 Warsaw, Poland*

Abstract. The aim of this study was to determine the effect of the support phase composition (Al_2O_3–TiO_2) and the addition of gold on the catalytic properties in low-temperature CO oxidation. In this paper, the physicochemical properties and results of catalytic measurements performed for the obtained samples were investigated. The catalyst carriers were prepared using the sol-gel method with alkoxide $Al(C_3H_7O)_3$ and $Ti(C_3H_7O)_4$ employed to obtain the corresponding oxides, in the case of which the molar ratio (Al_2O_3:TiO_2) was 0.75:0.25, 0.50:0.50 and 0.25:0.75, thus corresponding to the molar ratio of the Al:Ti elements, i.e. 1.5:0.25, 2:1 and 0.5:0.75 respectively. The gold catalysts were prepared through the deposition of gold by the deposition precipitation method using the theoretical loading of Au 2 wt.%. To examine the effect of the phase composition on the catalytic activity of the obtained samples, appropriate carriers were calcined at two different temperatures: 500°C and 1350°C. This made it possible to obtain the intended polymorphs of aluminum oxide and titanium dioxide (γ–Al_2O_3, α–Al_2O_3, anatase and rutile). For certain samples, calcined at a high temperature, the aluminum titanium oxide (Al_2O_5Ti) phase was also observed. The prepared samples were characterized by XRD, BET, SEM, and additionally both particle size distribution analysis and measurements of the catalytic activity were performed. The highest catalytic activity was shown by Au/75Al–25Ti_LT, where T_{90} was about 110°C.

1 Introduction

The increased interest in nanometric materials observed in recent years caused the discovery of nanocrystalline gold, which proved to be a material with a high catalytic efficiency. The number of nano gold applications has continuously grown in the last few years and a great interest in gold in the nanometric form is still observable. Nanocrystalline gold is used in many reactions, including low temperature CO oxidation, H_2 synthesis in the presence of water vapor and carbon oxide (WGS - Water Gas Shift Reaction), and further purification of H_2 with CO for PEMs fuel cells (PROX - PReferential OXidation) [1-5]. The latest trend in the use of nano-sized gold focuses on creating hybrid materials used as various types of biosensors [6-8]. Devices like biosensors based on nano Au show a remarkably high sensitivity with high specificity. Another very interesting application is medicine and especially the rapid diagnosis of cancer or DNA damage [9-11]. Nano gold deposited on various support structures is also used for the selective oxidation of alkenes, alcohols, and even alkanes [12,13]. For nanomaterials, including oxides, a relationship between the size of the crystallite or grain powder and the exhibited surface free energy that is different for a nanopowder and a material with large grains was found. It was discovered that the less thermodynamically stable structure can be a stable substrate for the nano-sized metal particles through the reduction of the surface free energy. This relationship was studied for the first time for various crystalline phases of alumina (γ-Al_2O_3 and α-Al_2O_3) [14]. Similar research also focused on the polymorphic forms of TiO_2 [15,16].

The selection of an appropriate material to be used as a support is important for the preparation of catalysts with a high efficiency in CO oxidation, or oxidation of the volatile organic compounds [17]. Among the oxides used for oxidation there is titanium dioxide (anatase), which stabilizes the size of the nanoparticles of gold, thus preventing them from excessive expansion during heat treatment, and provides a uniform distribution of Au, and, in addition, participates in the oxidation of CO [18,19]. The effect of the TiO_2 phase was studied also in the case of the hydrogenation reaction [20,21]. Due to the crystal structure of TiO_2, it can be divided into three polymorph forms (anatase, brookite and rutile) having various structural and physical properties [22]. For catalytic applications, anatase is the most often used solution due to its semiconductor properties and inconsiderable defection from stoichiometry [23]. Anatase grains usually do not exceed 50 nm and are used in catalysis as adsorbents for organic compounds. Rutile grains are larger, usually exceeding 200 nm [24]. TiO_2 in the rutile form due to its stable structure allowing the use of higher temperatures,

[a] Corresponding author: rajska@agh.edu.pl

is used for instance in the catalytic oxidation of hydrogen chloride to chlorine [25]. Gold can be added to the catalytic system by various methods, including precipitation methods, sputtering or CVD, which enables obtaining Au particles <10 nm. However, the deposition - precipitation allows the most efficient preparation of highly active catalysts with Au sizes between 2 and 6 nm [26].

In this work, the effect of the phase composition of selected oxides (TiO_2 and Al_2O_3) and the addition of gold in the catalytic activity in response to the low temperature oxidation of CO was analyzed. In addition, the influence of the phase composition on the crystallite size of Au and the distribution of gold particles on the surface of the sample were determined.

2 Experimental

The samples were obtained by the sol-gel method, using the following, appropriately chosen alkoxides: aluminum and titanium ($Al(C_3H_7O)_3$, $Ti(C_3H_7O)_4$) (p.a., Avantor Performance Materials Poland S.A.) with a concentration of 1 M in anhydrous ethanol [27]. The alkoxides of aluminum and titanium were used for the preparation of the corresponding oxides, in the case of which the molar ratio (Al_2O_3:TiO_2) was determined to be 0.75:0.25; 0.50:0.50 and 0.25:0.75. After hydrolysis, the substrate precursors underwent aging for 2 days at room temperature and were subsequently dried at a temperature of 70°C for 24 hours. The samples for further study were calcined for 3 hours at two different temperatures: 500°C and 1350°C. A detailed description of the preparation of the catalyst carriers was described in the previous work [28]. This allowed obtaining the proper polymorphic phase of oxides, i.e. γ-Al_2O_3, α-Al_2O_3, anatase, rutile. Table 1 shows the composition of the outcome samples and their corresponding symbols.

Table 1. Symbols of samples.

Sample	Abbreviation	Sample	Abbreviation
0.75Al_2O_3-0.25TiO_2_500°C	75Al-25Ti_LT	Au/0.75Al_2O_3-0.25TiO_2_500°C	Au/75Al-25Ti_LT
0.50Al_2O_3-0.50TiO_2_500°C	50Al-50Ti_LT	Au/0.50Al_2O_3-0.50TiO_2_500°C	Au/50Al-50Ti_LT
0.25Al_2O_3-0.75TiO_2_500°C	25Al-75Ti_LT	Au/0.25Al_2O_3-0.75TiO_2_500°C	Au/25Al-75Ti_LT
0.75Al_2O_3-0.25TiO_2_1350°C	75Al-25Ti_HT	Au/0.75Al_2O_3-0.25TiO_2_1350°C	Au/75Al-25Ti_HT
0.50Al_2O_3-0.50TiO_2_1350°C	50Al-50Ti_HT	Au/0.50Al_2O_3-0.50TiO_2_1350°C	Au/50Al-50Ti_HT
0.25Al_2O_3-0.75TiO_2_1350°C	25Al-75Ti_HT	Au/0.25Al_2O_3-0.75TiO_2_1350°C	Au/25Al-75Ti_HT

All polymorphic forms of alumina above 1200°C are transformed into corundum. At temperatures higher than 1300°C, it is possible to obtain a permanent connection between corundum and rutile [29]. Rutile is a thermodynamically stable structure which is formed during the exothermic conversion of anatase in the temperature range of 500-1000°C, in the case of which the lower the transition temperature, the smaller the crystallites of TiO_2 [30,31]. The gold particles were deposited on the substrate by deposition precipitation to obtain 2% Au by weight in the final catalyst. A precursor of gold was chloroauric acid $HAuCl_4$ (p.a., Chempur) with a concentration of 0.088 M.

2.1 Methods

X-ray diffraction measurements were performed using the X'PERT Pro PANalytical diffractometer (CuKa radiation). The average crystallite size of gold was calculated using the Scherrer equation, and the quantitative phase composition of the samples calcined at 1350°C was calculated using the Rietveld method. For morphology observation, the samples calcined at 500°C were analyzed using the JEOL JSM 7500F scanning electron microscope, whereas for the observation of other samples (1350°C) a SEM NOVA NANO 200 scanning electron microscope from FEI was used. The studies of the particle size distribution were made employing the Malvern Mastersizer 2000. The measurements of the specific surface area of the samples were performed using the ASAP 2010 apparatus from Micromeritics Instrument Corporation.

2.2 Catalytic measurements

Catalytic tests were performed in a flow-through, quartz reactor connected with the quadrupole mass spectrometer (Thermostar QSD 300, BALZERS). The ionic currents were used to calculate the rate of conversion of CO to CO_2. Before the catalytic test, the samples (0.3 g) were standardized by their heating to 300°C under a constant air flow and kept at this temperature for 30 min. Catalytic investigations were performed in a gas flow of 100 cm^3h^{-1} in the temperature range from 25°C to 300°C. The reaction mixture contained 1 %vol. CO, 9 %vol. He, 17.8 %vol. O_2 and 72.2 %vol. N_2. Each catalytic measurement was repeated three times. For a better characterization of the catalytic activity, the temperature at which the conversion of CO to CO_2 was 50% (T_{50}) and 90% (T_{90}) was determined.

3 Results

3.1 XRD

The phase analysis of the samples calcined at 500°C showed that next to the crystalline phase was a certain amount of the amorphous phase. Reflections derived from TiO_2 correspond to the anatase phase (ICDD 01-073-1764). The reflections of crystalline Al_2O_3 were very extended and indicating the γ-Al_2O_3 phase (ICDD 00-010-0425) [32,33]. The XRD analysis of the samples with gold showed the large half-width reflections derived from the crystallites of gold (ICDD 04-003-5615) (Figure 1). The Au crystallite size was calculated from the Scherrer equation (see Table 2). The crystallite size of Au for samples was similar, i.e. 4 nm (Au/75Al-25Ti_LT, Au/25Al-75Ti_LT) and 5 nm (Au/50Al-50Ti_LT). This means that the difference in the phase composition in the sample does not significantly affect the size of the crystallites of gold. The calculated Au crystallite size of the samples calcined at a high temperature is frequently many times larger when compared to the same samples calcined at a lower temperature, D_{Au} = 50

 DOI: 10.1051/e3sconf/20161000131

nm (Au/75Al-25Ti_HT), D_{Au} = 57 nm (Au/50Al-50Ti_HT), D_{Au} = 97 nm (Au/25Al-75Ti_HT), (Figure 2). The gold crystallites of such large sizes no longer show catalytic properties. Their presence does not improve the catalytic effect for the whole system [1].

Table 2. Results of XRD and specific surface area measurements.

Sample	S_{BET} [m^{2}g^{-1}]	Crystallite size [nm]	Phase composition
75Al-25Ti_LT	339.01 ± 0,61	$D_{anatase}$ = 26 ± 1 $D_{\gamma\text{-}Al_2O_3}$ = 7 ± 1	TiO_2-anatase, γ-Al_2O_3
50Al-50Ti_LT	227.70 ± 1.79	$D_{anatase}$ = 29 ± 1 $D_{\gamma\text{-}Al_2O_3}$ = 8 ± 1	TiO_2-anatase, γ-Al_2O_3
25Al-75Ti_LT	284.79 ± 0.87	$D_{anatase}$ = 30 ± 1 $D_{\gamma\text{-}Al_2O_3}$ = 8 ± 1	TiO_2-anatase, γ-Al_2O_3
Au/75Al-25Ti_LT	346.26 ± 1.29	$D_{anatase}$ = 23 ± 1 $D_{\gamma\text{-}Al_2O_3}$ = 6 ± 1 D_{Au} = 4 ± 1	TiO_2-anatase, γ-Al_2O_3, Au
Au/50Al-50Ti_LT	288.37 ± 0.88	$D_{anatase}$ = 26 ± 1 $D_{\gamma\text{-}Al_2O_3}$ = 7 ± 1 D_{Au} = 5 ± 1	TiO_2-anatase, γ-Al_2O_3, Au
Au/25Al-75Ti_LT	305.76 ± 0.59	$D_{anatase}$ = 25 ± 1 $D_{\gamma\text{-}Al_2O_3}$ = 7 ± 1 D_{Au} = 4 ± 1	TiO_2-anatase, γ-Al_2O_3, Au
75Al-25Ti_HT	0.73 ± 0.05	D_{rutile} = 126 ± 1 $D_{\alpha\text{-}Al_2O_3}$ = 108 ± 1	TiO_2-rutile, α-Al_2O_3
50Al-50Ti_HT	1.29 ± 0.01	D_{rutile} = 129 ± 1 $D_{\alpha\text{-}Al_2O_3}$ = 108 ± 1 $D_{Al_2O_5Ti}$ = 126 ± 1	TiO_2-rutile, α-Al_2O_3 Al_2O_5Ti
25Al-75Ti_HT	1.60 ± 0.01	D_{rutile} = 138 ± 1 $D_{\alpha\text{-}Al_2O_3}$ = 122 ± 1 $D_{Al_2O_5Ti}$ = 180 ± 1	TiO_2-rutile, α-Al_2O_3 Al_2O_5Ti
Au/75Al-25Ti_HT	2.40 ± 0.01	D_{rutile} = 126 ± 1 $D_{\alpha\text{-}Al_2O_3}$ = 107 ± 1 D_{Au} = 50 ± 1	TiO_2-rutile, α-Al_2O_3, Au
Au/50Al-50Ti_HT	1.19 ± 0.02	D_{rutile} = 129 ± 1 $D_{\alpha\text{-}Al_2O_3}$ = 108 ± 1 $D_{Al_2O_5Ti}$ = 126 ± 1 D_{Au} = 57 ± 1	TiO_2-rutile, α-Al_2O_3, Al_2O_5Ti, Au
Au/25Al-75Ti_HT	1.68 ± 0.01	D_{rutile} = 137 ± 1 $D_{\alpha\text{-}Al_2O_3}$ = 123 ± 1 $D_{Al_2O_5Ti}$ = 180 ± 1 D_{Au} = 97 ± 1	TiO_2-rutile, α-Al_2O_3, Al_2O_5Ti, Au

For sample 75Al-25Ti_HT corundum (ICDD 04-007-9906) was 92.1 wt.% and rutile was 7.9 wt.% (ICDD 04-007-5987). In sample 50Al-50Ti_HT corundum was 48.4 wt.%, rutile 7.4 wt.% and a new phase aluminium titanium oxide of 44.2 wt.% appeared (ICDD 00-041-0258). In sample 25Al-75Ti_HT corundum was 40.3 wt.%, rutile 31.3 wt.% and aluminium titanium oxide 28.4 wt.%. A new phase containing aluminum and titanium was observed for the samples with a high ratio of titanium to aluminum, i.e. for sample 50Al-50Ti_HT where the Al:Ti molar ratio is 2:1, and for sample 25Al-75Ti_HT where the Al:Ti molar ratio is 0.5:0.75. For sample 75Al25Ti_HT the theoretical weight percent of titanium is 12%, for sample 50Al-50Ti_HT it is already 28 wt.% and 61 wt.% for sample 25Al75Ti_HT.

The calculated content of the new phase (Al_2O_5Ti) shows that it rises in the first place at the expense of decreasing the amount of rutile. The losses of the α-Al_2O_3 phase appear to be larger, but in fact, they are strongly correlated with predefined stoichiometry and the stoichiometry of the new phase.

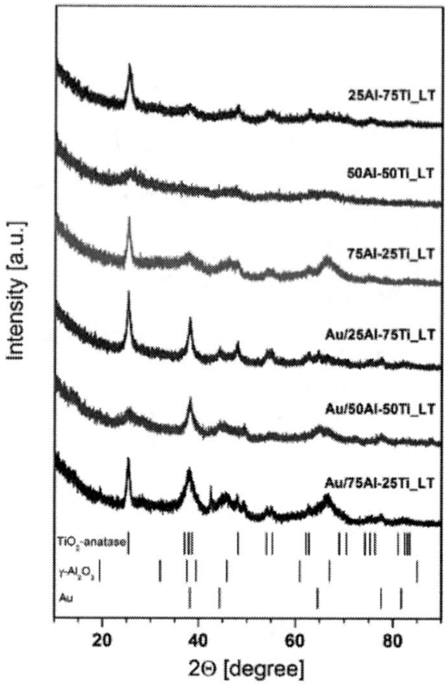

Figure 1. X-ray diffraction patterns of samples calcined at a low temperature.

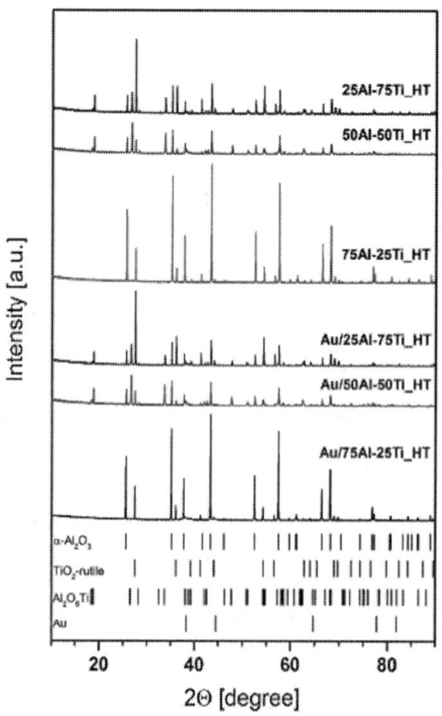

Figure 2. X-ray diffraction patterns of samples calcined at a high temperature.

The phase composition of the samples after the gold addition was not changed. A share of Au in the sum of the oxides was 1.5 by weight for Au/75Al-25Ti_HT, for Au/50Al-50Ti_HT it equalled 1.3 wt.% and for Au/25Al-75Ti_HT it reached 1.3 wt.%. The planned gold content is lower than the one obtained, i.e. 2 %wt., which is still within the optimum range of the Au content for this type of a catalyst [19].

3.2 Particle size distribution

The analysis of particle size distribution expressed as volume and quantity percentage indicates that the grains of powder form agglomerates. Most of the particles are between 0.3 and 10 microns; however, this distribution is changed after the gold addition (Figure 3 and Figure 4).

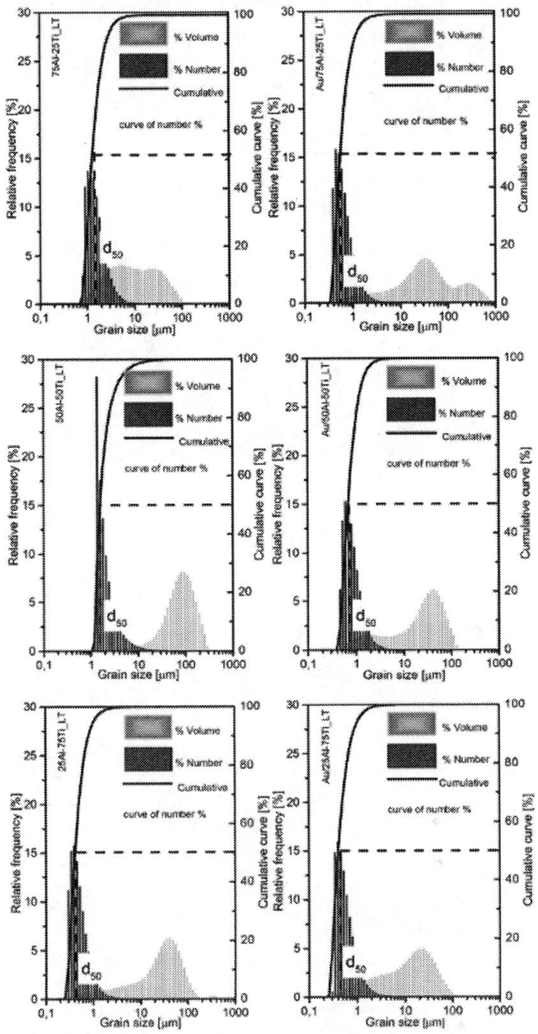

Figure 3. Grains distribution for samples calcined at a low temperature.

The average particle size of sample 75Al-25Ti_LT was 1.26 μm and after the application of Au it was reduced to 0.55 microns. For sample 50Al-50Ti_LT the average particle size was 1.55 μm and after gold deposition it was 0.68 μm. The situation was found to be similar for sample 25Al-75Ti_LT, for which the average grain size was 0.45 μm and 0.39 μm after the application of gold. In the case of samples 75Al-25Ti_LT and 50Al-50Ti_LT the average particle size was 2.3 times lower, and in the case of sample 25Al-75Ti_LT it was 1.15 times lower.

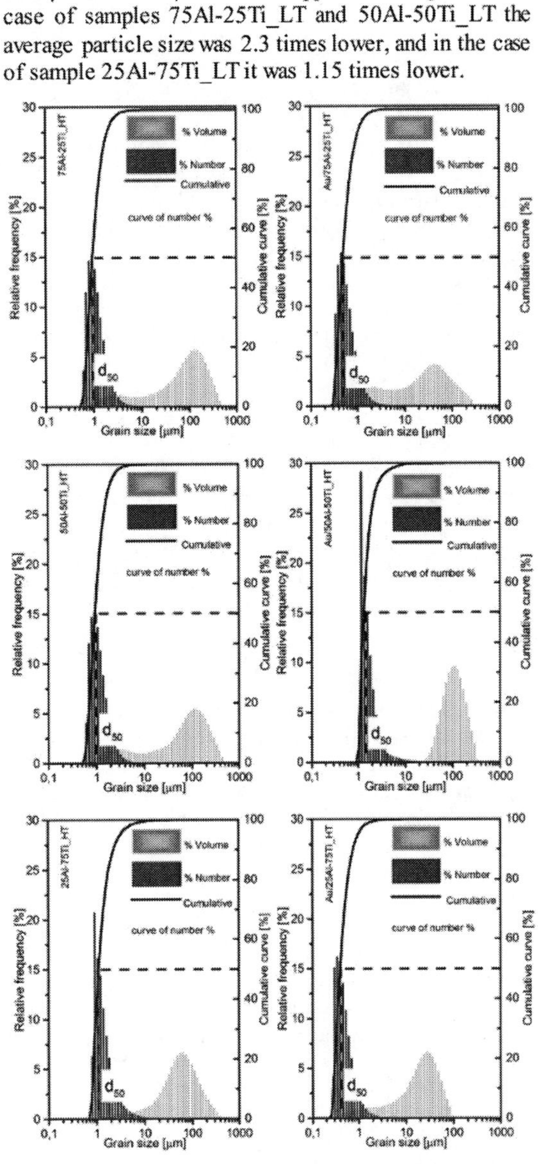

Figure 4. Grains distribution for samples calcined at a high temperature.

It was observed that the particle size was reduced and simultaneously the specific surface area increased. For samples 75Al-25Ti_LT, 50Al-50Ti_LT and 25Al-75Ti_LT the specific surface area was 339.01 m^2g^{-1} 227.70 m^2g^{-1}, 284.79 m^2g^{-1} respectively and increased to 346.26 m^2g^{-1}, 288.37 m^2g^{-1} and 305.76 m^2g^{-1}. The increase may be associated with the disintegration of agglomeration during the gold deposition-precipitation procedure. It is also probable that in the case of the sample containing γ-Al_2O_3 partial dissolving of a quantity of alumina which then recrystallized again could have taken place[34].

Similarly, changes occur in the case of the samples calcined at a higher temperature. For 75Al-25Ti_HT the mean grain size was 0.96 μm and after the addition of gold it reached 0.53 μm. The average grain size for samples 50Al-50Ti_HT and 25Al-75Ti_HT was 0.89 μm and 1.03 μm, whereas after gold deposition it changed to 1.35 μm and 0.39 μm respectively. For the samples calcined at a high temperature, after the application of gold the average particle size generally decreases: 1.8 times for sample Au/75Al-25Ti_HT and 3.5 times for sample Au/25Al-75Ti_HT. The only exception was sample Au/50Al-50Ti_HT with a small increase in the average particle size, suggesting that the effect of the addition of gold outweighs the potential effect of the disintegration of powder agglomerates. These results are in line with the measurements of the specific surface area showing that after the addition of gold the specific surface area increased for Au/75Al-25Ti_HT (0.73 m^2g^{-1} to 2.40 m^2g^{-1}) and Au/25Al-75Ti_HT (1.60 m^2g^{-1} to 1.68 m^2g^{-1}) and decreased for Au/50Al-50Ti_HT (1.29 m^2g^{-1} to 1.19 m^2g^{-1}).

The decrease in the specific surface area after Au deposition, described in the literature, was observed for the samples the carriers of which were calcined at 1350°C [18]. This effect is noticeable for samples coming from the carriers calcined at low temperatures. This could be most likely due to the dominant effect of the disintegration of agglomerates.

3.3 SEM

Grains of the carriers and the gold catalysts are presented in the micrographs shown in Figure 5 and Figure 6. All samples have irregular shapes and form agglomerates with different sizes. For the samples calcined at 500°C, agglomerates have a size of about 0.5 micron and for the samples calcined at higher temperatures agglomerates measure about a few microns. After the gold addition the carrier grains were smaller, which can be associated with the disintegration of the agglomerates presented in the beginning and secondly, the partial recrystallization of γ-Al_2O_3 which could have been possible for a sample calcined at a lower temperature (see Figure 5). In aqueous systems, γ-Al_2O_3 can again be transformed into aluminum hydroxide or aluminum oxyhydroxide, when the final heat treatment re-creates γ-Al_2O_3, but with a high specific surface area [35].

The distribution of the gold particles on the surface is uniform. For sample Au/75Al-25Ti_LT the gold particles have a small size, reaching about 10 nm (D_{Au} = 5 nm – XRD) (Figure 5).

As a result of the exothermic phase transition of γ-Al_2O_3 to corundum one observes the loss of porosity and reduction in the specific surface area as well as an increase in the grain size. α-Al_2O_3 powder has a specific surface area (S_{BET}) of less than 1 m^2g^{-1} and a micropore volume of less than 0.5 cm^3g^{-1}. It is also possible to obtain the polymorph of the aluminum oxide specific surface area of 66 m^2g^{-1} for decomposition and calcination of α-$Al(OH)_3$ at 530°C [36]. After Au deposition, the grains of the samples (from the carriers calcined at high temperature) did not show significant deformation or there was no reduction in the size of the agglomerates. The observed Au particles were evenly distributed over the surface of the sample.

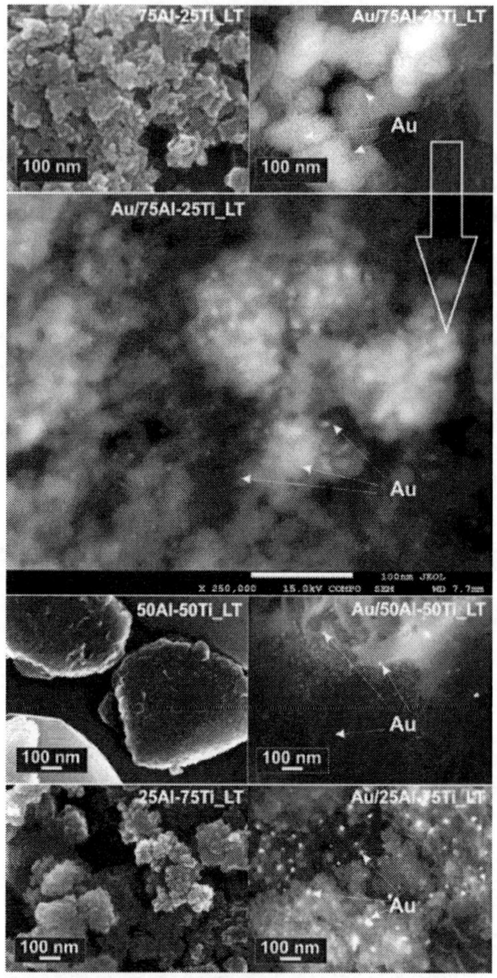

Figure 5. SEM images of the support and the gold supported catalysts for samples calcined at a low temperature.

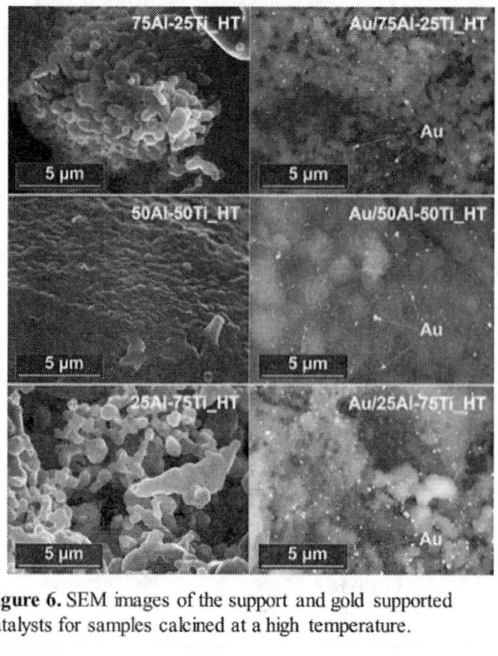

Figure 6. SEM images of the support and gold supported catalysts for samples calcined at a high temperature.

 DOI: 10.1051/e3sconf/20161000131

The average size of the gold particles determined from the microimages of sample Au/75Al-25Ti_HT was 80 nm (Figure 6). This is a value close to the result calculated by the Scherrer equation (D_{Au} = 50 nm). For the remaining samples and based on the microphotographs, the Au particles have the calculated size of 85 nm for Au/50Al-50Ti_HT (D_{Au} = 54 nm – XRD) and 110 nm for Au/25Al-75Ti_HT (D_{Au} = 97 nm – XRD).

3.4 Catalytic tests

The catalytic activity of the obtained carriers and Au/catalysts was determined on the basis of the following oxidation reaction of carbon monoxide: $2CO + O_2 \rightarrow 2CO_2$.

The CO oxidation test was carried out in the 25-300°C temperature range due to the fact that above a temperature of 320°C the phenomenon entailing the sintering of gold crystallites takes place, which results in a significant increase in their size and thus a loss of their catalytic properties [1]. For the samples that do not contain gold it can be observed that the carriers calcined at a higher temperature had better catalytic properties than the samples with gold (Figure 7).

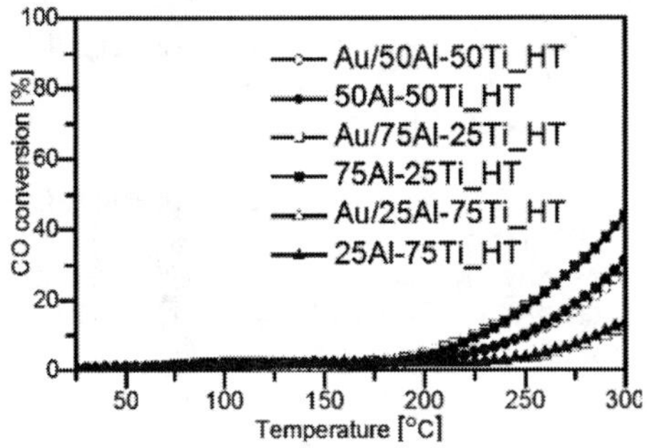

Figure 7. CO conversion for samples calcinated at high temperature.

This is probably related to the phase composition of the sample, i.e. the presence of the following phases: α-Al_2O_3 and TiO_2-rutile, especially when taking into account that a specific surface area for this material is small (about 0.3 m^2g^{-1}). The efficiencies of the system (30% conversion of CO) are still unsatisfactory from the point of view of its using as a catalyst in the low temperature oxidation of CO. After the gold addition, changes in the efficiency of the tested samples were observed.

The highest conversion rate of CO to CO_2 was obtained for sample Au/75Al-25Ti_HT, for which at a temperature of 300°C conversion reached about 35%. In this case, there were no positive changes when adding gold to the system. The calculated degree of conversion of the CO for media is low with respect to the systems containing gold in the low temperature range (Figure 8). The presence of gold brings a significant improvement in the catalytic properties of the samples calcined at a lower temperature. The CO oxidation reaction began at about 30°C for all samples and 90% conversion of CO to CO_2 was achieved at a temperature of about 110°C (Au/75Al-25Ti_LT), as shown in Table 3. The best carrier was sample 50Al-50Ti_LT with CO conversion similar to the conversion of 75Al-25Ti_HT. With a decrease in the percentage of the titanium dioxide degree, the conversion of CO (T_{50} and T_{90}) is moved towards lower temperatures.

Table 3. Temperatures of 50% CO conversion and 90% CO conversion.

Sample	T_{50} [°C]	T_{90} [°C]
Au/75Al-25Ti_LT	65	110
Au/50Al-50Ti_LT	100	155
Au/25Al-75Ti_LT	120	200
Au/75Al-25Ti_HT	>300	-
Au/50Al-50Ti_HT	>300	-
Au/25Al-75Ti_HT	>300	-

For the best sample (Au/75Al-25Ti_LT) lowering of the temperature at which the system shows a complete CO conversion from the high temperature range (>300°C) to the low temperature range (>110°C) was observed. This would mean that in the test reaction, a dominant role in obtaining a high catalytic activity for CO oxidation is played by the crystallite size of gold. For samples characterized by a high efficiency of CO oxidation, gold crystallites have a size of about 4 nm. The samples which are obtained from the carrier calcined at high temperatures have large gold crystallites (D_{Au} = 50 nm, D_{Au} = 57 nm, D_{Au} = 97 nm). Gold of this size does not exhibit catalytic properties.

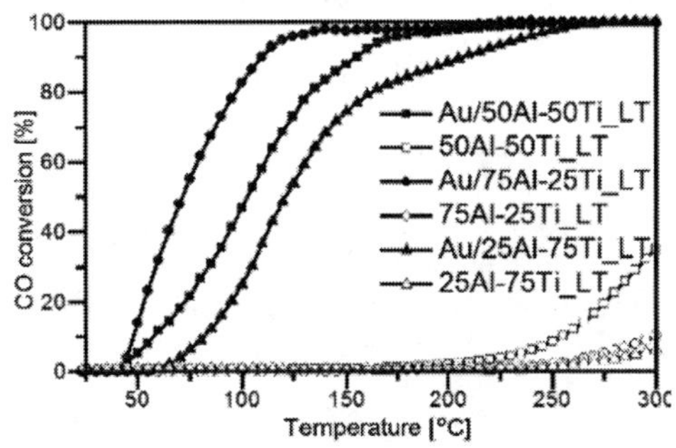

Figure 8. CO conversion for samples calcined at a low temperature.

Summary

The study is an attempt to explain the change in the catalytic activity in low-temperature CO oxidation, depending on the presence of different polymorphic forms of the base material. These research works led to the conclusion that the catalytic activity of this reaction is very intense for the catalysts in which there is TiO_2, which is in the form of anatase. For analogous systems containing rutile the catalytic activity was reported to be very low (T_{50}>300°C) and not depending on whether the sample with gold or a pure carrier was analyzed. The phase composition of the substrate significantly affects the size of the crystallites of gold and stabilizes their growth, thus significantly influencing the high catalytic activity of the samples the carriers of which were calcined at a low temperature. The size of gold crystallites was calculated using the Scherrer equation and was within the following two size ranges, i.e:4 to 5 nm and from 50 to 97 nm. Sample Au/75Al-25Ti_LT has the best catalytic activity . Samples, the carriers of which were calcined at a high temperature, were characterized by a low specific surface area. They contained α-Al_2O_3 and rutile and did not create proper conditions for the deposition of the gold particles of a nanometric size. The nanometer-sized gold is necessary to obtain the high catalytic activity in a low temperature CO oxidation reaction.

Acknowledgment

M. Rajska acknowledges the Statutory Project for Science (Grant for Young Scientists no.15.11.160.763) 2015 at the Department of Inorganic Chemistry, Faculty of Materials Science and Ceramics, AGH UST.
This work has been partially financed as a research project no. DEC-2014/12/S/ST8/00582 from the resources assigned to science by National Science Centre (NCN, Poland) and the National Centre for Research and Development (NCBR, Poland) under the project no. PBS3/A5/49/2015.

References

1. G.C. Bond, D.T. Thompson, Gold Bull. **33**, 41 (2000)
2. G.C. Bond, C. Louis, D.T. Thompson *Catalysis by Gold, Catalytic Science Series – Vol.6* (Imperial College Press, London 2006)
3. V.G. Milt, S. Ivanova, O. Sanz, M.I. Domínguez, A. Corrales, J.A. Odriozola, M.A. Centeno, Appl. Surf. Sci. **270**, 169 (2013)
4. A.T. Miah, B. Malakar, P. Saikia, Catal. Lett. **146**, 291 (2016)
5. T. Montini, M. Monai, A. Betram, I. Romero-Ocaña, P. Fornasiero, Mat. Sci. Semicon. Proc. **42**, 122 (2016)
6. J.E. Kim, J.H. Choi, M. Colas, D.H. Kim, H. Lee, Biosens. Bioelectron. **80**, 543, (2016)
7. Y.-M. Cheng, H.-B. Fa, W. Yin, C.-J. Hou, D.-Q.Huo, F.-M. Liu, Y. Zhang, C. Chen, J. Solid State Electrochem. **20**, 327 (2016)
8. W. Ma, L. Xu, L. Wang, H. Kuang, C. Xu, Biosns. Bioelectron. **79**, 220 (2016)
9. A. Barbasz, M. Oćwieja, J. Exp. Nanosci. **11**, 564 (2016)
10. J.H. Myung, K.A. Tam, S.-J. Park, A. Cha, S. Hong, WIREs. Nanomed. Nanobi. **8**, 223 (2016)
11. Y.-P. Kim, H.-S. Kim, Chembiochem, **17**, 275 (2016)
12. S. Rautiainen, O. Simakova, H. Guo, A.-R. Leino, K. Kordás, D. Murzin, M. Leskelä, T. Repo., Appl. Catal. A: Gen. **485**, 202 (2014)
13. P.-L. Zhu, X.-Y. Tang, M. Shi, Chemistry Open, **5**, 33 (2016)
14. J.M. McHale, A. Auroux, A.J. Perrott, A. Navrotsky, Science **277**, 788 (1997)
15. H. Zhang, B. Chen, J.F. Banfield, Phys. Chem. Chem. Phys. 11, 2553 (2009)
16. [c] S.-G. Shang, X. Li, L. Zhao, R. Lu, H.-F. Chen, Mater. Sci. Forum **852**, 238 (2016)
17. M. Hosseini, S. Siffert, H.L. Tidahy, R. Cousin, J.-F. Lamonier, A. Aboukais, A. Vantomme, M. Roussel, B.-L. Su, Catal. Today **122**, 391 (2007)
18. W.-C. Li, M. Comotti, F. Schüth, J. Catal. **237**, 190 (2006)
19. M.M. Schubert, S. Hackemberg, A.C. van Veen, M. Muhler, V. Plzak, R.J. Behm, J. Catal. **197**, 113 (2001)
20. Y. Li, B. Xu, Y. Fan, N. Feng, A. Qiu, J.M.J. He, H. Yang, Y. Chen, J. Mol. Catal. A: Chem. **216**, 107 (2004)
21. [B] X.-Z. Ding, X.-H. Liu, Y.-Z. He, J. Mater. Sci. Lett. **15**, 1789 (1996)
22. H. Iida, A. Igarashi, Appl. Catal. A: Gen. **298**, 152 (2006)
23. J.A. Moma, M.S. Scurrell, W.A. Jordaan, Top. Catal. **44**, 167 (2007)
24. M.Y. Kang, H.J. Yun, S. Yu, W. Kim, N.D. Kim, J. Yi, J. Mol. Catal. A: Chem. **368– 369**, 72 (2013)
25. K. Seki, Catal. Surv. Asia 14, 168 (2010)
26. G.C. Bond, D.T. Thompson, Catal. Rev.- Sci. Eng. **41**, 319 (1999)
27. D.A.H. Hanaor, I. Chironi, I. Karatchevtseva, G. Triani, C.C. Sorrell, Adv. Appl. Ceram. **111**, 149 (2012)
28. L.L. Hong, Y.J. Jong, M.L. Hyung, J. Mat. Sci. **32**, 5687 (1997)
29. M. Rajska, Materiały Ceramiczne/Ceramic Materials **67**, 393 (2015)
30. [A] J.M. Low, B. Curtain, M. Phillips, Z.Q. Liu, M. Ionescu, J. Aust. Ceram. Soc. **48**, 198 (2012)
31. [AA] J. Li, L. Liu, T.-K. Sham, Chem. Mater. **27**, 3021 (2015)
32. [D] S.R. Kumar, S.C. Pillai, U.S. Hareesh, P. Mukundan, K.G.K. Warrier, Mater. Lett. **43**, 286 (2000)
33. [E] R. Janes, L.J. Knightley, Dyes Pigm. **56**, 11 (2003)
34. L. Samain, A. Jaworski, M. Edén, D.M. Ladd, D.-K. Seo, F.J. Garcia-Garcia, U. Häussermann, J. Solid State Chem. 217, 1 (2014)
35. A.A. Tsyganenko, P.P. Mardilovich, J. Chem. Soc., Faraday Trans. **92**, 4843 (1996)
36. T. Tsuchida, Solid State Ionics **63–65**, 464 (1993)

 DOI: 10.1051/e3sconf/20161000132

The importance of LNG for natural gas consumption in the EU

Klaudia Metelska[1], Rafał Biały[2], Tomasz Cieślik[1,3], Tomasz Blacharski[4] and Adam Szurlej[1,a]

[1] AGH University of Science and Technology, Faculty of Drilling, Oil and Gas
[2] AGH University of Science and Technology, Faculty of Energy and Fuels
[3] Polish Academy of Science, Institute of Nuclear Physics, Aleja Radzikowskiego 152, 31-342 Kraków
[4]Polska Spółka Gazownictwa sp. z o.o. (Polish Gas Company), ul. Marcina Kasprzaka 25, 01-224 Warszawa

Abstract. The World market of liquefied natural gas (LNG) is growing rapidly. In 2015 LNG production exceeded 333 bcm with its predicted increase up to 450 bcm in 2019. The analysis of LNG role in natural gas import to the EU in recent years shows variability: LNG share in overall import reached 25% in 2011 and it went down to 15% in 2014. The smaller demand for natural gas including LNG in the EU can be due to, among others, a slower economic growth and a dynamic development of the use of renewable energy sources. The article shows the role of natural gas in the structure of consumption of primary energy as well as the changes in demand for natural gas in the years 2007-2014 for the main groups of end users: industry, energy production and individual households. The biggest fall in demand for natural gas has been observed in energy production sector in recent years. This publication continues to analyse the structure of natural gas supplies to the EU, with special emphasis on the directions of LNG import to the countries such as: The UK, Spain, France, Greece, Belgium, Portugal, Italy, Lithuania and The Netherlands. The significance of LNG in the balance of consumption of natural gas in these countries has been presented as well as the infrastructure connected with LNG and plans of development of regasification terminals. In the summary the most important conclusions have been drawn and a chance of the increase in significance of the role of LNG in the balance of natural gas supplies has been pointed out, which is due to the steep fall of LNG prices which has taken place in recent years.

1 Introduction

The main alternative for gas supplies through the gas pipeline network is its import in the form of liquefied natural gas (LNG) by sea in order to balance the demand for natural gas in the selected UE countries. The analysis of the significance of LNG in satisfying the demand for gas shows that it has changed rapidly all over the world in recent years. In so far as in 2007 the world production of LNG amounted to 230 bcm, it exceeded 333 bcm in 2015 with its predicted growth to at least 450 bcm by the end of 2019 [1]. In the years 2001-2014 the volume of natural gas trade in the liquefied form rose annually by 9.5% which is more than twice as much (4.5%) compared to the gas sent by pipeline networks. It needs to be stressed that the regasification capabilities exceed significantly liquefying capabilities (in 2001 the ratio of regasification capabilities to liquefying capabilities was 2.14 and in 2013 – 2.46). The increase in significance of LNG is connected, above all, with the increase in demand for natural gas as well as the high level of irregularity of the location of natural gas deposits which results in the need for LNG long distance transport – the average distance in 2015 was 14260 km (16 days). LNG is a form of condensed ecological energy and thanks to its properties it can be transported on long distances. One thing which is important is that this energy carrier is commonly accepted [2]. It is confirmed by the strategy proposed by the European Commission and accepted by The Council of Europe in March 2015 for stable energy union, which assumes strengthening the role of LNG on the European market of natural gas. The new EU energy policy aims at providing its consumers with safe, balanced and competitive energy at affordable prices. LNG plays an important role in this policy, especially in guaranteeing the security of supplies as well as raising the integration and competitiveness of the gas market [3]. As far as energy security is concerned, diversification of supplies of natural gas is to be based on creating in northern and central Europe LNG trade centres which will guarantee diversification of suppliers. These centres are supposed to ensure the possibility of buying gas from suppliers all over the world. The strategy of action which bases on the full use of the potential of access to the international LNG market falls within three areas [4]: infrastructure, which gives access to LNG directly or through other member states, completing the construction of the internal gas market which sends the right price alerts and the necessity of removing barriers in LNG trade and the promotion of the free market. The existing EU LNG terminals have sufficient regasification capabilities to satisfy the needs but they are not fully used (in 2013 the scale of use of terminals worldwide was around 35% while the EU figure was around 20%) [5].

[a] Corresponding author: szua@agh.edu.pl

2 The volume of consumption of natural gas in the years 2007-2014 and its structure of use

In 2004 in the EU primary gas consumption amounted to 213 thousand TWh, coming both from fossil fuels and renewable sources of energy while in 2014 energy consumption decreased to 211 thousand TWh. The drop in energy consumption is due to the economic slump in the years 2008/2009 whose effects are still visible in many EU countries and are characterised by low indexes of GDP growth. The declining trend regarding energy consumption is also connected with the introduction of the suitable regulations by the European Parliament in 2009 (so called climate and energy package) which assume a 20% reduction of greenhouse gases by 2020 (compared to the level in 1990), a 20% share of renewable sources of energy in the total energy consumption in the EU and a 20% increase in the energetic effectiveness [6][7]. A significant growth of the amount of renewable energy being used can be observed which is accompanied by the decrease in the use of fossil fuels which should be treated as the realisation of the adopted commitments of the EU member states. In 2004 the share of natural gas was 24% in the structure of primary energy consumption and in 2014 its share declined to 21%. For the sake of comparison the share of renewable energy rose from 6% in 2004 to 13% in 2014. Analysing natural gas consumption in the EU in absolute terms, in the years 2005, 2008 and 2010 it was at its highest at 500 bcm. To illustrate the volume of the decline of natural gas consumption, the consumption of natural gas in 2014 can be shown - 387 bcm [8][9].

In the EU natural gas is used in three main sectors: individual households, industry and energy production. A fall in gas consumption in the energy production sector can be observed (fig.1). It was caused, among other things, by the continuing relatively high prices of gas on the European market and by the appearance of cheap coal from The United States on the European market and by the focus of the majority of countries on the development of renewable energy use. [10]. In 2007 gas consumption by individual households was 34% with 33% in the industry and 26% in the energy production sector. In 2014 an increase in gas consumption by individual households to the level of 41% took place while in the industry it stayed on the same level and gas consumption in the energy production sector decreased to 23% [8]. In 2007 individual households gas consumption was 164 bcm. In Poland it was 5.13 bcm. In 2014 a fall in the consumption of gas by individual households took place to the level of 158 bcm whereas in Poland it rose to 5.40 bcm. It is worth mentioning that countries such as Germany, France, Italy, the UK, Spain and the Netherlands had a 82% share in the gas consumption by individual households in 2007. In 2014 this share was 80% which means that an increase in individual households gas consumption in countries such as Poland, Belgium, Romania and Portugal [8].

In 2007 gas consumption by the industry in the EU was 156 bcm. In Poland the consumption was 7.1 bcm. In 2009 a significant decrease in gas consumption to the level of 137 billion cubic metres took place, mainly because of the economic crisis. In 2011 while a fall in two other sectors was observed, namely, the energy production sector and individual households, an increase in the industry to the level of 149 bcm could be seen. In 2014 the EU industry gas consumption decreased to 128 bcm but in Poland it rose to 7.74 bcm. It needs to be observed that 72% of gas consumed by the industry in 2007 was in Germany, France, Italy, the UK, Spain and the Netherlands. In 2014 it was 71% which means a minimal increase in gas consumption by the industry in countries such as Poland, Finland and Estonia [8].

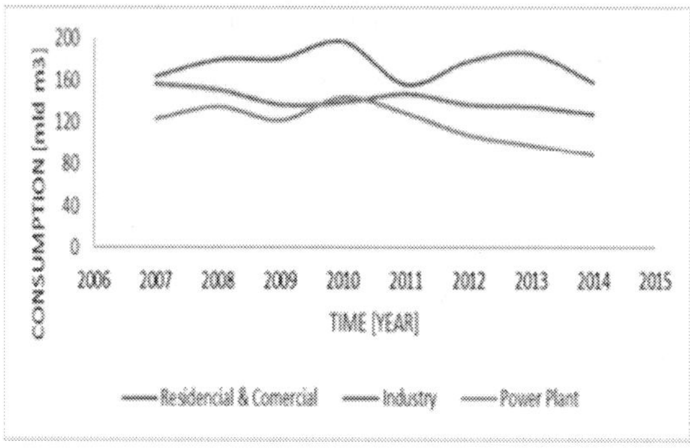

Figure 1. Gas consumption in different sectors (own study based on the data [8]).

In 2007 the EU gas consumption in the energy production sector was 124 billion cubic metres. In Poland it was 0.9 bcm. In 2014 a significant fall in consumption in energy production in the EU took place to the level of 88 bcm. In Poland, however, it rose to 1.35 bcm. It has to be said that 78% of gas consumed by the energy production sector was in countries such as Germany, France, Italy, the UK, Spain and the Netherlands but in 2014 it reached the level of 76%. In 2014 an increase in gas consumption in the energy production sector could be observed merely in countries such as Poland, France and Estonia [8]. However, as can be clearly seen, in recent years the biggest decrease in natural gas consumption was recorded in the energy production sector - in the years 2010 - 2014 it dropped by almost 40%.

3 The structure of gas supplies to the EU

In 2007 304 bcm of natural gas were imported to balance the supplies while the extraction was 187 bcm. In 2014 due to a smaller demand for natural gas, its import fell to 270 bcm and the extraction to 141 bcm. Countries such as Germany, France, Italy, the UK, Spain and the Netherlands extracted 85% of gas in the EU in 2007 but in 2014 – 79%. The analysis of the recent years indicates a declining trend of receiving gas from deposits in the EU, which is confirmed by a twofold decrease in the extraction of gas in the UK in the years 2007-2014 (37 billion cubic metres in 2014) [8][9]. Recent attempts to make use of unconventional natural gas deposits have been unsuccessful. Poland and Great Britain are the countries which have made the biggest effort in looking for these deposits [11].

At present natural gas is imported to the EU from Norway, Russia, Africa and The Middle East. The European Union has LNG terminals located in Great Britain, Poland, Spain, France, Greece, Belgium, Portugal, Italy, Lithuania and the Netherlands. It needs to be emphasized that 100% of supplies came from Russia in countries such as Estonia, Finland and Latvia. However, Bulgaria, Austria, Hungary, Slovakia, Slovenia and Poland received gas to satisfy from 89% to 59% of the demand (fig. 2) [8]. Therefore the Russian – Ukrainian gas crisis of early 2009 spread on many Central Europe countries [12].

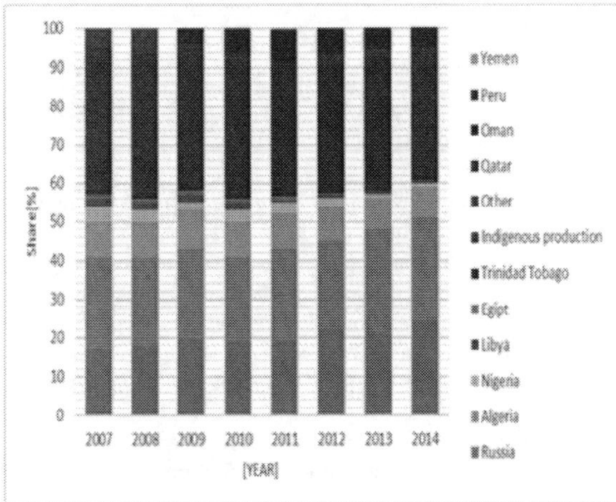

Figure 2. The structure of natural gas import to the EU (own study based on [8])

4 LNG share in balancing natural gas consumption in the European Union

LNG import to the EU countries possessing regasification terminals (with the re-export correction) in normal conditions was ca. 53 bcm, which covered 11.9% of the total demand for natural gas in the EU. In 2013 economic slump and the increase in the use of coal in the EU industry caused a reduction of import of LNG to 37.44 bcm (8.6% of the total demand for natural gas in the EU). In 2014 a further decline in the import of LNG was recorded to the level of 34.27 bcm, which resulted in the reduction of the share of gas imported to Europe in the world import of LNG to 13.9% (footnote – including Turkey). An increase was observed in 2015 when over 40 bcm of gas were imported to the EU. Attention should be drawn to the fact that in this period gas in the amount of 4.65 bcm was exported from the EU (mainly from Spain, the Netherlands and Belgium). The table 1 and fig. 3 show selected data with functioning LNG terminals in the EU countries as well as the directions of supplies of LNG. In order to illustrate the role of LNG, an analysis of the changes on the gas market has been carried out in countries such as the UK, Italy, France, the Netherlands, Spain, Poland, Belgium, Portugal, Greece, Lithuania and Sweden, namely, the countries which use LNG.

In 2012 natural gas consumption in the UK was 73.9 bcm and in 2014 this figure fell to 66.7 bcm. Similarly, the LNG import dropped from 13.1 bcm in 2012 to 10.6 bcm. The amount covered 16% of the total demand of the UK for natural gas in 2014. In 2015 an increase in the import of LNG to the UK in the amount of 12.7 bcm could be observed. Gas was imported from Algeria - 0.37 bcm, Norway - 0.17 bcm, Trinidad and Tobago - 0.22 bcm and from the key supplier Qatar - 12.3 bcm. The reception and regasification can be carried out in four terminals in the UK. The overall storage capacity of the British terminals in 2015 was 2.23 million cubic metres and their maximum regasification capability 53.6 bcm per year. Among the latest tasks concerning the development of the LNG infrastructure in the UK, one seems to be worth highlighting, namely, building work aiming at adjusting the terminal on The Isle of Grain for loading gas tankers [13].

In the years 2012-2014 natural gas consumption in Italy dropped by over 17% (from 68,7 to 56.8 bcm). At the same time a 35% decrease in the import of LNG was recorded from 6.49 bcm to 4.14 bcm. The amount of the LNG import to Italy could barely cover 7% of the total annual demand for natural gas. In 2015 a 30% increase in the amount of imported LNG was observed compared to 2014. Gas was imported from Qatar - 5.41 bcm, Algeria - 0.03 bcm and 0.05 bcm from re-export. Regasification of LNG can be carried out in three terminals in Italy located in: Toscana, Panigaglia and Rovigo with the overall storage capacity of 0.487 million cubic metres of liquefied LNG and regasification capability equalling 15.1 billion cubic metres per year. It appears that in recent years the most important building work connected with the development of LNG infrastructure in Italy has been carried out in the terminals Toscana and Panigaglia. In the latter a decision was made in 2013 to adjust it to unloading gas tankers of over 140 thousand cubic metres (including the rebuilding of the tanks, parking trens and building the cogeneration installation). Also, in 2015 preparations for the feasibility study were started for the project of building an installation for loading tankers and tank trucks in the existing terminal.

In France natural gas consumption in the years 2012 and 2014 dropped by ca.15% to 35.9 bcm. Similarly, LNG import declined by 36% (from 9.07 in 2012 to 5.8 bcm). That amount covered 16% of the demand for gas at that time. In 2015 a further 5% drop of LNG import took place (compared to 2014). The import was from (in bcm): Algeria - 4.19, Nigeria - 0.9, Norway - 0.42, Qatar - 0.24 and Peru - 0.25. Additionally, 0.46 bcm of gas were exported from France. LNG is received in three terminals in France with the overall storage capacity of 0.84 million cubic metres and the annual regasification capability of 21.3 bcm. Among a lot of work done on the development of LNG infrastructure in France, attention should be drawn to those in Montoir de Bretagne where the installation for loading tankers was modernised (direct transmission between two tankers) and in Fos Cavou where consultations for doubling the capacity took place.

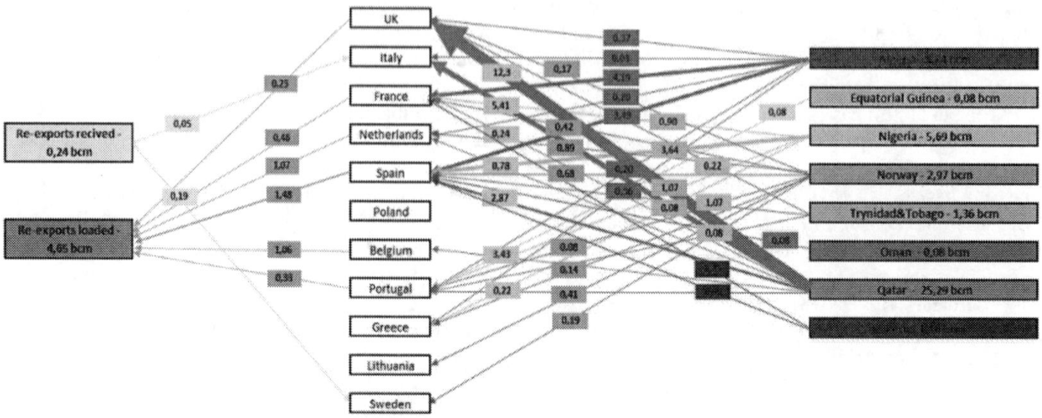

Figure 3. Directions of LNG import to the EU countries in 2015 (own study based on [13])

Also, in Fos Tonkin as of April 2015 terminal's regasification capability was reduced to 3 billion cubic metres per year. It is worth noticing that a terminal in Dunkirk is being built with the regasification capability of 13 billion cubic metres per year (expected start-up date 2016) [13].

In the Netherlands natural gas consumption in 2014 was 32.1 bcm. This amount was smaller by 11% compared to 2012. At the same time the amount of LNG import went down from 0.7 in 2012 to 0.55 bcm in 2014. The imported LNG covered merely 2% of the total demand for gas. In 2015 an increase in the import of LNG was recorded to 0.8 bcm (excluding the reloaded and re-exported gas in the amount of 1.07 bcm). Gas was imported from (bcm): Algeria - 0.2, Norway - 0.89 and Qatar - 0.78. Reception and regasification of LNG took place in the Netherlands in the terminal in Rotterdam with the storage capacity of 0.54 million cubic metres and regasification capability of 12 bcm per year.

Spain thanks to its geopolitical location possesses a well developed LNG infrastructure. There are seven terminals (the biggest number in the EU) with the overall regasification capability of 67.2 bcm per year and the total storage capacity of 3.61 million cubic metres (Barcelona boasts the oldest operating LNG terminal in the EU opened in 1969). The elaborate infrastructure allows a significant share of LNG in the structure of natural gas consumption which covered 40% in 2013 and 39% in 2014 of the overall demand for gas, which to provide an example, was 26.3 bcm in 2014 (17% less than in 2012). It had its impact on the amount of LNG import. In 2014 a 45% fall in the import was observed compared to 2012. The situation changed in 2015 when the import rose to 11.23 bcm and the well developed infrastructure allowed a strong diversification of supplies (bcm): Algeria - 3.49, Nigeria - 3.64, Norway - 0.68, Trinidad and Tobago - 1.07, the Oman - 0.08, Qatar – 2.87, Peru - 0.9 and export - 1.49. The modernisation of the regasification facilities located in Spain concerned, above all, building a new tank in the terminal in Bilbao in 2014 with the storage capacity of 150 thousand cubic metres and the installation for loading tank trucks.

Attention should also be drawn to the planned projects of building regasification terminals in Gran Canaria and Tenerife [13].

In Poland natural gas consumption has maintained the level of 16 bcm per year in the last three years. In order to realise the main objective of the energy policy of the Polish state - diversification of gas supplies, a decision was made to build the first LNG regasification terminal in Poland located in Świnoujście. Its main objective is raising Poland's energy security by gaining more independence of gas supplies from the import from the eastern directions. The terminal was officially opened in May 2016. In December 2015 it received the first supply of LNG for the installation start-up. The storage capacity of the terminal in the first stage will be 0.32 million cubic metres of liquefied LNG and the regsification capability 5 bcm per year, which will cover 30% of the domestic demand for gas once the terminal is fully used. There is a possibility of building a third tank, which will increase the regasification capability to 7.5 bcm per year and building an installation for reloading gas tankers and loading tank trucks. The deliveries will mostly come from Qatar within a long-term contract but also within spot contracts (first supplies in June 2016 from Norway). The operation of the terminal should have its impact on the prices of the import of natural gas to Poland [14].

In Belgium natural gas consumption in 2014 decreased by 12% compared to 2012 and the supplies in the amount of 1.21 bcm covered 8% of the domestic demand for gas. In 2015 a 95% increase in gas import was observed (compared to 2014) to the level of 2.32 bcm. The imported gas came entirely from Qatar - 3.43 bcm. Attention should be drawn to the amount of LNG export from Belgium in 2015 which was 1.06 bcm. Regasification and reloading of LNG takes place in the terminal in Zeebrugge, which has four tanks with the overall storage capacity of 0.38 million cubic metres. The annual regasification capability is 9 bcm. The important building work which is being/has been carried out within the development of LNG infrastructure in Belgium is building a second pier in Zeebrugge with the infrastructure which allows to receive small gas tankers up to 2 thousand cubic metres – which is to facilitate the shipment of LNG to Lysekill in Sweden. Besides, building a fifth storage tank of 0.18 million cubic metres is being planned.

Portugal, like Spain, owing to its geopolitical location in relation to the biggest deposits of natural gas requires a significant share of LNG in the structure of consumption of natural gas. The demand for it in 2014 was 3.8 bcm (16% less compared to 2012). The imported LNG in 2013 satisfied 45% and 33% in 2014 of the total demand for gas. In 2015 LNG in the amount of 1.38 bcm was imported to Portugal. The supplies were very diversified. They came from Algeria, Guinea Equatorial, Nigeria, Norway, Trinidad and Tobago and Qatar. Additionally, 0.33 bcm of LNG were exported. The reception and regasification of LNG in Portugal takes place at a terminal opened in 2004 located in Sines with the storage capacity of 0.39 million cubic metres and regasification capability of 7.6 bcm per year.

In Greece natural gas consumption in 2013 was 3.6 and in 2014 2.7 bcm. LNG imported to Greece in 2013 covered 16% of the demand and in 2014 18%. In 2015 LNG import rose to 0.57 bcm (18% more compared to 2014) and it was imported from (in bllion cubic metres) Algeria - 0.36, Nigeria - 0.08 and Norway - 0.14. In Greece LNG regasification is carried out at a terminal in Revithoussa, opened in 2000 with the storage capacity of 0.13 million cubic metres and regasification capability of 5 bcm per year. Within the development of the terminal, building a third tank with the storage capacity of 95 thousand cubic metres has been planned as well as building the infrastructure for loading small and medium size ships.

In Lithuania natural gas consumption in 2014 was 2.6 bcm. LNG import in 2014 was 0.14 bcm, which covered 5% of the domestic demand for gas. In 2015 it increased by over 290% to the level of 0.41 bcm (it was the first year of the operation of the terminal under full load since its opening in 2014). LNG was entirely imported from Norway. LNG regasification takes place at the terminal in Klaipeda with the rgasification capability of 4 billion cubic metres per year. It is a so-called floating terminal (FSRU - *Floating Storage Regasification Unit).*

It is worth mentioning the terminals in Sweden and Malta. In Sweden a terminal in Lysekil was opened in 2014 with the storage capacity of 30 thousand cubic metres. This facility is equipped with a tank, loading docks, LNG vaporisation installation and a pipeline leading to the refinery in Preem. In Malta, however, a terminal is being built, whose main priority will be supplying gas to two gas power stations on the island. The terminal will use a floating unit as a LNG storage facility (FSRU). Regasification installations will be located on the coast.

In conclusion, it needs to be stressed that the biggest share of the LNG market in the EU in 2014 belongs to Spain and Portugal. These two countries also have the most diversified structure of supplies of gas. What is equally important is the fact that the biggest storage capacities and LNG regasification capabilities in 2015 in the EU were located in the UK and Spain. It causes that these countries can, in all probability, increase LNG share in the structure of supplies in the future. It is also worth drawing attention to the numerous LNG projects in the EU, in the countries which already have terminals and in the countries which have started using LNG recently or which will start using it in 2016 i.e. Lithuania, Sweden, Poland and Malta.

Table 1. Characteristics of LNG regasification terminals in the EU countries (own study based on [13]).

Country	Site	Total capacity liquide LNG [cm liquid]	Nominal capacity [bcm/y gas]
UK	Dragon	320 000	7,6
	Isle of Grain	1 000 000	20,5
	South Hook LNG	775 000	21,3
	Teesside GasPort	138 000	4,2
Italy	Toscana	137 500	3,8
	Panigaglia	100 000	3,3
	Rovigo	250 000	8,0
France	Fos-Cavaou	330 000	8,3
	Fos Tonkin	150 000	3,0
	Montoir de Bretagne	360 000	10,0
Netherlands	Rotterdam	540 000	12,0
Spain	Barcelona	760 000	17,1
	Bilbao	450 000	7,0
	Cartagena	587 000	11,8
	El Musel	300 000	7,1
	Huelva	619 500	11,8
	Mugardos	300 000	3,6
	Sagunto	600 000	8,8
Poland	Świnoujscie	320 000	5,0
Belgium	Zeebrugge	380 000	9,0
Portugal	Sines	390 000	7,6
Greece	Revithoussa	130 000	5,0
Lithuania	Klaipeda	173 000	4,0

5 Summary

LNG share in relation to the overall import in the years 2008 - 2014 ranged from 13% to 25% (excluding re-export), which corresponds to the volume of supplies 41-81 bcm. Such a low level of LNG infrastructure use clearly informs about significant possibilities of increase in LNG import, which will undoubtedly be conditioned by the demand for natural gas, LNG prices (especially in relation to the import of gas through gas pipelines), accessibility of gas sent through pipeline networks as well as the political situation in the regions where the biggest deposits are. The analysis of the costs of natural gas import by the EU countries indicates that the UK had the lowest costs of gas import - ca. 35% lower in comparison with the average cost of gas import to the EU. It can be stated beyond any doubt that a well developed transmission infrastructure enabling the UK to diversify the supplies of gas has had a considerable impact. It is worth highlighting the fact that gas supplies can be realised by both gas pipelines and 4 LNG terminals. Furthermore, the UK also stood out from other EU countries in terms of low costs of LNG import [14].

As mentioned before, the range of use the regasification potential of LNG terminals in the EU has been relatively low in recent years. It seems that the situation may change in 2016. In 2015 alone an increase in LNG import was observed in comparison with 2014

 DOI: 10.1051/e3sconf/20161000132

due to the decrease in the prices of LNG. In the period between May 2013 and January 2016 the prices of LNG in the EU went down by ca. 50% on average [16]. It has been mainly caused by the falling prices of oil in recent years. Current situation on the LNG market and the low prices will probably support a rising trend of LNG supplies within spot contracts. The rising role of LNG in supplying the EU can also be connected with the start of LNG export from the USA to Europe (first supply from the Sabine Pass terminal in Louisiana reached Portugal in April 2016). One must not forget about the projects being carried out which will increase the world supply of LNG in the near future, especially about Australia's plans in this respect [17] [18].

Hopes for the increase in the significance of LNG in the EU are also connected with the use of this fuel in transport, particularly sea transport, after the introduction of rigorous regulations limiting the content of sulphur in fuels used by ships in SECA (*Sulphur Emission Control Areas*) regions [19].

References

1. M. Gałczyński, M. Ruszel, P. Turowski, R. Zajdler, A. Zawisza – *Globalny rynek LNG*, Wydawnictwo Rambler, Warszawa 2015
2. Rychlicki, J. Siemek, *Polityka Energetyczna – Energy Policy Journal,* **11**, 1 (2008)
3. M. Ruszel, *Polityka i Społeczeństwo*, **4,** 12 (2014), Uniwersytet Rzeszowski
4. K. Ficoń. W. Sokołowski, *Logistyka*, **5** (2012)
5. LNG SNAPSHOT (Analysis) http://lngsnapshot.com
6. M. Malec, J. Kamiński, P. Kaszyński, *Polityka Energetyczna – Energy Policy Journal*, **19,** 1 (2016)
7. L. Gawlik, A. Szurlej, A. Wyrwa, *ENERGY,* **92** (2015)
8. Eurogas Statistical Report, http://www.eurogas.org (2007-2015)
9. BP Statistical Review of World Energy, https://www.bp.com/ (2015)
10. A. Szurlej, J. Kamiński, P. Janusz, K. Iwicki, T. Mirowski, *Rynek Energii,* **6,** 115 (2014).
11. J. Siemek, S. Nagy, *Archives of Mining Sciences – Archiwum Górnictwa,* **57,** 2 (2012)
12. M. Kaliski, P. Janusz, A. Szurlej, *Gaz, Woda i Technika Sanitarna*, **83**, 7–8 (2009)
13. The LNG Industry in 2013-2016, Groupe International des Importateurs de Gaz Naturel Liquéfié, http://www.giignl.org
14. A. Szurlej, M. Ruszel, T. Olkuski, *Rynek Energii,* **5** (2015)
15. R. Biały, P. Janusz, M. Łoś, A. Szurlej: *Zeszyty Naukowe Instytutu Gospodarki Surowcami Mineralnymi i Energią PAN*, **87** (2014)
16. The Federal Energy *Regulatory Commission (FERC):* www.ferc.gov.
17. M. Kaliski, R. Wojciechowski, A. Szurlej, *Polityka Energetyczna-Energy Policy Journal*, **14**, 4 (2013)
18. P. Janusz, M. Kaliski, A. Szurlej, *Gospodarka Surowcami Mineralnymi/ Mineral Resources Management*, **31**, 3 (2015)
19. M. Krupa, M. Moskalewicz, A. P. Sikora, A. Szurlej, *Przemysł Chemiczny,* **94**, 12 (2015)

E3S Web of Conferences 10, 00133 (2016) DOI: 10.1051/e3sconf/20161000133
SEED 2016

Characterization of the wood combustion process based on the TG analysis, numerical modelling and measurements performed on the experimental stand

Mateusz Szubel[1,a], Mariusz Filipowicz[1], Wojciech Goryl[1] and Grzegorz Basista[1]

[1]*AGH University of Science and Technology, Faculty of Energy and Fuels, A. Mickiewicza 30 Av., 30-059 Kraków, Poland*

Abstract. The paper presents selected results of thermogravimetric (TG) analyses for softwood (pine) and hardwood (beech). The composition of the studied fuels has been defined and described. Both wood types used in the TG tests were studied in order to define their content of basic components such as lignin, cellulose and hemicellulose. Types of wood used in the TGA have been combusted on the experimental stand which is equipped with a set of temperature sensors and an exhaust analyser. A comparison of the TG analysis and the combustion in the heating unit has been performed to find relations between the kinetics of devolatilisation for different wood species and to determine the exhaust composition. Numerical modelling using computational fluid dynamics (CFD) has been performed for the process of carbon monoxide oxidation to supplement the tests results. The results of the comparisons of the performed analyses can be useful in all areas related to the process of optimisation and improvement of combustion, pyrolysis and devolatilisation process conditions in small scale heating units.

1 Introduction

Biomass, as an environment-friendly fuel, is gaining popularity. Local regulations often determine ranges of fuels which can be combusted in household heating devices. For this reason, utilization of wood in near future could rise rapidly.

Wood is the oldest form of biomass used for heat generation. Today, traditional fireplaces, furnaces and boilers have developed into advanced devices, equipped with expanded measurement instruments and control systems improving the efficiency of heat generation, but also producing a low emission of harmful combustion products [1].

A high quality of the device operation is possible if the details of the combustion process for each individual fuel are known, which allows to develop sets of operation recommendations, such as the process control algorithm, the appropriate load of the fuel, the position of the air nozzles in the combustion chamber, the construction materials etc. Changes of the air distribution in relation to the dynamics of temperature changes, but also to the current exhaust composition have to be taken into account.

Four fundamental stages of wood (and generally biomass) combustion need to be mentioned. The first phase includes the warming and drying of the fuel and there are no chemical reactions until the moment when the thermal decomposition starts (second stage). The moment of the beginning of this process depends on fuel properties, especially on the content of the major components. Long drying related to a high moisture content in the fuel can lead to delayed ignition and it has impact on the composition of the volatiles. Products of solid fuel thermal decomposition are reduced and oxidized in a series of chemical reactions constituting the gasification process [2]. Gasification residues are coke and mineral substance (ash). The combustion of fixed carbon is the last stage of biomass combustion.

Gasification of wood can start in a relatively low temperature, about 220°C, but different ingredients have different temperature of decomposition. The first important component, which undergoes gasification, is hemicellulose – in a temperature lower than 320°C. Then decomposition of cellulose begins, between 320 and 370°C. Lignin has the widest range of gasification temperature – between 320 and 500°C [3].

Effective combustion of wood is possible when the energy contained in the volatiles is efficiently converted to heat in the post-combustion process. The content of the total fuel calorific value which is contained in the volatiles could achieve even 70%, making the gasification process the most significant stage of wood combustion, but also combustion of the other types of solid biomass, such as straw.

In the research of the process of combustion and the gasification process, both experimental and numerical methods are used. Tenorio and Moya [4] have performed TGA analyses to investigate the thermal stability of different fast-growth wood species planted in the tropics. They have also examined mass consumption during the burning process in order to identify different

[a] Corresponding author: mszubel@agh.edu.pl

stages of the combustion process for these species. As a result of research, the most suitable species for the combustion process have been selected.

Garcia-Maraver et al. [5] have performed studies of the influence of different olive tree components on the combustion process of this type of biomass, using thermogravimetric analysis. As a result, a higher content of lignin in olive leaves has been identified as a factor responsible for the lower reaction rate of this component, which results in a higher ash content.

Vincente et al. [6] describe wood combustion experiments. The goal of study was to determine the effect of ignition technique, fuel load and cleavage as well as secondary air supply on carbon monoxide, total hydrocarbon, particulate matter (PM10) and particle number emissions from a wood stove. Pine wood and beech wood was selected to carry out the experimental tests.

According to Lamberg et al. [7], there is a significant difference in the particle emissions of samples collected from appliances representing different combustion situations in small-scale combustion. In the study described in the aforementioned paper, particle samples were collected from six appliances: five wood log fuelled stoves, including one stove equipped with modern combustion technology, three different conventional devices, one sauna stove and a modern pellet boiler. It was found that the combustion in the pellet boiler is the most efficient, which results in the smallest emissions, whereas inefficient batch combustion is always related to larger emissions (in this case, especially in the sauna stove).

Menghini et al. [8] have performed experimental and CFD analysis of a wood-open fireplace to determinate the possibility of the improvement of the total environmental impact of the device. A complete optimization cycle has been carried out and two configurations of heat exchanger were studied. It was established that it is difficult to improve combustion emission in domestic devices, where there is very little process control. However, significant improvements can be achieved as far as heat recovery efficiency. This can be done using suitable modifications of the heat exchanger.

2 Materials and methods

2.1 Thermogravimetric analysis

The thermogravimetric tests constituting a part of the research described in this paper were performed with use of TA Instruments SDT 2960 simultaneous TG – DTA apparatus. The device provides detection of the mass loss with a resolution of 0.1μg. The measurement of temperature is performed inside the sample holder, situated in the furnace (Fig. 1). The furnace is a radiant chamber that creates a uniformly heated zone where a quartz reactor is located [9].

The remaining major parts of the aforementioned apparatus are: a PID controller, a gas feeding system with flowmeters, a data acquisition set and a precision balance [9].

Figure 1. TA Instruments SDT 2960: a) general view, b) furnace.

The sample is exposed to thermal radiation by means of a stainless steel mesh screen the sides of which are wrapped on two stainless steel rods connected to a precision balance [9]. Inert gas flow, divided by two streams, connected with the furnace chamber and balance, establishes the proper reaction environment. An atmosphere of high purity nitrogen was used for the tests, at a flow rate of $1 \cdot 10^{-4} m^3/min$.

An additional series of measurements has been carried out using an oxygen atmosphere, which was necessary to perform the fit of the mass loss curves for the samples and the curves designated using the mathematical model, which has been described in the subsequent part of the chapter.

The fuels (beech hardwood, pine softwood) used in the tests were prepared by drying and milling to a particle size of <0.25mm. The accurate mass of the samples is presented in table 1.

Table 1. Masses of samples used in tests

No. of sample	Fuel	Mass [mg]
H1	Hardwood (beech)	53.4
H2	Hardwood (beech)	53.4
S1	Softwood (pine)	53.5
S2	Softwood (pine)	53.5

Each sample was heated with the heat rate $20 K \cdot min^{-1}$ to 110°C, then the temperature was maintained for 15 minutes (isothermal stage). Next, the sample was heated to 900°C (heat rate $20 K \cdot min^{-1}$). The last stage of the process was cooling them down to about 30°C.

To achieve accurate results, each sample was compared with the correction curve and a subtraction of both data series has been performed.

DTG curves always show partially overlapping peaks [10]. To separate the peaks related to the individual substances included in the samples, a mathematical model was used. It was assumed that three pseudo – components are included in the samples (hemicellulose, cellulose and lignin). The mathematical model used in the calculations is based on an approximation of the individual reaction steps by reaction order of one.

The equations are presented below:

$$\frac{-dm}{dt}=\sum_{i=0}^{3}C_i\frac{da_i}{dt} \tag{1}$$

where:
a – the volatile fractions released during devolatilisation
i – hemicellulose, cellulose, lignin
C_i – contribution of the partial decomposition process
da_i/dt – the mass loss rate for each component
t – time

$$\frac{da_i}{dt}=A_i\exp\left(\frac{-E_i}{RT}\right)(1-a_i) \tag{2}$$

where:
A_i – pre-exponential factor
E_i – activation energy
R – universal gas constant
T – temperature

2.2 Characterization of biomass combustion

To describe the combustion behaviour of the analysed biomass, the double parallel reaction random pore model (DRPM) was applied [11]. According to this model, two competing reactions occur simultaneously:

$$\text{wood\_volatile} + O_2 \rightarrow H_2O + CO_2 \tag{3}$$

$$\text{wood\_residue} + O_2 \rightarrow CO_2 + \text{Ash} \tag{4}$$

"wood_volatile" is considered as volatile matter in the sample and "wood_residue" is considered as carbon residue.

The DRPM model took into account the pore structure and its evolution during the reactions. According to the model, the pores of the particles are composed of randomly cylindrical pores of arbitrary size distribution, in which the reaction surface is given by the internal surface of the pores. It supposes that the reaction rate is proportional to its associated surface area. The combustion rate can be expressed as:

$$\frac{d\alpha}{dT}=\frac{A}{\beta}\exp\left(\frac{-E}{RT}\right)\alpha\sqrt{1-\psi \ln\alpha} \tag{5}$$

where α is non-converted fraction of the sample, β – the heating rate in the non-isothermal thermogravimetric method, here equal to 20 K/min, E – apparent activation energy and ψ is a parameter of particle structure.
The ψ parameter can be expressed as:

$$\psi=\frac{4\pi L_0(1-\varepsilon_0)}{S_0^2} \tag{6}$$

where S_0 is the pore initial surface area, L_0 – the total pore length per unit volume (m/m^3), ε_0 – the solid porosity [12].
After the integration of eq. (5), the non-converted fraction can be expressed in the following form [11]:

$$\alpha=\exp\left(-A\exp\left(\frac{-E}{RT}\right)\left(\frac{T-T_0}{\beta}\right)\left(1+\frac{A\exp\left(\frac{-E}{RT}\right)}{4\psi}\left(\frac{T-T_0}{\beta}\right)\right)\right) \tag{7}$$

where T_0 is the initial reaction temperature (373 K in the present study).

The DRPM is expressed as a linear combination of the two competing processes (3) & (4), each included by the appropriate fraction dependence α_1 and α_2.

So finally:

$$\alpha=\varepsilon\ \alpha_1(A_1,E_1,\psi)+(1-\varepsilon)\ \alpha_2(A_2,E_2,\psi) \tag{8}$$

To use this equation, it is necessary to determine six fitting parameters: $A_{1,2}$, $E_{1,2}$, ψ and ε. The non-linear fitting procedure was applied to that end.

2.3 Experimental analysis

To compare results from the laboratory scale experiments (TGA) and mathematical calculations (DRPM) with the experimental data, a measurement stand based on an accumulation fireplace, equipped with temperature sensors and an exhaust analyser (Fig. 2), as well as a control damper of the chimney draft has been applied (Fig. 3).

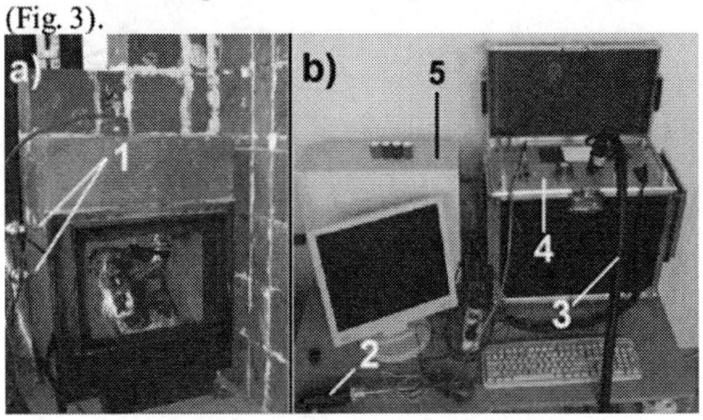

Figure 2. Experimental stand: a) general view of the accumulation stove with visible temperature sensors (1); b) main elements of the exhaust analysis system, including the probe (2), the gas duct (3), the measuring system (4) and the computer (5).

In case of the heating unit applied in the experimental works, ceramic accumulation material has been used to optimize the current power of the device and reduce the concentration of undesirable components of exhaust (mainly carbon monoxide). Details of the design and the operation of the accumulation stoves – fireplaces are described in [13].

For both pine and beech slivers of wood, two experiments have been performed. In view of the fact that during the measurement it was not possible to combust samples of wood mass in milligrams as it took place in the TG analysis, the mass of wood for hard and soft wood was 5.4 kg. The wood that was combusted in the fireplace was the same as in the TG analysis, but in form of slivers. Before each measurement series, pre-combustion has been carried out to warm up the combustion chamber. In both cases, the humidity of fuel was approximately 12-15%. The fuel load was 5.4 kg in each experiment. Each type of wood was combusted two times. The time of the measurement was about

DOI: 10.1051/e3sconf/20161000133

55 minutes. Due to the recommendations of the device producer, the control damper of the chimney draft was set on the -12 Pa.

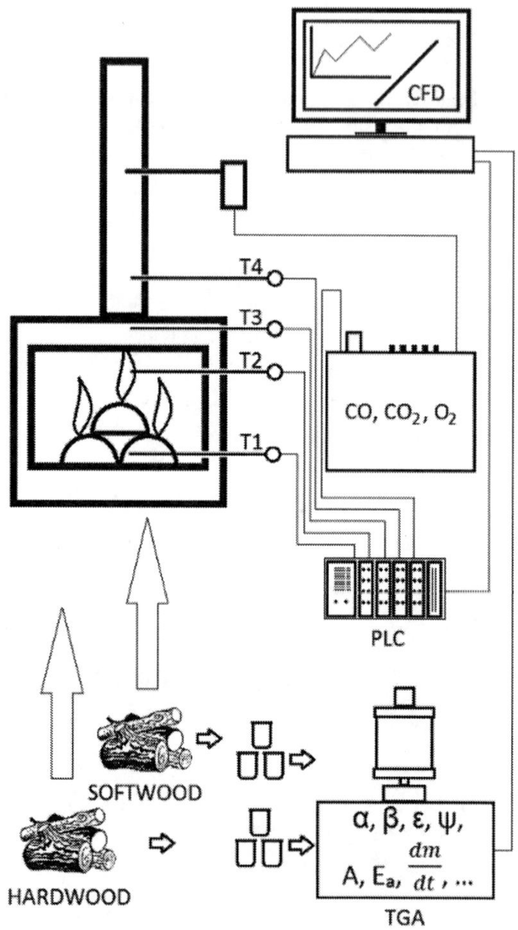

Figure 3. Schematic diagram of the experimental stand.

The air for the combustion process is provided by two parallel gaps located on the bottom and top of the front frame of the hearth. The charging of the air for both supplying degrees is realised through one pipe (150 mm diameter), installed under the body of the hearth.

During the experimental measurements, type K thermocouples were used. In Figs. 6 and 7 they are listed as T1, T2, T3 and T4 and were situated in the combustion chamber, above the deflector, above the smoke conduit and in the chimney, respectively. Moreover, the WAGO System has been used to connect thermocouples to the computer where the data was collected.

The MRU ECO 3000 exhaust analyser has been applied in the studies.

2.4 Numerical analysis

To analyse the distribution of gases and the reaction process inside the combustion chamber of the ceramic hearth, the CFD method has been applied. Spatial geometry of the hearth has been developed using the CAD tool and exported to the ANSYS Workbench 15. Then DesignModeler has been employed to define the computational domain of the combustion chamber interior (Fig. 4).

ANSYS Meshing has been applied for the domain discretisation. As a result of this process, $5 \cdot 10^5$ of the mesh elements were generated. Multizone method has been used to create mesh in most regions of the domain.

A set of significant process and device operation parameters has been defined as boundary conditions. Domain reference pressure was 101325 Pa. Inlet was defined as an opening boundary with relative pressure of 0 Pa and air temperature of 283.15 K. The outlet relative pressure, according to the experiment, was set as -12 Pa. The outlet opening temperature (after the outlet from the computational domain) was defined based on the average temperature on the outlet surface. The glass contact surface and the walls contact surfaces were characterised by heat transfer coefficients of 145 W/(m^{2}K) and 110 W/(m^{2}K) respectively. Based on the measurement carried out during the experimental tests, the assumed ambient temperature was 283.15 K.

The K-epsilon model of turbulence has been applied due to the lack of the near-wall layer analysis.

The goal of the study was to investigate the distribution of fuel decomposition products and the areas of efficient combustion, as well as the impact of the locations of air inlets. Gasification products have been limited to CO and H_2O. Top surfaces of the wood slivers have been defined as the volatiles inlet. The mixture composition as well as temperature of the gas fuel have been determined based on the DG and experimental tests.

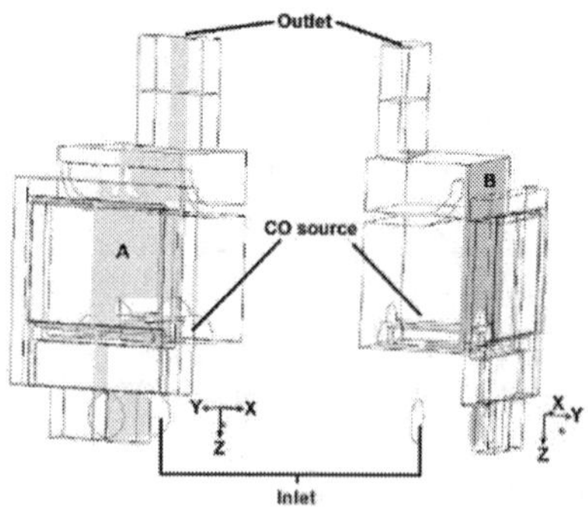

Figure 4. Computational domain of the combustion chamber, designed based on the experimental stand with marked inlet, outlet and CO source. The cross sections used in the presentation of the results are marked as "A" and "B" (on two different views of the domain).

Finite rate chemistry and the eddy dissipation model of combustion with the Arrhenius equation has been applied. Discrete transfer model of radiation with "surface to surface" mode has been used, including multigray mode of the mixture components properties.

3 Results and discussion

To perform an evaluation of the kinetic parameters concerning the thermal decomposition process and a determination of the fractions of components for hard and soft wood (table 1) (regarding thermal devolatilisation behaviour), dedicated software based on Matlab was employed. To define specific temperature points of thermogravimetric curves, first derivative of mass loss in samples has been calculated (Fig. 5). The curves in Fig. 3 present an average value from two samples H1, H2 and S1, S2 for hard and soft wood, respectively. Then, specific temperature points at the curves were defined (points of the curve trend changes – especially peaks, onsets/offsets, inflections etc.).

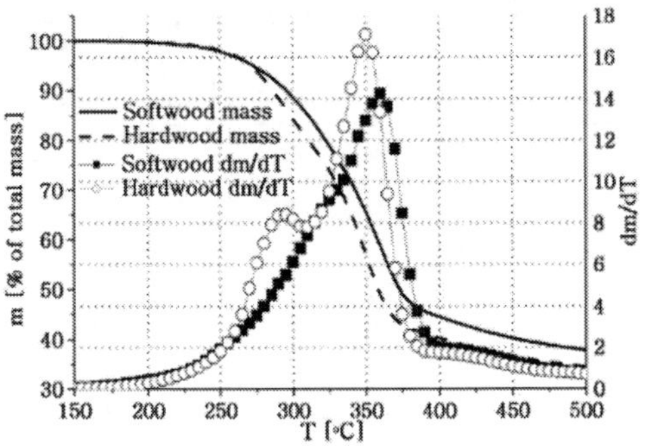

Figure 5. Thermogravimetric curves and first derivatives of mass loss for the considered wood species.

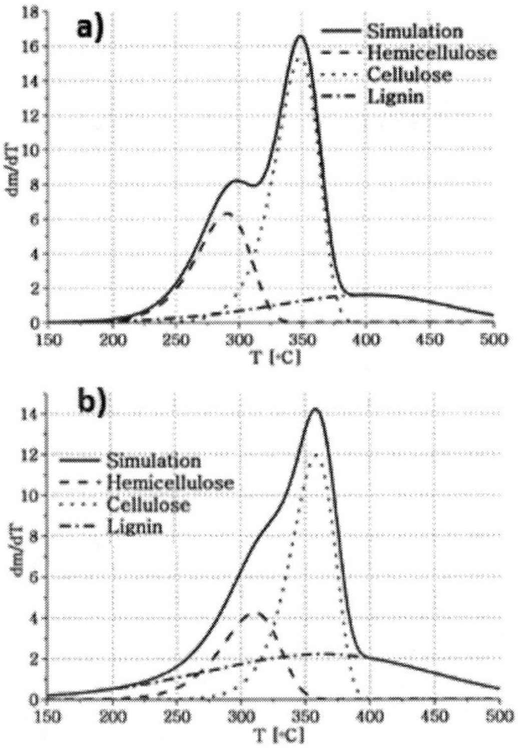

Figure 6. DTG for: a) hardwood, b) softwood.

The shoulders and tailings on the DTG curves of fuels (Fig. 6) indicate that the analysed biomass consist of components characterized by different pyrolysis reactivity. As in Fig. 5, curves in Fig. 6 present an average value from two samples H1, H2 and S1, S2 for hard and soft wood, respectively.

For hardwood, the point which can be understood as a beginning of hemicellulose decomposition was found for the temperature value of about 210°C. Because of the specifics of beech wood pyrolysis in the 250°C – 300°C temperature range, a second characteristic point of the process was found as a first peak on the *dm/dT* curve in Fig. 5. The point is related to the temperature of the maximum hemicellulose decomposition rate, which in this case was equal to about 6.3% of dry mass per minute in the temperature of 275°C. The second peak in the top region of the chart can be understood as the temperature of the maximum cellulose decomposition rate, equalling to about 15.4% per minute. The temperature for this point is 350°C. For hardwood, the average temperature of the end of cellulose decomposition point is equal to about 475°C. The moment when the curve in Fig. 5 representing mass loss returns to horizontal is related to the achievement of solid residue of sample (charcoal). In case of beech wood, it is about 35% of the initial sample mass.

Different decomposition characteristics were obtained as a result of the measurement of pine wood samples. First, the lower peak identified in case of beech wood is not present in this case. The temperature of the beginning of hemicellulose decomposition is higher than for beech samples, equalling to about 220°C. It is difficult to define the maximum hemicellulose decomposition rate because of the insignificant curve angle change in the range between 250°C and 360°C. This parameter has been established using simulation fit as a point of the curve inflection, in the temperature of 315°C, which was higher than for beech wood (about 40°C). A smaller difference between pine and beech wood has been observed for the temperature of the maximum cellulose decomposition rate – about 360°C.

Temperature values, decomposition rates and activation energy of decomposition of the considered pseudo components for beech and pine wood samples are included in table 2. Table 2 presents the average value of sample H1, H2 and S1, S2 for hard and soft wood, respectively. The point of maximum decomposition rate for lignin in comparison to beech samples is moved to the left. For pine wood, it is between 360°C and 370°C, whereas in case of hardwood it is about 400°C.

Some information related to fixed carbon (charcoal fraction) was obtained using the DRPM model. These results can be applied to the analysis of the last stage of biomass combustion, an aforementioned stage of fixed carbon combustion. The content of fixed carbon and its temperature dependence is important at the end of wood combustion in fireplaces.

The obtained results for two types of wood (hard and soft wood) using DRPM (eq. 7) are presented in Fig 7. The solid curves show results of the fitting procedure of the fractional conversion of the biomass samples.

 DOI: 10.1051/e3sconf/20161000133

Table 2. Results of the calculations concerning the content and decomposition rates of the pseudo components included in the wood samples

No.	Fuel	Component	*TGA*						*DRPM*		
			Mass fraction	Dec. start	Dec. end	$[-dm/dt]_{max}$	*Ea*	*ln(A)*	E_1	E_2	ψ
			[wt% d.b.]	[°C]	[°C]	[wt%/min]	[kJ/mol]	[s^{-1}]	[kJ/mol]	[kJ/mol]	
1.	Hardwood	Hemicellulose	16.7	210	337	6.3	123.3	22.4	92.8	2.9	2.2
		Cellulose	33.5	267	383	15.4	191.2	33			
		Lignin	15.3	<200	>500	1.7	44.5	2.6			
2.	Softwood	Hemicellulose	14.2	220	366	4.4	112	19.9	88.5	0.5	9.2
		Cellulose	26.7	276	395	12	193.8	34.2			
		Lignin	21.3	<150	>500	2.1	31.3	0.4			

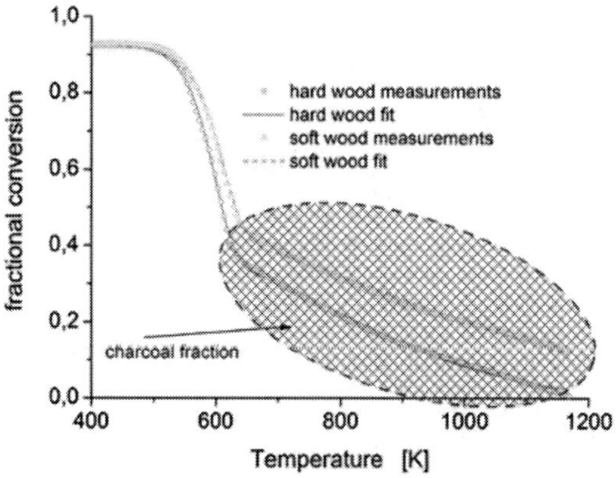

Figure 7. Fractional conversion data fitted by DRPM model for hard and soft wood.

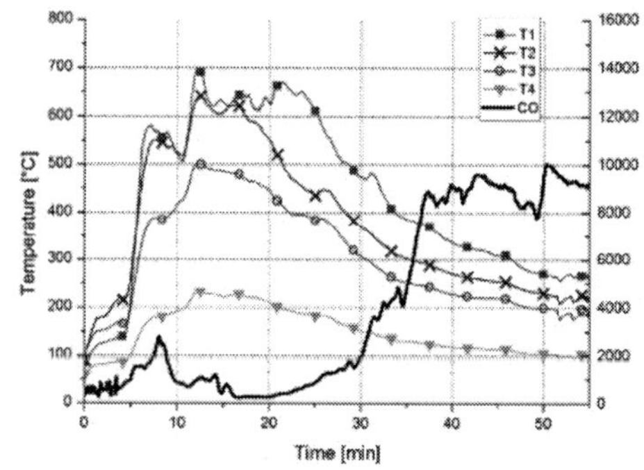

Figure 8. Average temperature and CO emission (13% O2) changes in time during hardwood combustion in the fireplace.

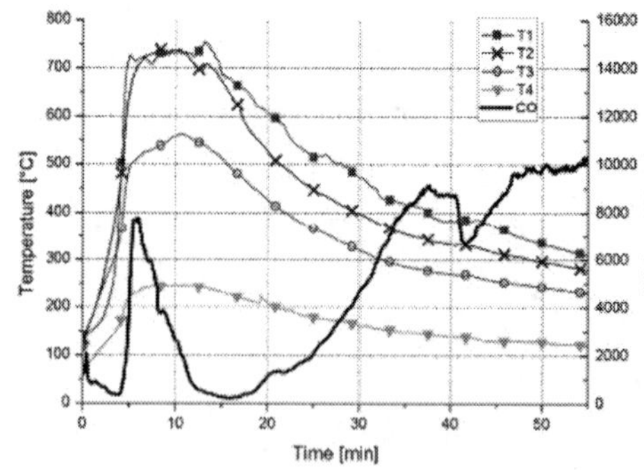

Figure 9. Average temperature and CO emission (13% O2) changes in time during softwood combustion in the fireplace.

Better fitting results were obtained for hardwood. It is visible that the charcoal fraction in the samples is much higher for softwood as compared to hardwood. In both cases the combustion of fixed carbon starts in a temperature a little higher than 600K.

According to eq. 8, the first activation energy E_1 is similar in both cases and equal to 92.8 kJ/(mol) for hardwood and 88.5 kJ/(mol) for softwood and the second activation energy E_2 is 2.9 and 0.5, respectively. The parameter of the particle structure ψ is equal to 2.2 for hardwood and 9.2 for softwood (see Table 2).

Fig. 8 shows the variation of temperature and CO emission during hardwood combustion in the fireplace. Two combustions of beech wood were performed and in Fig. 8 the average temperature and CO emission is presented. First, a rapid increase of temperature caused by intensive burning of volatiles is observed from the 3rd – 4th minute of measurements. Then, after about 15 minutes, the temperature in combustion chamber becomes relatively stable. Next, there is a rapid drop of temperature until the end of the combustion process. The emission of CO during beech wood combustion in the fireplace is relatively low until approximately 25 min. Then the emission increases rapidly up to 8000 – 10000 ppm, where the value is stabilized.

The curves shown in Fig. 9 present the changes of temperature and CO emission during pine wood combustion. Two combustions of pine wood were performed and in Fig. 9 the average temperature and CO emission is presented. Burning softwood is characterized by significantly different waveforms of temperature and CO emission. Pine wood, as compared to beech wood, reached much higher temperatures in the combustion chamber, which were caused by a higher resin content in the wood. The period of stable temperature, after the initial increase at the beginning of the burning, is shorter by 5 minutes. After 15 minutes from the start of the combustion process, the temperatures start to slowly fall

[a] Corresponding author: mszubel@agh.edu.pl

down to reach about 300°C. The curve showing CO emission is characterized by a sharp peak at approx. 5 min of combustion process. Mostly, it is caused by the high resin content of pine wood and its higher lignin content, which start to decompose in lower temperature as compared to beech wood.

The comparison of Figs. 8 and 9 shows that in the last stage of the process related to fixed carbon combustion, the higher temperature was obtained for softwood – ca. 300°C as compared to ca. 250°C for hardwood, and the rate of the temperature decrease is a little slower for softwood. It is related to the results presented in Fig. 7, where the higher content of fixed carbon for softwood is visible (45% vs. 35%). Moreover, the charcoal fraction from the softwood samples is burning a little slower.

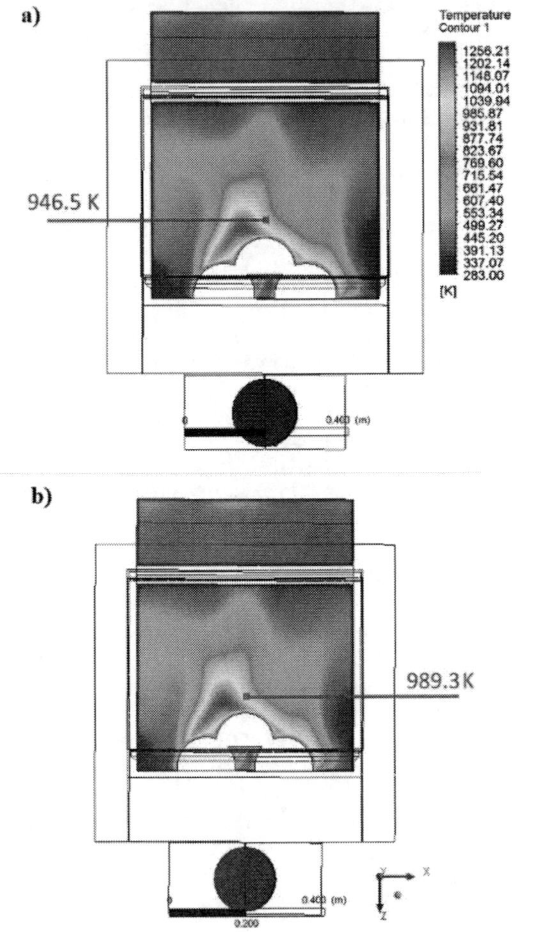

Figure 10. Temperature in the combustion chamber for: a) hardwood, b) softwood.

Figures 10 and 11 present selected results of the evaluated CFD modelling of the heating device. The simulations has been carried out to analyse spatial distribution of some exhaust components, and consequently to investigate impact of current design of the air distribution system on the process of the fuel combustion. Obtained results confirm, that impact of construction of the combustion chamber, including air staging system configuration is from the combustion process point of view equally important as the fuel properties. To show clearly impact of the gas streams (combustion air, gaseous fuel, reaction products) in the combustion chamber on the characteristics of the oxidation of the volatiles represented mainly by the CO, front view as well as side view of the combustion area are presented.

Observed different temperature levels and CO_2 concentration for soft and hard wood confirms effect of pseudocomponets composition decomposition rate. As mentioned in previous chapters, thermal decomposition of the pseudocomponents of the biomass fuel is strongly temperature-dependnd. Due to the fact, that in terms of the high temperature liquid products of the lignin destruction are secondary-converted to simple molecules such as CO, intensified phenomena of CO production can be noted in case of the fuel characterised by higher content of the lignin (softwood).

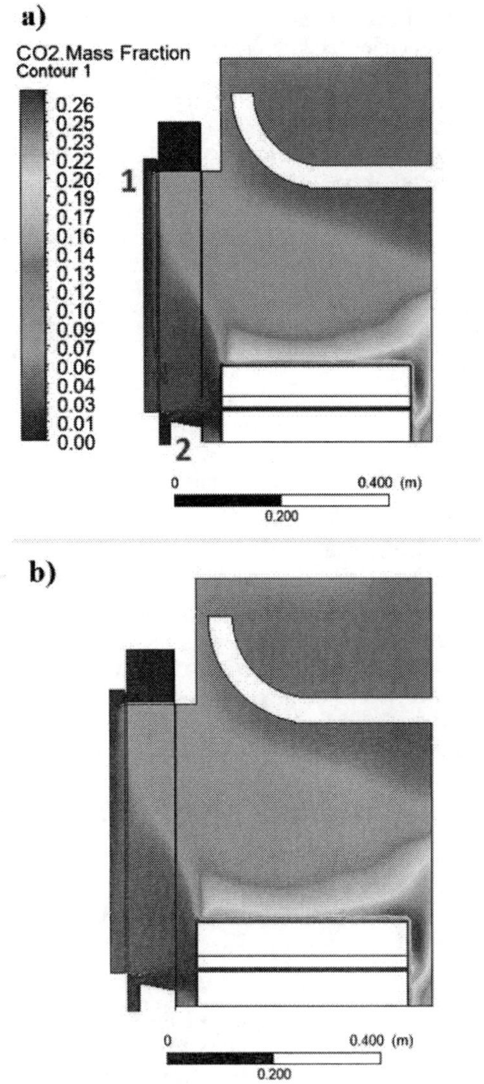

Figure 11. CO_2 mass fraction for: a) hardwood, b) softwood.

Despite the difference in range of the temperature level, profile of the flame is similar in both cases. It proves that significant factor which determines flame propagation is distribution of the gas streams in the combustion chamber. Air staging is achieved by the system of the channels situated in front housing of the glass door. First part of the combustion air flows from the bottom (below door, just above the bottom of the combustion chamber) through the fuel bed, to the rear wall of the chamber. Second part of the air is released from the slot above the door and it is distributed on the glass, mainly to avoid soot (see red arrows on Fig. 11 a): 1 – first part of the air, 2 – second part of the air). As a result of this approach to air feeding process, it is visible, that temperature level and the CO_2 concentration is much lower in regions in which free flow of the air is possible (near to glass and walls, as well as in the left and right part of the bottom of the combustion chamber). Moreover, due to different hydraulic resistance for streams 1 and 2, mass flow rate in the both inlets is different and not homogenous along the slots. In analysed cases, it causes deflection of the flame region to the left wall (Fig. 10).

Products of gasification are transported from the areas of syngas by relatively cool air released from the inlet slots and after heating, reactions can occur in rear part of the combustion chamber (Fig. 11). Inhomogenous temperature distribution related with the phenomena mentioned above can influence on the efficiency of the combustion of the gasification products. In case of the CO oxidation, presence of some radicals (i.eg. -OH) is equally important that presence of oxygen or high temperature [14]. Due to this fact, different rate of the stages of combustion in different areas of the chamber can disturb combustion process and increase concentration of CO in exhaust. This efect is partially responsible for the high CO emmision at the begining of each experimental test presented previously.

4 Conclusions

Two types of wood were compared as fuel in fireplaces. The first analysis was devoted to the detailed study of its decomposition using small samples of fuel, then the combustion of the appropriate fuel was performed in the fireplace. Results obtained during the laboratory analysis can be applied to the analysis of fuel combustion in the fireplace. Especially the different contents of three main components, ie. hemicellulose, cellulose, and lignin determine the ignition rate, the temperature growth rate and the maximal temperature. The fixed carbon contents determine the temperature decrease in last stage of the process. It has been demonstrated that all these parameters can be applied in CFD modelling of biomass combustion.

References

1. W. Rybak, *Spalanie i współspalanie biopaliw stałych Combustion and co-combustion of solid biofuels.* (Oficyna wydawnicza Politechniki Wrocławskiej, Wrocław, 2006)
2. B. Sørensen, *Renewable energy. Its physics, engineering, environmental impacts, economics & planning* (Academic Press, London, 2000)
3. W. Kordylewski, *Spalanie i Paliwa Combustion and Fuels.* (Oficyna wydawnicza Politechniki Wrocławskiej, Wrocław, 2008)
4. C. Tenorio, R. Moya 2013.. Thermochim. Acta **563**, 12–21 (2013)
5. A. Garcia-Maraver, D. Salvachua, M.J. Martinez, L.F. Diaz, M. Zamorano, Waste Manage., **33**, 11 (2013)
6. E.D. Vincente, M.A. Duarte, A.I. Calvo, T.F. Nunes, L. Tarelho, C.A. Alves, Fuel Process. Technol, **131**, 182–192 (2015)
7. H. Lamberg, K. Nuutinen, J. Tissari, J. Ruusunen, P. Yli-Pirila, O. Sippula, M. Tapanainen, P. Jalava, U. Makkonen, K. Teinila, K. Saarnio, R. Hillamo, M.-R. Hirvonen, Jokiniemi, Atmos. Environ., **45**, 7635–7643 (2011)
8. D. Menghini, T. Marchione, G. Martino, S.F. Marra, Ch. Allouis, F. Beretta, Exp. Therm. Fluid Sci., **21**, 477–482 (2007)
9. G.M. Ronli, G. Varhegyi, C. Di Balsi, Ind. Eng. Chem. Res., **41**, 4201–4208 (2002)
10. C. Branca, C. Di Barsa, Energ. Fuel., **17**, 1609–1615 (2003)
11. G. Wang, J. Zhang, J. Shao, S. Ren, Thermochim. Acta, **591**, 68–74 (2014)
12. H. Fei, L. Sun, S. Hu, J. Xiang, Y. Song, B. Wang, G . Chen, J. Anal. Appl. Pyrol., **91**, 251–256 (2014)
13. L. Kurcz, M. Filipowicz, K. Sornek, M. Szubel, K. Rzepka, J. Ręka, Ciepłownictwo Ogrzewnictwo Wentylacja, **46**, 104–111 (2015)
14. S. van Loo, J. Coppejan, The handbook of biomass combustion and co-firing (EARTHSCAN, London 2008)

Acknowledgements

The thermogravimetric analysis was carried out within the "Biofuels Research Infrastructure for Sharing Knowledge" project in Bioenergy 2020+ company (Graz, Austria). The studies performed on the experimental stand and the CFD analyses were carried out under contract (11.11.210.217), AGH-University of Science and Technology, Faculty of Fuels and Energy, Cracow, Poland.

The construction of the measurement stand with the accumulation stove was possible thanks to the CEBUD company (Krakow, Poland).

Combined Heat and Power Systems for the Provision of Sustainable Energy from Biomass in Buildings

Andreas Ortwein[1,a]

[1]*Deutsches Biomasseforschungszentrum gemeinnützige GmbH, Torgauer Straße 116, 04347 Leipzig, Germany*

Abstract. Against the background of greenhouse gases causing climate change, combined heat and power (CHP) systems fueled by biomass can efficiently supply energy with high flexibility. Such CHP systems will usually consist of one or more thermo-chemical conversion steps and at least one (the more or less separated) electric power generation unit. Depending on the main products of the previous conversion steps (e.g. combustible gases or liquids, but also flue gases with sensible heat), different technologies are available for the final power conversion step. This includes steam cycles with steam turbines or engines and different working fluids (water, organic fluids), but also combustion based systems like gas turbines or gas engines. Further promising technologies include fuel cells with high electric efficiency. When integrating such CHP systems in buildings, there are different strategies, especially concerning electric power generation. While some concepts are focusing on base load production, others are regulated either by thermal or by electric power demand. The paper will give a systematic overview on the combination of thermo-chemical conversion of biomass and combined heat and power production technologies. The mentioned building integration strategies will be discussed, leading to conclusions for further research and development in that field.

1 Introduction

Climate change caused by greenhouse gases is a major challenge for the world. At the world climate conference in Paris 2015, it was decided to keep the increase of global average temperature below 2 K. Substitution of fossil fuels like coal or natural gas by renewable energy is considered to be a key element of carbon dioxide emission reduction since energy consumption is one of the main contributors to greenhouse gas emissions.

Compared to solar and wind power, biomass has the potential to provide energy whenever it is needed. Especially for decentralized applications (e.g. in domestic and industrial buildings), combined heat and power (CHP) systems fueled by biomass are an interesting option and should be considered. Flexible operation of such systems in terms of power provision, heat-power-ratio and usage of advanced control strategies is a key issue for successful decentralized grid stabilization. This paper aims to give an overview on several CHP technologies and their application in buildings.

In section 2, a systematic overview on CHP technologies fueled by solid biomass will be given. Section 3 will present several technology examples for such systems including relevant research on their flexible operation. In section 4, several strategies on the implementation of such technologies into buildings will be discussed. Section 5 presents conclusions of the paper.

2 CHP Technologies for Solid Biomass

Combined heat and power systems for solid biomass usually consist of several process steps (see Figure 1).

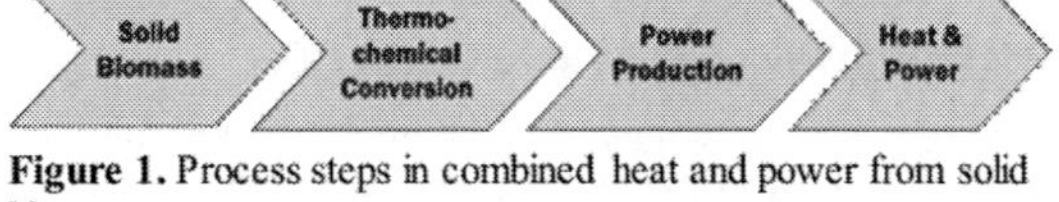

Figure 1. Process steps in combined heat and power from solid biomass.

In a first step, solid biomass is provided as a fuel for the process. Typical fuels to be considered are woody biomass, agricultural residues, and biogenic fractions of municipal solid waste (MSW).

In the next step, usually one or more thermo-chemical conversion processes (or primary conversion technologies) take place. This may include pyrolysis, gasification, combustion or more uncommon processes like hydro-thermal conversion [1]. Additionally, fuel preparation processes like chipping, pelletizing or briquetting might occur.

Pyrolysis is defined as the thermal decomposition of fuel without additional oxidation agent. The main products of pyrolysis include solid char and ash, liquid products, and gases (combustible and inert). The product composition depends on several parameters, including maximum conversion temperature, heating and cooling rate. [1]

Gasification is the process of under-stoichiometric oxidation of fuel. This is usually obtained by using

[a] Corresponding author: andreas.ortwein@dbfz.de

DOI: 10.1051/e3sconf/20161000134

a gasification agent like air, steam or carbon dioxide. Depending on the process design, gasification as an endothermal process is allothermal (external heat necessary) or autothermal (heat is provided by partial oxidation of fuel). The main products of gasification include carbon monoxide, hydrogen and methane. Depending on the process and gasification agent, further product gas components include carbon dioxide, water (steam), nitrogen and several trace components. [1]

Full oxidation converts the fuel completely to carbon dioxide and water (steam). This is conducted by using an oxidation agent containing enough oxygen for complete oxidation. Depending on the process conditions, the flue gas also contains nitrogen and oxygen as well as several trace components and byproducts (e.g. carbon monoxide from incomplete oxidation). All energy content is converted to sensible heat.

Hydrothermal conversion of fuel takes place in hot, pressurized water. The specific conditions of the reaction medium allows for the processing of wet biomass (e.g. from municipal waste). Depending on the process conditions, different products can be provided, including biochar or synthesis gas. [1]

Depending on the product of the previous conversion step(s), especially hot flue gas and combustible syngas, the following process step consists of one or more power production technologies (or secondary conversion technologies) [2]. Typical examples for these are steam turbines or engines, Organic Rankine cycles (ORC), gas turbine or engines or fuel cells. More uncommon technologies include externally fired gas turbines (EFGT) and thermo-electric generators (TEG).

Finally, heat and power is provided and has to be integrated into the system. For electric power, this is usually done by injecting the electricity into the grid. Heat is transported to the users by hot water or steam (for higher temperatures), or is stored in hot water storages.

3 Technology and research examples

One of the most common technologies to provide heat and power from solid fuels is based on the usage of a water-steam-cycle. The basic (simplified) principle is given in Figure 2. In this process, water is evaporated under pressure within a boiler, heated by a furnace. The pressurized steam is fed to a steam turbine or steam engine, where power is generated from its relaxation. Further heat is extracted from the steam in heat exchangers for heating purposes, or in condensers.

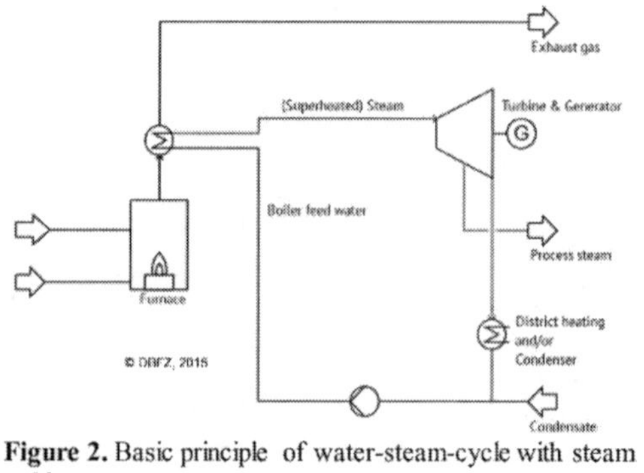

Figure 2. Basic principle of water-steam-cycle with steam turbine.

Decentralized applications of the steam-water-cycle have been developed within the last years, including a linear-piston based steam engine [3].

The basic principle for a combination of combustion and Organic Rankine Cycle (ORC) is given in Figure 3. Here, instead of water, an organic fluid is used for the thermal cycle. Thus, it is possible to optimize the process for specific conditions, e.g. combustion temperatures.

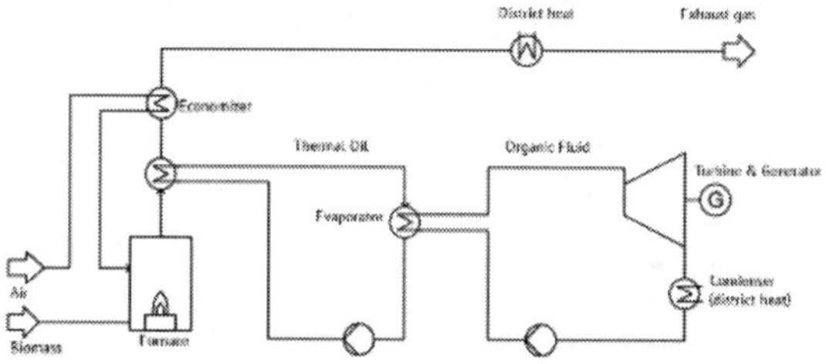

Figure 3. Basic principle of Organic Rankine Cycle.

Research on decentralized application of ORC on biomass combustion has been conducted during the last years, e.g. in [4]. In this specific case, instead of a turbine a modified compressor was used as a piston engine to overcome the technological and economic disadvantages of very small turbines.

Another example for CHP from solid biomass is the combination of gasification and gas engine. As one example, the flow scheme of a newly developed system for flexible power generation from solid biomass is given in Figure 4 [5]. Solid biomass is converted to a combustible gas within the gasifier. After a steam trap and filter, the product gas is mixed with air and fed to an adapted gas engine that can handle gases with relatively low heating values. Due to the changing gas composition, specific attention has been given to the management of ignition point [6]. Exhaust gas from the gas engine is cooled in a heat exchanger for further heat usage.

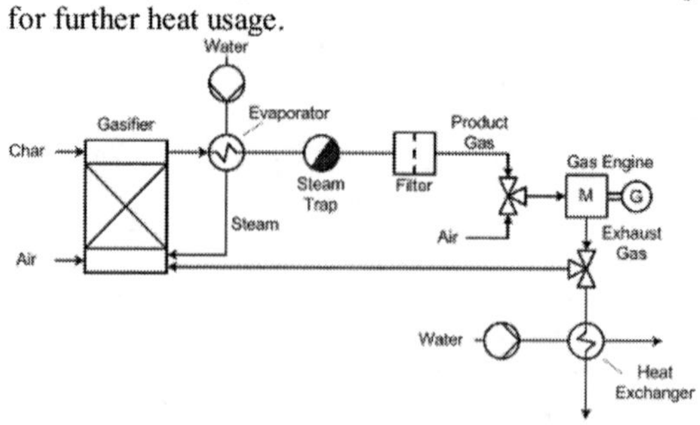

Figure 4. Micro-CHP system for solid biomass gasification in combination with gas engine.

The system shows high flexibility, allowing for high flexibility in power output while using solid biomass, see Figure 5 [5]. It was possible to change power output within less than one minute from 100% to 20%, and vice versa. Due to the good insulation of the gasification chamber, it was also possible to stop the system and restart within a certain period without long heat-up times.

Further research on small-scale CHP with gasifier and reciprocating engines is reported in literature, e.g. in [7–10].

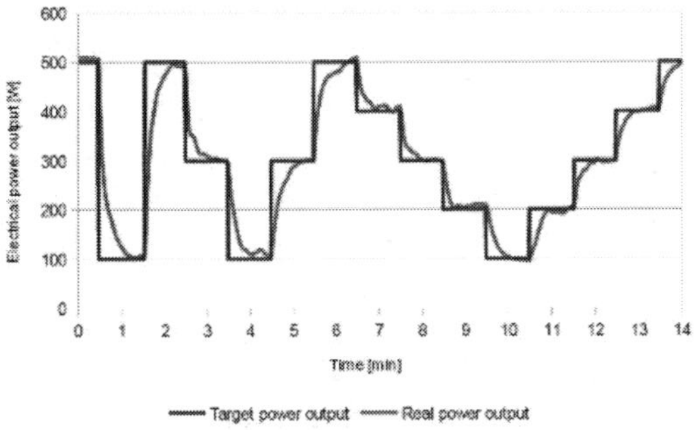

Figure 5. Fast alternation of power output in gasification based CHP.

4 Integration into buildings

To provide thermal energy and electrical power in a smart way for buildings, some criteria for the CHP system have to be fulfilled. Besides high efficiency and flexibility, it should be low in emissions (e.g. CO, NO_x, dust). Handling (fuel feeding, operation and cleaning) should be as easy as possible. A high degree in automation allows that several strategies can be considered depending on the frame conditions.

Heat-led operation is favored for many installations. Design of the system will usually be in such a way that it can provide a minimum heat load, with additional boiler or storage for peak load.

Base load or constant power production is also a common option and can be advantageous under certain conditions. By base load power production, a high number of full load hours of the system can be achieved, which leads to positive economic effects. For most systems, this also means continuous heat production. Depending on the building and its usage, this might require different heat usage strategies, including heat storage and seasonal operation.

Finally, electricity-led operation is the third main option for CHP integration. The system can be controlled to either produce power according to the demand of the local building, or in a way that it stabilizes the power grid. For both options, integration into communication systems and the implementation of advanced control technologies is important for optimized operation. This may not only include model predictive control of the CHP system itself, but also integration of weather forecast as well as self-learning algorithms that optimize system operation in the context of user behavior.

Against the background of increasing solar and wind power that bring stress to the electrical power grid, the last possibility has the potential to help stabilizing the grid with renewable energy.

For all options, proper transient simulation of the building [11] can help to design not only the technology, but also the control algorithms and the necessary communication infrastructure. Additionally, user satisfaction should be considered from the very beginning as one of the main promoters for the integration of renewable energies into buildings [12,13].

For owners of buildings, there are several aspects that have to be considered when integrating biomass-bases small-scale CHP technologies into buildings, including availability of technology and fuel [14].

5 Conclusions

There are numerous options to integrate combined heat and power systems fueled by solid biomass into buildings. This includes several thermo-chemical conversion processes like pyrolysis, gasification and oxidation as well as different options for electrical power generation like water-steam-cycles, ORC or gas engines.

Such systems can use different fuels including woody biomass, agricultural residues and biogenic fraction of municipal solid waste. By mechanical and thermal pretreatment (e.g. pyrolysis), fuel properties can be adapted to different CHP technologies. Products from gasification as first conversion step can be used in highly flexible technologies, while solid products are advantageous in terms of storability, grind ability and energy density.

Current examples show, that these technologies can provide thermal and electrical power in a flexible way to allow for different integration strategies (heat-led, base-load or electricity-led operation). For a small-scale gasification system with reciprocating engine, power output has been varied between 20% and 100% within less than one minute. Additional research is required to furtherly increase efficiency and flexibility.

Integration of CHP systems into building has to be conducted under consideration of thermal load, electricity consumption, user satisfaction and availability of technology and fuel.

References

1. A. Pollex, A. Ortwein, and M. Kaltschmitt, Biomass Convers. Biorefinery **2**, 21 (2011).
2. A. González, J.-R. Riba, R. Puig, and P. Navarro, Renew. Sustain. Energy Rev. **43**, 143 (2015).
3. J.-B. Bouvenot, B. Latour, M. Siroux, B. Flament, P. Stabat, and D. Marchio, Appl. Therm. Eng. **73**, 1041 (2014).
4. K. Sornek, M. Filipowicz, M. Szubel, E. Bożek, and K. Izdebski, MATEC Web Conf. **18**, 1010 (2014).
5. D. Krüger and A. Ortwein, in *Proc. Eur. Pellet Conf. 2016* (Wels, Austria, 2016).
6. D. Krüger and A. Ortwein, in *Tagungsband „Biomass Power Heat“* (Zittau, Germany, 2015), pp. 88–93.
7. J. Kalina, in *Proc. 25th Int. Conf. Effic. Cost Optim. Simul. Energy Convers. Syst. Process. ECOS 2012* (Perugia, Italy, 2012), pp. 331–349.
8. D. Prando, F. Patuzzi, G. Pernigotto, A. Gasparella, and M. Baratieri, Appl. Therm. Eng. **71**, 152 (2014).
9. D. Mertzis, P. Mitsakis, S. Tsiakmakis, P. Manara, A. Zabaniotou, and Z. Samaras, Energy **64**, 367 (2014).
10. A. Zabaniotou, P. Mitsakis, D. Mertzis, S. Tsiakmakis, P. Manara, and Z. Samaras, Energy Procedia **42**, 745 (2013).
11. J. Petzold, D. Büchner, and A. Ortwein, in *Proc. Ninth Int. Conf. Syst. Simul. Build. 2014* (Liège, Belgium, 2014), pp. P32, 1–15.

DOI: 10.1051/e3sconf/20161000134

12. D. Büchner, C. Schraube, E. Carlon, J. von Sonntag, M. Schwarz, V. K. Verma, and A. Ortwein, Appl. Energy **160**, 390 (2015).
13. G. Angrisani, C. Roselli, and M. Sasso, Prog. Energy Combust. Sci. **38**, 502 (2012).
14. K. Alanne, Renew. Sustain. Energy Rev. **8**, 401 (2004).

The political importance of energy cooperation between Germany and Denmark on the European Union energy market

Mariusz Ruszel[1]

[1]*Department of Economics, Faculty of Management, Rzeszow University of Technology, al. Powstancow Warszawy 8 street, 35-959 Rzeszow, Poland*

Keywords: *Germany, Denmark, EU internal energy market, interconnector, renewables*

Abstract. Denmark and Germany have similar goals of energy transition. Both states are going to implement a policy that is aimed at the transition from a fossil fuel-based system towards a renewable energy system with variable renewables generation. The main objective of the German energy transition is to cover more than 80% of the energy consumption with renewables by 2050. For comparison, Denmark is going to build a fossil fuel-free system, which would be able to cover 100% of energy consumption with renewables by 2050. To this end, stronger cooperation between Germany and Denmark could enhance the position of both countries on the EU energy market. The main aim of this paper is to analyse the impact of the energy cooperation between Germany and Denmark. It is crucial to answer the question of how both states could build a stronger energy position and which tools would be useful in this regard. Denmark is the world's leader in the deployment of wind power, which is also implemented in the northern part of Germany. For this reason, both states have similar challenges with respect to the stabilisation of electricity systems. The current and prospective integration of energy systems of both countries is also important for the improvement of energy security. It is also crucial to answer the questions whether the energy integration will have political implications for both countries, and whether stronger energy cooperation between Germany and Denmark will strengthen the competitive advantage of German economy in the EU.

1. Introduction

Denmark and Germany are two neighbouring countries with similar long-term goals of energy policies. Both countries have decided to carry out energy transition understood as gradual replacement of fossil fuels with renewable energy sources. Denmark currently has one of the most innovative energy policies worldwide. In the mid-1970s (after the 1973 oil crisis), a decision was made to develop cogeneration and extend the potential of renewable energy sources (RES), improve energy efficiency and develop dispersed generation [1]. In Germany the interest in the issue of renewable energy sources grew after the 1973 oil crisis [10]. Both countries strive to enhance their energy self-sufficiency and to reduce the dependence on imported energy raw materials, such as crude oil and natural gas. Denmark and Germany have social support for the implementation of energy policies resulting in high prices of energy for the community. Having similar objectives connected with energy transition also means that they have common challenges and goals. Therefore, the question arises of whether their mutual political cooperation will contribute to higher positions of these countries on the common EU energy market. How much will the existing and prospective natural gas and electricity interconnections help achieve flexible exchange and balance of the grids? This is especially important in the context of high participation of unstable renewable energy sources in energy balances of those countries. Another question is what will be the role of Norway in the integration of energy systems of Denmark and Germany. Are Denmark and Germany going to become electricity distribution centres in the EU?

2. Selected challenges

Energy transition in Denmark and Germany is associated with similar political and economic challenges for the countries. First, both of them are extensively developing wind energy systems, which need to be properly integrated with the electric system. This energy source is dependent on changeable atmospheric conditions, namely, wind speed. That is why the challenge is to stabilise the power grids with which wind power plants are integrated. In Denmark, these power plants satisfy nearly 40% of electricity demand (2014), and by 2020 their participation is to grow up to 50% [2]. The challenge faced by Denmark and Germany is to ensure the proper amount of energy in the system whenever the energy production is very high or whenever energy is not produced because of lack of wind [3]. Therefore, it is necessary to balance energy production in unstable weather conditions. Moreover, not all the elements of power transmission system in Germany and Denmark have enough capacity. This is especially noticeable in Germany, where it is necessary to further extend power transmission lines with energy storage projects [4]. Technical limitations caused by insufficient capacity of transmission networks lead to disturbances on the energy market and "loop flows", which upset Polish, Czech, and even Austrian power systems. A very important challenge is to improve the effectiveness of forecasting electric energy demand in certain regions of Germany and Denmark. Along with the increase of effectiveness of these activities, the risk of ordering the wrong amount of balance energy will decrease.

In Denmark and Germany, different measures to improve the flexibility of the power system are being considered. The effective use of electricity

[a] Corresponding author: Mariusz Ruszel PhD, mruszel@prz.edu.pl

 DOI: 10.1051/e3sconf/20161000135

interconnections is perceived to have huge potential. Investments are being realized that involve increasing the flexibility of production capacity, as well as systems designed to use electrical energy for the production of thermal energy. Introducing new services to the market and better management of supply and demand by means of implementation of intelligent grids is also discussed [5].

A significant political challenge is the growing price of electricity for citizens. It has made the social support for energy transition in Germany and Denmark slightly decrease in the 2013-2015 period. A study carried out by TNS market research company for the German Industrial Mining, Chemistry and Energy Union shows that in 2015, 57% of German citizens evaluated the decision on energy transition as good or very good. [6]. This is a drop by 11% in comparison with the year 2013 [6]. The situation is the result of growing prices of electricity for the citizens of the two countries. A significant political instrument preventing the social support for energy transition from decreasing even more quickly is the participation of both countries' local communities in RES sector investments, which makes the investing part of the community also benefit from it financially. However, whereas in Denmark energy production is one of the cheapest in Europe, ecotaxes imposed on the energy meant that the Danes paid the highest price for electricity in 2015. The situation is similar in Germany, where citizens also pay one of the highest prices for electrical energy. It is estimated that the total extra payments for renewable energy in an electricity bill of a German citizen are already more than double the price of electricity on the energy market [6]. In the past, it seemed that the German and Danish societies could afford it, since the share of energy costs in household expenses is 4-5%. The costs of electricity are particularly high for the people living in Denmark [7]. Research carried out in the Danish community showed that the Danish people identify with the energy policy of their country that aims at increasing energy self-sufficiency, the development of renewable energy sources, and the development of new technologies in the field of energy [7]. A study carried out by B.K. Sovocool and P. L. Blyth confirms that the Danish people perceive the technological development as the main solution to the problems related to energy [7]. At the same time, it is evident that the Danes' attitude and understanding of the goals of their country's energy policy is influenced by their education in the area of energy [8]. However, the current European challenges make particular communities expect lower electricity prices. In May 2016, the Danish government decided that the extra payment for green energy in electricity bills, amounting to 11%, would be abolished by the end of 2016 [9]. In June 2016, the Danish government made a political decision to withhold the work on wind power plants that were to be located along the coastline [11].

Comparing the two countries, it must be noted that both of them have very well developed energy infrastructure. Germany has a varied energy mixture thanks to carbon, gas, nuclear and renewable power plants. Proper use of the infrastructural potential makes it easier to achieve the goals connected with energy transition. In Denmark, ongoing development of dispersed generation is important. Denmark has decentralised its electrical energy system, developing small generations all over the country. Both countries also benefit from the export of technologies connected with renewable energy sources. The key role in RES promotion is not only played by the energy and climatic policy of the EU, but also by the International Renewable Energy Agency (IRENA), established at the cross-governmental conference in Bonn in 2009. IRENA facilitates the promotion of international use of RES and green technologies.

3. Political and economic benefits from the cooperation

Political and economic cooperation in the field of energy between Denmark and Germany gives mutual benefits in terms of energy security of both sides.

First of all, it strengthens the security of natural gas supplies. Denmark and Germany have the Ellund interconnector, which ensures transfer from Germany to Denmark of 2.26 billion m3/year, and from Denmark to Germany of 1.22 billion m3/year. Currently, this two-way connection is used to receive natural gas contracted by the DONG Energy concern (Denmark) from the Gazprom concern (Russia), supplied via the Nord Stream pipeline. Under the signed contracts, DONG Energy receives 2 bcm natural gas a year, and the German transmission system serves the transit function as part of performance of this gas contract. It must be pointed out that this connection can also be used to transfer natural gas to Germany.

Second, mutual electricity interconnections improve the flexibility of the electrical energy system and facilitate the commercial exchange of power in the case of market surpluses. Denmark and Germany have an electricity connection Kontek with the power of 600 MW, and another AC network with the power of 1,500 MW is planned. Furthermore, both countries are currently extending electricity interconnections with Norway, which has well developed water energy based on pumped storage power plants, which can be used to store electrical energy. Denmark already has electricity interconnectors with Norway: Skagerrak 1/2 – 500 MW, Skagerrak 3 – 500 MW, and Skagerrak 4 – 700 MW [14]. As for Germany, it is planning to build a 1,400 MW electricity interconnector Nord-Link, which is to be completed by 2020. It will be an undersea 380 kV direct current cable with the length of 623 km, connecting Tonstad in Norway with Wilster in Germany. It must be underscored that Germany has extended the production capacity of renewable energy in the north part of the country, but has not managed to construct relevant transmission infrastructure to connect the northern part of the country with the southern part. Therefore, Germany has limited possibilities to send power and needs to send electric power to the south through the Polish transmission system, causing so-called loop flows. This leads to significant disturbances of the electrical energy system in the neighbouring countries, e.g. in Poland and the Czech Republic. It seems that integrating the electrical energy systems of Norway, Denmark and Germany will contribute to the development of flexible exchange of electricity between these countries. Flow directions may differ in the case of small or large amount of energy produced by wind power plants. Moreover, it will help meet the assumed goals of energy transition in Denmark and Germany. It will also strengthen emergency response mechanisms. Besides, it will provide the basis for better use of Norwegian energy potential based on water power plants. Norway may become the EU "green storage". Integrating electrical energy systems and flexible exchange is an option to choose instead of extension of electrical energy storages or creating excessive power reserves in Germany. Especially that unstable wind energy requires power protection in order to stabilize the grid. It is also worth pointing out that further development of HVDC (high-voltage direct current) connections is planned between Denmark and: the Netherlands (COBRAcable – 700 MW by the year 2018), Great Britain (DK1-United Kingdom – 700 MW/1400 MW by 2020) and Germany (DK-1 – 3,500 MW by 2022 as well as Kriegers Flak – 400 MW by 2019) [12].

Figure 1. Electricity connections between Denmark and Germany – 2015 [12].

Third, electrical energy integration of Denmark and Germany will help strengthen the economic position on the common energy market and the political position in the EU. Comparing wholesale prices of electricity in EU countries, it is evident that in Denmark and Germany electricity is cheaper than in other member states. This results from the following factors: Surpluses of electricity produced by wind power plants in the northern part of Germany lead to the phenomenon of negative electricity prices, which affect electricity prices on the German stock exchange. This way, the price of electricity exported from Germany can be lower than the prices of electricity in the neighbouring countries. In the years 2014-2015, all the EU neighbours of Denmark and Germany had higher wholesale prices of electricity. Only in 2015, the wholesale price of electricity was the same in Germany and in Austria, whose electrical energy system is integrated with the German one. This results in increasing the possibility of exporting Danish, German and even Norwegian electricity. In the context of integration of common energy market in the EU and extension of capacity of electricity connections, the proper energy infrastructure will be developed to make it easier for the countries producing cheapest electrical energy to use their comparative advantage. It should be remembered that electricity bills for German or Danish citizens are higher than for industrial customers. Consequently, the retail prices of electricity in those countries are among the highest in the EU. This increases the competitiveness, not only of the energy sectors in Denmark and Germany, but also of the fields of industry that are dependent on electrical energy. This way, economic development is promoted in those countries, which leads to new workplaces. This mechanism can only work well if there is social approval for the fact that the price of electricity for German or Danish citizens is higher than for industrial customers. This way, production companies get discounts.

Fourth, electrical energy integration between Denmark, Germany and Norway is part of the European plans of establishing a common energy market. In 2006, the European Regulators' Group for Electricity and Gas (ERGEG) presented the Electricity Regional Initiative (ERI), including seven regional electrical energy markets in Europe: Baltic, Central-East, Central-South, Central-West, Northern, South-West, France-UK-Ireland [13]. It must be pointed out that Germany is the only country that belongs to four regions: Central-East (Austria, Czech Republic, Germany, Hungary, Poland, Slovakia, Slovenia), Central-South (Austria, France, Germany, Greece, Italy, Slovenia), Central-West (Belgium, France, Germany, Luxembourg, Netherlands), Northern (Denmark, Finland, Germany, Norway, Poland, Sweden) [13]. On the one hand, geographical conditions, i.e. Germany being located centrally in Europe, gives it the potential of becoming the electricity distribution centre. On the other hand, it is worth stressing that in all the four regions that Germany is in, wholesale prices of electricity are lowest in three countries: Norway, Denmark, and Germany. This means that developing the project of a common European energy market by developing regional markets and enhancing the capacity of interconnections and stock exchange instruments will provide the right conditions for the export of electricity produced in Norway, Denmark and Germany to the other European countries. Germany will serve a unique role in this process, as it can become a transit country for Norwegian and Danish electrical energy.

4. Summary

In the 1970s, Denmark and Germany decided to increase their energy self-sufficiency and decrease the dependence on imported fuels, i.e. crude oil and natural gas. These decisions have resulted in energy transition, currently being realized in both countries. Denmark and Germany have similar goals regarding energy transition, which is going to increase the use of renewable energy at the expense of conventional energy in the future. In order to meet these goals, they have to face challenges of integrating renewable energy with their national electrical systems. Among many possible options, it seems most prospective to integrate the energy industry with the heat industry to a greater extent. This will ensure better use of energy surpluses, lower energy losses and facilitate energy storage, since heat energy storage facilities are more efficient. A huge political challenge is the growing price of electricity for citizens, which in 2016 caused some adjustments in political decisions as a result of poor social support for energy transition. The model of transition adopted by Denmark and Germany requires appropriate expenditure, as by assumption it is to support the competitiveness of industries in these countries and become the instrument to achieve the goals of the economic dimension of foreign policy. Political cooperation between the countries strengthens their energy security, improves the flexibility of the energy system, facilitates the export of electricity, fits the European plans of establishing a common electricity market, and enhances their competitive advantage on the integrating market. Gradual development of regional markets – and ultimately, a common EU energy market – is going to increase competition between electricity producers. At the moment, Denmark and Germany have one of the lowest wholesale prices of electrical energy. Taking into account the plans of enhancing the integration of systems in these countries and the connections with Norway, where the water energy potential is huge, Denmark and Germany may definitely play the crucial role on the common energy market of the EU. They have all the instruments necessary to become electricity distribution centres, which will translate into economic development and more workplaces.

References

1. F. Hvelplund, Environ. Res. Eng. Manag., **66/4** (2014).
2. A. Estermann, *Interconnection and cross-border market integration*, Perspective of 50Hertz, Berlin 24.09.2015.
3. A. Kofoed-Wiufd, *The Danish Experience with Integration of Variable Renewable Energy*, Ea Energy Analyses, Berlin 24.09.2015.
4. M. Stark, *System and Market Integration - The Perspective of a German Energy Trader,* Berlin 24.09.2015.
5. T. Olkuski, E. Ciesielka, A. Szurlej, *Rynek Energii,* **2** (2015).
6. Niemcy płacą za OZE więcej, niż za samą energię *[Germany pays for RES more than for energy itself]*, http://wysokienapiecie.pl/oze/1302-cena-energii-elektrycznej-polsce-2015-niemczech [access: 10.06.2016].
7. B. K. Sovacool, P. L. Blyth, Environ. Sci. Policy, **54** (2015), pp. 304-315.
8. B. K. Sovacool, Soc. Stud. Sci. **40** (2010), pp. 1147-1157.
9. M. Druś, *Duńczycy nie będą dopłacać do „zielonej" energii [Danes are not going to pay extra for "green energy"], http://www.pb.pl/4432983,32820,dunczycy-nie-beda-doplacac-do-zielonej-energii* [access: 10.06.2016].
10. B. Morris, M. Pehnt, *Energy Transition – The German Energiewende,* Heinrich Boll Stiftung, Berlin 2014, pp. 57-59.
11. Dania rewiduje politykę klimatyczną. Promocja OZE była zbyt droga *[Denmark revises its climate policy. The promotion of RES proved to be too expensive]*, http://www.cire.pl/item,130388,1.html?utm_source=newsletter&utm_campaign=newsletter&utm_medium=link&apu=32706 [access: 10.06.2016].
12. G. Brønmo, *Interconnection as a flexibility option*, Energienet/dk, Berlin, 24.09.2015.
13. *Electricity Regional Initiative (ERI)*, http://www.ceer.eu/portal/page/portal/EER_HOME/EER_ACTIVITIES/EER_INITIATIVES/ERI [access: 10.06.2016].
14. M. Ruszel, Bezpieczeństwo. Teoria i Praktyka [Sc. Theory and Pract.] **1/XXII** (2016), pp. 71-72.

DOI: 10.1051/e3sconf/20161000136

The possibility of using flue gases as a medium for straw drying

Wojciech Goryl[1,a] and Mariusz Filipowicz[1]

[1]*AGH University of Science and Technology, Faculty of Energy and Fuels, A. Mickiewicza 30 Av., 30-059 Krakow, Poland*

Abstract. The paper presents the possibility of drying straw in dedicated and innovative straw dryers with a modified drying system. The basic problem behind the use of bales of straw as a fuel is their moisture content. The moisture content is mostly dependent on the time of harvest and the conditions of storage. The humidity level of dry straw may be as low as 10%, however harvesting the straw during unfavourable weather conditions may cause the moisture level to increase up to 60-70% of the relative humidity, a value often observed for fresh biomass. Experimental studies were conducted to examine the effectiveness of drying and heat transfer in the straw bale. The studies have shown that the inner layers of the bale heat up much faster and achieve significantly higher temperatures than the outer ones. With the application of dedicated straw dryers, a homogenous field of moisture content in the straw bale is achieved in a very cost effective way.

1 Introduction

Since the great energy crisis in the 70s of the 20th century, society has been increasingly interested in alternative energy sources and possibilities of using them for the generation of power, heat production or powering vehicles. In recent years, we have observed a rapid and dynamic development in the field of renewable energy sources. More and more companies manufacturing elements for photovoltaic or wind farms as well as biomass boilers are being established. This is caused mainly by two factors. The first one is the uncertainty in terms of supply of fossil fuels, namely oil and natural gas, due to the current political crisis between Russia and the European Union, and the political situation in the Middle East. The second reason for the development of technologies based on renewable energy sources are the Directives of the European Union aiming at the limitation of the emissions of greenhouse gases to the atmosphere [1,2,3,4]. An additional impetus for the development of renewable energy-based technologies of all kinds is provided by various subsidies for low-emission devices and the introduction of feed-in-tariff or e.g. green certificates.

There are many studies confirming the fact that biomass is one of the most important sources of renewable energy in Poland. Its technical potential fluctuates within 800-900 PJ per year [5,6]. In the estimation prepared for the purposes of the GHG Reduction Strategy report, the number of 100 PJ [5] is given. Biomass exhibits the highest potential in relation to other renewable energy sources. Moreover, it is the most predictable energy source and – as opposed to solar energy or wind – it can be used throughout the year. In addition, biomass may be easily stored and used in the periods of increased demand and transported to locations with deficits.

Poland, as an agricultural country with a large woodland area, has considerable biomass resources which may be used for generating electricity or in heat production. One of the basic problems regarding the use of biomass for energy production is bulk density – as well as calorific value. In a given biomass volume, there is approx. 4-8 times less energy than in a similar volume of coal. This causes a number of problems including the issue of transportation – the costs related to biomass transportation over long distances and emissions associated with its transportation from the place of harvesting to the place of utilisation [7,8].

Large biomass resources in the form of straw are available in Poland. It is estimated that the annual production of straw is approx. 20 million tons [9], while the technical potential of obtaining straw for energy purposes is approx. 7 million tons [10]. This potential is going to increase each year due to a decrease in the number of breeding animals [11], which will result in a decreased demand for straw for animal and bedding purposes. Such large surpluses of straw will popularise the use of straw for heating purposes by farmers or in public buildings. This may be exemplified by the Municipality of Człuchów, where straw is used for the heating of four school buildings. Straw is also used in district heating in towns such as Lubań in Lower Silesia.

Dry straw humidity may reach a level of as low as or even less than 10%, harvesting straw during unfavourable weather conditions, however, causes the moisture level to increase up to 60-70% of the relative humidity [12], a value often observed for fresh biomass [13]. In the Polish climatic conditions, as well as in many European countries, the heating season is relatively long and lasts from October to May, while straw is only

[a] Corresponding author: wgoryl@agh.edu.pl

DOI: 10.1051/e3sconf/20161000136

harvested in the summer period – in July and August. It is therefore necessary to store the straw that will subsequently be used in the heating period. It should be stored in closed barns or haystacks covered with foil, tarp or special waterproof materials. This allows to minimize the impact of weather conditions on the quality of the stored fuel. Excessive straw moisture level has a negative impact on the burning value of the fuel and results in problems related to the burning of wet straw in boilers [12, 14]. It is best for the straw used in batch boilers to exhibit humidity at a level of approx. 10-20%. This has a favourable effect on the combustion process due to the catalytic effect of water vapour on the combustion of the excessive amount of volatiles present in the biomass [12].

2 Experimental stand

The main elements of the dryer used in the study include the drying chamber, the transport table, the discharge fan, the fire prevention system, the regulation assembly and the system for the control of operational parameters, the system of pre-extraction and separation of sparks and the exhaust introduction system.

The heat used in the drying process comes from the process of biomass combustion in the boiler. Such utilised waste heat increases the efficiency of the entire system and helps to obtain a much more efficient fuel. The drying process involves the injection of waste gas characterized by a properly controlled temperature by the automatic system (Fig. 1) using a specially designed nozzle with a diameter of 80 cm into the bale of straw.

For safety reasons, the exhaust gases coming from the boiler are pre-dusted in the sedimentation chamber (cyclone) due to the presence of large, sometimes incandescent stalks of unburned straw. Subsequently, the gases are transported to a special mixer (Fig. 2), where, if necessary, the process of mixing of the exhaust gas with ambient air takes place in order to obtain the desired temperature of the drying medium. After the mixing process, the fan introduces the previously prepared mixture using the specially designed nozzle into the cylindrical bale of straw. The schematic diagram of the system has been presented in Figure 3.

The effectiveness of drying cylindrical straw bales (diameter: 160 cm, length: 140 cm) with exhaust gases was measured using a specially prepared recorder equipped with eight measuring probes. The probes were calibrated by the manufacturer using a sample of straw characterized by varying humidity. Capacitive relative humidity sensors and temperature sensors with the SHT11 device were installed.

The measuring probes varied in length (80 cm, 60 cm, 40 cm and 20 cm denoted respectively by lowercase "a", "b", "c" and "d"), so as to make it possible to measure the humidity and temperature in locations at a preset distance from the axis of the straw bale. Figure 4 presents the visualisation of the placement of the right probe in the bale during the measurement.

Figure 1. General view of the control system

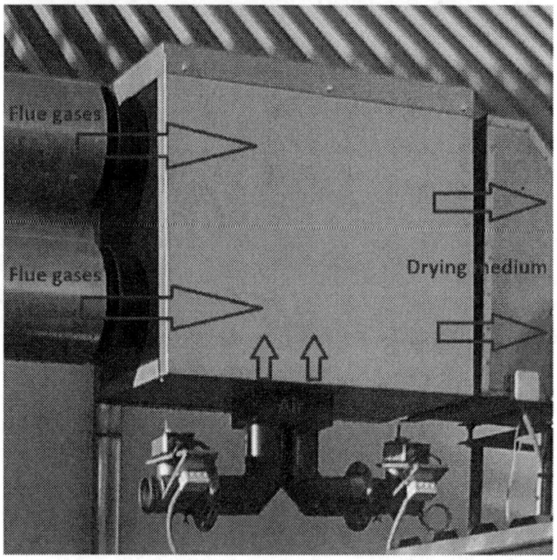

Figure 2. General view of the medium mixer

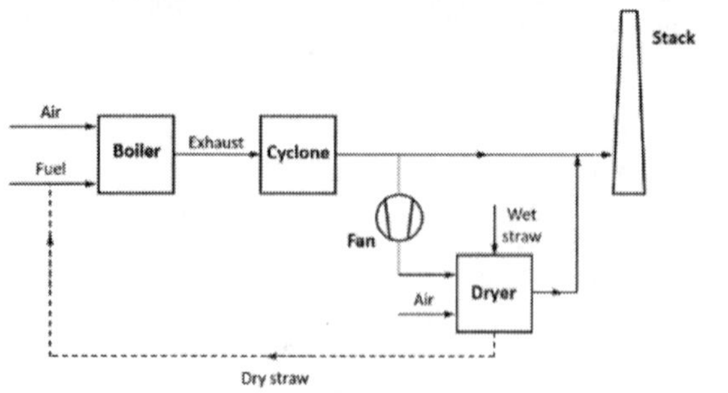

Figure 3. Schematic diagram of the installation used in the experiment

The measurements were always carried out in two planes. The measurement points were located at a distance (x=0 – the place where the drying mixture was pumped to the bale) of 40 cm (plane A, x=40) and 100 cm (plane B, x=100) from one end of the bale (Fig. 4).

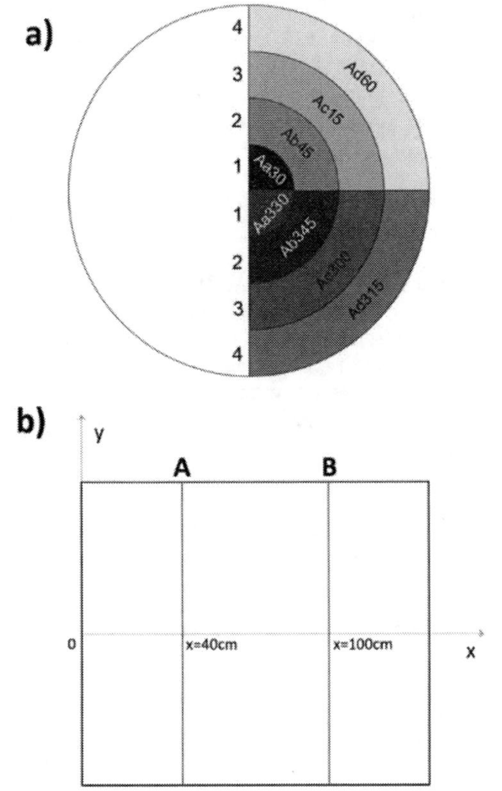

Figure 4. Visualisation of the locations of measurement probes: a) locations of individual probes, b) coordinates of the A and B planes.

The results of the experiment have been presented below. The curves show the change of temperature and humidity during the drying process. The prefixes "A" and "B" designating the probes refer to the measurement plane according to Figure 4b, while the prefixes "T" and "H" refer to the measurement of temperature and humidity, respectively.

This article presents the possibilities of drying straw using a basic and a modified drying process. After gaining some experience in the process of drying straw bales [15], some modifications of this process have been implemented, which involved preparing a duct along the bale axis with the diameter of 63mm and applying a metal plate with the diameter of 1,000 mm at the side opposite to the fitted drying medium. This modification was designed to allow for the highest radial propagation of the medium and to obtain a uniform temperature and humidity inside the bale.

3 Results and discussion

A study of changes in the humidity and temperature inside the bale of straw were conducted in the A plane (Fig. 4b). The measurements were carried out in two configurations (with and without three holes) to examine the effect of the presence of the holes inside the bale of straw on the dynamics of drying cylindrical straw bales and on the temperature changes inside. In addition, in each drying process the temperature of the drying medium injected into the bale was recorded. The exhaust gas temperature was presented in charts in the form of a dotted curve with the values given on the right axis. During the basic drying process, one bale of straw was dried in 280 min, and during the modified drying process one bale of straw was dried in 270 min.

3.1. The basic drying process

The first two graphs (Fig. 5 and 6) present the variation of humidity inside the bale of straw in the basic drying process (without the holes). Immediately after the completion of measurement no. 1, measurement no. 2 was performed. The initial humidity of the bale was in the range of 23-43%. The internal layers of the bale were drier, and the humidity increased along with the distance from the centre of the bale. This was caused by the fact that the bale had not been covered while storing.

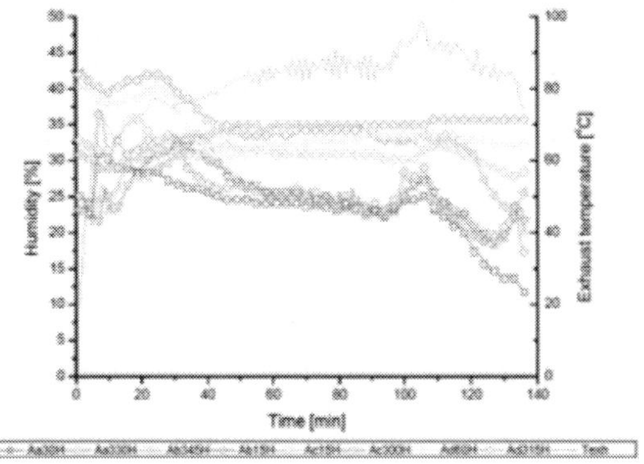

Figure 5. Basic drying process, humidity measurement No. 1

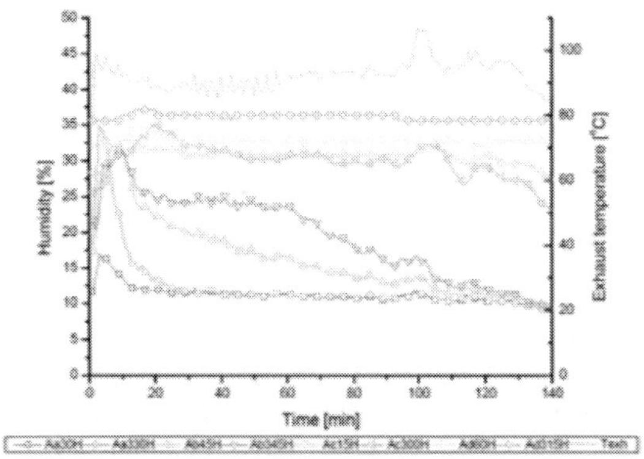

Figure 6. Basic drying process, humidity measurement No. 2

Measurement No. 1 was carried out for approx. 140 min. In the initial phase, one could note a sharp increase in the humidity at the inner portion of the bale represented by sensors Aa30H and Aa330H. This was due to the rapid temperature increase in this part of the bale, which contributed to the temporary increase in relative humidity. After the 30th minute of the measurement, a decrease in humidity in the first and second layer (the layers are presented in Fig. 4a) could be observed. This decrease continued until the end of the measurement. In case of planes 3 and 4, a stable humidity content oscillating in the range of 30-35% was observed. The final humidity value after the first measurement reached the value of 12-35%.

Measurement No. 2 was started 5 minutes after the completion of measurement No. 1. In this case, the drying time using the hot drying medium (the period of combustion in the boiler) was also approx. 140 min. In the initial drying phase, a similar increase of humidity as in the case of measurement No. 1 was observed. During the entire drying process, the humidity in planes 1 and 2 decreased until it reached 10%. No significant changes in humidity were observed in planes 3 and 4 and after approx. 280 minutes of drying, the humidity oscillated within the range similar to the initial humidity, reaching 30-35%.

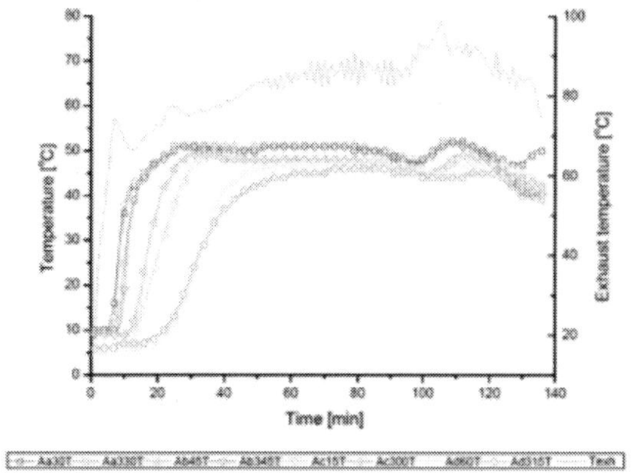

Figure 7. Basic drying process, temperature measurement No. 1

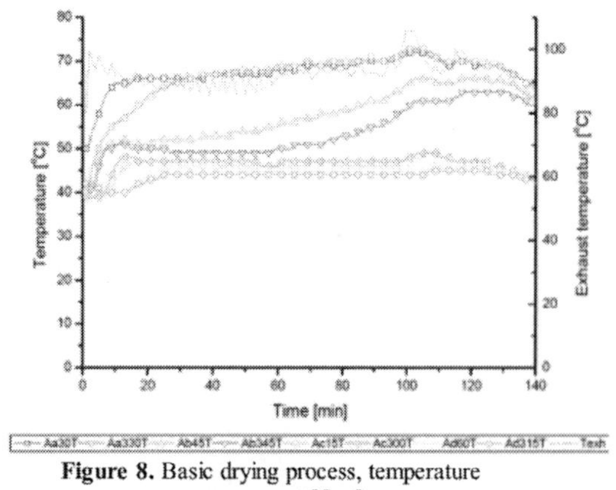

Figure 8. Basic drying process, temperature measurement No. 2

The two graphs above (Fig. 7 and 8) present the temperature variations inside the bale of straw in the respective measurement points during the basic drying process.

Figure 7 presents measurement No. 1 and the temperature changes inside the bale of straw in a period of approx. 140 minutes of the drying process. Five minutes after the drying process was started, a rapid temperature increase was observed in plane 1. Subsequently, after another 2 minutes have passed, rapid heating of plane 2 was observed. Planes 3 and 4, because of their distance from the centre of the bale, heated up much slower, and the beginning of the temperature increase in these places could be observed from the 15th and 20th minute, respectively. The maximum temperature for planes 1 and 2 was achieved in the 25th minute of the measurement and its value was approx. 52°C. In case of plane 3, the highest temperature of 50°C, was reached in the 35th minute. The extreme plane was heated at the lowest rate and reached the lowest temperature values, reaching the maximum of 46°C at approx. the 60th minute of the measurement. After reaching the temperature of 85°C, the flue gas temperature was maintained until the end of the combustion process.

Fig. 8 above presents the temperature changes during measurement No. 2. In this case, the highest temperatures were achieved only in plane 1, with values within the range of 65-73°C. These values were achieved after approx. 15 minutes from the start of the measurement. In plane 2, the maximum temperatures were achieved as late as 100 minutes from the beginning of the measurement and they reached approx. 60-65°C. During the entire drying process, no significant temperature changes were observed in planes 3 and 4 and the temperatures oscillated within the range of 44-48°C. In this case, the exhaust gas temperature was maintained at a relatively stable level within a range of 85-90°C.

3.2. The modified drying process

The following graphs (Fig. 9 and 10) present the changes in humidity for the bale with the application of the three holes and the metal plate.

The initial humidity of the bale which was subject to the measurement for the modified drying process oscillated within 30-45% and was comparable to the bale used in the basic drying process. The inner layers of the bale were drier, and the farther from the centre of the bale, the higher was the humidity. Measurement No. 1 was carried out for approx. 140 minutes. In this case, problems occurred with igniting the straw in the biomass boiler, and thus the drying mixture had a very low temperature. In the 18th minute of the measurement, it was possible to observe a sudden increase of the flue gas temperature, which proved that the fuel started burning in the combustion chamber. After a sudden increase in the drying mixture temperature, a sudden humidity increase was observed, and after approx. 5 minutes the humidity slowly declined. After 5 more minutes, the humidity in the whole bale was stabilised and reached 35%. Subsequently, until the 120th minute of the measurement, a sudden decrease of humidity was observed, and after another 20 minutes the humidity in planes 1 and 2 was stabilized at the level of 10%. Planes 3 and 4, just as in the previous measurements, were characterised by a similar humidity of approx. 35%.

Measurement No. 2 was carried out for 120 minutes and was initiated after approx. 5 minutes after the completion of measurement No. 1. In this case, a constant humidity value for planes 1 and 2 was observed during the entire period of the drying process, and the value reached approx. 10%. In addition, one of the sensors (Ac15H) in plane 3 recorded humidity values similar to planes 1 and 2. The second sensor located in plane 3 (Ac300H) showed a constant decline in humidity, until it reached a minimum (approx. 10%) in the 30th minute of the measurement. In plane 4, the humidity remained at a stable level until the 90th minute of the measurement, in which one of the

DOI: 10.1051/e3sconf/20161000136

sensors (Ad315H) started registering a decrease in the humidity value. At the end of the drying process, the bale of straw was characterised by a very low content of humidity of approx. 10%. Only one of the sensors (Ad60H) showed a higher value – 35%.

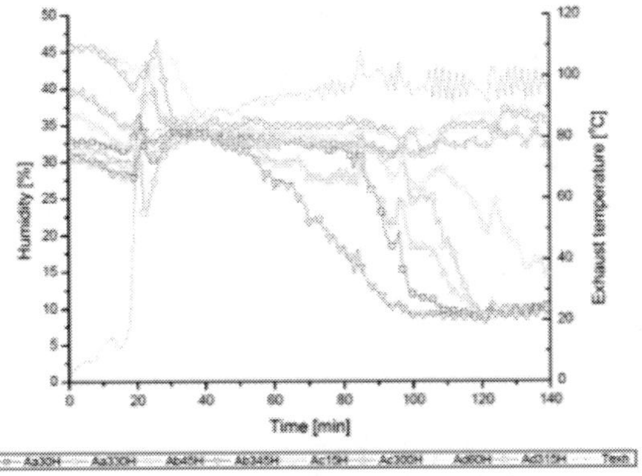

Figure 9. Modified drying process, humidity measurement No. 1

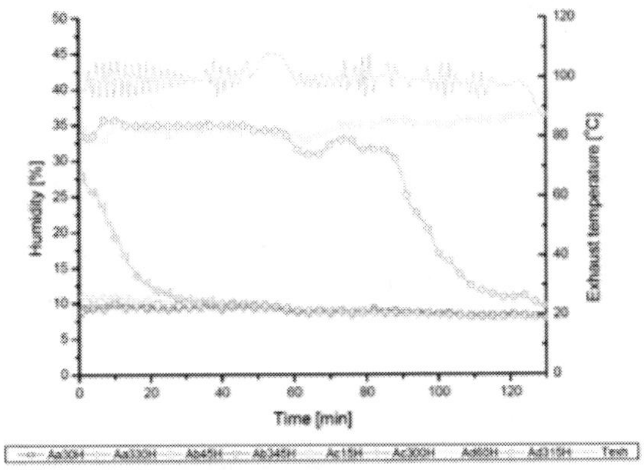

Figure 10. Modified drying process, humidity measurement no. 2.

The following two graphs (Fig. 11 and 12) present the temperature changes inside the bale of straw with the application of the three holes and the metal plate.

Figure 11 presents the temperature changes inside the bale of straw during measurement no. 1.

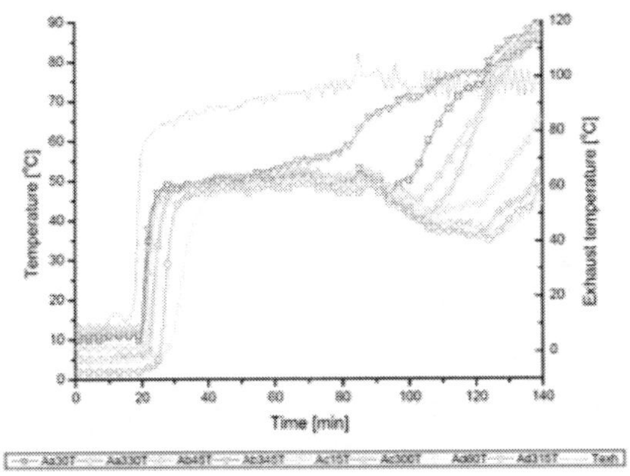

Figure 11. Modified drying process, temperature measurement No. 1

Until the ignition of the fuel in the combustion chamber, one may observe a similar temperature value throughout the entire bale of straw. In the 20th-25th minute of the drying process, a rapid rise in temperature in all measurement planes was observed, until the temperature reached 50°C. Subsequently, a stable temperature value may be observed in all planes, until the 90th minute. After this period, the temperature in planes 1 and 2 started to increase and it reached a maximum of 90°C. The maximum temperature reached at planes 3 and 4 was the range of 60°C. The flue gas temperature during the entire period of the drying process oscillated between 90 and 100°C.

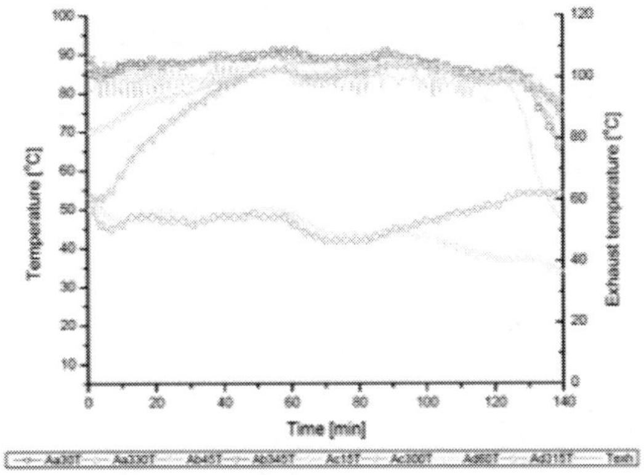

Figure 12. Modified drying process, temperature measurement No. 2

Fig. 12 above presents the temperature changes during measurement No. 2. During the entire drying process, the temperature in planes 1 and 2 remained at a constant level of approx. 90°C. In plane 3, the temperature was steadily growing until reaching approx. 85°C in the 50th minute of the measurement. At that point, it reached a stable level until the end of the measurement. The temperature in the outermost layer exhibited the lowest temperature values, which oscillated within the range of 40-60°C. The temperature of the exhaust gas was oscillating within the range of 95-100°C.

4 Conclusions

This article presents a study concerning the possibilities of drying cylindrical straw bales. Within the study, measurements were conducted for the basic and modified drying processes. In the modified drying process, the bale of straw had three holes made along the centre of the bale to achieve the best distribution of the drying medium. The measurement results confirm the validity of the implemented modifications. In this modified process, a satisfactory drying time was achieved. A bale with a relative humidity of 45% was dried to a level of 10% of moisture in 270 minutes. Within the basic drying process, however, a much worse coefficient of heat transfer was achieved, and thus the drying was insufficient.

Acknowledgments

This work was financially supported by the Statutory Grant of the Faculty of Energy and Fuels at AGH University of Science and Technology and it was carried out by BioORC Project within the KIC InnoEnergy Programme.

DOI: 10.1051/e3sconf/20161000136

References

1. Directive 2009/28/EC of the European Parliament and of the Council of April 23rd, 2009
2. Directive 2001/77/EC of the European Parliament and of the Council of September 27th, 2001
3. Directive 2003/30/EC of the European Parliament and of the Council of May 8th, 2003
4. Directive 2006/32/EC of the European Parliament and of the Council of April 5th, 2006
5. Report of the Ministry of Environmental Protection, *Strategy of the RES development*, Warsaw, 2000 [in Polish]
6. Report of the Ministry of Economy, *Posibilities of using RES in Poland until 2020*, Warsaw, 2007 [in Polish]
7. A. Guła, P. Wajss, W.Goryl, Przegląd Elektrotechniczny, **5a**, 198-203 (2012)
8. A. Guła, W. Goryl, Pol. J. Environ. Stud., **4**, 1377-1380 (2014)
9. Central Statistical Office of Poland, *Concise statistical yearbook of Poland 2015* (Warsaw, 2015).
10. W. Goryl, *Proceedings of 11th International Conference ENEF*, Banska Bystrica (2014)
11. A. Ludwicka, A. Grzybek, Prob. of Agri. Eng., **2**, 101-111 (2010) [In Polish]
12. A. Grzybek, *Straw – Implementation in Heating*, (Wydawnictwo ITP, Falenty, 2012)
13. V. Vladut, M. Chitoiu, A. Danciu, M. Militaru, C. Lehr, Bull. UASVM Agric., **67**, 292-300 (2010)
14. S. Kalinauskaite, A. Sakalauskas, E. Sarauskis, A. Jasinskas, Agron. Res., **11(2)**, 319-328 (2013)
15. W. Goryl, M. Filipowicz, EPJ Web of Conferences, **114**, 02033 (2016), DOI: http://dx.doi.org/10.1051/epjconf/201611402033

The comparison of solar water heating system operation parameters calculated using traditional method and dynamic simulations

Krzysztof Sornek[1,a]

[1] *AGH University, Faculty of Energy and Fuels, Mickiewicza Av. 30, 30-059 Krakow, Poland*

Abstract. The proper design of renewable energy based systems is really important to provide their efficient and safe operation. The aim of this paper is to compare the results obtained during traditional static calculations, with the results of dynamic simulations. For this reason, simulations of solar water heating (SWH) system, designed for a typical residential building, were conducted in the TRNSYS (Transient System Simulation Tool). Carried out calculations allowed to determine the heat generation in the discussed system as well as to estimate the efficiency of considered installation. Obtained results were compared with the results from other available tool based on the static calculations. It may be concluded, that using dynamic simulations at the designing stage of renewable energy based systems may help to avoid many exploitation problems (including low efficiency, overheating etc.) and allows to provide safe exploitation of such installations.

1 Introduction

The proper sizing of the components of a renewable energy based systems is a complex problem which includes both predictable (performance characteristics of individual elements) and unpredictable (weather data) parameters. Therefore, computer modelling presents many advantages including [1]:

- elimination of the expense of building prototypes,
- understandable organization of the complex systems,
- possibility to optimise the system components,
- possibility to estimate the amount of energy delivery from the system,
- possibility to estimate the design variable changes on system.

From software dedicated to computer modelling of energy systems, TRNSYS (Transient System Simulation Tool) was used to simulate operation of solar water heating (SWH) system designed for a typical residential building.

TRNSYS is worldwide known software allows to dynamic simulation of the energy systems operation. Primarily, it is used to simulate renewable energy based installations, including solar water heating (SWH) systems. From available references we can mention some investigations devoted to modelling and performance evaluation of solar domestic hot water systems [2], investigation of the effect of load profile [3], and simulation of the systems for hot water preparation but using e.g., heat pumps assisted by solar energy [4]. There are also many works devoted to individual elements of the solar water heating systems. For example, Morrison and Braun studied characteristics of horizontal and vertical tanks [5], Shariah and Ecevit studied the effect of different load temperatures on the performance of the solar hot water systems with auxiliary electric heaters [6], Shariah et al. optimized the tilt angle of the collectors [7] and Wongsuwan and Kumar studied the performance of forced circulation system [8].

These models aim to simulate the behavior of a solar thermal collector and are usually based on the energy transfer phenomena (such as radiation, convection and conduction) and on the thermophysical properties of materials [9, 10]. Other models are intended to serve as a basis for developing experimental test methods for identification of the characteristic parameters of the collector through nonintrusive means [11, 12].

The accuracy of TRNSYS simulations is high and close to the performance of solar water heating systems in real conditions. It was demonstrated e.g. in [13], where energy generated in thermal collectors in the TRNSYS simulation was only 1.2% higher than the measured one. On the other hand, on a seasonal basis the deviations had higher values and opposite trend (up to ±7-8% of average deviations weighted on the monthly transferred energy). In another case, comparison of the system energy yield, calculated using TRNSYS for periods higher than one month, with measured energy yield for these periods, shows a very good agreement with differences lower than ±3% [14].

Simulations are considered in general as powerful tool for modelling, design, prediction of performance and research and development of the energy systems. However, as there are limits to their use – a high level of skill and scientific judgement is required in order to produce correct and useful results [15].

[a] Corresponding author: krzysztof.sornek@agh.edu.pl

2 Methodology

The initial step in modelling is derivation of a structure to be used to represent the analysed system (it is always an imperfect copy of reality structure and should not be confused with the real system). The first established elements in the developing structures are boundaries conditions consistent with the considered problem. This is accomplished by specifying internal and external items, processes and effects. Simplified analysis methods have such advantages as computational speed, low costs and rapid turnaround. On the other hand, disadvantages include limited flexibility for design optimisation and lack of control over assumptions.

The following sections describe the assumptions of solar water heating system simulations using TRNSYS software and structure of the considered installation.

2.1 Assumptions of the dynamic simulations

The TRNSYS allows to simulate operation of the variety types of solar thermal collectors, including e.g. flat-plate and evacuated tube collectors (representing respectively by Type 1 and Type 71). These components model the thermal performance using a quadratic efficiency equation, which is a generalization from the Hottel-Whillier equation (eq. 1). Thermal performance of the collector array is determined by the number of modules in series and the characteristics of each module [16].

$$\eta = \frac{Q_u}{A \cdot I_T} = \frac{\dot{m} \cdot c_{pf} \cdot (T_O - T_i)}{A \cdot I_T} = F_R(\tau\alpha)_n - F_R \cdot U_L \frac{T_i - T_a}{I_T} \quad (1)$$

where:

Q_U - heat generated in collector array, kJ/h

A - total collector array aperture or gross area, m^2

I_T - global radiation incident on the solar collector, $kJ/(h \cdot m^2)$

m - flowrate at use conditions, kg/h

c_{pf} - specific heat of collector fluid, $kJ/(kg \cdot K)$

T_i - inlet temperature of fluid to collector, °C

T_o - outlet temperature of fluid from collector, °C

T_a - ambient (air) temperature, °C

F_R - overall collector heat removal efficiency factor, -

τ - short-wave transmittance of the collector cover(s), -

α - short-wave absorptance of the absorber plate, -

$\tau\alpha_n$ - $(\tau\alpha)$ at normal incidence, -

U_L - overall thermal loss coefficient of the collector per unit area, $kJ/(h \cdot m^2 \cdot K)$

The loss coefficient U_L is not exactly constant. Taking into account a linear dependency of U_L versus (T_i-T_a), the thermal performance may be expressed as:

$$\eta = \frac{Q_u}{A \cdot I_T} = F_R(\tau\alpha)_n - F_R \cdot U_L \frac{T_i - T_a}{I_T} - F_R \cdot U_{L/T} \frac{(T_i - T_a)^2}{I_T} \quad (2)$$

where:

$U_{L/T}$ - thermal loss coefficient dependency on T, $kJ/(h \cdot m^2 \cdot K^2)$

Equation (2) can be rewritten as:

$$\eta = a_0 - a_1 \frac{T_i - T_a}{I_T} - a_2 \frac{(T_i - T_a)^2}{I_T} \quad (3)$$

where:

a_0 - intercept (maximum) of the collector efficiency, -

a_1 - negative of the first-order coefficient in collector efficiency equation, $kJ/(h \cdot m^2 \cdot K)$

a_2 - negative of the second-order coefficient in collector efficiency equation, $kJ/(h \cdot m^2 \cdot K^2)$

The thermal efficiency is defined by three parameters defined as: a_0, a_1 and a_2. Those parameters are available for collectors tested according to ASHRAE (American Society of Heating, Refrigerating, and Air-Conditioning Engineers) standards and rated by SRCC (Solar Rating and Certification Corporation), as well as for collectors tested according to the another European Standards.

2.2 Structure of the solar water heating system

Considered solar water heating system has a typical configuration dedicated for single-family houses [17,18] and including flat-plate solar thermal collectors, water tank, circulating pump, diverter, tee piece, controller and other components.

The general assumptions of considered system are:

- location: Warsaw (Poland),
- daily hot water consumption: 60 dm^3 per person,
- number of residents: 5,
- solar fraction achieved (the ratio of solar heat yield to the total energy requirement for water heating): 60%,
- hot water temperature: 55°C,
- cold water temperature: 10°C,
- collectors' orientation: 0° (south),
- collectors' inclination: 35°,
- type of collectors: flat-plate,
- distance between the tank inlet and collectors: 10 m.

Calculations have been conducted using traditional method for designing SHW systems. The average global insolation for the Warsaw was taken as 1050 kWh/m^2 per year and time step was set as one month. The calculated number of connected in series solar thermal collectors (with parameters shown in Tab. 1) is five.

Table 1. The basic parameters of chosen solar thermal collectors.

Collector's parameter	Value, unit
Aperture area	1.85 m^2
Optical efficiency	0.791
1st order heat loss coefficient	3.782 $W/(m^2K)$
2nd order heat loss coefficient	0.015 $W/(m^2K^2)$
Power output (when T_m-T_a = 0 K)	1.46 kW*
Power output (when T_m-T_a = 30 K)	1.23 kW*
Power output (when T_m-T_a = 50 K)	1.05 kW*

*) tested according to EN 12975-2, assuming irradiance G = 1000 W/m^2 and flow rate 0.02 kg/s per m^2

The structure of the solar water heating system has been implemented into TRNSYS program (see Fig. 2). The load profile, based on predicted usage of hot water in residential buildings, was added. The weather data for the Warsaw has been taken from Meteonorm database. Time step for simulation was set to 15 minutes.

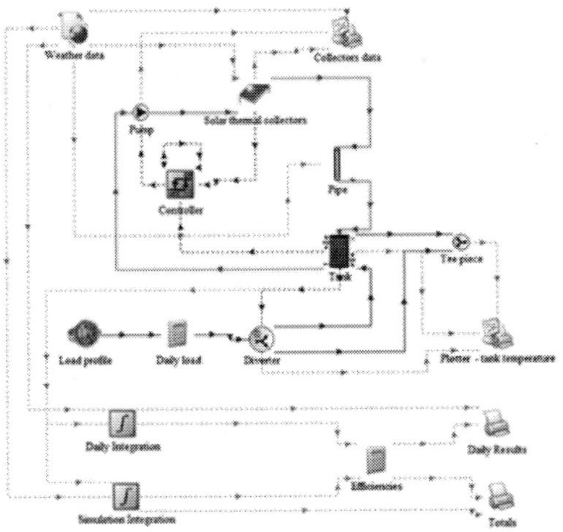

Figure 2. The structure of solar hot water system implemented in TRNSYS program.

3 Discussion of the results

The results obtained during traditional calculations (using one of the available tools) were compared with the results of dynamic simulations (conducted in TRNSYS).

3.1 Results from traditional calculations

The average yearly ratio of solar heat yield to the total energy requirement for hot water heating calculated based on the above listed assumptions, is 63.4%. On the other hand, monthly values of the solar fraction achieving varies from 9% in December to 117% in May. There are also high in June, July and August (equal and higher than 100%) as well as exceed the average yearly value still in April and September. The variations in solar fraction achieving during year for considered solar water heating installation is shown in. Fig. 4.

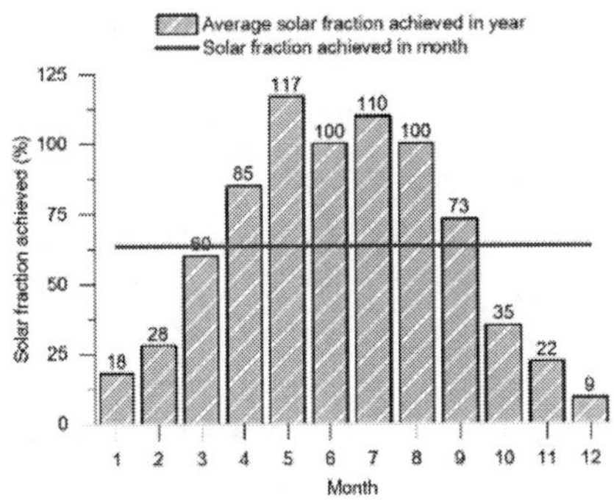

Figure 4. Solar fraction achieved in month calculated using traditional method

3.2 Results from dynamic simulations

Comparing solar fraction achieved (SFA) calculated using traditional method with results of dynamic simulations, we can see significant differences. The amount of heat generated in collectors is lower than expected for about 90% of time in the year and consequently the value of solar fraction achieved is close to 40%. What is also important, the SFA values strongly fluctuate in time (see Fig. 5).

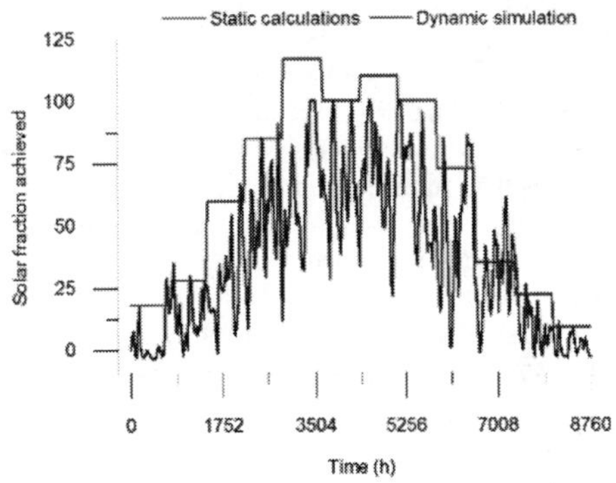

Figure 5. The comparison of solar fraction achieved calculated using traditional method and using dynamic simulations.

Taking into account high accuracy of simulations done in TRNSYS, it can be estimated the amount of additional energy delivered by auxiliary heater (e.g. electric heater) to meet the energy demands of SWH system (see. Fig. 6).

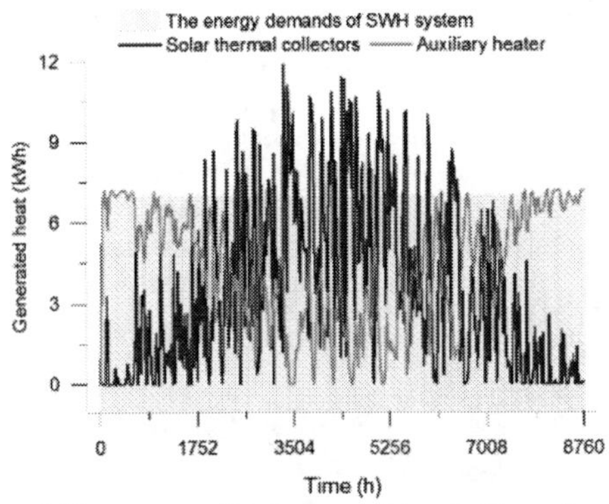

Figure 6. The amount of heat generated in solar thermal collectors and auxiliary heater.

As we can notice in Fig. 6, the total amount of energy generated in collectors and taken from the network in the whole year is respectively 1238 kWh and 1585 kWh (for assumed energy demands of solar water heating system). These values can be directly used to calculate the cost of electricity consumption for heating water. As we can conclude from dynamic simulations, real cost of auxiliary heater operation is significantly higher than expected basis on the conducted static calculations.

The real amount of heat generated in collectors may be different from the value calculated during carried out simulations. On the other hand, basis on the investigations available in literature, it can be expected, that average difference should be no more than 5%. Field tests (for validation calculated data) are planning as a next step of described research.

4 Conclusions

The analysis of results from traditional calculations and dynamic simulations shows significant differences in the amount of heat generated in solar thermal collectors. The solar fraction achieved during TRNSYS simulations fluctuates in time and its average yearly value is close to 40% (while SFA calculated using traditional method exceeds 60%). This difference makes, that more energy must be taken from network to power an auxiliary heater (~1585 kWh in comparison to ~1238 kWh, calculated using traditional methods). Consequently, the estimated costs of water heating are higher than expected.

Due to high accuracy, the use of dynamic simulations in the stage of designing solar water heating systems may help to avoid many further exploitation problems (such as low efficiency, overheating etc.). It may also help to provide safe exploitation of SWH installations.

The work has been completed as part of the statutory activities of the Faculty of Energy and Fuels at the AGH University "Studies concerning the conditions of sustainable energy development". It was realized using infrastructure of the Center of Energy, AGH University. Supervisor: prof. Mariusz Filipowicz

References

1. S.A. Kalogirou, C. Papamarcou, Renew Energ **21**, 471-493 (2000)
2. M. Oishi, T. Noguchi, Eurosun'2000 (2000)
3. U. Jordan, K. Vajen, Eurosun'2000 (2000)
4. B. Mette, H. Druck, S. Bachmann, H. Muller-Steinhagen, ISES Solar World Congress (2009)
5. G.L. Morrison, J.E. Braun, Sol Energy **34**, 389-405 (1985)
6. A. Shariah, A. Ecevit, Energ Convers Menage 36, 289-296 (1995)
7. A. Shariah, M.A. Al-Akhra, I.A. Al-Omari, Renew Energ **26**, 587-598 (2002)
8. W. Wongsuwan, S. Kumar, Int J Sustain Energ 24, 69-86 (2005)
9. J. Cadafalch, Sol Energy **83**, 2157-2164 (2009)
10. M.C. Rodrıguez-Hidalgo, P.A. Rodrıguez-Aumente, A. Lecuona, G.L. Gutierrez-Urueta, R. Ventas, Appl Therm Eng **31**, 2394-2404 (2011)
11. S. Fischer, H. Muller-Steinhagen, ISES World Congress 2007 (2009)
12. W. Kong, Z. Wang, J. Fan, P. Bacher, B. Perers, Z. Chen, S. Furbo, Energy **86**, 1838-1848 (2012)
13. F. Bava, S. Furbo, B. Perers, Energy Procedia **70**, 4-12 (2015)
14. P. Almeida, M.J. Carvalho, R. Amorim, J.F. Mendes, V. Lopes, Sol Energy **104**, 60-70 (2014)
15. S.A. Kalogirou, Prog Energ Combust **30**, 231-295 (2004)
16. W. DeSoto, S.A. Klein, W.A. Beckman, Sol Energy J (2005)
17. K. Sornek, K. Rzepka, T. Mirowski, Rynek Instalacyjny **3**, 47-52 (2015)
18. T. Mirowski, K. Sornek, Polityka Energetyczna **18**, 73-84 (2015)

Energy efficiency in the agricultural and food industry illustrated with the example of the feed production plant

Jacek Gembicki[1,a]

[1]*Szkoła Główna Gospodarstwa Wiejskiego w Warszawie Wydział Inżynierii Produkcji*

Abstract. Energy efficiency is an indicator specifying the amount of saved electric energy thanks to implementation of suitable systems and solutions aimed at reducing the energy consumption in a production plant. Effective use of electric energy or heat energy is intended to reduce the amount of energy required to manufacture products and provide services. Decreased demand for electric energy in the production plant by only a few percent's may result in considerable savings which in turn assure increased production profitability. If we reduce the energy consumption, it will translate into reduced pollution generated and emitted to the environment. Thanks to this, the plant may limit its negative impact on the surrounding. The feed industry is known to consume much amount of energy for the purposes of production. This energy is intended for pre-processing of substrates, actual production and preparation of ready product to be taken over by the customer. Farmers use fodders to feed their animals. Quality of fodders (feeds) and their ingredients determine health of farm animals, which has a direct impact on the quality of products we consume, and consequently on our health. An thorough analysis of feed production plants and reduction of their energy consumption should translate into improved effectiveness. Saved energy allows producing high-quality products and using ingredients of higher quality, which in turn may influence competitiveness of prices of ready products.

1 Introduction

The plants producing animal feed which belong to the group of agricultural and food processing plants form a manufacturing-technical-organizational structure. The structure of these production plants is dependent upon materials being processed and products these materials form. It entails application of various production technologies relevant to particular production facilities. Feed production plants mostly receive electric energy, and only a small percent of plants produces electric energy which is consumed to supply the entire plant, whereas the surplus of generated energy may be sold to other entities.

The issue of optimizing the energy consumption in these production plants has been popular for many years and implemented in certain production facilities. The main reason for optimizing these plants is the need to lower costs of particular energy carriers, which may directly cause the profits of the plant to rise. In addition, if we decrease the energy consumption, we will support the local eco-climate by reducing the emission of atmosphere pollutants, reducing the noise and producing green energy, which may well be resold to local residents.

Energy optimization process in feed production plants may be a multiple-step process and should engage the entire plant, starting from production halls, through technical rooms or storage rooms, ending up with offices.

2 Description

Energy efficiency and renewable energy are considered to be two pillars of well-balanced energy policy. Both strategies must be developed simultaneously in order to stabilize and reduce emissions of carbon dioxide and other pollutants.

Efficient energy consumption is essential and results in hindering the rise of the demand for energy so that rising supplies of pure energy may cause serious reductions of the use of fossil fuels.

Many countries acknowledge the fact that energy efficiency brings benefits for national safety because thanks to this we may limit importing energy from abroad and may hinder the pace of using the domestic sources of energy.

Energy efficiency stands for the amount of saved energy established by means of measuring or estimating the consumption before implementing the means intended to improve the energy efficiency and after implementation, with simultaneous assurance of normalization of external conditions affecting the consumption of energy (Resolution of the EU Commission no. 651/2014 of 17 June 2014 approving certain types of aid in accordance with the internal market pursuant to the art. 107 and 108 of the Treaty (CELEX: 32014R0651)

Effective use of energy is aimed at reducing the amount of energy required to provide products and services. When it comes to feed production plants, it refers to various

[a] Corresponding author: gembicki.jacek@gmail.com

fodders which can be divided into fodders of plant origin, animal origin or mineral feed.

There are many reasons encouraging us to improve energy efficiency. If we limit the energy consumption, we reduce operation costs and consumers may increase savings, provided that energy savings exceed additional expenses related to implementing energy-efficient technologies.

For example, the building thermal insulation assures the use of reduced amount of energy for the purpose of heating and cooling in order to reach and keep the comfortable temperature. The use of fluorescent lamps or skylights reduces the amount of energy but assures as strong illumination as traditional bulbs provide. The improvement of energy efficiency is frequently feasible thanks to more efficient technologies or production processes.

There are many reasons convincing us to improve energy efficiency. Limited use of energy results in reduced operation costs and may lead to savings in the production plant on condition that the energy-related savings exceed additional costs related to implementing energy-efficient technologies. Additionally, it serves to limit the emission of greenhouse gases and other pollutants, and in effect brings other benefits.

The figure 1 shows the impact of the feed production plant on the local agro-ecosystem. It includes products manufactured by the plant, as well as the influence on the neighbourhood through emission of waste water, noise and carbon dioxide.

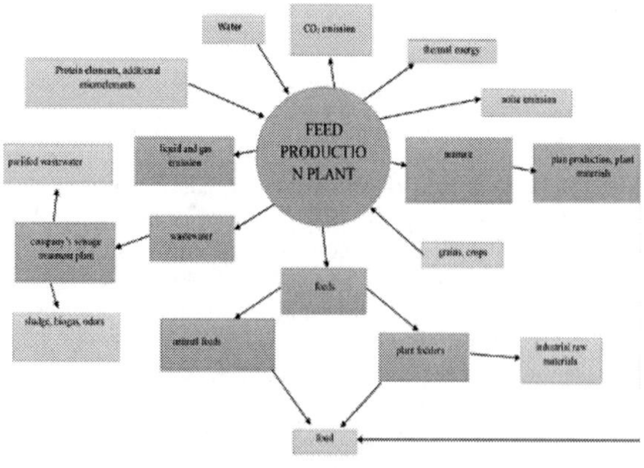

Figure 1. Feed production plant in Agro-ecosystem (prepared by the author)

3 Literature review

It has been proved that energy efficiency is a cost-effective economy building strategy and does not necessarily require increased energy consumption. Modern and energy-efficient equipment consumes a way less energy than other old-type devices. Advanced boilers and furnaces may operate at high temperatures but consume less fuel. They are more efficient and generate less substances which are harmful to the environment. Electric motors usually operate at the regular speed, but drives of variable speed allow setting the required velocity [1]. Thanks to this, it is possible to save from 3 to 60% of energy, depending how the engine is used. Engine elements made of superconductive material may also reduce energy losses. Similar effects assures voltage optimization. In the feed industry (but not only) are used many pumps and compressors of various shapes, sizes with a wide range of intended uses. Their efficiency depends on numerous factors, but it can be improved by implementing enhanced inspection of processes and improved maintenance [2]. Compressors are widely used to provide compressed air which is then applied on production lines to transport substrates or ready products. Feed production plants generate steam water and electricity for further own use [3]. When electric energy is generated, the heat, being a side effect of the process, may be collected and used to generate steam, to heat areas, or for other industrial purposes. The conventional generation of electric energy is efficient in ca. 30%, whereas a simultaneous generation of heat and energy (also referred to as cogeneration) processes up to 90% of fuel into usable energy [4].

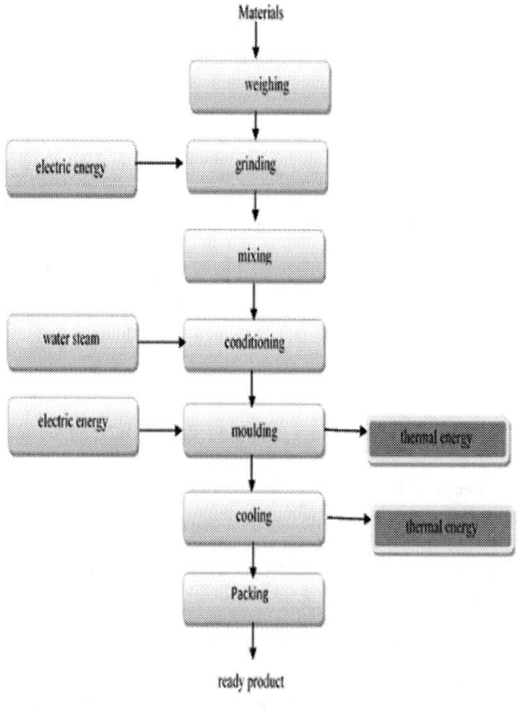

Figure 2. Energy-material flows in the feed production plant (prepared by the author)

The figure 2 shows the example of material flow in the feed production plant. A typical manufacturing process is showed there, starting from supply of substrates used to manufacture a fodder/feed, through particular processes related to the substrates, and ending up with ready product in the form of animal feed [5].

The feed industry uses large amounts of energy to supply various manufacturing processes. Many of these processes require large amounts of heat and mechanical energy which derive from natural gas, oil and electricity. Since manufacturing processes are so diverse, it is impossible to describe all available methods of assuring energy efficiency in the feed industry. A crucial role is played by specific technologies and

DOI: 10.1051/e3sconf/20161000138

mechanisms implemented in production plants [6]. Nevertheless, we may distinguish an array of processes and energy-related services which can be put into use in feed production plants.

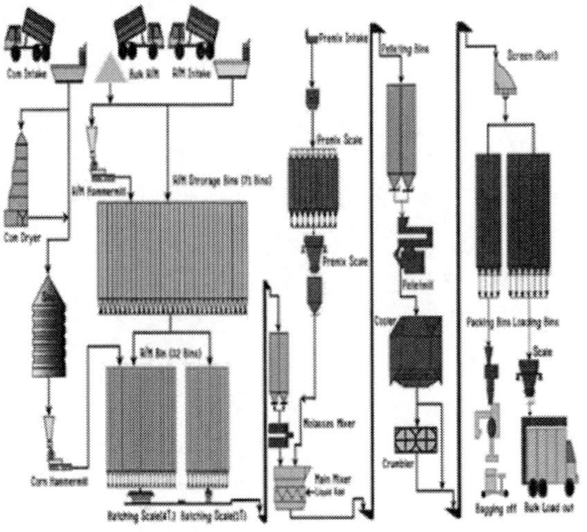

Figure 3. Feed processing system (http://www.scielo.br)

Energy forms and carriers:

- Energy in the form of heat (heating, thermal sterilization, evaporation),
- Energy of compressed air (pneumatic transport, packing),
- Electric energy used for motor drives,
- Mechanical energy (grinding, mixing, cleaning, transporting, separating, dosing).

The energy-manufacturing system of agricultural and food processing plants is a separate and internally arranged system of elements with a structure assuring specific production purposes. To examine this sort of system, it is necessary to get familiar with the manufacturing processes in a given plant and to see and measure places which energy is used in. Such assessment of energy management in the plant should provide information on energy carriers wear, method of use and tips on how to improve it. The knowledge of the energy carriers wear may be used in many ways, either to implement new solutions or modernize existing ones in order to make them more effective. The question arises of whether the existing testing methods, used to specify chances for reduction of energy carriers wear and specification of improvement actions, are sufficient and whether have been used in a proper way. Aside from basic features, the method should be justified from the scientific and technical point of view [7].

We distinguish two basic methods. The top-down method is concerned with the top-down approach and limits itself to analyzing input and output parameters. In this method processes occurring inside the system in question are not considered. The main advantages of this method is among others low costs, simplicity, easy monitoring and forecasting of energy consumption, as well as considerable flexibility and versatility. In turn, the bottom-up method focuses on a precise and in-depth analysis of the production process, thanks to which we can improve energy efficiency. Plus, it is possible to develop a detailed system model by means of which we can carry out a precise simulation showing weak links of the system and their improvement [8]. In order to indicate possible methods of assessing the energy management, in the first place it is necessary to isolate areas and then conduct their analysis to draw conclusion influencing the decisions related to improvement of energy efficiency in the production plant [9].

How to save energy:

- Suitable (full) use of electric motors in terms of power
- Increased use of production machines during continuous operation
- Reasonable use of lighting installation
- Use of thermal insulation of receivers and buildings
- Reduced number of energy conversions (e.g. replacing water steam with hot water),
- Heat recovery (recirculation)
- Shortening the length of transmission lines of heat carriers.

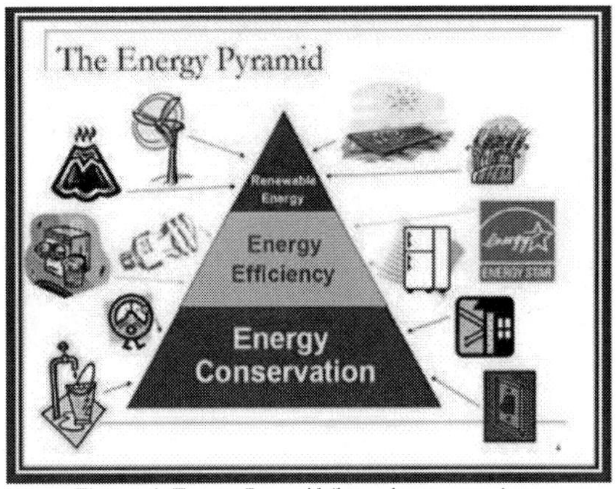

Figure 4. Energy Pyramid (legendpower.com)

4 Summary

To have an insight into energy consumption in animal feed production plants, it is necessary to choose relevant testing methods and specify which of them are the most effective, and provide credible and reliable results. Thanks to a suitable testing method, we are able to estimate the energy consumption and if it is too high, we may try to implement relevant processes in order to reduce the aforementioned consumption. If the process turns out to be effective, we may increase profits of the production plant, and thus become more competitive on the feed market. When discussing the issue of energy consumption, it is not only about the use of electric energy but also thermal energy. If we lower the demand for thermal energy or if we use it in a better way, we will be able to reduce the emission of greenhouse gases to the environment, and for this reason the production plant will have lower influence on the local eco-climate.

To sum up, if we test the energy consumption and attempt to reduce it, the feed production plant and its surrounding will benefit from it. We should remember however that it is of paramount importance to get familiar with the entire manufacturing process and the plant structure, as well as estimate potential improvement costs and forecasted profits in order to make sure

DOI: 10.1051/e3sconf/20161000138

the improvement pays off. Thanks to such balance, we will learn if the modernization is profitable or not.

References

1. Z. Narski, Przemysł Spożywczy, **8**, (1983)
2. N. Marks, Z. Sobol, Inżynieria Rolnicza, **3**, 78 (2006)
3. B. Dróżdż, Annals of Warsaw University of Life Sciences, **56**
4. B. Dróżdż, Metodyczne i utylitarne aspekty analizy energochłonności przetwórstwa rolno-spożywczego, p.20-25 (2013)
5. B. Dróżdż, MOTROL Motoryzacja i Energetyka Rolnictwa, Modelowanie zużycia energii w przemyśle rolno-spożywczym, p. 349-356 (2010)
6. R. Akinoso, K. Olatoye, International Food Research Journal Energy, **20** (2013)
7. K. Brodowicz, M. Markowski, Gospodarka Paliwami i Energią, **9** (1997)
8. J. Budny, Problemy gospodarki energią w przemyśle mleczarskim, Definicja i ogólna metodyka wyznaczania wskaźnika zużycia energii w przemyśle mleczarskim, p.25-26 (1988)
9. J. Budny, J. Turowski, Inżynieria przetwórstwa spożywczego, **4**, 4 (2012)
10. J. Grochowicz, K. Zawiślak, Inżynieria Przetwórstwa Spożywczego, **2**, 2 (2012)

E3S Web of Conferences **10**, 00139 (2016) DOI: 10.1051/e3sconf/20161000139
SEED 2016

The sizes of Flat Plate and Evacuated Tube Collectors with Heat Pipe area as a function of the share of solar system in the heat demand

Małgorzata Olek[1], Piotr Olczak[1,a] and Dominik Kryzia[2]

[1]Institute of Thermal Engineering and Air Protection, Faculty of Environmental Engineering, Cracow University of Technology, Warszawska 24, 31–155 Cracow, Poland
[2]Mineral and Energy Economy Research Institute, Polish Academy of Sciences, Wybickiego 7, 31–261 Cracow, Poland

Abstract. The popularity of solar collectors in Poland is still increasing. The correct location of the collectors and a relatively high density of solar radiation allow delivering heat even in spite of relatively low ambient temperature. Moreover, solar systems used for heating domestic heat water (DHW) in summer allow nearly complete elimination of conventional energy sources (e.g. gas, coal). That is why more and more house owners in Poland decide to install solar system installations. In Poland the most common types of solar collectors are flat plate collectors (FPC) and evacuated tube collectors with heat pipe (ETCHP); both were selected for the analysis. The heat demand related to the preparation of hot water, connected with the size of solar collectors' area, has been determined. The analysis includes FPC and ETCHP and heat demand of less than 10 000 kWh/year. Simulations were performed with the Matlab software and using data from a typical meteorological year (TMY). In addition, a 126–year period of measurements of insolation for Krakow has been taken into account. The HDKR model (Hay, Davis, Klucher, Reindl) was used for the calculation of solar radiation on the absorber surface. The monthly medium temperature of the absorber depends on the amount of solar system heat and on the heat demand. All the previously mentioned data were used to determine solar efficiency. Due to the fact that solar efficiency and solar system heat are connected, the calculations were made with the use of an iterative method. Additionally, the upper limit for monthly useful solar system heat is resulted from the heat demand and thus the authors prepared a model of statistical solar system heat deviations based on the Monte Carlo method. It has been found that an increase in the useful solar system heat in reference to the heat demand is associated with more than proportional increase in the sizes of the analyzed surfaces of solar collector types.

1 Introduction

Solar conditions in Poland, variable depending on time and place, are moderate when compared to other European countries. This is related to favorable conditions for obtaining solar energy in the spring and summer months (April – August) (about 7.5 kWh/m^2/day) and unfavorable in the autumn and winter (about 0.1 kWh/m^2/day) [1]. The highest annual insolation is in the eastern part of the Lublin province (>1048 kWh/m^2/year). In central Poland it is in the range from 1 022 to 1 048 kWh/m^2/year, while for the rest of the country insolation do not exceed 1 000 kWh/m^2/year [2, 3]. In spite of the diversified conditions the amount of objects with solar systems in Poland is constantly increasing. The total area of solar collectors installed in Poland at the end of 2004 and 2014 amounted to 99 000 m^2 and 1 744 000 m^2, respectively [4, 5]. The only producers and consumers of the energy from solar systems are trade and services sector and households (Fig 1.).

One of the important parameters of the solar system, especially for owners of single-family homes, is the selection of the optimum size of the collector area. Outside the local weather conditions and the amount of heat needed for domestic hot water (DHW) or low-temperature central heating systems (floor or wall-mounted), the collector area can be associated with the demand for DHW and the type of heat source [7]. The increased interest in solar energy in Poland has resulted in a number of studies regarding the selection of collector area but these are mainly focused on flat plate collectors [8-12].

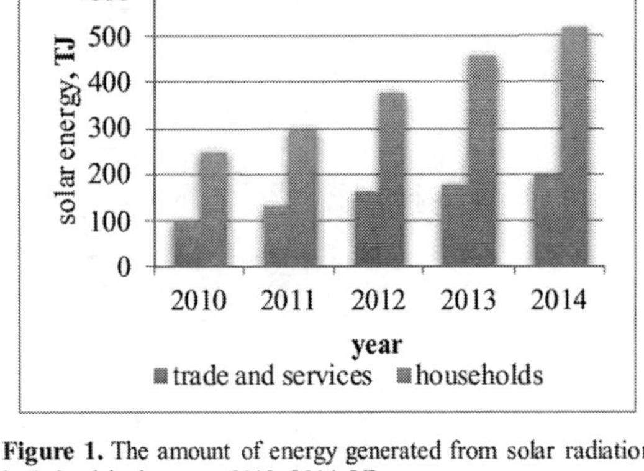

Figure 1. The amount of energy generated from solar radiation in Poland in the years 2010–2014 [6].

[a] Corresponding author: piotrolczak@hotmail.com

DOI: 10.1051/e3sconf/20161000139

The gross area (GA), absorber area (AA) and aperture area (AAp) for the variable, dependent on the number of residents, heat demand for the preparation of DHW and the variable degree of meeting this demand by the solar system, have been determined in the paper. The analysis has been conducted for flat plate collectors (FPC) and evacuated tube collectors with heat pipe (ETCHP). The paper has resulted in nomograms allowing the owner of a single-family building, knowing the amount of heat needed for DHW, to choose the collector area depending on the degree of meeting the heat demand.

2 Research object

2.1 Heat demand for domestic hot water

A residential building located in Cracow inhabited by a maximum of 6 people was selected as the object of the analysis. The annual heat demand for DHW ($Q_{w,a}$) was estimated from the relationship [13]:

$$Q_{w,a} = \frac{(V_{DHW} \cdot n_r \cdot c_w \cdot \rho_w \cdot (T_w - T_O) \cdot 365}{3600 \cdot \eta_{tot}} \quad (1)$$

where:

V_{DHW} - average daily demand for DHW, dm³/resident
n_r – number of residents,
c_w – specific heat of water, kJ/(kgK),
ρ_w – density of water, kg/dm³,
T_w – temperature of hot water, °C,
T_o – temperature of cold water, °C,
η_{tot} – domestic hot water system efficiency.

It was assumed that the differences in cold water inlet temperature correlate with seasonal fluctuations in demand for DHW and the perceived temperature of hot water. Under Cracow conditions, cold water temperature is in the range between 2 and 20°C [14]. The simulation assumed the cold water inlet temperature (T_O) and the tap water temperature (T_w) at 10°C and 55°C, respectively [15]. The assumed specific heat of water (c_W) and density of water (ρ_W) were 4.18 kJ/(kg·K) and 1 kg/dm³, respectively. Theoretical and actual daily DHW consumption is presented in Table 1. The analysis of the demand for DHW was performed for the range from 20 to 70 dm³/resident/day. The assumed efficiency of DHW system (η_{tot}) was 0.69.

Table 1. Daily water consumption per capita in residential buildings

Building	DHW, V_{DHW}
	dm³/resident /day
Multi-family house [16]	19.4 – 48
Single-family house [17]	37.5 – 39.3
EN 15316:3:2007 standard	36 – 66.7

An annual heat demand required ($Q_{w,a}$) for the preparation of DHW depending on the number of inhabitants and the daily demand for DHW is shown in Fig. 2.

Further analysis of the collector area was carried out for the heat demand for DHW in the range from 1000 to 10 000 kWh/year.

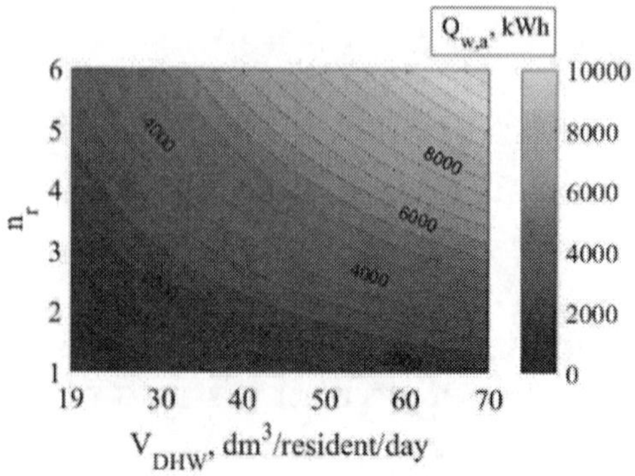

Figure 2. The annual heat demand for DHW as a function of the number of inhabitants and the average daily demand for DHW

2.2 Collectors

In Poland, flat plate collectors (74%), followed by evacuated tube collectors (26%), are the most popular types of solar systems [18]. Tube collectors can be divided into several subtypes: evacuated tube collectors with heat pipes or parabolic mirror collectors (CPC - Compound Parabolic Concentrators).

Flat plate collectors (FPC) are made of transparent protective cover, absorber, flow channels and insulation (mineral wool). The gross area of FPC depends on the surface of the absorber mounted in a rectangular metal housing (Fig. 3A). The aperture area of FPC depends on the way of mounting the glass protective panels and is usually smaller than the absorber area.

Evacuated tube collectors with heat pipes (ETCHP) usually consist of parallel rows of glass tubes. The outer tube has a protective function, while the inner tube is coated with an absorptive coating. Between tubes is formed a vacuum. The heat pipe is positioned coaxially to the glass tubes. It is filled with a liquid with a low boiling point. The heat between the absorber and the heat pipe is transmitted mainly by the heat exchange fin (Fig. 3B). A set of pipes is connected to the manifold tube where the heat exchange between the heat pipes and the solar fluid takes place. The manifold tube is not involved in the absorption of solar radiation. The gross area of ETCHP depends on the dimensions of the single outer tube (d_o), the number of tubes, the distance between them (d_{bp}) and the dimensions of the manifold tube. The distance between the tubes is optimized in order to minimize the risk of overlapping of the tubes and to maximize the absorber area relative to the gross area. The absorber area is calculated [19]:

$$AA = d_a \cdot L \cdot p \quad (2)$$

AA – absorber area, m²
d_a – outer diameter of inner tube (Fig. 3B), m,
L – length of tube, m,
p – numbers of pipes.

The aperture area for ETCHP is calculated as the product of the inner diameter of the outer tube, its length and numbers of pipes.

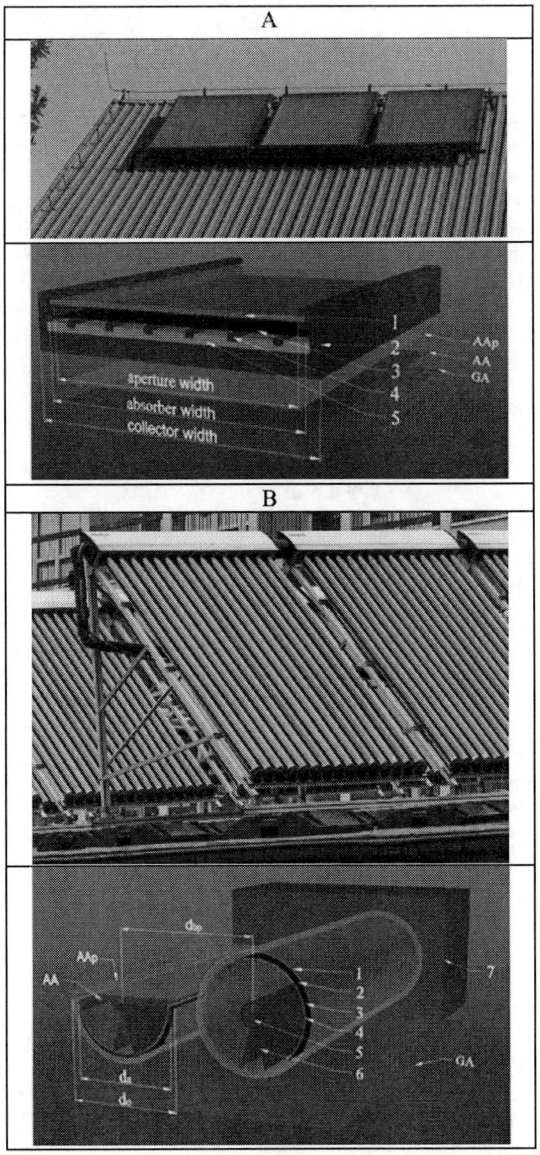

Figure 3. Solar collectors, GA–gross area, AAp – aperture area, AA – absorber area A) Flat plate collectors (1– transparent cover, 2– metal housing 3– absorber, 4– flow channels, 5– isolation); B) Evacuated tube collectors (1– outer glass tube, 2– vacuum, 3– absorber, 4– inner glass tube, 5– heat pipe, 6– heat exchange fin, 7– manifold tube).

The main difference between FPC and ETCHP is their efficiency (Table 2). The advantage of ETCHP is associated with the properties of insulation layer of the vacuum. The thermal conductivity coefficient for this layer is in the range from 0.001 to 0.0025 W/(m·K). In the case of the insulating layer in FPC, that is the air between the absorber and the glass panel, it amounts to 0.026 W/(m·K) [20]. The linear heat loss coefficient (a_1) for FPC depends on the model of collector and can vary from 2.741 W/(m^2·K) to 4.999 W/(m^2·K) [21-30], while for the ETCHP is in the range from 0.749 W/(m^2·K) to 1.771 W/(m^2·K) [31, 32]. For a selected day of May 23 the potential ranges of solar yields for boundary coefficients (a_1) are presented in Fig. 4.

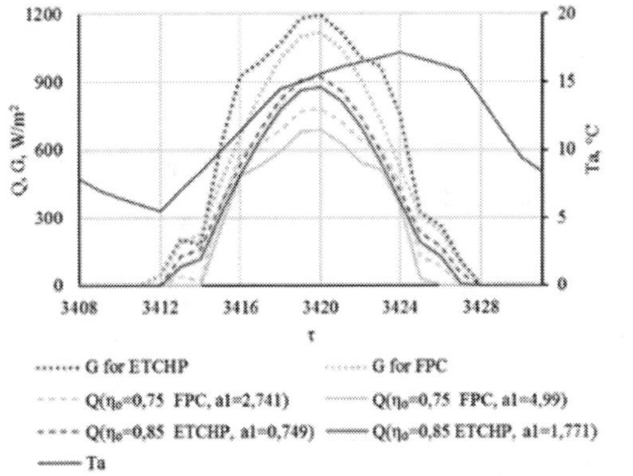

Figure 4. Comparison of the impact of the coefficient a_1 on the acquisition of solar energy for the analyzed collectors, $T_m = 50°C$.

The second difference is the access to solar radiation, affected by the shape of the absorber [19] and favorable for ETCHP (Fig. 5).

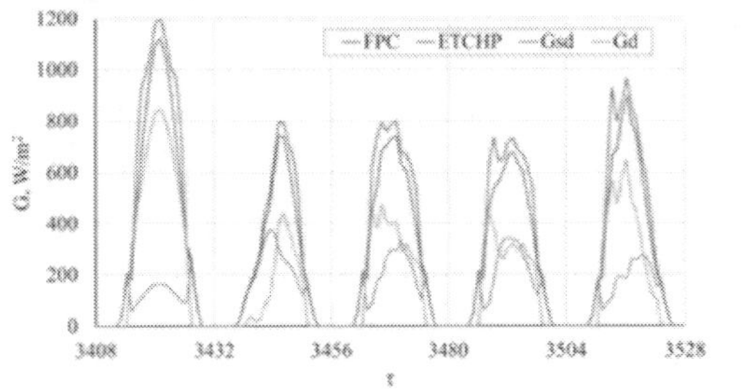

Figure 5. The intensity of solar radiation calculated on the surface of the FPC and ETCHP absorber for the selected days in a typical meteorological year and the available beam (G_d) and diffuse radiation (G_{sd}).

The paper assumed that the diameters of the analyzed vacuum tubes (d_o) and absorber (d_a) are 0.058 m and 0.0514 m, tube length at 1.8 m and the distance between the axes of adjacent tubes (d_{bp}) are 0.078 m. The simulated optical efficiency (η_0), linear (a_1) and nonlinear (a_2) heat loss coefficients for ETCHP correspond to the values typical for the Energosol collector (Table 1) tested in a large-scale installation [33, 34]. In the case of FPC, the Watt 2020 collector technical data has been used [25].

Table 2. The parameters of FPC and ETCHP collectors and the surface conversion [25,32,35].

	Unit	FPC	ETCHP
η_0	–	0.750	0.850
a_1	W/(m^2·K)	3.517	1.771
a_2	W/(m^2·K^2)	0.017	0.0192
AA/GA	–	1.88/2.04	1.21/2.42
AAp/GA	–	1.87/2.04	1.4/2.42

It was assumed that the collectors are oriented towards the south (azimuth 0 °) and are set at an angle β = 30 °

to the ground level [10]. The collectors are located on the roof covered with a reflectance coefficient (r_c) of 0.1. For ground surface, the assumed reflectance coefficient (w_o) was 0.2 [36].

Olczak et al. experimentally indicated that, even in the case of oversized collectors, there are periods of adverse weather conditions in the months of May or June, requiring to support the solar system with conventional source of energy eg. boiler [37, 38]. Therefore, the study has adopted a variant with a solar system delivering max. 90% of the monthly heat demand. The possible excess heat has not been included in the balance of useful energy.

Calculations of the potential solar yields for the investment located in Cracow have been performed for 1 m^2 of the absorber area. Assuming the demand and the conversions area factors (AA/GA and AAp/GA) for the particular type of solar collectors (Tab. 1), the total absorber area (AA), gross area (GA) and aperture area (AAp) have been estimated. In the case of FPC, the aperture and absorber areas weren't analyzed separately due to the fact that the difference between these surfaces is only about 0.6%.

3 Computational model of solar radiation

The intensity of solar radiation for each hour of the reference year (τ) has been determined using the HDKR (Hay, Davis, Klucher, Reindl) radiation model [36, 39]. The aforementioned model takes into account the beam solar radiation (G_d), diffuse circumsolar radiation (G_{dh}), diffuse isotropic radiation (G_{di}), diffuse horizontal radiation (G_{dlh}) and diffuse reflected radiation (G_{dr}) (Fig. 6) and its one of the most accurate models [40, 41]. For each hour (τ) of the year and the variable tilt the collector surface to the ground level (β), the HDKR model can be expressed with the equation (3):

$$G(\tau,\beta) = G_d(\tau,\beta) + G_{dh}(\tau,\beta) + G_{di}(\tau,\beta) + G_{dlh}(\tau,\beta) + G_{dr}(\tau,\beta) \quad (3)$$

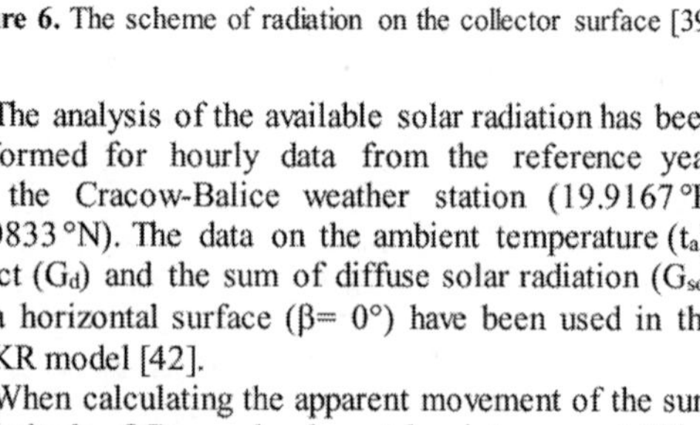

Figure 6. The scheme of radiation on the collector surface [39, 36].

The analysis of the available solar radiation has been performed for hourly data from the reference year for the Cracow-Balice weather station (19.9167°E, 50.0833°N). The data on the ambient temperature (t_a), direct (G_d) and the sum of diffuse solar radiation (G_{sd}) on a horizontal surface (β= 0°) have been used in the HDKR model [42].

When calculating the apparent movement of the sun, the latitude of Cracow has been taken into account. What is more, changes resulting from the difference between solar time for the meridian 19.9167° E and for the GMT time have also been included [42]. Due to the analysis of the average hourly data, the equation of time E has been omitted.

The angle of the collector to the ground level β = 30° for Cracow adopted in the model results in the highest yields per year assuming good heat collection efficiency [10]. In addition, for the adopted angle, the diffuse horizontal radiation [36, 39] and reflected radiation [36, 39] have a limited share in the intensity of radiation incident on the absorber.

The available solar radiation incident on the absorber of the FPC directed south at an angle β for each hour of reference year (τ) is calculated with the following formula:

$$G(\tau) = R_b(\tau)\cdot K_\theta(\tau)\cdot\left(G_d(\tau) + Ai(\tau)\cdot G_{sd}(\tau)\right) + R_d\cdot K_{\theta d}\cdot\left(1 - A_i(\tau)\right)\cdot\left(1 + R_{jh}(\tau)\right) + wo\cdot R_o\cdot K_{\theta d}\cdot\left(G_d(\tau) + G_{sd}(\tau)\right) \quad (4)$$

where :
G – available intensity of solar radiation incident on absorber dependent on time (for the average year), W/m^2
G_d – beam radiation on a horizontal surface, W/m^2
G_{sd} – diffuse radiation on a horizontal surface, W/m^2
A_i – anisotropy index,
R_b – geometric factor of beam radiation on tilted collector to that on a horizontal surface,
R_d – geometric factor for diffuse radiation,
R_{jh} – geometric factor for diffuse brightness horizontal radiation,
R_o – geometric factor for diffuse reflected radiation (reflected by surfaces in front of collector),
w_o – reflection factor from ground,
K_θ – incident angle modifier for beam radiation,
$K_{\theta d}$ – incident angle modifier for diffuse solar radiation.

The correction factor for diffuse horizontal radiation (R_d) and diffuse radiation (R_o) is dependent on angle of the collector to the ground level and for β=30 ° is 0.933 and 0.067, respectively. The incident angle modifier K_θ was determined using data for the Watt collector [25]– Fig. 7. For diffuse solar radiation, the $K_{\theta d}$ is assumed at 0.95 [43].

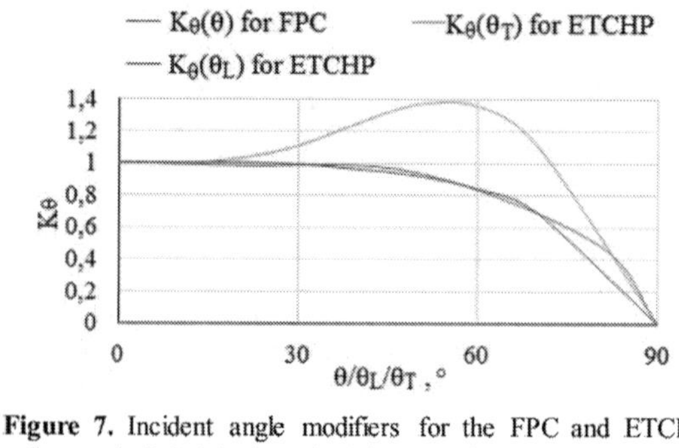

Figure 7. Incident angle modifiers for the FPC and ETCHP collectors included in the calculation model.

The available solar radiation incident for ETCHP was calculated with the following formula:

$$G(\tau) = R_b(\tau)\cdot K_\theta(\tau)\cdot\left(G_d(\tau) + Ai(\tau)\cdot G_{sd}(\tau)\right) + R_d\cdot \pi\cdot F_{cs}\cdot K_{\theta d}\cdot\left(1 - Ai(\tau)\right)\cdot\left(1 + R_{jh}(\tau)\right) + w_o\cdot R_o\cdot \pi\cdot$$

$$F_{cs} \cdot K_{\theta d} \cdot (G_d(\tau) + G_{sd}(\tau)) + r_c \cdot R_{ot} \cdot \pi \cdot F_{cs} \cdot K_{\theta d} \cdot (G_d(\tau) + G_{sd}(\tau)) \quad (5)$$

F_{cs} – view factor from the collector front side to sky
r_c – reflectance of radiation by surfaces behind collector.

The view factor from the collector front side to the sky (F_{cs}) associated with overlapping of the tubes, was determined on the basis of studies by Shah & Furbo [19] and data provided by the Energosol [32]. For the analyzed ETCHP, the F_{cs} is 0.3758. Due to the circular shape of the absorber, the diffuse radiation it is not available for its entire circumference. The area available (A_d) for the aforementioned radiation can be determined with the following equation:

$$A_d = \pi \cdot F_{cs} \cdot d_a \cdot L \cdot p \quad (6)$$

where:
d_a – diameter of absorber, m,
p – number of tubes,
L – length of tube, m.

The paper assumes that the area available for the diffuse radiation is not dependent on the number of tubes and their geometrical parameters. In Equation 5 for the diffuse radiation were used values referenced to the absorber area (AA=1 m^2) according to the relation:

$$\frac{A_d}{d_a \cdot L \cdot p} = \frac{A_d}{AA} = \pi \cdot F_{cs} \quad (7)$$

For the diffuse radiation, referred to half of the tube circumference, a constant correction factor $K_{\theta d}$ equal 0.88 - different than for the beam radiation - has been determined. The access to radiation reflected by surfaces in front of the collector has been adjusted by the F_{cs} factor [19]. Similarly, in the case of radiation reflected from the back, a 10% reflectance of radiation (r_c) by surfaces behind the tubes has been taken into account. This value is affected by both the transmission of radiation between the tubes and the surfaces behind the collector.

ETCHP are characterized by different availability of solar radiation when compared to FPC. This is related to a different shape of absorber so that different parts of the absorber area receive various types of solar radiation. Major importance for ETCHP is also passive solar tracking. In the case of ETCHP, the correction factors (K_θ) are considered in relation to the two angles: longitudinal incidence angle (θ_L) and transversal incidence angle (θ_T). Values of the angles for each hour of the reference year were determined according to formula from the EMD company guide [42]. The correction factor associated with transversal angle ($K_\theta(\theta_T)$), related to the tilt of tubes, depends on the distances between the axes of tubes (d_{bp}) and the diameter of the absorber (d_{ap}). On the basis of the data from Energosol company [32], a polynomial approximation of the correction factors (K_θ) for both types of angles (θ_L, θ_T) has been determined (Fig.7). The correction factor $K_\theta(\tau)$, used in eqn. 5, was established with the following equation [44, 45]:

$$K_\theta(\tau) = K_\theta(\theta_T(\tau)) \cdot K_\theta(\theta_L(\tau)) \quad (8)$$

The correction factor in relation to longitudinal angle ($K_\theta(\theta_L)$) is dependent on the difference between the tilt angle (β) and the angle of solar altitude above the horizon. Due to the achieved value of the θ_L angle (no larger than 60 °) for β = 30 °, the discussed correction factor is irrelevant for the calculations.

The monthly total insolation available for 1 m^2 of the absorber area and the gross area for both collectors was calculated according to the relationship (4) or (5) and (8) is presented in Fig. 8.

$$I(month) = \sum_{\tau 1}^{\tau n} G(\tau) \cdot \frac{1h}{1000} \quad (9)$$

where:
I – insolation, kWh/m^2
τ1 – first hour of a month,
τn – last hour of a month.

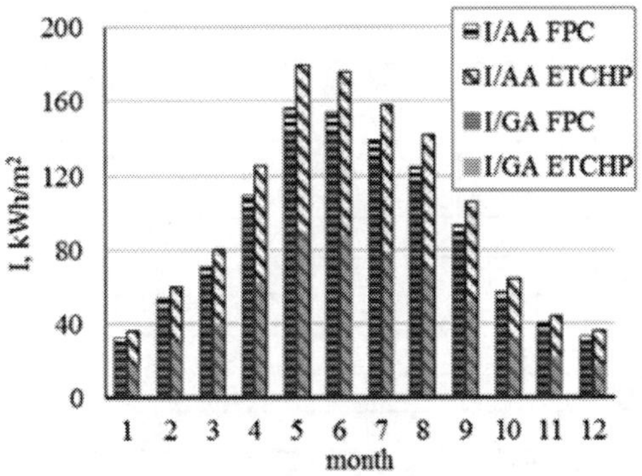

Figure 8. The comparison of available monthly insolation (I) for flat plate collectors (FPC) and evacuated tube collectors with heat pipe (ETCHP) calculated for the absorber area (AA) and the gross area (GA).

Generally, the available solar radiation incident values (G) determined from the data from the reference year (Equation 4, 5) are used to indicate the solar yield (Q):

$$Q = AA \cdot G \cdot \eta \quad (10)$$

where:
η – efficiency of the collectors depending on time and demand.

Thus, G value is determined on the basis of the average solar radiation from 30 years. In fact, solar yields for each month in subsequent years can significantly differ from the average values, which can result in oversized or undersized area of the collector.

In this paper, based on the known average solar radiation from 30 years (from the reference year) and the known standard deviations from 126 years of solar radiation measurements in Cracow (Table 3) [46], a model of standard deviations of solar yields has been created. The monthly insolation values for a period of 100 years have been determined using the Monte Carlo method. The values obtained allowed calculating the correction factor (σ) for the intensity of radiation taken into consideration when determining the average value of solar yields. A constant value from the reference year, usually assumed in the calculations, and the insulation values obtained using the Monte Carlo method for the month of July, is presented in Fig. 9.

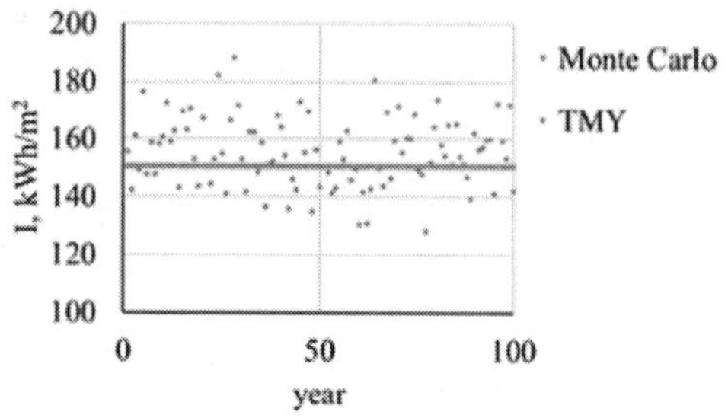

Figure 9. The comparison of values from Monte Carlo method with the values from the reference year (TMY – typical meteorological year) for July.

Table 3. TMY insolation and standard deviation percentage values [13, 46].

Month	Insolation, kWh/m²	Standard deviation, %
January	26	7.6
February	43	9.1
March	66	8.4
April	106	8.4
May	163	7.7
June	167	7.8
July	151	7.3
August	126	6.5
September	90	8.8
October	52	8.5
November	32	6.9
December	24	6.9

4 The calculation of solar yields

The amount of energy obtained (Q1h) for 1 m² of the absorber area (AA) over 1 h has been determined according to the relationship (11). The value has been determined for each hour of the year for a period of 100 years.

$$Q1h(\tau, y, m) = AA \cdot G(\tau, y, m)' \cdot \eta^i(\tau, T_m) \cdot 1\,h \quad (11)$$

where:
$\eta^i(\tau, r, m, T_m)$ – efficiency of the collectors depending on time and demand,
τ – hour of reference year,
y – calculation year,
m – month of reference year.

The intensity of all types of solar radiation (Fig. 6) available at a given time has been calculated using the HDKR model (4 and 5) and then modified by taking into account the estimates for a period of 100 years:

$$G(\tau, y, m)' = G(\tau) \cdot \sigma(y, m) \quad (12)$$

where:
$G(\tau)$ –available intensity of solar radiation incident on absorber dependent on time (for the average year), W/m²,
σ (y, m) – correction factor for the intensity of radiation in the subsequent calculation years (y) and months (m) determined using the Monte Carlo method.

In the case of solar systems, the temperature difference of the fluid at the inlet and outlet of the collector (T_m) depends on the degree of meeting the heat demand (shd_m). The average monthly value of T_m is determined by iteration. Due to the actual operating temperature of the solar system it is assumed that the monthly average temperature of the fluid is in the range from T_{min}=20 °C to T_{max}=50°C [33]. The efficiency of heat transfer from the collectors to the solar tank (η_{ts}) was 90%. The subsequent T_m values are calculated with the equation (12). It has been assumed that the monthly heat demand for domestic hot water ($Q_{W,m}$) is equal to 1/12 of the annual demand ($Q_{w,a}$) from eqn. (1).

$$T_m^i(y) = \frac{Q_m^i(y) \cdot \eta_{ts}}{Q_{W,m}} \cdot (T_{max} - T_{min}) + T_{min} \quad (13)$$

where:
$Q_m^i(y)$ – calculated theoretical monthly solar yields for a given calculation year, kWh,
$Q_{W,m}$ – monthly heat demand for DHW, kWh,
η_{ts} – efficiency of heat transfer from collectors to solar tank,
y – calculation year.

Theoretical monthly solar yields for a given calculation year have been determined according to the relationship:

$$Q_m^i(y) = \sum_{\tau 1}^{\tau n} Q1h(\tau, y, m) \quad (14)$$

For the available insolation shown in Fig. 9, adjusted for subsequent calculation years and determined according to the algorithm (12) T_m, the theoretical efficiency of collectors has been determined [47]:

$$\eta^i(\tau, y, m, Tm) = \eta_0 - a1 \cdot \frac{T_m^i - Ta(\tau)}{G(\tau,y,m)'} - a2 \cdot G(\tau, y, m)' \cdot \left(\frac{T_m^i - Ta(\tau)}{G(\tau,y,m)'}\right)^2 \quad (15)$$

where:
η^i– efficiency of the collectors depending on time and demand,
η_o– optical efficiency of collector,
a_1 – linear heat loss coefficient, W/(m²·K),
a_2 –nonlinear heat loss coefficient, W/(m²·K²),
T_m– average temperature between inlet and outlet of collector, °C,
T_a– ambient temperature, °C,
G (τ,y,m)'– available, corrected intensity of solar radiation incident on the absorber, W/m².

For the calculated annual heat demand eqn. (1) ranging from 1 000 to 10 000 kWh/year, the annual solar energy (Q_y) has been determined according to the eqn. (16). The simulation result is shown in Fig. 10.

$$Q_y = \sum_I^{XII} \frac{\sum_{y=1}^{100} Q_m(y)}{100} \quad (16)$$

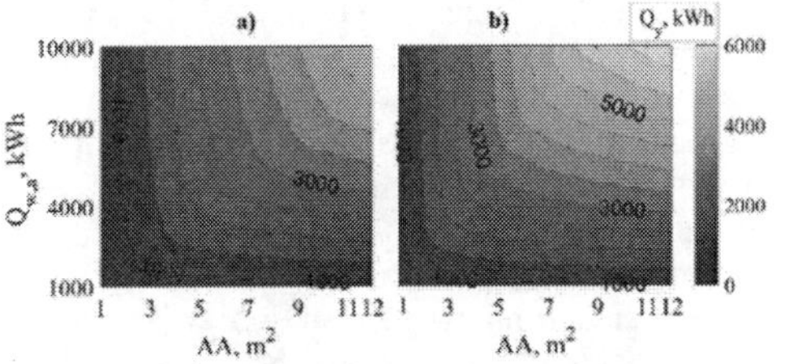

Figure 10. The theoretical solar yields (Q_y) for: a) FPC, b) ETCHP as a function of the absorber area (AA) and heat demand of the installation ($Q_{w,a}$).

The monthly degree of meeting the heat demand for domestic hot water (shd_m) is determined as the ratio of solar yields (Q_m) to the heat demand ($Q_{w,m}$). The values of solar yields determined on the basis of relationship (14) for a period of 100 years have been averaged.

$$shd_m = \frac{\sum_{y=1}^{100} Q_m(y)}{100 \cdot Q_{w,m}} \quad (17)$$

where:
$Q_m(y)$– calculated monthly solar yields for a given calculation year according to eqn. (14),
y– The calculation year (from a period of 100 simulated years).

The average share in annual heat demand for domestic hot water provided by solar collectors:

$$\mathrm{shd} = \frac{\sum_{I}^{XII} \mathrm{shd}_m}{12} \quad (18)$$

where:
shd – average share in meeting the annual heat demand calculated in months from I (January) to XII (December).

The share in annual heat demand, depending on the absorber, gross and aperture area, is presented in Figure11–15.

Knowing the number of residents of a single-family house and the amount of used hot water, the heat demand for DHW can be determined according to the eqn (1). The aforementioned heat demand for a given building can also be determined on the basis of the energy performance of building. With the known heat demand and degree of meeting the demand (ranging from 0 to 90%), it is possible to determine the required collectors area on the basis of nomograms presented in Figures 11–15. For example, the annual heat demand for DHW for an average family of 4 persons, 40 dm^3 per person per day, is 4800 kWh/year. The absorber areas of both types of the analyzed collectors, determined on the basis of Figures 11 and 14 and depending on the degree of meeting the heat demand, are shown in Figure 16.

Despite significantly lower access to sunlight calculated on the gross area of ETCHP, meeting 40% of the demand requires minimally larger surface when compared to FPC. An inverse relationship occurs in the case of coverage of 60% of the heat demand per year. Increasing the demand coverage from 40% to 60% for both types of collectors requires to double area of the absorber. In this case, especially in the summer, the yields are excessive compared to the needs. The excess energy can

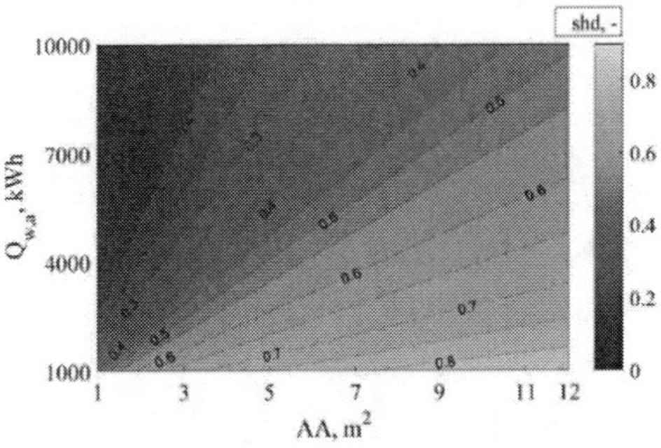

Figure 11. The share in meeting the annual heat demand (shd) for flat plate collectors as a function of the absorber area (AA) and heat demand ($Q_{w,a}$).

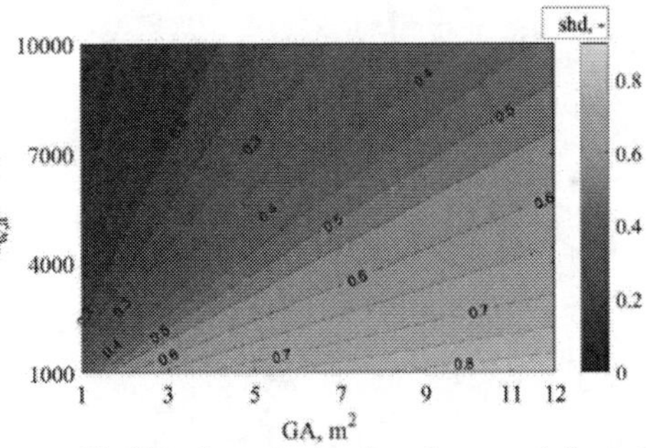

Figure 12. The share in meeting the annual heat demand (shd) for flat plate collectors as a function of the gross area of the collector (GA) and heat demand ($Q_{w,a}$).

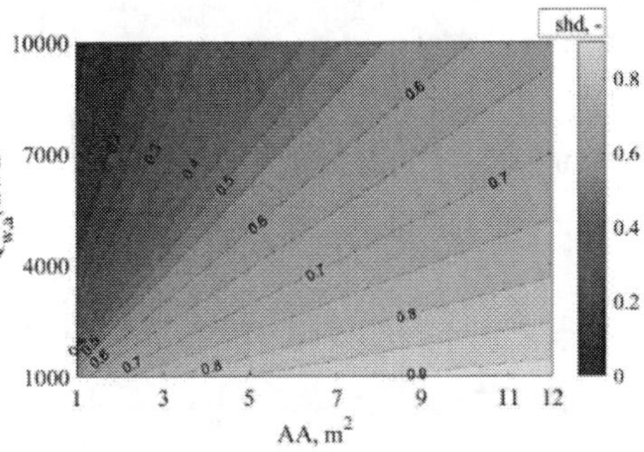

Figure 13. The share in meeting the annual heat demand (shd) for ETCHP as a function of the absorber area (AA) and heat demand ($Q_{w,a}$).

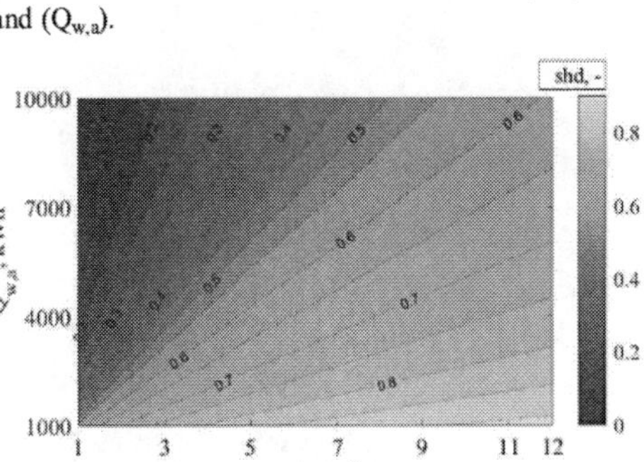

Figure 14. The share in meeting the annual heat demand (shd) for evacuated tube collectors with heat pipe as a function of the aperture area (AAp) and heat demand ($Q_{w,a}$).

DOI: 10.1051/e3sconf/20161000139

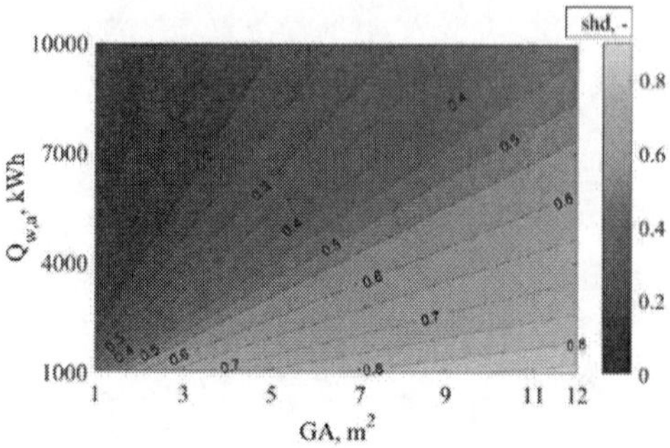

Figure 15. The share in meeting the annual heat demand (shd) for ETCHP as a function of the gross area of the collector (GA) and heat demand ($Q_{w,a}$).

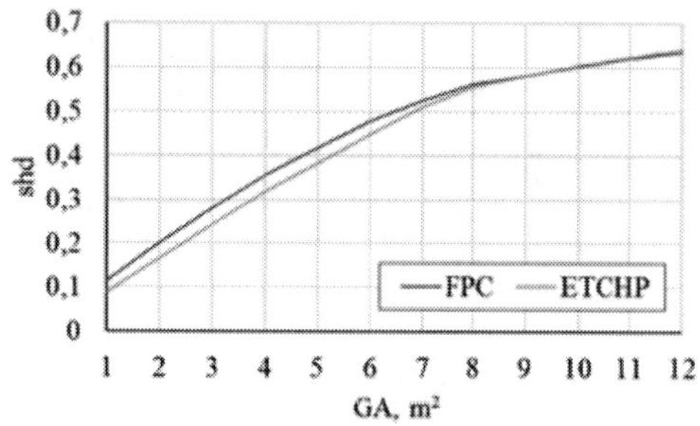

Figure 16. The summary results for heat demand, $Q_{w,a}$ =4800 kWh/year; shd – the share in meeting the annual heat demand, GA – the gross area of the collector

not be stored more than a few days without the use of additional technologies.

The degree of utilization of solar collectors was determined on the basis of the useful efficiency, calculated according to the relationship (19):

$$UE = \frac{Q_y}{I \cdot AA} \qquad (19)$$

where:
UE – useful efficiency.

The useful efficiency of solar collectors as a function of the absorber area and heat demand is shown in Figure 17.

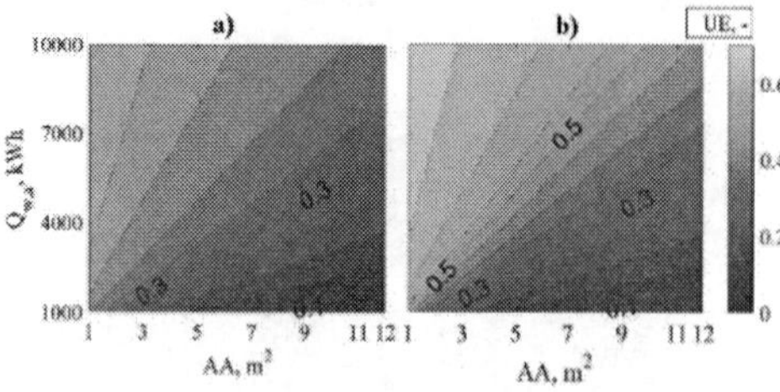

Figure 17. The useful efficiency of: a) FPC, b) ETCHP as a function of the absorber surface and heat demand.

For both types of collectors most useful efficiency (>50%) is obtained for small absorber area and a high heat demand. Due to the higher efficiency and better access to insolation (Figure 8.) ETCHP have greater useful efficiency regardless of heat demand and surface change.

The useful efficiency preseented in Figure 17 is affected, among others, by the average temperature of the absorber (T_m), depending on the coverage of heat demand on a monthly basis and the amount of unused energy, mainly in the summer months. In addition, exploiting the potential of the devices, and thus the costs, are of crucial importance. However, this is an optimization issue separated from the demand coverage and useful efficiency of solar energy generation.

5 Summary

Due to the variability in solar radiation over a year and the changing weather conditions, it is difficult to ensure a high level of coverage of the heat demand. By increasing the absorber area, it is possible to increase the share in heat demand coverage in the months of spring and summer. Further increasing of the absorber surface increases the possibility to cover the heat demand in the winter months.

Due to the lower heat loss rates, the use of ETCHP is more favorable in the annual balance sheet. The shares in meeting the heat demand suggest non-linear relationship between the coverage of heat demand and the analyzed surfaces. This is related to, among others, adverse working conditions of collectors in the winter months and surpluses, gained during the summer, which cannot be used. Therefore, the ongoing efforts to increase the share result in a low efficiency.

References

1. K. Neupauer, J. Magiera, Technical Transactions, **4**, 57 (2009) (in polish)
2. IEO for Mazowiecka Energy Agency, Available: http://www.mae.com.pl/files/poradnik-dla-uzytkownikow-instalacji-slonecznych-na-mazowszu_-mae.pdf, (07 04 2016) (in polish)
3. GeoModel Solar, Available: http://solargis.info/doc/_pics/freemaps/1000px/ghi/SolarGIS-Solar-map-Europe-en.png, (22 12 2015)
4. A. Głuszek, J. Magiera, Polityka energetyczna, **11**, 41-19 (2008), (in polish)
5. A. Więcka, Available: http://www.teraz-srodowisko.pl/aktualnosci/Zapomniane-kolektory-sloneczne-829.html, (07 04 2016), (in polish)
6. GUS, „*Energia ze źródeł odnawialnych w 2014* r." (GUS, Warszawa, 2015), (in polish)
7. D. Kryzia, P. Olczak, Logistyka, **4**, 4587 (2014), (in polish).
8. K. Żarski, Civil and Environmental Engineering, **59**, 905 (2012), (in polish)
9. K. Kurowski, Instal Reporter, **17**, 14 (2011), (in polish)
10. M. Żukowski, P. Radzajewska, District Heating, Heating, Ventilation, **46**, 138-142 (2015), (in polish)
11. J. Dziedzic, P. Olczak, D. Kryzia, Logistyka, **4**, 8832-8840 (2015), (in polish)
12. K. Pytel, A. Kłos, Civil and Environmental Engineering, **59**, 650 (2012), (in polish).
13. Minister Infrastruktury i Rozwoju, Dz.U. 2015, **376**, (2015), (in polish)
14. M. Kwietniewski, K. Miszta-Kruk, A. Piotrowska, Technical Transations Environmental Engineering, **108**, 113-127 (2011), (in polish)

15. Minister Infrastruktury i Rozwoju, Dz.U. 2015, **1422**, (2015), (in polish)
16. P. Bugajski, Infrastructure and Ecology of Rural Areas, 105-1132 (2009), (in polish)
17. P. Bugajski, G. Kaczor, Infrastructure and Ecology of Rural Areas, 17-26 (2005), (in polish)
18. EuroObservER,Available: http://www.energies-renouvelables.org/observ-er/stat_baro/observ/baro215.pdf. (07 04 2016)
19. L. Shah, S. Furbo, Applied Energy, **78**, 374–378 (2004)
20. L. Fen Su, L. Miao, S. Tanemura, G. Xu, Science and Technology of Advanced Materials, **13**, 4 (2012)
21. ET Polska, Available: http://www.etpolska.com/pl/oferta/kolektory-plaskie.html. (22 12 2015)
22. Hewalex, Available: http://www.hewalex.pl/oferta/plaskie-kolektory-sloneczne/kolektor-sloneczny-ks2000-tp-am.html. (22 12 2015)
23. Sonneko, Available: http://sonneko.pl/dat/attach/449_kolektor_plaski.pdf. (21 12 2015), (in polish)
24. Veelman, Available: http://www.veelman.com/pl/produkty/kolektory-solarne-plaskie (21 12 2015), (in polish)
25. Watt, Available: http://www.watt.pl/pl/produkty/kolektory-sloneczne/kolektory-plaskie/watt-2020-s.html. (21 12 2015), (in polish)
26. Watt, Available: http://www.watt.pl/pl/produkty/kolektory-sloneczne/kolektory-plaskie/watt-3020-s.html. (21 12 2015), (in polish)
27. Watt, Available: http://www.watt.pl/pl/produkty/kolektory-sloneczne/kolektory-plaskie/kolektor-sloneczny-watt-8020.html (21 12 2015), (in polish)
28. Watt sp. z o.o., *Watt CPC 9+* (ISFH, Emmerthal, 2012), (in polish)
29. Ensol, Available: http://www.ensol.pl/solartermia/kolektory-sloneczne/plaskie/z-absorberem-meandrycznym-aluminiowym/es2v20sal.html. (22 12 2015), (in polish)
30. Biawar, Available: http://www.biawar.com.pl/systemy-solarne/hevelius-wunder-kolektory-plaskie/hevelius-cls-3-bz. (02 01 2016), (in polish)
31. K. Neupauer, J. Magiera, Technical Transactions, **10**, 223-232 (2010), (in polish)
32. ENERGOSOL, Available: http://www.energosol.pl/oferta_kolektory_sloneczne_kolektory_prozniowe.html. (29 08 2015), (in polish)
33. P. Olczak, J. Zabagło, District Heating, Heating,Ventilation , **46**, 430 (2015), (in polish)
34. S. Kandefer i P. Olczak, District Heating, Heating,Ventilation , **46**, 11-13 (2015), (in polish).
35. Watt sp. z o. o., Available: http://www.watt.pl/pl/imagesdb_a0ff1f78dcaab2cb7b4b70db2a649c3a.pdf. (20 03 2016), (in polish)
36. D. Chwieduk, *Modelowanie i analiza pozyskiwania oraz konwersji termicznej energii promieniowania słonecznego w budynku* (Warszawa IPPT PAN, 2006), (in polish)
37. P. Olczak, J. Zabagło, S. Kandefer i J. Dziedzic, Między ewolucją a rewolucją – w poszukiwaniu strategii energetycznej, **II**, 739–752 (2015), (in polish)
38. OGIMET Available: http://ogimet.com/resynops.phtml.en. (10 11 2015)
39. J. A. Duffie i W. Beckman, Solar Engineering of Thermal Processes, 2006.
40. J. Karlsson, *Windows - Optical Performance and Energy Efficiency* (Uppsala, 2001)
41. P. Loutzenhiser, H. Manz, C. Felsmann, P. Strachan, G. Maxwell, Solar Energy, **81**, 254–267 (2007)
42. Ministerstwo Infrastruktury i Rozwoju, Available: http://www.mir.gov.pl/strony/zadania/budownictwo/dane-do-swiadectw-charakterystyki-energetycznej-budynkow (23 07 2015), (in polish)
43. J. Zimny, R. Brzegowy, S. Bielik, *Kolektory słoneczne* (Kraków 2013), (in polish)
44. J. P. Ryan, *Incidence angle modifiers in cylindrical solar collector design* (Fort Collins 1997)
45. EMD International A/S,, *Solar Collectors and Photovoltaic,* Available: http://www.emd.dk/files/energypro/Solar%20Collector%20and%20Photovoltaic%20in%20energyPRO.pdf (10 11 2015)
46. D. Matuszko, International Journal of Climatology, **34,** 228–234 (2014)
47. M. Rodríguez-Hidalgo, P. Rodríguez-Aumente, A. Lecuona, G. Gutiérrez-Urueta, R. Ventas, Applied Thermal Engineering, **31**, 2385 (2011)

AUTHOR INDEX

AUTHOR INDEX

AUTHOR INDEX